A Dictionary of

Agricultural and Allied Terminology

A Dictionary of Agricultural and Allied Terminology

John N. Winburne, Ph.D.
EDITOR-IN-CHIEF

Michigan State University Press

1962

MICHIGAN STATE UNIVERSITY PRESS

Library of Congress Card Catalog Number: 62-9169

The Lakeside Press

R. R. DONNELLEY & SONS COMPANY

Chicago, Illinois, and Crawfordsville, Indiana

Manufactured in the United States of America

PREFACE

PROSPECTUS

English, the common second language of the world, is a language of letters, the arts, business, diplomacy, science, medicine, education, sport, warfare, and agriculture.

In the last century, agriculture has become an accepted branch of knowledge. Educated men and the general public alike respect the agriculturist for his demonstrations of how science applies to the production of food, fiber, and related products, improves the product, increases the productivity of land and animals, and, as a result, improves man's lot.

As agriculture gained acceptance as a discipline, it utilized economics, entomology, botany, chemistry, physics, genetics, engineering, hydrology, climatology, sociology, geology, to create horticulture, agronomy, agricultural chemistry, agricultural engineering, rural sociology, land and water conservation, forestry, animal husbandry, and veterinary medicine. The technical language of agriculture overlaps and intertwines the jargon of practically every science, thus becoming much more inclusive than just *farming* or *husbandry*.

But the language of the researcher, scholar, and teacher by no means comprises the total language of agriculture. The speech of the practitioner, the dirt farmer, the sheep herder, the rancher, the implement manufacturer, the pesticide salesman, the fungicide distributor, the poultry producer, and the feedlot buyer, also forms a vocabulary of great scope.

These jargons, that is, the technical languages of trades, sciences, and professions, may be well known by those working in other fields, but often they are not. The floriculturist's *hard pinch* is certainly not clearly understood by the dairyman whose *milk-fat basis* may be meaningless to the agricultural engineer.

As the literature of agriculture increases in volume from the vigorous investigation of scholars; as more persons of different specialties become involved in agriculture; as more persons whose native language is not English need to learn the most recent discoveries of agriculture; and as more of our urban youth turn to one of the branches of agriculture, it became necessary to collect and define the words and idioms of agriculture.

BREAKING GROUND

The Disciplines. The problem of what constitutes the language of agriculture can not now, and probably never will be adequately solved, for what one man will include, another will discard. Certain disciplines had to be established, and yet the language of the practitioner, the tradesman, the manufacturer, and the scholar of those fields involving the production, handling, marketing, and utilization of food, fiber, flowers, wood, etc., had to be included.

This would encompass, as far as was physically possible in a single volume, general agriculture, agricultural chemistry, agricultural economics, agricultural engineering, animal husbandry, beekeeping, dairy, feeds, fiber crops, field crops, forage crops, forestry, horticulture, insects, meat products and processing, plant diseases, poultry husbandry, seeds, soils and fertilizers, veterinary medicine, weeds and weed killers, poisonous plants, and water.

Other limits immediately became necessary since each science or area of knowledge has its own extensive vocabulary, and not all of the words used in each could conceivably be collected. For example, botany is a fundamental science of agriculture, yet infrequently used botanical terms were bypassed in favor of those more widely used in agriculture in general.

Selection of Words. The problem of which words from the various sciences were to be collected was solved in a pragmatic fashion. Two-hundred-fifty books, magazines, pamphlets, and catalogs were selected as source material for the dictionary (See Appendix A).

Words and idioms (stereotyped phrases like *foreign matter, forest land, middlebuster*) found their way into the files, as did some highly technical terms and colorful colloquialisms appearing repeatedly in the source materials. In general, however, the narrowly technical terms of a particular science are omitted.

Source Material. The bibliography was prepared by the editorial board, whose membership (listed in Appendix B) represented the various large areas to be investigated. Each member of the board prepared a bibliography for his area in such a manner that items overlapped each other slightly, not being wholly mutually exclusive.

Preparing the Manuscript. Following the selection of the bibliography, the editor and five associates (See Appendix B) began reading the materials, selecting words and idioms of agricultural significance, and preparing manuscript. Of the 85,000 items collected, many were finally omitted from the dictionary.

Each man's work was criticized by the others, and from a first draft, a second was prepared for further criticism by the editorial board. After this was done, a third draft was prepared which was sent to members of the United States Department of Agriculture and to the Bureau of Primary Industries in Australia for further critical analysis. A thusly revised manuscript was then submitted to a separate editorial board for mechanical revision (See Appendix B).

BUILDING THE STRUCTURE

A definition is an explanation, a signification, of the meaning of a word or idiom in context. Dictionaries commonly use synonyms assuming the reader understands the meaning of the synonym. This volume, however, attempts for the most part to define by establishing either a scientific name followed by a brief description or statement of a well-known class followed by limitations designed to set the particular off from the other members of the class. The definitions may be encyclopedic in an effort to be clear. On the other hand, no definition is wholly satisfactory, for words, no matter how well chosen, are not always good symbols for things and notions. In the main, the definitions are free from highly technical words, being designed for the use of no particular group or intellectual level.

Order of Information. The alphabetical order of entries is literal: e.g., *Colored Dorking* and *colored fowl* precede *color gardening.* Words and phrases are alphabetized in the same manner, using each letter of each word as it appears: e.g., *color marking, color of fat, color of lean, color profile,* in that order.

The entries and definitions follow a general pattern: (a) the head word; (b) the scientific name, if any; (c) a general statement or description; (d) the agricultural implications; (e) synonyms; (f) cross-references.

Parts of Speech. Sometimes a word has a variety of meanings or implications, so that it was often necessary to give more than one definition for a head word. Again, the English word, generally having no inflectional ending, may be defined in any of several different parts of speech and/or agricultural applications. For example:

tail. 1. The posterior part of the vertebral column of animals. It is usually covered with hair, some of which may be quite long. Also called *brush.* 2. A fanlike row of rather stiff feathers on the posterior part of a bird. . . . 3. The lowest grade of flour. . . . 4. A wisp of hay not properly tucked into a bale. 5. The bottom layer of produce in a container. 6. The weakest and poorest sheep (Australia and New Zealand). 7. The woody cane of a plant which has been propagated by tip layerage. 8. To dock an animal's tail. . . . 9. To keep sheep together in a flock (Australia and New Zealand). 10. To remove the tail from a carcass. 11. To assist an undernourished cow to its feet by pulling on the tail.

Abbreviations. The abbreviations which appear are those used in standard practice, as *var.* for "variety," *sp.* and *spp.* for "species." Such common abbreviations as *i.e., e.g., q.v.* are used instead of *that is, for example,* and *which see.*

Use of Latin Names. The common names for plants and insects presented what seemed an insurmountable problem in that one plant or one insect may have a half dozen or more common names. To handle these items economically the plants are, for the most part, defined under the common name shown in *Standardized Plant Names.* Other names are entered but refer to the one: for example, *Lilium candidum* is known by at least two names—*Madonna lily,* and *Annunciation lily.* Since *Standardized Plant Names* gives *Madonna lily,* it is under *Madonna lily* that the term is defined. Under *Annunciation lily* is shown "See *Madonna lily.*" Similarly, the common names of insects are entered according to the *Bulletin of the Entomological Society of America.*

As a matter of editorial policy, few scientific names (*i.e.,* Latin names) appear as main entries, for it is common names which were collected and defined.

No Phonetic Spelling. The phonetic spelling is not included in this book, for which pronunciation would be given: American, English, or Australian? If American, which dialect would be used? Which system of phonetic transcription would be most clear and usable? Does a dictionary of this sort need pronunciation; most of its users will be native speakers of English, or those who have learned enough English to use this book. Finally, inclusion of pronunciation would have caused printing costs to soar.

AN APOLOGIA

It would be boastful to claim that this volume contains all the terms of agriculture, for it is beyond the ability of the lexicographer to collect the complete lexicon of any culture. In producing such a work as this, at a certain point in time the search for words must, necessarily, cease, and as a result, terms which enter the vocabulary later are not included.

Noah Webster in his *An American Dictionary of the English Language,* 1828, summarized the task best:

For all practical purposes, the defining part of a dictionary is by far the most important. Yet in this part, embracing, as it does, the whole circle of ideas embodied in the language of a people, the utmost efforts of the lexicographer are only an *approximation* towards the end in view. No single mind can enter, with perfect exactness, into all the multiplied distinctions of thought and action, among a highly civilized people. For his information on many subjects, the author of a dictionary must rely on the statements of others; and he is liable to be misled either by a want of accuracy in those statements, or by an erroneous conception of their meaning. Yet in the several departments of life there are those, to whom each of these distinctions is "familiar as household words;" and who, therefore, regard with wonder and contempt, the mistakes of a dictionary upon subjects of so plain a nature. Hence the lexicographer has ordinarily to accept much censure and little praise. His deficiencies and errors are sure to be detected; while accuracy of his information is lost on those, whose familiarity with the subject has led them to consider such accuracy as a matter of course.

East Lansing
November 9, 1961

A

Aaron's-beard. See *Kenilworth ivy; smoketree; strawberry saxifrage.*

Aaron's-beard St. John's-wort. *Hypericum calycinum,* a Eurasian evergreen undershrub, of the family Hypericaceae, which is grown in fairly warm areas as an ornamental ground cover. Also called *rose-of-Sharon.*

abaca banana. *Musa textilis,* a banana plant, of the family Musaceae, native to the Philippines. A fiber called (Manilla) hemp is produced from the leaves. Also called *abaca; abaka.*

abactor. Cow-thief.

abandoned land. Unoccupied and uncultivated land. The title may or may not have been forfeited.

abandoned tree. A tree, killed by insects, which no longer receives sprays or cultural treatment.

abattoir. A slaughter house.

Abbe refractometer. A device used to identify and describe a liquid by measuring the refraction of light passing through it. The liquid (in a thin film) is placed between two prisms, through which a beam of light is passed. The refraction of the beam is measured and the reading compared with those of similar liquids of known composition. It is used in food processing.

ABC soil. A soil having a completely differentiated profile. See *soil profile.*

abdomen. In mammals, that part of the body which contains the digestive organs; in lower orders of animals and insects that part of the body which roughly corresponds to the *abdomen; belly.*

abdominal dropsy. In animals, a condition caused by an accumulation of an excess of serous fluid in the abdomen; ascites.

abele. White poplar *q.v.*

abelmosk. See *musk mallow: 2.*

Aberdeen Angus. A breed of beef cattle originally from Scotland, which are black, hornless, blocky, low-set, and smooth. Also called *Angus; Black Angus; doddie.*

aberration. In genetics, an irregularity in chromosome distribution during cell division which results in deviation.

Abertam. A hard cheese made from ewe's milk.

abietene. An oily hydrocarbon consisting of heptane and its isomers. It is distilled from the resin or balsam of the Digger pine, *Pinus sabiniana.*

abietic acid. An organic acid obtained from the resins of certain firs and pines; used in the manufacture of driers, varnishes, soaps.

abiotic disease. A disease caused by an inanimate agent; a non-parasitic disease. Also called *physiogenic disease.*

ablactate. To wean.

abluent. A cleansing agent.

abnormal egg. A marketing term used in official grading to denote a misshapen egg, a shell faulty in strength or texture or an egg with a double yolk, no yolk, blood spots, meatspots, or off-colored yolk.

abnormal milk. 1. Any milk different in flavor, odor, appearance, or composition from fresh milk. 2. Milk from a diseased udder which is usually low in nonfat-solids, fat, lactose, and calcium, but high in sodium chloride.

abomasum. The fourth, or true, stomach division of a ruminant, or cud-chewing animal, like a cow or sheep. The compound stomach consists of: (1) rumen or paunch, (2) reticulum, (3) omasum, (4) abomasum. The herbage is swallowed nearly unchewed, and passes into the rumen or reticulum, from which it is regurgitated and masticated in the mouth while the animal is resting. It is then swallowed again and passed through the reticulum and omasum into the abomasum, where it is acted on by the gastric juices.

abort. 1. In animals, to expel the embryo or fetus from the womb prematurely. See *brucellosis; leptospirosis.* 2. In plants, to produce few or no seeds, or to drop fruit prematurely.

abortion tested. Animals tested for brucellosis.

abrasion. 1. Wear in farm implements caused by friction, such as *abrasion* of a plowshare by the soil. 2. A section of skin or mucous membrane from which the surface layers have been rubbed or worn off.

abscess. A localized collection of pus which causes a swelling under the skin, or in the organ where it is located.

abscission. In botany, the dropping off of a leaf, fruit or flower; shedding.

abscission layer. A narrow, transverse band of cells at the base of leaf, flower, stalk, fruit stalk, or a portion of stem, in which the cell walls dissolve, leaving the vascular bundles for support. These eventually break and the leaf, flower, fruit or even a portion of the stem falls. Also called *absciss layer; separation layer.*

absconding swarm. A bee colony which abandons a hive because of disease, wax moth, or other causes; unusual in modern apiaries.

absentee owner. A person, or corporation, who rents land, including housing, to a tenant farmer or hires a manager.

absolute forest land. Land utterly unsuited for any crop except trees.

absolute humidity. 1. The mass of water vapor per unit volume of space. 2. The gaseous pressure exerted by water vapor.

absorb. 1. To assimilate or incorporate one substance into another; as a blotter *absorbs* ink. 2. In botany, to assimilate water and nutrients.

absorbing complex. All that part of the soil capable of taking in liquids, solutes, gases, and radiant energy.

absorption. 1. The processes by which water enters the earth; atmospheric water-vapor, hygroscopic absorption, wetting, infiltration, gravity flow into sink-holes. 2. The passage of digested food from the alimentary tract into the circulatory system.

absorption loss. The loss of water through assimilation when a canal or reservoir is primed or flooded for the first time. Subsequent loss is *seepage* or *evaporation loss.*

absorption terrace. A ridge which slows the flow of run-off water so it can be absorbed by the soil.

abstract of title. A summary document which states the evidences of ownership of real estate and its encumbrances from the time it was taken from the Federal or other Government down to the present date. In most states the county office of the Register of Deeds office provides the *abstract of title.*

Abutilon. See *flowering maple.*

abutting. Adjoining.

Abyssinian oats. *Avena abyssinica,* one of several species of oat, Avena, which is found in northern Africa. The *Abyssinian oat* is grown to a limited extent in Abyssinia and neighboring regions. Probably it has had little influence in the development of oat varieties grown in Europe and America, these having derived from *A. fatua* from Africa and *A. sativa* from eastern Europe.

acacia. Any shrub or tree of the genus Acacia, family Leguminosae. Although basically tropical and short-lived, many species are grown in southern regions of the U.S. as ornamentals for their showy, dense clusters of small, usually yellow, flowers. See *kangaroo-thorn acacia; sweet acacia; Sydney acacia; blackwood acacia; pearl acacia; hairy-wattle acacia; goldwreath acacia; black locust.*

acanthella. A larval or immature thorn-headed worm found in the body cavity of the intermediate hosts, such as beetle grubs, insects, and other arthropods.

acanthor. The microscopic embryo or larva of the thorn-headed worms, which develop into the next larval stage, acanthella, in various intermediate hosts, such as insects and other arthropods.

acariasis. The condition of plants or animals infested with acarines: mites, ticks.

acaricide. A substance, solution, or paste which kills mites or ticks.

acarine. Any mite or tick.

acarine disease. 1. Any infestation with mites or ticks. 2. In Europe, a disease of the adult honey bee which is caused by *Acarapis woodi* Rennie. The mite spends its life in the trachea of the bee living on its body fluids. Also called *Isle of Wight disease.*

acarocecidium. On a plant, any gall caused by acarines.

acarology. That branch of zoology which deals with acarines.

accelerated erosion. Excessive wearing away of soil or rock brought about by changes in the natural cover or ground conditions, including changes due to human activity.

accented plant. Any plant with a natural, well defined form or one so pruned. Used as a decorative landscape feature.

acceptable character. Representative of a type or breed.

acclimation fever. Denoting any local disease which attacks domestic animals brought to a new climatic region. See *cattle tick fever.*

accommodation cell. An odd or irregularly shaped cell built by bees to attach combs which do not coincide or to attach the edge of a comb to another surface.

accredited herd. Dairy cattle certified to be free from tuberculosis as a result of two successive tests given under the direction of the U.S.D.A. The term is sometimes erroneously applied to a brucellosis-free herd.

accretion. 1. Increase of height, diameter, quality, and value of a tree or woods; increment; growth. 2. Increase of soil water. 3. Addition of soil by gradual deposition through natural causes. 4. Process of recovering land from the sea by diking and draining. See *relict.*

accretion borer. In forestry, a hollow auger to obtain a sample from the core of a tree to determine growth by counting annual rings.

accretion cutting. The thinning out of a forest to allow the proper growth of trees.

accrual method. One of the two main methods of accounting used in computing farm earnings in reporting Federal Income Tax. Increases and decreases in inventory items for the year are reflected in income. Farm income is included as income for the year in which it is earned regardless of when payment is received. Farm expenses are deductible in the taxable year in which incurred, whether paid or not. Complete inventories of livestock, crops, produce, feed, etc., are required. See *cash method.*

acequia. Irrigation canal (southwestern U.S.).

acequia madre. A main irrigation canal (southwestern U.S.).

acervulus. An erumpent mass of fungus hyphae bearing conidiophores and conidia, often with surrounding or intermingled setae. This is the typical asexual reproductive structure of Order Melanconiales of Class Fungi Imperfecti. See *anthracnose.*

acetarious plants. Salad plants.

acetic acid. An organic acid, CH_3COOH, which is the chief active component of vinegar. It is also important as a product in lactic acid fermentation and therefore, an important constituent of flavor in many milk products. Also called *pyroligneous acid.*

acetic bacteria. The bacteria which produce acetic acid (vinegar acid) from alcohol.

acetic fermentation. The process by means of which acetic acid is formed from ethyl alcohol in weak solution through the action of bacterial ferments: for example, the formation of vinegar from hard cider or from wine.

acetoarsenite of copper. See *Paris green.*

acetobacter. A bacterium, family Nitobacteriaceae, which grows in beer and wine.

acetoin. Acetyl methyl carbinol, mother substance of diacetyl, an important factor for aroma in good butter and cheese.

acetone. A ketone; a clear, rapidly-evaporating liquid which (1) in nature is obtained from fermentation of sugar and starch and is present in diabetic urine, breath, blood. Its presence in the breath of a lactating animal (cattle, sheep, milk goats) indicates her failure to oxidize the fatty material of her feed, making her deficient in carbohydrates; (2) is commercially obtained from distillation of wood and has wide application as a solvent.

acetone bodies. Chemical substances which are the intermediary products of the metabolism of fatty acids and are normally present in small quantities in the blood. See *ketone bodies.*

acetonemia. A disease characterized by the presence of excessive amounts of acetone bodies in the blood. Also called *ketosis; false milk fever; chronic milk fever; acidosis; acetonuria; pregnancy disease in ewes.*

achemon sphinx. *Pholus achemon* (Drury), family Sphingidae, a hawk moth, the larva of which is about three inches long and feeds on grape leaves and on the Virginia creeper.

achene. Any small, dry fruit having but one seed whose pericarp does not burst when the fruit is ripe.

achondroplasia. Skeletal malformation during prenatal development caused by a genetic factor; e.g., a bull-dog calf whose legs are very short, the head short and thick, the upper lip split, the lower jaw protruding, the nose flattened, and the tongue swollen. In some instances the fetus is carried to the full term of pregnancy, in which case the calf dies within a few days.

acid. A chemical term generally applied to hydrogen compounds which are water soluble. It is sour to the taste, and reacts with certain vegetable dyes, such as *litmus* (*q.v.*), changing its color from red to blue. See, for example, *acetic acid.*

acid board. A sheet of metal fastened to a wooden frame which is the same size as the top of the beehive. The underside of the sheet is covered with cloth, and the topside is painted black. The cloth is saturated with carbolic acid, the fumes of which drive the bees away from the honey in the super so that a person can work with the hive.

acid casein. The precipitate or curd of milk produced by the addition of acid or by developed acidity. Used for food, for making cold water paints and glue, wallpaper and fabrics, coating paper, and plastics.

acid-cut. A color defect which renders American Cheddar cheese opaque or transparent, bleached, dull, and lifeless. Sometimes it appears as a mottling effect around mechanical openings. The condition is associated with high moisture and high acid, although sometimes the cheese has dry and crumbly texture. Acid-cut differs only in intensity from *faded q.v.*

acid-fast. Bacteria not readily decolorized by mineral acids after being stained with dyes: the bacteria that cause tuberculosis and leprosy.

acid flavor. A sharp, sour taste.

acid-forming fertilizer. Fertilizer that increases the residual acidity of the soil.

acidified silage. Silage preserved by the addition of acid, such as commercial phosphoric acid, sulfuric acid or hydrochloric acid, or a combination of the latter two.

acidimeter. A device or chemical which measures the amount of acid present in a solution.

acid land. See *acid soil.*

acid lead arsenate. A common insecticide used on orchard trees and field crops. Also called *dibasic lead arsenate; diplumbic orthoarsenate; dilead orthoarsenate; diplumbic arsenate; lead hydrogen arsenate.*

acid lemon. *Citrus limon,* one of three types of *Citrus limon* used in fruit culture. See *rough lemon; sweet lemon.* Also called *lemon.*

acid lime. See *lime:* 2.

acid mercury treatment. A method of treating seed potatoes to prevent *Rhizoctonia disease of potato q.v.*

acidophilic. Of, or pertaining to, certain bacteria tolerant of, or able to live in, an acid solution. Also called *aciduric.*

acidophilus milk. The food which results from

souring milk with *Lactobacillus acidophilus.* Used for therapeutic reasons, since *Lactobacillus acidophilus,* established in the intestines with fermentable carbohydrates, causes a mild acid condition. Often prescribed for cases of intestinal autointoxication. Also called *acidophilus cultured buttermilk.*

acido-proteolyte. Bacteria which attack proteins in milk, break them down into simpler, soluble proteins, and produce acid.

acidosis. See *acetonemia.*

acidosis of pregnant ewes. See *pregnancy disease in ewes.*

acid phosphate. See *superphosphate.*

acid silica gravel. A medium for growing hothouse plants in gravel culture.

acid soil. A soil giving an acid reaction (precisely, below pH 7.0; practically, below pH 6.6) throughout most or all of the portion occupied by roots. A soil having a preponderance of hydrogen ions over hydroxyl ions in the soil solution. In common parlance, a "sour" soil, as opposed to a "sweet" soil.

acid tolerant crops. Crops, such as oats and corn, which will grow on moderately high acid soils.

acidulated fish scrap. A fertilizer of non-edible, steam-cooked fish from which the oil has been extracted and treated with sulfuric acid. Also called *acidulated fish tankage.*

aciduric. See *acidophilic.*

acid wood. Wood, usually of deciduous trees, harvested for the manufacture of wood alcohol, acetone, etc.

acid yellowing. A yellowish-green condition of melon plant leaves accompanied by poor growth which results from high soil acidity.

aciniform. 1. Having the shape of a bunch of grapes. 2. Containing small grape-like kernels.

aclang-parang. A timber tree (*Albizzia procera*), family Mimosaceae, of the Philippine Islands.

acle. 1. A tall tree, *Xylia xylocarpa,* family Mimosaceae, from which comes dark-colored wood, called *ironwood of Pegu.* Native to Asia. 2. A timber tree, *Albizzia acle,* family Mimosaceae, used in cabinet making. Native to the Philippine Islands.

acme harrow. A harrow constructed with long, thin, twisted, knife blades that project rearward. Used to cut clods and to stir the soil. Also called *knife harrow; pulverizer; pulverizer harrow; curved knife-tooth harrow; blade harrow.*

aconite buttercup. *Ranunculus aconitifolius,* a tuberous, perennial herb, of the family Ranunculaceae, grown for its white or yellow (sometimes double) flowers. Also called *garden buttercup; fair-maids-of-France.*

aconite monkshood. *Aconitum napellus,* a perennial herb, of the family Ranunculaceae, a native of Eurasia, grown for its showy, blue flowers and as a source of the drug aconite. Also called *Venus's chariot; face-in-the-hood.*

acorn. The fruit, or seed, of oaks, any species of the genus Quercus. *Acorns* are eaten by domesticated grazing animals, cattle, hogs, and goats; and by several kinds of wild life, both birds and mammals, and to a limited extent by man. They were an important source of food for some of the tribes of American Indians.

acorn fruit. A symptom of a virus disease, *Citrivir pertinaciae,* of citrus trees characterized by acorn-shaped fruit; as much as one-half the production of a tree may be so deformed. The fruit has very thin skin toward the stylar end. Also called *pink nose; blue nose.*

acorn squash. Hubbard group; characterized by a small, deeply fluted diamond or acorn shaped shell.

acquired character. A change in character, morphological or physiological, due to the environment, which is not passed on to the next generation; not a genetic change.

acquired immunity. Ability to resist a disease to which the individual would ordinarily be susceptible; opposed to natural immunity (*cf.*). It may be a result of antibodies built up through exposure to, or inoculation with, pathogens (*active immunity*) or it may result from the individual receiving antibodies from another individual who has become immune (*passive immunity*).

acre. A unit of land measure used in England and the United States which is equal to 43,560 square feet, or 1/640 of a square mile, or 160 square rods, or 4,840 square yards, or 4,047 square meters. The Scottish acre is 1.26 and the Irish acre is 1.62 times larger. One acre as used in England or the United States equals 0.4 hectare, 0.96 feddan (Egypt), 0.31 carreau (Haiti), 1.03 cuerdas (Puerto Rico), 0.03 caballerias (Cuba), 5.99 shih mou (China), 4.08 tan (Japan), 0.84 arpent de Paris (some sections of Canada), 0.37 dessiatine (Russia).

acreage. An indefinite quantity of land; a collective number of acres.

acreage allotment. The acreage given, under measures of control authorized by Congress, to specific crops (such as wheat, corn, tobacco, rice, and peanuts) designated as "commercial" areas. The U.S.D.A. determines the total number of acres which must be planted to enable the national supplies of crops to approximate their national demands. The primary purpose is to curtail production and thus raise farm prices and income.

acre foot. The quantity of water necessary to cover one U.S. or English acre to the depth of one foot: 43,560 cubic feet or 325,850 gallons of water.

acre inch. The quantity of water necessary to cover one U.S. or English acre to the depth of one inch: 3,630 cubic feet or 27,154.2 gallons of water.

acre return. Gross value of production per acre.

acre right. The share of common lands belonging to a citizen of a New England town. The value was fixed acreage but varied in different towns (obsolete).

acres in crops. That acreage on which field crops

are growing or have been harvested in any year.

acres in farm. The area which includes all the land in the farm.

acres of crops. Land actually given to crops. If double-cropping or inter-planting is practiced, the "acres of crops" will exceed the "acres in crops."

acre yield. Production per acre.

acriflavine. A dye with antiseptic properties that is especially effective against Gram-negative bacteria.

acroclinium. See *rose sunray.*

acromania. See *crazy top.*

acronecrosis. See *top necrosis.*

acropetal. In plants, designating a development from base to top; such as the blossoming pattern of gladioli.

acrospire. The young spiral shoot of germinating seeds.

acrostalagmus fungi. A fungus belonging to the family Dematiaceae. (a) *Acrostalagmus albus* Pr., a fungus which attacks brown scale on citrus. (b) *Acrostalagmus aphidum* Oud., a fungus which attacks the citrus-inhabiting aphid, cotton aphid, green peach aphid, and the black citrus aphid.

acrylonitrile. An insect fumigant of low vapor-pressure used in flour mills in dead spots. If mixed with carbon tetrachloride, a more effective fumigant with reduced flammable qualities results.

AC soil. A soil which has an incomplete profile; with A and C horizons but having no clearly developed B horizon. See *soil profile.*

ACTH. An abbreviation for the adrenocorticotrophic hormone secreted by the front lobe of the pituitary gland. This hormone stimulates the outer part or cortex of the adrenal gland to secrete its hormones into the blood.

actinobacillosis. A chronic, infectious, rarely fatal disease of cattle, sheep, and swine caused by *Actinobacillus lignièresi.* In cattle it is characterized by movable swellings up to the size of an egg, involving the soft tissues of the head and neck region. The lymph nodes of the neck may become enlarged. Sometimes the tongue is affected, and there may be ulcerations of the mucous membranes of the throat. It may spread to various other internal organs. In sows, it is responsible for udder abscesses. In sheep, it is usually a secondary infection. Also called *lumpy jaw, big jaw, wooden tongue.* (Sometimes it is mistaken for *actinomycosis.*)

actinomycosis. A chronic, infectious, rarely fatal disease caused by *Actinomyces bovis,* which affects the jawbones of cattle and swine and is characterized by difficulty in breathing, enlarged, spongy bones with cavities filled with pus, swollen gums and loosened teeth, bony growths through the tissues and skin, resulting in an abscess with a foul odor. It is sometimes seen in the udders of swine where it causes a tumor-like enlargement. Man is susceptible to a similar disease caused by *Actinomyces hominis.* Also called *lumpy jaw; big jaw; clams; clyers; wooden tongue.* (Sometimes it is mistaken for *actinobacillosis.*)

actinomycosis of potato. See *corky scab.*

actionable fire. A conflagration started or allowed to spread in violation of law or regulation.

activated sewage. An organic fertilizer made from sewage freed from grit and coarse solids, and aerated after being inoculated with microorganisms. The resulting flocculated organic matter is withdrawn from tanks, filtered with or without the aid of coagulants, dried in rotary kilns, ground, and screened. Also called *sludge.*

active agent. That component part of an insecticide, herbicide, or fungicide formula that has toxic properties.

active carbon charcoal. A catalytic agent used to purify sirups.

active immunity. See *acquired immunity, cf. natural immunity.*

actual use. The total number of livestock that have actually grazed on a range and the length of time that they have used it which is usually expressed in *animal unit months; cattle yearlong; cow months;* or *sheep months q.q.v.*

actual-yield method. Scientific determination of plant performance in which each plant is numbered, and its performance is carefully noted. This system is of value in plant breeding.

acute fowl cholera. See *fowl cholera.*

acute indigestion in horses. A severe pain in the abdomen (stomach or bowels) of horses due to exposure to bad weather; overeating; spoiled feed; ingestion of indigestible particles; drinking too much water when hot, accompanied immediately by eating a large amount of feed; fatigue; bad teeth; etc. Also called *colic; stomach bloat; intestinal bloat; wind colic; spasmodic colic; cramp colic; impaction of the large intestine* (depending on the location of the pain).

acute salt poisoning. A sometimes lethal poisoning of animals due to excessive ingestion of salt: it is characterized by thirst, vomiting, pain, diarrhea, muscular spasms, excessive urination, weak pulse.

acute water deficiency. A condition of plants in which there is a sudden wilting without previous stunting or discoloring, as in a transplanted plant. See *chronic water deficiency.*

adabe. See *adobe soil.*

Adam's-apple. See *crapejasmine tabernaemontana.*

Adam's method. An accurate method for determining the weight of the fat content of milk. A known quantity of milk is absorbed on fat-free blotting paper which is then dried; the paper is extracted by ether in a Soxhlet apparatus, and the residue is weighed.

Adam's-needle yucca. *Yucca filamentosa,* a semidesert plant, of the family Liliaceae, which is cultivated as an ornamental shrub. It is one of the species which is hardy where severe frosts occur. Also called *Eve's-thread.*

Adansonia. *A. digitata,* the baobab *(q.v.)* of Africa and India; *A. gregorii,* the cream-of-tartar *(q.v.)*

tree of Australia. Both species belong to the silk-cotton family Bombacaceae. Their leaves are palmately divided and they bear white flowers and capsular fruits.

adaubi. See *adobe soil.*

adaxial crack stem. See *brown check of celery.*

adder. 1. The harmless North American milk snake, *Lampropeltis triangulum,* so called because it is sometimes found in dairy buildings where it goes in search of mice rather than milk. 2. The harmless North American puffing adder, *Heterodan contortix;* also called *blowing adder; flat-headed adder; hognose snake; spreading adder; sand viper.* 3. The common venomous European snake, *Vipera berus.* 4. Any species of the venomous, European snake family Viperidae. 5. The Australasian venomous death adder, *Acanthophis antarctius.* 6. The Indian banded adder or cobra-like krait, *Bunganis coeruleus.* 7. The African venomous puff adder, *Bitus* or *Clotho arietans.* 8. The venomous Near-Eastern horned adder, *Cerastes cornutes.* 9. The venomous Indian asp or serpent, *Naje Haje.* 10. The sea adder, *Spinachia spinachia,* or pipefish of the genus Nerophis, family Syngnathidae.

adder's-tongue. 1. *Ophioglossum vulgatum,* a popularly cultivated fern, of the family Ophioglossaceae. 2. See *fawnlily.*

adder's violet. See *downy rattlesnake plantain.*

additive. Anything added to a diet or reaction mixture to increase its efficiency or to give it more desirable characteristics.

addled egg. An egg in which the yolk has become mixed with the white; sometimes a rotten egg, hence considered inedible. See *white rot: 2.*

adhesions. The adherence of separate tissues or organs of the body to each other as the result of an inflammation.

adhesive agent. A material that acts as a cementing agent between two solids. In spraying, it causes the dispersed material to cling to leaves and fruit.

adipose. Body fat.

adipose tissue. Body tissue in which fat is stored.

adjustable gate. A device which is used for varying the size of an opening, as in a seed drill, in an irrigation ditch, or in a "gated" irrigation pipe.

adjuvant. An additive which intensifies or modifies the action of a chemical or food. Carbon tetrachoride is often used as an *adjuvant* in insecticides.

adobe soil. Any clayey soil of the arid and semiarid regions of western U. S. Soils included under the term may vary widely in different localities. Also called *dobe; dobie; dobby; doby; adaubi; adabe; dogie.*

adobe lily. See *pink fritillary.*

adobe tick. See *fowl tick.*

Adoretus. A beetle genus, family Scarabaeidae, which attacks shrub roots. Found in Hawaii, and in Japan as *Adoretus tenuimaculatus.*

adrenal cortex. The outer part of the adrenal glands.

adrenal glands. A pair of ductless glands, each located on or adjacent to a kidney, which secrete adrenalin and cortin, whence cortisone; suprarenal glands.

adrenocorticotrophic hormone. See *ACTH.*

adsorbate bacterian. A suspension of killed bacteria which is adsorbed on an aluminum hydroxide gel.

adsorption. The adhesion of the molecules of a liquid or gas to another surface. Used in flavor control in the milk industry.

adsorption complex. The inorganic and organic colloids in a soil.

adulterate. To lower the quality of a product by mixing in another substance, as to *adulterate* milk by adding water to it. The Federal Government and most states legislate against the adulteration of food and drugs.

adustiosis. An external, physiological breakdown of lemon after picking, characterized by a reddish brown staining. Also called *red blotch.*

advance and final settlement plan. One method of handling products through farmers' cooperatives. The farmer receives an advance of money less than the market price for the products which are graded and sold by the cooperative. The money received by the cooperative less expenses and advances, is given to the farmer.

adventitious. Out of place; not in the normal, expected place. Arising from the secondary meristem. See *aerial root.*

adverse adaptation. Protection against an adverse factor of environment through functional or structural characteristics.

adz(e). A hatchet-like tool for dressing timber. Its bow-shaped blade is set at right angles to the handle.

adz hoe. In forestry, a fire-trenching tool which resembles a carpenter's adz, but which has a broader blade and lighter head; its tapered, rectangular eye is designed for a standard ship-carpenter's (double-curved) adz handle.

adzing and boring. 1. A method for conditioning unseasoned wood for preservative treatment. 2. The shaping of railroad ties to give a smooth bearing surface and to prevent loosening of spikes. This is often done before preservatives are forced into the wood.

adzuki bean. *Phaseolus angularis,* family Leguminosae, a bean, little known or grown in America, that is cultivated in North China and Japan.

aegagropila. See *hair ball.*

Aegilops. A genus of grasses, of the family Gramineae, native to the Mediterranean countries; related to wheat (*Triticum*). It has become a weed in the U. S.

aeolian (soil). Soil carried and deposited by wind.

aerate. 1. To force a thin layer of cooled air over milk to remove odors. 2. To cause air to pass throughout or around a substance.

aerial base map system. A system of mapping terrain by use of aerial photographs. Physical and cultural features of an area are shown to scale.

aerial bulblet. A bulb capable of producing a

plant, found in the leaf axils or in the place of flowers on certain plants, as in the tiger lily and onion. Also called *bulblet; brood bud; bulbil; bulbel.*

aerial drainage. See *air drainage.*

aerial dusting. Spreading insecticides, fungicides, herbicides in the form of powder from an airplane. Commonly called *crop dusting.*

aerial fertilization. The broadcast distribution of fertilizers from airplanes.

aerial potato. A small tuber produced aboveground in the leaf axil of the potato.

aerial propagation. The production of roots on a cutting, under moss or the like, without a covering of sand or soil. See *air layering.*

aerial root. 1. The root of an *epiphyte* (*q.v.*), or air plant, which anchors the plant and, in some cases, functions in photosynthesis. 2. A root growing from the stem which is exposed to the air. Also called *adventitious root.*

aerial seeding. Broadcast seeding from aircraft, especially used in areas such as rice lands.

Aerobacter. A genus of aerobic, Gram-negative bacteria, normally inhabiting the intestines of men and animals. It produces acid and gas from sugars; e.g., *Aerobacter aerogenes* produce lactic acid but also other acids which puncture the curd and cause difficulty in the manufacture of cheese.

aerobe. Bacteria or other organisms which live only in free oxygen.

aerobiology. The study of microorganism viruses carried in the air.

aerosol. Finely atomized spray or smoke with particles ranging between 0.1 and 50 microns in size. The particles are produced by blasts of heated air, or exhaust gases, or rapid volatilization of a liquefied gas containing a nonvolatile chemical solution. *Aerosols* include insecticides, antibiotics, germicides, and deodorants.

aerosol bomb. A cylinder containing finely atomized spray or smoke. See *aerosol.*

aerosol generator. A machine which breaks down an oil solution into tiny droplets by blasts of hot air. See *aerosol.*

aetiology. See *etiology.*

affinin. An insecticidal agent obtained from the roots of *Heliopsis longipes;* of the same order of toxicity and paralytic action as the pyrethrum powder *q.v.*

afforestation. Developing or re-developing a forest.

Afghan erysimum. *Erysimum perofskianum,* native to Central Asia, a hardy, annual herb of the family Cruciferae; favored for its orange or yellow flowers.

A-frame. A structure resembling the capital letter A. Used as a derrick and for opening or building irrigation ditches.

African agapanthus. *Agapanthus africanus,* a south African, tuberous, perennial herb, of the family Liliaceae, grown in tubs or large pots for its showy flowers. Also called *lily-of-the-Nile.*

African arctotis. *Arctotis stoechadifolia,* a South African, annual herb, of the family Compositae, grown as a flower-garden plant blossoming from June until frost. Also called *African daisy.*

African cane. See *pearl millet.*

African caper. *Capparis aphylla,* family Capparidaceae, a shrub bearing edible berries. Found in the desert regions of North Africa.

African daisy. See *African arctotis.*

African geese. A breed of domesticated fowl. The gander weighs about 20 pounds. It has a large knob on its forehead, a dewlap under the throat, and a dark stripe down the back. The body is light-brown turning almost white at the fluff.

African glanders. See *mycotic lymphangitis.*

African hair palm. See *Mediterraneanpalm.*

African juniper. *Juniperus procera,* an evergreen, timber tree, of the family Pinaceae, of the mountains of East Africa; sometimes grown as an ornamental. Also called *East African cedar.*

African marigold. See *Aztec marigold.*

African millet. See *pearl millet.*

African oil palm. *Elaeis guineensis,* a palm, of the family Palmaceae, important for the African oil trade, but an ornament in southern U.S.

African rose. See *corn poppy.*

African rue. See *harmel peganum.*

African valerian. *Fedia cornucopia,* an annual herb, of the family Valerianaceae, of the Mediterranean region, grown as a salad crop, but also for its red flowers.

African violet. *Saintpaulia ionantha,* an herbaceous plant, of the family Gesneriaceae, native to Africa. Grown outside the tropics as a house plant or in greenhouses. The flowers are usually deep violet-blue; in some varieties they are streaked with white or are entirely white. Not related to the true violet. See *Kew African violet.*

afterbirth. The matter which surrounds and attaches the fetus to the uterus. It is expelled after the fetus.

aftercrop. A second crop in one year on one piece of ground. See *acres of crops.*

afterfeed. Grass, used for feed, which grows in a field after harvest.

aftergrass. See *afterfeed.*

aftergrowth. The second growth of a crop within a single year.

afterings. The *strippings* (*q.v.*); the last milk drawn from a cow's udder at each milking. Also called *after milk.*

aftermath. The second mowing of the grass of a hayfield or meadow; or the growth produced after a cutting.

afternoon lady. See *four-o'clock.*

afterpains. The pains which accompany the expulsion of the afterbirth.

after-ripening. The biochemical and physical changes occurring in tubers, seeds, fruits, bulbs after harvest; necessary for germination.

afterswarm. The second or later swarm of bees to leave the hive.

afterwale. The padded underside of a draft horse-collar.

aftosa fever. See *foot and mouth disease.*

agalactia. A failure to secrete milk following the birth of offspring.

agarita. See *laredo mahonia.*

agathea. See *blue felicia.*

agave. 1. Any plant of the genus Agave, of the family Amaryllidaceae, including sisal agave, henequen, centuryplant agave, etc. 2. The fiber from such plants.

age class. Trees in a stand or forest which fall within stated age limits. The classes are usually divided into 20-year periods (5 years in coppice). Trees falling between the ages of 1 and 20 years would be referred to as Class I. Those falling between the ages of 21 and 40 would be referred to as Class II, etc. A normal distribution comprises all age classes to permit a given annual or periodic felling.

aged. 1. Stored to improve taste or texture. 2. A show term, designating dairy cattle calved prior to July 1 of third preceding year; or designating swine farrowed prior to August 1 of second preceding year; or designating sheep born prior to January 1 of previous year (or September 1 of second preceding year); or designating horses foaled prior to January 1 of fourth preceding year.

aged cheese. A cheese which has been properly stored to become mellow and to attain its proper texture and flavor.

aged flavor. An off-taste in butter resulting from improper storage.

aged herd. A show term designating one boar and three sows.

aged horse. In America, a horse over eight years old; in Australia, over six years.

ager. 1. A field; land generally; a portion of land enclosed by definite boundaries. 2. In old English law, an acre (obsolete).

ageratum. 1. A genus of tropical herbs, of the family Compositae. 2. See *mistflower eupatorium.*

agglutination. 1. A joining, as by adhesion. 2. The gathering together of the fat globules in milk or of bacteria suspended in immune sera. 3. The sticking together, of bacteria, other organisms, or cells suspended in a fluid by antibodies called agglutinins.

agglutination-lysis test. The clumping and dissolving of certain organisms when mixed with serum containing specific antibodies.

agglutinins. 1. Antibodies produced in the animal body in response to an infection or to the injection of microorganisms. They cause agglutination of the organisms responsible for their formation. 2. Any antibody capable of clumping the organism which stimulated its production in the animal body.

aggradation. Building up to a uniformity of grade or slope by the addition of material; especially the depositing of sediment in the beds of streams, and on the floors of ponds, lakes, etc.

aggregate. A single mass, or cluster, of separate soil particles; variable in shape and size, and degree of coherence, as granules, nuciform aggregates, clods, prisms. See *soil aggregate.*

aggregate fruit. The product of the fusion of several ovaries in the same flower: *e.g.,* blackberry, raspberry, strawberry, magnolia. See *collective fruit.*

aghanni. In Hindustan, the main and second of three annual rice crops.

agistment. 1. The rental of lands for pasture. 2. The price, rent, which is paid for pasturage. 3. The edible forage of a forest, and the rights to it (British).

agitating discontinuous pressure sterilizer. A sterilizer used primarily in the canning of milk. Cans of milk are placed in a cage and revolved to keep the milk in motion during sterilization to prevent scorching. See *continuous agitating cooker.*

agitating pressure cooker. Canning equipment used to increase heat penetration; consisting of a large tank in which hot water is kept in motion around the cans.

agitator. Revolving paddles which keep a liquid, powder, or gas in motion to maintain a proper mixture within a tank. Used in dairying, spraying, fertilizer drill, etc.

agoraphobia. A fear of open places; especially that occurring in animals kept in stalls for too long a period. Among horses, also called *picket-line-bound; barn rat.*

agostadero. A pasture used only in the summer (southwestern U.S.).

agrarian. 1. Pertaining to agriculture. 2. Pertaining to political action, or movements, for the benefit of farmers. See *rural; urban.*

agrarianism. The theory of equal, or equitable, distribution of productive land through redistribution; the establishment by law of favorable conditions for better use of land.

agrestal (agrestial). Uncultivated; growing wild.

agrestic. Rural, as opposed to urban.

agricolist. A farmer.

agricolous. Pertaining to a farmer.

agricoltor. A farmer.

Agricultural Adjustment Act (A.A.A.). This act was passed by Congress in 1933 to provide economic relief for farmers by the federal government's making benefit payments to farmers for limiting production (it being assumed that farm prices would then rise), and making rescue loans for several mortgages, as well as marketing loans and providing credit to cooperatives. The basic provisions of the A.A.A. are still in effect.

agricultural chemistry. That branch of the science of chemistry which is concerned with the composition and transformation of the plants and animals on which the economy of the farmer rests.

agricultural college. An educational institution devoted to study, research, and the dissemination of knowledge in agronomy, horticulture, animal husbandry, agricultural economics, etc. The U.S.D.A. works in conjunction with most *agricultural colleges* (sometimes called a *school* or *college* of agriculture within a university).

agricultural commodity. A general term; any product of agriculture.

agricultural economics. 1. The branch of economics concerned with farm management and production. 2. A department in a college or university.

agricultural education. 1. The general, formal knowledge of agriculture. 2. The course of study (in a college or university or a department of government) to prepare and assist teachers of agriculture in the secondary schools. See *vocational agricultural education.*

agricultural engineering. A course of study in the design, construction, and use of agricultural implements and buildings; soil and water management; rural use of electricity; and processing of agricultural products.

agricultural experiment station. See *Hatch Act.*

agricultural geology. The study of soils, and to a lesser extent, the study of the origin of mineral fertilizers.

agricultural gypsum. See *calcium sulfate.*

agricultural land. All the land devoted to raising crops and livestock, including farmstead, roadways, drainage and irrigation ditches, ponds, water supply, cropland, and grazing land of every kind. (The term is not strictly synonymous with *land in farms, cropland, pasture land, land suitable for crops or land suitable for farming.*)

agricultural lime (stone). See *calcium carbonate; lime: 1; marl: 1.*

agricultural planning. Policies proposed by governmental agencies, such as removal of land from cultivation, reclamation, resettlement, adjustment in crop acreages, reforestation, setting up river valley authorities, etc.

agricultural region. A classification of land according to its predominant feature, such as the crop most commonly grown or the type of farming employed.

agricultural sulfur. A coarsely ground mineral which increases the acidity of soil or corrects sulfur deficiency. Also called *brimstone; flowers of sulfur.* See *sulfur.*

agriculture. The production and processing of plants and animals for man's use and enjoyment.

agriculturist. In the broadest sense, a farmer; also ancillaries such as teachers of agriculture, farm editors, researchers, etc.

agrobiology. The science of plant nutrition and growth; crop production and soil management.

agroecology. The study of the relation of agricultural crops and environment.

agrogeology. See *agricultural geology.*

agrology. The science of soils. Also called *pedology; edaphology.*

agronomy. The theory of crop production and soil management; management of farm lands.

Agropyron mosaic. A virus disease of wheatgrass (*Agropyron* spp.) characterized by leaf mottling and striping, necrosis, and stunting.

agrostography. Delineation of grasses.

agrostology. The study of grasses.

agrotechny. The conversion on the farm of agricultural commodities into usable products.

A harrow. A spike-tooth harrow, shaped like the capital letter *A*, which is dragged from a hitch at the apex.

A horizon. The leached upper member of a soil profile; the eluvial layer. It may be further subdivided as *A_1 horizon,* which contains a relatively high amount of organic matter and *A_2 horizon* which has had more leaching and contains less organic material. See *soil profile.*

A_0 horizon. A top part of a soil profile; a layer composed mainly of partially decomposed organic matter. See *humus: 1; soil profile.*

A_{00} horizon. The topmost layer of undecomposed organic matter on soil. See *litter: 1; soil profile.*

aigues mortes. Standing water in an abandoned river channel.

Ailanthus. A tree genus, of the family Simaroubaceae, grown as street trees because of their smoke-resisting foliage; the pinnated leaves are food for silkworms. It has yellowish-green leaves and red showy fruit. Native to the East Indies, China, and southern Europe and now a pest in California. Also called *Moluccas; tree of heaven.*

ailanthus webworm. *Atteva aurea* (Fitch), family Hyponomeutidae; a caterpillar which feeds with others within a web on tree of heaven ailanthus.

air blast. 1. A rapidly moving stream of air blown by a fan. Used on threshing machines, combines, seed cleaners, corn shellers, etc., to clean grain, seeds, etc. 2. The circulation of air at high velocity in storage, freezing, pre-cooling rooms and compartments. 3. High velocity air used to spray trees and field crops.

air-borne. Carried, distributed, by air and winds. See *wind-borne.*

air bubble. An abnormality formed within the white of the egg, usually as a result of a crack in the shell.

air cell. Usually the large end of an egg which is filled with air resulting from shrinkage of its contents. See *shell membranes.*

air chamber. Part of a sprayer or other hydraulic apparatus, in which air, under pressure, reduces the pulsation or shock caused by the pump or pressure regulator.

air cure. A method of drying and curing tobacco either in the open or in a shed with natural air. It is often used to prevent pole rot *q.v.* See *flue cure.*

air drainage. The flow of cold air down a slope due to gravity rather than to wind.

air-dry. To dry or dehumidify forage, hay, wood, etc., by means of natural air movement. It is often performed in the open.

air-drying. The seasoning of wood out-of-doors or in an unheated shed.

air dry soil. A soil that has lost its free moisture by evaporation or transpiration; and which therefore contains only hygroscopic moisture and chemically combined water.

air hose. A tube used to transmit air under pressure *e.g.,* a tube of a milking machine which transmits the suction from the vacuum pump to the teat cup.

air layering. A method of propagation by rooting

branches of woody plants. An incision is made in the stem to be rooted and sphagnum moss is wrapped around the wounded stem. The wounded area is moistened and covered tightly with an oil paper, cheesecloth, wire screen, or plastic film, and left until sufficient roots have established themselves in the sphagnum moss. Then the branch is cut away from the main stem and transplanted. Also called *Chinese layerage.*

air-line cleaner. Used in a gin to clean the cotton as it passes through large pipes where it is beaten or agitated causing the dirt and foreign matter to drop. (Being replaced by the *unit cleaner q.v.*)

air pipe. The tube, about one foot in diameter, connected to the fan in a gin, which draws cotton from a truck or wagon into the gin. See *cotton pipe.*

airplane duster. A device which is attached to an airplane to expel and to distribute insecticides, fungicides, herbicides.

airplane plant. See *plantain; bracted plantain.*

airplane seeding. See *aerial seeding.*

airplant kalanchoe. *Kalanchoe pinnata,* an Asiatic, perennial herb, of the family Crassulaceae, grown in hothouses, as a pot plant and in desert gardens for its fragrant, pink flowers.

air pocket. Air or gas interrupting the flow of liquids or solids in a pipe.

airpotato yam. *Dioscorea bulbifera,* an East Indian, tender, climbing, herbaceous vine, of the family Dioscoreaceae, grown for ornament and for its 8 to 12 inch long edible tubers, which are formed in the axils of the leaves.

air-sac mite. *Cytodites* (*Cytoleichus*) *nudus,* Order Acarina, a mite that invades the lungs, windpipe, air sacs, and air-containing cavities of the bone in fowls. Also called *pulmonary mite.*

air sacs. 1. In plants, air cavities in the stems. 2. In animals, the terminal air cells of the bronchioles.

air seasoning. See *air-drying.*

air under the skin. Subcutaneous emphysema; a condition in which the skin is distended by air. Caused in fowls by leakage from the lungs or by the rupture of an air sac. It can occur in other animals following an operation, an injury, or in an infection, as in *blackleg.* Also called *windpuff.*

air vesicle. 1. An air cell in the lungs. 2. An inflated air cavity sometimes found in aquatic plants to maintain buoyancy.

air-way super. A frame placed on top of the hive proper for storage of honey. There is ventilation along all four sides between the sections and the sides of the super. Also called *ventilated T-super.*

aitchbone. 1. The bone of the rump; the floor of the pelvis. 2. The retail cut of meat which contains the aitchbone. Also called *edge bone.*

ajowan. Fruit of the Eurasian plant, *Carum copticum,* family Ammiaceae, which is valued as a seasoning and for medicinal properties. Antiseptic ajowan oil is made from it. Also called *Javanese seed; ajava.*

akaakai. A European bulrush, *Scirpus lacustris,* family Cyperaceae, found in Hawaii, where baskets and mats are made from it. Native to Europe.

akala. *Rubus macraei,* family Rosaceae, a shrub grown in the Hawaiian Islands for its large, red, edible fruits.

akaroa. The ribbon tree (*q.v.*) of New Zealand.

akeake. See *clammy hopseedbush.*

akebi. *Akebia quinata,* family Lardizabalaceae, a woody vine cultivated for its oily seeds and used in basketmaking. Native to Asia.

akepiro. *Olearia furfuracea,* family Carduaceae, an ornamental tree grown in New Zealand.

alamo. The cottonwood tree or poplar (southwest U. S.).

alang grass. *Imperata cylindrica,* a grass used in paper-making and thatching roofs. Grown in southern Europe.

alang-ilang de China. See *fragrant tail-grape.*

alang-ilang song-song. See *fragrant tail-grape.*

Alaska cedar. See *Nootka falsecypress.*

Alaska cypress. The yellow cedar, *q.v.*

Alaska fringecup. *Tellima grandiflora,* a hairy, hardy, perennial herb, of the family Saxifragaceae, native to western North America. It has greenish flowers that become red or pink, one variety has red leaves and yellow flowers. Also called *false alumroot.*

Alaska goldthread. *Coptis trifolia,* a low, perennial herb, of the family Ranunculaceae, native to Alaska. Sometimes confused with *C. groenlandica,* the *common goldthread q.v.*

Alaska sable. The fur of black skunk or raccoon, which resembles sable.

Alaska spruce beetle. *Dendroctonus borealis* Hopk., family Scolytidae, a beetle which attacks the white and Englemann spruce in Alaska and northwestern Canada. It bores into the inner bark, and its larvae bore into the wood of the tree.

alatern. See *Italian buckthorn.*

albedo browning. An abnormal condition of stored citrus fruits in which the white, spongy portion of the peel becomes discolored. Caused by storage at low temperatures and poor ventilation.

albino. 1. A man or animal with pink eyes, white skin and hair caused by a congenital deficiency of pigment in the skin. 2. In botany, plants with a chlorophyll deficiency.

albino cherry. A virus disease of sweet cherry found in the Rogue River Valley, Oregon. The disease spreads rapidly and kills trees: the foliage turns bronze-green; the fruit turns white and fails to mature.

Albizzia. 1. A genus of trees belonging to the family Mimosaceae. These mimosa trees have twice-pinnated leaves and bear clusters of pink and white flowers. Native to southern Europe. 2. Silktree albizzia, *q.v.*

albopruinose. A powdery, white bloom which sometimes covers grapes and plums.

albumen. The white of egg.

albumin. Proteins consisting primarily of amino acids. Albumin contains about two percent sulfur;

it is soluble in water and coagulated by heat; it is found in milk, blood, egg white, muscle, and vegetables.

alburnum. The part of a tree trunk between the bark and the wood, consisting of the cambium and the phloem cells. The alburnum of some species of trees: *e.g.*, the *Pinus ponderosa* of the Northwest, was eaten by American Indians during times of famine. See *sapwood.*

Alcaligenes. A bacterial genus of the family Bacteriaceae found in human and animal intestines. They may not cause disease but *Alcaligenes viscosum* does make milk glutinous or ropy.

Alcock spruce. *Picea bicolor,* a hardy, pyramidal, Japanese evergreen tree, of the family Pinaceae. An ornamental.

alcohol-alizarin test. A test in which alcohol and alizarin are added to milk. A color reaction of lilac-red indicates normal milk, yellow-brown indicates sour milk, and violet indicates mastitis milk.

alcoholic fermentation. The transformation of simple hexose sugars, especially glucose, into alcohol and carbon dioxide. Useful as a method of preservation.

alcoholic slime flux. A white, frothy seepage from the bark or sapwood near the base of a tree trunk. Rich in starches, sugars and proteins. Over long periods, it is fatal to the tree. See *slime flux; brown slime flux.*

alcohol test. A test in which equal parts of milk and ethyl alcohol are mixed to discover abnormal milk or milk with unusual salt balances. Normal milk breaks clear of the test tube, while the curd of abnormal milk clings to the glass. The test also detects milk which is apt to coagulate during sterilization of condensed milk.

alder. Any tree or shrub of the genus *Alnus,* of the family Betulaceae, some species are called tag alders. See *European alder; hazel alder; speckled alder.*

alder bark beetle. *Alniphagus aspericollis* (Lec.), family Scolytidae. It infests alders in western U.S.

alder blight aphid. See *woolly alder aphid.*

alder buckthorn. *Rhamnus frangula,* a shrub cultivated for its lustrous green leaves which turn bright yellow in autumn.

alder catkin disease. *Taphrina robinsonian* in eastern North America. *Taphrina occidentalis* in western North America. It affects the bracts of the *Alnus* spp., causing enlargement and projection from the cat-like tongues. It also affects the leaves.

alder flea beetle. *Altica ambiens* (Lec.), family Chrysomelidae. It feeds on the leaves of alder, willow, poplar trees.

alder leaf blister. *Taphrina tosquinetii* (West) Tul. A fungous disease of alder leaves.

alder leaf miner. See *European alder leaf miner.*

Alderney. A breed of cattle in effect identical to the Guernsey from the Channel Island of Alderney.

alder witches' broom. A European fungous disease, *Taphrina epiphylla,* family Taphrinaceae. It causes profuse branching.

aldrin. A white crystalline, chlorinated hydrocarbon insecticide.

Alectryon. A New Zealand tree belonging to the soapberry family, Sapindaceae, which is grown for ornament.

alehoof. See *ground ivy.*

Alemtejo. A soft, ripened cheese made from ewe's milk, sometimes with goat's milk added.

Aleppo gall. A hard, spherical swelling of the tissues of Aleppo oak (*Quercus infectoria*), and Portuguese oak (*Q. lusitanica*), caused by an insect *Cynips tinctoria,* family Cynipidae; used in the Middle East as a source of an astringent, tonic, and antidote; also used as a source of tannic and gallic acids for dye for leathers, and wools, and for special inks. Also called *gallnut; nutgall.* See *oak gall.*

Aleppo grass. See *Johnsongrass.*

Aleppo pine. *Pinus halepensis,* of the family Pinaceae, native to the Mediterranean region. It is a tall ornamental evergreen.

alerce. Incensecedar, genus Libocedrus, family Pinaceae. A compact cone-shaped tree.

aleurone. Protein grains found in the endosperm of ripe seeds.

aleurone layer. In wheat and maize, the outer differentiated layer of cells of the endosperm.

Alexander. A plant, *Smyrnium olusatrum,* of the family Umbelliferae, eaten by man until about the end of the seventeenth century, when it was replaced, in Italy, by celery.

Alexandrine rat. See *roof rat.*

alfaje. *Trichilia tuberculata,* family Meliaceae, a timber tree found in Panama.

alfalfa. *Medicago sativa,* a perennial, leguminous herb, of the family Leguminosae, a native of central Asia. Used as forage, green manure, and hay. Also called *lucerne; purple medic; purple alfalfa.* See *Turkestan alfalfa; variegated alfalfa; yellow-flowered alfalfa.*

alfalfa caterpillar. *Colias philodice eurytheme* Bdv., family Pieridae, an insect pest in the southwestern U.S., which feeds principally on leaves.

alfalfa dehydrator. A large drying oven, usually of the rotating cylinder type, used in preparing alfalfa meal.

alfalfa dodder. See *bigseed alfalfa dodder.*

alfalfa drill. A drill (*q.v.*), having furrow openers about four inches apart, used for sowing small seeds, such as alfalfa, clover, redtop, bluegrass, rape, etc.

alfalfa dwarf. A virus disease of alfalfa characterized by dwarfing of the plant.

alfalfa gall midge. *Asphondylia websteri* Felt, family Itonididae, an insect pest of alfalfa in Arizona and New Mexico which attacks the floral head, causing the seed pods to become swollen.

alfalfa leaf meal. A stock feed consisting chiefly of ground, leafy materials separated from alfalfa hay or meal. Reasonably free from other crop plants and weeds, and does not contain more than 18 percent of crude fiber.

alfalfa leaf spot. Caused by *Pseudopeziza medicaginis* (Lib.) Sacc., family Mollisiaceae, a world-wide, fungous disease of minor importance in most areas. It causes defoliation. Also called *common leafspot of alfalfa.* See *alfalfa yellow leaf blotch.*

alfalfa looper. *Autographa californica* (Speyer), family Noctuidae, the caterpillar of which feeds on alfalfa in western Canada and along the U. S. Pacific coast.

alfalfa meal. A stock feed of ground alfalfa hay. Reasonably free from other crop plants and weeds, and does not contain more than 33 percent of crude fiber.

alfalfa-molasses feed. A stock feed of alfalfa meal and cane or beet molasses, in which the molasses makes up from 20 to 40 percent of the mixture. Used as replacement to supply proteins and vitamins in the rations of dairy cows, beef cattle, and sheep.

alfalfa mosaic. A virus disease characterized by spotting or mottling of foliage.

alfalfa root rot. *Rhizoctonia crocorum,* the imperfect stage of the more rarely found *Helicobasidium purpureum,* family Auriculariaceae, a fungous, killing disease. The affected plants become yellow, wilt, and the top and roots die.

alfalfa rust. *Uromyces striatus* var. *medicaginis,* family Pucciniaceae, a leaf disease of alfalfa which causes rust fungus, premature shedding.

alfalfa seed chalcid. See *clover seed chalcid.*

alfalfa sifter. A machine containing a sieve, which removes extraneous matter from alfalfa meal.

alfalfa snout beetle. *Brachyrhinus ligustici* (L.), family Curculionidae, a flightless beetle, ½ inch long. The adults feed on foliage, the larvae feed on roots.

alfalfa stem meal. A ground stock feed reasonably free from other plants and weeds.

alfalfa tea. Made from alfalfa leaves; credited with medicinal value and possibly a slight tonic effect.

alfalfa teeth. Detachable, narrow points for a spring-tooth harrow.

alfalfa treehopper. See *three-cornered alfalfa hopper.*

alfalfa webworm. *Loxostege commixtalis* (Wlk.), family Pyraustidae; the insect covers the leaves of a plant with a light web under which the larvae feed.

alfalfa weevil. *Hypera postica* (Gyll), family Curculionidae, a common insect pest in western and many eastern states of U.S. It destroys first cuttings and delays the subsequent cuttings.

alfalfa white spot. Spotting of leaves because of dry weather followed by heavy rains.

alfalfa wilt. *Corynebacterium insidiosum,* family Bacteriaceae. Diseased plants are stunted and have excessively weak stems; the leaves curl up, turn yellow, and die; the taproot has a wet, yellow zone under the bark. Causes serious damage in U.S. Other organisms, such as *Fusarium* spp. and *Cylindrocarpon* spp. also cause wilt.

alfalfa worm. See *fall armyworm.*

alfalfa yellow leaf blotch. *Pseudopeziza jonesi,* (*Pyrenopeziza medicaginis*) family Mollisiaceae. A world-wide, fungous disease which causes defoliation.

alfalfa yellows. A boron deficiency disease characterized by pale leaflets or bronzed leaves with green midribs, or, in some cases, a stopping of terminal growth. The margins of the leaves may dry up and shrivel. Also called *alfalfa mosaic.*

alfilaria. See *alfileria.*

alfileria. *Erodium cicutarium,* an annual herb, of the family Geraniaceae, native to Southern Europe, an important forage plant in California and southwest U.S. Also called *storksbill; pin grass; filaree; alfilaria, alfillerilla; red-stem filaree.*

alfillerilla. See *alfileria.*

algarroba. 1. Carob, *q.v.* 2. Mesquite, *q.v.* 3. West Indian locust, *q.v.* 4. Rain tree, *q.v.*

alga spot. A disease of citrus trees caused by the alga *Cephaleuros virescens* Kunze, characterized by spots on the leaves and twigs, which may be greenish-grey and velvety or covered with reddish-brown, hair-like structures. The bark may become swollen.

Algerian iris. *Iris unguicularis,* a perennial herb, of the family Iridaceae, a beardless iris. It is grown in the southern U.S. for its early, lilac and white flowers.

Algerian ivy. *Hedera canariensis,* family Araliaceae, an evergreen, climbing, woody vine from the Canary Islands; grown in warm areas as a climber on brick or masonry walls.

algerita (algireta, alderita). See *laredo mahonia.*

algology. Study of seaweeds.

alhenna. See *henna.*

Alice's fern. See *Hartford fern.*

alienated land. Passing of title to property by the owner as distinguished from transfer by law, as in the case of inheritance of property.

alizarin-alcohol test. See *alcohol-alizarin test.*

alkali. 1. A chemical compound of oxygen and hydrogen with one element, such as sodium, potassium, calcium, magnesium, or the ammonium radical, capable of neutralizing acids. 2. A general term denoting salts of sodium, calcium, potassium, and magnesium which injure plant growth. 3. A general term for caustic soda, sodium hydroxide, caustic potash, sodium carbonate, etc., all cleansing agents in food processing plants.

alkali chlorosis. A yellowing or blanching of the plant leaves caused by excessive alkali.

alkali disease. Selenium poisoning (*q.v.*) caused by grazing on plants containing excessive amounts of selenium. Characterized by emaciation, loss of hair (mane and tail among horses), deformed hoofs, and blind staggers.

alkali dropseed. See *alkali sacaton.*

alkali dust. An irritant: particles of alkali from the surface crust, carried by wind in arid regions.

alkali flat. Usually a depression, as a dry lake bed, with a strongly alkalied surface. Found in the arid regions of western U.S.

alkali heath. *Frankenia grandifolia,* family Frankeniaceae, a low-growing shrub bearing clusters of pink flowers. Found in California.

alkali heliotrope. See *salt heliotrope.*

alkali mallow. See *alkali sida.*

alkali muhly. *Muhlenbergia asperifolia,* family Gramineae, a native grass of northern and western U.S., sometimes an undesirable weed pest in grazing lands. (Frequently referred to as *Sporobolus asperifolius.*) Also called *fine-top salt grass; scratchgrass.*

alkaline. Containing alkali; having a pH greater than 7.

alkaline detergent. A cleansing agent that yields a distinct alkaline reaction in water solution. Trisodium phosphate (Na_3PO_4), washing soda (Na_2CO_3), and borax ($Na_2B_4O_7 \cdot 10H_2O$) are examples.

alkaline flavor. A serious off-flavor in butter from using too much neutralizer in the sour cream.

alkaline goods. A fertilizer of superphosphate and potash salts.

alkaline soil. A soil that gives an alkaline reaction, precisely a reading above pH 7.0, but in practice readings as high as pH 7.3 may be considered neutral. In humid regions alkalinity most commonly comes from calcium and magnesium carbonates; in drier regions nitrates and salts of sodium and potassium as well as calcium and magnesium may be the source. See *alkali soil; acid soil; pH.*

alkali plant. Plants which grow principally or more luxuriantly on alkali soils.

alkali sacaton. *Sporobolus airoides,* a native grass to U.S., of the family Gramineae, used in reseeding grazing lands. Also called *alkali dropseed; tussock grass.*

alkali sida. *Sida hederacea,* of the family Malvaceae, a weed pest.

alkali soil. 1. A soil containing alkali salts, usually sodium carbonate (with a pH value of 8.5 and higher). The term frequently includes both *alkali* and saline soil as here defined. *White alkali* is applied to saline soil in some localities, and *black alkali* to *alkali soils.* 2. Soils with harmful concentrations of mineral salts. In general, *black alkali* consists of sodium carbonate. Water containing it dissolves humus, leaving a black residue. Generally *white alkali* consists of sodium sulfate. It is less injurious to plants than the black alkali. See *alkaline soil.*

allaeanthus. *Allaeanthus zeylanithus,* family Moraceae, grown for fiber, from which ropes are made. Native to Ceylon.

allantiasis. See *botulism.*

allée (alley). A purely ornamental, highly developed garden walk maintained primarily for its beauty.

Allegheny (Allegany, Allegeny). In the U.S., the humid transition zone stretching from the New England states to eastern North and South Dakota, including most of Pennsylvania and mountainous Georgia.

Allegheny barberry. *Berberis canadensis,* a shrub, family Berberidaceae, with leaves grayish color on the underside and red berries. An ornamental, it is native to Virginia and Georgia and westward to Missouri. Also called *American barberry; Canada barberry.*

Allegheny blackberry. *Rubus allegheniensis,* an erect blackberry, of the family Rosaceae, common in the northern U.S. The source of many cultivated varieties of blackberries.

Allegheny foamflower. *Tiarella cordifolia,* a perennial plant, family Saxifragaceae, native to eastern U.S., grown for its clusters of white flowers. Also called *false miterwort; foamflower.*

Allegheny menziesia. *Menziesia pilosa,* a hardy shrub, family Ericaceae, native to the U.S., common in rock gardens. Also called *minniebush.*

Allegheny mound ant. *Formica exsectoides* Forel, a mound-building ant, family Formicidae. Common in the northeastern U.S. It kills trees, shrubs, etc., within a radius of 20 to 30 feet, that threaten to shade its mound.

Allegheny pachysandra. *Pachysandra procumbens,* an evergreen (deciduous in the north), perennial, North American herb, family Buxaceae, grown for foliage and white or purple flowers. Also called *Allegheny spurge.*

Allegheny spurge. See *Allegheny pachysandra.*

Allegheny-vine. See *mountainfringe.*

allel(e). 1. The alternative forms of genes having the same place in homologous chromosomes which influence the development of alternative traits or characters. 2. A pair of Mendelian genes at the same locus of a pair of homologous chromosomes. Also called *allelomorph.*

allelomorph. See *allel.*

allergy. An exaggerated susceptibility to a substance harmless to most members of the same species.

allethrin. A synthetic insecticide similar to pyrethrins.

Allgäuer Rundkäse. A type of Swiss cheese.

all-heart. Lumber which is wholly composed of heartwood.

allicin. An antibiotic derived from garlic.

alligator. 1. In logging, a boat for transporting timber, which is equipped with a windlass and cable for portaging. 2. See *crotch: 2.*

alligator head. See *rough buttonweed.*

alligator pear. See *American avocado.*

alligator tree. See *American sweetgum.*

alligator wampee. See *pickerelweed.*

allite. Soils characterized by a silica-alumina ratio less than 2 to 1 in the hydrochloric acid extract. A soil from which silica has been removed during formation, and in whose clay fractions aluminum is in the form of Al_2O_3, as in bauxite and laterite.

all mash. Poultry feed composed of ground grains and soybean meal, meat scraps, dried skimmed milk, alfalfa meal, salt, limestone, and fish oil or vitamins or other fortifying materials. A whole and complete ration.

allodial land. Land held in absolute ownership; fee simple, as opposed to feudal land. Land in New York State is by statute allodial; also in Louisiana.

allogamy. Cross-fertilization.

allopolyploid. A polyploid containing genetically different chromosome sets: for example, from two or more species.

allosomes. Chromosomes distinguished by peculiarities of behavior or sometimes by a difference in size or shape.

allosyndesis. The pairing of chromosomes in different basic sets, particularly in allopolyploids. See *autosyndesis.*

allowable burned area. For any area, the maximum acceptable acreage of forest burned in one year. Used as a standard to determine success of fire-control methods.

allowable range use. See *range utilization standards.*

allspice. *Pimenta officinalis,* a tropical American tree, of the family Myrtaceae, so named because its spice resembles the flavor of a mixture of cinnamon, nutmeg, and cloves. Grown in the U.S. only in Florida, and there as an ornament. Also called *pimento allspice.*

alluvial cone. A delta-form feature composed of soil and rock detritus, deposited as storm wash at the mouths of streams. Similar to *alluvial fan,* except that the angle of slope is greater.

alluvial fan. Soil and rock fragments moved by swiftly flowing water and deposited at the mouth of a stream in the form of fans or deltas. Alluvial fans are most common in arid regions, at the base of mountain slopes. Several fans may coalesce to form terraces and piedmont plains. The agricultural value of the soils varies.

alluvial soil. Soil which consists of, or is derived from, alluvium; generally regarded as highly fertile.

alluvio maris. Soil formed by the washing-up of the sea; sand or any other material adjoining land near the sea.

alluvion. Land added by accretion, as that built up by deposition from sea, lake, or river, or that left by recession of water or by the natural shifting of the channel of a river. It always belongs to the owner of the land to which it is an accretion. The word is often erroneously used as the equivalent of alluvium.

alluvium. Matter transported and deposited by streams; usually composed of mixed inorganic (mineral) and organic particles. The word most commonly refers to the recent deposits on the flood plains; more inclusively it embraces deposits on *second bottoms* or river valley terraces, with the detritus spread out as fans. See *alluvial fan.*

all-wire excluder. A wood frame with round wire and metal crossbars forming a screen to prohibit the queen bee from entering or leaving a section of a hive.

almendron. 1. The Brazilnut *(q.v.)* tree. 2. Malabar almond tree *q.v.*

almond. *Prunus amygdalus (Amygdalus communis),* an Asiatic, deciduous, small-nut tree, of the family Rosaceae, cultivated for its popular edible kernel nut, the almond of commerce.

almond bud failure. A minor disease of unknown cause in which many almond buds fail to develop. Also called *shatter bud, crazy top, mule tail.*

almond calico. A virus disease, characterized by chlorotic, mottled, or necrotic patterns in the leaves.

almond cherry. *Prunus glandulosa,* an Asiatic, deciduous shrub, of the family Rosaceae, grown for its clusters of white or pinkish-white flowers. Also called *flowering almond.*

almond moth. *Ephestia cautella* (Wlk.), family Phycitidae, a serious insect pest of dried fruits, dried vegetables, and grain products. Also called *fig moth.*

alodium. Unrestricted ownership of land and now synonymous with *fee simple* in the United Kingdom and U.S.

aloe. See *true aloe.*

aloewood. *Sebesten sebestena,* family Boraginaceae, a tree bearing scarlet flowers. Native to the West Indies.

aloe yucca. See *yucca.*

alpaca. 1. *Lama pacos,* a llama, domesticated in Peru and adjoining countries. 2. The long wool of the alpaca.

alpestrine. Denoting plants growing on mountains below the timber line.

Alpin. A soft, ripened cheese of the Mont d'Or *(q.v.)* type.

alpine. Growing and flourishing in high, mountainous areas.

alpine aster. *Aster alpinus,* a Eurasian, and Rocky Mountain, perennial herb, of the family Compositae, grown in rock gardens for its blue or violet flowers. Also called *rock aster.*

alpine azalea. *Loiseleuria procumbens,* a prostrate shrub of the family Ericaceae, used as a mat-forming evergreen in rock gardens of northern Canada and New England. Also called *trailing azalea.*

alpine beardtongue. *Pentstemon alpinus,* family Scrophulariaceae, a perennial herb having bluish-purple flowers which grow in terminal clusters. Found on the Rocky Mountains.

alpine catchfly. See *alpine silene.*

alpine columbine. *Aquilegia alpina,* a perennial, European herb, of the family Ranunculaceae, grown for its large blue flowers.

alpine currant. *Ribes alpinum,* an ornamental shrub, of the family Saxifragaceae, native to Europe. Its fruits are insipid to the taste. Also called *mountain currant.*

alpine fescue. *Festuca ovina* var. *brachyphylla,* family Gramineae, a variety of sheep fescue, found from the Arctic to California, in the Rocky Mountains to New Mexico, and in the higher mountains of New England and New

York. A good forage plant, but smaller than *sheep fescue q.v.*

alpine fir. *Abies lasiocarpa,* an ornamental, pyramid shaped evergreen tree, of the family Pinaceae, of western America. Also called *Arizona fir; mountain balsam; Rocky Mountain fir.*

alpine forget-me-not. *Eritrichium howardi,* a perennial herb having blue flowers and yellowish foliage. Found on the Rocky Mountains.

alpine hemlock. *Tsuga mertensiana,* family Pinaceae, a tall tree valued for its timber. Found from Alaska to Montana, Idaho and California.

alpine laurel. See *bog kalmia.*

alpine meadow grassland. Summer pasturage (meadows or parks) near or above timberline.

alpine meadowrue. *Thalictrum alpinum,* a low, perennial, flowering plant, of the family Ranunculaceae, native to Eurasia and North America, grown in rock gardens for its greenish flowers with yellow stamens. Also called *arctic meadowrue.*

alpine meadow soil. Intrazonal dark-colored soils near and above the timberline. Also called *mountain meadow soil.*

alpine poppy. *Papaver alpinum,* a hardy, European, perennial herb, of the family Papaveraceae, grown for its fragrant, white or yellow flowers.

alpine savory. *Satureia alpina,* a perennial, southern European herb, of the family Labiatae, used for flavoring and grown for its aromatic foliage and its purple flowers.

alpine silene. *Silene alpestris,* a tender, European, spreading, perennial herb, of the family Caryophyllaceae, grown in rock gardens for its small, white flowers. Also called *alpine catchfly.*

alpine timothy. *Phleum alpinum,* family Gramineae, a smaller grass than cultivated *timothy,* common in alpine meadows from Alaska to northern New England and northern Michigan. Good forage.

alpine wallflower. See *flaxleaf erysimum.*

alsike clover. *Trifolium hybridum,* a common, domesticated, Eurasian hay and pasture legume, of the family Leguminosae, with pinkish-white blossoms. Sprawling when mature, resembling common red and white clovers. Also called *Swedish clover; bastard clover.*

alsike clover rust. A disease caused by *Uromyces trifolii* var. *hybridi,* family Pucciniaceae, characterized by pale-brown pustules, and a torn epidermis on the lower surfaces of leaves and sometimes on the petioles and stems.

alta fescue. See *meadow fescue.*

altai columbine. *Aquilegia glandulosa,* a Siberian, perennial herb, of the family Ranunculaceae, grown for its lilac-blue flowers.

Altenburger. A cheese made from goat's milk.

alter. 1. To castrate. Also called *cut; geld; emasculate; spay.* 2. To prune. 3. To remove the comb from a beehive.

Alternaria. A common saprophytic fungus of the family Dematiaceae; causes early blight and rot.

Alternaria blight. See *early blight of potato.*

Alternaria blight of carrot. A fungous disease caused by *Alternaria dauci*; characterized by irregular brown spots with yellow centers and with yellow halos near the margins of the leaves. The leaflet tip or leaflet may turn brown.

Alternaria core rot. A fungous disease of apples in storage caused by *Alternaria tenuis,* characterized by a dark growth over the seed and carpel walls and often by rotting of the flesh.

Alternaria leafspot. 1. Any leafspot caused by any species of *Alternaria.* 2. A fungous disease of carnations caused by *Alternaria dianthi,* characterized by a white spot centered with a black or purplish spot. The growing tips often bend to one side, and leaves seem to be stuck together. When the axil of the leaf becomes yellow, it dries up.

Alternaria mold. A mold of butter caused by a species of *Alternaria,* in which dark, smoky, sometimes greenish areas appear on the surface.

Alternaria rot of lemon. A fungous disease caused by *Alternaria citri,* characterized by softening of the fruit at the stem end and progressing to the core. It principally infects fruit of low vitality which is exposed to low temperatures, sunburning, dry winds, internal decline, or membranous stain. Also called *center rot.*

Alternaria rot of orange. See *black rot.*

Alternaria spot. A disease of citrus trees caused by certain species of *Alternaria,* characterized by spots, pits, and rot on mature fruit. The spots change in color from light brown to dark brown to black. The fungus sometimes may penetrate the skin, spread, and even reach the center of the fruit.

alternate bearing. The bearing of heavier and lighter crops of flowers or fruits, or both, in successive years. See *annual bearing; biennial bearing.*

alternate grazing. Changing pastures so that the forage grows back before it is grazed again. Also called *rotational grazing.*

alternate host. A plant or animal upon which a disease organism exists for only a part of its life cycle.

alternate husbandry. Rotation of planting grass and crops.

alternation of generations. Reproduction in which common characteristics are only found in every second generation; one generation reproduces sexually and the next asexually.

alum. 1. Potassium aluminum sulfate, used to acidify alkaline soil in greenhouse culture. 2. Aluminum sulfate *q.v.*

alumina. Aluminum trioxide (Al_2O_3), an impurity in phosphate rock which causes superphosphate to be wet and sticky.

aluminum sulfate. Alum used in tanning leather, sizing paper, and purifying water, and to acidify greenhouse soils. Also called *alum; pearl alum; pickle alum.*

alumroot. Any plant of the genus Heuchera, of the family Saxifragaceae, especially the *American alumroot* and *coralbells q.v.*

alveloz. Sap from the plant *Euphorbia heterodoxa,* family Euphorbiaceae, which is used by natives in Brazil to treat epithelial cancer.

alveolus (plural **alveoli**). A small cavity or sac: (1) A tiny, thin-walled air sac of the lung, in the walls of which gases are exchanged between the air and the blood; (2) An acinus or terminal lobule of a racemose gland; (3) One of the honeycomb pits in the wall of the stomach; (4) A tooth socket; (5) A cell of the honeycomb; (6) The tiny, balloon-like structure of the udder lined with epithelial cells which are the active principal in the secretion of milk; (7) A pore or pitted perithecium in a fungus; (8) A pit of a carduaceous plant.

alyceclover. *Alysicarpus vaginalis,* an Asiatic, tropical, annual herb, of the family Leguminosae, grown for forage, hay, and green manure.

alyssum. Any plant of the genus Alyssum, of the family Cruciferae. Used mainly for bedding and rock gardens. See *goldentuft alyssum; sweet alyssum.*

amaranth feathers. See *fountain humea.*

amaranth pigweed. See *lambsquarters goosefoot.*

amarelle (cherry). A pomological *sour cherry* (*q.v.*) species. The fruit is light colored and yields colorless juice. Common varieties are *Montmorency, Early Richmond,* and *Dyehouse.*

amaryllis. The genus Amaryllis, or closely related genera, of the family Amaryllidaceae, some of which are raised as flowering pot plants; some are raised outdoors in frost free areas. The most commonly known *amaryllis* is the *Barbadoslily* *q.v.* See *belladonnalily.* Also called *equestrian star.*

amasesis. Incapable of masticating.

amateur pears. Small pear fruit, unattractive and unsuitable for market, but of fine flavor.

amatungula. See *Natalplum carissa.*

amaurosis. Total loss of vision without any discernible lesion in the eye structures or optic nerve. Also called *glass-eye; gutta-serena.*

Amazonlily eucharis. *Eucharis grandiflora,* a bulbous, South American herb, of the family Amaryllidaceae, with star-like white flowers. Used in corsages and as greenhouse and pot plants.

Amazonvine. See *fringed Amazonvine; Argentine Amazonvine.*

ambarella. See *mombin.*

amber. See *St. John's-wort: 2.*

amber durum. A grade of *durum wheat* (*q.v.*) which has 60 to 75 percent hard vitreous kernels. See *hard amber durum.*

amber seed. See *musk mallow: 2.*

Ambert. A semi-soft, ripened cheese made from cow's milk.

ambler. A horse whose best gait is pacing; a pacer.

amblotic. Likely to cause abortion.

Amboina pine. *Agathis alba,* a tall tree whose resin is used for making varnish. Native to the Philippines and the Moluccas.

ambosexous. Bisexual.

ambrosia. See *Jerusalemoak goosefoot.*

ambury. See *anbury.*

amelanchier. See *serviceberry.*

amelcorn. See *emmer.*

ameliorant. Soil enricher.

ameliorated wine. See *modified wine.*

amendment. Soil enricher.

American alder. See *hazel alder.*

American alkanet. See *bloodroot.*

American allspice. See *strawberry shrub: 1.*

American aloe. See *centuryplant agave.*

American alumroot. 1. *Heuchera americana,* a perennial herb, family Saxifragaceae, grown in the shade for its greenish-white flower. Also called *American sanicle; alumroot.* 2. See *spotted geranium.*

American arborvitae. See *northern white cedar; eastern arborvitae.*

American arrowwood. *Viburnum dentatum,* family Caprifoliaceae, a tall hardy shrub bearing white clusters of flowers and bluish-black fruit. Native to North America.

American avocado. *Persea americana* is native to tropical America and includes Mexican (*Persea americana drymifolia*), Australian, Guatemalan, and South African avocados. A pulpy, green or purple, pear-shaped fruit, used in salads. Also called *avocado; alligator pear; ahuacate; aguacate; avocat; avocato; abacate.*

American balm of Gilead. *Protium carana,* family Burseraceae, the tree from which caranna gum is gathered for its supposed medicinal properties. Native to South America.

American barberry. See *Allegheny barberry.*

American basswood. See *basswood.*

American Beauty. This rose is the floral symbol of the District of Columbia. It belongs to the perpetual hybrid group of the family Rosaceae.

American beautyberry. *Callicarpa americana,* a four to five foot shrub, of the family Verbenaceae, grown for its showy, violet berries.

American beech. *Fagus grandifolia,* a hardy, deciduous tree, of the family Fagaceae. Cultivated ornamentally and used commercially for hardwood distillation, crossties, pulp, slack-cooperage, veneer, etc.

American bittersweet. *Celastrus scandens,* a woody vine, of the family Celastraceae, which produces orange berries that remain on the vine throughout the winter; poisonous to livestock. Used as a winter decoration indoors. Found from eastern Canada to South Dakota and from North Carolina to New Mexico. Also called *false bittersweet; waxwork; stafftree; climbing bittersweet.*

American bladdernut. *Staphylea trifolia,* a North American, hardy, deciduous shrub, of the family Staphyleaceae, grown for its white or greenish-white flowers and large, inflated, fruit pods.

American broomrape. Squawroot: 1 *q.v.*

American cedar. Northern white cedar *q.v.*

American cheese. A term applied to all Cheddar cheeses, the most common type made in the U.S.

American cherry. *Prunus serotina.* The wilted foliage of this wild black cherry tree is poisonous. Also called *chokecherry.*

American cherry-laurel. See *Carolina laurel-cherry.*

American chestnut. *Castanea dentata,* a tree of the family Fagaceae, now nearly exterminated by chestnut blight; found from Maine to Alabama, west to Michigan. It bears a sweet, edible nut, and its wood is used in making tannin, semichemical pulp, poles, fence posts, crossties, slack cooperage, furniture, etc. Also called *chestnut.*

American chinaroot. *Smilax pseudochina,* family Smilacaceae, a herbaceous vine with tuberous roots which bears greenish-yellow flowers.

American cinnamon. *Nectandra cinnamomoides,* family Lauraceae, a South American forest tree which bears greenish-yellow flowers and aromatic foliage.

American cockroach. *Periplaneta americana* (L.), family Blattidae, a house-dwelling insect, active at night, feeding on book bindings, paper, garbage, food. Also inhabits greenhouses where they feed on seedlings and flowers, and girdle plants.

American Code of Botanical Nomenclature. The Latin nomenclature given to plants in U.S. from 1905 to 1930, when it was superseded by the adoption of the revised International Code of Botanical Nomenclature (*q.v.*) at Cambridge, England. The *American Code* stipulated that plants be known by their oldest published name, beginning with Linnaeus' *Species Plantanum* (1753); it rejected the idea of nomen conservandum (*q.v.*) and Latin diagnoses.

American colombo. A perennial herb, *Frasera carolinensis,* which bears greenish-white flowers and whose root has medicinal properties. Found on the Pacific coast of U.S.

American columbine. *Aquilegia canadensis,* a North American, perennial herb, of the family Ranunculaceae, grown for its yellow flowers with red spurs.

American cotton. See *sea-island cotton.*

American crab. See *wild sweet crabapple.*

American cranberrybush viburnum. *Viburnum trilobum,* a North American, deciduous shrub, of the family Caprifoliaceae, grown for its large, white or pinkish flower clusters. Its red berries remain on the bush during the winter. Also called *cranberry bush.*

American dagger moth. *Acronicta americana* Harr., one of the largest of the family Noctuidae, whose larvae feed on maple, elm etc.

American devilsclub. *Oplopanax horridus,* a very prickly shrub, of the family Araliaceae, native to northwest America and Japan, grown for its bright green leaves and scarlet flowers. Also native to Isle Royale in Lake Superior. Also called *fatsia.*

American dog tick. *Dermacentor variabilis* (Say), family Ixodidae, a blood-sucking acarid. It carries Rocky Mountain spotted fever, tularaemia, bovine anaplasmosis. Also called *wood tick.*

American dog violet. *Viola conspersa,* a perennial, flowering herb, of the family Violaceae, native to eastern North America, grown for its pale violet to white flowers. Also called *dog violet.*

American dogwood. See *flowering dogwood.*

American dwarf birch. *Betula glandulosa,* family Betulaceae, a North American shrub which bears short catkins and circular leaves.

American earthnut. See *peanut.*

American ebony. Granadilla tree *q.v.*

American elder. *Sambucus canadensis,* a shrub, of the family Caprifoliaceae, grown for its flowers and fruit, which are used in wine and jelly. The leaves, opening buds, and young shoots are lethal to cattle or sheep. Also called *elderberry; sweet elder.*

American elm. *Ulmus americana,* a beautiful, long-lived deciduous tree of the family Ulmaceae, a popular street tree and used in shelter belts. Used in slack-cooperage, crates, veneer, furniture, and wooden ware. Also called *white elm; water elm.*

American false elm. *Celtis occidentalis,* family Ulmaceae, a tree found in eastern North America which bears edible drupes.

American falsehellebore. *Veratrum viride,* a stout, perennial, flowering herb, of the family Liliaceae, native to eastern U.S., grown for its yellowish-green flowers. Also called *falsehellebore; earth gall; Indian poke; American hellebore; American white hellebore.*

American false manna. Exudation from sugarpine trees.

American false nettle. *Boehmeria cylindrica,* family Urticaceae, a stingless herb bearing white flowers. Found in eastern North America.

American Farm Bureau. An entirely voluntary organization founded in 1919 to protect the economic interests of farmers through legislation and which, in conjunction with the U.S. Extension Services, fosters agricultural education. The Bureau is organized on three levels: the national bureau is comprised of independent state bureaus, each of which is comprised of independent county bureaus.

American fir. In the timber trade, an American tree belonging to the genera Abies or Picea.

American foulbrood. An infectious disease of the honey bee larvae, caused by *Bacillus larvae;* characterized by sunken, discolored, perforated brood cappings. If widespread, the hive ceases to function. *Cf. European foulbrood.*

American galipot. Resin from the eastern white pine *Pinus strobus.*

American germander. *Teucrium canadense,* a hardy, perennial herb, of the family Labiatae, sometimes grown for its purple flowers.

American ginseng. *Panax quinquefolium,* a hardy, perennial herb, of the family Araliaceae, native to America, grown for its thickened, aromatic roots, the source of commercial ginseng.

American gooseberry. 1. See *gooseberry.* 2. Barbados-gooseberry *q.v.*

American gooseberry mildew. See *gooseberry mildew.*

American grades of wool. Twelve classifications established by the U.S.D.A. They are designated by numbers, according to the diameter of wool fibers. *Cf. Australian grades of wool.*

American grasshopper. *Schistocera americana* (Drury), family Acrididae, an insect pest in the southeastern U.S.

American ground laurel. Trailing arbutus *q.v.*

American hellebore. See *American falsehellebore.*

American holly. *Ilex opaca,* an evergreen tree, of the family Aquifoliaceae. A popular Christmas decoration. Used in inlay wood, engravings, scrollwork, carvings, scales, rules, piano keys and other wooden ware. Also called *white holly.*

American hophornbeam. *Ostrya virginana,* a North American, deciduous tree, of the family Betulaceae, grown for its fruiting clusters. Also called *ironwood.*

American hornbeam. *Carpinus caroliniana,* a native, small, deciduous tree, of the family Betulaceae, cultivated for ornament. Slow growing, but hardy, in shady and protected places. Used in handles, vehicle parts, and fuel wood. Also called *bluebeech; ironwood; water beech.*

American ipecac. See *Indian physic.*

American ironwood. 1. See *American hophornbeam; American hornbeam.* 2. Brumelia *q.v.*

American ivy. See *Virginia creeper.*

American Jack. A breed of ass (*Equus asinus*) developed largely from the Catalonian, black in color with white points, weighing about 1300 pounds, standing about 15 hands high. Also called *Mammoth Jack.*

American joy. Virginia creeper *q.v.*

American Judas-tree. Eastern redbud *q.v.*

American jute. See *chingma abutilon.*

American Landrace. An American breed of swine, established in 1950, largely from Danish, Norwegian, and Swedish Landrace hogs. The development met the need for a meat-type hog. Mature boars weigh up to 900 pounds, sows up to 750 pounds. The breed is solid white in color with long, drooping ears. See *Landrace.*

American larch. See *tamarack; eastern larch.*

American larkspur. *Delphinium exaltatum,* family Ranunculaceae, a perennial herb which bears showy purplish-blue flowers on tall stalks, and is poisonous to livestock. Found in the cooler parts of U.S.

American laurel. Mountain laurel *q.v.*

American leaf spot. A serious fungous disease of coffee in the Western Hemisphere caused by *Mycena citricolor* (*Omphalia flavida*), family Agaricaceae, characterized by a small, dark spot on the leaf, in the center of which is a yellow fruiting body. The lesion grows into a round, grayish spot, on which are produced small, yellow mushrooms with yellow, hair-like stalks. Results in excessive defoliation and fruit drop or even in the death of the tree.

American leg. A retail cut of leg of lamb for roasting; the shank is removed at the stifle joint; the flesh is folded back under the vellum and fastened with a skewer.

American licorice. See *wild licorice: 1.*

American linden. See *basswood.*

American live oak. See *live oak.*

American lotus. *Nelumbo lutea,* a hardy aquatic herb, of the family Nymphaeaceae. Grown in shallow pools for its lovely, large, pale yellow flowers; its tubers and seeds were eaten by the American Indians. Also called *water chinquapin.*

American lungwort. See *Virginia bluebells.*

American mandrake. Mayapple *q.v.*

American martagon. Turk's-cap lily *q.v.*

American mistletoe. Any epiphytic plant of the genus Phoradendron, especially the *Christmas American mistletoe q.v.*

American mountain ash. *Sorbus americana,* a short-lived deciduous tree or shrub, of the family Rosaceae, grown for its foliage and red, berry-like fruits. Found from Newfoundland to North Carolina and west to Michigan and Manitoba. Also called *American rowan tree; dogberry; fowlers-service; Indian mozemize; quickbean.*

American mountain holly. See *mountain holly.*

American mountain laurel. See *mountain laurel.*

American mountain tea. Checkerberry *q.v.*

American mulberry. See *red mulberry.*

American nightshade. Pokeberry *q.v.*

American persimmon. See *persimmon.*

American planetree. *Platanus occidentalis,* a deciduous tree, of the family Platanaceae, native from Maine and Ontario south to Florida and Texas. One of the tallest of the deciduous American trees; used for shade and street trees. It will not stand city cultivation as well as the *London planetree q.v.* Used in veneer, slack-cooperage, boxes, crates, furniture, millwork, laundry appliances. Also called *buttonwood; buttonball;* known in lumber trade as the *American sycamore.*

American plum. *Prunus americana,* a deciduous tree, of the family Rosaceae, a most prolific source of native plum varieties. Found east of the Rocky Mountains, it is used as an understock in the colder sections and is important for crossing with the Japanese plum (*P. salicina*). Its fruit is red or sometimes yellow. Also called *wild plum.*

American plum borer. *Euzophera semifuneralis* (Wlk.), family Phycitidae, an insect which feeds on the cambium tissue of plum trees, entering through injuries, knot holes, or crotches.

American potatobean. *Apios americana,* a tuberous-rooted vine, family Leguminosae, grown for its brownish flowers. Its roots, once food for the Indians, have sometimes been mistaken for the potato. Also called *ground nut; potatobean; wild bean; Dakota potato.*

American powdery mildew. See *gooseberry mildew.*

American quarter horse. A type of horse developed for work; popular as a cow horse in the southwest of the U.S. It weighs 1100 pounds or more and has a short back and good feet. It is active, quick, and given to sudden bursts of speed; popular in quarter-mile racing.

American redbud. *Cercis canadensis,* family Leguminosae, a small or medium sized tree native from Connecticut to Florida and west to Wisconsin and southwestward. Grown for its beautiful red-purple clusters of flowers, which appear before the leaves do. Also called *Judas tree.* See *Cercis.*

American rowan tree. American mountain ash *q.v.*

American saddle horse. A light breed of horse developed from the Narragansett pacers, the Canadian pacers, the Morgans, and Thoroughbred breeds. Popular for pleasure riding and in horse shows because of its beauty, style, high action, and ability to perform 3 to 5 gaits. It is from 14 to 16 hands in height, from 900 to 1200 pounds in weight, and of solid colors, usually chestnut.

American sanicle. See *American alumroot.*

American sarsaparilla. Wild sarsaparilla *q.v.*

American scab. See *corky scab of potato.*

American senna. See *wild senna.*

American silver fir. Balsam fir *q.v.*

American smoketree. *Cotinus americanus,* a small, deciduous tree, family Anacardiaceae, cultivated for its plumy, fruiting clusters which suggest smoke. Also called *chittamwood; Aaron's-beard.*

American spikenard. *Aralia racemosa,* a native, perennial herb, family Araliaceae, grown in shrub borders or as a specimen plant because of its beautiful foliage and brownish-purple flowers. Also called *spikenard.*

American Standardbred. See *Standardbred.*

American sweetgum. *Liquidambar styraciflua,* a hardy deciduous tree, family Hamamelidaceae, grown for ornament for its handsome, scarlet, autumnal foliage. The red colored heartwood is called *American satin walnut.* Used in veneer, slack-cooperage, pulp, furniture, planing-mill products, turned articles. Also called *red gum; sweetgum; white gum.*

American sycamore. See *American planetree.*

American tarbush. *Flourensia cernua,* an aromatic, poisonous, bitter shrub, family Compositae, common on the dry plains from western Texas and Arizona into Mexico. Sheep and goats that graze on the ripe fruits suffer inflammation, ulceration, and perforation of the gastrointestinal tract. Acute inflammation of the abomasum can bring death within a day and a half to five days. Also called *black brush; tarbush.*

American Thoroughbred. A branch of the *Thoroughbred (q.v.)* horse, which differs from the English Thoroughbred only in that in 1913 the English restricted the Thoroughbreds to descendants of certain mares in England.

American Trotter. See *Standardbred.*

American twinflower. *Linnaea borealis* var. *americana,* an acid-loving, woody, evergreen, trailing plant, family Caprifoliaceae, grown in North America for its fragrant, rose-pink or white flowers. Also called *deer vine; twin sisters.*

American type Ayrshire. Red or brownish-red *Ayrshire (q.v.)* cattle with only a little white on the body. See *Scotch type.*

American type cheese. Descriptive names given to *Cheddar, Colby, Washed-curd, Soaked-curd, Monterey (q.q.v.)* cheeses.

American umbrellaleaf. *Diphylleia cymosa,* a perennial herb, family Berberidaceae, grown for its foliage and white flowers. Native from Virginia to Georgia.

American vegetable wax. See *northern bayberry; southern bayberry.*

American waterlily. *Nymphaea odorata,* an aquatic, herbaceous plant, family Nymphaeaceae, grown for its large, white, fragrant flowers. Also called *white waterlily; toad lily.*

American waterplantain. *Alisma triviale,* an aquatic or marsh herb, family Alismataceae, grown along the water for its small, white flowers. (This name is confused with the European *A. plantago-aquatica* in old manuals.)

American wayfaring tree. See *hobblebush viburnum.*

American white cedar. Whitecedar falsecypress *q.v.*

American white hellebore. See *American false-hellebore.*

American white pine. Eastern white pine *q.v.*

American wood anemone. *Anemone quinquefolia,* a perennial herb, family Ranunculaceae, grown for its white, solitary flowers which bloom in early spring. Native to northeastern U.S. Also called *wood anemone.*

American wormseed oil. See *wormseed oil.*

American yellowwood. *Cladrastis lutea,* a deciduous tree, family Leguminosae, grown for its showy flowers. Native from Indiana to Georgia. Also called *Kentucky yellowwood; gopher-wood; yellow ash; Virgilia.*

American yew. See *Canada yew.*

amino acids. Chemical compounds containing nitrogen, which compose the ultimate units of proteins. Those important in nutrition are glycine, alanine, valine, leucine, isoleucine, phenylalanine, tyrosine, tryptophan, serine, threonine, aspartic acid, glutamic acid, lysine, histidine, arginine, cystine, cysteine, methionine, proline, hydroxyproline, iodogorgoic acid, thyroxine, hydroxyglutamic acid, hydroxylysine. Twenty-three are known to be protein "building stones." An "essential amino acid" is one which the body cannot synthesize.

aminosis. An abnormal condition of the body caused by excessive ingestion of amino acids.

amitosis. Cell division without formation and splitting of chromosomes. *Cf. mitosis.*

amli. Tamarind *q.v.*

amlong. *Pothos rumphii* or *Raphidophora mervilii,* vines which resemble the vanilla and are used in basketmaking in the Philippine Islands.

ammate. See *ammonium sulfamate.*

ammonia. A colorless gas (NH_3) soluble in water. 1. Liquid ammonia is used in fertilizer plants. 2. Sometimes added to livestock feed to increase nutritious value. 3. Used as a refrigerant. 4. A product of the decomposition of urine.

ammonia fixation. Adsorption of ammonium ions

by soils or minerals in such form that they are neither water soluble nor readily exchangeable.

ammonia liquor. An aqueous solution containing ammonium hydroxide, formed when ammonia is absorbed in water. Used, among other things, to supply nitrogen in fertilizers. See *ammonia: 1.*

ammonification. The formation of ammonia compounds in the soil. See ***nitrification.***

ammonium chloride. NH_4Cl, the ammonium salt of hydrochloric acid. Similar in fertilizer properties to ammonium sulfate, but more expensive and may acidify the soil more. Also called *sal-ammoniac; muriate of ammonia.*

ammonium dinitro-o-sec-butylphenate. A water-soluble herbicide used to kill broad-leaved weeds. Not effective on *purslane, lambsquarters,* weeds with smooth, waxy leaves.

ammonium nitrate. An ammonium salt with a nitrate radical, excellent as a source of nitrogen for fertilizers. Made in large quantities from the air nitrogen and contains 30 to 35 percent nitrogen. Dangerous to work with in the pure state, due to its explosive nature.

ammonium phosphate. See *monoammonium phosphate.*

ammonium polysulfide. An insecticide which contains ammonium hydroxide, hydrogen sulfide, sulfur. Soluble in water and effective for control of the red spider mite.

ammonium sulfamate. $(NH_4)_2O\ SO_2NH_2$, used as a powder or dissolved in water as a spray for control of undesirable broad-leaved plants.

ammonium sulfate-nitrate. A double salt of ammonium sulfate and ammonium nitrate, which is a synthetic, inorganic, nitrogen carrier containing about 26 percent nitrogen.

ammophilous. Denoting plants or insects in sand.

amoeba disease. A bee disease of minor importance in the U.S. caused by *Vahlkampfia (Malpighamoeba) mellificae* which affects the Malpighian tubules, part of the excretory organs. This organism belongs to Phylum Protozoa.

amoebiasis. A disease caused by single-celled, or protozoan, parasites belonging to the amoeba group, Phylum Protozoa. Also called *amebiasis; amebiosis.*

amoebic. Of, or pertaining to, protozoan organisms known as amoebae, Phylum Protozoa, or the disease condition or lesions produced by parasitic species. Also called *amebic; ameba.*

amole. 1. *Chlorogalum pomeridianum,* a California soap plant. 2. Mexican species of the families Manfreda and Prochnyanthes; and *Yucca gloriosa.*

amole soapplant. *Chlorogalum pomeridianum,* a bulbous herb, family Liliaceae. It is an important bee plant in California; once used by the Indians to make soap.

amoor corktree. See *Amur corktree.*

amorpha. Any shrub of the genus Amorpha, family Leguminosae. Also called *false indigo.* See *indigobush amorpha; leadplant amorpha.*

amorphophyte. Any plant which bears irregular flowers.

amorphous. A plant of indeterminate shape.

amorphus. A heartless, limbless, and headless fetus.

amortization. A specified plan to repay a loan in a specified period of time.

amotus. Denoting the hind toe of certain birds, which does not touch the ground.

amourette. Quaking grass *q.v.*

ampalaya. Balsam apple *q.v.*

amper. A blotch or swelling.

amphegen. A thallophyte.

ampherotoky. Production of both male and female from a parthenogenic ovum, seed, or spore.

amphibious. Capable of living in water as well as on land.

amphibryous. In botany, even growth over the entire periphery.

amphicarpogenous. Fruit, such as the peanut, which buries itself below ground before maturity.

amphicarpous. Bearing two varieties of fruit which may mature at the same time.

amphichrome. A plant of two colors.

amphidiploid. A plant which possesses the somatic chromosome of two species. In an amphidiploid the two species are definitely known, whereas in an allopolyploid they are not.

amphigean. In botany, the blooming of flowers from the rootstalk.

amphimixis. Reproduction through cross-fertilization, a normal fecundation process. The union of two germ plasms in sexual reproduction (syngamy).

amphipolyploid. A polyploid which arises through the addition of the complete somatic chromosome sets of two or more known species. See *amphidiploid.*

amphispermous. A plant, such as grass, in which the pericarp tightly encloses the seed.

amphivorous. Omnivorous *q.v.*

amphoteric. Capable of reacting as either an acid or a base, as *casein.*

amuguis. *Koordersiodendron pinnatum,* a tree grown for its reddish timber which resists water and ants. Native to New Guinea, the Philippines, Celebes.

amulla. *Myoporum debile,* family Myoporaceae, a shrub bearing an edible fruit which is similar to a berry. Native to Australia.

Amur ampelopsis. *Ampelopsis brevipedunculata,* a woody, vigorous vine, family Vitaceae, grown to cover walls, fences, etc. Produces an inedible fruit. Native to China.

Amur corktree. *Phellodendron amurense,* a picturesque, rapid-growing deciduous tree, family Rutaceae, used for lawn specimens.

Amur honeysuckle. *Lonicera maacki,* an Asiatic, hardy, tall, deciduous shrub, family Caprifoliaceae, grown for its white flowers and red fruit. One of the best honeysuckles.

amylase. An enzyme (present in milk, etc.) which breaks down starch and similar substances into sugars.

amylopsin. A pancreatic enzyme which breaks

down starch to sugars; the chief carbohydrase of pancreatic juice.

anabasine. An insecticide, more effective than nicotine for aphid control obtained from *Anabasis aphylla,* family Chenopodiaceae, and also from *Nicotiana glauca,* family Solanaceae. Also called *neonicotine; piperidine.*

anabiosis. Ability of plants, such as sorghums, to remain in a quiescent state when confronted with an unfavorable condition of growth.

anabo. *Abroma fastuosa,* a shrub whose fibers are used for making rope. Native to the Philippines.

anabolism. Building up of nutritive substances into living tissues. See *catabolism; metabolism.*

anacahuite. *Cordia boissieri,* a small tree bearing yellowish flowers and having medicinal properties. Native to Texas and Mexico.

Anacardiaceae. The cashew or sumac family which includes approximately 60 tree and shrub genera of which many are of economic value, *e.g.* mango *q.v.*

anacardium nut. Cashew nut *q.v.*

anacthartic. A medicine or substance used as an emetic.

anaerobe. Organisms (usually bacteria) which live and multiply without free oxygen.

anaerobic spore test. Determines if anaerobic bacteria are in milk by placing milk and melted paraffin in a test tube, heating the tube until the paraffin rises to the surface, and allowing the paraffin to harden. If gas is formed, the paraffin plug will be pushed up, indicating the presence of anaerobic bacteria.

anagap. *Pithecolobium scutiferum,* a hardwood tree found in the Philippines.

anagep. *Terminalia edulis,* a tree grown for its soft timber. Preserves are made from its red fruit. Native to the Philippines.

anagreeta. A term used in the U.S. in Revolutionary or pre-Revolutionary time for corn gathered before maturity and dried in an oven or hot sun.

anahau. *Livistona rotundifolia,* family Arecaceae, a tall palm tree. Golf clubs are made from its wood; thatch from its leaves; bowstrings from its fibers. Native to the Philippines.

anama. *Ehretia elliptica,* a shrub cultivated in Texas and Mexico for its edible fruit. Also called *anaqua.*

anapanapa. *Colubrina asiatica,* family Rhamnaceae. The bark of this tropical shrub serves as soap in the Hawaiian Islands.

anaphase. A stage of nuclear division during which chromosomes pass from the equatorial plate toward opposite poles of the cell (toward the ends of the spindle).

anaphylactic (shock). Designating an exaggerated reaction of an animal or man to a foreign agent. Usually related to the injection of a protein to which the body is highly sensitive.

anaplasm. An abbreviated name for Anaplasma, the generic name of the organism of Protozoa, which causes anaplasmosis.

anaplasmosis. A blood disease of cattle caused by the parasite, *Anaplasma marginale,* Phylum Protozoa, located in the red blood cells. Symptoms are: anemia, milk flow ceases, high fever, almost total collapse; in a day or two the hairless areas become pale-yellow with jaundice; death may occur in two to three days. Also called *yellow teat disease; gall sickness.* In Africa, a much milder form of the disease is caused by *Anaplasma centrale* or *Anaplasma marginale* var. *centrale.*

anbury. 1. See *clubroot of crucifers.* 2. A soft tumor or bloody wart on a horse or ox. Also called *ambury.*

Ancien Imperial. A soft French cheese.

Ancona. A breed of chickens originating in Italy, having black feathers with white tips. Excellent layers of white eggs. Mature cock weighs about 6 pounds, hen about 4½ pounds.

Ancylocladus. Rubber is produced from some species of this genus of the dogbane family (Apocynaceae). Native to Asia and the East Indies.

anda. *Joannesia princeps.* The seeds of this timber tree yield oil. Native to Brazil.

Andalusian. A Spanish breed of horse of medium size, 15 hands or less in height, gray, with well shaped legs, large bones and a good head.

Anderson larkspur. *Delphinium andersonii,* a perennial, poisonous herb, family Ranunculaceae, found in semi-desert conditions in Oregon, and from Nevada to California.

Andes berry. *Rubus glaucus,* family Rosaceae, a prickly shrub bearing raspberrylike fruit. Native to tropical America.

Andes cotton. See *sea-island cotton.*

Andira. One of the tree genera of the family Leguminosae. Some of its species yield valuable timber, *e.g.* the cabbage angelintree *(q.v.),* whose dark wood is used for flooring.

Andrews gentian. *Gentiana andrewsii,* a perennial herb, family Gentianaceae, of North America, grown in bog gardens for its purplish-blue, almost closed flowers. Also called *closed gentian; bottle gentian; dumb foxglove; dumb gentian; blind gentian; cloistered heart.*

androecium. (plural **androecia**). The male organs or stamens of a flower collectively. *Cf. gynoecium.*

androgametophore. A plant having male sexual organs.

androgenesis. Development of offspring with the paternal chromosomes only.

androgynous. In botany, a plant having stamens and pistils.

anemia. A blood condition in which there is a decrease of red blood cells, of hemoglobin, or of both.

anemia in suckling pigs. Low blood pressure in young pigs caused by a nutritional deficiency of iron or copper. Produced by housing the pigs indoors on wood or concrete floors during their first few weeks. Characterized by decrease of vigor, and paleness of the mouth.

anemochore (anemochorous). An ecological term

which denotes a plant distributed by wind.

anemometer. An instrument used to measure speed of wind. Also called *wind gauge.* The Dines anemometer measures pressure, which is translated to velocity.

anemone. Any plant of the genus Anemone, family Ranunculaceae. See *poppy anemone; Japanese anemone; European wood anemone; American wood anemone; snowdrop anemone.*

anemone clematis. *Clematis montana,* a hardy, Asiatic, woody vine, family Ranunculaceae, grown on trellises, fences, etc., for its early-blossoming, slender-stalked, white flowers.

anemone dahlia. A type of *dahlia* (*q.v.*) that has open-centered flowers and one row of ray florets with elongated tubular disk florets. It looks like a pin-cushion.

anemoscope. Any device used to indicate and/or record, wind direction.

anesthesia. Loss of feeling, especially loss of tactile and pain sensations, with or without loss of consciousness.

anestrous period. That time when the female is not in heat; the non-breeding season.

aneuploid. An organism or cell which has a chromosome number other than an exact multiple of the monoploid or basic number; *i.e.* Hyperploids =higher. Hypoploid=lower.

Anezeh. A horse bred by that Arab tribe in the Syrian desert.

angel. A slang term for an inexperienced buyer who bids on an unsound horse.

angel-eyes. 1. Hungarian speedwell *q.v.* 2. Bluets *q.v.*

angelica. Any plant of the genus Angelica, family Umbelliferae. Usually aromatic herbs. They are livestock forage throughout the western U.S. The garden angelica, *A. archangelica,* was formerly blanched and eaten like celery; the tender stalks were cut for candying.

angelica tree. See *devil's-walking-stick.*

angelintree. See *cabbage angelintree.*

angelique. *Dicorynia paraensis.* A timber tree native to South America.

Angelot. A rich Normandy cheese.

angelstears. 1. *Narcissus triandrus,* a European, bulbous herb, family Amaryllidaceae, grown for its cup-like, pure white flowers. 2. See *baby-tears.*

angel's trumpet. See *floripondio datura.*

angel tree. Horse-radish tree *q.v.*

Angers. A quince sometimes used in pear cultivation as a dwarfing stock.

angico. *Piptadenia rigida,* family Mimosaceae, this tree is the source of a brown gum used in tanning leather. Native to South America.

angili wood. The hard wood of an Indian tree, *Artocarpus hirsuta,* used in shipbuilding.

angle berry. An old name for a fleshy excrescence on the bellies of cattle, which hangs in a pendulous fashion. Possibly the same as *anbury q.v.*

angle of pull. The angle between a line of travel and a chain which connects the source of power and the load.

angle the harrow. To change the angle between the gangs of a disc harrow.

angleworm. See *earthworm.*

angleworm cast. The above ground accumulation of soil caused by the underground activity of the worm.

angling adjustment. A device which varies the angles of the gang of a plow.

Angola grass. See *paragrass.*

Angola pea. The edible seed of a shrub supposedly imported from Angola, Africa, and once cultivated in Louisiana, probably the species, *Cajanus cajan,* family Leguminosae. Also called *pigeon pea; pidgeon pea; Oregon pea.*

Angora goat. A breed of domestic goat from Asia Minor; bred especially for mohair.

Angora rabbit. A species whose mohair-like fur is valued. See *rabbit.*

Angoumois grain moth. *Sitotroga cerealella* (Oliv.), family Gelechiidae, the larvae of which feed on stored grain.

Anguilla seed. Seed of sea-island cotton (obsolete).

Anguillulidae. A family of extremely small worms found in the soil or in decomposing matter.

angular leaf spot. An infectious disease of cucumbers caused by the bacterium *Pseudomonas lachrymans,* in which the leaves look water-soaked and later turn brown. See *bacterial blight of cotton; blackfire.*

angular-winged katydid. *Microcentrum retinerve* (Burm.), family Tettigoniidae, a long-horned grasshopper which feeds on many plants.

anguria. See *West Indian gherkin; watermelon.*

Angus. See *Aberdeen Angus.*

anhidrosis. A deficiency of sweat in any animal. Also called *anidrosis.*

anhydrous liquid ammonia. Dry ammonia gas, NH_3, compressed into a liquid form. The fertilizer grade contains over 99 percent of actual ammonia.

anidrosis. See *anhidrosis.*

anil. *Indigofera suffruticosa,* family Leguminosae. This West Indian shrub is a source of indigo dye. Also called *anilla.*

anilau. A shrub, genus Grewia, family Tiliaceae; fibers used in rope.

animalcule. Minute animals such as those belonging to the Protozoa class.

animal dip. See *dip: 1, a; dipping vat.*

animal fat. Fat derived from animals; greases, tallows, lards, etc.

animal husbandry. Agriculture which deals with livestock production, usually limited to meat animals and horses, although it may include poultry and dairy cattle.

animal industry. Agriculture which deals with livestock and the processing of the products derived therefrom.

animal inoculation test. A method of diagnosis whereby tissue or other material from an infected animal is injected into a healthy animal to attempt to reproduce the suspected disease in the healthy animal.

animal kingdom. All animals collectively. In the Kingdom *Animalia* there are 12 branches or *Phyla* ranging from I, *Protozoa* to XII *Chordata* which includes the Class, *Mammalia* in the Subphylum *Vertebrata.*

animal liver and glandular meal. The animal products feedstuff obtained from drying and grinding liver and other glandular tissues. At least 50 percent of the dry weight must be derived from liver and the product must contain at least 18 milligrams of riboflavin per pound. (Different standards are in force in different countries.)

animal protein factor. The term which formerly designated unknown growth factors in feeds of animal and marine origin, such as tankage, meat scraps, fish meals, milk products, liver meals, etc. Commercial *A.P.F.* concentrates originally supplied both vitamin B_{12} and certain antibiotic residues (obsolete). Vitamin B_{12} is the chief entity in *animal protein factor.*

animal regulator. A nostrum; a patent medicine of little or no medicinal value.

animal science. See *animal industry.*

animal starch. See *glycogen.*

animal tankage. See *tankage.*

animal tonic. 1. Drugs which improve the health of an animal, especially after illness. 2. See *animal regulator.*

animal unit. Measurement based upon the amount of feed eaten, and manure produced by, an average mature horse, or cow, or the equivalent; as two heifers over one year, four calves under one year, seven ewes or bucks, two and one-half brood sows or boars, five hogs raised to 200 pounds each, or 100 hens.

animal unit month. The forage required to maintain a group of animals for 30 days. Also referred to as *A.U.M.* See *cattle yearlong.*

animated oat. *Avena sterilis,* family Gramineae, a wild red Mediterranean oat, gathered as a novelty because of long, twisted awns which, when wet, cause the grains to move.

anion. An ion carrying a negative charge of electricity. See *ion.*

anise. *Pimpinella anisum,* an annual herb, family Umbelliferae, grown for its aromatic seed which causes the expulsion of stomach and intestinal gas. Native to eastern Mediterranean lands.

anise hyssop. See *fennel gianthyssop.*

anise oil. An oil which improves the palatability of feeds.

anise plant. *Seseli harveyanum,* family Ammiaceae, an aromatic Australian herb.

anise-root. See *sweetanise.*

Anisomeles. Genus of the mint family Laminaceae. The herb bears purple flowers and is found in Asia and Australia.

Anisostichus. Cross vine *q.v.*

anisum. 1. Anise *q.v.* 2. Frequently erroneously referred to as *dill q.v.*

anjan. *Hardwickia binata,* family Caesalpiniaceae. An Indian tree of hard, dark red wood whose fibers are used in ropemaking.

Anjou pear scald. A disease which occurs after storage for three to four months. Characterized by a dark brown surface discoloration which affects only the skin of the fruit.

ankee. Barnyardgrass *q.v.*

ankle boot. Worn by a horse to protect its ankle.

ankle cutter. A horse which injures his legs by striking them with his hoofs.

ankle hobble. A restraining device which prevents straying, consisting of a short piece of rope or chain with straps at each end fastened to the legs of an animal just above the hoofs.

Ankoli. The animal of a cattle breed characterized by being humpless and long-horned. Found on the plains of Uganda.

annatto. A vegetable-coloring for butter, oils, cheese.

annatto tree. *Bixa orellana,* family Bixaceae, from which the dye *annatto* (*q.v.*) is derived. Sometimes incorrectly referred to as *arnatto.*

annual (plant). Any plant which completes its entire life cycle and dies within one year or less.

annual bearing. Producing fruit each year. See *biennial bearing; alternate bearing.*

annual blanket-flower. See *rose ring gaillardia.*

annual bluegrass. *Poa annua,* family Gramineae, native to Eurasia, an annual grass which may become a weed pest in turf. Also called *annual meadow grass; early meadow grass; low spear grass; annual poa; winter grass; goose grass.*

annual candytuft. See *globe candytuft.*

annual capemarigold. *Dimorphotheca annua,* a tender, South African, annual herb, family Compositae, grown for its white, yellow or orange (sometimes double) flowers.

annual chrysanthemum. See *crowndaisy chrysanthemum.*

annual clover. See *crimson clover; low hop clover; Persian clover; subterranean clover.*

annual cut. The forest products cut from a forest in one year.

annual downy chess. See *cheatgrass brome.*

annual larkspur. See *rocket larkspur.*

annual layer. The layer of wood which is produced by a tree or other perennial in one year. See *annual ring.*

annual lespedeza. See *lespedeza.*

annual marjoram. See *sweet marjoram.*

annual meadow grass. 1. Any annual grass used in a meadow as pasture or for hay. 2. See *annual bluegrass.*

annual pasture. Pasture seeded each year to take the place wholly or in part of permanent pasture. The pasture may include such crops as rye, Sudan grass, rape.

annual phlox. See *Drummond phlox.*

annual plant. See *annual.*

annual poa. See *annual bluegrass.*

annual poinsettia. See *painted euphorbia.*

annual precipitation. The water, usually expressed in inches of depth, deposited on the earth in one year by rain, snow, dew, hail, fog, mist, sleet, or other precipitation.

annual ring. The growth layer of a perennial plant

which represents the growth of one year's time, as viewed on the cross-section of a stem, branch, or root. See *annual layer*.

annual snowfall. The snow falling in one year, usually expressed in inches in the U.S. For conversion to its approximate water equivalent, the inches of depth is divided by 10. (In some other countries: for example, Australia, the ratio used is 12:1. However, the ratio can vary from 5 to 50:1 depending on many factors.)

annual vernalgrass. *Anthoranthum aristatum,* an aromatic grass, family Gramineae, sometimes a weed pest.

annual white sweetclover. See *Hubam sweetclover.*

annual wildrice. *Zizania aquatica,* an annual marsh grass, family Gramineae, cultivated for its edible grain and for ornament. Widely planted by sportsmen as feed for waterfowl.

annual yellow sweetclover. *Melilotus indica,* a Eurasian, perennial herb, family Leguminosae, useful for pasture, hay, silage, soiling, soil improvement, and as a bee-plant. Uncontrolled, it can become a weed pest. Also called *sourclover.*

annular budding. Removing a ring of bark from a stock to replace it with a budding ring.

annulation. The ring growths visible in the cross-section of a log or around the body of a worm, etc.

Annunciation lily. See *Madonna lily.*

anodendron. *Anodendron paniculatum,* family Apocynaceae, a large Indian woody vine which is the source of a very strong fiber used in ropemaking.

anoestrus. See *anestrous period.*

anonang. *Cordia myxa,* family Boraginaceae, this tree is the source of fiber used in ropemaking. Native to the Philippines.

anorexia. Lack or loss of appetite, a symptom of many diseases.

anounou. *Lepidium owaihiense,* family Brassicaceae, a peppergrass found in the Hawaiian Islands.

ansu. Apricot *q.v.*

ant. Any insect of the family Formicidae. Ants modify natural soils; certain species damage crops, and some are lawn pests.

antarctic beech. *Nothofagus antarctica,* an evergreen found in the extreme south of South America.

ant cattle. See *aphid.*

ant colony. The community of ants which live together in a close association.

ant cow. See *aphid.*

antelope brush. *Purshia tridentata,* family Rosaceae, which is good forage in the western arid areas of North America.

ante-mortem inspection. An examination of meat animals before slaughtering, in pens (or on the way to the scales).

ante-partum paralysis. An abnormal condition of the cow (seldom of sheep or goats and rare in mares), in which the hindquarters of the pregnant animal become paralyzed about a week to three weeks before parturition. A symptom of many diseases.

anterior pituitary. The front part, or lobe, of the pituitary gland, located at the base of the brain. It produces and secretes specific hormones, regulates growth of the body, and regulates and controls the thyroid, adrenal cortex, ovaries, and testicles.

anthelmintic. Any drug used for expelling or killing stomach or intestinal worms. See *vermifuge.*

-anthemum. A common suffix of flowering plant genera, *e.g. chrysanthemum.*

anther. That sac-like part of the stamen on seed-producing plants which contains the pollen.

Anthericum. A genus, family Liliaceae, of African lilies characterized by tuberous roots and strap-shaped leaves.

anther mold. A fungous disease of clovers, caused by *Botrytis anthophila,* family Moniliaceae, characterized by the destruction of the yellow pollen grains and replacement with gray spores of fungus. If widespread, it materially reduces seed production.

anther smut. 1. A smut, *Ustilago violaceae,* family Ustilaginaceae, which attacks the anthers of certain flowers of the pink family (Caryophyllaceae). 2. Infection of anthers of any plant by the invasion of a smut fungi.

-anthes. A suffix of flowering plant genera such as *zephyranthes.*

anthesis. Full blossom. The opening of a flower bud: by extension, the duration of life of a flower, from the opening of the bud to the setting of fruit.

ant hill. The mound of dirt thrown up by ants.

anthocyan. See *anthocyanin.*

anthocyanin. Pigments in sap responsible for scarlet to purple or blue coloration in plants. See *anthoxanthin.* Also called *anthocyan.*

anthoecology. Study of flowers in relation to their environment.

anthography. Flower description.

Antholyza. A genus, family Iridaceae, comprising 20 species of South African herbs characterized by their gladiolus-like appearance and the allegorical resemblance in the opening of their red and yellow flowers to the mouth of an angry beast.

Anthony pig. The smallest pig in a litter, sometimes dedicated to St. Anthony, the patron saint of swineherds.

anthopilous. Flower eating or living amongst flowers.

anthophorous. Bearing flowers.

anthotaxy. Arrangement of flower blooms.

-anthous. Adjectival suffix denoting *possession of, e.g., monanthous,* having one flower.

anthoxanthin. Yellow crystalline pigments in sap responsible for pale yellow to ivory white coloration in plants. See *anthocyanin.*

anthracnose. Plant diseases of fungus origin or their special lesions, the acervuli, where the spore-bearing fungus hyphae are produced in an erumpent, subepidermal mass. The slime-surrounded spores are distributed to other hosts by insects or rain. The causative fungi are mostly Fungi Imperfecti, producing only conidia in the acervuli. *Anthracnoses* are found on many species of host plants. See *apple a.; avocado a.* etc.

anthracnose seedling blight. See *sore shin.*

anthrax. An acute, specific, infectious, and rapidly lethal disease of men and animals, caused by *Bacillus anthracis,* which occurs in soil of certain areas throughout the world. It produces skin ulcerations and splenic enlargement. Also called *charbon; splenic fever; milzbrand; splenic apoplexy; Cumberland disease.* See *malignant pustule.*

anthropic soils. Soils that have been under cultivation for a long period of time.

anthurium. *Anthurium scherzerianum,* a tropical, perennial herb, family Araceae. Grown intensively in greenhouses for its bright scarlet spathe and its coiled yellow spadix. Many varieties of different sizes and colors are cultivated. Native to Central America. Also called *flamingo flower.*

antibiosis. An association between two organisms in which one harms the other.

antibiotic. Any chemotherapeutic agents produced by living organisms, notably fungi.

antibody. A factor in tissues, blood, or serum which reacts with and neutralizes specific foreign materials, such as toxins; or the organisms that produce toxins or other organisms. Agglutinin, antitoxins, and precipitin are antibodies. When foreign substances (antigens) are introduced into the blood stream, specific antibodies may be formed.

anticor. In horses, swelling of the chest, near the heart. It is caused by friction or pressure of the harness.

anticyclone. A center or area of high barometric pressure in which the air movement is downward and outward in a clockwise direction in the Northern hemisphere and counter clockwise in the Southern hemisphere. In the Northern hemisphere the general direction of movement is southeasterly. Its rate of movement varies from zero to 25 mph. Cool weather usually follows northerly circulation and warm weather usually follows southerly circulation. In the U.S., it is common in early fall for an anticyclone to remain stationary for up to 10 days. Also called *high; high pressure area.* When strong winds are accompanied by freezing temperatures the storm may be referred to locally as *blizzard* (U.S., U.K.); *wet northern* (southwestern U.S.); *buran* (central Asia); *purga* (Alaska, Siberia).

antidote lily. *Crinum asiaticum,* family Amaryllidaceae. The fluid from the bulb of this Polynesian plant is used as a vomiting agent.

antigen. Any substance that stimulates the formation of antibodies when it is introduced into the body. Pure cultures of organisms or their products may be employed as *antigens* in various biological testing methods to measure the antibody content of blood serum and other body fluids.

antimellin glucoside. See *Java plum.*

antisepsis. Prevention of infection by the exclusion, inhibition or destruction of the causative organisms.

antiserum. A serum containing antibodies to give temporary protection against certain specific infectious diseases. *Antiserum* is obtained from the blood of animals that possess a high degree of immunity against certain infectious diseases.

antitoxin. An antibody *q.v.*

antler moth. *Charaeas graminis.* The larvae of this moth ravages grassland. Found in Europe.

antrycine. A drug used to treat African animal trypanosomiasis, or nagana, a disease caused by a parasite carried by the tsetse fly.

ant's wood. *Bumelia angustifolia,* family Sapotaceae, a tough shrub which bears edible berries and which provides a haven for ants. Found in Florida and the Bahamas.

ant tree. Any tree or shrub of the genus Triplaris, family Polygonaceae, whose hollow stems provide habitation for poisonous ants. Native to South America.

antu. Alpha naphthyl thio urea, a raticide which is nearly as poisonous as strychnine.

Antwerp hollyhock. See *figleaf hollyhock.*

anus. Terminal opening of the alimentary canal.

anvil. A block of iron sometimes faced with steel on which metal is shaped by hammering.

Apache plume. *Fallugia paradoxa,* an evergreen shrub, family Rosaceae, good as browse for goats. Found in southwestern U.S. and Mexico.

Apalousa. See *Appaloosa.*

aparejo grass. *Muhlenbergia utilis,* (*Sporobolus depauperatus*), family Gramineae, a grass once used for stuffing the pads of the aparejo, a kind of packsaddle formerly used in southwestern U.S.

apatite. $(CaF)Ca_4(PO_4)_3$. Phosphate of lime with fluorine. Present as a rock mineral in very small amounts; but it is the primary source of phosphorus of soils and hence of secondary phosphate deposits which are mined as a source of phosphorus for fertilizers.

apetalous. Without petals. See *naked flower.*

A.P.F. See *animal protein factor.*

aphid. A small insect, family Aphidae, which sucks plant juices, often doing great damage. Many aphids secrete honeydew, of which ants are fond. Common agents for conveying virus diseases of plants. Also called *plant louse; ant cattle.*

aphis. See *aphid.*

aphis foot. Foot mange *q.v.*

aphosphorosis. Phosphorus deficiency.

aphrogala. An old name, recorded in Germany in the sixteenth century, for a preparation of milk and cream, or the cream itself, shaken to a froth and frozen. Probably a predecessor of modern ice cream as was the *melca* of Rome of the second century A.D.

aphtha. In veterinary science, a vesicle or sac, containing a thin fluid, on the udder, inside the mouth, or sometimes between the toes of a cloven-footed animal.

aphthous fever. See *foot-and-mouth disease.*

apiarist. A beekeeper.

apiary. Bee colonies, hives, and other bee equipment assembled in one location. Also called *bee yard.*

apical leaf roll. A type of rolling which persists only on upper leaves in succeeding generations.

apiculture. The science and art of raising honey bees. Also called *bee culture; beekeeping.*

Apidae. The bee family, including the subfamilies of the common bee, *Apinae,* and the stingless, *Meliponinae,* which have long tongues and lower lips or labial palpi.

apii. *Aloscasia macrorrhiza,* family Araceae. Cultivated in Pacific islands for its edible, starch-yielding stems.

apiology. The scientific study of bees, especially honey bees.

Apis. Family Apidae, the genus of bees which includes the common honeybee *q.v.*

apitong. *Dipterocarpus grandiflorus.* A tree important for its timber. The resinous extract is used for calking boats, as lamp fuel, and as varnish. Native to the Philippines.

apogamy. The development of a sporophyte from other cells of the gametophyte (embryo sac) instead of from a gamete (egg).

apomixis. Development of an individual from an unfertilized egg without sexual fusion, whether the egg be normally haploid, or abnormally diploid, through failure of reduction division.

apoplectic anthrax. A form of *anthrax* (*q.v.*) from which death may occur in a few minutes although in the case of sheep, cattle, and goats, death may not occur for three hours.

apoplectiform septicemia. An infectious, lethal disease of fowls caused by the streptococci, *Streptococcus capsulatus gallinarum* and *S. zymogenes,* characterized by listlessness, lack of appetite, staggering gait, drawn-back head, and ruffled feathers. There may be a discharge from the eyes. Also called *streptococcus infection; avian streptococcosis.*

apoplexy. A disease of grapevines often found with fungous infection, which causes the full grown plant to shrivel.

apospory. The development of a gametophyte from a vegetative cell, that is, from a cell other than a spore in plant groups normally developing gametophytes.

apothecaries measure. A standardized system for measuring fluids. In the U.S. the gallon contains 8 pints, the pint 16 fluid ounces, the ounce 8 fluid drams, the dram 16 minims. In the U.K. the gallon contains 8 pints, the pint 20 ounces, the ounce 8 drams, the dram 60 minims.

apothecium. The *ascocarp* (*q.v.*), a fruiting structure of fungi which consists usually of a fleshy—more rarely a hard—disk, cup, or salver-shaped or convex mass of fungus hyphae. The upper surface is made up of a fruiting layer (hymenium) which in turn is made up of a palisade of elongated sacs (asci) packed side by side or with intermingled sterile hyphae (paraphyses). At maturity the asci discharge ascospores from their apical openings, air currents transporting them. See *perithecium; cleistothecium.*

Appalachian tea. See *witherod viburnum.*

Appaloosa. A strain of western stock horse characterized by color spots, particularly across the croup. From 14 to 15 hands high, having a short back. Also called *Apalousa; rain drop; dollar spot; leopard spot; Colorado ranger; buttermilk horse.*

apparel wools. Any wool manufactured into cloth.

apparent density. See *bulk density.*

Appenzeller. A type of Swiss cheese often made from skim milk.

Appetitost. A Scandinavian cheese made from sour buttermilk.

apple. 1. The tree or fruit of species of the genus *Malus* (or *Pyrus*), family Rosaceae, especially *M. pumila,* variously designated as *P. malus, M. malus, M. sylvestris,* and *M. domestica.* Specifically the fruit of the tree which is cultivated throughout the temperate zones. 2. A name applied to other plants, the fruit of which resembles the common apple, such as custard apple, mayapple, pondapple, roseapple.

apple and thorn skeletonizer. *Anthophila pariana* (Clerck), family Glyphipterygidae, a caterpillar which attacks neglected apple, thorn, and pear trees, feeding under a web on the upper leaf surfaces.

apple anthracnose. Caused by *Neofabraea malicorticis* (Cord.) Jack., family Mollisiaceae; characterized by cankers on the younger branches, cracks in the canker, a wearing-away of the diseased portion, exposing wood, and by a pronounced callus layer around the edge of the wound. The fruit (apple, pear, quince) is infected while on the tree, but the disease develops slowly and decay does not become apparent until the fruit has been in storage several months. The decay is also called *bull's-eye-rot.* See *anthracnose.*

apple aphid. *Aphis pomi* DeG., family Aphidae, an insect which feeds on apple, pear, wild crab, hawthorn trees. Also called *green apple aphid; apple leaf aphid.*

apple banana. A variety of the banana plant bearing a small, good quality fruit.

apple bark miner. *Marmara elotella* (Busck), family Gracilariidae, an insect that burrows under the bark of twigs and under the skin of the apple.

appleberry. *Billardiera scandens,* family Pittosporaceae, a woody vine bearing flowers and berries. Native to Australia.

apple bitter rot. A destructive disease in the south of the U.S., caused by the fungus, *Glomerella cingulata,* family Gnomoniaceae, characterized by a light brown rot, causing the fruit to shrivel and become hard. It is also characterized by cankers on twigs, and by lesions on leaves. Sometimes it kills twigs. See *apple anthracnose.*

apple black rot. A disease of great significance east of the Rocky Mountains, occasionally occurring on quince and pear, and on other woody plants though not destructively. Caused by the fungus *Physalospora obtusa,* family Pleosporaceae. The imperfect stage, *Sphaeropsis malorum,* that causes the rots of apples in storage also causes leaf spots (frog-eye) that defoliate trees. The perfect stage causes cankers on twigs and limbs. See *New York apple tree canker.*

apple blight. See *fireblight.*

apple blister canker. A major disease from Nebraska east to Virginia, caused by the fungus, *Nummularia discreta,* family Xylariaceae. It infests the limbs of trees; characterized by cankers, shedding bark, dying limbs. Also called *nailhead canker; staghead.*

apple blotch. A serious disease east of the Rocky Mountains, especially in the southern midwest U.S. caused by the fungus *Phyllosticta solitaria,* family Sphaeropsidaceae. Characterized by numerous, tiny, white leaf spots, cankers on the limbs and twigs, and by lesions on the outer layers of the fruit. Also called *black scab of pome fruit.*

apple box. Species of the Australian *Eucalyptus* (*q.v.*), family Myrtaceae, which have gray bark.

apple brandy. Produced by distilling apple cider.

apple brown spot. See *bitter pit.*

apple bucculatrix. *Bucculatrix pomifoliella,* family Tineidae, a moth which feeds on appletree leaves.

apple butter. A sauce made from stewing apples in cider, or by allowing apple juice to thicken by evaporation.

apple canker. A general term for any of several apple diseases which produce cankers on the limbs or twigs. See *apple black rot.* Also called *apple tree canker.*

apple cheese. The apple pomace or cake-like residue found in a press after making cider.

apple cider. The juice pressed from apples; unless pasteurized it ferments rapidly. Unfermented juice is called *sweet cider* and the fermented liquid *hard cider.*

apple cider vinegar. Vinegar produced from long-fermented apple cider. Also called *cider vinegar.*

apple cider yeast. *Saccharomyces malei,* family Saccharomycetaceae, a yeast which occurs naturally on fruits; used in making sparkling hard cider.

apple curculio. *Tachypterellus quadrigibbus* (Say), family Curculionidae, a snout beetle which infests the fruit of apple, haw, wild crab, quince, pear, shadbush. It punctures the skin in many places, resulting in knotty, misshapen, undersized fruit, which often drops prematurely. Found east of the Mississippi River. Also called *apple weevil.*

apple essence. Apple oil *q.v.*

apple flea weevil. *Rhynchaenus pallicornis* (Say), family Curculionidae, a snout beetle, found from Missouri east to New York. It eats tiny holes in the leaves of apple, haw, wing elm, hazelnut, quince, wild crab, and blackberry. The larvae mine the leaves.

apple fly. *Drosophilia melonogaster,* a fruit fly. Also called *pomace fly.*

apple fruit miner. The larvae of a small moth, *Marmara pomonella* Busck, family Gracilariidae, which makes a white, winding mine on the apple fruit. Found in the Pacific Coast region. Also called *apple skin miner.*

apple fruit moth. *Argyresthia conjugella* Zell., family Hyponomeutidae, a moth whose larvae burrow in apples. Found from California to Alaska.

apple geranium. See *nutmeg pelargonium.*

apple-grain aphid. *Rhopalosiphum fitchii* (Sand.), family Aphidae, a minor insect pest that infests apples, pear, wild crab, and haws, but often attacks grains and grasses.

apple grater. Revolving knives fastened on a drum which grate apples, grapes, berries into very small particles, preparatory to expressing the fruit juice.

apple gum. *Eucalyptus stuartiana,* family Myrtaceae, a timber tree similar in appearance to the apple tree. Native to Australia.

apple jack. 1. Apple brandy *q.v.* 2. A drink made by freezing hard cider and then drawing off the unfrozen alcohol.

apple juice. See *apple cider.*

apple leafhopper. *Empoasca maligna* (Walsh), family Cicadellidae, a small, hopping, sucking insect found east of the Rocky Mountains (except in the lower Mississippi Valley). It causes infested foliage to turn pale in late summer; the upper surface of the leaves has many small white spots; the fruit is speckled with excrement. It also infests rose, currant, gooseberry, raspberry, potato, sugar beet, celery, among other plants.

apple leaf skeletonizer. The larvae of the moth, *Psorosina hammondi* (Riley), family Phycitidae, which eat the fleshy part of the leaves of apple, plum, quince, turning them brown. Found predominantly in the upper Mississippi Valley, but also in the eastern U.S.

apple leaf trumpet miner. The larvae of a minor moth, *Tischeria malifoliella* Clem., family Tischeriidae, which make trumpet-like mines in apple leaves. Found in Oregon and in eastern U.S.

apple maggot. The larvae of a fly, *Rhagoletis pomonella* (Walsh), family Tephritidae, which bore through the flesh of apples, making winding galleries. It infests wild crabs, blueberries, European plums, and cherries. Also called *railroad worm.*

apple mealybug. See *Comstock mealybug.*

apple mildew. See *powdery mildew.*

apple mint. *Mentha rotundifolia,* a European, aromatic, perennial herb, family Labiatae, grown for its purple flowers on dense spikes. Also called *horse mint.*

apple mosaic. A virus tree disease characterized by polygon-like spots, white in spring then turning light yellow.

apple moth. *Carpocapsa pomonella,* family Tortricidae, codling moth or any moth whose larvae eat their way into apples causing serious damage.

apple of Cain. See *Pacific madrone.*

apple of Hesperides. Quince *q.v.*

apple of love. Tomato *q.v.*

apple of Peru. *Nicandra physalodes,* a Peruvian, annual herb, family Solanaceae, grown for its solitary, tubular, blue flowers.

apple-of-Sodom. See *Carolina horsenettle.*

apple oil. 1. Believed to contain acetate, amyl formate, acetaldehyde, caproate and caprylate. 2. Amyl isovalerate, or any artificially prepared oil imitative of apple flavor or smell.

apple oyster-shell scale. See *oystershell scale.*

apple pine. The *eastern white pine* (*q.v.*). A colloquial term once used in parts of New England.

apple pole. Used in harvesting apples.

apple powdery mildew. A fungous disease; also attacks pears, quince, cherries, prunes. Caused by *Podosphaera leucotricha,* family Erysiphaceae. Characterized by white or pale gray, felt-like patches on the margins of leaves which eventually cover the leaves. Infected leaves are narrowed and folded longitudinally, and there is a blighting of the flower and a russeting of the fruit.

apple pox. Blister canker of apple *q.v.*

apple redbug. *Lygidea mendax* Reut., family Miridae, an important sucking insect that feeds on the sap of the leaves and the juice of apples, leaving the fruit misshapen and pitted with russeted spots. Found east of the Mississippi River, especially in New England, New York, and southeastern Canada.

appleroot. Floweringspurge euphorbia *q.v.*

apple rosette. 1. A boron deficiency disease; the leaves are clustered together very closely. 2. Symptoms of plant diseases characterized by clustering of leaves.

apple rot. See *downy mildew.*

apple rust. A very serious fungous disease caused by *Gymnosporangium juniperi-virginianae* Schw., family Pucciniaceae. Characterized by rust spots on the leaves, rust lesions on the twigs, lesions on the fruit; results in defoliation, dwarfing, and reduction in the quality of the fruit. Found east of Rocky Mountains. The fungus lives alternately on *Juniperus* spp. and on apples. Also called *cedar rust; cedar-rust disease of apples; cedar apple rust.* See *cedar apples; cedar galls; cedar flowers; hawthorn rust; quince rust.*

apple rust mite. *Vasates schlechtendali* (Nal.), family Eriophyidae, a mite which attacks the under surfaces of the apple, pear, and prune leaves, causing them to roll longitudinally.

apple scab. A widespread parasitic fungous disease caused by *Venturia inaequalis,* family Pleosporaceae. It attacks species of *Malus;* characterized by lesions of the leaves, blossoms, fruit; dwarfed or misshapen fruit, and premature dropping of fruit. The perfect stage develops on the dead, over-winter leaves. The imperfect stage develops on living leaves, fruits, and twigs and is called *Fusicladium dendriticum.* Also called *scab of apples; scurf; black spot; black spot fungus; Tasmanian black spot; black spot scab and rust.*

apple scald. A serious, physiological disorder caused by alcoholic esters emanating from the apples. Characterized by browning of the skin after harvest, in storage, or after storage. Usually associated with immaturity at harvest; more prevalent on green-skinned varieties.

apple scale. A scale insect which attacks apple trees, *e.g.* oystershell scale, *q.v.*

apple seed chalcid. *Torymus druparum* Boh., family Torymidae, an insect that lays its eggs in apple seeds and related plants, causing fruits to become dimpled. Found in the northeastern U.S. and southeastern Canada.

apple soft scald. A disease of stored apples; characterized by sharply defined spots of dead skin, under which the flesh is brown and dry. Fungi may invade the fruit at the dead spot. See *soggy breakdown of apple.*

apple sucker. *Psylla mali* (Schmdb.), a jumping plant louse whose feeding on apples and honeydew injures buds, flower clusters, and prevents fruit setting. Also called *apple psylla.*

apple tree. 1. Belonging to the genus Malus, family Rosaceae, bearing apples. 2. Apple gum *q.v.* 3. *Angophora subvelutina,* family Myrtaceae, an Australian tree which produces kino *q.v.*

apple tree anthracnose. See *apple anthracnose.*

apple tree aphid. Plant lice which attack apple trees, such as the woolly apple aphid (*q.v.*). It attacks the roots and bark. On the other hand, *Aphis pomi* and *Aphis sorbi* attack the leaves.

apple tree aphis. See *apple tree aphid.*

apple tree borer. See *roundheaded apple tree borer.*

apple tree tent caterpillar. See *eastern tent caterpillar.*

apple twig beetle. *Stephanoderes obscurus* (F.), family Scolytidae, a small bark insect whose larvae feed on twigs.

apple twig borer. *Amphicerus bicaudatus* (Say), family Bostrichidae, a small beetle active in early spring. Also bores into the twigs of grape, pear, etc.

apple vinegar. See *apple cider vinegar.*

apple wine. Another name for hard cider. See *apple cider.*

apple worm. An insect whose larvae infect apples. See *apple moth.*

applied insect control. The use of sprays, insecticides, and other methods.

approach channel. A small basin on the upstream side of a weir, which reduces eddies and swirls in the flow approaching the weir. Also called *weir basin.*

approaching. Inarching *q.v.*

appropriative right. A doctrine of rights to irrigation water. Rights are based on date of appropriation of water; the right to water is limited by beneficial use on the land.

appulsion. The tearing away by storm waters and wind of river banks, which are then carried downstream to become part of the bank elsewhere.

apricot. *Prunus armeniaca,* a deciduous popular Asiatic tree, family Rosaceae, which bears a drupe fruit, the apricot of commerce. Resembles the peach.

apricot brown rot. See *brown rot of stone fruits.*

apricot Coryneum blight. See *Coryneum blight of stone fruits.*

apricot leather. The sun-dried pulp made into sheets. A confection and used for cooking in Asia Minor.

apricot palm. *Butia eriospatha,* family Palmaceae, a pinnate-leaved tree which bears apricot-like fruit. Native to Brazil and grown in California.

apricot plum. *Prunus simonii,* family Rosaceae, a tree bearing a tart fruit; cultivated for cross-breeding. Native to China.

April fool. Spreading pasqueflower *q.v.*

apron. 1. Canvas, or rubber-like material, or of metal links made into a continuous belt for conveying fruits or vegetables into bins (especially on grading machines); silage, straw, manure, etc. 2. Folds of flesh on the neck of an animal or fowl or the fatty layer covering its belly. 3. A strip of metal placed to guide the sap into a cup which is suspended from it. 4. A slab of concrete or stone which connects adjoining areas, or over which water is directed to dissipate its energy.

apterium. A featherless portion on the body of a bird.

apteroid. A bird with undeveloped wings, *e.g. emu.*

apterous. Wingless.

apteryx. Any tailless bird of the genus Apteryx. It has a long bill, undeveloped wings, gray and brown hairlike feathers and stout legs. This almost extinct, hen-sized bird lives in a burrow and is native to New Zealand.

aquagium. A canal, ditch, or watercourse which runs through marshy ground.

aqua rosae. Fragrant solution made from rose flowers and water.

aquatic. 1. An aquatic plant (one that lives in water) which may be classified as submerged, emergent, or floating. 2. An aquatic animal.

aquatic rights. Ownership or permission necessary to navigate or fish rivers, lakes, etc. See *riparian doctrine.*

aqueous tissue. Plant tissue composed of thin-walled cells, having neither air spaces nor chloroplasts.

aquiculture (aquaculture). The culture of plant and animal life in natural seas, lakes, streams.

Arab(ian). A type of saddle horse from 14 to 15 hands high; from 850 to 1000 pounds; predominant colors are gray, bay, chestnut. It possesses endurance and docility and belongs to perhaps the oldest known pure breed.

Arabian coffee. *Coffea arabica,* coffee made from an evergreen shrub, family Rubiaceae, native to the Yemen; grown in Florida as a curiosity. Also called *mocha coffee.*

Arabian cotton. See *Levant cotton.*

Arabian jasmine. *Jasminum sambac,* a climbing shrub, family Oleaceae; usually grown indoors for its white, very fragrant flowers. Also called *zambac.*

Arabian millet. See *Johnsongrass.*

Arabian myrrh. *Commiphora myrrha,* family Burseraceae. This Southern Arabian tree produces an inferior myrrh.

Arabian star-of-Bethlehem. *Ornithogalum arabicum.* A Mediterranean bulbous herb, family Liliaceae. It has large flowers with attractive black pistils surrounded by creamy-white petals.

arable. Suitable for cultivation; tillable.

Arabprimrose arnebia. *Arnebia cornuta,* an annual herb, family Boraginaceae, grown for its orange and black-spotted flowers which turn to red and then yellow.

arachidonic acid. The most important unsaturated acid in butterfat; an essential fatty acid.

Araucaria. A genus of trees, family Araucariaceae, important for its timber and edible seeds. Native to South America and Australia.

arbitrage. Purchase in one market at a low price while selling at a high price in another.

arbor. 1. A bower; small structure of latticework to support vines and provide a shady retreat. 2. A tree as distinguished from a shrub. 3. A wooden platform used for sun-drying fibers. 4. The central or supporting beam or rod, as the *arbor* bolt on a disk gang harrow.

arboraceous. From Latin *arbor,* tree-like, or pertaining to trees.

arborator. A tree-tender.

Arbor day. In U.S., a ceremonial occasion for planting trees and shrubs, usually in early May.

arboretum. A garden where trees and shrubs are grown for study and exhibition.

arboriculture. Cultivation of woody plants; particularly those used for decoration and shade.

arborist. A specialist in the field of tree culture.

arborvitae. Any plant of the genus *Thuja,* family Pinaceae. The species are valuable evergreen trees used for ornament and in the timber industry. White cedar or Thuja oil is used as a vermifuge. See *northern white cedar; oriental arborvitae; giant arborvitae.*

arborvitae leaf miner. The larvae of the moth, *Argyresthia thuiella* (Pack.), family Hyponomeutidae, which burrow into foliage.

arborvitae weevil. *Phyllobius intrusus* Kono, a weevil which feeds on the roots of arborvitae, falsecypress, and juniper.

arbutus. See *Pacific madrone; strawberry madrone; trailing arbutus.*

arcadian. Simple, rustic; a term applied to a place of contentment.

Arcadian nitrate. A fertilizer (99.5 percent nitrate of soda) of synthetic ammonia and soda ash.

Arcadian sulfate of ammonia. A nitrogenous fertilizer of sulfate of ammonia especially neutralized, dried, and screened to remove lumps. It has a minimum of 20.56 percent nitrogen and less than 0.5 percent moisture.

arch. 1. A curve. 2. The anterior-posterior curvature of the backs of hogs. 3. A log-supporting device used with a tractor in skidding logs. 4. The curved portion of a dam, bridge, or other building.

archetype. In biology, the antecedent of a group of plants or animals, from which certain typical characteristics have been inherited; a progenitor.

arching layers. See *tip layering.*

arcifinious. Having natural boundaries.

arctic birch. *Betula nana,* family Betulaceae, an arctic shrub.

arctic chrysanthemum. *Chrysanthemum arcticum,* a hardy autumnal herb, family Compositae, used as a border plant. Native to Eurasia and America. Also called *aster daisy; arctic daisy.*

arctic daisy. See *arctic chrysanthemum.*

arctic meadowrue. See *alpine meadowrue.*

arctic timothy. *Alopecurus alpinus,* family Gramineae, a grass grown by Eskimos for stock feed.

Ard's Rover. A climbing hybrid perpetual rose of deep crimson. Developed in England, it is now used for breeding in U.S.

are. 100 square meters in area.

area-kill. Total kill of wildlife per unit of area in one year.

arenales. A Spanish term for sandy beaches.

areolate mildew. An imperfect fungus which causes cotton leaf spot.

arethusa. *Arethusa bulbosa,* a North American tuberous-rooted bog orchid, family Orchidaceae, cultivated in bog gardens for its rose-purple flowers fringed with yellow. Also called *dragon's-mouth; wild pink.*

argan tree. *Argania sideroxylon,* family Sapotaceae; the fruit is used as animal feed.

Argas. A genus of ticks, family Argasidae, which in warm climates infest poultry.

Argentine Amazonvine. *Stigmaphyllon littorale,* a delicate tropical American, climbing, woody vine, family Malpighiaceae, which produces yellow flowers.

Argentine ant. *Iridomyrmex humilis* Mayr, family Formicidae, a household and orchard insect. It also attacks shade trees, bees, their broods and hives. Found in the southeastern U.S. and California.

Argentine thistle. Scotch cottonthistle *q.v.*

Argentine trumpetvine. *Clytostoma callistergioides,* an evergreen, woody vine, family Bignoniaceae. Grown in California and Florida for its purple- or lavender-streaked flowers.

argol. Crude cream of tartar which crystallizes on the sides of wine barrels. Also called *wine stone.*

argus tortoise beetle. *Chelymorpha cassidea* (F.), family Chrysomelidae. Attacks the sweetpotato, the morning-glory, and bindweed by eating holes in the foliage or consuming the entire leaves.

-aria. In botany and zoology, a suffix used in forming the names of genera and other groups.

arid. A term denoting dry, barren, desert, or semi-desert land, or land capable of supporting only xerophytic vegetation (*q.v.*) in a warm or tropical zone. See *arid climate.*

arid climate. A dry desert, or semi-desert climate with only sparse vegetation. Precipitation varies, with an upper limit for cool regions of 10 inches or less and for tropical regions of as much as 15 or 20 inches. (The Thornthwaite precipitation effectiveness [P-E] index ranges between 0 and 16.) See *precipitation-effectiveness.*

Arizona black walnut. *Juglans major,* family Juglandaceae. Produces an edible, usually thick-shelled nut. Found in southwestern U.S. and Mexico.

Arizona buckthorn. *Bumelia rigida,* family Sapotaceae, a hardwood tree. Native to southwestern U.S.

Arizona cotton stainer. *Dysdercus mimulus* Hussey, family Pyrrhocoridae, a cotton insect which causes the lint to become stained. Found in the southwestern U.S.

Arizona cypress. *Cupressus arizonica,* an evergreen tree, family Cupressaceae; grown as a windbreak and an ornament in south and southwestern U.S.

Arizona fescue. *Festuca arizonica,* family Gramineae, an excellent forage grass growing at elevations of 6,000 to 10,000 feet in southwestern U.S.

Arizona fir. See *alpine fir.*

Arizona gourd. See *calabazilla.*

Arizona longleaf pine. *Pinus mayriana,* family Pinaceae, a tree distinguished by its clusters of three leaves. Native to Arizona and New Mexico. *Cf. ponderosa pine.*

Arizona mistletoe. *Phoradendron macrophyllum,* family Loranthaceae, a parasite on alder, ash, poplar, hackberry, sycamore, walnut, and willow. Also called *bigleaf mistletoe.*

Arizona palm. Mexican Washingtonpalm *q.v.*

Arizona pine. Ponderosa pine *q.v.*

Arizona pine beetle. *Dendroctonus arizonicus* Hopk., family Scolytidae, an insect which burrows under the inner bark of yellow pine. Found in Mexico, and southwestern U.S.

Arizona walnut. See *Arizona black walnut.*

Arkansas black. A winter apple with deep red skin.

Arkansas soft pine. Lumber from either the loblolly pine or the shortleaf pine *q.q.v.*

Arlington meal. Whole-wheat flour *q.v.*

arm. 1. The main branches or extensions of the trunk of a tree or shrub. 2. A branch of a grapevine which is two years old or older. 3. That portion of an animal between his shoulder and forearm. 4. Any mechanical device which resembles in action the human arm. 5. To sprout to the side; to develop many off-shoots.

Armand clematis. *Clematis armandi,* an evergreen, woody vine, family Ranunculaceae, native to China; grown for its white flowers that change to pink.

Armavir. A soft, ripened cheese made from ewe's milk to which is added sour buttermilk or whey.

arm board. A tool, made from the bark of the cork oak (*q.v.*). Used for softening and finishing leather.

arm chop. A cut of meat for broiling (particularly lamb) from the part of the shoulder containing the humerus.

armed tapeworm. *Taenia solium,* a dangerous

parasite transmitted to man through pork infected with larvae. See *tapeworm.*

Armenian grapehyacinth. *Muscari armeniacum,* family Liliaceae, grown under shrubs in rock gardens for its deep violet to pale blue flowers. Native to southwestern Asia.

Armillaria root rot. See *mushroom root rot.*

armor. A hard covering which protects a plant or animal.

arm renewal. Upright training of grapevines. The old arms are cut back to the trunk and one-year old canes are selected for the development of new arms.

Armsby feeding standard. A method, common in U.S., of measuring, in therms, the net nutritive value of livestock feeds.

Armstrong freesia. *Freesia armstrongi,* a South African, bulbous herb, of the family Iridaceae, grown as a pot plant for its fragrant flowers.

army ants. Ants which group together to find and capture prey.

army cutworm. The larva of the moth, *Chorizagrotis auxiliaris* (Grote), family Noctuidae, which infest the roots of many kinds of green vegetation. Found in western U.S.

armyworm. The larvae of a noctuid moth, *Pseudaletia unipuncta* (Haw.), family Phalaenidae, which moves in hordes, destroying grain and grass, among other crops. Found east of the Rocky Mountains.

arnatto. See *annatto tree.*

aromatic vinegar. Vinegar (*q.v.*) to which spices have been added.

aromatic water. Water distilled from a plant, such as the rose, which contains fragrant oil.

aromatic wintergreen. See *checkerberry; wintergreen.*

Aroostook. A small farm area characterized by rolling topography and by a mellow, silty, loam soil; famous for growing potatoes.

aroya. See *arroyo.*

arpent. A French measure of land. In Louisiana it is equal to 0.845 of an acre or a square of 192 feet, as compared to the square of approximately 209 feet for an acre. Also called *arpen; arpenus.*

arrent. To give license to use, as farmland, at a fee.

arrowarum. Any plant of the genus *Peltandra.* See *Virginia arrow arum.*

arrow crotalaria. *Crotalaria sagittalis,* a poisonous annual herb, family Leguminosae; eating it green or in hay may kill a horse. Found in southern and eastern U.S. Also called *rattlebox; rattleweed; wild pea.*

arrowgrass. See *shore podgrass.*

arrowhead. *Sagittaria latifolia,* a North American, perennial herb, family Alismataceae; grown in shallow pools, bog gardens, or aquariums for its pure white flowers. Escaped, it can be a weed pest in rice fields. Its roots were eaten by the Indians of the western U.S. Also called *tule potato.*

arrowleaf balsamroot. *Balsamorhiza sagittata,* a perennial herb, family Compositae, which has forage value on the spring range in the western U.S.

arrowleaf butterweed. See *arrowleaf groundsel.*

arrowleaf groundsel. *Senecio triangularis,* a spring plant, family Compositae; good forage for sheep but less so for cattle. Found from Colorado to California and northward to Alaska up to 9,000 feet. Also called *arrowleaf butterweed.*

arrowroot. See *Bermuda arrowroot; coontie.*

arroya. See *arroyo.*

arroyo. An incised channel; usually dry, occurring in alluvial slopes at the bases of mountains. Found in arid regions of southwestern U.S. Also called *arroya; aroya.*

arsanilic. A chemical compound of substances with one of three poisonous acids containing arsenic; it may have medicinal properties at low levels.

arsenate of lead. See *lead arsenate.*

arsenate of lime. See *calcium arsenate.*

arsenic. A chemical element, As, present in pesticides. It may occur in soils as a trace element.

arsenic acid. Common name for orthoarsenic acid, metarsenic acid or pyroarsenic acid, all derivatives of arsenic pentoxide *q.v.*

arsenic pentoxide. A compound, As_2O_5, which is used in the manufacture of poisons.

arsenic poisoning. A condition which results from ingestion, or from absorption through the skin, of an arsenical; acute cases are characterized by trembling, weakness, prostration, and diarrhea. In chronic poisoning, symptoms develop more slowly.

arsenic trioxide. A compound, As_2O_3, used for rodent poisons and insecticides.

arsenotherapy. Use of arsenic as a curative.

artar. *Zanthoxylum senegalense,* family Rutaceae, a tree with pinnate leaves and green flowers. The root has medicinal properties. Native to Africa.

artemisia. See *sagebrush.*

artery curing. The preservation of ham by forcing brine into the femoral artery with a small needle.

artesian well. A deep water well under hydrostatic pressure. The term *artesian* is often applied to non-flowing wells.

arthritis. Inflammation of the joints characterized by stiffness, swelling, soreness.

arthropod. A phylum or division of the animal kingdom, which includes insects, spiders and Crustacea; characterized by a coating which serves as an external skeleton, and by legs with distinct moveable segments or joints.

artichoke. *Cynara scolymus,* a perennial herb, family Compositae; the edible portions are the bracts of the unripe flower head. Native to southern Europe. Also called *bur artichoke; globe artichoke.*

artichoke betony. *Stachys sieboldi,* a perennial, Asiatic herb, family Labiatae; grown for its small, white, edible tuber. Also called *Chinese artichoke; Japanese artichoke; knotroot; chorogi.*

artichoke plume moth. *Platyptilia carduidactyla*

(Riley), family Pterophoridae, a moth, the caterpillar of which feeds on new foliage and mines the stems, stalks, and buds of the artichoke.

artichoke thistle. See *cardoon.*

articular. Pertaining to the joints.

articulation. 1. The point of attachment of a leaf stem to the main stalk. 2. Any thickened portion of the plant stem which resembles such a point.

artifact. 1. Any man-made product, especially from a primitive culture.

artificial breeding. The use of *artificial insemination* (*q.v.*) to enhance a breed of animals or poultry.

artificial brooding. Rearing chicks without a hen, usually in heated nests.

artificial buttermilk. See *cultured buttermilk.*

artificial cell cup. A cell of beeswax inserted into a beehive to induce bees to produce a queen.

artificial drying. Drying forage crops with blowers and heaters.

artificial incubation. Hatching chicks by artificial means. See *incubator.*

artificial insemination. Fertilization by means other than coitus. Also called *artificial service; artificial breeding.*

artificial manure. See *synthetic manure.*

artificial refrigeration. Maintaining consistently low temperature within a chamber to preserve food stuffs.

artificial selection. Scientific breeding practices.

artificial service. See *artificial insemination.*

artificial swarm. A new colony of bees which is made by dividing an established colony.

artificial vagina. In artificial insemination, a device used to collect semen consisting of a heavy rubber or metal cylinder, an inner liner of soft rubber, and a tube.

artillery clearweed. *Pilea microphylla,* an annual, tropical American herb, family Urticaceae; grown indoors and as an ornamental ground cover in California and Florida. Also called *artillery plant.*

artillery plant. See *artillery clearweed.*

Artiodactyla. The order of hoofed mammals which includes oxen, swine, sheep, goats.

Artocarpus. A genus of trees, family Moraceae, which includes the breadfruit (*q.v.*) tree. Native to Asia.

arviculture. In agriculture, the study of field crops.

arythmic pulse. A pulse with an indefinite rhythm.

arzun. 1. The Italian millet, *Setaria italica,* family Graminae, cultivated as forage and cereal. 2. The American millet, *Panicum miliaceum,* family Graminae.

Asadero. A white, whole-milk, Mexican cheese.

asa dulcis. A drug made from the herb *Thapsia garganica,* family Umbelliferae.

asafetida. An offensively-smelling gum obtained from *Ferula assafoetida,* family Umbelliferae. Used as a drug for spasmodics. The plant is native to India.

ascariasis. A chicken disease caused by a parasitic roundworm, family Ascarideae. The symptoms are diarrhetic.

Ascarid. Family Ascaridae. A parasitic roundworm that infests the intestines of man and animals.

ascites. An accumulation of fluid in the abdomen. See *abdominal dropsy.*

ascocarp. The fruiting structures or shield of class Ascomycetes, which contain the asci, such as apothecium, cleistothecium, perithecium, hysterothecium, etc.

Ascochyta blight of cotton. See *wet-weather blight.*

Ascochyta blight of pea. A serious composite of three diseases caused by one or more of the following: *Ascochyta pisi; A. pinodella,* both of family Sphaeropsidaceae; *Mycosphaerella pinodes,* family Mycosphaerellaceae. It is characterized by lesions on the leaves, pods, stems; by the shrivelling and drying up of the leaves; by the withering and dropping of the pods and blossoms; and by a root rot which may develop.

Ascochyta canker. See *wet-weather blight.*

Ascochyta leaf spot of alfalfa. See *spring black stem of alfalfa.*

Ascochyta leaf spot of cotton. See *wet-weather blight.*

Ascochyta spot of sorghum. A fungous disease caused by *Ascochyta sorghina* and characterized by lesions on the leaves, and sheaths. The lesions coalesce and appear as reddish-brown patches.

ascorbic acid. A chemical compound, $C_6H_8O_6$, vitamin C, which occurs in fruits and vegetables and prevents scurvy.

ascus. The spore producing cell of fungi of the Class Ascomycetes. It consists of a terminal or subterminal cell of a hypha, containing at first two nuclei (probably always haploid). These then unite to form the diploid ascus which then divides by meiosis (reduction division) to produce four haploid nuclei (or by further divisions, eight or sometimes more nuclei). Around each nucleus is gathered a mass of cytoplasm surrounded by a cell wall, thus becoming an ascospore.

asepsis. Being free from microorganisms.

asexual. Sexless. In botany, specifically flowers that have no stamens, pollen, pistil or ovules.

asexual reproduction. Process of propagation by budding, dividing of cells, grafting, etc. as opposed to sexual union.

ash. 1. Any tree of the genus *Fraxinus,* family Oleaceae; grown mainly for timber, sometimes for ornament. See *European ash; black ash; red ash; green ash; blue ash.* 2. Residue after combustion. See *wood ashes.* 3. Specifically, minerals left after combustion has freed food from carbon.

ash borer. *Podosesia syringae fraxini* (Lug.), family Aegeriidae. The larvae of this moth tunnel in the bark and wood of the ash and mountain ash, causing the tree to become stunted. Found from Minnesota to Montana.

ashes from leached wood. See *leached wood ashes.*

ash-gray blister beetle. *Epicauta fabricii* (Lec.), an insect which can destroy forage or crops in transit; commonly found in eastern U.S.

ash hopper. 1. A wedge-shaped wooden box, with a hole at the point, used by pioneer farmers to leach wood ash for lye. 2. A tilted barrel with a hole in the side near the bottom used for the same purpose. Also called *lye hopper.*

ash of milk. The residue remaining after volatization of water, carbon, and nitrogen compounds, which consists of the oxides and salts of potassium, calcium, sodium, magnesium, iron, phosphorus, chlorine, and sulfur, etc.

Asiago. A hard cheese made from partly skimmed milk; sometimes ripened.

Asian moluccabalm. *Molucella laevis,* family Labiatae, an aromatic herb, native to western Asia; grown for its fragrant white or pinkish flowers. Also called *shell-flower; molucca balm.*

Asiatic beetle. See *Oriental beetle.*

Asiatic class. Domesticated chickens including Brahmas, Cochins, and Langshans which are heavy, and which have feathered legs. They lay brown-shelled eggs.

Asiatic cotton. See *Levant cotton.*

Asiatic elm. See *Chinese elm.*

Asiatic garden beetle. *Autoserica castanea* (Arrow). This insect eats holes in leaves of many plants. The larva is a whitish grub which feeds on roots of grasses. Found in eastern U.S.

Asiatic globeflower. See *Siberian globeflower.*

Asiatic oak weevil. *Cyrtepistomus castaneus* (Roelofs). The young of this insect feed on the roots of oak and chestnut, and the adults on the foliage of hickory, filbert, beech, dogwood, raspberry, etc. Found in eastern U.S.

Asiatic redbud. See *Chinese redbud.*

Asiatic tree cotton. *Gossypium arboreum,* a tropical, perennial, fiber plant, family Malvaceae, which grows six to ten feet tall; it is less productive than the herbaceous species and is more expensive to harvest. Also called *tree cotton of India; nurma cotton.*

Asin. A soft, ripened, sour-milk, washed-curd cheese.

asparagus. 1. *Asparagus officinalis,* family Liliaceae, a perennial herb whose tender new shoots are edible. Native to Europe. Also called *garden asparagus.* 2. See *smilax asparagus.*

asparagus bean. See *yardlong cowpea.*

asparagus beetle. *Crioceris asparagi* (L.), family Chrysomelidae. It gnaws out tender asparagus shoots as they emerge, and lays eggs on the tips, making the asparagus shoots unsalable. It is found in northern U.S., Colorado and California. See *spotted asparagus beetle.*

asparagus broccoli. See *broccoli.*

asparagus buncher. Any device to hold asparagus stalks together for butt trimming and tying for marketing.

asparagus fern. See *fern asparagus.*

asparagus fern caterpillar. See *beet armyworm.*

asparagus garden lettuce. *Lactuca sativa* var. *asparagina,* family Compositae, one of the four types of *lettuce* (*q.v.*) grown for its thick edible stem. Uncommon in U.S.

asparagus knife. A knife, about 15 inches long with a hooked blade, used in harvesting.

asparagus miner. Maggots of this fly, *Melanagromyza simplex* (Loew), family Agromyzidae, burrow up and down the stems of asparagus near the base, causing the foliage to turn yellow and die. Found in northeastern U.S., California, and southern Canada.

asparagus rust. The only serious, fungous disease of asparagus in the U.S., caused by *Puccinia asparagi,* family Pucciniaceae; characterized by reddish-yellow spots that later cover the plant and turn to orange or brown. It weakens the stock and reduces the yield.

aspect. 1. Physical appearance of a plant type. 2. Seasonal appearance. 3. Direction in which a slope faces: also called *exposure.*

aspect society. In botany, a planting sequence which contrives the successive appearance of various plant groups.

aspen. See *quaking aspen.*

aspen leaf beetle. *Chrysomela tremulae* F., family Chrysomelidae; attacks willows and poplars, often defoliating them.

aspen peavine. *Lathyrus leucanthus,* family Leguminosae; a good forage herb found on the western range of the U.S.

aspergillosis. A respiratory disease of chickens and turkeys caused by the fungus, *Aspergillus fumigatus,* family Moniliaceae, in the lungs and air sacs. Characterized by listlessness, loss of appetite, a darkening of the comb, ruffled feathers, high temperature, thirst, diarrhea, gasping for breath. High mortality among chicks and poults. Also called *brooder pneumonia; pneumomycosis.*

asphalt. A solid, bituminous material or a petroleum residue occurring naturally; it is soluble in petroleum solvents but insoluble in water. Used in tree fillings, tree paints, wound dressing, and grafting compound.

asphyxiation. A systemic condition of plants, etc., brought on by oxygen deficiency; characterized by seeds not germinating, blighting of plants, and breakdown of fruits, tubers, etc.

aspidistra. *Aspidistra elatior,* a foliage plant, family Liliaceae. Because of its ability to thrive under adverse conditions, such as rough usage, heat, cold, and drought, it is sometimes incorrectly called *iron plant; cast-iron plant; saloon plant; lager beer plant.*

ass. 1. *Equus asinus,* the donkey; a domesticated beast of burden used in the breeding of mules *q.v.* 2. Any of several species of *Equus:* e.g., wild asses of Asia and North Africa.

Assamking begonia. See *rex begonia.*

assessed value. The worth of a property established by a governmental unit as the basis for estimating taxes.

assets. Property implements, stock, or right, such

as water right, etc., which has commercial value.

assignment. 1. A transfer of notes, mortgages, etc. 2. The gradual payment of a debt, as in the case of a farmer who authorizes a milk dealer to assign a certain portion of his milk check to a bank in payment of a bank loan to buy dairy cows.

assimilation. Process whereby nutrients become a part of living tissues.

association. 1. In botany, a grouping of plants having common growth requirements; the association is identified according to the dominant species such as *coniferous association*, etc. 2. A grouping of soils to form a land type. See *soil association.* 3. Persons united for a common end.

assurgent. The curving upward of a plant stem.

aster. 1. Any herb usually perennial, of the genus *Aster,* family Compositae. Also called *Michaelmas daisy; starwort.* See *Rhone aster; alpine aster; Italian aster; heartleaf aster; Hindukush aster; white wood aster; heath aster; great aster; New England aster; New York aster; Porters aster; white upland aster; seaside aster; East Indies aster; Tatarian aster.* 2. The commonly cultivated garden aster is the *China aster q.v.*

aster daisy. See *Arctic chrysanthemum.*

aster leaf miner. See *serpentine leaf miner.*

asteroid spot. A virus disease of stone fruits; characterized by chlorotic, mottled, or necrotic patterns on the leaves.

aster wilt. See *Fusarium wilt diseases.*

aster yellows. A virus disease of many seed plants spread by the six-spotted leafhopper, *Macrosteles fascifrons* (Stal.), family Cicadellidae; characterized by stunting, production of many secondary shoots, chlorosis, sterility, and by the blossoms turning green.

asthenia. Debility.

asthma-weed. See *Indiantobacco lobelia.*

astomous. Mouthless.

Astoria bentgrass. See *colonial bentgrass.*

Astrakhan. The lambskin of Karakul or Bokhara sheep which is less lustrous, has longer hair and more open curl than Persian lambskin.

astral. Frequently used when describing star-shaped flowers.

astrologer. A poetical name sometimes given to the cock as the harbinger of dawn.

astrometeorology. Study of the possible connection between the activity of stellar systems and the Earth's weather.

asymmetry. Lack of proportion.

asynapsis. The failure of normal pairing, resulting in the production of univalents at meiosis; failure of homologous chromosomes to occur in pairs during meiosis.

atamasco lily. *Zephyranthes atamasco,* a bulbous garden herb, family Amaryllidaceae. It is poisonous to sheep and goats causing swellhead and goat fever (*q.q.v.*). Found in wet lands from Pennsylvania to Florida. Also called *Easter lily* (not *Lilium longiflorum eximum*); *daffodil lily; fairy lily.*

atavism. Reversion; reappearance of a characteristic or disease after a lapse of one or more generations.

ataxia. Loss of ability to control bodily movements.

atelectasis. Failure of the pulmonary air sacs to function at birth.

athel tamarix. *Tamarix aphylla,* an Asiatic tree, family Tamaricaceae, whose blossoms provide nectar for bees. Grown in beach and desert gardens in southwestern U.S. Also called *evergreen tamarix.*

Atlantic leatherwood. *Dirca palustris,* a hardy, tough-wooded shrub, family Thymelaeaceae, grown for its early yellow flowers. The bark contains poisonous properties which can severely irritate and blister the skin. Used by the Indians for making thongs. Native to North America. Also called *moosewood; wicopy; leather bush; rope bark.*

Atlantic pigeonwings. *Clitoria mariana,* a perennial vine, family Leguminosae, grown outdoors for its pale blue flowers in southern California and Florida or inside greenhouses elsewhere. Also called *butterfly pea.*

Atlantic white cedar. See *whitecedar falsecypress.*

Atlas beetle. *Chalcosoma atlas,* a huge olive-green insect found in the Far East.

Atlas cedar. *Cedrus atlantica,* an evergreen tree, family Pinaceae, resembling the larch, one of four traditional cedars. It is grown on lawns and in parks. Native to the Atlas Mountains.

atlas joint. The first vertebra of the neck supporting the skull.

atmometer. Any instrument used to measure evaporation rates.

atmospheric concentrator. A drum-like device to reduce the water content of food products. It consists of a cylinder within a cylinder with a space between the two for steam. Fans within the inner cylinder maintain a partial vacuum and force out vapor caused by the evaporation.

atmospheric drought. Transpiration from a plant in excess of the plant's ability to obtain moisture from the soil: e.g., a plant may wilt under hot dry winds, although there is moisture in the soil sufficient under normal conditions to supply the needs of the plant.

atmospheric drum dryer. A dehydrator for liquid food products consisting of a large, steam-heated drum which is coated with the food product as it slowly revolves. The water content is rapidly evaporated and close-fitting knives scrape the dehydrated material off the drum.

atmospheric pressure. Weight equal to that of air at sea level, or 14.7 pounds.

atomize. To reduce a liquid into very fine particles in a sprayer.

atrophy. The wasting away or decreasing in size of muscle, fat, or any tissue or organ. May result from weakening disease or from disuse.

attaint. A wound on a horse's leg caused by its own hoofs.

attapulgite. Clay mineral, $(OH_2)_4(OH)_2Mg_5Si_8 \cdot 4H_2O$, used as a carrier for poison dusts.

Atteal. Species of wild duck found in the Orkneys and Shetland Isles.

attendant's contract. A contract made by a livestock shipper with a railroad. The shipper agrees to furnish someone to look after the animals to relieve the railroad from responsibility.

attenuate. To dilute, thin down, enfeeble or reduce. Specifically, to reduce the virulence of an organism; to destroy the power of disease-producing bacteria by means of chemical action, or viruses by passing them through unnatural hosts, etc. This principle is used to develop certain bacterins and vaccines.

attractant. Specifically, any substance that attracts or draws insects or other animals. A substance identified as gyptol, obtained from the terminal segments of the female gypsy moths, is an *attractant* for male gypsy moths. Also called *attrahent.*

attrahent. See *attractant.*

A-type Merino. A type variation in the *Merino breed* (*q.v.*) of sheep which has very wrinkled skin covered with a fine-wool fleece and many body folds. See *B-type Merino; C-type Merino; Merino.*

atypical. Disagreeing with the form, state, or situation usually found under similar circumstances; not typical.

aubergine. The edible, blackish-purple fruit of the eggplant *q.v.*

aubin. An irregular gait of a horse, slower than a gallop but resembling it.

aubrietia. 1. Any plant of the genus *Aubrieta,* family Cruciferae. 2. A low, Eurasian, perennial herb, *Aubrieta deltoidea,* which is spreading and mat-forming. Grown for its pink, rose-lilac flowers in rock gardens and wall crevices, and as an edging plant. Also called *false wall cress; purple rock cress.*

auction. A public sale of goods to the highest bidders.

auctioneer. The agent for the sellers at an auction who sells the goods at the highest price. After the sale is completed he becomes the agent for the buyer.

aucuba mosaic. A virus disease of tomatoes, potatoes, and other plants, family Solanaceae; characterized by round, yellowish patches on the upper leaf surface.

auger. 1. A conveyor or elevator made on the principle of Archimedes' screw: a spiral or broad-threaded screw in an open tube. 2. A tool for boring holes: especially in agriculture, a broad-threaded screw for boring fencepost holes, etc. See *soil auger.*

Aujeszky's disease. A lethal virus disease of cattle, less so of dogs, cats, and pigs; characterized by intense itching, paralysis and convulsions. Death occurs within 36 to 48 hours after symptoms become evident. Also called *pseudorabies; infectious bulbar paralysis; mad itch.*

A.U.M. See *animal unit month.*

aureomycin. An antibiotic derived from the organism *Streptomyces aureofaciens.* Used in livestock feeds as a growth stimulant and for treatment of certain diseases.

australene. The most important component of English turpentine-oils which is produced from the extract of *Pinus australis.*

Australian beech. See *redbox eucalyptus.*

Australian bluebellcreeper. *Sollya heterophylla,* an evergreen vine that has small blue flowers. Also called *Tasmanian bluebell.*

Australian brome. *Bromus arenarius,* winter grass, family Gramineae, which has some weed-like characteristics. Also called *Australian chess; sand brome.*

Australian brown rot of citrus. A mildew of citrus caused by *Phytophthora syringae* Kleb; a brown rot of citrus found in Australia. See *brown rot of citrus fruits.*

Australian chess. See *Australian brome.*

Australian citrus scab. See *Tryon's scab.*

Australian cockroach. *Periplaneta australasiae* (Fab.), family Blattidae, an insect found throughout the world; particularly in southern U.S.

Australian fan palm. *Livistona australis,* one of the most widely cultivated ornamental fan palms in America.

Australian fern. See *birdsnestfern.*

Australian finger-lime. *Microcitrus australasica,* family Rutaceae, a plant of interest in citrus hybridization because of its vigor. Found in the coastal regions of southern Australia.

Australian grades of wool. Classifications determined by specific manufacturing use. In general, there are eight Merino groups for combing wool, four Merino grades for clothing wool, six grades of comeback (*q.v.*) wool, and eight grades of crossbred wool. *Cf. American grades of wool.*

Australian lady beetle. See *vedalia.*

Australian nut. See *Queensland nut.*

Australian oat. See *rescue brome.*

Australian orange. Rank-growing trees of the Washington navel orange, bearing coarse or inferior fruit.

Australianpea dolichos. *Dolichos lignosus,* an Asiatic, evergreen, annual, herbaceous vine, family Leguminosae; grown for its white or rose-purple flowers.

Australian pine. See *horsetail beefwood.*

Australian-pine borer. *Chrysobothris tranquebarica* (Gmel.), an insect pest of mangrove and casuarina in Florida. The larva bores into the sapwood, constructing pupal cells.

Australian ryegrass. See *Italian ryegrass.*

Australian saltbush. *Atriplex semibaccata,* an alkaline and saline tolerant herb, family Chenopodiaceae. It is of some feed value to livestock. Native to Australia but has spread by escape along the coast of California. Also called *creeping saltbush.*

Australian teatree. See *Victoria teatree.*

Australian wool. Wool grown and sheared in Australia.

Australorp. A breed of chicken with black plumage developed for its egg production. The cock weighs up to 8½ pounds and the hen up to 6½ pounds.

Austrian brier rose. *Rosa foetida,* an Asiatic shrub, family Rosaceae, which grows seven to ten feet in height. Its deep yellow flowers are nearly three inches wide, and have an unpleasant odor.

Austrian brome grass. See *smooth brome.*

Austrian fieldcress. *Rorippa austriaca,* a perennial weed, family Cruciferae.

Austrian pine. *Pinus nigra,* an evergreen tree, family Pinaceae. One of the most widely cultivated of all ornamental pines; used in shelterbelt planting.

Austrian winter pea. See *garden pea.*

autecology. The study of the ecology of any individual species.

autochthonous flora. The native croflora, usually present in the soil in relatively constant numbers. They are presumed to subsist on the more resident soil organic matter and consequently to be little affected by the addition of fresh organic materials. See *zymogenic flora.*

autoclave. A pressure cooker; a container for cooking, sterilizing, etc., with steam under pressure.

autoecious rust. A rust having all stages of its life cycle on a single species of a plant.

autogamous. Self-fertilizing.

autoheader. An engine driven *header q.v.*

autointoxication. Literally, self-poisoning; the formation and absorption of a poison within one's body.

autolysis. Self-digestion. The natural softening process of fruits or vegetables after picking, or meat after slaughtering.

automatic topper-lifter-harvester. A machine, usually tractor drawn, which lifts beets from the soil, cuts off the leaves (tops), and conveys the beets to a wagon or trailer.

autopass. A device constructed to allow automobile traffic but to prevent cattle or sheep from getting out of a pasture.

autopsy. Examination, including dissection, of a carcass to learn the cause and nature of a disease or the cause of death. Also called *post mortem.*

autosome. Any chromosome (*q.v.*) except the sex chromosome (allosome).

autosyndesis. In genetics, the pairing during meiosis of homologous chromosomes derived from the same original species. This occurs in polyploids, particularly allopolyploids.

autotrophic. Capable of utilizing carbon dioxide as a source of carbon; of obtaining energy for the reduction of carbon dioxide, and other life processes from the oxidation of inorganic elements or compounds: e.g., sulfur, hydrogen, ammonium, and nitrite salts, or from light. See *heterotrophic.*

autumn. Fall; the season between summer and winter.

autumncrocus. 1. Any bulbous herb of the genus *Colchicum,* family Liliaceae. A source of colchicine; poisonous to cattle, horses, sheep, but very popular in autumn gardens. 2. *Colchicum autumnale,* a Mediterranean bulbous herb, family Liliaceae, grown for its autumnal, purple or white flowers. Also called *meadow saffron.* See *Bornmueller autumncrocus; showy autumncrocus.*

autumn lycoris. *Lycoris squamigera,* a Japanese, bulbous herb, family Amaryllidaceae. After its foliage has died, its lilac and pink flowers blossom.

autumn sun. *Rudbeckia nitida,* family Compositae, a perennial, flowering herb, grown for its yellow flowers. Native to southeastern U.S.

auximone. Any of several substances required in minute quantities for plant growth.

auxins. Plant hormones; substances which promote the growth of plants.

available energy. See *metabolizable energy.*

available moisture. The amount of water in soil that can be absorbed by the roots of plants. Technically, it is the difference in the weight of moisture held in a soil at field capacity and that held at permanent wilting. It may be expressed in inches of water per unit depth of soil.

avenaceous. Of, pertaining to, or like oats.

avian. Of, or pertaining to, birds.

avian diphtheria. See *fowl pox.*

avian encephalomyelitis. A virus disease of chickens which affects the nervous system. Found only in chicks from one to three weeks old; characterized by tremor of the head, lack of coordination, sitting on haunches.

avian leukosis complex. A widespread, virus, transmissible disease-complex of chickens which is more destructive than all other poultry diseases combined. See the various types: *lymphomatosis, neural lymphomatosis gallinarum, ocular lymphomatosis, visceral lymphomatosis, osteopetrotic lymphomatosis, erythroblastosis, granuloblastosis, myelocytomatosis.*

avian monocytosis. A fowl disease of unknown cause often found in pullets of five to seven months old. It is characterized by depression, lack of appetite, darkening of the comb, watery diarrhea, distended crop, dehydration, drop in egg production. Death loss is from five to ten percent. Also called *pullet disease; blue comb; X disease; contagious indigestion.*

avian pasteurellosis. See *fowl cholera.*

avian plague. A lethal, infectious, virus disease of fowls; characterized by dullness, drooping wings and tails, closed eyes, running nose and eyes, thirst, rapid breathing, high temperature, brownish combs and wattles, and sudden death. The whole range of symptoms occurs within 24 to 36 hours. Prevalent in Europe and the Middle East. Also called *bird pest; fowl plague; Brunswick bird plague; fowl pest.*

avian pneumoencephalitis. See *Newcastle disease.*

avian tuberculosis. A disease of fowls, particularly of chickens; caused by the acid-fast microorganism, *Mycobacterium avium.* It is characterized by loss of weight, listlessness, lameness, good appetite, diarrhea; the comb, wattle, and skin become pale; may result in death.

avian typhoid. A wide-spread, infectious disease of of fowls caused by the microorganism, *Salmonella gallinarum*; characterized by droopiness, listlessness, head drawn in, sagging wings, ruffled feather, pale comb and wattles, yellowish or greenish diarrhea. May result in death.

aviary. A place for keeping birds: pigeon loft, etc.

aviculture. The caring for and the raising of birds.

avirulent strain. A type of a normally pathogenic microorganism that has no disease-producing power on a particular host species or variety.

avitamic acid. See *vitamin C.*

avitaminosis. A condition or disease of an animal resulting from a vitamin deficiency.

avives. Strangles *q.v.*

avocado. See *American avocado.*

avocado anthracnose. Caused by imperfect fungus *Colletotrichum gloeosporioides,* family Melanconiaceae; characterized by sunken black spots up to ½ inch in diameter. Also called *black spot.* See *anthracnose.*

avocado red mite. *Oligonychus yothersi* (McG.), family Tetranychidae. The larvae of this mite feed on the leaves of avocado, camphor tree, Australian silky oak, eucalyptus, elm, willow, white oak, pecan. Found in California, Florida.

avocado root rot. See *Phytophthora.*

avocado scab. A fungous disease of avocado caused by *Sphaceloma perseae,* family Melanconiaceae; characterized by corky, raised, brown oval spots on the fruit. The fruit has a russeted appearance and cracks may develop in the skin. *Cf. black spot.*

avocado sun blotch. See *sun blotch.*

avocado whitefly. *Trialeurodes floridensis* (Quaint.), family Aleyrodidae. This sucking insect infests papaya, banana, guava, anona, citrus in Florida.

avoirdupois. In English-speaking countries, the established system of weighing materials other than medicines, precious stones, and precious metals. The ounce contains 16 drams; the pound, 16 ounces. But the hundredweight contains 100 pounds in the U.S., 112 pounds in the U.K.

avulsion. The sudden transference of land from the estate of one party to that of another without change of ownership. Caused by inundation, or by sudden change in location of the channel of a river or stream.

away-going crop. A crop sown before, but which cannot ripen until after the expiration of tenancy. The old tenant is entitled to it.

"A" white grease. Grease secured from inedible hog fat which contains less than two percent fatty acid.

awl nematode. *Dolichodorus heterocephalus,* Class Nematoda, an ectoparasite which infests the root tips of many plants causing stunting to the plants.

awn. A hair-like appendage found on glumes and lemmas in barley, oats, and many other grass plants. Also called *beard.*

awnless brome. See *smooth brome.*

axial feather. The short feather in the middle of the wing of a bird which separates the primary from the secondary feathers.

axil. The angle resulting from the divergence of a leaf from the stem.

axillary. Developing in the axil of a leaf, as a bud or flower.

axis. 1. The main stem of a plant. 2. The second cervical vertebra.

axunge. Rich fat from the center of the kidneys of pigs and geese.

Aylesbury. A white, domesticated breed of ducks weighing about nine pounds when mature.

Ayrshire. An important breed of dairy cattle originating in Scotland. The animals vary in color from red to brown, or mahogany interspersed with white spots. The Ayrshire is between the Jersey and Holstein in size; the bulls weigh from 1400 to 2000 pounds.

Ayrshire field rose. See *field rose: 2.*

azalea. Certain ornamental shrubs of the genus *Rhododendron* (*Azalea*), family Ericaceae, which do not have leathery leaves. See *sweet azalea; flame azalea; piedmont azalea; Ghent azalea; Japanese azalea; pinxterbloom azalea; hiryu azalea; roseshell azalea; royal azalea; pinkshell azalea; swamp azalea; yodogava azalea; Chinese azalea.*

azalea bark scale. *Eriococcus azaleae* Comst., family Dactylopiidae. This insect infests azaleas in greenhouses. Also called *azalea mealybug.*

azalea bud rot and twig blight. A disease of azaleas which is caused by the fungus *Briosia azaleae* (syn. *Periconia* or *Sporocybe*), family Dematiaceae. It kills the flowers, leaf-buds and twigs in the blooming period.

azalea gall. A swelling found on the leaves of azaleas, caused by the fungus *Exobasidium vaccinii,* family Exobasidiaceae.

azalea lace bug. *Stephanitis pyrioides* Scott, family Tingidae, a sucking insect which is a major pest, causing the foliage to turn brown.

azalea leaf miner. The caterpillar of the moth, *Gracilaria azaleella* Brants, family Gracilariidae, burrows into and rolls the leaf around itself. Usually found in greenhouses. Also called *azalea leaf roller.*

azalea mealybug. See *azalea bark scale.*

azalea pot. See *three-quarter pot.*

azalea whitefly. *Aleyrodes azaleae* B & M, family Aleyrodidae, a sucking insect frequently found in southern U.S.

azarole. *Crataegus azarolus,* family Rosaceae, a hawthorn-like shrub bearing sweet-flavored fruit. Native to southern Europe. Also called *Neapolitan medlar.*

azobenzene. $C_{12}H_{10}N_2$, mite or tick fumigant or spray used in greenhouses.

azonal soils. Any group of soils lacking well developed profile characteristics, owing to immaturity or conditions of parent material. In the U.S these groups include the following: alluvial soils, lithosoils (skeletal soils), and some dry sands.

azotemia. See *azoturia.*

azoturia. An ailment common to draft horses kept on full feed when idle. It is characterized by excessive sweating, stiff movements, lameness, reluctance to move, sitting down or lying on one side, stiffness of muscles, passing red or brown urine. Also called *azotemia, blackwater, holiday disease, lumbago, Monday morning disease, myohaemoglobinuria, paralytic haemoglobinaemia.*

Aztec dahlia. *Dahlia pinnata,* a tuberous-rooted herb, family Compositae, grown for its rose or lilac-colored blossoms. Native to Mexico.

Aztec lily. *Sprekelia formosissima,* bulbous herb, family Amaryllidaceae, grown for its erect bright red flowers. Native to Central America. Also called *St. James's-lily; Jacobean lily.*

Aztec marigold. *Tagetes erecta,* tender, annual herb, family Compositae, grown for its yellow or orange flowers. Native to Mexico. Also called *big marigold; African marigold.*

Aztec tobacco. *Nicotiana rustica,* family Solanaceae, a wild tobacco of high nicotine content. It is sometimes cultivated in flower gardens and used in the manufacture of insecticides. The tobacco first grown by the Indians in eastern U.S. Also called *emetic weed; Indian tobacco; wild tobacco; eyebright.*

B

B. A. 1-6. Buenos Aires or South American wool grades. *Cf. American grades of wool, Australian grades of wool.*

babassu. A huge palm, *Orbignya martiana,* family Palmaceae. The nuts of this tree are ground for livestock feed, and produce an oil used in making soap. Native to Brazil.

Babcock test. A fat test of milk, introduced by Dr. S. M. Babcock, University of Wisconsin in 1890. The test is based on the action of sulfuric acid centrifugation and may be modified for products other than milk.

Babesia. A genus of single-celled organisms, class Sporozoa, which parasitize cattle, sheep, and horses, causing diseases of the bloodstream.

babesiosis. See *piroplasmosis.*

baby beef. A market classification denoting a carcass 8 to 18 months old of a prime grade of beef breed, which weighs from 600 to 1,100 pounds on foot.

babyblue-eyes nemophila. *Nemophila menziesii,* a North American, annual herb, family Hydrophyllaceae, grown in rock gardens for its very small size and its blue flowers. Also called *babyblue eyes.*

baby combing. See *French combing.*

baby gardenpea. *Pisum sativum* var. *humile,* an edible green pea, family Leguminosae, which grows up to 24 inches high. Also called *early dwarf pea.*

baby lima (bean). *Phaseolus limensis* var. *limeanus,* a dwarf vegetable bush, family Leguminosae.

Babylon weeping willow. *Salix babylonica,* deciduous tree, family Salicaceae, having long, slender, pendulous branches; grown as an ornamental. Native to China. Also called *weeping willow.*

baby nucleus. A ripe queen bee cell which is brought to maturity by being placed in a hive between two honeycombs of bees.

baby rambler. A small rose (*q.v.*) resulting from crossing *Rosa chinensis* var. *minima* and *R. multiflora.*

baby's-breath. *Gypsophila paniculata,* a handsome, Eurasian, perennial herb, family Caryophyllaceae. It is grown in borders or beds for its white and pink flowers. Also called *gypsum pink mist; chalk plant.*

baby's-slippers. See *birdsfoot deervetch.*

babytears. *Helxine soleiroli,* a moss-like, pot plant or ornamental ground cover, family Urticaceae. Native to Corsica and Sardinia. Also called *angelstears; Corsican nettle.*

bacaba. *Oenocarpus bacaba or O. distichus,* family Phoemicaceae, a tree with pinnate leaves and edible fruit. Native to South America.

bacciferous. Producing berries.

bachelor's-button. See *English daisy; cornflower; globeamaranth.*

bacillary white diarrhea. See *pullorum disease.*

Bacillus. A genus of single-celled, rod-shaped organisms, family Bacillaceae. One species, *B. anthracis,* causes anthrax (*q.v.*) in animals. Although most other species are harmless, they do cause chemical changes in animal and vegetable matter.

bacitracin. An antibiotic derived from the organism, *Bacillus subtilis.*

backband. That part of a harness made of a piece of leather or heavy canvas which passes over the back of a horse and supports the shafts of a vehicle.

back break. A new growth caused by cutting through the pseudobulb of an orchid preparatory to severing and potting.

back-burn. A controlled fire set to burn against the wind. It is used as a firefighting measure in prairie and forest fires. See *backfire.*

backcross. The crossing of a hybrid with one of the parental types. The offspring are referred to as the backcross generation or progeny. See *testcross.*

backfire. A fire set intentionally ahead of an advancing forest or prairie fire to burn undergrowth, etc., and thus to stop or to control the main fire. See *backburn.* Also called *fire-break.*

back house. See *privy.*

back line. See *haul-back.*

back pad. Part of the harness for a horse: a wide leather strap which rests just behind the shoulders.

backset. A stick or rod placed inside a carcass to spread the carcass open to make it easier to dress for market.

Backsteiner. A soft, ripened brick-shaped type of Limburger cheese made from partly skimmed milk in northern Germany.

backstrap. A leather strap attached at one end to the collar and at the other to the hip strap of a harness.

backwash. To remove oil from wool after it has been combed.

backwoods. Woods lying beyond more settled areas. Uncleared wooded country.

backyarder. One who breeds animals or fowl in a small area.

bacon. 1. The cured or smoked back and sides of a pig carcass. 2. Formerly used to describe any part of a hog carcass.

bacon beetle. See *larder beetle.*

bacon square. A retail cut of meat from the jowl of a hog; used for seasoning and panbroiling.

bacon-type hog. A type of hog bred to produce a large amount of lean meat. See *Tamworth; Yorkshire.*

baconweed. See *lambsquarters goosefoot.*

bacteria (singular: **bacterium**). Microscopic, one-celled, chlorophyll-less organisms which multiply by fission and do not form noticeable fungous threads. Some are parasites upon other plants or upon animals, or may live saprophytically upon non-living, organic matter. Sometimes they are classified in Class Schizomycetes, Order Bacteriales or are divided among several orders. Also called *germ.*

bacteria count. The number and kind of bacteria per unit of volume found in a substance.

bacterial black rot of the potato. See *blackleg.* 2.

bacterial black spot of stone fruits. See *bacterial spot of stone fruits.*

bacterial blight of barley. A disease caused by the bacterium, *Xanthomonas translucens.* It is characterized by small, water-soaked translucent spots that elongate and dry up resulting in the death of the leaves. There is also a bacterial exudate.

bacterial blight of beans. A disease caused by *Xanthomonas phaseoli.* It is characterized by small, translucent, water-soaked leaf spots, the surrounding tissues turning yellow and dying; small spots on the pods turning reddish with age; and sometimes by water spots or streaks appearing on the petioles and stems. A brownish discoloration of the vascular system becomes evident as the disease progresses. A plant may break over at the first node due to surface cankers and the seeds may shrivel and become discolored. See *halo blight of beans; fuscous blight; bacterial brown spot of beans; bacterial wilt of beans.*

bacterial blight of carrot. A serious disease caused by *Xanthomonas carotae.* It is characterized by yellow spots on the tips of the leaves, which turn brown and appear water-soaked; bacterial exudate appears on the stem. The leaves dry up and die. Found in western U.S.

bacterial blight of celery. A disease caused by the bacterium, *Pseudomonas apii.* It is characterized by small, yellow spots on the leaves which turn brown and develop a yellow border.

bacterial blight of clover. A disease caused by *Pseudomonas syringae.* It is characterized by angular, dark brown spots on the leaves, stipules, petioles and stems; also there is some exudate.

bacterial blight of corn. See *bacterial streak; bacterial stripe of sorghum; bacterial wilt of corn.*

bacterial blight of cotton. A disease caused by *Xanthomonas malvacearum* which is prevalent in all cotton producing areas. It is characterized by small, circular, or angular, translucent, water-soaked spots on the leaves which darken to brown or black. It may result in the shedding of leaves, the shedding of bolls, stem blight, boll rot, and a reduction of the lint grade due to stains. The damage over the whole U.S. is from one to two percent of the total crop. Also called *angular leaf spot; bacterial boll rot; black arm; bacterial boll spot; vein blight.*

bacterial blight of cowpeas. A disease caused by the bacterium *Xanthomonas vignicola.* It is characterized by stem cankers that blight and sometimes kill the infected plant.

bacterial blight of filberts. A most destructive parasitic disease caused by *Xanthomonas coryli* which is present in nearly all filbert orchards in Washington and Oregon. It is characterized by small, pale, yellowish-green, water-soaked spots on the leaves which eventually turn brown and water-soaked spots on the bark which later turn reddish-brown, and which may girdle twigs and branches.

bacterial blight of grapes. A disease caused by *Erwinia vitivora.* It is characterized by pale, yellowish-green, water-soaked spots on the young shoots which eventually become grayish-black, extend along the shoot, and widen around the nodes; cracks appear on the lesion and the shoots are malformed. On the leaves there is a bright red, marginal discoloration and necrosis sets in; the bunch stems die quickly and the fruit withers. Found in Europe and South Africa.

bacterial blight of grasses. See *chocolate leaf spot.*

bacterial blight of peas. A disease caused by *Pseudomonas pisi.* It is characterized by small, translucent, water-soaked spots on the leaves and pods and by similar streaks on the stems. The leaf spots turn brown and coalesce causing the leaves to wither and die; the stem lesions remain water soaked. The bacteria also infect cowpeas, sweetpea, hyacinth bean, and the perennial pea.

bacterial blight of rice. A disease caused by *Xanthomonas oryzae.* It is characterized by water-soaked linear stripes on the leaf between the veins of the leaf blade and the sheath. The stripes become light brown or bleached and show an exudate. Found in the Philippines, Japan and Mexico.

bacterial blight of rye. A disease caused by *Xanthomonas translucens* f. sp. *secalis.* It is characterized by irregular shaped lesions on the leaves.

bacterial blight of soybeans. A widespread disease caused by the bacterium, *Pseudomonas glycinea.* It is characterized by small, angular, yellow, water-soaked spots (sometimes with a yellowish-green ring) on the leaves. The spots later turn brown or black and the adjacent leaf surface turns yellow. Defoliation can result.

bacterial blight of walnut. A most serious disease of Persian walnuts caused by the bacterium, *Xanthomonas juglandis.* It is characterized by reddish-brown spots on the margins of the leaves and by black depressed spots on the stems that may circle the stem and kill it; also by the young leaf and catkin bud turning black or dark brown and dying, and by black spots on the nuts.

bacterial boll rot. See *bacterial blight of cotton.*

bacterial boll spot. See *bacterial blight of cotton.*

bacterial brown spot of beans. A disease caused by the bacterium, *Pseudomonas syringae.* It is characterized by very small, water-soaked spots on the leaves and pods which eventually turn reddish-brown; sunken lesions may also appear on the stem.

bacterial canker of cherries. See *bacterial canker of stone fruits.*

bacterial canker of cowpeas. See *bacterial blight of cowpeas.*

bacterial canker of poinsettias. A disease caused by *Phytomonas poinsettiae.* It is characterized by elongated, water-soaked streaks on the stem and on the leafstalks resulting in the spotting of the leaves and defoliation. In the advanced stage there is an exudate from the stem and leaf lesions.

bacterial canker of stone fruits. A widespread disease caused by the bacterium, *Pseudomonas syringae.* It is characterized by abnormal oozing of gum or watery material from the affected branches; cankers on the branches and trunk; elongated, water-soaked or gum-soaked lesions in the bark or outer sapwood; blighting of dormant buds. It infects lilac, most stone fruits, pear, apple (causing blossom blast), citrus, avocado, clover, cowpea, lima bean, common bean, rose, sorghum and related grasses, corn, pearl millet, foxtail, hibiscus, and forsythia. Also called *bacterial gummosis; bacterial sour sap.*

bacterial canker of sugar beets. A disease caused by *Phytomonas beticola.* It is characterized by overgrowths on the leafstalk, the root, and the crown; the center of the overgrowth is yellow and water-soaked. It is found in Virginia, Michigan, Maryland, Wisconsin, Colorado, Wyoming, New Mexico, Utah.

bacterial canker of tomato. A widespread seed-borne disease caused by *Corynebacterium michiganense.* It is characterized by brown spots on mature green fruit which are raised and surrounded by a white ring; the leaves wilt on mature plants and light streaks appear on stem and leafstalks; the plant is stunted.

bacterial count. See ***bacteria count.***

bacterial gall. Hypertrophy, usually also hyperplasia caused by bacteria; a condition of overgrowth which results from a pathogen effect on a host. See *crown gall; olive knot; hairy root; cane gall of brambles; beet pocket rot; oleander knot.*

bacterial gummosis. Exudation of gum-like liquid resulting from bacterial infection, particularly *bacterial canker of stone fruits q.v.*

bacterial gummosis of sugar cane. A disease caused by *Pseudomonas vascularum.* It is characterized by slime in the vascular bundles, dwarfing of the plant, streaking of the leaves, and general decay.

bacterial leaf blight of corn. A sporadic disease caused by *Pseudomonas alboprecipitans.* It is characterized by lesions in the shape of narrow water-soaked and olive-green stripes which later turn to a light brown. The leaves shred and die as the infection spreads; the stalk begins to rot at the top of the plant; the plant is dwarfed, and tassels may not appear.

bacterial leaf spot of begonia. A disease caused by *Phytomonas begoniae.* It is characterized by dead spots with translucent margins appearing on the leaves.

bacterial leaf spot of carnation. A greenhouse disease caused by *Phytomonas woodsii.* It is characterized by brown spots surrounded by purple, eventually light green, rings, and exudate. Later, the leaves wither, twist, and turn completely brown, but the spots remain clearly visible.

bacterial leaf spot of celery. A disease caused by *Pseudomonas apii.* It is characterized by small reddish-brown spots on the leaves which remain small and sometimes have pale yellow rings.

bacterial leaf spot of clovers. A widespread, but relatively minor, disease caused by *Pseudomonas syringae.* It is characterized by lesions on the leaves and sometimes on the stems, petioles, and flower pedicels. The small spots appear first on the lower leaf surface and are translucent; as they enlarge, they fill the area between the veins and become black except for the margins which remain water-soaked; the tissue around the spots is yellowish-green. See *bacterial leaf spot of corn.*

bacterial leaf spot of corn. A relatively minor disease caused by *Pseudomonas syringae.* It is characterized by small, dark green, water-soaked lesions on the leaves; the spots develop into light brown with a dark brown border surrounded by a yellowish ring. See *bacterial leaf spot of clovers.*

bacterial leaf spot of maple. A disease caused by *Phytomonas aceris.* It is characterized by small, water-soaked dots on the leaves surrounded by yellow rings; the spots in time turn black or brown; the petioles may become infected, causing defoliation.

bacterial leaf spot of peaches. A disease caused

by *Xanthomonas pruni.* It is characterized by small irregular lesions on the leaves and small corky spots on the fruit. The infected area dies and falls out of the leaf giving it a shot-hole appearance; sometimes the tree may be defoliated. May attack plum trees or cherry trees. Also called *bacteriosis of peach, plum, cherry.*

bacterial leaf spot of tobacco. See *blackfire.*

bacterial ooze. A discharge, exudate, of sap or gum, etc., from a plant as a result of a bacterial disease.

bacterial pocket disease of sugar beets. A disease caused by the bacterium, *Pseudomonas beticola.* See *bacterial canker of sugar beets.*

bacterial pustule of soybeans. A widespread disease caused by *Xanthomonas phaseoli.* It is characterized by small, yellowish-green spots on the leaves and sometimes on the pods. The center of the spot is reddish-brown; pustules often develop on the underside of the leaf resulting in irregular brown areas encircled with yellow.

bacterial ring rot of potato. See *ring rot.*

bacterial rot of lettuce. See *slime: 1.*

bacterial soft rot. One of the most destructive diseases of vegetables in storage or in transit which is caused by one or by combinations of the following: *Erwinia carotovora, E. atroseptica, E. aroideae, Bacillus polymyxa, B. megatherium, Pseudomonas solanacearum, P. viridilivida, P. cepacia.* It is generally characterized by soft decay of the fleshy tissues which become watery or slimy. Some affected vegetables give off an offensive odor.

bacterial speck. A disease of tomatoes caused by *Pseudomonas tomato.* It is characterized by tiny raised dark brown specks. Decay does not extend into the flesh of the fruit.

bacterial spike blight of wheat. A disease caused by *Corynebacterium tritici.* It is characterized by a yellow exudate on the diseased inflorescence. Found in Australia, China, Egypt, and India.

bacterial spot of cucumber. A disease caused by *Pseudomonas lachrymans.* It is characterized by small, circular, water-soaked spots that eventually dry and crack. The centers of these spots become sunken and chalky white, and sometimes a gummy exudate is present. Often results in soft rot. Also called *angular leaf spot.*

bacterial spot of lima beans. A disease caused by *Pseudomonas syringae.* It is characterized by reddish-brown spots on all the above-ground parts. The most seriously affected part is the pod which often drops off the plant resulting in a reduction of yield.

bacterial spot of peach. See *bacterial spot of stone fruit.*

bacterial spot of peppers. A common disease caused by *Xanthomonas vesicatoria.* It is characterized by spots on the leaves and fruit.

bacterial spot of sorghum. A common disease caused by *Pseudomonas syringae.* It is characterized by spots which vary from dots to ½ inch spots. At first they are water-soaked; later they become dry with a red or brown border. The spots may run together to form stripes. May attack Sudangrass, Johnsongrass, pearlmillet, foxtail millet, broomcorn, corn.

bacterial spot of stone fruits. A serious disease especially prevalent on peach, plum, and occasionally on sweet cherry, caused by *Xanthomonas pruni.* It is characterized by small, circular, light brown, water-soaked spots on the leaves and fruit that soon enlarge, dry out, darken, and crack. Small cankers may result from succulent twig infections, and sometimes a translucent exudate is present which dries to a flaky crust or white powder.

bacterial spot of tomato. An important disease caused by *Xanthomonas vesicatoria.* It is characterized by raised dark brown scabby spots ⅛ to ¼ inch in diameter. Old spots are gray and paper-like. Decay does not penetrate the flesh of the fruit.

bacterial standard. The maximum number of bacteria allowed per milliliter of milk for a particular grade. It is usually stated in reference to plate or microscopic count.

bacterial stem blight of alfalfa. A minor disease caused by *Pseudomonas medicaginis.* It is characterized by lesions on the stems, petioles, leaves. At first, the lesions have a light brown bacterial exudate, and appear water-soaked. Later they turn dark brown or nearly black. The disease follows wet weather and low temperatures on the first crop. Found in western U.S.

bacterial streak. A disease of Johnsongrass, Sudangrass, and sorghums caused by *Xanthomonas holcicola.* It is characterized by long, narrow, water-soaked lesions which have a light brown exudate. Later, red or brown margins develop around the lesions, and the exudate dries to become white or cream-colored scales.

bacterial stripe. An elongated brown leaf lesion which is common to several different bacterial diseases of cereals and grasses. On grass, it is caused by *Xanthomonas translucens;* on oats by *Pseudomonas striafaciens;* on broom corn millet by *Xanthomonas panici.*

bacterial stripe of sorghums. A very common and destructive disease of Sudangrass, Johnsongrass and sorghums. It is caused by *Pseudomonas andropogoni* and is characterized by long, purplish-red, brown, or tan lesions with exudate that dries to form scales of the same color as the stripes.

bacterial wilt of alfalfa. One of the most destructive diseases of alfalfa caused by *Corynebacterium insidiosum.* It is characterized by yellowing and dwarfing of the stems and leaves; sometimes by a rot of the crown, a yellow ring just under the skin, and a yellowing of the roots. The disease is most serious in wet areas.

bacterial wilt of beans. A disease caused by *Corynebacterium flaccumfaciens.* It is characterized by water-soaked lesions on the leaves and stalks. The pods become slightly discolored along the suture, and the seeds are shrunken and yellow.

bacterial wilt of carnation. A disease caused by *Phytomonas caryophylli.* It is characterized by the tissue under the bark turning yellow; the roots rotting and becoming sticky. The bark on the stem also becomes soft and sticky and the plant dies quickly.

bacterial wilt of corn. A disease which is particularly troublesome on sweet corn. It is caused by *Bacterium stewartii* which infects the whole plant; characterized by long, wavy streaks on the leaves, which emit a yellow exudate. Plants that survive are stunted and produce abnormal ears. The disease attacks sweet, dent, and pop corn; teosinte, and Jobstears. It is more severe after a mild winter. Also called *Stewart's disease.*

bacterial wilt of cucurbits. A serious disease of cucumber and muskmelon (rarely of the other cucurbits) caused by *Erwinia tracheiphila.* It is characterized by the withering of first a single leaf followed by the complete withering and death of the plant. Freshly cut infected vines have a white exudate. It is found in the midwestern, northeastern, and north central U.S.

bacterial wilt of eggplant. A disease of solanaceous plants, caused by *Pseudomonas solanacearum.* It is a vascular disease resulting in withering and death. See *bacterial wilt of peppers, tomatoes.*

bacterial wilt of peanut. A widely distributed disease, caused by *Xanthomonas solanacearum.* It is a very serious enemy of the peanut in the more tropical parts of the world. This organism causes bacterial wilt or brown rot of many species of the family Solanaceae; e.g., potato, tomato, tobacco, eggplant, etc.

bacterial wilt of peppers. A disease caused by *Pseudomonas solanacearum.* It is characterized by drooping of the leaves followed by withering and death of the plant; the inner tissues of the bark having a dark, water-soaked appearance.

bacterial wilt of potato. See *brown rot of potato.*

bacterial wilt of tobacco. See *Granville wilt.*

bacterial wilt of tomatoes. A disease caused by *Pseudomonas solanacearum.* It is characterized by a sudden withering and a black discoloration of the stem. If cut, the stem has a gray exudate.

bactericide. Anything which destroys bacteria; a germicide. See *disinfectant.*

bacteriemia. The presence of bacteria in the blood.

bacterin. Bacterial vaccine injected into animals to increase their resistance to infection.

bacteriophage. A virus-like, bacteria-destroying agent which can propagate itself only in the presence of young, active, susceptible bacteria.

badger. 1. A nocturnal carnivorous mammal, *Taxidea taxus,* family Mustelidae, the symbol of the state of Wisconsin. It has a black and gray coat, with a white streak running from the muzzle to a point behind the ears, short legs, a heavy body, and a round back. Native to North America. 2. A similar European species, *Meles meles,* family Mustelidae.

badlands. Areas of clay, soft shale, sandstone, or limestone carved into gullies by the dissection of many dried-up watercourses which have entrenched themselves in the soft lands. A specific example on a large scale is the dissected region east and south of the Black Hills in western South Dakota.

bad quarter. See *mastitis.*

bad water. Natural water which is unfit for drinking, especially alkali water found in western U.S.

baelfruit. *Aegle marmelos,* a small, thorny tree, family Rutaceae, which has large, greenish-yellow flowers. Its fruits have a hard shell and a fragrant, mucilage-like pulp that is edible. The pulp is used for preparing drinks and is believed to have medicinal properties. It is native to India; occasionally grown in warmer parts of the U.S. Also called *bel apple; elephant apple.*

bag. 1. Udder. 2. Scrotum. 3. A sack. 4. The total kill in a hunter's possession.

bagasse. Residue from sugar canes or stalks after the juice has been extracted by crushing. Used in the manufacture of wall boards, insulating materials, and in mixtures with molasses for livestock feeds. Also denotes sugar beet, sorghum, and sisal residue.

Bagazzo. A hard, ripened, Parmesan-type cheese.

bagflower. See *glorybower.*

bagger. A hollow tube shaped like an inverted capital *J* through which grain, etc., is blown into bags for packaging.

bagger elevator. A tube or open metal chute by which potatoes, litter, grain, etc., are raised to another level.

bagging. 1. Cotton, jute, or other cloth material used in covering cotton bales. 2. Any cloth used for making bags.

baggy crop. See *pendulous crop.*

bag-type strainer. A cloth bag used to filter extraneous materials from honey, etc.

baguilumbang. See *soft lumbang.*

bagworm. *Thyridopteryx ephemeraeformis* (Haw.), family Psychidae. The larvae of this moth spin a silken bag or sack about themselves for protection. They infest nearly all trees. Found in eastern U.S.

Bahama grass. See *Bermudagrass.*

Bahama hemp. See *sisal.*

Bahiagrass. *Paspalum notatum,* family Gramineae, a palatable range grass whose seed is low in germination.

bail. 1. A half-ring or half-circle such as the handle of a bucket. 2. A *doubletree q.v.* 3. A portable or mobile *dairy house q.v.*

bait. A food lure to capture or destroy birds, beasts, fishes, or insects.

bait broadcaster. See *bait spreader.*

bait comb. A comb honey placed in the super of a beehive to stimulate the bees to begin work in comb foundation. Also called *bait section.*

bait dispenser. A device which is used to dispense or discharge poison bait from an airplane.

baited sirup. A rich sugar sirup in which specific blossoms are steeped. It is placed in a beehive in the evening so that the bees, having fed on it, will in the morning go out and visit similar blossoms.

bait insecticide. An appetizing mixture of poison and food which when eaten by an insect results in its death.

bait spreader. A device consisting of a large, rotating disk onto which poisoned bait is dispensed from a hopper. The disk distributes the bait 7 to 10 yards from the spreader.

bait the super. To entice bees to use the super of a beehive and not to swarm by means of placing drawn combs among the frames of the comb foundation in the super. See *bait comb.*

bajada. The slope of an alluvial fan or piedmont in the arid regions of the southwestern U.S. Also called *bahada.*

Bakanae disease of rice. See *cereal scab.*

baked soil. Soil which has a hard, or even-glazed, surface due to a wet condition followed by an intense drying action of the sun.

Baker mealybug. See *grape mealybug.*

Bakers' cheese. A skim milk cheese similar to cottage cheese, but finer grained and containing acid.

balance. 1. The even or proportionate distribution of parts or weight; symmetry. 2. Specifically an important farm organization term used to express the ratio between the resources of land, labor, capital and management which attains the optimum use of each resource in the production of crops and livestock to maximize financial returns and to maintain or improve soil productivity.

balanced fertilizer. A soil additive containing suitable proportions of each necessary mineral element to develop a plant or crop.

balanced ration. A daily allowance of livestock or fowl feed; mixed to contain suitable proportions of nutrients required to promote normal development.

balance frame. The frame of a horse-drawn, walking-wheel cultivator designed so that the tongue remains horizontal when the gangs are raised; as a result, very little weight hangs on the horses' necks.

balance sheet. A list of the assets (current and fixed) and the liabilities (current and other) of an individual or business. The difference between assets and liabilities is the owner's equity, net worth, or financial gain (or loss). See *financial statement.*

Balantidium coli. A parasitic genus which attacks the intestines of mammals, especially pigs. It can also cause severe dysentery in human beings.

bald. 1. Designating a small, treeless area in a forested hilly or mountainous region, especially a crest, as *bald* knob. 2. Beardless, as *bald* wheat.

baldcypress. *Taxodium distichum,* a deciduous swamp tree, family Pinaceae, which is very durable in water or moist conditions; grown as an ornamental and used in outdoor construction for silos, railway ties, in boat-building, etc. Found in southern U.S. Also called *taxodium; deciduous cypress.* See *pond baldcypress.*

bald eagle. *Haliaeetus leucocephalus,* the North American bird portrayed on the coins and in the coat of arms of the U.S.

bald face. A white face or a white streak on the face of an animal.

baldness. See *favus: 1.*

Baldwin. A good quality, large, red, winter apple grown in New York and New England.

Baldwin hoptree. *Ptelea baldwini,* a poisonous shrub, family Rutaceae, which grows along wooded slopes; contact with it may result in a severe dermatitis. Found in southwestern and sometimes in northeastern U.S. Also called *wafer ash.*

Baldwin spot. See *bitter pit.*

bale. A large, tightly pressed and bound bundle of produce, as a *bale* of cotton, straw, hay, wool.

bale bagging. See *bagging.*

bale breaker. A machine used to disentangle, spread, and fluff up cotton fibers after it has been removed from a bale and before cotton is spun into thread.

bale buster. A self-powered machine to break up and shake out tightly baled hay before it is fed to animals.

bale compress. See *compress.*

baled straw silo. A wall of baled straw or hay, enclosing a round pile or stack of ensilage, which is usually about eight to ten feet high and 20 to 25 feet in diameter.

bale loader. An elevator attached to a truck and used to lift bales of hay or straw.

baler. See *hay baler.*

baler twine. Cord used for tying bales of hay.

bale to the acre. Denoting cotton yield: usually 500 pounds of lint cotton per acre.

baling chamber. The rectangular compartment of a hay baler in which the hay is compressed into a bale.

baling press. See *press box.*

balisier. See *carib heliconia.*

balk. Specifically, a ridge of untilled land missed in plowing; forming a division between common land; preventing land erosion.

ball. 1. Specifically, the compact earth adhering to the roots of a plant in transplanting. See *balled and burlapped.* 2. A large pill. 3. A mass of bees formed tightly around a queen bee.

ball dahlia. A type of *dahlia* (*q.v.*), family Compositae, having double flowers that are ball-shaped. The floral rays are arranged spirally and are blunt or rounded at the tips; the margins of the floral rays are rolled toward the upper side. The flower is 3½ inches or more across.

balled and burlapped. The covering with sacking of the compact mass of earth left on the roots of a plant in transplanting. *Cf. bare-root.*

ballfern. *Davallia bullata,* family Polypodiaceae, an Asiatic, greenhouse fern.

ball-hooter. In forestry, a man who shouts a warning cry as he rolls logs down a slope.

balling. A measurement of sugar percentage in simple sirup at fixed temperature (20° C.). Also called *Brix.*

balling forcep. A pair of tongs for administering medicine, especially pills, capsules, etc., to animals. Used to place medicine on the back of the tongue. Also called *balling gun; balling iron.*

ball mustard. *Neslia paniculata,* family Cruciferae, a European weed.

balloonberry. See *strawberry raspberry.*

balloonflower. *Platycodon grandiflorum,* a Eurasian, perennial herb, family Campanulaceae, cultivated as a border plant for its dark blue, summer flowers. Also called *Chinese* or *Japanese balloonflower.*

balloon frame. A type of barn construction made from a wooden frame without using heavy timbers. See *post frame; pole frame.*

balloon hood. A light, wooden frame covered with heavy roofing paper, etc., so that it is air-tight. It is placed over stacked trays of sliced pears, peaches, or apricots which are to be treated with sulfur fumes before being placed in the sun for drying.

balloonvine heartseed. *Cardiospermum halicacabum,* family Sapindaceae, a tropical American, decorative, annual vine, grown to cover trellises, arbors, fences. Also called *heartseed; heart-pea.*

ball planting. A method of transplanting in which plants are taken from the ground with enough compact soil on the roots to provide nourishment and help their growth in a new location.

balm. 1. A kind of propolis or bee gum produced inside bees whenever they eat pollen. It is a transparent substance, golden-greenish in color, and contains pollen shells and little hairs which never appear in propolis gathered from outside the hive. Also called *bee balm.* 2. Any of certain aromatic, oily, resinous exudates of woody plants, especially of the Myrrhtree, which may be of medicinal value. 3. A general term for a plant yielding *balm.* 4. *Melissa officinalis,* a Eurasian, aromatic, perennial herb, family Labiatae; grown for its lemon-scented leaves. Also called *lemon balm; bee balm.*

balm-of-Gilead poplar. *Populus candicans,* a hardy, short-lived, deciduous tree, family Salicaceae, which is grown sometimes for ornament. Its native home is unknown, but possibly it is a hybrid of American origin. Also called *balsam poplar.*

balm-of-heaven. See *California laurel.*

balocanad. See *soft lumbang.*

balsam. 1. A resinous or oleoresinous aromatic exudate from plants consisting of a mixture of resins and essential oils, cinnamic and benzoic acid. 2. Any plant or tree which yields balsam. Also see *garden balsam.*

balsam apple. *Momordica balsamina,* a tropical, annual, ornamental climber, family Cucurbitaceae, producing a fruit supposed to be good for burns or sprains. Native to the West Indies. Also called *balsam pear; balsam cucumber; ampalaya.*

balsam fir. *Abies balsamea,* an evergreen tree, family Pinaceae, used as Christmas trees, and in making cooperage, boxes, etc. Native to northern U.S.

balsam fir sawfly. *Neodiprion abietis* (Harr.), an insect that lays its eggs within the tender shoots of the balsam fir and spruce and whose larvae feed on the foliage.

balsam fir sawyer. *Monochamus marmorator* (Kby.), family Cerambycidae, a common beetle which quickly infests felled trees, thus establishing its larvae in the heartwood within a few months. Found in Canada and Maine.

balsam pear. See *balsam apple.*

balsam poplar. See *southern poplar.*

balsamroot. See *arrowleaf balsamroot.*

balsam twig aphid. *Mindarus abietinus* Koch., a homopterous insect that infests conifers, often distorting and sometimes killing them.

Baltic hemp. See *flax: 1.*

Baltic rush. *Juncus balticus,* family Juncaceae, a fairly good forage plant; useful in controlling erosion due to its elaborate root system. Also called *wire rush.*

balucanag. See *soft lumbang.*

bamboo. Woody-stemmed perennials of the genus Bambuseae, family Gramineae, apparently the most primitive grass tribe known. More than 20 genera and 200 species are recognized, some growing 60 or more feet in height. The hollow stems may die down to the ground or may remain living for many years. They are often thick enough to be used in building construction, etc., in the tropics and subtropics, the succulent young shoots are used for food, some varieties bear a drupe-like fruit with a fleshy exterior. Among the more important genera are *Bambusa, Phyllostachys, Dendrocalamus,* and *Arundinaria.* The last one forms the cane-brakes of southern U.S.

bamboofern. *Coniogramme japonica,* family Polypodiaceae, a fern which is grown indoors as an ornament in the U.S. Native to Japan.

bamboo rake. A light garden rake with teeth made of strips of bamboo (*q.v.*); used for gathering leaves, cleaning flower beds, etc.

banana. An herbaceous tree-like perennial, *Musa paradisiaca* var. *sapientum,* bearing clusters of pulpy fruit ranging from four to nine inches in length and usually one inch in diameter. It is native to southeastern Asia, but there are also many varieties in other tropical zones cultivated for their high food value and agreeable flavor.

banana bean. See *velvet bean.*

banana nematode. *Pratylenchus musicola* (Cobb) Filipjev, a nematode which infests the roots of citrus trees in California.

banana shrub. *Michelia fuscata,* a tropical shrub, family Magnoliaceae, whose name derives from its banana-like smell. It is an excellent greenhouse plant in the north, and it grows outdoors in the extreme southern part of the United States. Native to China.

banco. A bank or deposit of sand or silt exposed due to shifts in location of river channels; a term used in southwestern U.S.

bandage. 1. Cloth, preferably sterilized, used for dressing wounds. 2. Cloth used to cover and compress a cheese.

band cap. A metal top with a rubber lining which is crimped to the beaded rim of a glass jar to form an air-tight seal.

band cutter. 1. An attachment on the feeder of a threshing machine to cut string binding sheaves before they enter the cylinder. 2. A person who manually performs the operation described above.

banded corky bark. A disease of citrus trees thought to be caused by a virus, which is characterized by raised bands of cork-like eruptions extending horizontally around the trunk or limb of a tree. See *corky bark.*

banded cucumber beetle. *Diabrotica balteata* Lec., family Chrysomelidae. A pest of beans, legumes, cucurbits. Also called *belted diabrotica; banded diabrotica; belted cucumber beetle.*

banded diabrotica. See *banded cucumber beetle.*

banded fruit. See *belted: 1.*

banded greenhouse thrips. *Hercinothrips femoralis* (Reut.), family Thripidae. A pest which infests sugar beets and cacti in California and date palms in Arizona; also found in greenhouses.

banded hickory borer. *Chion cinctus* (Drury), family Cerambycidae, a beetle which does much damage to hickory lumber during seasoning.

banded woollybear. *Isia isabella* (J. E. Smith), family Arctiidae, a hairy caterpillar which riddles garden foliage during the summer and fall. Found throughout the U.S.

banding. 1. Marking of animals or birds by fastening bands to their legs or ears or by painting their feathers for identification. 2. The placing of a collar or band around the stem of a plant or the trunk of a tree to poison, repel, or trap insects.

band mill. 1. A sawmill equipped with a band head saw. 2. A machine on which band saws are mounted.

band seeding. 1. Planting seed in close proximity to a row of fertilizer which has been drilled in at a shallow depth and covered over. 2. The surface sowing of legumes and grasses on drilled ground by using a special attachment to the grain drill.

bane. A general and unscientific name for sheep diseases. Also called *rot.*

baneberry. Perennial herbs, of the genus Actaea, family Ranunculaceae, bearing a poisonous fruit. Found throughout the north temperate zone of the U.S. See *white baneberry; red baneberry.*

banewort. See *belladonna.*

bangalay eucalyptus. *Eucalyptus botryoides,* a tall tree, family Myrtaceae, grown as an ornamental for its striking foliage and flowers and as a bee tree. Native to Australia and found in California and Florida. Also called *bastard mahogany* (the common name in Australia).

bang board. See *sideboard.*

bangle ear. A drooping ear of a horse.

Bang's disease (of cattle). See *brucellosis.*

bang-tail. A horse whose tail hair has been evenly trimmed.

bangtail muster. In Australia, a roundup of cattle whose tails are clipped after they have been counted.

banjo-shaped. A term used in meat judging, designating flatness; a lack of plumpness; extreme length of a cut of meat.

bank. 1. The acclivity of the sides of the channel of a river, or other natural watercourse, or of a canal or ditch; the rising ground, beginning at the margin of the water, bordering a lake. See *river bank.* 2. Any steep declivity of earth. 3. The crown or upper surface of a furrow slice. Also called *furrow crown.* 4. To pile earth around plantings.

bank and bank. A New Zealand term designating a river or stream full to its banks. See *bank full.*

bank barn. A barn built on the side of a hill so that the second story or loft is on ground level on one side.

bankcress. See *hedgemustard.*

bank erosion. Land or soil losses of the banks of rivers, lakes, ditches, or canals resulting from currents, wave cutting, undermining, caving, rilling, or gullying of slopes.

bank full. Denoting a stream or river which is almost in flood.

Bankhead-Jones Farm Tenant Act. A law of Congress in 1937, which authorized the Farm Security Administration (now called the Farmers Home Administration) to make loans to qualified farm tenants, sharecroppers and farm laborers to enable them to buy farms and engage in farming.

banking. The mounding of earth or soil to preserve moisture or coolness or to cover the crowns of cultivated plants.

Banksia. An Australian genus of honeysuckles, family Proteaceae. Its yellow flowers in dense spikes mature into woody cones which bear winged seeds.

Banks rose. *Rosa banksiae,* an evergreen perennial rose, family Rosaceae, which often grows up to 20 feet; it has few prickles and bears clusters of small white or yellow flowers. Native to China; found in southern U.S. Also called *Lady Banks' rose.*

bankul nut. See *candlenuttree.*

bankweed. See *hedgemustard.*

banner crop. A crop which is high in yield, quality, or both.

banquette. Specifically an earthen terrace added to the base of a high levee away from the water to prevent seepage and sloughing of the levee.

bantam. Any dwarf or miniature breed of domestic fowl.

banty. See *bantam.*

banucalag. See *soft lumbang.*

banuyo. *Wallaceodendron celebricum,* family Mimosaceae, the dark, hard timber of this Philippine tree is used in fine woodwork.

banyan tree. *Ficus benghalensis,* an Indian fig tree. Aerial roots grow from the branches to the ground and form additional trunks so that one tree may cover a very large area. The name is derived from the Hindu merchants called banians who traded under the shelter of the tree. Also called *banian tree.*

baobab tree. 1. *Adansonia digitata,* family Bombacacae, a huge tropical tree which bears a large edible fruit called monkey's bread. The fibers are used in making rope, cloth, and paper. Native to Africa and long cultivated in India. 2. *A. gregorii,* a species found in northern Australia, which sometimes attains a circumference of 80 feet. Also called *cream-of-tartar tree* because of the pleasantly acid taste of the fruit.

bar. 1. Mouthpiece of a bridle bit. 2. Landside of a plow. 3. The straight line of a cattle brand (western U.S.). 4. That part of a horse's hoof which is bent inward extending toward the center of the hoof. 5. (pl.) Ridges in the roof of a horse's mouth. 6. The space in front of a horse's molars where the bit is placed. 7. A pole used to close a fence gap.

barb. 1. (cap.) A breed of horses related to the Arab and native to Barbary. It was introduced into Spain by the Moors and was probably smaller and coarser than the Arab; its strain is evident in all known present breeds. 2. A hairlike side branch of a feather. 3. (pl.) Mucous membrane projections for the openings of submaxillary glands under the tongues of horses and cows. 4. The teeth or spines on the awns of grasses, especially barley. 5. A pointed projection on a fence wire. 6. In botany, a hooked hair or bristle.

barbadine. See *giant granadilla.*

Barbados aloe. See *true aloe.*

Barbados-gooseberry. *Pereskia aculeata,* family Cactaceae, a cactus vine which grows up to 30 feet; used as a root stock on which to grow various weak cacti. Native to tropical America. Also called *lemon vine; blade apple.*

Barbadoslily. *Hippeastrum vittatum,* a bulbous plant commonly called *amaryllis,* family Amaryllidaceae. The plants are hybrids having large red, pink, and white lily-like flowers; popular for forcing as indoor pot plants. Native to South America.

Barbadosnut. *Jatropha curcas,* a tropical shrub, family Euphorbiaceae, bearing a seed which yields a purgative oil poisonous in quantity. It is grown for ornament in southern U.S. Also called *French physic nut; purging nut.*

Barbados pride. See *sandal beadtree.*

Barbados royal palm. See *Carib royal palm.*

Barbara's buttons. *Marshallia trinervia,* a small perennial herb, family Compositae, grown for its purplish-white or pink flowers. Native to southern U.S.

Barbary fig. See *devil's tongue.*

barbed wire. Fencing made of twisted wires, to form one strand, and to which wire barbs are attached at short intervals.

barbels. See *barb: 3.*

barber pole worm. See *twisted stomach worm.*

Barberey. A soft cheese made from fresh warm milk coagulated with rennet; sometimes sold before ripening. Also called *Fromage de Troyes.*

barberry. Any shrub of the genus *Berberis,* family Berberidaceae, often grown for its ornamental foliage and for its berries, which are of great value to birds in the winter. *Berberis vulgaris,* the European, or common barberry, and a few other species, are the alternate hosts for the stem rusts (*Puccinia graminis*) of cereals, such as wheat, barley, rye, etc., as well as many other grasses. See *stem rust of grasses.* However, many species of barberry, including the Japanese barberry, *Berberis thunbergii,* are not hosts for this rust. See *Magellan barberry; Allegheny barberry; Darwin barberry; wintergreen barberry; Sargent barberry; Japanese barberry; verna barberry; warty barberry; European barberry; Wilson barberry.*

barberry rust. See *stem rust of grasses.*

barbfruit pavonia. *Pavonia spinifex,* an evergreen shrub, family Malvaceae, which is grown out-of-doors in very warm areas and elsewhere as a pot plant for its yellow flowers. Also called *escobadura.*

bar bit. A specific type of *bit* (*q.v.*) consisting of a solid metal piece that has no lever action.

bar clip. A bar shoe.

bareback. 1. A poultry term denoting a chicken which is not fully feathered out on the back. 2. To ride *bareback* is to ride a horse without a saddle.

barebelly. A New Zealand term to describe a sheep, usually a ewe, which has no wool on its stomach.

bare fallow. Land unsown for a season but kept in cultivation to keep down vegetation. Also called *bare summer fallow; naked fallow.*

bare-root. A method of transplanting in which plants are taken from the ground with little soil left on the roots. *Cf. balled and burlapped.*

bare root grafting. A method of joining a cion to a plant by grafting onto the bare root stock before growth has started.

bare summer fallow. See *bare fallow.*

barium antimonyl tartrate. An effective drug for treatment of fowl gapeworm (gapes). It is administered as a dust to the fowls while they are cooped up in a tightly enclosed box for a period of 15 to 20 minutes.

barium carbonate. $BaCO_3$. This solid is used as a rat poison and as a water purifier; known as witherite in its natural state.

barium fluosilicate. $BaSiF_6$, a compound used as an insecticide and a rodent poison. Also called *barium silicofluoride.*

barium poisoning. Livestock poisoning caused by excessive ingestion of preparations containing barium. Characterized by profuse salivation, sweating, convulsions, palpitations, and general paralysis.

bark. 1. Specifically, the exterior of a woody stem containing phloem (*q.v.*) tubes and usually some bast fibers. Cork cells may develop in the outer layer. See *cortex: 4.* 2. To *bark* is to remove the external layer from a tree or log. 3. To cure or dye with bark extract; to tan. See *tan: 3.* 4. To enclose or surround with bark.

bark beetle. Any of several beetles, family Scolytidae, which is destructive to standing or felled trees. They feed upon the cambium and adjacent tissues by mining just beneath the bark.

bark blazer. A tool used for marking trees by cutting into the bark.

bark blotch (of citrus). A fungous disease caused by *Ascochyta corticola,* family Sphaeropsidaceae. Affected bark areas are characterized by dark brown coloring with raised margins and by gumming. Uncommon in U.S.

bark bound. A tree or plant retarded in growth by the tightness of its bark.

bark disease. 1. Any bacterial fungous or virus infection of bark. 2. See *chestnut blight.*

bark fungus of stone fruit. A non-serious fungous disease of stone fruits caused by *Calosphaeria princeps* Tul., family Sphaeriaceae.

bark graft(ing). A method of joining plants in which the cion is inserted between the bark and the xylem of the stock.

barking bill. An instrument used in removing bark from trees.

barking squirrel. Prairie dog *q.v.*

bark patch grafting. See *bark grafting.*

bark rot of sour orange. A disease of the rootstock suspected of being caused by a rot of the bark between the root and the bud union. It is frequently associated with infections of the fungus *Phytophthora citrophthora,* family Pythiaceae. It may attack the bark only but can extend into the wood causing a gummy infiltration, bark degeneration, and surface gumming.

bark spudder. In logging operations, a worker who strips bark from logs.

barley. One of the cereals, *Hordeum* spp., family Graminae, cultivated since prehistoric times for human and animal consumption, for malting, or as nurse or smother crop. *Hordeum vulgare* is the commonly cultivated barley.

barley anthracnose. See *cereal anthracnose.*

barley corn. A grain of barley.

barley feed. A stock feed consisting of the entire by-product of the manufacture of pearl barley from clean barley.

barley fork. A long-handled fork having a guard at the handle end of the tines; used to pick up short-stemmed straw, etc.

barley hulls. Outer coverings of the barley grain.

barley hummeler. An instrument consisting of parallel iron plates once used in Scotland for threshing the grain from the awn.

barley leaf rust. An uncommon leaf rust of barley caused by *Puccinia anomala.*

barley malt. Fermented barley used in manufacturing beverages, food, and medicines.

barley mixed feed. Livestock feed consisting of the entire residue from the milling of barley.

barley net blotch. A leaf disease caused by the fungus *Pyrenophora teres,* family Pseudosphaeriaceae, whose asexual stage is frequently given as *Helminthosporium teres,* family Dematiaceae. The infected leaves bear elongated spots which may form a ladder-like pattern of dark veins. The *Helminthosporium* stage develops in abundance on the leaves but the perfect stage develops mostly on the straw and stubble.

barley reel. A seed-separating machine consisting of a rotating, corrugated, woven screen.

barley scab. See *scab: 3.*

barley scald. See *scald: 1.*

barley spot blotch. See *spot blotch.*

barley stripe. A disease caused by the fungus *Helminthosporium gramineum,* family Dematiaceae; characterized by yellow stripes and gray leaves. Also called *stripe disease of barley.*

barley stripe mosaic. A virus characterized by brown stripes, chlorotic mottling, spotting, and streaking. The chlorotic areas are white or ashy-gray. Found in the U.S. and in Canada.

barley yellow dwarf. A virus disease of barley and other cereals which is characterized by the leaves rapidly turning light green and yellow and by stunting of the plant.

barn. A farm building used for storage of hay, grain, farm implements, etc., for housing of domestic animals.

barnacles. Specifically, a restraining device for animals which grips their noses. See *twitch.*

barnacle scale. *Ceroplastes cirripediformis* Comst., family Coccidae, an insect that infests Australian pine, citrus, begonia, eupatorium, guava, jasmine, Jerusalem thorn, lignum-vitae, myrtle, pear, poinsettia, quince, persimmon. Found in Florida, Louisiana, Mississippi, California.

barn boards. Soft timber used for barn walls.

barn book. A record book in which details of breeding dates, births, etc., are kept.

barn cleaner. A mechanical chain device used to clean manure gutters.

barn-cured. Cured within a barn, as contrasted to sun- or air-cured.

barn-door fowl. See *barnyard fowl.*

barn fowl. See *barnyard fowl.*

barn gallon. 2.40 U.S. gallons or 2 imperial gallons.

barn grass. See *barnyardgrass.*

barn itch. A variety of skin conditions in livestock, namely cattle, causing severe irritation. See *scab mite; ringworm; biting louse.*

barn lot. See *barnyard.*

barn manure. See *manure: 1.*

barn owl. *Tyto alba,* family Tytonidae, an owl which frequents barns and other such buildings and feeds on rodents, etc.

barn-raising. The constructing of a barn. In parts of the U.S. often this implies a gathering of friends and neighbors to assist in the construction, at which time much food and drink is served for a pleasurable occasion.

barn rat. Specifically, a horse that will not readily leave the stable or picketline.

barn swallow. *Hirundo erythrogastra,* family Hirundinidae, a gregarious migratory bird, native to N. America, which nests in farm buildings. In Europe, the name is applied to *Hirundo rustica.*

barny. A term used in judging dairy products to designate an off-flavor characterized by cowstable odor and its persistence on the palate. Closely related off-flavors are: *smothered; cowy; unclean.*

barnyard. Land immediately adjacent to a barn. It is usually fenced in to enclose livestock or fowls. Also called *barn lot.*

barnyard fowl. Any commonly domesticated fowls: chickens, ducks, geese, turkeys, etc.

barnyardgrass. *Echinochloa crus-galli,* family Gramineae, an annual grass which is often a weed in cultivated grassland. It can be cut for hay if special care can be given to the curing, as it is a host for ergot. Also called *ankee; barn grass; cockspur-grass; water-grass; barnyard millet; Japanese barnyard millet.*

bar pad. A leather covering for a horse's hoof to prevent slipping. It extends to each heel over the bars of the hoof and is fitted as a part of a short shoe having a soft pad to cover the frog. See *bar shoe.*

bar plate. See *bar shoe.*

barrancas. Deep *U*-shaped gulches in alluvium formed by, or enlarged by, waterfall erosion. A term used in arid regions of southwestern U.S. Also called *barranco.*

barred. In poultry husbandry, a term used to describe striped markings on fowl.

Barred Plymouth Rock. A breed of chicken reared for both egg production and meat. The cock weighs about 9½ pounds and the hen about 7½ pounds. The feathers are striped horizontally with black or almost black alternating bars. Also called *Barred Rock.*

barrel. 1. The trunk of a domesticated animal. 2. A measure of corn in the ear: one *barrel* equals one bushel of shelled corn. 3. In the earlier period of flour milling in the U.S., the wooden *barrel,* net weight 196 pounds of flour, was a common container for shipment. Quotations of flour prices were also given by the *barrel.* 4. A cylindrical wooden or metal container.

barrel planter. An early planter made of a barrel with holes bored at intervals around the middle. The barrel was rolled down a row and the seeds were supposed to drop out and into the soil.

barrel pump. A sprayer which develops 200 to 300 pounds per square inch of pressure with a barrel or drum of 15-45 gallon capacity. It is often transported on a hand cart or sled. Also called *barrel sprayer.*

barrel sprayer. See *barrel pump.*

barrel-type heat exchanger. A compact heater used in the dairy industry. The milk passes through pipes surrounded with steam within a barrel-shaped container. It is usually rated at about 1,000 pounds of milk per hour per eight square feet of heating surface.

barren. Denoting an area unable to support vegetation; incapable of producing offspring, seed or fruit.

barren myrtle. See *bearberry.*

barrens. A name used by settlers and explorers in eastern U.S. to describe land characterized by sparse vegetation, open or scrubby tree growth. Frequently designated by the kind of growth, as *pine barrens, oak barrens,* etc.

barrenstrawberry. *Waldsteinia fragarioides,* a strawberry-like herb, family Rosaceae, grown for its white flowers. Native to North America. Also called *hairy Norwegian cinquefoil.*

barrenwort. *Vancouveria hexandra,* an evergreen herb, family Berberidaceae, grown as ornamental ground cover. It has small white or yellow flowers. Native to South America.

bar roller. A device made of heavy metal drums and usually pulled by a tractor; used to crush clods of soil. See *cast-iron drum roller; clod crusher; smooth roller; planker; treader.*

barrow. 1. A castrated male hog. 2. In New Zealand, to shear or partially shear a sheep. 3. A small cart. 4. A hill or mound.

bar share. A plowshare which has a landside (*q.v.*) attached to it.

bar shoe. A horseshoe which has a bar plate consisting of a flat piece of metal across the heel opening to protect the frog of the hoof. See *bar pad; frog: 2.*

bar spade. A fork-like spade having a cutting blade welded across its three prongs; used for digging trenches in wet, sticky soil.

Bartlett. A widely grown variety of pear (*q.v.*), genus Pyrus, family Rosaceae, introduced into the U.S. by Enoch Bartlett of Dorchester, Massachusetts, from England where it is known as *bon chrétien.*

barton. 1. A poultry yard or farmyard. 2. Farm buildings attached to a farmhouse. 3. Land belonging to an estate.

Bartram oak. *Quercus heterophylla,* family Fagaceae, a hybrid of the willow and red oak. Found in eastern U.S.

baru. Linden hibiscus *q.v.*

barwal (bharal). *Pseudois nahura,* a goatlike sheep characterized by backward-curving horns. Native to northern India where it was bred for fighting.

barwood. The hard, red wood of *Baphia nitida.* First timber of commercial value exported from Africa. Used in making dye, violin bows. Also called *camwood.*

basal area. The area of a cross section of a tree measured at breast height.

basal-area plot. A field area where vegetation is studied to ascertain species composition, reproduction, the area of ground covered by the stems of each plant in contrast to the full spread of its foliage.

basal canker. A fungous disease of Norway maples caused by *Phytophthora cambivora,* family Pythiaceae; characterized by a decrease in size and number of leaves and by cankers at the base of the trunk near the soil. Death of the tree results when the cankers girdle the tree or when the root system decays.

basal glume rot of wheat. A minor disease of the wheat husks caused by the bacterium, *Pseudomonas atrofaciens.*

basal growth. Development from the axis of a plant as opposed to growth at its extremities.

basal leaf. Grows close to ground. It is usually larger than higher leaves.

basal metabolism. The energy necessary to sustain the vital functions of cellular change, respiration, and circulation when an organism is at rest.

basal-nerved. A leaf which has veins emanating from the base.

basal rot. See *root rot.*

basal stem rot. See *black leg: 1.*

bascule pan. The evaporating pan formerly used in manufacturing sugar from sugar cane. Also called *tilt pan; see-saw pan.*

base. 1. In chemistry, a substance which reacts with an acid to form a salt; a substance which gives off hydroxyl ions when dissolved in water, as contrasted with an acid, which gives off hydrogen ions. 2. In botany, the part of a leaf or branch attached to a stem or trunk. 3. The portion of a machine on which the operating parts are mounted.

base broom. Woadwaxen *q.v.*

base exchange. 1. In soil science, denoting a physical-chemical process in which cations absorbed by soils can be replaced by other cations in chemically equivalent quantities. 2. A term used in medical science usually in relation to the acid-base exchanges of the blood necessary to keep the blood and tissues at a physiologic constant.

base-exchange capacity. 1. In soil science, the cations that a soil can absorb. It is expressed in milliequivalents per 100 grams of soil. 2. In medical science, the maximum capacity of the blood to exchange base (alkali) for acid in maintaining a physiologic acid-base balance.

base-exchange solid. Non-fat milk solid from which most of the calcium has been removed by ion-exchange; used in making ice cream.

base goods. 1. A dry mixture of fertilizer materials with other materials carrying either ammonia or potash or both. 2. A wet mixture of fertilizer material produced by treating phosphate rock and nitrogenous substance with sulfuric acid.

base horehound. White deadnettle. See *deadnettle: 4.*

base level. The point at which erosion by running water is halted, *i.e.* sea level.

basement barn. 1. A two-story barn whose lower story is built of stone, brick, tile, etc., and the upper story of timber. 2. A bank barn *q.v.* Also called *English-type barn.*

base rocket. 1. *Reseda luteola,* family Resedaceae, an herb from which yellow dye is produced. Also called *weld.* 2. *R. lutea,* a fragrant flowering herb of the mignonette family, Resedaceae. Also called *yellow mignonette.* Native to the Mediterranean area.

base-saturation. In soil science, designating a soil in which the cation exchange capacity is completely saturated wth basic ions.

basic crops. Crops which because of acreage, value, climate, etc., are considered most important in an agricultural economy; such as corn, wheat, rice, peanuts, cotton, tobacco, etc.

basic slag. A finely ground by-product of steel containing 12 to 25 percent P_2O_5, 40 to 50 percent CaO, and 5 to 15 percent SiO_2. It has some value as a fertilizer because of the phosphoric acid content.

basil. See *sweet basil.*

basin. 1. A hollow for holding water, as around a tree, made by forming a ridge or levee of earth on all sides. 2. The depression in a pome fruit such as apple and pear, in which calyx lobes are located. Also called *eye.*

basin attachment. See *basin lister.*

basin crotch. A depression formed by three or more branches at the same level on a tree trunk. Also called *water pocket.*

basin irrigation. An efficient system of irrigating in which a field or orchard is divided into basins which are filled with water.

basin land. 1. A natural or artificial feature: a hollow area, completely or partly enclosed by higher land. 2. The drainage area or watershed of a river system. 3. The bed of a lake.

basin lister. A plow with an attachment to form small basins to hold rainwater by automatically depositing an earth dam in the furrow every six to ten feet.

basis prime crude cottonseed oil. Oil which must refine to a specified color no higher than grade ACOS 12 or with a loss in weight not exceeding 20 percent.

basket. 1. A container usually made of thin strips of wood, reeds, canes, or other flexible materials in which commodities are picked, handled, packed, shipped, or sold. See *climax basket; splint basket.* 2. The contents of a basket, as a *basket* of oranges.

basketfern. *Nephrolepis pectinata,* a small plant, family Polypodiaceae, which is grown in a hanging basket. Native to tropical America.

basketflower centaurea. *Centaurea americana,* a tall, annual herb, family Compositae; grown for its rose or flesh-colored flowers. Native to North America.

basketflower hymenocallis. *Hymenocallis calathina,* a bulbous herb, family Amaryllidaceae; grown outdoors or in the greenhouse for its long white flowers. Native to South America. Also called *spiderlily.*

basket osier. Basket willow *q.v.*

basket plant. Any vine or plant that grows well in a hanging basket.

basket press. A press used for extracting juice from grapes. It consists of a cylinder made of heavily reinforced strips of wood in which the crushed fruit is squeezed.

basket willow. *Salix viminalis,* a deciduous Eurasian tree, family Salicaceae; grown for its flexible twigs used in basket weaving. Also called *osier willow; osier; basket osier.*

Bassia. A tropical tree genus of the family Sapotaceae, from which a butter-like oil is extracted.

basswood. *Tilia americana,* a beautiful, symmetrical, deciduous tree which is commonly grown as a shade or street tree or as a good bee plant and used for making veneer core for plywood, slack cooperage, excelsior, boxes, dairy and apiary supplies, piano keys, etc. Also called *American linden; lime; whitewood; linwood; American basswood.*

basswood leaf miner. *Baliosus ruber* (Web.), family Chrysomelidae, a common pest which also infests oak and apple. It burrows into the leaves and feeds on them.

basswood leaf roller. The caterpillar of *Pantographa limata* G. & R., family Pyranstidae, a common pest. They roll the leaves into tubes in which they spin cocoons.

bastard. In botany, a hybrid; an offspring between species which ordinarily do not mate.

bastard branch. Any shoot from the root of a plant which grows independently of the stem. Also called *bastard slip.*

bastard clover. See *alsike clover.*

bastard cress. See *field pennycress.*

bastard indigo. See *indigobush amorpha.*

bastard mahogany. See *bangalay eucalyptus.*

bastard-saw. To saw wood at a tangent to the tree or log. Also called *tangent saw.*

bastard speedwell. *Veronica spuria,* a perennial, Eurasian, hardy herb, family Scrophulariaceae; grown for its blue flowers.

bastard strangles. See *strangles.*

bastard trenching. Ditches are dug two feet wide and almost as deep; a layer of manure is placed in each ditch which is then refilled with soil from the next ditch so that the subsoil is uppermost. *Cf. trenching.*

bast fiber. The slender, flexible fibers which are produced in the bark (cortex) of woody plants. It is the source of commercial fibers of many sorts, such as hemp, flax.

bat. 1. Loosely matted cotton lint or other fiber. 2. Any one of the numerous, flying mammals of the Order Chiroptera.

batamoe. See *seepwillow baccharis.*

batch mixer. A mechanical device used to mix seeds, feeds, etc.

batch peeler. A mechanical device used to peel a limited quantity of fruit or vegetables.

batch processing. The handling or pasteurizing of a limited amount of milk or juices. Also called *discontinuous pasteurizer.*

bate. In tanning, to soften hides by immersing them in a bath of dung or pancreatic enzymes.

bat guano. Excreta.

bat manure. Dry, bat excrement.

batoko plum. See *ramontchi.*

Battalmatt. A soft Swiss-type cheese made from cow's milk and ripened for three to four months.

battery. In poultry husbandry, a series of cages used for raising chickens without a hen; for fattening chickens and broilers for market; increasing egg production in laying hens.

batting. Cotton, flax, or wool loosely matted and not spun.

batting hammer. A mallet or flail used in beating flax to prepare batting.

batts. See *colic.*

Baudouin's test. A test for sesame oil adulteration of butter in which the butter fat is melted and added to a solution of cane sugar in hydrochloric acid. If the acid turns pink to cherry-red, sesame oil is present.

Baur clip. See *split clip.*

bay. 1. A term frequently used to describe a horse of yellowish-tan, mahogany, or dark brown. 2. The laurel *q.v.* or its berry. 3. An indentation of a plain into a range of hills. 4. A shallow swamp supporting dense tree and shrub vegetation, and containing peat or muck soil. See *pocosin.* 5. A coastal inlet. 6. A compartment in a barn for storage or for a special use, as a horse *bay.*

bayberry. In U.S., fruit of northern and southern bayberry (*q.q.v.*); in Jamaica, fruit of bay-rum tree *q.v.*

bayberry wax. See *northern bayberry.*

baygall. See *redbay persea.*

bayleaf willow. See *laurel willow.*

bay leaves. The leaves of the bay-rum tree (*Pimenta acris*) which are used for flavoring and as a source for bay rum.

bayonet plant. Any of the yuccas (*q.v.*). See *Adamsneedle yucca.*

bayou. 1. A secondary creek or water course which is often sluggish with muddy water. 2. A channel of a creek or river closed or partially closed at both ends. 3. An abandoned channel of meandering rivers, especially the lower course of the Mississippi. 4. A narrow navigable channel connecting two rivers or lakes. 5. The *bayou* contains standing water and is sometimes an oxbow lake.

bay poplar. See *water tupelo.*

bay roan. A coat-color of a horse which is brown with gray.

bay-rum tree. *Pimenta acris,* a tropical, aromatic tree, family Myricaceae; grown for the essential oil yielded by the leaves which is used in the manufacture of perfume and bay rum. Found in the West Indies and South America.

bay tree. See *Grecian laurel; California laurel.*

bay willow. See *laurel willow.*

BCG. Bacillis Calmette-Guerin, originally a strain of bovine tubercle bacilli used in some countries for vaccinating men and animals against tuberculosis.

BC soil. A soil with a profile which has no A horizon (*q.v.*). See *soil horizon.*

beach aster. See *beach fleabane.*

beach erosion control. The application of various mechanical and vegetative measures to prevent the washing away of a beach by waves.

beach fleabane. *Erigeron glaucus,* an herb, family Compositae, which is grown for its solitary, violet or lilac flowers. Found on west coast of U.S. Also called *beach aster; seaside daisy.*

beach goldenrod. See *seaside goldenrod.*

beachgrass. See *European beachgrass.*

beach peavine. *Lathyrus littoralis,* a perennial herb, family Leguminosae; grown for its cluster of purple and white flowers. Found on west coast of U.S.

beach wormwood. *Artemisia stelleriana,* an aromatic herb, family Compositae; grown for its clusters of aromatic yellow flowers. Native to Asia. Also called *dustymiller; old woman.*

beadplant. *Nertera granadensis,* a tender, perennial herb, family Rubiaceae; grown as ornamental ground cover and indoors for its orange-colored persistent fruit. Native to the Andes, Australia, southeast Asia; found in California.

beadruby. See *Canada beadruby.*

beadtree. See *Chinaberry.*

beak. 1. The upper and lower mandible or nib of a fowl. 2. The awn on the outer chaff of wheat. 3. Any part of a plant resembling a bird's *beak.*

beaked. Rostrate; beaklike.

beaked filbert. *Corylus cornuta,* a hardy deciduous shrub, family Corylaceae; grown as an ornamental and for its fruit. Native to North America. Also called *beaked hazel.*

beaked nightshade. See *buffalobur nightshade.*

beam. 1. The central shaft of a plow, supporting all principal parts, and by which it may be drawn. 2. A heavy piece of timber used to support a building. 3. The balance bar on a scale. 4. In tanning, a sloping board on which hides are dressed.

beamage. A deduction made when weighing a freshly dressed carcass to allow for the loss of weight by evaporation during cooling: for a pig, usually two pounds per side.

bean. 1. A vegetable grown for its highly nutritive value to men and animals, and eaten when green, dried, or ground into meal. The kidney-shaped (or in the broad bean the flattened) seeds of leguminous annual or perennial herbs, which are more often climbing or trailing. The common field and garden beans of America belong to the native species of the genus *Phaseolus.* Two varieties are: climbing, or pole beans, and bush beans. *Phaseolus vulgaris* is the most common species; known in England as *kidney bean* and in France as *haricot.* Some of the small beans of this species are called in America *navy beans* or *pea beans.* The *Lima bean* (*Phaseolus limensis*) and the *Sieva bean* (*P. lunatus*), came originally from tropical America and are now common in the U.S. The *tepary bean* (*Phaseolus acutifolius* var. *latifolius*) is still grown by the Indians of southwestern U.S. The *frijoles* of Mexico and Spain are various forms of *Phaseolus vulgaris.* In India and southeastern Asia two species of small-seeded *beans* known as *gram* (*P. aureus* and *P. munga*) are extensively cultivated. The *bean* of classical literature is a southern European vetch (*Vicia faba*) known as *broadbean, horse bean, Windsor bean,* etc. The *cowpea* or *black-eyed pea* is an African bean, (*Vigna sinensis*). The *soybean,* native of extreme eastern Asia, is *Glycine max,* a species of *Phaseolus genera.* 2. Any plant bearing such seeds. 3. A protuberance on the upper part of the beak of certain water fowl.

bean anthracnose. Caused by imperfect fungus *Colletotrichum lindemuthianum,* family Melanconiaceae, which infests the above ground parts, being especially noticeable on the pods in the form of lesions and cankers. Also called *spot disease; speck; pod canker; pod spot; rust.* See *anthracnose.*

bean aphid. *Aphis fabae* Scop., family Aphidae. A pest which gathers in great numbers on the succulent parts of a plant causing plant debility and a yellowing of the foliage. It infests beans, beets, peas, celery, asparagus, onion, leek, rhubarb, horse-radish, turnips, parsnips, etc.

bean beetle. See *Mexican bean beetle.*

bean blight. See *bacterial blight of beans; halo blight of beans; fuscous blight; bacterial brown spot of beans; anthracnose of beans.*

bean dolphin. See *bean aphid.*

bean ladybird. See *Mexican bean beetle.*

bean leaf beetle. *Cerotoma trifurcata* (Forst.), family Chrysomelidae. An insect which feeds on the undersides of leaves making round holes in them, and eating stems of seedlings. The larvae feed on the underground portion of the plant. It attacks beans, peas, cowpeas, soybeans, corn, beggar-tick, bush clover, tick trefoil, quail pea, hog peanut, and other weeds.

bean leaf roller. *Urbanus proteus* (L.), family Gelechiidae, an insect that attacks beans in the southern U.S. The caterpillar burrows into the leaves and feeds on the plant.

bean mildew. See *powdery mildew.*

bean mosaic. Common virus diseases transmitted by aphids or by the seeds; characterized by various forms of chlorotic mottling on the leaves. See *mosaic disease; virus.*

bean pole. A stick usually six to ten feet tall used to support climbing beans.

bean pork. Fresh trimmed hog jowls.

bean rust. A fungous disease caused by *Uromyces phaseoli* var. *typica,* family Pucciniaceae, characterized by reddish-brown pustules on the leaves. It is most prevalent in humid areas.

bean thrips. *Hercothrips fasciatus* Perg., family Thripidae, a common sucking pest of avocado, olive, pear, orange, legumes, grasses, and deciduous trees.

bean tree. 1. *Castanospermum australe*; grown for its bright yellow flowers, edible nut-like fruit, and timber. Native to Australia. Also called *Moreton Bay chestnut.* 2. A name commonly applied to trees or shrubs bearing bean-like pods. See *coral bean; carob; goldenchain laburnum.*

bean weevil. *Acanthoscelides obtectus* (Say), family Bruchidae, a snout beetle that feeds on beans until the pods are formed and then lays its eggs in the seed; the larvae burrow into and devour the dried beans. It attacks the kidney bean, lima bean, cowpea, and beans, peas, and lentils in storage. Commonly found in U.S.

bearberry. *Arctostaphylos uva-ursi,* an evergreen, trailing shrub, family Ericaceae. It is grown to prevent erosion on steep exposed slopes because its branches quickly take root; also grown for its red berries in winter. Found in U.S. and Eurasia. Also called *kinnikinnick.*

bearberry honeysuckle. *Lonicera involucrata,* a small hardy deciduous shrub, family Caprifoliaceae; grown for its yellow flowers tinged with red and its shiny, black fruit. Also called *twinberry.* Native to North America.

beard. 1. Hairy appendages on the face of a man or animal, such as the whiskers of a goat; the mane of a lion. 2. A group of feathers hanging from the throat of certain breeds of chickens. 3. A tuft of bristly hair projecting from the upper part of a turkey's breast. 4. Awns of grasses. 5. That part of a horse's lower jaw which carries the curb of the bridle.

bearded bellflower. *Campanula barbata,* a perennial herb, family Campanulaceae, grown in rock gardens for its nodding, pale blue flowers.

bearded darnel. See *darnel ryegrass.*

bearded iris. The type of iris (*q.v.*) with bearded petals.

bearded wheatgrass. *Agropyron subsecundum,* family Gramineae, a grass used in range reseeding. Found in western U.S.

beardless bluebunch wheatgrass. *Agropyron inerme,* a good perennial forage plant; also grown to prevent soil erosion. Found in western U.S. Sometimes called *beardless wheatgrass.*

beardless iris. The type of iris (*q.v.*) having smooth petals. Commonly found throughout America, Europe, Africa, and Asia.

beardless pinegrass. See *pine dropseed.*

beardless wildrye. See *creeping wildrye.*

beardlip penstemon. *Penstemon barbatus,* a stout, perennial herb, family Scrophulariaceae; grown for its long, red flowers. Found in western U.S. and Mexico.

beardtongue. See *penstemon.*

beargrass. 1. Any herb of the genus Xerophyllum, family Liliaceae. 2. *Xerophyllum tenax,* grown for its violet stamens. Native to North America. Also called *fire lily; elk grass, turkey's beard.*

bear huckleberry. See *black huckleberry.*

bearing rein. Checkrein *q.v.*

bearing tree. See *witness tree.*

bearmat. *Chamaebatia foliolosa,* a perennial, woody plant, family Rosaceae, sometimes grown for its flower and fragrance, but a weed pest in ranges. Native to California. Also called *mountain-misery.*

bear's breech. See *soft acanthus.*

bear-scratches. Distortions in the grain of a plank resembling bird's eyes.

bearstail celsia. *Celsia arcturus,* a tall, perennial herb, family Scrophulariaceae; grown as a biennial for its yellow flowers. Native to Eurasia. Also called *cretan bear's tail.*

bear trap bridle. A homemade, bitless bridle which has a light chain across the horse's nose. See *bitless bridle.*

beast. A quadruped animal.

beastlings. See *colostrum.*

beast of burden. A domesticated animal used to carry, haul, or pull, etc., as a horse, ox, or ass.

beauty-berry. See *American beautyberry.*

beautybush. *Kolkwitzia amabilis,* a deciduous shrub, family Caprifoliaceae, grown for its pink flowers with yellow throats. It is commonly found in U.S. Native to China.

beauveria fungus. *Beauveria globulifera,* family Moniliaceae, a fungous parasite which attacks and kills over sixty species of insects in the U.S.

beaver management. 1. The prevention of overbreeding. 2. In wild-life conservation, protection from over-trapping. 3. In water conservation, the controlling of beaver dams.

beaver meadow. Grassland flooded by water from beaver dams.

beaver poison. See *spotted waterhemlock.*

beaver root. *Nuphar advenum,* a yellow pond lily.

beaver tree. See *sweetbay magnolia.*

Bechtel's crabapple. *Malus ioensis* var. *plena,* a deciduous tree, family Rosaceae, grown for its profuse, double, pink, rose-like flowers. It is possibly a hybrid, originating in U.S.

bed. 1. A small plot of soil used for growing seedlings, vegetables, flowers or shrubs, and often raised above the level of the surrounding soil. 2. A place for animals to sleep. See *bedding.*

bed-a-tree. In logging, to prepare the ground onto which a tree will be felled so as to prevent shattering or splitting of the trunk.

bed bug. *Cimex lectularius* L., family Cimicidae, an insect which feeds by sucking blood from humans and animals.

bedding. 1. Plowing or otherwise elevating surface soil into a series of crowned parallel beds or lands with shallow surface drains separating them. 2. A method of surface drainage consisting of narrow plow lands in which dead furrows run parallel to the prevailing slope; the land between adjacent dead furrows is called a *bed.* 3. Straw, leaves, sawdust, sand, peat moss, pine straw, etc., to make a bed for an animal. It is used where an animal may lie, to absorb its urine. 4. Flowers which are appropriate for growing in flower beds for massed decoration effect.

bedding pansy. See *horned violet.* Also called *bedding violet.*

bedrock. Unweathered hard rock which lies directly beneath the soil layers or beneath superficial geological deposits, such as glacial drift.

bed rot. See *damping-off.*

bed sore. Inflamed or ulcerated skin areas resulting from the continued pressure of a bed after prolonged lying-down; commonly found on sick animals. Also called *decubital ulcers.*

bedstraw. Any plants of the weed genus Galium, family Rubiaceae.

bedstraw milkweed. See *poison milkweed.*

bee. 1. Any insect of the families: Prospidae, Colletidae, Megachillidae, Xylocopidae, Ceratinidae, Nomadidae, Andrenidae, Anthophoridae, Bombidae, Apidae. They are generally beneficial, being insect parasites, predators, pollenizers, or producers of honey and beeswax. 2. See *honey bee; drone; worker; queen bee; bumblebee.* 3. A social gathering of neighbors, especially in rural areas, to husk corn, to raise a barn, or to quilt in the home, etc.

bee balm. 1. See *balm: 1 and 4.* 2. Any plant of the genus Monarda, family Labiatae. See *lemon beebalm; Oswego beebalm; wildbergamot beebalm; spotted beebalm.*

beebread. The flower pollen gathered by bees, which is mixed with a little nectar or honey, and deposited in the cells of the comb.

bee brush. A thin brush used to remove bees from the comb or other parts of the hive.

bee candy. Artifical food for bees made from sugar and another substance, such as honey.

bee cellar. A cave or cellar used for keeping beehives during the winter.

beech. Any tree of the genus Fagus, family Fagaceae. See *American beech; European beech.*

beech bark disease. An infection of the American and European beech caused by the combined attack of the beech scale, *Cryptococcus fagi,* family Dactylopiidae, and the fungus *Nectria coccinea,* family Hypocreaceae. It is a relatively new disease which is rapidly decimating beech stands in eastern Canada and northern New England.

beech blight aphid. *Prociphilus imbricator* (Fitch), family Aphidae, a bluish-white louse which infests the undersides of beech limbs, often killing twigs and young trees.

beechfern. A narrow beechfern, *Dryopteris phegopteris,* or *Dryopteris hexagonopteria,* a broad beechfern.

beech scale. *Cryptococcus fagi* (Baer), family Dactylopiidae, a sucking insect, a vector for *Nectria coccinea.* See *beech bark disease.* Also called *woolly beech scale.*

bee culture. See *apiculture.*

bee escape. A device used to remove bees from supers or buildings; permits only a one-way passage.

beef. 1. Any cow, bull, or steer, which is fattened for slaughter for food. 2. The meat derived from cattle nearly one year old or older.

bee-farming. The raising of bees for the honey and beeswax they produce, and for their pollinating of plants.

beef breed. Any breed of cattle which is especially developed for use as a source of meat, as Hereford, Angus, Shorthorn, Galloway. Also called *beef cattle.*

beef bung. See *casing: 2.*

beef cut. 1. In butchering, a portion of a beef carcass. 2. A group of cattle separated from the main herd.

beef ham. A cut of beef consisting of the top round, the bottom round with the eye muscle, and the sirloin tip used in making dried beef by curing, pickling, or smoking, etc. Also called *beef set.*

beefiness. A judging term denoting desirable beef characteristics.

beef middles. The mid-section of a cow's or steer's intestine used as sausage casing. See *caecum; sewed casings.*

beef pritch. A rod used to support a carcass during skinning. See *gambrel.*

beef set. See *beef ham.*

beefsteak begonia. See *Feasti begonia.*

beefsteak saxifrage. See *strawberry saxifrage.*

beef tree. A large hook used by butchers to suspend a beef, hog, or lamb carcass. Also called *double tree.*

beefwood. See *horsetail beefwood.*

beefy. 1. A term used in judging to designate characteristics of a beef animal as contrasted to a dairy animal. 2. A term used in poultry judging to designate: (a) that the combs are coarse and overgrown, or (b) that the birds are fat and coarse.

bee glue. See *propolis.*

bee gum. 1. A beehive originally made from a hollow log of the gum tree. 2. Beeswax.

beehive. See *hive.*

beekeeping. Apiculture *q.v.*

bee martin. Any birds of the genus Tyrannus, which catch and kill bees in small numbers. Also called *kingbird.*

beemaster. See *apiarist.*

bee milk. See *royal jelly.*

bee miller. See *greater wax moth.*

bee moth. See *greater wax moth.*

beennut. See *peanut.*

bee pasture. Vegetation, attractive to bees, which is within flying distance of a hive. Also called *bee pasturage.*

bee plant. Any plant to which bees are attracted for nectar and pollen.

bee repellent. A chemical spray for plants used to discourage bees from gathering pollen or nectar.

beer grains. See *brewers' dried grains.*

beer yeast. *Saccharomyces cerevisiae,* family Saccharmycetaceae. One of the most important yeasts used in baking and distilling.

bee sage. *Audibertia grandiflora* (syn. *Salvia grandiflora*), a perennial herb, family Labiatae; grown for its crimson-purple or red flowers and as a bee plant. Native to California. Also called *big-flower sage; crimson sage.*

bee smoker. A device consisting of a funnel, firepot, and bellows for blowing smoke into a beehive to force out the bee colony.

bee's-nest plant. See *wild carrot.*

bee space. A space which is large enough to permit free passage for a bee but too small to encourage bees to build comb and too large to induce depositing propolis.

beestings. See *colostrum.*

beeswax. A mixture of organic compounds, secreted by special glands on the last four visible segments on the ventral side of the worker bee's abdomen; used by bees for building comb. Its melting point is 143.6 to 147.2° F. It is used for making cosmetics, church candles, comb foundation, salves, ointments, dental wax, modeling compounds, patterns, polishes, adhesives, crayons, ink ingredients, etc.

beeswing. A thin particle of the wheat kernel which contains the germs and bran that is separated during flour milling.

beet. 1. Any plants of the genus Beta, family Chenopodiaceae. 2. *Beta vulgaris,* the garden beet, grown for its root food and for its young edible leaves. 3. See *sugar beet.*

beet and bean cultivator. A cultivator designed for crops grown in narrow rows.

beet and bean planter. A multi-row planter designed for crops grown in closely spaced rows.

beet armyworm. *Laphygma exigua* (Hbn.), family Phalaenidae, a worm which eats the tops and roots of sugar beets and feeds on cereal and truck crops. Found in western U.S.

beet blight. See *curly top.*

beet blocker. A machine for chopping out beet seedlings in the row leaving one or more plants in blocks at intervals of every 10 to 14 inches.

beet digger. A harvesting machine. See *beet lifter; automatic topper-lifter-harvester.*

beet downy mildew. A fungous disease of sugar beets and chard caused by *Peronospora schachtii,* family Peronosporaceae; characterized by a yellowing and downward curling of the leaves of the seedling plant; older plants show spots on the affected parts ringed with a darker pigmentation. It is serious only in cool, wet weather.

beet heart rot. A fungous disease of beets caused by *Mycosphaerella tabifica,* family Mycosphaerellaceae; characterized by decay of the heart of the root and by dying back of the leaves.

bee tight. Of, or pertaining to, a building, structure, etc., which is so constructed that bees cannot get in or out.

beetle. Any insect of the Order Coleoptera. The upper pair of its four wings form hard cases to protect the true wings when folded.

beet leaf beetle. *Erynephala puncticollis* (Say), family Chrysomelidae, a beetle which feeds on sugar beets, garden beets, and spinach. Found in the western U.S.

beet leafhopper. *Circulifer tenellus* (Baker), family Cicadellidae. It feeds on sugar beets, tomato, garden beets, cantaloupe, celery, cucumber, pepper, spinach, squash and other vegetables, as well as on geranium, nasturtium, pansy, zinnia, and other ornamentals. It is the vector of curly top disease. Found in western U.S.

beet leaf miner. See *spinach leaf miner.*

beetleweed. See *galax aphylla.*

beet lifter. A device consisting of two prongs that straddle the row to pull beets from the ground. Also called *beet puller.*

beet molasses. A stock feed, high in sugar and low in phosphorous, slightly laxative, which is a by-product of beet sugar manufacture.

beet pocket rot. One form of hypertrophy, a bacterial gall, in which the cells in beet root become abnormally large and disorganized.

beet puller. See *beet lifter.*

beet pulp. See *dried beet pulp; wet beet pulp.*

beetrave. See *beet: 2.*

bee tree. 1. A hollow tree which is occupied by a colony of bees. 2. A tree whose flowers are a source of nectar or pollen.

beet root tumor. A fungous disease of beet roots due to *Urophlyctis leproides,* family Physodermataceae; characterized by various sized galls formed on the roots. It is unknown in North America but is common in Europe and North Africa.

beet rot. See *beet heart rot.*

beet rust. A rust caused by *Uromyces betae,* family Pucciniaceae; characterized by brown-colored spore masses on leaves, but is rarely of much harm to garden beets. Found in Europe and in western U.S.

beet scab. A disease caused by the bacterial organism, *Streptomyces scabies,* which is characterized by scab-like lesions on the roots.

beet slop. See *beet sugar residue.*

beet sugar. Sucrose sugar produced from sugar beets. *Cf. cane sugar.*

beet-sugar residue. The liquid left after sugar has been extracted from beet roots. It contains 3 to 4 percent nitrogen and 8 to 10 percent potash. When used in fertilizers, it is acidulated with phosphate rock and sulfuric acid.

beet tops. The green leaves removed from sugar beets at harvest. They may be fed to stock fresh or as silage.

beet webworm. *Loxostege sticticalis* (L.), family Pyraustidae, a caterpillar, which feeds on cabbage, beets, sugar beets, beans, peas, potato, spinach, cucurbits, other vegetables, as well as on field crops. Found in western U.S.

bee veil. A cloth or wire veil used for protecting the head and face from bee stings.

beeves. See *beef.*

beeweed. See *poison milkweed.*

bee wolf. A solitary wasp *Philanthus triangulum* (F.), which captures honey bees and stores its nest with them as food for its larvae. Where plentiful, it can cause beekeepers considerable trouble. Found in Europe. Related species have been observed in the U.S. preying on native bees but not on honey bees.

bee yard. See *apiary.*

beget. To procreate, as a sire. See *get; get of sire.*

beggar's lice. The dried, prickly fruit of any of the species of *Lappula, Galium, Desmodium, Hackelia.*

beggarticks. Any weed plant of the genus Bidens, family Compositae.

begonia. Any plant of the genus Begonia, family Begoniaceae; grown for its foliage and flowers inside or outside in warm regions. See *perpetual begonia; tuberous begonia; rex begonia; Feasti* *begonia; socotra begonia.* Also called *elephant's ear; beefsteak geranium.*

begonia treebine. *Cissus discolor,* a greenhouse plant of the family Vitaceae; grown for its leaves which are green on the upper surfaces and white, pink, or purple underneath. Also called *trailing begonia.* Native to Java.

behen. See *cowbell silene.*

Belgian. 1. A breed of horses originating in Belgium from the Flemish horse. They are strong brown draft horses; a stallion may weigh over a ton. 2. A breed of rabbits raised for meat production.

Belgian cooked cheese. A cheese made of curdled skim milk; after ripening for a week or two, cream and salt are added, and the ripening continues.

Belgian elm. A hardy variety of the Dutch elm (*q.v.*) that holds its branches upright and is grown for ornament. See *Huntington elm.*

Belgian walnut. See *candlenuttree.*

belladonna. *Atropa belladonna* L., a perennial herb, family Solanaceae. Its berries are poisonous, and its sap yields alkaloid atropine from which the drug is made. Also called *death herb; doftberry; deadly nightshade.* Native to Eurasia; grown in the U.S. where it is sometimes found escaped, particularly in eastern U.S.

belladonnalily. *Amaryllis belladonna,* a bulbous, flowering herb, family Amaryllidaceae; grown for its fragrant, pink flowers. Native to South Africa and found in southern U.S.

belle honeysuckle. *Lonicera bella,* a hybrid, deciduous shrub, family Caprifoliaceae; grown for its white or pink flowers that fade to yellow.

Belle Isle cress. See *early wintercress.*

Bellelay. A soft, blue-veined, whole milk cheese similar to Gorgonzola, made of heated sweet milk with added rennet. It is usually ripened for 12 months in a moist cellar. Also called *Monks head; Tête de Moine.*

bell-ended sewer pipe. A sewer or drain pipe, usually of vitreous clay, which has one end enlarged into which the small end of the next section fits.

bell-glass. A glass covering which is sometimes used in greenhouses for covering special plants and cuttings. See *cloche: 1.*

bellows duster. A portable duster for spreading insecticide, consisting of a bellows, a discharge chamber, and a pipe. The bellows forces the pesticide out in puffs rather than in a continuous stream. Also called *puff duster.*

bell pepper. See *sweetbell redpepper.*

bell scraper. A butchering tool consisting of a short handle and a flared bell-shaped steel cutting edge. It is used to remove hair from a hog's skin after slaughter and scalding.

bellwether. A ram or wether, with a bell tied around his neck, which serves as a flock leader.

belly. The abdomen of a man or animal.

bellyband. See *girth: 2.*

belly-fleece. Wool on the abdomen of sheep.

Bel Paese. A soft, uncooked, sweet, fast-ripening, table cheese of Italian origin, made of cow's milk to which is added lactic starter with rennet; and then cured about three weeks.

belt. 1. A geographical term denoting an area of similar soil and climate particularly suited to specific crops or animals; such as the wheat *belt.* 2. A band of hair or skin of another color, often white, around an animal's body; as in Dutch *belted* cattle. 3. In machinery, a broad flexible band passed around two wheels to transmit motion from one to the other.

belted. 1. Any fruit which has a retarded growth forming a ring or band extending completely around its middle; often due to frost injury when the fruit is very young. This band is usually shrunken. Also called *banded fruit.* 2. Any animal having a band of different color around its body, such as Dutch *belted* cattle.

belted diabrotica. See *banded cucumber beetle.*

Beltsville Small White turkey. A variety of small turkey developed by USDA, Beltsville, Maryland; hens weigh eight to ten pounds, toms fifteen to seventeen pounds.

Belvedere summercypress. *Kochia scorparia,* a branched, tender, annual herb, family Chenopodiaceae; grown for its bushy red, green, or yellow foliage which turns purple-red in the autumn. Escaped, it is a persistent weed pest. Native to Eurasia. Also called *summer cypress; standing cypress; mock cypress.*

bench. In horticulture, the platform in a greenhouse on which plants are grown.

bench killing and picking. A method of killing and dressing poultry so that the carcass of a bird is kept clean. A bird is suspended by its legs over a slanted table top and its lower bill is attached to a hook on the table. As the carcass is drained, the blood runs through a hole in the table into a bucket.

bench lands. 1. Terraces or shelf-like land features representing former water levels or shore lines of lakes, rivers, seas. They are usually composed of alluvium or unconsolidated coarse sediments. 2. A foothill below a mountain.

bench-legged. Knock-kneed.

bench root. Abnormal root development in which the roots of a seedling are twisted or badly formed. Also called *gooseneck.*

bench terrace. A step-like embankment of earth with a flat top and a steep or vertical downhill face constructed along land contours to control run-off and erosion.

benders. Cotton staple of intermediate length.

bendy tree. See *Portiatree.*

benefit payment. Payment made by the U.S. Federal Government to farmers for compliance with programs and regulations for control of crop acreages and agricultural production.

Bengal rose. See *Chinese rose: 1.*

benign. In reference to disease or disease processes, mild or nonmalignant; such as a benign tumor or growth, in contrast to malignant or cancerous, tumor or growth.

Benjamin-bush. See *common spice bush.*

bentgrass. Any plant of the genus Agrostis, family Gramineae. See *velvet bentgrass; colonial bentgrass; creeping bentgrass; cloud bentgrass; redtop.*

bent hock. See *sickle-hocked.*

bentonite. A porous rock of clay minerals derived from weathered volcanic ash or tuff. It is used: (a) as a dust diluent and carrier for insecticides; (b) in clarifying or fining wines, fruit juices, etc.; (c) as a wetting agent; and (d) as a seepage retarder in the bottom of a water storage reservoir.

benzene. C_6H_6, a volatile and inflammable hydrocarbon which is a derivative of coal tar and is used: (a) in degreasing bones, etc., in fertilizer manufacture, and (b) for destroying larvae. It is particularly effective against the screw-worm fly.

benzene hexachloride. An effective insecticide. Also called *BHC.*

benzil. An insect repellent which is especially effective against chiggers.

benzoate of soda. See *sodium benzoate.*

benzoic acid. $C_6H_5CO_2H$, a chemical food preservative, antiseptic, and perfume additive.

benzol. See *benzene.*

benzyl benzoate. $C_6H_5CO_2C_7H_7$, an acaride used to repel chiggers.

bere. Barley (obsolete).

bergamot lime. See *limeberry.*

bergamot mint. *Mentha citrata,* a perennial herb, family Labiatae; grown occasionally for ornament but most frequently for its essential oil. Native to Europe.

bergamot orange. *Citrus bergamia,* family Rutaceae; the rind of the pearlike fruit yields an aromatic oil which is used in perfume manufacture. Native to Europe.

bergenia. Any plant of the genus Bergenia, family Saxifragaceae. See *heartleaf bergenia; leather bergenia.* Also called *elephant's ear.*

Berkshire. A popular breed of hogs originating from Berkshire, England. Of medium size, it is black with white feet, nose, and tail, and it has a very short, dished face and is shorter in the leg than other lard-type breeds.

Berlin poplar. *Populus berolinensis,* a hybrid of *Populus laurifolia* and *P. Nigra* var. *italica* (the Lombardy poplar), sometimes called *Populus certinensis.* A columnar poplar resembling the Lombardy poplar but hardier. It is grown on the prairies of the northwestern U.S. Also called *certinensis poplar.*

berm. 1. The distance from the top of the bank of a drainage ditch to the edge of the spoil bank. 2. The shoulder of a highway.

Bermuda arrowroot. *Maranta arundinacea,* an herb, family Marantaceae; grown for its fleshy roots which are the source of arrowroot of commerce. Native to western India. See *coontie.*

Bermuda buttercup. See *buttercup oxalis.*

Bermuda Easter lily. *Lilium longiflorum* var. *eximium,* a popular bulbous herb, sometimes four feet high, family Liliaceae, which is often forced in greenhouses for the Easter bloom.

Bermudagrass. *Cynodon dactylon,* family Gramineae, a perennial grass providing excellent forage, but becoming a weed in cultivated fields; also useful for making lawns. Commonly found in southern U.S. Also called *scutch grass; Bahama grass; devil grass; doob; couchgrass.*

Bermudagrass cutter bar. A cutter bar for a mowing machine having twice as many guards in the same length as a regular bar and ledger plates on each side of each guard. The knife passes through two guards on each stroke. It is used in cutting closely matted grass.

Bermuda onion. Any species of onion introduced into U.S. from Bermuda. Now commonly grown in southwestern U.S.

Bermuda palmetto. See *Blackburn palmetto.*

Bernarde. An Italian cheese made from cow's milk to which about ten percent of goat's milk is added. Saffron is introduced for color, and rennet is added to coagulate the heated milk before it is cured for approximately two months.

berry. 1. Any small pulpy fruit, as the raspberry, strawberry, etc.; usually edible. 2. Any fruit which has either a fleshy or pulpy pericarp, such as the banana, cranberry, tomato, etc. 3. The dry seed of certain plants, as the coffee *berry,* which is really a seed contained in the fleshy true *berry.*

berry bladderfern. *Cystopteris bulbifera,* family Polypodiaceae, grown for its delicate foliage. Native to North America. Also called *bladderfern.*

berry cone. The ripe fruit of certain conifers in which the scales of the cone are bound together so as to resemble a berry, as in the juniper.

berry grape. See *bullace grape.*

berry hook. A metal rod with a sharpened hooked blade used for pruning low-growing vines.

berry moth. See *grape berry moth.*

berry thinning. Thinning vinifera grapes by removing the end of the main stem of the cluster and several branches of the cluster to improve the quality when the cluster is tightly packed.

berry worm. See *grape berry moth.*

berseem clover. See *Egyptian clover.*

Bessey cherry. *Prunus besseyi,* family Rosaceae. A shrub bearing a small, sweet, juicy, edible fruit which is purplish-red. The plant is sometimes crossed with plum tree or other species of cherry. Also grown as an ornamental. Found in central U.S.

Bessonette bale (cotton). See *round bale: 1.*

beta naphthol. A fumigant used against the codling moth. It is placed on tree bands to kill larvae when they enter the bands to pupate.

betel. *Piper betle* or Chavica betel, an Indian shrub, family Piperaceae, from which a masticatory is made. The leaf of the plant is used to enclose shavings of an areca palm nut and a small quantity of lime.

betony. Any herb, of the genus Stachys, family Labiatae; sometimes grown in the garden. Native to Asia, Europe, North America. See *woolly betony; artichoke betony.*

bezoar. Stones, hair balls, wool balls, etc., found in the alimentary canals of ruminants and sometimes other animals. At one time these were thought to have antipoison properties. See *hair ball.*

Bgug-Panir. A cheese made partly or entirely of skimmed ewe's milk to which is added rennet as well as salt and herbs.

BHC. See *benzene hexachloride.*

B horizon. See *horizon B.*

bichloride of mercury. $HgCl_2$, a poisonous white powder chemical used as a disinfectant; water soluble. Also called *mercuric chloride; perchloride of mercury; sublimate of mercury; bichloride; corrosive sublimate.*

biennial. Any plant which completes its life cycle from germinating seedling to seed production and death in two growing seasons. See *annual; perennial.*

biennial bearing. Designating a tree producing fruit every two years. See *annual bearing; alternate bearing.*

biennial wormwood. *Artemisia biennis,* family Compositae, a weed pest.

biestings. Colostrum.

bigarade. See *sour orange.*

bigarreau cherry. Denoting varieties of the sweet cherry (*q.v.*) species (*Prunus avium*). Some common varieties are Duke cherry, Bing, Lambert, Napoleon, and Black Republican. See *mazzard cherry: 2; heart cherry.*

big bluegrass. *Poa ampla,* family Gramineae. This grass is useful for range reseeding. Found in the Yukon Territory and southward to New Mexico and California. See *bluegrass.*

big blue huckleberry. A name commonly but incorrectly given to the blueberry (*q.v.*). See *blueberry.*

bigblue liriope. *Liriope muscari,* a perennial herb, family Liliaceae; grown as an ornamental ground cover and for its lilac-purple flowers. Native to Asia. Also called *blue lily-turf.*

bigblue lobelia. See *great lobelia.*

big bluestem. *Andropogon furcatus,* family Gramineae; a tall grass, producing good forage. It grows in central and southwestern U.S. in dry places, on prairies, and in open woods.

bigbract verbena. *Verbena bracteata,* an herb, family Verbenaceae; a weed found on abandoned fields in the early and late weed stages. It is native to U.S.

big bud hickory. See *mockernut hickory.*

big bud mite. See *currant bud mite.*

big cerastium. *Cerastium vulgatum,* a short-lived perennial herb, family Caryophyllaceae, which is a weed pest of lawns. It is a small, white-flowered, prostrate plant which spreads from creeping stems on the soil surface. Native to Eurasia. Also called *mouse-ear chickweed.*

bigcone pine. See *Coulter pine.*

bigflower butterwort. *Pinguicula grandiflora,* an insectivorous herb, family Lentibulariaceae, which is grown as a curiosity. Native to Europe. See *butterwort.*

bigflower sage. See *bee sage.*

bighead. 1. A disease of sheep, caused by grazing on certain plants which act as photosensitizing agents; characterized by swelling of the tissues of the neck and head. 2. One of the manifestations of osteofibrosis and osteomalacia in the horse in which the bones of the head lose calcium and become spongy and enlarged out of proportion to the rest of the skeleton. It may be due to deficient calcium and excess phosphorous in the diet.

big heliotrope. *Heliotropium corymbosum,* a tropical, perennial herb, family Boraginaceae; grown as a tender annual for its narcissus-scented, violet flowers. Native to South America.

big heronbill. *Erodium botrys,* an annual herb, family Geraniaceae, which has taken the place of many of the native grasses, particularly on range lands of the western U.S. where overgrazing occurs and in the San Joaquin Valley of California.

big hickory. See *shellbark hickory.*

big jaw. See *actinobacillosis; actinomycosis.*

bigleaf honeybush. *Melianthus major,* an aromatic, evergreen shrub, family Melianthaceae; grown in warm areas for its foliage and its reddish-brown flowers. Native to South Africa. Also called *honey-flower.*

bigleaf hydrangea. *Hydrangea macrophylla,* a shrub, family Saxifragaceae; grown indoors for its pink, blue or white, globe-shaped, flower clusters. Native to Japan.

bigleaf magnolia. *Magnolia macrophylla,* family Magnoliaceae; grown for its large evergreen leaves and white, fragrant flowers with purple centers. It thrives in cities where it is grown as a street tree; particularly in southern U.S. Native to North America. Also called *cucumber tree.*

bigleaf maple. *Acer macrophyllum,* a deciduous tree, family Aceraceae; grown as a shade and as a park tree. Its large leaves turn bright orange in the fall. Native to North America. Also called *Oregon maple.*

bigleaf periwinkle. *Vinco major,* a tender, trailing, evergreen herb, family Apocynaceae; planted in window boxes for its blue flowers. Native to Europe.

bigleaf shagbark hickory. See *shellbark hickory.*

bigleaf wintercreeper euonymus. *Euonymus fortunei* var. *vegetus,* a spreading evergreen shrub, family Celastraceae, which grows to a height of five feet and climbs when supported; grown for its large clusters of greenish-white flowers, and handsome, orange-colored fruits. Native to Asia. Also called *evergreen bittersweet; wintercreeper.*

big liver disease. See *fowl paralysis.*

big mountain brome. See *mountain brome.*

big-neck. In piglets, lambs, and ewes, an enlargement of the thyroid glands due to a deficiency of iodine. See *goiter.*

big root. See *root knot.*

bigroot morningglory. *Ipomoea pandurata,* a hardy perennial vine, family Convolvulaceae; grown for its white, purple-throated flower. Escaped, it can become a serious weed due to its very large root. Also called *manroot; wild sweet potato; man-of-the-earth; scammony; hardy* or *perennial moonflower; Colorado manroot.*

big sagebrush. *Artemisia tridentata,* a shrub, family Compositae, which is good forage on the western ranges of U.S. Also called *black sage.*

bigseed alfalfa dodder. *Cuscuta indecora,* a parasitic weed. Also called *alfalfa dodder.*

big shagbark. See *shellbark hickory.*

big spike bentgrass. *Agrostis exarata,* a perennial grass, family Gramineae, which is good forage. Found in western North America.

bigsting nettle. See *great nettle.*

big valley sweep. A plowpoint with narrow wings used on a cultivator for weeding.

big vein of lettuce. A virus, soil-borne disease which causes severe loss. At first the veins of the lettuce leaves show slight yellow discoloring, and as the veins become more pronouncedly yellow, the leaf becomes thickened and crinkled. All of the leaves are affected but the plant is not killed; the head forms but is much smaller than normal, and the leaves have a poor flavor and texture. The virus is limited to the roots and remains in the soil.

big wandflower. *Sparaxis grandiflora,* a perennial herb, family Iridaceae; grown for its spring-blooming, yellow or purple flowers. Native to South America.

big whortleberry. *Vaccinium membranaceum,* a shrub, family Ericaceae. It is a drought resistant plant and bears one of the best flavors of the wild blueberries. Found in northwestern U.S. Also called *broadleaved huckleberry; mountain blueberry.*

bilateral horizontal cordon. A method of training vinifera grapes and dwarf fruit trees. When a tree has grown to about two feet in height, two horizontal arms or lateral branches are developed from the trunk. See *unilateral horizontal cordon; vertical cordon; espalier.*

bile. Yellow or greenish fluid secreted by the liver which aids digestion.

bilharziosis (bilharziasis). A disease caused by blood flukes (parasitic trematodes) belonging to the genus Schistosoma (syn. Bilharzia) in men and animals. The disease is more commonly known as schistosomiasis.

biliary fever. See *piroplasmosis.*

bill. 1. See *brush hook.* 2. A natural land feature: a narrow headland or small peninsula. 3. A bird's beak.

billbug. *Calendra* spp., family Curculionidae, a group of insects that feed on wheat and other grasses. See *maize billbug; clay-colored billbug.*

billet. 1. A short log or bolt of wood cut especially for cooperage, shingles, pulp, or cordwood. 2. On a harness, a strap which enters a buckle or a loop to hold the end of a strap.

bill hook. 1. A metal hook to grasp string in a twine knotter. 2. A cutting knife used in pruning.

billion-dollar grass. See *barnyardgrass.*

bill of sale. A formal legal paper for the transfer of goods, real estate, animals, etc. It is a written promise that title will be transferred when the conditions of the sale have been completed.

billy. A tin pot for boiling tea; standard equipment among Australian and New Zealand sheepmen.

billygoat. A male goat.

Biltmore stick. A rule graduated so that the height and diameter of a standing tree may be calculated; used to estimate board feet in a tree.

bin. A box, crib, or enclosure used for storage, particularly of grains.

bin-burn. The discoloration and deterioration of stored goods due to heat; as corn or grain in a bin.

binder. 1. A harvesting machine for cutting grain and tying it into bundles which is being rapidly replaced by the combine harvester. Also called *reaper and binder.* 2. A pole used to tighten a chain which holds together a load of logs. 3. A fiber connecting the staples of wool so as to form a fleece. 4. Material to improve cohesiveness of soils, feeds, etc.

binder canvas. A piece of slatted canvas several feet wide forming a belt used to convey cut grain from the platform of a binder to the binding attachment.

binder head. See *binding attachment.*

binder tobacco leaf. Tobacco wrapped around the filler leaves and segments of a cigar.

binder twine. A coarse, loosely twisted twine used for tying grain bundles after cutting. It is usually made from sisal, henequen, or abaca. See *baler twine.*

binding attachment. The part of a binder which ties grain into bundles (sheaves) after it has been cut and elevated.

bindweed. Twining or creeping species of *Convolvulus,* family Convolvulaceae. Also called *glorybind; wild morning-glory.*

bioassay. The determination of the potency of various substances, such as drugs, hormones, and vitamins, by their action upon animals or microorganisms. The term also refers to the standardization of such preparations, in which case animals or microorganisms are given minimal doses under controlled conditions.

biochemical activity. Chemical process in living organisms, tissues and cells.

bioclimatics (bioclimatology). The study of effects of climate on living organisms.

biological control. The destruction or suppression of undesirable insects, plants, or animals by the introduction or propagation and dissemination of predators, parasites, and diseases.

biological immobilization. See *immobilization.*

biological interchange. The interchange of elements between organic or inorganic states in a soil or other substrate through the agency of biological activity.

biological mineralization. The conversion of an element occurring in organic compounds to the inorganic form through biological decomposition.

biologicals. Products derived from a living process or living matter; such as serums, vaccines, bacterins, and antitoxins, etc. Also called *biologics.*

Biological Survey. Formerly a bureau in the United States Department of Interior which examined and reported on American animals and birds. In 1940 it was consolidated with the Bureau of Fisheries to form Fish and Wildlife Service *q.v.*

biologic erosion. Erosion by water or wind as a result of soil being exposed by the burrowing of rodents, destruction of vegetation by insects, etc.

biology. The science of life and living organisms.

biometry. Application of statistical methods to the study of biological problems.

bionomics. The study of the relations among organisms and the relationships between them and their environment. See *ecology.*

biopsy. The microscopic or chemical analysis of tissue removed from a living body.

biosequence. A sequence of soils whose properties are functionally related to differences in organisms, as a soil forming factor.

biotic. Pertaining to life; biological.

biotic influence. Biological influences on vegetable life as contrasted to climatic influences.

biotic potential. Inherent ability of an organism to reproduce and survive.

biotype. A group or race of people, fauna, or flora of identical genetic constitutions.

birch. Any tree of the genus Betula, family Betulaceae. Species are valuable as timber having a hard, close grain wood, and as ornamentals. See *white birch; river birch; yellow birch; paper birch; European white birch; sweet birch; gray birch; canoe birch.*

birch borer. See *bronze birch borer.*

birch casebearer. The larvae of the moth *Coleophora salmani* Heinr., family Coleophoridae. They attack all birches and the speckled alder from within a case, burrowing into leaves as far as possible without becoming detached from the case.

birching. The cutting or hauling out of birch timber.

birch red-leaf blister. A disease of the birches caused by the fungus, *Taphrina bacteriosperma,* family Taphrinaceae, which induces a curling of the leaves (leaf curl).

birch skeletonizer. *Bucculatrix canadensisella* Chamb., family Lyonetiidae, an insect that attacks and often defoliates birches. The larvae bore into the leaves cutting crescent-shaped openings on the lower side of the leaves.

birch witches'-broom. A fungous disease of the birches caused by *Taphrina betulina,* family Taphrinaceae; characterized by a brush-like development of weak shoots. It is only found in Europe. See *witches'-broom.*

birch yellow-leaf blister. A disease of birches caused by the fungus, *Taphrina flava,* family Taphrinaceae.

bird. 1. Any member of the class Aves of the phylum Vertebrata, the body of which is more or less covered with feathers. 2. Domesticated birds: chickens, turkeys, ducks, geese, guineas, etc.

bird banding. See *banding: 1.*

bird grape. *Vitis munsoniana,* family Vitaceae, a wild grape of Florida, the Key Reefs, and the Bahamas. Also called *mustang grape of Florida; everbearing grape; everlasting grape.*

bird guano. See *guano.*

bird mite. See *chicken mite.*

bird-of-paradise. See *Queen's bird-of-paradise flower.*

bird peck. A small hole or distortion of grain in wood resulting from sapsuckers' attacking a tree. It may be associated with discoloration, such as mineral stain *q.v.*

bird pest. See *fowl plague.*

bird plague. A serious fowl disease caused by a virus; characterized by symptoms extremely similar to Newcastle disease, avian typhoid, infectious laryngotracheitis, fowl plague and acute fowl cholera (*q.q.v.*). Also called *fowl pest.*

birdseye gilia. *Gilia tricolor,* an annual herb, family Polemoniaceae; widely grown for its violet or lilac-colored flowers with yellow aubes. Found in California.

birds-eye hickory. See *mockernut hickory.*

bird's-eye maple. The lumber of the sugar maple having grain markings resembling eyes.

bird's eye of grape. See *black rot of grape; grape anthracnose.*

birdsfoot clover. See *birdsfoot deervetch.*

birdsfoot deervetch. *Lotus corniculatus,* a perennial herb, family Leguminosae; of importance as forage. Native to Eurasia. Also called *birdsfoot clover; birdsfoot trefoil; brown honeysuckle; baby's slippers; bloom-fell; crosstoes.*

birdsfoot trefoil. See *Spanishclover deervetch; deervetch.*

birdsfoot violet. *Viola pedata,* a handsome, perennial herb, family Violaceae; grown for its lilac or blue-violet flowers. Native to eastern U.S.

bird's nest. See *wild carrot.*

birdsnestfern. *Asplenium nidus,* a tropical fern, family Polypodiaceae. Handsome, it forms a clump resembling a bird's nest. The bright green fronds are from one to three feet long; usually grown in greenhouses, but found wild in southern Florida. Also called *Australian fern.*

bird-weed. See *chickweed.*

birth. See *parturition.*

Biscay heath. *Erica mediterranea,* an evergreen shrub, family Ericaceae; grown in warm areas for its foliage and its deep-red flowers.

bisect. A profile of plants and soil showing the roots and shoots in normal position and indicating vertical and lateral relations.

bisexual. Of two sexes, such as a flower having both stamens and pistils.

bishop. Specifically, to drill small holes in a horse's teeth and to grind the surface of the front teeth so as to make the horse appear younger than he is. The term is derived from a horse dealer named *Bishop.*

bishop pine. *Pinus muricata,* an evergreen tree, family Pinaceae; well suited for shade and shelterbelt planting along the western coast of U.S. Also called *prickle-cone pine.*

bishop's goutweed. *Aegopodium podograria,* a perennial herb, family Umbelliferae; grown as edging or ornamental ground cover. Native to Eurasia. Also called *bishop's weed; goutweed; goatweed; farmer's plague.*

bismuth subsalicylate. A carboxylic acid used for the control of tobacco downy mildew.

bistournage. Castration by the twisting of the spermatic cord.

bistoury. A long, narrow, surgical knife, straight or curved, used for incising abscesses, opening fistula tracts, etc.

bit. 1. That part of a horse's bridle which is placed in the mouth; usually made of steel. 2. An earmark for cattle. 3. The blade of an axe or a similar tool.

bitch. Female of the canine species.

bitch chain. A short chain with hook and ring used to attach a gin pole to a lumber sled or car.

biting louse. An insect which infests livestock. See *horse biting louse; cattle biting louse; sheep biting louse; goat biting louse; dog biting louse.*

bitless bridle. A bridle that has shanks of light steel or aluminum connected by a leather strap which crosses over the horse's nose. The shanks are held together in the rear by a curb strap. Also called *bear trap bridle.*

bit ring. A metal ring on a bridle to which the bit and reins are fastened.

bit snap. A metal ring with a spring opening which fastens to the bit ring of a bridle.

bit strap. That piece of leather on a bridle which goes over the horse's nose.

bitter actinea. See *bitterweed actinea.*

bitter aloe. See *true aloe.*

bitterbrush. *Purshia tridentata,* a valuable forage plant, family Rosaceae; found in the high mountain areas of the western U.S.

bitter dock. *Rumex obtusifolius,* family Polygonaceae, a very common weed in the U.S. Native to Eurasia.

bittern. A concentrated solution of salts which remain after the common table salt has been refined out. Sometimes used as an herbicide.

bitter nightshade. *Solanum dulcamara,* a woody vine, family Solanaceae, which has escaped as a plant pest throughout the eastern U.S. Native to Europe. Also called *bittersweet nightshade; climbing nightshade; fellenwort; blue bindweed; bittersweet.*

bitternut hickory. *Carya cordiformis,* a deciduous tree, family Juglandaceae; grown for shade and for its wood which is used in making tool handles, ladders, furniture, sporting goods, implements, woodenware, and for fuel and smoking meat.

bitter orange. See *sour orange.*

bitter pecan. See *water hickory.*

bitter pit. A non-parasitic malady of apples connected with a disturbed water relation in the fruit. First characterized by slightly discolored spots on the skin, which later become sunken. These spots are circular and vary in size up to ¼ inch in diameter. The pits eventually become brown. The pulp directly beneath the spots consists of dead, brown, corky and spongy cells. Fungi sometimes establish themselves in the bitter pit lesions. Also called *fruit spot; apple brown spot; Baldwin spot; fruit pit.*

bitterroot. *Lewisia rediviva,* a perennial herb, family Portulacaceae; grown in rock gardens for its rose-pink or white flowers. Formerly its starchy roots were eaten by the Indians. Native to western U.S.

bitter rot. See *apple bitter rot.*

bitter rot of cranberry. A fungous disease of cranberry caused by *Glomerella cingulata,* family Gnomoniaceae; infected fruit develop brown spots about the time of harvest or just after, and a soft rot develops.

bitter rot of ripening grapes. This fungous disease affects American and European varieties, with lesions appearing on the shoots and peduncles. Caused by *Melanconium fuligineum,* family Melanconiaceae, especially on the white varieties. Also called *bird's-eye of grape.*

bitter rubberweed. See *bitterweed actinea.*

bitter sneezeweed. 1. *Helenium tenuifolium,* a common, annual herb, family Compositae; causes cows which graze on it to produce a bitter flavored milk. It is a widespread weed in southern U.S. Also called *bitterweed; fennel; wild garlic bitterweed.* 2. See *horseweed fleabane.*

bittersweet. See *American bittersweet; oriental bittersweet; bitter nightshade.*

bittersweet nightshade. See *bitter nightshade.*

bitterweed. See *bitter sneezeweed.*

bitterweed actinea. *Actinea odorata,* an annual herb, family Compositae; poisonous to sheep which may eat it during a shortage of forage. Poisoning is characterized by loss of appetite, cessation of chewing of the cud, abdominal pain, bloating, frothing at the mouth, and a green nasal discharge. Found on overgrazed ranges in western U.S. Also called *bitter rubberweed; Colorado rubberweed; bitterweed.*

bitter wintercress. *Barbarea vulgaris,* an herb, family Cruciferae; sometimes used as a substitute for watercress. Escaped, it can become a troublesome weed. Native to Europe. Also called *yellowrocket; herb barbara; wintercress; rock cress; wild mustard; upland cress.*

bitterwort. See *yellow gentian.*

bitting rig. A device for breaking in a horse; made of a padded back band with a crupper attached to it and used in conjunction with a bridle. The back band has rings on either side near the top and two more on either side near the middle.

Bitto. A firm, semi-cooked cheese similar to Swiss cheese; made from cow's milk, ewe's milk, or a mixture of goat's and cow's milk, with rennet added to coagulate the heated milk. Curing may take up to two years.

bivalent bacterin. A bacterin prepared from two different types of disease-producing bacteria so as to give protection against both diseases.

bivalent chromosome. One of a pair of homologous chromosomes present during meiosis.

biversale adaptation. An adaptation in which a plant or other organism utilizes the favorable factors of the environment to the fullest extent, protecting itself against an unfavorable factor working in excess.

bizarre tulip. A type or class of tulip of variegated colors; usually yellow splashed with red or brown.

black acacia. See *blackwood acacia.*

black adobe. A heavy, black, clay soil of California. See *adobe soil.*

black alder. See *European alder: 1.*

black alkali. A non-saline alkali soil in which the organic matter has a very dark color due to the caustic action of sodium carbonate. The alkali mainly consists of sodium carbonate. See *alkali soil.*

black arm. See *bacterial blight of cotton.*

black army cutworm. *Aotebia fennica* (Tausch), family Phalaenidae, a variegated cutworm; a pest of blueberries, fruit trees, garden crops, strawberries, grains, and pasture. Found in Canada and northern U.S.

black ash. *Fraxinus nigra,* a deciduous tree, family Oleaceae, which prefers a swampy, moist soil. It is used in cabinet-making and basket-making. Found in eastern North America. Also called *brown ash; hoop ash; basket ash; swamp ash.*

black ballota. *Ballota nigra,* a perennial herb, family Labiatae; grown for its decorative purple flowers. Native to Eurasia. Also called *black horehound; fetid horehound.*

black baneberry. *Actaea spicata,* a perennial herb, family Ranunculaceae, a poisonous plant, often grown for its ornamental berries. Native to Europe.

black barley. A market classification of barley which includes all varieties of black (glume) barley (whether two or six rowed); it must not include more than ten per cent of other barley classes.

black bead. See *catclaw apes-earring.*

black bean aphid. See *bean aphid.*

blackberry. Any of the upright plants of the genus Rubus, family Rosaceae; grown for the edible, black, sweet berries. The edible torus is picked with the berry. Of horticultural interest are the Allegheny blackberry (***Rubus allegheniensis***), cutleaf blackberry (***R. laciniatus***), highbush blackberry (*R. ostryifolius*), Himalaya blackberry (*R. procerus*). See *raspberry* in which the torus stays on the bush when picked.

blackberry cane and leaf rust. A fungous disease which attacks species of *Rubus;* caused by the rust *Kuehneola uredinis,* family Pucciniaceae. It produces large, bright yellow lesions on the canes, smaller lesions on the leaves and fruit. Found widely in U.S.

blackberry cane gall. A bacterial disease caused by *Agrobacterium rubi.* Characterized by knobby outgrowths on the canes which cause the canes to split open. All species of the genus Rubus may be affected.

blackberry crown gall. A bacterial disease caused by *Agrobacterium tumefaciens.* Characterized by knobby outgrowths in the crown and on the leaves of the plant.

blackberry leaf and cane spot. A fungous disease caused by *Septoria rubi.* Characterized by light-colored, small spots with red or purple borders on the leaves. The spots also appear on the canes and result in premature defoliation of fruiting cane. Found mostly on dewberry in Oregon and Idaho and in the coastal regions of California, and on other species of *Rubus* in eastern U.S. *Mycosphaerella rubi,* family Mycosphaerellaceae has been found associated with this fungus in several regions and may be its perfect stage.

blackberrylily. *Belamcanda chinensis,* a perennial herb, family Iridaceae; grown for its red-spotted, orange-colored flowers. Native to eastern Asia. Also called *leopard flower.*

blackberry yellow rust. A minor, fungous disease caused by *Kuehneola uredinis,* family Pucciniaceae. Characterized by numerous, very small, yellow uredia on the lower leaf surfaces in the spring. Most cultivated species are resistant to this disease. Found in Europe and eastern U.S.

black bindweed. See *dullseed cornbind.*

black birch. *Betula lenta,* a deciduous tree, family Betulaceae. Its twigs are distilled for oil of wintergreen and it is used for hardwood distillation. From its wood are made cooperage, railroad ties, furniture, boxes, interior trim, spools, butcher blocks, agricultural implements, musical and scientific implements. Found in eastern U.S. See *sweet birch.* Also called *cherry birch.*

black blight of sugar beets. A phosphate deficiency of sugar beets which causes the leaves to become susceptible to invasions of parasites and spots the leaves with black lesions.

black blister beetle. *Epicauta pennsylvanica* (DeG.), family Meloidae; feeds particularly on China asters and Japanese anemone. Commonly found in eastern U.S.

black blizzards. Clouds of dust consisting of organic matter and the finer particles of soil from bare fields and grazing land denuded of vegetation. Prevalent in southwestern U.S.

black blow fly. *Phormia regina* (Meig), family Calliphoridae, which breeds primarily on animal carcasses. It also attacks soiled sheep's wool, in which the maggots feed on the wool around the rump; and in attacking horns, it causes the adjacent skin to fester and the wool to loosen and become putrid.

black brood. See *European foulbrood.*

black brush. See *American tarbush.*

black bulb. Any of the old pseudobulbs of an orchid.

black bundle disease. A disease of corn caused by the fungus, *Cephalosporium acremonium* Cda., family Moniliaceae; characterized by blackening of the fibrovascular bundles.

Blackburn palmetto. *Sabal blackburniana,* family Palmaceae; a fan palm grown in very warm areas for ornament. Native to the West Indies. Also called *Bermuda palmetto.*

blackcalla arum. *Arum palaestinum,* a tender, indoor perennial herb grown for its foliage and flowers, which resembles the calla-lily. Native to Palestine. Also called *Solomon's lily.*

black canker of sweet cherry. A minor virus disease known to be transmitted from diseased propagated wood. Found on the northwestern coast of North America.

black canker of willow. A disease caused by the fungus, *Physalospora miyabeana,* family Pleosporaceae. Characterized by dark-brown spots on the upper leaf surface and by whitish-gray lesions with black borders on the twigs and stems. Black fruiting patches develop on the stem lesions.

blackcap raspberry. *Rubus occidentalis,* whose ripe fruit is black as contrasted to the red fruit of other raspberries. Also called *common blackcap; black raspberry.*

black carpenter ant. *Camponotus herculeanus pennsylvanicus* (DeG.), family Formicidae, the largest American common ant. It attacks tree stumps, tree trunks, telephone poles, window sills, and old timbers. Found in eastern U.S.

black casse. A non-bacterial defect of wine due to excessive amounts of iron which oxidize and cause the wine to become dull, acquiring a bluish or black tint. It is the result of the solvent action of the fresh juice on the iron of the crusher and other equipment. Also called *blue casse; ferric casse.*

black centaurea. *Centaurea nigra,* a perennial herb, family Compositae; grown for its spherical, rose purple flowers. Escaped, it can become a weed pest. Native to Europe. Also called *knapweed; hardhead; Spanish buttons.*

black chaff of wheat. A seed-borne, common bacterial disease caused by *Xanthomonas translucens* var. *undulosum.* It produces water-soaked lesions on the seedling leaves and dark stripes running lengthwise on the chaff.

black check. 1. A dark stain in sapwood and heartwood, surrounded by abnormal formation of wood. 2. A small resin-filled pocket in wood.

black cherry. *Prunus serotina,* a deciduous tree, family Rosaceae, an important timber tree. It also is a stock poisoning plant. Also called *wild black cherry; black choke.*

black cherry aphid. *Myzus cerasi* (F.), family Aphidae, a large black louse that commonly infests the sweet cherry tree, causing a curling and distortion of the leaves. It also infects the sour cherry, but leaf curling does not result.

black cherry fruit fly. *Rhagoletis fausta* (O.S.), a fly, family Tephritidae, that infests cherry, wild cherry, pear, and plum. It feeds by scraping the surface of leaves and fruit and by sucking plant juices. It lays eggs in small cuts made in the flesh of fruit. Infested trees produce misshapen and undersized cherries. Found widely in northern U.S.

black chicken. A marketing term denoting a non-white chicken whose legs are black or greenish brown.

black chokeberry. *Aronia melanocarpa,* a deciduous low shrub, family Rosaceae; grown for its white flowers and shiny-black fruit. Native to North America.

black chokecherry. *Prunus virginiana* var. *melanocarpa,* a deciduous shrub or small tree, family Rosaceae. Although it is moderately palatable to sheep and goats but little fed on by cattle, it is of poor forage value, and under certain conditions poisonous. Native to North America.

black citrus aphid. *Toxoptera aurantii* (Fonsc.), family Aphidae, a louse that infests citrus trees on the coasts of California and Florida.

black cohosh. See *cohosh bugbane.*

black comb. See *avian monocytosis.*

black cosmos. *Cosmos diversifolius,* a tender annual herb, family Compositae; grown for its red flowers. Native to Mexico.

black crowberry. *Empetrum nigrum,* an evergreen, alpine, prostrate shrub, family Empetraceae; grown in rock gardens to form mats and for its small, purple flowers. Native to northeastern America and to Eurasia.

black crown rot of celery. A rot caused by *Centrospora acerina,* family Moniliaceae.

black currant. See *European black currant.*

black cutworm. *Agrotis ypsilon* (Rott.), family Phalaenidae, a surface cutworm which attacks truck crops. Found throughout U.S.

black disease. See *infectious necrotic hepatitis.*

black dusters. Dust storms, associated with northwest winds, in the panhandle country of Texas and Oklahoma.

black earth. A term synonymous with Chernozem (*q.v.*); used in U.S., and sometimes in Australia, to describe self-mulching black clays.

black East India duck. A breed of small domesticated ducks having a rather short black bill with black shanks and toes, and having lustrous, greenish-black plumage.

black elder. See *American elder.*

black elm bark weevil. The young grubs of the beetle *Magdalis barbita* (Say), family Curculionidae, which burrow into the inner bark and sapwood of elm. It is found from New York south to Georgia, and west to South Dakota.

black end of pear. A non-parasitic disease perhaps connected with the use of Japanese rootstocks; it may be caused by an abnormal moisture relationship between parts of the tree. Characterized by a blackening of the skin and the flesh at the calyx end of the fruit.

black ergot horn. A sclerotium of the ergot fungus, *Claviceps purpurea,* family Clavicipitaceae, which infects the heads of rye and many other grasses. The sclerotia, a spur-like mass of fungous tissue, form in the place of the grains. The spurs constitute the drug known as ergot. Falling to the ground the sclerotia germinate and produce stalked heads from which are discharged the ascospores that infect the next season's crop of rye or other grasses. See *ergot: 1.*

black eye. A pineapple disease caused by the bacterium, *Phytomonas ananas;* characterized by dark-brown or black spots.

blackeyed clockvine. *Thunbergia alata,* a climbing perennial, herbaceous vine, family Acanthaceae; grown in very warm regions on small trellises, or as a window-box plant for its white, purple-throated flowers. Native to Africa. Also called *blackeyedsusan.*

blackeyed heath. *Erica melanthera,* a small evergreen shrub, family Ericaceae; grown in warm regions for its foliage and winter-blooming, rose-red flowers. Native to South Africa.

blackeyed peas. A variety of the cowpea, *Vigna sinensis;* grown mainly for livestock forage and for green manure. The seeds are eaten as a table vegetable. Found in southern U.S. See *cowpea: 2.*

blackeyedsusan. 1. *Rudbeckia hirta,* a perennial herb, family Compositae; grown for its golden-yellow flowers with a purplish-brown disk. Escaped, it can become a weed pest. Native to North America. 2. See *blackeyed clockvine.*

black eyes of strawberry blossoms. The condition resulting from freezing the receptacle of the strawberry flower. It turns brown or black leaving the leaves of the plant unaffected.

Black-faced Highland sheep. A small, black-faced hardy breed, originating in the Highlands of Scotland; noted for its long wool and yielding good mutton.

black-face wool. A British term designating Black-faced Highland sheep.

blackfire. A widespread, bacterial disease of tobacco caused by *Pseudomonas angulata;* characterized (a) in the seed bed by small, angular, dark-brown or black spots on the leaves, (b) in the field by similar spots which develop into larger lesions with an exudate. Also called *angular leaf spot.*

black fly. An insect, family Simulidae, which frequently bites with very irritating results. Some species are vectors of several diseases of men and animals. The larvae live in streams. See *citrus blackfly.*

black frost. A blackening of vegetation which occurs when the temperature drops low enough below freezing point to destroy vegetation, including staple crops, but there is no actual deposit of frost crystals.

black fungi of scale insects. Three species of Myriangium, family Myriangiaceae, which attack citrus scale insects: (a) *M. duriaei,* widely distributed on many scale insects; (b) *M. floridanum,* occurs on the snow scale, *Unaspis citri* (Comst.), and the purple scale, *Lepidosaphes beckii* (Newm.); (c) *M. montagnei* (Berk), occurs in New Zealand, Australia and Tasmania on the mussel scale, *Lepidosaphes* spp.

black garget. See *gangrenous mastitis in ewes.*

black grain of corn (rye). See *ergot: 1.*

black grain stem sawfly. *Cephus tabidus* (F.), family Cephidae, a wheat insect that also feeds on rye.

black gram. See *mung bean.*

black grama. *Bouteloua eriopoda,* family Gramineae. An excellent forage grass. Found on southern desert ranges of U.S.

black gravel soils. A colloquial name denoting soils characterized by aggregates, or concretions, of gravel size, of black, hydrated iron oxides, bog iron ore. Found in soils developed under swampy, or bog, conditions.

black greasewood. *Sarcobactus vermiculatus,* a shrub, family Chenopodiaceae, whose presence indicates a very high soil content of alkali salts. Valuable as a winter and spring forage plant for sheep, but poisonous to them when they eat too much in a short time. Symptoms of poisoning are dullness, weakness, weak pulse. Found in arid regions of North America.

black ground. A New Zealand term to describe land after a fall of snow has melted.

black gum. See *black tupelo.*

blackhaw viburnum. *Viburnum prunifolium,* a deciduous, hardy shrub or small tree, family Caprifoliaceae; grown for its large, white, fragrant flowers and for its edible, black fruit. Native to North America. Also called *blackhaw; sloe; stagbush.*

blackhead. A chronic disease of poultry, chiefly of turkeys, caused by a protozoan parasite, *Histomonas meleagridis,* in which the principal lesions are confined to the liver and spleen. The disease is also characterized by listlessness, diarrhea with droppings from light-green to brown, and darkening of the skin on the head. Chickens may harbor the parasite, although they are seldom seriously affected, whereas with turkeys, the disease is commonly fatal. Most old poultry yards are contaminated with the organism. Also called *enterohepatitis.*

blackhead disease. A banana disease caused by eelworms. Also called *eelworm black rot.*

black-headed ash sawfly. *Tethida cordigera* (P. deB.), a serious insect pest which feeds on the foliage of the red and white ash. Found in eastern and central U.S.

blackheaded fireworm. *Rhopobota naevana* (Hbn.), family Olethreutidae, a destructive, insect pest of cranberry. The larvae web leaves together and feed within the web.

blackhead of wheat. See *loose smut of wheat.*

blackheart. 1. An injury to woody stems of plants caused by extreme cold. Characterized by dark discolorations of pith or by one or more annual rings of the wood of a tree. The leaves of the trees may be silvered as an after-effect. Also called *internal necrosis.* 2. See *Fusarium wilt diseases.* 3. Any non-parasitic disease of plants characterized by a darkening or blackening of the central tissues.

blackheart of celery. A non-parasitic, wide-spread disease which appears suddenly about mid-season. Characterized by a yellowing or blanching of the outer leaves. The heart leaves shrivel and turn black, and soon the entire heart blackens leaving the plant unmarketable.

blackheart of pineapple. A soft rot of pineapple caused by *Thielaviopsis paradoxa,* family Ophiostomataceae. The perfect stage is *Ophiostoma paradoxum.* Also called *rot of pineapple.*

blackheart of potato. A non-parasitic disease characterized in the tuber by internal necrosis and blackening of the internal tissues. Caused by a lack of oxygen; frequently infects potatoes stored at a high temperature. Sometimes diseased potatoes appear perfectly normal externally but are brown or black inside, having a star-like radiation of blackening. Sometimes the disease may be noted by moist places on the skin and by shallow brown discolorations on the surfaces. Also called *heart rot; internal necrosis.*

black hellebore. See *Christmas rose.*

black henbane. *Hyoscyamus niger,* a poisonous annual or biennial herb, family Solanaceae. Animals usually find it unpalatable. Symptoms of poisoning are nausea, vertigo, thirst, burning skin, dilated pupils, blindness, loss of voluntary motion, convulsions. It can result in death. Native to southern Europe, it is now commonly found in Canada and northern U.S. Also called *henbane.*

black hickory. See *pignut hickory; mockernut hickory.*

Black Hills beetle. *Dendroctonus ponderosae* Hopk., family Scolytidae, a bark beetle that causes serious destruction of pines and spruces. Found in the southern Rocky Mountains and Black Hills of South Dakota.

black horehound. See *black ballota; hoarhound.*

black-horned tree cricket. *Oecanthus nigricornis nigricornis* Walkr., family Gryllidae. It injures tree and bush fruits by making holes in the twigs or brambles in which to lay its eggs. The insect eats the ripe fruit, and the canes it punctures frequently die. It infests the apple, prune, plum, peach, cherry, raspberry, loganberry, blackberry, among other plants.

black horse fly. *Tabanus atratus* F., family Tabanidae. It infests domesticated animals by sucking their blood; a vector of tularemia.

black huckleberry. *Gaylussacia baccata,* an evergreen shrub, family Ericaceae; grown for its edible, black fruit. Native to North America. Also called *bear huckleberry; highbush huckleberry.*

blackjack oak. *Quercus marilandica,* a common deciduous tree, family Fagaceae, which has a heavy black bark. It tends to form dense thickets, is useless for lumber, and a nuisance. Found in eastern U.S. Also called *jack oak.*

black joint of celery. A discoloring and scarring of the stems and leaf petioles of celery caused by the feeding of the tarnished plant bug *q.v.*

black knot. A disease of plums and cherries caused by the fungus *Dibotryon morbosum,* family Dothideaceae; characterized by rough, black, knot-like swellings on twigs and limbs. The first evidence of infection is a slight swelling of a twig or branch near an old knot. The bark ruptures and a yellow or yellowish-brown growth fills the crack. In the summer the surface of the swelling becomes olive-green, covered with a velvet-like pile. By late fall the velvet surface disappears and the swelling becomes black and rough. Small twigs are killed, but larger limbs do not succumb quickly. An important disease east of the Alleghenies from North Carolina to southern Ontario and Maine. Also called *plum wart.*

black knot of currant and gooseberry. A fungous disease caused by *Phragmodothella ribesia,* family Dothideaceae. (Syn. *Plowrightia ribesia*).

black knot of filbert. A fungous disease caused by *Apioporthe anomala,* family Diaporthaceae; characterized by black, wart-like excrescences on the bark.

black knot of grape. See *dead arm of grape; crown gall.*

black knot of plum and cherry. A fungous disease caused by *Plowrightia morbosa;* characterized by abnormal black growths on the branches.

blackland. Generally, areas having very dark-brown or black surface soils. Specifically, the prairies of central Texas, the soils of which are black, waxy, clayey.

blackland plow (bottom). A moldboard and share designed to penetrate and scour in sticky, black soils.

black-laurel. See *loblollybay gordonia.*

blacklaurel leucothoe. *Leucothoe davisiae* Torr., a poisonous shrub, family Ericaceae. One or two ounces of the leaves of this plant will kill a sheep. Symptoms of poisoning are salivation, excessive flow of tears, increase in nasal discharge, convulsions, paralysis of the limbs. Found in mountains of northern California.

black leaf speck of crucifers. A disorder of cabbage, cauliflower and crucifers, caused by lack of oxygen during storage or transportation; characterized by small, lead-gray to black specks.

black leaf spot. Any parasitic or abnormal leaf spot of a black color. See *elm leaf spot.*

black leaf spot of arborvitae. A leaf spot caused by *Didymascella thujina,* (syn. *Keithia thujina*), family Phacidiaceae.

blackleg. 1. A disease of cabbage and other crucifers caused by the fungus *Phoma lingam,* family Sphaerioidaceae. The disease persists in the soil and is carried by the seed. It is characterized by lesions on the stem near the ground or on the petioles and leaves, which soon become sunken and dark in color. Seedlings may collapse completely from it. 2. A potato disease caused by a complex of soft rot bacteria, including *Erwinia carotovora* and *E. atroseptica.* Bacteria cause decay from the seed to the stem, producing a black moist soft rot on immature plants. It is likely to be serious wherever soil conditions are unfavorable, that is where it is cold, wet or waterlogged. 3. A disease of geranium caused by *Botrytis* spp., *Rhizoctonia* spp., or a bacterial leaf spot; characterized by stems rotting and blackening at the base. 4. An acute, infectious disease of young cattle and sometimes of sheep, goats, and swine; caused by bacteria, *Clostridium chauvoei,* that occur in the soil. It infests the animals' legs and produces a high fever. Also called *black quarter; quarter ill; felon; emphysematous anthrax; symptomatic anthrax; emphysematous gangrene.*

black-legged tick. *Ixodes ricinus scapularis* Say, family Ixodidae. It infests mammals; a vector for British and red-water fever.

black-legged tortoise beetle. *Jonthonota nigripes* (Oliv.), family Chrysomelidae. It attacks the sweetpotato, morningglory, and glorybind by eating the foliage.

black line of Persian walnut. A disorder of grafted walnuts; results in inadequate growth of shoots and premature yellowing and shedding of leaves; affects limb-grafted and crown-grafted trees. A tree dies three to five years after symptoms become apparent. Small chocolate-colored, water-soaked spots appear on the bark at the point of the graft; they then spread horizontally and coalesce, penetrating the wood, and breaking the connective tissues. Eventually the layer goes around the tree and stops the transfer of water and minerals. Also called *girdle.*

black liquor. The oil and juice of freshly pressed olives.

black locust. *Robinia pseudoacacia,* a deciduous tree, family Leguminosae. Planted extensively in gullies and on slopes to prevent soil erosion. Its uses are: fence posts, mine timbers, poles, railroad ties, stakes, wooden-ware, novelties, boxes, crates. It is poisonous to range animals which eat the bark or young shoots. Also grown as an ornament. Native to North America. Also called *common locust; false acacia; yellow locust.*

black loose smut. A world-wide fungous disease of oats and related grasses caused by *Ustilago avenae* (Pers.) Rostr.; characterized by a black spore mass.

black mangrove. *Avicennia marina,* an evergreen tree, family Verbenaceae, whose flower is a good source of honey in Florida.

black mapau. See *tawhiwhi pittosporum.*

black maple. *Acer nigrum,* a hardy deciduous tree, family Aceraceae, one of the American sugar-maples. Grown in highway plantings and in city yards. Also used in hardwood distillation, and for railroad ties, veneer (especially good for birdseye), shoe lasts, bowling pins, fuel, furniture, flooring, boxes. Also called *black sugar maple.*

black medic. *Medicago lupulina,* an annual prostrate herb, family Leguminosae; regarded as a weed in lawns, but has value as a pasture plant. Also called *hop medic; nonesuch; blackseed; hop clover; yellow trefoil.*

black melanose. See *melanose.*

black mildew. A grapevine disease which dries up the vine suddenly at full growth. Found in California. Also called *apoplexy; black measles; Spanish measles.*

black millet. See *pearl millet.*

black mold. 1. A mold which has dark mycelium or spores. 2. The diseased or darkened condition of produce or plants resulting from such a mold or mold-like fungus. The name is mostly employed for the common mold of bread.

black mold of grafts. A disease of rose grafts caused by the fungus, *Chalaropsis thielavioides,* family Moniliaceae. It kills the tissue at the point of the graft union.

black mold of onions. A widespread, fungous disease of onions caused by *Aspergillus niger,* family Moniliaceae; characterized by a black, powdery mass of spores on the exterior of the bulb scales. The spores may be readily rubbed off the scales. The affected scales shrivel slowly and get a brittle texture.

black mustard. *Brassica nigra,* an annual herb, family Cruciferae; largely grown for its seed used in the manufacture of mustard (*q.v.*). Escaped, it is a weed pest. Native to Europe.

black nightshade. *Solanum nigrum,* an annual herb, family Solanaceae; a common weed pest. Poisonous to livestock and to bees. Poisoning of livestock is characterized by salivation, vomiting, diarrhea and bloating. Also called *poison berry; deadly nightshade; petty morel; stubbleberry.*

black oak. *Quercus velutina,* a hardy, columnar, deciduous tree, family Fagaceae. Used for railroad ties, cooperage, fence posts, mine timbers, poles, piling, veneers, firewood, flooring, mill products, furniture, railroad cars, boxes, agricultural implements. Native to North America. Also called *tanbark oak; quercitron; yellow-bark oak.*

black oats. A market classification based on the color of the lemma and palea of the matured grain.

black patch of clover. A disease of small economic importance; caused by an unidentified fungus. It is characterized by leaf lesions, varying in color from brown to dark-gray surrounded by concentric rings which sometimes kill the bottom leaves. The stems may be girdled beneath the flower head before the seeds develop.

black peach aphid. *Anuraphis persicae-niger* (Smith), family Aphidae; a louse that feeds on the leaves of the peach and infests the roots of small trees.

black pecan aphid. *Melanocallis caryaefoliae* (Davis), a louse that attacks leaves causing them to fall prematurely.

black pepper. *Piper nigrum,* a woody vine, family Piperaceae; grown for its fruits which yield white and black pepper. Native to the East Indies.

black pine leaf scale. *Aspidiotus californicus* Colm., family Diaspididae, an insect of pine, Douglas fir, hemlock, and other conifers. Found throughout the U.S.

black pinto. Used to describe a horse spotted with black and white.

black pit. A bacterial citrus disease, most noticeable on lemon fruits, caused by *Pseudomonas syringae* (Van Hall); characterized by spots or pits on the fruit which turn from light brown to black. The same bacteria causes citrus blast on the leaves and twigs. See *citrus blast; lilac blight.*

black poplar. *Populus nigra,* a fast-growing, hardy, deciduous tree, family Salicaceae, found throughout the U.S. A less popular ornamental than Lombardy poplar *q.v.*

black pox. A minor, fungous disease of apples and pears, caused by *Helminthosporium papulosum,* family Dematiaceae. Characterized by blackish pustules on the bark and by small sunken spots on the fruits.

black prairie. A range characterized by black soil. See *blackland.*

black raspberry. See *blackcap raspberry.*

black rat. *Rattus rattus.* It infests houses and barns. See *brown rat.*

black ring of grasses. A fungous disease caused by *Balansia strangulans,* family Hypocreaceae, which surrounds the grass culms with a tight girdle and strangles the stem and leaves. The seedhead becomes blighted and fails to form.

black roan. Used to describe a horse whose coat is black and white, rather uniformly mixed. Also called *blue.*

black root. See *cotton wilt; damping-off; Culversphysic.*

black root of beets. A seedling disease which causes damping-off; results from any number of fungi, particularly: *Rhizoctonia solani, Phoma betae, Aphanomyces cochlioides,* and *Pythium* spp. Characterized generally by the root becoming soft and black and the leaves wilting.

black root of lettuce. See *Botrytis blight of lettuce.*

black root of radish. A serious fungous root rot of radish, particularly the long-rooted, icicle type, caused by *Aphanomyces raphani,* family Saprolegniaceae.

black root of strawberry. See *cortical root rot of strawberry.*

black root rot of apple. A most common and destructive parasitic disease. It is caused by *Xylaria mali,* family Xylariaceae. Characterized by large, finger-like fruiting patches, near the base of a diseased tree, that at first are white but soon black. The roots of a tree are covered with a black, charcoal-like coating. The wood becomes brittle, and a tree in an advanced stage of the disease is easily uprooted because roots break off readily. The leaves become yellow, buds fail to grow, terminal growth is short, main branches may die, and fruit is small and of poor quality. Found in southeastern and southcentral U.S.

black root rot of tobacco. A parasitic fungous disease in old fields, caused by *Thielaviopsis basicola,* family Moniliaceae; characterized by a retarded growth of stems and leaves and wilting of plants on dry days.

black rot. 1. A fungous disease of apples, pears, quinces, and other hosts caused by *Physalospora malorum,* family Pleosporaceae; characterized by cankers on the twigs and limbs, by a leafspot, and by rotting of the fruit as it approaches maturity or when in storage. The fruit usually becomes shrivelled and mummified. The disease is sometimes called *frog-eye* because the leafspots are small and dirty grey. Widely found east of the Rocky Mountains. Also called *ring rot; blossom end rot; brown rot.* The symptoms on twigs and limbs are also known as *dieback; twig blight; apple canker; black rot canker; New York apple tree canker.* The leafspot symptom is also known as *leaf blight; brown spot.* 2. A fungous disease of citrus fruit, particularly oranges, caused by *Alternaria citri,* family Dematiaceae; the infected fruits turn deep orange. They appear sound but are decayed within near the stylar end. The inside is brown to black. In lemons the disease is known as *Alternaria rot; center rot.* 3. A general marketing term describing bad eggs.

black rot of crucifers. A bacterial disease caused by *Xanthomonas campestris.* Characterized, especially in cabbage, by yellowing of the leaves and blackening of the veins at the leaf margins and developing toward the petiole and the main stem. Once established, the disease rapidly affects the entire plant, causing the leaves to wilt, turn yellow, dry up, and become thin and parchment-like in texture. Also called *bacteriosis; bacterial rot; brown rot.*

black rot of grape. A disease caused by *Guignardia bidwelli,* family Mycosphaerellaceae. Characterized by circular brown spots on the foliage which have a definite margin. The spots may coalesce and affect large areas of the leaf. Lesions form on shoots, petioles, and tendrils and then on the fruit as small, tan spots which rapidly increase in size, causing the fruit to blacken, shrivel, and mummify. Found east of the Rocky Mountains and particularly in humid parts of southern U.S.

black rot of sweet potato. A fungous disease of sweet potato caused by *Endoconidiophora fimbriata,* family Ophiostomataceae; characterized by circular, brown or black spots up to two inches in diameter. The blackening extends to the vascular rings of the potato or further, and fruiting bodies may develop at the center of the spots. Small round spots on the underground stem of the sprout cause *black shank,* a blackening and decay of the stem.

black russet. A disease of citrus fruits resulting from infestation by the citrus rust mite, *Phyllocoptruta oleivora* (Ashm.), family Eriophyidae.

black rust. A term used to describe a stage of either leaf or stem rust of grains and grasses. See *stem rust of oats, barley, or wheat; potash hunger.*

black sage(brush). *Salvia mellifera,* a perennial

herb, family Labiatae; important as a bee plant and sometimes as forage. Found on the Pacific Coast of U.S.

blacksalsify serpentroot. *Scorzonera hispanica,* a perennial herb, family Compositae; grown as an annual for its long, black skinned, edible root and for its foliage, which is sometimes used in salads. Native to Europe. Also called *black salsify; scorzonera.*

black sampson. See *coneflower.*

black scab. See *Rhizoctonia disease of potato.*

black scab of pome fruits. See *apple blotch.*

black scale. *Saissetia oleae,* family Coccidae. It attacks citrus and neighboring plants, such as pepper trees (*Schinus molle*) and olive trees. It extracts the juices of the plant which then becomes covered by the sooty mold fungus, *Meliola camelliae* (Catt.) Sac., that grows in the insect's honeydew. Found throughout the world, especially in western U.S. and Spain.

black scurf. See *Rhizoctonia disease of potato.*

black seed. See *black medic.*

black semi-loose smut. A fungous disease of barley caused by *Ustilago nigra,* family Ustilaginaceae; characterized by dark brown or black spore masses that replace the flowers. Found throughout the world.

black shank of tobacco. A disease caused by the parasitic fungus, *Phytophthora parasitica* var. *nicotianae* Tucker, family Pythiaceae; characterized by blackened and dead roots, by decay at the base of the stalk, brown blotches on the leaves, and by damping-off of the seedlings.

blacksmith. A mechanic who welds and fashions iron and other metals, sharpens plowpoints, shoes horses, etc.

black smut. See *fig smut; date smut.*

black snake. A harmless, common snake, which is black on the dorsal side and light on the ventral; may exceed five feet in length. In North America, this name is applied to *Coluber constrictor;* in Jamaica, to *Ocyophis ater;* in Australia, to *Pseudechis porphyriaceus, Hoplocephalus superbus,* and *Notechis scutalus.*

black snakeroot. See *cohosh bugbane.*

black soils. A Canadian term for Black Earth or Chernozem (*q.v.*) zonal group of soils which includes Wiesenboden, and Solonetz.

Black Spanish. See *White-faced Black Spanish.*

black speck. See *Rhizoctonia disease of potato.*

black spot. 1. See *apple scab; avocado anthracnose.* 2. Bull's-eye rot; a disease of apple caused by *Neofabraea malicorticis,* family Mollisiaceae. 3. A fungous disease of citrus caused by *Phoma citricarpa,* family Sphaeropsidaceae; characterized by reddish-brown spots on the fruit. Reddish-brown halos form, and the brown centers become depressed. The spots then turn dark, sometimes black. 4. See *peach scab.* 5. A disease of potatoes. The tissue underneath the skin breaks down and turns black or blue-black. The blackening frequently does not show up until the potatoes have been removed from storage. Heavy fertilization with nitrogen and excessive soil moisture seem to increase blackening following storage. 6. A fungous disease of roses caused by *Diplocarpon rosae,* family Mollisiaceae; characterized by black spots having rays and fibrils which sometimes cover the leaves. In some cases entire leaves turn yellow and defoliation follows. Similar spots also appear on canes. Inconspicuous spots on the petioles, stipules, flower receptacles, sepals, and petals may cause distortion of the flower. 7. A minor bacterial disease of sugar beets caused by *Pseudomonas aptata* which blights plants that are grown for seed. Also called *black streak.* 8. A low-grade infection resulting from mechanical injury to the teat of a cow.

black-spot canker. See *apple anthracnose.*

black spot of drupe fruits. See *bacterial spot of stone fruits.*

black spot of elm. See *elm leaf spot.*

black spot scab. See *apple scab.*

black spruce. *Picea mariana,* a hardy, slow growing, often stunted, evergreen, coniferous tree, family Pinaceae. Grown in moist soils, it is a most common species in strongly acid peat bogs. Used for pulpwood, boxes, planing-mill products, furniture, woodenware, etc. Native to North America. Also called *double spruce; eastern spruce; bog spruce.*

black stele. See *strawberry red stele.*

black stem of alfalfa. A fungous disease of alfalfa caused by *Ascochyta imperfecta,* family Sphaeropsidaceae.

black stem of China aster. See *Fusarium wilt.*

black stem of clover. A widely distributed disease caused by the fungi, *Phoma* spp., *Mycosphaerella* spp., *Ascochyta* spp. Characterized by a blackening of the stem and repeated defoliation which weakens or destroys a stand of clover.

black stem of field pea. A destructive disease caused by *Ascochyta pinodella,* family Sphaeropsidaceae, and *Mycosphaerella pinodes,* family Mycosphaerellaceae. Characterized by a dark brown or black stem tissue above the ground. It often kills plants before they produce seed.

black stem rust. See *stem rust of barley.*

black stem spot of cranberry. A widespread, fungous disease caused by *Mycosphaerella nigromaculans,* family Mycosphaerellaceae; characterized by black spots on the stems. If girdling results the distal portions die.

blackstrap. The liquid by-product of crystallization of sugar from sugar-cane or sugar-beet juice; used as a feed for livestock having about 75 percent of the feed value of corn. Also called *blackstrap molasses.*

black sugar maple. See *black maple.*

black swallowtail. *Papilio polyxenes asterius* Stoll., family Papilionidae, a butterfly whose larvae feed on celery, carrots, dill, parsley, carroway, parsnips, and related plants. Found east of the Rocky Mountains. Also called *celeryworm; parsleyworm.*

black thorn. See *pear hawthorn.*

black thread scale. *Ischnaspis longirostis* (Sign.), family Diaspididae. An insect that sucks the sap of many plants including palms, lantana, citrus, oleander, ivy, hibiscus, rose, and other greenhouse plants.

blacktongue. 1. A disease of dogs caused by a deficiency of one of the B vitamins, nicotinic acid (niacin); characterized by ulcers in the mouth and inflammation of the alimentary canal and severe nervousness. Identical with pellagra in humans. 2. A variation of anthrax disease which attacks cattle and horses; characterized by ulcerations on the tongue.

black tupelo. *Nyssa sylvatica,* a deciduous tree, family Nyssaceae; grown for its fine, red, autumnal foliage. Used in making veneer, pulp, cabinet work, boxes, etc. Native to North America. Also called *black gum* (forestry name); *sour gum; pepperidge; tupelo.*

Black turkey. A breed which has black plumage and a slight, bronze tinge on the back and tail. The tom weighs up to 33 pounds and the hen up to 18 pounds. Also called *Norfolk.*

black turpentine beetle. *Dendroctonus terebrans* (Oliv.), family Scolytidae. It infests felled pines, excavating irregular egg galleries. Its larvae feed through the inner bark, making a cavity between the bark and the wood. It does not kill a tree, but weakens it so that it is susceptible to attacks by other insects or diseases. Found throughout the eastern and southeastern U.S.

black vine weevil. *Brachyrhinus sulcatus* (F.), family Curculionidae. A serious insect pest of the yew tree in Europe and of garden shrubs and greenhouse plants in North America, such as azalea, begonia, blackberry, cranberry, cyclamen, gloxinia, geranium, gardenia, maidenhair-fern, primrose, raspberry, rhododendron, spirea, strawberry, and wisteria. The grubs feed on roots, and the adults chew foliage at night. Also called *cyclamen weevil.*

black vomit. Emetic resulting from the presence of blood in an animal's stomach. The black masses of vomited blood may look like coffee grounds.

black walnut. See *eastern black walnut.*

black wart. See *potato wart.*

black water. See *azoturia; cattle tick fever.*

black-water fever. See *cattle tick fever.*

black wattle. An Australian term denoting several species of acacia (*q.v.*). See *Sydney acacia.*

black waxy land. A colloquial term for clayey prairie lands in Texas; black and extremely sticky when wet.

black weevil. See *rice weevil.*

blackwood acacia. *Acacia melanoxylon,* a tall, unarmed, tropical tree, family Leguminosae; cultivated as a street tree in California and for wood in Australia.

black wool. A term used to describe any color of fleece other than white.

bladder campion. See *cowbell silene.*

bladderfern. See *berry bladderfern; brittle bladderfern.*

bladder gall mite. See *maple bladder gall mite.*

bladdernut. A tree or shrub of the genus Staphylea, family Staphyleaceae; grown for ornament. See *American bladdernut; European bladdernut; colchis bladdernut.*

bladder plum. See *plum pockets.*

bladder pod. See *Indiantobacco lobelia.*

bladdersenna. *Colutea arborescens,* a shrub, family Leguminosae; grown for its foliage and pealike, yellow flowers, and as a stock for grafting cions of glorypea parrotbeak (species of the Australasian genus *Clianthus*). Native to Eurasia. Also called *false senna.*

bladder worm. A larval form of tapeworm in which a single head, or scolex, is enclosed in a bladderlike sac.

blade. 1. The expanded portion of a leaf. 2. A cutting tool, as the blade of an axe or the blade of a bulldozer. 3. A retail cut of meat from the forequarter or shoulder. 4. The hind part of a fowl's single comb *q.v.*

blade apple. See *Barbados-gooseberry.*

blade blight. See *halo blight of oats.*

blade harrow. See *acme harrow.*

blanch. 1. To bleach or whiten a vegetable as it is growing by wrapping the stalk and leaves with paper or by throwing soil around the portion to be whitened, as celery is *blanched.* Also called *etiolate.* 2. To heat vegetables and fruit in water, live steam, or dry heat, to inactivate enzymes, preparatory to processing.

bland apple sirup. Made from apple juice which has been treated with calcium hydroxide slurry to a pH of 8.0 to 8.5, heated to 175° F., filtered, reacidified, and concentrated in vacuum.

bland lard. A deodorized pork product, to which hydrogenated lard is added, as well as a stabilizing agent to retard rancidity.

blank. 1. The rough, sawed or split wood pieces from which a finished product is made, such as handles, chair rounds, patterns, etc. 2. An unstocked area in a tree plantation where few or no trees are growing. 3. An unassembled box or package.

blanket. See *saddle blanket.*

blanket application. The application of spray or dust over an entire area rather than on rows, beds, or middles.

blanketed cattle. See *Dutch belted cattle.*

blanket flower. See *perennial gaillardia.*

blank list. To furrow a field with a lister without planting seed in the furrow.

blast. 1. A disease of sheep characterized by flatulence. 2. A disease in which a plant dries up or shrivels because of electricity or some other agent in the atmosphere. 3. See *blast of rice; citrus blast.*

blast-furnace slag. The dross from the smelting of iron ore, which contains calcium silicate, an important source of lime for agricultural use in the southeastern U.S. It is also a carrier of phosphorus.

blasting of oats and sorghum. The failure of a plant to produce seed; seemingly because of high temperatures and deficient water at blossoming time.

blastoderm. The germ spot in an egg.

blast of rice. A serious, fungous disease caused by *Piricularia oryzae,* family Moniliaceae; characterized by lesions on the leaf sheath and at the neck of the culm. The lesions may cause the stem to break off. Also called *rotten neck.*

blastomycosis of geese. A yeast-like fungous disease, *Blastomyces (Cryptococcus) anseris,* family Torulopsidaceae, occurring in the liver of geese crammed for fattening. It is found in colonies on the edge of the liver. The disease makes the goose liver unfit for human consumption, as it causes thrush.

blastophaga. The fig wasp, especially *B. grossorum,* family Agaontidae, of the cultivated fig, *Ficus carica.* The common fig wasp undergoes its full life cycle in the wild fig called the capri fig *q.v.*

blaze. 1. A white mark, such as a star, on the face of an animal. 2. A cut or other mark on a tree which indicates a trail or timber to be felled.

blazing star. Plants with showy flowers: (a) species of *Liatris,* family Compositae, in the mid and western U.S.; (b) the *Crocus tritonia,* a hybrid of the genera *Crocosmia* and *Tritonia,* family Iridaceae; (c) *Mentzelia laevicaulis,* a perennial herb, family Loasaceae; grown for its light yellow flowers. Found in western U.S. See *Kansas gayfeather; fairy wand.*

bleach. To whiten by exposure to sun or by chemical action.

bleached. Designating a color defect in strawberry ice cream, which is accompanied by an oxidized flavor. Also called *faded.*

bleached hay. Hay faded and reduced in feeding value, palatability, and digestibility because of excessive rain exposure or sun during curing.

bleaching powder. Chlorinated lime.

bleat. The sound made by a sheep, calf, or goat.

bled timber. Turpentine-treated pine trees.

bleed. 1. To lose blood. 2. To drain blood from a slaughtered animal. 3. For any wounded plant, to exude sap, water, resin, etc., and in trees, sometimes softening or discoloring the wood. 4. In the manufacture of ice cream and sherbets, for the unfrozen sirup to settle.

bleeder. 1. An animal that has hemophilia *q.v.* 2. A wounded or pruned plant that exudes sap excessively.

bleeding canker. A fungous disease of maples and American beech caused by *Phytophthora cactorum,* family Pythiaceae; characterized by exuding sap from fissures which overlie the canker in the bark. The inner bark cambium and sapwood become reddish-brown with an olive-green margin. The leaves wilt and the branches die back.

bleedingheart. 1. *Dicentra spectabilis,* a perennial herb, family Fumariaceae; grown for its red flowers. Native to Asia. 2. Any plant of the genus Dicentra. See *squirrel corn; Dutchman's-britches; fringed bleedingheart; Pacific bleedingheart; golden eardrops bleedingheart.*

bleedingheart glorybower. *Clerodendron thomsonae,* a woody vine, family Verbenaceae; grown outdoors in warm areas and in greenhouses for its almost continuous blooming of bright red flowers with pure white calyxes. Native to West Africa. Also called *bagflower.*

bleeding necrosis of sweetgum. A fungous disease believed to be caused by *Dothiorella berengeriana,* family Sphaeropsidaceae. There is much bleeding of the bark at the soil line. The exudation looks like motor oil and makes the bark dark reddish-brown and the outer layers of sapwood brown or olive-green. Terminal branches die back and foliage is underdeveloped.

bleeding pit. A pit in which the blood from a hanging slaughtered animal is collected. Also called *bleeding trough.*

bleeding wart. *Verruca vulgaris.* A contagious filterable virus common to young cattle. It attacks the sides of the neck, head, and shoulders in the form of growths from the size of a pinhead to ten inches in diameter which damage the hides. Also called *common wart of cattle; verruca vulgaris.*

bleicherde. Light colored, leached A_2 horizon of Podzol soils. See *soil profile.*

blemish. A spot or stain which affects appearance but does not necessarily limit usefulness. See *defect.*

blemished hock. Disfigurement of an animal, particularly a dairy cow, which disqualifies it for showing.

blemished yolk. Deep mottling which causes an egg to be classed as abnormal.

blend. 1. To cross-breed animals or plants to derive an offspring having characteristics of each. 2. To mix two or more ingredients to obtain a standardized product.

blended flour. A mixture made from different wheats.

blended inheritance. Transmission of characteristics intermediate to those of the parents.

blessed milkthistle. *Silybum marianum,* an annual herb, family Compositae; grown for its edible roots, leaves, and flower heads and as an ornamental for its silvery leaves and purplish-red flowers. Native to southern Europe.

blight. Disease symptoms, parasites, fungi, bacteria, viruses or unfavorable atmospheric conditions which cause the withering and death of a plant.

blight canker. A canker on pear or apple trees occurring from infection by the fire blight, *Erwinia amylovora.*

blighted flower. A horse suffering from St. Vitus's dance.

blind. 1. Swiss cheese that does not develop holes because of the absence of *Bacterium acidi propionici.* 2. The failure of a stem to develop a flower on the apical end. 3. Growth of leaves rather than flowers.

blind area. In forestry, an area which cannot be seen from a lookout station.

blind cane. See *blind wood.*

blind check. An egg which has an almost invisible crack.

blinded. Covered by 6 to 12 inches of loose top soil; usually pertaining to drain tiles.

blinders. Small pieces of material, usually leather, fastened to a bridle to keep a horse from seeing anything not straight ahead. Also called *blinkers.*

blind furrow. A dead or double furrow caused by plowing two adjacent furrows and throwing the soil in opposite directions.

blind gut. The cecum. Also called *bung.*

blind knot. An overgrown knot that forms a pronounced swelling on the trunk of a tree. Also called *blind conk.* See *knot: 4.*

blindness. Failure of vision in man and animals.

blind node. A node at which a bud does not develop in the axil of the leaf.

blind pocket. A virus disease of citrus caused by *Citrivir psorosis* var. *alveatum.* It has two forms—the blind pocket and the eruptive. The pocket of this disease form has narrow, abruptly depressed regions, often with two convex or concave sides. At the bottom of the depression are cores of salmon-colored tissue which become impregnated with a gummy substance. The eruptive form may occur with the scaling of bark inside an area of blind pockets.

blind quarter. A quarter of an udder that will not secrete milk during lactation or one that has an obstruction in the teat. It may be a common hereditary defect in swine and first-calf heifers. A serious defect. Also called *blind teat; nipples; inverted nipples.*

blind seed disease of grasses. A fungous disease, especially of ryegrass, caused by *Phialea temulenta,* family Helotiaceae; characterized by a pink slime on the surface of the pericarp and by a reduction in the yield of viable seed. Found in Oregon, northern Europe, New Zealand and Tasmania.

blind spavin. A hidden spavin among the bones at the hock joint. See *bone spavin.*

blind staggers. A disease of horses, cattle and swine caused by an excessive intake of the mineral selenium. Characterized by the loss of hair from the mane and tail in horses, from the tail in cattle, and the general loss of hair in swine. The hoofs slough off, lameness occurs, food consumption decreases even to the point of starvation and vision becomes gradually impaired. The disease is prevalent on the plains of U.S., particularly in South Dakota. See *alkali disease.*

blind Swiss cheese. See *blind: 1.*

blind teat. See *blind quarter.*

blind tillage. Plowing for weed control after a crop is in its early stages of growth.

blind valley. In a karst region, a valley which ends suddenly where its stream disappears underground.

blind vine. See *blind wood.*

blind wood. A vine or stem which does not produce flowers or fruits, because of disease, lack of nourishment, or improper planting. Also called *blind cane; blind vine.* See *blind: 2.*

blinky. Sour or slightly sour milk.

blissom. A ewe in heat.

blister beetle. Beetles belonging to the family Meloidae, which feed on plants. The acid, cantharides, in their bodies causes painful burns to humans. See *ash-gray blister beetle; black blister beetle; clematis blister beetle; spotted blister beetle; striped blister beetle; old-fashioned potato bug.*

blister buttercup. *Ranunculus scleratus,* a poisonous perennial herb, family Ranunculaceae; causes severe dermatitis if it comes in contact with human's skin. If eaten, it inflames the intestinal tract of animals and causes cattle to produce bitter milk. Found in low, wet areas throughout northern U.S. and Canada. Also called *bitter buttercup.*

blister canker of apple. A fungous disease caused by *Nummularia discreta,* family Xylariaceae; characterized by rough, black cankers on the trunk and limbs of the apple tree. Also called *apple pox.*

blister cheese. A defect which occurs when paraffin does not adhere to the rind of the cheese.

blister cress. See *Erysimum.*

blister flower. See *tall buttercup.*

blister fly. See *blister beetle.*

blister rot. See *blue contact mold.*

blister rust of pine. See *white-pine blister rust.*

blister smut of grasses. A non-serious fungous disease caused by *Entyloma* spp., family Tilletiaceae; characterized by flat or slightly raised, black, round spots on the leaves.

blizzard. A severe storm of high wind, low temperatures, and heavy snow.

bloat. 1. A severe distension of the abdomen by gas; usually in ruminant animals. Caused by eating watery food or by eating too quickly. In horses, the condition is called *colic.* 2. See *huffed cheese.* 3. Distention of one end of a can because of gas pressure. Also called *swellhead.* 4. See *nemarot.*

bloater. 1. An immature or insufficiently sundried fruit. 2. An animal, especially a ruminant, prone to *bloat.*

block. 1. A piece of wood used as a divider between two bales of hay. 2. A pulley used to increase pulling or hoisting power. 3. In logging, an administrative division of a forest. 4. About five pounds of cotton hanks. 5. A portion of hay thrown into a baler. 6. To thin out plants, especially sugar beets, into smaller bunches which will be again thinned into single plants. 7. To trim fleece to enhance the appearance of a sheep. 8. See *cutting block.*

block beef. A side of beef good enough to be sold over a retail butcher's block. Also called *side beef.*

blockiness. Descriptive of animals, especially cattle and draft horses, having deep, wide, low-set, compact bodies.

blocking. Preparing tomatoes and other plants for transplanting by cutting the soil around the plant into blocks with a straight edged tool.

blocking hoe. A hoe with a straight shank and a blade which is in direct line with the handle.

block milk. A concentrated, solid milk containing 40 percent sugar; uncommon.
block out. To remove covering vegetation to control a plant disease.
block salt. A cube of about ten inches of compressed, sometimes medicated, salt for consumption by domesticated animals.
block setter. A protective device for placing a block (*q.v.*) in a hay baler.
blocky. See *blockiness.*
blocky structure. The arrangement of particles which indicates that soil units exceed one cubic centimeter.
blood. 1. The fluid that carries nourishment and oxygen to all parts of a body and carries away waste products. 2. In plant or animal breeding, descended from parents of registered breed; relationship by blood. 3. A market term referring to the fineness of wool. See *three-eighths blood; quarter blood; low quarter blood; half-blood; blood classification.* 4. A market classification for a type of Spanish orange.
blood agar. A medium for growing microorganisms which consists of nutrient agar (1 percent peptone, 1 percent lemco, 0.5 percent sodium chloride, 2 percent agar, pH 4, to which is added 5 percent defibrinated blood at 45° to 48° C.).
blood albumen glue. A water-resistant glue used in the manufacture of plywood.
blood-and-feather-dressed. See *New York dressed.*
blood bay. A horse of a red shade of bay.
bloodberry rougeplant. *Rivina humilis,* a tropical, perennial herb, family Phytolaccaceae; grown for its white flowers and red berries. Found in Florida and the Gulf states of the U.S. Also called *rouge plant.*
blood classification. Originally the grading of wool according to the amount of Merino blood in sheep. The term now describes any wool having Merino characteristics. See *American grades of wool; Australian grades of wool.*
blood cup. In poultry dressing, a cup attached to the bill to catch blood.
blood disease of banana. A vascular disease caused by *Pseudomonas celebensis* Gäumann, characterized by blighting of the leaves and rotting of the fruit.
blooded. Of recognized breeding.
blood flour. Dried blood reduced to fine powder. Principally fed to dairy calves on a minimum milk diet. See *blood meal.*
bloodflower milkweed. *Asclepias curassavica,* a tropical American, perennial herb, family Asclepiadaceae, one of the showiest of milkweeds. Grown for its brilliant, orange-red flowers. Escaped, it can be a weed pest.
blood horse. A purebred horse.
bloodleaf. See *linden bloodleaf; Herbst bloodleaf.*
blood line. A direct line of descent; pedigree; strain.
blood meal. Dried, ground animal blood which is heated to coagulation, drained, and then pressed to remove remaining water. The meal is 80 percent hard-to-digest protein, but it is mixed into calf starter feeds. See *blood flour.*
blood orange. Any sweet orange having dark red pulp.
blood(y) oval. A market classification for Spanish oranges.
blood poisoning. See *hemorrhagic septicemia.*
blood pudding. A very dark sausage, the principal ingredient of which is animal blood.
blood ring. In poultry, the ring formed during the death of an embryo.
bloodroot. *Sanguinaria canadensis,* a perennial, poisonous herb, family Papaveraceae. The rhizome is unpalatable; the red juice supposedly is an emetic and a narcotic and was used by Indians to paint their bodies. Symptoms of poisoning are nausea, burning of the mucous membrane, faintness, vertigo, loss of consciousness, great thirst and difficulty in breathing, dilation of pupils, muscular prostration and coldness of the skin surface. Found in northeastern U.S. Also called *corn root; tetterwort; redroot; Indian plant; red puccoon; American alkanet.*
blood spavin. In horses and mules, a soft swelling due to a distention of veins around the hock; does not affect the action of the animal. Also called *thoroughpin.*
blood spot. An egg defect which occurs most commonly at the beginning of the laying year. Usually detected during candling.
bloodstock. Purebred horses.
bloodsucking conenoses. The insect *Triatoma sanguisuga* (Lec.), family Pseudococcidae, whose intensely painful bite causes swelling over much of the body, faintness, and vomiting which may last for weeks or months. Also called *Mexican bed bug.*
bloodsucking insect. Any insect, –*e.g.*, mosquito, fly, tick–which draws the blood of its host.
blood sugar level. The amount of sugar, usually glucose, in blood. Abnormally high or low blood sugar levels may indicate the presence of a disease.
blood system. See *blood:* 3.
blood test. Any chemical or microscopic test which uses blood in determining presence or absence of disease, blood disorders, etc.
bloodtrumpet. See *Mexican bloodtrumpet.*
bloodtwig dogwood. *Cornus sanguinea,* a showy shrub, family Cornaceae; grown for its white flowers, black fruit, and blood-red twigs. Native to Eurasia. Also called *dog tree; red dogwood.*
bloodwood. Any tree having red juice or red wood, e.g., the dark bloodwood eucalyptus.
bloody butchers. See *toad trillium.*
bloody discharge. The passing of blood or blood-tinged material from an opening, such as the vulva, rectum, nose, or mouth, which may be symptomatic of injury or disease.
bloody egg. An egg having blood diffused throughout the white; an uncommon term. See *blood spot; meat spot of egg.*
bloody groin. Damage caused to poultry during soft-scalding, either by insufficient bleeding, or by hanging the fowl by one leg. Characterized by a dark streak of fat along the inside front edge of the second leg joint. Exposure causes

the streak to darken and extend over the back, making the poultry unmarketable.

bloody milk. Milk from an animal suffering from ruptured blood vessels and injured inner gland tissues. Such milk should always be discarded.

bloody murrain. See *cattle tick fever.*

bloody scours. A form of diarrhea that attacks small calves and pigs; characterized by bloody feces. It is highly contagious and often fatal. See *common scours; calf scours.*

bloody whites. See *blood spot; meat spot of egg.*

bloom. 1. A blossom; flower. 2. The state of flowering. 3. Freshness and vigor approaching perfection. 4. A surface coating, *e.g.*, a cloudy deposit of substance on milk bottles. 5. A powdery substance coating fruits or leaves. See *pruinose.*

bloomer. A plant which flowers.

blooming-sally. See *fireweed.*

bloomless buckwheat. See *Tartary buckwheat.*

bloom of fruit. See *bloom: 5.*

bloomy. Pruinose.

blossom. 1. The flower or bloom of a seed plant; to flower. 2. A horse with a sorrel or bay coat mixed with white.

blossom blast and fruit rot of squash. A fungous disease caused by *Choanephora cucurbitarum,* family Mucoraceae; characterized by blighting of blossoms and by rotting of young fruit. The flowers quickly become covered with conidiophores which soon become brown and purplish-black and the tissue turns brown and soft, looking water-soaked.

blossom blast of pears. A disease caused by the bacterium, *Pseudomonas syringae;* characterized by death of dormant buds, withering of blossom clusters, superficial cankers on young shoots, and spotting of leaves.

blossom blight. Any cause of the sudden death of blossoms. In citrus flowers *Botrytis cinerea* may cause abscission or deformity of young fruits.

blossom bud. A bud that produces flowers rather than leaves.

blossom drop of tomato. The shedding of blossoms during early growth when the weather is cool.

blossom-end decay. See *endoxerosis.*

blossom-end rot. A decay at the blossom end of fruit.

blossom-end rot of tomato. A non-infectious physiological derangement which kills the protoplasm in the cells of growing tomatoes. Characterized by decay (dry rot) at the blossom end. Caused by too-rapid growth of seedling and by transplanting to dry soil which lacks humus. See *blossom-end rot.* Also called *the rot; black rot; fruit rot; point rot; dry rot; dry weather rot.*

blossom-end rot of watermelon. A disease similar to *blossom-end rot of tomato.*

blossom spray. A spray applied during blooming.

blotch. Common name for the diseased area of a plant.

blotch of apple. A fungous disease of apple caused by *Phyllosticta solitaria,* family Sphaeropsidaceae, which produces dark spots on the fruit and leaf, and tree cankers.

blotch of mushroom caps. A bacterial disease caused by *Pseudomonas tolaasii.* Characterized by pale yellow spots which later form chocolate-brown patches that are water-soaked beneath. The infection, which has been found in England and Minnesota, penetrates 1 to 3 mm. in depth.

blotch of Persian walnuts. A fungous disease caused by *Gnomonia leptostyla* (*Marssonia juglandis*), family Gnomoniaceae, which also attacks *Hinds walnut, eastern black walnut* and *butternut.* Characterized by reddish-brown round spots with gray-brown centers on leaves; similar oval-shaped lesions appear on shoots. Irregular, circular, depressed, dead spot lesions on the husks of the nuts cause premature drop. Also called *walnut blotch.*

blotch of stone fruits. See *peach scab.*

blotch rot. A disease of cranberries caused by the fungus, *Acanthornynchus vaccinii,* family Fimetariaceae. Characterized by small, light-colored lesions or by dark-colored blotches on the skin of the fruit, which destroy it. Also called *cranberry rot.*

blow. 1. In New Zealand, in handshearing sheep, a stroke by the shearer as he makes a cut. 2. To elevate grain, etc., into a bin or barn by air blast. 3. To sprinkle tobacco with water before curing. 4. Of flowers: to blossom, to bloom.

blow-down (blowdown). See *windfall.*

blower. A machine which drives air; used to elevate or ventilate grain, silage, etc., to clean seed, to dust plants, to increase the draft of a fire, etc.

blower pipe. A tube for carrying silage, seed, hay, etc., under air pressure.

blowfly. Any fly that breeds in decaying flesh, or that infests wounds. Many brightly colored species are known as *blue bottle* and *green bottle flies.*

blow holes. Holes so large and numerous that the cheese between them is extremely thin.

blowing. 1. A sound made by horses; not a symptom of disease or injury. 2. The bulging or bursting of sealed cans caused by fermentation (the production of carbon dioxide).

blowing out. The exhaling of surplus air by a horse after the saddle girth has been tightened.

blowings. Cotton waste from gins.

blow-out. 1. An excavation, by the wind, of loose soil, usually sand. See *sandboil.* 2. A hole in soil above a tile line caused by excessive pressure in drain.

blowout grass. *Redfieldia flexuosa,* family Gramineae; used in reseeding ranges and establishing pastures and to control blow-out (*q.v.*). It grows naturally on deep, loose sands in the Great Plains of southwestern U.S.

blowup. An uncontrollable forest fire.

blue. 1. A market term for dark-gray cotton. 2. Long wool of high quality from the neck of a sheep. 3. See *Blue cheese.*

blue anemone. See *roundlobe hepatica.*

blue ash. *Fraxinus quadrangulata,* a deciduous tree, family Oleaceae; often grown in parks and lawns in central and southern U.S.

blueback. Blue or black discolorations on the backs of turkeys caused by feather picking.

blue bean. See *silvery lupine.*

bluebeech. See *American hornbeam.*

bluebell. 1. *Campanula rotundifolia,* the bluebell-of-Scotland. A perennial herb grown for its bluish-purple flowers. Also called *harebell.* 2. The wood hyacinth, *Scilla nonscripta* and grape hyacinth, *Muscari botryoides* of Europe. 3. American blue bell-shaped flowers: blue jasmine, *Climatis crispa;* Virginia creeper, *Mertensia virginica;* creeping polemonium, *Polemonium reptans;* columbine, *Aquilegia coerulea,* etc. 4. Species of the New Zealand family Wahlenbergia.

blueberry. About 50 species of *Vaccinium,* family Ericaceae, of which the most important are the highbush blueberry, *V. corymbosum,* the large-fruited blueberry of the northeastern and east north central U.S.; the rabbiteye blueberry, *V. virgatum,* of the southeastern U.S.; and the lowbush blueberry, *V. angustifolium,* of Canada and northeastern and northcentral U.S. (often incorrectly called *huckleberry*).

blueberry bud mite. *Aceria vaccinii* (Keif.), family Eriophyidae, an insect pest.

blueberry cane canker. A fungous disease caused by *Physalospora corticis,* family Pleosporaceae; characterized in the autumn by reddish, conical swellings on the fruiting canes. The cankers become roughened, black, and deeply fissured by the second year. On some varieties, however, the disease causes only a moderate, grayish swelling.

blueberry maggot. *Rhagoletis pomonella* (Walsh), family Tephritidae, a serious insect pest which also attacks huckleberries. Also called *apple maggot.*

blueberry stunt. A virus disease of highbush blueberries, *Vaccinium corymbosum,* which dwarfs the plant, reduces leaf size, yellows leaves, and then reddens them in late summer, and makes the berries small, giving them an unpleasant flavor. Found in New Jersey, North Carolina and Michigan.

blue bindweed. See *bitter nightshade.*

bluebirdvine. See *purplewreath petrea.*

blue blazing star. See *tall gayfeather.*

blueblossom ceanothus. *Ceanothus thyrsiflorus,* an evergreen shrub, family Rhamnaceae; grown for its foliage and clusters of blue flowers. Native to Pacific Coast; also grown in southern U.S.

blue boneset. See *mistflower eupatorium.*

bluebonnet. See *Texas lupine.*

bluebrush. See *deerbrush ceanothus.*

bluebug. See *fowl tick.*

blue bugleweed. See *bugleweed.*

bluebunch wheatgrass. See *beardless bluebunch wheatgrass.*

blue casse. See *black casse.*

Blue cheese. Any cheese ripened by a blue mold, thereby resembling Roquefort (*q.v.*). Made from cow's milk (in France from ewe's milk) and cured for two or three months. Also called *Blue-Mold cheese; Blue-Veined cheese.*

blue cohosh. *Caulophyllum thalictroides,* a perennial herb, 2 to 2½ feet high, family Berberidaceae. Its poisonous rootstock is seldom eaten because of its extremely bitter taste. Poisoning irritates the mucous membranes and causes dermatitis. Common in the wooded areas of North America. Also called *false cohosh; papoose root.*

blue comb. See *avian monocytosis.*

blue contact mold. Soft decay of citrus fruit caused by *Penicillium italicum* which spreads by contact with diseased fruit. It may develop inside the core or in the spongy part of the rind; resulting water-soaked, soft areas are easily punctured by pressure. As the infection enlarges, blue spores form, which are surrounded by white mycelium and a halo of water-soaked, faded tissue. Also called *blue mold of citrus; pinhole decay; blister rot.*

bluecrown passionflower. *Passiflora caerulea,* an ornamental climber, family Passifloraceae, which has large, slightly fragrant, pink, white, and purple flowers. It is the most common passionflower grown in greenhouses. Native to Brazil.

blue Cupid's-dart. *Catananche caerulea,* a perennial herb, family Compositae; grown for its showy, blue flowers. Native to southern Europe. Also called *Cupid's-dart; blue succory.*

blue curls. See *self-heal.*

blue daisy. See *blue felicia.*

blue dandelion. See *chicory.*

blue dawnflower. See *dawnflower morningglory.*

blue devil. See *vipersbugloss.*

bluedicks brodiea. *Brodiaea capitata,* a perennial herb, family Liliaceae; grown in warm areas, particularly California, for its showy, blue flowers.

blue dogwood. See *pagoda dogwood.*

blue dove. The wild pigeon, *Columba livia,* derived from the domestic pigeon. It is bluish-grey in color, having a purple breast and white rump. Found on rocky coasts.

blue Egyptianlotus. *Nymphaea caerulea,* a tropical aquatic herb, family Nymphaeaceae, which is day-blooming and has light-blue flowers with white centers. Native to Africa. Also called *Egyptian lotus; blue lotus of Egypt.*

blue erythea. *Erythea armata,* a small fan palm, family Palmaceae, which is grown in lower California for ornament. Also called *Mexican blue palm; blue palm.*

blue eye. In marketing, an apparent defect in the eye of a horse or mule. *Cf. smoky: 2; speck-in-eye.*

blue-eyed-Mary collinsia. *Collinsia verna,* an annual herb, family Scrophulariaceae; grown for its white or purple and blue flowers. Native to North America.

blue false indigo. See *blue wildindigo.*

blue felicia. *Felicia amelloides,* a subshrub, family Compositae; grown for its daisy-like flowers with blue rays. Native to South Africa. Also called *blue daisy; blue marguerite; agathea.*

blue fescue. See *sheep fescue.*

blueflag iris. *Iris versicolor,* a perennial herb

(beardless iris), family Iridaceae; grown in wet places for its purple flowers. Its rootstock is poisonous but is seldom eaten by grazing animals; contact may cause severe dermatitis in humans. Also called *wild iris; blue flag.*

bluefly. Any flesh fly *q.v.*

blue gall. See *Aleppo gall.*

blue gentian. See *stemless gentian.*

bluegill. *Helioperca incisor,* a perchlike fish found in the Mississippi Valley. This sunfish makes good eating.

blue grama. *Bouteloua gracilis,* a native American grass, family Gramineae, used for reseeding on the western ranges of U.S.

bluegrass. 1. Any grass of the genus Poa, family Gramineae. Valued for hay and lawns and as the most palatable range and pasture grasses. See *Kentucky bluegrass; Canada bluegrass; roughstalk bluegrass; bulbous bluegrass; annual bluegrass; big bluegrass; Texas bluegrass; Sandberg bluegrass; mutton bluegrass.* 2. Grasses of the genus Andropogon in Australia. 3. *Agropyron scabrum* in New Zealand.

bluegrass leaf rust. A common rust caused by *Puccinia poae-sudeticae* which infects bluegrasses, especially Merion bluegrass, and occasionally meadow fescue and alpine timothy. Characterized by many, small, orange-red pustules on blades and sometimes on stems.

bluegrass leaf spot. A fungous disease caused by *Helminthosporium vagans,* family Dematiaceae; characterized by small, brown spots on leaves. The spots sometimes have purple-brown halos. The infected tissues turn straw-color and die.

bluegrass webworm. *Crambus teterrellus* (Zinck), family Crambidae, an insect whose larva cuts grass at surface of the ground. A loose silken web leads to a silk-lined tunnel at the base of the plant.

blue-gray. 1. A calf of a white Shorthorn bull and an Aberdeen Angus or Galloway cow. 2. A horse having steely-blue hairs scattered through a gray coat.

blue gum. Australian timber trees belonging to the genus Eucalyptus, family Myrtaceae. See *eucalyptus.* The giant *E. globulus* is also grown in California. See *Tasmanian blue eucalyptus.*

blue huckleberry. See *dangleberry.*

blu(e)ing. See *blue stain.*

blue jasmine. See *curly clematis.*

blue jay. *Cyanocitta cristata.* This bird, common to the eastern U.S., has a fine crest and a blue back; uncrested blue jays in western U.S. may also be identified by this term.

blue John. Colloquial term for skimmed or watered milk having a bluish tint.

bluejoint reedgrass. *Calamagrostis canadensis,* family Gramineae, a perennial grass; fair forage. Found throughout U.S. up to 12,000 ft. and usually in marshy or wet soil. Also called *bluejoint; blue top.*

blue laceflower. See *skyblue laceflower.*

blueleaf honeysuckle. *Lonicera korolkowi,* a hardy deciduous shrub, family Caprifoliaceae; grown for its bluish-green foliage, pale-rose flowers, and red fruits. Native to Asia.

blue lily-turf. See *bigblue liriope.*

bluelips collinsia. *Collinsia grandiflora,* a hardy annual herb, family Scrophulariaceae; grown for its purple or white and blue or violet flowers. Native to North America.

blue lobelia. See *great lobelia.*

blue loco. *Astragalus diphysus,* a common poisonous herb, family Leguminosae; causes the death of cattle and horses. It is characterized by a shaggy coat and by loss of: muscular control in walking, nervous sensibility, weight, appetite. It grows extensively in Arizona and New Mexico during January and February. Also called *blue locoweed.* See *loco disease; locoweed.*

blue lotus of Egypt. See *blue Egyptianlotus.*

blue louse. An insect (*Anoplura*) that infests cattle. See *long-nose cattle louse.*

blue lupine. *Lupinus angustifolius,* an annual legume, family Leguminosae, sometimes grown to improve soil. Native to southern Europe where it is cultivated as fodder.

blue lupine anthracnose. Caused by the fungus, *Glomerella cingulata,* family Gnomoniaceae, which infests the aboveground parts, causing lesions and cankers on the leaves, pods, and stems. See *anthracnose; apple bitter rot; peach anthracnose.*

blue marguerite. See *blue felicia.*

blue marsh violet. See *bogbice violet.*

blue melilot. See *bluewhite trigonella.*

blue milk. Colloquially, skimmed, slightly sour milk. Also called *blue John.*

blue mold. See *blue contact mold; powdery mildew; downy mildew of spinach.*

blue mold of apple. A common fungous disease caused by *Penicillium expansum,* family Moniliaceae. It attacks stored apples through wounds and lenticels, causing core rot and soft rot.

blue mold of citrus. A rot of ripe citrus fruits, especially those bruised in packing, which is caused by *Penicillium italicum,* family Moniliaceae. See *blue contact mold.*

blue mold of tobacco. A disease in plant beds caused by *Peronospora tabicina,* family Peronosporaceae. It is characterized by cupped leaves with a whitish or violet-colored moldy growth over the lower surfaces. It kills young plants and the leaf tissues of older plants. Found in tobacco areas throughout U.S. except Wisconsin.

blue myrtle. See *blueblossom ceanothus.*

blue nightshade. See *bitter nightshade.*

blue nose. A symptom of stubborn disease (*q.v.*) of citrus fruits, especially the grapefruit; characterized by a bluish color on the inside of the rind; sometimes the pulp in the stylar end has a sour, disagreeable taste and odor. See *acorn fruit.*

blue oak. 1. *Quercus oblongifolia,* an oak growing in southwestern U.S. 2. *Quercus macrocarpa,* the bur oak *q.v.* 3. *Quercus douglasi,* the California blue oak *q.v.* The leaf is nutritional but the acorn is low in fiber and fat. Therefore, the shrub is good for fattening but poor for growth.

blue oxalis. See *shamrock pea.*

blue paloverde. *Cercidium floridum,* a tropical American shrub, family Leguminosae; grown in desert gardens for its showy, yellow flowers. Also called *green-barked acacia; paloverde.*

blue pea. See *lupine.*

Blue Pearmain. A large winter apple with a purplish skin and yellow flesh.

blue phlox. See *sweet William phlox.*

blue racer. *Coluber constrictor flaviventris,* a subspecies of the harmless black snake *q.v.* Found in southern U.S.

blue ribbon. A prize or award for the best in a class; made of inscribed blue cloth.

blue roan. A horse with black hair interspersed with white, giving an impression of bluish or grayish color.

blue rocket. See *monkshood.*

blue rot. See *blue stain.*

blue sailor. See *chicory.*

blue skin. Bluish spots on the skin of purebred sheep; a fault or disqualification.

blue spruce. See *Colorado spruce.*

blue squill. *Scilla nonscripta,* a late-spring blooming, bulbous herb, family Liliaceae, having a raceme of drooping, bell-shaped, blue, purple, pink, or white flowers. Usually grown in masses in the flower garden. Also called *English bluebell; wood hyacinth; wild hyacinth.*

blue stain. A condition affecting all conifers caused by the fungus, *Ophiostoma piliferum* (syn. *Ceratostomella pilifera*), family Ophiostomataceae. Characterized by a bluish discoloration of the sapwood. Found on living, dying, and dead trees. It degrades lumber but does not rot or affect the strength of the wood.

bluestar bellflower. *Campanula ramosissima,* an annual herb, family Campanulaceae; grown for its solitary, violet flowers with white bases. Native to southeastern Europe.

bluestem. 1. Many grasses of the genus Andropogon, family Gramineae. Some species are good forage grasses. See *big bluestem; little bluestem.* 2. Verticillium wilt *q.v.*

bluestem joepyeweed. *Eupatorium purpureum,* a perennial herb, family Compositae, which grows up to 10 ft. tall; grown for its showy, purple, flower clusters.

bluestem of raspberry. See *Verticillium wilt.*

bluestone. See *copper sulfate.*

blue stripe wilt. See *Verticillium wilt.*

blue succory. See *blue Cupid's-dart.*

blue surface. An uncommon color defect of ice cream due to seeping of ammonia from the refrigeration lines.

blue thistle. See *vipersbugloss.*

blue tick. See *cattle tick.*

bluetongue. A virus disease of sheep and sometimes cattle. The affected animal runs a fever of about 105° for about four days; the nose and mouth mucous membranes redden, the animal salivates heavily, the lips and tongue swell, and the sternum becomes flushed.

blue torenia. *Torenia fournieri,* a tender, annual herb, family Scrophulariaceae; grown as a border plant or in hanging baskets for its blue, violet, or yellow flowers. Native to Asia.

bluets. *Houstonia caerulea,* a delicate, low, perennial herb, family Rubiaceae; grown as a rock garden plant for its violet, blue, or white flowers with yellow centers. Native to North America. Also called *angel-eyes; eyebright; innocence; Quaker ladies.*

Blue-Veined cheese. See *Blue cheese.*

Blue Veiny. See *Dorset: 2.*

blue verbena. *Verbena hastata,* a perennial herb, family Verbenaceae; grown in flower borders and as a pot plant for its small, blue flowers. Native to North America.

Blue Vinny. See *Dorset: 2.*

blue violet. See *palmate violet.*

blue vitriol. See *copper sulfate.*

blueweed. See *vipersbugloss.*

bluewhite trigonella. *Trigonella caerulea,* an annual herb, family Leguminosae; grown in flower borders for its blue and white flowers. Native to Europe. Also called *blue melilot.*

blue wildindigo. *Baptisia australis,* a perennial herb, family Leguminosae, which has handsome leaves and deep indigo-blue flowers. Grown in flower borders and wild gardens. Native to North America. Also called *false indigo; wildindigo.*

blue wildrye. *Elymus glaucus,* a fairly good forage grass, family Gramineae, with strong seed habits. Found up to 9000 feet in altitude from Ontario and Michigan to Alaska and south as far as South Dakota, New Mexico and California.

blue wood aster. See *heartleaf aster.*

bluewood condalia. *Condalia obovata,* a spiny evergreen shrub, family Rhamnaceae; a common chaparral plant in west Texas which is of some value as a browse plant. Also called *brasil.*

bluff. Steeply sloping land flanking rivers, lakes, or bottom-land; caused by erosion. Also called *bluffland.*

Blumea. A genus of shrubs and herbs, family Carduaceae, which bear yellow or purple flowers. Ngai camphor is produced from *B. balsamifera.* Native to Australasia and Africa.

boar. Uncastrated male swine.

board. 1. Lumber less than 2 inches thick and more than 8 inches wide. 2. Lumber 1½ inches thick, 6 or more than 6 inches wide and 8 or more than 8 feet long (British). 3. Lumber of all widths one-inch thick; widths of less than 6 inches are sometimes called *strips.* 4. Plank floor on which sheep are sheared (New Zealand).

boarder. An animal or fowl.

board foot. A lumber measurement: a board one foot long, one foot wide, one inch thick, based on the original cut before planing and surfacing. Also called *super foot* (Australia).

Boardman feeder. A feeder for bees made from a Mason jar, which has a perforation at the top; inverted over a box-like device, which is inserted in the entrance of the hive.

board mill. A sawmill that cuts 1- and 2-inch lumber, as compared to a timber mill that cuts thicker lumber.
board of hired labor. Meals provided by a farmer to a hired man in addition to his wages.
boar pig. Male swine under one year of age.
bob. 1. Veal from calves slaughtered when less than four to six weeks old. Also called *slunk.* 2. The docked tail of a horse. 3. In transporting lumber, a single pair of runners on which the forward ends of the logs are placed. 4. To transport logs on a *bob.*
bobbed wire. See *barbed wire.*
bobber. 1. A sunken log. 2. A horse with weak loins, which cause its hind quarters to wobble. Also called *jig back.*
bobby (calf). See *bob: 1.*
bobolink. *Dolichonyx oryzivorus,* family Icteridae, a North American migratory bird whose name is derived from its distinctive song.
bobsled. 1. Two coupled short sleds; a compound sled. 2. To ride a *bobsled.*
bobtail. An animal with a short tail.
bobtail barley. *Hordeum jubatum* var. *caespitosum,* family Gramineae, a subspecies of foxtail barley with short awns; of fair forage value. Found from North Dakota to Alaska, south to Kansas and Arizona.
bob veal. See *bob: 1.*
bob wire. See *barbed wire.*
bockwurst. A non-smoked sausage made of veal stuffed into hog casing.
bodewash. The dried dung of buffalo and cattle which was used for fuel on the treeless frontier parts of the Great Plains of the U.S.
body. 1. Physique. 2. The consistency of a substance, such as honey, butter, cheese, etc. 3. The fullness of flavor, especially of a beverage.
body blight. See *fire blight.*
body check. An egg with a healed crack in the shell. Also called *sealed check.*
body cloth. A horse blanket.
body down. Maturity in animals exemplified by acceptable body proportions.
body stock. A root to which branches from another stock have been grafted or budded.
body wood. Lumber or cordwood cut from the main stem of a tree.
Boer lovegrass. *Eragrostis chloromelas,* family Gramineae, a grass used for reseeding or forage on the western ranges of U.S. Native to South Africa.
bog. Wet, miry ground; swamp, marsh, fen, or moor. More specifically, wet, lake basin depressions filled with acid peat and covered by mosses, sedges, and ericaceous plants (such as sphagnum, black spruce bog, cranberries, etc.). Also called *bogue.*
bog asphodel. 1. The European bog herb, *Narthecium ossifragum.* 2. The American bog herb, *N. americanum,* found in New Jersey and Delaware.
bogbean. *Menyanthes trifoliata,* a perennial herb, family Gentianaceae, grown in bog gardens for its white flowers. Its leaves are said to be used in Germany as a substitute for hops in beer. Found all over North America. Also called *buckbean; marsh trefoil.*
bogbice violet. *Viola cucullata,* a perennial herb, family Violaceae, which grows well in poor soil. Its violet-colored flowers have dark throats. Native to eastern North America. Also called *blue marsh violet.*
bog butter. Hydrocarbon compound found in marshy land.
bog-down. In the U.K., any grass of the genus Eriophorum. Also called *cotton grass.*
bog garden. A marshy natural or artificial garden without pools but which retains soil moisture throughout summer.
boggy in the hocks. See *bog spavin.*
bogie. Tobacco leaves twisted into small rolls.
bog iron ore. Impure, ferruginous deposits in bogs or swamps caused by the oxidizing action of algae, bacteria, or atmosphere.
bog kalmia. *Kalmia polifolia,* a swamp shrub, family Ericaceae, which bears small purple flowers. It contains andromedotoxin which is poisonous to stock. Also called *swamp laurel; alpine laurel; pale laurel.*
bog lime. See *marl: 4.*
bog myrtle. See *sweet gale.*
bog orchid. See *arethusa.*
bog rider. A cowboy who rescues cows stuck in muddy places. A colloquial term of western U.S.
bog rosemary. See *moorwort.*
bogrush. See *rush.*
bog soil. Intrazonal soils with peaty surface soil underlain by a glei layer.
bog spavin. In horses, a distention of the joint caused by excess fluid on the inner surface of the hock. It makes a horse absolutely unsound. Also called *boggy in the hocks.*
bog violet. See *butterwort: 2.*
bog flowage. In cranberry cultivation, the water used in winter to overflow a bog to protect the plants from frost and insects.
boil down. To concentrate a liquid, such as a fruit juice or sap of a tree into a thick sirup.
boiled ham. A retail product. The outside fat and bone is removed from cured, unsmoked ham. The meat is then compressed, tied with cord, and simmered 30 minutes per pound.
boiled milk. Denoting an off-taste that milk acquires when heated to 180° F.
boiler. 1. A chamber in which steam is generated. 2. Any vegetable which may be boiled.
boiling chamber. A pan used to heat and evaporate or condense liquids, such as tree sap.
boiling point. Temperature at which a liquid is vaporized.
boil over. For bees in a hive to become excessively nervous and active.
boil smut. See *corn smut.*
bois côtelet. Fiddlewood *q.v.*
Bois d'arc. See *Osage orange.*
Bokhara clover. See *white sweetclover.*
Bokhara fleecevine. *Polygonum baldschuanicum,*

a perennial, climbing, woody vine, family Polygonaceae; grown for its foliage and rose-colored flowers; also useful as a bee plant. Native to Asia. See *silvervine fleeceflower.*

Bolander waterhemlock. *Cicuta bolanderi,* a perennial, California herb, family Umbelliferae. It is poisonous to stock, causing vomiting, diarrhea, dilated pupils, difficult respiration, frothing at the mouth, weak but rapid pulse, convulsions.

bolar roast. A retail cut of beef taken from the foreleg.

bold. 1. Designating the assured gait of a horse. 2. Designating assurance of character in an animal.

boldo. See *boldutree.*

boldutree. *Peumus boldus,* a tropical evergreen tree, family Monimiaceae; grown in California for its fragrant foliage. Native to Chile. Also called *Chilean boldo; boldo.*

bole. The main stem of a tree.

bole area. 1. The surface area of the main stem of a tree, which is calculated by multiplying circumference by height. 2. The sum of all the *bole areas* of trees in a stand.

bolewood. Black centaurea.

boll. 1. The leathery capsule which contains the seed and lint of the cotton plant. 2. The seed pod of flax.

bolleana poplar. *Populus alba* var. *bolleana,* a deciduous tree, family Salicaceae; grown as an ornamental for its upright habit. Native to Eurasia.

bollies. Unopened bolls of cotton which contain usable lint.

bolling. A tree whose branches have been removed.

boll period. That phase in the growth when cotton bolls have formed but not opened.

boll rot. See *cotton anthracnose.*

boll weevil. *Anthonomus grandis,* family Curculionidae, a beetle that causes cotton losses of more than 100 million dollars annually. The larvae feed only on cotton but the adults may feed on okra, hollyhock, and hibiscus. Native to Mexico and Central America, the boll weevil has gradually moved to the northern edge of the Cotton Belt in the U.S.

bollworm. *Heliothis zea* (Boddie), family Phalaenidae, an insect whose larvae feed on cotton bolls, corn, tobacco, tomatoes, beans, vetch, alfalfa, and other garden plants and flowers. Also called *corn earworm; tomato fruitworm.*

bolly. A boll of cotton which remains unopened after frost. See *bollies.*

bolly cotton. The lint extracted from the bollies.

bologna bull. Colloquially, a medium sized bull, the meat of which, since it carries little fat and absorbs water, is used in bologna.

bolson. A drainage basin surrounded by high land and flanked by alluvial fans in the arid regions of southwestern U.S. and Mexico.

bolster. 1. The padded undersection of a saddle. 2. The excrescence at the leafstalk where it joins the axis. 3. A cross member of a vehicle on which the bed, rack, or load rests.

bolt. 1. A section sawed or split from a short log from which blocks, shingles, staves, etc., are made. 2. A short log used in making pulpwood or veneer. 3. A strong, metal pin with a head at one end and threads at the other. 4. To sift coarse elements from fine, as bran from flour. 5. To flower or to produce seed stalks, often prematurely. 6. To eat rapidly. 7. To run away.

bolter. 1. A sieve. 2. A circular ripsaw and a mechanically driven carriage used for sawing round bolts into boards. 3. A plant, such as the sugar beet, which produces a seed stalk from the crown of the plant.

bolting cloth. A sieve cloth, usually silk.

Boltonia. A plant of the genus Boltonia, family Compositae. Species found in America are tall perennials with aster-like flowers. See *white boltonia; violet boltonia.*

bolus. 1. A large pill for dosing animals. 2. Regurgitated food that has been chewed and is ready to be swallowed.

bomb fly. See *northern cattle grub.*

bonavist. See *hyacinth dolichos.*

Bondon. A small, unripened, whole-milk French cheese of the Neufchatel type.

Bondost. A Swedish cheese made from cow's milk. Lactic starter and color may be added in the early stages when the milk is warm; later cumin or caraway seed is mixed with the curd. The cheese is cured six to eight weeks.

bone. A piece of the skeleton of a vertebrate.

bone ash. The highly phosphatic residue of bones burned in air; of limited use as a fertilizer. Also used in making assay cups and in cleaning jewelry.

bone black. The residue from heating bones in a closed retort similar to that of making coke. *Bone black* is used in refining sugar, oil, etc. After use, it is sold as spent bone black (*q.v.*) for fertilizer manufacture. Also called *bone charcoal; bone char; animal charcoal; animal black.*

bone char(coal). See *bone black.*

bone chewing. The eating, by animals, usually cattle and horses, of the bones of dead animals; indicative of a mineral deficiency, usually of phosphorus and calcium. Also called *pica; depraved appetite; licking disease.*

bone dust. See *bone meal.*

boned chicken. A dressed chicken from which all the skeleton has been removed, leaving the skin nearly intact.

bone earth. An obsolete term for phosphate of lime obtained from bones. See *bone ash.*

bone glue. One of three main types of glue which are by-products of packing houses.

boneless beef. Retail beef with the bones removed.

boneless chicken. See *boned chicken.*

bone marrow. See *marrow: 1.*

bone meal. The product of drying and grinding animal bones not previously steamed under pressure. The composition is nitrogen 3.3 to 4.1 percent, equivalent to ammonia 4 to 5 percent, bone phosphate 43 to 50 percent or phosphorus 10 to 12 percent, and calcium 21 to 24 percent. Used as a fertilizer and stock feed.

bone meal (steamed). A product of grinding animal bones previously steamed under pressure. The composition is nitrogen 1.65 to 2.5 percent, equivalent to ammonia 2 to 3 percent, bone phosphate 50 to 55 percent, or phosphorus 12 to 15 percent, and calcium 24 to 33 percent. Used as a fertilizer and stock feed.

bone out. To remove the bones from a carcass or cut of meat.

bone phosphate. See *tricalcium phosphate.*

boneset. *Eupatorium perfoliatus,* a very tall perennial herb, family Compositae. Grown for its white flowers and formerly for its medicinal properties as a persperative and tonic. Found in low-lying parts of North America. Also called *feverwort; thoroughwort.*

bone shears. A scissor-like device for cutting the bones of poultry.

bone spavin. A bony growth in the hock which lames a horse. Also called *jack spavin; spavin; spavined hock.* See *spavin.*

boning knife. A butcher knife for removing meat from bones.

Bonitierung. A German system of land classification in which lands are grouped according to normal yields of staple crops. Emphasis is on the quality of the land, not the skill of the producer.

book farmer. A contemptuous name for an advocate of scientific farming, as opposed to someone who farms by rule of thumb and by signs.

boom. 1. A long beam projecting from a mast; used for hoisting or stacking hay. 2. A barrier made by fastening logs end-to-end across a river to contain cut timber. 3. A group of floating logs. 4. The spar which projects from a log-loading machine to support logs. 5. An obstruction composed of floating logs. 6. A construction designed to float logs in a certain direction. 7. To rise in volume, as a river *booms* sufficiently to float logs.

boom sprayer. A long pipe to which several nozzles are attached to spray a wide area.

booster dose. The subsequent injection of vaccine which stimulates the production of antibodies and maintains resistance to disease.

boot. 1. The hollow metal casting on a planter or a drill through which the seed passes to be planted. 2. Profuse feathering on the shank and toes of fowls. 3. The sheath-like leaf structure on the upper end of grain or grass plants which encloses the inflorescence prior to its emergence. During inflorescence the plant is said to be in the *boot stage* or *in boot.* 4. In a grain elevator, the box which contains the lower pulley or sprocket.

boot laces. Narrow strips of skin cut from sheep by a careless shearer (New Zealand).

boot stage. See *boot: 3.*

borage. *Borago officinalis,* a hardy, annual herb, family Boraginaceae; grown for its blue flowers, as a bee plant, and as an additive in cordials. Native to southern Europe.

borax. Sodium tetraborate, $Na_2B_4O_7{\cdot}10\ H_2O$. It is sometimes applied directly to soil, or added to commercial fertilizer in very small quantities (toxic to plants in high concentrations) to supply boron. It is also used as an herbicide and an insecticide, especially to kill the maggots of flies in manure and refuse piles. Also called *borac; tincal.*

Bordeaux burning. See *Bordeaux injury.*

Bordeaux injury. Damage to orchard trees from excessive spraying of Bordeaux mixture (*q.v.*). Dark specks on the fruit coalesce and russet; deep cracks may be healed over by cork cells, but the fruit may be misshapen. Affected apples lose moisture and are easy prey to rotting fungi; affected cherries are small; affected peach trees spot and burn and defoliate and twigs redden. Also called *Bordeaux scald; spray injury; Bordeaux burning; spray russeting; yellow leaf.*

Bordeaux mixture. A popular spray of copper sulfate, lime, and water; a fungicide for leaf diseases. See *Burgundy mixture.*

Bordeaux nozzle. A nozzle which throws a flat, fan-shaped spray (obsolete).

Bordeaux paste. A thick, paint-like mixture of raw linseed oil and dry Bordeaux powder. It has antiseptic properties and hinders callous growths on trees and shrubs.

Bordeaux scald. See *Bordeaux injury.*

border. Trees or shrubs planted along a wall, hedge, etc.

border forsythia. *Forsythia intermedia,* a hardy, deciduous shrub, family Oleaceae; grown for its early-spring, yellow flowers. It is a hybrid between *F. suspensa* and *F. viridisaima,* of China.

border irrigation. A system of irrigating land which has a slope of 2 to 4 feet per 1,000 feet. Parallel levees are built and the area between levees is flooded.

Border Leicester. A type of the Leicester (*q.v.*) which, though a long-wooled breed of sheep, has a shorter fleece than the English Leicester.

boreal. The northern and cold mountainous parts of the world; also the north wind.

borecale. See *kale.*

bored well. A well sunk by means of auger-buckets (obsolete).

Borelli. A small Italian cheese made of buffalo's milk.

borer. An insect which bores into and feeds on the woody sections or bark of trees.

Borggreve thinning. See *selection thinning.*

Borna's disease. See *equine encephalomyelitis.*

Bornmueller autumncrocus. *Colchicum bornmuelleri,* a bulbous herb, family Liliaceae; grown for its autumnal, large, cup-shaped, mauve flowers. Native to Eurasia.

boron. The non-metallic chemical element B, which is not found in a pure state. Plant diseases such as yellows of alfalfa and heart rot of beets have been attributed to boron deficiency in the soil. Small amounts of *boron* are mixed into fertilizer. See *boron injury.*

boron deficiency. A soil deficiency. In vegetables, the terminal buds and phloem tissue are killed and the development of lateral buds is arrested. In citrus trees, the surface and main veins of the leaves cork or split and the leaves curl downward; long, narrow splits occur on the trunk bark of small citrus trees. In deciduous trees, an internal bark necrosis, which can often be detected on the surface, occurs.

boron injury. Slowed or arrested germination, killing or stunting of plants, fading of color, premature ripening, or small yields, caused by an excess of boron in the soil.

borrow pit. A hole formed by digging earth to be used for a levee, road, etc. Also called *bar pit; bar ditch.*

boscage. A thicket of shrubs.

bosque. A name given to a wooded thicket, usually of tamarisk, willow, and cottonwood, on the flood plains of southwestern U.S.

boss. 1. A lumpy formation on an animal or plant. 2. Metal stud on either side of the bit of a horse's bridle.

bossy. An affectionate name for a cow or calf.

Bostonfern. *Nephrolepis exaltata* var. *bostoniensis,* a tropical fern, family Polypodiaceae; grown in greenhouses for its foliage.

Boston ivy. See *Japanese creeper.*

Boston style butt. A retail roast of pork from the shoulder. Also called *cottage roll.*

Boston yellow daisy. See *Marguerite chrysanthemum.*

bot. Larva of the bot fly.

botanic garden. A garden of plants which have scientific interest.

botany. 1. The science of plants: anatomy, cytology, ecology, morphology, mycology, paleobotany, pathology, physiology, phytogeography, taxonomy, etc. 2. A general term for Australian wool.

bot fly. See *horse bot fly; nose bot fly; throat bot fly; sheep bot fly.*

bo tree. *Ficus religiosa,* a sacred tree of India, family Moraceae, which is sometimes grown in Florida and California. Also called *peepul tree.*

botryose. Used to describe a plant whose flowers grow in clusters.

Botryosphaeria canker and fruit rot of apple. A fungous disease caused by *Botryosphaeria ribis,* family Botryosphaeriaceae, resulting in an internal and external decay of fruit. It is characterized by small spots on the twigs which become depressed and have a watery exudate. The following year pimple-like elevations appear on the lesions; the thin outer-covering of the bark sloughs off easily, leaving a slimy underlying tissue.

Botrytis. A genus of parasitic fungi, family Moniliaceae, whose various species cause spots on flowers, leaves, and stems of plants. In wet weather, the spots spread and are covered by a gray, moldy, sporal growth of the fungus. See the following:

Botrytis blight of blueberry. Caused by *Botrytis cinerea* which blights the flower and stem of the plant, and spots its leaves. The fruit infected at harvest time may rot during marketing. Also called *blueberry blossom and twig blight.*

Botrytis blight of lettuce. Caused by *Botrytis cinerea,* which produces brown lesions near the base of the stem and on the bases of the lower leaves. These lesions progress upward and in wet weather become covered with gray mold. Also called *Botrytis rot of lettuce.*

Botrytis blight of lily. Caused by *Botrytis cinerea* or by *B. elliptica.* It produces small, brown, yellow-margined spots on leaves and flowers, which may rot along with the base of the stem.

Botrytis blight of peanut. Caused by *Botrytis cinerea* which destroys the leaves and stems with a gray moldy rot. Also called *Botrytis late blight of peanut.*

Botrytis blight of snapdragon. A serious disease of snapdragons caused by *Botrytis cinerea.* The flower spikes are wilted and spots may enlarge to girdle and kill the stem and spike.

Botrytis blight of tulip. Caused by *Botrytis tulipae,* which may kill the whole shoot from the bulb upwards. Adjacent plants then become infected with spots on the leaves and flowers.

Botrytis disease of gladiolus. Caused by the fungus *Botrytis gladiolorum,* which produces spots on the leaves, flowers and stems. These rapidly enlarge, becoming rotten in moist weather. The infected corms produce infected plants when transplanted. It is also a very serious disease on flower stalks shipped to the market.

Botrytis gummosis. A bark disease of citrus caused by the fungus *Botrytis cinerea* Pers.; characterized by softening of the invaded tissue and development of a gray, furry, fruiting body of the fungus. In the center of the spot the bark is entirely destroyed. It is often accompanied by shell bark *q.v.*

Botrytis rot of citrus. A fungous disease of citrus caused by *Botrytis cinerea* Pers.; characterized by gray fruiting stalks and spores of the fungus which cover the fruit. On lemons the color is cinnamon-brown and on oranges it is dark yellowish-brown. Also called *gray mold.*

Botrytis twig blight. A disease of citrus caused by the fungus *Botrytis cinerea;* characterized by a blight beginning at the extremities of twigs and spreading downward. The infection girdles the limb and causes it to wither and die. See *Botrytis gummosis.*

bots. See *sheep bot fly; horse bot fly; throat bot fly; nose bot fly.*

bottle brush. 1. Shrubs and trees of the genus Callistemon, family Myrtaceae, with cylindrically shaped clusters of bright red or yellow flowers with long, colored stamens. Native to Australia, and planted extensively in California and southern Florida. 2. See *field horsetail.*

bottlebrush Indianwheat. See *bracted plantain.*

bottle-butted. See *churn-butted.*

bottled gas. Popular name for liquefied petroleum gas kept under pressure in a metal container; usually butane (C_4H_{10}), propane (C_3H_8), or any mixture of the two. It is often called LP (liquefied petroleum) gas. Formerly used for home lighting systems, it is used now for cooking, heating, and for internal combustion engine fuels, particularly pumping engines and farm tractors. Butane is also widely used in many synthetic-producing industries.

bottle fern. See *brittle bladderfern.*

bottle garden. See *terrarium.*

bottle gentian. See *Andrews gentian.*

bottle grafting. See *side whip grafting.*

bottle jaw. A pendulous swelling of tissues underneath the lower jaws of cattle and sheep, caused by an infestation of stomach worms or other parasites.

bottle method. The self pollinating of corn. A tassel is cut from the stalk, its shank being placed in a two ounce bottle of water that is then hung on a corn stalk until the pollen is shed.

bottle palm. *Hyophorbe amaricaulis,* family Palmaceae; a feather palm, with a swelling near the base of the plant; grown for ornament. It thrives in shade and sunshine in a variety of soils. Native to Mascarene Islands; cultivated in Florida. Also called *bitter-stem palm.*

bottle tree. Any tree of the genus Brachychiton, family Sterculiaceae. See *flame bottletree; Kurrajong bottletree.*

bottom. 1. (often plural) Low-lying land adjacent to a river, usually rich in alluvial deposits, *e.g.,* Mississippi River *bottoms.* 2. A section of a plow consisting of the moldboard, share, frame, and landside. 3. Stamina in a horse.

bottom bar. The lower, wooden bar of a frame for a comb in a beehive.

bottom board. Floor of a beehive on which the hive body rests.

bottom chuck. A boned cut from a side of beef below the blade bone.

bottom heat. Heat applied beneath greenhouse benches to induce plant growth.

bottomline. A term used in judging stock to designate the line formed by an extension of the chest along the belly.

bottom round. A cut of beef composed of the two tender muscles of the leg; used either for steaks or roasts.

bottom rot. A disease of cabbage and lettuce caused by *Rhizoctonia solani,* after mid-season. The outer leaves are attacked by the fungus which advances from the main stem along the petioles and midrib, resulting in dark lesions and death of the leaves.

bottom sirloin. See *sirloin tip.*

bottom yeast. Yeast which produces sedimentary fermentation. See *yeast.*

botulism. A food poisoning caused by the bacterium, *Clostridium botulinum* in preserved foods or feed for animals. The poison can be fatal to men or animals. Called *limberneck* in poultry. Also called *food poisoning; allantiasis.*

Boudanne. A hard, French cheese made from cow's milk.

bougainvillea. Any tropical vine of the genus Bougainvillea, family Nyctaginaceae. Species are grown for their climbing habit and handsome flowers. See *lesser bougainvillea; Brazil bougainvillea.*

bougie. An instrument used for insertion into natural passages of a body for dilation or medication, *e.g.,* a teat *bougie.* It is usually constructed of steel, rubber, or plastic.

bouillon. 1. An abnormal growth on a horse's frog. 2. A beef extract or clear soup.

boulder (bowlder). (a) Large, rounded (or subangular), detached, or transported rock; larger than a cobble, *i.e.,* more than 10 or 12 inches in diameter; common in the soil of the glaciated and mountainous regions of the U.S. (b) In common usage *boulders* are often not distinguished from *stones.* (c) In soil and geologic literature, no precise definition as to size and shape has been made.

boulder clay. Massive, compact deposits of gritty clay, of glacial origin, containing imbedded gravel and boulders. It is the parent material for a number of highly productive types of soil.

boulder fern. See *hayscented fern.*

boulder flat. Level tract of land covered with boulders.

boulder raspberry. *Rubus deliciosus,* a perennial plant, family Rosaceae, which has profuse flowers for extended periods, but no prickles. Grown for ornament in warm areas. Native to North America. Also called *Rocky Mountain flowering raspberry.*

bouncing Bet. *Saponaria officinalis,* a hardy, perennial herb, family Caryophyllaceae; grown in rock gardens. Escaped, it can become a weed pest. It is toxic, but usually animals refuse to eat it. Poisoning results in irritation of digestive tract, vomiting, nausea, diarrhea, vertigo. Also called *soapwort; old maid's pink; London pride; lady-by-the-gate; hedge pink; farewell-summer.* See *cow soapwort.*

boundary keeper. A sheep herder who keeps sheep within an unfenced boundary (New Zealand).

boundary layer. Thin layer of water between the conduit surface and moving water in which the greatest variation in velocity is more or less concentrated.

boundary rider. A man responsible for the upkeep of boundary fences, as well as sheep and cattle, in remote parts of an Australian station.

boundary tree. An old, tall, distinguishable tree standing on a property line, usually blazed or otherwise marked.

bound moisture. Moisture intimately associated with the finer wood elements of the cell wall by absorption and held with sufficient force to reduce the vapor pressure significantly.

bound water. Water adsorbed by colloids and therefore not easily freezable. Also called *unfree water.*

bountyland. Land given as a bounty for military service. According to a U.S. statute of 1850, lands were granted to persons in military service or their dependents.

Bourbon cotton. *Gossypium purpurascens,* a perennial shrub or small tree, family Malvaceae; grown in the tropics as a source of commercial cotton. Also called *Puerto Rico cotton; Siam cotton.*

Bourbon Red. A breed of turkey with dark, reddish-brown feathers, except for the white tail and primary and secondary wing feathers. The standard market weight is 23 pounds for the young tom and 14 for the young hen.

Bourbon rose. *Rosa borboniana,* a hardy shrub, family Rosaceae, which is a famous hybrid of the Provence and China roses. It has prickly stems and purple, double flowers.

Bourgain. A low-fat, uncured, Neufchatel-type French cheese.

bourne (bourn). English name for a minor land feature; a small hollow containing a stream. (Uncommon in the U.S. or Australia.)

bout. One complete circuit with a plow, mower, combine. Also called *round.*

bouvardia. Any shrub of the genus Bouvardia, family Rubiaceae. Native to tropical and subtropical America. See *sweet bouvardia; scarlet bouvardia.*

bovine. An animal of the family Bovidae; a cow, bull, steer, calf or ox.

bovine farcy. A slow, chronic, and fatal disease of cattle, caused by *Actinomyces farcinicus.* It is characterized by small swellings on the legs, especially above the hocks.

bovine genital trichomoniasis. A venereal disease of cattle caused by a flagellated protozoan, *Trichomonas fetus;* characterized by difficulty in breeding, temporary sterility, and an accumulation of pus in the uterus without a fever. Also called *tric; bovine venereal trichomoniasis.*

bovine hemoglobinuria. See *British red-water.*

bovine infectious abortion. See *brucellosis.*

bovine malaria. See *cattle tick fever.*

bovine piroplasmosis. See *cattle tick fever.*

bow. 1. A contrivance of elastic wood used before the invention of the cotton gin to remove rubbish from lint cotton and downy fleece. 2. See *crook:* 2. 3. The curved parts at the top of the handle of sheep shears which serve as a spring to force the blade apart (New Zealand).

bowed. Describing an animal whose toes point outward.

bowed tendon. A sprain behind the cannon bones of a horse, which when healed results in the tendon being shortened, causing the leg to be bowed.

bower actinidia. *Actinidia arguta,* a hardy, high-climbing, woody vine, family Dilleniaceae; grown for its beautiful foliage and as a trellis cover. Native to Asia. Also called *tara vine.*

bower plant. See *jasmine pandorea.*

bowery. A farm in the early Dutch settlement of New York. Also called *bouwrie; bouwery.*

bowl. 1. A vat or tank in which wool is scoured. 2. A basin for holding liquids. See *stock tank.*

Bowman's root. *Gillenia trifoliata,* a perennial herb, family Rosaceae, cultivated for its white flowers. It grows well in partial shade and rich soil. Native to North America. Also called *false ipecac; Indian hippo.*

bowstring hemp. A group of tender, perennial plants, genus Sansevieria, family Liliaceae, with thick, short roots, and long, erect leaves, often blotched. The plants bear white or pale yellow flowers and are easily cultivated indoors. Native to India and Africa; grown in southern U.S. Also called *sansevieria.*

box. 1. *Buxus sempervirens,* a shrub or small tree, family Buxaceae, one of the most valuable broad-leaved evergreens. Since the days of the Romans it has been planted for hedges and topiary work. It is lethally poisonous to cattle, horses, sheep, and pigs, having emetic and purgative properties. Native to southern Europe. 2. The lowest grade of softwood lumber. 3. A cavity cut into the base of a pine tree to collect sap to make turpentine and rosin. 4. A system for bracing branches of trees. 5. To chip back on the underside of a cut in a tree to prevent the main stem of the tree from splitting when it falls. 6. An accidental mixing of two herds of sheep or cattle (Australia). 7. To corral animals. 8. A device for dividing water in an irrigation system into two or more ditches. 9. A canyon with one entrance.

box barberry. See *Japanese barberry.*

box-berry. See *checkerberry.*

box board. See *box:* 2.

box curing (without pressure). A method of curing pork in which the meat is rubbed with curing ingredients and placed in tight containers. The meat must be rearranged every three days.

box curing (with pressure). A method of curing bacon in tight boxes. The strips are placed in a box snugly, without overlapping. A dead weight or a screw jack is placed on a loose lid. Pressure causes the brine to rise and cover the meat.

box drier. A device for making egg powder in which heated air is forced through a chamber, the egg liquid being sprayed into the upper portion of the hot air stream. As the velocity of the air decreases, the dry egg powder falls to the bottom of the chamber.

box elder. *Acer negundo,* a deciduous shade tree, family Aceraceae. Its branches are brittle; the wood is used for paper pulp and woodenware. Also called *ashleaved maple.*

box-elder aphid. *Periphyllus negundinis* (Thoms.), family Aphidae, a louse that infests the box elder.

box-elder bug. *Leptocoris trivittatus* (Say), family Coreidae, a sucking insect that infests the leaves of the box elder; not especially destructive. It may enter homes during the late fall.

box-elder leaf miner. See *boxwood leaf miner.*

box-elder leaf roller. *Gracilaria negundella* Chamb., family Gracilariidae, an insect that infests the undersides of the leaves of the box elder in Colorado.

box-elder twig borer. *Proteoteras willingana* (Kearf.), family Olethreutidae, a moth whose larvae infest the twigs of the box elder in Utah.

Box (firm). A German cheese made of cow's milk and cured for three months.

box hive. Any bee hive in which the combs are not removable. (It is illegal to keep bees in such hives in Michigan and in most other states.)

box honeysuckle. *Lonicera nitida,* a tender, evergreen shrub, family Caprifoliaceae; grown for its glossy leaves, fragrant, white flowers, and purple fruit. Native to China.

box huckleberry. *Gaylussacia brachycera,* a hardy, evergreen, prostrate shrub, family Ericaceae; grown for its white or pink flowers. Also called *juniperberry.*

box-packed fowl. One or more packaged birds, dressed or ready to eat.

box sandmyrtle. *Leiophyllum buxifolium,* family Ericaceae, a neat, upright, perennial shrub about 18 inches high; grown for its glossy, dark-green leaves and white flowers. Native to North America.

Box (soft). A German cheese made of equal parts of skim and whole milk, to which saffron, rennet, and caraway seeds are added. Ripened for three months.

box stall. An enclosure in which an animal can be confined without being tied.

box the heart. To cut boards from the sides of a log so that the center or heartwood is left as one piece of timber.

boxthorn. See *matrimonyvine.*

boxtree. See *boxwood.*

boxwood. Any broad-leaved evergreen shrub or tree, genus Buxus, family Buxaceae. See *Harlands box; littleleaf box; box: 1.*

boxwood leaf miner. *Monarthropalpus buxi* (Laboulb.), family Itonididae, an insect which is not a serious pest. Originally native to Europe.

boxwood nematode. *Pratylenchus* spp., a genus of nematodes which infests, sometimes stunting or killing, the roots of the box, *Buxus* spp.

boxwood psyllid. *Psylla buxi* (L.), family Psyllidae, a jumping insect that infests new growth on the ornamental box, *Buxus* spp., causing leaf curl and stunting the twigs.

Boysenberry (Boysen). *Rubus* spp., family Rosaceae, a large, sweet, deep red, high quality berry, the hybrid of blackberry, loganberry, and raspberry. Commercially important on the West Coast of U.S.

bozal. A bridle-hitch on the mouth of a horse or mule. Also called *bosal* (southwestern U.S.).

B.P.L. Bone phosphate of lime. See *tricalcium phosphate.*

Bra. A hard, white Italian cheese, made of partly skimmed milk, to which rennet is added.

braccate. Feathered on the ankle and foot.

brace. 1. Any device designed to strengthen or support, such as a corner-post *brace,* wall *brace,* etc. 2. Specifically, a rod connected to the beam, which reinforces the landside of a plow. 3. A curved tool for holding and rotating bits. 4. A pair, as a *brace* of pheasants.

brace and bridge comb. A comb in a beehive which keeps other combs in position and which is a bridge for the bees to pass to other parts of a hive. It also serves as temporary storage for nectar during honey flow as well as for water in hot weather.

brace comb. A small piece of comb which connects combs or a comb and the hive.

brace roots. The supporting roots, of corn, etc., which spring from the first two or three nodes above the ground. Also called *aerial roots; prop roots.*

brachiate. Being branched in pairs which sprout from opposite sides of a stem, as on a box elder.

brachysm. A dwarfing of plants by shortening of the internodes.

bracken. *Pteridium aquilinum,* a widely distributed fern, family Polypodiaceae; sometimes grown in gardens, but which, if escaped, is a weed pest. The young curled fronds or fiddlenecks are edible. Also called *female fern; fern brake; ern fern.* See *eastern bracken; western bracken.*

bracken sickness. A livestock disorder caused by eating bracken. See *eastern bracken; western bracken.*

bract. A modified leaf that subtends a flower or belongs to a flower cluster.

bracted plantain. *Plantago aristata,* family Plantaginaceae, an annual or winter annual weed, which persists in poor soil. Native to Great Plains of U.S., now found eastward to New England. Also called *airplane plant; bottlebrush Indianwheat.*

bracted vervain. See *bigbract verbena.*

bradshot. See *braxy.*

bradsot. See *braxy.*

Bradford. A cross between Hereford and Zebu breeds of cattle.

brae. A steep bank; a slope; hillside.

Brahma. Large or bantam meat chickens of the Asiatic class, of which there are the Light (Columbian pattern), Dark (Silver Penciled), and Buff varieties. The *Brahma* has a pea comb, yellow skin, heavily feathered shanks and toes, and the eggs are brown. Cocks weigh as much as 12 lbs., hens 9½ lbs.

Brahman. See *Zebu; ranch cattle.*

braid. Market designation for the coarsest of American grades of wool (*q.v.*) comparable to the English 40's and 36's.

brail. A Y-shaped leather thong with a buckle which prevents the flying of captive game birds.

braining. 1. Fowl slaughter, in which a knife is inserted through the roof of the mouth to the rear of the skull, piercing the medulla oblongata. 2. Removing the brain from a carcass.

brains. 1. The nerve tissues enclosed in the skull or cranium of vertebrates. 2. Calf brains, sold for human consumption.

brain stem. The part remaining after the removal of the cerebrum and cerebellum.

brake. 1. An irrigation check. 2. A place overgrown with bracken, brush, wood, cane, etc., *e.g.*, a cane *brake;* a thicket.

bramble. 1. Any plant of the genus Rubus, family Rosaceae, as the blackberry, raspberry and dewberry (*q.q.v.*) called bramble fruits. 2. Any prickly, clinging shrub.

bramble anthracnose. See *cane fruit anthracnose.*

bran. The seed coat of wheat or other cereal grains which is separated from the flour and used as stock feed.

bran bait. A mixture of bran, sweetening, and a poison used to kill insects. See *bait.*

branch. 1. A lateral stem arising from the main stem of a plant, bough or limb. 2. Part of the horseshoe from the first nailhole to the end of the heel. 3. A tributary stream. 4. An ancillary pipe attached to a main pipe.

branch blight of grapefruit. A blight thought to be caused by sunburn, high water table, high salt content of the soil. Characterized by a decay of the tips of the branches. The lesions can extend and ultimately kill the entire branch. During the growing season there is a gum exudate near the margins of the affected tree. Occurs in California.

branched wheat. See *poulard wheat.*

branch gall. A poplar disease caused by the fungus *Macrophoma tumefaciens* Shear, family Sphaeropsidaceae; characterized by round galls up to 1½ inches in diameter at the base of the twigs. The twigs and branches may be killed, but the disease is rarely serious.

branching broccoli. See *broccoli.*

branching francoa. *Francoa ramosa,* a tender, perennial herb, family Saxifragaceae; grown in flower gardens in warm areas and in greenhouses for its showy, white flowers. Native to Chile. Also called *maiden's wreath; bridal wreath.*

branch ivy. See *drooping leucothoë.*

branch knot of hackberry. A witches'-broom of the hackberry tree, caused by a phytoptid mite.

branch of frog. The soft, rear part of a horse's foot which branches into two sections that are divided by the middle cleft of the frog.

branch rot. See *Fusarium rot; Fusarium wilt (diseases).*

branch wilt of walnut. A fungous disease of Persian walnuts (*Juglans regia*) caused by *Hendersonula toruloidea,* family Sphaeropsidaceae; characterized by sudden wilting of the leaves in summer. It appears on small twigs and branches, but eventually involves the larger limbs as well, elsewhere being most common on the south or southwest side of the tree. The leaves dry to a brownish color, remaining on the tree until late winter. Both the bark and wood of the infected areas turn almost black in color. Cankers usually appear on the uppersides of the branches at affected areas. The outer areas of corky bark over diseased areas loosen and slough off, exposing a black powdery deposit.

brand. 1. A mark of identification burned into the skin of cattle or horses. See *tag: 2; set brand.* 2. The steel or iron die used to mark the animal. 3. The design of the mark of identification commonly used to identify foods and feedstuffs. 4. (cap.) A German hard cheese made from sour milk curd and butter. Beer is used to moisten the cheese as it ripens in kegs.

brand artist. 1. A person who brands cattle. 2. A person skilled in forging brands.

brand book. An official record of ranch brands.

brand canker. A common disease of rambler roses caused by *Coniothyrium wernsdorffiae,* family Sphaeropsidaceae; characterized by reddish spots on the canes. As they enlarge, the centers turn light-brown, and lethal lesions girdle the cane.

brand fluid. A chemical used to put a brand on an animal.

branding bee. A festive, annual gathering at which animals born within that year are branded.

branding iron. An iron rod with a metal stamp used to burn a mark on an animal for purposes of identification.

branding paint. A paint, indelible or removable, used for marking animals, particularly sheep, for identification.

branding race. A narrow chute into which sheep are herded to be branded (New Zealand).

brand inspector. An official who records and recognizes brands of identification on animals.

brand tillage. A system of farming in which new land is periodically cleared for plowing, while abandoned fields are allowed to regain their natural cover. *Cf. milpa agriculture.*

bran duster. Used for sifting flour from bran after bolting.

brandy mint. See *peppermint.*

bran mash. Feed prepared by moistening bran with warm water.

brasada. Brush country or thicket vegetation (southwestern U.S.).

brashness. An abnormal condition of lumber in which a piece of wood is low in shock resistance and breaks without splintering.

brash oak. See *post oak.*

brashy soil. Soil composed almost entirely of coarse, angular rock fragments, with fine soil matter filling the interstices.

brasil. See *bluewood condalia.*

Brassica. A genus of plants, family Cruciferae, which includes the common cabbage, turnip, black mustard, etc.; characterized by smooth, greenish leaves and fruiting in long pods with conical tips. Native to Asia, Africa, and Europe.

brasswood. A common name for Osage orange *q.v.*

brassy beetle. See *corn flea beetle.*

bratwurst. A sausage made from lean pork, salt, white pepper, celery seed, mace, and rubbed sage.

braunschweiger. A pork sausage made from pork liver, pork, onion, salt, white pepper, ground celery seed, cardamon, and sweet marjoram.

Braunvieh. See *Brown Swiss.*

brawn. 1. Muscles, especially of the arm and leg. 2. Boar's flesh, especially the boiled and pickled meat. 3. A pig fattened for its meat. 4. Hardened skin.

braxy. An acute infection of Norwegian, Scottish, and Faroe Island sheep; characterized by hemorrhagic abomasitis. Apparently caused by *Clostridium septicum,* the bacterial organism which causes malignant edema in U.S. livestock. Also called *bradsot; bradshot.*

bray. The cry of a donkey.

Brazil bougainvillea. *Bougainvillea spectobilis,* a woody vine, family Nyctaginaceae, grown for its crimson-red bracts. Native to South America.

Brazilian glorybush. *Tibouchina semidecandra,* a shrub, family Melastomaceae, grown in very warm regions for its violet or reddish-purple flower clusters. Also called *spiderflower.*

Brazilian guava. *Psidium guineense,* a tropical shrub, family Myrtaceae, grown in very warm regions for its edible, greenish-yellow fruit. Native to South America. Also called *araça.*

Brazilnut. *Bertholletia excelsa,* a tree, family Myrtaceae, whose seeds are the *Brazilnuts* (niggertoes) of commerce. Native to South America. Also called *almendron.*

Brazil peppertree. *Schinus terebinthifolia,* an ornamental, evergreen tree, family Anacardiaceae, grown for its bright red fruits and as a popular shade tree. Native to South America; grown in southern California. Also called *Christmasberry tree; peppertree; California peppertree.*

breachy. Describing an animal who passes over, or through, fences.

Breadbasket of the nation. Colloquial name for the Corn and Wheat Belt regions of the north-central United States.

breadfruit. Fruit of the bread tree (*q.v.*). Its delicious, white pulp has the consistency of the common bread.

bread grain. A cereal, such as wheat or rye, which yields flour for leavened bread. Other cereals are called coarse grains or feed grains.

bread madness. See *ergotism.*

breadroot scurfpea. *Psoralea esculenta,* a perennial herb, family Leguminosae, which has large, tuberous roots. Native to western North America, and once extensively eaten by the American Indians. Also called *pomme de prairie; pomme blanche; prairie potato; meadow potato.*

breadstuffs. Various grains used for making bread.

bread tree. *Artocarpus communis,* a tropical tree, family Moraceae; it has heavy foliage of thick oval leaves, small flowers, and bears an edible fruit, called breadfruit (*q.v.*). Native to Polynesia; grown rarely in the extreme tip of Florida. Also called *breadfruit tree.*

break. 1. A weak spot in wool. 2. An opening in a bale of cotton for taking samples. 3. A sharp drop in prices. 4. A public sale of tobacco. 5. An uncontrolled forest fire. 6. The first process in the separation of the fiber from the stalks of flax or hemp. 7. A temporary division in a pasture established for pasture rotation (New Zealand). 8. To pick pineapples. 9. To tame an animal, especially a horse. 10. To plow fallow or virgin land. 11. To come into leaf or blossom prematurely. 12. To vary from the normal.

breakdown. 1. A disorganization of cellular tissues. 2. A two-wheeled log car. 3. To reduce logs to an acceptable size for a sawmill. 4. To cut the exterior part, including the bark, away from a log.

breaker. 1. A metal device, used to decrease water pressure, which is attached to a pipe or hose to prevent injury to plants. 2. A plow with a long, low moldboard used for turning sod land. Also called *breaking plow.*

breaker bar. A sprayer, used in aerial dusting, which consists of a boom of tubing with drilled holes in place of nozzles.

breaker roll. That part of a feed grinder which breaks the ears of corn, crushes the cob, and feeds the mixture to the grinding plates.

breakfast cream. See *substandard cream.*

breakfast style sausage. An all-pork, finely ground link sausage seasoned with sage, salt, and pepper. It is stuffed into sheep casings.

break ground. To plow land for the first time.

break in. To discipline; to train. See *break: 9.*

breaking plow. See *breaker: 2.*

breaking room. Where eggs are broken out of shells for freezing or dehydrating.

break joint. The epiphyseal, or temporary, cartilage above the hoof of a sheep. Used in grading mutton carcasses.

break out a row. 1. To cultivate new land in a row crop field. 2. To snap, husk, or harvest a row of corn or other row crop.

breaks. Bluff land or steeply sloping escarpments broken by stream dissection as in the eastern edge of the High Plains, south of the Arkansas River to the Texas Panhandle. This area is known as *Breaks of the Plain.*

break sod. To plow an uncultivated field.

break up. 1. To plow and harrow. 2. To prepare a seedbed, usually by plowing. 3. A spring thaw.

breast. 1. The front of the body between the neck and the abdomen. 2. The breast bone and the flesh surrounding it, especially in fowl, lamb, or veal. 3. The forepart of a moldboard on a plow. 4. The part of a cotton gin, consisting of a frame which contains the huller ribs, the picker rollers, and the roll box. 5. That part of a bulk milk tank which connects the inside lining to the outer wall.

breast band. A strap, used in place of a collar, which extends over the fore-chest of a horse. Also called *breast collar.*

breast blister. See *keel cyst.*

breast board. See *moldboard; breast chain.*

breast bone. The osseous sternum of higher animals.

breast chain. 1. A chain or strap which fastens the collar of a harness to the neckyoke. 2. A chain, strap, board, etc., by which a horse maintains its balance in a van. Also called *breast board; tailboard.*

breast high diameter. The point at which the diameter of a tree is measured. Custom has fixed this point at 4½ ft. above ground level. Also called *diameter at breast height.*

breast pack. A style of packing dressed poultry. Both sides of the breasts are in view from the top of the box. *Cf. double-layer side pack; single-layer side pack; squat pack.*

breast roll. Mutton consisting of boned and rolled breast of lamb for roasting.

breast strap. See *breast chain.*

breast wood. Espaliered shoots on the front of a fruit tree.

breather. A vertical section of pipe, which is an air vent, preventing pressure build-up. When installed after a steep grade, it is called a *relief wall.*

breathing pore. See *lenticel; stoma.*

breath-of-heaven. See *buchu diosma.*

bred heifer. A pregnant, young cow which has not previously borne a calf. *Cf. open heifer; heifer.*

breech. 1. That part of an animal from the base of the tail to the hocks. 2. Coarse wool from the *breech* and hind legs of a sheep.

breeches type duster. An aerial crop duster consisting of a large hopper with a W-shaped base.

breeching. A part of a harness, usually of leather, which is fastened around the hind quarters. It allows a horse to stop or back a wagon.

breed. 1. Animals having a common origin and distinguishing characteristics. 2. To improve, through control, characteristics in plants and animals.

breed association. A group organized to safeguard the purity of a breed and to promote the breed. It sets the standards, and maintains registers for the breed.

breed character. Details of conformation or color, *e.g.*, shape of horns in cattle, size and shape of ear in swine, etc., which distinguish a breed of animals.

breeder. 1. A specialist in breeding. 2. An animal or plant used for breeding. *Cf. breeder tom.*

breeder-feeder litter. A show term for swine: a barrow, a boar, or a gilt from the same litter.

breeder hatchery. An establishment which promotes a strain of poultry.

breeder house. See *breeding house.*

breeder's young herd. A show group consisting of one bull less than two years of age, two heifers less than one year of age, and two heifers between one and two years of age.

breeder tom. A male turkey used for breeding, which is marketed between April and August.

breeder tulip. A class of tulip with long stems and flowers of dull colors. The flowers, which blossom in mid-season, are globular and vari-colored. Also called *Dutch breeder tulip.*

breed in and in. See *in-and-in breeding.*

breeding and hog ringing crate. A wooden box to facilitate nose-ringing, and the mating of different-sized swine.

breeding chute. A wooden enclosure which holds a mare or cow, to facilitate mating. Also called *breeding rack.*

breeding class. Designating animals used for breeding purposes.

breeding herd. See *breeder: 2.*

breeding hobbles (hopples). A fetter fastened to the feet of a mare to prevent her from kicking the stallion.

breeding house. 1. A building, divided into pens 6x10 ft. in size, used for breeding poultry. One male and approximately a dozen females are placed in each pen. 2. Any structure used for breeding.

breeding pen. See *breeding house.*

breeding paralysis. See *dourine.*

breeding rack. See *breeding chute.*

breeding season. Period of the year when animals normally reproduce.

breeding shed. See *breeding house.*

breeding stock. See *breeder: 2.*

breeding unsoundness. Malformation of the genitals, cryptorchidism, scrotal rupture and other abnormalities which disqualify animals for procreation though not for other purposes. *Cf. serviceable unsoundness; temporary unsoundness; hereditary unsoundness.*

breed out. To eliminate undesirable characteristics. See *in-and-in breeding; out-breeding.*

breed trademark. The symbol of a distinguishing characteristic.

breed true. The ability to transmit a characteristic, uniformly, to an offspring.

breed type. See *breed character.*

Brewer oak. *Quercus oerstediana,* family Fagaceae, an oak used for forage on the western ranges of U.S.

brewers' dried grains. The dried residue left after wort (*q.v.*) has been extracted from barley malt. It is used in mixed feeds and equals corn gluten in content of digestible protein. It is unpalatable. Also called *dried brewers' grains; beer grains.*

brewers' dried yeast. A source of B complex vitamins from yeast filtered from fermented beer. It has roughly 50 percent protein and 1½ percent phosphorous.

briar. 1. A crosscut logging saw. 2. See *brier.*

brick. 1. Any food packaged in the shape of a brick. 2. A block of dry manure, at one time used to establish mushroom beds.

Brickbat. An English cheese made of fresh milk and cream and set with rennet.

Brick cheese. A high-moisture, American cheese which is similar to Limburger and Swiss.

bridal wreath. Branching francoa *q.v.*

bridalwreath spirea. *Spiraea prunifolia,* a shrub, family Rosaceae, growing up to 6 ft. tall; widely cultivated for its numerous, white flowers. Native to Asia.

bridge. 1. A part, usually 24 inches or less in width, which extends across a bulk milk tank. Usually in the center, it strengthens the tank and supports the agitator, gauge rods, etc. 2. To form an arch under pressure, as with fertilizer over a distributor in a spreader.

bridge graft(ing). A method of preserving trees which have suffered from winter injury, rodents, or disease. A split with loose ends is made in the bark above and below the wound. Then a cion longer than the affected spot is cut and inserted in the slit. The ends are finally covered with wax.

bridle. 1. The part of a harness which includes the bit, reins, and headstall. 2. A short rope with hooks on either end. 3. On logging sleds, a chain placed around the rear runners to control the speed.

bridle wise. Describing a horse accustomed to a bridle.

Brie. A soft, semi-ripened, French cheese, made from cow's whole, or skimmed milk, which is similar to Camembert, being ripened partly by surface-growing molds or bacteria.

brier (briar). 1. Any plant or bush bearing thorns or prickles on a woody stem, as *Rosa, Rubus, Smilax,* etc. 2. A thorn, twig, or branch of a *brier.* 3. *Erica arborea,* a shrub whose root is used in making pipes. Native to southern Europe.

bright. A term for a good grade of sorghum grain.

bright and shiny swords. Sheep shears (Australia and New Zealand).

bright bay. A horse with a light, glossy, tan coat.

bright sap. In lumbering, unstained sapwood. See *saps.*

bright tobacco. Cured light-yellow tobacco.

bright wool. Light-colored wool of light shrinkage, originating in the northern states east of the Mississippi River.

brimstone. See *sulfur.*

brindle. 1. An animal which is tawny, gray, or brown having irregular, dark streaks or spots. 2. Tobacco mosaic *q.v.*

brine. A salt solution of water used for food preservation, refrigeration, etc.

brine-pipe system. Refrigeration in cold-storage; cold brine is circulated in exposed pipes near the ceiling.

brine poisoning. Excessive consumption of salt by animals, resulting in severe vomiting, thirst, salivation, diarrhea, flatulence, dilation of the pupils, and hyper-sensitivity to touch. Convulsions and blindness occur in severe attacks. Also called *salt poisoning.*

brine-spray system. Refrigeration in cold-storage rooms; air is blown over coils, which are continuously sprayed with brine.

bring forth. To produce; to give birth.

Brioler. A German cheese, similar to Limburger, made from cow's whole milk curdled with rennet and cured for six weeks.

Brisbanebox tristania. *Tristania conferta,* a tree, family Myrtaceae, grown for its white flowers. Native to Australia; grown in California.

brisket. 1. In quadrupeds, the breast or lower chest between the forelegs. 2. The meat from these parts.

bristle. Stiff, sharp, hair or hairlike parts which grow on animals and plants.

bristly rose-slug. *Cladius isomerus* Nort., family Tenthredinidae, an insect pest of roses. The larvae, feeding on the undersides, skeletonize the leaves.

British bee. The brown bee *q.v.*

British red-water. A parasitic disease of cattle, caused by the protozoan, *Babesia bovis.* Characterized by fever, emaciation, diarrhea, anemia. It occurs in Europe, Africa, South America, and the East Indies. Also called *bovine hemoglobinuria.*

brittle. Describing crumbly fat particles in butter or meat which do not stick together. See *crumbly: 1.*

brittle bladderfern. *Cystopteris fragilis,* a cosmopolitan fern, family Polypodiaceae, of little significance except for the fern garden. Also called *brittlefern; bottle fern.*

brittle maidenhair. See *fan maidenhair.*

brittle thatchpalm. *Thrinax microcarpa,* family Palmaceae, a fan palm grown for ornament; its leaves are used in making hats and baskets. Native to Florida and the West Indies.

Brix. Balling *q.v.*

broadax(e). A wide-bladed ax used to cut logs or to make incisions in trees to obtain turpentine.

broad base terrace. An erosion prevention system for steep slopes. It consists of a horizontal ridge of earth 10 to 30 inches high, 15 to 30 feet in base width with gently sloping sides, a rounded crown, and a broad, shallow, water channel along the upper side. It may be level or have a slight fall toward one or both ends.

broadbean. *Vicia faba,* family Leguminosae, one of the most ancient cultivated food plants. It has a strong, upright stalk with small leaflets, and fruits in a pod, ½ to 1 ft. long, which contains edible seeds. Native to North Africa and southwestern Asia; seldom grown in the U.S. Also called *horsebean; Windsor bean.*

broadbean weevil. *Bruchus rufimanus* Boh., family Bruchidae, a beetle which attacks the broadbean, peas, and vetches in California.

broadbean wilt. A fungous disease of broadbean caused by *Fusarium oxysporum f. fabae,* family Tuberculariaceae; characterized by chlorosis, stunting, vascular discoloration, and wilt.

broad beech fern. *Dryopteris hexagonopteria,* a forest plant with nicely cut leaves and light yellow stripes. Native to North America.

Broad Breasted Bronze. A variety of the Bronze (*q.v.*) turkeys noted for its unusually meaty breast, thighs, and drum sticks.

broadcast. 1. To sow seed in all directions by scattering. 2. To scatter manure, lime, etc. 3. To plow, throwing the soil in one direction, so that the field is left level.

broadcast burning. Use of fire to destroy surface material. See *slash disposal.*

broadcaster. A device for scattering seeds, especially grasses, clovers. See *knapsack seeder.*

broadcast seeder. Broadcaster *q.v.*

broadcast seeding. Distributing seeds uniformly over an area, as contrasted with seeding in rows.

broadleaf. 1. Cigar tobaccos with broad leaves. 2. Tobacco leaf used for wrapping cigars. 3. A tree, or a plant, usually a weed, with wide leaves.

broadleaf filaree. Big heronbill *q.v.*

broadleaf ladypalm. *Rhapsis excelsa,* family Palmaceae, a palm, grown for decoration indoors and outdoors in very warm areas. Native to Asia. Also called *ground rattan.*

broadleaf uniola. *Uniola latifolia,* perennial grass, family Gramineae, growing 4 to 5 feet high; sometimes cultivated for its ornamental flower spikes. Native to North America.

broadleaved clover. Red clover *q.v.*

broadleaved endive. Escarole (*q.v.*); *cf. curled endive.*

broadleaved evergreen. Evergreen plants which are not coniferous, *e.g.,* the box, rhododendron, eucalyptus, wattles, etc.

broadleaved herb. Forb *q.v.*

broadleaved huckleberry. See *big whortleberry.*

broadleaved milkweed. Woollypod milkweed *q.v.*

broad-leaved plantain. See *rippleseed plantain.*

broad mite. *Hemitarsonemus latus* (Banks). Infests greenhouse plants, distorting the leaves and flowers.

broad-necked root borer. *Prionus laticollis* (Drury), family Cerambycidae, a grub which burrows in the roots of oaks, poplar, chestnut, pine, and grapes.

broads. In Canada, lumber 12 or more inches wide.

broadtail. Lambskin with a wavy pattern. Usually obtained from premature, or newly born, Karakul (*q.v.*) lambs.

broad-winged katydid. *Microcentrum rhombifolium* (Sauss.), family Tettigoniidae, a 1½ inch long grasshopper, widely distributed in the eastern U.S.

Broccio. A soft, highly perishable cheese made of fresh and cured whey.

broccoli. *Brassica oleracea italica,* an herb, family Cruciferae. The thickened flower branches are popular as a food. Also called *asparagus broccoli; sprouting broccoli; branching broccoli; Italian broccoli; calabrese.*

brod. A blacksmith's nail.

Brogdex process. Coating the skin of an apple with wax or oil to prevent apple scald *q.v.*

broiler. A chicken from 8 to 12 weeks old, weighing 2½ or more lbs., sufficiently tender to be broiled.

broiler mash. A complete mixed feed fed to broilers. Usually contains 20 percent or more protein with vitamins, minerals, and antibiotics added.

broke. Describing a gentled horse that can be harnessed or ridden (colloquial).

broken. 1. Describing land plowed for the first time. 2. Of plants or flowers, having irregular color distribution.

broken away udder. An udder which is abnormally pendulous, due to the stretching of tissues connecting it to the body. Common to cows with above-normal milk production.

broken crest. In stallions, a thick ridge of broken hair under the mane; may be a symptom of fistulous withers *q.v.*

broken down. 1. Describing a horse whose fetlock bends too near the ground. 2. Describing a horse whose toes turn up as a result of a tendon injury. 3. See *breakdown:1.*

broken flat. Designating land which has been plowed broadcast. See *broadcast:3.*

broken foot. A horse foot lacking part of its protective horn, which makes shoeing difficult.

broken knees. Of horses, scarred knees.

broken land. Terrain with numerous sharply cut valleys.

broken-mouthed. Designating an animal which has lost teeth. Also called *gummer.* See *smooth-mouthed sheep.*

broken rice. Rice kernels broken during milling.

broken tube retriever. In artificial insemination, a device, slightly larger than the insemination tube, which is slipped over a broken tube in removing the latter from a vagina.

broken wind. An incurable, respiratory disease of old horses, in which the air cells of the lungs are ruptured or dilated. The symptom is forced, double exhaling. Probably caused by hard exercise, or consumption of dusty feed. The disease is hereditary. Also called *heaves.*

broker. An agent entrusted to buy or sell products or commodities.

brokerage. The business of bringing buyers and sellers together and arranging contracts for a fee.

brome. Any plant of the genus Bromus, Family Gramineae. See *smooth brome; mountain brome; rescue brome; rattle brome.*

bromeliaceous. Having characteristics of the family Bromeliaceae, *e.g.,* a pineapple.

brome mosaic. A virus disease of smooth brome (*q.v.*) grass, caused by *Marmor graminis;* characterized by local lesions, yellow stripes on the foliage, and death of new growth. Occurs in Kansas and Nebraska.

brome pink snow mold. A late winter disease caused by *Fusarium nivale,* which destroys field and turf grasses. *e.g.,* cheatgrass brome. The dead leaves of affected plants form thin pink or straw-colored mats. Common on Pacific coast of northern U.S.

Brompton stock. See *stock:1.*

bronc. A wild or unbroken horse. Also called *bronco.*

bronc buster. 1. A trainer of wild horses. 2. In fairs and rodeos, a rider of a bucking horse.

bronchitis. Inflammation of mucous membranes in the bronchia, which are the tubes connecting the lungs to the windpipe. It affects horses, sheep, pigs, cattle, and domestic animals. The symptoms are hard, dry coughing, vomiting, rising temperature, nasal discharge, and difficulty of respiration. In horses, it sometimes results in broken wind *q.v.* Also called *infectious bronchitis; chick bronchitis; gaping disease.*

broncho-pneumonia. Inflammation of an animal's lungs, which usually occurs after another illness has become established. It is characterized by a bronchitis attack with severe rise in temperature; increased difficulty in breathing; short, moist cough; thick white catarrhal discharge from the nostrils; fast pulse; and general impression of severe illness. The appetite is lost, the head is extended, and depth of breathing is increased. In sheep and cattle, rumination ceases. In dogs, the lips are blown out and sucked in with each breath. The disease may be fatal within three or four days, but with proper treatment, most animals recover within two weeks.

bronchus. One of the forked branches of the trachea; a bronchial tube.

bronco. See *bronc.*

bronc spur. A pointed implement, secured to a cowboy's boot heel by strong straps. It is used to urge or goad a bucking horse; usually in rodeo arenas. Its short, dull rowels are bent to the inside of the boot heel, rather than projecting straight out, as in the ordinary spur.

Bronze. A popular breed of turkey with brilliant reddish-bronze plumage. It is the heaviest breed of turkeys; the adult tom may weigh more than 36 lbs., and the adult hen, as much as 20 lbs.

bronze birch borer. *Agrilus anxius* Gory, family Buprestidae, an insect which infests the white birch in northern U.S. The upper branches swell and turn brown. The *borer* burrows in the bark, making holes about ⅛ inch in diameter, leaving the tunnel tightly packed with sawdust. This pest has destroyed the white birch almost entirely in many areas.

bronze cane borer. See *rose stem girdler.*

bronzed cutworm. *Nephelodes emmedonia* (Cram.), family Phalaenidae, a cutworm (*q.v.*) which has dark brown and striped larvae. It infests corn, grains, and grasses.

brood. 1. The young of any animal. 2. Eggs, larvae, and pupae of the bee. 3. A single generation of insects. 4. To incubate eggs, or protect young from the weather. 5. Of, or pertaining to, an individual kept for breeding as a *brood* mare, *brood* sow, *brood* cow. 6. A hen inclined to set on eggs, as a *brood* hen.

brood bud. See *bulbel.*

brood capping. A porous substance with which honey bees seal the brood cells in a beehive.

brood cell. A cell in which the larvae of bees are reared.

brood chamber. That part of a hive in which the brood is reared. It includes one or more hive bodies and the combs within.

brood comb. A sheet of brood cells constructed in a wooden frame in a bee hive. Also called *brood nest.*

brooder house. A structure in which chicks or baby animals are raised without a mother. Also called *brooder.*

brooder pneumonia. See *aspergillosis.*

brood foundation. A base used to control the size of bee combs, consisting of a sheet of pure beeswax, embossed with the shape of cells. Bees follow this pattern to build up combs.

brood frame. Wooden frame containing the brood comb.

broodiness. Describing a hen which attempts constantly to sit on eggs.

brooding hen. A hen used to hatch eggs and rear young chicks.

brood mare. A mare reserved for breeding.

brood nest. Brood comb *q.v.*

brood sow. A sow reserved for breeding.

brood tree. A tree infested by insects, usually bark beetles.

brook euonymus. *Euonymus americanus,* a deciduous shrub, family Celastraceae, grown for its pinkish, warty fruit. Also called *burning bush; strawberry bush; skewerwood.*

brooklime. *Veronica baccabunga,* a species of speedwell (*q.v.*), with small blue flowers.

Brooks fruit spot. Phoma fruit spot of apple and quince blotch *q.v.*

broom. 1. A spineless shrub of the genus Cytisus, family Leguminosae, used for borders. Native to the eastern hemisphere. *Cf. Scotch broom; canary broom; butchers'-broom; weaversbroom.* 2. A sweeping brush made from broomcorn or other fibers. 3. Witches'-broom.

broomcorn. *Sorghum vulgare* var. *technicum,* a stiff grass, family Gramineae, used in making brooms and brushes. The mature seed is fed to poultry.

broomcorn millet. Proso *q.v.*

broomcrowberry. *Corema conradi,* a low evergreen shrub, family Empetraceae, grown in gardens. Native to North America.

broom deervetch. *Lotus scoparius,* family Leguminosae; a fair forage plant. Native to southern California. Also called *deerweed; tanglefoot.*

broomgrass. Broomsedge *q.v.*

broom gun. A device consisting of a spray rod three to five feet long with a shut-off valve at its hose end and a group of three to eight nozzles at the other. It produces a fog-like discharge. Also called *fog-driven gun.*

broom hickory. Pignut hickory *q.v.*

broomie. A boy who sweeps up locks, etc., at sheep shearing time (New Zealand).

broom millet. Proso *q.v.*

broom pine. Longleaf pine *q.v.*

broomrape. A root-parasitic herb of the genus Orobanche, family Orobanchaceae; subsists on herbaceous and woody plants. *Orobanche ramosa,* native to Eurasia, injures tobacco, tomato, muskmelon, etc.

broom rot. A disease of woody or herbaceous plants caused by the bacterium, *Phytomonas rhizogenes*; characterized by broomlike grouping of the fine roots at the end of a side root. *Cf. crown gall.*

broomsedge. A grass, *Andropogon virginicus,* family Gramineae, a widespread weed on poor land in southeastern U.S. It is an inferior livestock forage, but a covering of it protects poor soils from erosion. Also called *broom grass; yellowsedge bluestem.*

broom shoots. Willow shoots *q.v.*

broomtail. 1. Wild horse or a horse raised on the open range. 2. A pony with a short bushy tail (western U.S.).

broom tea-tree. *Leptospermum scoparium,* family

Myrtaceae, an evergreen shrub, grown for its myrtle-like leaves and many pink-to-red flowers. Native to Australia and New Zealand; grown in warm regions of the U.S. Also called *tea broom; manuka tea-tree.*

broomweed. *Scoparia dulcis,* a wild plant, family Scrophulariaceae, with small white flowers. Native to the tropical areas. Also called *sweet broomweed; sweet broomwort.*

brothers. A horse breeding term for colts of the same sire and mare.

brothers in blood. A horse breeding term for colts by the same sire out of full sisters, or by full brothers of the same mare.

brotherwort. 1. Mother-of-thyme *q.v.* 2. Pennyroyal mint *q.v.*

brow. A place where logs are collected for transportation.

browallia. Lovely browallia *q.v.*

brow band. That part of the bridle which goes around the brow of the horse's head.

brown apricot scale. European fruit lecanium *q.v.*

brown-banded roach. *Supella supellectilium* (Serv), family Blattidae, a small cockroach, less than ½ inch long, which infests houses. Found in northern and gulf coast states of U.S.

brown bark spots. A disease of pear and apple trees that flecks the bark with brown, circular spots: nonparasitic.

brown bee. The honey bee native to Europe; reputedly the first honey bee introduced into America in 1638.

brown bent. Velvet bentgrass *q.v.*

brown blight of lettuce. A soil-borne disease of lettuce, cause undetermined; characterized by small, light yellow spots in the leaves at the center of the plant. Then the spots become bigger, the leaves shrink, and the leaf area turns yellow-green, and the plant discolors. Finally the leaves turn brown and dry up.

brown blotch. Sooty blotch of clover *q.v.*

brown bush. Leatherleaf *q.v.*

brown canker of rose. A destructive fungous disease, caused by *Cryptosporella umbrina,* family Diaporthaceae. Lesions occur all over the plant; those on the cane are surrounded by a red-purple band; those girdling the stem kill everything above them. Lesions on the cane are raw-umber and have black spots; those on the leaves, petioles and stipules turn from purple to white or cinnamon. But those on petals turn cinnamon brown and have black spots in the center.

brown casse. The clouding and change of color of wine (red to brown, white to yellow) on exposure to air; caused by the enzyme oxidase in grapes. Common in France.

brown checked wood rot. A fungous disease of trees caused by *Polyporus sulphureus,* family Polyporaceae.

brown check of celery. A disease caused by boron deficiency and excess potash. Shallow, tan, greasy-looking spots, which appear on the leaf stalks, turn dark brown and crack horizontally. Variety Utah 10B is particularly susceptible to this disease. Also called *adaxial crack stem.*

brown chicken louse. *Goniodes dissimilis* Denny, family Linognathidae, a feather-chewing louse.

brown cress. Watercress *q.v.*

brown dog tick. *Rhipicephalus sanguineus* (Latr.), family Ixodidae, a hard-backed tick that infests dogs and ruminants; a carrier of piroplasmosis *q.v.* Also called *European dog tick; European brown tick.*

Brown Earth (Braunerde). One of the great soil groups of humid regions, which have a mull surface layer but no gray, highly leached or podzolized subsurface horizon. The use of the term is not rigidly restricted, being used interchangeably with Brown Forest soil (*q.v.*). It should not be confused with Brown soil *q.v.*

browned. Describing dry milk darkened with age. Also called *darkened.*

brown-eyed Susan. *Rudbeckia triloba,* a hardy, biennial herb, family Compositae, grown for its yellow flowers with brownish-black disk florets. Native to North America.

brown felt blight of conifers. A twig and leaf disease of the Englemann spruce, caused by the fungus *Herpotrichia nigra,* family Pseudosphaeriaceae.

brown felt blight of pine. A fungous disease caused by *Neopeckia coulteri,* family Pseudosphaeriaceae. Common in Europe.

Brown Forest soil. An intrazonal group of soils with some acidity, little iron and alumina, and some calcium in the soil colloids. The surface horizons are dark brown and relatively rich in humus (mull), but gradually shade into the lighter colored soil of the parent material. This soil group develops under deciduous forests in temperate humid regions.

brown fungus of citrus whitefly. *Aegerita webberi,* family Tuberculariaceae, destroys the citrus whitefly and other whiteflies in Florida, West Indies, and tropical regions of Asia.

brown grain sorghum. A market classification which includes all varieties of brown grain sorghum and very few other varieties.

brown grease. A grease obtained from inedible hog fat. Its oil is used in heavy lubricants, leather and illuminating oils, stearic acid, and in candle making.

brown hay. Hay discolored by fermentation, because of insufficient drying, which decreases its nutrient value.

brown-headed ash sawfly. *Tomostethus multicinctus* (Roh.), family Tenthredinidae, an insect pest which feeds on the leaves of red and white ash. Found in eastern and midwestern U.S.

brown heart. A common disease of beets and turnips due to a boron deficiency.

brown heart of pome fruits. A minor, nonparasitic disease, thought to be due to an accumulation of carbon dioxide. It causes the flesh to turn brown.

brown hickory. See *pignut hickory.*

brown horse. A blackish-bay horse with a black mane and tail.

browning. Abnormal browning of the foliage; when grape foliage is affected, it is called brunissure.

browning and stem break of flax. Caused by the fungus *Polyspora lini,* family Dematiaceae. Gray or brown circular flaws appear on infected seedlings; their stems break at the canker, droop, ripen early, and produce little seed. Found in Europe, New Zealand, Asia, and U.S.

browning of milk. Discoloration which occurs when milk is heated above 212° F.

browning root rot of cereals. A fungous disease of cereals caused by *Pythium arrhenomanes,* family Pythiaceae; characterized by rotting of the root. Common in central North America in marshy areas. The same organism attacks sugarcane in Louisiana and Hawaii.

brown lacewing. An insect, family Hemerobiidae, which preys on citrus mites.

brown leaf blight of grasses. Streak of grasses *q.v.*

brown leaf rust. A fungous disease of rye caused by *Puccinia dispersa,* family Pucciniaceae, a widespread, injurious rust of rye. Long, brown or reddish-brown, granular pustules burst through any part of a plant, but mostly on the stem and leaf sheath.

brown leaf spot of tobacco. A fungous disease occuring in periods of high temperature and humidity, which is caused by *Alternaria tenuis,* family Dematiaceae. Large, circular, brown spots appear on the mature leaves after harvesting.

brown lint. Brown fibers which occasionally appear on any variety of cotton.

brown patch. A fungous disease of grasses occuring during periods of high temperature and humidity, caused by *Pellicularia filamentosa,* family Thelephoraceae. Spots appear over the entire turf area; in time the turf turns brown and dies. *Cf. dollarspot.*

brown rat. *Rattus norvegicus,* family Muridae, the common rat. Also called *Norway rat; house rat; wharf rat; sewer rat. Cf. black rat.*

brown rice. Rice with the hull removed but not polished. It retains the bran layers normally removed by milling.

brown root rot of tobacco. Black root rot of tobacco *q.v.*

brown rot. A common disease of fruit, vegetable, and allied plants. The tissues turn brown and finally decay. See the following:

brown rot gummosis. See *foot rot of citrus.*

brown rot of cherry. See *brown rot of stone fruits.*

brown rot of citrus fruit. A fungous disease of citrus caused by *Phytophthora citrophthora, P. parasitica, P. syringae, P. boehmeriae, P. palmivora,* or *P. cactorum.* It turns the surface of the fruit purple or brown. The tissues become leathery and then soft after blue and green molds invade the fruit. In moist atmospheres, mycelium (*q.v.*) appears and spreads the decay rapidly.

brown rot of conifers. A fungous disease caused by *Trametes pini,* family Polyporaceae. Also called *pecky wood rot; root rot; ring shake; peckiness.*

brown rot of peach and plum. See *brown rot of stone fruits.*

brown rot of potato. A rot of potato stems and tubers, *Solanum tuberosum,* and of the peanut *Arachis hypogaea,* caused by the bacterial organism *Xanthomonas solanacearum.*

brown rot of stone fruits. A humid region disease of peaches, plums, prunes, cherries, almonds, and apricots, caused by *Monilinia fructicola* or *M. laxa,* family Sclerotiniaceae. Brown spots and spores appear on the fruit and rapidly destroy it. These spores also may blight the plants.

brown scab. Corky scab of potato *q.v.*

brown shorts. Wheat brown shorts *q.v.*

Browns lily. *Lilium brownii,* a perennial, bulbous herb, family Liliaceae, grown for its long, fragrant flowers, which are white inside and purple outside. Native to China.

brown slime flux. A foul-smelling seepage from cuts in shade trees, usually birch, elm, and maple of declining vigor. The clear sap, which contains nutrients, is fed upon by fungi, bacteria, and insects, and then turns brown and slimy.

Brown soil. Zonal soils having a brown surface horizon, which grades below into lighter colored soil and finally into a layer of carbonate accumulation. It develops under short grasses, bunch grasses, and shrubs in a temperate to cool, semiarid climate.

Browns peony. *Paeonia brownii,* family Ranunculaceae, a forage plant of secondary importance on the western ranges of the U.S.

brown spot. 1. A disease of corn caused by *Physoderma zeaemaydis* Shaw, family Physodermataceae; characterized by brown discolorations on the leaves and stalks. Also called *corn measles; corn pox; dropsy; spot disease.* 2. See *black rot.* 3. Fungous disease of pine needles caused by *Scirrhia acicola,* family Dothideaceae. Longleaf pine (*q.v.*) is especially susceptible to the disease.

brown spot of blue lupine. A fungous disease, in the form of black spots, caused by *Ceratophorum setosum,* family Dematiaceae; characterized by black spots on the plant, and defoliation in serious cases. The spots may form a cankerous girdle on the stem and kill the plant.

brown spot of celery. A fungous disease caused by *Cephalosporium apii,* family Moniliaceae, which attacks the parts aboveground, forming light or dark brown lesions. The lesions may eventually form a scurfy crack on the inner sides of the stalk and distort growth.

brown spot of corn. See *brown spot: 1.*

brown spot of rice. A fungous disease caused by *Helminthosporium oryzae,* family Dematiaceae, which kills the seedling plants, discolors leaves with reddish-brown spots, and rots the stalk at the base of the head. It pecks and specks the seeds, reducing the yield. Serious in the U.S. Also called *rotten neck of rice.*

brown spot of smooth brome. A common disease caused by *Helminthosporium bromi,* family Dematiaceae; characterized by small, dark brown, oblong spots on the leaves in spring. The spots coalesce into large, yellow areas, the leaves later turning brown from tip to base and dying. The disease attacks plants in U.S. in wet June weather.

brown spot of soybeans. A fungous disease caused by *Septoria glycines,* family Sphaeropsidaceae. It causes defoliation, especially in wet seasons, with eventual spotting of stems and pods. Widespread in Europe, Asia, and North America.

brown spot of sugarcane. A widespread, minor disease, caused by *Cercospora longipes,* family Dematiaceae. Narrow spots appear on the leaves, enlarging into ovals with straw-colored centers. The leaf blades die.

Browns rose. *Rosa brunoni,* a viny, perennial plant, family Rosaceae; grown in warm areas. Native to the Himalayan region of Asia. Also called *Himalayan musk rose.*

brown stain of oranges. A physiological breakdown, characterized by a superficial browning of the orange rind, which develops in storage when oranges are kept at 32° F. for more than two months.

brown stele. Strawberry red stele *q.v.*

brown stem. Rhizoctonia disease of potato *q.v.*

brown stem rot. A fungous disease of soybeans, caused by *Cephalosporium gregatum,* family Moniliaceae; characterized by a sudden blighting of leaves which resembles frost damage in fall. Leaf tissues become brown and the entire leaf soon withers. A badly infested field has a brownish appearance.

brown stripe of grasses. A fungous disease, caused by *Scoletotrichum graminis,* family Dematiaceae. The young leaves develop water-soaked, olive-gray lesions, which turn dull gray when dry, forming streaks. The lesions kill leaves slowly.

brown stripe rust of grasses. A fungous disease of wheatgrass, *Agropyron* spp., wildrye, *Elymus* spp., and coastrange melic, *Melica* spp., caused by *Puccinia montanensis,* family Pucciniaceae; characterized by long, narrow, reddish-brown stripes on the leaves.

Brown Swiss. Oldest of the purebred dairy cattle, very light to dark brown in color, with no off-color spots. The cows are strong and vigorous, but not coarse. Their minimum weight should be about 1400 lbs. while mature bulls weigh 2,000 lbs. or more. The breed originated in Switzerland, where it was raised for milk and meat, and used as a beast of burden. In U.S., it is strictly a dairy breed. Also called *Braunvieh.*

browntail moth. *Nygmia phaeorrhoea* (Donov.), family Lymantriidae, this moth is destructive to foliage of many trees. Its hair carries a poison which irritates the skin of humans.

brown ticked gray. The color of a horse with very small spots of brown scattered over a background of gray.

brown tip. A non-parasitic disease of lilies in which the leaf-tips turn brown and die back for a distance of ¼ to ½ inch.

brown top millet. *Panicum ramosum,* an annual herb, family Gramineae; grown for feeding game birds in southeastern U.S. Native to Asia.

browse. Shrubs or trees, whose sprouts, twigs, stems, vines, and leaves are cropped by livestock and wild life.

brucellosis. A highly contagious disease of cattle, goats, hogs, etc., caused by *Brucella abortus, B. melitensis,* and *B. suis.* The animal appears to be normal, but the disease brings about abortion, infertility, and low milk production. Easily communicable to humans. Also called *contagious abortion; Bang's disease of cattle; infectious abortion; epizootic abortion.*

brumby. A term for a wild horse in Australia. *Brumbies* are not native to the country, but are an escaped group of the imported domestic stock; found in thousands in Queensland, New South Wales, and South Australia.

Brumelia. A genus of the Sapodilla family. Shrubby plants of southeastern U.S. Also called *American ironwood.*

Brunigra. A term for the dark-colored Prairie group of soils of midwest U.S. See *prairie soils.*

Brunswick bird plague. Fowl plague *q.v.*

brush. 1. Shrub vegetation and trees that do not produce timber. 2. A revolving cylinder in a cotton gin, with bristles on its outer surface to remove cotton lint from the gin saw. 3. The lower part of the tail of the cow or horse. 4. The panicle and flower clusters of broomcorn. 5. The tuft of hair on the tip of a wheat kernel. 6. A device with set-in bristles for grooming the coat of an animal; used with a curry comb.

brush and bog plow. A heavy disk plow used to cut up vegetation and light brush.

brush bar. See *weed and brush bar.*

brush beef. Semi-wild cattle on brush country (southwest U.S.).

brush breaker. A large moldboard plow used for breaking land covered with brush.

brush burn. An area on which brush has been burned off.

brush control. The control, usually by herbicides, of herbaceous and low growing woody plants.

brush cover. 1. Small trees or bushes placed over seeded areas. 2. Small trees or brush growing on an area.

brush cutter. Weed and brush bar *q.v.*

brush drag. A makeshift harrow of short, tough branches attached to a log.

brushfield. Scrub vegetation and isolated clumps of trees; often left in the wake of a forest fire.

brush hand. A cowboy working in brush country of southwestern U.S.

brush harrow. Brush drag *q.v.*

brush hook. A heavy knife attached to a long handle or thick-bladed scythe; used to cut brush. Also called *bill; bush hook.*

brushing. 1. An injury to a horse in the region of the fetlock; caused by striking a fetlock with the opposite hoof while running. 2. Removing brush from an area.

brush killer. An herbicide for woody plants.

brushland. An area of little commercial value except perhaps for grazing goats and sheep.

brush matting. A covering of branches placed on badly eroded land to conserve moisture and reduce erosion while other vegetative cover is growing.

brush monkey. A lumber term for someone who cuts down small trees so that logs may be removed.

brush plow. Brush breaker *q.v.*

brush shredder. A machine for cutting up pruning-brush; used in orchards or along city streets.

brushy. Designating an abundance of brush and scrub vegetation.

Brussels sprouts. *Brassica oleracea* var. *gemmifera,* an annual or biennial herb, family Cruciferae, grown for its axillary buds which develop into small, edible heads. An important vegetable plant related to the cabbage.

brut. Describing dry wines.

bryology. That branch of botany dealing with moss and hepatica.

bryony. Redberry bryony *q.v.*

B. T. V. (Boynton's Tissue Vaccine). A vaccine used as a preventive against hog cholera.

B-type Merino. A variation of the *Merino* breed of sheep. It has a moderately wrinkled skin, including body folds. *Cf. A-type Merino; C-type Merino; Merino.*

bubble poppy. Cowbell silene *q.v.*

bubbly air cell. A defect of eggs; air bubbles form below the ruptured main air cell.

bubbly-jock. In Scotland, a male turkey.

bucare coralbean. *Erythrina poeppigiana,* a common coral tree, family Leguminosae; grown for its striking cinnabar-red flowers. Also used in the tropics to shade young coffee and cacao plants. Probably native to Peru; grown in Florida and California.

buccal capsule. An enlarged mouth cavity in some roundworms, which enables them to feed on tissues and blood of their hosts.

buccaneer palm. Sargent cherrypalm *q.v.*

buchu diosma. *Diosma ericoides,* an aromatic, small shrub, family Rutaceae; grown for its reddish flowers. Native to South Africa; grown in California and Florida. Also called *breath-of-heaven.*

buck. 1. A male goat, sheep (U.S.), rabbit, deer, or antelope. 2. Wool from rams. See *bucks.* 3. Buck rake *q.v.* 4. Sawbuck *q.v.* 5. To saw felled trees into logs. 6. The common beech (*q.v.*) or its nuts. 7. The quick leaps of a horse, with its back arched and head held low.

buckaroo. A cowboy (southwestern U.S.).

buckbean. Bogbean *q.v.*

buckbrush. See *Indiancurrent coralberry; western snowberry.*

buckbrush ceanothus. *Ceanothus cuneatus,* a shrub, family Rhamnaceae, which is a forage for livestock on the western ranges of the U.S. Also called *wedgeleaf ceanothus.*

bucked knees. Used to describe a horse or cow with knees bent inward.

bucket orchid. A plant of the genus Coryanthes, family Orchidaceae; it has bright yellow flowers. Native to Panama.

bucket pump. A lift pump clamped to a pail; used in spraying.

bucket sprayer. Bucket pump *q.v.*

bucket tree. A sugar maple producing sufficient sap to warrant hanging a bucket on it.

buckeye. Some trees or shrubs of genus Aesculus, family Hippocastanaceae. See *Ohio buckeye; yellow buckeye; red buckeye.*

buckeyed. Describing a horse's eye with a convex cornea, which appears to protrude beyond the eyelids. The horse is shortsighted and shies easily.

buckeye poisoning. Poisoning caused in livestock by eating plants belonging to the genus Aesculus. It causes inflammation of the mucous membranes, vomiting, depression, stupor, lack of coordination, twitching, and paralysis.

buckeye rot of tomato. A fungous disease, caused by *Phytophthora parasitica,* family Pythiaceae, (syn. *P. terrestris*); so named because of the speckled appearance of the fruit rot.

buck fat. Fat of a goat.

buck fleece. See *buck: 2.*

buck herd. A group of rams.

buckhorn plantain. *Plantago lanceolata,* family Plantaginaceae, a common, serious weed found in lawns and fields. Eaten with some relish by sheep. Native to Europe but widespread in the U.S. Also called *buckhorn; English plantain; narrow leaved plantain; ribwort; ripple-grass; rattail; ribgrass.*

bucking bronco. Usually an unbroken horse ridden at rodeos.

bucking chute. 1. In logging, a chute at a landing, in which logs are cut into equal lengths before loading. 2. An enclosure where broncos are saddled.

buck jacket. A water-proof canvas bag with an aluminum drain tied under the belly of a ram or buck to prevent breeding.

buck jumping. See *buck: 7.*

buck-knee. Bucked knees *q.v.*

buck moth. *Hemileuca maia* (Drury), family Saturniidae, a moth whose caterpillars infest oaks and willows in the autumn and feed voraciously on the leaves.

buckra. Denoting whiteness of certain vegetables.

buck rake. A rake with long, straight, horizontal teeth, for gathering dried hay and transporting it to a stack. Also called *bunching rake; sweep rake; bull rake.*

buckram. 1. Ramsons *q.v.* 2. Cuckoopint *q.v.*

bucks. The wool obtained from rams, distinguishable from ewe or wether wool by its odor and usually heavier shrinkage. It often sells at a lower price because of its shrinkage and coarseness.

bucksaw. A manual wood saw consisting of a narrow blade set in an H- or U-shaped frame.

buck-shinned. Describing a horse with a bulging front on the cannon (*q.v.*) bones.

buckshot soil. 1. Soils containing hard, round, iron and manganese concretions of buckshot size. Common in southern U.S. 2. Wet clay soil which upon drying disintegrates into aggregates about the size of buckshot. Descriptive of the soils of the Yazoo bottomland of the Mississippi River in southern U.S.

buckskin. 1. A disease of citrus, usually grapefruit and oranges, caused by the mite *Phyllocoptes oleivorous.* It is characterized by a crusted, leathery condition of the rind. 2. A term for a grayish-brown horse. 3. The skin of a buck deer. 4. A disease of sweet cherry, the symptoms of which vary according to the strain of virus and the stock. In general, the fruit loses its luster and drops or fails to ripen. Three recognized strains of the virus produce symptoms on peach and chokecherry similar to X-disease (of stone-fruits) *q.v.*

buckskin dun. See *dun.*

buckstrap. A saddle strap used for a hand-hold on a rearing horse.

buckthorn. 1. *Rhamnus cathartica,* family Rhamnaceae, a poisonous shrub once used for hedges but now escaped in fencerows and pastures in eastern U.S. Its bark, leaves, and berries contain purgative, but being bitter, are seldom eaten by livestock. Native to Eurasia. Also called *Hart's thorn; waythorn; Rhineberry.* 2. *Ceanothus sorediatus,* a thorny shrub, also called *jimbrush.* It is the alternate host for the fungus causing the crown rust of oats. 3. See *lotewood condalia.*

buckthorn aphid. *Aphis abbreviata* Patch, family Aphidae, a small insect which infests buckthorn (*Rhamnus*) and potato.

buckwheat. *Fagopyrum sagittatum,* family Polygonaceae, a fleshy, annual herb whose grain is sometimes erroneously considered a cereal. It is used in making griddle cakes and as a stock and poultry feed. It is harmful to unpigmented stock, which develop photosensitization (*q.v.*) after eating it. Native to Asia; a minor grain crop in U.S. Also called *fat hen; brank.*

buckwheater. An incompetent farmer; a novice (U.S.).

buckwheat feed. A by-product of buckwheat milling in which the middlings are mixed with hulls for livestock feed. It should not contain more than 30 percent crude fiber.

buckwheat hulls. The husks of buckwheat seed used as feed and bedding for livestock. They contain 29 percent protein and 7.4 percent fiber.

buckwheat middlings. Portions of buckwheat grain immediately inside the hull which are separated from the flour during milling. It is used with buckwheat hulls (*q.v.*) for dairy cattle feed, being about equal to corn gluten feed or brewer's dried grains in food value.

buckwheat straw. The harvested straw of the buckwheat; of little food value. It may cause digestive disturbances in livestock.

buck wool. Bucks *q.v.*

bucky. 1. Describing characteristic vigor, etc., in rams. 2. Pertaining to a tendency towards inelasticity in certain bread doughs during fermentation.

bud. 1. A protuberance, containing miniature leaves or flowers, located terminally or laterally on a stem. 2. An undeveloped shoot or stem. 3. To graft by inserting a bud in a slot in the stem of a different plant. It is economical as only a single bud, instead of a cion, is used. Also called *budding. Cf. shield budding; plate budding; patch budding; H-budding; June budding; prong budding.*

bud blast. 1. A disease of rhododendron and azalea caused by the fungus, *Sporocybe azalea,* family Dematiaceae; characterized by the buds turning dark brown and the external bracts becoming silvery, covered with black bristles. 2. In peony, a disease caused by *Botrytis paeoniae,* family Moniliaceae, which wilts and dries the small young buds. Larger buds turn brown and do not open; open flowers rot; stalks may turn brown or black. In damp weather, soft, gray-brown fungus spores cover the buds and may spot the leaves.

bud blight of soybean. A virus disease which curves, discolors, and dries bud tips. The leaf below the tip is flecked with rust, the plant is dwarfed and produces no seed. The pith in the stem may become brown. A later infection may produce underdeveloped or purple-blotched pods, which drop prematurely. The plant remains green after other plants have matured.

bud chip. A bud used in grafting.

bud cone. The cavity left in a stock when a shield bud is cut for grafting.

bud cutting. A method of quick propagation of plants. Parts of a stem are cut and split lengthwise so that each cutting has one bud. These are then buried under a shallow covering of soil to germinate. Also called *single eye cutting; leaf bud cutting.*

Budde's process. Preservation of milk by adding hydrogen peroxide and heating the mixture to 55° C.

Buddhist bauhinia. *Bauhinia variegata,* a small, tropical tree, family Leguminosae; grown for its purple flowers. Also called *orchid tree; mountain ebony.*

budding. See *bud:* 3.

budding cloth. A strong grade of muslin, saturated with a wax, which prevents buds from drying out after they have been grafted.

Buddleia. A genus of shrubs and small trees, family Loganiaceae; characterized by opposite-growing leaves and cup-shaped, yellow or violet flowers. The fruit is usually surrounded by dried flowers. Also called *butterflybush. Cf. orange-eye butterflybush; Farquhar; Lindley butterflybush; pale butterflybush.*

bud drop. The dropping of the buds from plants due to an unfavorable physiological condition, caused by excessive amounts of nitrogen in soil or fertilizer.

bud graft. The union of two different plants by inserting the bud of one into the stem of the other. *Cf. bud: 3.* See *shield budding; plate budding; prong budding; June budding; H-budding; patch budding; bud cutting.*

bud gum. The gluey exudate on a bud.

bud moth. A moth which, in its early stage, destroys buds on fruit trees.

bud mutation. A genetic change in the tissues of a shoot of a grafted bud. See *sport; bud variation.*

bud nematodes. Parasitic species of the genus Aphelenchoides, which infest the buds of strawberry (Fragaria), violet (Viola), African violet (Saintpaulia), etc., by sucking the tissue juices. Their larvae migrate to other buds or are washed by rain into the soil from where they infest other plants.

bud-parent tree. A tree from which buds are taken for propagation.

bud rot. Any disease characterized, in part, by decay of the buds.

bud rot of carnation. A fungous disease caused by *Fusarium poae,* family Tuberculariaceae, which arrests the development of the bud. The inner surface becomes discolored, rotted and moldy, and is infested by the grass mite, *Pediculopsis graminum,* which is a vector for the fungus.

bud rot of coconut. Caused by the fungus *Phytophthora palmivora,* family Pythiaceae, which decays the buds and the surrounding leaves, eventually destroying the whole tree. The buds may become infected by various bacteria, among which *Escherichia coli* (*Bacillus coli*) has been reported.

bud selection. Choosing buds from suitable plants or trees for grafting.

bud shield. See *shield budding.*

bud shoot wilt. In grafting, the wilting of shoots from inserted buds.

bud sport. See *sport.*

bud stick (budstick). A shoot whose buds are removed for propagation.

bud strain. In citrus fruit, a slight variation within the clone.

bud union. Section of a plant where a bud has been inserted.

bud variation. Abnormal variations in stems, leaves, or fruit which can be transmitted by asexual reproduction. See *bud mutation; sport.*

bud variety. Bud variation *q.v.*

budwood. Wood which bears buds desirable for grafting.

budworm. See *fall armyworm.*

buff. 1. A dull, golden brown color of animals and birds. 2. Hide of a cow, steer, or bull, not marked by a brand.

buffaloberry. A shrub of the genus Shepherdia, family Elaeagnaceae. See *silver buffaloberry; russet buffaloberry.*

buffalobur nightshade. *Solanum rostratum,* family Solanaceae; an annual herb commonly found in waste places with light, sandy soil. Its berries or herbage may poison animals. Native to North America. Also called *buffalobur; beaked nightshade; prickly nightshade; Texas thistle; Colorado bur; sand bur; yellowbloom sticker weed.*

buffalo cholera. A buffalo disease which may spread as an epidemic among domestic animals.

buffalo currant. See *clove currant.*

buffalo gnat. A small insect of the genus Simulium, found in the lower Mississippi valley. Innumerable bites may kill a domestic animal.

buffalo gourd. See *calabazilla.*

buffalograss. *Buchloë dactyloides,* family Gramineae, a short, valuable, native forage grass of the Great Plains of the West, occasionally used as a lawn grass in that area. Also called *early mesquite.*

buffalo treehopper. *Stictocephala bubalus* (F.), family Membracidae, an insect which lays eggs in the bark of fruit trees.

buffalo wallows. Shallow depressions lined with clay, on the Great Plains of U.S. Popularly thought to have been bathing holes for buffaloes.

buffer. 1. A substance in soil, such as organic matter or carbonate and phosphate compound, which preserves hydrogen-ion concentration and resists changes in pH value. 2. Animals which serve as food for predators, thus reducing danger to game. 3. A tool used to cut clinches from horseshoe nails before removing the shoe. Also called *clinch cutter.*

buffer capacity of soils. See *buffering action.*

buffer compound. See *buffer: 1.*

buffered. In plants or soils, a solution or system which requires more base or acid to neutralize it than the pH value indicates. See *lightly buffered.*

buffering action. Resistance to change in acidity or alkalinity. Also called *buffering.*

buffer species. See *buffer: 2.*

buffer strip. Rows of vegetation planted along contours to prevent erosion.

buffer zone. Public land surrounding and supplementing private range lands.

buffing. In horses, the striking of the inside of one hoof at the quarter with the opposite hoof. See *brushing.*

buffing boot. A protective covering strapped to a horse's fetlock to prevent buffing or brushing.

Buff Orpington. A variety of Orpington fowl. It is buff colored with red comb, face, and wattles. Raised for brown egg production and for meat. The cock may weigh 10 lbs. and the hen 8 lbs.

bug. 1. Any insect, specifically of the orders Hemiptera and Heteroptera. 2. A flaw in construction and operation of machinery.

bugbane. A plant of the genus Cimicifuga, family Ranunculaceae. See *cohosh bugbane; cordate bugbane; Kamchatka bugbane.*

buggy. In U.S., a light, four-wheeled, horse-drawn carriage. In U.K., the *buggy* usually has two wheels.

buginwillaea. See *bougainvillea.*

bugle land. In southeastern Georgia and northern Florida, low flatlands with wet, gray soil. Vegetation is limited chiefly to the pitcherplant.

buglelily. An herb of the genus Watsonia, family Iridaceae. Native to South Africa. See *taperpetal buglelily; irisleaf buglelily; rosy buglelily.*

bugleweed. *Lycopus virginicus,* family Labiatae, a perennial American herb, sometimes used as a remedy for lung hemorrhage.

bugloss. 1. A plant of the genus Anchusa, family Boraginaceae. See *Italian bugloss; early bugloss; cape bugloss; wild bugloss.* 2. *Anchusa officinalis,* a biennial or perennial herb, family Boraginaceae, grown for its blue, purple, or flesh-colored flowers. Native to Europe.

bug-tree. A tree infested with bark beetles *q.v.*

buhrstone. Any porous, siliceous rock, usually coarse sandstone, which has the required qualities for use as a millstone. Also called *burrstone, burrh-stone.*

build a plant. To develop an extensive root system and foliage in a young plant, so that it can withstand the removal of flower stems.

built-up litter system. See *deep-litter system.*

bulan. An organic insecticide, the nitro-alkyl analogue of DDT; namely, 1,1-bis-(p-chlorophenyl)-2 nitrobutane. Found to be more effective than DDT against certain insects.

bulb. 1. A subterranean bud in some plants, which has a short stem bearing overlapping, membrane-like leaf bases, as in onions and tulips. It functions as a storage of food for reproduction and represents the inactive stage of the plant. 2. Any plant or flower shaped like a *bulb.* 3. The upper part of a horse's heel.

bulb buttercup. *Ranunculus bulbosus,* family Ranunculaceae, a perennial herb grown for its yellow flowers. Native to Europe; a naturalized weed in the U.S. Also called *eggs-and-butter; bulbous buttercup; buttercup.*

bulb cellar. An insulated, subterranean enclosure for storing bulbs.

bulb crop. 1. A market term for vegetables such as onions, leeks, garlic, shallots, cibals (Welsh onion), chives, etc. 2. In floriculture, the blossoms and bulbs of lilies, narcissi, tulips, hyacinths, etc.

bulb disease. See *bulb nematode.*

bulbel. A small bulb produced above the ground, among the flowers or in the axil of a leaf. It is capable of reproducing the plant. Also called *bulbil; brood bud. Cf. bulb: 1.*

bulbil. Bulbel *q.v.*

bulbil lily. *Lilium bulbiferum,* a perennial bulbous herb, family Liliaceae, grown for its wide, orange-red, dark-spotted flowers with yellow centers. Native to Europe.

bulblet. A small bulb borne in the scale of a parent bulb. As the scale becomes thin, the bulblet may become attached to the parent bulb. See *bulbel.*

bulb mite. *Rhizoglyphus echinopus* (F. & R.), family Acaridae, a tiny insect that attacks many kinds of bulbous plants, giving them a sickly appearance. The leaves and the flowers become stunted and misshapen.

bulb nematode. *Ditylenchus dipsaci,* a species of parasitic worms, family Tylenchidae, which infest the living tissues of bulb-plants, *e.g.,* hemp (Cannabis), teasel (Dipsacus), etc. They make their way through the cell walls by stout stylets and at various places produce larvae. *Nematodes* of various ages thus may be found in the spaces between the scales of bulbs, in buds, and on various other fleshy, host tissues. See *nematode.*

bulbous begonia. *Begonia socotrana,* a plant of the family Begoniaceae. It has nearly round, half foot wide leaves and pink flowers in clusters. Native to the island of Socotra in the Arabian Sea.

bulbous bluegrass. *Poa bulbosa,* family Gramineae, a perennial grass with a bulb-like growth; occasionally found in pastures. Native to Europe. Also called *winter bluegrass.*

bulbous buttercup. Bulb buttercup *q.v.*

bulbous chervil. Turniproot chervil *q.v.*

bulbous iris. Species of *iris* (*q.v.*) that have a rootlike bulb, *e.g.,* the netted iris, the Spanish iris.

bulbous melic. Oniongrass *q.v.*

bulb pan. A clay pot about half the depth of a regular flower pot; used for shallow-rooted plants, bulbs, or seedlings. It is 5 to 12 inches in diameter.

bulb panicum. *Panicum bulbosum,* family Gramineae, a grass common in the southwest U.S.; relished by most livestock. Also called *bulbous panic.*

bulb rot. Any decay of bulbs, corms, etc.

bulb scale mite. *Steneotarsonemus laticeps* (Halb.), family Tarsonemidae, a pest of bulbous plants, which causes yellow-brown, scar-like streaks on the bulbs.

bulb waterhemlock. *Cicuta bulbifera,* a perennial herb, family Umbelliferae, which causes stock poisoning, the symptoms of which are nausea, difficult respiration, dilated pupils, frothing, weak and rapid pulse, and convulsions. Native to North America.

Bulgarian buttermilk. A fermented milk produced by growing a culture of *Lactobacillus bulgaricus* (Bulgarian bacteria) in sterile milk at a comparatively high temperature. It is generally thick with a distinctly acid taste. Also called *Bulgarian milk; Bulgarian sour milk; Bulgaricus milk; yoghurt.*

bulk comb foundation. A thick base for bee hives, 4½ by 16½ inches, which has heavier walls than an ordinary comb foundation. It is used for bulk production of honey. Also called *cut comb foundation.*

bulk comb honey. Honey produced from large combs, as compared to section comb honey.

bulk comb honey super. A detachable section, placed on top of a beehive, used for honey storage in bulk production.

bulk density. Mass per unit bulk volume of soil that has been dried to a constant weight at 105° C.

bulk milk cooler. A sanitary container for cooling

and storing milk or milk products, usually at a temperature between 33 and 40° F.

bulk (bulk down) of broomcorn. To remove cleansed brush from slats in the curing shed and place it in a compact pile.

bulk pruning. The removal, with few cuts, of large limbs from a tree. *Cf. fine pruning.*

bulk tank cooler. Bulk milk cooler *q.v.*

bulk volume. The volume of an arbitrary soil mass including the volume of the solid particles and the pores (interstices, voids).

bull. 1. An uncastrated male bovine. 2. A market term for a male bovine at sexual maturity. 3. The male elk, moose, etc. 4. The central bar of a harrow. 5. To be in heat; to desire the *bull* (used for a cow).

bullace grape. *Vitis rotundifolia,* commonly known as *muscadine grape.* A woody vine bearing edible, musky grapes; the principal species are cultivated in the warm areas of U.S. Native to North America. Also called *bullet grape; bull grape; everbearing grape; southern fox grape; berry grape.* See *scuppernong.*

bullace plum. *Prunus domestica insititia,* family Rosaceae, a deciduous tree, grown for its edible fruit, the *Damson plum.* Native to Eurasia. Also called *damask plum.*

bullamacow. Beef; beef cattle (S. Pacific islands).

bull bay. See *redbay persea; southern magnolia.*

bull beef. A federal meat grading for beef from a bull's carcass.

bull burn. See *granular vaginitis.*

bull cane. A very large cane on a vine.

bull chain. 1. A heavy chain to which short chains with hooks are attached for pulling logs. 2. On horse-drawn binders, a heavy drive chain which transmits power from the bull (main drive) wheel.

bull controller. A halter with a metal face plate to protect the eyes and forehead of a bull.

bull ditcher. A very heavy middlebuster (*q.v.*) plow.

bulldog. 1. A steel clip fitted into the nostrils of a bull, used to control him. Also called *bull holder.* 2. To throw a steer or cow to the ground by grasping its horns and twisting its neck (U.S.).

bull-dog calf. Achondroplasia *q.v.*

bull donkey. A large auxiliary steam engine with a drum and cable, used for pulling logs.

bulldozer. A crawler tractor with a blade mounted in front to move earth, snow, gravel, and underbrush.

buller. A cow in heat for a long time; a cow difficult to get with calf.

bullet grape. Bullace grape *q.v.*

bull exerciser. A revolving device with long booms, to which bulls are tied for exercising.

bull grape. Bullace grape *q.v.*

bullhead. A rose, with short or excessive petals, which fails to open properly because of thrips, mildew, or excessive shoots.

bull holder. See *bulldog.*

bulling heifer. A heifer in heat and ready to be bred.

Bulli soil. A black Australian soil used to surface turf cricket pitches. Obtained in the area of Bulli, New South Wales.

bull leader. A long rod used to control a bull. It is fastened to a ring in the animal's nose. Also called *bull staff.*

bull marker. A metal identification tag fitted around the horns of a bull and chained to the ring in his nose.

bull nettle. See *Carolina horsenettle; risky tread-softly.*

bull nose. In pigs, a swelling of the face and snout, caused by an infected wound. Also called *chronic rhinitis; necrotic rhinitis; sniffles.*

bull nose punch. A tool used to pierce the septum of a bull's nose and insert a ring.

bull nurse. In southwestern U.S., a man who tends cattle during shipment.

bullock. A castrated bull. See *ox.*

bullocksheart custardapple. *Annona reticulata,* family Annonaceae, a small tree with edible, heart-shaped fruit, grown in southern Florida. Native to West Indies. Also called *corazon; mamon; anonas.*

bull pen. 1. A small wood or metal enclosure for confining bulls. 2. The sale ring at a horse auction. 3. Living quarters for a group of men. 4. Storage room for machinery.

bull pine. See *ponderosa pine.*

bull rake. Buck rake *q.v.*

bull ring. A metal ring fastened in an animal's nose; used to control him.

bulls. Inferior bull hides.

bull seg. In Scotland and northern England, a castrated bull. See *stag: 1.*

bull's-eye rot. Apple anthracnose *q.v.*

bull's-eye spot of maples. A fungous disease of sycamore, red, sugar, and silver maples, caused by *Cristulariella depraedens,* family Moniliaceae. It appears as concentric rings on the leaves.

bull snake. A harmless snake of the genus Pituophis, which eats rodents. Also called *gopher snake; pine snake.*

bull staff. Bull leader *q.v.*

bull stag. See *stag: 1.*

bull thistle. *Cirsium vulgare,* family Compositae, a thistle with large heads and prickly leaves. In the U.S., it is a common weed; incorrectly called *Cirsium lanceolatura.* Native to Europe. Also called *down thistle; pasture thistle; button thistle.*

bull tie. A harness or halter made of chain and heavy leather straps.

bull-tongue. 1. A plow or cultivator with one shovel. 2. A narrow cultivator shovel which throws the soil to both sides.

bull wheel. 1. A large wheel which, while in motion, accumulates power from contact with the ground and transmits it to other parts of the machine by a belt or chain. 2. The main drive wheel of a machine.

bully. Beef preserved by pickling or canning.

bulrush. Any plant of the genus *Scirpus,* family Cyperaceae. Some species are grown as ornamental plants; *S. fluviatilis* and *S. mucronatus* are weed pests in rice fields. See *weeping bulrush; tabernaemontanus bulrush.*

bumalda spirea. *Spirea bumalda,* a hybrid shrub, family Rosaceae, grown for its pink flowers.

bumblebee. An insect of the genus Bombus; valued for flower pollination and honey, which it often stores in underground nests. Its colonies are small.

bumble flower beetle. *Euphoria inda* (L.), family Scarabaeidae, a beetle which infests the ears of ripening corn, causing damage similar to that of squirrels.

bumblefoot. In poultry, an abscess between the toes, caused by vitamin A deficiency or penetration of a hard substance such as glass or corn. The resulting lameness is long-lasting, as pus is dry and cannot be discharged.

bummer. A small two-wheeled dolly or truck used in transporting logs. Also called *drag car; self-loading skidder.*

bumpy ash. Bunji-bunji *q.v.*

bunch. 1. A cluster of plants or fruits, as a *bunch* of grapes. 2. In pigs, an improperly healed wound following castration. 3. To skid logs together for hauling. 4. To pile harvested crops in the field.

bunch bean. Kidney bean *q.v.*

bunchberry dogwood. *Cornus canadensis,* family Cornaceae, a shrub grown for its beautiful, white flowers usually in wild gardens. Native to North America and Asia. Also called *crackerberry.*

bunch disease of pecan. A virus disease which produces large broom-like formations on the lateral branches of the tree. Clusters of sprouts may also develop directly from the trunk or main limbs. The tree may not die, but it is often necessary to destroy it.

bunched up herding. Herding of sheep on the range by keeping them close together. Also called *close herding. Cf. open herding.*

bunchflower. *Melanthium virginicum,* family Liliaceae, a stout plant, native to U.S., grown in bog gardens for its greenish-white flowers. Escaped, it is poisonous to livestock. Symptoms are rapid and weak heartbeat, labored, difficult breathing, muscular weakness, stupor, and lack of appetite.

bunch grape. Summer grape *q.v.*

bunch grass. Grasses which, lacking stolons, form thick bunches: *e.g.*, sacaton, dropseed, feather grass, fescue, wheat grass *q.q.v.*

bunching attachment. On a mowing machine or harvester, a hood arrangement attached to a number of finger-like bars which extend to the rear of the cutter bar. It serves to bunch accumulated hay.

bunching machine. A device for tying asparagus in bunches.

bunching rake. Buck rake *q.v.*

bunch peanut. A type of peanut whose plant is bushy and erect; largely grown for forage. *Cf. runner peanut.*

bunch pink. See *sweetwilliam.*

bunchy top. A virus disease of banana, plantain, and Manila hemp which stunts the plant and bunches the leaves. Such plants fail to bear fruit.

bunding. A centuries-old soil erosion control method in India; building a low ridge of soil along the contour of a field's lower edge.

bundle browning. A disease of plants manifested by the browning of fibrovascular fascicles in stems or tubers.

bundle carrier. An attachment to a grain binder for accumulating and dumping bundles of harvested grain.

bundle corn. See *shock corn.*

bundle feed. Bundles of sorghum or corn, stored as stockfeed, usually without separating the grain from the fodder.

bung. Sausage casing.

Bunge desert candle. *Eremurus bungei,* family Liliaceae, a perennial herb, grown for its yellow or orange flowers. Native to Iran.

bunji-bunji. *Flindersia schottiana,* family Rutaceae, a tree with pinnate leaves and clusters of small, showy flowers growing in terminal panicles. Its bark is poisonous. Native to Australia. Also called *cudgerie; bumpy ash.*

bunk. 1. A crossbeam of heavy timber which supports logs during transportation. Also called *bolster.* 2. A car or truck used to transport logs. 3. A sleeping space in a lumber camp or a ranch.

bunker silo. A horizontal silo consisting of two, vertical side walls usually 8 to 10 feet high, 18 feet apart and as long as desired. Adjustable self-feeder sections may be used at either end.

bunk fence. A fence made of heavy logs.

bunkhouse. A building where rangehands or workers sleep.

bunt ball. An ovary of wheat containing the bunt fungus instead of the grain.

bunted. See *nematode disease of wheat.*

bunt of wheat. A fungous disease due to infection at germination by parasites *Tilletia caries* and *T. foetida.* The stricken plant and its head remain undersized. The glumes tend to be loose, the head becomes bluish green and never regains the normal color of ripened wheat. Individual grains in the head are filled with an oily black powder, mainly smut spores and, if broken, give an offensive odor: hence the name *stinking smut.* In bearded varieties, awns are often shed. Also called *high smut; low smut; pepper brand; smut ball.*

bunt smut. Bunt of wheat *q.v.*

bunya-bunya araucaria. *Araucaria bidwilii,* family Pinaceae, a tall, coniferous tree, endemic in Australia. Its tasty, roasted seeds are an important food for the aborigines. Grown in Florida and California as an ornamental. Also called *bunya.*

bur. 1. The rough, sticky, or prickly envelope of any fruit. 2. A mature hull on the cotton boll. 3. The cone on the hop plant at flowering. 4. A weed with burs. 5. To remove burs, as in wool cleansing.

bur artichoke. The flower head of artichoke; used as a vegetable along with its oval bracts.

bur bristlegrass. Hooked bristlegrass *q.v.*

bur cleaner. A revolving, carpet-covered roller on which peas and other small vegetables are placed for removing burs. The bristles stick to the roller and are later removed.

bur clover. See *California bur clover; spotted medic.*

burdekin plum. *Pleiogynium solandri* (*Spondias s.*), family Anacardiaceae, an evergreen tree with clusters of greenish flowers. It bears an edible, plum-like, red fruit. Native to Australia; grown in Florida and California.

burdekin vine. *Vitis opaca,* a trailing vine whose tubers are edible. Native to Australia.

Burdizzo emasculatome. Burdizzo forceps *q.v.*

Burdizzo forceps. A pincers-like instrument with wide jaws which is used for bloodless castration of animals with a pendant scrotum. Also used for bloodless killing of poultry and for docking lamb tails.

burdock. Plants of the genus Arctium, family Compositae, all of which are weeds. See *great burdock; smaller burdock; cotton burdock.*

burgeon. 1. To bring forth; to sprout buds. 2. To grow forth, as an appendage or a disease in an animal. 3. A young bud.

bur gherkin. West Indian gherkin *q.v.*

burglar. A horse, good in appearance, but with a slight defect that can be remedied temporarily when offered for sale. Also called *robber.*

bur grass. Mat sandbur *q.v.*

Burgundy. 1. A soft, white loaf-shaped cheese. Also called *Fromage de Bourgogne.* 2. A red wine from France.

Burgundy mixture. A fungicide which consists of water, blue vitriol, and washing soda. *Cf. Bordeaux mixture.*

burgundy trefoil. Alfalfa *q.v.*

buried soil. The soil of an original land surface which is buried to considerable depths by subsequent geologic deposition. Such soils have been observed in sections of thick alluvial deposits; soils of interglacial periods *e.g.,* found in glacial deposits.

buried soil profile. A buried soil recognizable by the succession of soil horizons as where deposits of a new soil overlie and bury an older soil profile with A or both A and B horizons represented. It is often found beneath Indian mounds and other artificial fills, and where valley bottom soils have been covered by an immature soil which consists of wash from the valley sides.

buripalm. *Corypha utan,* family Palmaceae, a large palm which yields a fiber called eruc (*q.v.*). Native to tropical Asia. Also called *talipot palm.*

burl. 1. A knot-like growth on the trunks of trees or plants. It is often produced with adventitious buds. 2. A distorted grain in lumber surrounding the pits of undeveloped buds. Considered valuable, if of sufficient size.

burlap. A coarse cloth made of jute, flax, etc., used for bags, bales, bundles, and for covering the roots of plants. See *balled and burlapped.*

Burley. A strain of tobacco, very light-colored when cured. Grown in Kentucky and Tennessee.

burn. 1. An area in which the trees and/or grass have been destroyed by fire. 2. A branding iron. 3. In plants, to give a scorched appearance by excessive manuring. 4. To blister the skin of an animal by friction with a rope or hobble, etc. 5. To become withered due to lack of moisture, as the foliage of a tree. 6. To become discolored because of insufficient drying, as in lumber.

burn beat. To cut and burn the rough turf on fallow land to prepare it for tillage.

burner. A flame thrower consisting of a tank, full of liquid fuel, and a pressure pump; used for destroying plant life.

burnet. Any perennial herb of the genus Sanguisorba, family Rosaceae. See *small burnet.*

Burnet rose. Scotch rose *q.v.*

burning bush. See *euonymus; Belvedere summer-cypress; gasplant dittany.*

burning index meter. A device for measuring inflammability and rate of spread of fire. It works by integrating the combined effects of the moisture content of fuel, herbaceous stage, wind velocity, and other factors.

burning period. Peak period of a forest or prairie fire.

burn of potato. A disease caused by heat, intense light, and a lack of moisture. The tips of potato leaflets become wilted, turn yellow, and later brown. The browning kills the tip tissues and spreads further in the vascular system of the leaf, destroying its various portions. *Cf. tip burn; hopper burn.*

burn out. To erase the brand on an animal by burning the hide.

burnt lime. Calcium oxide, a germicide. It is a white alkali which reacts with water to produce great heat. Also called *quicklime; live lime; caustic lime; unslaked lime. Cf. hydrated lime.*

burnt sole. Burns or blisters caused on the sensitive part of a horse's foot by placing a hot shoe on it.

burnt taste. Defective flavor of dairy products, caused by overheating or intense sunlight.

bur nut. Any wild plant of the genus Tribulus. See *puncturevine.*

bur oak. *Quercus macrocarpa,* family Fagaceae, a very hardy, massive oak, with its acorns nearly covered by big, fringed, scaly cups. Its wood is pest-free and very durable, and is commonly used for railroad ties, agricultural implements, furniture, etc. Also called *mossy-cup oak.*

burr. 1. Burl *q.v.* 2. Glanders *q.v.* 3. A seed capsule with hooks or spines.

burrawang. *Macrozamia spiralis,* family Cycadaceae, a small palm-like plant with stiff, pinnate leaves. Native to Australia, where its nuts are relished by the aborigines.

burr comb. Brace comb *q.v.*

burrel. 1. A variety of pear with soft, tasty flesh. 2. Blue colored wild sheep found in the Himalayas.

burr grinder. A feed grinder with flat, roughened, iron plates. It is not used for crushing feed that has an excessively high fiber content.

burr knot of pome fruits. Crown gall.

burro. A donkey; ass.

burroweed. *Suaeda fruticosa* (*moquini, Dondia f.*), family Chenopodiaceae, a common weed which has lowered the grazing capacity of large areas in western U.S.

burrowing carrion beetle. *Necrophorus* spp., family Silphidae, an insect whose larvae may infest young birds.

burrowing nematode. *Radopholus similis,* family Tylenchidae, a worm which attacks citrus, banana, sweet potato, sugarcane, edible canna, etc.

Burrows and Quinn method. A method of collecting semen for artificial insemination among birds; inducing a seminal discharge by massaging the soft part of the bird's abdomen.

burr stock. Low grade wool.

burry. Describing wool with excessive quantities of burs, birdseed, chaff, etc., in the fiber.

bur sage. Any shrub of the genus Franseria, family Compositae, the characteristic vegetation of the southwestern deserts in U.S.; often a cause of hay fever.

bursattae. A skin disease of horses caused by larvae of the large stomach worms, *Draschia megastoma.* The larvae migrate into pre-existing wounds or sores on the head and front of the body, making them larger and pussy. The afflicted areas become chronic and difficult to heal. Endemic in wet season in East Indies. Also called *summer sores; jack sores.*

burster. 1. A flower blossom, especially in carnations, in which the petals are increased far beyond the natural number. 2. Buster: 1.

bush. 1. A grove of sugar maple trees. See *sugar bush.* 2. A forest wilderness or vast scrubland (Canada, Alaska, New Zealand, and Australia; formerly used in northern U.S. also). 3. A shrub. 4. A method of training plants and fruit trees to assume a desired shape by means of stakes, ties, wires, and other supports. 5. The brush of an animal; the tail. 6. To force a seller to accept a lower price for a horse than was bid in the auction ring.

bush bean. Beans cultivated for food which grow in bush form and do not send out climbing runners. See *kidney bean.*

bush cinquefoil. *Potentilla fruticosa,* a small shrub, family Rosaceae, grown for its clusters of bright yellow flowers. Native to the cooler parts of northern Eurasia and North America. Also called *shrub cinquefoil.*

bush clover. Roundhead lespedeza *q.v.*

bush disease. A livestock disease caused by food deficiency. Its symptoms are a stiff gait, soft bones, and anemia (New Zealand and Australia). Also called *creeps.*

bushel. 1. A dry measure of 32 quarts: four pecks. 2. A weight of a product assumed to be the equivalent of a *bushel:* beans, 60 lbs.; barley, 48 lbs.; ear corn, 70 lbs.; seed corn, 56 lbs.; flaxseed, 56 lbs.; oats, 32 lbs.; peas, 60 lbs.; rye, 56 lbs.; wheat, 60 lbs. 3. Any basket, tub, or other vessel having the capacity of a bushel. See *imperial bushel.*

busher. One who clears the ground of underbrush, fallen trees, etc.

bush fallow. An Appalachian region term for growing a poor crop on cleared, worn out or eroded land before allowing wild vegetation to cover it.

bush fruit. Fruits produced on shrubs or bushes, as contrasted to those grown on trees and vines; *e.g.,* the blackberry, raspberry, gooseberry and currant.

bush honeysuckle. Any shrub-like plant of the genus Lonicera, family Caprifoliaceae. See *Tartarian honeysuckle; European fly honeysuckle.*

bush hook. Brush hook *q.v.*

bush lima. Any of the lima beans (*q.v.*) that grow as a shrub, as contrasted to those with climbing runners.

bushmint. Hoary rosemarymint *q.v.*

bush muhly. *Muhlenbergia porteri,* family Gramineae, a grass, with good forage rating, used in reseeding grazing lands. Native to southwestern U.S. Also called *mesquite grass.*

bush nasturtium. *Tropaeolum minus,* family Tropaeolaceae, a soft-stemmed herb whose flowers are used as a garnish; stems and leaves in salads, and seeds as a condiment. Native to South America. Also called *dwarf nasturtium.*

bush pea. Soft thermopsis *q.v.*

bushpoppy. See *stiff bushpoppy.*

bush pumpkin. *Cucurbita pepo* var. *melopepo,* family Cucurbitaceae, a summer maturing variety which includes the types: Cheyenne bush, pattypan, scallop, and summer crookneck squash. The plants are compact and have no tendrils.

bush redpepper. *Capsicum frutescens,* a tropical American herb, family Solanaceae, used as a salad plant and in condiments. Also called *chili pepper.* See *cherry redpepper; red cluster pepper; sweetbell redpepper; long redpepper.*

bush sickness. In New Zealand, a cobalt deficiency or nutritional anemia of ruminants. See *cobalt deficiency; salt sick; nutritional anemia.*

bush squash. Bush pumpkin *q.v.*

bushwhacker. 1. A hook for cutting brush. 2. A backwoodsman or frontiersman.

buster. 1. One who trains animals for saddle or work. 2. Middlebuster *q.v.*

butcher. 1. A person who slaughters animals or dresses meat. 2. One who sells meat. 3. Animals which are to be killed for their meat: heifers, cows, steers, or bulls suitable to be sold in carcass as block beef. 4. To slaughter or dress animals.

butchering tool. A knife, cleaver or saw used in slaughtering, skinning, or dressing an animal.

butcher's-broom. *Ruscus aculeatus,* a low, stiff evergreen shrub, family Liliaceae, grown for its foliage. It is often dried and colored for winter decorations. Native to Eurasia.

butcher's round. A wholesale cut of meat; about 20 percent of the hind quarter.

butcher tie. A method of trussing fowl. A string is drawn across the shoulders, over the wings and drumsticks and tied at the base of the tail.
butoxypolypropylene glycol. An insect repellent, which repels stable and horn flies.
butt. 1. The thick base or lower end of a tree or plant. 2. A stump, especially a walnut stump. 3. In butchering, the upper half of a ham or shoulder. 4. In Australia, a package of wool containing 196 lbs. greasy or 112 lbs. scoured, with a tare weight of more than 11 lbs. 5. To strike with the head or horns, as do cattle, goats, or rams.
butt adjuster. The vibrator or moving belt of a grain binder; it evens stalks and moves grain to the packers.
butt cut. 1. The first log to be cut above the stump. 2. Bark removed from the butt of a tree before it is felled; specifically for tanbark. Also called *butt log.*
butte. A high, flat-topped hill with steeply sloping sides, similar to but smaller than a mesa. Common in the arid and semi-arid regions of western U.S.
butt end sirloin. The first cut of steak taken from the rump end of the hip loin.
butter. 1. Solidified milk or cream fat to which salt and artificial coloring may be added. In commerce, non-fat constituents are limited to 20 percent by weight. 2. Butt adjuster *q.v.*
butter-and-eggs. See *butter-and-eggs toadflax.*
butter-and-eggs toadflax. *Linaria vulgaris,* a common perennial, family Scrophulariaceae, cultivated in the U.S. for its yellow and orange flowers. It grows well in poor soils and is a weed pest, when escaped. Native to Europe. Also called *ramstead; ramsted; eggs-and-bacon; butter-and-eggs; eggs-and-butter; toadflax; flaxweed.*
butter bean. 1. Small-seeded lima bean (*q.v.*) found in southern U.S. 2. Wax bean *q.v.*
butter board. Butt adjuster *q.v.*
butterbur. Any plants of the genus Petasites, family Compositae, used to cover unsightly or stony banks where other plants do not grow well. See *sweet butterbur.*
butterbush. *Pittosporum phillyraeoides,* family Pittosporaceae, a small Australian laurel tree with drooping branches and yellow flowers.
butter culture. Lactic acid bacterial culture used in butter manufacturing to produce a desirable aroma.
buttercup. A weed of the genus Ranunculus, family Ranunculaceae, which grows in wet soil. When dried, it may be fed to livestock, but if eaten green, it causes severe inflammation of the intestines. Some species are cultivated as garden plants; *Ranunculus asiaticus* is grown indoors. Also called *crowfoot buttercress; butterdaisy.* See *aconite buttercup; tall buttercup; Persian buttercup; bulb buttercup; creeping buttercup.*
buttercup comb. In poultry, a cup-shaped crest with regular serrations on the edges.
buttercup oxalis. *Oxalis cernua,* family Oxalidaceae, a perennial, bulbous herb grown for its showy yellow flowers. Native to South Africa; grown in Bermuda and Florida. Also called *Bermuda buttercup.*
buttercup winterhazel. *Corylopsis pauciflora,* a deciduous, Japanese shrub, family Hamamelidaceae, grown for its yellow flower clusters.
butterene. Oleomargarine to which a small amount of butter has been added.
butterfat. Pure milk fat; a mixture of butyrin, olein, palmitic acid and other glycerides *q.v.* It is the main constituent of butter.
butterfly. 1. An insect of the order Lepidoptera; noted for its bright colors. 2. A retail meat cut; the halves remain connected on one side and resemble *butterfly* wings when laid flat.
butterfly bauhinia. *Bauhinia monandra,* a tropical tree, family Leguminosae, grown for its foliage and pink flowers streaked with purple. Native to South America. Also called *Jerusalem date; butterfly flower.*
butterfly bush. See *Buddleia.*
butterfly gardenia. See *crapejasmine tabernaemontana.*
butterfly milkweed. *Asclepias tuberosa,* a perennial herb, family Asclepiadaceae, grown in sandy soil for its bright orange flower clusters; poisonous and unpalatable. Native to North America. Also called *butterfly weed; pleurisy root.*
butterfly orchid. *Oncidium papilio,* family Orchidaceae, a tropical epiphytic orchid. Its small flowers are yellow with brown spots. Cultivated in greenhouses in U.S. Native to Venezuela.
butterfly pea. See *Atlantic pigeonwings.*
butterflyvine. See *fringed Amazonvine.*
butterfly weed. See *butterfly milkweed.*
butter granule plug. Clustering of granules in defective flake buttermilk.
butteris. A chisel-like instrument used to pare horse's hoofs.
buttermilk. The fluid that remains when fat is removed from milk or cream in making butter. It contains at least 8.5 percent non-fat milk solids. *Cf. cultured buttermilk.* Also called *churnmilk; churned buttermilk.*
buttermilk cheese. A dairy product resembling cottage cheese, made from buttermilk curd.
buttermilk flavor. An undesirable taste in butter caused by highly acid or over-ripe cream, over-ripe culture, or buttermilk. Also called *acidy flavor.*
buttermilk horse. A roan with a coat color between red and blue.
buttermilk land. Marsh land.
buttermilk powder. A stock feed of dried buttermilk. See *semi-solid buttermilk.*
butternut. 1. *Juglans cinerea,* family Juglandaceae, a tall, deciduous, nut tree. The husk of the fruit is used to dye cloth yellow; the nut meat is oily, rich, and spicy. The wood is used for furniture, cabinet work, interior trim, and novelties. Found from New Brunswick to Georgia and west to the Dakotas and south to Arkansas. Also called *white walnut.* 2. In early settlement

period of eastern U.S., a backwoodsman who wore clothing dyed with butternut bark.

butter oil. Melted, clarified butterfat used in making process butter. It contains some water and non-fat solids and can be shipped without refrigeration.

butter rose. See *tall buttercup.*

butter trier. A long, two-edged, curved knife used to sample butter. *Cf. cheese trier; trier.*

butterweed. See *groundsel: 2.*

butterworker. A machine which kneads and adds salt to butter after it is churned.

butterwort. 1. Any insectivorous herb of the genus Pinguicula, family Lentibulariaceae, grown in rock gardens. The flowers are white, purple, or yellow; the leaves are covered with a sticky secretion which holds insects. There are several species, both European and American: mountain butterwort (*P. alpina*), tall butterwort (*P. elatior*), bigflower butterwort (*P. grandiflora*), yellow butterwort (*P. lutea*), etc. 2. A small herb, *Pinguicula vulgaris,* with violet or purple flowers, grown in rock gardens. Also called *eccle grass; earning grass; bog violet.*

buttery body. Poor texture in ice cream caused by incomplete homogenization, dull freezer blades, frozen ingredients, temperature extremes, high acidity, or excessive fat content.

butter yellow. Yellow coal-tar dyes used to color butter.

butt-headed. Hornless.

butt-off. 1. To square an end of a log. 2. To remove a defective piece of wood from the end of a log.

button. 1. An irregularly shaped berry. 2. A bud. 3. A round, seed vessel. 4. Any stunted or immature fruit. 5. A round, firm, cheesy curd of condensed milk; a defect of the body of the milk. 6. A nipple, especially of a hog. 7. A partially dismantled queen bee cell in a bee hive which resembles a small acorn cup. 8. Onion set. 9. A stunted or immature horn growth, as on a cow. 10. A metal clasp used to connect sections of a check row wire. 11. Cartilage on the chine bone of cattle. 12. Any shell-like bone construction of the body. Also called *concha.* 13. A leather ring for adjusting a horse's bridle.

buttonball. See *American planetree.*

buttonbush. *Cephalanthus occidentalis,* a hardy shrub, family Rubiaceae, grown for its small, white, tubular, flowers. Its leaves are poisonous when eaten by livestock. Native to North America. Also called *button willow.*

button clover. *Medicago orbicularis,* family Leguminosae, a forage plant introduced into the U.S. from Europe. Also called *button medic.*

button disease. A tropical, parasitic, fungous disease of chicks resulting in buttonlike excrescences at the corners of the mouth.

button flower. See *twining button flower.*

button pink. *Dianthus latifolius,* a cultigen, a perennial herb, family Caryophyllaceae, grown for its rose-pink and dark red flowers.

button rot. Shallow surface pitting of the potato, caused by the death and desiccation of areas of surface tissue. It may be followed by Fusarium (*q.v.*) decay.

button snakeroot eryngo. *Eryngium aquaticum,* family Umbelliferae, a stout, perennial herb cultivated for rock gardens in the wet soil of eastern U.S. Also called *rattlesnake master.* Some species of Eryngium are called *seaholly.*

buttonweed. See *chingma abutilon.*

button willow. See *buttonbush.*

buttonwood. See *American planetree.*

buttress. 1. The rear points on a horse's hoof. 2. The *buttress*-like growths at the base of large trees which give support in heavy gales. 3. The promontory of a hill or mountain.

butynorate. The chemical compound, di-n-butyrin dilaurate; used to treat certain intestinal parasites in fowls.

butyric acid. The normal *butyric acid,* $C_3H_7CO_2H$, is a colorless, fatty acid with a rancid smell, found in butter and other oils, especially cod-liver oil. Its second modification, *isobutyric acid,* $(CH_3)_2CHCO_2H$, which occurs free in some oils, resembles the normal acid.

butyrin. Any of the compounds derived from glycerine and butyric acid, especially tributyrin contained in the butterfat of milk and butter to the extent of about four percent.

butyrinase. An enzyme which can split butyrin into glycerol and butyric acid.

buzzard-wing sweep. A cotton cultivator point with very wide wings.

B. W. D. Bacillary white diarrhea. See *pullorum disease.*

"B" white grease. A marketing term for inedible hog fat in which there is less than 3.5 percent of free fatty acid.

by. Begotten by a male; sired. *Cf. out of.*

by-bidding. Bidding at an auction on behalf of the auctioneer or owner to raise the price.

bybloemen tulip. A class or type of tulip which has many varieties. The flowers are white with rose or violet-colored streaks.

by-product lime. See *waste lime.*

Byrsonima. 1. A genus of tropical trees, family Malpighiaceae, characterized by yellowish flowers. Native to America. 2. *Brysonima crassifolia,* a tropical American tree whose bark is used in tanning.

C

caballada. A herd of horses or mules (southwestern U.S.).

caballeria. A unit of land measurement in Spanish-speaking countries; equal to 33.2 U.S. acres in Cuba, 111.82 in Costa Rica, 111.51 in Guatemala, 111.13 in Honduras, 105.75 in Mexico, 112.41 in Nicaragua, 194.1 in Puerto Rico, and 95.48 in Spain.

caballo de oro. Palomino *q.v.*

cabbage. 1. *Brassica oleracea* var. *capitata,* family Cruciferae, a widely cultivated plant with a mass of enlarged leaves packed together around a short stem, forming a large bud-like structure, the *cabbage* head, an extremely valuable vegetable. Developed from an ancient coastal weed of Europe, *B. oleracea. Cf. Brussels sprouts; kale; collard; cauliflower; broccoli; kohlrabi.* 2. A cheese whose holes are so large and numerous that the cheese between is often paper thin. Its cross-section has a *cabbage*-like appearance. Also called *overset; blow holes.*

cabbage angelintree. *Andira inermis,* family Leguminosae, a tropical tree grown for its clusters of purple flowers. Native to America.

cabbage aphid. *Brevicoryne brassicae* (L), family Aphidae, a louse which infests cabbage, cauliflower, Brussels sprouts, kohlrabi, kale, turnips, and radishes, often covering the plants with clusters of whitish-green insects. It attacks from seed time through all the stages of growth.

cabbage black rot. Black rot of crucifers *q.v.*

cabbage butterfly. See *imported cabbage worm.*

cabbage caterpillar. See *imported cabbage worm.*

cabbage curculio. *Ceutorhynchus rapae* Gyll, family Curculionidae, a little weevil which gouges out the leaves and stems of crucifers.

cabbage downy mildew. See *downy mildew of crucifers.*

cabbage flea. Any of the numerous winged insects that infest cabbage.

cabbage fly. Adult stage of the *cabbage maggot q.v.*

cabbage hammock. A patch of land in marsh or pine forest containing a grove of the cabbage palmetto, *Sabal palmetto* (southeastern U.S.).

cabbage looper. *Trichoplusia ni* (Hbn.), family Phalaenidae, a larva which infests the leaves of crucifers, lettuce, spinach, beet, pea, celery, parsley, potato, tomato, carnation, nasturtium, and mignonette.

cabbage maggot. *Hylemya brassicae* (Bouche), family Muscidae, a larva which infests the roots of cabbage, cauliflower, broccoli, Brussels sprouts, radish, turnip, beets, cress, and celery. The adult, a two-winged fly, lays the eggs near the plant stem. After hatching, the larvae promptly eat into the roots, honeycombing them with tunnels, and thus rotting them. Also called *cabbage root maggot.*

cabbage mosaic. Diseases of the crucifers caused by different viruses, singly or in combination. The foliage becomes mottled, and there may be chlorosis along the veins, which results in defoliation and stunting of the plants. The viruses are transmitted by the cabbage aphid *q.v.*

cabbage palm. See *Carib royal palm.*

cabbage palmetto. *Sabal palmetto,* family Palmaceae, a palm tree grown as an ornamental in warm coastal areas of North America.

cabbage root maggot. Cabbage maggot *q.v.*

cabbage rose. *Rosa centifolia,* family Rosaceae, a hardy shrub grown for its large, fragrant, pink flowers.

cabbage seedpod weevil. *Ceutorhynchus assimilis* (Payk.), a pest of crucifers which lays its eggs in the seedpods.

cabbage seedstalk curculio. *Ceutorhynchus quadrideus* (Panz.). See *cabbage curculio.*

cabbage tree. See *giant dracena.*

cabbage webworm. *Hellula rogatalis* (Hulst), family Pyraustidae, a common caterpillar which feeds on the leaves of crucifers.

cabbage yellows. A fungous disease of crucifers caused by *Fusarium oxysporum f. conglutinans.* The leaves and the vascular system of the plants turn yellow and then brown. The stems become warped and the growth of the plant is retarded. The symptoms closely resemble those of black rot. Also called *cabbage wilt.* See *Fusarium.*

cab-horse disease. In old horses, new growth of bone on the long pastern bone of the front leg which results in a swelling and some lameness.

cable exhaust box. A sterilizing chamber for cans and their contents by heating them before sealing.

cable grader. Rope grader *q.v.*

cable out. To remove brush by attaching a chain or cable to two crawler tractors which, moving in the same direction, drag out the brush and underbrush.

cable stacker. An overhead cable supported by two sets of poles and used with a carrier to stack hay.

cableway. A method of skidding logs in the forest. It consists of a cable suspended between two trees on which runs a lead line to carry or drag the logs.

cabling. Supporting tree branches by means of flexible cables in the crown of a tree. Cables to provide additional support are also placed between branches with a common weak crotch.

cablish. Brushwood; or wood blown off by the wind.

cabresto. A lariat or halter of horse hair.

cacanapa. Nopal *q.v.*

cacao. *Theobroma cacao,* family Sterculiaceae, a tree from which the cocoa of commerce is harvested. Its yellow flowers grow on the trunk and larger branches of the tree and produce seed-containing oval pods several inches long. The fat extracted from these seeds is the cacao-butter and the remainder of the seed, dried and ground, is the cocoa. The seeds are also used to make chocolate. Native to America; cultivated extensively in Asia and Africa.

Cacciocavallo. An Italian hard cheese especially suitable for warm climates. It is made of cow's milk with rennet added, and is cured for three months.

Cacio Fiore. A soft, yellowish, Italian cheese made from ewe's or goat's milk. The milk is colored with saffron to which vegetable rennet mixture is added. It is then cured for about 10 days.

cactus. Any desert plant of the family Cactaceae. Mostly spiny, some are regarded as ornamentals and others as food for range livestock. Native to America, except the genus Rhipsalis (Hatiora), which is sometimes found in Madagascar.

cactus boomer. Wild brush cattle (southwestern U.S.).

cactus dahlia. *Dahlia juarezi,* family Compositae, a perennial, tuberous-rooted herb grown for its rose or lilac flowers; its rays have incurved margins.

cactus scale. *Diaspis echinocacti* (Bouche), family Diaspididae, a common armored insect which infests a wide variety of greenhouse plants.

cadastral surveys. Surveys which relate to land boundaries and subdivisions, and are made to create or define limitations of titles or ownership.

cade. *Juniperus oxycedrus,* family Pinaceae, a species of juniper which yields a thick, dark oil, called *cade* oil, used in veterinary surgery. Native to Europe. Also called *prickly juniper.*

cadelle. *Tenebroides mauritanicus* (L.), family Ostomatidae, a serious pest of stored grain and packaged food. It is sometimes a predator on other insects.

cadence. The rhythm of a horse's gait.

cadet. A young man learning sheep ranching (New Zealand).

caducous. Describing an early dropping off of parts of the plant, such as leaves. *Cf. deciduous; fugacious.*

caecum. In men and animals, the closed sac which is the beginning of the large intestine. Also called *cecum.*

Caerphilly. A Welsh, semi-soft, granular cheese made of cow's milk inoculated with a starter and ripened slightly. It is cured for three weeks and eaten soon afterwards.

Caesarean section. See *Cesarean birth.*

cage. A metal or wooden enclosure used to confine birds or animals, or to study the effects of grazing on plants by restricting the animals to a certain area.

cahiota. Chayote *q.v.*

cainito starapple. *Chrysophyllum cainito,* a small tree, family Sapotaceae, grown in warm areas for its showy, leathery, green leaves and its apple-shaped, greenish-purple, edible fruit. Native to the West Indies.

caja. See *mombin.*

caja manga. See *mombin.*

cajeput tree. *Melaleuca leucadendron,* family Myrtaceae, a large tree grown for ornament in the extreme southern parts of the U.S. and in southern California, in almost any type of soil for its creamy-white flowers. It is especially a good bee plant. Native to Australia. Also called *punk tree; paper bush.*

cajeta. A Mexican, sweetened, condensed milk produced by boiling milk and sugar together in open kettles.

cake. A residue from the pressed kernels of cottonseed, linseed, or soybean, etc., from which the oil has been removed. It is pressed into hard cake-like masses and is used either as a feed or a feed supplement.

cake breaker. A machine in a cottonseed oil mill which breaks cottonseed cakes into small pieces for grinding.

caked. 1. Solidified into a hard or dry mass. 2. A defect of dried whole milk in which the milk has lost its powdery consistency and has become a solid mass. Even though the solid mass may be broken, it never returns to its original state. It is a result of exposure to moisture and a crystallization of the lactose. 3. An udder condition. See *caked udder.*

caked hay. Hay that has been baled before it is sufficiently dried. When the bale is cut the sections do not spring apart. It is deficient in food value.

caked udder. 1. Physiological edema of the udder which is most often associated with recent freshening *q.v.* 2. A doughy, swollen udder.

cake poisoning. See *cottonseed meal poisoning.*

calabash gourd. *Lagenaria vulgaris,* a tropical vine, family Cucurbitaceae, planted widely because of its fruit which is known as the *dipper gourd, calabash pipe,* etc. The fruit is variable in size ranging from three inches to three feet in length and has a round, flattened, crooked neck, or dumbbell-shape with a hard rind. Used in making the necks of tobacco pipes. Also called *calabaya.*

calabash tree. *Crescentia cujete,* a tree, family Bignoniaceae, grown for its large, hard-rinded fruit whose shells or husks are used for dippers, utensils, etc. Native to tropical America.

calabazilla. *Cucurbita foetidissima,* an annual, trailing, herbaceous vine, family Cucurbitaceae, grown in warm areas for its fruits which when dried are used as ornamental gourds. Its roots were once used as a cleanser and its leaves medicinally. Native to North America. Also called *Arizona gourd; buffalo gourd; Missouri gourd; wild pumpkin; calibasa; calabazza; calabaza.*

calabrese. Broccoli *q.v.*

caladium. 1. Any tuberous-rooted herb, of the genus Caladium, family Araceae. Species are grown for their brightly colored leaves. See *mottled caladium.* 2. *Caladium bicolor,* a tropical American herb grown for its varied colored, large leaves.

cala ham. See *picnic ham.*
calamondin. *Citrus mitis,* family Rutaceae, a small, acidic orange which is widely cultivated in the Philippines, Hawaii, and the continental U.S. It produces an intensely acid fruit. Native to China. Also called *Panama orange.*
calandria. A common heating device used in vacuum evaporating of liquids. The liquid to be evaporated is circulated in pipes surrounded by steam.
Calcagno. A hard, Sicilian cheese made of ewe's milk; suitable for grating.
calcareous soil. Soil containing sufficient calcium carbonate (often with magnesium carbonate) to effervesce visibly when tested with cold, hydrochloric acid. A soil may be designated *calcareous* if a high content of lime is present at a shallow depth, but absent in the surface layer.
calciferol. Vitamin D_2: $C_{28}H_{43}OH$.
calciferous. Producing, or containing carbonate of lime.
calciferous glands. Glands that produce or secrete carbonate of lime (calcium carbonate, $CaCO_3$). Such glands are found near the esophagus of earthworms and in other lower animals.
calcify. To deposit or secrete lime salts which harden, as when gristle (cartilage) becomes bone.
calcimorphic soil. Suborder of intrazonal soils characterized by the high content of calcium available to plants (sometimes in the form of soft calcium carbonate) in the parent material, such as Brown Forest and Rendzina soils. Also called *calomorphic.*
calcined phosphate. Treated phosphate rock which has been heated under proper temperature with an alkali salt or in an atmosphere of steam so that the fluorine is expelled. It is suitable as an animal feed or phosphatic fertilizer. *Cf. fused phosphate.*
calcite. A mineral, $CaCO_3$, calcium carbonate, in crystalline form. Sometimes used as a diluent for insecticidal dusts.
calcium. Ca, a chemical element present in variable amounts in all soils. It is essential for plant and animal growth and is the principal mineral element in bones.
calcium arsenate. $Ca(AsO_4)_2$, an insecticide used in the control of the cotton boll weevil and cotton leafworm.
calcium carbonate. $CaCO_3$, a common substance, valuable for neutralizing acidity in soils and mixed fertilizers. Also used in livestock feeds as a source of calcium. Also called *agricultural lime.* See *liming; limestone; marl.*
calcium chloride. $CaCl_2$, a chemical used as a bleaching agent and for controlling dust.
calcium cyanide. $Ca(CN)_2$, a very effective fumigant for controlling insects and rodents.
calcium deficiency. A physiological condition of plants or animals which results from insufficient calcium in the body. In plants, new, fibrous roots die within four weeks, then the terminal bud, and ultimately the plant. A deficiency of vitamin D in birds and mammals makes it impossible for them to utilize the calcium in their diets. In such cases, hens may lay normal eggs at the expense of calcium needed for bone development, resulting in crooked breast bones. In animals it produces rickets, abnormal and fragile bones, and low blood calcium.
calcium hydroxide. $Ca(OH)_2$, a calcium compound, useful in raising the pH of soils. It is of greater neutralizing power than ground limestone. Also called *slaked lime; hydrated lime.*
calcium hydroxy-apatite. See *apatite.*
calcium hydroxy-phosphate. See *apatite.*
calcium metabolism. The physical and chemical processes by which calcium is utilized by the cells.
calcium meta-phosphate. $Ca(PO_3)_2$, a carrier of phosphorus for use in commercial fertilizers.
calcium nitrate. $Ca(NO_3)_2$, the calcium salt of nitric acid sometimes used as a fertilizer.
calcium oxide. CaO, a salt produced by heating limestone, marble, etc. In the hydrated form, it is used as a dust diluent, calcium carrier, and for raising the pH of soil. Also known under many other names. See *burnt lime.*
calcium phosphate. Tricalcium phosphate *q.v.*
calcium sulfate. $CaSO_4$, hydrated sulfate of calcium, commonly known as gypsum. Used for supplying calcium and correcting the alkaline content of soils. Also called *agricultural gypsum; land plaster.*
calculi. Stones which may form inside the bodies of animals due to stagnant fluids and other secretions.
calendula. See *potmarigold calendula.*
calf. 1. The young of certain large mammals. Of the bovine animals, specifically those more than 3 months old but sexually immature. 2. The hide of such a *calf.*
calf crop. Calves produced by a herd of cattle in one season.
calf (or lamb) crop percent. The ratio of calves or lambs, usually at weaning time, to the cows or ewes in the herd.
calf diphtheria. A herd infection of half-grown calves caused by the necrosis germ, *Spherophorus necrophorus,* (formerly *Actinomyces necrophorus*); characterized by difficulty in eating, drinking, and breathing. The mouth and throat are affected with ulcers and diphtheritic membranes to a varying degree. Also called *calf diphtheroid.*
calf feeder. A pail with a tube and a rubber nipple through which a calf sucks milk.
calf herd. A bull and 2 heifers, all under 1 year of age, bred by an exhibitor.
calfkill. See *lambkill kalmia.*
calf kneed. Describing a horse whose knees incline backward.
calf meal. A feed for calves to replace milk. It consists of ground cereal with the germ and inner kernel removed, or a mixture of cereal and by-products, usually fortified with growth-promoting substances. A calf starter *q.v.*

calf pellet. A small pellet made of compressed calf meal; used in feeding calves.

calf pneumonia. An infectious respiratory disease of calves, caused by faulty sanitation, inadequate ventilation, overcrowding, or debility from other ailments, such as diarrhea. Its symptoms are high temperature, rough coat, poor appetite, weakness, and coughing when the chest is tapped or compressed by the hand.

calf roping. A rodeo sport in which a rider lassoes a running calf, dismounts, throws the calf to the ground, and ties its feet together. The sport derives from old round-up practices.

calf scours. An acute, infectious disease, cyclic in occurrence, which reaches panzootic proportions for calves less than 10 days old. There is marked depression accompanied by severe diarrhea. The most lethal form appears at birth or 6 to 72 hours after birth. In these cases there may be no diarrhea; the calf is found in a cold, weak, and dying condition. Also called *white scours; infectious diarrhea; calf septicemia; 3-day calf disease.*

calf skin. A sub-class of packer hides and country hides. Used in making gloves, shoes, and in bookbinding, etc.

calf starter. Feeds used to replace milk in early weaning. See *calf meal.*

calf vaccination. Any vaccination, but especially one with a strain of low virulence of *Brucella abortus* to protect young females from infection of brucellosis (*q.v.*); injected between four and eight months of age.

calf weaner. Any device placed on the muzzle of a calf to prevent it from nursing and so wean it from its mother.

calibration. The process of determining gradations, and marking them on a gauge, *e.g.*, the volumetric *calibration* of a bulk milk tank.

calibration chart. A graph prepared for each bulk milk tank listing volumetric (gals.) and/or a gravimetric (lbs.) equivalent for each graduation on the gauge rod or surface gauge.

caliche. 1. In arid or semiarid regions, layers on or near the surface of carbonates of calcium or magnesium precipitated from the soil solution. Found in Arizona. Also called *croute calcaire.* It is not a geologic deposit. 2. Alluvium which is cemented with sodium nitrate, chloride, and other soluble salts in the nitrate deposits of Chile and Peru.

calico. 1. (a) Any cotton cloth from India. (b) Coarse printed cotton cloth from U.S. (c) Plain white cotton cloth from U.K. 2. A spotted horse. 3. A virus disease of potato which is characterized by irregular blotches of various shades on the leaflets. It is one of the potato mosaics. 4. Tobacco mosaic *q.v.*

calico back. See *harlequin bug.*

calico bush. See *mountain laurel.*

calico dutchmanspipe. *Aristolochia elegans,* a tropical vine, family Aristolochiaceae; widely grown for its purple and white flowers. Native to South America. Also called *calico flower.*

calico scale. *Lecanium cerasorum* Ckll., a scale insect which infests cherry, elm, maple, pear, Virginia creeper, Persian walnut in California.

California aralia. *Aralia californica,* a bush-like, herbaceous perennial, family Araliaceae; grown for its whitish or greenish flowers. Native to North America. Also called *elk clover.*

California barberry. See *cluster mahonia.*

California bay tree. See *California laurel.*

California beet disease. See *curly top.*

California black oak. *Quercus kelloggii,* family Fagaceae, a browse tree on the western range of U.S.

California black walnut. See *Hinds walnut.*

California blight. See *Coryneum blight of stone fruits.*

California bluebell. See *harebell phacelia.*

California blue oak. *Quercus douglasii,* family Fagaceae, an acorn-bearing tree with whitish bark and light colored timber.

California brome. See *mountain brome.*

California buckthorn. *Rhamnus californica,* a shrub, family Rhamnaceae, which is planted as a bee tree in California. Found in western U.S.

California bur clover. *Medicago hispida,* family Leguminosae. A Eurasian forage which is good grazing throughout most of the year in southwestern U.S. It may also be used as a cover crop and for green manure. See *spotted medic.* Also called *bur clover.*

California dandelion. See *spotted catsear.*

California danthonia. *Danthonia californica,* a perennial grass, family Gramineae, used for reseeding western ranges of U.S. Also called *California oatgrass.*

California false hellebore. *Veratrum californicum,* a hardy, perennial herb, family Liliaceae. The nectar and pollen of this shrub are poisonous to bees. Also called *western false hellebore.*

California fan palm. See *Mexican Washington-palm.*

California fawnlily. *Erythronium californicum,* a bulbous herb, family Liliaceae; grown in wild gardens for its leaves, which are mottled with brown or white spots, and for its white flowers. Native to North America. Also called *fawnlily.*

California flu. See *Newcastle disease.*

California fremontia. *Fremontia californica,* family Sterculiaceae, an evergreen shrub up to 10 feet high, or small tree up to 20 feet tall. It is free flowering with large, yellow blossoms. Also called *flannel bush* because of its woolly (stellately pubescent) leaves.

California goldfern. *Pityrogramma triangularis,* a tropical American fern, family Polypodiaceae, grown as a house plant.

California harvester ant. *Pogonomyrmex californicus* (Buckl.), an ant which collects seeds. Often bites man and animals severely. Found in southwestern U.S.

California holly. See *Christmasberry.*

California incensecedar. *Libocedrus decurrens,* family Pinaceae. An evergreen tree grown as an ornamental and for its valuable timber. Also called *white cedar.*

California larkspur. *Delphinium californicum,* family Ranunculaceae, a poisonous plant which causes loss of appetite, constipation, nausea, and bloating to livestock. Native to coastal regions of California. Also called *coast larkspur.*

California laurel. *Umbellularia californica,* an aromatic, evergreen tree, family Lauraceae; grown for ornament on the west coast of U.S. Also called *spice tree; bay tree; balm-of-heaven; mountain laurel; sassafras laurel; California bay tree.*

California lilac. See *blueblossom ceanothus.*

California live oak. *Quercus agrifolia,* family Fagaceae, a broad-leaved evergreen tree, grown for decoration or shade in warm areas. Native to North America. Also called *coast live oak; encina; evergreen oak.*

California melic. Coastrange melic *q.v.*

California nutmeg. California torreya *q.v.*

California oakworm. *Phryganidia californica* Pack., family Dioptidae, a moth whose young feed on the leaves of California live oak, California white oak, and a few other trees. Also called *California oak moth.*

California oatgrass. See *California danthonia.*

California peach blight. See *Coryneum blight of stone fruits.*

California peppertree. *Schinus molle,* family Anacardiaceae, a resinous evergreen tree with pinnate leaves and yellowish-white flowers. Native to South America but widely grown in California. Also called *peppertree; Peruvian mastictree; false pepper; pepper shrub.*

California picnic shoulder. In pork, a retail cut; the lower portion of the shoulder.

Californiapitcher. *Darlingtonia californica,* family Sarraceniaceae, an insectivorous plant with sparse yellow-purple flowers. Native to North America. Also called *Chrysamphora.*

Californiapoppy. *Eschscholtzia californica,* family Papaveraceae, a very popular, annual plant grown for its orange-yellow flowers. Native to western North America. Also called *cup of flame.*

California prionus. *Prionus californicus* Mots., family Cerambycidae, a beetle that infests citrus and ash trees in southwestern U.S. Its larvae bore into the trunks just below the surface of the ground.

California privet. *Ligustrum ovalifolium,* family Oleaceae, a hedge plant with small clusters of flowers, planted extensively in Asia and U.S. Native to Japan.

California red fir. Red fir *q.v.*

California red scale. *Aonidiella aurantii* (Mask.), family Diaspididae, a major pest of citrus which also infests the castorbean, nightshade, laurel, sumac, eucalyptus, camphor, willow, carob, mulberry, avocado, fig, Persian walnut, and rose.

California redwood. See *redwood.*

California rose. See *rose glorybind.*

California scrub oak. 1. *Quercus dumosa,* family Fagaceae, a North American evergreen shrub not over eight feet high. 2. See *Oregon white oak.*

California sorrel. Palomino *q.v.*

California standard. Weight per box of oranges; 70 lbs.

California standard flat. A cantaloup crate, 4½ x 13½ x 22⅛ inches, used for delivery to local markets. (Not accepted in Colorado).

California tent caterpillar. *Malacosoma californicum* (Pack.), family Lasiocampidae, a pest of oak trees in California. In spring the caterpillars construct webs on fruit and forest trees, such as the almond, apple, apricot, ash, California Christmasberry, California coffeeberry, ceanothus, cherry, cottonwood, madrona, plum, redbud, willow, etc.

California thistle. Canada thistle *q.v.*

California torreya. *Torreya californica,* family Taxaceae, an evergreen tree grown for its foliage, grayish-brown bark tinged with orange, and its green nutmeg-like fruits streaked with purple. Native to North America. Also called *California nutmeg.*

California tree poppy. Matilija-poppy *q.v.*

California tussock moth. Western tussock moth *q.v.*

California vine disease. Pierce's disease *q.v.*

California Washingtonpalm. Mexican Washingtonpalm *q.v.*

California waterhemlock. *Cicuta californica,* family Umbelliferae, a highly poisonous perennial herb found in wet areas and bogs in California. Its poisoning symptoms are stomach pain, nausea, violent vomiting, diarrhea, dilated pupils, labored breathing, frothing at the mouth, rapid and weak pulse, and convulsions.

California white oak. *Quercus lobata,* family Fagaceae, a tall, Californian, deciduous tree which grows well in adobe soils.

calk. A downward pointed part of a horse's shoe, which prevents slipping. Also called *caltrop.*

callalily. 1. Any tropical herb of the genus Zantedeschia, family Araceae. Species are grown in greenhouses for decoration. See *golden callalily; rose callalily.* 2. *Zantedeschia aethiopica,* family Araceae, the popular lily-like white-yellow flowers with thick rootstocks; sold by florists as the *callalily.* It is a tropical herb native to South Africa. Also called *Egyptian lily; Ethiopian calla.*

caller. A fowl suffering from infectious laryngotracheitis *q.v.*

Callicoma. 1. A small genus of opposite-leaved trees, family Cunoniaceae, which bear clusters of small flowers. Native to Australasia. 2. *Callicoma serratifolia,* family Cunoniaceae, an Australian tree with a black bark and dented leaves. Its saplings are used for making baskets. Also called *sawleaf callicoma.*

Callitriche. A genus of small waterweeds, family Callitrichaceae, characterized by opposite-growing leaves. Widely distributed in ponds and ditches. Also called *water starwort.*

Callitris. A genus of resinous, quick-growing evergreen trees, family Pinaceae, grown in Florida and California for their beautiful foliage. Native to Africa and Australia. Commonly called *cypress pines.*

callow. 1. The surface or humus soil overlying subsoil (England). 2. A marshy meadow (Ireland).

callow adult. A recently transformed insect whose body wall has not completely hardened.

callus. (a) In plants, protective covering which forms over a wounded surface. (b) In animals, an area of skin which becomes horny and thickens. (c) A hard exudate which forms at the end of a broken bone and eventually becomes part of it.

callusing mound. A mound of earth in which wounded cuttings are placed to hasten healing.

callusing room. A room for grafting of plants, in particular grapes, in which the grafts are first dipped into warm water to a depth of just below the unions. They are then sprinkled with warm water, and are placed in a well ventilated room where the temperature is held at 75° to 80° F. This hastens the callusing of the wounds.

calomel. Hg_2Cl_2, mercurous chloride. A purgative of tasteless, white powder. Used as a spray for turf diseases and to combat club root of crucifers. Also called *horn quicksilver* when found as a native mineral.

calomorphic soils. Soils which owe their chief characteristics to the high content of calcium available to plants (frequently, but not always, in the form of calcium carbonate) in the parent material, such as Brown Forest and Rendzina soils. Also called *calcimorphic.*

caltrop. See *puncturevine; calk.*

calumba. *Jateorhiza palmata,* family Menispermaceae, a climbing plant whose root is used in medicines. Native to Mozambique.

calumbang. See *soft lumbang.*

calycifloral. Describing flowers which have their leaves and stamens inserted in the calyx.

calyciform. Describing flowers shaped like a calyx; cup-shaped.

calycle. A group of co-axial leaves around the base of a calyx in some flowers. It has the appearance of an outer calyx.

calyx. The outer, floral leaves or sepals of a flower or the external part of the flower.

calyx split. See *splitting.*

cam. A piece of machinery that produces intermittant motion; a cogwheel.

camas. *Camassia quamash,* family Liliaceae, an ornamental, bulbous plant. Its stick-shaped leaves grow up to 3 feet from the base and are overtopped by foot-long clusters of white or blue flowers. Found in northwestern U.S. Also called *camass; commas; kamas; camash.*

cambial zone. In wood anatomy, that region of a stem which includes the terminal cambium and its undifferentiated derivatives.

cambium. Thin-walled, meristematic plant tissue which gives rise to the functional tissues of the plant. The term is commonly applied to the lateral meristem which produces secondary xylem and phloem in dicotyledonous plants. The cork cambium (phellogen) gives rise to the cork that forms the outer portion of the bark. The cells of such cambium layers divide parallel to the outer surface of the stem (periclinal division), the external new cells forming one kind of tissue and the inner cells another kind. See *meristem; cambium layer; phellogen.*

cambium insect. Any insect that lives in the cambium region of the stem throughout its developmental period.

cambium layer. A common term for the lateral meristem layer which produces secondary xylem and phloem of dicotyledonous stems. In woody plants it may be reduced to one layer of cells in periods of dormancy but during active growth may consist of several layers of cells. See *cambium.*

cambium-wood insect. Any insect that spends its developmental period in both the cambium and wood of a living or dying tree, or freshly cut logs. Pupation usually occurs in the wood.

Cambridge. A soft, English cheese made of cow's whole milk with enough rennet added for coagulation when the milk is set at 90° F. Also called *York.*

camel. See *ship-of-the-desert.*

camellia. *Camellia japonica,* an evergreen shrub or small tree, family Theaceae, grown in warm climates for its showy, red flowers, and waxy foliage. Native to Asia. *Cf. sasanqua camellia.*

camellia scale. *Lepidosaphes camelliae* Hoke, family Diaspididae, an armored insect common in greenhouses in the U.S. which infests various plants of the genus *Camellia.*

camelthorn. *Alhagi pseudalhagi,* family Leguminosae, a perennial, thorny plant that resprouts vigorously even when seriously damaged. It is a forage for camels. Also called *Alhagi camelorum.*

Camembert. A soft, surface-ripened, French cheese, made from cow's milk. It is set with the mold *Penicillium camemberti,* family Moniliaceae, and cured for 6 weeks.

cam engine. An engine in which the pistons are reciprocated by means of a cam and roller mechanism.

camera set. A fixed position for photographing a particular area within a season or in several seasons. Used in the study of plant ecology.

cam follower. The part of an engine which rides on the contour surface of a cam, being driven by cam lobes.

cam gear. A toothed wheel which drives a camshaft in an engine.

cam governor. A controlling device used in connection with Otto-cycle gas-engines: a stepped or differential cam is used giving three or four grades of valve lift; the action of the governor balls automatically slides the roller on to one or another of these cams depending on the centrifugal force of the balls, thus keeping the speed under a certain pitch.

cam lobes. The raised portions on the contour of a cam which operates the cam followers.

cammock. *Ononis repens,* family Leguminosae, a small, hardy, perennial shrub which bears rose-pink flowers. Native to Europe. Also called *rest-harrow; thorny ononis.*

camomile. Any plant of the genus Anthemis, family Compositae. See *Roman camomile.* Also called *chamomile.*

Camosun. A soft, textured American cheese prepared from cow's milk and cured for three months in a humid room.

Campanula. Any herb of the genus Campanula, family Campanulaceae. Species are grown for ornament. See *spurred bellflower; bearded bellflower; singlenod bellflower; Carpathian bellflower; Greek bellflower; elatines bellflower; Danesblood bellflower; Italian bellflower; milky bellflower; great bellflower; Canterbury bells; peachleaf bellflower; Dalmatian bellflower; spotted bellflower; chimney bellflower; radde bellflower; bluestar bellflower; creeping bellflower; saxifrage bellflower; Coventry bells.*

campeachy wood. Logwood *q.v.*

Camperdown elm. *Ulmus glabra* var. *camperdowni,* family Ulmaceae, a tall tree with short-stalked leaves, grown widely for its pendulous branches and fine foliage. It is a cultivated variety of Scotch elm *q.v.* Also called *umbrella elm.*

campernelle jonquil. *Narcissus odorus,* a bulbous herb, family Amaryllidaceae, widely grown for its fragrant, yellow flowers. Native to Europe.

camphor scale. *Pseudaonidia duplex* (Ckll.), family Diaspididae, an insect found in southern U.S. where it infests camphor and satsuma orange and over 200 other hosts.

camphor thrips. *Liothrips floridensis* (Watson), a small insect which feeds on buds and tender twigs of the camphor tree. Infestation results in blackening, cracking, and dieback of the bark, and deformation of the limbs.

camphor tree. *Cinnamomum camphora,* family Lauraceae, an aromatic evergreen tree with alternate leaves, whose wood contains the camphor gum of commerce, obtained by steaming and sublimating the bark. Native to China.

camphorweed. Vinegar bluecurls *q.v.*

Campine. A breed of fowls originating in Belgium. They are bred primarily for egg production and are non-setting, utility fowls. The cock weighs 6 lbs. and the hen 4 lbs.

campion. See *silene.*

campo pea. Royal peavine *q.v.*

camp tender. One whose job it is to locate grazing ground for sheep, and establish and provision a camp for the herder and himself.

Canada barberry. See *Allegheny barberry.*

Canada beadruby. *Maianthemum canadense,* family Liliaceae, a low, perennial herb with two leaves, small white flowers in a terminal cluster, and brownish-red fruits. Native to North America. Also called *false lily of the valley; wild lily of the valley; Canada mayflower.*

Canada blueberry. *Vaccinium canadense,* family Ericaceae, a species of wild blueberry (*q.v.*) with white flowers and dark blue fruits. Native to North America.

Canada bluegrass. *Poe compressa,* family Gramineae, a perennial European grass, found throughout temperate and colder zones. Used widely in permanent pastures and lawn. Also called *English bluegrass.*

Canada fleabane. Horseweed fleabane *q.v.*

Canada garlic. *Allium canadense,* family Liliaceae, an annual, bulbous weed. Cows feeding on it produce onion-flavored milk. Also called *wild onion.*

Canada gooseberry. *Ribes oxyacanthoides,* family Saxifragaceae, a low shrub with greenish-white flowers and red or purple fruits. Native to northern U.S. and Canada.

Canada hemlock. *Tsuga canadensis,* family Pinaceae, a hardy evergreen tree, grown as an ornamental and for its valuable lumber used in pulpwood and general construction. Native to North America. Also called *hemlock; eastern hemlock; hemlock spruce; tanbark tree.*

Canada lily. *Lilium canadense,* family Liliaceae, a North American, bulbous herb, grown for its orange-yellow or red, dark-spotted flowers. Also called *meadow lily.*

Canada mayflower. Canada beadruby *q.v.*

Canada parrotsfeather. *Myriophyllum verticillatum,* family Haloragidaceae, a fresh-water herb grown in pools for its interesting foliage of half-submerged, pinnate leaves. If not controlled, it can soon become a nuisance. Also called *water milfoil.*

Canada pest. Russell prairiegentian *q.v.*

Canada plum. *Prunus nigra,* family Rosaceae, a hardy, deciduous tree, grown for its edible fruit and for plant breeding. Also called *red plum.*

Canada potato. Jerusalem artichoke sunflower *q.v.*

Canada rice. Annual wildrice *q.v.*

Canada sagebrush. *Artemisia canadensis,* family Compositae, a perennial herb, grown for its grayish foliage and greenish flower clusters. Native to North America. Also called *sea wormwood; wild wormwood.*

Canada selected. A grade of dressed poultry established by the Canadian government. It includes all poultry not showing white color or evidence of milk feeding.

Canada thistle. *Cirsium arvense* (*Carduus arvensis*), family Compositae, an obnoxious perennial weed which is very widely distributed in meadows, orchards, cultivated fields, and roadsides in North America. Native to Europe. Also called *Canadian thistle; creeping thistle; cursed thistle; small-flowered thistle; corn thistle; California thistle.*

Canada violet. *Viola canadensis,* family Violaceae, a perennial plant grown for its white flowers tinged with purple and its long blossoming season. Native to mideastern North America.

Canada waterweed. *Elodea canadensis,* family Hydrocharitaceae, a submerged, perennial water plant, grown in pools. Uncontrolled, it becomes a pest. Native to northern U.S. Also called *water-weed; ditch moss.*

Canada wildginger. *Asarum canadense,* family Aristolochiaceae, a stemless, perennial, medicinal plant having large leaves and brownish-purple flowers. Grown in wild gardens and as a ground cover. Native to eastern North America. Also called *false coltsfoot; false crowfoot.*

Canada wildrye. *Elymus canadensis,* family Gramineae, a perennial forage grass, widely distributed throughout the U.S. and Canada.

Canada yew. *Taxus canadensis,* family Taxaceae, an evergreen, low-growing shrub, useful as a ground cover under evergreen trees. Highly poisonous. Also called *ground hemlock; American yew.*

Canadian bacon. Smoked pork cut from the back and loin. Also called *English bacon.*

Canadian bur. Jerusalem artichoke sunflower *q.v.*

Canadian pacer. The parent stock for the Standard breed of trotters and pacers.

Canadian singer. A hog whose carcass is suitable for export. See *singer.*

canafistula. See *golden shower senna.*

canaigre. *Rumex hymenosepalus,* family Polygonaceae, an herbaceous plant found in western U.S. which has clustered, tuberous roots from which tannin is obtained.

canal. 1. An artificial channel to carry water for irrigation or drain water-logged areas. *Cf. ditch; drainage ditch.* 2. A tubular conduit in the bodies of men and animals, *e.g.,* the alimentary *canal.* 3. A small channel in the tissues of plants.

canary bird flower. Canary nasturtium *q.v.*

canary broom. *Cytisus canariensis,* family Leguminosae, a spineless, evergreen shrub, grown in greenhouses for its profuse, fragrant, yellow flowers. Also called *genista.*

canary creeper. Canary nasturtium *q.v.*

canarygrass. 1. *Phalaris canariensis,* family Gramineae, an annual grass grown for its seed, the canary seed of commerce, the most commonly used bird seed. It has escaped as a weed in many states of the U.S. 2. See *reed canarygrass.*

canary nasturtium. *Tropaeolum peregrinum,* an annual herb, family Tropaeolaceae; grown for its pale yellow flowers. Native to Peru. Also called *canary bird flower; canary creeper.*

Canby pachistima. *Pachistima canbyi,* a low, evergreen shrub, family Celastraceae; grown in borders or in evergreen plantings for its foliage and small reddish flowers. Found in the Appalachian Mountains. Also called *rat-stripper; mountain lover; cliff green.*

cancer. 1. A malignant tumor of man and animals. 2. In plants, a growth similar to that caused by crown gall *q.v.*

cancer eye. A non-infectious, sporadic, malignant ailment of the eyeball and the contiguous tissues in cattle (particularly Hereford) and occasionally horses.

cancrosis A. See *citrus canker.*

candelilla wax. A wax produced from the *wax euphorbia* (*Euphorbia antisyphilitica*) which is used as a substitute for beeswax.

candelospora rot. A fruit rot of citrus of minor importance, caused by *Candelospora citri,* family Moniliaceae.

candle. To examine an egg in front of a light to observe characteristics associated with edible quality or hatchability, such as air cell size, yolk shadow position, presence of blood or meat spots and presence or lack of germ development.

candleberrytree. See *candlenuttree.*

candlenuttree. *Aleurites moluccana,* family Euphorbiaceae, a tree widely grown in the tropics and closely related to the tung oil tree (*q.v.*). The seeds yield an oil similar to linseed oil, and are used to some extent as food, and the wood is worked. Also called *bankul nut; Belgian walnut; candleberrytree; candlenut; Indian walnut; kukui; Tahiti walnut; varnish tree; tallow tree.*

candle out. To sort out eggs into grades by candling.

candlewick. See *flannel mullein.*

candlewood. See *ocotillo.*

candy. A paste-like mixture fed to queen bees in transit. Made by kneading confectioners sugar into honey or invert sugar. In Europe, candy for feeding bees during the winter is made by evaporating sugar sirup and recrystallizing it in blocks.

candytuft. Any plants of the genus Iberis, family Cruciferae. Species are very popular in flower gardens. See *rocket candytuft; evergreen candytuft; globe candytuft.*

cane. (a) Woody stems of any small fruits, as of the grape or raspberry. (b) Stems of reeds and large grasses, such as the bamboos and sorghums. Also sometimes applied to the stems of rosebushes and some small palms. (c) Any hollow or pithy stem which is flexible and slender. (d) Sugar cane *q.v.*

cane and leaf rust. A disease of the cutleaf blackberry on the northwest coast of U.S. caused by *Kuehneola uredinis,* family Pucciniaceae; characterized in spring by a lemon-yellow color, and by large lesions which split open the canes. In summer small, yellow pustules appear on the lower leaf surfaces and then on the fruit. In the fall the leaf pustules become buff-colored.

cane blight. A minor disease of raspberries, roses, blackberries, apples, caused by the fungus *Leptosphaeria coniothyrium,* family Pseudosphaeriaceae.

cane blight of currants. A fungous disease of currants caused by *Botryosphaeria ribis,* family Botryosphaeriaceae; characterized by a sudden wilting and dying of canes and bushes before the fruit ripens, and by wart-like growths in parallel lines on the canes. Sometimes there are brick-red fruiting bodies.

cane blight of raspberries. A fungous disease of raspberries caused by *Leptosphaeria coniothyrium,* family Pleosporaceae; characterized by a weak growth on the fruiting laterals, wilting of the leaves and flower clusters. In the black raspberry, the bark turns purple, often only on one side of the stem. Numerous black dots appear on older invaded areas.

canebrake. A thicket or dense growth of native bamboos, species of Arundinaria. Found in the southern states of U.S., especially Mississippi, Louisiana, Alabama.

cane fruit. Any small fruit, such as the blackberry, raspberry, dewberry, which bears its fruit on canes. See *bramble.*

cane fruit anthracnose. A common disease of black and purple raspberries; sometimes damaging to blackberries, red raspberries, and other cane fruits. Caused by the fungus *Elsinoë veneta,* it is characterized by lesions and cankers on the leaves, stems, and drupelets; dry and woody fruit. Also called *gray bark.* See *anthracnose.*

cane gall of brambles. Bacterial swellings on the fruiting canes of raspberries and blackberries. They are whitish at first but become darker and are caused by the organism *Agrobacterium rubi.* It is sometimes very harmful to the canes but does not infect other plants and is distinct from crown gall *q.v.* Also called *cane gall of Rubus.*

cane grind. A social gathering at a cane mill at crushing time in southern U.S. Similar to sugar off parties at maple sirup time in the north.

cane hay. Hay produced from sweet sorghum.

cane knife. A heavy, machete-like knife with a hooked blade; used for cutting sugarcane or cornstalks.

cane leafhopper. Sugarcane leafhopper *q.v.*

Canella. A genus with only one species, *Canella winterana,* family Canellaceae, an aromatic tree with white bark, whose pale orange inner bark is used in medicines and foods. Native to Florida and West Indies. Also called *wild cinnamon; cinnamon canella.*

cane molasses. Blackstrap *q.v.*

cane mosaic. Sugarcane mosaic *q.v.*

cane pruning. Systematic cutting back of grape vines in order to limit the fruit yield, improve the quality, and keep the plant shapely and manageable. The fruit-bearing branch is cut back to the last two buds at the end of the season. Of these, one bud is to produce the new shoot for the next year and the other is an emergency substitute. The new shoot, in its turn, will be similarly cut back after the season, and so on. Thus each year's branch forms a spur for the next year's shoots and the whole cane comes to have an elongated, crooked shape. In another system of *cane pruning,* the whole fruit-bearing branch is cut off, and a new shoot is encouraged to develop from a bud on the permanent vine or stalkhead.

cane reed. Giant reed *q.v.*

cane renewal. Replacement of two-year-old grape canes by new canes that will produce the next season's crop. Practiced in all systems of grape pruning.

cane spot of brambles. Cane fruit anthracnose *q.v.*

cane sugar. The carbohydrate sucrose or saccharose in the juices of many plants, especially sugar cane; extracted and refined to form the sugar of commerce.

canine. The dog family. It includes dogs, wolves, foxes, jackals, etc.

canine distemper. Distemper of dogs *q.v.*

canine madness. Rabies *q.v.*

canine tooth. In mammals, one of the four sharp teeth on both jaws between the incisors and bicuspids.

canistel lucuma. *Lucuma nervosa,* family Sapotaceae, a tropical tree bearing small, white flowers; grown in Florida and West Indies for its orange-yellow, edible, sweet fruit. Also called *canistel; eggfruit; ti-es.*

canker. 1. A diseased lesion of the bark and underlying tissue in woody plants. 2. An ulcer in the mouth. 3. A serious disease of the horse's hoof. The frog frequently discharges a stinking fluid and ultimately the sole and the frog rot.

canker of hickory. A fungous disease caused by *Rosellinia caryae,* which produces cankers on the plants.

cankerworm. See *fall cankerworm; spring cankerworm.*

canna. Any tropical herb of the genus Canna, Family Cannaceae. Distinguished by showy flower clusters and large, broad leaves. Cultivated *cannas* are of hybrid origin. Native to Asia and America. See *edible canna.*

canner. 1. A marketing classification for low grade beef. 2. A wild horse, from the western range of U.S., slaughtered for its meat. 3. Berries, etc., too soft for shipment but fresh enough for immediate table use or canning.

canners' alkali. Sodium carbonate-sodium hydroxide mixture; used to loosen fruit skin to make peeling easier, as in the peach.

cannibalism. A habit of some fowls of pecking at other fowls. Frequently blood is drawn and in severe cases the victim may be killed. See *pick out.*

canning factory silage. Husks, cobs, and undesirable kernels of sweet corn, pea vines, pear waste, etc., which are unfit for canning, but usable as a stock feed.

canning ripe. Fruit and vegetables of full size and good flavor; suitable for canning but not ripe enough for table use. *Cf. firm ripe; canner: 3.*

cannon. 1. The shank of a horse's leg, specifically the bone from the knee to the fetlock in the front legs or the hock to the fetlock in the rear legs. 2. A horse's bit. 3. To lift and swing a log crosswise to a load of logs.

cannula. A metal, rubber, or glass tube inserted into a body cavity to allow the escape of fluids or gas and through which liquids may be introduced into the body.

canoe birch. Paper birch *q.v.*

cañon. See *canyon.*

canopy. Covering of green leaves and branches formed by crowns of individual trees in a forest. See *crown cover.*

Canquillote. A skimmed milk cheese, made in eastern France.

cant. A log which has been cut and squared.

Cantal. A hard, yellow, French cheese, made from cow's milk. It is cured for four months or six months during which time it is frequently washed with salty water.

cantaloup(e). 1. *Cucumis melo* var. *cantalupensis,* an annual herb with trailing vines, family Cucurbitaceae; grown in Europe for its edible fruit with a hard rind. 2. For American *cantaloup,* see *muskmelon.*

canter. The gait of a horse in which the feet strike the ground one at a time. An abbreviation of *Canterbury gallop,* describing the gait of horses ridden by pilgrims to Canterbury. Also called *slow gallop.*

Canterbury bells. *Campanula medium,* family Campanulaceae, a popular, biennial herb grown for its violet-blue flowers. Native to southern Europe.

Canterbury umbrellapalm. *Hedyscepe canterburyana,* family Palmaceae, a feather palm grown in Florida and California as an ornament for its graceful foliage. Native to Lord Howe Island off the northeast coast of Australia. Also called *umbrellapalm.*

cant hook. A wooden lever with a variable iron hook on one end; used for handling logs.

cantle. The upward curving back end of a saddle.

canvas. The conveyor apron on a grain binder or combine.

canvas curtain. A curtain of canvas placed near the middle of a threshing machine to retard straw as it passes over the straw racks.

canvas hose irrigation. A method of irrigation which originated in Michigan. A porous canvas hose in which water flowing under pressure is placed on the ground alongside each row. The water seeps out of the canvas hose along its length and is absorbed by the soil.

canvas jacket. An apron made of canvas or other heavy material which is placed on the back of a female turkey to prevent injury during breeding. Also called *saddle; apron.*

canvas slapper. A heavy piece of canvas wrapped on the end of a stick; used to slap livestock to control them.

canyon. A steep-walled, deep valley formed by a stream cutting through soft rocks: a characteristic land feature of western U.S.

canyon live oak. *Quercus chrysolepis,* an evergreen tree, family Fagaceae, found in western U.S. Also called *maul oak; green cape oak.*

cap. 1. A piece of metal placed over the end of a log to make it skid over obstacles. 2. The top member of a trestle or similar support. 3. Milk tooth of an animal. 4. The crown of a kernel of corn. 5. The calyx of the strawberry. 6. To remove the calyx of the berry after it is picked. 7. To place a cover over a stack of hay, bound grain, etc. 8. To seal a cell in the honeycomb with wax, as by a bee.

Capacola. A sausage made from picnic hams seasoned with red pepper pods, etc.

cape. The short feathers underneath the hackle of chickens.

cape blue waterlily. Cape waterlily *q.v.*

cape bugloss. *Anchusa capensis,* family Boraginaceae, a biennial herb, popular as a pot plant for its red buds and blue flowers. Native to South Africa.

cape chinkerichee. *Ornithogalum thyrsoides,* family Liliaceae, a bulbous herb with white flowers; native to South Africa. A botanical variety (*O. t. aurem*) has golden yellow flowers. They are valued for pot and outdoor culture in warm regions. Also called *chinkerichee.*

cape forget-me-not. Cape bugloss *q.v.*

capefuchsia phygelius. *Phygelius capensis,* family Scrophulariaceae, a small, tender shrub with scarlet flowers, grown as an ornamental in the U.S. Native to South Africa.

cape gooseberry. Peruvian groundcherry *q.v.*

capejasmine. *Gardenia jasminoides,* family Rubiaceae, a popular plant with large, white, fragrant flowers, grown outdoors in warm areas as an ornamental.

cape nightphlox. *Zaluzianskya capensis,* family Scrophulariaceae, a small, night-blossoming plant whose flowers are white inside and black outside. Cultivated in gardens in summer. Native to South Africa.

cape plumbago. *Plumbago capensis,* family Plumbaginaceae, a tender shrub, grown for its azure-blue, flower clusters. Native to South Africa.

cape pondweed. Cape waterhawthorn *q.v.*

cape pittosporum. *Pittosporum viridiflorum,* family Pittosporaceae, a tall, evergreen shrub raised in warm climates for its clusters of green and yellow flowers. Native to South Africa.

caper. *Capparis spinosa,* family Capparidaceae, a Mediterranean shrub, grown for its young flower buds, the *capers* of commerce, which are pickled and used for seasoning.

caper euphorbia. *Euphorbia lathyrus,* family Euphorbiaceae, a poisonous, annual herb, contact with whose milky sap causes a dermatitis with blisters and inflammation. Animals will not eat the growing plant, but sometimes eat hay containing the dry plants, which causes swellings on the head and excessive scours and may eventually prove fatal. Native to Europe; occasionally grown in gardens. Its pods are sometimes used for pickles and seeds as a purgative. Also called *English capers; Capuchin capers; caper spurge; molewood; mole plant; false caper.*

caper spurge. Caper euphorbia *q.v.*

cape waterhawthorn. *Aponogeton distachyus,* a tuberous-rooted, aquatic plant, family Aponogetonaceae, widely cultivated for its floating leaves and tiny white flowers. Native to South Africa. Also called *cape pondweed; waterhawthorn.*

cape waterlily. *Nymphaea capensis,* family Nymphaeaceae, an aquatic herb, grown for its sky-blue flowers with whitish bases. Native to South Africa. Also called *cape blue waterlily.*

cape weed. Spotted catsear *q.v.*

Cape wool. Wool from South Africa.

capillaire. Creeping pearlberry *q.v.*

capillaria worms. Internal parasites of the genus Capillaria, which infest the crop, esophagus, small intestines, and ceca of chickens, ducks, turkeys, and some wild birds. Symptoms are unthriftiness, weakness, diarrhea, loss of weight, ruffled feathers, soiled vent, and a tendency to sit on the ground. Also called *hair worms; threadworms.*

capillary. Extremely narrow, microscopic blood vessels. The presence of blood in *capillaries* produces the pink color of the skin.

capillary action. The phenomenon by which water or other liquids are carried by minute pores throughout the soil or upwards into the plant.

capillary attraction. An assumed, apparent attraction between soil and water. It seems to cause the moisture to move towards the drier parts of the soil.

capillary capacity. The amount of moisture which is held in the soil by capillary attraction after excess or free water has drained away.

capillary conductivity. 1. (Qualitative) The physical property relating to the readiness with which unsaturated soils transmit water. 2. (Quantitative) The ratio of the water flow velocity to the driving force in unsaturated soil. The calculation is valid under conditions where flow velocity is proportional to driving force: for example, in practical units when the driving force is expressed in terms of the hydraulic gradient, *capillary conductivity* is the ratio of flow velocity to hydraulic gradient and has the dimensions of velocity. As saturation is approached *capillary conductivity* approaches the hydraulic conductivity.

capillary fringe. A belt of capillary interstices in the soil belt that overlie the zone of saturation, or water-table. Some or all of these are filled with water, held by capillary action against the force of gravity. The *fringe* water may be used by deep rooting or phreatic plants.

capillary potential. The attractive force of soil for water when the soil is at a particular moisture content.

capillary water. Water held by adhesion of surface tension as a film around soil particles and in the capillary spaces; used by plants in their growth.

capital gain. The excess of proceeds obtained from the sale or exchange of a non-inventory asset (or *capital asset* as defined under section 117a of the Federal Internal-Revenue Code) over its book value, cost, etc.

capital investment. Value of the inventory, often at the beginning of the fiscal year, of all farm properties; includes farm real estate (except owner's dwelling), machinery and equipment, livestock, feed crops, and supplies.

capitalization. Determining present value of future net money income.

capitalization rate. The ratio of net annual money income to the capitalized money value of property. The rate should probably be near the current rate of interest charged on first mortgages on similar farms in the area. It is influenced by the prevailing money market, the risk both physical and economic, the marketability of the farm in question, and the competition with other forms of investment.

capitalization value. The value of real estate based on the present value of future net money income. It is computed by dividing the annual net money income by the selected capitalization rate *q.v.*

capitalized earning value. The capitalization at a certain rate, as five percent of the net returns to real estate, in order better to estimate the earning value of the farm.

capital requirements. The amount of fixed and working capital which is needed to operate a farm or ranch. It includes capital for land, buildings, fences, water supply, drains, machinery, equipment, livestock, current operations, etc.

capital turnover. The ratio of the gross income to the total capital investment. It is expressed by the number of years required for the gross income to equal the capital investment.

capitular. The growing of fruits or flowers in a small cluster or head.

capon. A cockerel castrated to fatten it for the market. It is usually seven to ten months old and weighs 6 lbs.

caponette. A young meat chicken treated with an estrogenic compound during its growth.

capon style. A method of dressing fowl; feathers are left on the neck, wings, upper part of the back, the tail, and the lower part of the drumsticks.

capped elbow. A swelling on a horse's elbow, caused by irritation from lying on hard surfaces, or by constant rubbing from a shoe. It is painful and characterized by lameness. Also called *hygroma of the elbow; shoe boil.*

capped hip. A severe injury or fracture of the hip of an animal which makes the hip on one side appear lower than the other. It is a disqualification in cattle judging. Also called *knocked down hip.*

capped hock. A swelling on a horse's hock, similar to *capped elbow* (*q.v.*). It usually does not affect the usefulness of the horse.

capped stand. A large, upright pipe, with small outlets in its sides, which receives water under low pressure, and releases it through these outlets into different small irrigation ditches. It is free from debris, of low cost, and has small leakage loss.

capper. A device used for removing the calyx from strawberries.

capping drier. A device used to remove honey from wax cappings.

capping melter. A device used to melt the cappings on honey comb after the excess honey has been drained off.

cappy flavor. See *oxidized flavor.*

cap rail. The top rail of a fence.

capric acid. $C_9H_{19}CO_2H$, a fatty acid found in butter, fusel oil, etc., which gives a characteristic buttery flavor to cheese. Occurs more in goat's than in cow's milk. Also called *caprinic acid; decanoic acid.*

caprification. Artificial pollinization of certain figs by the fig wasp *Blastophaga psenes* (L.), family Agaontidae, by hanging the wasp-containing capri fig (*q.v.*) on the fig tree. The wasp forces its entrance through the terminal pore of the edible fruit. The term is also applied to the ancient practice of puncturing the green figs artificially to hasten ripening.

capri fig. *Ficus carica* var. *sylvestris,* family Moraceae, a wild variety of the common fig (*q.v.*), useful only for the artificial pollinization of certain domesticated figs. See *caprification.*

caprinic acid. Capric acid *q.v.*

caproic acid. $C_5H_{11}CO_2H$, a fatty acid found in butter along with capric acid to which it is similar but less peppery in taste and more unpleasant in odor. Also called *hexanoic acid.*

capsanthin of paprika. A carotinoid present in paprika.

cap screw. A screw-bolt intended to be used without a nut. It is so named because it was first used to secure the covers of small steam engines instead of studs.

cap seat. The inner ridge in the top of the neck of a milk bottle on which the cap or cardboard closure rests.

Capsicum. An important genus of tropical plants, family Solanaceae, which includes the common red pepper, long pepper, etc. The plants bear small, whitish flowers, and produce berry-like fruits which are used as vegetables and for pickling and flavoring. Native to America.

cap stem. The small appendage that attaches the berry to the main stem.

cap stemmer. A device consisting of two concentric truncated cones of heavy screen, one of which revolves rapidly. Raisins are placed between them and revolved and rotated to remove the stems. A blast of air separates the raisins from the removed stems.

capsulated. Surrounded by, or enclosed in, a sac of living tissue.

capsule. 1. A fruit (dry at maturity) of more than one carpel but not necessarily of more than one locule, which dehisces at maturity permitting the escape of the enclosed seeds. Typical capsules are found in the iris, lily, poppy, cotton. See *legume: 1; silique; follicle: 2; achene; pod: 1.* 2. Gelatine covering for medicaments.

capsule breeding. An older term for *artificial insemination q.v.*

captive bolt pistol. A pistol with a steel plunger in place of a bullet, used for stunning animals.

Capuchin capers. 1. Pods of nasturtium, sometimes used for pickles. 2. See *English capers; caper euphorbia.*

capybara. *Hydrochoerus capybara,* a quadruped, considered to be the largest extant rodent; lives exclusively on grass and vegetables. Native to tropical South America. See *cavy: 3.*

caracul. Karakul *q.v.*

Caragana. A genus of ornamental trees and shrubs, family Leguminosae, with pea-like flowers. Native to central Asia. See *Siberian peashrub.*

carambola. *Averrhoa carambola,* family Oxalidaceae, a small, tropical tree grown for its yellowish-brown edible fruit which smells like quince and is used fresh or in jellies. Native to southeast Asia.

caraway. 1. *Carum carvi,* family Umbelliferae, an annual or biennial herb, whose aromatic seeds are used as a condiment. 2. Any cheese spiced with caraway seeds.

carbohydrate. Any of certain organic chemical compounds of carbon, hydrogen and oxygen, which include sugars and starches. Formed in plants by photosynthesis, *carbohydrates* make up a large part of animal food.

carbohydrate-nitrogen relationship. The relative proportion of accumulated carbohydrates and nitrogen in stems and leaves of plants. It changes with the age, vigor, fruit-setting, etc., of the plant, and influences flower bud initiation in some species. Also called *carbohydrate-nitrogen ratio.*

carbolfuchsin. Staining fluid used in bringing out details of organisms.

carbolic acid poisoning (phenol poisoning). The poisoning of animals, externally or internally, by application or ingestion of carbolic acid (phenol). Symptoms of internal poisoning are burning of the tongue, throat, and pharynx, profuse salivation, characteristic acidic odor in the breath, muscular tremors, rapid respiration, staggering, and paralysis. Vomiting is frequent in the early stages in some animals. In external poisoning the skin is burned and blistered.

carbolic pad. A cloth-lined frame on which carbolic acid is sprinkled; used for driving bees out of a hive to remove supers full of honey.

carbon. An essential chemical element component in plants and animals. It is present in soils, in humus, plant residues, charcoal, and particles of coal, carbonaceous shale, etc.

carbonaceous. Pertaining to carbon; carbonous.

carbonate. A salt of carbonic acid.

carbonated cider. Aerated apple cider.

carbonate of lime. Calcium carbonate *q.v.*

carbonate of potash. K_2CO_3, a chemical obtained from beet residue, wood ashes, etc. Used in the manufacture of special glass, soaps, and fertilizers, especially for root crops.

carbonation. A chemical process of soil development in which carbon dioxide resulting from decomposition of organic matter in the soil combines with bases, in the presence of moisture, to form bicarbonates.

carbon bisulfide. CS_2, a clear, highly inflammable liquid, used as a fumigant for stored grains. It is destructive to mice and other rodents. Also called *carbon disulfide; high-life.*

carbon cycle. The sequence of transformation undergone by carbon utilized by organisms. Absorbed from plants and other sources by an organism, it is later liberated upon the organism's death and decomposition, and returned to its original form of carbon dioxide gas, to be again taken up by plants.

carbon dioxide. CO_2, a heavy gas, which combines with metals to form carbonates. It is the atmospheric gas which is necessary for plant growth and is dangerous to animals, if inhaled.

carbon dioxide evolution. The liberation of gaseous carbon dioxide from soil by biological processes.

carbon disulfide. Carbon bisulfide *q.v.*

carbonize. To reduce to carbon; to coat with carbon.

carbonizing. A process for extracting vegetable matter from wool by treating it with acid or some other chemical. It carbonizes the vegetable matter without harming the wool fiber. Wool thus treated is known as *carbonized wool.*

carbon monoxide. CO, a highly poisonous gas produced by the incomplete oxidation of carbonaceous products, such as fuels. Sometimes present in silos.

carbon monoxide poisoning. Poisoning by inhalation of carbon monoxide. It is characterized by stupor, drowsiness, labored breathing, staggering and spasms in fowl. Fatal in large doses.

carbon-nitrogen ratio. The relative proportion, by weight, of organic carbon to nitrogen in a soil, feed, etc.

carbon tetrachloride. CCl_4, a chemical that is an excellent solvent for grease. It is not inflammable and can be used as a fire extinguisher. Frequently used as a fumigant though not in large scale operations; toxic when inhaled.

carbon tool steel. Steel containing little or no alloying metals, with a carbon content of 1.10 to 1.30 percent, used in tools for cutting operations at low speeds, where a hard, keen cutting edge and a high quality finish is desired.

carcajon. A wolverine; also a name for a lynx or a badger (North America).

carcass. The body of a dead animal; especially one prepared for market.

card. A wire-toothed instrument used to groom animals or to clean and straighten wool, flax, and cotton fibers.

cardamon. *Elettaria cardamomum,* family Zingiberaceae, an herb grown in southern Florida. Its seed is used in medicine and as a condiment. Native to East Indies.

cardboard. In pasteurized milk, a defective flavor caused by the absorption of copper from storage equipment. Uncommon in raw milk. See *metallic flavor; oily: 2; oxidized flavor; stale: 1; tallowy flavor.*

carded woollen. A fabric made from carded, uncombed wool.

card fly. A classification of wool substitutes consisting of waste fibers after carding.

cardia. Juncture of the esophagus and stomach.

cardiac. 1. That which is related to the heart. 2. A heart stimulant.

cardinal climber. See *cardinal starglory.*

cardinal flower. *Lobelia cardinalis,* family Lobeliaceae, a perennial herb grown in moist, shady wild gardens for its scarlet flowers. Native to North America. Also called *scarlet lobelia.*

cardinal larkspur. *Delphinium cardinale,* a perennial herb, family Ranunculaceae, grown in decorative borders for its red flowers with yellow petals. Native to North America. Also called *scarlet larkspur.*

cardinal ladybird. See *vedalia.*

cardinal starglory. *Quamoclit sloteri,* an annual hybrid vine, family Convolvulaceae; grown for its ornamental scarlet flowers with white throats. Also called *cardinal climber.*

carding. One of the first steps in the preparation of scoured wool; fibers are separated from other fibers in the locks or bunches of wool.

cardoon. *Cynara cardunculus,* family Compositae, a stout perennial herb whose stalks and leaves are used in cooking or to curdle milk in making cheese. Native to Europe. A relative of the artichoke, it is also called *artichoke thistle.*

careless. Of, or pertaining to, a horse which stands with its knees sprung.

careless weed. See *redroot amaranth.*

Carex. A large genus of grassy-looking plants with hard stems, family Cyperaceae, commonly called *sedges,* a few of which are occasionally cultivated in gardens. Fiber from some of its species is used in matting.

Carey Act. An act of Congress, 1894, granting desert lands to states for irrigation and sale to farmers.

cargo mill. A sawmill with harbor facilities for shipping its products. Also called *tide-water mill.*

Carribbean flint corn. A Latin-American corn variety having chunky, tapering ears, enlarged at the base. The plants are large and vigorous.

Caribbean hymenocallis. *Hymenocallis caribaea,* a bulbous herb, family Amaryllidaceae, with fragrant, long, white flowers. Cultivated in flower gardens and greenhouses. Native to South America.

Carib heliconia. *Heliconia bihai,* a banana-like plant, family Musaceae, with orange-red flowers. In tropical America, its leaves are used to cover packages. Outside of the tropics, it is cultivated in greenhouses. Also called *false plantain; wild-plantain; balisier.*

caribou. The reindeer of North America.

Carib royal palm. *Roystonea oleracea,* family Palmaceae, a slender, tall palm. In tropical regions, used as a street tree. Also called *Barbados royal palm; palmiste; cabbage palm.*

caricature plant. *Graptophyllum pictum,* family Acanthaceae. A tall, tropical shrub; grown in greenhouses for its variegated foliage. Native to India.

carissa. Any spiny shrub of the genus Carissa, family Apocynaceae; grown for ornament and for its edible fruits. See *Natalplum carissa; Karanda carissa.*

cark. In old English law, a long ton of wool: 2,240 pounds.

car lot. The quantity of produce or number of animals which can be transported in a railroad car.
carminative. A chemical which expels gas from the alimentary canal to relieve colic, etc.
carmine crabapple. *Malus atrosanguinea,* family Rosaceae. A bushy, deciduous, hybrid shrub with deep rose flowers and red fruit.
carnallite. The mineral $KMgCl_3 \cdot 6H_2O$ used in Europe as a source of potash for fertilizers.
carnation. *Dianthus caryophyllus,* a hardy, perennial herb, family Carophyllaceae. Many varieties are grown in greenhouses and gardens for the fragrant blossoms in shades of red, yellow, and white. Native to Europe. Also called *clove pink; English gillyflower.*
carnation bud rot. See *bud rot of carnation.*
carnation clip. See *split clip.*
carnation rust. A common, greenhouse, fungous disease caused by *Uromyces caryophyllinus* (Schr.), Wint.; characterized by brown, powdery lesions which disfigure the leaves.
carnation split. See *splitting.*
carnation yellows. A virus disease characterized by mottling on the leaves and stems, by distortion and color breaking of the flowers, and by reduction of vigor of the plant. The method of transmission other than by division of the plants or cuttings is not demonstrated, but insect vectors are suspected.
carnauba wax. A hard wax derived from the leaves of a palm tree, *Copernicia cerifera,* family Palmaceae, of Brazil. Used in shoe polish, varnish, floor wax, etc.
carney. A disease of horses in which the mouth becomes so coated with dead tissue and mucous that the animal is unable to eat.
Carniolan bee. A variety of the honey bee which originated in Carniola, Austria. Popular in U.S. for their honey and gentleness.
Carnivora. A large order of mammals which customarily eat flesh; *e.g.,* dogs, cats, bears, and seals.
carnivorous. Flesh eating. In plants, those which subsist on protein obtained from entrapped insects. *Cf. herbivorous; omnivorous.*
caro. See *guanacaste earpodtree.*
carob. *Ceratonia siliqua,* a Mediterranean, evergreen tree, family Leguminosae, grown for its edible fruit which is used as fodder for farm animals. Thought to be the "locusts" eaten by John the Baptist, and the "husks" eaten by the prodigal son. Also called *St. John's-bread; algarroba; bean tree.*
caroba. Leaves of the *Jacaranda procera* or *Cybistax antisyphilitica,* family Bignoniaceae. Native to Brazil where they are used as purgative and in the treatment of syphilis.
Carolina allspice. See *strawberry shrub:2.*
Carolina buckthorn. *Rhamnus caroliniana,* family Rhamnaceae, a tender shrub grown for its red fruit which turns black. Native to North America. Also called *yellow bush; Indian cherry.*
Carolina fanwort. *Cabomba caroliniana,* family Nymphaeaceae, an aquatic plant with narrow, floating leaves and white flowers having two yellow spots at the base of each petal. Native to North America. Also called *fanwort; watershield; Washington plant; fish grass.*
Carolina geranium. *Geranium carolinianum,* an annual weed herb, family Geraniaceae, a pest of lawns in southern U.S.
Carolina grasshopper. *Dissosteira carolina* (L.), family Acrididae, a common grasshopper, two inches in length.
Carolina hemlock. *Tsuga caroliniana,* a beautiful, evergreen tree, family Pinaceae; commonly grown for ornament, especially in southern U.S. Native to North America. Also called *spruce pine.*
Carolina horsenettle. *Solanum carolinense,* a perennial weed herb, family Solanaceae, which may be poisonous to livestock feeding on its herbage or berries. Commonly found in southern U.S. Also called *sand brier; bull nettle; apple-of-Sodom.*
Carolina jasmine. See *Carolinajessamine.*
Carolinajessamine. *Gelsemium sempervirens,* family Loganiaceae, an evergreen, twining shrub, which has fragrant, bright yellow flowers; grown to cover banks and porches. Native to southeastern U.S. where it is the state flower of South Carolina. Also called *Carolina jasmine; evening trumpet flower; yellow jessamine; Carolina yellow jessamine; evergreen trumpet flower.*
Carolina laurelcherry. *Prunus caroliniana,* family Rosaceae, a small, evergreen tree, which has small cream-white flowers; planted for shade. Native to southeastern U.S. Also called *evergreen cherry; American cherry-laurel; wild-orange; mock orange; cherry laurel.*
Carolina mantis. *Stagmomantis carolina* (Johan.), family Mantidae, the praying mantis, so called because their front legs resemble hands folded in prayer. It preys on other insects but is not injurious to man or other animals. Also called *rearhorse.*
Carolina moonseed. See *Carolina snailseed.*
Carolina phlox. *Phlox carolina,* a perennial herb, family Polemoniaceae, grown for its purple, rose, or white blossoms. Native to North America. Also called *early perennial phlox; thick-leaf phlox.*
Carolina poplar. *Populus canadensis* var. *eugenei,* family Salicaceae, a soft, woody, columnar, deciduous tree; undesirable as a street tree because its roots clog drains and lift pavements and because it drops leaves throughout summer. Its timber is used for paper pulp and boxes. It is a hybrid of *P. canadensis* and *P. nigra.* Also called *Eugene poplar.*
Carolina rhododendron. *Rhododendron carolinianum,* family Eriaceae, a perennial evergreen shrub grown for its pale purple flowers. Native to North America.
Carolina rose. *Rosa carolina,* an erect shrub, family Rosaceae, grown for its large, rose-pink flowers. Native to North America. Also called *pasture rose.*

Carolina silverbell. *Halesia carolina,* a deciduous tree, family Styracaceae, grown for its profusion of early-spring, white flowers. Native to North America. Also called *silver-bell tree; snowdrop tree.*

Carolina snailseed. *Cocculus carolinus,* a woody vine, family Menispermaceae, grown for its greenish flowers and red fruit. Native to North America. Also called *Carolina moonseed.*

Carolina vanilla. See *vanilla trilisa.*

Carolina yellow jessamine. See *Carolinajessamine.*

carosella. See *fennel.*

carotene. An orange or red pigment found in such foods as carrots, sweet potatoes, milk fat, and egg yolk and also in green leaves. *Carotene* may be converted in the animal body into vitamin A; hence called the precursor of vitamin A.

carotenoid. Any of the yellow and red pigments found in plants and the fats of animals being related to carotene.

Carpathian bellflower. *Campanula carpatica,* a perennial herb, family Campanulaceae, grown in rock gardens for its solitary, blue flowers. Native to eastern Europe.

carpel. In seed plants, a megasporophyll which is a modified leaf. The simple pistil consists of one *carpel*; the compound pistil, of two or more *carpels*, which may be united or distinct.

carpenter ant. Ant species that bore into wood or a tree to build their nests. The American species belong to the type *Camponotus abdominalis* and C. *herculeanus.*

carpenter bee. Glossy-winged female bees of genus Xylocopa, which burrow into decaying timber to lay eggs.

carpenterworms. Larvae of carpenter moths, *Prionoxystus robiniae* (Peck) and *P. macmurtrei* (G.-M.), family Cossidae, which burrow into trees and weaken them. Their presence is marked by sawdust or dark-colored sap coming out of small holes and discoloring the trunk. They attack the black and red oak, elm, locust, poplar, willow, maple, ash, chestnut, etc., and are distributed throughout central North America.

carpet bedding. Arrangement of low, compact, foliage plants in patterns to resemble the design of carpets. See *bedding: 4.*

carpet grass. Tropical carpetgrass *q.v.*

carpet wool. Very low grade wool from unimproved sheep. Only small quantities of it are produced on Indian reservations in the southwestern U.S.

carpology. The study of structural anatomy of fruits and seeds.

carpoptosis. Abnormal fruit drop.

carrageen. Irishmoss *q.v.*

Carre. A rich, fresh, Neufchatel-type French cream cheese made from a rich milk and cream mixture coagulated with rennet. The curd is placed in a cloth to drain thoroughly and is then salted and put into molds.

Carre de l'Est. A French, Camembert-type cheese made from pasteurized or unpasteurized cow's whole milk.

carriage. 1. A frame on which a log is held. 2. Manner or bearing of an animal; its physical control and behavior.

carriage horse. Breeds of light harness horses, weighing up to 1300 lbs., formerly used primarily for pulling carriages, but now bred mostly for the show ring. See *Hackney; Yorkshire Coach; French Coach; German Coach; Cleveland Bay; Russian Orloff.*

carrier. 1. Any person, animal, or plant that has in its system the organisms of a disease to which it is immune or largely immune, but which it may transmit to other plants or animals. See *vector.* 2. Any substance added to a chemical compound to facilitate its storage, shipment, or use in the field by means of increasing the bulk. See *diluent.*

carrier-away. A person who carries the shorn fleece to the proper bin after classification (New Zealand).

carrion. Putrefying carcass.

carrionflower greenbrier. *Smilax herbacea,* family Liliaceae, an herbaceous vine which bears bluish-black fruit. Its flowers have a strong odor of carrion. Native to North America.

carrot. *Daucus carota* var. *sativa,* family Umbelliferae, an annual herb grown for its edible root used in salads and stews, and as a vegetable; a very important root crop. Native to Eurasia.

carrot beetle. *Ligyrus gibbosus* (DeG.), family Scarabaeidae, a black insect that feeds on the roots of carrot, parsnip, celery, beet, potato, cabbage, corn, cotton, sunflower, dahlia, amaranthus, and other plants. It is widespread throughout the U.S. except the northern states.

carrot rust fly. *Psila rosae* (F.), family Psilidae, an insect whose larvae attack carrots, parsnips, celery, parsley, etc., feeding on the fibrous roots of the plants, stunting the growth. Usually they eat the lower root-ends, but sometimes the entire root is riddled with burrows. Found over eastern North America.

carrot weevil. *Listronotus oregonesis* (Lec.), family Curculionidae, a pest of celery, parsnip, dill, and carrots, which makes zigzag grooves on the plant surface or burrows through the roots.

carrot yellows. A serious virus disease of carrots; characterized by a yellowing of the young leaves, side shoots, and a bronzing or reddening and twisting of older, outer leaves. Later the crown turns black and the plant dies. This is the same virus as that of aster yellows and is transmitted by leaf hoppers, especially *Macrosteles fascifrons* (Stal.), family Cicadellidae.

carry. 1. To keep an animal on a maintenance ration without obtaining any weight gain or any products from it. 2. To bear, as a pregnant cow *carries* a calf. 3. To sustain, as a farm *carries* a debt, or as a range *carries* stock. 4. To convey during transportation and marketing, as a container *carries* satisfactorily or a product *carries* well.

carryall. 1. A light-weight carriage, truck, wagon,

or the like, for transportation. 2. A heavy excavating device used to level fields and construct canals and waterways.

carrying capacity. Ability of pasture land to provide feed for a certain number of livestock at a given level of production.

carry over. Part of a crop stored from one season to the next.

carse. A stretch of rich land in a valley (Scotland).

cartilage. A firm but pliant type of tissue forming portions of the skeleton. Some of the *cartilage* becomes calcified, changing into bone. Hence in young animals the proportion of *cartilage* in the skeleton is greater than in mature animals. In certain fishes, sharks, and rays, it is permanent, and never becomes ossified. Also called *gristle.*

cartridge. A pointed cylindrical arrangement on a mole plow for digging drains.

carucate. In old English law, a unit of land used as a basis for taxation; variable from 60 to 120 acres.

caruncle. 1. Fleshy protuberance on the head of a turkey or Muscovy duck. 2. A protuberance on or about the ilium of a seed.

Caryocar. A small genus of hardy trees, family Caryocaraceae, whose wood is used in furniture and other woodwork. Their large, edible, nut-like seeds yield an oil used in cooking. Native to South America. Also called *souari; sawarri.*

caryopsis. Small, one-seeded, dry fruit with a thin pericarp surrounding and adhering to the seed, as in barley, wheat, etc.

Caryota. A small genus of tropical palms, family Palmaceae, some of which are cultivated for their small fruit and for decoration. Their leaf-endings are broad and jagged like a fish tail. Native to southeast Asia. Commonly called *fishtail palms.*

casaba melon. Winter melon *q.v.*

Cascade fir. *Abies amabilis,* family Pinaceae, a tall evergreen tree with silvery-white bark grown in U.S. as an ornamental. Native to the west coast of U.S. and Canada.

cascara buckthorn. *Rhamnus purshiana,* family Rhamnaceae, a small tree, bearing dark purple fruit, whose dried bark, *cascara sagrada,* is used for medicinal purposes. Native to western North America. Also called *coffeeberry; bearwood; chittamwood; shittimwood.*

case. 1. A box and sometimes its contents, *e.g.*, a *case* of eggs. 2. In tobacco leaves, to absorb moisture after curing so that the leaves become pliable enough for handling. 3. In mushroom culture, covering the prepared bed with its final layer of soil.

casease. An enzyme of type trypsin, which is sometimes added to fresh cheese to hasten its ripening.

casebearer. Any insect of the family Coleophoridae, whose larvae build case-like formations on leaves, fruits, flowers, or seeds, and feed on them. Most of them belong to the genus Coleophora.

case-count buyer. A country buyer who buys eggs in case-lots according to quality and grade.

casehardening. 1. A condition of drying wood in which the surface becomes hard and set. If the interior tensile stresses exceed the strength of the wood, honeycombing results. 2. A condition in the artificial drying of cone evergreens when the scales become overdried and hard, usually due to low humidity in the dehydrator. This retards the rate of dehydration of the major portion of the fruit or vegetable. 3. Treatment of iron or steel machine parts with carbon so that the outer surface becomes hard to withstand wear, and leaves a soft center which withstands shocks.

casein. 1. The chief protein of milk; a phosphoprotein which occurs in the milk of different animals. Also called *caseinogen.* 2. A protein produced in milk curdled by rennet; the chief protein of cheese and casein plastics. Also called *rennet casein; caseinogen.*

caseous. Cheeselike.

caseous enteritis. See *infectious enteritis in pigs.*

caseous lymphadenitis. A chronic, bacterial disease of the Ovidae, sheep and goats, caused by *Corynebacterium ovis,* characterized by pea-sized swellings in the lymph glands. The swelling may burst but the pus is often too viscid to flow out. Also called *pseudo-tuberculosis; cheesy broncho-pneumonia.*

Caserta peperoni. A dry or semi-dry sausage, containing 75 percent pork and 25 percent beef with spices, stuffed into hog casing. The sausages are linked together in pairs and weigh approximately 12 ounces each.

cash. Pertaining to commodities sold at a market and delivered upon immediate payment. See *futures.*

cash crop. A crop sold directly on the market contrasted to one which is fed to animals or is otherwise used but not sold. Formerly applied mainly to crops which were readily salable, *e.g.*, cotton, wheat, sugar cane, etc.

cashew. *Anacardium occidentale,* family Anacardiaceae, a tropical evergreen tree with small, sweet-smelling, pale pink flowers. It is chiefly known for its kidney-shaped fruit, the *cashew* nut of commerce, which is edible after roasting. Its bark has medicinal properties. Native to America.

cash farm expense. *Farm expenses* which include only cash items or items equivalent to cash: *e.g.*, purchases of animals, expenditures for breeding fees, taxes, farm improvements, machinery, feed, crops, hired labor, interest on farm debts, farm insurance, etc. Principal payments on mortgages and old accounts, and expenses on the farm dwelling are not included in the year's farm expenses.

cash farm receipts. Receipts which include sales and trade-in allowances from transactions on the farm; *e.g.*, concerning animals, poultry and dairy products, crops, farm improvements, machinery, work off the farm, etc.

cash grain receiver. A market expert thoroughly trained in handling country business. Familiar with quality and grades of grain, he checks the samples, studies freight rates and conditions for disposal of the grain, and handles claims, questions, etc.

cash income. Money received by a farmer from his operations. Also called *cash farm receipts.*

cash market. A method of trading in the standardized grades of certain products swiftly and economically on a monetary basis at the place and on the date the product is offered for sale. The cash price for the specified grade at the specified price will be almost the same as the futures price at maturity. *Cf. futures; hedging.*

Cashmere goat. A domestic goat raised in the Himalayas for its soft fleece, which is used in making rich shawls and cloths.

cash method. One of the two main methods of accounting used in computing farm earnings for Federal Income Tax purposes. All income is included as taxable income in the year it is actually received. Farm business expenses are deductible only in the taxable year in which they are paid. Inventories of livestock, crops, produce, feed, etc., are not used in determining income by the *cash method. Cf. accrual method.*

cash record. An account of farm cash receipts and cash expenses for the year.

cash rental. The rental received by an owner of, or paid by a tenant for, a tract of land for cash as contrasted with renting for a share of the crop.

cash return. Cash income *q.v.*

cash tenant. A farmer who rents a farm from an owner for a fixed amount of money.

Casigiolu. A Sardinian, plastic-curd cheese made from cow's milk or from a mixture of cow's milk and ewe's milk; very similar to Cacciocavallo *q.v.*

casing. 1. A pipe sunk into a well to prevent the walls from caving in and within which the pipe for pumping liquids is placed. 2. Cleaned sections of hog, cattle, or sheep intestine used as sausage skins.

casing head. In well-boring, a driverhead screwed into a casing pipe to take the force of the blows while the pipe is being driven down.

casing shoe. In well-boring, a ring or ferrule of hard steel with a sharp edge which is screwed on to the bottom of a string of casing to cut its way through the formation as the pipe is forced down.

casiri. Drinkable juice from cassava (*q.v.*) which is sometimes fermented to make beer.

cask. A barrel, made of tightly fitted wooden staves and hoops, for holding liquids. It is a general term for such items as barrel, pipe, keg, etc. Formerly commonly used as a container for marketed tobacco; and in this connection also, the weight allowed for container.

cassava. *Manihot esculenta,* a woody herb, family Euphorbiaceae, grown for its roots in the warmest parts of U.S. as a stock feed. It is of importance as a source of tapioca starch and as a food plant in tropical America. The rootstock contains a poison which must be dissolved or heated out before it is edible. Also called *tapioca plant; manioc; mandioca; bitter cassava; sweet potato tree.* See *casiri.*

Cassia. A large genus of trees and plants of the senna family (Caesalpinaceae) which grows in warm regions. The leaves of many species are used in medicinal tea. The pods of *Cassia fistula* contain a pulp used as a laxative. Also called *pudding-pipe tree; canafistula; drumstick tree.* Seeds of *C. occidentalis* called *coffee senna* and *Mogdad* are marketed as an adulterant of coffee. *C. alata* is used in treating ringworm.

cassiabarktree. *Cinnamomum cassia,* an Asiatic evergreen tree, family Lauraceae; sometimes grown in Florida and California. In Asia it is grown as a fraudulent adulterant of cinnamon. Also called *Chinese cinnamon.*

cassia bud. *Cinnamomum cassia,* or any dried bud from a flower of the genus Cinnamomum, family Lauraceae, found in Asia and Australia.

cassowary. A flightless bird of the family Casuariidae which resembles the emu except for its elongated claw of the inner toe and bony casque. It grows to 5 ft. and is found in the wooded areas of northeastern Australia, New Guinea, Aru Islands, Ceram.

cast. 1. The number of young produced. 2. A caterpillar which has had its entire larval body filled and displaced with a fungus which replaces any normal host structures, as by the fungus *Cordyceps.* 3. Device for immobilization of fractured bones. 4. A sheep which is unable to rise (New Zealand). 5. Wool which has fallen off a sheep before shearing (New Zealand). 6. Angleworm cast *q.v.* 7. Afterswarm *q.v.* 8. To throw off a horseshoe, as by a horse when the nails become worn, or loosened, etc. 9. To force to the ground, as to *cast* an animal to the ground to control it. 10. To broadcast seed or fertilizer.

Castelmagno. A blue-mold, Gorgonzola-type cheese made in Italy.

cast for age. Designating a sheep culled because of its old age (New Zealand).

castings. Excrement of earthworms which is an important factor in the modification of soils.

casting the withers. The eversion of the uterus or vagina at or following parturition. A colloquial term for prolapse of the uterus, prolapse of the vagina.

cast-iron drum roller. A long roller or clod crusher, usually about eight to ten feet in length, which consists of large cylinders made of cast-iron and held together by a long axle passing through the center. A frame supporting a seat is sometimes attached. In the middle of the frame is attached a pole or other device for hitching to a team or tractor. Also called *clod crusher.*

cast-iron plant. See *aspidistra.*

Castiza. A group of vinifera grape varieties used in Spain as pollinators for the Almeria grapes. They also produce marketable fruit.

castor. A callus on a horse's leg.

castor bean. *Ricinus communis,* family Euphorbiaceae, a tall, annual herb, with alternate leaves, the only species of genus Ricinus; grown for its large, striking foliage and seeds which yield oil. The plant itself is poisonous, causing nausea, gastric pain, diarrhea, thirst, and dullness of vision. Its ingestion in large amounts is lethal. Native to tropical Asia and Africa.

castor oil. The vegetable oil from the seeds of castor bean (*q.v.*) cultivated in the East generally. It is used as a purgative, a lubricant, and for dressing leather. The specific gravity is about 0.960 to 0.966 at 60° F.

castor-oil plant. Castor bean *q.v.*

castor pomace. Residue of castor bean left after the oil has been extracted; an organic fertilizer containing up to 7 percent nitrogen, 1½ percent phosphoric acid and 1½ percent potash.

cast plow. A moldboard plow with a bottom made of extremely hard cast-iron.

castrate. 1. To remove the testicles or to destroy their use; to geld. *Cf. spay.* 2. To remove the stamens from a flower.

castration. The surgical removal of a male's testicles, or a process whereby the testicles are rendered inactive, *e.g.,* by pinching the spermatic cord, or the use of hormones, etc.

Casuarina. The only genus of family Casuarinaceae, large trees and shrubs with leafless branches. The bark of some species is used as a medicine and dye by southwest Pacific Islanders. *C. equisetifolia,* sometimes called ironwood, forest oak, swamp oak or bull oak, is found in Asia and Africa but most of the other 40 species are native to Australia. They include *C. cunninghamiana,* the river oak; *C. glauca,* the swamp oak; *C. stricta,* the she-oak or beefwood; *C. torulasa,* the forest oak.

cat. 1. *Felis libyca domestica,* a lithe, carnivorous digitigrade, domesticated as a pet by man. Probably native to Egypt. 2. A general term for an animal of family Felidae, *e.g.,* the lion, leopard, tiger, etc.

catabolism. The breaking-down phase of metabolism as contrasted to the anabolic process. Nectar and other secretions in plants are exuded through this process. *Cf. anabolism.*

catalase. An enzyme that aids in the decomposition of hydrogen peroxide, liberating oxygen.

Catalina mariposa. *Calochortus catalinae,* family Liliaceae, a bulbous plant grown for its white or lilac flowers with deep red bases. Native to California and Catalina Island.

Cataline cherry moth. Filbertworm *q.v.*

catalo. A hardy crossbreed of the American bison and domestic cattle. Also called *cattalo.*

Catalonian. A quality breed of Spanish ass that has contributed to the American jack in the breeding of mules. They are black or brown with light spots.

catalpa. Any tree of the genus Catalpa, family Bignoniaceae. They are attractive flowering trees grown as street trees. Native to America and Asia. See *southern catalpa; northern catalpa; Chinese catalpa; umbrella southern catalpa.*

catalpa mealybug. Comstock mealybug *q.v.*

catalpa midge. *Itonida catalpae* (Comst.), family Itonididae, a fly whose maggots infest the undersides of catalpa leaves close to the midrib and large vein. Fungus-like dead, circular spots appear on the infected leaves, which wilt, turn brown, and fall. The seeds may also be destroyed.

catalpa sphinx. *Ceratomia catalpae* (Bdv.), family Sphingidae, an insect which is a very serious pest of the catalpa leaves. It is distributed over the U.S. from New York to Colorado.

catalyst. An agent that promotes interaction between or among other chemical substances without itself being changed.

catalytic. Relating to the action of a catalyst.

catamaran. 1. A raft for supporting a windlass and a grappling hook for recovering logs sunk in a river. 2. Any raft with at least two logs or floats fastened together.

catamount. Any wild animal of family Felidae; usually a panther, puma, or cougar.

Catananche. A small genus of five known species of perennial herbs, family Compositae, of which the blue Cupid's-dart (*q.v.*) is of horticultural interest. Probably native to the Mediterranean area.

cataplexy. A temporary, rigid, hypnotic state in man or animals, induced by certain stimulations or medicines. Commonly described as "shamming death."

catapult fruit. A fruit that throws out its seeds by the spring-like action of its peduncle.

cataract. An opaque or cloudy condition of the lens of the eye, or of the capsule for the lens, which obstructs vision.

catarrh. An inflammatory condition of any mucous membrane, in which the discharge is thin and watery.

catarrhal fever. See *equine influenza; bluetongue.*

catawba rhododendron. *Rhododendron catawbiense,* family Ericaceae, an evergreen shrub widely planted for its beautiful purple flowers. Native to southeastern U.S. Also called *mountain rose bay.*

catbrier. Greenbrier *q.v.*

catch. 1. A sheep whose fleece is cut after the day's work has stopped (New Zealand). 2. A successful establishment from a seedling, *e.g.,* a *catch* of grass or grain. 3. To germinate, sprout, and become established as a crop after sowing. 4. To conceive after breeding, as a mare.

catch basin. A chambered well, designed to minimize the admission of grit and detritus into a tile drainage system or sewer.

catch crop. a. A quick-growing crop, planted and harvested between two regular crops in consecutive seasons, or between two patches of regular crops in the same season. b. Such a crop planted after another has failed, or when the season is too late for the usual crops.

catchfly. Silene *q.v.*

catching coop. A small coop with a sliding gate at one end into which a limited number of chickens can be placed for treatment, grading, culling, or moving from one pen to another.

catching net. A net resembling a fishing dip-net, used for catching fowls.

catching panel. A slatted device, roughly 4 by 4 feet, used to corral fowl in a catching coop.

catching pen. A pen into which sheep are driven at shearing time.

catchment basin. 1. A watershed or drainage basin. 2. The area of such a basin. 3. The natural area drained by a stream. See *watershed.*

catchup. A thick sauce made from tomatoes, properly strained, and mixed with spices, vinegar, and sugar, sometimes mushrooms and walnuts. Also called *katsup; ketchup; catsup.*

catclaw acacia. *Acacia greggi,* family Leguminosae, a forage plant of poor pasture rating which grows in semi-arid areas of western U.S. It is not very palatable, but is utilized in dry years when other feed is scarce. Also called *devil's claw; cat's claw.*

catclaw apes-earring. *Pithecellobium unguis-cati,* family Leguminosae, a tropical shrub or small tree, grown as an ornamental for its greenish-yellow flowers and red, twisted pods. Native to America. Also called *cat's claw; Florida cat's claw; escambron; black bead.*

catclaw funnelcreeper. *Doxantha unguis-cati,* family Bignoniaceae, a beautiful, evergreen vine with claw-like tendrils. Native to tropical America. Also called *cat's claw; escambron.*

catclaw mimosa. *Mimosa biuncifera,* family Leguminosae, a browse plant growing on semi-arid areas in southwestern U.S.; ranked fair in feed value. Also called *cat's claw.*

category. Any one of the sub-divisions of the system of classification in which soils are arranged on the basis of their characteristics. Beginning with the lowest *category,* soil type, soils are classified on the basis of progressively fewer characteristics into groups of higher or more inclusive *categories;* namely, series, family, great soil group, suborder and order.

catena (soil). A group of soils within one zonal region, developed from similar parent material but differing in characteristics of the profile above the parent material due to differences in relief or drainage.

caterpillar. The worm-like larva of a butterfly, moth, or other insects. All larvae have several pairs of legs, short antennae, and strong cutting jaws. Some are covered with hairs, others not. They feed on the succulent parts of the plant and are known as *canker worm, army worm, cutworm,* etc.

catface. 1. A healing or healed wound on the trunk of a tree; a blemish. 2. Fruit abnormalities of tomato, attributed to various unknown growth disturbances, partially climatic. 3. A scar or irregular growth on an apple or peach which results from insect stings.

cat flea. *Ctenocephalides felis* (Bouche), family Pulicidae, a biting insect which infests man, dogs, and cats.

cat grape. *Vitis palmata* (*V. rubra*), family Vitaceae, a variety of grape from central and southern U.S. Also called *Missouri grape.*

catgut. 1. The tough string manufactured from the intestines of sheep and other animals, used in surgical sutures, musical instruments, etc. 2. See *Virginia tephrosia.*

cat ham. A thin and flat upper leg in domestic animals, *e.g.,* horses and cows.

Cathay Japanese rose. *Rosa multiflora* var. *cathayensis,* family Rosaceae, a climbing or trailing shrub; source of several horticultural varieties of rose. Native to Asia. Also called *Chinese rose.*

cathedralbells. See *purplebell cobaea.*

catheter. A slender, tube-like, metal, rubber, plastic, or glass instrument inserted into body passages for the insertion or removal of fluids, *e.g.,* removal of urine from the bladder or insertion of semen into the cervix in artificial insemination *q.v.*

cation. An ion carrying a positive charge of electricity.

cation exchange. The interchange between a cation in solution and another cation on the surface of a colloidal or other surface-active material.

cation exchange capacity. The sum total of exchangeable cations adsorbed by a soil, expressed in milliequivalents per 100 gms. of soil. Measured values of *cation exchange capacity* depend on the method used for the determination.

catjang cowpea. *Vigna catjang,* family Leguminosae, an herbaceous vine grown in warm areas as a cover crop, a green manure crop, and for forage. Native to Asia.

catkin. An inflorescence; a spike or raceme having a very flexible rachis, which often makes it appear to be hanging. It has a scaly appearance in plants that lack petals on the flowers. However, this scaliness is not noticeable when the petals are formed.

cat-mint. Catnip *q.v.*

catnip. *Nepeta cataria,* family Labiatae, a perennial herb grown for medicinal purposes. Its fresh or dried leaves are very attractive to cats, and the oil is used as a lure in trapping wild animals of the feline family. Also called *cat-mint.*

catpine. See *white spruce.*

cat's claw. See *catclaw funnelcreeper; catclaw apes-earring; catclaw mimosa; catclaw acacia.*

cat's-ear. See *spotted catsear.*

cat's-tail speedwell. See *spike speedwell.*

catsteps. Very small, irregular terraces, on steep hillsides, especially in pastures, formed by cattle tracks or slippage of saturated soil.

catsup. Catchup *q.v.*

cattail. 1. Any marshy reed of the genus Typha, family Typhaceae, notable for its long, cylindrical, brown spike containing innumerable, tiny, petalless flowers. Generally 5 or 6 feet high, the plant has an astringent rootstock and long, thin, stiff leaves. It is the principal plant of bogs and marshes and its young shoots are sometimes eaten in England and U.S.S.R. The floss from the spike is useful for insulating purposes and filling mattresses. The spike itself is sometimes used as a cut flower for decoration. The leaves are used in making mats, baskets and chair seats. 2. *Typha latifolia,* the common *cattail.* Also called *bulrush; cat's-tail reed; cattail flag; cooper's flag; cooper's fly; ditchdown; reed mace; cattail rush.*

cattail disease. A fungous disease of grasses caused by *Epichloe typhina,* family Clavicipitaceae, characterized by the whitish body of fungus which surrounds grass stems like the spike of a cattail plant.

cattail millet. See *pearl millet.*

cattail rush. Cattail: 2.

cattle. 1. Collectively, mature bovine animals. Calves and yearlings may sometimes be included. In a broad sense, the term may also include all domesticated quadrupeds. 2. Denoting beef carcasses in the meat trade.

cattle biting louse. *Bovicola bovis* (L.), family Trichodectidae, an insect that attacks cattle and deer. Infested animals rub against fences, stanchions, etc., causing irritation of the skin. In severe cases, the animals become weakened.

cattlebush. 1. *Atalaya hemiglauca,* family Sapindaceae, a short tree with scaly, brittle bark and small flowers. Native to Australia, where its foliage is used as cattle-fodder in dry seasons. Its gum-exudate and white timber are also useful commercially. Also called *whitewood.* 2. *Heterodendron oleifolium,* family Sapindaceae, a small, bushy tree with narrow, stiff leaves, and small, petalless flowers. Native to Australia; its seeds are eaten by the aborigines, and the foliage used as fodder. Also called *bullocky-bush; whitewood.*

cattle chip. Dried cow manure; formerly used as fuel. Also called *cow chip.*

cattle country. 1. The grazing area of the western plains and mountains in U.S. where cattle are one of the principal sources of income. 2. A country whose principal product is cattle, *e.g.,* Argentina.

cattle cover. Shrubbery and brush on a range which offer shelter for cattle.

cattle drive. Moving cattle on foot from one place to another.

cattle feeding. The finishing of range, or locally raised, cattle for market by feeding them mainly on corn and other feed concentrates for fattening. *Cattle feeding* is most extensive in the corn states of U.S.

cattle follicle mite. *Demodex bovis* Stiles, family Demodicidae, a tiny mite which burrows deep into the pores of the skin of cattle. It causes pea-sized lumps or pustules to form on the head, neck, and shoulders, injuring the hides.

cattle grub. See *common cattle grub; northern cattle grub.*

cattle knocker. A worker in a slaughter house who stuns an animal with a heavy hammer before slaughter.

cattle leader. A rod, or rod and rope, attached to a spring fastened in the nostril of a cow, bull, or steer for restraining or leading the animal.

cattle louse. Any of the lice which infest cattle. See *cattle biting louse; long-nose cattle louse; short-nosed cattle louse.*

cattle malaria. Cattle tick fever *q.v.*

cattle mange. Common scab (mange) of cattle *q.v.*

cattle marker. A metal or composition tag with identifying numbers or characters, fastened to the ear, neck or horns. Also called *cattle tag.*

cattle on feed. See *cattle feeding.*

cattle pass. A culvert constructed under a road for cattle to pass from one pasture to another without obstructing traffic.

cattle plague. Rinderpest *q.v.*

cattle proof. Designating a fence which will successfully restrain cattle (New Zealand).

cattle ranch. Usually a very large farm for the purpose of breeding and/or raising cattle.

cattle range. Particular sections of a country devoted to the grazing of cattle, *e.g.,* certain areas of the western states and Florida in the U.S.

cattle run. 1. The number of cattle sent to market within a stated period, *e.g.,* a day, week, etc. 2. An alley or passageway for cattle.

cattle saltbush. *Atriplex polycarpa,* family Chenopodiaceae, a forage plant found in the dried areas of desert shrub vegetation in North America. Also called *desert saltbush.*

cattle show. See *livestock show.*

cattle station. See *station: 2.*

cattle supplement. Extra feed given to cattle to balance their ration.

cattle tag. Cattle marker *q.v.*

cattle terrace. See *terracette.*

cattle tick. *Boophilus annulatus* (Say), family Ixodidae, an insect that sucks the blood of cattle, reducing their resistance to disease, and causing general unthriftiness. It is the vector of cattle tick fever and sometimes of anaplasmosis *q.q.v.*

cattle tick fever. An infectious disease of cattle due to the protozoan parasite, *Babesia bigemina* (*Piroplasma bigemina*), whose vector is the *cattle tick.* The protozoan enters the blood and destroys red corpuscles, causing high fever, an enlarged spleen, engorged liver, thick flaky bile, red-colored urine, jaundice, and emaciation. It is fatal in about 90 percent of acute cases. Also called *Texas fever; red water; black water; southern cattle fever; acclimation fever; murrain; bloody murrain; Mexican fever; Spanish fever; splenetic fever; hemoglobinuria; bovine piroplasmosis; bovine malaria; cattle malaria.*

cattle tight. See *cattle proof.*

cattle trail. The route on which cattle were driven for long distances in the U.S. in the 19th and early years of the 20th century.

cattle wire. Barbed wire.

cattle yak. A crossbreed of the yak and domesticated cattle. *Cf. catalo.*

cattle yard. The pen or pens in which cattle are assembled for transshipment.

cattle yearlong. The forage or feed required to maintain an animal unit for one year. See *animal unit month; animal unit.*

cattley guava. *Psidium littorale* (*Cattleianum*), family Myrtaceae, a small tree grown in warm areas for its round, purplish-red edible fruit used in jellies. Also called *strawberry guava.*

Caucasian bee. A variety of the honey bee from the Caucasus Mountains.

Caucasian bladdernut. See *Colchis bladdernut.*

Caucasian leopardbane. *Doronicum caucasicum,* family Compositae, a perennial herb grown for its yellow flowers. Native to Eurasia.

Caucasian lily. *Lilium monadelphum,* family Liliaceae, a bulbous herb grown in flower gardens for its tall, golden flowers, occasionally spotted with purple. Native to the Middle East.

Caucasian scabious. *Scabiosa caucasica,* family Dipsacaceae, a popular, perennial herb, with white or purple flowers, grown for decoration. Native to the Caucasus. Also called *pincushion flower.*

Caucasian valerian. *Valeriana phu,* family Valerianaceae, a perennial herb grown as a border plant for its whitish clusters of flowers. Native to the Caucasus. Also called *Cretan spikenard.*

cauchillo. *Lecythis ollaria,* family Lecythidaceae, a gigantic tropical tree whose red-brown wood is valued as timber. Native to America.

caudal. Near the tail.

caudal fold. Fold of skin at the junction of tail and body.

caudex. The upright stem of a fern, palm, etc.

caul. 1. The fat around the stomach of cattle, sheep, and swine. 2. In embryology, a part of the thin membrane or web surrounding the fetus.

caulescent. Describing a plant with a leafy stem growing above the ground.

caul fat. Fat obtained from the visceral cavities of beef cattle.

caulicle. The initial stem in a young seedling or an embryo.

cauliflory. The bearing of fruit and flowers on the trunk and large limbs, especially in many tropical trees like the redbud, carambola, etc.

cauliflower. *Brassica oleracea* var. *botrytis,* family Cruciferae, a stemless, annual or biennial herb, grown for its white or green, enlarged, edible inflorescence. It is an important vegetable crop. Native to Europe. *Cf. broccoli; cabbage.*

cauliflower disease. See *potato wart.*

cauliflower disease of strawberries. Caused by the nematode, *Aphelenchoides ritzema-bosi,* family Aphelenchidae, and the bacterium, *Corynebacterium fascians,* this disease results in a cauliflower-like development in the crown, which stunts the plant.

caulis. The stem, or the stalk, of an herbaceous plant.

caulote. *Guazuma ulmifolia,* family Sterculiaceae, a medium sized tropical tree with beautiful yellow flowers. Its leaves are used for forage, wood for timber, and bark for cordage. Native to America. Also called *bastard cedar.*

causal organism. An organism, such as fungus, bacterium, nematode, etc., which produces a particular disease.

causative agent (organism). Any substance, pathogen, etc., which produces a disease.

caustic lime. See *calcium oxide; burnt lime.*

cauterize. To burn the flesh or skin of an animal with a drug or heated metal.

cave. 1. A cellar or underground place which has the proper humidity and temperature for the growing of mushrooms. 2. An underground place used for the proper ripening and curing of special kinds of cheese.

cavenia acacia. *Acacia cavenia,* family Leguminosae, a quick-growing, spiny shrub grown in hedges. Its fruit contains tannin. Native to Chile. Also called *espino; espino cavan.*

cavesson. A noseband of iron, wood or leather, used to hold the head of a horse during training. Also called *caveson.*

cavity. 1. A hollow space within a body, *e.g.,* the buccal *cavity.* 2. Any hole in the trunk or branches of a tree.

cavity bracing. The insertion of metal rods across the cavity in a tree; used to support the sides and prevent the trunk from breaking.

cavy. 1. An American-Indian pony. 2. One or a group of stray animals (U.S.). 3.a. Any South American rodent of the family Caviidae of which the capybara and guinea pig are the main species. b. A guinea pig.

cayenne pepper. A condiment prepared by grinding the seeds of several plants of the genus Capsicum (*q.v.*), *e.g.,* redpepper, chili, etc.

Cayuga. A breed of duck with lustrous greenish-black plumage. The drake weighs up to 8 lbs., and the duck up to 7 lbs.

cayuse. A riding horse or pony of no particular breed (western U.S.). Also called *fuzz tail.* See *broom tail.*

cc. A cubic centimeter; a cube with an edge of one centimeter.

Ceanothus. A large genus of beautiful flowering plants, family Rhamnaceae, a few species of which are grown for their clusters of small blue or white flowers. Native to North America. Also called *wild lilac.* See *Jerseytea ceanothus; feltleaf ceanothus; squawcarpet ceanothus; redheart ceanothus; blueblossom ceanothus; deerbrush ceanothus.*

cecal. Pertaining to a caecum *q.v.*

cecal abligation. An operation on fowl in which the caecum is tied off to cause it to shrink and become severed from the main gut. It is effective in the treatment against blackhead *q.v.*

cecal coccidiosis. See *coccidiosis.*

cecidiology. The division of biology which studies galls on plants caused by insects, mites, and fungi.

cecidium. A gall.

Cecropia. A tree genus, family Moraceae, whose bast fiber is used in ropemaking and in bark tanning. Some species yield a milky juice containing caoutchouc. Native to tropical America.

cecropia moth. *Hyalophora cecropia* (L.), family Saturniidae, a moth whose larvae feed on fruit and forest trees. The adult moth has a wingspread of seven inches. Native to eastern U.S.

cecum. See *caecum.*

cedar. 1. Any tree of the genus Cedrus, family Pinaceae, especially the *deodar cedar, Atlas cedar, Cyprian cedar, cedar-of-Lebanon.* 2. Commonly but not accurately, any tree of the genus Juniperus, family Pinaceae, as the *southern redcedar, eastern redcedar, Rocky Mountain juniper.* 3. Commonly but incorrectly, certain trees of the genus Cedrela, family Meliaceae, as *cigarbox cedrela.* 4. Commonly but incorrectly, any tree of the genus Libocedrus, family Pinaceae, as *California incensecedar.* 5. Commonly but incorrectly, any tree of the genus Torreya, family Taxaceae, as *California torreya.* 6. Commonly but incorrectly, any tree of the genus Chamaecyparis, family Pinaceae, as *whitecedar falsecypress.*

cedar apples. Hard, brown, sphere-like excrescences of *Juniperus* spp. caused by the rust fungus *Gymnosporangium juniperi-virginianae,* family Pucciniaceae, up to 2 inches in diameter, scattered over the tree, which sometimes bend the branches with their weight. They represent the telial stage in the development of the fungus which may also spend part of its development upon the apple or other members of the family Rosaceae. Also called *cedar balls; cedar galls; cedar flowers.* See *apple rust.*

cedar apple rust. See *apple rust.*

cedar balls. See *cedar apples.*

cedar elm. *Ulmus crassifolia,* a medium-sized elm, family Ulmaceae, which bears stiff corky or scaly branches and blooms in autumn. Native to central southern U.S. where it is grown as an ornamental. Also called *evergreen elm.*

cedar flowers. Cedar apples (*q.v.*) which in spring develop elongated, gelatinous, orange-colored horns giving a flower-like appearance. See *apple rust.*

cedar galls. Small swellings on branches, twigs, and leaves of junipers and related trees and shrubs. See *cedar hawthorn rust.*

cedar hawthorn rust. A fungous disease of hawthorn and cedar caused by *Gymnosporangium globosum,* family Pucciniaceae, which alternately infests junipers and hawthorns, rarely apples. It produces small, irregularly shaped galls on the junipers which have wedge-shaped, gelatinous, orange spore masses. Its symptoms are similar to apple rust (*q.v.*) on the apple and hawthorn.

cedar moss. See *hornwort.*

cedar-of-Lebanon. *Cedrus libani,* family Pinaceae, the famous, evergreen tree of the Bible. It is one of the traditional cedars growing up to 100 feet in height and sometimes bearing branches 40 to 50 feet long. Native to Turkey; grown in southern U.S.

cedar pine. See *scrub pine: 2.*

cedar quince rust. See *eastern pear and apple rust.*

cedar rust. See *apple rust; cedar hawthorn rust.*

cedar tree borer. *Semanotus ligneus* (F.), family Cerambycidae, a beetle whose larva infests arborvitae, redwood, Douglasfir, and Monterey pine making winding galleries in the inner bark and sapwood.

cedrat. See *citron: 1.*

Cedrela. A genus of large mahogany trees, family Meliaceae, whose aromatic bark is used to relieve fever; its timber is used in making furniture. Found in tropical America, the West Indies, India and Australia.

ceibo. See *cockspur coralbean.*

ceiling price. The highest price allowed for a commodity by the U.S. federal government.

celandine. See *greater celandine.*

celandine poppy. *Stylophorum diphyllum,* a hardy, perennial herb, family Papaveraceae, grown in wild gardens for its deep yellow flowers. Escaped, it can be a weed pest. Native to North America.

Celastrus. A woody vine genus, family Celastraceae, to which belongs the sudorific American bittersweet (*q.v.*) and the East Indian *C. paniculatus* which is the source of the oil, *oleum nigrum.*

Celebes pepper. *Piper ornatum,* a climbing vine, family Piperaceae, frequently grown in greenhouses as an ornamental for its handsome white- or pink-spotted foliage.

celeriac. *Apium graveolens* var. *rapaceum,* a Eurasian, biennial herb, family Umbelliferae, grown for its edible, thickened, celery-flavored root. It is a minor vegetable. Also called *turnip-rooted celery; knob-rooted celery.*

celery. See *garden celery.*

celery blight. See *late blight of celery; early blight of celery; bacterial blight of celery.*

celery cabbage. See *petsai.*

celery harvester. A manual or tractor-attached device with two or more L-shaped teeth which turn inward toward either side of a row and cut the roots of the celery before harvesting.

celery hiller. A mechanical device for blanching celery. It throws soil up and around the growing plants.

celery juice. A food prepared by trimming, washing and steaming celery stalks. The juice is expressed by means of a screw-type press and then screened before salt and citric acid are added.

celery late blight. See *late blight of celery.*

celery leaf tier. *Udea rubigalis* (Guen.), family Pyraustidae, a major pest of celery, peas, spinach, and beets. It feeds on the tender growth just above the heart of the plant, covering the leaves with its web and excrement. The caterpillar usually feeds on the under surface of the leaves. It also attacks other soft-leaved plants, *e.g.,* snapdragon, cineraria, aster, sweetpea, ageratum, ivy, rose, etc. Also called *greenhouse leaf tier.*

celery mosaic. See *western celery mosaic.*

celery salt. A mixture of 25 percent ground celery seed, 73 percent common salt, and 2 percent calcium phosphate, used as a condiment.

celery seed. A condiment made of the dried fruit of wild celery, *Apium graveolens.* The seeds are also used in preparing oil of celery.

celeryworm. Black swallowtail *q.v.*

celery yellows. A fungous disease caused by *Fusarium oxysporum f. apii,* family Tuberculariaceae, which enters the roots and stems of the plant, discoloring the vascular bundles, and yellowing and stunting, sometimes killing, the infected plants.

cell. 1. A hexagonal unit compartment of a honey comb. 2. The ultimate functional unit of an organic structure, plant or animal. It consists of a microscopic mass of protoplasm which includes a nucleus surrounded by membrane. In most plants it is surrounded by a *cell* wall. 3. A single element of an electric battery, either primary or secondary, generally consisting of a jar filled with a liquid or a pasty electrolyte, into which the electrodes are inserted or connected. 4. A very small, enclosed compartment. 5. See *accommodation cell.*

cellar. See *storm cellar; cold cellar.*

cellar gardening. 1. The forcing of plants, *e.g.,* rhubarb, asparagus, French endive, etc., to grow in a cool room. 2. The indoor-growing of mushrooms, chard, parsley, etc., during the winter.

cellar wintering. Bringing beehives into a storeroom, especially a cellar, during cold winter months to protect the bees.

cell cup. An artificial cell made of beeswax for rearing queen bees.

cell drop. A device in mechanical seed planters consisting of a plate with cells on the outer edge. Seeds from a hopper drop into the cells, one by one, and from there into a spout and finally into the earth at regular intervals along the row. See *corn planter plate.*

cellular. Referring to a cell or cells.

cellular plant. A non-flower-producing plant without vascular tissues.

cellulose. An inert, complex carbohydrate which makes up the bulk of the cell walls of plants. Derived from stalks, straws, etc., it is an important farming by-product, and is used for making paper, fabrics, and building materials.

cell wall. The membranous covering of a cell secreted by the cytoplasm in growing plants. It consists largely of cellulose but may contain chitin in some fungi and silica in some algae.

Celsia. A genus of hardy herbs, family Scrophulariaceae, which bear large yellow flowers. Native to western Asia and southern Europe.

Celtic ox. *Bos taurus longifrons,* a fairly small, Neolithic ox with a high forehead; probably the progenitor of the Brown Swiss and Jersey breeds.

cement disease. Lung inflammation in pigs caused by living on new cement floors.

cement dust. A by-product from cement manufacture; contains six to nine percent potash. Used as a carrier for potash in the manufacture of commercial fertilizers.

cement-dust injury. The retarding or complete stopping of the setting of fruit as a result of dust from cement mills.

Cenangium twig blight. A fungous disease of certain species of pine, caused by *Cenangium ferruginosum.* Characterized by the death of terminal buds and the reddening of needles.

Cenchrus. A genus of shallow-rooted, spreading grass-weeds, distinguished by their prickly burs. Some species can be used as forage before the appearance of burs.

cenizo. A Spanish term for the fourwing saltbush *q.v.*

census plot. A small area of land used for vegetation studies, experiments, etc., to determine standard yields of crops. See *list quadrat; stocked quadrat; frequency-index quadrat.*

Centeotl. Aztec goddess of growing vegetation. See *Ceres.*

center brick mold. A block of ice cream with a special design in the center.

center cut. The tender parts near the center of a piece of meat, *e.g.,* pork chops.

center fire. A method of controlled burning in which initial fires are set in the center of an area to be broadcast-burned to create a strong draft. Later, additional fires are set near the outer control lines of the area to act as backfires to check outward spread.

center-fire rigging. One of the four main positions on the rigging of a saddle. The saddle has only one cinch, placed in the center.

center of draft. A point within the plow about which all the forces which act on the plow are balanced. Also called *center of load.*

center of load. See *center of resistance; center of draft.*

center of pull. The true point of hitch or center of power in machines. On a tractor, this is the point halfway between the wheels, at which the drawbar is attached.

center of resistance. A point where the vertical, horizontal cross-furrow, and longitudinal forces meet in a plow bottom. Also called *center of load.*

center rot. Alternaria rot of lemon *q.v.*

center-trip hay sling. See *hay sling.*

center windrow. A windrow attachment for a cutter bar so arranged that the windrow is made at the center of the cutter bar rather than at the end. Used in harvesting peas for canning.

centgener method of grain breeding. An obsolete method of breeding small grains whereby 100 seeds from each selected plant are planted in trial plots. The more promising plants are selected from these trial plots and the seed from them is planted. The process is repeated yearly.

centigrade thermometer. A thermometer on whose scale the interval between the freezing and boiling points of water is divided into 100 parts or degrees.

centipede. Any myriapods of the class Chilopoda, close relatives of the true insects. They are beneficial because they prey on insects.

centipede grass. *Eremochloa ophiuroides,* a creeping, perennial grass, family Gramineae, introduced from China. It has proved useful in stabilizing waterways in soil erosion work in southern U.S., and as a lawn grass because it quickly forms a dense turf.

centipede plant. See *ribbonbush.*

central. A large sugar mill or sugar refinery in the center of the sugar growing country.

Central American sugar. An unrefined grade of cane sugar which may contain as little as 84 percent sucrose. Once used in the canning of fruits.

central market. Public livestock market *q.v.*

central nervous system. The brain and spinal cord in vertebrates which receive sensory impulses and from which motor impulses issue.

centrals. The two central incisors of a horse. Also called *pincers.*

central service court. The yard around which the farm buildings are grouped and from which the work of the farm is carried on. Ideally, it should be possible to reach each farm building from this yard without going through the stockyard or opening gates. Also called *central service yard.*

centrifugal. 1. A principle of force; moving or flying away from the center. 2. In botany, growing outward from the center.

centrifugal clarifier. A high-speed centrifuge designed to throw out impurities or sediment from a liquid.

centrifugal cream separator. A mechanical device consisting of a bowl fitted with fast revolving discs which separate fat from milk for cream. This quick separation of fat is possible due to the relative low specific gravity of the fat.

centrifugal force. The force which acts upon a body revolving in a circular path, tending to force the body farther from the center of the circle. If the *centrifugal force* is just sufficient to balance the attraction of the mass around which it revolves, the moving body will continue in a uniform curved path. Should the *centrifugal force* increase, the body will either take up a larger path at a further distance from the center, or else tend to fly off in a straight line.

centrifugal pump. A water-lifting device that utilizes the centrifugal force imparted to the water by a rapidly rotating impeller. The water is admitted to the center of the impeller and discharged at its outer periphery.

centrifugal separation. The separation of substances of different densities by means of centrifugal force, as the separation of fat from milk.

centrifugal separator. See *centrifugal cream separator.*

centrifugal sugar. Sugar separated from liquid by a centrifuge.

centrifuge. A machine, such as a Babcock milk tester, used to separate the fat from acid-milk mixture.

century aloe. See *centuryplant agave.*

centuryplant agave. *Agave americana,* a succulent, fleshy-leaved, tropical American plant, family Amaryllidaceae, grown for its white or yellow flowers. Also called *century aloe;* incorrectly referred to as *American aloe.*

cepaceous. Smelling or tasting like onion or garlic.

cephalocaudal. To do with the head and the tail of an animal.

Cephalosporium fungus. *Cephalosporium lecanii,* family Moniliaceae, a parasite used effectively in the biological control of pyriform scale, brown scale, lemon scale, green scale, and cottony cushion scale.

Cephalosporium wilt. See *Dothiorella wilt.*

Cephus. A genus of sawflies, family Cephidae, whose larvae burrow into plant stems. *C. pygmaeus* infests and destroys wheat; *C. cinctus* is harmful to some grasses and grains.

Cerambycidae. The larvae of the many species of beetles belonging to this family burrow into the wood or roots of trees and shrubs, sometimes causing great harm. See *locust borer; twig pruner.*

ceramic oven. A type of dehydrator used for fruit and vegetables. The produce is placed on trays or trucks on the floor and the heat is radiated from the walls, ceiling, and floor which are made of heavy masonry and fire brick. There is also a forced draft to circulate the air and improve drying.

Ceratostomella. A fungus genus, family Ceratostomataceae. *C. pilifera* is the cause of blue stain *q.v.*

Cercis. A genus of redbud shrubs, family Leguminosae, grown for their showy red-purple or pink flowers which blossom in early spring. See *American Redbud.*

Cercospora. An imperfect fungus genus, family Dematiaceae, of which over 1,000 species have been given names, their distinguishing characters being largely based upon the species of the plants infected (host plants) and the type of lesions produced. The species produce elongated, slender spores or tufted conidiophores which emerge from the stomata of the infected spots. These spores are borne by air currents or are splashed by rain drops and carried to other leaves where germination of the sperm occurs. A few named species of *Cercospora* grown under controlled conditions on special culture media have produced the perfect stage of reproduction.

Cercospora blight of carrot. A disease caused by the fungus *Cercospora carotae,* which results in the spotting of leaves and petioles. The fungus can lie dormant in the refuse and seeds of the infected plant so that carrots planted in successive years in the same soil may also become infected. See *Cercospora.*

Cercospora leaf and stem spot of clover. Caused by *Cercospora zebrina,* on various species of clover. The leaflets, stems, petioles and seeds may become infected with small and linear or angular and reddish or woody-brown lesions. Stem infection may cause the death of the distal parts of the plants. The disease can lie dormant in residue and infected seeds. See *Cercospora.*

Cercospora leafspot of beets. The most common and destructive leaf disease of the sugar beet and garden beet, caused by the fungus *Cercospora beticola.* Characterized by dark-bordered circular spots, about ⅛ inch in diameter, on the leaves. The spots on the petioles are elongated with brown centers. Most prevalent in humid areas. See *Cercospora.*

Cercospora leafspot of cotton. A minor disease caused by *Cercospora gossypina.* It causes irregular-shaped spots with purple borders and white centers on the plant leaves, especially toward the close of the season. This fungus is one of the species of *Cercospora* of which the perfect reproduction stage, *Mycosphaerella gossypina,* family Hycosphaerellaceae has been determined. See *Cercospora.*

Cercospora leafspot of peppers. A fungous disease of redpepper (*Capsicum*) caused by *Cercospora capsici,* characterized by pale gray spots with wide, dark margins on the leaves and stems. Severely infested plants may defoliate, reducing the yield. See *Cercospora.*

Cercospora spot of rice. A disease caused by the fungus *Cercospora oryzae* forming narrow, elongated, colored lesions on the leaves and leaf sheaths of rice. Some varieties of rice are less susceptible than others. See *Cercospora.*

Cercosporella. A genus of imperfect fungi, family Moniliaceae, with characteristics similar to the fungi of genus Cercospora (*q.v.*) but having translucent spores.

Cercosporella foot rot of cereals. A plant disease of wheat and other cereals, usually caused by the fungus *Cercosporella herpotrichoides.* It produces eye-spot lesions at the base of the stem and in the root system, causing the plant to droop or break. In northern Europe, and sometimes in northern U.S., the fungus that causes this disease is *Leptosphaeria herpotrichoides,* perhaps the perfect stage of *C. herpotrichoides.* Also called *stem break.*

cereal. 1. Any grass grown for its edible grain. 2. A prepared foodstuff from grain, *e.g.,* oatmeal.

cereal anthracnose. A minor disease caused by *Colletotrichum graminicola,* family Melanconiaceae, and characterized by lesions and cankers on the stem, leaves, and grain. Infects rye, wheat, oats, sorghum, Johnsongrass, Sudan grass, corn, and barley. See *anthracnose.*

cereal cream. Substandard cream *q.v.*

cereal foot rot. Cercosporella foot rot of cereals.

cerealist. One who deals with or specializes in the study of cereals.

cereal milk. See *substandard cream.*

cereal rust. See *rusts; leaf rust of cereals; stem rust of barley, oats, rye and wheat; stripe rust; crown rust.*

cereal scab. A fungous disease of the heads of wheat, barley, rye and ears of corn, caused by *Gibberella zeae,* family Nectriaceae, which makes the grain unwholesome. The fungus may also bring about stem and root rots of these plants, so that the infected stems may become sufficiently weakened to break over in strong wind.

cerebrospinal. Referring to the brain and spinal cord, *e.g.,* the *cerebrospinal* canal.

cerebrospinal fluid. Liquid found in or around the brain and spinal cord filling the cerebrospinal canal, which runs through the length of the spinal cord.

cerebrospinal meningitis. Inflammation of the membranes covering the brain and spinal cord. The condition occurs in some types of forage poisoning and in some diseases of the central nervous system.

Ceres. Roman goddess of corn and growing vegetation, adapted from the Greek goddess Demeter in the 5th century B.C. Wheat was considered to be her special gift to mankind and she was yearly honored with sacrifices and other rituals just after the seed-time and before harvesting. The word *cereal* is derived from her name.

cereza. Spanish word for cherry, but specifically for a fruit of the plants of genera Cordia and Malpighia *q.q.v.*

ceriman. *Monstera deliciosa (Philodendron pertusum),* family Araceae, a climbing or trailing tropical vine grown for its large perforated leaves, and conelike edible fruits which taste like a mixture of banana and pineapple. Native to America. Also called *fruit-salad plant.*

ceriops. *Ceriops tazal,* family Rhizophoraceae, a large tree with thick, leathery leaves whose bark is valued for tanning purposes. Native to southeast Asia.

Ceroxylon. Genus name of very tall palms, family Palmaceae, with huge, featherlike leaves and small, round, purple fruit. Native to western South America.

cerrero. Describing a wild, untamed horse (western U.S.).

cerris. See *European Turkey oak.*

certation. The partial selective competition between pollen classes produced by a heterozygous plant which results in a greater percentage of fertilizations by one class, as revealed by the character of the offspring. It may result from differential germination of pollen, or differential pollen tube growth.

certification. See *plant certification.*

certification of seed. Seed production and marketing under the control of a certifying agency to maintain varietal purity and freedom from seed-borne pests.

certification standard. Rules and regulations of a certifying agency concerning the breeding, production, and cleaning of seeds along with the maximum amount of impurities allowed, etc.

certified meat sire (cms). A boar whose progeny in a specified number of litters has met certain requirements regarding carcass excellence, growth rate, and size, etc.

certinensis. Berlin poplar *q.v.*

cervelat. Summer sausage *q.v.*

cervical. Referring to the neck.

cervicitis. Inflammation of the narrow end of the uterus.

cervix. The neck or constricted part of an organ, especially the narrow end of the uterus.

Cesarean birth. Delivery of the fetus through an abdominal incision. Also called *Caesarean section.*

cespitis. A sod horizon, or surface soil layer held together by a tangle of roots, mostly grass.

cespitose. Turflike; having dense growth as in turf.

cess. A levy or tax, as land *cess,* usually exacted locally.

cesspit. A pit in which refuse, night soil, etc., is dumped.

cesspool. A receptacle into which household sewage or other liquid waste is discharged. The liquid leaches into the surrounding soil or is otherwise removed. See *septic tank.*

cestode. See *tapeworm.*

cevitamic acid. Vitamin C *q.v.*

Ceylon cinnamon. *Cinnamomum zeylanicum,* family Lauraceae, an evergreen tree of tropical Asia grown as a source of the spice, cinnamon, used as a condiment.

Ceylon gooseberry. Ketembilla *q.v.*

C.F.S. Cubic feet per second.

C.G.S. Centimeter-gram per second.

chacate. *Krameria grayi* (*Canescens* A. Gray), family Leguminosae, a shrub with alternate leaves and purplish flowers whose bark yields a dye. Native to southern U.S. and Mexico. Also called *Grays Krameria.*

chacte. *Caesalpinia mexicana,* family Leguminosae, a small tree whose wood is used in cabinet work. Native to Mexico.

Chaenactis. A genus, family Compositae, of annual or perennial herbs with alternate leaves. Native to western North America.

chafer. See *rose chafer.*

chaff. 1. The outer covering of grain, the glumes, husks, separated from the seed by threshing. 2. Finely cut hay or straw used as feed for cattle. 3. The scales associated with the seeds in the heads of some composite plants.

chaffer. A sieve on a threshing machine that separates the threshed grain from chaff.

chaff scale. *Parlatoria pergandii* Comst., family Diaspididae, an insect pest of citrus and other tropical plants in the Gulf states of U.S.

chaffy. Having too much chaff or being too light in weight.

chaffy corn. Immature corn cut and dried to 17 percent or less of moisture content.

chain. 1. A series of metal links forged or welded together and used for fastening, hauling, etc. 2. A 66 foot long tape or chainlike measure used in land surveying.

chain gage. A device for determining the elevation of a water surface. It consists of a tagged or indexed chain, tape, or other line attached to a weight lowered to touch the water surface whereupon the gage height is read on a graduated staff opposite an index. Especially suited to situations where the water is difficult to reach. The graduated staff is sometimes placed horizontally and the line run over a pulley.

chain harrow. A harrow having no rigid frame. It consists of interlocked round, square, or different-shaped links which make up a strong, flexible mesh to adjust to the inequalities of the ground surface. Used for rolling up and collecting weeds that have already been uprooted by other harrows. See *spiked flexible harrow.*

chain hoist. A mechanical device for lifting heavy weights, such as milk cans, engines, etc. It consists of a hook attached to a chain or rope and a set of pulleys, fastened to a track.

chaining. 1. The wrapping of a chain around logs, lumber, etc., for skidding or hauling. 2. Locking of a wheel with a chain to act as a brake when descending steep slopes.

chain-of-love. Mountainrose coralvine *q.v.*

chain pump. A machine consisting of an endless chain dipping into water in a shallow well at one end and passing over a revolving wheel at the other. Attached to the chain, at regular intervals, are cups or buckets which lift the water and drop it into a spout as they pass over the wheel at ground surface. A larger version of this mechanism, using animal power to activate the wheel, is a widely practised method of land irrigation in Asia. See *Persian wheel.*

chain saw. A portable, motor-driven saw which consists of an arm on which travels an endless chain with attached saw teeth. Used in felling trees, sawing logs, or in cutting underbrush.

chain tie. 1. A section of light chain attached to a manger in a barn that branches into a Y to go around both sides of a cow's neck and fasten at the top. It serves to restrain the cow in a manner similar to a stanchion. 2. A light wrapping chain thrown across the top of a load of logs to secure it for transportation.

chain whip. A large hook attached to a chain; used for pulling out tree stumps. The chain is tied around the stump to be uprooted and is pulled by animal or mechanical power.

chalaza. 1. Either of two cordlike, opalescent, albuminous strands, in the white of a fowl's egg, which are attached to the yolk and prolonged toward the ends of the egg. The *chalazas* aid in keeping the yolk in proper position. 2. The place in an ovule or seed at which the integuments diverge from the nucellus.

Chalcedonian lily. *Lilium chalcedonicum,* a bulbous herb, family Liliaceae, with ill-scented, deep-scarlet flowers. Also called *scarlet Turk's cap lily.* Native to Greece.

chalcid (wasp) fly. Any small insect of the superfamily Chalcidoidea which in its early stage exists in or on other larvae or pupae, thus destroying many undesirable pests.

chaldron. An old English measure; usually 36 bushels, but varying according to district.

chalicevine. See *goldcup chalicevine.*

chalk. 1. Soft white, fine grained limestone which is highly fossiliferous. As the parent rock, it lends distinctive character to soils. Locally impure chalks may be called *marl.* 2. To spread *chalk* over land as a manure.

chalk brood. A disease of brood combs in a bee hive caused by the fungus, *Pericystis apis,* family Pericystaceae, which usually attacks only the drone brood. Uncommon in U.S.

chalk maple. *Acer leucoderme,* a deciduous tree, family Aceraceae, grown as a shade tree and as an ornamental for its light gray bark. Native to North America.

chalone. An internal secretion which depresses or inhibits metabolism. Also called *colyone.*

Chamaecyparis. A genus of pine trees, family Pinaceae, which yields valuable timber. Native to North America, Japan.

Chamaedaphne. A genus, family Ericaceae, of evergreen shrubs grown in rock gardens. Found in bogs of northern Europe, northern Asia, North America. See *leatherleaf.*

Chamaelirium. A plant genus, family Liliaceae, of which the only known species is *C. luteum,* the fairy wand (*q.v.*). Native to eastern North America.

chamber. 1. A space enclosed by walls; a compartment. 2. In mining, to explode a small charge to create a cavity for placing a larger charge.

chamfer. a. The flat surface exposed after cutting off the right-angled corner of a block of wood, stone, etc. b. To bevel.

chamisal. A dense thicket of chamiso. See *chaparral.*

chamise lily. Lambstongue fawnlily *q.v.*

chamiso. 1. *Adenostoma fasciculatum,* family Rosaceae, an evergreen shrub grown in warm, humid regions for its needle-like foliage and clusters of white flowers. The most characteristic shrub of the California chaparral 2. *Atriplex canescens,* family Chenopodiaceae, a semidesert shrub found in New Mexico in U.S. Also called *greasewood chamise.*

chamiza. Fourwing saltbush *q.v.*

chamois. 1. *Rupicapra rupicapra,* a small antelope which resembles a goat. Found in the mountains of Europe and New Zealand. 2. Soft, pliable leather made by rubbing oil into the skins of the *chamois,* goat, sheep, etc. Also called *shammy.*

chamomile. Roman camomile *q.v.*

chamoudi orange. See *Shamouti orange.*

champ. To chew noisily, as a horse *champs* at the bit.

champaign. An expanse of flat, open country.

champignon. 1. A mushroom, especially *Marasmius oreades,* the fairy-ring mushroom. 2. Inflammation of a horse's spermatic cord.

champion brand Chilean nitrate. A commercial grade of nitrate of soda derived from caliche in Chile whose nitrate content ranges from 98.3 to 98.8 percent, being the equivalent of 16.2 percent nitrogen. *Cf. old style Chilean nitrate of soda.*

champion oak. See *northern red oak.*

champion pen. Winner as a group in a show (usually three animals per pen) rather than as individuals.

Champney rose. See *Noisette rose.*

Champoleon. A hard, skim milk French cheese similar to Canquillote.

chance. In logging, a place in a forest suitable for one operation, *e.g.,* logging *chance,* cutting *chance,* etc.

chance seedling. A fruit or nut tree which grows from seed scattered by wind, birds, etc.; sometimes bears fruit. Many varieties of fruits originated as *chance seedlings.*

chandul. A fiber from the upastree of India, *Antiaris toxicaria,* family Moraceae, used to make sacking.

changa. *Scapteriscus vicinus* Scudd., family Gryllidae, a mole cricket that attacks garden crops, peanuts, and tobacco by uprooting seedlings, cutting the roots off, and eating into the underground parts. Native to South America, the West Indies. Also called *Puerto Rican mole cricket; mole cricket.*

change agent. One who advocates or attempts to introduce another method; a teacher, an agricultural extension agent; the new system itself.

channa. An Indian confection resulting from the acid precipitation of the protein from heated milk; fat is removed from the milk with the coagulum and the whey is discarded. It is preserved by adding sugar and making it into sweetmeats.

channel. 1. A canal or small river. 2. The deeper part of a river or harbor usually caused by the main current. 3. A passage of water joining two bodies of water. 4. Any narrow passage through which liquid or air flows.

Channel Island breeds. See *Alderney; Guernsey; Jersey.*

channery. Small, thin, flattish fragments of limestone, sandstone and schist found in soils.

Chantecler. A Canadian breed of fowls with good winter laying qualities; capable of resisting a rigorous climate. Characterized by small cushion-shaped combs, small wattles, tight feathering and yellow skins; the eggs are brown. The cock weighs 8½ lbs., the hen 6½ lbs. There are two varieties—White and Partridge.

chanterelle. *Cantharellus cibarius,* a common, deep yellow mushroom.

Chaource. A soft, whole-milk French cheese, similar to Camembert.

chap. See *chop: 6.*

chaparajos. See *chaps.*

chaparral. A dense growth of shrubby vegetation or stunted and dwarf trees in arid and semi-arid regions of the southwestern U.S. Characteristic shrubs of the coast ranges of California are chamiza, ceanothus, and mazanita. In other places dense thickets of mesquite (*Prosopis* spp.) and thickets of oaks may be called *chaparral.*

chaparral broom. Kidneywort baccharis *q.v.*

chaparral lily. *Lilium rubescens,* a tall, bulbous herb, family Liliaceae, grown for its showy, pale lilac to rose-purple flowers. Native to North America.

chaparral snapdragon. *Antirrhinum coulterianum,* an annual herb, family Scrophulariaceae, grown for its purple to white flowers which have yellow hairs in the throat of the corolla. Native to North America.

chaparral whitehorn ceanothus. *Ceanothus leucodermis,* family Rhamnaceae, a browse shrub which is moderately palatable to livestock. Found from British Columbia to California, Idaho and Montana.

chapote. *Diospyros texana,* family Ebenaceae, the persimmon of Mexico.

chaps. Leather coverings from the waist to the feet, worn by cowboys to protect their legs from cacti, trees, or brush. Also called *chaparajos.*

character. 1. One of the details of structure, form, substance, or function which make up a species and distinguish it from others. 2. A judging term for desirable, necessary, or attractive qualities of an animal or product. 3. The evenness and distinctness of crimp in wool fibers.

characterize. To indicate, represent, describe; to form a characteristic of, to make distinct.

charbon. French name for the smut diseases of cereals and anthrax.

Charbray. A breed of beef cattle, about ¾ Charolaise and ¼ Brahman.

charco. A waterhole; or a standing pool of water, which remains following rains in adobe regions of southwestern U.S.

charcoal. A black, porous, carbonous substance, obtained by the partial burning of an animal or vegetable substance in the absence of air. A common fuel, it is also a mild antiseptic and deodorant.

charcoal rot. 1. A fungous disease caused by *Macrophomina phaseoli,* family Sphaeropsidaceae, which attacks over 30 different plants. Characterized by a rotting of the roots and the stalk just above, which stunts the plant and reduces the yield. 2. A decay of limes (*Citrus aurantifolia*) caused by *Diplodia natalensis,* family Sphaeropsidaceae, characterized by a brownish spot on the mature fruit. The fruit dries and becomes filled with a black, hard mass of mycelium. 3. Jet-black rot of the root system of sweet potatoes caused by the imperfect fungus, *Sclerotium bataticola.*

chard. Leaf beet *q.v.*

charge. 1. In electrostatics, the amount of electricity, measured in coulombs, present upon any substance which has accumulated electric energy. 2. In storage battery work, the amount of current absorbed by the battery during the operation of *charging.* It is measured in ampere hours. 3. In a gas engine, the amount of mixture taken into the cylinder during the suction stroke.

Charlier shoe. A narrow, small horseshoe without a toe-clip which is nailed into a prepared groove in the hoof. Sometimes used when the horse is put out to pasture for a period of time.

charlock. *Brassica kaber* (*arvensis*), family Cruciferae, a weed, bearing yellow flowers, found in grainfields and among other cereal crops. Feeds containing large amounts of its seed may result in chronic enteritis, hemorrhagic diarrhea, colic, abortion, nephritis, apathy, etc., in animals. Native to Eurasia. Also called *wild mustard; corn mustard.*

charming mariposa. *Calochortus venustus,* family Liliaceae, a bulbous herb, grown for its showy, tulip-like, pale-lilac flowers with a reddish-brown spot at the base. Native to western U.S. Also called *white mariposa (lily).*

Charolaise (Charollais). A breed of French beef cattle now also raised in the U.S.

char-pitting. A method of removing stumps by burning. A fire is built on the stump and kept covered until the wood is consumed.

charqui. "Jerked" beef; naturally dried strips of meat.

charred peat. Peat, dried at such a temperature as to cause partial decomposition.

charring. Burning partly; scorching.

chart quadrat. A small plot originally one meter square, used for making a detailed, graphic record of vegetation.

Chaschol. A hard, Swiss cheese made from cow's skim milk.

chase. A tree-lined path in fields.

chasmophyte. A plant growing in rock crevices.

chassis. The essential running gear in automobiles and trucks, consisting of the rectangular framework supported on springs attached to the front and rear axles.

chaste shrub. Virginia sweetspire *q.v.*

chaste tree. *Vitex agnus-castus,* family Verbenaceae, a deciduous shrub with some cultivated varieties which bear white or lilac flowers. Its fruit has medicinal properties. Native to Eurasia. Also called *hemp tree; monk's peppertree.*

chat. A general term for the seeds or inflorescence of various plants, *e.g.,* a catkin, a spike, a pine cone, etc.

chat potatoes. Small imperfect potatoes fit only for hog feed (old, English term).

chattel. Movable personal property, as contrasted with immovable real estate. Certain types of *chattel* are technically immovable, *e.g.,* a tenant's interest in a growing crop.

chausee. 1. A levee used for retaining water. 2. A term for a levee also used as a road in Louisiana.

Chautauqua system. A system of training grapes in which permanent arms are established and attached to low wires, the canes being tied in an upright position on the upper wires.

chaute. Cobblersthumb livingrock-cactus *q.v.*

chayote. *Sechium edule,* family Cucurbitaceae, an annual, tropical herb grown in coastal areas of U.S. for its edible tubers and pear-shaped fruit which has the taste of summer squash. Native to Mexico and Central America, where its foliage and young stems are also used as forage. Also called *cahiota; chocho; choko; Christophine.*

cheat. Chess brome *q.v.*

cheat chess. Chess brome *q.v.*

cheatgrass brome. *Bromus tectorum,* family Gramineae, an annual grass weed, palatable in the early stages of growth. Native to Europe; a troublesome weed in the forage fields of midwestern U.S. Also called *annual downy chess; downy brome; wild oats.*

checharron. A highly-flavored, brown lard; popular in Cuba.

check. 1. In irrigation: a. A basin into which the flow of water is regulated by levees and dykes. b. An adjustable gate on a canal to regulate the flow of water. c. Concrete blocks or wooden ties placed on a channel bottom to prevent erosion. d. A crack which appears in drying soil. 2. In lumbering, a lengthwise separation in the grain, caused by strains during seasoning, which extends across the annual rings. 3. A short crack within the body of a cheese. 4. A narrow crack in a rice grain that may cause it to break during milling. 5. An egg which has a cracked shell with the inner membrane intact. When the crack is naturally mended it is called a blind crack. Also called *crack; dent.* 6. To retard growth.

check bit. A small bit attached to the checkrein of a bridle. The checkrein is attached to the backpad of the harness, thus holding up the horse's head.

check board. A device of wood or metal in a thresher or a combine. Hinged at the upper edge and placed at the rear of the revolving blade mechanism, it prevents grain or straw from being thrown to the rear of the machine.

check dam. A small dam constructed in a gully or other watercourse; it minimizes erosion.

checked paraffin. Breaks or cracks in the covering of cheese, usually caused by a too heavy coating of paraffin. Also called *cracked paraffin.*

checkerberry. *Gaultheria procumbens,* family Ericaceae, an evergreen spreading shrub bearing white or pinkish, bell-shaped flowers and purplish berries, called *checkerberries.* Its fragrant foliage yields a medicinal oil, commonly known as *wintergreen oil* or *gaultheria oil.* Native to North America. Also called *wintergreen; teaberry; ground holly; spiceberry; aromatic wintergreen; partridge berry; box-berry; creeping wintergreen.*

checkered fritillary. *Fritillaria meleagris,* family Liliaceae, a bulbous herb grown for its bell-shaped, checkered, deep red flowers. Native to Eurasia. Also called *toad lily; checkered lily; snakes head; guinea-hen flower.*

checkered lily. See *checkered fritillary; riceroot fritillary.*

check fork. A device on a checkrow planter (*q.v.*) which strikes the button on the check wire to open the valve in the planter for dropping the seed.

check hook. A hook on the back pad of a harness to which the check rein is fastened.

check irrigation. See *check: 1. a.*

check plot. In agricultural experimentation, a small area treated or untreated, which is used as a standard for comparison. See *plot; census plot.*

checkrein. A leather strap fastened to the back pad of a harness to prevent a horse from lowering its head. Also called *bearing rein.*

checkrow. A row with seeds planted at regular intervals by a *checkrow* planter. Series of *checkrows* are established crosswise to form squares to facilitate cultivation.

checkrow planter. A machine equipped to permit the operator to drop corn seed in the planting hills in a checkrow.

check strap. A part of a harness which passes between a horse's forelegs. It is fastened to the collar on one side and bellyband on the other. It prevents the collar from rising when the horse is backing up or stopping.

check wire. A wire which is stretched across a field to trip the seed lever of a checkrow planter (*q.v.*) at definite intervals.

Cheddar. A popular English cheese, now made in many countries. It is made from cow's whole or skimmed milk which is then cured for about 6 months at 40 to 50° F. The finished product is hard and may be white or yellow. The term *Cheddar* cheese includes cheeses of the granular or stirred curd types.

cheddaring. In making Cheddar cheese, the cutting, piling, and repiling of matted curd to facilitate draining and the formation of body and texture.

cheddar pink. *Dianthus caesius,* a perennial, mat-forming herb, family Caryophyllaceae, grown in rock gardens for its fragrant, rose-colored flowers. (Syn. *D. gratianopolitanus*). Native to Europe.

cheek. 1. The fleshy side of the face in animals. 2. The part of the ventral side of a wheat kernel next to the crease.

cheeking. Grasping of the cheekpiece of a bridle to pull the head of the horse toward the rider.

cheekpiece. A leather strap, part of a bridle, which lies along a horse's cheek. Also called *cheekstrap.*

cheekstrap. Cheekpiece *q.v.*

cheek teeth. The premolar and molar teeth; the large teeth at the side of the mouth.

cheese. 1. Food made from consolidated curds which have been separated from milk by a coagulating agent. Cheese may be made from the whole milk, skim milk, or cream of any mammal. 2.The beeswax refuse, usually wrapped in burlap, from which beeswax is forced when the mass is submerged in a hot water press.

cheese bail. Cheese hoop *q.v.*

cheese board. The covering for a cheese vat.

cheesecloth. A thin, loosely woven cloth made of cotton, used in the pressing of cheese curds.

cheese fly. Cheese skipper *q.v.*

cheese group. Squash or pumpkin with a smooth, creamy-buff colored, thin rind.

cheese hoop. A cylinder in which curd is pressed in the making of cheese. See *cheese press.*

cheese knife. A heavy, long-bladed knife, similar to a machete, used in cutting American Cheddar cheese. See *curd mill; curd knives; Swiss harp.*

cheeselip. 1. Dried stomach from a calf. 2. See *cheese rennet: 1.*

cheese mite. *Tyroglyphus siro,* family Tyroglyphidae, a pest that infests ham, cheese, flour, and other stored grain and food products. It irritates the skin of handlers and causes grocer's itch *q.v.*

cheese press. A machine which extracts whey from curds. In one kind of press, a vacuum is created below the cheese hoop, and the whey is driven out by atmospheric pressure; in the other, screws, toggle joints, etc., are used to pressure out the whey.

cheese rennet. 1. A solution of the enzyme, rennin, used in setting milk for cheese-making. Also called *cheeselip.* 2. Yellow bedstraw *q.v.*

cheese rind. Cooked and partially defatted edible rind from cheese.

cheese running. Yellow bedstraw *q.v.*

cheeses. Running mallow *q.v.*

cheese skipper. *Piophila casei* (L.), family Piophilidae, a black insect whose larvae infest cheese and smoked or cured meat; so called because its maggots jump several inches. Also called *cheese fly; hopper.*

cheese spread. A food product of buttery consistency which contains cheese, as well as other ingredients: *e.g.*, milk, whey, emulsifying agents, relishes, seasonings, condiments, and coloring matter. It is similar to modified process cheese, *q.v.*

cheese trier. A sharpened, curved blade used to remove a piece from a cheese for examination.

cheese vat. A tub or similar container in which curds are formed and cut or broken.

cheese weed. Running mallow *q.v.*

cheesewood. See *orangeberry pittosporum.*

cheesiness. Spoiled butter, the putrid odor of which resembles that of various cheeses.

chemical. 1. A substance obtained by or used in a *chemical* process. 2. In dairying an uncommon flavor defect which occurs when cheese has been made from contaminated milk.

chemical brown stain. A brown discoloration of chemical origin that sometimes develops on wood during air-seasoning or kiln drying; probably due to the oxidation of extractives in the wood.

chemical caponizing. The injection or implantation of a hormone, Diethylstilbestrol (Stilbestrol) under the neck skin of a male fowl, including old roosters, to cause it to have feminine characteristics. After six weeks it results in loss of fighting instinct, paleness of the undeveloped comb and wattles, paleness of shanks and skin, and an increase in weight.

chemical dehorning. The application of chemicals to the horn buttons of young calves to stop horn growth.

chemical fertilizer. See *commercial fertilizer.*

chemical flavor. A flavor defect of butter made in a churn which has been improperly chlorinated, or made from milk which has been contaminated by the use of strong-scented chemicals on the teats of cows.

chemical gardening. See *hydroponics.*

chemically precipitated phosphorus. A form of fixed phosphorus compounds.

chemical seasoning. A seasoning process in which a hygroscopic chemical is applied to green wood before drying to reduce drying defects. The chemical may be applied dry to the surface of the lumber, or the lumber may be soaked in an aqueous solution.

chemical stimulation. In turpentining, the application of chemicals, usually acids, to increase the rate and duration of the oleoresin flow.

chemical toilets. A type of dry vault toilet in which the sewage is decomposed by adding caustic chemicals, such as quick lime.

chemical weed control. The application of herbicides as pre-emergence and post-emergence sprays for the control of weeds in crops.

chemical wood. Wood cut or prepared for the manufacture, distillation, or extraction of chemicals, charcoals, gases, or other products of a nonfibrous nature.

chemisorbed phosphorus. A chemically precipitated monolayer on surface of another crystalline species, formed by reaction of phosphorus compounds with solid phase constituents. A form of sorbed phosphorus, involving surface valence forces and chemical affinities between the phosphate ion and the lattice constituents on which sorption takes place.

chemolysis. Decomposition of organic matter brought about by chemical agents.

chemotherapeutic. Referring to a chemical used to treat disease.

chemotherapy. The treatment of diseases by means of chemicals.

chemuck. A food consisting mainly of ground acorns. Formerly used by Indians in northwestern U.S.

chemurgy. Chemical research which is designed to promote the use of agricultural products for industrial purposes, or to obtain new, valuable products from plants.

Chengtu lilac. *Syringa sweginzowi,* family Oleaceae, a hardy deciduous shrub grown for its fragrant, reddish or lilac flowers. Native to China.

cheniers. Low, narrow ridges in the coastal marshes of Louisiana in U.S. Usually 5 or 6 feet above the surrounding level, they were home sites and farm lands for the early settlers.

chenille copperleaf. *Acalypha hispida,* family Euphorbiaceae, a tropical herb used for bedding or hedges in southern California and Florida, U.S., for its drooping, reddish-purple spikes of flowers looking like bits of chenille. Native to East Indies.

chenopodium oil. Wormseed oil *q.v.*

cherimoya. *Annona cherimola,* family Annonaceae, a medium-sized, tropical tree which is grown for its oval, edible fruit. Native to Peru; occasionally grown in Florida, U.S.

Chernozem. A zonal group of soils having a rich black or very dark brown surface horizon which grades below into lighter colored soil and then into a horizon of lime accumulation. It was developed under mixed, tall and short grasses in a temperate to cool and subhumid to semiarid climate. Found particularly in European Russia. Variant spellings: Tschernozem, Tshernosem, Tchernayzem. See *Chestnut soils.*

Cherokee outlet. A strip of land in northwestern Oklahoma, U.S., ceded to the Cherokee tribe of Indians in 1828 by the federal government to provide them with an outlet to buffalo hunting grounds in the west.

Cherokee rose. *Rosa laevigata,* family Rosaceae, an evergreen, climbing shrub with white flowers, widely naturalized in southern U.S. Native to China.

Cherokee tickclover. *Desmodium tortuosum,* family Leguminosae, a tropical plant grown extensively in Florida and adjacent states of U.S. as a cover crop and for green manure. The plant is resistant to strains of the rootknot nematode. Native to U.S. Also called *Florida beggarweed.*

cherry. 1. Species of the genus Prunus grown for its edible fruit or for ornament. Cultivated *cherries* are native to Europe and are classified in 3 groups: sour *cherries,* sweet *cherries,* and dukes, which are hybrids between the first two. The sour varieties are derived from the species *Prunus cerasus;* the more common ones being Montmorency, Early Richmond and English Morello. The sweet varieties come from *P. avium.* The main ones are Seneca, Black Tartarian, Napoleon, and Windsor. Of the dukes, Royal Duke is a popular variety. Many species of cherry are grown for ornament: *e.g.,* the oriental cherry (*P. serrulata*), Siebold cherry (*P. sieboldi*), Higan cherry (*P. subhirtella*), Manchu cherry (*P. tomentosa*), etc. One wild variety, native to eastern North America, the black cherry (*P. serotina*) is a very large tree, the wood of which is used for fence posts, rails, crossties, furniture, and fuel. 2. Australian name for the edible fruit of a number of trees belonging to genera Antidesma, Eugenia, and Exocarpus.

cherry apple. Siberian crabapple *q.v.*

cherry bay. Laurel cherry: 1.

cherry birch. Black birch *q.v.*

cherry buckskin. See *buckskin: 4.*

cherry casebearer. *Coleophora pruniella* Clem., family Coleophoridae, an insect whose larvae damage the leaves of cherry.

cherry coryneum blight. See *Coryneum blight of stone fruits.*

cherry crab. Siberian crabapple *q.v.*

cherry elaeagnus. *Elaeagnus multiflora,* family Elaeagnaceae, a hardy deciduous shrub grown in windbreaks and as an ornamental for its foliage. Native to Asia. Also called *gumi.*

cherry fruit fly. *Rhagoletis cingulata* (Loew.), family Trypetidae, a fly that attacks cherries, resulting in misshapen, undersized fruit. Sometimes a closely related species, *R. fausta* (O. S.), is also present.

cherry fruit sawfly. *Hoplocampa cookei* (Clarke), family Tenthredinidae, a sawfly, the larvae of which attack cherries and plums.

cherry fruitworm. *Grapholitha packardi* Zell., family Olethreutidae, a moth whose larvae feed within the fruit of the cherry.

cherry gum. An exudation from the wood of the various species of *Prunus,* particularly after injury.

cherry laurel. See *Carolina laurelcherry; laurel cherry: 1.*

cherry leaf beetle. *Galerucella cavicollis* (Lec.), family Chrysomelidae, a small beetle; the larvae and adults feed on cherry leaves.

cherry leaf spot. A fungous disease caused by *Coccomyces* spp. (thought by some to be *Higginsia),* family Mollisiaceae, signalized by spots on leaves and frequently by lesions on fruit and pedicels. The spots are small and isolated, purple or reddish at first, soon turning brown. Later the spots may run together, resulting in large, dead areas, which may leave holes in the leaves, and white spore masses may appear on the under sides. The leaves turn yellow and fall; the defoliation depletes the carbohydrate reserve and brings on uneven ripening of fruits and a predisposition to winter injury. Also called *leaf blight; leaf spot; yellows; yellow leaf; shot hole disease; spot disease of cherry.*

cherry mildew. A powdery mildew caused by *Podosphaera oxyacanthae,* family Erysiphaceae; sometimes destructive; especially in nursery plots.

cherry mottle leaf. See *mottle leaf.*

cherry necrotic rust mottle. See *necrotic rusty mottle.*

cherry pie. A name for the garden heliotrope. See *heliotrope: 2; valerian: 2.*

cherry pit. The stone of the cherry, in which the seed is enclosed.

cherry plum. Myrobalan plum *q.v.*

cherry plum hybrid. Varieties developed from hybridizing the Bessey cherry (*q.v.*) with species of plums, *e.g.,* compass-cherry. Because of winter hardiness they are popular fruits in cold regions of the U.S.

cherry raspberry leaf. See *rasp leaf.*

cherry redpepper. *Capsicum frutescens* var. *cerasiforme,* family Solanaceae, a woody plant yielding a very pungent, edible fruit which is yellowish-red or purple. Grown as an ornament and for its fruit, the latter of which is a condiment. See *bush redpepper.*

cherry ring spot. See *ring spot of stone fruits.*

cherry rugose mosaic. A minor virus disease in California; characterized by a general chlorosis of the leaf midway between the midvein and the margin of the leaf, and by shriveling and premature drop of the affected leaves.

cherry slug. See *pear slug.*

cherry tomato. *Lycopersicon esculentum* var. *cerasiforme,* an annual herb, family Solanaceae, grown for its very small, edible, tomato-like fruits used in salads. Native to South America. Also called *plum tomato.*

cherry tree Tortrix. Ugly-nest caterpillar *q.v.*

cherry yellows. A virus disease of sour cherries; a portion of the leaf becomes mottled green and yellow and the leaf falls six weeks after bloom. Although total yield is reduced by 50 per cent or more, the fruit which develops is large and of good quality. The virus is frequently latent in sweet cherries.

chert. A very dense, cryptocrystalline, flint-like form of silica, which breaks with a splintery fracture. It is very resistant to decomposition, remaining as inert angular fragments in the residual mass of weathering, sometimes imparting a distinctive character to the soil.

chervil. Salad chervil *q.v.*

Cheshire. 1. A lard-type breed of white swine which originated in New York State from crossing the Yorkshire with White Suffolk hogs. 2. A popular old English cheese; less firm and compact than Cheddar, which has white or yellow curd. It is made of cow's milk, and annatto added for coloring. It is cured for 3 weeks (early ripening), for 2 months (medium ripening), or 10 weeks to 10 months (late ripening).

Cheshire-Stilton. An English cheese which combines the characteristics of Cheshire and Stilton cheeses.

chess-apple. Fruit of whitebeam mountainash *q.v.*

chess brome. *Bromus secalinus,* family Gramineae, a weed pest found in grain fields; especially troublesome in wheat fields when its seed becomes mixed with the threshed wheat, thus lowering the grade. Grown for hay in some regions of northwest Pacific. (The name has also been applied to darnel ryegrass, *Lolium temulentum.*)

chessel. A cheese vat or mold usually made of thick oak staves bound with iron to withstand pressure, and perforated with holes in the bottom and sides to drain the whey.

chessits. Cheese hoop (Scottish).

chest. The part of the body enclosed within the front ribs and breastbone.

Chester White. A large, lard-type breed of white swine which originated in Chester County, Pennsylvania, from crossing Yorkshire, Lincolnshire, and Cheshire hogs. Mature boars weigh 600-1000 lbs. and sows 500-700 lbs.

chest floor. That part of the body which lies along the breastbone.

chest founder. 1. An old term used by blacksmiths for a disease of horses, the symptom of which was inflammation about the chest and ribs. 2. A concavity of the muscles in the front of a horse's chest caused by insufficient exercising.

chestnut. 1. See *American chestnut; European chestnut.* 2. A horse color of brown with red-yellow hues which lacks brilliance. 3. The sweet, edible nut produced of *Castanea* spp., family Fagaceae. 4. The horny projection found on the lower inner part of the forearm and hock joint of a horse.

chestnut-bark disease. Chestnut blight *q.v.*

chestnut bean. Gram chick-pea *q.v.*

chestnut blight. A fungous disease caused by *Endothia parasitica,* family Diaporthaceae, which attacks the bark and cambium of the twigs, branches and main trunk of chestnuts. The cankers formed girdle the stems resulting eventually in the death of the tree. The new yellowish or yellow-brown patches of infection are slightly raised on the smooth-barked shoots, in marked contrast to the olive-green bark. On old wood infections are discolored, elongated, dead areas, slightly depressed with raised margins. The blight has almost completely destroyed the American chestnut, *Castanea dentata.* However, the East Asian varieties have shown some degree of resistance to it. Also called *endothia canker; chestnut canker; chestnut-bark disease.*

chestnut canker. Chestnut blight *q.v.*

chestnut oak. *Quercus montana,* family Fagaceae, a tall, deciduous tree found on dry hillsides in eastern U.S. It becomes crooked if not trained upright. The wood is used for railroad ties, railroad cars, fences, pilings, timbers, veneer, flooring, vehicles, planing-mill products, furniture, cooperage boxes, crates, agricultural implements, caskets, coffins, handles, etc. Native to America. Also called *tanbark oak; rock chestnut oak.*

chestnut roan. A horse on which white hairs are mixed with the basic chestnut color. Sometimes called *strawberry roan.*

Chestnut soils. A zonal group of soils with a dark brown surface horizon grading below into lighter colored soil and finally into a horizon of lime accumulation. The soils occur on the arid side of chernozem soils, into which they grade. *Chestnut soils* developed under mixed, tall and short grasses in a temperate to cool and subhumid to semiarid climate. See *Chernozem.*

chestnut timberworm. *Melittomma sericeum* (Harr.), family Lymexylidae, a beetle whose larva infests oaks and chestnuts, boring tunnels in the heartwood and sapwood.

chevaline. Horseflesh; or concerning the flesh of a horse.

Cheviot. A hardy, hornless, breed of English sheep; the rams weigh up to 200 lbs. and ewes up to 160 lbs. These sheep are small, thick set and well-proportioned, and are able to withstand severe climatic conditions, but they do not have good fleece weight.

chevon. Goat meat.

chewing louse. Any wingless, parasitic insect which feeds on birds and sometimes on mammals. See *horse biting louse; cattle biting louse; chicken head louse; chicken body louse; shaft louse; small pigeon louse; slender pigeon louse.*

Chewing's fescue. *Festuca rubra* var. *commutata,* family Gramineae, a perennial, fine-leaved grass, used for turf in New Zealand and U.S.

chewing the cud. The chewing by a ruminating animal of coarse food regurgitated into the mouth from the first stomach. The food is macerated into small pieces and mixed with saliva before final swallowing.

Chhana. An Asian sour cheese made from cow's whole milk.

chiasma. An apparent exchange of partners in a system of paired chromatids. In the diplotene stage (a late prophase stage) of meiosis the four chromatids of a bivalent are associated in pairs but in such a way that in one part of their length two chromatids are associated, but in the remainder of their length each is apparently associated with one of the other two chromatids. This point of apparent change of partner is the *chiasma.*

chiasmatype theory. That chiasmata are connected with crossover, either as a cause or as a consequence.

Chiavari. An Italian sour milk cheese made from cow's whole milk.

Chibol. *Allium fistulosum,* family Liliaceae, a bulbous herb, with the characteristic onion odor, used in seasoning, etc. The name is also applied to the fresh onion bulb with the green shoot attached. Native to Asia. Also called *Welsh onion; spring onion.*

Chicago style. A method of cutting a beef carcass into standard wholesale cuts (ribs, chucks, rounds, etc.); in general use in midwestern and western U.S.

Chickasaw plum. *Prunus angustifolia,* family Rosaceae, a small tree or shrub from which many cultivated varieties of plum in the southern U.S. have been derived. Native to North America.

chick bronchitis. See *infectious bronchitis.*

chick dermatosis. A deficiency of pantothenic acid in the diet of chicks, characterized by sores and incrustations on the corners of the eyes, the mouth, bottom of the feet and the joints of the toes. It also upsets the feather development and the general growth.

chicken. *Gallus domesticus,* family Phasianidae, the domestic fowl, whose ancestry, while obscured by antiquity, is attributed to *Gallus gallus* and other species of jungle fowl of India and the Malay Peninsula. It is one of the most widely distributed and most commonly kept of all farm animals.

chicken berry. Checkerberry *q.v.*

chicken body louse. *Menacanthus stramineus* (Nitz.), family Menoponidae, a biting pest that lives on the skin of chickens or turkeys, especially around the vent and under the wings. It feeds by nibbling on dried skin, scales and feathers, resulting in irritation to the skin of the infested birds.

chicken cholera. See *fowl cholera.*

chicken corn. 1. *Sorghum vulgare* var. *drummondi,* family Gramineae, an herbaceous plant, which has escaped from cultivation and is a troublesome weed in some places. Native to North America. 2. An old U.S. southern term for sorgo *q.v.*

chicken fat. The raw or rendered fat from chickens; an edible by-product of poultry; particularly in demand by the Jewish bakery trade.

chicken feet. A by-product of poultry processing, especially rich in gelatin. By removing the outer skin and boiling the inner flesh, a rich stock that jells readily is obtained.

chicken grape. See *frost grape.*

chicken head. A poultry by-product, a table delicacy. Some consider the edible portion of the head, that is, the brains, eyes and tongue served in the jelly of the feet, as a delicacy.

chicken head louse. *Cuclotogaster heterographus* (Nitz.), family Philopteridae, a biting pest which infests young chickens and turkeys. *Cf. chicken body louse.*

chicken horse. A wild horse in the western U.S. used for chicken and dog feed. See *canner: 2.*

chicken house. A structure for housing chickens.

chicken mite. *Dermanyssus gallinae* (DeG.), family Dermanyssidae, a blood-sucking pest that attacks fowl. It lives in the cracks about the roosts, nests, and walls of the poultry house in the daytime and upon nesting fowls at night. In severe cases it causes unthriftiness, lack of egg production, and at times even proves fatal. Also called *bird mite.*

chicken pest. See *fowl plague.*

chicken pox. See *fowl pox.*

chicken septicemia. See *fowl cholera.*

chicken sticker. A poultry knife with a short, thin blade, used for killing poultry. The blade is forced through the roof of the mouth into the brain.

chicken stomach worm. *Tetrameres americana,* an internal parasite which enters the gland of the proventriculus. Infested birds do not always show symptoms, but young chickens appear emaciated and sometimes die.

chick feathers. The feathers from chickens, a by-product of the poultry industry, gathered for use in the millinery trade.

chickling. Grass pea (vine) *q.v.*

chick-pea. See *gram chick-pea.*

chick starter. A balanced feed for the quick growth of baby chicks, consisting of ground grains, meat scraps, leaf meal, soybean meal, dried milk, limestone, iodized salt, vitamins, antibiotics, and other items.

chickweed. *Stellaria media,* family Caryophyllaceae, a weed pest found throughout the U.S. in cultivated fields, gardens, lawns, waste ground and around ornamental shrubs. Native to Eurasia. Also called *satin flower; tongue grass; starwort; starweed; winterweed; bird-weed; chickenwort.*

chickweed wintergreen. See *starflower.*

chico. See *Coville creosotebush.*

chicory. a. *Cichorium intybus,* family Compositae, a perennial herb cultivated for its roots; used as an adulterant of coffee, and for its leaves which are used as a potherb and salad green. It is an escaped weed along roadsides in U.S. Native to Europe. Also called *succory; blue dandelion; coffee weed; blue sailor.* b. A French variety of chicory with blanched leaves, used raw as a salad, is called *French endive* or *witloof.*

chigger. *Eutrombicula alfreddugesi* (Oud.), family Trombiculidae, a larva infesting men, domestic animals, some birds, snakes, turtles, and rodents. Its bite results in inflamed spots accompanied by intense itching. Also called *red bug; jigger.*

chigoe. *Tunga penetrans* (L.), family Tungidae, a small, reddish-brown flea which burrows into the skin, especially between the toes and under the toenail of men and animals, causing intense pain, itching, and sometimes chronic sores. Sometimes wrongly called chigger *q.v.*

chigoo. *Sarcopsylla penetrans,* type of the family Sacropsyllidae. The female of this flea genus attacks men and animals by getting under the skin of feet and hands causing severe sores. Also called *jigger.*

chilblain. A painful swelling on feet or hands due to exposure to cold, sometimes resulting in ulceration.

children's-bane. See *spotted waterhemlock.*

Chilean alstroemeria. *Alstroemeria chilensis,* family Amaryllidaceae, a garden herb grown for its rose-red or white flowers. Native to Chile.

Chilean arborvitae. Chilean incensecedar *q.v.*

Chilean bellflower. See *red Chilebells; dark blue nolana.*

Chilean clover. A variety of alfalfa grown in Chile. The name has been used as synonymous with alfalfa.

Chilean deervetch. *Lotus subpinnatus,* family Leguminosae, a valuable species of herbaceous forage plants found in California, U.S. Also called *fineleaf lotus.*

Chileanguava. *Myrtus ugni,* family Myrtaceae, a shrub or tree widely grown for ornament in the U.S. It bears a small, pleasantly flavored, edible fruit. Native to Chile.

Chilean incensecedar. *Libocedrus chilensus,* family Pinaceae, a medium-sized, evergreen tree, grown as an ornamental and for its valuable timber. Native to Chile. Also called *Chilean arborvitae.*

Chilean strawberry. Chiloe strawberry *q.v.*

Chilean tarweed. *Madia sativa,* family Compositae, an annual herb grown for its heavy-scented, brownish-yellow flowers. Native to America.

Chilebells. See *red Chilebells.*

Chile dodder. *Cuscuta suaveolens,* family Convolvulaceae, a parasitic plant which attacks both clover and alfalfa. Native to South America; common in red clover and alfalfa seed imported in U.S. from there.

Chilejasmine. *Mandevilla suaveolens,* family Apocynaceae, a tropical, climbing vine grown for its fragrant, showy, funnel-shaped, white or pinkish flowers. Native to America.

Chile parrotsfeather. *Myriophyllum proserpinacoides,* family Haloragidaceae, an aquatic herb grown in pools for its foliage. Escaped, it becomes a nuisance. Native to Chile. Also called *water feather.*

Chile saltpeter. See *nitrate of soda.*

chili pepper. Bush redpepper *q.v.*

chill. To reduce the temperature of a product, such as a meat-carcass or milk, to cure it, improve the flavor, prevent spoilage, and increase market life.

chilling. The development of a sweetish taste in stored potatoes kept at temperatures approaching the freezing point. It is not a serious injury since the tubers become normal upon exposure to higher temperatures.

chilling injury. The injury to fruits and vegetables resulting from temperatures of 32° to 50° F.

chill in the udder. See *mastitis.*

chill(ing) room. An insulated, refrigerating compartment especially constructed for the storage or transport of perishable food in which the temperature is maintained at a few degrees above freezing point. Meats and delicate fruits keep fresh and sweet in it for about a month; quality is not impaired as in a cold storage with temperatures below freezing point.

chiloe strawberry. *Fragaria chiloensis,* family Rosaceae, an American perennial herb which has served as a breeding stock for cultivated varieties of strawberry (*q.v.*) in the U.S. Also called *Chilean strawberry.*

chilopod. A centipede, Order Chilopoda, Class Myriopoda, whose two sets of anterior legs act as foot-jaws and whose body is segmented.

chilver. Ewe lamb.

chimera. A plant or part of it, composed of tissues of two or more genetically different types. If tissue of one genetic type is external to and surrounds another genetic type, it is a periclinal *chimera.* If the tissue extends from the surface into and is deeply seated in another genetic type, it is a sectorial *chimera.* When the tissues are greatly intermingled, it is hyperchimera.

chimney. A vertical opening, a foot or more wide, in a pile of lumber to facilitate circulation of air.

chimney bellflower. *Campanula pyramidalis,* family Campanulaceae, a perennial herb grown for its narrow clusters of flat, pale-blue flowers. Native to southern Europe. Also called *chimney plant.*

China aster. *Callistephus chinensis,* family Compositae. It is the common aster (*q.v.*) from which the various cultivated varieties with their single blooms of various colors are derived. Native to China and Japan. Also called *garden aster; fall rose.*

China bean. See *cowpea.*

Chinaberry. *Melia azedarach,* family Meliaceae, a semi-evergreen tree with a handsome foliage grown for shade in U.S. Its seeds are sometimes used for rosaries; the drupes are poisonous, producing complete paralysis, irregular respiration, and suffocation. Also called *China tree; beadtree; pride of India.*

Chinabox jasminorange. *Murraya exotica,* family Rutaceae, an evergreen tree-like shrub grown in very warm areas for its aromatic foliage and fragrant, white flowers. Also called *orange jasmine.*

Chinacypress. *Glyptostrobus pensilis,* family Pinaceae, a small, evergreen tree, grown as an ornamental in warm areas: commonly planted around rice fields in the Orient. Native to China. Also called *embossed cypress; water pine.*

Chinafir. *Cunninghamia lanceolata,* family Pinaceae, a tall, evergreen tree with drooping branches grown as an ornamental in warm areas. Native to China.

Chinagreen. *Aglaonema simplex,* family Araceae, a tropical herb, popular as a house plant for its variegated leaves. Native to China. Also called *Chinese evergreen.*

Chinaman's redberry. See *redroot amaranth.*

China rose. See *Chinese rose: 1; Chinese hibiscus.*

China tree. See *panicled goldraintree; Chinaberry.*

China wood oil tree. See *tung oil tree.*

chincapin. Chinkapin *q.v.*

chinch bug. *Blissus leucopterus* (Say), family Lygaeidae, a black and white insect, destructive to small grains, corn, and sorghum; found all over North America below central Canada, particularly in central U.S.

chincherichee. See *cape chinkerichee.*

Chincoteague. A small, spirited, wild pony found on the tidal marsh flats of Chincoteague Bay islands off the southeastern coast of U.S. It is a durable, active and good-looking animal, suitable for children's use.

chine. An animal's backbone.

Chinese. A breed of medium-sized, white or brown, domestic geese from China, having long arched necks and large knobs at the base of the beak. The adult gander weighs as much as 12 lbs., the goose as much as 10 lbs.

Chinese angelica tree. Chinese aralia *q.v.*

Chinese anise. Truestar anisetree *q.v.*

Chinese aralia. *Aralia chinensis,* family Araliaceae, a shrub or small tree, with large, compound leaves and prickly stems, grown for its greenish or whitish flowers. Native to China. Also called *Chinese angelica tree.*

Chinese artichoke. See *artichoke betony.*

Chinese azalea. *Rhododendron molle,* family Ericaceae, a low, deciduous shrub, grown for its golden-yellow flowers. Native to China.

Chinese boxorange. *Severinia buxifolia,* family Rutaceae, a spiny shrub or dwarf tree with small white flowers, black shiny berries, and leaves that resemble the common box (*Buxus*); sometimes grown as an ornamental or for hedges in subtropical areas. Native to China and Formosa.

Chinese cabbage. See *petsai.*

Chinese catalpa. *Catalpa ovata,* family Bignoniaceae, a deciduous tree, grown for its yellowish-white orange-striped flowers with violet spots inside. Native to Asia.

Chinese cork tree. *Phellodendron Chinense,* family Rutaceae, a small tree with clusters of yellow-green flowers and berry-like black fruit, grown in U.S. as an ornamental for its rapid growth. Native to China.

Chinese date. See *jujube.*

Chinese dogwood. Kousa dogwood *q.v.*

Chinese elm. *Ulmus parvifolia,* family Ulmaceae, an autumn-blossoming, deciduous tree, grown for its fine foliage, and rapid growth. It is the most useful of all the elms. Native to eastern Asia. Also called *evergreen elm; dwarf Asiatic elm.*

Chinese evergreen. Chinagreen *q.v.*

Chinese fan palm. *Livistona chinensis,* family Palmaceae, a very widely cultivated plant, grown in U.S. as an ornamental. Native to southern China.

Chinese fleece-vine. See *silvervine fleeceflower.*

Chinese flowering crabapple. *Malus spectabilis,* family Rosaceae, a small, deciduous tree, grown in U.S. for its rose-pink flowers. Native to China.

Chinese flowering quince. *Chaenomeles sinensis,* family Rosaceae, a small shrub or tree, grown for its bright-pink flowers; not as popular as the Japanese quince. Native to China. Also called *Chinese quince.*

Chinese forget-me-not. See *Chinese houndstongue.*

Chinese fringetree. *Chionanthus retusus,* family Oleaceae, a shrub grown in U.S. for its showy, white flowers. Native to China.

Chinese hackberry. *Celtis sinensis,* family Ulmaceae, a medium-sized tree grown for shade, as it is very tolerant of city conditions. Native to China.

Chinese hat-plant. *Holmskioldia sanguinea,* family Verbenaceae, an evergreen, tender shrub, grown in the greenhouses in U.S. for its red flowers. Native to the Himalayas.

Chinese hibiscus. *Hibiscus rosa-sinensis,* family Malvaceae, a tropical shrub, grown in frost-free areas for its rose-red flowers. Its sap is used as a hair dye and in shoe blacking. Native to Asia. Also called *China rose; shoeblack plant.*

Chinese holly. *Ilex cornuta,* family Aquifoliaceae, a broad-leaved, evergreen tree, grown for its lustrous, green, spiny leaves, and red, marble-sized fruit. Native to China.

Chinese houndstongue. *Cynoglossum amabile,* family Boraginaceae, a biennial or annual herb, grown for its clusters of beautiful, blue or pink blossoms. Native to China. Also called *Chinese forget-me-not.*

Chinese houses. Pagoda collinsia, *q.v.*

Chinese hull-less. A variety of oats grown infrequently in the U.S. In the threshing process, its kernel separates easily from the hull.

Chinese insect wax. A yellowish-white wax deposited on trees by a scale insect, *Ericerus pela,* family Coccidae, common in China and India. It is used for candles, polishing wax, sizing paper, etc., in place of beeswax. Also called *insect wax; vegetable spermaceti.*

Chinese jujube. See *jujube.*

Chinese juniper. See *pyramid Chinese juniper.*

Chinese layerage. See *air layering.*

Chinese lespedeza. *Lespedeza cuneata,* family Leguminosae, a coarse, perennial herb which, if cut young, makes fair hay for cattle, but because of its tannin content, becomes extremely unpalatable later on. It is grown in southern U.S. for pasture, hay, and soil improvement. Native to eastern Asia. Also called *sericea lespedeza.*

Chinese loquat. See *loquat.*

Chinese mantis. *Tenodera aridifolia sinensis* Sauss., family Mantidae, a large, important, insectivorous insect which devours immense numbers of insects. It was introduced in the U.S. from Asia. Also called *preying mantis.*

Chinese monthly rose. Everblooming Chinese rose.

Chinese parasoltree. *Firmiana simplex,* family Sterculiaceae, an ornamental tree, grown in warm areas for its showy foliage. Native to Asia. Also called *Phoenix tree; Japanese varnishtree.*

Chinese pear. See *sand pear.*

Chinese peony. *Paeonia albiflora,* family Ranunculaceae, a beautiful, perennial herb grown for it white flowers. Native to northeastern Asia. See *doublecrimson Chinese peony.*

Chinese pink. *Dianthus chinensis,* family Caryophyllaceae, an annual or biennial herb, grown for its fragrant, red, white or lilac flowers. Native to Eurasia. Also called *Indian pink.*

Chinese pistache. *Pistacia chinensis,* family Anacardiaceae, a medium-sized tree grown in Florida and California as an ornamental, and as a stock to graft the true pistache, *P. vera.* Native to China.

Chinese pistachio. See *Chinese pistache.*

Chinese potato. Cinnamon vine *q.v.*

Chinese preserving melon. Chinese waxgourd *q.v.*

Chinese primrose. Fairy primrose *q.v.*

Chinese quince. See *Chinese flowering quince.*

Chinese radish. *Raphanus sativus* var. *longipinnatus,* family Cruciferae, an annual herb which bears long, hard, edible roots, eaten raw or cooked. Native to eastern Asia.

Chinese redbud. *Cercis chinensis,* family Leguminosae, a deciduous, shrubby tree, grown for its early-spring, rosy-purple flowers. Also called *Judas-tree.*

Chinese rose. 1. *Rosa chinensis,* family Rosaceae, a low, semi-evergreen shrub, from which many cultivated varieties of rose have been derived. Native to China. Also called *Bengal rose.* 2. See *Cathay Japanese rose.*

Chinese rose beetle. *Adoretus sinicus* Burm., family Scarabaeidae, a beetle which infests roses and beans.

Chinese sacred bamboo. Nandina *q.v.*

Chinese silvergrass. *Miscanthus sinensis,* family Gramineae, a tall, perennial grass, grown in lawns and flower-beds for its leaves with whitish center lines and its feathery flowers. Native to China and Japan.

Chinese starjasmine. *Trachelospermum jasminoides,* family Apocynaceae, a climbing vine, grown for its evergreen foliage and fragrant, star-like, white flowers. Native to China and Malaya. Also called *Confederate jasmine.*

Chinese sumac. Tree of heaven ailanthus *q.v.*

Chinese tallowtree. *Sapium sebiferum,* family Euphorbiaceae, a tree whose wood is used for engraving work and the waxy covering of whose seeds yields a tallow-like substance used in soap and candle making. Native to China and Japan; grown in U.S.

Chinese trumpetcreeper. *Campsis grandiflora,* family Bignoniaceae, a climbing vine, grown for its large, scarlet flowers. Native to China; grown in U.S. where it is not so hardy nor so large as the American species, the common *trumpetcreeper.*

Chinese velvetbean. *Stizolobium niveum,* family Leguminosae, an annual herb grown in warm areas for forage. Native to Asia. Also called *Lyon bean.*

Chinese water chestnut. See *water nut: 2.*

Chinese water evergreen. Chinagreen *q.v.*

Chinese watermelon. Chinese waxgourd *q.v.*

Chinese wax. See *Chinese insect wax.*

Chinese waxgourd. *Benincasa hispida,* family Cucurbitaceae, a twining plant, grown for its large, edible fruit used largely for pickles and conserves. Native to Southeast Asia. Also called *tallow gourd; wax gourd; Chinese preserving melon; Chinese watermelon.*

Chinese wingnut. *Pterocarya stenoptera,* family Juglandaceae, a tall, deciduous tree, grown for its attractive leaves and winged fruit. Native to China.

Chinese wisteria. *Wistaria sinensis,* family Leguminosae, a high climbing vine, grown for its fragrant, bluish flowers.

Chinese yam. Cinnamon vine *q.v.*

Chinese yeast. A yeast isolated from food products of the Orient, used in fermenting sugary liquids for distillation of brandy.

chin fly. Throat bot fly *q.v.*

chingma abutilon. *Abutilon theophrasti,* family Malvaceae, an annual herb, grown for its yellow flowers and beautiful foliage. In India, its coarse fiber is used for making ropes and bags. Escaped, it can be a pernicious weed. Native to India; naturalized in North America. Also called *American jute; Indian mallow; buttonweed; velvetleaf.*

chinkapin. a. Certain trees of the genus Castanea, family Fagaceae, and their edible nuts. b. See *giant evergreen chinkapin.*

chinkapin oak. Either of the two plants of family Fagaceae: a. The tall tree, *Quercus muhlenbergii,* also called *yellow chestnut oak.* b. The shrubby oak, *Quercus prinoides,* whose sweet acorns are edible. Native to North America.

chinkerichee. See *cape chinkerichee.*

chinook wind. A northwestern U.S. term for a hot and dry wind which, with its moisture precipitated, descends from the Rocky Mountain crests and spreads over the lower land eastward, evaporating the snow rapidly. It is beneficial in winter, but may desiccate crops during the growing season.

chinquapin. See *chinkapin.*

chip. 1. A small piece of wood or stone. 2. To cut wood or stone into small pieces. 3. To rewound the face of a scar on a pine tree to renew the flow of oleoresin in the turpentining industry. 4. To hoe or cultivate (Australia).

chip budding. A cion, beveled at both ends, which consists of one bud, bark and a small amount of wood. It is inserted into a mortise in the stock and is usually held in place by tying.

chip hat palm. Brittle thatchpalm *q.v.*

chipped beef. Beef which has been smoked, cured, and sliced into paper-thin slices.

chipper notcher. In logging, the head of a number of saw crews who undercuts (*q.v.*) timber and keeps a tally of the logs cut by each crew.

chips. 1. Small pieces of wood or bark which have fallen into the cup or box of a turpentined tree, or which have been removed with the scrape from the scarified face above the box. 2. Pieces of dried dung used for fuel.

chipstuff. See *shipstuff.*

chisel. 1. A machine with penetrating points that are drawn at a depth of 12 inches to loosen subsoil. 2. A hard steel cutting tool with a sharp-ended blade.

chisel cultivator. See *field cultivator.*

chiseling. Breaking or loosening subsoil with a chisel cultivator or chisel plow.

chisel mouth. See *failing-mouth.*

chisel plow. See *chisel: 1.*

chit. 1. Second or third grade rice. 2. A young animal. 3. A plant shoot.

chittamwood. See *cascara buckthorn; American smoketree.*

chitterlings. Edible intestines of a hog which are emptied, rinsed but not scraped, and fried. Also called *chittlings.*

chive. *Allium schoenoprasum,* a perennial, bulbous herb, family Liliaceae, grown for its onion-flavored leaves. Used in salads and as a flavoring. Native to Eurasia. Also called *cive.*

chloranthus. Describing a plant with green flowers. *Chloranthy* describes a plant whose flowers have returned to leaves.

chlorate of soda. An herbicide which when used in excess may poison livestock.

chlordane. A chlorinated hydrocarbon insecticide. It is a viscous amber liquid, with a boiling point of 175° C. at 2mm. pressure.

chlorellin. An antibiotic derived from a fresh water alga of the genus Chlorella.

chloride content of milk. Chloride normally found in milk, usually about 0.07 percent but ranging from 0.05 to 0.12 percent.

chloride of lime. See *chlorinated lime.*

chloride of soda. Common salt.

chlorinate. To treat with chlorine; a chlorinated product.

chlorinated lime. Calcium hypochlorite; a powerful disinfectant and deodorant. Also called *bleaching powder; bleach; chloride of lime; calcium chloride.*

chlorination. Disinfecting a substance, retarding decomposition, or oxidizing organic matter with chlorine or bleaching powder.

chlorine. The chemical element, Cl, a greenish-yellow, poisonous gas. *Chlorine* compounds are used in disinfectants and deodorants.

chlorophyll. A substance present in all growing plants; it evidences itself as the green coloring in leaves. *Chlorophyll* transforms light energy from the sun into chemical energy for the manufacture of plant food from carbon dioxide and water, which process is called photosynthesis, *q.v.*

chloropicrin. A tear gas used as a soil fumigant, a nematocide, rodenticide, and an herbicide; poisonous to humans.

chloroplast. Minute structure within plant cells which contains the green pigment, chlorophyll.

chloroplast plastid. A spongelike structure in which chlorophyll is enmeshed. The plastids perform specific functions, such as manufacturing food, building up starch grains, and producing colors.

chlorosis. A yellowing between veins on upper foliage which results from chlorophyll deficiency. Many factors, including heredity cause *chlorosis.*

chlorotic phyllody of strawberries. A minor, virus disease; characterized by chlorosis (*q.v.*) and by metamorphosis of floral organs into leaves.

chlorotic ring spot. A circular, chlorotic area on plants; a common symptom of a virus disease.

chlorotic streak. A world-wide, virus disease of sugarcane and other grasses; characterized by white, wide, continuous, *chlorotic streaks* with irregular margins. If the plant is infected when it is young, dwarfing results.

chock (block). A blocking device used as a wedge to keep a log, truck, etc., from moving.

chocolate. Cocoa *q.v.*

chocolate flower. Spotted geranium *q.v.*

chocolate leaf spot. A bacterial disease of grasses caused by *Pseudomonas coronafaciens* var. *atropurpurea;* characterized by water-soaked lesions which turn brown and coalesce, forming purplish-brown areas on the leaf and sheath. In severe cases the upper nodes are killed. Also called *bacterial blight of grasses.*

chocolate tree. See *cacao.*

choctaw root. See *hemp dogbane.*

choice. A market designation of high quality, usually second from highest; *e.g., choice* beef as opposed to prime beef.

choice (fat type) grade slaughter barrows and gilts. A market grade of swine, denoting superior conformation, quality and finish, for the production of lard, rather than high quality cuts. The yield of major lean cuts is usually less than 47 percent of cold carcass weight, and at 200 to 225 lb. live weight, the backfat thickness averages over 1.9 inches.

choice (meat type) grade slaughter barrows and gilts. A market grade of swine, denoting the minimum degree of finish required to produce high quality cuts. The yield must be more than 50 percent of carcass weight in major lean cuts of hams, loins, picnics and shoulder butts. At a live weight of 200 to 225 lb., the backfat thickness ranges from 1.6 to 1.9 inches.

choice hand-picked. 1. Designating products chosen from a conveyer belt, by hand, for high quality. 2. A grade of dry, edible beans containing less than 1.5 percent of split, or damaged beans, and other foreign material.

choice summer yellow cottonseed oil. An oil free from visible foreign material, clear and brilliant at temperatures sufficient to melt the stearine, sweet in flavor, of a color no higher than AOCS 7.6, and containing in excess of 0.10 percent moisture and volatile matter.

choke. 1. An acute condition in livestock brought about by a food mass lodged in the esophagus. 2. Hairy or filamentous undeveloped scales at the base of globe artichoke heads which are removed before eating. 3. In engine carburation, to increase the ratio of gasoline to air in the fuel mixture when starting a cold engine. 4. In plants, to kill, dwarf, or stunt through excessive competition for space and nutrients, *e.g.*, the *choking* of dodder or bindweed by a climbing vine. 5. In plowing, the gathering of wet straw, weeds, etc., about the plowshare, which reduces its efficiency.

choke apple. See *crabapple: 2.*

chokeberry. See *red chokeberry; purplefruit chokeberry; black chokeberry.*

chokecherry. *Prunus virginiana* (var. *demissa*), family Rosaceae, a treelike, deciduous shrub of little decorative value, but dangerously poisonous to men and animals. The poisoned animal becomes uneasy, staggers, collapses, has convulsions, breathes with difficulty; its eyes roll, and finally, within an hour, it bloats and dies. Its reddish-black fruit has a high content of astringent substance, largely tannin. The plant is a host of the "X" disease of stone fruits (*q.v.*). Native to North America.

choke down. To break in or control an animal by putting a rope around its neck so that it chokes if it struggles.

choke of grasses. Epichloë head blight *q.v.*

choke pear. A pear that has an astringent taste.

choke pondweed. See *Canada waterweed.*

choker. A rope with a noose for pulling logs.

choker-man. One who puts chokers around logs for pulling.

chokestrap. A part of a harness which connects the bottom of the collar to the ring at the end of the neck yoke or tongue chain; used in backing or in preventing a vehicle from running on to a team when going down hill.

choke weed. Any weed or plant that crowds out other plants, especially the broomrape *q.v.*

choko. See *chayote.*

cholera. Any of a number of diseases, as hog cholera, fowl cholera, *q.q.v.*

cholera gallinarum. See *fowl cholera.*

cholesterol. A crystalline sterol found in all animal tissues but concentrated in bile, gallstones, egg yolk, nerve tissue, and milk.

choline. An organic compound present in animals and plants; essential for growth.

cholla. Any of the several cylindrical-stemmed species of cactus, family Cactaceae; serious pests of the stock ranges in U.S.

chomophyte. Any plant which grows in crevices and fissures and in rock detritus or debris.

chop. 1. Animal feed of coarsely crushed or finely ground cereal grains. 2. A small cut of meat which usually includes a rib: *e.g.*, mutton *chop.* 3. To cut hay into small portions for easy storage without baling. 4. To hoe a row crop, especially cotton; often the first hoeing is called *chopping.* 5. To crush grain. 6. Jaw; used generally in the plural to denote the jaws forming the mouth. Also called *chap.*

chopped alfalfa. A livestock feedstuff, consisting of the entire alfalfa hay, chopped finely enough to become a meal, and an admixture of alfalfa straw or other foreign matter. Also called *cut alfalfa.*

chopped hay. Hay cut into short pieces for easy handling and convenient storage.

chopper. a. A cordwood cutter. b. A machine for cutting forage into small pieces.

chord-foot. A unit of measurement for stave bolts; measured across the inner edges of the sapwood.

chore. A regular or odd job around a house or farm.

chorion. a. The outermost membrane that encloses the unborn fetus in mammals. b. The tender, fleshy substance of the original nucleus of a plant seed.

C horizon. See *horizon C.*

chorizos. A Spanish pork sausage coarsely ground, seasoned, and stuffed into hog casing and lightly smoked, after which it is dried.

chorogi. See *artichoke betony.*

chowchow. A preserve which is a mixture of green tomato, peppers, cauliflower, onions, pickles, etc.

chow sum. Dried ginseng roots which are treated with sugar and honey.

chresard. The total amount of water in soil available for plant use.

Christalinna. A hard Swiss cheese made from cow's milk.

Christian IX. A Danish cheese that differs from Kuminost (*q.v.*) in size and shape.

Christmas American mistletoe. *Phoradendron flavescens,* family Loranthaceae, a woody plant parasitic on various deciduous trees of the eastern U.S., which causes the limbs to become malformed. Its branches, preferably bearing berries, are a Christmas decoration. Also called *eastern mistletoe.*

Christmas begonia. See *Socotra begonia.*

Christmasberry. *Heteromeles arbutifolia,* family

Rosaceae, an evergreen shrub widely planted in western U.S. for its dense, white flower clusters, and bright red berries. Poisonous to stock because its leaves contain prussic acid. Poisoned animals become uneasy, stagger, have difficulty in breathing, roll their eyes and hang out their tongues. An hour after ingestion, the animal becomes quiet, bloats, and dies. Native to California. Also called *toyon; redberry; California holly.*

Christmasberry tree. Brazil peppertree *q.v.*

Christmascactus. *Schlumbergera truncata,* family Cactaceae, a small, graceful house plant with arching, flattened stems, which bears many small, pink, rose, white, or purple flowers at ends of its branches. Native to Brazil. Also called *crab cactus.*

Christmas fern. *Polystichum acrostichoides,* family Polypodiaceae, a leathery evergreen fern, grown as winter decoration. Native to North America. Also called *evergreen wood fern; dagger fern.*

Christmas pinch. The pruning of greenhouse roses in U.S. six weeks before Christmas to force blossoms.

Christmas rose. *Helleborus niger,* family Ranunculaceae, an evergreen, perennial herb, grown for its late fall or winter, white or purple with green flowers. The plant is poisonous to stock, and contact with its bruised parts may also result in severe dermatitis in human skin. Also called *black hellebore; felon grass.*

Christmasvine porana. *Porana paniculata,* family Convolvulaceae, a high-climbing herbaceous vine, grown in very warm areas for its showy, white flowers. Also called *mountain creeper; horsetail creeper.*

Christophine. See *chayote.*

Christ's-thorn. *Paliurus spina-christi,* family Rhamnaceae, a prickly shrub with dented, alternate leaves. According to Christian legend, Jesus' crown of thorns was made from this plant. Native to Eurasia. Also called *Jerusalem thorn.*

chromatid (pl. chromatids). A longitudinal half chromosome as seen from early prophase through metaphase of somatic mitosis and from Pachytene through second metaphase of the meiotic mitoses.

chromatin. Portions of the chromonemata, especially in interphase nuclei, which are Feulgen positive and therefore presumed to contain relatively large amounts of desoxyribose nucleic acid (DNA).

chromatography. A separating of closely related compounds by allowing a solution of them to seep through an absorbent, so that the different compounds become absorbed in separate colored layers comprising a chromatogram.

chromocenter. Comparatively large and usually spherical concentrations of chromatin presumed to be "condensed" portions of the chromonema. Several such centers may fuse to form a large one, frequently referred to as a false nucleolus. Hence a body produced by fusion of the heterochromatic kinetochore regions of the salivary gland type chromosomes in certain *Diptera* including *Drosophila.*

chromomere. A deep staining granule on the chromonema. The chromosome of early prophase of the first division of meiosis appears to be made up of a number of such bodies joined by thinner and usually lighter staining interchromomeric strands. Modern cytology interprets these bodies as regions where submicroscopic fibrils are tightly folded or coiled.

chromonema (pl. chromonemata). The smallest optically resolvable linear subdivision of the chromosome. Also the helically coiled thread in an anaphase chromosome or metaphase chromatid.

chromoseres. The smallest particles identifiable by characteristic size and position in the chromosome thread. They are minute subdivisions of chromatin arranged in a linear, bead-like manner on the chromosome.

chromosome. A microscopic, dark-staining body, visible in the nucleus of the cell at the time of nuclear-division, which carries the genes, arranged in linear order. Its number in any species is usually constant and it serves as the bridge of inheritance, *i.e.,* the sole connecting link between two succeeding generations.

chronic alveolar emphysema. See *broken wind; heaves.*

chronic bacillary diarrhea. See *Johne's disease.*

chronic bacterial dysentery. See *Johne's disease.*

chronic coccidiosis. See *coccidiosis.*

chronic disease. Any disease which persists for long in the host.

chronic inflammation of the knee joint. Knee spavin *q.v.*

chronicity. Disease conditions of long standing.

chronic mastitis. See *mastitis.*

chronic milk fever. See *ketosis; acetonemia.*

chronic molybdenum poisoning. A condition which develops when animals constantly take more molybdenum with their food than they can absorb. It causes profuse diarrhea, abnormal bone formation, rapid emaciation, swollen genitals, marked anemia, general weakness, stiffness, and fading of coat color. Molybdenum is a trace element in soils, a lack of which limits plant growth.

chronic rhinitis. See *bull nose.*

chronic water deficiency. A plant condition due to water deficient soil. The plant is stunted, the roots are undersized, and the leaves become discolored. In cereals, the grain is shriveled. In fruit crops, the fruit is spotted, deformed and shriveled. Excess light and heat injury may accompany the deficiency.

chronosequence. A sequence of soils whose properties are functionally related to time as a soil formation factor.

chrysalis. The pupa, or the resting stage (intermediate between larva and adult) of butterflies.

Chrysamphora. Californiapitcher *q.v.*

chrysanthemum. Any plant of the genus Chrysanthemum, family Compositae, whose annual or perennial species originated in Eurasia, but which are now found in most parts of the world.

Some species are grown as ornamentals, some as medicinal plants, while others are pernicious weeds. They flower in all colors except true purple and blue. Among the weeds, the most common is the European *C. leucanthemum,* the oxeye daisy or white weed. *C. coccineum,* florist's pyrethrum, and *C. cinerariaefolium,* Dalmatian pyrethrum, yield insecticides; and *C. parthenium,* the feverfew chrysanthemum, *C. maximum,* the Shasta daisy, *C. morifolium,* florist's chrysanthemum, and *C. indicum,* mother chrysanthemum, are the important ornamentals. The last two species have produced over 3,000 hybrid varieties in U.S. See *arctic chrysanthemum; costmary chrysanthemum; florists pyrethrum; crowndaisy chrysanthemum; Marguerite chrysanthemum; oxeye daisy; Pyrenees daisy; florists chrysanthemum; Nippon oxeye daisy; feverfew chrysanthemum; corn chrysanthemum; giant daisy; Dalmatian pyrethrum; mother chrysanthemum; Shasta daisy.*

chrysanthemum aphid. *Macrosiphoniella sanborni* (Gill.), family Aphidae, an insect pest in greenhouses which causes stunting.

chrysanthemum gall midge. *Diarthronomyia chrysanthemi* Ahlb., family Itonididae, an insect that infests chrysanthemums in greenhouses. Plants have misshapen leaves, and in severe attacks the flowers become distorted, the stems bend, and numerous galls appear.

chrysanthemum lace bug. *Corythucha marmorata* (Uhl.), family Tingidae, an insect pest of chrysanthemums in greenhouses which feeds on the leaves, yellowing them and causing premature drop.

chrysanthemum leaf miner. *Phytomyza atricornis* Meig., family Agromyzidae, a greenhouse insect whose larvae mine chrysanthemum leaves. Heavily infested plants are stunted and have inferior flowers; the leaves dry up but remain on the plant.

chrysanthemum mosaic. A virus disease which causes foliage to mottle. Transmitted by grafting, no insect vector has been reported. Some chrysanthemums carry the disease without displaying any visible symptoms.

chrysanthemum nematode. *Aphelenchoides ritzema-bosi,* family Aphelenchidae, a microscopic, eel-shaped worm which rapidly spreads through and kills the tissues of a chrysanthemum. The symptoms are dark spots. When the leaves fall, the nematodes spread through the soil to other hosts.

chrysanthemum powdery mildew. A fungus, *Erysiphe cichoracearum,* family Erysiphaceae, that grows on the outside of chrysanthemum leaves, forming a white, powdery, mycelial growth by the production of the conidia. The fungus is external to the host plant, except for the numerous haustoria, which may enter the epidermal cells in their search for nutriment for the fungus and seriously damage the foliage. Other plants, *e.g.,* lettuce, dandelions, are also subject to the disease.

chrysanthemum ray blight. A fungous disease caused by *Ascochyta chrysanthemi,* the imperfect stage of *Mycosphaerella ligulicola,* family Mycosphaerellaceae. The corolla of the ray flowers is blighted, resulting in poorly developed, discolored, and one-sided heads. Sometimes the flower bud fails to open.

chrysanthemum rosette. A virus disease in which the infected plant may show crinkling, yellowing, dwarfing, and yellow vein banding. This disease can be transmitted manually, but no insect vectors are known. The visible symptoms may be absent in some varieties.

chrysanthemum rust. A fungous disease of the foliage caused by *Puccinia chrysanthemi,* family Pucciniaceae.

chrysanthemum thrips. *Thrips nigropilosus* Uzel, family Thripidae, an insect pest which feeds on chrysanthemums.

chuchu. See *chayote.*

chuck. A cut of meat from the neck including the part around the shoulder blades and the upper three ribs.

chuckies. See *red columbine.*

chucking. Strong, coarse hemp used for making rope.

chuckle. To cluck or cackle, as a hen *chuckles* to the chicks.

chuck rib roast. A roast cut from the five upper ribs of a beef carcass.

chuck steak. A steak cut from the upper five ribs of a beef carcass.

chuck wagon. A wagon that carries food and cooking and eating utensils for cattlemen or livestock in U.S.

chufa flatsedge. *Cyperus esculentus,* family Cyperaceae; a grass-like, perennial herb, which is often a weed in sandy fields, but sometimes grown for its edible tubers for chickens and hogs. Native to North Temperate zone. Also called *earth almond; edible galingale; rush nut; edible rush; earth nut; yellow nut grass.*

chunk. 1. A strong, thickset horse which is smaller than a draft horse. 2. To pile up brush after logging.

chunk honey. Comb honey packed in jars with liquid honey.

churchmouse threeawn. *Aristida dichotoma,* family Gramineae, a perennial, tufted grass of dry, sandy soils. Its sharp awns are injurious to grazing animals. Native to U.S. Also called *poverty grass.*

churka. A manual spinning wheel used in India.

churn. A vessel in which cream is agitated vigorously to obtain butter.

churn-butted. Of, or pertaining to, a tree trunk much enlarged at the base. Also called *bottle-butted; swell-butted.*

churned body. Buttery body *q.v.*

churned buttermilk. 1. See *buttermilk.* 2. Cultured buttermilk to which churned cream, containing small lumps of butter, is added.

churning. The beating, shaking, or stirring of whole milk or cream to make butter.

churnmilk. See *buttermilk.*

chute. 1. A narrow passage through which animals are moved for branding, spraying, or loading, or through which grain slides to a lower level. 2. A trough constructed of round timbers in which logs are slid up or down a grade. 3. A stampede of men or animals (western U. S.). 4. A high-velocity conduit for carrying water. 5. An inclined drop or fall.

chyle. Lymph fluid; usually high in lipide content.

chyme. The partly digested food passed from the stomach into the duodenum.

chytrid diseases. Plant diseases caused by one of the Chytridiales, an order of fungi, *e.g.*, clubroot of cabbage; powdery scab of potato; Physoderma brown spot; potato wart; cranberry gall; crown wart; beet root tumor.

cibol (ciboule). See *Chibol.*

cicada. An insect of the family Cicadidae. See *periodical cicada.*

-cide. A suffix which indicates a killer, *e.g.*, amoebi*cide,* that which kills amoeba.

cider. See *apple cider.*

cider vinegar. The product of alcoholic and subsequent acetous fermentation of apple juice.

cider vinegar stock. Fermented apple juice intended for conversion into vinegar.

cider yeast. *Saccharomyces malei,* family Saccharomycetaceae, a yeast which causes rather slow fermentation; useful in the manufacture of sparkling hard cider.

cienaga. Mud flats or a swamp.

cigarbox cedrela. *Cedrela odorata,* family Meliaceae, a tropical tree. Its fragrant, colored wood is used for cigar boxes. Native to America; grown for ornamental purposes in southern Florida.

cigar casebearer. *Coleophora occidentis* Zell, family Coleophoridae, a small insect whose larvae weave tough, silken cases about themselves which they drag about while feeding on leaves, buds, and fruit of apple, pear, quince, plum, and cherry.

cigarette beetle. *Lasioderma serricorne* (F.), family Anobiidae, a major pest of stored tobacco. It cuts holes in the plant leaves and builds nests in them; also destructive to upholstered furniture, seeds, and dried spices, etc.

cigar filler. Tobacco leaves placed in the core of a cigar. The leaf fragments are as long as the cigar in superior brands, or short or shredded in low-priced cigars.

cigarflower cuphea. *Cuphea platycentra,* family Lythraceae, a popular shrub grown in U.S. for its solitary, petalless, variegated flower. Native to Mexico.

cigar-leaf states. States of the U.S. where leaf tobacco for cigars is grown: Connecticut, Florida, Georgia, Wisconsin, Pennsylvania, Ohio, and New York.

cigar wrapper. A tobacco leaf (usually imported or shade-grown domestic) wrapped spirally around the filler.

cilia. 1. Minute hairlike formations on the cells of many animals. Constantly vibrating, they serve as means of locomotion in aquatic unicellular animals and help remove mucus and fluid residue in higher forms. 2. Tender, sensitive hair forming a fringe along the edges of leaves.

ciliated. Equipped with cilia.

ciliate. A protozoan that moves by means of cilia.

Cilician fir. *Abies cilicica,* family Pinaceae, a handsome evergreen tree grown in U.S. as an ornamental. Native to Turkey and Syria. Also called *Taurian fir.*

cinch. A band of leather or heavy canvas attached to the saddle. It is tied around the horse's chest to hold the saddle in place.

Cinchona. A large genus of trees and shrubs, family Rubiaceae, characterized by opposite leaves and small panicled flowers. The bark from several of its species yields the commercial quinine. Native to South America; grown in southeast Asia and West Indies. See *medicinal cinchona.*

cinder soil. Soil composed of the ejecta of volcanic eruptions, generally fragmental matter of cinder size. It forms considerable aggregate area in parts of western U.S.

cineraria. *Senecio cruentus,* family Compositae, a low-growing perennial herb, grown in greenhouses for its blue flowers. Native to Canary Islands.

cinnamon. 1. See *Ceylon cinnamon.* 2. A flavoring made from the fragrant dried bark of cultivated varieties of *cinnamon.* The bark is stripped from the trees, piled up and permitted to undergo fermentation to soften the outer skin for removal. It is then dried in the sun.

cinnamon apple. Sugar apple *q.v.*

cinnamon fern. *Osmunda cinnamomea,* family Osmundaceae, a coarse, easily grown fern, found in wet, low woods. Its fibrous root is the source of most orchid peats used in potting. Native to Asia and America.

cinnamon oil. The oil obtained by steam distilling the inner bark of Ceylon cinnamon *q.v.*

cinnamon rose. *Rosa cinnamomea,* family Rosaceae, a hardy, erect shrub grown for its fragrant, purplish-red flowers. Native to Eurasia.

cinnamon vine. *Dioscorea batatus,* family Dioscoreaceae, a high-climbing, herbaceous vine, grown for its edible tubers. Native to China; cultivated in U.S. for its white, cinnamon-scented flowers. Also called *Chinese yam; Chinese potato.*

cinnamonwood. See *sassafras: 1.*

cion. A branch or portion of a branch having one or more buds. Detached from a woody plant, it is then used in grafting or budding. Also called *scion.*

Cipoletti weir. A trapezoidal measuring weir (*q.v.*) in which each side of the notch has a slope of one horizontal to four vertical. It does not require correction for end contractions.

Circassian walnut. Persian walnut *q.v.*

circle rider. A horseback rider who rounds up cattle for branding, etc. (western U.S.).

circling disease. An infectious disease of sheep, swine, cattle, and goats caused by the bacterium

Listeria monocytogenes. So named because it affects the central nervous system, causing the animal to move in circles. Also called *listeriosis.*

circular exhaust box. In canning, a box through which cans follow a circular or spiral path as air is removed.

circular gang mill. A machine with a battery of circular saws, up to 26 inches in diameter, which are fitted to the same shaft. They are usually spaced to cut 1-inch flooring strips from 4-inch or 6-inch cants. The saws are sometimes mounted on an edger arbor.

circular level. A device to indicate a level plane in a bulk milk tank. A bubble in alcohol is coincident or concentric with a permanent reference circle on the glass face of the device.

circular purple scale. See *Florida red scale.*

circular saw. A thin, circular steel disc with teeth cut on the edge. It turns on a motion-driven axle; used in cotton gins and for cutting logs. See *circular gang mill; gin saw.*

circular-spot corky bark. A disease of citrus, probably caused by a virus. Cork-like eruptions, with gummy exudation, appear on the bark, affecting the wood underneath. The lesions are circular and resemble psorosis *q.v.*

circulation. a. The pulsatory movement of blood in the body. b. The flow of cytoplasm in the cells of a plant.

circulatory. Referring to the circulation of blood in the bodies of men and animals.

circumneutral. Describing soil with a pH value slightly above or below 7.00; neither strongly acid nor alkaline.

circumposition. See *air layering.*

cirrhosis. A diffuse fibrosis (hardening) of the liver and some other organs.

ciruela. See *mombin.*

cistern. A natural or artificial reservoir or receptacle for holding rain water or other liquids.

citradia. *Poncirus trifoliata* x *Citrus aurantium,* family Rutaceae, a hybrid citrus tree resembling citranges but more vigorous and resistant to cold. It is used as a stock in the propagation of common citrus fruits.

citrange. *Poncirus trifoliata* x *Citrus sinensis,* family Rutaceae, a hybrid tree which bears a fruit, more acid and with a more pronounced odor than the orange; is mainly used as a stock for citrus breeders.

citrangequat. Citrange x kumquat (*Fortunella* spp.), family Rutaceae, a hardy hybrid tree, useful as a root stock. Its fruit is used in marmalades and citrus beverages.

citrate of lime. The dried citrate obtained from the lemon pulp after the oil has been removed. It usually contains more than 60 percent citric acid.

citrate soluble phosphoric acid. See *reverted phosphoric acid.*

citric acid. An acidic white, crystalline substance present in lemons, limes, currants, gooseberries, raspberries, etc. It is usually obtained from citrate of lime by decomposition and filtration. Used in artificial lemonade, medicines, and dyeing.

citricola scale. *Coccus pseudomagnoliarum* (Kuw.), family Coccidae, an important insect pest of citrus in California.

citron. 1. *Citrus medica,* family Rutaceae, a large, thorny shrub grown in very warm areas for its highly aromatic fruit, the citron of commerce. The thick fruit rind is candied and used as a preserve. It is one of the oldest citrus fruits: native to eastern Asia. Also called *cedrat.* 2. See *citronmelon.*

citron daylily. *Hemerocallis citrina,* family Liliaceae, a summer plant which is grown for its fragrant, large, light yellow flowers. Native to China.

citronellagrass. *Cymbopogon nardus,* family Gramineae, a perennial grass grown in warm areas for its fragrant oil which is used in perfumes and insecticides. Native to Asia.

citronella horsebalm. *Collinsonia canadensis,* family Labiatae, an aromatic, perennial herb grown for its lemon-scented, yellow flowers. Native to North America. Also called *horse balm; stoneroot; rickweed; citronella.*

citronmelon. *Citrullus vulgaris* var. *citroides,* family Cucurbitaceae, a tropical, annual, herbaceous vine, grown for its small fruit, the hard flesh of which is used in making preserves. Native to Africa. Also called *citron; piemelon.*

citrophilus mealybug. *Pseudococcus gahani* Green, family Pseudococcidae, an important pest of citrus trees in California and some other parts of U.S.

citrumelo. *Poncirus trifoliata* x *Citrus paradisi,* family Rutaceae, a citrus tree, a hybrid between the grapefruit and the trifoliate-orange; used as a root stock.

citrumquat. *Fortunella* spp., x *Poncirus trifoliata,* family Rutaceae, a hybrid citrus tree, a cross between kumquat and the trifoliate-orange.

Citrus. An important genus of trees, family Rutaceae, extensively cultivated in tropical and subtropical areas for its fruits, such as the lime, orange, bergamot, lemon, citron, grapefruit. Native to Asia.

citrus anthracnose. A disease of oranges, lemons, grapefruits, etc., caused by the fungus *Colletotrichum gloeosporioides,* family Melanconiaceae; characterized by rot, spot, and tear stain of the fruit, wither-top of the twigs, and occasional spotting of the leaves. Also called *anthracnose rot of citrus; anthracnose spot of citrus; anthracnose tear-stain of citrus.* See *anthracnose.*

citrus belt. The citrus producing regions of the world lying generally within 35 degrees north and south of the equator.

citrus blackfly. *Aleurocanthus woglumi* Ashby, family Aleyrodidae, an important pest of citrus and coffee plants in tropical areas. The insect sucks the plant juices, causing the tree to become stunted and yield poor fruit. It also harms the plant by producing sooty mold on the honeydew.

citrus blast. A bacterial disease of citrus trees caused by *Pseudomonas syringae.* Small, dark lesions appear on the petiole and extend to the base of the leaf blade and to the twig at the place of attachment. As a callus forms on the diseased spot, the color changes from reddish-brown to chestnut. The leaf withers rapidly on the stalk or falls. It is not considered a dangerous disease. See *black pit; lilac blight.*

citrus blight. A disease which attacks citrus trees 12 to 14 years of age in Florida; characterized by a progressive wilting of the foliage. New growth is reduced and only small, highly acid fruit is produced. The tree appears to be suffering from lack of water, but wilting continues even through damp weather. The cause is unknown.

citrus bud mite. *Aceria sheldoni* (Ewing), family Eriophyidae, infests citrus, particularly lemons, in California. It attacks the buds and blossoms beneath the calyx, and causes deformities of the fruit, leaves, blossoms, buds, and twigs.

citrus canker. A bacterial disease caused by *Xanthomonas citri,* formerly present in Louisiana and Mississippi, but now occurs only in South America and the Orient. It is characterized by spongy eruptions of the fruit tissue. These lesions show a glazed, oily, watery margin. On the leaves there are small, spongy, white eruptions becoming tan. The margins around the leaf lesions become watery and greasy being yellowish-brown or green with a yellow halo. The disease attacks the grapefruit, trifoliate-orange, lemons, sweet oranges, limes, Satsuma orange. Also called *cancrosis A.*

citrus knot. A fungous disease of citrus caused by *Sphaeropsis tumefaciens,* family Sphaeropsidaceae, characterized by galls up to 3 inches in diameter on the twigs and branches which may extend parallel to the axis of a branch for several inches. At first they are smooth but later become rough and fissured. The disease may result in witches'-broom. Occurs in Jamaica and Florida.

citrus mealybug. *Pseudococcus citri* (Risso), Order Hemiptera (Homoptera), family Pseudococcidae, a serious, widespread pest which attacks all citrus, many ornamentals, and greenhouse plants by sucking the plant juices.

citrus mite. See *citrus bud mite; citrus red mite; six-spotted mite.*

citrus molasses. A sirup prepared from the juice, rind, and waste of dehydrated oranges and grapefruit.

citrus nematode. *Tylenchulus simipenetrans* Cobb., family Tylenchidae, a microsopic worm which infests the roots of citrus. Soil particles cling to infested roots and the bark separates readily from the woody part of the root.

citrus psorosis. See *psorosis.*

citrus pulp. See *dried citrus pulp.*

citrus red mite. *Metatetranychus citri* (McG.), family Tetranychidae, one of the most serious pests of citrus; extracts chlorophyll from the leaves, which become silver or brown and fall. The fruit turns gray or yellow. Also called *purple mite.*

citrus root weevil. *Pachnaeus litus* (Germ.), family Curculionidae. The adult feeds on citrus blossoms and leaves, and the larva on the roots, particularly limes on the Florida Keys of U.S.

citrus rust mite. *Phyllocoptruta oleivora* (Ashm.), family Eriophyidae, a pest which extracts juices from citrus fruit skin and leaves, which causes rusting of oranges and silvering of lemons. It occurs in California and the Gulf States of U.S.

citrus scab. See *sour orange scab; sweet orange fruit scab; Tryon's scab.*

citrus stem-end rot. A rot during storage caused by *Phomopsis citri* and *Diplodia natalensis* fungi. In U.S., serious in the Gulf States but of minor significance on fruit grown in California.

citrus thrips. *Scirtothrips citri* (Moult.), family Thripidae, an American pest of deciduous fruits and other plants in California and Arizona. It infests the buds, new growth, and the fruit.

citrus white fly. *Dialeurodes citri* (Ashm.), family Aleyrodidae, a pest which sucks the sap on the underside of leaves of citrus and other plants, especially the camellia, causing stunting of the trees and undersized fruit of poor color. Found in the Gulf States of U.S., including Florida, where it attacks the Chinaberry. See *brown, red,* and *yellow fungus of citrus white fly,* as these are natural means for the biological control of this pest.

city chicken. A cut of meat made of veal shoulder in which small pieces of meat are fastened on a skewer. The pieces then are shaped to resemble a chicken's leg and breaded.

city crop. See *loose cotton.*

city forestry. The preservation and protection of shade trees in streets and parks.

cive. See *chive.*

civy. A sauce or soup eaten with hare. Also called *civey.*

clabber. Curdled milk in which whey has not separated from curd. See *curdled milk.*

clack valve. A hinged pump valve which makes a clacking sound when working.

claim. 1. A right or supposed right to unsettled land, or the mineral or oil discovered in land within the public domain. 2. To cost.

claiming pen. A pen or small enclosure within a pen in which a ewe and her newborn lamb are placed until the ewe accepts the lamb.

clambering monkshood. *Aconitum uncinatum,* a poisonous, perennial herb, family Ranunculaceae, found along streams in shady places from Pennsylvania to Georgia in U.S. Symptoms of poisoning are weakness, irregular and difficult breathing, weak pulse, bloat, attempted swallowing, belching. Also called *wild monkshood.*

clammy campion. *Lychnis viscaria,* a perennial, Eurasian herb, family Caryophyllaceae, grown for its red or purple flowers. Varieties have white and double flowers. Also called *German catchfly.*

clammy cuphea. *Cuphea petiolata,* an annual weed pest, family Lythraceae, bearing purple flowers. Found in fields and on roadsides of eastern North America. Also called *waxweed.*

clammy hopseedbush. *Dodonaea viscosa,* a shrub or small tree, family Sapindaceae. Native to Australasia and South Africa, but planted in Florida, Arizona, Mexico, Hawaii for ornament. Also called *akeake.*

clammy locust. *Robinia viscosa,* family Leguminosae, a deciduous tree grown for its ornamental foliage and pink and yellow flowers. The leaves and bark, covered with sticky glands, are poisonous (sometimes fatally) to horses, cattle, sheep and chickens.

clamp. The mound formed in a field or a farmyard by covering a harvested root crop with straw and a layer of earth. It protects the crop from winter weather.

clams. See *actinomycosis.*

clan. A plant colony composed of secondary species occurring locally or in a few scattered areas. It is a permanent feature of climax communities or of consocies which persist for a long time.

clarified milk. Milk that has been cleared of solid impurities by passing through a centrifugal separator.

clarifier slime. The residual solid matter left after centrifugal clarification of milk. It consists of body cells, bacteria, etc.

clarify. To remove undesirable, solid substances from a liquid, such as milk or fruit juice, by ordinary or by centrifugal filtration.

clarin. See *Mexican bloodtrumpet.*

Clark dwarf. A dwarfing rootstock for apples, believed by some to be Malling VIII, which has been used as an inter-stock. See *Malling (root) stock.* Also called *Clark rootstock.*

Clarke-McNary Act. An Act of U.S. Congress (1924) providing for cooperative farm forestry work. It authorized the federal Secy. of Agriculture to seek the help of appropriate state officials in assisting farm owners to establish, renew and improve woodlots, shelterbelts and other valuable forest growth. Under its provisions, the federal appropriations for it were to be matched by the state or other cooperating agency.

clarkia. Any annual herb of the genus Clarkia, family Onagraceae. Species are grown for their beautiful flowers. Native to western North America. See *rose clarkia.*

Clark rootstock. Clark dwarf *q.v.*

clary. Referring to clary sage and other plants considered beneficial to the eyes.

clary sage. *Salvia sclarea,* family Labiatae, a hardy, biennial medicinal herb with bluish-white flowers and grayish-haired leaves. The leaves are used to flavor preserves and omelets and to perfume sachets. Tea and a cordial called clary water are made from the flowers, and an oil is extracted from the plant for use in perfumes. Native to southern Europe and western Asia.

clasping mullein. *Verbascum phlomoides,* family Scrophulariaceae, a biennial herb, grown for its night-scented, yellow flowers. Native to Eurasia; naturalized in eastern North America. Also called *tall mullein.*

class. 1. a. A group of soils with a definite range in a property or attribute, such as acidity, slope, texture, structure, land capability, degree of erosion, and drainage. b. A group of animals, plants, fibers, etc., which have common characteristics or attributes. 2. To sort individuals into groups which have common characteristics; especially, to sort for marketing according to grade, colors, etc.

Class I. An old tariff term for wool from Merino sheep or from sheep having Merino blood.

Class II. An old tariff term for medium and long wools from British sheep.

Class III. An old tariff term for native, unimproved wools from all over the world. Also called *carpet wool.*

classer. A person who sorts individuals into groups of common characteristics or attributes, as a cotton *classer.*

classification. The systematic arrangement of units according to respective groups; *e.g.,* taxonomy, the arrangement of plants, animals, rocks, minerals, soils into categories according to a definite plan.

classify. 1. To systematically categorize plants or animals according to a set scheme. 2. To sort individuals together into groups having common characteristics or attributes.

classing table. A long table on which samples of lint cotton are placed for classification.

class of seedling stock. The age of nursery stock denoted by two or more figures, as 2-0, 2-1, or 1-1-1. The first figure shows the number of years in the seedbed and the succeeding figure or figures the number of years in the transplant bed or beds. The number of figures following the first indicates the number of times the stock has been transplanted.

class ratio. The relation between the numbers of different kinds of livestock, game, or rodents which a given area can support.

class use. See *selective grazing.*

clavel de l'Inde. Crapejasmine tabernaemontana.

clavus. See *ergot: 1.*

claw. 1. The sharp nail on the toe of an animal or bird. 2. The slender, extended lower part of the petal, as in the iris, lily, etc. 3. A device on a milking machine to which the stanchion tubes are connected.

claw-fern. See *Japanese clawfern.*

claw sickness. See *foot rot (of animals).*

claw weeder. A manual implement with claw-like metal projections from one end which is used for weeding flower beds, gardens, etc.

clay. 1. A hydrous aluminum silicate; a widely distributed substance in the earth which is hard when dry but becomes pliable when wet. The varieties depend upon variations in amount of combined water, included elements (other than silicon and aluminum) and lattice structure of crystals. 2. A soil particle less than .002 mm. in diameter. 3. A textural class of soil; one that contains 40 percent or more of particles of clay size.

clay-backed cutworm. *Feltia gladiara,* family Phalaenidae, a larval pest cutworm that infests plants near the soil line. It weakens a plant by eating part of the stalk, causing it to fall over.

claybark dun. A coat-color of animals, especially horses. It is dun imposed on a sorrel chestnut background. Also called *copper dun.*

clay-block silo. A silo constructed of fired-clay bricks.

clay-colored billbug. *Calendra aequalis* (Gyll.), family Curculionidae, a snout beetle infesting corn and grasses. It makes small holes in the stem through which it consumes the plant tissue, and in which it lays its eggs. It also feeds on the unripened kernels of wheat and millet.

clayey. Containing or resembling clay.

clay land. An area of heavy soils.

clay loam. A textural class of soil with 27 to 40 percent clay, 15 to 53 percent silt, and 20 to 45 percent sand. See *soil texture.*

claypan. A subsurface soil horizon; an alluvial layer of stiff, compact, and relatively impervious clay, especially when wet. Like hardpan (*q.v.*), it prevents water movement and root growth.

clay soils. See *clay: 3.*

Clayton fern. Interrupted-fern *q.v.*

clayweed. Coltsfoot *q.v.*

cleamer. Smeared wool from the shearing board clinging to fleece.

clean. 1. Free from flaws, blemishes or foreign matter. 2. To remove undesirable matter from a substance. 3. To hoe and cultivate to remove weeds from a field.

clean basis. Quotations of prices based on the estimated weight of fiber that a lot of grease wool may contain.

clean-boled. Free of branches, as a tree. Used to designate timber with a satisfactory length of clear bole.

clean-bred. Thoroughbred, as a horse.

clean-burning. A backfire set close to the fire edge to control a forest fire.

clean cotton. Seed cotton, free from extraneous hulls, bolls or trash.

clean culture. Intensive cultivation of a field so as to remove all weeds, etc. Also called *clean cultivation; clean tillage.*

clean rice. Milled rice with hulls removed.

cleaner. See *air-line cleaner.*

cleaner feeder. A device which connects the elevating and distributing system of a cotton gin. It also beats dirt, dust, and trash out of the seed cotton.

cleaning. In a nursery of young trees, the weeding out of undesirable forms or species which might injure those of greater prospective value.

cleaning shoe. A part of a threshing machine consisting of a sieve and fan which separates chaff, weed seeds, and other impurities.

cleansing flight. A flight of honey bees during which they are able to void their feces, following a period of confinement to the hive, caused by inclement weather.

clean tillage. Clean culture *q.v.*

clear. 1. In milling, the residual wheat flour left after the separation of patent flour *q.v.* 2. In land reclamation, to remove trees, brush, stones or other obstacles to tillage. 3. See *clean: 1.* 4. Clean-boled *q.v.*

clearance. 1. In marketing, the number of animals received and sold at the end of the day. 2. See *throat of a plow.*

clear belly. The square-trimmed bacon slab, ready for marketing. It is free from mammary gland pigmentation.

clear-boled. Clean-boled *q.v.*

clear-cut shoulder. The heavy muscle back of the arm bone in a beef carcass used for pot roasts.

clear-cutting. Removal of the entire timber stand from an area.

cleared. Land from which trees and underbrush have been removed.

clear egg. An infertile egg.

clear-eye. See *clary sage.*

clear face. Clear, usable lumber.

clearing. A piece of land cleared in preparation for cultivation.

clearing and grubbing. In a field, the removal of trees and brush that interfere with seedbed preparation, cultivation, mowing, and other cultural operations for the production of crops and pasturage.

clear length. The limb-free portion of the stem of a tree, *i.e.,* from the ground to the lowest branch or branch stub.

clear lumber. Clean lumber.

clearly visible germ development. The development of a germ spot in the yolk of a fertile egg which is plainly visible as a definite, circular area or spot with no blood in evidence.

clear money. Profit.

clears. Barren, peaty land on the Chatham Islands (New Zealand).

clear-span roof construction. A roof supported by rafters placed on the walls without interior posts.

clear the board. An order to stop shearing for the day (New Zealand).

clear winged grasshopper. *Camnula pellucida* (Scudd.), family Acrididae, found throughout the esatern U.S. and in the Rocky Mountain area at high elevations. It feeds on a wide variety of vegetation.

cleat. 1. A strip of wood or iron fastened crosswise to something for strength, or to prevent warping. 2. A wooden or metal device having two arms around which turns may be taken with a rope to hold securely and yet be readily released. 3. In mining coal, joints in seams at which the coal breaks.

cleavage. 1. In animal or plant reproduction, the splitting of one cell into two identical parts. Each resulting daughter cell matures and may divide again. 2. Tendency of certain minerals or woods to split along particular planes or angles. 3. The weight which is required to cause splitting in a standard piece of wood three inches long, expressed in pounds per inch of width.

cleaver. A heavy, wide butcher's knife for cutting bones.

cleft-footed. An animal having a divided hoof or foot, as a cow or hog. Also called *cloven-footed.*

cleft graft. A simple method of grafting in which large trees are used for stock. The branch is sawed squarely across and split lengthwise and two cions are inserted into the cleft. A thin coat of wax is spread over the cut surface and down both sides of the graft.
clefts. In blacksmithing, cracks in the heels of horses.
clefty. Pertaining to flat, clean, fine quality, cannon bones. Also called *clifty.*
cleistothecium. A closed, often hollow ascocarp of a fungus, Class Ascomycetes, which differs from the open apothecium or from the nearly closed perithecium. Often the distinction is not made sharply between *cleistothecium* and perithecium, *q.v.*
Clematis. A genus of perennial herbs or shrubby vines, family Ranunculaceae. Species are grown for ornament, generally as vines, for their bright colored flowers. Numerous species are found in North America and Asia. See *Armand clematis; curly clematis; Douglas clematis; plume clematis; Jackman clematis; anemone clematis; Oriental clematis; sweetautumn clematis; golden clematis; scarlet clematis; leatherflower clematis; virginsbower; travelersjoy; Italian clematis.*
clematis blister beetle. *Epicauta cinerea* (Forst.), family Meloidae, a pest which infests clematis, verbena, and other ornamental plants.
cletch. See *clutch: 1.*
cleuch. A cleft or gorge; a ravine. Also called *cleve.*
cleve. A steep hillside (England). Also called *cleuch.*
Cleveland Bay. A heavy carriage breed of horse in Yorkshire, which is the oldest existing, one-purpose breed of horse being 16 to 17 hands tall and weighing 1,250 to 1,550 pounds. The color is bay with a black mane and tail; it has a long back, and it is active, fast walking, and stands summer heat well. See *carriage horse.*
Cleveland cherry. See *Jerusalem cherry.*
clevis. A coupling device made of a U-shaped piece of iron whose ends are perforated and through which a pin or pole is placed.
click beetle. The adult of a wireworm *q.v.*
clicking. See *forging.*
cliff rose. See *Stansbury cliffrose.*
clifty. See *clefty.*
climacteric. Of, or pertaining to, the climacteric peak *q.v.*
climacteric peak. The maximum point of the respiration rate of mature fruit, especially apples. The term is usually applied to harvested fruit, but it may apply to fruit still on the tree. See *preclimacteric stage; post-climacteric stage.*
climate year. A year which is selected for presentation of data on streamflow, precipitation, etc. The *climate year* of the United States Geological Survey called a *water year,* extends from October 1 to September 30 following. Also called *climatic year.*
climath. See *poison ivy.*
climatic index. An index number which condenses climatic data into a simplified expression.
climatic stratification. The adaptation of a species to a particular climate.
climatography. The description of climates.
climatology. The science of climates.
climax. The ultimate or culminating stage of possible development of the vegetation (plant succession) for a given natural environment.
climax basket. Any wooden-veneer, generally rectangular, basket having a solid wood bottom and a wire handle which is often used for retail sale of grapes. In size, they are one pound, two quart, four quart, and twelve quart.
climax dominant. Any species which is dominant in the more or less stable biotic community.
climax vegetation. See *climax.*
climber. Any plant having long, relatively slender branches which, in order to support its branches, may attach itself to other plants, fences, walls, etc., by means of tendrils, disks, entwining, etc.
climbing bittersweet. American bittersweet, *q.v.*
climbing cutworm. Dark-sided cutworm, *q.v.*
climbing dahlia. *Hidalgoa wercklei,* family Compositae, a tender, herbaceous vine grown in warm areas for its dahlia-like, orange-scarlet flowers. Native to Central America.
climbing fern. See *Hartford fern.*
climbing fig. *Ficus pumila,* family Moraceae, a tender, woody vine grown in northern U.S. in greenhouses and in warm areas out-of-doors as an ornamental and a basket plant. Native to Asia and Australia.
climbing fumitory. Mountainfringe, *q.v.*
climbing hydrangea. 1. *Hydrangea petiolaris,* family Saxifragaceae, a high-climbing, woody vine with white flowers, useful as an ornamental. Native to Japan. 2. See *southeast decumaria; Japanese hydrangeavine.*
climbing ivy. Poison ivy, *q.v.*
climbing lily. Glory-lily, *q.v.*
climbing nightshade. See *bitter nightshade.*
climbing plumbago. *Plumbago scandens,* family Plumbaginaceae, a tropical, tender, spreading shrub with white flowers. Its bitter leaves and roots are sometimes used to cure toothache. Native to America. Also called *toothwort.*
climbing sailor. Kenilworth ivy, *q.v.*
climbing spur. A device consisting of a metal strap shaped like an "L" with a heavy iron spike at the apex of the angle. It is strapped to the boots of a person to enable him to climb trees and poles.
climbing ylang-ylang. See *fragrant tail-grape.*
climograph. A graph to show the influence of climatic conditions on man.
clinch. Part of a nail bent over to keep it from being pulled out, as in horseshoeing, box manufacture, etc.
clinching tong. Pliers which are used to bend the end of the horseshoe nail that extends through the hoof of a horse.
cling. 1. The tendency of cotton fibers in a sample to stick together. 2. Diarrhea.
clingstone. a. Any fruit, especially varieties of the peach, whose flesh adheres strongly to the seed,

or stone, even at maturity. b. The stone of such a fruit. *Cf. freestone.*

clinical. Concerning the investigation of disease on a living subject by observation, as contrasted to a controlled experiment.

clinical abnormalities. Unusual condition or behavior observed in a diseased animal and used as an aid in a diagnosis; a symptom.

clinical evidence. Any symptom of disease that can be determined by direct observation, such as fever, lack of appetite, swellings, paralysis, etc.

clinical form. A form or stage of a disease in which clinical symptoms are observed.

clinical infection. An infection in which clinical symptoms develop.

clinical symptoms. Clinical evidence, *q.v.*

clinosequence. A sequence of soils whose properties are functionally related to the amount of slope on which they are formed.

Clinton woodfern. *Dryopteris cristata* var. *clintoniana,* family Polypodiaceae, a hardy fern grown as an ornamental in U.S. Native to Europe and North America.

clip. 1. A semi-circular metal piece extending from the outer surface of the horseshoe at the toe or side to prevent the shoe from shifting on the hoof. 2. Shears. 3. a. An inclusive term for shorn wool. b. The process of shearing wool. c. The year's production of wool. 4. a. To shear the hair of an animal close to the skin, as to *clip* a dog, or a cow's flank and udder. b. To remove fleece from a sheep or goat. Also called *shear.* c. To cut the feathers from a fowl's wing·to prevent it from flying. 5. a. To trim a plant. b. To cut off the tops of a crop, as clover or alfalfa, at an early stage of growth when the crop is to be harvested for seed. Also called *preclipping.*

clipped board. A board square trimmed on one end (eastern U.S.).

clipped oats. Oats that have the tip ends cut off to increase their test weight and groat percentage.

clipped oat by-product. A feed from the residue of clipped oats, containing light, chaffy hull ends, and immature oats.

clipper. 1. See *strawberry weevil.* 2. A device for clipping the coats of animals.

clip plot. A small area of land used to determine the weight of forage produced under different conditions or to measure potential forage yield on different sites.

Clitocybe root rot. A fungous disease of pear, peach, apple, and other woody plants caused by *Clitocybe tabescens,* family Agaricaceae. Found on citrus in Florida and other trees throughout the southern U.S. Its characteristics are very similar to Armillaria root rot and mushroom root rot *q.v.*

cloaca. The common passage for intestinal, urinary and generative discharges in fowls. It is the pouch of which the vent is the outside opening.

cloacal fluke. *Prosthogonimus macrorchis,* a leaf-like, parasitic trematode which infests the cloacal region of chickens. The symptoms are a loss of weight, decrease in egg production, and general pallor. This parasite has an interesting life history. It lays eggs in the cloaca of the fowl from where they sometimes escape into water through feces, etc., subsequently hatching into minute ovoid worms. The worm enters the body of an aquatic snail and, developing there, produces a series of tailformed larvae, which again escape and individually enter the body of a dragon-fly larva where they become encysted. When a chicken eats the dragon-fly, these develop into adult flukes in the new host and settle in the cloaca, thus completing the life cycle.

cloacal lips. See *vent: 2.*

cloacitis. A foul-smelling inflammative condition of the cloaca and vent caused by a non-infectious unknown organism. Also called *vent gleet; vent disease of fowl.*

clobber. See *curdled milk.*

cloche. 1. A French term for a bell-glass *q.v.* 2. A miniature greenhouse used outdoors to protect plants against low temperatures. They are often placed close together along a row of plants to be protected.

clod. 1. A lump of soil produced by plowing or digging in contrast to a ped or natural aggregation of soil particles. The soil particles may be loosely or firmly bound together but in time the *clod* is subject to slaking, and is transient in nature. 2. Shoulder clod *q.v.*

clod crusher. A roller consisting of numerous uni-axled, circular, cast iron sections, about 30 inches in diameter, which is used to break up clods after plowing. See *bar roller; land roller; float: 3.*

cloddy. Referring to soil having clods.

clodhopper. 1. A heavy, high-topped shoe. 2. A farmer (slang).

clog. 1. A heavy, wooden block attached to a horse's hind pasterns by a strap to prevent it from kicking when put out to pasture. 2. An impediment, encumbrance, or restraint. 3. A wooden shoe. 4. To stop a machine, as a harvester, by feeding material in too rapidly; to plug up; to choke up.

clogging. The thickening of lubricating oils due to the absorption of oxygen and the presence of dust. Machinery is said to *clog* when its lubrication oil becomes thick and dry.

cloistered heart. See *Andrews gentian.*

clone. A group of plants originating from a single plant, propagated vegetatively rather than by seeds, as cuttings from a rose.

close. Generally, a tract of land enclosed by a fence or other boundary, but legally it may apply to a particular piece of land whether enclosed or not.

close breeding. The breeding together of closely related plants or animals. See *in-and-in breeding.*

close-centered tree. In pruning, a globe-shaped tree in which the trunk is contained as a leader in the center of the tree, produced by heading back the side branches and the trunk. Seldom used in orchards. See *open-center tree.*

close cheese. A Cheddar cheese without mechanical or gas openings, *i.e.,* with a close-knit, solid body.

close-coupled. 1. Describing an animal having a short back. Also called *short-coupled. Cf. long-coupled.* 2. In tube wells, the mounting of the pump and its power source on the same axle; *i.e.*, the engine crankshaft is extended to be the impeller shaft of the pump.

close cropping. The grazing by animals of grasses and other forage dangerously close to the crown of the plant which can ruin grazing lands. Also called *close grazing.*

closed drain. Conduits laid to grade beneath the surface to remove excess water from wet land. The joints are sealed to make a closed line on steep slopes or in areas where tree roots can penetrate and clog the line.

closed-end. Designating a harrow which has the teeth fixed to the guard rails attached to the ends of the bars.

closed flock. A flock of fowl or sheep bred from one strain only. See *close breeding.*

closed formula. The statement of ingredients on a tag attached to a sack of feed which gives the guaranteed percentages of protein, fat, and fiber. It does not give the amount of each ingredient. *Cf. open formula.*

closed gentian. See *Andrews gentian.*

closed herd. A herd in which no outside blood is introduced.

closed pedigree. Hybrid combinations of unrevealed inbreds which have been put together by privately owned organizations in the seed business. Also called *unpublished pedigree.*

closed tire wheel. A wheel with a solid tread used on corn planters. *Cf. open-center tire wheel; double wheel.*

closed womb. An imaginary cause of sterility in mares and cows. Charlatans practice dilating the cervix which may ruin the female for breeding purposes. Also called *imperforate neck of the uterus.* See *hollow tail; hollow horn.*

close-fertilization. See *self-fertile; self-pollination.*

close grain. Wood grain characterized by narrow and inconspicuous growth rings.

close grazing. See *close cropping.*

close-growing. Pertaining to plants which have many stems to the square foot.

close herding. See *bunched up herding.*

closh. See *laminitis.*

closing price. The price of the last contract on an item in a market on a particular day.

clostridial organism. Bacteria of the genus Clostridium which cause diseases such as tetanus, malignant edema, and blackleg.

clot. 1. A coagulum; a semisolid mass, as a *clot* of cream, or a blood *clot.* 2. To coagulate, as blood.

clotbur. Cocklebur *q.v.*

cloth-covered coldframe. A cloth-covered frame used for growing seedling tobacco and most vegetable crops that are to be transplanted. It provides protection from winds, evaporation, intense light, low temperature, and insects. See *cold frame.*

cloth-covered house. See *cloth house.*

clothesline hay drying. See *rack hay drying.*

clothes moth. Several species of moths of Order Lepidoptera, family Tineidae, especially *Tineola bisselliella* (Hummel), the webbing clothes moth, and *Tinea pellionella* (L.), the casemaking clothes moth. The female lays the eggs in clothes and rugs, etc., which hatch in about 5 days. The larvae, depending upon the food supply, temperature, and humidity, may require six weeks to four years to reach maturity and, in the meantime feed on hair fiber, wool, silk, felt, feathers, etc., in clothing, carpets or upholstery.

cloth house. A structure resembling a greenhouse, but covered with light cloth instead of glass. It is used for growing certain flower crops, such as asters and chrysanthemums to maturity and provides reduced light, reduced temperature, protection from wind, slightly higher humidity, and protection from insects to the plants. Also used for growing tobacco to maturity when a certain quality is desired.

clothing wool. Short staple wool, too short to be combed, suitable for woolens *q.v.* Also called *carding wool.*

clothofgold crocus. *Crocus susianus,* family Iridaceae, a perennial herb grown for its early-spring, orange-yellow flowers touched with brown. Native to Crimea.

clotted cream. See *Devonshire cream.*

clotted milk. Formation of thick globules in milk; a defect.

cloud bentgrass. *Agrostis nebulosa,* family Gramineae, an annual grass grown for its white cluster of tiny spikelets used in dry bouquets. Native to Spain.

cloudburst. An intense rain storm which usually lasts for a short time.

clouding. An undesirable translucent condition of liquids.

clouding (vinegar). The formation of a metal haze in white-wine vinegar due to the absorption of salts of iron, tin, and copper in its manufacture.

cloudland pink. *Dianthus alpestris,* family Caryophyllaceae, a perennial herb grown in rock gardens for its deep pink or rose flowers. Native to Europe.

cloud of pome fruits. See *sooty blotch of pome fruits.*

cloud on a title. A legal defect in the title to property.

cloud seeding. The production of rain by use of artificial nuclei. Favorable clouds are seeded with dry ice, silver iodide, or other substances to produce nucleation and thereby precipitation. The success of such procedures remains controversial.

cloudy-winged whitefly. *Dialeurodes citrifolii* (Morg.), family Aleyrodidae, an important pest of citrus trees in Florida; sometimes of yams and some species of fig also. It stunts the plants and reduces the fruit yield; like the citrus whitefly *q.v.*

clouted cream. See *Devonshire cream.*

clove. 1. A small division of a separable bulb, as a *clove* of garlic. 2. A weight unit of 8 lbs. of wool, cheese, etc., (England). 3. The dried flower bud of *Syzygium aromaticum* (*Eugenia aromatica; E. caryophyllata*), family Myrtaceae, a tropical tree native to the East Indies, but now extensively cultivated throughout the tropical regions. It is used in flavoring, perfumes and medicine.

clove currant. *Ribes odoratum,* family Saxifragaceae, an ornamental shrub, which is one of the finest of the currants grown for its fragrant yellow flowers and black fruit. Native to western U.S. Also called *buffalo currant.*

clove gillyflower. See *carnation.*

cloven-footed. See *cleft-footed.*

clove pink. See *carnation.*

clover. 1. A general term for low herbs of genus Trifolium, family Leguminosae, with compound, usually trifoliolate, leaves. *Clovers* are excellent forage, and serve also as good bee plants and for nitrogen fixation in soil. 2. A name applied with a qualifying word to plants of genera Medicago (bur clovers), Lespedeza (bush clovers), Petalostemum (prairie clovers), Melilotus (sweet clovers), and Alysicarpus (alyceclovers), etc. See *alyceclover; sweet clover; lespedeza; prairie clover; hop clover; rabbitfoot clover; alsike clover; crimson clover; red clover; white clover.*

clover anthracnose. A fungous disease caused by *Colletotrichum trifolii,* family Melanconiaceae. Lesions and cankers are formed on the aboveground parts of the plant, which sometimes kill it. Also called *southern anthracnose.*

clover aphid. *Anuraphis bakeri* (Cowen), family Aphidae, a louse that has on occasions caused total loss of seed crops of red and alsike clovers. Its honeydew causes the seeds to stick together in storage.

clover big vein. A virus disease of clovers, some vegetables, and common weeds in which the leaflets are distorted by the enlargement of veins on lower side. In some plants, tumors are also produced on the veins, stems and roots. Also called *wound tumor disease of legumes; clover club leaf.*

clover broom. Yellow wildindigo, *q.v.*

clover dodder. *Cuscuta epithymum,* family Convolvulaceae, a yellow, threadlike plant, parasitic upon various species of clover, alfalfa, and some other legumes. It is an introduced species occurring throughout North America. Also called *epithyme; fairies' hair; thyme dodder.*

clover hayworm. A larva of a moth, *Hypsopygia costalis* (F.), family Pyralididae. It attacks bases of weathered hay stacks, destroying large quantities of hay.

clover head caterpillar. A larva of the insect, *Grapholitha interstinctana* (Clem.), family Olethreutidae. It infests clovers, causing irregularly shaped heads and flowers.

clover huller. A threshing machine for small-seeded legumes.

clover leafhopper. *Aceratagallia sanguinolenta* (Prov.), family Cicadellidae, an insect that infests many plants, especially the clovers, alfalfa, grains, and grasses. Heavily infested plants show a lack of thriftiness; the leaves become whitened and mottled, and may turn yellow, red, or brown. It is especially destructive to clover during the seedling stage and after cutting.

clover-leaf style pruning. See *four-leaf clover system of pruning; wedge-style pruning.*

clover leaf weevil. *Hypera punctata* (F.), family Curculionidae, a snout beetle whose larvae eat out smooth-edged notches from the leaves of clover or alfalfa. The adults feed on various flowers and on the heads of wheat.

clover mite. *Bryobia praetiosa* Koch., family Tetranychidae, a mite that feeds on clover, grass, and weeds. In winter it may invade houses and become a nuisance, although it does not bite humans, transmit disease, or damage household furnishings.

clover root borer. *Hylastinus obscurus* (Marsh.), family Scolytidae, a small beetle which infests the roots of clover, sweet clover, alfalfa, peas, and vetch. Infested plants turn brown, wilt, and die. Found in northern U.S. and eastern Canada.

clover root curculio. *Sitona hispidula* (Fab.), family Curculionidae, an insect that infests clover, alfalfa, soybeans, cowpea, and other legumes, by scoring and furrowing the roots.

clover rusts. Fungi of the genus Uromyces, family Pucciniaceae, which attack clovers. The species concerned are *Uromyces elegans* on Carolina clover in southern U.S. *U. Trifolii* var. *fallens* on red clover, *U. Trifolii* var. *hybridi* on alsike clover, and *U. trifolii* var. *trifolii-repentis* and *U. nervi-philus* on white clover. These rust species are autoecious, *i.e.,* all known stages of the rust occur on the same host. The telial stage (and often the uredial stage) of the rust form rounded, brown pustules on the lower side of the leaf and on petioles and stems, sometimes leading to defoliation.

clover seed chalcid. *Bruchophagus gibbus* (Boh.), family Eurytomidae, an important insect pest of alfalfa and clover seed. It lays its eggs in the seed before it is hardened, and the larvae consume the entire inside of the seed. It does not affect the production of hay. Also called *alfalfa seed chalcid.*

clover-seed midge. *Dasyneura leguminicola* (Lint.), family Itonididae, an insect pest of red and white clovers. Infested plants are stunted and have irregular blooms; the heads remain green. If seeds form, they become shriveled.

clover-sick. Designating soils on which red clover fails because of infestation by parasitic fungi or exhaustion of the soil.

clover silk. The clover dodder, *q.v.*

clover stem borer. *Languria mozardi* Latr., family Languriidae, an insect pest of red and sweet clover. It eats the pith from the stems often causing them to break off.

clover taint. An off-flavor of milk produced from cows who have grazed on clover prior to milking.

cloves. See *clove: 3.*

club. See *clubroot of cabbage.*

clubbing. See *clubroot of cabbage.*

club foot. 1. See *clubroot of cabbage.* 2. A deformity in an animal's foot in which the foot turns inward to resemble a club.

club gourd. Edible snakegourd, *q.v.*
club grass. Cattail, *q.v.*
club hair mariposa. *Calochortus clavatus,* family Liliaceae, a bulbous herb grown for its brown-tinged, deep yellow flowers with thick hair at the bases. Native to California. Also called *mariposa tulip; mariposa lily.*
clubroot of cabbage. A fungous disease of crucifers caused by *Plasmodiophora brassicae,* family Plasmodiophoraceae. Because it attacks the underground parts of the plant, the disease may run its course before host symptoms appear. Root tissues are infested with soft-rot organisms, filling the root with gall and deforming it. This results in a slow reduction in growth, wilting, and premature death of the plant. Also called *clump foot; club foot; clubbing; finger and toe disease; anbury; club; dactylorhiza.*
club steak. A cut of beef from the loin.
club wheat. *Triticum compactum,* family Gramineae, an important annual cereal plant with flat leaves and club-shaped spikes. Grown for its grains which are a principal source of common bread. Probably native to Western Asia. Also called *cluster wheat.*
cluck. The call of a brooding hen.
clump. 1. A cluster of plants. 2. A mass of roots or rootstocks, as a *clump* of iris.
clump count. In bacterial counts, a cluster of bacteria as a single unit.
clump foot. See *clubroot of cabbage.*
cluster. 1. The form in which bees cling together in the hive after swarming or during winter. Also called *clustering.* 2. A large number of plants grouped, arranged, or growing in close proximity to one another. 3. An inflorescence.
cluster-cup rust. The aecial stage of a rust fungus, so called because the spores formed are always grouped in cuplike clusters. See *rusts.*
cluster-cup rust of gooseberries and currants. A minor disease caused by *Puccinia caricis* var. *grossulariata,* family Pucciniaceae; characterized by the appearance of cluster-cup rust formations on leaves. The spots become red and the leaves are holed and distorted. Spots may also appear on the shoots, petioles, and fruit. This rust is a heteroecious species with the cluster-cup stage on gooseberries and currants and the uredial and telial stages on sedges of the genus Carex, *q.v.*
cluster-cup stage. See *cluster-cup rust.*
clustering. See *cluster: 1.*
cluster mahonia. *Mahonia pinnata,* family Berberidaceae, an evergreen, tender shrub, grown for its foliage and pale yellow flowers. Native to North America. Also called *California barberry.*
cluster mallow. *Malva verticillata,* family Malvaceae, an annual or perennial herb, grown for its pink, stalkless flowers. Native to Eurasia. Also called *curled mallow.*
cluster palm. See *MacArthur clusterpalm.*
cluster primordium. The rudimentary growth which develops into a cluster of grapes.
cluster redpepper. See *red cluster pepper.*
cluster thinning. The selective removal of grape clusters after bloom, soon after the berries have set, to improve quality by influencing uniformity of cluster size, size of berry, and coloring.
cluster wheat. Club wheat, *q.v.*
clutch. 1. a. A nest of eggs in a hatchery. b. A brood of chickens. Also called *cletch.* 2. A device to engage or disengage the power from various working parts of machinery.
Clydesdale. A quality breed of Scottish draft horse, distinguished by an abundance of fetlock hair. Usually bay-colored with white points, sometimes roans and browns are also found among them. A stallion weighs up to 2,000 lbs. and a mare up to 1,800 lbs.
clyers. Actinomycosis, *q.v.*
coach breed. Any of several breeds of horses adapted for drawing coaches as the Cleveland Bay, the Yorkshire Coach, and German Coach Horse. They are usually heavier than riding horses.
coachwhip. Ocotillo, *q.v.*
coagulability. Ability to coagulate; the property of normal blood to form a clot.
coagulant. a. A substance which acts on a liquid to coagulate it, as rennin. b. An agent which aids the clotting of blood.
coagulate. To curdle, clot, as rennet *coagulates* milk.
coagulated whey protein. The precipitate of heated acidified whey.
coal ashes. Residue from burned coal. The ashes have no fertilizer value, but are sometimes used as a soil antiseptic and a conditioner for heavy, or clay, soils.
coal-oil brush. See *littleleaf horsebrush.*
coal-slack. Small bits of coal sometimes used in the diet for swine.
coal-tar creosote. A dark green or brown distillate oil frequently containing naphthalene and anthracene, which is used as a wood preservative and disinfectant. It is poisonous and a skin-irritant.
coal-tar dip. An emulsified coal-tar disinfectant used in a dipping vat to rid animals of external parasites. It consists of coal-tar oils, tar acids, and soap mixed with water to form a milky emulsion.
coarse. 1. Designating large, disproportionate features in meat and animals. *Cf. smooth.* 2. Referring to the roughness in appearance or the large size of a grain or particle, as *coarse* sand.
coarse-acid flavor. A defect in butter flavor caused by a slow formation of lactic acid in cream.
coarse flavor. A butter flavor defect characterized by a lack of sweetness.
coarse-grained wood. Wood with wide and conconspicuous annual rings.
coarse grit. The texture of an emery wheel or grindstone suitable for rough grinding.
coarse sands. A sandy soil containing over 35 percent fine gravel and rough sand, less than 50 percent fine sand. See *soil separate; soil texture.*
coarse sandy loam. A soil which contains over 45 percent fine gravel and rough sand.

coarse strain. A bud variant strain of the Valencia orange characterized by a change in the skin texture of the fruit.

coarse texture. 1. In soils, the sand, loamy sand, and sandy loam (except the very fine sandy loam) textural classes. 2. a. A body defect of sherbet and ice due to insufficient sugar or stabilizer, bad refrigeration, or failure to put them in the hardening room immediately after removal from a freezer. b. A body and textural defect of ice cream, characterized by a lack of smoothness. 3. A textural condition of a fruit, vegetable, or other plant or plant produce.

coarse wool. Referring to a breed of sheep which produces a coarse wool, as the Cotswold, Leicester, Lincoln, Romney Marsh. See *fine-wool; medium wools.*

coast disease. In South Australia, a copper-deficiency in sheep which causes a noticeable loss of crimp in the wool.

coaster. The long-horned cattle found along the Texas coast of U.S.

coast larkspur. California larkspur, *q.v.*

coast live oak. California live oak, *q.v.*

coastrange melic. *Melica imperfecta,* family Gramineae, a perennial grass, with fair forage rating, found in California. Native to North America. Also called *California melic.*

coast redwood. See *redwood.*

coast trillium. Pacific trillium, *q.v.*

coat. The outer covering of an animal, *i.e.,* the fur of a beaver, the feathers of a fowl.

coat flower. Saxifrage tunicflower, *q.v.*

cob. 1. The chaffy axis upon which the kernels of corn grow. Also called *corn cob.* 2. A stocky, short-legged horse used for pulling light carts. 3. Male swan.

cobaea pentstemon. *Pentstemon cobaea,* family Scrophulariaceae, a perennial herb, used as a border plant for its purplish flowers. Native to North America. Also called *dewflower.*

cobalt. Co, a trace element that occurs in soils in quantities up to 15 parts per million. It is a constituent of vitamin B_{12}, and is more significant in animal than in plant nutrition.

cobalt deficiency. Chiefly a vitamin B_{12} deficiency in ruminant animals due to a reduced synthesis of this vitamin which results in anemia and a thin, emaciated appearance in the animal. Also called *enzootic marasmus.*

cobbler. A sheep which is very difficult to shear (New Zealand).

cobblersthumb livingrock-cactus. *Ariocarpus retusus,* family Cactaceae, a low, flat-topped cactus, three to six inches wide and much cut up into tubercules, grown for its rose or whitish-pink flowers. Native to Mexico. Also called *chaute; seven-sisters; seven-stars.*

cobbles. Rounded rock fragments in soils or in wash deposits and alluvial debris, intermediate in size between pebbles and boulders; arbitrarily defined as three to ten inches in diameter.

Cobb's bacterial wilt of sugarcane. See *gumming disease of sugarcane.*

Cobb scale. A comparative scale consisting of a series of six strips on which are printed dots resembling the cereal rust lesions. The concentration of dots increases from the first strip to the sixth. Used to estimate the concentration of cereal rusts on leaves and stems.

Cobb's disease. See *bacterial gummosis of sugar cane.*

cob end. Small pieces of corn cob which may remain attached to the seeds of corn after shelling.

cob filbert. *Corylus avellana* var. *grandis,* family Corylaceae, a deciduous shrub, grown for its large, edible nuts. Native to Eurasia. Also called *cobnut.*

cob meal. See *corn-and-cob meal.*

cobra. A variety of highly venomous snakes found in Asia and Africa, which, when excited, expand the neck and sides of the head to give the appearance of a hood.

cob stacker. A part of a corn sheller. It consists of an open, metal channel with an endless belt with attached lugs running inside. It is used to elevate corn cobs after shelling to a truck bed, or stack.

cobweb houseleek. See *spiderweb houseleek.*

coca. Huanuco cocaintree, *q.v.*

cocaine plant. Huanuco cocainetree, *q.v.*

cocaine tree. Any shrub or small tree of the genus *Erythroxylum* (Erythroxylon), family Erythroxylaceae. Native to South America. See *Huanuco cocainetree.*

coccidia. A group of microscopic, parasitic protozoa that invade organs of the digestive tract of animals. They pass through a complicated life cycle within the host and produce minute, egg-like cysts, called oöcysts, which are usually discharged from the body of the host with the droppings.

coccidioidomycosis. An infection causing swelling of the lymph nodes in men and animals. It is produced by the fungus *Coccidioides immitis,* probably related to the yeasts. The characteristics may be vague unless the affected lymph node ruptures, in which case septic infection follows. The disease is sometimes confused with *actinomycosis, q.v.*

coccidiosis. A disease caused by one or more of the different kinds of protozoa belonging to the group coccidia (*q.v.*). In most domestic animals, *coccidiosis* is a disease of the digestive tract and causes diarrhea.

coccidiostat. An agent which gives practical control of coccidiosis by arresting its development.

coccidium. Parasitic protozoan of the group coccidia, *q.v.*

cocculiferous. Referring to plants which bear berries.

coccus. A spherical bacterium, as of the family Coccaceae.

cochil sapota. White sapote, *q.v.*

Cochin. An Asiatic breed of very large, egg-producing chickens with profuse, soft plumage and fluff, and full feathered shanks and feet. The cock weighs up to 11 lbs. and the hen up to 8½. Varieties are Buff Cochin; Partridge Cochin; White

Cochin; Black Cochin. Also called *Chinese Shanghai fowl; Cochin China.*

cochineal. A carmine-red dye prepared from the dried bodies of the females of the *cochineal* insect, *Dactylopius coccus* Costa, family Dactylopiidae.

cochineal nopalcactus. *Nopalea cochenillifer,* family Cactaceae, a cactus of frost-free areas, grown as an ornamental for its scarlet flowers and red, edible, juicy fruit. It had great economic importance at one time, being the main food of the cochineal-dye insect. Native to Central America.

cock. 1. The adult male fowl. Also called *rooster; stag.* See *cockerel.* 2. A small conically-shaped pile of hay, manure, etc. 3. A device for regulating the flow of fluids through a pipe. It consists of a conical plug working in a shell of iron or brass bored out to receive it, which is lodged in the pipe. The passage of fluids is controlled by rotation of the plug controls.

cockatoo bush. *Myoporum serratum,* family Myoporaceae, a shrub or tree with long, narrow leaves: bears edible berries. Native to Australia. Also called *native juniper.*

cockatoo farmer. A farmer who takes up land on a squatter's station. Also called *cocky* (Australia and New Zealand).

cockatoo fence. A rough wooden fence erected by cockatoo farmers, *q.v.* (Australia).

cocked ankle. A dislocation of the fetlock in a horse due to contraction of the tendon. Also called *knuckling over.*

cockerel. A male chicken less than a year old.

cockle. 1. A small gall on wheat which results from the attack of a nematode, *Anguina tritici,* family Tylenchidae. 2. A kiln for drying hops. 3. See *corn cockle; cow soapwort.*

cocklebur. Any plant of the genus Xanthium, family Compositae. Species are troublesome, poisonous weeds having heads with spiny hooked barbs, and large leaves. Animals after eating the plant become depressed, nauseated, stagger and have rapid and weak pulse with a drop in temperature. Also called *clotbur.*

cocklebur weevil. *Rhodobaenus tredecimpunctatus* (Ill.), family Curculionidae, a beetle which infests sunflower, evening primrose, dahlia, and chrysanthemum. Its larvae hollow out the lower portion of the stalk.

cockroach. An insect pest of the family Blattidae. Species are found in homes and stores, etc. See *croton bug; American cockroach; Australian cockroach; brown-banded roach; Surinam cockroach.*

cocksclaw. See *Japanese raisintree.*

cockscomb. 1. A fleshy protuberance on a cock's head; the comb or crest. 2. Any plant of the family Amaranthaceae, genus Celosia, an herb grown for its heavy-crested, deep red, purple, or yellow floral head. Native to warm countries. See *feather cockscomb, cockspur coralbean.*

cockscomb gall. On elm leaves, elongated galls resembling a rooster's comb. Caused by an aphid, *Colopha ulmicola* (Fitch.), family Aphidae, feeding on the leaves.

cockseggs. *Salpichroa rhomboidea,* family Solanaceae, a tropical, perennial, climbing vine with white urn-shaped flowers and white, or yellow, edible berries; grown as an ornamental in warm areas. Native to South America.

cocksfoot. See *orchardgrass.*

cockspur. See *ergot: 3.*

cockspur coralbean. *Erythrina crista-galli,* family Leguminosae, a handsome shrub grown in warm areas for its clusters of crimson flowers. Native to South America. Sometimes called *ceibo; cockscomb.*

cockspur-grass. See *barnyardgrass.*

cockspur hawthorn. *Crataegus crusgalli,* family Rosaceae, a small thorny tree whose wood is useful for fenceposts and fuel. Native to North America. Also called *cockspur thorn.*

cocoa. A food product obtained from the beans of cacao (*q.v.*). The cacao fruit pod resembles a small melon with well-defined segments. The seeds from the pods are fermented, which destroys the spongy pulp about the bean and eliminates a bitter taste, and are then dried at temperatures less than 125°. The dried beans are screened, roasted, crushed, and separated from shells by winnowing. The crushed product is ground, made into a thick paste, and cooled to set into solid slabs (bitter chocolate). To make *cocoa,* the slabs are warmed and pulverized by hydraulic presses to remove cocoa butter (fat). Also called *chocolate.*

cocoa meal. A feed by-product of secondary importance made from the cacao bean.

cocoa shell meal. A fertilizer obtained by grinding cacao bean husks. It contains 2.5 percent nitrogen, 1.0 percent phosphoric acid, 2.5 percent potash.

cocoa shell. The residual hull of the cacao beans, after the cocoa has been extracted. Used for animal feed and bedding, and to make cocoa tea.

coco grass. Nutgrass flatsedge, *q.v.*

coconut. The *coconut palm (q.v.)* fruit. Its fibrous skin yields coir (*q.v.*), and the meat inside, after drying, yields commercial copra.

coconut bud rot. A fungous disease caused by *Phytophthora palmivora,* family Pythiaceae, which destroys the terminal buds and adjacent leaves, and eventually the entire tree. See *bud rot of coconut.*

coconut cake. See *coconut oil cake.*

coconut meal. See *coconut oil cake.*

coconut mealybug. *Pseudococcus nipae* (Mask.), family Coccidae, a soft scale insect which infests coconut plants.

coconut oil. The oil extracted from the dried fruit of the coconut (copra); used in making soap and coconut butter. Also called *copra oil.*

coconut oil cake. A stock feed, a by-product of dried coconut meat after coconut oil has been extracted. It averages 21.3 percent protein; it equals corn gluten in feed value. Also called *copra oil cake.*

coconut palm. *Cocos nucifera,* family Palmaceae, a widely distributed, pinnate-leaved tropical tree, grown for its drupe fruit, the *coconut* of commerce. Native to Asia and Pacific Islands.

coconut palm nematode. *Aphelenchoides cocophilus,* family Aphelenchidae, a nematode which causes rindwing disease of tropical American coconut palms.

coconut scale. *Aspidiotus destructor* Sign., family Diaspididae, a scale insect which infests palms in Florida, U.S.

cocoon. The silky outer covering which the larvae of certain insects spin around themselves before passing to the pupa stage.

coco sedge. Nutgrass flatsedge, *q.v.*

cod. The part of the scrotum left after castration.

codling moth. *Carpocapsa pomonella* (L.), family Olethreutidae, a destructive insect whose larva chews its way into apples. It also attacks the pear, quince, wild haw, crab and English walnut.

cod liver oil. Oil obtained from the cod, *Gadus morrhuae,* or from other species of *Gadus.* It is sometimes mixed with other fish liver oils or synthetic vitamins to increase its food value.

codominant. 1. A tree which grows close enough to other trees as to receive very little sunlight from the sides. 2. Any plant or animal species which shares dominance with other species in an area.

codominant crown class. Codominant trees in a forest which form the general level of the forest canopy.

coefficient of digestibility. The percentage of digestibility of a nutrient.

coefficient of discharge. a. A coefficient by which the theoretical discharge of water through orifices, weirs, and other hydraulic structures must be multiplied to obtain the actual discharges. *Coefficients of discharge* are determined by experiments on scale model structures. b. The ratio between the actual and theoretical flow of water.

coefficient of roughness. A variable factor used to estimate the flow of water in rivers and canals, determined by the roughness or smoothness of channel lining, the mean hydraulic radius, mean velocity of the flow, and slope.

coffee. Any evergreen shrub or tree of genus Coffea, family Rubiaceae, which has opposite leaves, cream-colored flowers, and a berry-like fruit. The seeds inside the berries are freed of the pulp, dried in the sun, and used green or are roasted and ground to make coffee. Native to Asia and Africa.

coffee anthracnose. A fungous disease, caused by *Colletotrichum coffeanum,* family Melanconiaceae. Dark lesions appear on the leaves, which spread to the branches and stem of the plant, killing these parts.

coffeeberry. A name which designates several plants bearing coffee-like fruit. See *California buckthorn; cascara buckthorn; hairy ceanothus; shiny coprosma.*

coffeebush. Coffeeberry, *q.v.*

coffee cream. Cream that contains 18 to 22 percent milk fat. Also called *table cream.*

coffee huller. A machine consisting of serrated surfaces on a belt, which removes the husk or sac covering the coffee bean. Rubbing, brushing, dusting, and winnowing complete the process of hulling.

coffee nut. See *Kentucky coffeetree.*

coffee pea. See *gram chick-pea.*

coffee substitute. A food product which contains no coffee; generally prepared from a roasted and ground mixture of bran, wheat, and molasses.

coffee-tree. Kentucky coffeetree, *q.v.*

coffin bone. The bone enclosed within the hoof of horses and other animals with similar foot construction.

coffin joint. The joint just above the coffin bone, *q.v.*

cohosh bugbane. *Cimicifuga racemosa,* family Ranunculaceae, an herbaceous perennial, grown in borders and wild gardens for its double row of oval follicles extending upward from lateral branches. Native to eastern U.S. Also called *richweed; black snakeroot; black cohosh; bugbane.*

cohune palm. *Orbignya cohune (Attadea c.),* family Palmaceae, a tropical feather palm sometimes grown in southern Florida, U.S., as an ornamental. Native to Central America, where its wood is used in construction. Its nuts yield an oil which is used in soaps and cooking.

coiled rose worm. See *curled rose sawfly.*

coil-type heat exchanger. A device for rapid heating or cooling of liquids, such as milk. It consists of a tank which contains coils through which the heating or cooling medium is pumped.

coincidence. In genetics, a term used in estimating the distance between two genes. It is the ratio between the actual percentage of double crossing over and the percentage expected on the assumption that each crossing over is an independent event.

coir. The fiber from the husks of the coconut; used for cordage and coarse matting.

coital vesicular exanthema. An acute virus disease of horses; characterized by vesicles on the mucous membranes of the vagina or penis. Not common in the U.S. Also called *vesicular venereal disease.* See *granular vaginitis.*

coitus. Sexual intercourse; copulation; service; mating.

col. A saddle-like depression in the crest of a ridge.

Colby. An open-textured cheese made similarly to Cheddar, except that the curd is kept from matting by stirring.

colchicine. A poisonous alkaloid extracted from the common autumncrocus; used in medicine and in plant breeding to serve as an intracellular poison to block anaphase separation during cellular division, thereby producing polyploidy and doubling the number of chromosomes. It often increases plant size.

colchicum poisoning. See *autumncrocus.*

Colchis bladdernut. *Staphylea colchica,* family

Staphyleaceae, a hardy, deciduous shrub, grown for its yellowish-white flowers. Native to Caucasus. Also called *Caucasian bladdernut.*

Colchis ivy. *Hedera colchica,* family Araliaceae, an evergreen woody vine grown in warm areas as a climber. Native to Caucasus and Turkey.

cold. 1. See *infectious coryza.* 2. Temperatures near or below freezing. (An inaccurate expression, as scientifically, it can only be a greater or less degree of heat at any point above absolute zero.) 3. Designating a wire which is not electrified.

cold-blooded. 1. Referring to a horse of unknown breeding or one which does not have an ancestor of pure blood. See *hot-blooded.* 2. Referring to certain animals whose blood temperature depends on the environment, *e.g.,* fish and snakes.

cold cellar. An underground room for the storage of vegetables, fruits, etc. The temperature should remain about 38 to 40°F.

cold chisel. A steel tool strong enough to cut cold metal. Also called *cold cutter.*

cold collar. 1. A freshly harnessed horse (southwestern U.S.). 2. A balky horse.

cold deck. In lumbering, logs stacked in a large pile before transportation.

cold frame. A bottomless box with a removable glazed top; used to protect, propagate, or harden off plants. No heating device or fresh manure is used. See *hotbed; heated frame; frame; garden frame; cloth-covered coldframe.*

coldframe sash. See *hotbed sash.*

cold front. The face of swift-moving polar air which moves like a wedge through warm air. Such contact usually results in brief, heavy precipitation.

cold graded. Designating meat graded after the carcass has been cooled.

cold-hold tank. A low temperature milk storage tank.

cold jawed. Describing a horse difficult to control with the bit and bridle. Also called *hard mouthed.*

cold manure. Farmyard manure which does not ferment and heat unduly, while in storage. See *hot manure.*

Cold Pack cheese. A food product made from mixing various unheated fine ground cheeses; vinegar, or lactic, citric, acetic, or phosphoric acid, water, salt, color, and spices may be added. The pH should not be below 4.5. Also called *Club Cheese; Comminuted Cheese; Potted Cheese.*

cold-pack process. 1. A method of canning in which vegetables or fruits are scalded and placed immediately in hot containers. Hot sirup or water fills each container before it is closed. The containers are next sterilized or pressure cooked. See *hot-pack process.* 2. The preservation of fruits by storage in bulk containers at freezing temperatures.

cold pit. A deep pit with a sash cover which is used for the winter storage of bulbs and hardy plants.

cold pocket. A low area into which cold air settles from adjoining slopes. See *frost pocket*

cold pressed. See *crude cottonseed oil.*

cold-pressed cottonseed cake. See *whole pressed cottonseed.*

cold room. A storage room with heavily insulated walls, floors, and ceiling, which may be refrigerated. See *storage.*

cold soap. A semi-boiled soap in which the glycerine remains in contrast to its separation as a by-product in boiled or hard soap.

cold sterilization. 1. The use of cathode ray or electron beam gun in food processing to kill bacterial or insect life. 2. Chemical sterilization of instruments.

cold-storage butter. Butter stored at a temperature of about 0° F. for up to 6 months.

cold storage contest. A competition in which butter is judged when fresh and again after 6 months storage at 0° F. It determines who can make a high quality product that stores well.

cold storage eggs. Eggs held in storage at a temperature of 45° F. or less for more than 30 days.

cold test. A test for viability in which sample seeds are placed on wet soil, sand, cloth, or blotter at a constant temperature of 40° F. for 3 days or longer. The sample is then placed in a warm place and the percentage of germination recorded.

cole. Any plant of the genus Brassica, family Cruciferae. See *cabbage; cauliflower; broccoli; Brussels sprouts; kohlrabi; collard.*

Coleus. 1. Genus of herbs bearing blue or lilac flowers, family Labiatae; grown for their beautiful foliage. Native to tropical Asia and Africa. 2. *Coleus blumei,* a tender herb grown for its variegated leaves. Native to Indonesia. Also called *foliage plant.*

colibacillosis. A septicemic disease of chickens; probably caused by the bacterium *Escherichia coli.*

colic. 1. Pertaining to the colon. 2. A pain in the abdomen caused by irregular muscular contractions, obstruction, spasm or distension of the viscera. See *acute indigestion in horses.*

coliform. Referring to aerobic and facultative anaerobic, gram-negative, non-spore-forming bacilli in the colon, which ferment lactose to form acid and gas.

coliseum ivy. See *Kenilworth ivy.*

colitis. Inflammation of the mucous membrane of the colon.

collapse. 1. A breakdown. 2. The corrugation on the surface of wood caused by the irregular drying of cell walls in the early stages of drying hardwood. Also called *crimps.*

collar. 1. The padded and reinforced part of a harness which girdles the neck. Straps for pulling are attached to it by means of hames. 2. A closely bound retail cut of rolled meat. 3. The part of a tree at the line of union of the roots and the trunk. *Cf. crown: 1.*

collar blight. The canker of fire blight of pear and apple which encircles the trunk of a tree near the ground level.

collar boil. A sore resulting from an ill fitting horse collar. Also called *collar gall; sitfast.*

collar burn. A lesion caused by a collar on the neck or shoulders of a horse.
collard. A large kale.
collar disease. See *collar rot; collar blight.*
collarette dahlia. A dahlia that has open-centered flowers, and one row of ray florets. The central disk is surrounded by several rows of petaloids, usually of a different color, which form a collar.
collar pad. A harness pad which fits under the collar to protect the horse from chafing.
collar rot. Any of several plant diseases in which lesions girdle the stem or trunk near the ground.
collateral. Property pledged as part or full security for a debt.
collateral relationship. The relationship of two plants or animals of common ancestry. See *direct relationship.*
collecting tube. See *collector tube.*
collection. 1. The balance of a horse and rider which makes for instinctive coordinated movement. 2. The accumulation of sperm for use in artificial insemination. 3. In flower shows, an exhibit of several varieties of plants, flowers, fruits, or vegetables. A *collection* differs from a group in that the number of varieties is an important factor. Artistic arrangement is not necessarily mandatory.
collective fruit. The product of the fusion of several distinct flowers into a fruit: *e.g.*, mulberry, fig. See *aggregate fruit.*
Collectivism. An ideology or social theory in agricultural economics: the grouping or collectivization of farms, or other land, under unified management, usually with control vested in the State, as opposed to individual ownership and free management of single units.
collector tube. A tube fastened to the artificial vagina to collect semen from the male animal.
Collins dynamometer. A device to measure the tractive pull of draft horses. It permits the movement of the vehicle to which the horses are hitched only when the tractive pull equals the weight set on the machine. A distance of 27½ feet is most frequently used to determine horsepower.
Collins rule. A guide for the effect of marketing time on the fat content of the milk; the percentage alters by 0.2 percent for each hour's change in the marketing interval.
colloid. a. An insoluble substance consisting of particles small enough to remain suspended indefinitely in a medium. b. A mineral particle less than 0.0002 mm. in diameter. c. A substance which does not form a true solution in water and does not diffuse readily through animal or vegetable membranes. Its presence does not effect the freezing point or vapor tension of the solution.
colloidal. Pertaining to a state of matter in which particles are larger than those which ordinarily form crystals. The *colloidal* state of matter as a solid is generally amorphous: *e.g.*, starch, gelatin, and gums. The complex organic compounds in mucus, blood, lymph, and some of those in the urine are in the *colloidal* state.
colloidal matter. In soil, the fine-sized material, organic or inorganic, which is smaller than clay particles and has more accentuated qualities, such as being sticky, gelatinous, absorptive, gummy or plastic.
colloidal peat. The unctuous, gelatinous mass of organic residue formed as a basal or bottom deposit in lakes. It may lie at the base of a peat deposit, but does not conform in texture and appearance to the material commonly designated as peat.
colloidal phosphate. See *soft phosphate with colloidal clay.*
collop. A unit of grazing area which can support a full grown horse or cow for one year.
colluvial soil. Soil consisting of colluvium, *q.v.*
colluvium. Heterogeneous, non-stratified soil material, or transported regolith, which has accumulated slowly at the base of slopes as a result of rainfall runoff. It is not transported by streams nor is it stratified like alluvium; it is more earthy and less brashy, or restricted to formation at the base of very steep slopes than talus.
colmatage. The practice of spreading silt or mud sediments over poor land for reclamation or improvement which originated in Italy. See *warping.*
Colombo agent. See *Solomon Islands ivyarum.*
colonche. A fermented drink made from prickly pear cactus (southwestern U.S.).
colonial bentgrass. *Agrostis tenuis,* family Gramineae, a perennial grass which is used for golf putting greens and lawns. Native to North America.
colonial pine. See *Cunningham araucaria.*
colony. 1. An aggregate of worker bees, drones, and a queen bee, living together in a hive as a unit. 2. A cluster of bacteria or fungi grown on a culture medium, usually originating from a single bacterium, spore or inoculation transfer. 3. A group of people bound by communal or religious ideals that occupies tracts of land, often wild, for permanent settlement, especially in the early development of the U.S. and Canada.
colony brooder house. A small, movable structure used to house young chicks, poults, etc., during the early growing period.
colony house. A small, movable structure used to house a sow and her litter when the piglets are large enough to be cold resistant.
colony morale. The zeal of bees in rearing a brood and accumulating stores; determined by the resistance of the colonies to various factors of environment.
colophony. See *rosin.*
color. 1. In exhibitions of flowers and fruits, the natural *color* of the product as contrasted with the variation or change caused by rubbing or polishing, except for apples, which may be polished. 2. a. A substance with a definite tint or hue which, when added to another substance, gives the whole its own *color.* b. To change the *color* of a substance by adding a coloring matter or material.
Colorado bean. See *pinto bean.*

Colorado blue spruce. See *Colorado spruce.*
Colorado columbine. *Aquilegia coerulea (A. caerulea)*, family Ranunculaceae, a hardy, perennial herb grown for its bluish-purple flowers. Native to the Rocky Mountains, U.S. It is the state flower of Colorado.
Colorado corn rootworm. See *western corn rootworm.*
Colorado fir. See *white fir.*
Colorado grass. See *Texas panicum.*
Colorado hemp. See *hemp sesbania.*
Colorado loco vetch. See *Lambert crazyweed.*
Colorado manroot. See *bigroot morningglory.*
Colorado pine beetle. *Dendroctonus approximatus* Dietz, family Scolytidae, a beetle which infests Arizona pine, ponderosa pine, and the Chihuahua pine. Its egg galleries in the inner bark are longitudinal, winding, and branched. Found in western U.S.
Colorado potato beetle. *Leptinotarsa decemlineata* (Say), family Chrysomelidae, a striped beetle whose adult and larvae feed on the leaves and retard growth of the potato. It also infests the tomato, tobacco, pepper, ground cherry, henbane, horsenettle, belladonna, thistle, and mullein. Found east of the Rocky Mountains, U.S.
Colorado ranger. Appaloosa horse.
Colorado red cedar. See *Rocky Mountain juniper.*
Colorado river hemp. See *hemp sesbania.*
Colorado rubberweed. See *bitterweed actinea.*
Colorados. Sheep which were bred in Colorado, U.S.
Colorado spruce. *Picea pungens,* family Pinaceae, a hardy evergreen tree grown as an ornamental. The most widely grown variety is the blue spruce.
Colored Dorking. A variety of the Dorking chicken.
colored fowl. A market term for a non-white hen of a heavy breed which is too mature for a roaster.
color gardening. The systematic planting of flowers for color effect as contrasted with themselves and the background. One color predominates, and the garden is known by that color: *e.g.*, the blue garden, red garden, white garden, etc.
colorin. *Erythrina setosa,* family Leguminosae, an herbaceous shrub grown for its bright red flowers. Native to Mexico.
color marking. The color design in the coat of an animal or feather covering of fowl.
color of fat. In meat judging, the hardness, density and whiteness of the meat fat characteristic of the better grades.
color of lean. In meat judging, the bright red color of the lean meat, assumed to be an index of quality.
color profile. A soil profile described according to the colors of its separate horizons.
color scoring system. An arbitrary system for grading breads, by comparison with standard loaves. Bread made from ordinary sound wheat will score 90 points or better, exceptional bread 100 points.
colostrum. Milk of high protein and mineral content produced for a few days after parturition. It is especially suitable for calves, as it acts as a laxative and stimulant to the digestive organs, contains many antibodies, and increases resistance to disease. Also called *vistings; beastlings; beestings.*
colpindach. A young heifer.
colt. A young male horse, ass, or mule under four years of age. *Cf. filly.*
colt evil (ill). See *navel-ill.*
colter (coulter). A variously-shaped metal device attached to the beam of a plow. It eases the strain on the plowshare by cutting the sward ahead of it. See *fin colter; hanging colter; knife colter; notched colter; rolling colter; skim colter.*
coltsfoot. *Tussilago farfara,* family Compositae, a perennial herb found as a weed in clay soils in northern U.S. Its leaves are used in medicines for curing lung diseases. Native to Eurasia. Also called *clayweed; coughwort; bullfoot.*
Columbia. A breed of sheep developed in the U.S. from the Lincoln and Rambouillet breeds. It is large, vigorous, heavy-boned and long legged, with the rams averaging 275 lbs. and ewes up to 135. It yields long-stapled fleece of quarter-blood wool.
Columbia Basin wireworms. The larvae of a beetle, *Limonius subauratus* Lec., family Elateriadae, which infests the roots of potato, corn, cereals.
Columbia lily. *Lilium columbianum,* family Liliaceae, a perennial bulbous herb grown for its drooping orange-red, dark-spotted flowers. Native to North America.
Columbia monkshood. *Aconitum columbianum,* family Ranunculaceae, a perennial herb which is highly poisonous to men and animals, especially just before flowering. Characteristics of poisoning are weakness, difficult breathing, weak pulse, bloating, belching, swallowing, and contracted or dilated pupils. Native to northwestern U.S. and western Canada.
Columbian. A plumage color pattern in chickens. The main body plumage is white, but certain feathers in the neck, tail and wing sections are variously tinged with black. It is the plumage pattern of the Light Brahma, Columbian Plymouth Rock, and the Columbian Wyandotte.
Columbia River larkspur. *Delphinium trolliifolium,* family Ranunculaceae, a perennial herb bearing small blue flowers, which is an important poisonous plant. It causes loss of appetite, general uneasiness, staggering, constipation, nausea and bloating in animals. Native to the Pacific coast of the U.S. Also called *cow poison.*
columbine. Any plant of the genus Aquilegia, family Ranunculaceae. The species are popular garden perennials. See *alpine columbine; American columbine; golden columbine; Colorado columbine; fan columbine; Sitka columbine; altai columbine; longspur columbine; Pyrenees columbine; Skinner columbine; European columbine.*
columbine borer. *Papaipema purpurifascia* (G. & R.), family Noctuidae, the larva of a moth which feeds on columbines by boring within the stem and roots.

columbine leaf miner. *Phytomyza minuscula* Gour., family Agromyzidae, a maggot that mines the leaves of columbines and larkspurs.
columbine meadowrue. *Thalictrum aquilegifolium,* family Ranunculaceae, a deciduous, perennial garden herb, grown for its columbine-like, purple, pink, orange, or white flowers. Native to Eurasia. Also called *feather(ed) columbine.*
columnar. Describing a plant whose outline resembles a column; *i.e.,* one formed by relatively perpendicular sides.
columnar Norway maple. See *erect Norway maple.*
columnar structure. In soil structure, describing particles in which the units are prismatic with rounded tops.
column red maple. *Acer rubrum* var. *columnare,* family Aceraceae, a horticultural variety of the red maple with upright limbs like the Lombardy poplar, grown as an ornamental. Also called *upright red maple.*
colyone. The secretion of one of the ductless glands in animals which inhibits metabolism. Also called *chalone.*
colza oil. A pale yellow oil, specific gravity 0.912 to 0.920 at 60° F., pressed from the seeds of the winter rape. It is used in making soaps and blackening, as a cooking medium, an illuminant and a lubricant. For the latter purpose, air is blown through it while it is heating; the blown oil may be used alone or mixed with mineral oil. The residue of crushed seeds is made into oil cake. Also called *rapeseed oil.*
comate seed. A seed covered with long silky hairs or bristles.
comatose. Unconscious; as in a coma.
comb. 1. A fleshy crest on the head of fowls. 2. A back to back arrangement of two series of hexagonal cells in a beehive. Worker cells are approximately five cells to a linear inch and drone cells about four cells. 3. a. An instrument consisting principally of metal, bone, plastic, etc., in which long, narrow, thin teeth are cut for the purpose of arranging and cleaning hair on an animal. Also to clean and arrange the hair of an animal with such an instrument. See *currycomb.* b. A device for cleaning and removing buds, stems, etc., from produce.
comb disease. See *favus: 1.*
comb duck. Any large duck, the male of which has a stiff knob at the base of the bill.
combflower. See *sunflower.*
comb foundation. A thin sheet of embossed or stamped beeswax on which the bees will construct combs.
comb honey. Honey preserved in the comb.
comb honey foundation. Special, lightweight, white-colored foundation used in the production of comb honey.
comb honey section. A square or rectangular wooden container in which section comb honey is produced; usually made of sanded basswood ⅛ inch thick which is grooved for folding and dovetailed on the ends for fitting together. Recently, plastic sections have been developed.
comb honey super. One of several types of hive body in which section comb honey is produced.
combination class. A subdivision of the American saddle horse division.
combination horse. In exhibitions, a horse shown both in the harness and the three-gaited saddle classes.
combination plow. A moldboard plow with interchangeable bottom parts.
combination spray. A spray consisting of two or more insecticides, herbicides, fungicides, etc.
combine. A self-propelled or tractor-drawn machine which cuts, threshes, and cleans the standing crop while moving across the field. It is adapted to harvesting all the small grains, soybeans, grain sorghums, peanuts, rice, beans, etc. In some areas the heads are cut and placed in windrows by a swather (windrower), and a *combine* with a pickup attachment gathers the grain and threshes it at a later date.
combined harvester-thresher. See *combine.*
combing. An operation in the manufacture of worsted yarn by which the long fibers are separated from the short and arranged parallel to each other.
combing wool. Wool long enough (at least two inches), and strong enough for combing.
combless package. A quantity of bees (2 to 5 lbs.) with or without a queen, contained in a shipping cage.
comb out. To remove lint from the cotton seed.
comb package. See *nuclei package.*
comb rack. A rack placed over a tank in which the honeycomb may be kept before it goes to the extractor.
come-along. A clamp, fastened to a block and tackle, which is used to maintain tension in stretching wires.
comeback. 1. a. A sheep having a cross-bred dam and a Merino sire. b. Wool from the *comeback* sheep. 2. A guinea fowl.
come in(to) heat. To be in estrus.
comfrey. 1. a. Any plant of the genus Symphytum, family Boraginaceae. The species are hardy, perennial herbs bearing purple, blue or yellow flowers. Native to Eurasia. b. *Symphytum officinale,* the common *comfrey,* grown for its yellowish, rose, or white, flower clusters. The plant is eaten by stock, the root is used in medicines, and the young leaves and blanched stalks and roots have been used also as human food.
coming in order. The absorbing of enough moisture by flue-cured tobacco through humidity or by steaming to make it pliable.
commensal. A non-parasitic organism which lives attached to a host and shares its food.
commensurability. The extent to which competitive ranch outfits should share in grazing privileges on companionate, public, or cooperatively controlled range, which is determined by the gross forage resources of privately controlled base property.
commensurable property. See *commensurability.*
commercial. A low grade market meat.

commercial butter. a. Average grade creamery-made butter. b. Butter for a contest or show which is prepared, with or without culture, from the regular cream. See *culture butter.*

commercial buttermilk. See *cultured buttermilk.*

commercial culture. Any bacterial culture which is prepared and sold by a special culture producing laboratory. See *culture: 3, 4.*

commercial fertilizer. Plant food containing a single mineral nutrient or a mixture of mineral nutrients or organic materials. See *green manure; manure; fertilizer: 1.*

commercial pine tar. An insect repellent and disinfectant especially effective against the spinose ear tick, *Argas megnini.*

commercial round. A wholesale cut of beef consisting of the upper part of the hindleg.

commercial sheep. Grade or scrub sheep, as contrasted to purebred sheep.

commercial sour cream. See *cultured cream.*

commercial sweet cream. A high quality, sweet cream.

commercial variety. See *cultivar.*

commercial wheat germ meal. A by-product of flour milling which consists of the wheat germ and some bran and middlings. It contains 31.1 percent protein and 9.7 percent fat, and is used as a supplement in poultry feeds and calf meals. Also called *wheat germ oil meal.*

comminutor. An agricultural implement for breaking down the soil after plowing.

Commission. A Dutch cheese made like Edam from lightly skimmed milk.

commission company. A business concern which buys or sells farmers' produce or livestock on commission.

commission florist. A business concern which sells flowers to retailers on commission.

commission house. See *commission company.*

commission merchant. A person entrusted with goods for sale on commission. He may possess and sell the goods in his own name, as distinguished from a broker. See *factor: 2.*

Commodity Credit Corporation. An agency of the U.S. Federal Government which is authorized to purchase and make loans on farm products, dispose of surplus commodities, and engage in other operations with a view to stabilizing the agricultural market. Organized in 1933 as an affiliate to the Reconstruction Finance Corporation, it was transferred to the U.S. Department of Agriculture in 1939.

commodity drain. The annual or periodic loss in growing stock on commercial forest land by the removal of merchantable material by any cause.

Commodity Exchange Authority. An agency of the U.S. Department of Agriculture, established on February 1, 1947, to supervise trading on the 17 markets. Formally composed of the Commodity Exchange Administration and the Grain Futures Administration.

Commodity Stabilization Service. An agency of the U.S. Department of Agriculture, established on November 2, 1953. It is responsible for (a) acreage allotments and marketing quotas, (b) soil bank, (c) price support, (d) disposal of government-owned surplus farm products, (e) International Wheat Agreement Act, (f) storage, shipping and related service activities, (g) administration of the Sugar Act, and (h) assigned mobilization planning.

commodores. See *junior twins.*

common. 1. A joint pasture in a village or community to which all members have access for their herds. 2. Legal right of a person in sharing the profit or use of another's land. 3. A low, market grade of meat animals. 4. The American wool grade equivalent to the English 44's; a low grade only better than braid. 5. A defective grade of lumber useful only in framing, etc.

common board. See *common: 5.*

common cattle grub. The larva of *Hypoderma lineatum* (De Vill.), family Hypodermatidae, a fly which lays its eggs on the hides of animals. The maggots tunnel through the skin, migrate through the body, and lodge in the back. In each swelling of the skin there is a small breathing hole. Infestation results in localized dermatitis lowering the quality of meat and depreciation of the value of the hide. Also called *cattle grub; ox warble fly; cattle maggots.*

common everlasting. See *fragrant cudweed.*

common fig. The fig (*q.v.*), *Ficus carica,* from southwest Asia.

common garget. See *mastitis.*

common goldthread. *Coptis groenlandica,* family Ranunculaceae, a small, perennial herb, bearing white flowers. Its yellow fibrous roots yield a dye and are also used medicinally. Found in bogs and marshes in northern North America. Also called *yellowroot; cankerroot.*

common green mold. A soft decay of citrus fruits caused by *Penicillium digitatum,* family Moniliaceae, characterized by a water-soaked, soft area on the rind, and later by olive-green spores. The spots enlarge and are surrounded by a pasty and wrinkled white halo.

common leafspot of strawberry. See *leaf spot of strawberry.*

common malaria mosquito. *Anopheles quadrimaculatus* Say, family Culicidae, an important insect vector of malaria in southern U.S.

common mild mosaic of gladiolus. A virus disease characterized by angular, light and dark green mottling in young leaves and by a stripe break in the flowers.

common miterwort. *Mitella diphylla,* family Saxifragaceae, a small, perennial herb, bearing white flowers; grown as an ornamental. Its fruit capsule looks like a bishop's miter. Native to east-central North America. Also called *bishops-cap.*

common salt. NaCl, sodium chloride, a chemical used in livestock feeding, in seasoning, and in preservation of food products. It is also an effective herbicide.

common scab (mange) of cattle. A contagious skin disease caused by the parasite *Psoroptes* *bovis,* family Psoroptidae; characterized by intense itching on the withers, on the top of the neck, and at the tailhead. Lesions eventually spread over the entire body. Rare in the U.S. Also called *psoroptic scab; barn itch; cattle mange.*

common scab of Irish potato. See *corky scab of potato.*

common scours. A looseness of bowels in bucket-fed calves from overfeeding, dirty pails, irregular feeding, lack of proper sanitation, or cold, sour, or too rich milk. It is characterized by listlessness, loss of appetite, bloating, diarrhea, and foul smelling feces. See *bloody scours; calf scours.*

common spice bush. *Lindera benzoin* (*Benzoin aestivale*), family Lauraceae, a deciduous, aromatic shrub, grown for its small clusters of greenish-yellow flowers. Native to U.S. Also called *Benjamin-bush; spicewood.*

common storage. Storage of products in an air-cooled room, without artificial refrigeration. *Cf. cold room.*

common-use range. Range containing grass, forbs, and browse which allows two or more kinds of stock to graze together during the entire season or separately during part of a season.

common warts of cattle. See *bleeding wart.*

common wheat. See *wheat: 2.*

common winterberry. *Ilex verticillata,* family Aquifoliaceae, a deciduous shrub grown for its bright red berries which last through winter. Native to the U.S. Also called *black alder.*

common wool. A U.S. wool grade. Next to the coarsest, slightly finer than braid.

communalism. The biotic relations among individual species of plant and animal life, both beneficial and harmful, at a given place.

community nest. A large nest without partitions for use by several laying hens at the same time. It is usually about 24 inches wide and 4 feet long, with one or two 8x8 inch entrances, and is provided with a sloping top hinged for egg collection.

community pens. A system of goat rearing in which kids of about the same age are placed together in a pen during the day, and individually with the does at night.

commutator. A contrivance for reversing the direction of an electric current in a circuit, dynamo, etc.

compact. 1. To press soil particles together so as to firm the soil before or after planting. 2. Describing a densely packed, firmly massed soil.

compacter. See *clod crusher.*

compactness. In animal judging, a closely knit, firmly united animal.

compactoids. Mutants which occur in cultures of common wheat and possess characters of club wheat.

companion crop. A crop sown along with another crop on the same land; generally it is a secondary crop grown to increase returns from land. Also called *nurse crop. Cf. catch crop.*

comparison approach. A method of appraising the value of a farm by comparing it in all respects with a previously evaluated farm in the community. The values of the reference farms are established by recent sales or by a group of appraisers.

compartment. A unit of forest area, permanently established for orientation, administration, and silvicultural operations.

compass plant. 1. Any plant whose leaves are arranged in such a way as to indicate the points of the compass. See *prickly lettuce; Spanish-clover deervetch.* 2. *Silphium laciniatum,* family Compositae, a coarse, herbaceous, perennial, grown as background in flower borders. The lower leaves incline toward north and south if the plant grows in the open sunshine. Native to midwestern U.S. Also called *rosin weed; pilot-weed.*

compatibility. In fruit plants, a condition in which the male gamete will fertilize the female gamete. *Cf. incompatibility.*

compatible. Capable of compatibility, *q.v.*

compensatory payment. A method of supporting agricultural farm prices in which the government pays the farmer directly the difference between the market price and the support price of a particular commodity.

competing crops or livestock. Any two crops or livestock whose production demands the use of the same resources at the same time.

competitive crop. a. A crop planted on a piece of land to force out other forms of plant life. Also called *smother crop.* b. Crops which compete with each other for the same time in a farmer's work schedule.

competitive enterprise. Those enterprises, *e.g.,* crops or livestock which compete for the use of one or more of the farmer's productive resources, and thus tend to crowd each other out of the farming system.

competitive plant. A plant which is capable of forcing out plants of other species from a field.

complement. A ferment-like, heat-labile substance, existing normally in many sera and cell protoplasms, which helps destroy bacteria and foreign matter.

complemental feed. Fodder given to livestock in place of, or as a supplement to, range feed.

complementary. Referring to either of the two factors of production used in a specified proportion to form a certain product: *e.g.,* hydrogen and oxygen are *complementary* in the formation of water.

complementary crop or livestock. Any crop or livestock yielding a product which contributes to the success of another.

complementary enterprise. Any of those enterprises which are mutually contributive, one to the other, to farm income.

complementary genes. Genes which are similar in their individual effect but which together interact to produce a new character; *e.g.,* in the sweet pea, either of two dominant factors, C and

P, individually produce white flowers, whereas, when present together, the same two factors produce purple-flowered plants.

complement fixation. A biological reaction which is the basis of many tests for infection. It utilizes the three-phase system: (1) antigen, (2) complement, and (3) antibody. When a specific antigen and antibody are mixed in the presence of complement, a reaction occurs which fixes or renders the complement inactive. Therefore, if the presence of a specific antigen or antibody in a mixture is known, the status of the other can be determined by the mixture's reaction on the complement. Such reactions are used in the diagnosis of venereal and other infectious diseases.

complement fixing. The ability of a material, usually an antigen in serum, to fix or bind the complement.

complete clearance. Sale of all animals by the end of a market day. *Cf. incomplete clearance; holdover: 1.*

complete fertilizer. A fertilizer which contains nitrogen, phosphorus, and potash.

complete flower. A flower which has calyx, corolla, stamens, and pistil. *Cf. flower: 1; incomplete flower.*

completely saturated. Designating clay containing all the basic ions that it is capable of holding.

complete soiling. The forage portion of an animal's ration; green feed carted to the animal. *Cf. partial soiling; soiling.*

complex practice. Complicated farming practices related to or dependent upon other methods requiring on-site technical assistance for successful application.

complex virus. Two or more viruses or virus strains which together cause a disease in a host with symptoms that are not predictable from knowledge of the individual component.

Compositae. A very large family group of plants, herbs, and trees, which includes flowers, such as marigold, dahlia, chrysanthemum; vegetables such as chicory and lettuce; and weeds as cotton burdock, etc.

composite disease. An illness caused by a virus complex.

composite forest. A forest of saplings and sprouts which have grown up together forming the overstory and understory.

composition. 1. The relative production of various plant species in a given area. 2. The components used in any mixture of feed, fertilizers, hay, seeds, etc.

compost. 1. Organic residues, or their mixture, such as peat, manure, or discarded plant material and soil, placed in a pit, moistened, and allowed to become decomposed. Sometimes lime is also added. Used as a fertilizer. 2. To cause vegetable matter to become decomposed as a fertilizer.

compound flavor. A combination of natural and artificial flavoring in ice cream.

compound layering. See *serpentine layering.*

compound pistil. See *flower: 1.*

compound solution of cresol. A coal tar derivative used as a disinfectant and deodorant. It is less poisonous and more germicidal than carbolic acid, and is especially effective against the hog cholera virus.

compound structure. The size, shape and distinctiveness of the aggregate of individual particles belonging to more than one type of soil.

compress. 1. a. A hydraulic press used to reduce a bale of cotton to about one-third its original size for ease in handling and shipping. b. The business or building for such a press. c. To reduce the size of a cotton bale by a hydraulic press. 2. A pad of gauze or other material applied to put pressure upon any part of the body to reduce swelling or control hemorrhage.

compressed-air sprayer. A sprayer consisting of a tank and an air pump. Herbicides, insecticides, etc., are placed in solution in the tank and air above the liquid is compressed by the pump so that the spray material is forced out under pressure through nozzles. See *power sprayer; barrel pump; knapsack sprayer.*

compressed bale. a. A bale of cotton which has been reduced in volume to a density of 22½ lbs. per cubic ft. b. A bale of hay reduced in size by extra pressure.

compression process. A refrigeration system which employs a volatile agent with a low boiling point, such as ammonia, carbon dioxide, methylic ether, sulphurous acid, etc., for chilling. The volatile agent is forced by the compressor into a condenser or liquefier where, on contact with cool surfaces, it gives up its latent heat and becomes liquid under pressure. It is then released through a regulating valve into the expansion pipes (coils of pipes within an insulated space) and, encountering a low pressure, volatilizes into gas, absorbing the heat necessary for the transformation from the surroundings. The abstraction of heat chills or freezes the pipes, in turn cooling the insulated compartment. The expanded gas is then exhausted by the suction stroke of the compressor, thus completing the cycle, which is repeated continuously.

compression wood. Abnormal wood that forms on the lower side of branches and leaning trunks of softwood trees; characterized by relatively wide eccentric, annual rings with large amounts of summer wood. It shrinks excessively lengthwise as compared with normal wood.

compressor. The part of a refrigeration system which sucks in vapor from the evaporator and, after compression, forces it into the condenser.

Compton vine lilac. *Hardenbergia comptoniana,* family Leguminosae, a handsome, woody vine, grown in the warmer sections of U.S. for its blue or violet-blue flowers. Native to Western Australia.

Comstock mealybug. *Pseudococcus comstocki* (Kuw.), family Pseudococcidae, an insect whose larvae suck the sap of the catalpa, apple, boxwood, holly, horsechestnut, magnolia, maple, Osage orange, and poplar.

concave(s). A concave, stationary part of a threshing machine or combine over which the threshing cylinder revolves. A large number of grains are separated between the cylinder and the *concave* as the unthreshed material passes through. The remaining material passes on to the principal separating mechanism.

concave gum psorosis. A virus disease of orange and lemon trees caused by *Citrivir psorosis* var. *concavum.* It is characterized by broad, cracked concavities on the trunk and limbs from which gum may ooze in summer and autumn. Also called *concave gum disease.*

concave shoe. A horseshoe, with beveled inside edge, used for hunters and other fast moving horses given to overreaching.

concentrate. 1. a. Any feed high in energy; sometimes used with reference to other nutrients, such as protein *concentrate,* etc. b. Stock feed low in fiber content and high in digestible nutrients. 2. Whole fruit juice, thickened to the consistency of heavy sirup. 3. To increase in strength by removing dilutents.

concentrated buttermilk. See *evaporated buttermilk.*

concentrated by-product spread. A by-product of milk, prepared by concentrating skim milk, pure milk, or whey to a paste; used as a substitute for butter.

concentrated cultured skimmed milk. See *evaporated cultured skimmed milk.*

concentrated egg. A food product prepared by vacuum-concentrating whole eggs to which egg yolk has been added to increase the solids to 30 percent.

concentrated feed. See *concentrate: 1.*

concentrated milk. See *evaporated milk; sweetened condensed skimmed milk; sweetened condensed whole milk; plain condensed milk; plain condensed skim milk.*

concentrated pure vanilla. A vanilla extract containing the extractive material from more than 13.34 ounces of vanilla beans per gallon of extract.

concentrated sour skim milk. See *evaporated cultured skimmed milk.*

concentrated superphosphate. See *double superphosphate.*

concentrated tankage. Slaughterhouse garbage, or fish tankage, etc., containing about 55 percent solids for use as an organic fertilizer. See *stick: 2.*

concentrated tomato paste. Strained tomatoes concentrated by evaporation to contain more than 33 percent solids. Salt, or pure sodium carbonate, may be added to the paste to minimize acidity.

concentrated tomato paste from trimmings. Tomato paste prepared from strained trimming stock.

concentrated vinegar. A flavoring product, developed during World War II, which contains 25 percent acetic acid. It is prepared by freezing ordinary vinegar and then removing its water content (ice) in a centrifuge.

concentrating plant. 1. A plant for preparing cheese for marketing. 2. A milk condensery.

concentration. The amount of active material in a diluent. Specifications for *concentrations* of herbicides, insecticides, etc., are on the basis of lbs. per unit volume of diluent.

concentration area. 1. Overgrazed land, indicating an overpopulation of wild animals. 2. Any area where livestock or products are concentrated prior to shipping.

concentration yard. A livestock-buying station located in a cattle producing area.

concentrator. Any device used for the evaporation of liquids by heating: *e.g.,* the sorghum pan or iron kettle for the concentration of sirup.

conception. The beginning of pregnancy or gestation; first stage in the development of the embryo.

conception rate. In cattle breeding, the percentage of first services that conceive. It should be 70 percent in healthy herds.

conch. The bracket-shaped fruiting body of certain fungi on the bark, branches, or the crown of a tree. Also called *punk; conk.*

concha. See *button: 12.*

concho grass. Texas panicum, *q.v.*

conchuela. *Chlorochroa ligata* (Say), Order Hemiptera (Heteroptera), family Pentatomidae, a sucking insect which infests cotton bolls and squares causing the squares to shed. It is destructive to cotton in Texas and Mexico.

concrete frost. A type of frost in which the soil becomes virtually solid.

concrete frost structure. Extremely dense structure of frozen soil; usually soil frozen to a great depth.

concrete silo. A vertical silo constructed of concrete blocks, staves, or solid concrete walls.

concretion. An inorganic body formed in a human or animal body cavity, as a kidney stone.

condemnation value. Compensation paid to a land owner whose real estate is taken over under the powers of eminent domain. It includes the full justifiable market value of the property, plus the value of that property for the owner's particular purposes.

condemned. 1. Describing an animal, carcass, or food which has been declared unfit for human consumption. 2. Referring to real estate property acquired for public purposes under the right of eminent domain.

condensation. The process of turning a vapor into a liquid: *e.g.,* the carburetor in an automobile takes in raw gasoline and air, thus forming a gaseous mixture. If this mixture now comes into contact with a cold inlet manifold, it turns into small drops of raw gasoline. This process is called *condensation.*

condensed buttermilk. A stock feed prepared by evaporating buttermilk to one-third its original weight. Also called *semi-solid buttermilk.*

condensed cultured whey. The product remaining after the removal of a considerable portion of water from clean, sound, cultured whey.

condensed fermentation solubles. The product remaining after the removal of a considerable portion of the liquid by-product from the fermented grain, molasses, whey, etc. It should contain more than 18 milligrams of riboflavin per lb. on a moisture free basis.

condensed fish soluble. A stock feed obtained by condensing the water remaining from the hydrated extraction of oil from fish. It contains approximately 50 percent solids and is a rich source of vitamin B_{12}.

condensed milk. See *sweetened condensed whole milk; plain condensed milk; evaporated milk.*

condensed molasses distillers' solubles. A stock feed obtained by condensing the residue from the yeast fermentation of molasses after the removal of alcohol by distillation. It has sirupy consistency, and is a source of B vitamins.

condensed skim milk. A dairy by-product from the removal of much of the water from clean, sound skim milk. It contains more than 27 percent solids.

condensed whey. A dairy by-product from the removal of much of the water from clean, sound whey. It usually contains more than 62 percent whey solids.

condensed whey by-product. A dairy by-product containing at least 50 percent of whey solids which usually consist of 50 to 55 percent lactose.

condensed whey solubles. A stock feed produced by the removal of albumin and the partial removal of milk sugar from clean, sound whey.

condenser. 1. In a cotton gin, a device consisting of two different-sized rollers, the larger of which is covered with a wire screen. The lint is blown against the screen-covered roller where it is packed into bats by the smaller roller. 2. A part of a refrigeration unit which changes the vaporous refrigerant into a liquid. The *condenser* may be cooled by air or water or by both. See *condensation.*

condensing pan. A retort for vacuum condensing of liquids.

condiment plant. A plant which yields spices or other seasoning material.

condition. 1. The state of wool regarding the amount of yolk and other foreign matter it contains. See *heavy shrink: 1; light in condition.* 2. The general appearance and/or state of health of an animal, seed, fruit, or flower at a show. 3. To get an animal in good health and appearance by proper feeding and grooming.

condition class. A unit of forest land which is fairly homogeneous with respect to diameter class (*q.v.*) distribution, economic or silvicultural conditions, etc.; a term used in the classification of forests of mixed species for management purposes.

conditioner. 1. A nostrum; a patent medicine. 2. A chemical or other substance whose application to soils, direct or mixed with fertilizers, improves the soil structure and prevents caking or coalescence of fine soil particles.

conditioning. 1. Any process for evenly distributing moisture in timber. 2. Any process in which animals, plants, or their products are treated prior to processing or ultimate use. 3. The process of controlling the moisture of stored grain by passing air at a low rate through the material.

conditioning house. 1. An official testing laboratory in some European countries where wool is tested and certified regarding its moisture content in proportion to bone-dry weight. See *regain.* 2. Any building in which goods are conditioned for sale or consumption.

condition wool. To determine the amount of moisture in wool.

conducting tissue. The plant tissue which is primarily concerned with the movement of water and food. Xylem and phloem are usually considered as *conducting tissues.*

conduction. In plants, the movement of liquid nutriments, water, etc., from one part to another.

conduit. 1. Any channel, open or closed, intended for the conveyance of water; any container for flowing water. 2. A metal pipe in which wiring is installed.

cone. 1. a. The carpellate or the staminate strobilus of plants of the families Pinaceae, Taxaceae, Cycadaceae and Ginkgoaceae. See *conifer.* b. The *cone*-shaped inflorescence of plants, as in the hop (*Humulus lupulus*), or the *cone*-shaped gynoecium, as in the flowers of Magnolia. 2. A part of the artificial vagina, attached to the collector tube, through which the semen enters the collector tube.

cone bearer. A coniferous tree or shrub. See *coniferous.*

cone beetle. Any of the small, dark, shiny beetles, family Scolytidae, which bore into the supporting stem or base of a cone to lay eggs. The legless grubs feed upon the cones.

cone delta. See *alluvial fan; alluvial cone.*

coneflower. a. Any of several plants of the genus Rudbeckia, family Compositae. See *autumn sun; blackeyedsusan: 1; brown-eyed Susan; niggerhead: 4; orange coneflower; cutleaf coneflower; great coneflower; pinewoods coneflower; showy coneflower; sweet coneflower; golden glow.* b. Also a common name for plants of genera Echinacea (Brauneria) and Retibida (Lepachys) of family Compositae. See *prairie coneflower; purple coneflower.*

cone gall. A cone-shaped gall found on the tips of the willow branches which hinders bud development. It is caused by a tiny maggot, *Rhabdophaga strobiloides* Walsh, family Itonididae.

cone moths. Small moths whose caterpillars feed upon bracts, scales, and seeds of tender growing cones.

cone of depression. The inverted-bell-shaped, continuously increasing downward slope of the water table surface around a well as a result of pumping.

cone pepper. See *tabasco redpepper.*
cone shaker. A screened box or cylinder in which dried or partially opened cones are rotated against baffles to free the seeds.
cone tree. Any coniferous tree. See *coniferous.*
cone wheat. See *poulard wheat.*
Confederate jasmine. Chinese starjasmine, *q.v.*
Confederate rose. Cottonrose hibiscus, *q.v.*
Confederate vine. Mountainrose coralvine, *q.v.*
Confederate violet. *Viola papilionacea* forma *albiflora* (*V. priceana*), family Violaceae, a hardy, perennial herb with silvery-white, blue-centered flowers, sometimes grown in the South as a symbol of the Confederacy. Native to North America.
confined water. Underground water trapped as petroleum is in geologic structures. It is usually highly mineralized and once drawn out is not renewable since such water is not a part of the general hydrologic circulatory system.
confine livestock. To keep livestock in a certain area by means of a surrounding barrier.
conformation. In judging contests, the type, form, and shape of the live animal, usually with reference to some performance characteristic.
confused flour beetle. *Tribolium confusum,* family Tenebrionidae, a very common insect of stored products such as grains, flour, beans, peas, baking powder, ginger, dried plant roots, dried fruits, bran, nuts, chocolate, snuff, cayenne pepper, etc. Also called *bran bug.*
congenial. Able to cross-fertilize readily; to unite easily, as with stocks and cions.
congenital loco in poultry. A lethal genetic character of poultry in which the chicks are unable to stand when hatched. The head is held back so that the beak points toward one side, and the chick loses its balance and falls. About 88 percent die by the end of one month.
congestion of the udder. An enlargement of the udder of a lactating animal which is accompanied by tenderness and excessive warmth, usually just after calving. The swelling may extend some distance along the abdomen. Generally, it occurs in a high producer. Also called *edema of the udder; caked-udder.*
conical valve. A lift valve for a pump whose sides are cone-shaped sections. It fits on an annular seating and has no wings. It is sometimes used in place of the common lift valve with parallel wings.
conidium. Any asexual reproductive spore of fungi, excluding the zoospores (*q.v.*), which has its own surrounding cell wall and no means of self-locomotion. *Conidia* are either dry and distributed by currents of air, or sticky and adhere to passing insects or other animals, or scattered by splashing rain drops. They may be one-celled or, by the formation of septa, may become pluricellular. Upon reaching a suitable medium they germinate and establish a new growth of the fungus.
conifer. Any plant belonging to the Coniferae, a large group of trees and shrubs, which includes the families Taxaceae, Pinaceae, Cycadaceae, and Ginkgoaceae. Conifers bear cones or strobili, and because many of the species produce new leaves before all the old leaves are shed they are looked upon as synonymous with evergreens.
coniferous. Pertaining to trees or shrubs which bear cones.
coniferous rust. Any of several rusts of coniferous trees: *e.g.,* the needle rust of pine. See *rusts.*
conilignosa. Vegetation which is dominated by trees and shrubs with needlelike foliage.
conjugation. Side-by-side association (synapsis) of homologous chromosomes, as in meiosis.
conjunctivitis. Inflammation of the membrane that lines the eyelids and covers the forepart of the eyeball.
conk. See *conch.*
conky. Describing a log or tree bearing a wood-rotting fungi. See *conch.*
connate water. Water trapped in sediments at the time of deposition. Such water tapped by wells may be highly mineralized and may be a brine or salt water. See *confined water.*
Connecticut group. A group of squash or pumpkins, *Cucurbita pepo,* family Cucurbitaceae, which have fruits with flattened ends and grooved surfaces. Some fruits may also be oblong in shape. They are popular for use in pies and for stock feed.
Connemara pony. An Irish pony which stands 13 to 14 hands high and may be gray over brown or bay. It has a short back and strong quarters. Its action is free, straight, and easy.
conservancy district. A legal subdivision of a state, embracing the drainage basin of a river, which is established primarily for the control of floods. The governing body may have the right of eminent domain, power to levy taxes, and regulate the use of surface waters within the drainage basin.
conservation. The control and preservation of natural resources for present and future use. In agriculture, the maintenance and improvement of soil fertility and productivity and the control of erosion.
conservation pool. The unusable, undrainable space below the outlet in a reservoir. It is often allocated to silt storage, fish, waterfowl, recreation, etc.
conservation survey. A survey to determine and record the nature of the soil, slope, erosion conditions, and present land use of an area.
conservative grazing. a. The practice of limiting the number of livestock on a grazing range in such proportion as would not exhaust the range in successive seasons. b. A degree of grazing that causes little or no soil disturbance.
conservatory. A glass house constructed for the accommodation of plants for display as contrasted to a structure for propagation.
conserve. 1. A spread usually consisting of a mixture of fruits, nut meats, and sugar cooked to a consistency. 2. To control and improve soil and water and to use them most logically and intelligently.

consign. To entrust produce or animals to another person for sale on a commission basis.
consignee. A person entrusted with the sale of produce or animals on commission.
consignment. a. The entrusting of a product or animal to a person for sale on commission. b. Animals or produce consigned for sale.
consignment sale. A sale of produce or animals through a commissioned agent.
consignor. A person who puts produce or animals for sale in charge of a commissioned agent.
consistence. Degree of adhesion or cohesion of the soil mass. *Consistence* of wet soils is expressed by such terms as sticky, plastic; of dry soils, by loose, friable, firm.
consociation. A unit of climax vegetation within an association which has a single dominant species.
consocies. A developing consociation.
consolidated. Referring to a soil material which is made solid by pressure or other natural process from a previous fluid or loosely aggregated condition.
consolidated school. A school which is supported financially by a combination of several school districts according to state law. The students from the several districts attend the school tuition free. Such a school is often found in rural areas.
constant flow chicken waterer. A device for watering chickens in which fresh water is added to a trough and surface water and trash are carried away from it in a continuous cycle.
constant level method. A method of watering plants in a greenhouse by a channel having a "V" bottom (or a flat bottom covered by half tile). This is covered with pea gravel. A layer of sand, topped by a soil layer, is placed over the gravel. The water table is held constant, usually in the sand layer, by use of a float valve.
constipation. The retention of feces in the intestines for a longer than normal period.
constituent. The components or elements of a substance: *e.g.*, the *constituents* of soil.
constitution. A system of related parts. body, frame of temperament. The term is also used to describe the characteristics of an animal which determine in part its ability to put on flesh, reproduce, and maintain health, vigor, and longevity.
constructed stock trail. A passage constructed or cleared to facilitate the movements of livestock to and from different grazing areas.
consumer packaged. Referring to fruits, vegetables, meats, flowers, etc., which are usually packaged attractively for retail sale.
consumptive use. The total amount of water used by vegetation in transpiration or building of plant tissue and evaporated from adjacent soil, snow, or intercepted precipitation in a given area in any specified time.
contact frost injury. An injury to potatoes in storage or in transit resulting from localized freezing sufficient to cause necrosis; characterized by darkened threadlike areas radiating from the stem end of the tubers.
contact herbicide. An herbicide that kills a plant primarily by contact with plant tissue rather than by internal absorption. *Cf. translocated herbicide.*
contact insecticide. Any substance that kills insects by contact in contrast to a stomach poison which must be ingested. *Cf. stomach insecticide.*
contact spray. A solution containing an herbicide or insecticide that kills on contact.
contagious. Transmissable by contact. See *contagious disease.*
contagious abortion. See *brucellosis.*
contagious bovine pleuropneumonia. A highly contagious pneumonia of cattle which is generally accompanied by pleurisy. The causal organism is *Asterococcus mycoides*, which in acute cases, causes death within a day or two. In less acute cases symptoms of varying intensity exist for three or four weeks, followed by a lingering death or eventual recovery. Extinct in U.S. since 1892. Also called *lung plague.*
contagious catarrh of fowls. See *roup of fowls: 1; fowl cholera.*
contagious disease. A disease transmitted or spread from animal to animal or from plant to plant, by direct or indirect contact with the diseased.
contagious ecthyma. A highly contagious, virus disease of sheep and goats, under one year of age, which is characterized by pus-filled lesions on the lips and sometimes on the face and ears. Fatalities result from the animals not being able to eat when normal gains should occur. It may cause secondary infections. Also called *sore mouth; contagious pustular dermatitis.*
contagious epithelioma. See *fowl pox.*
contagious equine abortion. A contagious disease among mares caused by *Salmonella abortivo-equina*, characterized by abortion at any stage of pregnancy, but usually during the 8th to 10th month.
contagious equine pleuropneumonia. An infectious disease of horses, thought to be caused by a virus. Its symptoms are those of pneumonia. Severe cases rarely recover. Autopsy reveals fibrinous deposits and gangrene in the lungs. Frequently confused with equine influenza and strangles. This disease is rare in the U.S. Also called *contagious pneumonia of horses; contagious gangrenous pneumonia; hemorrhagic septicemia.*
contagious gangrenous pneumonia. Contagious equine pleuropneumonia, *q.v.*
contagious granular vaginitis. See *granular vaginitis.*
contagious indigestion. See *avian monocytosis.*
contagious ophthalmia in cattle. See *pinkeye.*
contagious pneumonia of horses. Contagious equine pleuropneumonia, *q.v.*
contagious pustular dermatitis. Contagious ecthyma, *q.v.*
contaminant. An agent that causes contamination.
contaminate. To make impure by contact or admixture of harmful bacteria, fungi, or dangerous chemicals, etc.; to render unfit for use.

contamination. a. Pollution; the process of being contaminated. b. Specifically, the addition of bacteria or other foreign substance to milk or other products by means of utensils, containers, exposure to air, etc.

contentment. In animal judging, a characteristic of a stable, as contrasted to a nervous animal.

contiguous houses. Two or more adjoining greenhouses which have a common gutter to drain the roofs and a partition under the gutter. See *house: 1; ridge and furrow houses; lean-to house.*

continental climate. Climate typical of great land masses. It is characterized by a great range of temperature occurring in parts of a continent which are not affected materially by nearness to the sea.

continuous-action atomizer. A type of hand spray consisting of a pump in a chamber which holds the spray liquid. The pump builds up enough pressure after a few strokes to force out a continuous stream of spray.

continuous agitating cooker. A device for processing fruits and vegetables in cans. The cans enter a small porthole at one end of the sterilizer and, rolling continuously, travel along a spiral inside. The contents are thus thoroughly agitated. *Cf. agitating discontinuous pressure sterilizer; agitating pressure cooker.*

continuous agitating pressure sterilizer. A device for sterilizing canned fruits, vegetables and evaporated milk under pressure. It differs from the continuous agitating cooker in that the cans enter and leave the sterilizer through vestibules permitting the maintenance of pressure within the device.

continuous apron. An endless belt which lifts harvested potatoes, etc., from the ground into a truck or wagon.

continuous cropping. The growing of the same crop on the same land year after year.

continuous cultivation. Continuous cropping, *q.v.*

continuous-flow irrigation. A system of irrigation in which each farmer receives his allotted quantity of water at a continuous rate.

continuous grain drying. Drying of grain by running it through a continuous stream of air.

continuous grazing. Grazing limited to the same land throughout the season.

continuous irrigation. A method of applying disinfectants on animals. The disinfectant is placed in a container and held above the wound surface. A rubber tube is fastened to the container and its lower end is placed on the wound area. Thus, the antiseptic trickles continuously into the wound.

continuous layerage. A method of propagation of woody plants by rooting the branches which are still attached to the parent plant. The whole shoot is bent down flat in a shallow trench, running radially from the plant, and is covered with soil. However, the end remains exposed. This method is used for the shrubs and vines whose buds will grow readily even though covered with soil. The snowball European viburnum is often propagated this way.

continuous manger. A long manger for feeding many animals simultaneously. Sometimes partitions are used to separate the animals.

continuous spread. The spread of plant disease into adjacent fields by one or more vectors, such as wind, wind-driven rain, water, insects, animals, agricultural machines, man, etc. *Cf. discontinuous spread.*

contour. A line joining the points of the same elevation above a datum. A series of contours delineates the relief or topography of the land.

contour basin. A basin made by levees (*q.v.*), constructed along contours, with occasional cross levees. Also called *contour checks.*

contour border irrigation. A method of irrigating gently sloping fields. The whole area is divided into strips by ridges along the contours and cross ridges. The ridges confine the water to a particular strip until it is completely full, before letting it flow to the next lower strip. Also called *contour check irrigation.*

contour check. A series of small levees or ridges constructed along contour lines to restrict the application of water to a particular strip.

contour cultivation. See *contour farming.*

contour farming. A method of cultivation, following the contour of the land, which is a deterrent to erosion.

contour feathers. Plumage on the body of a bird, including the wing feathers, tail feathers, exposed body feathers, and ear coverts which determine its outer contour.

contour furrow. 1. A furrow plowed along a contour or at a uniform grade. 2. A level furrow made in a field or pasture to prevent water runoff and soil loss, and to increase water infiltration in soil.

contour interval. On a contour map, the difference in elevation, or the vertical distance, between contours.

contour line. See *contour.*

contour liner. A device on a tractor which, by enabling the operator to maintain an approximately constant level, assists in plowing along a contour line.

contour planting. The planting or drilling of crops in rows along contour lines in contrast to plantings which run parallel to field boundaries or up and down hill on sloping land.

contour plowing. Plowing a field on a contour or line of equal level.

contour row. a. A row, all points of which have the same elevation within a given tolerance. b. A level row which runs at right angles to the line of slope regardless of the irregularities of the landscape.

contour strip cropping. The production of crops in long, variable width strips which are placed approximately on the contour, and crosswise to the line of slope.

contour tillage. A system of farming in which the various operations are performed on the contour.

contract buster. A person who contracts to break horses (western U.S.).

contracted feet. A condition in which a part of the horse's foot, usually the heel, shrinks to less than normal size.

contracted heel. See *contracted feet.*

contracted tendon. a. A thickening and shortening of the tendon in the leg of a horse, resulting from a severe strain or injury. The fetlock joint knuckles forward, and in severe cases the heels do not touch the ground. b. A knuckling over of the fetlock joint in new-born foals which often renders the animal useless. Can be remedied by surgery.

contracted weir. A device for measuring the flow of water, whose sides produce a contraction in the cross-sectional width of the water channel.

contract farming. An agreement about price and designated products between a farmer and a processor, usually made prior to the growing season. See *participating contract; vertical integration.*

contract feeding. An arrangement for finishing cattle or swine for the market. The contribution of each participant depends on a written contract; in the case of cattle the stock raiser usually furnishes the cattle. The feeder furnishes the feed, equipment, and labor, and usually receives his income on the basis of the increase in weight of the cattle, etc.

contraction. The decrease in the cross-sectional area of a jet of water after passing an orifice, weir, or notch. See *vena contracta.*

contract labor. Labor hired for farm operations, whereby the farmer contracts with a family or group leader to care for a certain acreage. Often the laborer's income is increased as the yield per acre goes above a given standard.

control. 1. Prevention of losses from plant or animal diseases, insect pests, weeds, etc., by any method. 2. A section of an open water channel where conditions exist that make the water level above it an index of the discharge. 3. A standard entity used for comparative purposes in experimentation. Also called *check.*

control burning. Setting fire to land cover under conditions that presumably will accomplish specific silvicultural, wildlife, grazing, or fire hazard reduction purposes.

control flume. An open conduit or artificial channel arranged for measuring the flow of water. See *Parshall flume; Venturi flume.*

controller. See *electric fence controller.*

control line. All the constructed or natural barriers which are used to control a fire.

control method. The method of regulating the cutting of trees on the basis of periodic inventories of the growing stock.

control of a fire. The surrounding of a fire with control lines and backfiring of any unburned surfaces adjacent to the inner edge of the control lines. This brings the fire *under control.* A fire is *safe* when it has been restricted to a controlled area. It is *out* after all known fire in the controlled area has been extinguished.

control rope. 1. A rope attached to a plow or harrow by which the operator on a tractor or sulky controls its action. 2. A rope tied to an animal for restraining it.

convalescent serum. The blood serum of an animal or man that is recovering, or has recovered from an attack of disease.

conval lily. See *lily-of-the-valley.*

convergent improvement. In animal husbandry, a system of backcrossing progenies of one parent with another parent in order to produce a new strain of desired characteristics.

convergent lady beetle. *Hippodamia convergens* G.-M., family Coccinellidae, a small, beautifully colored beetle which feeds on certain insects and their eggs. It is helpful in the destruction of various plant insects and scales.

converse adaptation. Adaptation of a plant or other organism to its habitat to such an extent that it gains advantage over other competing species.

conversion. 1. Sawing or cutting timber to any shape. 2. The change from one reproduction method to another. 3. The transformation of a forest from one dominant type to another.

conversion factor. A numerical value which indicates the ratio of exchange for different classes of grazing animals for a range.

converted rice. Rice steamed to cause the vitamins from the hulls to penetrate the endosperm. This changes the characteristics and appearance of milled rice, but when cooked by boiling, the kernels appear as white as raw rice and retain their shape better.

convertible husbandry. See *rotation: 1.*

convex sole. A condition of a horse's sole which, instead of being concave or ridged as normally, projects to a lower level than the outer rim of the hoof. Frequently the result of laminitis.

conveyance loss. The loss of water from a conduit due to leakage, seepage, and evaporation.

conveyor. Any mechanical device, as an endless belt for moving milk bottles, cans, grain, etc., from one place to another.

convulsions. A clinical symptom usually of a deranged function of the brain or spinal cord in an animal. The muscles contract and relax violently, producing aimless body movements.

cooked cheese. A cheese prepared by cooking or heating the curd.

cooked cream. See *Devonshire cream.*

cooked flavor. An off-flavor of butter, ice cream, or cheese which suggests heated milk. Also called *custard flavor.*

cooking eggs. Eggs as good as, or better than, U.S. grade C; satisfactory for cooking purposes.

cooking grade. A U.S. butter grade scoring 86 to 88 points. It is next to the lowest grade.

cooking sirup. A commercial sirup; a by-product of the sugar industry. See *molasses: 1; table sirup; sorghum molasses.*

cool. a. To reduce the temperature to a desirable level as for preservation or curing of milk or meat products. b. Describing a lower temperature.

cooler. Any chest-like device for reducing the temperature of products, such as milk.

cooler shrinkage. A loss in weight of an animal's carcass during storage in a cooler which usually amounts to 2.5 percent of the carcass weight.

cooler space. Storage space held at temperatures of 29° to 50° F.

Cooley spruce gall aphid. *Chermes cooleyi* (Gill.), family Chermidae, an aphid whose nymphs feed at the bases of new growths of Norway, Englemann, Sitka, and Colorado spruce, producing galls which envelop them.

cool greenhouse. See *cool house.*

cool house. A house (*q.v.*), in a range of greenhouses which is maintained at a day temperature of about 55° to 60° F. and a night temperature of about 50° to 55°. Chrysanthemum, carnation, sweet pea, snapdragon, and other plants do well in these temperatures. Also called *cool greenhouse.* See *warm house.*

cooling rack. A rack with several parallel, horizontal iron bars on which dressed poultry is hung to cool.

cool season. Designating plants which thrive best in cool weather: *e.g.*, garden peas, radishes, etc. See *warm season.*

coolwort. Allegheny foamflower, *q.v.*

Coon. A Cheddar cheese cured by a patented process so that it becomes dark-colored and crumbly, with a sharp and tangy flavor.

coontail. Hornwort, *q.v.*

coontie. *Zamia integrifolia,* family Cycadaceae, a tough, woody plant grown as an ornamental and for its roots used to make arrowroot-like starch. Native to West Indies and Florida.

coop. 1. A small light enclosure, usually made of wire netting, used to confine fowls. 2. A heap, usually of manure.

co-op. Cooperative.

cooper. A person who makes barrels, kegs, etc.

cooperage. 1. The barrels, kegs, etc., produced by a cooper. See *tight cooperage; slack cooperage.* 2. The lining of boxcars to prevent leakage of bulk products.

cooperative. A voluntarily organized business association, democratically controlled by its members or patrons. Its primary function is to secure fairness in marketing: *e.g.*, consumers' *cooperatives,* farmers' marketing *cooperatives,* and farmers' purchasing *cooperatives.*

Cooperative Extension Service. Field work and research carried on by the Land-Grant colleges and universities in the U.S. in cooperation with the U.S. Department of Agriculture.

cooper's flag. Cattail, *q.v.*

cooper's fly. Cattail, *q.v.*

coot. The ankle-joint of a horse; the fetlock.

cootie. Referring to feather-covered legs, as in a chicken.

coppa picante. A square-shaped sausage consisting of a cala butt, cured and coated with black pepper. It weighs about 3 lbs.

copper. Cu, a metallic element found in soils at 1 to 50 parts per million, and in plants up to 100 parts per million. It is functional in animal and plant life, especially in the production of commercial crops.

copper acetoarsenite. Paris green, *q.v.*

copper A compound. A powder, based on tetracopper calcium oxychloride, used as a fungicide.

copper arsenite. An inorganic compound of copper, used as an insecticide.

copperas. See *ferrous sulfate.*

copper beech. See *purple beech.*

copper carbonate. A compound used as a seed disinfectant.

copper 8-hydroxyquinoline. A standard bacteria-inhibiting agent, used in the treatment of vascular diseases.

copper iris. *Iris fulva,* family Iridaceae, a rhizomatous perennial herb grown for its copper-colored, variegated flowers. It is a beardless iris. Native to midcentral and southeastern U.S.

copper-lime. A dust sometimes used in place of Bordeaux spray.

copper nitrate. A highly toxic, selective herbicide.

copper oxide. A chemical used in the seed treatment sprays or dusts for sugar beets and vegetables. It prevents damping off, rose leaf diseases, celery blight, tobacco and hops downy mildew. Also called *cuprocide.*

copper oxychloride sulfate. A chemical used for the control of beet leaf spot, cucurbit anthracnose, rose leaf diseases, celery blight, potato and tomato foliage diseases, apple fire blight, cucumber downy mildew, grape black rot, and cherry leaf spot.

copper phosphate. A chemical used for the same purposes as copper oxychloride sulfate, *q.v.*

copper poisoning. A livestock poisoning usually in sheep, from consuming too much salt to which copper sulfate has been added for control of stomach worms. Symptoms are loss of appetite, jaundice, and often bloody urine.

copper spot. A fungous disease of grasses caused by *Gloeocercospora sorghi,* family Tuberculariaceae, which produces copper- or orange-colored spots on grass blades.

copper spray. A plant spray consisting mainly of a copper salt.

copper sulfate. $CuSO_4$, a chemical of many agricultural uses: *e.g.*, in fertilizers and Bordeaux mixtures, as a selective herbicide, an anthelmintic, and as a disinfestant and disinfectant. Also called *blue vitriol; bluestone.*

coppertip. Golden coppertip, *q.v.*

coppice. A recently cleared area in which new shoots have sprouted forth from the stumps of woody plants.

coppice forest. A forest of second-growth sprouts. Also called *sprout forest.*

copra. The dried coconut kernel from which coconut oil (*q.v.*) is obtained. Also used for ceremonials in India and southeast Asia.

copra oil. Coconut oil, *q.v.*

coprogenous. Designating the influence of animal excrement, as of the earthworm, in forming soil, especially humus.

coprolites. 1. Round lumps of fossilized excrements of prehistoric animals, rich in phosphate of lime. 2. Phosphate nodules which are not fossil excrement.

coprose. See *corn poppy.*

copse laurel. Spurgelaurel daphne, *q.v.*

copsy highbush blackberry. See *Allegheny blackberry.*

copulation. See *coitus.*

coquito palm. Syrup jubea, *q.v.*

coral bean. a. The bright colored seeds of the species of coral tree. b. *Erythina corallodendrum,* family Leguminosae, the *common coralbean,* a small tree grown for its scarlet flowers. Native to tropical America. Also called *coral plant.* See *coral tree: a.*

coralbells. *Heuchera sanguinea,* family Saxifragaceae, a hardy, perennial herb grown for its red, bell-shaped flowers. Native to North America.

coral berry. See *Indiancurrant coralberry.*

coral drop. See *showy coraldrops.*

coral evergreen. See *running pine.*

coralgem deervetch. *Lotus bertholeti,* family Leguminosae, a prostrate shrub grown in warm areas or greenhouses for its scarlet flowers. Native to Canary and Cape Verde Islands.

coral honeysuckle. See *trumpet honeysuckle.*

coral jasmine. Purple cestrum, *q.v.*

coral lily. *Lilium pumilum,* family Liliaceae, a bulbous herb grown for its scarlet flowers. Native to eastern Asia.

coralline honeysuckle. *Lonicera chrysantha,* family Caprifoliaceae, a deciduous shrub grown for its foliage, yellow flowers, and red berries, which are a favorite bird food. Native to Asia.

corallita. Mountainrose coralvine, *q.v.*

coral plant. a. *Russelia equisetiformis,* family Scrophulariaceae, a leafless, tender shrub grown in greenhouses or outdoors in warm areas for its showy, red flower clusters. Native to Mexico. b. See *coral bean: b.*

coral spot (or dieback) of woody plants. A fungous disease caused by *Nectria cinnabarina,* family Hypocreaceae, which produces lesions and bright red fruiting pustules on the twigs and branches.

coral sumac. Florida poisontree, *q.v.*

coral tree. a. Any of the several trees and shrubs of genus Erythrina, family Leguminosae, which bear long bean pods, and are grown for their gay-colored flowers in Florida and for shade in tropical America. Native to the tropical regions of the world. Also called *bean tree.* b. See *Japanese raisintree.*

coralvine. Mountainrose coralvine, *q.v.*

cord. 1. a. The spermatic cord. b. The umbilical cord. 2. A unit of measurement of timber. It contains 128 cubic feet, usually 8′ x 4′ x 4′. The long *cord* has 160 cubic feet, the short *cord* less than 128. 3. Twine. 4. To stack in rectangular shaped tiers.

cordate bugbane. *Cimicifuga cordifolia,* family Ranunculaceae, a tall, perennial herb, grown for its white, graceful flowers. Native to America. Also called *bugbane.*

Cordia. A large genus of tropical trees and shrubs, family Boraginaceae. The species are grown for ornament, timber, or their fleshy, edible berries.

cordon. The training of vinifera grapes and dwarf fruit trees in such a way that the main stem is 2 ft. high and from it only one or two horizontal branches extend opposite or parallel to each other. See *unilateral horizontal cordon; bilateral horizontal cordon; vertical cordon; espalier.*

corduroy road. A passage made by laying saplings or small poles close together across the road. Used in swampy or wet areas.

cordwood. Wood cut to 4 ft. lengths for use in furnaces, or fireplaces. It is sold in cords. Also called *firewood.*

core. 1. The innermost part of an apple, pear, etc., which contains the seeds. 2. The receptacle in certain plants, as the raspberry. 3. To remove the central, seed-containing portion of fruits for canning or preserving.

coreopsis. Any plant of the genus Coreopsis, family Compositae. Species are grown for their beautiful flower heads. See *goldenwave coreopsis; Pacific coreopsis; rose coreopsis; plains coreopsis.*

coreopsis yellows. See *aster yellows.*

core wall. A wall of masonry, sheet piling, or puddled clay built inside an earth dam or embankment to reduce percolation.

coriander. *Coriandrum sativum,* family Umbelliferae, an annual herb grown for its seeds which are used as a condiment, and in liqueurs. Its fruit is used in medicines and for seasoning. The oil obtained from the seeds is used in toilet waters.

cork. 1. The thick external suberized tissue of some woody plants, especially the cork oak, produced by the activity of the phellogen, used commercially for floats, stoppers, etc. 2. Plant tissue similar to oak bark in composition or structure. 3. A spot disease of apples which causes internal tissue to become dry and punky, but does not show any external malformations such as depressions or ridges. The large dry rot areas may become hollow, reducing the size of the apple. Also called *punky disease; York spot; hollow apple; crinkle; confluent bitter pit.*

cork elm. Rock elm, *q.v.*

cork fir. Alpine fir, *q.v.*

cork oak. *Quercus suber,* family Fagaceae, a fairly small tree, whose thick and corky bark is the commercial source of cork. Native to southern Europe.

cork of apples. See *cork: 3.*

cork russeting. See *Bordeaux injury.*

corkscrew flower. Snail bean, *q.v.*

cork tree. Cork oak, *q.v.*

corkwood. Light, porous wood of cork-like structure.

corky. 1. Describing dry, coarse, and crumbly cheese. It is springy and does not mold together. 2. Designating cork, *q.v.*

corky bark. A possible virus disease of citrus in which cork-like eruptions appear on the bark. See *necrotic cavity corky bark; banded corky bark; tattoo netted corky bark.*

corky end. See *powdery scab of potato.*

corky scab of potato. A bacterial disease caused by *Actinomyces scabies.* Scab-like areas appear on the surface of the tuber. Also called *brown scab; Oöspora scab; American scab; deep scab; actinomycosis of the potato.*

corm. A solid, enlarged, fleshy, underground stem which resembles a bulb, as in the crocus and gladiolus. Also called *solid bulbs; bulbil tuber.*

cormel. A small, leafless corm produced by an old corm.

cormlet. A cormel.

cormophyte. An old term referring to all plants with root and stem; a primary division of the vegetable kingdom (Endlicher). Also called *axophyte. Cf. thallophyte.*

cormous. Describing a plant which produces corms.

corm rot. Any rot of corms.

cormus. A corm.

corn. 1. *Zea mays,* family Gramineae, an annual herb, a cereal grain which is one of the most important agricultural crops in the U.S.; grown for human and animal consumption. It is also a source of vegetable oil. Also called *Indian corn; maize.* See *dent corn; flint corn; sweet corn; waxy corn; pod corn.* 2. The seed or the plant of any of the cereal crops, as wheat, rye, etc., (U.K.). 3. A growth on the foot of a fowl afflicted with bumblefoot. 4. An afflicted part of a horse's forehoof. An injury to the soft tissue of the sole causes blood to diffuse into the horny sole between the bar and the outside wall. 5. To preserve beef by seasoning with salt, sugar, baking soda, saltpeter.

corn-and-cob meal. A stock feed consisting of the grain and cob of corn ground together. Also called *ground ear corn.*

corn and oat chop. Corn-and-oat feed.

corn-and-oat feed. A ground feed, particularly for dairy cows and horses which contains various proportions of good grade corn and oats, often with a large amount of low grade materials, such as hulls and ground corn cobs.

Corn and Winter Wheat Belt. That area of the midwestern U.S. where soft red winter wheat is grown in rotation with corn. Includes Ohio, Indiana, and Illinois.

corn beef. See *corned beef.*

Corn Belt. Central part of the U.S. where corn is extensively cultivated. Because of increasing diversification of crops, the area is becoming less distinctive and the term is losing some of its value as a geographic referrent.

corn billbug. See *maize billbug.*

corn binder. A machine for harvesting rows of corn or sorghum. It cuts and bundles the standing stalks with twine prior to ensilage.

corn blotch leaf miner. *Agromyza parvicornis* Loew, family Agromyzidae, a black fly whose larvae mine leaves of corn and grasses.

cornborer. See *European cornborer.*

corn bran. The outer coating of the corn kernel having little or none of the starchy part of the seed germ, used as a stock and poultry feed.

corn buttercup. *Ranunculus arvensis,* family Ranunculaceae, a weed pest with light yellow flowers, usually associated with clay soils. Native to Europe. Also called *corn crowfoot.*

corn campion. Corn cockle, *q.v.*

corn catchfly. See *sweet William silene.*

corn chop. A fine or coarse stock feed made by grinding or chopping grains of corn. It contains less than four percent of foreign material. Also called *ground corn; cracked corn.*

corn chrysanthemum. *Chrysanthemum segetum,* family Compositae, a very popular annual herb, grown for its white or yellowish flowers. Escaped, it becomes a weed pest. Native to Eurasia. Also called *cornmarigold; fat hen; marigold.* See *chrysanthemum.*

corn cob. The woody core of the corn ears; the highly fibrous axis on which the kernels are borne; used alone or with the grains as ground cattle feed. It is a source of furfurals and is used in the production of synthetic rubber, etc.

corn cob meal. Pulverized corn cobs, used for cattle feed, cleaning furs, burnishing metal, removing oil from tin and metal, sweeping compounds, etc.

corn cockle. *Agrostemma githago,* family Caryophyllaceae, a common weed pest found in areas where winter wheat is grown. Animals eating feed which is heavily contaminated with its seeds become seriously poisoned resulting in vomiting, nausea, vertigo, diarrhea, and depressed breathing. Native to Europe. Also called *corn campion; corn rose; crown-of-the-field; purple cockle.*

corn cracker. A mill used to coarse-grind corn kernels.

corn crib. a. A building for the storage of ear corn. The roof is watertight, but the sides are usually slatted for ventilation. b. A circular or rectangular temporary storage, with or without a roof, of woven wire or slats. See *single crib; double crib; double crib with overhead bin.*

corn crowfoot. See *corn buttercup.*

corn cutter. See *stalk cutter.*

Corn Dance. A ceremonial dance held by American Indians at the planting and harvesting of corn.

corn distillers' dried grains. A stock feed consisting of the dried residue from the distillation of alcohol from corn or from a grain mixture in which corn predominates.

corn distillers' dried grains with solubles. A stock feed of corn distillers' dried grains containing a large portion of condensed, screened, dried stillage.

corn earworm. *Heliothis zea* (Boddie), Order Lepidoptera, family Noctuidae, an insect pest which infests corn, cotton, tobacco, tomatoes, sorghum, vetch, alfalfa, etc. Also called *tomato fruitworm; bollworm.*

corned beef. Cured beef; prepared by packing fresh beef into vats and weighting it down to keep it submerged in the pickling solution of salt, sugar, spices, and water. The treatment requires about 25 days, during which the meat is turned to insure thorough pickling. Sometimes pickling is carried out in tierces; the barrels are then headed up and allowed to stand, with occasional rolling to insure mixing.

Corneliancherry dogwood. *Cornus mas,* family Cornaceae, a hardy perennial tree, widely grown for its handsome, small, yellow flowers. Its red edible berries make excellent jelly. Native to Eurasia.

corn enamel. A can enamel containing a small amount of zinc oxide, used in metal cans in which sweetcorn is packed. It prevents a blackening or discoloration of the can. Also called *C-enamel.*

corner. 1. The outer pair of incisor teeth in the upper and lower jaws of a horse. 2. The junction point of boundary lines. 3. To tie up or control all available items of produce for speculation. 4. In lumbering, to cut through the sapwood on all sides of the tree to prevent it from splitting when the tree is felled.

corner brace. A reinforcement to keep a corner fence post erect so that the tension of the sides does not pull it out of line.

corner post. The fencepost at the apex of an angle which bears the strain of the fence wire from two directions.

corner tree. A tree, at or nearly adjacent to a land-survey corner, blazed by the original survey party to indicate the corner.

corn-fed. Designating an animal or fowl fattened on corn prior to marketing; hence, well finished and of high grade.

corn feed meal. A stock feed of fine siftings obtained in the manufacture of screened corn chop, screened ground corn, or screened cracked corn.

corn field. A grain field. In the U.S., the crop is maize. In England it is usually wheat, barley or rye.

cornfield ant. *Lasius alienus americanus* Emery, family Formicidae, an ant that nests in open places in soil or rotten wood. The worker ants often foster and attend honeydew-excreting insects. Distributed over most of the U.S.

cornfield horsetail. Field horsetail, *q.v.*

cornfield pea. See *cowpea: 2.*

corn flakes. The treated, rolled and toasted endosperm of corn; ready-to-eat breakfast cereal.

corn flea beetle. *Chaetocnema pulicaria* Melsh., family Chrysomelidae, a beetle that infests corn, making small holes in the leaves which retard their growth. It is a vector for bacterial wilt of corn, *q.v.* Also called *brassy beetle.*

corn flour. a. A flour produced by grinding and bolting corn or as a by-product of the manufacture of grits or hominy. b. Cornstarch.

cornflower. *Centaurea cyanus,* family Compositae, a very popular, annual herb, grown for its pale-blue, pink, white, or purple flowers. Native to southern Europe. Also called *bachelor's button; blue bottle; ragged sailor; blue bonnet; cyaniflower.*

corn fly. See *seed-corn maggot.*

corn fodder. The entire corn plant cut and fed green or harvested when mature and dried for future feeding. The term is used also for the stalk remaining after ears have been husked from the plant. It is often confused with *corn stover.*

corn following corn. A system of crop production in which corn is planted successively for the second year.

corn forage. A stock feed prepared from the entire corn plant harvested in the milk stage or earlier.

corn germ cake. A stock feed consisting of the corn germ with other parts of the corn kernel from which oil has been partially pressed. Obtained in the dry milling process manufacture of corn meal, corn grits, hominy feed, and other corn products.

corn germ meal. A livestock feedstuff which is ground *corn germ cake.*

corn gluten feed. A stock feed consisting of the residue which remains after the extraction of the larger part of the starch and germ from commercial shelled corn. It may contain corn solubles and corn oil meal.

corn gluten meal. A stock feed from the residue of commercial shelled corn which remains after the extraction of the larger part of the starch and germ, from which the bran has been separated. It may contain corn solubles and corn oil meal. Also called *gluten meal.*

corn grader. A machine used to separate various sizes and shapes of kernels for the seed trade. The size and shape of seed conforms to the size of the holes in the corn planter plate. See *cell drop.*

corn grits. See *hominy grits.*

corn gromwell. *Lithospermum arvense,* family Boraginaceae, an annual or biennial weed pest of fields. Native to Europe.

corn grower. A farmer whose principal crop is corn.

corn harvester. See *corn binder; corn picker.*

corn hill. Any of the several spots in a cornfield where the stalks of corn are growing.

corn-hog ratio. See *hog-corn price ratio.*

cornhusk. The bract or outer, leafy covering which encloses the corn ear. Also called *corn shuck.*

cornhusker. a. A power-driven machine which snaps ear corn from the stalks and strips the husks from the ears. See *corn picker.* b. A person who shucks corn by hand.

Cornhusker. 1. A cured cheese, made from cow's whole milk, in the U.S. It is similar to Cheddar and Colby. 2. A resident of Nebraska, U.S.

Cornhusker State. State of Nebraska, U.S., where a large quantity of Indian corn is grown.

cornhusking. The removal of husks from the ears of corn.

cornicle. An abortive spur on the leg of a hen which hardens with age. It does not develop into a regular spur, as in the cock.

Cornish. A very heavy, meat-producing breed of chickens originating in Cornwall, England. They have compact bodies, close-fitting plumage, pea comb, yellow skin, and unusually well-fleshed breasts, thighs, and drumsticks. Mature males weigh about 10½ lbs., females about 8. The

Cornish is sometimes crossed with other breeds for producing market poultry, as *Cornish-Hamp.*, a cross-breed between a Dark Cornish and a New Hampshire.

Cornish cream. Devonshire cream, *q.v.*

Cornish heath. *Erica vagans*, family Ericaceae, a low, spreading shrub, grown for its pinkish-purple flowers. Native to western Europe.

corn knife. A long-bladed knife with a wide, blunt end; used for cutting corn stalks by hand.

cornland. Land especially suitable for growing corn.

corn lanternfly. See *corn planthopper.*

corn leaf aphid. *Rhopalosiphum maidis* (Fitch.), family Aphidae, an aphid which sucks juices from the leaves, stems and flowering parts of corn, barley, sugarcane, millet, broomcorn, sorghum, Sudangrass and other grasses, and coats them with honeydew. It is especially destructive to sorghum. Common throughout the Corn Belt, but more abundant in southern U.S.

corn leaf hopper. See *corn planthopper.*

corn leaf spot. See *southern corn-leaf blight.*

cornlet. A small or immature ear of corn.

corn maggot. Seed-corn maggot, *q.v.*

corn marigold. Corn chrysanthemum, *q.v.*

corn meal. Finely ground but unbolted corn, used in corn bread, mush, etc. See *degerminated corn meal.*

corn measles. See *brown spot: 1.*

corn mildew. See *Diplodia disease of corn.*

corn mill. a. A machine which grinds corn on the cob for feed. b. A roller mill for producing corn flour.

corn mosaic of Hawaii. A disease of corn and sorghum caused by the virus *Fractilinea zeae*, which produces chlorotic markings on the foliage similar to corn streak. This virus also causes the corn stripe disease of Cuba.

corn mustard. See *charlock.*

corn oil. A yellow, edible oil obtained from the germ of corn. Used as a salad oil, in cooking oil, and soap manufacture.

corn oil cake. A stock feed consisting of the corn germ from which oil has been partially pressed.

corn oil meal. A ground corn oil cake.

corn on the cob. a. Kernels on the cob which are still in the soft milk or dough stage. b. Ear corn, or unshelled corn.

corn picker. A corn harvester. It is a single- or double-row machine equipped with snapping rolls to remove ears of corn from the standing stalk. Of the several different types, there are the snapper which does not remove the husk, the picker-husker which removes the husks from the ears, the picker-sheller which after picking corn, husks and shells it, and the picker-shredder which shreds the stalks after the ears have been picked.

corn planter. Any mechanical device for planting corn. It is designated according to the manner in which the seed is dropped: *e.g.*, checkrow planter, lister planter *q.q.v.* See *drill: 3.*

corn planter plate. The metal bottom plate of the hopper on a corn planter which has, around its circumference, a series of notches of a size to hold one or several grains of corn depending upon the method of planting. This plate rotates, dropping the kernels into the seed tube, so that they are planted at regular intervals. See *cell drop.*

corn planthopper. *Peregrinus maidis* (Ashm.), family Delphacidae, a leafhopper-like insect which infests the buds and leaves of late-planted corn. It is found in southern U.S.

corn poppy. *Papaver rhoeas*, family Papaveraceae, an annual herb, grown for its red, deep purple, scarlet, or white flowers. Escaped, it is considered a weed. It is the famous Flemish poppy of World War I. Also called *African rose; coprose.*

corn pox. See *brown spot: 1.*

corn root. See *bloodroot.*

corn root aphid. *Anuraphis maidi-radicis* (Forbes), family Aphidae, an insect that infests the roots of corn, cotton and a number of grasses and weeds. It sucks the plant juices.

corn root webworm. *Crambus caliginosellus* Clem., family Crambidae, an insect whose larvae cut corn off near the surface of the ground especially on spring-plowed sod land. It also infests timothy, tobacco, grasses and certain weeds.

corn rootworm. See *northern corn rootworm.*

corn rose. See *corn poppy; corn cockle.*

corn rust. A minor, fungous disease of corn caused by *Puccinia sorghi*, family Pucciniaceae, which produces circular or elongated, cinnamon-brown pustules over the leaf surface. As the corn matures, black pustules are formed on all parts of the plant above the ground.

cornsalad. Any of the several annual herbs of the genus Valerianella, family Valerianaceae, which are grown as potherbs, salad plants, or for ornament. See *European cornsalad; Italian cornsalad.*

corn sap beetle. *Carpophilus dimidiatus* (F.), family Nitidulidae, a beetle that infests corn and dates.

corn screenings. A stock feed consisting of small and broken grains of corn, obtained by screening shelled corn and other material having feed value.

corn shock. A bunch of bound or unbound corn stalks placed upright on the ground for curing or drying.

corn shredder. A farm implement which husks the ears and tears the dried stalk into bits for feed or to assist in decay.

corn shuck. Cornhusk, *q.v.*

corn silage. An excellent stock feed prepared from the entire corn plant. The plant is cut while still green but when the ears have begun to dent. It is then chopped and placed in a silo for fermentation and preservation.

corn silk. The pistillate floral part of the corn plant, protruding from the husk, which catches the pollen and serves as the avenue for the fertilization of the egg cell which develops into the kernel. Each silk is the much elongated style and stigma of the flower.

corn sirup. Commercial glucose, obtained chiefly by acid hydrolysis of corn starch. It is an uncrystallizable sirup, used as a sweetening agent.

cornsled. A crude, wooden sled provided with cutting knives on one or both sides; used in harvesting corn before the advent of the corn binder.

corn smut. A fungous disease caused by *Ustilago maydis,* family Ustilaginaceae, which accounts for up to two percent of the crop losses over wide areas in the U.S. It produces galls on the aboveground parts of the plant, usually pea-sized on leaves, but up to about six inches in diameter on other parts. The galls appear greenish-white, but on enlargement, they rupture and expose a powdery, black spore mass. The young galls while still firm are often cooked and eaten. Also called *boil smut.*

corn snapper. A corn picker which detaches the ears of corn from the stalk but does not remove the husk. See *corn picker.*

corn sow thistle. Field sow thistle, *q.v.*

corn speedwell. The common *speedwell, Veronica arvensis,* family Compositae, an herbaceous plant bearing small, blue flowers, found in the open field. Native to Europe and America.

corn spurry. *Spergula arvensis,* family Caryophyllaceae, an annual weed pest which flourishes on acid soils. Native to Europe.

cornstalk disease of horses. A lethal disease probably caused by the eating of toxic, moldy corn stalks. The affected animal exhibits nervous symptoms as dullness, excitement, local or general paresis, circling, staggering and tremors. Jaundice is a common side effect.

cornstalk pine. See *loblolly pine.*

cornstarch. Starch extracted from the corn kernel; used for puddings, thickening of foods, etc.

Corn State. State of Iowa, U.S.

cornstook. A corn shock.

corn stover. The dried corn stalk from which the ears have been removed; used as a roughage for livestock.

corn streak. A virus disease of corn, sugarcane, and certain other grasses in Africa and India. It produces narrow, interrupted, yellow to translucent streaks, small, chlorotic spots, and severe dwarfing.

corn streak of wheat. A virus disease which causes fine, linear, chlorotic streaks; shortened, curled leaves; excessive tillering; shortened culms and spikes; and sterility.

corn stripe. A virus disease characterized by green and yellow stripes on the leaves.

corn stripe disease of Cuba. A virus disease caused by the virus, *Fractilinea zeae,* characterized by chlorotic markings on the foliage similar to corn streak. See *corn mosaic of Hawaii.*

corn stubble. The basal portion of the stems of corn left after the plants are cut.

corn stunt. A virus disease of corn and teosinte characterized by yellowing, stunting, necrosis, bud proliferation, excessive tillering, and poor development of the tassel and ear.

corn sugar. See *corn sirup; dextrose.*

corn tassel. The terminal inflorescence of the corn plant, borne at the top of the stalk, which bears the male pollen grains.

corn thistle. See *Canada thistle.*

corn wallaby-ear. A virus disease of corn in Australia which produces dark-green, faint mottles and inward curving of the leaves. Narrow, elongated, white galls also develop on the underside of leaf veins.

corn wireworm. See *sand wireworm.*

corn worm. See *corn earworm.*

corn-yield test. A procedure for determining the comparative yields per acre of corn hybrids.

coroica. A strain of corn grown in Bolivia, which produces ears with an odd number of rows, usually 9.

corolla. The petals of a flower considered as a whole.

corollita. See *mountainrose coralvine.*

corona. An appendage or series of appendages on the inner side of the corolla of some flowers, such as the daffodil. Also called *cup; trumpet; coronet.*

coronal root. A root arising just at or slightly above the soil surface from the nodes at the crown of a plant of the grass family. The *coronal roots* comprise the main root system of such plants.

coronamen. See *coronet.*

coronet. The part of a horse's hoof where it joins the skin. Also called *coronamen; coronary band.*

corporate farm. A farm owned and operated by a corporation which may be made up of the same family. See *family farm; tenant farm.*

corporation. A group of persons legally associated for transacting business. Sometimes a large or complicated farm business is incorporated to attain certain advantages of the corporate structure. Share owners are not individually responsible for debts or other obligations of the *corporation.*

corporation farming. Large scale farming carried on by hired managers and labor who apply the efficiency procedures developed in business and manufacturing.

corpus luteum. A yellow mass in the ovary which develops after the egg has been discharged. Also called *yellow body.*

corral. A small enclosure in which cattle are confined for handling, branding, identification, etc.

corrasion. The wearing away of rock material by running water, glaciers, winds, waves, or mass movements. The effectiveness of these agencies is determined in part by the abundance and character of the abrading materials they carry.

correction strip. In strip cropping, a strip of irregular width placed between uniformly wide contour strips of cultivated crops to reduce soil erosion.

Corriedale. A popular crossbreed of sheep developed in New Zealand and Australia from the Romney, Lincoln, or Leicester rams and native or Merino ewes. It is a good meat producer and carries a good fleece in grade of three-eighths to one-half blood. The mature rams weigh up to 250 lbs., and mature ewes up to 185 lbs.

corrosion. 1. The process whereby surface or ground waters, by their own solvent action and by the help of solutes they carry, dissolve or chemically alter rock materials with which they come in contact. 2. The action of an agricultural chemical on the metal parts of distributors and containers.

corrosive sublimate. $HgCl_2$, bichloride of mercury, a highly poisonous salt available commercially in powder or tablet form. It is used agriculturally as an insecticide and fungicide: *e.g.*, for the control of onion maggot and the cabbage maggot, and as a dip for corms, bulbs, and seeds. It also may be used for controlling fungus gnat larvae and earthworms in greenhouses, and to preserve composite milk samples.

corrugated iron. A furrowed sheet of iron used for roofs or walls of farm buildings, or for the manufacture of large tubes, irrigation pipes or flumes.

corrugated roller. See *clod crusher.*

corrugation system. A method of irrigation used primarily for small grain and hay crops. Small irrigation furrows are placed 18 to 36 inches apart to wet the soil between them.

Corsican. A breed of horse. Also called *Sardinian.*

Corsican fir. Corsican pine, *q.v.*

Corsican heath. *Erica stricta,* family Ericaceae, an evergreen shrub, grown in warm areas for its small rose-purple flowers. Native to southern Europe.

Corsican larch. Corsican pine, *q.v.*

Corsican mint. See *requien mint.*

Corsican nettle. See *babytears.*

Corsican pine. *Pinus nigra* var. *poiretiana,* family Pinaceae, a variety of the Austrian pine having a narrow crown. It yields large amounts of resin and is also a source of Burgundy pitch. Also called *Corsican fir; Corsican larch.*

Corsican sandwort. *Arenaria balearica,* family Caryophyllaceae, a very low growing herb, grown for its solitary, white flowers. Native to Balearic Islands.

cortex. 1. The portion of the primary body which surrounds the central cylinder, limited externally by the epidermis and internally by the endodermis. 2. The surface or gray matter of the cerebellum. 3. Groups of cells lying immediately below the connective tissue capsule of the adrenal glands. 4. A pharmaceutical name for the bark and peel of certain plants used for medicinal purposes.

cortical root rot of strawberry. A fungous complex caused by *Coniothyrium fuckelii, Hainesia lythri, Ramularia* spp., *Fusarium orthoceras, Pachybasium candidum, Rhizoctonia solani,* and *Pythium* spp. It produces brown areas on light-colored roots in early spring. The lesions spread, blackening and weakening the entire root system. The plant is dwarfed and, in some cases, killed.

cortisone. A hormone essential in regulating vital functions in the body. It is produced and secreted into the blood by the outer part of the adrenal gland (adrenal cortex) and aids in the mobilization of body proteins and fats and effects carbohydrate metabolism. It also has been synthesized as a chemical compound.

corymb. A flower cluster (inflorescence) in which the flower stalks, around the axial stalk, grow to different lengths producing flat-topped clusters. The marginal flowers are the first to open and the central ones the last.

Coryneum blight of stone fruits. A fungous disease of apricots, peaches, and cherries caused by *Coryneum beijerinckii,* class Fungi Imperfecti, order Melanconiales, family Melanconiaceae. Cream-colored spots, surrounded by red rings, form on the leaves, and on dropping out, leave clean-edged holes in the leaves. Brown lesions, accompanied by profuse gumming, also appear on the twigs, frequently killing them. Severe defoliation occurs, the buds are killed, and the fruit becomes spotted. Endemic in Australia, New Zealand, Africa, and western coast of North America. Also called *corynosis.*

coryza. In fowls, a cold in the head, which causes an acute catarrhal condition and a mucous discharge. See *infectious coryza.*

cos lettuce. *Lactuca sativa* var. *longifolia,* family Compositae, a horticultural variety of lettuce (*q.v.*), with long erect leaves forming a head, which is grown as a salad plant used in regions too warm to grow other types of lettuce. Also called *romaine lettuce.*

cosmos. a. Any plant of the genus Cosmos, family Compositae. Some species are grown as ornamentals in U.S. in summer and fall. Native to Mexico. See *black cosmos; yellow cosmos.* b. *Cosmos bipinnatus,* an annual herb, grown for its white, pink, or red flower rays with yellow disks.

cosset. A lamb raised without the help of its dam.

cost account. A system of accounting in which records are kept of every cash and non-cash cost chargeable to a crop or livestock enterprise to prepare an account showing costs of production, returns, and net profit or loss. It includes records of labor, power and machinery use, building's use, feed, interest charges, etc. Types of *cost accounts* are (a) enterprise *cost accounts,* for a single crop or livestock; and (b) complete *cost accounts,* which under a double entry system of accounting, cover all aspects of the farm business.

Costa Rican nightshade. *Solanum wenlandi,* family Solanaceae, a prickly, climbing shrub grown in very warm areas for its lilac-blue flowers. Native to Costa Rica. Also called *paradise flower.*

costmary chrysanthemum. *Chrysanthemum majus* (*C. balsamita*), family Compositae, a hardy, perennial herb grown for its aromatic foliage and as a salad plant. Sometimes used in flavoring liquors. Native to western Asia. Also called *mint geranium; lavender.*

cote. A shed for sheltering small animals: *e.g.,* a sheep-cote.

coteau. 1. An elevated land, or morainic ridge which may be a divide between two valleys. 2. A low, dry ridge in a swampy area (Louisiana, U.S.).

Cotherstone. A blue-veined cheese similar to Stilton, made from cow's milk in England. Also called *Yorkshire-Stilton.*

cotoneaster. Any shrub or small tree of the genus Cotoneaster, family Rosaceae. Species are grown for ornament. See *sharpleaf cotoneaster; Peking cotoneaster; creeping cotoneaster; cranberry cotoneaster; Diels cotoneaster; spreading cotoneaster; glossy cotoneaster; Franchet cotoneaster; rock cotoneaster; Hupeh cotoneaster; rockspray cotoneaster; silverleaf cotoneaster; redbox cotoneaster; Simons cotoneaster; Himalayan cotoneaster.*

Cotronese. An Italian cheese made from ewe's milk.

Cotswold. A white-faced, long-wool breed of sheep from Gloucestershire, England. Its coarse wool is finer and softer than the Lincoln wool, but lacks the luster of the Leicester wool. Its former usefulness as a good cross-breeder has declined due to the popularity of the Lincoln. The mature ewes weigh up to 225 lbs.

Cottage cheese. A soft cheese prepared by coagulating raw or pasteurized skimmed milk by lactic culture with or without rennet. The broken coagulum is heated sufficiently to form a curd and drain off the whey. The cheese is then washed and salted and used without aging.

cottage lily. See *Madonna lily.*

cottage roll. See *Boston style butt.*

cottagers. See *foxglove.*

cottage tulip. One of the principal classes of tulips which is tall-stemmed, May-flowering, self-colored in a wide variety of colors, with oblong shaped flowers. Also called *May-flowering tulip.*

cotted fleece. Cotted wool.

cotted wool. Tangled or matted wool on a sheep's back.

cotton. 1. Any plant of the genus Gossypium, family Malvaceae. Species are the most important fiber plants in the world. Some are native to Asia, others to Central and South America. Its seed yields an edible oil and the residue is an important stock feed. See *sea-island cotton; Egyptian cotton, upland cotton; Levant cotton.* 2. The fiber from the cotton plant, used extensively in textiles, surgery and crafts.

cotton abroma. *Abroma angusta,* family Sterculiaceae, a shrub which yields a fiber used for cordage. Native to East Indies. Also called *devil's cotton.*

cotton afloat. An obsolete practice of buying and selling futures in which samples of cotton bales were sent from U.S. to England on fast sailing ships which arrived there much earlier than the freighters carrying the sampled bales.

cotton anthracnose. A serious disease caused by the fungus *Glomerella gossypii,* family Gnomoneaceae, which infests all the above-ground parts of the plant. It causes lesions and cankers on the leaves, stem, and bolls, and is sometimes lethal. Also called *boll rot.*

cotton aphid. *Aphis gossypii* Glov., family Aphidae, an aphid that sucks the juice of cotton, cucurbits, okra, and citrus plants. Also called *melon aphid.*

cotton bale. A bound package of compressed cotton lint. It averages 500 lbs. in gross weight. See *bale; compressed bale: a.*

cotton-ball of cranberry. A fungous disease caused by *Sclerotinia oxycocci,* family Sclerotiniaceae, which forms a white, cottony mass of mycelium in the center of the berries. See *Sclerotinia tip blight.*

Cotton Belt. The main cotton-growing region in the U.S. It comprises the states of Florida, South Carolina, Georgia, Alabama, Arkansas, Texas, Louisiana, Missouri, Tennessee, North Carolina, Mississippi, Oklahoma, New Mexico, Arizona, and California.

cotton boll. The pod of the cotton plant in which the lint and seeds form. It is the enlarged, ripened ovary of the plant.

cotton bollworm. See *corn earworm.*

cotton burdock. *Aectium tomentosum,* family Compositae, a biennial plant bearing prickly flower heads.

cotton burs. Bits of dried bolls sticking to seed cotton after it is picked. They are extracted from the lint before ginning.

cotton buyer. An agent who buys cotton in bales, direct from the producer in competitive bidding.

cotton cake. See *cottonseed cake.*

cotton caterpillar. See *cotton leafworm.*

cotton chopper. A person employed on wages or as a share cropper to hoe out weeds and thin out the cotton plants.

cotton classer. A person who grades cotton by estimating the length of cotton fibers in a sample. A sample of lint is drawn from the bale and the fibers are pulled to get a typical fiber length. The price of a bale depends, in part, on the length of the fiber.

cotton cutter. See *stalk cutter.*

cotton cutworm. See *yellow-striped armyworm.*

cotton factor. An agent who buys and sells cotton and lends money to the cotton farmers on behalf of another person or firm.

cotton field. A field or tract of land planted to cotton. Also called *cotton patch.*

cotton flea. Cotton fleahopper.

cotton fleahopper. *Psallus seriatus* (Reut.), family Miridae, an insect that damages cotton plants, especially along the coast of Texas, by puncturing the young squares which turn brown or black and fall from the plant. It also causes abnormal growth and excessive branching.

cotton fuzz. See *linters.*

cotton gin. a. A machine used to separate the cotton seed from the lint. b. A plant where the lint cotton is compressed into bales. Also called *gin; ginhouse.*

cotton ginner. A person who operates and maintains a cotton gin.

cotton gum. Water tupelo, *q.v.*

cotton hopper. See *cotton fleahopper.*

cotton house. A shed or bin for storing cottonseed.

cotton hull ash. The potash-rich residue of burned cottonseed hulls used as a fertilizer.

cotton leaf caterpillar. Cotton leafworm, *q.v.*

cotton leaf curl. A virus disease which causes

curling and mottling of the leaves and shedding of the blossoms.

cotton leaf louse. Cotton aphid, *q.v.*

cotton leaf perforator. *Bucculatrix thurberiella* Busck., family Lyonetiidae, a small moth whose larvae mine the leaves of cotton. Found in Texas, Arizona, and California.

cotton leafworm. *Alabama argillacea* (Hbn.), family Noctuidae, the larva of a moth which infests cotton leaves. The adult sometimes feeds destructively on grapes, peaches, etc.

cotton lint. See *lint.*

cotton louse. Cotton aphid, *q.v.*

cotton mule. A fairly small mule, weighing 750 to 1,110 lbs., used on cotton farms because its small hoofs do not damage the plants.

cotton of immature staple. Cotton lint picked and baled before the fiber has matured. It has a weak staple of low value.

cotton of perished staple. Cotton lint with the usual length of fiber, but weakened by exposure to weather or other causes.

cotton oil. See *cottonseed oil.*

cotton patch. A field in which cotton is growing, often a small field.

cotton pipe. A tube which conveys seed cotton from a wagon bed to the cotton gin by vacuum.

cotton press. A machine for baling cotton for transportation and storage. The various types of *presses* are the screw, toggle, beater, revolving, hydraulic, portable, double acting windlass, rack and pinion, repressing, and rolling pressure presses.

cotton rat. *Sigmodon hispidus,* family Musidae, a long-haired rat, found in southern U.S. and Central America, which damages the citrus trees by its burrowing.

cotton root rot. An important fungous disease caused by *Phymatotrichum omnivorum,* family Moniliceae, which attacks over 2,000 species of plants, especially cotton. The leaves become yellow, and later, wilt and turn brown. In the early stage, white or tan-colored threads of the fungus mycelium appear on the roots which break down the outer cells. As the disease progresses, it destroys the cortical tissues and penetrates into the central cylinder giving it a tan or buff color. In the advanced stage the cortical tissues are completely disintegrated and the central cylinder becomes red or brown as contrasted with the whitish color of the healthy tissue. Common in southwestern U.S. and Mexico. Also called *Texas root rot.*

cottonrose hibiscus. *Hibiscus mutabilis,* family Malvaceae, a shrub, grown in warm areas for its beautiful white or pink blossoms which change to deep-red at night. Native to China. Also called *Confederate rose; cotton rose.*

cotton rust. A potash deficiency in cotton which causes bronzing and marginal browning of the leaves. Later, the leaves become reddish-brown and die; the bolls are dwarfed, do not open, and contain inferior fiber.

cottonseed. The seed of cotton after the lint has been removed. It is a source of cottonseed oil, the residue being used as a stock feed.

cottonseed cake. The solid residue left after the extraction of oil from cotton seeds. It should contain more than 36 percent protein. It is sold according to its nitrogen or protein content. See *nut-size cottonseed cake; sheep-size cottonseed cake; pea-size cottonseed cake; pebble-size cottonseed cake.*

cottonseed cake, off quality. Cottonseed cake which does not fulfill the quality requirements of *cottonseed cake, prime quality* as to color, odor, and texture.

cottonseed cake, prime quality. A livestock feedstuff that must be firm, but not flinty in texture, of sweet odor, free of mold, and when ground into meal, produces cottonseed meal, prime quality. It is designated and sold according to its protein content which must be not less than 36 percent.

cottonseed cake screenings. That prime quality product consisting of fine particles and small pieces of cottonseed cake which have passed through a ⅜ inch round perforation. Protein percentage is designated.

cottonseed cubes. Prime quality cottonseed cake formed into cubes or pellets.

cottonseed feed. A stock and poultry feed consisting of cottonseed meal and cottonseed hulls. It is sold according to its protein content which is usually less than 36 percent, but not less than 22.

cottonseed flour. A food product prepared from cleaned, de-linted and hulled seeds. The seeds are rolled thin and cooked at about 225° F. for 1½ hours which renders the flakes plastic and facilitates pressing out the oil. The pressed cakes are cooled and aged for about 30 days, before grinding in hammer mills. The flour is then screened through a 200-mesh screen. It is light colored with a nutty taste and contains about 57.5 percent protein.

cottonseed hulls. The outer covering of the cottonseed. It is part of the residue after the extraction of the oil and used extensively for a stock feed. *Cottonseed hulls* supply no digestible protein but contain 43.7 percent digestible nutrients; they are fed with protein rich feeds, often cottonseed meal. See *prime cottonseed hulls; off cottonseed hulls.*

cotton seedling blight. See *sore shin.*

cottonseed meal. The residue of cottonseed kernels from which oil has been pressed. Usually containing the portions of the fiber, hull, and oil, left after processing the seed, it is used as a stockfeed and fertilizer. See *solvent extracted cottonseed meal, degossypolized cottonseed meal.*

cottonseed meal, off quality. A livestock feedstuff not fulfilling the requirements for prime quality.

cottonseed meal poisoning. A livestock poisoning, especially in pigs, caused by excessive use of cottonseed meal high in free gossypol. The animal has difficulty in respiration and/or a "potted" belly due to accumulation of fluid in the abdomen and general weakness. Death results apparently from edema of the heart and lungs or liver. Also called *cake poisoning.*

cottonseed meal, prime quality. A finely ground meal of sweet odor, bright yellowish in color, free from excessive lint, and containing not less than 36 percent protein. It is designated and sold according to its protein content.

cottonseed oil. The oil extracted from cottonseed graded and used for human and animal consumption. Within the grades of crude, prime, and choice, there are, *e.g.*, crude cottonseed oil, prime summer white cottonseed oil, and choice summer yellow cottonseed oil.

cottonseed oil mill. A plant at which cottonseed oils and stock feeds are processed from the cotton seed.

cottonseed pellet. See *solvent extracted cottonseed flakes.*

cottonseed tree. See *eastern baccharis.*

cotton sick. Designating land of low productivity because of serious erosion, or infestation by cotton insects, nematodes, or fungi.

cotton sled. A machine with a V-shaped throat which, drawn along the rows of cotton, picks the bolls and directs them into a container: an early, unsuccessful type of cotton-picking machine.

cotton smut. A general, misleading term for the cotton seedling blights.

cotton square. The unopened cotton flower bud with its surrounding bracts.

cotton square borer. *Strymon melinus* (Hbn.), family Lycaenidae, a rare pest of cotton in Texas, whose larvae bore into and eat the cotton squares.

cotton stainer. *Dysdercus suturellus* (H.S.), family Pyrrhocoridae, an insect that punctures the cotton seed inside the growing boll to suck the juices, staining the developing lint in the process.

cotton sweep. See *sweep: 1.*

cottontail. The common American rabbit, *Sylvilagus floridamus,* family Leporidae, a pest which often injures crops, shrubs, bushes, etc., by feeding on them. It is also a carrier of tularemia.

cotton thistle. See *Scotch cottonthistle.*

cotton tree. Black poplar, *q.v.*

cotton wax. A wax obtained from raw cotton fiber; a cheap substitute for beeswax.

cotton weed. See *milkweed: 2.*

cotton wilt. A fungous disease of cotton caused by *Fusarium vasinfectum* which yellows and blights the foliage and kills the plant. Also called *black root; black heart; frenching.* See *Fusarium.*

cottonwood. See *southern poplar.*

cottonwood borer. *Plectrodera scalator* (F.), family Cerambycidae, a beetle that infests the willow and poplar. Common in midwestern U.S.

cottonwood leaf beetle. *Chrysomela scripta* F., family Chrysomelidae, a beetle that infests poplar. Found in midwestern U.S.

cotton wool. Raw cotton, baled but not otherwise processed.

cotton worm. See *cotton leafworm.*

cottony-cushion scale. *Icerya purchasi* Mask., family Margarodidae, a scale insect of Australian origin which infests citrus and other plants in U.S.

cottony jujube. See *India jujube.*

cottony leak of cucumbers. A watery rot of cucumbers, caused by the fungus *Pythium aphanidermatum.*

cottony maple scale. *Pulvinaria innumerabilis* (Rathvy.), family Coccidae, a destructive insect which infests the soft maple in U.S. and Canada. Cottony masses of scale appear along the undersides of twigs and branches of the infested trees in May and June. The entire foliage of the tree turns yellow and the heavily infested branches die. It also attacks the linden, Norway maple, willow, apple, pear, poplar, grape, hackmore, sycamore, honey locust, beech, elm, plum, peach, gooseberry, Virginia creeper, currant, and sumac.

cottony peach scale. *Pulvinaria amygdali* Ckll., family Coccidae, a scale insect which infests peach, prune, plum, and apple in southern U.S.

cottony rot. A contagious disease of stored citrus caused by the fungus *Sclerotinia sclerotiorum.* Infected fruit develops a leathery but pliable skin over which a cottony mass later appears. Finally the entire fruit collapses. Also called *cottony mold; twig blight.*

cotyledon. 1. The first leaf to be developed by the embryo in seed plants. Also called *seed leaf.* 2. The areas of attachment of the fetal placenta to the maternal placenta (carunde) in certain types of ruminants.

couch grass. See *quack grass.*

couch wheat. See *quack grass.*

coughwort. See *coltsfoot.*

coulee. 1. A small creek or run; a sharply cut narrow valley carrying only storm water. Also called *arroyo* (southwestern U.S.) 2. A steep-walled, dry, glacial stream valley; a dry canyon as in the scablands of the Columbia Plateau region of northwestern U.S.

coulee cricket. *Peranabrus scabricollis* (Thos.), family Tettigoniidae, a wingless insect which looks like a field cricket and is very destructive to vegetation in Montana and Washington in U.S.

Coulommiers. A soft, mold-ripened, unwashed cheese made from cow's fresh milk. It is usually eaten fresh or with only a few days' aging.

coulter. See *colter.*

Coulter pine. *Pinus coulteri,* family Pinaceae, a large, evergreen, coniferous tree, grown as an ornamental. Native to California, U.S. Also called *bigcone pine.*

coumarin. $C_9H_6O_2$, a bitter organic chemical present in most sweet clovers, associated with a hemorrhagic disease of livestock that eat spoiled sweetclover hay. It is also found in tonkabeans and is used in scenting tobacco.

counterfeit. Referring to deceptive characteristics of good breeding or physique in an animal which it does not actually possess.

counting out pen. A narrow pen outside the woolshed in which the sheep assigned to a shearer are kept by him for tallying (New Zealand).

country borage. See *Spanishthyme coleus.*

country elevator. An elevator which buys grain from the farmers in a local area. The *country elevator* is the first step in the marketing of grain, and offers such services as weighing, grading, conditioning, drying and cleaning, and providing commercial storage of the grain. It also provides the farmer with market information.

country hide. A hide not uniformly removed from the carcass. It often contains numerous cuts and scores. The curing is not well done resulting in undersalting, hair slips, and maggot infestation. *Cf. packer hide.*

Country Life Commission. A commission organized in 1908 in U.S. to study farm problems and make recommendations for the betterment of rural standards of living.

country life movement. The working out of the desire to make rural civilization effective and satisfying. Its inception in the U.S. dates from the work of the Country Life Commission appointed by President Theodore Roosevelt in 1908.

country packer. A small plant which processes meat, poultry, eggs, etc., on the farm or in small towns for the market.

country-style. a. Describing a method of processing a farm product, such as *country-style* sausage. b. Referring to products processed or prepared on a farm.

county. A political unit in the United States established by a state legislature. A *county* usually contains 16 townships. Many *counties* are irregular in shape and may be smaller or larger than the usual size. The corresponding unit in Louisiana is called a parish.

county agent. An official employed jointly by the county, state, and federal governments in U.S. under the Smith-Lever Act of Congress approved May 8, 1914, to help diffuse practical information on agricultural and allied subjects among farming communities in a county. Men are usually known as *county agricultural agents* and women as *home demonstration agents,* or for brevity: *county agent* and *home agent.*

county brand. A designated, common brand for cattle, used by cowmen to show the place of origin of an animal.

county drain. A constructed drainage system consisting of an open ditch, a closed conduit, or a combination of both that provides a drainage outlet for agricultural and urban areas. It is administered according to the laws of the individual state. See *drainage district; judicial ditch.*

county extension agent. See *county agent.*

county farm. A home for aged indigents, often maintained by a county on a farm.

county seat. The city, town, or village in which all or most of the governmental offices of the county are located.

couple. To attach; to hitch: as to *couple* a trailer to a tractor.

coupled. Referring to the loin of an animal. See *close-coupled; long-coupled.*

coupler. Any device used to connect two or more devices.

coupling. 1. The loin. 2. Mating; breeding; copulation. 3. A coupler.

coupling pole. In a wagon, the shaft which connects the rear axle to the front bolster. Also called *reach.*

coupling rein. The rein or harness strap, used with a team, which connects the draft rein to the inside bridle ring. See *draft rein.*

coupling strap. A checkstrap used to fasten together horses' bits in a team; often used with nervous horses.

course. A horizontal layer, as a layer of lumber in a pile or grain bundles in a stack.

court. See *infection court; central service court.*

cove. a. A small, flat-bottomed valley with a steep head (eastern U.S.). b. A narrow, limestone valley, associated with mountainous ridges of less valuable land in the Appalachian Valley of eastern U.S. c. A narrow strip of prairie land, a finger or re-entrant from a prairie, which extends into a forest. d. A small inlet, or creek, protruding into a shore.

Coventry bells. *Campanula trachelium,* family Campanulaceae, a perennial herb grown for its nodding, bluish-purple flowers. Escaped, it becomes a weed pest. Native to Eurasia.

cover. 1. A lid placed over something for protection. 2. Woods, underbrush, etc., which may conceal game. 3. Plant life, such as grass, small shrubs, herbs, etc., used to protect soil from erosion. 4. The flesh, hide, and fat on a fattened animal. 5. To buy back future contracts. 6. To copulate with a female, as a bull *covers* a cow. 7. To incubate, to hatch eggs, as a hen.

cover boards. Low-grade lumber placed over a pile of lumber during drying for protection of the pile.

cover crop. A crop grown between periods of regular crops for adding organic matter to soil, and/or protection against erosion.

cover cropping. The establishment of close growing crops.

covered kernel smut. A fungous disease of sorghum caused by *Sphacelotheca sorghi,* family Ustilaginaceae, which usually affects the kernels on the smutted plant. The disease produces smutted heads and enlarged, cylindrical or cone-shaped, smutted galls in place of the kernels. The smut galls are light-gray or brown at first, but on breaking, they release dark-brown spores.

covered smut of oats and barley. A fungous disease caused by *Ustilago hordei,* family Ustilaginaceae, which produces spores in balls enclosed by the glumes. The covered smut panicles ripen prematurely, but the spore balls remain intact until shattered in harvesting. Also called *kernel smut of oats.*

covered smut of wheat. See *bunt of wheat.*

covering. 1. See *cover: 4.* 2. The distribution of fleece on a sheep.

covering shovel. See *shovel: 2.*

covering wire. A piece of wire attached to a moldboard plow which drags behind the plow and covers trash and vegetation by the furrow slice.

covering yard. A yard or court used primarily for breeding livestock.
cover pickle. The pickling or saline solution used for curing meats.
cover spray. a. Usually, a spray applied to fruit trees after blossom parts other than the ovary have fallen off. b. Any spray applied to small fruits after the leaves have formed.
covert. 1. A unit of wildlife cover: *e.g.*, a thicket. 2. One of the small feathers covering the bases of large feathers in fowls.
covert escape. Vegetation which serves as a refuge for animals.
covey. A flock of birds: quail, partridge.
Coville creosotebush. *Larrea tridentata,* family Zygophyllaceae, a heavily-scented evergreen shrub grown in desert gardens. Native to southwestern U.S. and Mexico: it has become a nuisance on desert ranges in U.S. Sometimes wrongly called *chico.*
cow. a. Any bovine. b. The mature, female bovine. c. A market term for designating a female bovine which has produced a calf.
cowage velvetbean. *Stizolobium pruritum,* family Leguminosae, an Old World, tropical vine, grown for forage. It is little grown in U.S. except in southern Florida. Also called *cowhage.*
cow alley. The entrance and exit area within the milking room.
cow and calf ranch. A cattle ranch where a breeding herd is maintained and the calves are sold either as vealers or at a slightly older age. See *mixed-herd ranch.*
cowbane. See *spotted waterhemlock.*
cow barn. A barn for housing the dairy cows, usually equipped with stanchions, mangers, gutters, and calf pens, with a storage space for hay and other feed in a loft overhead. See *dairy barn.*
cowbean. See *cowpea: 2.*
cowbell. A bell, usually of brass, strapped around a cow's neck to indicate its location by the tinkling of the bell.
cowbell silene. *Silene cucubalus,* var. *latifolia,* family Caryophyllaceae, a common weed pest found along roadsides and buildings in U.S. It is especially troublesome in clover fields where the clover is raised for seed production. Native to Eurasia. Also called *bladder campion; behen; white bottle; maiden's tears; cowbell; bubble poppy; fairy potatoes.*
cowberry. *Vaccinium vitis-idaea,* family Ericaceae, a prostrate, evergreen shrub grown as an ornamental. Native to Eurasia.
cowboy. 1. A man who works directly with cattle and horses; specifically, the cattle herders on the large ranches of western U.S. and Canada. Sometimes called a *cowpoke* or a *cowpuncher.* 2. The boy who milks cows (New Zealand).
cow brute. Cow, steer, bull, calf (southwestern U.S.).
cow cabbage. See *cowparsnip.*
cow cockle. See *cow soapwort.*
cow college. A sarcastic term for an agricultural college or university (U.S.).
cow country. See *cattle country.*
cow crack. A crack in the front of the hoof of a horse.
cow cress. See *field pepperweed.*
cow crowd. A herd of cattle (southwestern U.S.).
cow drive. See *round-up.*
cow-eyed. Describing a condition of poultry indicating that they are being properly fed. The eyes of the fowls are large, soft, and luminous.
cow grass. See *zigzag clover.*
cowhand. See *cowboy.*
cow heifer. A cow that has calved only once.
cowherb. See *cow soapwort.*
cowherd(er). A person who tends cattle; *e.g.*, a cowboy (seldom used in the U.S.).
cowhide. 1. A hide from a cow, steer, or bull. 2. A coarse, leather whip. 3. To whip an animal with *cowhide.*
cow-hocked. Describing a condition of a horse or cow in which the hocks are close together and the fetlocks wide apart.
cow horse. A horse or pony trained in rounding up cattle, etc., usually on the western range in U.S. See *stock horse: 1.*
cow hunt. See *round-up.*
cowitch. See *cowage velvetbean.*
cowl. A curved, flaring top fitted to a ventilator, turned by a wind vane in a down wind direction to improve drafts.
cowlily. See *spatterdock cow lily.*
cow month. A unit of measurement of the feed or forage acreage necessary to maintain a mature cow for 30 days.
cow mumble. See *cowparsnip.*
cow oak. See *chestnut oak.*
cowparsnip. *Heracleum lanatum,* family Umbelliferae, a perennial herb which is a weed suspected of causing dermatitis. Native to North America. Also called *hogweed; cow cabbage.*
cow-pat. Cow dung.
cow-path. A narrow path used by cattle, horses, etc., in going to and from the pasture.
cowpea. 1. Any bean of the genus *Vigna,* family Leguminosae. 2. *Vigna sinensis,* a hot-weather legume grown for food, forage, and green manure. Two of the several edible groups are the blackeye pea group and the cream crowder pea group. The latter are crowded in the pod resulting in blunted ends and are about one-third the size of the former. Also called *cornfield pea; blackeyed pea; blackeyed bean; blackeyed cowpea; cowbean.*
cowpea aphid. *Aphis medicaginis* Koch., family Aphidae, a plant louse which infests the cowpea and sometimes cotton.
cowpea curculio. *Chalcodermus aeneus* Boh., family Curculionidae, an insect pest which lays its eggs in the pods of cowpeas, string beans, and lima beans, causing wormy peas and beans. It also infests cotton and strawberry. Found in southern U.S.
cowpea huller. A special device for removing the cowpea seed from the hulls or pods. Also called *cowpea hulling machine.*

cowpea mosaic. See *crotalaria mosaic.*

cowpea pod weevil. See *cowpea curculio.*

cowpea weevil. *Callosobruchus maculatus* (F.), family Bruchidae, a weevil which is very destructive to cowpeas. The female deposits her eggs in the cowpeas in the field, and the larvae eat out the insides of the seed. It also attacks other beans and peas in storage.

cowpea wilt. A fungous disease caused by *Fusarium tracheiphilum* which affects the vascular system of the plant, causing it to wilt and die.

cowpenning. An old practice of manuring a patch of land by penning cattle on it. The cattle are moved to a new location when the land is sufficiently manured for cropping. The practice is still carried on in parts of southeastern U.S.

cow poison. See *Columbia River larkspur.*

cow poke. 1. A metal or wooden device used to discourage fence breaking by cattle. It is fastened around the animal's neck and has two hooked arms extending upward and downward. A barb on each of the arms points toward the animal so that when it attempts to put its head through a fence, it receives a sharp prick from the barb. 2. See *cowboy: 1.*

cow pony. Cow horse, *q.v.*

cow-pox. A contagious eruption appearing on the teats and udders of cows caused by a filterable virus, similar to the smallpox virus in man. It spreads through contact with contaminated premises, cattle, or milkers' hands. Cows have been known to contract the disease from people recently vaccinated for smallpox. Man sometimes contracts a pox-like disease from the afflicted cows.

cowpuncher. See *cowboy: 1.*

cow ranch. A cattle ranch running many cattle, largely cows and calves, which grows most of the crops consumed by the animals.

cow sense. The ability or intelligence in some horses which show unusual dexterity in working with cattle.

cowslip. See *marshmarigold; cowslip primrose.*

cowslip lungwort. *Pulmonaria angustifolia,* family Boraginaceae, a perennial herb grown as an ornamental for its blue flowers. Native to Europe.

cowslip primrose. *Primula varis,* family Primulaceae, a low, perennial plant with clusters of fragrant yellow flowers, grown as an ornamental, a salad plant, and a potherb in England. Native to Europe. Also called *cowslip; English cowslip; English primrose; fairy cup.*

cow soapwort. *Saponaria vaccaria* (*Vaccaria vulgaris*), family Caryophyllaceae, an annual herb bearing deep pink flowers, grown as an ornamental. Escaped, it becomes a weed. Its seeds are toxic and most animals refuse to eat grain containing these seeds. Native to Europe. Also called *cowherb; cow cockle; cockle; soapwort.*

cow sorrel. See *sheep sorrel.*

cowspanker. A dairy farmer or cow milker (New Zealand).

cowtail. Coarse grade wool sheared from the hind legs of the sheep.

cow tongue. *Clintonia borealis,* family Liliaceae, a common spreading herb bearing greenish-yellow flowers and blue berries, grown as an ornamental in U.S. Native to eastern North America. Also called *yellow clintonia; yellow beadlily.*

cow town. A village or small town in western U.S. in the last century and in the early days of the 20th century serving principally as a trading point for cattle ranchers and cowhands.

cow vetch. *Vicia cracca,* family Leguminosae, a perennial, climbing herb, with violet-purple or white flowers, grown as an ornamental cover crop, or for green manure and for forage. Also called *bird vetch; crow vetch.* Native to Eurasia: naturalized in North America.

cow waddie. See *cowboy.*

cowy. 1. See *barny.* 2. A bull which lacks masculinity.

cowyard. See *barnyard.*

coyote. *Canis latrans,* a species of wolf which is a predator of sheep, goats and chickens. However, it also keeps the rabbit population in check.

coyote brush. See *kidneywort baccharis.*

coyote tobacco. *Nicotiana attenuata,* family Solanaceae, a poisonous plant widespread in western U.S. in dry, sandy soils and along stream beds. Animals eating it suffer from vomiting, retching, salivation, bloating, diarrhea, colic, weakness, staggering, falling, lameness, spasms, paleness, stupor and violent palpitation of the heart. Also called *wild tobacco.*

crab. 1. A short shaft or axle with squared ends to receive hand cranks. It is mounted on a frame and used to raise a load by winding up the rope tied to it. 2. See *crabapple: 2.*

crabapple. 1. Any of the several species of genus Malus, family Rosaceae, cultivated as ornamentals for their uncommonly beautiful flowers. Next to the Japanese flowering cherries, they are considered the most beautiful ornamentals. See *Bechtel's crabapple; southern crabapple; carmine crabapple; Siberian crabapple; Manchurian crabapple; wild sweet crabapple; Japanese flowering crabapple; Halls crabapple; midget crabapple; Sargent crabapple; Toringo crabapple; Chinese flowering crabapple.* 2. A group of apples with small fruits that are of two types: Siberian, descended wholly or in part from the Siberian crabapple (*Malus baccata*), and the native types, descended in part from American species, principally the prairie crabapple (*Malus ioensis*). They are acid in flavor, and their high pectin content renders them especially useful in making preserves. Also called *crab; choke apple.*

crab cactus. See *Christmascactus.*

crab-eye (vine). See *jequirity rosarypea.*

crabgrass. Any of the several species of genus Digitaria, family Gramineae. Species are serious weed pests of lawns and fields.

crab meal. A marine poultry feed derived from undecomposed, dried waste of the crab industry. The contents are the shell, viscera, and all of the flesh. The product is more than 25 percent protein and contains up to 3 percent common salt.

Salt exceeding this amount shall be so labeled; in no case shall the salt content exceed 7 percent.

crab stock. A seedling apple plant used as grafting stock.

crab tree. See *crabapple: 1.*

cracked. A market designation for an egg with a broken shell but whose membrane is intact and whose contents do not leak.

cracked corn. See *corn chop.*

cracked heel. See *scratches.*

cracked oat groats. See *cut oat groats.*

cracked paraffin. See *checked paraffin.*

cracked rind. A defect in the finish and appearance of a cheese in which cracks or checks appear on the sides. Such openings allow the cheese to dry, permit the entrance of mold, and harbor pests.

cracked stem of celery. A boron deficiency disease which is characterized by brown mottling of the leaves, brown stripes on the stalks, breaking and curling of the epidermis, and browning and dying back of the roots.

cracker. 1. A rustic, a backwoodsman, or poor white (southeastern U.S.). 2. A short piece of rawhide, flax, silk, etc., attached to the end of a whip.

cracking. 1. A fissure which develops as wet soil dries and bakes in the sun. 2. The development of fissures in the skin of fruits which usually occurs during rainy periods at or just before ripening. Also called *fruit cracking.*

crackling. A food delicacy consisting of the pressed, rendered, pork fat left after extracting the lard.

cracks. See *check: 5.*

cradle. 1. A framework of several slender wooden fingers about 2½ feet long attached to a scythe to gather the grain into bunches and to protect it from shattering as it is cut. 2. A large, smooth area leveled out for felling large trees without breaking. 3. Any framework used as a support. 4. A framework of ten or twelve rigid sticks tied to an animal's neck to prevent it from turning its head sideways. It prevents self-sucking and tearing off of bandages, etc.

cradle knoll. A mound or ridge of earth on a forest floor often no more than a foot or two high; probably caused by the overthrow of large trees and consequent accumulation of the soil in the root mat through rainfall, frost action, and decay of the wood (Lake States, U.S.).

cradle swinger. A man who cuts grain with a cradle.

cradling. The cutting of standing stalks of grain with a cradle.

crambe. *Crambe maritima,* family Cruciferae, a hardy, perennial plant whose young, blanched, edible shoots are used like asparagus. Native to Europe. Also called *sea kale.*

cramberry. Cranberry.

cramming. The force feeding of poultry for two or three weeks to produce a superfat carcass; a common practice in Europe.

cramming machine. A device for force feeding of poultry which consists of a tripod and a tank to hold the ration. The feed flows into a pipe connected to the tank and is forced into the throat of the bird by a foot- or machine-operated plunger.

cramp colic. See *acute indigestion in horses.*

crampy. a. A stringhalted horse. b. Any stiff-limbed animal. c. A joint-type lameness in bulls, probably due to arthritis.

cranberry. *Vaccinium macrocarpum,* family Ericaceae, a bog plant grown for its edible fruit, used in jellies, jams, fruit juice drinks, and salads. Native to North America. See *small cranberry.*

cranberry barrel. A standardized container of 5,826 cubic ins. capacity, used for marketing cranberries.

cranberry bean. A mottled-red colored variety of kidney bean (*Phaseolus vulgaris*).

cranberry bog. A low-lying, wet area of peat, muck, or gray, sandy, acid soil capable of being flooded and drained periodically which is used for the production of cranberries.

cranberry bush. See *American cranberrybush viburnum.*

cranberry cotoneaster. *Cotoneaster apiculata,* family Rosaceae, a deciduous shrub, with beautiful foliage, pink flowers, and scarlet fruit, grown as an ornamental in U.S. Native to China.

cranberry false blossom. See *false blossom.*

cranberry fruitworm. *Acrobasis vaccinii* (Riley), family Phycitidae, the larva of a moth which eats its way into the cranberry fruit in midsummer, closing the hole with silk. Infested fruit is shriveled, reducing the yield.

cranberry gall. *Synchytrium vaccinii,* family Synchytriaceae, a fungous disease in which small, reddish galls are formed on the leaves, stems, flowers, and fruit of the plant. It also attacks some related species.

cranberry juice. A food product prepared by hot or cold extraction of cranberry juice. In the former process, the berries, after addition of water, are heated for 10 minutes. The pulp is then pressed in a cider press, strained, and diluted with water. Sugar is then added to give a sugar content of 15 to 20 percent.

cranberry rootworm. *Rhabdopterus picipes* (Oliv.), family Chrysomelidae, a beetle's larvae which feed at night on young foliage of rhododendron, camellia, and other ornamentals, making holes. It is a less serious pest of cranberry.

cranberry rot. See *blotch rot.*

cranberry scoop. A comblike device for raking cranberries from the bushes. See *Wisconsin rake.*

cranberry weevil. *Anthonomus musculussay,* family Curculionidae, an insect pest which eats out the fruit buds of cranberry. The few injured buds that develop form a rosette of misshapen leaves. The crop may be reduced up to 50 percent.

cranberry tree. See *European cranberrybush viburnum.*

cranberry worm. See *yellow-headed fireworm.*

cranesbill. See *geranium.*

crank duster. A small, hand-cranked duster for spreading poison dust for insect or disease control.

crapejasmine tabernaemontana. *Tabernaemontana coronaria* (*Ervatamia c.*), family Apocynaceae, a tropical shrub grown in warm areas for its white flowers. Native to India. Also called *East Indian rose bay; Adam's-apple; Nero's crown; clavel de l'Inde.*

crape myrtle. *Lagerstroemia indica,* family Lythraceae, a deciduous shrub grown in warm areas for its clusters of long-blooming, pink flowers. Native to China.

crape myrtle aphid. *Myzocallis kahawaluokalani* Kirk, family Aphidae, a plant louse which infests the crape myrtle. Honeydew from the aphid's action encourages fungous diseases.

crate. A wooden box; a framework of slats or a box without a top, used for transporting or measuring various farm products.

crate storage. A type of mound storage in which vegetables are put in crates and covered with straw and soil.

craw. See *crop: 4.*

crawfoot. See *tall buttercup.*

crawler. 1. A newly hatched insect. 2. A large earthworm. 3. A tractor equipped with tracks, as compared to one equipped with wheels.

crawler-type cane wagon. A wagon with track-covered wheels used for transporting heavy loads over soft ground, used in distributing seed cane in the Florida Everglades region in U.S.

crayfish (crawfish) land. Flat, wet land, underlaid by gray, clayey soil, which is characterized by holes and tiny mounds made by crayfish, the fresh water crustacean. (Used loosely as a term for poor agricultural land.)

crazy chick disease. A vitamin E deficiency of chicks up to 8 weeks of age, which causes nervousness and tremors. Affected chicks sit on the hocks, move backwards by pushing their feet, wheel in circles, fall frequently, and sometimes lie prostrate. Also called *nutritional encephalomalacia; epidemic tremors; encephalomalacia.*

crazy top. A non-parasitic disorder of cotton grown on calcareous soils. The plant growth, checked by water shortage, becomes normal when water is abundant. Abnormal branching and fruiting occurs on the upper parts of the affected plants. Fruiting branches usually are replaced by straight-growing, abnormal vegetative branches. The leaves, internodes, bracts, and floral parts are distorted. Flowers produced are misshapen and small. Leaves become round, small, thickened, and indented. The bolls that develop are small and malformed. Also called *acromania.*

crazyweed. See *hemp; two-grooved loco; timber poisonvetch; blue loco; Earl loco; specklepod loco; woolly loco; sheep loco; Hayden poisonvetch; fourwing poisonvetch; Wooton loco; Lambert crazyweed; Rocky Mountain crazyweed; spike crazyweed.*

cream. The sweet, fatty liquid or semi-liquid separated from cows' milk, containing more than 18 percent of milk fat. Also called *sweet cream.*

creambush rockspirea. *Holodiscus discolor,* family Rosaceae, a deciduous shrub grown for its clusters of creamy white flowers. Native to western North America.

Cream Cheese. A soft, uncured cheese prepared from milk or a mixture of cream and milk, used as a spread and a constituent in salads. A popular soft cheese in the U.S.

cream crowder pea. See *cowpea: 2.*

creamed buttermilk. Cultured buttermilk with cream added before heating or at the time of breaking of the curd to increase the fat content from 1½ to 2 percent.

creamed honey. Extracted honey which has become granulated but keeps a soft, creamy texture.

creamer. Any device used for separating cream from milk.

creamery. A commercial establishment which buys milk and/or cream and prepares finished dairy products for the market.

creamery butter. Butter produced in a factory or creamery from cream or milk gathered from many dairy farms, as contrasted with home-churned or farm-made butter.

cream layer. The fatty portion of milk which rises to the top, after milk has set for several hours in a bottle or other receptacle.

cream line. The line of demarcation between the layer of cream and the skim milk.

cream-of-tartar (tree). *Adansonia gregorii,* family Bombaceae, a large tree grown for shade and for its edible acid fruit. Native to Australia. Also called *Australian baobab.*

creamometer. A graduated glass tube used to make a rough estimate of the fat content of milk.

cream plug. A defect of bottled milk or cream in which the partially churned cream adheres to the bottle at the closure making pouring difficult.

cream ripening. A butter-making process consisting of adding a pure, exclusively developed culture of certain organisms to the cream to be churned. It improves the aroma and flavor of butter.

cream screw. An orifice type set screw on a cream separator bowl which controls the outward flow of the cream, and adjusts the ratio of cream to skim milk.

cream separator. A centrifugal machine which is used for the separation of cream from milk. See *centrifugal: 1.*

cream stage. A stage in the ripening of corn in which the kernel is developed but when mashed, has a creamy consistency. See *milk stage.*

cream-style corn. A method of canning corn in which the kernels are cut off the cob and the portions remaining on the cob are scraped off and added to the cut kernels.

cream test. See *fat test.*

cream tube. Creamometer.

crease. 1. The lengthwise indentation on the kernel of wheat or rye. 2. A slot or countersink in the nail hole of a horseshoe into which the head of the nail fits.

creaser. 1. A tool used in horseshoeing to put a shallow depression in a horseshoe in which to countersink the nails. See *crease: 2.* 2. A method of capturing a wild horse used in the U.S. in the last century. A bullet is fired into the ridge of the horse's neck temporarily stunning it.

creasing. Capturing a wild horse by stunning it through shooting into the crest of the neck (western U.S.).

creasing and puffing. A physiological malady of citrus fruits characterized by irregular grooves on the fruit, probably due to climatic or soil conditions.

creatine test. The test for acetylmethyl carbinol plus diacetyl to determine the relative amounts of aroma-producing substances in butter starters.

credit insurance. Insurance taken out and paid for by the borrower to provide for the payment of the loan in case of the death of the insured borrower. Of the two major types of this insurance, one pays the total amount of the loan and the other pays the unpaid balance.

creek. A small stream or river. The term is relative, since in one section of the U.S. a stream of water called a *creek* may be called a river in another.

creek bottom. See *bottom: 1.*

creek bottom land. The floor of a valley occupied by a creek. Usually alluvium on a flood-plain.

creep. 1. See *creep feeder.* 2. A phosphorus deficiency disease of animals which is characterized by anemia, softening of the bones and a slow gait. Also called *bush disease.* 3. Moving of the check wire in conjunction with the corn planter as the planter moves across the field. 4. To spread, as a plant, by means of creeping stems which root at the nodes.

creeper. 1. A plant which spreads over or under the ground surface by roots or branch stems. 2. A vine. 3. Any plant which is prostrate or trailing in its habit of growth. 4. A chicken with a lethal genetic factor resulting in short, deformed legs and wings. It usually dies within four days of hatching.

creep erosion. A form of natural erosion. The slow gravitational movement of detritus of weathering downslope, mainly as a result of freezing and thawing, micromovement from saturation, and other causes. Distinguished from mass slippage avalanche movement, and wash from rapid rainfall run-off erosion.

creep feeder. A specially constructed feeding pen which permits only young animals to enter and feed.

creep feed(ing). A method of feeding young livestock in which the feed is placed inside an enclosure large enough to admit only young animals.

creeping. Referring to plants with prostrate or trailing habits of growth.

creeping baby's breath. Creeping gypsophila.

creeping barberry. Creeping mahonia.

creeping bellflower. *Campanula rapunculoides,* family Campanulaceae, a perennial herb grown for its terminally-hanging, bell-shaped, blue or violet-purple flowers. Escaped, it becomes a weed pest. Native to Eurasia. Also called *European bellflower; false rampion; rover bellflower.*

creeping bentgrass. *Agrostis palustris,* family Gramineae, a perennial grass used on putting greens of golf courses and, mixed with other grasses, for lawns and turf. Native to North America.

creeping bur. Running pine.

creeping buttercup. *Ranunculus repens,* family Ranunculaceae, a poisonous plant with yellow flowers which may cause a dermatitis on touch, and when eaten, may result in a severe inflammation of the digestive tract. The poisonous principle is volatile and does not remain in the dried herbage: hay containing the *creeping buttercup* is safe for livestock. Native to Europe and North America.

creeping chalk plant. Creeping gypsophila.

creeping Charlie. A common name for several low-growing plants such as European glorybind, ground ivy, moneywort, etc.

creeping cotoneaster. *Cotoneaster adpressa,* family Rosaceae, a deciduous, prostrate shrub grown as an ornamental for its pinkish flowers and red fruit. Native to China.

creeping crowfoot. Creeping buttercup.

creeping fescue. See *red fescue.*

creeping fever. See *equine infectious anemia.*

creeping fig. Climbing fig.

creeping fire. A slow-spreading fire.

creeping gypsophila. *Gypsophila repens,* family Caryophyllaceae, a trailing, perennial herb grown as an ornamental ground cover for its habit and its white to reddish flowers. Native to Europe.

creeping forget-me-not. See *creeping navelseed.*

creeping Jennie (Jenny). Common name for the low-growing plants balsam apple, European glorybind, and moneywort.

creeping juniper. *Juniperus horizontalis,* family Pinaceae, a prostrate, evergreen shrub with blue-green leaves; grown as an ornamental. Native to North America.

creeping lily-turf. Creeping liriope.

creeping liriope. *Liriope spicata,* family Liliaceae, a perennial herb grown as an ornamental ground cover in warm regions. Native to Asia. Also called *creeping lily-turf.*

creeping loosestrife. Moneywort.

creeping mahonia. *Mahonia repens,* family Berberidaceae, a small evergreen shrub often planted around houses as shrubbery borders. Native to western North America. Also called *dwarf hollygrape.*

creeping myrtle. See *periwinkle.*

creeping navelseed. *Omphalodes verna,* family Boraginaceae, a perennial plant with blue flowers, often used as a ground cover. Native to Europe. Also called *creeping forget-me-not.*

creeping oxalis. *Oxalis corniculata,* family Oxalidaceae, a creeping, perennial herb with yellow flowers which is a weed pest in America. Native to Europe. Also called *yellow oxalis; yellow wood sorrel.*

creeping pearlberry. *Chiogenes hispidula,* family Ericaceae, a small, evergreen, trailing plant with bright green leaves and white flowers. Native to North America. Also called *teaberry; creeping snowberry; running birch; moxieberry; moxie plum; capillaire.*

creeping St. Johnswort. *Hypericum repens,* family Hypericaceae, a prostrate, perennial herb with yellow flowers in terminal clusters, grown in warm areas as an ornamental. Native to Turkey and the Balkans.

creeping Sally. Moneywort.

creeping saltbush. Australian saltbush.

creeping skyflower. *Duranta repens,* family Verbenaceae, a large shrub grown for its trailing habit and lilac flower clusters. Native to the warm regions of America. Also called *tropical lilac; pigeonberry; golden dewdrop; skyflower.*

creeping snowberry. Creeping pearlberry.

creeping speedwell. *Veronica repens,* family Scrophulariaceae, a prostrate, perennial herb grown for its bluish or rosy flowers. Native to Corsica.

creeping thistle. Canada thistle.

creeping thyme. See *mother-of-thyme.*

creeping wildrye. *Elymus triticoides,* family Gramineae, a forage grass used for reseeding western ranges in U.S. Also called *beardless wildrye.*

creeping wintergreen. Checkerberry.

creeps. See *rickets.*

cremation. The burning or incineration of dead bodies.

crenate leaves. Leaves with short, rounded teeth, or scalloped edges.

crenshaw melon. A horticultural variety of winter melon. See *muskmelon.*

Creole. A soft, unripened, cottage cheese prepared by mixing equal parts of cottage-type curd and rich cream.

Creole flint. A yellow variety of tropical flint corn.

creosote. An antiseptic liquid, a wood preservative, obtained by the distillation of wood tar or coal tar. It is also a mild insecticide. See *coal-tar creosote.*

creosote bush. See *Coville creosotebush.*

creosote emulsion. A wood preservative in emulsified form composed of creosote and water.

creosote paint. A liquid paint containing creosote. It is used to paint areas or cavities in a tree where fungi may have become established.

creosote-petroleum. A preservative mixture of petroleum and coal-tar creosote commonly used for treating crossties.

crepe myrtle. See *crape myrtle.*

crescent. The condition of a horse's hoof in which the enclosed bone slips downward causing spreading of the hoof.

crescent-marked lily aphid. *Myzus circumflexus* (Buck.), family Aphidae, a plant louse which infests greenhouse lilies, asparagus fern, aster, California laurel, callalily, crocus, cyclamen, and many other plants indoors and out. It is a vector of mosaic and other virus diseases.

Crescenza. An uncooked, soft, creamy, mildly sweet, fast ripening, yellowish cheese similar to Bel Paese, made of cow's whole milk coagulated with rennet.

cresols. Oily liquids or solids derived from coal tar or wood commonly used as disinfectants and insecticides. Classified chemically as metacresols, paracresols, and orthocresols. Obtained commercially from the coal tar and petroleum industry.

cress. A name commonly applied to certain members of the family Cruciferae which have pungent, edible leaves; occasionally used in reference to other plants having the same characteristics. See *watercress; gardencress peppergrass; early wintercress; bitter wintercress; wall rockcress; Kellerer rockcress; downy rockcress; crambe; Jericho resurrectionmustard.*

crest. 1. The tuft of feathers on the heads of certain fowls. 2. A cock's comb. 3. The ridge of an animal's neck. 4. The top of a dam, dike, spillway, or weir. 5. The peak or high water mark of a flood. 6. A toothed or irregular ridge or appendage on the petals or flower cluster of certain plants.

crested. 1. Pertaining to flowers which bear a crest. 2. Describing animals with ridged necks. 3. Referring to the feathered tuft on the heads of certain fowls.

crested iris. *Iris cristata,* family Iridaceae, a perennial herb, grown in rock gardens for its terminal, light lilac-colored flowers. Native to southeastern U.S.

crested pricklepoppy. *Argemone platyceras,* family Papaveraceae, a tropical, annual herb grown for its white flowers. Native to America.

crested shield fern. See *crested woodfern.*

crested wheat grass. *Agropyron cristatum,* family Gramineae, a hardy, perennial bunchgrass which produces an abundance of leaves. It grows well in the higher elevations of western U.S., its palatability and high feed value making it one of the most valuable of the forage grasses. Native to Russia.

crested woodfern. *Dryopteris cristata,* family Polypodiaceae, a beautiful fern for gardens. Native from Maine to Minnesota and southward to Louisiana in U.S.

crest-feathering whorl. An inherited defect of swine in which featherlike whorls or spirals of hair occur along the topline.

creston. A ridge running across a slope (southwestern U.S.).

cresylic acid. An unpurified mixture of the three cresols, used as an insecticide for the control of aphids and as a component in sheep and cattle dips.

cresylic type. A disinfectant that depends for effect mainly on its content of cresylic acid, although it usually contains some xylenol, *e.g.,* lysol, creolin, and cresolin.

Cretan bear's-tail. Bearstail celsia.

Cretan celsia. *Celsia cretica,* family Scrophulariaceae, a biennial herb grown for its yellow, purple-marked flowers. Also called *Cretan mullein.*

Cretan mullein. Cretan celsia.

Cretan spikenard. Caucasian valerian.

cretinism. 1. A severe deficiency of thyroid secretion dating from birth or early life which results in arrested physical and mental development, dwarfism, sluggishness, thickened tongue, etc. 2. Used to describe the physical characteristics of the deficiency. See *hypothyroidism.*

Creuse. A dry, firm cheese made with skim milk and rennet, usually aged for more than a year.

crevasse. A break in a levee.

Crevecoeur. A breed of chicken with greenish-black plumage and white skin which produces white eggs. It originated in Normandy. The cock weighs up to 8 lbs. and the hen up to 7.

crib. 1. A granary for the storage of ear corn; a *corn crib.* See *single crib; double crib; double crib with overhead bin.* Also, the act of placing ear corn in a crib. 2. An orderly pile of lumber, the boards of which are laid flat so that they form a triangle with only the ends of the boards touching. 3. A container from which animals feed; a manger.

cribber. A horse which has the nervous habit of biting some stationary object, usually the crib, while holding its neck muscles rigid and noisily sucking air into its stomach. The condition is referred to as *wind sucking; cribbing; crib biting.*

crib dam. A water barrier made of a timber framework containing stone, etc.

crib-wall. 1. A type of wall construction for grain bins; 2 in. boards are laid flat, one on top of the other, and spiked together. 2. A slatted wall used to provide ventilation in ear corn storage cribs.

cricket. A large group of leaping insects, of the family Gryllidae, some of which are nocturnal and known for their chirping. The most common is the black fieldcricket.

Crimean iris. *Iris chamaeiris,* family Iridaceae, a perennial herb grown for its various colored flowers. It is a bearded iris.

Crimean linden. *Tilia euchlora,* family Tiliaceae, a deciduous, hybrid tree grown as a shade tree, street tree, and ornamental.

crimp. The natural curl or waviness in wool fiber.

crimped oats. Oat chop.

crimp of strawberries. See *spring dwarf of strawberry; summer dwarf of strawberry.*

crimps. See *collapse: 2.*

crimp-wired comb foundation. A non-sagging brood foundation for beehives which has steel wires imbedded in each sheet of wax to hold the combs straight. It reduces the number of drone cells.

crimson clover. *Trifolium incarnatum,* family Leguminosae, an annual herb grown chiefly in the southeastern U.S. and along the Atlantic seaboard as a cover crop and green manure crop. Also called *annual clover; scarlet clover; incarnate clover; Italian clover; German clover.*

crimson rambler. *Rosa multiflora* var. *cathayensis,* family Rosaceae, a hardy, climbing or trailing shrub grown for its deep pink flowers. Native to Asia. Also called *Cathay Japanese rose; Chinese rose.*

crimson sage. *Salvia spathacea,* family Labiatae, a perennial herb cultivated for its use as a bee plant and an ornamental. Also called *bee sage; Andibertia grandiflora.*

crimson weigela. *Weigela floribunda,* family Caprifoliaceae, a shrub grown as an ornamental for its dark crimson flowers. Native to Japan.

crinkle. 1. See *cork: 3.* 2. A widespread, virus disease of strawberries characterized by wrinkling of the leaves and pinhole spots which enlarge. The veins become translucent and the leaf takes on a yellowish color. 3. A virus disease of chrysanthemums characterized by crinkled leaves and stunted growth. Sometimes there is a yellowish-green vein banding which is followed by a mottling and reduction of leaf size. There may be a blast of buds and leaves as well as a dieback of young shoots. 4. A virus disease of potatoes characterized by the mottling and yellowing of the leaves, dwarfing of the plant and, frequently, a wavy leaf margin. The leaves roll downward and there is a stem and leaf necrosis.

crinkled flannel moth. *Megalopyge crispata* (Pack.), family Megalopygidae, a moth, which, in the caterpillar stage, feeds on the apple, bayberry, birch, cherry, locust, oak, raspberry and sweetfern.

crinkled garden lettuce. *Lactuca sativa* var. *crispa,* family Compositae, an annual herb grown popularly as a salad plant. Also called *garden lettuce; leaf lettuce.*

crinkle leaf. A nonparasitic disorder of cotton characterized by a puckering, mottling, chlorosis, and distortion of the leaves. The branches may be twisted and the floral buds, flowers, and bolls distorted. The lint is weak and worthless.

crinkle of geranium. A graft-transmissible virus disease of greenhouse geraniums characterized by crinkled or curled leaves. The young leaves are wrinkled and marked with hyaline spots composed of white hypertrophic cells. Apparently also transmissible by whiteflies.

crinkleroot toothwort. *Dentaria diphylla,* family Cruciferae, a perennial herb cultivated in shady places for its white flowers, pinkish inside. Native to eastern Canada and southeastern U.S. Also called *pepper-root.*

crinkly leaf. A virus disease of citrus fruits, principally lemons, which causes the leaves to become pocketed and warped and may cause bark lesions.

Criollo. An Argentinian breed of horse which originated from Spain. A horse of this breed may weigh up to 990 lbs. and be 60 ins. in height and close-coupled.

cripple ail. An old, common name for a disease of animals which may be the same as osteomalacia *q.v.*

crippled. 1. Pertaining to an animal having difficulty in movement caused by a physiological condition, disease, or injury. 2. An imperfect flower.

crips. 1. A lame animal. 2. An animal injured in transit (U.S.).

Crismer test. A test to determine the chemical nature of butter and fats.

crissum. a. The area surrounding the cloaca on a bird. b. The feathers around the cloaca.

critical age. The time when decay begins to increase alarmingly in a stand of trees.

critical day length. Critical photoperiod.

critical depth. 1. The depth of water in a channel which corresponds to one of the recognized critical velocities, ordinarily, Belanger's *critical depth:* the depth at which a given flow would attain its minimum energy with respect to the bottom of the channel. 2. The depth at which in a given channel, a given quantity of water flows with minimum energy.

critical element. Essential element.

critical flow. A condition of flow for which the mean velocity is at one of the critical values; *i.e.,* ordinarily, flow at Belanger's critical depth and velocity. Another important usage is in reference to Reynold's critical velocities, *i.e.,* the point at which the flow changes from stream line or nonturbulent flow to turbulent flow. See *critical velocity; supercritical flow.*

critical light period. The minimum amount of daily light required by certain plants for growth and flowering.

critical photoperiod. Critical light period for plants. It is shorter in the tropics and longer in the arctic region.

critical size. The maximum grain size of sand that can be transported by a given velocity of water.

critical temperature. The environmental temperature below which the heat production in a warm-blooded animal increases to prevent a lowering of body temperature.

critical velocity. The velocity at which some basic change occurs in the nature of flow of liquids. *Critical velocities* are commonly determined with reference to Belanger's *critical velocity,* a velocity at which the flowing liquid possesses the minimum possible energy.

critter. 1. Cattle. 2. Any animal or fowl.

crocus. a. Any plant of the genus Crocus, family Iridaceae. Species are spring- or autumn-blooming garden plants. Native to southern Europe and southwestern Asia. b. *Crocus vernus,* the common crocus, grown for its early-spring, reddish-blue flowers.

crocus tritonia. *Tritonia crocosmaeflora,* family Iridaceae, a bulbous, herbaceous hybrid developed in France from the African plants Potts tritonia and golden coppertip. It is grown as an ornamental for its deep orange flowers. Also called *montbretia.*

croker sack. Gunny sack.

Crolo. Criollo *q.v.*

crook. 1. An abrupt bend in a log; a defect. 2. An edgewise deviation from a straight line in a board; a defect. 3. A panicle of broomcorn in which the fibers are curved near the base. 4. Any implement having a bent end.

crook cucumber. A cucumber which fails to develop fully at the blossom end so that the fruit has a stunted end. Also called *nubbin.*

crooked-foot. A deformed hoof of a horse, congenital or caused by poor trimming and shoeing.

crooked-necked. Describing a weak-stemmed flower.

crookneck group. A group of squash or pumpkin, *Cucurbita pepo,* family Cucurbitaceae, which has edible, golden-yellow or white, warted fruit with long curving necks; a popular vegetable.

crop. 1. Any product of the soil. In a narrow sense, the product of a harvest obtained by labor, as distinguished from natural production or wild growth. 2. 10,000 boxes of turpentine, a unit of turpentine orcharding. 3. The young fowls or animals bred on a farm. 4. Craw; a saclike enlargement in the esophagus of many birds to store food. 5. Part of a cow's body just behind the shoulder blades. 6. A whole tanned hide. 7. The honey sac of the honey bee. 8. An identification notch or earmark on an animal. 9. To cut: (a) to clip the ears or tail of an animal, especially a dog; (b) to bite off, as a cow *crops* grass; (c) to cut off the wattles of a bird; (d) to clip the hair of an animal; to shear. 10. To raise plants usually during one season.

crop acres. Acres in crops.

crop adaptation. The harmonious use of the hereditary and environmental factors of plants in particular areas and seasons so that the plants will grow to maturity. See *ecology.*

crop area. See *cropland.*

crop back. 1. In grazing, the cattle biting off the plant near its crown. Also called *crop down.* 2. See *head back.*

crop boom. An extension pipe of a pressure sprayer. It has nozzles attached to increase the coverage of the sprayer.

crop bound. Referring to a domestic fowl with a distended crop from which food does not pass out in the regular manner. Also called *impacted crop.*

crop down. See *crop back: 1.*

crop dusting. Aerial dusting *q.v.*

crop-eared. 1. Describing an animal with its ears cut. 2. Designating a small-eared horse.

crop ecology. A science which deals with the study of agricultural crops in relation to their environment, such as soil, climate, plant reaction, etc.

crop expense. Denoting the amount spent for seed, seed treatment, fertilizer, lime, marl, spray material, crop insurance, twine, and sometimes custom expenses.

cropland. That portion of tillable land on a farm.

crop mycosis. See *thrush: 2.*

crop-outlook. An official prediction by the U.S. Department of Agriculture, which attempts to forecast the total amount of the national or state harvest for a particular crop.

crop pasture. A sown crop which is normally harvested but may be used for pasture if necessary, as oats, wheat, soybeans, etc.

cropper. 1. A laborer to whom a farmer assigns a definite tract of land together with all or most of the necessary farming equipment, seed, and fertilizer and a share of the crops grown as remuneration for his labor. He is closely supervised by the farmer, receives free use of a tenant house, and ordinarily is given some opportunity to work for wages when his farming operations do not require his attention. Also called *sharecropper.* 2. A thrown rider.

cropping. See *crop: 10.*

cropping plan. The scheme of growing different crops in succession on the same land, in contrast to a one crop system or a haphazard change of crops. See *rotation: 1.*

cropping station. The place in a poultry dressing plant where a bird is cleaned. Food is removed from a bird's crop by massage, which forces the food out its mouth or out an incision made in the neck of the bird.

crop plant. A plant cultivated by man for its fruit, fiber, root, flower, etc.

crop production. See *agronomy.*

croppy. Pertaining to a domestic fowl which has a stuffed or excessively full crop.

crop reporter. A person, usually a farmer, who, during the growing season, submits a monthly report to the U.S. Department of Agriculture, estimating the crop outlook in an assigned area, making estimates in percentages of the normal, and often providing other related farm information, such as number of sows farrowing, price of farm land, etc.

crop residue. That part of a plant left in the field after harvest: leaves, stubble, roots, straw, etc.

crop residue management. The system of retaining crop residue on land between harvest and replanting to prevent erosion, insuring future crop production.

crop rotation. A systematic changing of crops grown on the same land to help prevent soil exhaustion. A cropping plan.

crop surplus. The portion of a particular crop for national consumption which remains unsold after the normal period for selling, usually at the time of the next harvest.

crop tree. An individual tree which is best adapted for a crop, usually as a saw log because of species, form, and condition.

crop value per tillable acre. The total value of crop production of a farm computed on a tillable acre basis. It is established by multiplying the production of each crop by the average price per bushel or ton and dividing the result by the number of tillable acres. Pasture value is computed by multiplying the estimated yield in tons on pasture fields by the price or value of hay per ton, etc.

crop weed. See *spotted centaurea.*

crop with. To grow one crop after another, as in rotation.

crop year. 1. The span of time from the planting to the harvesting of a crop. 2. The span of time from one harvest to the next.

crop yield. The amount of harvest per acre, or other land measure, for a particular crop.

crop yield index. A percentage index showing the relation of the combined crop yield on a farm to the combined average yield of the same crops for the region. The total production of each crop is divided by the average yield per acre for the same crop for that region. The quotients are added and divided by the total of acres of the same crops on the farm.

cross. 1. A plant or animal which is the result of mating individuals belonging to different species, races, breeds, varieties, etc. See *crossbred; hybrid.* 2. To mate individuals belonging to different species, varieties, races, breeds, etc. See *single cross; double cross: 1; backcross; top cross; inbred: 2; self-pollination.* 3. To plow a field at right angles to the first plowing. Also called *cross-cultivation.*

cross-alley. An alley in a barn which runs from side to side crossing the main alley at right angles.

cross break. A cross-grain break in a piece of lumber resulting from unequal shrinkage or mechanical stress.

crossbred. Pertaining to a plant or animal which is the result of crossing different species, races, breeds, varieties, etc. Specifically pertaining to a horse whose sire and dam are of different purebred breeds; a chick which results from mating two or more distinct varieties or breeds of fowl.

crossbred wool. Wool produced from sheep which results from crossing any of the Merino subbreeds with any of the English long-wool breeds. It is as coarse or coarser than half-blood (*q.v.*) grade (U.S. and Australia).

crossbreed. See *cross: 2.*

crossbreeding. The pollinating or breeding of plants or animals which belong to different species, races, breeds, varieties, etc.

cross-conveyor. An attachment on a windrow pickup hay baler which carries the hay laterally from the pick-up attachment to the baling device.

cross-cultivation. The working of land by plowing the second time at right angles to the first.

crosscut. To cut across the grain of wood.

crosscut saw. A saw with teeth especially designed for cutting across the grain of wood.

crosser. See *sticker: 2.*

cross-eye. An inherent physiological condition of some Holsteins in which the eyes are turned inward.

cross-fenced. A farm or ranch enclosed by a fence with fields or pastures fenced off within the enclosure.

cross-fertility. Interfertility *q.v.*

cross-field. Ray crossing *q.v.*

cross-firing. In a pacing horse, the striking of the near-fore and off-hind foot in the air.

cross-footed. Describing a horse whose toes turn inward. Also called *parrot-toed.*

cross-fruitfulness. Interfruitfulness. *Cf. interunfruitfulness.*

cross grain. 1. Wood with a grain deviating from the longitudinal one parallel to the wood; a grain that interweaves. 2. Wood cut across the grain.

cross gutter. A short channel in the barn floor running at right angles to the main gutter, used in connection with mechanical gutter cleaners.

crosshaul. A method of loading logs on a vehicle: one end of a line is made fast to the vehicle, passed over the bed and around the log to be loaded, power is applied to the line, and the log is rolled onto the vehicle.

cross-head gang. See *parallel shift.*

cross-immunity. The immunity of a host animal to one species of parasite or bacterium which occurs as a result of active infection with a related species.

crossing. See *cross: 2.*

crossing-over. The interchange of segments between the chromatids and paired (homologous) chromosomes prior to the formation of gametes. The result is an exchange of linked characters in combinations different from those shown by the parents. *Cf. linkage; recombination.*

crossing permit. A permit issued to a rancher to drive livestock through part of a national forest.

crossleaf heath. *Erica tetralix,* family Ericaceae, a nearly prostrate, evergreen shrub, grown for its rose-red flower clusters. Native to Europe.

cross-over. 1. An attachment for a four-row corn planter which prevents the check wire from moving sideways as the planter moves across the field. 2. An occurrence of crossing-over *q.v.* 3. An individual having characters which are the result of crossing-over.

cross-over unit. A frequency of exchange of one percent between two pairs of linked genes.

cross-pollinated. Pertaining to a plant which has undergone cross pollination.

cross pollination. The transference of pollen from one flower plant to another.

cross protection. The action of viruses, strains, or isolates in preventing the manifestations of symptoms of other viruses, strains, or isolates.

cross section. 1. A cut made at right angles to ground level to permit examination of earth layers, underlying strata, etc. 2. In surveying, a grid system linking points of elevation and thus determining land contours.

cross-slope bench. A bench terrace running along or near the contour which has a steep drop on the downhill side with the space between terraces being nearly flat. See *bench terrace.*

cross-sterility. Intersterility *q.v.*

cross-striped cabbageworm. *Evergestis rimosalis* (Guen.), family Pyraustidae, one of the many caterpillars which infests cabbage.

crosstie. 1. In the U.S., a hewn or sawed piece of timber placed beneath railroad rails for support. Also called *sleeper; railroad tie.* 2. To tie together 3 of the feet of an animal as it lies on the ground.

crosstoes. Birdsfoot deervetch *q.v.*

cross unfruitfulness. Interunfruitfulness. *Cf. interfruitfulness.*

cross vine. *Bignonia capreolata,* family Bignoniaceae, a handsome, climbing vine and ground cover grown as an ornamental for its habit and its funnel-shaped, orange-red flowers. Native to North America. Also called *Anisostichus.*

cross with. See *cross: 2.*

crosswort. *Crucianella stylosa,* family Rubiaceae, a prostrate herb grown in shady places in the rock garden for its dark pink or rose-red flowers.

crotalaria. Any plant of the genus Crotalaria, family Leguminosae, some species of which are used for improving soils in southern U.S. Native to southeastern U.S. and the tropics. See *striped crotalaria; showy crotalaria.*

crotalaria mosaic. A virus disease of *Crotalaria* spp., cowpea, and other legumes, characterized by vein clearing, light-green mosaic pattern, puckering of the leaflets, rosetting, and diversity of leaf shape and coloring.

crotalaria poisoning. See *showy crotalaria.*

crotch. 1. A fork formed by the separation of two or more branches of a plant or tree. 2. A small sled made from the fork of a tree used to skid logs. Also called *alligator; crazy dray; go-devil; travois; lizard.* 3. The area between the hind legs of an animal.

crotch bracing. Artificial support for heavy tree limbs provided to prevent them from breaking at the crotch.

crotch disease of tangerines. A frequently lethal disease of tangerine trees, probably caused by a virus, characterized by dark, water-soaked areas in the crotches. The bark at such places is injured or killed, splits open, and rots, exuding a white foam with an alcoholic odor.

crotch injury of fruit trees. The killing of bark in the crotch area resulting from severe cold weather. See *winter injury.*

croton. See *woolly croton; Texas croton; variegated leafcroton.*

croton bug. Name given to members of the cockroach family; specifically, *Blattella germanica,* family Blattidae, the common cockroach which is dark brown, short winged, and has a life span of about one year. Omnivorous household pests, cockroaches are especially fond of starchy foods.

crotovines. Burrows of animals, especially rodents, which, open or by their natural filling after abandonment, modify the soil. Also called *krotowinas.*

croup. 1. That part of an animal's back from the hips to the tailhead. 2. A respiratory infection.

croute calcaire. See *caliche: 1.*

crow. *Corvus brachyrhynchos,* family Cornidae, a large, black bird which often damages crops, especially corn.

crowberry. Black crowberry *q.v.*

crowd. To press against one another, as plants competing for space.

crowder (pea). See *cowpea: 2.*

crowding pen. A pen used for sorting or handling animals; a corral.

crowd out. To compete so vigorously for space, as

a plant, as to kill other vegetation. See *competitive crop: 1.*

crowdy. Mash.

crowfoot. See *buttercup; geranium.*

crowfoot grass. 1. Any member of the genus Dactyloctenium, family Gramineae, especially Durban crowfootgrass. 2. See *hairy crabgrass.*

crow foot wire fence. A wire fence with small diamond-shaped mesh which restrains animals and trespassers.

crow garlic. Field garlic.

crow-hop. A mild form of bucking by a horse (western U.S.).

crown. 1. a. The upper part of a tree which bears branches and leaves. b. To remove the top of a plant or tree. Sometimes called *topping.* 2. The place at which the stem and root join in a seed plant: the *crown* or top of the root. See *collar: 3.* 3. The upper surface of a furrow slice. 4. The width of the top of a levee. 5. The neck of a sweep or cultivator point used to fasten the sweep to the cultivator shank. 6. The top of an animal's head. 7. The dent in the cap of a kernel of corn. 8. That part of a bridle which passes over the top of a horse's head. 9. The top of a hill. 10. The higher, center portion of a road.

crown and trunk canker. A disease of deciduous fruit trees and the black walnut caused by several species of the fungus Phytophthora, family Pythiaceae, characterized by cankers on the trunks of the trees. See *Phytophthora; crown canker.*

crownbeard. *Actinomeris squarrosa,* family Compositae, a late-flowering perennial herb considered a good bee plant because of its high nectar yield. Native to North America.

crown borer. See *strawberry crown borer.*

crown bud. Any growing point located at the transition zone between the stem and root of a plant. Sometimes used in connection with trees and shrubs but more commonly with plants propagated by division of the root crown, such as peony and rhubarb.

crown canker. 1. A fungous disease of greenhouse roses caused by *Cylindrocladium scoparium,* family Moniliaceae. Characterized by slowly developing, black, watersoaked, soft, canker areas that are just above or at the graft union, which result in weak plants producing few and poor blossoms. 2. The most serious disease of flowering dogwood in the northeastern U.S. caused by the fungus, *Phytophthora cactorum;* characterized by an unthrifty appearance of the tree, weak top growth, small leaves of a lighter green color which redden prematurely in late summer, and curl and shrivel in dry spells. May be fatal to twigs and branches. See *Phytophthora.*

crown canopy. See *crown cover.*

crown class. Any of several classes of trees in a crown cover which are classified by relative height, position, and condition. See *dominant crown class; codominant crown class; intermediate crown class; overtopped crown class.*

crown closure. a. The closing together of the crowns of trees in a forest, as in a maturing stand, or after thinning. b. Any stage in this process: by extension, the enlarging proportion of the ground area covered by the aggregate vertical projection of all the crowns in the crown cover.

crown cover. The canopy formed by the crowns of all the trees in a forest, or in an uneven-aged forest, by the crowns of all trees in a specified crown class. Also called *crown canopy; leaf canopy.*

crowndaisy chrysanthemum. *Chrysanthemum coronarium,* family Compositae, a hardy, annual herb grown for its yellowish-white flowers. Native to Europe. Also called *annual chrysanthemum.* See *chrysanthemum.*

crown density. An expression of the crown closure in relation to the land area involved, measuring the extent of shading exercised by the crowns, with regard to the habit of the species, site, and age. It is usually expressed in decimal fraction of complete cover.

crown division. A method of reproduction of crown budding plants, such as the peony and the chrysanthemum, by dividing the crown into several separate plants.

crowned cover plate. An arched plate in the bottom of a seed hopper of a corn planter which causes the kernels to slide to the outer edge of the hopper where they fall into the cell drop.

crown fire. A fire which runs through the tops of living trees or brush.

crown gall. A destructive bacterial growth on the juncture of the stem and root, at various points on the stem, or even on the leaves of a variety of woody and herbaceous plants, produced by the action of *Agrobacterium tumefaciens.* A distinct, but related, species of the genus Agrobacterium causes the galls known as cane gall and hairy root. Also called *crown knot; root knot; root tumor; burr knot of pome fruits; bacterial gall. Cf. broom rot.*

crown girdler. Strawberry root weevil.

crown grafting. Bark graft(ing).

crown imperial. Imperial fritillary.

crown injury. The killing of bark areas near the root crown which results from low winter temperatures or ice pressure.

crown knot. Crown gall.

crown land. a. In a monarchy, land which the reigning sovereign possesses. b. In certain British domains, *crown lands* are public lands.

crown-of-the-field. Corn cockle.

crownofthorns euphorbia. *Euphorbia mili,* family Euphorbiaceae, a creeping, spiny, greenhouse plant, of subtropical origin, grown for its habits and its brilliant red bracts.

crown roast. A fancy cut of trimmed ribs from two loins, circularly fastened with the ribs inside.

crown rot. See *mushroom root rot; crown canker.*

crown rot of alfalfa. A disease caused by any of several fungi: *Leptosphaeria pratensis; Sclerotinia trifoliorum; Sclerotium rolfsii; Ascochyta* *imperfecta; Colletotrichum trifolii; Rhizoctonia solani; R. crocorum.*

crown rot of grasses. A disease caused by any of several fungi: *Fusarium* spp., *Helminthosporium* spp., *Pythium graminicola.*

crown rot of hollyhock. A disease caused by *Phytophthora megasperma.*

crown rot of house plants. A disease caused by any of the fungi: *Pythium* spp., *Rhizoctonia* spp., *Botrytis* spp., *Fusarium* spp., *Phytophthora* spp.

crown rot of rhododendron. A disease caused by *Phytophthora* spp.

crown rot of rhubarb. A disease caused by *Phytophthora parasitica.*

crown rot of trees. An early form of injury to trees characterized by lesions on the bark at the base of the trunk or on adjacent roots, which may be due to cold, drought, or fungi. The bark may split and decay. If the rot girdles the trunk, it may seriously affect the growth of the tree and/or kill it. Also called *collar rot.*

crown rot of walnut. A disease of the Persian walnut caused by *Phytophthora cactorum.* It attacks below the ground, producing brown or black cankers.

crown rot of wheat. A disease associated with the infection of the crown of the immature wheat and barley plants by the fungus *Helminthosporium sativum,* family Dematiaceae, and sometimes other species of Helminthosporium and various species of Fusarium. The affected plant is stunted and a conical area of necrotic tissue is visible in a section through the crown. Also called *dryland root rot.*

crown rust. *Puccinia coronata,* family Pucciniaceae, a leaf rust of grasses, and one of the most destructive diseases of oats, which is characterized by red pustules which rupture the epidermis, causing it to turn black. The alternate host is buckthorn, *Rhamnus* spp. Also called *leaf rust; orange leaf rust of oats.*

crown scab. An ulcer at the crown of a horse's hoof.

crown thinning. a. In the upper crown classes, the cutting of dominant and codominant trees that are crowding out the most promising trees. b. In the lower crown classes, the cutting of those trees that will die and be wasted before the next cutting is made.

crownvetch coronilla. *Coronilla varia,* family Leguminosae, a sprawling, perennial herb grown for its dense clusters of pink flowers. Sometimes poisonous in old fields and along roadsides in northeastern U.S., but little is known about its toxic action. Native to Europe. Also called *axseed.*

crown wart. A fungous disease of alfalfa caused by *Urophylctis alfalfae* (more properly *Physoderma alfalfae*), family Physodermataceae. It attacks the scales and leaf tissues of buds developing at the crown and results in their becoming large, rough white galls which in time turn brown and disintegrate, scattering the fungous spores.

crown onion. Field garlic.

crow pea. *Vicia sepium,* family Leguminosae. Also called *bush vetch; hedge vetch.*

crowpoison. *Amianthium muscaetoxicum,* family Liliaceae (sometimes called *Zigadenus muscaetoxicus*), an herbaceous, perennial plant from a coated bulb with a raceme of white to greenish flowers. At one time the poisonous bulbs were used in the manufacture of a fly poison. Native to North America, it grows in low sandy grounds, bogs, etc., in the Atlantic Coastal states from Florida to New York and westward to Missouri and Oklahoma. (Sometimes misspelled *Amianthemum.*) Also called *stagger grass; fly poison.*

croy. Marshland (Old English law term).

crucifer. Any plant of the family Cruciferae (Brassicaceae), such as cabbage, mustard, cress, radish, candytuft, sweetalyssum, honesty, stocks, cauliflower, horse-radish, broccoli, Brussels sprouts, turnip, kohlrabi, etc.

crude cottonseed oil. Unrefined oil, produced from cottonseed by either the hydraulic, expeller or screw press, prepress solvent or solvent extraction process.

crude fiber. The tough, woody, fibrous portion of plants and feedstuffs, largely carbohydrate, which is made up principally of cellulose, lignin, and other relatively insoluble and indigestible carbohydrates.

crude lees. First lees.

crude nitrogen solution. A mixture of 10 parts water, 45 parts anhydrous ammonia, and 45 parts sodium nitrate, used in making fertilizer mixtures.

crude petroleum. An effective constituent of contact poison used in cattle and sheep dipping vats to control lice, mites, and ticks.

crude protein. See *protein.*

crude sap. The stream of sap traveling upwards in a plant, carrying soluble salts which have been absorbed by the roots. *Cf. elaborated sap.*

cruise. A forest survey made to estimate quantity and quality of timber.

cruiser. A forest timber estimator. Also called *land looker; valuer.*

cruising ax. A light-weight ax, either single or double bitted, weighing about 2½ pounds. Also called *cedar ax.*

crumb. A natural structure in soil: small, spheroidal, very porous, easily crushed aggregates.

crumber. A device on a machine for digging trenches for tile. It follows behind the digger unit and cleans out and smooths the indentation into which the tile will be placed.

crumbles. Crushed pellets of feed. See *pelleted feed.*

crumbly. 1. A body or texture defect of butter in which the fat particles lack cohesion. Also called *brittle.* 2. A defect of ice cream which is dry, open, friable, and tends to fall apart when dipped. 3. Describing raspberries which have not matured properly because of injury to the blossom or immature fruit. 4. A body defect of cheese which tends to fall apart when sliced.

crumbly-gum corky bark. A virus disease of orange trees characterized by irregular, gall-like, eruptions or ridges from which light-brown, crumbly masses protrude. See *corky bark.*

crumb structure. See *crumb.*

crupper. 1. A leather loop; a part of a horse's harness which passes under the tail and is attached to the saddle, back band, etc. 2. See *croup: 1.*

crush. 1. A cattle chute or narrow passageway in which cattle are branded, etc. (Australia and New Zealand). 2. To mash, as grapes or seeds, so as to extract juice, oil, etc.

crushed ear corn. Ground kernels and cob, used as a livestock feed.

crushed oats. Oat chop.

crusher. 1. A machine or device used to extract oils, juices, etc., from seed, fruit, etc. 2. A machine used to crush the stems of newly mowed hay to facilitate drying.

crusher boards. Boards attached at right angles to the runners on a sled lister cultivator used to smooth the soil.

crushing. The smashing of the stems of harvested herbaceous plants to reduce the curing time.

crust. 1. a. A thin, brittle layer on the bare surface of many dried soils, which sometimes prevents plant emergence. b. To form a hard, brittle surface. 2. The outer wall of an animal's hoof. 3. Rough, tanned hide of a sheep or a goat. 4. On desert soils, a hard layer which contains calcium carbonate, gypsum, or other binding material often exposed at the surface.

crust over. Crust: 1. b.

crutch. a. The area between the hind legs of a sheep. Sometimes called *crotch.* b. To clip or shear wool from around the *crutch* of a sheep removing stained or dungy wool.

crutching. Wool shorn from around the crutch of sheep before the regular shearing time (Australia).

cryolac number. An index relating to the freezing point of milk which varies from 393 to 435 with a mean of 413.

cryolite. Na_3AlF_6, sodium fluoalumite or sodium aluminum fluoride, a mineral found naturally and synthetized. Used as an insecticide; particularly effective against the codling moth and orange worm.

cryopedology. The study of frozen ground, including soils of arctic and subarctic regions.

cryoscope. An instrument used to determine the freezing point of milk.

cryptogam. A lower plant such as fungus, alga, lichen, fern, moss, etc., which reproduces by spores and does not produce a proper flower. *Cf. phanerogam.*

cryptomeria. *Cryptomeria japonica,* family Pinaceae, an evergreen tree grown as an ornamental. Native to Japan. Also called *Japan cedar; sugi.*

cryptophyte. A plant, specifically a cryptogam, whose buds, shoot tips, or spores survive the unfavorable season in the ground, or at the water level if it is growing in water.

cryptorchid. A male whose testicles fail to descend into the scrotum.

cryptorchidism. The condition of being cryptorchid. In the horse it is popularly referred to as *original* or *ridgling.*

cryptostegia. A genus of the family Asclepiadaceae (milkweed), whose woody vines are a small source of rubber. Native to tropical Africa. See *palay rubbervine.*

cryptoxanthin. A yellow carotenoid pigment found in yellow corn and other grains.

crystal garden. Terrarium.

crystalliferous cell. In wood anatomy, a crystal-bearing cell.

crystallization. Sclerocystosis.

crystalloidal. Pertaining to a state of matter which will form crystals. Common substances that exist in this state are usually simple compounds such as salts, acids, and bases.

crystaltea ledum. *Ledum palustre,* family Ericaceae, an evergreen shrub grown in northern U.S. bog gardens for its white flowers. Also called *wild rosemary.*

crystal violet vaccine. A preventive vaccine for hog cholera prepared from the blood of cholera-infected swine. The vaccine contains killed hog cholera virus and is therefore non-infectious.

C-type Merino. The largest type of *Merino* sheep that is free from folds of skin on the body and that has only two or three small folds on the neck. These sheep produce more mutton, although the wool is less dense on the legs and head than on other types. Also called *Delaine Merino; Delaine type. Cf. A-type Merino; B-type Merino.*

cuadra. A unit of land measurement which in Argentina equals 4.17 acres; in Chile, 3.88 acres; in Paraguay, 1.85 acres; in Uruguay, 1.82 acres; and in Brazil, 0.92 acres.

Cubalaya. A breed of medium-size chickens which originated in Cuba from fowls of Oriental breeding. The three varieties, White Cubalaya, Black Cubalaya, and Black-Breasted Red Cubalaya, have short beaks, broad, short heads, and projecting eyebrows.

Cuban bast. Jamaica linden hibiscus.

Cuban frangipani. *Plumeria emarginata,* family Apocynaceae, a tropical tree grown in southern Florida, U.S., for its fragrant, white flowers.

Cuban lily. Peruvian squill.

Cuban pine. Slash pine.

Cuban royal palm. *Roystonea regia,* family Palmaceae, a tropical street palm, suitable only to Florida, U.S.

cube. 1. A sale package of butter which ranges in size from 63 to 80 lbs. 2. Small pellets of compressed stock feed. 3. (cubé) *Lonchocarpus* spp. (especially *L. nicou*), family Leguminosae, a tropical shrub whose roots are a source of rotenone.

cubeb pepper. *Piper cubeba,* family Piperaceae, an East Indian, woody, treelike vine grown mainly in greenhouses, U.S. It yields the cubeb of commerce used in manufacturing cigarettes which are smoked to relieve catarrh. Oleoresin from cubeb is used to treat gonorrhea.

cubed steak. A retail cut of meat for broiling prepared from any of several of the tougher cuts and macerated to tenderize it.

cubic foot per second. The standard unit of measurement of flow in irrigation, which is one cubic foot of water flowing past a given point in one second.

cubic nitre. An obsolete term for nitrate of soda, obtained from Chile and Peru.

cubing machine. Any device which macerates meat to make it tender.

Cuboni test. A chemical, urine test to determine pregnancy in a mare.

cuckoo bees. Bees of families, particularly the Nomadidae, whose larvae are parasites in others' hives.

cuckoo-flower. Ragged robin.

cuckoo lamb. A lamb born after the usual lambing season.

cuckoopint. The tuberous root of the plant, *Arum maculatum,* family Araceae. In the past, during periods of famine, its starchy root was boiled, dried, and ground and used in making bread. Native to Europe. Also called *cuckowpit; root of Aaron; buckram.*

cuckoo spit(tle). The frothy, white secretion from spittle insects, found on the stems of low-growing plants or on the grass. Also called *frog spittle.*

cucullate. Hooded.

cucumber. *Cucumis sativus,* family Cucurbitaceae, a sprawling, tendril-bearing vine which produces the commercial *cucumber* used in salads, pickles, and relishes. Also called *earth apple.*

cucumber angular leaf spot. See *angular leaf spot.*

cucumber beetle. See *southern corn rootworm; striped cucumber beetle.*

cucumber flea beetle. See *potato flea beetle.*

cucumber magnolia. See *cucumbertree magnolia.*

cucumber mosaic. A disease caused by the virus *Marmor cucumeris,* spread by aphids, which attacks a wide range of horticultural crops and is widespread throughout the U.S. On cucurbits, characteristics are a mottled leaf and fruit pattern of yellow and green, especially in the terminal leaves, dwarfed vine, and imperfectly formed fruit. Also called *white pickle.*

cucumber scab. See *scab: 3.*

cucumber tree. See *bigleaf magnolia; cucumbertree magnolia.*

cucumbertree magnolia. *Magnolia acuminata,* family Magnoliaceae, a deciduous tree grown for its greenish-yellow flowers and pink or red fruits, and for its wood used in furniture, Venetian blinds, interior finish, doors, boxes, and millwork. Also called *cucumber tree; cucumber magnolia.*

cucurbit anthracnose. A serious disease of muskmelon, watermelon, cucumber, and gourd caused by the fungus *Colletotrichum lagenarium,* family Melanconiaceae, characterized by lesions and cankers on the leaves, stems, and fruit.

cucurbit. Any plant of the family Cucurbitaceae, such as the cucumber, squash, pumpkin, watermelon, citron melon, gourd.

cucurbit mosaic. Cucumber mosaic.

cucurbit wilt. See *bacterial wilt of cucurbits.*

cud. A small wad of regurgitated food in a ruminant's mouth, which is rechewed.

cud-chewing. A ruminant.

cuerda. A unit of land measurement in Puerto Rico equal to 0.971 acre.

cuff. To strike, as a horse strikes itself with one of its feet in running.

cuke. Slang for cucumber.

cull. 1. a. Anything worthless or non-conforming which is separated from other similar and better items. b. The act of removing the inferior items; to *cull* out. 2. The lowest marketing grade of meat carcasses or dressed poultry. 3. a. Any animal or fowl eliminated from the herd or flock because of unthriftiness, disease, poor conformation, etc.; a reject. b. A dairy cow removed from the herd because of unprofitable production. 4. a. A lumbering term for defective or low grade timber. b. A term applied to pecky cypress and other hardwoods below grade #2 common. c. To cut a few of the best trees in a stand without regard to silviculture. d. See *woods cull.* 5. Any fruit that fails to meet grading specifications because of defects, maturity, conditions, etc., *e.g.,* ripe berries which have green tips.

culled forest. A forest from which only certain of the better trees have been cut and removed.

culler. Scaler.

cull grade slaughter barrows and gilts. A marketing grade of hogs having a very low degree of finish.

culling. See *cull: 1. b.*

culling chute. A chute through which sheep pass in single file so that their fleeces and other qualities can be judged prior to removal of the poorest animals.

cull out. See *cull: 1. b.*

cull stock. Seedlings or other plants not available for planting, usually because of poor development or injury.

cull tree. A living tree, of merchantable size, but unmerchantable because of defects or decay.

culm. The jointed grass stalks which, excluding the nodes, are usually hollow, or pith-filled at maturity.

culm smut of grasses. Smut sori on culms and portions of the inflorescence of various grasses, caused by *Ustilago hypodytes* and *U. spegazzinii* and its variations. Among the many host grasses are various species of Agropyron, Elymus, Melica, Poa, Stipa, etc. It is widespread in South America and in western North America.

culti-cutter. An implement with heavy spade-like blades used for cultivating sod cover in orchards.

cultigen. Any horticultural variety not known in the wild state.

cultipacker. A clod crusher and soil packer with a corrugated roller.

cultivar. The name officially given for all cultivated variants of plants. Also called *horticultural* or *agricultural varieties,* but distinguished from the botanical use of the term. See *variety: a.*
cultivate. 1. To till the soil. 2. To loosen the soil and remove weeds among growing plants. 3. To plant, tend, harvest, and improve by plant breeding.
cultivated. 1. Plants which have been especially bred or improved. 2. Land which has been used to grow plants. 3. Soil which has been tilled, usually by a cultivator, hoe, etc.
cultivated variety. Cultivar.
cultivated forage. Forage especially grown for livestock consumption.
cultivating shovel. See *shovel: 2.*
cultivation. 1. The planting, tending, harvesting, and improving of plants. 2. Tillage of the soil to promote crop growth after the plant has germinated and appeared above ground. 3. Loosening the soil and removing weeds from among desirable plants.
cultivator. 1. A farm implement used to break the surface of soil in which plants are growing, and remove weeds, consisting of a frame to which several shovels, teeth, disks, or blades are attached. 2. A person who plants, tends, harvests, and improves plants.
cultivator duckfoot. A cultivator shovel which has a wing on each side for cutting through plant roots.
cultivator shield. See *shield: 1.*
cultivator shovel. See *shovel: 2.*
cultivator sweep. See *sweep: 1.*
cultivator teeth. See *cultivator duckfoot.*
cultural. 1. Pertaining to, or produced by, plant breeding. 2. Pertaining to tillage practices. 3. Pertaining to equipment for tilling.
cultural area. A geographical area with a number of similar cultures, usually derived from a single cultural tradition and adapted to a specific type of environment. See *culture: 6.*
culturalism. A biotic factor; the conversion of native cover to cultivated crops by man. See *communalism; symbiosis.*
cultural practice. Either the sum total of, or the component items of, the tillage, tending, harvesting, and improving of plants.
cultural variety. See *cultivar.*
culture. 1. The working of the ground in order to raise crops; cultivation; tillage. 2. Attention and labor given to the growth or propagation of plants or animals, especially with a view to improvement. 3. a. The growing of microorganisms in a special medium. See *subculture.* b. The microorganisms which are so grown. 4. Bacteria used in making dairy and other products. See *mother culture.* 5. Man-made features of an area. 6. The specific way of life of a given society (community, tribe, nation).
culture butter. Butter manufactured from cream to which a culture of bacteria has been added.
cultured buttermilk. The product obtained by souring pasteurized skimmed or partially skimmed milk, which contains not less than 8.5 percent of nonfat milk solids, by means of a suitable culture of lactic bacteria. This produces a thicker buttermilk than that which is a by-product of butter making.
cultured cream. A heavy, smooth, viscous, creamy product which results from ripening pasteurized sweet cream with a lactic culture.
cultured milk. Milk fermented by inoculation with a culture. See *cultured buttermilk; acidophilus milk; yoghurt.*
cultured sour cream. Cultured cream.
culture media. Liquid or solid preparations which are used for the cultivation of microorganisms.
culture yeast. Any yeast used in industry for fermentation. See *yeast: 1. Cf. yeast: 2; pseudo yeast.*
culturing. Artificial propagation of pathogenic or nonpathogenic organisms on nutrient media or living plants.
Culversphysic. *Veronicastrum virginicum* (syn. *Veronica virginica*), family Scrophulariaceae, a small perennial herb, grown for its white or pale blue flowers. It contains leptandrina, a violent emetic and cathartic. Also called *black root; Culver's root.* Native to North America.
culvert. A conduit to convey water underneath a road, through an embankment, etc.
culverwort. European columbine.
Cumberland disease. An Australian term for anthrax.
Cumberland side. A cut of pork, cut off square on the ham end, including the side and shoulder, with the leg cut off at the knee. A popular export cut from U.S. to U.K., Canada and Denmark. See *regular long clear (pork).*
cumin. *Cuminum cyminum,* family Umbelliferae, an annual herb grown for its aromatic fruit, whose seeds are used to flavor cheese, bread, kraut, and as an ingredient of curry powder. Oil distilled from the seeds is used to flavor liqueurs. Native to the Mediterranean.
cumulative humidity. The algebraic sum of the departures from, or differences between, daily relative humidity readings and a fixed base.
cumulose soil. Sedentary soil formed by the accumulation of more or less decayed organic remains; organic soils. See *peat; muck soil.*
Cunningham araucaria. *Araucaria cunninghami,* family Pinaceae, a tall, ornamental tree grown in southern California, U.S. Native to Australia. Also called *hoop pine; Moreton Bay pine.*
cup. 1. The receptacle attached to a tree to collect resin in turpentine orcharding. 2. See *crown: 7.* 3. A notch in a tree trunk made by two downward cuts of an ax for inserting a fatal herbicide. 4. A curve across the face of a piece of board. 5. A mechanical object resembling a drinking *cup:* (a) in lubrication, a vessel or small funnel for receiving oil, etc., and conveying it to a machinery part, an oil *cup;* (b) in grain elevators, a bucket or receptacle with a more or less curved outline.
cup-and-saucer vine. Purplebell cobaea.

cup drill. See *grass-seed attachment.*
cup fern. Hayscented fern.
cup flats. A flexible piece of molded pulp with cup-like indentations for each egg, used to separate egg layers in cases.
cupflower. Any plant of the genus Nierembergia, family Solanaceae. Also called *nierembergia.* See *dwarf cupflower; tall cupflower; white cupflower.*
cupid's-dart. See *blue cupid's-dart.*
cupid's-flower. Cypressvine starglory.
cup of flame. Californiapoppy.
cupping. 1. A turning upward of the edges of a leaf. *Cf. curling.* 2. In turpentine orcharding, attaching a cup to a tree to collect the resin.
cupping ax. A concave bitted ax used in turpentine orcharding for slabbing the face and seating the apron.
cup-plant. Cup rosinweed.
cup protector. In turpentining, a wire or cloth device which covers the cup to keep the cup and gutter free from trash and chips.
cupressus canker. A fungous disease of cypress caused by *Coryneum cardinale,* characterized by sunken, dark, resinous cankers at the base of lateral twigs.
cuprocide. Copper oxide.
cup rosinweed. *Silphium perfoliatum,* family Compositae, a perennial herb grown for its very tall stalk and large, yellow flowers. Native to North America.
cup shake. A frost crack in a limb or the trunk of a tree along an annual ring involving a small or extended circumference. *Cf. longitudinal crack.*
curative. a. A remedy used in the cure of diseases. b. Relating to the cure of diseases.
curb. 1. A thickening of the ligaments or tendons on the back of a horse's hind leg extending from the hock downward for a distance of 5 or 6 ins. 2. To restrain an animal.
curb bit. A bridle bit so designed that a slight pull on the reins will exert great pressure on the horse's tongue or jaw.
curb strap. A part of a check bridle which passes over the nostrils of a horse and connects to the cheek strap just above the bit.
curby hock. See *curb: 1.*
curculio. See *apple curculio; plum curculio.*
curd(s). 1. The coagulated part of milk which results when milk is clotted by rennet, by natural souring, or by the addition of a starter. 2. The edible white head of cauliflower and broccoli.
curd knives. A taut, piano-wire tool used in cutting coagulated or curdled milk in cheese making. See *curd mill; Swiss harp; cheese knife.*
curdle. To coagulate, as milk, so as to form curds.
curdled milk. The coagulation or thickening of milk resulting from the development of acids by bacterial action or the addition of rennet. Sometimes called *clabber.*
curd mill. A multiple series of knives used for cutting curd in making Cheddar cheese. See *curd knives; Swiss harp; cheese knife.*
curd scoop. In cheese making, a shovel-like piece of equipment for stirring during cooking and for picking up the drained curd and piling it onto the sides of a cheese vat.
curd tension. A measure, expressed in grams, of the pressure necessary to cut the curd of milk; the hardness of milk curd.
curdwort. Yellow bedstraw.
curdy. 1. Pertaining to a body defect of cheese characteristic of freshly made, uncured cheese which is almost hard. 2. Pertaining to a body defect of fermented milk in which the texture lacks uniformity and smoothness; the curds may be large or small.
curdy melt down. A body texture defect, characterized by flocculant curds in melting ice cream.
curdy-texture buttermilk. A defect of buttermilk, characterized by a rough, coarse body.
cure. 1. To preserve a product by drying, smoking, pickling, etc., such as hay, meat, or tobacco. 2. To heal; to restore to health.
cured forage. 1. Dry range grasses, harvested or standing, slightly weathered, nutritious, and palatable stock feed. 2. Any roughage preserved by drying.
cured meat. Meat preserved by smoking, and/or soaking in brine or other solutions.
curing of fertilizers. Mixing and storing fertilizer in piles to facilitate certain chemical reactions before it is ground, screened, and bagged.
curing plant. A storage place designed for the curing and aging of cheese.
curiosity. Any plant grown for its peculiar properties and not for ornamental or economic reasons.
curl. See *peach leaf curl.*
curled dock. Curly dock.
curled endive. A type of *endive* having fringe-edged leaves; a popular salad plant. Also called *frilled-leaved endive. Cf. escarole.*
curled mallow. Cluster mallow.
curled mustard. *Brassica juncea crispifolia,* family Cruciferae, an edible herb with crinkled foliage. Native to Asia.
curled rose sawfly. *Allantus cinctus* (L.), family Tenthredinidal, an insect pest of all roses, found from Maine to Virginia and west to Minnesota, U.S.
curled toe paralysis. A vitamin B_2 deficiency of riboflavin in chicks which causes twisted or flexed toes.
curlew bug. A maize billbug.
curling. Abnormal bending of the leafblade downward along the main vein. *Cf. cupping: 1.*
curling comb. A comb used to untangle and curl an animal's hair coat for a stock show.
curl of brambles. A virus disease which causes a downward curling of the leaves.
curly clematis. *Clematis crispa,* family Ranunculaceae, a woody vine grown for its bell-shaped, blue or bluish-purple, fragrant flowers. Native to North America. Also called *blue jasmine; bluebell.*
curly dock. *Rumex crispus,* family Polygonaceae, a perennial, herbaceous weed pest in grain fields

and elsewhere. Native to Europe. Also called *yellow dock; narrow-leaved dock; curled dock.*
curly dwarf. A virus disease of potatoes characterized by plant dwarfing and wrinkling and cupping of the leaves.
curly garden parsley. *Petroselinum crispum,* family Umbelliferae, a biennial or short-lived perennial herb grown for its edible leaves and used as a flavoring and garnish.
curly indigo. Sensitive jointvetch.
curly leaf. See *peach leaf curl; curly top.*
curlymesquite. *Hilaria belangeri,* family Gramineae, a leafy, wiry, perennial grass of excellent forage value which is drought resistant and withstands close grazing. Found from Texas and Arizona, U.S., to Central America, between altitudes of 1500-5500 ft. Also called *mesquite grass.*
curly palm. Sentry palm.
curly toe paralysis. Curled toe paralysis.
curly top. A virus disease destructive mainly to the sugar beet crop. It causes curling, retarded growth, yellow, blighted leaves, foliage rigidity, and prematurely ripened fruit, with blister-like elevations on the leaves, transparent venation of young leaves, wart-like swellings on the veins, and a sweetish exudate. Also called *California beet disease; western blight; whiskered beets; hairy root; curly leaf; woolly knot; woolly root.* See *beet leafhopper.*
curly wood. Lumber with a wavy, but uncrossing, grain caused by submerged adventitious buds, disease, or early injury to the tree.
currant. 1. Any of certain plants of the genus Ribes, family Saxifragaceae, popular as ornamentals and for the fruit they produce. See *red currant; Houghton currant; northern red currant; European black currant; gooseberry: 1.* 2. Dried seedless raisins, varieties of *Vitis vinifera,* family Vitaceae, used mainly in breads and pastries. Largely produced in Corinth, Greece.
currant anthracnose. A serious, fungous disease caused by *Pseudopeziza ribis* Kleb., family Mollisiaceae; characterized by a spotting of leaves, petioles, new canes, fruits, and fruit pedicels. Also called *leaf blight of currants; leaf spot of currants.*
currant aphid. *Capitophorus ribis* (L.), family Aphidae, a plant louse which infests currants and gooseberries throughout the U.S. and Canada.
currant borer. *Ramosia tipuliformis* (Clerck), family Aegeriidae, a boring insect which infests gooseberry, black elder, sumac, and currant; especially destructive of the black currant.
currant bud mite. *Cecidophyes ribis* (Nal.), family Eriophyidae, a mite which infests buds of currants, causing them to swell and die.
currant cane blight. A fungous disease of currants caused by *Botryosphaeria ribis,* family Botryosphaeriaceae, characterized by dieback and death of fruiting canes as the berries are coloring.
currant fruit fly. *Epochra canadensis* (Roew), family Trypetidae, a pest of currants and gooseberries, especially in western U.S. The adult lays her eggs in the fruit and the maggots cause the fruit to turn red and drop prematurely.
currant rust. Any rust attacking currants. One important variety is white-pine blister rust.
currant sawfly. *Pteronidea ribesii,* the most harmful of the insect larvae which attacks the currant.
currant spanworm. *Itame ribearia* (Fitch), family Geometridae, an insect which eats the fruit and leaves of currants, gooseberries, and sometimes blueberries.
currant tomato. *Lycopersicon pimpinellifolium,* family Solanaceae, a perennial, weak plant grown for its small, red, tomato fruit, which are used in salads. Native to Peru.
current annual forest percent. Rate of interest, expressed in percentage, of a stand of trees which represents the annual increment on the value of a forest.
current annual increment. The growth of a stand of trees for a specific year.
current cross. A plant or animal offspring developed in the first season of cross-mating.
current meter. A device used for measuring the velocity of flow of water in irrigation streams, ditches, and rivers.
curry. To comb and dress the coat of an animal to improve its appearance.
currycomb. A comb with rows of metallic teeth used in dressing the coats of horses and cattle.
cursed crowfoot. Blister buttercup.
cursed thistle. Canada thistle.
curva palm. *Attalea spectabilis,* family Palmaceae, a short palm tree with oil-rich fruit. Native to South America.
curved-eave house. A greenhouse which, instead of having eaves, has curved pieces of glass joining the roof and side walls.
curved knife-tooth harrow. Acme harrow.
curved-roof house. A greenhouse having a curved roof, most often used for conservatories and show houses for its attractive appearance; an expensive construction. Also called *curvilinear roof house; curvilinear house.*
curved runner. A furrow opener on a corn planter which arches upward to slice gradually through the soil.
curvilinear house. Curved-roof house.
curvularia leaf spot. A fungous disease of gladioli, caused by *Curvularia lunata,* family Dematiaceae, characterized by oval, tan spots with dark brown margins on the leaves, and large, brown, often round spots on the petals. Found in eastern U.S.
cusec. In India, a cubic foot of water per second.
cushaw. *Cucurbita moschata,* family Cucurbitaceae, a crookneck squash. Also called *Canada crookneck squash; winter crookneck squash.*
cushcush yam. *Dioscorea trifida,* family Dioscoreaceae, an herbaceous, twining vine, cultivated for its fine flavored, edible tubers. Grown only in very warm areas. Native to South America. Also called *yampee.*
cushion. 1. The inside face of a ham. 2. The fleshy pads on a horse's hips. 3. The frog of a

horse's hoof. 4. The mass of feathers on the rear or back of a bird, especially in the Cochin hen.

cushion comb. The comb of a chicken which is solid, low-set, and free from points and wrinkles.

cushion euphorbia. *Euphorbia epithymoides,* family Euphorbiaceae, a hardy, perennial herb grown for its yellow flowers. Native to eastern Europe. Also called *cushion spurge.*

cushioning. The moving of a potted plant to a larger pot just before selling it, to insure a clean pot and additional soil around the **root ball.**

cushion lamb shoulder. A retail, square-cut, boneless lamb shoulder in which the bone cavity is filled with meat or dressing. The edges of the meat are sewed to form a shape like a pillow.

cushion pink. Moss silene.

cushion style picnic shoulder. A retail cut of pork consisting of the boneless picnic shoulder in which the center part is filled with meat or dressing. The edges of the cut are sewed to form a shape like a pillow.

custard. A cooked milk product which contains generous amounts of eggs and spices.

custard-apple. *Annona reticulata,* family Annonaceae, a medium-sized tree with edible fruits 3 to 5 ins. in diameter, custard-like in the interior. Native to tropical America. Also called *bullock's heart custardapple.*

custard flavor. See *cooked flavor.*

custom gardener. One engaged in the business of custom gardening.

custom gardening. The maintenance of landscape plantings on private estates and around homes or industrial buildings. See *grounds superintendent.*

cut. 1. A piece of meat prepared for retail or wholesale trade. 2. A slash wound. 3. The opening made by an ax, saw, etc., on a tree. 4. An excavation in the earth, either man-made or natural. 5. The action of a horse's hoofs striking its legs or other hoofs in walking or running, interfering with its gait. 6. The yield of certain crops, as wool or lumber. 7. a. An animal separated from the main herd. b. To separate an animal from the main herd. 8. A severing of the stem or a part of a plant. 9. The output of a sawmill for a given length of time. 10. a. A reduction in numbers, amount, size, etc. b. To reduce in numbers, amount, size, etc. 11. To mow. 12. To castrate; emasculate. 13. To sever the jugular vein of an animal or fowl for slaughter.

cut alfalfa. Chopped alfalfa.

cut and bind. To mow and tie grain into bundles.

cutaneous. Pertaining to the skin.

cutaneous abscess. A pocket of pus and the surrounding diseased tissues on the skin.

cutaneous form. A form of disease confined to and primarily affecting the skin, such as cutaneous anthrax.

cutaneous myiasis. An infestation of the skin of animals, particularly sheep, by larvae of species of blow flies, of the family Calliphoridae, such as *Phormia regina, Phaenicia sericata, P. cuprina, Callitroga macellaria,* etc. The condition is seen in various parts of the world, especially in warm climates. Mules' operation (*q.v.*) is sometimes performed on sheep to prevent outbreaks. Also called *dermal myiasis; strike.* See *myiasis.*

cut-away colter. See *notched colter; rolling colter.*

cut-away disk. A disk harrow with a serrated edge used for deep penetration of the soil. Also called *spading disk.*

cut back. 1. a. Animals which are removed from a higher to a lower classification in sorting, for a specific purpose. b. Animals which are refused by a buyer because of their failure to meet certain specifications. 2. To prune a plant by removing terminal ends of shoots or branches, as contrasted to thinning. 3. To reduce to a former or a lower amount, number, size, etc.

cutbank. The bank of an arroyo which has steep sides resulting from erosion (southwestern U.S.).

cut bean. 1. A marketing classification of Mexican vanilla beans. 2. A classification of green or wax beans cut to designated lengths or thicknesses before freezing or canning.

cut cattle. To separate cattle from a herd.

cutch. A commercial source of tannin being the extract of Bengal gambirplant, *Uncaria gambir,* family Rubiaceae.

cut comb honey. Bulk comb honey which is cut into pieces, drained, and wrapped for sale.

cut flower. Any flower separated from the plant and usable in bouquets and corsages. *Cf. potted plant.*

cut hay. 1. Hay chopped into fairly small pieces for storage and feeding. 2. Hay that has been mowed but not raked.

cuticle. 1. A thin layer of cutin which covers the epidermis of plants above the ground, except where cork has replaced the epidermis. 2. One of the four major parts of an eggshell.

cutin. A noncellular, waxy material secreted in or on the walls of epidermal cells and, less frequently, on subepidermal cells of plants.

cut in. To mix one ingredient with another by a short, chopping action.

cutleaf blackberry. *Rubus laciniatus,* family Rosaceae, a trailing shrub which is evergreen in mild climates. There are several varieties, one of which is raised for its fruit on the Pacific Coast, U.S. Also called *evergreen blackberry; Oregon evergreen blackberry.*

cutleaf coneflower. *Rudbeckia laciniata,* family Compositae, a branching, perennial herb. The golden glow, commonly raised in borders, having double, yellow-orange flowers, is a horticultural variety. Native to eastern and southern U.S. Also called *tall coneflower; golden glow.*

cut-leaf nightshade. *Solanum triflorum,* family Solanaceae, a poisonous plant found from Ontario to British Columbia, Canada, and south to Kansas and Arizona, U.S.

cutleaf toothwort. *Dentaria laciniata,* family Cruciferae, a perennial herb grown for its small, purplish or white flowers. Also called *pepperroot; crow's-toes.*

cut oat groats. A livestock feed of cut, cracked, or ground oat groats. Also called *cracked oat groats.*

cut-off. 1. An artificial channel for a stream constructed to straighten the channel or to reduce the possibility of flooding. 2. A natural channel made when a stream cuts through the neck of an ox-bow meander. 3. A wall, collar, etc., constructed to reduce the seepage of water along otherwise smooth surfaces or through porous strata.

cut-off cover plate. A plate in the seed hopper of a corn planter which prevents more than one kernel at a time from dropping through the seed tube.

cut-off pawl. A part of a corn planter hopper which rides above the rotating seed plate, and, acting under spring pressure, pushes back extra seed kernels which tend to pass under the plate cover. At the same time it presses the kernel firmly into the seed cell. As the plate revolves further the seed cell is placed over the seed tube where a knockout pawl, also under spring pressure, comes in contact with the kernel, knocking it into the seed tube where it falls into the soil.

cut-off (rainfall). An obsolete term for that part of the total rainfall which penetrates into the earth beyond the reach of plant roots. Also called *ground water. Cf. runoff: 1; fly-off.*

cut-off system. (tile drainage). The location of tile on a slope or hillside to intercept water from seeps or springs which would otherwise flow onto adjacent bottom land.

cut-off trench. An excavation in the base of a dam or other structure which is filled with relatively impervious material to reduce percolation.

cut-off wall. A wall or diaphragm, made of concrete, steel, wood or earth, through the center of a fill to reduce seepage.

cut out. 1. The soluble solids concentration, on a percentage basis, in the drained juice of canned fruit. 2. See *drained weight.* 3. To separate an animal from the herd. 4. To dress an animal. 5. To remove cuts of meat from a carcass. 6. To finish shearing (New Zealand).

cut-over. An area from which some or all of the trees have been removed.

cutover forest. A forest from which some or all of the most usable trees have been removed. Also called *logged-over forest.*

cutover land. Forest land that has been logged over, leaving only stumps, unmarketable trees, slashings, and other evidence of recent lumbering operations. The term may be applied several years after lumbering, even though considerable new growth has appeared and evidence of original lumbering is no longer apparent.

cutover pasture. Land from which trees have been removed, as in lumbering, leaving the stumps. Both new and old tree growth may be present.

cuts. 1. Small groups of animals which have been separated from the main herd. 2. A four-wheeled timber cart with the axles connected by a long pole which forms the bed of the cart.

cut sample. A sample of ginned cotton taken from the press box before pressure has been applied to make a bale.

cuttage. The making and the tending of cuttings. See *cutting: 1.*

cutter. 1. A low grade of beef just above canner. 2. See *stalk cutter.* 3. A part of a bean harvester consisting of two broad blades set in a wheeled frame at a 60° angle, so as to cut two adjacent rows about 2 ins. below the surface. 4. In turpentining, a three-cornered steel tool used to sharpen hacks and pullers. 5. The device which chops forage or other plant material preparatory to placing the item in a silo. 6. A sleigh drawn by one horse.

cutter bar. The cutting mechanism on a mowing machine, binder, or combine which consists principally of knives and shear plates.

cutterhead. The device on a silage harvester which cuts forage into suitable lengths for silage. Also called *cutter; cutting head.*

cutter shredder. A machine used to shred corn stalks or other crop residue in the field.

cutting. 1. Any part which can be severed from a plant and be capable of regeneration. 2. Mowing. 3. See *brushing: 2.* 4. Removing a horse or cow from the main herd. 5. Felling a tree. 6. An area on which trees have been or are to be cut. Also called *cutting area; felling area; cut-over.* 7. The striking of a horse's feet against its joints while running.

cutting bench. A bench, usually in a greenhouse, containing a medium, such as sand, peat moss, or vermiculite, in which cuttings are rooted. Also called *cutting table.*

cutting block. A large block of wood standing so that the grain of the wood is vertical; used in butchering.

cutting board. A board or table on which something is placed to be cut, such as bulk honey comb.

cutting box. 1. A device which is used to chop hay and straw. 2. An open-ended, rectangular box without a top, similar to a miter box, used to cut foundations to the proper size for combs in beehives.

cutting chute. A narrow passageway into which animals are driven to remove certain ones from the main herd.

cutting cycle. The planned lapse of time between major felling of trees in a forest.

cutting edge. The sharp edge of a tool which actually does the cutting, *e.g.,* the *cutting edge* of an ax.

cutting fat. Fat removed during the trimming or cutting of pork carcasses, as contrasted with offal fat.

cutting-graftage. A method of plant propagation in which a cion is grafted on a cutting which in turn is allowed to form roots.

cutting head. A cutterhead.

cutting horse. A trained horse used to separate animals from the main herd.

cutting knife. A knife used to castrate animals.

cutting plan. A written program under which the cut of trees in a given area is limited or regulated.

cutting rot. Decay of a cutting caused when the

sand in which it is placed is too fine, too compact, too wet, or when a fungus attacks the stem.

cutting shoe. A specially designed shoe for a cutting horse. It is nailed only at the toe, and the inner branch is narrower than on a regular shoe. See *cut: 5.*

cutting table. 1. A table on which a carcass is placed to be cut. See *cutting block.* 2. Cutting bench.

cutting wood. That part of a plant which is to be used as a cutting. See *cutting: 1.*

cuttoo. A flap over a wheel that protects the axle from flying mud; a mudguard.

cut-up fowl. Pieces of poultry, such as wings, thighs, etc., which are packaged together for retail sale.

cut weight. To lose weight.

cutworm. The larval stage of moths of the family Noctuidae, which are soft-skinned caterpillars up to 1½ ins. long. Some climb up plants and devour the leaves, some chew their way into fruits, and some feed at or below ground level. They infest most garden vegetables, flowers, field crops, and fruit trees.

cyanamid. A commercial product which may be used as a fertilizer. It contains about 50 percent cyanamide (CH_2N_2) and 25 percent calcium hydroxide. In its free state, *cyanamid* contains about 35 percent nitrogen, but impure materials bring the content down to 21 or 22 percent. It may be toxic to plants unless used with care. It is sometimes mixed with other fertilizers.

cyanamide. CH_2N_2, a white crystalline compound made by the action of cyanogen chloride on ammonia. See *cyanamid.*

cyaniflower. Cornflower.

cyanogenetic plants. Plants containing glucosides which, under certain conditions, may be hydrolyzed to yield hydrocyanic (prussic) acid as one of the end products. A number of such plants, including the sorghums, Sudangrass and Johnsongrass, may, under these conditions, cause hydrocyanic or prussic acid poisoning in the animals which feed upon them.

cyclamen. Any plant of the genus Cyclamen, family Primulaceae. Species are very popular florist flowers. See *florists cyclamen; ivyleaf cyclamen; Neapolitan cyclamen.*

cyclamen mite. *Steneotarsonemus pallidus* (Banks), family Tarsonemidae, a greenhouse pest of cyclamen, snapdragon, geranium, chrysanthemum, larkspur, begonia, fuchsia, and petunia; and of strawberries grown out-of-doors. Infested plants have distorted leaves, and buds which either fail to open or, on opening, produce distorted flowers.

cyclamen pot. A clay flower pot with five holes in the bottom of the pot instead of one.

cyclamen weevil. Black vine weevil *q.v.*

cycle. A systematic, recurring fluctuation, as in the density of a species.

cyclic salt. Soluble salt carried onto the soil by salt sea winds and returned by rivers to the sea or inland salt lakes.

cyclone. 1. A device for reducing tomatoes to pulp. Also called *pulper.* 2. Popularly, but incorrectly, a tornado. 3. A low atmospheric pressure area, frequently several hundred miles in diameter, which advances at approximately 25 miles per hour with winds blowing in a counterclockwise direction in the Northern Hemisphere and clockwise in the Southern; associated with warm and cold fronts in the middle latitudes. Also called *low pressure area.* See *anticyclone.*

cyclone cellar. An earthen-covered pit, formerly used for refuge from severe storms. Also called *storm cellar.*

cyclone juice. Tomato juice obtained by separation in a cyclone. See *cyclone: 1.*

cylinder. The part of a combine, whose projecting, revolving teeth beat out or loosen the grain between the stationary concave teeth.

cylinder cleaner and grader. A device with an indented cylinder which cleans and grades seed according to size, separating the unwanted seeds.

cylinder hay loader. A device with flexible fingers attached to a rotating belt which gathers and elevates mown hay onto a moving vehicle. Also called *windrow loader; endless apron; elevator belt; web; drum; hay loader.*

cylinder plow. One-way disk plow *q.v.*

cylinder sheller. A device consisting of a cylindrical series of adjustable staves, which enclose revolving shelling rings attached to a main shaft, that remove kernels from corn.

cylinder teeth. The teeth attached to the cylinder in a grain thresher which shatter the grain from the heads.

cylinder type side-delivery rake. A cylinder consisting of three rake bars placed at an angle of about 45° to the direction of travel which windrow the hay as the rake moves.

Cylindrosporium fruit spot. See *Phoma fruit spot of apple and quince blotch.*

cymling. *Cucurbita pepo* var. *melopepo*, family Cucurbitaceae, a scalloped summer squash.

Cynips. The small gall wasps of the family Cynipidae which form galls on oak trees.

Cynthia moth. *Samia cynthia* (Drury), family Saturniidae, a giant silkworm moth whose larvae feed on the tree-of-heaven. The mature larva is about 3½ ins. long, light green, and covered with a glaucous bloom. The adult moth has a crescent-shaped white marking in the center of its grayish-brown wings and a wing spread of 6 to 8 ins.

cypress. 1. Any tree of the genus Cupressus, family Pinaceae; grown as an ornamental in warm regions. See *Arizona cypress; mourning cypress; Mexican cypress; MacNab cypress; Monterey cypress; Italian cypress.* 2. See *baldcypress; Hinoki falsecypress.*

cypress lavendercotton. *Santolina chamaecyparissus,* family Compositae, an aromatic, evergreen shrub grown for its foliage and solitary, yellow flowers. Native to the Mediterranean.

cypress oil. An oil obtained by distillation from the cypress, *Cupressus sempervirens;* medically useful as a whooping cough inhalant.

cypress pine. See *Callitris.*
cypressvine starglory. *Quamoclit pennata,* family Convolvulaceae, an annual vine grown for its funnel-shaped, scarlet flowers. Native to tropical America. Also called *cupid's-flower; starglory; star ipomae; cardinal climber.*
Cyprian bee. A honey bee used in bee breeding in the U.S. Native to Cyprus.
Cyprian cedar. *Cedrus brevifolia,* family Pinaceae, one of the traditional evergreen cedars, grown in warm areas as an ornamental. Native to Cyprus.
cypriere. Swamp (Louisiana, U.S.).
cyrtosis of cotton. A virus disease characterized by dwarfed leaves and proliferation of the apical buds.
cyst. An abnormally developed, closed sac containing a living organism, a foreign body or fluid; formed to protect the foreign matter, or to protect the plant or animal in which the sac grows.
cystacanth. A young, thorn-headed worm in the intermediate or transport host, which develops to maturity in the final host.
cysticercoid. A form of tapeworm larva composed of a small, double-walled bladder and a retractable head, or scolex.
cystic fluke. *Collyriclum faba,* a parasite found in the skin of a chicken or a turkey which blisters the skin and disfigures the fowl, reducing its market value.
cystic ovary. An ovary with a benign cyst, often congenital.
cystic stage. The time in the development of a parasite during which it is enclosed in a sac which separates it from the tissues of the host.
cystine. $C_6H_{12}N_2O_4S_1$, a white crystalline amino acid from proteins which is necessary to an animal's diet to develop horn, hair, etc.
cystolith. In a plant cell, a hardened deposit of calcium carbonate.
cytase. An enzyme which dissolves cellulose, found especially in grass seeds.
cytology. The scientific study of the structure and function of plant and animal cells, often with particular reference to the chromosomes in the cell.
cytophagy. The engulfing of cells by other cells.
cytoplasm. The living substance within a plant or animal cell excluding the nucleus: the protoplasm.
Cytospora canker. A fungous disease caused by *Cytospora kunzei,* family Sphaeropsidaceae, which girdles and kills many trees, including the poplar and the willow. Characterized by brown sunken cankers covered with numerous red pustules.

D

D activated animal sterol. A livestock feed supplement supplying vitamin D_3, obtained by activation of a sterol fraction of animal origin with ultra-violet light or other means.
D activated plant sterol. A livestock feed supplement supplying vitamin D_2, obtained by activation of plant sterol with ultra-violet light or other means.
D activated yeast sterol. See *irradiated yeast.*
dactylorhiza. Clubroot of cabbage *q.v.*
Dadant hive. A beehive, developed by Charles and C. P. Dadant, which contained 11 frames having an inside dimension of 10 x 17⅞ ins., and which was one of the earliest well-devised hives.
daddock. Rotten wood.
dadhi. See *yoghurt.*
daffodil. *Narcissus pseudonarcissus,* family Amaryllidaceae, a bulbous herb grown for its beautiful, pale yellow flowers. Also called *trumpet narcissus.* See *petticoat daffodil; nonesuch daffodil.*
daffodil garlic. Naples onion.
daffodil lily. Atamasco lily.
daffodilly. See *purple trillium; daffodil.*
daffy. Daffodil.
daft lamb. A lamb afflicted with cerebellar atrophy. The disease has been found in northern England and Scotland.
dag(s). 1. The dirtied wool around the crotch of a sheep. Also called *tags.* 2. To cut or shear such wool from a sheep (Australia, New Zealand).
dag-boy. Dag-picker *q.v.*
dagger fern. Christmas fern.
dagger plant. Yucca.
daggers. Yellowflag iris.
dagging. See *tagging: 1.*
daglock. See *tags: 2.*
dag-picker. One who sorts tags for usable wool (Australia, New Zealand). Also called *dag-boy.*
dag-tailed. Tag-tailed.
dahlia. Any herb of the genus Dahlia, family Compositae. Species are very popular garden plants. See *cactus dahlia; Aztec dahlia.*
dahlia sugar. Levulose produced from the root of the dahlia.
dahlia sunflower. Thinleaf sunflower.
dahoon. *Ilex cassine,* family Aquifoliaceae, an evergreen tree or shrub grown in warm areas for its beautiful foliage and dull red or yellowish berries. Native to North America. Also called *yaupon; dahoon holly.*
Dahurian larch. *Larix gmelinii,* family Pinaceae, a hardy, deciduous tree, an ornamental growing to about 90 ft. Native to Asia.
Dahurian lily. *Lilium dauricum,* family Liliaceae, a bulbous herb grown for its orange-red, dark-spotted flowers. Native to Siberia. Also called *candlestick lily.*
daikon. Japanese name for Chinese radish.
dairy. 1. a. A plant in which milk is processed

and where dairy products are manufactured and sold. b. Pertaining to that which is related to the production, processing, or distribution of milk and its products. 2. A place where milk is kept. See *dairy house.*

dairy bacteriology. The study and practice of bacteriology as applied to the dairy industry.

dairy barn. Any barn in which dairy cows are housed and milked.

dairy-bred. Pertaining to a cow or bull of dairy breeding.

dairy breed. Any breed of cattle especially developed for milk production: Holstein, Ayrshire, Jersey, Guernsey, Brown Swiss, Dutch Belted, Kerry, French Canadian, or Red Dane.

dairy bull. A bull of one of the dairy breeds used for breeding.

dairy butter. Butter made on a farm, as contrasted to creamery butter.

dairy calf. A calf of one of the dairy breeds.

dairy character. Those characteristics of conformation desirable in a dairy cow.

dairy cheese press. A device in which cheese is placed under pressure to compact it and remove whey.

dairy code. The laws of a city, county, or state which relate to or govern the cleanliness and inspection of dairies and dairy products.

dairy conformation. The desirable body characteristics of dairy cows. See *dairy type.*

dairy cow. A milk producing cow which shows predominant characteristics of one of the dairy breeds.

dairy farm. A farm on which milk production is the principal source of income.

dairy farmer. A farmer who raises and tends dairy cattle and derives his living from the sale of milk.

dairy goat. Milk goat.

dairy herd. 1. A herd of milk-producing cows. 2. In judging dairy cattle, a group class owned by the exhibitor which consists of 4 cows over 2 years old that are in milk, or are freshening.

Dairy Herd Improvement Association. A cooperative organization of approximately any 25 dairy farmers whose purpose is the testing of dairy cows for milk and fat production and the recording of feed consumed. Each farmer receives a monthly record for each of his cows and a complete, yearly summary of production and feed costs.

dairy house. A one-room structure usually attached to or adjacent to a dairy barn, in which the dairy farmer stores and cools milk, and cleans and keeps milk implements. In many countries, the construction and cleanliness of the room are regulated by law. Also called *milking house.*

dairy husbandry. 1. The care, breeding, feeding, and milking of dairy cattle, and the production and sale of milk. 2. The formal study of milk production.

dairy industry. The production of milk and its products.

dairyman. 1. A dairy farmer. 2. Any person who works in or manages a dairy.

dairy pasture. Pasture whose forage is consumed by dairy cattle.

dairy plant. A place where milk is pasteurized, bottled, and distributed for sale, and where milk products are made.

dairy product. Any product derived from milk, *e.g.,* cream, homogenized milk, buttermilk, butter, yoghurt, cheese, etc.

dairy ration. Any feed or feed mixture which is adapted to the daily needs of dairy herds.

dairy region. That section of the country in which dairy farming is a principal agricultural activity, as the Lake States and New England, U.S.

dairy scale. A scale used to weigh milk and its products which is usually devised to show net and gross weight.

Dairy Shorthorn. A type of the Shorthorn breed of cattle which originated in northeastern England. A typical cow weighs from 1,200 to 1,350 lbs. and is red, white, or roan. It is the principal dairy cow of England. Also called *Milking Shorthorn.*

dairy sire. Dairy bull.

dairy state. Any state whose principal agricultural activity is dairy farming, as Wisconsin, U.S.

dairy stock heifer. A market classification denoting a young dairy cow who has not calved and who is to be sold for dairy production.

dairy store. A retail shop in which milk and its products and sometimes eggs, etc., are sold.

dairy temperament. The inherent ability of a cow to utilize her feed for production of milk, rather than to put on body weight. See *dairy type.*

dairy type. A cow who indicates an ideal conformation for dairy production. The cow is angular; carries no surplus flesh, but shows evidence of good feeding; has a good development of the udder and milk veins; and shows a marked development of the barrel in proportion to its size. See *dairy breed.*

daisy. 1. Any plant of the genus Bellis, family Compositae, especially the English daisy. 2. Certain plants of the genera Chrysanthemum, Townsendia, Arctotis, Aster, Rudbeckia. See *arctic chrysanthemum; crowndaisy chrysanthemum; Marguerite chrysanthemum; oxeye daisy; Pyrenees chrysanthemum; giant daisy; stemless townsendia; African arctotis; blackeyedsusan.* 3. A cylindrical Cheddar cheese having a diameter of 13½ ins., a height of 4¼ ins., and a weight of 20 to 22 lbs.

daisy clipping. The low, skimming action of a horse whose hooves barely appear to clear the ground while trotting.

daisy cutter. A horse which does not allow its feet much clearance above the ground as it trots.

daisy cutting. See *daisy clipping; daisy cutter.*

daisy fleabane. Various species of the genus Erigeron, family Compositae, annual or biennial weed pests, especially in meadows and waste ground, over large areas of the U.S. One species, *E. canadensis,* is widespread in Europe and Asia. Native to North America.

daisy ham. A boned and smoked piece of pork from the shoulder.

daisy tree. a. *Olearia stellulata,* family Compositae, a shrub which bears white, daisy-like flowers. Native to Australia. b. Any member of the genus Olearia, family Compositae. Native to Australia and New Zealand. Also called *daisy bush; tree aster.*

Dakota potato. American potatobean *q.v.*

dale. 1. A valley. In common usage there is no distinction between *dale,* valley, vale, and dell, except that *dale* is not applied to a large river valley. 2. A piece or share of common land.

dale-backed. A sway-backed horse.

dale land. Valley land.

Dale's pony. A breed of pony which has a sturdy body and a short back, stands about 14.2 hands high, and is used to pack heavy loads in coal mines. Found in Northumberland, England.

dale vuelta. See *dally.*

Dallisgrass. *Paspalum dilatatum,* family Gramineae, a deep-rooted, perennial bunchgrass of excellent forage rating, grown in southern and western U.S. Native to South America.

Dallisgrass poisoning. Cattle poisoning caused by grazing on Dallisgrass pastures which have ergot growing on the grass seeds. See *ergot: 1; ergotism.*

dally. To loop the end of a lariat around the horn of a saddle so that the horse may serve as an anchor in roping animals.

Dalmatian bellflower. *Campanula portenschlagiana,* family Campanulaceae, a perennial herb grown for its bluish-purple flowers. Native to Dalmatia.

Dalmatian cherry. Marasca sour cherry *q.v.*

Dalmatian insect powder. See *pyrethrum extract.*

Dalmatian pyrethrum. *Chrysanthemum cinerariaefolium,* family Compositae, a perennial herb whose dried flower heads are a source of insecticide. See *pyrethrum extract.*

dalus. A measure of land; a strip of pasture or unplowed land between strips of plowed or arable land (England). Also called *dale.*

dam. 1. A quadruped, female parent. 2. a. An artificial structure which obstructs a stream of water for the purpose of water storage, conservation, water power, flood control, irrigation, recreation, etc. See *reservoir: 2.* b. To obstruct the flow of water.

damask plum. Bullace plum.

damask rose. *Rosa damascena,* family Rosaceae, an erect shrub grown for its fragrant, pink or red flowers. It is cultivated in Europe as a source of an expensive perfume, attar of roses. Native to Eurasia.

Damen. A soft, uncured Hungarian cheese made of cow's milk.

dame's-rocket. *Hesperis matronalis,* family Cruciferae, a coarse, perennial herb widely grown for its large, fragrant, lilac, or purple flowers. Native to Eurasia. Also called *Eveweed; garden rocket; dame's-violet; sweet rocket; Whitson gillyflower; damewort.*

dame's-violet. Dame's-rocket.

damewort. Dame's-rocket.

Damiana goldenweed. *Aplopappus venetus,* family Compositae, a weed pest which grows with great luxuriance on alkali soils in arid and semiarid soils in the U.S.

damming attachment. See *basin lister.*

damming lister. See *basin lister.*

damping-off. Any of several fungous diseases which attack and cause rotting of seedlings and cuttings. Pre-emergence *damping-off* is the killing of seedlings after germination but before the seedlings appear above soil; post-emergence *damping-off* is the killing of the seedlings after the seedlings appear above soil. In the latter and on cuttings, rotting occurs near the soil and the seedlings usually topple over and die. The fungi concerned are more often various species of Pythium or of Phytophthora, family Pythiaceae, or of Rhizoctonia, or of Pellicularia, family Thelephoraceae. See *Rhizoctonia disease of potatoes.*

damping-off of beets. Black root of beets.

damping-off of celery. A fungous disease of celery seedlings usually caused by Pythium spp., family Pythiaceae. See *damping-off.*

damping-off of cotton seedlings. A fungous disease usually caused by Pythium spp., family Pythiaceae. See *damping-off.*

damping-off of grasses. A fungous disease usually caused by *Pythium debaryanum,* family Pythiaceae, or *Rhizoctonia solani.* See *damping-off.*

damping-off of lettuce. A fungous disease of lettuce seedlings frequently caused by *Rhizoctonia solani,* or *Pythium ultimum,* family Pythiaceae. See *damping-off.*

damping-off of sugar beets. A fungous disease of sugar beet seedlings frequently caused by *Pythium aphanidermatum,* family Pythiaceae. See *damping-off.*

dampness. Moderate to high relative humidity; moisture; fogginess.

damped-off. To be attacked by damping-off *q.v.*

damp spot. A blue demarkation between the heart and sapwood in the yew.

damsel. Bullace plum.

Damson plum. Bullace plum.

dance. A stylized performance given by the worker bee upon returning to the hive after an initial search for pollen or nectar, to indicate to other worker bees the location of the pollen producing plants.

dandelion. *Taraxacum officinale,* family Compositae, a common, yard and field, weed pest. A perennial herb, it is sometimes used as a potherb, especially improved forms which may be of commercial value. Native to Europe. Also called *dindle.*

Danesblood bellflower. *Campanula glomerata,* family Campanulaceae, a hardy, perennial herb grown for its dense clusters of blue or white flowers. Native to Eurasia.

dangleberry. *Gaylussacia frondosa,* family Ericaceae, an evergreen, spreading shrub which bears a blue, edible fruit. Native to North America, and found from New Hampshire to Louisiana.

U.S. Also called *tangleberry; blue huckleberry; blue tangle.*

Danish Export. A small, flat, ripened, Danish cheese made of skim milk and buttermilk.

dank. Pertaining to plants which grow in moist areas.

Danubian geese. Sebastopol geese *q.v.*

Danzig fir. Scotch pine.

dapple. A circular pattern in an animal's coat color in which the outer portion is darker than the center.

dapple bay. A term used to describe a horse that has a black mane and tail and light chestnut-colored body, or a light chestnut-colored body covered in part by small rings of darker color.

dapple gray. A coat-color pattern of some animals, in which the gray color is overlaid with spots of lighter or darker tones.

dariloid. A sodium alginate, a gel, obtained from the giant Pacific Coast kelp, *Macrocystis pyrifera,* which has remarkable water absorbing properties. It is used as a constituent in ice cream mixes to stabilize the frozen product against large ice crystal formations.

dark air-cured. A tobacco classification denoting its method of curing in which the leaves have been air-cured and are dark in color.

dark blue nolana. *Nolana atriplicifolia,* family Nolanaceae, a prostrate, perennial herb grown for its blue flowers with white or yellow throats. Native to Chile. Also called *Chilean bellflower.*

dark center. See *brown heart.*

dark-chestnut. A term used to describe a brownish-black, mahogany, or liver colored horse.

dark cranberry. A variety of the kidney bean *q.v.*

dark crop. A displeasing defect of dressed poultry in which the crop appears dark, caused by blood swallowed by the bird during killing.

dark-dun. A term used to describe the mouse color of an animal's coat.

darkening. Any browning or darkening in color which occurs in a product during processing, storing, aging, or as a result of disease, frost, etc.

dark fire-cured. A tobacco classification denoting its method of curing in which heated air is used, darkening the leaves.

dark friable rot. A wood rot caused by a fungus in which the rotting wood is dark colored and easily broken.

dark mealworm. *Tenebrio obscurus* F., family Tenebrionidae, the larva of a black beetle that infests cereals and cereal products, meat scraps, and feathers.

dark meat. The legs and thighs of cooked fowls.

dark northern spring wheat. A subclass of hard red spring wheat which contains above 75 percent dark, hard, vitreous kernels.

dark of the moon. That period of the lunar month during which the moon does not shine. Superstition has it that certain crops should or should not be planted during this period.

dark pine. Sturdy cypresspine *q.v.*

dark-sided cutworm. *Euxoa messoria* (Harr.), family Noctuidae, a cutworm that feeds on newly set plants in the spring, attacking the stems at ground level, or climbing the leaves to attack older plants.

dark slate. A color term denoting a dark gray which is almost black.

darkstripe bloomeria. *Bloomeria crocea,* family Liliaceae, a perennial, bulbous herb grown for its golden-orange, crocus-like flowers. Native to North America. Also called *golden star.*

dark tobacco. Tobacco which is partially air-cured and partially heat-cured to give special aroma to the leaf. See *dark air-cured; dark fire-cured.*

dark winter wheat. A subclass of hard red winter wheat in which there is above 75 percent dark, hard, vitreous kernels. It is used for bread making.

Darling pea. Goat's-rue swainsonpea *q.v.*

darnel ryegrass. *Lolium temulentum,* family Gramineae, a weed pest found in grain fields and on waste land. It has been thought to be poisonous to animals although experiments have shown no such result. Native to Eurasia. Also called *tares; poison ryegrass; ivary; bearded darnel.*

dartars. 1. A sheep mange caused by the mite, *Sarcoptes scabiei,* family Sarcoptidae. 2. A scab or ulceration on the chin of lambs. Also called *black muzzle.*

dart grass. Velvetgrass *q.v.*

Dartmoor pony. A breed of hardy ponies developed in Devonshire, England.

Dartmoor sheep. An English breed having long wool, no horns, and a long face similar to Leicester sheep.

dartre. See *dartars.*

dartrose. See *Rhizoctonia disease of potato.*

Darwin barberry. *Berberis darwini,* family Berberidaceae, an evergreen shrub grown in warm areas for its handsome foliage, its golden-yellow, drooping flower clusters, and its dark-purple fruit. Native to Chile.

Darwin tulip. A type or class of tulips with many forms which has tall stems and bright-colored flowers. The blossoms are large, broad, triangular or cup-shaped and crimson, red, or purple. The flat base inside the flower is usually blue, white, or both.

dash. 1. See *dashboard.* 2. See *dasher.*

dashboard. 1. A wooden or leather shield placed on the front of a buggy or other vehicle to protect the passengers from mire, etc., thrown up by the horses' feet. 2. The instrument panel of an automobile.

dash churn. An earthen or wooden container, generally cylindrical, in which cream is churned into butter by working the dasher up and down. See *dasher: 2.*

dasheen. *Colocasia esculenta,* family Araceae, a tropical herb which produces a starchy, edible tuber. It is widely grown in the Pacific Islands and occasionally in southern U.S. Native to tropical Asia. Also called *taro; eddo; tanier.*

dasher. 1. A series of panels extending at right

angles from an axle used in an ice cream freezer to agitate the mixture during freezing. 2. A device used in a dash churn to agitate the cream in making butter, consisting of two flat, wooden cross pieces mounted on a long wooden handle.

data. Plural of datum. See *datum: 2.*

datana caterpillar. Yellow-necked caterpillar.

date. a. *Phoenix dactylifera,* family Palmaceae, an important tropical palm tree. b. The edible fruit of the date palm. It is an important food; oil is extracted from the grooved seed and the ground seeds are sometimes used to make a coffee-like brew. Native to western Asia and northern Africa.

date mite. *Oligonychus pratensis* (Banks), family Tetranychidae, an important pest of the date palm as well as other palms, maple, cypress, and ornamentals, which webs leaves and fruit together, so that the fruit is scurfy, cracked, and yellow.

date of delivery plan. A method employed in certain areas for improving the quality of cream. The cream can is dated, and if the cream is delivered before the expiration of a certain date, the cream is given a specific grade.

date palm. See *date: a.*

dateplum persimmon. *Diospyros lotus,* family Ebenaceae, a hardy, round-headed, deciduous tree grown for its reddish-white flowers and yellow or purplish fruit which become black and edible when ripe. Native to Asia. Also called *European lotus.*

date smut. A smutty fungus of dates and figs caused by *Aspergillus niger,* family Moniliaceae. Also called *dark smut.*

date sugar. Sugar prepared from the fruit of the date palm.

datum. 1. Any level surface taken as a surface of reference from which to measure elevations. In surveying, sea level is the basic reference. 2. A figure indicating a fact, as the number, quantity, or weight of an item, used principally in the plural, *data.*

daughter. 1. The female offspring. 2. The primary division or first generation offspring of any plant, regardless of sex.

David astilbe. *Astilbe davidii,* family Saxifragaceae, a perennial herb grown in flower gardens for its clusters of white to purple flowers. Native to China.

David lily. *Lilium davidii,* family Liliaceae, a bulbous herb grown in flower gardens for its red, black-spotted flowers on tall stalks. Native to China.

dawnflower morningglory. *Ipomoea leari,* family Convolvulaceae, a tropical, perennial vine grown for its large, violet-blue flowers with a white tube that turns pink. Also called *big dawnflower; blue dawnflower.*

day cestrum. *Cestrum diurnum,* family Solanaceae, a shrub grown in warm areas for its daytime fragrant, white flowers. Native to the West Indies. Also called *day jasmine.*

dayherd. To ride cattle during the day; to care for cattle on a range during the day. *Cf. nightherd.*

day jasmine. Day cestrum.

day length. The length of daylight in hours during 24 hrs. which is an important factor in the growing of plants. See *long-day plant; short-day plant; indifferent plant; day-neutral plant.*

daylily. Any perennial herb of the genus Hemerocallis, family Liliaceae. Species are important flowering plants. See *orange daylily; citron daylily; early daylily; lemon daylily; tawny daylily; Middendorff daylily; grassleaf daylily; manyflower daylily; Thunberg daylily.*

day-neutral plant. A plant in which the flowering period or some other process is not influenced by length of daily exposure to light. Also called *neutral-day plant.* See *photoperiodism.*

day nursery. A shed which has pens roughly 4 x 4 ft., a feeding alley, and feeding racks in which ewes with new lambs are placed and where the lambs are kept until they are old enough to join the flock. It may or may not have a glass front.

day-old chick. Designating a chick which is less than a day old. Chicks are commonly bought at this age by a farmer from a hatchery.

days of work. A computed figure estimating the number of productive days of work for the farm determined by multiplying assigned factors by each acre and head of livestock, plus days of work off the farm.

day's work. The amount of work performed by one laborer in one working day. The length of the working day may vary in different sections of any given country.

D. C. See *dry cure.*

DDT. An insecticide used to control a large number of insects. Its chemical name is 2, 2-bis (P-chlorophenyl)-1, 1, 1-trichloroethane.

deacon. A veal calf which is marketed before it is a week old. Also called *bob; bob veal.*

deacon skin. The hide taken from a veal calf less than a week old.

dead-and-down. Describing fallen timber or dead, standing timber.

dead arm of grape. A fungous disease caused by *Cryptosporella viticola,* family Sphaeriaceae, characterized by ragged margins on the leaves, and a yellowing, dwarfing, and crimping of leaves and arms. A canker appears on the arm or on the trunk below the attachment of the arm, from which a dry rot spreads in the woody part of the trunk in all directions killing the arm. Fruit may be infected and become dark-gray, shriveled mummies.

dead-born. Still-born.

dead cap. A fungous rot of strawberries which occurs in the field, in transit, and in the market resulting in the drying and shrivelling of the cap, the attached stem, and sepals, giving the berries an unattractive appearance. The fungus frequently enters the fruit affecting the flavor.

dead cotton. The unripe, almost worthless fibers of cotton which have very thin walls and lack twist or curliness.

deadening. An area of standing trees which are killed by girdling and fire preparatory to clearing land for crop use (in common use during the early period of settlement of the forested region of eastern U.S.).

dead eye. See *dull glossy.*

deadfall. 1. A tree which has fallen as a result of decay or age. *Cf. windfall: 1.* 2. A trap used in catching fur animals.

dead furrow. A double furrow, or trench made by throwing two furrow slices in opposite directions.

dead head. 1. A log which is sunken or partially sunken in a stream or body of water. 2. Blighted or dwarfed seed stalks which produce reduced yields of poor seed.

deadlines. A term used in western U.S. in late 19th century and early 20th century by stockmen for boundaries on the free range, marking areas that could be grazed by each rancher.

deadly hemlock. Poison hemlock *q.v.*

deadly nightshade. Black nightshade *q.v.*

dead man. A log or stone which is partially or completely buried in the ground to serve as an anchor.

dead-man's bells. See *foxglove: 2.*

dead mouth. Describing a horse whose mouth is no longer sensitive to direction by rein and bit.

deadnettle. 1. All plants of the genus Lamium, family Labiatae, the mint family, which resemble, but do not have the sting, of the nettle. 2. Purple deadnettle, *L. purpureum,* an annual, sprawling herb grown for its purple-red flowers. Native to Eurasia. Also called *dog nettle; red deadnettle; French nettle.* 3. Spotted deadnettle, *L. maculatum,* a perennial herb which can be increased by division, grown for its purple-red, or sometimes white, flowers. Often there are white blotches along the midrib of the leaf. Native to Eurasia. 4. White deadnettle, *L. album,* a perennial herb bearing white flowers. Native to Europe. Also called *base horehound; dumb nettle.*

dead ripe. Describing fruit, grain, etc., which is absolutely ripe.

dead spavin. A cured lameness in the hock joint of a horse which results from firing or other such treatment of a spavin.

dead wood. That portion of a stem, trunk, branch, etc., which has died.

dead-wood fence. A fence or barrier constructed of dead stumps, brush, etc.

dead wool. That wool which has been clipped from sheep shortly after death. Also called *fallen wool.*

deaerate. To remove air or gasses from food products.

deaf ear. 1. A scab of cereals. 2. The fold in the skin of a fowl just below the ear.

deal. A term for pine or fir boards: (a) in U.S., cut yellow pine lumber usually more than 9 ins. wide and up to 5 ins. thick; (b) in U.K., softwood board up to 4 ins. thick and 11 ins. wide.

deal apple. A white pine or fir cone.

dealer. Any person who distributes merchandise of various sorts, as a trader or middleman buying and selling goods.

dealkalization. The removal of alkali from the soil, usually by amendments and by leaching. Technically, it is the replacement of monovalent metallic ions, such as sodium, by alkaline earth cations, *e.g.,* calcium, or by hydrogen ions.

deal pine. Eastern white pine.

death camass. Any perennial, poisonous, bulbous herb, of the genus Zigadenus, family Liliaceae, especially Z. *venenosus,* found mostly in western U.S. Cattle rarely eat these plants, but sheep often do. The characteristics of poisoning are salivation, nausea, vomiting, lowered temperature, staggering, prostration, and finally, a coma in which an animal may remain for days, followed by death. Also called *onion poisoning.*

deathcup amanita. *Amanita phalloides,* a poisonous mushroom which is very often lethal when eaten.

death herb. Belladonna.

death-ring treatment. The placing of a layer of insecticide around the base of a tree.

death root. Purple trillium.

death-temperature. a. The temperature at which life ceases. b. The temperature at which spores of certain bacteria are killed.

deathweed. Poverty sumpweed.

debark. 1. To remove bark from the trunk or a limb of a tree. 2. To remove the vocal cords from a dog.

debeaker. A foot-operated device with an electrically heated blade used to burn off a part of the upper beak of chickens or turkeys and thus prevent picking, cannibalism, and egg eating among fowls.

debilitating. Weakening, enfeebling, or impairing strength.

debility. Weakness.

deblossom. To remove blossoms from a plant to increase its vigor.

debris. 1. The coarse material resulting from the decay and disintegration of rocks occurring either in the place where it was produced, or the place of deposition after transportation by streams of water or glaciers. 2. Dead stalks, leaves and other rubbish left on a field after harvest. 3. A mixture of straw particles, dirt, shattered pupal cases, etc., found in the wool of sheep infested with sheep keds.

debris cone. A fan-shaped deposit of soil, sand, gravel, and boulders which is built up at the point where a mountain stream flows into a valley, or where its velocity is otherwise reduced sufficiently to cause such deposits. See *alluvial cone.*

debris dam. A barrier built across a stream channel to store debris, such as sand, silt, driftwood, etc.

debt adjustment. The examination and selling of a person's debts by a *debt adjustment* committee, without court action if possible, in cases in which the indebtedness is clearly in excess of the debtor's ability to pay and foreclosure action is threatened. Debt Adjustment Committees were common in U.S. during the Depression of the 1930's.

decalcification. 1. Removal of calcium carbonate by leaching. A natural process in soil formation. Technically, it is the replacement of calcium ions by monovalent cations. 2. The removal of calcium from bones of animals in negative calcium balance.

decaliter. Ten liters, equivalent to 610.25 cubic inches, or 0.284 bushel, (U.S.).

decapitation. A method of pruning trees for rejuvenation in which most of the branches are severely pruned. Also called *deheading.*

decare. A unit of measurement which is the area equivalent of 10 ares, 1,000 square meters, or 0.247 acre.

decastere. A unit of measurement which is the equivalent of 10 cubic meters, or 13.08 cubic yards.

decay. 1. a. The decomposition of organic matter by aerobic bacteria or fungi in which the products are completely oxidized. See *putrefaction.* b. To decompose by aerobic bacteria, fungi, etc., whereby the products are completely oxidized. 2. General disaggregation of rocks which includes the effects of both chemical and mechanical agents of weathering with stress on the chemical effects. 3. Any chemical or physical process which causes deterioration or disintegration.

decayed knot. In timber, a knot that is softer than the surrounding wood because of decay.

decay increment. The increase in decaying of the wood in a tree or stand of trees.

Deccan hemp. Kenaf hibiscus.

December. A futures trading designation for contracts for grains and cotton to be delivered or repurchased on or before December 31 of the current year.

deciduous. 1. Pertaining to a woody plant whose leaves fall at the end of the growing season. *Cf. evergreen; coniferous.* 2. Pertaining to the falling of leaves, fruits, and petals of flowers. 3. Pertaining to the falling of the calyx lobes of a pome fruit, especially pear, before the fruit is ripe. *Cf. persistent.* 4. Pertaining to teeth, horns, hair, etc., which are shed at maturity of an animal, or at certain seasons, as in a molting animal.

deciduous cypress. Baldcypress.

deciduous-fruit marmalade. A preserve which contains pieces of fruit from a deciduous plant.

deciduous tree. A tree which sheds its leaves each year at the end of the growing season.

decimal rule. A log rule in which the volumes are stated in units of ten board feet.

decistere. A unit of measurement containing 0.1 of a cubic meter or 3.5315 cubic feet, used in measuring capacity.

deck. A platform, as a *deck* of brooders in a brooder house; a double *deck* railroad car or truck for transporting livestock.

deck feather. Either of the two central feathers in a bird's tail.

decline. 1. A gradual wasting away; a gradual loss of vigor. 2. To diminish or decrease in volume. 3. A physiological condition of citrus trees of more than 7 years in which there is a dieback of small branches and twigs, a yellowing of the foliage, poor fruit yield, and a lack of feeder roots. The cause is unknown. See *quick decline; spreading decline.*

decline age. The age beyond which a species of tree begins to show extensive decay.

decline disease. See *dry root rot: 1; quick decline; spreading decline.*

declivity. A downward sloping land surface.

decoct. To condense by boiling, as in food preparation.

decoction. The residue remaining after a substance has been boiled down.

decompose. To decay; to rot.

decomposition. 1. The breaking down of complex, chemical compounds to form simpler ones. 2. In soils, the breaking down of rock by physical and chemical change.

decontrol. To remove price controls or rigid conditions for producing a product.

decore. To remove the core of a fruit.

decorticate. To remove bark, husk, peel, etc.

decortication. The removal of the bark, husk, rind, etc.

decorticosis. Shell bark.

decoy hive. A hive placed to attract stray swarms of honey bees.

decoy sow. See *dummy: 2.*

decrease in inventory. A lessening of the value of a farm inventory at the end of a year from its value at the beginning of the year. *Cf. increase in inventory.*

decreasers. Range plants which gradually are replaced by other species in a stand. *Cf. increaser; invading plant.*

decumbent. Denoting a plant whose stems lie horizontally on the ground with only the extremities rising vertically.

deed. a. A legal document asserting right of ownership of a property. b. To transfer legally the ownership of a property.

deep. 1. Describing an animal which has a well-developed chest cavity. 2. Describing an animal which has well-developed muscles, *e.g., deep* shoulders.

deep feeder. Any plant which roots deeply, as alfalfa.

deep freeze. 1. Freezing conditions which penetrate unusually far into the ground. 2. See *freezer: 1.*

deep-furrow opener. See *lister.*

deep-grown. Designating long-stapled wool.

deep-litter nest. A poultry nest especially prepared to keep eggs clean and protect them from breaking, consisting of straw or other material 6 to 10 ins. deep.

deep-litter system. The placing of wood shavings, sawdust, straw, etc., mixed with hydrated lime, on the floor of a chicken house to about the depth of 6 ins. The droppings are not removed, although at intervals the litter is stirred and new material added. Normally the litter is changed only in the springtime. Also called *built-up litter system.*

deep percolation loss. Water that permeates downward through the soil beyond the reach of plant roots, as a result of inefficient irrigation practice.

deep plunging. The placing of cut flowers and greens in deep containers full of water so that the distance the water must move by capillary action and plant processes is materially lowered. This reduces the exposed leaf surface, preventing water loss by leaves, thereby maintaining or increasing turgidity.

deep-rooted. Designating a plant which roots deeply into the soil, as alfalfa. *Cf. shallow-rooted.*

deep scab. Corky scab of potato *q.v.*

deep scald. See *apple soft scald.*

deep-setting system. An outdated system of cream separation in which whole milk was placed in a tall, straight-sided can with a small diameter which was sometimes equipped with a glass gauge on the side and a faucet at the bottom. The cans of milk were set in cold water or in a cool place for 24 to 36 hours to permit the cream to rise. The skim milk was then drawn off through the faucet, or the cream was removed by a skimmer.

deep soil. 1. A thick, humus, surface layer of soil. 2. A very considerable thickness of soil and soil parent material over bed rock. *Cf. thin land.*

deep suck. A designation of suction of a plowshare; the bottom of the share is arched 3/8 in. and used in soil that is dry and hard.

deep suture. A virus-like disease of Bing cherries of western U.S. and Canada, characterized by narrow, elongated, roughened, leaf surfaces, and by fruits with a pronounced depression on the suture side.

deep-tillage. The tilling of the earth at greater depths than is possible or usual with ordinary plows.

deep well. Any well whose water level during pumping is below the practical suction lift of 25 ft. (under sea-level conditions). *Cf. shallow well.*

deerbrush ceanothus. *Ceanothus integerrimus,* family Rhamnaceae, a flowering shrub browsed by livestock. The flowers are usually white, rarely pale blue or pink. Native to North America. The blue-flowered form is sometimes called *bluebrush.* Also called *deerbrush; mountain lilac.*

deerfern. *Blechnum spicant,* family Polypodiaceae, a hardy fern which may be grown inside or outside for ornament. Found in Eurasia and North America. Also called *hard fern.*

deergrass. *Muhlenbergia rigens,* family Gramineae, a forage bunchgrass of southwestern U.S.

deerhorn cactus. *Peniocereus greggi,* family Cactaceae, a cactus grown in desert gardens for its curious habits and its nocturnal, white flowers. Native to southwestern U.S.

Deering velvetbean. *Stizolobium deeringianum,* family Leguminosae, a tropical vine grown for its clusters of purple flowers, as a cover crop, as a green manure crop, and for stock feed. Native to Asia. Also called *Florida velvetbean.*

deer laurel. Rosebay rhododendron.

deer's-tongue. Showy frasera.

deervetch. Any herb or undershrub of the genus Lotus, family Leguminosae. See *birdsfoot deervetch; St. James deervetch; squarepod deervetch.*

deer vine. American twinflower.

deerweed. Broom deervetch.

deeryard. An area where deer gather in the winter for feed and protection from the weather.

defatted cocoa. A food product of cocoa manufacture nearly devoid of chocolate taste, containing from 0.5 to 3.0 percent fat, used as dusting powder and in the preparation of imitation chocolates.

defatted milk solids. See *nonfat dry milk.*

defecation. Voiding of excrement; movement of the bowels.

defecator. An apparatus for purifying cane juice in the manufacture of sugar and sorghum.

defect. 1. Any blemish, fault, irregularity, imperfection in an animal, fowl, or farm product that reduces its usability or impairs its value. 2. In animals or fowls, a departure from breed or variety specification. Because of a tendency to be inheritable, a serious *defect* is also a disqualification *q.v.*

defective wool. Wool having a blemish or fault which will injure its value after scouring, *e.g.,* burry wool in the U.S.

deferred grazing. The keeping of livestock from a pasture until there is enough vegetation to support the animals, or in the western range of U.S., until the seeds of the herbage have matured.

deferred-rotation grazing. The system of range management in which grazing is postponed on various parts of a range during succeeding years, allowing each part successively to rest during the growing season to permit seed production and better vegetative spread.

deferred use. See *deferred grazing.*

defibrinated. Deprived of fibrin *q.v.*

deficiency. 1. An insufficiency in reference to amount, volume, proportion, etc.; a lack; a state of incompleteness. The measure of the *deficiency* can be useful: *e.g.,* the *deficiency* of the natural flow of a stream in meeting a given irrigation demand determines the storage necessary, the additional supply necessary, or the limitation of the irrigable area. 2. Absence, deletion, or inactivation of a segment of a chromosome.

deficiency disease. A pathological condition in plants or animals which results from a deficiency of a nutrient, mineral, or other necessary element in the food supply: *e.g.,* the kind of goiter in man or animals which results from a deficiency of iodine. Also called *hidden hunger.*

deficient. 1. Designating a flavor defect of ice cream which is characterized by a flat taste. Also described as lacking flavor. 2. Designating a lack or omission of some characteristic or physical development in a plant or animal.

definite. In cheese grading, an identifiable but not intense flavor.

definitely acidy. In butter grading, an acid flavor in the butter. Such butter is rated B.

definitely aged. In butter grading, an aged flavor in the butter. Such butter is rated B.

definitely storage. In butter grading, butter with a flavor resulting from or suggesting storage. Such butter is rated B.

definitely utensil. In butter grading, butter with a flavor suggesting the use of unclean utensils by the manufacturer. Such butter is rated B.

definitely weedy (common). In butter grading, butter with a weedy flavor. Such butter is rated C.

definitely yeasty. In butter grading, butter with a greasy or yeasty flavor. Such butter is rated C.

definition. Distinctness and clarity of outline and form which characterizes desirable traits of breeding in animals.

definitive host. The animal or host in which a parasite attains the adult stage and in which sexual reproduction takes place: *e.g.*, sheep are the *definitive hosts* of the sheep tapeworm. *Cf. intermediate host.*

deflation. 1. A fall or drop in the general price level. 2. The removal of loose material, such as soil, by the wind, thus leaving a bare surface.

deflea. To remove fleas from an animal, frequently by using an insecticide.

deflector. Any device which causes an object striking it to change course: *e.g.*, a *deflector* on a grain harvesting machine.

deflocculate. To separate or break down soil aggregates of clay into their individual particles: *e.g.*, the dispersion of the particles of a granulated colloid to form a clay which tends to run or puddle.

deflorate. To remove the flowers or blossoms from a plant.

defloration. The removal of flowers or blossoms from a plant.

defluorinated phosphate. 1. A carrier of phosphorus for use in fertilizers or in livestock feeds. It is produced from rock phosphate by heating with water vapor, or by other treatment, to remove fluorine. 2. A phosphorus feed supplement which is ranked below bone meal in value.

defoliant. Any substance which, when applied to a plant, causes the foliage to drop off.

defoliate. To remove the foliage from a plant.

defoliation. The removal of the foliage from a plant.

defoliator. 1. Any insect which destroys the foliage of a plant. 2. See *defoliant.*

deforest. To remove or destroy most or all of the trees in a stand so that natural reproduction is not possible.

deforestation. The denuding of a forest area.

deformed. Denoting a plant or animal which is misshapen.

deforming leg weakness. Perosis *q.v.*

deformity. Any physical deviation from the normal in animals or plants caused by injury or disease.

degeneration (degenerative) disease. Any disease of plants or animals which results in a progressive deterioration.

degerm. To separate germ cells from cereal grains such as corn and wheat.

degermed wheat product. Any wheat or wheat by-product feedstuff which has had all or most of the germ cells removed.

degerminated corn meal. That meal which remains after grinding corn and removing the germ, hull, and fine screenings. It has low feed value due to its low protein and fat content. *Cf. corn meal.*

degerminator. A machine which splits cereal grains and removes germ cells.

deglutinate. To remove gluten from a cereal.

degossypolized cottonseed meal. Cottonseed meal in which the gossypol (toxic principle of cottonseed) has been removed or deactivated so as to contain not more than 0.04 percent free gossypol. It must be so designated at the time of sale and must meet the prescribed quality specifications of cottonseed meal, prime quality.

degradation of soils. 1. Change of one soil type to a more highly leached one: *e.g.*, the podzolization (*q.v.*) of a soil which originally developed under the calcification process, as in the formation of a Podzol from a Chernozem. 2. Sometimes, incorrectly, a decrease of soil fertility.

degrade. To reduce lumber in grade.

degraded Chernozem. A zonal group of soils having a very dark-brown to black surface horizon underlaid by a dark to light gray leached horizon resting upon a brown horizon. It developed in the region between Chernozem and Podzolic soils where forest vegetation had encroached upon grassland. See *Chernozem; soil horizon; leaching; podzolic soil; zonal soil.*

degras. Crude wool grease.

degrease. To remove the wool oil from wool fiber and to obtain the fat from which lanolin is derived.

deguelin. $C_{23}H_{22}O_6$, an insecticide derived from derris and cubé roots which is closely related to rotenone.

dehair. To remove the hair from a dead animal's hide, especially swine, preparatory to tanning the hide or processing the carcass.

deheading. See *decapitation.*

dehiscence. 1. The bursting of certain dry fruits at maturity. 2. The opening and discharge of pollen from the anther.

dehorn. 1. To remove the horns from cattle, sheep, and goats or to treat young animals so that horns will not develop. 2. To cut back drastically the larger limbs of a tree. See *pollard: 1.*

dehorner. Any device which is used to remove the horns from animals.

dehorning. The removal of horns from cattle, goats, and sheep.

dehorning saw. A saw especially designed for cutting off animals' horns.

dehorning shears (clippers). A pair of shears especially designed for cutting off animals' horns.

dehull. To separate hulls from grains, seeds, etc.

dehusk. To separate husks from grains, seeds, etc.

dehydrate. To remove all or most of the moisture from a substance particularly, for the purpose of preservation.

dehydrated alfalfa leaf meal. A ground feedstuff consisting chiefly of alfalfa leaves. It must be reasonably free from other crop plants and weeds and must not contain more than 18 percent crude fiber. The freshly cut leaves are artificially dried in such a manner that a temperature of at least 100° C. is attained for a period of not more than 40 minutes.

dehydrated alfalfa meal. A feedstuff obtained by drying and grinding freshly cut alfalfa hay, including the leaves but not any alfalfa stems, alfalfa straw or foreign material. It must be reasonably free from other crop plants and weeds and must not contain more than 33 percent crude fiber. It must have been artificially dried in such a manner that a temperature of at least 100° C. is attained.

dehydrated alfalfa stem meal. A feedstuff which is the product remaining after the separation of the leafy material from alfalfa hay or meal. It must be reasonably free from other crop plants and weeds. The freshly cut stems are artificially dried in such a manner that a temperature of at least 100° C. is attained.

dehydrated chopped alfalfa. A feedstuff consisting of the entire alfalfa plant, chopped or cut but not ground finely enough to become a meal. It must not contain an admixture of alfalfa straw or other foreign material. The freshly cut plants are artificially dried in such a manner that a temperature of at least 100° C. is attained.

dehydrated potatoes. A foodstuff made from chopped or thinly sliced potatoes which have been subjected to heat for the removal of moisture, used as human food and a stockfeed.

dehydrated soybean hay meal. A feedstuff which is the product obtained from the artificial drying and grinding of the entire soybean plant, including the leaves and beans, but not any stems, straw, or foreign material. It must be reasonably free from other crop plants and weeds and must not contain more than 33 percent crude fiber. It must have been artificially dried when freshly cut.

dehydrater(or). 1. Any device which artificially removes moisture from a product for preservation or for the reduction of weight and bulk. Also called *drier; evaporator.* 2. In refrigeration, a vessel in which ammonia gas is passed through trays containing quicklime to remove any traces of aqueous vapor.

dehydration. The removal of moisture, as in the drying of foods for preservation or for reduction of weight and bulk.

dehydrofreezing. Preservation of food products by partial dehydration followed by freezing.

delactation. a. Cessation of giving milk. b. Cessation of suckling young; weaning.

Delaine Merino. C-type Merino *q.v.*

Delaine wool. Any long-stapled wool from midwestern U.S., not necessarily from Merino sheep, which is similar in length and fineness to fine strictly combing. See *strictly combing.*

delayed dormant spray. A delayed action spray used in the control of fungous diseases and insects on fruit trees. It is applied when the fruit buds show green leaf tips about ¼ in. long.

delayed germination. See *dormancy: 1.*

delayed settling. See *slow breeding.*

deletion. Absence of a segment of a chromosome involving one or more genes.

Delft. A spiced cheese of the Netherlands, very similar to Leyden (*q.v.*), which is made from partly skimmed cow's milk.

Delicious. A well-known, American variety of late dark red eating apple.

delignification. The removal of lignin from wood or other cellulose material.

deliming. The soaking of a hide in a sulfuric acid bath to remove the lime used in dehairing the hide.

delint. To remove the linters from cottonseed.

delinter. A device, usually in a cotton oil mill, which removes the linters from the cotton seed.

deliquesce. 1. To absorb moisture from the air, thus becoming liquid. 2. To become solute, as certain parts of fungi melting away in the action of growth or decay. 3. To dissolve into ramifications, as tree trunks or leaf veinings.

deliquescence. a. Absorption of moisture from the air by a substance and its tendency to become a liquid. b. The resulting solution.

deliquescent. Of, or pertaining to, the ability of a substance to absorb moisture from the air, thus being reduced to a liquid state.

delivery month. The month of delivery for a futures contract.

delivery rate (watershed sediment). The ratio of sediment yield to gross erosion in a watershed or drainage basin.

dell. A small, wooded valley.

delousing. The extermination of lice by insecticides.

delphinium. Larkspur *q.v.*

delta. 1. An alluvial deposit at the mouth of a river which is usually fan-like or triangular. These soils are generally high in fertility, and those of some great rivers, *e.g.*, the Nile of Egypt and the Hwangho of China, support a dense agricultural population. 2. Deposits which were spread out at the mouths of streams during the Glacial Period.

delta maidenhair. *Adiantum cuneatum,* family Polypodiaceae, a sturdy fern grown by florists for decoration. Native to Brazil.

deluge. A very heavy and prolonged downpour of rain sometimes followed by floods.

demand. The consumer requirements for a product in its processed form or in the raw state.

demarcate. To limit; to mark off; to separate from.

Dematophora root rot. A root rot of apple, fig, grape, walnut, poplar, and other plants, caused by *Rosellinia necatrix,* family Sphaeriaceae, which is characterized by sparse foliage, little or no growth, and wilting of leaves, followed by twig, branch, and leaf necrosis. The dead leaves may stay on the plant for months. When moisture and temperature are favorable the fungus may appear an inch or so above the ground line of the bark of the dead plant as a dark, velvety mat with white-headed, spore-bearing structures. Early in the disease white, cottony mycelium is found throughout the cortex, wood, and soil next to the tree. Later the root dies and its surface is covered by a dark, hard mat from which a cottony white growth may develop.

Demeter. Ceres *q.v.*

demisang. 1. Progeny resulting from the mating of purebred parents of different strains. 2. See *French Coach.*

demodectic mange of cattle. A mange caused by the mite, *Demodex bovis* Stiles, family Demodicidae, which is characterized by lesion-like nodules or small lumps, usually on the neck, shoulders, breast, and dewlap. Also called *follicular mange of cattle.*

demoid. A paleontological term for a plant or animal by which a place or period of time is characterized, because of its abundance, pervasiveness, etc. Also referred to as a *demoid type.*

demonstration agent. County agent *q.v.*

demonstration area. A tract of land on which various agricultural practices are tested to evidence the validity of agricultural procedures. See *test field.*

demoralized market. In marketing, a particular time when prices fall so that there is no certainty as to the monetary value of a product.

denature. 1. To make a product unfit for human consumption without destroying its value for other purposes: *e.g.*, denatured alcohol. 2. To change the properties of a protein, as to coagulate eggwhite.

dendriform. Tree-like in structure; branched.

dendrite. A tree-like pattern on a stone or mineral caused by reaction with a foreign substance.

dendritic. 1. Marked by a branching habit resembling that of a shrub or tree; usually said of river systems, various plants, and of the leaf veins of many higher plants. 2. Marked by the pattern presented by certain apple scab lesions; having a tree-like marking.

dendrium. An orchard community.

dendrocola. A swelling on a tree.

dendrograph. An apparatus which measures and records the diameters of trees.

dendrology. The study of trees including the identification of trees by external features.

dendrometer. An instrument which estimates the diameter of a standing tree at any given distance from the ground.

denitrification. The reduction of nitrates to nitrites, with the formation of ammonia and free nitrogen, as in nitrate reduction in soil by soil organisms, particularly anaerobic organisms under certain conditions.

dense. 1. Any substance whose component parts are closely packed together, as soil, baled hay, etc. 2. Designating a soil mass in which a relatively small proportion of the total volume consists of pore spaces, with an absence of any large pores or cracks. 3. In judging horses, designating a thick, heavy hoof.

dense-unproductive strain. A strain of the Eureka or Lisbon lemons which produces an abundance of foliage but has a low fruit yield and a low commercial value.

densimeter. A device which measures the density of a material or substance.

density. 1. Mass per unit volume. 2. The number of wild animals per unit of area. 3. The degree of closeness with which wool fibers are packed together.

density of browse. That part of the browse (*q.v.*), estimated from the ground surface, which is readily available to livestock.

density rule. A rule for estimating the density of wood which is based on the rate of growth, *i.e.*, width of the annual rings, as applied to southern pine and Douglas fir; a grading specification.

dent. See *dent corn.*

dental cup. A small dark-colored depression in the wearing surface of the incisor teeth of horses, having a rim of hard, glistening, white enamel which disappears as the teeth wear down.

dental pad. The very firm gums in the upper jaw of cattle, sheep, goats, and other ruminants.

dental star. On the incisor teeth of a horse, a yellow, transverse line on the wearing surface just behind the anterior border of the teeth, which first appears in the central, lower incisors at about 8 years. With advancing years this mark appears on the other incisors.

dentate. Of, or pertaining to, a leaf with regular serrations on the margins.

dentated sill. A notched sill at the end of an apron used to augment the apron's ability to check the force of flowing water, thus reducing erosion below the apron. See *apron: 4.*

dent corn. *Zea mays* var. *indentata*, family Gramineae, an Indian corn whose kernels are typically dented on the cap or crown at maturity.

dentition. The character, arrangement, and number of teeth in an animal; the formation and growth of teeth.

den tree. 1. See *redbox eucalyptus.* 2. A hollow tree which is used as a home by a mammal.

dents. See *check: 5.*

denudation. 1. The sum of the natural processes which wear down the surface of the earth, including natural or normal erosion. 2. Removal of all vegetative cover from an area.

denude. To remove the vegetative cover from soil.

denuded plot. A field plot used to study range vegetation which, having all growth removed is used in studying trends in succession, viability of seeds, and the survival of seedlings.

denutrition. Lack or withdrawal of nutrition; the failure to transform food elements into nutritional substances.

deodar. *Cedrus deodara,* family Pinaceae, a graceful, pyramid-shaped evergreen tree, one of the traditional cedars used as an ornamental in warm areas. Native to the Himalayas.

deodar weevil. *Pissodes nemorensis* Germ., family Curculionidae, a snout beetle which infests pines by feeding on the cambium. Found throughout southern U.S. and as far north as New Jersey.

deodorant. Any substance which is used to destroy, mask, or combat offensive odors.

depasture. To graze a pasture; to consume the pasture crop.

depauperate. A plant which does not achieve its natural size and has a starved or undeveloped appearance.

dependency. In range management, the year-round need for a forage or feed that is seasonal, which must be stored or otherwise preserved to achieve a balanced operation.

dependency zone. In range administration, a specific area of public land surrounding certain private lands which may be used to supplement private range land.

dependent plant. A plant that does not make its own food but lives on other living things or on dead things. See *parasitic seed plant; symbiosis.*

dependent property. Privately owned or controlled land or water which is deemed to have special claims for companionate use with certain public or cooperatively controlled range lands.

depilate. To remove the hair from the skin.

depilatory. 1. A substance which is applied to the flesh side of pelts to loosen fibers from the skin. 2. An agent for removing hairs from the living body.

depleted soil. Any soil which has been so used, or misused, that the resulting loss of fertility has impaired its economic utility for crop production.

depletion. Exhaustion of, or the loss of, certain minerals, chemicals, etc., from the soil, water, body, etc. See *soil depletion; soil deterioration.*

deplume. To remove the feathers from a fowl, as by disease, mites, etc.

depluming itch. An infestation of poultry by the depluming mite.

depluming mite. *Knemidokoptes gallinae* (Raill.), family Sarcoptidae, a mite that burrows into the skin of fowls near the base of the feathers, causing severe irritation.

depluming scabies. See *depluming itch.*

deposition. 1. The addition of fresh soil, as by floods. 2. The process of accumulation of rock material or other debris when dropped due to slackening of transporting agencies, as water, waves, winds, glaciers, or gravitational mass movement. Alluvial fans, off-shore bars, dunes, glacial moraines, mudflow debris are such depositional features. 3. In pesticide application, the dust or spray which remains on the plant surface.

depot. 1. A place of storage for supplies; a depository. 2. In lumbering, the center of operations; headquarters.

depot fat. Fat that is accumulated or stored in the body.

depouille. An obsolete term for thick layers of fat under the skins of cattle or buffalo. Also called *depuis.*

depraved appetite. A craving in animals and fowls for items not normally eaten by them which may be caused by a diet deficiency.

depreciation. The decrease in value of buildings, other improvements, and machinery which is caused by wear, tear, and obsolescence.

depressed. Reduced; flattened.

depression. 1. A slight hollowing in the flesh of an animal. 2. A hollow in the surface of the land. 3. A severe drop in income and prices for an area or for a nation. 4. The *Depression* denotes the economic reversal that took place in the U.S. in the 1930's, which was precipitated by the Stock Market Crash, October, 1929.

depth. The thickness of an animal's body from the center of its topline to the underline. Also called *depth of body.*

depth crank. A crank which, when turned, varies the depth of cut made by a plow or disc. See *depth lever.*

depth gauge. A little-used attachment on a plow, planter, etc., which indicates or measures the depth of the cut. See *depth crank; depth lever.*

depth lever. A lever on a plow or disk which, when raised or lowered, varies the depth of the cut. See *depth crank.*

depth of air cell. In egg grading, the distance from the top to the bottom of the air cell when the egg is held so that the air cell is upward. See *air cell; candle.*

depth of carcass. The distance from the top of the backbone to the underline of a carcass.

depth regulator. Any device on a disk harrow or plow which regulates or varies the depth of cut. See *depth crank; depth lever.*

depulp. To remove the fleshy substance from around the seeds of berries or fruits. Also called *macerate.* See *macerator.*

dequeen. To deprive a colony of its queen bee.

Derby. A hard, sweet-cured cheese made of cow's milk which is cured from 1 to 4 months. Also called *Derbyshire.*

Derbyshire. See *Derby.*

derived. Designating an invading plant, or one coming into a plant community from another plant community.

dermatitis. A non-specific skin condition affecting both man and animals, which is characterized by inflammation or infection of the skin. Animals with light skin are most sensitive, although *dermatitis* may result from external parasites, irritating chemicals or plants, sunlight or nutritional deficiencies. See *eczema.*

dermatomycosis. Any skin infection caused by a fungus, *e.g.,* ringworm.

dermatophytes. Fungi which infect only the skin.

dermatosis. Any skin disease.

derrick. An apparatus for lifting and moving heavy weights which is similar to a crane but has the boom pivoted at the lower end so that it may take different inclinations.

derrick stacker. A device for stacking hay which consists of a large fork or sling attached to a derrick boom by means of a block and tackle.

derris. A powder or liquid derived from the roots and stems of the plant, *Derris elliptica,* family Leguminosae, which contains rotenone, an insecticide.

desalinization. The removal of harmful salts from the soil, usually by leaching.

descaling. The removal of milk stone, rust, or other corrosive oxides from metals by means of a weak acid.

descriptive grade. Any color description of a product which is used in a grade term: *e.g.,* middling white cotton.

deseret. The honey bee (Book of Mormon).

desert. A land area having an arid climate (*q.v.*) which is either absolutely barren or sparsely vegetated with shrubs and herbs; occasionally, grasses or trees may be present. See *xerophytic vegetation.*

desert candle. Any plant of the genus Eremurus, family Liliaceae. Species are perennial herbs grown as ornamentals in warm areas. Native to Asia. Also called *foxtail lily.* See *Bunge desert candle; Elwes desert candle; Himalayan desert candle; giant desert candle; Tubergen desert candle.*

desert corn flea beetle. *Chaetocnema ectypa* Horn, family Chrysomelidae, a beetle which infests corn, sorghum, and small grains. Found in southwestern U.S.

desert crust. A hardpan (*q.v.*) of calcium carbonate, gypsum, or other binding material which is exposed in desert regions by wind or water erosion.

desert garden. A flower garden in which plants adapted to desert areas are planted.

desert gum. Moitch eucalyptus *q.v.*

desert-horse. See *barb: 1.*

desert Indianwheat. *Plantago fastigiata,* family Plantaginaceae, a range plant which is fairly palatable to livestock. Found in western U.S.

desertion limit. In wildlife, the number of days from the beginning of incubation of bird's eggs to the stage when normal disturbance will not cause desertion of the nest.

desert pavement. A space in a desert where the soil has been blown or washed away, leaving a covering of stones on the surface.

desert peach. *Prunus andersonii,* family Rosaceae, an extremely drought resistant shrub, which is fairly good sheep and goat browse. Native to western U.S.

desert plant. Any plant which is adapted to environments of little rainfall.

desert plantain. See *desert Indianwheat.*

desert saltbush. Cattle saltbush *q.v.*

Desert soil. A zonal group of soils which has a light-colored surface soil due to absence of humus, and which is usually underlain by calcareous material, frequently by a hardpan. It develops under an extremely scant shrub vegetation in hot to cool, arid climates.

desertthorn. See *wolfberry: 1.*

desert tobacco. *Nicotiana trigonophylla,* family Solanaceae, a widely distributed, poisonous herb of arid and semiarid areas. Symptoms of stock poisoning are vomiting, salivation, bloating, diarrhea, colic, weakness, staggering, falling, spasms, contraction of the eye muscles, stupor, and irregular pulse. Native to western and southwestern U.S.

desert varnish. A glossy covering or coating of dark-colored compounds, probably composed largely of iron and manganese, found on exposed pebbles, stones and large rock surfaces in hot deserts.

desert waste. A large acreage in arid areas which may provide a little grazing, timber, and watershed protection.

desert wheat corn. See *shallu.*

desert willow. *Chilopsis linearis,* family Bignoniaceae, a small, deciduous tree or shrub grown as an ornamental, for shade, and for shelterbelt planting. Also called *flowering willow; mimbre.* Native to Mexico and southwestern U.S.

desert wind. Periodic, desiccating, strong wind of arid regions.

desi. Jute fiber which is prepared from potherb jute *q.v.*

desiccate. To dry out; to exhaust of water or moisture content.

desiccated milk. See *dried milk.*

desiccation. A drying out; exhaustion of water or moisture.

dessert fruit. Any fruit, berries, etc., which may be served after a meal. See *fruit: 2.*

dessert wine. A sweet wine to which high-proof, grape brandy has been added.

destructive distillation. The decomposition of organic compounds by heat in a closed container, and the collection of resulting volatile liquids and oils, as in the production of wood tar or coal tar.

destructive erosion. Accelerated erosion *q.v.*

detachable points. Spring-steel teeth of different widths and shapes which may be used on spring-toothed harrows and cultivators for particular uses.

detachable shin. A replaceable cutting edge of a moldboard plow which is especially practical for use in gravel soil where wear is excessive and replacement is frequently necessary.

detached house. A greenhouse which is a complete unit in itself, as contrasted with ridge and furrow houses, lean-to house, contiguous houses *q.q.v.*

detailed-reconnaissance soil map. See *soil map: b.*

detailed soil map. A map which shows the boundaries of soil types and phases, all plotted upon a base map from precisely located points, and in sufficient detail to indicate those differences of significance in the practical use of small bodies of land. See *soil map: a.*

detassel. To remove the tassels from the plants of corn, a practice prevalent in corn-breeding and seed corn production.

detasseling machine. A device which carries the workers along the rows in a corn field while detasseling.
detention reservoir. See *retention reservoir.*
detergent. Any chemical compound that has the property of cleaning, but, more commonly, a foaming, soaplike compound made from alkyl sulfates, etc.
determiner. See *gene.*
detin. To remove the tin lining from the inside of a can by the chemical action of the canned product.
detoxified. Having the toxic, or poisonous quality of a substance removed.
detritus. 1. A heterogeneous mass of disintegrated rock fragments. 2. The deposits of disintegrated material or rubbish carried off mountain slopes by torrential rains.
deutectomy. The removal of the yolk sac from newly hatched chicks.
deutoplasm. The nutritional substance in the yolk of an egg. See *yolk: 1.*
deutzia. Any shrub of the genus Deutzia, family Saxifragaceae. Species are grown as ornamentals. See *slender deutzia; kalmia deutzia; Lemoine deutzia; Mongolian deutzia; redbud deutzia; rosepanicle deutzia; fuzzy deutzia.*
developed pressure. The force or thrust of a soil mass per unit area which results from the resistance of the soil mass to displacement.
developed water. Underground water brought to the surface and used for some purpose, as for irrigation.
development. The process of growth and change which takes place in an individual animal or plant from its embryonic state to maturity.
devil grass. Bermudagrass.
devil-in-a-mist. Love-in-a-mist.
devil-in-the-bush. See *love-in-a-mist; fennelflower.*
devil's-apple. See *Jimsonweed datura; Mayapple.*
devil's-bit. See *spike gayfeather; fairy wand; orange hawkweed.*
devil's bite. See *falsehellebore.*
devil's-claw. *Proboscidea jussieui,* family Martyniaceae, a tender, annual sprawling herb grown in home gardens for its immature seed pods used in pickling and as a curiosity. Native to southern U.S. Also called *unicorn plant; proboscis flower.*
devil's-club. American devilsclub.
devil's cotton. Cotton abroma.
devil's-daisy. Oxeye daisy.
devil's-darning-needle. Dragonfly.
devil's-fig. Pricklypear.
devil's-fingers. Black poplar.
devil's-flax. Toadflax.
devil's-garter. Hedge glorybind.
devil's-grass. Bermudagrass.
devil's-guts. Dodder.
devil's-hair. Dodder.
devil's-hopvine. Greenbrier.
devil's-paintbrush. Orange hawkweed.
devil's-pincushion. Spikespine coryphantha.
devil's-pitchfork. Beggarticks.
devil's-plague. Wild carrot.
devil's ringlet. Dodder.
devil's-tether. Dullseed cornbind.
devil's-tongue. *Opuntia compressa,* family Cactaceae, a pricklypear which grows wild from Massachusetts to Georgia, U.S. Also called *Barbary fig.*
devil's-trumpet. Jimsonweed datura.
devil's-turnip. Redberry bryony.
devil's-walking-stick. *Aralia spinosa,* family Araliaceae, a small, spiny tree grown for its large, white flower clusters. Native to North America. Also called *Hercules'-club; tree aralia; angelica tree.*
devilwood osmanthus. *Osmanthus americana,* family Oleaceae, an evergreen grown in warm areas for its foliage and its fragrant, greenish flowers. Native to North America.
devitalize. To take life from; to reduce vitality.
devitalized. Not viable.
Devon. A red, dual-purpose breed of cattle, originating in Devonshire, which is one of the oldest English breeds. Bulls average 1,370 lbs. and cows, 1,000 lbs. The *Devon* is rarely used in the U.S., and then, principally for beef or as work oxen. In U.S., found primarily in New England.
Devonshire clotted cream. See *Devonshire cream.*
Devonshire cream. A table delicacy obtained by skimming cream from shallow pans of scalded, whole milk by hand. After the milk stands for 10 to 12 hours, it is heated at 190° F. (87.8° C.) for 20 to 30 minutes. It is then cooled, held for 24 hours, and skimmed. Also called *clotted cream; Cornish cream; scalded cream; cooked cream; clouted cream.*
Devonshire Cream. A soft, English cheese which is made of cream.
Devonshire myrtle. Sweet gale *q.v.*
dew. The condensed, atmospheric moisture which has collected, usually at night, on plants, soil, and other surfaces which have cooled below the condensation point of the moisture in the atmosphere. The amount of water contributed by *dew,* though small, is potentially significant in the growth of plants especially in semiarid, or arid, regions. See *dewpoint.*
dewberry. Certain trailing prostrate plants of the genus Rubus, family Rosaceae, cultivated for their blackberry-like fruits. The Lucretia *dewberry* (*R. flagellaris* var. *roribaccus*) is raised in northeastern U.S. and varieties of the southern *dewberry* (*R. trivialis*) are raised in southeastern U.S. The grapeleaf California *dewberry* (*R. ursinus* var. *vitifolius*) is believed to be one parent of the loganberry *q.v.* Native to North America. See *swamp blackberry.*
dew bill. Snood *q.v.*
dew-blown. Designating a bloat which results from an animal eating too much dew-covered herbage; the adverse effect stemming from the extra moisture on the herbage. See *bloat: 1.*
dew-boine. Dew-blown *q.v.*
dewclaw. A vestigial digit on, or just above, the foot of a quadruped which does not touch the

ground as the animal stands or walks. Also called *false hoof*.

dewfall. The condensation of atmospheric moisture as dew.

dewflower. Cobaea penstemon.

dewing. To detach the wings from seed.

dewlap. The pendulous skin fold hanging from the throat of any animal, particularly a member of the ox tribe, and certain fowl.

dew leaf. Any leaf so cupped that it retains dew.

dew plant. Roundleaf sundew.

dew point. Under field conditions, the temperature at which condensation begins in a cooling air mass.

dew pond. A man-made pond on the downs used for watering livestock, so called because the only apparent source of water seems to be dew. However, it must be artificially filled first in order to function (England).

dew retting. The exposing of hemp and flax to the sun, dew, rain, etc., to free the fibers. Also called *dew rot; dew ret.* See *ret.*

dewtry. Jimsonweed datura *q.v.*

dewy. Covered with, moist with, or characterized by dew.

Dexter-Kerry. A dual-purpose breed of cattle originating in southern Ireland from the Kerry breed. The animals are usually black but occasionally red with a little white. Mature live weight is about 760 lbs. Also called *Dexter.* See *Kerry.*

dextrin. A gummy, but soluble carbohydrate; an intermediate product in the hydrolysis of starch to sugar.

dextrorotatory honey. Any honey which contains honeydew. See *honeydew honey.*

dextrorse. Designating a vine which spirals to the right as viewed from without. *Cf. sinistrorse.*

dextrose. $C_6H_{12}O_6$, a dextrorotatory sugar of crystalline form, found in many plants and all animals. Also called *glucose; corn sugar; grape sugar.*

dhauri. *Woodfordia floribunda,* family Lythraceae, a shrub bearing red flowers and which produces a gummy exudate sometimes used in preparing medicines. Native to the East Indies.

dhava. *Anogeissus latifolia,* family Combretaceae, a tree valuable for its timber and for a gum it yields used in printing calico. Also called *Bakligum axlewood.* Native to the East Indies.

D.H.I.A. records. Production records of cows tested under the supervision of the tester of a Dairy Herd Improvement Association *q.v.*

D-hoe. A hoe which has a blade shaped like the letter *D.* The handle is attached to the curved side, the straight edge being used for weeding.

D-horizon. In soil profile description, a subsurface layer or rock stratum which is neither parent material nor the source of parent material for the soil. Also called *D-layer.* See *soil profile.*

diacetyl. The aromatic, chemical compound developed in the ripening of cream for butter which is the chief constituent in butter aroma.

diagnosis. The process of identifying a disease by examination and study of its symptoms.

diagnostic. Pertaining to the identifying of a disease.

diagnostic agent. A substance or principle used in testing to determine if a disease is present.

diagnostic antigens. Biological agents which are prepared for use in the diagnosis of a specific disease: *e.g.,* tuberculin, which is injected under the skin of cattle to determine if they have tuberculosis. It has little or no effect on healthy subjects, but in an infected animal, it will cause inflammation of the skin around the injection and a possible rise in temperature.

diagnostician. Any person, especially one trained in diagnosis, who determines the nature and cause of a disease or abnormality in plants or animals and prescribes a treatment.

diagnostics. A group of biological agents used in detecting certain specific conditions or diseases in animals.

diagonal gait. Trot *q.v.*

diagonal grain. Annual rings at an angle with the axis of a piece of wood as a result of sawing at an angle of less than 90° to the bark of the tree.

diakinesis. A stage of meiosis just before metaphase of the first division: the homologous chromosomes are associated in pairs near the nucleus and have undergone most of the decreases in length.

diallel crossing. In animals, a method of progeny-testing in which two males, bred at different times to the same females, are compared to the two sets of progeny.

diameter above bottle neck. In turpentine orcharding, the diameter of a tree measured at a distance of 10 ft. above the ground.

diameter at breast height. The accepted diameter of a tree, taken at a distance of 4½ ft. above the ground. Abbreviated *DBH.* Also called *breast high diameter.*

diameter class. All trees which fall within specified size limits according to diameter at breast height, usually stated in 1 or 2 in. limits. In the 1 in. limit, all trees with diameters 0.0 in. below and 0.5 in. above the stated limit are in the same *diameter class.* In the 2 in. limit, trees 1.0 below and 0.9 in. above the stated limit are included in the same class: *e.g.,* the 10 in. class would include all trees with breast high diameters from 9 in. through 10.9 in.

diameter distribution. The manner in which the number of trees in a stand is distributed over diameter classes.

diameter limit. The smallest, and occasionally the largest, size to which trees or logs are to be measured, cut or used. The points to which the limit usually refers are stump, breast height, or top.

diammonium phosphate. See *monoammonium phosphate.*

diamondback moth. *Plutella maculipennis* (Curt.), family Hyponomeutidae, a minor cabbage pest whose worm-like larvae feed on the foliage of crucifers, ornamentals, and greenhouse plants.

diamond canker. A virus disease of French prune

that causes necrotic pockets in the bark which grow in size as the bark gets older.

diamond drag. A type of harrow consisting of slightly-corrugated, heavy wire screening which is used to crush clods or level the soil in the preparation of a fine seed bed.

diamondflower. Portugal diamondflower.

diamonding. The warping of a piece of lumber during drying that causes the cross-section of the piece to change from a square or rectangle to a diamond shape.

diamondleaf laurel. Diamondleaf pittosporum.

diamondleaf pittosporum. *Pittosporum rhombifolium,* family Pittosporaceae, an evergreen tree grown for its white flowers and red fruits. Native to Australia. Also called *diamondleaf laurel; Queensland pittosporum; mignonette wood.*

diamond plow. Any moldboard plow or cultivator point which is diamond-shaped.

diamond shoe. A diamond-shaped horseshoe used on the rear hooves of horses that have a tendency to overreach *q.v.*

diamond skin disease. See *swine erysipelas.*

diamond-tooth harrow. See *diamond plow.*

diapause. A temporary halt in the development of an insect embryo or larva.

diapersant. A chemical product that acts to lower the surface tension between a solid and a liquid.

diaphragm. The muscular wall between the thoracic cavity and the abdomen which, on contracting, enlarges the thorax, aiding respiration.

Diaporthe blight. A fungous blighting of plants caused by species of the genus Diaporthe, family Valsaceae, characterized by a drying and browning of the foliage and a brown rot at the base of the stem. When the stem is girdled, the plant dies.

diarrhea. Frequent and profuse fluid defecation commonly caused by an infection of the gastrointestinal tract.

diastase. An enzyme which splits starch into sugars.

diatomaceous earth. Geological deposits which consist mainly of the siliceous skeletal remains of diatoms.

diatomite. A light siliceous, earthy material, derived from the remains of diatoms, sometimes used as a poison dust carrier or abrasive.

diatoms. Microscopic algae which belong to the Class Bacillarieae. They are found in both fresh and salt water and are distinguished by their siliceous cell walls, which sink to the bottom of the water after the living organism dies, forming diatomaceous earth.

dibasic lead arsenate. See *acid lead arsenate.*

dibber. See *dibble.*

dibble. a. A small, hand tool used to make holes in the soil for planting bulbs and seeds, or for transplanting plants. It usually has a wooden handle and a steel or brass point. Also called *dibber.* See *trowel: 1; planting bar; spud: 1.* b. To transplant with a *dibble.*

dibble board. A board to which are attached wooden or metal spikes arranged in regular rows that, when pressed down on soil, produce several holes at once for transplanting. Frequently used with planting in flats. Also called *spotting board.*

dibble in. To plant seed or young trees in holes made with a dibble.

dibbler. A two-wheeled vehicle which, when moving across a field, makes regularly spaced holes for transplanting by means of long, rounded spikes on the wheel rims.

dibenzo-Γ-pyrone. See *xanthone.*

dibutyl phthalate. A chemical which is used as an insect repellent.

dicalcium phosphate. $CaHPO_4$, a manufactured product consisting chiefly of a dicalcium salt of phosphoric acid, used as a supplement to stock feed and as a constituent of commercial fertilizers.

dicaryon cells. In some groups of fungi, especially in the Ascomycetes and Basidiomycetes, a cell which contains 2 haploid nuclei (of separate sexual origin), the result being a cell with the 2 sets of genes in separate nuclei, instead of one diploid nucleus as in the majority of plants undergoing sexual reproduction.

dice. a. To cut into small cubes, as to *dice* carrots. b. Used in the past tense to denote that which has been cut into small cubes: *e.g.,* diced beets.

Dicentra. A genus of plants of the family Fumariaceae. See *Dutchman's-breeches; bleedingheart: 1; squirrel corn; fringed bleedingheart; Pacific bleedingheart; golden eardrops bleedingheart.*

dichloroethyl ether. A soil fumigant used for the control of sod webworms, the peach tree borer, and nematodes.

dichloropropylene. A soil fumigant and nemacide.

dichogamous. Of, or pertaining to, dichogamy.

dichogamy. The maturation of male and female elements of a flower at different times to insure cross-pollination. *Cf. homogamy: 1.* See *protandry; protogyny.*

dickey. A slang term for a donkey or a small bird.

Dicksonia. A genus of tree ferns, family Cyatheaceae, which have marginal spore cases and fronds that are bipinnate or tripinnate.

diclinous. See *flower: 1.*

dicot. See *dicotyledon.*

dicotyledon. A plant whose embryo has two cotyledons or seed leaves. Also called *dicot.*

dicoumarol. A chemical compound found in spoiled sweet clover and lespedeza hay. It is an anti-coagulant and may cause internal hemorrhages, which are sometimes fatal, if eaten by cattle.

dictyospermum scale. *Chrysomphalus dictyospermi* (Morg.), family Diaspididae, an armored scale insect which infests citrus and greenhouse plants.

didle. A triangular-shaped, ditching spade.

dieback. The progressive dying, from the tip downward, of terminal twigs or branches of woody or herbaceous plants, which may be caused by disease, winter injury, etc. See *exanthema.*

dieback of coffee. See *coffee anthracnose.*

dieldrin. A widely used chlorinated hydrocarbon insecticide. It is a white crystalline material which has a melting point of 172°-175° C. It has been used as a seed dressing on wheat crops in England and has been found to be poisonous to foxes.

Diels cotoneaster. *Cotoneaster dielsiana,* family Rosaceae, a semi-erect shrub grown for its pinkish flowers and red fruit. Native to China.

die off. 1. To cease to exist. 2. To die and fall to the ground, as a leaf.

die out. To cease to exist.

diesel oil. A petroleum derivative used principally as a motor fuel for diesel engines, but which is also an effective herbicide, used along ditches, banks, and other uncultivated areas.

diet. The type and amount of food and drink habitually ingested by a given animal. See *ration.*

diethylstilbestrol. $C_{18}H_{20}O_2$, a synthetic hormone and source of several estrogens, used in causing a female animal to come into heat, in chemical caponizing, and to pacify tom turkeys. Also called *stilbestrol.* See *estrogen.*

die well. To yield many choice cuts of meat from a carcass (British).

differential. 1. The difference in price paid for grades which are higher or lower than the basic grade, usually fixed by contract. 2. A system of gears in a car which permits one wheel of a pair to travel independently of the other when necessary, at the same time causing each wheel to receive its share of the power from the engine.

differential grasshopper. *Melanoplus differentialis* (Thos.), family Locustidae, a large, destructive grasshopper which infests all kinds of crops, but which is especially destructive to corn.

differential host. A plant that produces distinctive reactions to a specific virus (or viruses) allowing for differentiation of viruses.

differential locust. Differential grasshopper.

differential pulley. A system of 3 pulleys, 2 of which, having different diameters, are fixed on a single axis, the other pulley being attached to these by an endless chain.

diffused runoff. Rainfall which runs off simultaneously over the whole area of a slope, in contrast to runoff concentrated in a single channel.

diffuse-porous wood. That wood in which the pores are fairly uniform in size, usually quite small, and well distributed throughout the wood. See *nonporous wood; ring-porous wood.*

dig. 1. To turn up the earth. 2. To excavate, *e.g.,* to *dig* a ditch. 3. To remove from the earth, *e.g.,* to *dig* potatoes; to *dig* strawberry plants for transplanting.

digest. 1. To convert food within the body into a form that can be assimilated. 2. To soften, dissolve, or alter any substance by heat and chemical action.

digester. A closed vessel or retort in which a substance may be cooked, softened, or decomposed by heat, pressure, and chemical action.

digester tankage. A feed supplement or fertilizer made from finely ground, dried residue of animal tissues which has been placed in a tank under live steam. It does not include hair, horn, hoof, manure, or the contents of an animal's stomach, except in small, unavoidable traces not in excess of 0.05 percent of any of these. When this product contains more than 4.4 percent phosphorus, the word *bone* must appear as part of the brand name. If more than 5 percent of this product cannot be passed through a 2 mm. screen, the words *coarsely ground* must be published. Also called *meat meal tankage; feeding tankage.*

digester tankage with bone. Digester tankage that contains more than 4.4 percent phosphorus. Also called *meat and bone meal digester tankage; feeding tankage with bone.*

digester tankage with paunch contents. Digester tankage to which the processed contents of an animal's stomach have been added.

digestibility. 1. The readiness with which a substance may be converted into an absorbable form within the body. 2. The rate or amount of a nutrient digester, *i.e.,* the difference between the amount of a nutrient fed and the amount found in the feces, but since some of the nutrient voided in the feces may not have been a part of the nutrient fed, this is considered *apparent digestibility.* When the amount of a nutrient digested is converted to percentage it is called the *digestion coefficient.*

digestible. a. That feed consumed and digested by an animal as opposed to that which is evacuated by the animal. b. Feed which can be converted into absorbable form within the body.

digestible crude protein. The total nitrogenous compound protein fed to livestock less the nitrogenous compounds eliminated in the feces.

digestible non-nitrogenous nutrient. The total digestible nutrients less the digestible protein in a feedstuff.

digestible nutrient. That portion of a nutrient which may be digested and absorbed into the body.

digestion. 1. The changes which food undergoes within the digestive tract to prepare it for absorption and use in the body. 2. The reduction of organic matter by biochemical action into more stable organic matter; *e.g.,* the manufacture of sludge from sewage.

digestion coefficient. The amount of a given foodstuff which is digested by an animal expressed as percentage of the gross amount eaten.

digestion trial. An experiment with any feedstuff by an animal or fowl to determine the amount of any substance that can be digested.

digestive tract. The mouth, the esophagus, the digestive organs (stomach or stomachs, crop, gizzard, and the small and large intestines), and anus; all of the organs of an animal or fowl through which food passes.

digger. 1. A plow which has a short, acutely curved moldboard. 2. A device for digging potatoes, peanuts, etc., for harvest which resembles an ordinary plow but has a V-shaped blade with one edge sharpened and so mounted as to run underneath the plants.

Digger pine. *Pinus sabiniana,* family Pinaceae, an evergreen tree which produces an edible nut, and which is often a pest on California foothills invading grazing lands. Also called *grayleaf pine.*

dig in. To cover or mix with earth; *e.g.,* to *dig in* manure.

digitalis. A valuable drug, having diuretic properties, made from the dried leaves of the foxglove and used in the treatment of heart diseases.

digitalis anthracnose. A disease of foxglove caused by the fungus *Colletotrichum fuscum,* family Melanconiaceae; characterized by lesions and cankers on the leaves and stems. Infected plants may be dwarfed or killed.

digitate. Similar to, or resembling, the fingers on a hand, as the leaflets of a horsechestnut.

di-hybrid. An individual which is heterozygous with respect to two pairs of genes.

di-hybrid cross. A cross between individuals differing in two single gene-pair characters.

dike. 1. A ridge of earth which is thrown up to impound water, as in irrigation, or to divert water, as in soil erosion control. 2. An embankment or levee which is constructed to prevent inundation of low land, as on the sea-coast or in the flood-plain of a river valley. Also called *dyke.* See *levee.*

dilan. An organic insecticide composed of an active mixture of the nitropropane and nitrobutane of DDT, namely 2-nitro-1, 1-bis (P-Chlorophenyl) propane, and 2-nitro-1, 1-bis (P-Chlorophenyl) butane. Effective in the control of certain insects, especially the Mexican bean beetle *Epilachna varivestis,* where the toxicity of DDT is inadequate. More toxic to a wider range of insect pests than bulan *q.v.*

dilead orthoarsenate. Acid lead arsenate *q.v.*

dill. *Anethum graveolens,* family Umbelliferae, an annual or biennial herb whose seed is used for seasoning and medicinal purposes. Native to Europe.

dill pickle. A cucumber pickle prepared by fermentation in a diluted brine solution to which has been added dill seeds and spices. It is marketed in this solution rather than in vinegar.

dilly. See *daffodil.*

diluent. Any gaseous, liquid, or solid inert material which serves to dilute or carry an active ingredient, as in an insecticide or fungicide. Also called *carrier.*

dilute. 1. To make more liquid by mixing with water, alcohol, etc. 2. To weaken in flavor, brilliancy, force, strength, etc., by mixing with another element.

diluter. See *diluent.*

diluvium. 1. An unconsolidated deposit of glacial drift. 2. Any superficial deposit other than river alluvium. 3. Any deposit which is attributed to the Deluge of Noah. (It has no precise meaning either as a geologic or pedologic term.)

dimethyl carbate. A chemical used as a repellent for bloodsucking insects, primarily against chiggers and mosquitoes.

dimethyl dilauryl ammonium chloride. A chemical used as a spray for brown spot and copper spot of turf.

diminishing returns. See *law of diminishing returns.*

dimite. An acaricide. Also called *DMC.*

dimple. A point on the hip of a horse which has been slightly deformed by an accident.

dindle. See *dandelion; hawkweed; sow thistle: 2.*

dingle. A small narrow valley.

dingo. *Canus antarticus,* a wild dog, found in Australia, which is a predator of cattle, sheep, and fowls. It is usually yellowish-red and has erect, triangular ears, a bushy tail with a white tip, and white, splayed feet.

dingy cutworm. *Feltia subgothica* (Haw.), family Noctuidae, a cutworm which infests all plants except those with hard, woody stems.

dingy wool. Any wool which is dark, discolored, and heavy in condition.

dinitro compounds. A group of chemicals used as herbicides, and as insecticides which are used to control eggs of apple aphids and leaf rollers. They are also used as fungicides. Also called *DN.*

dinmont. A term for a sheep which has been sheared only once and is between 1 and 2 years old.

dioestrum. That part of female's cycle between periods of heat.

dioicous. Referring to a plant having imperfect flowers that are either staminate or pistillate, but not both on the same plant. See *flower: 1.*

dip. 1. a. Any chemical preparation into which livestock or poultry are submerged briefly to rid them of insects, mites, ticks, etc. See *dipping vat.* b. Any preservative preparation in which produce is briefly submerged. 2. To remove ice cream from a container with a dipper. 3. To collect oleoresin from a cup in turpentine orcharding.

dip barrel. A tight barrel with a removable head used to transport crude turpentine gum from the dip bucket to the still.

dip grain. Wood characterized by undulations of the longitudinal fibers.

dip grained. Denoting lumber which has undulations in the grain around the knots.

diphenyl. A mild fungicide.

diphtheria. See *calf diphtheria; fowl pox.*

Diplodia boll rot. A fungous disease of cotton caused by *Diplodia* spp., family Sphaerioidaceae, characterized by the diseased bolls turning black and becoming covered with pycnidia so as to appear smutty.

Diplodia disease of citrus. A fungous disease caused by *Diplodia natalensis,* family Sphaerioidaceae, characterized by the death of the bark on branches of all sizes, an oozing of the gum, and by a stem-end rot of the fruit.

Diplodia disease of corn. A fungous disease caused by *Diplodia zeae,* family Sphaerioidaceae, characterized by a seedling blight and by a dry rot of the ears. The dry rot of the ears reduces the yield, the market value, and the quality of the corn for seed purposes. Also called *mold; moldy corn; mildew rot; ear rot; dry rot; Diplodia dry rot; Diplodia ear rot.*

Diplodia dry rot. See *Diplodia disease of corn.*

Diplodia ear rot. See *Diplodia disease of corn.*

Diplodia gummosis. See *Diplodia disease of citrus.*

Diplodia-phomopsis gummosis. See *Diplodia disease of citrus.*

Diplodia root necrosis. See *Diplodia disease of corn.*

Diplodia root rot. See *Diplodia disease of corn.*

Diplodia rot. See *Diplodia disease of citrus.*

Diplodia stalk rot. See *Diplodia disease of corn.*

Diplodia tip blight. A fungous disease of pine, fir, and spruce caused by *Diplodia pinea,* family Sphaerioidaceae, characterized by stunting of new growth and browning of the needles. Lower branches are heavily affected.

diploid. Having one genome comprised of two sets of chromosomes. Somatic tissues of higher plants and animals are ordinarily *diploid* in chromosome constitution in contrast with the haploid (monoploid) gametes. *Cf. haploid.*

diplotene. That stage in the prophase of meiosis following pachytene, but preceding diakinesis, in which the chromosomes are visibly double.

dipper. 1. A cup which has a long handle. 2. One who collects the oleoresin from turpentine cups, or the tool which is used to remove oleoresin from the cups.

dipper dredge. A power shovel or bucket suspended from an arm attached to an A-frame (*q.v.*) set on a floating platform. It is used in constructing ditches to drain swamp land.

dipper gourd. Calabash gourd.

dipping. 1. a. A method of preserving seasoned wood by submersion in an open tank of creosote or similar preservative. b. The submersion of any product for the purposes of preservation, addition of a substance, etc. 2. The submerging of animals or fowls in a liquid bath which contains insecticides, ovicides, repellents, etc.

dipping hook. A stafflike device used for immersing sheep in a dipping vat and for lifting them out of the vat.

dipping method. Jar method *q.v.*

dipping plant. A place where animals are dipped, consisting of a system of corrals or pens, a dipping vat, a draining pen, legging pen, etc.

dipping vat. A pit filled with a liquid containing insecticides, ovicides, repellents, etc., through which animals are forced to pass for disinfestation. It is concrete-lined or otherwise waterproofed, rectangular in shape, but so narrow that an animal is unable to turn in it. At the entrance it has a vertical drop of 5 or more feet while at the exit it has a cleated ramp.

dip siruper. A tank containing sirup into which cans filled with fruit, etc., are dipped to add sirup.

dip slope. A slope of a land surface which conforms approximately to the depression of the underlying rocks.

dipteron. An insect of the Order Diptera which usually has 2 wings. Many, such as the mosquitoe, tsetse, etc., are pests because of their ability to bite and to spread diseases.

direct expansion. A refrigeration system in which the heat is transferred directly to the evaporating or primary refrigerant of a compression refrigerated system. It is the opposite of indirect refrigeration which utilizes a secondary refrigerant (brine, water, alcohol, gycol, etc.) to carry the heat from the product to the primary or evaporative refrigerant.

direct-expansion surface coolers. Surface cooling units in which the cooling medium is a gas expanded directly within the tubes. This type of cooler is contrasted with those using a refrigerated liquid, such as water, as the cooling medium.

direct-expansion system. A method of the distribution of refrigeration in cold-storage rooms in which a refrigerant is circulated in exposed pipes near the ceiling. Fans may be used to assist the gravity circulation. See *direct expansion.*

direct-fired kettle. A device which consists of a cast-iron kettle placed over a fire, used to concentrate juice, sap, etc., to a sirupy consistency. It may be used in maple sirup or sorghum sirup, etc., production. Sirup so produced is often caramelized.

directional pruning. Removal of small branches of trees, usually ornamentals, to allow for passage of wires, etc.

direct marketing. 1. The sale of produce by a farmer directly to the consumer. 2. The sale of animals to slaughterers at places other than a central livestock market.

direct microscopic count. See *microscopic count.*

direct purchases. A method of price control in which the government competes with other buyers in purchasing an over-abundant commodity from the processors, in an effort to deplete the over-supply, thus returning the price to a desired level. The processors are, in turn, required to pass the supported price back to the producers.

direct relationship. In breeding, a parent-child relationship between individuals. See *collateral relationship.*

dirt. 1. Soil. 2. In excavation engineering, any unconsolidated, geologic deposit.

dirt band. A square, open-ended, heavy, paper box placed in a flat along with others. Soil is put in the boxes and seedlings are planted in it. On transplanting, the boxes are not removed so that the roots are undisturbed.

dirt farmer. A farmer who does all or part of his own work in the fields and stables, as contrasted to one who hires all labor or rents his land, or to an absentee farm owner who employs a manager to operate his farm.

dirt that will not wash off. See *Rhizoctonia disease of potato.*

dirty. In marketing, an egg which has dirt adhering to the shell, or a shell with stained, soiled spots covering more than ⅛ of the shell surface but with no adhering dirt.

disbud. a. To remove some of the buds from a plant for the benefit of the remaining ones. b. To remove shoot buds from a plant so as to prevent the growth of unwanted shoots.

disbudding. The removal of flower buds and/or shoot buds from a plant.

disc. See *disk.*

discharge. 1. The quantity of water, silt, or other mobile substances passing along a conduit per unit of time; rate of flows. 2. An exudate or abnormal material coming from a wound or from any of the body openings: *e.g.*, a bloody *discharge* from the nose.

discharge arms. A part of the binder attachment on a grain binder which moves the tied bundle from binder attachment to the bundle carrier or to the ground.

discharge bay. An open-ended, concrete basin into which water is discharged, used principally in pumping drainage from low areas of peat or mineral soils where an outlet of sufficient depth is not available.

discharge curve. A rating curve showing the relation between stage (depth) and discharge of a stream.

discharge hose (pipe). A hose or pipe attached to the discharge side of a pump, or to a forced system, through which the water or spray moves away from the pump or tank toward the nozzles or outlets.

disclimax. A climax (*q.v.*) that is a consequence of the disturbance of the natural landscape by man or domestic animals.

disc nozzle. A common nozzle, often used with garden hoses, that consists of a long, narrow chamber inside of which is a lens-shaped disc which may be moved by an adjustable plunger. Altering the position of the disk in the nozzle controls the angle of the spray.

discolored. In marketing, a grade designating grain sorghums of a color not true to type.

discontinuance. In zoning ordinances, abandonment.

discontinuous growth ring. A growth layer which is formed on only a portion of the stem of a tree and apparent when viewed in cross section.

discontinuous spread. The dissemination of similar or related diseases or organisms in widely isolated areas. *Cf. continuous spread.*

discount. 1. The amount allowed as a deduction for the settlement or payment of a debt before it is due to be paid. 2. The amount of money which is deducted as a fee by banks or brokerage houses from the sale of an issue of securities. 3. The excess money over the par value or face value of a land contract, or other security.

disc wheel. A solid or plated wheel as contrasted to a spoked wheel. The central portion may be flat, as in chain wheels and the lighter classes of toothed wheel, or it may be dished, as in the wheels of trucks and trolleys. The latter form is strong, because it will accommodate expansion and contraction due to changes in temperature, or to internal and external stresses.

disease. Any deviation from a normal state of health in plants or animals which temporarily impairs vital functions. It may be caused by viruses, pathogenic bacteria, parasites, poor nutrition, congenital or inherent deficiencies, unfavorable environment, or any combination of these.

disease control. Any procedure which tends to inhibit the activity or effect of disease-causing organisms, or which modifies conditions favorable to disease.

disease cycle. The complete chain of events in the manifestation of a disease, beginning with the infection of the normal organism and ending with the return to normalcy after the disease has run its course.

diseased. Designating a plant or animal that has a disease *q.v.*

disease development. That part of the disease cycle (*q.v.*) from the time of infection to the complete expression of the disease.

diseased plant. Any plant in which the normal physiological and morphological development has become inhibited or adversely altered.

disease outbreak. The onset of an epidemic, epiphytotic or epizootic.

disease resistant. Designating plants or animals which are not readily susceptible to, or are able to withstand, a particular disease.

disease X. See *X-disease: 1.*

dish. An indentation or concavity, as in the face, back, sides, etc., of an animal.

dishcloth gourd. Towelgourd.

dished. Designating an animal which has a concave-shaped face, side, back, etc.

dish-faced. Designating an animal which has a facial concavity. Also called *stag-faced.*

dish garden. Vegetable garden *q.v.*

Dishley Merino. A crossbred sheep from the Leiscester and Merino.

dish of the wheel. The degree of concavity of a wheel.

dishorning. Dehorning.

dishrag flavor. See *utensil flavor.*

dishrag gourd. Towelgourd.

disinfect. To destroy or render inert disease-producing germs, harmful microorganisms, and to destroy parasites.

disinfectant. 1. Any substance whose property is to disinfect; that which destroys disease-causing germs or frees a plant or animal from infection. 2. See *foreign flavor.*

disinfected. Designating any plant, animal, building, plot, etc., that has been treated with a substance which renders it free from infection.

disinfection. The freeing of a plant, animal, building, plot, etc., from infection by means of a disinfectant.

disinfective. See *disinfectant.*

disinfest. To rid, destroy, free from insects, parasites, rodents, etc.

disinfestant. 1. Any substance which destroys or rids an animal, fowl, building, etc., of insects, parasites, rodents, etc. 2. Any substance which destroys or inactivates disease organisms in the soil or on seed surfaces before these organisms can cause infection.

disinfestation. 1. The destruction of insects, parasites, rodents, etc. 2. The destruction or inactivation of disease organisms, in the soil or on seed surfaces before the organisms can cause infection.

disintegrator. See *size reduction.*

disintegration. See *rock disintegration.*

disjointed. Designating dressed poultry which has had the joints severed for convenience in canning or cooking.

disjunction. Separation of homologous chromosomes or of parts of chromosomes during mitosis: (a) in somatic mitosis, separation of chromosome halves, one going to each daughter cell; (b) in meiosis, separation of the synapsed maternal and paternal chromosomes, one going to each daughter cell.

disjunctive symbiosis. 1. The living together of two distinct, unconnected organisms without apparent contribution of benefit of one upon the other, as the living of ants in hollow thorns of some species of Acacia. 2. That symbiosis involving only temporary association of the symbionts, as in the cross-pollination of flowers by insects.

disk. 1. A round, usually sharp-edged, slightly dished, steel plate which cuts the soil as it revolves on a center axis and moves the soil to one side. Several *disks* are mounted and spaced on a horizontal shaft to make the disk harrow. Also called *disk blade.* 2. See *disc nozzle; disk harrow.* 3. One of a series of metal plates in a centrifugal separator bowl which increases the efficiency of the machine. 4. In botany, (a) an adhesive surface on the tendril ends of creeper plants which enables them to climb along flat surfaces; (b) an enlargement of a flower's receptacle; (c) the center of a composite flower; (d) an organ's surface; (e) the circular valve of a diatom.

disk blade. See *disk: 1.*

disk cleaner and grader. A device used to clean and grade seeds which consists of a series of disks on each of which are pockets deeper or shallower than the length of the seed which is being cleaned. Unwanted seeds are ejected at the end of the disk chamber. Also called *pocket cleaner.*

disk coverer. Two disks which are attached to a lister planter immediately behind the seed spout, whose purpose is to cover the planted seed with soil.

disk cultivator. A tiller which has two or more gangs of disks for cultivating. It is useful on ridges or in a field infested with vines.

disked. 1. Designating soil which has been tilled with a disk harrow. 2. Pertaining to chopping debris into small bits and putting it under the soil with a disk harrow.

disk exhaust box. A device used in canning to heat a can and its contents before sealing. It consists of a metal box in which are several revolving disks which force cans through the heated chamber.

disk floret. Those flowers of the head of a plant, of the family Compositae, in which the corolla is tubular, as contrasted with the marginal flowers where the corolla is ligulate or strap-shaped.

disk flower. See *disk floret.*

disk furrower. A planter attachment consisting of 2 disks which open a furrow for planting seed. See *double-disk furrow opener.*

disk gang. Several disks on a disk harrow rigidly placed about 6 ins. apart on a gang or arbor bolt.

disk go-devil. See *ridge buster.*

disk harrow. A harrow consisting of 2 or more gangs of disks which is used: (a) to cut debris, especially vegetable matter on the surface of the soil; (b) to pulverize the top layer of the soil before or after plowing; (c) to prepare plowed ground for planting; (d) to cultivate crops growing in widely spaced rows; (e) to cover seed sown broadcast; (f) for summer fallowing.

disk hiller. An attachment for a cultivator which consists of two disks so set that they throw soil onto the hill. It is frequently used in cotton cultivation for barring-off before chopping.

disking. The cultivation of soil by a disk.

disk nozzle. See *disc nozzle.*

disk plow. A plow which uses a disk rather than a share and a moldboard to cut and turn the soil. It is preferable to the moldboard in very loose soil, hard ground, or land covered with much plant residue or debris, as it will not scour. Also, it is possible to change the angle of the disk.

disk plow bottom. See *disk: 1.*

disk plow state. Texas.

disk tiller. Disk cultivator *q.v.*

disk waterhyssop. *Bacopa rotundifolia*, family Scrophulariaceae, a perennial marsh herb which is a serious weed pest of rice fields in California, U.S. Native to North America.

disk weeder. Disk cultivator *q.v.*

dislocating. A method of killing poultry in which the fowl's head is pulled and twisted until the neck separates from the skull and the spinal cord is severed. It is more commonly used in Europe than in the U.S.

dismount. To get down, alight, from a horse's back.

disodium ethylene bisdithiocarbamate. See *nabam.*

disorder. An unwholesome or unnatural physical condition of a plant or animal. See *disease.*

dispatcher's meter. A device which by integrating current local fire burning indexes, resistance to line-construction factors, fatigue, and travel time, indicates for specific fuel types the strength of attack required to control a forest fire within specified speed of attack standards.

dispersal. The spreading of plants or animals.

dispersed farmstead. A farmhouse, farm buildings, and land constituting an isolated unit, in contrast to a farm within a village community.

dispersion. 1. In reference to the gregariousness

of plants, the comparative isolation of certain plants. 2. In sod, the destruction of a soil structure so that each soil particle behaves as a unit.

dispersion failure. Failure to assemble animals in one place due to their scattering.

display. An exhibition of cut flowers, fruits, vegetables, foliage plants, etc., and other agricultural activities, which may be held and judged at a county or state fair.

disposition. The temperament, spirit, of an animal.

dispossess. To deprive a person of his possession or occupancy of real estate by legal action.

disqualification. In animal husbandry, a defect of characters of form or breed types which disqualifies an animal from an exhibition or from breed registration.

dissect. To cut or divide a plant or animal into pieces for examination.

dissected. Used to describe a leaf, etc., which is divided into many divisions.

dissected land. Surfaces that have been cut into valleys and ridges by stream erosion. Sharpness of sculpturing and close spacing of ravines and valleys is usually implied. Dissection may be deep or shallow. In some plateau regions the interstream areas or ridges may be flat-topped.

dissection. 1. That aspect presented by a land surface which has been cut into segments by valleys and ravines by the work of stream erosion. 2. The cutting or dividing of a plant or animal into pieces or the separating of parts for examination.

disseminate. 1. To sow broadcast. 2. To spread or scatter indiscriminately, as a disease or the seeds of plants by wind, water, animal carriers, etc.

dissemination. 1. The spreading or transporting of inoculum from a diseased individual to a healthy one. 2. The broadcast sowing of seed. 3. The natural scattering or distribution of weed seeds.

dissolved bone. A fertilizer material which consists of ground bone or of bone meal that has been treated with sulphuric acid.

distal. Away from the main part of the body, as the *distal* part of a leaf.

distemper of dogs. An acute, widespread, very contagious disease of dogs, especially of young ones, caused by a filterable virus which may cause death. Also called *dog disease; canine distemper.*

distemper of horses. Strangles *q.v.*

distill. To drive off gas or vapor from a substance by means of heat in a still and to condense the gas by cooling.

distillate. 1. The substance which is obtained by distillation. 2. The common name for one of the heavier, less volatile, petroleum products which was, at one time, widely used as a tractor fuel.

distillation. The converting of a substance into a gas or vapor by heat in a still and later the condensing of it.

distilled spirits. Any alcoholic beverage which has been distilled, *e.g.*, corn whiskey.

distilled vinegar. Vinegar which in its preparation has been subjected to distillation.

distilled water. Water which has been boiled, converted into steam, and then condensed by cooling.

distillers' condensed solubles. See *distillers' semi-solid solubles.*

distillers' corn dried grain. See *corn distillers' dried grains.*

distillers' dried grains. See *corn distillers' dried grains.*

distillers' dried grains with solubles. See *corn distillers' dried grains with solubles; rye distillers' dried grains with solubles; wheat distillers' dried grains with solubles.*

distillers' dried solubles. A by-product livestock feed which is obtained from the water-soluble fraction of grain fermentation mashes from which the alcohol has been distilled.

distillers' dried yeast molasses. The properly dried yeast resulting from the fermentation of molasses and yeast which is separated from the medium, either before or after distillation.

distillers' grains. A by-product livestock feed which is obtained as a result of the manufacture of alcohol and distilled liquors from corn, rye, and sometimes, a mixture of rice and other cereals.

distillers' rye dried grains. See *rye distillers' dried grains.*

distillers' rye solubles. See *rye distillers' dried grain with solubles.*

distillers' semi-solid solubles. A by-product livestock feedstuff obtained from the water-soluble fraction of grain fermentation mashes from which the alcohol has been distilled. It is partially dried and contains about 60 percent moisture.

distillers' soluble molasses. The product obtained by drying the residue from the yeast fermentation of molasses after the removal of alcohol by distillation.

distillery waste. The residue resulting from the fermentation and distillation of beer and distilled spirits which contains considerable potash and some ammonia, and which may be recovered by evaporation to be used as a fertilizer material.

distressed commodity. Any agricultural product sold at a price such that the producer receives little or no profit. In U.S., the Secretary of Agriculture may designate the product as a *distressed commodity.*

distributaries. 1. Any system of secondary conduits, especially laterals which carry irrigation water to farmlands. 2. The network of channels on a river delta.

distributing stand. Any device used in irrigating practices from which water is distributed into furrows and basins. Usually it is a part of a riser outlet from an underground pipe system. See *capped stand; valve-controlled hydrant.*

distributor. 1. Any device which is used to move produce from one place and to scatter it in another: *e.g.*, a fertilizer *distributor.* 2. An agent or a wholesaler who sells goods in quantity. 3. A device in some bulk milk tanks which spreads the milk over a cooling surface. 4. That part of a motor engine which conducts the secondary current to the spark plugs.

ditch. An artificial excavation, as a trench or channel dug to carry irrigation or drainage water. See *drain: 1.*

ditch bank. The side slope or shoulder of a ditch.

ditchbur. *Xanthium strumarium,* family Compositae, an annual herb, found as a weed pest from Massachusetts to Pennsylvania, U.S.

ditchdown. See *cattail: 2.*

ditcher. One who digs ditches or a machine used for that purpose. See *middlebuster.*

ditch fern. Royal fern *q.v.*

ditch-grader. A tool made in various sizes and usually with a mounted angle blade which is used to construct terraces, grade dirt roads, and cut ditches.

ditching machine. Used for digging ditches, tile trenches, and making excavations, the machine is powered by steam or internal combustion and has a bucket with a cutting edge attached to a steel frame. It may also have a continuous belt or wheel to which buckets are attached to dig out the soil and deposit it at the side of the trench.

ditching plow. An old-style plow which was specially designed with a deep, narrow share for cutting trenches and drains. It had boards on each side of the share for depositing the soil on either side of the excavation.

ditching scoop. A shovel used for digging ditches.

ditch millet. India pappusgrass.

ditch moss. Canada waterweed.

ditch rider. A person appointed by an irrigation company or a governmental unit to examine irrigation ditches, weirs, etc., and to adjust the measuring devices so that the correct amount of water is delivered to a farmer.

ditch stonecrop. Virginia penthorum.

dithane. See *nabam.*

dittany. Any plant of the genus Dactamnus, family Rutaceae, especially the gasplant dittany *q.v.*

diurnal. Designating a plant whose blossoms open during the day and close at night. *Cf. nocturnal.*

diverging-belt grader. A machine which grades fruits and vegetables according to size.

diversification. The practice of raising a variety of crops or animals on a farm.

diversified farming. See *diversification.*

diversified ranch. See *diversification.*

diversion chamber. A chamber that contains a device for diverting all or a part of the flow of water for irrigation.

diversion channel. A channel constructed along the edge of a hill which minimizes erosion and prevents excess runoff water from flowing onto low lying areas. Such a channel is deeper and wider than a diversion terrace, and is supported by a ridge.

diversion dam. A barrier built to divert part or all of the water from a stream into a different course: *e.g.,* into an irrigation ditch.

diversion ditch. A ditch constructed to change a flow of water.

diversion terrace. 1. A wide, relatively shallow channel of low gradient with gentle side slopes and ample water capacity, which is constructed across the slope of a field to intercept and change the direction of flow and to reduce the velocity of run-off water. 2. An artificial ditch used to divert water from its natural course.

divert. 1. To turn a river or stream aside from its natural channel into a new one. 2. To take water from a stream and direct it into another channel as by gravity channel or pumping.

divide. 1. The ridge or other land area which separates or lies between two watersheds or drainage systems. See *Great Divide.* 2. To separate a colony of bees so as to produce two or more colonies. 3. To cut the crown or underground parts of a plant into two or more portions so as to obtain more plants.

divided-apron system. A way of sorting fruits and vegetables by a machine which has a wide continuous belt with three lengthwise partitions over the belt. The products to be sorted are placed in the two outside alleys and the sorters select and place the sound products in the central alley.

divider. 1. In grain and hay harvesting, a metal or wooden device attached to the end of a cutter bar which separates tangled crops and prevents stems from becoming entangled in the reel shaft. 2. A board in the feeder of a threshing machine which helps to keep the bundles separated and following each other.

divider board. See *divider: 2.*

divider stick. See *divider: 1.*

dividivi. *Caesalpinia coriaria,* family Leguminosae, a tropical tree whose pods yield tannic and gallic acids.

divining rod (stick, wand). A rod or switch, often a forked willow or peach twig, which when held tightly in the hands of a dowser (water witch) is supposed to be irresistibly drawn to the earth over an underground supply of water, and hence, is used in locating a place to drill a water well. See *water witching; dowser.*

division. A method of plant propagation by cutting the crown or other parts of a plant so as to obtain more plants.

division block. A square piece of wood placed in the chamber of a hay baler to divide the bales.

division board. A thin board placed inside a beehive to reduce the size of the chamber.

division box. A structure for dividing and diverting all or part of a flow of water into channels. See *weir.*

division fence. A fence erected on a range to facilitate the distribution of animals.

division gate. A structure that divides the flow of water between two or more laterals.

divot. A small, irregular piece of turf.

Dixie iris. *Iris hexagona,* family Iridaceae, a perennial herb grown for its lavender flowers. Native to North America.

D-layer. D-horizon *q.v.*

DMC. See *dimite.*

DN. See *dinitro compounds.*

doab. 1. A tract of land which lies between and immediately above the junction of two streams (India). Also called *duab.* 2. Dark, sandy clay which is associated with bogs (Ireland).

dobbin. An affectionate designation for a gentle horse.

dobe (dobie, dobby, doby). See *adobe soil.*

dobe house. A house made of sun-dried mud cakes or crude bricks. Also, one whose walls are made of mud, sometimes mixed with straw and cast in a frame or form to dry and harden in the sun. It is a very ancient form of building structure of the Middle East and was also used in arid regions of southwestern U.S. by early settlers. See *adobe soil.*

dock. 1. A plant of the genus Rumex, family Polygonaceae. Species are for the most part pests. See *garden sorrel.* 2. That fleshy part of an animal's tail as distinguished from the brush. 3. To cut off or to shorten the tail of an animal.

dockage. 1. Weed seeds and foreign matter which may be in commercial seeds. 2. Foreign material in marketable grain which is easily removed by normal cleaning methods.

dockage sieve. A set of sieves, usually found at a grain elevator, into which a sample of grain is placed and shaken through to determine the amount of dockage that the sample contains.

docked. An animal which has had part of its tail amputated. See *dock: 3.*

docked and set up. A horse which carries its shortened tail high as a result of a docking operation.

docking. The removal of the tail a short distance from the body of an animal; cutting off most of the solid part of an animal's tail.

docking pincers. A pair of pincers which, when heated to a dull-red glow, are used to snip the tails off of lambs.

dockmackie. Mapleleaf viburnum.

dock nettle. See *dog nettle: 1.*

dock sorrel. See *garden sorrel.*

dod(d). A name sometimes given to cattail in England.

dodder. Any plant of the genus Cuscuta, family Convolvulaceae. These are parasitic vines whose seeds germinate in the ground and, as soon as they establish themselves on a host plant, draw all their food from the host. They have a reputation of being poisonous to livestock and to bees. Also called *strangleweed; love vine; goldthread; pull down; clover dodder; devil's-hair; hell bind; devil's-guts; hairweed.*

doddie. A polled cow.

dodrans. A full span, about 9 inches or 23 cm.

doe. The adult female rabbit, goat, or deer.

doer. Any animal or fowl which has the proper facility for producing, such as gaining weight.

doffing rolls. A part of the cotton gin which forms lint cotton into bats preparatory to baling.

doftberry. See *belladonna.*

dog. 1. *Canis familiaris,* family Canidae, a domesticated, carnivorous animal which is used as a household pet, a watchdog, a herder for sheep, cows, etc. 2. A device, usually consisting of a steel hook and chain, used in skidding logs. 3. A sort of iron hook or bar with one or more sharp fangs which may be fastened into a piece of wood or other heavy article to move it. 4. Any part of a machine acting as a claw or clutch, as an adjustable stop to change the direction of a machine tool. 5. A low grade, beef animal.

dogbane. Any plant of the genus Apocynum, family Apocynaceae. Species are perennial herbs grown sparingly for ornament and formerly used by the American Indians for the fiber contained in the cortex. See *hemp dogbane; spreading dogbane.*

dog bee. See *drone.*

dog bent. See *velvet bentgrass.*

dogberry. American mountain ash.

dog biting louse. *Trichodectes canis* (DeG.), family Trichodectidae, a louse which infests dogs. *Cf. goat biting louse.*

dog bur. Houndstongue.

dog button. See *nuxvomica poisonnut.*

dog disease. See *distemper of dogs.*

dog distemper. See *distemper of dogs.*

dog fennel. Mayweed camomile.

dog flea. *Ctenocephalides canis* (Curt.), family Pulicidae, an insect which infests man, rodents, dogs, cats, etc. It is a possible vector for bubonic plague and is an alternate host for the dog tapeworm and the rodent tapeworm which occasionally parasitizes man.

dog flower. Purple trillium.

dog follicle mite. *Demodex canis* Leydig, family Demodicidae, a mite infesting dogs which causes red mange.

dogfoot. See *orchardgrass.*

dog grass. An old name for quack grass (*q.v.*). The grass roots were once dried, ground, and used as a substitute for flour or mixed with flour or meal in the making of bread during periods of food scarcity.

dog hair. Coarse wool from the hind legs of sheep.

dog hobble. Drooping leucothoë.

dogie (dogy, dogey, doge). A term used on western ranges, U.S., for a motherless calf.

dog laurel. Drooping leucothoë.

dogleg fence. In Australia, a fence made of logs and saplings.

dogleg tobacco. An old term for an inferior grade of tobacco.

dog lilly. Spatterdock cow lily.

dog nettle. 1. *Urtica urens,* family Urticaceae, an annual herb which is a weed found on waste ground in U.S. Also called *dock nettle.* 2. See *deadnettle: 2.*

dog plum. Speckled beadlily *q.v.*

dog rose. *Rosa canina,* family Rosaceae, a shrub which bears a single white or pinkish flower. It is much used as an understock for roses and has become naturalized in parts of eastern U.S. Native to Europe. Also called *eglantine.*

dog rowan tree. Snowball European viburnum.

dog's finger. See *foxglove: 6.*

dog's mouth. See *snapdragon: 2.*

dog's thistle. Canada thistle *q.v.*
dog's tongue. Houndstongue.
dog sucking louse. *Linognathus setosus* (Olf.), family Linognathidae, a louse which infests dogs.
dog tick. See *American dog tick.*
dog tooth. Canine tooth *q.v.*
dog-tooth star grass. Bermudagrass *q.v.*
dogtooth violet. Fawnlily.
dog town. A community of prairie dogs (*Cynomys* spp., especially *C. ludovicianus*) prevalent in western U.S. Also called *prairie-dog town.*
dog tree. See *bloodtwig dogwood; European euonymus; European elder; snowball European viburnum; European alder: 1.*
dogtrot. A slow, gentle trot *q.v.*
dog violet. See *American dog violet.*
dogwood. 1. Any of the shrubs and trees of the genus Cornus, family Cornaceae. Species are grown as ornamentals. See *Tatarian dogwood; pagoda dogwood; silky dogwood; bunchberry dogwood; flowering dogwood; Corneliancherry dogwood; Pacific dogwood; gray dogwood; roundleaf dogwood; bloodtwig dogwood; red osier dogwood.* 2. See *snowball European viburnum; hobblebush viburnum; striped maple.*
dogwood borer. *Thamnosphecia scitula* (Harr.), family Aegeriidae, a whitish caterpillar which feeds in the cambium of dogwood and pecan. Small trees may be girdled and die as a result of the infestation.
dogwood club-gall midge. *Mycodiplosis alternata* Felt, family Itonididae, a midge that causes a gall on the small twigs of dogwood.
dogwood scale. *Chionaspis corni* Cooley, family Diaspididae, a scale insect which infests dogwood.
dogwood spittlebug. *Clastoptera proteus* Fitch, family Cercopidae, a sucking insect that infests the terminal shoots of the dogwood and pecan.
dogwood twig borer. *Oberea tripunctata* (Swed.), family Cerambycidae, the larva of a beetle which bores into dogwood, elm, viburnum, and many fruit-tree twigs.
Dokuchaiev formula. The formula given for soil formation by the Russian pedologist, Dokuchaiev, in 1899, as follows: P=f(K,O,G,V,) when P is soil; K, climate; O, organisms; G, "subsoil" or parent material; V, age or time factor.
dollar plague. Dourine *q.v.*
dollarplant. *Lunaria annua,* family Cruciferae, an annual herb grown for its dried pods which are used in bouquets. Native to Eurasia. Also called *honesty; moonwort; satinpod; satin flower.*
dollarspot. 1. A fungous disease, particularly of the bent grasses, caused by *Sclerotinia homeocarpa,* family Sclerotiniaceae, which is characterized by dark, water-soaked spots on the turf. These small spots then turn brown and later become straw-colored. 2. Used to describe the piebald coat of a horse.
dollop. A clump of grass (England).
dolly. A platform supported by rollers on which heavy loads may be moved, usually in a packing plant or warehouse.
dolomite. A natural mineral, $MgCO_3CaCO_3$, a double carbonate of lime and magnesia. See *dolomitic limestone.*
dolomitic limestone. A limestone which contains dolomite; a magnesium limestone. The rock is significant because soils derived from it may be markedly different in fertility from those derived from purer calcium carbonate limestones. Finely ground magnesium limestones are also used as soil amendments.
dolphin fly. Bean aphid *q.v.*
domestic. 1. Native wool produced in a country in contrast with foreign-grown wool. 2. Wool from eastern and midwest U.S. in contrast with that from the western areas. 3. Designating a species of animal or plant which has been brought under control by man and improved by selection and/or breeding so that its products and services are of use.
domesticate. To bring a wild animal, fowl, or plant under control and to improve it by careful breeding, selection, and handling, so that its products or services are more useful to man.
domestic fowl. *Gallus domesticus,* family Phasianidae, the chicken.
domestic market. A market within one's own country, as opposed to a foreign market.
domestic ryegrass. See *ryegrass.*
domestic sausage. A kind of sausage which includes fresh pork, frankfurter, and bologna. After being put into casings, the sausage is cooked and smoked. *Cf. dry sausage.*
domestic wool. Wool produced east of the range country, U.S.
Domiati. A so-called pickled cheese made in Egypt from whole or partly skimmed cow's or buffalo's milk. It is soft and mild when fresh, but acid when cured.
dominance. In genetics, the predominance of one of any pair of opposite Mendelian characters.
dominant. 1. In genetics, designating one of any pair of opposite Mendelian characters which dominates the other, when both are present in the germ plasm, and appears in the resultant organism. *Cf. recessive.* 2. Designating a plant species which characterizes an area.
dominant crown class. Trees with crowns which extend above the general level of the forest canopy and receive full light from above and some from the side. These trees are larger than the average tree in the stand with well developed crowns but are possibly somewhat crowded on the sides.
dominant trees. Those trees in a forest that are tallest.
Dominicker. A name formerly applied to the Barred Plymouth Rock *q.v.*
Dominique. An American breed of medium weight, dual-purpose chickens. The plumage is slate colored, crossed by dark and light bars. The cock weighs up to 7 lbs.; the hen up to 5 lbs. The egg shell is brown.
donga. A ravine or deep channel which results from water erosion (South Africa).

donkey. 1. A portable engine. 2. An ass.

donkey oats. Garden sorrel *q.v.*

doob (doub). Bermudagrass.

doodle. One of several small bunches of hay left in a field to cure.

doodlebug. 1. The larva of several small insects. 2. A homemade tractor constructed from a used automobile.

dool. A boundary or landmark which may be a strip of unplowed land between furrows (England). Also called *dole.*

Doolittle division-board feeder. A container for feeding bees which is the size and shape of a beehive frame containing floats on a U-shaped piece of screen wire to enable the bees to take the sirup without drowning in it.

doom palm. See *doum palm.*

door test. A test for moisture, quality, number per pound, etc., which is given to produce by a buyer at his receiving station.

door-weed. Prostrate knotweed *q.v.*

dooryard plantain. Rippleseed plantain *q.v.*

Dorking. A five-toed breed of chicken believed to have been brought to England by the Romans. They are dual-purpose birds which lay a white-shelled egg. The cock weighs up to 7½ lbs.; the hen up to 6 lbs. Varieties are White Dorking; Silver-Gray Dorking; Colored Dorking.

dormancy. 1. A physiological condition of plants, animals, and insects in which development ceases until more favorable conditions, such as of moisture or temperature exist. See *sleepy seeds.* Also called *quiescence; delayed germination.* 2. The inability of a plant or seed to resume growth even though environmental conditions are favorable for growth. Also called *rest period.*

dormant. Designating a period in a plant, animal, or insect's life when physiological activities are slowed to a minimum. Also called *resting; sleeping; inactive; quiescent.*

dormant bud. 1. A bud which is ordinarily inactive but which may begin to develop as a result of certain conditions: *e.g.*, when the stem beyond it is removed. 2. A bud which is inactive during the winter due to unfavorable conditions for growth.

dormant-budded. Designating a plant which has been bud-grafted when it is physiologically inactive.

dormant grafting. Grafting performed on a plant during the period the plant is physiologically inactive.

dormant pruning. Pruning carried on during the period of a plant's physiological inactivity.

dormant spray. An insecticide, fungicide, etc., applied to a plant during its period of physiological inactivity.

dornick. A piece of rock, or a stone, which is not too large to be thrown by a man. Also, a rock of iron ore.

dorsal. Referring to the back, *e.g.*, the *dorsal* fin of a fish, the underside of a leaf.

Dorset. 1. A medium-sized breed of sheep originating in southern England; a ram weighs up to 250 lbs. and a ewe up to 175 lbs. An individual produces about 7 lbs. of wool per shearing of three-eighths blood and quarter blood (*q.q.v.*) grade. 2. A hard, blue-veined, English cheese made from partly skimmed cow's milk cured with rennet. Also called *Dorset Blue; Blue Vinny; Blue Veiny.*

Dorset Horn. See *Dorset: 1.*

dosage response curve. In veterinary medicine, the measure of medicine containing a given chemical as plotted against the killing power of a chemical.

dose syringe. A syringe with a long pipe or nozzle which is used to force medicine down the throats of animals and poultry and to wash out the throats or crops of fowls.

dote. An early decay of wood characterized by patches or streaks of lighter or darker color than normal. Also called *doze.*

doted. Designating an early decay of wood. See *dote.*

Dothichiza canker. A fungous disease of the Lombardy poplar, balsam, California poplar, and eastern poplar caused by *Dothichiza populea,* family Sphaerioidaceae, characterized by elongated, dark, sunken lesions on the trunk, limbs, and twigs. The bark and cambium are destroyed, and when the lesions girdle the trunk or limbs, the distal portion dies.

Dothiorella gummosis. A fungous disease of citrus trees caused by *Botryosphaeria ribis,* family Dothioraceae, which is characterized by gum oozing out onto the surface of the bark. In mild cases the injury may heal over, while in severe cases large sections of the bark may be killed and the wood beneath discolored.

Dothiorella rot. A citrus fruit rot caused by the fungus, *Botryosphaeria ribis,* family Dothioraceae, usually characterized by a rot beginning at the style end of the fruit, resulting in a leathery, drab-brown, pliable decay. An amber juice is evident and as the rot continues, the color becomes dark-olive to black. It occurs on fruit in transit or in storage.

Dothiorella wilt. A fungous disease of elms caused by *Dothiorella ulmi,* family Sphaerioidaceae, characterized by a drooping and wilting of the leaves. The leaves become yellowed, then brown, roll upward to the midrib and drop off. There is an extensive formation of suckers below the point of infection. Also called *Verticillium wilt; Cephalosporium wilt.*

dotted hawthorn. *Crataegus punctata,* family Rosaceae, a deciduous, thorny shrub grown as a hedge plant or for its white flowers.

dotted melaleuca. *Melaleuca hypericifolia,* family Myrtaceae, a tall shrub grown for its red flowers. Native to Australia. Also called *hillock tree.*

double. 1. Designating a blossom which has more petals and sepals than normal, usually as a result of plant breeding. *Double* flowers are often highly prized. *Cf. single.* 2. Designating a team of horses: *e.g.*, to work horses *double.*

double-acting tank pump. A type of sprayer pump

which forces an ejection on both the downward and upward strokes.

double-action harrow. Double-disk harrow *q.v.*

double-bail plow. A frame plow which has two U-shaped iron pieces that attach the beam to the frame of the plow. This allows the shares to be raised to varying heights above the surface of the ground. *Cf. single-bail plow.*

double-band mill. A sawmill which has two band head saws and two carriages.

double-battery gin. A cotton gin which has two sets of gin stands.

double bit. A set of two bits attached to a horse's bridle: one for guiding the animal, the other for controlling it. The set consists of a curb bit and a straight bit.

double-bitted ax(e). An ax whose head has two cutting edges.

double-blossom disease. A fungous disease of blackberries and dewberries caused by *Cercosporella rubi,* family Moniliaceae, characterized by witches' broom; abnormally large buds; twisted, misshapen, pink blossoms; a reduction in yield; and by weakened plants. Common in southern U.S.

double-breaking cart. A cart which usually consists of the hind wheels and gears of a large wagon and a platform-like seat extending out behind the wheels. Its tongue is often made from a log up to 6 ins. in diameter, and it is used to train horses and mules to work in teams.

double breast. A part of certain cotton gins which is the boxed-in contrivance where the seed cotton comes into contact with the saws. One compartment receives the hulls, bolls, etc. Such gins are valuable for ginning bolly cotton or cotton which contains extraneous matter.

double capping. Sealing milk bottles with flat lids set in the mouths of the bottles and placing over these separate protective caps which are usually crimped or fastened below the outside lips of the bottles.

double cover. Coitus of a mare, cow, or other female animal with a male on successive days to insure conception. See *double mating: 2.*

double cream. Whipping cream *q.v.*

Double-Cream cheese. A soft, uncured cheese prepared from thick cream.

double crib. Two single corn cribs that have a roofed-over driveway between them.

double crib with overhead bin. A double crib that has an overhead storage space for small grains or shelled corn.

doublecrimson Chinese peony. *Paeonia albiflora* var. *sinensis,* family Ranunculaceae, an erect, tuberous-rooted, perennial herb widely grown for its very large, double, crimson flowers. Native to Asia.

double cropping. The growing of two crops in one year on the same field.

double cross. 1. The resultant seed secured from mating two single cross corn hybrids. It is planted extensively in hybrid corn production. 2. The cutting of corn kernels from the cob and the chopping of the same kernels to prepare a creamed corn for canning.

double cut (clover). Red clover which gives two cuttings or two hay crops in one season.

double-decked car. A railway car or truck having two main floors, one above the other. It is used for the shipment of sheep, hogs, calves, etc.

double deep suck. See *deep suck.*

double disk. To cultivate a field with a double-disk harrow (*q.v.*), or to cultivate twice with a single-disk harrow.

double-disk furrow opener. A furrow opener attachment on a row planter which consists of two disks so arranged as to throw the soil outward to open a furrow for seed planting. Also called *double-disk opener.*

double-disk harrow. A harrow (*q.v.*) in which two single-disk harrows are arranged in tandem. The front gangs are set to throw the soil outward, while the rear gangs are set to throw the soil inward to leave the surface level. Also called *double-action harrow; tandem disk harrow.*

double-disk marker. A row marker attachment for a planter which consists of a long rod fastened to the planter at one end. On the other end are two disks so arranged as to make a small, marking furrow in the soil. See *double-folding marker.*

double-disk opener. See *double-disk furrow opener.*

double drained. Designating an area of land which is drained by both the laterals and a main tile drain.

double dray. A type of tongueless, double logging sled which is used for short distance hauling.

double eye. Designating a plant cutting for vegetative reproduction which has two buds on it.

double fertilization. The usual fertilization in plants in which the egg nucleus is fertilized by one sperm nucleus and the two fused polar nuclei by another sperm nucleus.

double fleecer. Designating a sheep sheared at every other shearing (New Zealand).

double flowers. See *double: 1.*

double-folding marker. A type of row marker for a planter which consists of two separate rods attached to either side of the planter. On the end of each rod is a disk or spike tooth for marking the soil. One marker is used as the planter moves across the field, the other on the return trip.

double hacking. The girdling of a tree by making parallel, overlapping, downward, ax cuts and the removal of the chips between the cuts.

double ham-butt slice. A retail cut of ham in which the butt is cut off along the aitch bone. This piece is so sliced that only every other cut is complete. The double piece is then opened like a book and resembles a whole slice.

double-harpoon fork. A hayfork which has two, long, iron forks with barbs on the ends which, when forced to a desired depth in a load or pile of hay, may be sprung so as to grasp the hay for lifting. *Cf. single-harpoon fork.*

double immune. Designating a hog which has

been vaccinated with both live virus and anti-hog cholera serum for protection against hog cholera.

double-layer side pack. A method of packing dressed poultry in which 12 birds are packed in two layers on their sides. *Cf. breast pack; single-layer side pack; squat pack.*

double lift. The special effort made by a horse who suffers from broken wind to expel the air from its lungs.

double-main system. A system of tile drainage in which two main lines are laid parallel. It is used (a) where the depression between two slopes is broad and flat, and (b) for rehabilitating a drainage system which has been partially washed away.

double manure salt. Sulfate of potash-magnesia *q.v.*

double mating. 1. A rarely used breeding practice with Barred Plymouth Rocks to produce birds of exhibition type. 2. The mating of female livestock twice during a given estrous period to the same or different males to insure conception. See *double cover.*

double-neck slice. A retail cut of mutton in which a cut is made through the neck bone leaving two slices attached at the bottom which open like a book.

doubleness. The condition of a blossom which has more than the normal number of petals and sepals. See *double: 1; single.*

double-nosed. Designating a type of narcissus bulb which has two growing points present in the bulb.

double orange daisy. Orange fleabane *q.v.*

double planting. See *interplant.*

double-pointed shovel. A cultivator shovel that is sharpened at both ends so that it may be reversed when one end becomes dull. Also called *reversible point.*

double-post slip chain tie. A type of stanchion for dairy cows.

double reins. Two sets of reins attached to a horse's double bit *q.v.*

double-rib huller breast. See *double breast.*

double-rigged saddle. A saddle which has two cinches.

double-rod agitator. Two parallel axles placed within the seed box of a grain drill, to each of which power can be applied. Attached at intervals on each are paddles. As the axles turn, the paddles agitate the seeds to keep them flowing evenly into the feed cups.

double row. 1. A planting system for small vegetables and fruits in which the seeds or plants are set closely together in two parallel lines with a wider spacing between the pairs of rows. 2. A farm machine used on two rows of a crop simultaneously, as a *double-row* corn harvester. Also called *two-row.*

double-snaffle bit. A severe bit for a horse's bridle which consists of two bar bits, each hinged on opposite sides of the middle.

double spot. A fungous disease of highbush blueberries caused by *Dothichiza caroliniana,* family Sphaerioidaceae, characterized by tan spots with a purplish halo on the leaves. After midsummer the spots grow larger and a light brown band surrounds the original spot. Found in southern U.S.

double spruce. Black spruce *q.v.*

double superphosphate. A commercial fertilizer which contains 2½ to 3 times as much available phosphoric acid as superphosphate. The phosphate rock is treated with phosphoric acid, and in the manufacturing process the gypsum is removed. Also called *triple superphosphate; treble superphosphate.*

double-tine fork. A two-tined pitchfork which was commonly used to handle grain bundles.

double tree. Beef tree *q.v.*

doubletree. A crosspiece, usually made of hard, durable wood, up to 4 ins. across, which is fastened through a pivot hole by a steel pin to the tongue of a wagon, plow, etc. Two singletrees, for hitching animals, usually horses or mules, are attached near or at each end of and parallel to the *doubletree.* Also called *bail. Cf. singletree.*

double-tube heat exchanger. Internal-tube heat exchanger *q.v.*

double-turn plow. Two turn plows which are attached side by side to open a furrow similar to a lister.

double-virus streak. A disease of tomatoes caused by latent potato mosaic virus and tobacco mosaic virus which is characterized by small, irregular, angular spots on the surface of the fruit. The spots are originally glossy but in time turn brown, become sunken, and increase as the season passes. The fruit is unattractive, and seriously affected fruits may be a total loss.

double wheel. A type of wheel used on a planter which consists of two separate wheels having smaller diameters on one side, which are set with the smaller sides touching, forming an acute angle. The wheels serve to cover the seed and compress the soil.

double-wing. Designating a plow which throws soil both to its left and right. See *lister; middle-buster.*

double-worked tree. A tree which has been subjected to double-working.

double-working. Re-budding or re-grafting on a previously established graft on a plant. This system is used when two varieties or species of stock and cion do not unite except through an intermediary.

double-yolk egg. An egg which contains two yolks.

dough. See *dough stage.*

dough drying system. A method of drying milk in which the fluid milk is heated in a vacuum until most of the moisture is driven off. The partially dried milk is then put in a drying chamber, and the moisture content is reduced to the point desired.

dough stage. That stage in the development of a cereal grain in which the inside of the kernel has a consistency like dough.

Douglas clematis. *Clematis douglasi,* family Ranunculaceae, a hardy herb grown for its solitary, purple flower. Native to North America. Also called *sugar loaf.*

Douglas fir. Any tree of the genus Pseudotsuga, family Pinaceae, but specifically *Pseudotsuga taxifolia,* sometimes referred to as *P. douglasi,* the only commonly cultivated species of this genus. It sometimes reaches a height of 300 ft. and is a timber tree of great commercial value in northwestern U.S. It is also grown as a shade and shelterbelt tree and certain horticultural varieties are grown as ornamentals east of the Rocky Mountains, U.S. Also called *Douglas spruce; red fir.*

Douglas-fir beetle. *Dendroctonus pseudotsugae* Hopk., family Scolytidae, an insect pest of the Douglas fir and western larch which infests injured or dying trees and occasionally healthy trees.

Douglas-fir engraver. *Scolytus unispinosus* Lec., family Scolytidae, an insect pest of the Douglas fir which may also infest the western larch and the Engelmann spruce. It makes galleries in the bark of trees.

Douglas-fir tussock moth. *Hemerocampa pseudotsugata* McD., family Lymantriidae, an insect pest of the Douglas fir and fir in northwestern U.S. The larvae, distributed by the wind, defoliate the trees, usually from the top downward.

Douglas hemlock. See *Douglas fir.*

Douglas pine. See *Douglas fir.*

Douglas spirea. *Spiraea douglasi,* family Rosaceae, a shrub grown for its deep, rose-red, flowers. Native to North America.

Douglas spruce. See *Douglas fir.*

Douglas waterhemlock. *Cicuta douglasi,* family Umbelliferae, a poisonous, perennial herb. In grazing animals, the symptoms of poisoning are stomach pain, diarrhea, dilated pupils, labored breathing, frothing at the mouth, a weak but rapid pulse, and violent convulsions. Death results from respiratory failure. Roots and rootstocks are the most poisonous parts of the plant; the leaves and fruits may be eaten in hay without danger. Found from Alaska to California, U.S.

doum palm. *Hyphaene thebaica,* family Palmaceae, a fan palm tree which bears an edible fruit having a flavor similar to gingerbread. Fibers from the leafstocks are used to make ropes. Native to Egypt. Also called *doom palm; gingerbread tree.*

dour. Designating poor or barren soil (Scotland).

dourine. A parasitic, contagious disease of the horse and ass caused by *Trypanosoma equiperdum* as a result of infection during coitus. It is characterized in the stallion or jack by a swelling of the prepuce which in time spreads to the scrotum and abdomen, and the testicles becoming enlarged and sensitive. The penis extends from the sheath, is swollen, and discharges a yellowish fluid. Blisters on penis and sheath rupture and result in raw ulcers. There is repeated desire to urinate. In the mare or jenny, it is characterized by swollen, external genitals, a gaping vulva, and the clitoris being constantly erect. The mammary glands may be swollen. The female is uneasy, switches her tail, voids small quantities of urine, and appears to be in heat. A discharge similar to the male's is evident. Blisters on the inner and outer mucous membranes of the genitals rupture to leave raw ulcers. Both sexes become nervous. The skin lesions appear in the form of urticarial wheals similar to those in hives. The hind quarters may become paralyzed. Also called *dollar plague; breeding paralysis; genital glanders; equine syphilis; mal(adie) du coit.*

dousing rod. Divining rod *q.v.*

dover grass. Meadow fescue *q.v.*

dovetail. A downward, triangular projection of the posterior edge of the upper, corner teeth of a horse which occurs when the animal is about 7 years old. Also called *swallowtail.*

dove tick. Fowl tick *q.v.*

dovetree. *Davidia involucrata,* family Cornaceae, a tender tree grown for its handsome foliage and large, white flower clusters. Native to Asia.

dove-weed. Turkeymullein *q.v.*

dower right. The widow's right in her husband's real estate. In most states of the U.S. it is a one-third share.

Dow-Law planter. A cotton planter, invented about 1874, which has given way to more efficient planters.

down. 1. A sandhill. See *dune.* 2. The soft, furry or feathery covering of young animals and birds which occurs on certain adult birds under the outer feathers and which is used for stuffing pillows, furniture, and comforters, its most important source being the eider duck. 3. The pappus (*q.v.*) of a plant. 4. Used to describe the individual crop plants which have been blown or forced to the ground by rain, hail, wind, etc.: *e.g., down* rice.

Down breeds. One of two English type of sheep which produce a small sheep of the medium wool class. It includes the common breeds: Southdown, Shropshire, Hampshire, Suffolk, Oxford Down, and Dorset Horn.

down calving. Designating a cow which is about to calve (British).

down clothing wool. Wool from Down breeds or sheep which is too short for combing. It is a class I blood classification wool.

Down crossbreed. A sheep which results from crossing a Merino with any of the Down breeds.

downer. 1. An animal which, in transit in a truck or railroad car has fallen or lain down. 2. A diseased animal unable to get up.

down grain. Ripening grain that has been blown to the ground by a storm or that has fallen because of over-ripening, weakness of the straw, or heaviness of the ripened heads of grain.

downland. See *downs.*

Downland. The temperate grasslands of Australia which are similar to Steppes. See *Steppe: 2.*

downpour. A heavy rainstorm.

downs. Undulating grazing, or pasture land, especially the chalk lands of southeastern England.
Down sheep. See *Down breeds.*
downsman. A person who lives on the downs *q.v.*
downstream. Towards the mouth of a river or stream from the point of reference. *Cf. upstream.*
down suction. The downward bend of the point of a share which helps the plow penetrate the soil to the proper depth as it moves forward.
down thistle. Bull thistle.
down wool. Wool which comes from the Down breeds of sheep. See *down clothing wool.*
downy andromeda. *Andromeda glaucophylla,* family Ericaceae, a poisonous shrub which is sometimes eaten by sheep when other browse is unavailable. Symptoms of poisoning are salivation, flow of tears, a flow from the nostrils, convulsions, and paralysis. Found in acid bogs and uplands of northeastern U.S. and southeastern Canada.
downy ash. Red ash.
downy brome. See *cheatgrass brome.*
downy chess. See *cheatgrass brome.*
downy groundcherry. *Physalis pubescens,* family Solanaceae, a low, annual herb grown for its edible, round, yellow fruit. Native to North America. Also called ***husk tomato.***
downy hawthorn. *Crataegus mollis,* family Rosaceae, a deciduous tree grown for its large, bright green leaves, its white flowers with red disks, and its scarlet fruit. It is also used as an almost impenetrable hedge. Native to North America. Also called *red haw.*
downy leopardbane. *Doronicum clusi,* family Compositae, a perennial herb grown for its yellow flowers. Native to southern Europe.
downy mildew. Various plant diseases caused by fungi of the families Pythiaceae and Peronosporaceae. These fungi infect the leaves, herbaceous stems, flowers, and fruits of the host plants which are mostly herbaceous, although the herbaceous parts of some woody plants may be affected (rose, grape, hackberry). The rather coarse mycelium of the fungus grows between the host cells, sending into them the organs called haustoria which are the means of absorption of food by the fungus. From this internal mycelium, spore-bearing branches emerge with a downy or felty appearance, hence the name *downy mildew.* Within the diseased host, tissue may be produced through thick-walled oöspores which serve the purpose of overwintering by the fungus. The chief genera which are called *downy mildews* are Peronospora, Plasmopara, Rhysotheca, Bremia, Bremiella, Pseudoperonospora, Basidiophora and Sclerospora, all belonging to the family Peronosporaceae; and Phytophthora, of family Pythiaceae, but the genus Pythium belonging to this family rarely produces the external *downy-mildew* type of surface growth, although closely related to Phytophthora, so is not usually called *downy mildew.* Various *downy mildews* are described briefly below, including the names of the fungus and of the host plant.
downy mildew of alfalfa and clover. A fungous disease of alfalfa and clover caused by *Peronospora trifoliorum* which is characterized by a distortion and yellowing of the foliage, sometimes by death of the tender shoots, and by a purplish coating of the fungus on the underside of the leaf. The disease occurs widely in temperate regions. See *downy mildew.*
downy mildew of beets. A fungous disease of beets on the Pacific Coast and elsewhere in the U.S. and Europe caused by *Peronospora schachtii* which is characterized by a yellowing and curling of the leaves of plants in all stages of growth. In dry weather after infection, the spots of infection dry up; in wet weather, however, the underside of the lesion becomes covered with spore-bearing growth of fungus. See *downy mildew.*
downy mildew of clover. See *downy mildew of alfalfa and clover.*
downy mildew of corn. See *downy mildew of sugar cane.*
downy mildew of crucifers. A fungous disease of cabbage and other cultivated crucifers which is caused by *Peronospora parasitica.* Also called *cabbage downy mildew.* See *downy mildew.*
downy mildew of cucurbits. A destructive, fungous disease of cucumber, muskmelon, and watermelon caused by *Pseudoperonospora cubensis* which is prevalent in areas where there is rainfall during the growing season. It is characterized by angular, yellowish spots of the leaves. Severely infected leaves turn brown and shrivel, and maturing fruits do not have normal color or taste. It is present in northern U.S. on cucumbers raised under glass and in southern U.S. in field grown muskmelon and cucumber. See *downy mildew.*
downy mildew of grapes. A fungous disease of grapes in most eastern grape-growing areas caused by *Plasmopara viticola* which is characterized by a rot of the fruit bunches. Affected leaves show water-soaked lesions which soon become circular and light-yellow with no halo, and as the lesions enlarge, they become brown. A white, downy growth develops on the underside of the leaf during periods of rainfall and high humidity. Infected clusters may remain small and become covered with the downy, white fungous growth if infected early after blossoming. Berries infected later have dark, sunken areas on the surface of the decaying fruit. Some become hard, leathery, and reddish. (Also called *Rhysotheca viticola.*) See *downy mildew.*
downy mildew of grasses. A serious, fungous disease of some grasses caused by *Sclerospora graminicola,* which is characterized by dwarfing of the plants, excessive tillering, development of branches from the axillary buds, leaflike malformation of the floral bracts, failure of the kernel to develop, production of a downy mass of spores during the night, a browning and killing of the leaves, and shredding of the infected tis-

sue. Especially abundant in the fox-tail or pigeon grasses (genus Setaria). This same species or closely related downy mildews are also found on some of the cereal grasses: *e.g.*, barley, millet, oats, sorghum, and wheat, especially in subtropical or tropical regions. See *downy mildew.*

downy mildew of hops. A fungous disease of hops caused by *Pseudoperonspora humili* which is widely distributed in all hop-growing areas. It is characterized by thickened shoots, leaf spots, death of any of the plant parts, and reduction in yield. Losses vary from discoloration of the cones to total destruction of the vines. See *downy mildew.*

downy mildew of lettuce. A fungous disease of lettuce caused by *Bremia lactucae* which is widespread on the Pacific coast, U.S. where it occurs throughout the year. It also attacks greenhouse lettuce throughout the U.S. and is characterized by light-green to yellow spots on the leaves. In a few days a downy-white mass appears on the undersides of the leaves and the spots turn brown. The fungus may continue to develop on refrigerated lettuce in transit. See *downy mildew.*

downy mildew of lima beans. A serious fungous disease of lima beans caused by *Phytophthora phaseoli* in the eastern coastal states, U.S., which is characterized by a white, cottony mass on the pods surrounded by a purplish border. The pods shrivel, wilt, and die. The disease less frequently attacks leaves and floral parts. Infected leaves have irregular spots near the veins, with little or no fungous growth. See *downy mildew.*

downy mildew of muskmelon. See *downy mildew of cucurbits.*

downy mildew of onion. An important, fungous disease of onion caused by *Peronospora destructor* which is characterized by yellowish spots on the upper halves of the leaves and by bluish-gray fungous felts on the spots when the humidity is high. In high humidity the spots increase rapidly, and the tops of the plants die back. Plants seldom die, but the bulbs do not mature and bulb tissue may be spongy, resulting in the bulb not storing well. The mildew also attacks the Welsh onion, leeks, shallots, chives, and garlic. (Also known by the incorrect name *Peronospora chleidenii.*) See *downy mildew.*

downy mildew of pea. A fungous disease of pea caused by either of two fungi, *Peronospora pisi* or *P. viciae* which is characterized by fluffy white to blue cottony patches of mold on the underside of the leaf. The area on the top of the leaf opposite to the infection becomes brown. See *downy mildew.*

downy mildew of potato and tomato. A fungous disease of solanaceous plants, but principally of the potato and tomato, caused by *Phytophthora infestans* which is characterized on potatoes by black spots on the leaves, a withering of the leaves, an offensive odor, and a decay of the tubers; on tomatoes, by a rapid dying of the leaves after a dark spot appears on them. Large brownish lesions appear on the green fruit, and the affected areas do not ripen. This disease was largely responsible for the Irish famine in 1845-1846. Also called *late blight of potato and tomato.* See *downy mildew.*

downy mildew of rose. A fungous disease of roses which is caused by *Peronospora sparsa.* See *downy mildew.*

downy mildew of snapdragons. A fungous disease of snapdragons caused by *Peronospora antirrhini* which is characterized by a paling and curling of the affected leaves. Terminal growth is stopped, and death of the aerial parts of the plant can result. It is most common on seedlings. See *downy mildew.*

downy mildew of soybeans. A fungous disease of soybeans caused by *Peronospora manshurica* which is characterized by indefinite yellowish-green areas on the upper leaf surfaces. In time, these turn grayish-brown or dark-brown and have a yellowish-green halo. Infected leaves may drop prematurely. A grayish mass develops on the undersides of the lesions. See *downy mildew.*

downy mildew of spinach. A most serious fungous disease of spinach caused by *Peronospora effusa* which is characterized by yellow areas on the leaves. A downy-white (later purple) mass appears on the underside of the lesion. The disease develops when the weather is cool and the humidity is high. See *downy mildew.*

downy mildew of sugar beets. See *downy mildew of beets.*

downy mildew of sugar cane. A fungous disease of sugar cane caused by *Sclerospora sacchari* (*S. philippinensis, S. spontanea*), which is characterized by yellowish-green streaks on the leaves between the veins, and a downy white mass on the underside of the leaf. The elongated part of the stem is thin, and the leaves are few and stunted. Affected parts are shredded, and the yield is materially reduced. It may also attack corn and teosinte. Found in the Eastern hemisphere. See *downy mildew.*

downy mildew of sunflower and Jerusalem artichoke. A less serious, fungous disease of sunflowers and Jerusalem artichoke caused by *Plasmopora halstedii.* See *downy mildew.*

downy mildew of tobacco. A fungous disease of tobacco caused by *Peronospora tabicina* which is characterized by a bluish-gray mold on the seedlings in seedbeds. It may, in a very short time, kill thousands of seedlings. The disease is favored by cool weather and high humidity. Also called *tobacco blue mold.* See *downy mildew.*

downy mildew of tomato. See *downy mildew of potato and tomato.*

downy mildew of violets. A fungous disease of violets caused by *Bremiella megasperma* which is characterized by discolored spots on leaves and stems of cultivated violets and pansies. (Sometimes the fungus is called *Peronospora violae.*) See *downy mildew.*

downy mildew of watermelon. See *downy mildew of cucurbits.*

downy myrtle. See *downy rosemyrtle.*

downy rattlesnake plantain. *Epipactis pubescens,* family Orchidaceae, a mild herb distinguished by its green-white flowers. Native to North America. Also called *adder's violet.*

downy rockcress. *Arabis mollis,* family Cruciferae, a perennial herb grown for its clusters of white flowers. Native to the Caucasus.

downy rosemyrtle. *Rhodomyrtus tomentosa,* family Myrtaceae, a tropical, tender, handsome, evergreen shrub grown for ornament and for its edible berry. Found in Asia and Australia. Also called *downy myrtle; hill gooseberry.*

downy spot. A fungous disease of pecan foliage caused by *Mycosphaerella caryigena,* family Mycosphaerellaceae, which occurs in trees which are crowded or otherwise grown under unfavorable conditions. The spots are pale-yellow at first, but turn yellow-brown, brown, or black. Diseased tissue may drop out of the leaf.

downy violet. *Viola pubescens,* family Violaceae, a hardy, perennial herb which is grown in wild gardens for its yellow flowers. Also called *yellow violet.*

dowser. One who uses a divining rod to locate underground water and indicate the place to drive, dig, or drill a well. Also called *water witch; water diviner.*

dowsing rod (stick, wand). See *divining rod.*

Doyle rule. In the lumber trade, a simple method of estimating the amount of timber that can be obtained from a small log. The footage of the length is multiplied by the diameter of the log from which 4 ins. has been subtracted, and the result divided by 4 and squared.

Doyle-Scribner rule. A combination rule using the Doyle rule (*q.v.*) for scaling logs up to 28 ins. and the Scribner rule (*q.v.*) for logs larger than 28 ins. which is used extensively by private operators in southern U.S.

doze. See *dote.*

dozie. An area in which sheep lose condition and do not thrive (New Zealand).

dracaena. Any plant of the genus Dracaena, family Liliaceae. Species of this genus are large, tropical, woody-stemmed plants, some of which are used as pot plants in homes and greenhouses, others are trees. Plants of this genus are distinguished from those horticulturally referred to as *dracena,* of the genus Cordyline, by the occurrence of a single ovule in each cell of the ovary, whereas the *dracena* may have several. See *dragon dracaena; fragrant dracaena; Goldie dracaena; Sanders dracaena.*

dracena. See *ti.*

dracena palm. Giant dracena *q.v.*

draff. Any feedstuff obtained as a by-product of the distillation of grains.

draft. 1. A package deduction, 1 lb. per hundredweight, allowed a buyer of wool. 2. The horizontal component of pull of an implement parallel to the line of motion. Also called *directional pull.* 3. To divide livestock into classes (Australia, New Zealand).

draft animal. An animal, *e.g.,* a horse, mule, or ox, used to pull heavy loads.

draft breed. Any breed of horse which has been developed for the purpose of pulling heavy loads. See *Percheron; Belgian: 1; Clydesdale; Shire; Suffolk.* Also called *heavy breed.*

drafter. Draft mule *q.v.*

draft gelding. Any gelding of the draft breeds.

draft gelding class. A show-ring classification of draft horses.

draft hole. A hole in the center of an evener, singletree, or doubletree for the hitchpin.

draft horse. Any horse which, by reason of its size and weight, is used for pulling heavy loads, and which usually weighs over 1,500 lbs. and stands up to 17½ hands high. See *Percheron; Belgian: 1; Clydesdale; Shire; Suffolk: 1.*

draftiness. The character and ability of a horse to pull heavy loads.

draft mule. A mule weighing over 1,200 lbs. and standing up to 17½ hands high which is used for pulling heavy loads. *Cf. farm: 3; sugar mule; cotton mule; mining mule.*

draft rein. A long guiding rein which runs on the outside of a draft horse. *Cf. coupling rein.*

draft team. A show-ring classification for horses which are exhibited as a team. Also called *farm team.*

drafty. 1. Designating an animal which has the size and ability for pulling heavy loads. 2. Exposed to currents of air.

drag. 1. A type of harrow, usually homemade, consisting of several pieces of 2 x 8 or 2 x 10 in. planks fastened together, which is used in the preparation of seedbeds to crush clods, to level the soil, and to firm soil around the seeds. Also called *float; planker; slicker; clod crusher.* See *spike-tooth harrow.* 2. A sledlike device made of wood used to transport heavy or awkward loads. See *stone boat.* 3. See *cling: 1.* 4. To crush clods, to level soil surface, or to firm soil around seeds by use of a *drag.*

drag car. See *bummer.*

drag chain. 1. A series of short lengths of chain which are fastened to the seed tubes of small-grain drills to cover the seed without packing the soil. 2. An endless belt of chains which is used to move corn, etc., as from a loading platform to the silage cutter.

drag grader. A device which consists of heavy wood timbers fastened together so as to form a parallelogram. It is used to smooth and level the surface of dirt roads, fills, etc.

drag harrow. Spike-tooth harrow.

drag horse. Draft horse *q.v.*

drag in. 1. To drag logs from the place of felling to the skidway or to the landing. 2. To cover broadcast-sown seed with a drag or harrow.

drag iron. See *rudder blade: 2.*

drag-line excavator. An excavating machine used in digging large ditches and constructing levees. It consists of a power unit, a boom from the upper end of which a bucket or scraper is suspended, and a dragline which pulls the bucket toward the machine. The whole machine may be rotated on a turntable and is usually mounted on crawler treads.

dragon arum. Jack-in-the-pulpit.

dragon dracaena. *Dracaena draco,* family Liliaceae, a gigantic, slow-growing tree grown for its foliage, greenish flowers, and scarlet fruit. Native to the Canary Islands. Also called *dragon tree.*

dragonfly. A large, predaceous insect of the Order Odonata which feeds on mosquitoes, flies, gnats, bees, and other insects it can catch on the wing. The nymphs feed on mosquito larvae. Also called *devil's-darning-needle; snake feeder; snake doctor.*

dragonroot. Jack-in-the-pulpit.

dragon's-mouth. See *snapdragon: 2; arethusa.*

dragon's-tongue. Striped pipsissewa.

drag scraper. An excavating device which consists of a scoop with a cutting edge which has rear handles for guiding and dumping, and a bail in the front to which a team is hitched. Also called *slip scraper.*

dragstaff. A trailing pole attached loosely to the underside at the rear of a horse-drawn vehicle to prevent it from rolling backward.

drail. See *clevis.*

drain. 1. An artificial or natural channel which receives and carries off excess or unwanted water from an area of land. It does not carry sewage and is therefore distinguished from a sewer. 2. The loss of growing trees in a forest due to cutting, fire, or natural causes.

drainage. 1. The removal of surplus ground or surface water by artificial means. 2. The manner in which the waters of an area are removed. 3. The area from which waters are drained; a *drainage* basin.

drainage area. 1. The area (square meters, acres, etc.) of a drainage basin. 2. A catchment area; drainage basin.

drainage basin. The area from which water is carried off by a drainage system; a watershed or catchment area.

drainage coefficient. Drainage modulus *q.v.*

drainage district. An organization which operates under legal regulations for financing, constructing, and operating a drainage system.

drainage ditch. An open ditch used to carry off surplus ground and surface water. *Cf. canal: 1.*

drainage furrow. A furrow which is plowed for carrying off excess surface water. See *bedding: 2.*

drainage modulus. The run-off of surplus surface and ground water from a particular area, the depth of which is expressed in inches per 24 hours. Also called *drainage coefficient.*

drainage survey. A study performed to determine the need for and requirements of a drainage system.

drainage system. 1. An artificial network of furrows, ditches, tile drains, or combinations of these that provide drainage for farm land. 2. The natural means by which a river and its tributaries carry water from the land.

drainage terrace. A terrace constructed with a graded channel for run-off interception which is provided with a gentle grade so that the velocity of the water flow is non-erosive.

drainage water. 1. That water which the soil is unable to retain against the force of gravity. Also called *gravitational water; free water; excess water.* 2. Water flowing in a drain derived from ground, surface, or storm water.

drained weight. The weight of a canned fruit or vegetable after removing the liquid.

draining pen. A pen with a sloping concrete floor adjoining a dipping vat where dipped animals stand while excess fluid drains back into the vat.

drain rack. A wooden or metal rack on which bottles, milk cans, etc., may be inverted to drain.

drain tiles. Pipes of burned clay or concrete, in short lengths, which are laid with open joints to collect and remove drainage water.

drain well. A well into which drainage water is directed. The excavation is deep enough to reach a pervious substratum, such as sand and gravel, or a porous or cavernous limestone.

drake. The adult male duck, family Anatidae.

Drake almond bud failure. A virus disease found on the Drake almond in which buds fail to grow and twigs die. Found in California, U.S.

drape. A term applied to a cow or ewe incapable of bearing offspring, especially to such an animal selected for slaughter.

draper. 1. A machine used in cleaning sugar-beet seed. 2. The conveyor apron on a grain binder or combine.

draught. Draft *q.v.*

draw. 1. A small natural head tributary, drainage hollow, without a permanent stream in its bottom. (In local usage it is not always sharply differentiated from ravine, hollow, gulch, or coulee.) 2. A slip or shoot of a sweet potato.

drawdown. The temporary lowering of the water-table around a well during pumping. See *cone of depression.*

drawgear. Harness used on a draft horse.

draw hoe. See *garden hoe.*

drawing. 1. The removal of cuttings from a propagation bed. 2. The eviscerating of poultry.

drawing knife. 1. A tool used to make a shallow cut on the surface of wood along which the saw follows, preventing surface splintering or tearing. Also called *draw knife.* 2. A knife designed for paring hooves, especially horses' hooves.

draw line. In well boring, a rope used with a pole or rotary drills to lift the tools out of the hole.

drawn comb. Comb built up by honey bees to the point at which the cells are of the required depth for rearing brood and storing honey.

drawn poultry. Poultry which is ready for cooking. It has been bled, plucked, and eviscerated, and the head and feet have been removed. *Cf. dressed: 1.*

drawn weight. The weight of a bird prepared for cooking: feather dressed, eviscerated, with head and feet removed. *Cf. dressed weight: 2.*

draw off. To extract by distillation or by removing a portion of a liquid.

draw reins. To pull at a horse's reins to cause him to stop.

draw well. A well in which a bucket is used to lift the water to the ground surface.

dray. 1. A wagon used for heavy hauling. 2. A single sled used in hauling logs in which one end of the log rides on the sled and the other drags on the ground.

dredge. 1. Oat and barley seed sown together, cultivated to be used for making malt. 2. Any of several different types of machines to deepen channels and clean out ditches, etc. 3. To excavate or deepen and clean stream beds.

dreen. A swamp which separates marginal islands from the mainland.

drench (drensh). A fluid dose of medication given by introducing it into the mouth, usually with a dose syringe. Frequently used when attempting to remove parasites from the stomach and intestines of man and animals.

dress. 1. To cut the awns off oats (British). 2. To curry and brush an animal. 3. To remove the feathers and blood from a bird which has been killed. 4. To plane the surface of a board.

dressage. Designating a horse which has been trained to perform on almost unnoticeable signals by the rider. Also called *high school; haute école.*

dress a sample. See *trim a sample.*

dressed. 1. Designating a killed fowl which has had the feathers and blood removed. *Cf. drawn poultry.* 2. Designating a meat carcass which has had the excess fat and bone removed.

dressed size lumber. The dimensions of lumber after planing, usually 3/8 ins. less than the normal or rough size.

dressed weight. 1. The weight of a sheep, cow, or hog carcass. 2. The weight of a bird which has had its blood and feathers removed but not its intestines, head, or feet. Also called *New York weight. Cf. drawn weight.*

dressing. 1. Any application of medicine or bandage made to a wound on a man, animal, fowl, or plant. 2. An application of manure, fertilizer, mulch, etc. See *top dressing; side dress.* 3. The preparation of nursery stock for cutting, budding, grafting, etc. 4. The removal of feathers and blood from a bird. 5. The trimming of excess fat and bone from a meat carcass. 6. The external treatment of cheese, including: (a) the assistance given in forming a coat or rind; (b) improvement given to the external appearance of a cheese; (c) the prevention of loss of moisture from a cheese; (d) protection from external microorganisms.

dressing comb. A special comb used to prepare an animal's coat for a show. Also called *currycomb.*

dressing defect. A defect of dressed poultry characterized by one or more of the following: pin feathers left on the carcass; incomplete bleeding; a cut, tear, or abrasion in the skin longer than 2 ins.; a broken bone; feed left in the crop; dirty feet; dirty body; dirty vent.

dressing loss. The loss in weight between a live animal and its dressed carcass.

dressing percentage. The estimated loss in weight between an animal on the hoof and its dressed carcass which is determined by dividing the carcass weight by the live weight and multiplying by 100.

dressing tool. See *sticking knife; boning knife; hotel parer; bone shears; blood cup; shackle: 1.*

dressing weight. The weight of a dressed animal or bird in contrast to its live weight.

dried. Designating a product which, for purposes of preservation, has had all or most of the moisture removed, as dried apples, raisins, etc.

dried agave pulp. The residue obtained after extraction of the fiber from the leafy portion of the agave plant in the manufacture of sisal fiber.

dried apple pectin pulp. See *dried apple pomace.*

dried apple pomace. A feedstuff for livestock which is the sound, dried residue obtained by the removing of juice or cider from apples and subsequently drying the moist pomace.

dried apples. A food product prepared from apples. The fruit is peeled and cored by machines, washed, inspected, sliced, halved or quartered, and placed in sulfur houses for one hour for sulfuring to prevent browning while drying. The sulfured fruit is then loaded into drying kilns where heat, supplied from beneath the floor, comes through slats and drives the moisture from the fruit. After the fruit has been dried it is treated with fine sprays of water, again sulfured, and packed into containers.

dried apricots. A food product prepared from apricots. The fruit is halved, pitted, and placed cup up on trays which are then put on trucks and shunted to the sulfuring rooms. Here the fruit is exposed to the fumes of burning sulfur for 3 to 7 hrs. to fix the natural color of the fruit. The fruit is then placed on trays to dry in a sunny exposure or drying yard or in a mechanical or artificial dehydrator.

dried beef. A food product prepared by slowly drying meat which has been previously cured by pickling. The drying process, which removes about 25 percent of the moisture, is slow and expensive; hence, *dried beef* is sold at a high price. *Dried beef* is trimmed and automatically sliced, graded and packed by hand.

dried beet pulp. A feedstuff for livestock which is the dried residue left after sugar extraction from sugar beets; a by-product in the manufacture of sugar.

dried blood. The collected blood of slaughtered animals, dried and ground, which is sold as a fertilizer and which contains 8 to 14 percent nitrogen in organic form (50 to 87.5 percent crude protein). The better grades are used as an ingredient of feedstuffs. See *blood meal.*

dried brewers' grains. Brewers' dried grains *q.v.*

dried buttermilk. A product resulting from the removal of water from clean, sound buttermilk derived from natural cream to which no foreign substances have been added, excepting such as are necessary and permitted in the manufacture of butter. It contains not more than 8 percent moisture, not more than 13 percent mineral matter (ash), and not less than 5 percent butterfat, as determined by the Roese-Gottlieb method (32 to 33 percent protein). It is used both for human food and animal feed, depending upon the quality of the product.

dried casein. A dairy product which is the dried precipitate resulting from the treatment of skimmed milk with acid, which is used in the manufacture of paper, glue, paints, etc.

dried citrus pulp. A livestock feedstuff which is the dried and ground peel, residue of the inside portions, and occasional cull fruits of the citrus family with or without the extraction of part of the oil of the peel. If calcium oxide or calcium hydrate is added as an aid in processing, the maximum percentage present, expressed as calcium, must be stated.

dried comminuted shortening. A shortening used in prepared cake, biscuit, and doughnut mixes which is a by-product of milk. It is made (a) by homogenizing and spray drying mixtures composed of a suitable fat dispersed in milk, skim milk, whey, buttermilk. The milk by-product remains water soluble so that the fat globules are quickly dispersed when the mix is converted to dough (b) by the use of edible, hydrogenated, vegetable oil to which sufficient milk product is added and treated as in (a) above.

dried corn distillers' solubles. A livestock feedstuff which is the product obtained by drying *semisolid corn distillers' grain with solubles* (*q.v.*) which contains 28 percent protein.

dried cream. Dry cream.

dried currant. See *currant: 2.*

dried distillers' solubles. See *dried grain sorghum distillers' solubles; dried corn distillers' solubles; dried rye distillers' solubles; dried wheat distillers' solubles.*

dried egg albumen. Preserved egg albumen obtained by the separation of the albumen from the yolk, followed by the bacterial fermentation of the albumen and the drying of the albumen by heat.

dried eggs. Preserved eggs obtained by mixing the white with the yolk and removing all but 3 to 5 percent of the moisture by heat.

dried fermentation solubles. The product obtained by the concentration and dehydration of the liquid by-product which results from the action of the ferment on the basic medium of grain, molasses, whey, or other media; it must contain not less than 18 mg. riboflavin per lb. on a moisture-free basis. For label identification the source must be indicated.

dried fruit. Any preserved fruits, such as apples, apricots, figs, peaches, pears, prunes, raisins, dates, etc. which have had enough moisture removed by semi-drying or by artificial dehydration so that they no longer support the growth of decay microorganisms.

dried-fruit beetle. *Carpophilus hemipterus* (L.), family Nitidulidae, an insect pest of dried foods, especially figs and dates.

dried grain sorghum distillers' solubles. A livestock feedstuff which is the product obtained by drying *semisolid grain sorghum distillers' solubles q.v.*

dried honey-skim milk mixture. A miscellaneous by-product from milk, consisting of equal parts of honey and concentrated spray-dried skim milk, which is used in the manufacture of candy and in bakery goods.

dried ice cream mix. A dry dairy product consisting of cream and milk to which stabilizers and sugars are added. The original mixture is condensed and spray-dried so that the dry mix contains not more than 3 percent moisture.

dried kelp. A livestock feedstuff supplement which contains over 30 percent mineral matter and from 0.15 to 0.20 percent of iodine consisting of dried and comminuted brown seaweeds, of the genus Macrocystis or related genera. Also called *kelp meal.*

dried liver. See *animal liver and glandular meal.*

dried milk. The product resulting from the removal of water from milk. Also called *desiccated milk.* See *modified dried skim milk; dried sweetened milk; dried buttermilk; malted milk; dried whey; dried ice cream mix; nonfat dry milk.*

dried milk by-product. A designation which may be used in the list of ingredients of a mixed feed to indicate the presence of dried skimmed milk, dried buttermilk, or dried whey, or a blend of two or more of these products.

dried molasses-beet pulp. A livestock feedstuff which consists of a dehydrated mixture of beet pulp (the residue from sugar beets that have had the sugar extracted) and molasses.

dried molasses distillers' solubles. A livestock feedstuff which is the product obtained by drying the residue from the yeast fermentation of molasses after the removal of the alcohol by distillation.

dried rye distillers' solubles. A livestock feedstuff which is the product obtained by drying *semisolid rye distillers' solubles q.v.*

dried skimmed milk (feeding). A livestock feedstuff, especially suited for calves, piglets, and chickens, which is produced by evaporating water from clean, sound skimmed milk, and which contains not more than 8 percent moisture (approx. 34 percent protein).

dried skim milk. Nonfat dry milk.

dried spent hops. A livestock feedstuff which is produced by drying the material filtered from a dilute solution of sugars boiled with hops. Also called *hopped wort.*

dried sweetened condensed milk. Dried sweetened milk *q.v.*

dried sweetened milk. A dehydrated milk product, consisting of sweetened milk ranging from fat-free to full-fat, used in cooking and baking. Also called *dried sweetened condensed milk.*

dried sweet potatoes. A livestock feedstuff, rich in nitrogen-free extract (carbohydrate) and low in fiber, which is made from shredded and dehydrated sweet potatoes. See *sweetpotato pulp.*

dried tomato pomace. A livestock feedstuff consisting of a dried mixture of tomato skins, pulp, and crushed seeds, produced by extracting the juice from tomatoes.

dried wheat distillers' solubles. A livestock feedstuff which is the product obtained by drying wheat distillers' dried grains with solubles.

dried whey. A by-product from the manufacture of cheese or casein or both, which consists of at least 65 percent lactose (milk sugar) and approx. 12 percent protein; used as a livestock feedstuff.

dried whey by-product. A livestock feedstuff which is the dried product resulting from the partial removal of milk sugar from clean, sound whey to which no foreign substances have been added. It contains at least 55 percent lactose (milk sugar).

dried whey solubles. A livestock feedstuff which is the dried product resulting from the removal of albumen and the partial removal of milk sugar from clean, sound whey, to which no foreign substances have been added except such as are necessary in the manufacture of milk sugar.

dried whole-egg. The dehydrated albumen and yolk of eggs which, after removal from the shell, have been thoroughly mixed together. See *dried eggs.*

drier. Any device used for removing moisture from a product, usually consisting of a heating unit and a large fan for forcing the hot, dry air through the product being dried. Alfalfa is sometimes dried in such a way before being stored. See *dehydrater: 1.*

drift. 1. Detrital matter that has been transported by wind, ice, and water. Most commonly the deposition during the Glacial period is implied, but the term may be used with reference to some kinds of loess and other wind deposition. See *glacial drift.* 2. Piles of snow which are built by the wind. 3. In turpentine orcharding, a convenient working unit which is approx. 2,000 cups. 4. The movement of a herd or band of livestock not influenced by human action. *Cf. drive: 1.* 5. The movement of certain species of wildlife which is influenced directly by human action. 6. The carrying of spray or dust by the wind from the place for which it was intended.

drifter. 1. A term once used on the open ranges of western U.S. for a shepherd, or owner of a band of sheep, who moved about grazing his sheep. 2. A hired hand who moves from place to place with the season or job and has no permanent location.

drift fence. A line of fence, open at both ends, used to direct the movement of grazing animals on the ranges of western U.S.

drifting. The moving of bees from one hive to another because of loss of direction caused by wind or other circumstances.

driftway. A trail or driveway over which cattle are driven or drift in a definite direction.

drill. 1. A small furrow in which seeds are planted. 2. A row of seeds which have been planted by dribbling a small furrow. 3. An implement for planting seeds which forms a small furrow, deposits the seed in dribbles, covers the seed, and packs soil over it. It can also deposit fertilizer, lime, or other soil preparations into the soil, alone or with the seed. See *plain drill; fertilizer drill; alfalfa drill; grain drill; lister planter; checkrow planter.* 4. A very small, trickling stream or rill.

drill barrow. An early device intended to serve as a row planter which consisted of a barrel with holes on the circumference. It was filled with seed and rolled along a row, the seeds dropping out of holes as the barrel was rolled.

drilled. Designating those seeds or crops which have been planted in rows by means of a drill.

drilled well. A well which is sunk by percussion tools, by rotating bits, or by jets of water; the most common type of water well. *Cf. dug well; radial well; driven well; horizontal well; bored well.*

drill in. To plant seeds with a drill. See *drill: 3.*

drill planter (seeder). See *drill: 3.*

drimad. A plant which is tolerant of an alkali or salty soil.

D ring. In harness making, a piece of metal shaped like the letter *D* or *O* to form a link between two separate bands, straps, or other harness devices.

drinking cup. A metal or porcelain cup just large enough for an animal's muzzle, within which is a valve or float, activated when the animal pushes its muzzle against it. It is a handier method of watering animals than by a common watering trough, and it provides water for confined animals whenever they desire it.

drinking hole. Waterhole.

drinking pan. A relatively flat, shallow pan in which drinking water is placed for poultry, swine, etc.

drinking trough. A long vessel of wood, metal, concrete, etc., which is used to hold drinking water for livestock.

driodad. A plant of a dry thicket.

drip. Moisture from fog or cloud which condenses on leaves or twigs of plants, and drips onto the ground.

drip culture. The continuous application of nutrients in solution to the medium in which plants are growing. *Cf. slop culture: 1.*

drip gutter. A small gutter inside a greenhouse at the point where the roof and walls meet which carries off water formed by condensation inside the house.

drip loss. The loss in weight owing to exuding and dripping away of tissue juices during thawing of frozen products as affected by rapidity of freezing, composition of the liquid surrounding the product while frozen, etc.

drip of a tree. The place at which much water runs off the tree during rains, the periphery of the tree.

drive. 1. a. The moving of livestock under human direction. *Cf. drift: 4.* b. To herd livestock. 2. In wildlife management, the herding of wild animals past a particular point for counting or shooting. 3. The process of floating logs down a river from a forest area to a mill or to a shipping point. 4. The means for giving motion to a machine part or machine, as a *drive* chain, a *drive* wheel.

drive chain. A chain for power transmission which is composed of uniform links of malleable iron, working over suitable sprocket wheels. Also called *link belting.*

drive-ins. Designating those cattle which are herded-in to market, as contrasted to those transported there by truck, etc. See *truck-ins.*

drive nozzle. A high-pressure nozzle used in spraying trees, so designed that the jet breaks into a mist at the desired height.

driven well. A well which is sunk by forcing a metal pipe into the ground until it extends below the surface of the water table.

driver. 1. One who operates a vehicle. 2. That which communicates motion to something else, especially a wheel that transmits power to another wheel.

drive-through section. A section of a barn, usually the center, with door openings on both sides or ends to allow vehicles, tractors, and implements to move in and out without backing.

drive-through shed. A lean-to shed, without sides, which is attached to a barn or similar building to allow vehicles, tractors, and implements to move in and out without backing.

driveway. 1. In the range country, that land which is set aside for the movement of livestock from place to place: *e.g.*, from the home range to the shipping point. Also called *stock driveway; stock route.* 2. A path or passage, sometimes paved, for the movement of vehicles and/or livestock. 3. The farm road from the building site to the gate at the highway. 4. Vehicle passage into or through a barn.

driving. See *drive: 1.*

driving hammer. 1. A tool used to drive horseshoe nails into a horse's hoof. 2. A piece of pipe capped and weighted at one end which is used to drive steel posts into the ground. 3. A guided drop weight which is used to drive a post or piling into the ground.

driving iron. A sharp-pointed, steel rod used to force holes in the earth, into stumps, etc., in preparation for blasting.

driving rain. Rain accompanied by a high wind.

drizzle. Precipitation from stratiform clouds in the form of numerous, minute droplets, which appears to float in the air. It may amount to as much as 0.04 ins. per hr. and is heavier than a mist but less intense than a shower.

drone. The male bee hatched from an unfertilized egg. It is larger than the workers, does not gather honey, and has no sting.

drone brood. The male bee brood which, because of its larger size, is reared in larger cells than those of the worker bees.

drone comb. That comb in a beehive which has four cells to the linear inch on each side of the comb.

drone layer. A queen bee that has exhausted her supply of spermatozoa so that her eggs produce only drones.

drone trap. A device placed in the entrance of a hive which allows worker bees to pass but traps any queen or drone.

dronkgrass. *Melica decumbens,* family Gramineae, a grass which, when eaten by cattle, causes a condition similar to drunkenness. Native to South Africa.

droop. To hang downward from exhaustion, lack of water, or nourishment, etc., as a plant droops in the hot sun.

droopbead holly. *Ilex geniculata,* family Aquifoliaceae, a deciduous shrub grown for its red fruits and its yellow, autumnal foliage. Native to Japan.

droop croup. A horse with a short, steep rump, and a low-set tail.

drooping leucothoë. *Leucothoë catesbaei,* family Ericaceae, an evergreen shrub grown for its foliage and attractive, white, flower clusters. It has been reported as poisonous when browsed by sheep. Native to North America. Also called *fetter bush; dog hobble; branch ivy; dog laurel.*

drooping method. A system of training grape vines in which the growing shoots are not tied up on a trellis, but are allowed to hang down throughout the growing season.

drooping mockorange. *Philadelphus laxus,* family Saxifragaceae, a tender, deciduous shrub grown for its white flowers. Native to North America.

drooping wing. A natural, loose carriage of the wing of a bird in which the outer edge of the folded wing drops below the horizontal.

drop. 1. A structure in an open or closed conduit which is used for dropping the water to a lower level and dissipating its kinetic energy. 2. A decrease in height or elevation. 3. Any fruit which falls from a tree to the ground because of wind, or other conditions: *e.g.*, diseased, immature, and usually unfertilized fruit; also normal but ripe fruit. 4. A fungous disease of vegetables caused by *Sclerotinia sclerotiorum, S. minor,* or *S. intermedia,* family Sclerotiniaceae, which is characterized by a sudden collapse of the plant, or a slow stunting and premature ripening. The host tissue becomes soft and watery as the disease progresses. Sometimes the superficial mycelium grows profusely and binds stored vegetables together. Also called *watery soft rot; nesting; windfall.* 5. To give birth to young, as to *drop* a calf. 6. To shoot an animal and to cause it to fall.

drop band. A group of ewes or nanny goats which is managed separately at the season of bearing young. Also called *drop herd.*

drop crop. Pendulous crop *q.v.*

drop-crotching. The cutting back of the large branches of trees to the laterals which encourages growth to the sides.

drop herd. Drop band *q.v.*

drop jaw. Dumb rabies.

drop lever. The part of a planter which regulates the number and spacing of seeds.

drop method. A method of producing flake or pseudo-churned buttermilk, in which melted, high grade butter is dropped into cultured buttermilk. Also called *pellet method.*

drop off. 1. To fall to the ground, as a leaf or blossom. 2. To decrease in number, quality, demand, etc.

drop pan. See *dropping pan.*

dropped sole. Convex sole.

dropper. 1. A vegetative shoot in certain bulbous plants which grows downward to form a new bulb at the apex. 2. The young bulb of a tulip.

drop pin. Hitch pin.

dropping board. Boards which are nailed together and placed under a roost to catch the poultry droppings.

dropping pan. A flat, metal pan placed under the wire-mesh floor of poultry cages to catch the excrement.

dropping pit. A pit dug beneath a roost to catch the poultry droppings.

dropping plate. Cell drop *q.v.*

droppings. The excrement of animals and birds. See *manure: 1; guano.*

dropseed. Any plant of the genus Sporobolus, family Gramineae. Species are perennial, rarely annual grasses. See *alkali sacaton; sand dropseed.*

dropsy. 1. An excessive, intensive, watery enlargement of parts of a plant sometimes due to a superabundance of moisture in the soil. Also called *edema.* 2. See *brown spot: 1.* 3. Widespread edema throughout the body of an animal.

drop tongue. The tongue of a wagon, plow, etc., which is so fastened as to allow it to be raised and lowered.

drop-type waterer. A poultry waterer in which the bird pecks at a hanging drop of water and, in so doing, pushes a tiny valve which releases another drop.

drop wagon. A wagon with about ten separate compartments which follows the band of sheep in range country so that as lambs are born they can be placed in individual compartments with their mothers.

dropwort. *Filipendula hexapetala,* family Rosaceae, a perennial herb grown for its fern-like leaves and white flowers. It is sometimes sold as spirea. Native to Eurasia. Also called *meadowsweet.*

drosometer. An instrument which measures the quantity of moisture precipitated as dew.

drought. 1. A period of insufficient rainfall for normal plant growth which begins when soil moisture is so diminished that vegetation roots cannot absorb enough water to replace that lost by transpiration. A *drought* can be arbitrarily defined on the basis of amount of precipitation during a certain period although such a measure has no uniform significance because of variable amounts of moisture in soils when the measurement period begins. The intensity or damage from *drought* varies with the plant since some plants are much more tolerant of dry conditions than others. 2. Any dry spell. 3. A period when precipitation is not adequate to meet the needs of the human population. 4. In an arid area, a period when there is an especial shortage of water for irrigation. Also called *drouth.*

drought-enduring plants. Species of plants that require little moisture to grow and reproduce. Also called *xerophytes.*

drought form. The aspect which a plant attains under drought conditions.

drought necrosis. A nonparasitic disorder of potato, resulting from lack of sufficient water, which is characterized by a yellowing and browning of the vascular ring and the more external tissues of the tuber.

drought-resistant. Designating a plant which is adapted to growing under drought conditions.

drought ring. A zone of abnormal tissue noted in the annual rings of a tree induced by water deficiency.

drought spot. 1. A physiological condition of apple and prune fruits caused by a sudden and continued shortage of water following a normal supply earlier in the season. It is characterized by large, irregular, depressed, water-soaked lesions, usually near the calyx end. In prunes, the skin of the fruit may burst, and a gum ooze to the surface. 2. A boron-deficiency disease of apples characterized by russeting and cracking of the fruit.

droughty. 1. Designating periods of time or areas in which there is subnormal precipitation. 2. Dry. See *drought.*

drouth. See *drought.*

drouthy. Designating a soil which has low water-holding capacity: *e.g.,* one with a sand or gravel subsoil.

drove. A collection or mass of animals of one species, as a *drove* of cattle.

drover. 1. A person who drives cattle, sheep, etc., from one place to another. 2. A person who is in charge of shipment of animals and who accompanies animals to market.

drown. To deprive a plant or animal of life by immersion in water so that the supply of oxygen is cut off.

drown out. To kill a plant by drowning.

dru. A thicket in a valley.

drug fumitory. *Fumaria officinalis,* family Fumariaceae, an annual or biennial herb once widely used as a remedy for scurvy, but now grown for its small, purple, crimson-tipped flowers. Native to Europe. Also called *fumitory; earth smoke; hedge fumitory; wax dolls.*

drug lionsear. *Leonotis leonurus,* family Labiatae, a perennial herb grown in very warm areas and in greenhouses for its yellow, orange-red, or white flowers. Native to Africa.

drug of choice. The best of several medicaments that are available for a specific purpose when judged from standpoints of safety, efficiency, simplicity, and cost.

drug plant. Any plant from which a medicinal product is obtained.

drug sabadilla. *Schoenocaulon officinale,* family

Liliaceae, a tropical plant grown for its seeds which are ground into a powder and used as an insecticide and in medicine as a counter-irritant and nerve poison. The crude mixture of alkaloids from sabadilla is called veratrine. Native to Mexico. Also called *sabadilla.* See *sabadilla seed.*

drug speedwell. *Veronica officinalis,* family Scrophulariaceae, a prostrate, perennial herb which is a weed pest in lawns. Found in Eurasia and North America.

drugstore beetle. *Stegobium paniceum* (L.), family Anobiidae, a light brown beetle that infests foodstuffs and tobacco. Also called *biscuit beetle; bread beetle.*

drum. 1. A cylinder-shaped package or product, as a *drum* of cheese. 2. A revolving cylinder, as the *drum* of a seed cleaner. 3. To knock on the sides of a hive to drive the bees upward when transferring them from one hive to another.

drum-dried skim milk. A method of preparing nonfat dry milk in which milk is sprayed on heated, revolving drums for dehydrating. *Cf. spray-dried skim milk.*

drum drier. A type of dehydrator which consists of a revolving, heated drum on whose surface the materials are dried and scraped off after dehydration.

drumhead cabbage. A type of cabbage with heads that are broad but elliptical in cross section. Also called *tallow loaf.*

drumlin. An elongated, oval hill or ridge which is composed of glacial drift, normally compact and unstratified, usually with its longer axis conforming to the direction of the movement of the ice responsible for its deposition. *Drumlins* are present in parts of the glaciated region of the U.S. especially in parts of New York, Michigan, and Wisconsin where they are not only striking relief features, but also have agricultural significance.

Drummond pepperwort. *Marsilea drummondi,* family Marsileaceae, an aquatic or marsh herb grown for its foliage. If uncontrolled, it can completely cover the surface of a pool. Native to Australia. Also called *nardoo.*

Drummond phlox. *Phlox drummondi,* family Polemoniaceae, an annual herb grown in flower borders or rock gardens for its white to purple flowers. Native to North America. Also called *annual phlox; Texan pride.*

drum roller. Cast-iron drum roller *q.v.*

drumstick. The leg of a fowl from the hock joint to the joint next above.

drumstick tree. Golden shower senna *q.v.*

drum-type hay loader. Cylinder hay loader.

drum-type heat exchanger. A heat exchanger for pasteurizing or for ice-cream freezing in which a drum is jacketed with steam or refrigerant. The material to be treated is placed within the drum where agitator blades force the material around the periphery at high speeds.

drunken forest. A forest in which the trees have leaning positions due to displacement, caused by landslides or from freezing and thawing of soil.

drupe. A single-seeded, fleshy fruit which does not split open, *e.g.,* cherry, peach, plum, or olive. Also called *stone fruit.*

drupel. See *drupelet.*

drupelet. A small drupe, often one of many in a cluster or group, *e.g.,* the pulpy grain of the fruit of the dewberry, raspberry, blackberry, etc.

dry. 1. a. To cause a pregnant cow to stop giving milk shortly before she drops her calf. b. Designating a cow which has ceased to give milk shortly before she drops her calf. 2. To preserve a product by dehydration. 3. Designating dressed poultry packed for shipment without ice. 4. Designating a time or place of abnormal lack of precipitation, *e.g.,* a dry year. 5. Describing a well which yields no water.

dryad. A shade plant.

dry aggregate. A soil aggregate which is not broken down by dry sieving.

dry band. 1. A group of sheep or goats which is composed entirely of mature animals. Also called *dry herd.* 2. During lambing or kidding, that part of a herd whose members have not yet dropped their young.

dry bark. A disease of Eureka and Lisbon varieties of lemon, in coastal regions, U.S., which is characterized by lesions on or near the bud union or crotches of the large branches. The bark is penetrated from ⅔ through the cambium layer to all the way and dies. There is a yellowing of the leaves and defoliation. It is thought to follow injury by sunburn, cold, etc., when the tree is attacked by any of certain fungi; *Diaporthe citri, Botryosphaeria ribis, Botrytis cinerea, Diplodia* spp.

dry basis. Designating a product, *e.g.,* fertilizer or feed, which is analyzed for its constituents calculated to the basis of dry material.

dry beans. The dry, mature seeds of the bean (*q.v.*) varieties.

dry bible. A condition in which the omasum (one of the stomachs) of the ruminant becomes unusually dry or impacted.

dry body. A desirable characteristic of ice cream as compared to soggy or wet body which shows poor freezing quality.

dry bog. A cranberry bog that is not covered by water at any season.

dry brown rot. Apple soft scald *q.v.*

dry-bulb temperature. The temperature of the air which is shown by any type of thermometer not affected by the water vapor content or relative humidity of the air.

dry buttermilk solids. The product which results from drying buttermilk obtained from churning sweet cream.

dry cheese. A cheese made of soured, skim milk which is then cooked until firm. Salt and caraway seeds are added, and the cheese is aged until it becomes very hard.

dry-coil bunker system. A method of distributing refrigeration in cold-storage rooms in which coils of pipes carrying a refrigerant are placed in a separate room. Air forced by a large fan is blown over them and distributed by ducts to the storage room.

dry corn fodder. Corn fodder which is cured in the shock.

Dry Cottage cheese. Cottage cheese sold in bulk, to which no cream has been added.

dry cow. A cow which has ceased to give milk.

dry cream. Dehydrated, sweet cream which is used in confections and in bakery goods.

dry culture. A culture consisting of selected organisms in a media of skim milk solids, etc., which is dried so as not to reduce the vitality of the organisms; used in the production of commercial or lactic buttermilk.

dry cure. A method of curing pork by rubbing the meat with salt, sugar, and saltpeter, and allowing it to age. See *shelf curing; box curing (without pressure); box curing (with pressure).*

dry disease. A fungous disease of the roots of sugar cane caused by *Marasmius sacchan.*

dry end. Sclerocystosis *q.v.*

dryer. See *drier.*

dry-face. In turpentine trees, an abnormal or disease condition at or above the turpentine faces which stops the flow of oleoresin.

dry farming. A system of producing crops in arid and semiarid regions without the use of irrigation. Frequently, alternate years of fallow and cropping are practiced.

dry fog. A haze which is due to presence of dust or smoke in the air; not a true fog.

dry food. Livestock feedstuff to which no liquid has been added. *Cf. wet mash.*

dry frost. Black frost *q.v.*

dry fruit. A fruit with a hard pericarp.

dry gangrene. A physiological condition of grazing animals caused by ingestion of ergot which contracts the small blood vessels in the extremities of the body so that the parts receive no nutrition and die. See *ergotism.*

dry grinding. Grinding of grain without water. *Cf. water ground: 1.*

dry heat. Hot air which has a low relative humidity.

dry ice. Solid carbon dioxide which is purified, liquefied, expanded to form snow, and finally pressed into blocks. It is used as a refrigerant.

dry ice cream mix. See *dried ice cream mix.*

dry icing. The sprinkling or seeding of clouds with dry ice to cause rain, dispel clouds, or prevent the formation of a lightning storm.

drying. Reducing the moisture content of a product to the point at which the concentration of the dissolved solids is so high that osmotic pressure will prevent the growth of microorganisms.

drying by sublimation. The removal of water from a product while it is frozen to lock the molecular structure so that changes cannot occur during dehydration and to prevent bacteriological action.

drying off. 1. The gradual reduction of water, as from a bulb, tuber, corm, or potted plant until it becomes dormant and fit to store. 2. The reduction of the milk flow in a cow until it ceases.

drying oil. An oil which, either naturally or after being boiled with oxide of lead, absorbs oxygen from the air and dries rapidly. *Drying oils* are used as a basis for many paints and varnishes.

drying wind. A wind, the air of which has a very low humidity and does not usually portend snow or rain.

drying yard. The site on which a stock of lumber is piled for air-seasoning.

dryki. In logging, those trees which have been killed by backwater.

dry kiln. A structure in which lumber is seasoned artificially or in which cones are dried and opened.

dry land. 1. Farm land in arid or semiarid regions. See *dry farming.* 2. Land that normally has a dry surface, as contrasted to marsh or swamp land that is usually saturated.

dry-land dipper dredge. A power-driven dredge mounted on a platform which in turn is mounted on tracks. It consists of an open bucket with a cutting edge and a hinged-flap bottom, a boom, the dipper handle. It is used to dredge channels and to dig drainage or irrigation ditches.

dry-land farming. See *dry farming.*

dry-land foot rot. Root rot of wheat *q.v.*

dry-land wireworm. *Ctenicera inflata* (Say), family Elateridae, one of the destructive species of wireworms.

dry leaf spot. Gray speck *q.v.*

dry lightning. Lightning in a thunderstorm accompanied by insufficient rain to moisten fuels and prevent fire from a stroke of lightning.

dry lot. A bare, fenced-in area used as a place to keep livestock for feeding and fattening.

dry-lot fed. Designating livestock which have been fed and fattened in a dry lot in contrast to those fed on pasture.

dry mash. A mixture of ground cereal or cereal products with supplements which is fed to poultry without the addition of any liquid. See *laying mash; all mash; growing mash.*

dry masonry. Stone work which is laid without mortar.

dry matter. The total amount of matter, as in a feedstuff, less the moisture it contains. Dry feeds, such as cereal grains, usually contain about 9 percent water and 91 percent *dry matter.*

dry measure. A system for measuring dry products, such as grain.

dry milking. A complete milking to ascertain that the udder of the cow is completely empty or dry before starting an official test. Also called *dry-hand milking.*

dry milk solids. Any of the dehydrated milk products.

dry mill. A sawmill that has no log ponds.

dry off. 1. See *dry: 1.* 2. To reduce the amount of water given a plant, bulb, tuber, corm, until it becomes dormant and ready to store. 3. To bring the lactation period of a cow to an end.

dry out. To cause to become dry, as alfalfa has the capacity to *dry out* soil.

dry pack. 1. A method of preserving fruits by freezing without water or sugar. 2. A method of preserving cut flowers by packing them dry and placing them under very humid conditions in closed containers.

dry-packed poultry. A method of packing dressed poultry in boxes in which no ice is allowed to touch the skin of the bird. *Cf. ice-packed poultry.*

dry pea. A mature pea which is used as a vegetable or seed.

dry period. 1. That time just before parturition when a cow or other lactating animal ceases to give milk. 2. See *dry season; drought.*

dry pick. To dress a bird without scalding it. It may be bled by cutting the jugular vein, and its medulla oblongata punctured to relax the feather follicles to facilitate plucking. *Cf. hard scald: 2; slack scald.*

dry pod. See *capsule: 1.*

dry pox. Fowl pox (*q.v.*) in which there are no complications with coryza or roup.

dry-processed rendered lard. Lard extracted from pork by being melted by steam. The meat is placed in a vacuum container above a steam tank.

dry-rendered lard. Dry-processed rendered lard.

dry-rendered tankage. Meat meal *q.v.*

dry-rendering. The cooking of meat scraps, etc., in a vacuum above steam-jacketed tanks.

dry root rot. 1. A root rot of citrus, thought to be associated with *Fusarium* spp., family Tuberculariaceae, which is characterized by a thin appearance of the tree. This is followed by a yellowing and dropping of the foliage, and often by the setting of a large crop. The foliage may suddenly wilt. 2. A fungous disease of beans and peas caused by *Fusarium solani f. phaseoli* which is characterized by a dwarfing of the plants, failure of some varieties to produce vines, reduction in yield, a yellowing, and dropping of the foliage.

dry rot. 1. Any of many types of decay but especially of that which, when in an advanced stage, permits timber to be easily crushed to a dry powder. The term is actually a misnomer for any decay, since all fungi require considerable moisture for growth. 2. See *late blight and rot of potato; Diplodia disease of corn; cork: 3; inspissosis; blackleg: 2.* 3. A metabolic defect which causes ill health in animals.

dry rot of peanuts. A fungous disease caused by *Rhizoctonia solani;* characterized by lesions developing on the main stem near the soil line. If girdled, the plant will die; if not, the plant may partially recover.

dry rot of strawberries. Gray mold of strawberries *q.v.*

dry salt cure. A method of preserving pork, especially fat backs, in which salt is sprinkled heavily over the pieces of meat that are then piled one on top of another. The salted meat is cured from three weeks to a month. *Cf. dry cure; shelf curing; sweet pickle cure.*

dry salt pork. A preserved cut of pork which has been cured by dry salt cure (*q.v.*) with or without the use of one or more of the following: sodium nitrate, sodium nitrite, potassium nitrate, sugar, sirup, honey, spice.

dry sands. 1. An azonal group of soils consisting of well-drained, sandy deposits which have no clearly expressed soil profile horizons. See *azonal soils.* 2. Any kind of sand soil, especially in the humid regions, that is very dry because of free drainage and low moisture-holding capacity.

dry sausage. A sausage which may be eaten with or without cooking and which is low in moisture content and firm to the touch. *Cf. domestic sausage.*

dry season. That part of the year in which there is the least precipitation.

dry sheep. 1. Generally, a mixed band of wethers and lambless ewes. 2. Specifically, lambless ewes. 3. Pregnant ewes during the period from weaning to parturition.

dry shrinkage. The shrinkage of timber, which has been dried in a kiln, after its surface has been removed.

dry side. See *on the dry side.*

dry skim milk. Nonfat dry milk *q.v.*

dry spell. In humid regions, a period longer than usual with little or no rainfall which is less serious in consequence than a drought.

dry storage. A method of storage in which the temperature and humidity may or may not be controlled: *e.g.*, refrigerated storage of milk; low humidity storage of unformed paper cartons or packaging materials.

dry storm. A lightning storm during which less than 0.10 in. of precipitation reaches the ground.

dry strawberry. Barrenstrawberry.

dry tip. Endoxerosis.

dry top rot. 1. A disease of sugar cane ascribed to the fungus *Plasmodiophora vascularum,* family Plasmodiophoraceae, which is characterized by drying, wilting, and death of the plant.

dry up. 1. To become deficient in water, as a well, stream, or spring. 2. Gradually or abruptly ceasing to give milk for a time, as a cow prior to calving. See *dry: 1.*

dry valley. A valley in which the stream that once ran through it has dried up or sunk into the alluvium of the floor.

dry wash. A term in western U.S. for a dry, stony bed of a stream, often containing alluvium deposits, which is usually full of water only at times of flooding.

dry weather. A period during any part of the year which has relatively little or no precipitation.

dry weather legumes. Legumes, such as alfalfa, white sweetclover, yellow sweetclover, and red clover, which are sufficiently deep rooted to withstand drought conditions.

dry weight. The weight of a product or material less the weight of the moisture it contains. See *dry measure.*

dry well. 1. A well which contains little or no water. 2. A covered pit with open-jointed linings through which drainage from roofs, basement floors, or areaways may seep or leach into the surrounding porous soil.

dry whey solids. The residue resulting from dehydrating clean, wholesome whey which is used in confections and in bakery goods.

dry yard. The trays of drying fruit, the buildings, and the equipment used in sun-drying fruit.

dual-purpose. Designating a domestic fowl or animal which is bred for two purposes: *e.g.*, a chicken raised for both meat and egg production.

dual-purpose breeds of cattle. Cattle raised for both milk and meat production. The principal breeds in the U.S. are Milking Shorthorn, Red Polled, Polled Durham, and Devon.

dual use. 1. The use of a range by more than one kind of livestock at the same time. 2. The use of any area for two or more purposes, such as recreation and forestry, mining and forestry, etc.

dual-use range. A range containing a forage combination of grass, forbs, and browse that allows two or more kinds of stock to graze the area to advantage at the same time through the entire season, or separately, during a part of the season.

dub. To cut the combs and wattles from cockerels when they are 12 to 16 wks. old, to prevent injury from freezing or from fighting.

Duchartre lily. *Lilium duchartrei,* family Liliaceae, a bulbous herb grown for its white, purple-spotted flowers. Native to western China.

duck. 1. A bird, family Anatidae, especially the domesticated *duck* raised for meat and eider down. 2. The female *duck.* See *drake.* 3. A heavy cotton or linen cloth.

duckfoot cultivator. A field cultivator which is equipped with duckfoot shovels for subsurface tillage.

duck-footed. Designating a domestic fowl which carries its hind toe forward.

duckfoot shovel. A spear-headed type shovel which has swept-back wings normally inclined at a very flat angle to the ground so that it undercuts close to the surface of the soil.

duckfoot sweep. Duckfoot shovel.

duckfoot weeder. See *duckfoot cultivator.*

duckling. A young duck which still has downy plumage.

duck malaria. A protozoan, parasitic disease caused by *Leucocytozoon anatis* which is characterized by droopiness, loss of appetite, and a tendency to lie down. The blackfly is the vector. It particularly attacks ducklings.

duck potato. The tuber of the aquatic plant, *Sagitarria latifolia,* family Aliamaceae, which is a food for the wild duck. See *arrowhead.*

duckweed. 1. Any plant of the genus Lemna, family Lemnaceae. Species are very small, aquatic plants used in aquariums but which may become pests in garden pools. 2. Purslane.

duck wheat. Tartary buckwheat *q.v.*

ductless glands. Those glands of the body whose secretions pass directly into the bloodstream or into the lymph, as one of the endocrine glands.

dudaim melon. *Cucumis melo* var. *dudaim,* family Cucurbitaceae, an annual, viny, tendril-bearing herb sometimes grown for its orange-like, perfumed fruit. Also called *pomegranate melon; Queen Anne's pocket melon.*

dudder grass. Southern maidenhair *q.v.*

Duel cheese. A soft, cured, Austrian cheese, made from cow's milk.

duff. The surface horizon of forest soils which consists of matted, dead, organic matter which is only slightly decomposed, as well as humus.

duff hygrometer. An instrument for measuring the moisture content of litter or duff.

dug-out. 1. A crude pioneer dwelling in western U.S. which was often little more than an excavation in the side of a hill with a door and a hole for a chimney. *Cf. sod house.* 2. An excavated reservoir.

dug well. A well which is sunk by men digging with picks and shovels to the water table or just below it. Such a well is usually 4 ft. in diameter and walled with bricks.

Duke cherry. Any variety of a group of hybrid cherries which originated as natural crosses between the mazzard cherry (*Prunus avium*) and the sour cherry (*Prunus cerasus*). Some representative varieties are May Duke, Empress Eugenia, Rime Hortense, and Royal Duke.

Dulce de Leche. South American condensed milk which is prepared by boiling milk and sugar together in open kettles.

dull glossy. A defect in the eye of Swiss cheese in which the inner lining of the eye has the dull appearance of a nut shell. Also called *shell eye; dead eye.*

dull market. A term used to describe a time when produce is available for sale but there is little buying.

dullseed cornbind. *Polygonum convolvulus,* family Polygonaceae, an annual, twining or trailing vine which is a weed pest in annual crops and waste ground. Native to Europe. Also called *black bindweed; wild buckwheat; knot bindweed; devil's-tether.*

dumba. The fur or skin of the Karakul lamb.

dumb foxglove. Andrews gentian *q.v.*

dumb gentian. Andrews gentian *q.v.*

dumb rabies. Paralysis that develops in the last stages of rabies. Also called *drop jaw.*

dumb well. See *dry well: 2.*

dumetose. Bushy or relating to bushes.

dummy. 1. A sleepy or stupid horse, especially one that has suffered inflammation of the brain. 2. A strongly built frame suggesting the shape of a female animal, often covered with the hide of such a female, which is used to excite a male animal sexually so that it will mount the dummy and accept the use of an artificial vagina so that the semen may be collected for later use in artificial insemination.

dummy spoiled. Designating a male animal which has become so accustomed to the artificial vagina that it refuses coitus.

dump. 1. The storage yard of a portable sawmill. 2. A place for refuse. 3. To unload a truck or wagon by raising the bed and allowing the loaded material to fall to the ground. 4. To discard unwanted materials or refuse. 5. To place large quantities of a product on the market so as to depress the price.

dumpling. A whole, peeled and cored apple, which is ready for drying.

dump rake. A power or horse-drawn rake which consists of two wheels, an axle, a seat for the operator, and a series of sharply curved tines attached to the axle which curve outward and inward toward the ground. This rake is used to gather hay, etc., in rows or heaps. As sufficient hay is collected, the tines can be raised and the hay released.

dump truck. A truck having a specially built bed which can be raised at the front end so that material on the truck can be unloaded by sliding off.

dumpy. Designating an animal or fowl which does not feel very well, but which is not seriously ill.

dun. An animal's coat color characterized by a dark dorsal stripe over the withers and shoulders. Recognized shades are mouse dun, buckskin dun, and claybark dun.

duncecap larkspur. *Delphinium occidentale,* family Ranunculaceae, a poisonous plant found in U.S. on the high mountain slopes in Montana, Colorado, Utah, and Idaho. Characteristics of livestock poisoning are loss of appetite, constipation, staggering, uneasiness, lying with legs rigid, and sometimes bloat.

dune. A mound or ridge of loose sand piled up by the wind which is common where loose, bare sand is abundant, as along seashores, shores of the Great Lakes, in river valleys, and in desert areas. Under drought conditions *dunes* may form from the topsoil of cultivated land lying unprotected by vegetal cover.

dune and beach erosion control. The application of various mechanical and vegetative measures designed to stabilize coastal and inland sand dune or blow sand areas.

dune fixation. Dune and beach erosion control.

dune grass. Grass which is adapted to sand dunes, *e.g.,* beach grass, *Ammophila arenaria.*

dune heath. An area of low shrub-like vegetation on stretches of sandy waste.

dune sand. An area of wind-drifted sand in dunes, hummocks, and ridges, which is usually free from vegetation and which is undergoing active erosion and redeposition by winds.

dung. Manure; the feces or excrement of animals and birds. See *long dung; short dung; guano.*

dung hill. 1. A heap or pile of manure or farmyard excrement. 2. A scrub horse.

dung tag. See *tag: 1.*

dunk-in. Designating a chest-type freezer, as contrasted to an upright one.

Dunlop. A rich, white, pressed Scottish cheese similar to Cheddar made of cow's milk and cured for at least 10 weeks.

dun roan. A coat color combination of dun and roan in horses.

duodenal coccidiosis. See *coccidiosis.*

duodenum. 1. In birds, the part of the small intestine nearest the gizzard. 2. In mammals, the part of the small intestine nearest the stomach.

duplex. Designating a homologous pair of chromosomes, both members of which are present.

duplicate genes. Factors with the same recessive phenotypic expression.

duplication. The occurrence of a segment more than once in the same chromosome or genome.

durability. The quality of a product which gives it a comparative permanence with regard to decay or resistance to wear.

duramen. Heartwood *q.v.*

Durban crowfootgrass. *Dactyloctenium aegyptium,* family Gramineae, a tropical grass which provides forage for wildlife and which is found in lawns and cultivated fields in eastern U.S. Its habit is creeping and the spikes are finger-like. Also called *Egyptian finger grass; Egyptian grass; crowfoot grass.*

Durham cattle. See *Polled Shorthorn.*

Duroc. A prolific breed of hogs which grow rapidly. It is lardy in type and red in color. Boars weigh from 800 to 1,100 lbs.; sows up to 850 lbs. Also formerly called *Duroc-Jersey.*

durra. *Sorghum vulgare* var. *durra,* a variety of sorghum which is widely grown throughout the world for its grain and for forage. See *Kafir; milo.*

durum wheat. *Triticum durum,* family Gramineae, an annual, spring-planted grain grown as a livestock feed, and for the high quality flour obtained from it and used in the manufacture of macaroni, etc. Also called *macaroni wheat; hard amber durum; amber durum.*

dusky birch sawfly. *Croesus latitarsus* Nort., family Tenthredinidae, an insect pest of birches whose larvae feed in groups around the margins of the leaves, sometimes causing defoliation of saplings. Found in U.S. in New England and Great Lakes states.

dust. 1. Fine, dry particles of earth, or other matter, so attenuated that they may be wafted by the wind. 2. Particles less than 0.1 mm. in diameter which may be bounced off the ground by sand grains moving in saltation. 3. An insecticide, fungicide, etc., which is applied in a dry state, or the application of these to plants, animals, or fowls.

dust barrier. A barrier of pulverulent earth used as protection against migrating cinch bugs.

dust bath. An obsolete method of ridding fowls of parasitic insects, in which road dust, ashes, fine sand, or the like containing an admixed insecticide was placed in a receptacle so that poultry could wallow or bathe in it and thus rid themselves of insects.

Dust Bowl. A part of the Great Plains region, U.S., which is subject to severe droughts, during which dry soil is picked up by strong winds to form great clouds of dust. It includes parts of southeastern Colorado, and adjacent parts of Kansas, Oklahoma, and northwestern Texas, and is actually an elevated plain.

dust carrier. An inert, dry substance with which a dry insecticide, fungicide, etc., is mixed.

dust devil. A violent, dust-laden whirlwind.

dust diluent. See *dust carrier.*

duster. A mechanical device, consisting of a blower, a dust reservoir, and a nozzle, which is used in the application of an insecticidal or fungicidal dust so that the dust is emitted in a blast of air.

dust flue. In a cotton gin, a large, sheet-metal pipe which is used to carry off dust and dirt from lint cotton just before the lint cotton is pressed into bales.

dusting. The application of insecticides, fungicides, etc., in dust form. See *aerial dusting.*

dust mulch. A shallow layer of loose dry surface soil, which may be natural or created by tillage. Theoretically such a covering lessens losses of soil moisture resulting from direct evaporation, but in practice, it is of questionable value as a soil conservation measure.

dustproof. Describing anything constructed or protected so that an accumulation of dust will not interfere with its successful operation.

dust respirator. A mask to filter out harmful or irritating substances which is worn when applying poisons, working in dusty areas, etc.

dust soils. Soils, observed most commonly in arid regions, which are normally pulverulent. They are penetrated by water with extreme slowness and thus constitute a problem in management under irrigation.

dust storm. Huge clouds of dust, sometimes covering up to 1,000 sq. mi. which result from wind erosion of soils, especially in the western part of the Great Plains, U.S. See *Dust Bowl.*

dust tight. Describing anything constructed so that dust cannot enter.

dusty clover. Roundhead lespedeza *q.v.*

dusty cotton. Cotton which, after being baled, shows evidence of containing dust; as a result, it is reduced in grade.

dusty meadowrue. *Thalictrum glaucum,* family Ranunculaceae, a perennial herb grown for its fragrant, yellow flowers. Native to Europe.

dustymiller. 1. *Centaurea cineraria,* family Compositae, a perennial herb grown for its white felty foliage. Native to Eurasia. 2. A common name for beach wormwood; silver groundsel; rose campion.

Dutch ashes. Ashes which result from the burning of peat.

Dutch beech. White poplar *q.v.*

Dutch belted cattle. A dairy breed of cattle originating in the north of Holland which is black with a white belt circling the middle of the body. Bulls weigh up to 2,000 lbs. and cows up to 1,300. A minor breed, it is of little importance except as an odd, show animal.

Dutch breeder tulip. Breeder tulip *q.v.*

Dutch bulbs. Bulbs raised in Holland, *e.g.,* tulip, hyacinth, crocus, etc.

Dutch cattle. Loosely, all black and white dairy cattle. See *Holstein-Friesian.*

Dutch cheese. See *Edam; Gouda.*

Dutch clover. White clover *q.v.*

Dutch elm. *Ulmus hollandica,* family Ulmaceae, a hardy, hybrid, deciduous tree grown as a shade tree and as an ornamental. Also called *Holland elm.*

Dutch elm disease. A serious, fungous disease of the elm, caused by *Ceratostomella ulmi,* family Ceratostomataceae, which is characterized by a yellowing, curling, browning, wilting of the leaves and defoliation. The progress of the disease is more rapid on young trees than on old ones, and diagnosis is possible only in the laboratory. Also called *elm blight.*

Dutch hoe. See *D-hoe.*

Dutchman's-breeches. *Dicentra cucullaria,* family Fumariaceae, a delicate, stemless herb grown in deep shade in wild gardens for its white flowers tipped with yellow. It is poisonous to livestock. Symptoms of poisoning are: sudden trembling, frothing at the mouth, vomiting, convulsions, and sometimes death. Found in North America. Also called *little blue stagger; white eardrops; butterfly banners.*

Dutchman's pipe. *Aristolochia durior,* family Aristolochiaceae, a high-climbing vine widely grown for its foliage and U-shaped, yellowish-brown flowers. Also called *pipevine; siphonflower.*

Dutch medlar. See *medlar.*

Dutch myrtle. See *sweet gale; true myrtle.*

Dutch rush. See *scouringrush.*

Dutch woodbine. Woodbine honeysuckle *q.v.*

duty of water. The relation between the area of land served and the quantity of irrigation water used under different standards of practice which will vary from a large use under crude practice to small use under good practice. It is the measure of the use of water and may be distinguished as head gate or gross duty, lateral duty, duty at the farm, or net duty, and crop duty for different crops. It may be expressed in depth of water on the land, as a rate of flow per acre for a given time, or as the area per unit flow for a given time. Strictly speaking, *duty* should be expressed in terms of the crops that a given quantity of water will help produce or the number of acres that a given flow will irrigate. A high duty corresponds to an economical use of water; a low duty indicates small returns for the water used. Usage has broadened the meaning until it may mean exactly the opposite, that is, high duty signifying the use of abundant water, and vice versa. See *consumptive use; transpiration ratio.*

dwarf. 1. A serious virus disease of the trailing blackberry characterized by a yellowing of the plant and a shortening of the canes. The canes, spindly at first, become stiff, unusually upright, and thick. The leaves are dwarfed, crinkled, and mottled with bronze and light green patches. Found on U.S. Pacific Coast. 2. A plant, especially one that has been grafted, or animal which is much smaller when mature than others of its species. This condition is caused by disease, lack of water, mineral deficiency, etc. 3. See *dwarf (midget) cattle.* 4. Designating any plant disease, one of whose symptoms is a stopping or retarding of growth.

dwarf alder. Dwarf fothergilla *q.v.*

dwarf Asiatic elm. Chinese elm *q.v.*

dwarf banana. *Musa nana,* family Musaceae, a very small tree which produces edible fruit, but which is chiefly grown for its reddish-brown and yellowish-white flowers. Native to southern China. Also called *Cavendish banana; Chinese banana.*

dwarf bean. Any low growing bush bean plant. *Cf. pole bean.*

dwarf bunt. A fungous disease of wheat caused by *Tilletia caries,* family Tilletiaceae, which is characterized by severe stunting of the plant to ½ to ¼ its normal height, by the glumes having a feathery appearance, and by the diseased plants showing an excessive tillering, 40 tillers not being uncommon. It spreads by soil infestation.

dwarf bushhoneysuckle. *Diervilla lonicera,* family Caprifoliaceae, a hardy, low, deciduous shrub grown for its yellow flowers. Native to North America. Also called *gravelweed.*

dwarf cupflower. *Nierembergia hippomanica,* family Solanaceae, a tender, perennial herb grown in the flower garden or as a pot plant for its lavender flowers. Native to tropical America. Also called *nierembergia; cupflower.*

dwarfed. See *dwarf.*

dwarf elm. See *Siberian elm.*

dwarf Essex rape. See *winter rape.*

dwarf fan palm. See *Mediterraneanpalm: 2.*

dwarf fothergilla. *Fothergilla gardeni,* family Hamamelidaceae, a low shrub grown in warm areas for its white flowers. Native to North America. Also called *dwarf alder.*

dwarf glorybind. *Convolvulus tricolor,* family Convolvulaceae, an annual, semiprostrate herb grown for its blue flowers with yellow throats and white margins. Native to southern Europe. Also called *dwarf morningglory.*

dwarf hollygrape. Creeping mahonia *q.v.*

dwarfing. The stunting of a plant's growth by disease, lack of water, propagation techniques, etc. See *dwarf.*

dwarf iris. *Iris pumila,* family Iridaceae, a perennial herb grown for its varicolored flowers. It is a bearded iris. Native to Eurasia.

dwarf Japanese quince. Japanese flowering quince *q.v.*

dwarf laurel. Lambkill kalmia *q.v.*

dwarf lilyturf. *Ophiopogon japonicus* (*Mondo japanicum*), family Liliaceae, an herb grown as a sod-forming ground cover and for its light lilac flowers. Native to eastern Asia.

dwarf (midget) cattle. A genetic condition of increasing frequency in most breeds of cattle, especially beef cattle. Calves are of less than normal size at birth, remain small throughout life, and often have abnormal body conformation such as stubby legs, a dished or bulldog face, and pop eyes. Dwarfism is now considered an autosomal recessive character.

dwarf mistletoe. *Arceuthobium pusillum,* family Loranthaceae, a destructive, epiphytic pest of ponderosa pine.

dwarf morningglory. Dwarf glorybind *q.v.*

dwarf nasturtium. Bush nasturtium *q.v.*

dwarf of alfalfa. Alfalfa dwarf *q.v.*

dwarf of blackberry. See *dwarf: 1.*

dwarf palmetto. *Rhapidophyllum hystrix,* family Palmaceae, a small palm bearing leaves with long stalks. Also called *blue palmetto; needle palm.*

dwarf pinkray boltonia. Violet boltonia *q.v.*

dwarf Tasmanian blue eucalyptus. *Eucalyptus globulus* var. *compacta,* family Myrtaceae, a densely branched tree grown in California as a windbreak to protect citrus groves. Native to Australia.

dwarf trillium. *Trillium nivale,* family Liliaceae, a low, perennial herb grown in rock gardens for its white flowers, and as an ornamental ground cover. Native to North America. Also called *early trillium; early wakerobin; dwarf white trillium; snow trillium.*

dwelling. 1. A barely perceptible pause in the stride of a horse just before it places one of its feet to the ground when approaching a jump. 2. A farm home or family residence.

dwindle. To die off rapidly in the spring, as old bees.

dyad. The univalent chromosome, composed of two chromatids, at meiosis. The pair of cells formed at the end of the first meiotic division.

dyeing corn. A highly colored sweet corn which was used in making a dye in Colonial U.S. It is probably the corn known as papoon *q.v.*

dyersweed goldenrod. *Solidago nemoralis,* family Compositae, a perennial herb grown on dry sandy soils for its yellow flowers. Native to North America. Also called *dwarf goldenrod; gray goldenrod.*

dyers woodruff. *Asperula tinctoria,* family Rubiaceae, a perennial herb grown for its red or pinkish-white flowers. Native to Europe.

dying-back. See *dieback.*

dynamometer. A device which measures power: *e.g.,* an absorption dynamometer measures the power of a motor or engine and at the same time transforms the output energy into heat. A transmission dynamometer measures the engine power while at the same time transmitting it, practically undiminished, to some driven machine.

dysentery. 1. Any of the various intestinal diseases characterized by chronic diarrhea. 2. A condition of adult bees resulting from an accumulation of feces. It usually occurs during winter and is caused by unfavorable wintering conditions and low quality food. Its presence is detected by small spots of feces around the entrance and within the hive.

dysfunction. A lessening in the proper action of a part of the body.

dysgenesis. A defect in breeding in which hybrids cannot mate between themselves, but may produce offspring with members of either families of their parents. Such offspring are sterile.

dyspnea. Labored breathing.

Dzierzon's theory. The theory that a queen honey bee lays the male or drone eggs by parthenogenesis (*q.v.*) or without fertilization.

E

eagle claw. A cultivator gang with four shovels which are arranged in the shape of an eagle's claw.

eagle fern. Eastern bracken *q.v.*

eagle flower. See *snapweed: 1.*

ear. 1. The grain bearing spike of a cereal plant, *e.g.*, corn, rye, wheat, etc. 2. To produce or form fruiting spikes.

ear button. Ear tag *q.v.*

earcockle. Nematode disease of wheat *q.v.*

ear corn. Unshelled corn.

ear corn chops. A livestock feedstuff of chopped corn and cob, without the husk, and with no greater proportion of cob than occurs on the ear corn in its natural state. Also called *corn and cob meal.*

ear corn chops with husks. A livestock feedstuff of chopped corn, cob, and husks with no greater proportion of cob and husks than occurs on the ear corn in its natural state. Also called *ear corn chopped with husks.*

ear-covert. The very small feathers of a bird which cover the ear.

eared. 1. Designating the presence of ears or spikes on a plant. 2. Pertaining to an animal restrained by its ears. 3. Designating the presence of earlike tufts of feathers in some birds.

earflower. See *sacred ear.*

ear fly. Any of several small gadflies or deerflies of the genus Chrysops which infest the ears of livestock and humans. These flies have been known to transmit tularemia from rabbits to humans in western U.S.

ear leaf. A cotyledon or seed leaf.

earleaf tasselflower. *Emilia sagittata,* family Compositae, a tropical, annual herb grown for its scarlet-orange flower heads. Native to tropical America. Also called *tasselflower; Flora's-paintbrush.*

ear-leaved. Designating a plant which has earlike leaves.

earlet. A small, ear-shaped appendage of a plant part, such as a leaf. See *ear.*

Earl loco. *Astragalus earlei,* family Leguminosae, a perennial, poisonous herb. Symptoms of poisoning are dullness, irregularity of gait, lack of appetite, dragging of the feet, a solitary habit, loss of flesh, and a shaggy coat. The animal ceases to eat and dies. Found in western Texas, U.S. Also called *locoweed.*

ear lobe. The bare skin just below the ear of a bird.

early. 1. Designating a variety which regularly produces mature fruit, leaves, etc., earlier than normal. 2. Designating a period for planting or harvesting which comes before the normal or usual time.

early almond. Peanut *q.v.*

early aster. White wood aster *q.v.*

early blight of celery. A widespread, fungous disease of celery and celeriac caused by *Cercospora apii,* family Dematiaceae, characterized by small pale green or yellow spots on the leaves which enlarge and turn brown to slate-gray. The plant is stunted, and the yield is reduced. It attacks late winter crops from March to May in U.S.

early blight of potato. A troublesome, fungous disease of the potato and tomato caused by *Alternaria solani,* family Dematiaceae, characterized by dark brown or black, circular, dead areas on the leaflets appearing as a series of concentric rings. As the disease progresses, these areas enlarge, darken, and coalesce, so that the leaves have extended dead areas and ragged margins. Affected leaves may curl or roll. The stalks may show lesions and turn yellow. Also called *Alternaria blight; leaf curl; tuber rot.*

early blight of tomato. See *early blight of potato.*

early blueberry. Lowbush blueberry *q.v.*

early bugloss. *Anchusa barrelieri,* family Boraginaceae, a perennial herb grown for its blue flowers with white tubers and yellow throats. Native to Eurasia.

early cress. Early wintercress *q.v.*

early daylily. *Hemerocallis dumortieri,* family Liliaceae, a lily-like herb grown for its pale-orange flowers. Native to Japan.

early double tulip. A type or class of tulip, grown in flower borders and as pot plants, which blossoms early in the spring and retains its blossoms for a long time.

early dwarf pea. Baby gardenpea *q.v.*

early forsythia. *Forsythia ovata,* family Oleaceae, a hardy, deciduous shrub grown for its arching branches covered with early-spring, yellow flowers. Native to Korea.

early frost. A frost which occurs earlier in the fall than usual.

early hawkweed. See *poor-robin's-hawkweed; cohosh bugbane.*

early lamb. Any lamb born prior to the normal lambing season.

early leafspot. A fungous disease of peanuts caused by *Cercospora arachidicola,* family Dematiaceae, characterized by light tan leaf spots which change to reddish-brown to black on the under surface and light brown on the upper surface.

early lilac. *Syringa oblata,* family Oleaceae, a hardy, deciduous shrub grown for its early, pale lilac flowers. Native to China.

early maturity. The completion of development in a plant or animal in form and function sooner than others of its species. See *early: 1.*

early meadow grass. Annual bluegrass *q.v.*

early meadow parsnip. See *golden zizia.*

early meadowrue. *Thalictrum dioicum,* family Ranunculaceae, a delicate, perennial herb having purplish or greenish-white flowers, which is especially adapted to the rock garden and to acidic soil. Native to North America.

early mesquite. See *buffalograss.*
early parsnip. See *golden zizia.*
early perennial phlox. Carolina phlox *q.v.*
early rot of cranberry. A fungous disease of cranberries caused by *Guignardia vaccinii,* family Mycosphaerellaceae, characterized by very small, light-colored lesions on the half-grown fruit. Later, concentric, dark-colored rings appear and the fruit may shrivel and become covered with black, fungous, fruiting bodies.
early saxifrage. Virginia saxifrage *q.v.*
early single tulip. A class or type of tulip which blossoms early in the spring and whose varieties give a range of soft and brilliant colors.
early soil. A soil that warms up early in the spring. It is usually well drained with a sand or sandy-loam texture.
early spring. 1. In the northern hemisphere, March to May, in the southern hemisphere, Oct. to Dec., depending on the latitude. 2. A growing season which starts earlier than is usual for the latitude or location.
early sweet blueberry. Lowbush blueberry *q.v.*
early testing. A breeding procedure to discover the probable usefulness of an inbred line at the beginning of the inbreeding period.
early trading. The buying and selling which occurs just after the daily opening of the market.
early trillium. Dwarf trillium *q.v.*
early tulip. See *early single tulip; early double tulip.*
early wakerobin. Dwarf trillium *q.v.*
early wintercress. *Barbarea verna,* family Cruciferae, a biennial or perennial herb which is sometimes grown as a substitute for watercress. Native to Europe. Also called *early cress; Belle Isle cress; scurvy cress; scurvy grass.*
early wood. In a woody plant, the less dense, large-celled, first-formed part of a growth layer.
ear mange. A mange of cats, dogs, rabbits, and foxes caused by *Otodectes cynotis* Hering (ear mange mite), family Demodicidae, which infests the skin inside the ear, causing great irritation, impaired hearing or deafness, inability to coordinate movements, and sometimes causing the ear to become packed with dry scabs.
ear mold. A fungus that is moldlike and infests ears of cereals, especially corn.
earned income. Money received for salaries, wages, fees, etc., for services rendered.
ear notcher. A punch used to cut notches in the ears of animals for identification. *Cf. ear tag.*
ear of corn. The full-kerneled corn cob with or without the husks. See *ear corn.*
earrings. Flowers of species of the genus Laburnum, family Leguminosae, especially the goldenchain laburnum (*q.v.*), which are often used for Easter decorations.
ear rot. Molding or decay of corn which is caused by fungi or bacteria. See *Diplodia disease of corn.*
ear stage. The stage of development of a corn plant in which the ear is forming.
ear tag. A tag securely attached to the ear of an animal for identification. Also called *ear tab; ear button.*
earth. 1. Soil, as black *earth.* 2. Ground. 3. Soil-like material, as ocher. 4. Certain minerals, the rare earths, thorium, zirconium, etc. 5. Land, or land areas. 6. The underground hole of a burrowing animal. 7. To place *earth* over the roots or around a plant.
earth almond. See *chufa flatsedge; peanut.*
earth apple. See *potato; cucumber; Jerusalem artichoke sunflower.*
earth ball. The mass of earth which is dug up intact with the roots of a plant. See *balled and burlapped.*
earthboard. Moldboard *q.v.*
earth bob. See *grub: 2; maggot.*
earth dam. A barrier composed of earth, clay, silt, sand, sand and gravel, or a combination of earth and rock, as contrasted to a dam made of concrete, masonry, etc.
earth fall. Landslide *q.v.*
earth flow. The moderately rapid flowage of large masses of earth materials which is common on the edge of clay terraces, as in certain glaciated valleys and in some mountainous regions of shale bedrock, as parts of West Virginia, Pennsylvania, and Ohio, U.S. Also called *flow slides; mud-rock flow.*
earth gall. American falsehellebore *q.v.*
earthing. Banking up the soil, as in blanching celery.
earthing-up. See *earth up.*
earth ivy. See *ground ivy.*
earth louse. Any of several aphids that attack the roots of plants.
earth mover. Any large machine or device, such as a bulldozer or carryall which moves large quantities of earth.
earth nut. See *peanut; chufa flatsedge.*
earthnut oil. Peanut oil.
earth pea. See *southern hogpeanut; peanut.*
earth smoke. Drug fumitory.
earth spring. See *spring: 1.*
earth tank. A structure which is formed by excavation, as an earthen dam across a drainage area, used for impounding drinking water for livestock. Also called *livestock reservoir; farm pond.*
earth up. To cultivate in such a manner that soil is thrown toward plants, as in the hilling of potatoes.
earthworm. Any of the several species of the genus Lumbricus, family Lumbricidae, which are commonly, and often abundantly, represented in the fauna of surface soils, especially in the humid regions. They are generally regarded as beneficial because they improve the structure and fertility of soil; however, they may do minor damage by eating plants.
ear tick. *Otobius megnini* (Duges), family Argasidae, a tick that infests the ears of livestock, especially cattle, sheep, and horses.
ear-to-row. Designating an obsolete method of corn breeding in which the progeny of individual ears were tested and the seed of the better ones used for propagation.
ear wagon. A wagon used in corn picking, into

which the husked ears are thrown. It has a high sideboard which deflects the ears into the wagon bed.

earworm. 1. Eczema of an animal's external ear. 2. Corn earworm *q.v.*

easement. 1. The right of access or right-of-way through land which is usually attained by purchase for a specific purpose, such as for the construction and maintenance of a gas line, power line, oil line, or telephone line. 2. A vested or acquired right to use land, other than as a tenant or the owner, for a specific purpose.

ease off. To reduce slowly in price, quantity, demand, etc.

East African cedar. African juniper.

East African cherryorange. *Citropsis schweinfurthi,* family Rutaceae, a spiny tree grown in warm sections for its fragrant, white flowers and as a rootstock for grafting various species of citrus. Native to Africa.

East Coast fever (tick). A disease of cattle prevalent in East Africa caused by a protozoan transmitted by the biting of the brown dog tick *Rhipicephalus sanguineus* (Latr.).

easter. A high wind or rain storm coming from an easterly direction.

Easter anemone. European pasqueflower.

Easterbell starwort. *Stellaria holostea,* family Caryophyllaceae, a hardy, perennial herb grown for its showy, white flowers. Native to Eurasia. Also called *Easter bell; starwort; greater stitchwort; Easter flower.*

Eastercactus. *Schlumbergera gaertneri,* family Cactaceae, a cactus grown in desert gardens for its 3 in. scarlet flowers. Native to South America.

Easter daisy. Stemless townsendia.

Easter flower. A common name for poinsettia; European pasqueflower; Easterbell starwort; daffodil; spreading pasqueflower.

Easter giant. European bistort.

Easter heraldtrumpet. *Beaumontia grandiflora,* family Apocynaceae, a tall, woody vine grown in very warm areas for its habit and its large, fragrant, white flowers. Native to India. Also called *Easterlily-vine; Nepal trumpetflower.*

Easter ledges. European bistort.

Easter lily. 1. *Lilium longiflorum,* family Liliaceae, a bulbous herb grown in flower gardens and forced by florists, for its fragrant, trumpet-shaped, pure-white flowers. Native to Japan. Also called *white-trumpet lily.* 2. Any lily placed on the market principally during Easter week, although it may be sold at other times. The most popular is the Easter lily. Others include the speciosum lily, the Madonna lily, and the atamasco lily *q.q.v.*

Easterlily-vine. Easter heraldtrumpet.

eastern arborvitae. *Thuja occidentalis,* family Pinaceae, an evergreen tree from which innumerable horticultural ornamental varieties have been derived. Found in eastern North America. Also called *American arborvitae; white cedar; Queen Victoria arborvitae;* and among foresters: *northern white cedar.*

eastern baccharis. *Baccharis halimifolia,* family Compositae, a shrub grown in seaside gardens for its white fruiting heads. Native to North America. Also called *groundsel tree; seamyrtle.*

eastern black walnut. *Juglans nigra,* family Juglandaceae, a hardy, deciduous tree grown for its fruit, the black walnut of commerce, which is used in confections. Its lumber is one of the finest of the domestic cabinet woods, used also for veneer, interior paneling, furniture, etc. Native to North America.

eastern bluestem. A virus disease of raspberries in eastern U.S.

eastern bracken. *Pteridium latiusculum,* family Polypodiaceae, a fern poisonous to livestock at certain stages of its development although it is sometimes used as fodder or bedding. Cattle become sick with very high temperature, have difficulty in breathing, salivate, and have nosebleed and hemorrhage. Found chiefly in northeastern U.S. and Canada. Also called *brake fern; eagle fern; bracken.*

eastern burningbush. Eastern wahoo.

eastern field wireworm. *Limonius agonus* (Say), family Elateridae, the larva of a click beetle which infests the seed, underground stems, and roots of tobacco, corn, potato, beets, carrots, radishes, and onions. Commonly found in eastern U.S.

eastern larch. *Larix laricina,* family Pinaceae, a hardy, deciduous tree grown for its valuable timber and as an ornamental. Native to North America. Also called *American larch; hackmatack; epinette;* and among foresters: *tamarack.*

eastern larch beetle. *Dendroctonus simplex* Lec., family Scolytidae, a brown beetle which makes galleries in the bark of live or dead larches.

eastern lubber grasshopper. *Romalea microptera* (P. deB.), family Acrididae, a short-winged locust which infests grass, flowers, and ornamentals in southern U.S.

eastern mistletoe. Christmas American mistletoe.

eastern pear and apple rust. A leaf rust caused by *Gymnosporangium globosum,* family Pucciniaceae, which is common to apple, pear, and hawthorn, with alternate stages in its life cycle on pomaceous species (aecial stage) and on junipers (telial stage). It is common in North America where eastern redcedars are adjacent to susceptible pomaceous species or varieties. Also called *cedar quince rust.* See *rusts.*

eastern raspberry fruitworm. *Byturus rubi* Barber, family Byturidae, a grub which bores into raspberry and loganberry fruits, rendering them unfit to eat.

eastern redbud. *Cercis canadensis,* family Leguminosae, a deciduous tree grown for its early spring, rosy-pink flowers. Native to North America. Also called *American Judas-tree; American redbud; Judas-tree.*

eastern redcedar. *Juniperus virginiana,* family Pinaceae, a very hardy evergreen tree grown as an ornamental and as the source of cedar wood. Its valuable timber is used for posts, chests, closet linings, millwork, pencil slats, woodenware, and water buckets. Native to North America. Also called *red cedar; Tennessee red cedar.*

eastern red oak. *Quercus borealis* var. *maxima,* family Fagaceae, a tall, relatively quick-growing tree grown as an ornamental shade tree and for its wood which is used in furniture and interior finishing. Native to North America.

eastern snow mold. A fungous disease of fine turf caused by *Fusarium nivale,* family Tuberculariaceae. It appears soon after snow melts and is distinguished by whitish-gray dead areas of irregular circular outline. *Cf. snow mold.*

eastern spruce. Black spruce.

eastern spruce beetle. *Dendroctonus piceaperda* Hopk., family Scolytidae, a bark beetle which is a serious pest as it bores galleries into red, white, and black spruces.

eastern spruce gall aphid. *Chermes abietis* L., family Chermidae, a sucking insect which infests Norway and white spruce in eastern U.S. Infestation results in pineapple shaped galls at the base of new shoots.

eastern subterannean termite. *Reticulitermes flavipes* (Kollar), family Rhinotermitidae, an insect that infests woodwork of buildings, sometimes weakening joists and floors.

eastern tent caterpillar. *Malacosoma americanum* (F.), family Lasiocampidae, an insect whose larvae gather in the forks of the limbs of various fruit trees, spinning a large weblike nest from which they emerge during the day to feed on the foliage. They are serious pests found in North America from the Rocky Mountains eastward. Also called *apple-tree tent caterpillar.*

eastern wahoo. *Euonymus atropurpureus,* family Celastraceae, a shrub or small tree grown for its foliage and scarlet fruit. It is poisonous to sheep and goats, causing nausea, prostration, and cold sweat. Found in eastern U.S. and as far west as Texas. Also called *burningbush; wahoo; eastern burningbush.*

eastern white pine. *Pinus strobus,* family Pinaceae, an evergreen, coniferous tree which is one of the outstanding timber pines and ornamentals. It sometimes grows to 200 ft. at maturity and is often referred to as the "Queen of the forest." Its lumber is soft and is put to a wide variety of uses. Native to North America. Also called *white pine; northern white pine; American white pine; apple pine; deal pine.*

eastern X disease. See *X-disease (of stone fruits); buckskin: 4.*

East India(n) arrowroot. See *tikor: 2; Fiji-arrowroot tacca.*

East India bluestem. See *Turkestan bluestem.*

East Indian hawthorn. India raphiolepis.

East Indian lotus. Hindu lotus.

East Indian millet. Pearl millet.

East Indian rose bay. Crapejasmine tabernaemontana.

East Indian walnut. Lebbek albizzia.

East Indian wax. Ghedda wax.

East India thyme. Spanishthyme coleus.

East Indies aster. *Aster subcoeruleus,* family Compositae, a perennial herb grown in rock gardens for its blue or pale-blue flowers. Native to the Himalayas.

easy. 1. A command to a horse to move slower. 2. Designating an animal which has a low or weak back or pasterns, as *easy* in the back.

easy bit. A horse's bit which restrains the horse lightly. See *curb bit.*

easy gait. Any horse's gait which is conveniently slow for the rider. See *rack: 1; running walk; fox trot; slow gait.*

easy hitch. Any hitch for a tractor which is simple and quick to operate.

easy keeper. An animal which thrives with a minimum of feed, attention, and care.

easy market. A market in which prices decline under very little pressure.

eatage. 1. An edible growth of grass for livestock, especially the second mowing. 2. The privilege of using grassland for pasture.

eatberry. See *gooseberry: 1.*

eating. Designating food in regard to its quality, as an *eating* apple is one that has sufficient quality for *eating* out-of-hand.

eating ripe. Designating fruit which is somewhat riper than that suitable for canning purposes; a fruit that has attained a near optimum in degree of ripeness. *Cf. dead ripe.*

eau celeste. Cupric ammonium sulfate, a useful fungicide having nonstaining properties.

eaver. Any of certain species of ryegrass, especially darnel ryegrass and perennial ryegrass (England).

ebony spleenwort. *Asplenium platyneuron,* family Polypodiaceae, a fern grown in the wild gardens and other parts of home grounds for its dainty fronds with purplish-brown leaf stalks. Native to eastern and south central U.S.

ebrillade. The training of a horse by jerking one rein when he refuses to turn.

ecad. A plant or species of plant that has been changed morphologically from the normal by environment.

Ecballium. A vine genus, family Curcurbitaceae. See *squirting cucumber.*

eccaleobion. An apparatus once used in the artificial hatching of eggs in which steam was used as a source of heat.

eccentric. Used to describe an irregular or unsymmetrical tree trunk.

eccle grass. See *butterwort: 2.*

ecesis. The adjustment of a plant to a new habitat, involving germination, growth, and reproduction.

echard. Soil water which is unavailable for plant use. See *wilting coefficient; hygroscopic moisture.*

ecological competition. The competition of two or more organisms for a limited supply of an environmental resource.

ecology. The science of the interrelationships of organisms with their environment.

economic diameter limit. The minimum utilization of logs or trees which will produce the largest net income during the life of the operation.

economic entomology. That branch of insect study directed toward preventing man's losses and increasing his gains through manipulation of insect populations. Examples of *economic entomology* include the methods of protecting plants, man, and other animals from insect borne diseases by insecticides, drainage, crop rotation, the culture of silkworm, honeybee, and various beneficial parasitic insects.

economic maturity. The age at which a tree or stand of trees will no longer increase in value fast enough to earn a satisfactory rate of interest. Also called *financial maturity.*

economic poisons. Any of the insecticides, herbicides, rodenticides, etc.

economic size. The optimum size of a farm, ranch, or other agricultural operation which will yield the highest net return for the operator's labor, capital, and management.

economy of gain. The amount of feed required per unit of gain in the weight of an animal or fowl which is expressed in cash value, *i.e.,* the feed per unit of gain converted to dollars.

economy of scarcity. The theory that limited production and relative scarcity of a commodity will result in greater good for a country's economy than abundance, and consequently lower prices.

ecotone. A transitional area between two types of vegetation, as between forest and prairie.

ecotype. A variation in a plant brought about by a particular type of environment.

ecraseur. A surgical instrument consisting of a fine chain or cord that is looped around a diseased part and gradually tightened, thus severing the enclosed part.

ectodynamic soil. A soil whose dominant character has been determined by external forces, such as climate and vegetation.

ectoparasite. A parasite which lives on the outside of the body of its host.

eczema. A nonparasitic, skin disease of man and animals characterized by a red inflammation of the skin, development of papules, vesicles, and pustules, serous discharge, formation of crusts, severe itching, and loss of hair. It may be acute or chronic. It is thought to result from allergy, from hypersensitivity to chemicals, from photosensitization, from poor environment, or from faulty feeding. See *mange; parakeratosis; dermatitis.*

Edam. A hard, solid, spherical, Dutch cheese prepared from cow's milk containing about 2.5 percent fat. Usually the cheese is highly salted. Before marketing it is dipped in red wax.

edaphic. Pertaining to the influence of soil upon plant growth.

edaphology. That science which is concerned with the study of the relationships between plants and soils.

eddish. Aftermath *q.v.*

eddo. a. The root of the dasheen. b. The edible root or stem of certain other members of the same family Araceae.

eddy-chamber. A chamber in a nozzle in which liquid is atomized.

edelweiss. *Leontopodium alpinum,* family Compositae, a small, perennial, alpine herb grown in rock gardens and as a pot plant for its yellow flowers and silvery bracts. Native to Eurasia.

edema. 1. An accumulation, usually abnormal, of serous fluids within the intercellular tissue spaces of the body. If the *edema* is in the subcutaneous tissue, the affected area will be swollen and will pit with pressure. Also called *dropsy.* 2. Abnormal swelling of plant parts due to an excessive intake of water.

edema of the wattles. A condition in birds, in which the wattles are filled with inflammatory liquid so that they become very much enlarged. The cause is unknown, but *Pasteurella multocida* has been suggested. Also called *wattle edema.*

edematous. Pertaining to or affected by edema.

edge. The cover afforded by the vegetative margin between different types of wildlife habitats, as between fields and woodland.

edgebone. Corrupted form of aitchbone. See *aitchbone: 1.*

edge-drop. A type of seed plate used in a corn planter which carries the seed on edge in the cell of the plate.

edge firing. A method of controlled burning in which fires are set around the perimeter of the area to be burned and allowed to spread inward.

edger. A tool used for cutting the sod along walks, the edges of flower borders, shrub borders, and beds. Also called *edge iron.*

edge runner. A device used in extracting oil from olives which consists of a cast-steel bowl within which two heavy stone or steel wheels revolve, crushing the pulp and breaking some of the seeds.

edging. 1. A border of flowers, stones, metal, or wooden strips, etc., outlining the boundary of a flower bed, lawn, drive, etc. See *edging plant.* 2. Waste strips of wood which result from squaring boards.

edging candytuft. Evergreen candytuft.

edging lobelia. *Lobelia erinus,* family Lobeliaceae, a tender, annual herb grown as an edging plant for its blue flowers. Native to South Africa.

edging plant. Any low-growing plant which is used along the front of a flower border or around the outside of a flower bed. See *edging: 1.*

edible arum. Dasheen *q.v.*

edible bird. Any bird suitable for human consumption, being free from disease or any other condition which might render it unwholesome.

edible canna. *Canna edulis,* family Cannaceae, a large herb cultivated for its starchy rootstocks that are known as *tous-les-mois.* The extracted starch is sometimes sold as arrowroot and is used in the preparation of baby foods, as well as being an adulterant of cocoa. Native to the West Indies and South America.

edible galingale. Chufa flatsedge.

edible oil. Any vegetable oil which is fit for humans to eat.

edible pine. Any of several pines which bear large, edible seeds, especially the pinyon (pinon), as well as the digger pine and the Italian stone pine. Some of these are cultivated as a food crop. See *Digger pine; Italian stone pine; Mexican pinyon pine; nut pine.*

edible-podded pea. A type of garden pea whose seed and pod are edible. There are several varieties, the most common being ***Pisum sativum*** var. *macrocarpon,* family Leguminosae. Also called *sugar pea.*

edible rush. Chufa flatsedge.

edible snakegourd. *Trichosanthes anguina,* family Cucurbitaceae, an annual vine which bears white flowers and greenish, club-shaped fruit, sometimes used as food in India. It is grown as an ornamental in U.S. Also called *serpent gourd; club gourd.*

EDTA. Ethylenediamine tetraacetic acid, a chelate used to correct iron chlorosis.

eelworm. Nematode *q.v.*

eelworm disease. Root knot *q.v.*

effete. Designating an exhaustion of the ability in animals to produce young or in plants to bear fruit. See *impotence.*

efficiency. 1. Relative effectiveness of a process or causal agent in producing specified results from given means, specifically, the ratio of designated outputs to designated inputs in a given transformation process. In mechanics, *e.g.*, the ratio of energy recovered from a machine to the energy put in. 2. By extension, effectiveness in yielding private profits or socially desired products.

effloresce. To produce blossoms.

efflorescence. 1. The period or state of full bloom. 2. The crusts or coatings of any soluble salts, such as the sulfates and chlorides of magnesium and sodium which appear on the bare surface of soil, on clods, in cracks of natural soils, or on heavily fertilized soils as a result of the evaporation of moisture in the soil.

efwatakala grass. Molassesgrass.

egagropilus. Hair ball *q.v.*

egg. a. The reproductive body produced by a female organism: in animals, the ovum; in plants, the germ cell, which, after fertilization, develops into the embryo. b. The more or less oval reproductive body produced by females of birds, reptiles, and certain other animal species, enclosed in a calcareous shell or strong membrane within which the young develop.

egg albumen. The white of an egg, a mixture of proteins, which is used in confections, bakery goods, in manufacturing adhesives, etc.

egg apple. Eggplant.

egg basket. A specially designed, wire basket used in collecting eggs from the nests and as a container for cooling eggs.

egg bean. Kidney bean.

egg beater. A horse which places his feet carefully and smoothly.

eggberry. See *hackberry: 1; mazzard cherry: 1.*

egg bonnet. Schreber watershield.

egg bound. A physiological condition of a hen or other female bird that renders the bird unable to pass a formed egg in the normal manner which thus obstructs the cloaca. Symptoms are restlessness, frequent visits to a nest, prolapse of the oviduct.

egg-breaking. A preliminary step in the production of frozen or dried eggs in which the previously candled, clean eggs are broken into cups for inspection as to their wholesomeness.

egg candler. A person who candles eggs. See *egg candling.*

egg candling. The examination of poultry eggs in the shell by means of a bright light, for determining market grade and for the detection of meat spot of egg, blood spot, or air bubble *q.q.v.*

egg carton. A cardboard carton used in the retail sale of eggs to prevent breakage and to expedite handling. It usually contains 12 eggs and has a separate cell for each egg.

egg case. A case made of wood, fiberboard, etc., used for the shipment of eggs, which has cardboard layers with individual egg compartments for a capacity of 30 doz. eggs.

egg case shook. An unassembled egg case.

egg cell. The female reproductive cell of animals and plants. See *egg: a.*

Egg cheese. A cheese made from a mixture of fresh milk and fresh eggs.

egg cleaner. A device used to clean the surface of egg shells rendering the eggs more attractive for sale. It consists of a cushion base covered with sandpaper or cloth.

egg cleavage. Any mitotic division or the series of mitotic divisions of an egg cell after fertilization.

egg crate. Egg case.

egg cup. The indented or cuplike division of an egg carton or egg case divider into which individual eggs are placed.

egg drying. See *dried eggs.*

egg exchange. A market where eggs are bought and sold in large quantities and where eggs are traded on futures.

egg farm. A farm or establishment on which the principal source of income is from poultry eggs. Also called *chicken farm; chicken ranch.*

egg-feed price ratio. The pounds of poultry ration equal in value to one dozen eggs at prices existing at a particular time and place.

egg flat. A piece of cardboard or papier-mâché used between the fillers in an egg case to prevent breakage.

egg flavor. A mild flavor defect of ice cream characterized by a taste similar to that of egg yolk, sometimes accompanied by a musty flavor persisting in the mouth.

eggfruit. 1. The edible fruit of canistel lucuma *q.v.* 2. The fruit of the eggplant *q.v.*

egg grader. A mechanical device used for grading eggs according to weight. As the ungraded eggs are carried along a moving belt, they pass over a weighing device which automatically sorts them into separate bins.

egg industry. The industry which specializes in the production, manufacture, and distribution of eggs and egg products.

egg laying. The production of eggs specifically by hens and queen bees.

egg-laying stimulator. Any of various patented medicines purported to stimulate egg production in hens. Most often it is of no particular value.
eggler. One who buys and sells eggs (British).
egg mash. A mixture of ground cereals or cereal products to which has been added essential protein, vitamin, and mineral supplements in sufficient amounts to provide a balanced ration for laying hens.
egg meat. The contents of a fowl's egg.
egg membrane. The innermost parchment-like membrane of an egg shell.
egg money. Income derived from the sale of eggs from a general farm, an enterprise usually supervised by the farmer's wife. It is usually set aside for emergencies or for the purchase of personal luxuries.
egg nucleus. The nucleus of an egg cell.
egg oil. An oil which is obtained from the yolk of a hen's eggs.
eggplant. *Solanum melongena,* family Solanaceae, a tropical, perennial herb or small shrub grown as an annual for its edible, blackish-purple fruit. Native to the East Indies. Also called *garden eggplant; egg apple; eggfruit; aubergine; brinjal; mad-apple; Jew's-apple.*
eggplant blight. A fungous disease caused by *Phomopsis vexans,* family Sphaerioidaceae, characterized by a damping off or seedling blight, brown leaf spots, and circular, depressed, discolored fruit lesions of rotted tissue, which may cause the entire fruit to rot.
eggplant flea beetle. *Epitrix fuscula* Crotch, family Chrysomelidae, an insect that infests the foliage of the eggplant.
eggplant lace bug. *Gargaphia solani* Heid., family Tingidae, a sucking insect that infests eggplant and other solanaceous hosts.
eggplant leaf miner. *Keiferia glochinella* (Zell.), family Gelechiidae, a moth whose larvae mine the leaves of solanaceous plants in southern U.S.; however, they seldom become seriously destructive.
egg powder. Dried eggs.
egg-powder flavor. A flavor defect of ice cream characterized by its resemblance to the taste of dried eggs. See *egg flavor.*
egg producing. Designating egg production *q.v.*
egg product. Any product derived or manufactured from eggs or egg parts: *e.g.,* dried albumen.
egg production. 1. The tending, feeding, etc., of a flock of hens principally for the eggs they lay. 2. The number of eggs laid by a hen.
egg room. A room in which eggs are candled and graded and/or stored.
eggs-and-bacon. See *nonesuch daffodil; birdsfoot deervetch; butter-and-eggs toadflax.*
eggs-and-butter. 1. Butter-and-eggs toadflax. 2. Any herb of the genus Ranunculus, especially the tall buttercup and bulb buttercup *q.q.v.*
egg scale. A balance device for weighing individual eggs.
egg shell. The thin, hard, rather brittle, porous, calciferous, protective, outer covering of an egg.
egg sorter. A person who grades or sorts eggs according to size and condition.
egg tester. A person who candles eggs. See *candle.*
egg tooth. A scale on the tip of the upper mandible of the embryo chick, used as a reinforcement for the beak for breaking open the shell at hatching. Also called *pip.*
egg tube. The oviduct.
egg turner. A device used for changing the position of eggs in an incubator.
egg white. The albumen of an egg.
eggy. Of, or pertaining to eggs; being full of, or containing, eggs; being egg stained.
egg yolk. The central, yellow portion of a hen's egg which contains the egg cell.
egg yolk-buffer. A preparation derived from egg yolks which withstands variations in the hydrogen-ion concentration and serves as a diluter for animal semen in the practice of artificial insemination. See *egg yolk-phosphate diluter; egg yolk-citrate diluter.*
egg yolk-citrate diluter. An agent for diluting semen in the practice of artificial insemination which is made from crystalline sodium citrate, distilled water, and egg yolk.
egg yolk-phosphate diluter. An agent for diluting semen in the practice of artificial insemination which consists of distilled water, egg yolk, potassium dihydrogen phosphate, and disodium hydrogen phosphate dodecahydrate.
egilops. Valonia oak *q.v.*
eglantine. 1. Usually the sweetbrier rose. 2. Rarely the Austrian brier rose and dog rose *q.q.v.* 3. Woodbine honeysuckle.
eglatere. Eglantine.
egmont buttercup. *Ranunculus nivicola,* family Ranunculaceae, a bulbous herb grown for its yellow flowers. Native to New Zealand.
Egyptian arum. Dasheen.
Egyptian bath sponge. Towelgourd.
Egyptian bean. See *Hindu lotus; hyacinth dolichos.*
Egyptian bee. *Apis fasciata,* a honey bee that is nervous and ill-tempered, but prolific and useful in cross-breeding. It is a good honey producer but not suited to comb-honey production. Native to Egypt.
Egyptian clover. *Trifolium alexandrinum,* family Leguminosae, an annual herb grown in arid and semiarid regions for soil improvement and for forage. It is tolerant of white alkali in soils. Native to Egypt and Syria. Also called *berseem; berseem clover; winter lucerne.*
Egyptian corn. A variety of the durra grain sorghums.
Egyptian cotton. An agricultural variety of sea-island cotton (*q.v.*) noted for its long staple. Native to the western hemisphere.
Egyptian cucumber. Mango melon.
Egyptian finger grass. Durban crowfootgrass.
Egyptian ginger. The root of the dasheen *q.v.*
Egyptian grass. Durban crowfootgrass.
Egyptian henna. Henna.
Egyptian indigo. *Tephrosia apolinea,* family Le-

guminosae, a shrub grown in southern Europe, which yields an indigo dye.

Egyptian lily. See *callalily: 2.*

Egyptian lotus. 1. Usually, the Hindu lotus *q.v.* 2. Originally the blue Egyptianlotus and white Egyptianlotus *q.q.v.*

Egyptian melon. Mango melon.

Egyptian millet. See *Johnsongrass; pearl millet.*

Egyptian onion. Top onion.

Egyptian paperplant. Papyrus.

Egyptian papyrus. Papyrus.

Egyptian pea. Gram chick-pea.

Egyptian privet. Henna.

Egyptian reed. Papyrus.

Egyptian rose. Any of several species of the genus Scabiosa. See *scabious.*

Egyptian soaproot. Any of several southern European species of the genus Gypsophila whose roots are used as a soap substitute.

Egyptian sycamore. Sycomore fig *q.v.*

Egyptian thorn. Firethorn.

Egyptian uppers. Southern Egyptian cotton of a relatively short staple; *e.g.*, Ashmoni, Zagora cottons.

Egyptian wheat. See *pearl millet; poulard wheat; shallu.*

Egyptian white lotus. White Egyptianlotus.

ehuawa. *Cyperus laevigatus,* family Cyperaceae, a flatsedge, a grasslike herb cultivated in Hawaii for fiber.

80's, 70's, 64's (English). British wool grades which are the equivalent of the American grade, fine.

eight-spotted forester. *Alypia octomaculata* (F.), family Noctuidae, the larva of a moth which feeds on the foliage of grape, Virginia creeper, and Boston ivy, often defoliating the plant. It is found in U.S. from New England to Colorado and south to Texas.

eight-tooth sheep. A four-and-a-half-year-old sheep. Also called *full-mouthed sheep.*

einkorn. *Triticum monococcum,* family Gramineae, perhaps the most primitive of the wheats, which usually has only one kernel to the spikelet.

ejaculate. 1. Semen which is forced out of the body; specifically that semen taken from a male by use of the artificial vagina or methods other than by coitus for use in artificial insemination. 2. To eject semen from the body.

ejaculation. See *ejaculate: 1.*

ekaha. The Hawaiian name of birdsnestfern.

eker. See *watercress.*

elaborated sap. A watery solution containing dissolved foodstuffs such as sugars, organic acids, etc., produced by the green parts of a plant and carried mainly in the phloem tissues to those regions where these foods are to be used or stored.

elastration. Bloodless castration in which rubber bands constrict the spermatic cord causing it and the testicles to wither away. It is, however, not always successful.

elastrator. An instrument used for applying the rubber bands used in elastration.

elatines bellflower. *Campanula elatines,* family Campanulaceae, a vine-like, perennial herb grown for its sky-blue flowers. Native to southern Europe.

elb. See *jujube: 2.*

Elbinger. Werder *q.v.*

elbow. The joint in a front leg of an animal which corresponds to the elbow joint of a man's arm.

elbowbush. See *buttonbush.*

elder. Any plant of the genus Sambucus, family Caprifoliaceae. Species are occasionally cultivated for their flowers, and the fruit of wild and cultivated plants is used in making wine, pies, and jelly. Also called *elderberry; elder blow.*

elderberry. See *elder; American elder; European elder.*

elder blow. The blossoms of elder or the plant itself.

elder shoot borer. *Achatodes zeae* (Harr.), family Noctuidae, an insect pest of corn whose larvae work in the tassels of corn planted near elderberry.

elecampane inula. *Inula helenium,* family Compositae, a perennial herb having showy orange, daisy-like flowers. It is often grown in herb gardens for its root which is used in veterinary medicine, in candy, and in puddings. Native to Eurasia; naturalized in eastern U.S. Also called *elecampane; elf dock; elfwort; enula campana; horseheal.*

election pink. Pinxterbloom azalea.

electrical-resistance moisture meter. A device used for determining moisture content, directly or indirectly, by measuring the electrical resistance of water, or that of the material itself, as soil, in relation to its content of water.

electrical seasoning. Submitting wood to heat by placing it in an electrostatic field, to equalize drying.

electrical stimulation. A method of collecting semen from small animals by means of an electric current. An electrode is inserted about 4 ins. into the male's rectum and another electrode in held against moistened skin near the fourth lumbar vertebra. About ten stimuli of 30 volts in 5 second periods result in a satisfactory ejaculation.

electric brooder. A brooder that is heated by electricity. See *brooder house.*

electric-capacity moisture meter. A meter used for determining the moisture content of wood by utilizing the variation in dielectric capacity of wood with changing moisture content.

electric exerciser. A device, similar to a merry-go-round, consisting of a steel framework which is turned by a motor. Bulls are chained to the frame at the outer edge so that as the device revolves they are forced to walk in a circle for exercise.

electric fence. A fence or enclosing device which consists of a single wire supported by insulators on widely separated posts. The wire is attached to a controller which emits electric current for 1/10 of a second or less, 45 to 55 times per minute. When an animal touches the wire, it receives a sharp, short, but harmless, electrical shock. The fence, when properly installed, is not injurious to human beings.

electric fence controller. An electric device which times the duration and lapse of electric current in an electric fence.

electric-fly screen. A window screen so charged with electricity that when an insect alights on the screen it is electrocuted.

electric prod(der). A portable, battery-powered, canelike device used to control or drive livestock by giving them a slight electrical shock.

electuary. A soft, sweetened, medicated paste used as a throat medication for animals.

elephant apple. Baelfruit.

elephant-ear fern. *Hymenodium crinitum,* family Polypodiaceae, a tropical fern which has thick, tonguelike fronds. Because of its striking appearance it is sometimes raised in greenhouses. Native to tropical America.

elephantgrass. 1. *Typha elephantina,* family Typhaceae, a marsh plant whose leaves are used in basket-making and whose pollen is used in making bread in India. Also called *elephant's-grass.* Native to southern Eurasia. 2. See *Napiergrass.*

elephant-hedge Kafir-bean tree. *Schotia latifolia,* family Leguminosae, a tropical tree grown for its clusters of pinkish flowers. Native to Africa.

elephantiasis. The excessive thickening and swelling of an animal's legs (usually a horse) which results from various infections or disturbances in circulation.

elephant's-ear. 1. Any of several plants of the genus Colocasia, family Araceae. Species are widely grown for their ornamental foliage and edible tubers. See *dasheen; taro: 1.* 2. A name given to any of several other plants having large, ear-shaped leaves or appendages. See *begonia; bergenia; guanacaste earpodtree.*

elephant's-foot. Hottentotbread tortoiseplant.

elevating chain. See *gathering chain.*

elevating grader. A power-driven, earth-moving device used in the construction of terraces. It consists of a plow which turns the earth onto a belt conveyor which carries it to the other side of the machine, where it is deposited.

elevating pipe. A pipe or hose used in the deposition or removal of grains, etc., in a grain elevator. See *elevator: 2.*

elevating terracer. Elevating grader *q.v.*

elevation. 1. Height above sea level or height above any selected base point. 2. A land feature, such as a hill, which is of greater height than the surrounding land area.

elevator. 1. Any of a number of devices, most often consisting of endless belts, chains and buckets, or screws, which is used to raise materials from a lower to a higher level. 2. A building designed for the handling and storage of cereal grains, dried beans, and other seed crops. It may also be used for processing and selling products.

elevator bucket. In grain machinery, etc., a bucket of various designs which is fastened on a belt or chain or linked together with other similar buckets to raise materials in bulk.

elevator canvas. That part of a grain binder or threshing device which carries the cut grain from the platform to the binder or threshing deck.

elevator digger. A potato harvester which digs the potatoes from the ground, shakes the earth loose from the tubers on an open, elevating, endless chain, and arranges the plants so that the leafy parts and the potatoes are in separate rows.

elevator hay loader. See *cylinder hay loader.*

elf dock. Elecampane inula.

elfwort. Elecampane inula.

eligible producer. In government crop loans, any individual, partnership, association, corporation, or other legal entity which produces a commodity which has marketing quotas.

elegible storage. A term designating cribs or bins which are of such substantial and permanent construction as to afford protection against rodents, other animals, thieves, and weather, and are therefore eligible for use in storing grain for government loan.

eligible to registry. Designating an animal whose sire and dam are registered in the same breed registry and which meets other rules as to age, color, etc., of that breed.

elixir-of-love. Letterleaf.

elk-bark. See *bigleaf magnolia.*

elk clover. California aralia.

elk grass. See *beargrass: 2.*

elkhorn fern. See *staghorn fern: 2.*

Elkington system. A system of drainage in which wells are dug at regular intervals in low areas to allow collected water to drain away. First used by Joseph Elkington of Warwickshire, England.

elkslip. Elkslip marshmarigold.

elkslip marshmarigold. *Caltha leptosepala,* family Ranunculaceae, a perennial herb grown in pools and moist borders for its white-tinged flowers with blue on the outside. Native to western North America. Also called *elkslip.*

elk tree. Sourwood.

elk wood. 1. Sourwood. 2. Umbrella magnolia.

elm. Any tree of the genus Ulmus, family Ulmaceae. Species are very important for shade, street planting, and for lumber. See *American elm; smoothleaf elm; slippery elm; Scotch elm; Camperdown elm; Dutch elm; Chinese elm; Siberian elm; rock elm.*

elm bark beetle. See *native elm bark beetle; smaller European elm bark beetle.*

elm blight. Dutch elm disease.

elm borer. *Saperda tridentata* Oliv., family Cerambycidae, a beetle that infests the inner bark and sapwood of ornamental and shade trees, causing the flow of sap to slow or stop. Infested trees are characterized by moist spots on the bark, undersized leaves, and dying branches.

elm calligrapha. *Calligrapha scalaris* (Lec.), family Chrysomelidae, an insect pest of which both the larvae and adult beetles feed on the elm, linden, alder, and willow foliage. Also called *linden leaf beetle.*

elm casebearer. *Coleophora limosipennella* (Dup.), family Coleophoridae, an insect that infests the elms of northeastern U.S. by mining between the principal veins of the leaves.

elm cockscomb gall. *Colopha ulmicola* (Fitch), family Aphidae, an aphid that infests elms throughout the U.S. The infested trees have red-tipped galls filled with aphids growing out of the leaves, from which honeydew falls.

elm lace bug. *Corythucha ulmi* O. & D., family Tingidae, an insect that infests the American elm causing the leaves to turn yellowish in the early summer.

elm leaf aphid. *Myzocallis ulmifolii* (Monell), family Aphidae, an insect that infests elm leaves and produces a great quantity of honeydew.

elm leaf beetle. *Galerucella xanthomelaena* (Schr.), family Chrysomelidae, an insect pest of American, Scotch, English, and Camperdown elms throughout the U.S., which skeletonizes the leaves on which it feeds, sometimes defoliating the tree.

elm leaf miner. *Fenusa ulmi* Sund., family Tenthredinidae, an insect that infests English, Scotch, Camperdown, and American elms causing the leaves to drop prematurely.

elm leaf spot. A fungous disease of the American and Asiatic elms caused by *Gnomonia ulmea*, family Gnomoniaceae, characterized by black, white-encircled, raised leaf spots. In severe attacks defoliation results and twig terminals die. Also called *black leaf spot; black spot of elms.*

elm peeler. An old, colloquial name designating a scrub hog living on wild forage.

elm phloem necrosis. A serious, virus disease of elms, found in central U.S., which kills the phloem cells, causing wilting and rapid death of the trees.

elm sawfly. *Cimbex americana* Leach, family Cimbicidae, an insect that lays eggs in the leaf tissue, forming blisters on the underside of leaves. The larvae feed on the foliage, defoliating twigs and branches. They infest elm, willow, alder, basswood, birch, poplar, and maple. Common throughout northern U.S. and west to Colorado. Also called *giant American sawfly.*

elm scale. See *European elm scale.*

elm scurfy scale. *Chionaspis americana* Johns., family Diaspididae, a scale insect which infests the elm and the hackberry, killing young trees and the branches of older trees. Found in eastern U.S. and west to Texas. See *scale insect.*

elm spanworm. *Ennomos subsignarius* (Hbn.), family Geometridae, a moth whose larva feeds on the foliage of elm, horsechestnut, maple, linden, beech, and other trees. Found in eastern U.S. and west to Colorado.

elm twig borer. See *dogwood twig borer.*

eluvial. In soil horizons, designating those horizons of a soil profile in which eluviation (*q.v.*) has been dominant or very appreciable in their formation.

eluviated horizon. Eluvial *q.v.*

eluviation. The movement of soil material from one place to another within the soil, in solution or in suspension, when there is an excess of rainfall over evaporation. Horizons that have lost material through eluvation are referred to as eluvial and those that have received material as illuvial. *Eluviation* may take place downward or sidewise according to the direction of water movement.

eluvium. The residual deposits of rock weathering; *eluvium* can be the soil itself or the parent material of soils.

Elwes desert candle. *Eremurus elwesi,* family Liliaceae, a perennial herb grown in the flower border for its 10 to 12 ft. flower stalk and its pink flowers. Native to Asia.

emaciation. Excessive leanness, or a wasted condition of the body; loss of flesh.

emajagua. See *linden hibiscus; Jamaica linden hibiscus.*

emasculate. 1. To castrate. 2. To remove the androecium (the stamens and their appendages) from a flower for cross-breeding purposes.

emasculatome. A pincer-like instrument used for bloodless castrating, which severs or breaks the spermatic cord, causing the testicles to wither away. It may also be used in docking the tails of lambs. See *elastrator.*

emasculator. A scissors-like instrument used in castrating which severs the spermatic cord, simultaneously crushing it to prevent bleeding. See *elastrator.*

embargo. A governmental trade restriction which prohibits a particular commodity from entering a country.

embarras. A natural jam or lodged raft of trees in a river channel, which, upon reaching a certain size, may dam up the river and cause floods in the surrounding area. Such obstructions have created lakes, deadened bottom-land forests, and changed the character of soils and vegetation.

Embden. Emden *q.v.*

Embden groats. Crushed or coarsely ground oats without husks. Formerly used for making a gruel. See *groats.*

emblement. In law, the profits from sown land.

emblic leafflower. *Phyllanthus emblica,* family Euphorbiaceae, a tropical shrub or small tree sometimes raised in greenhouses for its handsome, hemlock-like foliage and for its edible fruits. Native to Asia. Also called *emblic; myrobalan.*

embossed cypress. Chinacypress.

embryo. 1. Any organism in its earliest stages of development. 2. The young sporophyte which results from the union of male and female cells in a seed plant. Also called *seed-germ.*

embryonated. In parasitology, a term used in reference to a parasite egg that contains a small, immature parasite, *i.e.*, an egg containing an embryo.

embryonic. 1. Pertaining to the embryo (*q.v.*) or its development. 2. Underdeveloped; immature. See *rudimentary.*

embryonic cells. A class of undeveloped cells

which make up the structure of embryos. In the very early development, such cells are non-specific in function, but as the embryo matures, they become more specific, making up tissues and organs, such as muscle, liver, and bone.

embryo sac. 1. The mature female gametophyte in higher plants. 2. A sac which contains the embryo in its very early life in animals. Also called *blastodermic vesicle.*

Emden. A domestic breed of geese which originated in Prussia, having white plumage and orange bill and feet. The gander weighs up to 20 lbs., the goose up to 18. Also called *Embden.*

emerald feather. Sprenger asparagus *q.v.*

emergence. 1. The appearance of the first leaves of the crop plant above the ground. 2. The escaping of an insect from its cocoon, pupal case, etc.

emergence hole. In forestry, a hole made by an insect through the outer bark, wood surface, or fruit, for its emergence.

emergence period. The interval between the free appearance of the first and last individuals of a generation or brood of insects.

emergency crop. See *catch crop: b.*

emergency tillage. The cultivation of the soil by listing, duckfooting, chiseling, or pitting in order to roughen the soil or to bring clods to the surface for the reduction of wind erosion.

emetic holly. Yaupon *q.v.*

emetic root. Floweringspurge euphorbia.

emetic weed. 1. *Lobelia inflata,* family Lobeliaceae, a medicinal plant which yields an emetic. Also called *Indian tobacco.* 2. Aztec tobacco *q.v.*

Emiliano. A very hard, Parmesan-like cheese made in Italy, which is mild to sharp in taste and cured from one to two years.

eminent domain. The right of the federal government or a state to take private property for public use upon payment of reasonable compensation.

Emmental cheese (Emmenthaler). See *Swiss.*

emmer. *Triticum dicoccum,* family Gramineae, an annual cereal herb, a wheat, which is grown as a source of bread flour in southern Europe, but is grown only as a stock feed in the U.S. Native to western Asia and the Mediterranean area. Also called *amelcorn.*

emphysema. An abnormal presence of air in the tissue of some part of the body, generally the lungs, where a stretching and rupturing of the walls of the smallest air sacs produces abnormal air space.

empress tree. Royal paulownia.

emptins. A kind of liquid leaven formerly much used in rural U.S. in place of yeast. Fermentations were made from hops, potatoes, or milk.

empty-cell process. Any process for impregnating wood with preservatives or chemicals, in which air is imprisoned in the wood under the pressure of the entering preservative which expands when the pressure is released, driving out part of the injected preservative.

empty glume. See *glume.*

emulsifiable concentrate. A liquid formulation, usually containing a pesticide, which will form an emulsion when mixed with water or certain other liquids.

emulsifier. A substance used in mixing materials so that they remain in suspension in water or other liquid: *e.g.,* oil and water, when shaken together and allowed to stand, quickly separate, but by adding bile, which the liver secretes, the oil is broken up into such fine particles that separation from the water takes much longer. Thus, bile is an *emulsifier* of oils and fats.

emulsion. 1. An oily or resinous substance held in suspension in an aqueous liquid. 2. A mixture of liquids that do not dissolve in each other.

emulsion process. A process of removing wool fat from wool fiber by washing with water in which the fat is emulsified.

emulsive oil. An oil, containing an emulsifier, which is to be mixed with water for use as an insecticide.

enaena. Purple cudweed.

enation. An outgrowth from the surface of an organ of a plant usually caused by the parasitic invasion of fungi, bacteria, or viruses.

encapsulation. Enclosure in a capsule or sheath.

encased knot. A knot in which the layers of annual growth are not intergrown with those of the adjacent wood. See *knot: 4.*

encephalartos. Kafirbread *q.v.*

encephalitis. Inflammation of the brain which results in various central nervous system disorders, characterized by excitement and irregular movements, depression, paralysis, and death. It may be a symptom of various diseases or it may occur as a primary disease.

encephalomalacia. 1. Softening of the brain. 2. Crazy chick disease *q.v.*

encephalomyelitis. Inflammation of the brain and spinal cord. See *equine encephalomyelitis.*

enchanter's herb. European verbena.

encina. California live oak.

encinal. A parklike growth of oaks (southwestern U.S.).

enclosure. a. A fenced area, usually of limited extent, within which animals of any kind may be held for a specific purpose. b. A small fenced area enclosing a portion of a range from which livestock or other animals are or may be excluded.

enclosure plot. A relatively small, fenced area (from 1 square rod to about 25 acres), used to study the effect of different grazing systems on vegetation and soil, as well as the grazing capacity of range communities.

encomienda. A grant of land to colonists in Spanish America which carried the right to subject the aboriginal inhabitants to forced labor in return for the obligation of evangelizing and governing them.

encrustation. A crust or hard coating on or in a body.

encumbrance. Any lien or liability attached to real property.

encyst. To become enclosed in a sac, bladder, or cyst.

encysted. Designating parasite, pus, fluid, or foreign particle which is shut off from the surrounding parts of the body by dense, fibrous tissue.

end board. Tailboard.

end check. A split which occurs on the end of a board during seasoning.

Endeavor River pear. *Eugenia eucalyptoides*, family Myrtaceae, a small tree or shrub which bears fruit used in preserves. Native to Australia.

endemic. a. Designating that which is prevalent in a particular district or region. *Cf. exotic.* b. Occurring frequently in a certain area, such as an *endemic* disease. *Cf. pandemic; epidemic.*

endgate. The board or gate at the rear end of a farm wagon box or bed or of a manure spreader which is made in a number of different types so that it may be raised or easily removed or adjusted for different loading or unloading conditions. Also called *tailboard.*

endgate lime spreader. A device fastened to the rear part of a wagon box or truck bed, consisting of a hopper and two rapidly rotating disks, used to spread lime on a field.

endgate rod. An iron rod used to brace the sides of a wagon box at the rear end.

endgate seeder. An attachment on the rear of a wagon box or truck bed for broadcast sowing. It consists of a hopper feeding device and two rapidly revolving disks which scatter the seed.

endive. *Cichorium endivia*, family Compositae, an annual or biennial herb grown for its edible leaves used in salads. Native to Asia. *Cf. French endive.* See *curled endive; escarole.*

endless. Designating that which is without end, such as a looped length of chain, belt, rope, etc., which has the ends fastened together. The loop is passed taut around a fixed set of pulleys or rollers. Power applied to the device communicates motion to the loop, which can then be used to move produce, material, etc., from one point to another, or to transmit power.

endless belt. A belt, whose ends have been united, which passes over two or more pulleys transmitting power, as from a tractor to a cornsheller.

endless chain. A chain whose end links have been united.

endless rope. A rope with both ends joined together, often used in the transmission of power.

endless saw. A band saw.

endocarp. The inner layer of a multiple layered pericarp (*q.v.*), *e.g.*, the shell of a cherry stone.

endocrine gland. Any gland of the body which secretes a substance or hormone thereby controlling certain bodily processes: *e.g.*, pituitary, pineal, thyroid, parathyroid, thymus, adrenal, pancreas, etc. See *ductless glands.*

endodynamorphic soil. A soil whose character has been determined largely by or inherited from the parent material or underlying rock with less influence from the external enrivonment.

endogen. The internal growth of new wood in a plant stem.

endoparasite. Any of the various parasites which lives within the body of its host. *Cf. ectoparasite.*

end-opening. Designating a barn or shed whose main doors are at either end rather than on a side.

endorsement. An additional signature on a promissory note which indicates that the signer pledges his unsecured property and is willing to pay the note should the borrower fail to do so.

endosperm. The nutritive portion in some seeds which originates in the embryo sac, but which is outside of the embryo. It usually follows the fertilization of the two primary endosperm nuclei of the embryo sac by one of the two male sperms. In a diploid organism, the *endosperm* is triploid. It serves as food for the growing plant during seed germination, often being entirely consumed in the process. The cereals have seeds which are referred to as "*endosperm* type" seeds wherein some of this nutritive material remains within the seed. Also called *nurse embryo.*

endothal. A chemical compound used as an herbicide, a defoliant, a drying agent, and as a pre-emergence herbicide.

endothia canker. Chestnut blight.

endoxerosis. An internal necrosis of the lemon characterized by a loss of yellow at the stylar end of the fruit, a depressed nipple, and pinkish to rust-brown masses in the albedo of the rind. It is thought to be related to drying within the tree and fruit brought about by inadequate irrigation. Also called *internal decline; blossom-end decline; blossom-end decay; dry tip; yellow tips; pink tip; tip deterioration.*

end plug. A cap used to shut off the flow of water from the end of an irrigation pipe. Also called *end cap.*

end point. 1. In jelly making, the time and condition of boiling fruit juice which, upon cooling, will form the desired gel-like consistency. 2. The neutral point in titrating acids and alkali.

end post. The post at the end of a row of posts, as in a vineyard or fence. See *corner post.*

endrin. The endo-endo isomer of the insecticide dieldrin *q.v.*

end rot of cranberries. A fungous disease caused by *Fusicoccum putrefaciens*, family Sphaerioidaceae, characterized by a rot beginning at either end of the fruit. At first there is slight discoloration of the fruit, but in time, affected fruits turn into dry, shriveled mummies.

endurance trial. A performance test for horses in which a horse and rider must travel 300 miles in 5 days. 60 percent of the resulting score depends on the condition of the horse at the finish and 40 percent on its speed.

energy head. The energy of a unit height of a stream which takes into account the elevation of the hydraulic grade line at any section plus the velocity head of the mean velocity of the water in that section. The *energy head* may be referred to any datum, or to an inclined plane, such as the bed of a conduit.

energy of position. Potential energy, or the energy possessed by a body in virtue of its position as distinguished from the energy of motion or kinetic energy: *e.g.*, water stored in an elevated reservoir represents potential energy, since its release to a lower level may be utilized to effect work, as in a hydroelectric power plant.

Engadine. A cheese made in Switzerland from cow's whole milk.

Engelmann oak. *Quercus engelmanni,* family Fagaceae, an evergreen tree used as a shade tree; the wood has a value only as a fuel. Native to southern California, U.S. Also called *evergreen white oak.*

Engelmann spruce. *Picea engelmanni,* family Pinaceae, an evergreen tree, one of the hardiest spruces, which grows up to a height of 150 ft. It is grown as an ornamental for its bluish-green to steel-blue needles and for its lumber which has a wide variety of uses including pulpwood. Native to the Rocky Mountains of western North America.

Engelmann spruce beetle. *Dendroctonus engelmanni* Hopk., family Scolytidae, an insect that is periodically destructive to spruce in the Rocky Mountain area of North America.

Engelmann spruce weevil. *Pissodes engelmanni* Hopk., family Curculionidae, an insect which attacks the Engelmann spruce in western North America, killing or seriously injuring terminal shoots, causing the infested trees to have worthless, crooked trunks.

engineer's chain. See *chain: 2.*

engine-oil emulsion. An emulsion made with lubricating oils used to carry insecticides, etc.

English bacon. Canadian bacon.

English bindweed. European glorybind.

English bluebell. Blue squill.

English bluegrass. 1. Canada bluegrass. 2. Meadow fescue.

English camomile. 1. Roman camomile. 2. Mayweed camomile.

English capers. The seed pods of caper euphorbia *q.v.*

English chop. A retail cut of meat; a double loin or rib chop which is made by cutting across the unsplit loin or rack.

English class. A standard class of domestic chickens which includes breeds of English origin and their derivatives. Included in this class are the Dorkings, Redcaps, Orpingtons, Cornish, Sussex, and Australorps. They are mainly general-purpose, white-skinned fowls, except for the Cornish which has yellow skin. All have nonfeathered shanks.

English clover. Mammoth red clover or American single-cut clover.

English cost. See *tansy: 2.*

English cowslip. Cowslip primrose.

English cut. A retail cut of beef for braising (pot roast) which comes from the thin end of the fourth and fifth ribs.

English Dairy. A very hard, English, Cheddar-like cheese used chiefly in cooking.

English daisy. *Bellis perennis,* family Compositae, a perennial herb which has white, pink, or rose flowers in heads 2 ins. across. It is very popular for bedding and for borders. Native to Europe. Also called *daisy; ewe-gowan; true daisy; bachelor's button.*

English elm. *Ulmus procera,* (*U. campestris*), family Ulmaceae, a tall, deciduous tree whose several varieties are popular as shade trees in northeastern U.S. Native to England and southwestern Europe. Also called *European elm.*

English gillyflower. Carnation.

English gooseberry. See *European gooseberry.*

English grain aphid. *Macrosiphum granarium* (Kby.), family Aphidae, a sucking insect pest of small grains and cultivated grasses which clusters in large numbers on the bracts of cereals, especially wheat. Infested plants have shriveled seed and, if infested early in the growing season, may die.

English grass. Kentucky bluegrass.

English Hackney. See *Hackney.*

English hawthorn. *Crataegus oxyacantha,* family Rosaceae, a spiny tree or small shrub grown as an ornamental and as a hedge plant for its white flowers and scarlet fruits. Native to Eurasia. Also called *espino; evergreen thorn; may.*

English holly. *Ilex aquifolium,* family Aquifoliaceae, a broad-leaved, evergreen tree grown for its lustrous green leaves with spiny points and for its red, pea-sized fruits. Native to Eurasia.

English iris. *Iris xiphioides,* family Iridaceae, a perennial, bulbous herb grown for its purple and yellow flowers. Native to the Pyrenees.

English ivy. *Hedera helix,* family Araliaceae, a climbing, evergreen vine, a variable species popularly used out-of-doors to cover masonry walls, in greenhouses as a pot plant and for cut greens, and in homes as a pot plant. Also called *ivy; evergreen ivy.*

English lamb chop. A retail cut of lamb for broiling which is a chop in which the loin is not split.

English laurel. See *laurel cherry: 1.*

English Leicester. A type of the Leicester breed of sheep, the smallest of the long-wool breeds, which has a coarse braid wool 7 to 8 ins. long, and a tuft of wool on the head. Rams weigh up to 250 lbs. and ewes up to 180. See *Border Leicester.*

English maidenhair. Maidenhair spleenwort.

English masterwort. See *bishop's goutweed.*

English meadow grass. Perennial ryegrass.

English mercury. Good King Henry goosefoot.

English oak. *Quercus robur,* family Fagaceae, a hardy, deciduous tree having many varieties. It is valuable for its timber and as a shade tree. Its foliage and young buds when browsed in large quantities are poisonous to cattle, sheep, and goats. Symptoms of poisoning are constipation, emaciation, edema, blood in the feces, and subnormal temperature. Native to Europe.

English pea. Garden pea.

English pepper-grass. Field pepperweed.

English plantain. Buckhorn plantain.

English poppy. Welshpoppy.

English primrose. Cowslip primrose.

English ryegrass. Perennial ryegrass.

English saddle. A lightweight, riding saddle which has a slightly raised pommel and cantle, and one cinch strap.

English scuffle hoe. See *D-hoe.*

English serpentary. European bistort.

English sparrow. *Passer domesticus,* a small bird that feeds on seeds, buds, insects, etc., and is often a pest. Native to Europe. Also called *house sparrow.*

English Stilton. See *Stilton.*

English thistle. Venuscup teasel.

English-type barn. See *bank barn.*

English violet. Sweet violet.

English wallflower. See *wallflower.*

English walnut. Persian walnut.

English yew. *Taxus baccata,* family Taxaceae, a magnificent evergreen tree having many forms and varieties, grown for ornament. It is poisonous when browsed in large quantities by cattle, horses, sheep, goats, and wild animals.

engorgement colic. See *acute indigestion in horses.*

engraft. See *graft: 1. b.*

engraver beetle. Bark beetle.

enheptin. A chemical, 2-amino-5-nitrothiazole, used in controlling blackhead, an infectious poultry disease.

enlarged and flattened. Designating an egg defect in which the yolk membranes have weakened and moisture from the white has been absorbed so much that the yolk appears larger and flatter than normal.

enlarged hocks. A poultry disease attributed to lack of vitamin E in feeds.

enology. The art and science of wine-making.

enphytotic. Designating a plant disease peculiar to a locality: *e.g.,* an *enphytotic* fungous disease. *Cf. endemic.*

enrich. 1. To add fertilizer or manure to soil. 2. To add a substance or vitamin to food products.

enriched cereals. Processed cereals or cereal products to which have been added certain minerals, proteins, and vitamins which were removed in the ordinary processing.

enriched grass. Grass grown on very heavily fertilized fields with especially heavy applications of nitrogen.

enriched rice. Rice to which vitamin nutrients have been added to compensate for the many nutrients lost in milling.

ensilage. Any green crop preserved by fermentation in a silo, pit, or stack, usually in chopped form. *Ensilage* can be made from practically any green crop having the proper moisture content. Also called *silage.*

ensilage blower. A device which consists of an auger or slotted chain conveyor, a power driven fan, and a stand-pipe. Green-chopped material is fed on to the conveyor which carries it to the blower or fan. The blower then forces the material up the pipe and into the silo.

ensilage cutter. A machine which chops green crop materials into short lengths suitable for storing in a silo. Also called *silo filler; silage cutter.*

ensilage fork. A pitchfork having 4 or more 15 to 16 in. tines, used in handling ensilage.

ensilage poisoning. See *forage poisoning: 2.*

ensile. To place green plant material, such as green crops of grain, grasses, cornstalks, etc., in a silo in such a manner as to bring about proper fermentation for preservation and storage.

ensiling. The process of placing green plant material in a silo, pit, trench, or stack for fermentation and storage. See *ensile.*

enter. To place an animal, fowl, or plant in competition in a show or exhibition.

enteralgia. See *colic.*

enteric. Intestinal.

enteritis. Any inflammatory condition of the lining of the intestines of animals or man. Characteristics of *enteritis* are frequent evacuations of a liquid or very thin, foul-smelling stool which may or may not contain blood, straining, lethargy and anorexia. In acute cases there is a rise in body temperature. The condition is seen as a symptom of a number of infectious diseases or it may be caused by specific bacteria or viruses. Other common causes include plant and chemical poisons, parasites, overeating, faulty nutrition and poor environmental factors. See *ulcerative enteritis; infectious enteritis in pigs; feline enteritis; hexamitiasis.*

enterohepatitis of turkey. See *blackhead.*

enterotoxemia. A disease of calves, sheep, and goats, which results in high mortality and is associated with *Clostridium perfringens.* It is characterized by a lack of coordination and sudden death. Also called *pulpy kidney disease.* See *hemorrhagic enteritis.*

enterprise. A project on a farm, such as the production of any crop or livestock.

enterprise cost accounts (records). Detailed physical and value records of all inputs and outputs which pertain to the production of an indivdual enterprise, rather than to all enterprises, on the farm.

entire. Designating an uncastrated male animal. See *stallion; bull: 1; ram; boar.*

entire leaf. Any leaf which is in no way divided and which is not toothed or serrated.

entire-tree variation. A sport or mutation commonly found in citrus orchards which is distinguished from the normal by its fruit, leaf characters, and/or habits of growth, and which involves the whole tree and not merely a single twig or branch.

entire-wheat flour. Whole-wheat flour.

entomologist. A person who specializes in the study of insects.

entomology. That branch of zoology which deals with insects. See *economic entomology.*

entomophagous insect. Any insect that eats other insects.

entomophagous parasite. An insect or fungus that parasitizes insects.

entomophilous. Designating flowers that are fertilized by insects which carry the pollen from plant to plant.

Entomophthora fungus. A fungus of the entomophagous genus Entomophthora, as *Entomophthora fumosa,* which infects and effectively controls the citrus mealy-bug in Florida, U.S., especially during summer rains.
Entomosporium leaf spot. See *leaf blight and fruit spot of pear and quince.*
entozoa. Internal animal parasites: *e.g.,* stomach worms.
entrails. The inner organs of animals, specifically the intestines.
entrainment. The moving of a liquid in a spray or gas: *e.g.,* in the evaporation of milk, or the carrying over of droplets of the solution in the process of distillation.
entrance. An opening in the hive which allows bees to pass in and out.
entrance switch. A wiring device for interrupting all electrical connection between the farmstead wiring and the service wires leading from the power company's source of energy supply.
entry. An animal, fowl, or plant which is placed in competition in a show or exhibition.
entyloma. A fungus which causes leaf smut of rice.
enucleation. The removal of an organ, tumor, or other body in such a way that it comes out clean and whole as a nut from its shell.
enula campana. Elecampane inula.
envelope. Any pod, leaves, cover, etc., of a plant part.
environment. The sum total of all external conditions which affect the life of an organism, plant, or animal. (May be used as synonymous with habitat.)
environmental. Of, or pertaining to, environment.
environmental resistance. The physical and/or biological factors which oppose survival of a species.
enzootic. Designating a disease which affects animals in a limited or confined area, or one which persists more or less regularly in such an area. *Cf. enphytotic; endemic.*
enzootic marasmus. Cobalt deficiency *q.v.*
enzootic meningo-encephalomyelitis. See *equine encephalomyelitis.*
enzymatic. Referring to a reaction or process which is catalyzed by an enzyme or group of enzymes.
enzyme. A chemical substance that can produce transformation of some other compound or compounds; digestive enzymes, *e.g.,* pepsin, break up complex food substances into relatively simpler ones, which can be readily absorbed into the blood stream.
eolation. All the direct, geologic activities of the wind which either tear down or build up: the process by which wind modifies land surfaces, both directly by transportation of dust and sand, by the work of sand blasts, and indirectly by wave action on shores.
eola-weed. See *St. John's-wort: 2.*
eolian. 1. Pertaining to action of wind, as soil erosion due to wind, or soil deposits transported by wind. 2. Loosely designating soils which are derived from geologic deposits that are wind borne in origin. Also called *aeolian.*
Ephedra. A genus of plants of the family Gnetaceae, a source of ephedrine (*q.v.*). Tea made from the plants was supposed to have medicinal value. Species browsed on the range are rated as good. Also called *Mormon tea; Brigham tea; Canatillo; teamster's tea.* See *Nevada ephedra.*
ephedrine. $C_{10}H_{15}NO$, an alkaloid used in testing hay fever and asthma.
ephemer. An introduced plant which is unable to persist and thus soon disappears.
ephemeral. Designating a very short-lived plant or animal.
ephemeral fever. A slight fever lasting a day or two.
epicarp. The outermost layer, or exocarp, of a fruit.
Epichloë head blight. A minor, fungous disease of grasses caused by *Epichloë typhina,* characterized by a mycelium in the crown buds, a systemic invasion of primordial tissues, and a white, felt-like layer over the tillers. Also called *choke of grasses.*
epicormic branch. A shoot developing from a dormant bud which is exposed to air and light.
epidemic. a. A widespread invasion or dispersion by an insect or a disease. *Cf. endemic; epiphytotic.* b. Designating a sudden and widespread attack by a disease or insect infestation. (Most scientists seem to prefer *epidemic* when referring to all types of disease: man, animal, and plant.)
epidemic tremor. See *avian encephalomyelitis; crazy chick disease.*
epiderm. Epidermis.
epidermal. Of, or pertaining to, the epidermis.
epidermis. The outer layer of skin or tissue.
epigent weathering. The rock weathering of the surface crust of the earth, including such end products as soils.
epilepsy. A chronic nervous disorder of man and animals, particularly of dogs, which is characterized by a sudden loss of consciousness and convulsions, a stiffened neck, chomping of the jaws, dilated pupils, salivation, distressed breathing, evacuation of the bladder and bowels. Cause is unknown. Also called *fits; falling sickness.*
epilobe (epilobium). Willow-weed.
epinasty. Stronger growth on the upper surface than on the under surface of a plant member which causes the member, such as a leaf, to curl downward.
epinephrin. $C_9H_{13}NO_3$, a drug used to arrest hemorrhage and to stimulate heart action which is a slaughterhouse by-product obtained from the adrenal glands. Also called *adrenaline.*
epinette. 1. Any of several spruces (*q.v.*) of North America which belong to the genus Picea. 2. Eastern larch.
epiphyte. A plant which germinates and grows on other plants, but is not parasitic; an air plant. Also called *air plant.*
epiphytic. Designating an epiphyte.
epiphytotic. a. That which is common or preva-

lent among plants. b. Designating a sudden and widespread attack of a plant disease. *Cf. epidemic; endemic; epizootic.*

epistasis. 1. The state or condition of being epistatic *q.v.* 2. The checking of any discharge, secretion, excretion, etc.; *e.g.,* blood.

epistatic. Designating a condition of genetics in which one factor prevents a factor other than its allelomorph from exhibiting its normal effect on the development of the individual.

epithelial. Pertaining to, or composed of, epithelium *q.v.*

epitheliogenesis imperfecta. A condition of newborn animals resulting from a recessive genetic factor. There are skin defects on the lower parts of the legs and on the hairless parts of the body. The mucous membranes are defective, and the ears and claws deformed. The animal so affected dies soon after birth.

epithelium. The dense, cellular tissue which covers all body surfaces and lines all body cavities.

epithyme. Clover dodder.

epizootic. Designating a disease of animals which spreads rapidly and affects many individuals of a kind at the same time, thus corresponding to an epidemic in man. *Cf. epiphytotic: b; epidemic; endemic.*

epizootic abortion. Brucellosis *q.v.*

epizootic aphtha. Foot-and-mouth disease.

epizootic cellulitis. Equine influenza.

epizootic cerebro-spinal meningitis. See *equine encephalomyelitis.*

epizootic enteritis. Feline enteritis.

epizootic lymphangitis. Mycotic lymphangitis *q.v.*

epizootiologist. A scientist who specializes in the study of epizootic diseases of animals.

epizootiology. The study of factors influencing or involved in the occurrence and spread of disease among animals.

epizooty. Equine influenza.

Epoisse. A soft, French cheese prepared from whole or partly skimmed milk.

Epsom salts. $MgSO_4 \cdot 7H_2O$, hydrated magnesium sulphate which is used as a laxative or purgative. Originally obtained from the mineral waters at Epsom, England, but now principally from kierserite *q.v.*

equalizing bar. See *evener.*

equestrian. One who rides horseback.

equestrian star. Amaryllis.

equestrienne. A woman who rides horseback.

equilibrium moisture content. 1. The equilibrium established between the vapor pressures of air and a product, such as grain, hay, popcorn, etc., at identical temperatures. 2. The moisture content at which wood neither gains nor loses moisture when surrounded by air at a given relative humidity and temperature.

equine. 1. A horse. 2. Pertaining to, or resembling, a horse or other member of the family Equidae. Horses, mules, and asses are referred to as *equines* or *equine* animals.

equine encephalomyelitis. A virus disease of horses and man characterized by fever, grinding of the teeth, sleepiness, wobbly gait, difficulty in chewing and swallowing, and frequently death. Also called *Borna's disease; horse sleeping sickness.*

equine infectious abortion. A contagious disease of horses which causes abortion usually in the latter stages of gestation. The causative agent is the microorganism, *Salmonella abortivoequina.* Relatively uncommon in the U.S.

equine infectious anemia. A widespread, contagious, virus disease of horses and mules characterized by dullness, rapid breathing, weakness of the hind legs, loss of appetite, swollen eyelids, intermittent fever, dropsy, frequent urination, diarrhea, general weakness, loss of the blood's ability to coagulate, emaciation, anemia, and death. Also called *pernicious equine anemia; swamp fever; horse malaria; malarial fever; slow fever; mountain fever; creeping fever.*

equine influenza. The most contagious and most widely spread disease of horses, mules, and asses, caused by a filterable virus with complications attributed to miscellaneous bacteria. It is characterized by fever, extreme weakness, depression, rapid respiration, coughing, watery nasal and eye discharges, a pinkish swelling of the eyelids, edema of the abdomen, legs, and head, and loss of appetite accompanied by excessive thirst. Also called *shipping fever; pinkeye; catarrhal fever; epizootic cellulitis; epizooty; stockyard fever.*

equine osteomalacia. See *osteomalacia.*

equine piroplasmosis. See *piroplasmosis.*

equine sleeping sickness. See *equine encephalomyelitis; encephalitis.*

equine syphilis. Dourine *q.v.*

equine variola. Horsepox; a contagious, virus disease of horses characterized by fever and pustular eruptions or pox, especially on the pasterns, fetlocks, and mouth. Also called *contagious pustular stomatitis.*

equinia. See *glanders.*

equinoctial plant. A plant whose flowers open and close at a particular hour of the day.

equipment. Material, fittings, devices, machinery, appliances, fixtures, apparatus, and the like, as well as separate items, such as milk coolers, hot water heaters, milk cans, can racks, water tanks, etc., which are used as a part of, or in connection with, farm machinery.

equisetum poisoning. A poisoning of horses and sheep which results from their feeding on *Equisetum* spp., especially *E. arvense,* the field horsetail (*q.v.*). It is characterized by unthriftiness, loss of weight and muscular control, nervousness, falling, lack of ability to eat, and death.

equitant. Designating leaves which overlap each other at the bases, as the iris.

equitation. Horsemanship.

equity. The net ownership of a farmer in his business which is the difference between the assets and liabilities of an individual or business as shown on the balance sheet or financial statement. It is comparable to net worth *q.v.*

eradicant. 1. Any agent which destroys the causative organism of a disease after the disease is established. 2. Any chemical or compound which is used to destroy plants.

eradicate. To destroy or abolish, as a disease or weed pest.

eradication. 1. Control of a disease by eliminating the pathogen after it is already established. 2. To control pests, insects, weeds, etc.

eradicator. 1. Any agent used for the destruction of weeds, insect pests, etc. 2. A device used for scraping the edible flesh from the pineapple shell after the central portion has been removed.

erect. 1. Designating upright plants, in contrast to prostrate ones. 2. To build or construct buildings, etc.

erection. 1. The assembling of the parts of a building and the making of the necessary permanent connections. 2. Physiological engorgement of certain organs with blood.

erect Norway maple. A horticultural variety of the Norway maple that is often planted as a street tree on narrow streets for its habit of upright limb growth.

eremacausis. The natural burning out (oxidation of organic matter), which results in a very low content, or nearly complete absence of humus in soils of arid regions.

eremium. A desert plant community.

ergosterol. A sterol which when activated by ultraviolet rays becomes vitamin D_2. Also called *provitamin* D_2; *ergosterin.*

ergot. 1. A fungous disease of cereals and wild grasses caused by *Claviceps purpurea,* family Clavicepitaceae, which attacks the inflorescence, replacing the grains with black or dark purple, club-shaped, horny structures (sclerotia) which are harvested with the grain and must be removed before the grain is used for flour or feed. *Ergot* causes a lowered grain yield and ergotism (*q.v.*) in men and animals. Also called *clavus.* 2. A drug obtained from such diseased plants. 3. A horny growth behind the fetlock joint of a horse.

ergotism. A disease of man and lower animals caused from eating grain or grain products contaminated with ergot which is characterized by excessive salivation, redness and blistering of the mouth epithelium, vomiting, colic, diarrhea, and constipation. Also called *holy fire; St. Anthony's fire; bread madness.* See *gangrenous ergotism; dry gangrene.*

ergoty. Contaminated with ergot.

erica. Heath *q.v.*

erinose. A plant disease, especially of the grape, walnut, and mountain maple, caused by mites of the genus Eriophyes, family Eriophyidae, which is characterized by an abnormal development of hairs or trichomes from the surface of the leaves, suggesting a feltlike patch.

eriogonum. A genus of plants of the family Polygonaceae. Species consists of herbs and subshrubs which are sometimes used as ornamentals. Native to western North America. See *sulfur eriogonum.*

ern(e) fern. Bracken *q.v.*

erode. 1. To wear away or carry away the surface of the earth, as water erodes a rock formation. See *erosion; erodible.* 2. To wear away, eat into the cells or layer of cells.

eroded area. A place that is worn away. In the body this may be mechanical, as the wearing away of the teeth or outer skin, or by the action of chemicals, or germs that destroy living cells.

erodible. Susceptible to erosion, as *erodible* soil.

erose. Designating a leaf or flower which is irregularly notched on its outer margin, as if it were partially eaten away.

erosion. The wearing away of the land surface by raindrop splash, flowing water, wind, or other geological agents, including such processes as gravitational creep. Natural *erosion* is normal or geologic, in contrast to that induced by the activity of man. See *accelerated erosion.*

erosion class. An arbitrary division of erosion conditions according to type and degree used in technical mapping of soil erosion in Soil Conservation Surveys. It is intended to be an appraisal of the effect of accelerated erosion at a specified time for the area mapped. Examples of erosion classes are moderate sheet erosion and severe or gully erosion. See *gully erosion; rill erosion; sheet erosion; splash erosion.*

erosion pavement. A residue of pebbles and stones on a land surface formed by the removal of the finer surface particles by wind or water.

erosion-resistant crop. Those crops which, because of dense foliage, root system, and/or litter, provide effective protection against soil loss by erosion.

erosive. Tending to produce erosion; *e.g.,* wind and water are *erosive.*

erosiveness. The tendency to cause erosion.

erosivity. The degree to which the land suffers from splash, sheet, rill, or gully erosion.

eruc. The fiber obtained from the buripalm (*q.v.*), used in making cordage.

eructation. The act of belching, or of casting up gas from the stomach.

eruption of horn. The appearance of horns on a calf.

eruptive blind pocket. See *blind pocket.*

Ervy. A soft cheese, similar to Camembert, which is made in France.

eryngo. Any plant of the genus Eryngium, family Umbelliferae. Species consist mostly of perennial herbs which are planted in flower borders and rock gardens. Also called *eryngium.* See *seaholly eryngo.*

erysimum. A genus of plants of the family Cruciferae grown for their yellow, lilac, or blue flowers. Also called *blister cress.*

erysipelas. See *swine erysipelas.*

erythea. Any palm of the genus Erythea, family Palmaceae. Species consist of low to medium-height palms which are sometimes planted in Southern California, U.S.

erythema. Redness or inflammation of the skin caused by congested capillaries.

erythroblastosis. One manifestation of avian leukosis complex, a virus disease of chickens characterized by paleness or a yellowish discoloration of the comb and wattles, a rapid loss of weight in spite of a good appetite, weakness, prostration, diarrhea, and tightly shut eyes. The infected fowl soon dies. Also called *erythroleukosis; erythroblastic leukosis.*

erythroleukosis. Erythroblastosis.

escallonia. Any shrub or small tree of the genus Escallonia, family Saxifragaceae. Species bear clusters of white, pink, or red flowers and are grown as ornamentals in southern and southwestern U.S. See *Montevideo escallonia; Organ escallonia.*

escambron. See *catclaw funnelcreeper; catclaw apes-earring.*

escape. 1. A fowl or animal which has got out of its enclosure. 2. a. Botanically, a cultivated plant which is found growing wild. b. To become wild after having been in cultivation.

escape board. A board, having one or more bee-escapes in it, used to remove bees from the supers.

escarole. A variety of endive having broad, thick, entire-edged leaves used as a salad plant, as a potherb, and in soups and stews.

eschallot. Shallot.

escobadura. Barbfruit pavonia.

escobita owlclover. *Orthocarpus purpurascens,* family Scrophulariaceae, an annual herb grown for its purple or crimson flowers tipped with white and streaked with yellow or purple. Native to California, U.S. Also called *owl's clover.*

esculent. a. Edibility, especially where it modifies plants or plant parts. b. A plant suitable for human food.

escutcheon. 1. The cion used in shield budding. 2. That part of a cow, etc., which extends upward just above and back of the udder where the hair turns upward in contrast to the normal downward direction of the hair. Also called *milk shield; milk mirror.*

esker. A minor land feature of the glaciated region of the U.S.; *i.e.,* a low, narrow ridge with curved or rounded slopes composed of gravel and sand showing evidence of deposition in water. Also called *osar; hogback.*

esophagus. Gullet; the tube that connects the throat or pharynx with the stomach. It varies greatly in the vertebrates; *e.g.,* in the crop of a bird, it is distended for retention of food.

espalier. a. A method of training a fruit tree in which the tree is usually planted against a wall and the main branches trained in a plane parallel to the wall in a geometric design. Plums, apples, pears, cherries, and peaches are suited to this type of training. b. A tree so trained. c. The trellis upon which an *espalier* tree is trained.

esparsette (esparcet). An old name for sainfoin.

espartograss. *Lygeum spartium,* family Gramineae, a grass which yields material for the manufacture of cordage, shoes, baskets, paper, etc. Also called *Spanish grass.*

espino. Any of many thorny shrubs and trees of South America, especially those of the genus Zanthoxylum, family Rutaceae. See *pricklyash: 2; Hercules'-club pricklyash; cavenia acacia; English hawthorn.*

espino cavan. Cavenia acacia.

esplees. The yield from farm land; *e.g.,* profit from rent, crops, or remuneration for such.

essence. A substance which contains to a high degree the essential oils or other stored products of a plant, etc., as *essence* of peppermint.

essential amino acid. Any of the amino acids which cannot be synthesized in the body from other amino acids or substances or which cannot be made in sufficient quantities for the body's use.

essential element. Any element necessary for growth and development of a plant or animal. Also called *critical element. Cf. trace element.*

essential floral organs. Those organs primarily concerned with reproduction; *i.e.,* stamens and pistils.

essential host. A host for one stage in the development of a parasite without which the parasite cannot develop to maturity.

essential oil. Any volatile, aromatic oil derived from plants which is characteristic of the plant in odor and sometimes in other properties. It is used in essences, perfumes, attractants, repellents, etc. It is distinguished from the so-called fatty oils in plant tissue by the fact that it evaporates or volatilizes in contact with air and gives off an aromatic odor or flavor such as peppermint, camphor, etc. Also called *ethereal oil; volatile oil.*

Essex. A small, easily fattened breed of swine which originated in England. It has a black body with a white belt and white feet.

established. 1. Designating a plant or plant community growing naturally in healthy adjustment to the environment. 2. Designating a pest or disease which is thriving. 3. Designating a plantation of trees which requires little or no cultivation, weeding, etc., and in which the young trees are almost certain to survive. 4. Designating a seeding of meadow or pasture plants that is thriving.

establishment. 1. The adjustment of a plant to a new site, consisting of three processes: germination, growth, and reproduction. Also called *acesis.* 2. A farmstead.

estancia. A small farm or small cattle ranch (southwestern U.S.).

estate. 1. A relatively large acreage on which are a residence and other buildings, lawns, flower gardens, etc. It connotes wealth and is not usually cultivated for profit. 2. One's entire property. 3. The property left by one after one's death.

ester. An organic salt formed by the interaction of an alcohol (base) and an organic acid.

esthetic woods. A woodland established and maintained for its beauty and esthetic values, as contrasted with forests or woods maintained for commercial purposes or for windbreaks. See *pinetum; arboretum.*

estimate. A judgment of the approximate volume or yield of a crop or of a timber stand made from incomplete data.

estimate-yield method. A method of recording individual tree yields of fruit by guess. *Cf. actual-yield method.*

estivate. A term of zoological reference: to become and remain dormant in summer. *Cf. hibernate.*

estivation. 1. A zoological term designating dormancy in the summer months. 2. The arrangement of the parts of a flower in the bud.

estragon. Tarragon.

estrapade. The bucking of a horse.

estriol. An estrogenic substance derived from the urine of pregnant animals and women which, when administered to female animals, causes them to come into heat. Also called *theelol.*

estrogen. Any hormonal substance which is capable of producing heat (estrus) in a female animal. Certain forms are fed to animals, and are used for producing estrus in the female and for chemical caponizing.

estrogenic. Of, or pertaining to, estrogen.

estrogenic effect. Production of estrus, or heat periods by an estrogen.

estrone. A product derived from the urine of pregnant animals and women or from the ovary which may be administered to female animals to cause them to come into heat. Also called *theelin; folliculin; follicular hormone.*

estrous. Pertaining to estrus.

estrous cycle. Estrual cycle.

estrual cycle. The interval of time between periods of estrus in the nonpregnant female. The definite physiological functional rhythm of the female reproductive system. Also called *estrous cycle; estrous.*

estrus. The period of sexual excitement (heat) at which time the female will accept coitus with the male.

eternal flower. See *everlasting.*

ethereal oil. Essential oil.

ether extract. In feedstuff analysis, that part which will dissolve out in ether, including fat and many nonfat substances, such as waxes, sterols, carotenoid pigments, etc. Also called *fat; crude fats.*

etherization. The subjecting of a plant to one part of ether to 400 or 500 parts of air for the purpose of shortening the dormant period and hastening germination.

ethide. A fumigant and insecticide.

Ethiopian calla. See *callalily: 2.*

Ethiopian eggplant. *Solanum integrifolium,* family Solanaceae, a tropical, tender, annual herb grown for its ornamental fruit. Native to Africa. Also called *tomato eggplant; scarlet eggplant.*

ethnic group. A segment of a population set aside because of distinct cultural and racial characteristics which is usually of a separate historical origin than the rest of the population. Recently, the term has tended to be used in reference to such a segment with distinct cultural characteristics whether or not racial differences are evident.

ethrog citron. *Citrus medica* var. *ethrog,* family Rutaceae, an evergreen tree whose fruit is used by orthodox Jews in the Feast of the Tabernacle. Native to the Mediterranean area. See *citron: 1.*

ethyl acetate. $CH_3COOC_2H_5$, an inflammable fumigant which leaves a characteristic odor on treated grain.

ethylene. C_2H_4, a colorless, inflammable, unsaturated hydrocarbon gas used to hasten ripening of harvested fruits, especially pears and green tomatoes, and to hull Persian walnuts, etc. It is used as a coloring agent for harvested citrus fruits, especially oranges, and as a defoliation agent for fall-dug nursery stock, especially rose bushes. It is also used for blowtorch welding and cutting.

ethylene dibromide. A heavy liquid used as a soil fumigant, especially for the control of wireworms in the irrigated lands of western U.S. Emulsified, it is effective against Japanese beetle grubs in grasslands and against the European chafer in soils.

ethylene dichloride. A soil fumigant and nematocide.

ethylene oxide. C_2H_4O, a chemical used as a fumigant.

ethyl formate. A fumigant.

ethylmercuric chloride. A fungicide for seed and bulb treatment.

ethylmercuric phosphate. A fungicide for seed and bulb treatment.

ethylmercuric-p-toluene sulfonilimide. A fungicide for seed and bulb treatment.

ethyl mercury phosphate. A fungicide.

etiolate. To prevent a plant or plant part from developing chlorophyll by shielding it from light; to blanch.

etiolated. Designating a plant or plant part which lacks chlorophyll because it was grown in the absence of light.

etiolation. The failure of a plant or plant part to develop chlorophyll due to the absence of light; the condition of being blanched.

etiological. Pertaining to the causes of diseases.

etiology. The science that deals with the origins and causes of disease. Also called *aetiology.*

etrog citron. Ethrog citron.

etua tree. Sausagetree.

eucalypt. Eucalyptus.

eucalyptus. Any tree of the genus Eucalyptus, family Myrtaceae, of Australia. Some species are gigantic, some highly aromatic, and some are evergreen. They are grown as ornamentals and as bee trees in warm regions. Also called *eucalypt; gum tree; stringy bark.* See *leaf eucalyptus; bangalay eucalyptus; sugar eucalyptus; yate eucalyptus; scarlet eucalyptus; Tasmanian blue eucalyptus; spotted eucalyptus; messmate stringybark eucalyptus; redbox eucalyptus; kino eucalyptus; Moitch eucalyptus; horncap eucalyptus; ribbon eucalyptus.*

eucharis. Any plant of the genus Eucharis, family Amaryllidaceae. Species are bulbous herbs useful in greenhouses, as pot plants, and in flower gardens in warm areas. See *Amazonlily eucharis.*

Eugene poplar. Carolina poplar.

eugenesis. Any fertility between hybrids.

eugenia. Any plant of the tropical genus Eugenia,

family Myrtaceae. Species are evergreen trees and shrubs used as ornamentals in Florida and California, U.S., and other places with a similar climate. In the tropics some species are grown for their fruit and for the preparation of some spices. See *pitanga; roseapple.*

eugenics. The application of knowledge of heredity to the improvement of the human race; the study of agencies under social control that may improve or impair the racial qualities of future generations, either physically or mentally.

eulalia. Silvergrass.

euonymus. Any plant of the genus Euonymus, family Celastraceae. Species are trees, shrubs, and vines of great garden importance. Also called *evonymus; spindletree; burning bush.* See *winged euonymus; brook euonymus; eastern wahoo; winterberry euonymus; European euonymus; evergreen euonymus; running euonymus; wintercreeper euonymus; yeddo euonymus.*

euonymus scale. *Unaspis euonymi* (Comst.), family Diaspididae, an insect that infests euonymus, particularly the viny species, causing the foliage to yellow and the twigs to die back.

eupatorium. Any plant of the genus Eupatorium, family Compositae. Species are largely herbaceous perennials raised in greenhouses, others, in flower borders and wild gardens. Also called *thoroughwort; boneset; eupatory; conoclinium.* See *mistflower eupatorium; boneset; bluestem joepyeweed; white snakeroot.*

eupatory. Eupatorium.

euphorbia. Any plant of the genus Euphorbia, family Euphorbiaceae. Species consist of herbs, shrubs, and succulent trees. The succulent or cactus-like forms are grown in greenhouses and are used as specimen plants and as hedge plants in warm areas. The hardy species are used in flower borders and rock gardens. Also called *spurge.* See *floweringspurge euphorbia; cushion euphorbia; scarletplume euphorbia; snow-on-the-mountain euphorbia; crownofthorns euphorbia.*

euploid. An organism or cell in which the chromosome number is the exact multiple of the monoploid or haploid number. Terms used for an *euploid* series are: haploid, diploid, triploid, tetraploid, etc.

Eurasian Solomonseal. *Polygonatum multiflorum,* family Liliaceae, a perennial herb grown for its greenish-white flowers and foliage and as a pot plant for spring forcing. Native to Eurasia. Also called *white root.*

European alder. 1. *Alnus glutinosa,* family Corylaceae, a deciduous tree easily grown as an ornamental, in swampy locations. Native to Eurasia and North Africa. Also called *black alder; dog tree.* 2. See *speckled alder: 1.*

European alder leaf miner. *Fenusa dohrnii* (Tisch.), family Tenthredinidae, an insect that infests the alder in northeastern U.S.

European annual grass. Annual vernalgrass.

European apple sawfly. *Hoplocampa testudinae* (Klug), family Tenthredinidae, an insect that infests apples in Connecticut and New York, U.S. The larvae bore into the fruit and leave brown "sawdust" on the surface.

European ash. *Fraxinus excelsior,* family Oleaceae, a deciduous tree growing to a height of 140 ft., which has 7 to 11 dark green leaflets in each compound leaf. The many varieties vary in growth habits and are desirable as ornamentals. Native to Eurasia.

European aspen. *Populus tremula,* family Salicaceae, a hardy, roundheaded, softwood tree, with quaking leaves which may be used for pulpwood and/or reforestation. Native to Asia.

European barberry. *Berberis vulgaris,* family Berberidaceae, a deciduous shrub grown as an ornamental. It is an alternate host of wheat rust and is often eradicated in wheat growing areas. There are several horticultural varieties. Native to Europe; naturalized in eastern North America. Also called *common barberry.*

European bark beetle. See *smaller European elm bark beetle.*

European beachgrass. *Ammophila arenaria,* family Gramineae, an excellent sand-binder (*q.v.*). Also called *beechgrass; marram grass.*

European beech. *Fagus sylvatica,* family Fagaceae, a tall, hardy, deciduous tree which has many varieties grown as ornamentals. Native to Europe.

European bellflower. Creeping bellflower.

European bindweed. European glorybind.

European birch. European white birch.

European bistort. *Polygonum bistorta,* family Polygonaceae, a perennial herb grown for its white, pink, or red flowers in spikes. Native to Europe. Also called *English serpentary; Easter giant; Easter ledges; snakeweed.*

European bittersweet. Bitter nightshade.

European black currant. *Ribes nigrum,* family Saxifragaceae, an unarmed shrub which is the source of many varieties of the black currant, bearing fruits of fine flavor. Native to Eurasia. Also called *black currant.*

European bladdernut. *Staphylea pinnata,* family Staphyleaceae, a hardy, deciduous shrub grown for its greenish-white flowers and for its large, inflated fruit pods. Native to Europe.

European brown snail. *Helix aspersa* Muller, family Helicidae, a snail (*q.v.*) which seriously infests citrus and avocado orchards, feeding on the foliage, blossoms, and young fruit. It also attacks many other plants. Native to Europe; thought to have been introduced into California, U.S., from France as a table delicacy. Also called *brown garden snail.*

European canker. A fungous disease of black walnut, butternut, pignut hickory, red maple, tuliptree, red oak, white oak, black oak, pear, and apple caused by *Nectria galligena,* family Hypocreaceae. The disease is characterized by rough, sunken, or flattened, red and black cankers having concentric prominent ridges. Girdling results in death to the distal portions of the tree.

European canker of poplar. A fungous disease of poplar caused by *Dothichiza populea,* family

Sphaerioidaceae, characterized by cankers on the trunk and branches. Girdling is lethal to distal portions of the tree.

European chestnut. *Castanea sativa,* family Fagaceae, a deciduous tree which grows up to 100 ft. or more with a spreading but compact head. It produces very large, spiny burs containing edible nuts enclosed in a dark, mahogany-brown shell. The nuts are larger than the American chestnut, and in Spain, France, and Italy they have constituted an important article of food since early days. There are several horticultural varieties. Native to the Mediterranean area. Also called *Spanish chestnut.*

European columbine. *Aquilegia vulgaris,* family Ranunculaceae, a short-spurred, perennial herb grown for its violet flowers. Several improved varieties have been developed by crossing with other species. Native to Eurasia. Also called *culverwort.*

European cornborer. *Pyrausta nubilalis* (Hbn.), family Pyraustidae, an insect pest of corn whose larvae bore into the corn stalk and ear shank. Infested plants show broken tassels, bent stalks, and "sawdust" outside small holes.

European cornsalad. *Valerianella locusta* var. *olitoria,* family Valerianaceae, an annual herb which is raised as an early-spring and fall salad crop in Europe and occasionally in North America. Native to Eurasia. Also called *cornsalad; lamb's lettuce; fetticus; vetticost; field salad.* See *Italian cornsalad.*

European cranberry. Small cranberry.

European cranberrybush viburnum. *Viburnum opulus,* family Caprifoliaceae, a hardy, deciduous shrub grown for its white flower clusters and red fruit. Native to Eurasia and northern Africa. Also called *cranberry tree.*

European earwing. *Forficula auricularia* L., family Forficulidae, an insect whose nymphs feed on the shoots of plants and eat holes in the leaves, while the adults feed on blossoms and ripening fruit. The adults often enter houses. Found in northern U.S.

European elder. *Sambucus nigra,* family Caprifoliaceae, a large shrub or small tree widely grown for its fragrant, yellowish-white flowers. Its many varieties give different forms, leaf colors, etc. Native to Eurasia and northern Africa. Also called *dog tree.*

European elm. The English elm, *Ulmus procera,* according to English authors; continental European authors classify it as *Ulmus carpinifolia.* See *English elm; smoothleaf elm.*

European elm bark beetle. See *smaller European elm bark beetle.*

European elm scale. *Gossyparia spuria* (Mod.), family Dactylopiidae, an insect that infests elms causing a yellowed foliage, premature defoliation, and death of smaller branches. Sometimes the entire tree dies.

European euonymus. *Euonymus europaeus,* family Celastraceae, a deciduous shrub or small tree grown in warm areas for its foliage and red or pink fruit. Native to Eurasia. Also called *dog tree; spindletree; prick timber.*

European evening primrose. See *evening primrose: b.*

European fan palm. See *Mediterraneanpalm: 2.*

European filbert. *Corylus avellana,* family Betulaceae, a hardy shrub grown in the U.S. principally for ornament, and in Europe for its edible nuts. It is the source of several varieties. Native to Europe. Also called *European hazel.*

European fly honeysuckle. *Lonicera xylosteum,* family Caprifoliaceae, a deciduous, ornamental shrub which has yellowish-white flowers and attractive dark red fruits. Native to Eurasia. Also called *European honeysuckle; bush honeysuckle.*

European foulbrood. An infectious, brood disease of bees considered to be caused by *Bacillus pluton.* Also called *black brood. Cf. American foulbrood.*

European fruit lecanium. *Lecanium corni* Bouche, family Coccidae, a widely distributed scale insect that infests fruit trees, ornamental trees, and shrubs. The nymphs feed mostly on the veins of leaves. Also called *brown apricot scale; brown elm scale.*

European fruit scale. *Aspidiotus ostreaeformis* Curt., family Diaspididae, an insect that infests fruit and ornamental trees.

European globeflower. Globeflower *q.v.*

European glorybind. *Convolvulus arvensis,* family Convolvulaceae, a slender, herbaceous vine, an extremely competitive weed pest over the northern half of U.S., which twines itself around other plants. Native to Europe. Also called *field bindweed; small-flowered morning-glory; small bindweed; hedge-bells; lap-love; creeping Charlie; creeping Jenny; wild morning-glory; English bindweed; lesser bindweed; bindweed.*

European gloryvine. European glorybind.

European goldilocks. *Linosyris vulgaris,* family Compositae, a perennial herb grown in flower gardens for its late-blooming, pale yellow flowers. Native to Europe.

European gooseberry. *Ribes grossularia* var. *uvacrispa,* family Saxifragaceae, a spiny shrub grown in Europe for its edible fruit used in jams, pies, etc. Native to Eurasia and northern Africa.

European gooseberry mildew. A powdery mildew of currants and gooseberries caused by *Microsphaera grossulariae,* family Erysiphaceae, which occurs mainly on the upper surfaces of the leaves.

European grape. *Vitis vinifera,* family Vitaceae, a woody vine which is the source of all the best wine grapes, commonly called vinifera grapes. Grown in Europe, Asia, and principally in California in the U.S., it is of vast economic importance. Native to Eurasia.

European green alder. *Alnus viridis,* family Betulaceae, a low shrub grown as an ornamental, as a specimen, or in a shrub border for its foliage and catkins. Native to Europe.

European hazel. European filbert.

European honeysuckle. See *woodbine honeysuckle; European fly honeysuckle.*

European honeysuckle leaf roller. *Harpipteryx xylostella* (L.), family Hyponomeutidae, an insect that infests the Tatarian honeysuckle.

European hornbeam. *Carpinus betulus,* family Corylaceae, a hardy, deciduous tree grown in hedges. Native to Eurasia.

European larch. *Larix decidua,* family Pinaceae, a hardy, deciduous, pyramidal tree grown for its timber, for its use in windbreaks, and for ornament. Native to Europe.

European larch canker. A fungous disease caused by *Dasyscypha willkommii,* family Helotiaceae, which is characterized by sudden death of small stems and branches girdled by cankers. There is a flow of resin from each canker, and affected trunks and limbs are distorted.

European linden. *Tilia europaea,* family Tiliaceae, a hardy, hybrid, deciduous tree grown in Europe, and sometimes in U.S. as a street tree. Also called *common lime; linden.*

European lotus. Dateplum persimmon.

European mallow. Hollyhock mallow.

European mantis. *Mantis religiosa* L., family Mantidae, an insectivorous insect, imported from Europe and established in the U.S., which devours large numbers of harmful insects. Also called *praying mantis; preying mantis.* See *Carolina mantis.*

European mountainash. *Sorbus aucuparia,* family Rosaceae, a deciduous tree grown as an ornamental for its bright red, small clusters of fruit which remain on the tree during the winter. Native to Eurasia. Also called *Roman tree.*

European pasqueflower. *Anemone pulsatilla,* family Ranunculaceae, a perennial herb which is especially grown in rock gardens for its blue to reddish-purple, spring flowers. Also called *Easter anemone; Easter flower.*

European peach scale. *Lecanium persicae* (F.), family Coccidae, an insect that infests English ivy, ginkgo, grape, holly, Japanese flowering quince, mulberry, nectarine, peach, and oleander.

European pellitory. Sneezewort yarrow.

European pine shoot moth. *Rhyacionia buoliana* (Schiff.), family Olethreutidae, a moth that lays its eggs on twigs, needle sheaths, or bark of red, mugho, Scotch, and perhaps other pines. Its larvae bore into the needle base and winter in the buds destroying them. It is found in U.S. from Massachusetts to Virginia and west to Illinois.

European privet. *Ligustrum vulgare,* family Oleaceae, a shrub which is grown as a hedge plant in Europe and North America.

European raspberry. See *red raspberry: 1.*

European red elder. *Sambucus racemosa,* family Caprifoliaceae, a shrub grown as an ornamental and in hedges for its yellowish-white flowers and scarlet fruit. Native to Eurasia. Also called *red-berried elder.*

European red mite. *Metatetranychus ulmi* (Koch), family Tetranychidae, a serious fruit pest in the U.S. that infests many deciduous trees and shrubs, especially plums, prunes, and apples.

European red raspberry. See *red raspberry: 1.*

European spruce sawfly. *Diprion hercyniae* (Htg.), family Diprionidae, an insect that infests white, red, black and Norway spruce in New England, New York, and New Jersey, U.S. The larvae feed on the foliage.

European strawberry. *Fragaria vesca,* family Rosaceae, a perennial herb which has been widely used in breeding strawberries, especially late-bearing varieties.

European sycamore. See *London plane tree.*

European Turkey oak. *Quercus cerris,* family Fagaceae, a deciduous tree grown as an ornamental, for its handsome, pyramidal shape, in warm regions of Eurasia. Also called *cerris.*

European-type grape. See *European grape.*

European verbena. *Verbena officinalis,* family Verbenaceae, an annual herb grown as a garden plant, a pot plant, or a greenhouse plant for its purple flowers. Native to Europe. Also called *European vervain; enchanter's herb.*

European walnut. Persian walnut.

European wheat stem sawfly. *Cephus pygmaeus* (L.), family Cephidae, an insect that infests grasses and small grains in U.S., in New York and Pennsylvania.

European white birch. *Betula pendula,* family Betulaceae, a hardy, deciduous tree, one of the several varieties of popular ornamentals, grown for its white bark that peels off in layers and its drooping branches. It reaches 60 ft. in height. Native to Eurasia. Also called *European birch.*

European white hellebore. See *white falsehellebore.*

European white waterlily. *Nymphea alba,* family Nymphaeaceae, a hardy, aquatic, herbaceous plant planted in out-of-door pools for its spring to autumn, pure white blossoms. Native to Europe and North Africa.

European wood anemone. *Anemone nemorosa,* family Ranunculaceae, a perennial herb grown for its early-spring, white or purple flowers. Native to Europe.

European yellowflag iris. Yellowflag iris.

European yellow violet. *Viola lutea,* family Violaceae, a perennial flowering herb grown for its usually yellow flowers (rarely violet colored). It is believed to be one of the parents from which the viola is a descendant, the other being the horned violet. Native to central and western Europe.

eutropic. Pertaining to the turning of a plant into the sun's rays.

eutropic peat. Peat or muck nearly neutral or alkaline in reaction which is relatively high in bases and essential mineral plant nutrients.

evacuated containers. Glass or metal containers from which most of the air has been removed.

evaginated. Turned inside out. Tapeworm larvae often have the head, or scolex, *invaginated;* when the larvae reach the intestines of the final host, the head is *evaginated,* or turned inside out, so the suckers can attach to the wall of the intestines.

evaporate. 1. To convert from a liquid or solid into a gaseous state. 2. To reduce the sap of a maple tree to maple sirup.

evaporated. Designating a product which has had most of the moisture driven off by boiling or other application of heat.

evaporated buttermilk. The product which results from the removal of a considerable portion of the moisture from clean, sound buttermilk, made from natural cream to which no foreign substance has been added. The product must contain not less than 27 percent total solids, not less than 0.055 percent butterfat for each percent of solids, and not more than 0.14 percent of ash for each percent of solids. Also called *concentrated buttermilk; condensed buttermilk.*

evaporated cultured skim milk. A product which results from the removal of a considerable portion of moisture from clean, sound skim milk which has been made from a suitable culture of lactic bacteria. This product must contain not less than 27 percent of total solids. Also called *concentrated cultured skimmed milk; condensed cultured skimmed milk; concentrated sour skim milk.*

evaporated milk. A liquid food product made by evaporating sweet milk to such a point that it contains not less than 7.9 percent of milk fat and not less than 25.9 percent of total milk solids. It may contain one or both of the following optional ingredients: (a) disodium phosphate or sodium citrate or both, or calcium chloride, added in a total quantity of not more than 0.1 percent by weight of the finished *evaporated milk;* (b) vitamin D in such quantity as increases the total vitamin D content to not less than 7.5 U.S.P. units per avoirdupois ounce of finished *evaporated milk.* The product is then pasteurized, homogenized, and sealed in a container which is sterilized by heat under pressure. Also called *concentrated milk.*

evaporated skim milk. A food product resulting from the removal of a considerable portion of the moisture from skim milk so that it contains 20 to 22 percent total solids. It is then hermetically sealed in a can and sterilized.

evaporated soured skim milk. See *evaporated cultured skimmed milk.*

evaporation. The changing of a liquid into a gas. Industrially it may mean merely drying or driving off the moisture of a product, and not complete vaporization. It may be loosely used as the equivalent of dehydration *q.v.* See *transpiration.*

evaporation discharge. Discharge of water, by evaporation, into the atmosphere from the zone of saturation which may be by transpiration from vegetation, or directly from the soil.

evaporator. That part of the refrigeration system in which the refrigerant is changed from a liquid to a vapor. The heat required to vaporize the liquid refrigerant is obtained from the product to be cooled. See *dehydrater: 2.*

evaporimeter. Atmometer *q.v.*

evapo-transpiration. That part of the root zone moisture which is consumed by evaporation and transpiration combined, including all water consumed by plants plus the water evaporated from bare land and water surface.

even-aged. Designating a stand of trees, all of which have an age difference of no more than 10 to 20 years.

even distribution. 1. In stock judging, an animal well fleshed over the entire body. 2. In sowing seeds, an even dispersal of the seeds over the entire area being sown or planted.

evener. A piece of iron or wood used to place the center of pull to the center of draft (*q.q.v.*) in hitching more than one draft animal to a plow, wagon, or farm implement. Also called *equalizing bar.*

evener pin. a. Any hard steel pin, either smooth or threaded, with a head at one end used to attach an evener to a tongue, clevis, or other point of attachment to a farm implement. b. Wagon hammer *q.v.*

even grain. Designating wood whose grain shows little contrast between early and late growth rings.

evening-bloom. Designating a flower which blooms at night. Also called *nocturnal.*

evening campion. *Lychnis alba,* family Caryophyllaceae, a biennial or perennial herb which has white, fragrant, night-blooming flowers. Native to Europe, it is widely naturalized in eastern U.S. where it is considered a weed. Also called *white campion; evening lychnis; white robin; white cockle.*

evening glory. Large moonflower *q.v.*

evening lychnis. Evening campion.

evening milk. Milk obtained from milking cows in the late afternoon or evening.

evening primrose. a. Certain plants of the genus Oenothera, family Onagraceae; others are called sundrops. Also called *evening rose.* b. *O. biennis,* the tall, coarse, weedy, biennial herb which has yellow, night-blooming flowers. It is occasionally grown in flower gardens, but is considered a weed in North America. Native to North America; naturalized in Europe. Also called *European evening primrose; night willowherb; field primrose.*

evening rose. See *evening primrose: 1.*

evening snow. Eveningsnow gilia.

eveningsnow gilia. *Gilia dichotoma,* family Polemoniaceae, an annual flowering herb which has white flowers with brownish-purple throats. Native to California, U.S. Also called *evening snow.*

eveningstar rainlily. *Cooperia drummondi,* family Amaryllidaceae, a small bulbous herb which has grasslike leaves and white, star-shaped, night-blooming flowers. Native to Texas, U.S. Also called *evening star.*

evening stock. Grecian stock.

evening trumpet flower. Carolinajessamine.

evenly balanced. In stock judging, designating an animal which stands well on its feet.

even-running. Designating bales of cotton of similar properties which conform to a single sample.

even-span house. A greenhouse which has both sides or spans of the roof equal in length.

even topline. In stock judging, straight along the backbone.

everbearer. Any small fruit plant which produces ripe fruit throughout most of the season, in contrast to those which produce the bulk of the crop in a short period of time.
everbearing. Designating any small fruit plant which produces fruit throughout most of the season.
everbearing grape. Either of two species, the bullace grape or the bird grape *q.q.v.*
everbloomer. A plant which produces flowers more or less continuously during the season.
everblooming. Producing flowers more or less continuously during the season.
everblooming Chinese rose. *Rosa chinensis* var. *semperflorens,* family Rosaceae, an erect, partly evergreen shrub grown in the warm regions for its solitary crimson or deep pink flowers. Native to China. Also called *Chinese monthly rose.*
everblooming honeysuckle. *Lonicera heckrotti,* family Caprifoliaceae, a hardy, low, deciduous shrub grown for its long-blooming, purple and yellow flowers.
everglade palm. Saw paurotis.
Everglades. A 4,000 to 5,000 sq. mi. tract of low-lying marshland in Florida, U.S., that extends from Lake Okeechobee to the southern tip of Florida. The area, less than 20 ft. above sea level, is covered with water the greater part of the year, and its characteristic vegetation is saw grass (*Cladium effusum*). The soil is mainly peat but extensive marl prairies occur in the south. Only a small percentage of the area is used for agriculture, while a large tract has been set aside as a national park.
evergreen. 1. A plant that retains its leaves or needles longer than one growing season so that leaves are present throughout the whole year. *Cf. deciduous: 1.* 2. (plural) The leafy branches of *evergreen* plants that are used for decoration. Also called *greens.* See *liveforever stonecrop; pipsissewa; Carolina laurelcherry.*
evergreen bittersweet. Bigleaf wintercreeper euonymus.
evergreen blackberry. Cutleaf blackberry.
evergreen blueberry. Ground blueberry.
evergreen candytuft. *Iberis sempervirens,* family Cruciferae, an evergreen, perennial herb grown as an edging or for its clusters of white flowers. Native to Eurasia. Also called *edging candytuft.*
evergreen cherry. See *Carolina laurelcherry; hollyleaf cherry.*
evergreen clematis. Vernal clematis.
evergreen elm. See *Chinese elm; cedar elm.*
evergreenery. See *evergreen: 2.*
evergreen euonymus. *Euonymus japonicus,* family Celastraceae, an evergreen shrub or small tree grown for its lustrous green leaves. Native to Japan.
evergreen grape. Evergreen treebine.
evergreen grass. Tall oatgrass.
evergreen ivy. English ivy.
evergreen magnolia. Southern magnolia.
evergreen millet. See *Johnsongrass.*
evergreen oak. See *holly oak; California live oak.*
evergreen privet. Italian buckthorn.
evergreen tamarix. See *athel tamarix.*
evergreen thorn. See *scarlet firethorn; English hawthorn.*
evergreen treebine. *Cissus capensis,* family Vitaceae, a tropical, strong-growing, evergreen vine grown for its handsome foliage. Native to South Africa. Also called *evergreen grape.*
evergreen trumpet flower. Carolinajessamine.
evergreen white oak. Engelmann oak.
evergreen winterberry. Inkberry.
evergreen wood fern. See *leather woodfern; Christmas fern.*
everlasting. 1. Any plant whose flowers retain their form and color when dry. 2. Any plant of the genus Helichrysum, family Compositae, which is useful as an *everlasting.* See *strawflower; fragrant cudweed.*
everlasting flower. The flower of an everlasting. See *strawflower.*
everlasting goldfields. *Baeria aristata,* family Compositae, a low, hardy annual grown in the flower garden for edging and for its yellow flowers. Native to North America.
everlasting grape. Bird grape.
everlasting peavine. *Lathyrus grandiflorus,* family Leguminosae, a perennial, climbing, vinelike plant grown for its rose-purple, slightly fragrant flowers. Native to Europe. See *perennial peavine.*
everlastings. Everlasting flower.
everlasting thorn. Scarlet firethorn.
ever-normal granary. A program designed to stabilize supplies of crops (a term used in the U.S. in the 1930's by Henry A. Wallace, Secretary of Agriculture). The government would finance storage of surplus crops, which were to be released during poor crop years. Under the Agricultural Adjustment Act of 1938, the Commodity Credit Corporation was authorized to make loans under certain supply and price conditions on wheat, cotton, and other crops.
eversion of the oviduct. The turning inside out and protrusion from the anus of the lower portion of the oviduct of a female bird. The condition renders the bird worthless for egg laying but not for human consumption.
eversion of the uterus. See *prolapse of the uterus.*
everted. Turned inside out.
ever-white. Pearlyeverlasting.
every-day-pickup. Collection of milk from the bulk milk cooling tank once each day; often abbreviated *ED pickup.*
every-other-day-pickup. Collection of milk from the bulk milk cooling tank once every two days; often abbreviated *EOD pickup.*
Eve's-cup. Trumpet pitcherplant.
Eve's-thread. See *tawny daylily; Adam's-needle yucca.*
Eveweed. Dame's-rocket.
evi. See *mombin.*
eviscerate. To remove the entrails, lungs, heart, etc., from a fowl or animal when preparing the carcass for human consumption. See *drawn poultry.*

evonymus. Euonymus *q.v.*
ewe. A female adult sheep.
ewe neck. A horse which has a straight or sagging neck rather than a gracefully arched one.
ewe's milk. The whole, fresh lacteal secretion which is obtained by the complete milking of ewes.
exanthema. Dieback of citrus trees characterized by gummy tissues and brown-stained leaves. This disease is caused by malnutrition.
excambiator. An exchanger of lands; a broker.
excellent. The highest grade or classification which is used in placing or grading dairy cattle in type classification.
excess water. 1. Legally, surplus water of a stream, that which is not adjudicated, or that which is in excess of the needs of those who have prior rights to its use. 2. Any water, particularly rainfall, over and above that needed for plant growth which results in ponded water or runoff. See *free water.*
exchangeable potassium. The potassium which is held mainly by the colloidal portion of the soil and which is easily exchanged with the cation of neutral salt solutions, less the water-soluble potassium. It is readily available to growing plant roots.
exclosure. An area from which animals are excluded by fences, etc. *Cf. inclosure.*
excrescence. An unnatural or wartlike growth, as on the stem of a tree.
excreta. Matter that is excreted; waste matter discharged from the body; materials cast out by the body. Excrement; waste matters discharged from the bowels.
excretion. See *excreta.*
exempt livestock. Livestock which is grazed on public ranges under proper authority without any charge for such use.
exhaust fan. An air propeller used to create a vacuum; the opposite of a blowing fan. See *fan: 1.*
exhaustion. Loss of fertility elements by crop removal or by leaching, etc., to such a degree that crop yields are scant or no longer profitable.
existent corner. A corner in field surveying whose position is evidenced by a monument or its accessories as described in the field note record, or whose location can be identified with the aid of acceptable testimony.
exocortis. A virus disease of the trifoliate-orange characterized by scaling bark, a severe stunting of the cion or the top of the tree. It is seen only on topworked trees.
exomphalos. A hernia which has escaped through the umbilicus. See *umbilical hernia.*
exostosis. An outgrowth of a bone: *e.g.,* splints, bone spavins, etc.
exotic. Introduced; alien.
expanded polyptyrene. A plastic used as an insulation material.
expansion joint. A device used in connecting long lines of piping, pavement slabs, etc., to permit linear expansion or contraction as the temperature rises or falls.
expectancy record. An index figure used in milk production record formulae to indicate what should be the expected record at maturity. It is ascertained from the production record of a first calf heifer.
expeller cottonseed oil. See *crude cottonseed oil.*
experimental conditions. Those conditions of environment, feeding, and breeding of plants or animals which are controlled or premeditated by experiments or trials, rather than by natural or field conditions.
Experiment Station. An institution for conducting organized research. See *Forest Experiment Station.*
exploitive farming. Crop production carried on for immediate profit and in disregard of fertility of soil.
exsiccation. The drying up of an area, a water body, or swamp, from causes other than a decrease in rainfall; evaporation.
extended family. A unit of human organization which includes only those males and females related by blood or believed to be so related, in terms of some cultural definition of what constitutes a blood relationship. See *family: 1; nuclear family.*
Extension Service. An agency of the U.S. Department of Agriculture which represents that department in the conduct of cooperative extension work with the various Land-Grant colleges and universities. See *Cooperative Extension Service.*
extensive farming. Farming in which a comparatively small amount of labor or capital is used per acre, usually implying a large acreage: *e.g.,* large cattle ranches and wheat farms in the Great Plains region, U.S. *Cf. intensive farming.*
extracted honey. Honey that has been removed from the cells of a comb, usually strained to remove impurities.
extracted meals. Animal feeds which are fermentation products.
extract wood. Wood which is suitable for the manufacture of tannin extract: *e.g.,* oak bark, sumac twigs, etc.
extrahepatic tissues. Tissues outside the liver; tissues or organs other than the liver.
extra-period fire. In national forest practice, a fire which remains uncontrolled within a specified time following discovery.
extravasate. 1. In geology, to erupt. 2. In physiology, to force out blood, etc., or to permit it to be let out or be infiltrated from a body channel.
exudate. A discharge deposited in or on an organ through pores, injured areas, or natural openings, as a bloody discharge from wounds of animals infested with screw-worms, or the gummy discharge from a wound in a tree.
eye. In gardening, an undeveloped bud, embryo, or growing point of a plant.
eyespot. A fungous disease of grasses caused by *Rhizoctonia solani* or *Cercosporella herpotrichoides,* family Moniliaceae, which is characterized by brown-bordered, elliptical lesions at the base of the stem.

eyespot mosaic. A virus disease of oats in southeastern U.S. caused by *Marmor terrestre* var. *oculatum,* characterized by light green or yellow spots with green centers on the foliage. Some plants develop a rosette.

eyespot of pearl millet. A fungous disease caused by *Helminthosporium sacchari,* family Dematiaceae, which is characterized by a conspicuous leaf spot and a blackening of the seed.

eyespot of sugar cane. A fungous disease caused by *Helminthosporium sacchari,* family Dematiaceae, which is characterized by elongated, oval, water-soaked spots which become yellowish halos. Long runners develop from the ends of the lesions. If the lesions coalesce, the leaf appears fired. It is found in Florida and Hawaii, U.S.

eyespot of timothy. A fungous disease of timothy and related species, caused by *Heterosporium phlei,* family Dematiaceae, which is characterized by small, oval, light-colored spots with narrow, violet halos. If a leaf is heavily infested, it appears fired and soon drops. It is found in the U.S., in Europe, and in Japan.

eye-spotted bud moth. *Spilonota ocella* (D. and S.), family Olethreutidae, a moth whose larvae are destructive to fruit trees.

eyetooth. Either of the canine teeth in the upper jaw. Also called *dog tooth.* See *canine tooth.*

F

F_1. The first filial generation; the first generation of a given mating.

F_2. The second filial generation, produced by crossing *inter se* or by self-pollinating the F_1. The term is loosely used to indicate any grandchildren of a given mating, but in controlled genetic experimentation inbreeding of the F_1 or equivalent is implied.

fabaceous. 1. Belonging to the legume or pea family, family Leguminosae. 2. Of, or pertaining to, the nature of a bean.

fabiform. Having the shape of a bean.

Fabraea leaf blight. See *leaf blight and fruit spot of pear and quince.*

face. 1. a. The bare skin on a fowl's head around and below its eyes. b. The front part of the head of an animal, including the eyes, nose, and mouth. 2. The side of a hill or furrow. 3. a. The top or bottom layer of produce, especially fruit, which is arranged in a container for display purposes when the container is opened. b. To arrange a layer of produce in a container for display purposes. 4. In turpentining, the exposed portion of the tree from which the oleoresin exudes. 5. In logging, ¼ of the circumference of a log for its whole length. 6. In lumber, the side of a board from which it is graded.

face cord. A pile of wood 8 ft. long and 4 ft. high, regardless of the length of the pieces. See *cord: 2.*

faced. See *face: 3. a.*

faced-in. Designating a method of arranging feeding alleys in a barn so that the animals stand in their stalls with their heads toward the middle of the barn. See *faced-out.*

faced-out. Designating a method of arranging feeding alleys in a barn so that the animals stand in their stalls with their heads toward the outside walls of the barn. See *faced-in.*

face-in-the-hood. Aconite monkshood.

face-out barn. See *faced-out.*

face piece. A part of the harness of a horse; a leather strap which is fastened to the headband.

face side. The side of a board from which the grade is determined. Softwoods are most frequently graded from the better side; hardwoods, the poorer side.

face veneer. Veneer used on exposed or finished surfaces of plywood and veneered panels for furniture, etc.

faciation. A phase, or geographic part of a plant association, in which one or more plant dominants has dropped out.

facies. The developmental equivalent of the faciation, *i.e.,* a disclimax.

factor. 1. A unit of inheritance occupying a definite locus on one or both members of a definite chromosome pair whose presence is responsible for the development of a certain character or modification of a character of the individual which possesses that genotype; a determiner or gene. 2. An agent, as one who buys and sells a commodity on commission for others. 3. An item in the analysis of a farm business; *e.g.,* labor efficiency, etc. 4. Inherent characteristics of the climatic, nutritional, cultural, or biological environment responsible for the specific performance of plants or animals.

factor of production or input. A good or service which is capable of contributing to production.

factory farm. A term applied to a farm or type of farming which is usually operated on a large scale according to modern business efficiency standards, solely for monetary profit, as contrasted to a so-called family farm, or farming as a way of life.

facultative. Designating an organism which is capable of living under more than one condition: *e.g.,* as a saprophyte and as a parasite, as an aerobic or anaerobic organism.

facultative parasite. Any organism which is normally and usually saprophytic but which may become parasitic; an organism that can grow on living or dead organic matter.

faded. 1. Designating a color defect of American Cheddar cheese in which there is an unnatural color associated with a dry body and a crumbly texture. 2. See *bleached.*

faeces. See *feces.*

fag. 1. Any tick or fly which attacks sheep. 2. Long coarse grass of the preceding season. 3. See *fagot: 2.*

fagopyrismus. See *bighead: 1.*

fagot. 1. A pork sausage composed of hog livers, hearts, fresh pork, onions, peppers, and sweet marjoram, molded into a 6 oz. ball which is encased in hog caul fat. 2. A small bundle of twigs which is to be used for fuel.

failing-mouth. Designating an animal whose teeth have just begun to decline. Also called *chisel mouth; fullmouth; broken-mouthed; smooth-mouthed sheep.*

fail spot. An area in a young stand of trees in which trees have not become established.

fair. 1. An exhibition of farm products, etc., on a competitive basis, usually with premiums offered for excellence. 2. The next to lowest grade of Mexican vanilla beans, hothouse lamb carcasses, and other products. 3. Absolute, in an unfavorable sense (Australian and New Zealand).

fair condition. 1. A range-condition class; a range which is producing only 25 to 50 percent of its potential. The cover consists of early maturing plants which are of low value for forage or for soil protection. Woody vegetation has begun to appear. The soil has a baked look, and gullies are beginning to form. The runoff water carries much silt. 2. Denoting a medium condition of plants, plant products, or animals.

fair grounds. An area where fairs are held. See *fair: 1.*

fairies' hair. Clover dodder.

fairies' horse. Ragwort groundsel *q.v.*

fairies' paintbrush. Orange hawkweed.

fairies' petticoat. See *foxglove: 2.*

fairies' table. Mushroom.

fairly well centered. Designating a yolk of an egg which is not more than ¼ the distance from the normal central position.

fairly well fleshed. 1. Designating a fowl which may not be fully developed but which has some fat over the thighs and back and which has all its bones covered fairly well with flesh. 2. Designating a reasonably well-fattened animal.

fair-maid-of-February. See *snowdrop: 2.*

fair-maids-of-France. See *aconite buttercup; sneezewort yarrow; meadow saxifrage; ragged robin.*

fair-maids-of-Kent. Aconite buttercup.

fair market value. In appraising, the summation of the price and terms for which a property should or does exchange within a reasonable time between a seller and an able, willing, and informed buyer with a reasonable cash payment. Also called *market value; normal sale value.*

fairy bell. See *foxglove: 2; hairy fairybells.*

fairybells. Any herb of the genus Disporum, family Liliaceae. Species are sometimes grown in wild gardens. See *hairy fairybells.*

fairy candle. Cohosh bugbane.

fairy cap. See *foxglove: 2.*

fairy circle. See *fairy ring: 1.*

fairy creeper. Mountainfringe.

fairy cup. See *cowslip primrose; miterwort.*

fairy finger. See *foxglove: 2.*

fairy glove. See *foxglove: 2.*

fairy grass. Quaking grass.

fairy green. See *fairy ring: 1.*

fairy lantern. White mariposa.

fairy lily. See *zephyrlily; redtinge rainlily.*

fairy potatoes. Cowbell silene.

fairy primrose. *Primula malacoides,* family Primulaceae, a tender, perennial herb grown as a temporary pot plant and out-of-doors in warm areas for its rose and lilac flowers. Native to China. Also called *Chinese primrose.*

fairy ring. 1. A darker green, more luxuriant growth which occurs as a ring or circle in a turf area or meadow, due to the presence of any fungus of the Class Basidiomycetes. At certain seasons of the year mushrooms appear on the periphery. It was once believed to be a dancing place for fairies. Also called *fairy circle; fairy green.* 2. A fungous disease of carnations caused by *Heterosporium echinulatum,* family Dematiaceae, which is characterized by round, bleached spots on the foliage. Black spores appear in concentric rings within the spots.

fairy-ring mushroom. 1. Any mushroom producing or causing fairy rings *q.v.* 2. The fungus *Marasmius oreades.*

fairy-ring spot. See *fairy ring: 2.*

fairy rose. Roulett Chinese rose.

fairy thimbles. The flowers of foxglove *q.v.*

fairy wallflower. See *wallflower: 2.*

fairy wand. *Chamaelirium luteum,* family Liliaceae, a perennial, flowering herb grown for its yellowish-white flowers in a spikelike raceme. Native to North America. Also called *blazing star; devil's-bit; false unicorn root; rattlesnake root.*

faitours grass. Leafy euphorbia.

faking. Any attempt on the part of an exhibitor at a show to deceive a judge.

Falconers mockorange. *Philadelphus falconeri,* family Saxifragaceae, a deciduous shrub grown for its fragrant, pure white flowers.

fall. 1. The dropping of a plant part, such as leaves or fruits. 2. The slope of land. 3. The amount of precipitation. 4. a. The quantity of trees cut. b. Felling trees. 5. a. The number of lambs born. b. Giving birth, as of lambs. 6. One of the three outer segments of an iris flower which is often drooping (usually used in the plural). 7. Autumn. 8. Of, or pertaining to, a plant or fruit which matures in the autumn, as a *fall* apple. 9. A plain strip of leather on a whip to which the cracker is fastened (New Zealand).

fallage. a. The act of cutting trees. b. Timber that has been felled.

fall armyworm. *Laphygma frugiperda* (J. E. Smith), family Noctuidae, the larva of a moth which travels in great numbers, consuming alfalfa, grass, corn, and other plants. It appears in northern U.S. in autumn. Also called *budworm.*

fall aster. Any species of the genus Aster, family Compositae, that blooms in early fall. Also called *flowering aster.*

fall barley. Barley sown in the autumn. Also called *winter barley.*
fall-bearing. Designating a plant or species which produces fruit in the autumn. *Cf. everbearing.*
fall-budded. Designating a plant which has been dormant-budded *q.v.*
fall calf. A calf which is born in autumn.
fall cankerworm. *Alsophila pometaria* (Harr.), family Geometridae, a moth that reaches the adult stage in late autumn. The moth feeds on the foliage of fruit and shade trees. *Cf. spring cankerworm.*
fall crop. A harvest in the autumn, especially of products which usually mature earlier, as a *fall crop* of raspberries.
fall daffodil. *Sternbergia lutea,* family Amaryllidaceae, a bulbous herb which is planted in the flower border for its autumnal, yellow flowers. Native to Eurasia. Also called *winter daffodil; lily-of-the-field.*
fall dandelion. Fall hawkbit.
fall disk. To cultivate land with a disk harrow in the autumn.
fall down. See *lodge.*
fallen. Designating a plant which has dropped to the earth as a result of storm, disease, insect infestation, etc. See *lodge.*
fallen hide. a. The hide of an animal which died of disease, accident, or natural causes. b. Any hide which is delayed in removal.
fallen wool. 1. Wool which is taken from sheep that have died. Also called *dead wool.* 2. Wool collected from fences or the ground which has been rubbed off the sheep.
faller. One who cuts down trees.
fall farrow. A pig born in the autumn.
fall hawkbit. *Leontodon autumnalis,* family Compositae, a weedy, hardy perennial herb, which has late-flowering, yellow heads; occasionally raised in flower gardens. Native to Europe; naturalized in eastern North America. Also called *fall dandelion.*
falling ax(e). A double-bitted ax designed for felling trees, having a long handle and narrow bits.
falling disease. A livestock disease in Australia believed to result from a deficiency of iron, copper, or cobalt. It is characterized by uncoordinated gait, sway back, abnormal wool production, and degenerative changes in the nervous system. Also called *sudden death.*
falling sickness. Epilepsy.
falling wedge. In lumbering, a wedge which is driven into the cut in a tree behind the saw to direct the fall of the tree.
fall in the market. A sharp decline in prices.
fall lamb. A lamb which is born in the autumn.
fall meadowrue. Tall meadowrue.
fallow. Cultivated land which may be kept free of vegetation by plowing, disking, etc., in order to destroy weeds or to conserve a supply of moisture for a succeeding crop.
fall planting. Crops which are planted in the autumn in preparation for a crop the following summer: *e.g.,* wheat and rye.
fall plowed. Designating land which is plowed in the autumn.
fall plowing. The autumnal tilling of land which is not to be planted until the following spring. The reasons for this practice are: (a) the operator may have more time available in the fall; (b) the moisture conditions may be more favorable; (c) a better tilth is obtained: *e.g.,* in soil that has a clay base.
fall rose. China aster.
fall rye. Rye planted in the autumn. Also called *winter rye.*
falls. The outer whorl of the perianth of the flower of an iris. The segments are often drooping.
fall-shorn wool. Fall wool.
fall short. Not to attain an expected or desirable yield.
fall sow. A sow born in the fall.
fall-sown. Designating seed sown in the autumn.
falltime. Autumn.
fall webworm. *Hyphantria cunea* (Drury), family Arctiidae, an insect that infests a wide variety of plants, feeding in May and June, but more destructively from July through September. The insects build unsightly, weblike nests and defoliate a whole branch.
fall wheat. See *winter wheat.*
fall wool. That wool which is sheared in the autumn. (The term is used in those sections where sheep are sheared twice a year.) Also called *six-months wool; fall-shorn wool.*
fally. Of, or pertaining to, autumn.
false acacia. Black locust.
false alder. See *winterberry: 1.*
false aloe. *Manfreda virginica,* family Amaryllidaceae, a bulbous herb grown for its greenish-yellow, mace-scented flowers. Native to North America.
false alumroot. Alaska fringecup.
false annual ring. In forestry, one of the growth rings in a multiple annual ring.
false arborvitae. Hiba false arborvitae.
false aster. White boltonia.
false banana. See *pawpaw: 2.*
false blossom. A virus disease of cranberries which is characterized by the flowers being enlarged or erect, instead of hanging down. The petals are shortened, broadened, and are reddish or greenish. Infected plants produce no fruit.
false box(wood). Flowering dogwood.
false budworm. See *corn earworm.*
false cabbage aphid. See *turnip aphid.*
false camomile. Boltonia.
false caper. Caper euphorbia.
false chinch bug. *Nysius ericae* (Schill.), family Lygaeidae, an insect pest which feeds on the blossoms of avocados, sugar beets, crucifers, and corn. Badly infested plants may die.
false cohosh. Blue cohosh.
false colic. Of a horse, any abdominal pain or distress not of the stomach or bowels.
false coltsfoot. Canada wildginger.
false cowslip. Shootingstar.
false crowfoot. See *geranium; Canada wildginger.*

false dittany. Gasplant dittany.
false dogwood. Striped maple.
false elm. See *hackberry: 1.*
falseflax. Any weed of the genus Camelina, family Cruciferae. The two species, littlepod *falseflax* (*C. microcarpa*) and bigseed *falseflax* (*C. sativa*), are common weeds in North America. Also called *gold-of-pleasure.*
false fringe tree. See *smoketree: 2.*
false grape. Virginia creeper.
false guinea grass. Johnsongrass.
false heartwood. That wood which is abnormally colored as a result of the action of fungi or bacteria or by the deposition of oxidation products.
false heat. The displaying of estrus by any female animal when she is pregnant or is out of season. It can occur in a healthy animal or may be caused by diseased ovaries.
false heather. Peru falseheath.
false heliotrope. See *hyacinth: 2.*
falsehellebore. 1. Any plant of the genus Veratrum, family Liliaceae. Species are hardy, perennial, poisonous herbs which are sometimes used in the wild garden. Derivatives of the roots have been used as insecticides for over a century. See *American falsehellebore.* 2. Pheasant's-eye adonis.
false honeysuckle. Pinxterbloom azalea.
false hoof. Dewclaw.
false ice plant. Mountain stonecrop.
false indigo. See *amorpha; indigobush amorpha; wildindigo; blue wildindigo; yellow wildindigo.*
false jasmine. Carolinajessamine.
false Jerusalemcherry. *Solanum capsicastrum,* family Solanaceae, an unarmed shrub grown as a greenhouse pot plant for its grayish foliage and persistent, orange-red or scarlet fruits. Native to Brazil.
false jessamine. Carolinajessamine.
false kernel. Bunt of wheat *q.v.*
false larch. Lovely goldenlarch.
false lily-of-the-valley. Canada beadruby.
false loose smut. See *nigra loose smut of barley.*
false loosestrife. Water seedbox.
false lote tree. Dateplum persimmon.
false mercury. Good King Henry goosefoot.
false mildew. See *downy mildew.*
false milk fever. Acetonemia.
false miterwort. Allegheny foamflower.
false molt. The shedding of feathers by a bird due to unnatural causes. See *molt.*
false nettle. See *ramie.*
false nutmeg. Torreya.
false oat grass. Tall oatgrass.
false-packed cotton. Cotton in a bale: (a) containing entirely foreign substances; (b) containing damaged cotton in the interior with or without any indication of such damage upon the exterior; (c) composed of good cotton upon the exterior and decidedly inferior cotton in the interior in such a manner as not to be detected by customary examination; or (d) containing pickings or linters worked into the bale.
false Paraguay tree. Witherod viburnum.
false pepper. California peppertree.
false plantain. Carib heliconia.
false quarter. A foot trouble in horses in which the coronary band is broken and the hoof fills in irregularly.
false rabies. See *Aujeszky's disease.*
false ragweed. Rag sumpweed.
false rampion. Creeping bellflower.
false redtop. Fowl bluegrass.
false ring. In forestry, an incomplete annual ring.
false sanicle. Miterwort.
false senna. Bladdersenna.
false smut. A fungous disease of range grasses caused by *Helminthosporium ravenelii,* family Dematiaceae, characterized by a velvety, brownish-olive colored mantle that grows over the seed head and, in time, becomes black and crusted. Also called *head mold; green smut.*
false smut of rice. A minor, fungous disease of rice caused by *Ustilaginoidea virens,* family Dematiaceae, characterized by the sclerotia developing and replacing the flowers; similar to ergot. See *ergot: 1.*
false Solomon's-seal. Solomonplume.
false spider mite. Privet mite.
false stable fly. *Muscina stabulans* (Fall.), family Muscidae, a fly which resembles the stable fly and the house fly but is of less importance as a pest.
false starwort. White boltonia.
false sting of pome fruits. A disease of apples in Nova Scotia, perhaps caused by a virus; characterized by small, depressed areas on young fruit which result in marked distortion of the mature fruit.
false sunflower. See *sunflower heliopsis; sneezeweed: 2; oxeye daisy.*
false sweet flag. Yellowflag iris.
false sycamore. Chinaberry.
false syringa. Mockorange.
falsetamarix. Any shrub of the genus Myricaria, family Tamaricaceae. Species are occasionally grown for ornament. See *German falsetamarix.*
false tea plant. Matrimonyvine.
false unicorn root. Fairy wand.
false varnish tree. Tree of heaven ailanthus.
false wall cress. See *aubrietia: 2.*
false walnut. Wingnut.
false wheat (grass). See *quack grass.*
false wild oats. Fatuoids.
false wormwood. Ragweed.
famble crop. In a ruminant, the first stomach.
family. 1. A unit of human organization which varies widely according to the cultural method employed in counting kinship. Generally, it is a local residence unit, such as a household which includes male(s) and female(s) of a parent generation and the children for which they are socially responsible. Under other definitions it may include nonlocally resident persons related or assumed to be related by blood through several generations. See *nuclear family.* 2. Any group of related plants or animals which make up a category for classification. It is usually more inclusive than a genus but smaller than an order.

3. See *Soil Family.* 4. An isolated group of organisms belonging to a single species; often descendents of a single plant.

family broken. Describing a horse which is gentle and safe for family use.

family characteristic. Particular parts of the conformation and/or the temperament which exists among certain families of dairy cattle, etc. Some families are consistently good or bad in straightness of legs, fore udders, depth of body, shape of head, etc. In temperament, there are families which are consistently nervous while others are normally docile. Those identifying marks or characters tend to display relationship.

family cow. Any cow maintained by a family and whose milk is largely consumed by one family. The keeping of a single cow for milk is now obsolescent in the U.S. because of costs, city ordinances, and disease.

family earnings. Compensation for a farm operator's labor and management and the labor of unpaid members of the operator's family. It includes the value of family living from the farm. The sum is computed by deducting an interest charge for the use of farm capital investment from net family farm income. *Cf. family labor earnings.*

family farm. A farm of such a size and character that it may be operated by the members of the farm family living on it with little if any hired labor and which under good management can return the family a satisfactory living for their effort and investment. (This definition may be considered as establishing the minimum size for a *family farm.* A *family farm* may also be described as a 2-man farm operated by a father and his son or by the farmer with the aid of one full-time hired man.) *Cf. factory farm.*

family farm operating agreement. The operation of a farm by the active participation of father and son(s) in respect to contributions of each one of land, buildings, machinery, livestock, labor, and management, and in sharing expenses and income, making settlements, handling finances, etc.

family farm-transfer arrangement. An agreement to transfer a farm to the heir(s) of a farm operator in a manner and at a time most appropriate to the families concerned.

family labor. Farm labor which is available within the farm family excluding the farm operator. Its value is determined by the amount of additional labor the operator would have to employ to carry on an equivalent amount of business if family labor was unavailable.

family labor earnings. The income from the labor and management of a farm operator and his family, as distinguished from total net farm income. *Cf. family earnings.*

family living. The farm value of home-grown food, fuel, and other products which are used by the farm family, plus the value of the use of the farm dwelling.

family twins. Junior twins *q.v.*

famine. A widespread, acute scarcity of food.

fan. 1. A piece of equipment which will deliver a large volume of air or other gases at low pressures up to 1 lb. per sq. in. for cooling, heating, etc. A blower or an exhauster is designed to deliver small volumes with pressures up to 35 lbs. per sq. in. A compressor will deliver small volumes at even higher heads of pressure. In some cases reciprocating compressors may go up as high as 5,000 lbs. per sq. in. by either rotary or reciprocating methods. There is considerable overlapping among the pressure of fans, blowers, and compressors. Fans may be classified into two general classes; those that operate by centrifugal action (radial), and those that operate by axial flow. The centrifugal fan type may be constructed with blades which are straight, forward-curved, or backward-curved. The latter group of fans may consist of propeller fans or axial flow fans with an impeller. Fan and blower are frequently used interchangeably. 2. Any winnowing machine; specifically, an old term for a kind of basket which, when filled with wheat, was tossed in the air to separate the chaff from the wheat.

fancier. A breeder who shows particular interest in, and the development of, a particular breed or type of animals or plants: *e.g.*, a rose *fancier,* dog *fancier,* Hereford *fancier.*

fan columbine. *Aquilegia flabellata,* family Ranunculaceae, a perennial herb grown for its lilac flowers. Native to Japan.

fancy. A top quality grade for many vegetables, fruits, flowers, poultry, and livestock.

fancy-bred. Designating an animal of a desirable pedigree.

fancy center. A special design used for the center or central portion of brick ice cream.

fancy points. Indications of purity of breeding: *e.g.*, the set of the horns, the carriage of the ear, the color of the hair on certain parts of the body, and the general style of an animal; such characteristics which are given special attention by some breeders are not necessarily an indication of purity of breed.

fancy price. 1. An unusually high price. 2. Prices which are paid for top grade products.

fan delta. See *alluvial fan.*

fanega. 1. A measure of land, used in Spanish settlements in America, which consisted of 8.81 acres. 2. A dry measure of Spanish origin, once used in southern U.S., equivalent to about 2½ American bushels.

fan feather. One of several feathers which remain on the wings of a fowl that has been mechanically picked; such feathers are removed by hand.

fang. 1. To form a forked root, especially of root crops, such as sugar beets. 2. a. Any tooth that is long and pointed. b. The root process of any tooth.

fanleaf. A minor, virus disease of grapes in California, U.S., characterized by dwarfing of the vine and spring shoots, a deforming of the shoots and leaves, a mottling of the leaves, and the setting of almost seedless fruit.

fanleaf palm. Mexican Washingtonpalm.
fan maidenhair. *Adiantum tenerum,* family Polypodiaceae, a tropical fern which has broad blades. Several horticultural forms are raised in greenhouses. Native to America. Also called *brittle maidenhair.*
fanning mill. In seed cleaning, a device equipped with screens to sift out foreign material and a power fan to separate and blow out chaff and other light material. See *seed cleaner.*
fan palm. Any palm having simple, fan-shaped leaves. Also called *fan tree.*
fantastic. A novelty type of chrysanthemum.
fan training. A method of training grapevines and fruit trees, especially dwarfs, in which there is practically no trunk. Branches from near the ground are trained obliquely upright so that the plant grows in one plane, fanlike in shape.
fan tree. 1. A tree which has been developed by fan training *q.v.* 2. Fan palm *q.v.*
fan weed. See *field pennycress.*
fanwort. Any plant of the aquatic genus Cabomba, family Nymphaeaceae. Species are grown in aquariums. See *Carolina fanwort.*
far. 1. In horsemanship, the right side of a horse. 2. *Triticum zea,* family Gramineae, a species of wheat recognized as one of the spelts *q.v.* It was formerly cultivated on inferior soils in Europe.
farcy. Glanders *q.v.*
farding bag. The first stomach, the rumen, of a cow or other ruminant.
farding deal. The fourth part of an acre of land.
farewell summer. See *bouncing Bet; heath aster.*
farewell-to-spring godetia. *Godetia amoena,* family Onagraceae, an annual herb grown for its rose to white flowers. Native to North America. Also called *farewell to spring.*
farina. 1. A food product which is the middlings of hard wheat, except durum. 2. Potato starch. 3. Originally, a flour or meal from any grain, seed, or starchy root.
farinaceous. Starchy.
farinaceous ipecacuanha. The emetic root of Mexican clover *q.v.*
farinose. Yielding farina *q.v.*
farinous. Designating a plant which is covered with a white, mealy substance.
farkleberry. *Vaccinium arboreum,* family Ericaceae, a blueberry-like shrub, seldom used as an ornamental except in masses, because of its straggling growth. Native to southeastern U.S. Also called *sparkleberry; tree huckleberry.*
farm. 1. a. A single tract of land, or separate parcels of land combined as an operational unit, used for the production of crops and/or animals. It all may be owned or all rented, or partly owned and partly rented, etc. The U.S. Census Bureau in 1954 arbitrarily defined a *farm* as a place three or more acres in size, producing an annual profit of $150 or more, exclusive of home garden products. A place of less than three acres, but producing $150 or more annually, may also be classified as a *farm.* b. A basic unit consisting of an agricultural firm and usually including a household. c. To cultivate a tract of land for agricultural purposes. d. Designating anything produced on a *farm,* such as *farm* sausage. 2. A French cottage cheese. 3. A market class of mules in midwestern U.S., which has great variation in quality and type.
farm accounts. Either of two types of accounts: (a) financial accounts, which are primarily records of inventories and financial transactions plus the production records, usually on a single farm enterprise. (b) cost accounts, which include the following in addition to the financial accounts: feed records; time records of labor, power, and machinery; building records; interest charges; etc., allocated to the different farm enterprises. They may be kept on single enterprises as well as on the entire farm business.
farm adviser. See *county agent.*
farm agent. 1. A person appointed by an absentee landlord to look after his interests and collect the rents, etc., from his tenant farmers (British). 2. See *county agent.*
farm animal. Any of the domestic animals which are commonly associated with farming, as a horse, cow, etc.
farm appraisal. A definite, written, detailed opinion of the value of a farm which is prepared by a qualified appraiser *q.v.*
farm area. See *acres in farm.*
farm bloc. A bipartisan group in the U.S. Congress, organized in 1921 to secure favorable legislation for the farming industry and farmers as a class.
farm budget. a. A plan for the organization and operation of a farm for a specified period of time which includes a detailed statement of the anticipated gross income, expenses, and net income. b. A plan for the future use of land, man labor, power, other resources, and which shows the crops to be grown, the livestock to be kept, and the estimated production, receipts, expenses, and farm income.
farm building. Any service building on a farm: *e.g.,* the barn, poultry house, tool shed, etc.
Farm Bureau. A free, independent, nongovernmental, voluntary organization of farm and ranch families in the U.S. and Puerto Rico, united for the purpose of analyzing their problems and formulating action to achieve educational improvement, economic opportunity, and social advancement; dedicated to the preservation of individual opportunity and freedom. It is local, national, and international in its scope and influence, and is nonpartisan, nonsectarian, and nonsecret in character. *Farm Bureau* policies are determined and implemented in the counties by the members; on the state level by a delegate body comprised of representatives chosen by the county *Farm Bureaus* in proportion to their respective memberships; at the national level, an annual board of delegates representing the state *Farm Bureaus.*
farm business. The series of production and maintenance activities and the financial transactions

which are performed in connection with the operation of a farm.

farm business record. A record of a year's business on a farm which includes an inventory of all farm property at the beginning and at the end of the farm year; a record of the number of livestock raised; quantities of livestock products produced; the acreage and production of each crop; and a record of all farm financial transactions, both sales and purchases, made during the year.

farm butter. Any butter which is manufactured on a farm. It is usually a highly salted butter made from raw, sour cream. See *creamery butter.*

farm capital. The value of the land, with such permanent improvements as are ordinarily transferred with the title to the land, livestock, machinery and equipment, feed crops, and other supplies which are included in a farm operating unit.

farm capital earnings. Compensation for farm capital which is computed by deducting the value of the operator's labor from net farm income. It is usually expressed as a percentage figure; percent earnings of farm capital; or rate of earnings on the investment.

farm chunk. A draft horse standing 15 to 16 hands high and weighing from 1,500 to 1,650 lbs.; used mainly for general farm work. See *southern chunk.*

farm conservation plan. A plan which states a cropping system, livestock system, and land use practices that are designed to reduce the rate of soil erosion and to maintain or improve soil fertility.

Farm Cooperative Service. An agency of the U.S. Department of Agriculture which was authorized by the Cooperative Marketing Act of July 2, 1926 and established December 4, 1953. The Service performs research, educational, and advisory service to assist farmers who belong to marketing, farm supply, and related service cooperatives, and works closely with college extension services and state agencies.

farm cost accounts. Detailed records of all economic facts pertaining to the operation of a farm, including all information in the farm business record (*q.v.*), plus a detailed record of the quantities and values of labor, feed, buildings, machinery, and all other resources used for each enterprise.

Farm Credit Administration. The supervisory body for the farm credit system in the U.S., created by Executive Order, March 27, 1933. It consists of 12 Federal Land Banks (created July 17, 1916 by the Federal Farm Loan Act); and 1,052 national farm loan associations, both of which became completely farmer-owned in 1947; 498 production credit associations, most of which are completely member-owned; 12 Federal Intermediate Credit Banks (created March 4, 1923); and 13 Banks for Cooperatives (created June 16, 1933). Borrowers from the latter two are now in the process of replacing government-owned capital and attaining member ownership.

farm crops. Economic crops, corn, wheat, alfalfa, oats, barley, etc., which are grown on farms on a field scale and differentiated from truck crops, fruit crops, etc.

farm depression. A period in which the returns to the factors of production used by the commercial farmers are not high enough to provide incomes which they consider to be satisfactory in relation to the incomes of other groups in the economy.

farm drainage. The removal, or exclusion, of excess water from farm land by open or closed drains alone or in combination with pumping plants.

farm drain tile. A low-quality grade of drain tile 15 ins. or less in diameter, for burying in a trench 4 ft. or less in depth, formerly considered suitable for farm drainage. (This grade was removed from the American Society for Testing Materials specifications in 1950.)

farm drier. See *drier.*

farm duck. Any duck, regardless of grade.

farm economics. See *agricultural economics.*

farm economy. The practical adjustment, organization, and administration of the affairs and resources of a farm.

farm elevator. See *elevator: 2.*

farm-engro. A term for a farm hand (Romany).

farm enterprise. A unit of farm business, as a crop or a class of livestock.

farmer. One who operates a farm, or is engaged in the business of farming; a person who produces agricultural products on a farm. Since there are many different types of farms, a *farmer* may be designated as a dairy farmer, a cattle farmer, fruit farmer, wheat farmer, sheep farmer, etc. Primarily, a *farmer* is one who devotes the major portion of his time to farming; *i.e.*, a full-time *farmer.* The 1954 U.S. Census of Agriculture defines a part-time *farmer* as a person operating a farm with a value of sales of farm products of $250 to $1,199 and who reported 100 or more days work off the farm in 1954 or whose nonfarm income was greater than the value of farm products sold. See *husbandman; agriculturist; rancher.*

farmer sausage. A sausage prepared from 65 percent beef and 35 percent pork which is chopped, seasoned, stuffed in beef middles, and heavily smoked.

farmer's cooperative elevator. A single elevator or a chain of elevators, operated by a cooperative organization of farmers, which handles grain and farmers' supplies. See *cooperative; elevator: 2.*

Farmers Home Administration. An agency of the U.S. Department of Agriculture which aids farmers financially. The agency also provides emergency loans to farmers in districts suffering from drought, flood, or other disasters. It operates through local offices located in 1500 counties throughout the U.S. Formerly called *Farm Security Administration.*

Farmers Institutes. A lyceum type of organization similar to the early teachers institutes which was organized to bring scientific information for better agricultural practices directly to farmers in U.S. They were first suggested by the Massachusetts State Board of Agriculture in 1852. By 1899 over 2,000 institutes were held annually in 47 states, with an attendance of more than 500,000. They were the forerunners of the Cooperative Agricultural Extension Service which offers a year-round educational service to all farmers.

farmer's lung. A peculiar respiratory illness among farmers; believed to be caused by exposure to moldy forage or by handling dusty materials.

farmers' market. A building, open space, etc., usually located in a town or city, where farmers may bring their produce for direct sale to the consumer.

farmer's plague. See *bishop's goutweed.*

farmer's square. Farmers' market.

farmer's style. Designating a hog carcass which is split on either side of the backbone.

Farmers' Union. An abbreviation for the Farmers' Educational and Cooperative Union of America, an organization founded in Texas in 1902. The *Farmers' Union* is organized in local community unions, county unions, state and territorial unions, and a national union. Among its objectives are those of maintaining and protecting the family-type farm, of expanding local and regional farmers' cooperatives, and of educating farm families in the economic, social, and cultural problems of agriculture. Also called the *National Farmers' Union.*

farm expense. The sum of the annual cash operating expenses, value of unpaid family labor excluding the operator, net decrease in farm inventory including depreciation, and the wages and cost of board furnished to hired laborers.

farm flock. Those animals or fowls, collectively by species, which a farmer raises on his farm, especially a small flock of sheep or chickens.

farm forestry. The practice of forestry on part of the farm land which is carried on with other farm operations.

farm game. That small wildlife which reproduces in agricultural areas and which is hunted for sport, food, or fur: *e.g.*, the rabbit, squirrel, etc.

farm hand. A man who is hired to work on a farm by the hour, day, week, month, or year. Also called *hired hand.*

farm implement. Any mechanical tool used on a farm: *e.g.*, a plow, a tractor, etc.

farm improvements. Those items of improvement which are conveyed with the land as real estate: *e.g.*, buildings, tiling systems, fences, and permanent water systems.

farm income. 1. The net income a farmer receives in operating a farm as a return for the total capital investment and for his own labor and management. 2. The collective income of farmers in a county, state, or nation.

farming land. See *farm land.*

farming operation. 1. The sum total of the activities of a farmer in operating a farm. 2. Any one single activity on the farm: *e.g.*, harvesting, planting, cultivating, fencing, etc.

farm inventory. A list of the amounts and values of all items of farm property as of a given date.

farm irrigation structure. a. Any structure which is necessary for the proper conveyance and application of irrigation water: *e.g.*, a flume, siphon, suspension, culvert, debris basin, detention dam, etc. These structures require special design, detailed staking, and/or on-site assistance. b. Any small farm irrigation structure for which standard designs are available: *e.g.*, a turnout, check, drop, or measuring device.

farm labor. a. The sum total of all the labor which is used in the operation of a farm. b. Those people who work on farms as hired hands.

farm laborer. A person gainfully employed by a farm operator to assist with the farm work who may be a wage worker, a family member, or a person working for perquisites. He may be regular, seasonal, local, migratory, full-time, or part-time. A person who, by experience or usual occupation, may be classed as a *farm laborer* even though unemployed or temporarily employed in another occupation.

farm land. Land which is, or is capable of, being used for raising farm crops without further clearing or draining.

farm lateral and field ditching. The construction, relocation, rehabilitation, or care of the permanent farm ditches or the small, temporary, field ditches which are often obliterated and rebuilt during tillage operations each year.

farm leader. A person who is influential among farmers and agriculturists, but who is not necessarily a farmer.

farm level. A simple type of surveyor's level, usually consisting of a telescope with cross-hairs and a tripod, used in the solution of simple problems of drainage, terracing, irrigation, etc.

farm livestock. Livestock kept on farms for family use or for commercial production.

farm machine. Any machine or implement used on a farm: *e.g.*, a tractor, grain combine, etc.

farm management. 1. The science and art of the organization and operation of farms so as to obtain the maximum amount of continuous net income. It considers the effectiveness of different sizes of operating units and of combinations of productive resources, enterprises, and practices for operating units; programs of adjustment for agricultural areas; and the impact of public policies and programs on economic activities and income on farms. 2. The practical adjustment, organization, and administration of affairs and resources of the farm.

farm management service. An individual or an organization which provides landlords such services as procuring tenants or hired workers and managers for farms, keeping farm records, advising the tenants, workers, or managers, handling the financial accounts on the farm, and serving as the land owner's agent in the operation of his farm. Such services are particularly prevalent in the Corn Belt of U.S. Also called *Commercial Farm Management Services.*

farm-management specialist. An expert in farm management *q.v.*

farm manager. A trained individual who directs the operation and management of a farm enterprise.

farm manure. The excrement from all or certain of the farm animals, including litter or bedding materials, such as straw. Also called *farmyard manure.*

farm map. An outline map of a farm showing roads, building sites, lanes, and field arrangement, used in farm planning and as a record of the tile lines, lime application, soil tests, etc.

farm mechanics. The unspecialized, mechanical activities which are performed on the farm and in the home, concerning farm power and machinery, farm building and conveniences, rural electrification, preparing farm products for storage or use, and soil and water management.

farm mixer. A mechanical device consisting of a drum and paddles attached to a power unit, used to mix feeds, etc.

farm mortgage middleman. A person (or agency) who brings together an investor and a borrower, usually to negotiate a farm mortgage loan between life insurance companies, distant banks, and private investors.

farm mule. See *farm: 3.*

farm operation. The consideration and adoption of cropping practices, methods of handling livestock, time and method of marketing, and other factors necessary to conducting a farm business.

farm operator. Any person who manages and directs the operation of a farm; one who applies capital and labor to land to carry on one or more agricultural enterprises. He may be the owner-operator, a tenant, or a hired manager.

farm organization. The selection of land, labor, crops, livestock, machinery, and equipment, and the proportions in which these elements are combined in a farm unit.

farm personal property. The personal, movable property on a farm which is involved in the farm business: *e.g.*, machinery, equipment, livestock, feed, growing crops, crops held for sale, and miscellaneous supplies; also a brooder house or any small building merely resting on the ground which is movable.

farm pond. A small reservoir of water on a farm usually formed by constructing a dam across a watercourse or by excavation.

farm population. All those persons living on farms, regardless of occupation.

farm practice. The techniques of farming which include methods, correlation, and timing of the various farming operations.

farm price. The price a farmer receives for his products at the farm.

farm problem. 1. Any problem of farmers collectively on the state or national level: *e.g.*, farm surplus. 2. Any particular problem on a farm: drainage, fertilizer use, marketing, etc.

farm product. Any product produced on a farm.

farm program. 1. A plan related to agriculture which is based upon, or resulting from, the action of the state or national government. 2. A planned program for the operation or management of an individual farm.

farm prosperity. A period in which the returns to the farmers are large enough to give them incomes which compare satisfactorily to the incomes of other groups in the economy.

farm real estate. The land in a farm including such improvements as the buildings, fences, windmills, tile drains, and the permanent features of a water system.

farm refrigeration. The preserving of meats, fruits, vegetables, and processed foods on the farm for the farm family's use by refrigeration and freezing.

Farm Security Administration. See *Farmers Home Administration.*

farm sheep. Sheep which are bred and raised on a farm, as contrasted to sheep raised on the range.

farm shop. 1. The building on a farm in which a farmer keeps tools, makes repairs, and maintains farm equipment. 2. A training program in agricultural high schools which trains students in the maintenance of equipment and in the production of home-made equipment.

farmstead. That part of the farm which is occupied by the main set of buildings, adjacent yards or corrals, kitchen garden, and family orchard.

farm team. Draft team.

farm tenancy. See *tenancy; tenant farmer.*

farm-to-market road. A hard-surfaced road which serves a farming area and leads to a town or city at which farmers can sell their produce.

farm town. A town or small city whose principal source of income is derived from agriculture rather than from industry or manufacturing.

farm tractor. Tractor *q.v.*

farm trailer. A large, wagon-like box mounted on rubber-tired wheels which is used to haul produce, animals, etc.

farm truck. 1. That part of a wagon consisting of the front and rear wheels which are the same size, the tongue, and coupling pole. 2. A truck powered by a gasoline engine used principally for hauling agricultural products and supplies.

farm wagon. A four-wheeled, horse-drawn vehicle used to transport produce, fertilizer, seeds, etc.

farm woodland. The wooded portion of a farm or ranch, or the wooded land operated in connection with a farm or ranch.

farm work. Any work which is performed at or on a farm.

farmyard. A small, enclosed area which is adjacent to a barn and in which animals are kept when not at pasture, or at work, or inside the barn. Also called *barnyard.*

farmyard manure. Farm manure *q.v.*

farolito. See *balloonvine heartseed.*

Farquhar. *Buddleia farquhari,* family Loganiaceae, a hybrid between *B. officinalis* and *B. asiatica,* a shrub grown for its rusty yellow foliage and pale pink flower clusters. Also called *butterflybush.*

farrier. 1. A person who shoes horses. 2. An obsolete term for a horse and cattle veterinarian.

farrier's knife. A curved, stout-bladed knife used by blacksmiths for trimming horses' hooves before shoeing.

farrow. a. To give birth to a litter of pigs. b. A litter of pigs.

farrowing. Denoting a sow which is about to give birth to a litter of pigs.

farrowing house. Any of several types of structures especially designed for a sow and her litter of pigs.

farrowing pen. A small pen or enclosure, usually in a farm building, especially designed for a sow that is ready to farrow or has given birth to a litter of pigs. This pen usually includes protective fenders (guard rails) and a brooder for the piglets.

farthing-dale. A rood *q.v.* Also called *farding-dale; farthingdeal.*

Far West. That section of the U.S. lying west of the Great Plains and including California, Oregon, Washington, Arizona, Nevada, Montana, Utah, Idaho, Wyoming, Colorado, New Mexico, and parts of Texas.

fasciate(d). 1. Fascicled. See *fascicle.* 2. See *fasciation: 2.*

fasciation. 1. A bacterial disease causing abnormal growths on asparagus, carnation, chrysanthemum, sweet pea, and other ornamental plants, caused by a species of the genus Phytomonas, especially *P. fasciens.* The stems of infected plants are thickened and shortened at or below the soil line or become flattened and the leaves distorted, turning brown or black along the midribs. 2. A flattening and enlargement of a branch as if several stems were fused, often accompanied by curving. It is believed to be caused by injury to the cells of the bud or by multiple terminal buds arranged in a single plane.

fascicle. A close or compact cluster of leaves, blossoms, or fruits.

fasciolasis. See *liver rot.*

fasciolinid. Pertaining to flukes belonging to the subfamily Fasciolinae, genus Fasciola which are commonly known as liver flukes *q.v.*

fasels. See *gram chick-pea; kidney bean.*

fast. Designating a heart beat in which the pulse is much more rapid than normal. See *thin: 3; thready; arythmic pulse.*

fast curdling. Ripening milk too rapidly as a result of setting at too high a temperature or by the use of too much rennet or culture.

fast-feathering. The maturing of first feathers by certain breeds and strains of chickens at a relatively young age. It is an inherent characteristic which is especially desirable in chickens sold as market broilers.

fastigiate tree. Any tree whose limbs are in an upright position: *e.g.,* the Lombardy poplar.

fast land. An uncommon term for coastal upland, or land not subject to flooding.

fast milking. The practice of udder stimulation followed by rapid systematic milking to increase total milk flow.

fast reining. Designating a horse which responds readily and quickly to directions given by the reins.

fast-stepping trot. One form of the trot (*q.v.*), a horse's gait, characterized by long rapid steps. See *high-stepping trot.*

fast walk. One form of the walk (*q.v.*), a horse's gait, which is somewhat more rapid than the walk but slower than the running walk.

fat. 1. a. The oily or greasy-substance bearing tissues of an animal. b. Designating any animal or fowl which abounds in *fat.* 2. The oily substance of milk; the chief constituent of butter. See *butterfat.* 3. The oily or greasy substances found in certain plants: *e.g.,* peanut oil, cottonseed oil. 4. Any food product: *e.g.,* lard or vegetable shortening, which is derived from animal or vegetable *fats.* 5. Those substances which can be extracted from dry feedstuff with ether. See *ether extract.* 6. Fattened cattle ready to market. 7. Of, or pertaining to, a prosperous year, as a *fat* year.

fat back. A retail cut of pork which is the upper half of a side after the loin, shoulder, ham, and belly have been removed.

fat cake. White ironbark eucalyptus.

fat calf. In marketing, a calf 8 to 9 months old weighing from 600 to 700 lbs. See *baby beef.*

fat cattle. Purebred grades and crossbred cattle that have been fed sufficiently to make them fat.

fat content. The amount of butterfat in milk, usually stated in percentages.

fat globules. The naturally occurring spheres of butterfat in milk.

fat hen. See *goosefoot; ground ivy; buckwheat; shepherdspurse.*

Father Hugo rose. *Rosa hugonis,* family Rosaceae, an armored, hardy shrub, grown for its yellow flowers. It is one of the best single, yellow roses. Native to China. Also called *Hugo rose; golden rose of China.*

fatigue. 1. Loss of productive capacity of soil. 2. A weakness of wood or metal which results from a repeated reversal of the load.

fatling. A young farm animal which is fattened for slaughter.

fatliquoring. In tanning, the rubbing of damp leather with fat or oil to prevent it from drying out and becoming stiff and hard.

fatmilking. The early practice of lubricating the teats with fat or milk prior to milking. Also called *wet milking.*

fatness. Having much fat, as an animal.

fat pine. 1. A term used in many parts of southern U.S., to denote branches and knots of pine that are rich in resinous material, and not to denote any particular species of pine. 2. See *longleaf pine.*

fat plus. A method of buying or selling market milk in which a given price for skim milk is constant but in which the seller receives an additional amount for the butterfat content.

fat-plus-two method. A method for arriving at a price for milk sold to cheese manufacturers in which two is automatically added to the butterfat percentage.

fat record. A record which is kept of the butterfat production of a dairy cow.

fat-rumped sheep. A class of sheep found throughout central Asia which is characterized by two great cushions of fat on the rump. See *long-tailed sheep; fat-tailed sheep; short-tailed sheep.*

fat sheep. A show-ring classification of mutton-type sheep which is limited to wether lambs. Also called *market class.* See *wether lamb: 1.*

fat show lamb. See *fat sheep.*

fatsia. See *American devilsclub; Japanese aralia.*

fat stock show. A show in which finished (fattened) market animals are exhibited.

fattable. Of, or pertaining to, an animal's capability of being fattened.

fat-tailed sheep. A class of sheep, typified by the Karakul breed, which is characterized by a long, flat, broad, tapering tail. See *long-tailed sheep; short-tailed sheep; fat-rumped sheep.*

fatted. Designating an animal or fowl which has been fattened.

fatten. 1. To add or put on fat, as an animal. 2. To cause an animal or fowl to put on fat.

fattened. Designating an animal or fowl which has put on a desirable amount of fat.

fattening. a. The feeding of animals or fowls so that they put on fat. b. Designating a feed which when fed to livestock causes them to add fat, as *fattening* grain.

fat test. A test sometimes applied to determine the fat content of a product. See *Babcock test.*

fat type hog. A hog that yields a high percentage of fat (lard), not necessarily synonymous with the lard type hog *q.v.* See *meat type hog.*

fatuoids. Mutants which occur in cultures of cultivated oats and possess the characters of *Avena fatua,* wild oat. Also called *false wild oat.*

fat-weight-gain. Fat which is put on a hog in preparation for marketing, as contrasted to its natural growth-weight-gain *q.v.*

fatwood. See *longleaf pine; lightwood: 1.*

fat yearling. See *baby beef.*

fault. To reduce an animal's score or rating in a show because of a malformation, etc.

fauna. The aggregate of wild animals which inhabit a given area or which exist within a stated period of time.

faunal. Of, or pertaining to, animal life.

Faverolle. A breed of dual-purpose chickens of French origin. The Salmon Faverolle is recognized as a standard variety and is included in the French Class. The skin color is white and the egg shell color light brown.

favor. To protect; to use carefully; as a horse *favors* a lame leg.

favus. 1. A fungous disease of poultry caused by *Achorion gallinae,* characterized by yellowish-white, scaly lesions on the unfeathered portion of the head. If it spreads to the neck, the feathers become brittle and break off. 2. A fungous disease of young cats, and sometimes of dogs, caused by *Achorion schoenleini,* characterized by circular, yellowish, or grayish patches which appear on the paws near the claws, the head, and the face. Thick, crusty layers develop. Both of these diseases are transmissible to man.

fawn. 1. A coat-color for an animal which is a soft, grayish-tan. 2. A young deer.

fawnlily. Any plant of the genus Erythronium, family Liliaceae. Species are spring-blooming, bulbous herbs grown in wild gardens and rock gardens. Also called *adder's-tongue; dogtooth violet.* See *California fawnlily.*

Fayal lily. Arabian star-of-Bethlehem.

fazenda. A coffee plantation (Brazil).

FCM. Fat-corrected milk which is based upon 4 percent butterfat.

F.D.A. An abbreviation for the Food and Drug Administration, a federal agency, U.S., which enforces the provisions of the Federal Drug and Cosmetic Act.

feaberry. An English colloquial term for the gooseberry.

Feasti begonia. A horticultural, fibrous-rooted variety of *Begonia feastii,* family Begoniaceae, a polybred house and conservatory plant grown for its light-pink flowers. Also called *beefsteak begonia.*

feather. 1. The epidermal structure partly embedded in follicles of the skin of birds. Feathers vary greatly in color, size, and shape, and generally cover the body of a bird. See *down: 2; pinfeather; primaries.* 2. (plural) Plumage. 3. The long hair of certain horses which grows below the knees and hocks.

featherball mammillaria. *Mammillaria plumosa,* family Cactaceae, a very small, clustering, globe-shaped cactus, which is grown in cactus gardens and as a house plant. Native to North America.

feather cockscomb. *Celosia argentea,* family Amaranthaceae, a tropical herb grown for its colorful, plumy, feathery flowers of a variety of colors. Native to Asia.

feather columbine. See *columbine meadowrue.*

feather eating. The pulling out of one another's feathers. It is a vice among poultry and caged birds resulting from irritation by lice and quill mites, from lack of exercise, or from faulty nutrition. If blood is drawn, cannibalism results. Also called *feather pulling.*

feathered. Designating a horse which has feathers. See *feather: 3.*

feathered columbine. See *columbine meadowrue; early meadowrue.*

feather-edge. 1. One edge of a board which is thinner than the other edge. 2. A tool which is sharpened to a thin but uneven edge.

feathered gillyflower. Grass pink.

feathered hyacinth. Tassel grapehyacinth.

feather fern. Japanese astilbe.

featherfew. Feverfew chrysanthemum.

feather fingergrass. Showy chloris.

featherfoil. 1. Any water plant of the small genus Hottonia, family Primulaceae. Also called *feather leaf.* Two species, the American featherfoil (*H. inflata*) and the European featherfoil (*H. palus-*

tris), are sometimes grown in small aquaria. 2. See *feverfew chrysanthemum.*

feather follicle. The depression in the skin of a bird from which the feather grows.

feather geranium. Jerusalemoak goosefoot.

feather grass. Certain species of the genus Chloris, family Gramineae, which have conspicuously feathery awns. Australian feathergrass (*C. elegantissima*) and European feathergrass (*C. pennata*) are grown for their feathery appearance to be used cut and dried, as ornaments.

feather hyacinth. Tassel grapehyacinth.

feathering. 1. A defect of coffee cream which is characterized by a lack of homogeneity, causing it to rise to the surface of coffee in flocculent masses and form a light, serrated scum. 2. A rough-edged hole which has been bored in wood in which wood fibers remain to project around the perimeter. It usually results from using a dull bit. It is especially bothersome in maple trees used for collection of sap. 3. The scuffing of the tender skin on an early potato in harvesting.

feather in the eye. A scar on a horse's eyeball which does not necessarily impair its vision.

feather leaf. See *featherfoil.*

featherleaf cedar. See *eastern arborvitae.*

feather-legged. Designating the breeds of chickens which have feathers on the shanks and toes.

feather louse. See *slender pigeon louse.*

feather marking. Fastening bright-colored feathers to the feather shafts of birds as a means of identification.

feather meal. A feed made from dehydrated, ground feathers.

feather mite. See *tropical fowl mite; northern fowl mite.*

feather palm. Any palm tree which has pinnate leaves.

feather picking. The plucking or removing of feathers from a fowl carcass in preparing it for human consumption. See *feather eating; dressing: 4.*

feather plucking. Feather picking.

feather pulling. Feather eating.

feather Solomonplume. *Smilacina racemosa,* family Liliaceae, a shade-loving, perennial herb, grown in borders and wild gardens for its white flowers and red fruits. Native to North America. Also called *false spikenard; false Solomon's-seal; treacleberry; wild spikenard.*

feather tip. A thin-ended piece of lumber: *e.g.,* that used in some shingles.

feather tree. See *smoketree: 2.*

feathery. Designating a novelty type of chrysanthemum in which the petals resemble feathers.

feathery wattle. Any of several species of Acacia *q.v.* See *Sydney acacia.*

febrifuge plant. See *feverfew chrysanthemum.*

febrile. Pertaining to fever or a rise in body temperature: *e.g., febrile* period (period of fever), *febrile* disease (a malady accompanied by fever).

February daphne. *Daphne mezereum,* family Thymelaeaceae, a small, deciduous shrub widely grown for its showy, lilac or rosy-purple flowers. It is a poisonous plant rarely browsed. Native to Eurasia. Also called *mezereon.*

fecal. Pertaining to excrement, manure, waste material passed from the bowels.

fecal contamination. Pollution with feces, manure, or excrement: *e.g., fecal contamination* of pastures, food, etc.

feces. Excrement discharged from the bowels. Also called *faeces.*

fecund. Fruitful; fertile; prolific.

fecundate. To pollinate; to fertilize.

fecundation. Pollination or fertilization.

fecundity. The ability to reproduce regularly and easily.

feddan. A measure of land somewhat larger than an acre (Egypt).

Federal Crop Insurance Act. A part of the Agricultural Adjustment Act of 1933 which was designed to provide crop insurance. The first insurance corporation under the act was empowered to provide insurance for loss of wheat crops.

Federal Extension Service. A U.S. agency created February 26, 1923, to assume leadership of all general educational programs in cooperation with the Land-Grant colleges of the states and territories and the county governments.

Federal Food Stamp Plan. A plan which was established in 1939 by the U.S. Department of Agriculture to dispose of surplus agricultural commodities and to aid low income families.

federal inspection. The inspection by U.S. Department of Agriculture appointed inspectors of slaughter animals and carcasses, fruits, vegetables, and processed products which are intended for human consumption to determine if the animals or products meet established specifications. See *ante-mortem inspection; post-mortem.*

Federal Land Bank. Any of several banks, which is a part of the Farm Credit Administration (*q.v.*) of the U.S. The *Federal Land Banks* through local national farm loan associations make long term (5 to 40 years), amortized, first-mortgage, farm real-estate loans. The Farm Credit Administration divides the U.S. into 12 districts and each district has a *Federal Land Bank.* In turn, each district has a number of national farm loan associations. The system is entirely member-owned and does not use government funds in its operations. Money for the loans is obtained from the sale of Federal Land Bank bonds which are secured by the real estate covered by the first mortgages. Members of the Board of Directors of the Farm Credit Administration and of the different districts and associations are usually elected by the member-borrowers.

Federal Surplus Relief Corporation. A U.S. corporation established in 1933 for the diversion of surplus farm products, acquired in price support operations, into domestic relief channels. In 1935 FSRC became the Federal Surplus Commodities Corporation. See *Section 32 Program.*

fed lamb. A market lamb which has been fed some grain, in contrast to one which has subsisted entirely on milk.

fee. 1. A charge made for livestock grazing on the basis of a certain rate per head for a certain period of time. 2. A charge made for any services as stud *fee,* marketing charge, loan *fee,* etc.

feed. 1. Harvested forage, such as hay, fodder, grain, or other processed food for livestock. See *forage.* 2. The quantity of *feed* in one portion. 3. To furnish with essential nutrients. 4. To deliver, carry, transport, as the canvas *feeds* grain to the thresher on a combine.

feed alley. The open, lengthwise space in a barn alongside of which are mangers, etc., in which livestock feed is placed for animal consumption.

feed analysis. The chemical or material analysis of a commercially prepared feed, printed on a tag and fastened to the bag in which the feed is to be sold.

feed apron. The shelf or apron on a hammer mill on which the material to be ground is placed and from which it moves into the grinders.

feed bag. Any bag in which feed is sacked. See *nosebag.*

feed bin. A bin in which feed is stored.

feed box. 1. A wooden or metal box or pan in which feed is placed for animal consumption. 2. A receptacle in which material is placed and from which it is delivered to a blower, grinder, etc.

feed bunk. A roughage feeding station.

feed cart. A hand-drawn cart used to transport feed within the barn or within a relatively small area.

feed collector. A generally inverted, conical, metal receptacle into which material is blown after it passes through a hammer mill and from which it passes into the bagger spout.

feed cooker. A stove used in cooking mashes, etc.

feed crop. Any crop grown as a feed for livestock, as hay, corn, oats, etc.

feed cutter. Any of a variety of agricultural implements used to cut, chop, or otherwise prepare provender for livestock: *e.g.,* compressing chaff cutters, root slivers and pulpers, and shredders for various roots and cereals.

feeder. 1. A young animal which does not have a high finish but shows evidence of ability to add weight economically. 2. Any device which carries material to or into a machine, as a *feeder* in a threshing machine or in a cotton gin. 3. A person who fattens livestock for slaughter. 4. A thing which feeds, as a plant. 5. Any of several types of appliances used for feeding sugar sirup to bees. 6. A hopper in which feed is placed for consumption.

feeder buyer. 1. One who buys cattle, sheep, or horses, fattens them and offers them for sale. 2. A buyer of feeder livestock.

feeder cattle. In marketing, cattle which are feeders. See *feeder: 1.*

feeder ewe. In marketing, a ewe which is a feeder. See *feeder: 1.*

feeder lamb. In marketing or stock judging, a lamb which is a feeder. See *feeder: 1.*

feeder pig. In marketing, a pig weighing less than 120 lbs. which is a feeder. See *feeder: 1.*

feeder root. Feeding root.

feeder sheep. In marketing, a sheep which is a feeder. See *feeder: 1.*

feeder wether. In marketing, a wether lamb that is a feeder. See *feeder: 1; wether lamb: 1.*

feed flavor. A flavor defect of milk or milk products which suggests the taste of cattle feed.

feed gate. A door or trap in the casing surrounding the conveyor in a grain elevator used for regulating the passage of grain to or from the various bins.

feed grain. Any cereal grain used as a livestock feed.

feedgrinder. See *hammer mill.*

feed hopper. See *feeder: 6.*

feeding. 1. Providing animals with the desirable quality and quantity of feeds; making feed available to animals. 2. The ingestion of feed by an animal or plant.

feeding alley. See *feed alley.*

feeding area. An area of a barn, shed, or open lot where cows are fed roughages, water, and sometimes concentrates. This area may or may not include feed storages.

feeding barley. That barley to be fed to animals. See *malting barley.*

feeding battery. A battery (*q.v.*) that is constructed for housing poultry to be fattened.

feeding beet molasses. A livestock feedstuff which is a by-product of the manufacture of sugar from sugar beets. It shall contain 48 percent or more of total sugars expressed as invert sugar. Its solution in an equal weight of water shall test not less than 39.75° Brix.

feeding cattle. See *feeder cattle.*

feeding corn sugar molasses. A livestock feedstuff which is a by-product of the manufacture of corn sugar from corn. It shall contain 48 percent or more of reducing sugars expressed as dextrose and shall contain 60 percent or more of total carbohydrates. Its solution in an equal weight of water shall test not less than 39.00° Brix.

feeding fence. A fence so constructed that animals can reach through and eat feed placed along the fence.

feeding floor. The ground surface of a feeding lot.

feeding gun. See *feeding needle.*

feeding needle. A long nozzle attached to a tank by a hose used to place fertilizer in the ground around trees and shrubs.

feeding oat meal. A livestock feedstuff which is the product obtained in the manufacture of rolled oat groats or rolled oats. It consists of broken rolled oat groats, oat-groat chips, and floury portions of the oat groats, with only such quantity of finely ground oat hulls as is unavoidable in the usual process of commercial milling. It must not contain more than 4 percent of crude fiber.

feeding oil. Any vegetable oil used as a feed supplement or additive, as cottonseed oil.

feeding pen. An enclosure in which animals or fowls are fed.

feeding practices. The method of supplying feed, the nature of the feed, the proper amounts of feed, and the timing of feeding.

feeding quality. Characteristics which indicate a good meat-producing animal: *e.g.*, broad, deep chest; emple middle; general ruggedness, etc.

feeding rack. A wooden, slatted rack in which hay is placed for animal consumption. The animal pulls the hay through the slats.

feeding root. A young root, or rootlet, bearing the root hairs which absorb water and mineral nutrients from the soil. Also called *feeder root.*

feeding standard. Established standards which state the amounts of nutrients that should be provided in rations for farm animals of various ages and classes in an attempt more nearly to attain the optimum economy of growth, gain, or production.

feeding station. A department of a poultry packing plant where feeding batteries (*q.v.*) are placed. Poultry are placed in these batteries and fed before killing to improve quality and flavor.

feeding stuff. See *feedstuff.*

feeding tankage. Digester tankage.

feeding tankage with bone. See *meat and bone meal digester tankage.*

feeding time. The regular time of the day at which animals and fowls are fed.

feed lot. The enclosed area in which animals are fed for fattening and finishing. Also called *feed yard.*

feed man. A person who manufactures, distributes, or sells feeds.

feed mill. See *hammer mill.*

feed mixer. A device for mixing various feeds consisting of an inverted cone of sheet metal within which are paddles or augers.

feed oats. 1. Any oats used as feed for livestock. 2. Any grain which consists of either (a) 30 percent or more but less than 80 percent of cultivated and wild oats combined; or (b) 80 percent or more cultivated oats and not more than 10 percent of wild oats. It may not contain more than 25 percent of other grains and not more than 10 percent of foreign materials including not more than 5 percent of fine seeds.

feed off. To harvest a crop by pasturing livestock on it. See *hog off; hog down.*

feed out. To prepare animals for market by fattening.

feed plate. Corn-planter plate.

feed production. The production of feedstuffs.

feed rack. See *feeding rack.*

feed rake. An implement resembling a garden rake that is used to stir feed in troughs.

feed room. A room in a barn used for the storage of livestock feeds.

feed saver. Any patent medicine with a secret formula supposed to cause an animal to gain weight with less feed than normal.

feed scoop. A shovel with a short handle used to lift or move feeds.

feed stable. A livery stable.

feedstuff. One or a mixture of the substances which form the nutrients; namely, proteins, carbohydrates, fats, vitamins, minerals, and water.

feed supplement. See *supplement.*

feed table. A feed apron.

feed tag. See *feed analysis.*

feed trough. A wooden, metal, or concrete basin in which feed is placed for animal consumption.

feed truck. Feed cart *q.v.*

feed unit. One pound of corn, or its equivalent in feed value in other feeds which is fed to cattle under normal farm conditions.

feed way. Feed alley.

feed wheat. Wheat used for livestock feed.

feed wheel. A fluted wheel which stirs the seed, fertilizer or other material in the drill box and carries it to the feed cups.

feed yard. Feed lot.

feedy (flavor). Feed flavor.

feeling their oats. Over-eager persons or animals.

feer. To mark off land with furrows for subsequent plowing (British).

feering. Land which is furrowed (British).

fee simple. Unrestricted ownership of real property.

feet. See *foot.*

Feijoa. 1. A genus of shrubs or small trees, family Myrtaceae. Native to South America. 2. *Feijoa sellowiana,* family Myrtaceae, a shrub raised in the tropical or subtropical regions for its white and purple flowers and edible fruit. Also called *pineapple guava.*

feldspars. A group of silicate minerals present in soils, igneous rocks, granites, etc. The *feldspars* are the primary source of potassium, sodium and kaolnitic clay of soils. Its two principal groups are orthoclase, potash feldspars; and plagioclase, soda-lime feldspars.

Felicia. A genus of herbs, family Compositae. Native to South Africa. See *blue felicia; kingfisher felicia.*

feline distemper. Feline enteritis.

feline enteritis. A highly contagious disease of cats caused by a filterable virus. It causes a high fever, loss of appetite, rapid loss of weight, depression, diarrhea, vomiting, soreness of the abdomen, and sometimes a nasal and eye discharge. Also called *infectious gastroenteritis; feline distemper; croupous enteritis; epizootic enteritis; malignant panleucopenia; infectious feline agranulocytosis; feline typhus; feline influenza.* See *enteritis.*

feline influenza. Feline enteritis.

feline typhus. Feline enteritis.

fell. 1. The elastic tissue just under the hide of an animal which is attached to its flesh; facia. 2. To cut down a tree.

fellenwort. Bitter nightshade.

feller. A person who fells trees. See *fell: 2.*

felling. A tract cleared, or to be cleared, of trees.

felling wedge. Falling wedge.

fellmonger. A dealer in sheepskins and pulled wool.

felloe. See *felly.*

Fell Pony. A breed of small, black, brown, or bay,

English horses, 13.2 to 14 hands high. They are good for carrying weight.

felly. The outer part or the rim of a wheel to which spokes are attached.

felon grass. See *Christmas rose; herbrobert geranium.*

felon herb. Mugwort wormwood.

felon quarter. Mastitis.

felonweed. Ragwort groundsel.

felt. 1. See *pelt.* 2. A non-serious, fungous disease of citrus caused by *Septobasidium pseudopedicillatum* and S. spp., family Septobasidiaceae, which produces a soft, feltlike, leathery covering of the twigs. 3. To cause to mat and cling together, as in wool.

feltleaf ceanothus. *Ceanothus arboreus,* family Rhamnaceae, an evergreen, treelike shrub grown in warm regions for its fragrant, light blue flowers. Native to North America. Also called *tree myrtle; island myrtle.*

fen. Wet, marshy land around fresh water lakes or in estuaries. It is rich in humus or muck.

fence. 1. A hedge or barrier of wood, metal, stone, or plants erected to enclose an area to prevent trespassing or the straying of animals. 2. To enclose an area with a *fence.* See *drift fence; snow fence.*

fence breaker. An animal, as a cow, horse, or hog, which habitually breaks through fences.

fence corner. The point at which there is an abrupt change in direction of a fence.

fence cutter. A stockman of the Old West, U.S., who cut farmer's fences to prevent the establishment of enclosed farms on the western range.

fence insulator. An insulator on a fence post or stake to which the live wire of an electric fence is attached.

fence line. 1. The live wire of an electric fence. 2. The dividing line between fields or farms on which a fence is usually built.

fence line feed bunk. A multipurpose structure designed for feeding roughage along a fence in the open. It is primarily a feed bunk with openings through which the animals may feed. The feed may be distributed by dump trucks or feed carriers operated outside of the enclosed area.

fence off. To enclose an area with a fence.

fence post. Any wooden, metal, or concrete post to which the fence, wire or other barrier is attached.

fence post rot. Any decay of buried wood.

fencer. 1. A person who builds or repairs fences (New Zealand). 2. A horse trained to jump obstructions, as a fence.

fence row. The uncultivated strip of land lying on both sides of a field fence.

fence stretcher. A device consisting of a long clamp, which holds the fence wire and a block and tackle. It is used to draw the fence wire taut before it is fastened to the posts.

fence tester. A device used to test the amount of amperage in electric fences.

fence viewer. In some states in the U.S., an individual or body of individuals, having legal status, whose duty is to determine the condition, value, ownership, and maintenance sharing of division fences in cases of dispute.

fencing. 1. Material used in the construction of fences. 2. The use of fences.

fender. 1. A broad piece of leather, attached to the stirrup strap of a saddle, used to protect the rider's legs from the sweat, etc., of the horse. Also called *sudadero; rosaderos.* 2. A device on a plow used to clear away weeds, vines, etc. Also called *trash spring.* 3. A rail set in a farrowing pen to keep the sow from rolling over on the pigs and killing them. 4. See *shield: 1.*

Fendler ceanothus. *Ceanothus fendleri,* family Rhamnaceae, a hardy, half-evergreen, woody plant. It is an important range browse plant for goats and is useful for covering dry, sandy banks. Native to North America.

Fendler Woods rose. *Rosa woodsii* var. *fendleri,* family Rosaceae, a thorny shrub browsed by sheep and goats on the western range. Native to North America.

fen lentil. See *duckweed.*

fennel. *Foeniculum vulgare,* family Umbelliferae, a perennial herb raised for its edible, fragrant seeds, young stems, and enlarged, thickened leaf bases. Also called *common fennel.* Native to Europe. The variety Florence, also known as Florence *fennel* and finocchio, is a low-growing, condensed plant with large edible leaf bases. The variety Italian, also known as Italian *fennel* and carosella, has stems which are edible when young.

fennelflower. Certain species of the genus Nigella, family Ranunculaceae. Also called *devil-in-a-bush.* See *love-in-a-mist; garden fennelflower.*

fennel giant. Giant fennel.

fennel gianthyssop. *Agastache foeniculum,* family Labiatae, an aromatic herb which is a very good bee plant. Also called *anise hyssop.* Native to North America.

fennel seed. See *fennel.*

fenugreek. Fenugreek trigonella.

fenugreek trigonella. *Trigonella foenum-graecum,* family Leguminosae, an annual herb raised for forage and for its seeds used in medicine, in making curry, and sometimes added as a flavoring to mixed feeds for horses and dogs. Also called *fenugreek.*

feracious. Fruitful; bearing abundantly.

feral. Wild, undomesticated.

ferbam. Ferric dimethyl dithiocarbamate, a fungicide.

fericulture. The propagation of game and fur bearing animals as a separate commercial enterprise or as an incidental or supplemental one to farming.

fermate. See *ferbam.*

ferment. 1. To undergo catalytic decomposition of complex compounds into simpler ones, as in the souring of milk, the formation of vinegar from cider, etc. 2. Any agent which can cause fermentation. See *enzyme.*

fermentable. Capable of being fermented.

fermentation. 1. The processing of food by means of yeasts, molds, or bacteria; the catalytic decomposition of complex compounds into simple ones, as in the souring of milk, the production of vinegar from cider, etc. The most important products of fermentation are alcohol, acetic acid, lactic acid, butyl acid, citric acid, and acetone. 2. An enzymatic curing of stored tobacco leaves which improves the quality of the product. 3. Digestion in the rumen by microorganisms.

fermented. Designating a flavor defect of cheese which is sweet and suggestive of fermentation or overripe fruit. Also called *fruity.* See *ferment: 1.*

fermented milk. Any milk to which a culture has been added to cause fermentation. See *cultured buttermilk; cultured cream; culture: 4.*

fern. 1. Any flowerless plant of the Order Filicales. Species resemble seed plants by having roots, stems, and leaves, but reproduce by spores instead of seeds. They are widely grown by florists and gardeners as ornamentals. 2. Any of various plants with fernlike characteristics.

fern asparagus. *Asparagus plumosus,* family Liliaceae, a fernlike, climbing vine commonly raised in greenhouses to be cut and used as greens in bouquets, etc. The long stems are also cut and used in decorations. Native to South America. Also referred to as *asparagus fern.*

fern brake. 1. Bracken. 2. A dense, wild growth of fern plants. Also called *fern scrub.*

fernery. A collection of growing ferns. See *terrarium.*

fernleaf ligusticum. *Ligusticum filicinum,* family Umbelliferae, a perennial, mountain plant palatable to livestock and wild animals. Its general forage rating is good for sheep and goats and fair for cattle and horses. Native to the northern hemisphere.

fern leaf of tomato. A disease of tomato caused by the cucumber mosaic virus which is characterized by an extreme discoloration of the leaves. See *cucumber mosaic.*

fern-leaved beech. A clon of the European beech (*q.v.*), family Fagaceae, grown as an ornamental for its sharply cut, fernlike leaves.

fernleaf peony. *Paeonia tenuifolia,* family Ranunculaceae, a perennial herb widely grown in flower borders, gardens, etc., for its crimson or purple flowers. Native to Eurasia.

fern nematode. *Aphelenchus olesistus,* a foliar nematode (*q.v.*) that infests many ornamentals.

fern nozzle. A nozzle used to apply sprays. See *disc nozzle.*

fern poisoning. See *eastern bracken.*

fern scale. *Pinnaspis aspidistrae* (Sign.), family Diaspididae, a scale insect (*q.v.*) that infests aspidistra, ferns, and other plants indoors and out-of-doors in mild climates.

fern scrub. See *fern brake: 2.*

fern seed. The asexual spores of ferns.

ferny. Denoting ferns; abounding in ferns.

ferric phosphate. A chemical compound containing 25 percent iron and 13.9 percent phosphorus used in mineral feeds for livestock. Also known as *iron phosphate.*

ferrous carbonate. A chemical compound containing about 50 percent iron used in feeds to prevent or treat iron deficiencies in animals.

ferrous sulfate. $FeSO_4 \cdot 7H_2O$, a compound used as an herbicide and soil acidifier in greenhouses. It is also used as a source of iron in livestock feed. Also known as *copperas.*

ferruginous soil. Soil containing a large amount of iron minerals, especially limonite and hematite, characterized by a vivid red color or various shades of yellow and brown.

fertile. a. Productive; producing plants in abundance, as *fertile* land. b. Capable of growing or of development, as a *fertile* egg. c. Capable of reproducing viable offspring, as a *fertile* seed. d. Able to produce fruit, as a *fertile* flower.

fertile egg. A fertilized, avian egg capable of embryonic development.

fertile queen. A queen bee which has mated and is capable of laying fertile eggs.

fertility. 1. The ability of a plant to mature viable seeds. 2. The ability of an animal or fowl to produce offspring. 3. The quality that enables a soil to provide the proper compounds, in the proper amounts and in the proper balance for the growth of specified plants, when other factors, *e.g.,* light, temperature, and the physical condition of the soil are favorable. See *sterile: 2.*

fertilization. 1. The union of two gametes. 2. The application of fertilizer to soil.

fertilize. 1. To supply the necessary mineral and organic nutrients to soil or water to aid the growth and development of plants. 2. To fecundate the egg of an animal or plant, or to pollenize the pistil of a flower.

fertilized pond. A pond to which fertilizer is regularly applied to promote the growth of algae or other aquatic plants as part of fish management.

fertilizer. 1. Any organic or inorganic material added to soil or water to provide plant nutrients and to increase the growth, yield, quantity, or nutritive value of the plants grown therein. 2. The insect or other organism that brings about the fertilization of a flower, as bees.

fertilizer analysis. The statement, usually on the label of the fertilizer container, which states the percentage of nitrogen, phosphoric acid, and potash contained, and more recently includes the minor elements which have also been found to be essential: *e.g.,* boron, copper, zinc, manganese, etc. See *fertilizer formula.*

fertilizer application. The use of organic or inorganic materials containing plant nutrients on soil. The compound may be spread on the surface or drilled into the ground with the seed.

fertilizer attachment. Any device that may be attached to agricultural machinery, especially to a drill or planter, to facilitate the distribution of fertilizer in or upon the soil. See *fertilizer distributor; fertilizer drill; fertilizer spreader.*

fertilizer distributor. Any of several types of devices and implements used to spread fertilizer.

fertilizer drill. An implement which places fertilizer in or on soil. It may be attached to a seed drill.

fertilizer feed. A device on a fertilizer distributor which regulates the distribution of fertilizer.

fertilizer formula. The quantity and grade of crude stock materials used in making a fertilizer mixture.

fertilizer grade. The guaranteed minimum plant food expressed in percentage terms of nitrogen, available phosphoric acid, and water-soluble potash in a fertilizer. See *fertilizer analysis; fertilizer formula.*

fertilizer spreader. A mechanical device used for broadcasting manure or coarse fertilizer in a field.

fertilizer tube. The part of a fertilizer spreader into which measured amounts of fertilizer are introduced to the soil for application.

fescue. Any species of grass of the genus Festuca, family Gramineae, grown for pasture, hay, and turf. Native to Eurasia and America. See *meadow fescue; red fescue; sheep fescue.*

fescue lameness. A disease condition of cattle attributed to vitamin A starvation and excessive eating of fully matured, tall fescue grass. It is characterized by lameness and apparent circulatory disturbances of ears, tail, and hind feet. Also called *fescue foot.*

Feta. A white, pickled, salted, Greek cheese made from ewes', or sometimes goats' milk.

fetal feces. Manure of the fetus.

fetal resorption. A pathological condition in animals in which the fertilized egg or the early fetus dies in the womb and is usually resorbed, leaving only a more or less unidentifiable, mummified or decomposed mass.

feterita. Durra.

feterita chop. A chop made from durra.

fetid. Designating an offensive odor.

fetid horehound. Black ballota.

fetid marigold. Prairie dogweed.

fetlock. The joint of a horse's leg just above the hoof; anatomically, the metatarso-phalangeal articulation. Also the tuft of hair growing there.

fetlock clipper. A hand or machine operated clipper, or shears, used for trimming the fetlock or other hair on a horse.

fetter. A shackle for the feet of horses.

fetticus. European cornsalad.

fetus. The unborn young of animals (usually vertebrates) which give birth to living offspring. The unborn offspring of human beings is usually called a *fetus* after the third month of gestation; before the third month, it is usually referred to as an *embryo.* Also called *foetus.*

feudal system. The system of land tenure and social arrangements prevalent in Europe during the Middle Ages. A man who held land from another was a dependent and subordinate of the latter and was subject to his order.

fever. A temperature higher than normal in animals which may be caused by disease organisms, poisonous plants, etc. In some cotton plant diseases, as Texas root rot, affected plants have a so-called *fever* or a higher temperature than normal, just prior to showing severe symptoms, which can be distinctly perceived by feeling the leaves in early morning.

feverfew chrysanthemum. *Chrysanthemum parthenium,* family Compositae, a perennial herb grown for its small, white flowers. It was formerly supposed to have febrifugal quality. Native to Europe. Also called *feverfew; featherfoil; featherfew.* See *chrysanthemum.*

feverish. 1. Referring to milk or stored cream that has an odor due to poor ventilation of the cow stable. Also called *smothered; barny; cowy.* 2. Pertaining to animals which have a temperature higher than normal because of disease, etc.

fever tick. Cattle tick *q.v.*

fever twig. American bittersweet *q.v.*

feverwort. Boneset.

FFA. Future Farmers of America.

fiador. The cord which fastens the bozal to the headstall in a halter. Also called *Theodore.*

fiber. 1. a. In plants, the elongated, elastic cells, often tapered at each end, which serve various functions, such as support of the stem, as in the case of cotton fibers on the seed; protection, as in fiber-tracheids, etc. b. Similar cells or structures in animals. 2. a. Commercially, true fiber, a single cell, or aggregation of such cells, as in flax, or even fibrovascular bundles used in the production of textiles. b. Similar synthetic structures, as in rayon, nylon, etc. 3. In the analysis of animal foods, the insoluble residue left after the sugars, acids, proteins, and ash are accounted for. It consists mainly of cellulose and lignified cell walls.

fiber-board. Building material made of compressed fibers, as of wood. It is light in weight and comparatively strong.

fiber crop. Crop grown for its fiber, as cotton and flax.

fiber flax. Flax grown for its fiber, as contrasted to flax grown for its oil-bearing seed. See *seedflax.*

fiberize. To separate wood or woody material into fibers.

fiber-saturation point. The stage in the drying or wetting of wood at which the cell walls are saturated and the cell cavities free from water. It usually occurs at a moisture content of 25 to 30 percent.

fiber sorter. An apparatus in which cotton seed with combed lint is sorted by length of lint and by the weight ratio of lint to the seed.

fiber-tracheid. 1. In plants, a fiberlike tracheid, commonly thick-walled, with a small lumen, pointed ends, and small bordered pits having lenticular or slitlike apertures. 2. The late-wood tracheids of certain gymnosperms as well as the fibrous tracheids of dicotyledons. A *fiber-tracheid* combines the functions of fibers for support and tracheids for transport of water. See *wood fiber; fiber.*

fibrillating current. Excessive current in an electric fence which on contact either stops an animal's breathing by paralyzing the chest muscles or stops the flow of blood.

fibrin. A whitish, insoluble protein formed from fibrinogen by the action of a ferment, thrombin.

fibrinogen. A soluble protein present in the blood and body fluids of animals.
fibrinous. Pertaining to, or composed of, fibrin.
fibrinous exudate. An exudate composed of fibrin of the blood.
fibrosis. The pathological formation of tissues composed of fibers on the body organs.
fibrotic. Relating to the deposition of fibrous connective tissue in diseased tissues or organs.
fibrous peat. See *peat.*
fibrous root system. A root system which is composed of profusely branched roots with many lateral rootlets.
fibrovascular bundles. Strands of water-conducting cells (tracheary tissue) and food-conducting cells (sieve cells) with some associated protective and supporting wood or bast fibers, which are united into branching (elongated bundles extending from the roots) through the stem and into the leaves and other parts of the plant. They serve to conduct food from the leaves or from storage organs to all living parts of the plant and to carry from the roots the water absorbed from the soil. In the leaves they are called *veins.* In some plants where they are easily separable from the leaf or stem tissues (by the process called retting) the whole *fibrovascular bundles* are popularly called *fibers,* as in sisal hemp.
fiddle-butt. A large, spruce, butt log which is suitable for the manufacture of musical instruments.
fiddleleaf dock. *Rumex pulcher,* family Polygonaceae, a perennial herb, which is an uncommon weed in waste places. Native to Eurasia.
fiddleleaf fig. *Ficus pandurata,* family Moraceae, a tropical tree, of unknown origin, grown in tubs or pots for its fiddle-shaped leaves.
fiddleneck. 1. Various species of weeds of the genus Amsinckia, family Boraginaceae, some of which are poisonous when their seeds are eaten by stock. They are mostly native to the states on the Pacific Coast of U.S. Also called *tarweed; fingerweed; fire-weed.* 2. The young curled fronds of the bracken and cinnamon ferns.
fiddlewood. The hard wood of a tree belonging to genus Citharexylum, family Verbenaceae, which is native to tropical America. The name is also given to other trees of the same family: *Petita dominigensis, Vetex umbrosa,* etc., and to *Tecoma pentaphylla,* family Bignoniaceae of British Honduras which is also called *bois côtelet.*
field. An area of agricultural land which is devoted to the production of farm crops, *e.g.,* cereals, vegetables, etc., or an area in which crop-growing has been postponed or abandoned.
field baler. A machine that bales hay or straw directly from the windrow.
field balm. Ground ivy.
field bean. 1. A term applied in the U.S., to varieties of the kidney bean, *Phaseolus vulgaris,* family Leguminosae, *e.g.,* navy, kidney, marrow, pea beans, etc. They are grown for human consumption of their dry seeds. 2. A term in southern and central Europe applied to varieties of the broadbean, *Vicia faba,* family Leguminosae, *e.g.,* the Windsor or flat bean. See *bean: 1; kidney bean.*
field bee. A worker bee which is about 16 or more days old and which works in the field to collect nectar, pollen, water, and propolis.
field beet. A large-rooted variety of the common species of beet grown mainly for stock feed. Also called *mangelwurzel; mangel.*
field bindweed. European glorybind.
field book. A notebook in which are kept the records of surveys, field experiments, field plots, expenses, income, etc.
field budding. The practice of budding as well as grafting, upon stocks already rooted in the field. See *bud: 3.*
field buttercup. Tall buttercup.
field capacity. The amount of water held in the soil after the excess or gravitational water has drained away. Also called *capillary capacity.*
field chopper. Field forage harvester *q.v.*
field chopping. The harvesting of forage, corn, other green crops, and straw by mechanical harvesters. Forage may be chopped directly as standing material or picked up from windrows and chopped and blown into wagons. See *green chopping.*
field conditions. The conditions for plant growth in the open field as contrasted with those which exist in the greenhouse or laboratory.
field corn. Any variety of corn that is grown extensively in large fields primarily for livestock feed, as contrasted with the horticultural varieties, such as sweet corn or popcorn. Most *field corn* is of the dent variety.
field cricket. *Acheta assimilis* F., family Gryllidae, an insect which may infest seedling plants and continue infestation until the plants are mature.
field crop. Any herbaceous plant grown in cultivated fields under a more or less extensive system of culture.
field cultivator. An implement which has several sweeps so mounted as to provide subsurface tillage. The standards which carry the sweeps are staggered to permit trash to flow around and between them. Also called *chisel cultivator.*
field cured. Designating hay or other produce that has undergone its curing process in the field before being placed in storage.
field curing. See *field cured.*
field daisy. Oxeye daisy.
field ditch. One of the small ditches from an irrigation canal which is used for carrying the irrigation water to a field.
field dodder. *Cuscuta pentagona,* family Cuscutaceae, a native species of dodder (*q.v.*) that is found on wild and cultivated plants, especially alfalfa, throughout North America, but most commonly east of the Mississippi River. (This common name is also applied to *Cuscuta campestris.*)
field drill. A various-sized drill used for planting seeds, as wheat or cotton seeds.
field ell. A right-angled pipe joint for main or lateral pipe lines to change the direction of the water used in irrigation.

fielder. Field bee.

field fence. A fence which surrounds a farm field.

field forage chopper. Field forage harvester.

field forage harvester. A harvesting machine, usually tractor drawn and powered, which is used for field chopping of corn, legumes, and grasses into suitable lengths for either silo or mow storage. Also called *field forage chopper.*

field forget-me-not. *Myosotis arvensis,* family Boraginaceae, an annual or biennial herb grown for its blue or white flowers. Escaped, it is a weed, especially in damp places. Native to Eurasia.

field garlic. *Allium vineale,* family Liliaceae, a pernicious, perennial, bulbous herb which is a pest of fields and pastures. Native to southern Europe. Also called *wild garlic; wild onion; crown onion; crow garlic.*

field hand. A man employed to work in the fields on a farm. Also called *hired man; farm hand.*

field harvester. See *field forage harvester.*

field hawkweed. *Hieracium pratense,* family Compositae, a hardy, perennial herb which is a serious weed pest of eastern U.S. Native to Europe. Also called *yellow paintbrush; king devil.*

field highbush blackberry. Highbrush blackberry.

field horsetail. *Equisetum arvense,* family Equisetaceae, a perennial plant which possesses poisonous characteristics. Horses feeding on it fresh or in hay will show signs of poisoning: *e.g.,* unthriftiness, loss of weight and muscular control, staggering, falling, inability to rise, lack of ability to eat, extreme nervousness, death. Native to America and Eurasia. Also called *cornfield horsetail; horsetail fern; meadow pine; bottlebrush; snake pipes; mare's-tail.*

field mice. Various kinds of mice that live among grasses and weeds, mostly of the genus Microtus.

field mint. *Mentha arvensis,* family Labiatae, an aromatic, perennial herb, grown for its essential oil present in all the plant parts. Native to Eurasia and North America. Also called *corn mint; wild pennyroyal; water calamint.*

field mower. A machine used for mowing fields, as contrasted to a lawnmower. See *mower: 1.*

field pea. *Pisum sativum* var. *arvense* (*Pisum arvense*), family Leguminosae, a tall, annual herb, grown for fodder, as a vegetable, and for its seeds used dried for split-pea soup. Native to Europe.

field pennycress. *Thlaspi arvense,* family Cruciferae, an annual weed pest whose garlic-flavored seeds damage the value of grain crops. Also called *French weed; stinkweed; bastard cress.*

field pepper-grass. Field pepperweed.

field pepperweed. *Lepidium campestre,* family Cruciferae, an annual or biennial herb which is a weed occurring in waste places, in pastures, and in lawns. When mature, the ripe seeds contain an acrid substance which may be poisonous to stock feeding on it. Native to Eurasia. Also called *field pepper-grass; English pepper-grass.*

field pit. A pit dug in or near fields of potatoes or cabbages in which these products are placed for storage, covered by soil or layers of soil and straw to protect them from the winter cold.

field practice. Any agricultural practice used in the field, in contrast with that used in laboratory or experimental plots.

field primrose. Evening primrose.

field-proved. 1. Designating features or machines that have been proved by years of field tests. 2. Designating any new variety of plants which has been proven satisfactory by a field test. See *field test: 1.*

field rose. 1. Any rose grown in the open air during the producing season, without protection, as in the San Francisco area and western portions of Washington and Oregon, U.S. 2. *Rosa arvensis,* family Rosaceae, a trailing shrub, grown for its white flowers. A variety, Ayrshire field rose, *Rosa arvensis* var. *ayreshirea,* has double, white or pink flowers and is a more vigorous plant. Also called *rosebuck rambler.* Native to Europe.

field run. Designating products (potatoes, onions, etc.) harvested in the field and ungraded as to size or quality.

field salad. European cornsalad.

field seed. The seeds of field crops, as contrasted with vegetable and flower seeds.

field soil. The soil found in the farm fields, as contrasted with that found in special plots or gardens.

field sorrel. Sheep sorrel.

field sow thistle. *Sonchus arvensis,* family Compositae, a perennial herb, a weed pest that spreads extensively by means of its subterranean rootstalks, as well as by its windblown seeds. Native to Europe.

field storage. The storage of produce in trenches and pits in mounds on the surface of the ground, and in above-ground, horizontal silos. See *pit storage; mound storage.*

field strip cropping. A system of strip cropping in which a field is divided into parallel strips crosswise of the slope, not necessarily following contours to reduce soil erosion. See *strip cropping.*

field tee. A pipe junction for connecting three lengths of tubing used in irrigation where a branch of the main line or lateral line is taken off.

field test (trial). 1. In plant breeding or plant introduction, the planting of vegetation, previously tested in control plots, under normal field conditions. 2. Extensive *field tests,* usually in various localities, of new insecticides or fungicides, as a final step before their commercial introduction.

field windbreak. A barrier of trees and shrubs grown to check the force of wind for protection of orchards, feedlots, and fields, and to check drifting snow and retain it to increase soil moisture. See *windbreak.*

field work. The farm work that is mainly associated with farm fields in fitting, plowing, sowing, cultivating, and harvesting during the growing season, as contrasted with the work indoors.

fiery. Designating a reddish tinge on the outside surface, as of the meat carcass.

fiery hunter. *Calosoma calidum,* F., family Carabidae, a black, ground beetle, which feeds voraciously on cutworms, larvae of potato beetles, and other succulent larvae.

fiery thorn. Scarlet firethorn.

fifth wheel. A horizontal wheel-like bearing or pivot which partially supports and is supported by the front axle of a vehicle, enabling it to be revolved at an angle for turning corners. By the arrangement of gearing with worm gears the *fifth wheel* can be made to serve as steering gear.

fifty-eight's (58's). An English grade of wool equivalent to the American half-blood.

fifty's and forty-eight's (50's and 48's). English grades of wool equivalent to the American quarter-blood.

fifty-six's (56's). An English grade of wool equivalent to the American three-eighths blood.

fig. *Ficus carica,* family Moraceae, a tropical tree grown for its edible fruit. It is one of the oldest domesticated plants whose fruit is an important crop in almost all tropical and subtropical regions. Native to Southwestern Asia. See *capri fig.*

figleaf hollyhock. *Althaea ficifolia,* family Malvaceae, a biennial herb grown against walls, fences, etc., as an ornamental with lemon-yellow and orange flowers. Native to Europe. Also called *Antwerp hollyhock.*

figmarigold. See *mesembryanthemum.*

fig moth. Almond moth.

fig rust. A fungous disease caused by *Physopella fici,* belonging to the group of rusts which sometimes causes defoliation of the fig.

fig scale. *Lepidosaphes ficus* (Sign.), family Diaspididae, a scale insect that feeds on the foliage and other tender parts of the fig.

fig smut. A fungous disease caused by *Aspergillus niger,* family Moniliaceae, introduced into the young fig fruit by insects entering the apical opening. It produces a black mass of fungus spores, popularly called *smut.*

figure. The pattern produced on a wood surface by irregular coloration, growth layers, rays, knots, and such deviations from regular grain as bird's-eye, interlocked, and wavy grain.

fig wasp. *Blastophaga psenes* (L.), family Agaontidae, a wasp which develops in the galled, infertile flowers on the capri fig and carries pollen to the Smyrna fig. It is the agent in caprification.

Fiji-arrowroot tacca. *Tacca pinnatifida,* family Taccaceae, a perennial herb grown in the tropics for the starch obtained from its roots. Native to Asia and Africa. Also called *East Indian arrowroot; otaheite arrowroot; pia; Madagascar arrowroot.*

Fiji disease. A disease of sugar cane believed to be caused by a virus probably insect-borne, but which can also spread from plant to plant by the cutting knife. Endemic in southwestern Pacific regions.

filament. 1. The part of the stamen of a flower that is below the anther and supports it. 2. A series of cells, especially of some algae, bacteria, and fungi, attached to each other forming a slender object or a long cylindrical cell.

filaree. See *heronbill; alfileria.*

filarial. Pertaining to or caused by filariae, the roundworm parasites occurring outside the alimentary canal.

filbert. a. Any tree or shrub of the genus Corylus, family Corylaceae. Species are extensively grown in the Pacific Coast states, U.S. for their edible nuts. Also called *hazel; hazelnut.* b. The thick-shelled edible nut of such trees. See *European filbert; Turkish filbert; beaked filbert; giant filbert.*

filbert bacterial blight. See *bacterial blight of filberts.*

filbert bud mite. *Phytoptus avellanae* Nal., family Eriophyidae, a mite which infests filbert buds. Infested buds become swollen but do not develop.

filbert weevil. *Curculio uniformis* (Lec.), family Curculionidae, a beetle that infests the developing nuts of the filbert.

filbertworm. *Melissopus latiferreanus* (Wlshm.), family Olethreutidae, a moth whose larvae infest acorns, beechnuts, filberts, and chestnuts on the western coast of U.S. Also called *Cataline cherry moth.*

file. A steel instrument whose surface is covered with sharp-edged furrows or teeth, used for sharpening, abrading, or smoothing other substances, as metals, wood, tools, etc.

file blade. A tooth float.

filiform leaf. An extreme condition of the mosaic disease of tomato, sweetpotato, and other plants in which the leaves develop in sinuous ribbon or thread shapes. See *mosaic disease.*

fill. 1. The soil, sand, gravel, etc., used to fill in a depression in a field, or to build up a terrace or embankment. 2. The substances used in filling tree cavities: *e.g.,* asphalt, concrete, wooden or rubber blocks, etc. Also called *filler.* 3. The increase in weight and form of livestock that have been watered and fed after arriving at their destination. 4. The shaft of a vehicle. See *thill.* 5. To level a depression in a field or cavities in trees, or to build up an embankment or terrace. 6. To feed and water livestock at the end of the shipment to make up for the loss of weight enroute. 7. To enlarge with the enclosed seeds, as the pods of leguminous plants; or to be plump and shrivelled when approaching maturity, as the seeds of cereals. (Cereal grains before harvest are often referred to as poorly- or well-filled.)

Filled. A cheese made of whole or skimmed milk with extra fat added in the manufacturing process.

filled hocks. A swelling of the joints of a horse which may include bog spavin or thoroughpin.

filler. 1. The part, usually removable, of a container in which the individual products are placed, as the *filler* in which the eggs are placed in a crate. 2. An extra row intercultivated between two regular rows of a crop. 3. An extra, short-lived plant grown between slow-growing, larger plants and removed when the latter approach maturity. 4. Any of the various types of appliances for filling special receptacles, as bottle-*filler,* silo-*filler,* etc. 5. Any material, active or inert, added to a mixed fertilizer to increase bulk.

6. The nonessential matter in a manufactured or mixed feed, such as high fiber materials, oat hulls, screenings, etc. 7. See *fill: 2.*

filler leaf. The portion of tobacco that makes up the central part of a cigar. Also the special type of tobacco leaf used for this purpose.

filler row. See *filler: 2.*

fillet of beef. The tenderloin muscle, the small muscle on the inside of the backbone of a beef carcass. The so-called "eye muscle" of the round is often mistaken for *fillet of beef* or tenderloin by the average customer.

fill-in crop. See *filler: 2.*

filling. The weft; the yarn that runs the width of a woven fabric and is usually made of thicker, softer yarns than the warp.

filling equipment. See *filler: 4.*

filling machine. See *filler: 4.*

fill line. An imaginary line near the top of a milk bottle to which it is filled.

fill of container. The degree of fullness of a can with reference to standards established by the U.S. Department of Agriculture under the Federal Food and Drug Act. The standard *fill of container* is the maximum quantity of optional ingredients that can be sealed in the container and processed without crushing of product.

fill section. That part of the surface of an earth fill which is higher in elevation than the original ground level.

filly. A female colt.

film. The layer of dried milk which sticks to the surface of a dairy utensil after use.

film (drum) process. A method of drying milk by spraying or otherwise placing a thin film of the fluid milk on a revolving steam-heated drum and scraping off the dried layer. See *drum drier.*

film water. Tightly bound water which is absorbed on the surface of particles in unsaturated soil.

filter. 1. A porous substance or device which is used to strain out the particles suspended in a liquid or gas. See *filter disk.* 2. To pass a liquid or gas through a porous substance which strains out the particles suspended in it.

filterable (filtrable). Capable of passing through the pores of a filter, as the Pasteur filter which is made of clay.

filterable virus. A virus that is capable of passing through the pores of a filter which does not allow passage of the ordinary bacteria. See *virus.*

filter crop. Any close-growing crop planted across a slope to retard runoff and cause deposition of the sediment load. See *filter strip; buffer strip.*

filter disk. A disk-shaped pad through which a liquid is passed to filter it. In the dairy industry, this is usually a cotton pad which is destroyed when the process is completed. Also called *filter disc; filter pad.*

filter host. Screening host.

filtering. The process of purifying a liquid by passing it through a filter. Also called *filtration.*

filter pad. A filter disk.

filter press. A filtering device consisting of a series of metal or wooden plates between which pieces of canvas are placed on which a filter medium, such as diatomaceous earth, may be deposited to collect suspended materials in liquid forced through the filter pad.

filter strip. A strip of permanent vegetation of sufficient width and vegetative density above farm ponds, diversion terraces, and other structures which retards the flow of run-off water causing the flowing water to deposit soil, thereby reducing the rate of silting of structure or reservoir below.

filtration. The process in which solid matter is removed from juices or saps using jelly bags, filter paper, filter presses employing cloth and other mediums.

fimetarius. Growing on dung or manure heaps.

fin. A flat plate used as a colter.

final cut. The amount or volume of material removed in harvesting the main crop of trees at the end of the rotation in even-aged management.

final host. The animal in which a parasite develops to maturity.

final stage (decay). A stage of wood rot in which destruction of the wood substance has progressed very far with partial or complete collapse of the wood structure. Also called *final rot; complete stage; late stage.*

financial accounts. Records of inventories and financial transactions, sales and purchases, plus the production records. *Financial accounts* are primarily single-entry accounts which are often referred to as simple farm accounts.

financial maturity. The period in the life cycle of a tree or stand of trees in which there is the greatest excess of gross conversion value over cost of growing and conversion.

financial records. Financial accounts which show the profit or loss from the entire farm business as a unit, but which do not show the profit or loss of the individual farm enterprises.

financial statement. A statement usually required by a lender from an applicant for credit. It is a list of assets and liabilities of an individual or a business at a specific time. The difference between the assets and liabilities is the net worth at that particular time. The difference between the net worth at the beginning of the year and at the end of the year represents the financial gain or loss for the year. See *balance sheet.*

fin colter. An irregular-shaped, sharp piece of steel bolted to the land face or gunnel of a plow share and extended upward to cut loose the furrow slice; it is used principally in sod land. See *colter; hanging colter; rolling colter; skim colter; notched colter; knife colter.*

find. To give birth to young: *e.g.,* a cow *finds* a calf.

fine. 1. The finest grade of wool, of which very little is produced in the U.S. It corresponds to English 80's, 70's and 64's (the latter numbers indicate the number of hanks of 560 yards each that it takes to make one pound, the finer the wool the more hanks needed to make a pound).

2. Denoting the fine-grainedness in meat. 3. The *fine* material from grinding or mixing feeds. 4. To break up the soil by tillage until it is reduced to small granules suitable for planting seeds and of a texture that is favorable for growth of the crop. 5. To clarify fruit juices or other liquids, often by the addition of some substance, which in the course of settling, carries down impurities in the liquid.

fine and fine-medium. A label designating *fine and fine-medium* wool placed in one grade.

fine, clean bone throughout. Designating those bone features associated with horses, and to a lesser extent with dairy cows, which are considered to be a measure of quality.

fine cotton. See *sea-island cotton.*

fine grain. Grain of wood that is characterized by narrow and inconspicuous growth rings. See *grain : 4.*

fine granular. Designating an aggregate that is under 5 mm. in size.

fine gravel. A soil separate which includes particles 2 mm. to ½ in. in diameter. According to earlier definitions, the particle size was 1 to 2 mm.; this size limit is now classed as very coarse sand.

fine ground limestone. Pulverized limestone.

fine, highly pleasing. Designating the highest, official flavor classification for creamery butter: *e.g.*, AA.

fineleaf actinea. *Actinea linearifolia,* family Compositae, an annual herb, sometimes used for fattening cattle in southwestern U.S. Also called *tallowweed.*

fineleaf lotus. Chilean deervetch.

finely ground (slag fertilizer). Basic phosphate slag; the particle size is determined by its ability to pass through standard sieves as follows: 90 percent or more should pass a 50-mesh sieve, 70 percent or more should pass a 100-mesh sieve.

fineness. 1. One of the several properties of cotton which determines the grade in which it is classified. *Fineness* refers to the smallness of the cross diameter of the fibers (or lint). 2. One of several properties of dry milk. Usually the powders prepared by the spray process are extremely fine whereas those prepared by the cylinder or drum process are coarser. 3. One of the several properties of high quality hay which help in the determination of the grade of hay as feed for livestock. 4. The relative smallness of the wool fiber.

fine pruning. The removal of many small branches and spurs of a tree by making many cuts on the plant. *Cf. bulk pruning.*

fine sand. In the mechanical analysis of soils, a soil separate, or mineral particles, ranging in size from .25 to .10 mm. (U.S. Department of Agriculture scheme). See *soil separate.*

fine sandy loam. A class of soil which contains 30 percent or more of fine sand or less than 30 percent of very fine sand (or) between 15 or 30 percent very coarse, coarse, and medium sands. The clay content is 20 percent or less and the silt content less than 50. See *sandy loam.*

fine, smooth withers. A feature associated with cows of pronounced dairy temperament and large productive ability in which the withers are refined.

finetooth holly. *Ilex serrata,* family Aquifoliaceae, a deciduous shrub grown for its foliage and its bright red fruits. Native to Japan.

fine-top salt grass. See *alkali muhly; alkali sacaton.*

fine-wool. a. Designating wool from any of the following breeds: Merino (Spanish), Merino (American) A, B, and C types, Delaine Merino (C types), Rambouillet. b. See *Merino.*

Fine Wool Breed of Sheep. The principal breeds of fine wool sheep in the U.S. are Rambouillet, C-type Merino, and Merino *q.q.v.*

finger and toe disease. Clubroot of cabbage.

finger citron. *Citrus medica* var. *sarcodactylis,* family Rutaceae, an evergreen, tropical tree grown as an ornamental. Native to Asia.

finger flower. See *foxglove: b.*

finger-grass. See *crabgrass.*

finger plate. A revolving plate with fingers at the edge which catch the fertilizer and carry it out to the opening at the back of the box where it is dropped into the spout of the fertilizer spreader. Also called *star feed; wizard feed.*

finger root. See *foxglove: b.*

finger(s). Finger-like parts of machines that serve different functions according to the machine: (a) in a grain combine, special parts that pick up the cut grain from the windrow and start it on its way to the threshing cylinder; (b) in a fertilizer spreader, certain parts that feed the fertilizer in the box or hopper to the outlets; (c) in mechanical pickers or roughing machines, the rubber tubes or strips which pull off most of the feathers from the fowl passing through the machine. See *finger plate.*

fining. 1. The operations of tillage that reduce the size of the soil particles. 2. The substances which when added to a liquid clarify the liquid.

finish. 1. The degree of fatness or the proper distribution of the fat in an animal. Also called *finishing.* 2. The desired degree of smoothness of lumber. 3. Reflecting the care in molding, processing, and dressing cheese.

finished milk. Milk which has been pasteurized and is awaiting bottling.

finisher. 1. In the manufacture of tomato and other fruit and vegetable purées, the appliance which breaks up the various-sized pieces of pulp by the action of rapidly revolving, heavy brushes so that the final product is homogeneous and will pass through a fine sieve. 2. In poultry keeping, a battery of several cages within which broilers may be fed to the desired size and degree of fatness. Also called *finishing battery.* 3. The formulated mixture of feedstuffs which is used in the finishing of animals. See *finish: 1.*

finishing. See *finish: 1.*

finishing battery. See *finisher: 2.*

finishing pot. A flower pot in which a plant achieves maturity.

finishing unit. The final, largest brooder for growing chickens.

Finland pink. *Dianthus arenarius,* family Caryophyllaceae, a perennial herb, grown for its white flowers. Native to Europe.

finn hoe. In forestry, a digging tool with a thin, heart-shaped blade with either a pointed or square cutting edge and a long handle that is usually firmly bolted to the tang of the hoe blade.

finochio. See *fennel.*

Fiore Sardo. A hard Italian cheese made of ewes' milk.

fiorin. Redtop *q.v.*

fir. Cone-bearing trees (conifers) of various genera, family Pinaceae: *Abies pectinata,* silver fir of central Europe; *Abies balsamea,* silver fir of Canada; *Pinus sylvestris,* Scotch pine of arctic Asia and Europe; *Picea excelsa,* spruce fir of Norway.

fire. 1. In several different plant diseases caused by bacteria, fungi, or nutritional deficiency, the final result which gives a burning or scorching appearance. See *Botrytis blight of tulip; fire blight; wild-fire: 1.* Potash deficiency or leaf scorch is sometimes referred to as firing. 2. The burning off of vegetation on lands of various types. This is apparently beneficial in some cases but is most often decidedly harmful. 3. To treat a spavin or ringbone on a horse with a strong liniment in an attempt to cure or alleviate lameness.

fire ant. *Solenopsis geminata* (F.), family Formicidae, a species of ants which is harmful to plants and whose bite is very painful to man. See *southern fire ant; little fire ant; imported fire ant.*

fire bean. Scarlet runner bean.

fire behavior. The action of fire. See *fire danger.*

fire blight. A serious, bacterial disease of susceptible varieties of apple, pear, quince, and many other rosaceous hosts caused by *Erwina amylovorus* (syn. *Bacillus amylovorus*), family Bacteriaceae, characterized by the rapid invasion and death of blossoms or succulent leaves and shoots. The bacteria enter and invade the cambium killing twigs, branches, or entire limbs or trees by invasion or girdling until bacterial multiplication is checked by maturity of the tissues. The dead blossoms and leaves persist giving the tree a fire-scorched appearance. Native to North America and has spread to Japan, Italy, and New Zealand. Also called *apple blight; blossom blight; twig blight; blight canker; body blight.*

fire break. An existing barrier, or one constructed before a fire occurs, from which all or most of the inflammable materials have been removed. It is designed to stop or check creeping or running fire but not to stop crown fires, to serve as a line from which to work, and to facilitate the movement of men and equipment in fire suppression.

fire bug. Harlequin bug *q.v.*

firebush. Orange streptosolen.

fire cherry. Pin cherry *q.v.*

fire concentration. The total number of fires that occur per unit of area over a given period of time; the rate of fire occurrence.

fire control. All activities which are concerned with the protection of wild land and forest growth from fire including presuppression, prevention, detection, and suppression.

fire control strips. 1. The borders of highways on which the weeds have been destroyed by chemical sprays, forming fire breaks. 2. Strips of open areas through a forest which are made to form fire breaks.

fire-cured. Designating tobacco that has been hung in the smoke and heat of fires built on the curing-barn floor; a process that dries out the tobacco and halts further enzymatic decomposition.

fire curing. A method used in curing export types of tobacco. See *fire-cured.*

fire damage. The loss caused by fire which may be direct, through immediate injury to or destruction of property and life; or indirect, *i.e.,* all other losses, not necessarily restricted to the area on which the fire occurs.

fire danger. In forestry, the result of both constant and variable factors which determine whether fires will start, spread, and damage, determining as well, the difficulty of control. Constant factors are those which are relatively unchanging: *e.g.,* normal risk of ignition, topography, all fuels, and exposure to prevailing wind. Variable factors change from day to day, season to season, and year to year: *e.g.,* all weather elements, moisture content of fuel, and variable risks of ignition.

fire danger station. In forestry, a location where the variable factors of fire danger are determined daily.

fire detection. In forestry, the act or system of discovering and locating fires.

fire disease. A disease of tulips caused by species of the genus Botrytis, family Moniliaceae, characterized by small, black spots the size of a pinhead on outer husks of bulbs.

fired leaves. The drying of the lower leaves of plants which may be caused by a moisture or nitrogen deficiency or other unfavorable soil conditions.

fire-dried. Designating the rapid drying of hay by artificial heat. If properly done, the hay retains vitamin A, but frequently the vitamin D content is reduced.

fire effects. In forestry, any consequences, neutral, detrimental, or beneficial, which result from fire.

firefanging. The heating of manure by the action of the contained bacteria and fungi which causes loss in nutrient value.

fire farming. Clearing patches of land by destroying wild cover by fire, cropping for a few years, and then allowing the area to remain fallow for revegetation and restoration. Practiced mainly in humid, tropical regions. Also called *milpa* and *canuca* in Latin America; *langland* in Indo-China; *caingin* in the Philippines.

firefinder. In forestry, an instrument used by lookouts to determine the location of fires sighted.

fire hazard. Any practice which increases the danger of fire: *e.g.,* spraying with sodium chlorate for the destruction of weeds; allowing inflammable materials to accumulate where they may become ignited; the use of lighted tobacco near inflammable material, etc.

fire insurance tree. A seed tree left as a precaution against destruction of reproduction by fire.

fire lily. See *beargrass: 2.*

fire line. In forestry, the narrow portion of a control line from which inflammable materials have been removed by scraping or digging down to mineral soil.

fireman. In forestry, a fire guard whose principal function is suppression of fires, and who usually stays at a fixed position awaiting orders to go to a fire. Also called *smokechaser.*

fir engraver. *Scolytus ventralis* Lec., family Scolytidae, a bark beetle (*q.v.*) that infests fir trees.

fire occurrence. In forestry, the number of fires which occur on a protection unit for a given period.

fire-occurrence map. In forestry, a map that shows by suitable symbols the starting points of all fires of various classes or causes for a given period.

fire-on-the-mountain. Painted euphorbia.

fire pink. Firepink silene.

firepink silene. *Silene virginica,* family Caryophyllaceae, a perennial herb grown in the rock garden for its deep crimson flower clusters. Native to North America.

fire plow. A heavy duty, usually specialized machine, either of the share or disk type, which is designed solely for abusive work in the woods and is used with either horses or tractors to construct fire breaks and fire lines.

fire presuppression. Those forest fire control activities concerned with the organization, training, instruction, and management of the fire-control organization, and with the inspection and maintenance of fire control improvements, equipment and supplies to insure effective fire suppression.

fire prevention. Those fire control activities concerned with the attempt to reduce the number of forest fires through education, hazard reduction, and law enforcement.

fire-progress map. In forestry, a map maintained on a large fire to show the progress of controlling actions or the spread of fire in the fire area.

fire-resistant. Possessing considerable resistance to fire.

fire retardant chemical. Any chemical which injected into or applied on wood will reduce inflammability.

fire season. In forestry, the period or periods of the year during which fires are likely to occur, spread, and do sufficient damage to warrant organized fire control.

firethorn. Any thorny, evergreen or nearly evergreen shrub of the genus Pyracantha, family Rosaceae. Species bear clusters of white flowers and orange or scarlet fruits which attract birds. Native to Eurasia. Also called *Egyptian thorn; pyracantha.* See *scarlet firethorn; Laland firethorn; Rogers firethorn.*

fire trail. In forests, a trail constructed primarily to facilitate travel to and from or along, the edge of a fire.

fire trap. In forests, an accumulation of inflammable material near which it is dangerous to fight fire.

fire weather station. A forest meteorological station specially equipped and operated for measuring weather elements that have an important effect on fire control.

fireweed. *Epilobium angustifolium,* family Onagraceae, a perennial herb, which appears on burned-over land. It is grown in flower gardens for its rose-purple flower cluster and as a bee plant. Native to the North Temperate Zone. Also called *giant willow herb; French willow; blooming sally; rose bay; buckweed.*

firewood. Logs cut short enough and split into convenient sizes for use as fuel in fireplaces or wood-burning stoves.

fireworm. *Rhopobota naevana,* (Hbn.), family Olethreutidae, a moth, the larva of which infests cranberries, eating and killing the leaves. The afflicted plants appear to be scorched by fire.

fire wound. The fresh or healed injury of the cambium of a tree, caused by fire.

firing. 1. A physiological condition of tobacco in which the leaves are scorched due to the plants being grown in soil containing an excess of boron compounds. 2. A change of color of the lower leaves of green corn usually attributed to a lack of plant nutrients or moisture deficiency. 3. A form of therapeutic cautery used with horses to strengthen certain structures of the legs: *e.g.*, the blistering of sprains, ringbones, sidebones, etc., to stop ossification around the joints. It is done either with a hot iron or with chemicals in ointments or salves. Also called *scoring.*

firing of corn leaves. See *firing: 2.*

firkin. 1. A measure of weight, used for butter and cheese, of 56 lbs. avoirdupois. 2. A wooden cask or tub used for holding butter, fish, etc.

firm. 1. An economic unit recognized to be engaged primarily in production. 2. a. To compact the soil, crushing and pulverizing the lumps to facilitate capillary movement. b. Designating well-compacted soil which is not lumpy or powdery. 3. In marketing, designating optimistic conditions. 4. Designating a cheese which feels solid. 5. Designating whites of eggs which are sufficiently viscous to prevent free movement of the yolk. 6. a. Designating meat which is not soft or soggy. b. Designating a fruit or vegetable that is not overripe or shriveled.

firm body. See *firm: 4.*

firm knot. A knot that is solid across its face but which contains potential decay. See *knot: 4.*

firmness. See *firm: 6.*

firm ripe. Designating fruit picked when mature but firm enough to withstand usual or customary handling without bruising.

first. 1. The highest grade of lumber. 2. The second grade, next below extra, for butter. 3. The primary occurrence or the beginning of a series: *e.g., first* milk (colostrum); *first* cutting in haymaking; *first* prices (opening prices); etc.

first bottom. In a river valley, the first alluvial plain level above the channel of the stream which is most subject to inundation during flood stages of the river.

first lees. A sediment which forms after fermentation ceases in which the yeast and suspended particles, such as of skins and pulp, settle rapidly. Also called *crude lees.*

first lock. The portion of a horse's mane immediately behind the foretop and head.

first-year cane. Primocane *q.v.*

fiscal year. Any accounting period which consists of twelve successive calendar months, fifty-two weeks, or thirteen four-week periods; a twelve-month period which ends with the last day of any month other than December.

fish. Pork tenderloin (not common).

Fish and Wildlife Service. A division in the U.S. Department of the Interior which conducts research studies on fish and wildlife. It is also concerned with rodents and predators, farm fishponds and other conservation programs affecting agriculture.

fish back. 1. The transversely ridged risers in the straw racks of a threshing machine which toss the threshed straw to sift the grain mixed with it and direct it to the grain pan. 2. Ridged back of swine which usually lacks spring of rib.

fishbone. A heavy, central rib of wax found in honeycomb built on a too heavy foundation.

fisherman-farmer. A person who combines fishing and farming for a living (New England).

fisheye rot of pome fruits. A fungous disease caused by *Corticum centrifugum,* family Thelephoraceae, characterized by dry, spongy rot. Also called *hypochnus.*

fish eyes. Gas holes in cheese which result from contamination with yeast. They may be round but as the cheese undergoes curing they flatten out.

fisheye spoilage. Gas blisters formed beneath the skin of pickled ripe olives through the action of contaminating bacteria.

fish geranium. Fish pelargonium.

fish grass. Carolina fanwort.

fishguano. A fertilizer prepared from fish scraps cooked with steam and dried and ground after extraction of oil.

fishiness. A flavor defect of dairy products which suggests a fishy taste.

fishing. 1. Joining two parts in the same straight line by butting their ends and securing them together by means of two short side pieces covering the joint. 2. The operation of recoupling a section of pipe dropped into a well with another section controlled at ground surface.

fish ladder. An inclined trough carrying water down a dam at a velocity against which fish can easily swim upstream to reach their spawning grounds.

fish liver and glandular meal. A poultry feed obtained by drying and grinding the offal of fish.

fish management. The art of cultivating fish populations to make them most useful to man.

fish manure. An organic fertilizer made from fish.

fish meal. A commercial feed for poultry and other farm animals which consists of the clean, dried, ground tissues of undecomposed whole fish or fish cuttings, with or without the oil extracted.

fish moth. Silverfish.

fish oil. 1. Oil from domestic fish which is used as an adherent for lead arsenate sprays in the control of codling moth and gypsy moth. 2. Oil, such as obtained from the cod, sardine, halibut, etc., used in preparing feed mixtures containing vitamin A and vitamin D.

fish oil soap. A soap, with a fish oil as its fatty base, used in some spray formulas as an adherer or spreader for insecticides or fungicides. Also called *sticker.*

fish pelargonium. *Pelargonium hortorum,* family Geraniaceae, a hybrid, herbaceous plant commonly known as "geranium" and grown for its red, salmon-pink, or white flowers. The plant has a fishy odor. Also called *fish geranium; budding geranium; geranium.*

fish pomace. A fertilizer, the fish refuse after the oil has been extracted.

fish pond. A small body of water in which the fish population is managed. On a farm, the pond impounds rainfall runoff, reduces erosion, provides a watering place for livestock, and furnishes a place where fish can be grown.

fish pond fertilizer. Inorganic or organic fertilizer applied to a pond to promote the growth of plankton and the aquatic plants on which the fish depend directly or indirectly for food.

fish residue meal. Clean, dried, undecomposed residue from the manufacture of glue from non-oily fish, containing less than 3 percent of salt, which is used for feed for farm animals. The salt content, in no case, should exceed 7 percent.

fish-rice rotation. See *rice-fish rotation.*

fish scrap. See *fish tankage.*

fish screen. A barrier across the inlet or outlet of a pond to prevent the ingress and egress of fish.

fishtail outlet. A nozzle shaped like a fishtail, used in spraying or dusting insecticides and fungicides, which causes the spray or dust to spread out in a band that covers the plants.

fishtail palm. Caryota *q.v.*

fish tankage. The dried, ground product derived from rendered or unrendered fish. Also called *fish scrap; dry ground fish; fish meal.*

fishway. A fish ladder.

fishy flavor. A disagreeable fishy flavor which is a defect of butter or cooked poultry. In butter, the flavor results from oxidation of trimethylamine. In dairy products it is sometimes due to feeding fish meal; in poultry, it is sometimes due to feed containing fish scraps or tankage.

fission. A form of reproduction, common among bacteria and protozoa, in which a unit or organism splits into two or more whole units.

fission fungi. Fungi that reproduce only by fission.

fist. To separate the pelt of the sheep or lamb from the carcass by pushing the hand against the inside of the pelt.

fistula. a. An unnatural body passage linking a hollow space or abscess to the skin surface or a surface membrane or joining two such abscesses. b. Surgically established openings between a hollow organ and the skin or between two hollow organs for experimental purposes. *Fistulas* that communicate with the outside are closed with a mechanical plug.

fistula of the teat. A fistula that results frequently from barbed-wire cuts on the body of the teat so that the milk flows out of the wound.

fistulous withers. A disease of the withers in horses in which abscess-like structures form, usually connected with the outside by tubelike fistulas.

fit. 1. To notch a tree for felling, and to mark it into log lengths after it is felled. 2. To ring, slit, and peel tanbark. 3. To file and set a saw. 4. To condition livestock for use, sale, or exhibit. 5. To prepare land for sowing, *i.e.,* to perform plowing (or disking), harrowing, and rolling. Land so fitted should have no large, hard clods so that the seeds may be placed at a uniform depth.

fits. Epilepsy.

fitting cow ration. A ration fed to cows which are being prepared for exhibition, sale, or calving.

fittings. Coupling joints, valves, tees, etc., for pipes, used for various purposes; *i.e.,* a sprinkler irrigation system, dairy apparatus, high pressure fittings for lubrication of wheels or other moving parts.

fittonia. Any evergreen, herbaceous plant of dwarf habit, genus Fittonia, family Acanthaceae. Species are grown in greenhouses and are hardy out-of-doors in tropical or subtropical regions. *F. verschaffelti* has dark green leaves with red veins, and *F. argyroneura* has lighter green leaves and white veins. Native to Peru.

fitty. Referring to a horse which has fits when overheated.

fitweed corydalis. *Corydalis caseana,* family Fumariaceae, a biennial herb relished by sheep, grown in regions where they are grazed extensively. Sheep and sometimes cattle are fatally poisoned when they eat considerable quantities of this plant.

five acre yield test. In the U.S., any competitive test made under the supervision of corn-growers' associations on the farms of the entrants. It reveals the capabilities of the land used, of the variety planted, and of the production practices followed.

five-gaited saddle horse. A saddle horse which is trained to use the following gaits: walk, trot, rack, pace, and canter.

five-horse hitch. A hitch involving five horses.

five-needled pine. Any of the so-called soft pines, belonging to the subgenus Cembra, genus Pinus, family Pinaceae, in which the soft needles occur in clusters of five and the wood is white and soft with only a little resin: *e.g.,* eastern white pine, Swiss stone pine, whitebark pine, limber pine, western white pine, sugar pine, bristlecone pine.

fivestamen tamarisk. *Tamarix pentandra,* family Tamaricaceae, a shrub or small tree, grown for its feathery foliage, purplish branches, and pink flowers in long clusters. Native to Eurasia.

five-tooth weeder. An assembly of five cultivator teeth that can be attached to a garden cultivator used for weeding row crops.

fixation of nitrogen. Fixation of atmospheric nitrogen in the soil by bacteria both symbiotic and nonsymbiotic. Symbiotic bacteria grow in nodules on the roots of legumes: *e.g.,* alfalfa and clovers; the nitrogen fixed by them may become available to the host plant, may be released directly to the soil, or through decomposition of the legume may become available to other plants.

fixation of phosphorus. The chemical conversion of soluble phosphorus fertilizers in the soil into less soluble forms rendering it less available as a plant nutrient.

fixed ammonium. Ammonium, held in a soil, which is neither water-soluble nor readily exchangeable.

fixed capital. Capital that is invested in land, buildings, fences, tile drains, wells, and other relatively permanent fixtures.

fixed copper. One of several dried copper salts of low solubility used for spraying.

fixed cost. That cost which does not change with volume of production.

fixed farm capital. See *fixed capital.*

fixed nicotine. Any compound or combination of nicotine that is insoluble or nonvolatile and may be retained on the plant foliage for several days or weeks, thereby being an effective stomach poison for chewing insects.

fixed oil. Any oil: *e.g.,* linseed, soybean, castor, neat's-foot, fish oil, etc., the product of plant seeds or animals and is essentially a glyceride.

fixed phosphorus. Phosphorus in soil which is unavailable to plants; soluble phosphorus of fertilizers which, as a result of reactions taking place in the soil, has become relatively insoluble and unavailable as a plant nutrient.

fixed potassium. Potassium, held by a soil, which is neither water-soluble nor readily exchangeable.

fixed price supports. Governmental supports of prices which are paid for agricultural products to bring them up to a definite percentage of parity: *i.e.,* of the average prices for these products current in a specified period: *e.g.,* 1910 to 1914, 1947 to 1949. The products selected for such *fixed price supports* are mostly those few major crops in which the current supply greatly exceeds demand (U.S.).

fix free nitrogen. See *fixation of nitrogen.*

fjeld. An Arctic, stony desert.

flag. 1. A very popular flower garden plant of the genus Iris, family Iridaceae, or many other monocotyledonous plants that have similar ensiform (sword-shaped) leaves. See *iris: 1.* 2. A branch with dead or discolored leaves on an otherwise green tree. 3. In plowing, a slice of turf or sod. 4. To droop, as a plant or plant part.

flagella. Whiplike appendages of certain single-celled aquatic animals and plants, including some bacteria, the rapid movement of which produces motility in water or liquid media.

flagellates. Certain one-celled aquatic animals and plants which move about by means of flagella *q.v.*

flageolet. A green variety of the kidney bean (France).

flagging. 1. Dead or discolored, scattered, single leaves of a seedling, as in clubroot of cabbage. 2. See *flag: 2.*

flaggy. Designating a soil characterized by coarse rock fragments that are flat, thin, and angular, with dimensions (arbitrarily assigned) of 6 to 15 ins.

flag leaf. a. The first leaf immediately below the inflorescence of the cereals. b. In corn, a large flaglike leaf at the top of the plant just prior to tassel emergence.

flag smut. A fungous disease of wheat, rye, and other grasses, caused by smut fungi of the genus Urocystis, family Tilletiaceae, characterized by elongated spore masses in the upper-leaf sheaths and in the stalk of the spike.

flail. a. A wooden-handled implement of many designs at the end of which is fastened a free swinging, metal or wooden piece called a swingle or swipple, used to thresh the grain by hand from harvested plants by beating the seeds free of the straw. b. To thresh the grain from the stem by the use of the *flail.*

flake. A variety of carnation (*q.v.*) which has a single stripe in each petal.

flake albumen. A food product consisting of large pieces of dried eggwhite which have undergone a bacterial fermentation for some hours, cleared with lactic acid, and then dried until a rubbery stage is attained. The dried albumen then is graded according to size of the pieces into flake, granular, and powdered albumen.

flake buttermilk. Cultured buttermilk containing butter granules; produced by churning cream or adding drops of melted butter to the buttermilk.

flaked corn. A food product consisting of thin flakes obtained by running cracked corn, aspirated and properly tempered, over or between smooth, heated rollers. These are then dried and cooled.

flaking mill. A piece of machinery for rolling or pressing grain, such as corn, into thin flakes.

flaky. a. Describing a body defect of ice cream which has a tendency to fall apart when dipped or melted and to show a feathery, light-colored scum on the surface. b. Describing a body defect of cheese, characterized by a lack of meatiness and a tendency to break easily. When worked between thumb and forefinger, there is a mealiness.

flamboyanttree. *Delonix regia,* family Leguminosae, a medium-sized tree grown in very warm areas for its scarlet or scarlet-striped flowers. Native to Madagascar. Also called *royal poinciana; peacock flower.*

flame. To destroy weeds in fields by using flame throwers.

flame azalea. *Rhododendron calendulaceum (Azalea calendulaceae),* family Ericaceae, a deciduous shrub grown for its yellow, orange, or scarlet flowers. It is the source of many horticultural varieties. Native to North America. Also called *yellow azalea.*

flame bottletree. *Brachychiton acerifolium,* family Sterculiaceae, an ornamental tree grown in warm areas for its showy clusters of scarlet or yellowish flowers. Native to Australia. Also called *flametree.*

flame cultivation. See *flame; flame thrower.*

flame cultivator. Flame thrower.

flameleaf sumac. *Rhus copallina,* family Anacardiaceae, a deciduous shrub or small tree, grown for its autumnal, brilliant red foliage. Native to North America.

flame thrower. A device used in the cultivation of row crops, consisting of a tank, pump, and special nozzle or tip, that blows a flaming gas on weeds and destroys them.

flame-vine. *Pyrostegia venusta,* family Bignoniaceae, an evergreen, climbing, woody vine grown in greenhouses and in very warm areas for its foliage and clusters of reddish-orange flowers. (Often incorrectly called *Bignonia venusta.*) Native to Brazil.

flaming poppy. *Meconopsis heterophylla (Stylomecon h.),* family Papaveraceae, an annual herb, grown for its foliage and brick-red flowers. Native to North America. Also called *wind poppy.*

Flanders. A breed of dairy cattle which originated in the Low Countries.

flange. 1. The downward-turned edge of a metal intended to give stiffness, often called *flanged edge.* 2. The outward-extending margin of a pulley wheel which prevents the slipping off of the belt or rope running over it; hence the name *flange* pulley.

flanged edges. See *flange: 1.*

flange pulley. See *flange: 2.*

flank. 1. The lower portion of the side immediately behind the ribs and above the hip of an animal. 2. Any retail cut of meat which is taken from the *flank.*

flank cinch. A cinch which is separate from the saddle and pulled tight in front of the hips and under the flanks. It is used in rodeos to cause the horse to buck.

flank girth. See *flank cinch.*

flannel bush. California fremontia.

flannel-leaf. Flannel mullein.

flannel mullein. *Verbascum thapsus,* family Scrophulariaceae, a biennial herb with woolly leaves and yellow flowers which is widely distributed all over the temperate regions; often a troublesome weed in meadows and pastures. Native to Eurasia. Also called *flannel-leaf; flannel plant; velvet plant; candlewick; mullein dock; Aaron's rod; mullein.*

flashboard. A temporary barrier, usually of wood, placed along the crest of dams to impound increased volumes of water. It can be carried away during floods to lower the water level in the reservoir. A permanent structure is called *stop planks.*

flash-heating. The process of heating a liquid, such as fruit juice, milk, or honey, by running it quickly through a coil container in a hot water or steam bath with the temperature and rate of flow controlled to prevent overheating or scorching.

flash pasteurization. The process of bringing a liquid rapidly to a moderate heat to kill growth of yeasts or molds. Also called *flash method; flash process.* See *high-temperature, short-time method.*

flat. 1. A shallow box containing soil, in which seeds are sown or to which seedlings are transplanted from the seedbed. 2. A square piece of strawboard or papier mâché, placed between layers of eggs in a standard egg case. 3. A level, treeless prairie, especially between hills or mountains. 4. Any natural area of nearly level lowland or upland, which is in contrast with associated land forms because of lack of relief. 5. Referring to a defect in the flavor of butter, cheese, or milk, due to errors in processing, as lack of sufficient salt, uncleanliness, etc. The flavor may be insipid or lacking in the usual characteristics of the product.

flat bale. See *ginner's bale.*

flat-board seat. A temporary seat consisting of a flat board across a wagon box, etc.

flat bone sirloin. A retail cut of beef for broiling, one of the cuts of steak from the rump end of the beef loin.

flat break(ing). See *flat furrow plowing.*

flat-chested. Designating an animal too narrow in the forequarters.

flat fern. Any fern whose large, flat fronds are collected and shipped fresh in bales to florists in large cities to be used in making up floral displays, wreaths, etc. The ones most often used are species of the genus Polystichum, family Polypodiaceae, especially *P. acrostichoides* and *P. munitum.*

flat-footed. A condition of an animal's foot; the arch is flattened, allowing the entire sole to rest on the ground. Also called *splay-footed.*

flat furrow plowing. Plowing in such a way that the furrow slices are inverted and lie flat on the bottom of the adjacent furrow.

flat grain. Designating a board in which the grain runs parallel or approximately parallel to the face.

flat grain beetle. *Laemophloeus pusillus* (Schönh.), family Cucujidae, a grain beetle that feeds on ground grain.

flatheaded apple tree borer. *Chrysobothris femorata* (Oliv.), family Buprestidae, a beetle whose larvae feed on the sap wood of deciduous trees and shrubs.

flatheaded fir borer. *Melanophila drummondi* (Kby.), family Buprestidae, a beetle whose larvae bore in the wood of conifers.

flat limb of apple. A virus disease of the apple variety Gravenstein, and perhaps others, characterized by a slight, linear depression along a branch about half way around. The flattening becomes increasingly pronounced with age of the branch, and the branch becomes brittle.

flat pea. *Lathyrus sylvestris* var. *wagneri,* family Leguminosae, a perennial, herbaceous, climbing or straggling herb with flowers which resemble the sweetpea (*Lathyrus odoratus*), but lack its fragrance. Sometimes used as a forage plant or as green manure. Native to Europe. Also called *Wagner flat pea.*

flat ribbed. Designating a bovine which has little spring of ribs.

flat runner point. The teeth of a hayrake which are flattened near the tip so as to pass freely under the mown grass and rake cleanly.

flat saddle. See *English saddle.*

flat saddle riding. The equitation class which uses English saddles.

flat sour spoilage. Designating a can of food (usually low-acid canned vegetables), which has undergone spoilage by microorganisms without gas formation and which, from every outward appearance, seems good. The product usually tastes sour and has a sour odor.

flat-spray nozzle. A nozzle which produces a spreading, flat spray.

flat-spring snap. A harness snap that locks by a flat spring, but which can be opened by pressing the spring with the thumb.

flat-tool wheel. A type of grindstone which has a flat surface for sharpening flat-faced tools.

flat-surface type heat exchanger. A type of heat exchanger used mainly in dairies for the heating or cooling of liquids. The flat-surfaced unit is set in a tank, the liquid to be heated or cooled is added and is kept in continuous motion to insure uniform heat exchange.

flatulence. A digestive disturbance in which there is an often painful collection of gas in the stomach or bowels.

flatulent colic. A form of colic, in which gas collects in, and distends, the bowels; characterized by a bloated appearance and a drumlike sound which occurs when the swollen sides of the animal are slapped by the hand. Also called *tympanites; wind colic; bloat.*

flat walk. A horse's gait which is square on four corners, bold, easy and graceful. Also called *flat-footed walk.*

flatwoods. Any of various types of flat, wooded land usually of inferior quality for agriculture. The land may be wet, or intermittently wet and dry, and it is generally either a clay or a sand soil. *Flatwoods* may be designated by the kind of native vegetation as post-oak *flatwoods,* saw-palmetto *flatwoods,* etc. (Florida and other parts of southern U.S.)

flavedo. See *chlorosis.*

flavescence. See *chlorosis.*

flavone. $C_{15}H_{10}O_2$, a colorless, crystalline substance, found as a dusty film on the leaves, seed pods, etc., of the primrose; also prepared synthetically, but of uncertain function.

flavonol. A yellow to colorless, plant compound closely related to the anthocyanins, differing from flavone by the presence of an extra -OH group. It is of uncertain function.

flavor. 1. Odor and taste combined with the feeling of the substance in the mouth. 2. a. The material added to foods to gain a desired taste.

b. To give taste or *flavor* to a product by the addition of spices, sugar, etc.

flavor mixing tank. A supplementary tank designed for adding flavoring to the ice cream mix in continuous-type ice cream freezers.

flax. 1. *Linum usitatissimum,* family Linaceae, an annual, herbaceous, crop plant, grown for its fiber from which linen is made; for its seed from which is extracted a valuable oil (linseed oil) used extensively in the paint and other industries; to a very limited extent, as a human food; and for the oil-cake (seeds with the oil expressed) which is a valuable livestock feed. Native to Eurasia. Also called *Baltic hemp.* 2. Wild species of the genus Linum, common in western U.S.; sometimes grazed by sheep and cattle. Some species, (New Mexican flax and stiffstem flax), contain a cyanogenetic glucoside known as linamarin, and are poisonous to animals eating them.

flax anthracnose. A fungous disease caused by *Colletotrichum linicola,* family Melanconiaceae, characterized by lesions and cankers on the leaves and stems.

flax asterocystis. A fungus, *Olpidiaster radicis* (sometimes called *Asterocystis radicis*), that attacks the roots of flax and of some other crops, sometimes causing serious damage. The fungus belongs to the rather primitive fungi classified in the Order Chytridiales.

flax bollworm. *Heliothis ononis* (D. & S.), family Noctuidae, a lepidopterous insect which infests flax.

flax browning and stem break. A disease caused by the fungus, *Polyspora lini,* family Melanconiaceae, characterized by a browning and breaking of the plant stem.

flax buncher. A flax bunching attachment on a grain binder.

flax dodder. *Cuscuta epilinum,* family Cuscutaceae, a twining, pale green, parasitic vine which feeds on and is destructive to flax. Native to Europe.

flaxleaf erysimum. *Erysimum linifolium,* family Cruciferae, a perennial herb grown for its lilac-colored flowers. Native to Spain. Also called *alpine wallflower.*

flax lily. New Zealand flax.

flax pasmo. A disease caused by the fungus, *Mycosphaerella linorum,* family Mycosphaerellaceae, characterized by greenish-yellow to dark brown lesions on the cotyledons and lower leaves of the seedling. In time the lesions develop on the stems and the stems turn brown. In its imperfect stage the fungus is known as *Septoria linicola.*

flax plant by-product. The portion of the flax plant remaining after the separation of the seed, the bast fiber and a portion of the shives. It consists of flax pods, broken and immature flaxseed, and the cortical tissues of the stem and the remaining shives.

flax rust. A disease of flax caused by the fungus *Melampsora lini,* family Melampsoraceae, often causing serious damage to the crop. It is characterized by light yellow to orange-yellow sori on the leaves and stem. Reddish-yellow uredinia occur on the stem, leaves, and capsules.

flaxseed. The seed of flax, known as linseed, which is a source of linseed oil used mainly as a drying agent for paints and varnishes. The residual oil cake is used for livestock feed. See *flax: 1.*

flaxseed meal. See *ground flaxseed.*

flaxseed plantain. Fleawort.

flaxseed screenings oil feed. The ground product obtained after the partial extraction of the oil from the smaller, imperfect flaxseed, weed seed, and other foreign materials that are separated in cleaning flaxseed. It is used as a supplement in feeds for livestock.

flaxseed stage. A stage in the life cycle of the Hessian fly in which the puparium roughly resembles in size, shape, and color, a flaxseed as seen from the edge. See *Hessian fly.*

flax-sick. Designating soil which has become unusable for the growth of flax due to the presence of the flax wilt fungus, *Fusarium lini,* family Tuberculariaceae.

flax straw. The straw left after threshing the flax seed crop; used in the manufacture of various types of paper, including cigarette paper. *Flax straw* of good quality can be used as a substitute for oat straw roughage in wintering cattle. Also called *cattle roughage.*

flaxweed. Butter-and-eggs toadflax.

flax wilt. A disease caused by the fungus, *Fusarium lini* (*Fusarium oxysporum* f. *lini*), family Tuberculariaceae, which multiplies in soil where flax is grown for several years. The soil is called flax-sick. The only control is to plant wilt-resistant varieties of flax.

flay. To remove the skin from a carcass.

F layer. See *humus: 2.*

flaymaster. A type of motorized knife used to remove the hide from the beef carcass in slaughterhouse operation.

flea. Certain active, leaping insects of the family Pulicidae, which feed upon the blood of warm-blooded animals. Their legless larvae (maggots) eat organic debris. The most common *fleas* are found on cats and dogs. Some species attack birds and rats. The human *flea, Pulex irritans* L., is a nuisance in eastern Europe and many other regions.

fleabane. Any plant of the genus Erigeron, family Compositae. Species resemble wild asters and are raised in flower and rock gardens; escaped, they become weed pests. Also called *erigeron; daisy fleabane.* See *orange fleabane; beach fleabane; Himalayan fleabane; poor-robins-plantain.*

flea beetle. Any small, jumping beetle, of the family Chrysomelidae. It eats small holes in the leaves of plants. Some species are vectors of bacterial wilt of sweet corn.

flea-bitten gray. A coat color of horses which is gray with darker hairs throughout.

flea-bitten white. A coat color of horses which is white with numerous, tiny, dark spots.

fleahopper. See *garden fleahopper.*

fleawort. *Plantago psyllium,* family Plantaginaceae, an herb whose seeds are used medicinally. Native to Eurasia. Also called *flaxseed plantain.*

fleck. See *necrotic fleck.*

Fleckvieh. A leading breed of triple-purpose cattle in Germany used for milk, beef, and work. They resemble the Brown Swiss in body conformation but are broken in color. Also called *Spotted Highland.*

fleece. 1. The wool from all parts of a single sheep, which consists of the crinkly hair up to 12 ins. in length. This waviness enables the wool to be matted together into felt or spun into yarn, twine, or thread. 2. The fluffy mass of cotton which remains after the seeds have been removed by ginning. See *lint cotton.* 3. To shear sheep.

fleeceflower. See *silvervine fleeceflower; Bokhara fleecevine.*

fleece-vine. See *silvervine fleeceflower; Bokhara fleecevine.*

fleece wool. Wool from the states east of the Mississippi in the U.S.

fleecy. A fleece-picker; the boy who assists the sheep shearer (New Zealand).

Fleischmann's formula. A formula connecting the specific quantity of milk with the specific quantities of fat and the solids-not-fat it contains.

flesh. 1. The portion of an animal body which consists mainly of muscle. 2. Plumpness or corpulence, especially in such phrases as good *flesh,* etc. 3. The pulpy or juicy portion of fruits or of storage organs of plants, such as potatoes, etc. 4. To remove adhering fat, *flesh,* and membrane from the pelt of a butchered animal.

flesh-colored. A coat color of animals in which the hair color is the same as the skin.

fleshed. 1. Designating muscles or lean meat. 2. Designating a pelt with the flesh or fatty pieces removed from the inner surface.

flesh fly. Any of a large group of flies which lay their eggs, or place the larvae, in flesh.

fleshing. The distribution of flesh on animals.

flesh layer. The innermost layer of a sheepskin, next to the flesh, used for manufacturing chamois leather.

flesh-pink skin. In stock judging, a pink shade of skin which denotes good health in sheep.

flesh side. The side of leather which formed the internal surface of the hide.

fleshy. 1. Designating fruit, leaves, and storage organs of plants with juicy or pulpy tissues. 2. Fat or corpulent. 3. Designating the soft or edible portions of meat.

fleshy frog. The deep, fleshy part of a horse's foot. See *horny frog.*

fleshy fruit. A fruit which is pulpy or soft, such as the berry, pome, and drupe.

Fletcher scale. *Lecanium fletcheri* Ckll., family Coccidae, a soft scale insect, parasitic upon arborvitae and a number of other plants.

fleur-de-lis. The iris.

flexibility. 1. The quality of bending without breaking. 2. Adjustment to meet changing conditions.

flexible bracing. The installation of wire cables to reduce the load on certain members, especially of weak tree limbs.

flexible grass harrow. A harrow that serves the same purpose as a chain harrow.

flexible hog supplement. A hog feed formula which is modified in some details by the substitution of certain constituents without essential modification of nutrient content.

flexible support. A policy of the U.S. Government which favors adjustment of subsidies or price supports for agricultural products as conditions of supply change, in contrast to a policy of fixed or rigid supports of prices.

flexor tendon. The tendon on the posterior surface of the leg on men and animals that straightens the ankle, acting in opposition to the extensor tendon.

flicker roll. A roller composed of coarse, circular saws which removes the extracted seeds from the seeder roll in commercial raisin seeding operations.

flight feathers. Primaries.

flight limit. In forestry, the maximum distance which can be traversed by seed in one continuous movement.

flights. Primaries.

flinders grass. A pasture grass, native to Australia.

flindosa. *Flindersia australis,* family Rutaceae, a large timber tree whose wood has a high reputation for general building purposes. Native to Australia. Also called *ballroom flindersia; crow's ash.*

flint. A dense, amorphous, or cryptocrystalline rock composed essentially of silica (SiO_2) and not different in composition from chert. It remains as a residue in the weathering of chalks and limestones and hence is locally abundant in soils, thus characterizing land, as flint hills, flint ridge, flinty soil. See *chert.*

flint corn. *Zea mays* var. *indurata,* family Gramineae, a type of Indian corn having a soft, starchy endosperm, completely surrounded by a very hard, horny, vitreous, outer layer. The kernels are somewhat flat and rounded rather than pointed. Some varieties are of early maturity, others medium, and the tropical flints are late maturing. See *dent corn.*

flintwood. *Eucalyptus pilularis,* family Myrtaceae, a tall, evergreen tree whose wood is valuable for timber and oil. Its bark on the lower parts has a permanent burnt look. Native to Australia. Also called *blackbutt.*

flinty. Consisting of the hard, translucent endosperm of corn or wheat kernels in contrast to the softer, more mealy portions. In wheat, the various degrees of flintiness and mealy character may be due to plant nutritional factors.

flipper. The hinged flap, on some power dusters, which oscillates back and forth to force dust onto the leaves and branches of the trees.

flitch. 1. A portion of log, sawed on two or more sides, which is intended for sliced or sawed veneer. 2. A pile or bundle of veneer sheets from the same bolt laid together in the sequence of cutting. 3. A side or portion of meat, as a *flitch* of bacon.

flixweed tansymustard. *Descurainia sophia (Sophia parviflora),* family Cruciferae, a weed that invades denuded sagebrush land in Idaho, U.S. Native to Eurasia.

float. 1. An instrument used for filing animals' teeth. 2. A valve in the cream separator to regulate the flow of milk into the cream-separator bowl; or a similar contrivance for maintaining the desired level of water in a tank. 3. A drag or device for leveling soil. 4. To come to the top of the ground, as with a certain type of plow which automatically adjusts if the bottom strikes an obstruction. 5. To irrigate, as in floating a meadow (rare in U.S.).

float gauge. A device used for determining the elevation of the surface of a liquid whose actuating element is a float partially submerged in the liquid.

floating. 1. The preparation of a field for seeding by the use of a float or drag. 2. The incipient fermentation, by bacteria and yeast, of olives being processed in lye solutions, which causes them to rise to the surface of the solution, probably due to the formation of gas. 3. Filing an animal's teeth. 4. Irrigating.

floating arum. Goldenclub.

floating dipper dredge. A dredge mounted on a floating barge which uses a bucket type of digger to dig and deposit the excavated soil at one side of the ditch. It is suited for excavation of drainage ditches or canals in swampy, timbered regions.

floating flakes. A body defect of churned buttermilk caused by overchurning or by low viscosity of the product. It results in the formation of large butter granules.

floating grapple dredge. A dredge mounted on a floating barge similar to the floating dipper dredge but which has a dipper of the clamshell or orange-peel type. It is especially suitable for levee work.

floats. Finely ground rock, rich in phosphate, used as a fertilizer.

flocculation. The aggregation of suspended collodial materials into aggregates as contrasted to their dispersion.

floccule. A small cluster of fine particles which may appear in a soil that has been flocculated by addition of lime or other substances.

flock. 1. Several birds or domestic mammals, such as sheep, which are tended as a unit. Also called *herd; band.* 2. Short fibers sheared from the face of cloth or produced in milling or finishing cloth or obtained by shredding rags to almost a powder. 3. In stock judging, one ram, of any age, and four ewes of varying ages as designated by the show.

flock book. The record of breeding and ancestry of sheep, kept privately by the flock owner or officially by the Sheep Breed Association, to register purebred sheep. See *herd book.*

flock feeder. A trough in which feed is placed for fowls.

flocking tendency. The habit of congregating in large flocks inherent in Spanish Merino sheep.

flockmaster. The man in charge of a flock of sheep. See *shepherd.*

flock mating. The indiscriminate breeding of fowls.

flood. 1. The overflow by water from natural or human causes into fields, farms, cities, etc. 2. To run water upon a field to the depth of a few inches to supply sufficient moisture for the growth of the crop, as rice, cranberries, etc., to control weeds, or to protect the crop from injury by frost. (Probably the earliest and simplest method of irrigation.) 3. To overflow the natural banks or levee so as to inundate the adjacent or even more distant regions.

flood control. Any of the various engineering practices used to prevent a flood or damage resulting from overflowing rivers and streams. It includes deepening the existing channels, digging diversion canals, building levees and dams to impound the water in numerous small ponds or retention basins in the upper parts of the watershed, or by large detention basins farther downstream.

Flood Control Act. An act of U.S. Congress passed in 1944 to control floods, raise the ground-water level, and help land reclamation in the Missouri River basin.

flooding. 1. The inundation of land by water in excess of that confined within the stream banks. 2. See *flood: 2.*

flood irrigation. See *flood: 2.*

flood plain. The nearly flat land in a stream valley that is subject to flooding at times of high water.

flood prevention. Flood control.

flood protection. Flood control.

flood relief. 1. Assistance given by the International Red Cross, local governments, etc., to persons who have suffered loss in floods. 2. Flood control.

flood runoff. Water in excess of the carrying capacity of the stream channels.

flood stage. The stage in which the level of the water rises above the tops of stream banks or dikes. Frequently, an arbitrary level established, beyond which the stream or river is considered to be at *flood stage.*

flood water. The water that has passed beyond the stream bank or dikes and spreads out on the surrounding land.

floodway. The space between a river bank and a levee or other natural elevation on either side of a stream. It acts as an overflow channel.

floor. All dead vegetable matter on the soil surface including litter and unincorporated humus. See *duff.*

floor sweepings. The broken tobacco leaves, not stems, which are swept up from the floor and put into scrap-grade tobacco.

floor-type manger. A manger whose front edge is level with the floor.

floor-type milker. A type of milking unit which rests on the floor during the mechanical milking process.

flora. 1. The aggregate of plants which grow without cultivation in a given area within a stated period of time. 2. The bacteria in or on an animal, a plant, the soil, etc. Usually referred to as bacterial *flora*.

floral. 1. Referring to the structure and character of a flower. 2. Referring to the flora of a region. 3. Composed of, or characterized by, flowers.

floral bud. Any flower that has not opened.

floral nectary. The secretory glands of a flower which assist in pollination by producing the nectar which attracts the insects. The glands may be united into a complete ring near the base of the corolla or calyx, or may occur in special parts of the flower, as in the floral spurs of columbine, some orchids, etc.

Flora's-paintbrush. Earleaf tasselflower.

flor de la oreja. See *sacred ear.*

Florence fennel. See *fennel.*

floret. A single, small flower of a head of flowers, such as those in the family Compositae; other types are found in the grasses, sedges, and many other families of plants.

floribunda. Designating a plant that blooms abundantly, as the Japanese flowering crabapple and the *floribunda* rose.

floricane. In the bramble fruits, the second year's growth which bears the flowers and fruits and then dies. Sometimes called *second-year cane.* See *primocane.*

floricultural. Referring to floriculture.

floriculture. The production and cultivation of flowers and other plants which are grown as ornamentals.

Florida beggarweed. See *Cherokee tickclover.*

Florida cat's claw. Catclaw apes-earring.

Florida Everglades. See *Everglades.*

Florida fern caterpillar. *Callopistria floridensis* (Guen.), family Noctuidae, a caterpillar which infests ferns, stripping them of leaflets. Originally native to tropical America, it is now widely distributed throughout North America.

Florida gummosis. A disease of orange and grapefruit trees in Florida, U.S., possibly caused by a virus; characterized by longitudinal cracks in the bark of the trunk and large limbs. A gum exudes from the cracks, runs down the bark, and hardens in a mass. The bark on either side of the crack dies and sloughs off, leaving light-colored to brownish lesions.

Florida harvester ant. *Pogonomyrmex badius* (Latr.), family Formicidae, a troublesome, stinging pest which infests groves and gardens in southeastern U.S., feeding on seeds.

Florida lubber. See *eastern lubber grasshopper.*

Florida paspalum. *Paspalum floridanum,* family Gramineae, a native forage grass of moderate value. Native to Florida, U.S., found from Kansas and Texas eastward to the Atlantic coast.

Florida poisontree. *Metopium toxiferum* (*Rhus metopium*), family Anacardiaceae, a shrub or small tree, which, on contact, causes severe dermatitis to susceptible persons. Also called *poison wood; doctor gum; coral sumac.*

Florida red scale. *Chrysomphalus aonidum* (L.), family Diaspididae, a scale insect which is harmful to citrus twigs, leaves, and fruits, to many other cultivated plants in the Gulf States, U.S., and to greenhouse plants in cooler regions.

Florida shade-grown wrapper leaf. A type of tobacco grown for its large leaves used for wrappers in cigar making.

Florida silverpalm. *Coccothrinax argentea,* family Palmaceae, a small, sometimes practically stemless, palm tree often grown in tropical and subtropical regions and sometimes in greenhouses as an ornamental. Native to southern Florida, U.S., and the West Indies. Also called *silver-palm; thatchpalm; silvertop palmetto.*

Florida velvetbean. Deering velvetbean.

Florida wax scale. *Ceroplastes floridensis* Comst., family Coccidae, an important insect pest of citrus fruit trees in various parts of the world, whose chief damage is the profuse production of honeydew on which sooty mold develops.

Florida yellowtrumpet. *Stenolobium stans,* family Bignoniaceae, a tropical, large shrub or small tree, grown in very warm areas for its clusters of late-blooming, yellow flowers. Native to tropical America. Also called *yellow elder.*

floripondio datura. *Datura arborea,* family Solanaceae, a showy, medium-sized, tropical tree which bears green-striped, white flowers. Native to Peru. Also called *angel's-trumpet; maikoa.*

florist. A person who sells plants and cut flowers, often growing these plants and flowers himself. Sales of accessories, such as fungicides, insecticides, seeds, fertilizers, pots, pot labels, etc., are often a part of the florist's service.

florist crop. The plants grown and sold by a florist.

floristically. With reference to the types of plants growing naturally in certain regions. Thus, the Mediterranean region of Europe and northern Africa are quite similar *floristically* to southern California, U.S.

florist's chrysanthemum. *Chrysanthemum morifolium,* family Compositae, a perennial herb, grown in flower gardens and greenhouses for its flowers of many different colors. Probably native to China. See *chrysanthemum.*

florist's cyclamen. *Cyclamen africanum,* family Primulaceae, a popular, tuberous, stemless, perennial herb. The *florist's cyclamen* is so much a product of crossing of various species native to southeast Europe and Asia Minor, that the name of the original species is confused. The *florist's cyclamen* is often called *C. indicum, C. persicum,* or *C. macrophyllum.*

florist's pyrethrum. *Chrysanthemum coccineum,* family Compositae, a perennial, tender herb, grown in the flower garden and as a pot plant for its red, pink, lilac, or white flowers. Native to Eurasia. Also called *pyrethrum; painted daisy.* See *chrysanthemum.*

florist's smilax. Smilax asparagus.

florist's violet. Sweet violet.

floss. The lint of cotton and other plants, resembling fine wool, which consists of the longer hairs from the surface of the seed intermingled with short fuzz.

floss flower. See *mistflower eupatorium.*

flotation. Separation of materials by mixing with a liquid of a different specific gravity; one of which will float, the other sink. Overmature peas may be separated from green peas in this manner; frozen oranges may be separated from sound fruit in running water.

flotation concentrates. The high-grade phosphate fertilizer that has been separated by the flotation process from the clay, sand, etc., present in crushed phosphate rock.

flotation phosphate rock. See *flotation concentrates.*

flotation sulfur. An extremely fine powdered sulfur, more active chemically than flowers of sulfur; produced by the oxidation of H_2S recovered in aqueous alkali from coal gas fumes.

flotation sulfur paste. A paste containing flotation sulfur with substances admixed that assist in the dispersion of the sulfur when the paste is put in water. It is used as a fungicide.

Flotost. A boiled whey cheese made in Norway, containing not less than 20 percent fat.

flour. The fine-ground product obtained in the commercial milling of wheat, which consists essentially of the starch and gluten of the endosperm. It is used for baking bread, pastry, cakes, etc.

flour beetles. Beetles of the genus Tribolium (*T. confusum* and *T. castaneum*), family Tenebrionidae, which infest stored grains and other food products.

flour color. The chief factor in determining the grade of flour and its selling price. The most desirable is pure white.

flour corn. A type of corn in which nearly all the starch is soft, with only the shell being the more vitreous material. It was widely grown by the Indians in the U.S. as it could be easily made into corn meal in their primitive mills. The blue and variegated types are called *squaw corn.*

flour middlings. See *wheat standard middlings; wheat flour middlings; wheat red dog.*

flour mite. Any of the several species of mites of the genus Acarus (Tyroglyphus), family Acaridae, that infest flour and other cereal products. These mites may also infest sugar, dried fruits, cheese, etc. Also called *cheese mite.* See *grocer's itch.*

flour moths. Any of several types of small moths including the Angoumois grain moth, *Sitotroga cerealella,* family Gelechiidae. They lay their eggs on starchy grains, breakfast food, flour, and ripe grains of corn on the cob, etc. Endemic in Europe and America.

flour strength. The quality in flour which enables the baker to produce a loaf of bread of a particular volume and texture by the use of proper ingredients, together with proper mixing, fermentation, and baking. High-strength flour is used in making ordinary fermented bread. Low-strength flour is preferred in making crackers and pastry, as it lacks toughness and yields a flaky, brittle, or crumbly product.

flour sulfur. Finely-powdered sulfur, in contrast to flowers of sulfur, formed by sublimation.

floury. Designating a fine-textured soil, consisting primarily of silt or flocculated clay with aggregates of silt size which when dry is incoherent, smooth, and dustlike.

flour yield. The amount of flour that can be produced from a given amount of wheat.

flow. 1. The quantity of liquid that passes through a pipe, gate, channel, or other conveyance for a given unit of time under given conditions of head, or pressure, roughness, etc. Units of measure may be cu. ft. per second, acre ft. per day, gallons per minute, etc. 2. The movement of silt, sand, etc., in a channel. 3. Discharge from a pipe, etc. 4. The amount of milk produced per cow herd etc., at a specified time. 5. The ease or difficulty with which a product can be moved from one place to another.

flowage. The flooding of a field for irrigation, control of weeds, or preventing frost injury.

flowage line. A contour line around a reservoir, pond, lake, or along a stream which corresponds to some definite water level (maximum, mean, low, spillway crest, etc.), and is generally used in connection with the acquisition of rights to flood lands for storage purposes.

flow diversion. In the pasteurization of milk, the system by which the milk that has not been heated to the proper temperature, 161° F., is turned back into the raw balance tank.

flower. 1. The reproductive structure of a seed-bearing plant, consisting of the male and/or female organs which are surrounded by one or two series of outer coverings (calyx and corolla). Calyx, the outermost series, consists of sepals, generally greenish, which cover the unopened bud. The inner series, corolla, consists of petals which give the flower its characteristic color. These two enclose the essential reproductive system. The female organ, the pistil, consists of one or more carpels, and contains ovules, each of which produces an egg. Around it are the stamens, the male element, which produce the pollen grains (or microspores). On coming into contact with the pistil, the pollen grains send out a tube which grows down to an ovule and carries the enclosed male cell whose nucleus unites with the egg and fertilizes it. The fertilized egg enlarges and undergoes repeated divisions to form the embryo. *Flowers* may have both pistil and stamens in the same *flower* or in separate *flowers.* A perfect *flower* contains both corolla and calyx, although some flowers have only colored sepals. 2. A soft, cured, cows' milk, English cheese containing the petals of various flowers, such as roses or marigolds. 3. To develop blossoms; to flourish.

flower arranging. The art of arranging flowers and leaves in a vase or bowl to give the most pleasing effect. Also called *flower designing.*

flower bed. An area in a garden which is devoted to the growth of special types of flowers in sufficient number to form a unit, usually to be viewed from all sides.

flower border. A setting of flowering plants arranged to be viewed from one side which has a fence, building, or shrubs as a background.

flower box. A box, usually placed just outside a window or on a porch railing, in which ferns, flowering plants, etc., are grown. It is useful for interior or exterior house decoration, and where land is unavailable. See *porch box; window box.*

flower bud. A bud which, when mature and open, reveals a flower. On woody plants, the bud may be initiated in the preceding summer or fall and then, surrounded by tough bud scales, lie dormant until the next spring. In some bulbous plants, as some lilies, the bud is formed inside the subterranean bulb in the fall, but does not emerge from the ground until spring. In most of the other plants, the flower buds do not show until shortly before blossom time.

flower cluster thinning. In the culture of grapes, the removal of certain flower clusters shortly after the flowers have set their fruits, while the latter are still small, so that the food and water that would otherwise go to these clusters may go to the remaining ones, thus producing larger fruits.

flower designing. Flower arranging.

flower head. An inflorescence in which the usually small or medium-sized flowers are clustered together, as the heads of clover, or the heads of plants of the family Compositae. Each head is popularly called a flower although it consists of numerous small flowers, some of which (the outer series usually) may be large and showy.

flowering. Designating the stage of development of a plant in which the production of flowers occurs.

flowering almond. A double-flowered, horticultural variety of the almond cherry (*q.v.*), grown as an ornamental for its white or pink flowers.

flowering ash. *Fraxinus ornus,* family Oleaceae, a deciduous tree grown in warm areas for its whitish, fragrant flowers. The sweetish exudate from this tree, manna, is sometimes collected and used for medicinal purposes. Also called *manna tree.* Native to Europe.

flowering bean. Scarlet runner bean.

flowering bulb. a. A subterranean bud; a short, enlarged stem, variously shaped, which emits roots below and bears a number of overlapped, fleshy or membrane-like scales. Sometimes called *tunicated bulb.* Usually corms are miscalled bulbs, as are some tubers. See *bulb: 1. Cf. corm; tuber.* b. Any of several species of flowering plants, mainly of the families Liliaceae and Amaryllidaceae, which produce true bulbs; grown as ornamentals. In general practice, any plant producing a bulbous root, whether a true bulb, a corm, or a tuber, grown for its flowers, is called a *flowering bulb.*

flowering cherry. Any of several cherry trees grown for their large showy, often double and infertile flowers, horticultural varieties, mostly small trees, a number of native Japanese species of the genus Prunus, family Rosaceae, including *P. incisa, P. nipponica, P. serrulata, P. sieboldi, P. subhirtella,* and *P. yedoensis.* In the U.S. they are hardy in the southern states and in sheltered places further north. See *Fuji cherry; Higan cherry; Takane cherry; Oriental cherry; Siebold cherry; Yoshino cherry; almond cherry.*

flowering crabapple. Any of several small-fruited, small to medium-sized trees, of the genus Malus, family Rosaceae, grown for its usually profuse, showy blossoms. Most of the more showy varieties are descended from long established eastern Asiatic species, but some have been crossed with native American species. *Malus baccata,* the Siberian crabapple, has been used in hybridization to produce winter-hardy varieties. See *Bechtel's crabapple; carmine crabapple; Chinese flowering crabapple; Japanese flowering crabapple; wild sweet crabapple.*

flowering currant. Either of two hardy species of currants of the genus Ribes, family Saxifragaceae, which has long-tubed, golden-yellow flowers with a clovelike fragrance, grown for their attractive foliage as well as for their flowers. Native to North America. See *clove currant; golden currant.*

flowering dogwood. *Cornus florida,* family Cornaceae, a small tree which sometimes attains a height of 30 to 40 ft., grown for its white flowers. (Occasionally a pink variety occurs wild.) Its scarlet, drupe fruit is much eaten by game birds: *e.g.,* bob-white, ruffed grouse, wild turkey. Native to eastern U.S. Also called *American dogwood; boxwood; American cornelian tree.*

flowering maple. Any of various species of the tropical and subtropical genus Abutilon, family Malvaceae, grown in greenhouses in southern Florida, U.S., and elsewhere where the climate permits. The more commonly used greenhouse varieties are often called *A. hybridum* and are grown for their red, pinkish-purple, yellow, or white flowers. These plants are not related to the maple but gain their name from the maple-shaped leaves. Also called *Chinese bellflower; abutilon.*

flowering merrillia. *Merrillia caloxylon,* family Rutaceae, a hardwood tree which reaches a height of up to 100 ft. Native to the Malay Peninsula, Sumatra, and adjacent regions. Also called *Ketenggah.*

flowering moss. See *Texas stonecrop; sand pyxiemoss.*

flowering plum. *Prunus triloba* (*Amygdalus triloba*), family Rosaceae, a deciduous shrub or small tree grown for its pinkish flower clusters. Native to China. Also called *flowering almond; flowering cherry.*

flowering quince. 1. Any shrub of the genus Chaenomeles, family Rosaceae. Species are grown for ornament. See *Japanese flowering quince; Chinese flowering quince.* 2. *Chaenomeles lagenaria,* a hardy shrub grown for its scarlet-red flowers and yellowish-green fruit. Native to China. Also called *Japanese quince.*

flowering raspberry. Fragrant thimbleberry.

flowering shrub. Any shrub cultivated for its beautiful flowers.

floweringspurge euphorbia. *Euphorbia corollata,* family Euphorbiaceae, a small, perennial, herbaceous plant, grown for its showy flower clusters with white, petal-like appendages. Its saplings on cutting exude milky juice. Also called *appleroot; emetic root; flowering spurge; milk purslane.*

flowering stage. The stage of development of a plant when it begins to produce flowers.

flowering time. The date at which a plant comes into flower. This depends chiefly upon: temperature, length of daylight, exposure to sun, soil moisture and soil fertility.

flowering tobacco. Winged tobacco.

flowering wintergreen. Fringed polygala.

flowering wood. See *softwood cutting.*

flower of an hour. *Hibiscus trionum,* family Malvaceae, an annual, herbaceous plant, grown in the U.S. for its purple-centered, yellow flowers. The calyx enlarges to form a dark-striped, bladder-like structure about the yellowish-white seed capsule. Escaped, it is a weed. Native to Africa. Also called *bladder ketmia.*

flower of Tigris. Tigerflower.

flower pot. A container, usually of clay, for growing plants. The standard *flower pots* are as wide at the top as their height and are made in various sizes. Three-quarter pots, sometimes called azalia pots, are ¾ as tall as wide. Rose pots are taller than wide. Special types of pots are available for some plants. See *thimble pot; thumb pot; lily pot; three-quarter pot; cyclamen pot; pan: 4; rose pot.*

flower shears. A special type of shears for cutting flower stalks. It saves the hands from thorn-wounds and gives a clean cut.

flowers in the open. Out-of-doors growing of flowers as contrasted with growing in greenhouses.

flowers of sulfur. A very light, floury form of sulfur produced by sublimation and not by crushing or grinding, as in flour of sulfur or sulfur flour, to which it is superior for many purposes. It is used as a fungicide, especially against the powdery mildews and also in the control of certain insects or mites. Also called *sublimed sulfur.*

flower spot of azalea. An infectious fungous disease characterized by small, rapidly enlarging spots which eventually ruin the whole flower. The perfect stage of this fungus develops on the fallen, over-wintered petals of the previous year's infected flowers and produces its spores (ascospores) in cup-shaped structures (apothecia) at the time the new blossoms open. The infected petals carry the imperfect stage of the fungus, producing large numbers of asexual spores (conidia) which rapidly spread the disease. It is injurious to azaleas and rhododendrons in southern U.S.

flowerstalk. See *peduncle; pedicel.*

flowerstem. Pedicel.

flower thrips. *Frankliniella tritici* (Fitch), family Thripidae, small insects which feed on portions of the flower in many cultivated plants, such as wheat, strawberry, cotton, citrus, etc.

flowing. See *flow: 2.*

flow line. 1. The position of the water surface of a running stream for a normal or specified rate of discharge. See *flowage line.* 2. The bed of a stream or culvert.

flow of milk. 1. The passage of milk through a tube or a nipple. 2. See *flow: 4.*

flow of sap. The out-flow of the sweet, watery fluid from an accidental or intended wound, break, or a hole in the trunk of a sugar maple.

flow rate. The rate at which the cow lets down milk.

flow slides. See *earth flow.*

flu. Infectious laryngotracheitis *q.v.*

flue. 1. A vertical space, usually 6 ins. or less in width, between two adjacent tiers of stock in a lumber pile. 2. A square or round ventilating passageway. 3. The passage in a chimney for conducting smoke, flame, or hot air. 4. The special passageways for heated air in a barn for curing tobacco. See *flue cure.*

flue cure. A method of curing cigarette-type tobacco used in southeastern U.S. Ripe tobacco leaves are tied lightly together in bundles, hung on poles over flues in a barn, and cured with regulated furnace heat conducted through the flues. The method is also used in curing red pepper. In some modern tobacco barns, the flues are replaced with burners distributed over the barn floor, which are fired with L-P (liquid petroleum) gas. *Cf. air cure.*

flue-cured tobacco. Designating tobacco that has been subjected to flue cure.

fluff. a. The downy part of a feather. b. The soft feathers on the thighs and posteriors of birds.

fluffed out. Designating the condition of a newly hatched chick in which its wet down has lost its moisture and become loose and fluffy.

fluff louse. *Goniocotes gallinae* (De G.), family Philopteridae, a louse found on poultry.

fluffy. 1. Describing a body defect of ice cream characterized by large air cells and open texture. 2. Describing a soil in which the aggregates of the surface are loose, light, and fine.

fluidity. The property of flowing possessed by gases and liquids. *Fluidity* and viscosity are inverse characteristics.

fluid milk. The fluid product of a dairy farm or factory in contrast with the more solid products, such as cream, cheese, and butter.

flukes. Flat worms of the Class Trematoda which are internal parasites of vertebrate animals and man at maturity, but which usually have intermediate stages in snails. The mature worms are usually seed-shaped and found in the liver, alimentary canal, and other body cavities, attached usually by two suckers. The eggs of the adults are discharged from the host and hatch under favorable conditions. This resulting intermediate stage enters a snail, passes through several stages of its life cycle, and is finally deposited on grasses. Domestic animals and man, involved in the food chain from the grasses, are reinfected by the adult flukes. The liver fluke, *Fasciola hepat-*

ica may cause a disease fatal to sheep and cattle.

flume. An inclined, open conduit or troughlike structure for the conduction of water and floating objects, as logs, sugar cane, etc. An aqueduct is a special-purpose *flume.*

fluor-apatite. $Ca_{10}F_2(PO_4)_6$, a mineral which is a source of phosphorus, but because of its fluorine content is not recommended as a stock feed supplement. See *apatite.*

fluorescent bacteria. Species of the genus Pseudomonas, family Bacteriaceae, which produce a fluorescent pigment.

fluoride. A salt of hydrofluoric acid with a base of sodium, calcium, or another element or a radical. See *fluoroacetate of sodium.*

fluorine. A pale greenish, gaseous, acid-forming element, having the symbol F, small quantities of which are detrimental to plants and to animals since it is a cumulative poison. Occasionally tested as a fumigant of plants against insects, it is dangerous to handle without extreme care. Many of the compounds of *fluorine* are used for insecticides or rodenticides. In sufficiently dilute solutions in drinking water, some *fluorine* compounds reduce the occurrence of dental caries, but care must be taken not to exceed certain limits or the teeth may be harmed. Minute amounts of *fluorine* are present in normal animal tissues, bones, teeth and skin, but it has never been known to be essential to plant life. It is widely distributed in rocks, soils, and water, and is present in the rock phosphates which are mixed in feed rations. Soluble fluorides are toxic to germinating seeds, and chronic fluorosis may occur in man and animals. See *fluoride; fluoroacetate of sodium; fluorosis; fluosilicate of soda; cryolite.*

fluorine poisoning. See *fluorine.*

fluoroacetate of sodium. A chemical used as a raticide.

fluorosis. A toxic condition due to an excess of fluorine, usually in the drinking water, which manifests itself mainly in mottling of teeth.

fluosilicate of soda. A rodenticide which may be spread on floors where rats and mice run and which enters their alimentary canal when the animals lick their feet.

flush. 1. To irrigate a field with just enough water to soften the crust. 2. To increase the feed allowance to ewes with protein-rich supplement or to place them on green feed a short time before and during the breeding season. 3. A vigorous or abundant, sudden, new growth. 4. To send out vigorous growth on the twigs of a tree. Thus, in the tropical fruit tree, the mango, most of the twigs *flush* several times a year. 5. To empty, clean, or wash out any material with a quick, heavy supply of water. 6. To introduce and shortly afterwards withdraw an irrigating solution of mild antiseptic of medicinal value, as to *flush* the vagina or uterus.

flush cut. To prune a branch so that the cut is approximately even and parallel with the branch or trunk from which it was removed.

flushed. See *flush: 4.*

flushes. The periods of rapid growth of a plant. See *flush: 4.*

flushing. 1. Broadcast plowing. 2. Washing out by throwing or forcing water into a container, organ, etc.

flushing ewes. The feeding of a high protein ration to ewes at breeding time for the purpose of increasing the lamb crop. See *flush: 2.*

flush season. The season in which the egg production is at its height.

fluted. Designating an unevenness of the rind of citrus fruits in which there are more or less easily distinguishable longitudinal ridges.

fluted scale. Cottony-cushion scale *q.v.*

fluted strain. A bud strain of citrus fruits characterized by the fluted shape of fruit.

fluted-wheel feed roll. The fluted shell or wheel placed in a feed cup below the seed box of a grain drill which, as it revolves, measures out the seed by the volume of the recesses in the fluted surface.

fluted-wheel grain drill. A grain drill which uses the fluted wheel type of forced feed mechanism.

flutter. Gill flirt *q.v.*

fluviatile. Produced by river or other stream action. *Fluviatile* sediments are those deposited in the channels, flood plains, deltas of rivers, or in alluvial fans.

fluviraption. Breaking, cutting, and movement of rock and loose earth by force of running water.

fly. Any winged insect, such as a moth, bee, gnat, etc. Specifically, a two-winged insect of the family Muscidae. Many *flies* are blood-sucking pests of men and animals, such as the mosquitoes, horse and deer *flies,* black *flies,* punkies or no-see-ums, and some sand *flies.* Some are vectors of diseases, such as the stable *flies,* etc. Some *flies* destroy other insects; some are parasites on plants, as the Hessian fly; others are valuable scavengers. Many *flies,* as the housefly, pass their larval stage in manure and garbage and upon attaining maturity carry with them the bacteria of filth, thus spreading diseases such as typhoid fever, etc. See *horse bot fly; nose bot fly; sheep bot fly; throat bot fly; black fly; black horse fly; sandfly; screw-worm; tsetse fly.*

fly-free date. The date after which it is safe to plant wheat to avoid serious infestation by the Hessian fly.

flying-dragon trifoliate orange. *Poncirus trifoliata* var. *monstruosa,* family Rutaceae, a dwarf variety of the trifoliate orange, with crooked, slender leaves and spines; grown occasionally as a curiosity.

flying drive. A log drive whose main portion is dispatched without stopping to gather stray logs in the stream.

flyings. A trade term for the cured, lowermost leaves of the tobacco plant.

flying stall. A portable stall for a horse in which the animal remains while being loaded or unloaded from a vessel or while in transit.

fly net. A meshed covering made of strips of leather or cords of fabric and placed over an animal, usually a horse, to keep the flies away.

fly netting. A fabric woven in small enough meshes to exclude flies.

fly-off. The part of the total rainfall which passes into the air through direct evaporation and transpiration of plants (obsolete). *Cf. cut-off (rainfall); runoff: 1.*

fly poison. See *crowpoison.*

fly repellent. See *repellent: 1.*

fly speck. 1. A common disease of the apple fruit, caused by the imperfect stage of the fungus, *Leptothyrium pomi,* family Leptostromataceae, characterized by minute fruiting bodies of fungus on the fruit which resemble the spots made by flies on defecation. 2. The small, dark specks which are dried fecal excrementa of flies.

flyspeck fungus. See *fly speck: 1.*

fly spray. 1. Aerosol spray, usually containing rotenone or pyrethrum, which is expelled under pressure from spray-bombs into the air. Flies and other insects coming into contact with the minute droplets are poisoned; often used in buildings to free halls and rooms from pests. 2. Any spray containing DDT, or any other effective insecticide, which is sprayed upon screens, walls, furniture, etc., or upon farm animals, where it remains toxic for some time. The insects that alight upon the sprayed surface are poisoned. 3. Any of various types of insecticides sprayed upon accumulations of manure or garbage, which contain insecticides that kill the fly maggots.

fly time. The summer season when flies are abundant.

fly trap. A device of wire screen with easily accessible entrance for flies but from which exit is difficult so that the flies accumulate in it in large numbers. Bait is often provided to attract the flies. Used in places where flies are numerous.

fly weevil. Angoumois grain moth.

fly wheel. A heavy wheel, attached to the rotating shaft of a machine, which by its inertia tends to make the rate of rotation more uniform.

foal. 1. The unweaned young of the horse. 2. In stock shows, a horse foaled on or after January 1 of the year shown.

foal heat. The estrous period that normally occurs a few days after foaling. *Cf. false heat.*

foaling. Giving birth to a foal.

foaling barn. The barn in which mares are placed for foaling.

foamflower. Allegheny foamflower.

foaming. The production of foam or froth in a liquid. It is troublesome in milk in various dairy processes.

foamy. Designating a body defect of ice cream which is characterized by the presence of large air cells in the ice cream and a foamy, frothy appearance. Also called *foamy melt-down.*

foamy appearance. A defect of cultured milk in which too much air has been incorporated, largely due to faults in preparation.

foamy melt-down. See *foamy.*

fobbing. Frothing produced in bottle-washing machines.

fodder. Feed for livestock, specifically the dry, cured stalks and leaves of corn and the sorghums. In the case of corn, the ears may be removed from the stalk leaving the stover, or the leaves of corn and sorghums may be pulled and cured to constitute *fodder.* See *forage: 1; roughage: 1.*

fodder bundle. A quantity of corn stalks or other fodder tied in a bundle for convenience in handling. See *corn shock.*

fodder corn. Corn grown primarily for fresh or cured forage.

fodder variety. A usually tall, freely suckering variety of corn grown for fodder and silage.

foehn wind. A wind descending down the Alps which becomes warm and dry in its descent. A similar wind in the U.S. Rocky Mountains is called *chinook.*

foetus. Fetus *q.v.*

fog. 1. Almost microscopically small water drops suspended in air which reduce horizontal visibility to less than 1 mile. 2. Heat-generated and gas-propelled aerosols, in which the particles of the liquid are 50 microns or less in diameter. See *aerosol.* 3. *(fogge)* The rather luxuriant regrowth of grass following the first mowing (northern England, Scotland). 4. Rotting and decaying grass in winter.

fog belt. A narrow belt approximately a mile or two wide, immediately adjacent to certain coasts, in which fogs abound due to the proximity of the ocean, the direction of the prevailing winds, and the atmospheric humidity.

fog drip. The moisture from fog which has collected on trees, shrubs, etc., and which drips to the ground. In some regions the amount may be equivalent to a light rain, and thus is a considerable factor in supplying moisture to plants. Also called *wet fog.*

fog-drive gun. A broom gun.

fog fever. An illness that sometimes occurs in cattle which have been grazing on fog grass. Its cause is unknown; sometimes it causes death of the affected animals. See *fog: 3.*

Foggiano. An Italian cheese resembling Cotronese or Molitarno, made from ewes' milk.

fogging. The practice of grazing cattle on fog grasses. See *fog: 3.*

fog nozzle. A special nozzle in which streams of water strike each other and are broken up into a fine mist. It is useful for wetting the foliage of greenhouse plants and maintaining the humidity inside.

fold. An enclosure, or pen for sheep or cattle.

folded rib roast. A retail cut of beef for roasting. It is a beef rib with a two-inch piece of the rib removed midway between the rib end and the backbone to permit the end to be folded over.

folding boom. The boom for a trailer-type sprayer on which are attached the nozzles. It has hinges so that the end portion can be drawn up or folded back to avoid obstacles.

fold removal. An operation to remove folds of skin of a ewe from below the vulva towards the crotch to prevent their being soiled by urine and becoming inflamed. It prevents infestation by blowflies. Also called *Mules' operation.*

foliaceous. Leafy or leaflike.

foliage. The leaves, collectively, of a plant; leafage.

foliage burn. A browning and dying of leaves, caused by deficiency of water, injury from harmful chemicals from spray materials, or the poisonous secretions from the mouth parts of sucking insects such as the insects causing tip burn.

foliage dieback. A condition in which the tips of leaves wilt progressing towards the stem.

foliage plant. A plant grown for the color and/or shape of its foliage: *e.g.,* coleus.

foliar nematode. *Aphelenchoides olesistus,* family Anguillulinidae, and related species, known as nematodes or eel worms. These almost microscopic worms gain entry to the leaves of many greenhouse plants through the stomata (breathing pores), or they enter the roots from infested soil and work their way up through the stem into the leaves. Necrotic areas are produced in the affected part which sometimes involve the whole leaf.

foliar spray. 1. Any spray applied to plants when leaves are present, as contrasted to dormant spray *q.v.* 2. A spray which supplies fertilizer to plants by spraying liquid fertilizer directly onto the leaves where it is absorbed.

foliation. The process of producing foliage by a plant; the coming into leaf.

folic acid. A vitamin found in the leaves of leguminous and other plants, in yeast, liver meal, and wheat. *Folic acid* is needed in hemoglobin formation and for growth. Also called *pteroylglutamic acid.*

foliocellosis. A mottling of leaves of Citrus in which the areas between the lateral nerves are paler giving the leaf a uniform lateral mottling; probably caused by zinc deficiency.

follicle. 1. A small sac, gland, or pit for secretion or excretion. The hairs of an animal grow out of pits called *follicles.* 2. A one-celled, monocarpellary, dry seed vessel or fruit that splits on the ventral edge.

follicle mite. *Demodex folliculorum* (Simon), family Demodicidae, a mite that infests the hair follicles of men and domestic animals, sometimes causing intense itching. The infested follicles may become infected.

follicular. Of a follicle, or consisting of follicles; pertaining to pits, sacs, cavities, etc.

follicular mange. A type of mange caused by the follicle mite.

follicular mange of cattle. Demodectic mange of cattle *q.v.*

follow. To plant or raise crops in sequence, as corn *follows* soybeans.

follower boards. The two thin boards at either end of the frames in a beehive for the commercial production of comb honey.

follow up. 1. In forestry, the reinforcing of fire fighters by sending additional man-power to facilitate either suppression or mop-up work. 2. The subsequent investigation made to check on progress, recuperation, diagnosis, etc., after a preliminary investigation.

fomentation. A poultice; the external application of warm moist cloths or other objects to ease pain.

Fontina. A cooked, cured, semi-soft, yellowish cheese with a nutty flavor which is made in Italy from whole milk.

food. Any non-noxious nutrient material which is used by an organism as a source of energy, for building, or for repairing tissues. See *feed: 1.*

Food and Agriculture Organization. An agency of the United Nations, formally organized in 1945, for supporting world-wide studies of agricultural production and food supplies. It is also empowered to lend technical assistance to nations in the solution of problems in agricultural production and human nutrition.

Food and Drug Administration. A division of the U.S. Department of Health, Education, and Welfare, charged with the enforcement of Federal laws and inspection services pertaining to adulteration and misbranding of foods, feedstuffs, and drugs in interstate commerce.

food chain. A series of plant or animal species in a community, each of which is related to the next as a source of food.

food chamber. That portion of the beehive in which the winter food supply of the bees is stored.

food chopper. A machine for chopping food (meats, etc.) for cooking or feeding.

food constituent. Any essential nutrient (protein, carbohydrate, fat, etc.) contained in a food.

food crop. A crop grown primarily for human consumption.

food-infesting insects. Insects that infest fresh or dried foods in storage, in the home or while still in the growing stage.

food intoxication. A food poisoning in lambs and sometimes calves which are receiving a heavy ration of concentrates to fatten for early marketing. Possibly due to the presence of bacteria, such as *Clostridium* spp., whose poisons are absorbed, causing the symptoms of enterotoxemia *q.v.*

food item. A product used as food.

food poisoning. Any sudden, painful, intestinal disorder in man and animals caused by some harmful bacteria or fungal organism taken into the body with the food. The organism multiplies in the alimentary canal and produces poisonous substances which may bring about death. *Clostridium botulinum,* one of the bacteria, will cause the disease known as botulism *q.v.*

food poisoning in lamb. See *food intoxication; enterotoxemia.*

food preservation. The treatment of foodstuffs in various ways to keep them wholesome and free from decomposition: *e.g.,* as refrigeration, canning, smoking, adding of certain antiseptic chemicals, drying, exclusion of air, etc.

food producing. 1. Designating regions in which

farming is the principal industry. 2. Designating plants of which certain parts, or the whole, may be used for human food. 3. Designating special technical processes which produce foods.

food reserves. Food substances over and above those already used in the growth and activities of the plant which are stored in various organs of the plant for future use. The storage products may be starch, as in the pith of apple twigs, in tubers of the potato, in roots of the sweet potato, etc.; sugar, as in the sap of sugar maple, or sugar cane; organic acids, as in many succulent plants; almost always certain proteins or amino acids accompany the other *food reserves.* With the resumption of growth and developmental activity, these *food reserves* are used for respiration, energy, and building material.

Food Stamp Program. A U.S. federal program, in operation 1939-43, to supply needy families with additional purchasing power. Those eligible to receive various forms of public aid were entitled to buy orange-colored food stamps for the value of money they were spending for food. These stamps were to be used for any food stuff. In addition, for each dollar's worth of orange stamps purchased, the family received 50¢ worth of free blue-colored food stamps. The blue stamps were for buying surplus foods on the Secretary of Agriculture's list.

food stamps. Stamps, issued by the U.S. government through certain surplus agricultural commodity programs to persons in the low income group, which were exchanged at stores for certain surplus foods. The stamps were redeemed from the merchants by the Federal Government at fixed prices.

food storage. The necessary practices for the preservation of foods stored for periods of time. These involve storage space; protection from cold and moisture, heat and drying, and dust and dirt; exclusion of harmful insects and rodents; and such prestorage treatment as may be necessary to prevent decay. See *food preservation.*

food technology. The science of food qualities, values, preparation, processing, and marketing.

fool-hay. Witch grass.

fools. See *plum pockets.*

foolsparsley aethusa. *Aethusa cynapium,* family Umbelliferae, a strong-smelling, herbaceous plant, resembling parsley, which is poisonous to stock and to persons who mistakenly eat the foilage or roots for parsley or radishes. Native to Europe, a naturalized weed in U.S.

foot. 1. In man and quadrupeds, the terminal portion of the leg which rests upon the ground. 2. The portion of a cultivator to which the sweep is attached. 3. The organs of locomotion of various invertebrates, as the *feet* of a caterpillar, the *foot* of a clam or a snail, etc. 4. The base of a tree, tower, mountain, wall, hill, etc. 5. A unit of linear measure, 12 ins.

foot-and-mouth disease. An acute, highly communicable disease of cloven-footed animals caused by a filterable virus. It is characterized by a short incubation period, high fever, and vesicles in the mouth and on the feet which break to form erosions. Also called *aftosa fever; aphthous fever; epizootic aphtha.*

foot-foul in cattle. Foot rot *q.v.*

foothill checkermallow. *Sidalcea malvaeflora,* family Malvaceae, a perennial herb grown for its rose-colored flowers on many spikes. Native to North America. Also called *checkerbloom; wild hollyhock.*

foothill needlegrass. *Stipa lepida,* family Gramineae, a perennial grass which is an important forage. Native to California, U.S.

foothills. The hills or dissected, sloping land at the base of a mountain or mountain range. Such land differs in native vegetation, potentialities for agriculture, grazing, and other uses from the higher slopes and crests of the mountains.

footing. 1. The low elements of a foundation, the part that bears on the soil. 2. The condition of the ground surface as it relates to the pulling capacity of a draft animal without slipping.

foot lift. A foot lever in a mower by which the operator can lift the cutter bar to avoid obstructions.

foot-lift sulky plow. A riding plow in which a lever and a single or double bail may be used to lift the plow bottom to avoid stones, roots, etc.

foot maggot. The larva of some species of the blowfly, family Calliphoridae, which infests the feet of sheep and feeds on wounds. The secondary screw worm, *Callitroga macellaria,* is also sometimes implicated in this trouble.

foot mange. A foot-skin disease of animals, especially sheep and cattle, caused by the mite *Chorioptes bovis.* Also called *aphis foot.*

foot pad. The cushions on the bottom of the feet of such animals as cat, dog, etc.

foot pedal. A lever in agricultural machines which is operated by the foot. It may be attached to the beam of a gang plow, to brakes, or to the cutter bar of a mower, etc.

foot rest. An attachment on agricultural machines on which the rider rests his feet to keep them out of danger during operation.

foot rot. A frequently occurring inflammation in animals' feet, usually followed by pus formation on the soft tissues between the toes. It occurs especially in sheep and cattle living in or using wet and filthy places and is caused by infection, sometimes through bruises or small wounds, by various kinds of bacteria present in the filth. One type of the disease is infectious pododermatitis, caused by *Spherophorus necrophorus.*

foot rot of cereals. See *Cercosporella foot rot of cereals.*

foot rot of citrus. A disease in which the basal portion of the trunk and the crown roots are infected with species of Phytophthora, especially *P. citrophthora,* family Pythiaceae. Patches of the bark are killed down to the wood and cambium. The invaded tissues are darkened, and the cambium along the margins of the lesions turns yellow-ochre. Affected parts exude gum. The same fungus may attack the fruit and cause brown rot gummosis.

foot rot of gerbera. A disease of the Transvaal daisy which is caused by the fungus *Phytophthora cryptogea.* See *foot rot of plants.*

foot rot of lily. A foot rot caused by the fungus *Phytophthora cactorum,* family Pythiaceae. See *foot rot of plants.*

foot rot of plants. A fungus-induced disease of the roots or basal portions of the stems of many different plants. Usually various tissues of the roots and the cambium at the base of the stem are destroyed, resulting in the ultimate death of the whole plant. If not woody, the plant may topple over as the roots and stem bases die and lose their turgor, a phenomenon called damping off. Several types of fungi are involved in the disease: *e.g.,* various species of the genera Phytophthora and Pythium, family Pythiaceae, in woody plants, and possibly the fungus, *Cercosporella herpotrichoides,* family Moniliaceae, in the cereals.

foot rot of sweet potato. A minor rot of the roots of the sweet potato caused by the fungus, *Plenodomus destruens,* family Sphaerioidaceae. See *foot rot of plants.*

foot-rot shears. Sharp-pointed, double-bladed shears used to dig out infected areas of the foot of an animal affected with foot rot.

foots. Settlings or sediment, as in oil or molasses.

foot-treadle grindstone. A grindstone so mounted that the operator sits on a seat attached to the frame and turns the stone by means of a foot treadle.

forage. 1. That portion of the feed for animals which is secured largely from the leaves and stalks of plants, such as the grasses and legumes used as hays, straws, fodders, etc. It may either be for grazing or be cut and preserved as dry fodder. See *roughage: 1; fodder.* 2. To search for, spread out or seek, for food.

forage acre. A hypothetical rangeland acre completely covered by vegetation that can be totally utilized by grazing animals to the limits of the physiological endurance of the vegetation.

forage acre factor. In range management, a numerical figure which indicates the amount of forage of a given type in each surface acre, obtained by multiplying the weighted average proper-use factor of the type by the average density. This figure multiplied by the surface acres gives total available forage.

forage acre requirement. The number of forage acres, or fraction thereof, required to support a mature grazing animal for a specified period without injury to the range resources.

forage cover. 1. A crop which is grown primarily as forage for livestock: *e.g.,* hay, pasture grass, legumes, etc. 2. In range management, the amount of edible vegetation for livestock or game produced seasonally or annually on a given range area. See *grazing capacity.*

forage crops anther mold. A parasitic fungus, *Sclerotinia spermophila,* family Sclerotiniaceae, which infests the stem and leaves of Ladino clover (*q.v.*) without apparent injury to the foliage, but when it reaches the blossom is harmful to the anther and pistils, often rendering the seeds incapable of germination. It is not at present a serious disease.

forage crops blind seed disease. A fungous disease of perennial ryegrass caused by *Phialea temulenta,* family Helotiaceae, whose small, stalked apothecia grow out of the diseased seeds lying on or shallowly buried in the soil and discharge their ascospores into the air, thus infecting the flower of the ryegrass and sterilizing the developed seeds.

forage crops ergot. See *ergot: 1.*

forage density. The percentage of ground surface, viewed directly from above, which appears to be completely covered by vegetation within easy reach of livestock.

forage drier. A type of equipment frequently consisting of a large volume fan, with or without a source of heat, which is designed to quickly and effectively reduce the moisture content of freshly cut or partly cured, forage material.

forage feeds. 1. Bulky type feeds composed largely of pasture grasses, hays, silage, etc. 2. A mixture of ground or processed feeds which is composed largely of forages.

forage grass. Any grass that is especially adapted for hay, range cover, or pasture for domestic livestock feed.

forage legumes. Any of the legume plants, that are grown or used largely as forage for livestock, such as alfalfa, clover, etc.

forage looper. *Caenurgina erechtea* (Cram.), family Noctuidae, a small caterpillar which feeds on clover and alfalfa, doing only minor damage.

forage plant. Any plant cultivated in meadows or pastures, or growing wild, used as forage for livestock.

forage poisoning. 1. Any poisoning of an animal (or man) from eating food, usually forage, contaminated by the presence of some organism which was not destroyed in the usual processing. Such an organism may produce toxins which are fatal to the animal consuming even small quantities of the forage. The most serious cases are those in which the contaminating organism is *Clostridium botulinum* (*Bacillus botulinus*) which can withstand repeated heatings at the temperatures ordinarily used in the home canning of vegetables. 2. Poisoning of animals grazing or eating forage containing a plant chemical such as the glucoside of hydrocyanic acid in Sudangrass or sorghum, or by forages improperly cured, such as sweet clover, where the breakdown of the alkaloid, coumarin, results in rupture of the small blood capillaries in the animal.

forager. A bee which goes out in search of food; a foraging bee. Also called *worker bee.* See *worker.*

forage seed. The seed of plants that are grown mainly for forage, such as alfalfa, red clover, lespedeza, timothy, brome, etc.

forage seeding. The seeding of grasses or other

forage crops alone or with regular grain crops such as oats, barley, or wheat, to be used for forage after the grain is harvested.

forage sorghum. Any variety or type of sorghum that is grown primarily for forage, rather than for grain or sirup production.

forage value. a. The relative importance for grazing purposes of a range plant or plants, as a whole, on a range. b. The rank of a range plant or plant type for grazing animals under proper management, preferably expressed as proper-use factor. See *grazing value.* c. The comparative value of the forage portion of an animal's ration compared to the other feeds in the ration.

forage variety. That type or variety of plant which is grown or recognized primarily for its forage use or value rather than for seed or grain production.

forage volume. 1. That portion of a plant, above ground and within reach of grazing animals, considered in three dimensions. 2. The measure of the forage crops; *i.e.,* the aggregate amount of forage produced on a range area during any year.

forb. Any annual, biennial, or perennial, herbaceous plant that is not grass-like, its leaves being broader and shorter than those of the true grasses and sedges. Also called *broadleaved herb.*

Forbes primrose. *Primula forbesi,* family Primulaceae, a tender, perennial herb, grown in greenhouses for its lilac or rose-colored flowers. Native to China and Burma. Also called *baby primrose.*

Forbes scale. *Aspidiotus forbesi* Johns., family Diaspididae, a scale insect which occurs on fruit trees and bushes in eastern U.S. It is the chief apple scale in some regions.

force. 1. Powerful effect: *e.g.,* the power of electricity; the power of a current or tide. 2. To move a body or to accelerate a body already in motion. 3. To cause plants to develop to maturity more rapidly than by normal growth. See *forcing.* 4. To guide or lead a herd or flock of animals, specifically, the ability of dogs to direct sheep.

forced air circulation. Air which is compelled to pass through some apparatus or enclosure by the use of blowers (electric fans, etc.) for drying, cooling, etc.

forced burning. In forestry, a technique in broadcast or spot burning: the selection of a time when the lower layers of duff and flashy fuels are damp but which can be made to burn by setting a large number of fires in heavier concentrations of debris over the entire area. Sometimes slash disposal is accomplished by progressive burning.

forced supercedure. The construction by worker bees, especially in those colonies in which the queen is old and enfeebled, of queen brood cells within which the larvae develop to mature queens. The brood is protected from destruction by the kin queen by the workers who lead off a swarm, leaving the newly-emerged queen in the colony.

forceps. In veterinary medicine, a pliers-like instrument used for grasping, pulling, and compressing.

force pump. A pump employed to force water and other liquids above the range of atmospheric pressure. It is distinguished from a lift pump, in which the liquid is elevated to run out of a spout. In a *force pump* the water is forced out by a piston or plunger working against a pressure corresponding to the head, or elevation (above the inlet valves) to which the water is pumped.

forcing. Bringing a plant to a specific stage in its development or to maturity earlier in the season than normal, by growing it in a greenhouse, under artificial light, etc.

forcing box. A small coldframe used to cover a single plant after it has been planted in the field or bed to expedite growth of the plant. Also called *plant forcers.*

forcing house. A greenhouse used primarily for forcing (*q.v.*) foliage, flowering, and vegetable plants.

forcing pen. A pen, adjacent to the chute in a dipping vat, in which sheep are placed prior to dipping. Animals are guided down the chute and into the vat.

fore. See *forequarters.*

forearm. See *foreleg.*

forecarriage. A truck, consisting of an axle and one or two wheels, which is placed under the front and unsupported part of a vehicle, plow, etc., for support.

forecast. 1. A statement of the anticipated weather conditions in a given region, usually for a period of 12, 24, or 36 hours; 5 days or 30 days. 2. The predicted future price of a product.

foreclosure. A procedure whereby a lender takes steps toward attaining ownership of property given as security for a loan. The usual steps are: default on the mortgage or contract by the buyer; court proceedings; and sale of the land by the sheriff in case of a real estate mortgage after expiration of the redemption period.

forecooler. In refrigeration, a supplementary condenser which consists of a coil or series of coils through which compressed ammonia is made to pass before it enters the condenser proper. It is cooled by the overflow water of the condenser.

foreflank. That section of the body of a hog just behind the lower shoulder or foreleg. It provides the leanest section of bacon.

forefoot. To rope the front legs of a running animal usually from horseback. The pursued animal is thrown hard when it reaches the end of the rope, allowing the rider to dismount and tie the animal before it fully recovers.

forehopple. A strap or rope which is tied around the forelegs of an animal to prevent straying.

Foreign Agricultural Service. An agency of the U.S. Department of Agriculture which is responsible for developing foreign markets for U.S. farm products.

foreign color speck. A color defect of American Cheddar cheese which appears as specks, varying from occasional white or black specks to rust spots or red blotches.

foreign flavor. A flavor defect of milk or its products which is any flavor not commonly developed in or associated with milk or milk products.

foreign material. See *foreign matter.*

foreign matter. Any material, substance, etc., which is unnatural to, or not commonly developed in, a product.

foreign object. A body, mass, or particle of physical material which is not normal or natural to the place where it is found.

foreign odor. Any odor which is not natural to a product.

foreleg. The lower portion, above the pastern, of either front leg of four-legged animals. It is commonly used in reference to horses.

forelock. A lock of hair growing above a horse's forehead; the forepart of the mane.

foremilk. The first 25 to 50 milliliters of milk to be withdrawn from the udder at the beginning of milking in contrast to middle milk and strippings. This milk is of poor quality chemically and bacteriologically and should be rejected.

forepastern. A term commonly used to describe the portion of the front leg next to the hoof.

forepunch. A kind of counterpunch used to make a place for the nailheads to be set into a horseshoe.

forequarters. The front two quarters of an animal.

foreribs. The ribs immediately behind the shoulder blade of domestic animals. Sometimes the area is called *heart girth* in judging or evaluating animals.

foresaddle. The part of the carcass anterior to the twelfth rib or the two unsplit forequarters of a meat carcass.

foreshank. The lower portion of front leg next to the pastern and hoof of a meat carcass.

forest. A plant association predominantly of trees and other woody vegetation which occupies an extensive area of land. *Cf. woodlot; forest land.*

forest armyworm. See *forest tent caterpillar.*

forestation. The establishment or reestablishment of forests. Also called *afforestation; reforestation.*

forest capital. The lands, timber, reproduction, and physical improvement, etc., which make up a forest property.

forest cover. All trees and associated plants which cover the ground in a forest.

forest economics. The study of forest management which deals with the needs, values, and policies of forest lands in relation to public or individual gains.

forest economy. A local, regional, or national economy concerned wholly or in part with forest production.

forest entomology. The branch of zoology that deals with insects in their relation to forests and forest products.

forester. A person professionally trained in forestry.

Forest Experiment Station. A U.S. forest research organization which consists of 9 stations, plus branch stations, in regional locations. It conducts investigations and scientific research on the protection, management, use, and survey of forest resources, with particular application to each regional area. Created within the McNary-McSweeney Act, 1928, and administered by the Secretary of the U.S. Department of Agriculture.

forest finance. The application of principles of valuation, accounting, and investment analysis to the practice of forestry.

forest fire. On forest land, any fire not being used as a tool of an authorized plan in forest management.

forest influence. The effect upon water, soil, climate, and health which results from the presence of forests.

forestit dust. Drug sabadilla.

forest land. Land which bears forest growth or shows evidence of past forest occupancy and which is not now in other use. To qualify as forest, an area must be at least 100 ft. wide and one acre in area, have a sufficient number of trees to provide 10 percent crown coverage, or lacking 10 percent crown coverage, be likely to remain in forest use.

forest litter. Leaves, dead branches, bits of bark, rotting wood, etc., found on the forest floor.

forest management. The application of business methods and technical principles to the operation of a forest property. It involves the computation of income from forest lands; the establishment of cutting cycles; the conservation of cover, land and water; and the formulation and conduct of long-range plans of operations.

forestomachs. The three nonglandular stomachs found in ruminants: specifically, the rumen, reticulum, and omasum.

forest pathology. The science which pertains to diseases of forest trees or stands and to the deterioration of forest products by organisms.

forest regulations. 1. A branch of forestry concerned with the technical (as opposed to the business and administration) aspects of organization and maintaining a forest for sustained yield. 2. Control of private forest management by exercise of public authority.

forest reserve. See *National Forest.*

forestripping. Removal of a small amount of milk by hand from each teat prior to the milking operation; *forestrippings* are usually discarded because of high bacterial and low fat content.

Forestry. The art or science concerned with the management of forests.

Forest Service. A unit of the U.S. Department of Agriculture created in 1905. Its responsibilities are to administer the National Forests in the public interest, to cooperate with state and private owners for protection and management of forest lands, to conduct research pertaining to forest resources and to make results of research available to the public.

forest soil. In pedology, a soil whose characteristics have been influenced or largely determined by forest vegetation.

forest survey. Inventory of forest land to determine area, condition, and timber volume and species for scientific purposes: *e.g.*, timber purchase, forest management, forest policies and programs.

forest tent caterpillar. *Malacosoma disstria* Hbn., family Lasiocampidae, a caterpillar that feeds voraciously on many shade and forest trees and also on field and truck crops.

forest tree nursery. Any plantation upon which young trees are grown for transplanting in a forest.

forest type. A stand of trees similar in composition and development, due to certain ecological factors, which differentiates it from other stands.

forest utilization. In forestry, the operation of harvesting and marketing the forest crop and other products of the forest.

foretop. The long, coarse hair on the top of a horse's head.

fore udder. The front two quarters of the udder of a cow.

forewarmer. A device used to heat milk as it is received at a milk processing plant or creamery.

forewarming. The process of preheating milk before it is pasteurized or further processed in dairy manufacture.

Forez. A crude, French cheese cured in cellars by covering with dirt and trickling water over it. Good quality *Forez* resembles Roquefort.

forget-me-not. Any herb of the genus Myosotis, family Boraginaceae. State flower of Alaska, U.S. See *true forget-me-not; field forget-me-not.*

forging. The noisy striking of the foreshoe with the toe of the hind shoe by a horse when walking, trotting, or running. Also called *clicking; striking.*

fork. 1. Any of various tools or implements consisting of a handle, shank, and two or more tines which is used for digging soil, grasping or lifting hay, manure, etc. 2. The juncture of two roads, rivers or tree branches, etc. 3. See *swell: 1.*

forking agitator. In cheese making, a motor-driven agitator which serves to mix and cool the curd after cutting.

forking larkspur. *Delphinium consolida,* family Ranunculaceae, an annual herb, grown for its blue or violet flowers. Native to Europe. Also called *field larkspur.*

fork loader. A hay loader which loads hay onto a wagon or truck bed from the ground without windrowing. It consists of rakelike attachments which gather the hay and deliver it to an elevator.

fork-tailed bush katydid. *Scudderia furcata* Brunn., family Tettigoniidae, a widely distributed long-horned grasshopper which is a general plant feeder and occasionally a citrus pest.

form. 1. The shape and structure of anything, as distinguished from the material of which it is composed. 2. In mensuration, the shape of a tree or log. 3. Denoting the normal type or shape of a variety of a flower in a show. 4. See *conformation.*

Formagelle. A small, soft, fresh, Italian cheese made from ewes' or goats' milk.

formal decorative dahlia. A type of dahlia which has fully double flowers in which the margins of the floral rays are sometimes slightly rolled toward the lower side. The rays are usually broad and pointed or rounded at the tips. The outer rays tend to recurve, and the central rays to be cupped. Most of the floral rays are regular in arrangement.

formaldehyde. A gas, readily soluble in water, which is generally used in aqueous solutions as an insecticide, a fungicide, a soil fumigant, a tool disinfectant, a storage house disinfestant, in disease treatment, and as a disinfestant for seeds, tubers, and bulbs.

formaldehyde resins. Synthetic resins which are condensation products of formaldehyde with phenols, urea, etc.

formalin. A 40 percent aqueous solution of the gas formaldehyde *q.v.*

for market. Designating an animal or farm product which is for sale.

formation. 1. Any igneous, sedimentary, or metamorphic rock which is represented as a unit in geological mapping. 2. A major division of the climax vegetation of a continent.

form board. A device used to assist in the insertion of foundations into frames of beehives.

form constant. The numerical difference, approximately constant for any given tree species, between the form quotient and the form factor *q.q.v.*

form factor. In forestry, the ratio between a tree's volume and that of a geometric solid, usually a cylinder, having the same diameter and height.

Formosa juniper. *Juniperus formosana,* family Pinaceae, an evergreen tree grown as an ornamental for its several trunks and its upright limbs. Native to Asia.

form quotient. The ratio of the diameter of a tree stem at one-half height inside bark to the basal diameter.

formula. Any general equation: a rule, or principle expressed in algebraic symbols: *e.g.,* the statement of the quantity and kind of stock materials which are used in making a fertilizer, feed, or other mixture.

formula 62. A preparation of diphenylamine, benzol, turkey red oil, and lamp black which is almost a specific for screw-worm (*q.v.*) control. Also called *smear 62.*

forsythia. Any plant of the genus Forsythia, family Oleaceae. Species are very popular and handsome, spring-blooming shrubs. Also called *golden bell.* See *border forsythia; early forsythia; weeping forsythia; greenstem forsythia.*

fortified. 1. Designating a product to which has been added amounts of a vitamin, as vitamin A or vitamin D. 2. Designating a wine to which additional alcohol (brandy) has been added.

Fortin's barometer. A pattern of mercury barometer suited for accurate readings of the pressure of the atmosphere. The zero of the scale is indicated by a pointer inside the mercury cistern, the bottom of which is flexible and may be moved for correction by an adjusting screw until the mercury surface just touches the pointer.

Fortunes plantain-lily. *Hosta fortunei,* family Liliaceae, a perennial herb, grown for its pale purple or white flowers. Native to Asia.

Fortunes windmill palm. *Trachycarpus fortunei,* family Palmaceae, a low-growing, fan palm, grown in warm areas as a tubbed plant ornamental. Native to Asia. Also called *hemp palm.*

forty. Forty acres of land, as a *forty.*

forty-eight's (48's). An English grade of wool equivalent to quarter-blood *q.v.*

forty-four's (44's). An English grade of wool equivalent to low quarter-blood *q.v.*

forty's (40's). An English grade of wool equivalent to braid *q.v.*

forty-six's (46's). An English grade of wool equivalent to low-quarter blood *q.v.*

forward pricing. A method of price support designed to announce a guaranteed minimum price far enough in advance so that production can be adjusted. As contrasted with parity prices, *forward prices* would be set close to the expected free market price under reasonably full employment conditions.

fothergilla. Any shrub of the genus Fothergilla, family Hamamelidaceae. Species are grown as ornamentals. See *dwarf fothergilla; large fothergilla.*

foulbrood. See *American foulbrood; European foulbrood.*

foul-in-the-foot. Any of a number of different suppurating or necrotic lesions of the feet of cattle (England). Called *foot rot* in U.S.

found. Room, board, and laundry, as well as necessary equipment for a hired hand which may be offered in addition to, or as part of, his salary.

foundation. 1. Designating those animals, birds, or plants used in the developmental breeding. 2. See *comb foundation.*

foundation planting. Any plant used decoratively to conceal the juncture of a building with the ground.

foundation seed. Seed stock from which seed increases are grown; *i.e.,* seed stock which is so handled as to most nearly maintain specific genetic identity and purity and that may be designated by or distributed through an agricultural experiment station or foundation seed association.

founder. 1. Laminitis *q.v.* 2. To feed on one type of feed in such volumes as to cause the feed to lose its palatability to a particular animal.

fount. Fountain.

fountain. a. Any of several types of devices which furnishes a continuous supply of drinking water. b. A jet or stream of water used in pools, with statuary, etc., as decoration.

fountain humea. *Humea elegans,* family Compositae, a biennial herb grown as a pot flower and sometimes placed in a border during the summer for its reddish flowers. Native to Australia. Also called *amaranth-feathers.*

fountain plant. *Russelia lemoinei,* family Scrophulariaceae, an herb grown for its showy, red flowers. Native to tropical America.

four-abreast hitch. The hitching of four draft animals abreast rather than in tandem.

four-arm system. See *rope sling: 2.*

four-cane Kniffin. Single trunk, four-cane Kniffin system *q.v.*

four-cornered gait. A horse's gait in which each foot is placed on the ground individually. Also called *single-foot; rack.*

four footing. Throwing an animal by means of a rope around the feet.

4-H. See *4-H Club.*

4-H Club. An organization for boys and girls between the ages of 10 and 21, organized under the guidance of Cooperative Extension Service workers and volunteer, adult local *4-H Club* leaders. The program is the youth phase of the national system of Cooperative Extension work in which the U.S. Department of Agriculture, the Land-Grant colleges and universities, and the counties participate. Within this organization the youth engage in various individual projects in the fields of home-making, animal husbandry, handicrafts, plant breeding, etc.; a process of "learning by doing." Each project, activity, and experience in *4-H Club* work is directed toward the development of the members into mature, competent, responsible adults. The emblem of the club is a green four-leaf clover outlining the aims of the club: the improvement of head, heart, hands, and health.

4-H Club agent. See *county agent.*

four-horse team. A team of four horses.

four-leaf clover system of pruning. A little-used system of training apple trees in which four primary scaffolds are chosen and the leader removed when the tree is young. Four channels or wedges about 2 ft. in width between the four quarters of the tree are kept free of branches. Also called *clover-leaf style pruning. Cf. wedge-style pruning.*

four-lined leaf bug. See *four-lined plant bug.*

four-lined plant bug. *Poeciloapsus lineatus* (F.), family Miridae, a bug that infests currant, gooseberry, rose shrubs, and herbaceous plants, generally east of the Rocky Mountains, U.S. Leaves of infested plants show white or dark-colored spots.

four-o'clock. *Mirabilis jalapa,* family Nyctaginaceae, a perennial herb grown as a tender annual in flower gardens for its red, yellow, or white flowers which do not open until late afternoon. Native to tropical America. Also called *marvel-of-Peru; afternoon lady.*

four-spotted tree cricket. *Oecanthus nigricornis quadripunctatus* Beut., family Gryllidae, a cricket that infests tree and bush fruits, drilling pinholes in the twigs in which to deposit eggs. These punctures often serve as openings for diseases.

four-tooth sheep. A two-year-old sheep.

four-way hybrid. The hybrid which results from mating two single crosses. Also called *double cross.* See *single cross.*

four-wheel trailer. Any trailer which is supported

by two axles and four wheels, as contrasted to a two-wheel or one-wheel trailer.

fourwing poisonvetch. *Astragalus tetrapterus,* family Leguminosae, a plant in which the foliage, green or dried, is poisonous. Symptoms of poisoning are dullness, irregularity of gait, lack of appetite, dragging of the feet, a solitary habit, loss of flesh, and a shaggy coat. As the animal ceases to eat, it dies. Found in Nevada and southern Utah, U.S. Also erroneously called *locoweed.*

fourwing saltbush. *Atriplex canescens,* family Chenopodiaceae, an alkali-tolerant, evergreen shrub extensively browsed by cattle, sheep, and goats on arid land of western ranges, U.S. Also called *cenizo; chamiza.*

four-wire horizontal trellis. A method of supporting semitrailing blackberries in which two pairs of horizontal wires are set, the one about two feet above the other. New canes are usually supported by the lower pair of wires; the fruiting canes on the upper pair.

four-year-old. In stock shows, a horse which was foaled on or after January 1 and on or before December 31 of the fourth preceding year.

fowl. 1. Any mature chicken. 2. Live poultry generally. 3. A hen of any age or weight. 4. Meat of a bird or poultry, as distinguished from other meats. 5. In its broadest sense, any bird, but more commonly a large bird, as wild *fowl,* water *fowl.*

fowl bluegrass. *Poa palustris,* family Gramineae, a slender grass sometimes used as a pasture grass. Of Eurasian origin, it is found across northern U.S. Also called *false redtop; fowl grass; fowl meadow grass.*

fowl catcher. A long rod with a hook at one end used to catch fowls by the leg or foot. Also called *chicken catcher.*

fowl cholera. An acute, infectious, septicemic disease caused by the bacterium, *Pasteurella multocida.* It is characterized in very acute form by sudden death, in less acute forms by greenish or yellowish diarrhea, listlessness, sleepiness, ruffled feathers, stationary habit, impaired appetite, increased thirst, rapid respiration, accumulation of mucous in upper nasal passages, fever, and a purpling of wattles and comb; in chronic forms by emaciation, depression, paleness of wattles, comb, and membranes of the head, and lameness. Also called *cholera of fowls; roup of fowls; hemorrhagic septicemia of fowls; avian pasteurellosis; cholera gallinarium; pasteurellosis; chicken septicemia; acute fowl cholera.*

fowl diphtheria. Fowl pox.

fowlers-servie. American mountain ash.

fowl flu. Infectious laryngotracheitis.

fowl gapeworm. *Syngamus trachea,* a small roundworm which infests the trachea of birds. Infestation of birds under eight weeks old is characterized by a stretching of the neck with open mouth, sneezing, coughing, shaking of the head, loss of appetite, emaciation, prostration, closed eyes, head drawn back against the body. Also called *gapes.*

fowl grass. Fowl bluegrass.

fowl leukemia. The avian leukosis complex (*q.v.*) that attacks the blood cells.

fowl mannagrass. *Glyceria striata,* family Gramineae, a strong-growing bunchgrass rated as good to excellent forage for horses and cattle but fair for sheep. Widely distributed over the U.S.

fowl meadow grass. Fowl bluegrass.

fowl paralysis. A type of the avian leukosis complex (*q.v.*) probably caused by a filterable virus; characterized by lameness or drooping of a wing, prostration, legs held in peculiar positions, limpness of legs and wings, no motor control, withering of the muscles of the affected part and by the feathers in the region of the crop being damp, dark, and permanently discolored. Also called *neural lymphomatosis; neuritis of chickens; range paralysis; big liver disease.*

fowl pest. Fowl plague.

fowl plague. A highly acute, infectious, virus disease characterized by a sudden onset, depression, droopiness, standing still, head drawn in, eyes closed, irregular gait, difficult respiration, shaking of the head, prostration, and bluish-red, swollen wattles and comb. Also called *fowl pest; bird plague; chicken pest; bird pest; Brunswick bird plague.*

fowl pox. A highly infectious, virus disease of birds, especially chickens. Characterized in the skin type, by small, raised, grayish blister-like spots on the comb or wattles or other unfeathered parts which later rupture and discharge a sticky fluid. As these sores dry, a dark brownish scab forms. Characterized in the diphtheric type, by raised, yellowish patches in the mouth and throat which prevent closure of the mouth, difficult breathing, and loss of weight. Also called *chicken pox; contagious epithelioma; canker; avian diphtheria; sore head; fowl diphtheria.*

fowl tick. *Argas persicus* (Oken), family Argasidae, an ectoparasite of poultry in the south and southwestern parts of U.S. It feeds only at night, hiding during the day and is a powerful blood sucker, sometimes killing the bird. Infested birds are characterized by weakness of the legs, droopy wings, paleness of comb and wattles, and cessation of egg laying. Also called *adobe tick; chicken tick; dove tick.*

fowl tuberculosis. Avian tuberculosis.

fowl typhoid. Avian typhoid.

fox. *Vulpes vulpes, V. fulva, Urocyon cinereoargenteus,* etc., any of several doglike animals of the family Canidae, which under certain conditions are farm predators, especially on fowls.

foxberry. Cowberry.

fox farm. An enclosure, usually on a small tract of land, where foxes are bred and raised for their pelts.

foxglove. a. Any herb of the genus Digitalis, family Scrophulariaceae. b. *Digitalis purpurea,* a biennial herb grown for its purple, white, or yellow flowers, particularly in Michigan, U.S., U.K., and Germany. Also grown commercially as a source of the drug, digitalis *q.v.* Native to

Europe and central Asia. Also called *fairy finger; fairy cap; finger flower; finger root; fairy bell; fairies' petticoat; common foxglove; fairy glove; dog's finger; deadman's bells.* See *foxglove poisoning.* c. A common name for several other plants. See *trumpetcreeper: 2; pitcherplant.*

foxglove anthracnose. Digitalis anthracnose.

foxglove aphid. *Myzus solani* (Kltb.), family Aphidae, an aphid that infests many garden plants, particularly potato and foxglove. Infested plants often have a chlorotic condition.

foxglove penstemon. Smooth penstemon.

foxglove poisoning. A poisoning of animals feeding on foxglove, characterized by gastro-intestinal irritation, dullness, loss of appetite, nausea, contracted pupils, and slow but steady pulse.

fox grape. *Vitis labrusca,* family Vitaceae, a hardy, woody vine which produces an edible, thick-skinned, sweet, musky berry. It is the origin of many American grape varieties. Also called *northern fox grape.*

foxtail. See *foxtail barley; mouse barley; foxtail millet.*

foxtail barley. *Hordeum jubatum,* family Gramineae, a biennial or perennial grass which is a weed pest in meadows, pastures, roadsides, and waste places, western U.S. It is harmful to animals feeding on it when the awns get under the tongue or gums or into the eyes. Also called *squirreltail grass; wild barley; skunk barley.*

foxtail fescue. *Festuca megalura,* family Gramineae, an annual grass which has a fair grazing value when consumed green. Native to U.S.

foxtail lily. Desert candle.

foxtail millet. *Setaria italica,* family Gramineae, an annual grass grown in the U.S. for its grain to be fed to hogs, poultry, and birds, etc., but grown in the Orient for human food.

fox trot. An easy, horse's gait between the walk and trot; a short, broken trot.

foxy oats. Oats, stored before they are sufficiently dry, which have steamed, changed to a deeper yellow color, and taken on an acrid odor.

fragipan. A compact subsurface layer in a soil profile which is relatively low in clay but is dense enough to interfere with free movement of water and full root penetration. When dry the soil appears to be cemented and hard, but when wet the apparent induration disappears.

fragmented structure. Aggregation of soil particles into very small, irregular, blocky shapes.

fragrance. Sweetness or agreeableness of smell; a pleasant odor.

fragrant cudweed. *Gnaphalium obtusifolium,* family Compositae, a fragrant, annual herb which is a weed pest of thin soils and waste places. Native east of the Rocky Mountains, U.S. and Canada. Also called *common everlasting; sweet everlasting; sweet balsam; white balsam; old-field balsam.*

fragrant dracaena. *Dracaena fragrans,* family Liliaceae, an arborescent plant grown as a greenhouse plant for its handsome green and white striped leaves, sometimes with a yellow margin, and for its yellowish, fragrant flowers. Native to central Africa.

fragrant mimosa. *Mimosa borealis,* family Leguminosae, a tropical shrub with a fair forage rating which is browsed in winter and spring in southwestern U.S.

fragrant piqueria. *Piqueria trinervia,* family Compositae, an herbaceous, shrubby, perennial plant grown by florists for its fragrant, white flowers produced on long stems. Native to tropical America. Also called *stevia.*

fragrant plantain-lily. *Hosta plantaginea,* family Liliaceae, a perennial herb grown for its fragrant, waxy, white flowers. Native to Asia.

fragrant snowbell. *Styrax obassia,* family Styracaceae, a large, woody plant grown for its fragrant, white flowers. Native to Japan.

fragrant sumac. *Rhus aromatica,* family Anacardiaceae, a hardy, perennial shrub grown for its yellow flowers and aromatic foliage. Native to North America.

fragrant tail-grape. *Artabotrys uncinatus,* family Annonaceae, an evergreen, climbing shrub grown for its habit and its fragrant, brownish-red flowers. Native to East Indies and Madagascar.

fragrant thimbleberry. *Rubus odoratus,* family Rosaceae, a perennial, bristly shrub up to 6 ft. tall, often planted for its showy, purplish flowers and red, flat, insipid fruits. Native to eastern U.S. Also called *thimbleberry; flowering raspberry;* (erroneously) *mulberry.*

frail. See *tap: 2.*

frailer. A person who harvests tree nuts, *e.g.,* pecans, by shaking and flailing the limbs with poles (southern Illinois, U.S., and southward).

frame. 1. The skeleton of a structure; essential parts in their proper places, as the *frame* of a house, the *frame* of a tree, etc. 2. A structure supporting or enclosing things, as that which contains a door. 3. See *cold frame; garden frame.* 4. A unit designed to hold honeycomb which consists of one top bar with shoulders, one bottom bar, and two end bars.

frame scraping. The dark beeswax obtained from scraping the wooden frames and old honeycombs. See *frame: 4.*

framework. The basic branch and leader structure of a tree. See *frame: 1.*

frameworking. The grafting and budding of trees in such a manner that the original framework of the tree remains. Also called *porcupine-grafting.*

Franchet cotoneaster. *Cotoneaster francheti,* family Rosaceae, a partly evergreen shrub grown for its pinkish flowers and its orange-red fruit. Native to China.

Franciscan raintree. *Brunfelsia americana,* family Solanaceae, a shrub or small tree grown as a pot plant for its white to yellow, night-fragrant flowers. Native to tropical America. Also called *Franciscan nightshade; lady-of-the-night.*

frangipani. Any shrub or tree of the genus Plumeria, family Apocynaceae. Native to tropical America. See *white frangipani; nosegay frangipani; Cuban frangipani.*

frank. See *frankfurter.*

frankfurter. A popular, smoked sausage made of beef, pork, and various seasonings, such as pepper, salt, mace, sugar, saltpeter, and coriander.

franklinia. *Franklinia alatamaha,* family Theaceae, a deciduous shrub or small tree up to 25 ft. tall, grown for its large, late-blooming, white flowers. Native to Asia and U.S. Also called *Franklin tree.*

Franklin tree. Franklinia.

Fraser balsam fir. *Abies fraseri,* family Pinaceae, an evergreen tree grown as an ornamental, for its lumber used in pulp, cooperage, boxes, millwork, general construction, woodenware, and for oleoresin. Native to central eastern U.S. Also called *southern balsam fir.*

Fraser magnolia. *Magnolia fraseri,* family Magnoliaceae, a deciduous or half-evergreen tree grown for its fragrant, creamy-white flowers. Native to North America.

Fraser thimbleberry. *Rubus fraseri,* family Rosaceae, a shrub grown occasionally as an ornamental. Native to eastern North America. See *fragrant thimbleberry; western thimbleberry* (both probable ancestors).

frass. Excrement of larvae and/or refuse left by boring insects.

fraxinella. See *gasplant dittany.*

freak. An unusual type of plant sometimes found in breeding, caused by a Mendelian unit character which makes the plant unsuitable to produce seed for reproduction.

freckle. Any dark-colored plant lesion. See *peach scab.*

free acid. H_3PO_4, free phosphoric acid in fertilizers.

free air. Atmospheric air unrestricted in any way.

free air cell. An air cell which moves toward the uppermost point in the egg as the egg is rotated slowly. This results in a reduction in grade for the egg.

freeboard. The distance between the water surface and the top of retaining banks or structures.

freeburning. The part of a forest or prairie fire, or portion of the perimeter of a fire, on which no work has been done to hinder or stop its spread.

free choice. A method of feeding in which the animal is free to make its own choice of feeds, medicants, etc. Also called *cafeteria method of feeding.*

free flow. In hydraulic structures, designating a direct flow, not diverted by submerged or backwater.

free from defects. 1. Designating a yolk of an egg that shows no spots or areas on its surface indicating the presence of germ development or other defects. 2. Designating a condition of near perfection of products or animals.

free from tuberculosis. In selling cattle, a guarantee by the seller that an animal or animals are known by test to be free from bovine tuberculosis.

free grass. Grass that could be used by any cowman for grazing purposes (western U.S., 19th century).

free grazing. In the early history of western U.S., the unlimited, unbounded, and unrestricted grazing on lands.

freehold. An estate in land held for life, or in fees.

free-living larvae. Immature stages of worm parasites living independently of a host in direct contact with the soil, grass, etc.

free-martin. A female, twin calf, born with a male. Barrenness occurs in 90 percent of the females born co-twin to a male.

free of excess flesh. In judging or grading livestock, especially cattle, designating a desirable condition of flesh and quality or character of body conformation and finish.

free permit. A type of usage permit issued by the U.S. National Park and Forest Service for limited or controlled use of nationally owned lands. It is issued to bona fide settlers for a limited number of domestic livestock for grazing and to prospectors, campers, etc., for the actual number of pack animals.

free range. Unrestricted grazing which is not limited by fencing or yardage controls.

freesia. 1. Any bulbous herb of the genus Freesia, family Iridaceae. Species are grown for ornament. Native to South Africa. 2. *F. refracta,* common freesia, a very beautiful, fragrant, pot plant, is grown for its yellow or greenish-yellow flowers. See *Armstrong freesia.*

freestone. a. Any of several varieties of peaches and nectarines characterized by the easy separation of the fleshy part of the fruit from the stone, *i.e.,* the mesocarp separates readily from the endocarp. *Cf. clingstone.* b. Designating the stone of such a fruit.

free water. That water which the soil is unable to retain against the force of gravity. Also called *gravitational water; excess water; drainage water.*

free weir. An unsubmerged weir. The downstream water surface is lower than the weir crest. See *weir.*

freeze. 1. a. A condition of the weather in which the air temperature at plant level falls to 32° F. or 0° C. or lower, with the result that tender plants are frozen. See *frost: 1.* b. To reach the temperature of freezing, 32° F. or 0° C. 2. Of a plant, to die or be impaired as a result of cold. 3. To preserve food products by rapidly reducing the temperature to about 0° F. and maintaining the temperature well below 32° F.

freeze-drying. The evaporation of volatile material from a frozen product by high vacuum.

freeze out. 1. To lose a characteristic flavor, odor, color, etc., because of cold, as when a food product is stored in a freezer. 2. To lose a crop of fruit by cold temperatures which kill the blossoms.

freezer. 1. A refrigerated storage cabinet, room, locker, etc., in which food products may be frozen quickly and stored indefinitely without appreciable loss of flavor, color, odor, etc. See *locker.* 2. Any of many types of devices for the manufacture of ice cream. 3. A palsied horse. 4. Export mutton sheep (Australia).

freezer burn. An undesirable and unattractive condition frequently found with dressed poultry which is kept or stored in quick or deep freezers; caused by an improper control of temperatures and humidity within the freezer.

freezer-locker plant. A commercial enterprise in which the operator furnishes storage space in lockers within a large refrigerated area for storage of food products by individuals for a fee.

freezer space. Storage space for food products which is maintained at a temperature below 29° F.

freezer-storage plant. See *freezer-locker plant.*

freezing. See *freeze.*

freezing injury. 1. Any damage to living plant or animal tissue which is caused by low temperature. 2. Damage to fruit and vegetable tissues which are exposed to temperatures at or below their freezing points, usually occurring after harvest. See *contact frost injury.*

freezing mixture. A snow or ice and salt mixture used to lower temperatures below 32° F. Such a mixture is used in home freezing of ice cream. The rapidly melting ice takes up heat from surrounding objects, thus lowering the temperatures of the ice cream mix to freezing.

freezing temperature. 32° F. or 0° C.

freezing tunnel. A room in a commercial, food-products freezing concern in which the products are rapidly brought to a temperature of about 0° F.

freezing weather. An outside temperature of 32° F. or 0° C.

Fremont screwbean. *Strombocarpa odorata,* family Leguminosae, a shrub or small tree, which is closely related to the mesquites and is an aggressive invader of range lands. In the spring its leaves and twigs are cropped by livestock; in late summer its large, sweet, pulpy clusters of twisted seed pods are also eaten by stock. Its overall rating as a food plant, however, is poor. Found in southwestern U.S.

French Alpine. A breed of excellent, domestic milk goats which originated in the European Alps. Males weigh up to 170 lbs., females up to 125 lbs. No distinct color has been established for the breed.

French artichoke. See *artichoke.*

French bean. 1. A green bean which is sliced into long, slender pieces preparatory to cooking or canning. 2. See *kidney bean.*

French-Canadian. A breed of dairy cattle found in eastern Canada, especially in Quebec, which was brought in by the early French settlers from Normandy and Brittany. These cattle are much like Guernseys and Jerseys with which they have a common early origin.

French cherry. See *pitanga.*

French clover. See *crimson clover.*

French Coach. A medium-weight breed of horse which was originated in France for army use. Thoroughbred stallions were mated to native, Normandy mares of Flemish descent; the resulting offspring called demisangs.

French combing. That wool which is longer than clothing wool and shorter than strictly combing, but long enough to comb on French combs. Also called *baby combing.*

frenched. 1. Designating a plant affected by frenching. See *frenching: 2.* 2. Designating a retail cut of meat in which a rib chop, leg of lamb, etc., has had the meat cut away from the bone.

French endive. A variety of chicory (*q.v.*) grown for a winter salad, whose roots are dug from the ground in the autumn and stored as a root crop. As they are needed, they are placed in pots and forced. Short, crisp, flavorful leaves develop which are served in salads. Also called *witloof.*

frenching. 1. Removing the flesh a distance of 1½ to 2 ins. from the end of the bone in preparation of retail cuts of meat. 2. a. A nonparasitical disease of crops, especially corn, cotton, and tobacco, characterized by a thickening, narrowing, and curling of leaves. It is ascribed to deficiency of some nutritional element. b. See *mottle-leaf of citrus.*

french lands. Certain kinds of soils or land on which crops such as corn and tobacco are subject to the disease frenching (Piedmont Plain of eastern U.S.). See *frenching: 2.*

French lavender. True lavender.

French lilac. Any of a group of varieties of lilacs which are hybrids having as one parent the common lilac, *Syringa vulgaris.* Also called *French hybrid lilac.*

French marigold. *Tagetes patula,* family Compositae, a tender, annual herb grown for its yellow flowers with red markings. Native to Mexico. Also called *marigold.*

French mulberry. See *American beautyberry.*

French pear. See *pear: 2.*

French physic nut. Barbadosnut.

French rose. *Rosa gallica,* family Rosaceae, a hardy, upright, armed shrub grown for its large, solitary, pink or crimson flowers, for its petals which are used in jams, and for the oil distilled from the blossom for use in perfumes and condiments. Native to Eurasia. Also called *Provence rose.*

French spinach. Garden orach.

French tamarisk. *Tamarix gallica,* family Tamaricaceae, a shrub or small tree growing to 30 ft. in height; grown for its feathery foliage and white or pink blossoms. Native to Europe. Also called *tamarisk; salt cedar.*

French weed. Field pennycress.

French wheat. See *buckwheat.*

French willow. Fireweed.

frequency-index quadrat. A vegetation index which provides a rapid method of determining species abundance in a plant association by means of a series of plots.

Fresa. A mild, sweet, soft, cows' milk cheese made in Sardinia.

fresh. 1. Designating a cow which has recently dropped a calf. 2. Designating very recently harvested or gathered food products. 3. Designating an egg of good quality. See *fresh eggs.* 4. Designating sweet water; *i.e.,* not salty water.

fresh butter. Recently manufactured butter which has not undergone a period of storage.
fresh cheese. Any unripened cheese.
fresh-chilled. In marketing, designating a beef carcass which has been held at temperatures ranging from 34° to 38° F. from 24 to 36 hours.
fresh concentrated milk. A milk concentrated to the fat and solids-not-fat requirements of evaporated milk but which has not been sterilized.
fresh cow. A cow that has only recently calved.
fresh-dressed. In marketing, designating a dressed chicken which may have been dry- or ice-cooled, but has not been frozen.
fresh eggs. In marketing, denoting those eggs of U.S. quality A or better which have not been held or delayed in their movement to market. *Cf. cold storage eggs; storage-packed eggs; held eggs; undergrades.*
freshening. To renew milk production in cows after normal calving.
fresh frozen. In marketing, designating a fresh, dressed beef carcass which has been subjected to a temperature of —10° to —20° F. and stored at 0° F.
fresh ham. Any ham which has not been salted, smoked, or cured in any manner. See *ham.*
fresh hard-chilled. In marketing, designating a dressed chicken which has been hard chilled or frozen but which has not developed any characteristics typical of products in cold storage or any evidence of deterioration from freezing, and has not been held at low temperatures for more than 60 days.
fresh manure. Recently excreted animal dung whose application can be harmful to plant tissues because of rapid chemical and fermentive changes which take place.
fresh market. Designating fresh produce, as contrasted to canned, dried, or preserved produce.
fresh marsh. See *fresh-water marsh.*
fresh meadow swamp. A grassy swamp in which the water is fresh.
fresh meat. In marketing, designating a carcass which has undergone no substantial degenerative change since slaughter.
fresh milk. Sweet, wholesome milk recently produced by the cow; no deleterious changes which would lower its quality have taken place.
fresh mulch. The upper, coarse, leafy layer of only slightly decomposed plant material of the A soil horizon, or surface, of grasslands, which corresponds to litter of forest soils.
fresh-packaged. In marketing, designating food products which are cleaned, graded, washed, and placed in transparent containers for the retail trade soon after picking, slaughtering, etc.
fresh vegetable. Vegetables which have been recently harvested, as contrasted to canned, dehydrated, or frozen vegetables.
fresh water. 1. Clean, cool, newly pumped, or spring water. 2. Sweet water; *i.e.*, water having a minimum of salts in solution, as contrasted to ocean water which is high in salt concentration.
fresh-water marsh. Coastal marsh lowland, on which standing water is fresh, in contrast to the brackish or salt-water marsh on the seaward side. See *salt-water marsh; tidal marsh.*
fresh whole fruit. 1. In marketing, designating one food product, such as ice cream, which contains fresh, uncut fruit. 2. Fruit as it is harvested from tree, bush, or vine before processing.
fresno. See *fresno scraper.*
fresno scraper. 1. A horse-drawn, adjustable scraper designed for scraping down mounds and filling hollows where the distance of the haul is not great enough to require wheels to carry the load. It has a long, narrow scoop or bucket. 2. A tractor-drawn scraper of the rotary type for small scale cutting and filling.
friable soil. A soil with aggregates that can be readily ruptured and crushed with application of moderate force; one easily pulverized or reduced to crumb or granular structure.
fribby. Designating fribs *q.v.*
Fribourg. A cooked curd, Swiss-type cheese made in Switzerland and northern Italy.
fribs. Short, dirty bits of wool; those tufts of wool that result from the staple's being cut twice in shearing.
friction head. The energy which is lost as a result of disturbances set up by contact between a moving stream of water and its containing conduit. For convenience, friction losses are best distinguished from losses due to bends, expansions, obstructions, impacts, etc., even though all such losses are often included in *friction losses.*
friction loss. See *friction head.*
friction pulley. A friction wheel in which the surface of the rim is in contact with the rim surface of another pulley; usually the driving pulley is faced with wood, the driven pulley being smooth iron. Such pulleys are often used in lumber working machinery.
friction trip. A device, frequently a stiff spring, on the shovel shank of cultivators, planters, etc., which allows the shovel to pass over an obstinate or hard object without bending or breaking.
friction wheels. Wheels which transmit power by frictional contact.
Friday hoe. Grape hoe *q.v.*
Friedman test. A test for pregnancy in which a small amount of the urine of the tested animal is injected into the blood stream of a virgin rabbit. Pregnancy is indicated by certain changes in the ovaries of the rabbit.
frier. See *fryer.*
Friesian. A name sometimes incorrectly applied in the U.S. to the black and white dairy cattle, the Holstein-Friesian *q.v.*
Friesian Clove. A clove-spiced, whole or partly skimmed cows' milk cheese made in the Netherlands.
frijole. Any cultivated bean of the genus Phaseolus, but especially a black-seeded variety of the kidney bean (*q.v.*) that is grown for its edible, dried seed (Mexico and southwestern U.S.).
frijo-lillo. See *mescalbean sophora.*

frill. A series of overlapping ax cuts, made through the bark and into the sapwood, that encircles a tree trunk. Also called *ring; girdle.* See *girdle.*

frilled-leaved endive. Curled endive.

frill ring. A frill in which the chips are left in place on the lower edge of the girdle.

fringebell. *Schizocodon soldanelloides,* family Diapensiaceae, a hardy, low, evergreen shrub grown for its habit and its flowers, deep rose shading to white at the edges. Native to Japan. Also called *fringed galax.*

fringecup. Any perennial herb of the genus Tellima, family Saxifragaceae. See *Alaska fringecup.*

fringed Amazonvine. *Stigmaphyllon ciliatum,* family Malpighiaceae, a woody vine grown for its habit and its golden yellow flowers. Native to tropical America. Also called *butterflyvine; orchidvine.*

fringed bleedingheart. *Dicentra eximia,* family Fumariaceae, a slender herb grown for its white flowers tipped with yellow. Native to North America. Also called *wild bleedingheart.*

fringed galax. Fringebell.

fringed gentian. *Gentiana crinita,* family Gentianaceae, a biennial herb grown for its lance-shaped leaves and its bright blue, solitary, terminal, fringed flowers. Native to North America.

fringed heath. *Erica ciliaris,* family Ericaceae, a prostrate shrub grown in warm areas for its foliage and its rosy-purple flowers. Native to Europe.

fringed loosestrife. Fringed steironema.

fringed pink. See *lilac pink.*

fringed polygala. *Polygala paucifolia,* family Polygalaceae, a perennial plant of open, hardwood forests which spreads by underground rhizomes and stolons and is grown in the wild garden for its showy, rose purple flowers. It also produces small, inconspicuous, cleistogamous, subterranean flowers. Found from New Brunswick to Manitoba in Canada and south to the mountains of Georgia, U.S. Also called *fringed milkwort; gay wings; bird-on-the-wing; flowering wintergreen.*

fringed sagebrush. *Artemisia frigida,* family Compositae, a perennial herb grown for its silvery-white, aromatic foliage. It is found wild over a large part of western U.S., where it is browsed by sheep, goats, and cattle. Its forage rating is good for goats and sheep; fair for cattle. Also called *mountain sage; wormwood sage; wild sage; mountain fringe; fringed wormwood.*

fringed steironema. *Steironema ciliatum,* family Primulaceae, a perennial herb grown for its yellow flowers. Native to North America. Also called *fringed loosestrife.*

fringed tapeworm. *Thysanosoma actiniodes,* a parasite of domestic sheep and certain related wild ruminants in western North America and Central and South America. The tapeworm lives in the bile ducts of the liver and in the small intestine. It causes thickening, or hypertrophy, of the bile ducts. In the U.S., livers infected with this tapeworm are condemned as unfit for human consumption. The life history of this tapeworm is unknown, but is very likely transmitted by some small invertebrate animal.

fringed wormwood. Fringed sagebrush.

fringetree. Any shrub of the genus Chionanthus, family Oleaceae. Species are grown for ornament. See *Chinese fringetree; white fringetree.*

fringewater. Mobile water, suspended in the capillary fringe above the water table completely filling the smaller interstices, which moves with the fluctuations of the water table. When the water table is shallow, water moves upward from it to replace loss by evaporation and transpiration being actuated by the resultant force of unbalanced film pressure minus the downward pull of gravity.

frit fly. *Oscinella frit* (L.), family Chloropidae, a widely-distributed fly, the larvae of which feed within the stems of small grains and grasses.

Frizzle. An odd breed of chicken raised mainly for exhibition purposes; the feathers curl backwards and primary wing feathers and tail feathers are imperfect.

froe. A wedge-shaped cleaving tool which is used in making shingles or shakes by hand. See *frow.*

frog. 1. That part which holds the plow bottom parts together; an irregularly-shaped piece of metal to which the share, landslide, and moldboard are attached. 2. The triangular-shaped, horny pad, located on the posterior-ventral part of the hooves of horses, mules, etc.

frogbit. *Hydrocharis morsus-ranae,* family Hydrocharitaceae, an aquatic, floating herb grown in pools and aquaria. Native to Eurasia.

frogeye. 1. More or less rounded leaf spots of various plants in which the necrotic, central portion is darker colored than the outer portion; caused by any of several different fungi. *Frogeye* on apple leaves is due to infection by *Physalospora obtusa,* family Pleosporaceae, which, if it attacks the fruit, causes black rot. On tobacco leaves *frogeye* spots are due to *Cercospora nicotianae,* family Dematiaceae. In periods of variable moisture and temperature, the fungus produces concentric or zonate lesions resulting in *frogeye* or target board. 2. See *black rot: 1.*

frog flower. See *buttercup.*

frog mouth. A defect in the finish and appearance of Swiss cheese which is associated with a cracked open edge.

frog spittle. Cuckoo spit(tle).

frog wort. See *bulb buttercup.*

Fromage á la Creme. A soft, rich, uncured, whole milk cheese made in France.

Fromage Fort. A cooked, skimmed milk cheese, ripened in dry ashes, made in France.

frond. 1. The leaves of true ferns (Order Filicales); less often, the large leaves of palm trees. 2. Sometimes, the rounded, flattened stems of duckweed *q.v.*

front. See *swell: 1; breast: 1-3.*

frontage. That portion of real estate which lies immediately adjacent to a stream, lake, street, or highway.

front country. The hills and runs adjoining the plains (New Zealand).

frontfooting. See *forefoot.*

front furrow wheel. An attachment to the plow which is placed on the front and connected to the hitch to aid in guiding and turning.

front gather. The proper adjustment and forward inclination of the axle journals which keeps the wheels in line, preventing them from working outward.

frontier. In the settlement of the U.S., the margin between occupied land and the wilderness; the sparsely settled land that formed a transition between the more densely settled land and the wilderness.

frontlands. In wide alluvial bottoms, the strip bordering the channel of a river which includes natural levees, as opposed to backland which is the outer part of the bottoms, generally more clayey and often swampy.

frost. 1. A condition of the weather in which the air temperature at plant level falls below 32° F. or 0° C. due to a local cooling of the air during a calm, cloudless night, causing the ground and other objects to be coated by frozen vapor or dew. See *glaze; light frost; heavy frost; killing frost; hoar frost; black frost.* 2. A freeze. 3. Of, or pertaining to, a frozen condition of soil, as in the expression, "*frost* still in the ground." See *frost line.*

frost alarm. An alarm which rings by an electric mechanism when the temperature falls to a predetermined level near freezing. It is used as a warning of impending frost.

frost blistered. Designating a leaf in which the lower epidermis is raised in spots from the underlying mesophyll as a result of low temperatures at the time the leaf was unfolding from the bud.

frost canker. A canker (on the bark of trees) of various sizes and shapes which results from localized killing of the bark or bark and cambium layer by low temperatures.

frost comes out. A thaw.

frost control. Various measures taken to prevent frost damage to more valuable crops, such as citrus, or certain other orchard or truck crops. Preventive measures are: the use of any of several types of heaters; irrigating or sprinkling water on the plants; using propeller blades to mix the upper warm with the lower cold air; use of coverings of paper, cloth, or plastics.

Frost count. A procedure used to estimate the number of colonies of bacteria per milliliter of milk.

frost crack. A longitudinal crack or split in the trunk or large limbs of a tree, extending from the bark into the sapwood, which results from a sudden extreme drop in temperature during winter. Also called *trunk splitting.* See *cup shake.*

frosted. Pertaining to plants which have been damaged or changed physiologically by light or heavy frosts.

frosted food. Frozen food.

frosted teat. An udder injury of cows which results from exposure to severe cold weather.

frostflower. Mexicanstar.

frost-free. 1. Designating that period or season of the year between the last killing frost of spring and the first killing frost of autumn. See *killing frost.* 2. Designating a region in which frosts or temperatures below 32° F. are not experienced.

frost grape. *Vitis vulpina,* family Vitaceae, a hardy, woody vine which is a common wild grape from New England to Colorado and Texas, U.S. It produces an edible, black, sweet-after-a-frost fruit. It has been used as one parent in several horticulture varieties of grape. Also called *riverbank grape; chicken grape; winter grape; raccoon grape.*

frost heaving. Heaving *q.v.*

frosting. The marginal edging of color on the feathers of birds.

frost injury. Any injury to a plant which results from low temperatures after plants have started into growth in the spring, during the period of vegetative activity, or before the plant has matured and entered the period of winter dormancy. *Cf. winter injury.*

frost line. The maximum depth to which soil is frozen in the winter; this may be an inch or two, or may be several feet. The soil thaws completely in temperate regions but may remain permanently frozen a short distance below the surface in the Arctic regions. See *permafrost.*

Frost little-plate method. A quick method of making a plate count for bacteria of milk (named after the man who introduced the technique).

frost necrosis. See *contact frost injury; freezing injury: 2.*

frost-nipped. Frost-bitten; designating a frost injury *q.v.*

frost pocket. A depression or hollow in either a hilly terrain or a pitted plain into which cold night air descends creating a frost hazard for sensitive crops.

frost-proof. Designating a variety, plant species, or a product which is not harmed, injured, or impaired by freezing conditions.

frost rib. A pronounced linear protrusion of callus tissue which is formed by the repeated healing over of a frost crack *q.v.*

frost ring. A brownish line or zone of abnormal tissue induced by freezing, extending partly or entirely around a stem within a growth ring of a tree.

frost split(ting). See *frost crack.*

frothy. See *foamy.*

frow. A cleaving tool with a wedge-shaped blade used by coopers in splitting staves for casks, etc., and in making shingles and clapboards. The handle is at right angles to the back of the blade and is held in the left hand, while the mallet which drives the wedge into the log is held in the right hand.

frowsy wool. See *mushy wool.*

frozen. 1. Designating a frozen food. 2. Designating a condition of congealing, killing, etc., as a result of exposure to temperatures of less than 32° F. or 0° C.

frozen cream. A frozen-food milk product which contains 50 to 75 percent cream.

frozen eggs. A frozen food product which usually consists of whole eggs without the shells; however, either the yolks or whites may be missing also.

frozen food. Any food product which has been preserved by subjecting it to a very low temperature, usually about 0° F. to –20° F., and storing it at less than 32° F.

frozen food locker. See *freezer-locker plant.*

frozen milk. Milk which is processed and packaged for freezing and subsequent sale in the frozen state.

frozen pack. See *frozen food.*

fructification. Production of spores by fungi, algae, and other plants on the organ producing the spores or seeds. See *fruiting body.*

fructose. Fruit sugar found in all sweet fruits.

Frühstück. A small, Limburger-like, whole or partly skimmed, cows' milk cheese made in Germany.

fruit. 1. Botanically, the matured ovary of a flower and its contents including any external part which is an integral portion of it. 2. In a popular sense, the fleshy, ripened ovary of a woody plant, tree, shrub, or vine, used as a cooked or raw food; it is not always completely and satisfactorily distinguished from a vegetable. The latter may also include edible leaves, roots and tubers. A *fruit* may be considered a dessert and not the principal part of a meal as a vegetable is so considered (such a meaning has accepted usage and has also been confirmed by a court decision). See *vegetable.* 3. To bear or produce *fruit* in any of the senses.

fruit belt. Any section of a country throughout which the raising of fruit is a principal and commercially important industry.

fruit blight. One symptom of fireblight (*q.v.*), resulting from invasion of the fruit by bacteria, in which sunken, necrotic canker areas develop and gradually involve the entire fruit. There is a bacterial ooze from the surface in wet weather. Also called *fruit spot.*

fruit blotch. One symptom of apple blotch (*q.v.*) in which irregular, brown patches appear on the fruit. The superficial lesions are studded with many small, black pycnidia.

fruit branch. A branch of a plant which bears fruit.

fruit bud. See *blossom bud.*

fruit butter. A food product prepared by boiling screened fruit pulp, with or without the addition of sugar, fruit juices, and spices, to a semisolid mass of homogenous consistency.

fruit cellar. Cold cellar.

fruit cracking. See *cracking: 2.*

fruit crown gall. See *crown gall.*

fruit-cutting knife. Any knife used to cut fruit, but specifically, one used to cut the cling peach around the suture from the surface to the pit.

fruit drop. Any substantial shedding of immature fruit or of fruit as it approaches maturity.

fruit farm. A farm on which the raising of fruits is the principal enterprise. Also called *fruit ranch.*

fruit fly. See *Mediterranean fruit fly; West Indian fruit fly.*

fruit gumming. 1. A physiological disease of citrus, resulting from a variety of causes, in which a gum forms on the surface or within the fruit. 2. Gum exuding from any type of fruit resulting from various conditions.

fruiting. 1. The production or bearing of fruit by a plant. 2. Designating a branch, cane, etc., on which fruit is borne, as a *fruiting* branch, *fruiting* cane, *fruiting* wood.

fruiting body. A complex fungus structure containing or bearing spores. Important types are apothecia, perithecia, conidiophores, coremia, sporangia, pycnia, aecia, pycnidia, acervuli, and sporodochia. At maturity the spores are disseminated by the *fruiting body.*

fruiting shoot. In bramble fruit, a lateral growing from a second-year cane.

fruit jam. See *fruit preserve.*

fruit jar. A glass jar with a wide mouth and an airtight lid or seal used as a container in preserving or canning fruits, vegetables, meats, etc.

fruit juice. The expressed, unfermented juice of fresh fruits used as a food product, flavoring, etc.

fruit leather. A product which is the same as fruit paste (*q.v.*), except that it is further concentrated and cooled in sheets, has a leathery consistency, and may be eaten out-of-hand.

fruit marmalade. A clear jelly preserve in which are suspended slices of a particular fruit or peel.

fruit mix. A mixture of fruits which is canned or frozen in a sirup.

fruit of paradise. Grapefruit.

fruit paste. The screened fruit pulp, with or without the addition of sugar, of a paste consistency. See *fruit leather.*

fruit picker. A person who harvests fruit. In the U.S., he is often a migratory worker *q.v.*

fruit pit. The endocarp and seed of fruits of the genus Prunus, such as those of prunes and peaches.

fruit preserve. Preserves which are made by cooking prepared fruit in a sugar sirup until the concentration of sugar reaches 55 to 70 percent. The pieces of fruit must retain their shapes, be crisp, and be permeated with sugar.

fruit press. See *basket press; rack and cloth press; screw press.*

fruit purée. Fruit in the form of a purée *q.v.*

fruit rot. Any decay of fruits which is usually caused by bacteria or fungi.

fruit rot of eggplant. A serious, fungous disease of eggplant, caused by *Phomopsis vexans,* family Sphaerioidaceae, characterized by circular or oblong, brown lesions on all above-ground plant parts, becoming grayish with a black halo. On the fruits there are grayish or light-brown blotches which become soft red.

fruit-salad plant. Ceriman *q.v.*

fruit setting. See *set: 3.*

fruit spot. 1. See *fireblight; bitter pit.* 2. Any distinctive, small, fruit lesion caused by infection or physiological disturbances. 3. See *leaf blight and fruit spot of pear and quince.*

fruit spot of apple. A fungous disease caused by *Phoma pomi,* family Sphaerioidaceae, characterized by minute spots of the same color as the fruit, but darker at the fruit lenticels. In time the lesions become sunken and dark red, brown, or black. The lesions are generally superficial.

fruit spur. A short, stout branch on which the flowers and fruit of a fruit tree are borne.

fruit tree. Any tree which is cultivated for the edible fruits it bears, as the apple, pear, peach, cherry, plum, apricot, orange, etc.

fruit tree leaf roller. *Archips argyrospila* (Wlk.), family Tortricidae, a caterpillar which infests nearly all deciduous fruits and certain forest trees, bush fruits, and herbaceous plants. It feeds on the leaves, buds, and small fruits, weaving a light web around several leaves enclosing a fruit.

fruit tree pulvinaria. See *cottony peach scale.*

fruity. In butter grading, designating a butter with a flavor suggesting certain fruit. Such butter is rated C.

frutescent. Shrub-like in appearance or habit.

fruticetum. A growing collection of shrubs similar to an arboretum *q.v.*

fruticose. Designating a woody plant which has the characteristics of a shrub.

fryer. Any young chicken approximately 12 to 20 weeks old, of either sex, weighing more than 2½ lbs. but not more than 3½ lbs., which is sufficiently soft-meated to be cooked tender by frying.

fry out. To cook fat pork scraps so as to extract the lard. Also called *rendering.*

fuchsia. 1. Any shrub of the genus Fuchsia, family Onagraceae. Species are grown as outdoor ornamentals in warm areas, as greenhouse plants, as summer bedding plants, and as window-box plants. See *Magellan fuchsia; trailing fuchsia.* 2. *F. hybrida,* a shrubby plant usually grown as a house plant for its white, rose, or purple flowers with a crimson calyx.

fuchsia-flowered gooseberry. Fuchsia gooseberry.

fuchsia gooseberry. *Ribes speciosum,* family Saxifragaceae, a showy, ornamental, evergreen shrub, grown in warm areas for its long, pendulous, bright red flowers and red fruits. Native to North America.

fuel. Any substance which will, by burning, produce heat and light.

fuel oil. A petroleum derivative, most often used as a fuel, which has a higher flash point than kerosene and which is sometimes used as a herbicide. It is being replaced by furnace oil and diesel fuel.

fuel wood. 1. Rough wood for use as fuel which is prepared from waste material, such as branchwood, tops, defective or small logs, waste lumber, or from trees especially cut into lengths suited to the fire box. Also called *firewood.* 2. The portion of a stand less than saw log size.

fugacious. Soon perishing, as the flower of a plant.

Fuji cherry. *Prunus incisa,* family Rosaceae, a deciduous shrub or small tree grown as an ornamental for its white to pale pink flowers. Native to Japan. Also called *Mame cherry.*

full blood. Designating animals of unmixed breed.

full bloom. Designating a plant at the period during which most of its flowers are blossoming.

full-cell process. Any process for impregnating wood with preservatives or chemicals in which a vacuum is created to remove air from the wood before admitting the preservative.

full-cream cheese. Any cheese manufactured from whole milk.

Fuller rose beetle. *Pantomorus godmani* (Crotch), family Curculionidae, a beetle whose larva feeds on the roots of vegetables and ornamental plants; the adult feeds on the leaves of deciduous trees.

fuller's earth. A massive or shalelike clay deposit. The clay mineral is mainly montmorillonite, and the earth itself is relatively nonplastic. It is employed in fulling or degreasing wool and in clarifying mineral and vegetable oils.

fuller's teasel. *Dipsacus fullonum,* family Dipsacaceae, a thistle-like herb which is sometimes raised in flower gardens, but most often as a special crop. The prickly head is used for raising the nap on woollen cloth. Native to Europe.

full feed. A feed or ration being fed to the limit of the animal's appetite.

full fleece. Designating a sheep with a fully developed fleece long enough for shearing.

full-fleshed. Designating an animal or bird which is fully developed and meaty.

fulling. Shrinking and felting woolen cloth so that it becomes thicker and denser and has a face produced upon it which covers up the weave pattern.

full in the hock. Designating a horse afflicted with spavin or other ailments of the hock joint.

full load. The normal maximum load under which an engine or machine is designed to operate continuously.

full moon. The instant when the moon is directly opposite the sun, in elevation 180°, and therefore fully illuminated, appearing as a bright circular disc. Some people erroneously believe that certain plants should and others should not be planted in the full of the moon.

fullmouth. Designating an animal which has a complete set of incisor teeth; a mature animal.

full-mouthed sheep. See *eight-tooth sheep; failing-mouth.*

full pink. Designating that stage of blossom development in apples just before most of the buds expand to full bloom.

full-slip. In harvesting melons, the easy separation of the fruit from the vine in which a cleavage layer develops around the peduncle at the base of the fruit so that as the fruit is removed from the vine, the fruit slips off easily and leaves a large, clean scar. See *half-slip.*

full water capacity. See *field capacity.*

fully-stocked stand. A forest stand in which all growing space is effectively occupied but which has ample room for the development of the crop trees.

fulvic acid. In soil chemistry, the complex of organic substances remaining in solution upon acidification of a dilute alkali extract of soil.

fulvous daylily. Tawny daylily.

fume damage. The injury to the crown of a tree which results from exposure to injurious industrial chemical gases.

fumigant. Any liquid or solid substance which forms vapors that destroy pathogens, insects, and rodents. It is usually used in soils or in closed structures.

fumigate. To destroy pathogens, insects, etc., by the use of certain poisonous liquids or solids which form vapors. See *fumigant; soil fumigant.*

fumigation. The use of poisonous gases for destruction of pests, mainly, rodents and insects. *Fumigation* can also destroy microorganisms, but may be less effective since not all gases which kill animals, such as rats, are toxic to bacteria or other microorganisms. See *fumigant.*

fumitory. See *drug fumitory.*

functional. Designed or adapted primarily to perform some operation or duty.

functional analysis. A mathematical concept of soil in which the aim is to express soil knowledge in the form of a curve or an equation.

fundi finger-grass. *Digitaria exilis,* family Gramineae, a grass grown for its seed which is similar to millet. Native to tropical Africa. Also called *fundi.*

funeral cypress. Italian cypress.

fungaceous. Fungoid or fungus-like.

fungal disease. Any disease caused by a fungus.

fungi. The plural of fungus.

fungicidal. Designating any chemical substance capable of killing fungi.

fungicide. Any chemical which kills or inhibits fungi: *e.g.,* Bordeaux mixture, lime-sulfur, ferbam.

fungistat. An agent or chemical material which prevents the growth and reproduction of, but does not kill, fungi.

fungous. Pertaining to a fungus, as a *fungous* disease.

fungous brown stain. A deep brown, fungous discoloration of the sapwood, found most often in pines.

fungus. A lower order of plant organisms, excluding bacteria, which contains no chlorophyll, has no vascular system, and is not differentiated into roots, stems, or leaves. They are classified in the plant kingdom division, Thallophyta, and vary in size from single-celled forms to the huge puffballs of the meadows. *Fungi* are familiar as molds, rusts, smuts, rots, and mushrooms. For the large part they reproduce prolifically by means of single or multicelled spores of various longevities, which are disseminated by air or water. They are either saprophytic or parasitic and many are considered useful in breaking down dead vegetation and organic matter into humus and as agents of fermentation, as in yeasts; others are also destructive in rotting structural timbers, posts, cloth, leathers, etc. The parasitic forms cause destructive plant diseases; a few are human and animal parasites.

fungus gnat. Various members of the family Mycetophilidae, especially *Sciara* spp.; gnats which infest mushrooms and various greenhouse plants, among them, sweet peas, injuring the roots and weakening the plants.

fungus root rot. Mushroom root rot.

fungus spores. The reproductive cells of a fungus.

funnel. A metal or glass implement with a wide mouth and tapering spout used for transferring liquids or powdered material from one vessel to another.

funnelcreeper. See *catclaw funnelcreeper.*

funnel-shaped udder. A pendulous or poorly attached cow's udder.

fur ball. A digestive illness found in domestic cats caused by the ingestion of fur from the body forming an undigestible ball in the stomach. It is similar to hair balls found in cattle. See *hair ball; wool-ball.*

fur bearing sheep. The Karakul (*q.v.*); the only breed of sheep in the U.S. kept primarily for the fur pelts of its young.

fur farm. A specialized type of farm where fur is produced as the chief source of income, from animals kept in captivity, such as the fox and mink.

furious rabies. One form or stage of rabies (*q.v.*) characterized by the affected animal becoming morose, violent, and indifferent to authority. The animal has dilated pupils and a vacant stare, displays excitability, indifference to food, difficulty in swallowing, and a deranged appetite; these symptoms are accompanied by snapping, biting, and running away. See *dumb rabies.*

furlong. A measure ⅛ of a mile in length.

furnish. To supply a farmer, especially a sharecropper (*q.v.*), with food for his family, seeds, fertilizer, feed, etc., to enable him to raise a crop. The various items are charged to his account, and after the crop is harvested, the account with interest must be paid before any money received from the crop is paid to the farmer. It has generally been proven an unsatisfactory system and has largely fallen into disuse.

furrow. 1. a. The opening left in the soil after the furrow slice has been turned by the plow bottom. b. To make a *furrow* with a plow. 2. A depression in the ground surface, plowed along a prescribed line, for controlling surface water and soil loss.

furrow bank. Furrow crown.

furrow crown. The upper surface of the furrow slice (*q.v.*) after it is turned. Also called *furrow bank.*

furrow drain. To drain surface water from a field by means of open furrows made with a plow.

furrow drill. A planter to which are attached furrow openers *q.v.*

furrow face. The vertical side of the cut in the soil made by a plow, away from which the turned soil is thrown.

furrow irrigation. A method of irrigating in which water is run in small ditches, furrows, or corrugations, usually spaced close enough together to afford lateral penetration between them.

furrow opener. A disk or shoelike attachment placed ahead of the planting mechanism on a seed planter, such as a corn planter, to open a furrow for satisfactory placement of seed.

furrow slice. The soil that is cut, raised, and inverted by the share and moldboard.

furrow sole. The bottom of the furrow on which the plow bottom slides.

furrow wall. The unplowed land next to a furrow.

furrow wheel. The wheel of a plow which rolls along in a furrow as land is being plowed or disked.

furry jasmine. *Jasminum multiflorum,* family Oleaceae, a climbing shrub grown in the greenhouse for its fragrant, white flowers. Native to India.

fur-type sheep. See *Karakul.*

furuncle. An abscess; a boil.

furze. Gorse.

Fusarium. A genus of fungi of the family Tuberculariaceae, of which many species have been studied and named. Most of these are saprophytes, but many are important plant parasites. The hyphae of some parasitic species of *Fusarium* may invade the water-conducting tissues of plants resulting in wilting. Other species may cause rots of fruits, stems, roots, leaves, etc. Still others cause symptoms called "yellows," perhaps to be considered a form of wilt. Some species are limited to special hosts, while others are able to attack many different plants. Since those fungi largely gain entrance into the host plant through sound or injured roots, ordinary means of spraying are ineffective in controlling the disease. The most efficient means of control are by planting crop varieties which have been obtained by selection or by breeding for resistance to the disease. For *Fusarium* blight of the various cereal grains, such as barley, oats, rice, rye, wheat, etc., see *cereal scab.* In the definitions following are listed and named various specific diseases of plants due to *Fusarium.*

Fusarium kernel rot. See *Gibberella rot of corn.*

Fusarium rots. Fungous diseases that attack plants, leaves, stems, fruits, roots, bulbs, corms, etc., often through wounds. The fungi bring about an immediate rot of the infected area, or a rot developing in storage. See *Chart A.*

Fusarium wilt (diseases). Fungous diseases that enter the sound roots or those injured by nematodes of the host plant. The fungi soon enter the water-conducting tissues and spread through the plant which shows various symptoms such as yellowing and wilting of the leaves, branch rot, and eventual death. See *Chart B.*

Chart A

FUSARIUM ROTS	SPECIES
carnation, bud rot	*Fusarium poae*
corn, pink rot of ears	*Fusarium moniliforme*
cotton, boll rot	*Fusarium roseum* and others
cucurbit, root rot	*Fusarium solani* f. *cucurbitae*
gladiolus, corm rot in storage	*Fusarium oxysporum* f. *gladioli*
muskmelon, fruit rot	*Fusarium* spp.
peas, root rot	*Fusarium solani* f. *pisi*

Chart B

FUSARIUM WILTS	SPECIES
wilt of alfalfa	*F. oxysporum* f. *medicaginis*
wilt of China aster	*F. oxysporum* f. *callistephi*
wilt of cowpea	*F. oxysporum* f. *tracheiphilum*
wilt of cotton	*F. oxysporum* f. *vasinfectum*
wilt of Dianthus	*F. oxysporum* f. *dianthi*
wilt of flax	*F. oxysporum* f. *lini*
wilt of muskmelon	*F. oxysporum* f. *melonis*
wilt of pea	*F. oxysporum* f. *pisi*
wilt of soybean	*F. oxysporum* f. *tracheiphilum*
wilt of spinach	*F. oxysporum* f. *spinaciae*
wilt of sweet potato	*F. oxysporum* f. *batatatis*
wilt of tobacco	*F. oxysporum* f. *nicotianae*
wilt of tomato	*F. oxysporum* f. *lycopersici*
wilt of watermelon	*F. oxysporum* f. *niveum*

Fusarium yellows diseases. Fungous diseases that enter the sound roots or those injured by nematodes of the host plant. *F. oxysporum* f. *conglutinans* causes cabbage yellows; *F. oxysporum* f. *appi* causes celery yellows. See *Fusarium wilt (diseases).*

fuscous blight. A blight of beans caused by a variety of *Corynebacterium flaccumfaciens.* For characteristics see *bacterial blight of beans.*

fused phosphate. Phosphatic fertilizer material with a variable percentage of phosphate which is prepared by heating rock phosphate with alkali or another agent designed to make the phosphate more available. *Cf. calcined phosphate.*

Fusicoccum canker of peach. A fungous disease caused by *Fusicoccum amygdali,* family Sphaerioidaceae, characterized by large, irregular or circular, often brown lesions on the leaves, with black pycnidia near the center of the lesions. Cankers appear around the buds of the previous season's growth.

futtah. A storehouse or storage shed built on stilts (obsolete, New Zealand).

future contract. See *futures.*

future delivery. See *futures.*

Future Farmers of America. In the U.S., a national organization established in 1929 for high school students who are studying farming under the Smith-Hughes Vocational Agriculture program.

future market. See *futures.*

futures. Agreements to buy and receive or to sell and deliver a farm commodity at a future date, with these distinguishing characteristics: all trades in the same contract, such as the New York Cotton Exchange contract, have the same unit of trading: *e.g.,* 100 bales of cotton; the terms of all trades are identical; a trade, therefore, may be offset later by a contra trade through the same broker; all trades must be made by open outcry around a ring within the hours prescribed; the contract has a basic grade where more than one grade is deliverable; delivery is made during a single calendar month; the trades are cleared through a Clearing House daily. (Traders in cash or spot goods usually refer to sales for shipment or delivery in the future as deferred or forward sales. Such sales however, are not standardized as are futures contracts just described above.)

futures contract. An agreement between two people, one who sells and agrees to deliver, and one who buys and agrees to receive, a certain kind and quantity of product to be delivered during a specified delivery month at a specified price, and under the terms of the rules and regulations of the Exchange.

futures market. A market which provides the mechanics wherein trading in standardized grades of certain products can be done swiftly and economically in standardized promises, or contracts, to deliver or receive products at a specified time in the future. This organized market is complimentary to the buying and selling of actual products on the cash or spot market *q.q.v.*

futurity. A type of competition for future judging of offspring in certain classes of livestock in which the sire and dam are nominated at the time of mating and if the offspring looks promising it is kept eligible for future competition by the payment of certain eligibility fees to the breed association in which the animal is registered.

fuzz. Short, soft, epidermal hairs.

fuzz tail. Cayuse.

fuzzy deutzia. *Deutzia scabra,* family Saxifragaceae, a widely cultivated shrub grown in warm areas for its white or pinkish flowers. Florists also used the plant for forcing. Native to eastern Asia.

G

gad. Any implement used to urge horses, mules, or oxen into greater effort, which often consists of a rod with a sharp point on one end. Also called *goad.*

gadfly. Generally, any of the horseflies of the family Tabanidae.

gage. See *gauge.*

gag rein. A type of rein which pulls the bit in the mouth of a horse upward.

gagroot. See *Indiantobacco lobelia.*

gaillardia. Any herb of the genus Gaillardia, family Compositae. Species are very popular flower garden plants. Native to North America. See *maroon gaillardia; rose ring gaillardia.*

gain. An increase in the weight of an animal for a specified period.

Gaiskäsli. A soft, German and Swiss, goats' milk cheese which is ripened about three weeks.

gait. The particular style of an animal's manner of traveling, especially of a horse's walking or running. See *trot; pace: 2; gallop; walk; canter; rack: 1; fox trot; stepping pace.*

galactagogue. Any substance which promotes the flow of milk.

galax aphylla. *Galax aphylla,* family Diapensiaceae, a perennial, evergreen herb widely used in floral arrangements, especially for funerals. Native to southeastern U.S. Also called *beetleweed; galaxy.*

gale. A wind which has a velocity of about 40 miles per hour or more at the height of 32 ft. above the ground. On the Beaufort scale a *gale* is a wind of force 8.

galena. An obsolete name for salt pork (*q.v.*), taken from Galena, Ill., U.S., which was once a pork-packing center.

gall. 1. A plant tumor which may be found on practically any part of a plant. It results from irritation by a foreign organism, usually an insect, a fungus, or bacterium. 2. A sore spot on the shoulder of a horse produced by the rubbing or chafing of harness or a saddle.

gallberry. Inkberry.

gallberry flat. A tract of wet, sandy land, characterized by a dense growth of the shrub, inkberry (*q.v.*). The soil is acid, low in fertility, and generally regarded as inferior agriculturally (coastal plain of southeastern U.S.).

galled. 1. Designating an animal having a sore caused by chafing or rubbing. 2. Designating a plant which has galls growing on it.

galled spot. 1. A sterile area in a field which results from removal of the top soil by erosion. 2. A small area on a horse where friction has severely abrased or removed the skin.

gallery. A relatively small, usually circular, passageway or burrow made in bark or in wood by an insect for feeding or oviposition. Also called *tunnel.*

galleta. *Hilaria jamesi,* family Gramineae, an erect, perennial grass, a range species palatable to stock, whose forage rating is excellent. Native to North America; found from Wyoming to Texas and west to California, U.S.

gallinaceous. Resembling, or of the nature of, domestic fowls and pheasants; belonging to the group or order Galliformes, which includes chickens, turkeys, pheasants, partridges, quail, etc.

gall-making maple borer. *Xylotrechus aceris* Fisher, family Cerambycidae, a small beetle which infests several species of the maple, mining the trunks and large limbs, often killing distal portions.

gallnut. Aleppo gall.

gallon. 1. A liquid measure: (a) in the U.S., 8.3359 lbs. of water at 4° C., 231 cu. ins., or 3.7853 liters; (b) in the U.K., 10 lbs. of water at 62° F., or 277.3 cu. ins., or 4.546 liters. 2. A dry measure, in some areas, for the sale of berries, cherries, etc., of 231 cu. ins.

gallonage chart. A chart showing milk tank capacity in gallons only as related to the graduations on the gauge rod or surface gauge.

gallons per minute. An index or unit of flow which is used to indicate the rate of flow of water, especially in pump drainage and in irrigation, as 1,000 *gallons per minute.*

gallop. A fast, three-beat gait of a horse, in which two diagonal legs are paired, their single beat falling between the successive beats of the other two legs, the hind one of which makes the first beat of the three. A hind foot makes the first beat in the series, the other hind foot and diagonal fore foot make the second beat simultaneously and the remaining fore foot makes the third beat in the series. Then the body is projected clear of the ground and the hind foot makes the first beat in a new series.

gallop-depart. The change from another gait to the gallop.

Galloway. A polled breed of beef cattle originating in Scotland which has a black, long, curly coat; somewhat smaller than the Aberdeen Angus.

Galloway pony. A breed of hardy, small horses originating in Galloway, Scotland; used mostly as a small cart horse.

gallows-grass. Hemp.

gall sickness. Anaplasmosis.

galvayne. A system used in breaking a horse in which the halter is tied to the tail in such a way as to bend the neck toward the tail, restricting free movement.

gama grass. *Tripsacum dactyloides,* family Gramineae, a tall, coarse forage grass. Native to North America. Also called *eastern gama grass.*

Gambel oak. *Quercus gambelii,* family Fagaceae, a shrub or small tree which is a palatable browse. Cattle subsisting entirely on it, however, may be poisoned. Native to North America; found from Wyoming southward.

gamble (gambel). See *gambrel (stick).*

gambrel roof. A roof, each side of which is made up of two slopes. The roof truss approximates an inverted catenary curve. The lower rafter is cut to set at approximately 60° and the upper rafter at approximately 30° from the horizontal.

gambrel roof house. A type of greenhouse roof in which the lower half of the roof slopes at approximately 60° while the upper half slopes at approximately 30° from the horizontal so as to better utilize the heat and light of the sun during the winter months.

gambrel (stick). A wooden or metal rod whose ends are inserted in the hocks of hogs, and which is used to support animals while butchering. Also called *gamble.* See *beef pritch.*

game. Species of animal life which are hunted for food, sport, or recreation.

game bird. 1. Any chicken which is bred for fighting purposes. See *Modern Game; Old English Game.* 2. Any wild fowl, such as quail, turkey, partridge, pheasant, etc., which is hunted for food or sport.

game management. The manipulation of game populations and their food supply, habitat, competitors, etc., so that a given supply is available for harvest.

game refuge. An area which is closed to hunting in order that its excess game populations may flow out and restock surrounding areas.

gamete. A mature male or female reproductive cell (sperm or egg).

gamma globulin. A specific protein fraction of the fluid part of the blood in which are included the bodies which protect against certain infections (immune bodies).

Gammelost. A semisoft, ripened, blue-mold, table cheese which is made in Norway from sour skim milk.

gammon. 1. A cured ham. 2. A side or flitch of bacon.

gander. A full-grown male goose.

gang. 1. A combination of implements or tools, as nozzles, disks, etc., which are to be used simultaneously so as to increase productivity and reduce labor. 2. A group of workers.

gang bolt. See *arbor: 4.*
gang cheese press. A cheese press in which several cheeses are compressed simultaneously.
gang disk plow. A disk plow which has two or more bottoms.
gang lever. A lever which is used to raise and lower, or adjust, a particular gang. See *gang: 1.*
gang plow. A plow which has two or more bottoms.
gangrene. Massive death of tissue caused by bacterial infection or interference by injury, infection, freezing, etc., with blood circulation. Dry *gangrene,* or mummification, may occur without bacterial action; moist *gangrene* occurs as a result of the action of bacteria.
gangrenous ergotism. A chronic ergotism (*q.v.*) resulting from consuming food or feed contaminated with ergot through a prolonged period; characterized in horses and cattle by a loosening and sloughing off of hooves, tips of the ears, and tail, and shedding of the teeth, emaciation, and death.
gangrenous mastitis in ewes. Mastitis (*q.v.*) of sheep; an inflammation of the udder of ewes, frequently caused by infection with the filth bacteria, *Pasteurella mastitidis, Staphylococcus aureus,* etc. The condition is characterized by loss of appetite; depression; solitary habit; refusal to suckle her lamb; straddled gait; hot, painful udder; a hardening and caking of the udder; dark, bluish-violet spots on the udder. Also called *blue bag; black garget; garget; mammitis.*
gap. 1. A break in a fence or wall which may be used as a gate. 2. A depression forming a break in the continuity of the crest of a mountain ridge.
gaper. A bird infested with the fowl gapeworm *q.v.*
gapes. An infestation by the fowl gapeworm *q.v.*
gapeworm. Fowl gapeworm.
garambullo. *Myrtillocactus geometrizans,* family Cactaceae, a treelike, very spiny cactus grown in the desert garden as an oddity. Native to Mexico.
garawi. See *Sudan grass.*
garbage. Animal offal and/or vegetable scraps and table refuse which is sometimes fed to hogs and chickens.
garbage tankage. The rendered, dried, and ground product derived from waste household food materials which is used as a fertilizer. It contains 2.5 to 3.3 percent nitrogen (3 to 4 percent ammonia), 2 to 5 percent phosphoric acid, and 0.5 to 1 percent potash.
garbanzo. Gram chick-pea.
garden. 1. A plot of ground, usually less than an acre, devoted to the growing of vegetables, flowers, herbs, fruits, etc., which is often adjacent to the home. See *botanic garden.* 2. The plants in a *garden* collectively. 3. To cultivate plants in a *garden.*
garden aster. China aster.
garden balsam. *Impatiens balsamina,* family Balsaminaceae, a tender, annual herb grown for its flowers of a variety of brilliant colors. Native to Asia. Also called *balsam.*
garden celery. *Apium graveolens* var. *dulce,* family Umbelliferae, a biennial herb, an important truck crop, grown as an annual for its edible leafstalks. Native to Eurasia. Also called *celery.*
garden centipede. Garden symphylid.
garden chamomile. Roman camomile.
garden cress. Gardencress peppergrass.
gardencress peppergrass. *Lepidium sativum,* family Cruciferae, a weedy, annual herb sometimes grown for a sharp-flavored, salad plant. Native to western Asia. Also called *garden cress.*
garden crop. 1. Any crop which is raised in a garden. 2. Any vegetable crop as contrasted to a field crop.
garden cultivator. Any of various, small, man-powered devices for cultivating a garden or a similar tract, which usually consists of two handles attached to single or double wheels, a beam for attaching various tools (hiller hoe, cultivator teeth, rake teeth) and a plow bottom. See *garden tractor.*
garden currant. See *red currant.*
garden dibble. See *dibble: a.*
garden eggplant. See *eggplant.*
garden escape. See *escape: 2.*
garden fennelflower. *Nigella sativa,* family Ranunculaceae, an annual herb grown in the herb garden for its fragrance and its seeds used in seasoning. Native to Europe. Also called *fennelflower.*
garden fleahopper. *Halticus bracteatus* (Say), family Miridae, an insect which infests many kinds of plants in northeastern U.S., sucking sap from the leaves, producing discolored areas.
garden fly honeysuckle. Tartarian honeysuckle.
garden frame. A boxlike structure, with no bottom and with a glass, plastic, or glazed cover; used for protecting, propagating, or growing plants. See *cold frame; heated frame; hotbed.*
garden geranium. See *pelargonium.*
garden gooseberry. See *European gooseberry.*
garden heliotrope. See *valerian: 2.*
garden hoe. A hand tool which consists of a long, wooden handle with a steel cutting blade attached at right angles at one end. It is used in weeding, in cultivating, to make small furrows for garden seeds, etc.
garden hose. Rubber or plastic tubing, ½ to ¾ ins. inside diameter, used to carry water from an outside tap or a hydrant to the point of use on lawns, flower beds, gardens, etc.
gardenia. Any tropical, evergreen shrub or tree of the genus Gardenia, family Rubiaceae. Species are grown as ornamentals. See *capejasmine.*
gardening. See *garden: 1.*
garden lemon. Mango melon.
garden lettuce. See *crinkled garden lettuce.*
garden lovage. *Levisticum officinale,* family Umbelliferae, a perennial herb grown for its aromatic fruits used in confections. Native to southern Europe. Also called *lovage.*
garden melon. See *muskmelon.*
garden mint. See *spearmint.*
garden nasturtium. See *nasturtium: 2.*

garden onion. See *onion: 2.*

garden orach. *Atriplex hortensis,* family Chenopodiaceae, an annual herb grown as a potherb. It is tolerant of alkaline soils and is drought resistant. Native to Eurasia. Also called *French spinach; mountain spinach; sea purslane; butter leaves.*

garden pansy. *Viola tricolor* var. *hortensis,* family Violaceae, an annual or short-lived perennial, flowering plant, the common pansy, grown in flower gardens as a biennial. The flowers are of various colors, including purple, blue, yellow, and white. Native to Europe. Also called *pansy; heartsease.*

garden parsnip. *Pastinaca sativa,* family Umbelliferae, a biennial or perennial herb grown as an annual for its edible tap root. Native to Eurasia. Also called *parsnip.*

garden pea. *Pisum sativum,* family Leguminosae, an annual herb grown for its edible seed; it is a very important and popular vegetable. Native to Eurasia. Also called *English pea.* See *field pea; baby gardenpea; sugarpod gardenpea.*

garden petunia. See *petunia: 2.*

garden phlox. Summer phlox.

garden pink. Grass pink.

garden plant. Any plant, flower, or fruit grown in a garden, in contrast to field plants, trees, etc.

garden plow. See *garden cultivator.*

garden plum. *Prunus domestica,* family Rosaceae, a deciduous tree popularly cultivated for its edible fruit eaten fresh or used in jams, preserves, and jellies. Native to Eurasia. Also called *European plum; common plum.*

garden portulaca. See *portulaca: 2.*

garden radish. *Raphanus sativus,* family Cruciferae, an annual or biennial herb grown as an annual for its edible root. Native to Eurasia. Also called *radish.*

garden rake. A hand tool which consists of a long, wooden handle to which a rod with projecting fingers is attached at right angles. It is used to pull leaves, grass, stalks, etc., together and to smooth seed beds, etc. See *rake: 1.*

garden rhubarb. *Rheum rhaponticum,* family Polygonaceae, a perennial herb grown for its edible, acid leaf petioles used in pies, sauces, etc. Native to Asia. Also called *pie plant; wine plant; rhubarb.*

garden rocket. Dame's-rocket.

garden sage. *Salvia officinalis,* family Labiatae, a hardy shrub grown for its leaves used as seasoning and for the oil extracted from the whole plant used in perfuming soaps. Native to the Mediterranean area.

garden seed. Any flower or vegetable seed which can be planted in a garden.

garden serviceberry. *Amelanchier ovalis,* family Rosaceae, a deciduous shrub grown as an ornamental for its white flowers and bluish-black fruit. Native to Europe.

garden snail. Any of several species of the genus Helix, family Helicidae, particularly *H. aspera* and *H. hortensis,* snails which are destructive to garden crops.

garden sorrel. *Rumex acetosa,* family Polygonaceae, a perennial herb sometimes grown as a potherb, but which is also a weed pest of eastern U.S. Native to Eurasia. Also called *sorrel; donkey oats.*

garden spot. A locality or region in which the soil is highly fertile or has been made highly productive.

garden springtail. *Bourletiella hortensis* (Fitch), family Sminthuridae, a very small, jumping insect which chews holes in thin leaves and in cotyledons of beans, etc.

Garden State. A nickname which has been applied to several states in the U.S.: Kansas, Illinois, and New Jersey, especially to the latter, because of its highly specialized gardening and truck crop industries.

garden symphylid. *Scutigerella immaculata* (Newp.), family Scutigerellidae, a tiny, white myriapod which has 12 pairs of legs. It lives in the soil and feeds on the roots of many plants. Also called *garden centipede.*

garden tool. Any tool normally used in a garden: *e.g.,* a hoe, rake, dibble, garden cultivator, spade, fork, shovel, trowel, wheelbarrow, weeder, etc.

garden tractor. Any gasoline-powered tractor, usually not over five horsepower, which is used to plow and cultivate gardens. It may also furnish power for sprayers, pumps, weed bars, snow plows, carts, etc.

garden vegetable. Any edible plant which is raised in gardens or in truck gardens.

garden verbena. A general term, sometimes shown as *Verbena hortensis,* loosely used for hybrids derived from several species of verbena *q.v.*

garden violet. See *sweet violet.*

garden webworm. *Loxostege similalis* (Guen.), family Pyraustidae, an insect pest of clover, alfalfa, sugar beets, and many other plants with succulent growth. It weaves a light web over a leaf and feeds on the leaf beneath the web. It is found widely over the U.S., Canada, and Mexico.

garden weed. Any weed that is found in cultivated flower and vegetable gardens. See *weed: 2.*

gare. Straight, coarse, glossy wool fibers which are sometimes found mixed with finer wool. Also incorrectly called *kemp.*

garget. 1. The abnormal changes in the udder and its secretion as a result of mastitis *q.v.* 2. See *pokeberry: 2.* 3. A hog and cattle disease in which the head or throat becomes inflamed.

gargety milk. An abnormal condition of milk as it comes from the udder of the cow in which the milk appears flaky and lumpy due to the masses of fibrin and leucocytes produced by the udder tissues. The milk also has an objectionable taste. *Gargety milk* results from udder injury and infection and is associated with stubborn cases of acute mastitis *q.v.*

garland flower. Rose daphne.

garland heath. *Erica subdivaricata,* family Ericaceae, a low plant grown as an undershrub for its foliage and its rose-red or white flowers. Native to South Africa.

Garland's flower. See *gingerlily: 2.*

garland spirea. *Spiraea arguta,* family Rosaceae, a hybrid shrub grown for its profuse, early, white flowers.

garlic. *Allium sativum,* family Liliaceae, a perennial herb grown for its strong, pungent, distinctive flavored bulb used in seasoning. Native to Eurasia.

garlic flavor. A flavor defect of milk or milk products which suggests the odor and flavor of garlic.

Garnetted stock. See *Garnetted thread waste.*

Garnetted thread waste. Waste yarn and thread that have been reduced to fiber by a Garnett machine *q.v.*

Garnett machine. A machine which, by means of pin-like teeth on a card cloth, reduces wool cloth trimmings, rags, and waste into usable wool fiber.

garter. The feathers immediately about the hock joint of a fowl that are not completely removed by a mechanical picker and must be plucked by hand.

gas gangrene. See *blackleg: 4; enterotoxemia; malignant edema.*

gas hole. A defect in cheese, particularly Cheddar, resulting from the growth of undesirable bacteria; characterized by numerous, uniform, round openings.

gashouse tankage. That residue fertilizer material which has been used to remove sulfur compounds from illuminating gas. It is rich in nitrogen: 5 to 10 percent (equivalent to about 6 to 12 percent ammonia) and, as it contains cyanide, it must be used with great care. Also called *spent oxide.*

gas injury. Injury to plants which results from natural gas, illuminating gas, industrial gases, etc.

gaskin. The lower thigh of a horse's (or other quadruped's) hind leg; the area between the hock and the stifle.

gasoline. A highly volatile product, formed by the distillation of crude petroleum or by the distillation of bituminous coal, which is used as a fuel mainly for internal combustion engines.

gas plant. Gasplant dittany.

gasplant dittany. *Dictamnus albus,* family Rutaceae, a strong-smelling, perennial herb grown for its white flowers. A small flash of light will sometimes result if a burning match is held beneath the flower cluster on a still, warm night. Also called *false dittany; fraxinella; gas plant; burning bush; dittany.*

gassy fermentation. The development of gas in milk, cream, etc., through action of organisms.

gastritis. Stomach inflammation, especially of the lining or mucous membrane.

gastroenteritis. Inflammation of the stomach and intestines. There is an infectious type of *gastroenteritis* in swine probably caused by a virus. See *enteritis; feline enteritis.*

gastrointestinal. Pertaining to the stomach and intestine.

gas vent. A passage to permit the escape of gases of decomposition.

gate. An open frame or a frame covered with pickets, woven wire fencing, etc., hinged on one side to a fence post, wall, etc., and latching on the other; it opens and closes to allow or restrict passage.

gatepost. A sturdy, large fence post to which a gate is hinged.

gate valve. A sluice valve; a valve which has two inclined seats between which the valve wedges down in closing. The valve, when opened, is drawn up into a dome or recess, thus leaving a straight passage the full diameter of the pipe.

gather. 1. To harvest a crop. 2. To collect eggs. 3. To follow in a track, as a wagon's wheel. See *front gather.*

gathered cotton. Picked cotton in contrast to cotton still in the boll on the plant.

gathering board. Either of two inclined pieces on the corn harvester or picker to which the gathering chains are attached. They incline upwards at about a 45° angle.

gathering chain. Either of two endless chains which are attached to the gathering boards of a corn binder or picker. These chains move inward and upward, carrying the corn stalks to a binder, truck bed, or to the snapping rolls.

gauge. 1. A device, graduated to indicate the elevation of a water surface, for registering water levels, flow, velocity, pressure, etc. 2. Any instrument used to measure quantity of materials, etc.

gauge bracket. A substantial, rigid element, permanently affixed to a bulk milk cooling tank, designed for supporting and positioning a gauge rod or surface gauge.

gauge height. The elevation of a water surface above or below a figure corresponding to zero on a gauge.

gauge rod. A graduated measuring device designed to be partially immersed in a liquid and read at the point where the liquid's surface crosses the rod. Also called *dip stick.*

gauge rod support. Gauge bracket.

gauge shoe. A shoe attached to a small plow or furrow opener to maintain uniform depth of a furrow.

gauge wheel. A wheel attached to a plow beam near the clevis to secure uniform depth and width of a furrow.

gauging station. A selected station in a stream channel which is equipped with facilities for measuring the flow of water; a place on a stream where data are gathered for continuous discharge records.

Gautrias. A Port du Salut type of cheese made in the Department of Mayenne, France.

Gavot. A French cheese made from cows', ewes', or goats' milk in the Department of Hautes-Alpes, France.

gayfeather. Any perennial herb of the genus Liatris, family Compositae. Species are grown in the wild garden or in the informal border. Native to North America. See *Kansas gayfeather; tall gayfeather; spike gayfeather.*

gean. Heart cherry *q.v.*

gear sack. A type of duffle bag used by cowboys

to carry their personal belongings when traveling on horseback. It is usually tied behind the saddle.

gear wheel pump. A rotary pump in which two meshing gear wheels revolve in a casing.

gee. A command which directs a horse, mule, or ox to turn to the right. *Cf. haw: 3.*

geese. The plural of goose.

geest. Residual soil, or the loose mass of rock weathering (obsolete).

Geheimrath. A deep yellow, Gouda-type cheese made in the Netherlands.

geigertree cordia. *Cordia sebestena,* family Boraginaceae, an evergreen tree grown in frost-free areas for its scarlet or orange-red flowers. Native to tropical America.

gel. A colloidal solution in which the particles have precipitated: *e.g.,* the coagulum in cheese-making.

gelatin(e). A by-product of the meat-packing industry made from the connective tissues of skin and bone of beef and pork; used as a food, as a medium for bacteria, as a stabilizer for ice cream, etc.

gelatin lump. A rather rare, texture defect of ice cream characterized by small lumps of undissolved gelatin in the product.

geld. a. Designating an animal which is sterile. b. To render sterile, as in castration.

gelding. A castrated male horse.

gene. The hypothetical unit of inheritance located in the chromosome which, by interaction with the other genes and the environment, controls the development of a character. *Genes* are believed to be arranged linearly on the chromosomes.

genera. The plural of genus.

general disqualification. A disqualification (*q.v.*) which has general application and is not limited to a specific breed or variety.

general farmer. One who raises a number of crops and sells a number of farm products, in contrast to the farmer who specializes in a single crop or agricultural product.

general infestation. An insect outbreak which is widely and more or less evenly distributed over a large area or region.

generalized soil map. See *soil map.*

General Land Office. Formerly, a division of the U.S. Department of the Interior, constituted by Congress in 1812, which had charge of public lands. In 1946 it was combined with the U.S. Grazing Service to form the Bureau of Land Management.

general-purpose. Designating an animal, tool, implement, etc., which may be used for several purposes.

general-purpose moldboard. A plow which is intermediate between the sod and stubble types and can be used for either sod or stubble land. It has less curvature than the stubble moldboard and more than the sod moldboard.

General Stud Book (English). The first official record book of purebred livestock, originating in England, 1791; thoroughbred horses were first recorded in it.

generating plant. The equipment necessary for the generation of electrical energy, including the electrical equipment as well as the prime movers. At one time it was a common source of electricity on a farm.

genesis. Origin, or evolutionary development, as of a soil, plant, or animal. (Also a combining form of *genesis,* to indicate manner or kind of origin as parthenogenesis, biogenesis, etc.)

genetic. Of, or pertaining to, genetics.

genetics. 1. The science that deals with the laws and processes of inheritance in plants and animals. 2. The study of the ancestry of some special organism or variety of plant or animal. Also called *breeding.*

genetic selection. The selection of breeding stock for reasons which are not visible to the eye: egg production, milk production, fertility, etc.

genista. Canary broom.

genista caterpillar. *Tholeria reversalis* (Guen.), family Pyraustidae, a caterpillar which webs the foliage of the canary broom (*q.v.*) and feeds on the leaves, sometimes defoliating the plant.

genital. Pertaining to the organs of reproduction.

genital eminence. In sexing chicks, a very small, shiny or glistening projection which is the rudimentary, male, copulatory organ. Also called *male process.*

genital glanders. Dourine.

genital organs. The parts of the animal body that are functional in reproduction.

genitol 923. The 2,4-dichlorophenyl ester of benzenesulphonic acid which is an effective residual acaricide.

genome. A complete set of chromosomes (hence of genes) which is inherited as a unit from one parent.

genotype. The genetic constitution (gene make-up), expressed and latent, of an organism. Individuals of the same *genotype* breed alike. *Cf. phenotype.*

genotypic. Of, or pertaining to, genotype.

gentian. Any herb of the genus Gentiana, family Gentianaceae. Some species are choice plants for the rock garden, for the border, or for the wild garden. See *Andrews gentian; fringed gentian; yellow gentian; stemless gentian.*

gentian speedwell. *Veronica gentianoides,* family Scrophulariaceae, a perennial herb grown for its blue-veined flowers. Native to southeastern Europe.

gentle. 1. To tame an unbroken horse, mule, ass, etc., so that it can be handled and worked. 2. Designating an even-tempered, docile, quiet animal.

gentleman farmer. A person of wealth or social position who owns a farm but ordinarily does not expect the farm to be a profitable enterprise. The term was more current in usage in the U.S. during Revolutionary and pre-Civil War days when it applied to men of distinction or wealth who lived on country estates.

genus (plural: **genera**). One of the kinds of groups used in classifying organisms which consists of a number of similar species (sometimes of only one species). Similar *genera* are grouped in a family.

geographic environment. The natural phenomena (seasons, rainfall, temperature range, floods, drought, soil, topography, etc.) which may effect the origin or development of physical or social life.

geography. The science of the earth and its life: the description of land, sea, air, and the distribution of plant and animal life, including man and his industries, with relationships of these diverse elements.

geologic erosion. The natural process by which land surfaces are worn down by water, ice or frost, and wind.

geology. The science which treats of the history of the earth and its life as recorded in the rocks.

geomorphology. That department of physical geography which deals with the form of the earth, the general configuration of its surface, the distribution of land and water, and the changes that take place in the evolution of land forms.

geophagous. Pertaining to organisms feeding on soil.

geophyte. Any plant which is buried in the ground, but little exposed at the surface and which may fruit underground.

geoponics. The science of agriculture; the tillage of the ground of the earth. *Cf. hydroponics.*

Georgia pine. See *longleaf pine.*

Georgia stock. A walking, one-horse, one-man plow which consists of a beam, handle, and a shank to which the plow share or shovel is attached.

geotropism. A growth of shoots and roots of plants in response to the stimulus of gravity which is positive when the growth curvature is toward the source of the stimulus (*i.e.,* towards the center of the earth) or negative when the direction of growth curvature is opposed to the stimulus.

geranium. Any herb of the genus Geranium, family Geraniaceae. Some species are used in the border or wild garden. Also called *crowfoot; cranesbill.* See *pelargonium* (for the commonly used garden geranium); *herbrobert geranium; spotted geranium.*

germ. 1. In reference to animal disease, a small organism, microbe or bacterium that can cause disease. 2. The embryo of a seed.

German catchfly. Clammy campion.

German chamomile. *Matricaria chamomilla,* family Compositae, an annual herb grown for its seeds used for flavoring. Escaped, it can become a weed pest. Native to Europe. Also called *sweet false chamomile.*

German clover. Crimson clover.

German Coach. A German breed of horse which is large, rather coarse, and may be black, brown, or bay.

germander speedwell. Hungarian speedwell.

germander spirea. *Spiraea chamaedryfolia,* family Rosaceae, a stoloniferous shrub grown for its white flowers. Native to Asia.

German falsetamarix. *Myricaria germanica,* family Tamaricaceae, a deciduous shrub grown for its tamarix-like foliage and its pink or white flowers in terminal racemes. Native to Eurasia. Also called *German tamarisk; false tamarix.*

German iris. *Iris germanica,* family Iridaceae, a perennial herb, a source of many varieties, grown for its purple flowers with yellow beards. See *orrisroot iris.*

German ivy. Ivy groundsel.

German tamarisk. German falsetamarix.

germ cell. A cell which is capable of reproduction or of sharing in the reproduction of an organism, as contrasted with the somatic or body cells which may divide to produce new cells in the same organism.

germ free. Designating a product which is free from harmful organisms.

germicide. A substance that kills disease-causing microorganisms; a germ killer.

germinate. Of a seed or embryo, to begin to grow and develop.

germination. Of a seed or embryo, the beginning of growth and development.

germination percent. The percentage of seeds which germinate under specified conditions.

germination test. Any of several different types of tests conducted to determine the percentage of viable seed in a particular lot.

germinative energy. The relative strength of sprouts and roots at the time of germination. One lot of seeds may exhibit strong, healthy sprouts whereas another of the same variety may have weak ones.

germinator. Any of several devices used to test seed for germination prior to planting.

germ meal. See *corn germ meal.*

germ plasm. The material or physical basis of heredity.

germ spot. The germinal disc of an avian egg showing as a small, circular area on the surface of the yolk which is the area of the blastoderm of a fertilized egg and the point at which embryonic development starts.

germ trap. Any of several devices used to prevent the spread of infectious diseases which may consist of a pan holding a germicidal solution into which a person or animal must walk before entering another area.

germ tube. 1. The hypha which is produced by a germinated fungus spore. 2. The tube which is produced by a germinating pollen grain by means of which the sperm nuclei are brought into contact with the egg and endosperm nuclei.

Gerome. A French or Swiss, soft, cows' milk cheese, to which some goats' milk may be added, flavored with anise and cured for 6 weeks to 4 months.

gestate. To carry young in the uterus from conception to birth.

gestation. Pregnancy.

gestation period. The length of time from conception to birth of young in a particular species. Usual gestation periods for farm animals are: mare, 330-340 days; cow, 280-285 days; ewe,

145-150 days; sow, 112-115 days; goat, 148-152 days; jennet (female ass), 360-365 days.

get. The offspring of a male animal. Also called *get of sire.*

get in lamb. To cause a ewe to conceive.

get of sire. A show ring classification of a group of four animals which are either sex, any age, and the get of a single sire. Some shows have a junior get for those under 3 yrs. of age and a senior get for those over 3 yrs. (a) For sheep, four animals, the get of one sire, both sexes represented (some shows specify lambs). (b) For swine, four breeding animals, the get of one boar. (c) For dairy cattle, four animals, the get of one sire, both sexes represented, but usually not more than two bulls can be shown. (d) For beef cattle, four animals, the get of one sire, both sexes represented. (e) For horses, three animals, the offspring of one sire.

get up. A command which is given to a horse to move forward. See *haw: 3; gee; whoa.*

Gex. A hard, cows' milk, blue-mold cheese, cured 3 to 4 months; made in France.

Geyer larkspur. *Delphinium geyeri,* family Ranunculaceae, an herb which is a very poisonous plant on range lands. Symptoms of poisoning of animals eating it are loss of appetite, staggering, prostration with rigid limbs, constipation, nausea, and bloat. Found on the plains and foothills from Montana to Colorado, U.S.

ghedda wax. Beeswax which is obtained from *Apis dorsata* and other Indian species of bees. Also called *East Indian wax.*

Ghee. An Indian, semi-fluid butter prepared from the milk of the buffalo, cow, sheep, or goat. After the butter has been extracted from the milk, it is heated to drive off excess water, cooled, and only the more-liquid, oily portion used. It is virtually 100 percent butterfat.

Ghent azalea. *Rhododendron gandavense,* family Ericaceae, a hybrid shrub which has numerous varieties; grown for its handsome flowers of various colors.

gherkin. Any small cucumber used for pickling. See *West Indian gherkin.*

ghostweed. See *snow-on-the-mountain euphorbia.*

giant arborvitae. *Thuja plicata,* family Pinaceae, an evergreen tree grown as an ornamental and a valuable timber tree. Native to North America. Also called *western red cedar; red cedar; Washington cedar.*

giant arrowhead. *Sagittaria montevidensis,* family Alismaceae, a tender, tall, perennial, aquatic herb, grown in shallow water, bogs, or aquaria for its greenish-white flowers with brownish spots at the bases of the petals. Native to South America.

giant bristlegrass. *Setaria magna,* family Gramineae, a coarse, annual grass up to 12 ft. tall, with its spikelike panicle up to 20 ins. long and 1½ ins. thick, which is a weed in marshy lands and in wet pastures. Native to the coastal states from New Jersey to Florida and Texas, U.S.

giant brunsvigia. *Brunsvigia gigantea,* family Amaryllidaceae, a perennial, bulbous herb grown in the greenhouse for its long, red flowers. Native to South Africa.

giant cactus. Saguaro.

giant cane. *Arundinaria gigantea,* family Gramineae, a tall, bamboo-like grass with stems growing 12 to 25 ft. high and leaves nearly 12 ins. long with rough and cutting edges. It is the chief plant of the canebreaks from Virginia southward, U.S. Native to North America. See *giant reed.*

giant daisy. *Chrysanthemum uliginosum,* family Compositae, a tall, hardy, perennial herb grown in the flower border for its white flowers. Native to Europe. See *chrysanthemum.*

giant desert candle. *Eremurus robustus,* family Liliaceae, a perennial herb grown as a border plant for its bright pink flowers. Native to central Asia.

giant dracena. *Cordyline australis,* family Liliaceae, a tropical tree grown out-of-doors in very warm areas for its handsome foliage (ribbon-like leaves) and its clusters of waxy-white flowers, as a house plant, a window-box plant, a porch plant, and as a summer bedding plant. Native to New Zealand. Also called *tuft tree; cabbage tree; Ti tree.*

giant evergreen chinkapin. *Castanopsis chrysophylla,* family Fagaceae, an evergreen tree grown in warm areas for its leaves green above and golden underneath. Native to Asia. Also called *golden chinquapin; giant chinquapin.*

giant fennel. *Ferula communis,* family Umbelliferae, a tall perennial grown in the flower garden for its finely cut leaves and its profuse clusters of yellow flowers. Native to Eurasia. Also called *fennel giant; common giant fennel.*

giant filbert. *Corylus maxima,* family Betulaceae (Corylaceae), a deciduous shrub or small tree grown as an ornamental and as one parent of many varieties producing its edible nuts, the *filberts* of commerce. Native to Eurasia.

giant foxtail. See *giant bristlegrass.*

giant garlic. *Allium scorodoprasum,* family Liliaceae, a perennial herb grown for its garlic-like bulbs used in flavoring. Native to Asia Minor and southeastern Europe. Also called *rocambole.*

giant granadilla. *Passiflora quadrangularis,* family Passifloraceae, a woody vine grown in very warm areas for its fragrant, large, white and purple flowers and its greenish-yellow, edible fruit used in soft drinks, sherbets, jams, and jellies. Native to tropical America. Also called *granadilla real; barbadine; maracuja melao.*

giant horsetail. *Equisetum telmateia,* family Equisetaceae, a poisonous, perennial herb which is found in wet meadows in the Pacific Northwest. Native to North America. Also called *ivory horsetail.* See *field horsetail.*

giantism. See *gigantism: 2.*

giantleaf philodendron. *Philodendron giganteum,* family Araceae, a tropical plant grown in the greenhouse for its high climbing habit and its decorative leaves which grow up to 3 ft. in length. Native to the West Indies.

giant lily. Piteira furcrea *q.v.*
giant ragweed. *Ambrosia trifida,* family Compositae, an annual herb which is a weed pest in bottom lands. Native to North America. Also called *great ragweed; king head; horse head.*
giant reed. *Arundo donax,* family Gramineae, a 10 to 15 ft. tall grass which is a persistent weed pest of cultivated fields and waste places. Found in U.S. from New York to Nebraska and south to Arkansas. Also called *cane reed.*
giant sequoia. *Sequoia gigantea* (Syn. *Sequoiadendron giganteum*), family Pinaceae, which is perhaps the world's largest coniferous tree. Its growth is limited to small areas. Native to California, U.S. Also called *big tree.* See *redwood.*
giant snowdrop. *Galanthus elwesi,* family Amaryllidaceae, a bulbous herb grown for its white flowers on tall flowering stalks. Native to Asia Minor. Also called *summer hyacinth.*
giant star tulip. A group of species of mariposa, genus Calochortus, family Lilaceae. The most popular species is the Tolmie mariposa *q.v.*
giant sunflower. *Helianthus giganteus,* family Compositae, a perennial herb grown in moist places for its large yellow flowers. Native to North America. Also called *tall sunflower; wild sunflower; Indian potato.*
giant wildrye. *Elymus condensatus,* family Gramineae, a perennial grass growing to a height of 12 ft., which is grazed closely by cattle and horses but only to a limited extent by sheep. Its forage value is fair. Found from Manitoba to British Columbia, Canada, and south to New Mexico, U.S.
Gibberella ear rot. See *Gibberella rot of corn.*
Gibberella rot of corn. A disease of corn which is caused by *Gibberella zeae* (Syn. *G. saubinetti*), family Hypocreaceae, producing an ear rot with pink to brick-red fungus growths on the ear, husk and kernel; a stalk rot; and a root rot. It may also be caused by *G. fujikuroi,* resulting in discoloration of random kernels on the ear. This same fungus is reported as causing a root rot of rice in Japan. The imperfect stage of Gibberella is Fusarium *q.v.*
gibberellic acid. $C_{19}H_{22}O_6$, tetra-cyclic-di-hydroxy-lactonic acid, one of the gibberellins (*q.v.*). It is a plant growth hormone, applied to plants locally and as a root treatment by drenching the soil.
gibberellins. A group of substances originally isolated from *Gibberella fujikuroi,* an organism which causes elongation of rice shoots, and later from *Fusarium moniliforme.* Three chemically different substances have been isolated. These have been designated as gibberellin A_1 ($C_{19}H_{24}O_6$), A_2 ($C_{19}H_{26}O_6$), and A_3 also known as gibberellic acid (*q.v.*) and gibberellin X. Certain plants treated with these compounds increase in height much greater than normally; some treated biennial plants produce seed in one season; some plants show other responses.
giblets. Any of the internal organs of a fowl used as food, particularly the heart, liver, and gizzard.
gid. A disease of sheep and goats, but rarely of cattle, which is caused by the gid bladderworm (*Coenurus cerebralis*), the larval stage of *Multiceps multiceps,* a tapeworm of dogs. The internal parasites are carried by the blood stream to various parts of the body, and those that reach the brain complete their full development there in a cyst reaching the size of a hen's egg. Infestation is characterized by staggering, loss of control of limbs when excited, weaving gait, excitation, walking in circles, prostration, in some cases also by holding the head lowered or elevated, by resting the head on an object, turning backward somersaults, exhaustion, and death. Also called *sturdy; turn sickness; staggers.*
Giemsa's stain. A complex polychrome stain or dye, used in staining preparations of blood cells and tissues, which is prepared from Azur, methylene blue, and eosin; the eosinates thus formed are the active substances in *Giemsa's stain.*
gigantism. 1. The production of luxuriant vegetative growth which is usually accompanied by a delay of flowering or fruiting. Also called *gone to stalk; gone to weed.* 2. In animals, abnormal overgrowth of a part or all of the body. Also called *giantism.*
gigot. A leg of mutton, venison, or veal which is trimmed ready for the table.
gilgai. A succession of microbasins and microknolls on flat lands; similar to hog-wallow land (adapted from Australia).
gill flirt. A condition of a mare, which usually results from foaling, in which the perineum is so lacerated that the rectum and vagina are united into a common opening. Also called *flutter.*
grill-over-the-ground. See *ground ivy.*
gillyflower. See *carnation.*
Gilpin sulky. A popular sulky plow which was patented by Gilpin Moore in 1875.
gilt. A young female hog that has not farrowed a litter of pigs.
gimmer. A ewe from its first to second shearing. See *gimmer lamb; gimmer hog.*
gimmer hog. A ewe lamb from weaning until its first shearing.
gimmer lamb. A ewe lamb from birth to weaning.
gin. 1. a. The entire machine processes, considered as a unit, which removes cotton lint from the cottonseed and bales lint cotton. b. The grounds, buildings, management, etc., of a *gin.* Also called *cotton gin.* 2. The alcoholic beverage which is obtained by distilling pure spirits over various flavoring materials, principally juniper berries, but also coriander seed, angelica root, and other plant products.
Ginaca machine. A machine used in the canning of pineapples which automatically cuts a cylinder from the central portion of the fruit, removes the shell, cuts off the shell at each end of the fruit, and scrapes out the edible flesh remaining on the shell.
gin bale. A bale of compressed lint cotton as it comes from a gin, weighing about 500 lbs.
gin breast. See *breast: 4.*
gin compress. A compress adapted for use in a cotton gin. See *compress: 1.*

gin cut. Designating cotton that shows damage in ginning through cutting by the saws to an extent that its value is reduced more than two grades.

ginger. 1. *Zingiber officinale,* family Zingiberaceae, a perennial herb grown in India, East Indies, and China, as a source of the *ginger* of commerce. In the U.S. it is grown only in very warm sections out-of-doors as an ornamental for its yellowish-green flowers and as a pot plant. Native to southeastern Asia. 2. The dried or peeled rhizome of *ginger* which is grown in Jamaica, China, Malay, West Coast of Africa, and South America. The roots are dried after boiling in water for about ten minutes to prevent them from sprouting. The flavor of *ginger* is due to a volatile oil, the pungency to gingerol.

gingerbread tree. Doum palm.

gingerlily. Any tropical herb of the genus Hedychium, family Zingiberaceae. Species are grown out-of-doors for ornament and as pot plants. 2. *H. coronarium,* a perennial herb grown out-of-doors in warm areas and as a hot-house plant for its ornamental foliage and its very fragrant, white flowers. Native to tropical Asia. Also called *Garland's flower.*

gin hood. An enclosed conduit which is fitted to the rear of the cotton gin. As lint is removed from the seeds it is blown to the *gin hood* and thence to the lint flu.

ginkgo. *Ginkgo biloba,* family Ginkgoaceae, a deciduous tree with fan-shaped leaves, the female of which produces a fleshy, unpleasant-smelling fruit. The male is grown as a street tree. Native to China. Also called *maidenhair tree.*

ginner. One who owns or operates a cotton gin.

ginner's bale. A low density bale of cotton weighing 12 lbs. per cubic foot, which is 27 ins. thick, 54 ins. long, averaging 45 ins. in width. Average weight is 500 lbs. with 6 steel ties or bands.

ginner's compass bale. See *compress: 1.*

ginnery. See *gin: 1.*

ginning. The process of removing lint cotton from seed cotton. See *gin: 1.*

ginning outturn. The percentage of lint cotton to seed cotton in the same batch or of the same variety.

ginny. A stable attendant.

gin pole. A contrivance for raising or moving heavy weights which consists of a strong pole with four ropes fastened at the top end to guide it and a block and fall fastened on the top of the pole.

gin-run. Designating lint or seed cotton as it comes from the gin without further refinement.

gin saw. A saw, 10 to 12 ins. in diameter, one of 70 to 80 in a gin stand, which is used to draw the lint cotton from the seed.

ginseng. Any hardy perennial herb of the genus Panax, family Araliaceae, which yields *ginseng.* Red berries follow greenish flowers. Native to China and North America. See *American ginseng.*

gin stand. One complete unit of a gin.

gioddu. Yoghurt.

gipsy flower. Houndstongue.

giraffe. A mechanism used for elevating a man to do pruning in an orchard. Also called *orchard giraffe; steel squirrel.*

girasole. See *Jerusalem artichoke sunflower.*

girdle. A wound encircling a living plant which involves the absence of tissues as deep or deeper than the cambium layer. The encircling cuts are deep enough to check the flow of elaborated foods or to cause the death of the plant above the point of the *girdle.* A simple cut by man or animals, or even the removal of a narrow strip of bark during dormancy or very early in the season will check the rate of growth above that point; if done after the first flush of growth, it will result in an accumulation of elaborated foods above the *girdle.* Wide girdling may cause death of the plant above that point. See *injury by mice; injury by rabbits; ringing; ringing of grapes.*

girdling root. A large root which grows so close to the main trunk or another large root of a plant that it chokes or constricts the other roots or a portion of the crown.

girth. 1. The circumference of the body of an animal behind the shoulders. 2. A band or strip of heavy leather or webbing which encircles a pack animal's body; used to fasten a saddle or pack on its back. Also called *cinch.*

Gislev. A hard, cows' milk cheese made in Denmark.

giving milk. The act of yielding milk by a cow or other milk producing animal; lactating. Milk is produced within the udder tissues when these tissues are stimulated by the milking process or by the sucking of the young.

gizzard. The muscular posterior stomach of birds, which has muscular walls and a thick, horny lining; its principal function is the grinding or crushing of coarse feed particles. The presence of grit increases the efficiency of the grinding process. See *giblets.*

gizzard erosion. A poultry disease, probably the result of dietary deficiency, characterized by softening and thickening of the gizzard and by separation of the gizzard lining from the glandular layer.

Gjetost. A boiled, whey cheese made from a mixture of cows' milk and goats' milk in Scandinavian countries.

glabrous. Smooth, not hairy, as a plant part.

glacial drift. Coarse deposits left as a result of glaciation with particular reference to regions covered by continental ice sheets during the last Glacial Period.

glacial soil. Soil which consists of or is formed from materials transported and deposited by glacial action: *e.g.,* that of till plains, moraines, drumlins, kames, and eskers.

glacial till. The unconsolidated, heterogeneous mass of clay, sand, pebbles, and boulders deposited by receding ice sheets. See *boulder clay.*

glaciation. The geological action of glaciers; the overspreading of a large land surface with glacial ice and the consequent production of glacial phenomena: *e.g.*, glacial drift, grooves and striae on rocks, moraines, drainage modifications, etc.

glacis. A very gentle slope from a natural levee, or higher land along the bank of a large river, to the back land or back swamp of a wide alluvial river plain.

glade. 1. An open, grassy tract in a forest. 2. An open wet tract, generally covered with grasses and sedges.

gladiola. A gladiolus.

gladiolus. Showy, popular summer and autumn flowering herbs of the genus Gladiolus, family Iridaceae. Species have sword-shaped leaves, flowers in spikes, and corms. Most of the improved garden forms, sometimes indicated as *G. hortulanus,* have been derived from South African species and modified by variation, hybridizing, and selection. Also called *gladiola.*

gladiolus Botrytis disease. A fungous disease caused by *Botrytis gladiolorum,* family Moniliaceae; characterized by a spotting and rotting of all parts of the plant.

gladiolus corm rot. A fungous disease caused by *Fusarium oxysporum* f. *gladioli,* family Tuberculariaceae; characterized by a rot of all the underground parts of the plant, a yellowing of the leaves, and a distortion of the leaf, stem, and flower growths.

gladiolus cucumber mosaic. See *cucumber mosaic.*

gladiolus Fusarium disease. See *gladiolus corm rot.*

gladiolus grassy top. See *grassy top.*

gladiolus mosaic. See *common mild mosaic of gladiolus.*

gladiolus stunt. A virus disease of gladiolus which is characterized by a dwarfing of the plant.

gladiolus thrips. *Taeniothrips simplex* (Mor.), family Thripidae, a very small insect which is a pest of gladiolus and other ornamental plants. Infested plants have a silvery foliage, which later turns brown, and deformed flowers with whitish flecks or streaks.

gladiolus white break. See *white break.*

gladiolus white pitting. See *white pitting.*

glaesler. A serious body defect of Swiss cheese in which there are large, parallel, clean-cut cracks. Also called *glassler; glass.*

gland. 1. In animals, an organ that secretes substances for the body's use or that excretes waste matter. 2. In plants, any special secreting organ.

gland cistern. The internal portion of the mammary gland immediately above each teat into which milk collects as it is secreted by the milk producing glands of the udder. Each *gland cistern* is very irregular in shape and capacity.

glanders. A contagious disease of horses, mules, etc., communicable to man, caused by *Malleomyces mallei* (*Bacillus mallei*). Pulmonary *glanders* is characterized by unthriftiness and difficult breathing. Nasal *glanders* is characterized by ulcers in the mucous membranes of the nose which discharge a blood-streaked fluid, a swelling of the glands under the jaw, and unthriftiness. Skin *glanders* is characterized by hard nodules under the skin which break to form ulcers. There is a thick discharge from the ulcers which become confluent. Also called *burr; farcy.*

Glanrich. A prominent breed of cattle in southern Germany; a triple-purpose breed used for milk, beef, and work. They resemble Brown Swiss in conformation and color. Also called *Lowland.*

glass. 1. A general term for coldframes, hotbeds, greenhouses, conservatories, etc. 2. A body defect of cheese in which there are parallel, clean-cut cracks in the body of the cheese. Also called *glaesler.*

glass container. Fruit jar.

glass-eye. An animal's eye in which the iris is pearly white.

glass garden. See *terrarium.*

glasshouse. See *greenhouse; conservatory.*

glassler. See *glaesler.*

glass wicking. A wick made of fiber glass used to convey water from a reservoir to the soil around the roots of potted plants.

glassy cutworm. *Crymodes devastator* (Brace), family Noctuidae, a widely distributed cutworm that infests crops planted in soil where grass had been growing.

glassy shell. In marketing, an egg with a thin shell which gives a metallic sound when handled.

Glauber salt. $Na_2SO_4 \cdot 10H_2O$, sodium sulfate decahydrate; used as a laxative.

glauconite. A dull green, natural mineral, hydrated aluminum silicate of iron and potash, which has been used to a small extent as a fertilizer chiefly for its potash and lime. Found abundantly in greensand beds *q.v.* Also called *green earth.*

glaucous. Designating a plant or plant part covered with a whitish bloom that rubs off.

glaze. A smooth coating of ice deposited by rain or drizzle on terrestrial objects which have a temperature below 32° F. Also called *glazed frost.*

glazed frost. See *glaze.*

glazed stage. One of the stages of ripening in corn which occurs about 45 days after fertilization when the covering of the corn kernels changes from a soft, dull, semitranslucent appearance to a hard, smooth, glossy appearance. See *milk stage; soft dough stage; hard dough stage; ripe.*

glean. To gather the useful bits left in the field after harvest.

Gleaners (Ancient Order of). A rural, fraternal organization founded in Caro, Michigan, in 1894. One of its principal purposes was the promotion of cooperatives for the benefit of farmers.

glebe. An archaic and poetic term for soil, cultivated land, or a piece of land assigned to a clergyman as a benefice.

glei horizon. A soil horizon in which the material ordinarily is bluish-gray or olive-gray, more or less sticky, compact, and often structureless. It is developed under saturation in the presence of organic matter. The iron compounds are ferrous.

glen. A narrow valley in a mountainous or hilly terrain which usually contains a stream.

gleyboden. A soil which is characterized by a glei horizon *q.v.*

gley horizon. See *glei horizon.*
globe amaranth. *Gomphrena globosa,* family Amaranthaceae, an annual herb grown in the flower garden for its spherical, red, pink, white, or yellow flowers used as everlastings. Sometimes called *bachelor's button.*
globe artichoke. See *artichoke.*
globe candytuft. *Iberis umbellata,* family Cruciferae, a hardy, annual herb grown for its pink, red, lilac, or violet flowers. Native to Europe. Also called *annual candytuft.*
globe daisy. Any herb of the genus Globularia, family Globulariaceae. See *Syrian globe daisy.*
globeflower. *Trollius europaeus,* family Ranunculaceae, a hardy, annual herb grown in moist places for its unusual leaves and lemon, orange, whitish, or purple flowers. Native to Europe.
globe mallow. Any plant of the genus Sphaeralcea, family Malvaceae. See *rosy globe mallow; Mexican globe mallow.*
globe thistle. Any herb of the genus Echinops, family Compositae, grown for its handsome foliage and blue flowers. See *small globe thistle.*
globe tulip. Any bulbous herb of the genus Calochortus, family Liliaceae; better known as mariposa lily. Native to western U.S. See *mariposa; white mariposa: 1.*
globule. See *fat globules.*
gloomy scale. *Chrysomphalus tenebricosus* (Comst.), family Diaspididae, an insect that infests maples, hackberry, elm, and boxelder.
glorybind. Any herb of the genus Convolvulus, family Convolvulaceae. Some species are garden plants, some are pernicious weeds. See *European glorybind; dwarf glorybind; Morocco glorybind.*
glorybower. Any tropical shrub, vine, or tree of the genus Clerodendron, family Verbenaceae. Also called *bagflower.* See *bleedingheart glorybower; Java glorybower.*
glory-lily. Any showy vine of the genus Gloriosa, family Liliaceae. Native to tropical Africa and Asia. See *Malabar glory-lily.*
glory-of-the-snow. *Chionodoxa luciliae,* family Liliaceae, a low, bulbous herb grown in the rock garden for its early-blooming, blue flowers with white centers. Native to Eurasia.
glory-of-the-sun Chilestar. *Leucocoryne ixioides,* family Liliaceae, a bulbous herb grown out-of-doors in California, U.S., for its blue or whitish flowers. Native to Chile.
glory-pea. *Clianthus dampieri,* family Leguminosae, a viny plant grown out-of-doors in very warm areas and in greenhouses for its lovely, red, pealike flowers. Native to Australia.
glory-rider. A person who rides horses for the satisfaction of mastering an outlaw horse.
gloryvine. Glory-pea.
gloryvine grape. *Vitis coignetiae,* family Vitaceae, a woody, hardy, climbing vine grown for its autumnal, crimson foliage. Its dark fruit is inedible. Native to Asia.
glossanthrax. A disease of horses and cattle in which the oral cavity, especially the tongue, becomes ulcerated and gangrenous.
glossy cotoneaster. *Cotoneaster foveolata,* family Rosaceae, a tall shrub grown for its pinkish flowers and black fruit. Native to central China.
glossy hawthorn. *Crataegus nitida,* family Rosaceae, a small, deciduous thorny tree grown for its white flowers, its autumnal, orange or scarlet foliage, and its dull red fruit. Native to North America.
glossy privet. *Ligustrum lucidum,* family Oleaceae, a tender, evergreen shrub or small tree grown as a hedge plant. Native to Japan and China.
glottis. a. The opening into the windpipe. b. The vocal cords and structures that surround the opening.
Gloucester. A hard, English, blue-mold cheese which is very similar to Derby.
Gloucester Old Spot. A breed of black and white swine of southern England which influenced the American development of the Spotted Poland China swine.
Glover scale. *Lepidosaphes gloverii* (Pack.), family Diaspididae, an insect that is a serious pest of citrus and various other plants.
gloxinia. *Sinningia speciosa,* family Gesneriaceae, a tuberous-rooted herb grown as a pot plant for its violet, purple, or reddish and white-spotted flowers. Native to Brazil.
gloxinia penstemon. *Penstemon gloxinioides,* family Scrophulariaceae, a hybrid, perennial herb grown as a border plant for its flowers of various colors. Native to North America.
glucocorticoids. Hormones which are secreted by the tissues of the outer layers of the adrenal gland. When these hormones are carried by the blood stream to tissues of certain organs of the body, they increase the rate of formation of glucose from proteins and fats. *Glucocorticoids* can normally be used in maintaining the glucose level of the blood in instances when there is a lack of carbohydrates.
glucose. $C_6H_{12}O_6$, a monosaccharide hexose sugar, the end product of the hydrolysis of many polysaccharides; thus starch, cellulose, dextrins, raffinose, sucrose, maltose, and lactose are composed partially or wholly of *glucose* residues. The *glucose* of animal feeds is a sirup obtained by the incomplete hydrolysis of starch from corn and other plant sources. *Glucose* is also used as a supplementary injection in the treatment of many disease conditions. Also called *dextrose.*
glue. A viscous liquid, having strong adhesive qualities. Commercial *glue* is obtained from hides, hoofs, gelatin, starch, resins, etc.
glume. Either of the two lowest, chaffy bracts of a grass spikelet, especially the empty one.
glume blotch. A blight disease of wheat caused by *Septoria nudorum,* family Sphaerioidaceae, which affects the leaves, culms, and heads.
Glumse. A Cottage-cheese type cheese which is made of skimmed milk in Germany.
gluten. The protein-rich product derived from cereal grains. *Gluten* is spoken of loosely as the protein constituent obtained in separating starch from corn in the wet-milling process. For wheat,

the term is more specifically applied to the viscous and semi-elastic substance that gives adhesiveness and the rising quality to bread dough.

gluten feed. See *corn gluten feed.*

gluten meal. See *corn gluten meal.*

glutenous. Of, pertaining to, or designating, gluten.

glut on the market. Designating an oversupply of a product on the market.

glutted market. An oversupplied market.

glut the market. To oversupply the market reducing prices.

glyceride. Natural fats and oils formed in plants and animals by the chemical union of glycerin and fatty acids.

glycogen. $C_6H_{10}O_5$, a carbohydrate similar to starch, found abundantly in the liver and stored in lesser amounts in other tissues and organs. Also called *animal starch.*

glycophyte. A nonhalophytic plant; a plant that does not grow well in a salty soil solution when the osmotic pressure of the solution rises above two atmospheres.

gnat. Any dipterous, biting insect.

gnaur. A knotty excrescence on a tree trunk or root.

goad. A long, pointed stick or rod used to urge draft animals onward. Also called *goadstick; gad.*

go a little tender. To walk as if favoring a lame foot, as a horse.

goat. Any horned, ruminant of the genus Capra, family Bovidae, especially the domestic goat, *C. hircus,* which is bred as a source of milk, meat, and wool or hair. See *Angora goat; French Alpine; Nubian; Saanen; Toggenburg.*

goat biting louse. *Bovicola caprae* (Gurlt), family Trichodectidae, a louse that infests sheep and goats, eating wool fibers and tangling and soiling the fleece. It is very irritating to the infested animals. *Cf. goat sucking louse.*

goat fever. A photosensitization of sheep and goats resulting from ingestion of *Agave lophantha* var. *poselgeri,* family Amaryllidaceae, or the atamasco lily (*q.v.*), and exposure to bright sunlight; characterized by jaundice, liver and kidney lesions, and sometimes by edematous swelling of the face and ears. Also called *swell head; big head.*

goat herd. Several goats which are tended as a unit.

goat herder. A person who tends, drives, feeds, waters, and generally cares for a herd of goats.

goatroot ononis. *Ononis hircina,* family Leguminosae, a shrubby, perennial herb grown for its paired, butterfly-shaped, red and white flowers. Native to Europe. Also called *restharrow.*

goatsbeard. See *vegetable oyster; meadow salsify.*

goatsbeard astilbe. *Astilbe astilboides,* family Saxifragaceae, a perennial, spirea-like herb grown for its clusters of white flowers. Native to Japan.

goat's milk. The whole, fresh, lacteal secretion obtained from a female goat.

goat's-rue. 1. *Galega officinalis,* family Leguminosae, a perennial, erect herb grown for its purplish-blue, pealike flowers. Native to Eurasia. 2. Virginia tephrosia *q.v.*

goat's-rue swainsonpea. *Swainsona galegifolia,* family Leguminosae, a small shrub grown as a greenhouse plant for its cluster of red flowers; poisonous to grazing animals in Australia. Native to Australia.

goat sucking louse. *Linognathus stenopsis* (Burm.), family Linognathidae, a louse that sucks the blood of sheep and goats. It robs the animal of nutrition and stains the wool with fecal spots.

goatweed. See *bishop's goutweed; St. John's-wort:* 2.

goat willow. *Salix caprea,* family Salicaceae, a deciduous shrub or small tree grown for its bright yellow catkins. It is one of the common pussy willows. Native to Eurasia and northern Persia. Also called *sallow.*

go back. 1. To decline in vigor, as a tree which has been propagated on uncongenial or unsatisfactory stock. 2. To revert to an ancestral type.

gobby. A lumpy, unattractive condition of the fat covering the body of an animal, such as a sheep or beef animal.

goblet pruning. Head pruning.

go blind. To fail to develop a flower.

gobo. Great burdock.

godetia. Any plant of the genus Godetia, family Onagraceae. Species are mainly annual flowering plants. See *farewell-to-spring godetia; Whitney godetia.*

go-devil. 1. Any of several types of sled cultivators or of various other similar implements used on farms. 2. A rough sled used for hauling a log out of a woods; one end of the log is placed on the sled and the other allowed to drag on the ground. 3. A stoneboat. 4. Any old car made over into a tractor-like machine. 5. Any homemade device which is particularly handy for a given job.

go-down. A cut or break in the bank of a river or creek through which animals may pass to get water or to cross the stream (western U.S.).

going back. 1. A cessation of initial vigor of normal tree growth which is a phenomenon observed on fertile but very shallow soils, or on those underlaid at shallow depths by some unfavorable horizon, such as a hardpan. 2. Reverting to an ancestral type.

going crazy. A condition of animals attributed to the feed, ammoniated molasses.

going light. 1. The losing of weight by an animal or fowl. 2. Hauling a small load or using a minimum of equipment.

goiter (goitre). The enlargement of the thyroid gland which results from a deficiency of iodine in the diet.

go lame. To become lame; a condition often noticeable in the foot and leg action of horses. Causes are sidebones, tenderness of feet, sprains, etc.

goldaster. See *golden aster.*

goldband lily. *Lilium auratum,* family Liliaceae, a perennial, bulbous herb grown in the flower garden for its fragrant, white flowers which are crimson spotted with a yellow band. Native to Japan.

gold chain. See *goldmoss stonecrop.*

goldcup chalicevine. *Solandra guttata,* family Solanaceae, a woody vine grown in very warm areas or in the greenhouse for its winter-blooming, fragrant, solitary, yellowish-white flowers. Native to tropical America. Also called *chalice vine.*

gold dust. See *goldmoss stonecrop; goldentuft alyssum.*

golden. In marketing, the degree of injury on the peel of citrus fruit caused by the rust mite.

golden Alexanders. See *golden zizia.*

golden aster. Any herb of the genus Chrysopsis, family Compositae; sometimes grown in the flower garden for its yellow flower heads. See *sickleleaf golden aster; Maryland golden aster; hairy golden aster.*

golden bamboo. *Phyllostrachys aurea,* family Gramineae, a woody-stemmed bamboo grown as an ornamental for its erect, yellow stems which reach up to 15 ft. in height. Native to China and Japan.

Golden Bantam. A famous variety of sweet corn extensively grown for its flavorful kernels. This variety was open-pollinated and was one of the varieties used in developing the hybrid Golden Cross Bantam.

golden bay. The rich, yellowish-red of an animal's coat.

golden bee. A honeybee that was developed in America, probably from Cyprian or Syrian and Italian races.

golden bell. See *forsythia; yellow whispering bells.*

golden blue-eyed grass. *Sisyrinchium californicum,* family Iridaceae, a half-hardy, perennial herb grown in clumps as a border plant for its foliage and its bright yellow flowers with brown markings. Also called *satin flower.*

golden callalily. *Zantedeschia elliottiana,* family Araceae, a tropical herb, the common callalily, widely used as decoration for its golden-yellow, trumpet-shaped flowers. Native to South Africa. Also called *yellow calla.*

golden camomile. *Anthemis tinctoria,* family Compositae, a perennial herb grown in the flower border for its golden-yellow flowers. Native to Eurasia. Also called *golden marguerite; yellow camomile; oxeye camomile.*

goldenchain. Goldenchain laburnum.

goldenchain laburnum. *Laburnum anagyroides,* family Leguminosae, a large, poisonous, deciduous shrub or small tree grown for its attractive racemes of yellow flowers; sometimes used for Easter decorations. Also called *goldenchain; bean tree; base tree.*

golden clematis. *Clematis tangutica,* family Ranunculaceae, a hardy, woody vine grown on trellises, fences, etc., for its spring (and sometimes autumn), yellow flowers. Native to Asia. Also called *Russian clematis.*

goldenclub. *Orontium aquaticum,* family Araceae, a hardy, aquatic herb grown in a shallow pool or a pond for its white scapes and yellow spadices. Native to North America. Also called *tawkee; tawkin; floating arum; water dock.*

golden columbine. *Aquilegia chrysantha,* family Ranunculaceae, a hardy, perennial herb grown for its yellow flowers. Native to North America. Also called *golden-spurred columbine.*

golden coppertip. *Crocosmia aurea,* family Iridaceae, a tender, perennial herb grown for its bright, orange-yellow flowers. Native to South America.

golden coreopsis. Plains coreopsis.

golden corydalis. *Corydalis aurea,* family Fumariaceae, a biennial herb which is poisonous to sheep. Symptoms of poisoning are rapid respiration, rapid pulse, staggering, convulsions, collapse, and death. Native to North America.

goldencup. *Hunnemannia fumariaefolia,* family Papaveraceae, a perennial herb grown as a hardy annual for its bluish-green leaves and yellow flowers. Native to Mexico. Also called *Mexican tulip-poppy.*

goldencup oak. See *canyon live oak.*

golden currant. *Ribes aureum,* family Saxifragaceae, an unarmed, deciduous shrub grown for its numerous, yellow, clove-scented flowers. Native to North America. See *clove currant.*

golden dewdrop. Creeping skyflower.

golden eardrops bleedingheart. *Dicentra chrysantha,* family Fumariaceae, a perennial herb found on the ridges of the Coast Range of California and the foothills of the Sierra Nevadas, U.S. It is poisonous to livestock. Symptoms of poisoning are trembling, frothing, convulsions, collapse. Native to western North America.

golden-eyed grass. See *golden blue-eyed grass.*

golden-eye lacewing. *Chrysopa oculata* Say, family Chrysopidae, an insect whose larvae are predators to aphids, mealybugs, cottony-cushion scale, and sometimes on thrips and mites.

golden flax. *Linum flavum,* family Linaceae, a perennial herb grown for its golden-yellow flowers; sometimes confused with Indian yellowflax (*q.v.*). Native to Europe.

golden glow. *Rudbeckia laciniata* var. *hortensia,* family Compositae, a perennial herb which grows to a height of 12 ft. and bears yellow, double flowers.

goldenglow aphid. *Macrosiphum rudbeckiae* (Fitch), family Aphidae, a large, red aphid which infests the cutleaf coneflower, chrysanthemum, goldenrod, sunflower, and lettuce.

golden gram. See *mung bean.*

golden groundsel. *Senecio aureus,* family Compositae, a hardy, perennial herb grown in the flower border for its clusters of yellow flowers. It is found wild in wet meadows of northwestern U.S., and it may be poisonous to livestock ingesting large amounts. Native to North America. Also called *golden ragwort.*

golden honey plant. *Actinomeris squarrosa* (*A. alternifolia*), family Compositae, a hardy, perennial herb sometimes grown in the wild garden for its yellow flowers. It is a very good bee plant. Native to North America.

Golden Horse. See *Palomino: 1.*

golden larch. Lovely goldenlarch.

golden loosestrife. *Lysimachia vulgaris,* family Primulaceae, a bushy herb grown for its yellow flowers. Native to Eurasia. Also called *willow-wort.*

golden marguerite. Golden camomile.

golden meadow parsnip. Golden zizia.

golden moss. See *goldmoss stonecrop.*

golden-nematode disease of potato. A pernicious disease prevalent in Europe for many years; caused by the golden nematode, *Heterodera rostochiensis* (Wr.), family Heteroderidae. Now established in a limited area in the U.S., it is characterized by delay in sprouting, stunting, unthriftiness, and the production of small tubers. Also called *potato sickness.*

golden net. A virus disease of peach in Colorado, U.S., characterized by marginal yellowing of leaf veins. It results in minor economic losses on peach and plum crops and serious losses from inoculated apricots. There has been no natural spread.

golden oak scale. *Asterolecanium variolosum* (Ratz.), family Asterolecaniidae, a greenish-gold insect that infests oaks, sometimes causing death to an infested tree.

golden osier. Yellowstem white willow.

golden peony. *Paeonia lutea,* family Ranunculaceae, an erect, shrubby, perennial herb grown for its large, golden-yellow flowers. Native to China.

golden polypody. *Polypodium aureum,* family Polypodiaceae, a coarse fern grown as a greenhouse fern and as a house plant. Native to tropical America. Also called *hare's-foot fern.*

golden queen. Ledebour globeflower.

golden ragwort. Golden groundsel.

goldenrod. Any herbs of the genus Solidago, family Compositae. Some species are grown as ornamentals: silver goldenrod; dyersweed goldenrod; seaside goldenrod (*q.q.v.*), and some are pernicious weeds.

golden rose of China. Father Hugo rose.

goldenseal. *Hydrastis canadensis,* family Ranunculaceae, a perennial herb grown in the wild garden for its showy leaves and red fruit and for its rootstock used in tonics. It is poisonous to stock. Native to North America. Also called *yellow puccoon; yellowroot; orangeroot; Indian dye.*

golden shower senna. *Cassia fistula,* family Leguminosae, a tropical tree grown only in very warm areas for its foot-long yellow flower clusters which appear before the leaves. Native to Asia. Also called *pudding-pipe tree; canafistula; drumstick tree.*

golden star. 1. Darkstripe bloomeria. 2. Maryland golden aster.

goldentop. *Lamarckia aurea,* family Gramineae, an annual grass grown for its golden-yellow or violet spikelets. Native to southern Europe.

golden tortoise beetle. *Metriona bicolor* (F.), family Chrysomelidae, a small, golden-colored, nearly hemispherical beetle which infests plants of the family Convolvulaceae, eating rounded holes in the leaves or completely devouring them.

goldentrumpet. *Anemopaegma chamberlayni,* family Bignoniaceae, an evergreen, climbing vine grown for its habit and for its pale yellow flowers. Native to Brazil.

goldentuft alyssum. *Alyssum saxatile,* family Cruciferae, a perennial herb widely grown in the rock garden, in the flower border for edging, and for cuttings of its golden-yellow flower clusters. Native to Eurasia. Also called *gold dust; basket of gold.*

golden wattle. Sydney acacia.

goldenwave coreopsis. *Coreopsis drummondi,* family Compositae, a tender, annual herb widely grown in the flower garden for its yellow flowers, brownish-purple at the bases. Native to North America.

golden willow. See *yellowstem white willow.*

golden wreath. See *goldwreath acacia.*

golden zizia. *Zizia aurea,* family Umbelliferae, a perennial herb grown for its small, yellow flower clusters. Native to North America. Also called *early meadow parsnip; golden meadow parsnip; early parsnip; golden Alexanders.*

goldfields. Any herb of the genus Baeria, family Compositae. Species are grown in flower gardens. See *everlasting goldfields; tuber goldfields.*

gold-flower. *Hypericum moserianum,* family Hypericaceae, a hybrid undershrub grown as an ornamental for its yellow flowers and reddish stems.

Goldie dracaena. *Dracaena goldieana,* family Liliaceae, an arborescent plant grown in greenhouses for its green leaves cross-banded with strips of light green. Native to central Africa.

Goldie's fern. Goldie woodfern.

Goldie woodfern. *Dryopteris goldieana,* family Polypodiaceae, a hardy, outdoor fern grown in the flower or fern garden. Native to North America.

goldilocks. European goldilocks.

goldmoss stonecrop. *Sedum acre,* family Crassulaceae, a low, creeping, evergreen herb grown in the flower garden and in cemeteries. Escaped, it becomes a weed pest. Native to Eurasia. Also called *mossy stonecrop; wall pepper; gold chain; tangletail.*

gold-of-pleasure. Falseflax.

goldpoppy. Any plant of the genus Eschscholtzia, family Papaveraceae. See *Californiapoppy.*

goldthread. See *Alaska goldthread; dodder.*

goldwreath acacia. *Acacia saligna,* family Leguminosae, a tropical shrub grown as an ornamental for its racemes of yellow flowers. Native to Australia.

Gomost. A Norwegian cheese which is very similar to Mysost (*q.v.*), made from cows' or goats' whole milk.

gomuti sugarpalm. *Arenga pinnata,* family Palmaceae, a palm tree grown as an ornamental in frost-free areas, India, and the East Indies for the sugar made from its sap. Native to Malaya. Also called *jaggery palm; sugar palm.*

gonad. The gland of the male or female which produces the reproductive cells; the testicle or ovary.

gone to sugar. Designating honey, sirup, etc., in which the sugar has crystallized.
goober. Peanut *q.v.*
good color. 1. In stock judging, designating a bright wool fleece, free from yellow tints and from off-color fibers. 2. In judging, designating a bright, acceptable color for the product.
good cottonseed meal. See *cottonseed meal, prime quality.*
good drainage. The removal of excess surface or subsurface water from land at a rate which prevents damage to growing crops or other desirable vegetation.
good feeder head. A steer that will feed and fatten satisfactorily; one that is clean cut, strong in muzzle and clear in the eye.
good finish. The desired appearance and indication of proper fleshing of a fat steer or meat animal.
good going. The desired, efficient gait or leg action of horses.
Goodhue drier. Box drier *q.v.*
Good King Henry goosefoot. *Chenopodium bonus-henricus,* family Chenopodiaceae, a perennial herb sometimes grown as a potherb. Native to Europe; naturalized as a weed in North America. Also called *Good King Henry; weak mercury; mercury; false mercury; fat hen; English mercury.*
good-legged horse. A horse which shows legs of proper shape, size, and quality for proper action.
good middling. A U.S. grade of lint cotton, No. 3, which is within the range of a set of samples established by the U.S. Department of Agriculture, which shall be American upland cotton in color, leaf, and preparation.
good middling extra white. A U.S. grade of lint cotton which shall be American upland cotton which in leaf and preparation is similar to good middling (*q.v.*), but which is whiter than good middling.
good middling gray. A U.S. grade of lint cotton which shall be American upland cotton which in leaf and preparation is similar to good middling (*q.v.*), but which is more gray in color than good middling, and no darker in color than the dullest bale in No. 6 or strict low middling.
good middling spotted. A U.S. grade of lint cotton which shall be American upland cotton which in leaf and preparation is similar to good middling (*q.v.*), but which in color is between good middling and good middling tinged.
good middling tinged. A U.S. grade of lint cotton which shall be American upland cotton that in leaf and preparation is similar to good middling, but which in color is within the range represented by the set of samples labeled No. 3 tinged or good middling tinged; it has a slight yellow cast.
good middling yellow stained. A U.S. grade of lint cotton which shall be American upland cotton that in leaf and preparation is similar to good middling, but which in color is deeper than good middling tinged.
good-morning spring. Virginia springbeauty.
good mouth. In marketing, designating an animal whose teeth are in good shape.
good off summer yellow cottonseed oil. Cottonseed oil (the third class of summer yellow oil) which may be off in flavor and odor, must be free from visible foreign material, of a color no higher than ACCS 7.6; it contains not more than 0.25 percent free fatty acid and not more than 0.10 percent moisture and volatile matter.
good ordinary. A U.S. grade of lint cotton which shall be American upland cotton that in color, leaf, and preparation is within the range represented by the set of samples labeled No. 9 or good ordinary.
good ordinary extra white. A U.S. grade of lint cotton which shall be American upland cotton which in leaf and preparation is similar to good ordinary, but which is whiter than good ordinary.
good tilth. Physical condition of a soil favorable to seed germination and plant growth.
go off feed. 1. To cease feeding with a healthy and normal appetite. 2. To refuse feed in the given amounts and kind previously fed.
gools. Marshmarigold.
gooranut. Sudan colanut.
goose. 1. Any large, web-footed bird (intermediate in size between swans and ducks) of the subfamily Anserinae (family Anatidae) including the genus Anser and related genera. The domestic goose, *Anser domesticus,* includes a number of breeds that are kept for their flesh and feathers. 2. The female goose as distinguished from the male, or gander.
gooseberry. 1. The plant, and fruit of a bush, raised for its large green, purplish, black, and red fruits which are usually used green for pies, preserves, etc. European varieties belong to the European gooseberry (*Ribes uva-crispa*), family Saxifragaceae. Most American varieties are hybrids between the European gooseberry or other species and the hairystem or American gooseberry (*Ribes hirtellum*). 2. Any of several species infrequently used as ornamentals. See *fuchsia gooseberry; Canada gooseberry.*
gooseberry fruitworm. *Zophodia convolutella* (Hbn.), family Phycitidae, the larva of a moth that feeds within the fruit of gooseberries and currants.
gooseberry mildew. A powdery mildew (*q.v.*) of gooseberries caused by *Sphaerotheca mors-uvae,* family Erysiphaceae.
gooseberry tree. Otaheite gooseberry leafflower.
goose feather. Any feather (or feathers) from a goose. Especially prized are the mature, soft, downy feathers from the breast and abdomen which are used as a filling for pillows, sleeping bags, etc. These soft feathers are often plucked from live geese in the summer prior to molting, by goose growers. The larger quill-feathers obtained from dressing plants are commonly stripped for their webbing.
goosefoot. Any plant of the genus Chenopodium, family Chenopodiaceae. The fleshy-leaved species are sometimes locally called *fat hen.* See

wormseed goosefoot; lambsquarters goosefoot; Good King Henry goosefoot; Jerusalemoak goosefoot.

goose grass. 1.*Eleusine indica,* family Gramineae, an annual grass which is a weed pest. Native to Europe; naturalized in tropical and North America. Also called *wire grass.* 2. Annual bluegrass. 3. Shore podgrass.

goose nest land. That part of Kentucky, U.S., characterized by a high frequency of sinkholes *q.v.*

goose rump. A horse with a short, steep croup that is narrow at the point of the buttocks.

goosestep. A peculiar walk or body action, locomotor incoordination (spastic gait) of swine which is caused by a nutritional deficiency of pantothenic acid.

goosey. Of, or pertaining to, a horse that is nervous in the stall or while handling.

gopher. 1. A rodent, any species of the genera Geomys, Cittellus, Thomomys. It feeds on the roots, stems, and bark of plants and may do considerable damage to cultivated crops. By its habit of burrowing, it is a factor in modifying natural soils. Sometimes called *ground squirrel.* 2. The land turtle, *Gopherus (Testudo) polyphemus. Gopher* is a common name applied to this reptile, especially in Florida, U.S. It makes deep burrows in dry sand soils, sometimes damaging the trees in citrus groves.

gopher (bomb, gun). A bomb or explosive trap used, not too successfully, to destroy gophers.

gopher-wood. American yellowwood.

Gordon mockorange. *Philadelphus gordonianus,* family Saxifragaceae, a tender, deciduous shrub grown for its fragrant, white flowers. Native to North America.

gore. To pierce the body with an animal's horns.

gorge. 1. A deep valley with nearly vertical rock sides. 2. To overeat.

Gorgonzola. A blue-green, cured, veined, cows' milk cheese which is made at Gorgonzola, near Milan, Italy.

gorse. *Ulex europaeus,* family Leguminosae, a branched, spiny shrub, 2 to 4 ft. tall; grown for its fragrant, yellow flowers. Escaped, it can become a pest. Native to Eurasia. Also called *furze; whin.*

go short and stubby (stilted). To walk, as a horse, with a stilted or halting front leg or foot action as a result of short, steep pasterns and steep shoulders.

gosling. Any young goose before its sex can be determined.

gosmore. Spotted catsear.

gossypol. A bright yellow crystalline, toxic substance found in cottonseed.

go stale. 1. To suffer sperm deterioration, as a bull. 2. To go off flavor, as a product. 3. As an animal, not to work at normal standards of work or production. 4. As an animal, to go off feed and lose its bloom.

go straight at both the walk and trot. To walk or trot, as a horse, gracefully and true at different speeds.

Goteborg. A summer sausage composed of 60 percent beef and 40 percent pork which is coarsely chopped, seasoned, stuffed in beef casings, and smoked.

go too close. To walk or run, as a horse, so that the front feet cross over and are placed one before the other.

go to pieces. To deteriorate, as a cotton variety changes from year to year in cultivation.

go to seed. 1. To produce seed, especially as a weed, a garden plant, etc. 2. In plants, to mature beyond a usable condition.

go to stalk. To grow vegetatively at the expense of seed or fruit.

Gouda. A semisoft to hard, sweet-curd, cows' cured whole or skimmed milk cheese made in Holland. The Baby *Gouda* which weighs about 1 lb. is popular in the U.S.

gourd. Any cucurbitous fruit of various shapes and sizes, of the genera Lagenaria and Cucurbita, family Cucurbitaceae, that is hard-shelled and used as an ornamental or for making domestic utensils. 2. In England, a generic name for species of genus Cucurbita.

gourdseed corn. A late-maturing variety of corn (*Zea mays*) which was grown by early southern settlers, U.S. The gourdseed kernels are white, soft, deeply dented, large, and pointed, resembling gourd seeds. The dent varieties of today may be the result of intentional or accidental crossing with the old *gourdseed corns.* Also called *Virginia gourdseed corn.*

Gournay. A soft, fresh, Cottage cheese type of cheese made in France.

gout. An uncommon, nutritional disease of mature poultry, characterized by internal deposits of sodium urate in the viscera or joints. Also called *gowt.*

goutweed. Any plant of the genus Aegopodium, family Umbelliferae. Also called *English masterwort.* See *bishop's goutweed; silveredge bishop's goutweed.*

government aid. See *subsidy; price supports; parity; flexible support; compensatory payment; incentive payment; Soil Bank Plan.*

governmental controls. See *acreage allotment; marketing quotas.*

government farm credit. Farm credit provided by the federal government, U.S., particularly through an agency such as the Farmers' Home Administration or the Commodity Credit Corporation. It differs from credit provided by a government sponsored farm credit agency, such as the Farm Credit Administration, in that units of the agency are either member-owned or are attaining a member-owned status and obtain their loan funds from the sale of bonds and debentures on the open money market.

governor's-plum. Ramontchi.

Goya. A skimmed or partly skimmed milk, medium cured, Argentinian cheese.

grab. A piece of metal about 3/16 x 3/8 x 1 in. brazed on to a horseshoe, used in place of heels, which tends to help a horse travel straight.

grabbin' the apple. Catching the horn of the saddle while riding horseback.

grabbling (potatoes). Harvesting the larger tubers from a potato hill without killing the plant; the soil is replaced, permitting the remaining tubers to mature. It is an old practice by home gardeners to obtain early or new potatoes.

grabbot. Inferior lint cotton.

grade. 1. The slope of a road, channel, or natural ground. 2. The finished slope of a prepared surface of a canal bed, roadbed, top of embankment, or bottom of excavation. 3. Any surface which is prepared for the support of a conduit, paving, ties, rails, etc. 4. a. Any animal which has one purebred parent and one of unknown or mixed breeding. See *grade up.* b. Designating a herd, flock, band, etc., of such animals. 5. The classification of a product, animal, etc., by standards of uniformity, size, trueness to type, freedom from blemish or disease, fineness, quality, etc. 6. To smooth the surface of a road. 7. To raise the level of a piece of ground by the addition of earth, gravel, etc.

grade A pasteurized milk. Grade A raw milk (*q.v.*) that has been pasteurized, cooled, and bottled, in an approved milk plant, which has at no time after pasteurization until delivery an average bacteria plate count of over 30,000 per ml.

grade A raw milk. Raw milk produced on dairy farms conforming with certain standards of sanitation, in which the average bacterial plate count does not exceed 50,000 per ml. if clumps are counted or 200,000 per cc. if individual organisms are counted, or the average reduction time of which is not less than 8 hours.

grade B pasteurized milk. Pasteurized milk which violates the bacterial standard established for grade A pasteurized milk, and/or the provision of lip-cover caps, and/or the requirements that grade A raw milk be used, but which conforms to all other requirements for grade A pasteurized milk; it has been made from raw milk of not less than grade B quality and has an average bacterial plate count after pasteurization and before delivery not exceeding 50,000 per ml.

grade B raw milk. Raw milk which violates the bacterial standard and/or the abortion testing requirement for grade A raw milk, but which conforms with all other requirements for grade A raw milk; it has an average bacterial plate count not exceeding 1,000,000 per ml. or an average direct microscopic count not exceeding 1,000,000 per ml. if clumps are counted, or 4,000,000 per ml. if individual organisms are counted, or an average reduction time of not less than 3½ hours.

grade C pasteurized milk. Pasteurized milk which violates any of the requirements for grade B pasteurized milk.

grade C raw milk. Raw milk which violates any of the requirements for grade B raw milk.

graded eggs. Eggs which have been sorted and labeled according to size and quality.

graded herd. In stock judging, a herd of beef cattle which usually consists of a bull from the older two classes and four females of varying ages, as designated by the show.

graded terrace. A terrace which has a constant lengthwise slope sufficient to cause runoff to flow at a nonerosive velocity.

grade I milk. Milk with a clean, pleasant flavor, and a methylene blue reduction time over 5½ hours, which shows practically no soiling of the sediment disk. Tolerance should be shown for such off-flavors as slightly feed, flat, or salty and for a few specks on the sediment disk. The sediment score, using the U.S. Department of Agriculture sediment chart shall be 9.0 or above.

grade II milk. Milk which does not meet standards for grade I and which has off-flavors such as definitely feed, flat, or salt, slightly barny, bitter, foreign, malty, metallic, musty, oxidized, rancid, or very slightly weedy; it has a methylene blue reduction time between 2½ and 5½ hours or shows a medium amount of sediment. The rating of sediment, using the U.S. Department of Agriculture sediment chart, shall be 7.0 or above.

grade III milk. Milk which has off-flavors such as definitely barny, bitter, foreign, malty, metallic, musty, oxidized, rancid, or slightly high-acid; it has a methylene blue reduction time between 20 minutes and 2½ hours or is definitely high in sediment. The sediment score, using the U.S. Department of Agriculture sediment chart, shall be 5.0 or above.

grade of cotton. See *grading cotton.*

grader. 1. A person who classifies products into grades. See *grade:* 5. 2. Any mechanical device for separating material into several different sizes, qualities, or grades. 3. Any of several blade devices which move soil or smooth land, such as a road *grader.*

grade up. To improve a herd, band, flock, etc., by breeding to a purebred sire.

gradient. 1. The rate of change of elevation, velocity, pressure, temperature, or other characteristics per unit length. 2. Slope.

grading. 1. The classification of products, animals, etc., into grades. See *grade:* 5. 2. The mating of a purebred animal with one of mixed or unknown breeding. 3. The smoothing of the land surface.

grading board. A board, with markings designating inches, used to measure stems of cut roses.

grading cotton. A system of classifying cotton according to grade, composed of three factors: color, foreign matter, and ginning preparation. By means of variation in the color factor, the grade gives certain information concerning the weathering or exposure to which the open cotton has been subjected and in some instances, information concerning attacks by insects, fungi, and the like, both before and after the boll is opened. By means of variation in foreign matter, the grade gives information concerning the methods of harvesting and the effects of weather conditions before and during picking. Because of variations in preparation, the grade gives information concerning storage or handling conditions previous to and during ginning.

grading up. See *grade up.*

gradual cut-back. A system of pruning rose bushes in greenhouses, to prevent the bushes from growing too tall, in which the stems are pruned as flowers are harvested. Also called *knife pruning; gradual pruning.*

gradual pruning. Gradual cut-back.

graft. a. A cion or bud which is taken from one plant and inserted into a cut on a root, stem, or branch of another or the same plant on which it continues to grow. b. To make a *graft.* c. The growth which results from this process.

graftage. See *graft: c.*

graft hybrid. A stem arising from an adventitious bud formed at a graft union which contains the genetically different tissues of both the stock and cion. See *chimera.*

grafting. 1. The inserting of a piece of one plant into another or the same plant with the intention that it shall grow there. See *graft.* 2. A process of removing a worker bee larva from its cell and placing it in an artificial queen cup, for the purpose of having it reared into a queen.

grafting bench. A work bench in a greenhouse or propagation house on which plants are grafted.

grafting by approach. See *inarching.*

grafting case. An airtight frame placed on top of a propagating bench, usually 12 to 15 ins. deep, equipped with sash and sometimes a heating cable; used for the grafting of roses, azaleas, and gypsophila and for the rooting of gardenias and other tropical plants.

grafting compound. A water soluble asphalt compound used on graft unions and as a wound covering.

grafting knife. A knife having one straight-edged blade used in grafting.

grafting tool. Any of several types of tools used to make the cleft in the stock for a cleft graft *q.v.*

grafting wax. A wax consisting of a mixture of beeswax, tallow, and rosin, used to seal a graft, making it airtight and waterproof. It may also be used to seal up a wound on a plant.

graft union. The point or place on the root, stem, or branch of a plant at which a graft has been made.

grain. 1. The seed of the cereal crops. See *small grain; cereal: 1.* 2. Commercially, or as listed on boards of trade, buckwheat, soybeans, and flaxseed, in addition to the cereals. 3. In buttermaking, the butter particles after the emulsion has broken. 4. The arrangement, direction, quality, etc., of the fibers in wood. See *annual ring.*

grain bin. Usually one of several verminproof, moistureproof, small compartments in the granary building, used for the storage of grains, such as wheat, oats, etc.

grain binder. A horse- or power-drawn harvesting machine designed to cut, bundle, and bind harvested grain in a single operation. This machine consists of a large central wheel, called the bull wheel, which, in the horse-drawn machines supplies the power to the three principal parts that are built over and about it. These are the cutting and generating parts, the elevating and bundling mechanism, and the binding and tying equipment. Many binders derive power directly from the tractor.

grain box. That part of the seed drill or seeding machine in which the seed is stored before planting. Also called *seed hopper.*

grain cleaner. A device used for removing weeds, seeds, and trash from grain during or after harvest.

grain consuming animal unit. An index figure which represents an average mature horse, or its equivalent in other livestock, based on the grain consumption of 2,500 lbs. of corn or its equivalent in one year.

grain corn. Shelled corn, as contrasted to silage corn or corn on the cob.

grain crop. Any of several of the cereals or grains, such as wheat, oats, rye, etc., grown as the principal crop.

grain diameter. In soil science, the size of grain, usually expressed in millimeters; determined by sieve analysis or wet mechanical analysis; hence not a true grain size except for spherical grains. Also called *grain size; particle size.*

grain distillers' dried yeast. A livestock feedstuff, a by-product of the distilling industry, which is the properly dried yeast resulting from the fermentation of grains and yeast, separated from the mash, either before or after distillation.

grain divider. 1. A finger-like projection, placed at right angles to the end of the cutting bar of a grain binder or harvesting machine, which divides the standing grain and guides it into the cutting mechanism. 2. An individual who allocates threshed grain to the land owner and renter in accordance with the rental share schedules. 3. A sampler; a laboratory device that mixes samples of grain and divides them into equal parts prior to grading the grain.

grain drill. A mechanical planting device or farm machine used in the proper placing of grain seed into the ground at planting time.

grain-drill fertilizer attachment. An attachment for a grain drill which consists of a specially constructed hopper for dispensing fertilizer at a metered rate while simultaneously seeding grain.

grain dryer. A mechanical device used to remove excess moisture from harvested grain, usually by forcing unheated air through the grain.

grain elevator. 1. A mechanical device especially adapted to the moving or elevation of grain from a wagon or truck to storage bins. 2. A central facility for the handling and storage of grain from many farms.

grain exchange. A market place where buyers and sellers of grain, in person or through representatives, meet to trade.

grain fed. Designating animals, such as cattle, which are being or have been fattened largely by the use of grain feeds.

grain head. In all cereal grains except corn, the

spike or panicle at the top of the stalk which carries the grain.

grain header. A harvesting device which cuts off the heads of the grain and elevates them to a wagon. Now largely replaced by the combine, it is used somewhat in grain sorghum harvesting where the heads are further dried before threshing.

grain inspector. One who officially inspects grain to determine its market grade, quality, or fitness for food. He is usually an official of a governmental or recognized marketing agency.

grain layer. A longitudinal layer of a sheep skin, the wool side of the skin, which is used for making hat bands, bookbinding, etc. See *flesh layer.*

grain mill products. Any cereal preparation, rice refuse and polishings, flour and meal, and flour preparations used as animal feed.

grain mixture. A livestock feed which consists of two or more different grains with an indication of the pounds or percentage of each grain content in such a mixture.

grain pan. A metal pan on a grain thresher or combine onto which all grain falls to be shattered by the feeder, threshed, and separated by the cylinder concaves, grates, beater, and strawrocker. The grain is conveyed backward by an oscillating motion to the cleaning shoe.

grain (pea, bean) lifter. An attachment for a mowing machine, binder, or combine in which guards that project forward are attached over the regular guards. These guards lift the lodged plants so that the cutter bar slides underneath and cuts the stems close to the base.

grain probe. A sampling device commonly used by elevator operators when purchasing grain to obtain a sample of a load, bin, or bag in order to determine grain quality and dockage content. It consists of a long, hollow tube which is inserted into grain to obtain a representative sample of the load. See *grain trier.*

grain room. The room in a barn in which feed grain is stored.

grain sampler. See *grain trier; grain probe.*

grain scale. A balance scale used for the accurate weighing of small quantities of seeds, grains, etc., for grade determination, weight per bushel, purity, or experimental purposes.

grain screenings. The inferior, light, or broken grain or seed or other impurities separated in the process of cleaning grain. These screenings, after recleaning, are often used in livestock feed.

grain separator. See *grain threshing machine.*

grain size. Grain diameter *q.v.*

grain sorghum. Any variety of the species sorghum (*q.v.*) that is grown principally for the grain it produces as contrasted with the sorghum grown for forage or sirup.

grain sorghum distillers' dried grains. A livestock feedstuff which is the dried residue obtained in the manufacture of alcohol and distilled liquors from grain sorghums, or from a grain mixture in which grain sorghums predominate, and contains the major portion of the condensed screened stillage dried therewith.

grain sorghum head chop. A livestock feedstuff which includes the entire head, grain, and stem of grain sorghums, chopped. If the name of the variety is given, it must be true to name.

grain-sorghum header. See *header: 2.*

grain sorghum meal. The finely ground, unbolted meal made from kafir, milo, or any other variety of grain sorghum. If the name of the variety is given, it must be true to name.

grain storage bin. Any of many different types of facilities used for the storage of harvested grain. A common type is the galvanized, all-steel, cylindrical bin which is fire and vermin proof. See *grain bin.*

grain thresher. Grain threshing machine.

grain threshing machine. Any mechanical device which consists principally of a rapidly rotating cylinder with teeth, rasp bar, or rubber-edged angle bar on its surface rotating in a concave or shell where it knocks the grain from the heads, a system of separating sieves, and a blower or shaker which separates the chaff and straw from the threshed grain.

grain trier. An instrument or device usually made of brass and consisting of two tubes from 6 ins. to 8 ft. in length. These tubes are slotted on the sides and one is closely fitted but freely moving within the other. When closed this instrument may be inserted in a bag or bin of grain and then opened to admit a representative sample from each and every layer of the grain. After filling, the trier is closed and withdrawn and the sample discharged upon an examination cloth for inspection or grading. Also called *grain sampler.*

grain vinegar. Spirit vinegar.

grain weed. Any weed which is often found in grain fields.

grain wheel. The wheel located on the inner side of the binder, corn picker, etc., near the rear side of the divider. It is canted inward slightly at the bottom.

grainy. Designating a body defect of cheese in which there are irregular shaped, hard particles.

gram. 1. A unit of weight used in the metric system; one *gram* being equivalent to .035 ounces avoirdupois, or to .032 apothecary weight. 2. Mung bean. 3. An Indian gram.

grama. Any grass of the genus Bouteloua, family Gramineae. A large number of species are native on the Great Plains and elsewhere in western U.S. and are rated as good to excellent forage on ranges and pasture land. See *blue grama; hairy grama; sideoats grama; black grama; Rothrock grama; sprucetop grama; sixweeks grama; Parry grama; needle grama.*

gram chick-pea. *Cicer arietinum,* family Leguminosae, an annual herb which has long been cultivated in Europe for its edible seeds used for food and livestock feed. Grown in California, U.S., the herbage is said to cause livestock poisoning. Native to Asia. Also called *chestnut bean; chick-pea; garbanzo; Egyptian pea; fasels; coffee pea.*

gramineus. Like, or pertaining to, the grass tribe.

Gram-negative bacillus. See *Gram stain.*

Gram-positive. See *Gram stain.*

Gram stain. A staining method devised by a Danish physician, Hans Gram, to aid in the identification of bacteria. Bacteria either resist discolorization with alcohol and retain the initial deep violet stain (Gram-positive) or can be decolorized by alcohol and are stained with a contrast stain (Gram-negative).

Grana. See *Parmesan; Reggiano.*

granadilla. See *giant granadilla; purple granadilla.*

granadilla real. Giant granadilla.

granadilla tree. *Brya ebenus,* family Fagaceae, a hardwood tree, the source of a high quality green ebony. Native to the West Indies. Also called *American ebony; Jamaica ebony; ebony cocuswood.*

granary. 1. A storehouse, especially for threshed or husked grain. 2. Any store room where food or supplies are kept or stored. 3. A large food producing area: *e.g.,* the central west of the U.S. is sometimes called the food *granary* of the world.

granary weevil. *Sitophilus granarius* (L.), family Curculionidae, an insect which infests stored grain and flour.

grand champion. In stock judging, the best animal of the entire show selected from the junior and senior champions.

grandmother's bluebells. See *creeping bellflower.*

Grand Rapids disease. See *bacterial canker of tomato.*

Grand Traverse disease. See *cobalt deficiency.*

grange. See *granary: 1.*

Grange (The National). An organization dedicated primarily to the improvement of rural life, founded in the U.S., 1867, as The Order of the Patrons of Husbandry; membership included both men and women. The *Grange* has exerted a strong influence on agricultural legislation and rural education and has promoted cooperative enterprises designed to protect the farmer from economic exploitation.

granger. 1. A member of the Grange. 2. A farmer (obsolete).

granger movement. The development and program of the Grange.

granite chips. Small particles of granite which are used as a medium in gravel culture *q.v.*

granite pink. *Dianthus graniticus,* family Caryophyllaceae, a perennial herb grown in the rock garden for its purplish flowers. Native to the Pyrenees.

granite soil. Soil which overlies granite rocks or is derived from the residue of weathering of such rocks. It has no single distinctive universal character but is variable depending upon local environmental factors.

granular frost structure. Granules of ice which are scattered through porous and easily spaded soil. It is common in frozen soils high in organic matter to a shallow depth.

granular soil. A soil composed largely of small, rounded, or subangular aggregates, in contrast to one composed of larger, blocky fragments; one composed of single grains, as sand.

granular vaginitis. A disease of cattle characterized by the formation of small nodules having a granular texture on the vulvar and vaginal mucous membranes of the cow and prepuce of the bull. The affected mucous membrane may be slightly swollen and sensitive and may bleed easily. Frequent urination and some straining may be seen in the cow. The condition may affect young calves and heifers as well. The incidence is generally quite high, but the seriousness of the condition is debatable. Many organisms have been suspected but none has definitely been proven to be the causal agent. It may spread by coitus, by the hands of attendants, or perhaps by grooming tools. Also called *granular venereal disease; nodular vaginitis; infectious vaginitis; contagious granular vaginitis; bull burn.*

granulate. To form into grains, granules, or small masses.

granulated honey. Honey in which the levulose portion remains liquid but is surrounded by dextrose crystals. Also called *solidified honey.*

granulation. 1. The formation of dextrose hydrate crystals in honey and crystals in other sugar sirups. 2. The cementation of particles into masses, as grains, aggregates, or clumps. 3. Sclerocystosis *q.v.*

granule. See *aggregate.*

granuled buttermilk. Cultured buttermilk which contains butter particles.

granuloblastosis. One type of avian leukosis complex affecting fowl over 6 months of age; characterized by a yellowish discoloration of the comb and unfeathered parts, weakness, emaciation, sleepiness, and diarrhea.

granuloma. A tumorlike mass of chronically inflamed tissue.

Granville wilt. A bacterial disease of tobacco and other solanaceous plants caused by *Xanthomonas solanacearum;* characterized by sudden wilting of the leaves which turn yellow and brown, and by death of the plant. Also called *southern wilt; brown rot; bacterial wilt of tobacco.*

grape. Any woody, climbing vine of the genus Vitis, family Vitaceae. Some species are grown as ornamentals; others are extremely important for their clustered fruits used as dessert, in wine making, for raisins, in jams, jellies, and juices. See *summer grape; gloryvine grape; frost grape; fox grape; bullace grape; European grape.*

grape anthracnose. A minor fungous disease caused by *Elsinoe ampelina,* family Myriangraceae; characterized by lesions on the leaves, fruit, shoots, and stems, often causing the fruit to become dry and wrinkled. Also called *bird's-eye rot.*

grape berry moth. *Paralobesia viteana* (Clem.), family Olethreutidae, an insect pest whose larvae infest the grape clusters and web the berries together, causing them to drop prematurely. The larvae also may eat small holes in nearly ripened berries.

grape blossom midge. *Contarinia johnsoni* (Sling.),

family Itonididae, an insect pest of grape which lays its eggs in the buds, preventing fruit development.

grape cane gall maker. *Ampeloglypter sesostris* (Lec.), family Curculionidae, an insect pest of grapes which punctures the canes to lay eggs. The larvae feed on the pith within the cane forming gall-like swellings on the cane.

grape clipper. A specially designed, scissors-like instrument used to cut the peduncle in order to remove the clusters of grapes in harvesting.

grape colaspis. *Colaspis flavida* (Say), family Chrysomelidae, a beetle whose larva infests corn, timothy, grapes, strawberries, beans, clover, buckwheat, potatoes, cowpeas, muskmelon, and apples. It is especially injurious to corn, killing or stunting the plants.

grape concentrate. A concentrate made from grapes which is used as flavoring for beverages, etc.

grape crown gall. See *crown gall.*

grape curculio. *Craponius inaequalis* (Say), family Curculionidae, a beetle which infests grapes, laying its eggs under the skin of the fruit. The larva feeds on the flesh and seeds.

grape dead arm. See *dead arm of grape.*

grape downy mildew. See *downy mildew of grapes.*

grape fanleaf. See *fanleaf.*

grape flea beetle. *Altica chalybea* (Ill.), family Chrysomelidae, an insect pest of grapes which feeds on the buds as they unfold in the spring, resulting in ragged, tattered foliage. It also infests plum, apple, quince, beech, elm, and Virginia creeper.

grapefruit. *Citrus paradisi,* family Rutaceae, a tender, subtropical, evergreen tree which bears the *grapefruit* of commerce, one of the important citrus fruits. Also called *pomelo.*

grape hoe. A machine which is used for mechanical hoeing between blueberry plants in the row. Also called *Friday hoe.*

grapehyacinth. 1. Any bulbous, perennial herb of the genus Muscari, family Liliaceae. Species are grown for their early-spring flowers. See *tassel grapehyacinth; musk grapehyacinth; Armenian grapehyacinth.* 2. *M. botryoides,* an early-spring herb widely grown for its blue flowers. Native to Europe. Also called **bluebells; starch hyacinth.**

grape juice. The unfermented juice expressed from fresh grapes.

grapeleaf California dewberry. See *dewberry.*

grape leaf folder. *Desmia funeralis* (Hbn.), family Pyraustidae, a caterpillar pest of wild and cultivated grapes, the Virginia creeper, and redbud which rolls a leaf tightly and feeds inside the rolled leaf.

grape leaf skeletonizer. *Harrisina americana* (G.-M.), family Zygaenidae, a caterpillar which feeds on upper leaf surfaces of wild and cultivated grapes.

grape mealybug. *Pseudococcus maritimus* (Ehrh.), family Pseudococcidae, an insect pest of buckwheat, clover, willow, apple, Californiapoppy, coleus, columbine, lemon, orange, passionflower, pear, potato, Japanese flowering quince, strawberry, Persian walnut, and Japanese yew, which sucks plant juices and secretes honeydew in which disease may become established. Also called *Baker mealybug.*

grape plume moth. *Pterophorus periscelidactylus* Fitch, family Pterophoridae, a moth whose larva enters buds on grapes and webs the newly unfolding leaves.

grape pruning. See *cane pruning.*

grape root borer. *Vitacea polistiformis* (Harr.), family Aegeriidae, a caterpillar which bores into the roots of grapes.

grape rootworm. *Fidia viticida* Walsh, family Chrysomelidae, the larva of a beetle which feeds on the bark of large grape roots or on the root hairs. The adult feeds on grape leaves leaving a series of small, chainlike holes. Infested plants are characterized by a loss of vigor, yellowing of the leaves, and little new growth.

grape scale. *Aspidiotus uvae* Comst., family Diaspididae, a scale insect that generally infests grapes and sometimes peach, hickory, and sycamore.

grape shears. See *grape clipper.*

grape sugar. A sugar extracted from grapes and other sweet fruits which is less sweet than cane sugar. It is also produced artificially from starch, dextrin, and cellulose, by the action of acids, ferments, and reagents. Also called *glucose; dextrose.*

grape twine. Binder twine or other string used to tie canes of vines to a wire trellis.

grapevine. See *grape.*

grapevine aphid. *Aphis illinoisensis* Shimer, family Aphidae, an aphid which infests grapes east of the Mississippi River, U.S., often causing premature fruit drop.

grape vinegar. Vinegar made by alcoholic and subsequent acetous fermentation of grape juice. It must contain, at 20° C., more than 1 gram of grape solids, more than 0.13 grams of grape ash, and at least 4 grams of acetic acid per 100 cc.

grapevine looper. *Lygris diversilineata* (Hbn.), family Geometridae, a caterpillar which feeds on grape leaves.

grape whitefly. *Trialeurodes vittatus* (Quaint.), family Aleyrodidae, a sucking insect which infests grapes in California, U.S.

grapple excavator. An earth excavating machine using the opening and closing type of bucket, the grapple, instead of the usual shovel.

grapple fork. A type of mechanical tool consisting of two, large, hinged, curved forks which open and close by means of an automatic trip operated by the ropes and pulleys. It is used to lift hay from a wagon and convey it into the hay loft or barn.

grass. 1. In its collective usage, *grass* includes 169 genera, 1,398 different species, and an uncounted number of varieties, family Gramineae, of annual and perennial herbs (rarely woody, such as bamboo) with hollow or solid stems, 2-ranked, usually parallel-veined leaves which consist of two parts: a flat bladelike portion and a sheath which encloses the stem. At the junction between the two parts of the leaf, and on the inside, is an appendage called the ligule which is either hyaline or hairy and membranaceous. The flowers are perfect (rarely unisexual) with no distinct perianth, borne in spikelets, and aggregated into spikes and panicles at the end of the stems. The flowers may have 1 to 6 stamens, but usually 3 with 2-celled anthers. One pistil, 1-ovulated ovary, and the fruit is a caryopsis with starchy endosperm. The grasses include the cereals, sugar cane, sorghum, millets, and bamboo which produce much of our human food and, in addition, furnish the bulk of forage for domestic animals. This group of plants is important not only as food for man and animals, but also in industrial arts, for turf, soil-holding, and ornament. 2. Unscientifically, any low-growing, herbaceous, grasslike plant, as a rich green pasture, the forage in a meadow, the herbage of the range, a well-kept lawn, or golf course.

grass anthracnose. See *cereal anthracnose.*

grass board. A board attached to the outer shoe on a cutter bar of a mowing machine whose purpose is to divide and rake away the cut from the uncut grass so as to give a clean place for the inside shoe on the next round. Also called *swathboard.*

grass cover. A good stand of grass or low herbaceous plants on a field, meadow, range, etc.

grass disease. A horse disease of unknown cause, characterized by paralysis of part of the digestive tract. It occurs when horses are put out to pasture in the summer months (Scotland and Wales).

grassed waterway. A vegetated, natural or artificial waterway used to conduct the accumulated runoff from cultivated and uncultivated land. It is designed to prevent gully erosion.

grass egg. In marketing, an egg with an olive-colored yolk.

grass family. See *grass: 1.*

grass farming. The cultivation of grasses and legumes primarily for forage purposes.

grass-fattened. Designating an animal which has been fattened on pasture, in contrast to one fattened on grain or other feed concentrate.

grass fire. A fire which burns a meadow or field. It once was a common practice to set such fires to burn old, dry grass.

grass-flower. Virginia springbeauty.

grasshopper. Any insect of the order Orthoptera, family Acrididae (Locustidae). Species have long legs for jumping and may be voracious feeders on many kinds of foliage.

grasshopper-catching machine. See *hopperdozer.*

grass killer. See *herbicide.*

grassland. 1. Land which is in grass but not under cultivation. (As commonly used, it is not restricted to true grasses, but may include legumes and other nongrasses.) 2. Land that produced grass just previous to being planted to a particular crop. 3. Any of the great natural regions of the world in which the vegetation is mainly grass or herbaceous plants. Climatic conditions are intermediate between those of forest regions and of deserts. Prairie, steppe, savanna, and pampas have been recognized as subdivisions of grasslands. The Great Plains constitute the major grassland region of the U.S.

grasslanding. Growing grasses and improving pasture land as a major part of farming or utilization of land.

grassland lupine. Spur lupine.

grassland soil. A division of soils in relation to the vegetative cover: a soil which supports grasses; a soil which developed under and has characteristics related to a grass vegetation.

grassland weed. A weed found in association with grasses.

grassleaf daylily. *Hemerocallis minor,* family Liliaceae, a lily-like herb grown for its fragrant, yellow flowers. Native to Asia. Also called *dwarf yellow daylily.*

grassleaf scabious. *Scabiosa graminifolia,* family Dipsaceae, a perennial herb grown as an ornamental on rock walls for its violet flowers and grasslike, silvery leaves. Native to Europe. Also called *pincushion; mourning bride.*

grasslike plant. A plant which resembles a true grass, family Gramineae, superficially, but which belongs to the Cyperaceae (sedge) or Juncaceae (rush) families.

grassnut brodiaea. *Brodiaea laxa,* family Liliaceae, a tender, perennial herb grown for its purple or white flowers. Native to California, U.S. Also called *triplet lily; Ithuriel's spear.*

grass pea(vine). *Lathyrus sativus,* family Leguminosae, an annual herb used as forage in Europe and Asia. Its seeds, used for human food in India and elsewhere for stock feed, are poisonous if consumed over a period of time. A symptom of poisoning is paralysis of the legs. Also called *vetchling; grass vetch; chickling; chickling vetch.*

grass pink. *Dianthus plumarius,* family Caryophyllaceae, a perennial herb grown as an edging plant for its low-growing, spreading habit and for its fragrant, rose-pink, purplish, or white flowers. Native to Eurasia. Also called *Scotch pink; feathered gillyflower; cottage pink; garden pink; pheasant's-eye pink.*

grass plantain. *Alisma plantago,* family Alismaceae, an aquatic herb which is a rice field weed pest.

grass-seed attachment. A small seed box usually attached in front of the grain box of a drill which is used to drill grass and legume seed. Also called *cup drill.*

grass-seed nematode disease. A disease affecting grass seed caused by the nematode, *Anguina agrostis,* and others; characterized by the seeds being transformed into purple or black galls which are poisonous to livestock.

grass shears. Hand shears used for trimming grass in lawns, around walks, flower beds, etc.

grass silage. Any silage prepared from forage

plants. Corn and sorghums are not ordinarily included.

grass sorghum. A class of sorghums which includes Sudan grass, Tunisgrass and Johnsongrass. Sudan grass is the most important and is used for hay, pasture, and sometimes silage. See *sorgo; grain sorghum; broomcorn.*

grass tetany. A magnesium deficiency disease of cattle characterized by hyperirritability, muscular spasms of the legs, and convulsions. In sheep, it is apparently associated with a calcium and magnesium deficiency. The disease is seen when the animals are turned out to lush spring pastures in some areas. Also called *grass staggers.*

grass thrips. *Anaphothrips obscurus* (Müll.), family Thripidae, an insect pest of grains and grasses which sometimes destroys the softer parts of the leaves, flowers, and kernels.

grass vetch. Grass pea(vine).

grassworm. See *fall armyworm.*

grassy. Designating grass or grasslike plants or plant parts.

grassy flavor. An off flavor of milk or butter which is considered a normal feed flavor, especially of butter made when cows are first turned to spring grass pasture.

grassy top. A virus disease of gladiolus which is characterized by weak, yellow shoots and thin, weak roots.

grate. 1. Parallel steel bars or tines placed just below and behind the concaves on grain threshers or combines which separate most of the grain from the straw. 2. Parallel bars to screen out trash at a surface inlet to a tile drainage line.

Graubraunes Girbirgsvieh. See *Brown Swiss.*

gravel. 1. A suppurating corn on a horse's foot. 2. The formation of a calculus or stone in the kidney or urinary system of bulls, steers, rams, or wethers. Also called *water belly.* 3. Aggregate which is composed of hard, durable, more or less rounded stones or pebbles, crushed or uncrushed, often intermixed with sand. 4. In the mechanical analysis of soil, rounded or angular particles of rocks or minerals which are greater in size than coarse sand and up to 3 ins. in maximum diameter.

gravel culture. Growing plants in artificial nutrient solutions with gravel as the medium for supporting the roots.

gravelled. See *gravel: 1.*

gravelly loam. A loam which contains from 25 to 50 percent of gravel, much sand, and little silt.

gravelly soil. A soil which contains a high proportion of rock fragments of gravel size: coarse, rounded or subangular particles less than 3 ins. in diameter.

gravelweed. Dwarf bushhoneysuckle.

gravid. In regard to higher animals, designating the state of pregnancy or containing a fetus. A parasite or part of a parasite filled with eggs is also referred to as being *gravid.*

gravid segments. Segments or parts of a tapeworm that are filled with eggs. *Gravid segments* of tapeworms usually break away from the main body of the tapeworm in the intestine and are eliminated with droppings.

gravitational water. Free water *q.v.*

gravity conveyor. A conveyor in which the weight of the articles handled is sufficient to effect their transport from a higher to a lower point, as when they are allowed to slide down an inclined runway.

gravity dam. A dam which depends on its shape and weight distribution to resist the overturning, sliding, or pivoting forces of the water and soil load behind it.

gravity water. 1. Water that moves through soil under the influence of gravity. 2. A gravity supply of water as distinguished from a pumped supply.

Gray. A sour, cured, skimmed milk cheese made in the Tyrol.

gray. 1. A coat color of an animal's coat which has white hairs mixed with black. 2. A cotton-lint color designation which is the darkest in chroma.

gray bark. See *cane fruit anthracnose.*

gray berteroa. Hoary falsealyssum.

gray birch. *Betula populifolia,* family Betulaceae, a deciduous tree which is scrubby and of little value either as a forest tree or horticulturally, but will grow in very poor soil. Native to North America.

Gray-Brown Podzolic soil. A zonal group of soils having a comparatively thin organic covering and organic-mineral layers over a grayish-brown leached layer which rests upon an illuvial, brownish horizon. It developed under deciduous forests in temperate, moist climates.

gray dogwood. *Cornus racemosa,* family Cornaceae, a hardy deciduous shrub grown for its white flowers and white fruit. Native to North America.

gray ear rot. A fungous disease of corn caused by *Physalospora zeae,* family Mycosphaerellaceae; characterized by bleached husks adhering to one another, by the ear's remaining upright till harvest, by tiny, black specks through the pith of the cob, and by black specks or stripes beneath the seed coats of kernels.

gray eye. Ocular lymphomatosis.

gray goldenrod. Dyersweed goldenrod.

gray gum. See *horncap eucalyptus.*

gray horsebrush. *Tetradymia canescens,* family Compositae, a low shrub which is poisonous to stock. Symptoms of poisoning are depression, weakness, swelling of the head, neck, ears, eyelids, and nose, and in some cases sudden death results. Native to North America; found on ranges from Washington to Mexico.

grayleaf pine. Digger pine.

gray leaf spot of grasses. A fungous disease of foxtail millet caused by *Piricularia grisea,* family Moniliaceae; characterized by severe spotting, blighting, or blasting of the foliage.

gray middlings. See *wheat gray shorts.*

gray mold. 1. A symptom of downy mildew of grapes (*q.v.*) in which the fruit is covered with gray fruiting stalks and spores of the fungus. 2. The covering of any of a large number of fruits by gray fruiting stalks and spores of any of several fungi. 3. Botrytis rot of citrus *q.v.* 4. Any mold with gray spore masses.

gray mold neck rot of onion. A fungous disease caused by *Botrytis allii,* family Moniliaceae; characterized by softening of the affected tissue which appears sunken and cooked and by a grayish color of the affected areas which may be covered with a gray mycelial mat.

gray mold of castorbean. A fungous disease of castorbean caused by *Sclerotinia ricini,* family Sclerotiniaceae; characterized by a gray mold.

gray mold of lettuce. See *Botrytis blight of lettuce.*

gray mold of strawberries. A fungous disease caused by *Botrytis cinerea,* family Moniliaceae; characterized by a blossom blight, and a rot of the fruit showing as a light brown area. In moist weather the fruits are coated with fine, gray, fuzzy, fruiting stalks and spores of the fungus.

gray oats. A market classification of oats based on the grayish color of the lemma and palea of the matured grain. Other classifications are: white oats; red oats; black oats; mixed oats.

gray orchis. Showy orchis.

gray rat. See *roof rat.*

gray roan. A coat color for a horse which is roan in combination with gray. See *steel gray.*

gray rot. See *gray mold.*

gray shorts. See *wheat gray shorts.*

gray speck. A magnesium deficiency disease of oats characterized by light green to gray, irregularly-shaped flecks on the leaf blades that enlarge, dry out, and turn brown or buff-colored; dwarfing; and reduction of yield. Also called *dry leaf spot.*

graze. 1. To consume any kind of standing vegetation, as by domestic livestock or wild animals. See *browse.* 2. To cause domestic animals to *graze.*

graze off. To cause animals to feed on and almost completely consume the top growth of herbaceous vegetation.

graze on. To feed, as domestic ruminants, on plant vegetation.

graze out. To allow domestic animals to feed abusively on certain palatable grasses or forbs alone or in combination until the vegetation ceases to exist on a particular pasture, range, etc.

grazier. A rancher; a person who produces livestock on grazing land.

grazing. 1. Feed available to animals on ranges and pastures. 2. The process of feeding by livestock on live or standing plants other than browse.

grazing bit. A snaffle or easy curb bit which does not prevent a horse from grazing.

grazing capacity. In range or pasture management, the ability of a range unit, exclusive of severe drought years, to give adequate support to a constant number of livestock for a stated period each year without deteriorating with respect to this or other proper land use. It is expressed in number of livestock per acre of given kind or kinds, or in number of acres per specified animal.

grazing district. 1. In the U.S., an administrative unit of the federal range which was established by the Secretary of the Interior under the provisions of the Taylor Grazing Act of 1934, as amended. 2. An administrative unit of state, private, or other range lands which is established under certain state laws.

grazing fee. A charge which is made for livestock grazing on a range on the basis of a certain rate per head for a certain period of time, as distinguished from lease or rental of the land on which animals may be grazed.

grazing habit. The distinctive manner of grazing for a given kind of livestock.

grazing land. Land regularly used for grazing, except cropland and pasture, which is used in connection with a system of farm crop rotation. See *pasture land; range: 1, 2.*

grazing management. The care and handling of both vegetation and livestock on the range or pasture.

grazing period. The time livestock are grazed or are permitted to graze on a range or pasture. Also called *grazing season.*

grazing permit. A document authorizing the use of public or other lands for grazing purposes under specified conditions which is issued to the livestock operator by the agency administering the lands.

grazing preference. 1. In the administration of public lands, the basis upon which permits and licenses are issued for grazing use. 2. The preference animals exhibit in selective grazing of certain plants in a mixed plant population.

grazing privilege. An authorized use of public land for grazing purposes which is granted without giving any permanent or vested interest in or to the public land concerned.

grazing (range-management) plan. A specific grazing program for a private range or public grazing allotment which is preferably based upon range-inventory information. It is usually graphic, accompanied by written directions, and should include the year-long operation.

grazing season. Grazing period.

grazing survey. The systematic collection of data which pertain to forage and other resources pertinent to proper range management on a range allotment or management unit.

grazing unit. 1. The quantity of pasturage which is used by an average, mature cow or its equivalent in other livestock in a grazing season in a given region. 2. Any division of the range which is used to facilitate range administration or the handling of livestock.

grazing value. The worth of a plant or cover for livestock and/or game which is determined by its palatability, nutritional rating, amount of forage produced, longevity, and area of distribution.

grease. 1. See *fat: 1; lard; lanolin.* 2. A thick petroleum derivative which is used for lubrication. 3. Hog fat as distinguished from tallow which is the fat of cattle and sheep; commercially differentiated by temperature of solidification which is below 40° C. for *grease* and above

40° C. for tallow. 4. To lubricate a machine. 5. To apply salve to a wound, irritation, etc.

grease heel. See *scratches.*

greasewood. 1. Any low-growing bush that burns easily (western U.S.). 2. See *white sage; black greasewood; chamiso: 2.*

greasewood chamise. See *chamiso: 2.*

grease wool. Wool as it is shorn before washing or scouring.

greasy. In butter grading, designating a butter with a greasy flavor; such butter is rated B.

greasy cutworm. See *black cutworm.*

greasy flavor. A flavor defect of milk characterized by an oiliness.

greasy-heel. See *scratches.*

greasy pod. A virus disease of beans caused by a strain of the common bean mosaic virus; characterized by a shiny or greasy appearance of the pods.

greasy spot. See *melanose.*

greasy wool. See *grease wool.*

great aster. *Aster grandiflorus,* family Compositae, a perennial herb grown for its deep purple or violet flowers. Native to North America.

Great Basin region. A vast, geographic region of the U.S., lying between the Rocky Mountains and the Sierra Nevada Mountains, which is mostly arid or semiarid; a region of mountains and wide, intermountain, alluvial plains or valleys, which sustains mostly bunchgrasses and sagebrush on the semiarid land. It is largely sheep and cattle grazing land, and in part nearly barren desert shrub and alkali and salt flat wastes, but a small part is cultivated either under irrigation or dryland farming.

Great Basin wireworm. *Ctenicera pruinina noxia* (Hyslop), family Elateridae, a wireworm which is an insect pest of the roots of potato, corn, and small grains. The adult beetle infests the wild rose and apple.

great bellflower. *Campanula latifolia,* family Campanulaceae, a perennial herb grown in the flower border for its solitary purplish-blue flowers. Native to Eurasia.

great burdock. *Arctium lappa,* family Compositae, a biennial herb which is a common weed pest resembling a thistle. Also called *burdock; gobo.* Native to Europe; naturalized in North America.

great coneflower. *Rudbeckia maxima,* family Compositae, a hardy, perennial herb grown for its yellow flowers with brown centers. Native to North America.

Great Divide. In the U.S., the hypothetical line which is drawn along the highest points of the Rocky Mountains. Precipitation falling generally to the east of it flows into the Gulf of Mexico, while that falling generally to the west may flow into the Gulf of California or into the Pacific Ocean.

greater celandine. *Chelidonium majus,* family Papaveraceae, a perennial, weedy herb grown for its yellow flowers. Escaped, it can become a weed pest. Native to Eurasia. Also called *celandine; killwort; sightwort.*

greater stitchwort. Easterbell starwort.

greater wax moth. *Galleria mellonella* (L.), family Galleriidae, a moth whose larvae damage brood combs in beehives. Also called *bee moth; bee miller.*

great laurel. Rosebay rhododendron.

great lobelia. *Lobelia siphilitica,* family Lobeliaceae, a perennial herb grown for its deep blue to purplish flowers. Found in U.S. from Maine to Louisiana. Also called *blue lobelia; bigblue lobelia.*

great mullein. See *flannel mullein.*

great nettle. *Urtica dioica,* family Urticaceae, a perennial weed pest which bears stinging hairs. Native to Eurasia; naturalized in North America. Also called *bigsting nettle.*

Great Northern. A white variety of the kidney bean that is grown for its edible, dried seeds. It is distinguished from the navy or pea bean by its larger size.

Great Plains. A vast, treeless region of plains (prairie and steppe) which occupies central-western U.S. It may be somewhat arbitrarily delimited on the east by the ninety-eighth meridian and thence westward with increasing elevation to the foot-hills of the Rocky Mountains. North to south, it extends from the Canadian border of Montana and North Dakota to Texas. The climate is semiarid to subhumid and the region is subject to periods of disastrous drought. In use, it is grazing land; in part dry farming, mostly wheat and sorghums; and a very small (in relative acreage) part is under irrigation.

great ragweed. Giant ragweed.

great soil group. In soil classification, a group of soils having common internal soil characteristics which includes one or more families of soils. Among the zonal soils, each *great soil group* includes the soils having common, internal characteristics developed through the influence of environmental forces of broad geographic significance, especially of vegetation and climate. Among the intrazonal soils, each *great soil group* includes the soils having common, internal characteristics developed through the influence of environmental forces of both broad and local significance. Among the azonal soils, each *great soil group* includes similar soils that are without developed characteristics, because of some local condition of parent material or relief.

great starwort. See *chickweed.*

great willow herb. Fireweed.

Grecian laurel. *Laurus nobilis,* family Lauraceae, an evergreen tree grown as an ornamental in tubs. It can be trimmed easily and is hardy out-of-doors only in warm areas. Native to southern Europe. Also called *bay; bay tree; sweet bay; true bay; laurel.* See *laurel leaves.*

Grecian silkvine. *Periploca graeca,* family Asclepiadaceae, a woody, climbing vine grown for its habit and its brownish-purple, slightly fragrant flowers. Native to southern Europe.

Grecian stock. *Mathiola bicornis,* family Cruciferae, an annual or biennial herb grown for its fragrant, purple flowers which open in the evening. Native to Eurasia. Also called *evening stock; perfume plant.*

greedy scale. *Aspidiotus camelliae* Sign., family Diaspididae, an insect that infests greenhouse plants and a host of out-door plants in warm areas, among which are citrus and deciduous fruits and many ornamental trees and shrubs.

Greek bellflower. *Campanula drabifolia,* family Campanulaceae, an annual herb grown in the flower border or in the rock garden for its solitary, blue flowers. Native to Eurasia.

Greek fir. *Abies cephalonica,* family Pinaceae, an evergreen, coniferous tree grown as an ornamental. Native to Eurasia.

Greek honeywort. *Cerinthe retorta,* family Boraginaceae, a hardy, annual herb grown for its yellow flowers tipped with purple. Native to Eurasia.

Greek juniper. *Juniperus excelsa,* family Cupressaceae, an evergreen, pyramidal tree grown as an ornamental or specimen in warm areas. Native to Eurasia.

Greekvalerian polemonium. *Polemonium caeruleum,* family Polemoniaceae, a hardy, perennial herb, grown in the flower border for its large clusters of blue flowers. Native to Europe. Also called *green valerian; Jacob's ladder.*

green. 1. A small piece of land which is covered with grass, as a golf *green,* or a common, as a village *green.* 2. Unmatured; immature, as a fruit, seed, honey. 3. Uncured, as *green* lumber, *green* cheese. 4. Untrained, as a *green* colt. 5. Designating a plant whose foliage is a shade of *green,* as *green* grass. 6. Designating a flavor defect of milk, characterized by a weak or acid taste and by a repulsive odor.

green alder. European green alder.

green amaranth. Redroot amaranth.

green arrow arum. Virginia arrow arum.

green ash. *Fraxinus pennsylvanica* var. *lanceolata,* family Oleaceae, a hardy, deciduous tree grown in shelterbelts in the Great Plains, U.S., as a shade tree and for its lumber used in handles, furniture, sporting goods, boxes, implements, boats, woodenware, cooperage, etc. Native to North America.

green asparagus. Asparagus that has not been blanched.

green-barked acacia. Blue paloverde.

green beans. The immature, green, succulent pods of the kidney bean which are used as a potherb. Also called *snapbeans.* See *kidney bean.*

greenbelt. 1. A strip of land of varying size, used for recreation or farming, which encircles a community and protects it from the intrusion of objectionable real property uses from adjacent communities and/or provides the community with space for orderly expansion. 2. A similar strip around a commercial enterprise into which it can expand without the invasion of residential property.

green-boll. Designating a stage of growth of the cotton plant in which the boll has been formed but is still immature.

greenbrier. *Smilax rotundifolia,* family Liliaceae, a prickly, green-stemmed, wiry, climbing vine which is often a nuisance to eradicate once it has become established. Native to North America. Also called *catbrier; horse brier; roundleaf; bullbrier; devil's-hopvine.*

green bristlegrass. *Setaria viridis,* family Gramineae, an annual grass which is a weed pest of cultivated fields, meadows, barnyards, and roadsides. Native to North America. Also called *foxtail grass; green pigeon-grass.*

green broke(n) horse. A horse which has recently been broken but is not wholly trained as yet.

greenbug. *Toxoptera graminum* (Rond.), family Aphidae, an aphid which infests small grains, rice, corn, and sorghum and is a vector of sugarcane mosaic.

green cheese. An uncured cheese. Also called *fresh cheese.*

green chop. Green forage which is cut with a field chopper and hauled to lots or barns for livestock feed in lieu of pasturing. See *green chopping.*

green chopping. The practice of chopping and hauling green forage to dairy and beef cattle during the pasture period. The most common practice is to chop and haul forage once or twice daily. Green chopped forage may be fed in self-feeding wagons filled directly by the field chopper or in bunks. Also called *zero grazing.*

green citrus aphid. Spirea aphid.

green cloverworm. *Plathypena scabra* (F.), family Noctuidae, a caterpillar which infests clover, alfalfa, kidney beans, soybeans, cowpeas, and other leguminous plants in eastern U.S.

green corn. See *roasting ear.*

green duck. A young duckling grown rapidly nine or ten weeks which weighs about 5 lbs.

green-ear disease. See *downy mildew of grasses.*

green fallowing. Green manuring (obsolete).

green feed. Uncured forage.

green fescue. *Festuca viridula,* family Gramineae, a perennial grass found from British Columbia to Alberta, Canada, and south to California and Idaho, U.S., at 6,500 to 10,000 ft. altitude, whose forage rating is good. Native to North America. Also called *greenleaf fescue; mountain bunchgrass.*

green foxtail. Green bristlegrass.

green fruitworm. *Lithophane antennata* (Wlk.), family Noctuidae, an insect pest of several deciduous trees and of the green fruits of apple, cherry, and pear. It eats holes in the side of the immature fruits.

greengage. A horticultural variety of the common plum, *Prunus domestica,* family Rosaceae, developed in southern Europe. The fruit is green and has a fine flavor.

green geese. Young geese produced for the market which weigh 10 to 14 lbs. at 10 to 15 weeks of age.

green hay. 1. Uncured hay. 2. That hay which, on being cured, retains a green color.

green herbage. Range forage which is all live

green plant material and dead tips of growing grasses.

green honey. That honey which is in unsealed cells and has not evaporated enough to be properly thick. Also called *unripe honey.*

green horse. See *green broke(n) horse.*

greenhouse. Any of several different types of heated, glass or plastic covered structures used for the growing of plants. See *forcing house; conservatory; propagating house; greenhouse range.*

greenhouse leaf tier. Celery leaf tier *q.v.*

greenhouse man. One who owns, operates, manages, or works in a greenhouse.

greenhouse orthezia. *Orthezia insignis* Browne, family Ortheziidae, a greenhouse insect pest which sucks plant juices.

greenhouse range. Two or more greenhouses, physically connected in some way, under the same management, which are usually heated by one heating plant.

greenhouse thrips. *Heliothrips haemorrhoidalis* (Bouché), family Thripidae, a rasping, sucking insect pest of greenhouse plants, especially of roses, carnations, cucumbers, fuchsias, chrysanthemums, crotons, and cinerarias. Out-of-doors, it is a serious pest of citrus in California, U.S. On infested plants, the surface of leaves is whitened and becomes flecked; tips of leaves wither, curl up, and die; the undersides of leaves are spotted with black specks.

greenhouse whitefly. *Trialeurodes vaporariorum* (Westw.), family Aleyrodidae, a sucking insect pest of greenhouse plants. Infested plants lack vigor, wilt, turn yellow, and die.

greening. 1. The turning green of the potato tuber due to exposure to sunlight. Also called *sunburn.* 2. Any of several varieties of green-colored apples.

greenish-black discoloration. A serious, color defect of chocolate ice cream which is the result of the chemical reaction between the tannins in cocoa or chocolate and exposed iron or rust spots on the containers.

green June beetle. *Cotinus nitida* (L.), family Scarabaeidae, an insect pest, the adults of which feed on the foliage and fruit of a number of trees and plants; the larvae damage lawns, vegetable roots, and tobacco in seedbeds.

green lacewing. *Chrysopa californica* (Cog.), family Chrysopidae, an important insect predator on the citrophilus mealybug.

Greenland sandwort. *Arenaria groenlandica,* family Caryophyllaceae, a low, perennial herb planted in the rock garden for its tufted habit and white flowers. Native to North America. Also called *mountain sandwort; mountain starwort.*

greenleaf fescue. Green fescue.

green lumber. Newly sawed, uncured lumber.

green manure. A crop which is plowed under green for its beneficial effect on the soil.

green manuring. The use of green manure.

green meat. Vetches alone or with oats and/or rye which serve as green feed for stock.

green mint. Spearmint.

green mold. See *common green mold.*

green mosaic. See *wheat mosaic-rosette; take-all: 2; green mottle mosaic.*

green mottle mosaic. A virus disease of bramble fruits which has a wide variety of symptoms. Generally, it is characterized by dark green, mottled areas on the leaflets which may be accompanied by blistering.

green needlegrass. *Stipa viridula,* family Gramineae, a perennial bunchgrass widely distributed in North America from New York to Alberta and south to New Mexico on dry slopes. Its forage rating is good.

green olives. Olives which are harvested while green, and pickled. They are first soaked in a lye solution, washed, and placed in oak or chestnut barrels with a brine solution for fermentation.

green onion. An immature garden onion (*Allium cepa*) which is harvested before the bulb has begun to take on a spherical shape; a very popular salad vegetable. See *top onion; multiplier onion; onion: 2.*

green osier. Roundleaf dogwood.

green peach aphid. *Myzus persicae* (Sulz.) family Aphidae, a sucking insect pest of peach, plum, apricot, cherry, citrus, as well as of many ornamental shrubs, garden plants, and flowering plants. Also called *spinach aphid.*

green peas. The immature seeds of the garden pea (*Pisum sativum*) which are a very important human food.

green pepper. The sweetbell redpepper (*q.v.*) picked after it has attained full growth, but before it turns red; used in salads, cooking, etc.

green pigeon-grass. Green bristlegrass.

green ring mottle. A virus disease of the sour cherry (*Prunus cerasus*) characterized by dark green rings on otherwise clear yellow leaves. A portion of the leaves shows symptoms and the tree defoliates about six to eight weeks after blossom time in cool seasons. There is some reduction in yield but infected trees live many years. It differs from sour cherry yellows (*q.v.*) in type of leaf pattern and time of symptom appearance.

green-ripe olives. See *home-process olives.*

greens. 1. The green leaves of various plants, both wild and domesticated, which are boiled for human food, usually with fat meats, such as bacon or hog jowls. Wild plants, such as lambsquarter, docks, and young shoots of pokeweed and the wild blackberry were mixed together and boiled, especially in spring, by early settlers of the U.S. The leaves of cultivated crops such as turnips and spinach are also considered as *greens.* Also called *potherb.* 2. Any foliage which is added to a floral arrangement, such as bouquets, funeral pieces, etc.

greensand beds. Geological formations, mostly Cretaceous and Tertiary, which contain glauconite (*q.v.*). When used as natural soil amendment or fertilizer, it is sometimes called *greensand marl.*

green scale. *Coccus viridis* (Green), family Coccidae, a scale insect that is very injurious to coffee plants and citrus in parts of tropical Asia.

green shield scale. *Pulvinaria psidii* Mask., family Coccidae, a scale insect which infests avocado, citrus, guava, apple, mango, persimmon, fig, gardenia, hibiscus, jasmine, and palm in Florida, U.S.

green smut. False smut.

green sprouting. 1. The practice of allowing potatoes which are to be used for seed to be exposed to the air and light until small, sturdy, green sprouts appear at the eyes. 2. The practice of keeping soaked oats in trays under warm and humid conditions until a heavy vigorous growth of green sprouts covers the surface. This sprouted material is used as chick feed and also as an appetizer for cows in very heavy milk production.

greenstem forsythia. *Forsythia viridissima,* family Oleaceae, a hardy, deciduous, upright shrub, grown for its bright yellow flowers. Native to China.

green stink bug. *Acrosternum hilare* (Say), family Pentatomidae, an insect which infests peaches, nectarines, apple, cabbage, catalpa, corn, dogwood, elderberry, linden, maple, okra, orange, pea, tomato, turnip, and cotton.

green-striped mapleworm. *Anisota rubicunda* (F.), family Citheroniidae, a caterpillar which infests maple, box elder, and oak, often defoliating maples.

green tying. The tying of the current season's grape shoots in the spring with soft twine, etc., to the wires of a trellis. Also called *summer tying; straw tying.*

green up. To become green again following the winter period or in one season, as a plant after a period of drought followed by rain.

green valerian. Greekvalerian polemonium.

green vegetable. Any vegetable whose edible portions are green, such as spinach, garden peas, mustard, lettuce, broccoli, etc. See *yellow vegetable.*

green vitriol. See *ferrous sulfate.*

green wood. Uncured, unseasoned wood.

greenwood cutting. Softwood cutting.

gregarious oak leaf miner. *Cameraria cincinnatiella* (Chamb.), family Gracilariidae, an insect which mines the leaves of oak, especially white oak.

grenadine (sirup). A sirup which is prepared from the juice of the pomegranate fruit.

gridiron. 1. See *espalier.* 2. A system of parallel lateral under-drains which enter a submain or main from one side only.

gridiron system. 1. One system used in range mapping in which a General Land Office (U.S.) survey grid is used as a base and the surveyor moves from a known corner on compass or straight lines. He maps in the type and subtype plant associations, topography, streams, ridges, culture, etc. 2. A system of topographic mapping in which a field is subdivided into small squares; elevations are taken at each corner for basic information on which to prepare a contour map.

grill. 1. A perforated covering for an air inlet or outlet which is usually made of wire screen, pressed steel, cast iron, or plaster. 2. A small, portable stove for camp or yard use.

Grimes. A golden-yellow, fine-flavored, late variety of eating apple. Also called *Grimes golden.*

Grimes collar rot of pome fruits. A root and trunk rot which appears near the ground level of pome fruit trees, especially of the Grimes variety of apple. It is thought to be caused by the fungus *Phytophthora cactorum,* family Pythiaceae.

Grim-the-collier. Orange hawkweed.

grind. 1. To crush or reduce to a powder or to small bits, as to *grind* grain to make flour or feed. 2. To sharpen a cutting instrument by filing a cutting edge with a grindstone, a carborundum, etc. 3. See *Rhizoctonia disease of potatoes.*

grinder. 1. Any of several different mechanical devices which crush grain into small particles. 2. A motor-driven wheel of aluminum oxide or other abrasives used for sharpening cutting tools.

grinders. 1. The molars; molar teeth. 2. American Cheddar cheese which is intended for processed cheese manufacture.

grinding. See *grind: 1, 2.*

grindstone. A circular disk of sandstone, mounted on an axle, turned by a crank or treadle which is used to sharpen cutting tools. A *grindstone* is made from natural sandstone, in which quartz comprises the abrasive medium. Formerly it was widely used for sharpening tools, finishing of cutlery, snagging of castings, etc., but it has largely been displaced by artificial, vitrified abrasive wheels and by polishing wheels.

gripes. See *colic: 2.*

grip hooks. Hooks which are attached to the traveling apron of a silage cutter to facilitate the conveying of stalks to the cutter.

grist. 1. A quantity of grain which is brought to a grist mill for grinding. 2. The product from such grinding.

gristle. Cartilage.

gristle on the heel. In auction terminology, a horse defect of one or more side-bones. See *side-bone.*

grist mill. A grain grinding mill equipped with large, heavy grinding stones or burrs usually run by water wheel power. Grains, such as wheat, rye, and corn, are ground into meal and flour. Such mills were common in early colonial days in the U.S.

grit. Hard, gravelly substances, such as small, rough-edged stones, which are fed to chickens to be used in the gizzard to grind feed.

grits. 1. A corn food product which consists of the hulled, ground grain of even-sized particles larger than corn meal. Also called *hominy grits.* See *grist.* 2. See *groats: 1.*

gritty. See *grainy.*

groats. 1. The hulled seeds of cereals and buckwheat which are ground into particles of even size but larger than flour particles. 2. Unground oat kernels.

grocer's itch. An infestation of a person's skin by flour mites, *Sarcoptes scabiei* (DeG.), family Sarcoptidae, and cheese mites (*q.v.*), which is very similar to an infestation by chiggers. Persons who work with flour, meal, sugar, dried fruits, copra, cheese, hams, etc., are apt to be affected.

grog. See *navicular disease.*

groin. 1. A structure built from the shore into the water to protect the bank against erosion. 2. The part of the body of man or other animals where the thigh joins the body trunk.

grood. An aggregation of soil particles formed into a nutlike structure.

groom. 1. A person who curries, combs, washes, etc., an animal and cares for it generally. 2. To wash, curry, brush, and generally care for an animal. 3. To trim grass and to make a yard and flower garden neat and trim.

gros-pied. See *clubroot of cabbage.*

gross change. Any change that can be observed by the normal human eye without the use of microscopic lenses.

gross farm earnings. The sum of the farm receipts plus the value of the family living from the farm.

gross (farm) income. The receipts and net increases from the farm business for the year which include the cash receipts from the sale of dairy products, eggs, forest products, machine work off farm, labor off farm, and other receipts of a minor nature and net increases from dairy cattle, beef cattle, swine, sheep, poultry, crops, agriculture payments, and orchards, etc. (Net decreases in any of the foregoing items are subtracted from the total gross income.)

gross pathological change. An alteration in the appearance of an organ or tissue, as a result of disease, which can be seen with the unaided eye.

ground. 1. The surface of the earth. 2. The soil; land. 3. a. An electrical connection to the earth. b. To make such an electrical connection.

ground barley. A livestock feedstuff which is the entire product obtained by grinding clean, sound barley, containing not less than 90 percent of pure barley and not more than 10 percent of other grains (which should not exceed 6 percent of the total), weed seeds, and other foreign material. Weed seeds or other foreign material shall not be intentionally added.

ground beef. Retail meat which is usually ground, lean beef. Frequently called *hamburg(er).*

ground blueberry. *Vaccinium myrsinites,* family Ericaceae, an evergreen shrub grown in borders and wild gardens for its foliage. Also called *evergreen blueberry.*

groundcedar. *Lycopodium complanatum* var. *flabelliforme,* family Lycopodiaceae, a trailing, evergreen, mosslike herb grown in shady places in the wild garden for its foliage. Native to the North Temperate Zone.

groundcherry. Any herb of the genus Physalis, family Solanaceae. Also called *husk tomato.* See *strawberry groundcherry; tomatillo groundcherry.*

ground color. 1. The predominant color of the web of a feather. See *surface color: 2.* 2. The greenish or yellow color of the skin of ripening and ripe fruits. Also called *underlying color. Cf. surface color: 1.*

ground corn. Corn chop *q.v.*

ground corn and oats. See *corn-and-oat feed.*

ground cover. 1. Any of many different plants, usually perennials, which grow well on sites on which grass does not thrive, as on banks, terraces, in the shade, etc. 2. All herbaceous plants and low-growing shrubs in a forest. 3. Any vegetation which grows close to the ground producing protection for the soil. See *cover crop.*

ground-driven. Designating a machine which, when pulled by horses or tractor, receives its power from one of its own wheels in contact with the ground.

ground ear corn. See *ear corn chops.*

ground extracted penicillium presscake. A livestock feedstuff which is the filtered and dried mycelium (mycelia) obtained as a by-product from the manufacture of penicillin.

ground feed. See *corn-and-oat feed.*

ground fire. A fire that not only consumes all the organic materials on the soil surface but also burns into organic materials of the underlying soil itself: *e.g.,* a peat fire (usually combined with, but not to be confused with, a surface fire).

ground flaxseed. A livestock feedstuff obtained by grinding flaxseed previously screened and cleaned of weed seeds and other foreign materials by the most improved commercial process. The final product must contain less than 6 percent of weed seeds and other foreign materials, no portion of which shall have been deliberately added. Also called *flaxseed meal.*

ground gold-flower. Sickleleaf golden aster.

ground hemlock. Canada yew.

ground holly. Checkerberry *q.v.*

ground honeysuckle. See *birdsfoot deervetch.*

ground ivy. *Glecoma hederaceae* (*Nepeta h.*), family Labiatae, a creeping, perennial herb sometimes grown as a ground cover for its habit and its blue flowers. Escaped, it can become a weed pest. It was once used medicinally and as an infusion for ale and beer. Also called *alehoof; cat's foot; gill; gill-ale; gill-over-the-ground; hayhofe; haymaids; honey tunhoof; creeping Charlie; robin-run away; gill-runover; crow victuals; wild snakeroot; hedge maids; fat hen; field balm; earth ivy.*

ground laurel. See *trailing arbutus: 2.*

ground lily. Nodding trillium.

ground limestone. A commercial soil amendment; the product which is obtained by grinding either high calcic or dolomitic limestone so that all materials will pass an 8-mesh sieve and at least 25 percent or more will pass a 100-mesh sieve. Also called *coarse ground limestone.* See *lime: 1.*

ground mealybug. *Rhizoecus falcifer* Kunck., family Coccidae, an insect pest which feeds on the roots of potted plants, grasses, acacia, box, chrysanthemum, currant, gooseberry, grape, larkspur, orange, peach, pepper, petunia, plum, and thyme in California, U.S.

ground moraine. See *till plain.*

ground nut. See *peanut; chufa flatsedge; American potatobean.*

ground oat groats. Cut oat groats.
ground oats. Oat chop.
ground pea. Peanut.
ground peanut skins and hulls. A livestock feedstuff which consists of peanut skins with only sufficient hulls added to facilitate the grinding of the mixture. See *peanut skins.*
ground peanut vine hay. Ground peanut vine stem hay.
ground peanut vine stem hay. A livestock feedstuff which is the ground product obtained from only vines and stems of the peanut plant.
ground pine. Running pine.
ground pink. Moss phlox.
ground rattan. Broadleaf ladypalm.
ground rot. A rot of melons, originating at the point of contact with the ground, caused by the fungus, *Sclerotium rolfsii;* characterized by a heavy growth of white mold and the formation of many brown bodies suggesting buckshot.
ground rough rice. A livestock feedstuff which is ground paddy rice or ground rice from which the hulls have not been removed.
grounds. 1. The land surrounding a house, building, etc., in lawns, flowers, parks, which is usually carefully tended. 2. Sediment.
groundsel. 1. Any herb, shrub, or tree of the genus Senecio, family Compositae. See *golden groundsel; silver groundsel; leopard's-bane groundsel; ragwort groundsel; ivy groundsel.* 2. *Senecio vulgaris,* a weed pest. Also called *butterweed.*
groundsel bush (tree). Eastern baccharis.
ground shell. A commercial soil amendment; the product which is obtained by grinding the shells of mollusks so that 50 percent shall pass a 100-mesh sieve. The product shall carry the name of the mollusk from which it is made. See *lime: 1.*
ground shelled corn. Corn chop.
ground shell marl. A commercial soil amendment; the product which is obtained by grinding natural deposits of shell marl so that not less than 75 percent shall pass a 100-mesh sieve. See *lime: 1.*
ground skidder. In forestry, a power skidder that skids logs on the ground.
grounds-man. Grounds superintendent.
ground snapped corn. See *ear corn chops with husks.*
ground soybeans. A livestock feedstuff obtained by grinding whole soybeans without cooking or removing any of the oil.
ground squirrel. Any of several different burrowing rodents. See *gopher: 1.*
grounds superintendent. A person who is hired to maintain a single, large, landscaped, planted area, such as an estate, park, cemetery, institutional grounds, etc. Also called *grounds-man; head gardener; gardener.*
ground velvet bean and pod. See *velvet bean meal.*
ground water. 1. Technically, the water of the zone of saturation as distinguished from the water of the zone of aeration. See *phreatic water; vadose (water); water table.* 2. Well water; subsurface or underground water whether it is in the soil or deeper underlying formation.
ground-water flow. The movement of ground water under hydraulic gradient; the water that comes to the surface in springs or causes streams and rivers to flow after periods of surface runoff have ended. See *ground water.*
Ground-water Laterite soil. An intrazonal group of soils with bleached A horizons containing some concretions and subsurface horizons which are thick, more or less cellular hardpans composed largely of iron and aluminum compounds. It is formed under an alternating high and low water table in warm-temperate to tropical climates.
ground-water level (line). See *water table.*
Ground-water Podzol soil. An intrazonal group of soils which developed from imperfectly drained, sandy deposits in humid regions. They have a thin organic layer over a light gray, sandy, leached layer, resting upon a dark brown B horizon irregularly cemented with iron or organic compounds or both. The B horizon is called ortstein (*q.v.*) when cemented.
ground whole pressed cottonseed, prime quality. A livestock feedstuff which is whole pressed cottonseed, ground. It is produced from sound cottonseed and it must be firm in texture, must not have a sour, musty, or burnt odor, and shall not contain less than 22 percent of crude protein. It shall be sold according to its protein content.
groundwort. See *American potatobean.*
group action. Action by a group of farmers or ranchers in planning the application and maintenance of land use, soil and water conservation measures, etc., under the direction of a technician or leader.
group enterprise. The application or installation and the maintenance of a conservation measure which serves more than one farm or ranch through joint effort: *e.g.*, construction of a main drainage ditch, realignment of the irrigation system of an irrigation company, etc.
grout. Mortar, composed of sand, cement, and water, of such consistency that it can be worked easily; used to seal the seams in rock encountered in tunneling water through mountains and in dam construction.
grove. 1. A stand of trees, smaller than a forest, which grows naturally and has little undergrowth. 2. A small woods. 3. An orchard, or a planting of trees, as an orange *grove,* pecan *grove.*
grow. 1. To live and to increase in stature and girth toward maturity. 2. To cultivate plants. 3. To raise animals.
grower. See *farmer.*
growing conditions. The environmental conditions of soil, precipitation, sunlight, temperature, cultivation, etc., which affect the growth of a plant or animal.
growing mash. A complete, ground ration which is fed to growing birds.
growing-on-house. A greenhouse used for growing potted plants from the small plant stage to maturity.

growing point. Meristem *q.v.*
growing season. In the North Temperate Zone, the number of days from the average date of the last killing frost in the spring to the average date of the first killing frost in the fall. The length of the period varies according to the hardiness of the plant and geographic location.
growing stock. The sum of all the trees in a forest.
grow out. To feed cattle so that the cattle get a certain desired amount of growth without much, if any, fattening.
growth. 1. The increment in size of a living organism. 2. Plants or plant parts. 3. A tumor, gall, etc. 4. The development of an organism from its earliest stage to maturity. 5. The development or increase of an enterprise or an organization.
growth layer. See *annual layer.*
growth percent. The ratio of current or periodic annual increment of standing timber, usually stated in terms of volume, to the quantity at the beginning or middle of the period in question.
growth regulator. Any synthetic or natural organic compound, such as indoleacetic acid, 2-4-D, naphthalene acetic acid, etc., which in dilute amount will promote, inhibit, or modify plant growth processes. Natural growth regulators are sometimes called *auxins*. Also called *plant growth regulator; plant hormone; phytohormone.*
growth ring. See *annual ring.*
growth ring boundary. The outer limit of an annual ring *q.v.*
growth stages. Of plants, any definite period in the growth or development, *e.g.*: (a) tillering stage, that period when a plant produces additional shoots from a single crown, as in wheat; (b) jointing stage, that period when the internodes of the stem are elongating; (c) boot stage, that period when the seed head of a plant begins to emerge from the sheath—usually applied to the grain crops.
growth substance. A natural or synthetic chemical (*e.g.*: a hormone, vitamin, or other substance) found within or applied to plants, which noticeably affects the growth behavior or other plant functions.
growth-weight-gain. The gain in weight due to growth or aging of an animal, such as a steer, as against the fat-weight-gain or the increase in flesh or fattening due to feeding. *Growth-weight-gains* are always the less expensive of the two gains.
growthy. Designating an animal that is big, rugged, and well-developed for its age.
grow under glass. To grow a plant in a greenhouse, conservatory, etc.
grow under lath. To grow a plant in a lath house.
grow up. To develop into a full-sized individual plant or animal. (The term does not necessarily imply maturity.)
grow wild. To grow uncultivated or untended.
grub. 1. Roots and crown which remain in the soil after a tree or shrub has been cut off or killed in clearing land for farm fields. 2. The larva of any of several beetles which are common in soil. 3. See *common cattle grub.* 4. To dig out a plant by the roots, as to *grub* out persimmon crowns, etc.
grub ax. A tool for digging up roots which has two cutting edges: one a broad, chisel-shaped point and the other a flat, adzelike blade.
grubber. A mechanical, plowlike device which digs out plants by the roots.
grubbing hoe. An implement which is provided with a wide, heavy, adzelike cutting blade and a wood handle; a mattock which is used for digging and cutting grubs or roots.
grubensland. A small, Indian field which contains clumps of knotty oak roots in the soil due to repeated burnings of oak sprouts (southeastern Pennsylvania, U.S.).
grub hole. A hole or wound in the hide of an animal caused by the larva of the common cattle grub *q.v.*
grub-in-the-head. See *sheep bot fly.*
grub out. See *grub: 4.*
grub prairie. Forest land that has become grassland due to annual burning-over.
grub puller. See *stump puller.*
grubs in sheep. See *sheep bot fly.*
grubs in the back. See *common cattle grub.*
grub wagon. A horse- or mule-drawn wagon which carried food supplies for cowboys on roundups (western U.S.).
gruel. A food or feed prepared by mixing a ground feed with hot or cold water.
grulla. A coat color of some animals, especially the American quarter horse. It is a slate blue bordering on a sooty black.
grumusols. Tropical and subtropical soils, analogous to Chernozems of temperate regions, which are commonly clay soils and have physical properties unfavorable for tillage or plant growth.
grupon. The most valuable portion of cattle hides, that part which extends from the shoulders to the tailhead and downward to the belly line.
Gruyere. A type of Swiss cheese, named after the village, Gruyere, Canton of Fribourg, Switzerland; characterized by a sharper flavor and smaller eyes than Swiss cheese.
Guadalupe erythea. *Erythea edulis,* family Palmaceae, a fan palm grown in very warm areas as an ornamental. Native to Mexico and southern California, U.S. Also called *Guadalupe palm.*
Guadalupe palm. Guadalupe erythea.
guamachil apes-earring. *Pithecellobium dulce,* family Leguminosae, a tropical tree grown in very warm areas for its white flower clusters. Native to Mexico. Also called *guaymochil; huamuchil; Manila tamarind; Madras thorn.*
guanacaste earpodtree. *Enterolobium cyclocarpum,* family Leguminosae, a tropical tree grown in very warm areas as an ornamental, windbreak, or shadetree. Its pods are used for cattle feed. Native to tropical America. Also called *elephant's ear; caro.*
guano. The dry excrement of marine birds which is a fertilizer rich in nitrogen and phosphoric acid. *Guano* was first imported from Peru where it was obtained from a number of arid islands

off the coast. The term now has no specific meaning and has been used in some parts of the world for any kind of fertilizer, organic or inorganic. See *Peruvian guano; whale guano; superphosphate; commercial fertilizer.*

guano horn. A funnel-shaped, metal tube used in the hand distribution of fertilizer.

guar. *Cyanopsis proraloides* (*C. tetragonoloba*), family Leguminosae, an annual herb cultivated for its edible beans. Grown experimentally or only to a very limited extent in the U.S., the seeds yield a starch which has some potentiality for use in the adhesive industry. Native to the East Indies.

guard. 1. One of several identical attachments for the cutter bar of a mowing machine, combine, etc., which serves to protect the cutting knives, to provide a place for the ledger plates, and to divide the material being cut. 2. Any protective device which keeps people or animals from being injured by the moving components of a machine.

guard plate. See *guard: 1.*

Guatemala grass. See *Mexican teosinte.*

guatemote. Mule fat baccharis.

guava. Any tropical shrub or tree of the genus Psidium, family Myrtaceae. See *cattley guava; Brazilian guava.* 2. *P. guajava,* a shrub or small tree cultivated for its edible, egg-shaped, yellow fruit, the *guava* of commerce. Native to tropical America.

guayule parthenium. *Parthenium argentatum,* family Compositae, a shrub which contains a latex and is a possible source of commercial rubber. It is a semidesert shrub, somewhat resembling sagebrush. Native to North America, it grows wild in north central Mexico and in the Big Bend section of Texas, U.S.

Guernsey. A breed of cattle which originated on the island of Guernsey in the English Channel and is highly regarded for its dairy characteristics and qualities. It is widely distributed, with the largest numbers in the U.S. *Guernseys* are fawn or reddish-fawn and white, medium in size, weighing about 1,000 to 1,100 lbs. at maturity. The milk of *Guernseys* is high in color and percentage of butterfat.

Guernseylily nerine. *Nerine sarniensis,* family Amaryllidaceae, a bulbous herb grown for its crimson flowers. Native to South Africa.

Guiana-chestnut pachira. *Pachira aquatica,* family Bombacaceae, a tree grown in frost-free areas for its fringed, pink or purple flowers. Native to tropical America.

guide line. A chalk line between two fixed points which serves as a guide for ditching, excavating, etc.

guinea. A domesticated bird, about the size of a small adult chicken, which descended from a wild, North African species, *Numida meleagris.* The most popular variety is the Pearl, which has purplish-gray plumage regularly dotted with white. Other varieties are white and lavender. They are not raised extensively, but some farmers keep a few *guineas* because they utter sharp cries if hawks, crows, or other natural enemies appear near the premises. *Guinea* meat has a gamey flavor.

Guineagold-vine. Twining button flower.

Guineagrass. *Panicum maximum,* family Gramineae, a drought-resistant, coarse, perennial bunch grass; an important forage grass in tropical America grown only to limited extent in the U.S. in parts of Florida and Texas. Native to Africa; introduced at an early period into the West Indies.

guinea-hen flower. Checkered fritillary.

guinsensis. See *sweet sansevieria.*

guitarbush lomatia. *Lomatia tinctoria,* family Proteaceae, a shrub, the seeds of which yield a pink dye. Native to Australia. Also called *guitar plant.*

gulch. A deep, narrow, sharply cut trench or gorgelike feature which may or may not contain a stream in its bed. Deeper than a gully or arroyo, it has less magnitude than a canyon or gorge, although local usage of the term is not consistent in this respect (western U.S.).

Gulf Coast tick. *Amblyomma maculatum* Koch, family Ixodidae, an insect pest which sucks blood from man and other animals. It has been ranked second to cattle tick (*q.v.*) in cattle destruction.

gulf wireworm. *Conoderus amplicollis* (Gyll.), family Elateridae, an insect pest of sweet potatoes which infests the underground parts of the plant.

gullet. Esophagus.

gulley. Gully.

gully. 1. A gash cut into a slope of soil or loose, unconsolidated sediments by concentration of rainfall runoff into a channel. *Gullies* usually have a transverse V-shaped profile and generally are relatively shallow, although locally the cutting may extend to a depth of more than 100 ft. and have wall-like sides and flat floors. One of the most destructive forms of erosion, it affects cultivated, range, and forest lands. 2. A furrow, channel, or miniature valley which is cut by running water during and immediately after rains or during the melting of snow. At other times it usually does not contain water. Special forms may be described as: dendritic arborescent, resembling the branching of a tree or shrub; linear, resembling a line, rather long, narrow, and of uniform width. The distinction between *gully* and rill is primarily one of depth. A *gully* is a channel of sufficient depth that it would not be obliterated by normal tillage operation, whereas a rill would be smoothed by ordinary tillage. 3. To make *gullies;* to form small drainage channels by the action of running water.

gully control. Checking gully erosion by mechanical or vegetal means.

gully erosion. That type of accelerated erosion by water that produces definite channels too deep to be obliterated by normal tillage.

gully head. The upper extremity of a gully which is usually the point of water overfall.

gullying. See *gully erosion.*

gully washer. A very heavy rain that produces a large volume of runoff and erosion. See *trash mover.*

gum. See *oleoresin.*

gum arabic. An ice cream mix stabilizer, now rarely used, which is obtained from any of several species of the genus Acacia, family Leguminosae, especially *A. senegal* and *A. arabica.*

gumbo. 1. Okra *q.v.* 2. a. Any heavy soil which is extremely sticky when wet but baked and cracked when very dry. b. A colloquial term for soil of the alluvial bottomlands of the Mississippi and Missouri rivers and midwestern prairies (U.S.).

gumbo lily. Ten-petal mentzelia.

gumbotil. Till; a massive deposit of glacial clay which weathered into a gray, sticky, tenacious clay during the interglacial period.

gum disease. Any disease of plants which has a gummy exudate as a symptom. See *gummosis; Diplodia disease of citrus; Dothiorella gummosis; Botrytis gummosis.*

gumi. Cherry elaeagnus.

gummer. Smooth-mouthed sheep.

gumming. See *gummosis.*

gumming disease of sugar cane. A destructive, bacterial disease caused by *Xanthomonas vasculorum;* characterized by pale green to yellow stripes on the leaves which are flecked with reddish dots diffusing with age. In time the lesions turn red to brown and the affected tissues die.

gummosis. The formation of clear or amber-colored exudates which set into solid masses upon the surface of affected plant parts and which may be caused by mechanical injury, pathogenic fungi, insects, excessive nitrogen in the soil, etc. See *Florida gummosis.*

gummy. 1. Grease wool that has an excessive amount of yolk, or scoured wool that still has some yolk in it. 2. See *smooth-mouthed sheep.* 3. Designating a body defect of butter which is characterized by difficulty in spreading and by sticking to the roof of the mouth. 4. Designating a body defect of ice cream which is characterized by a heavy, soggy body.

gummy body. A defect of ice cream in which the ice cream resists cutting by tending to hold together in a solid mass.

gummy leg. A condition of a horse's legs in which the tendons do not stand out clearly.

gum myrtle. Any shrub or tree of the genus Angophora, family Myrtaceae. Species are grown for ornament. See *rusty gum myrtle.*

gum spot. An accumulation of gummosis as a small patch on a tree which may occur in conjunction with a bird-peck or other injury to the growing wood.

gum swamp. A swamp which is characterized by gum trees, especially the black tupelo and the American sweetgum *q.q.v.*

gum tragacanth. An ice cream mix stabilizer which is obtained from the exudate of certain species of the genus Astragalus, family Leguminosae, especially tragacanth milk vetch (*A. gummifer*).

gum tree. See *eucalyptus; American sweetgum; tupelo.*

gum turpentine. The essential oil which is obtained from oleoresin exuded from living pine trees; the turpentine of commerce.

gun. a. A device used to determine the direction of the fall of a tree. b. To aim the fall of a tree.

gun harvest. The total number of game animals shot in a particular period of time.

Gunnison mariposa. *Calochortus gunnisoni,* family Liliaceae, a bulbous herb grown in protected spots in the flower garden for its white, purple-streaked flowers. Native to North America.

gunny sack. A coarse fabric bag manufactured from jute fiber which is commonly used as a container for feeds, seeds, potatoes, and various other products. Also called *toe sack; burlap sack.*

Güssing. A skimmed milk, brick cheese made in Austria.

gut. The digestive tract, or sometimes only the intestines, of animals.

gut cart. A cart used in slaughter houses to collect offal.

guttation. The forcing out of water or cell sap upon free surfaces of plants, a normal physiological process in many leaves.

gutter. 1. A channel in the floor behind cows in stanchion barns used to catch the manure and to simplify cleaning. 2. A ditch, dug in the path of a forest fire into which burning materials fall, designed to arrest the fire. 3. In turpentine orcharding, the groove or channel along which the oleoresin travels from the tree to the cup.

gutter cleaner. A power-driven mechanical device used to remove manure from gutters in barns and dropping pits in poultry houses.

gut tie. 1. A condition of a young animal in which a part of the intestines becomes entangled with the spermatic cord, resulting from faulty castration. 2. A condition in animals in which the intestine becomes twisted, causing an obstruction.

gut weed. See *field sow thistle.*

guy. a. To anchor a post, plant, limb, tower, etc., by the installation of cables or wires from the member to a firmly established object. b. Designating a cable or wire so used.

Guzerat. A strain of Zebu *q.v.*

gymnosperm. 1. a. Naked seed not enclosed in an ovary. b. A plant having such seeds. 2. Any of the seed plant species.

gynandromorph. An individual of which one part of the body exhibits female characteristics and another part male.

gynecium. Gynoecium.

gynoecium. The pistil or pistils of a flower collectively. Also called *gynecium. Cf. androecium.*

gynomonoecious. Having pistillate and perfect flowers on the same plant.

gyp. Gypsum (southwestern U.S.). See *calcium sulfate.*

gyppo. A timber contractor.

gyp-rope. The rope used by a trainer to rope or to exercise his horse.

gypseous marl. Gypsum beds which are naturally crumbly or not firmly indurated.

gypsophily. The phenomenon of the abundance or restriction of certain plants to soils strongly influenced by gypsum.

gypsum. Calcium sulfate *q.v.*

gypsum pink. Baby's-breath.

gypsy moth. *Porthetria dispar* (L.), family Lymantriidae, an important insect whose larva defoliates apple, pear, linden, birch, hawthorn, oak, poplar, willow, cherry, elm, hickory, hornbeam, larch, maple, and sassafras.

Gyr. A strain of Zebu *q.v.*

gyrotiller. A machine used in England to pulverize subsurface soil below depths of ordinary plowing.

gyttja. The ooze or gelatinous deposit, largely organic in composition, which is formed on the bottoms of lakes. Such material is a soil for aquatic plants in lakes and becomes a dryland soil following the recession or exsiccation of lakes.

H

habit. See *habit of growth.*

habitant. 1. Any plant or animal that is a permanent resident of a place or section. 2. A settler or a descendant of settlers of French descent in Canada or Louisiana, U.S., who belongs to the farming class.

habitat. 1. The place where a plant or animal normally lives and grows under natural conditions, such as a forest *habitat,* marsh *habitat,* grassland *habitat,* etc. 2. In parasitism, the host.

habitat factor. Any influence or condition of the habitat that affects an organism.

habit of growth. 1. The general form or arrangement of stems, roots, and branches, or of the entire plant which is possessed in common by a species in a given habitat. 2. The erectness of the stems of young plants of the small grains as a means of distinguishing between winter and spring varieties.

hacendado. A ranch owner; cattleman (southwestern U.S.).

hacienda. 1. The principal dwelling or headquarters of a ranch. 2. The ranch itself. (Southwestern U.S.)

hack. 1. In the eastern U.S., a horse that is rented out for riding by the hour. 2. See *three-gaited horse.* 3. A poor horse, as an old *hack.* 4. A U-shaped tool used in turpentining to cut into the tree to induce the flow of oleoresin. 5. A carriage or coach. 6. A tool for shallow digging, such as a hoe, pick, mattock, or grub hoe.

hackamore. A type of head restraint for horses similar to a halter but provided with a loop or noose that may be placed about the nose of the horse to give additional restraint in handling unbroken horses. Also called *jaquima; headstall.*

hackberry. 1. Any tree or shrub of the genus Celtis, family Ulmaceae. Species bear a thin-fleshed, sweet drupe or berry. Some species are hardy, drought resistant, and have value as shade trees, plantings for windbreaks, and prevention of soil erosion. They are not highly valued for lumber, the wood having about the same uses as elm. The common tree species in eastern and southern U.S. are *C. occidentalis* and *C. laevigata.* Also called *sugarberry; nettletree; false elm; hoop ash; eggberry.* 2. *C. occidentalis,* a hardy tree grown for its easy culture as a shade or lawn tree. Native to North America.

hackberry nipple gall. *Pachypsylla celtidis-mamma* Riley, family Psyllidae, an insect pest of hackberry which causes small, round galls on the upper surface of leaves.

hackle. A board set with sharp iron spikes which was formerly used for combing out the fibers of hemp and flax.

hackles. 1. The long, narrow, neck plumage of male birds. 2. The erectile hairs on the backs of certain animals.

hackmatack. Eastern larch.

Hackney. A breed of light harness or carriage horses which originated on the east coast of England, averaging from 15 to 16 hands high and having considerable depth through the chest and body. It lifts its forefeet in an exaggerated manner. The preferred coat colors are chestnut and bay, with white on the face and legs not being uncommon.

Hackney pony. A breed of harness ponies which resulted from crossing a Hackney stallion and a pony mare of Welsh breeding. In general, it is very similar to the Hackney except for size, being 14 hands or less high.

hacksawfern. 1. Any fern of the genus Doodia, family Polypodiaceae. 2. *D. aspera,* a dwarfed, thickly-growing fern which is popularly grown as a house plant or as a window-box plant. Native to Australia.

haemonchosis. The infection of ruminants with the large stomach worm, species of the genus Haemonchus, or the disease caused by this parasite. *Haemonchosis* is usually accompanied by loss of blood and anemia caused by the activities of the worms, and it is one of the most serious helminthic diseases of ruminants.

haggerel. A male lamb from weaning time until its first shearing (New Zealand).

haggis. A Scottish food made from hearts, lungs, and livers of sheep or calves, highly seasoned, mixed with oatmeal, and boiled in a sheep stomach.

hag moth. *Phobetron pithecium* (J. E. Smith), family Limacodidae, a caterpillar with stinging hairs that feeds generally but not destructively on a wide variety of trees and shrubs.

ha-ha fence. 1. A sunken wall which encloses a lawn or pasture; one that does not obscure the view. 2. A boundary trench; a ditch or moat.

hail. Precipitation in the form of hard pellets of ice, called hailstones, which often falls from cumulo-nimbus clouds and accompanies thunderstorms. Hailstones are formed when raindrops are swept up by strong air currents into regions where the temperature is below the freezing point. In falling, the hailstone grows by condensation from the warm, moist air which it encounters and may increase greatly in size and cause severe damage to greenhouses, buildings, livestock, and standing crops. See *sleet: 1.*

hail insurance. Insurance which covers damage caused by hail.

hail storm. A storm in which hail falls.

hail weed. See *dodder.*

hair. The outgrowth of a cell in the epidermis of a plant or animal. In vast numbers it forms the coat of an animal and is frequently used as a fiber, such as wool. See *pubescence; root hair.*

hair ball. Hair that an animal has swallowed which has gathered in the stomach in the form of balls; common in the stomach of cats and rumen of ruminants. Also called *aegagropila; egagropilus; piliconcretion; trichobezoar.* See *bezoar; stomach ball.*

hairgrass. 1. Any grass of the genus Aria, family Gramineae. 2. Any grass of the genus Deschampsia, family Gramineae. See *tufted hairgrass.*

hairlessness. 1. In genetics, a lethal factor characterized in the Holstein breed by calves being born without hairs except around natural openings and at the end of the tail. 2. An iodine deficiency disease which causes pigs to be born hairless. See *big-neck.*

hair slips. An animal hide that has been improperly salted and cured, allowing some decomposition to take place as indicated by slipping patches of hair.

hairweed. Dodder.

hairy. Designating an abundance of hairs.

hairy cat's ear. See *spotted catsear.*

hairy ceanothus. *Ceanothus oliganthus,* family Rhamnaceae, a shrub which grows to a height of 12 ft.; grown for its attractive foliage and its deep blue or purplish flowers.

hairy chinch bug. *Blissus leucopterus hirtus* Montd., family Lygaeidae, an insect pest of turf which often kills grass in patches, northeastern U.S.

hairy crabgrass. *Digitaria sanguinalis,* family Gramineae, an annual grass which is a weed pest particularly of turf.

hairy fairybells. *Disporum lanuginosum,* family Liliaceae, a hairy, perennial herb grown for its greenish flowers and red fruits. Native to North America. Also called *fairy bell; liverberry.*

hairy golden aster. *Chrysopsis villosa,* family Compositae, a perennial herb grown for its yellow flowers. Native to North America. Also called *rosinweed.*

hairy grama. *Bouteloua hirsuta,* family Gramineae, a perennial bunchgrass which has a good forage rating on the western range. Found from Wisconsin to South Dakota, U.S., and south to Mexico.

hairy honeysuckle. *Lonicera hirsuta,* family Caprifoliaceae, a twining, woody vine grown for its habit and for its bright yellow flower clusters. Native to North America. Also called *rough woodbine.*

hairy lilac. *Syringa pubescens,* family Oleaceae, a deciduous, hardy shrub grown for its fragrant, pale lilac flowers. Native to China.

hairy nightphlox. *Zaluzianskya villosa,* family Scrophulariaceae, an herbaceous plant grown as a tender annual for its night-blooming, dark purple flowers which are white or pale lilac inside. Native to South Africa.

hairy root. 1. A stage in one of the types of crown gall disease in which an infected plant develops abnormal, hair-like, fibrous roots; especially common in the apple. See *crown gall.* 2. See *curly top.*

hairy toadlily. *Tricyrtis hirta,* family Liliaceae, a half-hardy, perennial herb grown for its white, purple-spotted flowers. Native to Japan.

hairy vetch. *Vicia villosa,* family Leguminosae, a hardy, annual or biennial, climbing herb grown for forage, as a cover crop, and for green manure. Native to Eurasia. Also called *winter vetch; sand vetch; Russian vetch.*

hairy-wattle acacia. *Acacia pubescens,* family Leguminosae, a tropical shrub grown in greenhouses for its yellow flowers. Native to Australia.

halfa. A name applied to various species of espartograss (*q.v.*) which are used in paper manufacturing (North Africa).

half-and-half. See *substandard cream.*

half-blood. In marketing, fine wool, 2¼ ins. long in combing, (1¼ to 2¼ French combing) which is next to the best grade and equivalent to English fifty-eight's.

Half Bog soil. An intrazonal group of soils with mucky or peaty surface soil underlain at shallow depths by gray mineral soil which was developed largely under swamp-forest types of vegetation mostly in a humid or subhumid climate. See *bog soil.*

half-bred. Designating a horse that has a Thoroughbred as one parent and a draft horse as the other. A popular, medium-heavyweight horse with considerable action.

halfbreed. 1. A cross between two botanical varieties of the same species. 2. A cross between two races.

halfbreed bit. An old Spanish spade bit for a horse that has a spoon in the center of the mouth piece.

half broke. Designating a partially broken horse. See *green broke(n) horse.*

half-(brother, sister). Animals from the same mother but by different sires, or by the same sire from different mothers.

half hardy. Designating a plant or plant part which is able to withstand a moderately low temperature but which is injured by severe freezing.

half-hardy annual flowering plant. Any annual flowering plant whose seeds will survive out-of-doors during a mild winter.

half-hardy perennial flowering plant. Any perennial flowering plant whose roots will survive a mild winter out-of-doors if a mulch is placed on the soil over the roots.

half hoist. In dressing a slaughtered animal, the lifting of the carcass from the floor to a convenient height for removing the hide.

half lands. Tracts of land reserved by the U.S. government for half-breed Indians.

half-loop system. A system of training bramble fruit canes which is similar to the loop system (*q.v.*) except that the canes are brought down to the lower wire and interlocked with the canes of the next plant. The portion of the canes left at the point of interlocking is cut off. Also called *interlocking system.*

half mealy. See *yellow berry of wheat.*

half-moon hoe. A type of scuffle hoe (*q.v.*) whose blade is half-moon shaped.

half-section. Three hundred and twenty acres of land.

half-shrub. A perennial plant having stems which are more or less woody at the base.

half-slip. In harvesting melons, a stage of ripeness in which, as the fruit is pulled from the vine, only a portion of the stem separates easily from the base of the fruit leaving a smaller scar on the fruit than when harvested at the full-slip *q.v.*

half-standard cone. The next to largest type of cone which marks feed in a fertilizer spreader; sometimes used with a cone.

half sweep. A sweep that has only one wing so that it can be used to cultivate close to plants. See *sweep: 1.*

Hallock box. A box used as a container in the harvesting and marketing of berries which is square with the bottom so raised that one box can be stacked on another without bruising the berries.

Halls crabapple. *Malus halliana,* family Rosaceae, a deciduous tree grown for its deep rose flowers with purple calyces. Native to western China.

Halls panicum. *Panicum halli,* family Gramineae, a grass which produces abundant forage on bottomlands and in irrigated fields and whose forage rating is fair. Native to North America; found in Texas and Arizona, U.S.

halm. See *haulm.*

halo. 1. A bright ring or system of rings sometimes seen surrounding the sun or moon. The reflection from the ice crystals of cirrus clouds, it is an indicator of a probable change in the weather. Also called *sun dog; moon dog.* 2. A narrow, light colored band of diseased tissue which sometimes surrounds a lesion on a plant.

halo blight of beans. A bacterial disease caused by *Pseudomonas phaseolicola,* family Pseudomonadaceae, characterized in cool weather by a halo-like zone of greenish-yellow tissue around a water-soaked spot on leaves. *Cf. bacterial blight of beans.*

halo blight of oats. A widespread, bacterial disease of oats, wheatgrass, and brome caused by *Pseudomonas coronafaciens,* family Pseudomonadaceae, characterized by egg-shaped, small, water-soaked spots on the leaf sheaths and floral bracts. The spots change from green to light brown, a light yellow halo forms around the spot, and the spots coalesce to form irregular halo areas.

halogens. A group of elements, including fluorine, chlorine, bromine, and iodine, which are powerful oxidizing agents and have disinfectant properties. An alcoholic solution of iodine, or tincture of iodine, is an example of an iodine-containing disinfectant; chlorinated lime is an example of a chlorine-containing disinfectant.

halomorphic soil. A suborder of intrazonal soils whose properties are determined by the presence of neutral and/or alkali salts.

halophyte. A plant which grows in an alkaline habitat.

halophytic vegetation. Salt-loving or salt-tolerant vegetation which usually has fleshy leaves or thorns and resembles desert vegetation.

halter. A leather, rope, or chain device which is essentially a loop over an animal's nose and another loop behind the ears for leading or restraining.

halter broke(n). Designating a horse or colt which has become accustomed to the halter and its uses. See *broke.*

halter class. Show horses at the halter in breeding classes that are shown where breed, character, conformation, and action are emphasized.

halter puller. A horse or colt which continually tries to pull away when fastened to hitching posts, etc.

halve. To divide into two equal portions, as to *halve* a peach preparatory to canning.

ham. A cut of pork which consists of the hindquarters of a swine from the hock to the hip, including the thigh and buttock, specifically, one that has been salt-cured and smoked. See *fresh ham; picnic ham.*

Hambletonian. The most important family of Standardbred horses which originated in the U.S.

Hamburg. A Dutch breed of light-weight chickens of graceful carriage and attractive color patterns which are commonly bred for exhibition purposes. The breed is composed of six varieties: Golden Spangled, Silver Spangled, Golden Penciled, Silver Penciled, Black, and White.

hamburg(er). Retail meat which consists of ground lean and fat beef (except that no heart, liver, kidney, etc., is used).

hamburg parsley. Turnip garden parsley.

ham butt. A retail cut of pork which consists of the thigh end of a ham. *Cf. ham shank.*

hame. The wood or metal parts of the harness of a draft animal that fit about the collar to which the traces are attached, as well as certain other straps.

hame ring. A ring on the hame of the harness

used for the proper alignment of the reins or spreader straps, choke straps, and other parts of the harness.

hame strap. A strap which fastens the hames together under the horse collar. Also called *hame string.*

hame tug. See *trace: 1.*

hammer. One of several swiveled, hard-steel bars which are attached to a rotor in a hammer mill to shatter grain, hay, etc.

hammer mill. A feed grinding device or mill in which hammer-like projections are mounted on the surface of a cylinder which revolves at a high speed within a heavy perforated metal enclosure and shatters the feed material by beating it to pieces.

hammer mill magnet. A magnet which is placed on the apron of a hammer mill to attract bits of iron, wire, etc., to prevent them from entering the mill and damaging the hammer.

hammock. 1. A vegetational land feature, recognized in peninsular Florida and westward along the Gulf Coast to Mississippi, U.S., whose tree vegetation is, or was originally, in contrast to that of the prevailing pine land. 2. A small isolated tree-covered area in extensive marshes. *Hammocks* may be designated specifically according to tree growth, as oak *hammocks,* palm *hammocks,* etc. 3. In some parts of Mississippi, alluvial second bottoms.

hammock seat. A seat on a riding cultivator which is suspended between two bars so that the operator has an unobstructed view of the plants.

hammock soils. The soils of hammocks. They are, or were originally, generally regarded as more limy, more fertile, better supplied with humus, or more suitable for agricultural use than associated soils, but in many cases the opposite may be true. See *hammock: 1.*

hamper. A basket in which vegetables are shipped.

Hamprace. An American breed of meat-type swine developed from a cross-bred foundation of Danish Landrace and Black Hampshire swine. Also called *Montana No. 1.*

Hampshire. A medium-wool, black-faced breed of sheep which originated in England. It is a popular mutton breed whose ewes weigh up to 200 lbs.

Hampshire swine. An American, lard-type breed of black, white-belted swine.

ham shank. The hock end of a ham. *Cf. ham butt.*

ham skipper. See *cheese skipper.*

ham string. The large tendon which is above and behind the hock in the hind leg of quadrupeds.

hamstrung. Designating an injury to the tendon behind the cannon in the hind leg.

hand. 1. A laborer who is either permanently employed or migratory, as harvest *hand,* hired *hand.* 2. A unit of measurement equal to four ins. which is used to measure the height of horses from the ground to a point at the shoulder. 3. A bunch of tobacco leaves of the same grade which are tied together for easier handling. 4. The near horse in a team used for plowing. 5. A half-whorl-like cluster of bananas attached to the rachis of the spike or bunch. 6. Designating any tool, implement, etc., which is manually operated. 7. Designating any manual labor, as *hand* chopping.

hand atomizer. A hand operated device used to reduce a liquid into fine particles.

hand breeding. A system of animal breeding in which the breeder controls the number of times coitus is performed.

hand capper. A manually operated device which places bottle caps on milk bottles.

hand card. See *card.*

Hand cheese. A small, sour-milk, surface-ripened cheese which is popular in Germany.

hand cultivator. Any toothed implement drawn or pushed by hand which is used for stirring the soil and killing weeds in vegetable gardens and flower beds.

hand duster. A manually operated device for applying insecticides or fungicidal dusts to plants, etc.

hand-feeding. A method of feeding poultry or other livestock in which a fixed amount of feed is supplied to the animals according to a feeding schedule or according to the judgment of the feeder, as contrasted with self-feeding *q.v.*

hand gallop. A restrained or slow gallop of horses. See *canter.*

hand gun. A spray gun which is held in the hands during operation. See *hand duster.*

hand harvest. To pick or gather by hand as contrasted with mechanical harvesting by combine, corn picker, etc.

hand hoe. See *garden hoe.*

hand husk. To remove the husks from ear corn manually.

handiness. A characteristic of a horse with good manners which has good coordination and is not clumsy or awkward in movement or action.

hand labor. Any labor or production effort which is produced entirely by the hands or use of the human body. (The term often denotes any unskilled farm laborer working with his hands or with simple tools, such as a hoe.)

handle. 1. The woody cane of a bramble plant which has been propagated by layering. Also called *tail.* 2. The specially designed part of a tool, implement, container, etc., which is grasped by the hands. 3. The stem or portion of stem attached to a broomcorn brush or panicle. 4. To be docile and obedient, as a horse *handles* well.

handle basket. A rectangular, woven basket with a wooden, U-shaped handle used for carrying harvested vegetables, fruits, etc.

handler. A person or company which buys, sells, sorts, packages, or stores a product. See *country packer.*

hand level. A leveling instrument without tripod or stand that is held in the hand while making a horizontal line of sight to establish relative elevations.

handling nicely. 1. Designating a horse that is well-trained or broken. 2. Designating a meat animal that shows good fleshing and finish when examined by the feel of the hand.

hand log. To fell, lop, and saw trees into logs and move the logs by human effort without the aid of animal or mechanical power.

hand logger. 1. A person who logs without the use of animals or power. 2. In northwestern U.S., one who uses animals instead of a power skidder.

hand mating. See *hand breeding.*

hand milker. A person who milks a cow manually.

hand-pack filler. In canning, a machine which semi-automatically fills cans with vegetables, fruits, etc.

handpick. 1. To harvest by hand, as contrasted to harvesting by machine. 2. To pluck the feathers of a fowl manually.

hand pollination. The pollination of flowers by hand, especially for the production of hybrid seed.

hand pump. A pump which is generally a small suction pump with a bucket and foot valve operated by one person manually.

hand rake. 1. A type of dump rake (*q.v.*) which has a hand lever for raising the tines. 2. A long-handled garden rake.

hand saw. Any of several different kinds of manually operated saws, in contrast to power saws, used for sawing wood, or lumber.

hand skid. To move timber by hand to a point at which it can be reached by transport.

hand skimming. The separation of cream from milk by means of ladles, spoons, etc., as contrasted to the use of mechanical separators.

hand sprayer. A sprayer in which the pressure is usually generated by hand pumping and which is small enough to be manually operated.

hand staking. The flexing of a tanned hide or skin over a rounded metal blade.

hand stick. A device used in measuring the height of a horse.

hand strip. 1. To take the last bit of milk from a cow's udder, usually following machine milking. 2. To harvest a seed crop by hand.

hand tool. Any tool or implement which is used manually as distinguished from a machine tool.

hand-turned shoe. A horseshoe which is made by hand from a bar of metal.

hand weeding. The pulling of weeds by hand.

hang-down neck. A broken crest on a horse.

hanger. 1. A woodland which is on the side of a steep hill. 2. A smaller branch growing downward from a scaffold branch *q.v.*

hanging. Catching accidentally, as a plow point *hanging* on a root.

hanging colter. A colter which consists of a straight piece of metal sharpened on one edge, attached to the beam of a plow, hanging down, which is set to cut a loose furrow slice. See *fin colter; knife colter.*

hanging crop. Pendulous crop.

hang-over fire. A forest fire, started usually by lightning, which remains dormant and undetected for a considerable period before it becomes active.

hank. 1. A measure of length for yarn (a *hank* of worsted yarn contains 560 yds.). 2. A bundle of animal casings.

Hansen squirreltail. *Sitanion hanseni,* family Gramineae, a species of squirreltail with culms growing 24 to 36 ins. tall, occurring in open woods and rocky slopes from Wyoming to eastern Washington, Utah, and California, U.S., which affords fair forage for livestock early in the season.

Hanson lily. *Lilium hansoni,* family Liliaceae, a bulbous herb grown in the flower border for its tall flower stalk with fragrant, drooping, orange-yellow, purple-spotted flowers. Native to Japan.

haploid. An organism or cell having only one chromosome of each pair of chromosomes usually present in the somatic cells of the species. *Cf. diploid.*

harbinger-of-spring. *Erigenia bulbosa,* family Umbelliferae, a perennial, tuberous herb grown for its early-spring, small, white flowers. Native to North America.

harbor. To serve as a habitat for, as to *harbor* insects.

hard amber durum. In marketing, a subclass of durum wheat which contains 75 percent or more of hard, vitreous kernels. See *amber durum; durum wheat.*

hard-bodied cotton. Cotton fiber which a classer thinks has good spinning qualities.

hard body. A body defect of sherbets or ices characterized by the body's being too hard to cut properly.

hard bottom. In cranberry culture, sand as a soil medium in contrast to peat or muck.

hard breeder. Designating a female animal that is difficult to breed. See *shy breeder.*

hard cheese. Any cheese that is solid or firm in texture and low in moisture content.

hard cider. Fermented juice, as of apple, pear, etc.

hard cot(t)s. Fleece wool which is very badly tangled or matted. See *soft cot(t)s.*

hard crown gall. A symptom of crown gall (*q.v.*) in which tumors develop a covering of bark and an interior woody structure. *Cf. soft crown gall.*

hard dough stage. An intermediate stage between the milk and hard or ripe stage of ripening grain, especially wheat.

harden. 1. To treat plants to make them more resistant to adverse environmental conditions, usually by increasing the cold resistance of tender greenhouse plants by placing them in cold frames before transplanting to the open field. The hardening is thought to be due to a combination of many morphological and physiological changes, as reduction in cell size, reduction in the amount of free water, increase in the amount of hydrophilic colloids, etc. 2. To firm the body of ice cream, sherbet, etc., by placing it in a special compartment after freezing.

hard end. See *black end of pear.*

hardening. The process of tempering steel-edge tools. The tool, heated to a cherry red, is plunged into brine or clear water, the sudden quenching rendering it hard enough to scratch glass. Subsequent reheating to a point determined by experience is necessary to draw the temper to the requisite hardness.

hardening room (cabinet). 1. A room, cabinet, etc.,

for firming the body of ice cream, sherbet, etc., after freezing to reduce the temperature. 2. A room kept at about 65° F. for hardening plants, grafted plants, etc. See *harden: 1.*

harden off. See *harden: 1.*

hard feeder. An animal that stays in a thin condition even though well fed. Sometimes called *hard keeper.*

hard fern. Deerfern.

hard fruit. A boron deficiency disease of citrus fruits characterized by hard, misshapen, undersized fruit with gum spots in the flesh and albedo, water-soaked spots on the leaves, premature defoliation, and dieback. Detected in South Africa.

hardback spirea. *Spiraea tomentosa,* family Rosaceae, a hardy deciduous shrub grown as an ornamental for its dense, spirelike, rose or purple, flower clusters. Native to North America. Also called *steeplebush.*

hardhead. Black centaurea.

hardiness. The state of being hardy. See *hardy: 2.*

hard keeper. An animal that is unthrifty and grows or fattens slowly regardless of the quantity and quality of feed.

hard land. Land which consists of clayey soil in contrast to more loamy or sandy soil (Great Plains, U.S.).

hard lumps in the udder. A diseased condition of the cow's udder causing excess or hardened tissues to develop. It is an aftermath of mastitis *q.v.*

hard milker. A cow that milks slowly due to hardened or constricted sphincter muscles in the end of the teat, or a fleshy udder with limited space for the free or fast accumulation of milk.

hard mouthed. Cold jawed.

hardness of feather. A condition of a feather which possesses little fluff and is webbed closely.

hardness of wood. A property of wood measured by the load applied at a standard rate required to imbed a 0.444-inch steel ball half-way.

hard on the heel. See *gristle on the heel.*

hard on the land. Designating a crop which draws heavily on or exhausts the fertility of the soil or leaves it in a bad physical condition.

hardpan. A slightly to strongly indurated, or very compact nonindurated, subsurface layer in a soil profile. It can interfere with root penetration and the free, downward movement of water.

hard pinch. The removal of the end of a shoot by pinching it off near the second or third node below its tip. See *soft pinch.*

hard red spring wheat. In marketing, a wheat (*q.v.*) producing hard, flinty kernels rich in protein and possessing the quality for making good bread. See *dark northern spring wheat; northern spring wheat; red spring wheat.*

hard red winter wheat. A red-berried wheat sown in the fall which matures the following year producing hard, flinty kernels and yielding a flour of superior baking qualities. See *hard winter wheat; yellow hard winter wheat.*

hard root rot. A fungous disease of citrus in Southern Rhodesia caused by *Rhizoctonia lamellifera;* characterized by a crack in the bark just above the bud union, a sloughing of the bark exposing light buff wood, dry, brittle, light colored roots, and loose bark.

hard rot disease of gladiolus. A fungous disease caused by *Septoria gladioli,* family Sphaerioidaceae; characterized by reddish-brown, circular lesions on the leaves developing a light gray center surrounded by a zone of purplish-brown or black. Sunken, dark spots appear on the corms and some corms are mummified.

hard rot of cranberries. See *Sclerotinia tip blight.*

hardsalt. A mineral, a mixture of muriate of potash, sulfate of magnesia, and sodium chloride, containing about 16 percent potash, K_2O, used in fertilizers; mined in Germany and France. See *kainit(e); sylvinite.*

hard scald. 1. A severe case of apple scald (*q.v.*). 2. A method of removing feathers from a bird's carcass in which it is dipped in water from 150° to 200° F. and doused up and down. (Obsolete except for home use.) Also called *hot scald.*

hard scrabble land. Land that has thin, stony soil or is so unproductive that it barely yields a poor living for the occupant.

hard smut. See *nematode disease of wheat.*

hard soap. A cleansing agent made from stearin fats treated with an alkali sodium compound.

hard starch. The compact starch cells that occupy the sides and back of the kernel of grain.

hardware cloth. Galvanized, square-meshed, wire screening made in several widths with openings of various sizes up to about 1 in. square; used in the construction of poultry and small animal cages, incubator and dehydrator trays, various types of screens, etc. See *screen: 1, 2.*

hardware disease. A condition most frequently found in cattle and other ruminants caused by foreign bodies, such as bits of iron, nails, baling wire, etc., in the food; sometimes these bits lodge in the reticulum or rumen, causing a penetration with abscesses, peritonitis, and frequently death. Also called *traumatic gastritis; traumatic reticulitis; traumatic pericarditis.*

hard wart. On an animal, a wart found on the skin which is covered over with a layer of horny cells. See *wart: 1; soft wart.*

hard water. That water which contains certain minerals in solution, usually calcium and magnesium sulfates, chlorides, or carbonates, in such amounts as to cause a curd or precipitate rather than a lather when soap is added.

hard wheat. See *hard red spring wheat; hard red winter wheat; hard winter wheat.*

hard wheat mixed feed. A livestock feedstuff which consists of wheat bran and wheat flour middlings combined in the proportions obtained in the usual process of commercial milling. This product must not contain more than 9.5 percent of crude fiber. Also called *mill-run wheat feed.*

hard white wheat. In marketing, a white wheat (*q.v.*) that contains 75 percent or more of hard, not soft and chalky, kernels. *Cf. soft white wheat; white club wheat; western white wheat.*

hard winter. A winter which is characterized by

severe temperatures and/or heavy precipitation in the form of snow.

hard winter wheat. A market grade of hard red winter wheat (*q.v.*) that contains more than 25 percent and less than 75 percent of dark, hard, and vitreous kernels. See *yellow hard winter wheat.*

hardwood. 1. a. A term used by lumbermen in southeastern U.S. designating any tree which has broad leaves, in contrast to the conifers. b. The wood or lumber of such a tree. 2. Designating several kinds of timber which resemble teak, used in buildings, etc. (Australia).

hardwood ashes. The ashes resulting from the burning of hardwood which contain 2 to 8 percent or more of potash (K_2O), principally as a carbonate; often used as fertilizer. See *wood ashes.*

hardwood creosote. See *creosote; coal-tar creosote.*

hardwood cutting. A mature shoot of the last season's growth which is removed from the plant after the leaves have fallen to be used in propagating new plants. See *softwood cutting.*

hardwood land. Land which has, or originally supported, a native growth of deciduous forest trees, as distinguished from the pine lands (north-central and eastern U.S.).

hardy. 1. A wedge-shaped tool which is mounted on a blacksmith's anvil, serving as the bottom shear. 2. Designating a plant which withstands various environmental factors, especially the low temperatures of a given region. See *tender: 1; half hardy.*

hardy ageratum. Mistflower eupatorium.

hardy annual. 1. Any annual plant which may be sown where it is to grow, as contrasted to a plant which must be transplanted. 2. An annual flowering plant whose seeds will live outdoors over winter.

hardy catalpa. Northern catalpa.

hardy garden pink. *Dianthus knappii,* family Caryophyllaceae, a perennial herb grown for its yellow flowers. Native to southern Europe.

hardy moonflower. Bigroot morningglory.

hardy orange. Trifoliate orange.

hardy tree fruits. Trees, such as apple, pear, cherry, peach, plum, and apricot, which are temperate zone plants as contrasted to tropical and subtropical fruit plants.

hare. *Lepus* spp., an animal related to, but larger than, the rabbit (*q.v.*). Its young have fur and are able to care for themselves from birth. Some important species are the Varying Hare, *Lepus americanus,* also called the *snowshoe hare;* European hare, *Lepus europas,* and jack rabbit of the western plains. The hare may be a pest of growing plants and trees and shrubs in the winter.

harebell. See *grapehyacinth: 2.*

harebell phacelia. *Phacelia campanularia,* family Hydrophyllaceae, an annual herb grown for its bell-shaped, deep blue flowers. It is an important bee plant. Native to North America. Also called *California bluebell.*

hare's-foot fern. Golden polypody.

harestail. *Lagurus oratus,* family Gramineae, an annual grass which is grown to be dried and used in bouquets. Native to southern Europe. Also called *rabbit's-tail grass.*

haricot. See *kidney bean.*

Harlands box. *Buxus harlandi,* family Buxaceae, a broadleaved, evergreen shrub grown in warm areas in hedges and for topiary work. Native to Asia.

harlequin bug. *Murgantia histrionica* (Hahn), family Pentatomidae, the most important insect enemy of cabbage and related crops in the southern half of the U.S. It also infests many other plants occurring as far north as Michigan. Also called *harlequin cabbage bug; fire bug; calico back.*

harlequin flower. See *wandflower.*

harmel peganum. *Peganum harmala,* family Zygophyllaceae, a poisonous plant which is seldom eaten by livestock except under starvation conditions. Native to Eurasia; now established in New Mexico, U.S. Also called *African rue.*

harmonious blending and correlation of parts. In judging dairy cattle, symmetry of form and conformation.

harness. The straps, bands, collars, hames, lines, and attachments which are necessary to equip a draft animal, such as a horse, properly to pull or move a load. The *harness,* of many types, is usually largely made of leather. Also called *gear.*

harness horse. A lighter and faster horse than the draft horse which is especially suited to carriage and light driving use. There are two recognized types: the carriage or heavy *harness horse* and the roadster or light *harness horse.*

harness room. A storeroom in a barn in which harnesses are kept when not in use, and which has usually served as a catchall (obsolete).

harp. A hand tool for cutting the curd of cheese consisting of a frame in which taut wires are placed so that as the tool is passed through the curd in the cheese vat it cuts it into small lumps of uniform size.

harrow. An implement used to level the ground, stir the soil, and eradicate weeds. There are many types of *harrows,* ranging from the simple hand garden rake to the large, many sectioned spike or spring tooth and the disk type *harrows.*

harrow cart. An attachment for a horse-drawn harrow in which a seat is mounted between two wheels for the convenience of the operator.

harrowed. Designating a field or plot of land which has been treated with a harrow.

harrow plow. One-way disk plow *q.v.*

harrow teeth. Teeth made of steel which may be any of several different shapes, such as square, round, oval, etc. The teeth are held rigid to the harrow frame, but the frame may be adjusted to give the teeth a wide range of contact with the soil.

harsh. 1. Designating a fleece which lacks character, as rough hair. 2. Designating vegetation which is rough, hard, or which has some physical characteristic objectionable to livestock.

harsh wool. See *mushy wool.*

Hartford fern. *Lygodium palmatum,* family Schizaeaceae, a twining, vine-like, hardy fern grown in partly shaded places for its attractive foliage. Native to North America. Also called *Alice's fern.*

harts-tongue. *Phyllitis scolopendrium,* family Polypodiaceae, a hardy fern grown in the fern garden for ornament. Native to Europe.

harvest. 1. To cut, reap, pick, or gather any crop or product of value, as grain, fruit, or vegetables. 2. The crop or product so harvested.

harvest cut. In forestry, the cutting of a crop or stand of financially or physically mature trees as a final cut in even-aged management or the removal of mature elements of a stand in uneven-aged management.

harvester. 1. An agricultural implement, a further development of the reaper, in which the grain is cut, gathered into sheaves, bound with string, and discharged clear of the machine. 2. Any machine used in the harvest of a crop.

harvest hand. A person, usually a temporary or unskilled worker, who assists in harvesting.

harvesting. The gathering of the harvest.

harvest mite. Straw itch mite.

harvest moon. The full moon occurring nearest to the autumn equinox which rises on several successive nights almost at the same hour, instead of progressively later, and was formerly therefore of use to harvesters working after sunset.

harvest season. The time of the year when the matured crop is gathered, especially in the autumn months.

harvest time. Harvest season.

Harzkäse. A small Hand cheese made in Germany.

hashish. Hemp.

hassock. A separate tuft of coarse grass which grows somewhat dispersed in a bog or meadow and is frequently elevated above the soil surface by an accumulation of dead leaves and stems. Also called *tussock.*

hat carludovica. *Carludovica palmata,* family Cyclanthaceae, a palmlike plant sometimes grown in the greenhouse as an ornamental and in Ecuador as the source of the material used in the manufacture of Panama hats. Native to tropical America. Also called *Panama hat palm; jipi-japa.*

hatch. 1. a. To bring forth young from the egg by natural or artificial incubation. b. The young produced from one incubation. 2. The access entryway to cellars, attics, haymows, etc.

hatchability. 1. In poultry farming, that quality of fertilized eggs that makes possible normal embryonic development and the emergence of normal young, when incubated. 2. In incubation practice, the percentage of fertile eggs which hatch.

Hatch Act. An act of the U.S. Congress in 1887 which allocated $15,000 annually to each state and territory for the development and operation of an agricultural experiment station. This act was amended in 1955 to include subsequent acts of Congress that appropriated funds for the operation of these experiment stations. The consolidation or amended Hatch Act includes the original Hatch Act, the Adams Act of 1906, the Purnell Act of 1925, the Bankhead-Jones Act of 1935, and title I, section 9 of the Amendment of 1946 to the Bankhead-Jones Act.

hatchery. A place, building, company, etc., where eggs are incubated, usually a commercial establishment where newly-hatched young (chicks, poults, ducklings, etc.) are sold.

hatcheryman. One who operates a hatchery.

hatchet. A small ax used for splitting or rough dressing timber.

hatching egg. An egg of good form and quality produced by a breeding flock which may be used for hatching.

hatch out. To emerge from an egg, as a fully developed chick comes forth from the shell.

haul. To transport usually in a truck or wagon from one place to another.

haul-back. In logging, a small, wire rope traveling between the power skidder and a pulley set near the logs, used to return the main cable with tongs, chokers, or hooks, to the next log. Also called *back line.*

hauler. 1. An individual who drives a bulk milk truck, collects the milk from the various producers, and transports it to the milk plant or buyer. 2. A truck driver who picks up farm produce at regular intervals on a route.

haulm. 1. The culm of a grass. 2. The stem of an herb. 3. The stalk of beans and peas which is cured for fodder or hay. Also called *halm; haum.*

haunch. 1. A pivot on the hind feet. It is commonly observed in stock horses used on ranches for culling or cutting. 2. A hindquarter of an animal.

Hauskäse. A disk-shaped, Limburger-type cheese having a 10-inch diameter, made in Germany.

hautbois strawberry. *Fragaria moschata,* family Rosaceae, a perennial herb, sometimes used in strawberry breeding, grown for its edible, musky, dull red fruit. Native to Europe.

haute école. See *high school horse; dressage.*

haver. Oats or an oat product (Scotland).

haw. 1. The third eyelid of animals. Also called *hook.* 2. See *hawthorn.* 3. A command to a horse to turn left. *Cf. gee; whoa.*

Hawaiian beet webworm. *Hymenia recurvalis* (F.), family Pyraustidae, an insect pest of garden crops, especially in warm areas.

hawksbeards. Any herb of the genus Crepis, family Compositae. Species are useful forbs on the western range, U.S. See *red hawksbeard.*

hawkweed. Any herb of the genus Hieracium, family Compositae. Species are often weed pests, but some are valuable forbs on the western ranges. See *orange hawkweed; mouse-ear hawkweed.*

hawthorn. Any shrub or tree of the genus Crataegus, family Rosaceae. Species are grown as ornamentals for their (most often) white flowers. Also called *haw; thorn; thornapple.* See *cockspur hawthorn; Lavalle hawthorn; downy hawthorn; singleseed hawthorn; English hawthorn; Wash-*

ington hawthorn; pear hawthorn; glossy hawthorn; dotted hawthorn.

hawthorn lace bug. *Corythucha cydoniae* (Fitch), family Tingidae, a sucking insect pest of hawthorn, cotoneaster, firethorn, and flowering quince. Infested plants are characterized by loss of color on the upper surface of leaves and by dark specks on the undersurfaces.

hawthorn rust. A minor, fungous disease of apples, pears, and hawthorn caused by *Gymnosporangium globosum,* family Pucciniaceae; characterized by small lesions on the leaves, rarely on the fruit. Species of the genus Juniperus are the alternate hosts for this fungus. *Cf. apple rust.*

hay. Any leafy plant material, usually clover, fine-stemmed grasses and sedges, alfalfa, and other legumes, that has been cut and dried principally for livestock feeding. See *fodder; marsh hay.*

Hay. A skim milk cheese, made in France, which is cured 6 weeks to 3 months in hay in a curing cellar.

hay and dairy region. A region of the country where grasses and legumes are naturally abundant or easily grown, and where the feed supply and barn and market sizes are generally more suitable for dairy cattle and milk production than for other types of farming.

hay bale. A quantity of loose hay which is compressed usually into a rectangular bale about 3 ft. by 18 ins. by 14 ins., containing from 40 to 125 lbs. depending on the kind of hay and moisture content. The compressed hay is held in the bale by baling wire, light metal strips, or heavy twine. Less common is the cylindrical bale.

hay baler. A machine used for compressing loose hay into compact bales. There are two principal types, the stationary and the field or pick-up baler, the latter being the most extensively used. See *hay press.*

hay barn. 1. Any structure used to store hay and protect it from the weather. It may be a permanent structure, such as the conventional farm livestock barn, or a structure of a temporary nature, such as a shed. 2. A structure at one time used to store harvested hay that was later baled and sold.

hay barrack. A roof, supported on four posts, capable of being raised or lowered, used to protect an outdoor stack of hay.

hay blower. An attachment used with a hay chopper to blow the chopped hay into the hauling truck or into the storage.

hay carrier. A device with pulleys and wheels used to hoist hay from a wagon into the loft of a barn. See *hay sling.*

hay chopper. A device which chops the hay as it is cut.

haycock. A small, conical pile of hay in the field used for curing and protection of the hay prior to transporting it to a barn or haystack.

hay conditioner. 1. A mechanical device consisting of two closely spaced, parallel, smooth surface rollers which crush the fresh cut stems of hay to facilitate drying and curing. 2. A mechanical device consisting of two closely spaced, parallel rollers with corrugations resembling gear teeth paralleling the axle that kinks the stems of hay to break them open. Both methods result in a more even and rapid drying of the hay and less loss of leaves than conventional swath curing methods. 3. Any chemical which is supposed to be beneficial in preventing storage losses and/or increasing palatability.

hay crop. Any wild or cultivated crop which is cut and cured for hay.

hay crusher. See *hay conditioner.*

Hayden poisonvetch. *Astragalus haydenianus,* family Leguminosae, a plant whose green and dried foliage are poisonous. Symptoms of poisoning are dullness, irregularity of gait, lack of appetite, dragging of feet, a solitary habit, loss of flesh, and a shaggy coat. Found in southern Wyoming, U.S. Also called *milk vetch.*

haydite. Shale and clay fused at high temperatures, then ground or crushed, and used as an inert material in gravel culture *q.v.*

hay egg. Musty egg.

hay elevator. 1. An implement used for stacking cut hay; usually called *hay stacker.* 2. Any implement resembling an enlarged grain elevator which is used to elevate baled hay into the barn.

hayfield. Any field or area used principally for the production of a hay crop.

hay fork. 1. A retractable harpoon device of single, double, or grapple design, which is inserted into a load of hay and then lifted by means of ropes, pulleys, and a track into the hay mow where the load is discharged. 2. A three-tined pitchfork, a manually operated tool, designed for handling hay.

hay grass. Any of the comparatively tall, finer-stemmed grasses which may be cut, cured, and stored for winter feed for livestock, such as timothy, smooth brome, or orchard grass.

hay grinder. A combination of a cutting and chopping device with either a burr or hammer mechanism, used for reducing roughages, such as hay, to a ground meal.

hay harvest. The cutting, curing, and handling of the hay crop in such a manner as to make it a suitable roughage feed for livestock. See *haymaking.*

hay hook. A small hand tool made from a short, curved, steel rod sharpened at one end and with a hand grip on the other. It is inserted into the end or side of a bale of hay to facilitate handling.

hay keeper. A special type of storage building similar to a silo in which chopped hay is stored. It is often equipped with self-feeding arrangements.

hay knife. A long, scalloped, sharp-bladed knife used in cutting hay from a stack outdoors, or from a tangled mass of hay in the mow.

hay loader. A light machine drawn behind a hay wagon which picks up the hay from the swath or windrow and elevates it onto the wagon. There are two common types: the fork type and the cylinder type.

hay loft. See *hay mow.*

hay maker. See *hay conditioner.*

haymaking. The process of cutting, drying, and handling the hay crop to render it a satisfactory feed for livestock.

hay meadow. A field on which hay is grown.

hay mow. 1. A loft or space in the upper part of a barn where hay is stored. Also called a *hay loft.* 2. A mass of hay stored in a barn.

hay-mower-and-crusher. A power-drawn machine, combining the cutting mechanism of the mowing machine with a set of rollers which crimp or crush the stems and heavy parts of the hay as it is cut, which facilitates drying and curing and reduces the loss of the valuable leaves.

hay pick-up. A pick-up cylinder; an elevating unit of either canvas or lugged chains and a cross-conveyor which receives the hay from the elevator and carries it to the self-feeder on the baler.

hay plant. Any herbaceous plant which can be used for hay.

hay press. An agricultural implement, usually horse-drawn, which presses loose hay into bales (becoming obsolete). See *hay baler.*

hay rack. A slotted, wooden frame usually about 8 ft. wide and 16 to 18 ft. long placed on the running gear of a farm wagon, used to haul loose hay and grain bundles from the fields to the hay barn, stack or threshing machine. Sometimes called *hay basket.*

hay rake. Any of several different kinds of rakes. See *side-delivery rake; buck rake; sweep rake; dump rake.*

hayrick. 1. An outdoor pile of hay, usually of smaller dimension than a stack. See *haystack.* 2. A stack of hay, usually thatched for protection against the weather (chiefly British).

hayscented fern. *Dennstaedtia punctilobula,* family Polypodiaceae, a hardy fern grown as an ornamental in the fern garden. Native to North America. Also called *cup fern; boulder fern.*

hayseed. 1. Bits of straw, chaff, and seeds which cling to the garments in haymaking. 2. A rustic farmer or countryman (obsolete).

hay shock. See *haycock.*

hay sling. A device which is especially adapted to unloading hay from the hay rack or wagon to the hay mow by means of pulleys and ropes. See *sling.* Also called *center-trip hay sling.*

haystack. Any pile of hay larger than a haycock. *Haystacks* may be round or oblong and of a size and height varying with the quantity of hay to be stacked and the stacking equipment available. The purpose of the stack is to bring the hay into a compact mass and protect it from the weather and deterioration.

hay tedder. A light, field tool with arms and teeth which kick up and stir the hay in the swath to aid in the drying and curing.

hayward. 1. In old English law, an officer whose duty it was to keep cattle grazing on commons from straying into enclosed fields. 2. In the U.S., one who impounds straying cattle.

hazel alder. *Alnus rugosa,* family Betulaceae, a deciduous shrub grown in damp soils for its foliage and catkins. Native to North America. Also called *American alder; smooth alder; speckled alder; tag alder.*

hazel hoe. In forestry, a fire-trenching or digging tool which resembles a grub hoe but which has a shorter, broader, and lighter blade, a round or oval eye, and usually a straight, picklike head.

hazelnut. Filbert.

hazer. In rodeos, the assistant to the bulldogger who attempts to keep the animal running in a straight line and endeavors to protect the bulldogger from being gored.

H-bone. Aitchbone *q.v.*

H-budding. A double-rectangular cut, resembling the letter H, made in the bark of the stock. The flaps are lifted in the center for insertion of the bud. *Cf. plate budding; bud: 3.*

head. 1. Any tightly-formed flower cluster, as in members of the family Compositae, or any tightly-formed fruit cluster, as the *head* of wheat or sunflower. 2. A compact, orderly mass of leaves, as a *head* of lettuce. 3. On a tree, the point or region at which the trunk divides into limbs. 4. The height of water above any point of reference, (elevation *head*). The energy of a given nature possessed by each unit weight of a liquid expressed as the vertical height through which a unit weight would have to fall to release the average energy possessed, used in various compounds, as pressure *head,* velocity *head,* lost *head,* etc. 5. Cows, asses, horses, collectively, as ten *head* of horses. 6. The part of the body which includes the face, ears, brain, etc. 7. The source of a stream; specifically the highest point upstream at which there is a continuous flow of water, although a channel with an intermittent flow may extend farther. 8. The upstream terminus of a gully. 9. To prune a tree severely. 10. To get in front of a band of sheep, herd of cattle, etc., so as to stop their forward movement. 11. To place a top on a barrel.

head back. To prune off the end of a branch or stem by cutting a bud or side branch on the distal side.

head blight of wheat. Cereal scab.

head cheese. A meat product made from meaty parts of the head of a hog excluding the skin. jawbones, eyes, and ears, subjected to prolonged cooking, then cut or picked into small bits, seasoned, and compressed into a cheeselike mold or loaf.

head chop. See *grain sorghum head chop.*

head cold. See *rhinitis.*

header. 1. A framing member used to support a load across an opening. 2. A machine used to cut the heads from the grain straw instead of cutting the straw near the base. The heads are then transferred to a wagon or truck. 3. A structure installed at the head of a gully to prevent erosion.

header binder. A header to which is attached a device for binding cut grain. The machine is set to cut far enough down the stalk so that binding is possible. See *header: 2.*

head face. A channel used to direct and control the amount of water to a water wheel or other device actuated by flowing water.
head flume. A flume, chute, trough, or lined channel used at the head of a gulley or at the lower end of a terrace outlet to prevent soil erosion.
head gardener. Grounds superintendent.
head garden lettuce. *Lactuca sativa* var. *capitata,* family Compositae, a commercially important annual herb grown for its edible, compact head of leaves. It is popular as a salad plant. Native to Eurasia. Also called *head lettuce; lettuce.*
headgate. a. The control works, or the gate itself, on the front of a cattle squeeze. b. The control works at the point where irrigation water is withdrawn from a canal into a service ditch or conduit.
head in. See *head back.*
heading. 1. The removal of the hide from the skull. 2. The piece of lumber from which a top of a barrel is cut.
heading back. See *head back.*
heading broccoli. See *broccoli.*
heading stage (time). The period in the development of cereal plants when the heads begin to emerge from the sheaths.
headland. 1. A type of farm road which is established at frequent intervals in sugarcane land to permit the best possible movement of heavy loads of cane with as little travel over the rows as possible. It also may be used in connection with the production of other crops which require the removal of several tons of growth per acre at harvest. 2. That portion at the end of the field which is reserved for travel and for turning around the tillage implement after completing a furrow or row.
head lettuce. See *head garden lettuce.*
head maggot fly. Sheep bot fly.
head mold. False smut.
head out. To form a flower head.
head pruning. The pruning of grape vines so that the mature stem bears at its summit a ring of arms ascending in the form of an inverted cone. Also called *vase pruning; goblet pruning.*
head race. A channel which leads water to a water wheel; a forebay.
head rice. In commercial milling, that portion of the milled rice which is made up of plump, whole, unbroken kernels.
head row. 1. A row which is drilled or seeded at right angles to the regular rows across the end of the field. 2. A progeny row in a plant breeding nursery sown from the seed of a single panicle or head. See *head-to-row.*
head shy. Designating a horse on which it is difficult to put a bridle, to lead, or to work around its head.
head smut. A fungous disease of corn and sorghums, caused by *Sphacelotheca reiliana,* characterized by large smut sori replacing the tassel and ear in corn and the panicle in sorghums.
head smut of grasses. A smutting of the heads of cereal grains caused by several fungi, most important of which are Ustilago spp., family Ustilaginaceae, Sorosporium spp., family Ustilaginaceae; and Sphacelotheca.
head smut of millet. A fungous disease caused by *Sphacelotheca destruens* (Schlect), family Ustilaginaceae. The disease is evident as the panicles emerge, the entire inflorescence becoming enclosed by a grayish-white false membrane which ruptures as the plant matures exposing the dark brown spore mass.
head space. The space which is left at the top of the fruit jar or can after filling with food.
headstall. A type of halter used to lead or restrain horses which is similar to the hackamore or Spanish jaquina.
headstrong. Designating an animal which tends to be stubborn.
head-to-row. Pedigree selection (*q.v.*) for such crops as wheat and oats in which a number of heads are selected from different plants throughout a field. The seeds from each head are planted in one row. Every tenth row is planted to a high-yielding standard variety for comparison. Each row is harvested and threshed separately, and those rows showing promise are planted in definite amounts. Again each row is harvested and threshed separately. The procedure is carried on for three years or more to see if a superior variety has been developed.
head up. 1. To place a top on a barrel. 2. To lead, as by natural leaders in a flock of sheep, etc. 3. Indicating an alert horse.
headwall. a. A vertical wall, usually of concrete, at the end of a culvert which is used to prevent earth from spilling into the channel. b. The vertical wall, usually of reinforced concrete, over which the water falls in a gully head control structure.
headwater. 1. The small tributary streams that join to form a river; the water source of a river. 2. The water upstream from a structure, such as a dam.
headwork. The diversion structure at the inlet of a conduit; an intake heading.
head wrap. A piece of paper used to enclose the head of a fowl dressed New York style.
heal-all. Self-heal.
heal on. To grow to a stock, as a bud in bud grafting.
hear corn grow. On hot, humid, summer nights, to hear popping noises from growing corn, indicating optimum growing conditions.
heart. The organ of the body which by its rhythmical contractions circulates the blood. It is an edible by-product of slaughter animals and fowls.
heart center. That wood immediately surrounding the pith which is characterized by discoloration, small checks, shake, or pin knots.
heart check. A check that extends across the growth layers in one or more directions from the pith toward, but not to, the surface of a piece of wood. See *check: 2.*
heart cherry. A pomological group of varieties of the sweet or mazzard cherry species. The fruit of these varieties is heart-shaped, soft-fleshed, and light or dark colored. Two rather common varie-

ties are Governor Wood and Black Tartarian. In France, varieties of this group are known as *Guignes;* in England, *geans.* See *mazzard cherry: 1.*

heart-girth. A measurement taken around the body just back of the shoulders of an animal, used to indicate fullness of the chest and lung capacity.

heartleaf aster. *Aster cordifolius,* family Compositae, a perennial herb grown for its late-summer, violet or blue flowers. Native to North America. Also called *blue wood aster.*

heartleaf bergenia. *Bergenia cordifolia* (syn. *Saxifraga cordifolia*), family Saxifragaceae, a perennial herb grown in the flower border for its heart-shaped leaves and its rose-pink flower clusters. Native to Asia.

heartleaf philodendron. *Philodendron cordatum,* family Araceae, a plant which is grown as a house plant for its trailing habit and its heart-shaped, decorative leaves. Native to tropical America.

heart rot. A symptom of a number of plant diseases in which the center of the stem or a plant part rots, as in blackheart of potato *q.v.*

heartsease. Garden pansy.

heartseed. Any tropical, herbaceous vine of the genus Cardiospermum, family Sapindaceae. See *balloonvine heartseed.*

hearts-on-strings. Woods ceropegia.

heart substance. In wood anatomy, the various tannins, gums, resins, and other materials that are deposited in the cell cavities and walls when transformation from sapwood to heartwood takes place. It serves to differentiate sapwood and heartwood.

heartwater disease. A disease found in various parts of Africa, affecting cattle, sheep, and goats, caused by a very small organism that can be seen under the highest power of the microscope. The name *heartwater* is derived from the appearance of the heart when seen at autopsy; the thin sac that surrounds the heart is filled with an excessive amount of watery fluid.

heartweed. Spotted lady's-thumb.

heartwood. The inner core of a woody stem which is wholly composed of nonliving cells and is usually differentiated from the outer, enveloping layer (sapwood) by its darker color. Also called *duramen.*

heat. 1. To ferment as a result of stored grains, such as wheat, corn, or barley and forages, becoming wet internally or through exposure to moisture. 2. See *estrus.*

heat canker of flax. A nonparasitic injury to flax caused by high surface temperatures in dark-colored soils; characterized by cankers on the stem before the plant is high enough to shade the soil surface, by brown lesions on the stems, or by death of the seedling. In plants that survive, there is an enlargement of the stem above the canker.

heated. 1. Designating a condition of stored hay, grain, and other products in which, due to improper storage, the temperature rises resulting in damage and spoilage of the product. See *heating.* 2. Designating a flavor defect of cheese which suggests uncleanliness and a whey taint.

heated cream. A flavor defect of butter characterized by an odor resembling that of high-heated, acid cream.

heated frame. Any garden frame (*q.v.*) that has a device for heating the air in the frame. See *cold frame; hotbed.*

heater. See *orchard heater.*

heath. 1. Any plant of the genus Erica, family Ericaceae. Species are evergreen shrubs and small trees grown in greenhouses and out-of-doors. Also called *erica.* See *tree heath; spring heath; fringed heath; twisted heath; Biscay heath; Corsican heath; garland heath; crossleaf heath; Cornish heath; blackeyed heath.* 2. A natural land feature, an extensive tract of uncultivated land, treeless or nearly so, which is covered by a dense growth of shrubby, ericaceous plants. It may be nearly the same as a high moor. *Heaths* are generally sandy and the soils strongly acid.

heath aster. *Aster ericoides,* family Compositae, a perennial herb grown for its late-summer clusters of white or pinkish flowers. Native to North America. Also called *farewell summer.*

heather. See *Scotch heather; heath: 1.*

heath pea. *Lathyrus tuberosus,* family Leguminosae, a perennial, climbing or trailing herb bearing tubers, parts of which are edible. Native to Eurasia. Also called *groundnut pea.*

heat increment. The additional amount of heat produced in the body of an animal following, and because of, the consumption of food. Also called *work of digestion.*

heating. A condition of stored grains and forages in which the product increases in temperature due to respiration or to associated microorganisms. The product may become unfit for use as seed or feed.

heat injury (necrosis). Injury to a plant which results from excessive heat. The usual symptoms are retarded growth, undersized plants, localized burns on leaves or stems, defoliation, premature ripening of fruits, and death.

heat kill. The death of all or part of the foliage of a plant caused by the heat from a fire.

heat-of-wetting. A laboratory procedure which has been used to determine the colloid content of soils; the heat evolved when a dry soil is immersed in water.

heat prostration (stroke). A condition of an animal resulting from excessively hot weather; characterized by lethargy, inability to work, staggering gait, convulsions, and high temperature. Death often occurs.

heat-resistant. 1. Designating a variety or a species which grows under comparatively high temperature conditions. 2. Any material which is resistant to high temperatures.

heat spot. A defect of a fertile egg which results from alternating high and low temperatures; characterized by the beginning of the development of the embryo without blood showing.

heat spot of plums. Kelsey spot of plums.

heave line. A noticeable tightening or constriction in the muscles in the flank of horses having the heaves or broken wind defect. See *heaves.*

heavenly bamboo. Nandina.

heaves. An incurable, respiratory disease of horses characterized by difficult breathing in which the air cells of the lungs are dilated or ruptured. It is thought to be caused by hard exercise, by consumption of dusty feed, or by heredity. It is found most frequently in old horses and is characterized by forced, double exhaling. Also called *broken wind; chronic alveolar emphysema.*

heavier yearling. In marketing, a baby beef, 18 to 24 months of age, which weighs over 1,000 lbs. *Cf. fat calf; super baby beef.*

heaving. An upward movement of soil caused by freezing and thawing of free water in the soil, thus involving expansion and contraction. Damage, and sometimes destruction of plants, may result from the lifting action and fence posts may also be pushed upward and topple. Also called *frost heaving; frost lifting. Cf. soil heaving.*

heaving out. The forcing of a plant crown above the soil line by heaving *q.v.*

heavy. 1. Designating any material or product which exhibits a comparatively high weight per unit volume. 2. Designating a clay or clayey type of soil that is difficult to plow. 3. In marketing, designating an abundant supply of a product for sale on one day at one market. 4. Designating a subclass of country hide *q.v.*

heavy assorted chicks. Mixed chicks of heavy breeds, or off-color chicks of one heavy breed.

heavy boned. Designating an animal which has comparatively large bones for its breed or type.

heavy broiler. A broiling chicken that weighs 2¼ to 3¼ lbs.

heavy burning. Range firing during the dry, hot season to insure a fire which will destroy the existing cover, facilitate travel and livestock handling, increase forage for game, and enhance hunting.

heavy capon. A capon that weighs over 7 lbs.

heavy fat calf. See *super baby beef.*

heavy fowl. A female chicken of any heavy breed too mature to be a roaster.

heavy frost. A more severe deposit of frost than light frost which does not cause serious injury to staple products. See *frost; black frost; killing frost.*

heavy grazing. The practice of keeping a large number of animal units on a pasture or grazing area so that the grass or herbage may be closely grazed.

heavy-harness horse. The carriage or coach type of horse, as distinguished from the roadsters or light harness type, which is largely an English horse.

heavy milking cow. A cow that has demonstrated an ability much above the average as a milk producer.

heavy over the shoulder. In judging dairy cattle, designating undue coarseness or fleshing over the front quarters of the animal.

heavy pruning. Removal of much wood, usually by thinning branches and heading back.

heavy pullet. A pullet of a heavy breed which weighs over 4 lbs. before it comes into egg production.

heavy shrink. 1. Designating sheared wool in which there is much shrinkage in weight from scouring. 2. Designating a comparatively large loss in weight of an animal which generally occurs between the shipping point and place of slaughter.

heavy sirup. A sirup of 20° to 23.9° balling *q.v.*

heavy soil. Generally, a clayey soil in contrast with a sandy one. (The reference is to resistance and need for greater power in pulling a plow rather than to actual weight or specific gravity which is actually less for clay than for sand.)

heavy textured. Designating a soil of fine texture, in which clay and fine silt predominate, which usually has a dense structure and is firm to compact in consistency.

heavy purée. See *purée.*

heavy veal. A market grade of veal calf of a wide range of age but weighing between 150 and 300 lbs.

heavyweight calf. A slaughter calf which weighs over 300 lbs.

Heavyweight Hunter. A classification of riding horse used for hunting, that can carry a rider weighing 190 lbs. or more.

Hebrew cream. Cultured cream.

hectare. A European unit of land measurement equal to 100 ares, 10,000 square meters, or 2.47 acres.

hedge. 1. A fence or barrier formed by bushes, shrubs, or small trees growing close together in a line, sometimes with interwoven branches. 2. A managed belt of shrubs or small trees usually placed across fields or along field or property boundaries, but which is not used for wind control. 3. To buy or sell futures to protect from loss due to a rise or fall in prices.

hedge-bells. Hedge glorybind.

hedge bindweed. Hedge glorybind.

hedged. 1. Enclosed with a hedge. 2. Designating plants trimmed uniformly across the top, sometimes by the grazing or browsing of livestock.

hedge fumitory. Drug fumitory.

hedge glorybind. *Convolvulus sepium,* family Convolvulaceae, a weed pest. Also called *devil's-garter; hedge-bells; hedge bindweed.*

hedge laurel. Tarata pittosporum.

hedge maple. *Acer campestre,* family Aceraceae, a deciduous hardy tree grown as an ornamental for its autumnal, yellow foliage. Native to Eurasia.

hedgemustard. *Sisymbrium officinale,* family Cruciferae, an annual herb which is a weed pest in cultivated fields. Also called *bankcress; bankweed.*

hedge pink. Bouncing Bet.

hedgerow. A barrier of bushes, shrubs, or small trees which grow close together in a line.

hedgerow system. 1. A method of training strawberries by a definite placement in the row of the runners from each mother plant. See *matted row system.* 2. The widened row used for some bramble fruits in which sucker plants are permitted to grow between the mother plants in the row.

hedge shears. A pair of shears with long, heavy blades and long handles, used to trim woody plants, hedges, etc.

hedging. 1. Protection against the risk of a price change which consists either of buying futures to prevent losses from spot selling, or selling futures to prevent losses from spot buying. It is simultaneous but opposite to transactions on the futures and spot markets. *Cf. cash market.* 2. Designating a plant which is grown in a hedge.

heel. 1. The basal end of a plant stem cutting along with a piece of the older stem. 2. See *hock: 1.* 3. The end of the branches of a horseshoe. 4. The rear end of the foot.

heel calk. A downward, right-angled turning of a horseshoe at the heel.

heel clearance. The elevation of the rear of the landside of the wheel-mounted plow above the bottom of the floor to give proper suction.

heel cutting. See *heel: 1.*

heeled-in. See *heel-in.*

heel fly. A name applied to the common cattle grub (*q.v.*) because it enters the body through the feet.

heel-in. To cover temporarily the roots of transplants, which are awaiting permanent planting, with earth.

heel of the landside. The rear end of the landside on a plow, one of the points of bearing.

heel of the round. A retail cut of beef: the heavy fleshing at the rear of the stifle joint which is cut as a roast.

heel weight shoe. A horseshoe with weights on the heel which makes a horse fold his legs high up under him before extending them forward.

hegira. A grain sorghum.

he-huckleberry. *Lyonia ligustrina,* family Ericaceae, a deciduous shrub grown in the informal border for its foliage and bell-shaped, white flowers. Native to North America. Also called *swamp andromeda; male berry; privet andromeda.*

heifer. A female bovine which has not borne a calf. See *bred heifer; open heifer.*

heiferette. In marketing, a good grade of young cows of beef breeding which approach heifers in form.

height class. A classification of forest trees by height.

height of head. The distance of the lowest scaffold limb of a tree from the ground.

height to a fixed top. In forestry, the length of tree stem from the stump to a point in the top of specified diameter.

height to a variable top. In forestry, the length of tree stem from the stump to a point in the crown at which merchantability is limited by branches, defect, or breakage.

height to base of crown. In forestry, the length of tree stem from the average ground level to the junction with the bole of the first line branches composing the main crown.

held. 1. The lower end of an ax or other tool handle which is encased by the metal portion of the tool. 2. Designating the controlling of an animal in a small space or by other controlling devices.

held eggs. Eggs which have been delayed sufficiently in their movement to market so that they are of distinctly lower grade than fresh (lower in quality than U.S. grade A), and have not been held under refrigeration long enough to be labeled cold-storage eggs.

heliotrope. 1. Any plant of the genus Heliotropium, family Boraginaceae. See *big heliotrope.* 2. *H. arborescens,* a tropical, perennial herb grown as an annual in the flower garden or in the greenhouse for its fragrant, purple or violet colored flowers. Native to South America. Also called *cherry pie.*

heliotrope ehretia. *Ehretia thrysiflora,* family Boraginaceae, a tropical tree grown as an ornamental for its showy, white flowers and orange, changing to brownish-black, fruit. Native to Asia.

heliotropism. See *phototropism.*

hell bind. Dodder.

hellebore. 1. A perennial herb of the genus Helleborus, family Ranunculaceae. See *Christmas rose; Lenten rose.* 2. See *falsehellebore: 1.*

hell fence. A fence made of stumps laid upside down or on edge.

helminthic. Of, or pertaining to, helminths, or worm parasites.

Helminthosporium blight of oats. A fungous disease of oats of Victoria descent caused by *Helminthosporium victoriae,* family Dematiaceae; characterized by seedling blight and culm necrosis, and root and crown rot. Dark, olive-colored conidia are produced on the necrotic areas of the culm.

Helminthosporium blight of rice. A worldwide, fungous disease caused by *Helminthosporium oryzae,* family Dematiaceae; characterized by a seedling blight, or by brown spots on the leaves that spread and turn reddish-brown with a gray center, shriveled kernels, and broken panicles.

Helminthosporium brown stripe of sugar cane. A fungous disease caused by *Helminthosporium stenospilum,* family Dematiaceae; characterized by yellow or brown spots on the leaves that elongate and turn reddish-brown or red. There is a gray spot in the center of the lesion.

Helminthosporium eye spot of sugar cane. A fungous disease of sugar cane, lemongrass, and Napiergrass caused by *Helminthosporium sacchari,* family Dematiaceae; characterized by yellow or brown spots on the leaves which spread, become oblong or linear, turn reddish-brown or red, and develop a gray center. Streaks extend from the spot to the leaf tip.

Helminthosporium leaf blight of grasses. A fungous disease of sorghums, Sudan grass, corn, and

Johnsongrass caused by *Helminthosporium turcicum,* family Dematiaceae.

Helminthosporium leaf blotch of oats. A minor, fungous disease of oats, caused by *Helminthosporium avenae,* family Dematiaceae, the conidal stage of which is characterized by oblong or linear, reddish-brown, sunken-centered blotches on the leaves. The leaf turns yellow and dries out.

Helminthosporium leaf spot of corn. A fungous disease caused by *Helminthosporium carbonum,* family Dematiaceae; characterized by numerous, elongated spots, buff or reddish-brown, or by irregular, chocolate-brown spots on the leaves, and by black, moldy growth on the kernels resulting in a charred-appearing ear.

Helminthosporium leaf spot of millet. A fungous disease of millet and millet-like grasses caused by several species of the genus Helminthosporium, family Dematiaceae; characterized by oblong or linear, reddish-brown blotches on the leaves.

helminths. Worm parasites, such as the flatworms (flukes and tapeworms) and roundworms (hookworms and lungworms).

help. 1. Any skilled or unskilled laborer. 2. Labor, generally and collectively.

hematite. Fe_2O_3, anhydrous sesquioxide of iron, a mineral which occurs as deposits of iron ore and is widely distributed in rocks and soils where, in an amorphous condition, it causes a reddish color in the soil.

hemicryptophytes. Perennial bunchgrasses and similar plants whose shoots die back to the ground surface at the beginning of the unfavorable growth season leaving the reproductive buds protected by the upper soil layer and the remains of the weathered plants.

hemiparasite. A partial parasite.

hemispherical scale. *Saissetia hemisphaerica* (Targ.), family Coccidae, a sucking insect pest of oleander, bay, vinca, croton, cyclamen, fern, palm, figs, citrus, and other plants.

hemizygous. The condition in which only one allel of a pair of genes is present in the cells of an individual plant or animal, the other one being absent, as in deletion.

hemlock. 1. Any evergreen tree of the genus Tsuga, family Pinaceae. Some species are very valuable ornamentals and some are good timber trees. See *Canada hemlock; Carolina hemlock; Japanese hemlock; Pacific hemlock; mountain hemlock; Siebold hemlock.* 2. See **poison hemlock**; *waterhemlock; Canada yew.*

hemlock borer. *Melanophila fulvoguttata* (Harr.), family Buprestidae, the larva of a beetle which infests hemlock and sometimes spruce, separating the bark from the wood and making wide, shallow galleries. Also called *spotted hemlock borer; eastern flathead hemlock borer.*

hemlock looper. *Lambdina fiscellaria* (Guen.), family Geometridae, a caterpillar which infests hemlock, balsam fir, arborvitae, beech, birch, blueberry, elm, maple, oak, pine, spruce, and willow. Infestation may result in defoliation.

hemlock poisoning. Poisoning of animals resulting from browsing poison hemlock (*Conium maculatum*); characterized by cessation of digestion, gas, salivation, dilation of the pupils, rapid pulse and breath, paralysis, unsteady gait, difficult breathing, and death.

hemlock sawfly. *Neodiprion tsugae* Midd., family Diprionidae, an insect whose larva infests hemlock in the Pacific Northwest, U.S., by feeding on the foliage.

hemlock scale. *Aspidiotus ithacae* (Ferris), family Diaspididae, a sucking insect which infests hemlock, Douglasfir, and pine in western U.S.

hemlock spruce. See *Pacific hemlock; Canada hemlock.*

hemlock waterparsnip. *Sium suave,* family Umbelliferae, a widespread, perennial herb suspected of being a poisonous plant. Also called *water parsnip.*

hemoglobin. The red pigment in the red blood cells of animals and man which carries oxygen from the lungs to other parts of the body. *Hemoglobin* is a complex chemical compound, made up of iron, carbon, hydrogen, and oxygen, and is essential to life in red-blooded animals.

hemoglobinuria. The presence of hemoglobin in the urine which is a symptom of some diseases: *e.g.,* cattle tick fever and azoturia. Bacillary *hemoglobinuria* is a specific, acute, infectious disease of cattle and sheep which is known to occur in western and parts of southern U.S. The causative organism is *Clostridium hemolyticum.*

hemolysis. The destruction of red blood cells and the resulting escape of hemoglobin.

hemophilia. A hereditary condition in man or animals in which the blood does not coagulate readily and the slightest wound or bruise can cause considerable loss of blood. An animal so affected is often called a *bleeder.*

hemorrhage. 1. Any escape or discharge of blood from the blood vessels. 2. To bleed.

hemorrhagic. Bloody.

hemorrhagic enteritis. An inflammatory condition of the intestines in which the small vessels (capillaries) in the lining become engorged and eventually rupture allowing blood to seep into the intestine. The evacuations are watery and bloody or very dark colored due to decomposition of the blood present in them.

hemorrhagic ova. Immature eggs, or ova, which contain blood, in the diseased ovary of poultry suffering from pullorum disease or avian typhoid.

hemorrhagic septicemia. An infectious disease of livestock characterized by numerous, small hemorrhages in the tissues and the presence of bacteria of the pasteurella group and their associated poisons in the blood. Sometimes called *shipping fever; stockyard fever; blood poisoning.*

hemorrhagic septicemia of fowls. See *fowl cholera.*

hemp. *Cannabis sativa,* family Cannabinaceae (Moraceae), an annual herb which yields hemp fiber (not Manila hemp), hashish, and bird seed. It is the oldest cultivated fiber plant which, escaped, is a weed pest. It also presents serious problems because of its use by drug addicts; the stems and leaves when smoked have a narcotic effect. It is an important crop in Russia and India. Native to Asia. Also called *Indian hemp; marijuana; red-root; gallows-grass; hashish; locoweed.*

hemp dogbane. *Apocynum cannabinum,* family Apocynaceae, a widespread, hardy, perennial herb grown for its roots yielding a cathartic and an emetic. It is poisonous to horses and cattle. Symptoms of poisoning are fever, rapid pulse, coldness of the legs, dilation of the pupils, discoloration of the mouth and nostrils, sweating, refusal of feed or drink, and death. Native to North America. Also called *Indian physic; American hemp; chocktaw root* (a regional name).

hemp palm. Fortunes windmill palm.

hemp sesbania. *Sesbania exaltata,* family Leguminosae, an annual, tropical herb grown in warm areas as a cover crop and for green manure. Escaped, it can become a weed pest. Found from Missouri to Louisiana, U.S. Also called *tall indigo; Colorado river hemp.*

hemp tree. Chaste tree.

hen. A female fowl; specifically, the female domestic fowl valued for its egg production.

hen-and-chickens. *Sempervivum tectorum,* family Crasulaceae, a perennial, hardy herb grown in the rock garden for its tightly-clustered leaves. Native to Eurasia. Also called *houseleek.*

henbane. Black henbane.

hen battery. A number of individual hen-cage units arranged in a group, usually single or multiple decks, which have provision for watering, feeding, collecting eggs, and disposing of droppings. See *hen cage.*

hen cage. A cage of wire or slats in which one or more hens is confined and in which provision is made to eat, sleep, and lay. See *hen battery.*

henequen agave. *Agave fourcroydes,* family Amaryllidaceae, a plant grown in Central America as a source of a fiber resembling sisal. Native to tropical America.

hen-feathered. Designating a male bird whose feather structure and marking resemble that of a hen.

hen house. A building, usually of light or shed type construction, which is suitable for housing poultry.

hen manure. The excreta of the chicken. See *poultry manure.*

henna. *Lawsonia inermis,* family Lythraceae, a spiny shrub widely grown in warm countries, especially Egypt, as an ornamental for its heavily fragrant, white, pink, or rose flowers, and as a source of a reddish-orange dye obtained from its leaves. Native to Africa and Asia. Also called *Egyptian henna; alhenna; Egyptian privet; mignonette tree.*

hennery. 1. A poultry farm, particularly one which specializes in the production of market eggs. 2. A building or enclosure where laying hens are kept.

hennery egg. An egg produced on a specialized poultry farm, thus presumed to be of high quality.

henny. See *hinny.*

Henry honeysuckle. *Lonicera henryi,* family Caprifoliaceae, a hardy, half-evergreen, woody vine grown for its habit and its purple-red flowers and black fruit. Native to China.

Henry lily. *Lilium henryi,* family Liliaceae, a bulbous herb grown in the flower border for its very tall flower stalk with drooping, orange, brown-spotted, flowers. Native to China.

Henry spirea. *Spiraea henryi,* family Rosaceae, a handsome, flowering, deciduous shrub grown for its white flowers. Native to China.

hepatica. Any perennial herb of the genus Hepatica, family Ranunculaceae. See *roundlobe hepatica.*

heptachlor. A chlorinated hydrocarbon insecticide which is closely related to chlordane.

heptane. C_7H_{16}, a colorless liquid obtained from the oleoresin of the Jeffrey pine, the Digger pine, and from petroleum, used as a constituent of test fuel, as solvent, insecticide, and drug.

heraldtrumpet. See *Easter heraldtrumpet.*

herb. 1. A plant with leaves, stems, roots (and often with flowers), not markedly woody, which dies down to the root crown at the end of its growing season. The roots may or may not survive for another season. 2. A plant which is usually grown in a garden for culinary seasoning or medicinal use.

herbaceous. Referring to a plant or to a portion of a plant that lacks a pronounced woody structure. See *herb: 1.*

herbaceous stage. In fire control, the current state of inflammability of herbaceous vegetation.

herbage. 1. Herbaceous plants, collectively, especially with reference to their value as animal food. 2. Nonwoody plants in contrast with trees and shrubs.

herbal. A book or manuscript which describes plants, often with illustrations, written mostly by the early herbalists *q.v.*

herbalist. One who studies plants for their character, virtues, uses, etc. (the term is now almost obsolete but was prevalent especially from the 10th to the early 18th centuries).

herbarium (plural **herbaria**). a. A collection of dried or otherwise preserved plant specimens used for reference and for the verification of names of plants. It is usually an adjunct to a college or university where botany is studied and taught. Such a collection of dried, pressed plants formerly was called a *hortus siccus.* b. The building in which such a collection is displayed.

herb barbara. Bitter wintercress.

herb garden. A garden in which plants are grown for use in flavoring, for fragrance, for historic, legendary, or other purposes.

herbicidal. Designating the effect or properties of an herbicide *q.v.*

herbicide. Any chemical which is used for killing plants.

herbivorous. Designating an animal (herbivore) that feeds, in the native state, on grass and other plants, as cattle, horses, sheep, goats, deer, elk, etc. *Cf. carnivorous.*

herbmercury. *Mercurialis annua,* family Euphorbiaceae, an annual herb which is poisonous to cattle, sheep, horses, and pigs. Native to Europe. Also called *mercury.*

herb Robert. Herbrobert geranium.

herbrobert geranium. *Geranium robertianum,* family Geraniaceae, a prostrate, annual or biennial herb grown in the rock garden for its small, bright crimson, geranium-like flowers. Native to North Africa. Also called *felon grass; red robin; herb Robert.*

Herbst bloodleaf. *Iresine herbsti,* family Amaranthaceae, a tropical, tender plant grown as a summer bedding plant for its purplish-red and green leaves with yellow veins. Native to South America.

Hercules'-club. Devils-walking-stick.

Hercules'-club pricklyash. *Zanthoxylum clavaherculis,* family Rutaceae, an aromatic, prickly tree grown as an ornamental for its yellowish–green flowers. Native to North America. Also called *pricklyash; toothache tree.*

herd. 1. A group of animals (especially cattle, horses, swine), collectively, considered as a unit in farming or grazing practice. See *flock: 1.* 2. To tend animals in *herds.* See *cowboy: 1; herder; shepherd.*

herd book. The recognized, official record of the ancestry of a purebred animal which is kept by the particular breed association. The first *herd book* for keeping thoroughbred pedigrees was started in England in 1791. This was followed by Coates *Herd Book* for Shorthorns started in England in 1822.

herd-bound. Designating a horse that cannot be ridden away from the corral without a great deal of effort by the rider.

herd-broken. Designating an animal tamed to know the herd and stay with it. See *broke.*

herd bull. See *herd sire.*

herder. One who cares for sheep on the open range. He may or may not be the owner or operator or be as skilled as a shepherd or flock master.

herd header. See *herd sire.*

herd improvement registry. A type of registry maintained by certain purebred cattle breeder associations to record the production and yearly records of all producing cows of that breed in a given breeder's herd.

herd-improvement test. A type of production test for dairy cows in which the whole herd, rather than the individual cow, is placed under test on a yearly basis.

herding. The control of animals on the range by guiding their direction and movements to procure grazing where and when desired.

herd-man. That person who is in charge of animals, usually cattle. See *herder.*

herd number. The official number given an animal when it is registered in the official herd book of the breed association.

herd's grass. See *timothy; redtop.*

herd sire. The male of the species which is kept for the sole purpose of reproduction. (The term is most commonly applied to cattle and horses but is also used with sheep and swine.) See *bull: 1; stallion; ram.*

herdsman. One who cares for and manages a herd of livestock, usually cattle.

herd test. A type of semiofficial testing for milk production in which the whole herd of cows of milking age are included. This test was first started by the Ayrshire Breeders Association in 1925 and places the emphasis on the entire herd as a unit rather than on the individual cow.

hereditary. Designating factors or characters transmitted from parent to offspring, the expression of which is controlled by specific genetic factors.

hereditary causes. Phenomena attributable to genetic factors controlling the development or manifestation of a character.

hereditary unsoundness. A physical defect of a horse largely due to heredity, such as certain kinds of sway-back, narrow joints, sickle-hocked, calf-kneed, knee-sprung, bowlegs, etc. *Cf. breeding unsoundness; temporary unsoundness.*

heredity. The hereditary transmission of genetic or physical traits of parents to their offspring.

Hereford. A leading breed of beef cattle which originated in England. They are blocky, heavy, and light red with a characteristic white face and underline. Certain strains of the breed are polled. Cows weigh up to 1,900 and bulls up to 2,000 lbs. Also called *white face.* See *Polled Hereford.*

Hereford swine. An American-developed lard-type breed of swine which has color markings similar to Hereford cattle.

hermaphrodite. An individual which possesses both male and female sex organs. In some species such individuals are capable of reproducing while in others they are sterile.

hermaphroditic. Designating the occurrence of both male and female reproductive organs in one individual.

hermetically-sealed. Designating an airtight closure of a container.

hermit. A sheep which will not live with the flock (New Zealand).

hernia. 1. The protrusion of an organ or part through some opening in the walls of its cavity. In domestic animals the most common *hernia* is one involving a rent or abnormally large opening in the abdominal musculature; *i.e.,* inguinal region and umbilicus. Also called *rupture.* 2. See *clubroot of cabbage.*

herniated. Designating a hernia condition.

heronbill. Any of several species of Erodium, family Geraniaceae, which have a value as forage plants, but some have become weed pests. Native to the Mediterranean area; established in southwestern U.S. Also called *erodium; storksbill; cranesbill; filaree.* See *alfileria.*

Herrgardsost. A medium-firm, mild, sweet, cured, cows' skim milk cheese made in Sweden.

herring-bone system. A system of laying tile lateral underdrains in which the main lies in a shallow depression and the laterals enter it at acute angles from both sides.

herring-gutted. Designating a horse with a slim belly.

herring oil. A livestock feedstuff supplement which is the product obtained by extraction of part of the oil from the whole or part of the herring. It is used as a source of vitamins.

Herve. A Limburger-like cheese made in Belgium.

Hessian fly. *Phytophaga destructor* (Say), family Cecidomyiidae, an important insect pest of wheat, barley, and rye, but especially of wheat. Its larva sucks plant juices and causes the infested plants to lodge. See *fly-free date.*

heteroecious rust. A rust having different stages of its life cycle on two unlike types of host plants, as stem rust on cereals and barberry.

heteroecism. See *heteroecious rust.*

heterogametic. Producing unlike gametes, particularly with regard to the sex chromosome. In species in which the male is of the "X-Y" type, the male is *heterogametic,* the female homogametic.

heterogamous. Of flowering plants, not having the male and female parts (stamen and pistil) regularly present.

heterogen. A variable group of plants or animals which arise as hybrids, sports, mutations, etc., certain types of which may or may not breed true.

heterogeneity. 1. The property of one or more samples implying their origin from different populations. 2. Heterogeneous state or quality.

heterogeneous. Designating elements having unlike qualities.

heteroploid. An organism characterized by a chromosome number other than the true haploid (monoploid) or diploid number.

heterosis. Hybrid vigor; the greater vigor, sometimes displayed by crossbred plants or animals, as compared with that of their parents.

Heterosporium leaf spot of spinach. A fungous disease caused by *Heterosporium variable,* family Dematiaceae; characterized by small, brown spots which enlarge so that most of the leaf surface is covered and the leaf dies. A greenish-black mold appears on the spots.

heterostyly. A difference in length of styles and filaments in flowers of a species or variety. Some flowers have short styles and long filaments, others, long styles and short filaments. This results in much cross-pollination between plants.

heterotic hybrid. A hybrid (*q.v.*) that has an advantage or is an improvement over the parents.

heterotrophic. Designating a plant found on soils different from its native habitat and/or nourishing upon external substances, particularly animals and certain plants which dispense with photosynthesis entirely or in part.

heteroxenous parasite. A parasite requiring several or different hosts for its complete development.

heterozygote. An organism possessing unlike genes with respect to any given allelomorphic pair governing the particular character. Such individuals produce two kinds of germ cells with respect to the genes involved.

heterozygous. Designating the condition in which an individual has both genes of an allelomorphic pair or two unlike genes of an allelomorphic series.

HETP. Hexacthyl tetraphosphate.

hewn timber. Timber finished to size by ax or adz although the ends are commonly sawn.

hexachloride of benzene. See *benzene hexachloride.*

hexachlorophene. A chemical compound which is an exceptionally persistent insecticide.

hexacthyl tetraphosphate. A chemical compound used as a pesticide for the control of aphids, leaf hoppers, thrips, and spider mites. Abbreviated *HETP.*

hexagonal planting. A system of planting trees in a grove or orchard at the angles of equilateral triangles.

hexamitiasis. An acute, infectious disease of turkey poults caused by the protozoan organism, *Hexamita meleagridis,* characterized by listlessness, droopiness, stilted gait, lowered temperature, watery and foamy diarrhea, emaciation, and death. Birds that survive the disease are stunted and emaciated.

hexenbesen. Witches'-broom.

Hiba arborvitae. Hiba false arborvitae.

Hiba false arborvitae. *Thujopsis dolabrata,* family Pinaceae, an evergreen tree grown as an ornamental. Native to Japan.

hibernate. To pass the winter or part of it in a dormant or quiescent state, as a bear. *Cf. estivate.*

hibiscus. Certain plants of the genus Hibiscus, family Malvaceae. See *cottonrose hibiscus; Chinese hibiscus.*

hick. Uncouth or rural backward person.

hickory. Any deciduous tree of the genus Carya, family Juglandaceae. Species are valuable as lumber, nut, and shade trees. The wood is frequently used where resiliency and toughness are desired, as in handles, etc. See *shellbark hickory; shagbark hickory; pecan; red hickory; mockernut hickory; bitternut hickory; nutmeg hickory; water hickory; pignut hickory.*

hickory bark beetle. *Scolytus quadrispinosus* Say, family Scolytidae, one of the most destructive insect pests of hardwoods in America. The adult beetles feed on twigs, stems, and leaves. The larvae mine under the cambium. It is found from Quebec to Georgia and west to Utah, U.S.

hickory horned devil. *Citheronia regalis* (F.), family Citheroniidae, a caterpillar which infests hickory, black walnut, butternut, sycamore, tupelo, ash, persimmon, sassafras, sumac, and cotton from Massachusetts to Texas, U.S.

hickory land. Land on which hickory (*Carya* spp.) was a dominant or characteristic growth and was valued accordingly as agricultural land by early settlers of eastern U.S.

hickory leaf roller. *Argyrotaenia juglandana* (Fern.), family Tortricidae, a caterpillar which rolls the leaves of hickory and feeds inside the rolled leaf.

hickory shuckworm. *Laspeyresia caryana* (Fitch), family Olethreutidae, an insect pest of hickory and pecan whose larva bores into the young nut causing a premature fruit drop. Also called *pecan shuckworm.*

hickory tiger moth. Hickory tussock moth.

hickory tussock moth. *Halisidota caryae* (Harr.), family Arctiidae, a moth which, in the caterpillar stage, infests deciduous trees and shrubs, especially walnut, butternut, apple, pear, and hickory.

hidden defect. A defect in a log or tree that cannot be detected prior to milling.

hidden hunger. Deficiency disease *q.v.*

hide. The tanned or untanned skins of animals, especially those of cattle, horses, sheep, and goats. See *kip; pelt: 1.*

hidebound. Designating an animal whose skin is very tightly fastened to its body, often resulting from poor feeding and emaciation.

hide cellar. A storage room in which hides are kept at temperatures of about 60° F. or less.

hide crop. The hides which are produced in one year.

hide of land. 1. In old English law, a measure of land that could be worked with one plow. 2. In old English law, the land which would support a family, the land belonging to a mansion house, or the mansion house itself.

hide peeler. A person who removes the hides from cattle that have died on the range.

hidrosis. Excessive sweating.

Higan cherry. *Prunus subhirtella,* family Rosaceae, a deciduous, small tree or shrub grown as an ornamental for its light pink flowers. Native to Japan. Also called *rosebud cherry.*

high. 1. Designating butter which has a very rich yellow color. 2. Designating wool that is finer than the average of its grade. 3. Designating a strong flavor or odor resulting from decomposition.

high acid flavor. 1. A flavor defect of milk characterized by a feeling of drastic cleanliness in the mouth. 2. A flavor defect of cheese characterized by a quick flavor sensation which rapidly disappears. 3. A flavor defect of ices and sherbets characterized by a tart flavor, sometimes strong enough to cause pain, which persists in the mouth for some time. 4. A flavor defect of ice cream characterized by a quick taste sensation and by a clean, fresh, acid taste in the mouth. 5. A flavor defect of butter characterized by a sharp, sour taste.

high action. The leg action in horses in which the feet are lifted a reasonable distance to prevent stumbling or foot interference.

high at the tailhead. Designating an objectionable type of conformation in cattle in which the tailsetting is higher than the rest of the back outline. When found on a cow it is sometimes thought to indicate barrenness or breeding difficulty.

high blower. 1. A horse which has broken wind. 2. A horse which snorts at each exhalation while galloping.

high blueberry. Highbush blueberry.

high bunt. See *bunt of wheat.*

highbush blackberry. *Rubus ostryifolius,* family Rosaceae, an erect shrub grown for its edible, black fruits. Also called *field highbush blackberry.* See *blackberry.*

highbush blueberry. *Vaccinium corymbosum,* family Ericaceae, a bushy shrub found in swamps and low woods. It produces a large, edible berry in the wild state and is the source of horticultural varieties in commercial production of blueberries. Found in northeastern U.S. Also called *high blueberry; swamp blueberry; tall blueberry; tall bilberry.*

highbush huckleberry. Black huckleberry.

high calcic product. Any fertilizer material in which 90 percent or more of the total calcium and magnesium oxide content consists of calcium oxide. See *lime: 1.*

high-climber. A person who tops and rigs spar trees or poles.

high-climbing beans. Pole beans.

high collected trot. See *high action.*

high color. A deep yellow butter color.

high condition. See *highly fitted.*

high count. In milk examination or grading, designating milk which has a high bacterial contamination and which should be graded down or rejected.

high cranberry. See *American cranberrybush viburnum.*

high-crested. Designating a coarse or roached condition of the neck of a horse which is especially undesirable in mares.

high degree of finish. See *highly finished.*

high-density bale. 1. A bale of cotton similar to standard-density bale (*q.v.*) except the average density is held at 36 lbs. per cu.ft. 2. A highly compacted bale of hay with a density of 12-15 lbs. per cu.ft.

high edge. A workmanship defect of cheese in which the finished cheese lacks the square edges of a well-finished cheese.

highest profit combination. In the production of a single product with varying and fixed resources, the combination of the factors of production in which (a) additional units of the variable factors exactly pay for themselves, *i.e.,* when marginal factor cost (*q.v.*) equals the marginal value product (*q.v.*) and (b) the revenue derived from the added units is declining.

high forest. A forest which originated from seed.

high grade. Designating a grade animal which has a large proportion of pure blood. See *grade: 4. a.*

high-grading. The removal of only the best trees from a stand.

high-headed. Designating an animal which holds its head higher than normal for its type or breed.

high-heeled boots. Boots, worn by western riders, U.S., which have no laces and a high heel for gripping the stirrup. They are often decorated with fancy stitching and/or embossed patterns. Also called *cowboy boots.*

high in the withers. An undesirable, coarse or heavy shoulder conformation found in some animals, such as cows.

Highland. Some cattle of Alpine Europe which are offshoots of the Brown Swiss.

Highland breed. Any breed of livestock which originated in the highlands of Scotland, such as Highland cattle and Black Faced Highland sheep (often incorrectly used for breeds from other mountain sections of the world).

high-life. Carbon bisulfide.

high-lime soil. A soil which contains lime in sufficient amount to cause effervescence when the soil is tested with an acid. Loosely used in humid regions for any soil that gives an alkaline reaction, presumably because of its lime content.

highly buffered. Designating a soil which has a high concentration of some buffering material. See *buffer: 1; lightly buffered.*

highly finished. Designating meat animals that have been heavily fed until they have a smooth, even covering of fat over the body and an excellent condition of hide and hair or wool covering. (Animals fitted for show or exhibit come in this class.)

highly fitted. Designating cattle that have been heavily fed until they have a very heavy layer of fat covering the natural fleshing.

high magnesic products. Any material in which more than 10 percent of the total calcium and magnesium oxide content consists of magnesium oxide. See *lime: 1.*

high-moor peat. A group of peat soils which are raw or undecomposed and usually strongly acidic, as opposed to low-moor peat which includes the decomposed and less acidic or alkaline, peats. Also called *hochmoor peat.*

high-protein. Designating a food or feed rich in protein.

high renewal system. A system of training grape vines on a trellis in which 3 wires are used: the lowest wire is 20 ins. from the ground and there are 20 ins. between wires. The trunk reaches the lowest wire and canes are trained to the right and left on it. As the shoots grow in the spring, they are loosely tied to the two upper wires.

high-roller. Designating a horse whose bucking action is higher than usual.

high school horse. A horse trained for performing certain relatively highly complicated routines.

high smut. Bunt of wheat.

high stepper. See *high action.*

high-stepping trot. A trot characterized by height and spring of stride.

high-temperature, short-time method. A method of pasteurization in which the liquid being treated is exposed to a temperature of not less than 160° F. for at least 15 seconds after which it is rapidly cooled. Also called *flash pasteurization.*

high-unnatural. Designating a color defect of American Cheddar cheese characterized by a very deep yellow color which may have an orange-yellow hue.

high viscosity. A buttermilk defect: thick body.

high water. A flood or a body of water approaching flood stage.

high-water mark. The level reached at the highest stage of a flood.

highway erosion control. The control of erosion on slopes of cuts, in ditches, and at cross drains within a highway right-of-way.

high wind. A windstorm; a strong wind.

high yields. Large yields.

hill. 1. A natural relief feature, a rounded elevation a few feet to a few hundred feet above the adjacent lowland which is distinguished from other features such as mesa, butte, and peak by its form and from mountain by its lesser magnitude. Locally without consistency, relief features in a prevailingly flat region which are little more than hummocks received the name *hill,* and features no more than a hundred feet high the name mountain. On the other hand, relief features of magnitude in a mountainous region may receive the name *hills,* as the Black Hills of South Dakota, U.S. 2. An artificially raised elevation: *e.g.,* a row in a field or garden which is slightly raised above the surrounding soil or a small mound of soil in which seed may be planted. 3. A plant or a cluster of plants, as a *hill* of potatoes, a *hill* of beans. 4. To make soil into a *hill* or to form soil up around a plant.

hillbilly. A person who lives in rural, hill, or mountainous areas (jocular and derogatory).

hill clematis. Western virginsbower.

hill country. Hilly or mountainous regions.

hill culture. The management of trees, shrubs, and perennial plants for protection and the economic utilization of slopes, especially for the prevention of destructive erosion.

hill-drop. 1. A planting method in which several seeds are dropped at definite intervals so that plants grow in clusters or hills. 2. A planter plate designed to collect seed for hill planting.

hiller hoe. A cultivator point which serves to heap soil around plants in the row.

hill gooseberry. Downy rosemyrtle.

hill indexing. Tuber indexing.

hill-land. Land characterized by strong slopes and relief in comparison with associated plains or relatively flat land; rolling upland in contrast to flat, bottom land.

hillock tree. Dotted melaleuca.

hill planter. See *checkrow planter.*

hillrow system. See *hill system.*

hillside. The side slope of a hill or ridge.

hillside ditch. A shallow ditch across a slope in a field which is used to control water movement and thus reduce scour soil erosion. See *graded terrace; diversion ditch.*

hillside plow. A plow in which the moldboard and share are hinged underneath the frame to be reversed either to the right or to the left. All the furrow slices can be thrown in one direction: *e.g.,* upslope.

hillside-type combine. A combine which has a leveling device so that the thresher part can be kept essentially level on sides of hills.

hill system. A system of planting and growing strawberries, raspberries, and other small fruits in which the plants are spaced apart in the row and daughter plants are not permitted to develop by means of runners of suckers. Also called *hill-row system.* See *hedgerow system.*

hilly. Designating a land surface that is broken in aspect and exhibits considerable differences in elevations of local relief features. Greater relief is implied than in the terms undulating and rolling and less relief than in mountainous. It is usually distinguished from sharply dissected and badlands types of topography.

hilo grass. Sour paspalum.

Himalaya berry. *Rubus procerus,* family Rosaceae, a woody, perennial-stemmed, creeping, evergreen shrub grown on trellises in the Pacific Coast states, U.S., for its edible, blackberry-like fruit. Native to Europe.

Himalayan cotoneaster. *Cotoneaster frigida,* family Rosaceae, a tall shrub or small tree grown for its white flowers and showy, red fruits. It is often used for foundation planting. Native to the Himalayas.

Himalayan desert candle. *Eremurus himalaicus,* family Liliaceae, a perennial herb grown for its long, white flower clusters. Native to the Himalayas.

Himalayan fleabane. *Erigeron multiradiatus,* family Compositae, a perennial herb grown in the rock garden for its very showy, purple flowers. Native to the Himalayas.

Himalayan musk rose. Browns rose.

Himalayan pine. *Pinus griffithi,* family Pinaceae, a tall evergreen tree grown as an ornamental. Native to the Himalayas.

Hinckley system. A system of cultivation for orchards which uses the same irrigation furrows year after year, only very shallow hoeing for weed control, only concentrated inorganic fertilizers, and in which the dead leaves are allowed to accumulate on the ground.

hind. See *hindquarter.*

hind cinch. The rear cinch strap that encircles a horse to keep the saddle from tipping up when roping. In normal use, saddles have only one cinch that is fastened just behind the forelegs.

hindgate. See *endgate.*

hind leg. In quadrupeds, either of the rear legs.

hindquarter. Either of the two rear quarters of a meat carcass starting with, or behind, the thirteenth rib. Sometimes called the *hind saddle.*

hind saddle. Hindquarter.

Hinds walnut. *Juglans hindsi,* family Juglandaceae, a deciduous tree grown as a street tree and as a stock for the English (Persian) walnut. Native to California, U.S., where it is often called *California black walnut.*

hind shank. In a meat carcass, the lower portion of the rear leg next to the pastern and hoof.

Hindukush aster. *Aster diplostephioides,* family Compositae, a perennial herb grown for its blue or pale purple flowers. Native to the Himalayas.

Hindu lotus. *Nelumbium nelumbo,* family Nymphaeaceae, an aquatic herb grown in a shallow pool for its stately leaves and its large, fragrant, pink or white flowers. Native to Asia. Also called *East Indian lotus; Egyptian lotus; Egyptian bean; Indian lotus.*

hinny. The offspring of a stallion and a jennet. *Cf. mule: 1.*

Hinoki falsecypress. *Chamaecyparis obtusa,* family Cupressaceae, a hardy, evergreen tree grown for ornament. It has many varieties. Native to Japan.

hip. 1. Cynarrhodion; a fruit which is fleshy, hollow, and enclosing achenes. 2. The external angle formed by the meeting of two sloping sides of a roof. 3. That region of one of the rear quarters of four legged animals where the hind leg joins the pelvic region. In an animal such as a cow the *hips* should be wide, approximately level laterally with the back, and free from coarseness or excess tissue.

hip bone. See *hip: 3.*

hipdown. A fracture of the prominence of the hip and a falling away in that area. Also called *hip shod; knocked-down hip.*

hip loin. A cut of meat from the rear end of the loin.

hipped. Designating a horse which has one hip lower than the other.

hippy. Designating an animal which has angular, prominent hipbones.

hip roof. A roof so constructed that rain falling on it drains off all four sides. If the building is square the four hip rafters meet at the center.

hip roof house. See *Gambrel roof house.*

hips. See *hip: 3.*

hips and loin broad and full. In stock judging, good dairy type about the rear quarters.

hips are neatly tucked in. In livestock judging, well-conformed rear quarters on an animal, not coarse or rough.

hip shod. Hipdown.

hip sweeny. A horse with atrophy of the muscles of the hip.

hired help. One or several unskilled or semiskilled persons.

hired man. A man who works on a farm for wages.

hire out. To get oneself employed.

hirsute. Covered with coarse hairs.

hiryu azalea. *Rhododendron obtusum,* family Ericaceae, a tender half-evergreen shrub which has been widely used in breeding and hybridization. Native to Japan.

histamine. An alkaline-type of chemical compound composed of carbon, hydrogen, and nitrogen which is probably formed in the animal body, mainly in the intestinal tract, by the action of bacteria on the amino acid, histidine. If it is absorbed into the blood stream and circulated through the body, it may produce a fall in blood pressure, spastic contraction of the smooth muscles, and other symptoms, some of which may be relieved by the antihistamine drugs.

histology. The science of the microscopic structure of plant and animal tissue.

histomoniasis. See *blackhead.*

histopathologic. Designating abnormal changes in body structures, as observed by microscopic examination of sections of abnormal or diseased tissues.

historical area. An area set aside by a governmental agency to preserve interesting evidences of the life and activities of people who have lived at an earlier period.

hitch. 1. A catch; anything that holds, as a hook; a knot or noose in a rope, which can be readily undone, intended for a temporary fastening. 2. a. The connecting of an implement, vehicle, etc., to a source of power, as a tractor, team, etc. b. The device which is used to make such a connection. 3. A horse or horses used to pull an implement or vehicle. 4. The stride of a horse when one of the hind legs is shorter than the other. 5. To fasten an animal to a post, rail, etc.

hitching strap. A leather strap attached to the bridle which is used to fasten a horse to a post, rail, etc.

hitch pin. The pin which fastens the doubletree to the tongue of an implement or vehicle. Also called a *drop pin.* See *evener pin.*

hitch rack. A railing to which a horse is tied.

hitch your horse to the ground. To let one's reins drop to the ground for hitching purposes. Horses are trained to stay at a place when the reins are dropped on the ground.

hive. A home for honeybees which is provided by man; it usually consists of a base, removable supers, and a top.

hive bee. A honeybee which readily lives and produces honey in a hive.

hive body. That part of the hive which contains the honeycombs and which may be used as a brood chamber or a honey super. See *brood chamber.*

hive of bees. A colony of honeybees living in a hive.

hiverman. A person who stayed the winter through in a certain locality (southwest U.S.).

hive stand. A low platform on which a beehive is placed.

hive tool. A tool, used by beekeepers, made of steel having one end straight and sharpened to permit it to slip readily between supers or bodies in order to separate them. The other end, bent at a right angle and sharpened, is used for prying the frames apart. Usually, it has a V-shaped hole for pulling nails. Either end can be used for scraping combs or propolis from the hive parts, cleaning the bottom boards, etc.

hoar frost. Frost which is accompanied by a visible deposit of frozen moisture on trees as well as on ground vegetation. See *frost; black frost.*

hoarhound. 1. *Marrubium vulgare,* family Labiatae, a perennial, aromatic herb grown as a source of aromatic oil. Native to Eurasia. Also called *hound's bane; horehound; common hoarhound.* 2. See *black ballota.*

hoary alder. See *speckled alder: 1.*

hoary alyssum. Hoary falsealyssum.

hoary falsealyssum. *Berteroa incana,* family Cruciferae, an annual or biennial herb which is a weed pest of lawns, roadsides, meadows, and pastures. Native to Europe. Also called *hoary alyssum; gray berteroa.*

hoary rosemarymint. *Poliomintha incana,* family Labiatae, a shrub grown for stabilizing drifting sand because of its dense surface root system. It has some value as a browse plant. Native to North America. Also called *bushmint.*

hobble. 1. A thong which couples an animal's forelegs together to restrict its movements. 2. To tie an animal's forelegs so as to prevent its straying.

hobblebush. Hobblebush viburnum.

hobblebush viburnum. *Viburnum alnifolium,* family Caprifoliaceae, a hardy deciduous shrub grown for its fine, autumnal, claret-red foliage and its white flowers. Native to North America. Also called *American wayfaring tree; witch-hobble; tangle-legs.*

hobbled stirrup. One of a pair of stirrups which is tied down under the belly of the horse.

hock. 1. The region of the tarsal joint in the hind leg of a horse or other quadruped, corresponding to the ankle in man. 2. The lower joint section of a ham.

hock boot. A boot made of leather faced with felt to be strapped on the hocks of animals to prevent injury when traveling in railroad cars or trailers.

hock disease. Perosis *q.v.*

hock hobbles. Hobbles that are used to fasten the hocks of the hind legs of a mare to prevent kicking preparatory to coitus.

hoe. 1. A long-handled hand tool to one end of which is attached, at right angles, a light cutting blade. It is used to control weeds, to cultivate garden and field crops, and to perform other tillage operations. 2. To use a *hoe* in a garden or field. See *chop: 4.*

hoe culture. A primitive type of agriculture in which the hoe was the principal implement or tool.

hoedown. An old-fashioned, riotous country dance. Also called *breakdown; barn dance.*

hoe fork. A short-handled hand tool resembling a hoe, but which has the cutting blade divided into about 6 flat-pointed tines. It is used to dig roots or tubers from the hill or row. Also called *potato fork.*

hoe furrow opener. A part of a planter which consists of a single- or double-pointed shovel fastened to the lower part of the planter boot.

hoe hand. A hired laborer whose employment consists of hoeing row crops, especially in cotton culture. Also called *hoer* (jocular).

hoeing. The removal of weeds from a row crop with a hoe.

hoe-man. Farmer (southwestern U.S.).
hoe-out. 1. To complete the hoeing of a field. 2. To remove weeds from row crops with a hoe.
hoer. See *hoe hand.*
Hoffman frame. A frame for a beehive with end bars wide enough for a short distance at the top so that they touch and are automatically spaced.
hog. See *swine.*
hogal. A turpentining tool with a strong, elliptical blade used to slab off bark preparatory to seating a cup and to cut the first streak.
hogback. 1. A low, narrow, but often rounded, crested ridge, such as an esker in the glaciated region of the U.S. 2. In western U.S., any hill or mountain ridge which has a sharp, rocky crest.
hog back. Of, or pertaining to, a horse with an arched or convex back. *Cf. sway-back: 2.*
hog bristles. The coarse, stiff hairs on swine which are used in the manufacture of brushes.
hog cabbage palm. Sargent cherrypalm.
hog casing. See *casing: 2.*
hog cholera. An acute, contagious, virus disease of swine characterized by sudden onset, fever, and high morbidity and mortality.
hog-corn price ratio. The ratio between the price of corn and the price of live hogs determined by the number of bushels of corn which is equal in price to 100 lbs. of live hogs. Also called *corn-hog ratio.*
hog down. To pasture hogs on a crop grown for stock feed, thus eliminating the harvesting process. Also called *hog off.*
hog-dressed. Designating a dressed lamb or calf carcass with the head and pelt left on but with the feet and viscera removed.
hog flu. Influenza in swine *q.v.*
hog follicle mite. *Demodex phylloides* Csokor, family Demodicidae, the demodex mange mite, a minute, skin parasite of swine which lives in hair follicles and sebaceous glands on the face, base of the tail, and the inner sides of the legs; it produces an inflammation of the skin which appears as small, hard pimples in size up to a marble. Upon breaking, the pimples discharge a yellowish-cheesy pus. Demodectic mange is rare in hogs in the U.S.
hogged-mane. A closely-clipped horse's mane.
hogget. A sheep from weaning until its first shearing.
hogg(et) wool. The first fleece of a full year's growth that is sheared from the sheep (British).
hogging down. See *hog down.*
hogging off. See *hog down.*
hog hair. Hog bristles.
hog holder. A tool which consists of a short iron handle on either end of which is a fixed metal ring. The ring is pushed over the swine's snout and into its mouth and leverage is applied to control the animal.
hog hook. A tool which consists of a short, iron shank with a ring or handgrip on one end and a sharpened hook on the other. It is used in hog slaughtering and dressing to lift, move, and hold the carcass.
hog house. Any structure used in farrowing or for shelter for swine.
hog lard. See *lard.*
hog lot. An enclosed area in which hogs are kept when not on pasture.
hog louse. *Haematopinus suis* (L.), family Haematopinidae, the louse of the European wild boar, also found on domesticated swine in central Europe. *H. adventicius chinensis* Fahrenholz, 1916, is the common sucking louse of domesticated swine in North America.
hog millet. Proso *q.v.*
hog off. See *hog down.*
hog oiler. Any of several different devices which contain oil and which, when rubbed by a hog, release the oil onto its skin, destroying skin parasites.
hog pasture. A pasture on which swine feed.
hog-plum. See *mombin.*
hog-price cycle. Fluctuation in hog prices associated with a cycle in hog production that averages four to five years from one low to the next low.
hog ring. A metal ring fastened in the cartilage of a hog's nostrils which prevents the hog from rooting up a field, under fences, buildings, etc.
hog ringer. A pincers-like device used for inserting a hog ring in a swine's snout.
hog ringing and breeding crate. A narrow chute which has swinging gates at one end. The hog is driven into the chute and as it attempts to go through the gates at the other end its head is caught between them. It is used to confine swine for ringing, castrating, breeding, treating, etc.
hog-scalding kettle. A large cast-iron kettle used on farms to hold boiling water for scalding the hog carcass.
hog scraper. See *bell scraper.*
hogshead. 1. A liquid measure which contains 63 U.S. gallons or 52½ imperial gallons. 2. A cask which contains from 100 to 140 gallons, as used for molasses, etc. 3. A very large cask or barrel often containing as much as 1,000 to 2,000 lbs. of cured tobacco. It was used in marketing tobacco especially in early and pre-Civil War U.S.
hog snare. A hand tool for controlling a hog which consists of a hollow, metal handle through which a wire loop is passed. One end of the wire is fastened to the handle, and the other goes through the handle to a grip. The loop is placed around a swine's snout, the grip is pulled, and the wire tightens around the snout.
hog style carcass. See *hog-dressed.*
hog tag. A metal tag fastened to a hog's ear on which is stamped identification information.
hog-tight fence. Any type of fence which is sufficiently strong and tight enough to confine or restrain hogs.
hog trap. See *hog ringing and breeding crate.*
hog trough. 1. A rack into which a slaughtered hog is placed to hold the carcass on its back to facilitate skinning. Also called *skinning rack.* 2. A trough set low to the ground used for feeding and watering swine.
hog-wallow land. Flat land characterized by nat-

ural microbasins and mounds, frequently no more than two or three ft. in relief. (A vernacular term, mostly southern U.S., without a very precise meaning; used especially in Texas to describe parts of black clay land prairies.) See *gilgai.*

hogweed. See *ragweed; cowparsnip.*

hog wire. A special type of barbed wire having four- rather than two-pointed barbs at intervals along its length; it weighs about 400 lbs. per mile.

hog wool. See *hogg(et) wool.*

hogwort. See *woolly croton.*

Hohenheim grazing system. A European system of pasture management which involves (a) dividing the pasture into several areas, (b) dividing the grazing herd according to production, (c) frequent rotation of the animals, and (d) intensified fertilization of the pastures.

hoist. 1. Any of several different types of hand- or power-operated devices which may be used to lift heavy weights, as the block and tackle and hydraulic pistons common to tractor-powered hoists. 2. To lift heavy weights.

holard. The total water content of the soil regardless of whether it is available to plants or not.

holcus spot. A bacterial disease of corn, sorghum, and Sudan grass caused by *Pseudomonas syringae,* family Pseudomonadaceae; characterized by tan, red-bordered, rough to elliptical lesions on the leaves. In time the lesions coalesce to form irregular blotches.

hold. 1. To restrain animals to a particular place. 2. To remain on the tree until mature, as a fruit. 3. To retain, as certain soils *hold* moisture. 4. To maintain condition, as a steer which *holds* flesh. 5. To store or retain in storage, as to *hold* eggs. 6. Not to market at harvest time or when an animal is fat, but to wait to sell for a better price.

hold back. See *hold: 6.*

hold-back strap. The strap of a harness used with wagons, buggies, etc., that is hooked to the shaft and serves to hold the vehicle off the animal and to aid in backing. Also called *britchin.*

Holderness. An early type of English cattle that had some influence in the early development of Shorthorns.

hold his top up well. A term applied to sheep in stock judging: to have a strong and sturdy back formation.

holding. An indefinite amount of land which is usually considered large in proportion to the average size of farms or ranches in its vicinity.

holding alley. A passageway in a milking parlor system or a section of a barn, shed, or open lot, where cows may wait their turns to be milked. Also called *holding area.*

holding board. A board to which cords and weights are attached, used as an operating table in caponizing.

holding chute. A chute in which an animal may be restrained.

holding funnel. A funnel-like device sometimes used to hold a chicken to prevent bruising while killing. The bird is placed head downward in the device so that its head protrudes through a narrow opening at the bottom.

holding method. A method of pasteurization in which milk is preheated to about 144° F. and then pumped to a holding tank where this temperature is maintained for 30 minutes. From the holding tank it flows over a surface cooler consisting of pipes through which brine flows where the temperature is reduced to 50° F. or lower. See *low-temperature, long time.*

holding paddock. Holding pen.

holding pen. A small pen close to the yards, sheds, etc., in which animals are restrained but not fed.

holding storage. Refrigerated or otherwise cooled storage for plants used to prevent blossoming before the desirable time.

holding temperature. The optimum temperature for the successful storage of a product.

holdover. 1. Designating a product which is not sold on the market on one day and is retained for sale the next day. 2. Designating an insect brood which fails to emerge normally in the late summer but emerges the following spring. 3. Designating a tree which escapes logging and occupies a dominant position in the stand. 4. Designating seed or produce which is neither sold nor used but held in storage until some future date.

holdover canker. A canker on a plant in which enough bacteria may remain alive through the winter to resume activity and spread the following spring when growth starts.

hold up milk. In a cow, to cause a cessation in milk secretion as a result of undue excitement at milking time. See *let down milk.*

hole. 1. In western U.S., a depression which results from a natural widening or expansion of a deep valley. 2. A small bay, inlet, or creek. 3. See *sinkhole.*

holiday disease. Azoturia.

Holland bulbs. Dutch bulbs.

Holland cattle. Holstein-Friesian *q.v.*

Holland elm. Dutch elm.

Hollands acre. A land measure of slightly more than two acres; at one time in use in New York, U.S.

hollow. 1. A relatively small, open depression between hills, or a small re-entrant tributary valley in hill country which may or may not contain a stream. 2. Any of variously shaped, shallow concavities in a plain surface.

hollow apple. See *cork: 3.*

hollow-back. A weak back conformation which is objectionable in most animals. See *sway-back: 2.*

hollow-cone nozzle. A nozzle which emits water in a rotating, hollow cone appearing first as a sheet, then, farther from the nozzle the spray breaks into threads and, finally, droplets.

hollow face. Concave profile, as applied to horses. Also called *dish face.*

hollow heart of potato. A nonparasitic condition of the potato tuber characterized by an internal split or cavity or a central cavity lined with dark brown dead cells. It results from rapid growth induced by an abundance of moisture and plant food.

hollow horn. An imaginary disease arising from the erroneous belief that loss of appetite and listlessness in a cow was due to *hollow horns.* The remedy was supposed to be (a) boring a hole in each horn just above the horn line, (b) filling the cavity with salt, sugar, and pepper, and (c) plugging the hole with a wooden peg. The belief was that if the cow had *hollow horn* this remedy would cure her, and if she did not have *hollow horn,* the remedy would prevent her getting it. See *hollow tail.*

hollow knot. A knot that is apparently sound but which has a hole larger than ¼ in. in diameter. See *knot: 4.*

hollow stalk. A nonparasitic condition of celery which results in pithiness of the stalk due to a check in growth, poor seed, or too rapid growth.

hollow tail. An imaginary disease of a cow similar to hollow horn (*q.v.*) but which affected the tail. The remedy was supposed to be the same as for hollow horn, except that the hole was made in the tail. Also called *wolf in the tail.*

holly. Any shrub or tree of the genus Ilex, family Aquifoliaceae. Species are grown for ornament. See *American holly; Chinese holly; dahoon; droopbead holly; English holly; finetooth holly; inkberry; Japanese holly; mountain winterberry; smooth winterberry; winterberry: 1; yaupon.*

holly fern. See *house-hollyfern.*

hollyhock. *Althaea rosea,* family Malvaceae, a perennial, tall herb grown as a biennial for its terminal flower clusters of many different colors. Native to China.

hollyhock mallow. *Malva alcea,* family Malvaceae, a perennial herb grown for its pink or white flowers. Escaped, it can be a weed pest. Native to Europe. Also called *European mallow; vervain mallow.*

hollyhock rust. A rust disease caused by *Puccinia malvacearum,* family Pucciniaceae; characterized by the undersurfaces of the leaves being almost covered by tan or purplish, spore-bearing pustules.

holly laurel. Hollyleaf cherry.

hollyleaf cherry. *Prunus ilicifolia,* family Rosaceae, a broad-leaved, evergreen shrub or small tree grown for its holly-like leaves, its small, white flowers, and its dark red to nearly black, edible fruit. Native to North America. Also called *evergreen cherry; holly laurel; islay.*

holly leaf miner. *Phytomyza ilicis* (Curt.), family Agromyzidae, an insect which infests holly causing yellow or brown serpentine mines or blotches on the leaves.

holly oak. *Quercus ilex,* family Fagaceae, a broad-leaved, evergreen tree grown in warm areas for its ornamental foliage. Native to southern Europe. Also called *holm oak; evergreen oak.*

holly olive. Holly osmanthus.

holly osmanthus. *Osmanthus ilicifolius,* family Oleaceae, a broad-leaved, evergreen shrub grown for its foliage and its fragrant, white flowers. Native to Japan. Also called *holly olive.*

holly rancher. One who raises holly for the market (Pacific Coast, U.S.).

holly scale. *Aspidiotus britannicus* Newst., family Diaspididae, a scale insect which infests holly on the West Coast, U.S. Heavily infested trees are weakened.

holm oak. Holly oak.

Holstein. See *Holstein-Friesian.*

Holsteiner. A dry sausage which is composed of 65 percent beef and 35 percent pork chopped medium fine, seasoned, and stuffed into beef casings. The ends of each sausage are tied together. Also called *horseshoe sausage.*

Holstein-Friesian. A widely used dairy breed of cattle. The name is American in origin but applies to cattle originally imported largely from the province of Friesland in Holland. They are black and white and the mature cows weigh from 1,100 to 1,800 lbs. They are noted for their high production of milk with a low butterfat content.

Holstein Health. A sour, skimmed milk, cooked cheese made in Germany.

Holstein Skim-Milk. A skimmed milk cheese, cured five to six months, made in Germany.

Holsts snapweed. One species of snapweed. See *snapweed: 1.*

holy clover. Sainfoin.

holy fire. Ergotism.

holy grass. See *sweetgrass: 2.*

holy thistle. See *blessed milkthistle.*

home. 1. Designating a person's permanent place of residence, as *home* county. 2. Designating a product made in the *home,* as contrasted to a factory-made product. 3. Designating a product used in the *home,* as a *home* freezer.

home brand. See *brand: 1.*

home demonstration agent. See *county agent.*

home freezer. 1. A freezer used for home storage of frozen products. See *freezer: 1.* 2. A small, hand-turned ice cream freezer.

home garden. A relatively small vegetable garden which is on the grounds of one's residence.

home-grown. 1. Designating crops, vegetables, or animals which are raised on one's property, as contrasted to purchased products. 2. Designating produce raised locally in contrast to that shipped in from a distance.

homemade soap. Soap made on the farm or in the home from waste fats, such as tallow and lard, treated with the water extract of hardwood ashes (lye) which under heat produces a satisfactory soap. Also called *soft soap.*

home-process olives. Ripe manzanillo or Mission varieties of olives which are pink when harvested. They are soaked in a succession of lye solutions, washed, and soaked in brine. At the end of the process they are golden-yellow.

home ranch. The headquarters for a ranch (western U.S.).

home site. The plot of land on which one's house is situated or a suitable location for a home.

homestall. A farm yard (England).

homestead. 1. A home place; especially a farm dwelling and the surrounding ground. 2. In the U.S., a unit of public land, usually not more than 160 acres, acquired by an individual under the Homestead Act of Congress, 1862, and subsequent acts. See *homesteading.*

homesteading. In the U.S., the process of securing a unit of virgin, undeveloped federal land under the regulations of the Federal Homestead Act of 1862. This was a war measure adopted following the Civil War to help the jobless and displaced people secure land and make a living. Certain states later passed laws making state owned lands similarly available.

homestead rights. The widow's or widower's rights in the family dwelling.

home use. In appraising, the plus or minus values of a farm above or below its capitalized net income value; the human satisfactions and conveniences which are derived from the use of the property as a home.

hominy. See *hominy feed; hominy grits.*

hominy chop. Hominy feed.

hominy feed. A livestock feedstuff which is a mixture of corn bran, corn germ, and part of the starchy portion of either white or yellow corn kernels or a mixture thereof as produced in the manufacture of pearl hominy, hominy grits, or table meal. It shall contain not less than 5 percent of crude fat. If prefixed with the words white or yellow the product must correspond thereto. Also called *hominy chop; hominy meal.*

hominy grits. A stockfeed and human food which consists of the fine or medium-sized, hard, flinty portions of sound corn containing little or none of the bran or germ. Also called *corn grits.*

hominy meal. Hominy feed.

homogamous. Designating a condition of homogamy *q.v.*

homogamy. 1. The simultaneous maturation of male and female elements of the same flower. *Cf. dichogamy.* 2. In flowering plants, bearing one kind of flower throughout, as members of the family Cichoriaceae.

homogeneity. The state of being homogeneous.

homogeneous. Being of uniform character or nature throughout.

homogenized milk. Milk which has been subjected to high pressure in a special, triple-piston machine to break up the fat globules, thus producing a completely nonseparating product, uniform throughout.

homogenizer. A high-pressure, triple-pump machine which renders products, such as milk, homogeneous by breaking up the fat globules.

homogenizer valve. The unit assembly of a homogenizer in which fat globules are broken up during passage under high pressure.

homologous chromosomes. Chromosomes occurring in somatic cells in pairs which are similar in size, shape, and supposedly in function, one being derived from the male and one from the female parent.

homothallic. Designating a fungus whose mycelium produces hyphae functionally male or female.

homozygote. An organism whose chromosomes carry identical membranes of any given pair of allelomorphs. The germ cells are, therefore, all alike with respect to this locus, and the individual will breed true.

homozygous. Possessing identical genes with respect to any given pair or series of alleles.

honda. A rope, rawhide, or metal loop at one end of a lariat through which the other end of the lariat is pulled to make a noose or large loop.

hone. 1. The whetstone, carborundum stone, metal, or other device which is used in sharpening knives, scissors, or other cutting tools. 2. To sharpen a knife.

honesty. Either herb of the genus Lunaria, family Cruciferae. See *dollarplant; perennial honesty.*

honey. 1. An aromatic, viscid, sweet food material derived from the nectar of plants through collection by honeybees; modified by the bees into a denser liquid and finally stored in honeycombs. Of acid reaction, liquid in its original state, it becomes crystalline on standing. *Honey* consists chiefly of two simple sugars, dextrose and levulose, with occasionally more complex carbohydrates, with levulose usually predominant, and always contains mineral, plant coloring materials, several enzymes, and pollen grains. 2. Legally, the nectar and saccharine exudation of plants, gathered, modified, and stored in the comb by honeybees, which is levorotatory, and contains not more than 25 percent of water, not more than 0.25 percent of ash, and not more than 8 percent sucrose.

honey agaric. The fungus, *Armillaria mellea,* which causes mushroom root rot *q.v.*

honey ball melon. Honeydew melon.

honeybee. *Apis mellifera* L., family Apidae, a very important economic insect which produces honey and acts as a pollinator of the flowers of many plants both wild and cultivated.

honey bell. *Mahernia verticillata,* family Sterculiaceae, a shrubby plant grown as a hanging basket plant for its sprawling habit and its fragrant, yellow flowers. Native to South Africa.

honeyberry. *Melicocca bijuga,* family Sapindaceae, a tree grown in frost-free areas for its edible, juicy, sweet fruit. Native to tropical America. Also called *mamoncillo; genip; Spanish lime.*

honey-bloom. Spreading dogbane.

honey bound. A condition under which the queen in a beehive is restricted in her egg-laying by the fact that all or most of the cells in the brood chamber are filled with honey.

honey butter. A mixture of creamery butter and 20 to 30 percent table quality, liquid honey used as a spread.

honeycomb. 1. The waxy structure in which honey is stored by bees. See *comb:* 2. 2. A natural arrangement of the soil mass in more or less regular five- or six-sided sections separated by narrow or hairline cracks which is usually found as a surface structure or arrangement. 3. See *reticulum.*

honeycomb filler. An egg-case filler which has 36 compartments or cells in rows of 6; the most common type of filler used in packing eggs in the standard 30 dozen case.

honeycomb frost structure. Frozen soil which has a granular structure or is easily broken into fragments, associated with shallow freezing and granular soils high in organic matter.

honeycombing. The development of checks in the interior of a piece of wood not often visible at the surface due to drying stresses usually along the wood rays.

honeycomb ringworm. A fungous skin disease of horses having the same manifestation as favus. See *favus: 2.*

honeycomb stomach. Reticulum *q.v.*

honey cream. A food product which is a mixture of fine quality extracted honey, previously heated to melt all crystals and to destroy yeasts and enzymes (to prevent fermentation and avoid rancidity), mixed hot in proportion of 42 percent honey to 58 percent high-test cream (75 to 80 percent butterfat).

honey crop. The amount of honey gathered from the hives.

honeydew. 1. A sweetish, sticky exudate which drops from the leaves and petioles of certain plants. It is attributed to infestation by sucking insects, excessive plant turgescence, or to a superabundance of sap. It often occurs during hot weather. 2. The excretion of certain homopterous insects, such as aphids, leafhoppers and scale insects.

honeydew honey. Honey which usually exhibits plus rotation at 20° C., a high plus rotation of the inverted solution at 87° C., an ash content much greater than 0.25 percent, and high nonsugar solids; usually characterized by a very dark color and a peculiar molasses-like flavor.

Honeydew melon. A very famous variety of the muskmelon *q.v.*

honey dipper. A person who removes human excrement from privies, cesspools, and septic tanks. In some sections of the world he is very important, serving as a supplier of night soil *q.v.*

honey extractor. A machine used for the removal of honey from the cells of honeycombs by centrifugal force.

honeyflow. The incoming of nectar to the beehive, used especially in reference to the periodic changes in quantity related to the blooming periods of dominant flowering plants of the area, such as the clovers.

honey-flower. Bigleaf honeybush.

honey gate. A metal faucet used for drawing honey from tanks and other receptacles.

honey house. A building where honey is extracted and supers are stored.

honey ice cream. A frozen food product; ice cream in which honey has been used to replace 50 to 100 percent of the sugar in the mix.

honeylocust. 1. Any thorny, deciduous tree of the genus Gleditsia, family Leguminosae. See *waterlocust; Japanese honeylocust.* 2. *G. triacanthos,* a thorny, deciduous tree grown as a shade tree, in shelterbelts, and for its wood used in fence posts, rails, furniture, and interior trim. Native to North America. Also called *three-thorned acacia.* 3. See *mesquite.*

honey-mushroom root rot. Mushroom root rot.

honey plant. Any plant from which honeybees gather nectar and pollen, especially one which either is in abundance or one which gives a distinctive flavor to honey. See *bee plant.*

Honey Rock. A variety of muskmelon *q.v.*

honey root rot. Mushroom root rot.

honey spoilage. Fermentation of honey.

honey stomach. An enlargement at the posterior end of a bee's esophagus; the sac in which a bee carries nectar or honey. Also called *honey sac.*

honeysuckle. Any shrub or woody climber of the genus Lonicera, family Caprifoliaceae. Species are grown as ornamentals and in hedges. See *bearberry honeysuckle; belle honeysuckle; coralline honeysuckle; European fly honeysuckle; everblooming honeysuckle; hairy honeysuckle; Henry honeysuckle; Ledebour honeysuckle; lilac honeysuckle; Manchurian honeysuckle; privet honeysuckle; sweet honeysuckle; Tartarian honeysuckle; Tibet honeysuckle; winter honeysuckle; woodbine honeysuckle.*

honeysuckle clover. See *red clover.*

honeysuckle sawfly. *Zaraea inflata* Nort., family Cimbicidae, an insect whose larvae feed on the foliage of the honeysuckle often causing defoliation.

honey tree. Japanese raisintree.

honey vinegar. Vinegar made from unmarketable honey or from honey washings.

honeywort. See *Greek honeywort.*

hooded barley. A type of barley (*q.v.*) in which a trifurcate structure called a hood appears in place of the awn.

hooded pail. A pail having a permanent metal cover across about two-thirds of the top, used in milking to protect the milk from falling foreign materials.

hood(ed) shield. A shield or fender attachment for a cultivator which consists of an open-ended, metal hood used to prevent young plants from being covered with soil during early cultivations.

hoof. The hard, horny, outer covering of the feet of horses, cattle, sheep, goats, and swine.

hoof and horn meal. A fertilizer product resulting from the processing, drying, and grinding of hoofs and horns which contains 10.7 to 15.6 percent nitrogen (equivalent to 13 to 19 percent ammonia).

hoof-and-mouth disease. See *foot-and-mouth disease.*

hoofbound. A condition of horses, mules, etc., in which the hoof is dry and contracted, causing pain and occasional lameness.

hoof gauge. An instrument used to measure the angle of a gaited horse's foot.

hoofhead. The attachment of the hard bony shell of the hoof to the skin and inner, softer parts of the foot of animals, such as horses. Also called *coronary band.*

hoof knife. A curved knife used to cut off excess sole and frog in horseshoeing or in trimming an animal's hoof. Also called *farrier's knife.*

hoof nipper. The cutters used to trim off excess horny material of the hoof in horseshoeing.

hoof rasp. A coarse file used for smoothing and removing hoof material from hooves, etc., especially in horseshoeing.

hoof shovel. A cultivator shovel which has a triangular point.

hook. 1. A piece of metal or other hard material, made in the shape of a curve, used for catching, holding, or sustaining, as a bale *hook,* gate *hook,* etc. 2. The hip of an animal. 3. See *haw: 1.* 4. The downward curve of the cranberry blossom. 5. The angle or jog in the stem or branch of a plant which results from pruning or pinching.

hooked bristlegrass. *Setaria verticillata,* family Gramineae, a grassy weed pest. Also called *bur bristlegrass.*

hook gauge. A pointed hook attached to a graduated staff or Vernier scale for accurately measuring the elevation of the surface of still water. The hook is submerged, and then raised until the point makes a pimple on the water surface.

hooks in the eye. See *haw: 1.*

hook-up. See *hitch.*

hookworm. Any of certain internal, parasitic worms of animals and man which belong to the family Ancylostomatidae. The larvae, which are voracious bloodsuckers of the small intestine, are capable of penetrating the skin to gain entrance to the body, but may also gain entry by ingestion.

hoolihaning. In bulldogging, a term which includes the action of the sport from the time a rider leaps from the horse until the steer is pinned. See *bulldog: 2.*

hoop. A circular, wooden or metal band which is used to form cheese, support or restrain barrel staves, etc.

hoop ash. See *hackberry: 1.*

hoop-coop plant. See *lespedeza: 2.*

hoop-petticoat narcissus. Petticoat daffodil.

hoop pine. Cunningham araucaria.

hoopwood. Smooth winterberry.

hoose. A verminous bronchitis of cattle, sheep, and goats due to a lungworm infestation (a) of cattle chiefly by *Dictyocaulus viviparus,* and (b) of sheep and goats chiefly by *D. filaria, Protostrongylus rufescens,* and *Muellerius capillaris.* Symptoms are a husky, spasmodic cough, emaciation, anemia, debility, thick nasal discharge, and breathing through the mouth. Also called *husk.* See *fowl gapeworm.*

hop. 1. Any vine of the genus Humulus, family Moraceae. See *Japanese hop.* 2. *H. lupulus,* a perennial, tall-growing vine grown for its ripened and dried pistillate cones, the *hops* of commerce, used in beer manufacture and in medicines. Native to Eurasia.

Hop. A German, sour milk cheese cured with hops.

hop ampelopsis. *Ampelopsis humulifolia,* family Vitaceae, a hardy, woody, climbing vine grown to cover trellises, fences, etc., and for its yellow or blue fruits. Native to Asia.

hop aphid. *Phorodon humuli* (Schr.), family Aphidae, a sucking insect pest of plum. Infested trees have split fruit, curled leaves, and stunted growth.

hop clover. 1. Certain members of the genus Trifolium, family Leguminosae, whose flower heads resemble those of the hop. See *low hop clover.* 2. Black medic.

hop flea beetle. *Psylloides punctulata* Melsh., family Chrysomelidae, a leaf beetle that feeds on the foliage of the hop.

hophornbeam. Any tree of the genus Ostrya, family Corylaceae (Betulaceae). See *American hophornbeam.*

hop medic. Black medic.

hopper. 1. The boxlike holding device on many farm machines into which materials are placed for processing or distribution or temporary accumulation: (a) the part of a hay baler in which the hay is placed before it is pressed as part of the hay bale; (b) the apparatus on the top of a grain planter which holds the seed; (c) the device into which feed materials are placed on a feed grinder. 2. A condition of bees brought about when poisonous pollen is carried into the beehive.

hopper burn. A disease of potatoes caused by the potato leafhopper, characterized by a yellowing of a portion of the affected leaves, curling, followed by a blackening, and drying out. *Cf. burn of potatoes.*

hopper-catching machine. Hopperdozer.

hopperdozer. A device for exterminating grasshoppers which consists of a shallow pan partly filled with a mixture of water and kerosene which is attached to the front of a tractor or automobile and pushed slowly over a field causing grasshoppers to fly up and fall back into the solution. It is less efficient than poisoning.

hopper system. See *horizontal batch ice cream freezer.*

hopple. See *hobble.*

hops. See *hop.*

hoptree. 1. Any tree or shrub of the genus Ptelea, family Rutaceae. 2. *P. trifoliata,* a hardy, deciduous shrub or small tree which is sometimes grown for ornament. Contact with the leaves may result in a severe dermatitis. Native to North America. Also called *wafer ash.*

hop trefoil. Low hop clover.

horehound. See *hoarhound: 1.*

horizon A. The upper horizon of the soil mass from which material has been removed by percolating waters; the eluviated horizon of the solum; the surface soil. This horizon is generally subdivided into two or more subhorizons of which A_0 is the accumulation of organic debris upon the soil surface. Other subhorizons are designated A_1, A_2, etc.

horizon B. The horizon of deposition to which materials have been added by percolating water; the illuviated horizon of the solum; the subsoil. This horizon may be divided into several subhorizons depending on the color, structure, consistency, or the character of the material. These subhorizons are designated as B_1, B_2, B_3, etc.

horizon C. The horizon of relatively unweathered material underlying the B horizon; the substratum. While often described as unweathered, it usually shows some modification in the upper portions and may have one or more subhorizons. In most cases it represents the parent material similar to that from which the soil was formed. In some cases a stratum or geological formation of material different from that of which the soil was formed may underlie the solum. With such conditions there is actually no true *C horizon,* the B horizons resting directly on unrelated material. In actual practice, however, this is usually called the *C horizon* (its character and lack of relation to the solum being brought out in the descriptions).

horizon of eluviation. See *horizon A.*

horizon of illuviation. See *horizon B.*

horizontal batch ice cream freezer. An early type of power-driven, brine-refrigerated, ice cream freezer. It had a horizontal cylinder or freezing unit with a hopper above it to hold the ice cream mix or batch. The freezing unit was two-walled with the brine circulating about it causing the ice cream to freeze on the inner walls. This was removed by a paddle or dasher that slowly revolved within the freezing cylinder. This type of freezer has been largely replaced by the continuous freezer.

horizontal bench terrace. A type of bench terrace which has no measurable slope from the back to the front of the bench.

horizontal silo. A silo which is built with its long dimension parallel to the ground surface rather than perpendicular as in the case of an upright silo. It may be built just below the ground level with openings at both ends to facilitate filling.

horizontal suction. The degree to which the point of the plow share is bent out of line with the landside toward the unplowed soil. The object of this suction is to make the plow take the proper amount of furrow width.

horizontal training. The training of plants to grow along a horizontal wire, trellis, etc., as the grape.

horizontal well. A well which consists of collection pipes and screens installed horizontally. It may be driven under a stream or trenched across shallow underground streamflows and then covered. See *radial well.*

hormone. A substance produced by one organ or by cells and carried to another organ or tissue on which it stimulates or inhibits activity, depending on its nature.

horn. 1. A natural, bonelike growth or projection on each side of the head of most breeds of cattle, sheep, and goats which is a natural weapon of defense. 2. The pointed end of a blacksmith's anvil which is used for shaping hot metal. 3. The front, upraised projection of a riding saddle, the snubbing *horn.* 4. The outer hard covering of the hoofs of horses, cattle, sheep, and swine.

hornbeam. Any tree or shrub of the genus Carpinus, family Corylaceae (Betulaceae). Species are grown as ornamentals. See *American hornbeam; European hornbeam.*

horn binding. The leather covering of the horn on a saddle.

horncap eucalyptus. *Eucalyptus umbellata,* family Myrtaceae, a large, aromatic, deciduous tree grown in very warm areas for its striking foliage and flowers. Native to Australia.

horned. Designating an animal which has horns. *Cf. polled.*

horned Hereford. See *Hereford.*

horned sash. A sash for a hotbed whose stiles extend several inches beyond the frame at each end.

horned squash bug. *Anasa armigera* (Say), family Coreidae, an insect pest of cucurbits and cabbage.

horned violet. *Viola cornuta,* family Violaceae, a perennial, flowering herb grown for its dark lilac flowers. Native to Europe. Also called *bedding pansy; tufted pansy; bedding violet; Pyrenees violet; Spanish violet.* See *viola: 1.*

hornet moth. *Aegeria apiformis* (Clerck), family Aegeriidae, a moth whose larvae bore into the limbs, trunks, and roots of poplars and willows.

horn fly. *Siphona irritans* (L.), family Muscidae, a bloodsucking fly which infests cattle and goats. It probably causes greater loss in livestock production in the U.S. than any other bloodsucking fly.

hornless. Designating a well-defined, polled condition of certain breeds of cattle, such as the Angus and Red Polled.

horn neck. That part of the saddle which connects the seat and the swell to the pommel. Also called *throat.*

horn poppy. See *yellow hornpoppy.*

horn scraper. A curved, knifelike tool used in scraping and smoothing the outer surface of the horns of cattle in preparation for showing.

horn socket. In well boring, an implement used to recover lost tools, especially broken drill poles, etc. It consists of a conical socket, the larger end downwards, which slides over the broken part; this trips a spring latch which grips the broken part. Frequently a flaring mouthpiece is riveted to the *horn socket,* making it a bell-mouth socket.

hornstub. A dwarfed horn or horny surface sometimes found on the heads of polled cattle. It is considered a hereditary unsoundness.

horn trainer. A metal device placed on the tips of the horns of cattle and so constructed that a steady pressure may be applied to the growing horns to cause them to take the shape or form desired. See *horn weight.*

horn weight. Ball-shaped weights, usually of lead, which are attached to the tips of the horns of cattle to assist in the training and shaping of the horn growth. See *horn trainer.*

horn weighting. See *horn trainer; horn weight.*

hornworm. See *tomato hornworm; tobacco hornworm.*

hornwort. *Ceratophyllum demersum,* family Ceratophyllaceae, a submerged, aquatic herb grown in the aquarium, pond, etc., for its finely dissected leaves. Escaped, it can be a pest. Native to North America. Also known as *coontail; fish blankets; cedar moss.*

horny frog. The semisoft, elastic, V-shaped structure in the sole of a horse's foot. Also called *foot pad.*

horny skin. See *X-disease.*

horse. 1. *Equus caballus,* a quadruped of very ancient domestication used as a beast of burden, a draft animal, and a pleasure animal for riding, and in some areas as a meat animal. Breeds and types vary greatly in size and color from the Shetland pony to the Belgian. It has been important as one of the parents of the mule, and as a source of meat and hides. See *mule: 1.* 2. A stallion. 3. Designating an implement which is drawn by a *horse.*

horseback work. Work done while seated on a horse.

horse balm. See *citronella horsebalm.*

horse barn. A barn designed with stalls, floors, mangers, and other equipment suitable for the stabling and handling of horses.

horse bean. Broadbean.

horse biting louse. *Bovicola equi* (L.), family Trichodectidae, an insect pest of horses, mules, and donkeys which bites dry skin and hair causing irritation.

horse bot fly. *Gasterophilus intestinalis* (DeG.), family Gasterophilidae, a fly which lays its eggs on the hairs of horses, usually on the front legs but sometimes on the shoulders, belly, or hind legs. When the horse licks himself the larvae cling to his lips or tongue. The larvae then burrow into the mucous membrane of the tongue working their way toward the base of the tongue. They pass to the stomach and remain there from nine to ten months, after which they release their hold on the walls of the stomach and pass to the ground in the excrement. They cause mechanical injury to the tongue, lips, lining of the stomach, and intestines, and they interfere with glandular action. They cause inflamed ulcerated conditions and absorb food which starves the host and may cause complete obstruction from the stomach to the intestines. Badly infested animals are run down, have digestive upsets, and rough coats. Also called *common horse bot fly.*

horse breaking. The breaking or taming of horses for riding, driving, or work.

horse brier. Greenbrier.

horsebrush. Any plant of the genus Tetradymia, family Compositae. See *gray horsebrush; littleleaf horsebrush.*

horse-camp. A place where workers live while working with or caring for horses, generally away from headquarters or on the open range (U.S.).

horse car. A special type of railroad freight car designed and arranged for the satisfactory and safe shipping of horses.

horsechestnut. 1. Certain trees of the genus Aesculus, family Hippocastanaceae. See *Japanese horsechestnut.* 2. *A. hippocastanum,* a deciduous tree popularly grown as a shade and street tree. It has clusters of white flowers tinged with red. Native to southern Europe. Also called *common horsechestnut.*

horse chewing louse. See *horse biting louse.*

horse-drawn. Designating an implement, vehicle, etc., which is pulled by a horse.

horse flu. Influenza in horses.

horse fly. In a broad sense, any member of the family Tabanidae, but usually restricted to the subfamily Tabaninae. See *black horse fly.*

horseguard. *Bembix carolina* (Fabricius), family Sphecidae, a black and yellow sand wasp found in southern U.S. where there are horses or cattle. It preys upon horseflies and bottleflies, which it feeds to its larvae.

horse herd. A band of horses.

horse hide. The pelt of a horse which is used as leather.

horse hoe. A horse-drawn implement, a forerunner of modern cultivators, which was invented by Jethro Tull and improved upon and first used in England during the latter half of the eighteenth century. It consisted of various kinds of plowshares, shovels, and sweeps attached to a frame. It was revolutionary at the time of its introduction in that it replaced hand hoeing in a large measure and was more efficient.

horse malaria. Equine infectious anemia.

horse-mane oat. See *side oat.*

horsemanship. The ability to show a high degree of skill in handling horses.

horse manure. Dried horse excrement used as fertilizer.

horse meat. The edible flesh of horses. In many countries the better grades are used for human consumption. In the U.S. it is used solely as the meat base for ready-mixed animal foods.

horse meat product. Any food product which contains horse meat. The Horse Meat Act of July 24, 1919, requires that all such articles be so labeled (U.S.).

horsemen. Persons whose main interest and work is in training and handling horses.

horse mint. See *spotted beebalm; apple mint.*

horse nettle. See *Carolina horsenettle.*

horse nursery. A farm for breeding horses.

horse plum. A name applied to the fruit of several species of the genus Prunus, family Rosaceae, but especially the American plum and the Canada plum *q.q.v.*

horse pole. See *tongue: 2.*

horse-pox. A virus pox disease of horses which is not prevalent in the U.S. See *cow-pox.*

horse-radish. *Armoracia lapathifolia* (*A. rusticana*), family Cruciferae, a perennial herb grown for its edible, pungent root used as a condiment and in flavoring. Native to Eurasia.

horse-radish flea beetle. *Phyllotreta armoraciae*

(Koch), family Chrysomelidae, an insect pest of horse-radish and mustard which feeds on the foliage.

horse-radish tree. *Moringa oleifera,* family Moringaceae, a tropical tree grown for its seeds, the source of ben oil which is used as a lubricant in watchmaking. It has a pungent, edible root. Native to the East Indies. Also called *angel tree.*

horse-radish type jar. A slender, glass jar which has a wide mouth.

horse raising. The breeding, feeding, and training of horses.

horse ranch. A ranch on which horses are kept and perhaps bred.

horse round-up. Collecting the scattered horses from the range for branding and checking.

horse salts. Large-animal laxatives.

horseshoe. 1. Any of several types of flat straps of iron curved to fit the shape of a horse's foot and held to the foot by the use of nails driven into the hard, horny, outer shell of the hoof. 2. To fit *horseshoes* to a horse or mule.

horse-shoer. One who follows the trade of fitting shoes on horses. (In the early frontier days of the U.S., the village blacksmith performed this service but now it is a separate profession with the trained *horse-shoer* going from show to show or from horse farm to horse farm to render this service.)

horseshoe sausage. Holsteiner.

horseshoe tile. Drain tile made of burned clay with the bottom flattened and left open so that a cross-section is in the form of a horseshoe (obsolete).

horse show. A popular type of exhibit or show where harness and riding horses are shown and horses and riders or drivers are judged on the correctness of training and performance.

horse sleeping sickness. Equine encephalomyelitis.

horse sucking louse. *Haematopinus asini* (L.), family Haematopinidae, a bloodsucking, insect pest of horses, mules, and asses. Infestation is characterized by scurfy skin, loss of vitality, scratching or rubbing, and raw spots on the skin.

horsetail beefwood. *Casuarina equisetifolia,* family Casuarinaceae, a tall tree grown as an ornamental in very warm areas for its seemingly leafless twigs which sway in the wind. Native to Australia. Also called *beefwood; Australian pine; horsetail tree.*

horsetail creeper. Christmasvine porana.

horsetail fern. Field horsetail.

horsetail milkweed. See *poison milkweed.*

horsetails. Field horsetail.

horse trader. A man who exchanges or buys and sells horses. (This term may carry a derogatory connotation.)

horse van. A large truck with an enclosed body especially equipped for transporting horses.

horseweed fleabane. *Erigeron canadensis,* family Compositae, an annual herb which is a common weed pest of fields and waste places. Native to North America. Also called *Canada fleabane; horseweed; mare's-tail.*

horse work unit. The average amount of work accomplished in 10 hours by a horse or mule or an equivalent amount of work done by mechanical power.

horse wrangler. 1. A hostler who kept the string of extra saddle ponies that accompanied every drive up the trail or were used during round-up. He kept them from straying and had them ready when they were needed (southwestern U.S.). 2. The rider on horseback who accompanies the rodeo rider and keeps the bucking broncho from injuring the contest rider before and after the riding has been done. 3. A person who works with horses during the breaking period.

hortensia. See *tea-of-heaven.*

horticultural. Of, pertaining to, or designating horticulture.

horticultural crop. A crop, usually considered as a garden or orchard crop, raised commercially or for home use.

horticultural variety. Cultivar.

horticulture. That department of the science of agriculture which relates to the cultivation of gardens or orchards, including the growing of vegetables, fruits, flowers, and ornamental shrubs and trees.

hose. A flexible pipe or tube, usually of rubber or plastic, used to carry liquids from some bulk supply to a point of distribution.

hose-in-hose. A color condition of a flower in which the calyx and the corolla are of the same color and in which one appears to be growing within the other, as in certain primroses and Canterburybells.

hoseport. An opening, usually with a self-closing door, in the exterior milkhouse wall to accommodate the transfer hose and electric motor cord during the milk transfer operation from the bulk tank to the tanker truck.

hoss. Horse (slang).

host. Any animal or plant upon or in which another organism lives as a parasite.

hostler. One who takes care of horses.

host plant. See *host.*

host range. The various kinds of plants that may be affected by a given pathogen.

host specific. Designating a parasite that can live in only one host, to which it is therefore said to be specific.

hot. 1. Designating a horse with a bad disposition. 2. Designating a fruit, such as a pepper, which has a pungent, strong, lasting flavor.

hotbed. A garden frame (*q.v.*) in which the soil is heated by decaying manure, electricity, steam or hot-water pipes.

hotbed sash. A frame containing glass, usually 3 ft. by 6 ft., used on either a hotbed or cold frame.

hot-blooded. 1. Designating a horse which has some Thoroughbred or Arab blood. 2. Designating a horse which is nervous and at times even vicious.

hotbox. See *hotbed.*

hot cutter. A wedge-shaped, tempered steel tool used to cut hot metals in a blacksmith's shop.
hotel parer. A boning knife used in drawing and boning poultry.
hotel rack. See *rib rack.*
hot gas method. A method of fumigation in which a tent is placed over a tree and poison gas is applied to the tree by a machine. A measured amount of poison is pumped into coils in a steam boiler in which it is vaporized then passes through a hose to the tented tree.
hot grading. The grading of a carcass immediately after it has been split and washed but before it has been cooled.
hothouse. See *greenhouse.*
hothouse broiler. A chicken that has been raised in a battery, under strict conditions of confinement, feed, temperature, etc.
hothouse lamb. 1. A very young lamb that has been raised in confined quarters and has received its mother's milk. 2. Lambs that are dropped out of the regular lambing season and marketed at live weights ranging from 25 to 60 lbs.
hot iron. The heated iron rod or stamp with a handle used in branding cattle, etc.
hot iron test. A test used to ascertain the desired acidity of Cheddar cheese curd before dipping and milling.
Hotis test. A test for the rapid detection of certain microorganisms in raw milk.
hot logging. In lumbering, avoiding delay between the cutting of the trees and utilization of the logs.
hot manure. Fresh manure which is going through the process of heating due to fermentation. Horse manure is designated as one of the *hot manures.* Cow manure is one of the cold manures.
hot-pack process. A method of canning in which the product is precooked and placed in the hot container, sealed, and processed either in boiling water or a pressure cooker. Formerly it was much used by processors as well as in home preservation. See *cold-pack process.*
hot pepper. Any of certain varieties of redpepper (*Capsicum* spp.) which has a distinctive, pungent, and persistent flavor. See *redpepper.*
hot-room callus. A callus which is formed on a graft packed in an inert medium and stored 3 to 4 weeks in a room in which the temperature is held between 75° and 80° F.
hot scald. See *hard scald: 2.*
hot-spotting. In forestry, the checking of the spread of a fire at hot spots alone or at salient points as an emergency measure used in advance of control-line construction.
Hottentotbread tortoiseplant. *Testudinaria elephantipes,* family Dioscoreaceae, a perennial, twining vine sometimes grown in the greenhouse as a curiosity and for its edible, above-ground rootstock. Native to South Africa. Also called *elephant's-foot; Hottentot bread.*
Hottentotfig. *Mesembryanthemum edule* (*Carpobrotus edulis*), family Aizoaceae, a perennial, sprawling herb grown in warm areas to cover banks and for its edible fruit. Native to South Africa.
hot walker. A person who leads horses in order to cool them slowly after a workout or race.
hot-weather. Designating a tender annual (*q.v.*) plant.
Houdan. An old French breed of chickens of medium weight which has a V-shaped comb, full crest and beard, and five toes instead of the usual four. There are two varieties: Mottled and Black.
Houghton currant. *Ribes houghtonianum,* family Saxifragaceae, a deciduous shrub, a hybrid resulting from a cross between *R. rubrum* and *R. sativum;* grown for its edible fruits used in jellies.
hound. 1. Either of a pair of bars used as braces to strengthen various parts of the running gear on vehicles. 2. A hunting dog.
houndstongue. *Cynoglossum officinale,* family Boraginaceae, a biennial herb which is a weed pest of pastures and waste places. Native to Europe. Also called *gipsy flower; dog bur; sheep lice; dog's tongue; western houndstongue.*
house. 1. One unit (greenhouse) of a range of greenhouses. 2. A dwelling.
house bee. A female honeybee which works inside of the hive. See *worker; bee: 1; field bee.*
house fly. *Musca domestica* L., family Muscidae, a world-wide insect pest of houses, man, and animals. It is the vector or suspected vector of more than 20 human diseases and intestinal worms.
household accounts. The amount expended for the care and upkeep of the farm dwelling, for food, furniture, clothing, recreation, travel, personal expenses etc., of the members of a family.
household milk. Whole milk consumed in the fluid state. Also called *market milk; beverage milk.*
house-hollyfern. *Cyrtomium falcatum,* family Polypodiaceae, an erect fern grown as a house plant. Native to Asia and South Africa.
houseleek. Hen-and-chickens.
house plant. Any decorative plant which may be grown successfully indoors as a potted plant.
house rat. Brown rat.
house sparrow. English sparrow.
hover. 1. The sheet metal canopy surrounding a heat source under which incubator chicks are kept warm. 2. To cover chicks, as a hen *hovers* her chicks.
Howard method for worm count. A quick test for fragments of insects in vegetable products.
Howard scale. *Aspidiotus howardi* Ckll., family Diaspididae, a scale insect which infests almond, apple, ash, peach, plum, prune, and other deciduous fruits and vegetables, but especially pear. It is found in Colorado and New Mexico, U.S.
HPAN. An abbreviation for a particular chemical, hydrolized polyacrilonite, a soil conditioner.
H-T, S-T method. See *high-temperature, short-time method.*
huamuchil. Guamachil apes-earring.
Huanuco cocaintree. *Erythroxylon coca,* family Erythroxylaceae, a shrub cultivated in the tropics for its cocaine yielding leaves. It is sometimes

grown in the greenhouse. Native to South America. Also called *coca; cocaine plant.*

Hubam sweetclover. *Melilotus alba* var. *annua,* family Leguminosae, an annual herb developed by Professor Hughes at Ames, Iowa, U.S., from plants found growing wild in Alabama, which is sometimes used for pasture, hay, silage, soiling or soil improvement, and as a bee plant. Also called *annual white sweetclover.* See *white sweetclover.*

hub and spoke. A flexible bracing system of trees in which cables radiate from a central limb or metal ring to the main branches, but which provides little direct or lateral support.

Hubbard group. A group of varieties of the winter squash that are generally ovoid in shape and pointed at both ends. Their surfaces vary from smooth to warted, and their colors range from deep green to orange and gray.

huckleberry. 1. Any of several wild species of the genus Gaylussacia, family Ericaceae. Species are seldom cultivated but are sometimes harvested from the wild for their edible fruits used at home or occasionally sold on markets. 2. Incorrectly, the lowbush blueberry *q.v.*

huckster. One who gathers up farm produce from the farmer and sells it in small lots or at retail on the market, such as a country egg buyer.

Hudson River umbrella system. A system of training grapes on a two-wire trellis in which the trunk is carried to the upper wire, and the shoots droop and attach themselves to the lower wire.

huerta. Irrigated land which produces two crops per year (Spanish).

huff. To swell or bloat, as a cheese in curing, due to undesirable fermentation.

huffed cheese. A Cheddar cheese that has become swollen or bloated in the curing process leaving pin holes, gas holes, or cracks in the body of the cheese as a result of undesirable gassy fermentation.

Hugo rose. Father Hugo rose.

huisquil. See *chayote.*

hull. 1. The outer shell or covering of any fruit. Also called *husk.* 2. The enlarged calyx of the fruit of the strawberry. 3. An excessively thin animal. 4. The outer shell or covering of any device or product. 5. To remove the *hulls* from seeds or calyxes from strawberries.

hulled oats. 1. A livestock feedstuff; the kernels produced from the undried grain in the process of hulling oats. 2. Oats which have been hulled in the threshing or cleaning process. Also called *undried oat groats.*

huller. 1. A device in a cottonseed oil mill which removes the hulls from cottonseed. 2. A machine for threshing clover seed.

hull fiber. The lowest grade of cotton linters *q.v.*

hull-less buckwheat. Tartary buckwheat.

hull-less oats. Naked oats.

human ecology. A discipline that draws upon ecological concepts originally developed in biology in order to describe and explain certain aspects of the form and development of human communities. Although it has been primarily developed by sociologists, the discipline is marginal to a number of fields, and overlaps portions of geography and economics as they are traditionally defined. The current objective of ecological research and theory is to describe and explain the manifold patterns of interconnection between functions (recurrent patterns of interdependent human activities) as they appear both within and between communities.

human saddle. A flat, English saddle.

humble plant. Sensitive plant.

Humboldt lily. *Lilium humboldti,* family Liliaceae, a bulbous herb grown in the flower border for its tall flower stalk with drooping, reddish-orange, dark-spotted flowers. Native to North America.

humic acid. 1. Any of certain substances related to humus which is formed by heating carbohydrates with water under pressure. These substances are weak acids which form salts. Part of the natural humus is acid in nature and is generally present in the soil in the form of a salt which, when dissolved by an alkali and the solution neutralized by an acid, precipitates *humic acid.* 2. Acid which is formed by the decomposition of organic matter.

humic-gley soil. A dark-colored soil which has a thin, organic surface layer, and an underlying gray, or gley, horizon. It develops in wet meadows or forested swamps. See *Wiesenboden.*

humic mulch. Decayed, disintegrated, organic residue of fresh mulch which is found on the soil surface.

humid climate. 1. Generally, a climate which has sufficient precipitation to support a forest vegetation. The lower limit of precipitation in cool regions may be as little as 20 ins., whereas in hot regions it may be as much as 60 ins. The Thornthwaite precipitation effectiveness index ranges between 64 and 128. 2. A climate in which the average relative humidity, as measured by the hygrometer, is very high. See *relative humidity.*

humidify. 1. To add water vapor to the atmosphere. 2. To add water vapor or moisture to any material.

humidity. 1. The mass of water vapor per unit volume of space. 2. The gaseous pressure exerted by water vapor present in space. See *relative humidity.*

humin. In reference to organic matter of the soil, that part of the organic matter which is not dissolved upon extraction of soil with dilute alkali.

hummock. 1. A minor feature of a land surface which is a low, rounded elevation of less magnitude than a hill or knoll: a common feature produced by wind-drifting of sand and other soils. 2. See *hammock.* 3. The small mound which is built up by grasses and sedges in a swamp or bog. See *tussock.*

hummocky. In topography, designating a land surface which consists of closely spaced hummocks, undulating, billowy, or low in relief, but which contains only a small proportion of flat land.

hump. A fleshy protuberance over the shoulders in the Zebu (*q.v.*) breed of cattle.

hump up. The attitude an animal takes in order to expose less body surface to rain, wind, cold, etc.: to pull the feet together and push the back up.

humus. 1. Organic matter in the soil that has reached an advanced stage of decomposition and has become colloidal in nature. It is usually characterized by a dark color, a considerable nitrogen content, and chemical properties such as a high base-exchange capacity. 2. Any organic matter in the surface layer of soil. 3. Commercially, peat and muck, regardless of degree of decomposition. 4. Leaf mold and duff of a forest floor.

hundredweight. A unit of weight: 112 lbs. avoirdupois in England and 100 lbs. in the U.S.

Hungarian brome grass. Smooth brome.

Hungarian iris. *Iris variegata,* family Iridaceae, a hardy, perennial herb grown for its yellow standards and brown and purple falls. A late, bearded iris, it is very important for breeding. Native to Eurasia.

Hungarian lilac. *Syringa josikaea,* family Oleaceae, a hardy, deciduous shrub grown in the shrub border and as an ornamental for its slightly fragrant, deep lilac to violet flowers. Native to eastern Europe.

Hungarian speedwell. *Veronica latifolia,* family Scrophulariaceae, a hardy herb grown for its blue, white, or pinkish flowers. Native to Eurasia. Also called *angel-eyes; germander speedwell.*

hunger fine. Designating wool fiber that is fine as a result of malnutrition or sickness of the sheep. It is weak and is classed as clothing wool.

huntaway. A dog which drives sheep (New Zealand).

hunt down. To search for sheep in the high mountains just before the winter snows begin (New Zealand).

hunter. A riding horse of quality conformation and size, especially adapted for the chase and riding to hounds; a Thoroughbred type of breeding, about 16 hands high and weighing from 1,100 to 1,250 lbs.

hunter's robe. Solomon Islands ivyarum.

Huntingdon elm. *Ulmus hollandica* var. *vegeta,* a generally hardy variety of Dutch elm, used for shade and ornament. See *Belgian elm.*

huntsman's-cup. See *pitcherplant: 2.*

Huonpine dacrydium. *Dacrydium franklini,* family Pedocarpaceae, an evergreen tree grown for its scalelike leaves resembling juniper leaves. Native to Tasmania.

Hupeh cotoneaster. *Cotoneaster hupehensis,* family Rosaceae, a shrub grown for its white flower clusters and its red fruit. Native to China.

hurdle plots. 1. A small fenced-in plot of land, from 4 x 4 to 20 x 20 ft., on which the herbage is clipped at predetermined intervals during the growing season to determine forage composition and yield. 2. Small plots grazed by means of hurdles.

hurdle system. A method of handling livestock by using hurdles, or movable fences, making it possible to change the area of immediate use daily or at frequent intervals.

hurricane. A tropical cyclone; a cyclonic whirl which has a calm central core or eye. Winds reach velocities ranging from 75 to 200 miles per hour. The central pressure may drop well below 28.50 ins. of mercury. It is accompanied by lightning and excessive rainfall. Called *typhoon* in the Pacific Ocean.

hursinghar. Night jasmine.

husbandman. A person engaged in the production of animals or crops by tillage or other use of land; a farmer.

husbandry. In its earlier usage, the skill, or art, of tillage, crop production, and rearing of farm animals. Today the word is occasionally used as a synonym for farming. More commonly used in combination, as animal *husbandry* and poultry *husbandry,* embracing the art, science, processing and business of production.

husk. 1. The dry, harsh, or rough, outer envelope of seeds or fruit, as a corn *husk.* 2. To remove the *husks* from anything, especially from mature corn. 3. See *hoose.*

husker-shredder. A machine which removes the ears of corn from the stalks, shreds the stalks, and blows the shredded stover into the barn or storage space. It is used to husk and shred previously shocked corn and is a stationary, power-operated machine.

husking bee. A gathering of neighbors and friends in which the men husked harvested corn for a particular farmer. The women, children, and sweethearts accompanied the men and prepared sumptuous meals. It was a delightful social gathering.

husking hook. A small, narrow, curved, metal strip, pointed at the end and shaped in the form of a small hook, which is attached to the hand by a metal plate and leather strap. The husks are hooked at the base of the ear to remove them and to snap the ear from the stalk.

husking machine. See *cornhusker: a.*

husking peg. A small, narrow, flat, metal strip with a curved, pointed end, or a sharpened, hardwood stick held in the hand by means of a leather thong or strap passing over the middle finger. It is used in husking corn by inserting the point into the husks on the tip of the ear and pulling them back so the ear is more easily cleaned of the shucks and snapped from the stalk. It is very similar to the husking pin.

husking pin. A small, straight rod, wooden pin, or narrow, flat, metal strip pointed on one end and held in the hand by means of a leather thong or strap. It is used in husking corn by inserting the point of the pin into the husks at the bottom of the ear so that the ear may be more easily snapped from the stalk. It is very similar to the husking peg.

husking rolls. The husking mechanism of the corn husker; either one or two pairs of corrugated rollers arranged in such a manner as to strip the husks from the ears by friction and pressure. One roll of each pair is usually made of rubber or is rubber coated.

husk tomato. See *downy groundcherry; groundcherry.*

hutch. A box-like cage or pen for a small animal: *e.g.*, a rabbit *hutch.*

Hutton Black. See *Hamprace.*

Hvid Gjetost. A Norwegian, goats' milk cheese.

hyacinth. 1. Any bulbous herb of the genus Hyacinthus, family Liliaceae. Species are grown in flower gardens. 2. *H. orientalis,* a reasonably hardy herb grown for its flowers of different colors. It is also a good pot plant and is used in forcing. Native to Eurasia.

hyacinth bean. Hyacinth dolichos.

hyacinth dolichos. *Dolichos lablab,* family Leguminosae, a tropical, rapid-growing, perennial vine grown as an annual in warm areas for its white or purple flowers and for its edible seed. Native to the tropics. Also called *hyacinth bean; bonavist; lablab.*

hyaluronidase. An enzyme that causes the breakdown of hyaluronic acid or its salts and so makes certain substances more fluid. This increased fluidity tends to increase the solubility of solids with which these substances are mixed.

hybernagium. In old English law, the season for sowing winter grain; also the grain sown and the land itself.

hybrid. A heterozygote or progeny of genetically unlike parents. In the science of genetics, it may be the progeny of two parents much alike but differing in genes at one or more loci on one or more pairs of chromosomes. On the other hand, it may be the progeny of parents that differ greatly, as of two species or of two genera. The former would be known as a species *hybrid;* the latter, as an intergeneric *hybrid.* A numerical *hybrid* is the progeny of two parents that differ in chromosome number. A structural *hybrid* is one whose parents differed in the structure of their chromosomes. See *hybrid vigor.*

hybrid chicks. Chicks which result from crossing two or more inbred lines of the same or different breeds, varieties, or strains.

hybridization. The production of hybrids by natural crossing or by manipulated crossing.

hybridize. To create a hybrid.

hybridizer. A breeder who creates hybrids, especially of plants, such as corn.

hybrid passionflower. *Passiflora alatocaerulea,* family Passifloraceae, a climbing vine, a hybrid resulting from crossing *P. alata* with *P. caerulea;* grown for its habit and its fragrant flowers, white outside and pink inside with a crown of purple, blue, and white.

hybrid sprayer. A sprayer which is a combination of an aerosol generator and spray blower.

hybrid vigor. The phenomenon in which the crossing of two species, genera, lines, or strains produces a hybrid that shows increased vigor over either parent.

hydatid. The larval stage of the dog tapeworm, *Echinococcus granulosus,* which usually develops in the liver and lungs of man, cattle, swine, goats, sheep, and some other mammals. *Hydatids* are cysts containing the larvae. They may develop to a very large size, displace parts of the organs in which they grow, and cause a disease known as *hydatid* disease, or echinococcosis.

hydrangea. 1. Any shrub or woody vine of the genus Hydrangea, family Saxifragaceae. Species are grown for ornament. See *bigleaf hydrangea; climbing hydrangea: 1; oakleaf hydrangea; panicle hydrangea; shaggy hydrangea; smooth hydrangea; tea-of-heaven.*

hydrated lime. $Ca(OH)_2$, a dry product which consists chiefly of the hydroxide of calcium and oxide-hydroxide of magnesium, sometimes used as a soil amendment. Also called *slaked lime; calcium hydroxide.* See *lime: 1.*

hydraulic. Designating fluids (usually water) which are moving or at rest under forces of gravity or pressure.

hydraulic dredge. A floating excavator consisting of a suction pipe to which is attached a cutting head, a high-pressure centrifugal pump, and a discharge pipe. Excavated material, consisting of mud, sand, gravel, or clay, is sucked from the bottom of the stream, bay, etc., and transported through a pipe to the shore.

hydraulic elements. Those factors which are involved in the flow of water in pipes and channels: *e.g.*, depth, area, wetted perimeter, hydraulic radius, slope, velocity, energy, roughness, and viscosity.

hydraulic fill dam. A dam composed of earth, sand, gravel, etc., sluiced into place. Generally the finer elements are washed toward the center allowing for greater imperviousness.

hydraulic friction. The combined effect of all sources of resistance to the flow of liquids; a hypothetical resisting force supposed to act at the boundaries of the liquid which can be overcome only by an expenditure of energy equal to that actually lost by the flowing liquid. With water, the major part of the resistance, under ordinary flow conditions, is brought about by the expenditure of energy in maintaining normal turbulence. The energy losses arising from excessive turbulence and impact at obstructions, curves, and sudden changes in channel cross section are not ordinarily ascribed to *hydraulic friction* but are considered separately in hydraulic calculations.

hydraulic grade line. 1. In a closed conduit, a line which joins the elevations to which water could stand in risers. 2. In an open channel, the free water surface.

hydraulic gradient. 1. The slope of the hydraulic grade line. 2. The slope of the free surface of water flowing in an open channel.

hydraulic jump. The sudden and usually turbulent passage of water from low stage below critical depth to high stage above critical depth, during which the velocity passes from supercritical to subcritical. It represents the limiting condition of the surface curve wherein it tends to become perpendicular to the stream bed.

hydraulic radius. 1. The cross-sectional area taken at right angles to the center line of a stream of water divided by the length of that part of its periphery in contact with its containing conduit. 2. The ratio of area to wetted perimeter.

hydraulic ram. A mechanical device which utilizes momentum of water flowing down an incline to raise a part of that water to a higher elevation.

hydraulics. 1. The mechanics of fluids; hydromechanics. Commonly used to designate that body of hydromechanical principles given practical application by the engineer in dealing with water. 2. That branch of a service or of engineering which treats of water or other fluid in motion.

hydraulic sprayer. Any sprayer in which the spray liquid is pressurized by means of a hydraulic pump and in which the pressure may be stabilized in an air chamber.

hydrochloric acid. HCl, a strong, mineral acid, used in solution for the volumetric analysis of alkalis.

hydroclimate. The climate of water bodies, lakes, etc. See *marine climates.*

hydrocyanic acid. HCN, one of the most valuable and widely used of the fumigants. Except that it is highly toxic to man, it approaches the ideal in a fumigant. It is found in some plants, particularly sorghums, and may under certain conditions cause poisoning and death to animals. Sometimes called *prussic acid.*

hydrofluoric acid. HF, a strong, mineral acid.

hydrogen. H, an element; a colorless gas; one of the essential elements for plant growth.

hydrogenated lard. Usually, steam-rendered lard to which extra atoms of hydrogen have been added. It is bland, has a high smoke point, is resistant to rancidity, and is whiter than the usual lard.

hydrogen cyanide. See *hydrocyanic acid.*

hydrogenic soil. Soil which was developed under the dominant influence of water.

hydrogen-ion concentration. The amount of hydrogen ions present in a liquid or soil, indicated by the symbol pH. It serves as an indication of acidity or alkalinity. A pH of 7 is neutral; below 7 is acid and above 7 is alkaline. See *pH; soil acidity.*

hydrogen swell. In canned goods, an expansion of the can caused by the formation of hydrogen as a result of corrosion of the tin plate.

hydrograph. A graph which shows the fluctuation of the flow of water with respect to time.

hydrography. The measuring, recording, and analyzing of the flow of water: *e.g.,* measuring and mapping water courses, shore lines, and navigable waters.

hydrologic cycle. The movement of water in nature through a complete cycle, commencing as atmospheric water vapor, passing into liquid and solid form as precipitation, then into the ground surface as infiltration or as surface runoff or ground water to larger bodies of water, from which, by transpiration and evaporation, it returns as vapor to the atmosphere and as runoff returns to the oceans (the basic source of moisture) for re-evaporation.

hydrology. The science which deals with the waters of the earth, their occurrence, distribution, and circulation through the unending hydrologic cycle of precipitation, runoff, infiltration and storage, eventual evaporation, transpiration and reprecipitation. A simple distinction between *hydrology* and hydraulics is that in hydraulics the source of the water is immaterial, attention being confined to its motion.

hydrolysis. A double decomposition reaction in which water is a reactant.

hydrometer. A float type of instrument constructed to ascertain the specific gravities of liquids.

hydromorphic soil. A soil that has been formed under a strong influence from water, as one developed in a bog or swamp.

hydrophobia. Rabies.

hydrophyte. A plant which grows under wet or moist conditions or one which requires a large amount of moisture. Also called *hygrophyte.*

hydroponics. The growing of plants with liquid nutrient solutions instead of in soil. *Cf. geoponics.*

hydroscopic. Designating the ability or tendency to absorb or condense moisture from the air.

hydroscopic water. Water remaining in the soil after gravitational and capillary water have been removed which is mostly held as thin films on soil grains, especially colloids.

Hydrosols. A group of natural soils in which water is the principal gross component. It is of the same category in classification as mineral soils and organic soils. The vegetation of such soils is aquatic or hydrophytic.

hydroxide of lime. See *hydrated lime.*

hydroxy-apatite. A natural mineral, a calcium phosphate–calcium hydroxide complex.

hydroxymercurichlorophenol. An organic mercury compound used as a seed and bulb disinfectant.

hydroxypentamethylflavan. A chemical used as a constituent of insecticides.

hydroxy-phosphate. A mineral or chemical compound, a calcium phosphate complex. It may be formed in the manufacture of some superphosphate, and there is some evidence that it exists in nature, but at present its significance to agriculture is academic.

hygiene. The science of health; the rules or principles of maintaining health; sanitation.

hygienic. Of, or pertaining to, hygiene.

hygroma. A swelling in connection with a joint (knee, hock, etc.) which results from repeated bruising.

hygrometer. An instrument used for determining the relative humidity of the air.

hygroscopic. See *hydroscopic.*

hygroscopic coefficient. The moisture, in percentage of dry weight, that a dry soil will absorb in saturated air at a given temperature.

hygroscopicity. The tendency to absorb moisture from the atmosphere.

hygroscopic moisture. Any amount of moisture a

soil contains up to and including the hygroscopic coefficient. Such water does not move in the soil, is not used by plants, and may exist in other than the liquid state.

hymen. A membranous fold located just anterior to the external urethral orifice, which partially or completely closes the vagina. If the *hymen* remains until breeding age, it is referred to as persistent. If the *hymen* completely occludes the vagina, it is called imperforate; one that partially occludes the vagina is called perforate.

hymenium. The spore-bearing layer of a fungous fruiting body.

hymenocallis. Showy, chiefly tropical herbs of the genus Hymenocallis, family Amaryllidaceae. Also called *spiderlily.* See *basketflower hymenocallis; Caribbean hymenocallis.*

hyperchimera. See *chimera.*

hyperesthesia. Excessive sensitiveness of the skin or of a sense organ or organs.

hyperimmune. Designating the state or quality of having a high degree of immunity which has been produced by repeated exposure to the same disease agent.

hyperimmune serum. The blood serum of an animal that has recovered from (or has been immunized to) a disease-producing agent by repeatedly exposing the animal to the same specific infectious agent.

hyperkeratosis. See *X-disease: 1.*

hyperparasite. An organism that is parasitic on another parasite.

hyperpituitarism. A condition brought about by excessive production of one or more hormones by the pituitary gland causing abnormal or excessive growth as in gigantism.

hyperplasia. The production of an abnormally large number of cells in a tissue.

hyperplastic. Designating a disease in which an abnormally large number of cells are produced in a certain tissue.

hypersensitivity. The violent reaction of an organism to attack by a pathogen, a condition in which the response to a stimulus is unusually prompt or excessive in degree.

hypersusceptible. Designating a condition of abnormal susceptibility to infection, or to a poison, to which a normal individual is resistant.

hyperthyroidism. A condition of man and lower animals in which the thyroid gland secretes excessively; characterized by rapid tissue change and restlessness.

hypertrophy. An abnormal increase in the size of cells or a general increase in bulk of a part or organ not due to tumor formation.

hypha. A single thread of fungus mycelium.

hyphal hole. A hole formed in a cell wall by the dissolving action of an advancing hypha of wood-inhabiting fungus.

hypocalcemia. A deficiency of calcium in the blood.

hypochlorites. Disinfectant compounds containing oxygen, chlorine, and a metal: *e.g.,* NaOCl, sodium hypochlorite; $CaOCl_2$, calcium hypochlorite. Their disinfectant activity depends upon the properties of freshly liberated oxygen atoms although their strength is usually measured by the amount of available chlorine.

hypochnus. Fisheye rot of pome fruits.

hypoglycemia. A deficiency of sugar in the blood. See *shivering in little pigs.*

hypomagnesemia. A deficiency of magnesium in the blood.

hypoplasia. Subnormal cell production.

hypoplastic. Designating a disease in which cell production in a tissue or organ is either arrested altogether or subnormal.

hypostasis. 1. The formation of a deposit, especially a settling of the blood due to a feeble blood flow. See *epistasis: 2.* 2. The state or condition of being hypostatic.

hypostatic. Designating a genetic factor which is prevented from exhibiting its normal effect in the development of the individual due to the presence of some factor other than its allelomorph. See *epistatic.*

hypothyroidism. A condition of man and the lower animals in which the thyroid gland secretes inadequately; characterized by cretinism.

Hypoxylon canker of poplar. A fungous disease of aspen and balsam poplar caused by *Hypoxylon pruinatum,* family Xylariaceae, characterized by various-sized, gray cankers on the trunk only. The cankers become black as the outer bark falls off and the sapwood around the canker is black.

hypsometer. Any instrument used to measure heights of trees, based either on geometric or trigonometric principles.

hyssop. *Hyssopus officinalis,* family Labiatae, an under-shrub grown for its spikes of blue flowers and for its leaves which are used in flavoring, as a source of tea, and as a source of an oil used in eau de Cologne. Native to Eurasia.

hysterectomy. Total or partial removal of the uterus.

I

iarovization. See *vernalization.*

IBMA. Isobutylene maleic acid, a synthetic soil conditioner used experimentally for improvement of soil structure for better tilth.

ibota privet. *Ligustrum ibota,* family Oleaceae, a hardy shrub grown as a hedge plant. Native to Japan.

ice. 1. Water that has been transformed to a solid state by low temperature; frozen water. 2. Any frozen food product made from water or sweet skimmed milk or whole milk and sugar, with or without eggs, fruit juices, or other natural flavoring.

ice-bank. A system of refrigeration in which there is a transfer of heat from water surrounding the evaporator coils. A bank of ice is thus formed round the coils. Water is then circulated across the ice and round the outside of the inner tank containing the milk or other product. Often abbreviated *IB.*

ice-cleft soils. Soils composed of the residual debris resulting from the splitting or disintegration of rocks by ice action on high mountains and in the Arctic region. Also called *Spaltfrostboden.*

ice cream. A popular food product made by freezing a combination of various ingredients, such as cream, sugar, flavoring, and stabilizers into a semisolid product.

ice cream freezer. The mechanical apparatus used in the making of ice cream which may be of many types, hand- and power-operated, etc.

ice cream improver. See *improver.*

ice cream salt. Rock salt.

iced. Designating a product stored in ice for preservation.

ice damage. 1. Breakage of tops, branches, or stems of plants by the action of ice storms. 2. Smothering of perennial plants, such as alfalfa, by ice sheet formation over the crown of the plants.

iced fowl. See *ice-packed poultry.*

ice house. A building that serves as a storage place for ice.

Iceland poppy. *Papaver nudicaule,* family Papaveraceae, a hardy, perennial herb grown for its sweetly fragrant, white, red, yellow, or orange flowers. Native to the Arctic.

ice milk. A frozen product which resembles ice cream except that the fat content is lower and the milk solids-not-fat content is usually higher. It contains 3 to 4 percent fat, 10 to 15 percent milk solids-not-fat, about 15 percent sugar, and a stabilizer.

ice-packed poultry. A method of packing dressed poultry in which alternate layers of crushed ice and fowls are placed in a shipping box or crate. *Cf. dry-packed poultry.*

ice pink. *Dianthus glacialis,* family Caryophyllaceae, a perennial herb grown in the rock garden for its reddish-purple flowers, yellowish underneath. Native to the mountains of southern Europe.

ice plant. An ice manufacturing concern, especially the building in which the ice is manufactured and stored. See *ice house.*

iceplant mesembryanthemum. *Mesembryanthemum crystallinum* (*Cryophytum c.*), family Aizoaceae, an annual, fleshy-leaved herb grown for its glistening foilage, its white or pale pink flowers, and its edible leaves which may be cooked like spinach. Native to South Africa. Also called *sea fig; sea marigold.*

ice scald. A physiological condition of wrapped peaches that have been stored at too low temperatures.

ice separation. A body defect of ice cream or sherbet in which small, ice crystals separate from the body of the product.

ice sheet. A condition in which a sheet of ice forms over the surface of the ground and on plants, as a result of freezing rain or thawing and freezing of snow.

ice smothering. A suffocation of a plant due to a covering of ice remaining on the plant and the surrounding soil for a prolonged period.

ice storm. A heavy deposit of glaze or frozen rain which causes much breakage in tops of trees and damage to other plants.

Ichang papeda. *Citrus ichangensis,* family Rutaceae, an evergreen tree, the hardiest evergreen citrus, used in breeding cold-resistant hybrids. Native to China.

icterus. 1. Jaundice. 2. In plants, a yellowing of the leaves due to cold, excessive moisture, etc.

Idaho fescue. *Festuca idahoensis,* family Gramineae, a perennial bunchgrass whose forage rating is excellent. Found from Alberta, Canada, south to Arizona and west to California, U.S., at elevations between 5,000 and 10,000 ft.

Idahos. 1. In marketing, sheep which are raised in Idaho, U.S. 2. Designating a certain strain of potato much raised in Idaho, U.S.

Idaho white pine. See *western white pine.*

identification tag. A tag of metal or other material attached to the ear of an animal for identification. See *ring: 6.*

idiopathic. Designating a disease or condition with no apparent cause.

idle land. 1. Land which has been managed or exploited for some particular use but is now in a state of disuse; abandoned land. 2. Farm land that is capable of producing but is not in use; by extension, any land that has potentiality, but is not being put to any social use. 3. Land which is being fallowed.

ileocecal valve. The valve at the junction of the lower end of the small intestine and the large intestine (the junction of ileum and cecum).

Ilha. A firm, cows' milk cheese made in Portugal.

Illinois apple tree canker. See *blister canker of apple.*
illite. A group of mica-like clay minerals which may be present in the clay constituent of soils, and which has chemical properties different from other clay groups, such as kaolinite and montmorillonite. The mineral is a hydrous silicate of aluminum, containing potassium, iron and magnesium, but is variable because of replacement of aluminum by iron and magnesium.
ill-shaped udder. A cow's udder that is pendulous, lacking in rear or forward attachment, and irregular in the size or the shape of the quarters.
illuvial horizon. A soil horizon that has received material from other portions of the soil profile through the process of eluviation. Usually called *horizon B q.v.*
imbibition. The process by which solids, especially organic substance in the gel state, take up liquids and in so doing, swell. The imbibing substance may develop great pressures during the swelling: *e.g.,* absorption of water by seeds, wood, gelatin; absorption of ether by rubber. It is important in the physiology of the plant as one of the means of absorbing water by roots and in the uptake of water by seeds and the splitting of seed coats.
imbricated snout beetle. *Epicaerus imbricatus* (Say), family Curculionidae, an insect pest which infests plants and is generally especially destructive to strawberries.
Imhoff sludge. Sewage sediment made by running sewage through tanks and settling with the aid of anaerobic bacteria. It is sometimes used as a low-grade fertilizer. Also called *ordinary sludge.* See *activated sewage.*
imitation spices. A mixture of spice oils in a flavorless base consisting of ground and powdered nut shells, soybean flour, or some cereal.
immature. Designating a plant or animal which has not reached full development.
immature soil. A soil in which the profile shows very definite and distinct development of the A and B horizons with their subdivisions, but which has not yet reached full equilibrium with the environment. 2. A young or imperfectly developed soil.
immature staple. That cotton which has been picked and baled before the fiber has reached a normal state of development resulting in a weakened staple of inferior value.
immature veal. See *bob: 1.*
immersion heater. A type of electric water heater used for warming small quantities of water in which the heating element is placed directly in the water to be heated.
immobilization. The action of rendering an object immovable, as the *immobilization* of an injured limb in an animal to allow for healing.
immortelle. See *everlasting: 1.*
immune. Designating the state of possessing immunity.
immunity. Having resistance to the action of something, such as a disease. It may be inborn, may result from exposure to a disease, from having had a disease, or from having received an injection of immune serum. Degree of *immunity* varies in each case. See *immunizing agent.*
immunization. The process of making an animal resistant to a disease.
immunize. To render an animal resistant to disease by vaccination or inoculation.
immunizing agent. A substance that, when introduced into the body of an animal, will build up antibodies in the blood that will resist or overcome an infection to which most of the same genus or species are susceptible.
immunogenicity. Ability to produce immunity or security against any particular disease or poison.
immunological. Of, or pertaining to immunology, the science of immunity.
impacted crop. Crop bound.
impaction of the bowels. Constipation.
impaction of the crop. Crop bound.
impaction of the large intestine. See *acute indigestion in horses.*
impact loss. 1. The pressure lost as a result of the impact of particles of water against other material. See *head: 4.* 2. The dispersal action exerted on the soil aggregates as a result of water falling on exposed soil.
impeller. The disc or rotor of a centrifugal pump.
imperfect drainage. Impeded removal of excess water from all or some portion of the soil profile so that there is some limitation in the ease of tillage or use of the land.
imperfect flower. A flower in which both stamen and pistil are not present. Also called *diclinous flower.* See *pistillate; staminate; perfect flower.*
imperfect fungi. In the classification of fungi, (*Fungi imperfecti*), that large group in which the sexual stage may be lacking or at least has not yet been discovered.
imperforate hymen. See *hymen.*
imperforate neck of the uterus. See *closed womb.*
imperial bushel. A dry measure used in Great Britain, slightly larger than the standard American bushel: 1 *imperial bushel* is equal to 1.0321 American bushels. See *bushel: 1.*
imperial fritillary. *Fritillaria imperialis,* family Liliaceae, a bulbous herb grown for its purplish, yellow-red, or terra-cotta colored flowers. Native to Iran. Also called *crown imperial.*
imperial moth. *Eacles imperialis* (Drury), family Citheroniidae, a moth whose caterpillar is a general feeder on forest and shade trees in eastern U.S.
Imperial Valley. A geographic, agricultural region in southern California, U.S., noted for its prolific production of truck crops, melons, fruits, etc. It was once a desert area, but with modern irrigation practices the valley was rendered fertile. The basin is filled with alluvium; the greater part of the land surface is flat and below sea level.
impermeability. 1. In ground-water hydrology, the property of rock or earth which does not permit water to move through it perceptibly under the pressure ordinarily found in subsurface wa-

ter. Certain highly porous materials are subcapillary in size. Bodies of impermeable material, such as clay, may lie within either the zone of saturation or the zone of aeration, but may not be an active, functional part of either zone. 2. Being impervious, not easily penetrated, as by oil or water, either because of low porosity, very small individual pores, or pores that are disconnected. 3. Having the property of not permitting the passage of liquids or gases.

imperviousness. That quality or condition of a material that minimizes percolation of water.

impervious soil. A subsurface soil layer that is resistant to the free, downward movement of water or to penetration by plant roots.

imphee. Sorgo *q.v.*

implement. Any tool which aids man to make his work and effort more productive and effective.

implement shed. A farm building used primarily as a storage place for implements.

imported cabbage worm. *Pieris rapae* (L.), family Pieridae, an insect pest whose larvae infest crucifers, nasturtiums, sweet alyssum, mignonette, and lettuce, riddling the leaves.

imported currantworm. *Nematus ribesii* (Scop.), family Tenthredinidae, an insect which infests currants and gooseberries. Severe infestation results in defoliation.

imported fire ant. *Solenopsis saevissima* var. *richteri* Forel, family Formicidae, an ant imported from South America; widespread in southern U.S. and rapidly moving northward. Its sting is very harmful to man, and it is a serious menace because of its destructive feeding on crop plants, its attacks on young animals, and the nuisance of its hills in fields.

imported willow leaf beetle. *Plagiodera versicolora* (Laich.), family Chrysomelidae, an insect which infests willows and Lombardy poplars. The adults eat holes in the upper surfaces of the leaves; the larvae skeletonize the leaves.

impotence. Temporary or permanent loss of reproductive power or virility.

impotent pollen. Pollen that is nonviable and incapable of fertilization.

impoverished soil. Soil which has been continuously cropped and has had little or no care so that it is incapable of satisfactorily supporting an economic crop.

impregnate. To fertilize a female animal or flower.

impregnator syringe. A syringe with a long neck used to insert semen into the vagina of females in artificial insemination. It is becoming obsolete.

improved breed. Any group of animals having a relatively uniform breeding value with respect to certain distinctive features of conformation, type, and productive ability.

improved land. Land that has been rendered more useful or valuable because it has been cleared, drained, and placed under cultivation or is occupied by buildings, yards, lanes, fences, etc.

improved pasture. Pasture which has been rendered more valuable through clearing or seeding without cultivation and the use of fertilizer, etc. See *tame pasture.*

improvement cutting. 1. Cutting in any woodland primarily to develop a satisfactory growing stock in accordance with the site and the objective of management, or to increase the utility of the forest for other purposes than wood production. While the principal products from such cuttings will come from trees of under rotation age, they may include occasional trees of crop size. 2. A cutting made in a stand of trees past the sapling stage to improve the composition and character by removing trees of undesirable species, form, and condition, occupying dominant positions in the main crown canopy.

improvement expense. The net expense for the year for farm improvements such as buildings, fences, tiling, water supply, erosion control, structures, etc. It is obtained by deducting the sum of the inventory of improvements at the end of the year and any sales of improvements during the year from the sum of the inventory of improvements at the beginning of the year plus all expenditures on farm improvements including capital expenditures, cost of repairs, and insurance.

improver. A food product sometimes included in the ice cream mix to give smoothness and stability to the frozen ice cream.

improve the land. To increase the productiveness of land by fertilization, better methods of tillage, and management.

in-and-in breeding. The breeding together of closely related plants or animals for a number of successive generations to improve or eliminate certain characteristics. Also called *breed in and in; line breeding; inbreeding.*

inappetence. Lack of appetite.

inarching. A method of plant propagation in which a plant, still attached to its own roots, is grafted to another such plant. The cambium regions of both plants are exposed and fastened together so that they will unite. After the union is accomplished, one plant may be severed from its roots. Also called *approaching.*

in bloom. Flowering, blossoming.

inbred. 1. An individual with parents which show 50 percent or more of common ancestry in their pedigree. 2. The offspring produced by self pollination in normally cross-pollinated plants. 3. Of, or pertaining to, a plant or animal which is produced by breeding between close relatives.

inbred-line chicks. Chicks which result from four or more generations of in-and-in breeding.

inbred poultry. Birds which result from a mating of poultry related as close or closer than first cousins.

inbreeding. In-and-in breeding.

incensecedar. Any evergreen tree of the genus Libocedrus, family Pinaceae. Species are grown as ornamentals and for their durable timber. See *Chilean incensecedar; California incensecedar.*

incensecedar mistletoe. *Phoradendron libocedri,* family Loranthaceae, a mistletoe (*q.v.*) found on incensecedars in Oregon, California, and Nevada, U.S.

incentive payment. A form of compensatory payment (*q.v.*) in which the support price is set at a level high enough to encourage the increase in production of a particular commodity to a desired level.

incentives (for labor). Compensation and considerations over and above the regular wage for a farm worker or manager which are intended to encourage more effective farm operation.

inchworm. Any larva of a moth of the family Geometridae. Also called *measuring worm; cankerworm.*

incidence. Range of occurrence of influence; extent of effects; frequency; as the *incidence* of disease.

incienso. See *white brittlebush.*

incipient erosion. The early stages of erosion, especially with reference to gullying.

incipient stage decay. An early stage of wood decay in which the wood may show discoloration but is not otherwise visibly altered.

incised. Cut into; made by cutting.

incision. A cut.

incisor. A cutting tooth; one of the front teeth. *Cf. molar; canine tooth.*

incitant. See *causal organism.*

inclement weather. Harsh, disagreeable weather.

inclosure. Any device, such as a fence, which restrains animals to a given place. *Cf. exclosure.*

inclusions in eggs. Foreign materials, such as blood spots and/or meat spots, present on the yolk or in the albumen of a poultry egg. They may be caused by a sloughing off of materials from the walls of the oviduct before the shell is formed. Also called *inclusion bodies.* See *blood spot; meat spot of egg.*

income cycle. In forestry, the period of time between successive major yields.

income per man. An index of income determined by dividing the gross income from an agricultural operation by the number of men engaged in that operation.

income per $100 expenses. An index determined by dividing the gross income from an agricultural operation by the total expenses of the operation and then multiplying the result by 100.

income valuation. In an appraisal, the value of a farm or other property on the basis of the income which is derived from it. See *capitalization value.*

incompatibility. In fruit plants, a condition in which the viable male gamete will not fertilize the viable female gamete. *Cf. compatibility.*

incomplete clearance. In market reports, designating that not all animals received on a particular day were sold by the end of that day's trading. *Cf. complete clearance.*

incomplete fertilizer. A commercial fertilizer that contains only one or two of the three elements essential to plant growth: nitrogen, phosphorus, and potash.

incomplete flower. A flower which lacks one or more of the four organs: sepals, petals, stamens, or pistils. *Cf. complete flower.*

increase in inventory. A condition in which the value of the farm inventory at the end of a year exceeds the total value of the inventory at the beginning of the year. *Cf. decrease in inventory.*

increaser. A forage plant on the range which under existing circumstances is spreading. *Cf. decreasers; invading plant.*

increment. The increase in volume, weight, etc., of an animal on pasture, of a tree growing in a stand, etc.

increment borer. A hollow auger that is bored into a tree at right angles, extracting a core showing the annual rings from which the rate of growth may be determined. Also called *accretion borer; pressure borer.*

increment core. That part of a cross-section of a tree extracted by an increment borer.

incubate. To keep eggs, embryos, bacteria, etc., under conditions favorable for hatching or developing.

incubation. 1. The period between inception and manifestation of a disease in plant or animal. 2. The keeping of a culture of bacteria or protozoa in an incubator to promote development. 3. The keeping of poultry eggs under suitable conditions of warmth, humidity, etc., for the development of embryos and the hatching of young. See *incubation period.*

incubation period. The period between the setting and hatching out of eggs. The following are average incubation periods for some common birds: chickens, 21 days; most ducks, 28 days; muscovy, 33 to 35 days; geese, 30 to 34 days; turkey, 28 days; guinea fowl, 26 to 28 days; pheasant, 23 to 25 days; pea fowl, 28 to 30 days; pigeons, 17 to 19 days.

incubator. An apparatus or chamber which provides favorable environmental conditions for the development of embryos, the hatching of eggs, or the growth of cultures.

incubous. Having leaves that overlap in such a way that the tip of each leaf lies flat against the base of the leaf next above it.

in cultivation. Being cultivated or farmed, as land.

incurved cactus dahlia. See *cactus dahlia.*

incurving horns. Horns on cattle which curve over the eyes.

Indalone. N-butyl mesityloxide oxalate, a chemical used in insect repellents for the control of the stable fly, buffalo gnat, and mosquito.

indehiscent. Designating certain dry fruits that do not split open when mature.

independent elevator. A privately owned and singly operated grain handling business, as contrasted with cooperative or chain-owned and operated elevators. See *elevator: 2.*

index feather. A single feather of a bird's wing which is of major rank (but slightly shorter than adjacent major feathers), interposed between the primary and secondary group of major wing feathers, thus preventing a gap between these groups at the wing joint when the wing is spread. Also called *thumb feather.*

indexing. 1. Determining of latent disease in plants by growing a sample of seeds or sprouts, or by grafting a portion of a plant to a susceptible variety under conditions favorable for symptom expression. 2. Determining the presence of disease in seed stocks, particularly in potatoes, by preplanting portions of the stock, or in fruit stocks, by graft inoculation to a susceptible variety.

indexing host. A susceptible plant used to demonstrate the presence or absence of the infectious principle of a disease; it may be present in a plant that is resistant or without symptoms.

indexing host range. A selected series of plants, consisting of varieties or individuals susceptible to each known virus, used to determine the virus content of individual plants that do not show symptoms.

index number. A comparative measure of magnitude of groups of data, relating to such things as prices and production, etc., at different times or places, expressed as a percentage of a specified base.

index of prices paid by farmers. The percentage that the average price of a selected group of commodities and services, purchased by farmers for production and living (including interest, taxes, and farm wage rates, weighted according to their relative importance, in a particular place and at a particular time), is of some previously selected base period (generally 1910-14). Also called *parity index.*

index species. An organism which is so exactly adapted to its habitat that it is unable to survive elsewhere.

India floatingheart. *Nymphoides indicum,* family Gentianaceae, an aquatic herb grown in a pool or tub for its white, yellowish-centered flowers. Native to the tropics. Also called *water snowflake.*

India ivy gourd. *Coccinia cordifolia,* family Cucurbitaceae, a tuberous-rooted, perennial vine grown in warm areas for its climbing habit and its cup-shaped white flowers. Native to Asia. Also called *ivy gourd.*

India jujube. *Zizyphus mauritiana,* family Rhamnaceae, an evergreen tree or shrub grown in very warm areas for its edible, round, red fruit. Native to India. Also called *cottony jujube.*

India lovegrass. *Eragrostis pilosa,* family Gramineae, a tufted, annual grass which is a weed pest. Native to Europe; naturalized in North America.

India mockstrawberry. *Duchesnea indica,* family Rosaceae, a trailing, perennial herb sometimes grown as an ornamental ground cover for its habit and for its solitary, yellow flowers. Also called *mock strawberry; Indian strawberry.*

Indian. An untamed horse (western U.S.). See *Indian pony.*

Indian almond. Tropical-almond terminalia.

Indian apple. Mayapple.

Indian bean. See *southern catalpa; silvery lupine.*

Indian blanket. Rose ring gaillardia.

Indian cattle. See *Zebu.*

Indian cherry. Carolina buckthorn.

Indian corn. See *maize.*

Indian cress. See *nasturtium:* 2.

Indiancurrant coralberry. *Symphoricarpos orbiculatus,* family Caprifoliaceae, a hardy, deciduous shrub grown for its white flowers, reddish-purple fruit, and its foliage which turns crimson in autumn. Native to North America. Also called *Indian currant; coral berry.*

Indian dye. Goldenseal.

Indianfig. *Opuntia ficus-indica,* family Cactaceae, a bushy or treelike cactus grown in arid, warm areas for its edible, red, juicy fruit. Native to Mexico.

Indian Game. Poultry resulting from crossbreeding native fowls from Sumatra and India with English Black-Breasted Red game fowl. The Cornish (*q.v.*) is a product of this crossing.

Indian gram. See *gram chick-pea.*

Indiangrass. Any grass of the genus Sorghastrum, family Gramineae. See *yellow Indiangrass.*

Indian hawthorne. India raphiolepis.

Indian hemp. See *hemp.*

Indian horse. See *Appaloosa.*

Indian Jack-in-the-pulpit. *Arisaema triphyllum,* family Araceae, a perennial herb grown in the wild garden for its spadix (Jack) arched over with a purplish spathe (pulpit). Its roots were eaten by the Indians. Native to North America. Also called *Indian turnip.*

Indian lettuce. Miner's lettuce.

Indian licorice. Jequirity rosarypea.

Indian lilac. See *Chinaberry.*

Indian lotus. Hindu lotus.

Indian mallow. Chingma abutilon.

Indian-meal moth. *Plodia interpunctella* (Hbn.), family Phycitidae, a widespread, troublesome insect pest of stored grains, grain products, dried fruits, nuts, dried meats, etc. Also called *flour moth.*

Indian millet. Pearl millet.

Indian mulberry. *Morinda citrifolia,* family Rubiaceae, a tropical tree grown in warm areas as a source of yellow and red dye. Native to Australia.

Indian physic. 1. *Gillenia stipulata,* family Rosaceae, a perennial herb grown in the wild garden for its loose clusters of white flowers and its emetic root. Native to North America. Also called *American ipecac.* 2. Hemp dogbane.

Indian pink. 1. Pinkroot spigelia. 2. Chinese pink.

Indian pitcher. See *pitcherplant.*

Indian plant. Bloodroot.

Indian poke. American falsehellebore.

Indian pony. 1. A small, riding pony used by the Indians of southwestern U.S., which originated from horses shipped by the Spaniards to Mexico in 1519. Many became wild and their offspring were captured by the Indians. They are the results of natural propagation rather than any selection or planned breeding. 2. Any piebald or pinto pony.

Indian posy. See *butterfly milkweed.*

Indian potato. Giant sunflower.

Indian rice. See *annual wildrice.*

Indian ricegrass. *Oryzopsis hymenoides,* family Gramineae, a perennial bunchgrass found from Manitoba to British Columbia in Canada, and south to Texas, U.S. Its general forage rating is good.

Indian sage. Boneset.

Indian salad. Virginia waterleaf.

Indian sanicle. White snakeroot.

Indian summer. A brief period in late autumn, usually after a killing frost, when the weather is unseasonably warm and the atmosphere is hazy.

Indian tea. Jerseytea ceanothus.

Indian tobacco. Aztec tobacco.

Indiantobacco lobelia. *Lobelia inflata,* family Lobeliaceae, an herb which is poisonous to livestock. Symptoms of poisoning are nausea, exhaustion, prostration, dilated pupils, stupor, coma, convulsions, and death. Found from Canada to Georgia and Nebraska to Arkansas, U.S.

Indian tree cotton. See *Asiatic tree cotton.*

Indian turnip. Indian Jack-in-the-pulpit.

Indian walnut. Candlenuttree.

Indian yellowflax. *Reinwardtia indica,* family Linaceae, an undershrub grown in the greenhouse or out-of-doors in warm areas for its yellow flowers. It is sometimes confused with golden flax (*q.v.*). Native to Indonesia.

India pappusgrass. *Paspalum scrobiculatum,* family Gramineae, a grass cultivated in the Orient and Australia. Also called *ditch millet.*

India raphiolepis. *Raphiolepis indica,* family Rosaceae, an evergreen shrub grown in very warm areas for its pinkish-white flowers in loose clusters and small, purple or black fruit. Native to China. Also called *Indian hawthorn.*

India rubber. See *latex.*

India-rubber fig. *Ficus elastica,* family Moraceae, a tree grown as a pot plant for its long, glossy, green leaves. It was once an important source of rubber. Native to India and Malaya. Also called *rubber plant.*

India-rubber vine. Palay rubbervine.

India wheat. Tartary buckwheat.

indicator. 1. A substance, used in testing soil for its reaction, that changes color according to the degree of acidity or alkalinity of the solution tested. 2. A gauge or measuring device on motor vehicles to indicate oil pressure, amount of gas in gas tank, water temperature, etc.

indicator host. A plant that will manifest symptoms for viruses that are latent in other hosts. There is no requirement that the manifested symptoms be specific for a particular virus, only that they indicate a virus content in the material being tested.

indicator plants. Plants which reflect the chemical or physical characteristics of soil or other peculiarity of habitat.

indifferent plant. Day-neutral plant.

indigenous. Endemic; designating species that are native in a region.

indigestion. A condition of the digestive system of man and other animals in which the normal digestive process is halted or disturbed. There are many causes, such as infection, spoiled food, over-eating, etc.

indigobush amorpha. *Amorpha fruticosa,* family Leguminosae, a hardy, deciduous shrub grown for its dull purple or blue flowers and purple pods. There are several botanical varieties of which *A. fruticosa* var. *angustifolia* is most often planted. Also called *false indigo; bastard indigo.*

indispensable element. Essential element.

individual claiming pen. On goat ranches, a pen in which an individual doe, that has disowned her kid, is kept until she reclaims her offspring.

individual drinking cup. A type of stock-watering equipment in wide use in modern stanchion dairy barns whereby water is supplied by a cup at the stanchion for each individual cow.

individual farrowing house. A separate building or quarters equipped so that sows that are to farrow may be kept isolated in warm, clean, and quiet individual pens at farrowing time.

individual selection. In breeding, selection of individual animals which have a desired character.

individual-stall barn. An ordinary barn equipped with an individual stall for each animal housed in it. It usually has storage rooms for feed and hay.

induction period. The number of times or periods of a given day-length to which a plant must be exposed to cause flowering. See *photoperiodism.*

indurated (soil). A soil horizon, cemented into a hard mass, that will not soften or lose its firmness when wet.

industrial forestry. The practice of forestry carried on to sustain a given industrial enterprise, as a saw mill, pulp mill, chemical plant, etc.

inedible egg(s). In marketing, eggs deemed unfit for human consumption as a result of such defects as stuck yolks, blood rings, visible embryos, dirt, decomposition, etc.

inert matter. Any material which is not active: *e.g.,* in seed analysis, any nonviable impurity; in feed analysis, grit or any matter that does not have nutritive value.

infect. To cause disease by the introduction of germs, parasites, or fungi. See *infection.*

infected. Harboring an infection.

infection. Invasion of the tissues of the body of a host by disease-producing organisms in such a way that injury results; the presence of multiplying parasites, bacteria, viruses, etc., within the body of a host.

infection age. The approximate age of inception of heartwood decay in a tree species.

infection court. The place where an infection may take place (leaf, fruit, etc.) on a new host.

infection stage. The period in the course of a disease during which the host responds, symptoms appear, and the disease develops.

infection thread. A fungous hypha that initiates infection of a host; a germ tube.

infectious. Designating a communicable disease.

infectious abortion. Brucellosis.

infectious anemia in horses. See *Equine infectious anemia.*

infectious (bovine) keratitis. A very common disease of cattle characterized by swollen, red, and congested eyelids, a watery discharge from the eyes which in time contains mucus and pus, cloudy eyeballs, and temporary or permanent blindness. The cause has not been established. Also called *pinkeye; contagious ophthalmia in cattle; infectious conjunctivitis.*

infectious bronchitis of chicks. A respiratory, virus disease characterized by sneezing, coughing, a peculiar breathing sound, nasal exudate, and watery eyes. Mortality rate is high in young chicks. See *bronchitis.*

infectious bulbar paralysis. Aujeszky's disease.

infectious catarrhal enteritis. See *hexamitiasis.*

infectious chlorosis. A graft-transmissible disease of green plants which causes irregular variegations, yellowing, or blanching of the leaves. The symptoms vary considerably among different species of plants; however, affected leaves may be stunted, and sometimes the plant dies. The disease is known to occur in the following families: Malvaceae, Celastraceae, Oleaceae, Leguminosae, Cornaceae, Rutaceae, and Rosaceae.

infectious conjunctivitis. Infectious (bovine) keratitis.

infectious coryza. A severe, catarrhal inflammation of the mucous membranes of the upper respiratory tract of poultry caused by *Hemophilus gallinarum.* Symptoms commonly include a nasal discharge, which becomes thick and sticky with an offensive odor; adjacent sinuses become filled with mucus which may change to a dry, cheesy form and cause bulging around the eyes. Affected fowls have difficulty in breathing, shake their heads frequently, are listless, have loss of appetite, and lose strength rapidly. Crowding, dampness, and lack of ventilation are predisposing factors. Also called *roup; cold; rhinitis.*

infectious diarrhea. Calf scours.

infectious disease. A disease caused by bacteria, protozoa, viruses, or fungi entering the body. It is not necessarily contagious or spread by contact.

infectious enteritis in pigs. A group of enteritis conditions in pigs caused by Salmonella bacteria (*Salmonella cholera suis*) and/or a group of other pathogenic enteric organisms such as *Spherophorus necrophorus,* coccidia, and viruses. The conditions are characterized by diarrhea and frequently by the presence of blood in the stool.

infectious enterohepatitis. See *blackhead.*

infectious feline agranulocytosis. Feline enteritis.

infectious gastroenteritis. Feline enteritis.

infectious laryngotracheitis. An acute, highly contagious, virus disease of fowls characterized by respiratory distress, coughing, watery eyes, sneezing, shaking of the head, sitting with eyes closed, gurgling breathing sound, and a stretching of the neck while inhaling. Also called *infectious tracheitis; flu.*

infectious necrotic enteritis. See *infectious enteritis in pigs.*

infectious necrotic hepatitis. An acute disease of sheep caused by *Clostridium novyi* in the presence of liver flukes. The onset is sudden and acute, and apparently healthy animals can be stricken and die overnight. Also called *black disease.*

infectious organism. An organism which may be transmitted from a diseased host to a healthy one and which is capable of inciting the disease under favorable environment.

infectious peritonitis of fowls. See *fowl plague.*

infectious pododermatitis. A type of foot rot (*q.v.*) caused by *Spherophorus necrophorus,* affecting the inner part of the foot of hoofed animals.

infectious sinusitis. A virus disease of turkeys characterized by a clear discharge from the nostrils, followed by a foamy condition of the eyes which are closed or partially closed, labored breathing, and loss of weight. Also called *swellhead; sinusitis.*

infectious tracheitis. Infectious laryngotracheitis.

infectious variegation. A form of psorosis virus disease of citrus characterized by crinkled, chlorotically variegated leaves (and misshapen fruit in lemons). See *infectious chlorosis; psorosis.*

infective. Capable of entering and establishing itself in a host; able to infect a susceptible plant or animal.

infective stage. The point in the development of an egg or larva at which it is capable of becoming established in the definitive host animal.

inferior. Designating the lowest or poorest grade of a product.

infertile. Designating that which is incapable of reproduction: *e.g.,* a barren female animal; a male animal with nonviable spermatozoa; an unfertilized egg; a flower that will not produce seed; a soil so low in basic elements that it will not produce a profitable crop.

infertile egg. 1. An ovum which has not united with a spermatozoon to form a zygote, or fertilized egg, hence, incapable of embryonic development. 2. An egg produced by an unmated female; a market egg produced by an unmated hen.

infest. To assail; attack; overrun; annoy; disturb, as ticks *infest* a cow, or as an insect *infests* a plant.

infestation. 1. Act of infesting, or state of being attacked, molested, vexed, or annoyed by large numbers of insects, etc., as an animal may be subject to an *infestation* of parasites, such as fleas, ticks, mites, etc. 2. Presence of disease in a population of plants, or of pathogens in a position, as in soil or on seed surfaces, where they have the possibility of producing disease. (Not to be confused with *infection* which can be applied only to living, diseased plants and animals.)

infested. Designating afflictions by such parasites as mites, ticks, etc. Sometimes used to designate the presence in an animal of such multicellular parasites as flukes, tapeworms, and roundworms.

infiltration. 1. The absorption of a liquid into a porous substance, as the entry of water into the soil, by way of openings such as natural pores, cracks, root and animal holes, and openings resulting from tillage. See *percolation.*

infiltration capacity. See *infiltration rate.*

infiltration complex. A group of similar areas which have comparable physical conditions of soil, slope, litter, and cover, and which possess a similar infiltration rate.

infiltration rate. The maximum rate at which a soil, in a given condition at a given time, can absorb water, commonly expressed in inches of depth per hour.

infiltrometer. A device used in measuring the rate of absorption into the soil of water applied uniformly over an area at a given rate or in a given volume.

inflammation. The reaction of a tissue to injury which tends to destroy or limit the spread of the injurious agent and to repair or replace the damaged tissues. The tissue is infiltrated by white blood cells, is red from capillary expansion and hot to the touch.

inflammation of the udder. See *mastitis.*

inflorescence. A cluster of flowers and the manner in which the individual flowers are arranged.

in flower. Flowering; blossoming; in bloom.

influence. The modifying effect of any environmental factor on plants or animals.

influenza in horses. A highly contagious virus disease characterized by a high temperature, general ill health, coughing, and nasal discharge. Also called *shipping fever; pinkeye of horses; horse flu.*

influenza in swine. An acute, highly contagious, infectious disease caused by the bacterium *Hemophilus influenzae suis* and a filterable virus; characterized by high temperature, thumpy coughing, discharge from the eyes and nose, and listlessness. The death rate is low. Also called *hog flu; swine influenza.*

in foal. Designating a pregnant mare.

informal decorative dahlia. A type of dahlia which has fully double flowers with the margins of most of the floral rays slightly rolled toward the lower side. The rays are usually long, twisted, or pointed and are irregular in arrangement.

infraorbital. Beneath or below the orbit or eye.

in full bloom. Designating a plant at the period during which most of its flowers are blossoming.

in full flow of milk. Designating a cow which is in heavy or maximum milk production, usually soon after freshening.

in full head. Designating the stage of development of a cereal or grass crop when the inflorescence has extended to its fullest length.

in full show condition. In stock judging, designating an animal which has been properly fitted and is in good condition for being exhibited.

infuse. 1. To steep in liquid so as to extract useful qualities, as tea. 2. To inject or introduce a liquid or medicinal agent, as to *infuse* the udder with a drug.

ingest. To eat, or take in food for digestion by way of the mouth.

ingesta. Any substance taken into the body through the mouth for nourishment.

ingestion. The taking of substances, usually food and water, into the body through the mouth.

ingluvies. The crop or craw of a chicken or other bird; also the rumen or first stomach of a ruminating animal, as a cow.

in good flesh. Designating an animal in a satisfactory condition of flesh showing good health and feeding.

ingrowth. The volume or number of trees that have grown past an adopted lower limit of measurement during a specified period.

inhalants. Medicinal preparations that are inhaled or drawn into the lungs.

in hand. Designating an animal which is under control or immediately available.

in harness. Designating horses at work in harness.

in heat. Designating a female animal at a period when she will accept coitus with a male. Also called *estrus.*

inheritance. 1. The transmission of genetic factors from parent to offspring. 2. The process or procedure of transferring property, both real and personal, from one generation to the next, either by will or by laws of descent and distribution.

inheritance tax value. The value of a farm or an estate used as a basis for the determination of state inheritance and federal estate taxes. It is established by a committee of three: one appointed by the probate court, one by the heirs or administrator of the estate, the third member being appointed by the first two committee members.

inherited characteristic. A character, the expression of which is determined by a particular gene or genes.

inherited soil characteristic. Any soil characteristic that is due directly to the nature of the parent material, as contrasted with those partly or wholly due to the processes of soil formation: *e.g.,* the red color of a soil is said to be inherited if it is due entirely to the fact that the parent material is red.

inhibit. To suppress; prevent; hinder; restrain.

inhibitor. Any agent, chemical, etc., which hinders, stops, suppresses, as a rust *inhibitor.*

in high condition. In stock judging, designating an animal that is in an excellent state of flesh and finish.

in high flesh. In stock judging, designating an animal which shows evidence of good feeding and has an excellent condition of flesh covering the body.

initial attack. The initial activity of a destructive, plant-eating insect (a term used by forest entomologists).

initial water deficiency. The amount by which the actual water-content of a given soil zone (usually the root zone) is less than field capacity at the beginning of a season or specified time period.

inject. To introduce a substance into the body of an animal or plant by mechanical means.

injurious insect. Any insect which causes injury of one sort or another to plants or animals.

injury by mice. The eating by mice of the bark and some of the xylem of many woody plants resulting in partial or complete girdling. The injury may be above, at, or below the surface of the soil and usually occurs in winter. Also called *mouse injury.*

injury by rabbits. The eating by rabbits of the bark and occasionally some of the xylem of woody plants, especially fruit trees. The injury usually occurs in winter and may be to the trunk, where girdling may occur, or to the lower, large branches. Small, low-hanging branches may be cut off. Also called *rabbit injury.*

inkberry. *Ilex glabra,* family Aquifoliaceae, a broad-leaved, evergreen or partly evergreen shrub grown for its foliage, turning rusty-green in the fall, and its pea-sized, black fruit. Native to North America. Also called *bear bush; winterberry; evergreen winterberry; gallberry.*

ink disease of chestnut. A destructive, fungous disease characterized by a black exudate or ooze on the trunk (Europe). Also called *black canker.*

ink smear. A bandage defect of cheese which results from carelessness with ink during branding.

inland saltgrass. *Distichlis stricta,* family Gramineae, a widely distributed grass of the western and midwestern states, U.S., whose general forage value is fair. Also called *desert saltgrass.*

inlaying. A graft which consists of removing an area of the stock and preparing a cion which will fit it exactly.

inlet. 1. A surface connection to a closed drain. 2. A structure at the diversion end of a conduit. 3. The upstream end of any structure through which water may flow. 4. A small indention in the bank or shore of a body of water.

in level. Designating that known and reproducible position of the bulk milk tank which results when a standard level-indicating means, such as a circular spirit level, two-way level, plumb-bob, etc., shows by its indication that the tank is in the normal, desired operating position.

in milk. 1. Designating a cow that is giving milk or milking. 2. Designating any animal during the period of milk secretion in contrast to the dry period.

inner bark. The physiologically active layer of tissues between the cambium and the last formed periderm in woody plants.

inner cover. A wooden cover normally used directly under the beehive cover to provide better air circulation and allow more ready access to the hive by preventing the hive cover from becoming stuck to the combs. It usually contains a center hole which may serve as a feed hole for sirup and also as a receptacle for a bee escape. It may also serve as a division board to separate the parts of a two-queen colony, for which a flight hole is made by cutting a small notch from the rim and covering the center hole.

in nice bloom. In stock judging, designating a healthy, vigorous animal with a good glossy hair coat.

innocence. Bluets.

inoculate. To introduce actual or suspected disease-producing organisms, such as fungi, bacteria, viruses, or toxins, into plants or animals to determine the pathogenicity of the organism, the resistance of the host, or to counteract other diseases, or influence the production of antibodies.

inoculation. 1. The act of inoculating. 2. The process of placing inoculum in an infection court. 3. The introduction of nodule-forming bacteria into soil, especially for the purpose of nitrogen fixation.

inoculum. The material that is introduced by inoculation.

inorganic. Not made up of or derived from plant or animal materials.

inorganic insecticides. Insecticide made from chemicals such as the arsenates of copper, lead and zinc.

inorganic soil. A mineral soil; a soil in which the solid matter is dominantly rock minerals in contrast to organic soils, such as peat and muck.

in season. 1. That part of the year when a product is normally harvested and, as a result, is cheaper, more plentiful, and more flavorful. 2. See *in heat.*

insect. An air-breathing animal (Phylum Arthropoda) which has a distinct head, thorax, and abdomen. *Insects* have one pair of antennae on the head, three pairs of legs, and usually two pairs of wings on the thorax. The opening of the reproductive organs is near the posterior end of the body. They may be harmful or useful depending upon their habits. Some infest plants and animals, some are insectivorous, some pollinate plants, and some produce edible products.

insectary. A laboratory in which insects are bred in large numbers. Also called *insectarium.*

insect control. The chemical or biological inhibition or killing of insect enemies.

insect enemy. 1. Any insect which is destructive or harmful to something desired by man. 2. An insect, bird, mammal, etc., which preys on insects.

insecticidal. Designating an insecticide or its lethal properties.

insecticidal cake. A briquette designed to deliver its insecticide slowly over a period of time.

insecticidal smoke. The fumes resulting from burning an insecticide mixed with some slow-burning material.

insecticide. A substance that kills insects by chemical action, as a stomach poison, contact poison, or fumigant.

insectivorous. Designating any animal that feeds on insects. Also applied to those plants which capture insects and absorb nutrients from them. The pitcherplant and the roundleaf sundew (*q.q.v.*) are sometimes so designated.

insect outbreak. A sudden increase in the number and destructiveness of insects which approaches the epidemic.

insect pests. See *insect enemy.*

insect powder. An insecticidal dust.

insect trap. Any mechanical device used to capture insects.

insect vector. An insect which carries a virus, bacterium, or the spores or mycelium of a pathogenic fungus and inoculates susceptible plants and animals.

insect wax. See *Chinese insect wax.*

inseminate. To place semen in the vagina of an animal during coitus, or, in the practice of artificial insemination, to introduce semen into the vagina by a method other than coitus.

inseminating tube. A rubber, glass, or metal tube, usually with syringe attachment, used in artificial insemination to introduce the semen into the vagina of a female animal. Such instruments vary in character depending on the species being bred.

insemination. The placing of semen in the reproductive tract of the female so that fertilization can take place.

insemination gun. A gunlike tube or instrument used in placing the semen capsule or unit into the cervix of the female animal. There are several different types depending on the semen unit used.

insemination tube. See *inseminating tube.*

insemination tube carrier. A cylindrical container with cap ends used for holding and transporting insemination tubes.

inseminator. The technician, in the employ of an artificial breeders unit, who brings the prepared semen to the farmer's herd and performs the technical service of inseminating the cows.

in service. Designating a sire being in condition for regular breeding service.

inside and outside quotation. In marketing, a price range between two figures, the low figure being the so-called inside figure and the high figure the outside quotation: *e.g.,* grade A eggs, 42 to 45 cents per dozen.

inside chuck. The bottom or inside cut of the chuck section of the forequarter of beef.

inside-lacquered can. A milk can lined with a special lacquer baked on the metal at a high temperature.

inside lining. That part of the bulk milk tank with which the milk comes in contact; usually made of 18-8 stainless steel.

inside shoe. A runner which supports the inner end of the cutter bar of a mowing machine when in operation.

insidious disease. A disease that develops slowly in a stealthy, subtle manner over a long period of time.

insipid. 1. Designating a flavor defect of cheese characterized by a lack of taste and odor. 2. Designating any such flavor.

insolation. 1. Solar radiation received by the earth. 2. The rate of delivery of solar radiation per unit of horizontal surface.

insoluble. Not soluble; designating a substance which does not dissolve in another. *Cf. soluble.*

insoluble ash. That portion of milk ash which is not soluble in water.

inspan. To hitch horses, oxen, etc., to a vehicle (South Africa).

inspissated. Designating sap or juice of plants thickened by evaporation, as maple sirup or molasses. Thickened, dried or rendered less fluid, as *inspissated* pus.

inspissosis. A fruit rot of citrus caused by either *Nematospora coryli* or *N. gosspii,* family Saccharomycetaceae, characterized by a thickening and drying of the internal tissues of the fruit with no external symptoms. Also called *dry rot.*

installment. 1. A partial payment of a debt often at regular intervals such as monthly or yearly *installments.* 2. A partial delivery of a given lot of produce, etc.

institute grade. One of a series of grades or numbers established by the American Meat Institute for rating the meat of an animal according to quality or economic worth.

instrumental insemination. The introduction of drone spermatozoa into the genital organs of a virgin queen bee by means of special instruments.

insufflation. The blowing of a powder or vapor into a cavity for medication.

insufflator. A specially designed blower for insufflation.

insulate. The protection of a material from heat or sound transfer or electrical potential.

insulated gate latch. In electric fences, an insulated latch which permits a person to open a gate without receiving a shock.

insulation. The product used to prevent heat or sound transfer.

insulator. 1. A substance which offers very high resistance to the passage of electricity; a nonconductor. 2. A device made of some nonconducting material, such as glass, used for fastening or supporting a conductor such as electric fence wire.

insulin. The hormone from a part of the pancreas that promotes the utilization of sugar in the organism and prevents its accumulation in the blood.

insurable value. The cost which would be required to replace a structure with identical materials and quality, less a deduction for depreciation.

insurance premium. The cost for contract which guarantees reimbursement for a property or business loss that might be caused by various types of events.

intake. The head works of a conduit; the place of diversion of water.

integrated farming. The combination of production, processing, and selling or distributing of agricultural crops as a unified business. Also called *vertical integration.*

integration method. A means of determining the mean velocity at a cross section station in a stream by noting the total number of revolutions of a current meter vane and the time consumed while the meter is slowly lowered at a uniform rate from the surface to the bed and returned to the surface. (Now considered a poor practice.)

integument. Any covering layer, as the skin of animals, the body wall of an insect, or the shell or outer layer of the ovules of plants.

intensity. The force, energy, and concentration with which a farm business is operated, usually based on the input of labor per acre. The higher the labor requirement per acre the greater the *intensity*.

intensity of rainfall. The rate at which precipitation is taking place at any given instant or time period, usually expressed in inches per hour.

intensive cropping. The practice of producing maximum yields of crops from a given area by special soil treatment and labor.

intensive farming. Farming in which a comparatively large amount of labor and working capital is used per tillable acre of farm land. Dairy, poultry, vegetable and fruit crops are examples of intensive enterprises. *Cf. extensive farming.*

intensive farm management. A method of operating a farm in such a manner as to increase the productivity of a given area by the expenditure of more capital and labor upon it; either or both crops and livestock may be involved.

intensive forestry. The practice of forestry to obtain the maximum in volume and quality of products per unit of area through the application of the best techniques of silviculture and management.

intensive grazing survey. A range survey which entails detailed mapping and data, usually including vegetation types of 5 acres or less, for a study of range conditions.

interbreeding. In livestock breeding, the breeding closely within a family or strain for the purpose of fixing type and desired characters.

interception. Precipitation caught by vegetation and evaporated before it reaches the ground.

interception channel. A channel excavated in the side or at the foot of slopes, or at other critical places, to intercept surface flow; a catch drain.

interception loss. The amount of water intercepted by vegetation from precipitation that eventually evaporates.

interchange. An exchange of segments of nonhomologous chromosomes.

intercrop. To grow two or more crops simultaneously, as in alternate rows in the same field or single tract of land. See *catch crop; companion crop.*

intercrossing. Crossbreeding.

interdigital. Between the fingers, toes, or claws.

interdigital pouch obstruction in sheep. A noninfectious disorder in sheep caused by the stoppage of a gland orifice between the toes, characterized by lameness, and swollen feet.

interference. In genetics, the process whereby crossing over in one region of a chromosome prevents or reduces crossing over in the next region.

interfering. The striking of the fetlock or cannon with the opposite foot as a horse walks or runs.

interfertility. The ability of a plant to produce viable seeds by cross pollination. Also called *cross fertility.*

interfluve. The area between two adjacent streams with the same direction of flow.

interfruitfulness. The ability of a variety of a fruit plant to mature fruit following pollination by another specific variety. Also called *cross fruitfulness. Cf. interunfruitfulness.*

intergeneric cross. A cross between species of different genera (rare).

intergrown knot. A knot whose growth layers are completely intergrown with those of the surrounding wood.

interior live oak. *Quercus wislizeni,* family Fagaceae, a broadleaved, evergreen tree browsed on the western range, U.S. Its browse rating is good. Native to California, U.S., and Mexico.

interior sap stain. Stain which is developed on the interior of sapwood boards and dimension stock during seasoning. There is no surface evidence of its presence.

interlocked grain. In lumber, grain in which the fibers incline in one direction in a number of growth layers, gradually reverse and incline in an opposite direction in succeeding growth layers, and then reverse again.

interlocking system. Half-loop system.

intermediate battery. A battery (*q.v.*) unit that is larger in size than the starting battery to allow for the growth of chickens. See *finisher: 2.*

intermediate crown class. Trees which have the crown below but still extending into the general level of the forest canopy which receive a little direct light from above but none from the sides. Usually they have small crowns considerably crowded on the sides.

intermediate cutting. Those cuttings made in a stand of trees between the time of its formation and its final harvest. They include various types of operations, the most important of which are cleaning, liberation cutting, thinning, and improvement cutting. *Intermediate cuttings* are not made with the aim of obtaining reproduction.

intermediate host. A host in which a parasite develops only in part, before getting into its final (or definitive) host, where it develops to maturity.

intermediates. The horse's incisors which lie between the corners and the centrals.

intermediate shelter belt. A narrow shelter belt which is grown to supplement a basic shelter belt. See *shelter belt.*

intermediate stock. A stock between cion and another stock.

intermediate-term credit. Credit extended for capital expenditures on a farm, particularly for the more expensive items with a relatively long life, such as tractors, bulldozers, combines, etc. The usual period of the loan is from one to five years. It is not a common type of loan at present. See *long-term credit; short-term credit.*

intermingling color. A coat color pattern of animals in which the separate colors merge where they meet. It is usually an objectionable color pattern.

intermittent attacks. The return of symptoms after

a period without symptoms: *e.g.*, recurrent attacks of malaria.

intermittent grazing. Grazing a pasture for indefinite periods with periods of rest between grazing.

intermittent parasites. Those parasites, such as mosquitoes or bedbugs, which approach the host only when in need of nourishment.

intermittent stream. 1. A stream which flows during the wet season, the flow being principally derived from surface runoff.

Intermountain Region. The great land area lying between the Rocky Mountains and the Sierra Nevadas, U.S. Its climate varies from hot and arid to cool and humid. It has regions of irrigated farmlands, orchards, dry-land farms, ranges, and desert wastes.

internal bark necrosis. A nontransmissible condition of pome fruit trees characterized by elevations scattered over the surface, brown necrotic areas in outer bark, stunted growth, and roughened appearance on the limbs. Studies indicate it may be due to a mineral imbalance.

internal black spot. A boron-deficiency disease of beets characterized by the young leaves becoming redder and narrower than normal, a downward rolling of the leaves, adventitious buds, and a spotting of the root with black, corky areas of dead cells.

internal breakdown. 1. A disease of cabbage in New York, U.S., of unknown cause, characterized by a purplish-gray discoloration of the veins of the leaves and translucency of the surrounding tissue. The affected portions die, becoming brown or brownish-black and papery. Usually only one leaf in a head is affected. 2. A breakdown of internal tissue in fruits, vegetables, etc.

internal browning of apples. A nonpathogenic, storage disease characterized by brown streaks in the flesh of normal appearing fruit. It is found in some varieties which are stored at or near the freezing point.

internal browning of cauliflower. A boron-deficiency disease characterized by small, concentric, water-soaked areas in the stem and central branches of the curd, poor growth, yellowing of the terminal growth, and death of the terminal buds.

internal browning of pome fruits. See *internal browning of apples.*

internal brown spot. A nonparasitic, necrotic disorder which develops, during seasons of drought and high temperatures, in potatoes grown on light, gravelly soil. Also called *sprain* (British).

internal cork of apples. A boron-deficiency disease characterized by poor growth, yellowing of the terminal growth, and death of the terminal buds. The terminal growth becomes brittle, breaks easily, and shows vascular discoloration. There are light brown spots of dead tissue within the fruit.

internal decline. Endoxerosis.

internal double-run force feed. A feeder for a drill planter consisting of a double-faced wheel having a small and large side. The small side is used for planting small seeds while the large side is used for planting larger seeds, such as oats, wheat, peas, and beans.

internal fat. Offal fat.

internal layer. A hen whose egg yolks repeatedly escape into the body cavity instead of entering the oviduct.

internal necrosis. Death of tissues within the body of a plant or animal.

internal-tube heat exchanger. A device which is used to cool or heat liquids, especially to cool milk, consisting of a tube within a tube. The liquid to be treated flows in the inner tube while the cooling or heating medium flows in the outer pipe. Also called *double-tube heat exchanger.*

International Code of Botanical Nomenclature. A system, adopted by European botanists and some American botanists, for applying the Latin names to plants. Established by a congress in Vienna in 1905, it was revised in 1931 when it was universally accepted, replacing, in the U.S., the American Code of Botanical Nomenclature (*q.v.*). It provided that most plants shall retain the name applied by Linnaeus in his *Species Plantarum* (1753). It accepted the idea of nomen conservandum (*q.v.*) and provided that after January 1, 1931, publication of all new species be accompanied by a Latin diagnosis. Also called *Vienna Code; Vienna Rules.*

international rule. In forestry, a formula rule allowing ½in. taper for each 4 ft. of length and ⅛in. shrinkage for each 1 in. board; closely approximating the green chain lumber tally, in one form it assumes a ⅛in. kerf; in modified form it assumes a ¼in. kerf. It is an official rule of the U.S. Forest Service.

International Unit. A unit of measurement for potent medications as defined and adopted by the International Conference For Unification of Formulas.

International Wheat Agreement. An international agreement regarding wheat which came into effect on August 1, 1949. The objectives of the agreement are "to assure supplies of wheat to importing countries and markets for wheat from exporting countries at equitable and stable prices." To accomplish these purposes the agreement is limited to guaranteeing exporting countries an assured market for a definite quantity of wheat at not less than the agreed minimum price and assuring importing countries a definite quantity at not more than the maximum price.

internode. That part of a stem of a plant between the joints or nodes.

interplant. 1. To set out young trees among other existing young trees, planted or natural. It is applicable also to planting land partly occupied by brushwood. 2. To plant seeds of one species between those of another species, as to plant soybeans with corn. 3. To plant one crop in rows and another between the rows: *e.g.*, peanuts grown between rows of cotton or corn.

interregional competition. The competition of one agricultural region with another for the market of the same or similar agricultural product or products: *e.g.*, Maine, Michigan, and Idaho potatoes competing in the same market. Production costs, yields, quality, prices, and transportation are important factors in determining the comparative advantage of one region over another.

interrow tillage. The tillage between the rows of growing plants to condition the soil and reduce weed growth.

interrupted. Designating a plant stalk not continuously furnished with leaves or flowers.

interrupted-fern. *Osmunda claytoniana,* family Osmundaceae, a coarse, upright fern grown for its foliage. Also called *Clayton fern.*

intersex. Designating an organism which displays primary and secondary sexual characteristics intermediate between male and female.

interspersion. The intermingling of different vegetation types.

intersterility. The inability of a variety of fruit to produce viable seeds if only the pollen of another variety is used. Also called *cross sterility.*

intertillage. *See interplant: 3.*

interunfruitfulness. The inability of a variety of plant to mature fruit if only the pollen of another variety is used. Also called *cross-unfruitfulness.*

intervales. The valley lands which occur in a hilly country; includes the terraces of river valleys.

interveinal. Between the veins, especially of a leaf.

interveinal mosaic. A virus disease of potatoes characterized by pale patches between the veins of the leaf, the tissues immediately adjacent to the veins remaining green, and undulations of the leaf margins.

intestinal bloat. Abnormal distention of the intestines with gas. See *acute indigestion in horses.*

intestinal coccidiosis. See *coccidiosis.*

intestinal mycosis. See *thrush: 2.*

intestinal worm. Any internal parasitic worm that inhabits the intestines.

intestine. The lower portion of the alimentary canal from the stomach to the anus. Also called *gut; bowels.*

in the bud. Designating a stage of growth of alfalfa when most of the plants have well-developed blossom buds, but few are in bloom; or the stage of growth of a fruit orchard just prior to blooming.

in the ear. Designating a stage of development of corn when the young ears are well developed on most of the stalks in the field.

in the milk. Designating a stage of maturity of the corn ear when the corn kernels may be easily broken open by a squeeze of the fingernail and will exude a white, milk-like juice.

in the rough. Designating a method of packing celery in crates in the field without any preparation except stripping the damaged and discolored leaves.

in the swath. Designating a newly mowed field of hay before it is cured enough to rake it into the windrow.

in the tassel. Designating the emergence of the male corn flower, the tassel, on the tops of corn stalks.

in thin flesh. Thin; not fleshed out.

into leaf. Leafing out; developing leaves.

intolerance. The inability of a woody plant to grow well in a particular environment; commonly used to denote trees that cannot endure heavy shade.

intoxication. The state of being poisoned or intoxicated.

intracaudally. Situated or applied within the tissues of the tail.

intracellular. Within, inside of, a cell.

intracellular fungus. A fungus that occurs or grows within a cell.

intracervical method. A method of artificial insemination whereby semen is placed directly in the cervix or uterus and not in the vagina.

intractability. The characteristic of a soil which causes it to be difficult to work.

intradermal. Within the layers of the skin.

intradermally. Situated or applied within the skin.

intradermal spore vaccine. A type of anthrax spore vaccine that is injected directly into the skin and not under the skin.

intradermal tuberculin test. A test, used to discover tuberculosis in animals, which consists of injecting into the skin of the tail (caudal fold), or elsewhere, a minimum of concentrated tuberculin. If the animal is infected, 72 hours after the injection there will be a swelling at the point of injection.

intradermic pullorum disease test. See *wattle test.*

intramuscular injection. Injection of a substance into a muscle.

intramuscularly. Within or into the muscle, in connection with injections of drugs or other treatments.

intranasal instillation. The placing of fluid or medicine inside the nose.

intraperitoneal. Within the cavity of the body that contains the stomach and intestines.

intratracheal. Within the trachea, or windpipe.

intra-uterine. Within the uterus.

intravenous. In, into, or from within, a vein or veins.

intravenous injection. The introduction of a substance into the blood stream by way of a vein.

Intrazonal Soil. Any of the great groups of soil with more or less well-developed characteristics that reflect the influence of some local factor of relief or parent material, or both, over the normal effect of climate and vegetation. Each *Intrazonal Soil,* such as Ground Water Podzol, Bog, Solonetz, or Rendzina may be found associated with two or more groups of zonal soils.

introduced. Designating a plant, animal, disease, etc., which is not indigenous to an area, but which is brought in purposely or accidentally.

introduced pine sawfly. *Diprion similis* (Htg.), family Diprionidae, a fly whose larva infests pines, feeding on the needles. Native to northeastern U.S.

introducing cage. A box made of wire screen and

wood used in shipping queen bees and for introducing a new queen to a colony.

introduction. The transfer of a plant, animal, or disease to a region in which it is not native.

intromittent. Designating the use, in copulation, of the external reproductive parts of many male animals.

intumescence. The knotlike or pustule-like enlargement on various organs, such as stems, leaves, or fruits, produced as a result of the enlargement of groups of cells.

in utero. See *intra-uterine.*

invader. See *invading plant.*

invading plant. A species of plant that comes in on lands after the more stable, climax plants have been diminished in the stand by drought, fire, or overgrazing.

invasion. The action or process by which plants or animals successfully establish themselves in a new area or environment in competition with the native occupants.

inventory. A list of property which shows the amounts and values of all physical assets involved; *e.g.*, in a farm business, a list of amounts and values of land, buildings, equipment, etc. In more strict accounting terms, farm *inventory* is restricted to livestock, feed growing crops, and supplies. Buildings, machinery, etc., are included in the depreciation schedule.

inventory control. The control of the annual cut from a forest area in which an inventory of the financially mature timber is made; the time required for the trees next in size to replace the volume of this mature timber is estimated, and the cut is spread over this period of time.

inventory records. A list of the amounts and values at the beginning and at the end of the accounting year of the land, farm improvements, machinery, livestock, feed, crops, and supplies.

inversion. 1. In meteorology, the reverse of the usual lapse rate of temperature in the vertical. Under normal conditions the temperature decreases with increased altitude. In the lower layers, under radiation frost conditions, the temperature, starting from the ground, may increase with elevation for some distance before changing to the normal lapse rate. 2. A rearrangement of a group of genes in a chromosome in such a way that their order in the chromosome is reversed. 3. The process in which cane sugar is changed into dextrose and levulose.

invertase. An enzyme produced by growing yeasts which converts sucrose to glucose and fructose.

invertebrate. Any animal which does not have a backbone or spinal column.

inverted T. In budding, a method of making an incision on the stock in which a vertical cut about 1½ ins. long is made at a point where the bud is to be inserted. At the base of this cut a horizontal cut is made so that the two cuts resemble an *inverted T.*

inverted teat. See *blind quarter.*

inverting. The turning of the soil by plowing.

inviscate. 1. To trap in a sticky substance or on a sticky surface, as insects on fly paper. 2. To render a substance more sticky by mixing it with viscous matter.

iodine. A trace element in soil, and an essential element for the health of man and animals. It is readily absorbed by plants from soil and water, but its value to them is problematical.

iodine number. 1. A measure of the degree of saturation of fats and oils. 2. The grade index for quality of linseed oil ranging from 175-201. The higher the *iodine number,* the better the quality.

iodized salt. That salt prepared by mixing salt with potassium iodide in the proportion of about one part of iodide to 5000 parts of salt. *Iodized salt* is of value in the prevention of simple goiter which is endemic in animals in certain parts of the U.S.

ion. An atom or a group of atoms carrying an electrical charge, which may be positive or negative. *Ions* are usually formed when salts, acids, or bases are dissolved in water. When common salt, sodium chloride, is dissolved in water, positive sodium *ions* and negative chloride *ions* are formed. See *cation; anion.*

ion-exchange. The replacement, in a colloidal system, of one ion by another with a charge of the same sign.

ion-exchanger. A water-softening system, such as the zeolite type, which functions through the exchange of ions, especially of sodium for calcium and magnesium; in water, the sodium salts of certain compounds readily release sodium and take in its place calcium and magnesium.

Iowa ear-corn probe. A probe used to obtain a sample of ear corn from the bin, consisting of a shaft and a handle attached at right angles at one end. A portion of the shaft is hollow and has saw teeth at an opening. When the shaft is twisted, kernels are shelled and fall into the hollow portion of the shaft for removal and inspection.

iris. Any perennial herb of the genus Iris, family Iridaceae. Species are grown in flower gardens for ornament. Also called *fleur-de-lis.* See *dwarf iris; Crimean iris; sweet iris; Hungarian iris; Trojan iris; Mardin iris; Siberian iris; yellowband iris; Japanese iris; rabbitear iris; yellowflag iris; blueflag iris; Dixie iris; Lamance iris; copper iris; Oregon iris; Rocky Mountain iris; Algerian iris; netted iris; orchid iris; crested iris, slender iris; roof iris; bearded iris; beardless iris; bulbous iris.*

iris borer. *Macronoctua onusta* Grote, family Noctuidae, an insect which infests iris. The larva enters the leaf by boring a small hole, eats its way down to the rhizome where it feeds on the internal tissue. Infested plants are characterized by water-soaked, ragged leaves.

Irish blight. See *late blight and rot of potato.*

Irish cobbler. An early-maturing variety of potato widely cultivated in the U.S. See *potato.*

Irish juniper. *Juniperus communis* var. *hibernica,* family Pinaceae, an upright, columnar, evergreen tree grown as an ornamental accent plant. Native to North America.

Irishmoss. *Chondrus crispus,* family Gigartinaceae, an edible seaweed. A jelly obtained from it is used in foods in making blancmange and as a stabilizer in ice cream mixes, also in laboratories as a nutrient medium for microorganisms. The marine alga, *Gigartina mamillosa,* is also sometimes called *Irishmoss* and may be mixed with this species as a source of the jelly or gum, carrageen. Found off the coasts of Ireland and New England. Also called *carrageen; carragheen; pearlmoss.*

Irish potato. See *potato.*

Irish (potato) blight. See *late blight and rot of potato.*

irisleaf buglelily. *Watsonia iridifolia,* family Iridaceae, a bulbous herb grown for its pink flowers. Native to South Africa.

iris leaf spot. A fungous disease caused by *Didymellina macrospora,* family Mycosphaerellaceae, characterized by small, brown spots with water-soaked margins on the leaves. The spots enlarge, turn yellow and brown, and dry up. The leaves dry prematurely and in severe attacks the whole plant dies.

iris root aphid. See *tulip bulb aphid.*

iris thrips. *Bregmatothrips iridis* Watson, family Thripidae, an insect pest of iris whose larvae feed on the inner leaves. Infested plants are russeted, stunted, with the tips of the leaves brown and dead.

iris weevil. *Mononychus vulpeculus* (F.), family Curculionidae, an insect which infests iris, especially blueflag iris. Its larva feeds on the seeds and pupates in the pods.

iron. Fe, the metallic element, which is essential for normal plant and animal development. In plants it is connected with the functioning of chlorophyll. It is present in abundance in most soils, but not always in available forms; in extreme cases the concentration in the soil solution may be so great as to be toxic to plants.

ironbark. Any of several timber trees of the genus Eucalyptus, family Myrtaceae, which have hard, gray bark. See *mulga ironbark eucalyptus; kino eucalyptus.*

iron chlorosis. A yellowing of the foliage of plants due to an iron deficiency.

iron deficiency. See *iron chlorosis.*

iron deficiency anemia. See *anemia.*

iron gray. A coat color term for a horse which (a) in the U.S., denotes a white coat with a high percentage of black hairs; (b) in England, a gray coat with reddish hairs throughout.

iron phosphate. $FePO_4$, a compound which occurs in phosphate rock. If it is present in large quantities, it has the tendency to cause super-phosphate made from the rock to be damp and sticky.

iron plant. See *aspidistra.*

iron sulfate. See *ferrous sulfate.*

iron tannate stain. A bluish-black stain of oak and other tannin-bearing woods caused by a chemical reaction between iron and tannic acids on surfaces in contact with saws, wedges, or mill machinery.

iron tie. A long, thin, narrow, iron band used to bind a bale of cotton.

ironweed. Any perennial herb of the genus Vernonia, family Compositae. Species are often tall weeds in pastures, neglected fields, etc.

ironwood. See *American hophornbeam; American hornbeam.*

ironwood of Pegu. See *acle: 1.*

irradiated. Designating a food or feed which has been treated with ultraviolet light to increase the vitamin D content.

irradiated milk. Market milk treated with ultraviolet light so that the ergosterol is changed to vitamin D.

irradiated yeast. Yeast subjected to ultraviolet rays in order to increase its anti-rachitic potency. It is used as an ingredient in feeds for its vitamin D and B-complex content.

irradiation. The subjecting of products to ultraviolet light to increase the vitamin D content.

irregular cotton. Poor lint cotton.

irregular eyes. An eye defect of Swiss cheese characterized by distorted, elongated, walnut-shell shaped eyes.

irregular market. A marketing condition in which prices for a particular grade of product on a particular day vary over a wide range.

irrigable area. The area under an irrigation system capable of being irrigated, as determined principally upon the quality of soil and elevation of land. It generally includes, in addition to farm land, roads, farm lots, building sites, and miscellaneous areas not actually irrigated.

irrigate. To furnish water to the soil for plant growth in place of, or in addition to, natural precipitation, by surface flooding or sprinkling and subirrigation methods, using surface water or that from underground sources.

irrigated. Designating land or crops which have received water from other sources as well as from natural precipitation.

irrigated land preparation. In surface irrigation, the reshaping of the ground surface to facilitate a more uniform and less wasteful application of irrigation water.

irrigating. The furnishing of water to soil for plant growth other than by natural precipitation.

irrigating head. 1. The measure of stored up water ready to be used in irrigation. 2. The depth of a body of water in covering land, as by flooding. The proper wetting of the whole ground. 3. The indication on a measuring device of the rate of flow of irrigation water.

irrigating stream. 1. The flow used for irrigation of a particular tract of land. 2. The flow of water distributed at a single irrigation or in a single farm lateral. 3. The flow which is rotated among a group of irrigators. (Sometimes called *irrigating head.*)

irrigation. The artificial application of water to soil for the purpose of plant production.

irrigation district. A cooperative, self-governing, public corporation set up as a subdivision of the state, with definite geographic boundaries, organized to obtain and distribute water for irrigation of lands within a district. A district may have taxing power under authority of the state legislature with the consent of a designated fraction of the landowners or citizens (U.S.).

irrigation ditch. A small irrigation canal.

irrigation efficiency. The ratio of the water consumed by crops of an irrigated farm or project to the water diverted from a river or other natural source into the farm or project canals. Also expressed as a percentage.

irrigation requirement. The quantity of water, exclusive of precipitation, required for crop production, including economically unavoidable wastes.

irrigation water. The quantity of water artificially applied in the process of irrigation. It does not include precipitation.

irrigator. One who applies water to land for growing crops.

is by. In horsebreeding, indication of the male parent, as Bird *is by* Captain Courageous. (Most often used in connection with purebred or race horses.) See *out of.*

Isigny. A Camembert-type cheese which is washed and rubbed during curing to check the growth of molds on the surface.

island myrtle. Feltleaf ceanothus.

island of Langerhans. The tissue lying in the framework of the pancreas which secretes the hormone insulin.

Island of Orleans. A soft, whole milk, cured cheese made in Canada.

islay. Hollyleaf cherry.

Isle of Wight disease. See *acarine disease: 2.*

isobar. A line on a chart or diagram which connects points having the same barometric pressure, most commonly drawn on weather maps to show relative pressure in terms of inches of mercury or millibars.

isobutyric acid. See *butyric acid.*

isodyne. A line drawn on a map of a cultivated field which connects points having equal dynamometer readings of the force required to pull a cultivating implement.

isogamy. Morphological similarity of fusing gametes.

isohyet. A line drawn on a map or chart to indicate places or points of equal rainfall.

isolate. To place an animal in confinement away from other animals to prevent breeding or spread of disease.

isolated form of settlement. That form of agricultural settlement in which the farmers' residences and farm buildings are located outside the village on the land farmed. This form is sometimes referred to as the dispersed or scattered settlement pattern. *Cf. village form of settlement; line village form of settlement.*

isolation strip. An area surrounding a sample plot treated the same as the plot.

isolation transect. A strip of fenced vegetation divided into rectangular plots opened or closed to grazing each year in succession.

isotherm. A line on a chart or diagram drawn through places or points having equal temperature.

isotopes. Atoms, of the same element, having different weights. Radioactive *isotopes* are used, among other things, in agricultural research.

isotype. A type or form of animal or plant common to different regions.

issue shoe. Keg shoe.

Italian aster. *Aster amellus,* family Compositae, a perennial herb grown for its purple flowers. Native to Eurasia.

Italian bellflower. *Campanula isophylla,* family Campanulaceae, a tender, trailing, perennial herb grown in a hanging basket for its habit and for its bell-shaped, pale blue flowers. Native to Italy.

Italian broccoli. See *broccoli.*

Italian buckthorn. *Rhamnus alaternus,* family Rhamnaceae, a large, bushy, ornamental, evergreen shrub, bearing bluish-black fruit, grown in warm climates. Native to the Mediterranean region. Also called *evergreen privet; alatern.*

Italian bugloss. *Anchusa azurea,* family Boraginaceae, a perennial herb, grown for its clusters of bright blue flowers. Native to southern Europe.

Italian clematis. *Clematis viticella,* family Ranunculaceae, a hardy, slender, woody vine, bearing blue or purple flowers, grown for ornament on a trellis, fence, etc. Native to Eurasia. Also called *vine bower.*

Italian clover. Crimson clover.

Italian cornsalad. *Valerianella eriocarpa,* family Valerianaceae, a salad crop which has lighter-colored leaves and thrives better in warm climates than the European cornsalad. Native to southern Europe.

Italian cypress. *Cupressus sempervirens,* family Pinaceae, an evergreen tree grown in warm areas. Native to southern Europe.

Italian fennel. See *hemp.*

Italian honeysuckle. See *sweet honeysuckle.*

Italian jasmine. *Jasminum humile,* family Oleaceae, a tropical, spreading, evergreen or semi-evergreen shrub grown in warm areas for its foliage and its fragrant, golden-yellow flowers. Native to Asia.

Italian kale. A variety of turnip with no tuberous thickening, grown as a potherb.

Italian millet. See *foxtail millet.*

Italian pear scale. *Epidiaspis piricola* (Del Guer.), family Diaspididae, a sucking insect which infests pear, peach, plum, prune, apple, Persian walnut, and especially Christmasberry in California, U.S.

Italian prune leaf spot. A leaf mottle and necrotic spot condition of Italian prune, a leading variety of prune (*q.v.*), bud-perpetuated but nontransmissible, responsible for serious losses of quantity and quality in Italian prunes and characterized by abnormally fewer leaves and small set of fruit.

Italian reed. Giant reed.

Italian ryegrass. *Lolium multiflorum,* family Gramineae, a short-lived perennial grass, used as a pasture grass and in lawn mixtures. Native to Europe. Also called *Lovitto ryegrass; Australian ryegrass.*

Italian stone pine. *Pinus pinea,* family Pinaceae, an evergreen tree with stiff, lustrous leaves, grown in warm areas as an ornamental and for its edible seeds. Native to southern Europe.

itch mite. *Sarcoptes scabiei* (DeG.), family Sarcoptidae, a mite which infests man and other animals. In man it causes itch and in animals mange.

Ithuriel's spear. Grassnut brodiaea.

-itis. A suffix denoting inflammation, as *dermatitis,* inflammation of the skin; *tonsillitis,* inflammation of the tonsils.

ivary. Darnel ryegrass.

ivory horsetail. Giant horsetail.

ivory nut meal. A livestock feedstuff left as ground waste material in the manufacture of buttons and similar articles from the vegetable ivory nut.

ivy. 1. Any plant of the genus Hedera, family Araliaceae. Species are evergreen, climbing, woody vines. See *Algerian ivy; Colchis ivy; English ivy.* 2. See *Virginia creeper; Japanese creeper; ivy groundsel; Kenilworth ivy; ivy treebine; Solomon Islands ivyarum; ivyvine pelargonium; India ivy gourd.*

ivy aphid. *Aphis hederae* (Kltb.), family Aphidae, an aphid which infests English ivy, European euonymous, cranberrybush viburnum, and aralia.

ivyarum. See *Solomon Islands ivyarum.*

ivy geranium. Ivyvine pelargonium.

ivy gourd. See *India ivy gourd.*

ivy groundsel. *Senecio mikanioides,* family Compositae, a tender, herbaceous, climbing perennial grown in the window box and as a house plant for its yellow flowers in small heads. Native to South Africa. Also called *German ivy.*

ivyleaf cyclamen. *Cyclamen indicum,* family Primulaceae, a tuberous, perennial, stemless herb grown by florists as a pot plant for its white or rose colored flowers, dark purple at the mouth. Native to Eurasia.

ivy scale. See *oleander scale.*

ivy treebine. *Cissus incisa,* family Vitaceae, a tender, woody, climbing vine grown for its handsome foliage. Native to North America. Also called *marine ivy.*

ivyvine pelargonium. *Pelargonium peltatum,* family Geraniaceae, a tender, green, perennial herb grown for its trailing stems and white to deep rose flowers. Native to South Africa. Also called *ivy geranium.*

J

jaboticaba. *Myrciaria cauliflora,* family Myrtaceae, a small, aromatic, tropical tree grown for its small, purple, edible fruit. Native to Brazil.

jaburan. See *white lilyturf.*

jacaranda. Any tropical shrub or tree of the genus Jacaranda, family Bignoniaceae. Species are popular ornamentals. Native to America. See *sharpleaf jacaranda; toothleaf jacaranda.*

jack. 1. The male, uncastrated ass. See *jennet.* 2. A mechanical, hydraulic device used for lifting heavy objects. 3. See *bone spavin.*

jackaroo. An apprentice on an Australian sheep station.

jackass. See *jack: 1.*

jackbean. 1. Any herb or vine of the genus Canavalia, family Leguminosae. Species are grown for human and animal consumption and for ornament. 2. *C. ensiformis,* an annual herb grown in warm areas for its purple flower clusters and edible seeds. Native to West Indies.

jackfruit. Jakfruit.

Jack-in-the-pulpit. Any perennial herb of the genus Arisaema, family Araceae. Also called *dragon arum.* See *Indian Jack-in-the-pulpit.*

Jackman clematis. *Clematis jackmani,* family Ranunculaceae, a hybrid, hardy, woody vine with light violet flowers, grown on the trellis, fence, etc., as an ornamental.

jack pine. *Pinus banksiana,* family Pinaceae, a scrubby, evergreen tree which usually grows on dry, sandy soils. Generally considered inferior in quality, its wood is used for pulp, ties, posts, fuel, rough construction, boxes, and planing-mill products. Native to northern U.S. Also called *scrub pine; gray pine.*

jack rabbit. Any of the long-eared, large hares of the genus Lepus, found in western U.S. Species have long hind legs and often feed on crops and damage citrus trees.

jack spavin. Bone spavin.

jackstock. A male ass used as a stud on a mule-breeding farm.

Jacobean lily. Aztec lily.

Jacob's ladder. Greekvalerian polemonium.

Jacobsrod. *Asphodeline lutea,* family Liliaceae, a perennial herb grown for its yellow flower clusters. Native to southern Europe. Also called *asphodel.*

Jacob's staff. Ocotillo.

jade plant. Silver crassula.

Jaffas. Shamouti orange.

jaggery palm. See *toddy fishtailpalm; gomuti sugarpalm.*

jakfruit. *Artocarpus heterophyllus* (*A. integra*), family Moraceae, a tropical tree grown for its large, edible fruit. Native to Asia. Also called *jak.*

jam. 1. A preserved fruit food prepared by boiling the whole fruit pulp with sugar (sucrose), in the ratio, 45:55, to a jelly-like consistency. 2. Fruit pulp cooked with sugar to jelly consistency (British).

Jamaica goldfern. *Piturogramma sulphurea,* family Polypodiaceae, a tropical, evergreen fern grown as a house plant for decoration.

Jamaica honeysuckle. Watermelon.

Jamaica linden hibiscus. *Hibiscus tiliaceus* var. *elatus,* family Malvaceae, a small, tropical tree grown for its pale pink flowers which change to deep red by evening. It is also the source of Cuban bast, a fiber used for tying cigars, etc. Native to West Indies. Also called *emajagua; mountain mahoe; Cuban bast.* See *linden hibiscus.*

Jamaica sorrel. Roselle.

jambo. Roseapple.

Jamestown weed. Jimsonweed datura.

January thaw. A period of warm weather which occurs regularly during the month of January, especially in eastern U.S.

Japan cedar. Cryptomeria.

Japan clover. See *lespedeza: 2.*

Japanese anemone. *Anemone japonica,* family Ranunculaceae, a perennial herb grown for its flowers of almost every color except yellow and blue. Native to eastern Asia.

Japanese angelica tree. Japanese aralia.

Japanese aralia. *Aralia elata,* family Araliaceae, a prickly shrub or small tree grown for its foliage, whitish flower clusters, and black fruit. Native to Japan. Also called *Japanese angelica tree; fatsia.*

Japanese artichoke. Artichoke betony.

Japanese astilbe. *Astilbe japonica,* family Saxifragaceae, a perennial herb grown for its terminal clusters of white flowers. Native to Japan. Also called *feather fern.*

Japanese aucuba. *Aucuba japonica,* family Cornaceae, an evergreen shrub grown by florists for its glossy foliage and scarlet fruit. Native to east-central Asia.

Japanese azalea. *Rhododendron japonicum,* family Ericaceae, a tender, deciduous shrub grown for its bell-shaped, orange to scarlet flowers. Native to Japan.

Japanese barberry. *Berberis thunbergi,* family Berberidaceae, a hardy shrub widely grown in the shrub border for its yellow, red-tinged flowers, bright red fruit and brilliant scarlet, autumnal foliage. It is resistant to wheat rust. Native to Japan. A dwarf variety, *B. t. minor,* is called *box barberry.*

Japanese barnyard millet. See *barnyardgrass.*

Japanese beetle. *Popillia japonica* New., family Scarabaeidae, a beetle which feeds on plants in eastern U.S. The adults attack the leaves and the fruit; the larvae feed on grass, vegetables, and nursery stock.

Japanese black pine. *Pinus thunbergi,* family Pinaceae, a hardy, evergreen tree grown as an ornamental. Native to Japan. Also called *Thunberg pine.*

Japanese cane. A variety of sugar cane used for fodder.

Japanese cherry. Oriental cherry.

Japanese chestnut. *Castanea crenata,* family Fagaceae, a deciduous tree, grown for its edible nuts and for use in breeding blight-free chestnut trees. Native to Japan.

Japanese clawfern. *Onychium japonicum,* family Polypodiaceae, a small fern grown as a house plant. Native to Japan.

Japanese clematis. See *sweetautumn clematis.*

Japanese creeper. *Parthenocissus tricuspidata,* family Vitaceae, a hardy, woody, climbing vine, grown as a wall covering for its autumnal foliage. Native to Asia. Also called *Boston ivy.*

Japanese evergreen oak. *Quercus acuta,* family Fagaceae, a small, broad-leaved, evergreen tree, grown in warm areas as an ornamental. Native to Japan.

Japanese farcy. Mycotic lymphangitis.

Japanese feltfern. *Cyclophorus lingua,* family Polypodiaceae, a tropical fern grown in greenhouses in the U.S., as an ornamental. Native to Asia. Also called *Japanese fern; felt fern; tongue fern.*

Japanese fern. Japanese feltfern.

Japanese field mint. *Mentha arvensis* var. *piperascens,* family Labiatae, a perennial herb grown as a source of peppermint oil. Native to Eurasia and North America.

Japanese fleeceflower. *Polygonum cuspidatum,* family Polygonaceae, a hardy, perennial herb grown as a hedge or screen plant. Escaped, it can become a pest. Native to Japan. Also called *Mexican bamboo; Japanese knotweed.*

Japanese flowering crabapple. *Malus floribunda,* family Rosaceae, a deciduous tree grown for its rose-red flowers which change to white. Native to Asia. Also called *showy crabapple.*

Japanese flowering quince. *Chaenomeles japonica,* family Rosaceae, a hardy shrub grown for its brick-red flowers and yellow fruit. Native to Japan. Also called *dwarf Japanese quince.*

Japanese hemlock. *Tsuga diversifolia,* family Pinaceae, a hardy, evergreen tree grown as an ornamental. Native to Japan.

Japanese holly. *Ilex crenata,* family Aquifoliaceae, a broad-leaved, evergreen shrub grown for its foliage and black fruit. Native to Japan.

Japanese honeylocust. *Gleditsia japonica,* family Leguminosae, a deciduous tree with thorns, usually grown for shade. Native to China and Japan.

Japanese honeysuckle. *Lonicera japonica,* family Caprifoliaceae, a hardy semi-evergreen, woody vine grown on trellis, fence, etc., for its habit, for its black fruit, and its fragrant, sweet-scented, white flowers tinged with purple. Escaped, it becomes a serious pest. Native to Asia.

Japanese hop. *Humulus japonicus,* family Cannabinaceae, an annual, climbing vine grown to cover a fence or an unsightly object. Native to Asia.

Japanese horsechestnut. *Aesculus turbinata,* family Hippocastanaceae, a hardy, deciduous tree grown as a street tree for its dense shade and yellowish-white flowers with red spots. Native to Japan.

Japanese hydrangeavine. *Schizophragma hydrangeoides,* family Saxifragaceae, a hardy, climbing, woody vine grown as a climber or creeping shrub for its white flowers. Native to Japan. Also called *climbing hydrangea.*

Japanese iris. *Iris kaempferi,* family Iridaceae, a perennial herb grown for its red-purple flowers. It is a beardless iris. Native to Asia.

Japanese ivy. Japanese creeper.

Japanese knotweed. Japanese fleeceflower.

Japanese lacquertree. *Rhus verniciflua* (*Toxicodendron v.*), family Anacardiaceae, a small tree grown in the Orient as a source of Japanese black lacquer. It is a poisonous plant. Native to Asia.

Japanese laurel. See *silver crassula; Japanese aucuba.*

Japanese leaf. Chinagreen.

Japanese lily. 1. *Lilium japonicum,* family Liliaceae, a bulbous herb grown for its fragrant, large, rose colored or pale pink flowers. Native to Japan. 2. See *speciosum lily.*

Japanese maple. *Acer palmatum,* family Aceraceae, a small, deciduous tree grown for shade and for its fine, autumnal foliage. Native to Korea and Japan.

Japanese method. A highly accurate method of determining the sex of day-old chicks which was developed in Japan. See *sexing baby chicks.*

Japanese millet. See *barnyardgrass.*

Japanese mint. Japanese field mint.

Japanese morning-glory. White-edge morning-glory.

Japanese mulberry. *Morus australis,* family Moraceae, a deciduous tree, grown for its edible berry-like fruit. Native to east Asia.

Japanese pachysandra. *Pachysandra terminalis,* family Buxaceae, an evergreen, perennial plant grown as a ground cover for its white flowers. Native to Japan.

Japanese pagodatree. *Sophora japonica,* family Leguminosae, a hardy, deciduous tree grown for its profuse, yellowish-white flowers. Native to Asia.

Japanese pear. Sand pear.

Japanese persimmon. Kaki persimmon.

Japanese pittosporum. Tobira pittosporum.

Japanese plantain lily. *Hosta japonica,* family Liliaceae, a perennial herb grown for its pale lavender or lilac flowers. Native to Asia.

Japanese plum. *Prunus salicina,* family Rosaceae, a hardy, deciduous tree grown for its edible fruit and for plant breeding. Native to China.

Japanese privet. *Ligustrum japonicum,* family Oleaceae, a tender, evergreen shrub, grown as a hedge plant. Native to Japan and Korea. Also called *wax privet.*

Japanese quince. See *flowering quince: 2.*

Japanese raisintree. *Hovenia dulcis,* family Rhamnaceae, a tree grown for its handsome foliage and its edible, fleshy, fruiting stalks. Native to China. Also called *coral-tree; cocksclaw; honey tree.*

Japanese red pine. *Pinus densiflora,* family Pinaceae, a hardy, evergreen tree grown for its bluish-green foliage. Native to Japan.

Japanese rose. *Rosa multiflora,* family Rosaceae, a deciduous shrub planted extensively as a wild-life cover, for the prevention of soil erosion, as a living fence, and very commonly used as bud stock for hybrid perpetual varieties. Also called *multiflora rose.*

Japanese rubber-plant. Silver crassula.

Japanese scabious. *Scabiosa japonica,* family Dipsacaceae, a perennial herb grown in the flower border for its clusters of white flowers. Also forced by florists and used as a house plant.

Japanese snowball. *Styrax japonica,* family Styracaceae, a shrub grown for its fragrant, white flowers. Native to Japan and China.

Japanese spirea. *Spiraea japonica,* family Rosaceae, a deciduous shrub grown as an ornamental for its clusters of pink flowers. Native to Japan.

Japanese spurge. Japanese pachysandra.

Japanese stauntonvine. *Stauntonia hexaphylla,* family Lardizabalaceae, a woody, tall-growing vine grown for its habit, evergreen leaves, and fragrant, white flowers tinged with violet. Native to Asia. Also called *stauntonvine.*

Japanese torreya. *Torreya nucifera,* family Taxaceae, an evergreen tree grown for its lustrous, dark green leaves, nutmeg-like fruit, and edible seeds. Native to Japan.

Japanese tree lilac. *Syringa amurensis* var. *japonica,* family Oleaceae, a deciduous shrub or small tree grown for its fragrant, white flowers. Native to Japan. Also called *Japanese lilac tree.*

Japanese varnish tree. Japanese lacquertree.

Japanese virginsbower. Sweetautumn clematis.

Japanese white pine. *Pinus parviflora,* family Pinaceae, a hardy, evergreen tree with bluish-green leaves, grown as an ornamental. Native to Japan.

Japanese white spirea. *Spiraea albiflora,* family Rosaceae, a deciduous shrub grown for its dense clusters of white flowers. Native to Japan.

Japanese wisteria. *Wisteria floribunda,* family Leguminosae, a woody vine grown for its habit and long clusters of violet-blue or violet flowers. Native to Japan.

Japanese yew. *Taxus cuspidata,* family Taxaceae, an evergreen shrub or tree widely grown in the shrub border, as a foundation plant or as an ornamental for its fine foliage and fast growth. Native to Asia.

Japan laurel. Japanese aucuba.

Japan plum. Loquat.

Japan wood oil tree. *Aleurites cordata,* family Euphorbiaceae, a small tropical tree grown for a drying oil obtained from its seed. Native to Asia. See *tung oil tree.*

japonica. See *flowering quince; camellia.*

jaquima. See *hackamore.*

jar. A glass or earthenware container, without spout or handles, usually cylindrical and having a wide mouth.

jar fount. A drinking fountain for poultry consisting of a quart or half-gallon jar with a pan screwed onto its mouth. The jar is filled with water and inverted. As the water is used from the pan, more runs in from the jar.

jar method. A method of applying a translocated herbicide. The tops of a plant to be killed are bent over and forced into a jar or tin can containing the herbicide.

jar ring. A flat ring of rubber used as a sealing gasket between the cap and the mouth of a fruit jar.

jasmine. 1. Any of the shrubs or vines of the genus Jasminum, family Oleaceae. Species are grown for their attractive flowers. See *showy jasmine; furry jasmine; Spanish jasmine; Italian jasmine; primrose jasmine; Arabian jasmine.* 2. *J. officinale,* a deciduous shrub grown widely as an ornamental and as a conservatory plant for its very fragrant, sweet-scented, white flowers. Native to Asia. Also called *white jasmine; poets jessamine.* 3. See *Carolinajessamine; capejasmine; Chilejasmine; Chinese starjasmine; Madagascar stephanotis; night jasmine.*

jasmine nightshade. *Solanum jasminoides,* family Solanaceae, a tender, woody vine grown in frost-free areas for its starlike, white, blue-tinged flowers. Native to Brazil. Also called *potato vine.*

jasmine pandorea. *Pandorea jasminoides,* family Bignoniaceae, a woody, climbing vine grown in very warm areas for its habit and for its white or pinkish flowers. Native to Australia. Also called *bower plant; wonga-wonga.*

jaundice. A syndrome seen in animals and man characterized by excess bile pigment in the blood and deposition of bile pigments in the mucous membranes and skin resulting in a yellow or yellowish-orange complexion. Also called *icterus.*

Java. One of the oldest breeds of chicken of the American class. A dual-purpose, yellow-skinned breed producing brown eggs. There are two varieties: the Black and the Mottled.

Java black rot. A sweet potato disease caused by the fungus, *Diplodia tubericola,* family Sphaerioidaceae; characterized by a dry rot of the roots and a brown decay of tissues. Black protuberances appear on the surface of the root.

Java glorybower. *Clerodendron speciosissimum,* family Verbenaceae, a shrub grown in greenhouses for its clusters of scarlet flowers. Native to Indonesia.

Java gum disease. Leaf scald.

Java plum. *Syzygium cumini* (*Eugenia jambolana*), family Myrtaceae, a large, tropical tree, which yields dark red, plumlike fruit used in medicines. Native to Burma and Indonesia. Also called *jambo.*

Java snapweed. A species of snapweed.

Java velvetplant. *Gynura aurantiaca,* family Compositae, a tropical, shrubby herb grown in greenhouses for its handsome foliage. Native to Indonesia.

jaw spreader. See *mouth speculum.*

Jeffrey pine. *Pinus jeffreyi,* family Pinaceae, an evergreen tree grown for lumber and as an ornamental. Native to North America.

Jeffrey shooting star. *Dodecatheon jeffreyi,* family Primulaceae, a perennial, stemless herb grown for its deep reddish-purple flowers. Native to North America. Also called *Sierra shooting star.*

jelling point. The stage in jelly making in which the boiling fruit sirup on cooling will form a jelly of the desired consistency.

jelly. A food product obtained by boiling fruit with or without water, expressing and straining the juice, adding sugar (sucrose), and concentrating it by boiling to such a consistency that gelatinization takes place on cooling. It is clear, sparkling, of attractive color, and retains the form when removed from its container.

jennet. The female ass. Also called *jenny. Cf. jack.*

jennet jack. A jack used in the breeding of asses. *Cf. mule jack; jackstock.*

jenny stonecrop. *Sedum reflexum,* family Crassulaceae, a creeping, evergreen, perennial herb grown for a mat effect in the rock garden and for its golden-yellow flowers. Escaped, it becomes a weed pest. Native to Europe. Also called *yellow stonecrop.*

jequirity bean. Jequirity rosarypea.

jequirity rosarypea. *Abrus precatorius,* family Leguminosae, a tropical, woody vine grown for its scarlet, black-spotted seeds, which are used in rosaries and in medicines. The seeds are poisonous; one seed can be fatal to a human. Also called *Indian licorice; weather plant; crabeye (vine); jequirity bean; prayer plant.*

Jericho resurrectionmustard. *Anastatica hierochuntina,* family Cruciferae, an annual plant grown in the desert garden for its habit of forming a tight ball when dry and reviving when wet. Native to Asia and Africa. Also called *rose-of-Jericho; resurrection plant.*

jerk. To cut meat, usually beef, into long, thin strips, and dry it in the sun.

jerked beef. Beef which has been naturally dried. Also called *charqui.* See *jerk.*

Jersey. A widely used breed of dairy cattle from the Isle of Jersey. They are of extreme dairy type, solid fawn, red, or nearly black in color, and their milk has the highest butterfat content of any breed. Weight at maturity varies from 800 to 1,000 lbs.

Jersey Buff. A small, well-fleshed, buff-colored variety of turkey. It is a recently developed variety from New Jersey, U.S.

Jersey Giant. A breed of large, meat-type chickens with a single comb, yellow skin, and non-feathered shanks. Mature males weigh about 13 lbs.; females about 10. The varieties are Black and White.

Jersey Red swine. A strain of red hogs developed in New Jersey from imported English swine. Large in size, coarse in hair and bone, prolific, and of rugged constitution, they are now incorporated in the Duroc breed.

Jerseytea ceanothus. *Ceanothus americanus,* family Rhamnaceae, a hardy, low, deciduous shrub grown for its clusters of white flowers. Its leaves were used as a substitute for tea during the

American revolution. Native to North America. Also called *New Jersey tea; Indian tea; Walpole tea; redroot shrub.*

Jerusalem artichoke. Jerusalem artichoke sunflower.

Jerusalem artichoke sunflower. *Helianthus tuberosus,* family Compositae, a perennial, strong-growing herb, with edible tubers containing levulose, which were used for food by American Indians. Grown in some localities for hog feed, it can become a weed. Native to North America. Also called *earth apple; girasole; Canada potato; Canadian bur.*

Jerusalem cherry. *Solanum pseudocapsicum,* family Solanaceae, a deciduous shrub, with poisonous, yellow or scarlet, berry-like fruit; grown as a greenhouse plant. Native to Asia. Also called *Natal cherry.*

Jerusalem cricket. *Stenopelmatus fuscus* Hald., family Gryllacrididae, a cricket which is a predator on other insects. It may damage potato tubers.

Jerusalem cross. Maltese cross campion.

Jerusalem date. Butterfly bauhinia.

Jerusalem oak. Jerusalemoak goosefoot.

Jerusalemoak goosefoot. *Chenopodium botrys,* family Chenopodiaceae, an annual herb, with oaklike leaves and small, white flowers, sometimes grown for its fragrance. Native to Eurasia. Also called *feather geranium; Jerusalem oak; ambrosia.*

Jerusalem sage. *Phlomis fruticosa,* family Labiatae, a perennial shrub, with numerous, yellow flowers, grown as an ornamental. Native to southern Europe.

Jerusalem thorn. Christ's-thorn.

jessamine. Jasmine.

jet black. A horse color which is a shiny black.

jet condenser. A chamber with baffle plates, connected to a vacuum pan, into which a stream or spray of cold water is forced. The water condenses the vapors from the pan on contact. It is used in the commercial production of fruit and vegetable sirups and concentrates.

jet nozzle. A flask-shaped nozzle, containing a removable inner core with spiral flanges, that imparts a swirling motion to the liquid and breaks it up into coarse droplets before it leaves the orifice.

jet pump. 1. *Shallow well type.* A combined centrifugal and jet pump with the jet located in the pump above ground. A single suction pipe connects the well to the pump. 2. *Deep well type.* It consists of a simple centrifugal pump above ground and a jet (or injector) in the well. When the impeller is operated at normal speed it draws water up the suction pipe. Some of this water is forced down the pressure pipe to a nozzle, or injector, where it passes through a restricted opening at high velocity which creates a vacuum that in turn causes atmospheric pressure to force water into the chamber from the well through a foot valve. The incoming water is mixed with this water coming through the jet and is carried up the suction pipe under pressure.

jetty. A dike built of piles, rock, or other material, extending into a stream or the sea at the mouths of rivers to induce scouring or build banks and levees.

jet-type washer. A bottle washer and sterilizer which sprays the bottles, inside and out, with washing solutions from a series of jets.

Jewish sour cream. Cultured cream.

Jew's harp. Nodding trillium.

Jew's-mallow. Potherb jute.

Jew's-myrtle. Butcher's-broom.

jib(ber). A balky horse.

jiffy-hoe. Garden cultivator.

jig. An uneven gait of a horse, closely associated with prancing and weaving.

jig back. See *bobber: 2.*

jigger. Chigger.

jimbling. Otaheite gooseberry leafflower.

Jim Hill mustard. Tumblemustard.

jimmies. A condition of sheep resulting from feeding on bulb cloak fern (*Notholaena sinuata*). Affected animals arch their backs, walk with a stilted movement of the hind legs, breathe rapidly, and tremble.

jimmyweed. *Aplopappus heterophyllus,* family Compositae, a perennial herb, poisonous to livestock. Symptoms of poisoning are trembling, depression, inactivity, constipation, nausea, labored breathing, and weakness. Native to North America.

jimsonweed datura. *Datura stramonium,* family Solanaceae, an annual herb, which is poisonous and is a weed pest. Livestock feeding on it have nausea, vertigo, extreme thirst, dilated pupils, blindness, loss of voluntary movement, convulsions, and finally die. Native to North America. Also called *Jamestown weed; thornapple; dewtry; devil's-apple.*

jipi-japa. Hat carludovica.

jobber. A middleman who sells to retailers in relatively small lots.

job lot. A form of contract which has a smaller unit of trading than the regular contract.

jobo. See *mombin.*

Job's-tears. *Coix lacryma-jobi,* family Gramineae, a perennial, tropical, tall grass, with white or gray, beadlike seeds, used as a cereal in some parts of Asia. Native to Asia.

Jochberg. A cow and goat milk cheese made in the Tyrol.

jockey stick. A stick fastened to the hame of a broken horse and to the bit of a green horse to prevent them from crowding and to keep the green horse in position while being broken.

joepyeweed. Bluestem joepyeweed.

Johne's disease. A chronic, infectious disease of cattle, sheep, and goats caused by *Mycobacterium paratuberculosis.* It causes loss of condition, unthriftiness, persistent diarrhea, reduction of milk yield, alternate refusal to eat and ravenous eating, general debility, prostration, and finally death. Also called *chronic bacillary diarrhea; pseudo-tubercular enteritis; paratuberculosis; chronic bacterial dysentery.*

Johnny Appleseed. The nickname of John Chapman, an eccentric in the U.S., who in the early part of the last century distributed apple and fennel seeds in his wanderings through the Ohio Valley.

Johnny-jump-up. Garden pansy.

Johnsongrass. *Sorghum halepense,* family Gramineae, a perennial grass growing up to 6 ft. in height which produces heavy yields on fertile soils. It is an important livestock feed in southern U.S., but a serious weed pest in cultivated fields. Native to Mediterranean area. Also called *Alabama Guinea grass; Aleppo grass; Arabian millet.*

joint. 1. The point or place at which two bones meet: the articulation between the two bones. 2. The point or meeting of two pipes. 3. A general term for a cut of meat (England). 4. See *node.*

joint-disease of sucklings. See *navel-ill.*

jointer. A miniature plow attached to the colter yoke whose purpose is to turn over a small, ribbon-like furrow slice directly before the main plow bottom to cover trash.

joint evil. Navel-ill.

jointfir. Any plant of the genera Gnetum and Ephedra, characterized by leafless, jointed stems.

joint ill. Navel-ill.

jointing stage. The stage when the internodes of grass stems begin elongating. See *growth stages.*

joint-tenancy (with rights of survivorship). A legal arrangement for ownership of real property by two or more persons in which, upon the death of one of the persons, the title to the property goes to the surviving party or parties of the *joint-tenancy* rather than to the estate of the deceased. A *joint-tenancy* may be dissolved by one of the parties during his lifetime.

joist. 1. A piece of dimension lumber 2 to 5 ins. thick and 4 or more ins. wide, commonly used to support the floor of a building. 2. A piece of lumber 8 ft. or more long, 1½ to 4½ ins. thick, and 6 to 12 ins. wide. (English markets.)

Jonathan. One of the older and better-known varieties of apple; ripens in late autumn; the fruit is bright red.

Jonathan freckle. Jonathan spot.

Jonathan spot. A skin disease of Jonathan, Rome Beauty, Wealthy, King David and some other varieties of apple found wherever these varieties are grown. A physiological disease, associated with metabolism as the apple approaches maturity, it is characterized by circular, deep brown to black, depressed spots usually more numerous on the red-colored side of the fruit and the stem end. In storage the spots increase in size and become irregular in outline. Also called *Jonathan freckle.*

jonquil. 1. *Narcissus jonquilla,* family Amaryllidaceae, a hardy, bulbous herb, with several varieties, grown for its fragrant, yellow flowers. Native to southern Europe and North Africa. 2. See *campernelle jonquil.* 3. See *narcissus.*

Josephine. A cows' whole milk, soft, cured cheese made in Silesia.

Josephine's lily. Giant brunsvigia.

Josephs-coat amaranth. *Amaranthus tricolor,* family Amaranthaceae, a tropical, annual herb grown for its clusters of varicolored flowers and colored leaves. Also called *fountain plant.*

Joshua tree. *Yucca brevifolia,* family Liliaceae, a plant grown in the desert garden for its grotesque branching habits. It has dense clusters of terminal leaves and greenish-white flowers. Native to North America.

jowl. Meat taken from the cheeks of a hog.

Judas goat. A trained goat used by a slaughter house to lead sheep to the killing pens.

Judas tree. *Cercis siliquastrum,* family Leguminosae, a tender tree grown for its early spring, rose-purple flowers. Native to Eurasia. See *eastern redbud; Chinese redbud.*

judge. 1. An experienced person selected to make an official placement of entries in exhibits at a show. 2. To consider the qualities, breed, conformation, etc., of animals, plants, products, etc., at a show or exhibit.

judging card. A list of selected characters to be considered in judging an animal or a product in a contest. Sometimes associated with the term, *score card.*

judging contest. An event at an exhibition of livestock or products in which organized groups such as judging teams, compete in judging various classes of animals or products.

judging team. A selected group of persons, usually youths, who represent an organization, such as a 4-H Club, Future Farmers of America, U.S., or a college of agriculture. They compete with similar groups at shows or exhibits in judging and placing classes of livestock or other products.

judicial ditch. An intercounty drainage ditch maintained under the jurisdiction of the circuit court (U.S.).

jug. 1. A small claiming pen where ewes with newborn lambs are kept until the ewe has nursed and claimed her young. 2. A glass bottle holding ½ gallon or a gallon used to bottle cider, sirup, molasses, etc. 3. A deep vessel of earthenware narrowed to a neck which is frequently fitted with a finger handle for carrying. (A popular container for carrying water to the farm field.) 4. A small pen, part of a corral, used to funnel animals down a narrow alley (chute) to facilitate sorting, loading, or catching; used primarily with cattle.

jughead. A horse with poor characteristics, little sense, and not amenable to training.

juice. The fluid in plant and animal tissue, especially the fluid of fruits and vegetables expressed and used in jellies and as a drink, such as orange *juice,* carrot *juice.*

juice extractor. Any of several different types of devices used to express juice from plant stems, roots, fruits, and vegetables.

jujube. 1. Any shrub or tree of the genus Zizyphus, family Rhamnaceae. Species are grown for their edible fruits. See ***India jujube.*** 2. ***Z. jujuba,*** a small, deciduous tree grown in warm areas for its edible, round or egg-shaped, brown fruit often made into preserves. Native to Europe and Asia. Also called *Chinese date; elb.* 3. A lozenge flavored with the fruit of the *jujube.*

jumbo crate. A standard crate for packing and shipping melons, which measures 13 x 13 x 22⅛ ins. inside.

jumbo frame. A frame for a beehive which measures 17⅝ by 11¼ ins.

jumbuck. Sheep (New Zealand and Australia).

jumper. 1. A class of horse, usually a Thoroughbred, able to jump high hurdles and used in the shows. 2. Any animal with a tendency to jump over fences and therefore somewhat difficult to restrain.

Juneberry. Serviceberry.

June bloom fruit. An orange fruit produced as a result of blossoming shortly after the beginning of the summer, rainy season in Florida, U.S. It ripens after the main crop is harvested.

June budding. Shield budding, usually in June, in warm areas in which it is possible to force the bud into growth in the same season it is set. The top of the stock is bent or partially broken about ten days after budding to force the growth.

June drop. A shedding of tree fruits during the early summer; believed often to be caused by, or associated with, embryo abortion.

June grass. See *prairie June grass; Kentucky bluegrass.*

June yellow-color butter. The high golden-yellow fat globules of butter from milk produced soon after the cows go on to spring pasture.

Jungherr's disease. Trichomoniasis of the upper digestive tract.

jungle. Any dense, nearly impenetrable growth of vegetation, especially in the tropics.

jungle flame. *Ixora coccinea,* family Rubiaceae, an evergreen shrub grown in very warm areas and in the greenhouse for its foliage and its dense clusters of red flowers. Native to Indonesia. Also called *jungle geranium; flame of the woods.*

jungle geranium. Jungle flame.

junior calf. In stock judging, a beef or dairy animal which was calved after January 1 of the year it is shown.

junior champion. In stock judging, the best animal of the younger classes selected from the first prize winners.

junior spring pig. In stock judging, a swine which was farrowed on or after March 15 of the year it is shown.

junior twins. A style of packaging cylindrically-shaped Cheddar cheese 9¾ ins. in diameter, 5 ins. high, and 11 to 12 lbs. in weight. Also called *commodores; family twins; pets.*

junior yearling. In stock judging: (a) a beef animal calved between January 1 and April 30 of the preceding year; (b) a dairy animal calved between January 1 and June 30 of the preceding year; (c) a swine farrowed between March 1 and August 31 of the preceding year.

juniper. 1. Any evergreen tree or shrub of the genus Juniperus, family Pinaceae. Species are valuable as ornamentals and for lumber. See *creeping juniper; eastern redcedar; Formosa juniper; Greek juniper; pyramid Chinese juniper; Rocky Mountain juniper; savin juniper; Sierra juniper; southern redcedar; Syrian juniper.* 2. *J. communis,* a very hardy evergreen tree grown as an ornamental. Native to North America.

juniper American-mistletoe. *Phoradendron juniperinum,* family Loranthaceae, a parasite plant which infests junipers in central North America.

juniperberry. Box huckleberry.

juniper blight. A fungous disease caused by *Phomopsis juniperovora,* family Sphaerioidaceae; characterized by a browning of the tips of branches and by a dieback of seedlings and young trees.

juniper scale. *Diaspis carueli* Targ., family Diaspididae, a scale insect which infests junipers.

juniper webworm. *Dichomeris marginella* (F.), family Gelechiidae, a larva which webs twigs and needles of junipers together and feeds within the web.

Jupiter's-beard. *Centranthus ruber,* family Valerianaceae, a perennial herb grown for its fragrant, dense, red flower clusters. Native to Eurasia. Also called *red valerian; scarlet lightning.*

jute. Any tropical, annual herb of the genus Corchorus, family Tiliaceae, especially roundpod jute (*C. capsularis*) or potherb jute (*C. olitorius*), sources of the jute fiber used in making cordage, fabrics, bagging, etc.

Jutland. The black-and-white breed of cattle, native to Denmark, which probably had early influence on the black-and-white cattle of the Netherlands.

K

K-1875. An organic compound, bis-(p-chlorophenoxy)-methane, used as a miticide. See *neotran.*

K-6451. A highly persistent mite ovicide.

kaelder milk. A cow's fermented milk from Norway which keeps well. It is a mixed fermentation in which lactic acid bacteria hydrolyze the lactose and the torulae ferment the hexose.

Kaffir-bean tree. Elephant-hedge Kafi-bean tree.

kaffir corn. Kafir.

kafir. *Sorghum vulgare* var. *caffrorum,* family Gramineae, a tall, annual grass, a sorghum grown principally in the semiarid sections of the U.S. for its grain, for forage, and silage. Also called *kaffir corn.*

Kafirbread. Any plant of the genus Encephalartos,

family Cycadaceae. Species are grown in the greenhouse and outdoors in warm climates. Native to Africa.

Kafir orange. *Strychnos spinosa,* family Loganiaceae, a tropical, spiny shrub grown in frost-free areas for its edible fruit. Native to central and southern Africa. Also called *Natal orange.*

kaido crabapple. Midget crabapple.

kainit(e). A potassic fertilizer which contains potassium and sodium chlorides and sometimes sulfate of magnesia, with not less than 12 percent of potash (K_2O).

Kajmak. A cream cheese made from ewes' milk in Serbia. Also called *Serbian butter.*

kak. A kind of saddle bag.

kaki persimmon. *Diospyros kaki,* family Ebenaceae, a deciduous tree grown in warm areas for its orange-colored, edible fruit. Native to China. Also called *Japanese persimmon.*

kale. *Brassica oleracea* var. *acephala,* family Cruciferae, an annual or biennial herb grown as a potherb. Collard is a form grown in warm areas. Native to Europe and Asia.

Kalmia deutzia. *Deutzia kalmiaeflora,* family Saxifragaceae, a hybrid shrub grown for its cup-shaped, white flowers tinged with carmine on the inside.

Kamchatka bugbane. *Cimicifuga foetida,* family Ranunculaceae, a perennial herb grown for its dense raceme of white flowers. Native to Siberia. Also called *bugbane.*

kame. A short, irregular ridge, hill, or hillock of stratified glacial drift. Most *kames* are interspersed with depressions, sometimes known as kettles, having no surface drainage.

kangaroo. An herbivorous marsupial, family Macropodidae, with a small head, upright ears, strong hind legs, and a thick, round tail. In large numbers, *kangaroos* may cause economic damage to grasslands. Found in Australia and surrounding islands.

kangaroo-thorn acacia. *Acacia armata,* family Leguminosae, a tropical, spiny shrub grown in the greenhouse for its bright yellow flowers. Native to Australia.

Kangra buckwheat. Tartary buckwheat.

Kansas gayfeather. *Liatris pycnostachya,* family Compositae, a perennial herb grown for its lance-shaped leaves and rose-purple, dense flower spikes. Native to North America. Also called *blazing star; prairie button snakeroot.*

kaphir. See *kefir.*

kapok ceiba. *Ceiba pentandra,* family Bombacaceae, a tropical tree grown for its cotton-like fiber, the kapok of commerce. Native to America. Also called *silk-cotton tree.*

kapuka. See *kupuka tree.*

Karakul. A small, light-shearing breed of broad-tailed sheep from west-central Asia, bred especially for the lambskins used in the fur pelt industry.

karanda. *Carissa carandas,* family Apocynaceae, a spiny shrub grown as a hedge plant for its white or pink flowers. Native to India.

karaya gum. An exudate obtained from any of several species of trees of the genus Sterculia, family Sterculiaceae, used as a stabilizer in ice cream mixes.

Kareish. A skimmed milk cheese made in the U.A.R.

karo. Karo pittosporum.

karo pittosporum. *Pittosporum crassifolium,* family Pittosporaceae, an evergreen shrub or small tree grown in warm areas in a windbreak and for shelter near the ocean. It has dark red or purple flowers. Native to New Zealand.

karst. A topographical formation; a plateau of limestone rock, interspersed with holes and ridges formed by water action.

Karut. A very dry, hard, skimmed milk cheese made in Afghanistan and northwestern India.

Kashaval. A partly skimmed, ewes' milk, cured cheese made in Rumania.

Kashgar tamarisk. *Tamarix hispida,* family Tamaricaceae, a shrub grown for its feathery foliage and its dense racemes of pink flowers. Native to U.S.S.R.

Kashmir larkspur. *Delphinium cashmerianum,* family Ranunculaceae, a perennial herb grown for its blue flowers. Native to the Himalayas.

kassaba melon. Casaba melon.

katabatic wind. Cool air which, impelled by gravity, flows downward. Also called *mountain wind; canyon wind; gravity wind.*

katabolism. Catabolism.

Katschkawalj. A plastic curd, cured, ewes' milk cheese made in the Balkans.

katsura tree. *Cercidiphyllum japonicum,* family Trochodendraceae, a hardy tree grown for its pyramidal habit and its foliage which turns from purple, when unfolding, to bright yellow and scarlet in the fall. Native to Asia.

katydid. A long-horned grasshopper of the family Tettigoniidae. See *broad-winged katydid; fork-tailed bush katydid.*

kauri dammarpine. *Agathis australis,* family Pinaceae, an evergreen tree grown for its valuable timber and as an ornamental in warm areas. Native to New Zealand.

kauri pine. Kauri dammarpine.

kea. *Nestor notabilis,* a greenish parrot which, when driven by hunger, kills sheep by digging into the back of the animal with its beak in order to get at the fat around the kidneys (New Zealand).

ked. Sheep ked.

kedassia. A certified, raw, cows' milk produced under the supervision of the shomar (*q.v.*) for orthodox Jews.

keel. 1. The part of a fowl's body which extends backward from the breast. Also called *breast bone.* 2. The two front, united petals of a pea-like flower. See *standard: 4; wing: 2.* 3. In ducks, the pendant fold of skin along the entire underside of the body. 4. In geese, the pendant fold of flesh from the legs forward on the underpart of the body.

keel-bone. The breast bone of birds; the sternum.

keel cyst. The cystlike formation between the skin of the breast and the keel bone of chickens and turkeys, especially in males of the heavy breeds. The cyst contains a cavity usually filled with a brownish, sterile, viscous fluid for which no specific cause is known. Also called *breast blister.*

keel disease. Avian typhoid.

keeled. Of, or pertaining to, a ridge (like the bottom of a boat) on an animal or plant part.

Keen tree classification. A system of classifying ponderosa pine trees into 16 divisions according to four age groups (stages of maturity) and four crown vigor groups (determined by crown size, dominance, and thrift).

keep. 1. Board and room furnished a hired hand. 2. Salary. 3. To maintain animals for use or profit, as to *keep* cows, bees, etc. 4. To sell or to have in a store for sale, as to *keep* tobacco. 5. To store; as apples *keep* well. 6. To withhold from sale, as to *keep* one's seed. 7. To maintain business records, as to *keep* books.

keep easy. See *easy keeper.*

keeping. Designating the storage qualities of a product.

keeping-quality test. A simple, sample-holding test in which a sample of the milk produced by a given herd of cows is taken to determine the possibility of maintaining its quality in storage.

keep (one's) seed. 1. To retain the seed of home-grown cotton after ginning. 2. To save home grown seed of any year's crop for next year's planting.

keep on feed. To maintain an animal on liberal feeding for heavy production.

keet. The young guinea fowl.

Kefalotyi. A hard, ewes' or goats' milk cheese made in Greece or Syria.

kefir. A self-carbonated, fermented cows' milk beverage, made from whole, partially skimmed, or skimmed milk, containing usually about 1.0 percent lactic acid and from 0.5 to 1.0 percent alcohol. Also called *kephir.*

kefir grains. Small, cauliflower-like clusters which consist of dried masses of lactobacilli, streptococci, yeasts, and very likely, other microorganisms, bound together in a matrix of coagulated milk protein; used to inoculate milk in making kefir.

keg gardening. See *tub gardening.*

keg shoe. A horseshoe, factory-made on a mass-production basis. (It was called an *issue shoe* by the U.S. Army.)

keiapple. *Dovyalis caffra,* family Flacourtiaceae, a tropical, thorny shrub grown for its edible, cranberry-flavored, yellow berry eaten only when cooked. Native to South Africa.

Kellerer rockcress. *Arabis kellereri,* family Cruciferae, a hybrid herb grown in the rock garden for its mat-forming habit and its small, white flowers.

kelp. Any of the seaweeds of the families Laminariaceae and Fucaceae, at one time used as a source of potash for fertilizers. Dried *kelp* contains 1.6 to 3.3 percent nitrogen, 1.0 to 2.0 percent phosphoric acid, and 15 to 20 percent potash. *Kelp* has been used in animal feeds as a source of iodine. See *dried kelp.*

Kelsey locust. *Robinia kelseyi,* family Leguminosae, a deciduous shrub or small tree grown for its feathery leaves and rose-colored flowers. Native to North America.

Kelsey spot of plums. A localized injury of fruits, due to temperatures above 100° F., over a period of several hours during the ripening period. Characterized by sharp-margined spots, mostly at the distal end, having a depressed center, often waxy, and purplish-red color with dead flesh beneath, sometimes extending to the pit. Also called *heat spot of plums.*

kemp. A chalky-white, brittle, weak, wool fiber which, when found mixed with normal fibers of a fleece, constitutes a very serious defect. *Kemp* will not take dyes.

Kenaf hibiscus. *Hibiscus cannabinus,* family Malvaceae, a fiber plant grown as a source of fiber for cordage and coarse cloth. Native to India and Indonesia. Also called *Deccan hemp; ambary.*

Kenilworth ivy. *Cymbalaria muralis,* family Scrophulariaceae, a perennial, herbaceous vine grown in a hanging basket and on an indoor trellis for its foliage and its lilac-blue flowers with the corolla yellowish inside. Native to Europe. Also called *Aaron's Beard.*

kentia palm. Sentry palm.

Kentucky bluegrass. *Poa pratensis,* family Gramineae, a perennial grass widely grown in the north-central states, U.S., as a pasture and turf grass. Native to Eurasia. Also called *English grass; June grass.*

Kentucky coffeetree. *Gymnocladus dioicus,* family Leguminosae, a deciduous tree sometimes grown for its feather-like foliage and its greenish-white flowers. Native to North America. Also called *coffee nut; coffee-tree; American coffee bean.*

Kentucky Saddle horse. See *American Saddle horse.*

Kentucky viburnum. *Viburnum molle,* family Caprifoliaceae, a deciduous shrub grown for its white flower clusters. Native to North America. Also called *poison haw.*

Kentucky wisteria. *Wisteria macrostachya,* family Leguminosae, a climbing, woody vine grown for its habit and for its lilac-purple, yellow-spotted flowers. Native to North America.

Kentucky wonder. A very well-known type of kidney bean.

Kentucky yellowwood. American yellowwood.

kephir. Kefir.

kepi. Kefir.

keratin. A complex protein distinguished by high insolubility. It is the substance of which hair, horns, claws, and feathers are composed.

keratitis. An inflammation of the cornea of the eye.

kerf. The cut of an ax, a saw, or other sharp instrument; the notch or slit made in wood by cutting or sawing.

kernel. 1. The portion of a seed contained within the seed coat. 2. The whole grain of corn, wheat, etc.

kernel mold of corn. A fungous rot caused by *Rhizopus, Aspergilus,* and *Penicillium* spp. It produces molding of the kernels with a characteristic moldy odor.

kernel rot of rice. A bacterial disease caused by *Xanthomonas itoana,* family Pseudomonadaceae; characterized by necrotic dark areas in the endosperm.

kernel smut of grasses. A fungous disease caused by *Tilletia* spp., family Tilletiaceae; characterized by the smut balls replacing the kernel in the spike or panicle. See *smut.*

kernel smut of millet. A fungous disease caused by *Ustilago crameri,* family Ustilaginaceae; characterized by masses of black, dusty spores in the kernels. See *smut.*

kernel smut of rice. A fungous disease caused by *Tilletia horrida,* family Tilletiaceae; characterized by dull gray-colored, smutted kernels of rice that are light in weight. See *smut.*

kernel smut of sorghum. See *loose kernel smut of sorghum.*

kerosene. A petroleum derivative with a higher flashing point than gasoline; used as fuel, a solvent for insecticides, and a pesticide.

Kerry. The smallest of all dairy cattle breeds. It originated in southern Ireland in Kerry County. The cows are black, sturdy, good milkers, and weigh from 700 to 900 lbs. See *Dexter-Kerry.*

ketembilla. *Dovyalis hebecarpa,* family Flacourtiaceae, a small tree grown for its purple, edible fruit. Native to India and Ceylon. Also called *Ceylon gooseberry.*

ketogenesis. The production of ketone bodies or ketones.

ketone bodies. Chemical substances such as diacetic acid, hydroxybutyric acid, and acetone, produced in the liver of animals. Excessive amounts of such substances in the body result in ketosis (*q.v.*). When *ketones* are eliminated in excessive amounts in the urine, the condition is called *ketonuria; acetonemia.*

ketonuria. See *ketone bodies.*

ketosis. A metabolic disease characterized by loss of appetite, markedly poor condition of skin and flesh, digestive disorders, sometimes nervous symptoms, and foul-smelling milk. It often occurs as a primary metabolic disturbance in cattle, in pregnancy disease of sheep, in diabetes, and certain other conditions. Also called *acetonemia; false milk fever; chronic milk fever; acetonuria.* See *pregnancy disease in ewes; ketone bodies.*

ketosis (in sheep). See *pregnancy disease in ewes.*

ketsup. Catchup.

kettle (hole). A constructional basin depression of variable size and depth, sometimes containing lakes and swamps. It is a feature of moraines in the glaciated regions of the U.S., especially in Wisconsin.

kettle-rendered lard. Lard made from the heavy, back fat and leaf fat of hogs in equal parts and open-kettle rendered at 240 to 260° F.

Keuka high renewal system. See *high renewal system.*

Kew African violet. *Saintpaulia kewensis,* family Gesneriaceae, a tropical, stemless, perennial herb, grown for its violet-colored flowers and white, hairy leaves. Native to Africa. *Cf. African violet.*

Kew cape-primrose. *Streptocarpus kewensis,* family Gesneriaceae, a tender, perennial herb grown for its bright purple flowers with striped, dark brown throats. Native to South Africa.

key area. A portion of a range, with representative characteristics, strategic location, or unusually high resource values, upon which the management of adjacent areas or of the entire unit is judged or based.

key-feather. The short feather which grows on the wing between the primaries and secondaries.

key fruit. A one-seeded fruit that does not split open and has a membranous wing attached, as in the maple; samara.

key species. Plant species that endure moderately close grazing and are fairly or potentially abundant.

khamseen (khamsin). A hot, southerly wind that blows intermittently for fifty days in spring in the Middle East.

khapon. See *Kefir.*

Kharif. One of the two main growing seasons in India: the summer and fall crop, including corn, cotton, some fodders, etc. *Cf. Rabi.*

Kharsgoard. A hard, skimmed cows' milk cheese made in Denmark.

khoa. An Indian confection prepared by slowly evaporating water from whole milk and adding sugar to the dough.

kicker. 1. An animal which habitually kicks. 2. Any mechanical device which operates in a kicking fashion, as the bundle *kickers* on a grain binder. See *oat kicker.*

kicking hobble. A rope or strap fastened to the rear legs or feet of a horse, ass, or cow, to prevent kicking.

kicking strap. A strap fastened to the crupper strap of the harness in breaking colts. It extends down each side to the shaft of the carriage and is used to prevent kicking.

kid. 1. A young goat. 2. To give birth to a *kid.*

kidding. The birth of a kid.

kid-glove orange. Tangerine.

kid Mohair. The finest quality of Mohair wool, taken only from young goats.

kidney. A bean-shaped organ of man and certain lower animals which excretes urine. The lamb *kidney* is edible, and *kidneys* from other animals are converted into edible products.

kidney bean. *Phaseolus vulgaris,* family Leguminosae, a tropical, annual herb widely and popularly cultivated for its edible seed or pod. The species has many varieties: *egg bean; flageolet; frijole; green beans; pole beans; bush beans; snap beans; string beans; wax beans; butter beans; fasels; haricot; Navy bean; bean; Great Northern; cranberry bean; Kentucky wonder; bunch bean.*

kidney faller. A horse which collapses in the hind quarters when it walks.
kidney veal chop. A retail cut of veal from the back, which includes a section of the kidney imbedded in the inner fat as a means of identification.
kidneywort baccharis. *Baccharis pilularis,* family Compositae, an evergreen shrub grown in dry, alkali areas as an ornamental. Native to North America. Also called *coyote brush; chaparral broom; squaw waterweed.*
kieserite. $MgSO_4 \cdot H_2O$, magnesium sulphate, one of the Stassfurt salts, a source of Epsom salts. Sometimes used alone as a soil amendment and in fertilizer mixtures as a source of magnesium.
kill. 1. Purposeful killing of insects, plants, or wild or domestic animals. 2. The total number of wild animals killed in a period. 3. To destroy life in an insect, herb, etc.
kill back. See *dieback.*
killed to the ground. Designating a plant whose above-ground portions have been killed by cold, fire, drought, etc.
killed virus. A virus whose infectious capabilities have been destroyed by chemical or physical treatment.
killer. 1. A person who slaughters animals. 2. A meat animal. 3. A pesticide. 4. A predator.
killing cattle. Cattle, usually grass-fattened on a ranch or range, of sufficiently good flesh and finish to be slaughtered profitably.
killing fat. See *offal fat.*
killing floor. The place in a slaughterhouse where a stunned animal is killed, bled, skinned, eviscerated, and where the carcass is split, washed, and wrapped.
killing frost. A frost or temperature condition sufficiently low to kill staple crops. See *frost; black frost; light frost; heavy frost.*
killing house. A slaughterhouse on a station (New Zealand).
kill off. To eradicate.
kill out. To eradicate.
kill out well. A satisfactory dressing out of a meat carcass.
kill strip. A longitudinal portion of the trunk of a living tree that has been killed by insects.
killwort. Greater celandine.
kiln. An oven, furnace, or large heated room for the curing of lumber, tile, bricks, etc.
kiln burn. A browning of the surfaces of lumber which occurs during kiln drying.
kiln charge. The total load of lumber, etc., for a single or progressive drying operation in a kiln.
kiln dried lumber. Lumber dried by artificial heat to a moisture content less than that obtained through natural, air seasoning.
kiln drier. Kiln.
kiln dry (kiln dried). Designating wood dried in a kiln.
kiln drying. The seasoning of wood, etc., by treatment with artificially supplied heat in a kiln.
kiln run. The drying of a single charge or kiln load of lumber.
kilt. To shear the legs and belly of a double fleecer in the autumn (New Zealand).
Kimberly horse disease. In Australia, a disease of horses, characterized by dullness, wasting, irritability, biting, gnawing on posts, yawning, uncontrolled galloping, slow staggering gait, and stiff neck. It is probably caused by feeding on some plant, at one time thought to be whitewood (*Atalaya hemiglauca*). Also called *walk-about disease.*
kind. 1. All the plants of the same type, accepted as a single vegetable or fruit, as tomato, cabbage, bean, apple, peach, etc. 2. A species, as a cow, sheep, etc.
kine. Cows (archaic).
king bee. Before the seventeenth century, it was popularly believed that the queen bee was a male who ruled the hive. See *queen bee.*
kingbird. Bee martin.
King Cotton. A phrase testifying to the importance of cotton in the U.S. for many years.
king cup. Marshmarigold.
kingdevil hawkweed. *Hieracium florentinum,* family Compositae, a perennial herb which is a troublesome weed pest.
kingfisher daisy. Kingfisher felicia.
kingfisher felicia. *Felicia bergeriana,* family Compositae, a hairy, low, annual herb grown for its daisy-like flowers with bright blue rays and yellow disks. Native to South Africa. Also called *kingfisher daisy.*
king head. Giant ragweed.
kingnut. Shellbark hickory.
King orange. *Citrus nobilis,* family Rutaceae, a small, hybrid, broadleaved, evergreen tree grown for its rough-skinned, edible, sweet fruit, and as stock for Mandarin and Satsuma oranges. Native to Asia.
King's clover. Sweet clover.
King ventilating system. A ventilating system for barns which introduces fresh air near the ceiling and removes foul air from a point near the floor.
Kinkaid Act. An act of U.S. Congress, passed in 1904, which increased the size of homesteads to 640 acres. It was restricted in application to the grazing lands of western Nebraska.
Kinkaider. A nickname for settlers who homesteaded western lands in U.S. under the Kinkaid Act.
kinnikinnick. See *silky dogwood; bearberry.*
kino. A brownish, astringent substance obtained from several tropical trees. Used in medicines and sometimes in tanning and cotton-dyeing.
kino eucalyptus. *Eucalyptus resinifera,* family Myrtaceae, an aromatic, tall tree grown as a source of kino, for its lumber, and as an ornamental. Native to Australia. Also called *red mahogany.*
kip. A hide taken from a very young calf.
kitambilla. See *ketembilla.*
kitchen garden. A small garden in which vegetables and fruits are grown for home use.
kitchen garden pepper. Bush redpepper.

kittul. A darkish, horsehair-like fiber from the leafstalks of toddy fishtail palm and gomuti sugarpalm; used in making polishing brushes for cloth.

kiwi. See *apteryx.*

Klamath weed. St. John's-wort.

Klein's disease. Avian typhoid.

klendusity. The ability of a susceptible plant to escape infection under conditions in which other susceptible varieties contract the disease.

Kloster. A soft, cows' milk, ripened, Romadur-type cheese, made in Germany.

Knaost. See *Pultost.*

knapsack duster. See *knapsack sprayer.*

knapsack seeder. A device for broadcast sowing which consists of a canvas sack fastened to a seeding mechanism. A crank, turned by hand, revolves a wheel having radial ribs which throw seeds to the front and sides. See *broadcaster.*

knapsack sprayer. A four-gallon, kidney-shaped tank carried on the back of the operator. A pump is used to build up pressure in the tank, and an extension tube connects a flexible hose from the tank to a nozzle for spraying.

knapweed. Black centaurea.

knawel. *Scleranthus annuus,* family Caryophyllaceae, an annual herb which is a weed pest. Native to North America.

knee. 1. The joint between the hip and the ankle in man and quadrupeds (hind leg); in birds, the tarsal joint. 2. The carpal joint (fore leg). 3. The spurlike, root growth of the baldcypress which develops when the tree grows in a swamp.

knee banger. A horse that strikes its knees with the opposite front foot while walking or running.

knee cushion. A pad which is fastened to a person's knee to enable him to work in some comfort while kneeling. Also called *knee pad.*

knee cutter. Knife colter.

knee spavin. A chronic inflammation of the small bones of the carpal joint of horses, characterized by lameness, bony enlargement, a dragging of the toe of the affected leg, and a fusion of the carpal bones.

knee splints. A condition of a horse: splints (*q.v.*) near the knee.

knee strap. A restraining device for animals, especially horses, which consists of a 2-in. wide, 3-ft. long belt with a buckle attached to one end. The belt is buckled to form a loop and is passed over the bent knee well up on the leg so that the hoof is held off the ground.

knife. 1. Any cutting tool with a long, relatively slender blade sharpened on one edge. 2. The steel strap to which the mower sickle or knife sections are attached in the mowing machine.

knife colter. A rearward-curving, sharpened blade attached to the plow beam to cut the soil just ahead of the point of the plow share. See *fin colter.*

knife coverer. An attachment on a planter which consists of rearward-curving blades which cover seed in the row.

knife harrow. Acme harrow.

knife head. The end part of the sickle bar of a mowing machine to which the pitman is attached.

knife-head strap. The strap of metal to which the knife sections of a mowing machine are attached.

knife pruning. See *gradual cut-back.*

knife section. One of several truncated, triangular, sharpened, metal pieces attached to the knife-head strap on the sickle bar of a mowing machine.

knifey. Designating an animal that is too sharp and thin over the withers and shoulders.

Kniffin system. A method of training grape vines; one or more trunks are carried to wires of a trellis along which fruiting canes are renewed and tied annually. See *single-trunk, four-cane Kniffin system; two-trunk Kniffin system; six-cane Kniffin system.*

knob. A kidney which shows in the inner fat of the loin cut.

knobby bark. A condition of the trunks and limbs of orange and lemon, suspected to be caused by a virus, characterized by the production of gall-like projections 1 or 2 ins. in diameter and ⅛ to ½ in. high covered with nearly normal bark. The wood of the knob is light brown and softer than normal.

knob-rooted celery. Celeriac.

knobwood. *Zanthoxylum capense,* family Rutaceae, a tree with hard, close-grained wood suitable for making implements. Native to South Africa. Also called *wild cardamom.*

knockdown. The killing ability and speed of a toxic material, especially an insecticide.

knocker. 1. On hand shears, a small, leather pad fixed near the heel of shears to prevent the blades from closing too far (New Zealand). 2. One who swings the knocking hammer.

knocking hammer. A heavy hammer used in slaughtering to stun an animal before it is killed and bled.

knocking pen. A narrow chute or pen on the killing floor of a slaughter house in which an animal is confined so that it may be easily and accurately hit on the head with the knocking hammer. It has sides which lift and the floor may be tilted to roll the stunned or killed animal out onto the dressing floor.

knock-knees. An undesirable conformation or weakness of the front legs of horses in which the legs are not straight due to the knees coming too close together.

knot. 1. The interlacing of string or rope pulled into a tie to fasten something. 2. A swelling or protuberance on the stem of a cut flower through which water moves very slowly to the flower. See *node.* 3. The knob attached to a check wire (*q.v.*). 4. That portion of a tree branch which has become incorporated in the body of the tree. See *blind knot; decayed knot; encased knot; firm knot; hollow knot; intergrown knot; loose knot; pin knot; punk knot; rusty knot; sound knot; unsound knot.* 5. See *root knot.*

knot bindweed. Dullseed cornbind.

knot cluster. Two or more knots grouped together

as a unit, the fibers of the wood being deflected around the entire unit. See *knot: 4.*

knot garden. A flower garden with beds arranged in intricate patterns.

knotgrass. 1. *Paspalum distichum,* family Gramineae, a grass with jointed stems, which grows in wet places. Native to Eurasia and North America. 2. Prostrate knotweed.

knot-hole. A hole in a piece of lumber or a tree caused by a knot slipping out.

knot of brambles. See *cane gall of brambles; root knot.*

knotroot. Artichoke betony.

knotter. An automatic device to tie knots, as on a grain binder.

knotty gut. An intestinal disease caused by the nodular worm *q.v.*

knotweed. Prostrate knotweed.

knuckle. 1. The joint of two bones. The *knuckles* of a quadruped are often used in making soup stock. 2. See *round tip.* 3. See *node.*

knuckle cut. The removal of a flower from a rose plant by a cut on an older stem below the node of the stem of the flower and ¼ to ½ in. above the next lower lateral branch on the older stem. It gives a longer-stemmed bloom with a hook at the base of the stem.

knuckling over. A condition of a horse in which the ankles or pasterns are weak and not properly extended. Also called *cocked ankle.*

ko. Thunberg kudzubean.

kochia. *Kochia scoparia,* family Chenopodiaceae, an annual, herbaceous plant frequently cultivated for its bright autumn color; locally a weed. It has been used as a forage crop during drouth periods in North and South Dakota, U.S.

kohlrabi. *Brassica oleracea* var. *gongylodes,* family Cruciferae, a biennial herb, related to the cabbage, grown for its edible, enlarged fleshy stem. Also called *turnip cabbage; turnip-rooted cabbage; stem cabbage.*

kohuhu. Tawhiwhi pittosporum.

kok-saghyz. *Taraxacum kok-saghyz,* family Compositae, an herb grown somewhat as a source of latex in the U.S.S.R. after 1931 and experimentally in U.S. in 1942. Also called *Russian dandelion.*

kolanut. Sudan colanut.

kolomikta actinidia. *Actinidia kolomikta,* family Dilleniaceae, a woody, climbing vine grown on a trellis, etc., for its white or pink-blotched leaves, white flowers, and greenish-yellow, edible fruit. Native to Asia.

Kopanisti. A cured, blue-mold, sharp-flavored cheese made in Greece.

Koppen. A cured, sharp-flavored cheese made in Bohemia and Silesia from sour, goats' milk.

Korean azalea. Yodogava azalea.

Korean lily. *Lilium amabile,* family Liliaceae, a bulbous herb grown for its red, dark-spotted flowers. Native to Korea.

Korean lespedeza. *Lespedeza stipulacea,* family Leguminosae, an annual herb widely grown in warm areas as a pasture, hay, and green manure crop. Native to Asia.

Korean pine. *Pinus koraiensis,* family Pinaceae, a pyramidal, evergreen tree grown as an ornamental. Native to Korea and Japan.

Koreanspice viburnum. *Viburnum carlesi,* family Caprifoliaceae, a deciduous shrub grown for its fragrant, white flowers. Native to Korea.

Korean spirea. *Spiraea trichocarpa,* family Rosaceae, a deciduous shrub grown for its numerous, white flowers. Native to Korea.

kosher. Designating any food produced, killed, or prepared according to Jewish dietary laws.

Koster's blue spruce. *Picea pungens Kosteriana,* family Pinaceae, one of the most widely grown ornamental, blue spruces, it is hardy, evergreen, and spreads radially. Native to North America.

koumiss. A limpid, slightly acid, alcoholic beverage made from mare's milk fermented with lactobacilli and yeasts. Used by some tribes in Central Asia.

kousa dogwood. *Cornus kousa,* family Cornaceae, a tender, deciduous tree grown for its showy, white flowers and pinkish fruit clusters. Native to Asia. Also called *Chinese dogwood.*

Kowhai-ngutu-kaka. Glorypea.

kraal. 1. In South Africa, a hut or collection of huts. 2. See *corral.*

kraut. Sauerkraut.

Kreis test. A test for rancidity in edible fats and oils. The substance to be tested is mixed with benzene and hydrochloric acid and treated with phloroglucinol in alcohol. It is shaken for a minute and centrifuged and the color is then matched in a Lovibond tintometer for determination of results.

kreosote. Creosote.

krotowinas. Crotovines.

Krutt. A skimmed milk cheese made from cows', goats', ewes', or camels' milk in Central Asia.

kudzu. Thunberg kudzubean.

Kuhbacher. A whole or partly skimmed, cows' milk, ripened cheese made in Bavaria.

kukui. Candlenuttree.

Kuminost. A spiced, whole or partly skimmed, cows' milk cheese made in Scandinavia.

kumquat. Any tree of the genus Fortunella, family Rutaceae. Species are grown for their edible fruit. See *Meiwa kumquat; Marumi kumquat; Nagami kumquat.*

kupuka tree. *Griselinia littoralis,* family Cornaceae, an evergreen tree with profuse foliage, grown as an ornamental in warm areas. Native to New Zealand.

kura clover. *Trifolium ambiguum,* family Leguminosae, a deep-rooted, perennial herb which is an important bee plant. Native to Caucasia.

kurrajong bottletree. *Brachychiton populneus,* family Sterculiaceae, a tree grown in very warm areas and in greenhouses for its dark-spotted yellowish-white flowers, reddish inside. Native to Australia.

Kurung oil tree. See *poonga oil pongamia.*

L

label. A tag of wood, metal, paper, or other material, fastened to a plant, animal, or product for identification, ownership, composition, etc.
labiate. 1. Any plant of the mint family. Species have divided calyx and corolla. 2. Designating the form of a flower: *e.g.*, the snapdragon, in which the corolla or calyx is divided into two parts of unequal size, one overhanging the other.
lablab. Hyacinth dolichos.
Labneh. A sour milk cheese made in the U.A.R.
laboratory. A room equipped to do testing and research work, such as a chemical or bacteriological *laboratory.*
labor earnings. 1. Net farm earnings less a charge for the use of total farm capital. 2. The operator's labor income plus the value of the family living from the farm which represents the earnings for the labor and management of the operator.
labor expense. The hired labor costs, plus the estimated value of unpaid family labor, and a charge for the operator's labor.
labor income. Farm income after all expenses, depreciation, unpaid family labor other than the operator, and an interest charge on the total capital investment have been deducted.
labor peak. The maximum requirement for farm labor at a particular season, usually at harvest time.
labor returns per man. A farm efficiency factor for the net returns per man for all man labor (*q.v.*). The total net returns from man labor is obtained by dividing labor income (*q.v.*) by the average number of men working on the farm.
labor-saving. Designating any tool, machine, method, device, or procedure which reduces the amount of human effort in the production of material things.
Labradortea ledum. *Ledum groenlandicum,* family Ericaceae, a hardy, cold habitat, evergreen shrub grown for its white flowers. Native to North America. Also called *black laurel.*
laced. Designating a feather edged with a band of color different from the ground color.
lace flower. See *wild carrot; Queen Anne's lace.*
lace-leaf. Lace waterhawthorn.
laceration. A wound, especially one caused by tearing.
lace vine. Silvervine fleeceflower.
lace waterhawthorn. *Aponogeton fenestralis,* family Aponogetonaceae, a profusely-veined, aquatic plant grown for its leaves which float just beneath the surface of the water. Native to Malagasy. Also called *lace leaf; lattice leaf.*
lace-winged fly. Any insect of the family Chrysopidae. Species are insectivorous.
lack balance. 1. To be pendulous, poor in attachment, or poorly quartered, as the udder of a dairy cow. 2. To be poorly proportioned in conformation; *e.g.*, a horse, cow, etc.
lack character. In dairy cows, a poor conformation or lack in breed characteristics.
lack depth and fullness in the heart girth. Designating a shallow or narrow-bodied animal.
lack development in the rear end. Designating an animal with a poor conformation in the rear quarters, especially undesirable in any meat animal.
lack flavoring. To have a taste or flavor defect of a processed product resulting from an insufficient amount of added flavoring, such as vanilla extract.
lack in breed type. Designating an animal defective in one or more of the important characters of the breed which it represents.
lacking depth of rib. See *lack depth and fullness in the heart girth.*
lacking desirable flavor. Designating products with a flat, unwholesome flavor rather than the desirable flavor characteristic of the product.
lack in general appearance. Designating an animal unattractive in conformation; rough, coarse, and unstylish.
lacking freshness. Designating products characterized by a stale or aged flavor.
lacking in condition. Designating an animal that shows lack of thrift, growth, or fleshing.
lacking in flavor. Designating a flavor defect of a food product in which there is no undesirable and very little, if any, desirable flavor development.
lacking somewhat in bone. Designating a horse with light bone structure and feet.
lack quality and refinement. See *lack in general appearance.*
lack snap and alertness. Designating an animal that is sluggish or lazy in action and disposition.
lack strength of top. Designating an animal with a sway back, an undesirable conformation defect.
lack style. Designating an animal of poor conformation, manners and alertness.
lack substance. Of an animal, to be small in size, particularly in bone, thin in fleshing or narrow, or to lack in depth of body conformation.
lack sweetness. Defect of a food product in which the proper sweetness is missing.
lack necessary bloom. Designating an underfed or unfattened animal.
lack thickness of loin. See *lack development in the rear end.*
lactalbumin fraction. The second chief protein of milk which is exceeded only by casein.
lactate. To produce and secrete milk.
lactating. Designating an animal which is actively producing and secreting milk.
lactation. The production or secretion of milk by animals.
lactation period. The number of days an animal produces milk after each calving.
lactic acid. $CH_3CH(OH)CO_2H$, *a*-hydroxypropionic acid, a sirupy liquid generated by the nat-

ural souring of milk or the induced fermentation of various food products, such as sugars, starches, etc., by any of various organisms. It is used in cultured milk products, pickles, soft drinks, etc.

lactic acid culture. Pure culture of lactic acid-producing organisms, such as *Lactobacillus delbrueckii,* used in creameries to insure the proper souring and ripening of cream and other dairy products, as cottage cheese, buttermilk, etc. Often called *buttermilk cultures.* See *souring.*

lactic acid fermentation. The action of organisms in milk which changes milk sugar into lactic acid.

lactic buttermilk. Cultured buttermilk.

lactic ferment. See *commercial culture.*

lactic fermentation. The preserving process which occurs in the making of sauerkraut, dill pickles, fermented string beans, silage, etc.

lactic ferment culture. See *commercial culture.*

lactic former. Any organism which changes sugars into lactic acid, as *Lactobacillus plantarum.*

Lacto. An ice cream which is made from frozen, sour, skim or whole cows' milk, with sugar, eggs, fruits, and flavoring. Also called *buttermilk sherbet.*

lactogen. The specific hormone that initiates and maintains lactation. Also called *gelactin.*

lactogenic hormone. Any hormone which stimulates lactation, especially prolactin *q.v.*

lactoglobulin fraction. That small portion of milk protein which contains more sulfur and less phosphorus than casein.

lactometer. An instrument which determines the density of milk. It acts as a specific gravity indicator, from which information the percentage of solids of milk may be calculated.

lactose. $C_{12}H_{22}O_{11}$, the sugar of milk; a valuable by-product of cheese making which is secured from the whey. It is the basis for much of the fermentation in milk.

lacustrine deposit. Sediments which are deposited in lake waters.

lacustrine plains. Lake-bed plains.

lacustrine soil. A soil which is composed of, or derived from, geological deposits formed in lakes.

Ladino. A violent, uncontrollable horse (U.S.).

Ladino white clover. A large, rapid-growing, perennial type of common white clover which spreads by creeping stems that root at the node; grown primarily as a pasture crop but also for hay, silage, and seed. Introduced to U.S. from Italy.

ladle. 1. A large, wooden paddle. 2. A market classification for retail lots of butter which result from putting several smaller lots together and mixing them into a uniform unit by means of a butter paddle or *ladle.* A common practice in the early days when farm butter was traded at the local store.

Lady Bank's rose. Banks rose.

lady beetle. Any of the insects of the family Coccinellidae, beneficial because they prey upon aphids, scale insects, and mealybugs. Sometimes they are collected and introduced into particular locations because of their value in biologic control. Also called *lady bug; ladybird beetle.*

ladybird beetle. Lady beetle.

lady bug. Lady beetle.

lady-by-the-gate. Bouncing Bet.

ladyfern. *Athyrium filixfemina,* family Polypodiaceae, a feathery fern widely grown as an ornamental. Native to North America. Also called *female fern.*

lady-of-the-night. Franciscan raintree.

lady's-mantle. *Alchemilla vulgaris,* family Rosaceae, a perennial herb grown for its silvery or grayish foliage. Escaped, it can become a weed pest. Native to Europe.

lady's-slipper. Garden balsam.

lady's-thistle. Blessed milkthistle.

lady's thumb. Spotted lady's-thumb.

lady toed. Designating a horse which stands with its hocks together and its toes turned out. See *cow-hocked.*

Lady Washington geranium. Lady Washington pelargonium.

Lady Washington pelargonium. *Pelargonium domesticum,* family Geraniaceae, a tender, shrub-like plant grown for its large, white, pink, or red flowers. Native to South Africa.

La Fleche. A general-purpose, French breed of chickens, characterized by black plumage, a V-shaped comb, and white skin. Mature males weigh about 8½ lbs.; mature females weigh about 7½. The egg-shell color is white.

lager beer plant. Aspidistra.

lagg. The depressed margin of a raised peat bog.

Laguiole. A cured, hard, whole or partly skimmed, cows' milk cheese made in France.

laid by. Designating the last cultivation or working of a row crop for the season.

lake-bed plains. The beds of extinct lakes, especially if geologically recent, such as former great lakes of the Glacial Period and the high levels of the predecessors of the present Great Lakes of North America. As land, the surfaces are, for the most part, relatively flat and the parent soil material, or lacustrine sediments, not greatly modified by weathering.

Lakenfelder cattle. The name for Dutch belted cattle (*q.v.*) in the Netherlands.

Lakenvelder. A German breed of Leghorn-like chickens, distinguished by the color of the surface plumage which is black at both extremities but white in the midsection, giving a belted effect. The skin and egg-shell color are white.

lake shore disease. See *cobalt deficiency.*

Lake States. States of the U.S. which border on at least one of the Great Lakes: Minnesota, Wisconsin, Illinois, Indiana, Michigan, Ohio, Pennsylvania, and New York.

Lake States tree classification. A system of classifying trees in even-aged stands according to position in the stand, crown density, soundness, form, and utility.

Laland firethorn. *Pyracantha coccinea* var. *lalandi,* family Rosaceae, a semi-evergreen, thorny shrub grown in the eastern U.S. for its persistent, orange-red fruits. Native to Asia. Also called *scarlet firethorn; firethorn.*

lamance iris. *Iris foliosa,* family Iridaceae, a beardless iris, a perennial herb grown for iris breeding and for its lavender flowers. Native to North America.

Lamarckism. In genetics, a theory of biological evolution which assumes that inheritance of acquired characteristics is the mechanism of evolutionary change and that mutation results from environmental factors.

lamb. 1. The young of sheep: specifically, (a) a young ovine that has not yet acquired the front pair of permanent incisor teeth; technically, (b) in stock judging, a sheep born in the same calendar year as shown. 2. To give birth to a *lamb.*

lamb breast. A cut of meat from the breast or rib section in the forequarters of a lamb. This cut is boned or rolled and sold for lamb roasts or stews.

lamb crop. The total of the living young produced by the breeding flock of ewes which, because of the prevalence of twinning, should be from 130 to 150 percent of the ewes in the flock.

lamb crop percent. See *calf (or lamb) crop percent.*

Lambert crazyweed. *Oxytropis lamberti,* family Leguminosae, a plant, the foliage of which is poisonous both green and dried. Symptoms of poisoning are dullness, irregularity of gait, lack of appetite, dragging of the feet, a solitary habit, loss of flesh, and a shaggy coat. As the animal ceases to eat, it dies. Found in the Great Plains area of the U.S., and south to Mexico. Also called *crazyweed; white locoweed; rattleweed; Colorado loco vetch.*

Lambert mottle. A virus disease of the severe rusty mottle type found on Lambert cherry, a commercial variety of sweet cherry, grown in Okanogan valley, British Columbia, and elsewhere. It is characterized by early mottle, later chocolate-brown, dead leaf spots; brown, torn areas in the leaf; late and irregular bud development; and reduced yields. See *necrotic rusty mottle.*

lamb flock. In stock judging, one ram lamb and three ewe lambs.

lamb hog. A male lamb from weaning time until it is shorn.

lambing. The dropping, or birth, of lambs (in sheep husbandry, a time or regular event in management). See *lambing time.*

lambing ground. See *range lambing.*

lambing paralysis. Pregnancy disease in ewes.

lambing pen. A specially equipped, isolated pen in the sheep barn in which a ewe is placed just before she gives birth to her young.

lambing season. Late winter and early spring; the season of the year when most lambs are born. See *lambing time.*

lambing shed. A shed or barnlike building in which ewes are placed just before they give birth to lambs.

lambing time. The season of the year, usually late winter or early spring, when sheep produce their young.

lamb joint. The joint in the foreleg of a lamb where the forefeet are removed from the carcass at the time of slaughter or when it is offered for sale. Often called the break joint (*q.v.*); its ease of breaking is taken as an indication of the age of the lamb.

lambkill kalmia. *Kalmia angustifolia,* family Ericaceae, an evergreen shrub of thin, open habit, usually growing 2 to 3 ft. high. It is considered to be injurious to sheep and other grazing animals and may occur as a troublesome wild growth in cranberry bogs. Also called *sheep laurel; wicky; narrow-leaved laurel; calfkill; dwarf laurel.*

lamb rack. A trough standing about 18 ins. high, which is 6 ins. wide at the bottom, and has sloping, 6-in. sides; used to hold a slaughtered sheep on its back during skinning. Also called *veal rack.*

lamb riblet. A retail cut of lamb for stewing or braising which consists of thin slices of breast along with ribs.

lamb's ears. Woolly betony.

lamb shank. A retail cut of lamb for braising which consists of the shank. Also called *trotter.*

lamb's lettuce. European cornsalad.

lambsquarters goosefoot. *Chenopodium album,* family Chenopodiaceae, an annual herb which is sometimes a weed pest and sometimes used as a potherb. Native to Europe. Also called *lamb's quarters; amaranth pigweed; fat hen; chou grass; poulette grass.*

lambstongue fawnlily. *Erythronium grandiflorum,* family Liliaceae, a bulbous herb grown for its early spring, bright yellow flowers. Native to North America. Also called *Adam and Eve; chamise lily.*

lamb's wool. Short wool taken from lambs not over 7 or 8 months old although some wool which is taken from lambs of 12 to 14 months of age is also graded as *lamb's wool.* It is soft and has less felting properties than sheep wool.

lameness. Any soreness, tenderness, or unsoundness which is indicated by the unnatural action of feet or legs of animals.

lamina (plural: laminae). The left blade; the expanded leaf.

laminar flow. A type of low velocity flow in conduits having parallel sides in which each particle moves in a direct path parallel to every other particle and in which the head loss is approximately proportional to the first power of the velocity. Also called *streamline flow; viscous flow.*

laminated. Designating soil arranged in very thin plates or layers, less than 1 mm. thick, which lie horizontally or parallel to the soil surface, usually fragile and of medium to soft consistency.

laminated wood. A piece of wood built up of layers of wood that have been joined with the grain parallel, either with glue or with mechanical fastenings. The term is most frequently applied where the layers are too thick to be classified as veneer.

laminitis. Inflammation or congestion of the sensitive tissues, the laminae, which lie immediately below the outer horny wall of a horse's foot. It is characterized in the acute stage by expression of pain, standing still, dilated pupils, bright red membranes of the eyes, and rapid breathing. The pulse at first is very strong and fast but becomes weaker and the temperature is above normal. Among the various causes are: ingestion of excessive amounts of grain (grain founder), drinking cold water while overheated, concussion during hard fast road work or from long truck or train rides, toxemias as a result of pneumonia or metritis, etc. Some cases may be an allergic reaction. Grain founder is usually the most common type. Also called *founder.*

Lamona. A breed of general purpose chickens of American origin developed about 1912 by the U.S. Government Experiment Station at Beltsville, Maryland, by crossing the Silver-Gray Dorking, White Plymouth Rock, and White Leghorn. The breed has white plumage, yellow skin and single comb; their egg-shell color is white.

lampas. A swelling of the mucous membranes in the hard palate immediately behind the upper incisors of the horse. Also called *palatitis.*

lampers. Lampas.

lamprey. See *Petromyzontidae; sea lamprey.*

Lanarkshire strawberry disease. See *red stele.*

Lancashire. A white, cured, cows' milk cheese made in England.

land. 1. The total natural and cultural environment within which production must take place. Its attributes include climate, surface configuration, soil, water supply, subsurface conditions, etc., together with its location with respect to centers of commerce and population. Oyster beds and even tracts or bodies of water, as where valuable fishing rights are involved, may be regarded as *land.* It is often convenient, in fact, to regard *land* as synonymous with all that nature supplies, external to man, which is valuable, durable and appropriable, thus including: *e.g.,* waterfalls and other sources of waterpower. 2. In a broad legal sense, any real part of the surface of the earth, including all appurtenances, anything in, on, above, or below the surface. 3. In plowing, a plowed or unplowed space between two furrows. 4. The total width of a strip of *land* tilled by a farmer, or some designated width, as a perch, 16½ ft. Also called a *stitch.* 5. Soil. 6. A natural part of the earth's surface characterized by any single factor, or combination, of topography, climate, soil, rocks, vegetation; the natural landscape. 7. Pertaining to agriculture, those areas which actually are in use or are capable of use for the production of farm crops and livestock.

Land Bank. Federal Land Bank.

land capability. The suitability of land for a specified use. As ordinarily used in the U.S., it is an expression of the effect of physical land conditions, including climate, on the total suitability for use, without damage, for crops that require regular tillage, for grazing, for woodland, and for wildlife. *Land capability* involves consideration of the risks or the difficulties in land use owing to physical land characteristics, including climate.

land capability class. One of several classes of land distinguished according to potentiality for agricultural use. See *land classification.*

land capability map. 1. A map which shows land-capability units, subclasses, and classes. 2. A soil conservation survey map which is colored to show land-capability classes.

land classification. 1. Classification of specific bodies of land according to their characteristics or their capabilities for use. (a) A natural *land classification* is one in which the natural land types are placed in categories according to their inherent characteristics. (b) A *land classification* according to use capabilities is one in which bodies of land are classified according to their capabilities for man's use with sufficient detail of categorical definition and cartographic expression to indicate those differences significant to man. 2. The application of the results of a soil survey is frequently made through the intermediate step of some sort of *land classification.* Such a classification may involve little more than a grouping of the soil units shown on the map into land classes according to their common natural characteristics or according to their productivity for specific crops or groups of crops. It may involve the grouping of soil units into classes according to their recommended use, considering present economic conditions and the fundamental, physical factors. Or it may involve such complete detail as that required for placing a rating on each unit (farm or ranch) for the purposes of tax assessment.

land contract. An agreement between buyer and seller which states the terms of the purchase and sale of an item of real estate. The initial down payment on the property involved is usually small, often from 5 to 20 percent of the purchase price. Subsequent payments are scheduled for regular stated dates. The seller often retains the deed and title of the property until the purchase price is entirely paid; or the deed and title may be transferred at some earlier period agreed to in the contract. It is an important method used by young farmers in attaining farm ownership. Sometimes called *instalment land contract.*

Land District. In the U.S., a division of a state or territory created by federal authority in which is located a U.S. Land Office with a register of the Land Office and a receiver of public money for the disposition of public lands within the district.

land division. The manner in which the land is divided among the users and/or owners. Boundaries of units may be established upon a system of astronomical observations, such as the so-called checker-board (rectangular survey) division of land in many parts of the U.S. Boundaries may be located and described in deeds in terms of metes and bounds based upon various land marks. Also river front and similar points of departure may be used in the system by which the land is divided.

land drainage. 1. The removal of surface water or excess ground water by artificial means, such as ditching or tiling. 2. Natural drainage, or removal of water by creeks, rivers, in natural water channels.

landed. Designating a person, or group, that owns a large estate of rural land, as *landed* gentry (archaic).

land fallow. The preparation and cultivation of the land to take in and retain the greatest possible amount of water. See *fallow*.

Land-Grant college. Originally, any public institution of higher learning which was financed with funds derived from the sale of public lands granted for the purpose. Now specifically, an institution operating under the provisions of the Morrill Act of the U.S. Congress in 1862, which authorized the establishment in each state of at least one college "where the leading object shall be, without excluding other scientific and classical studies and including military tactics, to teach such branches of learning as are related to agriculture and the mechanic arts . . . in order to promote the liberal and practical education of the industrial class of the several pursuits and professions of life." The Morrill Act granted to each state a tract of public land, or land script, to be sold, with the proceeds to be invested in perpetuity by the state and the income used to endow, support, and maintain such a college. Passage of the Morrill Act represented the final break with the educational traditions of Europe which catered to a social and intellectual elite in that it extended the scope of scholarly inquiry and teaching into all areas of human activity, placed heavy emphasis upon the sciences, cleared the way for the application of knowledge to the solution of the multifarious problems of mankind, and opened the door of educational opportunity to all classes. The *Land-Grant colleges* have been distinguished by their emphasis upon "the liberal and practical" in their curricula and their dedication to public service. There are 68 land-grant colleges and universities: Alabama Polytechnic Institute, Auburn, Alabama; Alabama Agricultural and Mechanical College, Normal, Alabama; University of Alaska, College, Alaska; University of Arizona, Tucson, Arizona; University of Arkansas, Fayetteville, Arkansas; Agricultural, Mechanical and Normal College, Pine Bluff, Arkansas; University of California, Berkeley, Los Angeles and other campuses in California; Colorado State University, Fort Collins, Colorado; University of Connecticut, Storrs, Connecticut; University of Delaware, Newark, Delaware; Delaware State College, Dover, Delaware; University of Florida, Gainesville, Florida; Florida Agricultural and Mechanical University, Tallahassee, Florida; University of Georgia, Athens, Georgia; Fort Valley State College, Fort Valley, Georgia; Georgia Institute of Technology, Atlanta, Georgia; University of Hawaii, Honolulu, Hawaii; University of Idaho, Moscow, Idaho; University of Illinois, Urbana, Illinois; Purdue University, Lafayette, Indiana; Iowa State College, Ames, Iowa; Kansas State College, Manhattan, Kansas; University of Kentucky, Lexington, Kentucky; Kentucky State College, Frankfort, Kentucky; Louisiana State University, Baton Rouge, Louisiana; Southern University, Baton Rouge, Louisiana; University of Maine, Orono, Maine; University of Maryland, College Park, Maryland; Maryland State College, Princess Anne, Maryland; University of Massachusetts, Amherst, Massachusetts; Massachusetts Institute of Technology, Cambridge, Massachusetts; Michigan State University, East Lansing, Michigan; University of Minnesota, Minneapolis, Minnesota; Mississippi State College, State College, Mississippi; Alcorn Agriculture and Mechanical College, Alcorn, Mississippi; University of Missouri, Columbia, Missouri; Lincoln University, Jefferson City, Missouri; Montana State College, Bozeman, Montana; University of Nebraska, Lincoln, Nebraska; University of Nevada, Reno, Nevada; University of New Hampshire, Durham, New Hampshire; Rutgers University (The State University of New Jersey), New Brunswick, New Jersey; New Mexico College of Agricultural and Mechanic Arts, State College, New Mexico; Cornell University, Ithaca, New York; State University of New York, Albany, New York; North Carolina State College, Raleigh, North Carolina; Agricultural and Technical College of North Carolina, Greensboro, North Carolina; North Dakota Agricultural College, Fargo, North Dakota; Ohio State University, Columbus, Ohio; Oklahoma State University of Agricultural and Applied Science, Stillwater, Oklahoma; Langston University, Langston, Oklahoma; Oregon State College, Corvallis, Oregon; Pennsylvania State University, University Park, Pennsylvania; University of Puerto Rico, Mayaguez, Puerto Rico; University of Rhode Island, Kingston, Rhode Island; Clemson Agricultural College, Clemson, South Carolina; South Carolina State College, Orangeburg, South Carolina; South Dakota State College, College Station, South Dakota; University of Tennessee, Knoxville, Tennessee; Tennessee Agricultural and Industrial State University, Nashville, Tennessee; Texas Agricultural and Mechanical College System, College Station, Texas; Prairie View Agricultural and Mechanical College, Prairie View, Texas; Utah State University, Logan, Utah; University of Vermont, Burlington, Vermont; Virginia Polytechnic Institute, Blacksburg, Virginia; Virginia State College, Petersburg, Virginia; State College of Washington, Pullman, Washington; West Virginia University, Morgantown, West Virginia; University of Wisconsin, Madison, Wisconsin; University of Wyoming, Laramie, Wyoming.

land grants. 1. Tracts of land given by the U.S. Government to states, colleges, railroads, canal companies, and others for the purposes of aid to education and promotion of economic development of the unoccupied public domain. 2. Grants of land, to individuals or groups, which were made by French, Spanish, and Mexican governments in North America before the territory became a part of the U.S. Such grants may still have a legal status. 3. Tracts of land awarded to individuals in the British Empire by the governments as recognition of their services.

land improvement. Any changes made on land to cause it to be more productive or more desirable.

landing. A place at which logs are assembled for transportation in loads or rafts.

landing of the beam. The angle that the plow beam makes with the line of the landside. It varies with the size of the plow. Wedges were forced between the beam and the frog, or landside, to obtain the desired landing.

landing the point of the share. See *horizontal suction.*

landlocked. Designating a piece of land belonging to one person which is surrounded by land belonging to other persons so that it cannot be approached except over their land.

landlord. One who owns land or property and leases it to a tenant. See *land owner.*

land manager. In western U.S., a person who is concerned with man and his animals, as well as predator and rodent control, management of big game, range administration, and grazing in relation to the welfare of the forest and other land resources.

land measure. 1. A small device on a grain drill so designed that it records the number of acres sown. 2. Any unit of land measurement. 3. Any method for determining the size, etc., of acres of land.

land owner. One who owns land, especially large tracts of rural or nonurban land.

land patent. A deed or record title issued by a government or state for the conveyance of some portion of the public domain.

land plane. A large, tractor-drawn machine designed for planing or smoothing land for more efficient use of irrigation water or for easier tillage of land not irrigated. It consists of a long steel frame mounted on wheels, near the center of which is attached a large, long, adjustable, combination steel blade and scraper to remove soil from high points and to convey it to depressions.

land plaster. See *calcium sulfate.*

land policy. The guiding attitude or political philosophy in relation to land use which is based upon analysis of facts and pressures arising from problems in land use. It is a consistently followed, fundamental course of action, though open to revision. (Policy is understood to be that of the U.S. government unless otherwise stated.)

land-poor. 1. Designating a person who owns a great deal of unproductive land. 2. Designating one who owns ample land but makes inefficient use of it or is obliged to borrow money to pay taxes, interest, etc.

Landrace. A popular, white, bacon-type hog, bred in Denmark, used in the U.S., to develop a better meat-type hog. See *American Landrace.*

land reclamation. Making land capable of more intensive use by changing its character and/or environment through operations requiring collective effort: *e.g.*, preparing land for irrigation by constructing reservoirs, dams, and canals. (The clearing of stumps, brush, and stones from land or simple techniques of erosion control which can be effected by the individual are not usually included in reclamation.)

land reclamation by drainage. The recovery of land from seas, lakes, and rivers, and the improvement of marsh, fen, and swampland for agricultural and other purposes.

land rent. The payment, by cash or part of the crop or services, for the use of land.

land roller. Any one of a number of implements used primarily for crushing clods and compacting or firming the soil in the preparation of seed beds. See *cultipacker; clod crusher; tube roller; cast-iron drum roller.*

landscape. 1. In soil geography, the sum total of the characteristics that distinguish a certain area of the earth's surface from other areas, as soil types, vegetation, rock formations, hills, valleys, streams, cultivated fields, roads, and buildings. These characteristics are the result not only of natural forces but also of human occupancy and use of the land, and together give the area its pattern. 2. In a broad sense, the complex pattern of an extensive area, such as the rural *landscape,* the mountain *landscape,* or the Chernozem *landscape.* 3. To beautify terrain as with plantings of trees, bushes, and flowering herbs; with ornamental features, as terraces, rock gardens, bog gardens, pools, walks, drives, etc.

landscape architect. A person who is by training and experience qualified for the professional practice of landscape architecture. The national professional organization, American Society of Landscape Architects, requires high standards and strict adherence to a code of professional practice.

landscape architecture. The art and science of developing land areas and the objects upon them for human use and enjoyment in a manner that combines the maximum of beauty and utility.

landscape gardening. 1. The practice of landscape architecture. 2. The caring for lawns, trees, plants, walks, etc.

land shark. A tricky or unscrupulous person who deals in land.

landside. That part of the plow bottom which slides along the face of the furrow wall. It helps to counteract the side pressure exerted by the furrow slice on the moldboard, and to steady the plow while being operated.

landside clearance. The adjusted space of ½ to ¾ ins. between the heel of the landside on the rear bottom of a trailing plow and the furrow sole out from the furrow wall. The rear wheel is adjusted to provide this clearance. On all moldboard plows mounted on wheels the heel of the landside does not touch the floor when properly set. The vertical suction in this case will be the amount the heel of the landside is elevated above the floor.

landslide. The sliding or falling of a mass of earth, rock, or mixture of the two. A small amount of water is usually present helping to lubricate the mass, but true *landslides* do not contain enough liquid to cause the mass to flow. Displacements may occur on moderate to steep slopes, irrespective of cover. During protracted rains, great masses of waterlogged soil slide and flow downhill, exposing raw subsoil and forming mounds, ridges, and benches where the soil piles up along the lower part of the slip. These conditions frequently result in accelerated runoff and gullying and usually make cultivation impracticable. See *mudflow; mud-rock flow.*

landslip. Landslide.

landslip terrace. A rough bench formed by the landslide of a segment of a hillside.

land slump. Landslide.

land stones. Loose stones which occur naturally on a land surface or in a field, as contrasted with those which are quarried, cut, or broken by man. Also called *fieldstones.*

land tenure. The holding of land and the rights that go with such holding which includes everything from fee simple title, embracing all possible rights within the general limits imposed by the government, down to the most restricted forms of tenancy (holding and/or operating land under the ownership of another).

land tortoise. See *gopher: 2.*

land type. A geographic division of land based upon some one or combination of natural factors as soil, relief, vegetation, and climate. The term is often loosely used to indicate man's use of the land such as agricultural, grazing, mining, forest, urban, etc. See *natural land type.*

land type mapping. A scheme of mapping in which soil associations, or natural land type, are employed as the unit of mapping. It differs from the prevailing type of soil mapping in which the soil type is the unit.

land use mapping. The mapping of land according to use: *e.g.*, cultivated land, nontillable pasture and grassland, idle land, woodland, and urban areas, homesites, etc.

land use pattern. The area design or arrangement of land uses, major and minor, and of operation units.

land use planning. The development of plans for the uses of land that will, over a long period, best serve the general welfare, together with the formulation of ways and means of achieving such uses.

land use survey. A survey of an area conducted to determine present use of land.

land water. Water (lakes, rivers, etc.) or sources of water (bodies of snow and ice) found on land.

land wheel. A large wheel that travels on the unplowed soil on the landslide of a sulky plow, helping to support the weight of the plow carriage.

lane. A passageway from barnyard to pasture or field areas; a narrow road confined between fences.

Langres. A soft, cured, cows' milk cheese, similar to Livarot, made in France.

Langshan. A breed of intermediate weight, general purpose chickens, of the Asiatic Class, characterized by a single comb, white skin, and moderately feathered shanks and toes. Cocks weigh about 9½ lbs.; hens weigh about 7½ lbs. The eggshell color is brown. The two varieties are White and Black.

Langstroth frame. A frame for a beehive which measures 17⅜ ins. by 9⅛ ins. deep.

lanolin. Wool fat; a yellowish, viscous mass of waxlike consistency, very resistant to acids and alkalis, which emulsifies easily with water and is used for making ointments. It consists of the palmitate, oleate, and stearate of cholesterol.

lantana. 1. Any plant of the genus Lantana, family Verbenaceae. 2. *L. camara,* a tropical shrub grown in warm areas for its yellow flowers turning orange or red. Native to America. Also called *red sage; yellow sage.*

Lapland. A hard, dumbbell-shaped, reindeer's milk cheese made by the Laplanders.

lap-love. European glorybind.

lap (wood). Tops of trees left in the forest after logging.

larch. Any deciduous tree of the genus Larix, family Pinaceae. Species are grown as ornamentals and for lumber. See *European larch; Dahurian larch; eastern larch; western larch.*

larch canker. European larch canker.

larch casebearer. Coleophora laricella (Hbn.), family Coleophoridae, a casebearer that infests eastern and European larches in the eastern half of the U. S.

larch sawfly. *Pristiphora erichsonii* (Htg.), family Tenthredinidae, a sawfly whose larvae infest larches, often defoliating the trees.

lard. The fat rendered from fresh, clean, sound, fatty tissues of hogs in good health at the time of slaughter, with or without lard stearin or hardened lard. It is used as a food product for frying, shortening, etc. See *fry out.*

lard breed. Lard-type breeds of swine.

larder beetle. *Dermestes lardarius* L., family Dermestidae, a beetle whose larva feeds on dried skins, feathers, meats, etc.

larding. The introduction of strips of pork fat into the body of lean roasts by means of a larding needle.

larding needle. A half-round, metal needle used for introducing fat into the internal portion of meat roasts.

lard oil. An extract of lard made from "A" white grease used for making a high grade lubricant for delicate running machine parts. See *"A" white grease; brown grease.*

lard press. A mechanical device, consisting of a hopper box and a screw press, used for pressing warm lard from the cooked fat of pork.

lard-type breeds of swine. The principal *lard-type breeds of swine* in the U. S. are: Duroc, Chester-White, Spotted Poland-China, Hampshire swine, and Berkshire (*q.q.v.*). The Kentucky Red Berkshire, Cheshire, Large Black, Essex, Victoria and Mule-Foot are of minor numbers and importance.

lard type hog. A fat type of hog which was developed in the U. S. Most of the breeds common in the U. S. were originally classified as this type. It is especially adapted to a high degree of quick-fattening, with heavy back and thick sides. It was once very popular in the Corn Belt of the U. S. but is now decreasing in popularity because of consumer demand for more lean meat and less fat.

lardy. Designating a thick, over-fat hog.

lareat. Lariat.

laredo mahonia. *Mahonia trifoliolata,* family Berberidaceae, an evergreen, thornless shrub grown in shade borders and foundation planting in a sheltered position for its fragrant, yellow flowers and its edible, red fruit. Native to North America. Also called *agarita; algerita.*

Large Black. A bacon-type, black, coarse breed of swine, developed in England; not generally bred in the U. S.

large chestnut weevil. *Curculio proboscideus* (F.), family Curculionidae, a weevil that infests chestnuts.

large chicken louse. *Goniodes gigas* (Tasch.), family Philopteridae, one of the less important lice which infest chicken.

large, clean-cut, wide and deep hocks. In stock judging, indicating good leg conformation on a horse.

large, coarse horns. In stock judging, designating an undesirable horn structure in cattle.

large crabgrass. Hairy crabgrass.

large egg. A U. S. grade of eggs having a minimum net weight of 24 ozs. per dozen, with a minimum weight for individual eggs at the rate of 23 ozs. per dozen.

large fothergilla. *Fothergilla major,* family Hamamelidaceae, a tender shrub grown for its white flowers. Native to North America.

large heart girth. A desirable conformation in an animal which indicates lung capacity and wide, deep body.

large hitch. A class of draft horses of either sex exhibited in groups of more than two as a team.

large in his feet. In stock judging, big feet in a draft horse, a desirable quality.

large-leaved cucumber-tree. Bigleaf magnolia.

large-leaved linden. *Tilia platyphyllos,* family Tiliaceae, a deciduous tree which grows to a height of up to 120 ft. grown as an ornamental shade tree. Native to Europe. Also called *bigleaf linden.*

large-leaved magnolia. Bigleaf magnolia.

large-leaved tomato. *Lycopersicon esculentum grandifolium,* family Solanaceae, a tender herb grown as an annual for its large, red fruit. Also called *bigleaf tomato; potato-leaved tomato.*

large milkweed bug. *Oncopeltus fasciatus* (Dall.), family Lygaeidae, an insect which may cause serious damage to grain crops.

large moonflower. *Calonyction aculeatum,* family Convolvulaceae, a tropical and subtropical twining, perennial vine grown out-of-doors in warm areas, and in greenhouses, for its habit, its large leaves, and large, white flowers. Native to America. Also called *evening glory; moonflower.*

large purple fringe-orchid. Meadow pink orchid.

larger canna leaf roller. *Calpodes ethlius* (Stoll), family Hesperiidae, a caterpillar which infests cannas, cutting a strip from the margin of the leaf, folding the cut portion over itself, and feeding within.

larger elm leaf beetle. *Monocesta coryli* (Say), family Chrysomelidae, an insect that infests elms, hazel, and hawthorns.

large, roomy ewe. A female sheep with a desirable large body formation.

large, round, well-balanced foot. A desirable foot on a horse, nicely spread at the heel and having an excellent hoofhead.

large roundworm. A common name referring to various species of roundworms belonging to the family Ascaridae, which parasitize man, domestic livestock, and birds. The scientific names of these large roundworms are: *Parascaris equorum* (equines); *Neoascaris vitulorum* (cattle); *Ascaris lumbricoides* (swine, man); *Toxocara canis, T. mystax, Toxascaris leonina* (dogs and cats); *Ascaridia galli, A. columbae, A. numidae* (domesticated birds).

larger shot-hole borer. *Scolytus mali* (Bechst.), family Scolytidae, an insect pest of deciduous trees.

larger yellow ant. *Lasius interjectus* Mayr, family Formicidae, an ant which fosters mealybugs and aphids.

large-seeded alfalfa dodder. Bigseed alfalfa dodder.

large-seeded legume. Any plant of the family Leguminosae which produces seeds of relatively large size, as kidney bean, peas, soybeans, cowpeas, peanuts, etc.

large-toothed aspen. *Populus grandidentata,* family Salicaceae, a deciduous tree growing to 60 ft. Found from the Carolinas to Tennessee, U. S. Also called *bigleaf aspen.*

large turkey louse. *Chelopistes meleagridis* (L.), family Philopteridae, a louse which infests turkeys.

lariat. Rawhide, horsehair, or hemp rope, generally 35 to 45 ft. long and arranged with a large loop on one end. It is carried in a coil on the saddle and used to lasso livestock for branding or restraining purposes.

larkspur. Any annual or perennial herb of the genus Delphinium, family Ranunculaceae. Species are very popular flower garden plants; poisonous to livestock. See *rocket larkspur; cardinal larkspur; Kashmir larkspur; forking larkspur; American larkspur; Siberian larkspur; orange larkspur.*

larkspur leaf miner. *Phytomyza delphiniae* Frost, family Agromyzidae, a fly whose larvae mine the leaves of larkspur and monkshood. Infested plants appear to have blighted foliage.

larva. (plural: *larvae*). The immature form of insects and other small animals, which is unlike the parent or parents and which must undergo considerable change of form and growth before reaching the adult stage: *e.g.*, white grubs in soil or decayed wood are *larvae* of beetles. Caterpillars and maggots also are *larvae.*

larval. Pertaining to a larva.

larval flukes. Immature stages of trematodes which may occur in water, in snails and insects, and on vegetation.

larval metamorphosis. See *metamorphosis.*

larvicide. A chemical used to kill the larval or preadult stages of parasites.

laryngeal paralysis. Roaring *q.v.*

laryngotracheitis. Infectious laryngotracheitis.

larynx. The upper portion of the windpipe (trachea).

lasher. A piece of rope used for binding or making one thing fast to another.

lasso. To throw a lariat (*q.v.*) in such a manner as to catch an animal by the horns, neck or legs in the loop at the end, generally for restraining purposes.

late. 1. Designating a variety of plant which flowers or fruits at a later time in the year than others of its species. 2. Designating an event which comes at a later time in the year than normal, as a *late* spring.

late blight and rot of potato. A widespread, serious, fungous disease of solanaceous plants caused by *Phytophthora infestans,* family Pythiaceae; characterized by a blight on the leaves following flowering, consisting of brown, dead areas at the tips or margins which spread downward. In heavy, damp soils the tuber is rotted (wet rot), while in other soils the rot may be superficial (dry rot).

late blight of celery. A widespread, fungous disease caused by *Septoria apii,* family Sphaerioidaceae; characterized by small, brown, leaf spots which coalesce and dry out the leaf; by elongated, similar lesions on the stems; by small, black, fruiting bodies protruding from the brown spot.

late dough stage. A stage in the ripening of cereals, about 10 days before harvest, when the kernels are beginning to harden.

late leaf rust. A fungous disease of red raspberries caused by *Pucciniastrum americanum,* family Melampsoraceae; characterized by a fine, light yellow, powdery mass of spores on the older leaves in midsummer, accompanied by a leaf curl and defoliation. Also called *late raspberry rust.*

late leaf spot. A fungous disease of peanuts caused by *Cercospora personata,* family Dematiaceae; characterized by dark brown or black spots on the leaves.

late lilac. *Syringa villosa,* family Oleaceae, a hardy, deciduous shrub grown for its lilac-pink flowers. Native to China.

late-maturing. The coming to maturity of a plant at a later time than is normal for the species.

late-milk stage. A late stage in the ripening of grains in which the grain kernel is fully formed but will exude a milk-like substance when crushed or broken open.

latent. Designating an infection that is present but which is not manifest in the host under consideration. *Cf. dormant.*

latent bud. A bud that remains dormant or semidormant in relation to those actively producing shoots.

latent mosaic. A virus disease of potato characterized by a mottle on the leaves, sometimes faint but sometimes with irregular, pale green to yellow chlorotic tissue contrasting with the dark green veins of the leaves.

latent period. See *preinfective period.*

latent virus. A virus which is present in a plant without the production of symptoms.

late pear. Sand pear.

late rain. A rain that comes out of the crop season in sections of the country having seasonal precipitation, such as the rainy and dry seasons of the western range, U. S.

lateral. 1. A part of a system that branches out from the main body of the system, such as the tile *lateral* drain that connects to a main drain in a drainage system. 2. A branch or twig of a tree.

lateral bud. A bud attached to the side of a branch or spur.

lateral canter. A horse's gait characterized by a simultaneous beat of corresponding front and rear feet, as opposed to diagonal front and rear feet.

lateral erosion. Sheet erosion.

lateral gait. See *pace: 2.*

lateral lacunae. The two grooves which border the horny frog of a horse's hoof.

lateral ringbone. A bony growth on the side of the pastern bone in a horse's leg. See *ringbone.*

Laterite (soil). The zonal group of soils having very thin organic and organic mineral layers over leached soil that rests upon highly weathered material, rich in alumina or iron-oxide, or both, and poor in silica. The soils are usually, but not necessarily, red in color. *Laterite soils* are developed under the tropical forest in a hot, moist or wet and dry climate with moderate to high rainfall: *e.g.*, India. This group of soils is now more frequently called *Latosol.*

laterization. The characteristic process in soil formation which tends toward the production of Laterites and lateritic soils. Essentially, it is the process of silica removal with consequent increase in the alumina and iron-oxide content, and decrease in base-exchange capacity of the soil.

latex. A milky, resinous fluid of variable composition formed in special secretory cells, usually the phloem and leaves, of certain tropical trees of the families Apocynaceae, Euphorbiaceae and Moraceae. It is the source of rubber, gutta-percha, and some other commercial products. The para rubbertree (*q.v.*) is the principal source of the world's supply of *latex.* Also called *caoutchouc; India rubber.*

latexosis. The production of latex by a plant.

lath house. A structure, for the propagating and raising of tropical plants or plants needing shade, which consists of a frame covered with slats or laths, with about 1-in. separations between pieces. This allows about ½ the amount of light received out-of-doors, and provides shelter from the wind.

lath screen. Trellis.

latigo. See *rigging ring.*

latitude. The distance on the earth's surface from the equator measured in degrees of the meridian.

Latosol. In pedology, a zonal group of soils which replaces that formerly designated as Laterite.

lattermath. Aftermath.

lattice-leaf. Lace waterhawthorn.

lattice structure. Orderly arrangement of atoms in crystalline material.

laurel. 1. Any tree or shrub of the genus Laurus, family Lauraceae. Species are grown as ornamentals and for their leaves used as spice. Also called *bay; bay tree; bay laurel.* See *Grecian laurel; laurel leaves.* 2. Any tree or shrub that is similar in appearance to the true *laurel.* See *laurel cherry; lambkill kalmia; mountain laurel; bog kalmia; rhododendron; California laurel; spurge-laurel daphne; laurel oak; laurel willow; Pacific madrone.*

laurel cherry. 1. *Prunus laurocerasus,* family Rosaceae, an evergreen tree grown for its lustrous, green foliage and its small, white flowers. Native to North America. Also called *cherry laurel; wild orange; cherry bay; mock-orange.* 2. See *Carolina laurelcherry.*

laurel leaves. The dried leaves of the Grecian laurel used as flavoring for soups, sauces, pickles, etc. Also called *bay leaves.*

laurel oak. *Quercus laurifolia,* family Fagaceae, a half-evergreen tree grown as an ornamental and as a shade tree. Native to North America.

laurel willow. *Salix pentandra,* family Salicaceae, a hardy, deciduous tree grown as an ornamental for its showy, golden-yellow catkins. Escaped, it can become a pest. Native to southeastern Europe. Also called *bay willow; bayleaf willow.*

laurelwood. Pacific madrone.

laurestinus viburnum. *Viburnum tinus,* family Caprifoliaceae, an evergreen shrub grown in warm areas for its foliage and its clusters of pinkish flowers. Native to the Mediterranean area.

lauryl rhodonate. Lauryl thiocyanate.

lauryl thiocyanate. A chemical used as an insecticide in the control of sucking insects.

lauseto-neu. An insecticide containing chloromethyl p-chlorophenyl-sulphone, particularly effective as a louse ovicide.

lavage. The process of washing out the stomach or intestines. In gastric *lavage* a double-way tube is passed down the esophagus into the stomach.

Lavalle hawthorn. *Crataegus lavallei,* family Rosaceae, a tender, hybrid, thorny, deciduous, small tree grown as an ornamental for its white flowers with red disks.

lavender. See *true lavender; costmary chrysanthemum; sea-lavender.*

lavender cotton. Cypress lavendercotton.

lawn. Any open-ground area about a house or building planted to fine grasses and maintained in good turf or sward especially for its aesthetic value.

lawn grass. Any of the grasses, alone or in combination (including certain legumes), planted in lawns.

lawnleaf. *Dichondra repens* var. *carolinensis,* family Convolvulaceae, a low, creeping, perennial herb grown as an ornamental ground cover, especially for lawns. Escaped it can be a weed pest. Native to North America. Also called *ponyfoot.*

lawn mixtures. A mixture of various grass and legume seeds used to establish a lawn.

lawn mower. Any of several mechanical devices used to cut lawn grass at an even height.

lawn roller. A device, frequently a large cylinder weighted with water, concrete, etc., which is rolled over a lawn, usually in spring, to even the surface of the soil.

lawn weed. Any weed which thrives under the maintenance practices used on lawn areas.

law of diminishing returns. An economic law which states that as an increasing amount of a variable factor of production, *e.g.*, manpower, is applied to a fixed factor of production, *e.g.*, land, the result is a decrease in average returns per variable unit applied.

Lawson falsecypress. *Chamaecyparis lawsoniana,* family Cupressaceae, an evergreen tree valued for lumber, and also grown as an ornamental. Native to North America. Also called *Port Orford cedar; Port Orford white cedar; Oregon white cedar; Oregon cedar.*

laxative. A mild cathartic.

lax heads. See *nodding.*

lay. 1. To ovulate and deposit an egg or eggs produced within the female generative organs, as a hen *lays* eggs. 2. A meadow (obsolete).

lay-by. To cease cultivating a crop, as corn, because of stage of growth, lack of weeds, etc.

lay eggs. See *lay.*

layer. 1. A mature female fowl which is kept for egg laying purposes, especially one in current egg production. 2. A course or stratum, as a *layer* of sand. 3. A plant twig or shoot, tied down and partially covered with earth, so that it can take root while remaining unsevered from the parent stock. 4. To reproduce by layerage *q.v.*

layerage. The method of propagating woody plants by covering portions of their stems or branches with moist soil or sphagnum moss so that they take root while still attached to the parent plant.

layering. Layerage.

layer society. In the plant community, low stature plants, such as grasses, which form a distinct vegetal layer under tall forest or brush.

lay heavily. Of poultry, to lay intensely; to lay an exceptional number of eggs in a given period of time.

laying ability. The ability of a particular hen to produce eggs, commonly measured by the number of eggs produced in a given time, as in a month or year. It is largely determined by five inherent qualities: (a) earliness of sexual maturity; (b) persistency of production; (c) intensity (or rate) of production; (d) broodiness; and (e) pauses *q.v.*

laying battery. A battery that consists of a number of hen cages, each housing an individual laying hen.

laying by. The last cultivation or tilling of a row crop.

laying flock. The unit of poultry kept for egg production.

laying hen. A hen which is currently laying eggs.

laying house. Any of several types of buildings equipped to house laying hens. Types may range from simple structures of small capacity to specially designed, insulated and air-conditioned, often multiple-storied buildings accommodating several thousand hens.

laying mash. Any of several special feed preparations, conducive to egg production, which is fed to laying flocks of poultry.

laying nest. Any of several types of nests provided for laying birds. For chickens it is commonly a 12 x 12 in. or 14 x 14 in. open-front, box-like nest; usually one of a group of similar nests which are arranged in tiers. See *trap nest; community nest.*

laying on meat. The actual fattening or deposition of lean and fat tissue on the body of a meat animal.

laying shelter. An inexpensive, low, open, shed-type or outside shelter for laying hens; it may be made of pole or light frame construction with heavy overhanging roof and poultry wire sides.

laying stock. The flock of chickens kept for egg production.

laying test. An official egg-laying test conducted under governmental supervision. Breeders submit an entry of pullets to obtain an official record of the performance of their line of breeding. The test extends over a period of 50 to 52 weeks, each bird or pen being rated on the basis of number as well as the size of eggs laid.

laying worker. A worker bee which lays eggs that produce drones. *Laying workers* usually develop in queenless colonies.

lay inspector. A practical, experienced person, with three or more years work in handling meat animals, who is employed by the government to inspect and grade meat in packing plants.

lay off rows. 1. To throw up hills in rows, as with a middlebuster (*q.v.*) 2. To lay out a field to a specific width for a specified number of crop rows.

lay-of-land. The topography of a tract of land, as level, sloping, rolling, hilly, etc.

lay on fat. Of an animal that is an easy keeper; to fatten easily.

La Ysabella. See *Palomino: 2.*

lea. Arable land under grass; a pasture. Also called *lay; ley.*

leach. 1. To remove alkali and/or salt, from soil by abundant irrigation combined with drainage. 2. To remove soluble materials in downward percolating water.

leached saline soils. Naturally saline soils which still possess the major physical characteristics of saline soils although the soluble salts have been leached, generally as a result of reclamation.

leached soil. A natural soil from which most of the more soluble constituents have been removed throughout the entire profile or removed from one part of the profile and accumulated in another part.

leached wood ashes. A fertilizer which is wood ashes with part of the plant food removed by artificial means or by exposure to fire, rains, snows, or other solvents, as in the extraction of dyes. *Cf. unleached wood ashes.*

leached zone. See *horizon A.*

leaching. 1. The removal of soluble constituents from soils or other materials by percolating water. 2. The removal of alkali and soluble salts from soils by abundant irrigation combined with drainage.

leaching basin. A drainage structure similar to a catch basin which has no outlet other than a sand-gravel or otherwise pervious bottom.

leach out. Leach.

lead. 1. In a mowing machine, the setting of the outer end of the cutter bar a little in advance of the inner end to offset the backward strain produced by the pressure of the cutting and the frictional drag of the ground and vegetation and to permit the knife and pitman to run in a straight line. 2. The head sheep herder (New Zealand). 3. The front of a band of sheep (New Zealand). 4. To follow obediently, as a cow.

lead arsenate. $Pb_3(AsO_4)_2$, a white, crystalline compound used as an insecticide.

lead cattle. The cattle at the head of a herd.

leader. 1. The main or dominant stem of a plant.

2. The front animals of a tandem hitch. See *wheel horse.*

lead hydrogen arsenate. Acid lead arsenate.

leading halter. Halter *q.v.*

lead line. A bluish line at the margin of the gums of an animal which indicates chronic lead poisoning, usually as a result of ingestion of paint or spray materials.

lead-out. An animal which has strayed from the herd (southwestern U.S.).

leadplant. Leadplant amorpha.

leadplant amorpha. *Amorpha canescens,* family Leguminosae, a shrub grown in the rock garden and in the shrubbery border for its long, dense spikes of blue flowers. Native to North America.

lead poisoning. Poisoning which results from ingestion of lead, usually of paint containing lead. It is characterized (a) in horses by convulsions, partial paralysis, colic, roaring, thirst, and increased urination; (b) in cattle, by staggering, impaired vision, and unusual postures. See *lead line.*

leadwort. See *plumbago.*

leaf. 1. A more or less flattened outgrowth from a plant stem, varying in size and shape, usually green in color, which is concerned primarily with the manufacture of carbohydrates by photosynthesis. 2. To put forth *leaves* as in the spring.

leafage. Foliage.

leaf and stem spot. A fungous disease of peonies caused by *Cladosporium paeoniae,* family Dematiaceae; characterized by dark purple leaf spots on the upper surface and dull chestnut on the lower. The infection begins at the margins and leaf pits. Also called *measles.*

leaf and twig blight. 1. A symptom of gray mold of strawberry (*q.v.*) in which leaves turn brown and the fruit clusters or peduncles are blighted. 2. A dieback of leaf and twigs, generally caused by parasitic organisms, in which the dead portions remain on the tree or plant.

leaf beet. *Beta cicla,* family Chenopodiaceae, a biennial herb popularly grown as an annual in the vegetable garden as a potherb. Native to Europe and Asia. Also called *chard; Swiss chard; sea kale beet; silver beet.*

leaf blade. See *blade: 1.*

leaf blight. 1. A symptom of any of several plant diseases generally characterized by a killing or partial killing of the leaves. A blight of twigs usually accompanies *leaf blight.* 2. A symptom of black rot in which many small purple specks appear on the leaves. As the spots enlarge, the margin remains purple but the center turns brown or yellowish-brown. Also called *frog-eye leaf spot.* See *black rot: 1.* 3. A symptom of fire blight (*q.v.*) in which the leaves are completely or partially killed. 4. A symptom of currant anthracnose (*q.v.*) in which small, dark brown, circular spots appear most obviously on the upper leaf surfaces. If numerous, the spots coalesce and result in larger, dead, leaf surfaces. There may be a chlorosis between the dead areas. 5. See *cherry leaf spot.* 6. A disease of sweet potato caused by *Phyllosticta batatas,* family Sphaerioidaceae, characterized by small, brownish spots with dark margins. Although widespread, it is not a serious disease.

leaf blight and fruit spot of pear and quince. A fungous disease caused by *Fabraea maculata,* family Dermateaceae, characterized by small, purple spots on the leaves which become larger, circular, and deep purple or dark brown. A small, black pimple appears in the center of the enlarged spot. Defoliation can result. Numerous black, slightly sunken spots appear on the fruit resulting in cracking. Similar spots appear on the current season's twigs.

leaf blight of arborvitae. A disease caused by *Didymascella thujina (Keithia thyjina),* family Stictidiaceae, which kills trees four years old or less but which is found on trees of all ages. It is characterized by a needle blight.

leaf blight of strawberries. A fungous disease caused by *Dendrophoma obscurans,* family Sphaerioidaceae, characterized by large, reddish-purple spots on the leaves.

leaf blight of white cedar. A fungous disease of falsecypress caused by *Didymascella chamaecyparissi (Keithia chamaecyparissi),* family Stictidiaceae, which is destructive in the southern U.S.

leaf blister. 1. Any of several fungous diseases of deciduous trees caused by *Taphrina* spp., family Taphrinaceae; characterized by circular, raised, yellow to brown blisters on the upper leaf surfaces and by corresponding sunken areas on the under leaf surfaces. Some of the causative organisms are: on oak, *T. caerulescens;* on maple, *T. carveri, T. dearnessii, T. sacchari;* on poplar, *T. johansonii, T. aurea, T. populi salicis, T. populina;* on birch, *T. americana, T. carnea, T. bacteriosperma.* 2. See *peach leaf curl; pear leaf blister mite.*

leaf blotch and kernel blight. Leaf stripe of oats.

leaf blotch of horsechestnut. A fungous disease caused by *Guignardia aesculi,* family Mycosphaerellaceae, characterized by small, irregular, discolored, water-soaked spots on the leaves which become reddish-brown with a yellow halo. Black spots appear in the lesions, and the lesions enlarge to cover the leaf.

leaf blotch of linden. A fungous disease of European linden caused by *Gloeosporium tiliae,* family Melanconiaceae, characterized by elongated, light brown spots along the leaf veins with a black band separating the lesion from the healthy tissue.

leaf breaking of geranium. A virus disease of the greenhouse geranium in which the normally almost round leaves with purple zones are transformed into fan-shaped leaves with the purple zone wanting or broken up into scattered purple spotting. Apparently it is transmissible by aphids, and is recognized so far only in the Pacific Northwest, U.S.

leaf bud. See *bud: 1.*

leaf bud cutting. Bud cutting.

leaf cactus. Barbados-gooseberry.

leaf canopy. Crown cover.

leaf-cast disease of conifers. The premature shedding of the leaves (needles) mostly of pine, larch, fir, spruce, etc., usually due to the effect of parasitic fungi growing within the leaves. These are most often species of the genera Hypoderma, Hypodermella, Lophodermium, etc., family Phacidiaceae.

leaf-comb. A comb of a chicken which is a combination of two single combs with serrated edges.

leaf crumple of cotton. A virus disease in California and Arizona, U.S., and Mexico, characterized by thickened or turned-under leaves, and by a hypertrophy of the interveinal tissue.

leaf crumpler. *Acrobasis indigenella* (Zell.), family Phycitidae, a caterpillar which infests various fruit trees, particularly the apple, feeding on the buds, shoots, and leaves.

leaf cupping of geranium. A disease of the greenhouse geranium in the Pacific Northwest, U.S., apparently of virus origin, in which the leaves are not crinkled but are strongly cupped and glabrous.

leaf-curl of alder. Enlargement and curling of the leaves of various species of alder caused by several species of the genus Taphrina, family Taphrinaceae. In some species the bracts in the catkins are also infected and grow out as tongue-like structures. It is found in temperate climates.

leaf curl of brambles. Septoria leafspot.

leaf curl of cotton. A virus-induced disease, very serious in parts of Africa. It is transmitted by the whitefly *Bemisia gossypiperda,* family Aleyrodidae.

leaf curl of ferns. A fungous disease caused by species of the genus Taphrina, family Taphrinaceae, characterized by thickened or discolored spots on fern leaves and sometimes galls or curled portions of the leaf. Found in tropical to cool temperate regions.

leaf curl of peach. A fungous disease caused by *Taphrina deformans,* family Taphrinaceae. The yeast-like fungus lives saprophytically and harmlessly on the waxy coating of peach twigs. In the spring under cold, slow growing conditions the fungus may infect the newly developed twigs, leaves, blossoms and fruits. The infected leaves become thickened and variously twisted and curled, usually turning red or pink, and falling prematurely. Infection of the fruit results in red, scar-like areas or malformations in which asci are produced.

leaf curl of plums and cherries. A fungous disease caused by *Taphrina* spp., family Taphrinaceae, characterized by thickening and curling of the leaves, or the formation of witches' broom. Upon infection, the developing fruits become enlarged and hollow. Also called *plum pockets; bladder plums.*

leaf curl of raspberries. Two types of virus disease of raspberries, characterized by downward curling of the deep green leaves: (a) the alpha type, spread by the insect *Aphis rubiphila,* family Aphidae, is common only on red raspberries; the fruiting shoots produce small berries and in two or three years the canes die; (b) the beta type occurs on black raspberries and is transmissible to red raspberries.

leaf-cutter. Any insect that eats or cuts out parts of plant leaves.

leaf cutting. A method of plant propagation in which a leaf is taken from a plant and placed in or on a rooting medium. Different treatment is preferable for each species; *leaf cutting* is especially adapted to gloxinia, begonia, and members of the family Geseriaceae.

leaf drop. Partial or complete defoliation; premature fall of leaves.

leaf drying. The loss of moisture through the leaves of felled trees, or those which have been girdled and left standing.

leaf eucalyptus. *Eucalyptus salicifolia,* family Myrtaceae, an aromatic, tall tree grown as an ornamental. Native to Australia. Also called *peppermint gum; black peppermint.*

leaf fat. The layer of fat which surrounds the kidney of an animal, especially of a hog.

leafflower. Any plant of the genus Phyllanthus, family Euphorbiaceae. See *emblic leafflower; otaheite gooseberry leafflower.*

leaf-footed bug. *Leptoglossus phyllopus* (L.), family Coreidae, a sucking insect which infests the pecan, potato, bean, cowpea, tomato, sunflower, Satsuma orange, and tangerine.

leaf gall. A fungous disease of a wide variety of plants caused by *Synchytrium globosum* and *S. aureum,* family Synchytriaceae; characterized by the growth of small galls on the leaves.

leafhopper. Any leaping insect of the family Cicadellidae; range in size from small to moderate and suck plant juices.

leafhopper injury. See *aster yellows.*

leaf lard. Leaf fat.

leaflet. 1. A small, immature leaf. 2. A separate division of a compound leaf.

leaf lettuce. Crinkled garden lettuce.

leaf miner. An insect that burrows into and feeds between the epidermal layers of leaves when in its larval stage.

leaf mold. 1. A fungous disease of carnations caused by *Heterosporium echinulatum,* family Dematiaceae, characterized by circular spots with concentric zones on the leaves, stems, and calyx. 2. The spongy humus, technically the F and H layers of the A horizon, of soil developed under cover of forest vegetation, especially of deciduous trees, in a temperate humid climate. See *mor; mull.*

leaf mold of spinach. Downy mildew of spinach.

leaf mold of tomato. A fungous disease caused by *Cladosporium fulvum,* family Dematiaceae, characterized by a rusty or cinnamon-brown mold.

leaf mottle. 1. A disease of beech of unknown cause characterized in spring by small, translucent leaf spots with yellowish-green to white halos. In time the spots turn brown and dry out so that the leaf appears scorched. There is a partial or complete defoliation followed by new leaves. 2. A symptom of certain virus infections; the pigmentation of green and yellow is not uniform on the leaf, and gives it a mottled appearance.

leaf nematode. Usually species of the genus Aphelenchoides, family Aphelenchidae, which live externally between the leaves and scales in the buds of the host, feeding on the host by means of a pointed, hollow buccal spear which pierces the plant cells. Some species penetrate the host tissues and find their way through the plant, producing discolored spots.

leaf-out. To produce leaves in the spring, as a plant.

leaf roll. 1. A virus disease of potatoes characterized by dwarfing, rolling of the leaves, uprightness, rigidity, and chlorosis. 2. A European virus disease of grapes. 3. See *Rhizoctonia disease of potatoes.* 4. A symptom of improper root functioning, or drought, in some plants, such as corn.

leaf roller. Any lepidopterous insect which in the larval stage rolls the leaves of plants in various ways and lives inside the rolls.

leaf-rolling mosaic. A symptom complex of a virus disease of potatoes characterized by slight dwarfing, diffused mottling, wrinkling, slight ruffling, rolling of the upper leaves, and reduction in the size of the tubers.

leaf roll of strawberries. A minor virus disease characterized by a downward rolling of the leaflets, by wrinkling and paleness with irregular yellowed areas.

leaf rust. 1. Any rust-fungous infection that occurs on leaves. 2. See *leaf rust of cereals.* 3. See *alfalfa rust.*

leaf rust of birch. A fungous disease caused by *Melampsoridium betulinum,* family Melampsoraceae; characterized by small, round, reddish-yellow pustules on the lower leaf surfaces, later turning dark brown to black.

leaf rust of blueberry. A fungous disease caused by *Pucciniastrum myrtilli,* family Melampsoraceae; characterized by discolored spots with yellow pustules on the lower surface of the leaf. See *rusts.*

leaf rust of cereals. A disease of cereals, caused in barley by *Puccinia anomala,* in corn by *P. sorghi,* in rye by *P. rubigo-vera,* and in wheat by *P. rubigo-vera tritici,* fungi of family Pucciniaceae; characterized by the appearance of orange-brown pustules on the leaf-blades and sheaths of the plants.

leaf rust of hemlock. A fungous disease caused by *Melampsora farlowii,* family Melampsoraceae, characterized by a yellowing and dropping of the needles.

leaf rust of poplar. A fungous disease commonly caused by *Melampsora medusae, M. albertensia,* or *M. abietis-canadensis,* family Melampsoraceae; which is characterized by yellowish-orange pustules on the lower leaf surfaces.

leaf scald. A bacterial disease of sugar cane in the South Pacific, caused by *Bacterium albilineans,* family Bacteriaceae, characterized by definitely margined, creamy-white leaf streaks which eventually turn brown or red and dry out, and development of weakened shoots from the basal axillary buds. Also called *Java gum disease.*

leaf scald of grasses. A fungous disease of grasses, rye, and barley caused by *Rhynchosporium secalis,* which causes a scald on the leaves.

leaf-scorch. A symptom of potash hunger of woody plants characterized by dead spots on the leaves, which sometimes appear to have been caused by drought.

leaf scorch of azalea. A fungous disease of rhododendrons caused by *Septoria azaleae,* family Sphaerioidaceae; characterized by yellow areas on the leaves enlarging, turning brown, and by premature leaf drop.

leaf sheath. In grass and grass-like plants, the lower part of the leaf which encircles the stem.

leaf skeletonizer. Any caterpillar which feeds by eating out the tissue of the leaves leaving the epidermis and the veins.

leaf smut of grasses. A widespread disease of noncereal grasses caused by *Ustilago striiformis,* family Ustilaginaceae, known to occur on over 75 grass species including 24 genera, among the most important of which are Agropyron, Agrostis, Bromus, Calamaguostis, Elymus, Festuca, Phleum, and Poa. The fungus fruits principally in the leaves, leaf sheaths, stalks, and sometimes the inflorescence. Life history studies of the fungus show great variability depending largely upon the host differences.

leaf smut of rice. A fungous disease caused by *Entyloma dactylidis,* family Tilletiaceae, characterized by small, black, angular to linear dots on the leaves.

leaf spot and blotch of rye. A fungous disease caused by *Helminthosporium secalis,* family Dematiaceae; characterized by a small, brown, leaf spot which enlarges until the entire leaf is yellow and dry.

leaf spot of alfalfa. Alfalfa leaf spot.

leaf spot of ash trees. One of a number of similar fungous diseases caused by *Gloeosporium aridum,* family Melanconiaceae, characterized by large areas of the leaf, especially along the edges, turning brown. The entire leaf may be affected, especially in rainy springs, and premature defoliation may result.

leaf spot of barley. Barley net blotch.

leaf spot of beets. Cercospora leafspot of beets.

leaf spot of brambles. Septoria leafspot.

leaf spot of cauliflower and cabbage. A bacterial disease caused by *Pseudomonas maculicola,* family Pseudomonadaceae, characterized by brownish or purplish-gray, water-soaked spots which coalesce, enlarge, and elongate giving the leaf a ragged appearance. A soft rot develops in the cauliflower.

leaf spot of cherry. Cherry leaf spot.

leaf spot of cotton. 1. A fungous disease of cotton caused by *Mycosphaerella gossypina,* family Mycosphaerellaceae, characterized by small, red spots on the leaves which enlarge, coalesce, turn brown; the leaves become ragged and full of holes. 2. See *bacterial blight of cotton.*

leaf spot of cyclamen. A fungous disease caused by *Phyllosticta* spp., family Sphaerioidaceae, characterized by circular, brown to black, water-soaked leaf spots.

leaf spot of eggplant. See *fruit rot of eggplant.*

leaf spot of elm. A fungous disease caused by (a) *Gloeosporium inconspicuum,* family Melanconiaceae, characterized by irregular, brown leaf spots; (b) *G. ulmicolum,* characterized by death of tissue along the margins and the affected veins; (c) *Gnomonia ulmea,* family Gnomoniaceae, characterized by the appearance, in the spring, of white or yellow flecks on the upper leaf surfaces, enlarging and turning brown with black centers. There may be a premature leaf drop.

leaf spot of gardenia. A fungous disease of capejasmine caused by *Rhizoctonia* spp., characterized by irregular, black, leaf spots.

leaf spot of hickory. Any of several fungous diseases caused by (a) *Gnomonia ovata,* family Gnomoniaceae, characterized by large, irregularly circular, reddish-brown leaf spots on the upper and lower surfaces; (b) *G. leptostyla,* characterized by irregularly circular, brown spots, which, if numerous, cause premature defoliation; (c) *Monochaetia desmazierii,* family Melanconiaceae, characterized by large, circular, yellow spots with a reddish-brown halo.

leaf spot of honeylocust. A fungous disease caused by *Linospora gleditsiae,* characterized by small, black structures on the lower leaf surfaces.

leaf spot of lily-of-the-valley. A fungous disease caused by *Gloeosporium convallariae,* family Melanconiaceae, characterized by purplish-brown leaf spots.

leaf spot of linden. A fungous disease of lindens (basswood) caused by *Cercospora microsora,* family Dematiaceae, characterized by small, circular, brown spots with darker halos. If the spots coalesce or become very numerous, the affected leaf may drop prematurely.

leaf spot of oak. Any of several fungous diseases caused by (a) *Marssonia ochraleuca,* family Melanconiaceae, characterized by large, circular spots with brown to yellow concentric markings; (b) *Monochaetia desmazierii,* family Melanconiaceae, characterized by large yellow spots with reddish-brown halos; (c) *Gloeosporium* spp., family Melanconiaceae and *Phyllosticta* and *Septoria* spp., family Sphaerioidaceae, characterized by leaf spotting.

leaf spot of peanuts. A widespread, fungous disease caused by *Mycosphaerella berkleyi,* family Mycosphaerellaceae, characterized by a brown to black leaf spot without a distinct halo. Lesions also appear on petioles, stems, and pods.

leaf spot of pear. A widespread, fungous disease caused by *Mycosphaerella sentina,* family Mycosphaerellaceae, characterized by small, light colored spots on the leaves enlarging to ¼ in. and becoming grayish-white with well-defined margins. If the lesions are numerous and coalesce, defoliation results.

leaf spot of pome fruits. See *Alternaria leafspot; frogeye.*

leaf spot of rice. See *Helminthosporium blight of rice; Cercospora spot of rice; blast of rice.*

leafspot of Saintpaulia. Ring spot of Saintpaulia.

leaf spot of strawberry. A widespread, fungous disease caused by *Mycosphaerella fragariae,* family Mycosphaerellaceae; characterized by indefinite, deep purple spots on the leaflets, petioles, fruit, and fruit stems. The spot enlarges and turns white or light brown with a purple halo. If the spots are numerous, the leaflet may be killed. Also called *common leaf spot; leaf blight; rust; spot disease.*

leaf spot of sugar beets. A fungous disease caused by *Cercospora beticola,* family Dematiaceae; characterized by a blight of the leaves which causes a dwarfed root and reduces the percentage of sucrose.

leaf spot of sweetpea. See *white blight.*

leaf spot of sweet potato. A fungous disease caused by *Septoria bataticola,* family Sphaerioidaceae; characterized by white leaf spots with brown halos. On the white area there are black specks.

leaf spot of tomato. A general, fungous disease caused by *Septoria lycopersici,* family Sphaerioidaceae; characterized by water-soaked leaf spots which turn brown; the leaf yellows, rolls, and drops. The disease attacks the lower leaves first.

leaf spot of walnut. Any of several fungous diseases of walnut and butternut caused by: (a) *Gnomonia leptostyla,* family Gnomoniaceae; characterized by irregular, brown leaf spots in early summer. Numerous lesions on a leaf may result in premature defoliation. (b) *Marssonia juglandis,* family Melanconiaceae; characterized by large, yellow blotches on the upper leaf surfaces and glistening white ones on the lower surfaces.

leaf stalk. Petiole.

leaf stem. Petiole.

leaf stripe of oats. A fungous disease of oats and some related grasses caused by *Helminthosporium avenal,* family Dematiaceae; characterized by small, brown spots spreading over the entire leaf. Defoliation may occur in southern U.S.

leaf tier. Any caterpillar which webs leaves together with silk and feeds within this shelter.

leaf tobacco. Harvested tobacco before the leaves have been cut or shredded; the chief form in which the crop is transported prior to manufacturing. The diverse classes and grades depend on variety, curing, country of origin, etc.

leaf-tyer. Leaf tier.
leafy. Designating any plant, usually herbaceous, with comparatively numerous or large leaves. Many such plants are grown for their leaves to be used in salads or as potherbs: *e.g.*, lettuce, spinach, etc.
leafy euphorbia. *Euphorbia esula,* family Euphorbiaceae, a perennial herb which is a weed pest found in waste spaces across northern U.S. Contact with its juice may cause dermatitis in man and death to fish. Native to North America. Also called *leafy spurge.*
leafy lily. *Lilium myriophyllum,* family Liliaceae, a bulbous herb grown for its nodding, fragrant, greenish-white flowers which are yellowish inside. Native to China.
leafy spurge. Leafy euphorbia.
leafy vegetable. See *leafy.*
league. A Spanish unit of land measurement, once used in California and Texas, U.S., varying in extent according to locality and official agreements. Some of the areas accepted were approximately 4426, 4340, and 4438 acres.
leak. See *Rhizopus rot of strawberries.*
leakage. The loss of fluid from any cause.
leaker. 1. An egg in which the shell and membranes are sufficiently broken so that the contents are exuding or are free to exude. 2. A sealed can whose contents are exuding. 3. See *leaking teat.*
leaking teat. A faulty condition of a cow's teat in which a relaxed sphincter muscle allows milk to escape from the udder.
leaky. Designating a faulty condition of butter in which moisture beads show on the butter trier or sample plug.
lean. 1. A piece of meat which consists largely of muscles lacking in fat or the proper proportion or distribution of fat. 2. Designating an animal lacking in condition of flesh or finish.
lean (milk). Milk that is abnormally low in percentage of butterfat.
lean-to. A building extension with a single pitched roof which projects from a high structure.
lean-to house. A greenhouse with a single pitch roof which is built against another greenhouse or a building. *Cf. ridge and furrow houses; even-span house; uneven-span house.*
lean trimming. In meat handling, the pieces trimmed from the primary cuts, such as hams, shoulders, and loin cuts, resulting in large portions of muscle or lean meat.
lease. 1. A legal, written agreement which is entered into between the owner and the tenant for the use of a given property or acreage of land for a given period of time. 2. To rent by written contract.
lease land. That land which is leased by its owner, sometimes a state or federal government, for farming or grazing purposes.
least cost combination. The cheapest combination of two factors of production capable of producing a given output. (On an iso-product line, it is the point at which the marginal rate of substitution between a pair of inputs is equal to the inverse ratio of their prices or, in the cases of non-homogenous and/or imperfectly priced factors, the inverse ratio of their marginal factor costs.)
least hop clover. Suckling clover.
leather. 1. The cured, tanned skins of animals, especially of the bovines. 2. A pad of leather which is placed between the shoe and the foot of a horse with a sensitive sole. 3. See *apricot leather; fruit leather.*
Leather. A cows' skim milk, cured cheese made in Germany.
leather bergenia. *Bergenia crassifolia,* family Saxifragaceae, a perennial herb grown in the flower border for its ornamental, heart-shaped, long leaves and its rose-pink or lilac flower clusters. Native to Asia. Also called *Siberian tea.*
leather bush. Atlantic leatherwood.
leatherflower clematis. *Clematis viorna,* family Ranunculaceae, a hardy, slender, woody vine grown on a trellis, fence, etc., for its ornamental urn-shaped, reddish-purple flowers. Native to North America.
leatherleaf. *Chamaedaphne calyculata,* family Ericaceae, an evergreen shrub grown in the bog garden for its leathery leaves and white, bell-shaped flowers. Native to the North Temperate Zone. Also called *brown bush.*
leather punch. A pliers-like tool, one jaw containing a free wheel, the spokes of which are hollow, sharp-edged cylinders of varying diameter; used to pierce holes of different sizes in leather and similar material.
leather rot. Leathery rot of strawberries.
leather tankage. See *nitrogenous tankage.*
leatherwood. See *California fremontia; Atlantic leatherwood.*
leather woodfern. *Dryopteris marginalis,* family Polypodiaceae, a fern grown in the shade and used as a floral decoration. Native to North America. Also called *evergreen wood fern; marginal shield fern.*
leathery rot of strawberries. A fungous disease caused by *Phytophthora cactorum,* family Pythiaceae, occurring only where berries touch the ground; characterized (a) on green fruit by a brown, water-soaked area; (b) on ripe fruit by a mushy condition of the fruit and a bitter flavor.
Lebanon Bologna. A sausage composed of 9½ parts of beef to ½ part of pork trimmings, ground and seasoned with brown sugar, pepper, mace, saltpeter, salt, nutmeg, and cloves. It is stuffed into a muslin casing, from 2½ to 4 ins. in diameter, and smoked.
lebbek albizzia. *Albizzia kalkora,* family Leguminosae, a tropical tree grown for its greenish-yellow flower and long pods. Native to Asia. Also called *East Indian walnut; lebbek; lebbek tree; siris tree.*
lebbek tree. Lebbek albizzia.
leben. An Arabian sour milk product, made from

whole curdled milk, high in acid and yeasty; somewhat similar to Koumiss or yoghurt. Also called *leban.*

lechuguilla agave. *Agave lophantha* var. *poselgeri,* family Amaryllidaceae, a succulent, desert plant which is a source of fiber, also grown as a pot plant for its bluish-green leaves, pale-banded along the midrib. It is poisonous to sheep and goats causing goat fever.

lectins. A group of protein substances, natural antibodies; agglutinins for type A red blood cells. *Lectin* may be obtained from lima beans.

Ledebour globeflower. *Trollius ledebouri,* family Ranunculaceae, a hardy, perennial herb grown in moist places in partial shade for its yellow flowers. Native to Siberia. Also called *golden queen.*

Ledebour honeysuckle. *Lonicera ledebouri,* family Caprifoliaceae, a tender, deciduous shrub grown in the shrub border for its orange-yellow flowers tinted red and its purplish-black fruit. Native to North America.

ledger. 1. A book in which a farmer keeps his accounts, registering all credits and debits, etc., under appropriate headings. 2. A horizontal board which forms the top rail of a simple fence, etc.

ledger plate. The nonmoving, hardened steel plate which is attached to the guards of the cutting bar of a mower or grain harvester and over which the cutting knives operate. It forms a part of the cutting mechanism.

leechee. Lychee.

leek. *Allium porrum,* family Liliaceae, a perennial herb grown as a mild, onion-flavored vegetable. Native to Europe and Asia.

left bank. The bank of a stream that is on one's left when one looks downstream.

left-handed bottom. A plow bottom which throws the furrow slice to the left. It has not generally been manufactured since World War II.

leg. 1. Any of the limbs of animals which support and move the body, as fore *leg,* front *leg,* hind *leg.* 2. A cut of meat which is that portion of the *leg* between the knee and ankle, as a chicken *leg, leg* of lamb. 3. To haul or drag a sheep from the pen to the shearing board by a hind *leg q.v.*

legal description. The exact, geographical, survey description of real estate used in abstracts of title and on the legal real estate tax rolls.

legal milk. Milk which has the required percentage of butterfat and total solids, as required by law or ordinance regulations.

leg band. A small celluloid or aluminum band with identifying numbers or symbols placed on the leg of a fowl.

leg cut ham. The lower third or shank of a cured ham. This is the least desirable section as it has a higher portion of legbone, skin, and less choice muscle meat, and is the lowest priced ham cut.

leg-feathers. Feathers which grow on the shanks and thighs of birds.

leg fender. Saddle fender.

leg finisher. A mechanical device used in a large poultry dressing plant to remove the leg-feathers of poultry.

legging. The skinning of the legs of a slaughtered animal. Also called *shanking.*

leggy. 1. Designating a plant which has unusually long stems. 2. Designating an animal, usually very young, as a colt, whose legs are disproportionally long in relation to its body size.

Leghorn. An egg-type breed of domestic chickens of the Mediterranean Class characterized by yellow skin, non-feathered shanks, nervous disposition, fast feathering, and relative nonbroodiness. Mature males weigh about 6 lbs., mature females about 4½. The egg-shell color is white. The Single Comb White Leghorn, the outstanding variety of the breed, universally raised, is one of the world's leading and most highly developed breeds for efficient production of market eggs.

Leghorn broiler. In marketing, any young chicken of the light-weight breeds.

Leghorn fowl. In marketing, any hen of the light-weight breeds.

leg of lamb. A cut of meat comprising that section of the leg in which the shank of meat is tucked into a pocket made under the vellum on the inside of the leg of a lamb.

leg of mutton. A retail cut of meat from a mature sheep that comes from the same portion of the carcass as the leg of lamb *q.v.*

legs filled down to the hocks. In livestock judging, heavy, well-meated, rearquarters of an animal.

legs go. In marketing horses, no guarantee that the legs of the animal being sold are sound.

legs out of one hole. In stock judging, a narrow-chested animal, usually a horse.

legumes. 1. Many plants of the family Leguminosae, as alfalfa, clovers, peas, vetches, soybeans, cowpeas, beans and others, characterized botanically by a fruit called a *legume* or pod that opens along two sutures when ripe. They make use of the free nitrogen of the air through symbiosis. *Legumes* provide a relatively greater amount of higher quality protein than non-*legumes.* They are of great agricultural importance as food for humans, hay and pasture for livestock, silage, green manuring, soil improvement, and honey production. 2. Any edible vegetable (obsolete).

legume inoculation. The application of a dust or liquid containing species of the bacterium, Rhizobium, to the seeds of legumes prior to planting so that nodules will form on the roots and the plant flourish.

legume roughage. The legumes, such as alfalfa, clover, and soybeans, which are extensively used as roughage for livestock, especially dairy cattle.

legume silage. Legume crops, such as alfalfa and Ladino clover, which make satisfactory silage especially if mixed with grasses and put into the silo in proper condition.

legume straw. The straw remaining from the threshing of legume crops for seed; a satisfactory feed for livestock if not damaged too much by weather.

leguminous. Designating any herb or woody plant of the family Leguminosae.

Lehmann lovegrass. *Eragrostis lehmanniana,* a grass of the family Gramineae, used for forage.

Leicester. 1. A long-wooled, white-faced, mutton-type breed of sheep which originated in England. 2. A hard, mild, cows' milk, cured cheese made in England.

Leichtlin camas. *Camasia leichtlini,* family Liliaceae, a bulbous herb grown for its showy, purplish-blue (sometimes white) flowers. Native to North America.

Leichtlin lily. *Lilium leichtlini,* family Liliaceae, a bulbous herb grown for its drooping, lemon-yellow, dark-spotted flowers. Native to Japan.

leishmaniasis. Any disease due to infection with microscopic protozoan parasites of the genus, Leishmania.

Lemma. The lower one of the two bracts subtending the flower in the spikelet of grasses.

Lemoine deutzia. *Deutzia lemoinei,* family Saxifragaceae, a hybrid, tender shrub grown for its pure white flowers.

Lemoine mockorange. *Philadelphus lemoinei,* family Saxifragaceae, a hybrid, hardy, deciduous shrub grown for its fragrant, white flowers.

lemon. *Citrus limonia,* family Rutaceae, a tropical, evergreen tree grown for its very valuable, edible fruit, the *lemon* of commerce. Native to southeast Asia.

lemonade sumac. Skunkbush sumac.

lemon balm. See *balm: 4.*

lemon beebalm. *Monarda citriodora,* family Labiatae, an annual herb grown in masses for its white or pinkish flowers and its lemon-scented leaves. Also called *lemon mint.*

lemon cucumber. 1. A variety of cucumber (*q.v.*) raised as a novelty and used for slicing in salads. 2. See *mango melon.*

lemon daylily. *Hemerocallis flava,* family Liliaceae, a lily-like herb grown in the flower garden for its yellow flowers. Native to Asia.

lemongrass. *Cymbopogon citratus,* family Gramineae, a perennial, tropical grass with lemon-scented foliage grown in southern Florida and southern California, U.S., and India; origin unknown. It yields *lemongrass* oil used in perfume.

lemongrass pulp. A livestock feedstuff which is the chopped, dried residue obtained by the removal of the volatile oil from lemongrass.

lemon juice. A food product prepared by deaerating and flash-pasteurizing the juice of lemons.

lemon juice method. An obsolete method of home canning in which vegetables were preserved in a mixture of one pint of lemon juice to seven pints of brine.

lemon mint. Lemon beebalm.

lemon verbena lippia. *Lippia citriodora,* family Verbenaceae, a shrub grown out-of-doors in warm areas as an ornamental and as a pot plant for its lemon-scented leaves and white flowers. Native to South America.

lemon vine. Barbados-gooseberry.

length of carcass. In butchering, the distance from the back of the rear quarter to the tip of the neck on the front quarter, from the aitch bone to the first rib.

length of fleece. The comparative length of wool fibers which influences both the quality and quantity of wool produced.

length-of-run. 1. The distance water must travel in furrows or over the surface of a field from one head ditch to another or to the end of a field. 2. The period of time that one man is allowed the irrigating head in the system of rotation deliveries. 3. The planning horizon of a firm or household. While the term implies a time dimension, a given *length-of-run* can best be identified in terms of the list of factors of production and products subject to the control of the management of the firm or household in making the plans under consideration.

length of staple. The normal measurement, without regard to quality or value, of a typical portion of lint cotton staple fibers under a relative humidity of 65 percent and a temperature of 70° F.

Lenten rose. *Helleborus niger,* family Ranunculaceae, a perennial herb grown for its late-blooming, white flowers. Native to Eurasia. Also called *Christmas rose.*

lenticel. An opening through the bark or outer covering of fruits, etc., which permits exchange of gases from the inner tissues with the surrounding air. It originates as a loose mass of rounded, corky cells formed beneath a stoma (*q.v.*) and appears as the light colored, lens-shaped structure visible in the outer bark of the cherry, the birch, and other woody plants, and on potatoes, carrots, etc.

lentil. *Lens culinaris,* family Leguminosae, an annual, herbaceous legume grown for its edible, pealike seeds. Native to Europe.

Lenzites dry rot. A dry rot of coniferous timber caused by the fungus, *Lenzites saepiaria,* family Polyporaceae.

leopardbane. Any perennial herb of the genus Doronicum, family Compositae. Species are grown in flower gardens. Native to Eurasia. Also called *leopard's-bane.* See *Caucasian leopardbane; downy leopardbane; plantain leopardbane.*

leopard flower. Blackberrylily.

leopard lily. *Lilium pardalinum,* family Liliaceae, a bulbous herb grown in the bog garden for its tall flower stalks and orange-red, purple-spotted, yellow-based flowers. Native to North America.

leopard moth. *Zeuzera pyrina* (L.), family Coccidae, a European moth with white, black spotted wings. Its larvae bore in fruit and shade trees.

leopard's-bane groundsel. *Senecio doronicum,* family Compositae, a hardy, perennial herb grown in the flower border for its showy, orange or yellow, spring flowers. Native to southern Europe.

leopard spot. Appaloosa.

leppies. Orphan calves (southwestern U.S.).

leprosis. A disease of citrus, especially of sweet oranges and less so of grapefruit and sour oranges, thought to be caused by a virus; characterized by spots on the leaves, twigs, and fruit. On the leaves the spots are chestnut to auburn, raised, and become glazed, hard, and brittle. On the fruits the spots are chestnut-brown with a lemon-yellow halo. Also called *nailhead rust; scaly bark.*

leptospirosis. A disease of animals and occasionally of man caused by species of the genus Leptospira. In dogs, it is caused by *L. canicola* and is characterized by loss of appetite, depression, fever, thirst, vomiting, loss of weight, nephritis, and death. In cattle, it is caused by *L. pomona* and is characterized by blood-tinged or yellow milk, fever, albuminuria, nephritis, and abortion. In pigs, it is caused by *L. pomona* or by *L. icterohemorrhagiae,* and is characterized by loss of appetite, fever, jaundice, abortion, and sometimes death.

Lescin. A ewes' milk, cured cheese made in the Caucasus.

lesion. Injury or diseased condition of tissues or organs.

Leslie box. An octagonal box with a raised bottom which is used to contain berries for shipment and sale.

lespedeza. 1. Any annual or perennial herb or shrub of the genus Lespedeza, family Leguminosae. Species are grown for hay, pasture, soil improvement, green manure, and ornament. See *shrub lespedeza; roundhead lespedeza; Thunberg lespedeza; Korean lespedeza; Chinese lespedeza.* 2. *L. striata,* an annual herb which is important as a hay, pasture, soil improvement, and green manure crop. Native to China and Japan. Also called *common lespedeza; Japan clover; hoop-coop plant.*

lespedeza meal. A livestock feedstuff obtained from the grinding of lespedeza hay without the addition of any lespedeza stems, lespedeza straw, foreign material, or the removal of leaves. It must be reasonably free from other crop plants and weeds and must not contain more than 28 percent of crude fiber.

lespedeza stem meal. A livestock feedstuff, the ground product remaining after the separation of the leafy material from lespedeza hay or meal. It must be reasonably free from other crop plants and weeds.

lespedeza straw meal. A livestock feedstuff, the ground product remaining after the separation of the seed from lespedeza, which must be reasonably free from other crop plants and weeds.

lesser appleworm. *Grapholitha prunivora* (Walsh), family Olethreutidae, an insect larva which infests apples making shallow, blotch mines on the surface of the fruits.

lesser bindweed. European glorybind.

lesser bougainvillea. *Bougainvillea glabra,* family Nyctaginaceae, a woody vine grown in very warm areas for its climbing habit and for its very showy, magenta or purple bracts. Native to South America.

lesser bud moth. *Recurvaria nanella* (Hbn.), family Gelechiidae, a moth whose larvae mine the leaves of apple, pear, and some stone fruits.

lesser bulb fly. *Eumerus tuberculatus* Rond., family Syrphidae, a fly whose larvae bore into, and feed on, the bulbs, corms, or rootstocks of various garden plants.

lesser canna leaf roller. *Geshna cannalis* (Quaint.), family Pyraustidae, a caterpillar which fastens the unfolding, young leaves of canna together. Infested plants have ragged, brown leaves.

lesser chicken louse. Fluff louse.

lesser clover leaf weevil. *Hypera nigrirostris* (F.), family Curculionidae, an insect pest of clover which feeds on the leaves, buds, and sometimes inside the stems.

lesser cornstalk borer. *Elasmopalpus lignosellus* (Zell.), family Phycitidae, a moth's larva which infests corn, boring into the stem of the plant just below the soil line.

lesser grain borer. *Rhyzopertha dominica* (F.), family Bostrichidae, an insect pest (both in the larval and adult stages) of grains, seeds, drugs, dry roots, and cork; especially destructive to wheat.

lesser peach tree borer. *Synanthedon pictipes* (G. & R.), family Aegeriidae, a moth's larva which infests peach, plum, cherry, wild plum, wild cherry, and Juneberry, by tunneling in the inner bark.

lesser sickles. The small, curved feathers (tail-coverts) in front of, and at the sides of, a male chicken's tail. See *sickles.*

lesser wax moth. *Achroia grisella* (F.), family Galleriidae, an insect pest of beehives which tunnels in and feeds on the combs.

lesser wintergreen. Waxflower pyrola.

let-alone slow process. A home process of vinegar manufacture in which fruit juice is allowed to undergo spontaneous fermentation. It is an unsatisfactory method.

let down milk. To accelerate greatly the secretion of milk at the regular milking time. The milk glands of the udder are stimulated by massaging the teats and udder or by the sucking of the calf. There is also some influence of stimulating hormones, the most important being prolactin. See *hold up milk.*

let down well in the ham. Designating a hog with heavy, meaty, rear quarters.

lethal. Deadly; causing death, as a *lethal* dose.

lethal gene. A gene which destroys an organism or a cell.

letter grass. *Grammatophyleum speciosum,* family Orchidaceae, a widespread, epiphytic plant having yellow, brown-spotted flowers 3 to 6 ins. across on 6 to 10 ft. stems. Native to Southeast Asia.

Letterman needlegrass. *Stipa lettermani,* family Gramineae, a perennial grass which is an important forage species on the western ranges, U.S. Native to North America.

letting down. The adjusting period, with some loss of flesh, which takes place when beef cattle are changed from heavy, dry lot feeding to open pastures or range feeding.

lettuce. 1. Any herb of the genus Lactuca, family Compositae. Most species are weedy plants but one is a very important salad plant. See *asparagus garden lettuce; crinkled garden lettuce; head garden lettuce.* 2. *L. sativa,* an annual herb widely grown for its leaves used in salads. Native to Eurasia.

lettuce drop. See *Sclerotinia drop.*

lettuce stunt. A fungous disease caused by *Pythium* spp., family Pythiaceae, characterized by dwarfing, darkening of the foliage, dropping of the central leaves, and by blackening of the vascular bundles.

leucite. $KAl(SiO_3)_2$, potassium aluminum silicate, a feldsphatoid mineral which is a minor source of potash for fertilizer. In the U.S., its principal occurrence is in the rock *Wyomingite* found in Wyoming.

leucocyte. The white cells of the blood that destroy disease germs and help remove foreign matter from the blood stream or tissue. Found predominantly in pus.

leucocytozoon infection. An acute and highly fatal disease of ducklings and poults caused by protozoan parasites: *Leucocytozoon simondi* in ducks, *L. smithi* in turkeys. Symptoms are loss of appetite, lethargy, thirst, rapid and labored breathing, excitement, loss of equilibrium, convulsions, coma, and foamy discharge from the nose and mouth.

leukosis. Avian leukosis complex.

Levant cotton. *Gossypium herbaceum,* family Malvaceae, an annual herb widely grown in Asia and the Near East as a fiber plant for its lint, the *cotton* of commerce. Native to Eurasia. Also called *Arabian cotton; Asian cotton.*

levee. An earthen dam placed at varying distances from the banks of a river to serve as a containing, protective barrier to adjacent low land during flood periods.

levee district. A geographical area, subject to flooding by a stream and its tributaries, in which the land owners have incorporated to build and maintain levees. It is a quasi-governmental unit which assesses fees by taxation.

level. 1. An instrument used to ascertain the horizontal lines at different altitudes as a guide in terracing, ditching, foundation laying, etc. 2. The amount of a component in a mixture, as the nitrate *level.* 3. The amount, usually expressed in parts per million of an element, of fertility in a soil solution. 4. To make a field approximately horizontal. 5. Designating a piece of land which is nearly on a horizontal plane. 6. See *water level.*

leveler. A drag or any other device used for smoothing land for irrigation or drainage.

level-indicating means. A sensitive device or other provision for determining the level position of the bulk milk cooling tank, as two-way or circular level, plumb-bob, scribe, or punch mark.

leveling lever. A lever on a disk cultivator which serves to level the gangs.

level of living. The level of current consumption or utilization of goods and services, both publicly furnished and private, which contribute to well-being or provide satisfactions to a family. For practical purposes, it is the same concept as the more current socio-economic status. See *standard of living.*

level top. In stock judging, a straight back.

level transit. An instrument used in running grade lines or elevations for laying tile and drainage ditches, designing terraces, etc. See *hand level; level: 1.*

lever lift cultivator. A cultivator whose gangs can be lifted by a lever. It facilitates the moving of the cultivator to and from the field.

lever press. In machinery, a press used for compression of substances into bales. It is set in motion by a long lever.

leverwood. Hophornbeam.

levis. Tight-fitting, blue denim pants with seams reinforced with stitching and copper rivets.

levorotatory honey. A honey which does not contain honeydew. It rotates to the left the beam of polarized light passed through it in a special optical tube.

levulose. Fructose or fruit sugar.

Lewis flax. *Linum lewisi,* family Linaceae, a perennial herb grown for its blue flowers. Native to North America. Also called *prairie flax.*

Lewis mockorange. *Philadelphus lewisi,* family Saxifragaceae, a deciduous shrub grown for its white flowers. Native to North America.

Leyden. A spiced, partly skimmed, cows' milk cheese made in the Netherlands.

liability. Money, goods, or services which a person owes to another.

liana. A high-climbing vine which produces long, serpentine stems.

Liberian coffee. *Coffea liberica,* family Rubiaceae, a tropical shrub sometimes grown for its white, fragrant flowers and red fruit. Native to Africa.

liberation cutting. An improvement cutting by which young trees are freed from suppression by removal of overhead shade.

lice. Small, nonflying, biting or sucking insects that are true parasites of man, animals, and birds. Biting *lice* belong to the Order Mallophaga; the sucking or true *lice* belong to the Order Anoplura.

lichen line. The edge of lichens on undisturbed rock surface; the clean portion of the rock indicating the extent to which the soil has been eroded from around its base.

lichens. Fungi, usually of the Class Ascomycetes (rarely of the Class Basidiomycetes) which grow in mutual symbiosis with algae, forming a joint structure in which the algal cells are embedded in the network of fungous hyphae, making a combination characteristic of neither component. The algae manufacture carbohydrates by means of photosynthesis while the surrounding fungus provides protection and furnishes the necessary water and dissolved nutrients. In exchange, the algae lose part of their dissolved food to the fungus. As the *lichen* weathers, the dying and decaying tissues form a substratum, the beginning of the formation of soil, on which spores of mosses and ferns and seeds of small plants may begin to grow. They form gray or bright-colored patches on soil, rocks, tree trunks, etc. The tundra of the Arctic and Sub-Arctic regions is composed largely of various species of the *lichen* genus Cladonia which form an important portion of the food of reindeer, caribou, muskox, and other Arctic animals. In somewhat warmer climates some rock-growing species of the genus Rocella furnish the dye, orcein, and species of Cetraria have food and pharmaceutical value. In the tropical regions, *lichens* may grow on the evergreen foliage of the tree, sometimes parasitizing the leaves which they enter through stomata. Large coatings of *lichens* are sometimes harmful to the growth of trees, such as oranges, etc.

lick. See *salt lick.*

licking disease. See *rickets; Aujeszky's disease.*

licorice. *Glycyrrhiza glabra,* family Leguminosae, a perennial herb grown as a source of the flavoring, *licorice,* used in cough remedies, tobacco, candy, etc. Native to the Mediterranean region.

lie bare. To remain unplanted, as improved or cultivated land left fallow or subject to the ill effects of wind and water erosion for a period of time without a crop cover.

Liebig's law of the minimum. A law formulated by the German chemist, Baron von Liebig, which states the minimal proportions of plant nutrients necessary for plant growth. If one nutrient is present in minimal proportions, the growth will be minimal; if absent, the plant will die.

lie dormant. Not to germinate, as seeds of certain plants which remain in the soil for several years before germinating; characteristic of many weed seeds.

lie idle. To allow improved or cultivated farm land to remain uncropped for one or more seasons.

lien. A legal claim on a property or asset for the security of a debt, such as a *lien* on a crop to secure payment of money advanced to grow it.

life cycle. Life history; the changes in form and mode of life that an organism goes through between recurrences of the same primary stage.

life form. The visible expression of the features in a plant which determine its adaptation to environment, as the features of form and structure by which a plant is adapted to cope with environmental peculiarities: *e.g.*, size and habit of growth, duration of the plant as a whole and of the stem above ground, duration and structure of the leaves, position and structure of the roots, and methods of vegetative production, etc.

life history. Life cycle.

life plant. Airplant kalanchoe.

life tenant. A person entitled to the use of, or income from, a property during his life. Essentially, the property, a life estate, is real estate which upon the death of the *life tenant* goes in fee simple to another person, the title holder.

lift. 1. A joint of meat, esp. beef, from the thigh. 2. Elevating or pulling power of a pump. 3. To loosen and remove seedlings or transplants from the seed bed or transplant bed prior to transplanting.

lifter. See *beet lifter.*

lifting. See *transplant.*

ligament. Any tough, dense, fibrous band which connects bones or supports viscera.

light. 1. The form of radiant energy consisting of wave lengths lying within the limits perceptible by the normal human eye, and, by extension, the shorter and longer wave lengths, the ultra violet and the infra-red *light,* invisible to the eye but which may be recorded photographically. *Light* can be absorbed by various substances and transformed into heat. Its excess may produce fading, the destruction of green color in plants, but an insufficiency can also cause lack of chlorophyll production. The coloration of fruits is dependent upon sufficiency of *light.* The growing parts of plants respond to the stimulus of the direction from which the *light* comes. See *phototropism; photosynthesis.* 2. To become ignited; to take fire. 3. Designating a deficiency or lack in degree, such as a *light* rain or a *light* crop.

light assorted chicks. In marketing, mixed or off-color chicks of one or more light breeds.

light-bay. A coat color of a horse which is a light tan with a black mane and tail.

light breed. A smaller and more refined type of a particular kind of livestock, as the Thoroughbred horse, the Leghorn chicken, etc.

light broiler. In marketing, a young chicken weighing 1 to 2½ lbs.

light brood. A honeycomb foundation, 10 x 16¾ ins., ten sheets of which weigh less than a pound.

light building. Photosynthesis.

light burning. 1. In forestry, a controlled burning to check disease, remove fire hazard, or to favor a certain type of growth. 2. The method of reducing the volume of inflammable material in forests. Fires are deliberately set at a time when inflammability is low and the likelihood of damage at a minimum; designed to lessen the danger of damage when fires run more freely.

light capon. In marketing, a capon which weighs less than 7 lbs.

light chestnut. A horse color; light, yellow-gold.

light cranberry. A mottled, reddish-seeded variety of the kidney bean.

light dirties. In marketing, slightly soiled eggs.

light dun. A horse color; a dun imposed on a sandy bay or sorrel.

light feeder. a. An animal that is being fed for maintenance and normal growth but not for quick finish or fattening. b. An animal that does not eat as much as most animals of its class.

light fore-arms. An undesirable front-leg conformation of horses: light-boned or poorly muscled.

light freeze. Light frost.

light frost. A frost which does not kill but which may cause some injury to the more tender and susceptible plants in exposed places. *Cf. black frost; killing frost.*

light grazing. Allowing livestock to partially graze an area, such as a pasture or range.

light hand. Designating a rider or animal handler with an easy or trained touch; gentle.

light-handed. Denoting a rider who handles a horse well with a minimum of tugging or jerking on the bridle.

light-harness horse. An American horse produced for speed and performance. Standardbred and American Trotter are breeds most often represented.

light horse. A horse not large enough to class as a draft horse or a chunk and too large to class as a pony; used for riding, driving, and racing.

light income. The amount of light which a plant receives.

light in condition. Designating wool which will shrink relatively little when it is scoured.

light in gaskin and stifles. Designating a horse with a poor pair of hind legs, especially in the stifle joints.

light in natural covering. Designating a meat animal that is deficient in fleshing and body conformation.

light in the middle. Designating an animal which is deficient in the spring of ribs or depth of body.

light in the timber. Designating a horse which has small bones below the knees and hocks.

light leg. Designating a horse that is trim in leg bone and quick in action; not clumsy or slow.

light leg of mutton. Designating a mutton sheep that is deficient in fleshing, especially in the rear quarters.

lightly buffered. Designating a soil which has a small cation exchange capacity; one which will have the characteristics of the colloidal fraction modified by the absorption of a relatively small amount of base or of H ion. See *buffer: 1; buffering action.*

lightness. A desirable quality of dry whole milk characterized by a lack of compaction.

lightning arrester. A device used across the circuit of a piece of electrical equipment to protect it or its operator from abnormal surges of high voltage, such as from lightning.

lightning injury. The injury to a plant, most often a tree, in which some bark is stripped off, limbs are broken, etc.; caused by lightning. Frequently herbaceous plants or crops in a circular area are killed by a single streak of lightning. See *lightning spot.*

lightning ring. A band of abnormal wood in a growth ring of a tree which results from mild lightning shock to the cambium.

lightning rod. A metallic rod set upright on the roof of a building, a tall chimney, etc., and grounded by a wire conductor. It serves to conduct the lightning to the earth, thus minimizing damage to the structure. Also called *lightning conductor.*

lightning spot. An area of dead plants which have been killed by lightning.

light of bone. Designating an animal that is small-boned or deficient in body build.

light of the sort. 1. Designating a horse that is lighter in color than most of the breed it represents. 2. Designating those animals of a group which are the lightest in weight when subject to sorting.

light oil. Any lubricating oil which has a viscosity number of ten or less, as graded by the Society of Automotive Engineers.

light period. See *photoperiodism.*

light quarter. 1. Designating a condition of a cow's udder in which one quarter is smaller than the other quarters or gives less or no milk. 2. Designating poor development in hind quarter or round section of beef cattle.

light range horse. A small stock horse. See *stock horse: 1.*

light ration. 1. A limited or scanty ration in contrast to a liberal ration. 2. A ration that is loose and bulky in relation to its weight.

light rider. A man who rides his mount in such a way as not to tire it or to cause saddle sores on its back (southwestern U.S.).

lights. The lungs of a dressed animal or fowl.

light sandy loam. A sandy soil which is near the lower limit in content of silt and clay based on mechanical analysis. See *sandy loam.*

light-seeded. 1. Designating those seeds which are relatively light in weight so that care must be taken in sowing or harvesting. 2. Designating a plant which produces few seeds.

light shearer. A medium-wool sheep which produces a light-weight fleece.

light shrink. Designating an animal or carlot of animals that loses little weight between ranch or farm and market place.

light sirup. A sugar sirup which is of 11 to 13.9 balling *q.v.*

light smutty. In marketing, wheat in which an 8 oz. sample contains 14 or more smut balls. See *smutty wheat.*

light soil. A soil that is easy to cultivate or one that requires relatively little power to plow because it lacks cohesiveness; *e.g.*, sands and sandy loams.

light spot. Designating a cheese which shows irregular or light and dark coloring throughout the body.

light veal. A veal which weighs 70 lbs. or less.

light weight hunter. A class of hunters (*q.v.*) that carry riders weighing from 135 to 165 lbs.

lightwood. 1. Conifer wood, especially pine and

Douglas fir, abnormally inflammable because of its high resin content. 2. The highly resinous heartwood, stumps, knots, and dead roots of pines, which are used mainly for kindling. Also called *lighterwood.* 3. Blackwood acacia.

ligneous. Like wood; woody.

lignification. The process in which plant cells become woody by conversion of certain constituents of the cell wall into lignin, generally considered to include the hardening, strengthening, and cementing of the cell walls in the formation of wood.

lignin. A principal constituent of wood, second in quantity to cellulose; it incrusts the cell walls and cements the cells together.

lignocellulose. The substance of which woody tissue is mainly composed.

lignocortosis. A necrosis of wood and bark of a strain of semi-dense Lisbon lemon and of limes; the cause is unknown. Also called *wood pocket of lemon.*

lignon-berry. Cowberry.

ligule. A small appendage of certain plants on the inside of the junction of the sheath and blade; found especially in grasses.

like produces like. A rule of thumb in breeding; the offspring will bear a close resemblance to the parents in essentials.

lilac. 1. Any shrub or tree of the genus Syringa, family Oleaceae. Species are grown as ornamentals. See *Chengtu lilac; early lilac; hairy lilac; Hungarian lilac; Japanese tree lilac; late lilac; little-leaf lilac; nodding lilac; Pekin lilac; Persian lilac.* 2. *S. vulgaris,* a shrub grown for its very fragrant, white, pink, purple, or lilac flowers. Native to southeastern Europe. Also called *common lilac.*

lilac blight. A blight caused by *Pseudomonas syringae,* family Pseudomonadaceae; or a seedling blight caused by *Phytophthora syringae,* family Phythiaceae. Characterized by a blackening and death of young shoots and tender parts.

lilac borer. *Podosesia syringae syringae* (Harr.), family Aegeriidae, a moth's larva which infests lilac, ash, mountainash, and privet, tunneling in the stems and branches. Found in the U.S. from Colorado eastward.

lilac chaste tree. Chaste tree.

lilac honeysuckle. *Lonicera syringantha,* family Caprifoliaceae, a hardy, deciduous shrub grown in the shrub border and as an ornamental for its fragrant, sweet-scented, rosy-lilac flowers and red fruit.

lilac leaf miner. *Gracilaria syringella* (F.), family Gracilariidae, a moth's larva which mines the leaves of lilac, ash, privet, euonymus, and deutzia.

lilac mariposa. *Calochortus splendens,* family Liliaceae, a cormous herb grown for its pale lavender or lilac flowers. Native to North America.

lilac pink. *Dianthus superbus,* family Caryophyllaceae, a perennial herb grown in the flower border for its fragrant, lilac or pale-rose flowers. Native to Europe and Asia.

lily. Any perennial, bulbous herb of the genus Lilium, family Liliaceae. Some species are important flowering plants. See *Bermuda Easter lily; Browns lily; bulbil lily; Canada lily; Caucasian lily; David lily; Duchartre lily; Easter lily; goldband lily; Hanson lily; Henry lily; Humboldt lily; Japanese lily; Korean lily; leaf lily; Leichtlin lily; leopard lily; Madonna lily; Nankeen lily; nodding lily; Parry lily; Philippine lily; Sierra lily; regal lily; slimstem lily; Szechwan David lily; Thunberg daylily; tiger lily; Turkscap lily; Washington lily; wood lily.*

lily bulb thrips. *Liothrips vaneekei* Pries., family Phlaeothripidae, a very small insect which infests lilies and orchids. Infested plants have rusty-brown, sunken spots.

lily-flowering tulip. A type or class of late-flowering tulip of many varieties, with attractive, long, slender flowers with curved, pointed, lily-like petals.

lilyleaf ladybell. *Adenophora lilifolia,* family Campanulaceae, a perennial herb grown for its fragrant, bell-shaped, pale blue flowers. Native to Europe and Asia.

lily leek. *Allium moly,* family Liliaceae, a perennial, bulbous herb grown in the rock garden for its showy, yellow flowers. Native to southern Europe.

lily-of-the-field. Fall daffodil.

lily-of-the-Nile. African agapanthus.

lily-of-the-valley. *Convallaria majalis,* family Liliaceae, a perennial herb grown in the shade for its very fragrant, white, nodding flowers. Also called *Maybells; conval lily.*

lily pot. A flower pot that is 5½ ins. deep and has an inside diameter of 5½ ins.

lily-turf. Any stemless ground-covering plant of the genus Mondo (Ophiopogan), family Liliaceae. See *white lily turf; dwarf lilyturf; bigblue liriope; creeping liriope.*

Lima bean. 1. *Phaseolus limensis,* family Leguminosae, a tropical, twining, perennial herb grown as an annual for its edible seeds, the *Lima beans* of commerce. Native to America. 2. See *Sieva bean.* (Sometimes the Sieva bean is also called *Lima bean.*)

limb. 1. A lateral branch of a tree or shrub. 2. The leg or wing of an animal. 3. To cut off a *limb* of a tree.

limber. One who cuts the limbs from felled trees.

limber honeysuckle. *Lonicera dioica,* family Caprifoliaceae, a deciduous shrub grown as an ornamental for its sprawling habit and its yellow flowers sometimes tinged with purple. Native to North America. Also called *small honeysuckle; small woodbine.*

limberneck. See *botulism.*

limber pine. *Pinus flexilis,* family Pinaceae, an evergreen tree sometimes grown as an ornamental. Native to western North America. Also called *white pine.*

Limburger. 1. A semisoft, surface and bacteria ripened, cows' whole milk cheese which originated in Limburg, Belgium. It has a strong odor and taste. 2. See *surface taint.*

lime. 1. Strictly, calcium oxide, CaO, but, as commonly used in agricultural terminology, calcium carbonate, $CaCO_3$, and calcium hydroxide, $Ca(OH)_2$, are included. Agricultural *lime* refers to any of these compounds, with or without magnesium carbonate, which are used as amendments chiefly for acid soils. 2. *Citrus aurantifolia,* family Rutaceae, a small, spiny, evergreen tree grown for its edible, very acid fruit, the *lime* of commerce. Native to Asia. Also called *acid lime.* 3. To apply *lime* to excrement to reduce odor and hasten decomposition. 4. See *liming.* 5. See *European linden; basswood.*

lime and sulfur dip. See *lime sulfur.*

lime anthracnose. A disease caused by the fungus *Gloeosporium limetticola,* family Melanconiaceae; characterized by cankers on the fruit and death of tender twigs. Also called *blossom blight.*

lime based superphosphate. A fertilizer used on acid soils. See *superphosphate.*

limeberry. *Triphasia trifolia,* family Rutaceae, an aromatic, spiny shrub grown in warm areas as an ornamental for its fragrant, white flowers, which resemble orange blossoms, and less so for its edible, red fruit. Native to Asia. Also called *bergamot lime; limoncito; orangeberry.*

lime chlorosis. A chlorosis (*q.v.*) that is associated with high lime soils.

lime-hater. Any plant which grows well only in an acid soil.

lime nitrate. See *calcium nitrate.*

lime pan. In soils, a subsurface layer which is cemented by calcium carbonate. See *caliche: 1.*

limequat. A hybrid tree resulting from crossing the lime (*Citrus aurantifolia*) with the kumquat (*Fortunella* spp.); grown for its edible fruits but more importantly for breeding purposes.

lime spreader. Any mechanical device used to spread lime on fields.

limestone. A rock which consists dominantly of calcium carbonate. Numerous kinds exist in accordance with impurities present, physical variations in texture, and hardness. *Limestones* are widely distributed and locally impart a distinctive character to soil and topography. The rock calcined or finely ground is used as a soil amendment and as a mineral supplement in livestock rations to supply calcium. *Cf. dolomitic limestone.*

limestone chips. Small chips or pieces of limestone used as a rooting medium for plants.

limestone grit. Small pieces of limestone furnished to fowls as a part of their diet.

limestone soil. A nonspecific soil having a great diversity in chemical and physical characters which is associated with or derived from limestone rock.

lime sulfur. A combination of calcium hydroxide, sulfur, and water obtained by boiling the mixture, preferably under pressure; it is an orange colored liquid with a specific gravity of 1.28 or more and is diluted variously for use as an insecticide, fungicide, and sheep dip.

lime-tree loiper. See *linden looper.*

liming. The application of lime to soil for the purpose of reducing acidity, improving structure, or supplying calcium as a nutrient.

limit. The top or bottom price which a commission house will allow its buyers to pay for a product at a certain place for a certain period of time.

limited feeding. 1. The feeding of livestock to maintain weight and growth but not to fatten or increase production. 2. Restricting an animal to less than maximum weight increase.

limited order. An order to buy on which the client sets a limit on the price, as contrasted with a market order. Sometimes called a *time order.*

limited partnership. A partnership in which one or more of the partners, but not all of them, has a limited liability to the partnership creditors.

limited use. 1. Designating a class of cheese in which serious defects restrict its marketability and its use for grinding purposes and immediate consumption. 2. Specific periods of grazing or grazing to less than capacity on a range.

limoncito. Limeberry.

limonite. $2Fe_2O_3 \cdot 3H_2O$, hydrous sesquioxide of iron. The minerals of the *limonite* group are widely distributed in rocks and occur in beds as an iron ore and in an earthy or ochreous condition. They are the main sources of the yellow, red, and brown colors in soils.

limy. 1. Designating a soil having a high content of lime, usually in the form of calcium carbonate. 2. See *alkaline flavor.*

linchet. An unplowed bank or narrow strip on the downhill side which separates two pieces of tilled land. It is supposed to be developed from continuous turning of the soil downhill in plowing and by sheet wash. *Linchets* date back to Celtic and pre-Roman agriculture in England. Also called *lynchet.*

Lincoln. A long-wooled breed of sheep from Lincolnshire, England. A large, white-faced, mutton breed, it is popular in New Zealand. The mature ewes weigh up to 250 lbs.

Lincoln Red Shorthorn. A dual-purpose breed of cattle developed from the older, red Shorthorn. Mature cows weigh 1,300 to 1,400 lbs. They are heavy milkers for a short lactation.

lindane. The pure gamma isomer of benzene hexachloride; used as an insecticide.

linden. Any tree of the genus Tilia, family Tiliaceae. Species are grown for shade, lumber, and as ornamentals. Also called *basswood.* See *basswood; Crimean linden; European linden; large-leaved linden; littleleaf linden; Moltke linden; silver linden.*

linden bloodleaf. *Iresine lindeni,* family Amaranthaceae, a tropical plant grown for its dark red leaves and stems. Native to South America. See *Herbst bloodleaf.*

linden borer. *Saperda vestita* Say, family Cerambycidae, the larva of a beetle which burrows through the bark and wood of linden and poplar.

linden hibiscus. *Hibiscus tiliaceus,* family Malvaceae, a tropical shrub or small tree which is a source of Cuban bast, a fiber for tying cigars, etc. It has yellow flowers. Probably native to Asia and Africa. Also called *emajagua; mahoe; majagua.*

linden leaf beetle. Elm calligrapha.

linden looper. *Erannis tiliaria* (Harr.), family Geometridae, an insect which infests oak, apple, birch, elm, hickory, linden, and maple, causing serious defoliation. Also called *lime-tree looper.*

Lindley butterflybush. *Buddleia lindleyana,* family Loganiaceae, a tropical shrub grown for its purplish-violet flowers. Native to China.

line. 1. The reins of a harness. 2. A rope, cable, string, wire, tube, etc., for tying or hanging objects, or conducting electricity, water, gas, etc., as a power *line,* gas *line,* milk *line.* See *guy.* 3. A boundary or limit, as a fence *line,* property *line.* 4. In marketing, the whole of a herd of sheep (New Zealand). 5. See *contour.*

linear plot. A sample plot or a transect laid out on a line usually following a fixed bearing.

linear programming. A mathematically formal type of budgeting in which the most profitable of all possible combinations and amounts of a limited number of sub-organizations (activities within an organization) are selected subject to resource limitations and under the assumption of constant returns to scale for each sub-organization.

linear system. A system of training raspberries in which the suckers are removed and only the original plants are kept for fruiting.

line back. A white spot or stripe on the back of a Hereford; a breed disqualification.

line breeding. See *in-and-in breeding.*

line camp. A camp away from headquarters; used by riders for temporary shelter, etc. (southwestern U.S.).

line clearing. The removal of trees or parts of trees along a right of way to reduce interference with public utility lines.

line elevator. An elevator for grain storage, operated by a large company or corporation. See *farmers' cooperative elevator; independent elevator; mill elevator.*

line fence. A boundary fence, usually of better quality and more permanent than the cross fences.

linen. A type of cloth made from the fibers of the flax plant.

line of breeding. Breeding from a certain line of descent.

line of credit. The amount of credit granted an individual or business by a lending institution that may or may not all be used at any one time by the borrower.

line of draft. Line of hitch.

line of hitch. An imaginary straight line from the center of load or resistance through the clevis or hitch to the center of power or where the hitch is attached to the power.

line of least cost combination. In the production of a single product, the locus of all combinations of the factors of production involved which satisfy the condition that the marginal rate of substitution (*q.v.*) between any two factors is equal to the inverse ratio of their prices.

line pattern. A mild, virus disease of stone fruits, characterized by chlorotic lines in leaves. The virus is widespread in plums.

line plot survey. A survey employing plots as sampling units in which plots of specified size are laid out, usually at regular intervals along parallel survey lines.

line post. Any of several fence posts used in a line fence.

line rider. A man who rode horseback along the outskirts or line fences of a cowman's range to see that all was in order (western U.S.).

line test. 1. A test in which a series of samples are taken from the milk supply of the whole herd during milking or milk processing; used in the bacteriological control examination of the milk supply. 2. The rapid agglutinin test used to detect abortion infection in herd milk. 3. A test used for the determination of the vitamin D content of milk. 4. A test to determine the degree of calcification of a growing bone: a measure for rickets.

line transect. A method of recording plant species in a range study in which a tape or chain is stretched on the ground and the plants touched by the tape are recorded.

line village form of settlement. A form of agricultural settlement in which the farmers' residences are arranged in a line or string fashion along a river or other natural barrier. The resulting land holdings are long and narrow. *Cf. isolated form of settlement; village form of settlement.*

liniment. A preparation used for bathing or rubbing sprains, bruises, etc., usually containing a counter-irritant.

linkage. Association of characters in inheritance due to the fact that the character-determining genes are physically located in the same chromosome. Such a group of linked genes is called a *linkage group.*

Linnaean classification of plants. The sporophyll-based system of plant classification, now outmoded, originally adopted from the proposal of the Swedish naturalist Von Linne. According to it, the plants are divided into 24 classes, which are subdivided into orders, on the basis of stamen-arrangement and the characteristics of the fruit or seed.

linseed. Flaxseed.

linseed cake. Linseed oil cake.

linseed cubes. A livestock feedstuff, consisting of a mixture of linseed oil meal with flaxseed screenings, oil feed, or other flaxseed by-products, or both.

linseed meal. Linseed oil meal.

linseed oil. The drying oil expressed from the seed of flax, used in paints, printer's ink, etc. See *flax: 1; flaxseed.*

linseed oil cake. A protein supplement for livestock feeds obtained after the partial removal of oil from ground flaxseed. It should not contain more than 0.5 percent of acid insoluble ash. If a name descriptive of the process of manufacture, such as hydraulic, mechanical, or solvent extracted be used, the product must correspond thereto.

linseed oil chips. See *linseed oil cake.*

linseed oil meal. A livestock feedstuff consisting of finely ground, linseed oil cake.

linseed pellets. See *linseed cubes.*

linsey-woolsey. A cloth made of a mixture of linen and wool.

lint. Cotton fiber; the epidermal hairs of the cotton seed.

lint cotton. The cotton fibers removed from the seed as contrasted to seed cotton *q.v.*

linters. Short cotton fibers which remain on the cotton seed after ginning. These are removed at cotton oil mills and are an important source of cellulose; used in manufacturing felts, upholstery, artificial fibers, etc.

lint flue. A part of the cotton gin through which lint cotton is blown after it is separated from the seed.

lint frequency. The weight in grams of cotton fiber of uniform length produced per square centimeter of seed surface.

lint hair. Lint.

lint index. The weight of lint cotton produced by 100 seeds.

lint percentage. See *ginning outturn.*

lint slide. An inclined chute down which lint cotton slides into the press box.

lion's-ear. Drug lionsear.

lion's-tail. 1. A peach shoot defoliated except for a few immature leaves at the tip; a symptom of X-disease. 2. See *motherwort; drug lionsear.*

lip. 1. In a dam, a small wall on the downstream end of the apron to break the flow from the apron. 2. One of the edges of a wound. 3. Either of the two external fleshy folds of the mouth opening. 4. Either one of the inner or outer fleshy folds of the vulva. 5. See *lip roll.*

lip and leg ulceration. A virus disease of sheep, characterized by lesions on the lips, face, lower portion of the legs, and between the toes. Also called *ulcerative dermatosis of sheep.*

lipase. A fat-splitting enzyme of milk responsible for its sometimes bitter, soapy flavor.

lip bot fly. Nose bot fly.

Lipizzan. A breed of riding horses once reserved for the Austrian royal family. They are mostly white, compact, with broad chest, and have a long, muscular back and powerful croup. The animals are active and have the capacity to learn intricate gaits. The breed originated in North Africa, was brought to Spain, and thence to Italy and the town of Lipizza near Trieste.

lip mold seam. The seam formed in the molding of the lip or cap seat on glass milk bottles.

lipoid. Concerning fat or fatty tissue.

lipolysis. The hydrolysis of fats by enzymes, acids, alkalis, or other means, to yield glycerol and fatty acids.

lipolytic. Referring to lipolysis.

lip roll. The roll of thick glass molded on the top of a glass milk bottle in which the milk cap is seated. It must be smooth and perfectly formed for the proper machine bottling.

Liptauer. A soft, pickled, ewes' milk, ripened cheese made in central Europe. Sometimes 10 percent cows' milk is also added.

liquefaction. The act or process of reducing a substance to a liquid state.

liquidation. The sale of a previously bought contract, otherwise known as *long liquidation.* It may also be the repurchase of a previously sold contract, generally referred to as *short covering.*

liquidation value. 1. In appraising, the price at which a property can be sold within a short period for cash or its practical equivalent and in which the seller forces the sale and the buyer dictates the terms. 2. The value of a forest estimated on the assumption that all merchantable trees will be cut.

liquid creosote. An oil obtained from coal-tar creosote by the extraction of part of the compounds that crystallize at normal atmospheric temperatures.

liquid culture. A culture of selected strains of desired organisms in a fluid state as contrasted to a dried culture.

liquid fertilizers. Water-soluble plant food elements supplied in liquid form and applied by pumping, air pressure, or gravitational flow from a spray applicator.

liquid manure. The liquid excrement from animals, mainly urine, collected from the gutters of barns into large tanks and hauled to the fields. (A dairy cow will produce about four tons of liquid manure per year.)

liquid sludge. Sludge (*q.v.*) that contains sufficient water (ordinarily above 80 percent) to permit it to flow by gravity.

liquid smoke. A liquid mixture applied to meats in curing, especially to ham and bacon, to replace curing by wood smoke.

liquid stick. See *stick: 2.*

liquor cresolis saponatus. Compound solution of cresol.

list. 1. A small ridge formed by throwing two furrows together, producing alternating beds and furrows. 2. To plow so as to form *lists.*

listed. Designating land with alternate ridges and furrows.

listed crop. Those crops which are planted in a listed furrow or trench or below the general level of the ground.

listed land. Cropland prepared by the use of the lister *q.v.*

lister. A double moldboard plow which throws the soil on either side; used in semiarid regions for planting crops in the bottom of the furrow. Also called *lister plow; middlebreaker; middlebuster.*

lister cultivator. See *sled lister cultivator.*

lister furrowing. A system of furrowing with a lister crosswise to the prevailing wind direction to retain moisture and to help control soil blowing or washing.

listeriosis. A sporadic, specific bacterial disease of ruminants primarily involving the central nervous system (brain and spinal cord). On occasions it may result in abortions in sheep and cattle. It is caused by *Listeria monocytogenes,* also infective for nonruminant mammals (including man) and chickens. Also called *listerellosis; circling disease.*

lister opener. See *lister.*

lister planter. A drill planter designed to place the seeds in the bottom of a deep furrow made by the lister plow. There are different types of planters for different types of seeds, such as corn, wheat, sorghum, etc.

listing. The opening of a furrow several inches deep with a lister prior to depositing the seed in the bottom of the furrow.

list quadrat. A small square census plot (*q.v.*) used for a range census study of plant species. The presence of each species is noted but not necessarily its total number.

litchee. Lychee.

litchi. Lychee.

liter. A standard measure for liquids denoting a capacity of 61.025 cu. ins. or 1.0567 liquid quarts.

lithochromic soil. A soil which owes its color to parent material, or rock from which the detrital material is derived, rather than to color developed in soil forming processes.

lithosequence. A sequence of soils whose properties are functionally related to differences in the parent rock.

Lithosol. An azonal group of soils having no clearly expressed soil morphology, consisting of solid rock or a mass of rock fragments incompletely weathered; found primarily on steeply sloping land and only recently differentiated from Regosols *q.v.*

lithosphere. That part of the earth which is composed predominantly of rocks whether coherent or incoherent, including the disintegrated rock materials: soils and subsoils, together with everything inside this rocky crust. Other materials (chiefly water and gases that are found in the air) are intermingled, but rocks and soil predominate.

litmus. An acid-alkali indicator derived from various species of lichens, principally *Lecanora subfusea* var. *variolosa;* the main constituents are azolitmin and erythrolitmin. The commercial indicator has a pH range of 4.5 to 8.3. The acid color is red and the alkaline color blue. Usually it is used as *litmus* paper or as an indicator in milk to show acid or alkaline reduction by microorganisms. Slips of *litmus* paper were formerly used to determine whether soils were acid or alkaline.

litre. See *liter.*

litter. 1. On a forest floor, the uppermost surface layer of debris, leaves, twigs, and other organic matter, undecomposed or but slightly altered. In a technical description of a soil profile, it is generally designated by the letter L. 2. Accumulation of leaves, fruits, twigs, branches, and other plant parts on the surface of the soil. See *mulch.* 3. A group of young animals born at a single birth, as a *litter* of pigs, etc. 4. See *bedding: 3.*

litter alley. The aisle behind the cow stanchions in dairy barns where litter and manure may collect. It also serves as a passageway for animals going in and out of the barn.

litter carrier. A container suspended from an overhead track behind the cows in a dairy barn which is used for removing manure and litter from the barn.

little barley. *Hordeum pusillum,* family Gramineae, an annual grass which is a plant pest on the western ranges, U.S. The dried awns penetrate the tissues of the mouth parts or eyes of grazing animals causing wounds and infections. Native to North America.

little black ant. *Monomorium minimum* (Buckl.), family Formicidae, a North American ant which is a general feeder on fruits, sweets, honeydew, vegetables, and meats.

little blue stagger. Dutchman's-breeches.

little bluestem. *Andropogon scoparius,* family Gramineae, a perennial grass which grows throughout most of the U.S. Locally valuable for grazing and wild hay, its general forage rating is fair. Native to North America.

little carpenterworm. *Prionoxystus macmurtrei* (G.-M.), family Cossidae, an insect pest which infests oaks, especially red oaks, tunneling in the heartwood.

little cherry. A virus disease of sweet cherry, important in British Columbia, whose symptoms are confined to the fruit which ripen abnormally and are small, pointed, dull, and lacking in flavor.

little coarse. Pertaining to a horse which is afflicted with spavin.

little fire ant. *Wasmannia auropunctata* (Roger), family Formicidae, a small ant pest living in citrus orchards on honeydew whose vicious sting can make harvest almost impossible.

littleleaf box. *Buxus microphylla,* family Buxaceae, a small, hardy, broadleaved, evergreen shrub grown in warm areas for topiary work and hedges. Native to Japan.

littleleaf buttercup. *Ranunculus abortivus,* family Ranunculaceae, an herb which, when grazed, is a poisonous plant. Hay containing the plant, however, may be fed without harm. Native to North America. Also called *small-flowered buttercup.*

little-leaf disease. Any of several diseases of trees and vines that result in small leaves. The causes are various, such as zinc deficiency, nematodes, root rots, and viruses.

little-leaf disease of citrus. A virus disease of citrus characterized by the formation of hard, knotty growths on the trees and small leaves; known in Palestine, Brazil, and Florida, U.S.

littleleaf horsebrush. *Tetradymia glabrata,* family

Compositae, a low, spreading shrub found from Utah to Oregon and California, U.S. It is poisonous to stock, especially to sheep. Symptoms of poisoning are depression, weakness, and swelling of the head, ear, and nose. Also called *horse-brush; spring rabbitbrush; coal-oil brush.*

littleleaf in pines. A destructive disease caused by the fungus *Phytophthora cinnamomi,* family Pythiaceae; associated with the death of fine rootlets and inability of the tree to absorb nitrogen.

littleleaf lilac. *Syringa microphylla,* family Oleaceae, a spreading, deciduous shrub grown for its fragrant, lilac flowers. Native to China.

littleleaf linden. *Tilia cordata,* family Tiliaceae, a deciduous tree grown as a shade and street tree and as an ornamental for its fragrant, yellowish flowers. Native to Europe.

littleleaf mockorange. *Philadelphus microphyllus,* family Saxifragaceae, a tender, deciduous shrub grown for its showy white flowers which smell like pineapple. Native to western North America.

little marsh cow. A very small bovine supposed to have been found in pre-historic times about the marshy shores of the lake regions in Switzerland.

little merrybells. *Uvularia sessilifolia,* family Liliaceae, a hardy, perennial herb grown in the border or wild garden for its spring blooming, greenish-yellow flowers. Native to eastern North America. Also called *wild oat.*

little nappy. Pertaining to a balky horse or a dummy. See *dummy: 1.*

little oat. *Avena brevis,* family Gramineae, an annual herb sometimes grown in Europe for its edible seeds. Native to Europe. Also called *short oats.*

little ouchy. Pertaining to a horse with laminitis, navicular disease, or any type of lameness.

little peach. A virus disease of peach closely related to peach yellows; characterized by small, late-ripening, insipid fruit, and by a clustering of foliage. Affected trees degenerate and usually die within 3 to 5 years. Also called *peach little.*

littlepickle othonna. *Othonna crassifolia,* family Compositae, a fast-growing, perennial, drooping herb grown in the greenhouse or in a hanging basket for its prostrate habit and yellow, solitary flowers. Native to South Africa.

little-pig anemia. See *anemia in suckling pigs.*

littlepod falseflax. *Camelina microcarpa,* family Cruciferae, an annual herb whose seeds if eaten by stock in large quantities may result in chronic enteritis, hemorrhagic diarrhea, colic, abortion, nephritis with hematuria, and paralysis of the heart and respiratory system. Also called *small-seeded falseflax.*

little potatoes. See *Rhizoctonia disease of potatoes.*

little reading on him. Designating a horse which has been branded.

little red louse. Cattle biting louse.

littleseed alfalfa dodder. *Cuscuta planiflora,* family Convolvulaceae, a species of dodder that infests alfalfa.

little strong in the mouth. Designating a tough-mouthed, unmanageable horse which pulls on the bit.

littoral. A shore and the land contiguous to it; the zone between high- and low-water marks.

Livarot. A soft, cows' milk, cured, French cheese, similar to Camembert.

liveability. The inherited stamina, strength, and ability to live and grow; important of all young animals, such as baby chicks, lambs, pigs, etc.

live delivery. A common guarantee by hatcheries, of delivering live chicks by the addition of two or more per order of 100.

live fence. A hedge made by planting closely-spaced thorny or other plants, such as Osage orange, and maintaining them in a living condition.

live foal guarantee. A common guarantee by stallion owners that a live foal will result from the breeding service. In case of failure, the breeding fee is not charged or, if already charged, is refunded.

liveforever stonecrop. *Sedum telephium,* family Crassulaceae, a succulent plant grown for its fleshy leaves and pink or purple flowers. Native to Europe. Also called *evergreen; orpine; live-forever.*

live lime. Burnt lime.

live market. A market for transactions in live animals, as the livestock yards and markets in Chicago and Detroit, U.S.

live oak. *Quercus virginiana,* family Fagaceae, a broadleaved, evergreen, tender tree grown for shade and ornament and for its wood used in boats and articles requiring strength and toughness. It also furnishes some browse in southwestern U.S. Also called *American live oak; encina.*

liver. A glandular organ in man and other animals that secretes bile and performs certain metabolic functions. *Liver* of various animals and fowls is edible, and its extracts are used medicinally, especially for anemia.

liver and gland meal. See *animal liver and glandular meal.*

liverberry. Hairy fairybells.

liver chestnut. A coat color of a horse: a dark shade of chestnut.

liver fluke. *Fasciola hepatica,* and *Fascioloides magna,* a macroscopic parasite of sheep, cattle, and swine which lodges in the liver, causing liver rot.

liverleaf. Hepatica.

liver rot. A disease of sheep, cattle, and swine caused by the liver fluke; characterized by a loss of flesh, soft swellings under the jaw and along the belly.

liver sausage. A domestic sausage which consists of ground pork, beef or veal, ground pork liver, and bread or whole wheat flour. Salt, pepper, and sometimes mace, celery seed, cardamon, and coriander are added as seasoning.

liverwort. Hepatica.

livery barn. Livery stable.

livery stable. A business concern which rented saddle or driving horses by the day and which fed and housed horses for others. In some instances, a *livery stable* operated in conjunction with an inn serving meals especially for farmers who remained in town overnight (obsolete).

live seed. Viable seed.

livestock. Any domestic animal produced or kept primarily for farm, ranch, or market purposes, including beef and dairy cattle, hogs, sheep, goats, and horses. Subdivided by class (use, purpose, age), as cows, steers, ewes, or drys; by kind (species or genus), as cattle, horses, mules, sheep, goats, or swine; by type (productive purposes or characteristic qualities), as wool or mutton-type of sheep. Also called *stock.*

livestock commission house. An organization of livestock buyers which acts as an agent for the sale of livestock of many different producers at a given market for a fee or commission. Most livestock marketing in the U.S. is handled by such houses.

livestock distribution. The apportionment of livestock over a range area.

livestock enterprise. A farm activity which consists of one phase of the livestock business, such as the dairy enterprise, laying flock enterprise, etc.

livestock exchange. An organization of commission houses and livestock buyers who transact business in a given market, such as in Chicago, U.S.

livestock farm. A farm on which the major interest or activity is the growing or handling of some type of livestock, such as beef cattle, hogs, or sheep, or a combination of such types.

livestock farmer. A farm operator whose major interest and income is in producing livestock.

livestock feed. The forage and concentrates commonly used in the feeding of livestock, such as hays, silage, corn, oats, and cottonseed meal.

livestock husbandry. The art and science of the production, management, and marketing of livestock. See *animal husbandry.*

livestock income. All cash income derived from sale of livestock and livestock products, plus or minus the increase or decrease in inventory values during the year, and less the cost of livestock purchases.

livestock index. See *livestock yield index.*

livestock industry. The breeding, raising, feeding, slaughtering, handling, and selling of animals and animal products. See *animal industry.*

livestock judging. The art and practice of the careful appraisal of the desirable qualities, type, and conformation of different classes of livestock.

livestock man. A person who breeds, raises, handles, and sells livestock.

livestock management. The application of business methods and technical principles to livestock production.

livestock market. A market which specializes in transactions of livestock for processing and distribution, as Chicago and Kansas City livestock markets or the numerous farmers' markets in the U.S.

livestock ranch. An area of land usually larger than a farm and largely in grass where livestock, such as beef cattle or sheep, are produced. See *livestock farm.*

livestock registry association. A purebred breeding association with which animals of known and recorded ancestry may be registered. See *breed association.*

livestock shipping association. An association of several producers who pool their livestock for the purpose of shipping and marketing; usually a cooperative.

livestock show. A show at which livestock are exhibited according to breed and class and are judged according to standards of perfection. Also called *cattle show.*

livestock truck. A motor truck with a large rack-type body especially suited for the handling and transporting of livestock from the farm or ranch to market.

livestock yield index. A comparative measure of the production per animal or for all classes of livestock on a given farm or group of farms expressed as a percentage of the average of all farms. In some cases, the index refers to output per unit of feed consumed.

live virus. A virus whose ability to infect has not been altered.

live weight. The gross weight of a live animal as compared to the dressed weight after slaughter.

lixiviated soils. Leached soils; soil layers from which soluble material and coloring matter have been removed by leaching under saturated water conditions. Lixiviation is a natural process in the formation of soils, especially in humid regions.

lizard-tail grass. *Hackelochloa granularis,* family Gramineae, a grass with flat blades and awnless spikelets in pairs; grows in warm regions.

llama. 1. *Auchenia llama,* a ruminant quadruped of South America, allied to the camel, but humpless, smaller (about 3 ft. high at the shoulder) and with a long, woolly coat; used in the Andes as a beast of burden. 2. *Llama's* wool or material made from it.

Llano Estacado. The level grassland comprising the High Plains division of the Great Plains in western Texas, U.S.

load. 1. The work sustained by a machine. 2. The resistance offered to a motor by the machinery it drives, apart from the friction of its own parts. 3. In the export trade, a unit of lumber measurement, used particularly in India; it comprises 40 cu. ft. of round timber or 50 cu. ft. of squared timber, equaling approximately a long ton. 4. The weight carried by a beam, girder, truss span, or structure of any sort, or any part of such structure, including its own weight. 5. A burden; a weight; as a heavy *load.* 6. The work done by a prime mover. 7. To place a *load* on a wagon, truck, etc.

loader. 1. Any of various types of elevators used to lift products and place them on a transportation unit. 2. A person who fashions the load on the vehicle. 3. Any mechanical device which aids in loading material for hauling.

loading. The act of placing loads on vehicles.

loading chute. A narrow, inclined, usually cleated walk-way which leads from the pen or yard floor to the elevated floor of the car or truck in which animals are to be loaded.

loaf. Designating a product regularly molded or shaped, as a *loaf* cheese, meat *loaf,* etc.

loafing area. An area of 60 to 100 sq. ft. per cow of well-sheltered and bedded space where a cow may exercise or lie in comfort. See *loafing shed.*

loafing barn. A light type of building usually attached to the main dairy barn or milking parlor where cows may be turned loose after milking for exercise and comfort. Hay and silage are usually supplied in side racks or bunkers. See *loafing shed.*

loafing pen. Loafing shed.

loafing shed. A light shed type of building usually attached to the main dairy barn for cows to rest or exercise after feeding and milking.

loam. A soil not definitely sandy or clayey; one not strongly coherent, mellow and well supplied with organic matter. Technically, a soil which contains 7 to 27 percent clay, 28 to 50 percent silt, and less than 52 percent sand.

loamy. Like, or made of loam.

loamy sand. Soil with a silt and clay content just below that required for sandy loam *q.v.*

loan value. In appraising, the value of real estate on which prudent and informed lenders base the amount of first mortgage credit they will advance.

lobelia. Any plant of the genus Lobelia, family Lobeliaceae. Some species are grown in borders and flower gardens for their showy flowers. Moist places are congenial to their growth. See *cardinal flower; edging lobelia; great lobelia; Indian-tobacco lobelia.*

loblollybay gordonia. *Gordonia lasianthus,* family Theaceae, a tender, evergreen tree grown for its foliage and its white, fragrant flowers. Native to North America. Also called *black laurel.*

loblolly pine. *Pinus taeda,* family Pinaceae, an evergreen tree, valuable for timber and pulpwood. Native to North America. Also called *old-field pine.*

local. Limited to one region, area, or place, rather than spread widely throughout: *e.g., local* pain, *local* anesthetic, *local* infection, *local* markets, *local* customs.

local infection. An infection restricted to a small area, as an ear infection.

localized disease. A disease which affects only one part of an organism in contrast to diseases which pervade the entire organism.

local lesion. 1. Roundish, nonsystemic dead spots produced on tissue inoculated with viruses. 2. Visible evidence of disease seen in a tissue at the side of injection of an irritant or infective agent or resulting from physical violence, etc.

local overgrazing. Overgrazing in localized areas on a range: *e.g.,* near watering places.

locate. To select the site of, or determine the bounds of, as to *locate* a barn.

lock. 1. A small tuft of cotton, wool, flax, etc., fibers. 2. Small bits of wool that are packed separately for market (Australia). 3. A locule or the cotton in a locule. 4. The single cavity in an ovary.

lockjaw. Tetanus.

locker. A unit in a cold storage plant which usually holds from 100 to 250 lbs. of quick frozen products, such as meats, fruits, and vegetables. It is usually rented by the user, but often is owned in a cooperative. See *locker plant.*

locker plant. A freezer storage business concern that rents locker units in a cold storage plant to people who wish to store frozen foods. *Locker plants* also kill and dress meat animals for customers or sell such meats wholesale for locker storage, as well as prepare fruits and vegetables for quick freezing and locker storage. Popular in the U.S.

loco. See *congenital loco in poultry; loco disease.*

loco disease. A poisoning of grazing animals which results from ingestion of certain species of Astragalus or Oxytropis, family Leguminosae. Symptoms of poisoning for horses and cattle are dullness, irregular gait, irregular feeding, dragging the feet, loss of muscular control, solitary habit, loss of nervous sensibility, loss of flesh, shaggy coat, and death. In sheep the symptoms are not so marked. See *Lambert crazyweed; Rocky Mountain crazyweed; spike crazyweed; blue loco; Earl loco; sheep loco; specklepod loco; two-grooved loco; woolly loco; Wooton loco; fourwing poisonvetch; Hayden poisonvetch; timber poisonvetch.*

locoed. Designating an animal which has the symptoms of loco disease.

loco poisoning. See *loco disease.*

locoweed. See *Earl loco; fourwing poisonvetch; Lambert crazyweed; Rocky Mountain crazyweed; Wooton loco.*

locule. 1. A small cavity, as of a pollen sac. 2. A single cavity of a compound ovary. Also called *lock.*

locust. 1. Any shrub or tree of the genus Robinia, family Leguminosae. Species are grown as ornamentals and for their hard, durable wood. See *black locust; clammy locust; Kelsey locust; New Mexico locust; roseacacia locust.* 2. See *grasshopper.* 3. See *periodical cicada.*

locust bean. See *carob.*

locust borer. *Megacyllene robiniae* (Forst.), family Cerambycidae, the larva of a beetle which tunnels into the sapwood and heartwood of the black locust. Infested trees show swollen areas on the trunk, cracked bark, and exposed tunnels.

locust leaf miner. *Chalepus dorsalis* Thunb., family Chrysomelidae, a beetle whose larva mines the leaves of black locust; infestation may result in defoliation.

locust leaf roller. *Nephopteryx subcaesiella* (Clem.), family Phycitidae, a moth's larva which infests locust and wisteria by rolling and binding leaves with silk and feeding within the roll.

locust twig borer. *Ecdytolopha insiticiana* Zell., family Olethreutidae, a caterpillar which infests black locust.

lodge. Pertaining to field crops, to break, bend over, or lie flat on the ground, sometimes forming a tangle. Lodging may be caused by high nitrogen level in the soil and lush growth, wind and heavy rainfall, and plant diseases.

lodgepole needle miner. *Recurvaria milleri* Busck., family Gelechiidae, a moth whose larva mines the needles of the lodgepole pine.

lodgepole pine. *Pinus contorta* var. *latifolia*, family Pinaceae, an evergreen tree which is valued for its lumber used in ties, mine timbers, poles, fuel, rough construction, and planing-mill products. Native to North America.

lodgepole sawfly. *Neodiprion burkei* Midd., family Diprionidae, a sawfly which infests lodgepole pines, defoliating and killing them.

lodging. The falling over of plants. See *lodge.*

Lodigiano. A hard, yellow, Parmesan-type, cured cheese made in Italy from the milk of a 3 to 5 year old cow.

loess. A massive deposit of tan or buff-colored silt whose particles are typically angular and uniform in size; it is usually calcareous, often contains concretions of calcium carbonate, and sometimes shows lamination or bedding. Where deeply eroded, *loess* areas have a distinctive topography characterized by vertical walls, sometimes many feet high. Some deposits in the U.S. are believed to have been formed from finely ground rock material produced during the Ice Age and transported by water and winds to surrounding highlands. There is also some evidence that the *loess* was deposited directly from river waters when they overflowed their channels in times of flood.

loessial. Pertaining to loess.

loft. The storage space under the roof in a building, such as a barn or house.

loft barn. A barn which has a second story under the roof available for storage of hay or straw. Also called *two-story barn.*

lofty. Designating wool or a woolen fabric that has springiness when compressed, is bulky in comparison with its weight, is light in condition, and has an even, distinct crimp.

log. 1. An unhewn, sawed or cut length of a trunk or large limb of a tree. See *saw log.* 2. In the preparation of chip steaks, the molded, frozen piece of meat from which the steaks are cut. 3. To fell trees for lumber.

loganberry. *Rubus loganobaccus,* family Rosaceae, a prickly, shrubby plant, of uncertain origin, grown for its edible, red fruit, the *loganberry* of commerce, eaten fresh or in jams, jellies, juices, etc.

loganberry juice. A food product prepared by crushing and screening loganberries, vacuum-sealing the juice in enamelled cans, and processing. Sugar is usually added to overcome the tartness of the juice.

log boat. A short, tongueless sled with wooden runners used for transporting logs for short distances.

log chute. 1. A by-pass around or through a dam for logs and drift. 2. A constructed channel or trough by which logs are moved endwise by gravity or by water.

log drive. A large collection of logs floating down a river.

logged. Designating an area from which trees have been cut for lumber.

logged-off. Designating land from which the heavy timber has been removed in the form of logs usually for lumber production.

loggering. A riding posture in which the rider holds the horn of the saddle rather than sitting free and erect in the saddle.

logging. The harvesting of trees for lumber.

logging wheels. A pair of wheels from 7 to 12 ft. in diameter used for transporting logs. Also called *katydid; sulky; timber wheels.*

log gum. A primitive type of beehive consisting of a hollow tree or log.

log haul-up. An inclined plane with a trough up which logs are drawn into a sawmill by means of an endless cable or chain.

log house (cabin). A building made from natural logs sometimes squared by the use of an adz, usually laid horizontally one on top of the other and notched at the ends to tie the walls together. It was a very common type of building construction in the timbered section of the U.S. in the pioneer days.

log mark. A letter or sign stamped or chopped on logs indicating ownership.

log off. To harvest nearly all trees from a stand for lumber production. See *clear: 2.*

log roller. A type of clod crusher or land roller consisting of a heavy log mounted on an axle, and which when pulled across a field crushes clods, levels and firms the soil.

log rule. A table showing the estimated amount of lumber which can be sawed from logs of given length and diameter.

log run. 1. In softwoods, all merchantable lumber as it comes from the saw. 2. In hardwoods, the full run of the log excluding lumber of the lowest grade.

log stain. A stain present in the log before it is sawed into lumber.

log tally. A record of the volume of logs estimated by a log rule.

logway. Log chute.

logwood. *Haematoxylon campechianum,* family Leguminosae, a tropical, spiny tree which is a source of a red dye. Native to America. Also called *campeachy wood.*

loin. 1. The part of the body of man and other animals on each side of the backbone that lies between the floating ribs and the hipbone. 2. A wholesale or retail cut of meat which comes from the loin of a carcass. See *tenderloin: 1; short loin; loin end.*

loin chop. A retail cut of meat taken from the loin section of the carcass of a lamb, veal, or hog.

loin disease of cattle. Botulism of cattle which results from eating bones or putrid flesh containing toxins produced by the organism *Clostridium botulinum.* (Cattle eat bones or putrid flesh because of a depraved appetite usually resulting from a phosphorus deficiency in the diet.) Symptoms of botulism are weakness in the hind quarters and loss of muscle control.

loin end. A wholesale cut of beef which is the lower portion of the loin including the hip; used for steaks and roasts. See *short loin.*

loin lamb chop. A retail cut of lamb in which a section of the tenderloin muscle remains attached to the underside of the lumbar vertebra to form the chop.

loin-rump coupling. That section of the back of an animal where the rump and the loin join.

loin veal chop. A retail cut of veal taken from the forward end of the loin section.

Lombardo. A Parmesan-type, Italian cheese, similar to Lodigiano, with a colored oil surface, cured 1½ to 2 years.

Lombardy poplar. *Populus nigra* var. *italica,* family Salicaceae, a hardy, short-lived, deciduous tree grown as an accent tree, as a screen, etc., for its quick growing habit and its columnar shape.

London plane tree. *Platanus acerifolia,* family Platanaceae, a hybrid, hardy, deciduous tree grown as a very valuable street and specimen tree resisting smoke, dust, wind, etc. Also called *European sycamore; Oriental plane tree.*

London pride. Bouncing bet.

Londonpride saxifrage. *Saxifraga umbrosa,* family Saxifragaceae, a perennial herb grown as an edging plant for its small, white flowers with pink spots. Native to Europe. Also called *St. Patrick's cabbage.*

London purple. An insecticide which is a mixture of calcium arsenite and calcium arsenate with other by-products from the manufacture of certain dyes and pigments.

lone star tick. *Amblyomma americanum* (L.), family Ixodidae, a blood-sucking tick which infests cattle, man, horses, dogs, goats, hogs and many wild mammals.

long. 1. Designating a person who has bought more contracts than he has sold. *Long* hedges are purchases of futures made as a hedge against the sale of the cash commodity. *Cf. short.* 2. Designating a pulse beat which is longer than normal.

long and deep sides. Designating an animal of good body conformation.

long-bodied, weak-backed, thin-necked ram. A male sheep of very undesirable and unattractive body conformation.

long butt. A section of the lower end of a butt log cut off to remove serious basal defect, such as deep fire scar and advanced butt rot.

long-coupled. Designating an animal with a relatively long back. *Cf. short-coupled.*

long day. In horticulture, a day with more hours of illumination than one at the critical photoperiod *q.v.*

long-day plant. A plant in which the flowering period or some other process is accelerated by a relatively long, daily exposure to light. See *photoperiodism.*

long dung. Fresh farmyard manure containing undecayed long straw, as contrasted with short dung which is more putrefied farmyard manure.

longe. To train or lead a horse with a long rope.

longer, more open, type. A large angular type of body conformation desired in some breeds of dairy cattle, such as Holstein-Friesians.

long feed. 1. Any coarse, unchopped feed for livestock, such as fodder, hay, straw, etc. 2. Feeding for a long period of time as contrasted with a short feed.

long-grain type. Varieties of rice with grains longer than normal, as Texas, Patra, Blubonnet, Prelude, Nira, etc.

long head. An undesirable characteristic, a longer than normal head in some breeds of cattle, such as Shorthorns and Jerseys.

Longhorn. A style of Cheddar cheese in a cylindrical package 6 ins. in diameter, 13 ins. high, and about 12 lbs. in weight.

longhorn cattle. 1. Cattle of the Spanish type first brought into Mexico in 1521; much of the early cattle of the western range, U.S., were derived from this type. 2. A breed of cattle in England probably of Spanish derivation. See *Highland.* 3. See *Texas longhorn.*

long horse. A horse that can travel far and fast (western U.S.).

Long Island duckling. A young duckling, having soft, tender flesh, well covered with pale yellow (nearly white) fat; produced to weigh 5 to 6 lbs., in 10 weeks or less, by growers in the vicinity of New York City. This fast growth produces meat of delicate flavor. Such ducklings are now produced under different trade names in various sections of the U.S.

longitudinal crack. An intense cold injury to a tree, in which a crack develops, extending radially from the bark through the sapwood.

long jointed. Designating a plant with long portions between joints.

long, large veins. The large mammary veins desired on the udders of good dairy cows.

longleaf pine. *Pinus palustris,* family Pinaceae, a coniferous, evergreen tree; an abundant native forest species in the Atlantic and Gulf Coastal plains of U.S., where it is an important source of lumber and naval stores. Also called *fat pine; southern pine; fatwood; broom pine; Georgia pine.*

long-legged. Long of leg.

long level rump. Designating an animal well conformed in the rear quarters.

long milk. Ropy milk.

long moss. Treebeard tillandria.

long, narrow head. An animal with an undesirable or badly balanced head conformation.

long-nose cattle louse. *Linognathus vituli* (L.), family Lingonathidae, a blood-sucking insect which infests cattle.

long of leg. Designating an ungainly or poorly conformed animal.

long on. In the wholesale nursery business, designating an abundant supply of stock of a certain variety for sale.

long paddock. The road right-of-way into which livestock are sometimes turned for grazing (New Zealand).

long pepper. Long redpepper.

longpod poppy. *Papaver dubium,* family Papaveraceae, an annual weed pest with pinnate leaves and rose or white, solitary flowers on long stems; scattered in the eastern U.S. in cultivated fields and waste places. Native to Europe. Also called *field poppy.*

long redpepper. *Capsicum frutescens* var. *longum,* family Solanaceae, a tropical, woody plant grown for its edible, long, red or yellow fruit with a sharp, pungent flavor, used in flavoring, condiments, etc. It is the source of cayenne pepper. See *bush redpepper.*

long saddle. A retail cut of the pork loin in which the forward cut is made between either the fourth and fifth or the fifth and sixth rib. See *loin: 2.*

long scale. Glover scale.

long-season. Designating the ripening characteristics of a crop, such as a late maturing variety.

longspur columbine. *Aquilegia longissima,* family Ranunculaceae, a perennial herb grown for its nodding, golden-yellow flowers with 5 in. long spurs. Native to North America.

long-staple. Designating cotton fibers 1⅛ ins. or more in length.

long, straight, and determined stride. Indicating the desirable leg action of a good horse.

long-tailed mealybug. *Pseudococcus adonidum* (L.), family Pseudococcidae, a sucking insect, a pest of greenhouse plants and citrus.

long-tailed sheep. An undocked sheep. Common in the U.S. and Europe.

long-term credit. Credit usually advanced for 5 to 33 years for the purchase of real estate. See *short-term credit; intermediate-term credit.*

long, thin-necked plain headed ewe. An unattractive and inferior type of sheep.

long-time feed. A system of fattening cattle for slaughter in which feeder cattle of 350 to 600 lbs. are fattened from 8 to 12 months. Usually they are purchased in the fall and sold the next summer or fall.

longtom. *Paspalum lividum,* family Gramineae, a marsh grass sometimes grown in southern U.S. as a tame pasture grass. Native to North America.

long whip grafting. A method of grafting in which the stock is cut at an angle of 15° to 25° and the cion at an angle of 65° to 75°.

long-wool breeds of sheep. The principal breeds of long-wool, mutton-type sheep: in the U.S., the breeds are, Lincoln, Cotswold, Romney Marsh, Black-Faced Highland, and Leicester.

longwool crossbreed. A sheep crossbred from two different breeds, one being a long wool type. A common sheep on western sheep ranches in the U.S.

long yearling. 1. An animal more than a year old. 2. A senior yearling of livestock-show class.

loofa. Towelgourd.

lookout. 1. A person employed to detect and report forest fires. 2. A station or post used primarily in the detection of forest fires.

lookout fireman. A forest fire guard at a lookout station whose duties include the functions of a lookout and fireman.

lookout point. A topographic vantage point selected for forest fire detection.

lookout station. See *lookout: 2.*

lookout tower. See *lookout: 2.*

loopers. Certain caterpillars of the family Geometridae. Some species have a looping motion when crawling. Also called *measuring worm; span worm; inch worm.*

loop system. A system of training trailing brambles in which the canes from a single plant are divided into two parts and each part is taken separately to the top wire of an arbor, twisted around it, brought down to the lower wire, and again twisted around it.

loose. 1. Designating a condition in the roll box of a cotton gin in which the lint cotton is kept fluffy as it comes from the saws. 2. Designating a texture defect of Cheddar cheese in which mechanical holes, gas holes, or fish eyes appear.

loose and open. See *loose: 2.*

loose bandage. A defect of Cheddar cheese, characterized by folds in the bandage and by heavy paraffin in and around the folds.

loose cotton. Repacked cotton from samples from cotton-classing rooms and brokers' offices.

loose-flower lupine. Spur lupine.

loose hay. Hay stored in the hay mow or stack without chopping, baling, or compressing.

loose housing. A management system for cattle wherein the adult animals are given unrestricted access to a feeding area, water, a resting area, and an adjoining open lot. In dairies, the lactating animals pass through a milking room at milking time. Other dairy animals may be in separate pens, lots, or buildings.

loose housing-barn. A barn designed for loose housing.

loose in front attachment. Designating the undesirable form of a cow's udder in its forward conformation and attachment.

loose kernel smut of sorghum. A fungous disease caused by *Sphacelotheca cruenta,* characterized by an early appearance of smutted heads, dwarfing, elongation and proliferation of floral bracts. The membranes rupture early and release a powdery, black spore mass.

loose knot. An unfirm knot; one that is not held firmly in place by growth or position.

loosely fitted in back and loin. Designating an animal with a weak back and loin region.

loose paraffin. A defect of Cheddar cheese in which the paraffin covering is loose or scaly. Also called *scaly paraffin.*

loose rein. A condition in driving in which the reins on the harness of a horse are generally relaxed.

loose rock dam. A dam built of rock without the use of mortar. The rock is usually confined by woven wire.

loose side. The left side of a beef carcass. *Cf. tight side.*

loose smut of barley. A fungous disease of barley and some grasses, caused by *Ustilago nuda,* family Ustilaginaceae, characterized by the appearance of smutted spikes earlier than the normal emergence of spikes. The smutted heads are conspicuous when barley is heading.

loose smut of wheat. A fungous disease caused by *Ustilago tritici,* family Ustilaginaceae, characterized by loose spore masses replacing the floral bracts and ovaries. The smutted heads are conspicuous when wheat is heading.

lop. 1. To chop branches, tops, or small trees after felling so that the slash will be close to the ground. 2. To cut the limbs from a tree or shrub. 3. To fall or spread over, as a plant *lops* over the space between rows.

lope. In a horse, a slow gallop.

lop ear. Referring to an animal with drooping ears.

lopped comb. A fowl comb which falls over to one side.

loppered milk. Curdled milk. Also called *clabbered milk.*

loppers. Lopping shears.

lopping. 1. A method of forcing budded trees; the top of the stock is cut almost through above the bud and left partially attached for some time. The cut is made on the bud side and the top is bent over to rest on the ground. 2. Pruning or cutting away parts of trees.

lopping shears. A long-handled pair of shears used to prune trees; one sharpened, half-moon shaped blade cuts against a heavy convex metal piece. Also called *loppers.*

lopsided. Designating a cheese not uniform in shape due to uneven distribution of curd in the hoops in the pressing process or gassy fermentation during curing.

lopstick. A tree with branches lopped to serve as a memorial or landmark (Canada).

loquat. *Eriobotrya japonica,* family Rosaceae, an evergreen tree grown in warm areas for its edible fruit eaten fresh or in jams and jellies. Native to Asia. Also called *Chinese loquat; Japanese medlar; Japan plum.*

lords-and-ladies. *Arum maculatum,* family Araceae, a perennial herb grown in the wild garden for its purple-spotted spathe; a poisonous plant. Native to southern Europe. Also called *cuckoo pint.*

Lorraine. A small, hard, German cheese made from soured, cows' milk, eaten comparatively fresh.

loss. In egg marketing, inedible, smashed, or broken eggs in which the contents leak, are contaminated, or show bloody whites, large blood spots, or large, unsightly meat spots.

loss of appetite. Failure of animals to eat, a reliable indication of some illness.

loss of cud. An old, false idea that a cow sometimes would lose her cud and would have to be supplied with another, such as placing a rag or other material in the mouth. Failure to regurgitate and chew cud is an indication of illness.

loss-off. In marketing eggs, a provision governing a purchase of eggs in which final settlement is made after inedible eggs are removed.

lost corner. In land surveying, a corner whose position cannot be definitely determined from traces of the monument or by reliable testimony relating to it; it can be restored only by surveying.

lost head. A measure of the energy expended by flowing water in overcoming frictional resistance to its motion.

lost its cud. See *loss of cud.*

lot. 1. A small piece of enclosed land usually adjacent to a barn or shed in which horses, mules, cows, etc., are allowed to exercise. 2. Any particular grouping of animals, plants, seeds, fertilizers, etc., without particular regard to number, as a seed *lot.* 3. A small tract of land, usually less than an acre, on which a house is constructed.

lotebush. Lotewood condalia.

lotewood condalia. *Condalia obtusifolia,* family Rhamnaceae, a stiff, spiny, much-branched shrub bearing edible, though inferior, black fruit. It is an obnoxious invader on the southwestern ranges, U.S. Native to warm regions of North America. Also called *Texas jujube; buckthorn; lotebush.*

lots of depth. Designating a blockily built animal; a very desirable characteristic in meat animals.

lots of finish. Designating an animal well fattened and in prime condition for the market or show ring.

lots of natural fleshing. Designating an animal with a deep, thick, natural flesh covering over the body frame. A desirable characteristic.

lots of spring in the ribs. Designating an animal with a deep, broad mid-body or barrel.

lots of strength to the back. Indicating a straight top line on the back of an animal.

lotus. Any aquatic herb of the genus *Nelumbium,* family Nymphaeaceae. Species have shield-like leaves and single, large flowers. See *Hindu lotus; American lotus.* 2. See *jujube; deervetch.* 3. See *waterlily.*

loudly spotted. Designating objectionable large or discordant color patterns in the skin of broken-colored animals, such as Spotted Poland China swine, Ayrshire cattle, etc.

lounge. Longe.

louping-ill. A tick-borne, virus disease of sheep in

Scotland, northern England, and northern Ireland, characterized by uneasiness, fever, twitching of the neck and jaw, staggering, frothing at the mouth, and spasmodic jumping. Also called *trembles; trembling; twitch; dizzy staggers; the jumps.*

louse. Singular of lice *q.v.*

louse fly. Sheep ked.

lousicide. Any chemical compound used as a spray or dust in the eradication or control of lice on animals or man.

louver. Slatted opening for ventilation in which the slats are so placed as to exclude rain, direct sunlight, or vision.

lovage. Garden lovage.

love apple. A name given the tomato (*q.v.*) before it was accepted as an edible vegetable.

lovecharm. Argentine trumpetvine.

love-entangle. Goldmoss stonecrop.

lovegrass. Any grass of the genus Eragrostis, family Gramineae. Species are used as range grasses in western U.S. See *alkali lovegrass; weeping lovegrass; sand lovegrass.*

love-in-a-mist. *Nigella damascena,* family Ranunculaceae, an annual herb grown for its finely cut, feathery leaves surrounding blue or white flowers. It is a good cut flower. Native to southern Europe. Also called *fennelflower; devil-in-a-mist; devil-in-the-bush.*

love-in-a-puff. Balloonvine heartseed.

love-lies-bleeding. *Amaranthus caudatus,* family Amaranthaceae, a tropical annual herb grown for its deep-red flower spikes. Also called *tasselflower.*

lovely browallia. *Browallia speciosa,* family Solanaceae, an annual herb grown for its purple flowers. Native to South America. Also called *tasselflower.*

lovely goldenlarch. *Pseudolarix amabilis,* family Pinaceae, a coniferous, deciduous tree grown for its feathery, light green leaves that turn bright yellow in autumn. Native to China. Also called *false larch; golden larch.*

lovely mariposa. *Calochortus amabilis,* family Liliaceae, a bulbous herb grown for its nodding, golden-yellow flowers. Native to North America. Also called *golden globetulip.*

love vine. Dodder.

Lovitto ryegrass. Italian ryegrass.

low. 1. To cry, as a cow; to moo. 2. Close to the ground. 3. Designating cheap prices. 4. Designating land which lies below the general level of surrounding or adjacent land, as *low* land, *low* ground. 5. Designating a stream level much below normal.

low at the pins. Designating a dairy animal with a sloping rump; a defect.

low back. Designating a sway-backed animal; considered a defect especially objectionable in breeding stock.

low bacteria count. Designating raw milk with a low bacterial content; indicative of high quality milk, cleanliness, and disease-free animals.

low behind the shoulder. Indicating an animal with a pinched or restricted heart girth.

lowbush blueberry. *Vaccinium angustifolium,* family Ericaceae, the lowest and earliest ripening blueberry with blackish, sweet fruit. Native to northeastern U.S. Also called *low sweet blueberry; early blueberry; early sweet blueberry.* Often erroneously called *huckleberry. Cf. highbush blueberry.*

low cutter. A poor quality of beef animals, almost in the canner class.

low down, thick one. Indicating a desirable body conformation of a young lamb or beef calf.

low dressing percent. Designating a meat animal that loses only a low percentage of the live weight in the dressed carcass.

lower middle. The lower part of the belly of an animal.

lower thigh. 1. In fowls, the section of the leg between the hock and the joint above. 2. See *gaskin.*

low grade crude cottonseed oil. Cottonseed oil that does not come up to the specifications for off-crude cottonseed oil or reddish off-crude cottonseed oil.

low-hanging udder. A pendulous udder.

low-headed. 1. Designating a horse that carries its head low; an indication of lack of alertness. Often associated with a short leg action or stride. 2. Designating a tree so pruned and trained that the first or lowest limb is 1½ to 3 ft. from the earth.

low hop clover. *Trifolium procumbens,* family Leguminosae, an annual herb grown for early spring pasturage. Native to Europe. Also called *annual clover; large hop clover; mignonette clover.*

low huckleberry. Lowbush blueberry.

low in the topline. Designating an animal with an undesirable sway in the back.

low in the withers. A deficient or restricted heart girth and shoulder conformation; a defect in some animals.

low-lactose concentrate. A skim milk derivative from which part or all of the milk sugar has been removed; used in ice creams to improve the body and texture. Also called *milk protein concentrate; low lactose skim milk.*

low-lactose skim milk. See *low-lactose concentrate.*

lowland. 1. That land which lies lower in elevation than surrounding or adjacent land and whose soil is often poorly drained. 2. Designating plants which grow in wet or moist soil, as *lowland* weeds. 3. Designating irrigated varieties of rice. *Cf. upland.*

Lowland cattle. Cattle from the low, flat countries, such as the Friesian cattle of the Netherlands, as contrasted with Highland cattle of Scotland.

lowland rice. See *lowland:* 3.

low-lift sulky plow. A frameless and tongueless, horse-drawn plow, upon which the operator rides. It may have one or more plow bottoms.

low meadowrue. *Thalictrum minus,* family Ranunculaceae, a perennial herb grown in the flower border for its greenish-yellow or cream-white flowers. One horticultural variety, maidenhair, has foliage like the maidenhair fern. Native to Europe and Asia.

low middling. A technical term for a grade of cotton which in color, leaf, and preparation is within the range represented by a set of samples in the custody of the U.S. Department of Agriculture designated as "Original official cotton standards of the United States, American uplands, No. 7 or low middling, effective August 20, 1936."

low middling extra white. A technical term for a grade of upland cotton which in leaf and preparation is whiter than No. 7 or low middling.

low middling spotted. A technical term for a grade of upland cotton which in leaf and preparation is No. 7 or low middling, but which in color is between it and No. 7 tinged or low middling tinged.

low middling tinged. A grade of upland cotton which in leaf and preparation is No. 7 or low middling, but in color is within the range represented by a set of samples in the custody of the U.S. Department of Agriculture, designated as "Original official cotton standards of the United States, American upland, No. 7 tinged or low middling tinged, effective August 20, 1936."

low-moor peat. More highly decomposed and less acid peat; a muck.

low quarter-blood. In wool grading, the next to the coarsest grade based on fineness of the fiber and a minimum fiber length of two inches. It is equal to English 46's. See *American grades of wool.*

low-set. Designating a short-legged animal, a condition generally considered desirable in sheep and cattle.

low-set, rugged and strong-bone. Designating a deep-bodied, short-legged animal of strong bone and rugged conformation.

low smut. Bunt of wheat.

low-sodium milk. Fresh, liquid, whole milk, processed to reduce its sodium content. The process involves passing the milk through an ion exchange resin in the cold state (35°– 40°F.) resulting in the replacement of sodium with an equivalent amount of potassium. It is of special use in sodium-restricted diets.

low spear grass. Annual bluegrass.

low sweet blueberry. Lowbush blueberry.

low-temperature, long-time. A method of pasteurization in which the material to be pasteurized is exposed to a temperature of 143°F. (61.7°C.) for at least 30 minutes. Also called *L-T, L-T.* See *high-temperature, short-time method.*

L-T, L-T method. Low-temperature, long-time.

lubricant. Any substance used to grease or oil machinery to reduce friction or wear. Most ordinary *lubricants* for farm machinery are petroleum compounds.

lubricating jelly. A semisolid petroleum compound used for dressing wounds and for certain types of lubrication, such as on a rubber glove before insertion into the vagina. It is better known for medicinal use as *white vaseline.*

lubricating oil. A type of refined oil usually of a petroleum base which is used for lubricating machinery and in emulsion sprays.

lucerne. Alfalfa.

luckynut thevetia. *Thevetia nereifolia,* family Apocynaceae, a tropical, evergreen shrub grown in very warm areas and in the greenhouse for its leathery leaves and clusters of yellow flowers. Native to America. Also called *yellow oleander; exile tree.*

Lucretia dewberry. See *dewberry.*

lug. 1. Any kind of a projection for traction or supporting, as a projection on tractor tires. 2. A wooden crate or box approximately 6 x 12 x 24 ins., inside measurement, used as a container for berries, grapes, etc., during harvesting or transportation to processing plants or to market. 3. A leather loop on a saddle.

lug box. See *lug: 2.*

lugger. A horse which pulls at the bit.

lumbago. Azoturia.

lumbar. Pertaining to the loin.

lumber. 1. The product of the saw and planing mill not further manufactured than by sawing, resawing, and passing lengthwise through a standard planing machine, cross-cutting to length, and matching. 2. Sawed or split timber. 3. To cut trees for logs. 4. To manufacture logs into *lumber.*

lumber-boards. Yard lumber less than 2 ins. thick and more than 8 ins. wide.

lumbering. The business of cutting or getting timber or logs from the forest for lumber.

lumberjack. A man who works at felling trees or delivering them to a lumber mill.

lumber tally. A record of lumber which gives the number of boards or pieces by sizes, grade, and species.

lumpiness. A body and texture defect of Cottage cheese characterized by large, uneven curds.

lump lime. CaO, calcium oxide, as it comes from the lime kiln. Also called *quicklime.*

lumpy. 1. Designating a body and texture defect of fermented milk characterized by lumps of firm curds interspersed throughout the milk. 2. Designating a texture defect of ice cream characterized by the presence of gelatin lumps. 3. Designating a body and texture defect of dry milk solids characterized by lumps larger than a grain of wheat, interspersed in the product. 4. See *grainy.*

lumpy cheese. An undesirable condition of the body texture of cheese as indicated by a piece showing small, hard or gritty particles when worked between the thumb and fore fingers.

lumpy jaw. See *actinomycosis; actinobacillosis.*

lumpy withers. Fistulous withers.

Luna cheese. A very large, cows' milk cheese which was made in ancient Italy. According to Pliny some weighed as much as 1,000 lbs. and were shipped mostly to Rome from the ancient seaport of Luna. The cheeses were branded with a picture of the moon.

Luneberg. A saffron-colored, cured, cows' milk cheese made in Switzerland. Its characteristics are between those of Swiss and Limburger.

lung fever. Hemorrhagic septicemia.

lungwort. 1. Any plant of the genus Pulmonaria, family Boraginaceae. Species are grown in flower gardens. See *cowslip lungwort.* 2. *P. officinalis,* an herb grown for its spring-blooming, purplish-red flowers. Native to Europe.

lunker. 1. A very big, awkward, heavy-boned horse. 2. Any animal which is considerably larger than the average.

lupine. Any plant of the genus Lupinus, family Leguminosae. Species are grown in the flower garden, as cover crops, for green manuring, and to a small extent, as a feed crop. Also called *blue pea.* See *tree lupine; sundial lupine; Washington lupine.*

lupulus. See *hop.*

lush. Designating succulent, luxurious vegetation.

lush growth. A fast, abundant growth of a plant or plants, such as the early spring growth of a rye field or early pasture.

luster. The property of wood, independent of color, which causes it to reflect light and to exhibit a sheen.

luster wool. Shining wool reflecting more than the usual amount of light; chiefly from the long-wool breeds.

lutein. A hormone, prepared from the dried corpus luteum (*q.v.*) of the sow; used to cause ovulation.

luxuriant. Designating a lush or abundant growth of plant life, such as a heavy yielding field of grass or hay.

Lyall angelica. *Angelica lyalli,* family Umbelliferae, a perennial herb found in western U.S. Its forage rating is good.

lychee. *Litchi chinensis,* family Sapindaceae, a tree grown in warm areas for its sweet, edible fruit, the litchinut of commerce. Native to China. Also called *litchi; litchee, leechee.*

lycopene. $C_{40}H_{56}$, a carotenoid responsible for the red color of tomatoes.

lye. Any strong liquid alkali, *e.g.*, hydroxide of soda, hydroxide of potash, or a mixture of the two, used in making homemade soap, for cleaning, peeling of fruits in the canning industry, etc.

lye hopper. Ash hopper.

lye peeling. The use of lye to remove the skin of fruits for canning or for the removal of the hard outer covering of the corn kernel in the preparation of hominy.

lyme grass. Wildrye.

lymph. A nearly colorless fluid made up of the liquid portion of the blood and the white corpuscles but free from the red corpuscles. It is derived from the blood by seeping through the walls of the capillaries (very minute blood vessels) which join the blood stream in the vicinity of the heart.

lymphangitis epizootica. Mycotic lymphangitis.

lymphatic. Referring to the system of vessels returning lymph from the tissues to the blood stream.

lymphoid. Referring to lymph.

lympholeukosis. Lymphomatosis.

lymphomatosis. A disease of fowls caused by the formation of tumors of lymphoid tissue in different parts of the body; a form of avian leukosis complex. See *fowl paralysis; ocular lymphomatosis; visceral lymphomatosis; osteopetrotic lymphomatosis.*

lynx flower. *Stanhopea tigrine,* family Orchidaceae, an orchid with yellow-spotted, deep red flowers. Native to Central America and northern South America.

Lyon bean. Chinese velvetbean.

lyophilization. The evaporation of water from a frozen product with the aid of high vacuum. The process is used for the preservation of stocks of bacteria and other microorganisms.

lysimeter. A vessel or container placed below the ground surface to intercept and collect water moving downward through the soil. It is used in research to determine the amounts of nutrients carried out of a soil by drainage water.

lysis. 1. The gradual disappearance of the symptoms of a disease. 2. The destruction or dissolving of cells.

M

Macabee gopher trap. A steel wire trap used to catch and kill gophers.

macadamia nut. See *Queensland nut.*

macan. A variety of lowland rice from the Philippines.

macaroni wheat. Durum wheat.

macaroon paste. A paste, prepared from almonds or apricot kernels, used in baking.

MacArthur clusterpalm. *Actinophloeus macarthurii,* family Palmaceae, a slender and graceful feather palm grown in very warm areas as an ornamental. In cool areas it is grown in a tub indoors. Native to Australia and New Guinea.

Macartney rose. *Rosa bracteata,* family Rosaceae, a tender, trailing, evergreen shrub grown for its white flowers. Native to China.

macaw palm. Any palm of the genus Acrocomie, family Palmaceae. Species produce soft nuts, some edible, which yield a heavily-scented oil used in perfumes. Native to tropical America.

macerate. To soften or separate the constituents,

as of food, by crushing, soaking or by digestion with or without heat.

macerator. A machine designed for extracting seeds of deciduous plants or for mashing fleshy fruits.

machete. A large, heavy knife used for cutting sugarcane and brush, especially in Latin America.

machine generation. A method of fumigating trees, especially citrus, in which a gas generating chamber is filled with sulfuric acid and water to which cyanide is added. The generated gas is forced by its own pressure into a tent covering the tree.

machine milking. The practice of milking dairy cows by a power-operated, mechanical device. Of the several different types, those employing a pulsating suction are most common. *Machine milking* is a great labor-saving development and has largely replaced hand milking on the dairy farms of the U.S.

machinery expense. The total operating cost, including depreciation and repairs, for farm machinery which includes trucks and the farm share of the family car.

machine shears. Power-operated shears used in the shearing of sheep which are much more efficient than the old-fashioned hand shears.

machine shed. A farm building, usually of open, light construction, designed especially for housing and storing farm machinery.

MacNab cypress. *Cupressus macnabiana,* family Pinaceae, an evergreen tree with heavy foliage, the hardiest of the cypresses, grown as an ornamental in the U.S. Native to the Pacific Coast of North America.

maco. *Gossypium vitifolium,* family Malvaceae, a cotton plant with leaves having hair underneath, but which otherwise resembles sea-island cotton (*q.v.*), through hybridization with which it has yielded the various varieties of Egyptian cotton. Native to tropical America.

Macomber. A strain of rutabaga.

Maçonnais. A goats' milk cheese made in France.

Macqueline. A soft, whole or partly skimmed, cows' milk, French cheese, similar to Camembert. It is cured for 20 days or more.

Macrophoma ripe rot. A fungous disease of muscadine grapes caused by *Macrophoma* spp., family Sphaerioidaceae. It produces small, circular, slightly sunken, tan or brown spots on the berries. A brown rot then extends over the berry which becomes dotted with black pycnidia.

macroscopic. Visible to the human eye without the aid of a microscope.

Madagascan feather palm. Yellow butterfly palm.

Madagascar arrowroot. Fiji-arrowroot tacca.

Madagascar jasmine. Madagascar stephanotis.

Madagascar raffiapalm. *Raphia pedunculata,* family Palmaceae, a tropical feather palm which yields raffia, a widely used tying fiber. Native to Madagascar and the southeastern coast of Africa. Also called *raffia.*

Madagascar rubbervine. *Cryptostegia madagascariensis,* family Asclepiadaceae, a woody vine, with red-veined leaves and white or pink flowers, grown in the warm greenhouse as an ornamental. Native to Madagascar.

Madagascar stephanotis. *Stephanotis floribunda,* family Asclepiadaceae, a woody, twining vine grown in very warm areas for its habit and its fragrant, waxy, white flowers. Native to Madagascar. Also called *Madagascar jasmine; wax flower.*

mad-apple. Eggplant.

Madeira Bay persea. *Persea indica, family* Lauraceae, a small evergreen tree grown in warm areas for its fine foliage. Native to the Azores, Canary, and Madeira Islands. Also called *vinhatico.*

Madeira vine. *Boussingaultia ramosa* (*B. baselloides*), family Basellaceae, a tropical vine with thick, shiny leaves and a profusion of white, fragrant, late-summer flowers, widely grown in warm areas as an ornamental. Native to America. Also called *mignonette vine.*

madderwort. Any of the several plants of the family Rubiaceae. Many species are tropical and grow showy clusters of flowers.

made land. An area consisting of artificial fills of earth material, or dumps of waste, refuse, debris of any kind, smoothed over and stabilized.

mad itch (of cattle). Aujeszky's disease.

madness. See *rabies.*

Madonna lily. *Lilium candidum,* family Liliaceae, a perennial, bulbous herb grown for its waxy, pure white flowers. Native to Europe and Asia. Also called *Annunciation lily.*

Madras thorn. Guamachil apes-earring.

madre de cacao. *Gliricidia sepium,* family Leguminosae, a tree grown in very warm areas for its clusters of pink flowers and to furnish shade for coffee trees. Native to tropical America.

madrona. Pacific madrone.

mad stone. Hair ball.

Magdalena mockorange. *Philadelphus magdalenae,* family Saxifragaceae, a deciduous shrub grown for its white flowers. Native to western China.

Magellan barberry. *Berberis buxifolia,* family Berberidaceae, an evergreen shrub grown for its foliage, its orange-yellow flowers, and its dark purple fruit. Native to southern Chile.

Magellan fuchsia. *Fuchsia magellanica,* family Onagraceae, a shrubby plant grown on a wall or trellis in warm areas, its branches sometimes growing to 20 ft. It usually has blue flowers with a red calyx, but varieties have different colored flowers. Native to South America.

maggot. The larva of a fly. Also called *earth bob.*

magnesia. MgO, magnesium oxide, obtained by heating or burning the hydroxide or the carbonate. A commercial product, containing a high percentage of MgO, is used as a liming material for soils. Magnesium oxide is also used in the treatment of grass tetany *q.v.*

magnesia hunger. See *magnesium deficiency: 1.*

magnesian limestone. A limestone which contains magnesium carbonate. See *dolomitic limestone.*

magnesium. Mg, a white metal, essential for animal growth and plant growth in that it is an element of chlorophyll.

magnesium arsenate. A white powder insecticide which has been used to control the Mexican bean beetle. It is not widely used now.

magnesium deficiency. 1. A physiological disease of plants grown on soil deficient in available magnesium, generally characterized by chlorosis and necrosis and a reduction of growth. In tomatoes, the veins remain green. In cabbages, the lower leaves pucker and turn white at the margin and center. Also called *sand drown.* 2. The cause of grass tetany (*q.v.*) in cattle.

magnesium oxide. See *magnesia.*

magnesium sulfate. $MgSO_4$, a soluble salt obtained from kieserite (*q.v.*) which is added to mixed fertilizers to supply the magnesium essential to plant growth.

magnolia. Any shrub or tree of the genus Magnolia, family Magnoliaceae. Species are grown as ornamentals and for their wood. See *cucumbertree magnolia; Yulan magnolia; Fraser magnolia; southern magnolia; bigleaf magnolia; whiteleaf Japanese magnolia; saucer magnolia; star magnolia; umbrella magnolia; sweetbay magnolia.*

magnolia scale. *Neolecanium cornuparvum* (Thro), family Coccidae, a soft scale insect which infests magnolias. Heavily infested trees have small leaves, and branches which appear to be covered with cotton. Sooty mold (*q.v.*), may become established in the honeydew.

magnolia waterlily. *Nymphaea tuberosa,* family Nymphaeaceae, a hardy, aquatic herb grown in a pool or pond for its floating foliage and pure white, 4 to 9 in. blossoms. Also called *tuberous waterlily.*

Mahala mat. Squawcarpet ceanothus.

mahaleb cherry. *Prunus mahaleb,* family Rosaceae, a deciduous tree grown for its wood used as a grafting stock and in cabinet-making, its flowers and young leaves used in perfumery, and its inedible fruit from which a cordial and a violet dye are extracted. Native to Europe and Asia. Also called *St. Lucie cherry.*

mahoe. Linden hibiscus.

mahogany. Any tropical tree of the genus Swietenia, family Meliaceae. Native to America. See *West Indies mahogany.*

mahogany bay. A coat color term for a horse: a brown shade of bay.

mahogany gum. *Eucalyptus robusta,* family Myrtaceae, an aromatic, evergreen tree, grown for its decorative fruit. Native to Australia. Also called *swamp mahogany; beakpod eucalyptus.*

maholtine. *Wissadula periplocifolia,* family Malvaceae, a tropical plant from which a valuable bast fiber is obtained.

maiden. 1. A year old, single-stemmed seedling fruit tree used for budding or grafting. See *whip: 2.* 2. An unbred animal.

maidencane. *Panicum hemitomon,* family Gramineae, a perennial marsh grass sometimes a weed pest, but also one of the better forage grasses of the fresh-water marshes of the Gulf Coast, U.S. Native to North America. Also called *paille firme grass; paille fine.*

maidenhair. Any fern of the genus Adiantum, family Polypodiaceae. Some species, having slender stalks and delicate fronds, are grown for decoration. See *southern maidenhair; delta maidenhair.*

maidenhair spleenwort. *Asplenium trichomanes,* family Polypodiaceae, a hardy fern grown as an ornamental. Native to the North Temperate Zone. Also called *English maidenhair.*

maidenhair tree. Ginkgo.

maidenhair-vine. Wirevine.

maiden heifer. See *maiden: 2.*

maiden mare. See *maiden: 2.*

maiden pink. *Dianthus deltoides,* family Caryophyllaceae, a mat-forming, perennial herb grown for its red or pink flowers. Native to Europe and Asia. Also called *meadow pink.*

maiden's tears. Cowbell silene.

maiden's wreath. Branching francoa.

Maile. A ewes' milk, brine-cured cheese made in the Crimea.

Maile Pener. A ewes' milk, brine-cured, crumbly cheese made in the Crimea.

mailing cage. A shipping container for a queen bee.

main. The principal pipeline in a water supply or drainage system.

Mainauer. A fresh, whole or partly skimmed milk cheese, similar to Münster, cured 4 to 5 months. It is made in Germany, Austria, and Switzerland.

main canal. In irrigation and other systems, the principal conduit feeding water from the supply source to the lateral conduits.

main crop. The principal or most important crop grown on a farm or in an area.

main duct. A tunnel which distributes or collects air to or from either two or more smaller (lateral) ducts, or to a perforated or slatted floor.

Maine-style corn. That corn for canning which is obtained by cutting through the kernels.

main herd. The single large group into which cattle were rounded up on the open ranges of western U.S. Smaller herds were made from this large group.

main tail feathers. The stiff feathers in the tail of a fowl.

maintaining feed reserves. The production, procurement, and reservation of feed, other than range and pasture forage, for emergency feeding of livestock during the normal grazing season when forage is insufficient or of inferior quality. The feed reserves may be either hay or concentrates and are in addition to feeds normally required annually.

maintaining forage reserves. The reservation of native forage supplies for such emergencies as drought, fire, or other unforeseen circumstances; obtained by the exclusion of livestock or by light use of grazing areas.

maintenance. 1. The servicing of facilities or

equipment of farms which is required for upkeep, replacement of worn or broken parts on machinery, lubrication, etc. 2. The upkeep of gardens and grounds.

maintenance ration. The amount of food needed to support an animal when it is doing no work, yielding no product, and gaining no weight.

main wheel. A large wheel equipped with lugs on a binder or harvester, which bears a high percentage of the weight of the machine and furnishes the power to operate the machine. Sometimes called *bull wheel.*

Mainzer Hand. A sour milk, small, round, German cheese which is cured 6 to 8 weeks.

maize. *Zea mays,* family Gramineae, an annual cereal grass of several varieties and strains, widely grown for its edible seed. Cultivated in North and South America for several thousand years, it was introduced into Europe and Africa around the 15th. century and its cultivation spread rapidly throughout the Old World. Also called *corn* (U.S. and Canada); *Indian corn.* See *dent corn; flint corn; pod corn; sweet corn; waxy corn.*

maize billbug. *Calendra maidis* (Chitt.), family Curculionidae, an insect pest of corn which feeds inside the stalk. Found in the southern U.S.

majagua. Linden hibiscus.

Majorca. A breed of Spanish asses which contributed to the American breed.

make. The daily production of butter in a creamery.

make a crop. To plow and fit land, to plant seed, to cultivate, to harvest, etc.

make a stand. To constitute a stand *q.v.*

make hay. To cut forage, and to cure, bale, stack, or store it.

makomako wineberry. *Aristotelia racemosa,* family Elaeocarpaceae, a shrub grown for its rose-colored flower clusters and small, purplish-black, edible fruit. Native to New Zealand. Also called *New Zealand wineberry.*

Malabar almond tree. *Terminalia catappa,* family Combretaceae, a deciduous tree bearing greenish or reddish fruit with edible, oil-bearing seed; grows in tropical regions. Native to India.

Malabar glory-lily. *Gloriosa superba,* family Liliaceae, a tuberous-rooted, tropical vine grown out-of-doors in very warm areas and in the greenhouse for its yellow flowers turning orange-brown. Native to Africa.

Malabargourd. *Cucurbita ficifolia,* family Cucurbitaceae, a climbing, tender, annual vine grown as an ornamental for its green, white-striped fruit.

malachite green. Copper carbonate.

mal(adie) du coit. Dourine.

malai. A clotted layer that forms on the surface of heated milk. Sometimes called *sar.*

Malakoff. A soft, Neufchâtel-type, French cheese which may be eaten fresh or after it is cured.

malanders. In horses, an eczema on the posterior portion of the foreleg knee. Pustules also appear on the neck.

malarial fever. See *equine infectious anemia.*

malaria mosquito. See *common malaria mosquito.*

malathion. An organic, phosphorus insecticide. Technical grade *malathion* (95–98 percent pure) is a viscous, dark brown liquid with a strong, offensive odor somewhat like that of garlic. Widely used to control pests, it is the least toxic of available phosphorus insecticides.

Malay. An old breed of chickens which originated in Asia. *Malays* were introduced into England about 1830 and somewhat later into America where they are raised mainly for exhibition. They are large, and upstanding, fierce-looking birds with short, broad heads and overhanging eyebrows. The Black-Breasted Red Malay is included in the Oriental class.

malde caderous. An infectious horse disease caused by *Trypanosoma equinum,* family Trypanosomatidae, a protozoan parasite that infests the blood plasma. It is characterized by rapid loss of flesh, red urine, anemia, partial paralysis, and edema. Endemic in South America.

male. 1. Designating an animal capable of producing spermatozoa, or male sex cells. 2. Designating the stamens of a flower. 3. Designating a staminate plant.

maleberry. He-huckleberry.

male fern. *Dryopteris filixmas,* family Polypodiaceae, a nearly evergreen fern with fronds growing in crowns; grown as an ornamental. Native to North America and Europe.

male process. Genital eminence.

malformation. Any unusual, abnormal growth, organ, or part of a plant or animal.

malice. Running mallow.

malienders. Malanders.

malignant. Tending to grow worse; exceedingly noxious, dangerous, infectious, or that condition which terminates in death, as a *malignant* growth.

malignant aphtha. Foot-and-mouth disease.

malignant catarrhal fever in cattle. An infectious disease of cattle probably caused by a virus; characterized by sudden onset, high fever, loss of appetite, extended head, painful cough, and thick nasal discharge. Also called *malignant head catarrh.*

malignant edema. An acute, fatal wound infection disease caused by *Clostridium septicum,* family Bacillaceae, which affects horses, cattle, sheep, swine, and man. Characterized by swelling and gangrene.

malignant oedema. Malignant edema.

malignant panleucopenia. Feline enteritis.

malignant pustule. Localized skin lesion of anthrax in man.

mallein. A diagnostic agent processed as an extract from *Malleomyces mallei* (glanders bacillus); used principally in the diagnosis of glanders in horses.

mallenders. Malanders.

mallet cutting. A hardwood stem cutting that has at its base a piece of the main stem to which it was attached.

Malling-Merton rootstock. Any of several clonal

understocks for apples developed in England mainly by crossing Northern Spy and rootstocks of the Malling series which produce trees of different sizes resistant to woolly apple aphids. See *Malling (root) stock.*

Malling (root) stock. Any of 16 clonal understocks for apples designated as Malling I, Malling II, etc., standardized at East Malling Research Station, East Malling, Kent, England. They are used to produce trees of different sizes, such as dwarf, semi-dwarf, and standard. See *Clark dwarf, Malling-Merton rootstock.*

mallow. Any herb of the genus Malva, family Malvaceae. Some species are grown in flower gardens and some are weed pests. See *curled mallow; hollyhock mallow; musk mallow: 1; running mallow; cluster mallow.*

malnutrition. An unhealthy condition resulting from either poor feed or lack of feed.

Malpighia. A genus of the family Malpighiaceae. Species are small trees or shrubs with opposite, simple leaves and red, purple, or white flowers. *Malpighia glabra* is grown for its acid fruit used in preserves, some of the other species being grown as ornamentals. Native to tropical America.

mal secco. A serious disease of citrus trees in Italy, Greece, Turkey, and Palestine caused by the fungus *Deuterophoma tracheiphila,* family Sphaerioidaceae, characterized by a sudden wilting and drying of leaves and twigs.

malt. A grain product, rich in protein and carbohydrates, made by allowing the grain to sprout for a sufficient length of time to produce adequate amounts of enzymes and then dried. The term *malt,* unqualified, implies barley malt *q.v.*

Malta fever. Undulant fever.

malt cleanings. A livestock feedstuff obtained from the cleaning of malted barley or from the recleaning of malt which does not meet the minimum protein standard of malt sprouts.

malted barley. See *malt.*

malted milk (dried). The food product obtained by combining whole milk with the liquid separated from a mash of ground barley malt and wheat flour, with or without the addition of sodium chloride, sodium bicarbonate, and potassium bicarbonate, in such a manner as to secure the full enzymatic action of the malt extract. Excess water is removed and the resulting product contains not less than 7.5 percent of butterfat and not more than 3.5 percent of moisture.

Maltese. A breed of asses, probably of Arabian origin, first imported into the U.S. from the Island of Malta at the end of the 18th century.

Maltese cross campion. *Lychnis chalcedonica,* family Caryophyllaceae, a perennial herb grown for its scarlet flower clusters. Native to Siberia. Also called *Jerusalem cross; scarlet lychnis; scarlet lightning.*

Maltese mushroom. *Cynomorium coccineum,* family Cynomoriaceae, a parasitic, fungus-like, fleshy plant of the Mediterranean region.

malt extract. A sugar substance obtained from the filtrate of a mixture of malted barley and grain.

malt hulls. A livestock feedstuff consisting almost entirely of hulls obtained in the cleaning of malted barley.

Malthusian theory. A theory or doctrine, expounded by Thomas R. Malthus (1766–1834), an English clergyman, according to which human population tends to increase at a greater rate than food supply. The corollary is that unless the birth rate is controlled, poverty and threat of famine are inevitable.

malting. The germinating of moistened barley grains especially for brewing or distilling.

malting barley. In marketing, a grade of barley which is six-rowed barley, of the class Barley, meeting the requirements of grades one to three, inclusive, which after the removal of dockage contains not more than 5 percent of two-rowed and/or other types or varieties of barley of unsuitable malting type, such as Trebi and Black; which contains not more than 15 percent of barley and other matter that will pass through a 20-gage metal sieve with slotted perforations 0.076 (4⅞/64) of an inch wide and ¾ in. long; which contains not more than 4 percent damaged barley; and shall not include Bleached barley. Barley of this class shall contain 75 percent or more of mellow barley kernels which kernels are not en masse, semi-steely. (Market standards set by the U.S. Department of Agriculture.)

maltose. $C_{12}H_{22}O_{11} \cdot H_2O$, a sugar formed by the action of diastase on starch. Also called *malt sugar.*

maltose process corn gluten feed. A livestock feedstuff which is the dried residue from degermed corn after the removal of the starch in the manufacture of malt sirup.

maltose processed corn. A livestock feedstuff which is the dried residue from cleaned, whole ground corn after the removal of the starch in the manufacture of malt sirup.

malt sprouts. A livestock feedstuff obtained by the removal of the sprouts from malted barley together with the malt hulls, and other parts of malt and foreign material unavoidably present. It should contain not less than 24 percent of protein. The term *malt sprouts* when applied to a corresponding portion of other malted cereals shall be used in qualified form: *e.g.,* "Rye Malt Sprouts," "Wheat Malt Sprouts," etc.

maltster. A person who makes, or buys and sells malt.

malt sugar. Maltose.

malt vinegar. A vinegar which is made from a mash of malt, barley, and oats. After fermentation, the wort may be treated with gelatin to precipitate suspended matter, with an enzyme to get rid of soluble proteins which may cause cloudiness in the finished product. The wort is then filtered and run through generators in the same manner as cider vinegar.

malt-worm. A weevil which infests malt.

malty flavor. A flavor sometimes present in milk

and cream caused by the growth of *Streptococcus lactis* var. *maltigenes.* The flavor defect may also be transferred into butter made from such milk or cream.

Mame cherry. Fuji cherry.

mamey. *Mammea americana,* family Guttiferae, a tree grown in very warm areas for its edible, sweet fruit. Native to the West Indies. Also called *mammee-apple.*

mammal. Any animal which suckles its young.

mammary gland. Udder or breast. In female mammals, a milk-secreting gland for the nourishment of the young. It is rudimentary in male mammals.

mammary system. The udder, blood vessels, and teats of an animal, especially of the dairy cow.

mammary vein. Any of the numerous, long, tortuous, prominent, large veins or blood vessels found just under the skin along the belly to the udder of a dairy cow.

mammee-apple. Mamey.

mammitis. Mastitis.

mammoth clover. A variety of red clover (*q.v.*) which is a very important forage crop.

Mammoth Jack. A distinct breed of ass developed in the U.S. in which the Catalonian Ass was an important influence. The *Mammoth Jack* is larger and heavier boned than other breeds of ass and for this reason is better suited for use in mule breeding. Also called *American Jack.*

mammoth wildrye. *Elymus giganteus,* family Gramineae, a grass, developed from a Siberian grass by the Soil Conservation Service (U.S.), and used to still sand in the Pacific Northwest. Also called *Volga wildrye.*

mamoncillo. Honeyberry.

manage. 1. Horse-training with regard to paces. 2. A trained horse's action and paces; specifically, a brief gallop at full speed.

managed milking. Systematized milking in which each cow is milked in the minimum time to make the best use of the regular let down period of milk flow.

management. The human factor within a production process (or firm) which delimits problems, accumulates information relevant to their solution, analyzes that information, reaches decisions, acts on those decisions, and bears responsibility for the consequences of those actions. See *farm management.*

management of odd areas. In wildlife management, the development, protection, and utilization of vegetation on odd portions of farms and ranches.

management of pastures. The handling of animals and plants in such a manner that the stand of desirable forage is not depleted and at the same time provides a sufficient supply of palatable, nutritious forage for the producing or growing animal for as long a time as is possible during the grazing period.

management of spoil banks. The establishment of some form of vegetative cover on raw materials which are exposed by strip mining, ditch construction, dredging, etc.

management plan. 1. In forestry, a written scheme of operations for forest property. It usually records data and prescribes measures for the optimum use of all forest resources. 2. A written plan providing for the most effective operation of a farm using accepted principles of organization and management.

management returns. Farm operator's labor earnings less a charge for the operator's labor.

management wage. The net farm income less the sum of the charges for the operator's labor and interest on the capital investment.

Manchineel. *Hippomane mancinella,* family Euphorbiaceae, a small tree whose milky sap is a very strong irritant. Contact with the sap may result in a dermatitis. Native to America. Also called *manzanillo.*

Manchurian crabapple. *Malus baccata* var. *mandshurica,* family Rosaceae, a hardy, deciduous tree grown for its fragrant, white flowers and yellow or red fruit. Native to China.

Manchurian honeysuckle. *Lonicera ruprechtiana,* family Caprifoliaceae, a hardy, tall, deciduous shrub grown in the shrub border or as a specimen plant for its white flowers fading to yellow and its red or orange-colored fruit. Native to Asia.

mandarin. The variable group of trees of the species *Citrus nobilis* var. *deliciosa* (*C. reticulata* var. *deliciosa*), family Rutaceae, which yield a yellow, loose-skinned, edible fruit, the mandarin orange of commerce. See *orange; tangerine; King orange.*

mandarin-lime. Either of the varieties Rangpur and Kusaie, hybrids from the mandarin and lime, commonly called limes, which if used at all are used as limes, but which probably do not properly belong with the lime species. The trees are frost-resistant and bear heavily.

mandible. The bone of the lower jaw; the lower jaw.

mandioca. Cassava.

mandrake. 1. *Mandragora officinarium,* family Solanaceae, a stemless, perennial herb grown for its greenish-yellow flowers. To the medieval European, the root resembled a man and superstitions of magic became attached to it. Native to Europe. 2. See *mayapple.*

mane. Long, heavy hair which grows about and on the upper side of the neck of the horse and some other animals.

man equivalent. The average number of men working on a farm for a year, obtained by dividing by 12 the total number of months that the operator and all paid and unpaid laborers have worked.

manettia. *Manettia bicolor,* family Rubiaceae, a woody vine grown in a greenhouse for its red, yellow-tipped flowers. Native to Brazil.

Manetti rose. See *Chinese rose: 1.*

manganese. Mn, a metallic element, found in soils (determined as manganous oxide) from a mere trace to as much as 15 percent, but when present it is often in forms unavailable to plants. It is regarded as essential to normal plant growth and is often applied, usually as manganous sulfate, to soils deficient in this element. It is also essential to growth, reproduction, and lactation in animals.

manganese arsenate. An inorganic, arsenical stomach poison containing about 40 percent arsenic pentoxide, used as an insecticide.

manganese chlorosis. A discoloring disease of plants caused by lack of manganese.

manganese deficiency. A nutrition deficiency which results in dwarfing of plants, chlorosis of the upper leaves often becoming spotted. Crops are frequently unmarketable. In animals, it causes perosis (slipped tendon) in poultry and impaired growth in swine.

manganese sulfate. $MnSO_4$, a chemical, sometimes used as an ingredient of mixed fertilizer. See *manganese.*

mange. A contagious skin disease caused by mange mites. See *scab mite.*

mangel wurzel. *Beta vulgaris* var. *macrorhiza,* family Chenopodiaceae, a biennial herb with large roots, grown for stock feed in northern U.S., Canada, and Europe. Native to Europe and Asia. Also called *mangel.*

mange mite. Scab mite.

manger. A trough or other receptacle of metal, wood, concrete, or stone, in which fodder is placed for cattle to eat.

mango. *Mangifera indica,* family Anacardiaceae, a fruit tree, grown throughout the tropics for its edible, large, usually red or yellowish-orange, aromatic, very juicy fruit. In the U.S. its culture is limited to California and southern Florida. Native to India and Southeast Asia.

mango melon. *Cucumis melo* var. *chito,* a variety of muskmelon, occasionally grown for its fruit useful for making pickles and preserves. It is not edible in its natural state. The fruit has the color and shape of an orange or lemon with a white or pale yellow, cucumber-like flesh and has no muskmelon odor. Also called *orange melon; melon apple; vine peach; garden lemon; vegetable orange.*

mango squash. Chayote.

mangosteen. *Garcinia mangostana,* family Guttiferae, a handsome, tropical tree grown for its leathery, shining green leaves, rose-pink flowers and reddish-purple fruit with juicy, white, orange-like segments. Native to India and Southeast Asia.

mangrove. See *black mangrove.*

mangrove borer. See *Australian-pine borer.*

man hour. The unit of work performed by one man in one hour.

Manila hemp. See *abaca banana.*

Manila rope. Rope manufactured from Manila hemp.

Manila tamarind. Guamachil apes-earring.

manilla. See *abaca banana.*

manioc. Cassava.

man-killer. A vicious horse which will attack its handlers.

man labor. Any work performed on the farm by the hired man, the operator himself, or members of his family.

man-land ratio. The ratio which exists between number of human beings and acres of land inhabited by them in a particular geographical area.

manmade group. A number of persons brought together for a specific purpose under a hired or elected leader.

manna. Sweet exudate of plants; honeydew, especially of the flowering ash *q.v.*

manna ash. Flowering ash.

mannagrass. Any member of the genus Glyceria, family Gramineae, tall, perennial marsh grasses fairly palatable to livestock but not widely distributed. Native to North America. See *fowl mannagrass.*

mannagrass rust. A rust caused by *Puccinia rubigo-vera agropyrina,* family Pucciniaceae, which appears as red pustules mainly on the leaves but sometimes on the stems. It may also attack wheat grass, brome, alkali-grass and other plants.

manna gum. Ribbon eucalyptus.

mannan. Any of the simple polysaccharides found as reserve carbohydrate in seeds and wood pulp.

man-of-the-earth. Bigroot morningglory.

manorial system. Feudal land holding including the exploitation of serf labor by feudal lords in the Middle Ages.

man-planted. In forestry, designating a tree planted by man, as contrasted with natural growth.

manroot. Bigroot morningglory.

Manson's eye worm. *Oxyspirura mansoni,* a slender, colorless roundworm about ¾ in. long. It is a parasite found under the transparent membrane over the eyeball of the fowl. Infestation causes puffiness around the eye, continuous winking, and sometimes blindness.

Manteca. An Italian butter (usually whey butter) enclosed in a plastic bag.

mantis. Any of the large insects, family Mantidae, predaceous on other insects, capturing their prey with their unusually well-developed front legs. Species are considered beneficial in gardens. See *praying mantis.*

manufacturing milk. Raw milk produced or used for the manufacture of dairy products, such as cheese, butter, powdered milk, etc. It may or may not be of lower quality than milk used for bottled milk and may sell for less.

manufacturing tobacco. One of the general classes of tobacco. It includes the Virginia sun cured, white burley, flue cured, and dark fire cured.

manuka. Broom tea-tree.

manuka tea-tree. Broom tea-tree.

Manur. A Serbian cheese made from cows' or ewes' milk to which a mixture of buttermilk, whey, and rennet is added.

manure. 1. Excreta of animals, dung and urine

(usually with some bedding), used to fertilize land. 2. In Europe, any material which contains the essential elements of plant food, as chemical fertilizers, excreta of animals, etc. See *compost.*

manure bucket. A hydraulic attachment for a tractor, used to lift and load manure. Also called *manure scoop.*

manure cellar. A cellar in a barn into which manure is dropped for storage through a trapdoor in the floor.

manure cistern. A large underground container used to receive liquid manure.

manure fork. A pitchfork, with 4 or 5 heavy tines, designed for handling barn manure.

manure pit. A large pit used for storage and preservation of barnyard manure.

manure salts. Crude salts in natural deposits, containing a high percentage of potash (K_2O) in the form of the chloride. Sometimes used as fertilizer. See *sylvanite.*

manure spreader. A wagon-type implement for carrying barnyard manure to the field, shredding it and spreading it uniformly on the land. The power for the spreading mechanism and conveyor is supplied from the rear wheels or from a tractor power take-off.

manurial. Concerning manure; manure-type.

man with dogs. An inefficient shepherd (New Zealand).

man work unit. The average amount of work accomplished by one man in 10 hours at the usual farm tasks under ordinary conditions.

manyflower daylily. *Hemerocallis multiflora,* family Liliaceae, a bulbous herb grown for its numerous, small, orange flowers. Native to China.

many leaves. The third stomach compartment of a ruminant; the omasum.

manyplies. See *omasum.*

manzanita. Any shrub or small tree of the genus Arctostaphylos, family Ericaceae. Species are used as ornamentals. See *woolly manzanita; bearberry.*

maori mite. Citrus rust mite.

map. A chart of any portion of the earth's surface which shows its location, shape, and extent of physical, political, and other features. See *soil map.*

maple. Any tree or shrub of the genus Acer, family Aceraceae. Species are grown as ornamentals, for shade, lumber, and for the sap which yields *maple* sirup and *maple* sugar. See *bigleaf maple; box elder; Japanese maple; Norway maple; planetree maple; red maple; silver maple; striped maple; sugar maple; vine maple.*

maple and oak twig pruner. *Hypermallus villosus,* a white, cylindrical, conspicuously segmented grub which infests maple and oak, girdling twigs.

maple bladder gall mite. *Vasates quadripedes* (Shimer), family Eriophydae, a mite which infests maples causing red (turning green and black) bladder-like galls on the leaves.

maple callus borer. *Sylvora acerini* (Clem.), family Aegeriidae, an insect which infests maples. It works in new callus tissue and girdling the tree, kills it.

maple cream. The sap of the sugar maple, processed to a consistency between sirup and crystalized sugar; used as a spread. Also called *maple butter.*

maple leaf aphid. Woolly alder aphid.

mapleleaf cutter. *Paraclemensia acerifoliella* (Fitch.), family Incurvariidae, a caterpillar that mines the leaves of maples.

mapleleaf viburnum. *Viburnum acerfolium,* family Caprifoliaceae, a hardy shrub grown for its white flowers and black-purple fruit. Native to North America. Also called *dockmackie.*

maple petiole borer. *Caulocampus acericaulis* (MacG.), family Tenthredinidae, a sawfly larva which tunnels in the leaf petiole of maples causing the leaf to drop. There is only one generation a year and the injury is mild.

maple sirup. A delicious, sweet, concentrated food product made in the spring by boiling the natural sap of the sugar maple tree (*Acer saccharum*) to the proper concentration. Further concentration will cause crystallization of the sirup into maple sugar.

maple sugar. A crystallized, sweet food product, largely saccharose in nature, made by evaporating the sweet sap of certain maple trees. A food delicacy now, at one time it was an essential human food in some parts of the U.S.

maple trumpet skeletonizer. *Epinotia aceriella* (Clem.), family Olethreutidae, a larva that infests maples, folding the leaves to form trumpet-like tubes, and living inside the fold near the skeletonized part of the leaf.

Maquée. A soft, cows' milk cheese made in Belgium.

maqui. *Aristotelia macqui,* family Elaeocarpaceae, an evergreen shrub with greenish flowers, grown in California, U.S., as an ornamental. Native to Chile, where a medicinal wine is obtained from its berries. Also called *Chilean wineberry.*

maracuja melao. Giant granadilla.

marasca sour cherry. *Prunus cerasus* var. *marasca,* a pomological group and a variety of the sour cherry (*q.v.*) that has small fruit with red colored juice; formerly used in making maraschino cherries. Also called *Dalmatian cherry.*

maraschino cherry. A firmed, artificially colored and flavored cherry usually made from light colored, sweet cherries, but which may be made from any variety. The cherries are sulfur-bleached and firmed with calcium, desulfured, sweetened, dyed and flavored.

marbled. Mottled or streaked, like certain kinds of marble.

marbling. The desired distribution of the fat in the muscular tissue of a cut of meat which gives it a spotted appearance.

marc. The insoluble residue left after processing fruits, sugar cane, and sugar beets.

marcescent. Designating leaves which wither on a plant without falling off.

Marches. A ewes' milk, hard cheese made in Italy.

marcott. A type of layerage (*q.v.*) practiced in sub-tropical Asia which consists of girdling a selected branch or twig and covering the girdled part with a ball of moist clay, moss, or coconut fiber. Also called *marcottage.*

Mardin iris. *Iris mesopotamica,* family Iridaceae, a perennial herb grown for its lavender-blue flowers. It is an important stock for garden varieties in mild climates. Native to Syria, U.A.R.

mare. A mature female horse.

mare and foal. In the show ring, a brood mare with foal at side.

mare class. In the show ring, draft mares 4 years of age or older.

mare mule. A female mule.

mare's-tail. Field horsetail.

mare with own foal. In the show ring, a mature female horse and her foal.

margarine. Oleomargarine.

Margarita spirea. *Spiraea margaritae,* family Rosaceae, a hybrid, deciduous shrub grown for its bright pink flowers.

margin. 1. The amount deposited by a client with his broker to protect the broker against losses on contracts being carried or to be carried by the broker. A *margin* call is a request to deposit either the original *margin* or to maintain *margin* with added cash. 2. The edge, border or border line, as *margin* of a leaf.

marginal analysis. The analysis of the effect on costs and returns associated with successive units of input applied to one or more fixed factors.

marginal factor cost (MFC). The increase in total cost associated with using an additional unit of a factor of production. When inputs are homogeneous and input markets are perfectly competitive, *MFC* is the price of the factor in question. Also called *marginal resource cost.*

marginal grower. One whose farm income from production over a period of years is not likely to exceed the farm costs for a particular enterprise or for the entire farm.

marginal land. Agricultural land definitely not first class, and near the margin between profitable and unprofitable use. See *submarginal land; supermarginal land.*

marginal leaf roll. A symptom of a plant disease in which the margins of the leaves or leaflets curl upward especially in the upper portion of the plant.

marginal physical product. The addition to total physical product occasioned by the increase in quantity of a variable factor of production by one unit. It may also be represented by the slope (first derivative) of a production function as the size of the factor or production unit approaches zero.

marginal rate of substitution. The amount of an input saved by the addition of one unit of a second input with production maintained at the previous level. It may also be represented by the slope (first derivative) of the iso-product relation.

marginal return. See *marginal value product; marginal physical product.*

marginal shield-fern. Leather woodfern.

marginal value product (MVP). The increase in the value of the total product associated with each unit of an input. Also called *marginal revenue product (MRP).*

marginal variegation. Narrow or broad zones of pale tissue which mark the edges of leaves or leaflets.

margined blister beetle. *Epicanta pestifera* Werner, family Meloidae, a black beetle which feeds on a wide variety of plants.

margin of profit. See *marginal analysis.*

Marguerite chrysanthemum. *Chrysanthemum frutescens,* family Compositae, a tender, very beautiful, branched herb grown for its daisy-like, white or pale yellow flower heads. Native to the Canary Islands. Also called *Boston yellow daisy.* See *chrysanthemum.*

Marguerite fly. Chrysanthemum leaf miner.

Marienhofer. A Limburger-type cheese made in Austria.

marigold. Any tender, annual herb, of the genus Tagetes, family Compositae. Species are grown in the flower garden. See *Aztec marigold; French marigold; sweet marigold; striped marigold.*

marijuana. See *hemp.*

marine climates. Climatic types determined by oceans. The climate of oceanic islands and of windward shores of continents, contrasted with interior or continental climates, which usually have a greater range in daily and annual temperatures, *i.e.,* colder winters and warmer summers.

mariposa. Any cormous plant of the genus Calochortus, family Liliaceae. Species are useful as border plants and occasionally are grown indoors. Native to western North America. Also called *mariposa lily; star tulip; globe tulip; mariposa tulip; butterfly lily.* See *white mariposa; skyblue mariposa; clubhair mariposa; yellow mariposa; sagebrush mariposa; Tolmie mariposa; charming mariposa; lilac mariposa; Gunnison mariposa.*

maritime pea. Maritime peavine.

maritime peavine. *Lathyrus japonicus* (*L. maritimus*), family Leguminosae, a fleshy, perennial herb grown in a seaside garden for its purple-violet to bluish flowers. Native to Japan. Also called *maritime pea.*

marjoram. Origanum.

mark. To place a sign on an animal, product, etc., for identification by attaching a light metal tag, notching the ears, etc. See *marking.*

marker. 1. In lumbering, one who performs any of the following: marking logs for identification; marking the boles of trees into log lengths for the buckers; scaling, marking and tallying sawn lumber at a portable sawmill; assessing, and marking by a symbol, the grades of pieces of lumber; on logging operations, marking those standing trees to be cut. 2. In range management, a colored or otherwise marked sheep in a range band. 3. A horse or man-drawn implement used to mark off rows before planting seeds by hand (obsolete).

market. 1. A place, usually public, where goods and commodities are offered for sale or exchange. 2. The demand for a product; or a country, region, etc., in which there is such a demand. 3. To buy and sell produce.

marketable. Designating a product, animal, etc., which is of sufficient quality and demand that it has a sale or exchange value.

market barrow. A castrated male hog finished or fattened for meat purposes.

market bird. Any fowl produced and fattened for meat purposes.

market cattle. In the show ring, classes of cattle fitted or fattened for marketing.

market classes and grades. Various market classes and grades established by the U.S. Department of Agriculture to sort livestock according to conformation, finish, quality, use, age, sex and weight.

market demand. The requirements or need of the market at a particular time.

market-fat. Sufficiently fat to meet the requirements of the market.

market fluctuation. Variation in prices which are received or paid for products placed on the market.

market garden. A farm which produces vegetables to be sold on a market in a nearby city. Also called *market garden farm.*

market gardener. Market garden farmer.

market garden farm. Market garden.

market garden farmer. A farmer engaged primarily in the production of vegetables to be sold on a nearby market. See *truck farmer.*

market garden grower. Market garden farmer.

market gardening. The raising of vegetables for local markets. See *market garden.*

market glut. An oversupply of a particular product to such an extent that the market has difficulty in absorbing it.

market grade. A set of descriptive terms, such as prime, choice, etc., used in livestock marketing to designate the comparative value of animals based on differences in type, conformation, degree of finish etc. See *grade: 5.*

market grades of carcass animals. See *prime; choice; utility: 1.*

market grades of cows and bulls. See *commercial; utility: 1; cutter: 1; canner: 1.*

market grades of lambs. See *prime; choice; utility: 1; canner: 1; cutter: 1; cull: 2.*

market grades of slaughter steers and heifers. See *prime; choice; utility: 1.*

market grades of stockers and feeder cattle. See *fancy; choice; medium: 4; common: 3.*

market grades of swine. Various market classes of swine established by the U.S. Department of Agriculture: U.S. Number 1, U.S. Number 2, U.S. Number 3, medium and cull.

market grain. Any grain offered on the open market meeting the requirements of one of the official grades.

market hay. Hay which has the quality to meet the requirements of one of the several market grades set for hay offered for sale on the market.

market hog. A hog grown and fattened for the market rather than for show or breeding purposes.

marketing. The art of selling or purchasing on a market.

Marketing Agreement Act. An act of the U.S. Congress authorizing the Secretary of Agriculture to establish orderly marketing conditions for agricultural commodities to help achieve parity. It gives the Secretary power to enter into agreements with processors and handlers of any product affecting interstate commerce.

marketing control. See *Marketing Agreement Act.*

marketing order. An order issued by the U.S. Secretary of Agriculture which makes a marketing agreement binding upon all the handlers of a commodity within a specified area, regardless of whether they sign the marketing agreement.

marketing quotas. Under the U.S. Federal Crop Control Acts, the quantity of a commodity a farmer is allowed to produce on his land under acreage allotments.

marketing specialist. An individual who by technical training and marketing experience is well informed on all phases of marketing and renders service to others in the field of marketing. He is usually associated with the extension service of an agricultural college or with a marketing organization.

market lamb. Lambs raised and marketed for meat purposes.

marketmen. Individuals who make a business of buying and selling on the market.

market milk. Whole milk sold to the consumer in its fluid state, usually as a bottled or packaged product.

market order. An order to buy or sell which is to be executed at the best possible price as soon as received in the trading ring or pit.

market poultry. Poultry produced primarily for sale on the market for food purposes.

market price. The price for a product, animal, etc., established by buyers and sellers in competition in the market place.

market standard. The official standard governing weights, grades, and qualities of products offered for sale on a market, such as No. 2 corn, grade A eggs, etc., set by the U.S. Department of Agriculture. See *grade: 5.*

market trend. The movement in commodity prices on the open market: *e.g.*, up or down, strong or weak, stable.

market unit. The unit of measure usually designated for a particular commodity on an established market, as a bushel of wheat.

market value. The price at which a commodity will sell on the open market.

marking. Castrating, docking, branding and earmarking.

marking corral. The place where lambs are marked. See *marking shelf.*

marking shelf. A plank 2" x 10" x 3', 3½' to 4' from the ground, level and secured firmly. The lamb to be marked is placed on the shelf on its rump and held securely by the holders while the marker operates.

marking (timber). Selection and indication, usually by blaze or paint spot, of trees which are to be cut or retained in a cutting operation.

Märkisch Hand. A Hand cheese (*q.v.*) that is salted, placed in a linen sack, and heavily pressed. Pieces are cured.

marks feed. On a fertilizer drill, an inverted cone with a stationary, projecting lip beneath which a revolving plate brings the fertilizer to the lip and to the center of the cone and drops it through the spout to the soil.

marl. 1. A natural, earthy deposit, consisting chiefly of calcium carbonate mixed with clay or other impurities in varying proportions. 2. Incoherent sands containing shells. 3. Soft, impure limestone. 4. Soft, mud-like deposits originating mostly in lakes and found underlying bogs widely distributed in the glaciated region of the U.S. Sometimes called *bog lime. Marls* are used as an amendment for soils deficient in lime. See *lime: 1.*

Marlatt whitefly. *Aleurolobus marlatti,* a tiny insect which extracts sap from trees and deposits honeydew. A sooty mold fungus, injurious to the leaves, forms on the deposits. The fly infests citrus fruit in Japan.

marl chlorosis. Marl frenching.

marl clover. Zigzag clover.

marl frenching. Certain disease symptoms of citrus trees characterized by yellowing of foliage; attributed to deficiencies of manganese and zinc. Also called *marl chlorosis.* See *frenching: 2.*

marmalade. A thick jam, most often made from oranges, lemons, or grapefruit, which includes the fleshy portions of the fruit and the usually finely-sliced rind or peel. Other fruits, such as the plum, quince, guava, etc., are sometimes used to make a preserve called *marmalade.*

marmaladebox genip. *Genipa americana,* family Rubiaceae, a shrub or small tree grown in very warm areas for its edible fruit used in preserves and beverages. Native to the West Indies. Also called *genip; genipap.*

marmalade plum. Sapodilla.

marmor. Marbled or veined, a type of chlorosis.

Marolles. A soft, cows' whole, skimmed, or partly skimmed milk, French cheese, cured from 3 to 5 months.

maroon gaillardia. *Gaillardia amblyodon,* family Compositae, an annual herb grown for its brownish-red flowers. Native to North America.

marram grass. European beachgrass.

marrow. 1. A soft, yellow or red, fatty tissue that fills the cavities of most bones. 2. The pith of plants. 3. See *vegetable marrow.*

marrowbone. Any bone containing marrow. See *marrow: 1.*

marrowfat. Any of the several varieties of cultivated, large-seeded peas.

marsh. A flat, miry, treeless area commonly covered by standing water, with a native growth of grasses, rushes, and sedges. Also called *merse* (Great Britain). See *fresh-water marsh; salt-water marsh.*

marsh elder. Rag sumpweed.

marshfern. *Dryopteris thelypteris,* family Polypodiaceae, a fern grown for its graceful, erect fronds. Native to North America.

marsh grass. Any water-tolerant plant which grows on a marsh, including sedges and rushes as well as true grasses. Also called *water-grass.*

marshland. See *marsh.*

marsh hay. Any of the swamp or marsh grasses and sedges, usually coarse in nature and low in feeding value, that are sometimes cut for hay. Also used for packing crockery, glassware, etc., and for making grass ropes. These were the only sources of hay for early settlers in the forested regions of the U.S.

marshmallow. *Althaea officinalis,* family Malvaceae, a perennial, marsh herb grown for its roots, source of a mucilaginous paste formerly used in medicines. Native to Europe. Also called *sweat weed.*

marsh management. Manipulation of vegetation of marshlands for grazing and for the production of wildlife, crops, etc.

marshmarigold. *Caltha palustris,* family Ranunculaceae, a perennial, swamp herb grown for its bright yellow flowers and rarely for food. Native to North America and Europe. Also called *bull-flower; cowslip; gools; king cup; May-blob.*

marsh muhly. *Muhlenbergia racemosa,* family Gramineae, a perennial grass found in moist places from Newfoundland to Arizona, U.S. Its forage rating is good. Native to North America.

marshpepper smartweed. *Polygonum hydropiper,* family Polygonaceae, an annual herb, a widely distributed weed pest.

Marsh Pony. See *Tackey Pony.*

marsh rosegentian. *Sabatia dodecandra,* family Gentianaceae, an attractive, little, salt marsh herb grown in moist, salty sand for its pink flowers. Native to North America. Also called *rose pink; sea pink.*

marsh trefoil. Bogbean.

martagon lily. Turk's-cap lily.

martingale. A leather strap attached to the girth of a horse's saddle or harness on one end, with the other divided and passing between the front legs and tied to the nose band of the bridle. It stops the horse from rearing.

martin heifer. See *free-martin.*

Martynia. An herbaceous genus, family Martyniaceae. Species are allied to devil's-claw (*q.v.*). Native to Southeast Asia and Central America.

Marumi kumquat. *Fortunella japonica,* family Rutaceae, a shrub or small tree grown for its small, oblong, edible fruit chiefly used for preserves. Native to China. Also called *round kumquat.*

marvel-of-Peru. See *four-o'clock.*

Maryland dittany. Maryland stonemint.

Maryland figwort. *Scrophularia marilandica,* family Scrophulariaceae, an herbaceous bee plant grown for its greenish-purple flowers. Escaped, it is a weed. Native to North America.

Maryland golden aster. *Chrysopsis mariana,* family Compositae, a perennial herb grown in sandy soils for its yellow flowers. Native to North America. Also called *golden star.*

Maryland stonemint. *Cunila origanoides,* family Labiatae, a perennial, branching herb grown for its purplish-pink flowers. Native to North America. Also called *Maryland dittany.*

Maryland-style corn. A method of commercial canning of corn in which the whole kernels are cut, and not scraped, from the cob and canned in brine.

Mascarpone. A soft, cream cheese made in Italy. It is eaten fresh.

masculine head. In cattle, the brightness of eye, strong muzzle, horn character, and neck attachment which shows the vigor, alertness, and strength associated with a male.

masculine sex character. In male animals, vigor, body build, alertness and aggressiveness; desirable characteristics in animals such as bulls and stallions.

masculine style. In stock judging, the desirable characteristics of a male animal: *i.e.*, up-headed, straight through the muzzle, wide between the ears, bold and clear in the eyes, wide in body conformation and vigorous in appearance.

mash. A ground feed of cereals and malt, etc., fed in a wet or dry form to livestock and poultry. Also called *crowdy.* See *dry mash; wet mash.*

mash concentrate. A poultry mash containing 20 to 40 percent protein.

mash-grain system. In poultry feeding, the allowing of free access to a mash and the controlling of the amount of a grain mixture in the daily ration. It is a compromise between the all-mash and cafeteria systems.

mash hopper. A special container for mash to feed poultry and livestock. It comes in different shapes and sizes.

mashing. The converting of starch in cereals, potatoes, etc., into sugar by the action of amylase enzymes of malt, and the accompanying breakdown of proteins by the protease enzymes of the malt, in the manufacture of alcohol or alcoholic beverages.

mash supplement. Protein, vitamin, and mineral substances used to balance the cereal base of the mash. See *mash.*

masked. Designating disease symptoms that are conditionally hidden.

masonry dam. A dam built of rock and mortar.

massive. Indicating a large-sized and deeply-muscled animal; designates an ideal quality in a draft horse.

massive (soil structure). Large uniform masses of cohesive soil which sometimes have irregular cleavage, as in the C horizons of many heavy clay soils.

mass movement. Unit movement of a portion of the land surface as in a landslide, earthflow, or subsidence.

mass selection. 1. In plant breeding, the selection of a large number of plants for propagation from which the off-type, low yielding, inferior, and disease-susceptible plants have been eliminated. 2. In animal breeding, the selection for breeding purposes of animals on the basis of their individual performances, type or conformation.

mast. Nuts collectively, such as acorns, beechnuts, chestnuts, and seeds of certain pines as piñon, especially when considered as food for domestic livestock and deer.

mastagium. An Old English term for the right granted to turn hogs into the King's forests to feed on acorns and beechmast.

Master Farmer. A recognition for outstanding farmers in the U.S., started by the journal, *Prairie Farmer,* and followed by cooperation from the Agricultural Extension Service of the Land-Grant colleges and universities. Certain farmers who have done notable work in farming and represent a high type of citizenship are honored as *Master Farmers.*

masterwort. See *cowparsnip.*

mastic. 1. *Pistacia lentiscus,* family Anacardiaceae, an evergreen, resinous shrub or small tree, grown for the gum it exudes, the *mastic* of commerce, used in astringents, in making varnish, etc. 2. A distilled liquor flavored with *mastic.* 3. See *California peppertree.*

mastica. A plastic material used in the place of putty for glazing greenhouse glass and hotbed sash.

masticate. To chew; to prepare food for swallowing and digestion.

mastic jungleplum. *Sideroxylon mastichodendron,* family Sapotaceae, one of a group of timber trees known as ironwood because of their hardness and strength. Native to the West Indies and Florida, U.S.

mastictree. See *California peppertree; mastic: 1.*

Mastigosporium leaf fleck of grasses. A fungous disease caused by *Mastigosporium rubricosum,* family Moniliaceae, characterized by purplish-brown flecks on the leaves of many grasses.

mastitis. An infectious or non-infectious inflammation of the udder. All domestic female animals are susceptible, but it is most common among cows, ewes, and goats. It causes serious economic loss. Symptoms include decreased production, varying degrees of abnormal milk, heat, pain, and swelling of the udder, followed by a permanent firmness (fibrosis) of affected parts. Also called *garget; felon quarter; weed; bad quarter; mammitis.*

mat bean. Moth bean.

match plank. In forestry, a plank from which match stick blocks are cut.

mate. 1. To pair off two animals of opposite sexes

for reproduction. Mating may be for a single season or for life. 2. In plants, to be cross pollinated.

maternal inheritance. Inheritance from mother to offspring unaffected by inheritance from the father.

maternal instinct. The natural instinct of all female mammals to nurse and protect their young.

maternity pen. A special pen in a barn where animals about to bring forth their young may be isolated from the rest of the herd. It should be of good size, well bedded, warm, and with good sanitation.

maternity stall. Maternity pen.

mat-grass. Prostrate knotweed.

matilija-poppy. *Romneya coulteri,* family Papaveraceae, a tender, perennial herb grown in the flower border for its white flowers. Native to North America. Also called *California tree poppy.*

mating. See *mate.*

mating flight. The flight of the queen bee during which she is mated for the one time in her life by one of the many accompanying drones.

mating season. The season of the year when animals naturally breed and when conception is normally high. It varies with different species.

mat lovegrass. Lehmann lovegrass.

matrimonyvine. *Lycium halimifolium,* family Solanaceae, a spreading shrub grown on a fence or arbor as an ornamental for its dull purple flowers and red fruit. Native to southeast Europe and western Asia. Also called *false tea plant; boxthorn.*

matroclinous. Resembling the female parent.

matron. A mare that has produced a foal.

matronly quality. The quality and refinement desired in a brood mare.

mat sandbur. *Cenchrus pauciflorus,* family Gramineae, an annual grass considered a weed. It produces burs which easily drop off or cling to clothing of man and to the hair of passing animals. Native to North America. Also called *loose bur; sandspear; bur grass.*

matted growth. An interlaced, tangled, dense growth of vegetation.

matted row system. A system of growing strawberries in which the plants are allowed to become interlaced within the row, as contrasted with the hill system.

matter. 1. Purulent discharge; pus. 2. To discharge *matter;* to generate pus.

mattock. A hand tool which consists of a short handle with a two-bladed steel head. One blade, which is perpendicular to the handle, is flat and is used for grubbing. The second blade may be pointed like a pick or it may be flat and ground to a sharp edge like an ax for cutting.

mattress. A blanket of brush or poles interwoven or otherwise lashed together and placed to cover an area subject to river bank erosion. It is weighted with rock, concrete blocks, or otherwise held in place.

maturation. 1. Becoming mature or ripe. 2. Changes in cell division, especially in reproductive elements, in which the number of chromosomes in the nucleus of the new cells is half that of the original.

mature. Designating any plant, plant product, or animal that has reached its full stage of development.

mature class. In the show ring, dairy cows 5 years of age or older that are in milk.

mature equivalent factor. An age conversion formula used to predict the expected mature milk record for a cow, based on a previous year's milk production. It is frequently used in cattle selection and breeding.

mature forest. A forest that has reached its age of utilization.

mature mutton. In marketing, a sheep carcass used for meat which comes from sheep 20 or more months old.

mature soil. A soil which has well developed characteristics produced by the natural processes of soil formation and which is in equilibrium with its environment.

maturity. 1. A state of full or complete growth development or ripeness. See *mature.* 2. The ability of a plant or plant part to withstand the cold.

maund. A basic weight generally equivalent to 82 lbs. avoirdupois, but varying from this standard in some states (India).

mauritius bitterorange. *Citrus hystrix,* family Rutaceae, a low tree or shrub which produces a fruit containing an insecticidal agent. Native to Asia. Also called *cabuyao.*

mauritius hemp. Piteira furcrea.

maverick. 1. Any unbranded animal, particularly a calf. 2. A motherless calf (western U.S.).

maverickers. Men who stole and put their brands on unbranded cattle (western U.S.).

maxilla. The jawbone, particularly the upper one in vertebrates. In insects and arthropods, the two accessory jaws or appendages immediately behind the mandibles.

maximum temperature. In plant physiology, the temperature beyond which growth ceases.

Mayapple. *Podophyllum peltatum,* family Berberidaceae, a perennial herb grown for its white flowers and edible, yellow fruit. Native to North America. Also called *American mandrake; Indian apple; mandrake.*

May beetles. Species of the genera Phyllophaga or Lachnosterna, family Scarabaeidae, the adults of the white grub (*q.v.*) which are brown or brownish black, appear May to July, fly at night, and are attracted by lights. They are widely distributed. Also called *June bugs; daw bugs.*

May-blob. Marshmarigold.

Mayflower. Roundlobe hepatica.

May-flowering tulip. Cottage tulip.

May lambing. Spring lambing which comes in early May.

maypop passionflower. *Passiflora incarnata,* family Passifloraceae, a tendril-climbing vine grown for its habit, its white flowers, and its yellow, edible fruit. Native to North America. Also called *wild passion flower.*

May troubles. The difficulties encountered in the spring by milk manufacturers, due to chemical and bacteriological changes in milk when the cows go from dry feeds of winter to pasture.

mayweed camomile. *Anthemis cotula,* family Compositae, an ill-smelling, yellowish-green, annual weed. Native to Europe and Asia. Also called *dog fennel; English camomile.*

Mazoe lemon. A lemon found growing wild along the streams by English settlers in Southern Rhodesia; identical with the Florida rough lemon which was introduced into Florida by the Spanish.

mazun. A fermented milk similar to yoghurt, leben, and giodder. Also called *mazoum; matzoon.*

mazus. *Mazus japonicus,* family Scrophulariaceae, a prostrate, perennial herb grown as an ornamental ground cover, for its habit and its blue, brown-spotted flowers.

mazzard cherry. 1. *Prunus avium,* family Rosaceae, a tree which bears edible, sweet fruit, more commonly known as the sweet cherry. Native to Europe and Asia. 2. A pomological group and a variety of *P. avium* with inferior fruit, whose seedling stock is used as an under-stock for horticultural varieties of cherries. Also called *mazzard stock.*

mazzard stock. See *mazzard cherry: 2.*

McCormick's reaper. One of the earliest machine reapers which caused a revolution in the harvesting of small grains. See *reaper.*

McIntosh. A hardy, aromatic, late variety of bright red, striped apple, with thin skin and tender, white flesh. Grown in Canada and northern states of U.S.

McLean County Sanitation System. A system developed in McLean County, Illinois, by veterinarians of the U.S. Bureau of Animal Industry for the prevention and control of filth-borne diseases of swine. It has particular reference to internal parasites.

McLeod tool. A combined hoe-rake (U.S.).

McNary-Haugen Bill. A proposed measure for assuring producers of certain staple farm crops a fair profit, whereby a government corporation would be created to buy surplus farm products and sell them abroad. Any losses thus sustained would be met from the proceeds of an equalization fee. The bill was twice passed by the U.S. Congress in 1927 and 1928, but vetoed by President Coolidge.

mead. An alcoholic beverage prepared from honey.

meadow. 1. A tract of land on which grasses and legumes are grown, primarily used for hay and pasture. 2. Natural areas of moist land which support grasses and other herbaceous plants in mountain regions. 3. Alluvial lowland which borders streams (not precisely distinguished from marsh and swamp). See *mountain meadow; beaver meadow.*

meadow barley. *Hordeum brachyantherum,* family Gramineae, a perennial bunchgrass with scant foliage and short awns, found in moist areas from Alaska to New Mexico. It is grazed closely by livestock until time of heading, but its forage value is only fair.

meadow brake. Sensitivefern.

meadow fescue. *Festuca elatior,* family Gramineae, an upright, tufted, perennial grass grown for pasture and hay. Also called *alta fescue; tall fescue; English bluegrass.*

meadow fescue crown rust. See *crown rust.*

meadow foxtail. *Alopecurus pratensis,* family Gramineae, an upright, perennial grass resembling timothy, used in wet areas, meadows and pastures, especially in the Pacific Northwest and some other parts of northern U.S. Native to Europe and Asia.

meadow grass. Any grass which grows in a meadow.

meadow grasshopper. See *grasshopper.*

meadow leafhopper. See *leaf hopper.*

meadow lily. See *Canada lily.*

meadow mixture. Any of several grass and clover seed mixtures sown to produce a meadow.

meadow mouse. A rodent of the genus Microtus which is larger than the common mouse but smaller than a rat. In fields it forms paths in the grass which connect holes and tunnels. It also damages many trees by girdling them.

meadow nematode. Species of the genus Pratylenchus, family Tylenchoidae, a widely-distributed, wormlike root nematode found in the cortical tissue of the roots causing sloughing off and the formation of fine roots near the surface of the ground. It infests many trees, such as apple, cherry, walnut, almond, and causes brown root rot of tobacco, alfalfa, cotton, and other plants.

meadow oat grass. Tall oatgrass.

meadow pine. Field horsetail.

meadow pink. Maiden pink.

meadow pink orchid. *Habenaria fimbriata,* family Orchidaceae, a wild orchid grown for its fragrant, purplish-lilac flowers. Native to North America. Also called *purple fringed orchid.*

meadow redtop. See *redtop.*

meadowrue. Any perennial herb of the genus Thalictrum, family Ranunculaceae. Species are grown for their handsome, unusually large, terminal flower clusters. Native to Northern Hemisphere. See *columbine meadowrue; dusty meadowrue; early meadowrue; low meadowrue; tall meadowrue; Yunnan meadowrue.*

meadow saffron. Autumncrocus.

meadow sage. *Salvia pratensis,* family Labiatae, a hardy, perennial herb, an important bee plant grown in the flower garden for its bright blue flowers. Native to Europe.

meadow salsify. *Tragopogon pratensis,* family Compositae, a yellow-flowered weed. Native to Europe. Also called *star-of-Jerusalem; yellow goatsbeard.*

Meadow saxifrage. *Saxifraga granulata,* family Saxifragaceae, a bulbous herb grown for its white flowers and kidney-shaped leaves. Native to Europe and northern Africa.

meadow snakegrass. Stinkgrass.

meadow soil. A mineral soil which developed under a grass-sedge or herbaceous vegetation in a wet habitat, such as a mountain meadow. Also called *Wiesenboden.*

meadow spittlebug. *Philaenus leucophthalmus* (L.), family Cercopidae, an insect pest which infests clover, alfalfa and other forage crops, strawberries, and flowers in gardens and greenhouses.

meadow strip. A sloping strip or field of grassed land which, in addition to yielding a hay crop, acts as a broad, shallow water channel during periods of runoff. Often used as a terrace outlet channel; usually much larger in extent than a grassed waterway.

meadowsweet. *Spiraea alba,* family Rosaceae, a handsome, hardy, flowering, upright shrub grown for its white flowers. Native to North America. Also called *narrow leaf meadowsweet spirea.*

meal. 1. A coarsely ground, often unbolted grain, as corn *meal,* oat *meal.* 2. Any dried and ground feed material, as fish *meal.*

mealiness. A defect of cheese in which the texture lacks smoothness upon working between the fingers. See *mealy: 2.*

meal snout moth. *Pyralis farinalis* (L.), family Pyralidae, a reddish-brown to clear yellow moth which infests ground grains.

meal worm. See *yellow meal worm.*

mealy. 1. Designating buff or red plumage in which the ground color is stippled with a lighter color. 2. Designating a gritty texture, as in coarse corn meal. 3. Designating a condition in the flesh of ripe apples in which the protopectin that cements the cells together has been hydrolyzed so that the cells slough apart when the apple is bitten giving a soft, less juicy texture. 4. See *mealy potato.*

mealy (mealies). The South African equivalent of maize (*q.v.*). See *corn: 1.*

mealy bay. The rust color of a horse's coat.

mealy bellwort. Wood merrybells.

mealybug. Species of the genus Pseudococcus, family Pseudococcidae, widely distributed, small, elliptical, waxy-covered insects which infest various fruit trees and greenhouse plants.

mealy gray. The rusty blue color of a horse's coat.

mealy plum aphid. *Hyalopterus pruni,* family Aphidae, a pale green aphid with a gritty coating which infests the plum, peach, apricot, etc., causing curling of the foliage, fruit splitting, spoilage by excrement, and sooty mold on the fruit.

mealy potato. A potato which produces a light, flaky, uniform product when cooked.

mealy (soil structure). A crumblike soil structure in which the aggregates are of soft to very soft consistency, usually less than 5 mm. in diameter.

mealyworm. See *meal worm.*

mean annual increment. In forestry, the total growth of a stand of trees divided by the total age.

mean annual precipitation. The mean depth of moisture received from the atmosphere in one year.

mean annual rainfall. The mean depth in inches of the yearly totals of rainfall for a period of years in a specific locality. See *rainfall.*

mean annual temperature. The mean of the yearly maximum and minimum means of temperature for a period of years in a specific locality.

mean depth. The cross-sectional area of an open water channel stream divided by its surface width.

meander. One of a series of turns or bends of more or less regular size and shape in the course of a stream, formed where the stream winds back and forth on a flood plain.

meander line. The water line or shore of a body of water, as a large lake, established at the time of an official land survey, as that of the U.S. General Land Office. Legally, it is important in determining the ownership and use of shoreland.

mean diameter. 1. Usually the diameter corresponding to the average basal area of all trees in a stand. 2. In frequency distribution analysis, the arithmetic average of all diameters of trees in a stand. 3. Occasionally the geometric mean of all diameters of trees in a stand.

mean sample tree. Any tree which represents the average of a group of trees in any given class or stand.

mean sea level. The mean level of the sea at a given station; *i.e.,* the mean elevation of the tidal curve which varies somewhat from place to place.

Means grass. Johnsongrass.

mean velocity. The velocity at a given section of a stream, obtained by dividing the discharge of the stream by a cross-sectional area at that section.

measles. See *brown spot: 1.*

measles in beef. The presence of one or more cyst stages, in the musculature of a beef carcass, of *Taenia saginata,* a parasite of man. Under U.S. Federal Meat Inspection regulations, carcasses extensively infested with these cysts are condemned.

measles in pork. The presence of one or more cyst stages, in the musculature of a hog carcass, of *Taenia solium,* a parasite of man. Under Federal Meat Inspection regulations, carcasses extensively infested with these cysts are condemned.

measles of apples. Small, raised areas in the bark of apples producing a rough, scabbed appearance. Three different diseases have been termed measles: (a) internal bark necrosis attributed to excess manganese; (b) black pox attributed to the fungus, *Helminthosporium papulosum,* family Dematiaceae; (c) a pustular spot attributed to *Bacterium populans* (Rose), family Bacteriaceae. A similar bark condition results from oviposition by tree hoppers.

measles of peony. A fungous disease characterized by dark purple spots on the upper surface of the leaf, dull chestnut on the under surface, caused by *Cladosporium paeoniae,* family Dematiaceae. Also called *leaf and stem spot.*

measly pork. See *measles in pork.*

measurable precipitation. Any precipitation equivalent to over 0.005 in. of rainfall. Less than 0.005 in. is considered a trace.

measuring weir. A notch of regular form, usually sharp crested, used for measuring the flow of water. See *weir.*

measuring worm. See *inchworm.*

meat. 1. The edible flesh of an animal. 2. The kernel of a grain or nut. 3. The entire egg except the shell.

meat and bone meal digester tankage. A livestock feedstuff obtained by the separation of fats, by either the wet rendering or dry rendering method, from the meat and bone substance obtained as a result of the slaughtering of animals, in the meat packing industry. See *digester tankage.*

meat-and-bone scrap. A livestock feedstuff which is the product obtained by the dry rendering of the meat and bone substance of slaughtered animals in the meat packing industry.

meat animal. Any animal which is produced primarily for its meat producing characteristics, whether for breeding or slaughter.

meat bird. A fowl produced primarily for its meat, in contrast to a fowl kept especially as an egg producer.

meat by-products. Any non-rendered, clean, wholesome part of the carcass of slaughtered mammals other than meat.

meat curing. Any of several methods of processing meats and meat products so they will keep in usable condition for a satisfactory period of time: salting, pickling, smoking, etc., are the common methods.

meat grinder. A machine for reducing meat to very small pieces.

meat hook. A hook, usually inside a cooler or refrigerator, upon which the carcass of a meat animal is hung for storage and curing.

meating. A desired characteristic of good, meat-type poultry which shows vigorous growth, early maturity and good flesh production. See *fleshing.*

meat inspector. A technically trained individual, usually a graduate veterinarian, engaged by regulatory authorities to inspect all meats as they pass through the slaughter house or packing plant, to certify that they are free from disease and are wholesome food products.

meat meal. A livestock feedstuff supplement which is the ground, dried, rendered residue from animal tissues exclusive of hoofs, horns, blood, manure, and stomach contents except in such traces as might occur unavoidably in good factory practice. Also called *dry-rendered tankage.*

meat meal tankage. Digester tankage.

meat packer. An individual or organization that slaughters meat-producing animals and processes, packs, and distributes meat and meat by-products. (Originally an individual or organization that placed meat and meat by-products in convenient form for distribution.)

meat packing. See *meat packer.*

meat-packing company. See *meat packer.*

meat packing establishment. Any plant engaged in meat packing. See *meat packer.*

meat packing industry. Those individuals and organizations which are engaged in meat packing. See *meat packer.*

meat processor. See *meat packer.*

meat production. The breeding and production of animals especially for meat, in contrast to producing animals for milk, egg production, or other uses.

meat quality. The quality of a meat carcass as indicated by color, texture, grain, degrees of marbling, and age of animal.

meat saw. A tool used in meat slaughter consisting of a thin steel saw blade set in a U-shaped handle, especially designed for sawing bone in meat carcasses.

meat scrap. Meat meal.

meat spot of egg. Any solid, foreign substance, which is dark red, brownish-red or deep brown, found between the vitelline and shell membranes of the egg.

meat type hog. A trim, smooth hog with good length and moderate width which shows evience of muscling, particularly in the hams and loin, and which yields a carcass with a high proportion of lean to fat.

meaty. 1. Designating a well-fleshed animal. 2. Designating a characteristic of good cheese showing flexibility. It will tear but not break short and is not dry or brittle. 3. Any cut of meat that shows a high percentage of good muscle structure. 4. Designating spareribs cut with a heavy layer of the bacon left on.

meaty udder. A cow's udder that is hard, lumpy, and deficient in soft, spongy, glandular tissue; an undesirable type.

mecate. A long, special kind of riding horse reins often made of horse hair (western U.S.).

mechanical buffer. A machine used in the final operation of plucking feathers from poultry slaughtered for market. Also called *buffer.*

mechanical choker. A horse which is otherwise sound but which makes a respiratory noise (roar) when pulling a heavy load.

mechanical control. Man-built structures which are used to control erosion, such as terraces, dams, retards, baffles, etc., in contrast to vegetative control.

mechanical corn shredder. A machine into which dry, harvested corn is fed, usually in bundles, which removes the husks and shreds the stalks.

mechanical corn picker. A harvesting machine which removes husks and ears from the corn stalks at harvest. Most pickers have husking rolls for removing the husks from the ears.

mechanical cotton picker. A machine which removes cotton from the bolls mechanically, thus replacing slow hand labor in harvesting the cotton crop.

mechanical cow. A machine in which dry milk

powder and water are mixed to produce liquid milk.

mechanical ducker. A machine designed to hold hog carcasses under water in the scalding vats of slaughtering plants.

mechanical egg washer. A washing machine which cleans soiled eggs to improve appearance for marketing.

mechanical gutter cleaner. A conveyor-type of machine used for cleaning gutters in dairy barns.

mechanical hole. A body defect of cheese in which there are large or small, irregularly shaped openings.

mechanical manipulation. Collecting semen for artificial breeding purposes by massaging the male's genital organs or by other use of the hands to produce ejaculation.

mechanical manure loader. A labor saving device usually mounted on the front of a farm tractor for loading manure from piles, yards, sheds or barns into manure spreaders for transport to fields. It is used extensively where loose housing of dairy cows is practiced. See *manure bucket.*

mechanical milker. A machine which replaces hand labor, completely or in part, in the milking of dairy cows. It consists of a vacuum pump, vacuum lines, pulsator, teat cups, etc. Also called *milking machine.*

mechanical mixtures. In plant breeding, the admixture of seeds of unwanted varieties which affect the purity of foundation-seed stocks.

mechanical openings. A body defect of cheese characterized by irregular-shaped openings (regardless of the number and size) caused by poor workmanship and not by gas fermentation.

mechanical poultry picking. The removal of feathers in dressing poultry for the market by the use of machines in contrast to removal by hand labor. See *mechanical buffer.*

mechanical refrigeration. Cold condition created by a machine compressing certain low boiling point gases which replaces cold cellars, the use of ice, and other older methods of preserving foods and products at low temperature.

mechanical separator. Cream separator.

mechanical snapper. A mechanical device which cuts the ears of corn from the stalks. It saves labor in harvesting but reduces the value of the corn stalks for fodder.

mechanical soil analysis. The separation by mechanical means of the particle groups (separates) and the determination of the percentage of each group in a given soil sample.

mechanical stone picker. A machine that collects stones from the surface of the soil and delivers them to a collection device for removal from the field.

mechanical tissue. Any of the various plant tissues which form the supporting framework of plants.

mechanical tramper. A device used in baling cotton: it may consist of a cylinder operated by steam pressure, which exerts intermittent pressure on the loose cotton in the press box.

mechanical tripping. Pollination brought about by mechanical means as a substitute for a lack of natural tripping in the flower: *e.g.*, in alfalfa seed production.

Mecklenburg Skim. A hard, skim milk cheese made in Germany.

meconium. 1. The first excreta of a new-born animal. 2. The juice of the opium poppy, *Papaver somniferum.*

mecron. The unhumified plant matter on a forest soil. See *litter: 1.*

media. 1. The middle coat of an artery. 2. The plural of medium (*q.v.*). 3. Material for artificial propagation of various microorganisms.

mediacid. Designating acid soils with a pH value of 4.5–5.0.

medialkaline. Designating alkali soils with a pH value of 9.0–9.5.

medic. See *medics.*

medicated stock food. A manufactured stock feed in which ingredients having a medicinal value are incorporated.

medication. 1. The application of medicines, salves, etc., to an injured or sick animal. 2. The forced introduction of a chemical, usually a water-soluble salt in solution, into the sapstream of a living tree to kill it or make it repellent to insect attack, or into a freshly felled tree to destroy barkbeetle and woodbeetle broods.

medicinal cinchona. *Cinchona officinalis,* family Rubiaceae, a tree which was cultivated as a source of quinine. Native to Peru and Bolivia.

medicinal flavor. A taste defect of ice cream which results from the use of synthetic vanillin.

medics. Certain plants of the genus Medicago, family Leguminosae. Some have a high forage value; some are regarded as weeds. Native to Europe. See *alfalfa; black medic; California bur clover; spotted medic.*

Mediterranean. A standard class of domestic chickens the original breeds of which originated in the Mediterranean region. It includes the Leghorns, Minorcas, Spanish, Blue Andalusians, Anconas, and Buttercups which are regarded as egg-type fowls. All have white ear lobes, nonfeathered shanks and rather nervous dispositions. Their egg-shell color is white.

Mediterranean barley. *Hordeum gussonianum,* family Gramineae, a grass usually regarded as a weed but which has some forage value.

Mediterranean climate. Any warm-temperature climate which is relatively dry in the summer and relatively moist in the winter.

Mediterranean fever. Undulant fever.

Mediterranean flour moth. *Anagasta kuhniella* (Zell.), family Phycitidae, an insect pest which infests stored grain and flour.

Mediterranean fruit fly. *Ceratitis capitata* (Wied.), family Tephritidae, a fly which infests a number of fruits, especially citrus fruits. Common in the Mediterranean region, it has invaded the U.S., notably the citrus groves of Florida.

Mediterranean orange. Any sweet orange originating in the Mediterranean area.

Mediterraneanpalm. 1. Any palm of the genus Chamaerops, family Palmaceae. 2. *C. humilis,* one of the hardiest and most widely cultivated of all palms. It is commonly grown outdoors throughout the warm sections of the United States and in the cool greenhouse for its bushy foliage. Also called *African hair palm; European fan palm; European palm; hair palm.*

Mediterranean sea-lavender. *Limonium vulgare,* family Plumbaginaceae, a salt marsh, perennial herb grown for its white and bluish-lilac flowers. It can be used as an everlasting. Native to Mediterranean region.

medium. 1. Any of a number of natural or artificial substances, paste-like or liquid, in or on which microorganisms, such as bacteria and fungi, can be cultured. 2. A soil or material, such as sand, peat moss, vermiculite, etc., in which plants are raised or cuttings are rooted, especially in the greenhouse. 3. A market grade of roses which have stems 12–18 ins. in length. 4. One of the six recognized grades of meat, lying between good and common.

medium acid. In measuring acidity of soil, a pH of approximately 5.0. See *pH.*

medium brood foundations. Comb foundations for beehives which are 5–8 sheets per lb.

medium cheese. A cheese that is semi-aged.

medium color cheese. A body color of cheese which is a deep straw or amber.

medium eggs. A U.S. size grade of eggs, between small and large, which specifies a minimum net weight of 21 ozs. per dozen with no individual egg under the rate of 20 ozs. per dozen.

medium grade (slaughter animals). In marketing, a grade of (a) cattle which have only a very slight covering of fat; (b) lambs that are somewhat rangy, with long legs and unsymmetrical conformation; (c) vealers that are leggy, hippy, rough in the shoulder, and light in the round; (d) lard type hogs which generally weigh less than 220 lbs. and are thin or under-finished.

medium grain rice. In marketing, a class of rice which is based on the length of the grains: the class intermediate between long grain and short grain rice.

medium sands. In soil analysis, a separate (*q.v.*) that includes particles ranging in size from 0.5 to 0.25 mm.

medium sandy loam. A soil class which contains 30 percent or more of very coarse, coarse, and medium sand combined, but less than 25 percent of very coarse sand, and less than 30 percent of very fine or fine sand. The clay content does not exceed 20 percent, and the silt content does not exceed 50 percent. See *sandy loam.*

medium shrink. A subgrade of wool based on shrinkage: intermediate between heavy and light shrinkage.

medium soil. A soil which has characteristics intermediate between the looseness of sand and the tightness of clay soil, as sandy loam or loam.

medium tomato purée. See *purée.*

medium-weight breed. Any medium-heavy breed of chickens, as represented by the Plymouth Rocks, Wyandottes, Rhode Island Reds, and New Hampshires. These breeds or their crosses furnish the greatest proportion of market poultry.

medium-weight calves. A market class for slaughter calves which weigh 200–300 lbs.

medium-weight veal. A market class for veal carcasses which weigh 71-110 lbs.

medium wools. Wools which for the most part come from certain of the Down breeds, as the Hampshires and Shropshires. See *Down breeds; medium wool sheep.*

medium wool sheep. Any of several breeds of sheep intermediate between the fine wool and long wool breeds, desirable for meat as well as wool. See *Cheviot; Columbia; Corriedale; Dorset; Hampshire; Oxford; Panama; Shropshire; Southdown; Suffolk; Tunis.*

medlar. *Mespilus germanica* (*Pyrus germanica*), family Rosaceae, a small tree grown for its fruit, edible after frost, sometimes used in preserves. Native to Europe and Asia.

medulla cell. A layer of globular cells in the center of the wool fiber.

medullary rays. Groups of parenchymatous cells, extending radially between the vascular bundles, which provide for radial transfer of the water and food of a plant.

Medusa's head. *Elymus caput-medusae,* family Gramineae, an annual grass which has become a dominant and serious local pest on western ranges (U.S.). Native to Europe.

Meeker harrow. A type of harrow consisting of a large number of disks uniformly spaced on an axle and mounted in a frame; designed to pulverize the soil, make a smooth surface and firm the seedbed, and especially useful in preparing fields for garden crops.

megaspore. A spore which has the property of giving rise to a gametophyte (embryo sac) bearing a female gamete. One of the four cells produced by two meiotic divisions of the megaspore-mother-cell (megasporocyte).

megaspore-mother-cell. The cell which undergoes two meiotic divisions to produce four megaspores.

megathermic tree. Any tree which requires a warm climate and long growing season, as palm, laurel, magnolia, southern oak, tupelo gum, redwood, bald cypress, slash pine, longleaf pine, and other southern pines.

megrims. Staggers.

Mehring press. In beekeeping, a comb formation invented by Johannes Mehring in 1857. It consisted of a flat press which printed the rudiments of cells on beeswax (obsolete).

meiosis. The process preceding the formation of gametes, in which the number of each chromosome pair separate, and the chromosome number in the resulting daughter cells is reduced to half the somatic number, as distinguished from ordinary cell division, or mitosis.

Meiwa kumquat. *Fortunella crassifolia,* family Rutaceae, an evergreen tree or shrub grown for its very small, highly aromatic fruit chiefly used for preserves. Native to China.

melada (malada). A thick, crude sugar.

Melanesia bitterorange. *Citrus macroptera,* family Rutaceae, a citrus tree having very long, narrow leaves and a nearly juiceless, bitter fruit. Introduced into Florida, it shows promise for rootstocks. Native to New Guinea, Luzon in the Philippines. Also called *Melanesian papeda.*

melanin. A dark brown or black pigment found in the hair or in the skin.

melanized soil. A soil which has a very dark color to a considerable depth in the surface layer, due to incorporation of humus.

melanose. 1. A disease of the grapevine caused by *Septoria ampelina,* a parasitic fungus which attacks the leaves and causes them to fall. 2. A fungous disease of citrus trees caused by *Phomopsis citri* and *Diaporthe citri,* characterized by small dots or pustules on fruit, leaves and small twigs.

melanosis. Dark brown pigmentation of skin surfaces.

melanotic. Designating melanosis *q.v.*

melassic acid. An acid generated in molasses and glucose by alkali action.

melaxuma. Any of several diseases causing dark to black bark cankers, as the fungous disease of the walnut caused by *Dothiorella gregaria.*

melene. $C_{30}H_{62}$, a white, waxy, crystalline hydrocarbon, extracted from certain types of paraffin, and from beeswax by dry distillation.

melezitose. $C_{18}H_{32}O_{16} \cdot 2H_2O$, a trisaccharide obtained from Persian manna, and from the manna of the Douglas fir and larch. It resembles, but is less sweet than, ordinary sugar.

melic. Oniongrass.

melilots. Sweet clover.

melittology. That branch of entomological science concerned with bees.

melliferous. Honeybearing.

mellit. A scabby, small sore on the coronet or heel of a horse's foot, resulting from injury.

mellow. Designating soft, tender, juicy, ripe fruit.

mellow soil. 1. A soil, especially one containing a high percentage of humus, that is soft and loosely coherent; *e.g.,* loam plow soil. 2. A soil that pulverizes easily into a good seed bed.

melon. The usually round, fleshy fruit of *Cucumis melo* and *Citrullus vulgaris,* family Cucurbitaceae, and their variations. See *cantaloup(e); citronmelon; muskmelon; watermelon.*

melon anthracnose. Cucurbit anthracnose.

melon aphid. *Aphis gossypii* Glov., family Aphidae, an aphid which infests melons and other greenhouse and outdoor garden plants. Also called *cotton aphid.*

melon apple. Mango melon.

melon fly. *Dacus cucurbitae* Coq., family Tephritidae, a fly which infests melons, squashes, beans, tomatoes, and other vegetables.

melongena. Eggplant.

melon louse. Melon aphid.

melon mold. An ice cream molded so that it resembles a half melon in size and appearance.

melon moth. The mature form of the melonworm *q.v.*

melon pear. See *pepino.*

melon shrub. See *pepino.*

melon tree. Papaya.

melonworm. *Diaphania hyalinata* (L.), family Pyraustidae, the larva of the melon moth which burrows into the rinds of muskmelons causing decay.

melt down. A sample of nearly melted ice cream which reveals its melting qualities.

melter. A device used in extraction of the honey crop from the comb.

melters. A plant disease caused by the soil fungus *Pythium debaryanum,* family Pythiaceae. Also called *watery soft rot; leak.*

melting quality. The quality of ice cream judged according to its resistance toward melting and smoothness upon melting when a sample is exposed to room temperature.

member vessel. In wood anatomy, one of the cellular components of a vessel. Also called *vessel element; vessel segment.* See *vessel: 2.*

membrane. 1. A thin, flexible sheet of vegetable or animal tissue; the thin protoplasmic tissue connecting, covering, or lining a structure, as a cell of a plant or animal. 2. The layer which surrounds fat globules of milk. Its nature is not definitely known: it may consist mainly of phospholipoids and a protein or proteins not completely identical with casein, albumin, or globulin. 3. A layer of low permeability material: *e.g.,* soil, chemicals, or plastics, placed in the bottom of a farm pond to reduce seepage losses.

membranosis. A membranous staining in lemons due to an internal breakdown which occurs in storage at temperatures below 55° F., particularly in old fruit and fruit of low vitality.

memorial rose. Wichura rose.

menadione. Vitamin K.

Mendel's Law. Either of the laws concerning hereditary transmission of genes or character determiners which resulted from the experiments in plant hybridization, by Gregor Johann Mendel, an Augustinian monk at Brno, Czechoslovakia, in 1866. The two laws, known as the principle of segregation and the principle of independent assortment, state that inheritance is not a hit-or-miss proposition but is subject to definite rules or laws. Offspring of parents with contrasting characters (the one dominant and the other recessive) will resemble the dominant parent. The second generation resulting from crossing will produce individuals, 25 percent resembling the dominant grandparent, 50 percent resembling the parents, and 25 percent resembling the recessive grandparent.

Mendel tulip. A class of hybrid tulips from the early variety Duc van Tholl and the Darwin tulips. Used for forcing and bedding, their flowers are of fine pastel shades blooming in midseason.

Mendelian genetics. The genetic principles which follow Mendel's Laws.

menhaden. a. A marine fish of the genus Brevoortia; abundant on the eastern coast of the U.S. In the early agriculture of this region, it was used as a soil fertilizer. b. The *menhaden* fish meal used as a protein supplement in animal feeds, especially for poultry and swine.

menhaden oil. The oil obtained from menhaden fish. A vitamin-rich supplement of feeds.

Menindie cloves. *Trigonella suavissima,* family Leguminosae, a clovelike, aromatic, perennial herb with small, light yellow flowers, grown for forage in Australia. Native to Australia. Also called *Darling clover.*

meninges. The membranes which cover the brain and spinal cord.

menominee. See *annual wildrice.*

mensuration. The science dealing with the measurement of volume, growth, and development of individual trees and stands, and the determination of various products obtainable from them.

Menzies larkspur. *Delphinium menziesi,* family Ranunculaceae, a perennial, poisonous herb, which, when eaten by cattle, causes uneasiness, staggering, constipation, nausea, bloat, and rigid limbs. Native to North America.

Menzies tolmiea. *Tolmiea menziesi,* family Saxifragaceae, a perennial grown in the wild garden, rock garden, or as a potted plant for its foliage. Native to western North America. Also called *pick-a-back.*

merchantable. 1. Designating the portion of a tree or stand which can be marketed under given economic conditions. *Merchantable* length pertains to the marketable length of a log in a tree. *Merchantable* volume pertains to the marketable volume in a tree. 2. Designating any product which possesses suitable quality for marketing.

merchantable height. The length of tree stem from the top of the stump to the top end of the last marketable section, usually expressed in feet, number of logs, or bolts of some standard length.

merchantable log. A log that will yield products of sufficient value to pay at least for conversion.

merchantable lumber. The entire cut of a sawmill except for culls.

merchantable timber. A tree or stand of trees which may be disposed of at a profit through conversion into salable products.

merchantable volume. The salable volume of wood in a single tree or stand.

mercurial ointment. A standard ointment, containing 30 percent of mercury, used for the control of the pubic louse or crab and the poultry head louse. Also called *blue ointment.*

mercuric chloride. Corrosive sublimate or bichloride of mercury, commercially available as a white powder, crystals, or in the form of tablets. It is used in animal medicines, as a disinfectant, and in solution for the control of fungi, gnats, earthworms, cabbage root maggot, etc.

mercurous chloride. Calomel *q.v.*

mercury. 1. Hg, a chemical element which, when incorporated in a heavy oil, such as vaseline, makes an effective ointment for lice on poultry. Also called *quicksilver.* 2. See *Good King Henry goosefoot; herbmercury; poison ivy; Pacific poisonoak.*

mercury toxicity. Poisoning from vapor or metallic mercury in a greenhouse, as in the rose.

mericap. Either of the two carpels of a fruit which splits away, seemingly as a separate fruit.

meridian line. A true north and south line; a line from which range lines are established in the General Land Office survey (U.S.).

Merino. 1. A breed of small sheep which descended from Spanish stock believed to have originated in Africa. Wool from the *Merino* is of excellent quality, very fine, and about 4 in. staple. Raised chiefly in the U.S. and Australia for its fleece, it has been crossed with other breeds, especially in New Zealand, to make it a better mutton sheep. Rams weigh up to 175 lbs., ewes up to 125. See *A-type Merino; B-type Merino; C-type Merino; Rambouillet.* 2. A term used in the textile industries to denote wool from the *Merino* or very high quality wool.

merinos. A grade of wool from Uruguay equal to fine medium (60's–64's).

meristem. The undifferentiated, thin-walled tissue of a plant from which, as development proceeds, the permanent tissues are produced. See *cambium.*

Merker grass. Napiergrass.

Merriam's life zones. Divisions of North and Central America on the basis of vegetation and animal life.

merry-go-round. A mechanical device for extracting honey from the comb. It consists of a round comb rack revolving above a stationary pan which slopes to the center. The honey, drained from the combs, is caught by the pan connected to the extractor or the honey sump.

merry-go-round dipper. A device for dipping raisin grapes into a boiling lye solution. Usually it consists of two or more hinged wire baskets suspended from the ends of levers which in turn are hinged to a central pivoted upright.

mersolite. Phenyl mercuric acetate used as a seed disinfectant, especially for cottonseed.

merthiolate. Sodium ethyl mercurithiosalicylate, a substance used to prevent citrus fruit decay and as a common skin antiseptic.

mesa. A land form: (a) an isolated, flat-topped hill, or mountain, with abrupt sides; or (b) a fairly extensive, relatively flat plateau (western U.S.).

mesa dropseed. *Sporobolus flexuosus,* family Gramineae, one of the less common, perennial, range grasses. Native to North America.

mesa oak. Englemann oak.

mescal. A Mexican liquor distilled from the cooked, fermented head and heart portion of the pulque agave (*q.v.*). Generally colorless, it contains about 44 to 55 percent alcohol.

mescalbean sophora. *Sophora secundiflora,* family Leguminosae, an evergreen shrub or small tree which causes poisoning of cattle, sheep, and goats when eaten in large quantities. It is sometimes planted for its fragrant, violet-blue flowers. Native to the southwestern ranges of the U.S. Also called *mescal bean.*

mescalbutton peyote. *Lophophora williamsi,* family Cactaceae, a small, globe-shaped cactus which yields the mescal-button, a bulb having narcotic properties. Native to North America. Also called *hikuli; peyote.*

mesembryanthemum. Succulent plants of the genus Mesembryanthemum, family Aizoaceae. Many species have been introduced for use in rock-gardens and other type gardens, some of which have escaped from cultivation and now grow wild. Native to South Africa. Also called *figmarigold.* See *iceplant mesembryanthemum; Hottentotfig.*

mesentery. A membrane which supports a visceral organ, particularly the intestine, and contains the vessels and nerves that supply that organ.

Mesitra. A soft ewes' milk cheese made in the Crimea and eaten fresh.

meslin. A crop of peas and oats sown together (New England, obsolete).

mesocarp. The layer between the endocarp and the epicarp of a multiple layered pericarp (*q.v.*). In stone-fruits, it is thick, fleshy and edible.

mesophyll collapse. A collapse of the soft interior tissue (mesophyll) of the leaf blade, especially the leaves of oranges, due presumably to an insufficient supply of water. Affected leaves lose their chlorophyll, turn gray and finally brown, and may be invaded by fungi.

mesophyte. Vegetation that grows under medium conditions of atmospheric or soil moisture, as contrasted with xerophytic vegetation which is able to withstand periodic or permanent conditions of low moisture and with hydrophytic or aquatic vegetation.

mesotherm. 1. Any plant which grows under intermediate conditions of temperature along with abundant moisture. 2. Any plant which grows under conditions of alternating high and low temperatures.

mesquite. Any plant of the genus Prosopis, family Leguminosae. Species are usually thorny shrubs, the foliage and seed pods of which have browse value in the desert, but the invasion of the plant on open grassland lowers the grazing capacity of the land. One of the more useful *mesquites* is *P. juliflora,* also called *honey locust* or *algarroba,* an important bee and forage plant. Also called *mesquit; musquit; muzquit.*

mesquite American mistletoe. *Phoradendron californicum,* family Loranthaceae, a parasitic shrub which grows on trees. Its berries are used for decoration chiefly at Christmas time. Native to North America. See *mistletoe.*

mesquite grass. See *curlymesquite; bush muhly; velvetgrass.*

mesquite-grass desert. A natural land division based on vegetation: desert areas characterized by an association of mesquite, catclaw acacia, Coville creosotebush, yucca, and grasses, such as Rothrock grama and curly mesquite, etc.

mesquite pod. The beanlike pod or seed of mesquite (*Prosopis* spp.) rich in sugar and a valuable feed for stock.

mesquite seed. Mesquite pod.

mess beef. Salted beef, mainly the chuck, flanks, and rump, packed in barrels.

messenger. Haul-back.

messmate stringybark eucalyptus. *Eucalyptus obliqua,* family Myrtaceae, an aromatic, very tall tree grown as an ornamental. Native to Australia.

mess pork. Salted pork in pieces of about 4 lbs. each, taken from the shoulders and sides of lightweight hogs and packed in barrels.

metabolic. Designating the chemical changes that take place in living cells whereby one compound is converted to one or more other compounds.

metabolism. The chemical changes in the cells, organs, and entire body which provide energy for the physiological functions and activities of living organisms, for growth, tissue repair, or maintenance. The building-up part of *metabolism* is called anabolism or synthesis and the breakdown processes, catabolism.

metabolite. Anything produced by the metabolism.

metabolizable energy. The total amount of energy in feed less the losses in feces, combustible gases, and urine. Also called *available energy.*

metabolized milk. Milk produced by cows that have been fed irradiated yeast.

metacercaria. The encysted larval stage of a trematode (fluke) which develops from another larval stage, the cercaria, and is infective to the final (definitive) host: *e.g.,* the *metacercaria* of the common liver fluke of sheep and cattle encysts on blades of grass and is swallowed by sheep or cattle while grazing. See *flukes.*

metagenesis. See *alternation of generations.*

metaldehyde. A crystalline polymer of acetaldehyde, used at two to three percent as an attractant in poison baits for snails and slugs.

metal cover. See *outer cover.*

metal haze. The clouding in white wine vinegar which results from iron dissolved from equipment, such as pumps and pipelines, used in its manufacture.

metallic flavor. An objectionable flavor in milk, butter, cheese, and ice cream, which results from metal contamination, particularly copper and iron. See *oxidized flavor.*

metamorphic rock. Any rock which has undergone profound changes by the combined action of pressure, heat, and water, as schist, slate, and marble.

metamorphosis. A process by which an organism changes in form and structure in the course of its development, as many insects do.

metaphase. That stage of cell division in which the chromosomes are arranged in an equatorial plate or plane. It precedes the anaphase stage.

metaphosphate fertilizer. A fertilizer which contains higher percentages of P_2O_5 than ordinary acid phosphate and superphosphate. Calcium metaphosphate, manufactured by the U.S. Tennessee Valley Authority for experimental use as a fertilizer, contains as much as 63 percent of P_2O_5. The metaphosphate salts of metaphosphoric acid, HPO_3, hydrolyze in the soil to form the more soluble orthophosphates, salts of orthophosphoric acid, H_3PO_4.

metate. The lower or bed stone used by Indians in southwestern U.S. for grinding corn and beans.

metaxenia. Direct effect of pollen on parts of the seed and fruit which lie outside the embryo and endosperm.

metayer. A tenant farmer who works a piece of land for a share in the crop.

Metcalfe bean. *Phaseolus metcalfei,* family Leguminosae, a perennial herb with a long trailing stem, grown mostly for forage. Native to North America.

meteorology. The study of the physical processes which occur in the atmosphere and of the connected processes of the lithosphere and hydrosphere. (Long usage has caused the term to be used almost solely in connection with phenomena of the weather.) 2. The science of the atmosphere and its phenomena.

metes and bounds. In U.S., a means of defining the legal boundaries of a parcel of land by giving the bearings and distances from points of reference. Often a tree, a stone, or some natural feature is used in the description; the line from one point to another is not necessarily straight but may be curved. Land descriptions were commonly made by *metes and bounds* prior to the establishment of the rectangular system of land surveying by the General Land Office.

methallyl chloride. $CH_2{=}C(CH_3){-}CH_2Cl$, a chemical compound, used as a fumigant for soil and grains.

methanesulphonyl fluoride. $CH_3\text{-}SO_2F$, a highly effective greenhouse fumigant.

methemoglobin. A rust-colored substance, formed when oxygen unites with an oxidized product of hemoglobin. The oxygen is not easily released from this combination, thus injuring the tissues. Its properties are in contrast to those of oxyhemoglobin, which is bright red and releases its oxygen readily. *Methemoglobin* results from blood decomposition or from poisoning of blood by nitrites and other substances.

methionine. A sulfur-containing essential amino acid; indispensable for animals.

methoxychlor. A chlorinated hydrocarbon insecticide, closely related to DDT and TDE.

methyl bromide. CH_3Br, an insect fumigant.

methyl chloride. CH_3Cl, a colorless gas, used in refrigerators.

methylene blue reduction test. A test for the bacteriological quality of milk which consists of placing methylene blue thiocyanate in the milk under controlled conditions. The more rapidly the methylene blue changes color, the greater the number of bacteria in the milk.

methyl formate. An insect fumigant.

methyl n-anyl. A ketone, produced along with other ketones by molds of the *Penicillium roqueforti* type. These are considered to be constituents of the Roquefort cheese flavor, which sometimes occurs in moldy butter. The ketones are presumably derived from the corresponding fatty acid.

methyl red. A chemical indicator; a dye stain used in determining the classification or taxonomy of certain bacteria such as coliform (*q.v.*) bacilli. See also *Gram stain.*

metritis. Uterine inflammation.

metrox. See *copper oxide.*

mett. A semi-dry, 100 percent pork sausage.

Mexican ageratum. *Ageratum houstonianum,* family Compositae, a tropical, annual herb grown as a border or bedding plant for its habit and its blue, sometimes white or pink, flowers. Native to Mexico.

Mexican bamboo. Japanese fleeceflower.

Mexican bean beetle. *Epilachna varivestis* Muls., family Coccinellidae, a brownish, spotted beetle whose larvae feed on kidney beans, soybeans, and cowpeas. Also called *bean ladybird.*

Mexican black bean. See *frijole.*

Mexican blindness. An animal disease caused by a filarial parasite, *Onchocerca volvulus,* which is carried by blackflies, buffalo gnats, and turkey gnats.

Mexican bloodtrumpet. *Phaedranthus buccinatorius,* family Bignoniaceae, an evergreen, woody vine grown in warm areas for its habit, its foliage, and its bright red flowers, yellow at the base. Native to Mexico. Also called *clarin.*

Mexican blue palm. Blue erythea.

Mexican boll weevil. Boll weevil.

Mexican chicken bug. See *poultry bug.*

Mexicanclover. *Richardia scabra,* family Rubiaceae, a tropical herb, whose root, known as *farinaceous ipecacuanha,* is used as an emetic. Native to America.

Mexican coraldrops. Showy caroldrops.

Mexican cotton boll weevil. Boll weevil.

Mexican-creeper. Mountainrose coralvine.

Mexican cypress. *Cupressus lusitanica,* family Pinaceae, an evergreen tree grown in warm areas for its bluish-green leaves. Native to Mexico. Also called *Portuguese cypress.*

Mexican earflower. Sacred ear *q.v.*

Mexican fire-plant. Painted euphorbia.

Mexican fruit fly. *Anastrepha ludens* (Loew.), family Tephritidae, a serious insect pest of citrus and other fruits in Mexico. At times it has appeared in the Rio Grande valley of the U.S.

Mexican globe mallow. *Sphaeralcea umbellata,* family Malvaceae, a tropical shrub grown for its scarlet flowers with white bases. Native to Mexico.

Mexican groundcherry. Tomatillo groundcherry.

Mexican group (lime). *Citrus aurantifolia,* family Rutaceae, a tender, small-fruited sour lime, whose varieties are grown very extensively in Mexico, the West Indies, and southern Florida. The trees are very spiny and susceptible to frost and to the withertip disease (*Gloeosporium limeticolum*). The rinds of the fruits are thin and closely adhering. Native to Asia. Also called *West Indian lime; Key lime.*

Mexican hog. A wild hog, the Texas peccary, at one time common in Texas and Arkansas, U.S.

Mexican horse. A breed of horse directly descended from the Andalusian and Arabian breeds brought to Mexico more than four hundred years ago by the Spanish conquerors.

Mexican ivy. Purplebell cobaea.

Mexican jack. The small, common burro, seldom used for breeding in the U.S. Also called *Spanish jack.*

Mexican jumping bean. The seeds of certain shrubs of the genera Sapium or Sebastiania, family Euphorbiaceae, which have been infested by a small moth, genus Carpocapsa, family Olethreutidae. The moth lays its eggs in the seeds and the action of the larvae spinning cocoons within cause the seeds to jump about.

Mexican marigold. Striped marigold.

Mexican mealybug. *Phenacoccus gossypii* T. and C., family Pseudococcidae, an insect which infests ornamentals in southern California.

Mexican milkweed. *Asclepias mexicana,* family Asclepiadaceae, a perennial herb, poisonous to sheep and poultry when eaten. Native to North America, found from Washington to Mexico. Also called *narrow-leaved milkweed; Mexican whorled milkweed.*

Mexican narrow ear. Any of the various varieties of corn widespread in western Mexico with ears long and somewhat tapering at both ends. The plants are tall and deep-rooted with slender leaves and wiry tassels.

Mexican orange. Ternate Mexicanorange.

Mexican pinyon pine. *Pinus cembroides,* family Pinaceae, a low tree which bears an edible seed. Native to North America. Also called *pinyon; Mexican stone pine; nut pine; piñon.* See *nut pine: 2.*

Mexican pricklepoppy. *Argemone mexicana,* family Papaveraceae, a low, prickly herb, growing wild in waste places in southern and southwestern U.S. It contains a substance reportedly poisonous to fowls. Native to Mexico. Also called *prickly poppy.*

Mexicanstar. *Milla biflora,* family Amaryllidaceae, a bulbous herb grown for its fragrant, white flowers. Native to U.S. and Mexico.

Mexican stone pine. Mexican pinyon pine.

Mexican tea. Wormseed goosefoot.

Mexican teosinte. *Euchlaena mexicana,* family Gramineae, a grass closely related to corn, grown in southern U.S. for forage. Native to Central America.

Mexican tulip-poppy. Goldencup.

Mexican vanilla. *Vanilla planifolia,* family Orchidaceae, a tropical tall-climbing, fleshy-stemmed vine grown in frost-free areas for its aromatic seed pods, the source of the vanilla of commerce used in flavoring. Native to Mexico.

Mexican Washingtonpalm. *Washingtonia robusta,* family Palmaceae, a palm grown in warm areas as an ornamental and as an avenue tree. Native to North America. Also called *California fan palm; fanleaf palm.*

Mexican weed. *Caperonia palustris,* a troublesome weed in rice fields in southern U.S.

Mexican whorled milkweed. Mexican milkweed.

mezcal. Mescal.

mezereon. February daphne.

mica. A common mineral in soils; muscovite, biotite, and a number of other, not easily soluble minerals of the mica group residual in soils derived from the weathering of granites, gneisses, and schistose rocks. When present in considerable quantity, *mica* imparts looseness and a greasy or soapy feel to a soil. See *vermiculite.*

micell. 1. Theoretically, a submicroscopic, crystalline, multimolecular unit of structure built from complete molecules which forms the foundation walls of cells. 2. The nucleus of a colloid which is usually negatively charged. Also called *micella; micelle.*

Michaelmas daisy. Certain late-summer- or autumn-flowering, perennial asters of the family Compositae, very popular in the flower garden. They are, for the most part, varieties of one of the following: Italian aster; New England aster; New York aster; heartleaf aster; heath aster. See *aster: 1.*

Michigan age. Of, or pertaining to, an old horse.

Michigan box. A standard box with one bushel capacity used for packaging apples.

Michigan holly. See *winterberry: 1.*

microbe. Loosely, any microscopic organism or microorganism.

microbial processes. The activities and influence of microorganisms in modifying soils or other substances.

microbiologist. A scientist concerned with the study of plant and animal microorganisms.

microclimate. 1. Purely local variations from the general or regional climate due to slight differences in elevation, direction of slope exposure, soil, density of vegetation, etc. 2. Climate within a few inches or feet of the ground surface.

Micrococcus. A specific genus of small, round, bacterial cells belonging to the family Coccaceae.

microconidia. The smaller spores of Fungi Imperfecti which are usually produced in a different sporocarp or at a different time than macroconidia, the larger spores of the species. In some cases they are known to be spermatia.

microelements. Trace elements.

microfibril. 1. A sub-microscopic, thread-like unit of the cellulose matrix of the cell walls in wood. 2. (plural) The small fibers cemented together to form the spindle cells of wool.

microfilaria (plural: **microfilariae**). Minute, immature forms of certain roundworms known as filarids, parasites in the tissue or blood of mammals.

microflora. 1. Microscopic plants, such as bacteria. 2. In plant geography, the vegetation of a small area in contrast to that of a region.

microfloral. Pertaining to, or of, microflora *q.v.*

micrometer. A device for measuring width, length, or thickness with precision, usually in conjunction with a microscope or telescope.

micron. A unit of measurement, 0.001 mm. or approximately 1/25,000 in. used for measuring spores, bacteria, fat globules, soil particles, and other microscopic objects.

micronized sulfur. Sulfur which is in an extremely fine state of division with a particle size mostly under 10 microns, used as dusts and sprays for plant diseases.

micronutrients. Nutrients essential to plants or animals but needed in only very small amounts. See *trace elements.*

microorganism. An organism so small that it cannot be seen clearly without the use of a microscope; a microscopic or sub-microscopic organism.

microparasite. A parasite so minute that it can be seen only with the aid of a microscope.

Micropedology. A branch of Pedology which deals with the study of the microscopic structure, arrangement and mineralogic composition of the particles of the soil mass, mainly by the use of the microscope.

micropyle. The minute necklike opening in the integuments of an ovule, where the sperms enter.

microscopic. Designating an organism so small that it can be seen only by the aid of a microscope.

microscopic count. The direct counting of bacteria in milk by the use of the microscope.

microscopic plant. A fungus or alga.

microspore. One of the four cells produced by the two meiotic divisions (mitoses) of the microspore-mother-cell (microsporocyte). It gives rise to a gametophyte bearing only male gametes. Also called *pollen grain.*

microspore-mother-cell. A cell from which other cells are formed. Also called *pollen-mother-cell.* See *mitosis.*

microthermic tree. A tree which tolerates a cold climate and short growing season: *e.g.*, aspen, birch, alder, willow, mountain ash, jack pine, Scotch pine, lodgepole pine, spruce, larch, and alpine fir.

mid-bloom. The middle period of blooming for trees, neither the very early nor very late.

Middendorff daylily. *Hemerocallis middendorffi*, family Liliaceae, a lily-like herb grown for its pale orange flowers. Native to Siberia.

middens. 1. The castings of organisms which live and work in the soil, especially of earthworms. 2. A pile of refuse; a manure pile.

middle. 1. In stock judging, the belly region. 2. The belly. See *middles.*

Middle Atlantic truck crop belt. One of the principal agricultural regions of the U.S.: the Coastal belt, extending from South Carolina to Long Island, N. Y., in which the production of truck or garden crops is a major industry.

middle breaker. Middlebuster.

middlebuster. A lister (*q.v.*) without a planting attachment. In tillage of soil, a double moldboard plow which is designed to make a bed, especially for cotton and corn. Also called *middle breaker.*

middle capacity. In animals, the dimension of the part between the thighs and shoulders; an indication of capacity to handle feed.

middle cleft of frog. A lengthwise depression in the posterior middle of the frog of a horse's foot.

middleman. A merchant, broker, jobber, or dealer who performs a marketing service between the producer and the retailer.

middle milk. The milk obtained from a cow during the middle of the milking period, as opposed to that obtained in the beginning and at the end. It has a higher fat content and a lower bacteriological count than the other two. See *foremilk; forestripping.*

middles. The spaces between the rows of trees or a crop which may be cultivated, planted to an interrow crop, or left in sod.

middle weight hunter. A hunter type horse which can carry riders weighing 165 to 190 lbs.

middling. A grade of lint cotton which shall be American upland cotton which in color, leaf, and preparation is within the range represented by a set of samples in the custody of the U.S. Department of Agriculture, designated: "Original official cotton standards of the United States, American upland, No. 5 or middling, effective August 20, 1936."

middling extra white. A grade of lint cotton which shall be American upland cotton which in leaf and preparation is No. 5 or middling, but which, in color, is whiter than it.

middling fair. A grade of lint cotton which shall be American upland cotton which in color, leaf, and preparation is better than No. 2, or strict good middling; it is graded No. 1.

middling gray. A grade of lint cotton which shall be American upland cotton which in leaf and preparation is No. 5 or middling, but which, in color, is more gray than No. 5 or middling but not darker than the dullest bale in No. 8 or strict good ordinary.

middling meat. The center side cuts between the hams and shoulders of a meat carcass, usually of a hog: salted and cured for home use by early settlers in south and western U.S.

middlings. The coarse particles separated in the milling of wheat which contain the germ of the wheat grain and fine particles of bran. Higher in protein and digestible nutrients than bran, the *middlings* are used as a feed chiefly for hogs, calves, and poultry. Also called *shorts.* See *wheat flour middlings; wheat red dog; wheat standard middlings.*

middling spotted. A grade of lint cotton which shall be American upland cotton which in leaf and preparation is No. 5 or middling, but which in color is between No. 5 or middling and No. 5 tinged or middling tinged.

middling tinged. A grade of lint cotton which shall be American upland cotton which in leaf and preparation is No. 5 or middling, but which in color is within the range represented by a set of samples in the custody of the U.S. Department of Agriculture, designated: "Original official cotton standards of the United States, American upland No. 5 tinged or middling tinged, effective August 20, 1936."

middling yellow stained. A grade of lint cotton which shall be American upland cotton which in leaf and preparation is No. 5 or middling, but which in color is deeper than No. 5 tinged or middling tinged.

midge. A loose name for any small, mosquito-like insect; specifically applied to those of family Chironomidae.

midget cattle. See *dwarf cattle.*

midget crabapple. *Malus micromalus,* family Rosaceae, a small, deciduous tree grown for its upright habit and for its deep pink flowers. Native to America. Also called *Kaido crabapple.*

mid-grass (range). An area on the western plains of U.S., where the prevailing growth height of native grasses is less than in the tall grass prairie or that characterized by species of the genus Andropogon, such as small and big bluestems. The *mid-grasses* include the wheat-grasses and Junegrass.

midrib. The main vein of a leaf; located halfway between the two edges. It is a continuation of the leaf stalk.

midseason oat. An oat of medium maturity.

midshipman's butter. American avocado.

Midwest eggs. Eggs produced on general farms in the U.S. Midwest, as a marketable surplus, as distinguished from specially produced hennery eggs from the same or other areas.

mignon dahlia. A type of dahlia which grows up to 18 ins., and has open-center flowers and one row of ray florets.

mignonette. 1. A plant of the genus Reseda, family Resedaceae. 2. *R. odovata,* an annual or biennial herb grown for its very fragrant, yellowish flowers. Native to Europe and Asia.

mignonette tree. Henna.

mignonette vine. Madeira vine.

mignonettewood. Diamondleaf pittosporum.

Mignot. A Livarot-type cheese made in France.

migrant. Designating a plant or animal which changes, or has changed, its natural location.

migrant labor. See *migratory worker.*

migratory worker. A farm laborer who moves from one locality to another according to variations in time of harvesting crops, or one who lives in a locality only for the time required to carry out operations in the production and harvesting of one or more crops. Usually, the family of the laborer is also employed on the farm.

milacre. A sample plot, 1/1000th of an acre (usually 1/10 chain square), used in plant reproduction or vegetation surveys.

Milano. A soft, sweet, fast-ripening, Italian cheese; cured for about 20 days.

milch. See *milk cow.*

mild. 1. Designating cheese which is not sharp or strong in flavor. 2. Designating a gentle animal.

milder in character. In cattle judging, a term referring to an animal's head that is neither bold nor outstanding.

mildew. 1. A plant disease in which the pathogen forms a fungous coating: *e.g.,* white, downy, or powdery, over the surface of plant parts. 2. See *Diplodia disease of corn; mold; downy mildew.*

mild mosaic. One of the mosaic diseases, carried by a virus, which affects potato plants. It causes slight dwarfing, distinct mottling, and some ruffling.

mild rusty mottle. A virus disease of cherry, prevalent in Oregon, U.S. There is a rusty or bronzed appearance of the foliage with varying degrees of mottling frequently with leaf casting. The plant slowly declines.

mild streak. See *streak.*

mild water pepper. Swamp smartweed.

milfoil. See *yarrow.*

military bounty. See *bountyland.*

milk. 1. The natural, whitish or cream-colored liquid discharged by the mammary glands of mammals. Unless otherwise stated, *milk* usually means cow's *milk.* 2. The juices or fluids secreted by certain plants, as the *milk* of the coconut or the *milk* of immature kernels of corn. 3. To draw *milk* from the udder of a cow.

milk area. An area in a dairy which includes the milking room and milkhouse. Utilities, such as toilet, boiler room and heating plant, locker room, and office may be adjacent to or included in this area.

milk-borne disease. Any human disease which results from the intake of pathogens in milk.

milk bottle. A round or square container made of special paper, or specially annealed glass, in different sizes: used in retailing milk. See *milk carton.*

milk can. A specially constructed, steel container lined with tin, provided with lifting handles, and either a plug or umbrella cover; used in the transport of milk from producer to distributor or processor. Its usual capacity is ten gallons.

milk carton. A plastic- or wax-coated, disposable, paper container used in retailing milk.

milk champagne. An effervescent, alcoholic, fermented milk, such as koumiss.

milk cistern. A part of a cow's udder; the cavity

holding about one-half pint which is drained by the teat ducts.

milk closure. A closure or cap for a milk bottle.

milk cooler. A device used for the sanitary cooling of milk before shipment or following pasteurization. It may consist of a stone urn etc., in which the milk is allowed to cool, or it may be a device in which the temperature is controlled by mechanical means. A *milk cooler* usually holds large quantities of milk.

milk cow. A cow bred and used for milk production, in contrast to one bred for beef production. Also called *milch cow.*

milk derivative. Any product obtained from milk, such as butter and cheese.

milk dry. To draw out the last of the milk from a cow's udder; to strip.

milk dryer (drier). Any mechanical device designed to remove most of the water from milk. The water may be evaporated from pans or from steam heated rollers; or the milk may be sprayed into a stream of hot air. See *dehydrater; drier.*

milk duct. The cavity in the teat which leads into the milk cistern. Also called *teat cistern.*

milker. 1. A milk-giving cow. 2. a. One who milks cows. b. A machine which milks cows.

milker pail. Part of a milking machine: the container which receives the milk from the tubes. Also called *milk pail.*

milk-fat basis. A method of paying for milk at receiving stations on the basis of price per pound of butter fat.

milk-fed. Designating animals fed largely on dairy products.

milk feed. In poultry husbandry, to feed chickens a fattening mash mixed with milk, six to ten days prior to marketing.

milk-feed price ratio. The pounds of dairy ration equal in value to one pound of milk, at prices existing at a particular time and place. The U.S. Department of Agriculture *milk-feed price ratio* is pounds of dairy concentrates, rather than the entire dairy ration.

milk fermentation. The process by which a change is produced in milk as a result of the activity of one or more species of microorganisms.

milk fever. A common febrile disease of cows usually occurring at termination of parturition characterized by a loss of consciousness and a generalized paretic condition. The most obvious and consistent abnormality displayed is a low calcium content of the blood. Also called *parturient paresis.*

milk fistula. A fistula or abnormal passage leading through the teat wall to the milk cistern or milk canal which may be caused by a wound to the teat, as a cut from barbed wire.

milk flow. The amount of milk produced by a cow or the entire herd in a day, week, season, etc.

milk flush. A flush, or purgative, sometimes used by poultrymen in the treatment of coccidiosis. It is made by adding from 20 to 30 percent of dried milk products (skim milk, buttermilk, or whey) to the mash.

milk goat. A goat bred primarily for milk production, as the Saanen, Toggenburg, and Schwartzenber-Guggisberger.

milk-handling room. A room separate from or attached to a modern dairy barn, and separate from the milking parlor in which milk is cooled and stored in preparation for shipment.

milkhouse. See *dairy house.*

milkiness. 1. Certain physical qualities of a dairy cow, associated with dairy temperament, indicating milk producing ability. 2. Denoting the appearance of any solution or emulsion which is milk-white in appearance.

milking. 1. The quantity or quality of milk removed from an animal's udder at one time. 2. The act of removing milk from an animal's mammary gland. 3. Pertaining to *milking,* as in *milking* machine, *milking* quality of a breed.

milking animal. Any animal which produces milk primarily for man's use.

milking barn. A barn constructed primarily for milking and housing of dairy cattle. More often called a *dairy barn.*

milking cow. A cow in milk; a lactating cow. Also called *milch cow.*

milking cup. In a milking machine, the part which fits onto a teat of a cow's udder.

milking heavy. Giving a large yield of milk.

milking heifers and cows. Dairy animals in lactation. Any heifer which has freshened and presently is in lactation.

milking house. See *dairy house.*

milking line. 1. The place where cows are milked regularly. 2. The vacuum pipe line running from the motor to the stalls where cows are milked. The milker is attached to it by a petcock or valve.

milking machine. A mechanical device replacing hand labor in the milking of cows. The essential parts are teat cups, a vacuum pump and a milker pail or milk line. Milk is drawn from the udder by application of alternate vacuum and atmospheric pressure.

milking parlor. In a dairy, an especially arranged and equipped room where cows are separately fed concentrates and milked by mechanical milking equipment.

milking ration. Any feed formula prepared especially for milk cows to maintain or increase milk production. More often called a *dairy ration.*

milking room. A room where the cows are milked but not housed. It is an essential part of loose housing but optional with a stall barn.

Milking Shorthorn. A breed of dairy cattle which originated from the Shorthorn in England. Heavy milk producers, they are popular in England and in the central U.S. Predominant colors are red, white, or roan. Individuals weigh from 1,200 to 1,400 lbs. See *dual purpose breed of cattle; dual purpose.*

milking space. In a dairy barn, the space which

is used for the milking of cows, but which may also be used for feeding and bedding.

milking stall. A partial enclosure in a dairy barn designed to restrict the movement of a cow during milking. It is usually about 8 ft. long including the manger and is provided with a stanchion or other means of securing the animals.

milking stanchion. Part of a milking stall; a frame clamped on both sides of the cow's neck to limit forward and backward motion.

milking time. 1. A regular or appointed time at which a herd of cows is milked. 2. The time required to complete a milking.

milking tube. An instrument used for insertion into the opening of a cow's teat in cases of hard milking or injury to the teats preventing the free flow of milk. Also called *teat tube; teat cannula.*

milk ipecac. Spreading dogbane.

milk irradiator. A device which uses an electric arc or mercury vapor lamp to convert ergosterol in milk into vitamin D.

milk line. 1. A visible layer of milk in the bottom of a bottle of cream. Sometimes called *serum separation.* 2. A pipe line which carries milk from the milking machine to the milkhouse.

milk of lime. A whitish suspension made by slaking calcium oxide (quicklime) in water; used as a disinfectant and for whitewash.

milk ordinance. A law which sets forth regulations concerning the handling and quality standards of milk sold for public consumption.

milk pail. A metal, open or partially covered bucket with bail handle, usually of 12 to 14 quart capacity, used as a container in hand milking.

milk plant. A building equipped for the handling or processing of milk.

milk plasma. The liquid part of milk as distinguished from its fat content.

milk powder. See *dry milk; non-fat dry milk solids.*

milk preservative. A compound used to preserve composite samples of milk for later testing. See *corrosive sublimate.*

milk producer. 1. One who is engaged in the production of milk. 2. A lactating cow. 3. Any of various tonics, conditioners, and patent medicines supposed to increase milk yield.

milk production. 1. The quantity of milk produced by a cow or herd. 2. The amount of milk produced by the dairy industry.

milk pump. A pump specially constructed for the sanitary handling of milk.

milk purslane. Spotted euphorbia.

milk record. A record of the amount of milk produced by a cow during a specified period: used for evaluation in terms of money, profit, culling, or for registry.

milkripe. Designating semimature seeds of a plant in which the seed contents are milklike in texture or appearance. See *dough stage; milk stage.*

milkroom. A room partitioned into sections for handling raw milk. It may be inside or outside the dairy house *q.v.*

milk sample. A small amount of milk, normally collected at each farm each time the milk is collected; used for making certain tests at the dairy plant to determine butterfat content, bacteria count, etc.

milk sanitarian. A professional worker who specializes in the supervision of sanitary milk regulations.

milk scale. A spring scale, graduated in tenths of pounds, constructed for weighing milk from each cow at the time of milking.

milk secreting tissue. The glandular tissue in the udder.

milk secretion. A physiological process of all mammals in the production of nourishment for their young; lactation.

milk-serum. The whey of milk. Milk which is free from fat and casein. See *milk plasma.*

milk shed. The rural area from which an urban community draws its fluid milk supply.

milk sheep. A sheep kept primarily for milk production.

milk sheet. A single sheet of a record form used by dairymen to keep records of milk production.

milk shield. See *escutcheon: 2.*

milk sickness. 1. A disease of cattle caused by eating white snakeroot. Also called *white snakeroot poisoning; trembles; milk sick.* 2. A disease in man caused by ingestion of contaminated milk or milk from a cow affected by *milk sickness.*

milk soil. See *milk stone.*

milk-solids. The total solids of milk after water is excluded, which includes fat and solids-not-fat.

milk souring organisms. Groups of organisms, such as lactic streptococci, and coliform bacteria, which are responsible for the souring of milk.

milk stage. The period in the development of a grain when the contents of the seed are still fluid and of the color of milk. See *dough stage.*

milk star-of-Bethlehem. *Ornithogalum lacteum,* family Liliaceae, a bulbous herb, with milk-white flowers in dense racemes, grown for pot culture and as an ornamental. Native to South Africa.

milk stone. In dairy manufacture, a grayish-white, thin, chalky deposit which sometimes accumulates on heating surfaces, coolers, utensils, and freezers when improper washing methods have been used. It is largely a mixture of the minerals of milk and water.

milk strainer. 1. A metal dish or funnel-like utensil with a wire sieve or perforated metal plate and cotton filter bottom, designed to fit into a milk pail, the milk tank of a cream separator, or the neck of a milk can for straining extraneous matter out of milk. 2. A sock-like cotton filter used in the line of a pipe line milker.

milk substitute. In animal or dairy husbandry, any of a number of gruel feeds or mixtures fed to calves or pigs as a substitute for milk.

milk sugar. Lactose.

milk-sugar feed. See *dried whey.*

milk tank. Any of the different sized and shaped receptacles, especially a metal or glass container, in which milk is transported on a truck from collection centers to distributors and processors.

milk teeth. The temporary teeth of a young animal which are much whiter and smaller than the permanent teeth.

milk the crop. To massage the crop of a slaughtered fowl so as to force feed remaining in its crop out of the mouth. It lessens the danger of spoilage of the dressed fowl.

milk thistle. Blessed milkthistle.

milk tube. A rubber or plastic tube which connects the teat cup and the milker pail of a milking machine.

milk tube brush. A brush used for cleaning a milk tube.

milk vein. The subcutaneous, abdominal veins which are a continuation of the mammary veins of the udder of a cow. The largeness and tortuousness of the veins is used in judging dairy cattle.

milk vetch. Hayden poisonvetch.

milkweed. 1. Any plant of the genus Asclepias, family Asclepiadaceae. The plants are weedy, perennial herbs characterized by milky juice. Some species are poisonous to stock, and some yield a good quality of light colored honey. The white silky floss of the mature seed pod has been harvested experimentally as a substitute for kapok and other uses. 2. *A. syriaca,* a perennial, weed pest of waste ground and untilled fields. Found in central and western U.S. and Canada. Also called *silkweed; cottonweed.*

milk well. The point at which the milk veins of a cow enter the abdominal wall.

milk willow-herb. Winged lythrum.

milkwort. Polygala.

milky. In judging dairy cattle, designating a well developed udder, as well as several other dairy characteristics.

milky bellflower. *Campanula lactiflora,* family Campanulaceae, a perennial herb grown in the flower border for its showy cluster of white or pale blue flowers.

milky disease. An insect disease caused by *Bacillus popillae* which is used in the biological control of the Japanese beetle.

milky disinfectant. See *coal-tar creosote.*

milky mirror. See *escutcheon: 2.*

mill. 1. A building which has machinery used in grinding, pressing, processing, etc.: *e.g.,* grain *mill,* cane *mill,* cider *mill,* saw *mill,* etc. 2. To process a product. 3. To move around in a circle, as a herd of cattle.

mill buyer. A manufacturer who buys wool or grain directly from the producer.

milled grain. Grain processed in a mill: *e.g.,* wheat that has been made into flour.

milled rice. Rice grains which have been run through hullers for removal of the bran coat.

mill elevator. An elevator that is operated by milling companies primarily for obtaining the grain needed for their own operation.

milleped(e)s. Arthropods of the division Diplopoda, similar to centipedes but having a greater number of legs. They are known to be abundant in some soils and may be a factor in the development of a mull. They seldom injure plants but occasionally enter houses. Also called *thousand legged worms; galleyworms.*

miller. 1. The operator of any kind of mill, especially of a flour mill. 2. Any moth with wings which appear dusty or powdery.

millerandage. In grape culture, production of small, seedless berries. The pistil does not develop into a normal berry but remains green and very small when the normal berries in the cluster are ripe, or the berry grows larger but is seedless and somewhat smaller than the normal berries.

millet. Any of certain annual grasses and cereals with little seeds of the family Gramineae. Species are grown as field crops for grain and forage. Native to Europe and Asia. The principal kinds grown in America are: proso; foxtail millet; and barnyardgrass *q.q.v.*

millet smut. A smut caused by *Ustilago crameri,* family Ustilaginaceae, which affects the yield of millets.

millet woodrush. *Luzula parviflora,* family Juncaceae, a grass-like plant with soft, flat leaves, which grows on very moist or wet sites; its forage rating is fair.

millimeter. A unit of measurement which is one-thousandth of a meter; about one twenty-fifth of an inch.

millimho. A measure of electrical conductivity in proportion to salinity, used in expressing the comparative salinity of soils and water.

milling. The processing of grain, or other products, by hulling, chopping, crushing, etc.: *e.g.,* the conversion of wheat into flour in rolling mills.

milling-in-transit. A privilege granted by railroads which allows shippers to stop loaded cars for intermediate manufacturing between the loading point and destination without losing advantage of the through rates.

milling quality. 1. The percentage of whole kernels and total milled rice that is obtained from a given quantity of rough rice. 2. Any wheat which will yield a high percentage of white, sound flour.

million-dollar weed. See *waterhyacinth: 2.*

mill pond. A small body of water retained by a dam above a mill which is driven by water power. See *millrace.*

millrace. A sluiceway through which water runs to drive the wheel of a water mill.

mill run. 1. All the lumber output of a sawmill which has a sale value. 2. A millrace.

mill-run mixed feed. A poultry feed which is wheat bran mixed with gray (wheat) shorts.

mill-run wheat feed. Hard wheat mixed feed.

millstone. A large circular stone made from layers of stone pieces cemented together and grooved from the outer rim to a hole in the center. It is used for grinding cereals into flour and meal and it was in fairly common use in water grist mills in the U.S. up to the end of the 19th century. The stones were quarried from a very hard, coarse sandstone formation or a cellular quartz conglomerate.

mill work. Remanufactured lumber products, such as sash, doors, and molding.

milo. *Sorghum vulgare* var. *subglabrescens,* family Gramineae, a drought-resistant, annual grass, one of the older varieties of grain sorghums, which is grown mainly in the southern Great Plains region of the U.S. mostly for livestock feed. Native to tropical Africa. See *grain sorghum; sorghum.*

milo bran. A livestock feedstuff consisting of the outer covering of the milo kernel obtained in milling milo for brewers' grits or other similar purposes.

milo chop. A livestock feedstuff consisting of the entire grain of milo sorghum removed from the head and chopped or ground.

milo disease. A destructive root rot affecting milo, probably caused by the soil-dwelling phycomycete, *Pythium arrhenomanes* and other fungi in conjunction.

milo grain. In stock feeding, the grain or seed of the milo sorghum, in contrast to the whole head cluster.

milo mill feed. A livestock feedstuff which is a mixture of milo bran, milo germ, and a part of the starchy portion of the grain, produced in the manufacture of grits from milo grain.

milo root rot. A fungous disease of milo caused by *Periconia circinata,* family Dematiaceae, characterized by rotted, blackened roots and dwarfed plants.

milpa. A cultivated field, specifically a cornfield (presumably originated from the Indians of southwestern U.S.).

milpa agriculture. A tropical system of agriculture under which land is cleared, cultivated for a few years and then abandoned and returned to jungle for restoration of fertility. *Cf. brand tillage.*

milt. A ductless gland near the stomach and intestine of fowls. It contributes to the formation of new blood cells. See *spleen.*

mimosa. 1. An obsolete name for the genus Acacia (*q.v.*); at present used to designate certain Australian species of this genus, the wattles, which yield a high quality tanbark. See *wattle: 4.* 2. Silktree albizzia.

mimosa wilt. A disease resulting in wilting of leaves and death of the silktree albizzia, caused by a soil-dwelling fungus, *Fusarium perniciosum,* family Tuberculariaceae.

mince. 1. To cut, chop, or grind (meat, etc.) into little pieces. 2. The mass of small pieces into which something, as meat, has been cut. See *ground beef.*

mine. To eat out a gallery or passage way, as a larva *mines* a fruit. See *gallery.*

mineral. 1. A chemical compound or element of inorganic origin. 2. Designating the inorganic nature of a substance.

mineral elements. The inorganic components of soil, plants, or agricultural products.

mineral feeder. Any of a number of box or bin devices which automatically supplies a mineral or mineral ration in the feeding of livestock.

mineralized salt. A salt compound for animal feeding which is composed of a common salt base with the trace mineral elements, such as manganese, copper, iron, iodine and zinc, added to it.

mineral land. A class of land valued chiefly for its subsurface or surface mineral resources; a use classification. See *agricultural land; forest land; rural land.*

mineral manure. Inorganic fertilizer.

mineral material. Particles of rocks and minerals, in contrast to the other major components of soils: organic matter, water, and gases.

mineral matter. Inorganic matter; in analysis of organic matter, the ash from burning.

mineral mixture. Any feed containing salt, limestone, phosphates, minor elements, etc.

mineral oil. Oil obtained from petroleum and other hydrocarbon minerals, as contrasted with vegetable oils.

mineral rights. Rights pertaining to ownership of subsurface land minerals and to access in their exploitation. An owner of land may sell only surface rights and retain ownership of *mineral rights.*

mineral soil. The common soil of land surfaces. Consisting of comminuted minerals and rocks, the solid matter is preponderantly inorganic. See *organic soil, hydrosols.*

mineral stain. In hardwoods, particularly hard maples, an olive to greenish-black or brown discoloration of undetermined cause and commonly associated with bird-pecks and other injuries. *Cf. bird peck.*

miner's inch. Given a certain head, the amount of water that will flow through a one inch square opening. (The amount of head varies in the U.S., in parts of California 50 miner's inches equal one second foot and in Colorado 38.4 miner's inches equal one second foot.)

miner's lettuce. *Claytonia perfoliata* (*Montia p.; Limnia p.*), family Portulacaceae, an annual herb grown for its small, white flowers and as a potherb or salad plant for its edible leaves. Native to North America. Also called *winter purslane; Indian lettuce.*

mine run. Designating a product (eggs, etc.) as produced and not sorted or graded.

miniature broiler. A fast-growing broiler chicken, which is produced in six to eight weeks, about 1½ lbs. in live weight and about 1 lb. in dressed weight. Usually marketed frozen and in individual packages, it is intended to be served whole and generally stuffed with dressing.

miniature dahlia. Any dahlia with flowers 4 ins. or less in diameter, not including pompon dahlia *q.v.*

minimum acquisition feeding period. The shortest length of time the vector must feed upon a virus source in order to acquire enough virus to become infective.

minimum temperature. 1. A degree of warmth at which growth first begins. 2. The lowest temperature recorded during a given period, as the minimum daily temperature.

minimum tillage. The least amount of tillage required for quick germination and a good stand. Several implements may be drawn behind the tractor to reduce the number of times it is driven over the field, but it does not imply that primary tillage, secondary tillage, fertilization and seeding must be done in one trip across the field.

mining mule. A market class of mules: individuals weigh 600 to 1,350 lbs. and stand 12 to 16 hands.

mining scale. *Howardia biclavis* (Comst.), family Diaspididae, an insect which infests a variety of plants and is common in greenhouses. One of the scale insects *q.v.*

mining the soil. Exploitation or detrimental use of soil for crops without any effort to replace or maintain inherent fertility.

ministry (ministerial) land. Land which was set aside in the past by the U.S. Congress or other agencies, for the use and maintenance of the Christian ministry.

mink. A semiaquatic, weasel-like mammal of the genus Mustela, especially *M. vison,* which has thick, soft, usually dark brown fur with lustrous outside hairs. In the fur farming industry, mink are bred in captivity in natural brown as well as various other shades of color.

mink farming. The business of raising minks for their valuable fur.

mink ration. A meat or other ration fed to mink raised as breeding stock or for fur.

Minnesota No. 1. A red breed of swine developed in the U.S., by the Minnesota Experiment Station from a foundation of Tamworth and Landrace.

Minnesota No. 2. A black and white spotted breed of hogs; developed in the U.S., by the Minnesota Experiment Station from a foundation of Poland China and Yorkshire.

Minnesota seed treated. Designating one of several types of treatments for disinfecting cereal seed, especially for the prevention of smut diseases.

minnie-bush. Allegheny menziesia.

Minorca. The largest breed of chickens of the Mediterranean Class. They have long, strong bodies, enamel-white earlobes, and white skin. The egg-shell color is white. Varieties are Single-Comb Black, Rose-Comb Black, Single-Comb White, Rose-Comb White, and Single-Comb Buff.

minor ear rot. An ear rot of corn produced by species of Rhizopus, Aspergillus, and Penicillium, weakly parasitic fungi that are not aggressive pathogens under field conditions but which occasionally are destructive to stored corn.

minor elements. Inorganic nutrients which, in trace or very small quantities in soil, are necessary for optimum plant growth: *e.g.*, copper, zinc, boron, manganese, iron, *etc.* See *trace element.*

minor plant food. See *minor elements.*

mint. Any herb of the genus Mentha, family Labiatae. Species are grown for their aromatic plant parts and as a field crop; source of the essential oil used in flavoring and for medicinal purposes. See *apple mint; bergamot mint; field mint; peppermint; pennyroyal mint; spearmint.*

mint flea beetle. *Longitarsus menthaphagus* Gentner, an insect which is especially destructive of growing crops of mint.

mint geranium. Costmary chrysanthemum.

mint rust. A common disease of spearmint caused by the organism *Puccinia menthae.*

Mintzitra. A soft, ewe's milk cheese, made in Macedonia.

mirabilite. Glauber salt (*q.v.*), a mineral of the white alkali soils in arid regions of U.S.

miracidium. The ciliated embryo or larva which hatches from the egg of a trematode (fluke).

mire. Wet, spongy ground, as in a bog or marsh.

mirliton. See *chayote.*

miscible. Designating two or more substances which, when mixed together, form a uniformly homogeneous solution.

miscible oil. Any commercial preparation in which the emulsifier is incorporated with the oil and a small quantity of water.

miscibility. The ability of two materials, especially liquids, to mix.

misplaced yolk. An egg yolk which has escaped into the body cavity of the hen. See *internal layer.*

misshapen. Designating a cheese characterized by nonparallel ends as a result of uneven distribution of the curd in the hoops.

Mississippi Delta. The fertile alluvial bottom lands which lie along the lower course of the Mississippi River northward from the delta proper to southeastern Missouri in the U.S.

Mississippi scraper. A type of plow designed especially for very shallow cultivation of cotton and used in southern U.S. till the middle of the 19th century (obsolete).

Missouri banana. Pawpaw.

Missouri gourd. Calabazilla.

Missouri grape. Cat grape.

Missouri mule. A large mule from Missouri, U.S., a state famous for mule production.

Missouri primrose. Ozark sundrops.

Missouri River-bottom disease. A disease of horses attributed to poisoning from eating arrow crotalaria, a legume found mostly in southern U.S.

mist. 1. Liquid particles in a fine state of division but of perceptible size which make up spray material. 2. Transparent water suspended in particles near the surface of the earth. See *fog: 1.* 3. See *baby's-breath.*

mist blower. A type of spray equipment constructed to break up and atomize concentrated insecticide and fungicide mixtures and to disperse them into an air current of high velocity. It is used to apply solutions or suspensions.

mistflower eupatorium. *Eupatorium coelestinum,* family Compositae, a late-blooming, perennial herb grown for its head of small, blue or violet flowers. Native to North America. Also called *blue boneset; hardy ageratum.*

mistletoe. 1. Any of a number of segmented, flowering plants of the family Loranthaceae. Species are parasitic on trees and other woody plants. The *mistletoes* are divided into: (a) A dwarf group including members of the genus Arceuthobium, all without apparent leaves, occurring only on conifers. Also called *lesser mistletoes.* (b) A leafy group including, in temperate North and South America, members of the genus Phoradendron, mostly with thick, flat, green leaves but sometimes without apparent leaves. This group usually occurs on hardwoods, but some species grow on conifers. See *Christmas American mistletoe.* 2. *Viscum album,* the common *mistletoe* of Europe. The plant bears small, yellow flowers and occurs on oak, apple, etc. Native to Europe and Asia.

mistletoe hook. A sickle blade attached to a long pole used to cut objectionable mistletoe from forest trees or to cut off the plant for forage, or for sale as a plant ornament.

mistletoe rhipsalis. *Rhipsalis cassutha,* family Cactaceae, a spineless, epiphytic cactus which resembles mistletoe. It is used in hanging baskets. Native to tropical regions of America. Also called *willow cactus.*

miterwort. *Mitella diphylla,* family Saxifragaceae, a delicate, perennial herb grown in the shady part of a rock garden for its clusters of white flowers. Native to North America. Also called *bishop's cap; fairy cup.*

mites. Arachnids, very small spiders belonging to the Order Acarina. These parasites constitute a very large group living in all parts of the world and are pests on both plants and animals.

miticide. Any chemical mixture or compound, such as parathion, which is used to kill mites.

mitosis. Cell division involving the formation of chromosomes, spindle fibers and the division of chromosomes by a process of longitudinal splitting. Each of the resulting daughter cells thus has a full set of chromosomes as distinguished from reduction division or meiosis in which the daughter cells have half the somatic number.

Mitzithra. A Greek cheese made by adding fresh ewes' milk to the by-product whey from Feta *q.v.*

mix. 1. A formula; a combination of two or more ingredients, blended together for a specific purpose, such as feedstuff for livestock. 2. To cross or interbreed animals, plant varieties, etc., more or less by chance.

Mixa hook. A specially designed hook used to hang chickens for boning in poultry slaughter.

mix cooler. A heat exchanger which can be used as a cooler for a mix in dairy manufacture.

mixed bale. A bale of cotton which is composed of different grades and classes of lint cotton.

mixed barley. A market class for barley, as black barley and western barley.

mixed breeds. Farm animals which result from crossing different breeds of livestock.

mixed calves. A market class for a group of slaughter calves which are not uniform in grade.

mixed colored eggs. Market eggs which are not uniform in color.

mixed corn. All corn that does not meet the color rules for yellow corn or white corn.

mixed dip. The hot solution of potassium carbonate, sodium hydroxide, olive oil, and water, into which seedless grapes are dipped preparatory to sun drying.

mixed farming. See *diversification.*

mixed feed. Any livestock feedstuff which consists of grains and various edible materials and by-products, and sometimes vitamins and/or antibiotics mixed to meet special feeding needs. It is used very largely by the dairy and poultry industries. *Mixed feeds* are mostly commercial products.

mixed feed barley. A market class of barley which is the entire product obtained by grinding country-run barley containing not less than 75 percent of pure barley and not more than 25 percent of other grains, weed seed, and other foreign matter and with nothing intentionally added.

mixed feed oats. A market class for oats which consists of less than 30 percent of cultivated oats, but either (a) not less than 65 percent of cultivated and wild oats combined or (b) not less than 65 percent of wild oats. It may not contain more than 25 percent of other grains, or more than 10 percent of foreign material, of which more than 5 percent is fine seeds.

mixed fertilizer. A formula, or mixture of fertilizer compounds, as those of nitrogen, phosphorus, and potassium in definite proportions, and as commercially produced, usually containing a filler which may be inert. It may also contain one or more of the minor elements.

mixed forest. A forest which is composed of trees of two or more species. In practice, usually a forest in which at least 20 percent are trees of other than the principal species.

mixed gait. Lack of uniformity or cadence in the stride of a horse which is often a permanent fault due to inheritance or conformation.

mixed grain sorghum. A market class which includes all grain sorghums not meeting the separate standards for white, yellow, red, and brown grain sorghums.

mixed hay. Hay which consists of two or more forage species, as a mixture of red clover and timothy.

mixed-herd ranch. A form of range organization where the calves are grazed generally beyond the first year of age when the steers, some heifers, and the culled cows and bulls are sold.

mixed oats. In marketing, oats in which different colors of the lemma and palea are present.

mixed-packed cotton. Baled cotton which, in the samples drawn therefrom, (a) shows a difference of more than two grades, if of the same color; or

(b) if of the same grade name but of different color, shows a difference of more than two color gradations; or (c) shows a difference of two or more grades and two or more color gradations; or (d) shows a difference in length of staple exceeding 3/32 of an inch.

mixed pickles. A mixture of pickled green tomatoes, cucumbers, peppers, cauliflower, and onions; chowchow.

mixed prairie. A phytogeographic region west of the tall-grass prairie in which mid-grasses form the upper story and short grasses the lower.

mixed rot. A rot of eggs, probably due to heterogeneous bacterial flora, in which the yolk and white are mixed giving an addled, murky appearance when candled.

mixed stand. A stand of trees in which less than 80 percent of the trees in the main crown canopy are of a single species.

mixed staple. A bale of cotton in which the staple grade is not uniform.

mixed wheats. In a quality classification or grading of wheats for milling purposes, a mixture of classes as hard winter and soft winter varieties.

mix filler. A machine used for placing ice cream in cartons.

mixing floor. In feeding barns, a floor space used for mixing livestock feeds.

mixing nozzle. A nozzle in which the extruded liquid is mixed with a stream of air before it finally leaves the orifice.

mixing pens. In range management, a large pen in which a number of animals are confined. In handling goats, the does and kids are placed together until such time as the kids are able to follow the mothers in grazing.

mixing plant. An establishment which primarily mixes feeds or fertilizers.

mix stabilizer. Any of various substances, such as pectin, gelatin, Irish moss, etc., which may be included in the ice cream mix to improve its texture and body.

mixture. 1. Two or more virus strains occurring together in a host which cause symptoms that are predictable from the symptoms caused by the individual components in the host. 2. A blending together of seeds or other materials of different kinds.

mix-whipping. Rapid agitating of an ice cream mix to incorporate air.

mob. A general term for a herd of sheep, horses, or cattle, in Australia.

mobile soil colloids. Soil colloids so dispersed that they may move in the soil with percolating waters.

mobile water. Suspended water in soils in excess of pellicular water (*q.v.*), which occurs as a film covering the surface of solid particles of soil or the walls of fractures. It is free to move in any direction under the pull of the resulting force of gravity and unbalanced film pressure. Also called *gravity water; free water.*

moccasin flower. *Cypripedium acaule,* family Orchidaceae, a hardy orchid which grows wild in bog areas and is sometimes grown in the bog garden for its pink flowers with yellowish petals. Native to North America. Also called *pink lady's slipper.*

Mocha coffee. Arabian coffee.

mochila. A large piece of skin or leather covering for a horse saddle. In the last century, *mochilas* fitted with pockets called continas were used in western U.S. by pony express riders.

mock chicken. A style of preparing retail meat in which alternate layers of veal and pork are placed on a skewer in such a way as to resemble a chicken leg. See *city chicken.*

mock duck. A retail cut of meat made from the wing or outside half of the shoulder of a lamb carcass.

mockernut hickory. *Carya tomentosa,* family Juglandaceae, a deciduous tree grown for shade and for its valuable wood used in tool handles, ladders, furniture, sporting goods, implements, woodenware, and for fuel and smoking meat. Also called *big bud hickory; bird's-eye hickory; black hickory; curly hickory; whiteheart hickory.*

mock feed. Horse feed that is devoid of grain.

mock-orange. Any of the several flower-bearing, deciduous shrubs of the genus Philadelphus, family Saxifragaceae. Species are grown as ornamentals. Also called *false syringa.* See *drooping mockorange; Falconers mockorange; Gordon mockorange; Lemoine mockorange; Lewis mockorange; littleleaf mockorange; Magdalena mockorange; Peking mockorange; scentless mockorange; snowbank mockorange; sweet mockorange; virginalis mockorange; Zeyher mockorange.* 2. See *Carolina laurelcherry; Osage orange.*

mock plums. A fungous disease of plums caused by *Taphrina* Spp., family Taphrinaceae, characterized by several whitish spots or blisters enlarging rapidly to cover the fruit. The fruit becomes greatly enlarged, withers, and leaves a hollow cavity. Also called *plum pockets*

Modena. A Parmesan-type, cows' milk cheese made in the U.S. during World War II.

moderate. Designating the sales activity in marketing that is neither very rapid nor very slow.

moderately dished face. In stock judging, a concave facial contour of livestock.

moderate temperature. In climatology, temperature that is neither extremely low nor high but is near the normal seasonal mean for a geographical region.

Modern Game. A breed of game fowl, developed by fanciers for exhibition purposes. The *Modern Game* is far removed from the Old English or pit game, bred in earlier times for cock fighting. They have short and hard feathers and a style and carriage peculiar to themselves. Their normal and bantam size varieties are: Black-Breasted Red, Brown-Red, Golden Duckwing, Silver Duckwing, Birchen, Red Pyle, Black and White. See *Old English Game.*

modernized parity prices. See *new parity prices.*

modified continuous layerage. A method of propagation of woody plants, usually vines, by rooting the branches while attached to the parent plant. The shoot or cane is laid in a trench about 5 ins. deep and, after laterals have grown from its buds, soil is filled in over the mother cane and around the laterals. See *continuous layerage.*

modified Dadant hive. A large beehive (17⅝ ins. long), containing eleven frames (11¼ ins. deep), spaced 1½ ins. apart.

modified dried skim milk. Dried skim milk which is altered in various ways for such uses as dried milk-pectin mixtures, rennet mixtures, ice cream, soft types of cheese, etc.

modified live-virus. A virus which has been changed by passage through an unnatural host, such as passing hog-cholera virus through rabbits, so that it no longer possesses the disease-producing characteristics but so that it will stimulate antibody production and immunity when injected into susceptible animals.

modified milk. 1. Cow's milk which is altered to a composition suitable for the special needs of infants. 2. A mineral-vitamin fortified milk for general beverage use. (Sometimes applied to homogenized milk.)

modified process cheese. A processed cheese to which certain food constituents have been added to make a specific cheese product, as bacon cheese, etc. See *cheese spread.*

modified stem. Any type of plant stem, such as a tuber, rhizome, runner, or stolon, which differs from the upright stem.

modified wine. A sparkling wine made from a mixture of the juice of sound, ripe grapes with sugar or a sirup in just sufficient quantity to raise the alcoholic strength after fermentation to 11 percent by volume. Also called *ameliorated wine.*

modifier. Any element which is added to, or taken from, a substance which alters its normal appearance or function.

modifying gene. A gene which changes the expression of the chief gene, or genes, controlling a character.

modiola. *Modiola procumbens*, family Malvaceae, a plant grown in California, U.S. for pasturage on alkali soils. Escaped it can become a weed. Native to Chile.

modoc. Squatter (southwestern U.S.).

modock. Wool which is sheared from range lambs that had been feedlot fed in the central states (U.S.). See *fedlamb.*

module. 1. A unit which is used as a standard or ratio of measurement. 2. In the design of gears in the United States, it is the pitch diameter in inches divided by the number of gear teeth.

mogote. A thick patch of shrubbery (southwestern U.S.).

mohair. 1. The long, lustrous hair from the Angora goat. 2. Cloth which is made from the hair of Angora goats.

mohair cincha. A Western cinch or saddle girth which is made of mohair.

Mohave Desert. A hot, extremely arid region of southern California, U.S., characterized by complete absence of rain during summer and only by scant winter rain. The natural vegetation is mostly desert shrub types. The area includes highly productive soils under irrigation and crops, such as alfalfa, citrus, cotton and dates are grown.

mohawkweed. See *wood merrybells.*

moist mash. A mixture of ground feeds and other ingredients (for poultry and livestock) moistened with water or milk.

moisture. 1. The total amount of water in any plant or animal product. 2. Any form of water. 3. The total amount of water (exclusive of that in chemical combinations) in the soil, both that which is available and that unavailable for plant growth. In common usage, either the available water or that which can be easily detected by the senses as wet (saturation), moist (normal), or dry. 4. The amount of water which is contained in wood usually expressed as: (a) a percentage of the weight of the oven-dry wood (used in the lumber industry); (b) a percentage of the total weight (used in the paper industry); (c) an absolute amount of water in an absolute amount of wood (by weight or volume). The moisture content of wood varies from 5 to 10 percent when it is kiln dry, and from 30 percent upward when it is green or unseasoned. 5. The water content of feeds.

moisture content tolerance. In wood technology, the allowable variation above and below the mean, usually in connection with moisture-content specifications.

moisture equivalent. In soil, the ratio of (a) the weight of water which the soil, after saturation, will retain against a centrifugal force 1000 times gravity, to (b) the weight of the soil when dry. It is expressed by the formula 100 $\frac{c}{w}$, where c is the weight of the water that remains in the soil after it has been subjected to the centrifugal force and w is the weight of the soil when dry.

moisture pan. An adjunct to egg incubators; a vessel that contains the water which provides the humidity necessary for incubation.

moisture penetration. 1. The depth to which moisture penetrates in soil following an irrigation or rain before the rate of downward movement becomes negligible. 2. Downward movement of water into a soil (a soil may be open, pervious and permit free penetration or percolation, or it may be compact and tight and thus prevent free downward movement).

moisture percentage (of soils). The ratio of the weight of water in a soil to the dry weight of the soil which is expressed as a percentage of dry weight; the standard basis for expressing soil moisture.

moisture-proofing. Any treatment that reduces the absorption or adsorption of moisture by wood. It is usually accomplished by the impregnation of wood so that some of its hygroscopic moisture is replaced by the moisture-repellent substance.

moisture-retentive soil. A soil that retains a high percentage of moisture or water. In mineral soils, retentiveness bears a relation to clay or colloid content: *e.g.*, a clayey soil is generally more retentive than a sandy soil.

moisture tension. The force at which water is held by soil, usually expressed as the equivalent of a unit column of water in centimeters: 1,000 cm. being equal to one atmospheric tension. *Moisture tension* increases with dryness and indicates the degree of work required to remove soil moisture for use by plants.

Moitch eucalyptus. *Eucalyptus rudis,* family Myrtaceae, an evergreen, aromatic tree, grown in very warm regions as a shade or shelterbelt tree and as a bee plant. Native to Australia. Also called *desert gum.*

moits. See *motes: 2.*

molar. In mammals, one of the large back teeth with ridged or rounded surface specially adapted for grinding.

molascuit. A cattle feed made from sugarcane bagasse (*q.v.*) and molasses.

molasses. 1. The dark brown sirup obtained as a by-product in the manufacture of sugar from sugar cane. 2. Loosely, a sirup obtained by evaporation of the sweet juices of any plant, as sorghum, corn, sugar beets, sugar maple, etc. See *blackstrap; beet molasses.*

molasses feed. An animal feed, usually a commercial mixture in which molasses is an ingredient; used as a substitute for grain because of its palatability and because it is a cheap source of readily digested carbohydrates.

molassesgrass. *Melinis minutiflora,* family Gramineae, a tall, perennial grass grown as a forage crop in the U.S. Native to Africa. Also called *efwatakala grass.*

molasses silage. Legume or grass silage to which specified amounts of molasses, such as blackstrap, are added to aid the proper fermentation, increase the carbohydrates, improve palatability, etc.

mold (mould). 1. A form containing a cavity into which material is poured or pressed to achieve a special shape and design: *e.g., molds* for Edam and Gouda cheese, butter, ice cream, etc. 2. The F layer and the humus of a soil on a forest floor. 3. Any soft, humous soil. 4. Fungi distinguished by the formation of a mycelium (a network of filaments or threads), or by spore masses; usually saprophytes. However, various kinds may do serious damage to fruits, hay, grain, growing crops, and ornamental plants. See *black mold: 2; blue contact mold; downy mildew; mildew: 2; powdery mildew.*

moldboard. The part of the plow just back of the share which receives the furrow slice from the share and turns it partially or completely over. Also called *breastboard; earthboard.*

moldboard hiller. A curved, metal, boardlike attachment for a cultivator, designed to bank or hill the soil around a growing crop.

moldboard plow. A plow equipped with a moldboard to turn the furrow slice.

moldiness. In judging and grading cheese, any defect due to the presence, or previous presence, of undesirable molds.

mold poisoning. Harmful effects upon animals from eating certain moldy feeds. The poisoning may be due to the degree of moldiness; or to prolonged feeding when less harmful molds are involved. With horses, the effect may be a nervous disease resulting in blindness and a staggering gait.

mold powder. Culture of mold spores used as a starter for blue-veined cheeses.

mold spot. In judging and grading cheese, a defect characterized by occasional, tiny, scattered mold areas on the surface. These are more generally on the ends than the sides of the cheese.

moldy. Infected by molds (fungi); designating objectionable quality due to appearance or odor, flavor, toxicity, caused by molds.

moldy butter. Butter affected by molds (fungi) causing it to become unpalatable or unmarketable.

moldy corn. See *Diplodia disease of corn.*

moldy egg. In marketing, an egg infected by molds which have entered the shell. It is classed as inedible.

moldy feed. See *bloat: 1.*

moldy hay. Hay affected by various kinds of molds observed on leaves and stalks. It is usually dusty and has a bitter taste.

mole. 1. One of several species of burrowing, insectivorous mammals, characterized by soft, dark fur, long nose, very small eyes, and large, strongly developed front feet. The moles belong mainly to the family Talpidae. A common species in eastern U.S. is *Scalopus aquaticus.* They can damage lawns and gardens with their underground burrows but otherwise are not serious pests. 2. A natural levee or depositional ridge on the bank of a large river. 3. A plow used in mole drainage *q.v.*

mole cricket. Any of the several insects of the family Gryllidae which eat plant roots, other insects, and earth worms.

mole drainage. A method of draining land by making open channels in the soil. These are constructed with a mole plow consisting of a two-wheel frame and a projectile-shaped ball varying in diameter from 2 to 8 ins. and chain linked behind a sharp pointed, sharp edged shank. The plow is pulled along the course of the desired drain with the ball at depths of as much as 2 to 3 ft.

mole plant. Caper euphorbia.

mole-tree. Caper euphorbia.

Moline elm. *Ulmus americana,* family Ulmaceae, a rapidly growing clon of the American elm that has dark green foliage. It must be carefully pruned to eliminate the formation of weak crotches to which it is subject. It is more upright and less spreading than the American elm.

Moliterno. A plastic-curd cheese made in Italy.

mollusk. Any animal of the phylum Mollusca, as a slug, snail, etc., characterized by a soft, unsegmented body protected wholly or in part by a calcareous shell.

molt. The casting off of old feathers, skin, horns, etc., before a new growth, as with the normal annual renewal of plumage of adult chickens, turkeys, etc. Also called *the castoff covering* (British); *moult.*

Moltke linden. *Tilia moltkei,* family Tiliaceae, a hybrid, deciduous tree grown as a shade and street tree.

molucca balm. Asian molluccabalm.

molybdenosis. See *molybdenum poisoning.*

molybdenum. Mo, a trace element (*q.v.*). Applied to soils in very small amounts in some available form, such as sodium, molybdate, under certain conditions, it may promote vigorous growth in a number of farm crops and prevent disease. Some plants may store up excessive amounts of *molybdenum* from the soil and thus become toxic.

molybdenum poisoning. The poisoning of cattle, sheep, and goats resulting from the presence of molybdenum in feeds or forage. Other animals appear to be quite resistant to the toxic effects of molybdenum. Symptoms are anemia, diarrhea, emaciation and reduction in milk production. Also called *molybdenosis.*

mombin. Any tree of the genus Spondias, family Anacardiaceae, grown in the tropics for their edible fruit or ornament. Also called *ambarella; caja manga; jobo; otaheitapple.*

monacillo. South American waxmallow.

Moncenisio. A blue-mold, Gorgonzola-type of cheese made in Italy.

Monday disease. Azoturia.

Monday morning disease. Azoturia.

Mondseer Schachtelkäse. An Austrian, whole or partly skimmed milk, Münster-type cheese, cured three to six weeks.

money crop. In a farm enterprise, the crop which is a major source of income, as cotton, wheat, potatoes. Other crops may be grown only as subsidiaries for producing feed for livestock or for maintaining or increasing the yield of the money crop. Also called *cash crop.*

moneywort. *Lysimachia nummularia,* family Primulaceae, a prostrate, perennial herb grown as an ornamental ground cover for its habit and its yellow flowers. Escaped, it can become a weed pest. Native to Europe. Also called *creeping Charlie; creeping Jennie; creeping loose-strife; creeping Sally.*

Mongolian deutzia. *Deutzia parviflora,* family Saxifragaceae, a shrub grown for its white flowers. Native to China.

Mongolian spirea. *Spiraea gemmata,* family Rosaceae, a deciduous shrub grown for its white flowers. Native to China.

mongrel. 1. A mosaic virus disease of tobacco characterized by leaf blotches. 2. Animals or plants of mixed or unknown breed.

mongrel seed. Seed used for planting which is a mixture of seeds of several strains or varieties, as opposed to pure seed or that of a single variety.

moniliasis. An infectious mycotic disease of the mouth and especially of the crop of domestic poultry and of the mucous membranes of dogs, calves, colts and man. The causative fungus is *Candida albicans.* Also called *thrush crop mycosis.*

monitor. Catamaran.

monkey-coconut. Syrup jubea.

monkeypuzzle araucaria. *Araucaria araucana,* family Araucariaceae, a tree grown as a curiously branched ornamental in warm regions. Native to Chile.

Monkshood. Any plant of the genus Aconitum, family Ranunculaceae. Species are poisonous, perennial herbs, sometimes semi-climbing, with blue, white, or yellow, hooded flowers. Native to the North Temperate Zone. Also called *blue rocket.* See *aconite monkshood.*

monkshoodvine. *Ampelopsis aconitifolia,* family Vitaceae, a hardy, woody, grape-like vine grown as an ornamental for covering a wall. Native to China.

monk's peppertree. Chastetree.

monoammonium phosphate. A chemical used in fertilizer mixtures. When pure, it contains 12.17 percent phosphoric acid.

monocarp. A plant that blooms and fruits only once and dies, as an annual.

monocephalous. 1. Designating a fruit or ovary having only one head or summit. 2. Designating a plant with flowers disposed in a single head or umbel.

monocotyledon. A plant characterized by its embryo having a single cotyledon, or seed leaf; by stems with woody fibers distributed through the ground tissue rather than laid down in layers showing secondary growth; by vascular bundles that are scattered through the stem tissue; by leaves typically parallel-veined and linear, and floral parts in threes or multiples of three. All grasses, cereal grains, sedges, etc., are *monocotyledons.*

monoculture. Cultivation of a single crop, as wheat or cotton, to the exclusion of other possible uses of the land.

monocytic. Characterized by, or having the nature of, monocytes, large, white blood cells with an oval or indented nucleus.

monoecious plant. A plant having both staminate and pistillate flowers. See *dioicous.*

monoestrous animal. An animal which has only one estrous (heat) cycle each year.

monogastric. Designating an animal with one stomach.

monoglucoside of cyanidin. Idaein, the pigment of cranberries.

monoglucoside of delphinidin dimethyl ether. A

very complex organic chemical to which is attributed the color of some fruits, as of the grape, *Vitis vinifera,* and of berries of the Virginia creeper.

monohybrid. A hybrid whose parents differ in a single character.

monolith. An undisturbed (not broken up) sample of a soil profile.

monolithic concrete silo. A concrete silo made as a poured casting, in contrast to one made of concrete staves.

monophagous parasite. A parasite restricted to one species of host.

monophyllous. Designating that plant part which is made up of a single leaf: *e.g.,* certain calyces. Sometimes used in reference to plants bearing only one leaflet.

monoploid. See *haploid.*

monopodial. Designating a plant in which the growth of the main stem is continued indefinitely by the terminal bud. *Cf. sympodial.*

monopterous. Designating a single-winged seed as in the ash tree. See *samara.*

monorchid male. A male animal which has only one testicle in the scrotum. Also called *ridgeling; rig; cryptorchid.*

monosomic. Designating a diploid organism which lacks one chromosome of its proper complement.

Monostorer. A ewes' milk, salt-washed, Rumanian cheese, cured from 8 to 10 weeks.

monotrophic. Designating bees which visit only one kind of flower.

monovoltine. See *voltine.*

monoxenous parasite. A parasite which requires only one host for its complete development.

monsoon. See *rainy season.*

Montadale. A new breed of sheep developed from a Columbian ram and a Cheviot ewe. A year old ewe weighs about 120 lbs., and rams should be proportionately heavier. They yield up to 12 lbs. of fleece each.

Montanas. In marketing, sheep which originate on the range or in the feed lots of Montana, U.S.

Montasio. A hard, Fontina-like, cows' and goats', or ewes' milk, Italian cheese which is eaten fresh or after curing.

Montavoner. A sour milk, herb flavored cheese, made in Austria.

montbretia. Crocus tritonia.

Mont Cenis. A hard, blue-mold, cows', ewes', and goats' milk, French cheese which is cured three to four months.

Mont d'Or. A soft, goats' or cows' milk, French cheese, salt-washed, and cured from one to three weeks.

monte. 1. A tract of dense mesquite brush. 2. The chapparal of the foothills of Southern California, U.S.

Monterey. A cows' skimmed or partly skimmed milk, American cheese which is cured up to six months.

Monterey cypress. *Cypressus macrocarpa,* family Pinaceae, an aromatic, tender, evergreen tree grown as an ornamental and for windbreaks. Native to California, U.S.

Monterey pine. *Pinus radiata,* family Pinaceae, an irregularly headed, picturesque, evergreen tree confined to the coastal region of California, U.S. Used to some extent as an ornamental and for shelterbelt plantings.

Monterey-pine weevil. *Pissodes ratiatae* Hopk., family Curculionidae, a weevil which infests pines on the West Coast of the U.S., mining twigs, stems, etc.

Montevideo escallonia. *Escallonia montevidensis,* family Saxifragaceae, a tender, evergreen shrub or small tree grown in warm areas for its foliage and white flowers. Native to South America.

Monthery. A soft, surface-ripened, blue-mold, skimmed or partly skimmed, cows' milk cheese, made in France.

Monti process. An Italian commercial method of concentrating fruit juice by freezing.

Montmorency. A famous variety of sour cherry that is very widely grown in commercial orchards in the U.S.

montmorillonite. A hydrous aluminum silicate, one of the clay minerals, found in the clay-colloid separate of soils. It is grayish or reddish in color and has a high cation exchange capacity.

mooley. See *muley.*

moonarian. A farmer who plants and cultivates crops according to phases of the moon.

moon-blind. Referring to a horse suffering from periodic ophthalmia *q.v.*

moonblindness. See *periodic ophthalmia.*

mooneyed. See *periodic ophthalmia.*

moonseed. *Menispermum canadense,* family Menispermaceae, a woody vine bearing grape-like fruit and attractive foliage, grown as an ornamental in the U.S. It is a poisonous plant: the rootstock contains bitter alkaloids, and cases of poisoning, attributed to wild grapes, have occurred from eating moonseed fruit by mistake. When eaten, the pits with their sharp ridges may also cause mechanical injury to the intestines. Native to North America. Also called *moon-creeper.*

moonwort. See *dollarplant; perennial honesty.*

moor. A bleak, treeless waste land. In Scotland and England, *moors* are generally characterized by peat soil and by heather (*Calluna* spp.), and in wetter spots by sedges and grasses. (There is no very similar natural landscape feature in the U.S., but the term *moor* is used in ecologic and pedologic literature. *Hochmoor* and *Niedermoor* from German refer to (a) moss bogs and very acid peats, and (b) less acid or more muck-like soil and reed marshes, respectively.

moorband. The siliceous pan cemented by iron oxides; a feature underlying peaty, moorland soils. Also called *heath pan.*

moorberry. Small cranberry.

moor myrtle. Sweet gale.

moor soils. See *moor.*

moorwort. *Andromeda polifolia,* family Ericaceae,

a low, evergreen shrub, bearing drooping clusters of small, pink flowers, that grows in acid soils. Native to the North Temperate Zone. Also called *andromeda; bog rosemary.*

moosewood. See *striped maple; Atlantic leatherwood.*

mopping up. In forestry, making a fire safe, after it is controlled, by extinguishing or removing burning material along or near the control line and by felling snags.

mor. A type of forest soil humus layer which consists of organic materials usually matted or compacted or both, distinctly delimited from the underlying mineral soil. It is not highly decomposed and granular and is nearly free from admixed mineral particles thus being differentiated from *mull* (*q.v.*). It may be equivalent to the F layer (*q.v.*) of soil profile.

moraine. The geological deposition formed on the margins of glaciers or beneath moving ice sheets. In the glaciated region of the U.S., *moraines* vary greatly according to kind: terminal, ground, etc., in rock composition relief and minor topographic features, but they are significant locally in soil and land classification.

moraine garden. A feature of rock gardens: a bed of coarse stones, 2 or 3 ft. thick, to which soil may be added. It may be watered either from above or below.

mora (mowra) meal. The meal or ground seeds with the fat removed, of the mowra buttertree of India, *Madhuca longifolia.* It contains about 2.7 percent nitrogen. When spread on lawns or golf greens and sprinkled with water, it causes worms to come to the surface.

morass. Any mirky land, bog, marsh, or swamp.

morbidity. 1. Being diseased. 2. The extent to which a disease exists.

morbus maculosus. Petechiale.

morel. An edible mushroom: the fruit body of *Morchella* spp., especially *M. esculenta,* family Helvellaceae.

Morello (cherry). 1. A pomological group of varieties in the sour cherry (*q.v.*) species. The fruit and juice of these varieties is dark colored. Some common varieties are English Morello and Wragg. 2. Any variety of the above group.

morelos orangeworm. See *Mexican fruit fly.*

Moreton Bay chestnut. *Castanospermum australe,* family Leguminosae, a tall tree grown in warm areas for its yellow-orange flowers, and for its seeds, edible when roasted. Native to Australia. Also called *black bean.*

Moreton Bay fig. *Ficus macrophylla,* family Moraceae, a large tree grown in warm areas as an ornamental. Native to Australia.

Moreton Bay pine. Cunningham araucaria.

Morgan. An American breed of light, general-purpose horses developed principally as a riding horse. Individuals stand 14.2 to 16 hands high and weigh 800 to 1,200 lbs. Standard colors are bay, brown, black, and chestnut with white marking not uncommon.

Mormon cricket. *Anabrus simplex* Hald., family Tettigoniidae, an insect found in Utah and other western states (U.S.), which can be very destructive to nearly all field, orchard, and garden crops.

Mormon tea. Ephedra.

Mormon weed. Chingma abutilon.

morning-glory. 1. Any of several plants of the genus Ipomoea, family Convolvulaceae. Species are mostly annual and perennial, twining vines grown as ornamentals on trellises, fences, etc. The sweet potato (*q.v.*), however, is grown for its edible tuber. See *bigroot morningglory; white-edge morning-glory.* 2. *I. purpurea,* an annual, twining, tender plant grown for its purple, blue, or pink flowers with pale tubes. Native to tropical America. Also called *common morning-glory.*

morning-glory inlet. An inlet structure to an underground pipe line whose weir is shaped like a morning-glory.

morning-glory sphinx. The larva of certain moths of the genus Herse, that feeds on plants of *Ipomea* spp. *H. convolvuli* attacks the morning-glories and *H. cingulata* attacks sweet potatoes (southern U.S.).

morningstar lily. *Lilium concolor,* family Liliaceae, a bulbous herb grown for its fragrant, bright red flowers. Native to Japan. Also called *star lily.*

Morocco glorybind. *Convolvulus mauritanicus,* family Convolvulaceae, a perennial, prostrate herb grown in a hanging basket and as a ground cover for its prostrate habit and for its blue or violet-purple flowers. Native to Africa.

morphogenesis. The developmental history of organisms or of their parts.

morphology. 1. A branch of biologic science which deals with the forms, rather than the functions, of plants and animals. 2. Pertaining to pedology or soil science, the study of soil horizons and their arrangement in profiles. 3. Pertaining to land surfaces, the shape or configuration of physical features.

Morrill Act. An Act of U.S. Congress approved July 2, 1862, which provided for the establishment of colleges of agriculture and mechanic arts in the various states. The bill was first introduced by Representative (later Senator) Morrill of Vermont in 1857. The main feature of this and later acts was the donation of public lands for the support of the land grant educational institutions in each state. A subsequent act (the second Morrill Act), passed in 1890, allowed an annual, Federal grant of $25,000 to each Land-Grant college and university. See *Land-Grant college.*

Morrison Standard. A feeding standard (*q.v.*) formulated by F. B. Morrison, author of *Feeds and Feeding,* which is adapted to practical feeding conditions and is used extensively.

Morrow honeysuckle. *Lonicera morrowii,* family Caprifoliaceae, a hardy, deciduous shrub grown for its white flowers and blood-red fruit. Native to Japan.

Mortadella. A sausage which originated in Bologna, Italy: 75 percent pork and 25 percent beef flavored with garlic. It is placed in beef bladders, steamed under pressure, and then dried.

mortality. Death; destruction of plants or animals due to disease, insect damage, drought, wind, fire, etc.

mortgage. A lien on land, buildings, personal property, etc., which is given by a borrower to the lender as security for a loan.

mortgagee. One to whom a mortgage is given as security for money he has loaned to another.

mortgage insurance. An insurance plan sponsored by the U.S. Federal Government for insuring the payment of real estate mortgage loans. First used by the Federal Housing Administration, it was later adapted to farm-ownership loans available through the Farmers Home Administration. In this program the farm purchaser must make a 10 percent down payment; the mortgage insurance is on the 90 percent balance of the reasonable value of the farm including necessary repairs and improvements. See *credit insurance.*

mortgagor. Giver of a mortgage as security for funds which he has borrowed.

mosaic disease. Any of the several virus diseases of plants so called because the characteristic symptom, mottling of the leaves, has the appearance of a mosaic pattern.

mosaic-rosette of wheat. A virus disease transmitted through soil and also mechanically; characterized by mottling, striping, and rosetting.

mose. A marsh.

mosquito. An insect belonging to the family Culicidae. The proboscis (*q.v.*) of the female has a needle-sharp tube in it with which it pierces the skin of man or animal to suck blood. Mosquitoes are pernicious pests and transmit diseases both to animal and man. See *common malaria mosquito.*

moss. A flowerless plant intermediate in character between algae and ferns. *Mosses* of the Tundra have a grazing value, some have value as ornamental plants, and some have a commercial value for packing and rooting plants.

mossberry. Small cranberry.

moss cabbage rose. *Rosa centifolia* var. *muscosa,* family Rosaceae, a shrub grown for its fragrant, pink flowers. Native to Caucasus. Also called *moss rose.*

moss campion. Schafta silene.

moss peat. Peat composed mainly of the identifiable remains of mosses, especially those of the genera Sphagnum, Polytrichum, and Hypnum. *Moss peats* are natural soils and are used commercially for mulches, pot and greenhouse soils, poultry litter, barn bedding, paper making, fuel, etc. Also called *highmoor peat; peat moss; Sphagnum peat.*

moss phlox. *Phlox subulata,* family Caryophyllaceae, an evergreen, creeping, herbaceous perennial herb grown for its habit and its bright, purple, pink, or white flowers. Native to North America. Also called *ground pink; flowering moss; mountain pink; moss pink.*

moss sandwort. *Arenaria verna* var. *caespitosa,* family Caryophyllaceae, a low, compact, mosslike herb grown for its needlelike, dark and yellow-green leaves and tiny, starry flowers. Also called *tufted sandwort.*

moss silene. *Silene acaulis,* family Caryophyllaceae, a perennial herb grown for its purplish-red flowers. Native to Europe and North America. Also called *cushion pink; moss campion.*

moss verbena. *Verbena laciniata,* family Verbenaceae, a tender, perennial herb grown in the flower garden and as a pot plant for its dense clusters of lilac flowers. Native to South America.

mossy. 1. Designating irregular, dark markings which spoil an otherwise desirable color contrast on the feathers of domestic birds. 2. Covered with moss.

mossy-coated. In the show ring, designating a beef animal which has hair of a solid color containing darker mottlings.

mossy-cup oak. Bur oak.

mossy horn. An animal horn which has become wrinkled and scaly from age.

mossy rose gall. *Rhodites rosae* L., family Cynipidae, an insect whose larvae cause galls on rose canes.

mossy stonecrop. Goldmoss stonecrop.

motes. 1. In cotton ginning immature seed and particles of trash. 2. Burs, etc., which are found in wool. Also called *moits.*

moth. An insect belonging to the Order Lepidoptera. The adults have wings covered with scales and differ from butterflies in being night fliers and in having heavy, hairy bodies. In the caterpillar stage, many *moths* are destructive; they may be injurious and do tremendous damage to cloth, crops, trees, and stored grain.

moth ball. See *naphthalene.*

moth bean. *Phaseolus aconitifolius,* family Leguminosae, a trailing, annual herb with yellow flowers, cultivated for its edible seeds and for forage. Probably native to India. Also called *mat bean.*

mother. In a chick hatchery, a shelter for newly-hatched chicks.

mother block. A group of plants known to be disease free or true to type from which cuttings are made, usually in the greenhouse, for obtaining plants for commercial production.

mother bulb. A bulb having several points or divisions from which splits or bulblets are obtained for propagation.

mother cane. A cane or branch, especially of grape, used in propagation by continuous or modified-continuous layerage.

mother chrysanthemum. *Chrysanthemum indicum,* family Compositae, a low perennial which bears flowers in yellow, showy clusters. Native to China.

mother clove. The mature dried fruit of the clove which are clove-like in appearance, but less aromatic. See *clove: 3.*

mother culture. In dairy bacteriology, milk or a

liquid milk product into which a control culture of actively growing bacteria, specific for the production of a definite fermented product, has been introduced.

mother ditch. A main canal which carries irrigation water.

mother-of-thousands. Strawberry saxifrage.

mother-of-thyme. *Thymus serpyllum,* family Labiatae, a woody, perennial, aromatic herb grown for its small, purplish flowers and for its foliage used in seasoning. Native to Europe, Asia, and North Africa. Also called *creeping thyme; wild thyme; brotherwort.*

mother-of-vinegar. The thick film of bacteria *Acetobacter* spp. which may form on the surface of vinegar during production.

mother plant. 1. In sugar-beet seed production, the steckling or first year root which is dug, stored over winter, and transplanted to the seed growing field the following spring. 2. Any plant from which parts are removed in vegetative propagation.

mother root. In the propagation of the sweet potato, the root which is bedded for propagation of the crop.

mother seed potato. The potato tuber or slice containing eyes which is planted for the propagation of the crop.

mother spleenwort. *Asplenium bulbiferum,* family Polypodiaceae, a fern with proliferating fronds, grown as an ornamental. Native to Africa and Australia.

mother tree. In plant breeding, the original tree from which varieties are developed.

motherwort. *Leonurus cardiaca,* family Labiatae, a wild-growing, coarse, perennial herb, with tooth-edged leaves and small purple flowers, which is a source of nectar and pollen for honey bees. The plant causes a dermatitis in susceptible individuals. Also called *lion's tail.*

moth mullein. *Verbascum blattaria,* family Scrophulariaceae, a strong-growing, biennial herb grown for its yellow flowers, lilac at the base. Escaped, it becomes a weed pest. Native to Europe and Asia.

motte. An isolated, small clump of trees or shrubs; a minor vegetational feature of the natural landscape in Texas and other states (U.S.). Also called *mott.*

mottle. A spot or blotch of a color different from the mass color of a surface, as a *mottle* caused by a virus disease of a plant.

mottled. 1. Designating bird feathers marked with white tips at the ends or spotted with colors or shades at variance with the ideal. 2. Irregularly marked with spots of different colors, as a *mottled* soil, butter, etc.

mottled caladium. *Caladium picturatum,* family Araceae, a tuberous-rooted herb grown as a summer bedding plant for its variegated leaves. Native to South America.

mottled color defect. A body defect in cheeses characterized by spots of different shades arising from microbiological action or from physical causes in manufacture.

mottle leaf. A virus disease of sweet cherry characterized by varying degrees of leaf mottling and rugosity depending on the cherry variety and strain of the virus. Endemic in western U.S.

mottle-leaf of citrus. A disease of citrus trees, ascribed to zinc deficiency and characterized by a yellowing of tissue between the main lateral veins of the foliage. Also called *frenching; small offs.*

mottling disease of sugar cane. Yellow spot.

mottling of leaves of cereals. A disease of cereal plants caused by a deficiency of magnesium in the soil.

mould. See *mold; molds.*

moult. Molt.

mound. 1. In tillage, to heap up or ridge the soil about a plant or row of plants. 2. To cover portions of branches in mound layering (*q.v.*).

mound layering. The rooting of branches of woody plants or shrubs by leaving the branches upright and mounding soil about the basal portions of the stems.

mound planting. A method of planting on wet ground, the seeds or young trees being planted on mounds, ridges, or hills; hill planting; ridge planting.

mound prairie. In some parts of the U.S., a presettlement land feature comprising a vast expanse of treeless grassland marked by enclosures, called mounds, believed to have been built by Indians.

mound storage. Storage of vegetables on the ground surface under a covering of straw and earth.

mount. 1. A horse for riding. 2. A riding seat. 3. To copulate, as a male animal. 4. To get into a saddle on the back of a riding horse.

mountain ash. American mountain ash.

mountain-ash sawfly. *Pristiphora geniculata* (Htg), family Tenthredinidae, an insect which infests the mountain ash sometimes defoliating it.

mountain azalea. Piedmont azalea.

mountain balm. 1. Yerba santa. 2. Oswego bee balm.

mountain balsam. Alpine fir.

mountain blackberry. Allegheny blackberry.

mountain bluebells. *Mertensia ciliata,* family Boraginaceae, a perennial herb of good forage value growing at altitudes of 5,000 to 12,000 ft.; also grown in the wild garden for its bright blue flowers. Native to western U.S.

mountain blueberry. Big whortleberry.

mountain brome. *Bromus carinatus,* family Gramineae, a bunchgrass of excellent forage value. Native to western U.S. Also called *California brome.*

mountain brome leaf spot. A fungous disease caused by *Selenophoma bromigena,* and endemic in the central and western states, U.S.

mountain brome smut. A smut, *Ustilago bromivora,* family Ustilaginaceae, which attacks the seed heads of the mountain brome.

mountain bunchgrass. Green fescue.

mountain butterwort. One species of butterwort *q.v.*

mountain camellia. Mountain stewartia.

mountain cranberry. Cowberry.

mountain creeper. Christmasvine porana *q.v.*

mountain currant. Alpine currant.

mountain ebony. Buddhist bauhinta.

mountain fetter-bush. Mountain pieris.

mountain fever. Equine infectious anemia.

mountain fleeceflower. *Polygonum amplexicaule,* family Polygonaceae, a perennial herb grown in the flower border for its rose or white flowers. Native to the Himalayas.

mountainfringe. *Adlumia fungosa,* family Fumariaceae, a very delicate, leaf-climbing, biennial vine grown in the wild garden for its white or purple flowers. Native to North America. Also called *climbing fumitory; Allegheny-vine; fairy creeper.*

mountain hemlock. *Tsuga mertensiana,* family Pinaceae, an evergreen tree with rounded, bluish-green leaves, grown as an ornamental. Native to North America.

mountain hickory. *Acacia penninervis,* family Leguminosae, a hardwood, timber tree growing up to a height of 40 ft. Native to Australia. Also called *hickory acacia.*

mountain holly. *Nemopanthus mucronata,* family Aquifoliaceae, a neat, deciduous shrub, with handsome fruit and autumn foliage, mostly growing in bogs and swamps. Native to eastern North America. Also called *American mountain holly.*

mountain land. The land surface of any physical feature or region with sufficient elevation and relief to be properly named mountain or mountainous. (The term usually implies very steep slopes, stoniness, adverse climate or other depreciating conditions, although all *mountain land* is by no means of poor quality or non-agricultural.)

mountain larkspur. Sierra larkspur.

mountain laurel. *Kalmia latifolia,* family Ericaceae, an evergreen shrub grown for its foliage and its rose or white flowers. It may be poisonous to bees and livestock. Native to eastern North America. Also called *American laurel; American mountain laurel; calico bush.*

mountain lilac. Deerbrush ceanothus.

mountain lily. See *starlily.*

mountain lover. Canby pachistima.

mountain lupine. *Lupinus alpestris,* family Leguminosae, an herb having wide distribution on the ranges of western U.S., which is highly palatable to cattle and sheep.

mountain mahoe. Jamaica linden hibiscus.

mountain mahogany. 1. Designating shrubs or trees of the genus Cercocarpus, family Rosaceae. Some species serve for browse and some are grown as ornamentals. Native to U.S. See *true mountain mahogany.* 2. See *sweet birch.*

mountain meadow. A glade or a moist or wet treeless tract which supports sedges, grasses, and other herbaceous vegetation, usually above the timberline at high altitudes.

mountain meadow soil. See *meadow soil.*

mountain-misery. Bearmat.

mountain oyster. The testicle of a lamb or pig which is used for human consumption.

mountain pieris. *Pieris floribunda,* family Ericaceae, a broad-leaved, evergreen shrub grown as an ornamental, as an accent plant, or in rock gardens. Native to North America. Also called *mountain fetter-bush.*

mountain pine beetle. *Dendroctonus monticolae,* Hopk., family Scolytidae, a bark beetle that is especially damaging to lodgepole and white pines in northwestern U.S.

mountain rose bay. Catawba rhododendron.

mountainrose coralvine. *Antigonon leptopus,* family Polygonaceae, a high-climbing vine, grown in frost-free areas on a fence, wall, etc., for its pink flowers. Native to Mexico. Also called *coralvine; corallita; pinkvine; Confederate vine; rosa-de-Montana; San Miguelito; chain-of-love; queen's jewels.*

mountain sage. Fringed sagebrush.

mountain sandwort. Greenland sandwort.

mountain sheep. Domestic sheep, such as the Cheviot, Black-faced Highland, Dartmoor, etc., native to the English and Scottish highlands.

mountain silverbell. *Halesia monticola,* family Styracaceae, a tender, deciduous tree grown for its profusion of white flowers. Native to southeastern U.S. Also called *tisswood.*

mountain snowberry. *Symphoricarpos oreophilus,* family Caprifoliaceae, a hardy shrub rated as fair to good as a browse for livestock, except horses. Found in the aspen-fir belt of the western mountains, U.S.

mountain spinach. Garden orach.

mountain starwort. Greenland sandwort.

mountain stewartia. *Stewartia ovata,* family Theaceae, a shrub grown for its white flowers. Native to North America. Also called *mountain camellia.*

mountain stonecrop. *Sedum ternatum,* family Crassulaceae, a white-flowered, perennial herb grown as a ground cover and as a plant for rock walls. Native to North America. Also called *false ice plant; wild stonecrop.*

mountain tea-tree. See *checkerberry.*

mountain timothy. Alpine timothy.

mountain winterberry. *Ilex montana,* family Aquifoliaceae, a tree grown for its foliage and for its red or purple fruits. Native to eastern U.S.

mounted corn picker. A mechanical corn picker which is mounted on a tractor.

mounted cultivator. A cultivator which is mounted on a tractor.

mounting block. Any block of wood, etc., which is designed as a step for easier mounting of a saddle horse.

Mount Morgan wattle. Pearl acacia.

mourat. Brown wool of some sheep (obsolete).

mourning bride. See *sweet scabious; grassleaf scabious.*

mourning-cloak butterfly. *Nymphalis antiopa* (L.), family Oecophoridae, an insect pest whose larvae feed on the foliage of elm, poplar, and willow. The larva is called *spiny elm caterpillar.*

mourning cypress. *Cupressus funebris,* family Pinaceae, an evergreen tree grown for its foliage and drooping branches. Native to China.

mourning widow. Sweet scabious.

mouse. Any of the numerous, smaller rodents of the family Muridae and related families, the most common of which is *Mus musculus,* the house mouse, which is of typical size, 6 ins. long including the tail.

mouse barley. *Hordeum murinum,* family Gramineae, an annual grass common in the foothills and valleys of California, U.S., which has a fair forage rating. Native to Europe.

mouse dun. A coat-color of some animals, especially horses. It is a dun color imposed on black, seal brown, dark mahogany bay, or dark liver chestnut giving a smoky effect. Sometimes called *smoky.*

mouse-ear chickweed. Big cerastium.

mouse-ear hawkweed. *Hieracium pilosella,* family Compositae, a weed resembling the orange hawkweed except that its flowers are more yellow. It is prolific but easily controlled in fields by cultivation. Native to Europe and Asia.

mousy. Designating the peculiar odor of fermented fruit juices damaged by the Tourne bacterium *q.v.*

mouth-borne disease. Any disease which gains entry to the body via the mouth: *e.g.,* on food.

mouth speculum. An instrument which is used by veterinarians for holding an animal's mouth open to facilitate inspection of the mouth, dental work, or the administering of medicaments. Also called *jaw spreader.*

mouton. The fur which is made by chemical treatment of wool on sheep skins.

movable dam. A water barrier that may be opened in whole or in part. The movable part may consist of gates, stop logs, flash boards, wickets, or any other device whereby the area for water flow through or over the dam may be controlled.

movable frame. A frame or comb based on the principle of bee space, which, when placed in a hive, remains separated from its surroundings by large pieces of burr-comb or heavy deposits of propolis and is easily removed for inspection.

movable hog house. A hog house, frequently mounted on skids, which may be easily moved for sanitary purposes or for better use of summer pastures.

movement. See *air blast.*

move out. In livestock marketing, to sell animals and to remove them from a pen.

moving screen. A screen used in moving beehives which consists of a rim the size of the top of the hive about 2 ins. in depth, covered with screen wire. It permits the bees to cluster in this space and provides additional ventilation.

mow. 1. The place in a barn where hay is stored. 2. A pile of hay. 3. To cut hay grasses or the grass of a lawn.

mow-burned. Designating hay that has been damaged or altered by heating and fermentation in the mow. Also called *mowburnt.* See *brown hay; tobacco-brown hay.*

mow-drying. The use of a blower and duct system in the hay mow to provide forced ventilation to reduce the moisture content of the hay. Hay having approximately 40 percent moisture may be put in the mow and the curing completed with the aid of the blower system. Normal outside air or artificially heated air may be used.

mower. 1. A machine with a mowing sickle cutting bar which is designed to cut forage for hay, weeds, etc. 2. A machine with a cutting reel or rotating blade for cutting the grass of a lawn.

mower attachment. Any special attachment on a mowing machine, such as one designed to windrow green forage and legume seed crops.

mower knife. The long, horizontal cutting bar to which mower sections are attached on a mowing machine.

mower pitman. A connecting rod from a crank wheel to the sickle bar of a mowing machine.

mower section. One of the triangular cutting blades on the mower knife of a mowing machine.

mower wheel roller. A land roller which is homemade from the wheels of worn-out mowing machines. It can be made by stringing the wheels on a shaft, or by using only two wheels fastened together to make a cylindrical roller.

mowing. See *mow: 3.*

mowing machine. Mower.

mowra meal. Mora meal.

mow storage. Storage of hay or unthreshed grain in the mow of a barn.

moxieberry. Creeping pearlberry.

moxie plum. Creeping pearlberry.

Moyes rose. *Rosa moyesii,* family Rosaceae, a hardy shrub grown for its dark red flowers. Native to China.

Mozarinelli. A cows' or buffalos' milk cheese made in Italy.

Mozzarella. A soft, plastic curd cheese first made in southern Italy, originally from buffalos' milk, but now from cows' milk. It resembles, and is made much in the same way, as Scamorze, and is eaten fresh with little or no ripening.

mucilage. A sticky or gelatinous substance manufactured by certain plants, as the marshmallow *q.v.*

mucilage cell. In wood anatomy, a specialized parenchyma cell which secretes mucilage.

muck. 1. Dung or manure in a wet, fermented or decomposed state, as in a barnyard. 2. See *muck soil.*

muck ditch. In flood control engineering, a trench excavated under the center of a levee and refilled with impervious material to form a cut-off wall and to prevent the seepage of water along the foundation plane of the levee. Also called *cut-off trench.*

muckland. A tract of land which is characterized by muck soil either in a virgin state or in crop use.

muck soil. 1. An organic soil which is derived mainly from plant matter, but one in contrast to peat in that the matter is more highly decomposed and has lost its botanical identity. A *muck* may be granular or amorphous but is humified and blackish. 2. Any organic soil in crop use even if technically peat. Peat with a high percent of mineral matter is sometimes designated *muck.*

mucopurulent. A body discharge which contains both mucus and pus.

Mucor. A widely distributed genus of molds of the family Mucoraceae, mainly saprophytic species, abundant in soil, decaying vegetable matter, dung, etc. Some are capable of converting starch into sugar, and thus important industrially. Probably the most frequently met and most troublesome species is the black bread-mold, *Rhizopus nigricans,* which appears wherever starchy substances are found, as on stale bread, and which causes a watery rot of strawberries (leak), and peaches (whiskers), a soft rot of roots of the sweet potato, and even rot in the tubers of the white potato when they have been weakened by cold or bruises or other unfavorable conditions. A few species of *Mucor* are weak parasites in animal tissues. Choanephora attacks and destroys the blossoms of squash and pumpkin (Cucurbita) and of Hibiscus.

mucor rot. A rot affecting citrus fruits caused by Mucor *q.v.*

mucosa. Mucous membrane.

mucous discharge. A discharge of mucus or one which has a mucuslike consistency.

mucous membrane. Membrane lining body cavities and canals connecting with the exterior, as the digestive tract.

mucus. A clear, viscid substance secreted by glands found in the mucous membranes.

mud. 1. Fine-textured soil so saturated with water as to be semifluid or in a sticky, plastic condition. 2. Freshly deposited alluvium of flood waters. 3. Flows of ash from the heavy rains occurring after volcanic eruptions.

mud boat. In lumbering, a sled with wide runners used in transporting logs out of swamps.

mud cap. A mass of mud placed on an explosive charge on top of an object to be shattered, as a large boulder. The mud tends to confine the explosive force.

mud cracks. Irregular or polygonal cracks, caused by shrinkage from drying, on the bare surface of clay soil or on flats of freshly deposited alluvium.

mudding. Dipping the roots of a transplant in a tub of fluid mud made from fine-textured soil. Also called *puddling.*

mud fence. A fence made of clay, adobe soil, etc.

mud fever. 1. A variety of erythema that attacks the heels and coronets of horses' feet, causing irritation. It may develop during wet weather from mud caked on the feet. 2. In some sections of the U.S., a popular term for avian monocytosis of turkey.

mud flat. Mud, or very fine sediments, built up under water at the mouth of an estuary along a sea-coast. It is usually bordered inland by salt water marsh and exposed only at low tide. Such land when reclaimed has a high agricultural potential. See *tidal flat.*

mudflow. The movement of water-saturated soil and other unconsolidated deposits downslope under the force of gravity. See *mud-rock flow.*

mud-in. 1. To seed grain, such as oats and corn, in excessively wet soil. 2. To transplant by dipping the bare roots of the plant in mud (rare).

mud-rock flow. A great mass of soupy earth and boulders carried down a steep slope by heavy rains or spewed in torrents from the mouth of a canyon. It is a phenomenon of mountainous regions. Also called *slumgullion.*

muermo. Certain hardwood trees of the genus Eucryphia, family Eucryphiaceae, grown for their timber. Native to Chile.

muffs. The whisker-like crop of feathers on either side of the face below the eyes of certain fowls.

mugging. Bulldogging or throwing a calf.

mugho pine. *Pinus mugo* var. *mughus,* family Pinaceae, a horticultural variety of the Swiss mountain pine (*q.v.*) with shorter cones than the parent stock. Native to southern Europe. Also called *mugho Swiss mountain pine.*

mugwort wormwood. *Artemisia vulgaris,* family Compositae, an herbaceous perennial with beautiful foliage; at one time used in medicines. Native to Europe and Asia. Also called *mugwort; felon herb; fat hen.*

Muhlenbergs smartweed. *Polygonum coccineum,* family Polygonaceae, a strong-growing, perennial weed troublesome on low, wet ground. Native to North America. Also called *marsh smartweed; swamp persicaria.*

muhly. Any grass of the genus Muhlenbergia, family Gramineae. Several species provide a moderate volume of forage on western ranges, U.S. See *aparejo grass; bush muhly; deergrass; nimblewill, spike muhly.*

Muir peach dwarf. A virus disease prevalent on Muir peach in California, U.S.; characterized by large, flattish, deep green leaves, closely appressed on short twigs. It is experimentally transmissible to other varieties of stone fruits.

mulatto land. An old term for land which contains brownish plow soil due to a mixing of humus layer and a red subsurface layer. (A non-technical term, it once was commonly used in some southern states, U.S., but now is largely in disuse.)

mulberry. Any of the several plants of the genus Morus, family Moraceae. Species are grown for their edible fruit and for foliage used as food for silkworms. See *Japanese mulberry; red mulberry; silkworm mulberry; white mulberry.*

mulberry whitefly. *Tetraleurodes mori* (Quaint.), family Aleyrodidae, an insect that infests mulberry, dogwood, ash, mountain laurel, linden, and maple.

mulch. 1. Soil, straw, peat or any other loose material placed on the ground to conserve soil moisture, or prevent undesirable plant growth or soil erosion. 2. Material, such as straw, placed over plants or plant parts to protect them from cold or heat, as the *mulch* for strawberries. 3. To modify soil, as by cultivation, so that it serves as a *mulch,* or to supply mulching materials to the soil.

mulched-basin irrigation. In soil irrigation, the building of permanent basins with substantial levees and covering each basin with a mulch of organic material, such as barnyard manure, bean straw, or alfalfa hay.

mulching. See *mulch: 3.*

muld. See *mull.*

mule. 1. The hybrid offspring of a jackass and a mare; used as a draft animal. *Mules* have advantages over horses in that they are more resistant to heat, require less care in feeding, have fewer digestive disturbances, are less nervous and more surefooted than the horses. *Cf. hinny.* 2. The spinning machine used for the manufacture of woolen yarn.

mule cart. A four-wheeled vehicle used in the Atlantic Coastal Plain region of the U.S. for hauling logs. The logs are suspended at one end under the axle of the rear wheels and are dragged over the ground.

mule fat baccharis. *Baccharis viminea,* family Compositae, a leafy shrub grown for its beautiful, yellowish flowers. Native to North America. Also called *guatemote.*

Mule-Foot hog. A very old breed of hogs with a solid hoof like that of a mule. Its origin is uncertain: possibly a native of Africa. It is bred locally in the U.S. but is not recognized as an official breed and is rarely seen.

mule jack. A jack bred to a mare in contrast to one used for jack stock perpetuation. See *jennet jack.*

mulesears wyethia. *Wyethia amplexicaulis,* family Compositae, a low, sunflower-like perennial with leaves shaped like the ears of a mule. Its forage rating is good, its leaves and flower heads are eaten by grazing animals. Native to North America.

mule skinner. 1. A mule-team driver, especially in the western U.S. Also called *muleteer.* 2. A whip used on a mule team.

Mules' operation. An operation involving removal of folds of the skin from the crotch of ewes. See *cutaneous myiasis.*

muley (cow). A cow without horns; generally, any polled or hornless beef or dairy cattle.

muley saw. A straight ripsaw used during the early lumbering days in the U.S.

mulga acacia. *Acacia aneura,* family Leguminosae, a small tree or shrub used as cattle fodder in times of drought. Native to Australia. Also called *mulga wattle.*

mulga grass. Certain grasses of the genera Danthonia and Neurachne, family Gramineae, that have high forage ratings, so named because they are often found growing among mulga acacias. Native to Australia.

mulga ironbark eucalyptus. *Eucalyptus sideroxylon,* family Myrtaceae, a medium-sized, evergreen tree with rough black or dark-red bark and white or yellowish flowers. Native to Australia. Also called *red ironbark.*

mull. A type of organic surface horizon of forested soils in which the organic matter is well decomposed and largely humus, granular in structure, relatively rich in bases, and medium acid to slightly alkaline in reaction.

mullein. Designating plants of the genus Verbascum, family Scrophulariaceae. Species are grown in the flower border. See *flannel mullein; purple mullein.*

mulley. Muley (cow).

mulsh. Mulch.

multiflora bean. Scarlet runner bean.

multiflora fence. See Japanese rose.

multiflora rose. Japanese rose.

multigerminal disk. A yolk of an egg which has more than one germinal disk, a somewhat rare occurrence.

multihitch. A hitch which consists of a combination of doubletrees, bars, braces, evener arms, and chain levers permitting successful use of a team of three or more horses in pulling a plow, etc. See *hitch: 2.*

multiparasitism. The condition of a host suffering simultaneously from the attacks of two or more species of primary parasites.

multiparous. 1. Giving birth to more than one offspring at a time. 2. Having had several offspring resulting from more than one pregnancy.

multiple annual ring. In tree growth, an annual ring which consists of two or more false annual rings.

multiple-arch dam. A water barrier which consists of a series of arches reinforced by buttresses or piers, through which the load is transferred to the foundation.

multiple-block board. A board of blocks used in the production of comb honey. The board consists of twelve 3⅝ in. square blocks in which the foundation sections are fastened.

multiple cropping. In favorable climates, the growing of two or more crops consecutively on the same field in a single year.

multiple factor inheritance. Type of inheritance in which a character is determined by many genes with similar effects.

multiple factors. In genetics, two or more factors which have similar effects on a particular character.

multiple land use. Use of land according to the principle that each resource is to be developed and used so far as possible without undue interference with other resources.

multiple plow. A plow so constructed that there are two or more bottoms. Also called *gang plow.*

multiple-price system. A proposal for variable prices, presumably set by the federal government (U.S.), for agricultural products to overcome inequities in production and to deal with surpluses: *e.g.*, one price for domestic consumption and another price when the same product is to be exported. Also called *two-price system.*

multiple-row seeder. Any planter that sows more than one row in a single operation. See *seeder; grain drill; planter: 1.*

multiple superphosphate. Double superphosphate.

multiplication plot. A field used for growing a crop to produce seed in large quantities from a comparatively small lot of selected seed.

multiplier onion. *Allium cepa* var. *solaninum,* family Liliaceae, a bulbous herb which produces separable bulbs used in propagation; grown for its edible bulb and leaves. Also called *potato onion.*

multiservice metal can. A metal can; a bulk container for ice cream.

multivoltine. See *voltine.*

multiwall heavy paper bag. A laminated paper bag which is used for packaging products, such as fertilizers, flour, and sugar. The bags are made of layers of heavy paper, asphalt, aluminum foil, or plastic to prevent moisture absorption.

mum. Chrysanthemum.

mummification. In animal reproduction, the drying up and shriveling of the unborn young.

mummified. Dried up and shriveled, as in fruit affected by the brown rot pathogen. See *stone brood.*

mummified fetus. In animal reproduction, a shriveled, dried fetus that has remained in the uterus instead of being aborted or expelled.

mummy. A dried, shriveled fruit which may hang onto the tree and resist further decay.

mummy-berry. A destructive disease of the cultivated blueberry caused by *Monilinia vacinii-corymbosa,* family Sclerotiniaceae, which results in a dry fruit rot and shedding of the diseased berries. Blossoms and shoots are also blighted. It can be very destructive in wet seasons.

mung bean. *Phaseolus aureus,* family Leguminosae, an annual herb whose small, green or yellow seeds are used for food, or to make bean-sprouts (*q.v.*). The plants are used for forage or green manure. Native to Asia. Also called *black gram; golden gram; green gram; gram; mungo bean.*

mungo. Shoddy or reworked wool.

Munro grass. *Panicum agrostoides,* family Gramineae, a common pasture grass. Native to North America. Also called *redtop panicum.*

Munsell color standards. A color designation system that specifies the relative degrees of the three variables of color: hue, value, and chroma. The standards may be used in precise comparison of colors of soils, or in standardizing agricultural products: *e.g.*, prime cottonseed cake is 10YR5/5, which means yellow-red with value = 5, and chroma = 5.

Munson system. A system of training grape vines in which 2 ft. long crosspieces are fastened at right angles to posts about 4 ft. above the ground. A wire passes over the ends of the crosspieces and another wire is fastened to the posts about 6 ins. lower. The trunk is brought up to the lower wire and two canes are fastened, one to the right and one to the left. The two upper wires support the growing shoots which develop from the canes and droop to the ground. The system protects the grapes from the sun.

Münster. A semisort, cows whole milk, German cheese; cured from 6 to 8 weeks.

muriate of potash. KCl, potassium chloride containing 50 to 62 percent of potash, K_2O. It may be used as a fertilizer, singly or in combinations with other materials.

murin. Wool from the carcasses of dead and decayed sheep. See *dead wool.*

muriti palm. *Mauritia vinifera,* family Palmaceae, a tall, fan palm grown for its edible nuts and useful fiber. Also called *wine mauritia.*

Murphy. 1. A potato. 2. An Irishman.

murrain. 1. A plague of potatoes, *murrain* is a term applied to the historically famous potato blight disease in Ireland. 2. See *cattle tick fever.*

murrina. An equine disease caused by *Trypanosoma hippicum,* a protozoan parasite in the blood. The affected animals are emaciated and feverish and may have anemia, conjunctivitis, edema, and paralysis of the hind legs.

muscadine grape. Bullace grape.

muscadine grape angular leaf spot. A disease of bullace (muscadine) grapes caused by the fungus, *Mycosphaerella angulata,* family Mycosphaerellaceae. It produces small, angular, black spots on the lower side of the leaves of the plant. Heavily infested leaves turn yellow and die.

muscadine grape black rot. A disease of bullace (muscadine) grapes caused by the fungus, *Guignardia bidwellii,* family Mycosphaerellaceae; characterized by conidia in tiny, black pycnidia, abundant in dead tissues of spots, cankers, and affected fruit during the growing season.

muscadine grape bitter rot. A serious disease of the bullace (muscadine) grape caused by the fungus, *Melanconium fuligineum,* family Melanconiaceae; characterized by a soft rot of the berries beginning at the pedicel. The surface becomes covered with a black crust, and the berry becomes mummified.

muscadine grape Macrophoma ripe rot. A berry rot of bullace (muscadine) grapes caused by a fungus of the genus Macrophoma, family Sphaeropsidaceae. It is characterized by circular, sunken, tan or brown spots which extend over the berry and shrink it.

muscatel. 1. A golden-amber dessert wine made from bullace (muscadine) grapes. 2. A raisin which is the dried bullace (muscadine) grape.

muscle contracture. A congenital, or lethal, disease of animals, characterized by stiff joints. The young pigs, lambs, and calves are sometimes affected in the womb and are dead at birth.

muscovado. Moist brown sugar, freshly separated from molasses (obsolete).

muscovite. See *mica.*

Muscovy. A domesticated, duck-like waterfowl which originated in South America. It is commonly classed as a duck but is of a distinct race and when crossed with other races of ducks its progeny are sterile. The period of incubation for eggs is 35 instead of 28 days, as with other ducks. They differ also in that the head and face of both male and female are covered with coruncles. There are two varieties: White and Colored.

muscular stomach. See *gizzard.*

mush. A food made of corn meal boiled in water to form a pudding.

mushmelon. Muskmelon.

mushroom. 1. A general term for the fruiting bodies of fungi of the Class Basidiomycetes. See *toadstool.* 2. More specifically, the fleshy, edible fruiting bodies of fungi such as the morels or cup fungi of Class Ascomycetes, Order Pezizales, truffles of Order Tuberales, of the same Class, puff-balls, of Class Basidiomycetes, Order Lycoperdales, and the sulfur mushroom of the Order Polyporales, of the same Class. 3. In the popular sense, *Agaricus campestris,* the extensively cultivated common *mushroom,* and other wild-growing, edible species of genus Agaricus. See *morel; puff-ball; sulfur shelf-mushroom.*

mushroom blotch disease. A disease of the mushroom cap caused by the bacterium, *Pseudomonas tolaasii,* family Pseudomonadaceae.

mushroom mite. *Tyrophagus lintneri* (Osb.), family Acaridae, a pest of mushrooms, and to a lesser extent of cheese and grain.

mushroom mummy disease. A drying up and shriveling of mushrooms; suspected to be caused by a virus.

mushroom olive mold. A weed mold in mushroom beds which appears following the sweating of beds to kill insects. It is controlled by a refinement in pasteurization; *i.e.,* by holding temperatures in the beds below 150°F.

mushroom root rot. An important, sometimes fatal, disease of fruit and forest trees caused by the fungus, *Armillaria mellea* (*Armillariella m.*), family Agaricaceae. It is characterized by a rot in the roots, by mycelium between bark and wood in the rotted portion, and by fungus-strands growing out into the soil. There is stunting of growth and yellowing and wilting of foliage. Also called *oak fungus disease; oak root fungus disease; honey-mushroom root rot; shoestring root rot; fungus root rot; crown rot; toadstool disease; Armillaria root rot.*

mushroom truffle disease. A disease, caused by a weed mold, which affects the production of mushroom spawn. It is controlled by growing the mushroom crop at temperatures below 65°F.

mushroom weed molds. Any undesirable fungi appearing in mushroom beds which competes with, and causes diseases affecting, mushroom spawn.

mushroom white plaster mold. A destructive, powdery mold, frequently encountered in the early days of mushroom culture. Controlled by the use of gypsum applied to the compost.

mush rot. See *soft rot of sweet potato.*

mushy-chick disease. Omphalitis.

mushy wool. Dry wool that has weathered on the sheep's back. It is wasty when combed.

musk daisybush. *Olearia argophylla,* family Compositae, a tree, 15 to 30 ft. tall, valued for its pleasant smelling wood used in cabinet-making. The leaves, when crushed, give out a musky odor. Native to Australia. Also called *musk tree.*

muskeg. A swamp or bog.

musk grapehyacinth. *Muscari moschatum,* family Liliaceae, a bulbous, perennial herb grown for its early-spring, purplish (later yellow or brownish) flowers. Native to Europe and Asia.

musk mallow. 1. *Malva moschata,* family Malvaceae, a perennial herb grown for its pink or white flowers. Native to Europe. Also called *musk rose.* 2. *Hibiscus abelmoschus,* family Malvaceae, a tropical, annual or biennial herb grown for its amber seeds used in perfume and for its yellow flowers with crimson eyes. Native to Asia. Also called *abelmosk.*

muskmelon. *Cucumis melo,* family Cucurbitaceae, an annual, trailing herb, of which there are several commercially valuable varieties bearing a round, fleshy, edible fruit. There are many cultivated forms of this species and their classification is difficult. The botanical variety *C. melo* var. *reticulatus* includes the more or less musk-scented types to which *muskmelon* is more restrictedly applied. They are also called *netted melon* or *nutmeg melon* and in the trade are commonly called *cantaloup.* The true cantaloup, however, belongs to *C. melo* var. *cantalupensis.* The fruits of *C. melo* var. *reticulatus* usually have a netted skin, shallow to moderate ribbing, and flesh from light green to reddish-orange in color. Common horticultural varieties include Hale's Best, Honey Rock, and Iroquois. The botanical variety *C. melo* var. *inodorus* is the wintermelon. The fruit is relatively non-odorous and has a smooth or ridged skin, ripens late, and keeps well. The commercial production of the wintermelon is largely confined to California and Arizona where Honeydew, Casaba, Crenshaw, and Persian are representative horticultural varieties. Native to Asia. See *mango melon.*

muskrat. *Ondatra zibethicus,* an aquatic rodent. Raising and trapping *muskrats* for their fur provides a supplemental source of income for some farmers in parts of the U.S. The propagation of this animal is one of the multiple uses of certain kinds of marsh lands. Also called *marsh rabbit.*

muskrat-weed. See *spotted waterhemlock.*

musk rose. See *musk mallow: 1.*

musk tree. Musk daisybush.

muskwood. 1. The aromatic, mahogany-like wood of the American muskwood, *Guara trichilioides,* and the musk bitterwood, *Trichilia moschata,* family Meliaceae. 2. See *musk daisybush.*

musquash. The dressed skin or prepared fur of the American muskrat. See *muskrat.*

musquash root. Spotted waterhemlock.

musquash weed. Tall meadowrue.

musrol. The nose band of a bridle.

must. The fresh juice of grapes before fermentation.

mustang. 1. In the southwestern U.S., a wild descendant of the Spanish horse, generally smaller and inferior to the domestic Spanish horses. 2. A small domestic pony that is largely of Spanish breeding. 3. A range horse that is somewhat wilder than the average. Also called *mestang.*

mustang cattle. Wild cattle (southwestern U.S.).

mustanger. A man who catches wild horses (obsolete).

mustang grape. *Vitis candicans,* family Vitaceae, a wild grape which grows very profusely in southern U.S. The fruit has a pungent taste and is gathered for making jelly. Also called *bird grape.*

mustang wine. Wine which is made from the mustang grape.

mustard. Common name for some of the herbs of the genus Brassica, family Cruciferae. Species are grown for use as potherbs and for their seeds used in the manufacture of the condiment, *mustard.* See *white mustard; potherb mustard; black mustard.*

mustard meal. The meal which remains after the oil has been extracted from the mustard seed. It contains about 5 percent of nitrogen and very small percentages of phosphorus and potassium. It has limited use as a fertilizer.

mustiness. An offensive, pungent odor caused by molds.

musty. Designating a defect of milk, cream, butter, or ice cream that has an offensive odor as that imparted by molds or by a damp, unventilated cellar or basement.

musty egg. A moldy egg; one that has a disagreeable odor and musty flavor. Also called *hay egg.*

mutage. The arresting of the fermentation process in must *q.v.*

mutant. An individual whose genotypic constitution differs from that of its parents because of a change in a gene or genes.

mutation. A sudden variation in an organism that is inherited. It may involve gene changes or chromosome changes.

mutterkorn. See *ergot: 1.*

mutton. The flesh of a grown sheep (at least one year old) as opposed to lamb.

mutton bluegrass. *Poa fendleriana,* family Gramineae, a perennial, bunch bluegrass, 1 to 2 ft. high, growing at high elevations in western U.S. It is a range plant of excellent forage value. Native to North America.

mutton bread. Mutton sheep.

mutton buck. A male mutton sheep.

mutton cane. *Panicum dichotomum,* family Gramineae, a slender grass browsed by sheep (eastern U.S.).

mutton corn. Ears of green corn, as roasting ears (southern U.S.).

mutton goat. A goat sold for its flesh.

mutton grass. Mutton bluegrass.

mutton joint. See *pastern.*

mutton sheep. Any of several breeds of sheep developed primarily for their meat. See *Shropshire; Southdown.*

mutualism. Dependency of two organisms upon each other. Among insects, an example is furnished by the cornfield ant and the corn-root aphid.

mutualistic. Designating a mutually beneficial relationship between organisms; symbiosis.

muzzle. The projecting part of an animal's head, comprising the nose, mouth, and jaws, as of a cow, horse, or dog.

mycelial fan. A thin layer of fungous growth composed of mycelial strands spread in a fanshaped pattern, occurring either in a layer over the surface of decaying timbers or as small patches between bark layers or between the bark and wood of trees.

mycelial neck rot. A rot affecting the neck of onion bulbs ascribed to the fungus *Botrytis byssoidea,* which is probably a form of *B. allii.*

mycelium. The hyphae or filaments of a fungus.

mycoderma. 1. The membrane or "mother" formed on fermenting wine, beer, vinegar, etc. 2. The genus of fermentation-fungi found in yeast etc., which produces this membrane. Also called *wine flowers.*

mycology. The science dealing with fungi.

mycorhiza. The morphological association, usually symbiotic, of fungi with the roots of seed plants. The association is referred to as *ectotrophic* in those cases in which the fungal hyphae occur on the root surface and penetrate only the intercellular spaces, and as *endotrophic* when the hyphae occur mainly within the cells of the host plant.

mycorhizae. Threads of hyphae of various fungi which grow in association with a higher plant, generally at the root.

mycorhizal fungus. A fungus which grows externally upon or internally within the feeding roots of some woody plants; apparently in some cases beneficial to the host plant but sometimes seeming to affect it adversely. See *mycorhiza.*

mycosis. A disease caused by the growth of fungi in plants or animals.

Mycosphaerella blight. 1. A blight of garden peas caused by *Mycosphaerella pinodes,* family Mycosphaerellaceae, characterized by small, purple leaf spots which may remain small or enlarge turning brown or black. The affected leaves may die and dry out but remain on the plant. Similar but elongated lesions appear on the stem and if they coalesce, the lower stem acquires a blue-black appearance. 2. Any blight resulting from infection by fungi of the genus Mycosphaerella.

Mycosphaerella fruit spot. A fungous disease of apple and quince caused by *Mycosphaerella* *pomi,* family Mycosphaerellaceae, characterized by dark red, brown and black, shrunken spots on the fruit. Also called *Phoma fruit spot.*

Mycosphaerella leaf spot. Any leaf spot disease caused by species of Mycosphaerella, family Mycosphaerellaceae, on cotton, citrus, and other plants.

Mycosphaerella leaf spot of gooseberries and currants. A fungous disease caused by *Mycosphaerella ribes,* family Mycosphaerellaceae, characterized by small, brown spots with grayish centers appearing on both sides of the leaves.

mycotic disease. A disease caused by a fungus.

mycotic lymphangitis. A glanders-like horse disease which affects the skin, superficial vessels and lymph nodes. It is attributed to a yeastlike fungus, *Cryptococcus farciminosus,* family Torulopsidaceae. Also called *epizootic lymphangitis; lymphangitis epizootica; African glanders; Japanese farcy; Neopolitan farcy.*

mycotic pneumonia. Pneumonia resulting from a fungous infection in the lungs. See *aspergillosis.*

myelocytomatosis. One of the manifestations of avian leukosis complex *q.v.*

myeloleukosis. One of the manifestations of avian leukosis complex *q.v.*

myiasis. A disease due to presence of fly's larvae in warm-blooded animals.

Mynot-Plymey process. A process for producing milk powder in which a single roller-drier is fed by a smaller, unheated roller partially immersed in milk beneath it. See *milk dryer; non-fat dry milk.*

myocardium. The heart muscle.

myriad-leaf. Canada parrotsfeather.

myricin. The waxy portion of beeswax remaining after the major portion has been dissolved in alcohol.

myrmecophilous. a. Designating insects, other than ants, that live in ant hills. b. Designating plants that live symbiotically with ants or are fertilized by them.

myrobalan. Emblic leafflower.

myrobalan plum. *Prunus cerasifera,* family Rosaceae, a tree grown for its seedlings used as stocks for stone fruits. Native to Europe and Asia. Also called *cherry plum.*

myrrh. An aromatic gum resin produced by several trees or shrubs of the genus Commiphora, family Burseraceae, especially *C. abyssinica* and *C. myrrha,* native to northeast Africa and Arabia. *Myrrh* is used in the manufacture of perfumes, dentifrices, and as an ingredient in medicinal tonics.

myrtle. Any plant of the genus Myrtus, family Myrtaceae. Species are trees or shrubs, generally aromatic and bearing rosy or white flowers and berry-like fruit. Native to tropical and subtropical regions. See *true myrtle; periwinkle: 2.*

myrtleleaf orange. *Citrus aurantium* var. *myrtifolia,* family Rutaceae, a spiny tree with thornless branches and small, closely-set leaves, grown as an ornamental. Native to Asia. Also called *dense-leaved box orange.*

Mysost. A cows' milk whey cheese, light brown and sweet; originally made in the Scandinavian countries.

myxa. *Cordia myxa,* family Boraginaceae, a deciduous tree growing to a height of 40 ft.; the mucilaginous, tan-colored fruit is used in medicines. Native to India and Australia. Also called *sebastanplum cordia; sebesten.*

myxedema. A disease resulting from insufficient secretion of the thyroid gland. Not common among animals, it is seen in some of the calves of the Dexter breed.

myxomatosis. A virus disease of the pox group, transmitted by vectors, nonlethal to the South American rabbit, its natural host, but lethal to most European varieties. Tested by the Commonwealth Scientific and Industrial Research Organization as a method of rabbit control and widely introduced (*e.g.,* into Australia and many European countries) in the 1950's, it made spectacular successes. In Australia the rabbit population, estimated to consume feed enough for 70 million sheep, was reduced by about 25 percent in 1951. Increasing natural resistance of rabbits to *myxomatosis* threatens to make this method of rabbit control obsolescent.

N

nabam. An organic fungicide used in the control of blights, rusts, mildews, and leaf spots.

nag. A horse or pony of nondescript breed.

Nagami kumquat. *Fortunella margarita,* family Rutaceae, a small, orange-like tree grown for its edible fruit used in preserves. Native to Japan. Also called *oval kumquat.*

Nageles. A cows' skim milk cheese spiced with cloves and cumin seed; made in the Netherlands.

nailhead canker. Blister canker of apple.

nailhead rust. A disease of the sweet orange tree, characterized at first by spots on leaves, twigs, and fruit. Later the bark breaks into scales. Not a true rust, it is possibly caused by mites. Also called *leprosis; scaly bark.*

nailhead spot. A fungous disease of tomato caused by *Alternaria tomato,* family Leptostromataceae, characterized by blight of the leaves and a spotting of the fruit.

naked fallow. Designating land not in crop use and resting, but plowed and harrowed, and so in a bare state, in contrast to land permitted to grow to grass or seeds. See *bare fallow.*

naked flower. A flower without petals (*apetalous*) or sepals.
naked jasmine. Winter jasmine.
naked ladies' poisoning. See *autumn crocus: 2.*
naked oats. *Avena nuda,* family Gramineae, a cereal which possesses a lemma and a palea readily separated from the kernel. Also called *hull-less oats.*
naked seed. Achene.
nandina. *Nandina domestica,* family Berberidaceae, a shrub grown for its brilliant foliage and red berries. Native to Asia. Also called *heavenly bamboo; Chinese sacred bamboo.*
nanism. Dwarf growth.
Nankeen lily. *Lilium testaceum,* family Liliaceae, a hybrid, showy, bulbous herb grown for its fragrant, apricot or yellowish flowers.
nannyberry viburnum. *Viburnum lentago,* family Caprifoliaceae, a large, hardy, shrub or small tree grown as an ornamental for its white flowers and bluish-black fruit eaten by grouse, pheasants, and rabbits. Native to and widely distributed in the eastern U.S. Also called *sweet viburnum; sheepberry wild raisin.*
nanny(goat). The female goat.
Nantucket pine moth. *Rhyacionia frustrana* (Comst.), family Olethreutidae, the larva of a moth which mines needles and buds of pines and causes deformation and occasional death of young pines.
nap. Matted masses of fibers that contribute to the rough appearance of ginned cotton.
nape. Back of the neck of an animal.
naphthalene. $C_{10}H_8$, a chemical derived from coal tar and widely used as a soil fumigant, as a repellent, and in the form of moth balls to fumigate clothes in the home.
napthalene acetic acid. One of the organic acids of hormone mixtures used by plant propagators for stimulating root development.
naphthol. $C_{10}H_7OH$, either of the two coal tar derivatives from naphthalene, alpha-*naphthol* and beta-*naphthol.* Alpha-*naphthol* is a pleasant smelling solid with a melting point of 122°F. It is the standard material for impregnating tree bands used to control the codling moth. The beta-*naphthol* is also an antiseptic.
Napiergrass. *Pennisetum purpureum,* family Gramineae, a perennial grass which yields an enormous amount of green fodder and is cut several times in a season. In the U.S. it is grown in Florida and Georgia. Native to Africa. Also called *Merker grass; elephant grass; Carter grass.*
napiform. Formed like a turnip.
Naples onion. *Allium neapolitanum,* family Liliaceae, an onion-scented, bulbous herb grown in warm areas for its white flower clusters. Native to Europe. Also called *daffodil garlic.*
nappe. The stream of water which overflows a weir or dam.
napped. Designating a matting or clumping of cotton fibers.
nappy. Coarse and sticky lint cotton.
narcissus. Any of the several herbs of the genus Narcissus, family Amaryllidaceae. Species are grown as ornamentals. See *angels tears; campernelle jonquil; daffodil; jonquil; nonesuch daffodil; petticoat daffodil; poets narcissus; polyanthus narcissus.*
narcissus basal rot. A rot of narcissus bulbs caused by the fungus *Fusarium oxysporum* f. *narcissi,* family Tuberculariaceae. The plant becomes stunted, the tips of the leaves brown and the base of the bulb becomes soft and rotten.
narcissus bulb and stem nematode. A species of Ditylenchus, family Tylenchidae, which infests the bulbs of narcissi and a large number of other plants. Infested plants show discoloration, especially of the bulb, and a swelling of stem.
narcissus bulb fly. *Lampetia equestris* (F.), family Syrphidae, a yellow and black fly that infests bulbs of narcissus, amaryllis, hyacinth, and other plants, causing yellow, stunted foliage and almost no growth.
narcissus fire. A fungous disease caused by *Sclerotinia polyblastis,* family Sclerotiniaceae, characterized by small, watery or light brown spots on the flowers. Bright, yellowish, elongated spots appear on the leaves which turn brown.
nardoo. Drummond pepperwort.
Narragansett. 1. A variety of turkey which resembles the Bronze in color pattern but has neither the irridescent, red-green sheen nor the characteristic copper-colored bronzing. The *Narragansett* colors are metallic-black, light silvery-gray, and white. In size it is slightly smaller than the Bronze (*q.v.*). 2. A hog of inferior breed. 3. An American breed of small and sturdy saddle horses, popular during the last century.
narrow-base terrace. A terrace for controlling soil erosion. It is similar to a broad-base terrace in all respects except the width of ridge and channel: the base of a narrow terrace is usually 4 to 8 ft. wide. It is subject to frequent failures and has not been widely accepted in the U.S.
narrow beech-fern. *Dryopteris phegopteris,* family Polypodiaceae, a fern with compound foliage, grown for its beautiful, triangular fronds. Native to North America, Europe, and Asia. Also called *long beech-fern.*
narrowleaf meadowsweet spirea. *Spiraea alba,* family Rosaceae, a deciduous shrub grown for its clusters of white flowers. Native to North America.
narrow-leaved dock. Curly dock.
narrow-leaved laurel. Lambkill kalmia.
narrow-leaved milkweed. Mexican milkweed.
narrow-leaved pittosporum. Willow pittosporum.
narrow-leaved plantain. Buckhorn plantain.
narrow nutritive ration. A feed with high protein content in proportion to its non-nitrogenous, fat and carbohydrate, nutrients.
narrow-track seeder. A seed drill which has the wheels closely spaced leaving the ends of the seeder box projecting beyond them on either side.
naseberry. Sapodilla.

nasturtium. 1. Any of the several herbs of the genus Tropaeolum, family Tropaeolaceae. Species are grown for their flowers, for salad plants, and for seeds used in sauces as a condiment. See *bush nasturtium; canary nasturtium.* 2. *T. majus,* an annual herb with red-spotted, yellow-orange flowers, grown as a covering for trellises and as an ornamental. Its young flower buds and fruits are used as seasoning. Native to South America.

Natal cherry. Jerusalem cherry.

Natal grass. *Tricholaena repens,* family Gramineae, a tall, short-lived, perennial grass grown in warm areas for forage and as an ornamental. Native to Africa. Also called *ruby grass.*

Natal orange. Kafir orange.

Natalplum carissa. *Carissa grandiflora,* family Apocynaceae, a spiny shrub grown for its white flowers and egg-shaped, red, edible berries. Native to South Africa. Also called *amatungula.*

National Council of Farmer Cooperatives. In U.S., a national federation dedicated to promote the interests of farmer cooperatives through its influence on various governmental and other agencies. The Council provides an avenue through which cooperatives are advised of current economic, technological, legal, and other developments, and also provides a forum through which better understanding among cooperatives can be attained.

National Farm Loan Associations. In U.S., local borrower-owned cooperative agencies that initiate and service farm mortgage loans made through the Federal Land Banks.

National Forest. A federal reservation dedicated to the production of timber, protection of watershed area, and equitable utilization and management of all natural resources within its boundaries. In the U.S., the *National Forests* are administered by the Forest Service of the Federal Department of Agriculture. The first law in relation to the establishment of *National Forests* in U.S. was passed by the Congress in 1897.

National Grange. See *Grange (The National).*

National Monument. In the U.S., an area or object of outstanding, historic, or scientific value which is set aside by the federal government and conserved unimpaired because of its national interest.

National Park. A large area of land owned by the federal government and managed in the public interest which is set aside for its scenic beauty, human interest, and value for recreation. Most of the parks were taken out of the public domain.

National Poultry Improvement Plan. See *U(nited) S(tates) National Poultry Improvement Plan.*

native. 1. Designating a plant that grows naturally in a country or region; one not introduced by man. 2. Designating animals, as cattle, hogs, and horses, which, though originally introduced into a region, have lost some of their original characteristics or have gone wild: a scrub or mongrel. 3. Designating an unbranded beef hide. 4. See *indigenous.*

native disease. A disease which is caused by an indigenous organism.

native elm bark beetle. *Hylurgopinus rufipes* (Eichh.), family Scolytidae, a beetle which makes galleries transversely across wood of the elm and linden. It is a vector of Dutch elm disease.

native forage. Indigenous vegetation suitable as food for livestock and game.

native orange. Australian finger-lime.

native pasture. Any pasture in which the forage species are wild or native in contrast to the seeded pastures of cultivated land.

native sheep. In marketing, sheep from the farms of the Midwest and East, as contrasted to those from the western ranges (U.S.).

native wool. Fleece wool.

Natural Area. In the U.S., an area permanently preserved in unmodified condition as representative of the virgin growth of a major forest or range type, primarily for the purposes of science, research, and education. Timber cutting and grazing are prohibited and general public use discouraged.

natural boundary. Any feature not made by man, as a river, or a mountain ridge, which separates states, countries, or tracts of land.

natural breed. A type of animal which results from natural conditions: *e.g.,* a breed may have been originally introduced into a region, but has been allowed to develop without interference from man.

natural cheese. A cheese which is not processed.

natural control. Nature's method of maintaining a biotic balance, as in the reduction of harmful insects through the action of heat, cold, rain, drought, parasites, predators, and disease.

natural cross. Interbreeding or hybridizing which takes place in nature without assistance from man.

natural-draft drier. A drier (*q.v.*) for fruits or other agricultural products which requires no fan to circulate the heat.

natural enemy. In nature, any organism which preys or feeds upon another. A *natural enemy* may be introduced by man in biological control. See *e.g., myxomatosis.*

natural flavor. In ice cream manufacture, flavor derived from a natural product: *e.g.,* the vanilla obtained from the vanilla bean, in contrast to an artificial vanilla flavor.

natural grape. In the drying of grapes, any grape that is not lye dipped or bleached.

natural group. A group of persons with similar attitudes, customs, interests, and modes of living, functioning under a natural leader.

natural group leader. The individual to whom members of a natural group defer.

natural habitat. The place which a plant or animal naturally occupies; natural environment.

natural immunity. Immunity to a disease, infestation, etc., which results from qualities inherent in a plant or animal. *Cf. acquired immunity.*

natural impurities of wool. The yolk of wool with the adhering mineral and vegetable matter.

natural increase. In range management, any increase in numbers within a herd by propagation alone.

naturalization. The process of successful adaptation by plants and animals introduced to an area where they are not indigenous.

naturalized disease. A disease whose causal organism is of foreign origin but which has become established in a new habitat and has reached a state of balance with native organisms.

naturalizing. 1. The planting of bulbs, shrubs, etc., in such quantity and manner as to suggest natural growth. 2. The process by which plants and animals adapt themselves to a new environment.

natural landscape. That aspect of the land surface resulting from the effect of relief, water, and vegetational features, produced by nature, which is essentially unmodified by man's activities. See *landscape.*

natural land type. A unit of land area capable of being mapped upon a combination or integration of natural characteristics, as soils, vegetation, relief, climate, as contrasted with types based upon use, money value, or modifications by man. See *land type.*

natural levee. A low ridge, generally of sands, on the rim of the bank of a large stream, formed by the initial deposition of sediment in the natural flooding of the bottomland.

natural pasture. Uncultivated land, suitable for grazing, occupied wholly or chiefly by native or naturally distributed plants.

natural refrigeration. The use of water, ice, or ice salt mixtures for cooling.

natural reproduction. Renewal of plants by self-sown seeds, sprouts, rhizomes, etc. See *artificial reproduction.*

natural reseeding. In range management, the restoration of depleted grazing land to permit natural revegetation by reseeding of desirable forage species.

natural resources. Assets supplied by nature, such as mineral deposits, soils, forests, waters, and wild life.

natural selection theory. A theory of evolution, propounded by Charles Darwin in the last century, which postulates that the forms best adapted to exist survive and that the distinctive characteristics of fitness can be inherited. Also called *the survival of the fittest theory.*

natural system. The system of land drainage in which the tile lines are laid in the natural drainage depressions. Also called *random system.*

natural thinning. The death of trees in a stand as a result of competition.

natural type. See *natural breed.*

natural water. Water (H_2O) as it occurs in nature with extremely variable proportions of solid, liquid and gaseous materials in solution and suspension. In common usage the term water refers to *natural water* rather than to chemically pure water.

natural water course. A natural stream which flows in a well-defined bed or channel. It is distinguished from an artificial water course, such as a ditch or canal.

nature trail. A footpath along which interesting plants, animal haunts, geological and other natural features are marked for educational and recreational purposes.

naval stores. Turpentine and rosin made from the sap of pines.

nave. The hub of a wheel to which spokes are attached.

navel. 1. In *Mammalia* spp. the point of connection between the umbilical cord and the fetus. 2. See *navel orange.*

navel cord. The umbilical cord.

navel hernia. See *umbilical hernia.*

navel-ill. A disease of new born animals which results from infection of the umbilical cord shortly after birth. It is characterized by temperature, weakness, listlessness, lameness, hot and painful swelling of the joints and loss of appetite. The affected animal finally dies.

navel infection. See *navel-ill; omphalitis.*

navel orange. A variety, usually seedless, of the sweet orange with a small navel-like formation at the apex of the fruit. It originated in Brazil as a budsport and was introduced in the U.S. in 1870. Widely grown in California.

navicular bone. A little oblong bone in the lower part of the horse's leg between the second and third phalanx; the sesamoid of the third phalanx.

navicular disease. In horses, an inflammatory disease of the navicular bone and bursa of the front foot. Affected animals go lame and have a short, stubby stride.

navy bean. White varieties of the kidney bean, grown for their edible seed, generally known as *pea bean.* See *field bean; 1.*

Neapolitan cyclamen. *Cyclamen neapolitanum,* family Primulaceae, a perennial herb grown in warm areas for its pink or white flowers. Native to Europe.

near side. The left side of a horse.

near wilt of pea. A fungous disease of the field pea caused by *Fusarium oxysporum* f. *pisi* race two, family Sphaerioidaceae, which remains indefinitely in the soil. The symptoms are similar to wilt, except that there is less stunting of the plant and the vascular discoloration is brick-red in color and occurs farther up the stem.

neat cattle. Horned cattle; generally not in use for the bovines.

neat's foot oil. An oil obtained by boiling the feet and bones of calves and other cattle; used as a fine lubricant, for dressing leather, and in hoof ointments.

Nebraska potash. Salts obtained from the brines of lakes in the semiarid regions of western Nebraska, U.S., which contain 20 to 30 percent of

K_2O in the form of carbonate and sulfate of potassium.

neck. In man and animals, the connecting link between the head and the body. 2. To tie cattle neck-to-neck (western U.S.).

neck ail. A cobalt deficiency disease of cattle which results in listlessness and an emaciated appearance.

neck finisher. A mechanical buffer or picking machine for finishing the dressing of poultry for the market. It cleans blood and feed from the head and removes feathers and pin feathers from the neck.

necklace orchid. *Coelogyne dayana,* family Orchidaceae, an epiphytic orchid grown for its white flowers with reddish-brown tips. Native to Southeast Asia.

neck (land). An elevated, narrow strip of land between two somewhat parallel streams, or water bodies; a promontory; a peninsula; an isthmus.

neck of the woods. Neighborhood; locality.

neck piece. Of a meat carcass, that part between the shoulders and head.

neck rein. To turn a saddle horse in a desired direction by particular pressures of the reins on the neck rather than on the bit.

neck rib. In a retail cut of pork, the rib next to the neck. See *Wiltshire side.*

neck roll. In a retail cut of beef, the neck meat from the carcass rolled with sheets of cod fat and tied.

neck rope. a. A rope tied around the neck of a horse being trained for use in lassoing cattle. b. A rope used in picketing horses.

neck rot of onion. A very destructive fungous disease which affects onion bulbs. It can be caused by any of the three species of Botrytis; the *B. allii,* family Moniliaceae, causing the common gray mold neck rot.

neckyoke. A bar, usually of wood, which connects two draft animals abreast by the neck and supports the forward end of the tongue of a wagon.

necro. 1. Ulcers of the intestine, necrotic enteritis. See *infectious enteritis in pigs.* 2. Swollen nose of hogs; bull nose.

necrobacilloses. Several disease entities in domestic animals, sometimes of uncertain etiology but involving some bacteria, usually the bacillus *Actinomyces,* and such predisposing factors as trauma, moisture or filth. These include such diseases as foot rot, bull nose of pigs, and necrotic stomatitis.

necrobiosis. A virus infection of plants; one of three types of phloem alteration characterized by gradual swelling of the cell walls in either healthy or diseased plants.

necro in pigs. Infectious enteritis in pigs.

necropsy. An examination of the internal organs of a dead body to determine the cause of death. Also called *autopsy; post-mortem.*

necrosis. Death of a cell or group of cells within living animal or plant tissue. See *necrobacillosis.*

necrosis of grapes. A physiological condition of grapes sometimes caused by a late-spring frost, characterized by localized death or splitting of tissues so that the leaves appear lacerated or torn.

necrosis of the beak. An inflammation of the beak of poultry chicks caused by particles of finely ground, all-mash feed remaining under the tongue and adhering to the edges of the upper and lower mandibles.

necrotic. Designating a necrosis.

necrotic cavity corky bark. A virus bark disease affecting trees, especially citrus, in which the lesions are in the form of necrotic cavities whose margins are composed of corky, raised tissues impregnated with hard, vitreous, gumlike material.

necrotic center. The degenerating and dying cells in the center of a lesion, such as may be observed in a skin lesion produced by an abscess.

necrotic diseases. See *necrosis.*

necrotic enteritis. Infectious enteritis in pigs.

necrotic fleck. A virus disease of Easter lilies in which flecks are produced on the leaves.

necrotic leaf spot. A virus disease of peaches differing from necrotic ring spot in that symptoms are recurrent.

necrotic ring spot. A virus disease of sour cherry caused by one or more viruses or strains of the same virus. The disease is widespread on *Prunus* spp. The symptoms are most severe after initial infection, consisting of necrotic rings or spots, and dieback of spurs and shoots. After this the virus becomes latent without symptoms. The effect on fruit production is controversial.

necrotic rusty mottle. A virus disease of cherry of major importance in Utah (U.S.), in which variable, brown, necrotic spots appear in the leaves 3 to 5 weeks after petals fall, followed by several waves of partial defoliation. Chlorotic areas, brown necrotic spots, and shot-holes are evident in late season. See *Lambert mottle.*

nectar. A sweet liquid secreted by plant nectaries usually located within the flowers, but in some species it may be exuded elsewhere, as from glands on leaf stalks.

nectar carrier. A worker bee actively engaged in carrying nectar from flowers.

nectar flow. The period or time during which the flow of nectar occurs in flowers.

nectarine. *Prunus persica* var. *nectarina,* family Rosaceae, a deciduous tree grown for its edible fruit, the *nectarine* of commerce. The fruit is quite similar to the peach except that the skin is fuzzless. Native to China.

nectarine Coryneum blight. See *Coryneum blight of stone fruits.*

nectary. The nectar-secreting gland of a flower or other plant part.

needle. 1. A timber set on end vertically, or inclined, to close an opening for the control of water; a form of top-plank. See *flashboard.* 2. The curved steel member which feeds twine around the bundle and places it in the twine holder disc, preceding the operation of tying the knot in the twine around the bundle.

needle-and-thread. *Stipa comata,* family Gramineae, a tufted, perennial grass widely distributed from Alaska to Texas and California (U.S.). It is rated as good forage on western ranges, but when mature and dry, the pointed, long-tailed fruits may cause serious injury by penetrating the skin. Native to North America.

needle and twig blight. 1. A fungous disease of balsam fir caused by *Rehmiellopsis bohemica,* family Sphaeriaceae, characterized by the needles of the current season's growth turning red, shriveling, and the new twigs being blackened and stunted. 2. Any of several diseases of needles and twigs of evergreens.

needle blight of Douglas fir. A blight caused by the cup fungus, *Rhabdocline pseudotsugae* Syd.

needle board. A device once used in California, U.S., in the preparation of prunes for the market; it consisted of a board containing sharp, metal points over which dipped prunes were passed to puncture the skins.

needle cast. A fungous disease of spruces and other conifers caused by *Lophodermium filiforme,* family Phacidiaceae, characterized by the lower branches of red and black spruces being completely defoliated.

needle grama. *Bouteloua aristidoides,* family Gramineae, an annual grass fairly common in southwestern U.S. It has a fair forage value in early spring. Native to Argentina.

needlegrass. Certain species of the genus Stipa, family Gramineae, perennial bunchgrasses with numerous, mostly basal, narrow or rolled leaves, open or narrow panicles, narrow, long-awned spikelets, and hardened lemmas ending in a long, bent, twisted, and sometimes feathery awn. Two species, green needlegrass and needle-and-thread (*q.q.v.*) are of good grazing value on western ranges (U.S.). Porcupine grass (*q.v.*) produces long-tailed, sharp-pointed fruits that may penetrate the skin of animals and cause their death.

needle pitman. A connecting rod on the binder attachment of a grain-harvesting machine.

needle rust. Any of several diseases of coniferous trees as pine, hemlock, fir and larch, caused by the fungi of the genus Coleosporium, family Melampsoraceae, and characterized by blister rust on the needles. It rarely does appreciable damage although some defoliation may occur.

needle trussing. In preparing poultry for roasting, tying the bird together compactly with a stout string and long needle.

neep. A turnip (Scotland).

negro bug. *Corimelaena pulicaria* (Germ.), family Pentatomidae, an insect that infests celery, corn, wheat, and other crops.

negro patch. On old Southern plantations in the U.S., a patch of land for growing crops for use of the negroes.

neigh. 1. The characteristic cry of the horse. 2. Of a horse, to utter its characteristic cry.

neighborhood. A small group of nuclear families living in a circumscribed area and having a high degree of mutual interaction. Members experience the *neighborhood* as a limited social group, tend to identify themselves with it and to view it as second in importance only to the family.

nellies. *See canner: 2.*

nemarot. A disease of onion bulbs and other plants caused by the nematode *Ditylenchus dipsaci,* family Tylenchidae.

nemas. Nematode.

nematocide. An agent that kills nematodes.

nematode. Any of the round, thread-like, unsegmented animal worms of the phylum Nematoda, ranging in size from microscopic to one metre long. *Nematodes* may be saprophytes or parasites of plants and animals. They are responsible for important animal and plant diseases resulting in much economic loss. The animal parasites are designated as roundworm of horses and swine, and hookworm of man. Also called *eelworms; roundworms; nemas.*

nematode disease of wheat. A disease caused by the nematode, *Anguina tritici,* family Tylenchidae. It invades the inflorescene and transforms the kernels into galls resembling smut balls. Also called *bunted; earcockle.*

nematode root galls. Root-knot.

nematodirus. The internal parasite (nematode) of sheep, a nematode of the genus Nematodirus. Also called *stomach worms; intestinal worms.*

nematology. A branch of zoology concerned with the study of nematodes usually including their relation to plants and animals.

nemoral. Of, or pertaining to, woods or forest, as *nemoral* culture.

neoeluvium. The soil or residue of weathering which results from the eluviation of recent deposition, such as alluvium and glacial drift.

neomycin. An antibiotic derived from the organism *Streptomyces fradiae.*

neonicotine. Anabasine.

Neopolitan farcy. Myootic lymphangitis.

neotran. A miticide.

nep. Catnip.

Nepal trumpetflower. See *Easter heraldtrumpet.*

neps. The small tangles of fibers visible as dots or specks in a thin web of fibers held to the light or against a dark background. Caused by irregular growth or processes like ginning. *Cf. nap.*

Nero's crown. Crapejasmine tabernaemontanta.

nerve poison. A poison which is soluble in tissue lipoids in contrast to respiratory poisons, such as cyanide gas, and physical poisons, such as oils and dusts.

nervous market. A market in which traders expect a sharp break momentarily and are easily influenced by rumors.

Nessel. A soft-cured, round and thin, English cheese made from cows' whole milk.

nest. 1. A farm secured by squatter rights (obsolete, western U.S.). 2. A cavity in which wild bees make their home to store honey and wax, often in a hollow tree. 3. Various kinds of constructions prepared by birds to lay and hatch their eggs. In poultry raising, any of various devices in which the hens lay their eggs. 4. In florists' shops and greenhouses, a stack of clay flower pots, one placed inside another. 5. A range of sashbeds.

nester. A homesteader or squatter who legally or illegally occupied land (western U.S.).

nesting. A phase of watery rot of vegetables, caused by the fungus *Sclerotinia* spp., family Sclerotiniaceae. It develops in the center of hampers of vegetables in transit, appearing as a cottony mass of mycelia.

net blotch. Barley net blotch.

net cash income. The difference between the total cash farm receipts and the total cash farm expenses for the year.

net duty of water. The amount of irrigation water delivered to the land to produce a crop. It is measured at the point of delivery to the field.

net-energy value. The amount of energy which remains after deducting from a feed's total energy value the amount of energy lost in feces, urine, combustible gases, and heat increment. Sometimes called *work of digestion.*

net farm earnings. The net cash income plus an increase (or minus a decrease) in total farm investment other than bare land, plus the value of farm products furnished to the farm family, and less the value of unpaid family labor (excluding operator's labor).

net farm income. The amount of money that the farmer earns as pay for his own labor, management, and the total capital investment in the farm business.

net height of terrace. The difference in elevation between the highest point in the terrace channel and the lowest point on a terrace crown as measured at proximate points along a terrace.

net increment. In forestry, the addition to tree growth which represents an increase in usable timber.

net livestock increase (or decrease). The clear profit or loss; a figure obtained by subtracting the total of the value of livestock at the beginning of the year and the cost of livestock purchased during the year from the total of the value of livestock at the end of the year and the receipts from livestock sold. If the result is negative, it is a net livestock decrease.

net necrosis. A disease of the potato caused by a virus, or sometimes a frost, in which cut tubers show a network of brown strands, especially near the surface and stem end; associated with leaf roll. See *leaf roll: 1.*

net russeting of apples. A discoloration of the skin of apples caused by the powdery mildew, *Podosphaera leucotricha,* family Erysiphaceae. The russeting may be spread all over the fruit or may be in the form of small patches, rings, or bands.

net sales. Gross sales less deductions for freight, handling, discounts, etc.

netted iris. *Iris reticulata,* family Iridaceae, a perennial herb grown for its red-purple flowers. It is a beardless iris. Native to the Caucasus.

nettletree. Hackberry.

nettling caterpillar. A caterpillar possessing nettle-like or poisonous hairs. A well-known type is the saddleback caterpillar *q.v.*

net type necrosis. See *net necrosis.*

net weight. 1. The weight of an article, or a mass of anything, after deducting from the gross weight the weight of the container or covering. 2. The weight of a slaughtered animal after removal of hide, viscera, etc.

net worth. The net ownership of a farmer or individual in his business; the difference between the assets and liabilities; the owner's equity (*q.v.*) in the business.

Neufchâtel. A soft, mild, French cheese, made either from whole or skim, cows' milk or a mixture of milk and cream. It is eaten fresh or after curing.

neural lymphomatosis. Fowl paralysis.

neuritis of chickens. Fowl paralysis.

neurolymphomatosis gallinarum. See *fowl paralysis.*

neurotoxicity. The state or condition of being poisonous to the brain and nerves of the body.

neuter. Of a plant, asexual; having neither stamen nor pistil. See, *e.g., zoospore.*

neutral. See *neuter.*

neutral breeding ground. A fenced-off plot, used for breeding to prevent the spread of venereal diseases.

neutral-day plant. See *day-neutral plant.*

neutralizer. An alkali, such as soda, used to neutralize a portion of the acidity of certain dairy products.

neutralizer flavor. A flavor defect in a dairy product, such as butter, suggestive of the alkali used in neutralizing the acidity of the milk or cream used in the manufacture of that product.

neutral lard. Leaf or back fat rendered in a water-jacketed kettle by slow melting at 126° F.

neutral soil. A soil that is not definitely acid or alkaline; specifically, a soil with a pH between 6.6 and 7.3. See *pH.*

Nevada bluegrass. *Poa nevadensis,* family Gramineae, a tufted, grayish-green, perennial grass which occurs on western ranges of the U.S. at elevations of 3000 to 11,000 ft. It has good forage value.

Nevada ephedra. *Ephedra nevadensis,* family Gnetaceae, a woody plant with very small leaves, growing in semiarid regions of Utah, California, and other states, U.S., which is good browse for cattle and sheep in late fall and winter. Native to the U.S. Also called *Nevada jointfir.*

Never-Never. A term applied to the northwest of Queensland and adjacent parts of the Northern Territory; the outback (*q.v.*), an expansive and lonely area having no definite limits (Australia).

new. 1. Designating a recently produced product which has not ripened or cured. 2. Designating any newly introduced breed, product, etc.

Newcastle disease. An acute, rapidly-spreading virus disease of poultry, caused by a filterable virus. It occurs all over the world. Also called *avian pneumoencephalitis; California flu.*

Newcastle thorn. Cockspur hawthorn.

new crop pip. In flower forcing, the pip (*q.v.*) of the lily-of-the-valley which is used for early forcing.

New England aster. *Aster novaeangliae,* family Compositae, a tall, branching herb grown for its deep purple flowers. Native to North America.

new ground. Land recently cleared of native, vegetal, woody cover, stones, stumps, etc., and put under cultivation for the first time.

new-ground plow. A plow especially designed for plowing new ground cleared of trees and brush but whose soil still contains roots.

New Hampshire. An American breed of intermediate-sized chicken, characterized by chestnut-red plumage, yellow skin, fast feathering, and rapid growth. It is of the same standard weight as the Rhode Island Red (*q.v.*) from which it was developed largely as a strain selection. The egg color is light brown.

New Jersey tea. Jerseytea ceanothus.

new land. Tillable agricultural land in a virgin state, or recently cleared or reclaimed land.

newly-mown sweet-clover flavor. An unnatural vanilla flavor of ice cream; a defect. See *vanilla-wafer flavor.*

New Mexico flax. *Linum neomexicanum,* family Linaceae, an annual herb whose ingestion is apparently poisonous to livestock. Native to southwestern U.S. Also called *yellow pine-flax.*

New Mexico locust. *Robinia neomexicana,* family Leguminosae, a small to large spiny shrub, grown as a browse plant for goats and horses. The pods after falling to the ground are eaten by sheep and cattle. Native to the U.S.

New Mexico range caterpillar. See *range caterpillar.*

New Orleans molasses. A light molasses obtained as a by-product in the manufacture of cane sugar used as a table sirup and in cooking.

new parity prices. Parity prices (*q.v.*) based upon the most recent 10 year period for an individual commodity and farm wages paid to labor. *Cf. old parity prices; transitional parity prices.*

new potato. 1. The tuber of the potato taken from a potato hill before full maturity. 2. The potato of the newly harvested crop in contrast to stored potatoes.

new strains test. In cotton plant breeding, a test conducted in the second year after the original selection is made on a portion of the seed from each of the choice progeny-row strains previously saved.

new wood. In horticulture and forestry, the current year's growth. It is softer and less liquefied in contrast to the wood of a previous year.

New York apple tree canker. A fungous disease of apple and other fruit trees caused by *Physalospora obtusa,* family Pleosporaceae, characterized by the bark becoming roughened or cracked until the limb is girdled and the wood killed. The fungus is often referred to by the name of its imperfect stage, *Sphaeropsis malorum.* Also called *blackrot canker.*

New York aster. *Aster novibelgi,* family Compositae, a perennial, fall-flowering plant grown for its bluish-violet flowers. Native to North America.

New York Cotton Exchange. A world-famous central exchange organized in 1870 in New York City, for trading, buying, selling, and speculating in cotton.

New York dressed. In marketing, designating a bird that has been bled and has had its feathers removed but is undrawn with head and feet on.

New York fern. *Dryopteris noveboracensis,* family Polypodiaceae, a woodfern with light green fronds, grown as an ornamental. Native to North America.

New York 15 mosaic virus. A variant of the bean mosaic (*q.v.*) virus. First recognized because it attacked bean varieties resistant to the other.

New York ice cream. A brand of ice cream which contains a generous amount of egg yolks. Also called *Philadelphia's Neapolitan; French ice cream.*

New York single stack system. A system of ventilation similar to the King ventilating system (*q.v.*), except that the outlet flues are all combined in one large stack or flue.

New York weevil. *Ithycerus noveboracensis* (Forst.), family Curculionidae, a chewing insect which infests the twigs and small branches of apples, especially of the young trees.

New Zealand flax. *Phormium tenax,* family Liliaceae, a fiber plant with long, broad leaves and red or yellow flowers. Its leaves yield a fiber used in ropes, matting, etc. Native to New Zealand. Also called *New Zealand hemp.*

New Zealand spinach. *Tetragonia expansa,* family Aizoaceae, an herbaceous annual grown as a potherb for its edible leaves. It is a hot weather substitute for spinach. Native to East Asia, Australasia, and South America. Also called *New Zealand ice plant.*

New Zealand wineberry. Makomako wineberry.

niacin. A vitamin of the B-complex group. Also called *nicotinic acid; antipellagra vitamin.*

niche. The ecological position of a particular animal species in the environmental complex.

Nichol's terrace. A small terrace for disposal of surplus water. It has a comparatively deep, narrow channel and a low, flat ridge with a slope that merges quite closely with the downhill side.

nicked. 1. Descriptive of a horse's tail in which some of the muscles have been severed so that it sets up. 2. Designating a mating of animals which results in an offspring superior to either of the parents.

nickel. Ni, a chemical element, a metal which is found in traces in soils. It appears to be deleterious rather than beneficial to plant growth.

nicker. To neigh.
nicking. Breeding progeny which are superior to their parents.
nicotiana. Any of several species of the genus Nicotiana, family Solanaceae. Species are grown as flowering plants. Also called *flowering tobacco.* See *tobacco; winged tobacco.*
nicotine. $C_{10}H_{14}N_2$, a poisonous alkaloid, extracted from tobacco by distillation and other procedures, which is used in insecticides and fungicides.
nicotine bentonite. An insecticide containing nicotine sulfate, bentonite, and soybean oil.
nicotine compound. A chemical compound, such as nicotine sulfate, nicotine tannate, and mixtures with clay, peat, or various pastes containing nicotine which is used in insecticides and repellents.
nicotine decoction. A home-made insecticide spray manufactured from tobacco stems or high-grade tobacco refuse brewed for 24 hours and stirred occasionally.
nicotine dip. Any of a number of dips containing nicotine in various amounts, and in combination with other substances, used in disinfesting livestock of ticks, for scabies, etc.
nicotine dust. A dust made by combining nicotine with a powder, used as a repellent and contact poison.
nicotine humate. The product which results from a combination of nicotine and peat. Also called *nicotine peat.*
nicotine peat. Nicotine humate.
nicotine-sulfate. An alkaloid extract of the leaves of the tobacco plant, used as an insecticide.
nicotine tannate. An old insecticide which is a compound of nicotine and tannic acid, a fixed compound used as an orchard spray.
nicotine thiocyanate. An aphicide produced by combining nicotine with a thiocyanate.
nicotinic acid. $C_5H_5NO_2$ (3-pyridine carboxylic acid), normally obtained by oxidation of nicotine. See *niacin.*
Nieheimer. A sour-milk, German cheese made from cows' milk and packed with hops for curing.
nierembergia. See *cupflower; dwarf cupflower.*
nievling. A method of milking cows by hand in which steady pressure is applied to the teat by the doubled hand. Also called *nieveling.*
niggerhead. 1. A rounded boulder or low, rounded protrusion of dark-colored rock, such as diabase or gabbro, common in cultivated fields in the Piedmont Plain region of eastern U.S. 2. A very strong, rank, black tobacco manufactured in the form of a twist. 3. A tussock of coarse grasses or sedges, especially in burned-over marshes and swamps. 4. *Rudbeckia occidentalis,* family Compositae, a tall herb, common in the Wasatch Mountains of Utah (U.S.) where the leafage and flowers are grazed by sheep. Also called *western cone flower.*
niggertoe. 1. An old variety of Irish potato once grown in Michigan, U.S., in the early period of agricultural settlement. 2. See *Brazilnut.*
niggerwool. Threadleaf sedge.
night blindness. One of several symptoms, in livestock, of Vitamin A deficiency. See *Vitamin A.*
night blooming cereus. *Hylocereus undatus,* family Cactaceae, a night-blooming, high-climbing cactus, grown for its showy, pure white flowers and edible red fruit.
night corral. An enclosure or pen on western U.S. ranges in which ewes may be placed at night or at lambing time.
night-dropping shed. A shed of special construction used on western U.S. sheep ranches for the more efficient handling of ewes during the lambing period.
night hawk. A night wrangler or herder of saddle horses.
nightherd. To ride herd at night; to keep cattle or sheep herded at night.
night horse. A special cow-horse that is sure-footed, has good eye-sight and a keen sense of direction.
night jasmine. *Nyctanthes arbor-tristis,* family Oleaceae, a shrub grown in warm areas for its fragrant, night-blooming, orange and white flowers. Native to India. Also called *tree-of-sadness; sad tree; hursinghar.*
night milk. The milk obtained at the evening or night milking.
night pasture. A pasture in which domestic animals graze during the night, usually close by the barn or dwelling.
nightphlox. Any plant of the genus Zaluzianskya, family Scrophulariaceae. Species are tender annuals grown in flower gardens. See *cape nightphlox; hairy nightphlox.*
nightshade. Any plant of the genus Solanum, family Solanaceae. Some species are flowers, some are shrubs, some vegetables, and some are poisonous plants. See *false Jerusalemcherry; Carolina horsenettle; bitter nightshade; Ethiopian eggplant; jasmine nightshade; pepino; black nightshade; Jerusalem cherry; potato; Costa Rican nightshade.*
nightshade poisoning. Poisoning of stock which results from eating species of nightshade. See *black nightshade; bitter nightshade.*
night soil. Human feces and urine, an unsanitary but potent manure for crop land. It is rarely used in the U.S. but remains in use in some other parts of the world, especially the Orient. See *honey dipper.*
night willow-herb. Evening primrose.
nigra loose smut of barley. A fungous disease caused by *Ustilago nigra,* family Ustilaginaceae, characterized at heading time, by dusty, smutty heads containing millions of dark brown microscopic spores.
Nigrospora cob rot. A fungous disease of corn caused by *Nigrospora oryzae,* family Dematiaceae, widely distributed throughout the U.S. Corn Belt. It is characterized by a shredding and gray discoloration of the cob. The pith of the cob may completely disintegrate.
Nigrospora dry rot. Nigrospora cob rot.
nikau rhopalostylis. *Rhopalostylis sapida,* family Palmaceae, a feather palm grown as an orna-

mental in very warm areas. Native to New Zealand.

Nikko fir. *Abies homolepis,* family Pinaceae, an evergreen tree grown for its beauty and hardiness. Native to Japan.

Nile blue sulfate. A dye used in media for detecting lipolysis by microorganisms.

nimblewill. *Muhlenbergia schreberi,* family Gramineae, a weed grass which spreads rapidly over uncultivated fields. Also called *nimble weed.*

nine-day pneumonia. See *pneumoencephalitis.*

ninety-day strain. A variety of corn which matures approximately 90 days after planting.

ninseng. An obsolete variety of ginseng *q.v.*

nip. 1. A knot in cotton fiber. 2. A sudden check in the plant or animal growth due to severe cold or frost.

nipa-palm. *Nipa fruticans,* family Palmaceae, an average-sized feather palm whose leaves are used to make mats and thatching for roofs. Its sap is made into wine. Native to southeast Asia.

nippers. 1. The two central incisor teeth of a horse. 2. Small pinchers used for holding, breaking, or cutting.

nipple. 1. A small, conical elevation. 2. The protuberance of the udder of a mammal which contains the mammary gland. See *teat.*

nipple pail. A pail with a tube and rubber nipple fastened at the bottom, used in calf feeding, weaning, etc.

Nippon oxeye daisy. *Chrysanthemum nipponicum,* family Compositae, a hardy, perennial herb grown for its white flowers. Native to Japan. See *chrysanthemum.*

niszler. A cheese which contains numerous small pin holes as a result of an abnormal gassy fermentation.

niter. See *nitre.*

niter burning of apple leaves. A browning of the margins of the leaves of apples; attributed to an excess of available nitrogen in the soil.

niter poisoning of apples. See *niter burning of apple leaves.*

nitragen. A fertilizer containing nitrogen-fixing bacteria *q.v.*

nitrated superphosphate. A mixed chemical fertilizer: a superphosphate treated with a nitrogen solution which contains 5 to 6 percent nitrogen and 16 to 17 percent available P_2O_5. See *nitrogen solution.*

nitrate of ammonia. Ammonium nitrate.

nitrate of lime. Calcium nitrate.

nitrate of potash. KNO_3, the potassium salt of nitric acid which occurs in nature only in small quantities. The pure compound contains about 46.6 percent potash. Used as a fertilizer or as a component in fertilizer mixtures, commercial grades contain about 13 percent of nitrogen and 44 percent potash.

nitrate of potash poisoning. Poisoning to cattle which results from ingestion of potassium nitrate in oat hay. Also called *saltpeter poisoning.*

nitrate of soda. $NaNO_3$, the sodium salt of nitric acid containing, when pure, 16.48 percent nitrogen. It is one of the chief sources of nitrogen for commercial fertilizers and is obtained in a natural form mainly from deposits in Chile. Also manufactured synthetically.

nitrate of soda potash. A natural combination of nitrate of soda and nitrate of potash from the caliche of northern Chile containing about 14 percent of nitrogen and 17 percent of potash. It is a source of nitrates for commercial fertilizer mixtures.

nitrates. Compounds of nitrogen and oxygen (NO_3) with some metallic element, such as sodium, potassium, calcium. The *nitrates* are the principal sources of the nitrogen of commercial fertilizers. At one time, *nitrates* were used in milk and butter as preservatives and deodorizers. The use of these and other preservatives is now prohibited in the U.S. by state and federal laws. Saltpeter, a *nitrate,* was used extensively in Europe and Canada for the prevention of blowing or formation of gas holes in cheese.

nitre. 1. A precipitate of malic acid formed in making maple sirup. 2. Nitrate of potash, (saltpeter), KNO_3, used as a fertilizer and in the manufacture of explosives. 3. Sodium nitrate, or Chile saltpeter.

nitrification. In the nitrogen cycle, the transformation of ammonia, NH_3, by bacteria, Nitrosomonas and Nitrosococcus to nitrites, NO_2, and of the nitrites, by bacteria of the genus Nitrobacter, to nitrates, NO_3, the soluble form used by plants.

nitrobacteria. Minute, rodlike bacilli which convert nitrogen into soluble nitrates in soils.

nitrofurans. A large group of closely related chemical compounds made up, in part, of substances obtained by the distillation of wood, as corn cobs, etc. Many *nitrofurans* have medicinal properties and act as a bactericide in the treatment of wounds and infections. The *nitrofurans* are used, alone or mixed with antibiotics and drugs such as the sulfas, in the treatment of a number of poultry and livestock diseases.

nitrogation. The method of supplying agricultural ammonia through or in irrigation water in which most of the ammonia combines with the irrigation water to form ammonium hydroxide.

nitrogen. N, a gaseous element which occurs in the atmosphere to the extent of 78 percent by volume. It occurs naturally in soil where it is converted into usable forms for plant use by bacteria and by other natural processes. A great number of compounds containing nitrogen, such as ammonium sulfate and sodium nitrate, are used in commercial fertilizers. It is an essential plant nutrient; all plant and animal proteins contain approximately 16 percent nitrogen.

nitrogen balance. The state of nitrogen equilibrium in the animal body.

nitrogen carrier. Any of a great number of inorganic or organic compounds which contain nitrogen and can be classed as a fertilizer or fertilizer material.

nitrogen deficiency. An insufficient amount of available nitrogen in a soil for the proper functioning of plant growth or for a desired yield of a crop. It is often characterized by a yellowing of the veins in the leaves of plants.

nitrogen excesses. The presence of nitrogen, as nitrates or ammonium salts in soils, in excessive amounts, which results in injury to plants, as burning or stunting. The excess may occur naturally or may result from excessive applications of nitrogenous fertilizers. In tree fruits, it may cause excessive vigor which results in reduced fruiting.

nitrogen fixation. The conversion, by nitrogen-fixing microorganisms, of elemental nitrogen to organic combinations or to forms readily utilizable in biological processes.

nitrogen-fixing bacteria. Species of the genus Rhizobium, family Rhizobiaceae, which live symbiotically in the root nodules of leguminous plants, upon which they are dependent. They are capable of extracting nitrogen from the air and converting it to a form utilizable by the plant. Species of the genus Azotobacter are free living and act independently (nonsymbiotically).

nitrogen-fixing plant. A plant which can assimilate and fix the free nitrogen of the atmosphere by the aid of bacteria living symbiotically in the root nodules.

nitrogen-free extract. Certain substances in a feedstuff which are determined by difference rather than by direct analysis. The extract includes the more soluble, digestible and valuable carbohydrates, and is the total dry matter in a feeding stuff minus the sum of the ash, crude protein, fiber, and fat.

nitrogenous material. A fertilizer or feeding stuff material which contains concentrations of nitrogen.

nitrogenous tankage. Crude, inert material, such as skins, wool, hair, feathers, etc., which is processed under steam pressure, with or without the use of acids, to make the nitrogen available. Such material, dried and ground, contains from 5.7 to 10 percent nitrogen and is used as a fertilizer. Also called *process tankage.*

nitrogen solution. A solution of 60 parts of ammonium nitrate dissolved in 40 parts of a 50 percent aqua ammonia. It is used for ammoniating superphosphate.

nitrojection. The method of applying agricultural ammonia by injection into the soil.

nitromethane. CH_3NO_2, a liquid, prepared from chloro-acetic acid and sodium nitrate, used as a solvent of pyrethrum.

nitrophilous. Designating a plant that is tolerant of alkali.

nitrosomonas. A specialized group of bacteria in soils that oxidize ammonia into nitrites. The nitrite is further oxidized to nitrate by other groups of bacteria such as nitrobacter. See *nitrobacteria.*

nits. The eggs of a louse.

noble cane. *Saccharum officinarum,* family Gramineae, and varieties, a widely cultivated, tall, stout sugar cane, having a chromosome number of 40. These originated in the Southwest Pacific Ocean region, where several other species of Saccharum are also native.

noble fir. *Abies procera,* family Pinaceae, a huge, evergreen tree grown on the U.S. West Coast as an ornamental, and for its lumber used in general construction, boxes, pulp, and planing mill products. Native to North America.

nocturnal. Of the night: *e.g.,* a *nocturnal* plant is one whose flowers open at night; a *nocturnal* parasite is one which is active at night. *Cf. diurnal.*

nodder. A horse that characteristically nods its head when walking, as the Tennessee Walking Horse.

nodding. Designating a flower, tree, plume, etc., which droops, bends, or inclines with a swaying motion.

nodding lilac. *Syringa reflexa,* family Oleaceae, a tall, hardy shrub grown for its nodding clusters of deep pink flowers. Native to China.

nodding lily. *Lilium cernuum,* family Liliaceae, a bulbous herb grown for its nodding, fragrant, lilac, purple-spotted flowers. Native to eastern Asia.

nodding trillium. *Trillium cernuum,* family Liliaceae, a hardy, perennial herb grown in shady places for its early-spring, white or pinkish flowers. Native to North America. Also called *ground lily; Jew's harp.*

node. A joint or knot, especially of a stem of a plant; the enlarged places from which leaves usually diverge.

nodular worm. A common name applied to nematodes found in cattle, sheep, goats, and swine, whose larvae normally penetrate into the tissues of the intestines and produce nodular lesions there; the adult worms live in the cavity of the large intestines. *Oesophagostomum radiatum* is the *nodular worm* of cattle; *O. columbianum,* the common *nodular worm,* and *O. venulosum,* the lesser *nodular worm,* occur in sheep and goats; four species of *nodular worms* have been reported in swine in the U.S.: *O. dentatum, O. longicaudum, O. brevicaudum,* and *O. georgianum.* The symptoms of infestation are called *pimply gut; knotty gut.*

nodulated. Designating plants which have nodules on their roots in which nitrogen-fixing bacteria live.

nodule. 1. A root tubercle or lump formation on certain leguminous plants produced by invasion of symbiotic, nitrogen-fixing bacteria. The bacteria furnish the plant with fixed nitrogen compounds and receive nutrient plant juices like carbohydrates. The genus Rhizobium and some species of Azotabacter and Clostridium fix free nitrogen. 2. A small knot, lump, or roundish mass of abnormal tissue.

Noekkelost. A Norwegian, spiced, cows' milk cheese similar to Kuminost and Leyden.

nogal. A walnut or pecan tree (southwestern U.S.).

no grade. 1. In the judging of dairy products designating a product which does not merit a standard grade classification. 2. Designating cheese that has deteriorated in quality as a result of defective finish, or one which has extremely serious defects in flavor, body, texture, or color. It may or may not be edible. 3. Designating butter which scores below cooking grade. 4. Designating milk which has high-acid, rancid, weedy, or foreign flavor. See *sample grade.*

noils. The short wool fibers which are removed from the long ones in the process of combing wool.

Noisette rose. *Rosa noisettiana,* family Rosaceae, a hybrid shrub, derived by crossing *R. chinensis* with *R. moschata;* it is widely grown as stock for tea and China roses. Also called *Champney rose.*

nomadism. A primitive unsettled pastoral type of existence, especially in steppe and desert regions; the life led by people who wander in search of grazing for their livestock and food for themselves.

nomen conservandum. In botanical and zoological nomenclature, a name of a species, genera, etc., that should be preserved in exception of the usual rules of nomenclature, because of wide popular usage, etc. The American Code of Botanical Nomenclature rejected this idea, while the International Code of Botanical Nomenclature, as revised in 1931 and superceding the American Code, accepts it.

nominal measure. In the European lumber import trade, the full measure of a board before it is surfaced.

nonacid forming fertilizer. A fertilizer containing ingredients which neutralize acidity or leave a basic residue, such as cyanamid and nitrates of soda, potash, and lime.

nonagricultural land. 1. Land that is nonarable and therefore not suitable for tilled or cultivated crops. 2. Any land not used for agricultural production, though potentially some or all of it might be so used.

noncalcareous. Designating soil with a very low calcium carbonate content; a soil with a complete absence of lime. See *calcareous soil.*

Noncalcic Brown soil. The zonal group of soils with slightly acid, light pinkish or light, reddish-brown A-horizons over light, reddish-brown or dull red B-horizons; developed under mixed grass and forest vegetation in a subhumid, wet-dry climate.

noncommercial forest land. Forest land withdrawn from its commercial use for timber because (a) it is utilized for such purposes as parks, game refuges, military reservations, or reservoir protection or (b) its poor growing conditions or inaccessibility render its commercial use unprofitable.

nonconforming uses. Uses of land permitted by zoning ordinances in areas in which such uses are not otherwise permissible.

nondisjunction. Failure of disjunction *q.v.*

nondisturbance agreement. A promise by one lender to another that he will not start any action against the borrower during a stated period of time.

none-so-pretty. Sweetwilliam silene.

nonesuch. Black medic.

nonesuch daffodil. *Narcissus incomparabilis,* family Amaryllidaceae, a group of natural or artificial bulbous, hybrid herbs of specific lineage, grown for their spring blossoms. Native to Europe. Also called *eggs-and-bacon.*

non-fat dry milk. The product obtained by the removal of fat and water from sweet, cows' milk. It contains lactose, milk proteins, and milk minerals in the same proportions as the fresh milk, maximum content of retained moisture being not more than 5 percent by weight and that of fat not over 1½ percent unless otherwise specified. Also called *defatted milk solids.*

nonhardy. 1. One of three broad groups of alfalfa: hardy, medium hardy, and *nonhardy,* recognized on the basis of their resistance to winter killing. 2. Any plant or animal not adapted to cold weather conditions. See *tender: 1.*

noninfectious chlorosis of strawberries. A physiological condition caused by a deficiency of nitrogen, magnesium, manganese, boron, or oxygen in a water-logged or alkaline soil; characterized by yellowing or loss of normal green color.

noninfectious plum shot hole. A nontransmissible leaf spot and shot hole condition apparently the result of a genetic weakness in the wildgoose plum, *Prunus munsoniana* and the garden plum, *P. domestica.*

nonleguminous. Designating plants that do not produce pods called legumes. See *legume.*

nonlignified. Designating the nonwoody type of crude fiber in feed; it contains little lignin.

nonmotile. Not capable of locomotion.

nonpalatable. 1. Designating a range plant species not grazed when the range or pasture is properly utilized. 2. Designating feedstuff not relished by animals.

nonparasitic. See *abiotic disease.*

Nonpartisan League. A nonpolitical organization of farmers formed in 1915 in North Dakota, U.S., to reform the system of marketing agricultural products. It also sponsored state ownership of public necessities.

nonpathogenic. Not capable of producing disease.

nonpersistent virus. Any virus whose vectors lose their ability to transmit a disease within one day of leaving infective plants.

nonporous wood. Wood devoid of pores or vessels, as the conifers and a few broadleaf species.

nonprofessional inspector. In official U.S. meat inspection, an inspector who passes certain Civil Service examinations but is not a veterinarian or graduate of an accredited veterinary college.

nonreacting infected animal. An animal which, though infected by a particular disease, does not react to standard tests for the disease.

nonrecourse loan. A loan on a commodity in approved storage in which the borrower is not under obligation to pay in cash; he can deliver the commodity mortgaged in payment of the loan when it comes due. Used primarily by the Commodity Credit Corporation (U.S.).

nonreturn. In artificial breeding, the breeding efficiency of bulls expressed as the percentage of cows that conceive on the first service.

nonrotating drag-line excavator. A type of excavator with only one platform which is stationary, and with a boom pivoted at its lower end; chiefly used in the construction of small drainage ditches.

nonruminant. An animal without a rumen. See *ruminant.*

nonsaccharine sorghum. The grain sorghum (*q.v.*), as milo, kafir, durra, as distinguished from the sweet sorghums which have a juicy pith and high sugar content.

nonsaline alkali soil. A group of soils distinguished from saline alkali soils by a larger amount of adsorbed sodium. Black-alkali soil belongs to this group.

nonselective contact spray. A contact spray that kills only the tissue to which it is applied.

nonselective herbicide. Any herbicide which kills all vegetation to which it is applied.

nonslip loop. A loop of rope so knotted that it will not slip or tighten when pulled; used on a foal being trained to lead.

nonsod-former. Any grass which does not form a turf, as bunchgrass or other herbaceous plants. See *sod.*

nonspecific dermatitis. A skin disease not due to any particular agent. It may affect various parts of an animal's body.

nonspecific mastitis. An abnormal condition in the udder or milk in the absence of specific mastitis-causing organisms. See *abnormal milk: 2; mastitis.*

nonstocked forest land. That land which is capable of producing trees but is generally lacking in tree growth.

nonsymbiotic nitrogen fixation. Fixation of nitrogen in soils by free-living microorganisms in contrast to that fixed through the agency of symbiotic organisms. See *nitrogen-fixing bacteria.*

nontoxic. Not poisonous to plant or animal. See *toxic; toxicity.*

noodles. A feed preparation used chiefly in Europe for fattening geese which consists of a mixture of ground barley and other cereals. After scalding it is formed into noodle-like rolls about 3 ins. long which are force fed to the geese. See *noodling.*

noodling. A method of goose fattening, based on the use of noodles (*q.v.*) for forced feeding.

Nootka falsecypress. *Chamaecyparis nootkatensis,* family Pinaceae, an evergreen tree grown as an ornamental and as a valuable timber tree. Its wood is used for poles, interior finish, furniture, caskets, hulls, oars, greenhouses, acid tanks, sounding boards, etc. Native to North America. Also called *Alaska cedar; Sitka falsecypress.*

nopal. *Opuntia lindheimeri,* family Cactaceae, a tree-like, flat, jointed cactus which grows to a height of 8 to 12 ft. Pioneers are reported to have used this cactus for clarifying drinking water. Native to North America. Also called *cacanapa.*

Nordmann fir. *Abies nordmanniana,* family Pinaceae, an evergreen tree grown as an ornamental. Native to southern Russia.

no return. In horse breeding, designating conditions of breeding in which a stallion is obligated to bring a mare in foal only during the season advertised. See *with return.*

Norfolk Island pine. *Araucaria excelsa,* family Pinaceae, an evergreen tree grown as a pot plant, as a florist's plant, and for its lumber. Native to Australia.

normal age class. The presence of a complete series of tree age classes corresponding to the rotation. See *age class.*

normal age class distribution. Such distribution of tree age classes that will permit annual or periodic felling of trees to be made without damage to adjoining stands. See *age class.*

normal agricultural value. The price at which it is estimated that a property will sell for agricultural purposes including: farm home advantages and location under forecasted normal farm commodity price conditions with a substantial down payment and with a reasonable amount of effort by a willing but not forced seller to a typical, desirous but not anxious purchaser experienced in the area.

normal crook. In lumbering, an allowable bend in a log for which no deduction is made.

normal diameter. In logging, the diameter of a tree at breast height after root swell has been eliminated.

normal erosion. The erosion which is characteristic of land surface in its natural environment, undisturbed by human activity including (a) rock erosion, consolidated or unconsolidated, as in stream channels, high mountains, and badlands, and (b) normal soil erosion, or the erosion characteristic of the soil type in its natural environment. Also called *geologic erosion.*

normal feed. In butter grading, designating a butter which has a flavor characteristic of good, natural forage, such as clover hay, as contrasted to that resulting from grazing on onions, garlic, etc. Such butter is rated A.

normal fire season. A forest fire season in which weather, rated fire danger, and number and distribution of fires are approximately average.

normal forest. A standard with which to compare an actual forest to bring out its deficiencies for sustained-yield management; an ideally regulated or organic forest with normal increment, normal age classes in area and distribution, and normal growing stock.

normal growing stock. The total volume of trees in a fully stocked forest which has an ideal distribution of age classes.

normal increment. In forestry, the best average increment which is attainable by a given species on given sites, as represented in normal yield tables.

normalizing. In the dairy industry, a heat treatment which is necessary to restore the fat globules and other constituents of milk to a standardized or normal condition. Usually the milk is heated to 140° F.

normal kill. In wildlife management, the average number of animals or birds that can be killed yearly on a particular area without diminishing future productivity.

normal moisture capacity. The amount of water which is normally held by a soil after the free or gravitational water has drained off. See *field capacity.*

normal soil. A soil having a profile in equilibrium with the forces of the environment, such as native vegetation and climate which is usually developed on gently undulating upland, with good drainage, from parent material (not of extreme texture or chemical composition) in place long enough for soil forming processes to exert their full effect.

normal stand. In forestry, a stand of trees fully stocked, in proper growing condition and having normal increment conforming to the yield table.

normal yield table. In forestry, an accepted standard yield table with which to compare actual yields. The statements of a *normal yield table* are derived as an average from the best producing, fully stocked areas for particular species and sites.

Norman. See *Percheron.*

Normandy fowl. See *Houdan.*

norms. Customs, ethics, moral codes, laws, and institutions. (Most sociologists regard *norms* as an ideal standard for behavior, but some would include statements about average or typical behavior patterns.)

nornicotine. An alkaloid, closely related to nicotine and anabasine which appears to be a better contact poison, but a poorer fumigant, than nicotine.

Norris-Doxey Act. An act of U.S. Congress, approved in 1937, which authorizes the Secretary of Agriculture to cooperate with Land-Grant colleges and universities and state forestry agencies for the development of farm forestry in the states and territories of the U.S. An important provision rules that these bodies produce or procure and distribute tree and shrub planting stock to farm owners on condition that the cooperator make available the land for planting without charge.

norther. A strong wind from the north which usually brings cold, sleet, rain, or dust, especially in Texas and the Plains of southwestern U.S.

northern anthracnose. A fungous disease of red clover in the cooler areas of the U.S., caused by *Kabatiella caulivora.* It produces brown lesions on the stems and petioles and a cracking of the stem tissues.

northern bayberry. *Myrica pensylvanica,* family Myriaceae, a deciduous shrub with fragrant foliage and grayish-white berries that last throughout the winter, grown for windbreaks and hedges. The wax covering of the fruits, called *American vegetable wax,* is used in making candles. Native to North America. Also called *bayberry.*

northern bedstraw. *Galium boreale,* family Rubiaceae, a perennial herb grown for its prostrate habit and white flowers. Native to North America.

northern catalpa. *Catalpa speciosa,* family Bignoniaceae, a deciduous tree bearing yellow-striped, white flowers, grown as an ornamental. Native to North America. Also called *hardy catalpa.*

northern cattle grub. *Hypoderma bovis* (L.), family Hypodermatidae, an insect which infests cattle in the U.S. The adult fly, called a *heel,* lays eggs on the heels of the animal and the larvae migrate to the back, where they cut small holes through the skin. The flies irritate the animal with consequent decreased production of milk and loss of meat. The grubs damage the hide. Also called *bomb fly.*

northern corn leaf blight. A fungous disease caused by *Helminthosporium turcicum,* family Dematiaceae; characterized by lesions beginning on the lower leaves and appearing progressively upward. Heavily infected fields may appear dry and fired.

northern corn rootworm. The larva of a beetle, *Diabrotica longicornis* (Say), family Chrysomelidae, which burrows through the roots of corn, leaving small, brown tunnels. It is a very serious corn pest in the upper Mississippi Valley of the U.S.

northern fowl mite. *Orithonyssus sylviarum* (C. and F.), family Dermanyssidae, a parasite of fowls, commonly found about the base of the tail and around the vent. It may cause scabs which damage the dressed carcass. Also called *northern feather mite.*

northern hornworm. Tomato hornworm.

northern leaf blight. A fungous disease of corn, Sudangrass, and sorghums caused by *Helminthosporium turticum,* family Dematiaceae; characterized by linear or irregular, elliptical, water-soaked lesions on the leaves. The lesions turn brown, and then black. Sometimes the entire leaf dies.

northern mole cricket. *Gryllotalpa hexadactyla* Perty, family Gryllotalpidae, a dark brown insect which digs tunnels, sometimes 2 or 3 ft. deep, occasionally doing some damage to potatoes and other root crops.

northern rat flea. *Nosopsyllus fasciatus* (Bosc.), family Dolichopsyllidae, a vector of diseases among men and animals, especially bubonic plague. See *European rat flea; rat flea.*

northern red currant. *Ribes rubrum,* family Saxifragaceae, a shrub with dull brown flowers, grown in warm areas of Europe and the U.S. for its edible, red fruit. Native to eastern Europe and western Asia.

northern red oak. *Quercus rubra* var. *borealis,* family Fagaceae, a northern variety of red oak. Also called *champion oak.*

northern spring wheat. A sub-class of hard red spring wheat. It contains more than 25 and less than 75 percent dark, hard, and vitreous kernels, as compared with dark northern and red spring wheat which contain respectively higher and lower percentages of such kernels.

Northern Spy. One of the older, very well known and popular varieties of apple, a juicy winter variety which originated in New York, U.S.

northern white cedar. *Thuja occidentalis,* family Pinaceae, an evergreen tree valuable for timber and as an ornamental. Native to North America. Also called *American arborvitae; American cedar; eastern arborvitae.* See *whitecedar falsecypress; Lawson falsecypress; giant arborvitae; California incensecedar; eastern arborvitae; Chinaberry.*

northern white pine. Eastern white pine.

northwestern anthracnose. Apple anthracnose.

northwestern box. A standard type of container for packaging apples which has the capacity of one bushel.

Norway maple. *Acer platanoides,* family Aceraceae, a deciduous, hardy tree grown as a shade and street tree and for its brilliant autumnal foliage. Native to Europe and Asia.

Norway-maple aphid. *Periphyllus lyropictus* (Kess.), family Aphidae, a honeydew-secreting insect that infests Norway maple, causing stunted, wrinkled, or blotched leaves.

Norway pine. Red pine.

Norway rat. *Rattus norvegicus,* family Muridae, a native rodent of northern Europe, now common in many civilized countries and harmful to man in various ways. Also called *house rat; brown rat; wharf rat; sewer rat.*

Norway spruce. *Picea abies,* family Pinaceae, a hardy, evergreen tree, probably the most widely grown ornamental evergreen in America; it has a large number of varieties differing in height, form, and foliage. Dwarf varieties are 1 to 2 ft. high while other varieties reach a height of 150 ft.

Norwegian cinquefoil. *Potentilla norvegica,* family Rosaceae, an annual or biennial herb which is a widely distributed weed pest. Native to North America.

nose. 1. The point of a plow share. 2. The part of the face of man and animals which covers the nostrils; snout; muzzle. 3. To round off the end of a log in order to facilitate snaking or skidding.

nosebag. A feed bag hung with the open end over a horse's nose and attached by straps over the head and back of the ears.

nose band. The bridle strap that passes over the nose of a horse; a part of a hackamore *q.v.*

nosebleed. Toad trillium.

nose bot fly. *Gasterophilus haemorrhoidalis* (L.), family Gasterophilidae, an insect which lays its eggs in hairs on the lips of the horse. The larvae migrate into the mouth and thence into the intestines. It causes great irritation and nervousness.

nose clamp. A device which may be fitted tightly on the nose of an animal and used for its control during shoeing, surgical operations, or various types of training. See *twitch; nose twitch.*

nose color. A distinguishing color of the skin and hair on the nose of different animals. It ranges from light flesh, cream, or buff, to mouse, smoky or black.

no-see-um. Punkie.

nose fly. Sheep bot fly.

nosegay frangipani. *Plumeria rubra,* family Apocynaceae, a tropical shrubby tree grown in frost-free areas for its showy, very fragrant, pink or reddish-purple flowers. Native to the West Indies. Also called *red jasmine.*

nose lead. A metal ring, clamp, or other device attached to the nose of an animal for control.

Nosema disease. An intestinal disorder of bees caused by the parasite *Nosema apis,* family Nosematidae.

nose ring. 1. A metal ring fastened through the cartilage of the nose of a bull for safe control of the animal. A staff, or metal rod, about 6 ft. long may be snapped into the ring for handling the animal. See *nose lead.* 2. A metal ring fastened in the nose of a hog to prevent it from rooting.

nose twitch. A looped, light rope or heavy cord attached to a short stick. The loop is placed over the end of the nose of the horse and when twisted acts as a control in breaking, training, or leading. See *nose clamp.*

Nostrale. A very old, hard or soft, cows' milk cheese made in northwestern Italy.

nostril. One of the two outer openings of the nose which serves as a passage for air in breathing.

nostrum. A quack remedy or worthless patent medicine.

notch. 1. The opening in a dam or spillway for the passage of water. 2. A gap, pass, or defile between mountains (chiefly in northeastern U.S.). 3. An undercut; in logging, a cut in the trunk of a tree to govern direction of fall.

notch buckwheat. *Fagopyrum emarginatum,* family Polygonaceae, the wing-seeded buckwheat (*q.v.*) which is of little importance except as an admixture.

notched colter. A rotating, circular attachment for the forward part of the plow which has a cutter edge with a serrated perimeter for cutting the soil surface and trash in front of the plowshare. Also called *cut-away disk; cut-away colter.* See *colter.*

notched-edged rolling colter. Notched colter.

notch grader. A metal or wooden strip with different-sized notches in the side which is used for grading grape cuttings into different sizes to facilitate their grafting to the European grape varieties.

notching. 1. Cutting dents on the ears of animals for identification. 2. Removing a V-shaped piece of bark from a branch just above or below a bud.

notchleaf sea-lavender. *Limonium sinuatum,* family Plumbaginaceae, a perennial or biennial, salt marsh herb, grown for its blue and yellowish-white flowers. Native to the Mediterranean area.

novale. Land newly plowed and brought under cultivation.

novalis. In civil law, land that has been fallow for a year after the first plowing.

novelty. 1. In horticultural exhibitions and shows, a recent, undisseminated variety or type which has not been previously exhibited. 2. In ice cream manufacture, a product which differs from those commonly sold.

November drop. The dropping of mature citrus fruit in South Africa.

noxious plant. Any grass or herb, usually designated a weed, which is injurious to the farm crops, to other useful plants, or to livestock.

nozzle. The pouring end of a spout; the orifice through which liquids are discharged in a stream or spray.

N.P.I.A. See *U(nited) S(tates) National Poultry Improvement Plan.*

N-P-K. Chemical symbols for nitrogen, phosphorus, and potassium, the chief ingredients of complete fertilizers.

N S quotient. The ratio between total annual precipitation and evaporation: it expresses the effectiveness of precipitation for plant growth.

nubbin. 1. A virus disease of the cucumber, characterized by yellow-green systemic mottling of the leaves with small, distorted plants, and mottled and misshapen fruit, called *crook cucumber.* The disease is also called *mottled-leaf mosaic; wart disease; white pickle.* 2. An underdeveloped ear of corn which is not fully covered with kernels. 3. A strawberry irregular in shape and smaller than normal.

Nubian. A breed of milk goats of Mediterranean origin with long drooping ears, a Roman nose and commonly colored. The black, red, or tan bucks weigh up to 175 lbs., and the does up to 135 lbs.

nubs. Crook cucumbers.

nuclear family. The fundamental unit of human organization, specifically a male, female, and their unmarried children, all of whom are coresident. See *family; extended family.*

nuclei package. In beekeeping, a packaged colony of bees established on combs. See *package bee.*

nucleus. 1. The central portion of the cell protoplast surrounded by a very thin membrane. It consists of nucleoplasm and includes within itself variously-arranged chromatin, nuclear sap, and nutritive substances. It is of crucial significance in metabolism, growth, reproduction and the transmission of the determiners of hereditary characters. See *cell. Cf. cytoplasm.* 2. A small colony of bees used in queen rearing or in pollination work in greenhouses. 3. A central core around which material collects or is grouped.

nucleus box. A box that can accommodate two or three frames of honeybees and brood.

nuda loose smut. A fungous disease of cereals caused by *Ustilago nuda* or *U. tritici,* family Ustilaginaceae; characterized by blossom infection and the destruction of the kernels of the plant.

nuisance birds. Starlings, sparrows, crows, red-winged blackbirds, pigeons, etc., which damage the farm, orchard, and garden crops, or become objectionable because of their roosts, or the grain they consume in the fields and around granaries.

Number (No.) 1 flax. A marketing grade for flaxseed weighing at least 49 lbs. per bushel and containing less than 30 percent damaged flaxseed.

Number (No.) 2 flax. A marketing grade for flaxseed weighing at least 47 lbs. per bushel and containing less than 30 percent of damaged flaxseed.

number of men. A size factor for farm operations which consists of the months of work done on the farm by the hired help, the operator, and unpaid members of his family, divided by 12 to put it on a one year basis. Also called *man equivalent; average number of men.*

numdah. A thick, belt blanket placed under a saddle to absorb sweat. Also called *namda; nammad.*

nuna. American potatobean.

nurse bee. A young honeybee that produces food for the brood.

nurse cow. A milk cow used to supply milk for nursing calves other than her own.

nurse crop. A companion crop grown to protect some other crop sown with it, as small grain is sometimes seeded with clover. The *nurse crop* can sometimes become competitive and harmful also.

nurse-grafted Y-cutting method. A method of obtaining cuttings of plants difficult to root. One limb of a Y-shaped cion, whose roots are held in a pot of soil, is inarched with a small nurse plant, a seedling, or a rooted cutting. When the graft is fused, the Y-branch is girdled below the graft. When the graft is complete, it is severed at the girdle. The top of the nurse plant is also cut back to the graft. See *inarching.*

nurse rig. A service rig, consisting of mobile tanks with pumps, which delivers mixed spray material, water, or other ingredients to the spray rig in an orchard.

nurse-root graft. In grafting, the cion of a root graft planted deeply to encourage its rooting. The original root is cut off afterwards.

nursery. 1. Any place where plants, shrubs, and trees are grown either for transplanting or as grafting stocks. 2. A group of young plants or trees in a plantation.

nursery dibble. A small, hand implement consisting of a triangular or rounded blade attached to a short handle which is pushed into the soil to make a hole suitable for planting seedlings.

nurseryman. One who operates a nursery. See *nursery: 1.*

nursery row. A row in a garden in which transplants, seedlings, and specimen plants are grown until they are planted in their permanent places.

nursery separator. Any small threshing machine which is so designed as to be efficient in threshing a large number of small samples of plant material from nurseries or small experimental plots.

nursery stock. Nursery-grown plants. See *nursery.*

nursery tree. A tree which is grown in a nursery to an age and size suitable for transplanting.

nurse tree. In forestry, a tree which protects or fosters another in its first year of growth.

nursing calf. A calf which runs with its mother and is not weaned.

nursing mare. A mare with an unweaned foal.

nursing visit. In beekeeping, the visits to cells by a nurse bee for the care and feeding of eggs and larvae.

nut. 1. A hard, indehiscent, one-seeded fruit. 2. In common usage and in commerce, fruits, such as almonds, peanuts, coconut and Brazil nuts, which are not true *nuts* botanically.

nutgall. Aleppo gall.

nutgrass. Chufa flatsedge.

nutgrass flatsedge. *Cyperus rotundus,* family Cyperaceae, a sedge which is widely distributed as an agricultural weed. The tubers were used as food by North American Indians. Native to tropical and subtropical regions.

nutmeg. *Myristica fragrans,* family Myristicaceae, a tree grown for its seed, the *nutmeg* of commerce used in flavoring. Native to Indonesia.

nutmeg hickory. *Carya myristicaeformis,* family Juglandaceae, a deciduous tree grown as a shade tree and for its valuable wood used in tool handles, ladders, furniture, sporting goods, implements, woodenware, and for fuel and smoking meat. Native to North America.

nutmeg liver. Chronic, venous congestion of the liver which is characterized by a finely mottled appearance; common in old or diseased animals.

nutmeg melon. See *muskmelon.*

nutmeg pelargonium. *Pelargonium odoratissimum,* family Geraniaceae, a tender, perennial herb grown for its nutmeg-scented, white flowers. Native to South Africa. Also called *apple geranium.*

nut pine. 1. Any of various pines which bear edible nuts. 2. *Pinus cembroides* var. *edulis,* family Pinaceae, a hardy variation of the Mexican pinyon pine (*q.v.*). This term may include other variations such as *P. c.* var. *monophylla.*

nutria. *Myocastor coypus,* an aquatic rodent valued for its pelt which is produced to a limited extent in fur farming in parts of the U.S. The animal exists in the wild state, as an escape, in the marshes of Louisiana and Texas, and is trapped annually in considerable numbers. Native to South America.

nutrient. 1. A substance which favorably affects the nutritive processes of the body; a food. 2. An element or compound in a soil which is essential for the growth of a plant. 3. In stock feeding, any food constituent or group of food constituents of the same general composition that aids in the support of life, as proteins, carbohydrates, fats, minerals, vitamins.

nutrient level. 1. In soils, the amounts or proportions of plant foods, such as phosphorus, potash, nitrogen in available forms especially in greenhouse culture. 2. Specifically, the concentrations of any particular nutrient in the ration of animals.

nutrient solution. A solution in which plants may be grown, made by dissolving salts containing the essential elements for growth in water, which is used in hydroponics (*q.v.*) and in experimental work in laboratories.

nutrient spray. Liquid fertilizer or plant food which is sprayed directly on the foliage and absorbed therefrom.

nutrition. The act or process by which organisms, whether vegetable or animal, transform food elements into body elements.

nutritional anemia. Anemia in animals which results from a nutritional deficiency, usually iron or copper. It may result from deficiency of some trace element in a soil, such as cobalt, and the consequent lack of the element in forage produced on that soil. See *bush sickness.*

nutritional blindness. Blindness of stock which results from the lack of vitamin A in the rations.

nutritional encephalomalacia. Crazy chick disease.

nutritional roup. Roup of fowls.

nutritive ratio. In animal feeds, a ratio or proportion between the digestible protein and the digestible non-nitrogenous nutrients which is found by adding the digestible carbohydrates, plus the digestible fat multiplied by 2.25, and dividing the sum by the digestible protein. The ratio is an expression of the energy value of a ration against its body building power.

nutritive value. The relative capacity of a given feed to furnish nutrition for livestock. (Usually prefixed by "high," "low," etc.)

nut-size cottonseed cake. A livestock feedstuff, a cottonseed product which will pass through 1½ in. round perforations and over ⅞ in. round perforations. It shall be free from meal, pea-size and pebble-size cake, and shall not contain in excess of 10 percent of sheep-size cake.

nut structure. The aggregation of fine particles of soil into small subangular, blocky, or roughly rounded, masses of definite size or shape.

Nuttall blister beetle. *Lytta nuttallii* Say, family Meloidae, an insect which feeds on leguminous crops; the larva feeds on grasshopper eggs.

nuttalliosis. A disease of equine animals caused by protozoan parasites of the genus Nuttallia. Not seen in the U.S. For symptoms see *piroplasmosis.*

Nuttall saltbush. *Atriplex nuttalli (A. gardneri),* family Chenopodiaceae, a useful browse plant found on alkaline or saline soils on western ranges (U.S.). Native to North America. Also called *Gardner saltbush.*

nutwood. The tree or the wood of nut-bearing trees, such as hickory and walnut.

nuxvomica poisonnut. *Strychnos nuxvomica,* family Loganiaceae, a tropical tree whose fruit yields strychnine. Native to India. Also called *dog button.*

nymph. A stage in the development of some insects and related forms, such as ticks, immediately preceding the adult stage.

O

O.A.C. honey strainer. A device for straining honey developed at the Ontario Agricultural College (Canada), which consists of a series of screens of different-sized meshes.

oak. Any deciduous, evergreen, or partly evergreen tree or shrub of the genus Quercus, family Fagaceae. Species are grown as ornamentals, as shade and lawn trees, for bark, lumber, acorns, and browse. See *black oak; blackjack oak; bur oak; California live oak; California scrub oak; California white oak; canyon live oak; cork oak; Engelmann oak; English oak; European turkey oak; holly oak; Japanese evergreen oak; laurel oak; live oak; northern red oak; pin oak; post oak; scarlet oak; scrub oak; swamp white oak; water oak; white oak; willow oak.*

oak apple. An English name for oak gall *q.v.*

oak barrens. In the eastern U.S., land which supports only a scrubby or sparse growth of oaks, especially scrub oak, *Quercus ilicifolia,* and blackjack oak, *Q. marilandica;* land of low agricultural value.

oak canker. A fungous disease of oaks caused by species of the genera Diaporthe, family Valsaceae, and Strumella, family Tuberculariaceae, characterized by a slow-growing wound in the bark and to a lesser extent in the wood.

oakfern. *Dryopteris disjuncta,* family Polypodiaceae, one of the shield ferns grown for ornament.

oak fungus disease. Mushroom root rot.

oak gall. Swelling of various shapes which commonly occurs on the leaves and stems of oaks; usually caused by insects.

oak hammock. Hammock land characterized by growth of oaks (Florida, U.S.). See *hammock: 2.*

oak-hickory land. Land on which the dominant native tree growth was oak and hickory. (In the early period of settlement of the forested regions of eastern and southern U.S. local distinctions based upon tree growth, implying quality or value of land, were in common use. To some extent such groupings remain convenient and useful in the geographic description of natural regions.)

oakleaf hydrangea. *Hydrangea quercifolia,* family Saxifragaceae, a shrub grown for its white flowers which turn purple. Native to North America.

oak-leaf pattern. A symptom of virus plant diseases: psorosis and necrosis in which cleared or lighter-colored spots on the leaves resemble the outline of an oak leaf.

oak-leaf poisoning. A digestive ailment, observed among cattle on the ranges of southwestern U.S.; attributed to excessive ingestion of oak leaves in the spring, before grass makes its appearance. In a few instances the ailment is fatal and otherwise may cause stunted growth.

oak mildew. A fungous disease caused by *Microsphaera quercina,* family Erysiphaceae, characterized by a delicate, white mycelial growth.

oak opening. A prairie area in an oak forest, or an area only very sparsely occupied by oaks. (Openings were a feature of the natural landscape of parts of midwestern U.S., particularly southern Michigan, recognized by early settlers and explorers.)

oak poisoning. Poisoning or digestive disturbances in cattle resulting from excessive feeding on the ripe acorns of various species of oaks. The symptoms are constipation, black feces, and rose-colored urine. Death may result.

oak root fungus disease. Mushroom root rot.

oak skeletonizer. *Bucculatrix ainsliella* Murtf., family Lyonetiidae, an insect whose larvae eat away oak leaves, leaving a skeletonized surface.

oak wilt. A very serious, fungous disease of various species of oak in the North Central states of the U.S. eastward to Pennsylvania, West Virginia and Michigan, caused by *Endoconidiophora fagacearum,* known in some publications by the name of its imperfect stage, *Chalara quercina,* family Ophiostomataceae. It is related to the fungus causing the Dutch elm disease, and like that disease is carried by various insects which feed on the pads of fungus mycelium which are formed under the bark and project outward through cracks. It is also spread by the natural grafting of the roots of diseased and healthy trees.

oast(house). A house heated by hot air, used in drying hops and malt.

oat. See *oats.*

oat chop. A livestock feedstuff which is the entire product made by chopping, cutting, crushing, and crimping whole oats. Also called *ground oats; pulverized oats; crushed oats; crimped oats.*

oat common root rot. A fungous disease caused by *Colletotrichum graminicola,* family Melanconiaceae, characterized by a black, cushion-like formation at the base of the stem of the oat plant near the surface of the soil.

oat-covered smut. A fungous disease caused by *Ustilago kolleri,* (*U. levis kands.* Magn), family Ustilaginaceae, characterized by smut-covered panicles of the plant ripening prematurely, and finally the oat kernel being destroyed. See *semi-loose smut of oats.*

oat feed. Oat mill feed.

oat flour. Flour of the oat grain; used in ice cream primarily to prevent the development of stale, metallic flavors.

oatgrasses. Any of the grasses of the genus Arrhenatherum, family Gramineae. See *tall oatgrass.*

oat groats. The kernels of cleaned and dried oats separated during the manufacture of oat meal.

oat gum. A gum-type stabilizer derived from oats; used in ice cream.

oat-hay poisoning. Poisoning, particularly of cattle, resulting from ingesting the oat plant which under certain conditions may contain a large amount of potassium nitrate. Also called *saltpeter poisoning.*

oat hulls. The outer coverings of the oat grain; used as an ingredient in certain mixed feeds and sometimes for bedding.

oat kicker. A machine for removing weed seeds from wheat. See *fanning mill.*

oat leaf blotch. A disease of the leaves, leaf sheaths and culms of oats caused by the fungus, *Helminthosporium avenae,* family Dematiaceae, which begins by producing spots on the leaves, spreads to the sheaths and finally blackens the stems, even filling the hollow of the stem with a white, wefty mycelium. In severe cases the stems break at the third or fourth internodal region.

oatmeal. Hulled oats boiled and crushed or rolled; used as a breakfast food and in feed mixtures. Also called *porridge* (British). See *feeding oat meal.*

oatmeal feed. Oat mill feed.

oat middlings. Oat mill feed.

oat mill feed. The entire by-product, comprising oat hulls, oat shorts, and oat middlings, left in the manufacture of oat groats; used in mixed feeds. Also called *oat hulls; oat shorts; oat middlings.*

oat root rots. See *root rots.*

oat rusts. See *rusts.*

oat(s). 1. Any grass of the genus Avena, family Gramineae. Species are grown for forage, as a cover crop, and for their seed used as food and feed. Native to temperate regions. 2. *A. sativa,* the common oat, an annual herb native to Europe and Asia. See *wild oat; animated oat; side oat.*

oats anthracnose. See *anthracnose.*

oats brome mosaic. A virus disease affecting a wide range of grasses and cereal crops, including oats, caused by *Marmor graminis* and characterized by leaf mottling, dwarfing, and chlorosis.

oat shorts. Oat mill feed.

oat smuts. 1. *Ustilago avenae,* family Ustilaginaceae, a loose smut, which has rough spores and destroys about all the tissues of the oat panicle, grain, chaff, etc., leaving nothing but a black mass of spores. 2. *Ustilago kolleri* (*U. levis*), which has smooth spores and leaves the chaff (glumes, etc.) almost intact but replaces the grain with a mass of spores.

oat straw. The dry stems and leaves of the mature oat plant left after threshing. It has a somewhat higher feed value than the straw of other small grains and is also used for bedding and industrial purposes.

oaxaca. Asadera.

obedient plant. Virginia lionsheart.

obligate parasite. Any parasite which cannot exist independently of its living host.

obligate saprophytes. Microorganisms not related to living cells which secure their nutrients from dead organic tissue or available inorganic materials.

oblique-banded leaf roller. *Archips rosaceana* (Harr.), family Tortricidae, an insect common in greenhouses. Its larvae build their nests by rolling up plant leaves.

obnoxious weed. In dairying, any weed which, ingested by a cow, imparts an obnoxious odor or flavor to milk, cream, or butter. See *noxious plant.*

obscure scale. *Chrysomphalus obscurus* (Comst.), family Diaspididae, a tiny insect which infests the bark of twigs and branches of oak and other trees.

observation area. An area reserved for special developments to facilitate the enjoyment of a view of outstanding beauty or interest, but itself not usually located in any wild region.

observation hive. A beehive made largely of glass to permit the observation of bees at work.

observatory lookout. In forestry, a small building on top of a tower, with walls largely of glass and occupied by a lookout man; designed for the observation of natural phenomena not for residence.

obsolescence. In appraisal, the impairment of desirability and usefulness of a structure, machine, or equipment consequent upon new developments in art, design, process, or any other circumstances.

obstetrical chain. A metal chain, 30 to 60 ins. long, used by veterinarians to assist animals having difficulty in the delivering of newborn (usually calves or foals). It is used to apply a small amount of traction to the unborn.

obstetrics. That branch of medicine which deals with the birth of the young and management practices during pregnancy.

occlusion. 1. The process of healing over or closing of the wound caused by cutting or breaking off of a limb in pruning. 2. The absorption of gases by solids: *e.g.,* the absorption of oxygen by milk powder.

occult spavin. A type of spavin that occurs between the bones at the hock joint of a horse. See *spavin.*

ocotillo. *Fouquieria splendens,* family Fouquieriaceae, a perennial, tall, woody, cactus-like plant producing leaves for short periods whenever moisture conditions are favorable and grown for its showy, scarlet flowers. Characteristic of certain soil conditions in the desert regions of the southwestern U.S. and adjacent Mexico. Native to North America. Also called *coachwhip; vine cactus; candlewood; Jacob's staff; Spanish bayonet.*

octamethyl pyrophosphoramide. The chemical name of the insecticide schradan. Also called *OMPA.*

ocular estimate. Estimate by sight only. (1) In forestry, the determination of the approximate volume and quantity of standing timber without the use of measuring instruments. (2) On a range, the qualitative procedures for determining the degree of cropping of forage plants. Observations of a general reconnaissance nature are made visually: (a) by examining small random plots; (b) by observing the cropping of individual plants; (c) by photographing grazed areas; and (d) by examining permanently established random plots at the end of each grazing season.

ocular lymphomatosis. One of the aspects of avian leukosis complex (*q.v.*). A part of the symptoms includes an absence of pigment in the iris. The affected birds ultimately lose their sight. Also called *gray eye; white eye.*

ocular roup. An advanced stage of coryza (*q.v.*) of poultry, mainly involving the eyes. It is characterized by a thick, cheesy mass of material that develops in the lower corner of the eye under the eyelid. The disease may result in blindness or death.

odd area. In farm planning, a small area, such as bare knob, fence corner, sink hole, blowout, borrow pit, or an irregularly-shaped area that is unsuitable for cultivated crops and is best used for wildlife crops.

odd lot. On livestock markets, animals that do not conform uniformly to some particular weight, age, or grade of quality class.

odometer. An instrument to measure the distance travelled by a wheeled vehicle. It is fixed on to the hub of the wheel and works by computing the number of its revolutions. Used in the early days of soil surveying by the U.S. Department of Agriculture.

odontomata. A tumor arising in a tissue which normally produces teeth; sometimes found in horses and cattle.

odorous house ant. *Tapinoma sessile* (Say), family Formicidae, a household ant which is a pest over much of the U.S.

oedema. See *edema.*

oedema of the wattles. See *edema of the wattles.*

oenin. Monoglucoside of delphinidin dimethyl ether.

oestrus. See *estrous.*

off . 1. In cotton transactions, designating grades below middling, the basic grade. 2. A low grade or inferior product. See *on.*

offal. 1. The inedible parts of a butchered animal, such as the digestive system, lungs, feet, etc. 2. Anything discarded as useless. 3. In grain milling, the by-products such as wheat bran, shorts, etc.

offal fat. In cattle slaughter, the internal fat. It is not a waste product. Also called *intestinal fat; internal fat; killing fat.*

off-bloom. See *off-season bloom.*

off-bloom fruits. Citrus fruits that develop from blossoms produced out of season. See *off-season bloom.*

off-center. Designating a yolk distinctly above, below, or to one side of center in a chicken egg.

off-color. In farm products, designating a departure from a standard or natural color, which usually implies inferior quality: *e.g.*, hay is *off-color* when it is not properly cured.

off cottonseed hulls. Low quality cottonseed hulls, which do not meet the specifications for prime cottonseed hulls (*q.v.*), or which are made from off-grade, heated, bolly or trashy cottonseed, or have been heated after being processed, or are otherwise damaged.

off-crude cottonseed oil. Cottonseed oil which refines to a color no higher than AOCS20 with or without a loss in weight up to 25 percent.

off-feed. a. Designating an animal which has digestive disturbances due to excessive or improper feeding. b. Designating an animal that fails, from any cause, to eat the normal feed.

off-flavor. Designating milk or any milk product which has a peculiar or unnatural flavor.

off-grade. Designating an agricultural product which fails to meet requirements of commonly accepted standards or legal or official standards in grading products for sale.

officer's mount. A horse, preferably a gelding, at least 15 hands tall, whose suitability depended on the height and weight of the rider (obsolete).

official testing. Certification of records of production of milk and butterfat of purebred dairy cows under the control of officials of agricultural experiment stations. The tests are an important factor in the culling, breeding, and development of registered dairy cattle.

off-odor. In egg inspection, designating egg meats that depart from natural odor, or are offensive.

off-quality cottonseed. Cottonseed that has been treated by either mechanical or chemical process other than the usual cleaning, drying, and ginning or is fermented or hot, or that upon analysis is found to contain 12.5 percent or more of free fatty acids in the oil, or more than 10.0 percent of foreign matter, more than 20.0 percent of moisture, or more than 25.0 percent of moisture and foreign matter combined.

off season. In farming, the season of the year when the production of a particular crop is difficult or impossible.

off-season bloom. Blossoms of citrus trees that are produced out of season. Also called *off-bloom.*

offset. A rather short, prostrate, many-noded, branch growing from the crown of a plant which has a somewhat fleshy scaly bud or a rosette of leaves located terminally. It often roots and tends to produce an individual plant.

offset-disk harrow. A harrow adapted for use in orchards and vineyards, which can be set to run to the side of the tractor and thus cultivate under branches too low for the tractor to pass under.

offshoot. A lateral shoot or branch which rises from one of the main stems of a plant.

offside. The right side of a horse.

offsorts. Fleeces or parts of fleeces rejected as unsuitable for the regular lots of graded wools as a result of being badly stained and having an undesirable color.

off-summer yellow cottonseed oil. That summer yellow cottonseed oil (*q.v.*) having the following characteristics: off in flavor and odor, free from visible foreign material, of a color no higher than AOCS 12, containing not more than 0.50 percent free fatty acid nor in excess of 0.10 percent moisture and volatile matter.

off-sweetening. A flavor defect of ice cream which is lacking in sweetness or is too sweet.

offtype. In plants or animals, any notable deviation from standard or normal.

off-type strain. A strain of a plant variety which is offtype *q.v.*

offwool. Any wool of inferior quality; a fleece that is otherwise known as discounts, rejections, or unmerchantable wools.

off-year. The year during which an alternate or biennial-bearing fruit tree produced little or no fruit. *Cf. on-year.*

ogee. The reversed curve shape of the face of an overflow dam.

Ogeechee tupelo. *Nyssa ogeche,* family Cornaceae, a deciduous tree grown in southern U.S., the acid fruit of which is sometimes made into a conserve. Native to North America.

Ohio buckeye. *Aesculus glabra,* family Hippocastanaceae, a deciduous tree grown as a shade tree and as a specimen for its clusters of yellowish-green flowers. Native to North America.

Ohio Importing Company (U.S.). A company which began, a few years prior to 1850, the business of importing farm animals and developing better breeds for Ohio and country westward. Also called *O.I.C.*

Ohio State System. A system of overhead, automatic watering of benches in a greenhouse in which water is spread from a special nozzle in the center of the bench and covers the soil uniformly without excessive moistening of foliage of plants.

oil. One of several kinds of fatty or greasy liquids that are lighter than water, burn easily, are not soluble in water and are composed principally, if not exclusively of carbon and hydrogen. See *fixed oil; mineral oil; essential oil.*

oil base. Oil used as a base or carrier for poison or insecticide sprays.

oil bearing crops. Any crop grown primarily for its oil content, such as soybeans, peanuts, cottonseed, and linseed.

oil cake. Stock feed which is a mass of compressed seed from which the oil has been largely extracted, as linseed cake, cottonseed cake.

oil cell. In wood anatomy, a specialized and usually much enlarged wood parenchyma or ray cell which secretes an essential oil.

oil-dipped raisin. Raisins prepared by dipping the fresh grapes in a cold solution of sodium bicarbonate on the surface of which is a thin film of olive oil.

oil drying. In forestry, artificial seasoning of wood wherein the heat required is supplied by immersing the wood in hot oil.

oiler. Any of several devices by which oil is deposited on an animal's body to kill parasites, such as lice and ticks. It may consist merely of posts, firmly set, and wrapped with oil-soaked burlap against which the animal rubs its body, or of mechanical devices, such as rollers set in a container of oil, against which the animal rubs itself.

oil gland. The gland at the base of the tail of a chicken which secretes an oil used by the bird in preening its feathers. The gland is removed in dressing chickens for the market. Also called *preen gland; uropygial gland.*

oiling machine. In the poultry industry, a machine used for dipping eggs in oil to form a protective coating and to retard loss of moisture and quality. It is used especially on eggs placed in storage. See *oil protecting eggs.*

oiling off. Designating a defect of cream or milk characterized by separation of free butter fat. When defective table cream is poured into hot coffee, droplets of butter fat appear on the surface.

oil meal. Linseed-oil meal, cottonseed-oil meal, etc.

oil mill. A mill or factory in which oils are extracted from vegetable seeds, such as cottonseed and soybean, by hydraulic, expeller, or solvent method of extraction.

oil nut. See *butternut.*

oil of chenopodium. Wormseed oil.

oil of lemon. An essential oil used in flavorings and in perfumes, extracted from the rind of the lemon.

oil of wintergreen. An aromatic, flavoring oil obtained from the checkerberry *q.v.*

oil protecting eggs. The process of protecting eggs by dipping them in a hot liquid for a very short time, partially to coagulate the albumen in the shell membranes and to seal the egg shell. See *shell protected.*

oils and fats. In foods and foodstuffs, the esters of glycerol and fatty acids. The lower members being liquid at room temperatures, are regarded as oils. See *fat: 1, 3.*

oil spray. A contact insecticide spray consisting of poisons mixed with an emulsion (*q.v.*) of oil.

oil-stained cotton. Cotton that has become more or less saturated with oil during the ginning process. It is usually recognized by its texture and odor.

oilstone. A fine granite stone used with oil for sharpening tools.

oily. 1. In butter grading, designating a butter with an oily flavor. Such butter is rated C. 2. Designating a flavor defect of milk which results from the oxidation of some of its fatty constituents.

oily cream. See *oiling off.*

oily flavor. See *oily.*

oily pork. A swine carcass that is soft or oily because of softening feeds or immaturity.

ointment. A salve; an unguent; a medicinal preparation which has a base of some kind of fat or soft unctuous substance, such as the petroleum product, petrolatum.

Oka. A pale yellow, semisoft, cows' milk cheese, made in a Trappist monastery at Oka, Canada.

okra. *Hibiscus esculentus,* family Malvaceae, a tropical annual, grown as a garden vegetable all over the world. The mucilaginous, immature pods of the plant are cooked and eaten by themselves or in soups. Native to Africa and Asia. Also called *gumbo.*

okra wilt. A fungous disease of okra and other species of the genus Hibiscus caused by *Fusarium oxysporum,* family Tuberculariaceae, characterized by the wilting of the plant.

old bird. A stewing chicken, hen or fowl, generally a year or more old.

old-cream flavor. In butter grading, a flavor suggestive of old cream. Such butter is rated B.

Old English Game. A breed of game fowls famous for their alertness, strength, and stamina. The present day *Old English Game* chicken is similar in type and form to the earlier Pit Game bred in England when cock fighting was in vogue. They are bred in the following varieties: Black-Breasted Red, Brown-Red, Golden Duckwing, Silver Duckwing, Birchen, Red Pyle, Black White and Spangled. There are also bantam sizes in most of these varieties. See *Modern Game.*

Old English sheep dog. A shaggy-coated, slate and white spotted, medium-sized dog believed to be a breed imported to England during the Roman occupation. It is stocky and square-nosed, with well-muscled hind legs; it may have a very short tail or none at all. Used to herd sheep and cattle.

old ewe disease. See *pregnancy disease in ewes.*

old-fashioned potato bug. Any of the blister beetles, *Epicauta* spp., family Meloidae, which were pests of the potato in the central and eastern U.S. before the Colorado potato beetle.

old fashioned weigela. *Weigela florida,* family Caprifoliaceae, an upright shrub bearing rose-pink flowers, grown as an ornamental. Native to China and Korea.

old-field balsam. Fragrant cudweed.

old-field clover. Rabbitfoot clover.

oldfield juniper. *Juniperus communis* var. *depressa,* family Pinaceae, an evergreen shrub grown as an ornamental. In the wild its berries furnish food for birds. Native to North America. Also called *prostrate juniper.*

old-field pine. Loblolly pine.

old field stand. A stand of trees which has grown up on land once used for crops or pasture.

old flavor. A flavor defect of dry whole milk, probably associated with the constituents other than fat; a characteristic age defect.

old geese. In marketing, geese over one year old.

old growth. A stand of mature trees. See *virgin forest.*

Old Heidelberg. A soft, surface ripened cows' milk cheese similar to Liederkranz.

old hen. In marketing, a female turkey over one year old.

old-house borer. *Hylotrupes bajulus* (L.), family Cerambycidae, an insect which infests well-seasoned wood in buildings.

old-ingredient flavor. A flavor defect in ice cream which suggests the use of a stale, musty, off-flavored product in the ice cream mix.

old maid. The defective kernels of popcorn which do not pop.

old maid's bonnet. Silvery lupine.

old maid's pink. Bouncing Bet.

old-man-and-woman. See *hen-and-chickens.*

old-man's-beard. White fringetree.

oldman wormwood. *Artemisia abrotanum,* family Compositae, a perennial herb grown for its leaves and stems which are supposed to repel moths and ants. Native to Europe. Also called *southernwood.*

old parity prices. The prices of certain farm commodities in terms of their buying or purchasing power in a selected base period, usually 1910 to 1914. Used as a base index of prices for calculating the cost of living and other statistical information of agricultural importance. During the years 1930 to 1950, the base period price was multiplied by the most recent index of prices paid by farmers. See *new parity prices; parity prices; transitional parity prices.*

old process. A method of extracting oil from soybeans in which the beans are ground, cooked, and subjected to hydraulic pressure.

old shell. Beef cows in very poor condition; classed as canners and cutters.

old skin. Referring to a decrepit horse.

old soils. In pedologic terminology, soils on flat land surfaces which have fully developed profiles but are stagnant, *i.e.,* undergoing changes with extreme slowness.

old style Chilean nitrate of soda. Natural nitrate of soda obtained by leaching caliche (*q.v.*) of Chilean mines with boiling water. The commercial product is about 97 percent nitrate of soda. *Cf. champion brand Chilean nitrate.*

old tom. A male turkey over one year old after molt is completed.

Old Virginia ham. Ham cured by a special method in which (a) the ham is thoroughly rubbed with salt and cured for seven weeks; (b) then rubbed with a mixture of New Orleans molasses, brown sugar, black pepper, cayenne pepper and saltpeter; and (c) allowed to cure for at least 30 days or for several months before being eaten.

old witch grass. Witch grass.

old woman. Beach wormwood.

old wood. A shoot or branch of a woody plant which is at least one year old.

oleander. 1. Any evergreen shrub or tree of the genus Nerium, family Apocynaceae. 2. *N. oleander,* a shrub or small tree grown as an ornamental in warm areas and as a tub plant in northern U.S. The leaves, both in green and dry condition, are poisonous to livestock. Native to northern Africa and southern Europe. Also called *rosebay.*

oleander knot. A bacterial gall which appears on the stems of the oleander and other plants. See *bacterial gall; gall: 1; hypertrophy.*

oleander scale. *Aspidiotus hederae* (Vallot), family Diaspididae, a scale insect which infests citrus throughout the world.

oleaster. Russian olive.

oleic acid. Colorless, oily liquid found in most fats; it has a pungent taste and is used in making soap.

olein. A colorless, oily compound of glycerine and oleic acid found in most fats and mixed oils.

oleocellosis. Spots most often occurring on lemon

fruits but also on oranges, caused by the action of oil freed from the oil glands of the rind by mechanical injuries, hot water, and ice formation. The area affected by the oil may remain green while the rest of the surface changes color. Also called *rind oil spot.*

oleomargarine. A product of certain animal or vegetable fats, sometimes fish and coloring matter, emulsified, churned, or mixed with cream, milk, water, or other liquid and made in the semblance of butter. Some of the vegetable fats used are cottonseed, coconut, corn, and soybean oil. Also called *margarine.*

oleo oil. A type of oil obtained in the rendering, at a temperature below 170° F., of certain of the inner or intestinal fats of beef or mutton; it is used in the manufacture of many animal fat mixtures and compounds.

oleoresin. A natural combination of resinous substances and essential oils which occurs in, or exudes from, plants. Upon distillation, the *oleoresins* obtained by chipping the living trees of *Pinus palustris* and *P. caribaea* yield turpentine and rosin. Also called *crude turpentine.*

oleoresin of pyrethrum. A product used in insecticides, obtained by treating finely ground pyrethrum with ethylene dichloride and then evaporating.

oleo stearin. The compound of fats and oils obtained in the rendering of beef and mutton fats at a temperature below 170° F.; it is used in the manufacture of compounds requiring a stearin base.

oleo stock. A compound of the fats and oils obtained in rendering beef and mutton fats at a temperature below 170° F.; it is used in the manufacture of oleomargarine.

olericulture. Vegetable culture, as distinguished from fruit and flower culture. A branch of horticulture. Also called *vegetable gardening.*

oligophagous parasite. A parasite capable of developing upon a few closely related host species.

oligotrophic peat. Peat soil that is strongly acid in reaction and low or wanting in bases, such as calcium, and other mineral plant nutrients.

olive. 1. Any shrub or tree of the genus Olea, family Oleaceae. 2. *O. europaea,* an evergreen tree which has been grown for centuries for its edible fruit, the *olive* of commerce, and as a source of *olive* oil (*q.v.*). Native to Europe and Asia.

olive-colored yolk. A greenish coloration of the yolks of eggs, believed to be caused by some particular feed eaten by hens. In egg grading it is regarded as a defect, although such eggs are edible. See *grass egg.*

olive fruit fly. *Dacus oleae* (Gmel.), family Tephritidae, a fly whose maggots may infest the olive fruit. It is the vector for olive knot.

olive knot. An irregular knot or excrescence of a spongy or cheesy character on aerial parts, especially the trunk or branches, of the olive tree; it is caused by the olive tubercle bacterium, *Pseudomonas savastanoi,* family Pseudomonadaceae. The vector is the olive fruit fly (*q.v.*). See *bacterial gall.*

olive oil. The edible oil derived from the fruit of the olive by crushing, pressing, and filtering. Chemically, *olive oil* consists of olein, palmitin, some stearin, and a small amount of linolein.

Oliver snapweed. One species of snapweed *q.v.*

olive scale. *Parlatoria oleae* (Colvee), family Diaspididae, a scale insect pest which infests and damages not only the olive, but a great number of other plant species in California, U.S.

Olivet. A soft, cows' milk cheese which derives its name from Olivet, a town in France. There are three types: unripened, half ripened (blue), and ripened.

olive tubercle. See *olive knot.*

Olmutzer Quargel. A sour milk, spiced Hand cheese containing caraway seeds, made in Austria and Bohemia.

Olney bulrush. *Scirpus olneyi,* family Cyperaceae, a perennial, grasslike herb which occurs in coastal marshes, where it is an important source of food for muskrats. Native to Europe. Also called *three-cornered grass; three square.*

Olympic horse. A horse that competes in Olympic games.

Olympic mullein. *Verbascum olympicum,* family Scrophulariaceae, a densely-white, woody biennial herb, which has large yellow flowers and is sometimes used as an ornamental for its bold tropical effect. Native to Greece.

Olympic poppy. *Papaver pilosum,* family Papaveraceae, a perennial herb grown for its brick-red to orange flowers. Native to Turkey.

omasum. The section between the reticulum and abomasum of the stomach of ruminating animals. The *omasum* of the cow consists of one hundred or more longitudinal folds known as laminae, leaves or manyplies. Also called *book; manyplies; manifold.* See *rumen.*

ombutree pokeberry. *Phytolacca dioica,* family Phytolaccaceae, a beautiful evergreen tree grown as an ornamental in warm areas for its foliage and its white flowers. Native to North America. Also called *umbra.*

omentectomy. The removal of a portion of one of the membranes attached to the stomach. One of these membranes is very large and may carry an excess of fat.

omentum. A fold of the peritoneum connecting the stomach to the adjacent organs. The fat deposited in the folds of the great *omentum* of cattle, sheep, and swine is in part the source of the suet of commerce.

omnivorous. Designating an animal which feeds on both flesh and plants, as man; as applied to insects, voracious, but not necessarily *omnivorous.* Also called *amphivorous. Cf. carnivorous; herbivorous.*

omnivorous leaf tier. *Cnephasia longana* (Haw.), family Tortricidae, a European insect which infests the fruit and foliage of strawberries and a large number of field crops and cultivated flowers; especially destructive in California and the northwestern U.S.

omnivorous looper. *Sabulodes caberata,* family Geometridae, an insect whose larvae are very destructive on avocado trees in California, U.S., often defoliating the entire tree. They also infest many other cultivated trees including the orange.

ompa. A systemic insecticide, a complex chemical compound, octa-methyl-pyro-phosphoramide, a selective killer designed to control species of aphids.

omphalitis. An infection, usually fatal, of the navel in baby chicks and turkeys largely due to failure of the navel to close properly and to unsanitary incubation conditions; characterized by listlessness, a puffed and mushy condition of the body, and a bluish appearance of the abdomen. Also called *mushy-chick disease.*

omphalophlebitis. An inflammation of the umbilical vein and other structures in the umbilicus of young animals caused by filth-borne bacteria. Also called *sleepy foal disease; pyosepticemia; navel-ill; joint ill.*

on. In cotton marketing in spot transactions involving grades other than middling, a premium over the grade. See *off: 1.*

on and off permit. A grazing permit issued only where movement of livestock is necessitated between National Forest land (U.S.) and adjoining outside range, or where private and forest lands intermingle.

onchocercosis. An ailment of animals characterized by fibrous nodules forming under the skin; caused by worms of the genus Onchocerca.

on clean basis. In marketing, designating clean wool offered for sale.

on contract. Designating a crop, as vegetables for processing, grown under an agreement whereby the producer sells to a processor usually at prices previously agreed upon.

oncosphere. The first larva of the tapeworm enclosed in the egg shell and armed with six hooks.

one bum lamp. A blind eye of a horse.

one-crop farming. A system of farming in which the producer grows a single crop as a source of income: *e.g.,* wheat or cotton.

one-half-row cultivator. A one-horse cultivator which cultivates one-half of the area between rows in one trip across a field; used in the eastern and southern U.S. where a farm often includes only a few acres, especially one not entirely cleared of stumps. It could be adjusted to cover the entire middle between the rows.

one-horse. 1. Designating a plow, planter, etc., drawn by one horse such as a *one-horse* wagon, *one-horse* drill, etc. 2. Designating a poor or neglected farm, as a *one-horse* farm.

one-horse drill. A small drill drawn by one horse; frequently useful in seeding small grains, peas, or soybeans between the rows of a growing row crop.

one-horse farm. A small, poorly equipped farm.

one-horse mower. A mower designed to cut grass for hay, similar in principle but smaller than a two-horse or a tractor drawn mower. See *mower: 1.*

one-horse planter. In cotton farming, a drill planter (*q.v.*) drawn by one animal.

one-horse wheat drill. See *one-horse drill.*

one-litter system. A swine production program in which each sow farrows only one litter of pigs, usually in summer or fall, and is then fattened and sold. Gilts are selected for breeding each time.

one-man type potato planter. A potato planter operated by one man.

one-night bedding-out. In sheep grazing on the western range, U.S., a system in which the sheep are bedded down on new ground each night.

one-night camp. See *one-night bedding-out.*

one-piece frame. In bee keeping, a single-box frame, in contrast to a divided or two-piece frame.

one-row four-wheel lister planter. A riding planter differing from the two-wheel type in that the bottom is carried on a frame which extends to the rear, being supported by two small wheels. See *lister planter.*

one-row ground-driven horse-drawn potato-harvesting machine. A riding potato digger, commonly called an elevator digger *q.v.*

one-row picker. A mechanical corn picker which harvests only one row at a time, in contrast to the two-row picker.

one-row riding cultivator. A one-row cultivator (*q.v.*) on which the operator rides, in contrast to one designed for the operator to walk.

one-row riding type of cotton planter. See *cotton planter: 2.*

one-row two-horse walking cultivator. An implement which cultivates both sides of the row at one trip across the field with the operator walking behind rather than riding.

one-row walking corn planter. A planter similar to the one-row walking cotton planter but differing in the size of the hopper and plate equipment.

one-row walking lister planter. A planter which consists of an ordinary walking middlebuster with a planting mechanism attached to the rear. See *lister.*

one-run process. A procedure in the manufacture of vinegar by the generator or German process. The alcoholic liquid is acidified by the addition of vinegar to increase the acidity to 3 to 3½ percent, representing about one gallon of hard cider to two gallons of vinegar. One passage of this liquid through the generator converts the remaining alcohol into acetic acid. See *two-run process.*

one-season pasture. Pasture composed of a forage plant, usually an annual, such as Sudan grass, which can furnish grazing only for a single season.

one-way disk plow. A widely-used plow which combines the principles of the disk plow and the disk harrow. It was designed initially as a disk harrow. The disks are set to throw the soil one way. There may be from four to thirty disks depending on the size of the plow. Also called *wheatland plow; cylinder plow; harrow plow; tiller plow.*

on feed. 1. Designating livestock kept on farms and ranches which are being fattened for the market. 2. Designating any animal being fed grain for milk or meat production.

on full feed. In feeding poultry and livestock, the feeding of the animal all the food which is necessary for optimum gain.

onion. 1. Any of several herbs of the genus Allium, family Liliaceae. Species include some of the very important vegetables. 2. *A. cepa,* a perennial herb grown as an annual salad plant and as a potherb for its strong-tasting bulb and leaves. It is a very ancient vegetable. Native to western Asia. Also called *garden onion.* 3. See *shallot; Welsh onion; leek; garlic; chive.*

onion black mold. A mold caused by the fungus *Aspergillus niger;* family Dematiaceae.

onion bulb and stem nematode. A nematode of the genus Ditylenchus, family Tylenchidae, which infests the bulbs of onions and of narcissus causing a softening of the bulbs.

onion downy mildew. A destructive, fungous disease caused by *Peronospora destructor,* family Pythiaceae. It affects the leaves of the plant when the humidity is high and appears as a bluish-gray, fuzzy mildew.

onion flavor. A flavor defect in milk and milk products which results from ingestion of the wild onion by dairy cows.

onion fly. See *onion maggot.*

oniongrass. Certain plants of the genus Melica, family Gramineae. Species are perennial bunchgrasses and the base of the stem is bulbous. They are regarded as fillers on the ranges and have only a secondary value as a forage. Native to Rocky Mountain regions of the U.S. Also called *melic.*

onion hoe. A small hand hoe with a small triangular blade and short handle which is used in weeding onions.

onion maggot. The larva of the onion fly, *Hylemya antiqua* (Meig.), family Muscidae, which tunnels into the bulb and crown of the onion; a destructive pest in southern U.S.

onion mildew. Onion downy mildew.

onion neck rot. See *neck rot of onion.*

onion or garlic flavor. In butter grading, a butter which has a flavor of onion or garlic. Such butter is rated CG (cooking grade).

onion pink rot. A fungous disease caused by *Pyrenochaeta terrestris,* family Sphaeropsidaceae, characterized by the affected roots turning pink, shriveling and dying, and a reduction in the size of the bulbs.

onion plant bug. *Labopidea allii* Knight, family Miridae, a pale green bug which infests the tips of leaves of onion plants causing them to turn white in spots and finally to die from the tip downward.

onion poisoning. Death camass.

onion purple blotch. A widespread, fungous disease which is serious on onions in warm, irrigated locations. The disease caused by *Alternaria porrii* (syn. *Macrosporium porrii*), family Dematiaceae, is characterized by purple lesions on the leaves and a soft, watery neck and bulb rot at or after harvest.

onion set. A small, immature onion bulblet which is grown from seed, harvested from a large bulb, or the bulbil of the top onion, for starting onions.

onion smudge. A fungous disease caused by *Colletotrichum circinans,* family Melanconiaceae. It normally attacks only white onions, and not those with yellow or red papery scales. In certain stages black, smudgy spots appear on the scales of the bulb and have a detrimental effect on the market value.

onion smut. A fungous disease caused by *Urocystis cepulae* Frost, family Tilletiaceae, an organism which can live in the soil for many years. Young seedlings are most affected and the disease appears to be restricted to northern states, U.S.

onion thrips. *Thrips tabaci* Lind., family Thripidae, a very minute, hardly visible insect that punctures the leaves or stems of onions and sucks up the exuding sap causing whitish spots on leaves and stems and weakening the plants.

onion yellow dwarf. See *yellow dwarf of onion.*

on pasture. Designating livestock that are grazing on pasture in contrast to those feeding in the barn or feedlot.

on the block. 1. Up for auction. 2. Designating carcasses ready for cutting for sale. *Cf. on the hoof.*

on the clean basis. In American wool market reports, referring to quotations given on the cost of the clean fiber which any lot of unscoured wool is estimated to contain.

on the cob. 1. Designating unshelled corn. 2. Designating green, or immature corn, which is cooked and eaten with the kernels attached to the cob.

on the contour. See *contour farming; contour row: b.*

on the dry side. In greenhouse culture, designating a reduction in the amount of water ordinarily given to plants to keep the soil comparatively dry when growth is inactive.

on the hoof. Designating a live meat animal.

ontogeny. The complete developmental history of an organism from the egg, spore, bud, etc., to the adult stage.

on-year. Referring to the year during which an alternate, or biennially fruit-bearing tree produces fruit. *Cf. off-year.*

oocyst. The encysted or encapsulated stage of parasites, known as coccidia, passed with droppings of infected animals.

oocyte-egg-mother-cell. The cell which undergoes two meiotic divisions, oogenesis, to form the egg cell, as primary oocyte, the stage before completion of the first meiotic division, and secondary oocyte, after completion of the first meiotic division.

oogenesis. The process by which germ cells are produced by the female.

oon egg. An egg, expelled from the vent of a hen without a shell.

Oöspora rot. A soft, slimy, contact decay of the

fruits of several species of Citrus caused by the fungus, *Oöspora citri-aurantii,* family Moniliaceae. Also called *sour rot.*

Oöspora scab. See *corky scab of potato.*

oöspore. A resting spore produced by sexual reproduction in the downy mildews and related fungi.

open. 1. Designating a female animal which is not pregnant. 2. Designating a body defect of cheese in which the cheese has many mechanical openings. 3. To unfold, as a flower *opens.* 4. Designating a rural or wilderness area in contrast to the congestion of cities and towns.

open-air system of milking. A method of milk production in which the cows are kept in the open and milked in a mobile house or bail. See *bail: 3.*

open bed. An uncovered box or frame containing soil in which seeds are planted to produce seedlings for transplanting. Used for starting tobacco plants in southern U.S. See *coldframe.*

open-behind. Designating a horse having hocks set far apart and the fetlocks close together.

open boll. The ripened cotton boll which has split open and from which locks of cotton hang.

open brand. A brand that has no surrounding or enclosing lines. See *brand: 1.*

open-center tire wheel. The wheel of a planter consisting of two rims with a space between them which serves both as a covering device and a press wheel.

open-center tree. An orchard tree pruned so as to have a vase-shaped top in contrast to one having a central leader.

open class. In agricultural shows, an exhibit for which no strict limitations are imposed either for the thing exhibited or for the exhibitor.

open concentrator. An uncovered vessel, such as a kettle, in which fruit juices and sap are concentrated by heat to a sirupy consistency.

open cotton. Cotton left unpicked and exposed to the weather in the field. It becomes dull and lower in grade.

open-ditch drainage. Drainage of excess water from land by open ditches as opposed to tile drainage.

open-end harrow. A harrow without guard rails.

open-faced. Designating sheep which have little wool over the face, especially those which are free from wool hanging over the eyes.

open field system (of farming). An old system of village or communal farming in which intermixed strips of unenclosed land were separately owned or cultivated by individual operators but subject in varying degrees to right of common. The system prevailed especially in England until the Enclosure Acts of the 18th and 19th centuries. It was used in Russia for centuries.

open-fire curing. One of the three common methods of curing tobaccos, in which the tobacco is hung on scaffolds in the curing barn and fires are built directly under it. *Cf. air cure; flue cure.*

open-fleeced. Designating the wool on a sheep which has a less dense coat and coarser wool than normal.

open-flume. An uncovered passageway for irrigation water.

open formula. On feed containers, the statement on the tag about the number of lbs. of each ingredient in a ton of mixed feed.

open formula fertilizer. A fertilizer which has both the analysis and the formula or materials used in it printed on a tag or on the bag container.

open heifer. A nonpregnant heifer. *Cf. bred heifer.*

open herding. Handling a band of sheep or goats in a way to allow the animals to spread rather freely and without trailing in grazing. *Cf. bunched up herding.*

opening. 1. A treeless or very sparsely timbered area in a forested region. See *oak opening.* 2. In cranberry culture, the first swelling of a terminal bud. 3. The unfolding of a flower or boll of cotton. 4. Designating the first price offered for a commodity when a market day begins.

open-kettle method. A method of canning fruits and vegetables in which the products are cooked in an open vessel to kill bacteria prior to placing them in sterilized cans or jars.

open-kettle one-period process. The preparation of fruit for preservation by boiling in steam-jacketed open kettles and impregnating with sirup, in a single operation. See *slow open-kettle process.*

open-kettle rendered leaf lard. The leaf fat (*q.v.*) of hogs rendered in uncovered steam-jacketed kettles provided with mechanical agitators. The system provides the highest grades of commercial lards outside of the neutral lards.

open pedigree. Among corn breeders, those pedigrees of inbred lines, which are published or available to the general public.

open-pollinated. Designating normal or natural pollination. Applied to corn, the term has two meanings: (a) all corn grown before there was any production of hybrid corn; (b) experimental fields where there is no control of the distribution of pollen from or to any of the plants within the field. See *pollination.*

open range. An extensive grazing area on which the movement of livestock is unrestricted.

open round-up. A round-up of cattle in which corrals or fences are not used (western U.S.).

open shield. A fender attachment for a cultivator which consists of spaced rods or wires, to protect young plants from being covered with soil.

open shoulder. In livestock judging, a depression between the shoulder blades.

open soil. Soil that is pervious, granular, or loosely aggregated.

open surface. A texture defect of cheese which shows many press or mechanical openings.

open-trough hopper. A feed hopper for poultry that does not require a grid to make it nonwasting of feed.

open weather. Pleasant weather with mild temperatures and an absence of fog or severe storms, especially during the winter.

operating capital. The value of the livestock,

equipment, feed and supplies, and the amount of cash necessary to run a farm business.

operation waste. The water which is wasted through spillways or otherwise discarded from an irrigation system after having been diverted into it.

operator. One who does the appropriate work and management necessary to conduct a business, as farming.

operator area. The area within the milking room used by the personnel performing the routine milking operations.

operator's earnings. The net earnings of the farm operator for his labor and management including the value of the family living obtained from the farm. It is derived by deducting from the net farm earnings an interest charge for the total capital invested in the farm business. See *operator's labor income.*

operator's labor earnings. See *operator's earnings.*

operator's labor income. Operator's earnings (*q.v.*) excluding the value of the family living obtained from the farm. It is one of the best measures of the degree of financial success on a farm.

ophthalmia. Inflammation of the eyeball or conjunctiva.

ophthalmic tuberculin test. A test for tuberculosis in which a small quantity of a concentrated tuberculin is placed within the lower eyelid of the animal to be tested. A discharge of mucopurulent material within six hours indicates a positive reaction. The test is commonly used on primates held in zoos: *e.g.,* gorillas, baboons, etc.

opium poisoning. The poisoning of animals which results from ingesting the opium poppy (*q.v.*) or from overdoses of opiates.

opium poppy. *Papaver somniferum,* family Papaveraceae, an annual herb grown as a source of opium, a drug, and for its seeds used on the crust of breads and rolls. Native to Asia and the Near East.

opportunities. 1. Production possibilities or isocost lines in product-product dimensions. A relation between a pair of products indicating all different combinations of the two products that may be produced with a given outlay. 2. A specific resource: *e.g.,* an iso-land cost line showing possible crop combinations on a given acreage.

opportunity cost. The return from the most profitable alternative use sacrified by employing a unit of a resource in a given manner.

optimal light income. The amount of light, measured by intensity and duration, which induces the best growth of a plant.

optimum condition. The ideal environment, with regard to nourishment, light, temperature, etc., for an organism's growth and reproduction.

optimum fruitfulness. That favorable condition for growth (especially in fruit trees) in which a plentiful supply of blossom buds is produced. The condition is associated with the carbohydrate-nitrogen relationship *q.v.*

optimum temperature. That certain temperature at which a particular plant or animal grows satisfactorily, other conditions being favorable for growth.

optimum water content. The amount of water in a soil needed by a plant for its optimum growth, varying from 40 to 60 percent of the moisture holding capacity.

optional payment. A term or condition in a loan contract whereby a borrower may pay any part or all of the loan at any time or at any interest bearing date before the loan matures. Often a great convenience to the borrower but an inconvenience to the lender.

orach. Garden orach.

orange. Certain tropical or subtropical trees of the genus Citrus, family Rutaceae. Species are grown for their edible fruits, the *oranges* of commerce, for their fragrant flowers, and as breeding stock. See *sour orange; sweet orange; calamondin; King orange; mandarin.*

orangeberry pittosporum. *Pittosporum undulatum,* family Pittosporaceae, an evergreen tree grown in frost-free areas for its fragrant, white flowers. Native to Australia. Also called *Victorian box; cheesewood; mockorange.*

orange bulbil lily. *Lilium bulbiferum* var. *croceum,* family Liliaceae, a bulbous herb grown for its orange, crimson-spotted flowers. Native to southern Europe. Also called *orange lily.*

orange concentrate. The oil-free juice of crushed oranges concentrated to a certain density, heated or pasteurized and canned. It is used in confections, baking, etc.

orange coneflower. *Rudbeckia fulgida,* family Compositae, a perennial herb grown in the flower border for its golden-yellow flowers with orange bases and purplish-black disk florets. Native to North America.

orange daylily. *Hemerocallis aurantiaca,* family Liliaceae, a lily-like herb grown for its burnt- or salmon-orange flowers. Native to Japan.

orange dog. *Papilo cresphontes* Cram., family Papilionidae, a butterfly which infests citrus feeding voraciously on the foliage, sometimes defoliating a young tree in a few days.

orange douce. See *sweet orange.*

orange-eye butterflybush. *Buddleia davidi,* family Loganiaceae, a shrub widely grown for its late-blooming, fragrant, lilac-colored flowers with orange throats. Native to China.

orange fleabane. *Erigeron aurantiacus,* family Compositae, a perennial herb grown in the rock garden for its orange-yellow flowers. Native to Eurasia. Also called *double orange daisy.*

orange-glowvine. *Senecio confusus,* family Compositae, an evergreen, twining vine grown in frost-free areas on an arbor, in a hanging basket, or as a pot plant for its large, orange-red, daisy-like flowers with lemon-yellow disks. Native to Mexico. Also called *Mexican trailing groundsel; Mexican firevine.*

orange-grapefruit buttermilk. Buttermilk flavored with the juices of orange and grapefruit.

orange grove. See *grove: 3.*

orange gum. Rusty gum myrtle.

orange hawkweed. *Hieracium aurantiacum,* family Compositae, a somewhat hairy, perennial herb grown in the rock garden for its orange-red flowers. Escaped, it can be a weed pest. Native to Europe. Also called *Burma weed; devil's-paintbrush; devil's-bit; fairie's paintbrush; Grim-the-collier.*

orange jasmine. Chinabox jasminorange.

orange larkspur. *Delphinium nudicaule,* family Ranunculaceae, a perennial herb grown for its red flowers with yellow petals. Native to North America. Also called *red larkspur.*

orange leaf rust of wheat. See *leaf rust.*

orange maggot. The maggot of the Mexican fruit-fly, *Anastrepha ludens* (Loew), family Tephritidae, which works inside the orange fruit.

orange melon. Mango melon.

orange milkweed. *Asclepias tuberosa,* family Asclepiadaceae, a perennial herb which causes poisoning in animals eating its leaves. Native to Mexico and Florida and Arizona in the U.S. Also called *butterfly weed; pleurisy-root; butterfly milkweed.*

orange oil. An edible oil produced (a) by crushing or grinding whole oranges and screening the juice centrifugally; or (b) by rasping the peel while sprays of water are directed against the fruit. The wash water is left in a tank. The oil and water emulsion rising is skimmed and centrifuged.

orange polygala. *Polygala lutea,* family Polygalaceae, a hardy, annual plant grown in the bog garden for its dense clusters of orange-yellow flowers. Native to southeastern U.S.

orange pulvinaria scale. *Pulvinaria aurantii,* family Coccidae, a common and serious scale insect pest which infests citrus in Japan.

orangeroot. Goldenseal.

orange rust. A fungous disease of blackberries and black raspberries caused by *Gymnoconia interstitalis,* family Pucciniaceae, characterized by a bright orange, powdery mass of spores on the lower leaf surfaces early in summer, stunting, production of little or no fruit or spines, tillering, and premature defoliation.

orangery. An artificially heated house, the predecessor of the greenhouse, used as early as the 14th or 15th century in Europe primarily for growing oranges and citrons in climates too cold for outdoor growth.

orange sirup. A concentrated solution of sugar and orange juice freed from peel and oil. Diluted with water, it is widely used as an ingredient and flavoring for bottled soft drinks.

orange sneezeweed. *Helenium hoopesi,* family Compositae, a perennial herb grown in the informal border for its yellow flowers. Found wild from Montana to New Mexico, U.S., it is poisonous to sheep. Symptoms of poisoning, known as spewing sickness, are depression, weakness, salivation, nausea, diarrhea, bloating, and vomiting. Also called *western sneezeweed; yellow-weed; owls-claws.*

orange spiny white fly. *Aleurocanthus spiniferus,* family Aleyrodidae, an insect pest, the most injurious of the several species that infest citrus in Japan.

orange streptosolen. *Streptosolen jamesoni,* family Solanaceae, a tropical, woody vine or pendulous shrub grown out-of-doors in warm areas to cover a wall, as a shrub, and as a pot plant in the colder areas. It has orange-red to bright red flowers. Native to southern U.S. Also called *yellow-heliotrope; firebush.*

orange-striped oakworm. The caterpillar of *Anisota senatoria* (J. E. Smith), family Citheroniidae, which infests oaks and occasionally other trees.

orange stubborn disease. A disease attributed to a virus, which affects citrus in California, U.S. It is characterized by slow growth of buds and by acorn-shaped or pink-nosed fruits.

orange sunflower. Rough heliopsis.

orange tortrix. The caterpillar of the moth, *Argyrotaenia citrana* (Fern.), family Tortricidae, which bores into the rind of the orange fruit. It also feeds on a large number of other plants.

orange-tree quick decline. A virus disease which affects sweet orange plants grafted on sour orange rootstocks. Young trees may collapse and the leaves and fruit suddenly dry up. Older trees decline more slowly.

orange tristeza. A virus disease, similar to orange-tree quick decline, which is very destructive in orange groves. It probably originated in South Africa but has spread to other countries including the U.S.

orange vinegar. Vinegar made from the fermented juice of oranges.

orange worms. The larvae of four species of small moths, *Argyrotaenia citrana, Pyroderces rileyi; Holcocera iceryaella,* and *Platynota sultana,* which infest the orange in certain areas of southern California, U.S., feeding on the fruit.

orchard. a. An acreage of fruit trees. b. A grove of nut trees, as pecans. c. An olive grove. d. A grove of sugar maple trees. e. In southern U.S., a tract of pine trees which is being bled for turpentine.

orchard bark beetle. See *bark beetle.*

orchard cover crop. Grasses, legumes, or any herbaceous plant, grown in an orchard to form a cover to prevent soil erosion, and to improve the soil through the addition of organic matter. It may also serve to use up available nitrogen so the trees will harden up for winter.

orchard cultivator. A cultivator which breaks up compact clay subsoils or hardpans without bringing them to the surface. Also called *subsoil tiller; chisel plow.*

orchard disk harrow. A disk harrow differing from the regular disk harrows by having a wide frame so that the gangs can be set at varying distances apart.

orchard disk plow. A specially constructed disk plow with the wheels set inside the frame to allow passing close to trees, and which has no long levers to catch overhanging branches. Its hitch is designed to allow the plow to be shifted to the right or left, making cultivating close to trees possible.

orchard duster. A power duster for applying insecticide and fungicide dusts to fruit trees. It is usually equipped with a flexible hose for applying the dust in any direction desired.

orchard giraffe. See *giraffe.*

orchardgrass. *Dactylis glomerata,* family Gramineae, a perennial bunchgrass, grown for hay, forage, and pasture. It is aggressive in growth and yields well, but the hay is inclined to be woody and of low value. Its pasture value is reduced when approaching maturity. Native to Europe and Asia. Also called *cocksfoot; dog foot.*

orchardgrass brown stripe. See *brown stripe of grasses.*

orchard harrow. A rigid tooth harrow used in extremely hard soils for subsoiling and preparing of seed beds.

orchard heater. Any of several types of heaters used in the orchards to prevent frost injury.

orchard heating. The use of stoves or other devices to prevent frost injury to orchards.

orchardist. A cultivator of orchards.

orchard mite. A spider mite, *Tetranychus pacificus* McG., family Tetranychidae, a serious pest along the Pacific Coast of the U.S. It infests fruit trees and other plants, heavily webbing the foliage. There may also be extensive defoliation. Also called *Pacific mite.*

orchard power sprayer. An insecticide and fungicide sprayer adapted for fruit trees. See *power sprayer.*

orchard weed. Any undesirable or noxious plant which grows in an orchard. See *weed: 1.*

orchid. Any of the plants belonging to the family Orchidaceae. Species are perennial epiphytic and terrestrial plants, valued for the beauty of their flowers. Native mainly to the tropical regions; grown in the greenhouses elsewhere.

orchid basket. A criblike basket filled with coarse fiber or orchid peat; designed for potting orchids, etc.

orchid-flowering dahlia. A type of single dahlia which has tubular rays or florets.

orchid fly. *Eurytoma orchidearum* (Westw.), family Eurytomidae, a small, black, wasplike fly; a serious orchid pest in greenhouses.

orchid iris. *Iris orchioides,* family Iridaceae, a perennial herb grown for its yellow or white flowers. It is a beardless iris. Native to Russian Turkestan.

orchid peat. A potting material composed of fern roots whose chopped pieces are soaked until they attain a spongelike consistency.

orchid plant bug. *Tenthecoris bicolor,* an insect from South America which causes irregular white spots on the under side of orchid leaves.

orchid thrips. *Chaetoanaphothrips orchidii* (Moult), family Thripidae, an insect which infests citrus in Florida, U.S., and does some minor damage.

orchid tree. Buddhist bauhinia.

orchidvine. Fringed Amazonvine.

orchitis. Inflammation of the testicles.

order. 1. In botanical classification of plants, a category in between class and family. 2. In soils, a higher category in classification: zonal, intrazonal, and azonal are *orders.*

order of terms. The order in which constituents of a mixed fertilizer appear on the printed formula: nitrogen, phosphorus, potash.

ordinary cut. In marketing, vanilla beans which are dry, broken beans, and windfalls.

ordinary buttermilk. Buttermilk which is the byproduct of the churning of cream for butter. *Cf. cultured buttermilk.*

oread. A sun plant or heliophyte.

Oregon cedar. Lawson falsecypress.

Oregon evergreen blackberry. Cutleaf blackberry.

Oregongrape. *Mahonia aquifolium,* family Berberidaceae, an evergreen shrub grown for foundation planting, and for its yellow flowers and edible fruit. The state flower of Oregon (U.S.). Native to North America.

Oregon iris. *Iris tenax,* family Iridaceae, a perennial herb grown for its lilac-purple flowers. It is a beardless iris. Native to North America.

Oregon lummox. A large, feather-legged stock horse (Pacific Coast states, U.S.).

Oregon maple. Bigleaf maple.

Oregon pine. Douglas fir.

Oregon tunnel drier. A type of drier used in dehydration of fruits, consisting of a series of parallel, sloping, narrow chambers above a furnace room. Heated air enters the lower end of the chambers through a sliding door which regulates the heat.

Oregon water-hemlock. Tuber waterhemlock.

Oregon white cedar. Lawson falsecypress.

Oregon white oak. *Quercus garryanna,* family Fagaceae, a deciduous tree valuable for browse on the western range (U.S.).

Oregon wireworm. *Melanotus oregonensis* (Lec.), family Elateridae, the larva of a click beetle, common to the West Coast area of the U.S., a destructive pest of grains, grasses, root crops, vegetables, and seeds.

orejuella. Sacred ear.

orf. A widespread disease of sheep and goats caused primarily by a virus and secondarily by a bacterial organism, *Fusiformis necrophorus,* family Mycobacteriaceae. In malignant form, ulcers appear inside the mouth and other parts of the body.

Organ escallonia. *Escallonia organensis,* family Saxifragaceae, a tender, evergreen shrub grown in warm areas for its foliage and its rose-red flower clusters. Native to Brazil.

organic. 1. Produced by plants and animals; of plant or animal origin. 2. More inclusively, designating chemical compounds which contain carbon.

organic disease. A disease which causes some actual alteration of an organ resulting in faulty action.

organic farming. A system of farming in which

strong emphasis is placed on the use of organic materials: plant and animal manures, and composts, to maintain fertility and productivity of soil. The use of chemical fertilizers, insecticides, fungicides, etc., is avoided or minimized.

organic fertilizer. Any fertilizer or manure which is of plant or animal origin.

organic fungicide. A fungicide composed of organic chemical compounds.

organic gardening. See *organic farming.*

organic insecticide. A preparation made from organic chemical compounds, used to kill or control insects, as rotenone, nicotine, etc.

organic manure. 1. Feces and urine of farm animals; barnyard manure. 2. Any fertilizer material which is mainly organic in composition, such as treated sewage, guano, etc.

organic matter. 1. Matter found in, or produced by, living animals and plants, which contains carbon, hydrogen, oxygen, and often nitrogen and sulfur. 2. Any matter, defined as organic (*q.v.*) incorporated in or on the surface of soil. It may include undecomposed plant matter as well as that which is highly humified. Sometimes used interchangeably with humus. 3. In farming practice, the animal manures and plant growth added, plowed under and thus incorporated with the soil.

organic mercury compound. A combination of an organic chemical with mercury, used in seed treatments or sprays, such as ethyl mercury.

organic phosphate insecticide. A chemical compound used as an insecticide, such as tetraethyl pyrophosphate, ethyl metaphosphate, and tetraethyl dithio-pyrophosphate.

organic phosphorus. Phosphorus present as a constituent of an organic compound or a group of organic compounds, as glycerophosphoric acid, inosital phosphoric acid, cytidylic acid.

organic soil. Generally, any soil whose solid part is predominantly organic matter as contrasted to mineral soil and water soil; muck or peat.

organic thiocyanates. Highly complex synthetic, organic chemical compounds used in insecticide sprays as a contact poison.

organism. A living thing, whether plant or animal.

organ of flower. Any of the parts of a flower, as sepals, petals, stamens, pistils. See *essential floral organs.*

organogen. Any of certain chemical elements without which organisms cannot exist: oxygen, carbon, nitrogen, phosphorus, etc.

organoleptic. Concerning the sensory impressions associated with eating and drinking.

oribatid mites. Small, oval mites, less than 1/16 in. long, of the family Oribatidae, found in soil, moss, humus, and on the blades of grass. Some species act as intermediate hosts or vectors of the common tapeworms of sheep, cattle, goats, and horses.

Oriental arborvitae. *Thuja orientalis,* family Pinaceae, an evergreen, low, bushy tree of several varieties, grown as an ornamental. Native to China and Korea.

Oriental beetle. *Anomala orientalis* Waterh., family Scarabaeidae, a beetle whose larvae infest the roots of plants, mostly grasses, and seriously injure turf. Also called *Asiatic beetle.*

Oriental bittersweet. *Celastrus orbiculata,* family Celastraceae, a woody vine grown on a wall, trellis, arbor, etc., for its beautiful foliage and orange-yellow fruit. Native to China.

Oriental cherry. *Prunus serrulata,* family Rosaceae, a deciduous tree grown for its pink, white, or red flower clusters. Native to Japan. Also called *flowering cherry; Japanese cherry.*

Oriental clematis. *Clematis orientalis,* family Ranunculaceae, a hardy, woody vine grown on a trellis, fence, etc., for its yellow flowers. Native to Asia.

Oriental cocklebur. *Xanthium orientale,* family Compositae, a common, noxious weed, which may cause poisoning in livestock, especially swine. Native to Europe and Asia. Also called *sheep-bur; cocklebur; clotbur.*

Oriental fruit fly. *Dacus dorsalis* Hendel, family Tephritidae, an insect pest, especially serious in areas of mild winters, which infests all citrus fruits as well as the banana, avocado, pineapple, and a host of other plants. See *Mediterranean fruit fly; West Indian fruit fly.*

Oriental fruit moth. *Grapholitha molesta* (Busck), family Olethreutidae, a most destructive insect pest whose larvae first bore into tender twigs of the peach and later generations attack the fruit making it wormy.

Oriental leaf disease. A fungous disease caused by *Hemileia vastatrix,* family Pucciniaceae, which seriously affects the coffee plant in Asia and Africa.

Oriental moth. *Cnidocampa flavescens* (Wlk.), family Limacodidae, a Japanese insect largely confined to Massachusetts in the U.S., whose larva feeds preferably on the Norway maple, sycamore maple, buckthorn, black birch, cherry, apple, pear, and plum, but may also infest other trees.

Oriental planetree. *Platanus orientalis,* family Platanaceae, a tree with a broad round head and a short trunk often divided into several stems near the base, much used in Europe as a shade tree. Native to southeastern Europe and western Asia. Also called *Chenar.*

Oriental poppy. *Papaver orientale,* family Papaveraceae, a perennial herb grown in the flower garden for its large, showy, scarlet flowers. Native to western Asia.

Oriental rat flea. *Xenopsylla cheopis* (Rothschild), family Pulicidae, a vector in the spread of plague from rats to man. It may also affect both domestic and wild animals.

Orientals. A standard class of certain breeds of chickens native to the Orient or largely of oriental type and breeding, as the Sumatras, Malays, and the Cubalayas.

Oriental sesame. *Sesamum indicum,* family Pedaliaceae, a tropical, perennial herb grown for its seeds, the benne of commerce, used as food and for oil, extracted from the seed, used in soaps. Native to the Old World.

Oriental yellow scale. *Aonidiella citrina* (Coq.), family Diaspididae, a scale insect which infests citrus.

orientation flight. The flight of a young bee. It often begins during the latter part of the nursing period, which may extend slightly beyond thirteen days in case of lack of nurse bees.

origanum. *Origanum vulgare,* family Labiatae, a hardy, perennial herb grown for its aromatic foliage used as a flavoring, in sachets, and with tobacco. Native to Europe. Also called *marjoram; pot marjoram; wild marjoram; winter sweet.*

original. A ridgeling *q.v.*

original bag wool. Wool that can be sold to the mills without grading. Also called *original.*

original forage appraisal method. A method for determining the grazing capacity of a new range area by comparing it with a similar range of known capacity.

originality. In horticultural exhibits in shows, unusual plant material or common plants displayed in an unusual manner.

orinoco. A tobacco variety of many strains which is prepared by flue-curing.

Orleans process. In the manufacture of vinegar, a process which results in aging and produces a vinegar superior in flavor. It employs *Acetobacter orleanenae,* and is used in Europe in the preparation of vinegar from wine.

ornamental bush redpepper. *Capsicum abbreviatum,* family Solanaceae, an annual or biennial herb grown as an ornamental and to some extent for pickling. Found in the tropical regions of America; probably originating from the Old World tropics.

ornamental orange. *Citrus taitensis,* family Rutaceae, one of the few species of orange, of unknown origin, bearing fruit when the plant is very small; used as a potted plant. Also called *Otaheite orange; Tahiti orange.*

ornamental pepper. Cherry redpepper.

ornamental tree. Any tree prized mainly for its beauty of form, foliage, flowers and decorative fruit rather than for food, flavoring, medicines, etc.

oromorphic soil. A soil which retains dominantly the characteristics of the parent material.

orpine. Liveforever stonecrop.

Orpington. A breed of general-purpose chickens of the English class. They are large, loose-feathered birds having single combs, white skin, and non-feathered shanks. Mature cocks weigh about 10 lbs., hens about 8 lbs. The egg-shell is brown. Varieties are the Buff, the Black, the White, and the Blue.

orrisroot iris. *Iris germanica* var. *florentina,* family Iridaceae, a perennial herb grown for its root stock, a source of orris root used in perfumery and medicine. Native to Europe. Also called *white flag.*

orterde. The aggregated, friable, non-cemented B, or subsurface, horizon of the profiles of Podzol soils. Organic matter or both organic matter and iron sesquioxide, accumulated here, impart the usual dark brown or yellowish coloring.

ortet. The original single plant from which a clon is derived.

orthodichlorobenzene. $C_6H_4Cl_2$; a liquid used in the control of termites.

orthoeluvium. The residual product from eluviation of the mass of weathering of primary or igneous rocks. It is the primary parent material of soil.

ortstein. An irregularly-cemented, usually sandy dark yellow to nearly black horizon; a characteristic of some Podzol soils.

Osage orange. *Maclura pomifera,* family Moraceae, a thorny tree grown as a hedge plant, for fence posts, and for its roots, a source of dye. Native to North America. Also called *Bois d'arc; bodark; brasswood; mockorange.*

Osage plum. A plant mentioned by Lewis in his account of the Lewis and Clark Expedition of 1804. The species identification of this wild fruit has not been determined. The name *Osage* is also sometimes given to the wildgoose plum of southwestern U.S.

Oschtjepek. A plastic, Cacciocavallo-like, curd cheese made from sheeps' milk in Slovakia.

oscillating pipe line. A water applicator used in overhead irrigation. A several hundred feet long line of pipe, with nozzles inserted in a straight line at approximate intervals of 3 ft., is supported on posts about 7 ft. above the ground. When in operation, a water driven oscillator causes the line to rotate back and forth to apply water to the soil first on one side of the pipe and then on the other. It is particularly well suited to fixed beds such as a nursery. Also called *nozzle line.*

oscillator. In nursery practice, a mechanical device, worked by hydraulic pressure, for automatically turning overhead sprinkler lines through an arc of varying degrees up to 180°. It produces an even spread of water.

osier willow. Basket willow.

osmanthus. Any evergreen shrub or small tree of the genus Osmanthus, family Oleaceae. Species are grown for their attractive foliage and fragrant blossoms. Some are also called *tea olive.* See *devilwood osmanthus; holly osmanthus.*

osmosis. The flow of a fluid through a semipermeable membrane separating two solutions which permits the passage of the solvent but not the dissolved substance. The liquid will flow from a weaker to a stronger solution, thus tending to equalize concentrations.

osmotic pressure. The hydrostatic pressure required to stop osmosis or to prevent diffusion of molecules of solvents of a dilute solution from passing through the walls of a semipermeable membrane into a more concentrated solution.

osmunda peat. A potting material made from the roots of ferns of the genus Osmunda. See *orchid peat.*

Ossetin. A cheese made in the Caucasus from

sheeps' or cows' milk. Also called *Tuschinsk; Kasach.*

ossolets. A soft, warm, sensitive swelling over the front and sometimes sides of the fetlock joint in the horse. It is an inflammation of the bone covering, periosteum, occurring usually only in the front legs. Also called *osslets.*

osteofibrosis. A loss of calcium salts from the bones which causes them to become fragile. A condition chiefly affecting the horse, it may also appear in pigs, goats, and dogs.

osteomalacia. A disease in which the bones of animals become softer, supposedly due to a deficiency of phosphorus, calcium, or both, in the diet. Also known as *creeping sickness; loin disease of cattle; down in the back; adult rickets.*

osteopetrotic lymphomatosis. A rare form of the avian leukosis complex (*q.v.*), in which the bones of the shanks become thickened, hot, hard, and insensitive.

osteoporosis. A bone disease caused by calcium deficiency which results in increased porosity and softness of the bone. In the equidae (horse, etc.), it is a specific disease marked by enlargement, softening, and increased porosity of the bones of the face. In poultry, the condition is commonly referred to as rickets, however, some specialists distinguish between *osteoporosis* and rickets or adult rickets.

ostiole. A porelike opening of fruiting bodies of algae and fungi, such as pycnidia and perithecia, etc., through which the spores are dispersed.

O-strain. A strain of the foot-and-mouth disease (*q.v.*) virus.

Oswego beebalm. *Monarda didyma,* family Labiatae, a perennial, aromatic herb grown for its dried leaves used in tea. Native to North America. Also called *bee balm; fragrant balm; red balm; Oswego tea.*

otaheitapple. Mombin.

otaheite arrowroot. Fiji-arrowroot tacca.

otaheite gooseberry. Otaheite gooseberry leafflower.

otaheite gooseberry leafflower. *Phyllanthus acidus,* family Euphorbiaceae, a shrub or small tree sometimes grown as an ornamental in the greenhouse and in gardens in Jamaica and the West Indies for its fruit used for pickles and preserves. Native to South Asia. Also called *jimbling; gooseberry tree; otaheite gooseberry; star gooseberry; Tahiti gooseberry; West India gooseberry.*

otaheite orange. Ornamental orange.

outage. A tax formerly imposed by the state of Maryland in the U.S. for inspecting and marking the hogsheads of tobacco intended for export.

out apiary. A subsidiary apiary maintained at a distance from the main apiary.

outback. A casual term referring to the farming and inland, arid areas of Australia. See *Never-Never.*

out-boundaries. In early Mexican land laws, certain boundaries within which grants of a smaller tract might be located by the grantee.

outbow-footed. Referring to a horse which stands with its toes turned outward; splayfooted.

out-breeding. The mating of unrelated plants or animals, selected for their individual qualities, to improve or restrain certain characteristics. *Cf. in-and-in breeding.*

outbuilding. See *outhouse: 1.*

outcrop. Natural exposure of bed rock at the surface.

outcross. A mating between individuals of the same breed or variety but belonging to different strains or families.

outdoor brooder. A portable brooder that may be used outdoors. See *brooder house.*

outdoor cellar. A cellar, especially one for winter storage of root crops, constructed outdoors. See *cold cellar.*

outer bark. The layer of dead tissue, of a dry corky nature, outside the last-formed periderm of a tree or woody plant.

outer cover. In beekeeping, usually a cover that telescopes over the top of the hive to a depth of an inch or more and is covered with galvanized iron or aluminum sheeting, to protect the hive from the weather.

outer shell. The exterior or outside portion of the bulk milk cooling tank. The outer tank is separated from the inside lining by insulation and/or integral cooling surfaces.

outer tank. See *outer shell.*

outfall. The point where water is discharged from a conduit; the mouth or outlet of a drain or sewer.

outfield. Land or a field distant from the farmstead (obsolete).

outfit. 1. A group of cowhands and their equipment (western U.S.). 2. To supply the necessary equipment for an expedition, a practice, a business, etc.

outgang. A lane used by cattle going to and from pasture.

outhouse. 1. Any small building on a farm, usually other than the barn, in which tools or small animals, etc., are kept. Also called *outbuilding.* 2. See *privy.*

outlaw. 1. A horse that refuses or fails to become tractable during the breaking process. 2. A wild, unbroken horse.

outlet channel. In soil conservation engineering, a waterway provided to receive and carry away the runoff discharge from terrace channels and diversions. Also called *terrace outlet channel.*

outlot. In early American land law, a lot or parcel of land which lay outside the corporate limits of a town but which was subject to its municipal jurisdiction or control.

out of. In breeding, mothered by. *Cf. get; by.*

out-of-air cleaner. Cleaners that act on cotton after it has been elevated from the wagon but before it reaches the gin stand.

out of condition. In stock judging, designating an animal which is not well-fed in appearance nor in normal good health.

out of heat. Designating a female animal not in estrus *q.v.*

out of season. Designating products, as fresh fruits and vegetables, available at a time at which they are not normally produced and marketed.

out-planting. The setting of trees in a plantation as contrasted with transplant(ing) *q.v.*

outrider. A cowboy whose duty it is to ride the range and protect his employer's interests (western U.S.).

outriding. Inspection trips over the grazing range of livestock (western U.S.).

outrigger. An extra horse hitched outside the shafts to a one-horse cart.

outrun. A pasture for cattle, etc., at some distance from the main farm buildings; an outlying pasture.

outside. See *cushion.*

outside branding. Branding away from the home range.

outside fruit. In an orange grove, that fruit which is on the outside of the tree canopy and thus more exposed to the weather.

outside man. 1. A man who attends round-ups on ranges away from those of the home ranch (western U.S.). 2. A farm laborer who works in the fields as contrasted to one who works in the barn, etc.

outside quotation. The highest price for a commodity on a particular market for a particular day.

outspan(n). To unyoke, as a pair of oxen, or a place where unyoking is carried out (South Africa).

outstanding crop. Legally, a crop which is not harvested. It is outstanding from the day it commences to grow until it is harvested and taken away.

outstanding in the class. In stock judging, designating a superior or outstanding individual in type and conformation.

out station. Any hut, including yards, on a sheep station (Australia).

out stock. Domestic livestock allowed to run at large on a range.

outwash plain. A physical land feature of glaciofluvial origin. The plains are smooth, or where pitted are usually low in relief. The deposits are mainly sandy or coarse textured. The plains constitute a large total area of land and a large number of separate soil and land types in the glaciated region of the U.S.

outyard. In beekeeping, an establishment of colonies away from or in another locality than the home yard.

ova. Plural of ovum *q.v.*

oval guinea pig louse. *Gyropus ovalis* Burm., family Gyropidae, a louse which infests the guinea pig in the U.S.

ovalhead sedge. *Carex festivella,* family Cyperaceae, a grasslike, herbaceous plant, found at elevations of 6,500 to 12,000 ft. from New Mexico to California (U.S.) and north to Alberta (Canada). Its forage value is good and its fibrous roots afford protection against soil erosion.

oval kumquat. Nagami kumquat.

oval pignut hickory. Red hickory.

ovarian extract. A by-product of packing houses, made from the ovaries of sheep, hogs, and cattle containing female sex hormones, useful in treating ovary disorders or hormone imbalances.

ovarian cyst. A benign cyst, often congenital.

ovarian follicle. The small cyst-like structures in ovaries which when fully developed contain a mature ovum. Also called *graafian follicle.*

ovary. 1. In animals, the female organs which give rise to the eggs, or ova. 2. In plants, the swollen part of the pistil, containing the ovules.

Ovcji Sir. A cheese made in the Slovenian Alps from ewes' milk.

oven canning. The processing of foods for preservation in glass jars, in an oven at temperatures from 250° to 275° F.

oven-dried. 1. Designating wood or other material dried by exposure to 100°C. or 212°F., or slightly higher temperatures, until ceasing to lose weight, usually 24 to 48 hrs. 2. Designating plant tissues usually *oven-dried* at about 150° F.

oven-dry soil. A soil dried at 105° to 110° C. until it is in moisture equilibrium at that temperature.

over. Designating one complete coverage of an entire field with a seed-bed fitting implement.

overall. A suit of clothes, usually one piece, made of heavy cotton; often worn by farmers over their regular clothes for added warmth or to protect their clothes when engaging in particularly messy work.

overbrowsing. Excessive cropping of shrubs or tree growth, usually by goats, sheep, or game animals.

overcheck bridle. A bridle used with an overcheck rein *q.v.*

overcheck rein. A checkrein (*q.v.*) which passes over a horse's head between the ears.

overcrop. To exhaust land by continuous cultivation and production of crops. See *overplow.*

overcutting. The cutting of a quantity of timber in excess of the annual or periodic cutting budget.

over-developed eye. A body defect of cheese which consists of large holes, usually larger than 2 ins. in diameter. See *blow holes.*

over-developed in the forequarters. In judging animals, designating a lack of proper body conformation.

overdone. Designating an animal that has been fed too long or too heavily, producing an overly fat or poorly finished condition.

overdone and soft. In stock judging, designating a lacking in firmness in the flanks and back of the shoulders.

overeating disease. Enterotoxemia *q.v.*

over-fall. 1. The part of a dam or weir over which the water pours. 2. The over-pouring water.

overfall dam. A dam constructed to allow water to flow over its crest.

overfall stabilization. The process of hardening the sides of a gully, where slight overfalls occur and where engineering structures are not deemed necessary, by group plantings of trees and shrubs,

overfat. In the wholesale meat trade, designating carcasses that have a very large amount of fat in proportion to lean. It may apply to either the live animal or the carcass. Also called *too fat; overfinished.*

overfeeding. Consuming excessive amounts of feedstuffs. It can cause various digestive disturbances, such as diarrhea, colic, bloat, and founder.

overflow. 1. In cotton gins with a belt distributor system, the surplus cotton carried on belts to the feeders, which drops to the floor when the feeders are full. 2. The flood water of swollen streams which spreads over bottom land.

overflow(ed) land. a. Land which is subject to overflow or flooding; generally flood-plains of rivers. b. Loosely, swampland. c. Legally, land that is covered by non-navigable waters, but not including land covered by the normal flow of tides.

overflowworm. See *fall armyworm.*

overgraze. To graze land in excess of or beyond its carrying capacity.

overgrazing. To graze land so heavily as to impair future forage production and to cause range deterioration through consequential damage to plants, soil, or both. Also called *overstocking.*

overgrown. Designating abundant growth on a field, usually of some undesirable vegetation as weeds or brush.

overgrowth. See *hyperplasia.*

overhaul. 1. The rehandling or repacking of ham during the pickling period to permit a more uniform distribution of pickle. 2. To repair and recondition tools, implements, machines, etc.

overhead. The general expenses of operating a farm, such as taxes, insurances, etc., which cannot be charged directly to any particular crop or livestock. In cost accounts, *overhead* expenses are prorated to the productive enterprises, such as crops and livestock, etc., in proportion to one of the major inputs or to the total debits of each enterprise.

overhead carrier. A container suspended from an overhead track for the more efficient carriage of feed to stalls and for removal of manure to storage pits; used especially in dairy barns.

overhead check. In horses, a rein which extends from the bit over the head of a horse to a hook on the harness back pad to prevent the animal from tossing or lowering its head.

overhead-frame type plow. A disk plow with an elevated overhead frame from which the disk bottoms can be suspended at varying heights to give more clearance.

overhead mow. A barn space for the storage of bedding, hay, and other feeds that is directly above the floor on which animals are kept.

overhead sprinkling system. A system of irrigating a small acreage by sprinkling from permanently installed overhead pipes.

overhead-training. Training canes, especially grapes, to spread on a horizontal trellis so as to form an overhead canopy.

overirrigation. Application of water in excess of the need of a crop and of the water-holding capacity of the soil: It damages the plants in addition to wasting the water.

overlander. A man who drives sheep or cattle over long distances (Australia); a drover.

overlapping furrow. A furrow whose slice is nearly on edge without much crumbling.

overlay. Any transparency, containing supplemental information, superimposed over a map or a data sheet to show this information more clearly.

overlength. In milling logs, an allowance for trim; an extra length left on a log so that it may be trimmed to an exact length.

overlying color. See *surface color: 1.*

overmature forest. An old forest in which growth has almost entirely ceased and decay and deterioration have begun.

overmaturity. The period in the life cycle of trees, fruits, and vegetables when growth or value is declining. See *maturity: 1; overripe.*

overplaced soil. Colluvial soil.

overplow. To till land excessively or so continuously as to cause deterioration of productivity.

overpotting. Growing a plant in an unnecessarily large pot.

overproduction. Crop or livestock production in excess of the market demand.

overreach. In a horse, to strike the heel of the forefoot with the front of the hind foot.

overripe. 1. Designating fruit which is mature, soft, or decayed to the degree that it is inedible or unmarketable. 2. Designating a degree of maturity, such as fruit, that is too ripe for canning or shipping but is not too ripe for table use.

overrun. 1. The pounds of butter churned over and above the pounds of fat which were in the cream put into the churn. 2. The increase in volume of ice cream caused by the incorporation of air during the freezing process. 3. The excess amount of lumber actually sawed from logs over the estimated volume or log scale, usually expressed as a percentage of log scale.

overrun tester. A weighing apparatus used in the dairy industry to test for the percentage overrun in an ice cream mix.

overseeding. 1. Seeding in a crop already established, such as seeding a forage crop in the spring on a field of winter wheat established the previous fall. 2. Using an amount of seed in excess of that necessary to assure a good stand.

overseer. 1. A manager or supervisor of a ranch, farm, etc., who has supervisory duties over the laborers. 2. In New Zealand, an undermanager.

overset. Designating a body defect of cheese in which the eyes are so numerous within the body of the cheese that they crowd each other leaving only a small portion of solid cheese.

overshot jaw. Malformed mouth of an animal, most commonly the horse, in which the upper jaw abnormally protrudes beyond the lower jaw. The incisor teeth of the upper and lower jaws do not come together. Also called *parrotmouth; overshot mouth.*

overshot stacker. A device used for lifting hay from the ground to the top of a haystack. The hay brought to the stacker teeth by the sweep rake is carried up over the stacker frame and delivered to the stack in much the same manner that a man would stack hay with a hay fork. The stacker usually has a rigid frame extending from the stack out to the stacker teeth. The whole is raised by means of ropes and pulleys.

oversow. See *overseeding: 1.*

overstocked. 1. Designating a condition of a stand of trees or of a forest, in which there are more trees than normal or full stocking would require. The overstocking may be to such a degree that growth is slowed down and many trees, including dominants, suppressed. 2. Designating a pasture range, or grazing game area which has more animals on it than the vegetation of the area will support. See *overgrazing.* 3. Designating a locality in which there are too many bees.

overstocked stand. See *overstocked: 1.*

overstory. That portion of trees in a forest stand which forms the upper crown cover. Also called *overwood.*

oversummer. To live through the summer.

over the block. 1. Beef suitable for block beef or to be sold by the retailer *over the block.* 2. Designating wholesale cuts of meat; *i.e.,* sides and quarters which can be cut by the butcher for retail sale.

overtopped crown class. Trees with the crowns entirely below the general forest canopy, which receive no direct light either from above or from the sides.

overtopping. In river flood control, the passage of water over the top of a levee during a flood.

overuse. The utilization of key forage species on a range to such an extent that they fail to maintain themselves over a period of time.

overwash plain. Outwash plain.

overwinter. To hibernate; to live in a dormant condition through the winter months.

overwood. Overstory.

ovicide. Any substance which kills parasites in the egg stage.

oviduct. The tube that leads from the ovary to the uterus or other organs where fertilization or further development of the ovum or egg cell occurs. Also called *fallopian tube; tubes.*

ovine. 1. An animal of the sub-family Ovidae; sheep, goats, etc. 2. Pertaining to such an animal, commonly to sheep.

ovine babesiasis. A disease of sheep which may be caused by one of two species of blood parasites, *Babesia motasi,* and *B. ovis.* The disease causes blood cell destruction, icterus and hemoglobinuria.

ovine balano-posthitis. A venereal disease of sheep causing ulcerations of the penis sheath and vulva, ulcerative vulvitis, thought to be caused by a filterable virus with bacteria as secondary invaders.

oviparous. Producing offspring from eggs which hatch outside of the body. *Cf. viviparous: 1.*

ovipositor. A tubular structure in female insects, used for depositing its eggs.

ovotestis. A gonad, part of which is composed of ovarian tissue and testicular tissue.

ovulate. To bear ovules; to release ovum (ova) from the ovary.

ovule. The macrosporange of flowering plants which consists of the integuments plus the nucellus, within which the macrogametophyte or embryo sac develops; the body which after fertilization becomes the seed. It is contained in the ovary.

ovum. The female germ cell.

owls-claws. Orange sneezeweed.

owner-operator. A farmer who owns the farm which he operates. See *part owner-operator.*

Owner-Samples D.H.I.A. plan. A system of records for dairy cows in which the owner weighs the milk and takes the samples which are later picked up by the tester who runs the tests and compiles the records. This system gives the owner the same information, is more economical, and the tester can handle many more herds each month, but is less official than the standard D.H.I.A. plan. See *Dairy Herd Improvement Association.*

own-root plant. A plant which has roots and tops of the same horticultural variety.

ox (plural: oxen). Any species of the bovine family of ruminants. Specifically, the domesticated and castrated male bovines which are used for work purposes, as distinguished from steers used for meat, or the uncastrated bulls used for breeding.

oxalacetate. An intermediate compound formed from sugars and sugar-producing (gluconeogenic) amino acids, essential for the oxidation of carbohydrates, proteins, and fats for energy.

oxalic acid. An organic acid which occurs in many plants. It is probably the acid secreted by crustaceous lichens having the power to dissolve minerals of hard rock surfaces: *e.g.,* granites. It is produced synthetically by the fusion of sawdust with caustic soda or potash.

oxalic acid poisoning. A toxic condition of sheep and cattle resulting from the ingestion of large quantities, over a short period of time, of dried plants in which the oxalic acid content is more than 10 percent.

oxalis. Any herb of the genus Oxalis, family Oxalidaceae. See *woodsorrel oxalis; creeping oxalis; violet woodsorrel oxalis.*

oxbane. 1. Any plant that is poisonous to cattle; specifically *Boophane disticha* (*Buphane toxicaria*), native to South Africa. Also called *cowbane lilly.* 2. See *spotted waterhemlock.*

ox boose. An ox stall (obsolete).

ox bot. Common cattle grub.

oxbow. 1. A U-shaped, wooden collar, part of the harness of an ox, placed on the neck of the animal with the open ends passing through the yoke. See *yoke: 1.* 2. A bend in the channel of a river resembling an *oxbow.* (A large river may cut across the narrow neck of a nearly circular meander and thus leave an abandoned channel, as a lake, swamp, or bayou.) Also called *oxbow meander*.

oxen. See *ox*.

oxeye camomile. Golden camomile.

oxeye daisy. *Chrysanthemum leucanthemum,* family Compositae, a perennial herb, a common troublesome farm weed in hayfields. It is grown in the flower garden for its flowers which have white petals and yellow disks. Native to Europe and Asia. Also called *white weed; white daisy; field daisy; devil's daisy; false sunflower.* See *chrysanthemum.*

Oxford sheep. A mutton and medium-wool breed of sheep which is a cross of several foundation breeds. Mature ewes weigh 175 to 250 lbs. First produced in Oxfordshire, England.

oxgang. In old English law, as much land as an ox could till. A measure of land of uncertain quantity: according to some, 15 acres; the Scottish *oxgang* contained 12 acres.

oxherd. See *cowherd(er).*

oxidase. Oxidising enzymes. See *enzyme.*

oxidasic casse. A defect of wine caused by an enzyme; characterized by clouding and changes of color of the wine on exposure to air: *e.g.*, a red wine may turn brown. See *Tourne bacterium.*

oxidation. Any chemical change which involves the addition of oxygen or its chemical equivalent. *Oxidation* may affect agricultural products adversely or even cause destruction: *e.g.*, it can be the cause of taints in milk or butter and it can cause spontaneous combustion of hay stored in barns.

oxidation-reduction potential. The balance between oxidized and reduced forms of chemical substances. The electrical determination (Eh) of the ratio of oxidized to reduced forms is useful in relation to the dairy industry, and is a laboratory procedure sometimes employed to determine the state of oxidation in soils which in turn bears a relation to plant growth.

oxidized flavor. A flavor defect of milk products which is suggestive of tallow, oil, paper, etc.

oxidized milk. An objectionable flavor (like wet cardboard) in milk in which the fatty constituents (and possibly some milk proteins) are adversely affected by (a) small amounts of copper, or iron absorbed from equipment and/or (b) to exposure of milk to direct or indirect sunlight.

oxlip primrose. *Primula elatior,* family Primulaceae, a hardy, perennial herb grown in the rock garden or flower border for its showy, yellow flowers. Native to Europe and Asia.

ox pull. A pulling contest often staged by early settlers (U.S.), and currently at some country fairs, to find the ox team in the community capable of pulling the greatest load.

oxtail. The tail of a beef carcass, one of the edible by-products of packing houses.

ox tongue. A food product which is prepared by pickling beef tongues.

ox warble fly. Common cattle grub.

oxygen. The chemical element, O, a colorless, odorless gas. The most abundant element, it is essential in the growth of all crops and for the respiration of most forms of life.

oxyhemoglobin. A substance formed by the union of oxygen with the hemoglobin of the blood, the union being of such a nature that the oxygen is readily given up at times and in places where it is needed.

oxyquinoline benzoate. A fungicide which is used as a control for Dutch elm disease and bleeding canker of maples and other hardwood trees. It is not used as a foliage spray but is either injected directly into the tree trunk or inserted into the soil over the root area at a depth of about 18 ins.

oyster. The tenderloin muscle of the poultry carcass. There are two, one on each side, located just in front of the hip bones.

oyster plant. See *vegetable oyster.*

oyster shells. Shells of the marine bivalve, genus Ostrea. They are nearly pure calcium carbonate in composition; when finely ground, they make good liming material for soils and a mineral feed for livestock and poultry as a source of calcium.

oystershell scale. *Lepidosaphes ulmi* (L.), family Diaspididae, a scale insect, about 1/10 in. long, which covers twigs and branches. The pests overwinter in the egg stage under the scales. The young crawling stage appears in late May. This scale is a pest of apples and a number of shade and ornamental plants.

ozaena. See *ozena.*

Ozarks. A high plateau region which embraces a large part of southern Missouri and northern Arkansas (U.S.). The terrain is generally hilly and deeply dissected by streams. The region contains fertile bottom lands and productive soils on slopes, but the greater part is inferior agriculturally because of steep slopes, cherty soils, and gray, unproductive soils on flat ridge crests. The greater part of the land is primarily suited for forestry and grazing.

Ozark sundrops. *Oenothera missouriensis,* family Onagraceae, a perennial herb grown for its yellow flowers. Native to North America. Also called *Missouri primrose.*

ozena. A fetid nasal discharge accompanied by chronic inflammation of the mucous membranes, associated with a disease of the bones in the nose.

P

pace. 1. A measure of length: the ordinary length of a step from heel to heel is 2½ ft. The geometric *pace* is 5 ft. Distances for land measure are usually stepped according to some definite distance per step, as 3 ft. Land measurements were often paced off, from some known mark as a base, in early land surveying in the U.S. 2. Applied to horses, a rapid, two-beat gait in which the lateral fore and hind legs work in pairs. It enables the pacers to get away at speed and involves a minimum of concussion, more or less side motion, an absence of much knee fold, and the necessity for smooth, hard footing and easy draft for its execution. Also called *amble; rack; trot.*

pacer. A horse, one of whose gaits is a pace. Also called *side-wheeler.*

pachymeningitis. Inflammation of the tough membrane enveloping the brain and spinal cord.

pachysandra. Any plant of the genus Pachysandra, family Buxaceae. Species are grown for ornament and ground cover in shady places. See *Allegheny pachysandra; Japanese pachysandra.*

pachytene. In cytology, a stage at about mid-prophase of the first meiotic division preceding diplotene when the chromosomes are paired but are somewhat shorter and thicker than in earlier prophase stages.

Pacific bleedingheart. *Dicentra formosa,* family Fumariaceae, a slender herb bearing heart-shaped, rose-purple or white flowers. Native to North America. Also called *California bleedingheart.*

Pacific bunchgrass association. One of the major grassland associations of the U.S. whose grasses have a characteristic bunch-growth. Regionally, it embraces Idaho, Montana, Nevada, Washington, Oregon, and California.

Pacific Coast canker. Apple anthracnose.

Pacific Coast tick. *Dermacentor occidentalis* Marx, family Ixodidae, a tick which is one of the vectors of Rocky Mountain spotted fever.

Pacific Coast wireworm. *Limonius canus* Lec., family Elateridae, a wireworm which infests bulbs and various root crops.

Pacific coreopsis. *Coreopsis maritima,* family Compositae, a perennial, tender herb grown for its large, yellow flowers. Native to North America; it is common to coastal areas of southern California and Mexico.

Pacific dogwood. *Cornus nutalli,* family Cornaceae, a deciduous, tender tree grown for its white or pinkish flowers. Native to the Pacific coast of Canada and U.S.

Pacific flatheaded borer. *Chrysobothris mali* Horn, family Buprestidae, a beetle which bores into the bark and wood of nearly all forest, shade, and fruit trees in western U.S.

Pacific hemlock. *Tsuga heterophylla,* family Pinaceae, a very tall, evergreen tree grown as an ornamental and for its lumber used as pulp, for general construction and planing-mill products. Native to the Pacific coast from Alaska to California. Also called *western hemlock* (forestry name); *West Coast hemlock; hemlock spruce.*

Pacific madrone. *Arbutus menziesi,* family Ericaceae, a tall, tender, evergreen tree grown in warm areas for its white flowers and orange-red fruit. Native to British Columbia, Canada, and California. Also called *apple of Cain; laurelwood; madrona; Oregon laurel.* See *strawberry madrone.*

Pacific mite. Orchard mite.

Pacific Northwest. A geographic subdivision of the U.S. Without any precisely fixed boundaries, it definitely includes the states of Washington and Oregon and sometimes Idaho and Montana.

Pacific poisonoak. *Toxicodendron diversilobum,* family Anacardiaceae, a woody shrub or vine which on contact with susceptible humans causes a severe dermatitis. Native to western U.S. and Canada. See *poison ivy.*

Pacific trillium. *Trillium ovatum,* family Liliaceae, a hardy, perennial herb grown in the wild garden for its early-spring, white flowers fading to rose. Native to British Columbia, Canada, and California. Also called *coast trillium.*

pacing plate. Running plate.

pack. 1. Fruit, vegetables, meat, etc., which is to be or is packaged, canned, frozen, etc., for the market. 2. The total amount of products which are processed in a season. 3. The manner in which produce is packaged. 4. The load which is carried by a *pack* animal. 5. To compress or firm soil with a special implement. 6. To damage soil structure by compacting or puddling heavy soils, as from the pressure exerted by the wheels of a tractor, by injudicious irrigation, or by excessive rainfall. 7. The sheep given to the shepherd as his share for tending the flock.

package bees. A colony of bees, weighing 2 lbs. or more, packaged in wood and wire cages for shipment for stocking or restocking hives.

package refrigeration unit. A mechanical cooling unit, an integral part of, and normally attached to, the frame of the bulk milk tank.

pack animal. A burro, mule, or horse used for carrying packs and equipment over rough areas and to places inaccessible by other means.

pack burro. A Spanish or Mexican burro, smaller than a mule or pack horse, used for carrying camping and mining equipment into rough or nearly inaccessible country. See *burro.*

packer. 1. One who operates a slaughter and meat processing business. 2. Pertaining to the business of packing fresh or processed fruits and vegetables or meats. 3. A field tool of the roller type, horse or tractor drawn, consisting of a set or series of rollers which pack the loose soil after plowing. See *cultipacker; land rollers.* 4. One who loads pack horses and leads or drives them from place to place (New Zealand). 5. One who makes a pack, as of vegetables, meat, etc. 6. Fingers or arms used on harvesting machinery, as grain binders and corn harvesters, to pack the cut material into compact bundles for tying or ease of handling.

packer buyer. A representative of a meat packing plant appointed to buy livestock for slaughter.

packer cap. A cork-lined, tin cap held in place by a turn-down wire bail used for sealing glass jugs, bottles of cider, vinegar, and other products.

packer hide. An animal hide prepared for the market as a by-product by meat packing companies as distinguished from country hides which are not uniformly processed.

packer's style. A method of dressing pork in the U.S. in which the carcass is split into two sides and the face removed but the jowls left attached. See *shipper-style dressing; farmer's style.*

pack horse. See *pack animal.*

packing. Arrangement and manner of packaging and preparing fruits, vegetables, meats, etc., for market and preserving.

packing house. 1. A slaughtering and meat distributing organization that buys, slaughters, processes, and distributes livestock products. 2. A separate building or shed adjacent to a range of greenhouses where flowers and greenhouse plants are packed for market. 3. A building which is equipped and arranged for grading and packing fresh fruit or vegetables.

packing-house by-product. Anything of value, other than dressed meat, which is produced in the slaughter of animals.

packing industry. The organization and development of the facilities and resources of the meat industry.

packing plant. See *packing house.*

packing shed. A roofed but otherwise open structure in which fresh fruits, vegetables, etc., are sorted, packaged, and graded for the market.

packing sow. A market classification for hogs, the grades based on age, condition, sex, and ratio of lean to fat quality of the animal.

pack mule. A small mule used by prospectors and miners for carrying camping and mining equipment and for hauling ore from mines in the isolated mining sections of western U.S.

pack rat. A bushy-tailed rat, *Neotoma cinerea,* which collects odds and ends in its burrow. Found in southwestern U.S., it sometimes damages citrus orchards adjacent to canyons with brushy cover.

packsaddle. A carrying unit for camping equipment or other material which is placed on the back of a horse or pack mule, either as separate units or attached to the rider's saddle. *Packsaddles* are of many types, much used in ranches, ranges and isolated areas.

pack stock. Horses or mules used to transport supplies and equipment, usually in mountainous or desert terrain.

pad. 1. Any soft material used under harness parts or saddle to prevent injury by direct chafing of the leather on the animal's skin. 2. In a double harness, that part to which the girths are attached. 3. A very gentle horse with an easy gait. 4. A hard, cushionlike surface on the sole of some animals.

paddle staff. A spud inserted into a long handle which is used to clean a ploughshare and to destroy weeds (England).

paddling. In the leg action of horses, an outward deviation in the direction of the stride of the foreleg, the result of a too narrow or pigeon-toed standing position; a defect.

paddlings and sishing. Horses that are cowhocked, splay-footed, or with tied-in elbows.

paddock. 1. A small inclosure or pasture used for grazing and exercising of a single large animal, such as a horse, herd bull, or for a few head of smaller animals, as sheep. More common in England, Australia and New Zealand than in the U.S. 2. Any fenced-in field or piece of land regardless of size or use (Australia).

paddy rice. Rice as harvested from the fields before it is milled.

paddy soil. The soil of a rice paddy: it is wet and subject partially to anaerobic conditions.

paddy whack. The wide yellow band of fat along the top of the shoulder of a beef carcass (Australia).

pad filter. A form of filter press used in the wine and brewing industries for the final or polishing filtration of wine and beer.

pad hook. A hook in a harness pad which supports the bearing rein.

paeony. See *peony: 1.*

paglia. A Gorgonzola-type of Swiss cheese.

Pago. A cheese made from ewes' milk on the island of Pag in Yugoslavia.

pagoda collinsia. *Collinsia bicolor,* family Scrophulariaceae, a hardy, annual herb grown for its flowers with white upper lips and rose-purple or violet lower lips. Native to North America. Also called *Chinese houses.*

pagoda dogwood. *Cornus alternifolia,* family Cornaceae, a deciduous shrub or small tree bearing white flowers and bluish fruit, grown as an ornamental in the U.S. Native to North America. Also called *blue dogwood; pigeonberry.*

pagoda tree. Japanese pagodatree.

Pahala blight. A manganese deficiency disease of sugar cane.

pail. See *milk pail.*

paille firme grass. Maidencane.

paint. See *paint horse.*

paintbrush. See *orange hawkweed.*

painted beauty. *Vanessa virginiensis* (Drury), family Oecophoridae, a butterfly whose caterpillar infests a number of garden plants. Also called *American painted lady; hunter's butterfly.*

painted daisy. Florist's pyrethrum.

painted euphorbia. *Euphorbia heterophylla,* family Euphorbiaceae, a showy, annual herb grown for its red or mottled-red and white floral bracts. Several horticultural varieties have been developed from it. Native to the U.S. and tropical

America. Also called *Mexican fire-plant; fire-on-the-mountain; annual poinsettia.*

painted hickory borer. *Megacyllene caryae* (Gahan), family Cerambycidae, an insect pest which infests hickory, walnut, butternut, Osage orange, and hackberry.

painted lady. 1. *Vanessa cardui* (L.), family Oecophoridae, probably the most widely distributed butterfly in the world whose caterpillar feeds on a large number of crop plants and garden flowers such as calendulas, hollyhocks, and lupines. 2. See *florist's pyrethrum.*

painted maple aphid. *Drepanaphis acerifoliae* (Thos.), family Aphidae, an aphid which infests the leaves of the sugar maple and silver maple.

painted tongue. See *scalloped salpiglossis.*

painted trillium. *Trillium undulatum,* family Liliaceae, a hardy, perennial herb grown for its white and crimson-marked flowers. Native to North America. Also called *wakerobin.*

paint horse. A horse, much favored by the American Indians in western U.S., which usually carries odd markings and strong color contrasts. The two common color patterns in *paint horses* are: piebald, or black and white, and skewbald, which is white and any other color than black. Also called *calico.*

paint pony. Paint horse.

paint roan. A coat-color of horses in which the roan color seems to be imposed on other colored areas.

pair. 1. A male and female of a species. 2. A team of horses frequently spoken of as a well-matched *pair* regardless of sex.

pair of calves. In the show ring, a bull and heifer from the senior or junior calf classes.

pair of yearlings. In the show ring, a bull and heifer from the junior yearling or summer yearling classes.

pakchoi. *Brassica chinensis,* family Cruciferae, an annual or biennial herb grown as a potherb and as a salad plant. Native to Asia. Also called *Chinese cabbage.*

palace stock car. In the railway transportation of livestock, the first railway car built to provide separate stalls for the animals, ventilation, and water troughs.

palaeosol. See *paleosol.*

palatability. The relative relish with which feeds are consumed by domestic livestock or other animals. On the range this may vary with composition of the plant cover or the season of grazing.

palatability-rating. A method of rating the condition of a range, from high to low, depending on the proportion of palatable plants carried by the range. The more such plants the higher the rating.

palatitis. A painfully swollen condition of the palate of a horse's mouth which may be caused by eating hard food. Also called *lampas; lampers.*

palay. Unhulled rice (Philippine Islands).

palay rubbervine. *Cryptostegia grandiflora,* family Asclepiadaceae, a tropical, woody vine grown in warm areas or in the greenhouse for its habit and its purplish flowers. It is also a source of small quantities of rubber. Native to Africa. Also called *India-rubber vine.*

pale agoseris. *Agoseris glauca,* family Compositae, a perennial, leafy herb widespread throughout northern U.S. Species of this genus contain poisonous alkaloids which may cause death to livestock, especially sheep. Native to North America.

pale butterflybush. *Buddleia (Buddleja) pallida,* family Loganiaceae, a tropical shrub grown in warm areas, or in the greenhouse, for its attractive flowers. Native to China.

pale face. A white egg.

pale juniper webworm. *Phalonia rutilana* (Hbn.), family Phaloniidae, a winged insect which infests juniper, tying the leaves together to form a tube in which the larva lives.

pale laurel. Bog kalmia.

Paleopedology. That branch of pedology which deals with buried soils, fossil soils, or those formed in past eras.

paleosol. A soil which was formed in a past climatic or geologic period. Such soils, where subsequently buried by sediments, may be exposed by natural erosion, or encountered in excavation or in well drilling.

pale-striped flea beetle. *Systena blanda* Melsh., family Chrysomelidae, the larva of a flea beetle which infests corn seed so that the seed often fails to sprout, or produces a pale, weak plant. The adult infests watermelon, pumpkin, pea, bean, and eggplant.

pale sweetshrub. *Calycanthus fertilis,* family Calycanthaceae, an aromatic shrub grown for its fragrant, brownish-purple flowers. Native to North America.

pale western cutworm. *Agrotis orthogonia* Morr., family Noctuidae, the larva of a moth which is very destructive to small grain, sugar beets, and alfalfa in western U.S.

paling. In strawberry culture, the development of leaves under mulch, resulting in low chlorophyll content and pale color.

paling fence. A light, modern fence made of narrow slats or poles. The palings may be sharpened and driven into the ground a few inches apart or supported at top and bottom by horizontal strips nailed to each pale. It is used for gardens, backyards, and small inclosures.

palisade layer. The cells of the mesophyll lying next to the upper epidermis of leaves. The process of food manufacture, or photosynthesis, is most active in these cells.

pallet. A platform of any size supported by two or three runners or stringers used as a base for stacking and transporting a *pallet* box (*q.v.*) or several packages as a unit.

pallet box. A box of from 20 to 40 bushel capacity with a pallet (*q.v.*) as its bottom, used for storing and transporting crops, such as carrots, potatoes, and apples.

palm. Any tropical or subtropical tree or shrub of the family Palmaceae. Species are grown for their wood, edible fruits, resins, fiber, and as a source of drugs.

palma christi. Castor bean.

palma dulce. Sweet braheapalm.

palmate. Arranged like the fingers of the hand.

palmate violet. *Viola palmata,* family Violaceae, a perennial, flowering herb grown for its rather large, violet-purple flowers. Native to North America. Also called *common blue violet; blue violet.*

palmerworm. *Dichomeris ligulella* Hbn., family Gelechiidae, a caterpillar, widespread from Maine to Texas (U.S.) which skeletonizes the leaves of trees and sometimes eats into fruit of apple, cherry, hazel, oak, pear, and plum.

palmetto. Any palm of the genus Sabal, family Palmaceae. See *Blackburn palmetto; cabbage palmetto.*

palmetto flatwoods. A kind of land, constituting extensive areas in Florida (U.S.), in which the ground cover or vegetation is characteristically sawpalmetto (*q.v.*), although various other plants are present and the trees are mainly pines. The land is periodically very wet and very dry. The soils are leached sands commonly underlaid by organic hardpans.

palmgrass. *Setaria palmifolia,* family Gramineae, a perennial, wide-leaved, ornamental grass grown for its grass-like leaves, up to 20 ins. long. Native to Indonesia.

palmitic acid. One of the fatty acids of butterfat.

palm kernel oil meal. A livestock feedstuff which is the ground residue from the extraction of part of the oil, by pressure or solvents, from the kernel of the fruit of the African oil palm *q.v.*

palm leaf skeletonizer. *Homaledra sabalella* (Chamb.), family Cosmopterygidae, a small moth whose larva skeletonizes the leaves of the saw-palmetto *q.v.*

palm mealybug. Coconut mealybug.

palm oil. The oil extracted from the kernel of the fruit of the African oil palm (*q.v.*), used as a foodstuff in Africa, and elsewhere in the manufacture of soap, candles, etc., and as a lubricant.

palm scale. *Chrysomphalus dictyospermi* (Morg.), family Diaspididae, a common greenhouse scale which infests palms and a number of other plants. Also called *dictyospermum scale.*

palmyra. *Borassus flabellifer,* family Palmaceae, a fan palm which bears large, round, drupe fruits. Cultivated in India for its timber, its leaves used in making matting, hats, paper, etc., and for its edible fruit. A wine is made from the sap.

Palomino. 1. A western U.S. horse, doubtfully a distinct breed, characterized by a cream, yellow, or golden color and light-colored mane. Averaging 15 to 16 hands high, the *Palominos* weigh 1,000 to 1,200 lbs., and are very popular riding horses. Also called *caballo de oro; California sorrel; Golden Horse.* 2. A coat-color of certain horses which is a golden chestnut, with flax-colored mane and tail. Ideal body color is that of a newly-minted gold coin. Purity of color and mane and tail is very important in *Palomino* selection.

palomino grape. European grape.

Palouse. A region of about 8,000,000 acres in eastern Washington adjacent to part of Idaho and the north central part of Oregon. It has a semiarid and subhumid climate, an undulating to hilly and dunelike topography, productive silty soil, mostly deep, but locally severely eroded and shallow over basalt rock. It was originally grassland characterized by bunch wheatgrass but is now largely in farms with wheat as the principal crop.

paloverde. See *blue paloverde.*

paludal. Designating wet or swampy land.

palustrine. 1. Designating swampy or boggy ground. 2. Designating a plant which grows in a swamp.

pampas. Vast, rolling plains without trees, especially in Argentina.

pampas grass. See *selloa pampasgrass.*

pan. 1. A compact, or indurated, subsurface layer in a soil profile. See *hardpan.* 2. A depression in a salt-water marsh. 3. In forestry, a wide, steel sheet used to support the end of a log in skidding. 4. A flower pot about one-half the depth of the standard flower pot; used for growing shallow-rooted plants, bulbs, and for pricking out seedlings. Also called *bulb pan; seed pan.*

Panama. A recently-developed breed of sheep, very similar to the Columbia. It is a cross of Lincoln ewes and Rambouillet rams. Developed in Idaho, U.S., especially for range conditions.

Panama gumtree. *Castilla elastica,* family Moraceae, a tree grown in the tropics as a source of latex and in the warmer sections of the U.S. as a curiosity. Native to Mexico and tropical America. Also called *ule; Mexican rubber-tree.*

Panama hat palm. See *hat carludovica.*

Panama orange. Calamondin.

panaschiering. A condition of plants in which the leaves are part green and part yellow, or spotted with white and green, or white-bordered.

pancake. The taffy-like deposit sometimes formed at the bottoms of cans containing bulk ice cream.

pancake land. Flat, prairie land with low-lime clay soils which tends to have a compact baked surface rather than a granular one (eastern Texas, U.S.).

pancheon. A large, shallow, earthenware basin in which fresh milk is placed to allow the cream to rise to the top.

pancreas. A gland below and behind the stomach which secretes pancreatic juice. The *pancreas* is commonly known as the *sweetbread* but should not be confused with the commercial sweetbreads of veal (thymus gland).

pancreatic juice. A secretion by the pancreas containing ferments which contribute to digestion of food.

pancreatin. The proteolytic enzyme, trypsin, of the pancreas. See *enzyme.*

pandemic. Designating a disease or organism of worldwide distribution; widely epidemic.

pandora moth. *Coloradia pandora* Blake., family Saturniidae, a moth whose caterpillar infests western pines and sometimes defoliates them.

American Indians have used this caterpillar for food.

panel. A plot, enclosed by wire netting, used in the study of forage species on grazing land: *e.g.*, in comparing the forage on protected and unprotected grazing areas.

panel plot. A small area protected from grazing during a part of the growing or grazing season by a movable fence. Also called *hurdle plot; panel.*

Pangolagrass. *Digitaria decumbens,* family Gramineae, a perennial, extensively stoloniferous or creeping, tender grass grown in Florida and southern California, U.S., as a pasture grass. It is tolerant to highly acid to neutral soils but not to saline soils, and is adapted to sandy soils when fertilized. Native to South Africa. See *spreading crabgrass.*

panhandle. See *Texas Panhandle.*

pan haus. A sausage in which meat from the split head (jaws and eyes removed), shoulders and feet, with kidney, heart, tongue, liver, is ground and cooked together to make a single product.

panic grass. 1. Any grass of the genus Panicum, family Gramineae. Some species are grown for their grain or as pasture grass, others are weeds. See *Guineagrass; paragrass; proso; Texas panicum; witch grass.* 2. Foxtail millet.

panicle. A type of inflorescence which has a main stem and subdivided branches. It may be compact and spikelike, as a timothy, or open and spreading, as in the oat.

panicled goldraintree. *Koelreuteria paniculata,* family Sapindaceae, a tree grown as an ornamental in the U.S. for its feathery leaves and yellow flower clusters. Native to East Asia. Also called *pride of India; China tree; varnish tree.*

panicle hydrangea. *Hydrangea paniculata,* family Saxifragaceae, a shrub grown as an ornamental in the U.S. for its white flowers turning pink and purple. Native to East Asia.

pan method. A method of grading strawberries which consists of emptying each box into a grading pan and removing the defective berries as the fruit is poured back into a box.

panmixy. Unrestricted random mating in plants.

pannage. a. Pasturing swine in a forest to feed on acorns, etc. b. The right or privilege to pasture swine in this manner. c. The payment made to the land owner for the privilege of pasturing swine in his forest.

Pannarone. A Gorgonzola-type Italian cheese, with white curd but without blue veining, which is fast ripening.

panoramic-photograph map. A panoramic photograph from a lookout point on which azimuths and vertical scales have been marked; used in connection with a fire finder to locate fires.

panoramic-profile map. A map drawn around the circumference of a fire-finder map to show the profile of the topography as it appears from the lookout point.

panphytotic. Designating world-wide distribution of a plant disease.

pansy. See *garden pansy.*

pantothenic acid. A vitamin of the B-complex group; required by poultry and swine. Also called *chick antidermatosis vitamin.*

panzootic. Designating a disease which affects all susceptible animals in an area.

pap. The teat of a cow.

papain. An enzyme present in the green papaya fruit used as a digestant.

papaw. See *pawpaw.*

papaya. *Carica papaya,* family Caricaceae, a tree grown in warm regions for its edible fruit, the *papaya* of commerce. Native to tropical America. Also called *papaw; papaia; papeya; papia; pawpaw; melon tree.*

paper birch. *Betula papyrifera,* family Betulaceae, a deciduous, average-sized tree grown as an ornamental in the U.S. for its white, paper-like bark. Native to North America. Also called *canoe birch; white birch.*

paper bush. Cajeput tree.

paper container. A single-service container for milk.

paper doll. A method of testing seed for germination in which seeds are placed on wet paper towels. The towels are then rolled and incubated.

paper fence. A strip of heavy paper placed on edge and treated with a repellent which serves as a barrier to check the advance of chinchbugs into a field.

paper flower. 1. *Bougainvillea glabra* var. *sandereana,* family Nyctaginaceae, a woody vine having oblong leaves and almost ever-blooming magenta flowers. Native to South America. 2. See *strawflower.*

paper grain. Grain which is purchased on contract on an Exchange or Board of Trade for speculation or hedging.

papermulberry. *Broussonetia papyrifera,* family Moraceae, a tender, ornamental grown as a street tree. Its bark is called tap or tappa. Native to Asia. Also called *Tahiti mulberry.*

paper mulch. Heavy paper rolled out in strips to cover the soil as a mulch between rows of plants, or squares of paper with slits through which individual plants grow. The paper may be quite as effective as, and in some instances preferably to, usual mulches of straw, manure, leaves, etc., in conserving moisture and preventing weed growth.

paper profit. The profit that would be realized if open contracts were liquidated at a certain time at a certain price.

paper pulp. The product used in the manufacture of paper which is made by mechanical and chemical treatment of wood or cellulose material.

papershell. Designating any variety of pecan which produces a thin shell and large nut.

papilionaceous. Having a butterfly-like corolla, as illustrated by the flower of the pea.

papilla. Any small protuberance on the surface of the body or an organ, such as the fleshy projections on the tongue.

papoon. A kind of sweet corn which was obtained from the Indians living along the Susquehanna

River in Pennsylvania, U.S., and introduced elsewhere. The theory that this early corn was the progenitor of present-day sweet corns has some acceptance.

papoose root. Blue cohosh.

pappus. Any appendage or tuft of such appendages forming the crown of an achene which aids in wind dissemination of the seed.

paprika. The powdered ripe fruit of sweet redpeppers which is used as coloring matter in ketchups and relishes, as a flavoring, and a garnish.

papule. A small elevation, usually of the skin, as a pimple.

papyrus. *Cyperus papyrus,* family Cyperaceae, a perennial, grasslike, aquatic herb, 10 to 15 ft. in height grown in a shallow pond as an ornamental. The pith was used by the Egyptians in making papyrus (paper). Also called *Egyptian paperplant; Egyptian papyrus; Egyptian reed.*

parabolic weir. A measuring weir, with a notch of parabolic shape which can be so proportioned to give a parabolic discharge equation.

paracasein. The protein which is precipitated when milk is treated with rennet. See *casein: 2; rennet.*

parachute fruit. A seed or achene with a spreading tuft of hairs, bristles, or scales at one end, which enables it to be carried by wind and become widely distributed: *e.g.,* a weed seed, such as that of the dandelion.

parade horse. A horse that has two gaits; the parade gait, a high prancing gait (4 to 6 miles per hour) and the parade walk.

paradichlorobenzene. A white crystalline compound which volatilizes slowly and has a distinctive odor. It is used as a fumigant to kill various insect pests, such as the peach tree borer, household moths, root aphids, etc. It is widely used for fumigation of clothes. Also known as *PDB.*

paradise flower. Costa Rican nightshade.

paradise poinciana. *Poinciana gilliesi,* family Leguminosae, a small tree or shrub with straggling, sticky-haired branches grown for its graceful, feathery foliage and showy, yellow flowers. Native to South America. Also called *bird-of-paradise bush.*

paraeluvium. The residual product which results from the eluviation of the weathering products of sedimentary rocks.

paraffin. A tasteless, odorless, waxlike substance obtained mostly as a residue in the distillation of petroleum; used in sealing the tops of jars of jellies, preserved fruits, as a coating for paper milk bottles, cheeses, and for various other similar purposes.

parafoulbrood. A disease of the brood of honeybees apparently caused by *Bacillus pluton* or *B. para alvei.* It is very similar to European foulbrood *q.v.*

paragrass. *Panicum purpurascens,* family Gramineae, a tropical, heavy, perennial grass grown in the southern U.S. for forage. It is adapted to wet sites. Native to Africa; introduced in the U.S. via Brazil. Also called *Angola grass.*

parakeratosis. A skin disease found in hogs which results from a zinc deficiency in the ration commonly associated with a high calcium intake. It is characterized by dry, hard, crusted proliferations of the superficial layers of the skin.

parallel shift. A method of control designed to prevent cultivators from plowing up plants in which the gangs are rigidly connected together at the front by bars. The *parallel shift* or crosshead gang prevents the shovels from trailing when shifted sidewise as a unit. Each gang is provided with lifting springs to aid in lifting and carrying the greater part of the weight. Also called *pivot gang; treadle guide.*

paralysis. 1. Abolition or impairment of function, especially loss of the power of voluntary motion or sensation. It is frequently a symptom or manifestation of various animal diseases. See *fowl paralysis.* 2. A disease afflicting honeybees which is characterized by trembling, sprawled legs and wings. 3. See *milk fever.*

paralysis of the facial nerve. In horses, a very common ailment due to bruising or permanent functional destruction of the motor nerve (facial or seventh cranial nerve) which is characterized by drooping of one side of the lower lip and a partial collapse of the nostril.

paralysis (of the hind parts in swine). Paralysis usually seen in pregnant or post-parturient sows which is generally believed due to a mineral or vitamin deficiency. Posterior paralysis in swine can also occur from brucellosis and injuries.

paralytic rabies. Dumb rabies.

Para-nut. Brazilnut.

paraoxone. A chemical used as an insecticide.

paraplegia. Paralysis of the posterior limbs of an animal which follows accidents or disease resulting in injury to the spinal cord.

para rubbertree. *Hevea brasiliensis,* family Euphorbiaceae, a tree grown as the chief source of the world's supply of crude rubber. Also called *caoutchouc tree.*

parasite. An organism that lives at least for a time on or in and at the expense of living animals or plants. Some diseases of man and animals are caused by *parasites* ordinarily classified as protozoan, helminthic, and arthropod species. There are also innumerable species of plant *parasites*

parasitic. a. Referring to a disease caused by a parasite. b. Designating a carrier of such diseases.

parasitic diseases. Any disturbances in the life and production of plants and animals which results from the attacks of animal or plant parasites.

parasitic fruit spot. See *phoma fruit spot of apple and quince blotch.*

parasiticide. An agent destructive to parasites.

parasitic seed plant. A plant that obtains part or all of its sustenance from other plants: *e.g.,* dodder and mistletoe.

parasitism. 1. The state of being a parasite. 2. The relationship between parasites and their hosts.

parasitized. Infested or infected with parasites.

parathion. A powerful insecticide compound.

parathomone. The hormone secreted by parathyroid glands.

parathyroid dysfunction. Faulty function of the parathyroid gland. The condition may result in a change in the bony structures and in abnormal calcium deposits in various tissues of the body, or in tetany and low blood calcium.

parathyroid glands. Small glands about the size of a grain of wheat located near the thyroid, which are mainly concerned with calcium metabolism.

paratuberculosis. Johne's disease, a chronic infectious disease of the digestive tract of cattle, rarely of sheep and goats, caused by the organism *Mycobacterium paratuberculosis.* It is closely related to tuberculosis. Also called *chronic bacterial dysentery.*

paratyphoid. See *Salmonellosis; infectious enteritis in pigs.*

parcel. 1. A unit or portion of land. 2. A lot.

parched corn. 1. Corn roasted for human consumption. (In early America it was eaten whole or ground into a meal.) 2. Popcorn and hominy (obsolete).

parchment paper. a. A superior paper made of rag and wood for writing. b. A paper used in food wrapping: *e.g.,* in the packaging of butter. It is made by processing paper through sulfuric acid.

parenchyma. 1. Simple, nonspecialized vegetative tissue composed of thin-walled, diametrically equal cells; especially such tissue as contains functional protoplasts. 2. In animals, the essential tissue of an organ, as a gland.

parenchyma disease. A disease in which the pathogene invades the soft or succulent parenchyma tissues of the host.

Parenica. A Cacciocavallo-type, ewes' milk cheese made in Hungary and Slovakia.

parent material. The rock material, or mass of weathering, from which the solum or soil proper of a mineral soil has been derived by the processes of soil formation.

parent seed. 1. In hybrid corn production, the single-cross seed used by a grower for producing hybrid seed corn. 2. Any seedstock from which an increase of seed is secured.

paresis. Slight or incomplete paralysis that affects the ability of an animal to move.

paretic. Referring to or affected with paresis.

parfait. An ice cream which is made from cream, sugar and egg yolks, with or without nuts or fruits and other natural flavoring.

paries. A wall of a cavity or hollow body organ in plants and animals.

parietal. Borne on or belonging to a wall or walls of a cavity.

paring and burning. A method of improving peaty and stiff clay lands at one time extensively practiced in England. Parings of turf and stubble were burned in heaps or over the whole surface. Probably some benefit resulted but on the whole, damage exceeded benefit.

Paris circaea. *Circaea lutetiana,* family Onagraceae, a soft,pubescent herb, having small, whitish flowers in racemes which is sometimes raised in the moist rock garden. It is the plant of Greek fable which the enchantress Circe used. Native to Asia and North America. Also called *enchanter's nightshade.*

Paris daisy. Marguerite chrysanthemum.

Paris green. Acetoarsenite of copper, one of the first arsenical compounds to be widely used in insect sprays as a stomach poison for the destruction of leaf-eating insects. It was almost exclusively used to control the potato beetle and is still extensively used for that purpose. It injures the foliage and fruit of the fruit trees and so is now largely rejected for orchard spraying.

parity. A legislative formula designed to maintain a just balance between the prices an agricultural producer or farmer receives for his products and the payment he makes for the things he buys. The base period most commonly used in determining the purchasing power of a commodity is August 1909 to July 1914.

parity prices. The prices which give a certain quantity of farm commodities the same buying or purchasing power that those commodities had in a selected base period, usually 1910 to 1914, in terms of (a) prices of commodities and services that farmers buy, (b) interest on farm indebtedness secured by farm real estate, (c) taxes on farm real estate, and (d) in the new parity price index, the cash wages paid to hired labor. See *new parity prices; old parity prices; transitional parity prices.*

parity ratio (all commodities). The ratio of the index of prices received by farmers to the index of prices paid by farmers.

park. a. A natural opening of grassland or a park-like area in a forested region. b. In the Rocky Mountain region, an enclosed valley which may be partly grassland. c. The land enclosed in the oxbow meander of a river. d. Land set apart to be used for public recreation.

park hack. A saddle horse which may have either three or five gaits.

parkland. A grassland region which contains trees either singly or in a group.

park nozzle. A type of nozzle throwing a solid stream of an insecticide solution; used in spraying tall trees.

park pace. A driving pace for harness horses.

parlatoria date scale. *Parlatoria blanchardi,* family Diaspididae, a scale insect which infests the date palm of California and Arizona in the U.S. Also called *date palm scale.*

parly. A small, weak lamb, taken out from a flock for better feeding (New Zealand).

Parmesan. A group of hard Italian cheeses made from skimmed, cow's milk and cured for about a year.

Parmigiano. One of the Parmesan cheeses.

parotid. A gland which secretes saliva. This is a paired gland located posterior to the mandible and somewhat below the ear.

parrotbeak. Any vine-like plant of the genus

Clianthus, family Leguminosae. See *glorypea parrotbeak; red parrotbeak.*

parrot fever. Psittacosis.

parrotmouth. Overshot jaw.

parrot's bill. Red parrotbeak.

parrot-toed. Of, or pertaining to, a horse whose toes turn inward. Also called *cross-footed; pigeon-toed.*

parrot tulip. A type or class of tulip which has many varieties. The flower petals are lacinated, serrated, or slashed at the edges and the blossoms are large and showy, being an intense red and yellow. It flowers late in the season and has weak stems.

Parry grama. *Bouteloua parryi,* family Gramineae, an annual grass, one of the common grama grasses on the ranges of southwestern U.S., Mexico, and South America, whose forage rating is fair. Native to the Americas.

Parry larkspur. *Delphinium parryi,* family Ranunculaceae, a very poisonous herb found in the coastal regions and foothills of California (U.S.). Symptoms of poisoning are loss of appetite, uneasiness, staggering, prostration with rigidly extended limbs, constipation, bloating, and nausea.

Parry lily. *Lilium parryi,* family Liliaceae, a bulbous herb grown for its fragrant, lemon-yellow flowers, spotted on the inside. Native to North America.

Parshall flume. A Venturi-type device commonly used in irrigation for measuring the discharge of small streams and ditches.

parsley. Curly garden parsley.

parsleyworm. *Papilio polyxenes asterius* Stoll, family Papilionidae, a caterpillar which infests celery, parsley, parsnip, carrot, and other plants.

parsnip. Garden parsnip.

parsnip webworm. *Depressaria heracliana* (L.), family Oecophoridae, a caterpillar which webs the leaves of parsnip and celery and feeds on the foliage.

parterre work. A geometric arrangement of ornamental-shaped flower beds which are separated by a pattern of walks or turf.

parthenocarpy. Production of fruit without fertilization.

parthenogenesis. The development of an individual from an unfertilized egg cell. Known to occur occasionally in turkeys and some lower forms in the animal kingdom but not in mammals. In honeybees, unfertilized eggs produce only drones.

partial etiolation. Low chlorophyll content of a plant or plant part resulting from insufficient light.

partial soiling. Green feeds cut and hauled to livestock to supplement pastures, for only part of the season. See *soiling.*

participating contract. A kind of share contract: *e.g.,* one which provides that growers and processors of sugar beets share essentially on a 50-50 basis.

parti-colored. Designating fowls which have feathers of two or more colors or shades.

partido system. In the southwestern U.S., a form of livestock operation in which sheep owned by the patron are let out on shares to a tenant who cares for them and returns part of the increase or income to the owner.

partnership. An association of two or more persons as co-owners of a profit-making business. Chief criteria of a *partnership* are participation in management, sharing profits, sharing losses, ownership of assets together, and having a firm name, a single joint bank account and a single set of farm records. No one factor is controlling. All are applied to the particular arrangement to determine whether in the sum it is a *partnership.* (In most states of the U.S., the law states that the *partnership* name must be filed with the county clerk in the court house. Most father and son farming arrangements are legal *partnerships* regardless of whether they are operating under a written or an oral agreement.)

part owner-operator. A farmer who owns part and rents, from another, part of the farm which he operates. See *owner-operator.*

partridge berry. Checkerberry.

partridge pea. Showy partridgepea.

partridge-wood. 1. A red hardwood, with darker parallel stripes, much valued for cabinetmaking, and also used for walking sticks. The wood is supposedly of the leguminous tree *Andira inermis,* and is from the West Indies. 2. Designating wood attacked by *Stereum foustulosum,* a saprophytic fungus, because of its speckled appearance.

parts per million. A measure of concentration, especially of chemicals in solution: one milligram per liter. Abbreviated *ppm.*

part-time farming. Farming in which the operator is employed much of his time in occupations other than farming and who derives a substantial part of his income from other occupations. As defined by the U.S. Census, farming in which the operator spends 100 or more days off the farm, or in which the nonfarm income received by him and members of his family is greater than the value of the farm products sold.

parturient paresis. Milk fever.

parturition. In mammals, the act or process of giving birth to young.

pasmo. A fungous disease of flax, caused by *Septoria linicola,* family Sphaerioidaceae, characterized by lesions on the cotyledons and foliage of plants, a seared appearance, and a serious reduction in yield.

pasonaria. Giant granadilla.

paspalum. Any grass of the genus Paspalum, family Gramineae. Species are most abundant in the tropics. A number are native in southeastern U.S. and have forage value. See *Dallisgrass; sour paspalum; Vaseygrass.*

pasqueflower. Spreading pasqueflower.

Passburg process. A process in the manufacture of milk powder involving the use of a vacuum roller-drier. This consists essentially of a steam-heated drum, partly immersed in the milk which is drawn into the vacuum chamber and evaporated, after which a knife removes the film of dried milk in the usual way. Its capacity is from 350 to 400 gals. per hr. The process is of German origin. See *milk dryer; nonfat dry milk.*

passionflower. Any of several ornamental plants of the genus Passiflora, family Passifloraceae, so named because each part of the flower has been associated with the passion and life of Christ. See *hybrid passionflower; bluecrown passionflower; wingstem passionflower.*

passion fruit. The edible fruit borne by any of several members of the genus Passiflora, family Passifloraceae. Also called *tacso.* See *giant granadilla; maypop passionflower; purple granadilla; waterlemon.*

passive immunity. Disease immunity given to an animal by injecting the blood serum from an individual already immune to that disease.

Pasta Filata. A group of Italian cheeses whose curd is immersed in hot water, or hot whey, and worked and molded while it is in a plastic condition. Some *Pasta Filata* cheeses, hard and soft, are Cacciocavallo and Mozzarella.

paste. A seasoned or unseasoned, highly concentrated fruit or vegetable juice, pulp, or purée of high solids content.

paste emulsion. An oil emulsion spray that has a very thick consistency; now superseded by more liquid emulsions.

pastern. The portion of a horse or other animal's leg which connects the cannon bone and the coffin bone in the hoof. There are two parts to the *pastern* joint, the long and short *pastern.* This flexible joint serves as a shock-absorbing mechanism in the action of the leg and foot.

paste up. The formation of pasty feces around or over the vent of young chicks which may be related to certain laxative feeds, such as soybean oil meal, etc.

pasteurellosis. Infectious disease caused by bacteria of the genus Pasteurella. See *shipping fever; fowl cholera.*

pasteurization. The process of destroying all or most of the vegetative bacteria in a substance, such as milk or fresh fruit juices, by application of heat of from 140° to 185° F. and then cooling. The word *pasteurization* is derived from the name of Louis Pasteur, the French scientist who first applied heat for the preservation of wine. See *flash pasteurization.*

pasteurize. To treat milk or other fresh liquids by the pasteurization process to prolong their freshness. See *pasteurization.*

pasteurized milk. Milk which has been subjected to a temperature over 142° F. for more than 30 minutes or at 160° F. for 16 seconds and then promptly cooled to 50° F. or lower. There are several systems of milk pasteurization. See *high-temperature, short-time method; low-temperature, long-time.*

pasteurizer. Any of several types of heat-treating equipment used in the pasteurizing process. See *pasteurization.*

pastor. A sheepherder (obsolete).

pastoral. 1. Rural; rustic; bucolic. 2. In farming, referring to a kind of livelihood dependent almost entirely upon pasturing or herding animals, as sheep, goats, cattle, reindeer, etc.

pastoralist. One who breeds and pastures sheep or cattle (Australia).

pastry-flour. Flour made from soft wheat which is lower in protein than hard wheat.

pasturage. All vegetation, grasses, and grasslike plants, forbs, and the fruits and twigs of trees and shrubs upon which grazing animals subsist.

pasture. 1. A fenced or unfenced tract of land on which farm animals feed by grazing. The pasturage is mainly grass but it may consist of various other herbs, brush, and trees. See *rotated pasture: 1; temporary pasture; permanent pasture; range: 1.* 2. Nectar and honey plants within flight range of bees of an apiary. 3. To place livestock on a field or area of grass to harvest the crop by grazing.

pasture corn. To turn livestock into a field of mature corn for grazing it off. See *hog down.*

pasture crop. A crop grown primarily for grazing. See *pasture plant.*

pasture culture. The technology and art of good pasture production and management.

pastured. Designating an area or crop that has been grazed off by livestock.

pasture fertilization. The proper application of fertilizer suitable to pasture crops and soils.

pasture field. A field used primarily for pasture. It may be a farm field in regular rotation and not necessarily a permanent pasture area.

pasture forage. A crop ordinarily grown for pasture but which may be cut for green feed, silage, or cured for roughage.

pasture grass. Any of the several grasses used for pasture, such as bluegrass, etc.

pasture husbandry. The art and practice of good pasture development and management. See *pasture culture.*

pasture improvement. The practice of grazing, clipping, fertilization, liming, seeding, contour furrowing, or other methods of management which improve the vegetation for grazing purposes.

pasture land. See *pasture: 1.*

pasture legumes. Plants of the family Leguminosae which are suitable for pasture, such as white clover, crimson clover, alfalfa, and lespedeza.

pasture management. The principles and practices that insure the greatest amount of satisfactory pasturage for a particular area or time. See *pasture husbandry.*

pasture mixture. Any mixture of several pasture grasses and plants that gives the most satisfactory pasture yield for a particular area, soil and climate.

pasture off. To remove a crop by grazing; a common practice on grass seedings and grain fields in late summer or fall.

pasture plant. Any plant, such as the grasses and legumes, grown primarily for pasturing livestock.

pasture rose. Carolina rose.
pasture rotation. The practice of moving the herd from one field or area to another after a few days of intensive pasturing.
pasture technician. A specialist in the science of soils and plants and in the technical problems of pasture production and management.
pasturing. The system of removing plant growth by allowing animals to graze it rather than harvesting by other methods.
pasty. Designating a body defect in cheese, characterized by its tendency to become distorted and sticky.
Patagras. A Gouda-like cheese made in Cuba.
patch. 1. A small piece of land of indefinite size used for cultivated plants, as garden *patch,* melon *patch.* 2. A relatively small island-like tract named after the vegetation growing on it, as a brier *patch, patch* of woods, etc.
patch budding. Budding by cutting out a rectangular piece of bark from the stock and replacing it with a similar piece of bark containing the bud from another plant.
patchwork ranches. A long, narrow or irregular acreage difficult to operate properly.
patchy at the tailhead. Designating a coarse, high or meaty tailhead setting; an undesirable characteristic in a cow.
patent. 1. A grant made by public authority to an inventor or discoverer entitling him to certain exclusive rights. In the U.S. *patents* may be granted by the U.S. Patent Office to originators of strains or some variation from existing varieties of plants. Many flowers, especially roses and some fruits, have been patented. 2. Open or unoccluded, as a *patent* duct or orifice.
patent flour. In the milling of wheat, the highest grade or most refined flour, in contrast to clear or second grade and to shorts or middlings.
pathobiology. The study of disease processes; biology of disease.
pathogen. A disease producing microorganism or virus.
pathogenesis. The development of morbid conditions of disease.
pathogenic. Capable of causing disease or morbid symptoms.
pathological heartwood. Sapwood which has prematurely assumed heartwood characteristics as a result of wounding or irritation.
pathological lesion. See *lesion.*
pathologically abnormal milk. See *abnormal milk.*
pathological rotation. The maximum rotation through which species may be grown without undue loss from disease.
pathologic alteration. A change in structure or behavior as a result of disease.
pathologic manifestation. A change in structure and activity of an animal or plant as a result of disease or injury.
pathology. Study of diseases, their nature and causes, and the structural and functional changes they produce in plants and animals.
pattern stock. Lumber suitable in quality for use in making models or as a guide for manufacturing.
pattypan group. A group of summer varieties of the pumpkin *Cucurbita pepo,* family Cucurbitaceae. Members are flat, with scalloped edges.
paulownia. Any tree belonging to the genus Paulownia, family Scrophulariaceae. Species are deciduous trees grown for their attractive foliage and white flowers. Native to eastern Asia. See *royal paulownia.*
Pauls betony. Thymeleaf speedwell.
paunch. See *rumen.*
paunchy. Designating a steer with a large middle or paunch which reduces its dressing weight when slaughtered.
pause. Of a laying hen, a temporary cessation of production during the laying year. An inheritable characteristic which, if of frequent occurrence and/or of long duration, seriously limits a hen's total annual production. See *laying ability.*
pavement ant. *Tetramorium caespitum* (L.), family Formicidae, a yellowish ant, a pest in lawns, gardens, and greenhouses.
pavie. Clingstone.
pawl. That part of the drive wheel of a machine, as a grain drill, that engages the ratchet and transmits power to the rest of the machine. *Pawls* consist of small pieces of cast iron or steel which fall in place in the ratchet as the machine moves forward.
pawpaw. 1. Any shrub or tree of the genus Asimina, family Annonaceae. Species are shrubs or small trees with white or purplish flowers. 2. *A. triloba,* a small, deciduous tree whose edible fruit is fleshy, pulpy, and banana-flavored when ripe. It is in limited cultivation for its foliage and green flowers changing to yellow and purple. Native to central and eastern U.S. Also called *custard apple; false banana; papaw.* 3. See *papaya.*
PDB. Paradichlorobenzene.
pea. Any annual or perennial herb of the genus Pisum, family Leguminosae. Species are grown for food and feed. See *garden pea; field pea; baby garden pea; sugarpod garden pea.*
pea anthracnose. A disease caused by the fungus, *Colletotrichum pisi,* family Melanconiaceae, characterized by lesions and cankers on the stems, leaves, flowers and seed pods. See *anthracnose.*
pea Aphanomyces root rot. A fungous root rot caused by *Aphanomyces euteiches,* family Saprolegniaceae, a water mold which is common where peas are grown and may do serious harm to canning peas.
pea aphid. *Macrosiphum pisi* (Harr.), family Aphidae, a large, green-colored aphid present in home gardens but which does its most serious damage in fields of canning peas.
pea Ascochyta blight. A blight caused by any one or all of the three fungi: *Ascochyta pisi; A. pinodella;* or *Mycosphaerella pinodes.*
pea bacterial blight. Blight caused chiefly by the bacterium *Pseudomonas pisi,* family Pseudomon-

adaceae, characterized by cankers and water-soaked spots surrounded by yellow, brown or black zones.

pea bacterial canker. Pea bacterial blight.

pea bean. Navy bean.

pea-cannery waste. The pea-vines and empty pods left after removal of the green peas at cannery stations. Stacked for silage, it has a strong odor but forms a nutritious feed for dairy cows, beef cattle, and sheep.

peach. *Prunus persica,* family Rosaceae, a deciduous tree grown for its edible, sweet fruit, the *peach* of commerce, eaten fresh, dehydrated, sun-dried, frozen, or canned. It forms one of the most widely distributed and valuable orchard crops in the U.S. Double-flowered and purple-leaved varieties are grown as ornamentals. Native to China. See *nectarine.*

peach anthracnose. A disease caused by the fungus, *Glomerella cingulata,* family Gnomoniaceae, characterized by lesions and cankers only on the fruit. See *anthracnose.*

peach Armillaria root rot. A destructive root rot caused by the fungus *Armillaria mellea,* family Agaricaceae, characterized by lesions generally near the union of the larger lateral roots with the main crown root. The symptoms may be confused with those of winter injury.

peach aphid. *Anuraphis persicae,* family Aphidae, an insect which is a vector of virus diseases such as peach mosaic and leaf roll of potato.

peach bacterial spot. A bacterial disease caused by *Xanthomonas pruni,* family Pseudomonadaceae, characterized by severe leaf drop. The fruit becomes unfit for market.

peach bark beetle. *Phloeotribus liminaris* (Harr.), family Scolytidae, a beetle which punctures the bark of the peach and other fruit trees. Beads of gum appear on the affected trunks and branches.

peach bells. Peachleaf bellflower.

peach borer. See *peach tree borer.*

peach blotch. A minor, virus disease of peach, characterized by well-defined leaf variegations.

peach brown rot. See *brown rot of stone fruits.*

peach calico. A mottle type of virus disease of peach leaves in which the veins are yellow and the leaf tissue grades from green to yellow or white. Leaf variegation (non-virus) differs with at least three different grades of green and affected areas are sharply delimited.

peach canker. See *bacterial canker of stone fruits.*

peach Clitocybe root rot. See *Clitocybe root rot.*

peach Coryneum blight. See *Coryneum blight of stone fruits.*

peach crown gall. See *crown gall.*

peach cutting and pitting knife. A knife with two blades to halve the peach and remove the pit.

peach freckle. See *peach scab.*

peachleaf bellflower. *Campanula persicifolia,* family Campanulaceae, a perennial herb grown for its terminal clusters of blue or white flowers. Native to Europe. Also called *peach bells.*

peach leaf casting yellows. See *X-disease (of stone fruits).*

peach leaf curl. A curl of peach leaves caused by *Taphrina deformans,* family Taphrinaceae. Affected leaves are noticeably reddened or paler than normal and become much curled, puckered, or distorted. Later the leaves acquire a grayish appearance, gradually turn brown, and finally wither and fall from the tree. Young twigs and fruit may also be involved. The disease is of world-wide distribution. Also called *curl; leaf blister.*

peach little. See *little peach.*

peach mosaic. A virus disease of peach prevalent in southwestern U.S. It is characterized by leaf mottling, retarded foliation, rosetting, and the breaking of color in the blossoms. The virus has a wide range of hosts in stone fruits and there are a number of variable strains. The disease causes large losses and quarantines have been set up for it.

peach mottle. A virus disease of peach occurring in western U.S. which causes varying degrees of mottle with no fruit symptoms. The virus is apparently transmissible to cherry and the disease is distinct from mottle leaf.

peach necrotic leaf spot. A virus disease that results in shock symptoms similar to ring spot of stone fruits (*q.v.*), but differs in that light brown, membranous, necrotic areas occur yearly on young peach leaves in midsummer. The disease causes serious losses in nursery stock in California.

peach phony. See *phony peach.*

peach rosette. A virus disease of peach which is indigenous to southern U.S. Infected trees turn yellow, followed by the formation of new shoots bearing many small leaves. Peach trees die shortly after symptoms appear. Some plums and apricots are symptomless carriers of the virus.

peach rust. A fungous disease of peach caused by the rust, *Tranzschelia discolor,* family Pucciniaceae; characterized by pale yellow spots on both surfaces of the leaf. Later the spots become bright yellow and brown. See *rusts.*

peach scab. A fungous disease of peach and apricot caused by *Cladosporium carpophilum,* family Dematiaceae, which on the fruit causes superficial, dark, sooty freckles. When numerous, they may merge into dark blotches beneath which the skin may dry and crack. Small lesions may be formed on twigs and on the underside of leaves. The disease is widely distributed and may occur on other stone fruits also. Also called *freckles; black spot.*

peach tree borer. *Sanninoidea exitiosa* Say, family Aegeriidae, a moth, whose larva is a very serious pest of the peach. It bores into the base of the tree from a foot above to 2 or 3 ins. below the surface of the soil causing masses of gum to exude. These borers produce dead areas in the bark and may eventually kill the tree.

peach twig borer. *Anarsia lineatella,* family Gelechiidae, a very small, gray moth whose larvae bore into the ends of tender new peach growths, causing a wilting and dying back of the twigs.

Generally a minor pest, it may be important in unsprayed or neglected orchards.

peach variegation. A leaf variegation disorder of peach which is regarded as a chimera (*q.v.*). It is of minor importance.

peach wart. Raised welts on peach fruits. Smooth wart is caused by a virus and is of rare occurrence. Peach foliage is unaffected but transmission of the wart virus to sweet cherry results in mottling. Crease wart, along the suture, is non-virus and inherent in certain varieties. Beady wart results from insect feeding.

peach western X-disease. See *western X-disease.*

peach yellow bud mosaic. A virus disease of peach in which pale yellow, oblong blotches appear along the midvein of the leaves resulting in distortion of the leaf blade. In advanced stages, defoliation occurs leaving bare the basal portion of affected shoots. The disease is known only in certain counties in California, U.S.

peach yellows. The first known peach virus disease indigenous to eastern and midwestern U.S. The fruit ripens prematurely and is insipid. Fine, thin shoots with small, yellow leaves are produced and the foliage generally turns yellow. Infected trees die within a few years. Orchards in entire districts have been periodically wiped out by this disease in the past 150 years.

peacock flower. Flamboyanttree.

peacock neck. The neck of a horse which is long and slim.

peacomb. A type of comb, common to certain breeds of chicken, which is composed of three small single-combs parallel to each other and joined at the base. The edges are serrated, but the middle row serrations are higher and thicker.

pea downy mildew. A mildew, caused by the fungus, *Peronospora pisi,* family Peronosporaceae, which attacks stems, leaves, and pods of peas and various other legumes.

pea harvester. A mechanical device for harvesting peas consisting of a winged drum revolving rapidly over a stationary, sharp-edged plate which knocks the pods back onto a platform from which they are elevated into a box or bag. Green peas are also harvested with a regular mowing machine and the cut vines are taken to a viner for hulling.

peak. 1. In turpentining, the upper point of the V-shaped streak at the top of the incision. 2. In hydrology, the maximum rate of flow recorded at a gauging station during a flood.

peaked at the dock. Designating a sheep with a narrow or sloping tail head.

peak out. To reach the top crest, as in prices, during a specified period.

peak year. The year of greatest production in the life of a fruit tree, a bed of berries, asparagus, etc.

pea lifter. See *grain, (pea, bean) lifter.*

pea louse. See *pea aphid.*

pea mosaic. A virus disease of garden pea which affects the pods and causes distortion, mottling, and sometimes dark green ridges with sunken, yellowish-green areas in between. Affected pods are poorly filled.

pea moth. *Laspeyresia nigricana* (Steph.), family Olethreutidae, a moth which emerges about the time the peas come into blossom and places its eggs on almost any part of the plant. The minute, active larvae locate the pods and bore into them, feeding on the developing seeds and spinning a very light web in the process. The insect attacks all varieties of garden peas, sweet peas, and vetch and is a serious pest in the northern pea-canning areas of the U.S.

pea near-wilt. A wilt of pea caused by the fungus, *Fusarium oxysporum* f. *pisi,* family Sphaerioidaceae, which resembles pea wilt but differs in that it progresses more slowly. The vascular discoloration is brick-red in color and occurs farther up the stem.

peanut. *Arachis hypogaea,* family Leguminosae, an annual herb whose pods ripen underground and contain usually two large, oily seeds, the *peanuts* of commerce, that are edible when roasted. The seeds are valuable as food, and serve as a source of peanut oil. The plant is widely cultivated as a cash crop in warm regions where it is harvested for nuts and forage and is popular for hog down. Also called *American earthnut; earth almond; earth pea; goober beenut; ground nut; pinda; pindar.*

peanut blue damage. A discoloration that often occurs on the seedcoat of Spanish peanuts cured during periods of warm, damp weather. The small spots usually have a bulls-eye appearance with colors ranging blue through black, and are a result of an interaction between pigments in the seedcoat and oxalic acid produced by the fungus, *Sclerotium rolfsii.*

peanut butter. A food made of roasted peanuts ground until soft and smooth. It is a wholesome and highly nutritious food, containing 2,825 calories per lb., and about 29 percent protein, 46 percent fat, and 17 percent carbohydrate.

peanut butter-whey mix. A mixture of equal parts of peanut butter and sweetened condensed whey; used as a sweet filling or candy center or as a sweet spread.

peanut charcoal rot. A seedling blight in peanuts caused by the fungus, *Sclerotium bataticola,* which turns the tissues dark gray or black on the inside. The initial infections are usually near the soil line. Under high temperature conditions, the lesions may girdle the stem and kill the plant.

peanut collar rot. A disease complex that may affect the runner peanuts when plants that have been wounded by cultivating implements, cankers, etc., are invaded by any of several species of fungi. The affected part wilts suddenly, the leaves turn dark brown or black, and the plant dies quickly.

peanut digger. Any of the various plowlike instruments for removing the peanut pods and the vine from the soil. These vary from a one-horse turning plow, with moldboard removed, to a machine potato digger, etc.

peanut dry rot. A form of peanut seedling blight caused by the fungus, *Rhizoctonia solani,* which is very much like peanut charcoal rot in the early stages. Lesions may develop near the soil line, but the plant partly recovers if the infection does not spread.

peanut early leaf spot. A disease of peanuts caused by the fungus, *Cercospora arachidicola,* family Moniliaceae; characterized by spots on the leaf that are light tan at first. With age, the spots become reddish-brown to black on the lower surface and light brown on the upper surface, and become surrounded by a yellow halo.

peanut hull. The shell of the peanut. It contains more than one half fiber and is less nutritious than straw; used chiefly as poultry litter, bedding, and fertilizer. See *peanut hull meal.*

peanut hull meal. Ground peanut hulls; a very light fertilizer containing 1.5 to 2.5 percent nitrogen.

peanut late leaf spot. A fungous disease of peanuts caused by *Cercospora personata,* family Moniliaceae; characterized by dark brown or black spots on both surfaces of the leaf. The spots are usually smaller than those in peanut early leaf spot, without the distinct halo.

peanut meal. Finely ground peanut oil cake. See *peanut oil meal.*

peanut meal and hulls. See *unhulled peanut oil feed.*

peanut oil. The oil expressed from peanuts; used primarily as a food. Also called *earthnut oil.*

peanut oil cake. A livestock feedstuff obtained as a by-product of the partial extraction of oil by pressure or solvents from peanut kernels under reasonable milling conditions.

peanut oil meal. A livestock feedstuff which is ground peanut oil cake.

peanut peg rot. All losses that result from rotting, sprouting, and other types of damage to developing nuts, or to the loss of nuts in the soil at harvest by breaking of the gynophores.

peanut root knot. Galls or knots which form on the feeding roots or fruits of peanut caused by any of several species of nematodes (*q.v.*). Symptoms of the above-ground parts are not characteristic. Plants are stunted, leaves are likely to show chronic symptoms of wilting, and the color is light green. Later, considerable leaf shedding results and the yield is reduced.

peanut seedling blight. See *peanut charcoal rot; peanut dry rot.*

peanut seed rot. A rot of peanut seed that occurs usually within the first week after planting, caused mainly by species of Fusarium, Rhizopus, Mucor, Diplodia, Penicillium, and Aspergillus fungi. Their entrance is facilitated by broken seedcoats and by other mechanical injuries produced by the sheller. The disease is most destructive under conditions that retard germination: cool, damp weather; abnormally deep planting; and waterlogged soils.

peanut skins. A livestock feedstuff which is the outer covering of the peanut kernel, exclusive of hulls, as obtained in the usual process of commercial milling. The product may contain broken peanut-kernels.

peanut southern blight. A blight of peanuts caused by the fungus, *Sclerotium rolfsii,* which may appear on plants at any time during the growing season but is more likely to cause damage in late summer or early fall. It causes wilting and eventual death of the plant above the point of infection.

peanut stem meal. A livestock feedstuff; the ground product obtained after the separation of peanuts and practically all leaves from the peanut vine, leaving only vines and stems.

pea powdery mildew. A mildew of pea caused by the fungus, *Erysiphe polygoni,* family Erysiphaceae, characterized by a white to grayish growth on the surface of the plant parts.

pear. 1. Any tree or shrub of the genus Pyrus, family Rosaceae. Species are grown for their edible fruits and as ornamentals. See *snow pear; sand pear.* 2. *P. communis,* a deciduous tree grown for its edible fruit, the *pear* of commerce. Native to Europe and Asia.

pear, apple powdery mildew. Apple powdery mildew.

pear blight. Fireblight.

pear borer. *Thamnosphecia pyri* (Harr.), family Aegeriidae, an insect whose larvae bore within the wood of the apple and pear tree.

pear bull's-eye rot. Apple anthracnose.

pear Clitocybe root rot. Clitocybe root rot.

pear common scald. A physiological breakdown of pear fruits which are held in storage at moderate or high temperatures under poorly ventilated conditions.

pear decline. A disease of pear trees of north-central Washington (U.S.), in which many of the finer feeding roots are infected with various fungi, the most abundant of which is *Pythium debaryanum,* family Pythiaceae. It occurs especially in the roots that have been held several days in water.

pear Dematophora root rot. Dematophora root rot.

pear fire blight. Fireblight.

pear gray mold. A serious disease of pear fruits in storage caused by the fungus, *Botrytis cinerea,* family Sclerotiniaceae, characterized by the appearance of a slate-gray mold on the surface of the skin and a rotting of the fruit. In advanced stages the fruit often has a sour odor.

pear hawthorn. *Crataegus calpodendron,* family Rosaceae, a thorny, deciduous shrub or small tree grown for its small, white flowers and small, pear-shaped, orange-red fruit. Native to North America. Also called *pear haw; black thorn.*

pearl acacia. *Acacia podalyriaefolia,* family Leguminosae, a shrub grown for its bluish-gray or silvery foliage. Native to North America. Also called *Mount Morgan wattle.*

pearl alum. Aluminum sulfate.

pearl-barley. Barley ground down to small rounded grains.

pearlbush. *Exochorda racemosa,* family Rosaceae, a shrub grown for its showy terminal clusters of wide, white flowers. Native to Asia.

pear leaf blight. A fungous disease of pears and quinces caused by *Fabraea maculata,* family Mollisiaceae, characterized by several purple dots extending to purple or dark brown lesions on the leaves, fruit, and shoots. Also called *leaf-spot; fruit spot; leaf scald.*

pear leaf blister mite. *Eriophyes pyri* (Pgst.), family Eriophyidae, an insect which infests pears, apple, mountain ash, and cotoneaster causing brownish blisters on the lower leaf surfaces.

pearleverlasting. Pearly everlasting.

pearl millet. *Pennisetum glaucum,* family Gramineae, a tall, annual, tropical grass grown in the tropics as grain for food, and in warm areas, generally, for forage. Also called *cattail millet; Indian millet; African millet; African cane; Egyptian millet; Egyptian wheat; East Indian millet; ragmillet.*

pearl millet bacterial spot. Holcus spot.

pearl popcorn. One of the types of popcorn whose kernels are rounded in contrast to the shoe-peg or rice-shaped kernels of other types.

pearls. A pearllike or amber-colored bacterial exudate which may appear on blighted blossom pedicels and on lesions on leaves of apples, pears, and other fruit trees as a result of attack by fire-blight *q.v.*

pearly everlasting. *Anaphalis margaritacea,* family Compositae, a perennial herb grown for its white flowers which are used as an everlasting, (*q.v.*). Native to the North Temperate Zone. Also called *ever white; American cudweed.*

pear midge. *Contarinia pyrivora* (Riley), family Itonididae, an insect pest which infests pear blossoms causing the fruit to become bloated or lopsided and drop prematurely.

pear plant bug. *Neolygus communis* Knight, family Miridae, an insect which infests pears and apples. The nymphs puncture the fruit causing deformities to develop and some fruit to fall prematurely. Also called *green apple bug; false tarnished plant bug.*

pear psylla. *Psylla pyricola* (Foerst), family Pysllidae, an insect which infests pears feeding on sap sucked from leaves, fruit buds, and the young fruit. Heavy infestation lowers the vitality of the tree and fruits do not develop properly.

pear rust. A fungous disease of pear caused by *Gymnosporangium sabinae,* family Pucciniaceae, of which the juniper tree is the alternate host. It can be controlled by the removal of junipers from within 500 ft. of the pear trees. See *rusts.*

pear rust mite. *Epitrimerus pyri* (Nal.), family Eriophyidae, a mite which infests the pear feeding on the surface of leaves and causing a russeting, followed by a blackening.

pear scab. A fungous disease of pear caused by *Venturia pyrina,* family Mycosphaerellaceae, characterized by lesions on the undersides of the leaves which are at first olive-green but which later turn almost black. There is an upward bulging of the scab, which curls and distorts the leaf. Similar spots appear on the fruit. The fungus also attacks the twigs of some pear varieties.

pear slug. *Caliroa cerasi* (L.), family Tenthredinidae, an insect pest which infests cherry, plum, pear, hawthorn, Juneberry, mountain ash, and quince, skeletonizing the foliage.

pear thrips. *Taeniothrips inconsequens* (Uzel), family Thripidae, a very small insect which infests the pear, cherry, apple, and other fruit trees. It attacks the buds of fruit trees early in the spring causing them to shrivel and turn brown. Mature fruit is very seriously damaged by scabbing, russeting, and deforming.

pear tomato. *Lycopersicon esculentum* var. *pyriforme,* a variety of the tomato whose fruit is pear-shaped.

pear wine. The fermented juice of pears. Also called *perry.*

pear woolly aphid. See *woolly pear aphid.*

pea shrub. Siberian peashrub.

pea-sick soil. A soil that is infested with fungi which cause the root rot of peas.

pea-size cottonseed cake. A livestock feedstuff consisting of small-sized solids left after the oil has been extracted from the cottonseed which can pass through perforations between ⅝ in. and ⅜ in. round. Technically, they are free from meal and nut-size cake and should not contain in excess of 10 percent of sheep-size and pebble-size cake.

peat. A geologic deposit, consisting predominantly of plant remains only very slightly decomposed, which accumulates in lakes, marshes and some swamps. Various kinds are recognized according to origin, texture, and plant composition. See *turf: 3; peat moss.*

peat bog. A bog (*q.v.*) in which the soil or cumulose deposit is peat, a common relief and vegetational feature in the northern glaciated region of the U.S. from Minnesota eastward. The bog may be treeless or it may support a growth of trees, such as black spruce.

peat-lover. A plant which grows in peat; a plant which is tolerant or adapted by nature to highly acid and other conditions found in natural peat soils.

peat moss. 1. Moss plants which grow on heath bogs, such as species of Sphagnum and Polytrichum. 2. See *moss peat.*

peat pot. A container made of compressed peat usually used for growing seedlings in preparation for transplanting without removal from the pot.

peat soil. Soil which consists of or is derived from peat as the parent material. It is spongy and lower in volume weight and higher in water-holding capacity than mineral soil. It is less balanced in plant nutrients and requires different management practices. Various mineral constituents, such as sand, may be mixed with the organic matter, and the peat may be of fresh-water origin and generally acid, or it may be from the saline marshes and swamps along the sea coast.

peavey. In lumbering, a stout lever, 5 to 7 ft. long, which is fitted with a socket, spike, and a curved steel hook that works on a bolt hinge. It is used in handling logs. Also called *peevy.*

peavine hay. The cured vines of peas.

peavine silage. Silage usually made from canning refuse consisting of the vines and pods of green peas left after the seeds have been removed for canning. It has a strong odor but is an excellent feed for dairy cows, beef cattle, and sheep.

pea weevil. *Bruchus pisorum* (L.), family Bruchidae, a beetle whose larvae bore into the pea seeds, feed, and later pupate there. It is one of the most serious insect pests of field peas.

pea wilt. A fungous disease of peas caused by *Fusarium oxysporum* f. *pisi,* family Tuberculariaceae, characterized by chlorosis of the plant. The fungus is confined to the lower four or five internodes of the plant and eventually kills it.

pebble-size cottonseed cake, prime quality. A livestock feedstuff which consists of cottonseed cake which can pass through perforations from 3/16 to 3/8 in. round. It shall be free from meal, nut-size and sheep-size cake and shall not contain more than 10 percent pea-size cake and cottonseed cake screenings.

pebrine. A destructive, contagious disease of silkworms caused by a microsporidian protozoan, *Nosema bombycis,* family Nosematidae; characterized by black spots and stunted growth.

pecan. *Carya illinoiensis,* family Juglandaceae, a tender, deciduous tree grown for its edible nut, the *pecan* of commerce, and for its lumber used in tool handles, ladders, furniture, sporting goods, woodenware, novelties, etc. Native to North America. Also called *Mississippi nut; Illinois nut.* See *paper shell.*

pecan borer. See *dogwood borer.*

pecan bud moth. *Gretchena bolliana* (Sling.), family Olethreutidae, the caterpillar of a moth which feeds on buds of the pecan sometimes causing excessive branching and stunting of the trees in commercial pecan orchards.

pecan carpenterworm. *Cossula magnifica* (Stkr.), family Cossidae, the larva of a moth which bores into the trunks and branches of the pecan, oak, and hickory in the southern U.S. Also called *hickory cossid.*

pecan casebearer. *Coleophora caryaefoliella* Clem., family Coleophoridae, the larva of a moth which mines the foliage of pecan, hickory, and black walnut, leaving holes and often causing defoliation. Also called *pecan cigar casebearer.*

pecan cotton root rot. See *cotton root rot.*

pecan downy spot. A leaf spot of pecan caused by the fungus, *Mycosphaerella caryigena,* family Mycosphaerellaceae.

pecan leaf phylloxera. *Phylloxera notabilis* Perg., family Chermidae, an aphid-like insect which causes galls on the stems of pecan.

pecan leafroll mite. *Aceria caryae* (Keif.), family Eriophyidae, a mite which feeds on leaflets of the pecan and causes them to roll parallel to the midvein.

pecan nut casebearer. *Acrobasis caryae* Grote, family Phycitidae, a moth whose larva does serious damage to the pecan crop by feeding on buds and nuts.

pecan phylloxera. *Phylloxera devastatrix* (Perg.), family Chermidae, an aphid-like insect which causes galls on stems of the pecan.

pecan rosette. A nutritional disease of pecans caused by a deficiency of zinc in the soil and characterized by a yellow mottling of the leaves in the tops of the trees, a dwarfing of the leaves, and death of the twigs and branches.

pecan scab. A fungous disease of pecans caused by *Cladosporium effusum,* family Dematiaceae, characterized by the blackening and shrivelling of the nuts.

pecan shuckworm. Hickory shuckworm.

pecan spittlebug. Dogwood spittlebug.

pecan sunscald. A sunscald (*q.v.*) of pecan characterized by dead or cankerous areas on the trunks and tops of the larger branches.

pecan weevil. *Curculio caryae* (Horn), family Curculionidae, a weevil which infests newly-formed pecans causing them to shrivel and drop. Also called *hickory nut weevil.*

peck. 1. A unit of measure: eight quarts or one-fourth of a bushel by volume of such materials as grains, fruits, or vegetables. 2. See *pocket rot.*

peck order. The order in which some birds within a flock may peck others without being pecked in return; hens at the top of the order can gain a place at feeders, etc., at will, and others according to rank. Males' social standing and mating order is largely determined by combat.

pecky rice. Rice with discolored hulls and kernels.

pecky wood. Wood having pockets of decay caused by fungi or pits or holes caused by insects and birds. It may be used for rustic interior paneling. Pecky cypress is used for greenhouse benches.

pecky wood rot. Brown rot of conifers.

Pecorino. Any of the different kinds of Italian cheeses made from ewes' milk.

pectic acid. A white, sour powder which is fairly soluble in water, and forming a clearer solution than pectin; obtained from pectin by hydrolysis with dilute alkalies.

pectin. Any of the fruit juice substances which form a colloidal solution with water and are derived from pectose (protopectin) in ripening processes or other forms of hydrolysis. *Pectin* is derived from citrus fruits and apple wastes and is used in jelly making to firm the body of the product.

pectinase. An enzyme capable of hydrolyzing pectin to pectic acid and alcohol. Also called *pectase.*

pectocellulose. Protopectin.

pectose. Protopectin.

ped. A natural aggregate or lump of soil in contrast to the clod produced by plowing or digging.

Pedalfer. A soil with a horizon of alumina and iron oxide accumulation in the profile with little or no carbonate accumulation; a category which includes mature soils developed in humid regions. See *Pedocal.*

peddler. 1. A petty, itinerant seller or merchant without a fixed place of business who carries his goods with him and sells from place to place. 2. The larva of the tortoise beetle, *Cassida bivittata* (Say), family Chrysomelidae, which looks like a moving bit of dirt or excrement.

pedicel. 1. A small stalk in plants, especially one of the several stalks attached to the peduncle, each supporting a flower in a branched inflorescence. 2. Any of the various small stalk-like structures in animals. Also called *peduncle.*

pediculicide. Lousicide.

pediculus. Infested with, or pertaining to, lice.

Pediculus. A genus of sucking lice infesting mammals, including man.

pedigree. 1. A list of an individual animal's ancestors, usually only those of the five closest generations. 2. A list of the ancestors of a crop plant, as the *pedigree* of corn.

pedigreed chicks. Chicks whose parents and female ancestors for at least two generations are exactly known. They are wing-banded for identification.

pedigreed stock. Animals which have a pedigree. (*Pedigreed stock,* pure-bred or registered stock are interchangeable terms.)

pedigree selection. In plant breeding and improvement, the selection of seed stock from healthy, high-yielding plants. The seeds from each plant are planted in one row so that yield, etc., may be carefully evaluated and comparisons with other similarly selected plants made.

Pedocal. A mature soil with a horizon of accumulated carbonates in the soil profile. *Pedocals* are soils of arid and semiarid regions in contrast to those of humid regions or Pedalfers.

pedogenic. Referring to processes of soil formation.

pedologic. Pertaining to pedology or soil science.

pedologist. One versed in pedology; a soil scientist.

pedology. The science which deals with the origin, fundamental nature, and classification of soils.

pedosphere. The outer, soil layer of the earth in and on which organic life exists.

peduncle. 1. The main or primary stalk of the flower cluster or the stem of a single flower. 2. See *pedicel: 2.*

peel. 1. The outer covering or skin of fruits or vegetables such as the apple, orange, etc. Also called *peeling; rind.* 2. To remove the skin from a fruit or vegetable, as to *peel* a banana.

peeler. 1. A rotary veneer machine that cuts logs into thin sheets. 2. Any of many devices which remove peels from fruit.

peeler log. A log suitable for the manufacture of rotary cut veneer.

peeling. See *peel.*

peg. The plant mechanism of the peanut by which the fertilized peduncle forces the ovary under the ground where it develops into the pod.

peg and pod rots of peanut. A rotting of various tissues of the pod and kernels during the various stages of growth of the peanut; caused by fungi of *Sclerotium* spp., *Rhizoctonia* spp., *Diplodia* spp., and *Rhizopus* spp.

pegged splint. A splint in which two splints are connected by a third splint around the cannon bone of a horse.

peg pony. A saddle horse trained to change direction rapidly.

peg-tooth harrow. See *spike-tooth harrow.*

Pekin. A heavily-meated breed of white ducks, excellent for market and table use. The breed is noted for its ability to make rapid and economical growth, young ducklings commonly attaining a weight of 5 lbs. or more in 8 to 10 weeks under commercial feeding. It is used almost exclusively for the production of market ducklings in the U.S. Also called *White Pekin.* See *Long Island duckling.*

Peking cotoneaster. *Cotoneaster acutifolia,* family Rosaceae, a shrub grown as an ornamental in the U.S. for its pinkish or whitish flowers. Native to China.

Peking mockorange. *Philadelphus pekinensis,* family Saxifragaceae, a deciduous shrub grown for its fragrant, creamy-white flowers. Native to China and Korea.

Pekin lilac. *Syringa pekinensis,* family Oleaceae, a deciduous shrub grown for its white flowers. Native to Asia.

pelargonium. Any perennial herb or shrub of the genus Pelargonium, family Geraniaceae. Species are the geranium of flower gardens, window boxes and pot plants. See *fish pelargonium; ivy-vine pelargonium; Lady Washington pelargonium; nutmeg pelargonium; rose pelargonium.*

pelham bit. A bit with two rings on each side for two controlling reins; used on polo ponies.

pelican dutchmanspipe. *Aristolochia grandiflora,* family Aristolochiaceae, a tropical, woody vine, grown in greenhouses for its heart-shaped leaves and very large, solitary, yellowish-green, purple-spotted flowers. Native to the West Indies. Also called *pelican flower.*

pelite. A rock composed of fine particles, such as clay and shale, which upon weathering yields silty and clayey soils.

pellagra. A deficiency disease of humans caused by restricted diet and lack of vitamin B_2. Black-tongue in dogs is also a vitamin deficiency disease of the *pellagra* type.

pellet. 1. A mass of indigestible hair and bones regurgitated by carnivorous birds or mammals. 2. See *pelleted feed.*

pelleted. Designating the practice of placing seeds, feeds, and fertilizers into uniform sized units so they may be handled in a more efficient or satisfactory manner. See *pelleted seed; pelleted feed.*

pelleted feed. A pill-like or cubical type of animal feed made by forcing the loose, bulky, or dusty feeds into small, uniform pellets by the use of grinding, molding, and compressing machinery. It is much used in the manufacture of specialized feeds, such as poultry, calf, rabbit, and dog feeds.

pelleted hay. Hay that has been highly compressed by passing through a pelleting machine. It is easy to handle and is free from dust.

pelleted seed. Seed which has been coated with an inert material, fertilizer, fungicide, etc., and shaped into a uniform unit like a pill or pellet. Pelleting enables better preservation of the seed, precision planting, and a great saving in seed.

pellet feeding. Feeding ration in the form of pellets, rather than in bulk form.

pelleting. 1. Coating seeds with a plastic substance in which a fungicide and/or fertilizer may be incorporated. 2. The process of making pelleted feeds.

pelletized fertilizer. Fertilizer compressed into small grains or pellets of uniform size and shape. These pellets can be more easily handled and more exactly placed by fertilizer machinery than the loose material, and there is less loss by deterioration or change in storage.

pelletized seeds. See *pelleted seed.*

pellet mill. A machine which compresses materials into pellets.

pellett clover. Kura clover.

pellicular water. Suspended water which adheres by molecular attraction to the surface of solid particles or to the walls of fractures, but in excess of hygroscopic water. It is fixed water capable of partial extraction by root absorption and evaporation but is not moved by balanced or the unbalanced film-forces resulting from localized evapoation and transpiration.

pelophyte. A plant which grows in clayey soil.

pelt. 1. The natural, whole skin covering, including the hair, wool, or fur of the smaller animals, such as sheep, foxes, etc. A large pelt is more often called a hide *q.v.* Also called *felt.* 2. To remove the whole skin or *pelt* from the body of an animal.

pemmican. A concentrated food used by explorers which consists of powdered, dried strips of meat mixed with dried fruits and vegetables.

pen. 1. A loose, rectangular stack of fuel wood or pulpwood in layers of two pieces each of varying height and width. 2. In poultry shows, a male and four female birds of the same variety. 3. A small space enclosed by any kind of fence, used for confining pigs, cows, and other animals.

pen barn. A structure providing shelter and freedom of movement of the milking herd between milkings, as contrasted to the stable wherein cows are stanchioned. See *loafing shed.*

pencil cedar. California incensecedar.

penciling. Small marks or stripes on a bird's feather.

pendulous crop. In chickens, an enlarged, loosely hanging crop. Also called *drop crop; water crop; baggy crop.*

pendulous udder. A low-hanging, poorly attached udder.

penetrability. A physical characteristic of soil measured by the work required to push a definite kind of probe a unit distance into the soil.

penetrometer. A device which measures the force required to push a probe rod into the soil. It can be used to measure the density or degree of compaction in a soil.

pen-fattening. Fattening poultry in which small flocks in pens are given full feed.

penicillin. An antibiotic derived from the mold, *Penicillium notatum,* family Moniliaceae.

Penicillium rot. A rot caused by a species of the mold genus, *Penicillium* spp.

penis. The male organ of sexual union.

pen lot. A number of animals in an enclosure. In livestock exhibitions, the animals are judged as a lot, or group, in contrast to an exhibition of single individuals.

penny cress. Field pennycress.

Penny-John. St. John's-wort.

pennyroyal mint. *Mentha pulegium,* family Labiatae, a perennial herb grown for its foliage and stems which yield the essential oil, pennyroyal. Native to Europe and Asia. Also called *brotherwort.*

pen of ewe lambs. In the show ring, three ewe lambs which are owned by one exhibitor.

pen of lambs. In the show ring, four lambs of both sexes, owned by the exhibitor.

penstemon. Any herb of the genus Penstemon, family Scrophulariaceae. Species are grown as ornamentals. Also called *beardtongue.* See *beardlip penstemon; gloxinia penstemon; scarletbugler penstemon; sharpleaf penstemon; smooth penstemon; type penstemon.*

penstock. A closed conduit which supplies water under pressure to a waterwheel or turbine.

pen system. A system of goat management on ranges in which the does and kids are penned, as contrasted to the toggle system *q.v.*

pentachloroethane. A soil fumigant; a nemacide.

pentachlorophenol. An organic chemical, which is a flaky, crystalline solid with a pronounced creosote odor used as a wood preservative, for protection from fungus decay and attack by termites or beetles.

pen-type barn. A modern barn designed for cattle in which the animals are not confined or stanchioned but have free run for exercise, loafing, and free access to roughage and water. The cows are moved a few at a time into a milking parlor arranged for the feeding of concentrates and for milking. See *pen barn.*

peony. 1. Any perennial herb of the genus Paeonia, family Ranunculaceae. Species are grown in flower gardens. See *Chinese peony; double-crimson Chinese peony; fernleaf peony; golden peony; tree peony.* 2. *P. officinalis,* a perennial herb widely grown for its large, crimson, white, or yellowish flowers. Native to Europe and Asia.

peony blight. A fungous disease of peonies caused by *Phytophthora paeoniae,* family Pythiaceae, characterized by the young shoots decaying at the soil line. The infection may start from the tip of the plant and progress downward.

peony dahlia. A type of dahlia which has open-centered flowers with two to five rows of ray florets. In addition, there may be curled or twisted floral rays around the disk.

Pepato. A Romano-type, cured, spiced cheese made in southern Italy and Sicily.

peperomia. Sanders peperomia.

pepino. 1. *Solanum muricatum,* family Solanaceae, a spiny herb grown in Chile for its yellow, somewhat acid, edible fruit. Native to Peru. 2. See *cucumber.*

pepper. 1. Any herb, shrub, woody vine, or tree of the genus Piper, family Piperaceae. One species yields the cubeb of commerce; one the black pepper; and one is grown as an ornamental. See *cubeb pepper; black pepper; Celebes pepper.* 2. See *redpepper.*

pepper anthracnose. An important wet weather disease caused by the fungus *Gloeosporium piperatum,* family Melanconiaceae, characterized by cankers on the green ripe fruit, and by lesions and cankers on the leaves and stems.

pepper-box. In honey production, a condition in which cells of healthy brood are irregularly intermingled with diseased cells either uncapped or with punctured or sunken cappings.

pepper brand. Bunt of wheat.

peppercorn. 1. The dried fruit of the black pepper (*Piper nigrum*), used whole or ground as seasoning. 2. In English law, a merely nominal rent for land.

peppered. Designating a feather of a bird which has small spots of gray or black.

peppergrass. A common name applied to many plants of the genus Lepidium, family Cruciferae. See *Virginia pepperweed; gardencress peppergrass.*

pepper maggot. *Zonosemata electa* (Say), family Tephritidae, an insect which feeds on the core of redpeppers and spoils the pod.

peppermint. *Mentha piperita,* family Labiatae, a perennial herb extensively grown in herb gardens and as a field crop. The leaves are used for flavoring as is the aromatic oil which the plant yields Also called *brandy mint.*

peppermint rust. A fungous disease of peppermint caused by *Puccinia menthae,* family Pucciniaceae, which attacks the leaves and stems of peppermint causing a serious loss in yields of oil.

peppermint Verticillium wilt. A fungous disease of peppermint, caused by *Verticillium albo-atrum* var. *menthae,* family Moniliaceae, which survives in the soil and is characterized by the leaves wilting and cankers forming on stems and rhizomes. It is the most serious disease affecting peppermint and spearmint in the U.S.

pepper Phytophthora blight. A fungous disease of peppers caused by *Phytophthora capsici,* family Pythiaceae, which occasionally occurs in southern U.S. Characterized by dark, water-soaked bands that girdle the stems and branches. The leaves show dark spots of irregular size and shape and dull water-soaked patches soon cover the entire surface of the fruits.

pepperidge. Black tupelo.

pepper ring spot. A virus disease of redpepper caused by the cucumber mosaic. Dark spots with greenish-brown rings appear on the pods and may seriously affect the market value of peppers.

pepperroot. Cutleaf toothwort.

pepper root rot. A fungous disease of redpepper caused by *Phytophthora capsici,* family Pythiaceae.

pepper shrub. California peppertree.

pepper spot. A fungous disease of clover caused by *Pseudoplea trifolii,* attacking clover throughout the U.S., especially Ladino white and white clover. It is characterized by very small, sunken, black flecks appearing on both the upper and lower surfaces of the leaves.

pepper tobacco etch virus. A virus disease of redpeppers caused by *Marmor erodens,* characterized by a mottling of the leaves.

peppertree. California peppertree.

peppervine. *Ampelopsis arborea,* family Vitaceae, a woody vine grown as an ornamental covering for a wall, etc. Native to North America.

pepperweed whitetop. *Cardaria draba,* family Cruciferae, a pernicious weed pest. Also called *hoary cress.*

pepper wilt. A bacterial disease caused by *Pseudomonas solanacearum,* family Pseudomonadaceae, which attacks redpeppers and other plants in the warmer sections of the U.S., characterized by a drooping of the leaves soon followed by wilting and death of the plant. The inner tissues of the stem have a dark, water-soaked appearance and when pressed there is a gray, slimy exudate.

pepsin. A digestive enzyme which is secreted by glands in the stomach. Commercially obtained from the lining of the pyloric end of a pig stomach, it is used in medicine as an aid to protein digestion.

peptones. Products obtained by the digestive action of enzymes on albuminous matter. *Peptones* of different kinds are produced from lean muscle tissue or by the action of yoghurt bacteria on lactalbumin in milk, and serve as easily assimilated forms of proteins or as cultural media.

peracute. Excessively acute: *e.g.,* when symptoms of a disease occur much earlier than usual and are well marked.

percentile taper. The relative taper of a tree in terms of diameter at regular intervals along the stem which is expressed in percent of diameter at breast height.

percent return on farm capital. Annual returns from farm capital expressed in percent of a farm capital.

perch. 1. In the U.S. and U.K., a unit of land measurement, 16½ ft. in length; in square measurement, a plot of land 16½ ft. on each side; a square *perch.* Also called *rod; pole.* 2. In a poultry house, a roosting pole or bar.

perched water table. A body of water which has been retarded in its downward course by an impermeable or nearly impermeable bed to such an extent that it forms an upper zone of saturation overlying but separated from a lower zone.

Percheron. A breed of draft horse: deep, short,

heavily muscled, and strong-boned, which originated in Normandy, France.

perchloride of mercury. $HgCl_2$, bichloride of mercury, a dangerous poison used in various forms as a fungicide and pesticide.

percolation. The process whereby water moves downward through the soil in response to the force of gravity.

perennial. A plant that lives for more than two years. There are two classes of *perennials;* herbaceous and woody. An herbaceous *perennial* is one having succulent above-ground parts that are usually killed by winter cold, the roots living more than two years. A woody *perennial* is one having above-ground parts that are woody and in the temperate zones normally withstand winter cold. Both roots and above-ground parts live more than two years.

perennial baby's-breath. See *baby's-breath.*

perennial canker. A canker in which the activity of the pathogen or other injurious factor prevents complete healing for a period of years.

perennial crop. A crop which continues to produce without annual or biennial reseeding or replanting, such as alfalfa, many grasses, asparagus, rhubarb, artichoke, Jerusalem artichoke sunflower and certain other varieties of sunflower, tree or shrub crops, etc.

perennial dollarplant. Perennial honesty.

perennial fruit-body. A persistent fruiting body (*q.v.*) that is seasonally active for two or more years.

perennial gaillardia. *Gaillardia aristata,* family Compositae, a hardy, perennial herb grown for its yellow flowers. Native to North America. Also called *blanket flower.*

perennial honesty. *Lunaria rediviva,* family Cruciferae, a perennial herb grown for its dried seed pod used in dried bouquets. Native to Europe and Asia. Also called *perennial dollarplant.*

perennial moonflower. Bigroot morningglory.

perennial mouse-ear chickweed. Big cerastium.

perennial peavine. *Lathyrus latifolius,* family Leguminosae, a perennial, climbing herb grown for its habit and its flowers of various colors. Native to Europe. Also called *everlasting pea; perennial pea.*

perennial phlox. Summer phlox.

perennial quakinggrass. *Briza media,* family Gramineae, a perennial grass grown for its slender leaves and purplish panicle. Native to Europe and Asia.

perennial ragweed. Western ragweed.

perennial ryegrass. *Lolium perenne,* family Gramineae, a tall, short-lived, perennial, rapid-growing grass used in permanent pasture mixtures in the humid areas of the northern states (U.S.), in turf mixtures, and as a seed or cover crop. Also called *English meadow grass; English ryegrass.*

perennial sow thistle. Field sow thistle.

perennial stream. A stream which flows throughout the year.

perennial tree onion. Top onion.

perennial wheat. Any of certain varieties of wheat that may produce crops for a period of years without annual reseeding. The breeding of such wheats for commercial cropping is still in the experimental stage.

Perez sea-lavender. *Limonium perezii,* family Plumbaginaceae, a salt marsh, shrubby plant grown for its bluish-purple and yellow flowers. It can be used as an everlasting. Native to the Canary Islands.

perfect flower. A flower which has both stamens and a pistil or pistils. *Cf. imperfect flower.*

perfect stage. Sexual reproductive stage of a fungus.

perfect head. 1. Each individual animal. 2. An individual head of lettuce, cabbage, etc.

perforate hymen. See *hymen.*

performance tested parent stock. See *U(nited) S(tates) Performance Tested Parent Stock.*

perianth. The floral envelopes of a flowering plant: the calyx, corolla, or both.

pericarditis. In cattle, inflammation of the membranes surrounding the heart which frequently results from hardware disease *q.v.*

pericardium. The membrane that encloses the cavity containing the heart.

pericarp. The outer layer of the ovary wall around the seed or seeds. It is sometimes constructed of three separate structural layers: the endocarp, the mesocarp and the epicarp. See *hull: 1.*

periclinal chimera. See *chimera.*

periclinal division. The division of tissue cells by the formation of walls parallel to the outer surface of the organ, as in the active division of the cells of the cambium layer or of the phellogen, etc.

periglacial soil. A soil on unglaciated areas adjacent to the fronts of ice sheets. Some soils in the U.S. on unglaciated land may retain traces of characteristics imparted by the ice environment during the last Glacial Period.

perilla oil meal. The ground residue obtained after the extraction of the edible oil from the seed of plants of the genus Perilla, family Labiatae. It contains approximately 38 percent protein, and is used in limited amounts in dairy and sheep feeds.

perimeter of a fire. In forestry: (a) the entire outer edge of the fire; (b) the length of the outer circumference or edge of the fire.

periodical cicada. *Magicicada septendecim* (L.), family Cicadidae, a cicada found in the northeastern part of the U.S., which has a life cycle of 17 years, with a number of broods overlapping. Injury to plants is caused by the female cicada depositing her eggs in roughened punctures, 1 to 4 ins. in length, in the twigs and branches of apple and many other trees. Native to North America. Also called *seventeen year locust.*

periodic annual increment. The growth of a stand of trees for any specified period, divided by the number of years in the period.

periodic increment. The growth of a stand of trees for any specified period, commonly 10 to 20 years.

periodic ophthalmia. An inflammatory disease of

the eyes of horses and mules characterized by recurrent attacks of acute symptoms. Repeated attacks result in blindness. The internal structures of the eye are primarily involved. According to ancient folklore its recurrence corresponds to changes of the moon giving rise to the term moon blindness which still persists in lay circles today. The cause of *periodic ophthalmia* has been in dispute for many years and the exact etiology is still not known. Some believe the disease to be intimately related to certain soils and climates; others believe it is due to nutritional deficiencies (namely riboflavin). Recent work shows that Leptospira infection may be involved in the condition. Also called *moon blindness; moon blind; mooneyed; recurrent ophthalmia.*

period of lactation. Period of milk production in an animal; the time between the beginning and end of the milk flow.

periople. The thin, outer layer of the hoof of a horse.

periosteum. The membrane that covers bone.

peripheral. Pertaining to the surface of the body or a plant, especially a tree.

perique. A kind of tobacco, black, strongly flavored, cured in its own juice and mixed in small amounts with some smoking tobaccos. In the U.S., it is grown only in Louisiana.

perishable. Designating any product which is liable to easy or quick destruction by rot, disease, or decomposition, such as fresh fruits, meats, and vegetables.

perished cotton. Cotton that has been seriously reduced in quality of fiber by exposure to weather, fire, dampness, etc.

peristalsis. a. The successive wavelike motion of the muscular fibers in the duct walls which forces the egg through the oviduct. b. The rhythmic contractions and movements by which the alimentary canal propels its contents.

perithecium. A round or oval, ascus-containing, fungus fruiting body.

peritoneum. The membrane which lines the abdominal cavity and invests the contained viscera (digestive organs).

peritonitis. Peritoneal inflammation. See *peritoneum.*

periwinkle. Any herb of the genus Vinca, family Apocynaceae, especially *V. minor,* a trailing, evergreen, hardy perennial grown as an ornamental ground cover for its blue flowers. Native to Europe. Also called *creeping myrtle; running myrtle; vinca.* See *bigleaf periwinkle.*

periwinkle phlox. *Phlox adsurgens,* family Polemoniaceae, a perennial herb grown for its rose or pale pink flower clusters. Native to North America.

perlite. A white and very porous volcanic mineral that is sometimes used as a medium for the rooting of cuttings.

permafrost. The permanently frozen subsoil of the tundra and other lands of the Arctic regions. (The Swedish is *tjäle* layer, and the Russian, *merslota.*)

permanent cover. Any vegetation which grows continuously on the land.

permanent dam. A dam made of concrete or masonry.

permanent grass. Any of the grasses which in nature reseed or renew themselves from year to year, as buffalograss, bluestem, and other prairie grasses.

permanent hay. Hay crops, such as alfalfa and grasses which occupy the land for a long period without intervening crops.

permanent meadow. A meadow planned and handled so as to produce a regular crop of grass for several years without plowing and reseeding.

permanent parasites. Parasites, such as bloodsucking lice, which spend all life stages on or in the body of the host.

permanent pasture. Grazing land in farms occupied by perennial grasses and legumes. It is not a part of a regular rotation of fields and usually remains unplowed for long periods or indefinitely.

permanent roots. In plants, the top or lateral roots which make up the bulk of the root system during the life of the plant in contrast to the primary roots which are temporary and function only in the early stage.

permanent soil sterilization. Sterilization of soil by chemical treatment lasting for more than a year to suppress all undesirable vegetation on roadsides and ditch banks. See *soil sterilization.*

permanent wilting point. The percentage of moisture in a soil at which plants wilt and fail to recover when placed in an atmosphere saturated with water vapor. Also called *wilting percentage.* See *wilting coefficient.*

permanganate of potash. $KMnO_4$, a chemical used as a mild fungicide, disinfectant, deodorant, and oxidizing agent. It was once used in determining the availability of nitrogen contained in organic materials.

permatodes. Roundworms which live in soil or water.

permeability. 1. A physical property of soil; the difficulty or the ease with which it permits water, air, or roots of plants to penetrate through its pores. 2. The rate at which a soil mass transmits fluids under specified conditions.

permissible velocity. The highest velocity at which water may be carried safely in a canal or other conduit without scouring the channel sides.

permittee. A person who is legally allowed to graze a certain number of livestock on a particular area of public range.

Pernambuco cotton. Sea-island cotton.

pernicious equine anemia. See *equine infectious anemia.*

perosis. A deforming leg weakness in poultry in which the Achilles tendon slips out of its natural groove at the hock joint causing the leg to become permanently bowed or badly twisted. There is usually an enlargement and flattening of the hock joint. It occurs most frequently in rapid growing chickens due to manganese deficiency and an excess of phosphorus in the diet. Also called *slipped tendon; hock disease.*

peroxidase. A heat resisting enzyme present in milk, especially in abnormal milk *q.v.*

perpelic. Referring to plants which inhabit or grow on clay soils.

perpetual begonia. *Begonia semperflorens,* family Begoniaceae, a fibrous-rooted herb grown as a pot plant and as a summer bedding plant for its white, pink, or red flowers. Native to Brazil.

perpetual (rose). A group of hardy hybrid roses that bear large flowers, are tall growing, and are permanent when once well-established.

perry. 1. A cordial made from pears. 2. Pear cider.

Persian buttercup. *Ranunculus asiaticus,* family Ranunculaceae, a bulbous herb grown for its yellow, often double, flowers. Native to Europe and Asia. Also called *turban buttercup.*

Persian clover. *Trifolium resupinatum,* family Leguminosae, an annual herb grown as a winter and spring pasture and hay crop. Native to southern Europe and Turkey. Also called *annual clover.*

Persian insect powder. A pyrethrum flower powder. Also called *Dalmatian powder.*

Persian lamb skin. The skin of the lamb of Karakul sheep. The wool is lustrous and tightly curled making the skin highly desirable for luxury coats and other apparel.

Persian lilac. 1. *Syringa persica,* family Oleaceae, a deciduous shrub grown for its pale lilac, fragrant flowers. Native to Asia. 2. See *Chinaberry.*

Persian melon. See *muskmelon.*

Persian nepeta. *Nepeta mussini,* family Labiatae, a perennial herb grown for its blue flowers and for its leaves, green above and whitish below, attractive to cats. Native to Iran. Sometimes incorrectly called *catnip.*

Persian walnut. *Juglans regia,* family Juglandaceae, a deciduous tree grown for its edible nut, the English walnut of commerce. Native to Europe and Asia. Also called *Circassian walnut; English walnut; European walnut.*

Persian walnut blotch. A fungous disease of Persian walnut caused by *Gnomonia leptostyla,* family Gnomoniaceae, which attacks leaves, young shoots, and nuts. Young nuts drop prematurely and the yields may be seriously reduced.

Persian walnut melaxuma. A fungous disease of Persian walnut caused by *Dothiorella gregaria,* family Sphaerioidaceae, which attacks branches and trunks and produces a black liquid which oozes out of infected areas.

Persian walnut mushroom root rot. A fungous disease of Persian walnut caused by *Armillaria mellea,* family Agaricaceae, which appears as mushroom clusters about infected areas. Roots and basal parts of the trunk are affected and a tree may die within two to five years.

Persian walnut ring spot. A fungous disease of Persian walnut caused by *Ascochyta juglandis,* family Sphaerioidaceae, characterized by round, brown spots on the foliage. It does not cause serious damage.

Persian wheel. An ancient device for land irrigation still widely used in parts of Asia. It consists of a various-sized, metal, rarely wooden, wheel with a horizontal axle that is worked to raise water from a lower to a higher level, usually from a well to the ground surface. Of the two very popular types of *Persian wheels,* one consists of a huge wheel with floats attached radially, or buckets on the circumference, which dips into water at its lower extremity. As the wheel rotates, water is picked up from the lower level and discharged at the other, higher end. It is a commonly practised mode of irrigation in the Middle East. In the other type, used more commonly in India, the wheel is smaller and is suspended over water. A hanging chain of buckets passes over the wheel and dips into the water at its other end, drawing it up as the wheel rotates. The circumference of the wheel tilts the passing buckets over at the upper level and the water is discharged into a conduit. Animal power is generally used to activate the wheel. See *chain pump.*

persicary. Spotted lady's-thumb.

persimmon. 1. Any tree of the genus Diospyros, family Ebenaceae. See *date-plum persimmon; kaki persimmon.* 2. *D. virginiana,* a hardy deciduous tree which bears an edible fruit and which is also a bee plant. Also called *American persimmon.*

persistent. 1. Of, or pertaining to, the presence of the calyx lobes on a ripened pome fruit, especially pear. 2. Designating fruits, flowers, leaves, etc., which remain on a plant after frost. *Cf. deciduous.* 3. Designating a disease which is difficult to control.

persistent hymen. See *hymen.*

persistent-style strain. A variation of the Valencia orange whose style remains attached to the fruit continuously instead of dropping early in the development.

persistent virus. Any virus whose vectors normally remain infective more than one day.

Peru falseheath. *Fabiana imbricata,* family Solanaceae, a small, tender, heathlike shrub grown for its white flowers. Native to Peru. Also called *false heather.*

Peruvian cotton. See *sea-island cotton.*

Peruvian groundcherry. *Physalis peruviana,* family Solanaceae, a tender, perennial herb grown for its edible, yellow berry. Native to Peru. Also called *cape gooseberry.*

Peruvian guano. The dry excrement of marine birds which is obtained from islands off the coast of Peru. See *guano.*

Peruvian lily. See *Chilean alstroemeria; yellow alstroemeria.*

Peruvian mastictree. California peppertree.

Peruvian squill. *Scilla peruviana,* family Liliaceae, a bulbous herb grown for its early-spring, purple, or reddish flowers. Native to the Mediterranean area. Also called *Cuban lily.*

pervious soil. An open soil; one with large pore spaces which permits a ready flow of water through it. It is generally a sandy or gravelly soil.

pest. Anything, such as an insect, animal, plant, or other organism that is obnoxious, *i.e.,* causes injury or loss to a growing crop, or injury or irritation to domestic animals or man: *e.g.*, weeds, flies, lice.

pest control. The use of disinfectant, herbicide, pesticide, insecticide, management or cultural practice which controls pests.

pesticide. A substance which is used to control insect, plant, or animal pests. *Pesticides* include insecticides, herbicides, fungicides, nematocides, and rodenticides.

pest resistant. Designating a plant developed through plant breeding which is resistant or more or less immune to infestation by various pests.

pet. Any animal, such as a cat, dog, lamb, bird, etc., which is kept for affection and companionship.

petal. A division of a flower inside the calyx; a unit of the corolla, consisting of petioles, which usually surrounds the pistil and stamens.

petal fall spray. A spray which is applied for the control of apple scab and codling moth immediately after the fall of the petals. Also called *calyx spray.*

petalody. The transformation of calyx bracts or other floral structures into petals; a morphologic change which may result from overnutrition of the plant.

peteca. A pitting or sinking of the rind of lemons in small discrete areas which may be caused by suboxidation. It is favored by cold weather, heavy oil sprays, and waxes.

petechiae. Small, purple or red spots in the skin caused by small hemorrhages in which blood is released from its normal channel in the blood vessels into very minute areas of the surrounding body tissues.

Peters snapweed. One species of snapweed. See *snapweed: 1.*

petiole. A thin stem supporting the blade of a leaf.

petite. A grade for canning peas of 9/32 in. or less in diameter.

Petit Suisse. A small, unripened cows' milk cheese made in France which is similar to Carré but which contains more fat.

Petri's infectious mottling. A virus disease of sour orange which is characterized by white, pale green, or irregular yellow areas of the leaves, narrow green bands along the midrib and at the bases of the secondary veins, blistering, and wilting.

Petromyzontidae. The major family of the group of vertebrates comprising the lampreys and borers, eel-like, scaleless, aquatic creatures found in both fresh and salt water. The fresh-water species, such as the brook lampreys, are harmless, nonparasitic and dwarfed, and furnish, along with their larvae, food for fresh-water game fish. The sea lamprey (*q.v.*) has become established in the Great Lakes in U.S., and is a destructive parasite of salt- and fresh-water fish.

pets. See *junior twins.*

petsai. *Brassica pekinensis,* family Cruciferae, a biennial herb grown as a potherb and salad plant for its loose, celery-like head. Native to China. Also called *celery cabbage; Chinese cabbage; Pekin cabbage.*

petticoat daffodil. *Narcissus bulbocodium,* family Amaryllidaceae, a bulbous herb grown for its yellow or white flowers. Native to southeastern Europe. Also called *hoop-petticoat narcissus.*

pettis frame. A seedbed frame with screened sides and top to insure thorough aeration of the seedbed and to give protection from birds and rodents.

petty euphorbia. *Euphorbia peplus,* family Euphorbiaceae, an annual herb which is a weed pest of cultivated spaces. Contact with its milky sap may result in a severe dermatitis. Native to Europe. Also called *petty spurge.*

petty-morel. Black nightshade.

petty spurge. Petty euphorbia.

petunia. 1. Any annual or perennial herb of the genus Petunia, family Solanaceae. 2. *P. hybrida,* a hybrid, annual herb of several varieties, grown in the flower border, for bedding, in a window box, and as a pot plant for its flowers of various colors, shapes, and sizes. Also called *common garden plant.*

peyote. Mescalbutton peyote.

Pfister. A cured, cows', fresh skim milk cheese resembling Swiss cheese; made in Switzerland.

pF value. The logarithm of the height, in centimeters, of a water column necessary to produce a force equal to the energy with which moisture is held by a soil. The *p* indicates a common logarithm, and the *F* suggests energy.

pH. An index designating relatively weak acidity and alkalinity, as encountered in soils and biological systems. Technically, it is the common logarithm of the reciprocal of the hydrogen-ion concentration of a system. A *pH* of 7.0 indicates precise neutrality; higher values indicate alkalinity, lower values acidity. See *soil acidity; alkalinity; hydrogen-ion concentration.*

phagocytic. Designating the ability of certain body cells to assimilate small objects by flowing around them, enclosing them completely, and digesting them, or to render them harmless by eradication. Certain of the white blood cells have this property and use it to destroy germs and worn-out cells and to eliminate foreign particles.

phanerogam. A plant with flowers in which stamens and pistils are developed. *Cf. cryptogam.*

Pharaoh ant. *Monomorium pharaonis* (L.), family Formicidae, a tiny ant, generally nesting in inaccessible places about the foundations and in the walls of buildings. It is annoying where food is handled and is a pest in bee yards.

Pharaonic bee. A honeybee kept by the ancient Egyptians and supposed to be the ancestor of today's Egyptian bee *q.v.*

pharynx. Throat; the part of the digestive and respiratory tracts between the mouth, nasal cavity, and esophagus.

phase. A subdivision of a soil mapping unit or soil type having minor variations from the characteristics of the type. The *phase* is more often differentiated on the basis of external characteristics, as relief, stoniness, or accelerated erosion, than on that of intrinsic properties.

phase inversion. The change of cream into butter; the emulsion of fat in water which becomes an emulsion of water in fat. See *butter: 1.*

pheasant. Any brightly colored, gallinaceous bird of the genus Phasianus. Some species, especially the ring-necked *pheasant* (*q.v.*), are farm game birds, and are also raised in limited numbers for the market as a luxury meat fowl. Locally, they may be regarded as a minor pest because of the damage they do to some truck and garden crops.

pheasants-eye adonis. *Adonis annua,* family Ranunculaceae, an annual herb grown for its red flowers with dark centers. Native to temperate regions of Europe and Asia.

pheasant's-eye pink. Grass pink.

phellogen. In growing woody plants, a layer of meristematic cells which by periclinal division produces corky tissue on its outer side. The *phellogen* may arise from the transformation of a layer of parenchyma tissue in the young cortex into cork-cell producing meristem. See *cambium.*

phenol. C_6H_5OH, carbolic acid, a colorless, crystalline compound, obtained by the distillation of coal tar; widely used as a disinfectant and as an ingredient in antiseptics. See *carbolic acid poisoning.*

phenol coefficient. An index showing the potency of a disinfectant in terms of a multiple of the potency of phenol.

phenology. The science that deals with the time of appearance of characteristic periodic phenomena in the life cycles of natural organisms, especially in relation to climate and other environmental factors.

phenospermy. The production of seeds that cannot reproduce.

phenothiazine. The compound produced by the cyclization of diphenylamine with sulfur. Highly toxic for many insects, it is used for the removal of parasitic worms on livestock.

phenotype. The observed character of an individual without reference to its genetic nature. Individuals of the same *phenotype* look alike but may not breed alike. *Cf. genotype.*

phenylmercuric acetate. A fungi-killing, chemical compound used in cottonseed treatment, for the mildew-proofing of textiles and as a spray for apple, strawberry, etc. Also used as an herbicide for control of crabgrass.

phenylmercuric nitrate. An antiseptic tree paint used to check sycamore canker.

phenylmercury triethanol ammonium lactate. A compound used in the control of fruit tree diseases, bean anthracnose, cucurbit anthracnose, as a dip for Irish and sweet potatoes, and for mildew-proofing textiles. Also called *puraturf.*

Philadelphia Cream Cheese. A fresh, soft cheese that resembles Neufchatel *q.v.*

Philadelphia dressed. In poultry slaughter, designating fowl bled and plucked but not eviscerated. Also called *New York dressed.*

Philadelphia ice cream. Vanilla ice cream which consists of cream, sugar, and flavoring with or without a stabilizer. No specific standards have been established for it. Also called *New York ice cream.*

Philippine downy mildew. A downy mildew of sugar cane (*q.v.*), caused by *Sclerospora philippinensis,* family Peronosporaceae.

Philippine lily. *Lilium philippinense,* family Liliaceae, a bulbous herb grown for its fragrant, large, greenish-tinged, white flowers. Native to the Philippine Islands.

philodendron. Any plant of the genus Philodendron, family Araceae. Horticultural species are climbing plants grown in homes and greenhouses for their ornamental leaves and stems. Native to tropical America. See *heartleaf philodendron; giantleaf philodendron; warty philodendron.*

phlobatannin. The iron greening tannins which are widely distributed in plants and fruits. Hemlock tannin is a *phlobatannin.*

phloem. A complex vascular tissue of plants which may include sieve tubes, companion cells, fibers, parenchyma, and secretory cells. In woody plants, usually that portion of the bark immediately adjacent to the cambium.

phloem necrosis. 1. A virus disease highly destructive to elms characterized by the phloem tissue being killed, the leaves drooping, curling upward at the margins, turning yellow, becoming stiff and brittle and falling off. 2. An internal disease of the potato characterized by a network of brown strands extending throughout the flesh of the tuber.

phloem parenchyma necrosis. A necrosis of the potato characterized by spotlike rather than blotchy necrotic areas in the tubers and showing no foliage symptoms. Also called *pseudo-net necrosis.*

phlox. Any herb of the genus Phlox, family Polemoniaceae. Species are grown for their attractive flowers. See *periwinkle phlox; sweet William phlox; Drummond phlox; trailing phlox; summer phlox; moss phlox; Carolina phlox.*

phlox plant bug. *Lopidea davisi* Knight, family Miridae, an insect which infests perennial phlox, feeding on tender leaves and buds.

Phoenixtree. Chinese parasoltree.

Phoma fruit spot of apple and quince blotch. A fungous disease of apple and quince caused by *Mycosphaerella pomi,* family Mycosphaerellaceae, characterized by scattered spots abundant on the stem end of fruit, deeper-red on the colored face of the fruit and darker-green on the lighter surface. The spots become sunken and the tissue browned. Also called *Brooks fruit spot.*

Phoma root rot. A fungous disease of celery and other truck crops, such as carrots and parsnips, caused by *Phoma apicola,* family Sphaerioidaceae.

phomopsis gummosis. A gummosis (*q.v.*) caused by Diplodia and Diaporthe fungi. On the lemon tree outer layers of bark are killed and a flow of gum induced.

phomopsis stem-end rot. A fungous disease of lemons and oranges caused by *Diaporthe citri,* family Valsaceae, characterized by a firm, pliable decay at the stem end of the fruit.

phomopsis twig blight. A twig blight of blueberries caused by the fungi, *Diplodia* spp. and *Diaporthe* spp. The fungus enters the tip of the succulent shoot and progresses rapidly toward the base. The tip wilts or turns dark brown or black within a few days, with a sharp line of demarcation separating it from the healthy, light green bark below.

phony disease. Phony peach.

phony peach. A virus disease of peaches which causes the infected trees to leaf out earlier and ripen fruit earlier than normal. The fruits are reduced in size and number but are more highly colored.

phosphatase. A class of enzymes that causes the breakdown of certain organic phosphorus compounds resulting from the interaction of phosphoric acid and alcohol-like compounds. The determination of the concentration of these enzymes is used as a diagnostic test for certain diseases.

phosphates. Salts of phosphoric acid, such as those made by combining phosphoric acid with sodium, potassium, and calcium. The *phosphate* most commonly used in fertilizers is made by treating phosphate rock (*q.v.*) with sulfuric acid.

phosphate rock. Any rock or mineral which contains phosphorus in sufficient quantity to make it useful directly, or indirectly through manufacture, as fertilizers. Most phosphate rock is tricalcium phosphate, $Ca_3(PO_4)_2$, with various impurities. See *apatite.*

phosphatic fertilizer. A fertilizer composed mainly of phosphates, which supplies the element, phosphorus, to soils and plants.

phosphatic clay. Soft phosphate with colloidal clay.

phosphatic sand. A low grade phosphate rock which has a high content of sand or silica.

phosphatide. A complex organic compound which contains choline phosphoric acid or amino-ethyl phosphoric acid. It occurs in milk, is easily destroyed by heat and oxidation, and is an important factor in the auto-oxidation of milk fat and the development of taints in milk.

phosphazote. A fertilizer product made by converting the nitrogen of cyanamid into urea with sulfuric acid, and neutralizing the urea with phosphate rock. It is not extensively used as a commercial fertilizer in the U.S.

phospholipid. Phosphatide.

phosphoric acid. H_3PO_4, orthophosphoric acid. To be made available to plants, it is converted into soluble acid salts by processing with sulfuric acid. These salts (phosphates) are important ingredients in fertilizers. Soluble *phosphoric acid* is referred to as phosphorus pentoxide (phosphoric anhydride) P_2O_5, and the amounts of available phosphorus in fertilizers are usually expressed in terms of percentage by weight of P_2O_5.

phosphoric acid silage. Legume, small grain, and grass silage to which a small amount of commercial phosphoric acid has been added. The advantages claimed for *phosphoric acid silage* are that there is little loss of green color or nutrients, and the available phosphorus in the ration is increased.

phosphorite. A variety of apatite (*q.v.*), which is fibrous and concretionary.

phosphorized limestone (feeds). A phosphatic limestone used as a mineral supplement in stock feeds. Fluorine is usually present in such phosphatic rocks.

phosphorus. The chemical element, P, which is a constituent of natural soils in various mineral forms both inorganic and organic. It is an essential element in plant and animal growth.

phosphorus deficiency. In soils, insufficient amounts of phosphorus in available form for optimal growth of plants, or in insufficient amounts to produce an economic or desired yield of a crop.

phosphorus paste. A poison containing yellow phosphorus used in killing cockroaches.

photocritical period. See *critical photoperiod.*

photolysis. 1. Chemical changes brought about by the absorption of light. 2. In botany, the grouping of the chloroplasts in relation to the amount of light the plant receives.

photoperiod. A plant's period of daily exposure to light. See *photoperiodism.*

photoperiodic adaptation. The adjustment of plants in their native or artificial habitat to a definite length of daily exposure to light. See *photoperiodism.*

photoperiodic after-effect. The continuing of a certain plant process during a period of daily exposure to light different from that which originally initiated the process. See *photoperiodism.*

photoperiodic induction. The stimulation of some plant processes by exposure to a certain amount of light daily. See *photoperiodism.*

photoperiodic inhibition. The retardation of the growth, primarily of the main axis of a plant, as a result of certain photoperiods.

photoperiodicity. The response of animals to the length of daily exposure to light.

photoperiodism. The reaction of plants to periods of daily exposure to light which is generally expressed in formation of blossoms, tubers, fleshy roots, runners, etc. See *short-day plant; long-day plant; day-neutral plant.*

photosensitization. A condition in which external parts of the body, usually the skin, become sensitive to light. This condition may occur in cattle after eating certain substances that have the ability to sensitize light-colored areas of the skin to sunlight. It can result in swellings and fluid-containing vesicles on unpigmented skins. The exudate from the vesicles forms a crust on the skin which may permanently harm the outer part of the affected skin. See *bighead: 1.*

photosynthesis. The assimilation process by which chlorophyll in plants constructs higher chemical compounds, such as sugars and starches, from simpler ones like carbon dioxide and water, under influence of sunlight. Also called *light building.*

photothermal induction. The inducing of a plant process, particularly reproduction, by a combination of photoperiods and temperature. Also called *thermo-photoperiodic induction.*

phototropism. The response of a plant to the stimulus of sunlight in which the plant or its parts seem to turn to face the light. The parts of the plant receiving the direct rays grow more slowly and the plant appears to turn.

phreatic water. That portion of the subsurface water within the zone of saturation. Also called *ground water. Cf. vadose (water).*

phreatophyte. Any plant which sends its roots deep into the soil to the level of ground water or to the capillary fringe above the water table. In semi-arid and arid regions, these plants may be objectionable because they rob more shallow-rooted, less hardy but more desirable plants of water. However, alfalfa, a useful plant, under some conditions may be considered a *phreatophyte.*

phylliform soil structure. A thin, leaflike layer of soil which is less distinct and thinner than platy (*q.v.*). Where this condition is present in the C horizon, as in soils developed from thin bedded sediments, the term *laminated* is used.

phyllody. The transformation of floral organs into leaflike structures.

phylogeny. The evolutionary history of a species, genus, or type.

phylum. One of the primary groupings used in classifying plants and animals which have a common lineage descent.

Phymatotrichum root rot of cotton. A fungous disease of cotton caused by *Phymatotrichum omnivorum,* family Moniliaceae, characterized by the leaves turning yellow or bronze-colored, wilting, and becoming dry. The bark and cambium of the roots turn brown and root rot appears.

physalis (mosaic). See *cucumber mosaic.*

Physalospora ear rot. A fungous ear rot of corn caused by *Physalospora zeicola,* family Mycosphaerellaceae, characterized by stalk lesions and the slate-gray color of the rotted ears. Endemic in the Gulf States of the U.S.

physical poison. Any poisonous material which exerts a physical rather than a biochemical effect, as heavy mineral oils and inert dusts.

physical weathering. In soil formation, the natural disintegration of rock and minerals by alternate heating and cooling, expansion by water freezing in crevices, corrosion by moving ice and water, splitting of rocks by roots of trees, and other physical factors, as contrasted with chemical weathering.

physic-nut. Barbadosnut.

Physiocratic system. A French politico-economic theory expounded by Quesnay during the 18th century. The physiocrats considered agriculture as the only creative occupation and espoused individual liberty and freedom from governmental interference for the agricultural producer.

physiogenic disease. Any disease caused by unfavorable environmental factors.

physiognomy. 1. The general, outward appearance of a plant community as determined by the life of the dominant species. 2. The physical appearance of a landscape created by a combination of land forms or relief features. 3. The external, facial features of a man or an animal.

physiological. 1. Referring to or concerning the science of physiology or the branch of biology which deals with life processes and functions. 2. Referring to the functions of the organs of plants and animals.

physiological drought. Inability of a plant to obtain water from soil although the water may be present in it, as when a soil is frozen, or by reason of weak osmotic force of plant roots.

physiological maturity. The period of advanced age in the cycle of a tree or stand when resistance to adverse influences is so low that death of a tree or net losses in volume of salable wood are likely to occur within a cutting cycle.

physiologic races. Pathogens of the same species and variety that are usually structurally indistinguishable but which differ in their physiologic behavior, particularly in their ability to parasitize varieties of a particular host.

Physoderma brown spot. A minor, fungous disease of corn in the southern areas of the Corn Belt of the U.S., caused by *Physoderma maydis,* family Cladochytriaceae. It produces oblong to round, water-soaked, light green lesions on the stalk and leaves. In time the lesions coalesce, turn brown, collapse, and form small pockets of dusty-brown sporangia. Also called *physoderma; brown spot of corn.*

phytobezoar. Undigestible plant fibers ingested by an animal which have gathered in the rumen or stomach in the form of balls.

phytoedaphon. The microflora of the soil.

phytogenesis. History of plant evolution; the origin of vegetation.

phytogenic soil. Phytomorphic soil.

phytohormone. See *growth regulator.*

phytome. The vegetable substance composing a plant.

phytometer. A plant used to measure the physiological activities of the habitat.

phytomorphic soil. In pedology, a group of soils which includes those developed under the predominant influence of vegetation.

phytopathogenic. Capable of causing plant disease.

phytopathology. The science of plant disease.

phytophagous parasite. Any parasite which lives upon plant hosts.

Phytophthora. A genus of fungi of the family Pythiaceae, that includes a number of plant-parasitic species many of which can also live as saprophytes. At the exterior of the infected host plants the fungus sends out a downy or felty mass of hyphae, which is the downy mildew phase of the *Phytophthora.* Within the host plant the advancing hyphae of the fungus cause the death and decay of the invaded tissues, the rot phase of the infection. If infections occur on young stem tissues or leaves, lesions of varying sizes are produced and the blight stage results. The stem may

FUNGUS SPECIES	PHYTOPHTHORIC PLANT DISEASES
Phytophthora cactorum	*Blossom blight and dieback of lilac, palm, and rhododendron*
	Seedling blight of pine and spruce
	Bleeding canker of beech, elm, and maple
	Stem or trunk canker of apricot, cherry, plum, and tomato
	Rots of cacti
	Collar rot of apple, avocado, chestnut, and walnut
	Fruit rot of apple, muskmelon, strawberry, tomato, and watermelon
	Root rot of rhubarb, sweet clover, and sweet pea
P. capsici	*Fruit rot of redpepper, watermelon, and plant genera Cucumis, Cucurbita*
	Wilt of redpepper and the genus Cucurbita
P. cinnamomi	*Collar rot of walnut*
	Root rot of pine, rhododendron, and tree genera Aleurites, Persea
	Seedling root rot of larch, oak, and spruce
P. citrophthora	*Trunk canker and dieback of almond, apricot*
	Brown gummosis of trunks of citrus trees
	Collar rot of honeylocust and some citrus trees
	Fruit rot of avocado, citrus, and watermelon
	Leaf rot of some citrus trees
P. drechsleri	*Root rot of onion*
P. erythroseptica	*Rot of sugar-cane seed*
	Pink rot of potato
P. fragariae	*Red stele of strawberry*
P. infestans	*Late blight or downy mildew of potato and tomato*
	Leaf blight of some nightshades
	Fruit rot of tomato
	Tuber rot of potato
P. megasperma	*Crown rot of hollyhock*
	Root rot of carrot and coles
P. palmivora	*Seedling blight of avocado*
	Bud rot of coconut, palmetto, and some other palm genera
	Fruit rot of fig
	Stem rot of castorbean
	Leaf spot of avocado
P. parasitica	*Boll rot of cotton*
	Collar rot of avocado and citrus
	Crown and root rots of rhubarb
	Dry rot of parsnip
	Fruit rot of muskmelon, redpepper, and tomato
	Pod rot of bean and pea
	Stem rot of bean and tomato
	Tuber rot of potato
P. parasitica var. *nicotianae*	*Black shank of tobacco*
P. phaseoli	*Downy mildew of Lima bean*
P. syringae	*Leaf blight of citrus trees*
	Shoot blight of lilac
	Brown rot of citrus fruit

be rotted off and if this occurs near or just below the soil surface, the condition is called damping-off (*q.v.*). If the conidia infect a fruit this may result in fruit rot. Storage regions underground, such as tubers, bulbs, corms, roots, may be infected by asexual spores carried into the depths of the soil by rain, as in the tuber-rot phase of late blight of potatoes. For the species of *Phytophthora* that cause various plant diseases see chart, p. 567.

Phytophthora collar rot of apple. A fungous disease of apples caused by *Phytophthora cactorum,* family Pythiaceae, characterized by the fungus growing both extra- and intercellularly through the bark tissue after girdling the tree.

Phytophthora root rot. See *Phytophthora.*

Phytophthora rot of sugar cane. A fungous disease of sugar cane caused by *Phytophthora erythroseptica,* family Pythiaceae, characterized by the soft rot of sugar cane cuttings in cold, wet soils.

Phytophthora trunk canker. See *Phytophthora.*

phytosterol. The sterol (*q.v.*) that is common in vegetable fats.

phytotomy. Plant anatomy; dissection of vegetation.

phytotoxic. Describing a substance poisonous to plants.

pia. Fiji-arrowroot tacca.

piaffe step. A high action step to which horses are trained for exhibition.

pica. Depraved or perverted appetite that animals sometimes show for substances not ordinarily considered as food. It is most commonly seen in animals suffering from a phosphorus deficiency.

pick. 1. The total amount of a crop harvested, or the yield of an individual tree, as the *pick* of oranges. 2. Small, irregular openings within the body of a cheese. 3. See *pickax.* 4. To pull or pluck ripe fruit, as berries, apples, cotton. 5. To pluck the feathers from a fowl in dressing it for the market. 6. To nibble at food.

pick-a-back. Menzies tolmiea.

pickax. A hand tool which has a short handle and a double blade. One part of the blade is narrow, curved, and sharpened on the face like an ax, the other is pointed. It is used in grubbing roots or working in stony or hard soil.

picked cotton. Cotton lint plucked from the boll, as contrasted with that obtained from bolls snapped off the stalk. *Cf. snapped cotton.*

picker. 1. One who harvests a crop by removing it from the plant: *e.g.,* an apple *picker.* 2. A mechanical device that removes the fruit from the plant. See *corn picker.* 3. A type of machine potato planter which is provided with metal forks (pickers) attached to a vertical revolving disk. The *picker* moves through the hopper and spears the seed potato which is then punched off by a special ejector into the seed spout.

pickerelweed. *Pontederia cordata,* family Pontederiaceae, a hardy, perennial, aquatic plant grown in a shallow pool, bog garden, on the edge of a pond, etc., for its attractive, funnel-shaped, blue flowers. Native to America. Also called *alligator wampee.*

picker roller. A part of the mechanism of a cotton gin; a roller, with spikes attached, which disposes of the hulls after the cotton lint is removed.

picker wheel. The rotating disk faced with sharp projections which is the basic shelling mechanism in the spring-type corn sheller.

picker-wheel drop. A type of cotton planter drop which consists of a horizontal, rotating, agitator plate with fins or fingers radiating outward from the body of the plate. The fingers press the cotton seed downward on the picker wheel, notched and about ¾ in. wide, which picks the seed from the mass dropped down on it.

pickery cotton. Cotton from a damaged bale which has had the marketable lint cotton separated out by hand.

picket. To tether or control the grazing range of an animal by a picket rope *q.v.*

picket fence. A light, wooden fence made of narrow pickets or slats which are usually woven into a fence by the use of wires at the top and bottom. It is often used for small enclosures, such as gardens, chicken yards, etc.

picket pin. A stake to which a horse or mule is tied in picketing (western U.S.).

picket rope. Any rope, with a swivel snap on one end to which an animal may be fastened, and with a picket pin on the other which may be driven into the ground to tether the animal while grazing.

picking machine. In poultry dressing, a machine consisting of a revolving drum equipped with flexible rubber fingers which picks the feathers from carcasses.

pickings. Cotton which has been picked or separated by hand from damaged bales. See *pickery cotton.*

picking season. The season of the year when a crop is harvested, especially such crops as cucumbers, string beans, cotton, etc., where the picking period extends over a number of days or weeks. The *picking season* for cotton may extend from August to January in some sections of southern U.S.

picking table. A part of the equipment of a fruit or vegetable grading machine: a roller table which carries the produce past sorters to facilitate the removal of undesirable fruit or vegetables.

pickle. 1. A solution of salt in water which is used for curing or preserving many different products, such as meat, vegetables, etc. There are many different types of pickling solutions with many different seasonings or condiments added for the particular product being pickled or preserved. 2. The cucumber, any vegetable, fruit, or meat which has been processed through a brine solution. 3. To place various products, such as cucumbers, fruits, or meat in a brine solution for preservation, one of the oldest methods of food preservation.

pickle alum. Aluminum sulfate.

pickle brine. A salt solution, generally about 10 percent salt, in which cucumbers, fruits, vegetables, etc., undergo a preliminary fermentation.

pickled. 1. Designating cheeses to which a considerable percentage of salt has been added to prolong keeping qualities. 2. Designating meats, as pig's feet, which have been prepared by placing them in a pickle for several weeks before cooking.

pickled pig's feet. The feet of the hog carcass that, after removal of the toenails, dew claws, and hair are cured in a special pickling solution, which is basically a brine, with certain seasoning ingredients added.

pickle pump. An instrument used in meat curing. The two methods of pumping are: (a) arterial, in which the curing solution is injected through the arterial system of the carcass; (b) stitch, in which the solution is injected into the cut of meat by means of a perforated needle.

pickler (onion). An onion bulb which is too large to be used as a set *q.v.*

pickleweed. *Allenrolfea occidentalis,* family Chenopodiaceae, a stiff shrub common in the San Joaquin Valley and in the Death Valley region of California (U.S.), whose presence in the native flora indicates a very high content of salts in the soil.

pickleworm. *Diaphania nitidalis* (Stoll), family Pyraustidae, a whitish to greenish caterpillar, common throughout the Gulf States of the U.S., which bores into ripening fruits of muskmelons, cucumbers, and squashes from the side next to the ground.

pickling. See *pickle: 3.*

picklock. A grade of wool in the wool markets of the U.S. which was at one time the finest grade of Merino wool produced in the country. It has now disappeared from commerce.

pick of land. A narrow strip of land running into a corner.

pick out. Of a chicken, to peck away the entire vent of another chicken so that the victim, generally a laying hen, dies.

pick-up. A light automobile truck usually with a load capacity of 0.5 to 1.0 ton which is adaptable to many different uses especially on farms.

pickup attachment. A special mechanical device which may be attached to a regular grain combine, baler, or field chopper to pick up cut or mowed crops from the windrow or swath.

pick-up baler. A power-driven, automatic, self-tying hay baler with a windrow pick-up and with or without cross conveyor attachment similar to that used on combines. It consists of a pick-up rotating cylinder, an elevator mechanism, and a cross conveyor to carry the hay to the self-feeder in the baler.

pick-up butter grading. See *pick-up contest.*

pick-up contest. A butter grading competition in which the exhibitor does not know when his sample is to be collected and entered for competition. Also called *surprise contest.*

pick-up guard. A special guard for an abnormally long cutter bar. It lifts fallen grain, etc., so that the cutter bar can cut off the stem below the head.

pick-up hay baler. See *pick-up baler.*

picnic ham. The cured and smoked lower half of a hog's shoulder. Also called *cala ham.*

picnic table. A wooden table designed for outdoor use, especially for parks, playgrounds, and recreational areas. Usually rectangular in shape, it has benches attached to seat 6 to 10 people.

picry. Poison ivy.

Pictou disease. An eastern Canadian cattle poisoning caused by ingestion of the plant, ragwort groundsel, *Senecio jacobaea,* either fresh or with hay. See *ragwort groundsel.*

piebald. Designating a horse with a black and white or dark-colored coat; a pinto. Also called *pied.*

piece. A variable short distance, as a *piece* to the creek.

piece out. To sow seed of another crop on the acreage remaining unsown after all the available seed of one crop has been planted.

piece-root graft. A graft with nursery stock in which seedling roots are cut into pieces 3 to 5 ins. long and cions of similar lengths and diameters are grafted onto them.

pieces. Separately packed skirtings or other pieces of wool from the main part of the fleece (Australia).

pie cheese. Any cheese, such as Cottage cheese, used in making cheese pie or cake, or other bakery goods.

piedmont azalea. *Rhododendron canescens,* family Ericaceae, a deciduous shrub grown for its pink or white flowers. Native to southern U.S. Also called *mountain azalea.*

Piedmont Plateau Region. A physiographic division of the eastern U.S.; the eastward-sloping, dissected, upland plain lying between the Appalachian mountains and the Atlantic Coastal plain. One of the older agricultural regions of eastern U.S., it is notable in its central and southern sections for tobacco, cotton, and fruit production; red soils; and for severe soil erosion and its attendant problems.

pie fruit. The trimmed, overripe, and otherwise low-quality fruit canned especially for use in pies. It is wholesome but not choice in quality for table use. Also known as *water grade.*

piemarker. Fruit of the plant chingma abutilon *q.v.*

pie melon. Citronmelon.

pie plant. Garden rhubarb.

Pierce's disease. A virus disease of grapes characterized by scaling, dwarfing, withering, drying, premature coloring of the berries, delayed leafing out, interveinal chlorotic mottling, failure of the canes to mature evenly, root death, and death of the plant. The causative virus is now considered identical with the one causing alfalfa dwarf. Also called *Anaheim disease; California vine disease.*

pig. 1. A young swine weighing less than 120 lbs. (A few markets, as Chicago, U.S., set 130 lbs. as the maximum weight for an animal of this class.) 2. Any young, unweaned swine.

pig brooder. A heated cover of light metal, wood, or canvas, placed in the corner of the farrowing pen to keep very young pigs warm. Usually heated with a large electric bulb, a *pig brooder* is similar in many respects to a small chick brooder.

pig-eating sow. A sow that devours her young at farrowing time presumably as a result of a ration deficiency, an unsuitable environmental situation, or exciting disturbances at farrowing time.

pigeon. Any bird of the genus Columba. In poultry, young *pigeons,* or squabs, are generally dressed for the market, and old *pigeons* are sold alive, generally for the Jewish trade. See *squab.*

pigeonberry. 1. See *pokeberry.* 2. See *creeping skyflower.*

pigeon fly. *Pseudolynchia canariensis* (Macq.), family Hippoboscidae, a blood-sucking fly, somewhat smaller than the housefly, which infests pigeons and related birds. Known in the U.S. since 1896.

pigeon horntail. *Tremex columba* (L.), family Siricidae, a wasp-like insect whose larvae bore into the trunks of diseased and dying elm, sugar maple, and other trees. Also called *pigeon tremen.*

pigeon loft. A dovecot, elevated house, or pen, atop buildings for breeding pigeons.

pigeon louse. Either of two species of lice, *Columbicola columbae,* or *Goniocotes bidentatus,* family Philopteridae. Each causes considerable annoyance both to old birds and to partially feathered squabs.

pigeon milk. A thick, creamy mixture produced in the crop of the pigeons and regurgitated by them to feed their squabs.

pigeonsbeak. Deervetch.

pigeon plum. 1. Any tree of the genus Coccolobin, especially *C. laurifolia,* family Polygonaceae, bearing edible, berry-like fruit. Native to the tropical and subtropical regions of North America. 2. *Chrysobalanus ellipticus,* family Rosaceae, a small tropical tree grown for its edible, drupaceous fruit. Native to Africa.

pigeon-toed. Designating a horse whose feet turn inward.

pigeon tremex. Pigeon horntail.

pigeonwings. Any vine of the genus Clitoria, family Leguminosae. Species are grown as ornamentals. See *Atlantic pigeonwings.*

pig eye. In a horse, a small, retracted eye which may cause imperfect vision.

piggin-string. A short rope used for tying down animals, as in calf roping.

pig-guard rail. A rail about 8 to 10 ins. above the floor and 8 to 10 ins. from the wall of a farrowing pen, beneath which recently farrowed pigs can move and thus be protected from crushing by the sow.

piggy sow. Any mature female hog that shows advanced pregnancy, as indicated by the swelling of the underline.

pignut hickory. *Carya glabra,* family Juglandaceae, a deciduous tree whose valuable wood is used in tool handles, ladders, furniture, sporting goods, implements, woodenware, and for fuel and smoking meat. Native to North America. Also called *black hickory; broom hickory; brown hickory.*

pigskin. The tanned hides of hogs.

pig-skin fluff. A food product made from skins removed from fat backs of hog carcasses. The skins are fried in deep fat, or roasted until brittle, and then salted and dried. See *crackling.*

pig snout. The snout of a hog carcass, used extensively in the manufacture of Philadelphia scrapple.

pig souse. Souse.

pig teeth nipper. A clipper-like instrument used for clipping the canine teeth (temporary or milk teeth) of suckling pigs.

pig typhus. Infectious enteritis in pigs.

pigweed. Lambsquarters goosefoot.

pilchard oil. The product obtained by extraction of part of the oil from the whole Pacific pilchard or from cannery refuse from this species of fish; it is used as a source of vitamins in animal feeds.

pilgrim. A worn-out, decrepit horse.

Pilgrim geese. A very old variety of geese said to have been brought to America by the early Pilgrim colonists. *Pilgrim geese* are of medium weight. The gander has pure white plumage; the goose has pure white neck feathers, soft dove-gray body plumage and brownish-gray tail feathers.

piling. Round timbers or steel cylinders driven into the ground to support structures such as buildings or piers. The steel cylinders, after being driven into place, are filled with concrete. These supports may extend below the water table in soils or through unstable soils into stable soils below, as through muck into mineral soil.

pillbug. *Armadillidium vulgare,* not an insect but a creature closely related to the crayfish which lives in damp places, as under rotting boards. Mainly a greenhouse pest, it may eat roots of seedlings and do other damage. See *sowbug.*

pilotweed. See *compass plant: 2.*

Pimento cheese. Any cheese to which pimentos have been added.

pimiento pepper. A type of sweetbell redpepper (*q.v.*) whose varieties are thick-walled and more or less pointed at the apex. It is raised primarily for canning. Also called *pimento.*

pimpernel. 1. Herbs of the genus Anagallis, family Primulaceae. Species are sometimes grown in the flower garden. See *scarlet pimpernel.* 2. See *saxifrage pimpinella.*

pimpernel root. The dried rhizome of the saxifrage pimpinella (*q.v.*) which is used for medicinal purposes.

pimple scab. Small, soft, pimple-like pustules, a symptom of corky scab of potato *q.v.*

pimply gut. An intestinal disease of sheep. See *nodular worm.*

pin. In poultry slaughter, to remove the pin feathers which are imbedded in the skin.

pin bone. The region on each side of the tail head on the hind quarters of a bovine. These should be wide apart and well-defined for good conformation.

pin-bone roast. A beef roast cut in the same manner as a pin-bone sirloin steak, but thicker. See *pin-bone sirloin.*

pin-bone sirloin. A steak cut from the sirloin having a tip of the ilium in it.

pincers. The incisor teeth of a horse. Also called *nippers.*

pinched at the dock. Designating an undesirable conformation of the rear end of a sheep; one that is tapering, narrow or pinched.

pinched in the heart girth. Designating an undesirable forequarter body conformation in livestock, indicating lack of development.

pin cherry. *Prunus pennsylvanica,* family Rosaceae, a deciduous tree or shrub which produces small, sour, bright red cherries on long slender stalks. Found in the U.S., from the Atlantic to the Rocky Mountains and from Canada to Virginia and Tennessee, it is especially abundant on burned-over land. The foliage is sometimes lethal when browsed by livestock. Native to North America. Also called *fire cherry.*

pinching. Removal of the terminal shoots of plants by pinching off, as with the fingers of the hand, to produce stocky, bushy plants or to produce long-stemmed roses for exhibition or market, and to prevent flower production in early stages of growth of a plant. See *hard pinch; soft pinch.*

pinclover. Alfileria.

pincushion cactus. Any plant of the genus Mammillaria, family Cactaceae. A globe-shaped cactus found in desert regions, it is often cultivated in the desert garden and as a house plant. Native to North America. See *featherball mammillaria.*

pincushion flower. See *scabious; Caucasian scabious.*

pindar. Old, colloquial name for peanut (*q.v.*), still recognized in parts of southern U.S. Also called *pinder.*

pindo palm. *Butia capitata,* family Palmaceae, a feather palm grown as an ornamental in very warm areas. Native to South America.

pine. Any cone-bearing, evergreen tree with needle-shaped leaves, of the genus Pinus, family Pinaceae. Species number about 50 in the U.S. and constitute valuable timber trees, a source of turpentine and rosin, ornamentals, and Christmas trees. See *Aleppo pine; Austrian pine; Coulter pine; eastern white pine; Himalayan pine; Italian stone pine; jack pine; Japanese black pine; Japanese red pine; Japanese white pine; Jeffrey pine; Korean pine; limber pine; loblolly pine; longleaf pine; Mexican pinyon pine; Monterey pine; mugho pine; pitch pine: 1; ponderosa pine; red pine; Scotch pine; shortleaf pine; slash pine; sugar pine; Swiss mountain pine; Swiss stone pine; western white pine.*

pineal gland. A reddish gland about one third the size of the pituitary which is located in a brain cavity behind and just above the pituitary. A by-product of the meat-packing industry, extracts of which are used in medicine to regulate growth.

pineapple. *Ananas comosus,* family Bromeliaceae, a plant grown in frost-free areas for its edible fruit, the *pineapple* of commerce. It is widely consumed fresh and canned as a dessert fruit. Grown to a very limited extent in Florida, most of that sold in the U.S. is produced in Hawaii and the West Indies. Native to tropical America.

pineapple-buttermilk. A pineapple-flavored buttermilk which contains about 15 percent pineapple juice.

Pineapple cheese. A cheese very similar to Cheddar which is made in the shape of a pineapple. It is supposed to have been first made about 1845 in Litchfield County, Connecticut.

pineapple disease of sugar cane. A seed-rotting disease of sugar cane caused by the fungus, *Ceratostomella paradoxa,* family Ceratostomataceae.

pineapple flower. *Eucomis punctata,* family Liliaceae, a bulbous herb grown in the greenhouse for its greenish-white flowers. Native to South Africa.

pineapple guava. See *feijoa: 2.*

pineapple mealybug. *Pseudococcus brevipes* (Ckll.), family Pseudococcidae, a tropical insect which sometimes infests pineapple, banana, and sugar cane, causing a condition known as pineapple wilt.

pineapple melon. See *muskmelon.*

pineapple scale. *Diaspis bromeliae* (Kern.), family Diaspididae, a tropical pest which infests pineapple. It invades greenhouses where it infests various tropical plants.

pineapple squash. A form of summer squash. See *pattypan group.*

pine bark aphid. *Pineus strobi* (Htg.), family Chermidae, a sucking insect which causes white, woolly patches on trunks and limbs, especially of white pine.

pinebarren deathcamas. *Zigadenus leimanthoides,* family Liliaceae, a bulbous herb which grows in bogs and on wet pine lands and is poisonous to livestock. Native to coastal areas of southeastern U.S. Also called *crowpoison.*

pine barrens. Land, embracing various soils, but generally sandy, and low in fertility, on which pines were the dominant growth. (A local colloquial term, especially during the period of earlier settlement, from New Jersey southward to Florida, U.S.)

pine beetle. Any of the tree-killing bark beetles of the genus Dendroctonus, family Scolytidae, which infests pines.

pine blister rust. A very serious rust disease of pines, especially the white pine, caused by the fungus, *Cronartium ribicola,* family Melampsoraceae. Cultivated and wild species of the genus Ribes, currants and gooseberries are alternate host plants. Pines are infected only from spore forms produced on *Ribes* spp., which in turn are infected only from spores produced on *Pinus* spp.

pine bluegrass. *Poa scrabrella,* family Gramineae, a highly palatable, bunchgrass of western ranges (U.S.), which grows at low to medium altitudes from Montana and southern Washington to California.

pine butterfly. *Neophasia menapia* (F. & F.), family Pieridae, a butterfly whose larvae are pests in the coniferous forests of western U.S., frequently defoliating large areas of pine and Douglas fir.

pine disease. See *nutritional anemia.*

pine dropseed. *Blepharoneuron tricholepis,* family Gramineae, a range grass of good forage rating found from Colorado and Utah to Arizona and western Texas. Native to western North America. Also called *beardless pinegrass.*

pine engraver. *Ips pini* (Say), family Scolytidae, a beetle which infests all species of pine, burrowing small circular holes into branches and trunks of trees. Native to northern U.S.

pine false webworm. *Acantholyda erythrocephala* (L.), family Pamphiliidae, an insect which infests red and white pines. The larvae living in webs cut off needles and can cause extensive defoliation. Native to northeastern U.S.

pinegrass. 1. *Calamagrostis rubescens,* family Gramineae, a tall, perennial, range bunchgrass which grows in the ponderosa and lodgepole pine belts of western U.S. It is rated as only fair in grazing value. Native to western North America. 2. *Carex pennsylvanica,* family Cyperaceae, a grasslike sedge with fair forage rating which grows in the coastal pine lands of southeastern U.S. Native to eastern North America.

pineland threeawn. *Aristida stricta,* family Gramineae, a perennial, erect grass characteristic of the open, sandy pine land of southern Georgia and northern Florida, U.S. It is extensively grazed by cattle but only the young growth is palatable. Also called *wiregrass.*

pine leaf scale. *Phenacaspis pinifoliae* (Fitch), family Diaspididae, a small, elongated, white insect which occurs generally over the U.S. on leaves of pines and sometimes on other evergreens. Pine needles may appear nearly white when heavily infested. Also called *pine needle scale.*

pine littleleaf. A fungous disease of shortleaf pines in southern U.S. caused by *Phytophthora cinnamomi,* family Mycosphaerellaceae, characterized by short, yellow needles at the ends of branches and death of the fine roots of the tree.

pinemat manzanita. *Arctostaphylos nevadensis,* family Ericaceae, a prostrate, woody plant grown to cover rockwork, sloping banks, etc. Native to California, U.S.

pine meadows. A vegetational land feature of southern Mississippi along the Gulf Coast (U.S.), characterized by flat topography, wetness, a scrubby growth of longleaf pine, cypress, and a ground cover of sedgy vegetation.

pine needle miner. *Exoteleia pinifoliella* (Chamb.), family Gelechiidae, a brown larva, about ⅓ in. long, which infests pines, entering the needles near the tips, excavating and causing the needles to turn yellow and dry up.

pine needle scale. Pine leaf scale.

pine oil disinfectant. A disinfectant widely used in animal sanitary procedures as a substitute for cresol preparations.

pine pitch borer. Any moth whose larvae burrow into the stems and branches of pines, causing exudation of pitch. See *pitch mass borer.*

pine rooter. In southern U.S., any hog that obtains its living in the forest from acorns, grubs, and the succulent roots of pines. Also called *razorback.*

pinery. A forest area which supports a heavy growth of pines. (A localism, especially in Wisconsin and Michigan, U.S.) See *pinetum.*

pine spittlebug. *Aphrophora parallela* (Say), family Cercopidae, an insect that infests a number of pines, especially injurious to Scotch pine. The insects feed on twigs, which are covered with spittle, and eject undigested sap.

pine straw. The fallen dead needles of pine trees. The loose, open mat at the base of trees may represent the fall of several years. It gradually becomes broken and decomposed at the bottom.

pine tar. Commercial pine tar.

pine tortoise scale. *Toumeyella numismaticum* (P. & M.), family Coccidae, an insect that infests Scotch, jack, and Austrian pines in northern U.S., producing a black sooty mold, and causing defoliation.

pine tree. A deformation of leaves of roses grown under gravel culture in greenhouses. It results from an insufficient supply of water containing essential nutrient solutions. The lower leaves have a *pine tree* appearance and fall off in a short time.

pine tube moth. *Argyrotaenia pinatubana* (Kearf.), family Tortricidae, an insect that infests white and lodgepole pines. The larvae make tubes of the needles by tying them together, and eat off the green ends.

pinetum. A small grove or planting of pine trees kept for its aesthetic and educational value rather than as a forest.

pine webworm. *Tetralopha robustella* Zell., family Epipaschiidae, an insect that infests pines. The larvae work near the end of terminal twigs, producing frass in silken webs.

pinewoods coneflower. *Rudbeckia bicolor,* family Compositae, an annual, flowering herb grown for its yellow flowers with black centers. It is valued as a cut flower. Native to North America. Also called *coneflower.*

pinewoods grape. *Vitis lincecumii,* family Vitaceae, a wild grape which has been used as one parent in developing hybrid grape varieties known as post oak hybrid grapes. Native to south central U.S. Also called *post oak grape.*

piney. Peony (colloquial).

piney woods. Native forest which consists predominantly of open stands of pine, especially longleaf, in contrast to deciduous and swamp forests. (Colloquial especially in the Gulf Coast region of U.S.)

piney woods cattle. Scrub cattle, or those nondescript in breed (southeastern U.S.).

pin feather. The young feather embedded in, or just emerging from, the skin of a fowl.

pingue. *Actinea richardsoni,* family Compositae, an herb which contains a toxic saponin (*q.v.*). It also contains a low percentage of rubber. Native to California (U.S.).

pinhole decay. A very early presporulation stage of the soft decay of citrus fruit caused by *Penicillium italicum,* family Moniliaceae.

pin holes. 1. Holes in timber or timber products caused by ambrosia beetles. Also called *black holes; shot holes.* 2. Small holes in leaves caused by defoliating insects or by a leaf disease. 3. The holes in the ends of an evener through which the singletrees are attached. The location of these holes determines the proportion of the load that each horse must pull. 4. Small round holes in cheese caused by gas formation from undesirable bacteria.

pining. 1. A disease of sheep in certain sections of England and Scotland which results from cobalt deficiency. 2. Designating any unthrifty or unhealthy animal.

pinion. 1. The outermost part of a bird's wing including the carpus, metacarpus and phalanges; the part of a wing, corresponding to the forearm, on which the primary flight feathers are borne. 2. To cut off the *pinion* on one (or both) wings to permanently prevent a bird from flying.

pinion feather. A feather (particularly a primary or flight feather) on the segment of the bird's wing most remote from the body. See *pinion: 1.*

pink. Certain annual or perennial herbs of the genus Dianthus, family Caryophyllaceae. Species are grown in flower gardens. See *cloudland pink; carnation; Chinese pink; Finland pink; ice pink; granite pink; maiden pink; hardy garden pink; button pink; rock pink; grass pink; lilac pink.*

pink and green aphid. Potato aphid.

pink boll rot. Cotton anthracnose.

pink boll worm. *Pectinophora gossypiella* (Saund.), family Gelechiidae, one of the most injurious insect pests of cotton. Eggs deposited on various parts of the plant hatch into small, pink caterpillars, which penetrate into the boll or square and later enter the seed.

pink callalily. Rose callalily.

pink color salt method. A method of testing seed corn for viability in which kernels of corn are soaked in lukewarm water for about twelve hours, then split the long way and perpendicularly to the germ side. These split kernels are then placed in a weak, colorless solution of a salt, tetrazolium chloride. The action of this salt on living germs in the soaked, split kernels turns the exposed parts pink in two hours or less.

pinkeye. An infectious disease of the eye in cattle, sheep, and goats which is characterized by an inflammation of the cornea and conjunctiva. Also called *keratitis; infectious keratitis; contagious ophthalmia in cattle.*

pinkeye of horses. Influenza in horses.

pink-eyed potato. A potato variety which has pink eyes.

pink-fleshed grapefruit. Any variety or strain of the grapefruit (*Citrus paradisi*), such as the Foster and Thompson, which has pink flesh. Extensively grown in Texas and Florida, U.S.

pink fringed orchid. Smallpurple fringeorchid.

pink fritillary. *Fritillaria pluriflora,* family Liliaceae, a bulbous herb grown for its pinkish-purple flowers. Native to North America. Also called *adobe lily.*

pink fruit. A virus disease of sour cherry which is closely similar to and possibly a strain of western X-disease (*q.v.*). It has caused serious losses in western Washington, U.S.

pink fungus. *Nectria diploa,* family Hypocreaceae, a fungus parasite which is employed in the biological control of scale insects on citrus trees.

pink joint. A fungous disease of the bush redpepper caused by *Sclerotinia sclerotiorum,* family Sclerotiniaceae, characterized by a stem and fruit rot.

pink locust. Roseacacia locust.

pink nose. See *acorn fruit.*

pin knot. A disease gall or knot that is less than ½ in. in diameter.

pink patch. A fungous disease of grasses, especially turf grasses, caused by *Corticium fuciforme,* family Thelephoraceae, characterized by the leaves having a water-soaked appearance, losing their green color, and the turf dying out in distinct isolated patches. The fungus bodies become bright, coral-red.

pink root. 1. A fungous disease of onions caused by *Pyrenochaeta terrestris,* family Sphaerioidaceae, characterized by the roots turning pink, shriveling, and dying.

pinkroot spigelia. *Spigelia marilandica,* family Loganiaceae, a perennial herb occasionally grown as a border plant in the flower garden for its red flowers with yellow throats. This plant yields a remedy for intestinal worms. Native to North America. Also called *worm grass; Indian pink; star bloom.*

pink rot. 1. A fungous disease of celery caused by *Sclerotinia sclerotiorum,* family Sclerotiniaceae, characterized by a watery rot at the base of the stem, and in storage, by pink tissues which later turn gray or brown. 2. See *Gibberella rot of corn.*

pink rot of pome fruits. A fungous disease caused by *Cephalothecium roseum,* family Moniliaceae, which is found on scab lesions causing a storage rot. It may appear in orchards under damp conditions.

pink sandverbena. *Abronia umbellata,* family Nyctaginaceae, a perennial, viny herb grown as an annual for its large, pink flowers. Native to North America.

pinkshell azalea. *Rhododendron vaseyi,* family Ericaceae, a tender shrub grown for its rose-colored, brown-spotted flowers. Native to North America.

pink spray. A spray for control of the apple scab, which is applied when the unopened blossoms are enlarged but before they are in full bloom.

pinkster flower. Pinxterbloom azalea.

pink sugarcane mealybug. *Trionymus sacchari* (Ckll.) family Pseudococcidae, an insect that infests both roots and above-ground parts of sugar cane, causing a general reduction in thrift and often killing small stalks.

pink tip. Endoxerosis.

pinkvine. Mountainrose coralvine.

pinna. A section of a pinnate leaf.

pinnate leaf. A compound leaf whose sections are arranged on both sides of the main petiole.

pinnate tansymustard. *Descurainia pinnata,* family Cruciferae, an annual herb which grows on the arid and semiarid range lands of western U.S. Eaten in the blossoming stage by stock, it is a cause of poisoning referred to by ranchmen as paralyzed tongue. Native to North America.

pinning knife. A dull, shortbladed knife which is used in removing pin feathers in the dressing of poultry.

pinny. In poultry marketing, designating a bird carcass which has an excessive amount of pin feathers, and is therefore of low grade and marketable only at a reduced price.

pin oak. *Quercus palustris,* family Fagaceae, a hardy, deciduous tree found from Massachusetts west to Michigan and Illinois and south to Arkansas. Easily transplanted, it is grown as a shade tree on city streets and in landscaping for its autumnal foliage. Its lumber is used for ties, tight cooperage, fuel, flooring, planing mill products, furniture, boats, coffins, etc. Native to North America.

pin oak sawfly. *Caliroa lineata,* family Tenthredinidae, an insect whose larvae skeletonize the leaves of the pin oak.

pinole. A food made from ground seeds of various kinds. A kind of hoe cake made from ground corn principally, but apparently other seeds, such as those from mesquite and honey locust were used by the Indians of southwestern U.S.

piñon. Mexican pinyon pine.

pin point colony. A very tiny colony of bacteria on the agar plate.

pin point scab. A late season infection of apple that shows on the fruit at harvest time or later in storage as small, scattered lesions.

pintado. Guinea.

pinto. Designating a horse which has a spotted or piebald coat-color.

pinto bean. A variety of the kidney bean (*q.v.*) characterized by mottled seeds. It is a medium-sized bean grown mainly in Colorado, Idaho, Nebraska, and New Mexico, U.S.

pinto leaf. A variegated leaf condition of sweet cherry caused by a virus which is of minor importance because of its limited occurrence.

pinworm. 1. A common name often applied to parasitic nematodes belonging to the family Oxyuridae, that infect man and animals. Some examples are: *Enterobius vermicularis,* the human pinworm; *Oxyuris equi,* the common horse pinworm; *Skrjabinema ovis,* the sheep and goat pinworm. 2. *Heterakis gallinae,* a cecal, grayish-white worm from ⅓ to ½ in. long which is found in the intestines of chickens, turkeys, and other fowls. It is involved in the transmission of blackhead disease.

pinxterbloom azalea. *Rhododendron nudiflorum,* family Ericaceae, a deciduous shrub grown for its clusters of pink or pinkish-white flowers. Native to North America. Also called *pinkster flower; election pink.*

pinyon pine. Mexican pinyon pine.

pioneer plant. A plant species which is capable of invading bare areas, often in large numbers over a considerable space, and of persisting until displaced as succession proceeds.

pioneer settlers. In the early settlement of the U.S., those settlers who were the first to enter new country or virgin wilderness to establish farm homes.

Piora. A hard, cows' milk cheese, similar to Tilsiter, made in the Canton of Tessin in the Swiss Alps.

pip. 1. A horny, dried condition of the tip of the tongue which develops in poultry in cases of infectious coryza when mouth breathing is long continued. 2. The raised crown or individual rootstock (*q.v.*) of a plant. 3. The specialized underground bud of lily-of-the-valley.

pipe. Round, hollow tubing made of metal, plastic, glass, wood, etc., used to convey water or other liquids in a closed conduit usually under pressure.

pipeline milking. A mechanical system so arranged that the milk flows directly from the milking machine to the bulk tank cooler without being lifted or handled by hand.

pipes. The casks used in the Italian method of brining citron.

pipevine. Dutchman's pipe.

pipevine swallowtail. *Battus philenor* (L.), family Papilionidae, a butterfly whose caterpillars feed on and defoliate the Dutchman's pipe.

piping. A series of shrill sounds made by queen bees.

pipped egg. An egg through which the chick has forced its beak in the first step of breaking out of the shell during incubation.

Pippin. A famous apple of which there are a number of varieties, such as the Newtown, Albemarle, Missouri.

pipsissewa. *Chimaphila umbellata,* family Pyrolaceae, a perennial, low, evergreen herb grown for its white flowers and evergreen leaves. Infusions, or tea, made from the leaves, have been used as a folk or home type of medicine. Native to Europe and Asia. Also called *prince's pine; wintergreen.*

piquia. 1. *Caryocar villosum,* family Caryocaraceae, a timber tree that bears edible, oily nuts. Native to South America. 2. Commonly, any tree of the above genus.

piroplasmosis. Any of a number of tick-borne diseases of domestic animals especially affecting cattle, horses, sheep, and dogs, caused by protozoan parasites of the genus Babesia, which invade and destroy red blood cells, giving rise to anemia, elevated body temperature, and sometimes bloody urine. These parasites are transmitted from host to host by ticks of the family Ixodidae. Commonly called *tick fever; Texas fever; splenetic fever; southern fever; babesiosis.*

piroplasms. Microscopic protozoan organisms belonging to the genus Babesia. They cause piroplasmosis *q.v.*

pirouette. A high-school exercise for horses in which the forelegs are held more or less in place while the horse moves his hindquarters around them.

piscary. The right of fishing. Common of *piscary* is the right of fishing in waters belonging to another person.

pisciculture. The production of fish in natural or artificial bodies of water under controlled conditions, such as stocking, feeding, and use of chemical fertilizers.

pistache. *Pistacia vera,* family Anacardiaceae, an aromatic tree grown in very warm areas for its fruit, the pistachio nut of commerce. Native to Europe and Asia.

pistachio. The fruit of the pistache.

pistil. That part of the flower which consists of ovary, style, and stigma.

pistillate. Designating a flower that has a pistil or pistils but lacks stamens; an imperfect flower. *Cf. staminate.*

pistillody. The change of stamens into carpels.

pistol casebearer. *Coleophora malivorella* Riley, family Coleophoridae, a minor insect pest of apples and other fruits in eastern U.S. The larvae live in tiny silken cases about ½ in. long, slightly curled resembling a pistol, and feed on the foliage.

pit. 1. The endocarp of a drupe; the seed-stone of a fruit, as the *pit* of a peach or prune. 2. An excavation in soil in which vegetables, such as potatoes, carrots, and parsnips are placed and covered over for storage. 3. A place on the floor of an exchange in which traders stand when dealing in wheat, cotton, and other commodities. 4. In botany, a small hollow or depression in a cell wall. Various types are recognized in wood anatomy as blind; bordered; primordial; simple. 5. That part of an upright silo located below ground surface.

pita. 1. Centuryplant agave (*Agave americana*) and related species. 2. The tough fiber obtained from the above species, used for cordage, etc. Also called *pita-fiber; pita-flax; pita-hemp.* 3. The pith-like wood of the tree piteira furcrea *q.v.*

pitahaya. *Lemaireocereus thurberi,* family Cactaceae, a tall, tree-like cactus of the arid region of Arizona, California (U.S.), and Mexico, which yields an edible fruit. Native to North America. Also called *sweet pitahaya; aya pitaya.*

pitanga. *Eugenia uniflora,* family Myrtaceae, a tropical, aromatic shrub grown for its fragrant, white flowers and its edible, ribbed, crimson fruit. Native to South America. Also called *Surinam cherry.*

pitch. 1. The resin which occurs in the wood of conifers, as the *pitch* from pines. 2. A heavy, dark-colored, viscous or more or less solid, fusible material which is obtained by distillation of the tar derived from coal, wood, rosin, and petroleum oils. It consists of many organic compounds, chiefly hydrocarbons, differing according to origin. 3. The jumping action of a horse in its attempt to unseat its rider.

pitcherplant. 1. Any herb of the genus Sarracenia, family Sarraceniaceae. Species are insectivorous. See *trumpet pitcherplant.* 2. *S. purpurea,* a perennial herb grown as a curiosity for its insectivorous habit. Native to North America. Also called *huntsman's-cup.*

pitcher pump. A hand-driven water pump with a cylinder that is part of the pump assembly. It has a short handle and a short suction stroke and is suitable for lifting water only a short vertical distance, as from a cistern or shallow well.

pitchfork. A farm implement consisting of a long handle usually with two or three long, curved, sharp-pointed tines used in handling loose hay, straw, bundles of unthreshed small grain, etc.

pitch hay. To handle hay with a pitchfork.

pitch mass borer. *Vespamima pini* (Kell.), family Aegeriidae, a moth devoid of scales whose larvae bore into pine trees, causing large amounts of pitch to exude from the bark.

pitch moth. Any moth whose larvae bore into the wood of conifers and cause an extensive discharge of pitch. See *pitch twig moth; sequoia pitch moth.*

pitch pine. 1. *Pinus rigida,* family Pinaceae, a hardy evergreen, rugged but beautiful tree sometimes grown as an ornamental. Native to North America. 2. Any pine tree with an abundance of pitch in the wood.

pitch pocket. A well-defined, lens-shaped opening between or within annual growth layers of conifer wood which usually contains solid or liquid pitch.

pitch streak. A well-defined accumulation of pitch (resin) which is in the form of a more or less regular streak in the wood of certain conifers.

pitch twig moth. *Petrova comstockiana* (Fern), family Olethreutidae, a moth whose larva infests small branches and twigs of pines leaving a thick mass of pitch at the mouth of the boring.

piteira furcrea. *Furcraea gigantea,* family Amaryllidaceae, a tropical plant grown in very warm areas as an ornamental in the desert garden and also for the fiber it produces, Mauritius hemp. Native to Brazil. Also called *giant lily; cabuya; Mauritius hemp.*

pith. The spongy center of exogenous plants. In the corn plant the *pith* fills the center cavity of the stalk and serves as a storehouse for moisture and food.

pith fleck. A small, localized, isolated patch of parenchymatous wound tissue, darker or lighter than the surrounding wood, which is produced in wood through injury to the cambium of certain hardwoods by the larvae of flies of the genus Agromyza, and to the cambium of conifers by the feeding of adult weevils. Also called *medullary spot; pith-ray; fleck.*

pithing. A method of animal slaughter in which the spinal cord is severed to cause death.

pithworm. See *wireworm.*

pit-making oak scale. See *golden oak scale.*

pitman. In machinery, a connecting rod that joins the crank of a rotary with a reciprocating part, as the knife of the cutter bar.

Pitman-Robertson Act. An act of U.S. Congress, passed in 1937, for aiding the states in the selection and improvement of land and water areas for wildlife restoration. Farm lands are included for improvement of wildlife habitats under the provisions of the Act.

pit silo. A shallow pit of variable size for storing silage, which is dug in well-drained soil and is frequently walled with lumber or concrete if of a permanent nature. It is very common in the western states of the U.S. See *horizontal silo; trench silo.*

pit storage. A shallow pit used for storage in which vegetables are covered with alternate layers of straw and earth.

pitted scab. One of the symptoms of potato scab (*q.v.*) in which a circular to irregular lesion becomes deep to form a pock or depression on the tuber.

pitter. A machine for removing the seeds or stones from drupes, such as peaches or cherries.

pitting. 1. The development of little cavities in metal, especially in aluminum cans, caused by the development of lactic acid by bacteria in milk residues. *Pitting* can also be caused from corrosion by brine. 2. In fruit processing, removal of seeds or stones, as from peaches and cherries. They may be removed by hand or machine pitters.

pittosporum. Any tree or shrub of the genus Pittosporum, family Pittosporaceae. Species are evergreens grown under glass in warmer climates as ornamentals. See *cape pittosporum; karo pittosporum; orangeberry pittosporum; tarata pittosporum; tawhiwhi pittosporum; tobira pittosporum.*

pituitary gland. A small endocrine gland located in the lower part of the brain. Among other functions, it secretes hormones into the blood stream that influence the growth of the body, stimulate the thyroid gland, the sex organs, and the mammary gland to initiate the secretion of milk after the birth of the young.

pituri. *Duboisia hopwoodi,* family Solanaceae, a poisonous shrub or small tree whose dried leaves and twigs are used by the aborigines in Australia as a narcotic and hunger-killer. Occasionally waterholes are poisoned with it to stupefy emus which then are easy to hunt. It is believed to hold some commercial possibilities because of its nicotine content. Native to Australia. Also called *pitchery; pitjuri.*

pivot. A leg action in which a horse pivots around his hindquarters while holding one leg in place and side stepping with the other hindfoot.

pivot axle. A mechanical device on a cultivator whereby the wheels may be pivoted either to the right or left by the use of foot pedals. It is useful especially on hillsides, among trees, and on stump land.

pivot frame. See *seat guide.*

pivot gang. See *parallel shift.*

pivot pole. See *seat guide.*

pivot wheel cultivator. A cultivator with a pivot axle *q.v.*

pizzle. The penis, especially of a bull.

place lands. In the U.S., lands granted in aid of a railroad company with certain limits on each side of the road, which became instantly fixed by the adoption of the line of the road.

placenta. 1. The membranous tissue that envelops the growing fetus in the uterus and establishes communication between the mother and the fetus by the umbilical cord. It is discharged from the uterus at the time of the birth of the young or shortly thereafter. Also called *afterbirth; calf bed.* 2. In plants, that portion of the ovary on which ovules are borne.

placental. Of, or pertaining to, the placenta.

placental retention. The undue retention of the placenta and other fetal membranes after the young is born.

plain. In marketing, designating an animal of poor quality, *i.e.,* one having rough, prominent shoulders, paunchy middle, bare ribs and loin, and small legs.

plain ankles. An auction term denoting swelling or puffiness about the ankles in horses and mules.

plain condensed milk. Unsterilized evaporated milk (*q.v.*) marketed in bulk. The concentration ratio is about 2:1 and no sugar is added. Also called *concentrated milk.*

plain condensed skim milk. Skim milk concentrated in the ratio of 3:1 to 4:1; generally used in ice cream manufacture or baking industry.

plain condensed whole milk. Evaporated milk.

plain drill. A grain drill used for sowing only, in contrast to one that has a fertilizer attachment.

plain (gin). Single-breast gin.

plain head. In stock judging, a very unattractive head on an animal. (A good type of head is highly desirable on all classes of livestock as it denotes breed type, character, and quality.)

plain ice. A confection made from water, sugar, with or without egg albumin, stabilizer and natural flavoring, frozen to a consistency of ice cream.

plain ice cream. A frozen confection made from cream and sugar, with or without a natural flavoring: *e.g.,* vanilla ice cream.

plain-leaved parsley. A variety of curly garden parsley (*q.v.*) that has a plain leaf in contrast to the more popular and common curled leaf variety.

plain lift. In the operation of a mowing machine, the lifting by hand of the outer end of the cutter bar and fastening it in a vertical position. See *vertical lift.*

plain quarters. An undesirable body conformation of an animal in the rear or front quarters.

plains. Flat or nearly level, extensive areas of land. Some major physiographic divisions named *plains* may have broken, hilly topography and may exhibit strong relief locally. They may be classified into various kinds, as alluvial, coastal, grassland, pine, desert, high, low, etc.

plains coreopsis. *Coreopsis tinctoria,* family Compositae, a tender, annual herb grown in the flower border for its long-blooming, showy, golden-yellow flowers. Native to North America.

plain section. In a beehive, a section with unscalloped sides for vertical passage of the bees: the common sizes are 4¼ by 4¼ x 1½ and 4 or 5 x 1⅜ ins.

plain throughout. In livestock shows, designating an animal, very ordinary in type and conformation.

plane. Planetree.

plane blight. A common fungous disease of plane trees in the U.S. caused by *Gnomonia veneta,* family Gnomoniaceae, characterized by the sudden browning and death of single leaves or of clusters of leaves in their spring-time expansion. Brown, dead areas appear along and between the veins of other leaves. The leaf falls prematurely when several lesions develop or when the leaf stem is infected.

planetree. Any tree of the genus Platanus, family Platanaceae. Species are valuable as street and shade trees. See *London planetree; American planetree; oriental planetree.*

planetree maple. *Acer pseudoplatanus,* family Aceraceae, a hardy, deciduous tree of different varieties grown for shade. Native to Europe and western Asia. Also called *sycamore maple.*

planimentric map. A base map of a specific area showing drainage courses and major topographic features.

planker. A drag made of planks, such as 2 x 8's or 2 x 10's fastened together. This type crushes clods but does not compact the soil appreciably.

plankton. Minute or microscopic flora and fauna that float or drift in the waters of seas, lakes, etc. *Plankton* includes protozoans, worms, coelenterates, lower chordates, and algae, and serve as an important source of food for many crustaceans, mollusks, small fish, etc. Some fish: *e.g.,* herring and mackerel, feed mainly on plankton their entire lives. Research is presently being carried on in Japan and the U.S.S.R. to determine whether *plankton* could be processed into foodstuff for man and/or animals. Fresh-water *plankton* is sometimes called *limnoplankton;* salt-water (or marine) *plankton, haliplankton.*

Planosols. An intrazonal group of soils with strongly eluviated surface horizons underlain by a B horizon more strongly illuviated, cemented, or compacted than associated normal soils. It developed upon nearly flat upland surfaces under grass or forest vegetation in a humid or subhumid climate.

plant. 1. An organism distinguished from the animals in that it takes nutrients entirely in liquid solution, rather than in solid form. (No completely non-controversial definition has yet been contrived.) 2. To set in the soil young *plants,* as seedling trees in forest plantings, cabbage *plants* in gardening. 3. To sow seed. 4. To set out roots, etc.

plantain. Various plants of the genus Plantago, family Plantaginaceae. Some species, as buckhorn, are common weeds in fields and lawns. *P. fastigiata,* desert Indianwheat, is a forage plant of some value on ranges in southwestern U.S. Also called *airplane plant; bracted plantain; tain.*

plantain-lily. Any perennial herb of the genus Hosta, family Liliaceae. Species are grown in flower gardens. See *Fortunes plantain-lily; wavyleaf plantain-lily.*

plantain leopardbane. *Doronicum plantagineum,* family Compositae, a perennial herb, grown in the flower border and garden for its yellow flowers. Native to Europe.

plantar cushion. A pad of fibro-fatty tissue in the foot to the rear of and under the navicular and coffin bones of animals. In horses it is pronounced and overlies the frog. Also called *digital cushion.*

plant association. A unit of natural vegetation essentially uniform in habitat, aspect, general appearance, ecological structure, and floristic composition; a plant community. An area may be dominated by some one of several species with minor species associated, such as a sugar maple-beech association or a bunchgrass association.

plantation. 1. A large-scale agricultural unit especially one devoted to the production of cotton and sugar. In pre-Civil war times in the U.S., a large manorial estate on which cotton, tobacco, and other crops were produced with slave labor. 2. An artificially reforested area established by planting or by direct seeding.

plantation clarified. Designating the grade of cane sugar centrifugally separated from the molasses.

plantation gait. A natural, horse gait, somewhat faster than a walk, which is easy on the horse and rider. Horses with this gait are sometimes called plantation horses, as they are popular on the plantations in the southern U.S.

plantation gin. A small cotton gin operated for individual plantation use as contrasted to one for custom ginning.

plantation horse. A horse that has a plantation gait (*q.v.*), as the Tennessee Walking Horse.

plant band. Short strip of heavy paper that may be folded or rolled into a potlike unit to replace clay pots for growing young plants, which are later transplanted direct to the field without the removal of the *plant band.* See *peat pot.*

plant bed. A specially prepared bed in which seeds are planted, the young plants being later transplanted into field areas: *e.g.,* tobacco and cabbage *plant beds.*

plant breeding. The improvement of plants by selection, inbreeding, hybridization, and other scientific procedures.

plant buffer. Plant tissues that tend to neutralize acid sprays used as herbicides. See *translocated herbicide.*

plant bug. Any insect that infests plants. See *bug: 1.*

plant cancer. See *crown gall.*

plant cell. See *cell: 2.*

plant certification. Certification by a public or private agency that a given plant conforms to certain standards as to varietal characteristics, freedom from disease, etc.

plant community. See *plant association.*

plant consumption (of water). Water used by plants in the process of growth which includes water stored in the plant body and that dissipated from its leaf and body surfaces by transpiration.

plant cover. 1. The vegetal mantle covering the soil. 2. All plants found on a particular range regardless of their availability, palatability, toxicity, or other characteristics.

plant density. Percentage of total ground area covered by vegetation.

plant disease. A condition which affects physiologic activities of a plant so as to check normal development, lead to abnormal formations or to premature death of all or a part of a plant.

plant doctor. A plant pathologist; a person skilled in the treatment of plant diseases.

plant-drawn. Designating poultry eviscerated in a dressing plant instead of a retail shop.

plant ecology. The science which deals with plants in relation to their environment. See *crop ecology; ecology.*

planter. 1. A mechanical device used for the rapid, efficient, and uniform planting of seeds. *Planters* are of many different kinds, from the simplest hand *planter* to the large highly mechanized, multi-row, power-driven machines. 2. A farmer, especially a cotton farmer. (In its earliest use, it generally implied wealth, and denoted the owner of a large cotton or tobacco plantation.) 3. A person who plants. 4. A container, usually rectangular in shape, containing soil or vermiculite, which is used for growing foliage plants indoors. It may be made of plastic, pottery, wood, metal, or other material and may be movable or a permanent part of the building.

Plant-Exploration Service. A very important activity of the New Crops Branch, Agricultural Research Service, U.S. Department of Agriculture. Plant explorers are sent all over the world to collect plants, seeds, or plant materials of desired species for breeding and development of new crops for agricultural and other uses.

plant explorer. An agent of the Plant-Exploration Service (*q.v.*), U.S. Department of Agriculture.

plant fertilizer. See *fertilizer: 1.*

plant flavor. An off-flavor, usually objectionable, imparted to milk by certain plants when eaten by the cow.

plant food. The organic compounds elaborated within the plant which nourish its cells. Sometimes the term is used loosely as an equivalent of plant nutrient (*q.v.*) or the mineral elements absorbed by plant roots and essential for plant growth.

plant forcer. See *forcing box.*

plant formation. A group of plants which live together under the same environment. The formation is delimited by climate and may include two or more associations. See *association: 1.*

plant growth regulator. See *growth regulator; hormone.*

plant hoppers. Any of a number of leaping insects with piercing-sucking mouths which belong to the Order Homoptera.

plant hybrid. A plant which results from cross pollination of two genetically different plants. See *hybrid.*

plant immunity. Resistance of a plant to a pathogen so that a disease does not cause appreciable damage.

plant indicator. An individual species of plant, or a community, which by its presence indicates a condition of soil, depth of moisture, climate, or other characteristics of the site: *e.g.*, a plant may indicate excessive alkali, fire, or overgrazing. See *virus indexing; virus indicator plant.*

planting. 1. An area seeded to a crop. 2. The sum of plants in a seeded area: *e.g.*, a strawberry *planting.* See *plant: 2.*

planting bar. A hand tool used in making a slit-like hole in the soil in which trees are planted.

planting board. 1. In planting orchard trees, a device used to gauge properly the exact location of each tree and its depth of planting with reference to the soil-surface level. 2. A board with evenly spaced holes which is used in space planting of seeds.

planting borer. A device used to remove a core of soil and make a hole for planting extensively rooted seedlings or transplants.

planting hoe. In forestry, a digging tool which consists of a rectangular steel blade with a pointed cutting edge firmly bolted to a standard adz handle.

planting machine. A mechanical device which opens a hole or furrow for a tree seedling, closes it again and firms the soil about the plant.

planting mattock. A specially constructed hand hoe used for tree planting.

planting plow. A special, heavy-duty machine, designed to plow furrows in reforestation planting.

planting rod. A hand tool which is used to make cylindrical holes in soil in which cuttings are planted.

planting season. The period of the year in which farm crops, trees, shrubs, or herbs may best be planted for successful growth.

planting stick. See *dibble: a.*

planting stock. Seedlings of which the kind, age, certification, source of seed, and other necessary and useful information is known.

planting stock age. The age of the planting stock from seed with distinction between periods spent in seedling and transplant beds, as 2-0 for two-year seedlings and 2-2 for stock two years in seed bed and two years in transplant bed.

planting time. Planting season.

planting wire. 1. A wire which is useful in setting out orchard trees at definite uniform intervals in rectangular blocks. The wire with solder or other visible attachments at definite intervals is drawn taut between sighted-in stakes and the trees are planted at the point of the attachment. 2. A wire with regularly spaced knots which is used in checkrowing corn and other crops at planting time. See *check wire.*

plant kingdom. One of the two great categories of living things, the other being animal kingdom.

plantlet. An embryonic or very young plant.

plant lice. See *aphid.*
plant life. Plant organisms of any kind from lowest to highest forms, as distinguished from animal organisms.
plant louse. See *aphid.*
plant nutrient. Any element or group of elements which is taken in by a plant from the air and soil. These are essential to growth and are used by the plant in the elaboration of its food and tissue.
plant parasite. A parasite (*q.v.*) which commonly affects or is peculiar to plants, as distinguished from animal parasites.
plant patent. A patent which is granted by the U.S. Patent Office to originators of varieties, strains, or some variation from existing varieties of asexually reproduced plants.
plant pathology. That branch of botanical science which deals with the diseases and disorders of plants.
plant physiology. The science which deals with the response of a plant to its environment including moisture, temperature, light, and nutrients.
plant propagation. The practice of producing new plants which may be by asexual means, such as division, graftage, cuttage, layerage, etc., or by the use of seeds.
plant protector. Any device used in protecting plants from nuisances such as rabbits and birds, from excessive sunlight, winter injury, etc.
plant quarantine. The isolation of or restriction placed on the movement of plant materials to prevent either the plant or plant-carried insects and diseases from contaminating an agricultural area which is free from them. The Plant Inspection Branch of the Agricultural Research Service in the U.S. Department of Agriculture processes all plant materials from foreign sources.
plant residue. Plant material, such as leaves, roots, straw, corn stover, grass, and weeds which is left after harvesting; any plant material remaining after any harvesting or other process.
plant selection. See *plant breeding.*
plant setter. A tool used for setting out young plants.
plant succession. The replacement of one community of plants by another which results in successive occupation of an area. It involves interrelated change in soil and other changes in factors of the environment.
plant symbol. An abbreviation used to designate a given plant species.
plant tissue. Material which consists of cells of similar origin and character. Tissues in turn form systems and organs of the plant.
plant-tissue tests. A test, usually rapid and not precisely quantitative, using various chemical reagents which are made upon tissue and sap of growing plants to determine the adequacy of plant nutrients. They are similar to and used in conjunction with and compared to such tests made upon the soil itself.
plant-to-row testing. The progeny of a single plant which is planted in a row the following year for more adequate testing.
plantule. A tiny seedling.
plant virus. See *virus.*
plashing. An English method of making or repairing a hedge by bending down shoots to the ground, severing them nearly in half with a knife to make them more pliable, and then interlacing them among the upright stems of the hedge plant. Also called *pleaching.*
plasma. 1. The liquid portion of blood or lymph. 2. The liquid which may be squeezed from muscle.
plasmolysis. A separation of the protoplasm from the cell wall by osmotic action when the cell is in contact with the liquids of stronger salt content. Wilting and in extreme cases death of a plant results.
plaster. See *calcium sulfate.*
plastic cream. A high-fat cream (above 50%, usually beyond 70%) which results from double or special separation. *Plastic cream* is a means of storing sweet cream in the frozen state.
plastic curd. A curd which is heat treated after whey removal to facilitate working and molding.
plastic (soil). A soil which can be readily molded or deformed without rupture when the content of water is optimum.
plate. 1. A cut or section of beef which is taken from the lower section of the front quarter just back of the brisket section. 2. In cotton marketing, a mass of inferior cotton which is concealed in the bale by higher grade lint on the exterior. 3. A flat, metal disk on a mechanical planter, containing holes at intervals through which seed falls or is pushed when planting. 4. A type of horseshoe.
plate budding. A method employed in the propagation of plants in which a rectangular incision of the bark is made by two vertical cuts and one cut across the top, and a rectangle of bark containing the bud to be united is cut to fit the face exposed on the stock. The flap of bark on the stock is then brought up over the bud and tied. *Cf. H-budding; patch budding; shield budding.*
plate count of milk. A method for the bacterial count of milk made by placing a measured sample on a sterilized media in a petri plate. After incubation, the colonies developing from each individual and/or group of bacteria in the sample may be counted.
plated bale. A bale of cotton composed of good cotton on the exterior and decidedly inferior cotton in the interior, concealed in such a manner as not to be detected by customary examination.
plate method. Plate count of milk.
plate roll. 1. A retail cut of beef taken from the plate section of the forequarter, boned, and made into a roll suitable for pot roasting or stewing. 2. The breast meat of a sheep carcass made into a roll for roasting or stewing.
plate-type pasteurizer. A milk pasteurizer made of a large number of plates clamped together with proper passageways for milk and water so that heating and holding, regenerating and cooling can be accomplished in one small apparatus.

platform cutter. A sled type of cutter, usually homemade, for harvesting corn as a silage crop, or for shocking. It is pulled by one horse or mule and has the advantage of cheapness of operation as compared with the corn binder (obsolete).

platform harvester. A type of corn harvester mounted on a wheeled or sled platform.

platform planter. A potato planter with a dropping mechanism which consists of a device to elevate the seed piece from the hopper to a horizontally revolving platform where the piece is carried to the opening over the seed spout through which it falls into the furrow.

platform scale. A weighing machine with a flat deck on which the objects to be weighed are placed.

platter pasture. See *soiling.*

platy. Designating laminated soil aggregates predominantly developed along the horizontal axes.

playa. A flat basin or sump area on the floor of a desert valley in the western U.S. The sediments of the *playa* left by flooding are generally fine or clayey, highly charged with salts or alkalis, and such areas are nearly devoid of vegetation.

play flight. The first flights of young bees. See *orientation flight.*

pleach. A method of pruning trees to produce a wall-like hedge.

pleasing in the head and neck. In stock judging, denoting a good type and conformation about the head and neck, an important aspect of desirable conformation in livestock.

pleasing type. In stock judging, an animal of good breed type and conformation.

pleasure horse. A classification for horses which includes those used for riding, driving, or racing. See *work horse; stock horse.*

Pleospora rot. A rot caused by the fungus, *Pleospora herbarium,* family Sphaeriaceae, which produces a firm rot either at wounds or at the stem end of citrus fruits.

plerocercoid. A worm-like larval stage of certain tapeworms.

pleura. The membrane that infolds the lungs, lines the thorax, or chest cavity, and is reflected upon the diaphragm.

pleurisy root. Butterfly milkweed.

plot. In agricultural research, a small parcel of land, usually rectangular in shape and of a definite size, used in comparing yields of crop varieties, testing different applications of fertilizers, comparing methods of tillage, etc. Also called *plat.* See *census plot; check plot.*

plottage. The value increase resulting from combining small plots of land to form larger ones.

ploughboy. A boy who leads the horse, or other animal, pulling a plow guided and manipulated by an older person also on foot (obsolete).

plow. 1. The whole implement, of various types, used to cut, break, or turn a soil layer in preparation for planting, seeding or other agricultural practices. More specifically, the removable metal point, share and moldboard or disk, attached to a *plow* stock or frame. 2. Any such implement not used primarily for agricultural purposes, such as a snow *plow.* 3. To make a furrow or to turn over a layer of soil. 4. To cultivate: *e.g., plow* corn. (This usage refers to that part of the cultivation process in which weeds are removed from the crop area by plowing, and does not mean plowing under the corn to destroy it.) Also spelled *plough.*

plowable. Designating land suitable, in its present condition, for plowing.

plowable pasture. Land ordinarily kept as pasture but which may be plowed and utilized for other crops.

plow beam. The wooden or metal part of a plow which extends out in front of the plow bottom and to which the pulling power is attached.

plow bottom. The real unit or base of the moldboard plow, composed of those parts necessary for the rigid structure to lift, turn, and invert the soil.

plow broadcast. To flat-break land.

plow down. To bury material lying on the surface of a field by plowing: *e.g.,* to plow in a surface application of fertilizer, or to plow under a cover crop.

plow draft. The amount of force required to pull a plow bottom through a soil under specified conditions. The draft may be measured by a dynamometer *q.v.*

plow horse. A work horse that responds readily to commands, has sturdiness and other qualifications for pulling a plow, especially a one-horse plow.

plowland. 1. Land that is regularly plowed for crop use. 2. Land that is capable of being plowed and tilled without excessive cost or labor.

plow layer. The depth: *e.g.,* 5 to 8 ins., to which soil is ordinarily broken or penetrated by a plow; the surface layer or top soil of cultivated land.

plowman. One who plows land or operates a plow. Also called *plowjogger.*

plow out. To lift out or remove, as to destroy young corn plants or to remove stalks of harvested corn in plowing operations.

plow pan. A compacted layer which is formed in the soil immediately below plow depth. It is attributed to the sliding action and weight pressure of the plow bottom. Also called *plow sole.*

plow planting. The plowing and planting of land in a single operation.

plow point. The hard steel point of the plowshare of the moldboard plow. The point is the first part of the plow to enter the soil in plowing. See *plow tip.*

plowshare. The cutting edge of a moldboard plow.

plow sole. Plow pan.

plowstaff. 1. A spade-like device used in cleaning the plowshare. 2. The handle of a plow.

plow stock. The beam, handle, and shank of a walking plow.

plowtail. The plow handle. Also called *plowstaff; plow tree.*

plow tip. The tip end of a plowshare. It is subject to the most severe wear and is replaceable by welding on a new tip in place of the worn tip on the old share. See *plow point.*

plow tree. Plowtail.

plow truck. A seat attached to a pair of wheels following behind the plow on which the farmer may ride while plowing.

plow under. To bury and incorporate in the soil a cover of living or dead plants by the use of the plow. See *plow down.*

plow up. 1. To cut a furrow slice and partly or completely to invert it. 2. To lift out of the ground unwanted plants more with the purpose of destroying than preserving. 3. To bring to the surface vegetation that was turned under during a previous plowing.

pluck. 1. In hog and cattle slaughtering, the organs that lie in the thoracic cavity consisting of the heart, lungs, gullet, and windpipe. 2. In poultry slaughter, to remove the feathers.

plucked wool. Pulled wool.

plucker. 1. A machine which, by means of spiked rollers, untangles long-fibered wool prior to the combing process. 2. In the meat-packing industry, one who removes the wool from dead sheep or feathers from poultry.

plug. 1. The mass removed by a trier or other special penetrating implement in sampling or testing an agricultural product, as a *plug* from a cheese, a bale of cotton, or from a watermelon. 2. An old, worn-out horse. 3. A horse with a poor conformation. 4. To repair a leak, as a dam or earth fill.

plug cover. A cover for a milk can which is pushed down inside of the rim of the can rather than fitted to the outside of the rim.

plum. 1. The tree and fruit of certain species of genus Prunus, family Rosaceae. Varieties of domesticated species are widely grown for their edible fruits, fresh and cured (as prunes); some species are grown for ornament; and some native wild species, such as the beach plum, *Prunus maritima,* which grows in dune areas of the east coast of U.S., especially Cape Cod, where the fruit is very popular for use in preserves. Four main groups are under cultivation; Domesticas or European plums; Damsons; Japanese; and varieties of native American species. See *American plum; Canada plum; Chickasaw plum; garden plum; Japanese plum; myrobalan plum.* 2. The fruit of species of several other plants. See *seagrape sapodilla.*

plum Armillaria root rot. See *mushroom root rot.*

plum bacterial canker. 1. A disease of plums and peaches more commonly known for leaf and fruit spot lesions caused by the bacterium, *Xanthomonas pruni,* family Pseudomonadaceae. This organism from overwintering cankers on twigs causes serious injury to the fruit and foliage. 2. A bacterial disease of plums and cherries, common in England, resulting in cankers on stems, trunks or branches, also in shoot wilt and leaf spot associated with two species of bacteria: *Pseudomonas mors-prunorum* and *P. prunicola.* A similar disease is common in western U.S. caused by *P. syringae.*

plumbago. Any plant of the genus Plumbago, family Plumbaginaceae. See *cape plumbago; climbing plumbago.*

plumbed. In stock judging, designating well-placed feet and legs on a horse.

plum blotch. A fungous disease of plum in which tiny, gray or brown, angular spots appear on the leaves. The fruit is blotched with brown or gray spots. Caused by *Phyllosticta congesta.*

plumcot. A hybrid of no commercial importance, originated by Luther Burbank by crossing a plum and an apricot.

plum curculio. *Conotrachelus nenuphar* (Hbst.), family Curculionidae, a snout beetle that infests plums, apples, peach, cherry, and apricots. The beetles first feed on leaves and petals and later feed upon, puncture, and lay eggs in the fruits making them grubby, misshapen, and unmarketable. Found east of the Rocky Mountains, U.S. Also called *plum weevil.*

plum Dematophora root rot. See *Dematophora root rot.*

plume clematis. *Clematis flammula,* family Ranunculaceae, a woody vine grown in warm areas on a trellis, fence, etc., for its fragrant, white flowers. Native to Europe and Asia.

plumed chrysanthemum. A novelty type of chrysanthemum (*q.v.*) that is grown by florists.

plume grass. Any grass of the tropical genus Erianthus, family Gramineae. Species are robust, tall-growing and are sometimes used as an ornamental garden plant, especially Ravennagrass, *E. ravennae,* which has a very showy, silky, flowering plume. Native to Europe and Asia. Also called *erianthus.*

plum gouger. *Anthonomus scutellaris* Lee., family Curculionidae, a snout beetle that feeds on the wild plum, but may infest cultivated plums and other fruits. Its larvae bore into the seed of the fruit to pupate.

plum leafhopper. *Macropsis trimaculata* (Fitch), family Cicadellidae, an insect vector of the virus that causes peach yellows and little peach disease, affecting plums, peaches, and other fruit trees.

plum line pattern virus. A mild reacting virus prevalent in plums and other stone fruits which is possibly the same as infectious chlorosis of apple. Also called *banded chlorosis.*

plumosa. The finely-cut, plumose leaves of the fern asparagus (*q.v.*), grown by florists and used mainly as greens in flower arrangements.

plump. 1. In poultry slaughter, to create an effect of plumpness by dipping the picked carcass in scalding water. It shrinks the skin and draws up the legs, wings, and neck closer to the body thus imparting a false plumpness. 2. In meat judging, designating a well-muscled carcass or wholesale cut.

plump grain. A quality designation for small grains: wheat, barley and oats, which denotes fullness of grain, better quality, and greater weight per bushel, as contrasted with a shrivelled grain.

plump leg but lacking in firmness. In stock judging, designating the hind leg of a sheep that is good-sized and well-fleshed but which has muscles that are soft and flabby.

plum pockets. A fungous disease of plums caused by *Taphrina pruni* and *T. communis,* family Taphrinaceae, which affects both wild and cultivated plums, causing the fruit to become greatly inflated and hollow. Also called *bladder plum; fools; pocket bladder.*

plum pox. A virus disease of plums which poxes and discolors the fruit and causes it to drop 30 to 40 days before ripening.

plum prune diamond canker. A disease caused by a virus, possibly the same as plum rough bark virus, which results in diamond shaped cankers on French prunes.

plum prune dwarf. A stunted growth of plum trees as a result of a virus which causes rough, narrow, thickened leaves, with symptoms most evident on Italian prune. The virus has several strains, which react variably on other Prunus hosts.

plump wheat. Wheat kernels of good size and well filled out, which are of heavy flour producing type.

plumule. In a germinating seed plant, the primary bud which develops into the primary stem. In a dycotyledon it is located between the cotyledons.

plum wart. See *black knot.*

plum-web-spinning sawfly. *Neurotoma inconspicua* (Nort.), family Pamphiliidae, an insect pest of the plum mainly in the Dakotas, U.S. Its larvae work in colonies, webbing the ends of branches together and sometimes defoliating trees.

plum white spot. A chlorotic leaf condition caused by a virus found on Santa Rosa plums. Of minor importance, the virus is not clearly distinguished from other plum mosaic viruses.

plumy coconut. Queenpalm.

plunge. To bury a plant-containing pot in soil. It is a common practice in floriculture as by this operation the plant can have some of the advantages of growing free in soil and can be lifted without repotting.

plurennial. A plant that can live and grow for many years. *Cf. annual (plant); biennial; perennial.*

Plymouth Rock. A popular breed of dual-purpose chickens of the American class, whose origin is attributed to early crosses of the Dominique and the Black variety of Java. They are characterized by a single comb, non-feathered shanks, yellow skin, and compact body. Cocks weigh about 9½ lbs., hens about 7½ lbs. Their egg-shell color is brown. The varieties are: Barred, White, Buff, Silver-Penciled, Partridge, Columbian, Blue. The White and the Barred varieties are used extensively for commercial egg and meat production. In earlier times, the Barred *Plymouth Rock* was commonly called *Dominicker* by farmers because of its similarity to the earlier Dominique breed.

PMS. See *pregnant mare serum.*

pneumatic chain saw. A light, portable saw designed to work at high speed in felling or cutting of trees. The power is commonly supplied by a light-weight gasoline engine. See *chain saw.*

pneumatic sprayer. A manual, compressed air sprayer in common use by gardeners for controlling insects and plant diseases. It is of a size that can be easily carried on the shoulder.

pneumoencephalitis. Newcastle disease.

pneumonia. Inflammation of the lungs in men and animals, characterized by fast and labored breathing, elevated body temperature, varying amounts of nasal discharge, and many times by a degree of consolidation of the lungs. *Pneumonia* is caused by various agents, such as bacteria, viruses, parasites, fungi, hot or cold air, liquids which might be taken into the lungs, and by dust.

pocket. 1. A relatively small basin-depression in a land surface. 2. A short cove-type valley or hollow in a mountainous terrain. Often used in connection with water drainage and air drainage (*q.v.*) of land.

pocket bladder. Plum pockets.

pocket gopher. A small, stout-bodied, short-legged rodent of the family Geomyidae, which builds nests underground in fields and orchards. Its burrows may constitute a nuisance as it has been known to girdle citrus trees below the ground surface, commit minor damage to corn, and destroy roots and bulbs in gardens. Commonly found in U.S. and Central America. Also called *ponched rat.* See *salamander: 1.*

pocket planting. Placing plants in separate holes of such size, depth, and spacing as may be required by each particular plant.

pocket rot. One of the several kinds of wood rots produced by fungi of the class Basidiomycetes.

pocosin. A minor, natural land feature of the flat Atlantic coastal plain of the eastern U.S. It is a swamp occupying shallow basins and is thus properly differentiated from creek swamps. (The term is largely restricted to the Carolinas.) Also called *pocoson.*

pod. 1. Technically, a dry, many-seeded fruit that splits open, such as a pea *pod* or bean *pod;* a legume or a silique. 2. A flock of animals, birds, etc. 3. To form pods.

pod and stem blight of soybeans. A fungous disease of soybeans caused by *Diaporthe phaseolorum* var. *sojae,* family Valsaceae, characterized by dark brown lesions on the stems and pods.

pod blight of Lima bean. A blight of Lima beans caused by the fungus, *Diaporthe phaseolorum,* family Valsaceae, characterized by brown patches on the leaves and older pods. Also called *pod spot.*

pod borer. The larva of the insect *Etiella zinckenella* (Treit.), family Phycitidae, which occasionally infests Lima beans in California (U.S.). Also called *Lima bean borer; Lima bean pod borer.*

pod canker. Bean anthracnose.

pod corn. *Zea mays* var. *tunicata,* family Gramineae, a curious variety of corn in which each kernel is enclosed in a pod or husk whose ear is also enclosed in husks. It is sometimes grown as a curiosity, and some mutants, which have striped and purple leaves, are cultivated as ornamental plants.

pod mottle. A virus disease of the kidney bean whose symptoms are similar to those of common and yellow mosaic. See *bean mosaic.*

pod spot. Pod blight of Lima bean.

podware. 1. Leguminous plants which bear pods. 2. Any agricultural product, as silage (obsolete).

Podzol. A zonal group of soils which has an organic mat and a very thin organic mineral layer above a gray or ashy, highly-leached layer resting upon an illuvial, dark brown or yellow horizon. The soils develop under coniferous or mixed forest, or under heath vegetation, in a temperate to cold, moist climate. The B horizon, often characterized by a very dark color from organic compounds, may be slightly to firmly cemented.

podzolic soil. A soil formed wholly or partly under the influence of the podzolization process.

podzolization. The natural process by which soils become depleted of bases, become acid, and develop eluvial (A) horizons and illuvial (B) horizons. Specifically, the process by which a Podzol is developed, including the more rapid removal of iron and alumina than silica, from the surface horizons. (The term is also used to include similar processes operative in the formation of other soils of humid regions.)

poets narcissus. *Narcissus poeticus,* family Amaryllidaceae, a bulbous herb grown for its very fragrant, white flowers with wavy, red margins. Native to southern Europe.

poic. Of, or pertaining to, grassland.

poinsettia. *Euphorbia pulcherrima,* family Euphorbiaceae, a shrub grown as a Christmas pot plant for its showy, usually vermilion bracts. Native to Mexico.

point. In the market, 1/10 of 1 cent; 1/1000 of a dollar.

pointing. An indication of pain in the foot of a horse in which a forefoot is rested forward to secure relief.

point men. Cowboys who ride toward the head of the herd to keep it going in the desired direction (western U.S.).

point-observation transect. A method developed in New Zealand for measuring the amount and quality of herbage produced by individual species of a plant community. A frame is used which consists of a wooden board having 10 holes bored 2 ins. apart. Through each hole is dropped an 18 in. long steel pin. The frame is held so that the points of the pins touch the ground and as each or any of the pins touch a plant, that plant is recorded.

point of buttock. The extreme rear end of an animal.

point of hock. The extreme back and top side of the hock on the hind leg of an animal.

point of shoulder. The point of the forequarter on an animal where the shoulder blade and arm of the leg join.

point row. A row which terminates without extending the full length of a field. When a field is irregular in shape or terraced, plowed and cultivated on the contour, some rows will necessarily be short. These *point rows* may terminate in a point in the terrace channel, at the base of the terrace ridge, or in between the ridges.

points. A condition of coat color in animals in which white or a lighter color appears about the muzzle, eyes, feet and tail.

poison. Any substance which causes lethal or injurious effects on plant or animal tissues.

poison ash. Poison sumac.

poison bait. A poison mixed with wheat bran, molasses, or other attractant used to control cutworms, grasshoppers, and other insects.

poison berry. Black nightshade.

poison chickweed. Scarlet pimpernel.

poison creeper. Poison ivy.

poison dogwood. Poison sumac.

poison haw. Kentucky viburnum.

poison hemlock. *Conium maculatum,* family Umbelliferae, a biennial, rank-growing herb, which is an escape and a weed in various sections of the U.S. It is dangerously poisonous to man and animals, being most poisonous to stock in the spring, when the herbage is fresh. Symptoms of poisoning of cattle are loss of appetite, salivation, bloating pain, feeble but rapid pulse, and loss of muscular power. Native to Europe and Asia. Also called *deadly hemlock; poison parsley; winter fern.*

poison ivy. *Toxicodendron radicans,* family Anacardiaceae, a small, erect shrub or climbing vine common throughout the U.S. in waste places, pastures, woodlands, and along old fence rows. Apparently more dangerous to man than to animals, contact with the plant among susceptible individuals causes severe blistering and inflammation, especially on the hands, arms, and face. Also called *climbing ivy; three-leaved ivy; climath; poisonoak; poison creeper; picry; Mercury.*

poison laurel. Mountainlaurel kalmia.

poison milkweed. *Asclepias galioides,* family Asclepiadaceae, a poisonous, perennial herb found on dry sites on the western ranges in the U.S. It is occasionally eaten by animals with fatal results. Sheep appear to be most susceptible to it. The symptoms of poisoning are loss of muscular control, staggering, falling, spasms, bloating, fever, weak but rapid pulse, and respiratory paralysis. The affected animal usually dies. Native to North America.

poisonoak. See *Pacific poisonoak; poison ivy.*

poisonous. Containing poison, as a *poisonous* plant.

poison parsley. Poison hemlock.

poison ryegrass. Darnel ryegrass.

poison soil. A soil that contains substances toxic to plant growth: *e.g.,* an excess of alkali (sodium carbonate), boron, sulfuric acid, etc.

poison suckleya. *Suckleya suckleyana,* family Chenopodiaceae, an annual herb which grows along the margins of water holes in the dry prairies of western U.S. Losses of cows and nursing calves are believed to have been caused by a poison from this plant. Native to North America.

poison sumac. *Toxicodendron vernix,* family Anacardiaceae, a shrub or small tree bearing greenish-white flowers, contact with which causes a severe blistering and inflammation to the skin of susceptible persons. Native to North America. Also called *poison ash; poison dogwood.*

poison vetch. Timber poisonvetch.

poison wood. Florida poisontree.

poke. 1. A yoke with an attached sharp spur pointing forward which is placed on the neck of an animal to prevent it from jumping a fence. 2. Pokeberry.

pokeberry. a. Any plant of the genus Phytolacca, family Phytolaccaceae. Also called *pokeweed.* b. *P. americana,* a tall, stout, perennial weed which has purplish or greenish flowers in racemes and black or dark wine colored fruit. Its young asparagus-like shoots are sometimes eaten as greens; the roots and seeds are poisonous. Native to North America. Also called *foxglove; pokeweed; common pokeberry; poke; scoke; garget; pigeon berry.*

pokeweed. Pokeberry.

pokkah-bong. A fungous disease of sugar cane caused by the fungi *Gibberella fujikuroi* and var. *subghetinans, G. moniliforme,* or its imperfect stage, *Fusarium moniliforme* and var. *subglutinans,* family Tuberculariaceae. It is characterized by light-colored, twisted leaves emerging from the terminal leaf whorl, and a rot of the top of the stalk.

Poland China. A large, formerly lard type, breed of swine which originated in Ohio, U.S., early in the 19th century, as the result of crossing the native swine with several breeds including Big China, Berkshire, and Irish Crozier swine. The coat color is black with white tips on the tail, white feet, and a white dash or spot on the forehead.

polar bodies. In female animals, the smaller nuclei produced at meiosis not developing into egg nuclei.

polder. A mud flat reclaimed from the sea for agricultural use through the construction of dikes and pump drainage of water (Holland).

pole. 1. A young tree 4 ins. or more in diameter at breast height. The maximum size of *poles* is usually, though not invariably, between 8 and 12 ins. See *sapling.* 2. A round timber used to support telephone and power lines. 3. A linear measurement of land, by statute, 16½ ft. In England, besides the statute pole, at one time there were other poles of 18 to 25 ft. Also called *perch; rod.*

pole ax. An ax head with a cutting edge on one end and squared off on the other with a point or claw bending downward, or projecting from the back of the head, which is fixed to a pole or handle. This type differs from the doublebit ax which has two cutting edges.

pole barn. A form of barn construction in which natural poles are used in the framing instead of the regular sawed lumber or dimension material generally used in such framing. See *pole frame.*

pole bean. Any variety of kidney beans (*q.v.*) or Lima beans (*q.v.*) that grows as a climbing vine and is supported in the garden on poles, strings, wires, etc.

pole fence. A fence similar to a zigzag rail fence but made of poles instead of split rails.

pole frame. The frame of a barn or building made from long poles of suitable size and length, set in the ground and joined together into a suitable frame with trusses or framing ties. *Cf. balloon frame.*

pole hook. In carriage and wagon making, the hook on the end of the tongue.

pole pruner. A tree pruning implement; a pole saw or clipper used for removing small limbs beyond the reach of a hand saw or hand shears.

pole rot (burn). A rot or burn of tobacco leaves, noticeable in the curing shed, which is possibly started in the field by the fungi *Sclerotinia sclerotiorum,* family Sclerotiniaceae, and *Botrytis cinerea,* family Moniliaceae, but is later overrun by secondary saprophytic fungi. Also called *pole sweat.* See *air cure; flue cure.*

pole saw. Pole pruner.

pole thinning. The removal of a portion of the fruits, as peaches, from a tree after the June drop by hitting scaffold branches with a length of garden hose which may or may not be attached to a pole.

pole timber stand. A stand of trees in which the bulk of the trees are of a diameter and height suitable for making poles.

Polish. A very old, light weight, handsome breed of chickens which originated in eastern Europe where its ancestry has been traced as far back as the 16th century. The most striking characteristic of the breed is the very large, profuse crest which adorns the head. The *Polish* is among the most ornamental and beautiful breeds of poultry and is highly prized for exhibition. There are a number of varieties, such as White Crested Black, Non-Bearded Golden, Non-Bearded Silver, Non-Bearded White, Non-Bearded Buff Faced, Bearded Golden, Bearded Silver, Bearded White, and Bearded Buff Faced.

Polish wheat. *Triticum polonicum,* family Gramineae, an annual herbaceous cereal, characterized by very long heads, often 6 or 7 ins. long and large, very hard, yellowish-white kernels. Its grain is used for macaroni manufacture but is not suitable for breadmaking. Not grown extensively in the U.S. Native to Mediterranean region and western Asia.

poll. 1. To cut back the crown of a tree. 2. To remove the horns of cattle.

pollard. 1. A tree whose crown has been cut back to invite the production of shoots from the top. Sometimes a tree is so pruned to induce a globe-like mass of foliage. 2. A hornless ox, sheep, or goat. 3. Wheat bran (English, obsolete).

polled. Designating animals, especially cattle, that normally do not develop horns. This is a character which can be produced in a pure form.

Polled Durham cattle. The original (U.S.) name for Polled Shorthorn cattle. The term was officially changed in 1919 by the Shorthorn Breeders Assn.

Polled Hereford. A strain of the Hereford breed of beef cattle which was developed in the U.S.; characterized by the absence of horns.

Polled Shorthorn. A strain of beef cattle within the Shorthorn breed, characterized by the absence of horns. See *Shorthorn.*

pollen. The male element which carries the spores in the fertilization of the egg nucleus in the ovule of a flower. The *pollen* is borne by the anthers and is usually a yellowish, dust-like mass of separate grains.

pollen basket. A flattened depression surrounded by curved spines located on the tibiae of a bee's hind legs which is adapted for carrying the pollen gathered from flowers to the hive.

pollen cake. See *pollen supplement.*

pollenizer. See *pollinizer.*

pollen packer. A house bee that manipulates and packs the pollen into cells delivered by the fielder bee.

pollen sterile. A variety, strain, or species of a plant whose pistils are fertile but the stamens sterile, so that inbreeding does not occur. The necessary pollen is secured from the desired male parent that produces viable pollen.

pollen supplement. A cake consisting of pollen, soybean flour in sirup, with or without other additives, fed to honey bees in early spring to stimulate brood rearing.

pollen trap. A grid placed at the entrance to a beehive, which removes pollen from the bees' legs as they enter; the pollen falls into a tray below the grid.

poll evil. A fistula of the poll between the ears which may follow a severe injury to that part, a common affliction of horses and mules.

pollinate. To bring about pollination *q.v.*

pollination. The transfer of the pollen from the anther to the stigma of a flower, the first step in producing a fruit or seed. See *cross pollination.*

pollinator. 1. Any of several different kinds of technical apparatus which may be used to collect pollen from one plant and transfer it to the flower of another plant for subsequent fertilization. 2. One who aids in the pollination of plants.

pollinizer. 1. An insect which carries pollen from flower to flower and thereby brings about the fertilization necessary for the production of the fruit or seed of the plant. 2. A plant (especially a tree) in a fruit planting which provides pollen for self-unfruitful varieties.

pollution. The contamination of clean surface or ground water, air, or land areas with harmful material, such as sewage, rubbish, gases from industrial plants, various chemicals, etc.

polo class. A show ring classification for small riding horses that have been trained in polo performance.

polo pony. Any small active horse with the ability to be trained to perform in polo.

polyacrolyte. A synthetic chemical used as a soil conditioner *q.v.*

polyantha rose. *Rosa polyantha,* family Rosaceae, a hybrid rose which is a low shrub with many horticultural forms grown as ornamentals.

polyanthus narcissus. *Narcissus tazetta,* family Amaryllidaceae, a bulbous herb grown for its fragrant, white flowers and as the source of an essential oil. Native to Europe, Asia, and the Canary Islands.

polybrid. A plant hybrid which originated from a cross between two particular species, varieties, or genera.

polychlor. Any of the group of the chlorinated class of insecticides, such as DDT.

polyestrous animal. An animal that experiences estrous cycles throughout the year, as cattle.

polygala. Any plant of the genus Polygala, family Polygalaceae. Species are annual and perennial herbs, subshrubs and treelike plants which are grown as ornamentals. Also called *milkwort.* See *fringed polygala; orange polygala; Seneca-snakeroot polygala.*

polygamous. Having both perfect and imperfect flowers borne on the same individual plant.

polygonal soil. A soil which is characterized by irregular (polygonal) prisms formed by intensive shrinkage of dispersed homogenous soil material. (This phenomenon may be observed both in torrid deserts and in the Arctic regions.)

polyhedral disease. A disease caused by a microorganism which attacks the gypsy moth and cabbage looper larvae and results in their decimation. However, man has not yet been able to spread the infection artificially for biologic control.

polymorphonuclear leucocytes. White blood cells in which the nuclei are constricted into irregular shapes.

polymyxin. Any of various antibiotics which are derived from cultures of the organism *Bacillus polymyxa.*

polyneuritis. An inflammation of many nerves at once.

polyp. A smooth, stalked, or projecting growth from a mucous membrane.

polyphagous parasite. A parasite which is capable of parasitizing a considerable number of host species.

polyphemus moth. *Antheraea polyphemus* (Cram.), family Saturniidae, one of the giant silkworm moths whose larvae feed on the foliage of a number of trees and shrubs and make silky cocoons. It is rarely injurious.

polyploid. An organism with more than two sets of the basic or haploid number of chromosomes: *e.g.,* triploid, tetraploid, pentaploid, hexaploid, heptaploid, octoploid.

polypody. 1. Any fern of the genus Polypodium, family Polypodiaceae. Species are grown as ornamentals. See *golden polypody; resurrection-fern.* 2. *P. vulgare,* a creeping fern grown in wild gardens as an ornamental. Native to Europe and Asia. Also called *common polypody; male polypody.*

polypody brake. Sensitivefern.

polysaccharides. A class of carbohydrates. Noncrystalline, tasteless solids, insoluble in water, they decompose on hydrolysis into two or more simple sugars or monosaccharides.

polyvalent colon bacterin. A bacterin made from several types of bacteria which inhabit the colon or intestinal tract of animals; sometimes useful in the prevention of calf scours.

polyvoltine. Designating an animal yielding several broods in a season. Especially applied to certain silkworms producing several broods of cocoons in a year. See *voltine.*

pomace. 1. The crushed apple pulp left after the extraction of juice in cider making. 2. Residue of fruit and plant parts after the expression of juices.

pomace fly. *Drosophila* spp., family Drosophilidae, an insect that infests bruised or decaying fruit. Also called *sour gnat.*

pomato. A horticultural curiosity; a chimera which originated by grafting a tomato cion on a potato stock. The fruit is fragrant, juicy, and tomato-like. Also called *potomato; topato.*

pome. A fleshy fruit having several seeds instead of a stone, as apple, pear, quince.

pomegranate. *Punica granatum,* family Punicaceae, a shrub or small tree grown in warm areas for its edible, red, orange, or yellow fruit about the size of an orange with astringent rind and juicy, slightly acid flesh. Native to Asia, it is occasionally grown outdoors in Florida, Georgia, and other parts of the U.S. for its fruit or ornamental value. Some purely decorative forms are grown in greenhouses. The juice of the fruit is used in making grenadine sirup.

pomegranate melon. See *dudaim melon; muskmelon.*

pomelo. Grapefruit.

pomes. 1. Lumps of fat on a mule (colloquial, auction). 2. See *pome.*

pomiculture. Fruit culture.

pomme cythere. Mombin.

pommel. 1. The knob, ball, or protuberant part serving as a means of grip on anything, as the high, forward part of a saddle. 2. A block of hard wood grooved like a crimping board and employed by curriers to render leather supple and impart a grain to it.

Pommel. A small, soft cheese made in France.

pomme-liane. Waterlemon.

pommy, Pommy. An Australian term for an immigrant from Britain. Of obscure origin, it is thought to have been introduced at the beginning of this century, probably by Australian soldiers returning from the Boer War (1899-1902). Popularly thought to be a corruption of pomegranate because of the frequently ruddy complexion of the British.

pomology. The art and science of growing and handling fruits, especially tree fruits.

Pomona. The Roman goddess of tree fruits.

pompelmous. Grapefruit.

pompin. Pumpkin (obsolete).

pompon. Any small, dense, globe-shaped flower.

pompon chrysanthemums. Any chrysanthemum (*q.v.*) having small, button-like flowers in clusters.

pompon dahlia. A type of dahlia with the same characteristics as the ball dahlia (*q.v.*) except that its flower never exceeds 2 ins. in diameter. See *miniature dahlia.*

pond. A small, sometimes stagnant, body of water. It may be a natural feature, or man-made, as a farm *pond* made by impounding runoff water by a dam or by excavating. (In some parts of the U.S., lake and *pond* are used for natural water bodies without any consistent distinction.)

pond baldcypress. *Taxodium ascendens,* family Pinaceae, a deciduous, coniferous tree grown for its valuable lumber used in shingles, buckets, window boxes, coffins, outdoor seats, etc., and sometimes as an ornamental. Native to southern U.S.

ponderosa pine. *Pinus ponderosa,* family Pinaceae, an evergreen tree, large or small depending upon the site, and valued chiefly for its timber. Native to British Columbia, Canada, and western U.S. Also called *western yellow pine; Arizona pine; bull pine. Cf. Arizona longleaf pine.*

pond lily. See *waterlily: 1.*

pond management. The adoption of suitable measures for the protection of ponds from siltation and for the production of fish and other wildlife.

pondweeds. A number of species of aquatic plants: *Potamogeton* spp., *Naias* spp., *Zannichellia* spp., *Ceratophyllum* spp., *Elodea* spp., etc., which may infest ponds, lakes (valued for recreation), millponds, and drainage ditches.

P_1. In genetics, the parental generation; parents used in a fruit cross.

pone. A handshaped, oval cake of corn bread.

Pont l'Eveque. A soft, cows' milk cheese, similar to Romadour and manufactured at Pont l'Eveque in Normandy, France.

pony. 1. Any small horse of nondescript breeding. 2. Any of certain recognized breeds of small horses, as the Shetland and Hackney pony *q.q.v.*

pony band mill. A sawmill where sawn logs or boards are resawn to different dimensions.

pony crate. A standard cantaloup crate with the inside dimensions of 11 x 11 x 22⅛ ins. See *standard crate.*

pony flat. A standard cantaloup crate with the inside dimensions of 4 x 12 x 22⅛ ins. See *standard crate.*

pony grass. *Calamagrostis neglecta,* family Gramineae, a perennial grass that is valuable forage in the Rocky Mountains, U.S. Native to the North Temperate Zone. Also called *slimstem reedgrass.*

pony harness class. In show ring classification of ponies, the class suitable for harness and carriage purposes.

pony-penning. The round-up of Chincoteague (*q.v.*) ponies.

pool. An arrangement by individuals or groups to assemble their separately owned products, such as wool or cotton, together for mutual advantage at a sale.

pool auction. A cooperative method of marketing in which individually owned products are pooled and sold to the highest bidder.

poolroot. Wild hoarhound.

Poona. A cows' whole milk, surface-ripened, soft cheese supposed to have originated in New York state (U.S.), which has a mild Limburger aroma. It is put up as a round, flat cake about 4 ins. in diameter and 1¾ ins. thick.

poonga oil pongamia. *Pongamia pinnata,* family Leguminosae, a woody plant grown frequently for rotenone (*q.v.*) derived from its bark and roots. Native to Asia. Also called *derris.*

poophyte. A meadow plant which grows under mesic or intermediate moisture conditions.

poor drainage. A natural condition of standing water on the surface of a saturated soil. (The term usually denotes that an excess of water is present for most desired agricultural uses of the land; for farm ponds the same drainage may be quite satisfactory.)

poor feeder. Any animal that does not fatten or grow in a satisfactory manner.

poor in uniformity. 1. An animal that is of undesirable conformation or type and appearance. 2. A quantity of fruit that lacks uniformity in size, color, or quality.

poor killer. An animal, such as a beef steer, that will dress out poorly or show a low percentage dressed weight in comparison to live weight.

poor land. Any land which is inferior in quality for some particular use, as for agriculture. (Commonly the term is loosely used: land may be poor either because of some unfavorable relief, or because of some physical or chemical condition of the soil. Often land and soil are used interchangeably.)

poorly bled. In poultry slaughter, designating a bird which shows red pin marks on its breast or thighs, or one whose skin is reddened from blood clots.

poorly fleshed. 1. In poultry slaughter, designating a bird which has a narrow breast and whose thighs and back are dark-colored from absence of fat. 2. Designating any meat animal in thin condition.

poor-man's-alfalfa. Lespedeza (*q.v.*) so named because of its capacity to make a fairly satisfactory growth on land unsuitable for alfalfa.

poor-man's-manure. Snow, which in some localities may be beneficial on plowed land, although it may not add appreciable amounts of the chemical elements which aid fertility.

poorman's pepper. Field pepperweed.

poorman's-weatherglass. Scarlet pimpernel.

poor-robin's-hawkweed. *Hieracium venosum,* family Compositae, a perennial herb which is a weed pest of dry woods and open sandy places generally throughout the U.S. Also called *early hawkweed; rattlesnake weed.*

poor-robin's-plantain. 1. *Erigeron pulchellus,* family Compositae, an annual herb grown for its showy, violet or violet-purple flowers. Native to North America. 2. See *poor-robin's-hawkweed.*

poor soil. A soil unsuitable for some particular plant growth because of deficiency in fertility or some unfavorable physical condition. A *poor soil,* however, may be valueless for one plant but quite suitable for the growth of another.

poor stand. A seeding of any crop which germinates poorly and produces an uneven stand of young plants or seedlings.

poor tilth. Poor physical condition of soil: hard or cloddy; usually restricted to top soil.

popcorn. A type of corn whose dried kernels explode into large, fluffy masses when heated. Most *popcorns* have small, hard, flinty kernels and small cobs. In some localities it is an important cash crop.

poplar. Any tree of the genus Populus, family Salicaceae. Species are mostly quick-growing, soft-wooded trees grown as ornamentals, for pulpwood, and other wood uses. See *balm-of-Gilead; Berlin poplar; black poplar; Carolina poplar; European poplar; large-toothed aspen; Lombardy poplar; quaking aspen; southern poplar; white poplar.*

poplar and willow canker. A fungous disease of poplar and willow caused by *Cytospora chrysosperma,* family Sphaeropsidaceae, characterized by girdling of the twigs, limbs, or trunk. There is a profusion of reddish-yellow sport horns from the cankers following rains.

poplar anthracnose. A disease caused by the fungus *Drepanopeziza populi-albae,* family Mollisiaceae, whose imperfect stage is *Marssonina castagnei,* family Melanconiaceae, (sometimes called *Trochile populorum*). It is characterized by lesions on the twigs and leaves.

poplar borer. *Saperda calcarata* Say, family Cerambycidae, an insect, the larva of which causes black and swollen scars on the limbs and trunks and sawdust at the base of poplars. Damage by this insect frequently makes it impossible to grow certain species of poplar in some localities.

poplar leaf-folding sawfly. *Nematus bozemani* (Cooley), family Tenthredinidae, a fly whose larvae eat holes through the leaves of poplar.

poplar-petiole gall aphid. *Pemphigus populi-transversus* Riley, family Aphidae, an insect which infests the roots of cabbages, turnips, and other crucifers in the Gulf States (U.S.). It winters on the roots of these plants and migrates in the summer to the poplars.

poplar rust. A fungous disease of poplars in eastern U.S. caused by *Melampsora* spp., family Melampsoraceae, characterized by yellowish-orange pustules on the lower leaf surfaces. It rarely causes serious damage.

poplar tent maker. *Ichthyura inclusa* Hbn., family Notodontidae, an insect, the larva of which chews the leaves of the poplars and makes silken nests on the twigs.

poplar yellow-leaf blister. A fungous disease of poplar caused by *Taphrina aurea,* family Taphrinaceae, characterized by yellow to brown blisters on the leaves.

poplar vagabond aphid. *Mordwilkoja vagabunda* (Walsh), family Aphidae, an insect that causes the peculiar convoluted galls seen at the tops of twigs.

popple. See *quaking aspen; large-toothed aspen.*

poppy. Any of various plants of the genus Papaver, family Papaveraceae. Species are herbs a few inches to as much as 4 ft. high mostly of Asiatic and European origin. Grown mainly as ornamentals in gardens for their bright red and other varicolored flowers. See *alpine poppy; corn poppy; Iceland poppy; Olympic poppy; opium poppy; Oriental poppy.*

poppy anemone. *Anemone coronaria,* family Ranunculaceae, a perennial herb grown indoors by florists and outdoors in warm areas for its red, blue, or white flowers. Native to Europe.

poppy-seed oil meal. A meal used as a minor stock feed. It consists of the residue left after extraction of oil from poppy seed.

population. In statistics, the sum of all units of individuals or organisms of any one kind.

population density. The number of units, persons, families, dwellings, etc., per given area. Rural density is usually expressed on a square mile basis.

population pressure. In wild life, the force of increasing numbers which results in dispersion and endangers perpetuation of a species on a particular range.

Populist movement. A political party, also designated as the People's Party, which was organized in the U.S. in 1891. It existed as a party only for a few years, ceasing after 1908, but many proposals which it supported were eventually enacted into law. Its membership came principally from west of the Mississippi River and included the support of agricultural and labor groups.

porcelain flower. Waxplant.

porch box. A wooden, concrete, or ceramic, rectangular, decorative container, placed on porches, stoops, etc., in which ornamental plants are grown.

porcine. Pertaining to swine.

porcupine-grafting. See *frameworking.*

porcupine grass. *Stipa spartea,* family Gramineae, a tall perennial grass, whose sharp awns cause severe injury to mouths and tongues of grazing animals. Native to the U.S.

pore. 1. In plant and animal membranous tissues, minute openings for absorption and transpiration of matter. 2. In wood anatomy, the cross section of a vessel or a vascular trachea. See *vessel: 2.*

pore spaces. The portion of a given volume of soil which is unfilled with solid matter: air spaces, irregular in shape and size.

pork. The meat of swine.

pork bird. A special type of pork chop made by cutting a pocket on the inside of the chop for stuffing.

porker. Any young hog.

pork goodie. A cured and smoked boneless Boston style butt of a pork loin. Also called *cottage roll.*

pork loin. A retail cut of pork made fairly close to the eye muscle from either side of the spinal column from the third rib down to the hip bone. *Pork loin,* as sold in retail, usually does not have more than ½ in. of fat back left covering the outside of the loin.

pork roast. A retail cut of pork for roasting made by removing the ribs and backbones from two blade end cuts and placing the inside cut surfaces together.

pork sausage. Any ground and seasoned pork product.

pork tenderloin. A retail cut of pork from the last rib to the hip joint which is the psoas major muscle located below the transverse spinal process in the lumbar region of the swine carcass.

porosity. The total pore space in a soil, expressed in terms of the ratio of void volume to total volume in percentages.

porous wood. Wood with pores or trachea (vessels) found in nearly all broadleaf species and a few gymnosperms.

porridge. Oatmeal.

portable bin. A bin (*q.v.*), usually made of sheet metal, with a removable top and bottom so it can be taken apart and moved from one place to another. A more temporary type of *portable bin* made of slatted snow fence is used largely for storage of ear corn.

portable mill. A small sawmill that can be readily moved from place to place.

portable vat. A portable dipping vat, frequently owned by a group of sheep owners, which can be moved from one location to another.

port bit. One of the many bits used with horse bridles. It is a light, smooth iron rod with a curved center and is used with a soft curb strap.

Port du Salut. A cured, semisoft, cows' milk, surface ripening cheese first made by the Trappist monks in the Department of Mayenne, France. Later it has been made under different names in many other parts of the world.

porterhouse steak. A retail cut of beef taken from the posterior end of the beef short loin containing a large section or area of the tenderloin muscle.

Porters aster. *Aster porteri,* family Compositae, a perennial herb grown for its numerous, small, white flowers. Native to North America.

Portiatree. *Thespesia populnea,* family Malvaceae, a tropical tree bearing bell-shaped, yellow flowers grown as an ornamental and as a shade tree. A dye is obtained from its flowers and its seeds yield a commercial oil. Native to Asia. Also called *bendy tree; seaside mahoe.*

Porton method of spraying. An airplane method of spraying for insect and plant disease control, developed at Porton Experiment Station, England, which employs wide, horizontal booms by which a swath of spray as much as 200 yards wide may be obtained.

Port Orford cedar. Lawson falsecypress.

Port Orford cedar root rot. A fungous disease of Lawson falsecypress in coastal Oregon, U.S., caused by *Phytophthora lateralis,* family Pythiaceae, which was known to cause only a serious root rot. Recently, however, it has spread epiphytotically so as to cause aerial infections, such as stem cankers and downy mildew of the foliage, sometimes sufficient to destroy the entire crown of the tree.

Portugal diamondflower. *Ionopsidium acaule,* family Cruciferae, a very small, herbaceous annual grown in the rock garden and in the greenhouse for its very small, violet or white flowers. Native to Portugal.

Portugal orange. See *sweet orange.* (The sweet oranges of Europe were believed to be derived from a tree imported into Portugal by voyagers, following Vasco da Gama, in the late 15th or early 16th century.)

Portuguese broom. *Cytisus albus,* family Leguminosae, a low shrub grown in warm areas for its profuse, yellowish-white flowers. Native to southeastern Europe. Also called *white broom.*

Portuguese cypress. Mexican cypress.

Portuguese laurelcherry. *Prunus lusitanica,* family Rosaceae, an evergreen shrub or small tree grown for its lustrous foliage and small white flowers. Native to Spain and Portugal. Also called *cherry laurel.*

portulaca. 1. Any annual or perennial herb of the genus Portulaca, family Portulacaceae. 2. *P. grandiflora,* an annual herb grown for its showy, white, pink, yellow, red, or purple flowers. Native to Brazil. Also called *rose moss; garden portulaca; sun moss; wax pink.* 3. See *purslane.*

pose. In the show ring, a special stance or position which a horse may be trained to assume. A posed horse stands with his front feet extended and his rear feet back.

position effect. The change in the phenotype of an individual, due to a change in the position of a gene or group of genes on the chromosomes.

positive ion. A cation; an ion which carries a positive charge of electricity.

post. 1. A short timber used in an upright position for supporting structures. 2. Any timber which supports fencing. It may be round, split, or sawn. (*Posts* may be of other materials, such as iron and concrete.) 3. The proper position and balance a rider assumes in the saddle in riding a horse while in the trot gait, rising with each second beat of the one-two rhythm of the trot.

post cenam. After feeding (in scientific writing often abbreviated: *p.c.*).

postclimacteric. Of, or pertaining to, the postclimacteric stage *q.v.*

post-climacteric stage. The stage of the respiration curve of mature fruit during which the respiration rate is decreasing after the climacteric peak. See *climacteric peak; preclimacteric stage.*

post-climax vegetation. A relic of a former community of plants which remains under favorable soil conditions.

post-emergence damping-off. See *damping-off.*

post-emergence treatment. Any treatment of plants made after they emerge from the soil.

posterior paralysis in pig. A disease characterized by lameness and sometimes complete paralysis of the rear part of the body. Causes are a lack of lime (calcium) in the diet, or an unbalanced ratio of calcium to phosphorus, especially if there is a deficiency in vitamin D. Other mineral and vitamin deficiencies may be involved.

post frame. A pole frame (*q.v.*) except that the posts which support the frame are set on a sill or foundation rather than being set in the ground. *Cf. balloon frame.*

post hole auger. A cylindrical auger or stirrup-shaped implement furnished either with a crown or a disk cutter, used for boring holes in the earth for the insertion of fence posts. It may be hand or power operated.

post hole digger. An implement for digging circular post holes which consists of two semi-cylindrical cutting blades, hinged together but each mounted on separate handles. The soil is cut or loosened by plunging the blades downward and removed by moving the handles outward and lifting.

post-legged. Designating a horse with too much set in the hocks, crooked hocks, or the other extreme of having hind legs that are too straight.

post-mortem. An examination of an animal carcass or human body after death.

post oak. *Quercus stellata,* family Fagaceae, a small to medium (rarely large) sized, deciduous tree, whose wood is suitable for posts, ties, tight cooperage, veneer, fuel, flooring, planing-mill products, etc. (Where abundant or characteristic in the cover, its name was applied to various kinds of land such as *post oak* flatwoods, prairies, glade, ridges, for the most part inferior agricultural land.) Native to central and southern U.S. Also called *brash oak.*

post oak grape. Pinewoods grape.

post oak hybrid grape. Any of several hybrid varieties of grape having the pinewoods grape (*q.v.*) as one parent. Some typical varieties are American, Bailey, and Beacon.

pot. 1. To place plants in various kinds of *pots* for growing indoors. 2. To grow seedlings in paper pots or other containers for transplanting without disturbing roots. 3. See *flower pot.*

potagerie. A kitchen garden in which various kinds of vegetables are collectively grown.

potash. Loosely, potassium and various compounds of potassium. *Potash* of chemical fertilizers is expressed as K_2O (potassium oxide) in the formula; the caustic potash, or lye, is potassium hydroxide (KOH). Old names for potash are *potassa; kali.* (*Potash* was made by early settlers in the forested regions of the U.S. by leaching the ashes of burned woods such as oak, beech, and maple. The liquid so obtained was evaporated in iron pots to a thick consistency and became black *potash,* an article of commerce and a source of supplementary income for the early settler. The black *potash* was further refined by heating, whitening and was known as *pearlash.*)

potash hunger. An abnormal condition of crop

plants due to a deficiency of available potassium in the soil. Symptoms may appear as: dwarfed growth of tubers or fleshy roots; weak development of terminal shoots of woody plants; yellow (chlorotic), brownish, or whitish spots on leaves; blight of foliage. Evidence of *potash hunger* has been observed especially on tobacco, potato, and cotton (rust hunger). See *rim-fire; dieback.*

potassium. K, the chemical element, an alkali metal which occurs widely in minerals: *e.g.*, in the orthoclase feldspars of granites and in salt deposits as the chlorides and sulfates. Regarded as an essential plant nutrient, *potassium* is present naturally in some form in all soils but in extremely variable amounts, and is likely to be in largest amounts in clay soils and in least amounts in highly siliceous soils and in peats.

potassium aluminum sulfate. $KAl(SO_4)_2 \cdot 12H_2O$, an alum sometimes used for lowering the pH of soils. Also called *potash alum.*

potassium ammonium nitrate. A combination fertilizer which is potassium chloride, KCl, and ammonium nitrate, NH_4NO_3. Its nitrogen content is about 16 percent and the potash (K_2O) about 27 percent.

potassium ammonium selenosulfide. A specific poison for red spider mites.

potassium antimonyl tartrate. $(KSbOC_4H_4O_6)_2 \cdot H_2O$, a water-soluble salt used medicinally as an ant poison and, in combination with sugar, as a stomach poison for various species of thrips and fruit flies. Also called *tartar emetic.*

potassium carbonate. K_2CO_3, a white, soluble salt obtained from wood ashes, sugar residue, cottonseed hulls, or other plant materials, and from potassium sulfate. It is used to some extent as a source of potash in chemical fertilizers.

potassium casse. A defect of wine caused by an excess of potash.

potassium chloride. KCl, the salt most commonly used as a source of potash in fertilizers. Its commercial form is the muriate of potash *q.v.*

potassium deficiency. See *potash hunger.*

potassium fixation. The process of converting exchangeable or water-soluble potassium to moderately available potassium, a form not easily exchanged from the soil complex, with a cation of a neutral salt solution.

potassium iodide. KI, a white, water-soluble chemical which may be given to farm animals as an iodine supplement.

potassium metaphosphate. KPO_3, a chemical containing 40 percent of potash (K_2O), which is used as a fertilizer. Commercial grades contain about 37 percent potash and 55 percent phosphoric acid.

potassium nitrate. See *nitrate of potash.*

potassium permanganate. $KMNO_4$, a powerful oxidizing compound used as a disinfectant, deodorant, and a reagent in analytical work, especially in the determination of available nitrogen in organic material. Sometimes also used as a fungicide in greenhouses.

potassium phosphate. Any of the various phosphates of potassium, especially potassium metaphosphate (*q.v.*), used as fertilizers.

potassium sulfate. K_2SO_4, one of the carriers of potash used in the manufacture of mixed fertilizers. The salt occurs in nature, but the commercial product is manufactured by treating potassium chloride with sulfate of magnesia or sulfuric acid. The product contains about 60 percent potash.

potassium sulfide. K_2S, a chemical compound used in sprays for red spider mite control.

potato. *Solanum tuberosum,* family Solanaceae, a perennial herb grown as an annual for its edible tuber, the *potato* of commerce. The *potato* is indigenous to South America and is believed to have been brought to England in 1586. It first became an important food crop in Ireland and was reintroduced into America from that country, hence the name *Irish potato.* It is now extensively grown throughout the world and is one of the important agricultural crops. Also called *earth apple; spud; white potato.*

potato aphid. *Macrosiphum solanifolii* (Ashm.), family Aphidae, a green or pinkish-winged or wingless aphid which infests the potato, tomato, eggplant, pepper and many other plants. It may occur in epidemic numbers and do tremendous damage as a sucking insect and as a vector of virus diseases. Also called *pink and green aphid.*

potato bacterial ring rot. See *potato ring rot.*

potatobean. American potatobean.

potato beetle. See *Colorado potato beetle.*

potato black leg. A rot of potatoes caused by the bacterium, *Erwinia atroseptica,* family Enterobacteriaceae, which invades the tops and tubers of the plant and produces black lesions at the base of the stem.

potato blight. See *potato early blight; potato late blight.*

Potato cheese. A cheese made in the state of Thuringia in Germany from cows', ewes', or goats' milk. Boiled potatoes are mixed with the curd before it is set.

potato curl. See *potato Verticillium wilt.*

potato digger. A machine which has a shovel-like bottom which runs beneath the potato plants in the field, lifts the tubers and plants out of the soil, and slides them onto the elevator where most of the clinging soil is shaken off. The potatoes are separated from the vine and windrowed on the ground, in separate rows from the vines, or elevated into trucks.

potato disease. See *potato late blight.*

potato distillers' dried residue. A livestock feedstuff which consists of the dried product obtained after the manufacture of alcohol and distilled liquors from potatoes or from mixtures in which potatoes predominate.

potato early blight. A fungous disease caused by *Alternaria solani* (Ell. & G. Martin) Sor., family Dematiaceae, developing ordinarily in mature plants, producing target-like lesions on leaves and stems and firm dark brown lesions on tubers. Tuber infection seldom causes decay and breakdown in storage. *Cf. potato late blight.*

potato flea beetle. *Epitrix cucumeris* (Harr.), family Chrysomelidae, an insect whose larvae feed on the roots and tubers of the potato, causing scurfy or pimply potatoes. It infests numerous other plants also and is one of the most destructive garden pests.

potato flour. A flour made from cooked and dried potatoes or from sliced raw potatoes, sulfured, dehydrated, ground, and bolted. Used in breads, it compares favorably with wheat flour in food value, but is lower in protein.

potato Fusarium tuber rot. A fungous rot of potato caused by species of the genus Fusarium, family Tuberculariaceae. The infection may enter through wounds in harvesting and thus cause very serious losses later when potatoes are in storage. See *Fusarium.*

potato hook. A four-pronged tool somewhat like a hoe which is used in hand digging of potatoes.

potato hopper burn. See *hopper burn.*

potato internal black spot. A blackening of internal tissues of the tuber which may follow bruising in handling.

potato late blight. A serious disease of the potato caused by the fungus, *Phytophthora infestans,* family Pythiaceae, which is characterized by a moldlike growth affecting the plant tissue and killing the tops. The fungus may also infect the tuber causing a rapid breakdown. *Cf. potato early blight.*

potato latent mosaic. A potato mosaic which is attributed to virus X. The virus is latent in a great many varieties of potatoes and may cause little or no damage. A very severe necrosis may be caused depending upon the particular strain of the virus.

potato leafhopper. *Empoasca fabae* (Harr.), family Cicadellidae, one of the most injurious insect pests of potatoes in the eastern half of the U.S. It also infests alfalfa, other legumes, and deciduous nursery stock.

potato leaf roll. One of the several degenerative or virus diseases of the potato characterized by a pronounced rolling of the leaves and reduction in size of plants and tubers. Apparently it is caused by more than one race of viruses. Aphids are vectors. Also called *potato phloem necrosis virus; potato virus I; Solanum virus 14; Corium solani.*

potato-leaved tomato. Any of several varieties of the tomato which has leaves resembling those of potato of which Red Jacket is the most common. See *large-leaved tomato.*

potato meal. An animal feed which is made from dried cull potatoes.

potato mosaic. Any mosaic disease of potato caused by any of several distinct viruses or virus complexes. Three important mosaics, latent mosaic, virus A, and vein banding mosaic are transmitted mainly by aphids.

potato onion. Multiplier onion.

potato pit. A shallow excavation in which potatoes, covered with straw and earth, are stored through the winter.

potato planter. One of several types of mechanical planters that opens a furrow and drops and spaces the seed pieces of the potato. See *picker: 3; platform planter.*

potato psyllid. *Paratrioza cockerelli* (Sulc.), family Psyllidae, an insect whose nymph causes the disease known as psyllid yellows (*q.v.*). Present from the Great Plains west to the Pacific Coast of the U.S. Also called *tomato psyllid.*

potato Rhizoctonia scurf. See *Rhizoctonia disease of potatoes.*

potato ring rot. A bacterial disease of potato caused by *Corynebacterium sepedonicum,* family Corynebacteriaceae, characterized by a wilt of the foliage and a rot of the vascular tissue of the tubers.

potato rugose mosaic. A composite infection of potatoes which is caused by both virus X and the vein banding mosaic virus. It may be more severe than either one of the above two virus infections. It is characterized by rugosity or wrinkled surface of the leaf, by necrotic spots and streaks, and by the dropping of the leaves in the later stages.

potato scab. See *corky scab of potato.*

potato scab gnat. *Pnyxia scabiei* (Hopk.), family Fungivoridae, an insect, the larvae of which feed on the surface tissue of the tuber in the ground and continue to feed upon stored potatoes, causing a scabby condition.

potato seed. Seed from the flower of the potato, as distinguished from a seed potato which is the tuber or part of a tuber normally used for planting and reproduction.

potato sickness. Golden-nematode disease of potato.

potato spindle tuber. A virus disease of the potato whose symptoms are an upward roll of the leaves and elongated tubers with numerous and conspicuous eyes. Also called *marginal leaf roll.*

potato stalk borer. *Trichobaris trinotata* (Say), family Curculionidae, a snout beetle whose grubs hollow out the stems of potato causing wilting and death. Occasionally it destroys entire fields of potatoes.

potato tuber necrosis. A relatively new disease of potatoes which results from infection by the same virus that causes alfalfa mosaic. The necrosis starts at the stem end of the tuber and thence may extend into all tissues.

potato tuber rot. A rot of potatoes caused by the fungus, *Pythium debaryanum,* family Pythiaceae, which secretes a toxin that kills the cells.

potato tuberworm. *Gnorimoschema operculella* (Zell.), family Gelechiidae, the larva of a moth which burrows into the potato tuber, often riddling it with slender, dirty-looking, silk-lined burrows. The larvae also infest tobacco leaves.

potato vein banding mosaic. A mosaic, caused by virus Y, which is distinguished by rugosity or wrinkled surfaces of the leaves in early stages, and necrotic spots and streaks in later stages. The disease is responsible for serious reductions in yields of potatoes. Also called *potato virus Y; Solanum virus 2; Marmor upsilon.*

potato Verticillium wilt. A fungous disease of potatoes caused by *Verticillium albo-atrum,* family

Moniliaceae, characterized by wilting of the tops and a vascular discoloration of the stems, tubers, and roots. It may affect one or all of the stems in a hill, often causing death of the plant.

potato vine. Jasmine nightshade.

potato virus A. An aphid-transmitted, mosaic disease of potato characterized by light green, rugose leaves similar to those induced by virus X.

potato virus X. The virus which causes potato latent mosaic *q.v.*

potato virus Y. The virus which causes potato vein banding mosaic *q.v.*

potato wart. A fungous disease of potatoes caused by *Synchytrium endobioticum,* family Synchytriaceae, characterized by abnormal wartlike protuberances in the eyes. Also called *black wart; cauliflower disease; potato canker.*

potato wilt. See *potato Verticillium wilt.*

potato witches'-broom. A virus disease of potato characterized by a more than normal upright, bushy clustering of stems, a very marked reduction in size, and great increase in numbers of tubers.

potato yellow dwarf. A virus disease of the potato transmitted by the common clover leaf hopper and characterized by a yellowing of the foliage, a decided dwarfing of the plant, and the production of ill-shaped, cracked, necrotic, and hence almost worthless, tubers.

pot-bellied. 1. Designating any animal that has developed an abnormally large abdomen, usually because of improper feeding or nutrition. 2. Designating a type of coal stove used for heating purposes in pioneer stores and homes, so named because of its large, round, lower half.

pot binding. The growing of a plant in a pot until its root system has become densely matted; a method practiced by florists to promote flowering.

pot-bound. Designating a plant grown in a pot until the root system has become densely matted.

Pot cheese. A soft, cows' skim milk cheese high in moisture content. Somewhat like Cottage cheese (*q.v.*), it is a popular cheese of southeastern Pennsylvania, U.S.

potential. The work required to move a unit mass against a force from one specified point to another. In soil physics, both gravitational and pressure *potentials* are used in dealing with saturated soils. For unsaturated soils within the capillary fringe, capillary *potential* is the work required to move a unit mass of water from the water-surface level against capillary forces in the soil column to any point of specified distance above the water surface. It is a negative *potential* and does not include the work done against the force of gravity.

potential soil acidity. Those hydrogen ions on the negatively-charged soil particles that are not dissociated, and therefore not active in the soil solution; the amount of exchangeable hydrogen ions in a soil that can be released by cation exchange. Also called *reserve soil acidity.*

potgarden. A garden for the cultivation of potherbs.

potherb. Greens; any plant yielding foliage which is edible when cooked, such as spinach, kale, chard.

potherb butterfly. *Pieris oleracea* Harris, family Pieridae, a butterfly whose larva feeds on cabbage and related plants.

potherb jute. *Corchorus olitorius,* family Tiliaceae, an annual herb which yields the jute fiber called desi. Young shoots are eaten as a potherb. Native to India and Egypt. Also called *Jew's-mallow.*

potherb mustard. *Brassica juncea* var. *crispifolia,* family Cruciferae, an annual herb grown for its leaves cooked for food. Native to the Old World. Also called *southern curled.*

pot label. Usually a flat, pointed piece of wood inserted in the soil of a pot to identify and to give pertinent information about the plant.

pot layerage. Air layering (*q.v.*) in which two halves of flowerpots are placed around the branch to be rooted, tied together, and filled with sphagnum moss or soil.

potlicker. The liquid left in the bottom of a pot after cooking greens with a fat or meat such as hog's jowls. It is used as a food, a sop, especially with corn bread.

potmarigold calendula. *Calendula officinalis,* family Compositae, a tender, annual herb grown in the flower garden and in the greenhouse for its whitish-yellow to deep-orange flowers. Native to southern Europe. Also called *calendula.*

pot marjoram. Origanum.

pot method. The propagation of plants in pots, as of strawberries to a limited extent for summer and fall planting in northern regions.

potomato. Pomato.

potometer. An instrument to measure the rate at which a growing plant absorbs moisture.

pot plant. A plant ordinarily grown in a flower pot as an ornamental.

pot planting. The field planting of trees in paper pots or cans in which they have been grown or transplanted; more often practiced in semiarid regions as it permits trees to be planted early and to grow in undisturbed, moist pot soil until rains occur.

pot roast. Any part of a meat carcass, cut, rolled, or prepared for moist cooking; usually applied to one of the less tender sections of meat.

pot rose. Any rose grown in a pot in the greenhouse, as hybrid teas, hybrid perpetuals, baby ramblers.

potsherd. A small fragment of broken, fired clay flower pot, placed over the bottom hole in a flower pot for drainage, aeration, and to prevent soil from being washed out. Also called *crock.* See *shard.*

pot system. An early method of fumigating citrus trees with hydrocyanic acid gas generated by placing water (three parts by volume), sulfuric acid (one part by volume), and potassium cyanide (one part by weight) in an earthenware vessel beneath each tented tree.

potted plant. An individual plant grown in a pot or small receptacle, usually in a greenhouse.

Such plants are of great economic importance to florists.

potter. An individual experienced in the practice of potting plants. See *potting.*

pot test. Any nutritional and varietal test of plant growth conducted in a pot indoors or in the greenhouse, in contrast to such tests in outdoor plots.

potting. The practice of transplanting very young plants from seed frames to individual flower pots. These are selected for size and provided with soil mixture and drainage according to the needs of the particular plant.

potting bench. In a greenhouse, a bench equipped with potting mediums, pots, tools, etc., for the easy potting of young plants. See *potting.*

potting medium. Soil mixture.

potting shed. A garden shed, usually small, used for rearing potted plants which are to be transplanted, at a later stage, into the garden (British).

potting soil. See *soil mixture; standard soil mixture.*

potting stick. A wood or aluminum stick approximately 16 ins. long and tapering toward both ends; used to firm soil around the roots of plants being potted.

Potts tritonia. *Tritonia Pottsi,* family Iridaceae, a cormous herb grown for its bright yellow-red flowers. Native to South Africa.

pouce. A colloquial term for flax dust.

poudrette. Dried night soil (*q.v.*) mixed with charcoal powder, gypsum, lime or peat; used as a fertilizer. Used in European countries from early times, probably it has never been used to any extent in the U.S.

poulard. A pullet which is fattened for table use by sterilization.

poulard wheat. *Triticum turgidum,* family Gramineae, an annual herb yielding large, boat-shaped, hard, light yellowish to nearly white kernels valuable only as feed for livestock. It is characterized by large heads and nearly solid stems. Also called *branched wheat; cone wheat; Egyptian wheat; rivet wheat.*

poult. The young turkey before its sex can be distinguished. Sometimes applied to the young of other fowls.

poulterer. A poultry dealer. Usually also applied to those dealing in hares or other small game.

poultry. Any or all domesticated fowls that are raised primarily for their meat, eggs, or feathers, as chickens, turkeys, ducks, and geese.

poultry band. A strip of metal, usually aluminum, marked with numbers, etc., attached to the leg or wing of a bird for identification.

poultry bug. *Haematosiphon inodorus* (Dugès), family Cimicidae, a blood-sucking insect pest of poultry.

poultry dip. See *dip: 1.*

poultry fattener. 1. Any of several feeds containing ingredients designed especially for fattening poultry for the market. 2. A person engaged in the business of fattening poultry for the market.

poultry feed. Any feed supplied to poultry. Commonly, specially formulated mashes designed to be fed either alone or in combination with cereal grains. (However, it may include any of various other poultry feedstuffs.) See *all mash; growing mash; laying mash; mash; mash concentrate; mash-grain system; mash supplement; pelleted feed; scratch mash; starting mash.*

poultry house. Any building equipped and used for housing and handling poultry. See *breeding house; brooder house; laying house.*

poultry husbandry. The science and art of the production and distribution of poultry and poultry products, including breeding, incubation, brooding, rearing, housing, feeding, marketing, and poultry farm management.

poultry inhalant. A remedy designed to relieve coughs due to colds, and bronchial irritations in poultry. It is usually sprayed in the house while birds are on roosts.

poultry knife. A specially designed knife with a very narrow curved blade, used to kill poultry and to sever arteries for bleeding. Also called *chicken sticker.*

poultry lice. Insects of the order Mallophaga, that infest fowls. See *chicken body louse; shaft louse.*

poultryman. A trained or experienced individual who operates a poultry farm or business.

poultry manure. The mixture of feces and urine voided by birds. When freshly produced it contains about 80 percent moisture, 1 percent nitrogen, 0.8 percent phosphoric acid and 0.5 percent potash. Poultry manure decomposes with a rapid and large loss of moisture and ammonia, especially during warm weather.

poultry mite. See *chicken mite.*

poultry netting. Wire netting with hexagonal meshes, made with 1, 1½, or 2 in. mesh openings, used for fencing, constructing wire partitions, cages, trellises, etc.

poultry picker. Any of various patented machines designed to remove feathers from poultry. It may consist of a revolving drum to which many rubber fingers are attached.

poultry pinner. A specially designed, flat-bladed knife used to remove pinfeathers from fowls in poultry dressing.

poultry range. Pasture which provides green feed and exercise for poultry.

poultry seasoning. Herbs and/or other plant products: sage, marjoram, thyme, savory, peppers, nutmeg, and allspice, used as flavoring in cooking poultry.

poultry shipping feed. Any of various feed mixtures specially prepared for feeding poultry in transit.

poultry tag. Poultry band.

pounce. Dust or powder applied to the gauge rod to obtain a water-level indication on the gauge.

pound. 1. A unit of weight: 16 ozs. avoirdupois, 12 ozs. troy. The standard British unit of weight equals 7000 grains, 1/2240 long ton and 453.59 grams, the weight of 27.692 cu. ins. of water at 4°C. 2. An enclosure in which stray animals are legally confined. 3. An enclosure in which groups of animals, as flocks of sheep, may be gathered for shelter, etc. 4. An enclosure used to trap wild animals.

pounding. In the movement of a horse, a heavy contact of the foot with the ground which usually accompanies a high stride.

pound of gain. In animal feeding, the net gain of weight in pounds derived from a particular number of pounds of feedstuff fed.

pound party. A social gathering formerly common in rural areas. Those invited were supposed to donate a pound of coffee, tea, or other provisions, usually for the benefit of the preacher of the local church.

pounds of butterfat. The production of butterfat by a cow for a certain period, such as a day, week, or lactation period, usually quoted in terms of total *pounds of butterfat.*

pounds of grain per acre of pasture. The feed value of an acre of pasture expressed in terms of pounds of dry grain feed.

pounds of milk. The output of milk by a cow, by weight, for a single milking, a day, or a lactation period.

pounds of milk per acre of pasture. A measure of pasture efficiency in a dairy region in which the pounds of milk produced while cows are on pasture are calculated on an acre basis.

pourridie. Mushroom root rot.

poverty sumpweed. *Iva axillaris,* family Compositae, an herbaceous weed present both on irrigated fields and on grazing lands, which is capable of growing in fertile soil, on alkaline, and saline soil. It has a rank, unpleasant odor and is avoided by grazing animals. Native to western U.S.

powder dip. A powdered rooting stimulant in which plant cuttings are dipped.

powdered buttermilk. Dry buttermilk.

powdered culture. A dry milk culture so dried and handled as to retain viability and purity of the desired microorganism.

powdered milk. Dry whole milk. See *nonfat dry milk.*

powdered potatoes. Potatoes which are dehydrated and ground to a powder; a commercial product used chiefly in making mashed potatoes for table use. See *potato flour.*

powdered skim milk. See *nonfat dry milk.*

powdered thalia. *Thalia dealbata,* family Marantaceae, an herbaceous marsh or aquatic perennial grown in warm areas and in the cool greenhouse for its dull-violet flowers. Native to the U.S. Also called *water canna.*

powder-post. Designating a type of damage in which dry wood is infested and reduced to powder by certain types of beetles.

powder-post beetle. Any of the beetles of the families Ptinidae, Anobiidae, Bostrichidae, or Lyctidae, which breed in old, dry wood and reduce it to powder. The greatest damage is done by species of Lyctus which confine their infestation to the seasoned sapwood of a number of hardwoods, especially oak, ash, and hickory.

powdery mildew. A common disease of many plants, caused by one of the several species of the fungus family Erysiphaceae, which are obligate parasites that grow on the surface of the leaf, stem, or fruit, obtaining their sustenance by means of root-like projections that penetrate the plant cells. Infections are local rather than systemic. Little moisture is necessary for infection and the disease often prevails under relatively dry conditions. The name is descriptive of the white or gray, powdery, dusty appearance of the infected areas. The disease may be destructive or of small consequence, depending on the fungus strain and the variety of the infected plant. Sulfur sprays or dusts and resistant varieties of plants have frequently been effective where control is desired. Sometimes also called *oidium disease.*

powdery scab of potato. A fungous disease of potatoes caused by *Spongospora subterranea,* family Plasmodiophoraceae, characterized by small to medium-sized, round pimples on the tubers which turn brown and burst open at maturity, allowing the powdery mass of spores to escape. The tubers and roots may be infected. Also called *corky scab; corky end of potato.*

power. 1. An exercise of control by a person or a group in a particular social system or situation which is based upon the ability of one or more participants to govern the actions of others regardless of their acceptance of such control. Control by possession or *power* is usually characterized by: (a) area and scope of control variable, often responding to situational expediency; (b) roles and relationships, sporadic and changeable; (c) rewards and punishments within the system distributed by whim or personal preference. *Power* is frequently vested in a person or group by means of such social processes as conquest and manipulation or in response to such personal qualities as character traits and physical attributes. Whatever criteria are used, the key fact is that social relations regulated by *power* are inclined to be inflexible and fraught with conflict and tension. 2. Energy which is produced by a *power* unit, such as an electric motor, gasoline engine, etc. 3. Physical ability of an organism to act or induce action, as the *power* of capillary attraction of a soil.

power-cart duster. A power dusting unit mounted on a homemade cart or a truck, in contrast to one mounted directly on a tractor. Power for the duster mechanism is furnished by a gasoline engine.

power-centered. Designating groups or social systems where action is initiated from a central office or position. Examples of *power-centered* groups are communities dominated by single large estates or haciendas. See *power: 1.*

power-diffuse. Designating groups or social systems which have no center for initiation of an action. Many unincorporated communities of family-sized, isolated farm holdings are *power-diffuse* in that there is no legal or formal community official who can determine action for the community.

power-dropped. Designating corn planting similar to drilling in which kernels are released by a power device operated either from the tractor or by the wheels of the planter.

power dump rake. A mechanical hay rake in which the teeth are raised by means of a small foot lever which engages pawls in the main wheels and supplies the power to trip the dumping mechanism.

power duster. A field dusting machine using the conventional principles of any duster which has its power supplied by a separate source, such as a gasoline engine or power take-off (*q.v.*). *Power dusters* may be very large when used in extensive orchard or plantation operations.

power lawn mower. Any of the several types of lawn mowers operated by power, usually a gasoline engine or electric motor, in contrast to the ordinary mower operated by the ground wheels when pushed by the operator.

power lift. A mechanical device which lifts a load or adjusts a part of the equipment by means of a hydraulic lift or power supplied by a power take-off *q.v.*

power-lift plow. A tractor-drawn plow in which the plow bottoms are lifted or lowered as a unit by means of a special clutch which can be engaged or disengaged by the operator of the tractor. This type has been now largely replaced by hydraulic power lifts.

power loader. A fork, or fork and scoop combination, mounted at the front or rear of a tractor and operated by hydraulic power to load manure, bales of cotton, baled hay, pallet boxes, soil, gravel, etc.

power press. A stationary hay baler operated by an internal combustion engine in contrast to earlier mechanisms in which horses supplied the power.

power sprayer. A conventional plant sprayer equipped with a source of power, such as a gasoline engine, to operate the pump and the spray equipment.

power station. Any building containing an installation of machinery for the conversion of energy from one form into another. *Power stations* are of several general classes: (a) central stations; (b) substations; (c) isolated plants.

power subsoiler. A tractor-drawn plow which has a long, narrow shank with a wedge-shaped point for deep penetration to break up stiff clay subsoils and hardpans.

power take-off. An attachment, usually consisting of a shaft and two or more universal joints, which is used to transmit power from a tractor to an attached unit, such as a combine, hay baler, mower, etc.

power trencher. A machine used on agricultural land in digging narrow trenches in which drain tiles are laid. Usually the power is supplied by an internal combustion engine.

power work unit. See *horse work unit.*

pox. 1. A minor fungous disease of corn. See *brown spot: 1.* 2. Shallow, sunken spots on the rinds of citrus fruits which are stored under improper conditions of temperature and humidity. The spots turn brown when exposed to warmer temperatures. 3. An infectious disease of animals. See *cow-pox.*

p.p.m. An abbreviation for parts per million *q.v.*

practically free from defects. In egg grading, designating a yolk showing no germ development which may show other very slight defects.

practically normal. In egg grading, designating a shell that approximates the usual shape, is of good even texture and strength, and is free from rough areas.

practically peeled peaches. In the process of preparing dried peaches, the aggregate in which the peel of the riper fruit is removed but not that of the unripe.

prairie. 1. The extensive, nearly treeless and dominantly grass-covered plains of the midwestern U.S. which lie east of the Rocky Mountains. In a more restricted sense, the tall grasslands with blackish soils; in a more general sense the semiarid short grass plains as well. Also called *savannah; steppe.* 2. In the generally forested eastern part of the U.S., any naturally treeless area which is generally dry or naturally well-drained. 3. Wet, treeless, marshy areas. 4. Prehistoric, treeless tracts which resulted from fires or from clearing by Indians.

prairie bird's-foot trefoil. Spanish-clover deervetch.

prairie breaker. A long, low, moldboard plow with a gradual twist that completely inverted the furrow slice with a minimum of breakup and complete coverage of vegetation. It was used especially to cut and turn the tough virgin sod of the prairies of the midwest.

prairie cat's-foot. *Antennaria campestris,* family Compositae, a perennial herb which is a common weed of grasslands and prairies of midwestern U.S.

prairie clover. Any of several plants of the genus Pentalostemon, family Leguminosae, perennial herbs, some trailing, flowers are white, pink or red. Native to midwest and southwestern areas of U.S. Also called *tasselflower.*

prairie coneflower. *Lepachys columnifera,* family Compositae, a hairy, rough, perennial herb with large yellow flowers. Native to central and southern U.S.

prairie cordgrass. *Spartina pectinata,* family Gramineae, a perennial, tall, marsh grass, sometimes used for thatching. Native to North America. Also called *tall marsh grass; slough grass.*

prairie crabapple. *Malus ioensis,* family Rosaceae, a deciduous tree grown as an ornamental for its white or rose-colored flowers. Native to North America. Bechtel's crabapple is a variety of this species.

prairie dog. Any animal of the genus *Cynomys.* Species are small, stubby, burrowing rodents found in abundance on the prairies and treeless plains of midwestern and western U.S. by early settlers and explorers. They are considered a nuisance by western cattlemen largely because their burrows constitute a danger to horses and their riders. They eat green crops and some insects, such as the Mormon cricket. They live in colonies or towns and through their extensive burrowing they have been a considerable factor in modifying the soil. Native to North America. Also called *barking squirrel.*

prairie dogweed. *Dyssodia papposa,* family Compositae, an herb which has become a weed pest. Also called *fetid marigold.*

prairie flax. Lewis flax.

prairie hay. Any hay which is made from the wild grasses of prairie land.

prairie June grass. *Koeleria cristata,* family Gramineae, a tufted, perennial grass, found from Ontario to British Columbia in Canada and south to Pennsylvania, Oklahoma, and California, U.S. whose forage rating is good. Native to North America.

prairie lily. Ten petal mentzelia.

prairie lotus. Spanish-clover deervetch.

prairie pine. Spike gayfeather.

prairie rose. *Rosa setigera,* family Rosaceae, a hardy shrub, grown for its rose-pink flowers fading to whitish. Native to North America.

prairie sandreed. *Calamovilfa longifolia,* family Gramineae, a perennial, drought-enduring grass, growing from 2 to 6 ft. tall, which is found from Michigan to Colorado, U.S., and Alberta, Canada. Important for winter pasture and for hay though grazed but lightly in the summer. It is an important, sand-binding grass on dunes and sand hills. Native to North America. Also called *prairie sandgrass.*

prairie schooner. A covered wagon which was used by early settlers in their westward movement and occupation of the prairies and plains of midwestern U.S.

Prairie soils. A zonal group of soils occupying the midwestern prairie regions of the U.S. which developed under grass vegetation and in a subhumid temperate climate. These soils are related to Chernozems but do not have a concentration of calcium carbonate in the solum of profiles. (*Brunigra* and *Brunizem* have been proposed for dark-colored soils of this group.)

prairie-type combine. A type of combine (*q.v.*) or harvester-thresher for small grain which may be either drawn by a tractor or self-propelled which is adapted for level land. *Cf. hillside-type combine.*

prance. To walk with a swagger or nervous action, as a horse.

Prato. A Gouda-type cheese made in Brazil which is similar to the Cuban cheese, Patagras.

Prattigau. A cows' milk cheese which was first made in Switzerland.

prayer plant. Jequirity rosarypea.

praying mantis. Insects of the family Mantidae, which prey upon other insects and are generally beneficial to man. They wait for their prey with front legs raised like hands in prayer, and seize it when it comes within reach. See *Carolina mantis; Chinese mantis; European mantis.*

prebloom. Designating a plant in bud: clover or alfalfa cut in *prebloom* makes high quality hay.

prechilling. In nursery practice, a seed pretreatment method to overcome dormancy. Seed is placed on moist blotting paper or comparable media in standard germinators and the whole apparatus is held at near freezing temperatures for a predetermined time.

precipitated bone. A by-product of glue manufacture from bones used as a fertilizer. It contains about 40 percent of phosphoric acid.

precipitated phosphate. A by-product from the manufacture of mono-calcium phosphate for use in baking powder which is sufficiently soluble in the soil solution to be of value as a fertilizer.

precipitation. Moisture condensed from the atmosphere and deposited upon land or water surfaces, as rain, dew, frost, snow, sleet, and hail. The amount is expressed as inches, or millimeters, of depth and designated as daily, weekly, monthly, seasonal, and annual. The term is often used synonymously and erroneously with rainfall.

precipitation-effectiveness (P-E) index. The sum of 12 monthly quotients of precipitation divided by evaporation.

precipitin. An antibody developed in the blood serum of an animal that has been injected previously with a foreign protein. *Precipitins,* present in the blood serum, have the property of bringing about a visible precipitate when brought into contact with a solution of the specific protein which induced their formation.

preclimacteric. Referring to the preclimacteric stage.

preclimacteric stage. The stage of respiration curve of mature fruit during which the respiration rate is increasing to the climacteric peak. See *climacteric peak; post-climacteric stage.*

preclimax. The stage of succession of an area that has never supported climax vegetation but is in the process of succession toward climax vegetation for a particular climate. Thus, an extremely shallow soil or soil material in the true prairie association may support *preclimax* blue grama rather than climax bluestem, because soil development has not yet reached a stage at which it will support the latter. See *climax; disclimax.*

preclipping. An agronomic practice, sometimes employed in growing clovers for seed production, in which tops of young plants are clipped before flowering to retard excessive growth and to induce flowering at a time when pollinating insects are numerous.

precondensing. See *sweetened condensed (skimmed/whole) milk; plain condensed milk; plain condensed skim milk; evaporated milk; milk dryer; nonfat dry milk.*

precook. In food preservation, to heat thoroughly before canning.

precooling. 1. Preliminary cooling of milk immediately after a milking to prevent spoilage from bacterial action. 2. The cooling of vegetables and fruits immediately after harvesting during periods of hot weather to retard ripening and deterioration. 3. The cooling of meats after slaughter and before cutting.

predatism. Intermittent parasitism, such as the attacks of mosquitoes and bedbugs upon man.

predator. Any animal, including insects, that preys upon and devours other animals: *e.g.*, a coyote or dog preying upon sheep. Some *predators:* such as a lady bug, may be beneficial also as they kill and eat parasites. A *predator* is distinguished from a parasite which lives only on one host at a time and usually does not destroy the host.

predial. Referring to agrarian land or landed property.

predisposition. The effect of one or more environmental factors which makes a plant vulnerable to infection by a pathogen.

preemergence damping-off. See *damping-off.*

preemergence spray. A spray applied to the ground to kill weeds after a crop is planted but before it comes up.

preemergence treatment. Any treatment made after a crop is planted but before it emerges, or before germination of the seed occurs. Some weeds may be chemically controlled by treatment of the soil before the seed germinates.

Preemption Act. A law passed by the U.S. Congress in 1841 providing for the sale of public lands at $1.25 per acre under certain conditions. It was employed in conjunction with the later Homestead Act in many western states to increase the acreage available to an individual and to secure land with water facilities by ranch owners. It recognized the vested interest of squatters on the public domain and their right to preempt or buy the land they occupied.

preen. Of a bird, to arrange or dress its feathers with its beak.

preen gland. An oil-secreting gland at the root of the tail in most birds whose secretion dresses their feathers. Also called *rump gland.*

preference permit. A license valid up to ten years which is issued by the U.S. Forest Service to permit domestic livestock grazing in a National Forest. The permit is renewable annually.

preformed terrace. A terrace or bench constructed or sloped before a planting, as in setting out an orchard.

pregermination. A special seed treatment applied prior to testing or sowing to increase the rapidity or completeness of germination. It usually involves seed exposure for a month or more to temperatures between 30° and 40° F. while in contact with moist peat or sand.

pregnancy. 1. The condition of a female animal having a living fetus in the uterus which occurs after the ovum has been fertilized by the male sperm cell. See *gestation period.* 2. Of soil, fruitfulness; fertility.

pregnancy disease in ewes. A usually fatal disease of pregnant ewes near term, associated with a disturbance of carbohydrate metabolism caused by deficient diet and lack of exercise. It most commonly occurs in ewes carrying twins or triplets. It is characterized by nervousness, inability to rise, pushing the head against an object, gnashing of the teeth. Also called *lambing paralysis; old ewe disease; ketosis; acidosis of pregnant ewes; twin lamb disease.*

pregnancy test. Any of several tests which determine whether or not a female is pregnant. Pregnancy can be determined by these tests in humans, mares, and rhesus monkeys because of two factors: pregnancy gonadotropins produced by the pregnant female from the 40th to the 150th day of pregnancy, and after that period, by the high amount of estrogens found in the urine. See, *e.g., Friedman test.*

pregnant mare serum. A gonadotrophic hormone, the secretion of the uterus, obtained from the blood of pregnant mares. It is used in treating certain types of shy-breeding animals. Its main value lies in its content of follicle stimulating hormone. Also called *P.M.S.*

preinfective period. The period before a viruliferous insect becomes infective after feeding on the virus source.

preinfective stage. Early, developmental stage (egg or larva) in the life cycle of a parasite during which it is incapable of becoming established in the host animals; the stage that precedes the infective stage.

preirrigation. In rice culture, the application of irrigation water prior to the usual irrigation after emergence of the plants.

Prejvalsky horse. The wild horse of Asia found in the steppes of Mongolia, used as a riding horse. See *tarpan.*

premature drop. The abnormal dropping of fruit from trees before maturity.

premature grazing. Grazing before important forage plants have reached a sufficient state of maturity and, on spring ranges, before the soil has become firm enough to prevent a cumulative injury to the range under the prevailing degree of stocking. See *range readiness.*

premature seeding. Bolting or the development of a seed stalk before the usual time or before maturity of other parts of a plant. It frequently results in losses especially in the growing of celery, sugar beets, spinach, and tobacco.

premature yellowing. A disease of head lettuce, of unknown cause, characterized by a yellowing of the leaves. It may appear in the field or during shipping and marketing.

premier breeder award. An award made to the breeder of animals who wins the most prizes, whether exhibited by himself or others, credit being given in all single classes but not in group classes.

premier exhibitor. In the livestock show ring, the exhibitor who wins the most prize money in the single animal classes.

premium. An added payment by the buyer to the producer over and above the market price of similar products of equal grade and quality, as for the first bale of cotton of the season.

premium price. A price for a product, above the average for the market, paid for prize livestock, the first bale of cotton, superior produce, etc.

premix. In animal feeding, a previously prepared mixture of small amount, as one containing a vitamin or medicine, which is added to the main feed mixture.

prenuptial flight. The flight made by the virgin queen bee supposedly to acquaint herself with landmarks which enable her to return to her own hive after her nuptial flight.

prepackaging. The packaging of agricultural products in convenient, attractive packages for display, ease of handling, and sale.

preparturient. Occurring before birth.

prepeeled potatoes. A food product; potatoes peeled in a flood type washer where they are agitated and rubbed against each other by action of compressed air and high pressure spraying, then conveyed to a pressure steamer-peeler where about 6 lbs. of them are charged continuously. Steam is maintained at 60 lbs. pressure and is condensed at the exit by spray water from water-jet condensers. The peeled potatoes are dipped in a sulfite solution of a controlled pH to prevent discoloration.

prepink. Designating the stage of growth of an apple blossom bud when the leaves have spread away from the blossoms but before the stems of the flower cluster have separated. Very little if any pink is showing in the blossoms at this time. *Cf. full pink.*

prepink spray. A spray applied to apples in the prepink state. (In England, it is called the spur-burst spray). See *prepink.*

preplanting treatment. Any treatment of soil, seed, or reproductive vegetative material before the crop is planted.

prepotent. Designating an animal which transmits its characters to its progeny to a marked or highly uniform degree.

prepuce. The sheath or foreskin covering the glans penis or glans clitoridis.

preputial. Pertaining to the prepuce.

prescribed burning. The application of fire to land under such conditions of weather, soil moisture, time of day, etc., as presumably will result in the intensity of heat and spread required to accomplish specific silivicultural, wildlife, grazing, or fire hazard reduction purposes.

precription. 1. Legal title to land obtained by long possession. See *squatter's right.* 2. A written direction for the preparation and administration of a remedy.

present market value. In farm appraisal, the price for which a property can be sold under present conditions with a substantial down payment, and a reasonable amount of effort, by a willing but not forced seller to a typical, desirous, but not anxious purchaser.

present value. In farm appraisals when applied to farm buildings, the replacement cost plus major improvements less depreciation and obsolescence.

preservation freezing. The process of preparing frozen foods.

preservative. 1. Any material, as salt or sugar, which delays or prevents spoilage and decay of food products, etc. 2. A substance, such as creosote, which when suitably applied to wood makes it resistant to infestation by fungi, insects, or marine borers.

preserve. 1. In wildlife management, a game-shooting area on which game species are propagated or released. 2. A tract of land set aside for preservation of natural conditions, and protected against exploitation or any commercial use. 3. To prepare foods by cooking with some preservative so as to prevent fermentation or decomposition.

preserves. A fruit product which has been processed by cooking with the addition of some preserving agent, such as sugar. Also called *jam.*

preserving melon. Citronmelon.

press. 1. A device for extracting juice from fruits or oil from seeds, etc. 2. Any device which compresses a loose or bulky product for packing or storage, such as a press box in a cotton gin.

press box. The part of a cotton gin in which the lint cotton is compressed into a bale. The size of the press box is standardized to make the bales of uniform size. Also called *baling press.*

press cake. Pomace, or the residue left after pressing the juices from fruits, olives, or tomatoes. Dried pomace has some value as animal feed.

press-cake meal. A by-product left after the separation of oils, by grinding and distillation, from the pits of peaches, apricots and cherries. It has some value as a stock feed.

pressing. 1. a. The extraction of the juice from fruit, as in cider making or in the production of orange juice. b. The solid material left after such a process. 2. The process of bringing the small loose pieces of curds together in a shaped solid body in cheese-making by means of a press.

Pressler cheese. A Swiss cheese that develops too many eyes or gas holes in the early stage of manufacture.

pressure atomizer. A type of atomizer used in evaporated milk manufacture which gives the fluid a rapid rotating motion as it comes from the nozzle.

pressure blower. A rotary blower which produces a definite and positive pressure.

pressure cooker. A small vessel or vat capable of being sealed to develop pressure by the formation of steam during the cooking process. It is used in the home and in food processing plants.

pressure necrosis. A sore back on a horse which develops from saddle pressure; an area of necrosis (dead tissue) resulting from excessive pressure.

pressure regulator. A device on a spray machine which adjusts and maintains the pressure desired and prevents excessive pressure from being built up especially when the nozzles are closed.

pressure system. In drying products, a system by which air is moved through air ducts at a pressure greater than atmospheric pressure and is, therefore, pushed through the product to be dried. Also called *pushing system; forcing system.* See *suction system.*

press wheels. Wheel attachments to field seed drills which follow the drills or tubes and press the loose soil over the seeds for better covering and compaction, resulting usually in increased germination.

Prestost. A cows' milk cheese made in Sweden. Also called *Saaland pfarr.*

presumptive diagnosis. Determination of the nature of a disease on the basis of probable evidence: *e.g.*, in a bacterial disease *presumptive diagnosis* may be made when symptoms appear to be characteristic of the disease.

presumptive standards. The regular standard for a product: *e.g.*, the *presumptive standards* for milk are the legal standards for fat and solids-not-fat. Milk which is below the legal limits is presumed adulterated.

prettyface brodiea. *Brodiaea ixiodes,* family Liliaceae, a perennial herb grown for its salmon or salmon-yellow flowers with dark-purple streaks. Native to North America.

pretty print bag. A cotton bag container for feedstuff which has attractive color patterns printed on it. The empty feed bags are usable for making aprons, children's dresses, etc.

pretty wool. In wool judging, fleeces that are soft, bright, and of light condition. *Pretty wool,* however, does not necessarily bring a higher price.

prevailing wind. The wind which comes most frequently in an area from a particular point of the compass, either seasonally or annually.

prevention of grazing damage. Providing protection from grazing by domestic or other animals to woodland or rangelands that are subject to damage by grazing.

preying mantis. A misnomer for praying mantis (*q.v.*), occasionally used because of the preying habit of the insect.

price cycle. Any regularly recurring movement of prices which extends over a period of more than a year.

price level. Weighted average of prices which arise from all business transactions during a specified period of time. The U.S. Bureau of Labor Statistics index of wholesale prices and consumer price index are commonly used as indicators of the general *price level* in the U.S.

price limit. The maximum fluctuation in price of a futures contract permitted during one trading session as fixed by the rules of a contract market.

price-spread. The difference between the price paid to the producer and that paid by the consumer for a product.

price supports. Programs, carried on by the U.S. Department of Agriculture through the Commodity Credit Corporation, which attempt to establish minimum prices for corn, cotton, wheat, rice, tobacco, peanuts, wool, mohair, tung, nuts, honey, milk, butterfat, and other farm products, by means of loans, purchase agreements, and purchases. The *price supports* may be rigid and fixed or may be flexible and based on parity *q.v.*

price trend. A persistent upward or downward movement of prices, usually over a relatively long period of time.

prick. To pierce or cut the tail of a horse so that it will be carried higher; an unlawful act in many states.

pricklyash. a. Any tree or shrub of the genus Zanthoxylum, family Rutaceae. Species are sometimes grown as ornamentals. See *Hercules'-club pricklyash.* b. *Z. americanum,* an aromatic, prickly shrub or small tree grown for its greenish flowers. Native to North America. Also called *angelica tree; toothache tree.*

prickly lettuce. *Lactuca serriola,* family Compositae, an annual weed which is sometimes troublesome in cultivated fields throughout the U.S. Also called *compass plant; wild lettuce.*

prickly nightshade. Buffalobur nightshade.

pricklypear. Any of various cacti of the genus Opuntia, family Cactaceae. Species are prostrate to treelike forms, which contain both spines and cushions or barblike hairs. Some species spread very rapidly and in some places have greatly depreciated the value of the land. With the spines and barbs burned off, the cacti have been eaten by cattle during periods of drought and scarcity of any other feed. The pearlike fruit when preserved is edible for humans. Also called *devil's-fig.* See *colonche.*

pricklepoppy. Various herbs of the genus Argemone, family Papaveraceae. See *crested pricklepoppy.*

pricklyseed spinach. *Spinacia oleracea,* family Chenopodiaceae, an annual herb grown for its edible leaves, the most important potherb grown in the U.S. Native to Asia. Also called *spinach.*

prickly thatch. Brittle thatchpalm.

prick off. In the greenhouse, to lift very small seedling plants from the seed beds and transfer them to transplant flats. Also called *prick out.*

prick out. Prick off.

prick timber. See *European euonymus; Oregon-grape.*

pride of India. See *Chinaberry; panicled goldraintree.*

primal cuts. The four major wholesale cuts of the pork carcass: ham, shoulder, loin, and belly (side). The yield of *primal cuts* is frequently used as a basis for classifying meat type hogs.

primaries. The outermost group of major wing-feathers (usually 10) located on the third joint of a bird's wing, hidden when the wing is folded. Also called *flight feathers; flights. Cf. secondaries.*

primary bud. The largest and strongest bud in the compound bud on the shoot or cane of a grape vine.

primary cotton market. In the Cotton Belt or southern states (U.S.), a small town, frequently a county seat, where cotton is first marketed from the fields.

primary cultures. The first bacterial cultures made from a sick animal.

primary drainage. Drainage in a watershed as a whole, in contrast to secondary, or that of an individual farm.

primary feather. Primaries.

primary fertilizer components. The nitrogen, phosphoric acid, and potash of commercial fertilizers. The three are sometimes called primary plant foods because they are needed by plants in relatively large amounts.

primary forage plants. Those plants on a pasture or range which bear the main grazing load.

primary infection. The first infection by a pathogen (inoculum). The infection responsible for the onset of a disease which may be followed by secondary infections.

primary inoculum. See *primary infection; secondary inoculum.*

primary insect. Any insect capable of attacking, injuring, or killing the growing parts of a living plant unaided by other insects or disease organisms.

primary leaf roll. The roll of all of the leaves of a plant in the potato leaf roll (*q.v.*) in contrast to roll of the apical leaves only.

primary particles. The discrete, individual particles of a soil mass.

primary plant food. Primary fertilizer components.

primary product. In slaughtering, the whole carcass of an animal.

primary production. The crop itself in contrast to a product derived from it: *e.g.*, corn is primary and dairy products or beef cattle are secondary products.

primary root. The main descending axis of a plant; the pole of the embryo opposite to the shoot.

primary scaffold. A large branch attached to the leader of a tree.

primary soil. A soil which developed through the disintegration and decomposition of rocks in place. See *orthoeluvium.*

primary succession. Plant succession on newly formed soils or upon surfaces exposed for the first time which have never borne vegetation.

primary xylem. The first woody tissue developed by a tree or shrub. See *xylem.*

primas. A grade of wool from Uruguay equal to half blood *q.v.*

prime. In the grading of various agricultural products, meat, fruits, beans, cottonseed oil, etc., of first class or high quality. *Cf. choice.*

prime bleachable summer yellow cottonseed oil. Cottonseed oil which is free from visible foreign material, clear at temperatures sufficiently high to melt the stearine, sweet in flavor and odor, and when bleached as provided in the rules of the National Cottonseed Products Association (U.S.) shall be of a color no higher than AOCS 2.5, and shall contain not more than 0.25 percent free fatty acid nor in excess of 0.10 percent moisture and volatile matter.

prime cattle. Slaughter cattle that have a high finish, desirable conformation, and yield prime meat cuts.

prime cottonseed hulls. Cottonseed hulls which shall be sound, not musty, and reasonably free from foreign substances and trash, and shall have no lint taken off except through the linters in the usual milling of cottonseed; a livestock feedstuff. Cotton burrs and stems, even though reground, also hull pepper, bran and loose lint reclaimed from a lint beater, shall be considered as excess foreign substances, if added to cottonseed hulls.

prime cottonseed meal. See *cottonseed meal, prime quality.*

prime crude cottonseed oil. That cottonseed oil which is not decidedly rancid, musty, sour, or bitter, does not have a foreign flavor and the fatty acid of which does not exceed 3.25 percent. It must refine to a color no higher than AOCS 7.6, with a loss in weight not exceeding 12 percent.

prime cut (vanilla beans). The best grade of Mexican cut vanilla beans. These beans are not adaptable to bundling and are inferior to the prime whole bean.

prime grade. A recognized market classification for meat animals being sold for slaughter. There is some difference in description for *prime grade* for the different types of livestock, such as beef cattle and lambs, but the principle of degree of quality is the same. See *market grade.*

prime log. A log free from defects, usually for export.

prime mover. The power source to which an implement is connected.

prime quality cottonseed. Cottonseed containing not more than 1.0 percent of foreign matter, not more than 12.0 percent of moisture, and not more than 1.8 percent of free fatty acids in its oil. Such cottonseed oil has a quality index of 100.

prime steam lard. Lard made by rendering hog fat in direct contact with steam in a closed vat under pressure of about 50 lbs. It does not contain fats obtained from cured meats or trimmings.

prime summer white cottonseed oil. Cottonseed oil which is brilliant, sweet in flavor and odor, free from visible foreign material, of a color no higher than AOCS 2.5, contains not more than .25 percent free fatty acid nor in excess of .10 percent moisture and volatile matter. See *cottonseed oil.*

prime summer yellow cottonseed oil. Refined cottonseed oil which is free from visible foreign materials, clear at temperatures sufficient to melt the stearin, sweet in flavor and odor, has a color no higher than AOCS 7.6, contains no more than 0.25 percent free fatty acid nor an excess of 0.10 percent moisture and volatile matter.

prime swarm. The first swarm of bees to issue from a hive usually with the old queen.

primeval forest. The forest as it existed before the advent of man.

prime winter white cottonseed oil. Cottonseed oil which is brilliant, sweet in flavor and odor, free from visible foreign matter, of a color no higher than AOCS 2.5, and does not contain more than 0.25 percent free fatty acid nor in excess of 0.10 percent moisture and volatile matter.

prime winter yellow cottonseed oil. Refined cottonseed oil which is sweet in odor and flavor, free from visible foreign material, has a color no higher than AOCS 7.6, contains not more than 0.25 percent free fatty acid nor in excess of 0.10 percent moisture and volatile matter. It must stand a test determining its resistance to crystallization which serves as an index of its fitness for winterizing or similar stearin removing processes.

priming method. A method of harvesting tobacco by picking the leaves from the stem, as they ripen.

primitive area. 1. In the U.S. Forest Service, wilderness area. 2. In the U.S. National Park Service, an area of indeterminate size in which no commercial development nor the construction of any roads for motorized transportation is permitted.

primocane. The first year's growth of bramble fruits (*Rubus*). Also called *first-year cane.* See *floricane; second-year cane.*

primordium. A member or organ of a plant in its earliest condition.

Primost. A kind of Scandinavian cheese which contains all the whey solids. It is buttery, light brown and has a mild sweet taste. Also known as *Mysost.*

primrose. Any of the various perennial herbs of the genus Primula, family Primulaceae. Many species are in cultivation. See *cowslip primrose; fairy primrose; Forbes primrose; oxlip primrose.*

primrose jasmine. *Jasminum mesnyi,* family Oleaceae, a tender, evergreen shrub, cultivated both in the greenhouse and outdoors for its yellow flowers. Native to China.

prince's pine. Pipsissewa.

princesplume ladysthumb. *Polygonum orientale,* family Polygonaceae, an annual herb grown in the flower border for its 6 ft. height and for its pink or rose flowers. Native to Australia.

principle of independent assortment. Mendel's genetic theory that differentiating characters (e.g., dominant and recessive characters), recombine at random. He proved by experiment that a certain mathematical relation exists among all the resulting combinations. See *Mendel's Law.*

principle of segregation. Mendel's first law of genetics, stating that alternatives in the reproductive cells remain separate. See *Mendel's Law.*

pringle ax. A turpentining tool.

print butter. Butter sold in prints or cartons of uniform style or size.

priority (water rights). The rights to special consideration and prior claim, over other individuals, to the use of a quantity of water.

prismatic soil structure. Blocky soil structure in which the vertical axis of a block is longer than the horizontal; the upper ends of the block, or column, are not rounded. See *columnar structure.*

pritch. A staff with sharp metal points on the ends; used in skinning operations in cattle slaughter. A pritch, stuck on either side of the withers, may be used in the place of a skinning rack to brace the body of the animal and to keep it on its back.

pritchel. A punch used to make or enlarge nail holes in a horse shoe.

private farm manager. One qualified by training and experience to manage farms efficiently and usually serving only one employer.

private general refrigerated warehouse. Any artificially cooled warehouse operated by a businessman to facilitate his main function as a producer, processor, or distributor but not to store commodities for others for pay.

private land permit. In the U.S., a permit issued when the government is compensated for use of U.S. forest lands by private lands of equal value, and when lands are released to the government for exclusive administrative control of the grazing use.

private treaty. In marketing livestock, a private sales agreement between different parties.

privet. Various shrubs of the genus Ligustrum, family Oleaceae. See *California privet; European privet; glossy privet; ibota privet; Japanese privet; Regels border privet.*

privet honeysuckle. *Lonicera pileata,* family Caprifoliaceae, a tender, low, evergreen or half-evergreen shrub, grown in the rock garden for its foliage, fragrant white flowers, and purple fruit. Native to China.

privet lippia. *Lippia ligustrina,* family Verbenaceae, an aromatic shrub, a troublesome invader in the grasslands of Southwestern U.S. Native to America. Also called *whitebush.*

privet mite. *Brevipalpus inornatus* (Banks), family Tenuipalpidae, a mite that infests privet and many other plants by feeding on the undersides of leaves or on the stems and petioles. Also called *false spider mite.*

privy. A gentilism (see Fowler), one of the terms applied to the outdoor water closet. In rural areas, often a building in a secluded corner of the house yard which is equipped with seats, with a pit underneath them in which human excrement is deposited. The *Oxford English Dictionary,* with rare good humor, refers to such a place as a necessary. Also called *backhouse; outhouse.*

probe. 1. A trier or sampling tube, of various kinds, used for taking samples of cheeses, grain, and other agricultural products. See *grain trier.* 2. A rod that is pushed into the soil to locate objects, such as pipe and drain tile.

proboscis. An elongated nose, such as the snout of a hog, or of some species of beetles.

proboscis flower. Devil's-claw.

process. To carry out the operations required to prepare agricultural products for sale and consumption, such as pasteurizing milk, curing meats, canning and preserving fruits, etc.

Process cheese. Cheese made by blending, grinding and melting Cheddar and various other kinds of cheese and emulsifying them with various salts. The mixture is pasteurized and wrapped in foil to keep it moist.

process dill pickles. Cucumber pickles made by soaking salt-stocked cucumbers in hot water to remove salt followed by storage in barrels with a solution in which dill is added.

processed eggs. See *shell protected.*

processed garbage. An animal feed composed of dried animal and vegetable waste from garbage collected before harmful decomposition has set in and from which harmful material, such as glass, has been separated.

processed milk. Milk which has been pasteurized, clarified, or changed in any way after production by the cow.

processor. 1. One who processes or prepares, by cooking, curing, etc., agricultural products. 2. The metal tank container, or cooker, in which cans of fruit and vegetables are cooked in hot water or steam in preparation for market.

process tankage. Organic fertilizer, made from nitrogenous waste material steamed in tanks under pressure. It contains 6.5 to 10.0 percent nitrogen.

procumbent. Designating a prostrate plant which differs from a creeper in that the parts that lie on the soil do not take root.

produce. 1. Articles produced from or grown in the soil. 2. In animal breeding, a female's offspring.

produce of dam. In the show ring: (a) for swine, four breeding animals, the offspring of one sow; (b) for horses, two animals, the offspring of one mare; (c) for dairy cattle, two animals, the offspring of one cow.

produce of mare. In the show ring class for horses, two offspring of either sex and any age of a mare.

produce of sow. In the show ring for swine, four offspring of either sex, and any age from one sow.

producer. One who grows farm crops or performs the farm or ranch operations which result in the product, in contrast to processor, distributor, and consumer.

producer-to-consumer. The different stages involved in the processing and distribution of a product.

producing cow. A cow in lactation.

product complementarity. Alternative combinations of two products that may be produced from a fixed set of resources. It describes the relationship in which an increase in production of one product can be accompanied by an increase in the quantity produced of another product without an increase in cost or outlay in the absence of technological change.

production. The total quantity or yield of plant and animal produce obtained by the operators of farms and ranches, singly or as a group.

production-bred poultry. Poultry, mainly chickens, bred especially for commercial purposes. See *Standard-bred poultry.*

production coefficient. The measurement of the production values of a certain unit of a feed, such as the yield value of a pound of oats as a feed for milk production.

production function. An expression of the relationship between the amount of product and the amounts of the factors of production used. Also called *input-output function.*

production get. In the show ring for dairy cattle, four females with official production records that are the get of one sire.

production index. The percentage comparison of the combined crop yield index and livestock index on a farm or group of farms with the average of all farms expressed as 100. In calculating *production index* the crop yield index and livestock index are weighted by the number of productive man work units on crops and livestock, respectively.

production records. In farming, a record of the crop raised, land utilized, numbers of livestock raised, and amounts of livestock products produced, during the year.

productive. Pertaining to abundance, such as a *productive* soil, a *productive* animal, or a *productive* worker, etc.

productive animal unit. An animal unit of productive livestock which includes all livestock except the work stock.

productive day's work. Amount of productive work that can be done by a man working at average labor efficiency in a ten hour day. The total *productive day's work* on a farm equals the number of ten hour days' work required to care for the crops and productive livestock on the farm with average labor efficiency. See *productive man work unit.*

productive energy. In feeds, the portion available for furnishing heat to the animal body or energy for life processes or work.

productive-energy value. Energy value of feeds expressed in therms.

productive flower growing. Raising flowers for commercial purposes.

productive labor. Work done on farm enterprises: e.g., field work on all kinds of crops and chores on cattle, hogs, sheep, and poultry. It does not include labor for repairing buildings, machinery, fences, terraces, and ditches, or the care of work stock.

productive land. Land from which abundant or comparatively high yields are obtained. The productivity may be due primarily to natural fertility of soil and favorable topography or lay of land, or it may result from skillful practices and management by the operator.

productive man work unit (PMWU). A standard of work performance; the amount of productive farm work that will be done by a man working at average labor efficiency in a ten hour day. Productive work includes work on crops, livestock, and labor off the farm. Work spent in caring for work stock, repairing machinery, buildings, fences, and other work connected with the farm is not included in calculating the number of *productive man work units* on a farm. This standard of work performance varies from region to region. Standards for *PMWU* may be obtained from any of the state colleges of agriculture in the U.S. As it is worked out, a standard of performance is determined for each acre of crops and head of livestock. For example, in Michigan (U.S.), a dairy cow requires 10 *productive man work units* (100 hrs.) to care for her for one year, as an average. On this basis 30 cows would require 300 *PMWU*. Grain crops require 0.6 *PMWU* an acre (6 hrs.). The *productive man work unit* is a very important factor in the analysis of a farm business. From it may be determined: the size of the farm business (the total *PMWU*); the intensity of the farm business (*PMWU* per tillable acre); and the efficiency of farm labor (*PMWU* per man). Often called *productive day's work.*

productive man work units per man. A measure of labor efficiency determined by dividing the total productive man work units by the man equivalent.

productivity. The volume of plant growth that a soil is capable of producing. *Productivity* may be regarded as a factor of management and therefore, to a degree, independent of fertility (*q.v.*). However, it is not uncommon for the two terms to be used interchangeably.

productivity rating. The productivity of the various soil types on a farm based on the expected yield of the major crop or crops. The rating is a percentage based on standards developed by research and statistical records of crop yields in the region. The best soils have a productivity rating of 100. In some instances, tables have been established showing the relation of crop yields to land values.

product-product competitiveness. The relationship between the outputs of two enterprises. If the output of one enterprise cannot be expanded for a given outlay without reducing the output of another, the two enterprises are said to be competing or competitive.

professional farm manager. One who is qualified by training and experience to manage farms efficiently for one or more farm owners. His services are available to the public in the same manner as other professional persons.

professional group leader. A professionally trained person who is hired for training and leading others: *e.g.*, in agricultural extension work.

professional inspector. A meat inspector who is a graduate of an accredited veterinary college and has passed the required civil service examinations for federal, state, or municipal meat inspection (U.S.).

profile. 1. The general outline of an animal's body. 2. See *soil profile.*

profile survey. A survey along a line or route in which elevations are determined usually at uniform intervals, as in staking out a terrace or drain line.

profits. The financial returns from a business, enterprise, or transaction above all costs, including both actual receipts and actual or estimated appreciation on the capital involved. The costs include the actual operating expense for all labor, materials, etc., used, all rents, interest and insurance, charges for unpaid family labor, and any depreciation on the capital involved. Net *profits* are the same as *profits*. Gross *profits* as defined in the Federal Farm Income Tax Report (U.S.) are the sum of the sale of livestock raised, the sale of produce raised, other farm income, and the gain (or loss) from the sale of purchased livestock and other purchased items. Other definitions consider gross *profits* to be the amount by which the receipts from the sale of purchased goods exceed the purchase price without deduction for other costs.

progenitor. An individual animal or plant that is recognized as the source of a certain type or character in its offspring.

progeny. The offspring of animals or plants.

progeny test. Evaluation of the genotype of an individual by a study of the progeny derived from it by controlled breeding. As applied to forest trees, the method usually involves knowledge of the female parent only because open-pollinated seed collected from individual trees is used.

progesterone. A hormone used in treating certain diseases of the ovary and uterus in domestic animals.

progestin. A female sex hormone. See *progesterone.*

proglottids. The segments or parts of a tapeworm other than the head (scolex) and neck region.

prognosis. The forecast of the probable course and outcome of a disease.

progressive. Of a disease, developing through successive stages, usually in a certain direction, whether improving or deteriorating.

progressive robbing. The depredation in which bees from one colony enter another hive to steal honey without antagonizing or destroying the robbed colony.

project. An undertaking designed to effect the improvement or reclamation of a given area of land: *e.g.*, by irrigation. Also the area of land itself.

prolactin. A specific hormone which initiates and maintains lactation.

prolapse. The dropping or sinking of an internal organ: *e.g.*, the uterus. See *eversion of the oviduct.*

prolapse of the uterus. Partial or complete turning inside out of the uterus, usually following parturition.

proliferation. 1. Exuberant or excessive growth by reproduction of similar cells. 2. The multiplication of morbid cysts. Also called *prolification.*

proliferous. 1. Pertaining to a new branch which arises from an older one, or one head or cluster of flowers from another. 2. Reproducing fully by offsets, buds, or other vegetative means, as in most species of cactus genus Echinopsis. 3. Developing shoots from flowers or fruit as in some pricklypears.

prolific. Capable of producing abundant offspring.

promethea moth. *Callosamia promethea* (Drury), family Saturniidae, the larva of an insect which feeds on the leaves of the sassafras, tulip, ash, wild cherry, and a number of other plants.

promissory note. The primary document in most credit transactions, the signed promise of the borrower to repay a loan. It states the date and amount of the loan, the interest rate, and terms of repayment.

prong budding. A kind of budding or plant grafting similar to shield budding (*q.v.*) except that a short spur or prong is used in place of a simple bud.

prong hoe. A hand implement having two or more curved prongs, used for cultivation or garden hoeing.

pronghorn. A section of a large branch which is left attached to a top or cull log (southwestern U.S.).

propagate. To cause to generate or to multiply by sexual or asexual means.

propagating bench. A bench container for soil or other rooting media, used by nurserymen and florists in greenhouses for multiplying desired plants by seed, cuttage, layering, etc.

propagating house. Any structure, often a greenhouse, used for the propagation of plants. It may or may not be provided with refinements, such as temperature controls, mist spray, artificial light, etc. Also called *propagation house.*

propagating part. In plants, a part, such as a tuber of the potato, the bulb of a tulip, a cutting, leaf, etc., used for asexual reproduction.

propagating sand. Sand, kept to the desired moisture content and in suitable containers, used for propagating plants in greenhouses and laboratories.

propagation house. Propagating house.

propagation method. Construction of drainage ditches by use of dynamite. Charges are closely spaced so that detonation of one charge causes the explosion of the next charge and so on until the complete series of charges is exploded. A water-filled soil is most desirable for this method.

propagator. An expert employed in greenhouses and nurseries to propagate plants.

proper range stocking. The utilization of a range at such an intensity that optimum utilization of forage is not exceeded.

proper stocking. Stocking of a range area on the basis of its true grazing capacity; *i.e.,* the number of animals that will graze the range to approximately proper utilization during the grazing period of the slightly below average year, and thus assure maintenance of forage production and improvement of deteriorated range.

proper use. The degree of grazing that an individual plant species, or the total palatable cover of a range area, may endure without damage to it or the soil.

proper use factor. As applied to individual range species, the estimated maximum percentage of the total vegetative production of the year within easy reach of the livestock to which a given range species may be grazed without damaging it or associated important palatable plants or the soil; the degree to which each species may be grazed when the range as a whole is properly grazed. The *proper use factor* for a range type is the weighted average of the *proper use factors* of the individual plants in the type. Also called *P. U. F.* See *palatability.*

prophase. The stage in mitosis or meiosis from the appearance of the chromosomes to metaphase.

prophylactic. Pertaining to that which prevents disease.

prophylaxis. The prevention of disease; preventive treatment; protection against disease by chemicals, drugs, serums, or mechanical means.

propionic acid. $C_2H_5 \cdot COOH$, a fatty acid resembling acetic acid in its odor and some of its properties. It is a flavor constituent of cheese, especially of Gruyère. It is also used to prevent mold growth.

propionic acid fermentation. The main fermentation (by propionibacterium) in Gruyère or Swiss cheese.

propolis. A resin collected from the buds of trees and other sources by bees and used to strengthen the comb, close up cracks, etc. Also called *bee glue.*

proportional weir. A weir so shaped that the flow rate varies directly with head. It has special application to irrigation.

proprietary. A medicine or food made by a person or company having the sole right to manufacture and sell it.

prop roots. Technically, the aerial roots of Rhizophora (*q.v.*). See *aerial root: 2; brace root.*

propylene dichloride. An organic chemical, less toxic and inflammable than ethylene dichloride, used as a grain fumigant, and in 15 percent emulsions as a soil fumigant.

propyl isomer. A chemical used as a synergist with pyrethrum to increase its effectiveness as an insecticide.

prosciutti. A dry-cured, black pepper flavored, Italian ham.

proso. *Panicum miliaceum,* family Gramineae, an annual herb grown for its seed used for birdseed, poultry and stock feed. Native to Asia. Also called *broom corn millet; broom millet; hog millet.*

prospective supply. The amount of a commodity available, including that in production and carry over.

prostrate amaranth. *Amaranthus blitoides,* family Amaranthaceae, a weed annual growing as a dense mat which is often one of the first weeds to appear on abandoned cultivated fields. Native to Europe. Also called *prostrate pigweed.*

prostrate juniper. Oldfield juniper.

prostrate knotweed. *Polygonum aviculare,* family Polygonaceae, an annual herb which is a weed pest in settled regions. Native to Europe, Asia, and North America. Also called *door-weed; knotgrass; mat grass.*

prostration. A symptom of an animal's ailment in which pronounced weakness and loss of strength, resulting in the animal lying in prone position unable to rise, is observed.

protandry. Pollen dehiscence before the stigma is receptive. See *dichogamy; protogyny.*

protease. An enzyme which attacks proteins. A proteolytic enzyme present in cow's milk, it is important in the ripening of certain types of cheese. See *enzyme; protein; proteolytic bacteria.*

protectants. Chemicals containing heavy metals, sulfur, or organic compounds used as sprays, dusts, or dips on seeds, stems, leaves, or wounds of living plants to prevent entrance of fungi or bacteria, and on fabric or wood products to prevent decay.

protection cut. In range management, a reduction in forage use to meet the requirements of good range use, *i.e.*, to prevent damage to soil and water resources. Also called *cuts for protection.*

protection forest. A wooded area managed primarily for its beneficial effects on water, climate, or soil, rather than for wood or other forest products or services.

protection forestry. The practice of forestry with the primary objectives of (a) conserving water supplies; (b) maintaining desirable streamflow; (c) increasing ground water storage; (d) preventing erosion and reducing sedimentation; (e) providing high quality water, and preventing pollution; (f) ameliorating adverse climatic conditions, especially wind.

protection land. Land set aside to protect other land, such as that furnishing a water supply or that used to protect more valuable land from the destructive effect of erosion.

protection strip. In white pine blister rust control, a belt or zone outside the pine stand which is included in the control unit and is freed of *Ribes* spp.

protective foods. Foods which furnish additional supplies of certain minerals, vitamins, and proteins for the normal growth of the body and the maintenance of good health.

protein. Any of a large number of complex, organic compounds which has a high molecular weight and is an essential part of all living organisms. *Proteins* consist largely of carbon, hydrogen, nitrogen, and oxygen; many contain sulfur, and some iron and phosphorus also. They constitute a large portion of the protoplasm, and are obtained from foods such as lean meats, or from vegetables such as beans. Chemically *proteins* are extremely complex combinations of amino acids *q.v.*

protein equivalent. An expression used in computing the protein of feedstuffs. In Great Britain, non-protein, nitrogenous compounds are considered to have one-half the value of true protein. *Protein equivalent* is considered to be the sum of the true protein and one-half the amount of non-protein nitrogen.

protein expeller soybean oil chips. A livestock feedstuff which is the product obtained after expressing part of the oil from soybeans by crushing, cooking, and mechanical pressure using an expeller, screw press, or any other mechanical press.

protein expeller soybean oil meal. The product which results from grinding protein expeller soybean oil chips.

protein extracted soybean feed. A livestock feedstuff which is the material resulting from the partial removal of protein and nitrogen-free extract from solvent extracted soybean oil meal.

protein hydraulic soybean oil cake. A livestock feedstuff which is the product obtained after expressing part of the oil from soybeans by crushing, cooking, and hydraulic pressure.

protein hydraulic soybean oil meal. The product which results from grinding protein hydraulic soybean oil cake.

protein palm nut oil meal. A livestock feedstuff which is the ground residue left after the extraction of part of the oil from the fruit of one or more species of palm.

protein solvent extracted cottonseed meal. A livestock feedstuff which is the product resulting from grinding solvent extracted cottonseed flakes.

protein solvent extracted soybean flakes. A livestock feedstuff which is the product obtained by expelling part of the oil from soybeans by the use of solvents.

protein solvent extracted soybean oil meal. The product which results from grinding protein solvent extracted soybean flakes.

proteolysis. The process by which casein or some insoluble casein derivative is broken down to water-soluble compounds through the action of organisms.

proteolytic. Referring to proteolysis or to an agent that produces proteolysis.

proteolytic bacteria. 1. Bacteria which produce a proteolytic action on the proteids in cheese and cause a strong odor and taste to develop. 2. Bacteria which act on proteins, breaking them down to simpler compounds.

proteolytic flavor. Any unpleasant flavor of butter and other milk products attributed to the breakdown of proteins and amino acids.

proteoses. The first stage of breakdown of protein molecules by proteolytic enzymes.

protogyny. Receptivity of the stigma of flowers before pollen dehiscence. *Cf. protandry.*

protopectin. The parent substance from which pectin *(q.v.)* is derived. Also called *pectose.*

protoplasm. The semifluid substance of which living cells or organisms are composed.

protoplasmic. Consisting of or pertaining to protoplasm.

Protozoa. The lowest group of the animal kingdom which consists of single-celled organisms.

protozoacide. a. Any chemical agent which kills protozoan parasites. b. Any agent that is employed for the prevention, suppression, or cure of infection by protozoan parasites.

protozoan parasite. A single-celled parasite of the coccidia, trypanosomes, and malarial organisms, which belong to the animal phylum Protozoa.

Provatura. A soft, plastic-curd cows' milk cheese that originated in southern Italy.

proved sire. A sire whose transmitting ability has

been measured by comparing the production performance of his daughters with that of the daughter's dam under uniform conditions.

Provence rose. French rose. Also called *Provincial rose.*

provender. All dry feed or fodder for domestic animals.

provenience. Origin; particularly the geographic origin of seed and inoculum.

proventriculitis. Inflammation of the glandular (or true) stomach which often occurs in growing chicks reared in confinement and occasionally in adult fowls.

proventriculus. The glandular or true stomach of birds which is a spindle-shaped organ between the esophagus and gizzard.

Providence. A cheese very similar to Port du Salut made in the monastery of Bricquebec in the Department of Manche, France.

Province (Soil). In the early period of the Soil Survey by the U.S. Department of Agriculture, the smaller units of mapping, series and types, were grouped and correlated according to the *Province,* a broad geographical division in which there was homogeneity in geologic formations or unity in physiography and it was therefore assumed that all soils, within the boundaries set would have common characteristics and values as agricultural land (obsolete in soil classification).

provision tree. Guiana-chestnut pachira.

Provole. A round, plastic-curd cheese made from buffalo milk in southern Italy.

Provolone. A plastic-curd cows' milk cheese made in southern Italy.

pruinose. Having a dusty or chalky appearance, as a leaf surface.

prune. 1. Any of the horticultural varieties of the plum *(q.v.)* which bear fruit that is sufficiently firm and sweet to be preserved by drying without fermenting. Cultivated extensively on the West Coast of the U.S., the most popular ones are: French (Prun d'Agen), Robe de Sargeant, Imperial, and Italian, all varieties of *Prunus domestica,* the garden plum. 2. To remove methodically parts of a plant so as to improve it in some respect for the purposes of the cultivator. See *root pruning.*

prune brown rot. A fungous disease of stone fruits caused by *Monilinia fructicola* and *M. laxa,* family Sclerotiniaceae, the latter especially on the Pacific Coast, U.S. Small, brown spots appear and then spread rapidly, soon covering the entire fruit. Blossoms may be blighted.

prune diamond canker. A virus disease of the French (Agen) prune characterized by a thickened and roughened, diamond-shaped area around a wound. The disease is slowly devitalizing and causes substantial losses in California, U.S.

prune dwarf. A virus disease of plum indicated on Italian prune by rough, strap-shaped leaves. There are varying degrees or no symptoms on other varieties and species of stone fruits. Similar symptoms may be caused by other viruses or the *prune dwarf* virus may be associated with other viruses in a complex mixture.

pruner. 1. A mechanical device used to prune roots of seedlings in the seedbed. 2. Any of several mechanical devices, such as a hooked cutting blade mounted on a pole, connected to an operating lever with a wire, or long handled shears for manual operation, used to prune trees. 3. A person who is engaged in pruning operations. 4. See *twig pruner.*

prunier d'Espagne. See *mombin.*

pruning. The removal of live or dead branches, roots, and other parts from trees, shrubs, vines, etc., for purposes of improvement. Natural *pruning* of trees results from deficiency of light, decay, snow, ice, etc. See *self-pruning; root.*

pruning compound. Any of various preparations, such as compounds of asphaltum, used to seal over wounds made in pruning to prevent infection and infestation.

pruning saw. Any manually operated saw having either a straight or curved blade, used for pruning.

pruning shears. Short-bladed shears which are used for small cuts on trees and pruning roots.

prussic acid poisoning. Poisoning of livestock from prussic acid or hydrocyanic acid (HCN) which may result from ingestion of sorghums, such as Johnson grass, under certain conditions. Hydrocyanic acid may be produced from the leaves of several species of wild cherries (*Prunus* spp.) which are occasionally browsed by livestock.

pseudo-albino. A coat-color of some animals which is a very light cream to white. In horses there is a silver mane and tail and often glass or blue eyes. It is due to the absence of pigment in hair and skin.

pseudobulb. The swollen, stemlike base of many orchids or an elongated above-ground, fleshy plant stem in which food and moisture is held.

pseudocarp. A fruit which consists of more than just the ripened ovary and its contents: *e.g.,* the apple.

pseudogamy. Parthenogenetic development of the female gamete which requires the stimulation of pollination but not complete syngamy *q.v.*

pseudolinkage. Apparent linkage of factors in individuals heterozygous for a chromosomal change: *e.g.,* interchange in which linkage no longer exists in individuals homozygous for the chromosomal change.

pseudo-net necrosis. Phloem parenchyma necrosis.

Pseudoplea leaf spot of clover. A fungous disease of clover caused by *Pseudoplea trifolii,* characterized by pepper spots on the leaves of clover and alfalfa. The spots enlarge often resulting in curled, light brown leaflets, and dead stem apexes. Also called *pepper spot.*

pseudorabies. Aujeszky's disease.

pseudo-tubercular enteritis. Johne's disease.

pseudo yeast. Any yeast which does not form spores. See *yeasts: 2.*

psittacosis. An acute or chronic virus disease of birds (domestic and wild) transmissible to man and characterized by systemic reactions and respiratory involvement. Also called *parrot fever.*

psoroptic scab. Common scab (mange) of cattle.

psorosis. A virus disease complex of citrus under six different forms (common *psorosis* formerly called type A. Type B is now thought a virulent, rapidly progressing form of A initiated from transmission from a bark lesion, concave gum disease, blind pocket, crinkly leaf, and infectious variegation). All have a common leaf flecking symptom. Also called *scalybark* (California, U.S.).

psychroenergetics. Science dealing with the effect of ambient temperature and humidity upon conversion of feed into bodily heat or energy. It is important in the study of housing for livestock and its effect on production.

psychrometer. An instrument used to measure a relative humidity of the atmosphere.

psyllid. Any jumping plant louse of the genus Psylla, family Chermidae (Psylliidae).

psyllid yellows. A disease of potatoes prevalent in western U.S., characterized by marked stunting of the plant, yellowing, and upward rolling of the leaves. The disease may be transmitted to other solanaceous plants, such as tomato and eggplant.

ptosis. The prolapse *(q.v.)* of a body organ, specifically the drooping of the upper eyelid.

ptyalin. An enzyme in saliva which digests starch.

pubescence. The downy covering of short, smooth hair found on leaf surfaces and some insects.

pubescent. Hairy; covered with a short, soft growth that resembles hair; as a *pubescent* leaf.

pubic bone. See *pubis.*

pubis. One of the paired bones comprising the pelvis which make up the floor of the pelvis. In poultry culling the spread of the pubic bones is sometimes used to determine whether a hen is a layer or non-layer.

public auction. See *auction.*

public ditch. A drainage ditch constructed by a county or a legally constituted drainage district.

public domain. All lands over which a national government exercises proprietary rights, including national parks, grazing lands, forests, and military and other reservations.

public general refrigerated warehouse. Any artificially cooled warehouse whose operator is engaged in storing perishable food commodities for others on a payment.

public lands. The general public domain; lands belonging to a national government which are subject to sale or disposal otherwise and which are not reserved or held back for any special governmental or public purpose.

public livestock market. An organized market system operating under certain federal and/or state regulations in which livestock are offered for sale. Also called *central market.*

public sale. An open auction of livestock, equipment, or other goods, to which the public is invited. Very common in rural areas. See *auction.*

puccoon. Bloodroot.

pudding-pipe tree. Golden shower senna.

puddin meat. See *liver sausage.*

puddle. 1. To deflocculate or destroy the granular structure of the soil which results in a hard or baked surface and therefore is an unfavorable tilth condition. 2. To make soil more dense, compact, and less pervious by working it in a wet condition, as in the bottom of an artificial pond, stock tank, or an irrigation canal. It prevents leakage.

puddle duck. Any duck of mixed breeding, generally small in size, not a good layer or particularly desirable as meat. These are sometimes kept on the farms mostly for decorative purposes.

puddling. A procedure in transplanting in which the roots of plants are dipped in liquid mud to coat the fine roots with soil just before planting.

puering. An incorrect term designating puring, the use of liquid dog manure in the bating of hides. See *bate.*

Puerto Rican mole cricket. Changa.

puff. A soft swelling which involves joints or tendons of animals.

puffballs. Certain globose-shaped fungi, of the family Lycoperdaceae, whose ripe fruit emits clouds of spores when disturbed. The spore fruits of most species of the family are edible when young.

puff duster. Bellows duster.

puffed cereals. Food products generally made from cleaned wheat or rice, placed in retorts and heated to a temperature of 560° F. from 30 to 40 minutes. Steam at a pressure of 175 lbs. per sq. in. is then introduced for about three minutes. The pressure is suddenly released from the retorts causing the grains to puff because of the rapid fall in pressure.

puffing and creasing. An abnormality of orange rind which results from the separation of the albedo in irregular lines with accompanying sinking of the yellow surface.

Pulaski tool. A combination chopping and cutting tool widely used in fire line construction for the protection of forests. It is a light, single-edged ax with a straight handle, having a narrow, adzlike trenching blade attached to its head.

pull. See *pulling.*

pullboat. A flatboat, carrying a steam skidder, or a donkey engine which is used in logging cypress and tupelo.

pull-down. Dodder.

pulled cotton. Cotton harvested by hand after frost by literally pulling the opened and partially opened bolls from the plants. *Pulled cotton* is usually associated with bollies or bolly cotton and is common in the High Plains Region of Texas and Oklahoma, U.S.

pulled flax. Flax harvested by pulling the whole plant from the ground and tying in bundles.

pulled wool. Wool removed from the skins of slaughtered sheep by means of a depilatory, sweating, etc.; a by-product of sheep slaughter. Also called *skin wool.*

puller. 1. In turpentining, a long-handled knife, with a strong, elliptical blade, used to chip timber when the face has advanced too far up the tree for chipping with a hack. See *hack: 4.* 2. In poultry slaughter, a hook tool used for drawing tendons. 3. In bean harvesting, the part of the harvester machine which consists of two broad blades set at an angle so that they cut off the bean plants just beneath the surface.

pullery. The wool house in the packing plants. It is the place where wool is pulled from the skins of slaughtered sheep.

pullet. An immature female chicken; in poultry shows, a young hen under one year of age.

pullet disease. Avian monocytosis.

pulling. A hand procedure for stapling cotton or determining length of fiber in which a tuft for a sample is pulled into two parts, of which one part is manipulated by fingers and thumbs properly to ready it for measurement.

pulling contest. An exhibit or public contest between two or more matched teams of horses to determine their pulling ability.

pulling face. In turpentining, a face advanced so high as to necessitate chipping with a puller.

pulling fodder. Stripping the leaves from growing corn plants just as they begin to turn brown. The pulled leaves or blades are usually bound together in small bundles and tied to the stalks to dry. Sometimes the tops of the plants are cut off just above the ear and handled in the same manner. Also called *stripping fodder.*

pulling leather. Designating a rider who hangs on to parts of the saddle with the hands to keep from being thrown from a bucking horse.

pullorin test. A test formerly employed in diagnosing pullorum disease of poultry. The test is performed by injecting into the skin of one wattle a small amount of the substance known as pullorin and noting the reaction obtained. This test is not sufficiently reliable for practical use in control and eradication of the disease. See *stained-antigen, rapid, whole-blood test.*

pullorum disease. A widespread, infectious, bacterial disease of poultry caused by the microorganism *Salmonella pullorum.* In baby chicks and poults the disease assumes an acute, septicemic form with a high degree of fatality, most losses occurring in the first two or three weeks of life. Adult fowls are less seriously affected but may harbor the organism and transmit the disease through infected eggs. The disease is largely controlled by the blood-testing of breeding stock to eliminate carriers of the infection, and by incubator and brooder house sanitation. *Pullorum disease* was formerly inappropriately called *bacillary white diarrhea (B.W.D.).* See *stained-antigen, rapid, whole-blood test.*

pullorum-typhoid clean. See *U(nited) S(tates) Pullorum-Typhoid Clean.*

pullorum-typhoid passed. See *U(nited) S(tates) Pullorum-Typhoid Passed.*

pullorum-typhoid test. See *stained-antigen, rapid, whole-blood test.*

pull plunger press. A hay baling machine so constructed that the hay is pulled rather than pushed into the compressing box. One or two horses are used on a sweep to supply power.

pull root. A root which contracts and thus pulls the stem of a plant deeper into the soil.

pull tail. To thin and shorten the tail of a horse.

pull weeds. To eradicate weeds by pulling out of the ground.

pulmonary mite. See *air-sac mite.*

pulmonary parasitism. The presence of parasites in the lungs.

pulp. 1. In paper making, the product obtained by digesting wood in a slightly alkaline or neutral sodium sulfite cooking liquor. 2. The juicy or fleshy tissue of a fruit.

pulpwood. Wood cut or prepared primarily for manufacture into wood pulp for subsequent manufacture into paper, fiber board, etc., depending largely on the species cut and the pulping process. The chief sources of pulpwood in the U.S. are the conifers, called soft woods; fir (*Abies* spp.); larch (*Larix* spp.); spruce (*Picea* spp.); pine (*Pinus* spp.); Douglas fir (*Pseudotsuga taxifolia*); and hemlock (*Tsuga* spp.). The more commonly used hardwoods are the maples (*Acer* spp.); birch (*Betula* spp.); poplar (*Populus* spp.); linden (*Tilia* spp.); elm (*Ulmus* spp.); American beech (*Fagus grandifolia*); and oak (*Quercus* spp.).

pulpy kidney. Enterotoxemia.

pulpy kidney disease. Enterotoxemia.

pulpy peat. Fine textured, non-fibrous peat derived from sediments and decay of lush aquatic plants. See *sedimentary peat: 1.*

pulque agave. *Agave atrovirens,* family Amaryllidaceae, a succulent, fleshy-leaved, tropical plant cultivated in Mexico as a source of the potent fermented drink, pulque. Native to America.

pulsator. A device on a milking machine which acts to break the steady suction on the teat cups and to produce a pulsating effect similar to the sucking of a calf.

pulse. 1. The edible seed of legumes, such as peas and beans. 2. The expansion and contraction of an artery associated with each heart beat, which may be felt with the finger.

Pultost. A sour, cows' milk cheese made in Norway. Also called *Ramost; Knaost.*

pulverized limestone. The product obtained by grinding either calcareous or dolomitic limestone so that all material will pass a 20-mesh sieve. It is applied to land to reduce soil acidity and to supply calcium as a nutrient, and it is mixed with commercial fertilizers to minimize their acid forming constituents. Also called *fine ground limestone.* See *lime: 1.*

pulverized oats. Oat chop.

pulverize (soil). To reduce clods, or peds, to a fine granular state.

pulverizer harrow. Acme harrow.

pulverulent. Powdery or dusty; of soil, designating a mass of soft crumbs and granules, but not necessarily dust or powder.

pummelo. *Citrus grandis,* family Rutaceae, a tree widely grown in sub-tropical regions for its large pale yellow, edible fruit similar to but larger and coarser than the grapefruit. Native to Asia.

pummice. 1. Those parts of the carcass of a sheep, or other animal, not of primary importance as meat food; the head, entrails, etc. 2. See *pomace: 1.*

pummy. 1. The residue from distillation of mint which may be used as a fertilizer. 2. Apple pulp, a by-product in cider making.

pump. A hydraulic machine, variously constructed for raising or transferring fluids. It consists essentially of a moving piece, moving pieces or a piston working in a hollow cylinder or other cavity, with valves properly placed for admitting or retaining the fluid as it is drawn or driven through them by the action of the pistons or blades, as an air *pump, a* rotary *pump,* centrifugal *pump,* etc. *Pumps* have many uses on the farm or ranch and in the production and processing of agricultural crops.

Pumpelly brome. *Bromus pumpellianus,* family Gramineae, a stout perennial grass, native at elevations of 5,000 to 9,000 ft. in the Black Hills of South Dakota, in Colorado, Idaho and other states (U.S.), where it furnishes good grazing for all kinds of livestock.

pumping pickle. A pickle forcibly introduced into meat for curing. It is usually five to ten degrees stronger than a cover pickle. See *pickle: 1.*

pump jack. A device for applying power to operate the pump rod of a deep well piston pump. Electric motors or internal combustion engines supply power through belts and/or gears to operate the device.

pumpkin. *Cucurbita pepo,* family Cucurbitaceae, an annual, prostrate, herbaceous vine of unknown origin grown for its edible fruit, the *pumpkin* of commerce. Varieties yield the squash *q.v.*

pumpkin bacterial wilt. A bacterial disease of pumpkin and squash caused by *Erwinia tracheiphila,* family Enterobacteriaceae, characterized by a wilting of the leaves. The plants are rarely severely damaged.

pumpkin flour. A dehydrated, finely-ground, coarsely-bolted mixture of the two vegetables, pumpkin and squash, used in cooking.

pumpkin wilt. Verticillium wilt.

pump-up type compressed-air sprayer. See *compressed-air sprayer.*

punch. 1. To herd cattle; to take care of cattle. 2. To forcibly drive sheep up hills when they are tired (New Zealand). 3. See *leather punch.*

punch-bar method. A method of fertilizing trees by punching holes in the ground, manually or without a soil auger, throughout the root area and filling them with fertilizer.

puncher. One who herds cattle; a cowboy. Also called *cowpuncher.*

puncturevine. *Tribulus terrestris,* family Zygophyllaceae, a pernicious, annual weed found in various parts of the U.S., especially in the southwest and California. It grows sometimes in fodder fields and the spiny fruits can cause both external and internal injury when fed with hay. Probably native to the Sahara region in Africa. Also called *bur nut; caltrop; Mexican sandbur.*

Punic bee. The Tunisian bee (*Apis nigra*) which is noted for its ferociousness.

punice foot. A bulging, convex sole on the foot of a horse.

punk. 1. A small, scrubby horse. 2. See *conch; knot: 4.*

punkies. A blood-sucking midge of the genus Culicoides, family Tendipedidae, which occurs in some wooded or marsh areas. Also called *no-see-um; sand fly.*

punk knot. A knot filled with fungous tissue.

punk tree. Cajeput tree.

punky. Designating a soft, weak, often spongy, wood condition which is caused by decay.

punky disease. See *cork: 3.*

pupa. The quiescent or inactive stage during which an immature insect or larva transforms to the adult. This period of metamorphosis is referred to as the pupal stage.

pupal stage. The period in the life history of insects between the caterpillar or grub stage and the mature or adult insect.

puparium. The hardened larval skin or protective case which encloses the living pupa.

pupate. To change, from an active, immature insect into the inactive pupal stage.

puratized apple spray. See *phenylmercuric acetate.*

puratized N5. Phenylmercury triethanol ammonium lactate *q.v.*

puratized nd-5. Phenylmercuri-w, 2,2-nitrilotrictanol; a chemical used in the prevention of citrus fruit decay.

purchase agreement. In the U.S., an agreement on the part of the Commodity Credit Corporation, an agency of the federal Department of Agriculture, to purchase from a producer, at the producer's option, not more than a stipulated quantity of a commodity at the support price. See *Commodity Credit Corporation.*

pure. A liquid manure, usually dog or pigeon dung in water, used in bating hides. See *bate.*

purebred. Designating an animal belonging to one of the recognized breeds of livestock. Such animals are registered or eligible for registry in the official herdbook of the breed. *Purebred,* registered, and pedigree stock, are often used interchangeably, and Thoroughbred is often improperly used for *purebred.*

purebreeding. The development of purebred livestock.

pure culture. 1. A crop of one species, variety, etc., grown in contrast to a customary mixture: *e.g.,* timothy grown from a pure seeding rather than mixed with clover. 2. A bacteriological culture which contains only a single species, or one which contains those desired for a particular purpose, as a yeast culture for making vinegar.

purée. The product obtained by straining sound, mature fruits or vegetables, raw or cooked, through a sieve that removes skins and seeds. It may be made with or without seasoning. A medium *purée* is a slightly concentrated *purée;* a heavy *purée* is more concentrated but not a paste *q.v.*

pure forest. A forest composed principally of trees of one species. In practice it is usually a forest in which 80 percent or more of the trees are of one species. In special instances, basal area or volume may be used as the basis.

pure line. A strain of organism that is comparatively pure genetically (homozygous) because of continued inbreeding, etc.

pure stand. A stand of trees in which at least 80 percent of the trees in the main crown canopy are of a single species.

pure wheat bran. The best grades of bran that have large clean flakes and contain no screenings.

purgative. A medicinal agent which actively empties the bowels. Aloes, arecoline, and calomel are examples of laxatives, drastic purgatives, and cholagogue purgatives.

purging nut. Barbadosnut.

purines. A group of organic chemical compounds, containing carbon, hydrogen and nitrogen, of which uric acid, the chief nitrogenous compound in the excrement of birds, is an example.

puring. The use of liquid manure to bate hides. See *bate; pure.*

purity. 1. Fleece that is all white wool; *i.e.*, free of dark fibers, kemp, or gare. 2. Freedom from extraneous or foreign materials.

purple-backed cabbageworm. *Evergestis pallidata* (Hbn.), family Pyraustidae, a caterpillar that usually feeds beneath a protecting silken web or burrows into the leaves of cabbages.

purple beech. A variety of the European beech (*q.v.*) grown for its purple foliage. Also called *copper beech.*

purplebell cobaea. *Cobaea scandens,* family Polemoniaceae, a woody vine grown as an annual in the greenhouse and in warm areas in a window box for its violet or greenish-purple flowers. Native to Mexico. Also called *cathedralbells; Mexican ivy.*

purple boneset. Bluestem joepyeweed.

purple brood. A disorder of the brood of bees that often appears in several of the southern states in the U.S. during May and June. It is not an infectious disease but a condition of the brood resulting from plant poisoning.

purple-cane raspberry. A hybrid between red and blackcap raspberries. Also called *purple.*

purple cestrum. *Cestrum elegans,* family Solanaceae, a shrub grown for its purple flowers. Native to tropical America.

purple clover. 1. *Trifolium wormskjoldi* (*T. involucratum*), family Leguminosae, a variety of clover which bears purple heads. Native to southwestern U.S. Also called *sierra clover.* 2. *Kennedia tabacina,* family Leguminosae (Papilionaceae), an herbaceous, perennial, woody vine, bearing large purple flowers, which has some forage value. Native to Australia.

purple cockle. Corn cockle.

purple coneflower. *Echinacea purpurea (Rudbeckia p.),* family Compositae, a perennial herb bearing light purple flower heads, sometimes grown as an ornamental. Native to the U.S. Also called *black sampson; hedgehog coneflower.*

purple cudweed. *Gnaphalium purpureum,* family Compositae, a woolly-foliaged herb found in eastern and central U.S. and valued as a wildlife plant. It is a lawn weed in the southern states. Also called *anaena* (Hawaii).

purple deadnettle. See *deadnettle: 2.*

purple eyespot. A fungous disease of maple caused by *Phyllosticta minima,* family Sphaerioidaceae, characterized by large, irregularly circular, brown spots surrounded by a broad, purple border.

purple foxglove. See *foxglove: a.*

purple fringed orchid. Meadow pink orchid.

purplefruit chokeberry. *Aronia prunifolia,* family Rosaceae, a hardy shrub which bears purple-black, berry-like fruit, grown for its white flowers. Native to North America.

purple granadilla. *Passiflora edulis,* family Passifloraceae, a climbing vine grown commercially in frost-free areas for its plum-shaped, dull purple, edible fruit. Native to Brazil. Also called *passion fruit.*

purple larkspur. *Delphinium bicolor,* family Ranunculaceae, an herbaceous perennial grown as an annual for its purple flowers which have brownish-yellow upper sepals. Native to the U.S. Also called *little larkspur.*

purple locoweed. Woolly loco.

purple lythrum. *Lythrum salicaria,* family Lythraceae, a shrub bearing profuse, purple-colored flowers, grown as an ornamental and as a bee plant since it yields large amounts of nectar. Native to Europe and Asia. Also called *purple loosestrife; red Sally; purple willow-herb.*

purple medic. Black medic.

purple mite. Citrus red mite.

purple mombin. See *mombin.*

purple mullein. *Verbascum phoeniceum,* family Scrophulariaceae, a biennial herb, the leaves of which form a green rosette on the soil surface. It is used as a border plant for its flowers which are soft shades of rose, mauve, and lavender. Native to Europe and Asia.

purple raspberry. Purple-cane raspberry.

purple rock cress. See *aubrietia: 2.*

purple scale. *Lepidosaphes beckii* (Newm.), family Diaspididae, an insect pest destructive of citrus.

purple-spotted lily aphid. *Macrosiphum lilii* (Monell), family Aphidae, an insect that infests lilies.

purple star thistle. *Centaurea calcitrapa,* family Compositae, a biennial weed found commonly in pastures. Native to eastern U.S. Also called *caltrops.*

purple stunt. A virus disease of loganberry in western Oregon, U.S. Canes are greatly stunted and remain purple throughout the growing season.

purple top wilt. A virus disease of potato and a number of vegetable crops caused by the aster yellow virus, characterized by stunting, chlorosis, sterility and the blossoms remaining green. See *aster yellows.*

purple trillium. *Trillium erectum,* family Liliaceae, a perennial, early-spring flowering herb grown for its white to pink or brownish-purple flowers. Native to North America. Also called *daffodilly; death root; dog flower; birthroot; bethroot; bumblebee root.*

purple vetch. *Vicia atropurpurea,* family Leguminosae, a viny, non-hardy, winter annual herb grown in the milder parts of California, U.S., for seed, as a green manure crop, and for hay. Native to southern Europe.

purple willow-herb. Purple lythrum.

purplewreath petrea. *Petrea volubilis,* family Verbenaceae, a woody, tropical vine grown in frost-free areas for its showy, blue flowers. Native to Mexico, Central America, and the West Indies. Also called *sandpapervine; bluebirdvine; queen's wreath.*

purpura hemorrhagica. An acute or subacute toxemic affection of horses marked by a rapid onset, small hemorrhages of the skin and body membranes, and swellings under the skin in many parts of the body.

purslane. *Portulaca oleracea,* family Portulacaceae, a fleshy-stemmed, low-growing, trailing, annual herb, a persistent garden weed. Relished by hogs, it is sometimes eaten as a salad and potherb by humans. Also called *pussley; duckweed.*

purslane sesuvium. *Sesuvium portulacastrum,* family Aizoaceae, a small, alkali-tolerant, prostrate, blood-red or purple, maritime herb with succulent stems, which is commonly found in the damp, coastal sands in warm areas of U.S., especially from Florida to Texas.

purslane speedwell. *Veronica peregrina,* family Scrophulariaceae, a common weed throughout the U.S. See *speedwell.*

purulent. Referring to pus, as a *purulent* discharge.

pus. A liquid inflammation product which consists of leucocytes, lymph, bacteria, dead tissue cells, and fluid derived from their disintegration.

push-in cage. A cage used to introduce a new queen bee into a colony. A 4-in. square of ordinary screen wire is bent along each edge and the corners clipped to form four sides, making a wire cage which is placed over the queen in an area of emerging brood and pushed into the face of the comb.

pushing. In plants, the swelling which precedes the initial opening stage of the bud.

push plunger press. A hay baler which has a long pitman that helps to push the hay into a box.

puss caterpillar. *Megalopyge opercularis* (J. E. Smith), family Megalopygidae, a stinging caterpillar that feeds on a large number of deciduous trees and shrubs.

pussies. Rabbitfoot clover.

pussley. Purslane.

pussy willow. *Salix discolor,* family Salicaceae, a deciduous shrub or small tree grown as an ornamental for the female catkins which are hairy and resemble a cat's paw. Native to North America.

pustular. Pertaining to, or covered with, pustules.

pustular spot. See *Coryneum blight of stone fruits.*

pustule. A small elevation on the skin that contains pus; a blister or pimple. On plants it may contain a spore mass.

putamen. The hard endocarp or pit of a stone fruit, as in cherries, peaches, and plums.

Putnam scale. *Aspidiotus ancylus* (Putn.), family Diaspididae, an insect that appears on orchard, shade, and forest trees.

put off. To lack the desire to copulate, as a bull.

putrefaction. The bacterial decomposition of proteins.

putrefactive. Causing decomposition or rotting.

putrefactive spoilage. Rotting of canned foods of low acidity shown by a distension of the can.

putrid. Decomposed, rotten; said of organic materials.

putty soil. Loosely, a soil puttylike when wet and extremely hard when dry yet containing a very small amount of clay. Some crayfish soils composed mainly of silt exhibit this phenomenon.

puzzolonic. Designating soils exhibiting the property of cohesion or cementation.

pycnidium. In certain fungi, the flask-like fruiting body containing conidia.

pyelonephritis. Inflammation of the kidney and the pelvis.

pyemia. A generalized infection in the blood stream caused by pyogenic organisms resulting in the occurrence of numerous abscesses in various parts of the body.

pygostyle. The flesh-covered, triangular bone plate formed by the union of vertebrae at the posterior extremity of a bird's body which supports the main tail feathers. Not found in rumpless fowls.

pyogenic. Pus producing.

pyometra. An accumulation of pus or mucopurulent matter in the uterus.

pyosepticemia. See *navel ill; omphalophlebitis.*

pyramidal. Designating a tree or shrub whose silhouette is generally triangular; like a pyramid or cone.

pyramid Chinese juniper. *Juniperus chinensis,* family Cupressaceae, an evergreen shrub or tree, of many varieties; grown as an ornamental. Native to Asia. Also called *Chinese juniper.*

Pyrenees columbine. *Aquilegia pyrenaica,* family Ranunculaceae, a perennial herb grown for its dark blue flowers. Native to the Pyrenees.

Pyrenees daisy. *Chrysanthemum maximum,* family Compositae, a perennial herb grown in the flower border as a biennial for its white flowers. The Shasta daisy was developed from this plant. Native to the Pyrenees. Also called *Pyrenees chrysanthemum.* See *chrysanthemum.*

Pyrenees violet. Horned violet.

pyrethrins. The active ingredients (*pyrethrins I* and *II*) of pyrethrum, which are derived from plants belonging to the genus Chrysanthemum, family Compositae. One of the oldest and most widely used insecticides, they control insects affecting man, animals, crops, and households. See *pyrethrum extract.*

pyrethrum. See *florist's pyrethrum; Dalmation pyrethrum.*

pyrethrum dust. A dust insecticide made by mixing pyrethrin extracts with talc, powdered charcoal, etc. See *pyrethrum powder.*

pyrethrum extract. An extract obtained by solvents from the ground flowers of species of chrysanthemum, chiefly *Chrysanthemum cinerariaefolium,* Dalmation pyrethrum, grown for commerce mainly in Kenya, Dalmatia, and Japan.

pyrethrum powder. An insecticide made from the ground flower heads of various species of chrysanthemums whose use as an insecticide is reported to have originated in the Trans-Caucasus region of Asia about 1800.

pyric. 1. Designating the fire factor in environment. 2. Designating modifications of soil and vegetation caused by fire.

pyriodoxine. Vitamin B_6, which appears to be necessary for the normal growth of chickens and pigs and for the prevention of a type of nerve disorder. Its distribution in ordinary feeds is fair.

pyroligneous acid. An acid liquor obtained by the destructive distillation of wood, especially of certain species of hardwoods. It is a complex mixture containing 80 to 90 percent water and many organic compounds including acetic acid and methanol. Also called *pyracetic acid; wood vinegar.*

pyrology. The scientific art of protecting forests from fire.

pyrophyllite. Hydrous aluminum silicate, a mineral used as a diluent or carrier for insecticide and fungicide dusts.

pyrophyte. A tree with a thick, fire-resisting bark, or one not subject to fatal damage from ordinary forest fires.

Pythium damping-off and stem rot of tobacco. See *Pythium seedling blight.*

Pythium root rot. One of the fungous root rots caused by species of the genus Pythium, family Pythiaceae. Usually more severe in moist soils, it infests wheat, oats, barley, corn, sorghums, etc.

Pythium root rot of millet. See *Pythium root rot.*

Pythium root rot of soybeans. See *Pythium root rot.*

Pythium root rot of sugar cane. See *Pythium root rot.*

Pythium seedling blight. A seedling blight and root rot of corn, tobacco, alfalfa, etc., caused by the fungus *Pythium arrhenomanes* Drechsler, family Pythiaceae.

Pythium seed rot of alfalfa and/or clover. See *Pythium seedling blight.*

Pythium wilt. A wilting of bean plants associated with infection of the roots and stem by the fungus *Pythium butleri,* family Pythiaceae.

Q

Quacheq. An ewes' milk cheese made in Macedonia.

quack grass. *Agropyron repens,* family Gramineae, a perennial grass, widespread in the northeastern and central states, U.S., it has become established on farm fields and is generally regarded as a pernicious weed. Spreading mainly by rootstocks or underground stems, it is difficult to eradicate. However, it has some value as a forage and hay plant and as a cover plant to prevent soil erosion. Native to Eurasia. Also called *couch grass; couch wheat; dog grass; false wheat (grass); quitch grass; quake grass.*

quack grass teeth. Sharp-pointed spring teeth for attachment to a spring-tooth harrow for use in eradication of quack grass.

quadrat. A small plot or sample area frequently one square meter or one milacre in size. In ecological field studies, a unit area in which vegetation and changes in species are recorded.

quadrivalent. In genetics, a group of four associated homologous chromosomes.

quagmire. Soft, miry ground, as a bog, marsh, or morass.

quail. Any of the various species of small upland, gallinaceous game birds belonging to the genus Coturnix and allied genera of the family Odontophoridae, found in many parts of the world and often erroneously called partridge. Adapted to agricultural lands, they are seldom regarded a nuisance and are frequently propagated for game purposes. The North American bobwhite *quail, Colinus virginianus,* is common in central and eastern U.S. The California *quail,* valley *quail* and mountain *quail* are common in western U.S.

quake grass. Quack grass.

Quaker bonnets. Sundial lupine.

Quaker ladies. Bluets.

quaking aspen. *Populus tremuloides,* family Salicaceae, a small to fairly tall, short-lived, deciduous tree widely distributed throughout the U.S., occurring in dense stands following logging or burning of forest lands. Its wood is now valued chiefly as a source of pulp for paper-making, and it furnishes browse for sheep, goats and deer and food for beaver. Native to North America.

quaking grass. Designating any grass belonging to the genus Briza, family Gramineae, of European origin naturalized in the U.S., sometimes grown as ornamentals. Also called *amourette; fairy grass.*

qualified appraiser. One who by education and experience has fitted himself to pass dependable judgment upon the value of properties, usually real estate.

quality. The composite of those characteristics that differentiate individual units of a product and have significance in determining the degree of acceptability of that unit to the user: (a) in wool, it is the fineness of wool fibers and sometimes strength and color; (b) in a horse, it is that refinement in bone and body which contributes to ability, beauty, and style; (c) in meat, it is color, texture, grain, and degree of marbling; (d) in milk, it may refer to flavor, chemical, or bacteriological analysis, and/or to organoleptic characteristics; (e) in fruits and vegetables, it is color, texture, maturity, and freedom from defects.

quamash. Camas.

quantity of wool. In sheep judging, the density, length and completeness of the fleece.

quarantine. 1. A regulation under police power for the exclusion or isolation of animal and plant pests or diseases and insects. (a) The isolation of an animal sick with a contagious disease. (b) A place where the sick are detained away from other animals until the danger of spread of a contagious disease has disappeared. In its wider application, the quarantine may be enforced against an individual animal, against all the animals, or all the animals of the same species, in a township, county, or state, and against those in a foreign country. 2. Prohibition to prevent the introduction or spread of any dangerous insects or plant diseases.

Quarg. Sour skim milk curd cheese which is usually not ripened.

quarter. 1. In slaughtering for meat, one half of the side of beef, as the forequarter or hindquarter. 2. Pertaining to a horseshoe, the branch between the last nail hole and the heel. 3. A unit of weight: (a) ¼ cwt. (avoirdupois); (b) eight bushels, formerly ¼ ton (especially of grain). 4. A quarter section of land or 160 acres. 5. A section of the bovine udder.

quarter blood. In grading wool or fleece of sheep, *quarter blood* length of fiber is 3½ ins. The American *quarter blood* compares with the English 50's and 48's, spinning grades.

quarter blood lamb. A one-quarter bred black-face lamb. Such lambs often show a mottle-colored face.

quarter boot. A leather piece fitted around a horse's forefoot to prevent injury from striking with the hind foot. See *shin boot.*

quarter clip. A clip on the shoe for the hind hoof of a horse. The clips are placed on the quarters of the shoe on the outside or inside to prevent the shoe from shifting laterally on the foot.

quarter crack. A vertical split in the wall of the hoof of a horse that results from improper hoof growth or shoeing. Also called *sand crack.*

quarter drain. In the sugar-cane system of surface drainage in Louisiana (U.S.), a small surface drain which carries water to the lateral drain or split ditch *q.v.*

quarter evil. See *blackleg: 4.*

Quarter Horse. A breed of horses developed in western U.S. for great endurance and high speed while being ridden over short distances. Especially useful as a cowpony on ranches, it has also been used for racing at a quarter mile.

quarters. 1. The parts of the body of a horse or other quadruped above the legs, as the breasts and hips. 2. A cabin or hut used to house workers on plantations, especially pre-Civil War times (southern U.S.). 3. Farm or ranch housing for domestic animals.

quarter section. According to the survey of public lands by the General Land Office (U.S.), one of four parts of a section of land, generally 160 acres. The square section of land was divided into four equal parts by a north-south and east-west line, and in a legal land description the quarters were designated by the points of the compass as N.E., N.W., S.E., and S.W.

quartet. A group of four spores formed by the two meiotic divisions.

Quartirolo. A soft, cows' milk Italian cheese made in Lombardy.

quartz. SiO_2, a mineral which is a crystalline form of silica. It is the most common mineral component of sand soils, especially in humid regions.

quassia. An insecticide derived from the wood of tropical trees of the family Simaroubaceae.

quaternary ammonium compounds. Very complex, organic chemical compounds used mainly as bactericides in sanitation, cleansing and sterilizing in the food and catering industries. Sometimes called *quats; Q.A.C.*

quayotto. Black grama.

Queen Anne's lace. See *wild carrot.*

Queen Anne's pocket melon. Dudaim melon.

queen bee. A fully developed, mated female bee, larger and longer than a worker bee whose function is to lay eggs. Formerly, it was believed that the *queen bee* was a male. See *drone; worker.*

queen breeder. One who breeds queen bees commercially or experimentally. Queen cells are removed from the cell building colony and introduced into the nuclei, which are placed in queen mating yards. See *baby nucleus.*

queen-cage candy. Candy made of powdered sugar, sugar sirup or honey, kneaded until it forms a stiff dough, used as a food in queen bee shipping cages.

queen cell. A special, elongated, wax cell resembling a peanut shell in which the queen bee is reared. It is usually at the bottom of the comb.

queen crapemyrtle. *Lagerstroemia speciosa,* family Lythraceae, a tropical tree, grown in frost-free areas for its showy, purplish flowers. Native to South Asia and Australia. Also called *queen's flower.*

queen excluder. A device made of wood and wire which permits worker bees to pass through but excludes the queen and drones. It is used to restrict the queen and her brood nest to certain parts of the hive.

queenlessness. A condition of a beehive which has lost its queen. This emergency results in supersedure or swarming.

queen-of-the-meadow. Willowleaf spirea.

queenpalm. *Arecastrum romanzoffianum,* family

Palmaceae, a feather palm widely cultivated by florists for its feathery, ornamental foliage. Native to South America.

queen-rearing yard. An outdoor yard in which nuclei are placed for the rearing of queen bees.

Queen's bird-of-paradise flower. *Strelitzia reginae,* family Musaceae, a trunkless, herbaceous perennial, grown for its beautiful yellow flowers with blue tongues and purple bracts. Native to South Africa.

queen's jewels. Mountainrose coralvine.

Queensland nut. The edible seeds of the trees *Macadamia integrifolia* and *M. tetraphylla,* family Proteaceae, considered by some to be the finest nuts in the world. Native to Australia; naturalized in Hawaii where its cultivation is estimated to be five times that given it in Australia. Also called *Australian nut; bopple nut; bopple-popple; bush nut; macadamia nut.*

Queensland pittosporum. Diamondleaf pittosporum.

queen's wreath. See *mountainrose coralvine; purplewreath petrea.*

Queen Victoria arborvitae. Eastern arborvitae.

quercitron. a. A yellow dye found in the inner bark of the black oak. b. Also, the black oak itself. See *black oak.*

Queso Anejo. A white, rather dry, skim milk Mexican cheese which has a crumbly texture.

Queso Blanco. The principal Latin-American cheese made from whole, cows' milk, partly skimmed, or skim milk, or from whole milk with cream or skim milk added.

Queso de Bola. A whole milk cheese similar to Edam made on a small scale in Mexico.

Queso de Cavallo. A pear-shaped cheese made in Venezuela.

Queso de Cincho. A sour-milk cheese made in Venezuela.

Queso de Crema. One of the principal cheeses of Costa Rica which resembles soft brick cheese.

Queso de Hoja. A cows' milk cheese made in Puerto Rico.

Queso del Pais. A white, pressed, semisoft, perishable cheese made in Puerto Rico for local consumption.

Queso de Prensa. A hard cheese made in Puerto Rico from cows' whole milk.

Queso de Puna. A cheese made in Puerto Rico which is like Cottage cheese except that it is molded in forms.

Queso Fresco. A rather dry cheese of the Cottage-cheese type which is made from skim milk in El Salvador and some other Latin American countries.

quick breaking emulsion. An emulsion in which the components separate rapidly.

quick coupler. A coupler used with a portable irrigation pipe that, through the use of split gaskets expanding under increased water pressure, effects a water seal between itself and the coupled section of pipe. Rapid coupling and uncoupling are possible with this special coupler.

quick cure. A pickling solution containing sodium nitrite or a combination of nitrite and nitrate.

quick decline. A virus disease of sweet orange or sour orange stock in California, U.S., characterized by the ashen color of foliage and curling of the leaves lengthwise and upward. A curling, wilting, and dropping of leaves, and a dying back of twigs takes place in the tops of the trees. See *decline: 3; tristeza.*

quick freezing. Freezing of products, fruits, vegetables, poultry, meat, dairy products, etc., for preservation under conditions in which the temperature of the product is lowered from 28° F. to −15° F. within 30 minutes. See *frozen food.*

quick-growing. Designating a plant in a particular climatic region that matures in a comparatively short period: *e.g.,* buckwheat and aspen are *quick-growing* among crops and trees.

quicklime. See *burnt lime; calcium oxide.*

quicksand. Any loose sand, usually fine grain, which is mobile or semifluid when supersaturated with water. It is encountered in excavations, especially in localities where soils are derived from, or the land surface underlaid by, lacustrine sediments or river alluvium. Also called *running sand.*

quick tests. Certain standard chemical tests devised for the very rapid determination, in the field or laboratory, of the amounts of nutrient elements in a soil and other chemical conditions. Such tests, although not complete or precise, may be useful, when properly interpreted, in making recommendations for fertilizers and soil amendments.

quick wet treatment. A rapid method of treating seed with fungicides in which the seed is thoroughly mixed with a small quantity of a concentrated liquid fungicide.

quiescence. See *dormancy: 1.*

quick heat. Lack of the usual nervousness, such as riding and other symptoms, that cows display when in heat.

quill. The hollow, horny part of a feather.

Quinby hive. A modern beehive that contains ten frames of the Quinby depth, 11¼ ins., and the Langstroth length, 17⅝ ins., spaced 1⅜ ins. from center to center. Also called *Jumbo hive.*

quince. *Cydonia oblonga* (*Pyrus cydonia*), family Rosaceae, a deciduous tree grown for its edible fruit, the *quince* of commerce, used in making preserves and jellies, and as an ornamental for its white flowers. Native to Asia. Also called *apple of Hesperides.* See *floweringquince: 1.*

quince blotch. See *Phoma fruit spot of apple and quince blotch.*

quince curculio. *Conotrachelus crataegi* Walsh, family Curculionidae, an insect which infests the quince, eating cavities in the fruit and causing it to be knotty and misshapen. The insect resembles the plum curculio but winters in the soil as a grub.

quince rust. A fungous disease of quince, apple, pear, and serviceberry caused by *Gymnosporangium germinale* (*G. claviceps*), family Pucciniaceae. The aecial stage is found on fruits or stems of the susceptible pome fruits and the telial stage infects stems and trunks of *Juniper* spp., particularly the lowbush juniper, *J. communis.* Control on pome fruits involves the destruction of nearby junipers. *Cf. apple rust.*

quincunx. Triangular planting.

quinoa. *Chenopodium quinoa,* family Chenopodiaceae, a plant used as a food in the Andes and the Mexican Cordilleras: the leaves as greens or a potherb, and the boiled seed as a substitute for rice. Also called *Peruvian rice.*

quintal. A measure of weight, 100 lbs.

quirt. A flexible, leather whip which is made of thongs of knotted rawhide mounted on a short stock about a foot long. It was used mainly by horseback riders in the early period of development of western U.S.

quitch-grass. Quack grass.

quit-claim deed. A document by which the seller conveys all his present rights to the real estate in question but does not give any further warranties as to the title. The use of the *quit-claim deed* is usually reserved for clearing title and settling the rights of persons who have some claim to the property being transferred.

quit rent. A rent paid by a tenant of the freehold by which he goes quit and free; *i.e.,* discharged from any other rent.

quittor. A discharging sore on the coronet of a horse's hoof; necrosis of the lateral cartilage of the third phalanx.

quota. 1. In the U.S., a means of regulating the production and marketing of commodities when supplies become excessive. It is always used in conjunction with acreage allotments and the *quota* is the quantity produced on that allotment. Following the proclamation of a need for a marketing *quota* for a specific crop by the Secretary of Agriculture, a referendum is held. If at least two-thirds of the farmers eligible to vote on the marketing *quota* do not approve it, the *quota* will not be put into effect. Under existing legislation in the U.S., the only commodities for which marketing *quotas* can be proclaimed are upland cotton, extra long staple cotton, wheat, rice, peanuts and tobacco. This includes all basics, except corn. 2. The amount of a commodity that may be imported into one country from a source outside that country. 3. A definite amount of any product, feed, etc.

quotation. The market price of a commodity.

Q virus. First reported as a casual agent of chrysanthemum mosaic (*q.v.*), now believed to be a complex disease caused by virulent mosaic virus and mild rosette virus.

R

rabbit. Any of certain mammals of the family Leporidae. Hare and *rabbit* are often used interchangeably. However, the biologist classifies the *rabbit* as having shorter and smaller legs and ears, and giving birth to naked, helpless young with eyes closed, while the newborn hare has fur and is quite able to care for itself. Species of the genus Sylvilagus, cottontail *rabbits* such as the eastern cottontail (*S. floridanus*), Rocky Mountain cottontail (*S. nuttali*), and western cottontail (*S. auduboni*), are considered rodents and may be injurious to gardens, shrubs and trees. The genus Oryctolagus includes the domesticated *rabbits* raised for their meat or fur, such as the Belgian and Angora *rabbits.* However, the San Juan or European *rabbit* (*O. cuniculus*), which originally came from England, was introduced into San Juan Island and thence to the mainland of the U.S., lives in colonies much like the prairie dog, and is considered a nuisance. See *Angora rabbit; Belgian: 2; cottontail.*

rabbit breeder. One whose work it is to improve rabbits through selection and breeding.

rabbit control. See *myxomatosis.*

rabbit coop. A wire cage used to contain rabbits for feeding, breeding, and show purposes. Also called *hutch.*

rabbit ear. A virus disease of several plants, as lettuce, celery, and asters, characterized by the foliage turning yellow. Also called *yellows.*

rabbitear iris. *Iris laevigata,* family Iridaceae, a perennial herb grown for its blue or white flowers. It is a beardless iris. Native to Asia.

rabbiteye blueberry. *Vaccinium virgatum,* family Ericaceae, a bushy shrub which is more heat and drought resistant and requires a shorter cold rest period than the highbush blueberry. It is grown in southeastern U.S. for its black, edible fruit and as a source of improved blueberry strains. Native to North America.

rabbit feed. A prepared feed which usually consists of ground corn, oats, bran, oil meal, beet pulp, molasses, minerals, and vitamins, usually pressed into pellets.

rabbitfoot clover. *Trifolium arvense,* family Leguminosae, a grayish-green, silky-haired, annual legume with slender upright stems and heads with very small flowers, found on poor soils throughout the U.S. The long hairs mat together to form hair balls in the stomach of animals grazing on it. Native to North America. Also called *stone clover; old-field clover; pussies.*

rabbit pen. Rabbit coop.

rabbitproof fence. A fence of special design which is placed around cultivated fields in New South Wales, Victoria, and Western Australia to keep out rabbits (*Oryctolagus cuniculus*) which have overrun the country and have become a very serious pest.

rabbit's-tail grass. Harestail.

rabbit test. 1. A pregnancy test in which virgin female rabbits are used. 2. A test for sweetclover poisoning in which rabbits are fed sweetclover hay and if no injurious effects appear the hay is considered safe for cattle.

rabbri. A confection which is made in India from the skin that forms when sweetened milk is boiled.

Rabi. In India, the important group of crops, including cereals (especially wheat), fodders, mustard, etc., sown in October and November and reaped in March and April. *Cf. Kharif.*

rabies. An infectious disease caused by a filterable virus which is communicable by means of the bite in which saliva containing the virus enters the wound. It occurs most frequently in dogs, but many other animals and man are quite susceptible. The virus has been found in the saliva at least five days before the animal showed recognizable symptoms which are melancholy, excitation, violence, and paralysis. Also called *hydrophobia; canine madness.*

raccoon. *Procyon lotor,* family Procyonidae, a nocturnal mammal, which is gray, with black and white facial markings. It eats and does some damage to corn especially when the kernels are well-developed but soft.

raccoon grape. Frost grape.

race. 1. A group of individual plants which have certain common characteristics because of ancestry. It is generally a subdivision of a species frequently used in plant breeding. 2. Pathogens of the same species and variety which are structurally indistinguishable but which differ in their physiological behavior, especially in their ability to parasitize varieties of a given host. 3. A human aggregate which is distinguishable from other aggregates in terms of genetically inherited characteristics, either morphological or metrical in nature. 4. The channel that leads water to or from a water wheel, the former is head *race,* the latter, tail *race.* See *millrace.* 5. A narrow passage or fenced land in a sheep yard for branding, dipping, etc. (Australia and New Zealand). 6. The elongated white mark on the face of a horse or dog. 7. To compete for top speed as in a horse *race.*

raceme. A flower cluster in which the pedicels or supports of the flowers are of equal length and attached along a common axis, the sequence of blooming being from below or upward.

racemised casein. Casein in a dilute alkaline solution which gradually loses its optical activity until a final low constant value is obtained. *Racemised casein* is not attacked by proteolytic enzymes or digested by animals.

race saddle. A very light saddle which weighs about 1½ lbs. used for racing.

race-way. Millrace.

rachilla. The axis to which the spikelet is attached in grasses and grains; a secondary rachis *q.v.*

rachis. 1. The elongated axis to which the pedicels of an inflorescence are attached. 2. The extension of the petiole to which the leaflets of a compound leaf are attached.

racing. The holding or competing in races as a business or sport, as in horse racing or dog racing.

racing snaffle bit. A bridle bit with one joint in the center of the mouthpiece which is especially designed for horse racing.

racing stable. One or more horses under one management which are used for racing.

rack. 1. The gait of a horse in which only one foot touches the ground at any one time, producing a four-beat gait. Characterized by a display of knee action and speed, it is exhausting, and thus rarely executed voluntarily. Also called *single-foot; four-cornered gait.* 2. A frame attached to a truck or wagon for the transportation of straw, hay, tobacco, etc.; any vehicle so fitted. 3. A framework for holding feed for cattle, swine, sheep, etc., with upright partitions so that the animal can insert its head between the partitions and have access to the feed. 4. A frame placed in a stream to prevent passage of fish. 5. A frame placed at the entrance to a pump sump to remove debris that would clog the pump.

rack and cloth press. Built-up layers comprising closely-woven, heavy cloths, folded to hold crushed fruit (as ground apples or grapes) between wooden racks constructed of hard slats. Several such layers are built prior to processing the fruit in a cider or juice press.

rack hay drying. A method of drying hay used in the humid climates of northern Europe in which the freshly-cut hay is placed on wooden racks built up off the ground. *Rack hay drying* is most often used only in small scale operations. Called *clothesline hay drying* when wires are used to support the hay rather than wooden racks.

rack meat. 1. Fodder or hay placed in racks as feed for horses. Also called *manger meat* (England). 2. The foresaddle of any carcass.

rack-vintage. Wines drawn from the lees.

radde bellflower. *Campanula raddeana,* family Campanulaceae, a perennial herb grown for its solitary, dark purple flowers. Native to the Caucasus.

raddle. 1. An attachment for a threshing machine consisting of an endless apron, used to convey straw from the thresher cylinder (now obsolete). 2. To mark or color highly.

radekrankheit. Nematode disease of wheat.

Radener. A hard cheese of Mecklenburg, Germany, made from cows' skim milk in much the same way as Swiss except that the curd is pressed less.

radial extractor. A device used for extracting liquid honey from the comb by means of centrifugal force.

radial well. Usually a large diameter, dug well, which has collection pipes and screens driven horizontally from the bottom of the excavation to increase the inflow of water.

radial wood surface. A longitudinal wood surface or plane which extends wholly or in part from the pith to the bark. Also called *quarter-sawed.*

radicle. 1. The rudimentary root of the embryo. 2. In chemistry, pertaining to a chemical radical.

radioactivated milk. Milk treated by cathode rays through special electrical equipment which sterilizes the milk without a sensible rise in the temperature. Also called *cold sterilization.*

radioactive soil. A soil which possesses a minute degree of radioactivity. Radioactive phosphorus and other plant food elements are sometimes added to soils in research work to discover more facts about their functions in plant growth.

radioisotope. A radioactive isotope of an element. *Radioisotopes* are being extensively used as tracers (*q.v.*) in many areas of scientific research including agriculture.

radish. Garden radish.

radish black rot. See *black rot of crucifers.*

radius cruising. The distance between locations at which an individual animal is found at various hours of the day, at various seasons, or at times during various years.

Radolfzeller Cream cheese. A semisoft cheese very similar to Munster made in the vicinity of Lake Constance (Switzerland) from fresh, whole cows' milk coagulated at 86° F.

raffia. Madagascar raffiapalm.

rafter. 1. A beam which gives slope to a roof and supports the roofing. 2. To plow land so that a space is left between furrows by turning the grass side of each furrow upon an unplowed ridge (England).

rag. A parchment-like membrane found in certain fruits: *e.g.*, in dates around the pit and in oranges around the sections.

rag doll. A method of germinating corn kernels by placing them on a strip of cloth which is then rolled, soaked in water, and stored at the proper temperature for germination. After a few days the cloth is unrolled and the percentage germination determined.

ragged. Of the fur, etc., of animals, shaggy, rough, and hanging in tufts.

ragged boring. A serious body defect of butter with a sticky crumb texture which gives it a ragged appearance.

ragged hips. Irregular or poorly conformed rear quarters of an animal.

ragged robin. *Lychnis floscuculi,* family Caryophyllaceae, a hairy, perennial herb grown for its red or pink flowers in loose clusters. Native to Europe and Asia. Also called *cuckoo flower.*

ragmillet. Pearl millet.

rag iron. The adjustable part of the corn sheller which permits the shelling of both large and small ears.

rag sumpweed. *Iva xanthifolia,* family Compositae, an annual weed with straight erect branches and drooping heads of greenish-white flowers. Ingested by cattle, its leaves can cause dermatitis and a bitter flavor in milk. The pollen may cause hay fever in humans. Native to North America. Also called *marsh elder.*

ragweed. *Ambrosia artemisifolia,* family Compositae, an annual weed with a bitter taste, which bears greenish-yellow flowers and produces an abundance of pollen during the late summer and early fall which causes hay fever. Native to North America. Also called *Roman wormwood; hogweed; bitter weed; sarriette.*

ragwort groundsel. *Senecio jacobaea,* family Compositae, a perennial herb, sometimes 4 ft. tall, grown for its yellow flowers. Native to Europe. Also called *tansy ragwort; stinking willie; ragwort.*

rail. See *rail fence.*

rail fence. A fence made of split logs, called rails, laid in a zigzag line, crossed at their ends, and sometimes held in place by stakes crossing slantingly. It was a very common way of farm fencing during the early settlement of the forested region of eastern U.S. In a second type of *rail fence,* the split logs were laid in a straight line, and mortised in posts or held together by stakes at the ends. Also called *worm fence; stake fence; rider fence.*

railroad tie. In railway construction, a wooden, beamlike, transverse support on which the steel rails are placed and fastened. The production of *ties* furnishes supplemental income for many farmers in U.S. Also called *tie; sleeper.*

railroad worm. Apple maggot.

rain. Precipitation in the form of water condensed from the atmosphere which falls in drops. See *precipitation.*

rain belt. A region of heavy or above average rainfall.

rainbow flower. Iris.

rain drop. See *rain.*

rainfall. The amount of precipitation, expressed in the form of water but including snow, sleet, and hail, which falls in a given time. In the U.S. it is usually measured in inches of water. See *precipitation.*

rainfall intensity. The rate at which rainfall occurs expressed in inches per hour.

rainfall interception. Interception of the fall of raindrops by a canopy of vegetation or vegetative residue, usually expressed as inches per depth retained.

rainfall penetration. The depth below the soil surface reached by a given quantity of rainfall.

rain forest. Properly, an equatorial forest where the annual rainfall is in excess of 80 to 100 ins. The Douglas fir forest of the U.S. Pacific Coast is also sometimes called a *rain forest.*

rainlily. Any of the several plants of the genus Cooperia, family Amaryllidaceae. Species are bulbous herbs some of which are grown for decoration. See *eveningstar rainlily; redtinge rainlily.*

rainmaker. 1. A rain-simulator used for applying water in drops upon a plot surface. See *infiltrometer.* 2. An individual who claims to be able to increase the rainfall of an area or locale. See *cloud seeding.*

rain-rot in sheep. A physiological condition of heavily fleeced sheep exposed to prolonged rains in which occasionally enough moisture penetrates to the skin to result in loosening of the wool so that patches start falling out.

rain shadow. The region of diminished rainfall on the lee side of a mountain mass where the rainfall is noticeably less than on the windward side.

rainstorm. A storm accompanied by rain.

rain-tight. Referring to a construction which, when exposed to rain, does not allow water to enter or penetrate through it.

rain tree. *Samanea samon* (*Pithecellobium s.*), family Leguminosae, a handsome, tropical annual with spineless branches, grown in warm areas as a shade tree and for its yellowish flowers in long, crimson stamens. Its sweet, pulpy pods are sometimes eaten by cattle. It is erroneously so named on the supposition that its leaflets exude water. Native to Central America and the West Indies. Also called *algarroba; monkey pod; zaman.*

rain water. Water that falls from the atmosphere in the form of rain, at times being caught and directed into cisterns for further use.

rainy. Designating a period in which rain falls more or less continually.

rainy season. In the tropics, the annual season of heavy or more than normal rainfall as contrasted to the dry season of little or no rainfall.

rainy spell. Frequent rainfall for a protracted period of time, usually several days.

rainy weather. A period of relatively short duration during which there are numerous showers of rain as contrasted to rainy season.

raise. To grow or produce, as to *raise* corn.

raised by hand. Designating young lambs, calves, or other livestock fed from birth out of bottles, specially-built pails, etc., in contrast to those nursed from birth by the dam.

raised grain. A roughened condition of the surface of dressed lumber in which the dense late wood is raised above the less dense early wood but not torn loose from it.

raisin. The dried fruit of certain varieties of grapes, such as Muscat, Thompson seedless, and Black Corinth. See *currant: 2.*

raising. Erecting the assembled framing of a barn or making the structure ready for putting on the roofing and siding. In former times in the U.S., neighbors came from miles around to help with the *raising.* It was quite a social event in a pioneer community and plenty of food and drinks were served by the farmer and his wife. Also called *barn raising.*

raisin moth. *Ephestia figulilella* Greg., family Phycitidae, an insect that infests grapes. Its larvae feed on the foliage and cover the clusters of developing fruit with a disfiguring web.

raisin variety of grape. See *raisin.*

raisin wine. The beverage made by the alcoholic fermentation of an infusion of dried or evaporated grapes.

rake. 1. A farm tool which consists of several bars or head pieces or wheels equipped with tines which are caused to rotate and move swath hay or other material into a windrow for ease in picking up with a loader or baler. 2. A hand tool consisting of teeth inserted in a head piece attached to a long handle; used for collecting hay and leaves and breaking and smoothing of the soil.

rake-bar hay loader. Fork loader.

rake teeth. The teeth or tines of a rake *q.v.*

raking. 1. The smoothing and pulverizing of soil, particularly of gardens, before seeding. 2. The collection of hay, grass, leaves, etc., with a rake *q.v.*

rales. Abnormal lung sounds in cases of pneumonia or lung inflammation.

ram. A mature male sheep usually used for breeding purposes.

rambler. A climbing rose of the multiflora type.

Rambouillet. A fine wool breed of sheep first developed in France from earlier Spanish Merinos. It is a good wool and mutton breed very popular in the sheep section of the western range in the U.S.

ramie. *Boehmeria nivea* (*Urtica utilis*), family Urticaceae, a plant growing 6 to 8 ft. and about ½ in. in diameter and bearing between the bark and the woody core a white, lustrous, strong fiber, the *ramie* of commerce, which can be spun and woven. Native to China and Japan, it is an important commercial crop there. Also grown to a very limited extent in southern Florida, U.S.

ramie leaf meal. A livestock feedstuff, the ground product consisting chiefly of leaves from the ramie plant. It must not contain more than 18 percent of crude fiber.

ram jacket. A jacket, designed to prevent coitus, placed on a ram to permit it to run with the flock before the breeding season.

ram lamb. A male sheep under one year of age.

ramontchi. *Flacourtia indica,* family Flacourtiaceae, a spiny tree grown in very warm areas for its round, red, edible fruit. Native to South Asia and Madagascar. Also called *governor's-plum; Batoko plum.*

ramoon. The foliage, used as fodder, of the tropical tree, *Trophis Americana,* family Moraceae, native to West Indies and Central America.

ramp. An inclined plane serving as a passage-way between two different levels; commonly used in loading livestock and machinery on trucks or trailers.

rampion. *Capanula rapunculus,* family Campunulaceae, a usually biennial herb bearing bell-shaped, lilac-like flowers. Its leaves and roots are sometimes used in salads. Native to Europe.

ramps. Ramsons.

ram service. The fee charged for the breeding of a ewe.

ram's-head ladyslipper. *Cypripedium arietinum,* family Orchidaceae, a hardy, slender stemmed herb, grown in the bog garden for its purple flowers whose lips are whitish and crimson streaked. Native to North America and China.

ramsons. *Allium ursinum,* family Liliaceae, a bulbous herb with a pungent taste and odor. Its growth is especially objectionable in pastures for milk cows. It is most common in the south central U.S. in the spring. Also called *buckram; ramps.* See ***wild leek.***

ramsted. Butter-and-eggs toadflax.

ranch. 1. A large tract of land including headquarters, pastures, corrals, and personnel, used for the raising of livestock, especially cattle (Western U.S.). 2. A very small tract of land with house regardless of use of land. (U.S. Census shows *ranch* and *farm* without distinction.)

ranch boss. The foreman of a large ranch.

ranch cattle. Any of the several crossbred varieties of cattle raised as beef animals. The most important of these are: the Santa Gertrudis (*q.v.*); the Beefmaster, a three-way cross of Brahman (Zebu), Hereford, and Shorthorn; and the Brangus, an animal which is about 37 percent Brahman (Zebu) and 63 percent Aberdeen Angus. Others include the Braford, a cross between the Brahman (Zebu) and the Hereford; and the Charbray, an animal 12.5 to 25 percent Brahman, the main strain being Charolaise. The purpose of this crossbreeding has been to produce beef animals better adapted to tropical climates and more resistant to tick fever.

ranch conservation plan. A statement of objectives in conservation and management on a given ranch or range together with the recommended course of action, supported by inventory data and analysis, to achieve them. It may consist of written, tabular, and graphic sections.

ranched. Designating an animal on a ranch.

rancher. One who owns, occupies, operates, or works on a ranch.

ranchero. A rancher.

ranch hand. A person hired to work on a ranch tending animals, making hay, mending fences, etc.

ranch headquarters. The usual place of residence of the rancher and his place of business.

ranch house. The main residential building on a ranch.

ranch land. The land being used or suitable for ranch purposes.

ranchman. A rancher.

ranch management. Plans and regulations for the operation of a ranch.

rancho. A ranch or large tract of grazing land for raising cattle and horses, as distinguished from a hacienda *q.v.*

ranch operations. See *ranch management.*

ranch unit. An operating unit of grazing land in the western U.S. consisting of individually owned land and permits to use public land issued by the U.S. Forest Service under the Taylor Grazing Act.

rancid. Designating an offensive smell or taste resulting from the chemical transformation or putrefaction of fat, butter, milk, ice cream, and other products.

rancidity. State of being rancid *q.v.*

Randolph plow. A hillside type of plow, named after Thomas Mann Randolph of Virginia, U.S. used in early American farming. Said to have been used on the farm of Thomas Jefferson, the third president of the U.S.

randomized block. An experimental design for testing crop varieties or strains by dividing them into groups and assigning treatments at random to the units within each group. The allotment is done so that all units in the same group or block are as nearly identical as possible.

random sample test. Under the U.S. National Poultry Improvement Plan (*q.v.*), an official test of the performance of progeny, conducted at a neutral location under the supervision of an official state agency and performed on two or more participants. See *random sampling; U.S. Performance Tested Parent Stock.*

random sampling. A method of sample selection in which each individual or experimental unit stands the same chance of being selected.

range. 1. Uncultivated land, including forest land, which produces forage suitable for livestock grazing. 2. Specifically, a unit of grazing land used by an integral herd of livestock. 3. Ecologically, the geographic area of natural occurrence of certain plants or animals. 4. In the U.S. land survey, a tier of townships, according to number east or west of a principal meridian line. See *township:* 2. 5. See *greenhouse range.* 6. The difference between extremes.

range administrator. A ranger.

range allotment. A specific range area to which certain livestock of a permittee or group of permittees are assigned. Also called *grazing allotment.*

range appraisal. A definite, written, detailed opinion by a qualified appraiser of the value of range land. Among the considerations involved are its grazing capacity, accessibility, facilities for handling livestock, availability of feed sources, and income producing ability.

range band. A large flock of sheep handled as one unit on the range.

range bull. A bull used for breeding purposes on a range.

range caterpillar. *Hemileuca oliviae* Ckll., family Saturniidae, a range pest on wild grasses in the southwestern U.S., which sometimes infests corn and other cultivated plants. Its larvae are covered with coarse, poisonous spines.

range cattle. Cattle raised under range conditions.

range classification. A system of classifying grazing areas, based on such items as cover, forage, season for use, and suitability for use by different kinds of livestock.

range condition. The state of health or productivity of both soil and forage of a given range in terms of what it could or should be under a normal climate and under the best practicable management.

range conservation plan. A statement of objectives in range conservation and management on a given ranch or range together with recommended course of action to achieve them, supported by inventory data and analysis.

range control. The handling of livestock on a range to bring about the desired production and use.

range count. A census made on a range of the animals using a grazing area as contrasted to feedlot, corral, driveway, or other similar counts.
range country. See *range.*
range crane fly. *Tipula simplex* Doane, family Tipulidae, an insect whose dark-colored, leathery maggots worm into and destroy plant roots. It is sometimes destructive on the ranges of southwestern U.S.
range ecology. The specialized branch of ecology which deals with vegetational response to environmental factors on range land, especially with the effects of grazing.
range economics. The scientific study of the relationship of the use of grazing lands by livestock to the material and social welfare of individuals, communities, and the nation.
range ewe. A ewe raised on a range.
range fattening. The fattening of steers or other livestock for market on a range.
range fire. On range land, any fire which is not being used as a tool in range management.
range grasses. Grass vegetation on the range areas. These are of three types recognized in western U.S.: (a) tall grasses, such as big bluestem; (b) medium grasses, such as wheatgrass; (c) short grasses, such as buffalograss and blue grama (*q.q.v.*). These grasses vary in intensity and character of stand depending on the nature of the range, whether tall grass prairie, short grass plains, or mesquite-grass desert.
range improvement. Increase in the utility, and ease of management, of the ranges secured by various means, such as development of water supply, fencing and other facilities, revegetation, and noxious plant control.
range indicator. Any plant community portraying the condition of its environment which can be used as an indicator for the condition of a range.
range inspection. A visit to a range for examining its condition to determine errors in the existing management and to gather data for remedial measures.
range inventory. The compilation of all important facts needed for perfecting a sound range management plan.
range lambing. Permitting ewes to drop their lambs on the range under approximately natural conditions of shelter and forage.
range land. See *range.*
range management. The scientific administration of a range for the continuous production of forage and livestock as consistent with use of the land for other purposes.
range of glass. See *greenhouse range.*
range paralysis. See *fowl paralysis.*
range plant. Any herbaceous or shrubby plant which grows on a range and forest land.
range plant cover. The ground vegetation composed of all herbaceous and shrubby plants which are within easy reach of livestock on a range area regardless of whether or not they constitute forage.
range-potential approach. Classification of range units to express their current production in terms of their ultimate potential.
ranger. An administrative officer in charge of a unit of forest land, usually a subdivision of a public forest or park. Various classifications are recognized, as a forest *ranger,* district *ranger,* county *ranger,* etc.
range-raised. Designating livestock raised on the range.
range readiness. The stage of growth of the important palatable plants on the range and the condition of soil which permits grazing without undue compacting of the soil or endangering the maintenance of the plants.
range reconnaissance. A range or grazing survey of an extensive or preliminary nature.
range rights. In the U.S. Old West, assumed rights to the range for grazing livestock by the ranchers in opposition to the farmer and small herd owner.
range sheep. Sheep handled in bands of one to two thousand on a range.
range sheep industry. See *range sheep.*
range shelter. See *summer-range shelter.*
range subtype. A subdivision of a range type for grouping range ares of similar character as regards composition, density, or condition due to ecolgical or use factors by which they may be differentiated from other range areas.
range survey. 1. A determination by inspection of the carrying capacity of extensive areas of natural vegetation, based on the density and palatability of the forage. 2. An inventory which assembles important facts needed for perfecting a sound management plan taking into account the number of animals grazed, grazing capacity, period grazed, salt plan, needed water development, areas in need of revegetation, better livestock distribution, and the location of special problem areas.
range technician. A person who assists farmers and ranchers in making the best use of their ranges by collecting and compiling all information pertaining to proper range management, analyzing it, and collaborating with the operator in the formulation of range use plans which are technically and economically sound.
range type. An area of range which differs from other areas primarily by the difference in the aspect of its plant cover, such as grassland, browse, or conifer. One vegetation group can be distinguished from another group by difference of dominating species.
range type description. The field notes taken for a range type or subtype in making a range survey including estimates of density and composition, record of proper use factors, appraisals of range and soil conditions, and other information pertinent to management. Also called *type writeup.*
range utilization. 1. For a single plant or species, the degree to which the foliage or herbage has been removed in percentage of the current growth within reach of livestock. 2. For an entire range or other unit, the relative amount eaten

by animals including domestic livestock, game, rodents, or insects. Range utilization is occasionally referred to in percentage of proper utilization but more often in general terms, such as lightly-grazed, heavily-grazed, or overgrazed. See *stocking: 1.*

range utilization standards. A yardstick or indicator guide used in judging the closeness of grazing allowable on a range area in the interest of livestock and the natural resources, such as forage, soil, and water.

range waste. Any portion of a range area, not including barren areas, which is unusable by livestock.

Rangiport. A cheese usually made from cows' milk in the Department of Seine-etoise, France, which is very similar to Port du Salut.

Rangoon creeper. Persian buttercup.

rangy. Designating an animal that is long, lean, leggy, and not too muscular in appearance.

rank. Designating vegetative growth that is luxuriant, vigorous, and often coarse.

ranny. A poor quality calf.

ranting. The restless activity of a boar as it matures sexually and as the breeding season approaches. It is usually accompanied by loss of appetite and weight.

Raoul's method. A scientific test used in the detection of water added to milk which is based on the principle that a solvent depresses the freezing point of a liquid.

rape. See *winter rape.*

rape-seed meal. A livestock feedstuff obtained from grinding the cake produced when winter rape seed is pressed to recover the oil. It contains approximately 5.7 percent nitrogen, 1.7 percent phosphoric acid and 1.2 percent potash.

rapeseed oil meal. Rape-seed meal.

rapid abnormality indicator. An electrical apparatus used for the rapid detection of abnormal milk.

rapid plant bug. *Adelphocoris rapidus* (Say), family Miridae, an insect that infests young cotton, alfalfa, and sweet clover.

rappee. An old name for snuff made by rasping tightly wound rolls of tobacco leaves known as *carrots.*

rareripe. Descriptive of a fruit or vegetable that matures early.

rasorial. Referring to the animals which scrape the ground to obtain food, as a fowl.

rasp. A filelike tool used to wear away excess when trimming an animal's hoof.

rasp-bar. A roughened, bar-type threshing cylinder used in combines for harvesting as compared to a threshing cylinder with teeth.

raspberry. Any of the various species of Rubus, family Rosaceae, consisting of brambly canes which yield red, black, purple, and yellow berry-like fruit used fresh or frozen for canning, jams and preserves. Of the Rubus spp., only the *raspberry* torus stays on the plant when the fruit is picked. See *blackcap raspberry; red raspberry.*

raspberry anthracnose. See *cane fruit anthracnose.*

raspberry cane blight. A fungous disease caused by *Leptosphaeria coniothyrium,* family Sphaeriaceae, which enters the bark of canes through insect wounds or mechanical injury causing brown dead areas to appear between blossoming and fruiting. The berries dry up and the canes become brittle.

raspberry cane borer. *Oberea bimaculata* (Oliv.), family Cerambycidae, a beetle whose larvae cut holes through the bark of raspberries and enter the pith, causing wilting of the young shoot and swelling and purple discoloration at point of entry. Canes frequently die and break at the point of swelling.

raspberry cane maggot. *Pegomya rubivora* (Coq.), family Muscidae, a fly whose larvae feed in the ring of wood outside the pith and girdle the canes of raspberries, causing them to wilt and break off.

raspberry fruit rot. A fruit rot of raspberries associated with excessive moisture at harvest time and infestation with several different fungi.

raspberry green mosaic. A virus disease of raspberries which causes a gradual loss of vigor, reduced yields, and lower quality of the fruit. Symptoms consist of a mottling of light and dark green areas on the foliage intermingled with dark green blisters.

raspberry late rust. A fungous disease of raspberry leaves caused by *Pucciniastrum americanum,* family Melampsoraceae, characterized by the infected leaves being covered with light yellow to orange masses of fungous spores turning brown to black due to the development of the teliospores. It occurs only on red raspberries.

raspberry leaf curl. A virus disease of raspberry which causes downward curling and crinkling of the leaves and a retardation of the growth of the veins. Leaves are bronzed in late summer with surface glistening. Berries are small and poor and winter kills canes easily.

raspberry leaf roller. *Exartema permundanum* Clem., family Olethreutidae, an insect whose larvae feed at first on the underside of the raspberry leaves under a silken cover and later migrate to the upper side, where they fold the leaf about themselves holding it with silken threads.

raspberry leaf spot. A fungous disease caused by *Septoria rubi,* family Sphaerioidaceae, characterized by brown or purple spots appearing on the lower leaves, then on the new cover and gradually involving all of the foliage. It weakens the plants and greatly reduces the production of fruit.

raspberry mosaic. A virus disease which causes a bright yellow discoloration of the leaves, a puckering of leaf tissue, and a gradual reduction in growth and fruit production.

raspberry orange rust. A fungous disease of black raspberries caused by *Gymnoconia interstitialis,* family Pucciniaceae. Conspicuous, bright orange spore masses show on the underside of leaves in the spring. The fungus becomes perennially established in the canes and roots. Shoots are numerous and slender. Infected plants are unfruitful.

raspberry powdery mildew. A disease caused by *Sphaerotheca humuli,* family Erysiphaceae, characterized by a white, powdery coating on the leaves and green shoots, and the berries becoming stunted, discolored and thus unmarketable.

raspberry ring spot. A virus disease characterized by the presence of circular, light green rings from ¼ to ½ in. in diameter bordering the nearly normal green tissue. It is not considered very destructive.

raspberry root borer. *Bembecia marginata* (Harr.), family Aegeriidae, an insect, the larvae of which enter the canes of raspberries near the ground and tunnel into the crown of the plant.

raspberry sawfly. *Monophadnoides geniculatus* (Htg.), family Tenthredinidae, an insect that lays eggs in raspberry leaf tissue. The larvae feed on the leaf surface and the damage may result in the loss of a crop of berries.

raspberry spur blight. A fungous disease caused by *Didymella applanata,* family Sphaeriaceae, characterized by reddish-brown areas on the stem and around the buds, weakening or killing them. Leaves may be killed and entire stalks may die.

raspberry Verticillium wilt. A soil-borne fungous disease caused by *Verticillium alboatrum,* family Moniliaceae, characterized by wilting, yellowing, and defoliation. Black streaks are found in the wood.

raspberry western yellow rust. A fungous disease caused by *Phragmidium rubi-idaei,* family Pucciniaceae, characterized by yellow flecks or pustules appearing on the upper surfaces of the leaves and light orange pustules breaking out 2 or 3 weeks later on the undersurface and on the canes. The canes weaken easily and fruit production is greatly decreased.

raspberry yellows. Any of certain virus diseases of raspberry characterized by mottling or yellowing of the leaves. See *raspberry mosaic.*

rasp leaf. A virus disease of sweet cherry reported from Colorado and other western U.S. regions, with a rapid spread only in Delta county, Colorado, characterized by leaflike outgrowths or enations on the underside of the leaves. Also called *cherry raspberry leaf.*

rat. *Rattus* spp., family Muridae, a long-tailed rodent which is much larger than a mouse. A serious pest, it is very destructive to stored food and may carry disease.

rata. *Metrosideros robusta,* family Myrtaceae, a tender, hard-wooded tree grown in frost-free areas for its showy, dark red flower clusters. Native to Australasia.

rat bait. Any food attractive to rats which contains a substance poisonous to them. There are many types.

rat bite fever. A disease caused by the flagellated organism, *Spierillum minus,* carried by rats and transmitted to man by their bites. Characterized by severe intermittent attacks, the ailment is not contagious. Especially prevalent in Japan.

rate earned on investment. An accounting figure determined by dividing the net income from the capital investment by the total investment in the farm business at the beginning of the year and multiplying the result by 100. This total represents the return for the capital and management expressed in percent.

rate of fall. The number of inches or feet of vertical drop in 100 ft. horizontal distance, as one in. drop per ft. in a tile line.

rate of growth. 1. The rate at which a tree has laid on wood, measured radially in the trunk or in timber cut from the trunk. (The unit of measure in use is the number of annual growth rings per inch.) 2. The rate at which a young animal increases weight and height.

rate of planting. Rate of seeding.

rate of seeding. The amount of seed sown or planted per unit area, usually an acre.

rate of spread. In forestry, the quantitative increase in area per unit of time, used to describe forest fires and serious new forest diseases.

rat flea. Oriental rat flea.

rather ripe. Maturing at an earlier date than the other fruit. Also called *rareripe* (England).

rating. Of a machine, apparatus or device, an arbitrary designation of an operating limit. A *rating* is arbitrary in that it must necessarily be established by definite fixed standards and cannot indicate the safe operating limit under all conditions that may occur in service.

ratio kill. The proportion or percentage of wildlife population which can be killed yearly without diminishing subsequent yield.

ration. The feed allowed an animal during a 24 hr. period regardless of whether it is fed at one time or at different times.

ration of maintenance. The feed which is necessary to maintain the body.

rat mange. A mange caused by an infestation of *Notoedres muris,* a very small mite just visible to the naked eye.

ratoon. A basal sucker used for propagation, as in sugar cane, pineapple, and banana.

ratoon stunting. A virus disease of sugar cane affecting the young shoots, characterized by a slow decline in yield. It was discovered in Australia in 1950 and shortly thereafter in Louisiana, U.S.

rat poison. 1. A poison for rats obtained from the seeds of a shrub, *Chailletia toxicaria,* native to West Africa. 2. *Hamelia erecta,* family Rubiaceae, a tall, evergreen shrub, bearing scarlet or orange flowers and small, purple-red fruit, sometimes grown as an ornamental in warm areas. Native to tropical America. Also called *scarlet bush.*

rat-stripper. See *Canby pachistima.*

rat-tail. 1. A slim, hairless tail of a horse. 2. A small, round, tapered file. 3. See *buckhorn plantain.*

rattail aporocactus. *Aporocactus flagelliformis,* family Cactaceae, a vine like cactus, grown as a window plant for its thin, rattail-like stems and crimson flowers. Native to Mexico.

rattail fescue. *Festuca myuros,* family Gramineae,

a slender, annual grass, distributed throughout the U.S. It is considered a grassy weed and is often found on over-grazed ranges. Native to Europe.

rattail radish. *Raphanus sativus* var. *caudatus,* family Cruciferae, an annual or biennial herb, grown for its edible, long, curved, often twisted pod, eaten either raw or pickled. Native to Europe and Asia.

rattle. A wholesale cut of meat made from the arm, shank, brisket, and short plate.

rattlebox. Arrow crotolaria.

rattle brome. *Bromus brizaeformis,* family Gramineae, an annual grass, with leaves up to 8 ins. long, and branches with many flattened spikelets, grown for use in dry bouquets. Native to Europe and Asia.

rattle-bush. Blue wildindigo.

rattlesnake fern. *Botrychium virginianum,* family Ophioglossaceae, a fern grown as an ornamental. Native to the North Temperate Zone.

rattlesnake master. See *button snakeroot eryngo.*

rattlesnake plantain. See *downy rattlesnake plantain.*

rattlesnake weed. See *cohosh bugbane; poor-robin's-hawkweed.*

rattletrap. Any rickety machine or vehicle that does not run smoothly.

rattleweed. See *blue loco; arrow crotolaria.*

rat typhus. A disease caused by the parasitic organism, *Rickettsia prowazekii,* whose symptoms are weakness, headache, chills and fever. It is transmitted to man by the oriental rat flea *Xenopsylla cheopis* (Rothsch.), family Pulicidae. Also called *murine typhus; flea typhus.*

rat virus. A food bait for killing rats which contains bacterial organisms of the genus *Salmonella;* its sale is prohibited by law in some states. Some authorities do not consider this bait effective.

Ravennagrass. *Erianthus ravennae,* family Gramineae, a stout, perennial, tall-growing grass with smooth and stiff stems, and a much-branched, silky, and showy flowering plume, grown for ornament. Native to Europe and Asia.

Raviggiolo. A soft, uncooked, creamy cheese which is made from ewes' milk in Tuscany, Italy.

ravine. A sharply cut, natural depression, V-shaped in transverse profile, usually without a permanent stream in its bottom. It has less width and size than a valley, but is larger than a gully. (In local usage, the meaning is not consistent, in relation to size, depth, and other attributes.)

ravison. Rape seed of inferior quality.

ravison meal. The meal produced by extracting the oil from ravison; very similar to rape-seed meal.

raw bone meal. The dried, ground product suitable for animal feeding obtained by cooking undecomposed bone in water at just enough atmospheric pressure to remove excess fat and meat. Because of incomplete cooking *raw bone meal* contains more than 23 percent protein and is lower in calcium and phosphorus than the steamed bone meal. It has limited use as a calcium and phosphorus supplement for livestock.

raw cane sugar. The crude, yellowish or brown unrefined cane sugar.

raw cotton. Unprocessed cotton.

raw-cream butter. Butter made from unpasteurized sweet or sour cream.

rawhide. Undressed skin of cattle.

rawhide belting. In millwrighting, belting which has not been subjected to the process of tanning except on the surface.

rawhide bit. A bit made of rawhide used in the breaking of horses to protect the mouth.

rawhide hackamore. A bitless bridle or halter made of rawhide, used chiefly in the breaking of horses.

rawhide quirt. A riding whip made of rawhide.

rawhumus (rohhumus). Soil organic matter which is completely or only slightly undecomposed and generally acid, such as the peaty material derived from Sphagnum mosses and ericaceous plants.

raw land. Virgin land; an uncleared tract which is not ready for cultivation.

raw milk. Fresh, untreated milk as it comes from the cow.

raw peat. See *peat; high-moor peat.*

raw-rock phosphate. Phosphate rock, finely ground but otherwise untreated, which is used as a fertilizer.

raw wool. Wool prior to the removal of the grease.

ray. 1. In wood anatomy, a ribbon-shaped strand of tissue formed by the cambium and extending in a radial direction across the grain which serves to store food and transport it horizontally in the tree. 2. In botany, one of the small flowers radiating out from the margin of a dense inflorescence.

ray crossing. In wood anatomy, the rectangular facet of a wall where a ray cell and a longitudinal tracheid or vessel member are contiguous, or where the walls of these cells appear to intersect, as viewed in radial section.

ray fleck. In wood anatomy, a portion of a ray, as it appears in radial section.

ray floret. The marginal flowers of an inflorescence distinct from the disk. These are the florets in a flower, such as a sunflower, which have well-developed petals.

ray fungus. *Actinomyces bovis,* an organism widely distributed in nature which may enter the tissue of the various organs of the animal causing swelling of either the bone or soft tissue. The infection usually reaches the surface so that fistulae are established and pus-producing organisms gain entrance. It is the cause of lumpy jaw in cattle.

ray grass. Rye grass.

rayless goldenrod. Jimmy weed.

Rayon. A special type of Swiss cheese used for grating, made in the Canton of Fribourg, Switzerland for export to Italy.

razorback. 1. A type of hog with long legs and snout, sharp narrow back and lean body; usually a half-wild mongrel breed (southern U.S.). 2. A sharp-ridged spur or hill.

razorback terrace. A narrow terrace constructed for the control of soil erosion. It may be only 3 to 5 ft. wide at the base and only 1 to 2 ft. high, and is either graded or level. The narrow terrace has serious deficiencies and is not in common use.

REA. Rural Electrification Administration.

reach. Coupling pole.

reaction. 1. A change in a market trend. 2. The degree of acidity or alkalinity (*e.g.*, of a soil mass) expressed in pH values and in words as follows: extremely acid, below 4.5; very strongly acid, 4.5–5.5; medium acid, 5.6–6.0; slightly acid 6.1–6.5; neutral, 6.6–7.3; (strictly, 7.0); mildly alkaline, 7.4–8.0; strongly alkaline, 8.1–9.0; very strongly alkaline, over 9.1.

reactor. Especially in testing for a disease, an animal which reacts positively to a foreign substance: *e.g.*, a tuberculous animal would be a *reactor* to tuberculin.

ready to cook. As applied to meat, poultry, and vegetables, designating a product which when purchased has been processed to the point at which it is ready to be cooked.

ready to eat ham. A ham cooked until the interior reaches 155° F. and held at that temperature or above for a further 2 hrs.

ready-whipped cream. Whipping cream, raw or pasteurized, flavored, whipped, and ready to be delivered to the consumer.

reagent. Any substance involved in a chemical action.

real estate. Real property, such as houses, land, buildings and attachments thereto.

real flank. See *flank.*

real property. Land and generally the improvements thereon.

reap. To cut or harvest, as a crop of grain.

reaper. An early machine for harvesting grain of which there have been several kinds: the first patented by Hussey 1833 and McCormick 1834. These have evolved into and have been replaced by the binder which has been replaced largely by the combine (*q.v.*). See *binder: 1.*

reaper file. A file, 7 to 10 ins. long, slightly tapering, with a trapezoidal section and two safe edges, used to sharpen the knives of reaping and mowing machines.

reap hook. Sickle.

reaping machine. Reaper.

rear attachment of udder. In the judging of a dairy cow, the rear development and conformation of the udder.

rear carrier. The carrier located at the rear of the threshing machine, used to carry away the straw.

rear hitch. A hitch at the rear or back of an implement, as rear-hitch sweep rake *q.v.*

rear-hitch sweep rake. A rake in which the team or tractor is hitched behind rather than forward of the rake head.

rear horse. Carolina mantis.

rear quarter. 1. The hind quarter of an animal. 2. A posterior section of the bovine udder.

rear shaker. The shakers at the back of the potato digger which separate the potatoes from the vines leaving them separated on the ground.

rear udder. The back part of a cow's udder.

reasonably firm cheese. Designating a cheese with a slightly weak body, but not to the extent of materially injuring the quality.

reasonably firm white of egg. The white of an egg that permits the yolk to move somewhat more freely from its normal position in the center of the egg and approach the shell more closely.

reata. Lariat; a kind of strong Mexican rope made by twisting thongs of hide together, used in roping cattle on western ranges (U.S.).

rebale cotton. To bale cotton the second time, usually under much greater pressure, producing a more compact bale for shipping.

Reblochon. A soft, French cheese made from fresh whole milk and cured for 4 or 5 weeks in a moist room at about 60° F.

reburn. Subsequent burning of an area over which fire has previously run.

receipts. The number of livestock, cattle, hogs, and sheep which arrive at a public market for a specified time, as by days or months.

receiving platform. The place of delivery of milk at a creamery or receiving station.

receiving room. The room at a creamery or receiving station where milk and other farm products are delivered.

receiving room equipment. Scales, vats, etc., used in the room receiving milk and other farm products at a receiving station or creamery.

receiving vat. A vat or container at a creamery or receiving station for the reception of milk.

receiving yard. A very large enclosure in which sheep are held when first brought into the sheep yards (Australia and New Zealand).

recent soil. A soil which has a profile without definite horizons of eluviation or illuviation: *e.g.*, recently deposited alluvium.

receptacle. 1. The swollen part of a plant stem which forms the base of a flower (floral *receptacle*). Also called *thalamus; torus.* 2. The stem of an inflorescence (*receptacle* of inflorescence).

receptive hypha. A fungous hypha in a pycnium of rusts, with a female sexual function, ready to be fertilized.

recessional moraine. A moraine which formed during the recessional stages of an ice sheet. See *moraine.*

recessive. In genetics, one of a pair of contrasting Mendelian factors, which, when both such factors are present, is subordinate to the other in its manifestations. In the homozygous state, a *recessive* factor gives rise to a *recessive* character in the resultant organism. *Cf. dominant: 1.*

reciprocating band-cutters. The type of band cutter of a threshing machine that moves back and forth rather than in circular fashion, for cutting the twine around the bundles of grain.

Recknagel phenomenon. The rise in specific gravity of milk which is allowed to stand several hours after milking, due to the solidification of the fat, probably also to the loss of gases, and to other changes.

reclamation. Land reclamation.

reclamation disease. A disease of oats appearing

when they are grown on reclaimed soils.

Reclamation Service. A bureau in the U.S. Department of Interior concerned mainly with the arid lands of western U.S.: classification, irrigation, improvement, administration, and with power development.

recleaned seed. Forage seed, as clover, alfalfa, and timothy seed, which is cleaned the second or third time to improve the quality.

recombination. The observed new combinations of characters which are different from those combinations exhibited by the parents. Percentage of *recombination* equals percentage of crossing over only when the genes are relatively close together. Cytological crossing over refers to the process, while *recombination* or genetic crossing over refers to the observed genetic result.

reconnaissance. 1. A cruise of forest property to obtain general information of the forest conditions. 2. An extensive range survey which is carried out to estimate average density and composition of range vegetation within a type or subtype without use of systematically established plots, 3. A type of survey in which land features are examined at wide intervals and are not delineated in detail. It is intended to furnish information primarily for extensive or over-all planning.

reconnaissance soil map. See *soil map.*

reconstituted cream. Cream which is made by the reemulsification of butter in milk with the aid of emulsifying agents.

reconstituted milk. The product which results from the recombining of milk fat and non fat dry milk or dried whole milk with water in proportions to yield the constituent percentages occurring in normal milk.

reconstituted milk powder. See *reconstituted milk.*

record form. See *pedigree: 1.*

record keeping. 1. Any of the different methods used for keeping a record of business operations on a farm which includes a record of cash income and cash expenses; inventories of real estate, machinery and equipment, livestock, crops, feed and supplies on hand; as well as of production records for crops and livestock. See *farm management.* 2. A method of recording the yield of every cow, preferably at every milking. At intervals an official visits the farm to check the record sheets and to sample the milk for butterfat and other tests.

Record of Performance. See *U.S. Record of Performance.*

recreational land. Land that is used primarily for recreational purposes, as hunting, fishing, parks, etc.; land which has potentialities for such use.

recruits. The volume or number of trees that have grown past an adopted lower limit of measurement during a specified period.

rectangular line. In describing beef cattle, the general, blocky conformation desired in a meat animal.

rectangular survey. A system of describing parcels of land in the U.S. by the section, township, and range as laid out in the original General Land Office survey.

rectangular system of land division. A system of land division which is based upon permanent meridianal and latitudinal lines yielding rectangular-shaped land areas as nearly as may be: 6 miles square divided into 36 secondary units called sections. See *land division.*

rectangular weir. A measuring weir which has a rectangular notch. A contracted weir, unless a suppressed weir is specified. See *weir.*

rectrix. A feather in the tail of a bird. See *main tail feathers.*

rectum. The terminal or lower part of the intestine which ends at the anus.

recurrent fever. An infectious virus disease of the horse which is characterized by extreme prostration, swelling in the lower parts of the body, jaundice, a pounding heart, and easy exhaustion. Also called *swamp fever; infectious anemia; creeping fever.*

recurrent selection. In corn breeding, a method to determine which plants of a composite make the best hybrids in combination with the particular single cross used as a tester.

recurved stamens. Stamens of some grape varieties which have reflexed stamens; *i.e.,* curved in the opposite or an unusual direction which produces pollen that is worthless for pollination.

red. A common coat-color for several species of animals which may range from dark red, deep rich-red, blood-red, golden-red, light red, or yellow-red. Shades of red are most common in cattle.

red-admiral. *Vanessa atalanta* (L.), family Oecaphoridae, a common and widely distributed butterfly whose larva feeds on nettles, false nettles, and hops.

red and black harlequin bug. See *harlequin bug.*

red and yellow podzolic soil. See *podzolic soil; Podzol.*

Red and Yellow soil. A major soil group in southeastern U.S. developed in an area of high temperature and rainfall, characterized by a small accumulation of surface organic matter and a deep horizon of eluviation in which a high rate of oxidation and hydration of iron produces bright red and yellow colors.

red apple leafhopper. *Erythroneura hartii* Gillette, family Cicadellidae, a leaf hopper which winters as an adult and produces two generations a year. Injury to apple shows to greatest extent in late summer and consists of spotting of fruit and leaf injury.

red aschersonia. *Aschersonia aleyrodis* Webber, a fungus parasitic on larvae of species of white flies in Florida, Brazil, the West Indies, South China, India and Ceylon. The perfect stage is now considered to be *Hypocrella libera.*

red ash. *Fraxinus pennsylvanica,* family Oleaceae, a hardy, deciduous tree, grown especially in the prairie states of the U.S. for shade, elsewhere for its lumber. Its wood is used in handles, furniture, oars, baseball bats, cabinets, boat buildings, implements, woodenware, novelties, cooperage, planing-mill products. Also called *river ash.*

red-backed cutworm. *Euxoa ochrogaster* (Guen.), family Noctuidae, an insect whose larvae feed on garden vegetables and other plants grown in rows or hills, cutting them off near the soil surface. Widely distributed in Canada and the northern states of the U.S.

red balm. Oswego beebalm.

red-banded leaf roller. *Argyrotaenia velutinana* (Wlk.), family Tortricidae, an insect whose larvae infest many small fruits and vegetables, and ornamental trees and flowers. It passes the winter in a silken cocoon under leaves on the ground.

red baneberry. *Actaea rubra,* family Ranunculaceae, a perennial herb grown in the wild garden for its white flowers and red fruits. The fruit is poisonous and the rootstock is a violent purgative. Native to North America.

redbay persea. *Persea borbonia,* family Lauraceae, a subtropical, evergreen tree, grown for its persistent blue or blue-black fruit on red stalks. Native to South America. Also called *bull bay; baygill.*

red beech. American beech.

red beet. See *beet:* 2.

red belt. Winter browning.

red-berried California holly. Christmasberry.

redberry. Christmasberry.

redberry bryony. *Bryonia diocia,* family Cucurbitaceae, a climbing herb, grown for its unusual spreading form. A cathartic is produced in Europe from its tuberous root. Native to Europe, Western Asia and North Africa. Also called *devil's-turnip.*

redberry disease of blackberries. A disease of blackberries caused by the redberry mite *Aceria essigi* (Hassan), family Eriophyidae, which feeds between the drupelets and the core of the berries causing them to become a brighter red, turn hard, and remain on the bushes.

redberry mite. See *redberry disease of blackberries.*

red birch. 1. Lumber taken from the heartwood of the yellow birch and the sweet birch *(q.q.v.).* 2. *Nothofagus fusca,* family Fagaceae, a valuable hardwood tree. Native to New Zealand.

red blotch. A reddish-brown discoloration of the surface of lemon fruit thought to be associated with immaturity, cold weather, forced curing with ethylene gas, and poor ventilation. Fruit from some orchards are consistently more susceptible than others, suggesting an inherited tendency to the weakness.

redbox cotoneaster. *Cotoneaster rotundifolia,* family Rosaceae, an evergreen shrub, grown for its foliage, white flowers, and scarlet fruit. Native to the Himalayas.

redbox eucalyptus. *Eucalyptus polyanthemos,* family Myrtaceae, an aromatic, evergreen tree, up to 150 ft. high, grown as an ornamental for its green leaves, and small flowers. Native to Australia. Also called *Australian beech; den tree.*

red buckeye. *Aesculus pavia,* family Hippocastanaceae, a tender shrub or small tree grown for its bright red flowers. Native to southeastern U.S.

redbud. Any of the small trees or shrubs of the genus Cercis, family Leguminosae. Species are grown as ornamentals. See *eastern redbud; Chinese redbud.*

redbud deutzia. *Deutzia purpurascens,* family Saxifragaceae, a tender shrub, grown for its star-shaped flowers purplish outside, white inside. Native to Western China.

red bugs. Chigger.

Red Cabbage group. One of the eight groups of cabbage distinguished by its purplish-red leaves.

Redcap. 1. A dual-purpose breed of chickens which originated in Derbyshire, England, characterized by red, brown and black plumage, a large rose comb, white skin, and bluish-slate shanks. Mature males weigh about 7½ lbs., mature females about 6 lbs. The egg-shell color is white. 2. (not capitalized) See *red raspberry.*

red carpenter ant. *Camponotus herculeanus pennsylvanicus ferrugineus* (F.), family Formicidae, an ant which lives in large colonies in the North Temperate Zone under stones, stumps, logs, and often far under ground.

red cedar. See *eastern redcedar; southern redcedar.*

red cedar aphid. *Lachnus sabinae* Gill., family Aphidae, a reddish-brown insect which secretes honeydew, a medium for the sooty blotch fungus often covering the leaves of the red cedar. Both the aphid and the fungus greatly weaken the tree.

red cherry. Cherry.

red chickweed. Scarlet pimpernel.

red Chilebells. *Lapageria rosea,* family Liliaceae, a high-climbing vine, grown on a trellis in frost-free areas for its white, trumpet-shaped, rose-colored flowers. Native to Chile.

red chokeberry. *Aronia arbutifolia,* family Rosaceae, a shrub grown for its white flowers, brilliant red fruit, and autumnal red foliage. Native to North America.

red clover. *Trifolium pratense,* family Leguminosae, a forage herb. Native to Europe, it is grown extensively in the northeastern quarter of the U.S. and locally in the northern Rocky Mountain and Pacific Coast states. A perennial, it behaves under cultivated conditions as a biennial. There are two types (a) the widely grown double cut, *T. pratense,* medium red or June clover, the first crop being used for hay and the second for either hay or seed, depending upon seasonal conditions; (b) the single cut, *T. pratense* var. *perenne,* called mammoth red, bull, sapling or peavine clover, producing only one crop a year which may be harvested for either hay or seed. The stems are coarser and the quality of forage usually lower than the double cut. The *red clover* is also called *broadleaved clover.*

red clover black patch. A disease of red clover caused by an unidentified seed-borne fungus which infests leaves, stems, flowers and seeds producing brown to grayish-black lesions usually with concentric rings.

red clover Cercospora spots. A fungous disease

of red clover caused by *Cercospora* spp., characterized by reddish or smoky-brown lesions appearing on the leaves, stems, petioles, and seeds. Seed-borne, the disease is most widespread in late summer when a silvery-gray down develops on mature lesions.

red clover hay. Hay or a hay mixture containing a high proportion of red clover. It supplies about two-thirds as much digestible protein and about the same total amount of digestible nutrients as alfalfa. If well cured and free from dust, it is satisfactory for all classes of livestock. One of the most important hays in the northeast quarter of the U.S.

red clover northern anthracnose. A fungous disease of red clover caused by *Kabatiella caulivora,* characterized by dark brown or black spots on the petioles, cracks in the center of the lesions, girdled stems and the field having the appearance of being scorched by fire because of blackened stems and leaves.

red clover powdery mildew. A fungous disease of red clover caused by *Erysiphe polygoni,* family Erysiphaceae, characterized by patches of fine, white, cobwebby mycelium developing on the upper surface of the leaves until the entire field appears white.

red clover Pseudopeziza leaf spot. A fungous disease of red clover caused by *Pseudopeziza trifolii,* family Mollisiaceae, characterized by olive to reddish-brown spots developing on either leaf surface, and amber globules or fruiting bodies appearing in the center of the spot. The disease, almost entirely limited to leaves, occurs during damp weather in the cooler humid regions of the red clover section of the U.S. and Europe.

red clover rust. A fungous disease of red clover caused by *Uromyces trifolii fallens,* family Pucciniaceae, characterized by irregular, pale brown pustules, surrounded by torn epidermis appearing on the lower surface of the leaves, petioles, and stems. The telial or black spore stage overwinters on diseased plants.

red clover sooty blotch. A fungous disease of red clover caused by *Cymadothea trifolii,* characterized by minute, olive-green dots appearing on the lower leaf surface enlarging and becoming thicker and darker and appearing as velvety, black, angular, elevated patches, causing the leaf to turn brown, die, and fall off.

red clover southern anthracnose. A fungous disease of red clover caused by *Colletotrichum trifolii,* family Melanconiaceae, characterized by irregular, dark brown spots on the leaves spreading to the stems and petioles as small water-soaked spots, lengthening to form depressed, black lesions, and the browning of the entire stem. It is a major disease in the southern red clover section of the U.S.

red clover Stemphylium leaf spot. A fungous disease of red clover caused by *Stemphylium sarcinaeforme,* family Dematiaceae, characterized by minute, light brown spots appearing on the leaflets followed by lesions with light and dark concentric rings. Defoliation may occur.

red clover vein-mosaic. A virus disease of red clover causing vein chlorosis, mild to severe mottling and chlorotic patches between the veins. It is more prevalent during the cooler growing season. Aphids are very important carriers. Damage ranges from little to severe.

red cluster pepper. *Capsicum frutescens* var. *fasciculatum,* family Solanaceae, a variety of bush redpepper, used as a condiment.

red cohosh. See *red baneberry.*

red columbine. See *American columbine.*

red copper oxide. See *copper oxide.*

red coralbush. *Templetonia retusa,* family Leguminosae, a shrub grown in very warm areas for its red flowers. Native to Australia.

red core. Strawberry red stele.

red currant. *Ribes sativum,* family Saxifragaceae, a hardy, popular shrub grown for its edible, red fruit used in jellies. Also called *rissel.*

red currant tomato. See *currant tomato.*

red cypress. See *baldcypress.*

Red Danish. The newest dairy breed of cattle to be developed whose origin is Denmark. In color it is red to brindle. Individuals are heavy milk producers.

red date scale. *Phoenicococcus marlatti* (Ckll.), family Dactylopiidae, an insect which firmly attaches itself to the bark and young branches of the date palm.

red dead nettle. See *deadnettle: 2.*

Red Desert Soil. A zonal group of soils having light, reddish-brown friable soil over a reddish-brown or dull red heavy horizon grading into an accumulation of carbonate of lime which is found in warm temperate and tropical deserts. It is characterized by scant desert-shrub vegetation.

Reddish-brown Lateritic Soil. A zonal group of soils with dark reddish-brown, granular surface soils, red friable clay B horizons, and red or reticulately mottled lateritic parent material which developed under humid tropical climate with wet-dry seasons and tropical forest vegetation.

Reddish Brown Soil. A zonal group of soils with a light brown surface horizon of whitish or pinkish lime accumulation. It developed under shrub and short-grass vegetation of warm temperate to tropical regions of semi arid climate.

Reddish Chestnut Soil. A zonal group of soils with dark brown, tinted pinkish or reddish surface soils up to 2 ft. thick over heavier reddish-brown soil over grayish or pinkish lime accumulation which developed under warm, temperate semi arid climate and mixed grass vegetation with some shrubs. It is approximately equivalent to southern Chernozem.

reddish off-crude cottonseed oil. The fourth grade of crude cottonseed oil as classified by the National Cottonseed Products Association. It must refine to a color no higher than AOCS 30 with a loss in weight not exceeding 40 percent.

reddish off-summer yellow cottonseed oil. The fifth grade of summer yellow cottonseed oil as classified by Interstate Cotton Seed Crushers'

Association. It may be off in flavor and odor, must be free from visible foreign material, of a color no higher than AOCS 20, and shall contain not more than 0.75 percent free fatty acid nor in excess of 0.10 percent moisture and volatile.

Reddish Prairie Soil. A zonal group of soils with dark reddish-brown, slightly to medium acid surface soils grading through somewhat heavier reddish material to the parent material which developed under warm-temperate humid to subhumid climate and tall-grass vegetation.

red dog flour. Wheat red dog.

red dogwood. Bloodtwig dogwood.

red durum wheat. A class of spring wheat grown primarily for livestock feed and not used for the manufacture of macaroni as are other durum wheats.

red dysentery. See *coccidiosis.*

red ear rot of corn. Gibberella rot of corn.

red elm bark weevil. *Magdalis armicollis* (Say), family Curculionidae, a reddish snout beetle which infests elms, especially those in an unthrifty condition.

red fermentation of milk. A red color of milk induced by bacteria or pink yeasts. See *red milk.*

red fescue. *Festuca rubra,* family Gramineae, a stoloniferous, perennial grass with weak underground stems, which is an excellent lawn grass especially on light soils and for shady areas. It is sometimes used in pasture mixtures on poorer soils. Native to Europe, Asia, and North America.

red fir. 1. *Abies magnifica,* family Pinaceae, an evergreen tree grown for its valuable lumber used for general construction, boxes, crates, pulpwood, and planing-mill products. Native to North America. Also called *California red fir.* 2. See *Douglas fir.*

red flour beetle. *Tribolium castaneum* (Hbst.), family Tenebrionidae, an insect whose larvae infest flour, stored grain, and other food products.

red fungus of citrus white fly. *Aschersonia aleyroidis,* family Zythiaceae, a fungus that under favorable conditions attacks and destroys the citrus white fly and related species in the tropics and subtropical regions. Its rarely found perfect stage is *Hypocrella libera,* family Hypocreaceae. This fungus has been cultured in pure culture on sweet potato strips and the spores sprayed on trees infested with the white fly.

red gall of cranberry A fungous disease caused by *Synchytrium vaccinii,* family Oynchytriaceae, characterized by red, globular galls on the buds, flowers, and young leaves.

red grain sorghum. A market class of grain sorghums as provided by the Official Grain Standards of the U.S. which includes all varieties of red grain sorghums and may include not more than 10 percent of grain sorghums of other colors.

red harvester ant. *Pogonomyrmex barbatus* (F. Smith), family Formicidae, an ant about 1/4 in. long, with a black head and legs, and red abdomen. It lives in large colonies and makes bare circles 2 to 12 ft. in diameter, with a deep nest. It is injurious to grasslands in southwestern U.S. because it feeds on grass seed. Also called *agricultural ant.*

red haw. Downy hawthorn.

red hawksbeard. *Crepis rubra,* family Compositae, an annual herb grown for its red or pink flowers. Native to Europe.

red-headed louse. Sheep biting louse.

red-headed pine sawfly. *Neodiprion lecontei* (Fitch), family Diprionidae, an insect whose young larvae destroy the needles of numerous kinds of pines. Found in eastern U.S., it is often very destructive.

red-headed scale fungi. *Sphaerostilbe* spp., a fungus, whose pustule of the imperfect form is orange-red to blood-red and blood-red in the perfect form. It is found on a large number of scale insects on citrus and other host plants.

red heart. Referring to a pronounced, reddish coloration in the heartwood, induced by fungi in conifers and a few hardwoods.

redheart ceanothus. *Ceanothus spinosus,* family Rhamnaceae, a tender shrub grown for its white to pale blue, dense, branched flower clusters. Native to North America.

redheart of lettuce. A physiological disease of lettuce characterized by chestnut-brown discoloration and breakdown of inner leaves, with outer leaves developing brown pits on the midrib and veins. It is caused by a lack of sufficient oxygen due to poor aeration or by prolonged exposure to low temperatures during shipment and storage.

red hickory. *Carya ovalis,* family Juglandaceae, a deciduous tree grown for its valuable wood used in tool handles, ladders, furniture, sporting goods, implements, woodenware, and for fuel and smoking meat. It is also grown as a shade tree. Native to North America. Also called *mockernut; oval pignut hickory.*

red honeysuckle. 1. *Hedysarum coronarium,* family Leguminosae, a perennial or biennial herb grown for its profuse, deep red, fragrant flowers. Native to Europe. Also called *sulla sweetvetch; French honeysuckle.* 2. *Banksia serrata,* family Proteaceae, an evergreen tree grown for its rust-colored leaves and beautiful, red flowers. Native to Australia. Also called *redwood banksia.* See *Banksia.* 3. Red clover.

red horse chestnut. *Aesculus carnea,* family Hippocastanaceae, a tree, hybrid between the horse chestnut and the red buckeye, grown for its scarlet or flesh-colored flowers.

red-hot-poker. See *torchlily: b.*

red-humped caterpillar. *Schizura concinna* (J. E. Smith), family Notodontidae, an insect whose larvae feed on both fruit and ornamental trees. Its moths lay eggs in masses on underside of leaves and may cause severe defoliation.

redia. A larval stage in the development of flukes. *Redia* of liver flukes of cattle, sheep and goats occurs in snails.

red ironbark. Mulga ironbark eucalyptus.

red jasmine. Nosegay frangipani.

red lady beetle. *Cycloneda munda,* family Coccinellidae, an insect whose larvae and adults feed on scale insects, aphids, and small soft-bodied insects.

red larkspur. See *orange larkspur.*

red leaf. 1. A disease of various plants as oak, pear, strawberry, caused by functional factors such as nutritional deficiency, etc., which produces red discoloration of the foliage and sometimes affects the yield. 2. See *marshpepper smartweed.*

red leaf blight. Cotton rust.

red leaf spot of cranberry. A fungous disease caused by *Exobasidium Vaccincii,* family Exobasidiaceae, characterized by bright red spots on the upper surface of the plant leaves.

red-legged flea beetle. *Derocrepis erythropus* (Melsh.), family Chrysomelidae, an early-season insect which infests young apple and peach orchards destroying the developing buds and sometimes killing the trees.

red-legged grasshopper. *Melanoplus femur-rubrum* (DeG.), family Acrididae, a grasshopper that infests soybeans, vegetables and fruit trees.

red-legged ham beetle. *Necrobia refipes* (DeG.), family Cleridae, an insect whose larvae are very destructive to cured and dried meats. Sometimes its adults are also found in infested meat.

red locust. *Nomadacris septemfasciata* Sert., family Acrididae (Locustidae), a migratory grasshopper that appears in swarms and feeds on the foliage, fruit, and bark of the twigs of citrus fruits.

red louse. *Haematopinus eurysternus* (Nitz.), family Haematopinidae, a louse which may be a serious pest on young cattle. Also called *short-nosed cattle louse.*

red maids rockpurslane. *Calandrinia caulescens* var. *menziesi,* family Portulacaceae, a fleshy herb grown as an annual for its rose or crimson flowers. Native to North America.

red mange. A skin disease of dogs caused by *Demodex canis* Leydig, family Demodicidae, a mite that lives deep in the hair follicles and produces bare, inflamed spots about the eyes, ears, and joints. Also called *demodectic mange; follicular mange.*

red mangel. Mangel wurzel.

red maple. *Acer rubrum,* family Aceraceae, a large tree, grown as an ornamental and as a shade tree for its brilliant scarlet and yellow, autumnal foliage. Native to North America.

red milk. A red color of milk caused by: (a) the bacterium *Serratia marcescens,* a common soil or dirt contaminant, which grows and produces a red pigment that appears first on the top of the milk and then develops throughout; (b) pink yeasts such as *Torula rosea* and *T. glutinis,* which produce spotty, red pigment in the cream layer; and (c) the presence of blood, as in acute mastitis or udder injury, which tends to settle to the bottom as the milk stands. *Red milk* caused by yeasts and bacteria is also referred to as *red fermentation of milk.*

red mite. One of the spider mites of the family Tetranychidae, which is a plant feeder and which may do serious damage to orchard trees, field crops, and greenhouse plants. See *two-spotted spider mite.*

red mite of poultry. See *chicken mite.*

red mombin. See *mombin.*

red mosaic of raspberry. See *red raspberry mosaic.*

red mulberry. *Morus rubra,* family Moraceae, a tree which produces red or purplish-red, edible fruit about 1 in. long, eaten by chickens, hogs, and young children. Native to North America. Also called *American mulberry.*

red-necked cane borer. *Agrilus ruficollis* (F.), family Buprestidae, a beetle that infests raspberry, blackberry, and dewberry. The young larvae burrow upward in the sapwood and around the cane causing formation of galls and swelling.

red nose. An infectious rhinotracheitis usually found in feeder cattle, characterized by light colored sometimes bloody nasal discharge, accompanied by drooling, coughing, and a *red nose.*

red oak. See *eastern red oak; northern red oak; southern red oak.*

red oats. A market class of oats designated by the color of the lemma and palea of the matured grain in the Handbook of Official Grain Standards of the U.S.

red onion. See *onion.*

red osier dogwood. *Cornus stolonifera,* family Cornaceae, a hardy shrub, seldom over 6 ft. high, grown as an ornamental for its erect and red branches, its white flowers, and bluish-white fruit. Native to North America. Also called *squaw bush.*

red parrotbeak. *Clianthus puniceus,* family Leguminosae, a vinelike plant grown in the greenhouse or out-of-doors in very warm areas as an ornamental for its habit and for its crimson, white-streaked flower cluster. Native to Australia. Also called *parrotbeak; parrot's bill.*

redpepper. Any plant of the genus Capsicum, family Solanaceae. The edible, usually red fruits are very pungent except in the so-called sweet peppers. In their tropical homes they are woody shrubs but in temperate zone cultivation they are treated as annuals being killed by heavy frosts. The dried fruits of the pungent kinds are ground and sold as *red pepper,* Cayenne pepper, capsicum, etc. Native to Central America. See *bush redpepper; ornamental bush redpepper; cherry redpepper; tabasco redpepper; red cluster pepper; sweetbell redpepper; long redpepper.*

red periwinkle. *Vinca rosea,* family Apocynaceae, a tender, everblooming perennial grown in pots or hothouse beds for its showy, pink or white flowers with red centers. Native to Madagascar. Also called *Madagascar periwinkle.*

red pine. *Pinus resinosa,* family Pinaceae, a quick-growing, hardy, evergreen tree, grown as an ornamental and for its valuable lumber. Native to North America. Also called *Norway pine.*

red plum. Canada plum.

Red Podzolic soil. A zonal group of soils having their organic and inorganic mineral layers over a yellowish-brown leached layer which rests upon an illuvial red horizon. It developed under a deciduous or mixed forest in a warm temperate moist climate.

Red Polled. A dual-purpose breed of cattle common to central U.S. which originated in Norfolk and Suffolk counties in England. Animals are red and polled, and mature cows weigh 1200 to 1400 lbs. with average milk production.

red raspberry. 1. *Rubus idaeus,* family Rosaceae, a shrubby plant, grown for its edible, red fruit. It is the parent of many cultivated varieties. Also called *redcap.* 2. Any variety of raspberry which produces a red fruit.

red raspberry mosaic. A virus disease, prevalent in mild or latent form in red raspberries, which produces more severe mottling and splotches of yellow and green on the leaves of black raspberries. Crops suffer loss of quality and yield. Also called *green mottle mosaic.*

red ribbons. *Clarkia concinna,* family Onagraceae, an annual herb grown for its showy, rose-purple flowers. Native to California, U.S.

red rice. *Oryza rufipogon,* family Gramineae, an annual cereal herb, origin unknown, which has a red color in the grain; a very serious weed pest in rice fields as it has many of the same characteristics as white rice and is very difficult to eradicate. Its presence lowers the grade and value of the threshed grain of rice.

red roan. A coat-color of horses and cattle, which is a bay, chestnut, or brown with a red ground color.

red robin. Herbrobert geranium.

redroot amaranth. *Amaranthus retroflexus,* family Amaranthaceae, an annual herb which is a weed pest of gardens, field crops, and waste ground. Cases of livestock being poisoned from feeding on the plant have been reported. Also called *careless weed; Chinaman's redberry; rough pigweed; green amaranth.*

red root-rot. See *strawberry red stele.*

redroot shrub. Jerseytea ceanothus.

red rot of cole crops. A physiological disease which apparently results from a boron deficiency in cole crops, as cauliflower and cabbage. First symptoms appear as a water-soaked area in the stem changing to a rusty-brown, causing thickening, brittling and downward curl of older leaves.

red rot of sugar cane. A fungous disease of sugar cane caused by *Physalospora tucumanensis,* family Mycosphaerellaceae, which attacks sugar cane stalks, reddening or rotting the interior and inverting sucrose to less desirable sugars.

red-rust. See *rusts.*

red sage. See *lantana: 2.*

red Sally. Purple lythrum.

red sandalwood. Sandal beadtree.

red scale. See *California red scale.*

red shorts. A livestock feedstuff; a by-product from winter wheat milling which consists of fine particles of bran, germ and a little of the wheat red dog *q.v.*

Red Sindhi. A milking strain of Brahman cattle of India used by the U.S. Department of Agriculture in breeding experiments to develop better dairy cattle for southern U.S. See *ranch cattle.*

red spider mite. *Tetranychus telarius* (L.), family Tetranychidae, a mite which makes mealy cobwebs on the undersides of the leaves of a large number of plants, sometimes covering an entire new shoot. Also called *two-spotted spider mite.*

red spot (rusty spot) in cheese. A defect caused by chromogenic varieties of two types of lactobacilli, *Lactobacillus plantarum* var. *rudensis* and *L. brevis* var. *rudensis,* which consists of very fine spots quite evenly distributed throughout the cheese.

red spring wheat. A subclass of hard red spring wheat established by the U.S. Grain Standards Act, U.S. Department of Agriculture, which contains 25 percent or less of dark, hard and vitreous kernels.

red spruce. *Picea rubens,* family Pinaceae, an evergreen, coniferous tree which is a chief source of paper pulp. Native to North America.

redstain of wood. A red staining and discoloration in timber, caused by fungi which may not impair the strength of the wood but which reduces its market grade.

red stele. See *strawberry red stele.*

redstem ceanothus. *Ceanothus sanguineus,* family Rhamnaceae, a shrub which provides sheep and goat browse in northwestern U.S. Native to North America.

red-stem filaree. See *alfileria.*

red-stemmed peavine. *Astragalus emoryanus,* family Leguminosae, a vetch-like herb with alternate compound leaves and small, pea-like flowers in clusters which sometimes causes poisoning of livestock. Native to Texas, U.S.

red stripe. See *sugarcane red stripe.*

red suture. A yellow-type peach virus disease in which the fruit ripens prematurely on the suture side. Foliage symptoms are somewhat like those in little peach *q.v.*

red tart cherry. Sour cherry.

red three-awn. *Aristida longiseta,* family Gramineae, a densely cushioned bunchgrass with sharp, pointed leaves, which has some forage rating depending upon season pastured. Native to southwestern U.S.

redtinge rainlily. *Cooperia pedunculata,* family Amaryllidaceae, a bulbous, night-blooming herb grown for its white flowers which are reddish outside. Native to Mexico and southwestern U.S. Also called *fairy lily.*

redtop. *Agrostis alba,* family Gramineae, a stoloniferous, creeping, perennial grass which produces a coarse, loose turf, and is tolerant to wet acid soils. It is used in mixtures for pasture, lawns, and golf courses. Native to Europe; widespread in the northeast quarter of the U.S. Also called *Burden's grass; fiorin; herd's grass; meadow redtop.*

redtop sorrel. Sheep sorrel.

red trillium. Toad trillium.

red turnip beetle. *Entomoscelis americana* Brown, family Chrysomelidae, an insect which feeds on turnips, cabbage, radishes, and other garden crops.

reducer. A coupling which links together pipes of

different sizes. It is commonly used in sprinkler irrigation and plumbing.

reducing mill. In vegetable oil manufacture, a machine which grinds up the rough edges of the oil cake cut off by the paring machine. These rough edges formed around the cake during pressing contain much oil and are ground so that they may be heated afresh in the kettle and repressed.

reductase. An enzyme (*q.v.*) which has the power of reducing or decolorizing methylene blue.

reduction hazard. In forestry, any treatment of inflammable material, such as wood or leaves, by methods other than a going fire that reduce the chances of fires starting or spreading through them.

reduction of acreage (of crops in the U.S.). A program for the control of crop surpluses carried out under the direction of the Agricultural Adjustment Administration of the U.S. Department of Agriculture in 1933 (in 1936 changed to the Agricultural Conservation Program), in which payments are made to farmers on the basis of planting crops classified as soil conserving rather than soil depleting. Additional changes were made with the ever-normal granary *q.v.*

reduction tankage. The product obtained from rendering scrap meat and bones from butcher shops for soap grease; used chiefly as a fertilizer but may be fed to livestock if of suitable quality.

red valerian. Jupiter's-beard.

red water. See *cattle tick fever.*

red wax scale. *Ceroplastes rubens* Mask., family Coccidae, an important citrus pest in Japan, China, and Australia, the female of which infests the young twigs and rarely the foliage and fruit of the plant.

red weevil. *Contarinia tritici* Kirby, family Itonididae, a midge whose larvae feed on the kernels of wheat and other small grains materially reducing the yield. See *wheat midge.*

red whortleberry. Cowberry.

red wing aphid. Violet aphid.

red-winged pine beetle. *Dendroctonus rufipennis* (Kby.), family Scolytidae, an insect pest of pine. The adult beetles lay eggs in tunnels between the bark and wood, and the very small larvae or grubs feed on the inner bark tissue. When present in very large numbers, they may eventually girdle the tree.

red winter wheat. A sub-class of the soft red winter wheat class, established by the U.S. Grain Standards Act, which produces the flour for the making of cake and biscuits.

redwood. *Sequoia sempervirens,* family Pinaceae, an evergreen tree which grows to 340 ft. and is valued as an important source of lumber. Native to California and Oregon, U.S.

redwood banksia. See *red honeysuckle: 2.*

red worms. *Strongylus* spp., family Strongylidae, nematodes which infest the large intestine of equines. Their red color is due to the presence of hemoglobin in the bodies of the worms, and not to blood sucked from their hosts.

Red-yellow Podzolic soil. A zonal group of soils having a thin organic mineral layer over a yellowish-brown leached layer which rests upon an illuvial red horizon. It developed under a deciduous or a mixed forest in a warm-temperate moist zone.

red ysabella. Designating the red, sorrel-like coat-color of some animals. Horses of this color have flax-colored manes and tails.

reed. 1. Any tall, slender plant, usually having coarse and jointed stems, including certain grasses and grasslike plants. 2. *Phragmites communis,* family Gramineae, the common *reed,* a widespread, perennial marsh grass with erect culms 10 to 15 ft. tall, and stout, creeping rhizomes; used for making mats, thatching, cordage, and carrying nets. Native to North Temperate Zone and South America. 3. See *abomasum.*

reed canarygrass. *Phalaris arundinacea,* family Gramineae, a perennial grass adapted to marshy places where it is valuable for hay and pasture. Native to the temperate regions of Europe and Asia.

reed fescue. *Festuca elatior* var. *arundinacea,* a variety of meadow fescue *q.v.*

reedgrass. Any grass of the genus Calamagrostis, family Gramineae. Species are tall, perennial grasses with creeping rhizomes, valuable as range grass and for wild hay.

reed mace. Cattail.

reed meadow grass. *Glyceria grandis,* family Gramineae, a tall, coarse, perennial pasture grass growing in wet areas. Native to North America.

reed peat. One of the classes of peat *q.v.*

reel. 1. A frame turning on an axis; used for winding rope, yarn, fiber, or other flexible material, as a hose or a fishing net. 2. A winch. 3. In a binder or combine, each of the revolving slats which throw the cut grain back toward the platform canvas. 4. A device for winding a surveyor's tape.

reel line. A hose reel so constructed that the hose can be played out or pulled in while water is being discharged through it under pressure.

reel slat. The horizontal part of a reel of a binder or combine which strikes the grain as it is cut and throws it back upon the platform canvas. See *reel: 3.*

reeve. To pass the end of a rope through a ring, a hole in a block, etc.

Reeves spirea. *Spiraea cantoniensis,* family Rosaceae, a tender, deciduous shrub grown for its dense clusters of white flowers. Native to China and Japan.

refinancing. Borrowing sufficient money to repay debts contracted previously. The procedure may involve the *refinancing* of debt now due or overdue or the consolidation of a number of debts or accounts payable.

refined bone. In stock judging, a small, high quality bone, as distinguished from coarse.

refined foods. Food products from which some of the minerals, vitamins, and/or other elements have been removed during processing and manufacturing, as white flour, degerminated corn products, and polished rice.

refined sugar. Sugar freed from impurities.

reforestation. The natural or artificial restocking with forest trees of an area previously under forest.

refractory wood. Wood possessing inherent tendencies to develop defects during seasoning.

refrigerant. 1. The volatile agent whose liquefaction and subsequent expansion furnishes the means of abstracting heat in mechanical refrigeration. 2. Anything that cools or refrigerates.

refrigerant compressor. The part of a mechanical refrigeration system which receives the refrigerant vapor at low pressure and compresses it at higher pressure.

refrigerant condenser. That part of a mechanical refrigerating system wherein the vaporized, condensed refrigerant is liquefied by removal of heat. It is of several types: submerged, evaporative, atmospheric, or air-cooled.

refrigerant evaporator. The part of a mechanical refrigeration system in which the refrigerant is vaporized.

refrigeration. Artificial cooling, either by the application of ice or by utilizing the principle of the latent heat of evaporation.

refrigeration machine. A device for producing low temperatures. A gas, such as air, ammonia, or carbon dioxide, is compressed, the heat generated being dissipated into air or into circulating water as the gas condenses. The cooled fluid is permitted to expand again to a gas in suitable chambers or pipes, abstracting the heat necessary for volatilizing from its surroundings, thus creating low temperature.

refrigerator. An insulated chamber, cupboard, or cabinet cooled by ice or mechanical refrigeration. See *refrigeration.*

refuge. a. A tract of land set aside for wildlife protection, particularly for game animals and wildfowl. In practice the protection afforded in a man-made *refuge* is more limited than in a sanctuary (*q.v.*). b. In nature, a place affording protection against enemies; a shelter; a covert.

refuge cover. Vegetation of sufficient density that can afford an unusual degree of protection to wildlife from hunters or predators.

refuse lime. See *lime: 1; limestone.*

regain. 1. In agricultural products, the percentage increase in weight due to increase in moisture as compared to the moisture-free weight. 2. The amount of moisture that wool or a fabric will naturally absorb from the air at a certain relative humidity.

regal lily. *Lilium regale,* family Liliaceae, a bulbous herb grown for its large, lilac or purplish flowers, white inside with yellow bases. Native to eastern China. Also called *royal lily.*

Regels border privet. *Ligustrum obtusifolium* var. *regelianum,* family Oleaceae, a variety of the privet with horizontally spreading branches, grown as a hedge plant in the U.S.

Regels threewingnut. *Tripterygium regeli,* family Celastraceae, a shrub grown as an ornamental for its profusion of terminal clusters of fragrant, white flowers and red stems. Native to Asia.

Reggiano. A hard, Italian cows' milk cheese used for grating. See *Parmesan.*

reginned cotton. Cotton that has passed through the ginning process more than once and afterwards has been subjected to a cleaning process before baling.

regional profile. A soil profile that owes its character largely to the effects of the climate and vegetation normal for the region in which it has been formed. See *soil profile.*

regional research laboratories. Laboratories established in certain geographical regions of the U.S. by the Federal Department of Agriculture to conduct research of special interest to agriculture.

regional (seasonal) feed flavor. Any flavor of feed commonly present in milk at particular times of the year or in certain regions.

registered seed. The progeny of foundation or registered seed, approved and certified by the certifying agency, that is so handled as to maintain satisfactory genetic identity and purity.

registered stock. An animal registered in the official herdbook of the breed. See *purebred.*

Register of Merit. 1. The plans of production for registration established by two different breed associations, the Jersey and the Shorthorn. 2. The plan of registration for male and female breeding chickens qualifying under the U.S. National Poultry Improvement Plan.

Register of Production. The plan of production for registration established by the breeders of Brown Swiss dairy cattle in the U.S.

Registration Certificate. A paper issued by a breed association which shows that a particular animal has been registered.

registration paper. Registration Certificate.

registry number. A number assigned to a particular purebred animal that has been registered with a breed association.

regolity. The mantle of loose material consisting of the product of weathering which overlies the solid rock of the earth.

Regosol. An azonal group of soils consisting of unconsolidated rock in which few if any clearly expressed soil horizons have developed. It is largely confined to recent sand dunes, loess, and glacial drift on steeply sloping lands.

regrassing. Reestablishing grass on areas where it was once prevalent but had been killed out by some environmental condition, such as drought.

regression. Destruction of the vegetation, as by fire, grazing, cutting, etc., usually with subsequent deterioration of the site, as by exposure, erosion, or loss of nutrients, to such extent as to give rise to a subsequent simpler vegetative type. It is not a true succession or development from forest to grassland, but a replacement as a consequence of more or less complete destruction of the trees, etc.

regrowth. Herbage that grows after grazing or after the plants have gone through a period of dormancy. *Cf. aftermath.*

regular ham. An unskinned ham.

regular head horseshoe nail. A horseshoe nail with a large head. The other type of nail with smaller head but the same length of shank is called *city head nail.*

regular long clear (pork). An export cut of pork made from the side of the carcass with the ham off and the backbone and ribs out, foreleg cut off close to the breast and the aitch and breast bones cut down smooth with the face of the side.

regular shoulder. A retail cut of pork made from the shoulder that has the neck bone, ribs, breast flap and foot removed but has all the fat left on.

regular vanilla bean. One of three subgrades of the beans of Mexican vanilla (*q.v.*): *i.e.*, good, regular, and inferior. The regular grade includes red, dry beans of fair quality which are not adaptable to bundling.

regulation cut. The determination, periodically or in advance of cutting, of the volume of timber to be felled under the objectives of a given management plan.

regulator. 1. Any chemical, other than a nutrient, which will increase or decrease the growth of a plant or a plant part. 2. A mechanical device which regulates an operation or condition, as a temperature *regulator.*

Regur. An intrazonal group of dark calcareous soils high in clay, mainly montmorillonitic in character, and formed mainly from rocks low in quartz. It occurs extensively on the Deccan Plateau of India.

regurgitate. To return undigested food from the stomach to the mouth, as by ruminants.

rein. To control, check, stop, guide, or back up a horse or horses with the reins *q.v.*

reindeer. Any of several species of the genus Rangifer, a type of large deer with branching horns used in Arctic regions to pull sleighs and for food. See *caribou.*

reinfect. To produce infection again in a previously infected plant or animal.

reins. The part of a horse's harness fastened to each side of the bit or curb by which the rider or driver directs and controls it.

reinsemination. To repeat the process of insemination *q.v.*

reject. Any product, such as lumber, milk, wool, etc., that does not measure up to recognized grade specifications.

relative advantage. The relative profitability of producing one product as compared to another in a particular area or firm; also, the marginal value productivity of a factor of production in producing different products within an area or firm relative to its marginal value productivity in producing the same product in different areas or firms. See *marginal value product.*

relative humidity. The ratio of the mass of moisture actually present in any volume of air of a given temperature to the maximum amount possible at that temperature and pressure, usually expressed in percentage.

release. 1. To free trees from competition by cutting or otherwise removing or killing nearby vegetation and branches, usually applied to young stands. 2. To cancel a claim, as to *release* a real estate or chattel mortgage.

relict. A plant or a community of plants that has survived some important change in natural conditions, as climate, etc.

relictland. Land left by the recession of the shore line of a lake or other body of water.

relief. 1. The variation in elevation of the ground surface. It is indicated on maps by hachures, shading, or more accurately, by contour lines. 2. The physical or geomorphological features of a land surface. 3. The local difference in elevation of land features expressed in feet or in such terms as strong, mild, low, rugged.

relish. 1. A substance added to food to make it more palatable. 2. To eat or feed on with enjoyment.

relisted. Listed (*q.v.*) land that is sometimes worked over a second time to mellow the soil and control the weeds.

remainderman. A person entitled to the remainder of an estate, usually real estate, upon the expiration of a prior claim or reservation, such as a life lease.

Rembrandt tulips. A type of Darwin tulip in which the color is broken; *i.e.*, the flowers are variously edged, feathered, streaked, or blotched with colors.

remiges. See *wing feather.*

remnant teeth. The first premolar teeth of the horse which usually remain embedded under the gum but which occasionally erupt. Also called *wolf teeth.*

remontant. 1. Having a second blooming period the same growing season, as with certain roses, delphiniums, and some other plants. It can be induced by special care, as feeding and pruning. 2. Describing certain hybrid perpetual roses.

remote refrigeration unit. A mechanical cooling unit which is not affixed to the frame of the bulk milk tank. The unit is connected to the tank proper only by water lines, refrigerant lines, and electric cable.

remount. A fresh horse used to substitute for another worn out or otherwise incapacitated horse.

Remount Service. Originally a branch of the Quartermaster Corps of the U.S. Army which was established to carry on a horse breeding program to provide sufficient horses of cavalry and artillery types for the Army. Transferred to the Department of Agriculture in 1948 and renamed the Agriculture Remount Service, it is now discontinued.

remuda. A collection or string of broken horses.

render. To extract, separate, or clarify by melting, as to *render* lard or fat.

rendered pork fat. Fat, other than lard, rendered from clean, sound carcasses, parts of carcasses, or edible organs from hogs in good health at the time of slaughter.

rendering kettle. A kettle used for extracting fat. See *render.*

rendering plant. A commercial plant for the extraction of fat. See *render.*

rendering-plant tankage. See *reduction tankage.*

rendering wax. Melting honeycombs and cappings to remove refuse from the wax.

Rendzina. A type of soil, belonging to the AC group, which consists mostly of a blackish or dark gray humous layer resting directly on limestone or some other highly calcareous rock formation. Also called *Rendsina.*

renew itself. To make new growth, as winter barley *renews itself* under grazing.

rennet. An extract of rennin; a digestive ferment secured from the fourth stomach of suckling calves which is used as a milk coagulant in cheese making. See *rennin.*

rennet curd. The curd produced in milk by rennet action either as a result of specific bacterial growth or by direct addition of a rennet solution.

rennet curd tension. See *curd tension.*

rennets. The salted or dried stomachs of suckling calves, pigs, or lambs.

rennet stomach. The abomasum.

rennin. A coagulant enzyme occurring particularly in the gastric juice of cows, and also in some plants and lower animals. See *rennet.*

renovate. To renew, as by pruning old wood to induce fresh growth; to fertilize and spray old lawns to produce new growth; to repair and again make usable as to *renovate* a machine or building.

renovated butter. Butter made from poorer grades by melting, clarifying and rechurning in skim milk, or fresh milk.

rent. 1. A return paid or received for the use of property. 2. To lease property for money or for a share of the produce.

rental. The amount for which a property will rent.

rental income. In value appraising, the income anticipated by a landlord based on a typical landlord-tenant arrangement under a particular cost-price relationship.

renter. One who rents farm or pasture land from an owner. See *rent.*

reorganization of fields. The necessary rearranging, laying out, or establishing of new fields; or the changing of the boundaries of old fields on farms and ranches. It is commonly used where soil and moisture conservation programs are put into practice making such changes in fields desirable or when the overall farm program is drastically changed.

repacked cotton. Lint cotton from two or more bales which is worked over and rebaled.

repeating spore. The urediospore, the only spore in the rust life cycle able to reinfect the same species of plant as that on which it is produced.

repellent. 1. Any substance obnoxious to insects, which prevents them from injuring their hosts or laying eggs for hatching future generations. 2. Any substance disliked by a nuisance animal, used to discourage its visits.

replacement cost. The total cost to a business of adding or replacing one unit of a factor of production, such as a new barn or a ton of hay. See *present value.*

replacement stock. Young animals used to replace older animals, usually for breeding purposes.

replacement value. The estimated cost at current prices of replacing an animal, a building, a tree, or some other object.

replant. To plant again, especially when the first crop is destroyed or fails to grow.

replenish. To fill again after having been diminished; to stock anew; hence, to refill completely.

replication (plant breeding). The systematic laying out of several test rows or plats to test soil fertility and other environmental factors.

repot. To place a growing plant in a larger pot to prevent roots from becoming pot-bound.

reprocessed wool. See *shoddy: 2.*

reproduce. 1. Of animals, to bring forth young. 2. Of plants, to bear fruit and seeds.

reproduction cost. In value appraising, the normal cost of exact duplication of farm improvements (*q.v.*) with the same or closely similar materials, as of a certain date or period. It is used primarily in reporting and analyzing the value of farm improvements.

reproduction period. The time required or normally decided upon for the renewal of a stand of trees by natural or artificial regeneration.

repullulate. To sprout or bud again.

repulsion. In genetics, the condition in which an individual heterozygous for two pairs of linked genes receives the dominant member of one pair from one parent and the dominant member of the second pair from the other parent: *e.g.,* AAbb x aaBB.

requeen. To introduce a new queen bee into an old colony to replace a malfunctioning old queen.

Requeijao. A cheese made in northern Brazil from skim milk.

requien mint. *Mentha requienii,* family Labiatae, a small, creeping herb, peppermint-scented, grown for its mauve or pale purple flowers. Native to Corsica.

rescue brome. *Bromus catharticus,* family Gramineae, an annual or biennial grass grown in the southern states (U.S.) for winter and spring pasture. Native to South America. Also called *Australian oat.*

reseed. To seed a second time on an area where the first seeding has failed due to unfavorable seasonal conditions, overgrazing, or other environmental conditions.

reserve. 1. Any tract of land, especially public land, set aside for some special use: *e.g.,* forest *reserve,* school *reserve.* Also called *reservation.* 2. A tree or group of trees left uncut on an area for a period, usually a second rotation. After the stand is reproduced, naturally or artificially, an active stand which is held for future utilization.

reserve champion. In the show ring, the best animal in a group of first prize animals after the champion has been selected.

reserve feed. A supply of roughage and concentrates in excess of normal requirements, set aside for emergencies, such as droughts, abnormally severe winters, etc.

reserve food. Plant food stored in the various parts of any plant, especially that stored in the roots and crowns of biennial and perennial grasses and legumes.

reserve grand champion. In most show rings, an individual awarded the second highest rating in the breed and of its sex.

reserve soil acidity. Potential soil acidity.

reservoir. 1. Any storage place. 2. A place where a quantity of water, usually large, is collected and stored for use when required, as to supply a city, for irrigation, etc. Such a body of water is often held from flow by a dam *q.v.*

reservoir host. An animal that harbors parasites without being particularly injured by them and which is capable of passing them on to other animals thereby injuring their health; a carrier of parasites: *e.g.*, a chicken that has recovered from coccidiosis may harbor the causative organisms and pass them on to susceptible chicks.

reset. Replant.

residual effect. The effect of a particular product or substance that remains after its initial use. Some insecticides and fertilizers remain effective for some time after they are applied.

residual fertilizer. That portion of a fertilizer left in the soil and not used by the crop to which it was applied.

residual soil. The residual mass of rock weathering, or soil derived therefrom, which remains in its place of origin. (Strictly, a term of geology rather than pedology.) Also called *sedentary; in situ; in place soil.*

residual stand. The stand of trees remaining after the completion of cutting.

residue. A deposit of an insecticide which persists on an exterior surface or which is absorbed by plant or animal tissue following treatment: *e.g.*, DDT sprays applied to barn walls leave surface deposits of the insecticide that will kill insects alighting on them. Also, DDT sprays applied to dairy cattle result in absorption of DDT and leave *residues* in the meat or milk.

resilient. Designating the tendency to recover or spring back to its original form when compressed, as the body of a cheese.

resin canal. In wood anatomy, an intercellular canal containing resin.

resin cell. In wood anatomy, wood parenchyma cells of coniferous woods.

resin-flux. Resinosis.

resinosis. An abnormal exudation of resin or pitch from conifers or the abnormal impregnation of their tissues by resin. Also called *resin-flux.*

resins. Solid or semisolid organic compounds obtained from the exudates of a number of plants, which are soluble in alcohol and ether but not in water. Natural and synthetic *resins* are used in varnishes, soaps, bonding for plywoods, etc. See *rosin.*

resistance. The property of not being easily influenced or affected. In pathology, the inherited or acquired properties by virtue of which an organism is able to resist an invasion by germs or the effects of poisonous or irritating substances. See *environmental resistance.*

resistant. Designating a plant or animal capable of withstanding disease, inclement weather, or other adverse environmental conditions.

respiration. 1. The oxidation of carbohydrates in living organisms and the attendant release of energy and liberation of carbon dioxide and water. 2. In animals, the act of breathing; the drawing of air into the lungs and its exhalation. In small organisms with no special breathing organs, the process takes place over a large part of the body surface.

respirator. 1. A device put over the mouth or nose as a protection against the inhalation of injurious substances, as dust or spray materials. 2. A device used in the application or inhalation of medicated vapors. 3. A device used for artificial respiration.

respiratory infection. Disease of the air passages or lungs.

respiratory poison. Any fumigant that blocks cellular respiration *q.v.*

respiratory system. In the body of an animal, the system concerned with respiration or breathing. It consists primarily of the trachea or windpipe and the lungs. It supplies oxygen, vitally necessary to carry on living processes in the body and eliminates carbon dioxide, an end product of the physiologic processes. See *respiration: 2.*

respire. To breath. See *respiration: 2.*

respond to. To show increased growth or favorable reaction, due to an application of fertilizer or any other treatment.

responsive mouth. Designating a horse so trained that it easily obeys commands given by means of the reins.

resprouting. Putting forth new growth, as sprouts.

restharrow. Cammock *q.v.*

resting. See *rest period; dormancy.*

resting area. A sheltered area of loose housing where cows are bedded but not fed. The manure is allowed to accumulate during all or a part of the year. Also called *bedded area; loafing area; lounging area.*

resting bodies. See *sclerotia.*

resting land. See *fallow.*

resting pasture. A pasture not grazed by livestock.

resting spore. A spore, often thick-walled, that can remain dormant for some time before germinating and becoming capable of initiating growth or infection.

restock. To stock again; to add new or additional livestock, game, or fish.

restocking. See *restock.*

restocking forest land. Cut-over, burned, or otherwise denuded areas which are in the process of being satisfactorily re-covered by a young forest.

rest period. A period of quiescence or inactivity in plants even though moisture and temperature conditions are favorable for growth. See *dormancy.*

restricted feeding. A system of feeding poultry

whereby feed is provided only during certain periods of the day. See *free choice.*

restriction. An encumbrance limiting the use of land imposed for mutual or community protection.

resurrectionfern. *Polypodium polypodioides,* family Polypodiaceae, a drought-resistant, epiphytic fern which grows on trees. Under dry conditions the fronds are coiled and the plant is ball-like but it opens and continues growth under moist conditions. Native to tropical regions of America.

resurrection selaginella. *Selaginella lepidophylla,* family Selaginellaceae, a desert, perennial herb growing about 4 ins. high in dense tufts, stems turning inward when dry, giving the plant a ball-like appearance. It opens and resumes growth in the presence of moisture. Native to America.

ret. To prepare flax, hemp, and some other fiber-producing plants in order to separate the fibers from the straw. In dew retting the flax is spread out in the field and wetted by dew, the moisture stimulating the action of bacteria and molds. Water retting is accomplished by soaking the flax in water at summer temperature for several days.

retail dairy. A type of dairy plant concerned primarily with the distribution and selling of milk and milk products to the consumer. See *dairy: 1; dairyman: 2.*

retained afterbirth. See *retained placenta.*

retained placenta. The fetal membranes or afterbirth which a mammal mother fails to expel normally within a few hours after her young is born.

retaining pen. A pen for holding cattle, sheep, and hogs at the time of dehorning, shearing, dipping or weighing.

retention. The difference between cumulative precipitation on and total runoff from a drainage area. It closely approximates consumptive use *(q.v.)* when considered for periods of a year or more.

retention of the afterbirth. See *retained placenta.*

retention pen. A pen in which new-born lambs are placed to separate them from their ewes.

retention reservoir. A reservoir used strictly for flood control. One or more openings at the base of the dam, open at all times, are of such size that the discharge is never more than the capacity of the stream below.

reticulate. A form of soil structure with more or less regular-sized fragments separated by a network of narrow or hairline cracks.

reticulate mottling. In laterite soils, a network of coarse streaks of different colors characteristic of parent materials. Also called *vermiculate mottling.*

reticuloendothelial system. A widely spread network of cells in the body concerned with blood cell formation, bile formation, and engulfing or trapping of foreign materials, which includes cells of bone marrow, lymph, spleen, and liver.

reticulum. The second chamber of the ruminant's stomach. The walls of this stomach have a honeycomb-like structure.

retired to pasture. Designating an animal whose productive life has ended and which is turned out to graze.

retired to the stud. Designating a horse which no longer races but is retained for breeding purposes.

retrogression. The retreating or deterioration of a plant community from its highest natural expression to a lower one which is usually brought about by biotic influences or influences of man.

retting. The loosening of the bast fibers of flax and hemp from the stem by utilizing bacteria in the decomposition of the pectin which binds the fibers together. See *ret.*

return on farm capital. The net farm income less the value of the operator's labor which represents the return on the farm capital.

reused container. Any produce-shipping container which is packed and transported to any destination, emptied, and reused for the packing and transporting of the same or a different type of produce, as a bushel basket.

revegetation. Reestablishment of vegetation which may take place naturally or be induced by man through seeding or transplanting.

reverse T-budding. See *shield budding.*

reversible-disk harrow. A single-disk harrow with the gangs constructed in such a manner that they can be reversed to throw the soil either in or out. The gangs are adjustable for cultivating ridges.

reversible-disk plow. A one-bottom disk plow which is so constructed that the disk can be reversed and the soil thrown in the same direction on each trip across the field.

reversible marker bar. A bar which is used to mark rows for a machine planter.

reversible plow. A plow which has a moldboard to turn the soil to the right and another to turn the soil to the left. It is sometimes used in contour farming or plowing land which is to be irrigated. Also called *2-way plow.*

reversible point. See *double-pointed shovel.*

reversion. 1. The interaction of a soluble plant nutrient with the soil which causes a precipitation of the nutrient in a less soluble form, usually restricted to the conversion of monocalcium phosphate to the less soluble dicalcium phosphate. 2. In animal husbandry, the appearance in an individual of a character possessed by some distant ancestor. Also called *reversion breeding.*

reverted phosphoric acid. In the fertilizer trade in the U.S., an increase of the citrate-insoluble phosphoric acid with a decrease in the available phosphoric acid.

revolving-coil-type agitator. An agitator which is used in the pasteurization of milk. See *pasteurization.*

revolving head cupola. A revolving ventilator head which is placed on the roof of a barn to provide ventilation.

revolving spike-cylinder seedstripper. A revolving cylinder set with spikes used to harvest Kentucky bluegrass seed.

revolving super. Changing supers as soon as the bees start to work so as to keep the bees spread throughout the entire stack of supers.

reworked wool. See *shoddy: 2.*

rex begonia. *Begonia rex,* family Begoniaceae, a popular indoor plant cultivated for its large, variegated leaves and showy, pink flowers. Native to Assam. Also called *beefsteak geranium.*

R. F. D. (rural free delivery). In the U.S., the free delivery of mail in rural areas which started in 1902. The term is now obsolete, and has been replaced by Route (R.) or Rural Route (R.R.).

rheology. The science of deformation and fluidity of matter as in cheese and butter.

rheum. See *catarrh.*

rheumatism root. Striped pipsissewa.

rhinitis. Inflammation of the membranes lining the nasal passages. See *infectious coryza.*

rhinitis atrophic. A disease affecting the upper respiratory tract of the pig which causes atrophy of the turbinate bone and frequently distortion of the snout. Affected pigs usually grow at a slower rate than normal.

Rhizobium. A genus of bacteria, Rhizobium, family Rhizobiaceae, several of whose species produce nodules upon the roots of legumes and supply them with some of their nitrogen. This phenomenon, of great agricultural importance, is called symbiosis. *R. japonicum* produces nodules upon soybeans; *R. leguminosarum* produces nodules upon garden and field peas, sweet peas, and vetches; *R. meliloti* produces nodules upon white sweetclover, yellow sweetclover, and alfalfa; *R. phaseoli* produces nodules upon kidney and navy beans; *R. trofolii* produces nodules upon red, alsike, mammoth, crimson and white Dutch clovers.

Rhizoctonia blight of wheat. A fungous disease of wheat caused by *Rhizoctonia solani,* and other species, which appears in patches, causing the plants to be stunted and a purple cast on the leaves. The roots show a tan cortical rot and tan lesions are present on the leaf sheaths. See *Rhizoctonia disease.*

Rhizoctonia crater spot of celery. Pit-like craters on the stalks of celery which are caused by the fungus *Rhizoctonia solani.* See *Rhizoctonia disease.*

Rhizoctonia disease. A disease with various symptoms occurring on many different plants, caused by the fungus usually called *Rhizoctonia solani,* which is the imperfect (asexual) stage classified as belonging to Class Fungi Imperfecti, Order Mycelia Sterilia. This may infect the underground growing tips of roots or germinating seeds or stems near the surface of the ground causing damping-off. Various above-ground parts of the plant may be infected, such as stems, leaves and fruits. Under certain conditions the fungus may cover parts of the stems and leaves with a grayish-white powdery growth on which may develop the perfect (sexual) stage, often called *Corticium vagum,* which then bears the name *Pellicularia filamentosa,* family Thelephoraceae. Following are the definitions of various of the more striking manifestations of the disease caused by this fungus on some of the more important hosts. A few other species of *Rhizoctonia* also cause diseases of plants but they are mostly of minor importance.

Rhizoctonia disease of cotton. See *Rhizoctonia disease.* Also called *sore shin.*

Rhizoctonia disease of golf links. Patches of dead grass in the putting green, etc., caused by *Rhizoctonia solani.* See *Rhizoctonia disease.*

Rhizoctonia disease of potato. A widespread fungous disease of the Irish potato caused by *Corticium vagum,* family Thelephoraceae, characterized by the production of stem lesions below the ground level, the appearance on the tubers of black specks that will not wash off; various minor injuries and secondary or accompanying effects of rather uncertain relationship. Also called *black scab; black scurf; black speck; brown stem; damping-off; dartrose; dirt that will not wash off; little potatoes; russet; stem canker.* See *Rhizoctonia disease.*

Rhizoctonia root and crown rot of grasses. A fungous disease of grasses caused by *Rhizoctonia* spp., characterized by the crown and leaf tissues being killed in localized spots early in the season. See *Rhizoctonia disease.*

Rhizoctonia root rot of alfalfa. A fungous disease of alfalfa caused by (a) *Rhizoctonia violacea,* characterized by a rotting and shredding of the roots with a dark brown to dark violet discoloration; (b) by *R. solani,* characterized by dark, sunken areas on the tap root and branches. See *Rhizoctonia disease.*

Rhizoctonia root rot of beans and Lima beans. A rot of beans and Lima beans caused by *Rhizoctonia solani.* See *Rhizoctonia disease.*

Rhizoctonia root rot of peas. A minor, fungous disease of seedling peas caused by the Rhizoctonia stage of *Pellicularia filamentosa,* characterized by a slightly eroded, yellowish-brown zone on the underground stem and root near the seed attachment. In time, the lesion turns brown and becomes sunken and heavily eroded. The growing tip may be killed before emergence, and a new shoot sent up which may also be killed. See *Rhizoctonia disease.*

Rhizoctonia root rot of strawberry. A root rot of strawberry caused by *Rhizoctonia solani.* See *Rhizoctonia disease.*

Rhizoctonia rot. See *Rhizoctonia disease of potato; Rhizoctonia disease.*

Rhizoctonia rot of citrus. Rot of roots of *Citrus* spp. caused by *Rhizoctonia solani.* See *Rhizoctonia disease.*

Rhizoctonia scab. See *Rhizoctonia disease of potato; Rhizoctonia disease.*

Rhizoctonia stem rot of China aster. A root and stem rot and leaf blight caused by *Rhizoctonia solani.* See *Rhizoctonia disease.*

rhizoid. A root-like structure which is capable of attaching itself to a surface.

rhizome. A horizontal stem which grows partly or entirely beneath the surface of the ground, often having scalelike leaves. It usually produces roots below and sends up aboveground shoots. Also called *rootstock.*

rhizomorph. The mycelial cables and strands especially of the fungus, *Armillaria mellea,* that ramify the soil, penetrate and cover the roots and trunks of decaying trees.

rhizomorphic root rot. Mushroom root rot.

rhizophora. The true mangroves, tropical and semi-tropical trees and shrubs of the family Rhizophoraceae. *R. mangle,* the common or American mangrove is native in marshes under sea influence, tidal shores and shoals from southern Florida (U.S.) through Central and South America. In such regions the mangrove acts as a protection for shores and as an aid in building land, and affords suitable background for the development of property.

Rhizopus rot of drupe fruits. A fungous disease of drupe fruits caused by the bread mold fungus *Rhizopus nigricans,* family Mucoraceae, characterized by rot of fruits in transit and storage. The fruit is covered with the mycelium of the fungus, giving a whisker-like appearance.

Rhizopus rot of strawberries. 1. A fungous disease mostly of harvested strawberries caused by *Rhizopus nigricans,* family Mucoraceae, characterized by soft, watery tissue on the surface of the fruit, easy exudation of the juice of the berry on slight pressure, a browning of the fruit, and in packages, by a white, fluffy growth across the space between the berries. Also called *leak; melters; leak and rot; watery soft rot.* 2. See *soft rot of sweet potato.*

rhizosphere. The root zone, the area where microorganisms are most active in increasing the availability of nutrients for plants and where injurious effects may take place under certain conditions.

Rhode Island bent. Colonial bentgrass.

Rhode Island Red. A breed of domestic chickens of the American Class. Single-comb *Rhode Island Reds* are hardy, dual-purpose birds, bred largely for egg production but with consideration for meat qualities. Individuals are oblong in appearance with relatively long, flat backs, rich red plumage, yellow skin, and yellow shanks tinged with reddish horn. Mature cocks weigh about 8½ lbs., mature hens about 6½ lbs. The eggshell color is brown. The Rose-Comb variety is less commonly raised.

Rhode Island White. A breed of medium-weight chickens of the American Class. Individuals have a rose comb and white plumage. The shape and weight specifications are the same as for the Rhode Island Red (*q.v.*). It is not a commonly raised breed.

Rhodesgrass. *Chloris gayana,* family Gramineae, an erect, perennial grass, propagated by seed and by running branches that root at the joints, grown for forage and as an ornamental in warm areas. Native to South Africa.

rhododendron. Certain North Temperate Zone, evergreen or deciduous shrubs of the genus Rhododendron, family Ericaceae. Species are grown as ornamentals. See *azalea; Carolina rhododendron; catawba rhododendron; rosebay rhododendron; Smirnow rhododendron.*

rhododendron borer. *Ramosia rhododendri* (Beut.), family Aegeriidae, a moth whose larvae bore in the sapwood of rhododendron causing the branches to break off.

rhododendron lace bug. *Stephanitis rhododendri* Horv., family Tingidae, an insect that infests rhododendrons.

rhododendron whitefly. *Dialeurodes chittendeni* Laing, family Aleyrodidae, an insect whose larvae cause a yellowish mottling on the upper side of rhododendron leaves with a rolling of leaf margins. Rhododendrons with leaves smooth on the underside are the only ones infested.

Rhone aster. *Aster acris,* family Compositae, a perennial herb, grown for its autumnal blue flowers. Native to southern Europe.

rhubarb. An herb of the genus Rheum, family Polygonaceae. Species are grown for their edible leaf petioles used in sauces, pies, etc., and as ornamentals for their striking foliage. See *garden rhubarb; sorrel rhubarb.*

rhubarb curculio. *Lixus concavus* Say, family Curculionidae, a large beetle which punctures the stalks of rhubarb to deposit its eggs. The larvae, however, seldom develop on the rhubarb plant but usually grow on curly dock. The injury on rhubarb results from punctures and a very small amount of feeding by the adult beetle. It is not considered to be a serious pest.

Rhynchosporium scald. A leaf scald primarily of barley and rye but also of many other grasses, caused by *Rhynchosporium secalis,* characterized on the leaves by oval to irregular areas, at first water-soaked and gray-green but later zonated and yellow, brown and gray-green. The pathogen produces colorless asexual spores scattered by rain and wind. Specialized races are restricted closely to the different small grain and grass species.

riata. Lariat.

rib. 1. a. A cut of meat made from five to eight ribs depending upon the method of cutting the forequarter. b. Any cut of meat which comes from the *rib* section, as a *rib* roast. 2. In judging of livestock, spring of *rib* indicates conformation and vigor. 3. That which resembles a *rib* in form or use, as a piece of timber to which boards are fastened. 4. To divide a side of beef into a fore and hind quarter.

rib back. A wholesale cut of veal which is the unsplit ribs or eight ribs on each side.

ribband (ribbon). In barn construction, a horizontal strip of wood notched into the studs in a balloon frame, which foims a support for joists above the first floor.

ribbed cocoon maker. Apple bucculatrix.

ribbing plow. The landside or moldboard plow. The reference is to the parallel ridges, or ribs, made by the turned furrow slice (England).

rib blight of lettuce. A disease of unknown cause occurring in head lettuce in the field and during shipping and marketing. It is characterized by a dark brown, necrotic strip of tissue along the mid-vein.

ribbonbush. *Homalocladium platycladum,* family Polygonaceae, an erect, tropical shrub with flat, ribbon-like branches, which is sometimes grown as a curiosity. Native to the Solomon Islands. Also called *tapeworm plant; centipede plant.*

ribbon cane. A variety of sugar cane, marked by a longitudinal reddish stripe on the stalk which is grown for sirup production in some southern states of the U.S.

ribbon eucalyptus. *Eucalyptus viminalis,* family Myrtaceae, an aromatic, very tall tree which is grown as an ornamental. Also called *manna gum.*

ribbon tree. *Plagianthus betulinus,* family Malvaceae, an average-sized tree bearing small, cream-colored flowers, whose bark yields a flaxlike fiber. Native to New Zealand. Also called *akaroa.*

ribby. Designating a pot-bellied meat carcass or a carcass in which the eye muscle is small and the rib is long and flat.

rib cotton gin. The part of a cotton gin which is so shaped that the partially stripped seeds are pressed forward and upward by the pressure of other seeds brought forward by the saw teeth as the lint is engaged. The ginning ribs separate the seed from the lint and permit the saws to engage the lint.

rib down. To separate the hind from the front quarter of a beef carcass, exposing the large muscles which lie on either side of the back bone.

Ribes eradication. In blister rust control, the removal of currant and gooseberry plants from stands of white or five-needled pines for protection against white-pine blister rust.

Ribes live stem. In blister rust control, the aggregate length of living stems and branches of a Ribes bush or of Ribes plants per acre; usually expressed in linear feet.

ribicide. A toxic chemical which is used to kill *Ribes* spp.

rib lamb chop. A retail cut of lamb made from the rib rack *q.v.*

riblet. A retail cut of lamb breast which is cut into individual ribs.

riboflavin. $C_{17}H_{20}N_4O_6$, a vitamin of the B-complex group. See *vitamin B complex.*

riboflavin supplement. A feed material used chiefly for its riboflavin content which shall contain not less than 1,000 milligrams of riboflavin per lb. according to the tentative method of analysis of the Association of Official Agricultural Chemists. The label shall bear a statement of origin.

rib rack. A retail cut of mutton which consists of the loin and eight ribs.

ribwort. Buckhorn plantain.

rice. 1. *Oryza sativa,* family Gramineae, an annual marsh grass, one of the oldest cultivated crops grown for its seeds, the *rice* of commerce, one of the most important starchy foods of the world. There are two types, lowland and upland, the first being irrigated. *Rice* is also classified as long, short, and medium grain depending upon the length of grain. Native to Indonesia. It has been grown in China for at least 4000 years and is now grown throughout the world, especially in eastern and southeastern Asia. Louisiana, Texas, California, and Arkansas are the principal *rice* producing states in the U.S. 2. A special variety of *rice* grown at the lower delta of the Tigris and Euphrates Rivers, which is extra long-grained and has a fine flavor. Also called *Euphrates rice.*

rice bean. *Phaseolus calcaratus,* family Leguminosae, a weakly climbing, annual herb grown for its edible seeds. Native to Asia.

rice binder. A binder designed to harvest rice. It differs from a grain binder only in that it is especially designed for operation on soft, muddy ground. See *binder: 1.*

rice bran. A livestock feedstuff consisting of the pericarp or bran layer of the rice and only such quantity of hull fragments as is unavoidable in the regular milling of rice.

rice combine. A combine (*q.v.*) for the harvesting and threshing of rice. It is gradually replacing the binder.

rice cutgrass. *Leersia oryzoides,* family Gramineae, a perennial grass with weak culms, often decumbent at the base and with creeping rhizomes. Native in marshy areas throughout most of the U.S., it is sometimes a weed in cultivated cranberry bogs.

rice-fish rotation. The alternation of rice and fish crops. Rice fields, which would ordinarily lie fallow and idle, are flooded and made to produce a crop of fish for one or two years. In addition to yielding income from the fish, the land is rested and enriched for the following rice crop.

rice hull. A product which consists of the outer covering of the rice. It is low in digestible nutrients and unpalatable but if well ground may be used as a low grade roughage for livestock.

rice huller bran. The product secured by the huller from brown rice and consisting mostly of rice bran and rice germs.

rice leaf smut. A fungous disease of rice caused by *Entyloma oryzae,* family Tilletiaceae, characterized by small, black, angular or linear spots on the leaves.

rice meal. A livestock feedstuff consisting of ground brown rice.

rice polish(ings). A livestock feedstuff consisting of the by-product obtained in the milling operation of brushing the grain to polish the kernels.

riceroot fritillary. *Fritillaria lanceolata,* family Liliaceae, a bulbous herb grown for its dark purple, greenish-yellow, mottled flowers. Native to the Pacific Coast of North America. Also called *checkered lily.*

rice root maggot. Rice water weevil.

rice stalk borer. *Chilo plejadellus* Zinck., family Crambidae, a caterpillar that burrows through the rice stalks causing them to break over, reducing thus the yield of the grain.

rice stem rot. Stem rot of rice.

rice stink bug. *Olbalus pugnax* (F.), family Pentatomidae, an insect that sucks the sap from the developing rice grains.

rice stone bran. A livestock feedstuff which consists of the siftings from the material obtained while removing hulls from rice. It contains rice germs, broken rice, and some rice hulls.

rice straw. The straw remaining after threshing rice. If well cured, it may be fed in the same manner as straw from the other cereals.
rice stripe. A virus disease of rice characterized by yellow stripes on the leaves and a dwarfing of the plant.
rice water. A product made by boiling rice grain to a pulp and mixing it with water; especially useful as a drink for horses suffering from diarrhea after exertion.
rice water weevil. *Lissorhoptrus oryzophilus* Ksl., family Curculionidae, a destructive rice insect. The white grubs first tunnel the roots and later chew them from the outside.
rice weevil. *Sitophilus oryza* (L.), family Curculionidae, a destructive grain insect, inhabiting farmer's bins, grain elevators, and ships where grain lies undisturbed for some time.
richardia. See *callalily: 1.*
Richmond's formula. A formula for calculating the percentage of total solids and solids-not-fat in milk from the fat content and specific gravity or density.
richweed. See *white snakeroot.*
ricinin. $C_8H_8N_2O_2$, a crystalline alkaloid, found in the castorbean, *Ricinus communis.*
rick. 1. Hay or grain carefully placed in a rectangular pile and ridged at the top which may be thatched for protection from the weather. 2. A pile of cordwood for use as firewood. It is 8 ft. long, 4 ft. high and has a variable width. 3. A pile of cordwood, stave bolts, etc., split from short logs.
rickets. A disease of man and animals due to a lack of calcium, phosphorus, or vitamin D and generally seen at a young age. In animals the clinical symptoms are soft and deformed bones, emaciation, poor appetite, and reproductive failure.
rickety. 1. Having rickets. Also called *rachitic.* 2. Designating an unstable building or old, infirm animal.
rickweed. See *citronella horsebalm.*
Ricotta. A small, white, salty, cows' milk cheese made from coagulable material, principally albumin, in the whey obtained in the manufacture of other cheeses such as Cheddar, Swiss, and Provolone. First made in Italy, it is now made in all the countries of Central Europe, and in Wisconsin and New York in the U.S.
riddle. A coarse-meshed sieve to separate coarser materials from finer. Usually made of wire.
ride. 1. To mount a horse for going from one place to another. 2. To push down a fence, as a cow. 3. To mount a cow, as another cow indicating heat.
ride bareback. To ride without a saddle.
ride fence. Regularly, to inspect range boundary fences for damage or for loss of cattle (western U.S.).
ride heavy. To ride in such a way as to tire one's mount and cause saddle sores.
ride herd. To attend cattle on the range.
ride light. See *light rider.*
ride slick. To ride a horse without securing a hand hold of any kind on any part of the saddle.
ridge. 1. A divide between two hollows; specifically, the strip of land thrown up by a plow or left between two furrows, or anything which resembles the elevation left by the plow. 2. A crest, as the summit of an abrupt ascent; a long hill. 3. The highest part of the roof of a building where the upper ends of the rafters meet. 4. The earth levee that is thrown up in terrace construction to confine runoff water.
ridge and furrow houses. A series of greenhouses built side by side, each sharing a common wall with another.
ridge buster. A disk or moldboard implement used for tearing down ridges and filling furrows. Also called *disk go-devil. Cf. middlebuster.*
ridge land. Any broad or narrow crested, elongated land feature which rises above or separates lowland plains or valley basins, sometimes constituting a drainage divide.
ridgeling. Any male animal whose testicles fail to descend normally into the scrotum. Also called *cryptorchid; original; ridgling.*
ridge planting. The practice of throwing two furrows together and planting on the ridge so formed. A practice on flat land which would otherwise be excessively wet for cropping.
ridging. Making small embankments or borders in fields to control irrigation water, conserve runoff from rainfall, or to assist in drainage.
riding academy. A business concern which rents saddle horses and teaches the art of riding.
riding crop. A short whip used by horseback riders.
riding down. Pushing over of small trees, shrubs, and fences by livestock in order to reach and browse on the foliage.
riding horse. A horse which is specially bred and trained to be ridden.
riding stable. A stable where saddle horses are maintained by the individual owners or are offered for hire by the management.
Riesengebirge. A soft cheese made from goats' milk in the mountains of northern Bohemia.
riffles. Shallow rapids in an open stream where the water surface is broken into waves by obstructions wholly or partly submerged.
rift crack. A longitudinal separation of wood through the pith which develops from a simple shake and with great age extends to or near the sapwood, separating by several inches at the center.
Rift Valley fever. A virus disease of domesticated animals which can be transmitted to man. Intranuclear inclusion bodies have been demonstrated in the liver of infected animals.
rig. 1. An apparatus or outfit of tools for any particular purpose, more especially for well boring. 2. The complete assembly of beam, shanks, and shovels of the corn cultivator. 3. The equipment used for the application of spray material for the control of diseases, insects and weeds. 4. The equipment used for threshing grain; threshing *rig* consisting of a farm tractor and a thresher. 5. A

carriage with its horses. 6. A ridge made in plowing. See *stitch: 1.*

Riga pine. Scotch pine.

rigging. That part of the saddle construction which has to do with the securing of the cinch around the horse.

rigging ring. The ring attached to the saddle tree for a cinch.

right bank. That bank of a stream which is on the right when one looks in the direction in which the current flows.

right-handed plow. A plow that throws the furrow slice to the right.

right-of-way. 1. A privilege which one person or persons may have for passage-way over the land of another. 2. A strip of land used for highways, railroads, pipe lines, etc. It may be an easement, or may have been purchased outright.

right side. One of the two sides of a beef carcass, the tight side in which the kidney hugs the carcass.

rigid bracing. The use of bolts and threaded rods to support weak or split crotches and long cracks in the trunk or branches of trees.

rigid drawbar. A steel bar rigidly attached to a tractor. Usually, it has several holes spaced horizontally to provide attachment points for implements.

rigid supports. The federal (U.S.) government's support of prices for farm commodities at 90 percent of parity, as contrasted to flexible supports at variable percentages of parity. See *parity.*

rigling. A male sheep or horse that has only one testicle in the scrotum (U.K.). Variously called *ridgel; riggil; riglen.* See *ridgeling.*

rigor mortis. The stiffness of the body muscles that is observed shortly after death.

rill. A very small stream, especially one with a rippling current.

rill erosion. Incipient gullies a few inches in depth and width which frequently develop on unprotected slopes. It is an erosional stage transitional between sheet and gully erosion.

rim. 1. In wood anatomy, the remnant of a perforation plate forming a border about a simple perforation. 2. A metal strip on the outside of a wheel. 3. The edge of a tank, gutter, natural depression, etc.

rim bound. In tobacco, the downward curve of the margins of leaves and especially of the leaf tips which is a symptom of potash hunger. Excessive applications of borax may also cause this condition.

rime. A white, icy coating on grass and leaves, formed by atmospheric pressure; hoar frost, white frost, etc.

rim-fire. A potash deficiency disease of plants, characterized by necrosis of the leaf along its margins.

rind. 1. A hard coating caused by the desiccation of the surface of the cheese. 2. Skin of an animal, especially of a hog, as a pork *rind.* 3. Skin of a fruit or vegetable. 4. Bark of a tree. 5. To remove the skin.

rind breakdown. Breakdown of the skin, especially of oranges, which may be caused by adverse weather conditions.

rinderpest. A specific, acute, lethal, and inoculable virus disease of cattle, characterized by an ulcerative inflammation of the mucous membranes, especially of the alimentary tract, fever, dullness, drooping head, loss of appetite, grinding the teeth, red eyes, and red nasal discharge. Not observed in the U.S. Also called *cattle plague.*

rind gall. A localized lack of cohesion between growth increments on woody plants caused by the growth of healing tissue over an area of dead cambium exposed by mechanical injury or fire.

rind-oil spot. See *oleocellosis.*

rind rot. Surface bacterial decomposition of cheese usually manifested by various-sized, smooth-edged, irregularly-shaped, brown-colored areas and usually intensely bitter to the taste.

ring. 1. A cut or girdle around the trunk, branches, or roots of a tree. See *girdle.* 2. Annual growth *ring* of a tree. See *annual ring.* 3. a. A circular band of metal or wood as the metal *ring* in the nose of a bull. b. To place a *ring* through the cartilage of the nose of an animal: *e.g.,* to prevent a hog from harmful rooting or to control a bull, etc. 4. a. A circular, metal or plastic band placed on the leg of a fowl for identification purposes. b. To place a *ring* on the leg of a fowl. Also called *ringing birds.* 5. A ridge which encircles the horns of a cow, the number increasing with age. 6. A circular exhibition place for the showing or sale of livestock or the racing of horses.

ring bit. A large ring that passes over the lower jaw of the horse which is used as a bit.

ring bologna. A sausage composed of a paste of beef, pork, salt and other seasonings, and ground, mixed and stuffed in casings. It is then smoked and has its two ends tied together.

ring bone. A bony enlargement, involving the pastern bones just above the hoof, which interferes with the action of the joints and tendons, thus causing lameness to an animal. Usually only seen in horses.

ring disease. A disease of plants caused by a stem and bulb nematode *Ditylenchus dipsaci,* family Tylenchinae, which infests ornamentals, strawberries, red clover, and alfalfa. The organism lives in the soil and survives between crops on crop refuse. The symptoms are discoloration or decay of underground parts and dwarfing of stems with a reduction in yield. It is spread chiefly by surface water.

ringer. 1. In horse races, a substitute race horse of higher or lower ability than the one it substitutes, which is entered surreptitiously to mislead persons who bet on races. It is a practice which entails severe penalties if discovered. 2. A substitute animal so similar in appearance that few differences are discernible.

ring-eyed. Designating the condition of an albino horse in which the iris of one or both eyes is devoid of the characteristic coloring matter and in which the dark shining lens appears to be surrounded by a ring of white.

ringing. To girdle a tree by making two parallel cuts in the bark around a trunk or branch as deep as the cambium and ⅛ to ¼ in. apart and removing the bark between the cuts. *Cf. scoring: 4.*

ringing of grapes. The removal of a narrow ring of bark from the fruiting shoot on the trunk side of the bunch which results in making larger berries. Usually done for exhibition purposes.

ringing of hogs. See *ring: 3, b.*

ring method of layering. A method of obtaining a number of rooted layers from the same seedling in which a year old seedling is cut back in early spring. New branches develop on the trunk and are prepared later for rooting by placing a ring of copper wire around each new stem and heaping earth above the ring. See *layering.*

ring-necked pheasant. *Phasianus colchicus torquatus,* family Phasianidae, a long-tailed, highly colored, gallinaceous bird with a white ring neck. A highly popular game bird, it is sometimes a nuisance on farms because it pulls young corn plants out of the ground and devours what is left of the kernel. It is also occasionally harmful to vegetable crops, such as tomatoes. Originally from China.

ring-porous wood. Any of a group of hardwoods in which the pores are comparatively large at the beginning of each annual ring and which decrease in size more or less abruptly toward the outer portion of the ring, thus forming a distinct inner zone of pores known as the spring wood and an outer zone with smaller pores known as the summer wood.

ring pox of apricot. A virus disease of apricots characterized by reddish-brown necrotic rings and blotches on the fruit at ripening time extending into the flesh of the fruit. Infested trees are commercially worthless.

ring rot. A very destructive, highly infectious, bacterial disease of the potato tuber caused by *Corynebacterium sepedonicum,* family Corynebacteriaceae. A black ring discoloration in the outer flesh often results in only a shell around the soft decayed center.

ring shake. Wind damage to hemlock in which the spring wood and summer wood separate. See *shake: 1.*

ring spot. 1. A fungous disease of cauliflower and other crucifers caused by *Mycosphaerella brassicaecola.* 2. A disease of agaves, including sisal, caused by the fungus *Colletotrichum agaves,* forming pale circular lesions. 3. See *fairy ring: 2; frogeye; ring spot of lettuce; sugar cane ring spot; tobacco ring spot.*

ring spot complex of drupe fruits. A virus disease of drupe fruits; a complex of many forms or strains with widely varying symptoms, but usually with some form of ring spot. See *ring spot of stone fruits.*

ring spot of lettuce. A disease caused by the fungus *Marssonina panattoniana,* family Melanconiaceae, characterized by brown spots on the ribs and petioles of the leaves and by dead spots in the leaf. These spots soon fall out leaving holes in the leaf. Also called *lettuce anthracnose; rust disease of lettuce.*

ring spot of Saintpaulia. A physiological disorder of African violet caused by cold water wetting of the foliage, characterized by a breakdown of some of the leaf tissue.

ring spot of stone fruits. A widespread, virus disease of stone fruits reported from Europe and North America, and frequently latent. The disease is characterized by chlorotic rings and necrotic spots on leaves and occasionally by dieback of shoots. Severe symptoms are usually limited to a time shortly following acquisition of the virus, after which they are rarely evident. There are numerous strains of the virus or similar viruses resulting in a gradient of symptoms from mild to severe. The virus may be mechanically transmitted to cucurbits and tobacco. The insect vector is unknown.

ring tail. A horse which switches its tail in a circular motion.

ring test. A milk test for brucellosis.

ring type necrosis. Internal frost necrosis of potato characterized by lesions in or adjacent to the vascular ring, making a continuous or broken ring, narrow and distinct or broader and more diffuse, occurring in potatoes which have been subjected to low temperatures not sufficient to freeze them solid.

ringworm. A skin disease of man and animals caused by parasitic fungi, usually marked by distinct, circular patches with a scaly appearance.

Rinnen. A sour milk, spiced cheese made in Poland.

Rinzgau. A German breed of cattle believed to have been used to some extent in the early development of certain strains of Brown Swiss cattle in Switzerland. It probably was an influence in the lighter color marking of some Brown Swiss cattle.

Rio Grande wood necrosis. Gum formation from cracks in the bark of citrus connecting with irregular bands of necrotic wood; probably caused by infectious organisms gaining entrance at unprotected wounds.

ripa. 1. A stream or river bank. 2. Land which borders the water line of a river or sea.

riparian. 1. Designating land which borders a stream or body of water. 2. Pertaining to the banks of a body of water; a *riparian* owner is one who owns to the edge of a body of water; a *riparian* right is the right to control and use water by virtue of the ownership to the bank or banks of a body of water.

riparian doctrine. Under common law (U.S.), the owner of land along a stream is entitled to have the streamflow to his land undiminished in quantity and unimpaired in quality by upstream riparian owners, except that such owners are entitled to use water for domestic purposes. In some states, it is held that the riparian right includes the right to make use of the water for irrigation purposes and other uses, and that such right is a property right entitled to protection. See *aquatic rights.*

riparian forest. Tree growth adjacent to streams or other water courses whose roots are in or close to the zone of saturation due to the proximity of surface or underground water.

riparian law. The common law which deals with the nature and extent of rights to the use of water by adjacent land owners. See *riparian doctrine.*

riparian rights. See *riparian doctrine.*

ripe. 1. Designating mature seeds which are fit for germination. 2. Designating fruit which has attained full development. 3. In plant propagation, designating wood that will root well. 4. In grafting, designating that wood which is ready for perfect union. 5. Designating the best condition for use, as *ripe* cheese, *ripe* wine.

ripe honey. That honey which has become properly thick through evaporation. *Cf. green honey.*

ripening. 1. Growing to maturity and being fit for food, as ripe fruit or ripe grain. 2. Bringing to a certain condition for use by keeping, as in wine. 3. Preparing milk or batch mixes for making butter or ice cream either by a natural souring or by the addition of starters. 4. Undergoing an aging process, as in meat.

ripe seed. See *ripe: 1.*

ripe wood. See *ripe: 4.*

ripe snow. Snow which during the process of melting and settling has attained its maximum power of water suspension without loss of water. Snow is overripe when it has exceeded its ability to hold water and is losing it. The difference in density between ripe and overripe snow does not exceed 10 percent.

ripgut brome. *Bromus rigidus,* family Gramineae, tufted annual grass which is a common weed in the western U.S. Native to Europe.

ripper. An unusually strong, large horse.

ripple-grass. Buckhorn plantain.

rippleseed plantain. *Plantago major,* family Plantaginaceae, an annual or perennial herb which is widely distributed throughout the U.S. as an ubiquitous weed. Native to Europe and Asia. Also called *plantain; dooryard plantain; whiteman's foot; broad-leaved plantain; roadweed.*

rippling. 1. The process of drawing flax or hemp through a kind of comb to remove the seed from the stalk. 2. The mechanics of continuous layering of purées, chocolate and other flavor substances in ice cream giving a rippled effect.

rip rap. 1. A bed of stones laid on a sloping river bank or side of a levee, etc., to prevent river bank cutting or flood water erosion. 2. A bed of stones laid in a swamp or miry place to make a roadway or effect a crossing. 3. A stone fence.

rise of alkali. Alkali, previously distributed in the soil, which is brought to the surface with the rise of capillary water and left as a surface incrustation when the moisture evaporates. This accumulation may cause crop injury, as in the San Luis Valley of Colorado, U.S. *Cf. waterlogging.*

rising two. Describing a horse, 1½ to 2 years old. Also called *coming two.*

risk rating. An estimate of the credit and responsibility of an individual or business concern.

risky treadsoftly. *Cnidoscolus stimulosus,* family Euphorbiaceae, an herbaceous plant with long perennial roots and stout, branched stems, whose bristly hair contains a strong irritant which upon contact produces a dermatitis. Native to North America. Also called *bull nettle; spurge nettle; stinging nettle.*

river. 1. A channelized, naturally-flowing body of water of substantial size. Generally larger than a creek, stream, brook, or run, it is formed by the union of these lesser hydrographic features. 2. In local usage, a narrow connecting body of water of short length or size which does not occupy a valley or drainage basin of its own making.

river ash. Red ash.

river bank. 1. The nearly vertical or steeply sloping ground, generally alluvium, which forms the side of the river channel. Generally, it is not as high as a bluff, although there is little distinction between the two. Banks are subject to continuous, slow corrosion by river flow, or to rapid undercutting and dissolution during flood periods. See *bank; bank erosion.* 2. The brim of a river or land bordering its edge.

riverbank grape. Frost grape.

river birch. *Betula nigra,* family Betulaceae, a deciduous tree of average size with reddish-brown, ragged bark, grown as an ornamental on moist sites. Native to North America. Also called *red birch.*

river bottomland. Low land which consists of alluvial deposits along a river. Generally, but not everywhere, it is composed of soil more fertile and potentially more productive than adjacent upland. See *bottom.*

river sunflower. Thinleaf sunflower.

riverwash. Alluvial deposition in the channel or channels of rivers, mainly sandy, barren, and exposed at low stream levels or during absence of stream water, but subject to inundation and shifting during flood periods.

roach. See *Australian cockroach; brown-banded roach; cockroach; Surinam cockroach.*

roach back. An arch in the loin section of the back, often caused by injury or disease.

road grader. A towed or self-propelled machine with an adjustable steel scraper slung under the main body, which is used in the building and maintaining of roads especially for moving and smoothing of earth and also in constructing waterways and terraces.

road hack class. A class of light breed horses usually of American Saddle Horse or Thoroughbred breeding.

road machine. A device largely used in road construction, having one or more scrapers or scoops mounted between the wheels so that it can scrape up earth and transfer it, as from the sides to the middle of a road or railway bank. See *carryall: 2.*

roadside erosion control. Any system of controlling and preventing erosion along highways.

Methods used vary from elaborate structural control to simple vegetative control.

roadside stand. A place which usually consists of a display shed along the roadside for the retail sale of farm produce by the farmer.

roadster breeds of horses. See *Standardbred.*

roan. a. A coat-color of a horse that is chestnut or bay or may be red or strawberry *roan,* blue *roan,* or chestnut *roan* depending upon the intermingling of the background colors. b. Designating the red-white color phase of Shorthorn cattle.

roarer. A horse afflicted with roaring.

roaring. A defect in the air passage of a horse which causes him to roar or whistle when respiration is forced.

roast. 1. To dry thoroughly by heating; *e.g.,* to *roast* peanuts. 2. To cook a large piece of meat, as a ham, roaster chicken, etc., in an oven.

roasted almonds. An unsalted food product prepared by roasting almond kernels without removing the pellicle. Used in ice cream and in confectionary products.

roaster. 1. That which is raised, prepared for and adapted to roasting whole: *e.g.,* a chicken. 2. A young chicken whose weight exceeds 4 lbs.

roasting ear. An ear of field corn having its kernels well filled, plump, soft or in the milky stage when it is ready to be eaten boiled or roasted; sometimes applied to sweet corn in the same stage of development.

roasting pig. A pig weighing from 15 to 50 lbs., dressed shipper style with the head on.

robber bee. A bee which robs food from another colony.

Robbiole. A soft, rich, fast-ripening cheese made in Italy in winter from cows' milk mixed with ewes' or goats' milk.

robust. 1. Designating Merino wool that is relatively coarse compared with other Merino wools. Also called *strong wool* (Australia). 2. Designating a strong vigorous animal or plant.

Rocamadur. A soft cheese made in southern France from ewes' milk.

rock. Any relatively homogeneous mineral mass; a combination of mineral materials that forms a portion of the earth's crust including loose, incoherent as well as solid or hard masses.

rock aster. Alpine aster.

rock cedar. *Juniperus mexicana,* family Cupressaceae, a tall, evergreen tree grown as an ornamental and for its aromatic, brownish lumber used for making boxes, pencils, etc. Native to southwestern U.S. and Mexico.

rock chestnut oak. Chestnut oak.

rock cotoneaster. *Cotoneaster horizontalis,* family Rosaceae, a half-evergreen shrub grown for its foliage, pinkish flowers, and red fruit. Native to western China.

rock cress. 1. Herbs of the genus Arabis, family Cruciferae, grown in wall gardens, rock gardens and in open borders for their white or purple flowers. Native to Europe and Asia. See *downy rockcress; Kellerer rockcress.* 2. Bitter wintercress.

rock disintegration. The natural process by which hard rock is disrupted and comminuted. The factors involved are: expansion and contraction due to change in temperature, the chemical and mechanical work of plant roots, movements of stone fragments down slopes through force of gravity, a partial chemical solution, abrasion by moving water and ice.

rock elm. *Ulmus thomasi,* family Ulmaceae, a deciduous tree sometimes grown as an ornamental for its autumnal, yellow foliage. Its wood is used in cooperage, crates, veneer, furniture, interior trim and implements. Native to North America. Also called *American cork elm.*

rocker toe shoe. A special horseshoe which is used on a horse that stumbles.

rockery. A rock garden.

rocket candytuft. *Iberis amara,* family Cruciferae, a hardy, annual herb grown for its fragrant white flowers. Native to Europe.

rocket larkspur. *Delphinium ajacis,* family Ranunculaceae, a hardy, annual herb grown for its violet, rose, pink, blue, or white flowers. Native to southern Europe. Also called *annual larkspur.*

rocketsalad. *Eruca sativa,* family Cruciferae, an annual herb sometimes grown as a potherb. Native to Europe and Asia. Also called *salad plant; roquet; rocket.*

rock-fill dam. A dam which is composed of loose rock usually dumped in place, often with the upstream part constructed of hand-placed, or derrick-placed rock and which is faced with rolled earth or with an impervious surface of concrete, timber, or steel.

rock garden. A garden which contains alpine and saxatile plants displayed in an appropriate rock setting.

rock land. Land which is characterized by a large proportion of rock outcrop with shallow soil covering. The U.S. Department of Agriculture may designate an area as such when it has from 25 to 90 percent rock outcrop. Such land is for the most part nonarable but may support tree growth and supply limited pasturage. See *rock outcrop.*

rock lifter. A tripod provided with a combination of differential pulleys and ratchet gear which is used to lift isolated rocks or boulders from the ground and which, through the multiplication of force of one or two men at the end of a 12 ft. lever, can move a great weight, as a rock or tree stump.

rock lily. A common name for many plants: (a) *Selaginella rupestris,* family Selaginaceae, a fern-like, flowerless plant used as ground cover in damp areas. Native to eastern U.S.; (b) *Dendrobium speciosum,* family Orchidaceae, an epiphytic plant cultivated in bog gardens in warm areas. Native to southeastern Asia; (c) American columbine *q.v.*

rock mat. Sweet alyssum.

rockmelon. See *muskmelon.*

rock outcrop. A surface exposure of bedrock which may be a feature of stony land but which is distinguished from boulders or detached fragments of rock.

rock phosphate. The natural mineral deposit essentially tri-calcium phosphate, $Ca_3(PO_4)_2$, which is the principal source of the phosphorus of commercial fertilizers. The raw rock finely ground is also applied to soils as a fertilizer. The principal mining (U.S.) at present is in Florida and Tennessee, but large deposits occur in Wyoming, Idaho, Utah, and Montana, and smaller deposits occur or have been mined in South Carolina, Kentucky, and Arkansas.

rock pink. *Dianthus petraeus,* family Caryophyllaceae, a perennial herb grown for its fragrant white flowers. Native to eastern Europe.

rock plants. Plants which will grow well in rock gardens.

rock purslane. Plants of the genus Calandrinia, family Portulacaceae, fleshy herbs grown in the rock garden for their bright rose-pink to purplish flowers. Native to California, U.S.

rock salt. Crude, common salt (sodium chloride).

rock soapwort. *Saponaria ocymoides,* family Caryophyllaceae, a perennial herb grown in the rock garden for its bright pink flowers with purple calyxes. Native to Europe. Also called *rosy soapwort.*

rockspray cotoneaster. *Cotoneaster microphylla,* family Rosaceae, a low, evergreen shrub grown for its foliage, its white flowers, and its scarlet fruit. Native to the Himalayas.

rock weathering. The natural processes which cause the physical and chemical disintegration and decomposition of rock; the residual mass of weathering in the parent material of mineral soils.

Rocky Mountain cedar. Rocky Mountain juniper.

Rocky Mountain crazyweed. *Oxytropis saximontana,* family Leguminosae, a poisonous plant found from Montana to Utah, U.S., whose foliage is poisonous to cattle. Symptoms of poisoning are dullness, irregularity of gait, lack of appetite, dragging of the feet, a solitary habit, loss of flesh, and shaggy coat. As the animal ceases to eat it dies. Native to North America. Also called *locoweed.*

Rocky Mountain flowering raspberry. Boulder raspberry.

Rocky Mountain grasshopper. See *grasshopper.*

Rocky Mountain iris. *Iris missouriensis,* family Iridaceae, a perennial herb grown for its pale blue flowers. Native to North America.

Rocky Mountain juniper. *Juniper scopulorum,* family Pinaceae, an evergreen tree grown as an ornamental. Native to North America.

Rocky Mountain locust. See *grasshopper.*

Rocky Mountain spotted fever. A disease of man and animals characterized by intermittent chills and fever, painful muscles and joints, and red blotches on the skin. Occurring in the Rocky Mountain area, U.S., the Rocky Mountain wood tick *(q.v.)* is the vector.

Rocky Mountain wood tick. *Dermacentor andersoni* Stiles, family Ixodidae, probably the most important tick vector of disease. This tick transmits Rocky Mountain spotted fever, tularemia, Colorado tick fever, American Q-fever, and encephalomyelitis and, experimentally, anaplasmosis. Most domestic animals, man, and numerous wild mammals are its hosts. The stages in the life cycle are egg, larva, nymph and adult. Larvae and nymphs usually live on small wild animals, mainly rodents; adults attack larger animals and man.

rod. 1. A linear measure of land equal to 16½ ft. In square measure, a plot of land 16½ ft. on each side; a square *rod.* Also called *perch; pole.* 2. An old, English measure of land which varied from 6 to 8 yds.

rodent. A classification of mammals, mostly vegetarians, characterized by their single pair of chisel-shaped, upper incisors. *Rodents* are members of the Order Rodentia (rats, mice, squirrels, etc.) and the Order Lagomorpha (rabbits, etc.).

rodenticide. Any poison which is lethal to rodents.

rodeo. 1. The rounding-up of cattle. 2. A public performance presenting features of a cattle roundup, as lariat throwing, bronco riding, and bulldogging.

rod-type moldboard. 1. A moldboard plow which consists of round rods attached to the plow in such a manner as to form a surface upon which the furrow slice will be turned without breaking up the soil; used on sticky soil. 2. An attachment for a mowing machine which turns the grass so as to clear a space for the next cut.

rod weeder. A machine used for the cultivation of summer fallow land. The weeds are pulled up by a slowly rotating, round or square rod.

roe. The eggs or testes of fish. Consisting of two types, the female eggs (hard *roe*) and the male testes (soft *roe*), they are widely used for human consumption: *e.g.,* the salted *roe* of sturgeon (caviar) is highly valued as a delicacy. See *sturgeon.*

Rogers firethorn. *Pyracantha crenulata* var. *rogersiana,* family Rosaceae, an evergreen, ornamental shrub with willowy, slim branches bearing many orange-yellow fruits, used for hedges especially in the Pacific northwest (U.S.). Native to Asia.

rogue. 1. A seedling or plant of inferior or objectionable quality; a variation from type. 2. A horse without inclination to work or cooperate with its handler. 3. To remove and destroy undesirable plants.

rolcut secateurs. An instrument for removing the folds of skin from the crotch of ewes, also used in the bloodless emasculation of calves and lambs.

role. 1. Anthropologically, the actual behavior which is performed in executing the right and duties of a status. 2. Sociologically, the performance of any or all of the activities properly associated with the complex position one has in an hierarchical socio-economic scale.

roll. 1. The revolving drum on which milk is dried in the drum or *roll* process. 2. To reduce grain to a flaky or gritty texture for feeding, by putting it through steel rollers. 3. To prepare a field for planting by use of a roller. See *roller: 1.* 4. To level mammoth clover plants in the field so that less plant growth will be handled when the late upright growth producing the seed heads is harvested.

roll box. The circular or elliptical compartment at the front of the cotton gin where the cotton comes in contact with the picker roll.

rolled. 1. Designating a retail cut of meat for roasting in which the bones have been removed and the piece rolled and tied with strings. 2. Designating grain that has been through steel rollers.

rolled oats. Oat groats which have been pressed between heated rollers.

rolled toe shoe. A type of horseshoe which is used to make the foot action of a horse break over easier by improving action at the pivotal point of the shoe.

roller. 1. A farm implement; a horse- or tractor-drawn device consisting of one or more large, heavy cylinders, used for crushing lumps, smoothing the surface, and firming the soil. 2. A mechanical device used for pressing the juice from canes for the production of molasses, which consists of two or more cylinders between which the canes are pressed. 3. A set of two or more heavy cylinders used for converting whole kernels of corn and other cereal grains into a flaky or gritty feed. 4. A worker in a wool shed who assists in the rolling and skirting of the fleece (New Zealand).

roller check. A crack in the wood structure caused by a piece of cupped lumber being flattened between machine rollers.

roller drier. A type of drier for milk in which the milk is dried in a thin film on the surface of a horizontal steel cylinder heated by steam. The dried milk is removed by a scraper knife.

roller foundation mill. A roller mill used for the manufacture of honey comb foundations used in the production of honey.

roller gin. A cotton gin of small capacity, not extensively used. The lint is removed from the seed by a roller, a fixed knife, and a moving knife.

roller grader. For the grading of fruit, two rollers which revolve from each other and are closer together at the upper end than at the lower. As the fruit passes along the rollers the small fruit drops through first and the large last. This type of grader is used for whole, unpeeled fruit.

roller wheel. A broad wheel attached at the rear of a cotton planter to regulate the depth of planting the seed.

rolling. 1. Excessive side motion of shoulders, common in horses with abnormally wide fronts or chests. 2. A part of seed-bed preparation in which the land is rolled to even out the surface. 3. See *roller: 2.*

rolling colter. A round, flat, steel disk, sharpened on the edge and suspended on a shank and yoke from the beam of a plow, so constructed that it can be adjusted up and down for depth and sideways for width of cut. It cuts the furrow slice loose from the furrow wall, especially when plowing under heavy plant growth with a moldboard plow. *Cf. cut-away colter; fin colter.*

rolling land. A land surface which has rounded undulations, and relatively low relief.

rolling landside. A metal wheel which revolves on an axle fitted to the plow bottom, and which serves the same purpose as the bar landside.

roll-less churn. A churn in which moisture is incorporated in the butter by allowing the soft, freshly churned butter to drop violently from one shelf of the churn to another. *Cf. roll-type churn.*

Roll of Honor. A division in the system of Advanced Registration of Ayrshire Cattle in which a cow must meet certain production requirements in 305 days and give birth to a calf within a specified period.

Rollot. A soft, ripened cows' milk cheese similar to Camembert, made in France.

roll-over terrace. A terrace which has little height and a broad base, hence gently sloping. Under certain conditions it is effective in conserving water and preventing erosion, and it has the advantage of offering no obstacle to the movement of farm equipment over fields. See *broad base terrace.*

roll powder. The dried milk which is obtained by drying a thin milk film upon the surface of a rotating, steam-heated drum.

roll tube method. A method of making colony counts of bacteria in milk in which a test tube is used instead of a plate count.

roll-type churn. A churn fitted with rollers to facilitate the working process. *Cf. roll-less churn.*

rollway. a. A platform at a lumber mill onto which logs are unloaded from cars and trucks. It may be built around the edge of a pond or along the bank of a stream to aid in dumping the logs into the water, or it may be so built that it is used as a place for dry land storage of logs. b. In the early days of logging in the Lake States of the U.S. a bank or bluff of a stream or lake, down which logs were rolled into the water for transportation to a saw mill.

Romadur. A soft, ripened cheese made in southern Germany from either whole or partly skimmed cows' milk which is similar to Limburger but has a milder aroma.

Roman camomile. *Anthemis nobilis,* family Compositae, a perennial herb with finely-cut, strong-scented foliage and flower heads, grown as a medicinal, and sometimes as a substitute for grass, especially on sunny banks and on paths in herb gardens. Native to Europe. Also called *English camomile; chamomile; garden chamomile.*

Romanello. A very hard Italian cheese, usually made from partly skimmed or skim milk, which is suitable for grating. It is made from cows' milk in the U.S.

Roman hyacinth. *Hyacinthus orientalis* var. *albulus,* a variety of hyacinth *(q.v.),* grown for its early-spring, white flowers.

Roman nettle. *Urtica pilulifera,* family Urticaceae, a small annual herb which is a weed with stinging hairs. Native to southern Europe.

Roman nose. A curved, hooked, prominent, or bold nose, like the beak of an eagle. Used to describe the face of certain breeds of livestock.

Romano. A very hard, popular Italian cheese made from cows' or ewes' milk. Also called *Pecorina.*

Roman profile. See *Roman nose.*

Roman wormweed. Ragweed.

Rome Beauty. A famous variety of apple whose fruit is large, red-striped, and of medium quality. The trees are small, very productive, and grow well southwest of New York, U.S.

Romney Marsh. An English breed of hornless, long wool sheep whose rams weigh 225 to 250 lbs. and ewes 190 to 215. The face, ears, and legs are white, the nose black. The wool is the finest of the long wool breeds. The body is relatively low set and blocky.

rood. One-fourth of an acre; 40 square rods (England).

roof garden. A garden site, often elaborate in design, on the roof, terrace, or set back of a high building.

roof houseleek. See *hen-and-chickens.*

roof iris. *Iris tectorum,* family Iridaceae, a perennial herb grown for its blue-purple flowers. Native to Asia.

roof rat. *Rattus rattus alexandrinus,* one of the three types of rats found in the U.S., about 15 ins. from tip of nose to end of tail. Its fur on the back and sides is gray to gray-brown and the undersurface of the body is almost white. It breeds four times a year with an average of six young in a litter. Also called *Alexandrine rat.* See *rat.*

rookery. A place where large numbers of wild birds or animals congregate and breed.

room temperature. In laboratory work, 68° to 70° F.

roomy middle. The ample middle portion of farm animals. A deep, wide middle section is considered desirable because it contributes to both feeding and breeding capacity.

roomy sow. A sow with ample development in the middle.

roost. a. A resting or lodging place for fowls. b. A group of roosting fowls. c. To rest upon a roost or perch.

rooster. Male chicken. Also called *cock; cockerel* (when under 1 year of age).

roosting shelter. A place that protects or shelters the resting place of poultry, usually a poultry house.

roosting space. Space necessary for each roosting bird, usually 8 to 10 ins. on the roost, depending upon the size of the bird.

roost mite. Chicken mite.

roost tumor. Keel cyst.

root. The lower portion of a plant bearing neither leaves nor reproductive organs which mostly develops underground and anchors the plant in the soil. It bears the root hairs which absorb water and mineral nutrients. Some *roots* are aerial or partly aerial. In aquatic plants, the *roots* do not necessarily anchor plants. See *aerial root; corm; rhizome; tuber.*

root absorption. Taking in of water and nutrient materials from the soil by the roots of a plant.

root antagonism. Injury to the roots of a species which are grown near certain other species or when one is grown after the other on the same plot, as *root antagonism* exists between tomato and cucumber roots.

root aphids. Certain sucking insects of the family Aphidae, which feed on the roots of many plants, as corn, cotton, grasses, and ornamentals, often causing serious damage. Ants collect the aphid eggs in the fall and store them over winter. In spring the young are carried to roots where, through a complex life cycle, many generations are produced. The aphids excrete honeydew which the ants use as food.

root bark beetle. Small cylindrical beetles, family Scolytidae, which sometimes infest young trees at the ground line and kill them by working downward into the roots.

root borer. See *borer.*

root cap. The extreme tip of the root consisting of a group of cells that slough off and are replaced as the tip moves through the soil. It protects the growing region of the root which is just behind it.

root cellar. A pit or underground room especially constructed for the storage of root crops during winter. See *cold cellar.*

root climber. Any plant, as the common ivy, fig, etc., which uses its unusually located (adventitious) roots for climbing.

root collar. The transition zone between stem and root usually recognizable in trees and seedlings by the presence of a slight swelling. See *collar:* 3.

root crops. Any of a number of field and garden crops whose underground roots are used as food for man and animals, as turnips, beets, carrots, and sweet potatoes.

root crown. The basal portion of a stem of a plant where the transition or change from stem to root tissue occurs. See *root collar.*

root cutting. The portion of the root of a plant used for propagation.

root development. Increased growth and development of roots which may be induced by cultural practices and which often increases crop yields.

root division. See *division.*

rooted cutting. A cutting *(q.v.)* that has developed roots.

root gall. See *root-knot; crown gall.*

root grafting. The grafting of a cion on a seedling root or a portion of a root. See *grafting.*

root growth. See *root development.*

root hair. A hair-like growth on an epidermal cell of the root. It absorbs water and mineral nutrients for the plant.

root hardy. Designating a plant whose roots can survive the winter.

rooting. 1. The production of roots by a plant. 2. Digging the earth with the snout, as hogs. 3. A root used for propagation. 4. The production of roots by a cutting.

rooting medium. The substance in which cuttings are placed to strike roots. It may consist of sand, a mixture of sand and peat, or other material.

root knot. A disease of plants caused by the penetration of nematodes, *Meloidogyne* spp., family Heteroderidae, which stimulates cell development, resulting in enlargement of the root system, and the formation of galls. Many crops, including potatoes, tomatoes, celery, spinach, cabbage, tobacco, cucumbers, sugar beets and cotton, are hosts. Infested plants are paler in color, wilt more easily, and die early. Also called *eelworm disease.*

root knot of drupe fruit. See *root knot.*

root knot of grapes. See *root knot.*

root knot of pome fruits. See *root knot.*

root knot of strawberries. See *root knot.*

root lesion nematode. *Pratylenchus* spp., nematodes which infest at least 100 plant species including strawberry, coffee, cotton, tobacco, beets, corn, various small grain crops, and many legumes. They attack the roots of the plants causing wilting, yellowing, poor yield, and often death.

rootlet. A small root.

root louse. See *root aphids.*

root maggot. Rice water weevil.

root mealy bug. See *ground mealy bug.*

root nematode. See *root knot; root lesion nematode.*

root nodule. See *Rhizobium.*

root-nodule bacteria. See *Rhizobium.*

root parasite. Any plant, as the broomrape *(q.v.),* which is completely parasitic on the root of another plant.

root pressure. The force with which water is pushed up through the intact stem of a plant or exuded from holes made in the stem or from fresh stumps of branches or stems. Also called *sap pressure; endodermal pressure; bleeding pressure; exudation pressure.*

root pruning. Any cutting back of the roots: (a) during potting of plants to contain roots in a smaller volume of soil; (b) before planting trees to reduce roots to the size of the hole already dug; (c) to reduce top growth; (d) to remove a root and promote branching for later removal of the tree with a ball of soil attached by undercutting the roots with a tree digger or power knife.

root pulper. Any of several hand- or power-operated machines for reducing root crops, usually rutabagas, to a form suitable for stock feed. The most common type consists of a hollow, rotating drum, lined with knives, into which the roots are fed. By controlling the direction of the drum's rotation, the operator can determine the form of the product. Slices are used for cattle; fingers for sheep.

root reserves. The storage of plant food in the form of proteins, carbohydrates, etc., in the root.

root rot. Decay or rotting of the roots of different plants caused by any of a large group of organisms.

root rot of alfalfa. A complex of diseases due to a number of different organisms usually associated with plants weakened by unfavorable weather conditions. Often the mycelium of a fungus appears around the roots and crowns and in ruptured and injured plants.

root rot of wheat. A fungous disease of wheat caused by *Pythium* spp., family Pythiaceae, characterized by a light brown, soft rot of rootlets and roots and a pale green, stunted growth of the tillers.

root spread. The extent of growth of the root system of a tree or other plant. Some trees have a *root spread* twice as great as the spread of the branches.

rootstock. 1. In grafting, the underground stock upon which a desirable variety may be grafted. 2. See *rhizome.*

root sucker. A sprout which rises from a root.

root swell. The gradual swelling at the base of a tree resulting from the buttressing of the stem to afford greater stability.

root system. All the roots of a plant, as a fibrous *root system* or a tap *root system.*

root tip. Root cap.

root tubercle. A small, rounded swelling on a root. See *Rhizobium.*

root tumor. See *root knot.*

root weevil. Any of various weevils of the genus Baryopadus, especially *B. squalidus,* whose larvae damage pear and apple trees by boring into the roots. Native to Australia.

rootworm. 1. Any insect larva, such as the northern corn rootworm *(q.v.),* that infests the roots of plants. 2. Any root-infesting nematode. See *root lesion nematode; root knot.*

root zone. The part of the soil occupied by the roots or plants.

R.O.P. See *U.S. Record of Performance.*

rope. 1. A thick, strong cord made by twisting or intertwining metal strands or fiber, as iron, steel, and manila hemp. 2. Lasso *q.v.*

rope bark. Atlantic leatherwood.

rope bridle. A bridle made of rope instead of leather.

rope burn. A skin irritation caused by pressure contact and speedy movement of a rope upon the surface of the skin of man or animal.

roped. Designating an animal which has been lassoed or tied.

rope grader. A grader for sizing olives which consists of almost parallel, traveling, continuous steel cables. The olive is carried along between the cables until space permits it to drop through. Also called *cable grader.*

rope halter. A halter made of rope instead of leather.

roper. A cowboy who is skilled with the lasso.

rope sling. 1. A large sling made of rope which is placed on the wagon or truck when loading hay in the field and which aids in unloading loose hay from the wagon to the mow. 2. A type of harness made from rope and used for lifting and supporting diseased animals when they are unable to rise and stand unaided.

rope surcingle. 1. A belt, band, or girth made of rope, passed under the horse and over a load on its back to hold the load fast. 2. A rope girth for a rider to hang on to when riding a bucking horse or steer.

rope walking. The leg motion in which the horse swings the striding leg around and in front of the supporting leg in walking or running.
ropiness. See *ropy milk.*
roping horse. A horse trained for use by a rider wielding a lasso.
ropy fermentation of milk. See *ropy milk.*
ropy milk. A defect of milk in which the growth of certain bacteria may result in a change in consistency so that the milk solids adhere and form threads or viscous masses.
Roquefort cheese. A blue-veined, semisoft to hard cheese made from raw, ewes' milk in Roquefort, France. In the U.S., a Roquefort type cheese made from cows' milk is designated Blue cheese.
Roquefort type. See *Roquefort cheese.*
rosary. A rose garden.
rosary pea. Jequirity rosarypea.
rosaderos. Saddle fender.
rosa solis. Roundleaf sundew.
rose. 1. Any of several prickly shrubs or vines of the genus Rosa, family Rosaceae. Species are grown for ornament, decoration, and for their beautiful, often fragrant flowers. See *Austrian brier rose; Banks rose; Bourbon rose; Browns rose; cabbage rose; Carolina rose; Cathay Japanese rose; Cherokee rose; Chinese rose; cinnamon rose; damask rose; dog rose; everblooming Chinese rose; Father Hugo rose; field rose; French rose; Japanese rose; Macartney rose; Noisette rose; polyantha rose; prairie rose; roulett Chinese rose; rugosa rose; Scotch rose; swamp rose; sweetbrier rose; tea rose; Wichura rose.* 2. A perforated metal cap attached to the spout of a watering-can, etc., for delivering water in fine jets.
roseacacia locust. *Robinia hispida,* family Leguminosae, a hardy shrub grown for its rose-colored flower clusters. Native to North America. Also called *pink locust; rose acacia.*
rose aphid. *Macrosiphum rosae* (L.), family Aphidae, a sucking insect which infests roses, especially the tender buds.
roseapple. *Eugenia jambos,* family Myrtaceae, a tropical, broad-headed tree grown in very warm areas for its showy, greenish-white flowers and greenish-yellow, edible fruit. Native to Asia. Also called *jambo.*
rosebay. See *oleander; rosebay rhododendron.*
rosebay rhododendron. *Rhododendron maximum,* family Ericaceae, an evergreen shrub grown for its foliage and rose-colored flowers. Native to North America. Also called *deer laurel; great laurel.*
rose black spot. See *black spot: 6.*
rose bloom of cranberry. A fungous disease of cranberry caused by *Exobasidium vaccinii,* family Exobasidiaceae, characterized by the buds in the axils of the old leaves growing out as abnormal shoots bearing greatly enlarged, rose-colored leaves. Small, round blisters appear on the leaves.
Rosebuck rambler. Field rose.
rose callalily. *Zantedeschia rehmannii,* family Araceae, a tropical, perennial herb grown as a pot plant for its trumpet-shaped, rose-red flowers. Native to South Africa. Also called *red callalily; pink callalily.*
rose campion. *Lychnis coronaria,* family Caryophyllaceae, a perennial herb grown for its crimson flowers. Native to southern Europe. Also called *dusty miller; mullein pink.*
rose canker. A fungous disease of roses caused by *Coniothyrium* spp., especially *C. fuckelii,* family Sphaerioidaceae, characterized by red or yellow cane lesions, turning brown and enlarging. If the stem is girdled, dieback to that point occurs.
rose chafer. *Macrodactylus subspinosus* (F.), family Scarabaeidae, a beetle that feeds on the leaves of the rose and many other plants including grape, apple, peach, cherry, corn and beans.
rose chafer poisoning. A poultry poisoning which is caused when rose chafers or rose beetles are eaten by chickens under 10 weeks old.
rose clarkia. *Clarkia elegans,* family Onagraceae, an annual herb grown for its rose-colored or purple flowers. Native to North America.
rose-comb. A chicken's comb which is low and solid having the upper surface covered with small, rounded points. There is no indentation in the center and there is a spike termination on the rear.
rose coreopsis. *Coreopsis rosea,* family Compositae, a perennial herb grown for its rose-pink, yellow-centered flowers. Native to North America. Also called *swamp tickseed.*
rose crown gall. See *crown gall.*
rose curculio. *Rhynchites bicolor* (F.), family Curculionidae, an insect pest that drills holes in the buds of roses, often destroying them.
rose daphne. *Daphne cneorum,* family Thymelaeaceae, a low, evergreen shrub which bears yellow-brown fruit, grown for its foliage and fragrant, pink flowers. Native to Europe. Also called *garland flower.*
rose-de-montana. Mountainrose coralvine.
rose geranium. 1. Any of the various scented-leafed geraniums, genus Pelargonium, family Geraniaceae, of which there are several species with their hybrids and derivatives. 2. See *rose pelargonium.*
rose glorybind. *Convolvulus japonicus,* family Convolvulaceae, a perennial, herbaceous, climbing vine grown for its habit and its pink flowers. Escaped, it can be a weed pest. Native to eastern Asia. Also called *California rose* in the U.S.
rose hip. In rose plants, the much enlarged, fleshy, berry-like stem end upon which the flower is borne. The exact nature of hips varies with the particular variety of rose.
rose hook. 1. The stub left after cutting a flower from a rose or after pruning. 2. The older stem attached to a cut rose flower stem which has been removed by a knuckle cut *q.v.*
rose leaf beetle. *Nodonota puncticollis* (Say), family Chrysomelidae, an insect that infests buds, blossoms, and foliage of roses and fruit trees as well as clover and other meadow plants in the spring.

rose leafhopper. *Edwardsiana rosae* (L.), family Cicadellidae, an insect that infests plants of the rose family, especially the apple and rose.

rosella. A sheep that has cast most of its wool before shearing (New Zealand).

roselle. *Hibiscus sabdariffa,* family Malvaceae, an annual herb grown in very warm areas for its thick, red calyx and bracts, which are used in acid jellies and as substitute for cranberries. Native to tropical Asia and Africa. Also called *red sorrel; Jamaica sorrel.*

Rosellinia root disease. A root rot caused by the fungi *Rosellinia pepo* (Pat.) and *R. bunodes,* family Sphaeriaceae, which are found on many host plants in the tropics, especially in proximity to dead and dying stumps. The lateral roots are covered with mats of fungi and fan-shaped mycelial masses appear underneath the bark.

rose mallow. Any of the various herbaceous plants of the genus Hibiscus, family Malvaceae: especially *H. palustris* (*H. moscheutos*), a hairy, perennial, marsh herb grown for its white, pink, or rose-colored flowers. Native to North America. Also called *marshmallow; sea hollyhock; swamp mallow; swamp rosemallow.*

rosemary. *Rosmarinus officinalis,* family Labiatae, a perennial shrub bearing evergreen, strongly-scented leaves and light blue flowers, grown as an ornamental and for use in seasonings and scents. Native to the Mediterranean area.

rose midge. *Dasyneura rhodophaga* (Coq.), family Cecidomyiidae, an insect pest of roses whose maggots feed at the bases of the flowers and on leaves, browning and killing them.

rose milkweed. Swamp milkweed.

rose mosaic. A virus disease of roses characterized by yellow areas in the leaflet next to the midrib, retarded growth, some distortion of the leaflet, and loss of vigor.

rose of heaven. *Lychnis coelirosa,* family Caryophyllaceae, an annual herb grown for its solitary, rose-pink flowers. Native to the Mediterranean area.

rose-of-Jericho. Jericho resurrectionmustard.

rose-of-Sharon. Shrub althea.

rosepanicle deutzia. *Deutzia rosea,* family Saxifragaceae, a tender shrub, hybrid between redbud deutzia and slender deutzia *(q.q.v.),* grown for its bell-shaped, pinkish flowers.

rose pelargonium. *Pelargonium graveolens,* family Geraniaceae, a tender, perennial herb grown as a pot plant, in window boxes, and in the herb garden for its leaves used in flavoring jellies and desserts. Its essential oil is used in perfumery. Native to South America. Also called *rose geranium.*

rosepink zephyrlily. *Zephyranthes grandiflora,* family Amaryllidaceae, a bulbous herb grown in warm areas for its red and pink flowers. Native to tropical America.

rose pogonia. *Pogonia ophioglossoides,* family Orchidaceae, a bog plant grown in the bog garden for its fragrant, rose-purple flowers. Native to North America. Also called *addersmouth; snakemouth.*

rose pot. A porous, unglazed clay pot with a hole in the bottom which has an inside diameter of from 2 to 2½ ins. and depth 2½ to 3¼ ins., and is used for growing small plants.

rose powdery mildew. A fungous disease of roses caused by *Sphaerotheca pannosa,* family Erysiphaceae, characterized by a slight leaf curl, and by a white, powdery growth on the foliage, canes, and buds. The flowers may become distorted or not bloom at all.

rose range. a. The part of a garden where roses are propagated. b. A series of two or more greenhouses used for the production of roses.

rose ring gaillardia. *Gaillardia pulchella,* family Compositae, an annual herb grown for its handsome flowers, with yellow rays at the tips and rose-purple bases. It is good for cut flowers. Native to North America. Also called *annual blanketflower; Indian blanket.*

rose root gall. A large swelling on the roots of roses caused by the larva of a wasp, *Rhodites radicum,* family Cynipidae.

rose rust. Any of the nine species of the fungus genus Phragmidium, family Pucciniaceae, which attack plants of the genus Rosa; however, only one *P. mucronatum* attacks cultivated roses. The stage appearing in the spring produces many bright orange or yellow spots. The spots turn black in the fall and are filled with black spores that live over the winter.

rose scale. *Aulacaspis rosae* (Bouché), family Diaspididae, an insect pest of rose, raspberry and blackberry canes that makes the plants appear as if they were whitewashed.

roseshell azalea. *Rhododendron roseum,* family Ericaceae, a shrub grown for its fragrant, bright pink flowers. Native to North America.

rose-slug. *Endelomyia aethiops* (F.), family Tenthredinidae, an insect that infests soft parts of rose leaves leaving a network of veins and one epidermis.

rose stem canker. A fungous disease of roses caused by *Coniothryium fuckelii,* family Sphaerioidaceae, characterized by red or yellow spots on the canes causing them to dry, turn brown, and enlarge.

rose stem girdler. *Agrilus rubicola* Abeille, family Buprestidae, an insect's larva which burrows upward in the sapwood and around the stem of a rose causing the formation of galls or swellings. Also called *bronze cane borer.*

rose streak. A virus disease of roses characterized by brownish rings and brown vein-banding in leaves, and brownish or greenish rings in canes. It invades roots as well as leaves and stems.

rose sunray. *Helipterum roseum,* family Compositae, a perennial herb grown as an everlasting and as a cut flower for its white or rose-colored flowers. Native to Australia. Also called *acroclinium.*

rose tickseed. *Coreopsis rosea,* family Compositae, a perennial herb grown for its rose-pink flowers. Native to eastern U.S.

rosette. 1. A radiating cluster of leaves. 2. An abnormal condition of leaf growth produced by a disease (due to a fungus, a virus, or malnutrition) in which the leaves are greatly stunted and very closely clustered, giving a rosette-like appearance. See *rosette-peach; rosette of pecan and apple tree.*

rosette disease of blackberries and dewberries. A fungous disease of blackberries and dewberries caused by *Cercosporella rubi,* family Moniliaceae, characterized by an abnormal production of multiple, short, bunchy shoots from the overwintering canes.

rosette disease of peanut. A virus disease of peanuts transmitted by *Aphis medicaginis, A. leguminosae,* and *A. craccivora,* characterized by mottling and chlorosis of the leaves and stems. Internode elongation suppression results in a rosette or witches' broom growth habit.

rosette mosaic of peach and plum. A soil-borne, virus disease of peaches and plums characterized by delayed foliation, chlorotic mottling of the leaves, and a shortening of the internodes which gives a rosette appearance to the shoots.

rosette of pecan and apple tree. A zinc deficiency disease of apple and pecan trees characterized by narrow and small, bunchy and erect leaves.

rosette of wheat. See *wheat mosaic-rosette.*

rosette-peach. A virus disease of peaches prevalent in southeastern U.S. The leaf buds develop compact tufts or rosettes which may contain several hundred small leaves on axes not more than 2 or 3 ins. long. Partial infection soon spreads over the entire tree and death results.

rosetum. A rose garden.

rose verbena. *Verbena canadensis,* family Verbenaceae, a perennial herb grown in the flower garden or as a pot plant for its purple, lilac, or white flowers. Native to North America.

rose weevil. Fuller rose beetle.

rosewood. Wood from any of several tropical trees of the genus Dalbergia, dark red to purplish in color with black streaks; used in cabinetmaking. The term is frequently used with country of origin; *e.g.,* African *rosewood,* Indian *rosewood.*

rose worm. Oblique-banded leaf roller.

rosin. A specific product of resin remaining after the distillation of turpentine, obtained only from the resin exudates of pine trees, especially the longleaf pine (*Pinus palustris*) and the slash pine (*P. caribaea*) of the southeastern part of the Atlantic Coastal Plain of the U.S.

rosin oil. A blue, fluorescent oil obtained in the destructive distillation of rosin and pitch. It is used as a lubricant.

rosin-rose. See *St. John's wort: 2.*

rosinweed. Any herb of the genus Silphium, family Compositae. Some species are grown as ornamentals. See *compass plant: 2; cup rosinweed.*

ross. To remove bark from, or to smooth a log.

Ross test. A test of urine for the presence of ketone bodies, used in detection of acetonemia in cattle and sheep.

rosy apple aphid. *Anuraphis roseus* Baker, family Aphidae, an insect pest of apples which feeds on the outside of the leaf bud and fruit clusters until the leaves begin to unfold. It then works its way down inside the cluster and sucks the sap from the stems and newly-formed fruits.

rosy buglelily. *Watsonia rosea,* family Iridaceae, a bulbous herb grown for its rose-red flowers. Native to South Africa.

rosy globemallow. *Sphaeralcea rosea,* family Malvaceae, a tropical shrub grown for its rose-colored flower clusters. Native to America.

rosy soapwort. Rock soapwort.

rot. 1. A state of decay caused by bacteria or fungi. See *decay.* 2. See *ret.* 3. See *liver fluke; liver rot.* 4. To decay.

rotary atomizer. A device used in the production of evaporated milk which consists of an extremely high-speed spinner plate upon which the product flows to be atomized. The rapid whirling action picks up the particles and throws them off by centrifugal force, the degree of atomization depending upon the speed of the atomizer.

rotary brush. 1. A metering device used in the application of insecticides and in sowing small seeds. 2. A cleaning device used in washing milk bottles and other containers.

rotary can washer. A device for washing milk cans by which the cans are washed on a large rotating table or carried rather than propelled by the ratchet drive straight-away type.

rotary cut veneer. Veneer which is cut in a continuous strip by rotating a log or bolt against a knife.

rotary duster. A dust applicator in which a rotary fan is used to supply the air flow needed for dust distribution.

rotary force pump. The general class of pumps in which rotating parts including the gear, roller, rotor, impeller, propeller, auger, and vane pump are used to cause displacement of a liquid.

rotary grass board. A revolving, spokeless wheel at the end of a mowing machine cutter bar which is used to divide and rake away the cut from the uncut grass.

rotary harrow. See *rotary hoe.*

rotary hoe. A ground-driven cultivating implement which consists of a series of curved, spikelike teeth, rotating on a main axle, the points of which dig into the ground destroying young weeds and breaking the crust of the soil.

rotary hog feeder. A round or many-sided feeder which the hog turns with its snout to make grain feed come down from the hopper to the animal for consumption.

rotary homogenizer. A machine in which the product to be homogenized is passed between two closely spaced rotating parts.

rotary knife cotton chopper. An implement which is equipped with rotary knives turning across the cotton row to thin or reduce the stand of cotton plants.

rotary mower. A machine that uses a rotary cutting mechanism for mowing forage, grain, weeds, lawns and other vegetation. Two common types are (a) the rotary knife blade which rotates rapidly in a horizontal plane having a vertical shaft;

(b) the cylinder type, in which knives attached to a horizontal shaft cut off the vegetation when passing over a horizontal shear plate.

rotary plow. Cutting knives (tines) or an auger mounted on a horizontal or vertical power-driven shaft whose revolving tines loosen the soil or destroy weeds. Also called *rotary tiller.*

rotary scraper. A scraper used for the handling of large quantities of earth which is constructed to tilt forward to load and to tilt backward to remove the blade from the ground for traveling. Also called *rotary fresno.*

rotary sprinkler. A revolving head spray sprinkler used on sprinkler irrigation systems.

rotary subsoiler. A subsoiler with a thin, strong shank that has a chisel-type point to loosen up the subsoil. A cylinder with fins to cause rotation is attached to the shank to aid breakup of the soil.

rotary tillage. Seed bed preparation with a rotary plow.

rotary tiller. Rotary plow.

rotary weed cutter. A large, power-driven blade rotating in a horizontal plane mounted on a tractor, used for cutting weeds.

rotary weeder. A rotary hoe attachment for a tractor cultivator.

rotated pasture. 1. A pasture in the regular crop rotation which is grazed for a few years, usually two or three, and then plowed for other crops. 2. A pasture which is divided into segments by use of fences: the livestock being confined to one segment at a time in a definite rotation pattern.

rotation. 1. In cropping, the growing of two or more crops on the same piece of land in different years in sequence and according to a definite plan. One of the most widely recommended crop rotations consists of (a) a tilled crop, often corn; (b) a small grain; (c) a legume or grass crop or a mixture of legumes and grasses often for more than one year. 2. In forestry, the period of years required to establish and grow timber crops to a specified condition of maturity.

rotational grazing. Dividing a pasture field into several areas usually by electric fences. The individual areas are grazed in a sequence which allows some forage to recover while other forage is being grazed.

rotation area. That portion of the farm upon which field crops, summer fallow, and pasture are changed from year to year.

rotation burning. The burning of brush and meadows every three years to provide more pasture.

rotation grazing. See *rotational grazing.*

rotation irrigation. A system of irrigation through which the irrigator receives his allotted quantity of water not at a continuous rate but as a large flow at stated intervals: *e.g.*, a number of irrigators receiving water from the same lateral may agree among themselves to rotate the water, each taking the entire flow in turn for a limited period.

rotenone. An insecticide which is obtained from the roots of derris, Virginia tephrosia, etc.

Rothenfusser test. A test for heat-treatment based on the fact that amylase, the starch-splitting enzyme of milk, is destroyed at temperatures above 60°C. Residual starch is detected by the iodine test.

Rothrock grama. *Bouteloua rothrocki,* family Gramineae, a short-lived, perennial grass which grows in small bunches and resembles blue grama. Occurring in southwestern U.S. on mesas and dry slopes, its forage value is fair to good. Native to North America.

rotifer. One of the Rotifera, a class of generally microscopic, many-celled animals abundant in stagnant and fresh water and in the soil.

rot in the ground. Not to germinate, as seeds, due to wet, cool weather.

rot off. See *damping-off.*

rot of pineapple. A fungous disease of pineapple caused by *Ceratostomella paradoxa,* family Synchytriaceae, characterized by a destructive rot beginning at the stem end. Also called *blackheart of pineapple.*

rotor hammer feed grinder. Hammer mill.

rotted manure. Animal dung which has undergone decay and is safe to use with growing plants.

rotten egg. An egg which has deteriorated as a result of bacterial invasion.

rotten neck of rice. Brown spot of rice *q.v.*

rotten. 1. Designating decomposed or putrid organic matter. 2. Designating ground or soil extremely soft and yielding because of decay, or rocks partially decomposed. 3. Designating sheep attacked by rot *q.v.*

rotten wood. Wood in a state of decay produced by an attack of bacteria or fungi.

rotting. See *rot: 1; damping-off.*

Rouen. A breed of ducks which takes its name from Rouen in Normandy, France. The plumage color is similar to the wild Mallard, the shape and weight very much like the Pekin. Although a good meat bird, it is not produced extensively for meat but is raised mainly by fanciers.

rough. 1. A market classification for hogs. See *roughs.* 2. To remove the major part of the plumage of a fowl leaving scattered feathers for the finisher. 3. In meat judging, designating uneven contour or uneven distribution of fat on a carcass. 4. Designating an uneven piece of land or a road. 5. Designating food of low quality. 6. Designating a horse not properly trained. 7. A calk for the horse's shoe.

roughage. 1. A dry fodder high in fiber content and therefore low in total digestible nutrients, as hay, corn fodder, straw, and silage. 2. In human nutrition, a coarse food containing considerable indigestible material usually in the form of cellulose, as uncooked celery and lettuce.

roughage cow. Any milk cow fed on hay and other roughage with a minimum of grain.

roughage cutter. A machine used for cutting or chopping roughages, as hay and straw.

rough ammoniate. Organic nitrogenous waste material, as scrap leather, felt, wool waste, fur and hair, which cannot be used as a fertilizer until suitably treated.

rough bark disease of the apple and pear. 1. A

condition of the bark caused by the fungus *Phomopsis mali.* Whole branches may be involved, the smaller ones being occasionally girdled and killed. The yellow Newtown variety is very susceptible and in the U.S., the disease is reported on York and Winesap varieties in the east and on pear in California. 2. Superficial bark canker; a fungous disease of the outer bark of apple and pear characterized by shallow cankers caused by *Myxosporium corticolum.* 3. The roughening, sometimes quite pronounced, of the normally smooth surface of the bark of apple and pear as a result of frost or spray action (non-parasitic).

rough bark of oriental flowering cherry. A virus disease of the flowering cherry that results in longitudinal splitting of bark with tissue regeneration.

rough-bearded barley. Any barley with tiny barbs projecting from the beards. The barbs are very irritating to both man and animals.

rough belly. On the cutting floor of a packing plant, the belly of the hog before it is trimmed into a bacon side for trade.

rough broken land. Land naturally so deeply and minutely dissected by stream cutting that it is nonarable. Technically, it is differentiated from badlands (*q.v.*) and gullied land. See *gully.*

rough buttonweed. *Diodia teres,* family Rubiaceae, a hairy or minutely pubescent, annual herb found as a weed in warm regions and often objectionable in rice fields. Native to North America. Also called *alligator head.*

rough cinquefoil. See *Norwegian cinquefoil.*

rough coat. Coarse, tangled, unkempt animal's coat which may indicate lack of thrift or its remaining in the open during the winter.

rougher. A person or a device which removes the major part of the plumage of a fowl in processing for market.

rougher and pinner system. A system of dressing poultry in which the rougher removes the major part of the plumage, the remaining pin and broken feathers being removed by pinners.

rough-fruited cinquefoil. See *sulfur cinquefoil.*

rough ground. An uneven, uncultivated land.

rough-headed cornstalk beetle. See *sugarcane beetle.*

rough heliopsis. *Heliopsis scabra,* family Compositae, a perennial herb grown in the flower border for its orange-yellow flowers. Native to North America. Also called *orange sunflower.*

roughing machine. A mechanical feather picker, usually consisting of a rapidly revolving drum equipped with flexible rubber fingers against which the previously lightly scalded fowl is held to remove feathers.

rough land. Nonarable land on steep, broken slopes. It may be either stony or highly dissected by erosion.

rough lemon. A horticultural variety of lemon (*q.v.*) introduced by the Spanish into Florida, U.S., where it is often used as a stock for other citrus fruits. Also called *Florida rough lemon; French lemon mazoe.*

rough lumber. Lumber as it comes from the saw and before it is dressed.

roughly finished. Designating a beef animal lacking in smoothness, uneven in covering, patchy, and hence having a low dressing percentage.

rough meadow grass. Roughstalk bluegrass.

roughness. See *roughly finished.*

roughness at the tail setting. See *roughly finished.*

rough over the rump. See *roughly finished.*

rough paraffin. A minor bandage defect of cheese characterized by a roughness of the paraffin covering.

rough plowed land. 1. Land which has been plowed but not harrowed or fitted. 2. Land that has been plowed so as to leave the surface rough as a deterrant to wind erosion.

rough post frame. See *pole frame.*

rough rice. The hull-covered rice as it leaves the thresher.

roughs. In marketing, coarse and rough, inferior hogs of all weights sorted out of various lots.

rough shoulder. An untrimmed shoulder of a hog carcass with the foot and jowl removed but neck bone left in.

rough spot. A fungous disease of Sudan grass, Johnson grass, and sorghums caused by *Ascochyta sorghina,* family Sphaerioidaceae, characterized by yellowish-brown or reddish-purple spots covered with small, black, fruiting bodies which roughen the texture of the affected areas.

roughstalk bluegrass. *Poa trivialis,* family Gramineae, a perennial grass grown in pastures and as a shady-lawn grass. Native to Europe. Also called *rough bluegrass; rough meadow grass.*

rough surface. Uneven surface of cheese, a body defect that may be due to unclean press cloths to which particles of dried curd have adhered, insufficient pressing, or rough, uneven shelf boards.

rough through. To feed a minimum of concentrates and a maximum of roughage to animals during the winter.

rough up. To make the surface of stubble land rough by cultivating to reduce soil erosion.

rough wood. Unpeeled cordwood.

rough woodbine. Hairy honeysuckle.

roulett Chinese rose. *Rosa chinensis* var. *minima,* family Rosaceae, a variety of the Chinese rose (*q.v.*), a dwarf shrub grown for its rose-red flowers and as breeding stock for Baby Rambler roses. Also called *fairy rose.*

round. 1. Rear quarter of beef animals. 2. See *bout.* 3. In stock judging, designating thick, broad, deep, full hindquarters carrying well down to the hocks. 4. To prevent a herd of cattle, etc., from scattering, or to reorganize an already scattered herd, by riding round it and forcing it together (U.S. and Australia).

round bale. 1. A bale of cotton made by winding the cotton bat around a central core in such a way that steel ties are not needed and wrappings of duck or hemp cloth hold the bale intact. Also called *Bessonette bale.* 2. See *hay bale.*

round barn. A circular or multi-sided barn with a litter alley around the wall and cows arranged in a single row round the periphery facing toward the center. A silo might be constructed in the center.

round bulb. In marketing, a bulb with only one growing point.

round dock. Running mallow.

round forty. In early lumbering days in the Great Lakes States, U.S., a tract of 40 acres which was enlarged by encroachment and cutting of timber on adjacent 40's not owned by the operator.

roundhead borer. Any of the large group of very small to very large, oblong beetles, called Longicorns, having extremely long antennae and belonging to the family Cerambycidae. The larvae are white, long, slender, usually legless grubs, with enlarged thoracic segments, with a horny plate on the top surface of the first segment. Most of these are wood borers or cambium miners in injured, felled, dying, or dead trees. Some, however, infest and develop in living trees. Also called *longhorn beetles; sawyers.*

roundheaded apple tree borer. *Saperda candida* F., family Cerambycidae, an insect larva which attacks the sapwood and heartwood of the apple, serviceberry, wild crab, hawthorn, mountain ash, quince, pear, and plum. Feeding on the cambium and heartwood, it weakens the tree and often kills it.

roundheaded fir borer. *Tetropium abietis* Fall, family Cerambycidae, a long, horned beetle whose grub works under the bark of felled and standing firs.

roundheaded pine beetle. *Dendroctonus convexifrons* Hopk., family Scolytidae, an insect that infests ponderosa pine.

roundhead lespedeza. *Lespedeza capitata,* family Leguminosae, a weedy herb which grows even in the poorest soils and is used for pasturage and soil improvement. Native to North America. Also called *bush clover; dusty clover.*

round hill. Hill country (New Zealand).

rounding hammer. A hammer with one face flat and the other convex; used for making horseshoes.

rounding rump. Rounding at the tail head which mars the general appearance and does not conform to the shape of the rump. Considered desirable in the good type cow.

rounding tool. In blacksmithing, a swage tool hollowed to a semicylindrical section and used for rounding and finishing iron rods.

round kumquat. Marumi kumquat.

roundleaf dogwood. *Cornus rugosa,* family Cornaceae, a hardy, deciduous shrub grown for its green, purple-spotted twigs and for its clusters of light blue or white flowers. Native to North America. Also called *green osier.*

roundleaf sundew. *Drosera rotundifolia,* family Droseraceae, a perennial herb grown in the bog garden or as a house plant for its insectivorous habit and its white or pinkish flowers. Native to the North Temperate Zone. Also called *dew plant; rosa solis.*

roundlobe hepatica. *Hepatica americana,* family Ranunculaceae, a low-growing, perennial herb grown for its lavender-blue, white, or rose-pink flowers. Also called *Mayflower; blue anemone.*

round-lumber. A single layer or course of piled lumber.

round of beef. A cut of meat; the upper part of the hind leg between the rump and the shank which provides several very desirable cuts of meat.

round-sausage casing. A casing made from the small intestines.

round shock. A method of shocking bundles of grain in the field which provides for curing and protection against damage by rain. The bundles are set up in a circular pattern around the central pair rather than in a rectangular pattern of paired bundles.

round shouldered. Designating a desirable characteristic of the saddle horse, smoothness through the neck and shoulders.

round steak. A retail cut of beef for frying, broiling, or braising which comes from the round of beef *q.v.*

round-timber. An untapped turpentine pine (southeastern U.S.).

round tip. A cut of beef taken from the upper tip of the round. See *round of beef.*

round turn. A purchase and its liquidating sale, or a sale and its liquidating purchase.

round-up. Bringing together scattered range stock for counting, branding, sorting, and sale.

round-up captain. The manager of a round-up *q.v.*

round worms. Parasites in man, animals and plants, which may cause disease and great economic loss. They vary in size from a fraction of an inch in length and as thin as a silk thread to over a foot in length and as thick as a lead pencil. In man and animals, they inhabit the intestine but in completing a complex life cycle may infest the blood stream, lungs, windpipe, liver, kidneys, etc. Symptoms vary but in general are those of unthriftiness. See *nematode.*

roup of fowls. 1. Infectious roup caused by *Hemophilus gallinarum,* characterized by swelling of the sinuses under the eyes and watery discharge from the nostrils. Also called *contagious catarrh of fowls.* See *infectious coryza.* 2. Nutritional roup, caused by an insufficient amount of vitamin A in the diet, characterized by a whitish exudate in the eye, which may cause serious loss before deficiency is noticed.

roustabout. 1. A man who does all kinds of work. 2. In Australia, a shed hand but not a skilled sheep shearer.

routine count. In the grading of milk and in the general control of milk supplies, a rapid count of bacteria.

routine laboratory test. 1. Any test of milk for chemical and bacteriological quality. 2. Any regular routine or recipe for determining certain constituents or nature of a product or condition in a laboratory, such as breaking strength of fiber, lime content of soils, organisms causing disease, etc.

rover bellflower. Creeping bellflower.

roving hog. Any hog with a tendency to stray and which is a poor feeder.

roving wool. Wool fibers drawn out to the proper thickness for spinning and needing only the twist to make it into yarn.

row. The arrangement of crop plants, shrubs, trees, or other objects planted in a line as opposed to broadcast *q.v.*

Rowan tree. European mountainash.

row application. An application of a substance, such as fertilizer, made in or immediately adjacent to the row of a crop. See *row.*

row binder. An implement used for harvesting and tying bundles of a crop grown in a row, as a corn binder.

row crop. A crop, as corn, beans, sugar beets, cotton, sorghum, etc., which is usually grown and cultivated in rows.

row crop drop extension. An attachment which converts boom sprayers from overtop spraying to spraying between the rows of row crops.

row crop field harvester. An implement used for harvesting and chopping row crops, especially corn. See *row binder.*

rowel. In horse-riding, the wheel of a spur with blunt to sharp projecting points.

row mulching. Mulch placed on the row so that the plants emerge through the mulch.

row planting. Row seeding.

row seeding. The planting of crops in a line as contrasted to broadcast. See *row.*

royal azalea. *Rhododendron schlippenbachi,* family Ericaceae, a deciduous, tender shrub grown for its pink, brown-spotted flowers. Native to eastern Asia.

Royal Brabant. A small, Limburger-type cheese made in Belgium from cows' whole milk.

royal cells. Those cells which produce queen bees Also called *queen cells.*

royal fern. *Osmunda regalis,* family Osmundaceae, a coarse, easily-grown fern found in damp places. Native to the North Temperate Zone. Also called *ditch fern.*

royal jelly. The food supplied by worker bees throughout the developmental period of the larvae destined to become queen bees. Also called *bee milk.*

royal lily. Regal lily.

royal palm. Any palm tree of the genus Roystonea, family Palmaceae. Species are grown as ornamentals. See *Carib royal palm; Cuban royal palm.*

royal paulownia. *Paulownia tomentosa,* family Scrophulariaceae, a hardy, deciduous tree grown for its pale violet flowers. Native to China. Also called *empress tree.*

royal peavine. *Lathyrus splendens,* family Leguminosae, a perennial, more or less shrubby herb grown for its rose-pink, violet, or magenta flowers. Native to North America. Also called *pride-of-California; campo pea.*

royal poinciana. Flamboyant tree.

royal snapweed. One species of snapweed. See *snapweed: 1.*

royal waterplatter. *Victoria regia,* family Nymphaeaceae, an aquatic plant grown in a pool as an ornamental for its large, floating leaves and its dull crimson flowers. Native to South America. Also called *royal waterlily.*

R-242. A residual acaricide.

rubber. See *latex.*

rubber finger. The hollow, rubber, fingerlike projection mounted in rows on a revolving steel drum which removes most of the feathers and hair from the slaughtered fowl.

rubber plant. India-rubber fig.

rubber rabbitbrush. *Chrysothamnus nauseosus,* family Carduaceae, a widely-distributed shrub whose presence is an indication of range depletion. Native to western North America.

rubber-seed oil meal. A livestock feedstuff which is a by-product of the manufacture of oil from the para rubbertree seed.

rubber sleeve. A protective, rubber covering attached to a rubber glove for the hand, arm and shoulder which is used by veterinarians and artificial inseminators in various phases of their work.

rubber tire. A tire made of rubber instead of steel which is used on various agricultural implements, as tractors, combines, wagons, etc.

rubber tree. Para rubbertree.

rubber vagina cylinder. A rubber tube which fits inside the artificial vagina *q.v.*

rubber vine. See *palay rubbervine.*

rubberweed. See *pingue.*

rubbery. Designating a body defect characteristic of freshly made, green or uncured cheese.

rubbery wood of apple. A virus disease of apples which causes the trees to show a lack of rigidity due to the incomplete lignification of the fibers and vessels of the wood.

rubbing machine. A device used to remove awns and glumes from the seed of native grasses so the seed will flow through a seed drill.

rubbing post. Any of several devices consisting mainly of a post set in the ground with a can of oil attached to the top, so arranged that the oil or medicament is allowed to slowly drip or run down the sides of the post so that when hogs or other livestock rub against it a small amount of oil will be deposited on the rubbed or infected area of the skin. Also called *hog oilers; livestock oilers.*

rubble. Small stones, chips from boulders, broken bricks, etc.

rubble dam. An unmortared dam, similar to a loose rock dam, which is made of broken stones, bricks, etc., and used in gully control in soil conservation.

rubble land. That land which has 90 percent or more of its surface covered by stones and boulders.

rubble masonry dam. A dam which is similar to a rubble dam *(q.v.)* except that reinforced concrete may be used for added protection and rocks are set in mortar.

rube. A rustic, or unsophisticated farmer (slang).

rubefaciant. Liniment, plaster, or any substance that produces redness of the skin when applied to it.

rubidium. A rare element known to appear in traces in soils and in plants. It probably enters into the same reactions in soil and plants as does potassium. In very small quantities it appears to stimulate plant growth.

ruby grass. Natal grass.

rudder blade. 1. An attachment for the bottom of a lister which serves to steady the plow. 2. An attachment which serves to steady a cultivator. Also called *drag iron.*

ruderal. Designating weeds which spring up on uncultivated or abandoned land.

rudiment. A part of an organ which is only partially developed; undeveloped or in the first stage.

rudimentary. Undeveloped.

rudimentary copulatory organ. A very small, shiny or glistening eminence, which is located in the median portion of the fold between the urodaeum and the proteodaeum. By examination of this organ the sex of day-old chicks can be determined.

rudimentary ear. The female flower of the corn plant before fertilization.

rudimentary teat. An undeveloped teat which may or may not be connected with milk-secreting tissue. It may be attached as a small nipple to the regular teats, an underdeveloped teat behind the regular ones or show up on the male of the species.

rue. *Ruta graveolens,* family Rutaceae, an evergreen, aromatic, perennial herb grown for its foliage and its dull yellow flowers. Native to southern Europe. Also called *herb-of-grace.*

ruffer. A rude kind of comb used in the first operations of hackling or combing flax by hand. It consists of a tin covered stock fastened to a board set slantwise to the bench, the teeth sloping away from the hackler, who operates it by pulling the flax through the teeth with a circular sweep.

ruffled hair. In stock judging, disordered, disturbed hair of an animal which may suggest an unhealthy condition.

ruffle fat. The fat that is located between the intestines of an animal.

ruffling. An unevenness of the surface of the leafblade caused by ridges that develop or become more pronounced from the midrib to the lateral margins.

rugged. Designating size, strength, and vigor in the body build or conformation of an animal.

rugged and straight in his bone. A strong, vigorous animal with good leg and body conformation.

ruggedness. See *rugged.*

rugose. In plants, designating a condition of wrinkling.

rugose mosaic. A mixed infection of potatoes caused by two virus entities characterized by distinct dwarfing, diffused mottling, a distinct type of wrinkling and brittleness.

rugosa rose. *Rosa rugosa,* family Rosaceae, a hardy shrub grown for its red or white flowers. Native to China and Japan. Also called *hedgerow rose.*

rumen. The first and largest compartment of the stomach of a cud-chewing animal. Also called *paunch.*

rumenal. Ruminal.

rumen fistula. A fistula *(q.v.)* of the rumen or first stomach of a cud-chewing animal.

rumen flora. The microorganisms of the rumen.

rumenology. A branch of animal science concerned with the study of the rumen *q.v.*

rumenotomy. The operation of cutting into the rumen to remove foreign bodies or to observe activity.

ruminal. Relating to, or having the properties of, ruminants, which are cud-chewing animals, such as cattle (rare).

ruminant. A cud-chewing animal with a stomach which contains four complete compartments through which the food passes in digestion. *Ruminants* include cattle, sheep, goats, deer, elk, antelopes, and camels.

rumination. The regurgitation of food from the stomach to the mouth for rechewing, a characteristic of ruminants.

rump. That part of the rear end of an animal which includes the buttocks or fleshy part of the rear quarters.

rump roast. A retail cut of meat taken from the rump section of the hind quarter.

run. 1. A period of time, as a maple sap *run.* 2. The amount of sap or sugar produced in a given time. 3. A swiftly moving tributary, rivulet or mountain stream (eastern U.S.). 4. The stream outlet of a large spring (Florida, U.S.). 5. Unrestricted movement, as the colts have the *run* of the pasture. 6. An area of land or a leasehold (New Zealand). 7. A small, often dry gully or channel carved by water. See *arroyo.* 8. A fenced-in pen used for the exercise of animals or poultry. 9. To feed; to graze, as steers *run* on the open range. 10. To operate, as a plow is set to *run* at a depth of 6 ins. 11. To cultivate, mow, combine, etc., as to *run* over a field with a weeder. 12. To maintain animals, as he *runs* sheep. 13. To work a dog with sheep or cattle. 14. To discharge pus, as a sore *runs.* 15. To be, as the prices *run* very high. 16. To move rapidly, as a horse. 17. To turn, as to *run* a wheel. 18. To grow, as when the vines begin to *run.* 19. To operate, as to *run* an engine.

run a level. To survey, as with a level to obtain topographic or profile information.

run a line. To survey or establish a line for terracing, grading, drainage, etc.

runaway horse. A horse that has the bad habit of running away with or without control or guidance when in harness or under the saddle.

run cutter. In forestry, one who clears narrow trails.

run-down. Designating anything that has been allowed to deteriorate, such as a fence, a set of farm buildings, or a field that has been poorly farmed and has lost productivity.

Runesten. A cows' milk cheese made in Denmark

and the U.S., similar to Swiss cheese but with smaller eyes. Each cheese weighs about 5 lbs. and is wrapped in red transparent film.

run free (loose). To be as animals that are not restrained and are allowed to move about freely in the barn, yard, or pasture.

run holder. Squatter, station owner or (recent) owner of a small grazing run (New Zealand).

runnel. A tree that has been pollarded or pruned so that all of its main branches are cut back to the trunk.

Runner. A breed of ducks of very distinctive type, having a long, narrow body and very erect carriage. The breed derives its name from its gait, which is a quick run, quite unlike the waddle of other ducks. The adult drake weighs about 4½ lbs.; the adult duck about 4 lbs. The *Runner* is noted as an egg-producing breed and has little or no value for meat production. Its three varieties are the White, the Fawn and White, and the Penciled. Also called *Indian Runner.*

runner. 1. A lateral, above-ground shoot of certain plants: *e.g.,* strawberries, which roots and forms young plants at some of the nodes, aiding in propagation. 2. A rope used to increase the mechanical power of a tackle. 3. The upper or rotating stone of a set of millstones. 4. A supporting attachment which slides along the ground, as a sled *runner.*

runner peanut. Any variety of peanut whose pods are scattered along procumbent rooting stems, as contrasted to the bunch type in which the pods are clustered at the base of the plant. *Cf. bunch peanut.*

runner plant. A plant which is capable of being propagated by runners (*q.v.*); a plant produced from a runner.

running birch. Creeping pearlberry.

running euonymus. *Euonymus obovatus,* family Celastraceae, a hardy, prostrate shrub grown for its foliage, habit, and crimson fruit. Native to North America. Also called *running strawberry bush.*

running fire. In forestry, a fire which spreads rapidly with a well-defined head but little or no spotting or crowning.

running fits. A disorder of dogs, of unknown cause, characterized ordinarily by the affected animal, for no apparent reason, suddenly starting to run and bark, then hiding in a secluded place, and sometimes by spasms and convulsions.

running gear. The wheels, axles, and frame of a vehicle, usually a wagon or trailer.

running horse. Any Thoroughbred *q.v.*

running iron. A short, iron rod which, when heated, can be used to cancel or rework a brand on an animal.

running mallow. *Malva rotundifolia,* family Malvaceae, an annual or biennial weed which is procumbent with dull green foliage, pubescent stems, and a deep tap root. It is common throughout U.S. Native to Europe and Asia. Also called *common mallow; cheeses; dwarf mallow; little mallow; malice; round dock.*

running martingale. See *martingale.*

running myrtle. Periwinkle.

running noose. A noose which has a running knot, as a lariat *q.v.*

running out. The condition in which an improved variety of plant or animal is reverting to a former and inferior type or is losing some of its desired qualities.

running pine. *Lycopodium clavatum,* family Lycopodiaceae, an evergreen, mosslike herb grown in the wild garden and used in Christmas wreaths. Native to North America. Also called *ground pine.*

running plate. The aluminum horseshoe worn by race horses.

running serviceberry. *Amelanchier stolonifera,* family Rosaceae, a deciduous shrub grown for its white flowers and black-purple fruit. The edible fruit is sometimes used in jelly. Native to North America.

running sheep. The handling and management of a flock of sheep under range conditions.

running strawberry bush. Running euonymus.

running walk. A slow, single-foot or four-beat horse's gait with the break in the impact or rhythm occurring between diagonal fore and hind feet. In the stepping pace which is also a slow, four-beat gait, the break in the impact occurs between lateral fore and rear feet.

running-walker. A horse which has a peculiar habit of nodding his head while bringing each front foot in contact with the ground a mere instant before the diagonally opposite hind foot.

run-of-the-hatchery. Chickens as they come from the hatchery without culling, sorting, or sexing.

runoff. 1. That portion of total precipitation that finds its way into drainage channels. It consists of ever varying proportions of both surface *runoff* and ground water *runoff.* 2. Water which is removed from soil by surface drainage or underdrainage. Also called *flowoff.* 3. Sirup which has been drawn off sugar crystals in the manufacture of sugar.

runoff coefficient. The percentage of rainfall in a storm which is of a specified magnitude that may be expected to flow from a watershed.

runoff modulus. The depth of water in inches over the under-drained area which must be removed by tile drains in 24 hours. It is a measure of the maximum rate at which the water will move through the soil to the laterals.

runoff water. See *runoff.*

run on. To graze or pasture on, as to *run on* the range.

run-out. See *running out.*

run-out fleece. Wool that is lacking in uniformity, being hairy or brittle and weak in fiber.

run to. To grow to, as to *run to* flower without branching.

runty. Designating a small or poorly nourished animal, especially a young pig.

run wild. To grow untamed, not handled by man.

run with. To be present with, as in the same pasture, such as the calves *run with* the cows.

rupture. 1. Forcible tearing or breaking of a part. 2. See *hernia: 1.*

rural. Concerning or like the country, as opposed to the town or city. *Cf. urban; agrarian.*

rural appraisal. A written opinion of a qualified appraiser as to the value of rural land or real estate.

rural community. A body of people having common interests and customs living in a rural area larger than a neighborhood.

Rural Electrification Administration. In the U.S., an agency which organizes loans for central station electric service to people who do not have it, for rural electric and telephone system loans, and loans to finance the wiring of farmsteads and installation of electric appliances and plumbing. Loans are not made directly to consumers but to R. E. A. borrowers for relending to members. Created by Executive Order May 11, 1935, and later reorganized and revised to become a part of the Department of Agriculture July 1, 1939.

rural family. A family which lives in an agricultural environment.

rural-farm. One of three residence categories used by the U.S. Bureau of the Census to encompass all persons living on farms without regard to occupation. See *rural-nonfarm; urban.*

rural-farm population. Persons who live on farms (as defined by U.S. Census) located in rural territory without regard to occupation.

rural folks. See *rural family.*

rural free delivery. See *R. F. D.*

rurality. The extent to which a community approximates a wholly rural status.

rural land. Land which is occupied by farmers or used for agricultural purposes as distinguished from urban land, park or recreational land and wilderness.

rural mail box. A metal or wooden box placed at the road side on a R. F. D. route in rural areas to receive mail.

rural mail carrier. An employee of the U.S. Post Office who distributes mail in the rural area.

rural-nonfarm. One of the three residence categories used by the U.S. Bureau of the Census which includes persons living outside urban areas (with certain exceptions) but not residing on farms. See *rural-farm; urban.*

rural non-farm population. All persons who live in rural areas (as defined by U.S. Census) but who are not engaged in farming as a business or who do not live on farms.

rural population. Those people who do not live in cities. (A city is arbitrarily defined in the U.S. Census statistics as a community of 2,500 persons or more.)

rural real estate. Agricultural and other non-urban land including improvements and rights appertaining thereto, such as mineral and riparian rights.

rural sociology. The science of rural human relationships.

rural zoning. In the U.S., enabling acts which permit land-use control measures under the police power of the state in rural areas. The zoning may be concerned with broad measures merely setting up districts in which the use of land is specified as primarily for agriculture, forestry, recreation, etc.; or in urban fringe areas the restrictions may be similar to those of urban zoning ordinances.

rurban. The transition zone between the corporate limits of a city, or thickly populated suburbs, and the open country. Occupied land which is neither definitely urban nor rural.

rush. 1. Any of the various marsh herbs of the genus Juncus, family Juncaceae. Some species have a fair value as forage for sheep and cattle. 2. *J. effusus,* a perennial, marsh herb whose stems are used in making mats. Native to the North Temperate Zone.

rush nut. See *chufa flatsedge.*

rush skeletonplant. *Lygodesmia juncea,* family Compositae, a perennial herb found on waste lands and dry soils of the Great Plains area (U.S.). The bitter, milky juice is thought by stockmen to be poisonous. Also called *skeleton weed.*

rush skeletonweed. *Chondrilla juncea,* family Compositae, a biennial, rush-like weed. Native to North America.

Russell prairiegentian. *Eustoma russellianum,* family Gentianaceae, a hardy, annual herb grown for its pale purple, bell-shaped flowers. Native to North America. Also called *blue bell; Canada pest.*

russet. 1. Brownish, roughened areas on the skins of fruit, tubers of potatoes, etc., resulting from abnormal production of cork tissue which may be caused by diseases, insects, other injury, or by a natural varietal character. 2. Any of the varieties of apples having natural, russet-colored skins.

russet buffaloberry. *Shepherdia canadensis,* family Elaeagnaceae, a very hardy, spreading shrub grown in areas of dry, stony, windswept soils for its egg-shaped, yellowish-red, sour fruit. Native to North America.

russet dwarf. A disease of the potato caused by a virus, characterized by shrinking, browning, and russeting of the leaves.

russeting. See *russet.*

russet scab. Rhizoctonia disease.

Russian centaurea. *Centaurea repens,* family Compositae, a perennial herb which is an objectionable weed with pale, stiff, creeping stems. Native to Europe and Asia. Also called *Russian knapweed.*

Russian clematis. Golden clematis.

Russian dandelion. Kok-saghyz.

Russian mulberry. *Morus alba* var. *tatarica,* family Moraceae, a very hardy tree with edible, red fruit. It is adapted for planting in bleak windswept regions as a windbreak. Native to Asia.

Russian olive. *Elaeagnus angustifolia,* family Elaeagnaceae, a hardy, wind-resistant, spiny shrub or small tree grown for its ornamental fruit and silvery leaves in hedges and windbreaks. Native to Europe and Asia. Also called *oleaster.*

Russian Orloff. A strain of trotting horses which was derived from an admixture of Dutch, Friesian, and Arabian breeds from the stud of Count Orloff, 1737-1809, of Russia.

Russian pigweed. *Axyris amaranthoides,* family Chenopodiaceae, an alternate-leaved, pubescent, annual weed. Native to Asia.

Russian sunflower. Sunflower.

Russian swine. An early strain of swine of uncertain origin used in Ohio and Kentucky (U.S.), in the early development of the Poland China breed of swine. They were of variable colors, fairly large in size, long bodied and long legged.

Russianthistle. *Salsola kali,* family Chenopodiaceae, an annual weed. When the plant dies it breaks away from the ground blowing about as a tumbleweed. Native to Europe and Asia. Also called *Russian tumbleweed.*

Russian tumbleweed. Russianthistle.

Russian vetch. Hairy vetch.

Russian wormwood. *Artemisia sacrorum,* family Compositae, a tall, stiff herb often 5 ft. high, grown for its ash-gray leaves and greenish-white, nodding heads. Native to Siberia.

rust. 1. A disease condition of some plants in which the leaves take on a red or rusty color and some of them drop off, due either to a deficiency or an excess of some particular nutrient, or from improper tillage practices. See *cotton rust.* 2. See *rusts.* 3. See *white rusts.*

rustic. 1. Pertaining to or of the country; rural. 2. A simple, unsophisticated countryman, a peasant (obsolete).

rustic borer. *Xylorechus colonus* (F.), family Cerambycidae, a black and gray beetle which infests oak and hickory in eastern North America.

rusting. See *rusts.*

rustle. 1. To hunt for food, especially with reference to domestic animals inadequately fed by their owner. 2. To brand or rework brands on animals belonging to another. 3. To steal livestock.

rustler. A cattle thief.

rustling. See *rustle.*

rust mite. Citrus rust mite.

rust of alfalfa. A fungous disease caused by *Uromyces striatus medicaginis,* family Pucciniaceae, characterized by reddish-brown spots on the leaves and stems. See *rusts.*

rust of flax. A fungous disease caused by *Melampsora lini,* family Melampsoraceae, whose life stages are all produced on the flax plant. See *rusts.*

rust of peanut. A fungous disease caused by *Puccinia arachidis,* family Pucciniaceae, which appears sporadically. See *rusts.*

rust of sorghum. A fungous disease caused by *Puccinia purpurea,* family Pucciniaceae, which occurs in limited amounts on both surfaces of the leaves of all the sorghums. See *rusts.*

rust of sugar cane. A fungous disease caused by *Puccinia kuehnii,* family Pucciniaceae, characterized by elongated, orange-brown spots on both surfaces of the leaves. See *rusts.*

rust-red flour beetle. See *red flour beetle.*

rust-resistant. Designating the quality possessed by certain varieties of plants to resist the attack of a particular rust fungus.

rusts. Fungi which are obligate parasites on the higher classes of plants (ferns and seed plants), *i.e.,* they are not capable of growth and development apart from their living hosts. The fungus is internal, consisting of fine, colorless, branching threads (hyphae) creeping between the cells of the host from where they draw nourishment by means of small suckers, haustoria (*q.v.*), which penetrate the walls of the host cells.

rust spot. See *cedar apples.*

rusty grain beetle. *Laemophloeus ferrugineus* (Steph.), family Cucujidae, a small beetle that usually feeds on ground grain and sometimes on grain products.

rusty-gray. A coat-color of a horse, with gray points, mane, and tail, and reddish hair through the coat.

rusty gum myrtle. *Angophora lanceolata,* family Myrtaceae, an average-sized tree which bears showy, white flowers and is grown as an ornamental in very warm regions. Native to Australia. Also called *gum myrtle; orange gum.*

rusty knot. A knot that has a rusty-red color characteristic of a heart-rot fungus attack. See *knot: 4.*

rusty menziesia. *Menziesia ferruginea,* family Ericaceae, a small, very hardy, deciduous shrub of no particular value, which has been known to cause poisoning in sheep. Native to western North America.

rusty mottle. A virus disease of sweet cherry which is localized but important in certain orchards in the U.S. The symptoms appear as chlorotic mottlings four to five weeks after bloom and later the leaves show yellow to red coloring and leaf casting. See *mild rusty mottle; necrotic rusty mottle.*

rusty plum aphid. *Hysteroneura setariae* (Thos.), family Aphidae, an insect that infests blossoms of plums preventing the setting of fruit.

rusty spot. A yellowish-red spot in the flesh of potato which indicates the presence of yellow dwarf of potato *q.v.*

rusty spot in Cheddar cheese. Red or rust-colored spots, very fine and quite evenly distributed throughout Cheddar cheese which are caused by two species of organisms belonging to the genus Lactobacillus.

rusty tussock moth. *Orgyia antiqua* (L.), family Lymantriidae, a moth whose larvae feed on many kinds of fruit trees as well as on beech, birch, oak, willow, etc.

rut. 1. The grooved track left by the wheels of vehicles in soft ground. 2. See *estrus; rutting season.*

rutabaga. *Brassica napobrassica,* family Cruciferae, a biennial herb grown for it yellowish, edible root used for human and animal consumption. Native to Europe and Asia. Also called *macomber; Swede; Swedish turnip.*

rutabaga blackleg. A fungous disease caused by *Phoma lingam,* family Sphaerioidaceae, characterized by leaf spots with ashen-gray centers. Black dots, smaller than a pin head, appear in these and serve as the fruiting bodies in which the spores are formed.

rutabaga black rot. A bacterial disease caused by *Xanthomonas campestris,* family Pseudomonadaceae, characterized by the invaded tissues turning black and the leaves turning brown and dropping prematurely. The organism enters the plant through the leaves, roots and injured places.

rutabaga clubroot. A fungous disease caused by *Plasmodiophora brassicae,* family Plasmodiophoraceae, characterized by the roots taking various shaped malformations which are soon invaded by soft-rotting organisms.

rutin. A yellow plant pigment isolated from buckwheat and flue-cured, bright tobacco. It is used by the medical profession in treating cases of fragility of blood capillaries.

rutter. 1. A form of plow used in the Lake States in the U.S. during the early days of logging to cut ruts in an iced logging road for the runners of the sled to fit. 2. A female mammal, such as a cow, which for some abnormality remains constantly in heat. Also called a *buller.*

rutting season. The recurring, usually annual, period when deer, cattle, etc., are in heat. See *estrual cycle.*

ryania. A contact and stomach poison insecticide obtained from the roots and stems of a South American plant, *Ryania speciosa,* family Flacourtiaceae.

Rydberg clover. *Trifolium rydbergii,* family Leguminosae, a wild clover of the Rocky Mountain region from Idaho to Arizona between elevations of 4,500 to 13,000 ft. whose forage value is excellent. Native to the U.S.

rye. *Secale cereale,* family Gramineae, a hardy, winter annual cereal grown as a cover crop, for forage, a bread grain, feed for livestock and for use in the manufacture of whiskey. It is one of the principal bread grains of northern Europe. Native to Europe and Asia.

rye anthracnose. See *cereal anthracnose.*

rye bran. A livestock feedstuff consisting of the coarse outer covering of the rye kernel separated from the cleaned and scoured rye.

rye buckwheat. Tartary buckwheat.

rye distillers' dried grains. A livestock feedstuff consisting of the dried residue obtained in the manufacture of alcohol and distilled liquors from rye, or from a grain mixture in which rye predominates.

rye distillers' dried grains with solubles. A livestock feedstuff consisting of the dried residue obtained in the manufacture of alcohol and distilled liquors from rye or a grain mixture in which rye predominates and contains the major portion of the condensed, screened stillage dried therewith.

rye ergot. A fungous disease of rye caused by *Claviceps purpurea,* family Hypocreaceae, characterized at first by the appearance of the honeydew stage when the exudate accumulates as droplets on the surface of the floral structure and later by the blue-black sclerotial body replacing the kernels in the infected flowers. Some of the chemical compounds of the sclerotia have medicinal value. See *ergot: 1.*

rye feed. A by-product obtained in the process of milling of rye flour from cleaned and scoured rye grain, consisting principally of the mill-run of the outer covering of the rye grain and the germs with small quantities of flour and aleurone, and containing not more than 9.5 percent of crude fiber.

rye flour middlings. A product derived from rye which consists of rye feed, rye red dog, and pure dark rye flour combined in the proportion obtained in the milling of rye flour and containing not more than 5 percent of crude fiber.

rye Fusarium head blight. A head blight of rye caused by *Fusarium* spp., characterized by infection of the flowers, which, spreading up and down the spike turns it to a straw color, causing the plant to produce shriveled kernels of a rough, scabby appearance with a very low food value.

ryegrass. Any grass of the genus Lolium, family Gramineae. Species are grown in lawns and pastures. Also called *ray grass.* See *darnel ryegrass; Italian ryegrass; perennial ryegrass.*

ryegrass blind seed disease. A fungous disease caused by *Phialea temulenta,* characterized by the infested seed becoming soft, pasty, and shriveled.

rye jointworm. *Harmolita secalis,* family Eurytomidae, a small, wasplike insect which lays its eggs in the tissues of the rye stems near the joints. The development of the larvae causes a gall-like thickening which interferes with conduction in the stems, reducing grain production and causing the stems to lodge or break over.

rye leaf rust. A fungous disease caused by *Puccinia rubigo-vera secalis,* family Pucciniaceae, characterized by dusty-rye to orange-brown rust spots on the leaves and leaf sheaths. The yield of grain is lowered.

rye lowgrade feed flour. A flour which consists principally of dark rye flour and small quantities of aleurone and fine bran particles and containing not more than 1.5 percent of crude fiber.

rye middlings. A by-product of rye milling which consists of rye feed and rye red dog combined in the proportions obtained in the usual process of milling rye flour and containing not more than 8.5 percent of crude fiber.

rye powdery mildew. A fungous disease caused by *Erysiphe graminis,* family Erysiphaceae, characterized by the cobweb-like growth on the surface of the leaves producing a powdery appearance.

rye red dog. A by-product obtained in the process of milling rye flour, consisting principally of aleurone with small quantities of flour and fine bran particles and containing not more than 3.5 percent of crude fiber.

rye Rhynchosporium leaf scald. A fungous disease caused by *Rhynchosporium secalis,* characterized by gray-green, water-soaked areas that later become zonated and yellow or brown. It affects the yield by reducing the number of kernels and the kernel weight.

rye rust. See *rusts.*

rye Septoria blight. A fungous disease caused by

Septoria secalis, family Sphaerioidaceae, characterized by leaf blotches and spots indefinite in color and shape but slowly turning straw color and brown, and resulting in shriveled kernels and lodging.

rye stalk smut. A fungous disease caused by *Urocystis occulta,* family Tilletiaceae, characterized by the sori forming in the stalk and flowers preventing the spike from emerging.

rye stem smut. See *rye stalk smut.*

rye straw. The mature stems and leaves of the rye plant after the grain has been removed. Hard and woody, it is usually used for bedding rather than feed.

rye stripe rust. A fungous disease caused by *Puccinia glumarum,* family Pucciniaceae, characterized by yellow or orange-yellow pustules arranged in rows of various lengths on the glumes, leaves, and leaf sheaths. The kernels are usually shriveled and yield reduced.

S

Saanen. A breed of milk goats which originated in the Saanen Valley of Switzerland. The animals are white or creamy-white, usually short-haired and hornless. The breed is noted for its persistence in production rather than for high production over short periods.

sabadilla. Drug sabadilla.

sabadilla seed. The seeds from certain species of Schoenocaulon, which grow in Mexico and contain cevadine, veratridine, sabadilline, sabadine, and the inactive cevine. See *drug sabadilla.*

sac. A baglike part of an animal. It may be a depressed area in a mucous membrane, small terminal outgrowths of the fine air passages of the lungs, or other baglike structures of various size.

sacahuista. *Nolina microcarpa,* family Liliaceae, a perennial plant with short woody stems found on certain Texas (U.S.) ranges. Sheep and goats grazing upon this plant have developed a disease similar to that produced by ingestion of lechuguilla agave (*q.v.*). The leaves can be used in cordage and basketry.

sacaton. *Sporobolus wrighti,* family Gramineae, a stout, perennial bunchgrass which grows from 2 to 8 ft. tall on alluvial fans and in arroyos from southern California to northern Texas (U.S.). It is grazed by all stock, especially when it is green, and has a fair forage rating.

sacbrood. An infectious disease of the brood of the honeybee caused by a filterable virus, characterized by decomposition of the larval tissue.

saccate fruit. A type of fruit which has an inflated, membranous envelope not common among weedy plants and not conducive to seed distribution.

saccharimeter. Any device used to determine the concentration of sugar in a solution, particularly some form of polarimeter. See *sirup hydrometer.*

saccharin(e). A coal-tar crystalline product used as a sugar substitute.

saccharine sorghum. Sorgo.

saccharometer. See *saccharimeter.*

saccharose. Sucrose.

sack. 1. A bag, usually large, oblong, and made of burlap or other stout fabric, used for holding grains, vegetables, etc. 2. To stroke or rub a horse. (A highly nervous horse is frequently sacked by the buster who, from a safe position, soothes the animal by stroking or rubbing with a burlap *sack* or broom.)

sacked. 1. Designating grain, produce, etc., which has been placed in sacks. 2. See *sack: 2.*

sacking spout. A spout to which sacks are attached in order to fill them with material, such as flour or grain, from a threshing machine, combine, mill, elevator, etc.

sacred bean. The seed of the Hindu lotus *q.v.*

sacred cotton. *Gossypium hirsutum* var. *religiosum,* family Malvaceae, a variety of upland cotton (*q.v.*) that bears rust-colored lint.

sacred ear. *Cymbopetalum penduliflorum,* family Annonaceae, a tropical shrub or small tree grown for its fragrant flowers whose dried petals are used medicinally and to flavor food. Native to America. Also called *earflower; xochinacaztli; teonacaztli; sacred earflower of the Aztecs; orejuella; Mexican earflower; flor de la oreja.*

sacred earflower of the Aztecs. Sacred ear.

sacrum. That composite bone structure which makes up the vertebral column between the coccygeal and lumbar vertebrae. It is connected directly with, and comprises part of, the pelvis.

saddle. 1. A seat designed to fit a horse's back and to make riding easy, comfortable, and safe. *Saddles* are of many types but are usually made of heavy leather over a well-padded frame. They are often raised in the rear to hold the body in place and they may have a firm, raised support or horn in front to attach ropes or gear. They are held to the horse's back by means of one or more wide straps or girths that pass around the body of the horse. See *English saddle; packsaddle; side saddle; western saddle.* 2. The whole upper back section of a meat carcass which in wholesale cuts is commonly divided into fore and hind *saddle.* In cookery, the rear upper back portion including the two loins. (Sheep, veal calves, goats, and deer.) 3. The rear part of the back of a male chicken extending to the tail, covered by *saddle* feathers. 4. An apron made of reasonably heavy canvas which is placed on the backs of valuable breeding hens to protect against injury during mating. *Saddles* are usually held in place by loops which fit under the wings. 5. The uncut stalks of four hills of corn from two rows, tied together, around which corn is shocked. 6. A saddlelike depression in the crest of a ridge. See *saddle-*

back: 1. 7. A specially designed transverse log used in a skid road to guide the moving logs. 8. To put a *saddle* upon an animal or thing.

saddleback. 1. A curved depression which connects two higher elevations in a ridge; a col. 2. Resembling or suggesting a saddle in shape or marking; saddlebacked.

saddleback caterpillar. *Sibine stimulea* (Clem.), family Limacodidae, a caterpillar with a green patch in the middle of the back, widely distributed through the Atlantic states of the U.S., which feeds on oak and cherry but sometimes on other trees and plants also.

saddlebag. A bag or pouch of leather or fabric carried at the saddle, especially one of a pair laid across the back of a horse behind the saddle.

saddle blanket. A blanket or pad of wool or cotton placed on the back of a horse beneath the saddle to prevent chafing; saddlecloth.

saddlebred. Of, pertaining to, or designating a type of horse which has the breeding and qualities desired in a saddle horse.

saddle breeds of horses. The principal breeds of saddle horses in the U.S. are: American Saddle horse; American quarter horse; Appaloosa; Arab(ian); Cleveland Bay; Morgan; Palomino; Tennessee Walking Horse; Thoroughbred *q.q.v.*

saddle-broken. Designating a horse trained to the saddle.

saddlecloth. See *saddle blanket.*

saddled prominent. *Heterocampa guttivitta* (Wlk.), family Notodontidae, a caterpillar, common in Atlantic and Gulf states of the U.S., which feeds on beech, maple, apple, oak, and other plants. Also called *antlered maple caterpillar.*

saddle duster. A manually operated insecticide duster which is placed on a horse or mule so as to permit the rider to dust a row on each side. Also called *saddle gun.*

saddle feather. One of a group of long, narrow feathers which grows out of the saddle of a male chicken.

saddle fender. A broad piece of leather attached to a stirrup strap to protect the rider's leg from the stirrup straps and perspiration from the side of the horse. Also called *sudadero; rosaderos; stirrup fender; leg fender.*

saddle gait. The foot movement or gait of a saddle horse; a definite manner of walking, running, etc., such as the walk, running walk, trot, canter, rack.

saddle gall. A sore or wound on the back of a horse which results from abrasion by a saddle.

saddle girth. In horses, a strong band or strap which encircles the body, usually just behind the front legs, and helps fasten a saddle to the back. With western-type saddles it usually consists of two latigo straps on the sides and a webbed band underneath with rings into which the latigo straps fit to tighten and fasten. Also called *cinch.*

saddle graft. A graft in which a wedge-shaped stock fits into a deep cleft cut in the cion. It is used more on herbaceous than on woody plants.

saddle gun. 1. A carbine or short rifle, which fits into a scabbard attached to the saddle. 2. See *saddle duster.*

saddle horn. A knoblike elevation at the front center of a western saddle, used in roping as an anchor for the lariat.

saddle horse. A type of horse especially suitable for riding because of its strong back, style, ease of action, and different gaits. It may include running horses, three and five gaited horses, stock horses, hunters, polo ponies, etc.

saddle leather. A special type or quality leather used in making saddles and other horse equipment.

saddle mule. A mule suitable for riding.

saddle pad. See *saddle blanket.*

saddler. 1. One who makes, repairs, or deals in saddles or saddlery. 2. A saddle horse.

saddle roof. A roof with two gables and a ridge.

saddle scabbard. A sheath or boot carried at the saddle to hold a carbine or other object.

saddle skirt. The leather part of a western-type saddle which extends beyond the sides and to the rear of the saddle seat. On English-type saddles, a broad piece of leather which extends under the saddle, down the side of a horse and under the stirrup strap.

saddle soap. A special soap used to clean and preserve saddles, boots, and leather goods.

saddle stock. Riding horses.

saddletree. 1. The framework of a saddle. 2. See *tulip tree.*

saddling. The placing of a saddle on a horse or other animal.

safety belt. A belt used to fasten a climber securely to a tree, pole, tower or seat so as to free his hands for work.

safety release hitch. An attachment designed to release a pulled implement, such as a plow, mower, etc., from a tractor when the implement strikes an obstruction.

safflower. *Carthamus tinctorius,* family Compositae, an herb, resembling a thistle, cultivated for its orange-colored flower heads which yield a drug and red dye, and for its seed, a source of a rapid-drying oil used mainly by the paint industry. Native to Europe and Asia. Also called *false saffron.*

safflower meal. The by-product obtained after oil has been pressed from safflower seeds.

saffron. A deep, orange-colored product consisting of the dried stigmas of the saffron crocus, used mainly to color and flavor foods.

saffron crocus. *Crocus sativus,* family Iridaceae, a perennial herb grown for its autumnal, white or lilac flowers and as a source of perfume, flavoring, and coloring. Native to Turkey. Also called *vegetable gold.*

saffronwood. *Elaeodendron croceum,* family Celastraceae, a semi-tropical tree whose yellowish wood is used in building and furniture-making. Native to South Africa. Also called *saffronhout.*

sag. A depression in a nearly flat or undulating land surface; a micro-relief feature.

sage. 1. *Salvia officinalis,* family Labiatae, a perennial herb grown for its grayish-green, dried leaves commonly used to flavor meats, in dressings, etc. Native to the Mediterranean area. Also called *garden sage.* 2. See *sagebrush.*

sagebrush. Any plant of the genus Artemisia, family Compositae. Species are herblike, bushy plants relatively important as browse for goats and sheep and to some extent for cattle especially in late fall, winter, and early spring when there is a scarcity of other forage. Growing wild in western and southwestern U.S., it often becomes a pest by invading better grazing lands that are overgrazed. Native to North America. See *big sagebrush; Canada sagebrush; fringed sagebrush; silverking sagebrush.*

sagebrush mariposa. *Calochortus macrocarpus,* family Liliaceae, a cormous herb grown for its purple flowers with a green stripe in the center of each segment. Native to western North America. Also called *mariposa lily; mariposa tulip.*

Sage cheese. An American-type, sage-flavored cheese usually made by the Cheddar process. It has a green, mottled appearance.

sage grass. Little bluestem.

sage tea. An infusion of the leaves of the sage, *Salvia officinalis,* used as a substitute for true tea during the American War of the Revolution. Supposed to have some medicinal value.

sage tick. Spotted fever tick.

sagging comb. Honeycomb in which the weight of the honey causes the wax cells to sag or droop.

sago. A foodstuff made from the starchy pith of the sago palm *(q.v.)* or some other plants, as cycads, jaggery, etc. In granulated or powdered form, the latter known as the *sago* flour, it is used as an item of diet or in making cakes, soups, and puddings. Also used as a stiffening agent in the textile industry.

sago palm. Any of the palms the pith of whose trunks is a source of sago, especially *Metroxylon rumphii,* family Palmaceae, which flourishes in marshy areas and is native to Indonesia. Some of the other *sago palms* are gomuti palm *(Arenga pinnata),* kittoolpalm *(Caryota urens),* and cabbage palm *(Corypha umbraculifera),* all native to Indonesia.

sagrain. One of several important hybrids between grain sorghums and sweet sorghums; grown for seed and forage.

saguaro. *Cereus giganteus,* family Cactaceae, a huge cactus which produces an edible fruit used in sweetmeats. Found in southeastern California, southern Arizona and Mexico. Also called *giant cactus; sahuaro; suwarro.*

sainfoin. *Onobrychis viciaefolia,* family Leguminosae, a perennial herb occasionally cultivated for fodder. Native to Europe and northern Asia. Also called *common sainfoin; holy clover; esparsette.*

St. Andrewscross. *Ascyrum hypericoides,* family Hypericaceae, a small, shrubby, perennial herb grown in the flower border for its bushy habit and its bright yellow flowers. Native to North America.

St. Anthony's-nut. *Conopodium denudatum,* family Umbelliferae, a slender, flexuous plant with a smooth stem and a brown, tuberlike, edible rootstock.

St. Augustinegrass. *Stenotaphrum secundatum,* family Gramineae, a tropical or subtropical, creeping, stoloniferous, perennial, lawn grass which grows in moist, especially mucky, soil mostly near the seashore from South Carolina to Florida and Texas. Native to North America.

St. Benoit. A soft cheese made in the Department of Loiret, France. Charcoal is added to the salt rubbed on the surface of the cheese.

St. Bernardlily. *Anthericum liliago,* family Liliaceae, an unbranched, perennial herb grown in warm areas for its small, white flowers. Native to Europe. Also called *St. Bernard's-lily.*

St. Bernard's-lily. St. Bernardlily.

St. Claude. A small, square, goats' milk cheese made in the vicinity of St. Claude in the Department of Jura, France.

saintfoin. Sainfoin.

St. James deervetch. *Lotus jacobaeus,* family Leguminosae, a perennial herb grown in frost-free areas for its black-purple, yellow, or yellow and black-purple flowers. Native to the Cape Verde Islands. Also called *St. Jame's-flower; St. Jame's-pea.*

St. Jame's-flower. St. James deervetch.

St. Jame's-pea. St. James deervetch.

St. John's-bread. Carob.

St. John's-wort. 1. One of the several plants of the genus Hypericum, family Hypericaceae. 2. *H. perforatum,* an erect, branching, mildly poisonous weed plant, 1 to 3 ft. tall, with abundant yellow flowers and runners at the base, now common in cultivated areas, range lands, waste ground, and along roadsides across the northern half of the U.S. and in Australia. Livestock usually avoid this plant, especially when in bloom, because of its acrid sap which may cause dermatitis or blistering of the unpigmented portion of the skin when the animal is exposed to bright sunlight. It may also produce undesirable flavor in milk and meat. Native to Europe. Also called *Klamath weed; Tipton-weed; goatweed; eolaweed; speckled-john; amber; rosin-rose.*

St. Lucie grass. One of the cultural varieties of Bermudagrass *q.v.*

St. Marcellin. A cheese made in the Department of Isère, France, from goats' milk to which either ewes' or cows' milk is sometimes added. Blue mold is cultivated on the surface of the cheese, but not in the interior.

Saintpaulia. See *Africanviolet; Kew Africanviolet.*

St. Regis. *Rubus idaeus,* family Rosaceae, a variety of red raspberry which bears fruit at frequent intervals; an everbearer. Native to Europe and Asia. Also called *Ranere.*

St. Stephano. A cheese, similar to Bel Paese, made from cows' whole milk in Germany.

St. Swithin's-apple. An apple which ripens prematurely due to the presence of worms.

sake. An alcoholic beverage produced in Japan by a process which involves steaming, cooling, and inoculating rice with spores of the mold,

Aspergillus oryzae, the mass being allowed to incubate and more steamed rice mixed with water being added at different progressive stages of the chemical action. The percentage of alcohol in *sake* is 18 to 24.

Sakellarides. *Gossypium barbadense,* family Malvaceae, a creamy-white cotton, a prominent variety, whose staple measures in length up to 1$^{11}/_{32}$ ins. Native to Egypt. Also called *Sak; Sakel.*

Sakhalin fir. *Abies sachalinensis,* family Pinaceae, a tall, evergreen tree grown for its bright green foliage and purple-green cones. Native to Sakhalin and northern Japan.

salad. 1. A combination of raw vegetables, cooked cold vegetables, or fresh ripe fruits. Such a preparation may be garnished with some type of salad oil, or complemented by the addition of cooked cold meats or eggs. Fruit salads may be garnished by adding lemon sirup or whipped cream. 2. Any vegetable or herb specifically grown for *salads.*

salad burnet. See *small burnet.*

salad chervil. *Anthriscus cerefolium,* family Umbelliferae, an annual herb grown for its edible leaves used as a garnish like parsley. Native to Europe and Asia.

salad plant. Any food plant grown primarily for its edible leaves, stalk, or other vegetative parts used in salads. Commonly included are such plants as lettuce, parsley, endive, cress, chicory, salad chervil, watercress, celery.

Salamana. A soft, ewes' milk cheese with a very pronounced flavor, made in southern Europe.

salamander. 1. *Geomys tuza,* a very common rodent in the dry sandy pine land of Florida (U.S.). Because of its burrowing, it is considered a factor modifying these sand soils. 2. A portable stove which is used in construction work to provide heat, in winter, to keep newly poured concrete from freezing while curing.

Salame. A large-style Provolone cheese made not only in Italy but in other countries, including the U.S.

salami. A dry sausage prepared from pork with 10 to 30 percent beef. The chopped meat is mixed with salt, sodium nitrate, and seasoning and then aged at 38° to 40° F. for about 4 days. The cured meat is stuffed into casings, hung in a dry room at 48° to 50° F. for 5 to 8 days and smoked at 50° to 75° F. for 24 to 48 hours. The smoked sausage is hung in a dry room at 60° F. and kept for about 30 days.

sal-ammoniac. Ammonium chloride.

sale catalog. A printed, descriptive list of the breeding records, performance, terms of sale, etc., of animals offered at a livestock sale.

saleratus. Literally, aerated salt, now commonly sodium bicarbonate (baking soda), but also potassium bicarbonate.

sale ring. The area which is occupied by the animal or group of animals being offered for sale by the auctioneer at a livestock sale.

sales pavilion. A building, large tent, or other covered structure in which livestock sales are held, especially auction sales.

saleyard. A yard used for the sale of livestock.

salicetum. An arboretum of willow trees.

salicylic acid. $C_6H_4(OH)CO_2H$, a colorless acid found in many plants and fruits, it may be obtained especially from oil of wintergreen. Manufactured commercially by processing dry sodium phenate (C_6H_5ONa), it is used as a preservative and as a fungicide.

saliferous. Producing, or containing a large proportion of salt.

salimeter. Salometer.

saline. Containing salt; salty.

saline-alkali soil. A soil having either a high degree of alkalinity or a high amount of exchangeable sodium, or both. The amount of soluble salts is sufficient to be harmful to most crop plants. See *alkali soil; saline soil.*

saline lake. A lake whose water contains a relatively high concentration of soluble salts. Certain *saline lakes* in western U.S. are a commercial source of potash.

saline soil. A soil which contains an excessively high concentration of soluble salts, such as sodium chloride and sodium sulfate. *Saline soils* usually occur in the arid and semiarid regions of western U.S. and in sea coast areas. See *alkaline soil; alkali soil.*

salinity. The degree of saltiness.

salinometer. Salometer.

saliva. The weakly alkaline fluid consisting of the secretions of the salivary glands; spit.

salivary secretion. See *saliva.*

salivation. A discharge of saliva. In large animals, excessive *salivation* usually results in the appearance of long strings of a clear, viscid material hanging from the mouth.

sallenders. An eczematous condition of the skin at the bend of the hock joint of a horse. *Cf. malanders.*

sallow. 1. Goat willow. 2. An unhealthy skin color.

salmonberry. *Rubus spectabilis,* family Rosaceae, a large raspberry having red flowers and salmon-colored fruit. Native to Pacific Coast area of North America.

salmonellosis. Infection by organisms of the genus Salmonella causing food poisoning in man and many diseases in domestic animals. See *food poisoning; paratyphoid; enteritis; pullorum disease; avian typhoid.*

salmon liver oil. A livestock feedstuff supplement; an oil obtained from the livers of salmon. It has properties similar to those of cod liver oil and has use as a vitamin A and D supplement in mixed feeds for livestock. *Cf. cod liver oil.*

salmon oil. A livestock feedstuff supplement which is the product obtained by extraction of part of the oil from the cannery refuse of salmon. It contains important amounts of vitamin A and D and has possible use as a supplement in livestock feeds.

Saloio. A small, cylindrical cheese made from cows' skim milk on farms in the region of Lisbon, Portugal.

salometer. A device, usually a hydrometer with

special graduations in degrees or percentages, used to determine the strength or salinity of a particular salt solution, such as a meat pickle or a pickle for cucumbers, etc. It is graduated so that 4 degrees on the *salometer* correspond to each 1 percent of sodium chloride in a solution at a temperature of 60° F.; thus, 70° of saltness would correspond to a solution having a strength of 17½ percent. When used for calcium chloride brine, the *salometer* shows a reading nearly 1/10 larger for equal percentages of strength. Also called *salimeter; salinometer.*

saloon plant. Aspidistra.

salpingitis. 1. Inflammation of a fallopian tube (oviduct). 2. An inflammation of the oviduct of poultry characterized by a discharge from the duct which causes irritation of the vent and a smeary appearance of the feathers below it.

salsa di pomodoro. A tomato paste used extensively as a flavoring for many Italian dishes.

salsify. See *vegetable oyster.*

sal soda. One form of sodium carbonate.

salt. 1. Sodium chloride (common salt), NaCL, a white crystalline compound occurring abundantly in nature as a solid or in solution. It is a requirement of all livestock and is generally deficient in all natural feed ingredients. It is commonly included in mixed feeds and is otherwise supplied as *salt* blocks in pastures, barn lots or stalls. It has wide usage for seasoning, preserving meats and vegetables, tanning hides, etc. Applied in small amounts to soils it probably has no direct value as a fertilizer. It may, however, liberate potassium and possibly phosphorus in the soil. It may improve or may injure the quality of plants depending upon the particular crop and other circumstances. In large amounts it is toxic and is frequently used to kill weeds and other undesirable plants. 2. To add *salt* as seasoning, preservative, supplement, etc.

salt alum tanning. A simple method of tanning sheep skins whereby the previously salted and fleshed skins are treated overnight with a 2 to 1 mixture of salt and alum and later on with neat's-foot oil or glycerine.

saltation. The bouncing, tumbling movement of soil particles along the soil surface which is caused by wind or water action.

salt block. A compressed block of common salt which is used to supply salt to livestock. Also called *saltera.*

salt bottom. A flat, alkali, relatively low piece of land (western U.S.).

saltbush. An herb or shrub of the genus Atriplex, family Chenopodiaceae. Species occupy large areas of the arid and alkaline soils of western and southwestern U.S. and Australia. They grow at low and intermediate elevations under desert conditions. The plants have a salty taste which adds to their palatability for stock. Some species, especially shadscale saltbush *(q.v.)*, have a fairly high rating as browse plants.

salt cedar. French tamarisk.

salt dome. Any anticlinal structure containing a core of rock salt developed during the normal folding and faulting by geologic forces. Valuable mineral deposits are frequently found with them as well as oil. They occur in areas all over the world: *e.g.*, the Texas-Louisiana coastal region of the U.S.; Colombia and Peru in South America; Germany, Denmark, Spain in Europe, as well as parts of Asia and Africa.

salt down. To add salt, especially for the preserving or curing of meats, hides, etc. See *salt: 2.*

salted. Treated, seasoned, supplied, or preserved with salt.

salted butter. A butter to which salt has been added. *Cf. sweet butter.*

saltgrass. Designating a grass of the genus Distichlis, family Gramineae. Species are low, leafy, perennial grasses with strong running rootstocks which occupy alkaline and saline lands and which are high in salt content. All *saltgrasses* are rather low in nutritive value and rate only fair for general forage purposes. See *inland saltgrass.*

salt grounds. The several locations on a range where salt is provided for the livestock usually to influence their movement and to bring about a more uniform utilization of the range.

salt hay. 1. A mulching material for plants which is cut from coastal marshes. 2. Hay which is made from grasses cut from salt marshes.

salt heliotrope. *Heliotropium curassavicum,* family Boraginaceae, an annual weed plant found on sandy seashores and borders of saline marshes. Also called *alkali heliotrope.*

salt lick. A place where salt occurs naturally on the surface of the ground and to which wild animals go to obtain salt needed in their diet.

salt log. A log with a hollow place in which salt is put for the use of livestock.

salt marsh. A flat, grass- and shrub-covered piece of land, occasionally inundated by salt water.

salt-marsh caterpillar. *Estigmene acrea* (Drury), family Arctiidae, the larva of a moth which is generally distributed and at times may be a severe pest infesting various types of vegetation.

salt-marsh peat. The peat occurring in salt marshes mainly along the sea coasts which typically consists of coarse, fibrous, only partially decomposed plants and large amounts of admixed mineral sediments.

saltpeter (saltpetre). 1. Potassium nitrate, a white crystalline salt of potassium and nitrogen whose principal agricultural use is as a fertilizer and as a preservative in curing and preserving of meats. 2. An incrustation of certain salts occasionally formed on tobacco leaves during processing.

salt poisoning. A condition that may result in the death of livestock and poultry caused by ingestion of an unusually large amount of salt. Farm animals that have been deprived of salt over a long period may consume it to excess if unlimited quantities are available.

salt pork. Usually pork belly or fat backs (*q.v.*) that have been cured by the dry salt method.

salt sick. A mineral deficiency disorder which results in loss of appetite, unthriftiness, anemia, and even death to cattle and other animals on certain poor, sandy soil areas of Florida and similar soil areas elsewhere. This condition is known to be the result of a critical deficiency in the forage of many of the trace mineral elements, as copper, iron, and cobalt. Also called *bush sickness*.

salt stock. In the manufacture of pickles, the cucumbers that have been salt-cured in brine.

salt tree. See *athel tamarix*.

salt-water marsh. Treeless, wet land along the seacoast and inland in which the free water and soil are saline, mainly from sodium chloride, in contrast to fresh-water marsh (*q.v.*). The vegetation consists largely of grasses and sedges, a number of which have a value for pasture, wild hay, and food and cover for wildlife. See *marsh*.

saltwort. See *Russianthistle*.

salty flavor. 1. A flavor appearing frequently in late-lactation. 2. A flavor defect of milk associated with mastitis.

salve. Usually a preparation, consisting of a fatty, greasy, or jelly-like base with added medications, which has properties of soothing, softening, healing, etc.

salvia. See *scarlet sage; sage: 1.*

samara. A single-seeded, winged fruit, whose wing helps in seed dissemination, like that of the maple, ash, and elm.

samp. A food product which is a corn mush consisting of coarse hominy prepared for use by boiling.

sample. 1. Anything taken or shown as representative of the whole, as *samples* of cotton from a bale, seed from a bag, etc. 2. In cotton buying, two parts taken from either side of the bale, which weigh about 6 ozs. and have a face 4 to 6 ins. wide. 3. To take a portion of some commodity and to test it, or offer it for test, as a representative of the whole.

sample bottle. In dairy herd testing, a bottle suitable for holding a sample for later inspection or testing. Also called *sample jar*.

sample dipper. A dipper especially designed or otherwise suitable for taking samples, as a milk *sample dipper*.

sample grade. In marketing, especially in grains, a substandard product that does not conform to the specifications of recognized standard grades and consequently is sold on the basis of samples which show the actual quality or condition of the particular offering. In grains, if inspection reveals that the product is heating, musty, sour, or otherwise of distinctly low quality, the shipment or lot is placed in *sample grade*.

sample jar. A sample bottle.

sample plot. An area laid out, permanently or for a short time, for experimentation or mensuration. Also called *study plot*.

sampler. 1. Any one of the several different devices used for taking a sample from a bag, a car of grain, a can of milk, a tub of butter, a cheese, etc. 2. One who takes samples for inspection or testing. See *sampling tube*.

sampling tube. a. Any of the several tubelike devices, commonly made of brass or stainless steel, of open or closable types used for extracting representative samples of grains, feeds, seeds, liquids, etc. b. Any of the several kinds of tubes used for collecting samples of soils.

sanctions. Social techniques such as influence, penalties, etc., that provide for conformance to norms, ethical standards, laws and institutions. *Sanctions* may be positive, as rewarding, or negative, penalizing.

sanctuary. A reservation where animals and birds are sheltered and protected from hunting or molestation. *Cf. refuge.*

sand. 1. Any of the small, loose, granular fragments which are the remains of disintegrated rocks. It may contain a great variety of minerals and rocks, but the most common mineral is quartz. 2. In soil science, a group of textural classes in which the particles are finer than gravel but coarser than silt, ranging in size from 2.00 to 0.5 mm. in diameter. It is the textural class of any soil that contains 85 percent or more of *sand* and not more than 10 percent of clay. 3. An area or deposit of this material, as a *sand* shore, a *sand* bank, etc. 4. In maple sugar processing, a granular deposit, mainly of silica and calcium malate, formed as the sap is condensed. 5. To cover with *sand*.

sandal beadtree. *Adenanthera pavonina,* family Leguminosae, a tropical tree grown in frost-free areas for its feathery foliage and white and yellowish flowers and for its seeds used in beadwork. Native to Asia and North Africa. Also called *Barbados pride; red sandalwood; bead-tree; peacock flower; flower fence.*

sandalwood. See *white sandalwood*.

sand-and-boll screen. A device for cleaning cottonseed which consists of an inclined reel covered with revolving screens of various-sized meshes to separate seeds and foreign matter according to size.

sand bag. A bag of burlap or other stout fabric which, when filled with sand or earth, is used in flood control to increase the height of or to reinforce a levee, etc.

sand bar. A ridge or narrow shoal of sand formed in a river or along a shore by the action of currents or tides.

Sandberg bluegrass. *Poa secunda,* family Gramineae, a perennial bunchgrass which occurs in lower elevations and exposed sites in western areas of the U.S. and Canada. It endures grazing better than most bluegrasses and is rated good in general forage value.

sand binder. Any plant which can hold sand in its place, and prevent it from shifting, by the tentacular grip of its rootstocks. Grasses of the genera Ammophila and Calamovilfa are commonly used for such a purpose. See *European beachgrass*.

sandboil. A bubbling spring carrying a volume of sand which occurs behind a levee from water seeping through a sand stratum under the levee. Also called *blow-outs*.

sand brier. Carolina horsenettle.

sand brome. Australian brome.

sand bur. Any of the various weeds, especially the annuals of the genus Cenchrus, which bear prickly fruit and grow in sandy or waste places. Also called *bur-grass; sand-spur.* See *buffalobur nightshade.*

sand cherry. 1. *Prunus pumila,* family Rosaceae, a low-growing bush grown as an ornamental and for its somewhat edible fruit. Native to northeastern U.S. and Canada. 2. See *Bessey cherry.*

sand-colic. A condition in horses which occurs when they graze on shallow-rooted vegetation and ingest a large quantity of dirt.

sand corn. *Zigadenus paniculatus,* family Liliaceae, an herbaceous plant, occasionally cultivated for its greenish-white, racemose, flower clusters. The leaves are poisonous to cattle. Native to western U.S. Also called *foothill deathcamas.*

sand crack. A vertical split which may develop at any part of the wall of a horse's hoof but is commonly seen at the toe in the hind foot and at the inside quarter of the forefoot. Also called *quarter crack.*

sand cricket. Jerusalem cricket.

sand disease. In horses and cattle, an inflammatory swelling of the stomach and small intestine due to the presence of sand in feed or water. Occurs frequently in flooded areas.

sand dropseed. *Sporobolus cryptandrus,* family Gramineae, a perennial bunchgrass that grows in small clumps and is recognized by the tufts of long, white hairs at the top of the sheath. It is adapted to sandy or gravelly soils at altitudes up to 5000 ft. Regarded as a fair forage grass, it is used sometimes in the west central states of the U.S. to prevent erosion along highways. Found in western U.S.

sand drown. A chlorophyll deficiency condition in tobacco and other plants. In tobacco, symptoms in mid-season are a progressive loss of normal green color on lower leaves which results in a lower market grade of affected leaves. It appears on sandy soils, especially in the coastal plain of South Carolina, U.S.

sand dune. Loose sand drifted by wind and piled up into hills and hummocks. It is commonly found on seashores, the shores of large inland lakes, and in deserts.

Sanders dracaena. *Dracaena sanderiana,* family Liliaceae, an arborescent plant grown as a house plant for its white-margined leaves. Native to central Africa.

Sanders peperomia. *Peperomia sandersi,* family Piperaceae, a tropical, fleshy herb grown as a pot plant for its foliage. Native to Brazil.

sand filter. A method of water purification by filtration in which the impure water is allowed to trickle downward through particles of sand which sift out the suspended impurities.

sandfly. *Culicoides* spp., family Tendipedidae, a very small, two-winged, blood-sucking midge which is a nuisance in certain areas because of its sharp sting. Also called *no-see-ums; punkies.*

sandgrass. See *sandreed.*

sand-hill. A hill-shaped mound of sand, usually on a sea or lake shore. It is not restricted to any particular geologic origin.

sandiness. 1. A condition of a product resembling or suggesting sand; grittiness. 2. The defect in foods experienced in tasting, when a food, such as ice cream or condensed milk, contains tiny crystals of sugar or ice.

sand lovegrass. *Eragrostis trichodes,* family Gramineae, a grass frequently used in the reseeding of sandy or semi-sandy upland pastures in the Great Plains region of the southwestern U.S. It provides useful forage and is used largely as a summer pasture grass.

sandmyrtle. See *box sandmyrtle.*

sandpapervine. Purplewreath petrea.

sand pear. *Pyrus pyrifolia,* family Rosaceae, a deciduous tree which is resistant to fire blight and has edible fruits with many grit cells. Its small-fruited forms were called late pears *(P. serotina);* while large-fruited forms were the *P. serotina culta* to which the name *sand pear* was applied. Horticultural varieties of the large-fruited forms are raised in Japan and formerly in America. Crossed with the common pear, *P. communis,* it has yielded the Kieffer, Leconte, and Garber varieties of hybrids. Native to China. Also called *Chinese pear; Japanese pear.*

sand pyxiemoss. *Pyxidanthera barbulata,* family Diapensiaceae, an evergreen plant grown for its habit, foliage, and bell-shaped, white or pinkish flowers. Native to North America. Also called *flowering moss; pyxie; pine barren beauty.*

sandreed. Any of a small group of rigid, tall, perennial grasses with creeping rhizomes, belonging to the genus Calamovilfa or Ammophila. The more important species are (a) the drought-enduring prairie *sandreed, Calamovilfa longifolia,* family Gramineae, a plant with tapering, long, slender leaves which grows from 2 to 6 ft. tall. It is grazed lightly in the summer; in the western U.S., it is important as winter pasture and for hay; (b) American beachgrass, *Ammophila breviligulata,* family Gramineae, found on the Atlantic shore and the Great Lakes area of the U.S. They are important sand-binding grasses on dunes and sand-hills.

sand sagebrush. *Artemisia filifolia,* family Compositae, a shrub that is a weed pest on grazing land in the southern Great Plains area of the U.S.

sand screen. a. A screen used in corn shellers, etc. for the removal of grit and sand. b. A screen used to separate out coarse aggregate from sand.

sand spreader. In forestry, a mechanical device used to facilitate spreading sand evenly over seedbeds sown broadcast.

sandstone. A sedimentary rock composed of sand particles, mainly quartz, whose particles are bound together by natural cementing materials as silica and lime carbonate. Soils resulting from the weathering of such rocks are generally sands or sandy loams in texture. *Sandstones* of certain textures are used in the manufacture of sharpening stones and grinding wheels.

sandstorm. A windstorm which carries such large amounts of sand that visibility is drastically reduced. See *dust storm.*

sand trap. 1. A device, often a simple enlargement of the sump, used in a ditch or conduit to arrest the heavier particles carried by the water. It often includes the means for the removal of such material also. 2. A constructed area on golf courses, filled with sand, to catch golf balls off line between the tee and the putting green. It serves to penalize poor play.

sand vetch. Hairy vetch.

sandwich. A graft made by inserting an intermediate cion, such as of a fungus-resisting variety, between a given root-stock and the cion that later develops into the tree top. It has possible use in the control of fungous diseases which work up from the roots in citrus trees, etc. See *double-working.*

Sandwich Nut. A cheese made by mixing chopped nuts with fresh Neufchâtel or Cream Cheese.

sand wireworm. *Horistonotus uhlerii* Horn, family Elateridae, an insect which infests corn, cotton, and potato plants below the ground surface. Also called *cotton wireworm; corn wireworm.*

sandwort. Any of the several plants of the genus Arenaria, family Caryophyllaceae. The *sandworts* comprise a large genus of low, usually tufted, herbs of the chickweed family. They are widely distributed in sandy, gravelly, or dry sterile soils. Only a few of over 100 species are of horticultural interest and these are largely limited to rock gardens and borders. See *Greenland sandwort.*

sandy. 1. Containing sand, as a *sandy* beach, soil, or desert. 2. Resembling or suggesting sand in color or texture. 3. Gritty, as ice cream or condensed milk in which tiny crystals of sugar or ice have formed.

sandy bay. The coat-color of a horse which is a light bay.

sandy clay. Soil which contains more than 45 percent sand, more than 35 percent clay, and less than 20 percent silt.

sandy clay loam. A soil which contains from 20 to 35 percent clay, less than 28 percent silt, and 45 percent or more of sand.

sandy cotton. Dusty cotton.

sandy loam. Gritty, loamy soil which contains less than 20 percent clay, less than 50 percent silt, and between 43 and 52 percent sand.

sandy soil. In common usage, the sand and sandy loam textural classes; any soil which contains more than 70 percent sand and less than 15 percent clay.

sandy textured ice cream. See *sandiness: 2.*

sanguivorous. Subsisting on blood, as mites, ticks, and certain other insects.

sanitary. 1. Concerning health or the conditions affecting health, especially freedom from filth and infection. 2. Favorable to health.

sanitary officer. An official, usually delegated by a governmental agency, whose duty is to inspect conditions which affect health and to enforce regulations designed to maintain or restore it: *e.g.,* a meat or dairy inspector, etc.

sanitary pump. Any of several kinds of pumps for the handling of milk and other liquids whose parts can be easily removed for cleaning and sterilization.

sanitary steam jets. A device for elevating unskimmed whey in the manufacture of cheese.

sanitary trap. A part of a vacuum pump installation in a milking machine which is designed for collecting and holding unwanted materials. It can be easily emptied and cleaned.

sanitary vat. A vat *(q.v.)* designed to maintain products treated therein free from any condition injurious to health.

sanitation. The developing and practical application of measures designed to maintain or restore healthful conditions, such as the treatment, removal, or destruction of contaminated or infested materials and possible sources of infection or infestation.

sanitize. To disinfect and make sanitary the utensils used in preparation of food products, as dairy utensils.

San José scale. *Aspidiotus perniciosus* Comst., family Diaspididae, an important scale insect of woody plants which may infest most of the cultivated fruits and a large number of ornamental trees and shrubs. It is often lethal if left unchecked. Trees lightly infested show small, round, grayish specks, about the size of a pin head, with the bark or fruit often reddened around each scale. On more heavily infested trees, the bark is covered by a gray layer of overlapping scales. See *scale insect.*

San Miguelito. Mountainrose coralvine.

Sansevieria. Bowstring hemp.

Santa Gertrudis. A breed of beef cattle developed at the King Ranch in Texas (U.S.) from crossbreeding Brahmans and Shorthorns. Cattle of this breed have about ⅜ Brahman blood and are said to be especially tolerant of sub-tropical conditions. See *ranch cattle.*

Santa Rosa plum. A member of the Japanese group of the genus Prunus. It is a large, reddish-purple plum of excellent flavor. Native to Japan.

sap. 1. The juices or fluid constituents of a plant, especially a woody plant. 2. The moisture and its soluble constituents in unseasoned wood.

sapling. A young tree which is usually between 2 and 4 ins. in diameter at breast height.

sapling clover. See *mammoth clover.*

sapodilla. *Achras zapota (Lucuma mammosa),* family Sapotaceae, a large, evergreen tree which bears an edible fruit, and whose latex yields chicle, the chief ingredient of chewing gum. Native to tropical areas of North America. Also called *Sapote; marmalade plum; naseberry.*

saponification. The chemical conversion of fats into soap.

saponification number. The number of milligrams of potassium hydroxide which is required to saponify one gram of a fat or similar substance.

saponify. To change a fat to soap by treating with an alkali.

saponin. Any of several glucosides or glycerides present in many plants: *e.g.*, soapwort, which has soap-like qualities. Used in detergents, to reduce the surface tension of water in emulsifying oils, and to produce foam in fire extinguishers and beverages.

saprogen. An organism capable of producing decay in non-living organic matter, such as dead wood, etc.

saprogenesis. That period in the life cycle of a pathogenic organism in which it is not directly associated with a living host and in which it may be either dormant or living as a saprophyte *q.v.*

saprolite. A kind of rock weathering, disintegrated or rotten rock, which remains in its place of origin.

sapropel. A slimy, fetid sediment on the bottoms of lakes and other bodies of water which is made up of the organic debris from aquatic plants and animals.

saprophyte. An organism, such as fungi, bacteria, etc., which lives on organic matter which is dead or rotting. Mushrooms, toadstools, various orchids, etc., are *saprophytes.*

saprophytic. Pertaining to or having the characteristics of a saprophyte.

saproplankton. A mass of decaying, aquatic plant and animal material (plankton) which may be found floating on or in stagnant water.

sap rot. Designating disintegration of the sapwood caused by parasitic, wood-destroying fungi which are, for the most part, wound parasites invading living trees through pruning wounds, mechanical injury, etc.

saps. Grades of lumber which apply to several sorts of sapwood, used with some modifying description, such as: kiln-dried *saps,* clear *sap,* bright *sap,* etc.

Sap Sago Cheese. A small, extremely hard, green cheese, of Swiss origin, made largely from skim, cows' milk, with some buttermilk and whey which is flavored and colored with the powdered, dried leaves of *Trigonella caerulea,* the bluewhite trigonella *(q.v.),* grown especially for this purpose.

sap stain. Any stain that predominantly affects sapwood.

sap sucker. 1. Any of several, small woodpeckers, *Sphyrapicus* spp., especially *S. varius,* the yellow-bellied sapsucker which makes a horizontal series of holes around trees, such as apple, birch, pine, etc. It returns from time to time to feed on the sap and insects from these holes. 2. Any insect that sucks sap from the roots, stem or leaves of plants, as the corn-root aphid, chinch bug, etc.

sapwood. That portion of the wood of a tree more recently formed which contains living cells intermingled with non-living, woody tissue. It is located between the bark and heartwood of trees. Also called *alburnum; splintwood.* See *heartwood.*

sapwood rot. The decay of sapwood caused by various fungi, as white spongy rot of the sapwood of horse chestnut caused by the toadstool *Collybia velutipes* Curt. The same fungus produces a similar rot in linden and American elm.

sapwood timberworm. *Hylecoetus lugubris* Say, family Lymexylidae, the grub of a beetle which bores into the sapwood of many trees.

sar. See *malai.*

Saratoga chop. A retail cut of lamb made from the large shoulder muscle taken from the rib half that is left when making a mock duck.

Saratoga spittlebug. *Aphrophora saratogensis* (Fitch), family Cercopidae, a spittlebug *q.v.*

sarcle. A hoe; the crude wooden implement first used by the ancient Egyptians which is conjectured to be the original tool from which modern chopping hoes were developed.

sarcocarp. See *mesocarp.*

sarcoptic mange. A persistent and contagious skin disease of man and animals, including dogs, cattle, sheep, goats, camels, rabbits and horses. It is caused by the itch mite *Sarcoptes scabiei,* a small, parasitic mite. See *common scab (mange) of cattle; foot mange; mange; scab mite.*

sarcoptic scab (mange) mite. See *itch mite.*

sarcous. Of, or pertaining to, flesh or muscle.

sardine meal. A fish meal resulting as a by-product of the oil extraction from sardines which is used as a protein supplement in animal feeds.

sardine oil. A product obtained by the extraction of oil from the whole sardine or pilchard or from cannery refuse of this species. It is used in animal feeds as a vitamin A and D supplement.

Sardo. A Romano-type cheese made from a mixture of cows' and ewes' milk on the Island of Sardinia.

Sargent barberry. *Berberis sargentiana,* family Berberidaceae, an evergreen shrub grown for its foliage, yellow flowers, and bluish-black fruit. Native to China.

Sargent cherrypalm. *Pseudophoenix sargentii,* family Palmaceae, a palm which grows up to 25 ft. high, with trunk about 12 ins. in diameter and usually bulged in the middle. It has large leaves in a terminal crown and orange-red, cherry-like fruits about ¾ in. in diameter. This palm grows well in sandy soils and is much cultivated, from Miami southward, as a substitute for the royal palm. Native to the Florida Keys (U.S.). Also called *hog cabbage palm; buccaneer palm; Sargent palm.*

Sargent crabapple. *Malus sargenti,* family Rosaceae, a hardy, deciduous tree grown for its white flowers. Native to Japan.

Sargent lily. *Lilium sargentiae,* family Liliaceae, a tall, bulbous herb grown for its fragrant, large, rose-purple flowers, white inside. Native to western China.

Sarrazin. A Roquefort-type cheese made in the Canton of Vaud in southern Switzerland.

sarsaparilla. The flavoring extract which is obtained from the roots of various tropical species of Smilax, family Liliaceae, especially S. *aristolochiifolia* and S. *officinalis.*

sasanqua camellia. *Camellia sasanqua,* family Theaceae, an evergreen shrub grown as a specimen plant, and for its foliage and its white, pink, red, or variegated flowers. Tea oil, used in cooking, is derived from its seeds. Native to China and Japan. Also called *tea oil tree.*

sash. 1. The frame which holds the glass in a window. 2. A glazed frame used to cover a hotbed or cold frame. See *hotbed sash.* 3. The frame in which gang and sash saws are stretched and which moves up and down in vertical slides by means of a connecting rod attached to it.

sash bar. A shaped wooden or metal bar used in the construction of a sash or frame and designed to hold and support the glass secured to it, as in a building window, greenhouse, or hotbed frame.

sashbed. See *hotbed; cold frame; cold pit.*

sash house. A frame structure which is covered with hotbed sash to provide an inexpensive, small greenhouse and which is commonly used only for the growing of plants for later transplanting.

sassafras. 1. Any of the several aromatic plants of the genus Sassafras, family Lauraceae. Also called *cinnamonwood; saxafras.* 2. *S. albidum,* a deciduous tree grown for its gorgeous, scarlet, fall foliage. Its dried root bark is used in flavoring and as a medicinal. The wood is used for fence posts, general millwork, rails, wooden pails, small boats. Escaped, it can be an invading pest. Native to North America.

sassafras laurel. California laurel.

sassafras oil. An oil obtained by steam-distilling the bark, wood and roots of sassafras. The bark contains 6 to 9 percent oil and the roots less than 1 percent. It is used as a flavoring and in perfumery.

Sassenage. A hard, blue-veined cheese about 12 ins. in diameter and 3 ins. thick made from cows' milk to which small quantities of goats' or ewes' milk are usually added. It is named for the village Sassenage in the Department of Isère, France.

satin flower. See *chickweed; dollarplant; golden blue-eyed grass.*

satin moth. *Stilpnotia salicis* (L.), family Lymantriidae, an insect found in New England, Washington, and Oregon (U.S.), whose larvae defoliate poplars and feed on willows and oaks.

Satsuma orange. *Citrus nobilis* var. *unshiu,* family Rutaceae, a hardy, tropical, evergreen tree grown for its edible fruits with loose, thin rinds. Native to the East Indies.

Satsuma orange anthracnose. A fungous disease of the Satsuma orange in Japan, caused by *Gloeosporium foliicola,* family Melanconiaceae. See *anthracnose.*

saturant. A substance used for saturating another substance.

saturate. To fill to the utmost capacity for absorption or containment.

saturated. Filled to the maximum capacity under existing conditions.

saturated air. Air containing as much water vapor as it can hold without condensation. In *saturated air,* the partial pressure of the water vapor is equal to the vapor pressure of water at the existing temperature.

saturated atmosphere. Saturated air.

saturated fat. A completely hydrogenated fat.

saturated soil. 1. A soil filled to its capacity with water. 2. A soil chemically saturated with cations or anions.

saturated solution. A solution, usually of water and some solid substance, such as salt, in which the solvent holds all it can contain of the substance in solution. When so much common salt is dissolved in water that no more salt will dissolve, the solution is said to be saturated.

saturation point. 1. The point beyond which no further additions can be held, utilized, or accommodated: *e.g.,* the water-holding capacity of a particular soil, the flow of goods to a market, or a population of plants. 2. In wildlife, the maximum density of population; the natural condition wherein a species is as abundant as living conditions permit.

saucer magnolia. *Magnolia soulangeana,* family Magnoliaceae, a hybrid, deciduous tree or shrub grown for its sometimes fragrant, purplish flowers.

sauerkraut. A food, the clean, sound product of characteristic flavor obtained by full fermentation, chiefly lactic, of properly prepared and shredded cabbage in the presence of not less than 2 percent nor more than 3 percent salt. It contains upon completion of fermentation, not less than 1.5 percent acid, expressed as lactic acid. *Sauerkraut* refined in the process of canning or packing contains not less than 1 percent of acid expressed as lactic acid.

sausage. 1. A ground or chopped meat product of many varieties which is made usually of meat trimmings of pork, beef, veal, etc., often highly seasoned with various blends of spices. It may be a bulk product or enclosed in a tubular casing, such as livestock intestines which have been specially prepared for this purpose. *Sausage* may be fresh, smoked, cooked, dry or semi-dry: *e.g.,* pork *sausage,* bologna, frankfurters, salami, etc. 2. A long, cylindrical object whose shape suggests a *sausage: e.g.,* the fruit of the sausagetree (*q.v.*), the bundles used in making a *sausage* dam, etc.

sausage bull. A market classification for mature bulls whose flesh is used mainly in the making of sausage.

sausage casing. The small and large intestines of hogs (or cattle and sheep) which are soaked, slimed, and scrubbed to be used as a covering for sausage meat.

sausage dam. A wire-wrapped, loose-rock dam whose resulting bundles are like cylinders and are laid in a horizontal or vertical position (obsolete).

sausage stuffer. A device consisting of a cylinder with a tubular outlet and a plunger, used for forcing prepared sausage meat into casings.

sausagetree. *Kigelia pinnata,* family Bignoniaceae, a tree grown in warm areas as a curiosity for its gourdlike fruits and its purplish-red, bell-shaped flowers. Native to tropical Africa. Also called *etua tree.*

savannah (savana). 1. An area of grassland usually flat and devoid of trees or containing only scattered trees, especially marshy or wet expanses in the southeastern U.S. 2. Dry grassland which contains isolated or scattered trees and shrubs.

saveloy. Summer sausage.

savin juniper. *Juniperus sabina,* family Pinaceae, a hardy, evergreen shrub grown extensively as an ornamental around buildings, etc. Native to Europe and western Asia. Also called *savin.*

savory. 1. Any aromatic annual or perennial herb of the genus Satureia, family Labiatae. See *summer savory.* 2. The dried leaves and flowering tops of summer savory used as a flavoring.

Savoy cabbage. A hardy variety of cabbage having wrinkled leaves.

saw. 1. A cutting device of various types, sizes, and shapes which consists essentially of a serrated metal blade; used for cutting wood, metal, bone, etc. 2. To cut with a *saw.*

saw arbor. The shaft and bearing on which a circular saw is mounted. Also called *mandrel.*

sawbuck. An X-shaped frame or saw horse used for supporting poles and small logs to be sawed.

saw cabbagepalm. Saw paurotis.

sawdust. Dust and small particles of wood produced by the cutting action of saws. A by-product of the lumber and wood industries, it was used for insulation in ice houses and is now used to a limited extent as a bedding material for livestock or as soil mulch mainly by gardeners. Mixed with soil, it has no fertilizer value until decomposed and may depress plant growth in its fresh state.

sawfly. Any of a large group of insects belonging to families Tenthredinidae, Pamphiliidae, Cimbicidae, Diprionidae, and certain other related families. The adult females have saw-like organs for puncturing foliage and twigs of plants to deposit their eggs in them.

saw gin. A cotton gin *(q.v.)* in which the cotton strands are pulled between parallel wires by the teeth of a series of revolving, circular saws. The wires are so closely spaced that the seeds cannot pass and thus are separated from the cotton. The tufts of ginned cotton are removed from the saw teeth by brushes or air blasts. The *saw gin* greatly aided the cotton industry. See *circular saw; gin saw.*

saw kerf. The width of a cut made by a saw.

sawleaf callicoma. See *Callicoma: 2.*

saw log. A log large enough to produce lumber or other products.

sawmill. A plant at which logs are sawed into salable products.

sawpalmetto. *Serenoa repens* (*S. serrulata*), family Palmaceae, a palm usually almost stemless, and having a prickly leaf stalk. In flatwood areas, especially in Florida, U.S., it is commonly an undergrowth with pines. Native to North America. Also called *scrub palmetto.*

saw paurotis. *Paurotis wrighti,* family Palmaceae, a fan palm grown as an ornamental. Its long leafstalk has upward curving prickles. Native to the West Indies. Also called *saw cabbagepalm; everglade palm.*

sawpetal waterhyacinth. *Eichhornia azurea,* family Pontederiaceae, an aquatic herb grown in a pool or pond for its lavender-blue, purple-centered flowers. Native to Brazil.

saw timber. Trees that will yield logs suitable in size and quality for lumber.

saw timber stand. A stand of trees in which most of the trees are of merchantable size or in which the merchantable volume per acre exceeds the minimum economic requirement.

saw-toothed grain beetle. *Oryzaephilus surinamensis* (L.), family Cucujidae, a small beetle which infests stored seeds, grain products, dried fruits, sugar, etc.

saxifrage bellflower. *Campanula saxifraga,* family Campanulaceae, a perennial herb grown in rock crevices in rock gardens for its tuft-forming habit and violet flowers. Native to Armenia.

saxifrage pimpinella. *Pimpinella saxifraga,* family Umbelliferae, an herb, bearing pinnate leaves and white flowers, grown for its aromatic root which is dried and used as a diuretic. Native to Europe. Also called *burnet saxifrage.*

saxifrage tunicflower. *Tunica saxifraga,* family Caryophyllaceae, a hardy, perennial herb grown for its tufted and spreading habit and pink or lilac flowers in terminal clusters. Native to Europe. Also called *saxifrage pink; tunic flower; coat flower.*

Say stink bug. *Chlorochroa sayi* Stål, family Pentatomidae, a bright green bug which infests grains, grasses, beans, peas, and some other plants.

scab. 1. The crust formed over a wound or sore. 2. A mangy disease of animals, especially sheep, caused by the itch or scab mite *q.v.* 3. Any of the numerous fungous or bacterial diseases of plants characterized by dark-colored or rough, scablike spots, as apple *scab,* cereal *scab,* cucumber *scab,* gladiolus *scab,* potato *scab.*

scabies. A skin disease caused by mites; mange. See *scab mite.*

scabiosa. Scabious.

scabious. Any of the several plants of the genus Scabiosa, family Dipsacaceae. Species are popular garden flowers. Also called *Egyptian rose; mourning rose; pincushion flower; scabiosa.* See *sweet scabious; Caucasian scabious; grassleaf scabious; Japanese scabious.*

scabland. 1. Land which is characterized by bare rock surface or a very thin covering of soil, especially the dry lava plateau country of southeastern Oregon (U.S.). 2. Land which has many bare spots of Solonetz or solodized-Solonetz soils. See *scoria: 2; solodized soil; Solonetz (soil).*

scab mite. Any of the certain, almost microscopic, parasitic mites which cause scab or mange of man and animals. Common *scab mites* of man and animals belong to the genera Sarcoptes, Psoroptes, and Chorioptes. These mites puncture the skin with their sharp mouth parts and feed on the raw skin, completely covered over by the scabs which are formed. Affected animals have an intense itchiness, become unthrifty, lose their hair or wool on infested areas, and may even die. See *common scab (mange) of cattle; foot mange; sarcoptic mange.*

scab of apples. Apple scab.

scab of barley. See *cereal scab.*

scab of citrus fruits. See *sour orange scab; Tryon's scab; sweet orange fruit scab.*

scab of wheat. See *cereal scab.*

scaffold branches. The lateral branches produced when young fruit trees are pruned to obtain the desired size, shape, and strength of framework of the top.

scald. 1. A serious burn of plant leaves, fruit, limbs, and trunks which may result from exposure to intense sunlight or extreme heat: *e.g.,* cowpeas, beans, and soybeans may develop tan blotches which eventually cause defoliation; exposed or nearly ripe tomatoes may *scald* and wither during extreme heat, etc. Winter sun *scald* of tree trunks is a variation of this condition due to cold temperature following a warm period. A condition resembling *scald* occurs in stored fruit, but is in fact caused by chemical esters. Also called *sun scald.* See *apple scald.* 2. An eroded spot on hillsides where subsoil is exposed. 3. A bare spot in freshwater coastal marshes caused by the killing of plant life by increases in salt content of the soil through encroachment of water from adjacent salt water marshes. 4. To dip in hot water or to pour on hot liquid, as to *scald* a killed bird to loosen feathers. 5. To heat a liquid to a temperature just short of the boiling point, as to *scald* milk.

scalded cream. Devonshire cream.

scalding vat. A tank, vat, or barrel used to hold very hot water for the dipping and scalding of slaughtered hogs or poultry to cause hair or feathers to loosen for easy removal.

scale. 1. Any instrument or device used to determine weight. 2. Either of the pans of a balance. 3. A measure of dimension, concentration, or intensity. 4. A graduated series of steps or degrees on measuring devices. 5. An instrument or device on which graduated spaces for measurement have been stamped or attached. 6. A graduated list of prices, wages, etc. 7. Any thin flake that peels from a surface, such as from skin. 8. One of the thin, flat, membranous plates which forms the outer, protective covering of certain animals or parts thereof, as the shanks of birds, etc. 9. One of several rudimentary, specialized leaves which protects the buds of plants and deciduous trees in cold climates. 10. A thin, dry, membranous part of a plant or flower. 11. See *scale insect.* 12. The estimated sound content of a log or group of logs scaled by use of a given log rule. Net *scale,* the *scale* after deducting for defects. 13. To form into *scales.* 14. To cover with *scales* or with an incrustation. 15. To remove in thin layers, as by scraping. 16. To come off in *scales.* 17. To make according to *scale.* 18. To estimate the sound dimensions (content) of a log or group of logs, using an accepted log rule.

scale caliper. A log scale stick fitted with calipers graduated for the volumes of logs of different lengths and middle diameters.

scalecide. A spray material especially formulated for the control of scale insects which is generally applied as a dormant spray in early spring before buds open.

scale insect. Any of the several members of a widely diverse group of usually small, sucking insects, so called because of the appearance of the more widely known members, such as the San José scale. The males sometimes have wings; the females are scalelike, gall-like, or grublike in form, and clothed in wax. The forms most commonly seen (adult female scales) generally remain fixed on the host. The families Coccidae, Diaspididae, and Pseudococcidae are typical of scale insects and their allies. *Scale insects* infest most cultivated fruits, a large number of ornamental trees, shrubs, and greenhouse plants, and feed by inserting their slender mouth parts into the plant tissue and sucking the juices. Heavy scale infestations often kill branches or entire trees or shrubs.

scale line (line of constant proportions). In the manufacture of a single product, the varying quantities of production expected from varying quantities of the factors of production, the factors always being combined in the same proportion.

scale of points. The values assigned to the various parts or sections of animals or birds by which ratings are arrived at in judging or in selection for breeding.

scaler. One who determines the volume of logs.

scale stick. A graduated stick used for measuring the diameters and contents of logs.

scallion. An onion with a swollen base which lacks a true bulb as a result of nutritional deficiency, disease, or insect infestation.

scalloped-edged rolling colter. See *notched colter.*

scalloped salpiglossis. *Salpiglossis sinuata,* family Solanaceae, a semi-hardy, annual herb grown for its funnel-shaped, variegated flowers. Native to Chile.

scalp. 1. To remove the top surface of soil containing the matted roots of grass and other plants. 2. To remove all vegetative material down to the soil surface. 3. To remove the straw and other coarse materials in cleaning harvested grain.

scalper. A machine or device used in flour or feed mills for the removal of the outer seed covering of grains.

scalping. 1. In flour mills, separating the fuzzy growth from the ends of wheat berries by attrition with or without the use of suction or blower fans. 2. In milling, partially separating the product from the break rolls into broken wheat and break flour by means of an arrangement of sieves, bolts, or screens of varying degrees of fineness. 3. In horses, a striking of the front of the hind coronet, pastern or cannon against the front toe when running. 4. See *scalp: 1.*

scaly bark. See *psorosis.*

scaly butt. See *exocortis.*

scaly leg. A condition in fowls caused by a very small itch mite, *Knemidokoptes mutans* (R. & L.), family Sarcoptidae, which burrows under the scales on the shanks and feeds on the leg tissue. The severe irritation set up results in the accumulation of grayish, powdery or crusty material under the scales, their loosening and lifting, and the enlarging of the shank, all of which creates a roughened appearance.

scammony. Bigroot morningglory.

Scamorze. A small, soft, mild, plastic-curd, Italian cheese, made from goats' or cows' milk, which is oval with lappets at the top for handling, usually tied in pairs. It is eaten while fresh. Also called *Scarmorze.*

Scandinavian feed-unit system. A feed evaluation system widely used in Scandinavian countries for measuring the relative values of different feeds. In this system, one pound of barley is taken as the standard. The feed-unit value for any other feed is the amount of that feed which is estimated to have the same productive value as a pound of barley.

Scanno. A cheese with a buttery consistency and a burned flavor made from ewes' milk in Abruzzi, Italy.

scantling. A piece of yard lumber 2 ins. thick and less than 8 ins. wide.

scapula. The shoulder blade; the principal bone of the shoulder girdle.

scar. A blemish which results from an injury to a plant or animal.

Scarborough lily. Vallota.

scarecrow. An object, usually a crude human representation, which is set up in fields to frighten crows and other birds away from crops.

scarf. 1. The tapered or notched ends of pieces of wood that fit together to form a scarf joint. 2. The beveled cut on a log or stump which results from undercutting a tree in felling. 3. The incision on a stock in which the cion is inserted in propagation by budding.

scarified seed. Seed with normally water impervious coats which are made pervious to water by abrasion or chemical treatment so as to facilitate or hasten germination.

scarified soil. Soil loosened to a shallow depth by some narrow-bladed implement.

scarify. 1. To abrade, scratch, or modify the surface for increasing water absorption, as to *scarify* an impervious seed coat. 2. To scratch, loosen to shallow depth or crush a hard uneven surface to provide a more tillable soil. 3. To make superficial incisions or punctures in the skin.

scarious. Having a wrinkled, dry appearance.

scarlet bouvardia. *Bouvardia ternifolia,* family Rubiaceae, a shrub grown in the greenhouse for its very showy, red flowers. Native to tropical America.

scarletbugler penstemon. *Penstemon centranthifolius,* family Scrophulariaceae, a small, bluish-green, perennial herb grown for its red, tubular flowers. Native to western U.S.

scarlet bush. See *rat poison: 2.*

scarlet clematis. *Clematis texensis,* family Ranunculaceae, a hardy, woody vine bearing vase-shaped, scarlet flowers, grown as an ornamental on a trellis, fence, etc. Native to North America.

scarlet clover. See *crimson clover.*

scarlet eggplant. See *Ethiopian eggplant.*

scarlet eucalyptus. *Eucalyptus ficifolia,* family Myrtaceae, an aromatic tree grown for its white, pink, or scarlet flowers. Native to Australia. Also called *scarlet-flowered gum.*

scarlet firethorn. *Pyracantha coccinea,* family Rosaceae, a partly evergreen, large shrub grown for its white flowers and bright red fruits. Native to Europe and Asia. Also called *evergreen thorn; firethorn; fiery thorn; everlasting thorn.*

scarlet gaura. *Gaura coccinea,* family Onagraceae, a perennial herb grown for its long, pink, red, or white flowers. Native to North America. Also called *wild honeysuckle.*

scarlet hamelia. See *rat poison: 2.*

scarlet oak. *Quercus coccinea,* family Fagaceae, a rather cylindrical tree grown as a shade tree and for its autumnal scarlet foliage. Native to North America.

scarlet pimpernel. *Anagallis arvensis,* family Primulaceae, a prostrate, annual herb sometimes grown in the herb garden but commonly considered a weed. It has scarlet or white flowers that close at the approach of bad weather. Native to Europe. Also called *eyebright; poor man's weatherglass; poison chickweed; red chickweed.*

scarlet plume. Scarletplume euphorbia.

scarletplume euphorbia. *Euphorbia fulgens,* family Euphorbiaceae, a small shrub grown for its slender, drooping branches and its conspicuous scarlet appendages. It is occasionally grown under glass farther north for cut-flower purposes. Native to Mexico. Also called *scarlet plume.*

scarlet runner bean. *Phaseolus coccineus,* family Leguminosae, a tropical, perennial, herbaceous vine grown on a trellis for its constant and profuse bloom of showy, scarlet flowers. A white-flowered form is raised for its edible seeds. Native to America. Also called *fire bean; multiflora bean; Dutch case-knife bean; flowering bean.*

scarlet sage. *Salvia splendens,* family Labiatae, a tender shrub grown as an annual for its scarlet flowers. Native to Brazil.

Scarmorze. Scamorze.

scavenger. A mammal, bird, or other organism that feeds on carrion or dead organic matter.

scentless mockorange. *Philadelphus inodorus,* family Saxifragaceae, a tender, deciduous shrub grown for its white flowers. Native to North America.

schafta silene. *Silene schafta,* family Caryophyllaceae, a small herb grown for its rose or purple flowers. Native to Europe and Asia. Also called *moss campion.*

Schamser. A cheese made from cows' skim milk in the Canton of Graubunden, Switzerland.

schizocarp. A dry fruit that splits at maturity into one or more closed carpels, each containing one or more seeds.

schizont. A stage in the life cycle of certain protozoan parasites in which a cell, without fertilization, gives rise to a number of new individuals, called merozoites.

Schizophyllum rot. A rot of various shade, nut, and

fruit trees, as well as of sugar cane, caused by the fungus *Schizophyllum commune,* family Agaricaceae, characterized by the formation, on the bark, of leathery, gray sporphores with split and recurving gills.

schlachter. The person, a rabbi, who performs the ritualistic killing in the kosher method of slaughtering cattle and other animals.

Schlesische Sauermilchkäse. A cheese made in Silesia in much the same way as Hand cheese.

Schloss. A small, soft, ripened cheese made in Germany and northern Austria.

schooled horse. A horse, usually of a light-weight breed, that is trained or drilled for some special performance. See *dressage; jumper: 1; high school horse.*

school lands. In the U.S., public lands of a state set apart by the state or by the federal government for the establishment and maintenance of public schools. See *Land-Grant college.*

School Lunch Program. In the U.S., a federal program which began as a means of supplementing the lunches of needy children. It became a permanent law after the passage by Congress of the National School Lunch Act of 1946. Under this act the following types of government assistance are authorized: (a) Cash reimbursement for a part of the school's expenditure for purchases of food from local sources. (b) Foods of special nutritive value purchased by the U.S. Department of Agriculture and distributed to participating schools in accordance with need. (c) Part of the food acquired by the U.S. Department of Agriculture under its surplus removal operations, distributed to all schools serving lunches on non-profit basis.

school marm. A school teacher (rural).

School Milk Programs. In the U.S., the special school milk programs which were established under the Agricultural Act of 1954, to move more milk directly into consumption channels by assisting states and local communities to increase the serving of milk in schools. This is accomplished by reimbursing schools for a portion of the cost of additional milk served. A base is established for each participating school representing former consumption of milk by children in the school. Schools are reimbursed up to a maximum of four cents for each half-pint served in addition to the base. Schools not serving lunches under the federal school lunch program are reimbursed up to a maximum of three cents per half-pint served to children.

school section. In the U.S., a section of land, 640 acres, given by the federal government to a state for the support of public schools. In the General Land Office Survey this was Section 16 of a township.

schorf. Corky scab of potato.

Schottengsied. A whey cheese made for home consumption by peasants in the Alps.

Schreber watershield. *Brasenia schreberi,* family Nymphaeaceae, a world-wide, aquatic herb grown in the garden and park pool for its purple flowers. Also called *watershield; egg bonnet.*

Schutzenkäse. A Romadur-type cheese made in Austria.

Schwarzenberger. A Limburger-type cheese made in southern Bohemia, western Hungary, and Austria.

Schwedler maple. A horticultural variety of Norway maple (*q.v.*) which has dark red foliage throughout the early growing period. It is grown as a shade tree and as an ornamental.

Schweitzer cheese. See *Swiss.*

scimitar boning knife. A knife of medium length with an upward curving (scimitar-shaped) blade which is adapted for boning, pelting, and skinning of slaughtered animals.

scion. Cion.

scion bud. Cion.

sciophilous. Designating a plant which thrives in the shade.

sciophyte. A plant which grows well on, or is adapted to, shady sites.

scirrhous cord. In castrated animals, abnormal swelling of the cut end of the spermatic cord caused by infection. It is usually accompanied by discharge of thick, white pus from the castration wound, and may be chronic.

sclera. The tough, white, supporting covering of the eyeball which encompasses all of the eyeball except the cornea.

sclerocystosis. An abnormal condition of citrus fruits which usually starts near the stem end and extends first to the region close to the core, causing the juice sacs to take on a clear-white, granular appearance and become hard, firm, and finally dry and woody. This condition, which is not fully understood, is found most often in Valencia oranges and with fruits picked late in the season. Also called *crystallization; granulation; dry end.*

sclerotia. Resting bodies of fungi consisting of hardened masses of mycelium from which fruiting bodies may develop.

sclerotial root, stem and crown blight of soybeans. A blight of soybean in which the roots, crown and basal portion of the stem are invaded by various fungi resulting in yellowing or wilting and the ultimate weakening or death of the plant.

Sclerotinia drop. A fungous disease of lettuce caused by *Sclerotinia sclerotiorum* and *S. minor,* family Sclerotiniaceae, characterized by a soft, watery rot of the stem and leaf base, and sudden collapse of the plant. The entire plant is a soft, watery mass and finally becomes brown and dry.

Sclerotinia root rot and crown rot of clover. A fungous disease of clover caused by *Sclerotinia trifoliorum,* family Sclerotiniaceae, characterized by brown leaf spots overrun by a mass of white mycelium. Spring symptoms are a soft rot of the crown and roots.

Sclerotinia rot. Any rot caused by *Sclerotinia* spp., family Sclerotiniaceae, including certain rots of fruits, garden vegetables, etc.

Sclerotinia tip blight. A fungous disease of cranberries caused by *Sclerotinia oxycocci,* family Sclerotiniaceae, characterized by the tender spring shoots turning brown and withering, a grayish powder on the dead stems, a white, cottony mass of mycelium in the berries which later forms a hard, black sclerotium.

Sclerotinia wilt. A disease of alfalfa caused by *Sclerotinia trifoliorum*, family Sclerotiniaceae, characterized by small, black, tough sclerotia in the diseased plant. Also called *wilt of alfalfa.*

Sclerotinia wilt of beans. A disease of kidney beans in northeastern Colorado (U.S.), especially in irrigated fields, caused by the fungus *Sclerotinia sclerotiorum*, family Sclerotiniaceae, in which the plants are killed off in scattered patches, and the ground is covered with a white, moldy growth on which the fruiting apothecia of the fungus are frequently found. The pods of the infected plants frequently rot in transit to the cannery. Also called *white mold of beans.*

scoke. Pokeberry.

Scolecotrichum leaf blotch of oats. A fungous disease of oats caused by *Scolecotrichum graminis* var. *avenae*, family Dematiaceae, characterized by grayish-brown to dark brown linear lesions on the leaf blade and sheath.

scolex. The anchoring or holdfast organ of a tapeworm which serves for the attachment of the parasite to the intestinal lining. Also called *head.*

scoliid wasp. Any wasp of the genus Scolia, all species of which lay their eggs on the larvae of various beetles. These wasps are important in the control of certain destructive beetles, such as the Asiatic or Oriental beetle.

scoop. 1. A large ladle; a bucket or vessel with a handle, used for dipping, gathering, skimming, etc. 2. A shovel-like implement used for dipping, gathering, digging out, or shoveling, as a grain *scoop*, cranberry *scoop*, curd *scoop*, etc. 3. A small implement (usually curved) used for cutting, gouging, or removing an amount of material, as an ice cream *scoop*. 4. A mechanical implement with a large bucket used for the removal or conveying of earth, liquids, etc. 5. The amount of material which is obtained with a *scoop*. 6. To remove, to take out from, to gather or to empty with a *scoop*.

scoop shovel. An implement consisting of a broad, flat, moderately deep ladle fitted to a D handle used for handling grain, coal, sand, gravel, etc.

scoop wheel. A wheel with scoops or buckets on its perimeter used for removing water, earth, etc.

scoot. 1. Hardwood lumber inferior in quality to No. 4 common. 2. A sled used in dragging logs. See *dray: 2.*

scooter. A narrow plow with a short V-point used for opening furrows or breaking up soil between crop rows. Also called *shovel; bull tongue; dolly.*

scorch. 1. The appearance of seemingly burnt or brown irregular patches on foliage, bark, or fruits as a result of heat, excessive sunburn, lack of water, action of bacteria or fungi, improper use of insecticides, fungicides, or some nutrients. 2. The brownish discoloration or the burning of foods, etc., in the process of cooking, drying, or by other use of heat. 3. See *anthracnose.* 4. To burn superficially; to give the appearance of a burn, as to brown with heat.

scorch disease. See *anthracnose.*

scorched. In butter grading, designating a butter with a *scorched* flavor. Such butter is rated B.

scorched flavor. An off-flavor characterized by a lightly burned or cooked taste. In dairy manufacture, also called *intensified cooked flavor.*

scorched neutralizer. A flavor as of being scorched; a defect of pasteurized cream or butter which may result from the cream not being properly standardized for acidity.

score. 1. A mark, scratch, line, incision, etc., especially one made for keeping tally. 2. A numerical valuation or rating as to the merit of an animal, a product, etc. 3. The record of points made by contestants in a contest or test. 4. To keep account by notches, lines, or figures on a tally or *score* card. 5. To keep *score* in a contest. 6. To mark with lines, notches, grooves, gashes, furrows, etc., as to *score* timber for hewing or squaring, hides in skinning, meat in cookery, etc.

score card. A numerical standard of excellence with definite values assigned to the various sections or aspects of an animal or product; used as a basis in judging.

scoria. 1. The slaglike clinker left after the smelting of metals. 2. A cellular or cinder-like rock of volcanic origin.

scoring. 1. Judging on the basis of a score card. 2. Marking with cuts, notches, slashes, etc. 3. Lining up horses preparatory to the start of a race (county and state fair races, etc.). 4. Making one or more cuts in the bark as deep as the cambium without removing the bark. The cuts may be directly or spirally around, or lengthwise on the trunk or branch. *Cf. ringing.*

scorpion. Any of the various arachnids of tropical and sub-tropical regions. They have a segmented body up to 4 or 5 ins. long, with a poisonous sting at the tip of the tail. Living on insects, they are probably more beneficial than harmful.

scorzonera. Blacksalsify serpentroot.

Scotch bottom shoe. A horseshoe used on draft horses.

Scotch broom. *Cytisus scoparius*, family Leguminosae, a hardy shrub grown for its bright yellow flowers. Native to Europe. Also called *common broom.*

Scotch chain harrow. See *chain harrow.*

Scotch cottonthistle. *Onopordum acanthium*, family Compositae, an annual or biennial herb which is the national floral emblem of Scotland. It bears pale purple flowers and is grown in flower gardens for its sub-tropical effect. Native to Europe and Asia. Also called *bull thistle; Scotch thistle; Argentine thistle.*

Scotch crocus. *Crocus biflorus*, family Iridaceae, a perennial herb grown in the rock garden for its early spring, purple-tinged flowers with yellow throats. Native to Europe and Asia.

Scotch elm. *Ulmus glabra*, family Ulmaceae, a smooth-barked, cultivated tree growing up to 120 ft. or more in height. There are many garden forms (including dwarf types) widely planted in lawns and parks. Native to Europe and Asia. Also called *wyche elm.*

Scotch ham. A fresh ham that is skinned, trimmed of most fat, boned, and mildly cured according to a particular formula and practice followed in some parts of Scotland.

Scotch heather. *Calluna vulgaris,* family Ericaceae, an evergreen, low shrub, of many horticultural varieties, grown for its foliage and white, crimson, golden-yellow, pink, deep red, or purple flowers. Native to Europe and Asia.

Scotch hobble. A restraining device consisting of a rope passed over a horse's neck in front of the shoulders and tied as a loosely fitting rope collar. The slack end of the rope is passed under a hind pastern and back through the loop, enough slack being taken up to pull the foot from the ground.

Scotch pine. *Pinus sylvestris,* family Pinaceae, a hardy, evergreen tree grown as an ornamental, a specimen, in windbreaks, for Christmas trees, and for its lumber. Pyramidal when young, it becomes irregular with age. It grows rapidly on acid, poor, sandy soil, and is resistant to cold, drought, and dry winds. Native to Europe and Asia. Also called *Danzig pine; Riza pine; Scots pine.*

Scotch pink. Grass pink.

Scotch rose. *Rosa spinosissima,* family Rosaceae, a hardy, erect shrub grown for its white, pink, or yellow flowers. Native to Europe and Asia. Also called *Burnet rose.*

Scotch thistle. Scotch cottonthistle.

Scotch type. A breed-type of cattle characteristic of Scotland and different from the same breed elsewhere: *e.g.,* the Scotch-type Shorthorn differs from the English-type Shorthorn; and the Scotch-type Ayrshire differs from the American-type Ayrshire.

Scotch whisky. An alcoholic beverage prepared from barley malt dried over a peat and coke fire, giving the grain its smoky flavor which passes into the mash and upon distillation, into the whisky. It is generally distilled in a pot still although some is rectified and blended with that from other pot stills. Some *Scotch whiskies* are distilled in patent stills and in this case the mash is composed of unmalted cereals and barley malt. These whiskies do not have the smoky taste.

Scots pine. See *Scotch pine.*

scour. 1. To abrade and flush as by the action of rapidly moving water on stream beds. 2. To cleanse dirt, grease, etc., by rubbing or scrubbing, as to *scour* wool. 3. To cleanse the bowels of an animal by purging. 4. In plowing, to pass through the ground cleanly without any soil sticking to the moldboard. 5. To rub off the flesh sticking to a hide. 6. See *scours.*

scoured basis. See *on the clean basis.*

scoured wool. Wool from which natural impurities and dirt have been removed.

scour erosion. Erosion caused by the scouring action of flowing water.

scouring. 1. The act of removing soil, etc., by water erosion. 2. The process of washing wool by scrubbing with soap, hot water, and alkali. 3. A diarrhea condition of an animal.

scouringrush. *Equisetum hyemale,* family Equisetaceae, an evergreen, perennial, rushlike herb, found wild on moist, springy grasslands, and along roadsides. The epidermis of its stems is very harsh and was formerly used for scouring purposes. It is poisonous to animals, especially when fed in hay. Native to the North Temperate Zone. See *field horsetail.*

scourings. Material removed in the scouring process; specifically that removed in scouring grain in milling. These are portions of the cuticle, brush, white caps, smut and other materials separated from the grain in the usual commercial process of scouring.

scours. A persistent diarrhea in animals. See *bloody scours; calf scours; common scours.*

scrag end. The nape or back part of the neck, especially in a sheep. Also called *scrag.*

scrambling rocket. Hedge mustard.

scrape. 1. A tillage implement; a sweep. 2. A spreading wing attachment for a cultivator. 3. In turpentine operations, oleoresin remaining on a tree from which volatile oils have evaporated. 4. To collect *scrape* from trees in turpentine operations.

scrape iron. In turpentining, an instrument used in gathering scrape from faces.

scraper. 1. A tool or device with which something is cleaned, leveled, etc., as a shaped piece of wood or metal used to remove sweat from a horse, an attachment which removes soil from a disk or roller while it is in operation, or a metal piece to clean butcher blocks, etc. 2. A large steel pan or scoop used in excavation or to even the surface, which is drawn, dragged, or pushed over a surface by horses or tractors. 3. See *bell scraper.*

scraper hiller. A blade or sweep attachment used on cultivators for hilling.

scrap grade. A grade of tobacco that is essentially a by-product from handling leaf tobacco including floor sweepings and other waste tobacco material except stems.

scrapie. A disease of sheep which produces twitching, intense itching, excessive thirst, weakness, and finally paralysis in the animal. The cause has not been determined.

scraping. The portion of the sweet corn kernel which is scraped from the cob after the main part of the kernel has been cut off. This is added to the previously cut kernel in the cannery process of producing cream style corn.

scrapple. A food product prepared from the edible parts of hog heads, hog skins, livers, corn meal, rye flour, salt and spices. The heads are cleaned by removing the eyes, ear tubes, teeth, etc., followed by subsequent washing and cooking until the meat falls off. The cooked meat is ground and corn meal added together with the spices and salt. The mixture is further cooked to a thick consistency, placed in molds and allowed to set.

scratches. An eczema-like condition of the skin

on the fold or hollow of the fetlock of horses which is accompanied by inflammation and tenderness of the affected part. It occurs on horses that are kept in damp, unsanitary stables, or are long subjected to mud and filth without proper cleaning. An advanced stage of this affection, in which fissures across the skin are formed with a discharge of a foul-smelling pus, is known as *cracked heel; grease; greasy-heel.*

scratch feed. The grain part of a ration for poultry which consists of cracked or whole grain (corn, wheat, oats, etc.) or a mixture of such grains. *Scratch feed* is fed in hoppers, in the litter or, with range poultry, sometimes on the ground. Also called *scratch grain.* See *all mash; mash concentrate.*

scratchgrass. Alkali muhly.

scratch up. To use a harrow lightly on a field, as to *scratch up* a light stand of alfalfa for reseeding.

scree. Hillside talus; a mass of stony detritus on the slope which results from landslips and gravitational or creep movement of soil. (More commonly used in the U.K. than in the U.S.)

screen. 1. A sieve or grating, such as a frame covered with meshed wire or fabric, or a perforated plate, etc., which is used for separating the finer or coarser parts of soil, gravel, grain products in milling, seeds in cleaning, etc. 2. Any form of grating used to prevent invasion or escape of insects, reptiles, fish, or other kinds of animals. 3. A construction or planting used to conceal an unpleasant sight. 4. To consider a group for the purpose of selecting a relatively small number of individuals. 5. To sift by shaking through a *screen.* 6. To conceal with, or as with a *screen.*

screened corn chop. A livestock feedstuff which is the coarse portion of corn chop, ground corn, or cracked corn from which most of the fine particles have been removed.

screened cracked corn. Screened corn chop.

screened ground corn. Screened corn chop.

screen filter. A filter made out of fine wire mesh. See *filter: 1.*

screening host. A plant into which the components of a virus complex or mixture are introduced in order that one or more of the components can be eliminated.

screenings. 1. A mixture of mill or elevator-run materials, or a combination of varying amounts of materials, obtained in the process of cleaning grain or seed. This is usually waste material unless recleaned. 2. Inferior, light, or broken grain or seed which is obtained in the process of cleaning grain or seed, or recleaning *screenings.* 3. The refuse which is screened from wheat in its preparation for milling. 4. Second grade or inferior rice.

screw cap. A screw-top closure of metal or molded plastic which is lined with lacquered paper, a rubber gasket, etc., to form a tight seal when screwed firmly onto bottles, jugs, or jars.

screw conveyor. A mechanical contrivance consisting of a single plate or double plate helix which is formed about a shaft in a spiral and whose auger-like channels push materials along a trough or through a tube, as with grain, silage, etc. Also called *spiral-auger conveyor.* See *auger: 1.*

screw crank depth regulator. A mechanical device consisting of a screw spindle with a crank on one end to permit the making of adjustment by turning, used on farm implements for adjusting working parts, such as for regulating depth of plows, harrows, tool gangs, etc.

screwpine. Designating a tree of the genus Pandanus, family Pandanaceae, comprising a large group of plants, ranging from small shrubs to tall trees. They are not at all related to the pines but are monocotyledonous subtropical to tropical species. *P. utilis* (common *screwpine*), which may attain the height of 60 ft. is very useful in the tropical regions where it grows wild or under cultivation, and some species produce edible fruits and leaves used for weaving seats, making hats, cordage, etc. A very few species are cultivated for ornament, and these are mainly grown as outdoor plants in southern U.S. *P. veitchi* (Veitch *screwpine*), which has striped leaves up to 2 or 3 ft. in length, is grown by florists as specimen foliage, for potted plants, and for accent points in window boxes. Native to India, Africa and Australia.

screw press. An apparatus used for applying pressure by means of a ram which is forced forward by the turning of a long screw spindle. It is used in pressing fruit to extract juice, in baling cotton, etc.

screw-press cottonseed oil. Crude cottonseed oil.

screw pump. A pump which consists of an inclined or vertical shaft with many propeller-like blades attached at the lower end of the shaft which rotates in a cylindrical casing. It is particularly adapted to low-lift conditions, such as exist in coastal drainage districts and organic soil areas. Also called *propellor pump.*

screw-top. See *screw cap.*

screw-worm. The larva of the fly *Callitroga hominivorax*, family Calliphoridae, occurring in the warmer regions of America which lays its eggs at the edges of sores and wounds of animals. The larvae, or maggots, are able to penetrate the flesh of the host by two hornlike mouthparts and, if left uncontrolled, will go deeply into living tissue causing serious results that are often fatal.

scribe marks. In a bulk milk tank, permanent marks placed on the inner or outer tank, usually at or near the four corners. The marks, if placed on the inner tank, are used to indicate a level plane of the bulk milk tank when the surface of a volume of water touches all four marks simultaneously. The marks are placed on the outside of the tank as a guide in placing a standard carpenter's spirit level.

scribe saw. A saw designed for cutting meat and bone which is used for cutting up carcasses, especially for making the longitudinal cut through the vertebrae.

Scribner decimal rule. In forestry, the Scribner rule which is modified by rounding off the last digit to the nearest 10 and dropping the cipher. (An official rule of the U.S. Forest Service). *Cf. Doyle rule.*

Scribner rule. In forestry, a diagram rule for determining the board measure of logs, one of the oldest in existence, which assumes 1-in. boards and ¼-in. kerf, makes a liberal allowance for slabs and disregards taper. (It is the official rule in many parts of the U.S.) *Cf. Doyle rule; Doyle-Scribner rule.*

scrotal hernia. A protrusion of loops of intestine or other viscera into the scrotum through an enlarged inguinal ring or through an accidental opening in the inguinal region which results in the scrotum being greatly enlarged.

scrotum. In most male mammals, the pouch of skin containing the testicles and related structures.

scrub. 1. Land which under natural conditions supports only shrubs or a dwarfed, stunted growth of trees. The soils are generally infertile sands, but *scrub* may appear on various kinds of sites. 2. The stunted trees or shrubs, often in dense stands, on such land. 3. An animal of inferior breeding or condition. 4. Designating an inferior product, animal, etc.

scrub bull. See *scrub: 3.*

scrub-growth. A poor, stunted growth of vegetation. See *scrub: 1.*

scrub oak. Any of the several dwarf oaks (*Quercus* spp.) which occur in various areas of the U.S., such as *Q. ilicifolia* of the northeastern states; *Q. myrtifolia,* or myrtle oak, of the southern states; and *Q. dumosa,* or California scrub oak, of the Rocky Mountain region. These species are usually of little value except for firewood and browse.

scrub pine. 1. *Callitris calcarata,* family Cupressaceae, a black or red pine which grows on stony hill-slopes and ridges. Twigs of this tree have been used as a vermifuge for horses. Native to Australia. 2. Any of several dwarf pines of the U.S. which often thrive on stony or arid land. See *jack pine; lodgepole pine; Virginia pine.*

scrub sire. See *scrub: 3.*

scud. In tanning, to scrape a dehaired and trimmed hide or pelt to remove any hairs, lime, or other materials remaining in the hair follicles.

scuffle hoe. A garden hand tool which consists of a straight, flat blade attached to a long handle. It is pushed back and forth so that the blade cuts just beneath the surface of the soil to break up the crust and destroy weeds.

scum. 1. Minute forms of lower plant life and impurities that gather on the surface of stagnant water. 2. Extraneous matter or impurities that rise to the surface of boiling or fermenting liquids. 3. The condensation of solids at the surface of boiled milk upon cooling.

scuppernong. A horticultural variety of the native bullace grape *(q.v.)* cultivated since early colonial times in southeastern U.S. This variety has a strong-growing vine, nearly round, triangular-toothed leaves, nonforked tendrils, and short, dense flower clusters. It bears an edible, amber-green fruit and is grown on arbors for its edible fruit and shade.

scur. A small, rounded, horny growth attached to the skin on the head of a polled animal; considered a hereditary unsoundness.

scurf. 1. Any scaly matter or incrustation on a surface. 2. A roughening or browning of a smooth plant surface which is more pronounced than russeting: *e.g.,* apple scab. 3. A disease of sweet potato, which results in brown discoloration on the surface of underground parts, caused by the fungus *Monilochaetes infuscans,* family Dematiaceae.

scurfy. Designating a condition of the skin in which the upper layer becomes dry and comes off in scales.

scurfy scale. *Chionaspis furfura* (Fitch), family Diaspididae, a scale insect which appears on the tender bark of a tree, sucking plant juices and causing serious injury. It infests apple, pear, quince, mountain ash, and other common deciduous trees and bush fruits.

scurvyweed. *Cochlearia officinalis,* family Cruciferae, a biennial herb grown as an annual for its edible leaves used in salads. It was used as a remedy for scurvy in earlier times. Native to the North Temperate Zone. Also called *scurvy grass.*

scutched fiber. The fiber, including tow, from which the woody tissue has been removed by scutching. See *scutching machine.*

scutch grass. Bermudagrass.

scutching machine. A machine or device which beats off the woody tissue (scutch) from flax or hemp and separates it from the long, unbroken fiber.

scutellum. The cotyledon of the grasses, sometimes referred to as the seed leaf.

scutum. The dorsal shield; a shieldlike structure covering all or a part of the dorsum (the whole dorsal surface of the body) of ticks and related animals.

scythe. A long, sharp, curved blade attached to a long bent handle, used for mowing or cutting grass and weeds by hand.

seagrape. 1. Any of the several plants of the genus Coccolobis, family Polygonaceae. 2. C. *uvifera,* a shrub or small tree grown for its edible, grapelike fruit used in jelly. Native to tropical America. Also called *shore-grape; uvadel mar.*

sea-holly. See *seaholly eryngo; soft acanthus.*

seaholly eryngo. *Eryngium maritimum,* family Umbelliferae, a perennial, flowering herb grown for its pale blue flowers. Native to Europe. Also called *sea-holly.*

sea-island cotton. *Gossypium barbadense,* family Malvaceae, a woody plant grown as an annual for its long staple lint cotton used in fine cotton fabrics. Native to tropical America. Also called *American cotton; Andes cotton.*

sea kale. Crambe.

sea kale beet. Leaf beet.

seal. To make a bottle, jar, can, etc., relatively impervious to air and water.

sea lamprey. *Petromyzon marinus,* family Petromyzontidae, an eel-like, external parasite of fish that reaches a maximum size of 3 ft. It attaches itself to the fish with its sucker mouth and drains the blood. Although by nature a marine creature, the *sea lamprey* has become established in the Great Lakes (U.S.), and has done considerable damage to the game and food fish of this area. Attempts have been made to control the increase of these parasites by trapping and destroying the adults as they enter fresh-water streams to spawn.

sea-lavender. Any annual or perennial herb of the genus Limonium, family Plumbaginaceae. Species are grown for their flowers. Also called *Algerian sea-lavender.* See *wideleaf sea-lavender; Perez sea-lavender; notchleaf sea-lavender; Mediterranean sea-lavender; Suworow sea-lavender.*

sealed bid sale. A sales method sometimes used with livestock, etc., in which sealed bids are submitted to the seller with the privilege of purchase going to the highest bidder if the terms of sale are met.

sealed bottle closure. See *tamperproof closure.*

sealed can. A sheet metal container tightly sealed by soldering, crimping, or other means, so as to prevent leakage or contamination of its contents.

sealed check. See *body check.*

sealed closure. A tamperproof closure or seal for bottles, etc., which cannot be removed and replaced without detection.

sealing. 1. The action of high intensity rainfall which causes disintegration of unprotected soil granules, a consequent separation of colloids which fill spaces between larger particles, thus producing a seal or relatively impervious crust. 2. See *seal.*

sealing wax. A mixture, as of shellac and turpentine with an added coloring matter that is fluid when heated but quickly solidifies on cooling. It is used for making seals, as on bottles, jars of preserved fruits, etc.

sea lyme grass. Wild rye.

seam. A log defect characterized by a scar running across the annual rings.

seaside aster. *Aster spectabilis,* family Compositae, a low, summer-blooming, perennial herb grown for its bright, purple-violet flowers. Native to eastern U.S.

seaside bent. A strain of creeping bentgrass (*q.v.*) with long stolons and narrow, stiff, oppressed blades, usually densely matted, which is used for lawns and putting greens of golf courses, etc. Native to the marshy areas of the Atlantic and Pacific coasts of U.S. and Canada.

seaside daisy. Beach fleabane.

seaside goldenrod. *Solidago sempervirens,* family Compositae, a perennial, thick-stemmed herb, 5 to 8 ft. tall, grown for its clusters of yellow flowers. Native to North America. Also called *beach goldenrod.*

seaside mahoe. Portiatree.

seaside plum. Seagrape.

sea silene. *Silene maritima,* family Caryophyllaceae, a perennial herb grown for its white flowers. Native to Europe.

season. 1. A portion of the year, as a growing *season* when plants grow, the fire *season* when danger of fire is greatest, the grazing *season* when grazing is possible. 2. To bring to or develop into a condition of greater usefulness, as to dry lumber by natural or artificial means. 3. To flavor food with spices, salt, etc.

seasonal. Referring to a season or seasons, or to something that follows the periodic cycle of seasons; as with work or with most fresh food products, a *seasonal* pattern or *seasonal* demand.

seasonal breeder. An animal that breeds naturally during a certain season of the year and rarely at other seasons, as sheep, goats, etc., are *seasonal breeders* in the fall.

seasonal distribution. 1. In pasturage, the progressive grazing in a sequence of moves from one part of a range to another as vegetation develops. 2. In meteorology, the amount and kind of precipitation, wind movement, etc., in different seasons.

seasonal grazing. The grazing of a range only during a certain period or periods of the year which roughly correspond to one or more of the four seasons.

seasonal labor. In farming, the labor required at a particular season, as at fruit picking, grain harvesting, sheep shearing, etc.

seasonal price variation. Periodic fluctuations of prices that recur with about the same pattern each year.

seasonally protected plot. In range management, the fenced area within an enclosed plot which is designed to regulate the use of grazing during different parts of the season.

seasoning. 1. The process of drying (curing) lumber or other forms of wood for better utilization. *Seasoning* is natural when the drying is done by air or other natural means; in artificial *seasoning,* the drying is carried out by means of a kiln, oils, or electrical appliances. 2. Adding salt, pepper, etc., to meat and other food products.

seasoning defects. Defects in lumber which occur during the curing process, as those due to unequal shrinkage as the wood dries, or staining which results from the activity of fungi.

sea squill. Spring squill.

seated land. Land under use, *i.e.,* occupied, cultivated, improved, reclaimed, farmed, or used as a place of residence.

seat guide. A method of controlling cultivator gangs on a riding cultivator so as to follow the row and prevent plowing up of plants. The seat support is pivoted to the rear end of the tongue, and gangs are controlled by shifting the seat sideways by an easy swaying motion of the body aided by pushing with the feet against the gangs. Also called *wiggletail; pivot pole; pivot frame.*

seaurchin hakea. *Hakea laurina,* family Proteaceae, a tall shrub or small tree grown for its crimson flowers in dense, stalkless clusters. Native to

Australia. Also called *pincushion flower; pincushion tree.*

seaweed. See *kelp.*

sea wormwood. Canada sagebrush.

sebaceous. Referring to anything that secretes or resembles fatty matter; oily, greasy, as the secretions of the *sebaceous* glands of the skin.

sebastanplum cordia. See *myxa.*

Sebastopol geese. A novelty breed of domestic geese which has the primary and secondary wing feathers, the main tail feathers, and the feathers over the back much curled and twisted. The plumage is white and the bill and shanks are orange-colored. The adult gander weighs about 14 lbs. and the goose about 12. Also called *Danubian geese.*

sebesten. The mucilaginous fruit of the myxa *q.v.*

seborrhea. A group of diseases of the skin involving the sebaceous glands, characterized by an accumulation of dry scurf or by the formation of an excessive oily deposit.

Sebright. A breed of ornamental bantams which originated in England. Males are unusual in that their feathering is like that of the female (hen-feathered) and free from the usual male hackle, saddle and sickle feathers. Each feather of the surface plumage of both sexes is evenly laced with a narrow edge of lustrous black. The two varieties are the Golden and the Silver.

sec. Dry or nearly dry, not sweet, wine. Also called *brut.*

second. See *second grade.*

secondaries. Of a bird's wing, the innermost group of major wing feathers which are located on the second wing joint. They contribute importantly to the wing surface of a bird in flight. See *primaries; index feather.*

secondary branch. A smaller branch which grows out of a main branch, as with a plant or tree.

secondary bud. 1. A small bud adjacent to a larger bud; the smaller bud of an eye, as with grapes. 2. A semidormant bud which becomes activated later in time than a primary bud.

secondary feather. See *secondaries.*

secondary fertilizer components. Those elements in a fertilizer (other than nitrogen, phosphoric acid, and potash) that are essential to the proper growth of plants and that may be needed by soils. Some of these components are: calcium, magnesium, sulfur, manganese, zinc, copper, and boron. Also called *minor element.* See *essential element.*

secondary growth. 1. Growth which may arise from a secondary bud if the first shoots are destroyed, as with shrubs or trees. 2. A shoot or spur that grows from a bud or growing point the same year that bud or growing point is produced, as with shrubs and trees. 3. Growth from a cambium.

secondary infection. 1. An invasion by a second and different organism after a first organism has become established in a host animal or plant. 2. An infection resulting from inoculum produced by an organism after it is first established in a host plant.

secondary inoculum. 1. Inoculum which is produced from the organism causing initial infection. 2. Inoculum consisting of spores or other reproducing cells which is produced by the primary infecting organism; inoculum consisting of conidia produced after, and as a result of, initial or primary infection.

secondary insect. Any insect which follows, is associated with a primary form, or feeds upon plant tissue but which is incapable of initiating injury by itself.

secondary pairing. The loose association without chiasmata of chromosomes of similar size into groups containing more numbers than normally expected at somatic mitosis or at meiosis.

secondary polyploidy. An allopolyploid in which certain of the chromosomes in the basic set are present more frequently than others.

secondary root. A lateral branch of a primary or main root.

secondary scaffold. A large branch which is attached to a primary scaffold *q.v.*

secondary screw-worm. The larva of the screw-worm fly, *Callitroga macellaria* F., family Calliphoridae, which normally infests unburied carcasses and transfers its attack to living animals only when it has become excessively abundant. It frequently attacks wounds already infested by the true screw-worm and, in southern U.S., may be the cause of primary invasion of wounds in animals.

secondary shoot. A lateral branch of a primary or main branch both of which are produced during one growing season.

secondary succession. Plant succession which is subsequent to the destruction of part or all of the original vegetation.

second bottom. The first terrace level of a stream valley which lies above the flood plain. It is rarely or never flooded.

second cuts. See *fribs.*

second foot day. An abbreviation for a flow rate of one cubic foot per second for 24 hours: 86,400 cubic feet; 1,983,471 acre feet; 646,317 gallons; the runoff of .0372 inch from one square mile.

second foot per square mile. An abbreviation of cubic feet per second per square mile; the average cubic feet of water flowing per second from each square mile of area drained assuming that the runoff is distributed uniformly both as regards time and area.

second grade. A grade of fruit, vegetable, meat, etc., which is slightly below the established standards for first grade products. It is commonly referred to as No. 2 grade.

second growth. 1. A forest which originates naturally after removal of a previous stand. 2. The smaller trees which remain after a cutting or the residual trees available for another logging on the same area.

second joint. The thigh or meaty portion of the leg of a fowl.

second lock. In wool sorting, a bunch or tuft of wool of shorter length or with a higher percentage of dirt than is tolerated for first locks.

second piece. The coarse heavy pieces of wool which come from the hind part of a fleece.

seconds. See *second grade.*

second stomach. Reticulum.

second-year cane. In bramble-fruits (*Rubus* spp.), the canes which in their second year make fruiting shoots, bloom, bear fruit, and die. Sometimes called *floricane.* See *primocane.*

secret bit. See *check bit.*

secretin. A hormone produced in the intestine (duodenum) which controls the secretion of the pancreatic juices or enzymes. See *hormone.*

secretion. The metabolic act or process of synthesis and liberation of substances from cells or glands of animals or plants, as saliva, milk, etc.

secretory cell. 1. In wood anatomy, a cell directly concerned with the secretion of essential oils, gums, resins, etc. 2. Any body or plant cell which secretes a substance concerned with digestion, metabolism, etc. See *secretion.*

section. 1. A distinct part of a town, country, region, people, etc., as a *section* of the population or of the country. 2. A unit of the township in the U.S. General Land Office Survey of the public domain. The full *section* is one square mile, or 640 acres, and the full township consists of 36 *sections.* 3. In forestry, a logical or natural length of the fire-control line which is handled as a unit during fire suppression. 4. In beekeeping, one of several small frames of basswood placed in a hive in which the bees build surplus honey. A filled frame weighs from 12 to 16 ozs. See *section comb honey; comb honey section.* 5. Segments making up the interior of a citrus fruit.

sectional variegation. A non-infectious chlorosis in which the yellow and green areas are distributed over the leaves and stems, or leaves only, in blotches, spots, bands or stripes.

section box. See *comb honey section; section: 4; section comb honey.*

section comb honey. Honey that is produced in square or rectangular basswood frames 4¼ x 4¼ x 1⅞ or 4 x 5 x 1⅜ ins. See *comb honey section; section: 4.*

section harrow. Spike-tooth harrow.

section holder. Any of several means of supporting the sections in a super *q.v.*

section press. A device to press comb honey sections together firmly and squarely. See *section comb honey.*

Section 32 Program. In the U.S., this section amended in 1935 the Agricultural Adjustment Act by allocating 30 percent of the collections from tariffs on all imports to the Secretary of Agriculture for his use to encourage the export of agricultural commodities and to increase domestic consumption by diverting surplus commodities to low income groups. The three primary subsidy programs developed under Section 32 have been Food Stamp Program, School Lunch Program, and the Relief and School Milk Distribution Program.

sectorial chimera. See *chimera.*

sector of a fire. A length of the fire-control line handled as a unit for suppression, normally under the supervision of one man.

secundines. The afterbirth; the placenta *q.v.*

sedentary. Remaining in one place or confined to one spot; not migratory.

sedentary soil. A soil remaining in its place of origin from parent material and without addition of any transported material.

sedge. Any member of the botanical family Cyperaceae which is related to the grasses but is generally inferior for forage. Sometimes erroneously applied to grasses, as broomsedge, *Andropogon* spp.

sedge peat. A brown to black peat, mostly the partially decomposed remains of sedge vegetation. It is generally less acid than moss peat and of lower water-absorbing capacity than moss peat.

sedging. An oat disease characterized by a thickened stem at the stalk base; supposedly caused by a grub.

sediment. 1. Matter that separates itself from a liquid and settles at the bottom. 2. Earthy or organic matter deposited by water.

sedimentary peat. 1. Organic material deposited in bodies of water, as lakes, usually as a basal layer of peat beds. Finer in texture than the ordinary peat, it is of minor significance as soil. 2. Admixed or alternate layers of peat and mineral sediments formed in coastal marshes and alluvial valleys.

sedimentary rock. A rock composed of particles transported or deposited by water. The chief groups of *sedimentary rocks* are conglomerates (from gravel), sandstones (from sands), shales (from clays), and limestones (from calcium carbonate deposits).

sedimentation. Deposition of sediment by natural or mechanical means.

sedimentation basin. In bottom land drainage districts, a basin or pond covering several acres of ground at the foot of the bluffs and at the upper end of the drainage ditch. It checks the velocity of the water of small streams entering the drainage system, causing the heavier suspended material carried in suspension to be deposited in the basin rather than in the drainage ditch.

sedimentation rate. The speed, measured in minutes, at which particles, usually red blood cells, settle down in a tube. The *sedimentation rate* of red blood cells is an index of the variations in the cell count, blood viscosity, and other chemical changes in the blood.

sedimentation test. A test for visible dirt in milk. A sample of milk is forced through a pad or disk of cotton wool and the sediment collected on the straining device.

sediment yield. The amount of sediment carried out of a watershed by streams. *Sediment yields* of a drainage system may also be determined by measuring the accumulation of sediment in reservoirs of known age.

seed. 1. The embryo of a plant; also kernels of

corn, wheat, etc., which botanically are seedlike fruits as they include the ovary wall. 2. Propagative portions of a plant other than true *seeds*, as tubers, bulbs, etc. 3. Offspring or progeny. 4. Sperm or semen. 5. To place *seed* on or in the soil for the production of plants of the same species; to sow *seed*. 6. Overseed; to sow *seed* on an area where a crop is established, as clover on wheat, lespedeza on oats. 7. To produce *seed*. 8. To extract *seed* as to *seed* cherries or grapes.

seed analysis. Determination by a seed analyst of the percentage of pure seed, and the approximate number per pound or percentage of each weed or other crop seed impurity, and inert material present in a given lot or sample of seed. See *seed label*.

seed analyst. A person who examines samples of seed to determine the percentage of pure seed, the kind and percentage of inert material and the seed viability.

seed association. An organized group of seed growers which produces seed of known ancestry, guaranteed to be pure, viable, and reasonably free from seed-borne diseases. Such seed is commonly offered for sale as registered or certified seed.

seed attachment. 1. The attachment on a grain drill for sowing small forage seeds. 2. A seeding device that can be attached to a farm implement ordinarily used for some other purpose, as to a weeder.

seed ball. A matured cluster of single-seeded fruits grown together in a single mass (glomerule) in which the seeds are enclosed: one seed for each flower concerned, as in the beet.

seedbed. The upper portion of the soil prepared by natural or artificial means to receive seed and to promote germination and growth.

seedbed leveler. A mechanical device, as a roller or land plane, used to level the surface of a seedbed preliminary to seeding.

seed-borne disease. A plant disease which may be transmitted by seed, as the fungous diseases of corn caused by Diplodia and Fusarium and some virus diseases as the common bean mosaic.

seed-borne infection. An infection produced by seed-borne organisms.

seed box. On a seed planter or drill, the box holding the seed. It is equipped with adjustable outlets to regulate the quantity of seed to be sown. Also called *seed hopper*.

seed capsule. A dry, dehiscent, seed vessel composed of two or more cells, usually containing many seeds.

seed case. Any pod, capsule, or other dry, hollow fruit that contains seed; a pericarp.

seed certification. See *certification of seed; certification standard; seed association*.

seed-certifying agency. See *seed association*.

seed chaff. Chaff with 5 to 50 percent desirable seed by weight; chaff with enough desirable seed to be used for seeding.

seed chalcid. See *apple seed chalcid; clover seed chalcid*.

seed cleaner. Any of several devices which remove unwanted material from seed, especially one with screens and a fan used for the separation and removal of chaff, weed seeds, shriveled seed, dirt, etc., from seed. Also called *fanning mill*.

seed coat. 1. The hard outer layer of a seed; the protective covering, or integument. Also called *testa*. 2. The bran of milled cereals.

seed corn. Ears or kernels of corn used as seed and usually tested for germination and for evidence of lack of seed-borne diseases prior to planting. See *certification of seed; corn grader; germination test*.

seed-corn beetle. *Agonoderus lecontei* Chaud., family Carabidae, an insect which eats out the contents of seed corn and is sometimes the cause of corn failing to sprout.

seed-corn maggot. *Hylemya cilicrura* (Rond.), family Muscidae, the maggot of the cornfly which burrows into seeds of peas, beans, corn, infests shoots of melon, cucumber, potato, young plants of cabbage, radish, onion, and several other garden vegetables. It is especially injurious in cold, wet seasons on land rich in organic matter.

seed cotton. Unginned cotton which consists of the seed with the attached lint as picked from the boll.

seed-cotton drier. Any of several devices which dries damp seed cotton so that it can be successfully ginned.

seed crop. A crop grown primarily for its seed, as seed grains or as clover, alfalfa, etc.

seed-delinter. A machine which removes the short-staple fiber still adhering to cotton seed after ginning.

seed drill. A machine that sows grass and grain seed in spaced rows. See *grain drill*.

seed ear. An ear of corn intended for use as seed either while on the stalk (as in hybridizing) or after harvesting.

seeded. 1. Planted or supplied with seed; sown. 2. Supplied or planted with spores, enzymes, cultures, etc. 3. Having the seed or seeds extracted; as *seeded* cherries.

seeder. Any device used for sowing grass and grain seed on the ground surface to be covered later if needed by a separate machine. See *broadcaster; endgate seeder; multiple-row seeder; seed flask*.

seeder roll. A roll, composed of very fine circular saws, which operates against a revolving rubber roll and extracts the seeds from previously softened unseeded raisins in commercial raisin-seeding operations. See *flicker roll*.

seed farm. A farm on which the production of seed is the most important part of the farm business.

seed flask. A seeding tool which distributes a uniform amount of seed in direct seeding. Also called *seed horn*.

seed flat. A receptacle, usually a shallow, flat, wooden box filled with soil in which seeds are planted for later transplanting.

seed flax. Any of several varieties of flax developed for producing flaxseed.

seed-flow reducer. A wire or cast metal device used in the seed cups of grain drills when seeding small seed, as alfalfa, to reduce the rate of seed outflow. Also called *wire reducer.*

seed fly. *Agromyza lantanae,* family Agromyzidae, a fly brought to the Hawaiian Islands from Mexico for the biological control of a thorny pest plant, *Lantana camara.* This *seed fly* infests only the lantana, and its larvae eat the berries so that birds do not spread the seed.

seed furrow. A furrow in which potatoes, etc., are planted and which is gradually filled by early cultivation.

seed-furrow opener. A double-pointed shovel, runner-shoe opener, or similar part of a planter, which opens up a furrow into which the seeds are deposited.

seed-germ. See *embryo: 2.*

seed germination. The sprouting of seed; the resumption of embryo growth.

seed grader. Any of several mechanical devices used to grade seed according to size, shape, weight, or any combination of these qualities, to eliminate inferior seeds and to obtain greater uniformity. See *fanning mill.*

seed hopper. A box or hopper holding seed on a corn planter, an endgate seeder, etc., which is equipped with a mechanical feeding device and a means of regulating the quantity of seed to be sown.

seed house. 1. A building used for the storage of seed and for the cleaning, grading, and preparing of seed for planting or for sale. 2. A company that deals in seed. 3. The place of business of a seed dealer.

seeding. 1. The act of sowing seed. 2. The kind or amount of seed sown or to be sown, as a light *seeding* of oats for a nurse crop, or the stand of a *seeding* of alfalfa, etc. 3. In a broad sense, all methods of planting regardless of whether the crop is reproduced by seed or by vegetative parts. 4. The act of extracting seed or seeds, as the *seeding* of raisins. 5. The addition of crystalline nuclei to induce mass crystallization, as the *seeding* of sweetened condensed milk.

seeding lath. A device on drills used to get an even distribution of seed in the seedbed.

seeding of range and pasture. The establishment of perennial vegetation on depleted range and pasture areas and on crop land suited for forage production, by drilling or broadcasting seed and covering it by such methods as brush dragging, harrowing, or trampling livestock. See *aerial seeding.*

seeding time. The time of the year at which seed of any given sort is sown. Spring is commonly regarded as *seeding time* in the temperate zones although seeds are sown at various times of the year, as forage crops, or the fall sowing of wheat, rye, etc.

seed inoculation. The treating of seeds of alfalfa, sweet clover, and certain other legumes with cultures of nitrogen-fixing bacteria. These beneficial organisms live on the roots of legumes and store up nitrogen gathered from the air in nodules on the roots of the host plant to the benefit of the host and subsequent crops.

seed label. A label or tag, provided by law for each lot or parcel of seed exposed for sale, which gives the purchaser pertinent information as to the kind of seed and the approximate percentage by weight of pure seed present. Most seed laws require the *seed label* to show the percentage by weight of inert matter, of seeds other than agricultural seeds, and to establish definite limits on noxious weed seeds that may be present.

seed leaf. A cotyledon, or a leaf-like part of a seed, which appears as the original leaf of an embryo.

seedleaf group. See *broadleaf: 2.*

seedless grape. Any of several varieties of grapes that have been developed by selection and breeding for seedlessness.

seedless raisin. A raisin which is produced from a variety of seedless grapes, such as Thompson Seedless.

seed level. The depth at which seed of a given sort is planted.

seedling. 1. The early growth stage of a plant grown from seed as it emerges above the ground surface. 2. In forestry, a tree which has not been transplanted in the nursery. 3. A tree which has not reached a height of 3 ft. 4. In horticulture, a fruit tree grown from seed which has not been budded or grafted.

seedling blight. A disease which attacks and kills germinating seeds and seedlings before or after emerging from the soil. *Seedling blight* may be caused by species of Fusarium, Diplodia, Gibberella and other fungi, and may at times result in serious damage to cereal grains, cotton, flax, garden and greenhouse plants, and other seedlings. Also called *damping-off.*

seedling blight and crown rot of grasses. A fungous disease of grasses caused by *Pythium* spp., family Pythiaceae, characterized by light brown soft rot of the rootlets and roots and a pale green, stunted growth of the tillers.

seedling disease. See *damping-off; smut; seedling blight.*

seedling for rootstock. A seedling raised in a nursery to be used as rootstock which has not been grafted or budded. See *rootstock: 1; seedling rootstock.*

seedling pasture. A field of seedling wheat, rye, etc., in which animals are placed to graze.

seedling rootstock. A rootstock grown from seed which has been grafted or budded. See *seedling for rootstock.*

seedling sprout. A tree of sprout origin which results from the cutting of, or injury to, a small seedling, usually less than 2 ins. in diameter at the ground line. It is differentiated from coppice sprout or shoot because of its subsequent behavior.

seed mixture. Two or more kinds of seed which have been mechanically mixed.

seed pan. A clay pot, about half the depth of standard flower pots, which is used to contain shallow-rooted plants and bulbs and for pricking-out seedlings. It may be in any diameter from 5 to 12 ins.

seed pellet. A mixture of clay, fertilizer, rodent and insect repellents, and a single seed which is pressed together to form a small hard unit like a shot for precision planting.

seed piece. Any plant segment including a node which is used for planting, as with sugar cane, potatoes, etc.

seed plant. A seed-bearing plant; a spermatophyte.

seed plant (parasitic). See *parasitic seed plant.*

seed plate. See *corn planter plate.*

seed plot. An area of ground which is used in the production of seed.

seed pod. 1. Technically, any dry, seed vessel or fruit that splits open, as a pea pod. 2. In common usage, any dry seed vessel or fruit that contains seed.

seed potato. A potato used for propagation. In planting, it is used either as a whole tuber or, more commonly, halved or quartered into seed pieces of 1½ to 2 ozs.

seed processing. 1. Treatment of seed for removal of awns or adhering stems or wings to facilitate seeding by machinery. 2. The act of cleaning, testing, treating, and packaging of seed for sale.

seed production. 1. The growing of a crop primarily for its seed. 2. The developing of seed by fruiting plants. With annual plants, such as corn, seed is produced and the life cycle is completed in one season. Biennial plants require two seasons to attain full development and bear seed, after which they die: *e.g.*, common sweet clover and sugar beet. Perennial plants may live for several years and may produce seed each year after attaining maturity.

seed protectant. A seed preparation used to treat disease, such as seedling blight and damping-off.

seed purity. The percentage of developed, whole seed of the desired species in a given lot or sample.

seed rot. A rotting of planted seeds which results from a fungous infection. Cold, wet soil conditions and low vitality seed may be contributing factors.

seed rot of peanut. A seed rot of peanuts caused by any of numerous fungi (*Fusarium* spp., *Rhizopus* spp., *Penicillium* spp., *Aspergillus* spp.), which cause seed to rot when poor quality or mechanically damaged seed is retarded in germination by cold, wet soils or deep planting.

seed scarification. See *scarified seed.*

seedsman. One dealing in seed who may perform such services as cleaning, grading, testing, packaging, etc.

seed staining. The staining of imported red clover and alfalfa seed at the port of entry as required by federal law.

seed stalk. A plant stem on which flowers and seeds are borne.

seed stem. The stem and seed of grasses and legumes at about the stage of maturity for seed harvest. (If left unharvested until the seed shed, the stem is an empty seed stem. A stem producing no seed is a barren seed stem.)

seed stock. 1. Pedigreed or well-bred livestock, which is maintained for breeding purposes. 2. A specially selected strain of plants, or seeds thereof, which are to be used as parents of future generations.

seed testing. See *germination test.*

seed-testing laboratory. A laboratory, usually operated by a federal or state agricultural department, in which samples of seed are tested for purity, viability, and for conformity to seed laws and regulations.

seed tick. The newly-hatched, six-legged larva of a tick, especially of the cattle tick *Boophilis annulatis,* a one-host tick in which the larva, the nymph, and the adult are all found on cattle. Newly-hatched larvae are found on the ground or on grass, weeds, and other objects in fields where infested cattle have pastured. See *cattle tick.*

seed treater. Any of several contrivances used for the treating of seeds with disinfectant in disease control, such as a mechanical cleaner-grader-duster, oil barrel treater, etc.

seed treatment. The treating of seeds with chemical dusts, chemical solutions, heat, etc., for the control of seed-borne infections, such as smuts of cereals, or scab of potato, and for the protection of seedlings from damping-off fungi, etc.

seed tree. 1. A tree that produces seed. 2. In forestry, a tree reserved, in a cutting operation, to supply seed for reforestation.

seed tube. One of the tubes on a planter or seeder which conducts the seed from the hopper or feed cup down through the boot and furrow opener into the furrow.

seed-value. Abbreviation for true seed value: seed purity x percent germination x 100.

seedy. 1. Abounding with seed. 2. Containing seed. 3. Bearing seed. 4. Having run to seed. 5. Suggesting seeds. 6. Lacking in vegetative characteristics.

seedy cotton. Cotton, undergoing ginning, in which a considerable number of whole seeds as well as parts of seeds have been left mixed with the lint.

seedy cut. The mammary portion of a hog belly which, in the trimming of a hog carcass, is cut off for rendering because of its texture.

seedy toe. A separation between the wall and the sensitive laminae of the hoof at the toe of a horse.

seedy wool. Wool which contains a large amount of seeds and other vegetable matter.

seep. 1. A spot where water oozes out slowly from the soil and gathers in a pool or produces merely a wet place, usually on a hillside or at a hill base. See *seepage.* 2. To soak through pores. 3. To lose liquid by drainage.

seepage. Infiltration of water into soil or substrata from a source of supply, as precipitation, a water reservoir, or an irrigation canal. Such water may reappear at the surface as wet spots or seeps. See *infiltration; percolation; absorption loss.*

seepweed. Burroweed.

seepwillow baccharis. *Baccharis glutinosa,* family Compositae, a deep-rooted shrub which grows in arid and semiarid regions of southwestern U.S. Because of its thicket-developing habit, it is supposed to be an indicator of water. Also called *water wally; botamote.*

segment. Any one of the parts into which an object can be naturally separated or divided; a section of an orange, a flower, or an insect.

sego. *Calochortus nuttallii,* family Liliaceae, a perennial herb bearing an edible bulb; also grown for its white, purple-lined flowers. The *sego* lily is Utah's state flower. Native to western U.S.

segregation. 1. In genetics, the separation of a pair of allelic genes in the formation of germ cells. *Segregation* of allelic genes is manifested in the distribution among related individuals of the contrasting characters controlled by individual gene pairs. Also called *Mendelian segregation.* 2. The separation of species or types into isolated groups.

seguin tuftroot. *Dieffenbachia seguine,* family Araceae, a foliage plant grown in the greenhouse for its yellow-blotched foliage. Ingestion of the leaves results in temporary speechlessness. Native to West Indies. Also called *dumb cane.*

select. A standard grade of hardwood lumber, being the highest grade with respect to the amount of clear usable material (clear face) that can be cut out of the pieces.

selected. Chosen out of a number of similar products, animals, etc.

selection. 1. Choosing certain individuals for breeding purposes in order to propagate or improve some desired quality or characteristic in the offspring. See *natural selection.* 2. A system of land settlement in Australia similar to the Homestead Acts in the U.S., whereby a prospective land owner (selector) was allowed to purchase a limited area, upon which he must live and make improvements. The aim of the legislation, to settle on the land a body of agricultural yeomen, was an almost total failure, due to loopholes in the legislation and the actions of the pastoral leaseholders. See *squatter: 2.*

selection forest. A forest in which many, or theoretically all, age classes are represented.

selection thinning. In forestry, the cutting of the largest dominant trees to prevent their becoming wolf trees of relatively low value, and of the poorest overtopped trees before they become a total loss, provided such trees possess a positive net conversion value. Also called *Borggreve thinning.*

selective breeding. The breeding of selected plants or animals chosen because of certain desirable qualities or fitness, as contrasted to random or chance breeding.

selective contact weedkiller. See *selective herbicide.*

selective cutting. In forestry, the removal of certain trees selected for cutting so as not to interfere with the growth and development of the remaining trees.

selective grazing. The tendency for livestock and other ruminants to prefer certain plants and to feed on these while grazing little on other species.

selective herbicide. Any of certain chemical toxicants (usually applied as a spray) that will kill weeds in a growing crop or turf and produce little or no injury on the crop plant or turf.

selective spray. See *selective herbicide.*

selector. One who purchased land under the selection acts (Australia). See *selection: 2.*

selects. A standard grade of hardwood lumber.

selenium. A nonmetallic element, related to sulfur, rarely present as more than a trace in soil. Where it is present it is absorbed by plants and is toxic to grazing animals. See *blind staggers.*

selenium poisoning. A disease of grazing animals and man caused by ingesting plants grown on poorly-irrigated soils containing more than very minute traces of selenium. In horses, it is characterized by mange of the mane and tail, overproduction of hoof tissue, and in severe cases it causes death. Recently, arsenic has been found to have some use as an inhibitor of the disease. See *alkali disease; blind staggers.*

Selenophoma leaf blotch of grasses. A fungous disease of grasses caused by *Selenophoma bromigena,* family Sphaerioidaceae, characterized by small, brown flecks on the leaves early in the spring. The flecks enlarge to cause blotches.

self-binding harvester. A farm implement which cuts the standing grain near the ground and mechanically binds the grain-bearing straw into twine-tied bundles before discharging it.

self-blanching. Any of certain varieties of celery that loses its chlorophyll as it matures.

self-boiled lime sulfur. A home-made spray mixture in which sulfur is added to hot slaking lime (CaO). It is less caustic than lime sulfur and was a favored peach spray before the advent of wettable sulfur. See *lime sulfur; sulfur lime spray.*

self color. 1. One color. See *solid color.* 2. Natural color.

self-colored flower. A flower whose petals are of one color. Also called *solid-colored flower.* See *variegated.*

selfed. Self-pollinated. See *selfing; self-pollination.*

selfed-line. An inbred strain of plants or animals which presupposes several generations of inbreeding until the parent lines are reasonably pure.

self-feeder. 1. In poultry or other livestock farming, any feeding device by means of which the animals can eat at will: *e.g.,* a hopper which supplies feed by gravity to a box-like trough. 2. Any automatic mechanism for supplying materials to a given point in an operation.

self-feeding. A feeding practice in which feed is kept before the animals at all times so that they can eat at will; a full-feeding method. See *hand-feeding.*

self-feeding hopper. See *self-feeder: 1.*

self-fertile. Designating a plant which can fertilize itself by its own pollen and produce viable seeds.

self-fertility. The ability of a variety of fruit to produce viable seeds without the aid of pollen from another variety.

self-fertilization. See *self-fertile; self-pollination.*

self-fruitfulness. The ability of a plant variety to mature fruit without the aid of pollen from another variety; fruit production resulting from self-pollination.

self-heal. *Prunella vulgaris,* family Labiatae, a low-growing, mat-forming weed found in waste ground, roadsides, and lawns. Native to Europe and Asia. Also called *heal-all; carpenter's weed; blue curls.*

selfing. In plant breeding, the pollination of a stigma with pollen from the anthers of the same flower, plant or variety. Also called *inbreeding.* See *self-pollination.*

self-mulching soil. A soil that cracks deeply and becomes so granular at the surface when very dry that the granular mulch washes into the cracks when rains begin, the whole soil swelling enough as it becomes moist to force material upward between the former cracks. 2. A soil in which the surface layer becomes so well aggregated that it does not crust and seal under the impact of rain.

self-pollinate. To perform self-pollination *q.v.*

self-pollination. The transfer of pollen from the anthers of a flower to its own stigmas or to those of another flower on the same plant or a plant of the same asexually propagated variety. In a broad sense, the transference of pollen to flowers of the same variety. See *cross pollination.*

self-propelled harvester. Designating a farm implement which contains within itself the means for its propulsion; self moving, as a self-propelled combine, etc.

self-pruning. Natural pruning; the natural death and fall of branches from live plants resulting from light and food deficiencies, decay, insect attack, snow, wind, and ice.

self-rake reaper. A reaper with a rake attachment for sweeping a quantity of mowed grain sufficient to make a sheaf off the platform. (It preceded the self-binding harvester for harvesting small grains.)

self-seed. To disseminate its seed and reestablish its kind, as some plants self-seed; to sow itself.

self-seeded. Designating a plant or crop seeded automatically by parent plants; self-sown.

self-seeding. 1. The act of disseminating its seed by a plant or plant species. 2. Designating a plant which scatters its seed and reestablishes its kind.

self-spacing frame. One of several beehive frames so made that when pushed together, they stand a bee-space apart.

self-sterile. Designating a plant variety unable to produce viable seeds if only the pollen of that very plant or variety is available for pollination.

self-sterility. The inability of a plant variety to produce viable seeds if only the pollen of that plant or variety is available for pollination.

self-sucker. A cow that has the habit of sucking its own teats.

self-sufficing farming. A type of farming in which the production is mainly for the use of the farm family. See *subsistence farmer.*

self-unfruitfulness. The inability of a plant variety to produce fruit if only the pollen of that very clonal variety is used for pollination.

sell short. See *hedge: 3.*

selloa pampasgrass. *Cortaderia selloana,* family Gramineae, a tall, reed-like grass grown in warm areas for its attractive leaves and for its feathery flowers. Native to Argentina.

semen. Secretion of the male reproductive system containing the male germ cells, or spermatozoa, and substances from associated glands.

semen collection funnel. A thin rubber cone used in the collection of semen for artificial insemination. It is attached to a rubber vagina cylinder and conveys the semen to a collecting tube.

semen collector. Any device or receptacle used for the collection of semen from males of domestic animals. It may consist of an artifical cloaca, or vagina, a test tube, a small beaker, etc. See *artificial vagina.*

semen plasma. The fluid portion of semen.

Semental cattle. The most commonly used type of cattle in Russia. They are dual-purpose in character, weigh from 1,600 pounds to a ton, are mouse-grey to light fawn in color, with a rugged, heavy bone, and resemble the Brown Swiss in body conformation. Originating in Switzerland, they are the second most popular breed there.

semen vial. A small vial which is used as a container for shipping or storing semen.

semiaged cheese. A cheese cured by holding under controlled conditions in a storage room until it has acquired a medium degree of flavor and texture, as a medium Cheddar as contrasted to a mild or sharp Cheddar.

semi-annual. 1. Of plants, those which bloom or grow for a half year. 2. Recurring every six months.

semiarid climate. The climate, characteristic of the regions intermediate between the true deserts and subhumid areas, under which precipitation effectiveness is such that a vegetation of scattered short grasses, bunchgrasses, or shrubs prevails. The maximum of average annual precipitation in the cool semiarid regions is as low as 15 ins. The Thornthwaite precipitation effectiveness (PE) index ranges between 16 and 32. See *precipitation-effectiveness (P-E index).*

semiboiled soap. A soft soap made by treating the more liquid fats with potash and boiling in an open vessel until the desired consistency is obtained. The boiling period is shorter than in the preparation of hard soap and the glycerine is not removed.

semibright. In wool grading or classing, designating a fleece that has a dark or stained color appearing dirty and unattractive but which will scour out white. Such wool is characteristic of certain highly-colored soil areas.

semicactus dahlia. A type of dahlia (*q.v.*) with margins of most of the floral rays rolled back for less than half their length, the rays being broad at the base.

semicircular roof and frame. A type of building

construction for barns in which each upright roofing support is bent into a semicircle on a radius equal to one-half the width of the desired structure.

semiclustered. Designating a type of upland cotton whose fruiting branches are irregularly jointed and whose bolls are moderately clustered.

semihard cheese. Any of the several varieties of cheese, such as brick or Roquefort, which is intermediate in texture between the hard cheeses, such as Cheddar or Swiss, and the soft cheeses, like Camembert or Cream Cheese varieties.

semihardy. Referring to cold resistance: designating a plant that will not endure as much cold temperature as the more hardy species do. It is a relative term designating plants between tender and hardy in a given climate.

Semillion. A golden-yellow variety of the European grape (*q.v.*) to which the famous sauterne wines of France largely owe their character. In California, U.S., the variety has been best naturalized in the north coast region.

semiloose smut of oats. A fungous disease caused by *Ustilago avenae,* family Ustilaginaceae, characterized by the dark brown to black spore mass on the inflorescence and the smutted spikes appearing later than in the case of loose smut. Also called *black smut.*

semimonitor house. A type of barn building with a gable-type roof on which the slope, usually on the south side is lowered to permit the vertical placement of a window sash at the peak for the better admittance of sunlight, particularly in winter. Formerly used as a small building for housing pigs and chickens but now replaced by other types because of excessive loss of heat entailed.

seminal fluid. Semen.

seminal root. One of the several small roots which grow laterally from a germinating seed of corn and other cereals.

semiprivate general refrigerated warehouse. Any artificially cooled warehouse whose operator uses part of the space for storing his own commodities, in addition to storing others' food commodities for pay.

semiscald. Designating a method in which previously bled birds are immersed in hot water, usually 125° to 130° F., before plucking. It is used for the easy removal of feathers in the preparation of dressed poultry. Also called *subscald.* See *hard scald; dry pick.*

semisolid buttermilk. Buttermilk condensed by evaporation to about ⅓ of its original weight to the consistency of a thick paste. It is used as a supplement in hog and poultry feeding. Also called *buttermilk powder.*

semisolid corn distillers' grain with solubles. The residue from the manufacture of alcohol and distilled liquors from corn, or from a grain mixture in which corn predominates, obtained by condensing to a sirupy consistency the entire stillage contained therein. Used as a protein and vitamin supplement in swine and poultry feeds.

semisolid grain sorghum distillers' solubles. A livestock feed product obtained in the manufacture of alcohol and distilled liquors from grain sorghums, or from a grain mixture in which grain sorghums predominate, by condensing to a sirupy consistency the screened stillage obtained therefrom.

semisolid processed fish. A livestock feed supplement which is obtained by the hydrolysis with acid of whole fish or fish trimmings, or both, at controlled temperatures and pressures and thereafter evaporated to a soluble consisting of at least 50 percent solids.

semisolid rye distillers' grain with solubles. A livestock feedstuff which is the residue from the manufacture of alcohol and distilled liquors from rye, or from a grain mixture in which rye predominates, obtained by condensing to a sirupy consistency the entire stillage contained therein.

semisolid rye distillers' solubles. A livestock feedstuff which is a by-product obtained in the manufacture of alcohol and distilled liquors from rye, or from a grain mixture in which rye predominates, by condensing to a sirupy consistency the screened stillage obtained therefrom.

semisolid sweetened condensed milk. See *block milk.*

semituberous begonia. Any of several species of the genus Begonia, family Begoniaceae. Species have no tubers but store food in the base of the stems giving a tuberous effect. They are propagated by leaf-petiole cuttings. See *begonia.*

semolina. A foodstuff made from that part of durum wheat (*q.v.*) which is resistant to millstone action. The purified, coarse middlings are used in making macaroni and comparable edible pastes. Also called *semola bran.*

Seneca grass. Sweetgrass.

Seneca-snakeroot polygala. *Polygala senega,* family Polygalaceae, a perennial herb, 12 to 18 ins. high, with lance-shaped leaves and small greenish-white flowers in terminal spikes, whose roots are used medicinally. Native to North America.

Senecterre. A whole-milk cheese, first made in St. Nectaire, in the Department of Puy-de-Dome, France.

senior. Designating an age classification of animals in livestock shows: *e.g., senior* yearling heifer, *senior* boar pig, etc.

senior calf. 1. In dairy cattle shows, a dairy calf born between July 1 and December 31, inclusive, of the year prior to the show. 2. In livestock shows, a beef calf born on or after September 1 and on or before December 31 of the preceding year.

senior champion. In stock judging, the best animal of the older classes selected from the first prize winners.

senior pig. In swine judging, a pig farrowed on or after September 1 of the preceding year and on or before February 28–29 of the year shown.

senior spring pig. In swine judging, a pig farrowed on or after February 1 and on or before March 14 of the year shown.

senior yearling. 1. In judging beef cattle, a bovine born on or after May 1 and on or before December 31 of the second preceding year. 2. In judging dairy cattle, a bovine born on or after July 1 and on or before December 31 of the second preceding year. 3. In swine judging, a pig which was farrowed on or after September 1 of the second preceding year and on or before February 28-29 of the preceding year.

senna. See *Cassia.*

senna-pea. Swainsonpea.

sensitivefern. *Onoclea sensibilis,* family Polypodiaceae, a hardy, perennial, rather coarse fern which grows to a height of 2 to 4 ft. It is widely distributed and is common in low, wet meadows, swamps, and marshes. Instances of poisoning to horses have been reported in New England (U.S.), where *sensitivefern* is frequently associated with hay containing brake fern and equisetum. Also called *meadow brake; polypody brake.*

sensitive jointvetch. *Aeschynomene virginica,* family Leguminosae, a common, very troublesome weed pest of southern U.S. Also called *curly indigo.*

sensitive plant. *Mimosa pudica,* family Leguminosae, a low-growing, woody, slightly spiny, perennial, with small sensitive leaflets and rose to lavender flowers in tight clusters. Native to tropical America; cultivated in Florida, U.S. Also called *humble plant.*

sensorium. The whole sensory apparatus of the body, principally sight, smell, hearing, taste, and touch (which includes pain). The *sensorium* gathers all of the external, and some internal, impressions and conveys them to the brain or spinal cord or both for proper evaluation and response by the organism.

sentry ginkgo. A horticultural variety of the ginkgo *(q.v.)* that holds its limbs upright like the Lombardy poplar. It is grown for ornament. Also called *upright ginkgo.*

sentry maple. A horticultural variety of the sugar maple *(q.v.)* that has upward-turning branches, parallel sides, and narrow form. It is suitable for planting on narrow streets.

sentry palm. *Howea belmoreana,* family Palmaceae, a feather tree reaching 40 ft., widely grown as a decorative plant. Native to Lord Howe Island (Australasia). Also called *curly palm.* Known horticulturally as *Kentia.*

sepal. One of the leaf-like divisions of the calyx of a flower.

separate. 1. A group of soil particles in the same diameter group. 2. To put apart, as to *separate* weanling calves from mothers. 3. To disunite, as cream from milk.

separated milk. Milk from which all but a minute fraction of the fat has been removed by skimming or by mechanical separation. Also called *skim milk.*

separation layer. Abscission layer.

separator. 1. Any machine or apparatus for separating: *e.g.,* a cream *separator (q.v.).* 2. The threshing machine which separates grain from the head in which it grows. See *grain threshing machine.*

separator bowl. A high-speed, revolving part of a cream separator; the metal casing or bowl with its enclosed series of conical plates which revolve at high speed.

separator dropper. In cotton ginning, the apparatus for taking the seed cotton from the air in the belt distributor system. In general, it consists of a wire-mesh screen through which the air may pass but not cotton. Also called *vacuum dropper.*

separator slime. The visible dirt, cell debris, etc., separated out and deposited on the inner walls of the separator bowl in the process of cream separation.

separator spoon. In commercial egg-breaking, a spoon, just large enough to hold the yolk of an egg, into which the opened egg is dropped. The white separates from the yolk and drops to the cup below.

sepsis. A state of contamination or poisoning by pathogenic bacteria.

septate. 1. Containing or divided by a septum or septa. 2. Fungus hypha or spores which have cross-walls. 3. Any fruit in which the ovary is divided into sections.

septic. 1. Relating to, or caused by, the presence of disease-producing organisms, or their toxins, in the blood or tissues. 2. Putrefactive; putrefying.

septicemia. Blood poisoning; a disease condition which results from the presence of toxins or poisons of microorganisms in the blood. See *hemorrhagic septicemia.*

septic sore throat. A diseased condition of the throat caused by certain strains of Streptococcus bacteria. Epidemics of *septic sore throat* are very largely milk-borne. Milk may be infected by cough-spray of milkers, other persons handling milk, or from the udder of a cow infected from a human source.

septic tank. A tank in which sewage is deposited and broken down by the action of anaerobic bacteria into carbon dioxide, soluble, and insoluble materials. A *septic tank* is used for single or individual sewage disposal systems. It is essentially a sewage settling tank in which the raw sewage flows slowly through the tank allowing fermentation and digestion to take place, with the lighter effluents passing out into the disposal field for absorption. See *cesspool.*

Septomoncel. A hard, blue-mold cheese made from cows' milk to which a small proportion of goats' milk is sometimes added. It is made near St. Claude, in the Department of Jura, France.

Septobasidium canker. A lesion of the bark of the limbs of apple, pear, and some forest trees caused by *Septobasidium pedicellatum* and related species of fungi.

Septoria blight of celery. See *late blight of celery.*

Septoria leaf blotch of barley. A fungous disease of barley caused by *Septoria passerinii,* family Sphaerioidaceae. For symptoms see *Septoria leaf blotch of wheat.*

Septoria leaf blotch of oats. A fungous disease of oats caused by *Septoria avena,* family Sphaerioidaceae, characterized by mottled, light and dark brown blotches on the leaves and floral bracts, frequently resulting in a brown discoloration of the enclosed kernels.

Septoria leaf blotch of rye. A fungous disease caused by *Septoria secalis,* family Sphaerioidaceae, which produces symptoms similar to the Septoria leaf blotch of wheat *q.v.*

Septoria leaf blotch of wheat. A fungous disease caused by *Septoria tritici* or *S. nodorum,* family Sphaerioidaceae characterized by light green to yellow spots between the veins of the leaves. The lesions eventually spread to form light brown, irregular blotches.

Septoria leaf spot. A fungous disease of the tomato and other plants caused by *Septoria lycopersici,* family Sphaerioidaceae, characterized by numerous, water-soaked spots that become circular with gray centers and later show dark specks.

Septoria spot of citrus. A spotting of citrus fruit and leaves by species of Septoria, usually by *S. citri* Pass, family Sphaerioidaceae. On fruits of lemon and grapefruit and to a lesser extent on orange, spots ranging from buff-brown specks to sunken, hard, dry, dark brown to black areas covering a considerable part of the fruit surface appear.

septum. Any dividing membrane or other layer in plants and animals: (a) a partition separating two cavities; (b) a division wall in a compound ovary; (c) a cross-wall in a fungus hypha or spore.

septuple planting. See *hexagonal planting.*

sequela. The sickly affection left after an attack of a disease.

sequoia. Either of the two tall trees, giant sequoia and redwood (*q.q.v.*), genus Sequoia, family Pinaceae, characterized by oval, drooping cones and woody scales.

sequoia pitch moth. *Vespamima sequoiae* (Hy. Edw.), family Aegeriidae, a moth whose dirty white larvae infest branches and trunks and mine the cambium layer of pines, firs, redwood, and other conifers in Montana, Washington, Oregon, and California, U.S.

seral development. The change in vegetation through successive recognizable stages of development (plant succession) from a pioneer stage to a more stable community, the climax (*q.v.*) stage, which is in equilibrium with local conditions. See *climax dominant.*

seral dominant. Referring to the plant species which are dominant in successional stages in the biotic community. See *climax dominant.*

serape. 1. A Mexican saddle cloth. 2. A small, often brightly-colored blanket, worn as an outer garment (Mexico and southwestern U.S.).

Serbian spruce. *Picea omorika,* family Pinaceae, a tall, evergreen tree grown for its beautiful foliage. Native to Europe.

sere. The series of plant communities that follow one another on any given area of the earth's surface.

sereh. A virus disease of sugar cane, of unknown cause, characterized by red gum, staining of the vascular tissues, and development of adventitious roots from nodes. Endemic in Indonesia.

sericea lespedeza. Chinese lespedeza.

sericeous. Closely covered with soft, silky hair as a *sericeous* leaf.

sericite. A variety of muscovite mica found in soils in regions underlaid by schistose rocks.

series. A group of related results or values with a tendency toward a characteristic pattern or distribution.

series cutting. A method of cutting stands of trees in a sequence so as to provide for reproduction and protection. Also called *felling series.*

sero-agglutination test. The examination of serum to determine the presence of agglutinins in it. Used for detection of certain diseases.

serological. Designating certain laboratory tests used in detecting disease which involve the use of the liquid portion of the blood.

serological diagnostic test. The use of blood serum of animals in various tests which aids in detecting different diseases.

serology. The science which deals with serums and their actions.

serotinous. a. Designating a late-flowering or late-fruiting plant. b. Referring to certain plants or trees which delay the dissemination of seed until long after time of maturing.

serotype. A type of organism differentiated from other organisms of the same genus by using serological tests. An organism of a specific *serotype* will show a greater reaction with serum containing antibodies specific for it than with serum containing antibodies specific for another type of organism and vice versa.

serous. a. Referring to or resembling serum. b. Designating a body that produces or contains serum.

serpent cucumber. Snake melon.

serpent gourd. Edible snakegourd.

serpentine. A secondary rock mineral, a hydrous magnesium silicate. Soils residual from the weathering of the rock have the reputation of being infertile and are locally known as *barrens.*

serpentine layering. Plant propagation in which several portions of a branch are covered with soil leaving the intervening portions above the soil until rooting takes place. Usually used with vines. Also called *compound layering.*

serpentine leaf miner. *Liriomyza pusilla* (Meig.), family Agromyzidae, a fly whose maggot mines the leaves and makes long, slender, white, winding trails under the epidermis of many garden plants and a number of field crops.

serpent melon. Snake melon.

serpentroot. Any plant of the genus Scorzonera, family Compositae. See *blacksalsify serpentroot.*

serradella. *Ornithopus sativus,* family Leguminosae, an annual, semivining, leguminous herb somewhat like vetch in habit of growth. It is used for forage and green manure but is not extensively grown in the U.S. Native to Europe and Asia.

Serra da Estrella. A popular, soft cheese with a pleasing, acid flavor made in Portugal.

serrated. 1. Having tooth-like notches on the edge, as a saw. 2. Referring to plant leaves which have sawlike edges.

serration. 1. The state of being notched or toothed. 2. The projection of the epidermal scales which give a toothed appearance to the surface of wool fiber.

serum. (plural **sera**) 1. The clear portion of any animal liquid which is separated from its cellular elements. It usually applies to the amber-colored liquid called blood *serum* which separates in the clotting of blood from the clot and corpuscles. 2. The whey of milk. (The term is applied often erroneously to skim milk.)

serum neutralization. A test employed with certain viral infections to determine the presence or absence of specific antibodies in the blood serum of animals by mixing varying amounts of serum with a constant amount of known virus and determining afterwards by animal inoculation if inactivation of the virus has taken place.

serum separation. A defect in cultured cream in which a watery, wheylike layer is separated in the bottom of a glass container of the cream. Such a defect may be accompanied by a grainy texture of the product. Also called *wheying off*.

serve. To copulate with a female; to cover, as a stallion, bull, or other male animal.

service. The act of mating, serving, or covering. See *serve*.

serviceable unsoundness. Any disease or disorder of a horse which does not impair its usefulness. See *breeding unsoundness*.

serviceably sound. Designating a horse which may have certain defects of such nature that they do not interfere with fitness for the service for which it is sold.

service alley. A passageway or aisle for servicing, such as an alley in front of mangers for feeding stock or for access to the rear of animals for milking, cleaning, bedding, etc.

serviceberry. 1. Any of several species of the genus Amelanchier, family Rosaceae. Species are deciduous shrubs or trees. Also called *shadbush*. See *garden serviceberry; running serviceberry; shadblow serviceberry*. 2. *A. laevis,* a tree reaching a height of 40 ft. grown for its early-spring, drooping clusters of white flowers. Also called *Allegheny serviceberry*. Native to eastern North America.

service court. The open work area or maneuvering space around farm buildings which has connections to fields and highways.

servicing. The rendering of service, especially such service as is given in the lubrication, fueling, adjustment, cleaning, or repair of automobiles, trucks, tractors, farm implements, etc.

sesame. Oriental sesame.

sesame oil. A fixed or non-volatile oil expressed from sesame seed whose chief uses are for food, and as the source of the product sesamin (*q.v.*). In certain countries the oil has been used for the adulteration of butter fat. Also called *benne oil*. See *oriental sesame*.

sesame oil meal. A livestock feedstuff which is the ground residue obtained after the extraction of part of the oil by pressure from oriental sesame seed as produced under reasonable milling conditions.

sesamin. A white, crystalline compound which is present in sesame oil (*q.v.*). *Sesamin* concentrates, obtained from sesame oil in the process of refining for use as an edible oil, are used in pyrethrum insecticides. *Sesamin* itself is not insecticidal, but it greatly increases the effectiveness of the pyrethrums.

sesamoid. A small bony or cartilaginous nodule developing in tendons, as at a joint, such as at the fetlock joint of a horse.

Sesbania. A genus of plants of the family Leguminosae. See *hemp sesbania*.

sessile. Of leaves, flowers, fruits, etc., attached directly by its base without a stem or stalk, such as leaves of grasses, sedges, and certain other plants.

set. 1. A small propagative part, a bulb, shoot, tuber, etc., which is suitable for setting out or planting, as an onion *set*. 2. A number of things usually used together, or forming a complete collection, as a *set* of tools, harness, etc. 3. The initial swelling of the ovary of a flower soon after petal fall. Also called *setting; fruit setting*. 4. To put in a particular place, position, condition, direction, adjustment, etc. 5. To put eggs under a fowl or in an incubator to hatch them. 6. Of the ovary of the flower, to swell appreciably soon after the petals fall. 7. To put a price or value on something. 8. To put in proper working condition. 9. To adjust the teeth of a saw. 10. To fix a trap or a net to catch animals or fish. 11. To fix in the ground, as a post, tree, or plant. 12. To become stiff, firm, or hard, as cement *sets*. 13. To adjust into proper apposition the ends of a broken bone, as to *set* a fracture.

seta. A stiff hair, bristle, or bristle-like body or appendage, as the stalk supporting the capsule of mosses; a bristle occurring on or in some fungus fruiting structures; or the hairs (setae) of a caterpillar, etc.

set brand. For range cattle, a brand made from a pattern print of iron. See *brand: 1*.

set hocks. In horse judging, an undesirable hock characteristic in which the hocks are set too close together (cow-hocked), too much angle (sickle-hocked), too far apart (bow-legged), or in which any other abnormal condition affects the correct conformation of the hind legs and the action of the horse.

set of horn. The manner in which the horns of cattle are set on the head. It is regarded as a more or less fancy point in the judging of horned breeds.

set of the legs. In horse judging, the conformation of the legs or the manner in which they are placed or set on a horse. The *set of the legs* has an influence on the action in movement as well as on the soundness of the legs.

setose. Designating a bristly, hairy texture or outer surface.

set out. To plant bulbs, shoots, cuttings, seedling plants, nursery stock, etc., in a garden, nursery plot, or other out-of-door, soil area. See *set: 11.*

set pod. To form or develop a seed pod.

set screw. A machine screw used to fasten one part to another, as to fasten gears, pulleys, etc., to a shaft.

set seed. Of plants, to make or develop seed from the flower or to produce a seed crop.

set squarely. In stock judging, beef cattle showing desirable placement of the hind legs. These should *set squarely* at the corners of the body, as such placement contributes to the form of a steer and the ultimate carcass.

setting. 1. The eggs placed, or the eggs intended for placement, in an incubator at one time. 2. The number of eggs which a fowl incubates at one time. 3. The position, depth, distance, etc., at which an implement, fence post, plant, or other object is set. 4. See *set: 3.*

setting hen. A broody hen in the act of incubating eggs.

setting the hair. In dressing hogs, a contraction of the skin around the base of the bristles resulting from overscalding which causes the bristles to be tightly held and thus affects their easy removal.

settle. 1. To sink gradually to a lower level, as of sediment, etc. 2. To cause a female to become pregnant, as a bull *settles* the female with which he has had coitus. 3. To occupy land usually for the purpose of farming, ranching, or homesteading.

settled. 1. Designating suspended particles or impurities in liquids dropped to the bottom. 2. Designating a pregnant cow or mare.

settler. A person who establishes residence, especially one who settles in a new country.

settling. The act or process of clearing a liquid of particles or impurities by precipitation, gravity, etc.

settling basin. An enlargement in a stream or conduit to permit the settling of debris carried in suspension; a form of sand trap.

settlings. Particles, in a liquid, that settle to the bottom: dregs, sediments, etc.

settling tank. A sedimentation tank; a container for holding liquids so that they may be cleared by precipitation, gravity, etc.

sevenbark. Smooth hydrangea.

7-dehydrocholesterol. A sterol transformed into vitamin D (*q.v.*) by the action of ultraviolet light. Through such action, vitamin D may arise in dead plant matter.

seven-sisters. Cobblersthumb livingrock-cactus.

seven-sisters Japanese rose. *Rosa multiflora* var. *platyphylla,* family Rosaceae, a climbing, trailing shrub grown for its deep pink flowers. Also called *crimson rambler.*

seven-stars. Cobblersthumb livingrock-cactus.

17-hydroxycorticosterone. A hormone produced by the adrenal cortex when stimulated by adrenocorticotrophic hormone (ACTH). It functions to increase the rate of breakdown of proteins and fats into smaller units which can be used for the immediate needs of the body.

seventeen western states. In irrigation, the 17 states of the U.S. where irrigation is commonly practiced: Washington, Oregon, California, Idaho, Nevada, Utah, Arizona, Montana, Wyoming, Colorado, New Mexico, North Dakota, South Dakota, Nebraska, Kansas, Oklahoma, and Texas.

seventeen-year locust. Periodical cicada.

seventies (70's). A British grade of wool equal to U.S. fine *q.v.*

severance tax. A tax upon the exploitation of timber or other natural resources.

severinia. Chinese boxorange.

Seville orange. Sour orange.

sewage-disposal system. A means for the treatment and disposal of sewage, such as a septic tank (*q.v.*) together with a disposal field for drain tile.

sewage sludge. An organic product, remaining after the treatment of sewage in sewage-disposal plants, sometimes used as a fertilizer. See *activated sewage.*

sewed casing. An animal intestine which has been cut longitudinally and sewed back together to be used as sausage casing.

sewer. A pipe or conduit, generally closed, but normally not flowing full, which is used for carrying off sewage and other liquid waste.

sex character. The peculiarity of appearance, other than the presence of sex organs, that distinguishes one sex from another, such as the thick neck and bold, rugged head of a bull as contrasted to the more finely-turned neck and head of a cow. Technically, these are secondary sex characteristics.

sex chromosomes. Chromosomes particularly connected with the determination of sex. Most animals have one pair of *sex chromosomes,* the remaining chromosome pair being known as autosomes.

sexed chicks. Baby chicks separated according to sex at hatching time, making possible the purchase of either pullets or cockerels.

sex feathers. The two top tail-feathers which characterize the mature males of most breeds of duck. These two tail-feathers at their outer ends turn upward and forward making a pronounced curl. This characteristic feathering is useful in sex determination.

sexing baby chicks. Any method for the determination of sex of baby chicks at hatching time. The method generally used in America was introduced by the Japanese who probably acquired it from the Chinese. Briefly, it consists of everting the cloaca of the chick by a slight pressure and in distinguishing differences in the parts thus exposed. With chicks of certain colored breeds, such as the Barred Plymouth Rock, and with certain cross-bred chicks, sex can be determined with reasonable accuracy by differences in color pattern.

sex-limited. Designating the expression of a character in only one sex: *e.g.*, milk production, horns in Rambouillet sheep, egg production in poultry, etc.

sex linkage. Association of character with sex due to the fact that the gene for that characteristic is in a sex chromosome.

sex ratio. The proportion of males to females in a group or population.

sexual reproduction. Reproduction brought about by the gametes, as contrasted with asexual or vegetative reproduction. See *asexual.*

sexupara. That form of female organism that produces new individuals without fertilization. See *parthenogenesis.*

shackle. 1. A device, such as a ring, chain, or band, used to restrict free movement of an animal. Particularly useful in slaughter houses. 2. A ring, clevis, chain, hobbles, etc., used for attaching or coupling, so as to leave some degree of movement.

shackling pen. The pen in a slaughter house in which the hind legs of hogs are secured by one end of a shackle, the other end of which is attached to a mechanical hoist.

shadbelly. A scrubby horse with a slight midsection.

shadblow. Serviceberry.

shadblow serviceberry. *Amelanchier canadensis,* family Rosaceae, a shrubby tree grown for its white flowers and red-purple fruit. Native to North America. Also called *downy serviceberry.*

shadbush. Serviceberry.

shaddock. See *pummelo.*

shade. 1. A spot or area protected from direct sunshine. 2. To protect seedlings and other plants from intense sunshine by screens, spaced strips on frames, etc.

shade board. A board placed over a beehive as a substitute for tree shade.

shade density. The degree of shade which in forest areas is in inverse relationship to the percentage of light passing through the crowns. Its determination is based on the assumption that unbroken light has at the time of measurement the value of 100 percent.

shade frame. A frame for the partial shading of a seedbed upon which laths, brush, or cloth can be arranged to admit light as desired. It is frequently combined with a wire screen to exclude birds or rodents.

shade-grown wrapper leaf. A leaf of the Cuban tobacco group used in cigar making which is produced on plants shaded by a screen of open-mesh cotton fabric during growth.

shade plant. 1. A plant that grows in shade or can tolerate shade, as moss, woodland flora, etc. 2. A plant grown to provide shade for crops that require it.

shade streak. In turpentining, a streak cut in such a way that the newly-exposed surface is shaded from the direct rays of the sun.

shade-tolerant. Designating a plant that grows well in open sunshine and can also show satisfactory growth in moderate shade.

shade tree. Any tree grown to provide shade; specifically, one of luxuriant foliage, such as a street tree or a park tree.

shade-up. Descriptive of livestock, which retire to the shade during the heat of the day and graze during the cool morning and evening hours.

shadfly. A horsefly.

shading. See *shade: 2.*

shadoof. A pivoted pole with a bucket attached at one end and a counterweight at the other, which facilitates the raising of water, used for irrigation purposes in the Near East.

shadow jumper. A nervous, skittish horse.

shadscale saltbrush. *Atriplex confertifolia,* family Saxifragaceae, a scurfy, gray shrub, which ranks well as a browse plant. Native to the western U.S. See *saltbush.*

shaft. 1. Either of the pair of long bars (thills) sticking out in front of a vehicle between which a horse or other animal is harnessed. 2. The stem of a feather to which the barbs are attached. 3. A connecting rod between a power source and a pulley, gear, or other contrivance to distribute such power.

shaft louse. *Menopon gallinae* (L.), family Menoponidae, an insect similar to the common chicken louse which is normally found along the shafts of bird feathers and around the vent or on the body and head. See *chicken body louse; chicken head louse.*

shafty wool. Wool that has been well grown and has fibers that are sound throughout their entire length. (England and Australia).

shagbark hickory. *Carya ovata,* family Juglandaceae, a deciduous tree so named because its shaggy, gray bark peels off in long strips. It is valued for its timber and edible nuts. Native to North America.

shaggy hydrangea. *Hydrangea bretschneideri,* family Saxifragaceae, a tender shrub grown for its clusters of white flowers. Native to China.

shake. 1. A separation, such as a crack or split, along the grain of the wood, frequently due to overstressing of standing timber by wind. 2. A crack or split in the trunk or limbs of a tree caused by very low temperatures. See *cup shake; ring shake.* 3. A section split from a block or short log of wood used unplaned for roofing or siding: *e.g.,* cedar *shakes.*

shake down. See *shakes.*

shaker screen. In flour mills, any of the various forms of screens provided with a mechanical means for agitating during a cleaning or separating operation. Commonly used as a part of grain or seed separators or cleaners and in machines used in milling cereals, grading devices, etc.

shakes. 1. In butter judging, samples of butter of very excellent quality (usually scoring 94 points or better) set aside for final scoring, after the other samples in the exhibit have been judged. 2. Trembling defect in baby pigs which may cause death due to inability to nurse.

shale. 1. A sedimentary rock which is composed of clay or silty mud finely laminated or fissile in structure. There are many variations in color and many gradations into hard, thinly-bedded sandstones and limestones and into slates. The residual product of weathering of the more typical shale rock is a silt or clay in texture. 2. The sliver-like lining of the carpel or bur which before the opening of the cotton boll serves to separate the locks. The presence of fine particles of *shale* in ginned cotton is highly objectionable as they are difficult to remove from the lint.

shallot. *Allium ascalonicum,* family Liliaceae, a perennial, bulbous herb grown for its edible, mild-flavored, white stalk and green leaves. Native to Europe and Asia.

shallow. 1. Not deep; lacking in depth. 2. In stock judging, designating an animal lacking depth of body.

shallow cultivation. Scraping the surface, scarifying, or cultivating to shallow depths, normally less than 5 ins.

shallow frog. Of a horse's hoof, a thin frog which has too little cortical depth, a condition commonly associated with flat-footedness.

shallow-pan. An obsolete gravity method of skimming milk.

shallow-rooted. Designating a plant which has all or nearly all of its roots near the surface of the soil. *Cf. deep-rooted.*

shallow soil. A few inches of soil over parent material or bed rock. Also called *thin soil.* See *soil depth.*

shallow super. A beehive super about half the depth of a full depth super, used in the production of section comb honey (*q.v.*) and bulk comb honey. See *comb honey super; bulk comb honey super.*

shallow well. A well from which a supply of water is obtained at little depth, commonly 20 ft. or less below the surface. *Cf. deep well.*

shallu. *Holcus sorghum* var. *roxburghii,* family Gramineae, a late-maturing variety of grain sorghum (*q.v.*). Also called *Egyptian wheat; dessert wheat corn.*

shaly soil. Soil derived from shale (*q.v.*), usually of small thickness, and containing a mass of very thin, flat fragments of this rock.

Shamouti orange. An important variety of sweet orange *(q.v.)* grown extensively in Palestine and known as Jaffas on European markets. The fruit is oblong, of medium size, orange to deep orange color, with rather thick skin and firm, solid flesh and medium juice. Also called *Chamoudi orange.*

shamrock. A plant having trifoliate leaves, traditionally believed to have been used by St. Patrick as illustrative of the doctrine of the Trinity, and hence adopted as Ireland's national emblem. The name *shamrock* is now commonly applied to the emblematic plant worn by the Irish on St. Patrick's day: the lesser yellow trefoil *(Trifolium minus).* The original plant is variously considered to have been the suckling clover *(Trifolium dubum),* the white clover *(T. repens),* or the wood-sorrel oxalis *(Oxalis acetosella).*

shamrock pea. *Parochetus communis,* family Leguminosae, a trailing, perennial herb grown in flower borders and hanging baskets for its cobalt-blue and pink flowers. Native to Asia and East Africa. Also called *blue oxalis.*

Shanghai. A breed of domestic fowls probably originating in Shanghai, China. Large in size, they have long legs and feathered shanks.

shank. 1. That part of the leg joining knee to ankle in man, or the corresponding part in various animals. 2. A market cut of meat from the lower part of the foreleg of a dressed carcass of beef, veal, mutton or lamb, or a corresponding cut from the hind leg. 3. The tarsus of birds: the part of the leg, below the hock joint, which is covered with scales. 4. That part of a tool, implement, etc., between the acting part and the part by which it is held, as a cultivator-shovel shank, etc. 5. Any connecting part of a plant; a foot stalk, as the part between the stalk and cob of a corn ear.

shank clamp. A metal clamp which holds the shank of a tool, implement, etc.

shank feathering. Those feathers which grow on the outer side of the shank of certain birds, as those on the Cochin, Brahma, and Langshan breeds of chicken.

shanking. The process of skinning out and removing the shanks of a slaughtered animal.

shape horns. To correct the shape or position of horns on show cattle by affixing metal weights to the tip of each horn or by otherwise changing their shape.

shard. 1. A piece of broken flowerpot which is placed over the hole in the bottom of another flowerpot to aid in providing drainage. Also called *crock.* 2. To provide or use a *shard.*

share. 1. That portion of the crop or livestock received by either the tenant or the landlord on a farm or from farm land. 2. That part which provides the front cutting-edge for a plow. The principal parts of the *share* are the point, the wing, and the cutting edge or throat.

share-crop. To cultivate farm land in return for a share of the crop produced. See *sharecropper.*

sharecropper. A tenant farmer who farms land for a share of the crop. The landlord commonly furnishes the land, living quarters, credit for food, the tools, and a part or all of the expenses, such as for seed, fertilizer, etc., while the *sharecropper* provides labor for his share of the returns, usually half the value of the crop after adjustments for fencing, drainage, and credit advances (southern U.S.).

shark-liver oil. An oil obtained from the liver of sharks. It has use as a vitamin A and D supplement in animal feeds.

sharkskin. The injury on the peel of citrus fruits caused by the rust mite.

sharp-crested weir. A notch cut in a relatively thin, sharp-edged plate, on the upstream side of the crest of a weir.

sharp flavor. Acid flavor.

sharp freezing. Freezing of products, fruits, vegetables, poultry, meats, dairy products, etc., at low temperatures (+5°F. to —20°F.).

sharpleaf cotoneaster. *Cotoneaster acuminata,* family Rosaceae, a shrub grown for its pinkish flowers and red fruit. Native to the Himalayas.

sharpleaf jacaranda. *Jacaranda acutifolia,* family Bignoniaceae, a popular tree grown in very warm areas for its foliage which remains on the tree almost until spring, and for its blue flower clusters. Native to tropical America.

sharpleaf penstemon. *Penstemon acuminatus,* family Scrophulariaceae, a perennial herb grown for its long, blue flowers. Native to North America.

sharp practice. Hard bargaining; a deal or transaction in which advantage is taken or sought through unscrupulous methods or trickery: *e.g.,* operating upon, or tampering with, an animal for the purpose of concealing faults.

Shasta daisy. A clon of the Pyrenees daisy (*q.v.*) grown for its showy, white, daisy-like flowers 2 to 4 ins. across. See *chrysanthemum.*

shatter. To scatter or disperse as leaves, grains, seeds, fruits, etc.: *e.g.,* leaves fall from dry hay, seeds drop out of husks or pods from over-ripeness, etc.

shea. *Bassia parki,* family Chenopodiaceae, a tropical tree whose seeds yield a fat, *shea* butter, used as a food in Africa and in soap manufacture in Europe. Native to West Africa and Sudan.

sheaf. A quantity of the seed-bearing stalks of cut grains which are tied into a bundle.

shear. To cut with shears or similar instruments, as to cut the wool or hair from sheep, goats, etc.

shear cut. The action of a rolling colter of a plow in cutting trash on the surface of the ground, using the ground as one edge of the shear.

sheared wool. Wool which is removed by shearing as contrasted to pulled wool *q.v.*

shearer. 1. A person who shears sheep. 2. A market lamb in full fleece. See *shearing lamb.*

shearing. 1. The act or operation of removing wool from a sheep or goat by means of shears. Shearing machines (*q.v.*) have largely taken the place of hand shears or blades with professional shearing outfits. 2. The thinning or heading-back of shrubs or trees by removing portions with trimming or pruning shears, as in the trimming of hedges, etc.

shearing comb. A comb which forms part of the cutting-head of a shearing machine *q.v.*

shearing lamb. In marketing, a subclass of lambs that are in full fleece and sold primarily for their wool.

shearing machine. A device with a mechanically operated cutter for shearing sheep and goats. The essential working parts are a metal comb and a cutter plate affixed to a handpiece.

shearing shed. A shed in which the sheep or goats are sheared.

shearling. 1. Short wool pulled from the skins of sheep that have been sheared just previous to being slaughtered. 2. Yearling sheep (England).

shearling gimmer. See *gimmer.*

shear plate. A metal cutter bar or plate which forms the cutting edge for the knives on a cutter bar. Some *shear plates* are reversible and may have as many as four cutting edges.

shears. Designating cutting instruments or devices consisting of opposed, metal cutting edges which have a cutting action when their edges are caused to pass closely to each other.

sheath. 1. The leaf base covering a stem or branch, such as the lower part of leaves of grasses which surround the stem. 2. The enclosing and protecting tubular structure within which the penis of a stallion and certain other male animals is retracted.

sheath and culm blight of rice. A fungous disease of rice caused by *Rhizoctonia oryzae,* characterized by lesions with reddish-brown margins and light yellow or greenish-yellow centers on the leaves and culms.

sheathing. 1. An undercovering of boards or waterproof material fitted to the framework of an outside wall or roof of a frame building. 2. Materials which are used for sheathing, such as lumber other than siding, sheets of waterproof material, etc.

shed. 1. A simple structure for shelter of livestock, storage of feeds, or both, open on one or more sides which may be separate from or attached to another structure. 2. To lose or cast off something, as a plant *sheds* leaves, pollen, seed, etc.; animals *shed* hair; birds molt or *shed* feathers; a roof *sheds* rain, etc.

shed barn. A one-story barn usually with limited feed-storage space, used for housing milk cows, beef cattle, or sheep.

shed lambing. Housing and feeding ewes at the time the lambs are dropped.

shed-roof house. See *lean-to house.*

sheep. Any of a variety of cud-chewing mammals, genus Ovis, related to goats, especially *Ovis aries,* which includes many breeds domesticated for their heavy wool, edible flesh (mutton), or skin.

sheep barn. A barn, or a one-story, shed-like structure with one or more open sides used as a shelter for sheep.

sheep bell. A bell hung about the neck of a sheep whose tinkling indicates the location of a flock.

sheep biting louse. *Bovicola ovis* (L.), family Trichodectidae, a small louse of sheep and goats which eats off wool fibers and skin scales, tangles and soils the hair, and causes severe irritation to the skin. These lice do not suck blood but when numerous may cause raw sores, scratching, and rubbing. Also called *red-headed louse.*

sheep bot fly. The maggot or larva of the sheep bot fly, *Oestrus ovis* L., family Oestridae. The eggs are hatched in the body of the female fly and the very small, active maggots are deposited in the nostrils of sheep, goats, and deer. The maggots work their way to the sinus cavities where they lacerate the tissues and feed, ultimately growing to ½ in. or more in length. Affected sheep shake their heads, stamp their feet, and hold their noses to the ground. Also called *head grub; nose fly; sheep grub.*

sheep bur. See *cocklebur.*

sheep camp. A place where sheep bed down at night; a camp serving as headquarters for sheep herders; an outstation.

sheep casing. See *casing: 2.*

sheep country. A sheep range area, especially an area in the western U.S., Australia, New Zealand, or South Africa where the raising of sheep is the major occupation.

sheep dip. A liquid mixture containing an arsenical, cresol, coal-tar, lime-sulfur, nicotine, etc., used in solutions for dipping sheep to destroy external parasites and disease germs. See *dipping: 2; dipping vat.*

sheep dog. A dog which is trained and used to tend or work sheep, such as an Old English sheep dog or similar dog of Shepherd or Collie breeding.

sheep drench. A liquid, medicinal preparation, usually a vermicide, which is given to sheep by mouth for the removal of stomach worms, tapeworms, hookworms, etc.

sheep fescue. *Festuca ovina,* family Gramineae, a perennial grass, some forms of which are native or introduced on western ranges (U.S.). It is sparingly used for pasture and in some turf grass mixtures where rough turf is desirable. Native to Europe and Asia; naturalized in North America.

sheep follicle mite. *Demodex canis* var. *ovis* Raill., family Demodicidae, a variety of the parasitic dog follicle mite that lives in follicles of sheep and causes a mange. It is not common in the U.S.

sheep grub. Sheep bot fly.

sheeping down. A method of harvesting and utilizing a crop, as corn, by the use of sheep which are turned into the field to feed. See *hog down.*

sheep ked. *Melophagus ovinus* (L.), family Hippoboscidae, a degenerate, wingless, louse-like fly that crawls about over the sheep's skin in the wool and feeds by thrusting its sharp mouth parts into the flesh and sucking blood. The irritation causes sheep to rub, bite, and scratch at the wool, thus spoiling the fleece. Also called *sheep tick.*

sheep-killing dog. A dog that has acquired the habit of worrying and killing sheep. See *dingo.*

sheep laurel. Lambkill kalmia.

sheep lice. 1. Lice that commonly infest sheep, primarily the sheep biting louse (*q.v.*), but also *Hinognathus pedalis,* a blood-sucking foot louse and *H. ovillis,* a blood-sucking body louse. (The latter two are more often pests of other animals, particularly goats.) 2. See *houndstongue.*

sheep loco. *Astragalus nothoxys,* family Leguminosae, a poisonous plant found in southern Arizona, U.S. Both its green and dried foliage are poisonous, causing dullness, irregularity of gait, lack of appetite, dragging of the feet, a solitary habit, loss of flesh, and shaggy coat in animals. As the animal ceases to eat, it dies. Also called *locoweed.*

sheep manure. The dung of sheep either moist or dried. Dried, it is sold for fertilizer. The approximate analysis is: nitrogen 1.4 percent, phosphoric acid 1.0 percent, and potash 3.0 percent.

sheep milk. The milk obtained from lactating ewes which is unusually high in fat and in solids-not-fat, totaling about 20 percent. In certain eastern Mediterranean and Asiatic countries it is used as a beverage, but its principal use is in the manufacture of Roquefort cheese *q.v.*

sheep nasal fly. Sheep bot fly.

sheep month. The amount of forage or feed which is necessary to maintain a mature sheep or a ewe and its suckling lamb for 30 days. It is commonly figured as one-fifth of a cow month *q.v.*

sheep panel. A lightweight, gate-like structure used in handling sheep. Several such panels may be used to form pens for confining sheep.

sheep pen. 1. A pen for confining or separating one group or lot of sheep from others, commonly made of adjustable panels. 2. In stockyards, any pen in which the different consignments of sheep are fed and held until sold. In the larger yards, the pen is roofed or enclosed in a building known as the sheep house.

sheep pest. See *cocklebur.*

sheep pox. A contagious sheep disease, similar to smallpox, causing fever and pock marks on the skin. Also called *ovinia; sheep pock* (British).

sheep ranching. The business of raising and grazing sheep, especially the handling and grazing of large flocks of sheep.

sheep rot. 1. A name for plants said to cause illness in sheep, as the marsh pennywort (*Hydrocotyle americana*). 2. See *liver fluke; liver rot.*

sheep run. The whole property of a sheep operation, including the residence and other buildings belonging to it, the stock, and freehold and leasehold country. Originally, the hut, yards, or other buildings where a squatter or his agent stationed himself. See *squatter: 2.*

sheep scab mite. *Psoroptes equi* var. *ovis* (Her.), family Psoroptidae, a scab mite that causes severe injury to sheep by puncturing the skin with its sharp mouth parts until the lymph exudes. A scab forms over the larvae, under which they feed, causing added scab formation which eventually lifts the hair out by the roots. Extreme irritation and itching are caused by the work of the mites, and the sheep loses bits of wool on weeds, fences, and other objects as a result of constant rubbing. Psoroptic mange is the most frequent type of mange affecting sheep and is also called *scab; scabies.* See *scab mite.*

sheep shearing. See *shearing: 1.*

sheep shearing machine. See *shearing machine.*

sheep shears. Any shears designed for the removal of the fleece from sheep, such as hand shears (blades) having a spring-opening action or the shears of a shearing machine (*q.v.*). See *shears; shearing: 1.*

sheep shed. A shed in which sheep are confined or in which they may go for shelter. See *shearing shed; sheep barn.*

sheep sickness. See *braxy.*

sheep-size cottonseed cake. A livestock feedstuff which is a cottonseed product which will pass through 7/8 in. round perforation and over 5/8 in. round perforation. It shall be free from meal and pebble-size cake and shall not contain in excess of a total of 10 percent of nut-size and pea-size cake.

sheepskin. 1. The skin of a sheep, especially one dressed with the fleece on it. Also called *sheep pelt.* 2. Parchment or leather made from the skin of a sheep, used for clothing, bookbinding, diplomas, etc.

sheep sorrel. 1. *Rumex acetosella,* family Polygonaceae, a perennial, sour-tasting herb which is a common weed in abandoned fields or waste ground and meadows, especially on dry, sandy, acid or depleted soils. Also called *cow sorrel; field sorrel; redtop sorrel; sourweed.* 2. Woodsorrel oxalis.

sheep station. A sheep run (or ranch) which includes the whole property of a sheep operation, such as the residence and other buildings belonging to it, the stock, and freehold and leasehold country. Originally the hut, yards, or other buildings where a squatter stationed himself to work his run (Australia, New Zealand).

sheep tag. 1. A heavy, dung-covered lock of wool which usually hangs from the crotch of a sheep. 2. A metal tag which is fastened to the ear of a sheep for identification.

sheep tick. See *sheep ked.*

sheep weed. See *butterwort: 2.*

sheet erosion. Removal by rainwater and snow melt runoff of a more or less uniform layer of soil or comminuted material from a portion of the land surface, as contrasted to gullying and stream erosion caused by surface water flowing in channels.

sheet piling. A diaphragm or partition, made up of meshing or interlocking members of wood, steel, concrete, or other material, driven individually, which may be used to form an obstruction to percolation, prevent movement of material, stabilize foundations, and build coffer dams.

shelf curing. A simple, meat-curing process whereby the meat is rubbed with the curing ingredients and placed to cure on shelves or in boxes or barrels with perforated bottoms that permit the juices to drain off.

shell. 1. A hard outer covering of certain animals, such as turtle, mollusk, or crustacean, which when ground may be used as a poultry feed supplement. 2. The hard outer covering of a fruit or seed, as a nut *shell,* etc. 3. The hard calcareous exterior covering of an egg. 4. To remove or separate from its *shell* or pod, as to *shell* nuts, peas, etc. 5. To drop out of a *shell,* or out of a cluster, as seeds, grapes, etc. 6. To separate seeds or kernels (of corn, wheat, oats, etc.) from cob, ear, or husk.

shell bark. A disease (decorticosis) of lemons, mainly of the trunk bark, whose cause is not known. With affected trees the bark dies, shrinks, cracks, and in advanced stages tends to loosen in vertical strips, leaving the inner bark and cambium alive.

shellbark hickory. *Carya laciniosa,* family Juglandaceae, a deciduous tree grown especially for its wood used in tool handles, ladders, furniture, sporting goods, implements, woodenware, for fuel and for smoking meats. It may be used as a shade tree, and its nuts are edible and sweet. Native to North America. Also called *bigleaf shagbark hickory; big hickory; thick shellbark hickory.*

shell bone. Side-bone.

shell box. A box or self-feeding hopper used as a means of supplying crushed oyster-shells or other suitable calcareous material to laying hens.

shelled corn. Corn kernels separated from the cob or ear. *Cf. ear corn.*

shell egg. In marketing, an entire egg with a shell as contrasted to an egg that has been broken out. *Cf. dried eggs; frozen eggs.*

sheller. Any device used for shelling, such as corn *sheller q.v.*

shell-flower. See *white turtlehead; Asian molucca-balm.*

shelling. Separating corn kernels from cob, peas from pod, grapes from cluster, etc.

shelling percentage. The weight of corn kernels (shelled corn), expressed as a percentage of the weight of the ear corn prior to shelling.

shell membranes. Either of the two membranes, the inner and outer shell membranes, surrounding the egg substance and located immediately inside the shell of the egg. At the large end of the egg, the membranes are separated to form the air cell *q.v.*

shell protected. Designating market eggs that have been dipped in mineral oil, usually heated paraffin oil, to retard evaporation and otherwise to maintain quality. Also called *shell treated; protected eggs.*

shell quality. The various characteristics of the shell of an egg, including thickness, porosity, texture, etc., which may vary widely between hens and may be improved through selective breeding.

shell treated. Shell protected.

shelly. In meat judging, designating a chuck or forequarter that lacks thickness.

shelter belt. An extended barrier of several rows of living trees and shrubs, established and maintained for protection of farm lands from wind erosion damage over an area larger than a single farm. Such plantings were initiated in the Great Plains (U.S.). See *windbreak.*

shelterbelt project. A U.S. government plan to plant trees on about 1,000,000 acres of land in a hundred mile wide belt extending from the Canadian boundary to the Texas Panhandle. A part of a huge program for relief and rehabilitation in the Great Plains region supported by funds appropriated by Congress, in 1934, the project has never been carried to completion.

shelterwood. A forest management method of securing natural reproduction under the partial shelter of seed trees which are gradually removed by successive cutting to admit the amount of light and heat that the seedlings will require.

Shenandoah Valley. A part of the Great Valley of the Appalachian Region of the U.S.; a great limestone belt of lowland plains and mountains lying directly west of the Blue Ridge in western Virginia, which is traversed in part by the Shenandoah River. It is notable for fertile soils, farm products, livestock, and apple orchards.

she-oak. A popular name of certain trees and shrubs of the genus Casuarina, family Casuarinaceae, (especially *C. stricta*), so designated because the grain of the wood resembled, but was inferior to, English oak. Native to Australia. Also called *Australian pine; beefwood.*

shepherd. A flockmaster or herdsman who feeds and guards a flock of sheep or goats.

shepherd's crook. The staff used by a shepherd; specifically a staff with a crook at one end which is used in catching and holding a sheep (used by Old World shepherds since ancient times).

shepherdspurse. *Capsella bursa-pastoris,* family Cruciferae, a very common, annual, introduced weed widely distributed in cultivated areas. Also called *fat hen.*

sherbet. A dessert made from water with the addition of milk or skim milk solids, stabilizer, sugar, with or without egg whites, and natural flavoring, frozen to the consistency of ice cream.

Shetland. A breed of small and stocky ponies (native to the Shetland Islands), noted for hardiness, patience, surefootedness, and strength as weight carriers. A desirable height is 39 to 42 ins. and the popular colors are black, chestnut, and pinto. The ponies are generally affectionate and sensitive, and because of their small size are popular for children.

shield. 1. A flat, metal plate or disk attached vertically or adjacent to the shovels of a cultivator to prevent soil from falling on small plants, or a similar part or parts used with shovel-openers to prevent soil from falling back into the furrow before the seeds are deposited. 2. The thick, hard, horny skin on the shoulders of a mature male hog. 3. The cion used in *shield* budding *q.v.*

shield budding. A bud graft made by inserting a bud of a specified variety, with an attached segment of bark and wood, cut in the shape of a shield, into an opening in the bark of a stem or branch of a different stock. The opening in the bark of the stem is a T-shaped cut. On some fruit trees the T is reversed for a better graft. Also called *T-budding; bud; budding; budgraft.*

shielded gauge. A rain and snow gauge which has a special collar so as to obtain a better catchment of light rain by stilling the air over it.

shield fern. See *woodfern.*

shield graft(ing). A side graft in which a T-shaped incision is made in the bark of the stock as for budding. Into this incision a wedge-shaped cion is inserted and tied in place. The stock is eventually cut off above a cion. Also called *sprig graft(ing).* See *bud:* 3.

Shigella infection. Omphalophlebitis.

shih mow. A Chinese unit of land measurement equal to 0.171 acre.

shin. 1. The front part of the leg between the knee and the ankle; the lower part of the foreleg in beef cattle. 2. The cutting edge of the moldboard of a plow just above the landside.

shin boot. A shaped, leather piece fitted to the shins and cannon bone of horses, especially saddle and running horses, to protect against injury from striking with a hoof when in motion.

shiner. Colorado blue spruce.

shingle. 1. a. A thin, oblong piece of wood sawed or split, with one end thinner than the other, which is lapped lengthwise in covering roofs and outer walls of buildings. b. A piece of any kind of material, as asbestos, used in the same way. 2. Loose, rather sharp shale or similar rock, common to certain rocky areas in New Zealand. 3. A mass of loose, coarse, rock fragments of cobble or small boulder size, rounded or flattish, found on the seashore and on the beaches of large inland lakes. 4. A thin slab of slate stone, cut or chipped into a section used as a *shingle.* 5. To cover with *shingles.*

shingle-tow. The fine, shred-like shavings produced as a by-product in the manufacture of cedar shingles. Where available, the shavings are used as a packing around roots of nursery stock, as chicken litter, etc.

shinnery. A dense thicket which, because the growth is relatively new or because growth conditions are unfavorable, is scarcely more than tall shrubbery. Although *shinnery* is usually made up of mixed woody species, it is generally thought of as oak *shinnery* because of the usual predominance of dwarf oak.

shin oak. A small tree, or mere shrub, in the semiarid parts of southwestern U.S. (The term has been applied locally to the following species: *Quercus gambelii, Q. undulata, Q. havardii, Q. vaseyana, Q. mohriana, Q. laceyi, Q. breviloba.*)

shin share. A share (*q.v.*) that has an extension to form the cutting edge or shin for the moldboard of a plow.

shin soupbone. A soupbone from the foreshank of a beef carcass.

shiny coprosma. *Coprosma lucida,* family Rubiaceae, a shrub grown for its bright, two-stoned, berry-like fruit. Native to the Pacific area. Also called *coffeeberry; coffeebush.*

shiny xylosma. *Xylosma senticosa,* family Flacourtiaceae, an evergreen shrub or small tree grown for its shiny, green foliage.

shipmast locust. A clon of the black locust (*q.v.*) that is supposed to be resistant to the black locust borer.

ship-of-the-desert. The camel; either of the two members of the genus Camelus, a large, herbivorous mammal which can travel great distances without water. Its large, flat feet enable it to walk over the desert sand without sinking. The one-humped, or Arabian camel (*C. dromedarius*) is the larger of the two; the two-humped, or Bactrian camel (*C. bactrianus*) is better suited to hilly, rocky country and cold climates. Both are used as workhorses throughout the arid areas of North Africa, Arabia, Asia, and northern Australia. Used in India and the Sudan for cavalry;

imported into the U.S. in the 19th century as experimental cavalry in Texas.

shipper. One who ships livestock and other farm products.

shipper-style dressing. An unsplit carcass of a pig with head on but entrails removed, a style of dressing often used with a whole carcass in the wholesale trade.

shipping cage. 1. A small, ventilated cage for shipping bees which may or may not have provisions for honeycomb or a queen. 2. A cage for shipping poultry.

shipping can. A strong, tin-coated, 5, 8, or 10 gallon-sized metal can, with handles and tight-fitting lid, used for shipping milk.

shipping fever. Any of several infections that may occur in cattle or other animals shipped to stock yards, or from place to place, if they are exposed to pathogenic organisms when their resistance is lowered by the nervous strain and rigors of being shipped. See *equine influenza; hemorrhagic septicemia; pneumonia.*

shipping season. That season of the year when an agricultural crop or commodity is shipped extensively.

shipstuff. Shorts or middlings: low grade flour, a by-product of wheat milling. Also called *chipstuff.*

Shire. An English breed of draft horses with a typical weight of 1,900 to 2,300 lbs. and a height of 16 to 17 hands which was reportedly developed from the English Great Horse or English Black Horse and used as a war horse by the armored knights of the Middle Ages. Bay is the most popular color, with some browns, blacks, chestnuts, grays, and roans. White is usually found in the form of a stripe or blaze in the face and it also occurs more or less below the knees and hocks. Long hair (feather) grows down the back of the legs from knees and hocks to pasterns.

shirley. 1. An obsolete variety of broom corn formerly grown in New England (U.S.). Also called *black brush.* 2. Any of certain varieties of the single garden poppy.

shive. 1. One of the pieces of the woody part of hemp or flax which is removed in breaking. 2. A fragment of plant material found in wool, exclusive of burrs.

shivering in little pigs. A blood sugar deficiency disease in newly-born pigs, characterized by shivering, dullness, no desire to nurse, upright hair and coma. See *hypoglycemia.*

shoat. See *shote.*

shock. 1. A quantity of a harvested grain crop or corn, etc., set together upright in a field. 2. A small pile of hay. 3. A state of collapse in men and animals, which results from severe loss of blood following surgery, accident, or sometimes under conditions of undue stress. 4. The severe symptoms in plants following acquisition of virus. Later symptoms are absent or milder. 5. A parcel or lot of 60 pieces or units. 6. Twelve sheaves of small grain. 7. To place grain, hay, etc., in *shocks.*

shock corn. Stalks or bundles of corn that have been placed in shocks to cure for later husking or for feeding unhusked. Also called *bundle corn.*

shocking horse. A wooden device 6 to 8 ft. long used as an aid in shocking corn. Its front end has spraddle legs and resembles a 3 to 4 ft. high carpenter's horse; the other end rests on the ground. The corn stalks are placed upright in each of four divisions formed by a removable 3 to 4 ft. stick which goes crosswise through a hole in back of the legs.

shock rows. Each of the more or less straight rows of shocks lined up in a field.

shock symptoms. See *shock: 3, 4.*

shod. Equipped with shoes, as a *shod* horse.

shoddy. 1. That leather made by grinding waste leather to a pulp and pressing it into solid sheets with or without the addition of a binding material. 2. Reworked wool that has been recovered from wool cloth trimmings, rags, etc. *Cf. virgin wool.*

shoe. 1. A plate or rim of metal, usually iron, nailed to the underside of a horse's or other such animal's hoof as a protection from injury or to assist in obtaining footholds. 2. A part or parts of certain farm implements, particularly parts that travel in or through the soil, such as the metal plate on the bottom of a subsoil plow, furrow openers on certain seeders, etc. 3. The shoelike runner which supports the outer end of the cutter bar of a mower. 4. Small pieces of the woody stalk of the hemp plant which are produced when the retted stalks are broken to remove fiber or lint. 5. To fit and attach a *shoe,* as on a horse's hoof.

shoeblack plant. Chinese hibiscus.

shoe boil. An abscess or swelling at the point of the elbow of a horse; a capped elbow.

shoe furrow-opener. A type of furrow-opener on certain grain drills, made from two flat pieces of steel welded together to make a cutting edge similar to the stub-runner furrow opener used on a corn planter.

shoepeg-shaped. Designating the shape of kernels of certain hull-less varieties of popcorn which are somewhat similar to a type of nail used by a cobbler in fixing shoes.

shoe puller. The tongs used to pull off old shoes before re-shoeing a horse.

shoe sole. The removable and adjustable steel plate which serves as the sole of the shoelike runner at the outer end of the cutter bar of a mower.

shoe string. See *tobacco frenching.*

shoestring of tomato. A virus disease of tomato characterized by the leaves becoming divided into long, narrow strips.

shoestring root rot. A fungous disease of trees and shrubs caused by *Armillaria mellea,* family Agaricaceae, characterized by long, round, brownish-black, stringlike strands of tissue (rhizomorphs) produced underneath infected bark and over infected roots. Also called *armillaria root rot; mushroom root rot; oak fungus disease; oak root fungus disease.*

shomar. The Jewish official who supervises the production and handling of special milk, such as Kosher; kedassia (*q.q.v.*) for orthodox Jews.

shoofly. Yellow wildindigo.

shook. 1. Pieces of lumber or veneer cut to size for box or crate manufacture, bundled together, but not yet assembled. 2. A set of staves and headings sufficient in number for one cask or barrel.

shoot. 1. A stem of a plant, including the leaves. 2. A young stem arising from or near the crown of a plant. 3. A new leafy growth developed from a bud. See *sprout.* 4. An area of land inhabited by game (grouse, etc.), and kept especially for the shooting of such game as a sport; the shooting rights to such an area. 5. To develop new stems. 6. To detonate a charge of dynamite or other explosive. 7. See *chute: 1.*

shoot bud. A bud containing a miniature shoot that becomes a shoot or spur. See *blossom bud.*

shootingstar. *Dodecatheon meadia,* family Primulaceae, a wild, herbaceous perennial grown in the wild garden for its rosy-pink flowers with reddish-yellow stamens, and purple anthers in a conelike collection. Native to North America. Also called *prairie pointer; false cowslip.*

shooting the red. The beginning of caruncle (*q.v.*) extension down the neck of a seven or eight weeks old turkey poult.

shore. The contact area between a body of water and the adjacent land; the land fronting on or directly bordering a large body of water, as the ocean or a lake.

shore erosion. Loss of sand, soil, or rock from the land area adjacent to oceans or large lakes which results from the action of waves and wind. Bank erosion may occur along streams as a result of undercutting, etc. Also called *beach erosion.*

shore erosion control. The control of erosion of beaches and bluffs which border on the oceans and large lakes. Also called *beach erosion control.*

shore podgrass. *Triglochin maritima,* family Juncaceae, a perennial, poisonous marsh herb. Symptoms of poisoning in cattle are abnormal breathing, trembling, spasms, and convulsions, causing death. Native to North America. Also called *goose grass; arrowgrass; sour-grass.*

shorn. a. Designating a sheep, lamb, or goat that has had its fleece removed by shearing. b. Referring to wool that has been sheared.

shorn hogget. A lamb after its first shearing and until it gets its first two teeth.

shorn wool. Wool that has been removed by shearing as contrasted to pulled wool *q.v.*

shorn wool clip. In market reports, the total quantity of shorn wool produced in a given year.

short. 1. Designating one who has sold a futures contract that does not liquidate a previously bought contract for the same delivery month. In other words, he sells in expectation of buying back at a lower price. *Short* hedges are sales of futures made as hedges against holdings of the spot commodity, or products thereof. *Short* interest is the sum total of all contracts or the owners of such contracts. 2. A body defect of ice cream. See *crumbly: 2.*

short-coupled. Designating a horse or cow whose top lines, particularly the loin section from the last rib to the hip, are short. Such animals have typically compact bodies and strong backs. See *rangy.*

short course. A course of study in agriculture of less time than a degree program which is usually offered by Land-Grant colleges (U.S.) for special groups of students. A non-degree training program involving a sequence of organized classroom instruction, it usually covers more than one subject matter field. Student registration and class preparation is required. Proficiency is evaluated by examination, grades are recorded, and certificates generally are awarded for acceptable achievement.

short crop. A crop which is below the average in total production.

short cut pork loin. A wholesale cut of pork which is similar to a pork loin except that the shoulder end has been removed behind the blade bone and the ham end has been removed next to the hip bone.

short day. In plants, a day with comparatively fewer hours of illumination than one at the critical photoperiod *q.v.*

short-day plant. A plant in which the flowering period or some other process is accelerated by a relatively short daily exposure to light. See *photoperiodism.*

short dung. Half-rotted or putrefied farmyard manure in contrast to fresh dung (obsolete). *Cf. long dung.*

shortening. Any fat which has the property of making pastry, cakes, etc., friable, easily crumbled, or short.

short-fed cattle. See *short-time feed.*

short grass. Grass or grasses of short stature, such as buffalo grass and blue grama, which are characteristic of and associated with certain range areas of the Great Plains of western U.S. These are commonly the dominant grasses of certain low rainfall regions west of the tall-grass prairies, and the area is often called or known as *short-grass country.*

short handed. Designating insufficient help, an inadequate labor force.

short haul. Carrying or transporting of commodities or produce for a short distance.

short-hold products. Products of a perishable nature which will permit only short-term storage without loss.

Shorthorn. A breed of beef cattle which has characteristically short horns, originating in England. These cattle vary in type from extreme beef conformation (Scotch type) to the dual-purpose (English type) which combines to some extent both beef and dairy qualities. The breed is characterized by various colors: red, white, roan, and red and white combined in different patterns. The Polled Shorthorns (*q.v.*) is a recognized strain.

shortleaf pine. *Pinus echinata,* family Pinaceae,

an evergreen tree used for pulp, poles, ties, slack cooperage, veneer, structural timber, planing-mill products, etc. Native to North America.

short leet. The best animals, usually horses, selected by judges from competing animals in a show ring. From these are selected the prize winners after further examination.

short-legged. Designating an animal that has the appearance of being set close to the ground. See *leggy: 2.*

short loin. A wholesale cut of beef being that part of a whole beef-loin which remains after the rear sirloin end has been removed. It is that part of the loin from which porterhouse, T-bone, and club steaks are cut.

short-nosed cattle louse. *Haematopinus eurysternus* (Nitz.), family Haematopinidae, a bloodsucking, cattle louse which pierces the skin, sucks blood, sets up irritation, and causes rubbing. They are found on the head, neck, or along the inner surface of the legs where they appear as blue patches on the skin. Heavily infested cattle are usually in poor condition.

short on. In the wholesale nursery business, designating a limited supply of stock for sale.

short ribs. A retail cut of beef for roasting or broiling which consists of short lengths of ribs cut from the plate of a forequarter.

shorts. 1. In the florist's trade, roses with short stems (approximately 9 ins. in length). 2. Lumber which is shorter than standard lengths. 3. See *wheat brown shorts; wheat gray shorts; wheat red dog; wheat standard middlings; wheat white shorts.*

short-staple. Lint cotton whose fiber length is ⅞ to 1 in.

short-tailed mealybug. Any of several species of mealybugs with short tails, family Pseudococcidae, such as the citrus mealybug *Pseudococcus citri;* the coconut mealybug *P. nipae;* and the Mexican mealybug *Phenacoccus gossypii;* none of which has a tail more than one-fourth the length of the body.

short-tailed sheep. Sheep with naturally short tails, such as the marsh or moorland sheep of northern Europe.

short-term credit. Credit which is designed for purchasing livestock, machinery, equipment, and supplies. Usual basis of security is the chattel mortgage. It is usually for a term varying from 90 days to 1 year. Also called *production credit.* See *long-term credit; intermediate-term credit.*

short-time feed. A system of fattening cattle for slaughter in which heavy stockers or feeders are fed a heavy grain ration for periods of 30 to 90 days and then sold.

short udder. A cow's udder which lacks length (not carried well forward or to the rear) and which is usually low-hanging and weakly attached.

short wool. 1. Wool which has short staple or fiber. 2. A sheep producing such wool, as the English Shropshire, Southdown, and Suffolk breeds.

short yearling. An animal which is slightly less than a year old.

shote. A young hog of either sex which is weaned but which usually weighs less than 150 lbs. Also called *shoat.*

shot hole. 1. A perforation of plant leaves which may result from localized lesions followed by the falling out of dead or diseased tissue leaving more or less circular perforations. It is not a diagnostic feature of diseases caused by certain species of bacteria or fungi but is more properly viewed as a host peculiarity. The leaves of stone fruits are prone to *shot hole* whenever localized areas of leaf tissue are killed. See *cherry leaf spot; Coryneum blight of stone fruit.* 2. A hole in timber or a timber-product which is caused by any of several wood-boring beetles. 3. In cheeses, a small gas hole (*q.v.*), more or less evenly distributed, which is formed by undesirable microorganisms found occasionally in Cheddar cheese.

shot-hole borer. *Scolytus rugulosus* (Ratz.), family Scolytidae, a small beetle which makes small holes in the bark of healthy fruit trees and in the branches and trunks of weakened trees. On stone fruits, the shot holes are usually covered with gum. The beetles and larvae also make galleries in the sapwood of trees which they infest. If the beetles are abundant, the foliage yellows and wilts, and the tree may die.

shot-hole disease. Shot hole that may occur as a result of any of several bacterial and fungous diseases, such as bacterial black spot caused by *Bacterium pruni,* the leaf spots or blights (*Coccomyces* spp.) of cherries and plums, and the California blight (*Coryneum beijerinckii*) of apricots, peaches, and cherries. See *cherry leaf spot; Coryneum blight of stone fruit; shot hole: 1.*

shot hole of cherry. Cherry leaf spot.

shot holing. A disease symptom in which small, roundish fragments drop out of leaves, giving them the appearance of having been riddled by shot. An irregular *shot holing* of the leaf blades of stone fruits in early spring may be the result of low-temperature injury to the tender leaves. See *shot hole: 1.*

shoulder. 1. Of livestock, the upper part of the fore-limb and/or the adjacent part of the back; the scapula region. 2. The upper foreleg and/or the adjacent part cut from the carcass of a hog, sheep, or other animal; the upper part of the carcass to which the foreleg is attached. 3. The part of a bird at which the wing is attached. 4. In turpentining, the uppermost corner of a face. See *face: 4; streak.* 5. A shoulder-like part or projection, as the projection around a tendon. 6. A sudden inward curvature in the outline of something, as the flaring part between side and neck of a milk bottle. 7. The edge of a road adjacent to the traveled way or pavement.

shoulder arm roast. A retail cut of meat which is a roast or roll made from the meat removed from in front of the arm bone of a foreleg of beef.

shoulder blade. The flat bone of the shoulder; scapula.

shoulder chuck. A cut of meat. See *chuck.*

shoulder clod. A retail cut of meat which is the heavy muscle back of the arm bone of a beef forequarter. This makes a very desirable pot roast or may be cut into steaks for Swissing. Also called *clear-cut shoulder.*

shovel. 1. A scoop or spade-like tool, consisting of a broad metal blade attached to a handle, used for lifting grain, soil, litter, etc. 2. A strap-shaped, V-pointed attachment which is used for various plowstocks, cultivators, grain drills, and soil scarifiers.

shovel cultivator. A cultivator which is equipped with shovels as contrasted to one equipped with disks or sweeps. See *shovel: 2.*

shovel plow. A double plow with moldboards set back to back which throws soil to both sides. It is used for opening furrows or for broadcast plowing. See *middlebuster.*

show. 1. A gathering together of livestock, other farm products, flowers, etc., for public exhibition, judging, etc. 2. To display livestock, farm products, etc., for rating by judges or for inspection by prospective buyers.

show barn. A building used for quartering and displaying show animals, as at a fair or other public exhibition.

show bull. A bull which is worthy of being exhibited because of excellent qualities of breeding, conformation, etc.

show cattle. Cattle of high quality as to breeding, conformation, finish, etc., that are on exhibit or are worthy of being shown.

shower. A local rain, usually of short duration.

show grounds. The site and buildings of an exhibition.

show halter. A fancy halter used on animals on display.

show horse. A sound, well-trained horse of good breeding, conformation, and performance.

showing. 1. The act of exhibiting or displaying. 2. That which is shown or displayed.

showman. One who exhibits or aids in exhibiting.

showmanship. The art or skill of showing or displaying products, animals, etc., to advantage.

show ring. The place at a show where animals are displayed or judged.

show string. The animals selected by an exhibitor for showing; the collection of animals taken to a show by an exhibitor.

showy autumncrocus. *Colchicum speciosum,* family Liliaceae, a bulbous herb grown for its autumnal, very large, cup-shaped, light pink to mauve flowers. Native to Europe and Asia.

showy chloris. *Chloris virgata,* family Gramineae, a grass widespread in the U.S., which is grazed on by livestock. Its forage rating is fair. In certain areas it is a weed. Native to tropical America. Also called *feather fingergrass.*

showy coneflower. *Rudbeckia speciosa,* family Compositae, a perennial herb grown for its golden flowers with purple-brown centers. Native to North America.

showy coraldrops. *Bessera elegans,* family Liliaceae, a tender, bulbous herb grown in very warm regions for its orange-red flowers, sometimes white-marked. Native to Mexico. Also called *Mexican coraldrops; coraldrops.*

showy crotalaria. *Crotalaria spectabilis,* family Leguminosae, a tropical annual herb grown as a cover crop and as a soil improving plant. Conditions of poisoning are reported with quail, chickens, swine, and cattle that have eaten the seeds. The green or dried plants are equally poisonous and, when eaten, frequently result in death. Native to tropical America.

showy frasera. *Frasera speciosa,* family Gentianaceae, a stout, perennial herb grown for its greenish-white, purple-spotted flowers. Native to North America. Also called *deer's-tongue.*

showy jasmine. *Jasminum floridum,* family Oleaceae, a half-evergreen, tender shrub grown for its golden-yellow flowers. Native to China.

showy lady-slipper. *Cypripedium reginae,* family Orchidaceae, a terrestrial orchid grown in the bog garden for its showy flowers with white petals and white lips strongly suffused with pink to purplish-pink. Native to North America.

showy milkweed. *Asclepias speciosa,* family Asclepiadaceae, a coarse herb which grows up to 6 ft. high, mostly in moist or somewhat shady situations, from Minnesota to British Columbia and south to Texas, Arizona, and California (U.S.). Although not generally eaten, instances of poisoning have been reported with sheep and cattle grazing on the leaves and stems.

showy orchis. *Orchis spectabilis,* family Orchidaceae, a hardy, woodland, tuberous-rooted herb grown for its purple-magenta flowers with a white, blotched lip. Native to North America. Also called *gray orchis.*

showy partridgepea. *Chamaecrista fasciculata,* family Leguminosae, a common herb which is a weed pest. Toward the end of the 18th century in Virginia and Maryland (U.S.), it was used as green manure. Also called *golden senna; partridge pea; prairie senna.*

showy pricklepoppy. *Argemone grandiflora,* family Papaveraceae, an annual herb grown for its large, showy, yellow or white flowers. Native to tropical America.

showy primrose. *Oenothera speciosa,* family Onagraceae, a perennial herb grown for its white flowers changing to pink. Escaped, it can become a weed pest. Native to North America.

showy tidytips. *Layia elegans,* family Compositae, a hardy, annual, hairy herb grown for its flower heads with yellow rays tipped with white. Native to California (U.S.).

shred. To tear or cut into small pieces, especially in long narrow pieces.

shredder. A machine which is used for tearing or cutting corn stalks and other forage crops into small pieces. See *husker-shredder.*

shredding. The process of reducing a thing to shreds.

shredding cylinder. The cylinder of a husker-shredder machine which has many knife-like blades on its perimeter and, when caused to revolve at high speed, shreds the corn stalks that are fed into it.

shrimp meal. A livestock feedstuff supplement produced from the undecomposed dried wastes of the shrimp industry. It contains the head, hull, or the whole shrimp, either singly or in mixture, and not more than 3 percent salt (NaCl), unless otherwise labeled, and in no case should it contain more than 7 percent salt.

shrink. To reduce in dimension, weight or volume, as with livestock, grain, wool, wood, etc.; to shrivel. See *shrinkage.*

shrinkage. 1. The process of reducing in dimension, weight, or volume. 2. With livestock, the loss in weight of an animal from the time it is taken off feed to the time it is marketed or slaughtered. 3. The loss in weight of grease wool due to scouring, expressed as a percentage of the original weight. 4. The loss in weight (moisture loss) of grain from harvest time to some later date. Such *shrinkage* is commonly reflected in the selling price. 5. The contraction of wood caused by drying, usually expressed as a percentage of some specific dimension (or volume) of the wood when green. 6. Loss in weight of either meat or carcass stored under refrigeration. Also called *cooler shrink; freezer shrink.*

shrinkage retarding treatment. A treatment of wood by impregnation with chemicals to reduce the normal shrinkage (and swelling) in service.

shrivel. To contract and wrinkle, as of leaves, grain, fruit, etc.; to dry up.

Shropshire. A widely distributed, English breed of hornless, brown-faced, medium-sized sheep. The breed possesses both mutton and wool qualities but is kept primarily for mutton and lamb production. Individuals are distinguished by having a complete covering of wool from the tip of nose to the feet. The wool is of medium length and fineness, and grades from (in U.S.) three-eighths to quarter blood. The average fleece yields between 8 and 10 lbs.

shroud. In butchery, a sheet of unbleached duck cloth that is immersed in warm water and pinned over the outside of a side of warm beef carcass before it is placed in the cooler so that after chilling the carcass will have a smoother appearance.

shrub. A perennial, woody plant, usually less than 10 ft. tall, which has several branches starting at or near the ground; a bush. No clear-cut distinction differentiates a tree from a *shrub.*

shrub althea. *Hibiscus syriacus,* family Malvaceae, a hardy shrub grown as an ornamental for its late-summer, red, purple, violet, or white flowers. Native to China. Also called *rose-of-Sharon.*

shrubbery. A group of shrubs considered as a whole; a shrub border or a planting of shrubs.

shrubby. Of the nature of, or consisting of, shrubs.

shrubby cinquefoil. Bush cinquefoil.

shrubby fern. Sweetfern.

shrub lespedeza. *Lespedeza bicolor,* family Leguminosae, a shrub grown as an ornamental for its late-summer, purple or rose-purple flowers. Native to Japan.

shrub live oak. *Quercus turbinella,* family Fagaceae, an evergreen shrub or small tree growing to 14 ft. in height at elevations up to 8,000 feet in Arizona, California, Colorado, New Mexico and northern Mexico. It is an abundant species of the chaparral and is browsed to some extent by livestock. Native to North America.

shrub yellowroot. See *yellowroot: 1.*

shrunken ice cream. A body defect of ice cream which may be associated with low solids, high over-run, and unfavorable storage conditions and which may manifest itself by open spaces between the ice cream and sides of the container.

shuck. 1. The outer covering of certain nuts, as of hickory, chestnut, etc., or the outer covering of an ear of corn; a husk or pod. 2. The calyx of the flower of stone fruits (as of the peach) which for an interval encloses the fruit but gradually dries, splits, and is ultimately pushed off by the expanding fruit. 3. To remove the shucks from corn, nuts, etc.

shucker. A person or machine which removes shucks usually from corn; a shucking machine.

shuck fall. The dropping of the shuck of stone fruits. Also called *shuck split.* See *shuck: 2.*

shucking. Removing shucks or husks.

shucking pin. See *husking pin.*

shuck split. Shuck fall.

shute. See *chute.*

shut-off nozzle. A spray nozzle equipped with a valve for restricting flow.

shut-off valve. A valve used as a means of restricting the flow of liquid or other substances.

shutter. A wooden cover of the same size as a cold-frame sash which is used as protection from cold for plants. They are either used alone or with soft, insulating mats.

shy. 1. Of, or pertaining to, a horse that is easily frightened or startled; timid; skittish. 2. To act in a startled manner, as a frightened horse may break gait and jump violently to one side, or otherwise show fright until he has a chance to observe the cause of his fear.

shy breeder. A male or female of any domesticated livestock that has a low reproductive efficiency. It is associated most often with males or females from highly developed breeds.

shy producer. Designating a tree or plant that bears a scant or very light crop.

siallite. A group of soils characterized by a silica-alumina ratio greater than two to one in the hydrochloric acid extract.

Siam cotton. Bourbon cotton.

Siamese coupling. A branched hose coupling used to divert the flow of water from one line into two lines or vice versa.

sib. In genetics, a brother or sister.

Siberian alfalfa. Sickle alfalfa.

Siberian crabapple. *Malus baccata,* family Rosaceae, a deciduous tree grown as an ornamental for its white flowers and yellow or red fruits. Native to Asia. Also called *cherry apple; cherry crab.*

Siberian elm. *Ulmus pumila,* family Ulmaceae, a hardy tree which is resistant to the Dutch elm disease and grows well in poor soils. It is much used in shelterbelts in the Plains states of the U.S. Native to Asia. Also called *dwarf elm.*

Siberian fir. *Abies sibirica,* family Pinaceae, an evergreen tree grown for its bright green foliage and bluish cones. The tree suffers greatly in hot, dry climates. Native to eastern Asia.

Siberian globeflower. *Trollius asiaticus,* family Ranunculaceae, a perennial herb grown in the moist flower garden for its orange-colored flowers. Native to Siberia. Also called *Asiatic globeflower.*

Siberian iris. *Iris sibirica,* family Iridaceae, a perennial herb grown for its tall flower stalk and blue flowers. It is a beardless iris. Native to central Europe and Asia.

Siberian larkspur. *Delphinium grandiflorum,* family Ranunculaceae, a perennial herb grown in the flower border for its blue or whitish flowers. Native to eastern Asia. Also called *bouquet larkspur.*

Siberian peashrub. *Caragana arborescens,* family Leguminosae, a spiny, hardy shrub or small tree grown for its pea-like, yellow flowers and as a windbreak shrub. Native to Asia. Also called *pea shrub; pea tree.*

Siberian squill. *Scilla sibirica,* family Liliaceae, a small, bulbous herb grown for its early-spring, deep blue flowers. Native to southwest Asia and Russia.

sibling. One of several sons and/or daughters of the same parents. See *sib.*

sib-mating. Mating between siblings, two or more individuals having one or both parents in common.

sick. 1. Suffering from illness of any kind; not sound or in fit condition. 2. Designating a soil unfit for the profitable production of certain crops as a result of being infected with disease organisms or because of the excess or deficiency of certain elements in its constitution.

sickle. 1. A sharp, curved metal blade fitted with a short handle; used for cutting weeds, grass, etc. It is one of the earliest hand implements used for harvesting small grain. Also called *reap hook.* 2. The cutting section of a hay or grain-harvesting machine.

sickle alfalfa. *Medicago sativa* var. *falcata,* family Leguminosae, a perennial, hardy herb which in its pure form has yellow flowers and sickle-shaped seed pods. It has been used extensively in hybridization for creating improved varieties more resistant to cold. Native to Europe. Also called *Siberian alfalfa.*

sickle bar. In a mowing or grain-harvesting machine the cutter bar on which the sickle with its cutting blades or sickle sections is mounted.

sickled. See *sickle-hocked.*

sickle-hocked. Designating a horse, cow, or sheep, having a crooked hock which causes the lower part of the leg to be bent forward out of a normal perpendicular straight line. It is a serious defect in the set of legs, especially of horses, and may result in curb *q.v.*

sickleleaf goldenaster. *Chrysopsis falcata,* family Compositae, a perennial low herb grown for its yellow flowers. Native to North America. Also called *ground gold-flower.*

sickles. The longer, curved tail feathers of a male chicken as distinct from the smaller, similarly curved tail-coverts, the lesser *sickles q.v.*

sicklethorn asparagus. *Asparagus falcatus,* family Liliaceae, a tall-climbing, leafless vine grown in warm areas for its dark green foliage, fragrant white flowers, and brown berries. Native to South Africa and tropical Asia.

sickly. In, or symptomatic of, a poor state of health; not robust or strong.

side. 1. In the body of an animal, either of the two lateral surfaces or parts extending from the shoulders to the hips. 2. An entire lateral half of a beef, veal, or pork carcass split lengthwise through the back bone and the corresponding mid under-surface. 3. Flesh taken from the lateral surface of an animal carcass, as a slab taken from the *side* of a hog carcass and used for preparing a *side* of bacon.

side arm. Dead arm of grape.

side arm of grape. Dead arm of grape.

side beef. Block beef.

sideboard. One of the boards which form the long side of a wagon or truck box; an additional board to raise the height of a wagon-box side, as those used in corn husking to prevent overthrow. Sometimes called *bang board.*

side-bone. Ossification of the lateral cartilages of a horse's foot, noticeable as hard, bony projections immediately above and toward the rear quarter of the hoof head. It is an unsound condition occurring, especially in draft horses, as a result of injury, conformation, excessive use, or perhaps old age. Also called *shell bone.*

side camp. In forestry, a small camp set up to accommodate a crew working away from a main camp or headquarters.

side check. A leather strap attached at the side of a horse's bridle to limit side movement of the head. *Cf. overhead check; checkrein.*

side clip. A metal piece, similar to a toe clip (*q.v.*), located on one or both sides of a horseshoe.

side-cut (timber). A timber cut from the side of a log.

side-delivery rake. A farm implement that rakes hay to one side into loose, continuous windrows for convenience in bunching or gathering by mechanical hay loaders. See *cylinder type side-delivery rake; dump rake; windrowing attachment.*

side draft. A sidewise pull produced when the center of a load or resistance is out of the true line of hitch or draft.

side-dress. To apply fertilizer to the soil at the side of a row, usually by a cultivator fitted with a fertilizer-distributing attachment or by a machine especially designed for this purpose.

side-dressing. Fertilizer used to side-dress growing plants.

side graft(ing). A method of grafting used outdoors and in the greenhouse. A rather long downward diagonal cut is made in the stock and a cion with the lower end shaped into a thin wedge is inserted within its bark, the cut side being kept close to the wood. Tying and waxing may or may not be done. See *shield graft(ing)*.

side-hitch sweep rake. A sweep rake (*q.v.*) with a horse hitched on either side of the implement rather than in front.

side hopples. Hopples (hobbles) arranged laterally to make a horse pace and prevent it from trotting or running.

side knife. One of the two large knives attached on either side of a sled-lister cultivator to destroy weeds and level middles.

side leak. A small, unnatural opening in the wall of a cow's teat; a defect in dairy cows.

sideline. In stock judging, the line formed by the extension of the lateral parts of an animal.

side oat. *Avena sativa* var. *orientalis,* family Gramineae, an oat very much like the common oat (*A. sativa*) except that the branches of its panicles tend to develop on one side. Not grown extensively because its yields are lower than those of the common oat.

sideoats grama. *Bouteloua curtipendula,* family Gramineae, a loosely-bunched, perennial, widely-distributed grass having a good forage rating on ranges and sometimes cut for hay. Native to western North America.

side of bacon. An entire slab of unsliced bacon.

side of beef. An entire lateral half of a beef carcass.

side-packed. Designating a method of packing dressed poultry in which fowls are placed on their sides in the shipping crates.

side saddle. A saddle for women so constructed that the rider sits with both feet on one (usually the left) side.

side slope. The slope of the sides of a hill, canal, dam, or embankment. Usually expressed as the ratio of the horizontal distance to the vertical such as two to one, four to one, etc.

side spur. A short, woody branch, commonly one which bears a blossom, bud or fruit, as in the apple tree.

side step. A stepping sideways; a foot movement expected of a show horse and desirable in a well-trained pleasure horse, as it assists in lining up horses in a show ring. Also called *transversal.*

side wall. The wall of a dam structure which extends from head-wall downstream the length of the apron, designed to prevent sloughing of banks or channels and to direct and confine overfall.

side wheeler. A horse which paces.

side whip grafting. A modification of whip graft(ing) (*q.v.*) in which the cion is placed on the side of a stock rather than on the end of it. Also called *bottle grafting.*

siding. 1. Boards used to cover the sides of buildings; one board usually being lapped over the upper edge of the board below. Also called *clapboard; weatherboard.* 2. Removing the hide from the side of an animal, as from a beef or veal carcass.

Siebold cherry. *Prunus sieboldii,* family Rosaceae, a small, deciduous tree grown for its pink, sometimes white, flowers. Native to Japan.

Siebold hemlock. *Tsuga sieboldii,* family Pinaceae, an evergreen, tender tree grown as an ornamental. Native to Asia.

Siebold plantainlily. *Hosta sieboldiana,* family Liliaceae, a perennial herb grown for its pale lilac flowers. Native to Asia.

Sierozem (soil). A zonal group of soils having a brownish-gray surface horizon that grades through lighter-colored material into a layer of carbonate accumulation and frequently into a hardpan layer. It developed under mixed shrub vegetation in a temperate to cool arid climate.

Sierra clover. See *purple clover: 1.*

Sierra juniper. *Juniperus occidentalis,* family Cupressaceae, an evergreen tree grown as an ornamental and for lumber. Native to North America. Also called *yellow cedar; western juniper.*

Sierra larkspur. *Delphinium scopulorum* var. *glaucum,* family Ranunculaceae, a variety of the tall mountain larkspur, a poisonous plant, which is found in the higher mountains of western Canada and Alberta, Washington to Nevada and California (U.S.). Also called *mountain larkspur.*

Sierra lily. *Lilium parvum,* family Liliaceae, a tall, bulbous herb grown for its orange-red flowers purple-spotted with yellow bases. Native to North America.

Sierra shooting star. Jeffrey shooting star.

sieva bean. *Phaseolus lunatus,* family Leguminosae, an herb usually of slender growth and not high climbing, which bears small, flat pods, 2 to 3 ins. long, with small, flat, edible seeds. It is widely cultivated in warm countries for its earliness and prolificacy, and has given rise to cultivated bush forms. Native to tropical America. Also called *civit bean; sivy bean; butter bean.*

sieve. A frame of metal or fabric with meshes through which the finer particles or substances are passed to separate them from the coarser forms. *Sieves* of special types are used in preparing many food products and also in the screening of seed. See *screen.*

sieved. Designating products, usually food products (fruits, vegetables, etc.) that have been reduced to a pulp and then passed through a fine-screened finisher to remove small pieces of fiber and to give a fine-grained, smooth purée. Such products are used chiefly as baby foods. Also called *strained.*

sieved fruit. See *sieved.*

sieved vegetable. See *sieved.*

sifter. Any device used for sifting, such as a mechanical sifter (sieve) used to separate small from large particles in the grain milling process.

sifting. The process of separating the finer from the coarser particles of a substance, as with a sieve.

sightwort. Greater celandine.

silage. Any material harvested in a green state and preserved in a succulent condition by partial fermentation in a container, as a silo. Also called *ensilage.* See *silo; silage crop.*

silage cart. A two- or three-wheeled hand cart with

a large box, used for hauling silage from the silo to the manger.

silage chute. A vertical chute down which silage is thrown to prevent scattering. It is a permanent part of a silo, located outside the silo door openings.

silage corn. Usually a vigorous-growing corn that is planted thicker than grain corn to produce a high yield of forage for ensiling. The kernels should be well dented and the ears should contain about 50 percent moisture at silo-filling time.

silage crop. Any of several crops grown and harvested for silage (*q.v.*), as corn, sunflowers, sorghum, small grains, legumes, grasses, etc., cut green and stored in a silo.

silage cutter. Ensilage cutter.

silage fork. Ensilage fork.

silage harvester. A field implement that harvests standing corn, cuts it into ensilage length, and elevates the chopped fodder into a truck or wagon box ready for transportation to and placement in the silo.

silene. Any herb of the genus Silene, family Caryophyllaceae. Species are grown in rock gardens and flower borders. Also called *campion; catchfly.* See *alpine silene; firepink silene; schafta silene; sea silene; sweet William silene.*

Silesian. A cheese made from cows' skim milk by a method similar to that used in making Hand cheese.

silica. SiO_2, silica dioxide, present in quartz, flint, etc.; one of the most abundant minerals in the earth's crust.

silica-alumina ratio. The molecular ratio of silica to aluminum oxide in a soil, clay, or other alumino-silicate mineral. It is the quotient obtained when the number of mol-fractions of silica are divided by the number of mol-fractions of alumina, both determined by standard fusion analysis of the soil or some part of it.

silicofluoride. See *sodium fluosilicate.*

silicon. Si, a nonmetallic element which always occurs in a combined form in nature, where it is exceeded in abundance only by oxygen with which it usually combines to form silica. Found in all soils, it is a constituent of plants but is not regarded as essential in fertilizers.

silique. A modified bicarpellary capsule in which a replum or so-called false septum separates the locule into two locules which are opened by marginal sutures.

silk. 1. The long, silky styles with stigmas of the corn plant, each of which is connected to an ovary on the ear and each of which must be pollinated to form a full ear of kernels. The *silks* are susceptible to pollination before they emerge from the husk and remain so for about two weeks afterwards. 2. The soft, fine, and shiny fiber which is produced by the silkworm.

silk-cotton tree. Kapok ceiba.

Silkie. An unusual breed of small, five-toed, crested, domestic fowls with an abundance of soft, fluffy plumage on their body, legs and toes, instead of the usual webbed feathers. The plumage is white, the skin is dark blue, and the face, wattles, and small, nearly round, flat comb are deep mulberry. The *Silkies* are believed to have originated in Japan. Their two varieties are the Bearded and the Non-bearded.

silking machine. A machine, equipped with revolving brushes and rolls, which removes nearly all of the silk from sweet-corn ears in the commercial process of preparing sweet corn for canning.

silkoak grevillea. *Grevillea robusta,* family Proteaceae, an ornamental tree grown for its showy, orange-colored flowers and handsome foliage with a fernlike appearance. It is a popular street tree in frost-free areas, and is grown by florists as a pot plant. Its timber is sometimes used in furniture making. Native to Australia.

silktassel. Any plant of the genus Garrya, family Garryaceae. Species are mostly ornamental, evergreen shrubs, rarely trees, which are grown for their attractive leaves and showy catkins. Also called *tassel tree.* See *tree silktassel.*

silk tassel tree. Tree silktassel.

silktree albizzia. *Albizzia julibrissin,* family Leguminosae, a tender tree grown in frost-free areas for its light pink flowers. Native to Asia. Often incorrectly called *mimosa.*

silkweed. Milkweed.

silkwood tree. Calabur tree.

silkworm. The larva of the moth *Bombyx mori* L., family Bombycidae, which feeds on mulberry leaves and in three or four weeks grows into a large caterpillar about 3 ins. long. It then constructs its cocoon by secreting a curious saliva that hardens upon exposure to the air to form soft, delicate threads of remarkable strength and pliability. The cocoons are skillfully unwound by expert operators and the threads are combined to form reels of raw silk. Silkworms that produce a single brood in a season are referred to as *monovoltine,* those producing more than one brood, *polyvoltine.* Native to China.

silkworm mulberry. *Morus alba* var. *multicaulis,* family Moraceae, a shrubby tree grown for its foliage which is fed to silkworms. Native to China.

silky camellia. Virginia stewartia.

silky cornel. Silky dogwood.

silky dogwood. *Cornus amomum,* family Cornaceae, a hardy, deciduous tree grown as an ornamental for its clusters of white flowers and its pale blue fruit. Native to North America. Also called *silky cornel.*

silky oak. Silkoak grevillea.

silky sophora. *Sophora sericea,* family Leguminosae, a small, perennial herb with white flowers and thick, leathery, hairy pods with few or several seeds. The seeds contain a toxic alkaloid and if eaten in large quantities may cause poisoning. Native to North America.

silo. A pit, trench, above-ground horizontal container, or vertical cylindrical structure of relatively air-tight construction into which green crops, such as corn, grass, legumes, or small grain and other feeds are placed and converted into silage (*q.v.*) for later use as a livestock feedstuff.

silo filler. A machine consisting of a power-driven conveyer, a cutter, blower, and a stand-pipe which is used to cut green crops into silage lengths and to elevate the material into the silo. See *ensilage blower; ensilage cutter.*

silt. 1. Small, mineral, soil particles, ranging 0.05 to 0.002 mm. or 0.02 to 0.002 mm. in the International system, in diameter. 2. A textural class of soils that contains 80 percent or more of *silt* and less than 12 percent clay. 3. Water-borne sediment with diameters of individual grains approaching that of *silt.* 4. To fill or obstruct a channel, etc., with water-borne sediment; to become filled or choked up with silt or sediment, as a channel, river-bed, etc. See *soil separate.*

silt basin. An enlarged construction or excavation to a deeper level in or beside a ditch or other drainage channel for the deposit of sediment carried by the water.

silt loam. A textural class of soil that contains 50 percent or more of silt and 12 to 27 percent clay, or 50 to 80 percent silt and less than 12 percent of clay.

silty clay. A textural class of soil that contains 40 percent or more of clay and 40 percent or more of silt.

silty clay loam. A textural class of soil that contains 27 to 40 percent of clay and less than 20 percent of sand.

silver Asiaglory. *Argyreia splendens,* family Convolvulaceae, a tender, tall-climbing vine grown in frost-free areas for its leaves, silvery-haired on the underside, and its funnel-shaped, rose-colored flowers. Native to Asia.

silver beardgrass. *Andropogon saccharoides,* family Gramineae, a coarse, perennial grass with dense, silvery-white, silky panicles, which is rated as good for grazing. Native to the U.S. Also called *silver bluestem.*

silver beet. Leaf beet.

silverbell tree. Carolina silverbell.

silverberry. *Elaeagnus commutata,* family Elaeagnaceae, a hardy, erect shrub, grown for its silvery-yellow, fragrant flowers and silvery berries. Native to North America. Also called *wolfberry.*

silver bluestem. Silver beardgrass.

silver buffaloberry. *Shepherdia argentea,* family Elaeagnaceae, a very hardy, spiny shrub or small tree, grown in a hedge and for its edible fruit used in jelly. Native to North America. Also called *wild oleaster; silverleaf.*

silver crassula. *Crassula argentea,* family Crassulaceae, a succulent shrub which is a very popular florist's plant for its very thick, shining leaves. It flowers only rarely, with rosy-red blossoms. Native to South Africa. Also called *jade plant; Japanese laurel; Japanese rubberplant.*

silveredge bishop's goutweed. *Aegopodium podograria* var. *variegatum,* family Umbelliferae, a stout, coarse, perennial herb grown as an edging plant and as an ornamental ground cover for its white-margined leaves. Native to Europe; naturalized in North America.

silver fir. *Abies alba,* family Pinaceae, an evergreen tree, grown as an ornamental. Native to central and southern Europe.

silverfish. *Lepisma saccharina* L., family Lepismatidae, a small, silvery or greenish-gray, wingless insect which prefers damp places, such as basement rooms, and feeds on papers, cards, wallpaper, or any starchy material. It may deface bookbindings, etc.

silver goldenrod. *Solidago bicolor,* family Compositae, a coarse, perennial herb grown for its white flowers. Native to North America. Also called *white goldenrod.*

silvergrass. Any grass of the genus Miscanthus, family Gramineae, grown for ornament. Also called *swordgrass; eulalia.* See *Chinese silvergrass.*

silver-gray. A coat color of a horse which is light grey with a white mane and tail.

silver groundsel. *Senecio cineraria,* family Compositae, a hardy, perennial, branching herb, grown for its silver-haired foliage and yellow or cream-colored flowers. Native to the Mediterranean area. Also called *dustymiller.*

silverking sagebrush. *Artemisia albula,* family Compositae, an herb grown for its white-felty leaves. Native to North America. Also called *silver-king artemisia.*

silver-lace vine. Silvervine fleeceflower.

silverleaf. Silver buffaloberry.

silverleaf cotoneaster. *Cotoneaster pannosa,* family Rosaceae, an evergreen or half-evergreen shrub grown as an ornamental for its foliage, profuse white flowers, and dull red fruit. Native to southwestern China.

silverleaf nightshade. *Solanum elaeagnifolium,* family Solanaceae, a perennial, poisonous herb which has silvery-haired stems and leaves and smooth, orange-yellow berries. Poisoning may result if the plant is ingested by livestock. Native to North America. Also called *white horse nettle; trompillo.*

silver leaf of fruit trees. A silvering of foliage of apple, plum, and other woody plants which follows injury during severe winters or as a result of invasion of trees by the fungus *Stereum purpureum,* family Thelephoraceae. The condition is commonly associated with blackheart. See *blackheart: 1.*

silver leaf of stone fruits. A fungous disease of stone fruits caused by *Stereum purpureum,* family Thelephoraceae, characterized by a metallic luster on the foliage and a dark discoloration of the heartwood where the fungus is growing.

silverleaf scurfpea. *Psoralea argophylla,* family Leguminosae, a perennial herb which is a poisonous plant. Native to North America.

silver-leaved poplar. White poplar.

silver linden. *Tilia tomentosa,* family Tiliaceae, a handsome, deciduous tree, pyramidal in form, grown as an ornamental shade tree. Native to Europe and Asia. Also called *white linden.* See *linden.*

silver maple. *Acer saccharinum,* family Aceraceae,

a deciduous tree grown as an ornamental for its silvery foliage. Native to North America. Also called *white maple.*

silver morning-glory. Silver Asiaglory.

silver-rod. Silver goldenrod.

silver sage. *Salvia argentea,* family Labiatae, a half-hardy, biennial herb grown for its whitish-yellow flowers. Native to the Mediterranean area.

silver scab. Silver scurf.

silver scurf. A fungous disease of potatoes caused by *Spondylocladium atrovirens,* family Dematiaceae, in which the skin is marked by irregular, silvery patches within which minute, dark specks are visible. Also called *silver scab.*

silver spruce. A clon of the Colorado spruce *(q.v.)* that is grown as an ornamental for its silvery foliage.

silvertop palmetto. Brittle thatchpalm.

silvervine actinidia. *Actinidia polygama,* family Dilleniaceae, a hardy, woody, climbing vine grown for its glossy, dark green foliage and white, fragrant flowers. Its sweet, edible fruits are greenish-yellow and as large as plums. Native to eastern Asia.

silvervine fleeceflower. *Polygonum auberti,* family Polygonaceae, a hardy, herbaceous, climbing vine grown for its foliage and dense spikes of greenish-white, fragrant flowers. Native to Asia. Also called *silver-lace vine; fleece-vine; Chinese fleece-vine; lace-vine.*

silverweed. See *early meadowrue.*

silvery cinquefoil. *Potentilla argentea,* family Rosaceae, a low-growing, perennial herb common throughout the northern U.S. east of the Rocky Mountains, which is sometimes a weed. Native to Europe, Asia and North America. Also called *silverweed.*

silvery lupine. *Lupinus argenteus,* family Leguminosae, a perennial herb whose pods and seeds (either green or dry) when eaten by livestock (especially by sheep) result in poisoning with symptoms of nausea, frothing at the mouth, nervous trembling, and convulsions. Native to North America.

silvicide. Any chemical which is capable of killing trees, usually by contact, as by spraying or injection.

silvicultural rotation. The rotation through which stands of timber should be grown to maintain maximum vigor of growth and reproduction.

silviculture. Forest culture; the application of the knowledge of silvics in the management of a forest.

silvics. The life history and general characteristics of forest trees and stands, with particular reference to environmental factors.

Simmenthaler. A large-boned, red-and-white spotted, Swiss breed of cattle, used for milk and beef as well as for work.

Simons cotoneaster. *Cotoneaster simonsii,* family Rosaceae, a partly-evergreen shrub grown for its foliage, white flowers, and scarlet fruit. Native to the Himalayas.

simple. A medicinal herb or the medicine made from it.

Sind. A breed of beef cattle which is under development in India. Probably of Arabian origin, it is adaptable to warm climates.

singe. To burn lightly, as to subject the carcass of a pig, fowl, etc., to a flame in order to remove bristles or hair, or to sear the ends of stems of flowers (which have a copious flow of sticky sap) to prolong freshness.

singer. A hog, singed after dressing, and suitable for making Wiltshire sides of bacon for export to England. Also called *Canadian singer.*

single. Designating a blossom having a more or less open center and one whorl each of sepals and petals. *Cf. double: 1.*

single-action atomizer. A type of hand spray consisting of a pump and a receptacle for holding spray liquid, which acts only on the stroke of the pump. *Cf. continuous-action atomizer.*

single-bail plow. A frame plow which has a single U-shaped iron piece attaching the beam to the frame of the plow. Its purpose is to allow the share to be raised to varying heights above the surface of the ground. *Cf. double-bail plow.*

single battery. A cotton gin which has one set of gin stands *q.v.*

single-bottom walking plow. One plow bottom *(q.v.)* that is attached to a beam with a clevis on its forward end for hitching to the pulling power, and a pair of handles at the rear for guiding.

single-breast gin. A type of cotton gin having a single gin breast, suitable for ginning only clean cotton that has no boll, hulls, or trash. *Cf. double breast.*

single comb. Of chickens, a comb consisting of a single, fleshy, serrated (usually five-pointed) formation which extends from the beak backward over the crown of the head.

single compound pistil. A single-seed-bearing organ (pistil) of a flower which is composed of two or more carpels *q.v.*

Single-Cream cheese. A soft, mild, uncured, cows' milk cheese prepared from thin cream (25 to 30 percent fat) and thickened with rennet. *Cf. Double-Cream cheese.*

single crib. A rectangular-shaped farm structure for storing ear corn, usually 5 to 8 ft. in width and of any desired length, which has open siding to permit circulation of air, and a roof usually sloping in one direction. *Cf. double crib.*

single-crop farming. Farming which depends largely upon one crop as a source of farm income. Also called *one-crop farming. Cf. diversification.*

single cross. The crossing of one strain, variety, inbred line, or breed of plants or animals with a different strain, variety, inbred line, or breed. The seed, plant, or offspring so produced is designated as a *single cross* or F_1. See *double cross: 1; three-way cross.*

single cut (clover). A clover, as mammoth clover, which gives only a single cutting of hay in a season.

single dahlia. A type of dahlia which has open-centered flowers and one row of ray florets.

single-disk. Designating a farm implement which is equipped with one disk: *e.g.,* a *single-disk* plow, or a *single-disk* furrow opener.

single-disk furrow opener. A furrow opener, used on certain grain drills, which consists of one disk securely fastened to the boot and set to run at a slight angle.

single-eye cutting. A stem or cion cut to include one eye or bud, used for rooting or grafting.

single-foot. See *rack: 1.*

single-grained structure. A soil in which there is no aggregation of particles, such as sand.

single-harpoon fork. A hay fork having one long, iron fork with an enclosed barb on the end which when forced to a desired depth in a load or stack of hay may be sprung so as to grasp the hay for lifting. *Cf. double-harpoon fork.*

single-layer side pack. A method of packing dressed poultry in which 12 birds are packed on their sides in one layer in each box. *Cf. double-layer side pack; breast pack; squat pack.*

singleleaf vinelilac. *Hardenbergia monophylla,* family Leguminosae, a woody vine, grown in frost-free areas for its small, pea-like, white, pink, or purple flowers in long, showy clusters. Native to Australia.

singlenod bellflower. *Campanula caespitosa,* family Campanulaceae, a perennial herb grown for its nodding, solitary, blue flowers. Native to Europe.

single-rib huller (gin). See *single-breast gin.*

single-rigged saddle. A saddle which has one cinch.

single-rod agitator. An axle, to which paddles are attached at intervals, placed within the seed box. As the axle turns, the paddles agitate the seeds to keep them flowing evenly into the feed cups. *Cf. double-rod agitator.*

single-row. 1. Designating a method of agriculture in which crops are planted in straight lines, with parallel, cultivated spaces of uniform width between all the lines. 2. Designating a machine which plants, cultivates, or harvests a *single-row* crop.

singleseed hawthorn. *Crataegus monogyna,* family Rosaceae, a thorny shrub or small tree grown as an ornamental for its white flowers and scarlet fruit. Often used as a hedge plant also. Native to Europe.

single-service container. A container for milk, cream, and other food products, made of stout, waxed paper. These containers are made in several shapes and with several types of closures and are meant to be used only once.

single-shovel plow. See *shovel plow; bull-tongue.*

singletree. In a horse-drawn carriage, the swinging bar usually made of wood or steel to which the traces of a harnessed horse or mule are hitched. Also called *swingletree; whiffletree; whippletree. Cf. doubletree.*

single-trunk, four-cane Kniffin system. A system for training grape canes in which a permanent trunk is trained to two horizontal wires, one above the other, 3 ft. and 5 to 6 ft. above the ground respectively. Four canes, two to the right and two to the left of the trunk, are selected and tied to wires. See *six-cane Kniffin system.*

sinistrorse. Designating plants whose stems twine or turn spirally from right to left as the bindweed. *Cf. dextrorse.*

sinker. A log whose specific gravity is greater than water.

sinker stock. Lumber made from logs whose specific gravity is greater than water.

sinkhole. A basin depression, often deep, circular, and well-like, present in many parts of the U.S., which is underlaid by limestone or gypsum. The sink results from the collapse of the roofs of subterranean solution caverns.

sinuate pear tree borer. *Agrilus sinuatus* (Oliv.), family Buprestidae, a beetle whose larvae work just beneath the bark and make narrow, winding galleries in trunks and branches of pear trees.

sinuate-striped flea beetle. *Phyllotreta Zimmermanni* (Crotch.), family Chrysomelidae, a striped flea beetle (*q.v.*) which is a garden pest.

sinus. 1. Any of the different cavities in men and some animals, some of which are enclosed by bone, others merely enlargements in a vessel or channel. Among the better known *sinuses,* because of the frequency with which they become infected in men and birds, are those located in the bones at the sides of the nose or nostrils and above the eyes. 2. A depression between two lobes in a leaf.

sinusitis. A sinus inflammation, usually of the bone-surrounded cavities of the head.

siphonflower. Dutchman's pipe.

Siraz. A Serbian, semisoft cheese usually made from whole milk.

sire. 1. The male parent. 2. To father or to beget; to become the *sire* of, as to *sire* a fine cow.

sire index. A mathematical index for measuring a dairy bull's transmitting ability in terms of milk and butterfat. It tends to measure what production may be expected of the daughter when a bull is mated to cows of known productivity.

siris tree. Lebbek albizzia.

Sir Iz Mjesine. A cheese made in Yugoslavia from ewes' skim milk and cured in the fresh skin of a sheep or goat.

sirloin. A loin or part of a loin of beef. In the U.S., a steak or roast cut from the rear end of a beef loin, next behind the porterhouse.

sirloin lamb roast. A roast prepared from meat cut from the rear end of a loin of lamb. Also called *sirloin roll.*

sirloin pork roast. A roast cut from the tenderloin end of a pork loin.

sirloin tip. A retail cut of beef made from the lower forward end of a full round of beef (cut U.S. National style) which includes the lower end of the sirloin section. Also called *bottom sirloin; round tip.*

sirloin veal cutlet. Sirloin veal steak.

sirloin veal steak. A veal steak cut from the large or rear end of a full loin of veal. It is a retail cut and corresponds to a sirloin steak in beef. Also called *sirloin veal cutlet.*

Sir Mastny. A cheese made in Yugoslavia from ewes' whole milk.

Sir Posny. A cheese made in Yugoslavia from ewes' skim milk.

sirup. The thick, sweet liquid made from water, the juice of fruits, etc., boiled with sugar; or the concentrated juice of certain saccharine plants, such as sugar cane, sugar sorghum, the sugar maple, etc. Also called *syrup.* See *corn sirup.*

siruper. 1. One who controls the boiling of sap to the proper consistency of maple sirup. 2. A person or device that fills bottles or cans with sirup, or adds sirup to food products, as in the commercial canning of fruits. See *siruping machine.*

sirup hydrometer. A floating instrument with a special scale for measuring the specific gravity of concentrated sugar solutions. Used to determine the strength of sirups, as in the making of maple sirup.

siruping machine. A machine that automatically adds a measured amount of sirup to each can in commercial fruit-canning operations. Also called *siruper.*

sisal. *Agave sisalana,* family Amaryllidaceae, a tropical plant with a short stem and long, succulent leaves which yield a strong, durable fiber used for making rope, twine, etc. Probably native to America. Also called *Bahama hemp; sisal hemp.* See *henequen agave.*

sissoo. *Dalbergia sissoo,* family Leguminosae, a tropical tree grown in frost-free areas for its small, pea-like, yellowish-white flowers in short clusters and for its durable lumber, called *rosewood,* used in railway construction, etc. Native to India.

site. 1. A location for a specific use. 2. A location for a forest planting or for the natural development of a forest. 3. An area, considered as to its ecological factors with reference to capacity to produce forests or other vegetation; the combination of biotic, climatic, and soil conditions of an area. 4. The situation of a fruit planting regarding elevation, topography, soil, and nearness to large bodies of water.

site index. An expression of forest site quality based on the height of the dominant stand of an arbitrarily chosen age.

sitfast. A piece of attached dead skin which prevents the healing of a sore on the back of a horse.

Sitka columbine. *Aquilegia formosa,* family Ranunculaceae, a perennial garden herb grown for its red and yellow flowers with straight red spurs. There are several horticultural varieties, some dwarf, others double-flowered, and one with nearly white flowers. Native to Siberia and North America.

Sitka falsecypress. Nootka falsecypress.

Sitka spruce. *Picea sitchensis,* family Pinaceae, an evergreen tree grown as an ornamental and as a timber tree. Native to North America.

Sitka-spruce weevil. *Pissodes sitchensis* Hopk., family Curculionidae, a small weevil whose larvae cause serious damage to Sitka spruce in the Pacific Northwest (U.S.), by killing or injuring terminal shoots of young trees and causing crooked trunks and forked worthless trees.

sitter. See *setting hen.*

sitting. See *setting: 1; 2.*

six-cane Kniffin system. A system for training grape vines in which a cane is trained to the right and left of the trunk along each of three parallel wires; otherwise it is similar to the more commonly used single-trunk, four-cane Kniffin system *q.v.*

six-months wool. Fall wool.

six-row barley. A variety of barley *(q.v.)* in which all three spikelets at each node are fertile so that the spike has six rows of kernels. It is the most important type grown in the U.S.

six-spotted leafhopper. *Macrosteles fascifrons* (Stål), family Cicadellidae, a leafhopper (*q.v.*) that transmits aster yellows.

six-spotted mite. *Eotetranychus sexmaculatus* (Riley), family Tetranychidae, a small, red spider mite that is a serious pest of citrus trees. It works with others in colonies on the underside of leaves, causing damage to foliage.

sixties (60's). An English wool grade which is the equivalent of the American grade, half-blood *q.v.*

six-tooth sheep. A three-year-old sheep.

sixty-four's (64's). An English wool grade which is the equivalent of the American grade, fine *q.v.*

sixweeks grama. *Bouteloua barbata,* family Gramineae, an annual grama grass of mesas and rocky hills of the southwest range country of New Mexico, Arizona (U.S.), and northern Mexico. It furnishes forage of value in the early spring.

size. To grade or classify according to bulk.

sizer. 1. A device which sorts seeds, fruits, etc., according to size. Also called *grader.* 2. A machine which surfaces timbers to exact dimensions.

size of business. The magnitude of the business conducted on a farm in a given year, commonly expressed in terms of the number of units of any important item of farm property, farm receipts, productive man work units, etc.

size reduction. Cutting, crushing, grinding, and/or milling. The reduction is brought about by means of a disintegrator or other machine, without change in chemical properties of the material. Such processes include cutting fruit or vegetables for canning, shredding sweet potatoes for drying, chopping fodder, grinding limestone, milling flour, and grinding grain.

sizing. That act of separating and arranging according to size. Also called *grading.*

skein. 1. A metal thimble which protects the spindle of a wooden axle. 2. A quantity of thread, yarn, silk, etc.

skeletal soil. A soil in which the mass of material consists mainly of coarse rock fragments. See *Lithosol.*

skeletonizer. Any insect that eats the green fleshy part of leaves, leaving only veins and a transparent membrane. See *apple leaf skeletonizer; birch skeletonizer.*

skeleton weed. See *rush skeletonplant; rush skeletonweed.*

skep. A beehive without movable frames, especially one which is made of twisted straw or willow twigs.

skewbald. A paint horse with some white and fawn (bay) markings.

skewer. A wooden or metal pin which fastens meat to keep it in form during cooking.

skewerwood. Brook euonymus.

skid. 1. A plank, roller, frame, etc., on which something heavy may rest or on which it may be slid or pushed along. 2. In forestry, logs or poles which are used in pairs to form a skidway (*q.v.*). 3. To pull logs from the felling site to a loading point or mill.

skidder. 1. A person who skids logs. 2. A yarding machine which skids logs by a cable.

skidding tongs. Tongs for grasping logs, used for skidding logs in logging operations.

skidway. A prepared place for piling logs prior to loading which consists of two parallel supports at right angles to the road generally raised at the end nearest the road.

skim. 1. To take something from the top, as to clear scum or matter floating upon the surface of a liquid. 2. To remove cream from milk which has stood for a period of time, by means of a special utensil or apparatus.

skim colter. A triangular-shaped, miniature plow attached to the colter yoke, which turns over a small, narrow furrow directly before the main plow for the purpose of covering trash and organic matter. Also called *jointer*.

skim milk (skimmed milk). That portion of milk which remains after removal of the fat. See *nonfat.*

skim milk cheese. A cheese made of skim milk, such as Sapsago, Gammelost, Hand cheese, and Cottage cheese.

skim milk concentrate. Skim milk condensed to concentrations between 20 and 40 percent solid content, used extensively in food manufacture as an excellent source of nonfat milk solids in ice cream, bakery goods, and confections. It may be used to standardize evaporated milk and is an intermediate form in the manufacture of dry skim milk. See *nonfat dry milk.*

skim milk powder. See *nonfat dry milk.*

skin. 1. The flexible integument which forms the external covering of an animal, especially of vertebrates. 2. Anything that resembles *skin* in nature or use, as the flexible outer covering or peel of fruits, etc. 3. The pelt of a small animal, such as calf, sheep, goat, fox, mink, etc., which is usually dressed, tanned, or intended for such treatment. 4. To strip the *skin* from. 5. To rub the *skin* off. 6. To peel.

skin abrasion. 1. A skin sore caused by rubbing or external friction. 2. In dressed poultry, reddened or bruised outer skin resulting from rubbing or rough handling.

skinner. 1. A person who cures or removes animal skins. 2. One who drives horses or mules.

Skinner columbine. *Aquilegia skinneri,* family Ranunculaceae, a perennial herb grown for its greenish-yellow flowers with red spurs. Native to tropical America.

skinning knife. A thin, sharp knife, with an upward-curving, 5 or 6 in. long blade, used for pelting and skinning slaughtered animals. See *scimitar boning knife.*

skinning rack. A trough or troughlike rack which holds a slaughtered sheep, hog, or beef animal on its back during skinning. See *beef pritch.*

skin off. Designating a manner of dressing a veal carcass for the wholesale trade in which the skin is removed. *Cf. skin on.*

skin on. Designating a manner of dressing a veal carcass for the wholesale trade in which the skin is left on to prevent the outer surface from becoming dark and dry. *Cf. skin off.* See *hog-dressed.*

skin out. To skin a slaughtered animal, especially to separate the skin from such parts as the face and shank.

skin spot. A superficial lesion of the potato, caused by the fungus *Oospora pustulans,* family Moniliaceae, which appears as a slightly raised, or depressed, dark, circular area. The skin is not generally ruptured. See *powdery scab of potato.*

skin tumor. A protuberant growth on the skin, such as a wart or a skin cancer.

skin wool. See *pulled wool.*

skin worm. 1. *Cacoecia franciscana,* family Tortricidae, a moth whose larvae burrow underneath the apple skin. Found in western U.S. 2. *Filaria medinensis,* family Filariidae, a thin, very long, parasitic nematode of white color which infests the skins of men and animals in West Africa and the Caribbean area. Also called *Guinea worm.*

skip. An inferior or cull lamb.

skipjack. See *click beetle; wireworm.*

skipper. Any of various skipping insects, especially the cheese skipper *q.v.*

skippy. In meat judging, lamb or veal that is thin and soft.

skirret waterparsnip. *Sium sisarum,* family Umbelliferae, a hardy, perennial plant grown as an annual for its tender, edible roots. Native to Asia.

skirt. 1. The diaphragm muscle, the muscular part of the membrane which separates the abdominal from the thoracic cavity. It is edible meat. See *skirt steak patty.* 2. To remove inferior grade wool from fleece. 3. See *saddle skirt.*

skirted. Designating wool from which belly-fleece and other inferior types have been removed. See *skirting.*

skirting. Inferior quality wool, especially that taken from the underside, the legs, etc. In sheared wool, it consists of inferior heavy-shrinkage parts which are removed from around the edges of a fleece and packed separately.

skirt steak patty. A steak patty which consists of slices of a rolled skirt of beef.

skite. To caper about in front of sheep, as a dog (New Zealand). It derives from the Australian slang word meaning "to boast."

skive. To cut leather or rubber into thin strips.

skiver. 1. A thin leather made from the grainy

side of a cut sheepskin, usually tanned and used for cheap bookbinding, purses, etc. 2. One who pares or splits leather. 3. A cutting device used in skiving leather.

skunk barley. Foxtail barley.

skunkbush sumac. *Rhus trilobata,* family Anacardiaceae, a hardy shrub occasionally grown in hedges for its brilliant red, autumnal foliage. Its foliage has a foul odor. Native to North America. Also called *squaw-bush; lemonade sumac.*

skunkcabbage. *Symplocarpus foetidus,* family Araceae, a foul-smelling, coarse marsh herb with large, deep roots and large, nearly round leaves. It flowers in very early spring and produces an aggregate of scarlet, berry-like fruits. Native to North America. Also called *polecat weed; skunkweed; swamp cabbage.*

skunk spruce. White spruce.

skyblue laceflower. *Trachymene coerulea,* family Umbelliferae, an annual herb grown in the flower garden and the greenhouse for its pale blue or lavender flowers which are very similar to flowers of wild carrot *(q.v.).* Native to Australia. Also called *blue lace flower.*

skyblue mariposa. *Calochortus coeruleus,* family Liliaceae, a cormous herb grown for its white flowers lined and dotted with blue. Native to North America. Also called *star tulip; skyblue pussyears.*

skyflower. Any of several tropical shrubs or trees of the genus Duranta, family Verbenaceae; native to America. See *creeping skyflower.*

skyline. In forestry, the cable suspended between the head spar tree and the trail tree on which the trolley travels in cableway logging.

sky pond. A pond constructed near the ridge or top of a hill or mountain. Since there is little watershed, it depends largely on direct rainfall or snowfall for filling it, and is sometimes supplemented by runoff from farm buildings.

slab. 1. The exterior portion of a log which is removed in sawing lumber. 2. An unattractive, overripe or broken, dried fruit (prune, apricot, etc.) which is sorted from a drying tray in commercial fruit drying. 3. A piece of unsliced bacon.

slab streak. In turpentining, a streak so chipped that the entire surface of the newly exposed wood is turned outward toward the direct rays of the sun.

slack at the heart. An objectionable body conformation of an animal, as a cow, in which there is a restriction in the spring or depth of ribs in the forequarters. Also called *shallow at the heart.*

slack coal. The fine screenings of coal sometimes fed to hogs as a mineral substitute. Hogs generally relish *slack coal* but its value as an adequate substitute for a balanced mineral mixture is doubtful.

slack cooperage. A barrel or keg suitable only for nonliquid products. *Cf. tight cooperage.*

slacker. A lumber-stacking machine with a drum-operated skyline *q.v.*

slack-filled. Designating a can in which the level of the contents is approximately 10 percent or more below the top. In fruit or vegetable products canning, a *slack-filled* container prevents damage to the contents from drastic changes in temperature.

slacking skyline. A skyline *(q.v.)* that can be raised and lowered by means of a drum.

slack in the loins. Indicating an animal, as a beef steer, which is thinly-fleshed or narrow in the back and rearquarters.

slack scald. See *semiscald.*

slack water. 1. The interval between high and low water when the tide is in a state of rest. 2. The apparently motionless water in the upper reaches of a dam. 3. In a log jam, the resulting sluggish flow of water.

slade. See *sole: 2.*

slag. The dross from iron ore furnaces and a by-product of the steel industry which can be converted into forms for agricultural use, as fertilizer and soil amendments. Some prepared material from *slag,* containing little or no phosphorus, but calcium in the form of silicates, may have value primarily as a liming material and a soil conditioner. Other *slags,* as basic *slag,* have some value as fertilizer because of the phosphoric acid which they contain.

slaked lime. 1. CaO, quicklime or burnt lime, treated with water to produce hydrated lime, which in its pure form has the composition $Ca(OH)_2$. It is marketed in a dry, powdered form and is used as a soil amendment in addition to other uses. Also called *calcium hydroxide; hydrated lime.* 2. Burnt lime which has been exposed to the open air and thereby lost its causticity.

slash. The branches, bark, tops, chunks, cull logs, uprooted stumps, and broken or uprooted trees which remain on the ground after logging.

slash disposal. Treatment or handling of slash to reduce the fire hazard or for other purposes which includes broadcast burning, strip burning, piling and burning, lapping, scattering, pulling tops away, etc.

slashing. Forest area which has been logged off and upon which the slash *(q.v.)* remains.

slash pine. *Pinus caribaea,* family Pinaceae, an evergreen tree grown for its lumber used in general construction and as a source of turpentine and rosin. Native to North America. Also called *Cuban pine; loblolly pine.*

slash (wind). See *windfall.*

slat. A thin, narrow strip of wood used in the manufacture of crates, etc. 2. A sheepskin after the wool has been pulled but before other treatment is given it.

Slate. A variety of turkey which has slaty-blue colored plumage. Its shanks and toes are pink.

slat moldboard. A type of moldboard *(q.v.)* that has lengthwise sections cut out leaving only about half of the surface to come in contact with the furrow slice.

slatted crib. A crib for storing ear corn whose sides are constructed of long, narrow boards, spaced to allow good ventilation.

slat tray. A shallow tray, about 3 by 3 ft. in size, which has spaced, narrow wooden strips attached to the bottom. It is used to hold fruit, etc., in forced-draft dehydrators.

slaughter. 1. The butchering of cattle, sheep, and other animals for food. See *slaughter laws.* 2. To kill cattle, sheep, and other livestock for food; to butcher. 3. Designating an animal suitable for *slaughter.*

slaughter animal. An animal intended for slaughter, especially one possessing sufficient fat or finish to meet the requirements of packers or consumers.

slaughter calf. In marketing, a calf less than 12 months but more than 3 months of age, suitable for slaughter.

slaughter cattle. Cattle which go immediately to meat processors as differentiated from feeder or breeder cattle.

slaughterer. One who slaughters livestock; a butcher.

slaughter ewe. In marketing, a mature female sheep sold for immediate slaughter, as differentiated from a ewe sold as a feeder or stocker.

slaughter house. A building or place where meat animals are butchered; an abattoir.

slaughter lamb. In marketing, a young ewe, wether, or ram (generally under 1 year of age), weighing 60 to 100 lbs. and having sufficient fat or finish to meet the requirements of packers and consumers.

slaughter laws. In the U.S., state and federal laws have been established dictating the methods to be used in stunning an animal prior to slaughter. Among the various humane methods is the use of carbon dioxide gas as an anaesthesia; a captive bolt which gives the animal an electric shock, stunning it; or a rifle bullet. The knocking hammer is not used in establishments under federal inspection, and tranquilizing drugs are not used as they leave a residue in the meat. Also called *humane slaughter laws.*

slaughter pig. In marketing, a young pig weighing under 120 lbs. which is slaughtered only when the slaughter demand exceeds that for feeders. See *feeder: 1.*

slaughter ram. An unaltered male sheep over one year of age that is sold for slaughter.

slaughter steer. In marketing, a steer suitable for immediate slaughter.

slaw. A type of salad made of fresh, shredded cabbage. Also called *coleslaw.*

sled. 1. A low platform or vehicle on runners, used to convey a load and usually pulled by two horses. It was much used on farms when horses were the common means of pulling farm implements. 2. To gather or harvest any crop by means of the field *sled,* such as cotton, corn for shocking, etc.

sled corn cutter. See *sled cutter.*

sled cutter. An implement which consists essentially of a sled, sometimes with low wheels, on one side of which is attached a long, sharp knife, sloping backward at an oblique angle to the row of the crop to be cut. The sled is driven close enough to the plant row for the slanting knife to cut the stalks with a sliding cut. There are many different types.

sledded cotton. Cotton mechanically harvested with the cotton sled *(q.v.).* Cotton harvested in this manner usually contains excessive amounts of trash.

sled harvester. Cotton sled.

sled lister cultivator. A sled cultivator which consists of two oak runners to which are attached crusher boards near the front end. Just behind these, on each side, are attached large knives to destroy weeds and level middles or ridges. Gangs of disks are coupled to the rear arch. This type of cultivator is adapted to the cultivation of a listed crop in its early stages of development.

sleeper. 1. A wooden cross-member supporting railroad rails; a tie. 2. An unbranded calf; a maverick. 3. A horse with equine encephalomyelitis *q.v.*

sleeper foal. A foal affected by omphalophlebitis *q.v.*

sleepering. Catching a calf before it has left its mother and placing a mark of ownership on it (southwestern U.S.).

sleeping sickness. Equine encephalomyelitis.

sleepy catchfly. Sleepy silene.

sleepy cream. Cream which will not make butter in the ordinary churning time.

sleepy Dick. Star-of-Bethlehem.

sleepy foal disease. Omphalophlebitis.

sleepygrass. *Stipa robusta,* family Gramineae, a coarse bunchgrass, growing 3 to 6 ft. tall, native to dry plains and open woodlands from Colorado and western Texas to southern Calfornia (U.S.), and to Mexico. It is said to produce sleepiness in horses and sheep and to a lesser extent in cattle. Its general forage rating is poor.

sleepy seeds. Seeds that are slow or irregular in sprouting, a condition of dormancy.

sleepy silene. *Silene antirrhina,* family Caryophyllaceae, a slender, erect, slightly branched, annual plant, with the upper portion of stem sticky at maturity. "Catchfly" refers to the retarding by this sticky stem of small insects approaching the flower from below. Also called *sleepy catchfly.*

sleet. 1. Fine ice particles, or partly frozen rain that freezes as it falls. 2. The coating of thin ice that sometimes forms on trees, buildings, etc., when rain or *sleet* falls at a low temperature. 3. A storm or shower of *sleet.*

slender deutzia. *Deutzia gracilis,* family Saxifragaceae, a shrub grown in warmer sections as an ornamental and forced by florists for its numerous, white flowers. Native to Japan.

slender duck louse. *Anaticola crassicornis* (Scop.), family Philopteridae, a louse sometimes found on ducks but rarely in sufficient numbers to cause serious injury.

slender goose louse. *Anaticola anseris* (L.), family Philopteridae, a louse which infests geese but seldom to the extent of causing serious injury.

slender iris. *Iris gracilipes,* family Iridaceae, a perennial herb grown for its lilac or pinkish-mauve flowers. It is a crested iris. Native to Japan.

slender oat. *Avena barbata,* family Gramineae, an annual herb which is a weed pest in the Pacific coast states of the U.S. It provides fair forage, especially when young. Native to Europe and Asia.

slender pigeon louse. *Columbicola columbae* (L.), family Philopteridae, a feather-chewing louse which infests pigeons. Also called *feather louse.*

slender turkey louse. *Oxylipeurus polytrapezius* (Burm.), family Philopteridae, a louse which infests the primary wing feathers of turkeys.

slender wheatgrass. *Agropyron pauciflorum,* family Gramineae, a widely-distributed wheatgrass which makes a choice forage for all kinds of livestock and is used in the reseeding of range lands. Native to western North America.

sliced veneer. Very thin sections of wood sliced off by moving a log or bolt of wood against a knife.

slick. Unmarked stock; unbranded calf (southwestern U.S.).

slick ears. Unbranded range horses or animals having no ear-slits to mark ownership. Also called *slicks.*

slicker. See *drag: 1.*

slick-seeded cotton. Any variety of cotton having smooth seeds as sea-island cotton. Also called *smooth-seeded cotton.*

slick spots. Small bodies of black alkali soil which, in a certain moisture condition, are slick and shiny when turned by a plow.

slide. 1. The slipping of a mass of earth or rock down the slope of a hill or mountain; a landslide. 2. A smooth surface for sliding; a chute *q.v.*

sliding poise. A movable weight mounted on the bar of a scale beam.

slight dressing defect. A minor dressing fault, as a few scattered pin feathers, a small abrasion, tear, or the like, which mars the appearance of poultry prepared for market.

slight grading. In egg sorting, a casual or loose method of grading; grading with consideration of only the more obvious deviations from grade specifications, as that done at many rural collection points.

slightly aged. Designating a condition of flavor in butter indicating a lack of freshness often caused by faulty or prolonged storage.

slightly flat. Designating a flavor defect in butter or cheese characterized by a lack of the desired flavors specific to the product.

slightly normal feed. Designating the mildly objectionable off-flavor in butter resulting from the feeding of sound, decay-free roughage, having an aromatic odor normal to the feed.

slightly obnoxious weeds. Designating an undesirable flavor in milk or butter frequently occurring in the spring or where grass is short and cows are consequently forced to graze on weeds.

slightly onion or garlic. Designating a mild, but very objectionable flavor frequently found in market milk, especially in the spring and in areas where wild onions or garlic are common.

slightly patchy at the pins. In judging, designating the rearquarters of a beef animal not evenly fleshed over the thurl or hip joint.

slightly rape, cabbage, turnip. A flavor defect of butter frequent in seasons of abnormal feed conditions.

slightly smothered. Designating an undesirable flavor in milk and cream caused by improper cooling and storage.

slightly soiled surface and surface mold. A body defect often associated with poor finish in manufacture and ripening of Cheddar cheese.

slightly storage. Designating a slightly objectionable flavor in dairy products. It is ascribed to staling during storage.

slightly weak. Designating an egg white lacking in thickness or viscosity to an extent that allows the yolk to move quite freely from its normal position.

slightly weedy. A flavor defect of market milk or butter resulting from the cows grazing on weedy pastures.

slightly woody. Designating an undesirable flavor of butter caused by old, worn, or poorly conditioned wooden churns or storage tubs.

slime. 1. A bacterial rot of lettuce resulting in a wet, slimy decay of the leaves and heads. It is caused by *Erwinia carotovora,* family Enterobacteriaceae, *Pseudomonas viridilivida,* family Pseudomonadaceae, *P. marginalis,* and other species of bacteria. *Slime* occurs in the field, during warm, muggy weather; in lettuce not refrigerated in transit, and at markets. Also called *bacterial rot of lettuce.* 2. In the clarification of milk, a collection or deposit of dirt, leucocytes, cells, and other viscous matter thrown to the walls of the clarifier bowl. 3. In the curing or storage of cheese, a condition frequently produced on the surface by yeasts, bacteria, or other contamination, or faulty handling, causing a slimy surface on the cheese. 4. To remove the slimy or viscous coating from animal intestines, as in the preparation of sausage casings.

slime flux. A persistent, fermenting exudate or seepage from trunk or branch wounds in deciduous trees. In some cases pathogenic organisms are associated with the flux. *Cf. alcoholic slime flux; brown slime flux.*

slime mold. Organisms which undergo a stage of development in which they are amoeba-like, eventually forming a definite slimy plasma, totally lacking cell walls and obtaining their food by engulfing it in the manner of animals. On spore formation, rounded bodies are formed each with a cell wall (the spores), which upon germination again become amoeba-like. They are usually called *Mycetozoa* or *Myxogastres,* depending on the idea of the student as to their true relationship. One order of this group is the Plasmodiophorales which are parasitic on plants or animals.

slime rot of lettuce. See *slime: 1.*

slimflower scurfpea. *Psoralea tenuiflora,* family Leguminosae, a bushy-branched herb found on dry prairies, in open woods, and on rocky banks of midwestern to southwestern U.S. Cattle and horses are reported to have been poisoned by grazing upon its herbage. Native to North America.

slimstem lily. *Lilium callosum,* family Liliaceae, a bulbous herb grown for its scarlet, dark-spotted flowers. Native to Japan.

slimy milk. An unwholesome condition sometimes noticed in sweet milk when delivered from the farm. It is caused by bacterial action, a contamination from unclean or poorly sanitized milk utensils. See *ropy milk.*

sling. A mechanical device consisting of ropes or cables, pulley and holding unit, used for lifting heavy loads.

slink. 1. The young of an animal brought forth prematurely or abortively, especially a calf. 2. The flesh or skin of such a calf or other animal. See *bob: 1.*

slink calf. See *slink: 1.*

slip. 1. A soft-wood or herbaceous cutting from a plant, used for propagation or grafting. 2. An incompletely castrated male. 3. Curdled milk. 4. To take a cutting or cuttings from a plant. 5. To abort.

Slipcote. A soft cheese made from cows' whole milk in the county of Rutland, England.

slipes. Pulled wool.

slip-nose share. A share *(q.v.)* whose point is detachable.

slipped tendon. A crippling in young growing birds which results from a nutritional disturbance. See *perosis.*

slipped wing. A wing (of a fowl) that is not held up in proper position but droops; a disqualification.

slippery elm. *Ulmus fulva,* family Ulmaceae, a tree grown for its wood used in slack cooperage, boxes, veneer, furniture, interior trim, etc. The mucilaginous inner bark is sometimes chewed by children, who like its taste. Native to North America.

slip scraper. A drag tool used for scooping up dirt and moving it by sliding. Also called *drag pan; slip pan.*

slip share. A share *(q.v.)* that has no extension to form the landside.

slip-tongue cart. A special form of logging wheels *(q.v.)* used for transporting logs.

Sliwowitz. A brandy made from plums (Yugoslavia).

slobber. The saliva dripping or drooling from the mouth of an animal, usually a condition found following extreme exertion or excitement, or after eating certain types of feed; sometimes a result of infection.

sloop. 1. A pair of long sled-runners equipped with a tongue and bunks on which short logs are hauled. 2. A pair of runners on which the forward ends of logs are loaded for hauling. 3. To haul logs down steep slopes on a *sloop* equipped with a tongue.

slop. A homemade, wet feed mixture made from flour mill by-products, grains, whole or crushed, and other waste or surplus farm products, such as kitchen waste, skim milk, etc. It has been much used with growing swine. See *swill.*

slop culture. 1. The application of nutrients in solution, at intervals, to the medium in which plants are growing. *Cf. drip culture.* 2. See *mud in: 2.*

slope. 1. Any land surface marked by an ascent or descent, as a hillside or terrace escarpment. 2. The *slope* of land in percent as determined by direct measurement with an Abney level; slope (percent) = vertical distance in feet × 100, divided by horizontal distance in feet.

slope of pile. The deviation from a horizontal plane of boards in a pile, from front to back; for air-seasoning, usually about 1 in. to 1 ft. in length.

sloping bench terrace. A bench having no fall from end to end but measurable slope from the back edge to the front edge. See *bench terrace.*

sloping pasterns. An undesirable weakness in the lower forelegs of a horse. Also called *coon-foot.*

sloping rump. An undesirable feature of the rump and rearquarters of cattle. The rump should be more nearly level and straight according to most scorecard descriptions.

sloppy attachment. An undesirable shape and attachment of the udder to the body of the cow.

slot grader. Notch grader.

slough. A narrow stretch of backwaters, frequently the abandoned channel of a river, in an alluvial plain, a narrow, marshy depression.

slow breaking cottonseed oil. That crude cottonseed oil that is designated slow breaking by a mechanical analysis.

slow breeding. 1. Delayed conception or settling after insemination. Also called *delayed settling.* 2. Designating an animal which lacks sexual desire. See *shy breeder.*

slow fever. Equine infectious anemia.

slow gait. One of the several forward movements or gaits of horses, faster than a walk, but slower than a canter. There are three slow gaits: the running walk, the fox trot and the slow pace.

slow gallop. Canter.

slow open-kettle process. The preparation of fruit preserves by heating in a steam-jacketed kettle for a short time on successive days in sirups of progressively increasing sugar concentrations. The method avoids undue injury to the color and flavor of the fruit. See *open-kettle one-period process.*

slow poke. The last cow in a herd to come up; a trailer.

sludge. 1. The residual solids which remain after sewage treatment; the sewage solids deposited by sedimentation in sewage treatment. It is often used as an organic fertilizer. See *activated sewage.* 2. Borings from a well.

slug. 1. Any of numerous soft, usually grayish or brownish, slimy creatures (Gastropods), up to 3 ins. or more in length, with unsegmented bodies, related to snails but without a shell. Chiefly terrestrial mollusks of the family Limnacidae, *slugs* are found in damp places in gardens, fields, and elsewhere under logs, rubbish, and in rotting vegetation. They feed at night by rasping holes in leaves and other plant material and leave slimy trails on surfaces over which they have moved. They may be serious pests in gardens and greenhouses. 2. A wholesale, forequarter cut of lamb or mutton which consists of the chuck, breast, neck and shanks: all of the forequarter except the rack. Also called *rattle.*

sluice. 1. A conduit which carries water at high velocity. 2. An opening in a structure for passing debris. 3. A water gate. 4. A channel which carries or drains off surplus water. 5. To cause water to flow at high velocities for wastage, for excavation, ejecting debris, etc. 6. To move or float logs over or through a dam from the water above into the stream below.

sluiceway. See *sluice.*

slumgum. The mixture of pupal cases and other refuse which remains after beeswax has been extracted from old honeycomb in a hot-water press.

slunk. See *bob: 1.*

slunk skin. The skin of a stillborn calf. See *slink.*

slurry. A thick suspension of chemicals used in coating seed with a fungicide.

small. A U.S. grade for eggs having a minimum net weight per dozen of 18 ozs. with a minimum weight for individual eggs at the rate of 17 ozs. per dozen.

small animal hospital. A veterinary hospital which is specially arranged to handle small animals, such as dogs and cats.

small beet webworm. Spotted beet webworm.

small bitter cherry. A minor virus disease of cherry found in Okanagan Valley in Washington, U.S. A variable portion of the cherries fail to mature and lack flavor.

small body louse. See *shaft louse.*

small brown patch. See *dollar spot: 1.*

small burnet. *Sanguisorba minor,* family Rosaceae, a perennial herb grown for its young, edible leaves having a cucumber flavor, used in salads. Native to Europe and Asia. Also called *salad burnet; burnet.*

small but typy heifer. In the show ring, a very promising calf or young heifer which may be somewhat undersized for age.

small chestnut weevil. *Curculio auriger* (Casey), family Curculionidae, a small weevil which penetrates the burs of the chestnut to lay its eggs. When these weevils are numerous, their larvae may destroy most of the nuts.

small crabgrass. Smooth crabgrass.

small cranberry. *Vaccinium oxycoccos,* family Ericaceae, a small, red-fruited shrub native to swamps in subarctic and alpine regions of the northern hemisphere which is seldom cultivated. Also called *European cranberry; moorberry.*

smaller burdock. *Arctium minus,* family Compositae, a coarse, vile-smelling weed with large, spade-shaped leaves and purplish-white flowers maturing to bur-covered fruits. The burs may snag passers-by and be carried along with them, thus spreading the seeds. A troublesome pest in grazing areas. Native to Europe and Asia. Also called *common burdock.*

smaller European elm bark beetle. *Scolytus multistriatus* (Marsh.), family Scolytidae, a beetle which, with its grubs, makes tunnels in the sapwood of elms and is considered the most common vector of Dutch elm disease *q.v.*

small farmer. Any farmer who farms a few acres deriving only a small yearly income from the farm.

small-flowered buttercup. Littleleaf buttercup.

small-flowered morning-glory. European glorybind.

small-flowered thistle. Canada thistle.

smallflower tamarisk. *Tamarix parviflora,* family Tamaricaceae, a shrub or small tree grown for its feathery foliage, slender, reddish-brown branches, and pink flowers. Native to southern Europe. Also called *tamarisk.*

small-frame nucleus. See *baby nucleus.*

small fruit. Fruit which is produced on low-growing plants, as the strawberry, grape, raspberry, blueberry, currant, etc. See *tree fruit.*

small game. Any of the smaller mammals or birds which are commonly hunted for sport.

small globe thistle. *Echinops ritro,* family Compositae, a thistle-like, perennial, hardy herb grown for its white-wooly foliage and blue flowers. Native to Europe and Asia.

small grain. Any of the cereal crops, such as wheat, oats, barley, rye, and rice.

small herd. Any herd of cows or swine of only a few in number, usually less than ten.

small horsebean. *Vicia faba* var. *minor,* family Leguminosae, a rapid-growing, rather coarse herb sometimes used as a winter cover crop and to control weeds in citrus orchards in the Pacific Coast area, U.S. Native to Asia and North Africa.

small nettle. See *dog nettle: 1.*

small offs. Mottle-leaf of citrus.

small pigeon louse. *Campanulotes bidentatus* (Scop.), family Philopteridae, a small louse which infests pigeons.

smallpox of pome fruits. See *bitter pit.*

smallpurple fringeorchid. *Habenaria psycodes,* family Orchidaceae, a perennial herb grown in the bog garden for its fragrant, lilac flowers. Native to North America. Also called *pink-fringed orchid; soldier's plume.*

small rabbitbrush. *Chrysothamnus stenophyllus,* family Compositae, a low-growing, shrubby plant which affords a shelter for jack rabbits in the Great Plains region, U.S. Native to western North America.

small ragweed. See *ragweed.*

small-sclerotial neck rot. See *gray mold neck rot of onion.*

small-seed broad bean. Small horsebean.

small-seeded alfalfa dodder. Littleseed alfalfa dodder.

small-seeded falseflax. Littlepod falseflax.

smartweed. Certain herbs of the genus Polygonum, family Polygonaceae. Also called *arsemart; waterpepper.* See *marshpepper smartweed.*

smear. Material smeared on a surface, usually a small piece of glass (slide), and examined under a microscope. *Smears* are generally stained with dyes before being examined. Materials suitable for examination from a *smear* include blood, milk, bacteria from a culture, pus, etc.

smear 62. Formula 62.

smear of cheese. A technique in the examination of cheese in which the surface of the cheese is pressed on a sterile slide for bacteriological study under the microscope.

smeary. Designating a cheese which has a very high moisture content, a weak texture and body.

smegma. A dried and hardened, fatty secretion found in the sheaths of stallions and geldings, bulls, etc. When not removed by washing, excessive accumulations may shut off the urinary passageway. Also called *bean.*

smelter injury. Toxic effects on plant life which are caused by flue gases or dusts discharged by smelters. It is similar to that caused by smoke in highly industrial areas.

Smierkase. See *Cottage cheese.*

smilax asparagus. *Asparagus asparagoides,* family Liliaceae, a branching, leafless vine with bright and lustrous green, leaflike oval branches, about 1 in. long, and sparse, greenish-white flowers, widely used for decoration. Native to South Africa. Also called *florist's smilax.*

smiling wakerobin. Painted trillium.

smilo. Smilograss.

smilograss. *Oryzopsis miliaceae,* family Gramineae, a perennial grass with relatively stout, erect culms, sometimes branching, growing 2 to 4 ft. tall, and sometimes used in the reseeding of range lands. Native to North America.

Smirnow rhododendron. *Rhododendron smirnowi,* family Ericaceae, an evergreen shrub grown for its foliage and its bell-shaped, rosy-red flowers. Native to the Caucasus.

Smith-Hughes Act. A U.S. federal law passed by Congress in 1917 which gives federal aid for vocational education in the secondary schools. It established a national system of vocational education of broad scope in agriculture, home economics, trades, and industries.

Smith-Lever Act. An Act of U.S. Congress which provided funds for extension work in agriculture cooperatively between the agricultural colleges of the several states and the U.S. Department of Agriculture. The Act was approved May 8, 1914.

smoke. To treat meat or meat products, fish, or other foods by prolonged exposure to the smoke of certain nonresinous woods, corn cobs, hardwood sawdust, etc., for preserving and imparting a desired flavor. Meat products commonly undergo a curing process before being smoked.

smoke chaser. A member of a forest-fire protective unit that goes to reported fires to extinguish them.

smoked. See *smoke.*

Smoked. A smoked cheese characterized by the flavor and aroma of smoke, usually American type or Cheddar cheese.

smoked ham. A ham which, in the process of curing, has been treated with smoke, as distinguished from an uncured fresh ham (*q.v.*). See *smoke.*

smoked picnic shoulder. A picnic ham (*q.v.*) that has been cured and smoked in the same manner as smoked ham (*q.v.*). See *smoke.*

smoked sausage. Any sausage that, in its preparation, has been treated with smoke *q.v.*

smoked turkey. A whole turkey or a turkey roll that has been treated with smoke *q.v.*

smokehouse. A small, separate building which is so arranged and equipped that meat to be smoke-cured may be hung over a low, smouldering fire producing a dense smoke.

smoke kiln. A primitive type of kiln in which unseasoned lumber is exposed to the heat and fumes of an open fire.

smoke machine. A device used to produce artificial smoke to determine the direction and speed of air circulation in a kiln.

smoker. See *bee smoker.*

smoke screen. A screen or blanket of smoke over an area produced by smudge fires of damp straw or other material. It was formerly erroneously believed to protect fruit crops or tender plants from frost.

smoketree. 1. Any tree or shrub of the genus Cotinus, family Anacardiaceae. See *American smoketree.* 2. *C. coggygria,* a deciduous shrub grown for its attractive foliage and feathery panicles producing a greenish-buff, smoke-like effect. Native to Europe and Asia. Also called *Aaron's-beard; common smoketree; false fringe tree; feather tree; Venetian sumac.*

smoky. 1. Of the color of smoke, a brownish or bluish shade of gray; dusky; cloudy. 2. An abnormality in the appearance of a horse's eye, which becomes cloudy, whitish, and pearly-colored. It is indicative of impaired vision. The condition is referred to as *smoky eye.*

smoky eye. See *smoky: 2.*

smoky nose. An animal's nose which has a dark, mouse color. Also called *mousy.*

smooth. 1. In stock judging, designating a favorable condition of the skin and/or hair; one that is smooth and silky and not dry and rough to the touch. 2. Designating an animal having a well filled-out, even contour that is devoid of rolls, roughness, or patchiness. *Cf. coarse: 1.*

smooth alder. Hazel alder.

smooth and nicely balanced. Designating an animal devoid of roughness and well proportioned.

smooth-awned. See *smooth-bearded.*

smooth-bearded. Designating a variety of barley with awns or beards practically free from barbs.

smooth behind. In stock judging, designating a beef animal with rearquarters well filled-out and free from rolls and coarseness.

smooth brome. *Bromus inermis,* family Gramineae, a perennial, awnless, drought-resistant grass which has a moderately spreading growth habit, forms a dense sod and grows up to 5 ft. in height. It is a palatable hay and pasture grass and is used extensively in reseeding range lands. Native to

Europe. Also called *Austrian brome grass; awnless brome; Hungarian brome grass.*

smooth crabgrass. *Digitaria ischaemum,* family Gramineae, a species of crabgrass (*q.v.*) distinguished from most of the other species by its smaller size, smoother leaves and shorter spikes.

smooth fingergrass. Smooth crabgrass.

smooth going. 1. Designating an easy gait or forward movement of a horse used for riding and driving. 2. Designating any practice which proceeds at a brisk, even pace without particular hindrance.

smooth hydrangea. *Hydrangea arborescens,* family Saxifragaceae, a small, deciduous, hardy shrub grown for its white flowers. Native to North America. Also called *sevenbark; wild hydrangea.*

smoothing harrow. A harrow used to smooth the surface of a field in preparation for seeding. Most common in use is the drag or spike-tooth harrow.

smoothleaf elm. *Ulmus carpinifolia,* family Ulmaceae, a deciduous, hardy tree grown as a shade and street tree and for its valuable lumber. Native to northern Africa, Europe and western Asia. It is a variable species with many garden forms. Also called *European elm.*

smoothly covered. Smoothly finished.

smoothly distributed flesh. See *smoothly finished.*

smoothly finished. Designating the desired condition of finish or fleshing found on livestock fed for slaughter.

smoothly laid shoulders. In stock judging, a desirable conformation or attachment of the shoulders to the back and forefront of the animal.

smooth mountain dandelion. Pale agoseris.

smooth-mouth. The mouth of a horse whose teeth have lost their natural cups and have become smooth by use and wear, indicating that the horse is ten or more years of age.

smooth-mouthed sheep. An aged sheep, usually a ewe, whose incisor teeth are missing or so badly worn out that it cannot eat normally; a gummer. Also called *broken-mouthed.*

smooth paspalum. *Paspalum pubiflorum,* family Gramineae, a good forage grass native to North America.

smooth penstemon. *Penstemon laevigatus,* family Scrophulariaceae, a perennial herb grown in the flower border for its long-toothed, purple flowers. Native to North America. Also called *fox-glove penstemon.*

smooth roller. A large, cast iron or steel roller with a smooth surface, used for compacting soil and crushing clods. It has been largely replaced by the culti-packer (*q.v.*), or corrugated roller.

smooth-seeded. Slick-seed cotton.

smooth-seeded pea. Any variety of garden pea having smooth (not wrinkled) seeds, as the Alaska variety.

smooth, shiny walls. In the examination of the texture and quality of Swiss-type cheese, the proper appearance of the wall structure of the eye.

smooth sumac. *Rhus glabra,* family Anacardiaceae, a hardy shrub or small tree grown as an ornamental for its showy, red, hairy fruits. Native to North America.

smooth-textured cheese. Any cheese which exhibits a fineness of grain and a continuous oily, silky, smooth film when the mass is worked between the thumb and fingers or is spread over the palm of the hand.

smooth winterberry. *Ilex laevigata,* family Aquifoliaceae, a hardy, deciduous shrub chiefly grown for its ornamental, orange-red fruits. Native to North America. Also called *hoopwood.*

smother. 1. To kill by suffocation, as panicky chickens or sheep are *smothered* when they pile up in a heap. 2. Of plants, to kill by covering thickly with dirt, straw, ice sheet, etc. 3. Of plants, to retard in growth or to kill by a dense or more vigorously growing crop. See *smother crop.*

smother crop. Any dense-growing crop that suppresses or stops the growth of less competitive plants; a quick-growing crop (buckwheat, rye, etc.) planted especially for the suppression of weeds.

smothered. Designating a characteristic, undesirable flavor of dairy products, especially cream, detected by smell or taste and caused by inadequate cooling and improper storage.

smothering. Designating that which smothers, as a *smothering* crop or a *smothering* layer of ice that stifles growth.

smothering disease of coniferous seedlings. A disease of coniferous seedlings caused by the smothering fungus *Thelephora laciniata,* family Thelephoraceae.

smouldering fire. A forest fire still smoking, not spreading appreciably, and burning with little or no flame.

smudge. A smoky fire or the dense smoke used to drive insects away. See *smoke screen.*

smudged type. A type of cheese molding with dark, smoky, or rarely, greenish areas suggesting soot or dirty finger marks.

smudge fire. A fire made from damp straw, smouldering manure, or any other fuel which produces a very heavy smoke that hangs low to the ground, erroneously believed to prevent frost injury to crops. See *smudge pot.*

smudge fungus. A fungous disease of white onions caused by *Colletotrichum circinans,* family Melanconiaceae.

smudge pot. A type of open pot or vessel in which an oil fire is maintained to heat the air in orchards and vegetable plots in hope of preventing frost damage.

smut. 1. The disease which is caused by the true smut fungi of class Basidiomycetes, order Ustilaginales, or the casual fungus. 2. Any disease which is caused by entirely different fungi agreeing with the true smuts only in the production in the diseased host plant of numerous dark-colored, mostly dusty spores, as fig smut (*q.v.*). In the true *smuts,* the spores (often erroneously called chlamydospores) accumulate in the host tissues as a dark, usually powdery mass. These spores germinate by producing a few-celled promycelium upon which normally arise the colorless sporidia that are in fact the cells that infect the next host plant. The dark spore mass may break up into dustlike powder which is distributed by the wind or may remain enclosed within the host tissues producing there the promycelia which emerge into the air and bear their sporidia. The true *smuts* are the cause of injury in cereals as well as in many other kinds of plants, more often destroying the floral parts but sometimes only found in the leaves or stem.

smut ball. A kernel of a cereal infected by a covered or kernel smut in which most of the inside tissue of the kernel has been replaced by smut spores, as in bunt of wheat (*q.v.*). The diseased kernels may appear darker and larger than normal.

smut of barley. See *covered smut of oats and barley; loose smut of barley.*

smut of corn. See *corn smut; head smut.*

smut of cotton. Cotton smut.

smut of fig. Fig smut.

smut of millet. See *millet smut; head smut of millet.*

smut of oats. See *oat smuts; covered smut of oats and barley.*

smut of onion. Onion smut.

smut of sorghum. Covered kernel smut.

smut of sugar cane. A fungous disease of sugar cane caused by *Ustilago scitaminea,* family Ustilaginaceae, characterized by the infection of terminal buds of the young shoots and the formation of spores in the inflorescences.

smut of wheat. See *bunt of wheat; loose smut of wheat; flag smut.*

smutty wheat. A U.S. federal grain-grading designation for wheat in which there is a considerable quantity of smut spores detected by the fishlike odor and/or by the presence of numerous smut balls *q.v.*

snaffle. Any type of mouth control used in handling horses, especially in breaking. See *snaffle bit.*

snaffle bit. A bridle bit; the mouthpiece of a horse bridle which is composed of two or more linked parts or sometimes made of a single chain which acts as a bit and serves as an effective control.

snag. 1. A sharp point of wood which results from a break. 2. A broken tree trunk more than 20 ft. tall.

snail. *Helix aspersa,* family Helicidae, designating a land mollusk characterized by a spiraled shell large enough to house the whole body of the *snail.* Generally to be found in warm, damp areas, it tends to come out only at night. Very prominent in Europe, the *snail* is very harmful to most vegetation. Sometimes erroneously called *slug.*

snail bean. *Phaseolus caracalla,* family Leguminosae, a tender, perennial, high-climbing vine grown in frost-free areas for its fragrant, white or purplish flowers resembling sweetpeas. Native to South America. Also called *snail flower; corkscrew flower; snailvine.*

snailseed. See *Carolina snailseed.*

snailvine. Snail bean.

snake. 1. Any of the numerous limbless vertebrates of the order Serpentes, characterized by a smooth-skinned, elongated body. The belly is marked by broad, transverse plates attached to the ribs. This feature enables the *snake* to move from place to place by dragging its body across uneven surfaces. The internal ears are very sensitive to sound. A transparent membrane coats the eye as the eyelids are immovable. The elastic ligament construction of the jaw and pointed recurved teeth enables the *snake* to ingest quite good-sized mammals. It is very helpful to farmers as it takes care of harmful mice or insects around the farm. Most *snakes* are harmless, only a very few being venomous or harmful to man. It gives birth to living young or in many cases lays eggs. As it tends to adjust its body temperature to that of its surroundings, the *snake* is found residing in warm, tropical areas. *Snakes* live in trees, in the ground, some both on land and in the water, and some occasionally are found to have adapted themselves completely to the water. See *adder.* 2. To drag or skid logs, etc., with a chain, cable, or rope.

snakebeard. See *lily-turf.*

snake cucumber. Snake melon.

snake doctor. Dragonfly.

snake feeder. Dragonfly.

snake-flower. Vipersbugloss.

snake gourd. See *edible snakegourd.*

snake grass. See *Virginia spiderwort.*

snakehead. White turtlehead.

snake melon. *Cucumis melo* var. *flexuosus,* a variety of the muskmelon (*q.v.*), which bears a long, slender, bent, twisted, furrowed, and edible fruit. It is raised largely as a curiosity. Also called *snake cucumber; serpent cucumber.*

snake pipes. Field horsetail.

snake plant. Sweet sansevieria.

snakeweed. European bistort.

snap. 1. A spring fastening or clasp which closes with a clicking sound, used as a fastener on harnesses, gate chains, dog chains, etc. 2. A brief period of cold weather; as a cold *snap.* 3. To harvest by pulling off with a quick motion, as the ears of corn, the bolls of cotton, etc. 4. In horses, especially in harness or saddle horses, to move briskly with animation; to pick up feet quickly and sharply.

snapbean. See *kidney bean.*

snap corn. Ear corn, with all or most of the husks adhering, which is harvested by breaking or snapping from the stalk.

snapdragon. 1. Any of the plants of the genus Antirrhinum, family Scrophulariaceae. Species are grown as ornamental plants. Also called *dog's mouth.* See *chaparral snapdragon.* 2. *A. majus,* a perennial herb which in its many horticultural forms is grown as an annual for its reddish-purple or white flowers. Also called *dragon's-mouth; toad's mouth.*

snapdragon anthracnose. A fungous disease of snapdragon caused by *Colletotrichum antirrhini,* family Melanconiaceae, characterized by lesions and cankers on the leaves and stems.

snapdragon rust. A fungous disease of the snapdragon caused by *Puccinia antirrhini,* family Pucciniaceae, characterized by dark brown, powdery sori on all green parts of the plant.

snapped. Designating an entire mature cotton boll, an unhusked corn ear, asparagus, a garden bean, etc., which has been harvested by breaking from the plant or stalk.

snapped boll. See *snapped cotton.*

snapped corn. Ear corn, with the husks adhering, that has been broken from the stalk in harvesting.

snapped cotton. Cotton harvested by hand in which the mature boll with burrs and lint still adhering is plucked from the plant; especially common in Texas and Oklahoma, U.S. See *pulled cotton.*

snapped sweet corn. Freshly gathered sweet corn, with husks adhering, that has been pulled or broken from the stalk.

snapping beetle. See *wireworm.*

snapping bug. See *wireworm.*

snapping rolls. The adjustable rolls on a mechanical corn picker or a husker-shredder that snap the corn ears from the stalk.

snapweed. 1. Any plant of the genus Impatiens, family Balsaminaceae. Species are herbs whose capsules or fruits snap open into spirally coiled valves and scatter the seeds. They are grown in flower gardens and greenhouses for their showy flowers. Also called *touch-me-not.* 2. See *garden balsam.*

snathe(e). The shaft or handle of a scythe; a snead.

sneezeweed. 1. Any of the several plants of the genus Helenium, family Compositae. Species are tall, coarse, mostly yellow-flowered herbs, somewhat resembling sunflowers; native to America. 2. *H. autumnale,* a stout, poisonous, branching, perennial herb with nearly two inch wide, yellow flowers, which is widespread in eastern North America and is common in open thickets, meadows, and along streams. Symptoms of poisoning are reported with sheep, cattle, and horses that have eaten large quantities of the tops of the plant. Also called *staggerwort; swamp sunflower.*

sneezewort. Sneezewort yarrow.

sneezewort yarrow. *Achillea ptarmica,* family Compositae, a perennial herb grown for its handsome, white flowers. Native to Europe, Asia, and North America. Also called *sneezewort; European pellitory.*

snig. To drag along the ground, especially logs or timbers (Australia and New Zealand).

snip. To remove by clipping, as to *snip* off a twig, flower, etc.

snipe. To round off the end of a log so that it will drag easily in snaking or skidding.

snippers. A small hand tool used for cutting, as for cutting wire.

snood. The fleshy protuberance at the base of the beak of a turkey. Also called *dew bill.*

snorting pole. See *teasing pole.*

snout. 1. The projecting part of an animal's head that contains the nose and jaws, as the *snout* of a hog. 2. An anterior elongation of the head; a proboscis; a piercing or sucking part of certain insects.

snout beetle. Any of the beetles of the family Curculionidae, to which belong the typical weevils which injure fruit and other crops. They are distinguished by the head which is prolonged forward and downward into a cylindrical snout, at the end of which are very small parts which function to chew plant tissues. The long snout is used for making a hole deep into the tissues of fruits or plants into which the beetle lays its eggs.

snow. Small ice crystals usually aggregated in the form of flakes which form during freezing temperature by the congealing of atmospheric vapor.

snowball aphid. Viburnum aphid.

snowball European viburnum. *Viburnum opulus* var. *roseum,* family Caprifoliaceae, a variety of the European cranberrybush viburnum *(q.v.)* which is a flowering shrub, and is grown for its ball-like, white flower clusters. Also called *dog rowan tree; Guelder rose; snowball.*

snowbank mockorange. *Philadelphus nivalis,* family Saxifragaceae, a hybrid, hardy, deciduous shrub grown for its white flowers.

snowbell. Any shrub or tree of the genus Styrax, family Styracaceae. Many species are handsome plants suited to tropical or warm regions where they are grown for their fragrant, bell-shaped flowers. Also called *styrax; storax.* See *fragrant snowbell; Japanese snowbell.*

snowberry. 1. A group of hardy, many-branched shrubs of the genus Symphoricarpos, family Caprifoliaceae. Some species are common weeds occurring in the Rocky Mountain region of U.S. where they are sometimes eaten as forage by sheep, goats, and cattle. 2. *S. albus,* a small shrub with slender branches and small, white berries, grown as an ornamental for its white fruit. Native to northern North America. Also called *wax berry.*

snowbreak. 1. The breaking of lumber trees by snow. 2. A planting of one or more rows of trees or shrubs to catch blowing snow and prevent drifts on roadways and other places; a snow fence.

snowbrush ceanothus. *Ceanothus velutinus,* family Rhamnaceae, a shrub which is moderately palatable and is browsed upon to some extent by livestock and deer. Native to North America.

snow catch. See *snowbreak: 2.*

snowcreeper. Christmasvine porana.

snowdrop. 1. Any of the herbs of the genus Galanthus, family Amaryllidaceae. Species are cultivated for their handsome flowers. Native to Europe and Asia. See *giant snowdrop.* 2. *G. nivalis,* a bulbous herb grown for its very early, white flowers. Native to Europe and Asia. Also called *fair-maid-of-February; Candlemas bells.*

snowdrop anemone. *Anemone sylvestris,* family Ranunculaceae, a perennial herb grown for its nodding, fragrant, white flowers. Native to

Europe and Asia. Also called *snowdrop windflower.*

snowdrop tree. Two-wing silverbell.

snowdrop windflower. Snowdrop anemone.

snoweater. A chinook wind.

snow fence. A slat and wire fence, used in winter to intercept drifting snow, which protects roadways, railways, and other areas from snow drifts.

snow-fence crib. A temporary corn crib made of snow fencing. It is very common in the midwestern U.S.

snow-fence silo. A temporary silo made of snow fence *(q.v.)*. Frequently it is lined with heavy paper for better protection of the silage.

snow hedge. See *snowbreak: 2.*

snow-in-summer. *Cerastium tomentosum,* family Caryophyllaceae, a prostrate, perennial herb grown for its mat-forming habit and its small, white flowers. Native to Europe.

snow line. On mountain slopes, a natural line above which snow lies during certain seasons of the year. In range lands, it marks the upper limits for effective grazing.

snow mold. A fungous disease of winter cereals in northern Europe and of grasses, grains, and alfalfa in northern U.S. and Canada, attributed to several fungi such as *Typhula itoana, Calonectria graminicola,* and *Pythium* spp. Its outstanding symptom is a conspicuous, white mold growth appearing on the plant as the snow melts. See *eastern snow mold.*

snow-on-the-mountain euphorbia. *Euphorbia marginata,* family Euphorbiaceae, an annual herb grown for its showy, white flowers and white-margined, green leaves. Contact with its milky sap may cause a dermatitis. Native to North America.

snow pear. *Pyrus nivalis,* family Rosaceae, a small, deciduous tree grown for its showy, white flowers and for its fruit used for making pear cider. Native to Europe and Asia.

snow plow. A bulldozer, or some other power-driven vehicle, with a V-shaped horizontal blade attached to its front, which is used for clearing snow from highways, railroads, parking areas, etc.

snow poppy. *Eomecon chionantha,* family Papaveraceae, a perennial herb grown for its white flowers. Native to China.

snow scale. *Chionaspis citri* (Comst.), family Coccidae, a scale insect (*q.v.*) which is an important pest of citrus in Australia, Asia, South America, Mexico, and to a limited degree in Florida, U.S. Also called *white louse; orange chionaspis.*

snowstorm. A weather condition in which falling snow flakes or particles are blown about by the wind. It usually implies a storm less severe than a blizzard.

snow trillium. Dwarf trillium.

snowy. Designating a body defect of ice cream which has a flaky, brittle texture with a tendency to fall apart when dipped.

snowy tree cricket. *Oecanthus niveus* (DeG.), family Gryllidae, a cricket that is generally beneficial, feeding on aphids, tree hoppers, and scale insects, but sometimes it feeds on flowers, fruits and leaves also. Widely distributed throughout North America.

snub. To control an animal by roping it and tying the rope to an adjacent post or other solid object so as to restrict its movement.

snubbing block. A part of the braking equipment on the rear gang of wheels on a wagon which consists of blocks placed in front of the wheels, controlled by a lever bar or brake handle. The blocks directly apply a retarding force to the rim of the wheels.

snubbing post. Any smooth, sturdy post to which animals may be fastened to restrict their movement.

snubbing ring. A heavy metal ring attached to the floor, or any other solid object, through which a control rope is passed and by means of which an animal's movment is restricted as in slaughtering or treating for ailment.

snub down. To place the hand on or over the nose of an animal to stop its breathing in order to control it.

snuffy ears. Heads of wheat destroyed by loose smut (Australia).

Soaked-curd. Washed-curd.

soaker-washer. A bottle washing machine in which the milk bottles are soaked in a caustic solution for a few minutes as they pass through the machine.

soap. A cleansing agent made usually by treating fats or fatty acids with sodium or potassium salts. There are two types, hard *soap;* soft *soap* (*q.v.*). *Soap* is sometimes used as a component of insect sprays for its insecticidal effect, as an emulsifier, wetter or spreader, and for enhancing the toxicity of nicotine sprays for aphids.

soap ashes. Wood ashes used by early settlers in the forested regions of the U.S. as a source of lye, which in turn was used in making soft soap.

soapless detergent. Any cleansing agent (detergent) which does not contain soap. See *detergent.*

soap plant. Amole soapplant.

soaptree yucca. *Yucca elata,* family Liliaceae, a semidesert plant found from western Texas to southern Arizona (U.S.), whose leaves and young flower stalks are grazed upon by cattle to some extent. The plant is used for silage during emergency feed periods.

soapwort. Any plant of the genus Saponaria, family Caryophyllaceae. See *bouncing Bet; cow soapwort.*

soapy. Designating an objectionable taste resembling that of soap (when soap is not the cause), sometimes found in dairy products. It is caused by the activity of organisms producing lipolysis, or the giving off of free fatty acids producing an alkaline reaction.

social change. Any alteration of a pre-existing set of social relationships which has consequences for common purposes or objectives including alterations of organizational structures and purposes.

social class. One type of social stratification (others being estate and caste) in which persons and groups are ranked in terms of economic criteria, such as property, occupation, income. Sometimes class membership is conscious and results in group solidarity (a "we" feeling) and in collective action. Marxists, *e.g.*, believe that class consciousness is a necessary ingredient for revolutionary success and those who are not members of the same class cannot gain legitimate identity with the revolutionary group (rejection of peasants, businessmen, etc.). Although some societies may disclaim or reject class as a method of classification an emphasis upon economic factors as key group values will result in a class-oriented type of stratification. A society in which a class-type of stratification is dominant will generally be relatively mobile, and will be characterized by power relationships, resulting in considerable conflict and tension.

social-cultural linkage. The process whereby the elements of at least two social systems come to be articulated so that in some ways they function as a unit. Thus, when a farm family borrows money from a credit agency *social-cultural linkage* is accomplished between two systems and the status-roles of debtor and creditor express this linkage.

social differentiation. A process by which units of a social group give up to other units the performance of some of the activities of group life in order to specialize in the remaining activities, thus creating a situation of increasing specialization and interdependence.

social insect. Any insect that lives with others of its kind in a more or less organized community, as ants, wasps, and bees.

social interaction. The process of relating two or more persons to each other by verbal and non-verbal communication.

social stratification. The classification of human groups in a society into a rank order based upon the degree to which such groups possess certain measurable characteristics. These characteristics may vary from ascribed elements, such as birth order, age levels, physique, or skin color to acquired elements such as education, business success, or war prowess. Whatever the constellation of factors may be, the members of the existent layers or categories have varying opportunities for vertical movement within the system, as a person achieving a recognized success goal may move into a higher group. In the broad sense, societies and groups are stratified according to the statuses and symbols which their members deem central, and by which they govern their behavior. Thus, a stratification system is intimately connected with whatever system of values has general acceptance in the group at large. Although people may not consciously apply the concept of stratification in their personal behavior, their adherence to a set of values and their judgments of others in terms of these values results in the kind of rank order concensus described above. See *status; social class.*

social system. 1. A set of on-going interpersonal relationships which differentiate one acting group from another. 2. Patterns of relationships which analytically differentiate types of interpersonal relationships.

society. 1. A group organized for the development of a division or subdivision of science, art, religion, or technology. Its objectives are less comprehensive than those of an association, and are not primarily political. 2. A group of animal or plant species living together or capable of living together under the same local environment. A plant community is characterized by one or more subdominants; that is, species of different life form from those of the regional dominants. Thus, a dense patch of daisy fleabane in a natural prairie is referred to as a *society. Societies* are usually conspicuous for only a part of the season. *Societies* that are a part of climax vegetation may properly be termed *sociations.* Those that are not climax are properly termed *socies.*

socio-economic status. The social prestige rank of a person or family primarily as it is manifested in material consumption standards. Chapin defines it as "the position that an individual or family occupies with reference to the prevailing standards of cultural possessions, effective income, material possessions, and participation in the group activities of the community." Indices have been constructed by Sewell, and by Belcher and Sharp to measure this concept among farm families.

Socotra begonia. *Begonia socotrana,* family Begoniaceae, a tuberous-rooted, tropical herb grown as a pot plant for its rose-pink flowers. Native to the island of Socotra (Indian Ocean). Also called *Christmas begonia.*

sod. 1. A tight, closely-knit growth of grass or other plant species which may or may not be closely cut at frequent intervals. 2. A surface layer of soil containing grass or other plant species with their matted root systems.

soda. The fertilizer, sodium nitrate (southern U.S.).

sod bottom. A type of plow bottom whose moldboard is so constructed as to have a long, auger-like twist to turn the slice completely upside down, covering all vegetable matter. It is designed to work in unbroken, prairie sod land.

sodbound. 1. Designating a condition in which the roots are so restricted and form such a dense mat that growth of the plant is impeded. 2. Designating a condition of a grass, sod, or turf in which the nitrogen supply in the soil has become depleted and which results in retarded, stunted vegetative growth.

sodbuster. A farmer or homesteader on the prairies and plains of the U.S. midwest during the early period of settlement.

sod crop. 1. Any crop or vegetative cover which quickly forms a heavy, close-knit, top growth over the surface of the soil and a root system which binds the soil particles together, thus forming a sod, as white clover, bluegrass, etc. 2. Any forage grasses or legumes.

sod culture. A soil management practice of maintaining permanent sod in orchards as opposed to annual tillage.

sod cutter. A tool for cutting a thin band of sod

free from the soil, used in connection with transplanting sod for sodding waterways, etc.

sodding. Removing sod from one area and placing it on a bare soil area in another location.

sod fence. A wall-like enclosure for gardens or the homestead yard which was made of blocks of grass sod cut from the virgin prairie (early agricultural settlement in the Prairie and Great Plains regions of western U.S.).

sod-former. See *sod crop: 1.*

sod-forming crop. See *cover crop; sod crop: 1.*

sod grass. Any grass that reproduces by stolons or rhizomes or otherwise forms a more or less continuous sod or turf.

sod house. A farm dwelling or other enclosure which was made of sod sections about 2 ft. long, 4 to 6 ins. thick, and 1 ft. wide, from firm grass sod. These were laid brick-fashion, forming the walls of the structure. Such walls were then roofed over with poles and willow thatch or adobe-mud covering. The inner surface of the walls was usually shaved smooth and plastered with a natural clay plaster. The floors were either of hard-packed clay or wood boards (western U.S. prairie in frontier days).

sodium. Na, the chemical element, a soft, waxy, silvery-white metal of the alkali group (lithium, sodium, potassium, etc.) which is one of the commonest and most widely distributed elements. It oxidizes rapidly in air and in the presence of air and water forms sodium hydroxide (caustic soda). Combined with chlorine it forms sodium chloride (common salt); other common forms are sodium nitrate and sodium carbonate. *Sodium* probably reacts in the soil to liberate potassium from insoluble compounds, thus increasing the available potash in the soil.

sodium alginate. A sodium salt of the seaweed, algae. It has excellent water binding properties and is used by some manufacturers to prevent water crystallization or iciness in ice cream.

sodium arsenate. A highly soluble form of arsenic which is used in poison baits.

sodium arsenite. A chemical compound which occurs in several distinct forms. Sodium meta-arsenite ($NaAsO_2$) is commonly used as a poisonous ingredient in the preparation of ant poisons, poison flypapers, animal dips, etc. Solutions of *sodium arsenite* are marketed for use as grasshopper baits and as weed killers.

sodium benzoate. $NaC_7H_5O_2$, a white, colorless, crystalline powder, very soluble in water, which is used as a food preservative. It is effective only in slightly acid media. Also called *benzoate of soda.*

sodium chlorate. $NaClO_3$, a colorless, crystalline salt widely used as a weed killer but which may make the soil sterile for a while.

sodium chloride. See *salt: 1.*

sodium dinitro-orthocresylate. A chemical compound sold under various trade names which is used as a dormant spray insecticide and fungicide, wood preservative, and selective weed killer.

sodium ethyl mercurithiosalicylate. A chemical compound sold under various trade names which is used as a spray in the prevention of citrus fruit decay.

sodium fluoracetate (Compound 1080). A powerful rodenticide developed by the U.S. Fish and Wild Life Service; a fine, white powder readily soluble in water which is used in food or drink to control rats, mice, and other rodents.

sodium fluoride. NaF, a white, powdered, chemical compound used for combating chewing lice on animals and poultry, commonly employed in roach and ant powders, as an anthelmintic used to eliminate ascarids (round worms) in swine, and as a fungicide.

sodium fluosilicate. A fluorine compound sometimes substituted for sodium fluoride for the control of lice, roaches, and certain other insects. Also called *silicofluoride.*

sodium hypochlorite. NaOCl, a chlorine compound which is commonly sold in a powdered form or in solutions under various trade names for use in the disinfection of processing equipment and as a household disinfectant and bleach.

sodium nitrate. $NaNO_3$, a white, deliquescent salt obtained in the natural form from deposits in Chile but also made synthetically. It contains about 16 percent nitrogen and is readily soluble in water. Its chief use is for fertilizing crops either singly or in fertilizer mixtures. Also called *nitrate of soda; Chili saltpeter.*

sodium orthophenylphenate. A chemical often used as a disinfectant.

sodium selenate. A highly poisonous, water-soluble salt sometimes applied in solution to greenhouse soil to control red spider mites and aphids infesting greenhouse plants. The compound is taken up from the soil by the plants, and the plant juices become toxic to the insects feeding on them.

sodium silicate. A silicate of sodium, commonly known as *water glass,* which is sometimes used to make a water solution in which eggs are stored.

sodium silicofluoride. Sodium fluosilicate.

sod land. Land that is covered with a heavy overcover of grass turf or other close vegetative growth.

sod mulch. A type of sod culture, composed of cut grass, sometimes complemented with straw or low-grade hay, which is placed around trees.

sod plow. A type of plow which is especially adapted by the shape and length of its moldboard for complete inversion of the furrow slice. Also called *sod bottom; breaker bottom.*

sod strip. 1. A band or narrow strip of sod used for checking erosion in waterways. 2. A narrow band of grass or other close-rooted crop placed across the channel of a gully to spread and retard the flow of water. 3. In strip cropping, a strip that is in sod.

sod waterway. See *sod strip: 1.*

sod webworm. Bluegrass webworm.

soft acanthus. *Acanthus mollis,* family Acanthaceae, a perennial herb grown for its white, lilac, or rose-colored flowers. Native to southern Europe. Also called *bear's breech; brank-ursine; sea-holly.*

soft brome. *Bromus mollis,* family Gramineae, a widely-distributed, annual grass growing 1 to 1½ ft. tall with velvety foliage, which provides fair forage on range lands. Also called *soft chess.*

soft brown scale. *Coccus hesperidum* L., family Coccidae, a small sucking insect of world-wide distribution which infests a large variety of plants and is found commonly on citrus. In Australia and South Africa, it is considered a major pest of citrus. Also called *soft scale.*

soft center share. A plowshare made from a three layer steel plate having a soft center and an outer layer of high carbon steel which will take a high polish so that the share will scour well in any kind of soil.

soft cheese. Any of certain types of cheese which has a high moisture content (more than 55 percent) and a short ripening process, as Cottage cheese, Cream cheese, and Neufchâtel.

soft chess. Soft brome.

soft corn. Corn that is immature or has been frosted before maturing. It has a high water content and poor keeping qualities. See *flour corn; soft starch corn.*

soft cot(t)s. Fleece wool which has a light degree of tangle or mat in the fibers. See *hard cot(t)s.*

soft crown gall. A crown gall (*q.v.*) which is made up of succulent and imperfectly vascularized tissue. *Cf. hard crown gall.*

soft-curd milk. A milk which upon specific curdling yields a cutting tension not exceeding 30 grams. It is especially desirable for certain invalids and infant feeding formulas.

soft cutting. See *slip: 1.*

soft dough stage. A stage of maturity of corn and other cereals in which the kernel when broken open shows a soft starchy or cheesy condition.

soft hog. Any hog that is fattened on peanuts, soybeans, rice bran and chufas, producing soft or oily meat carcasses (*i.e.,* soft or oily fat). Also called *oily pork.*

soft lumbang. *Aleurites trisperma,* family Euphorbiaceae, a tree closely related to the tung oil tree (*q.v.*) from the seeds of which a drying oil is obtained. Native to the Philippine Islands. Also called *banucalag; balocanad; baguilumbang; calumbang; balucanag.*

soft pasty texture. A defect of the body or texture of Cottage cheese.

soft phosphate with colloidal clay. A soft, clayey, low-grade, phosphatic rock, not suitable for use in the manufacture of superphosphate. It is the fine-grained waste which results from the washing of hard rock and pebble phosphate, collected in ponds, subsequently dried and sold as a low-grade fertilizer or used as a filler for mixed fertilizers. Also called *colloidal phosphate; waste-pond phosphate; phosphatic clay.*

soft pinch. Removal of the very tip of a shoot by pinching it off. *Cf. hard pinch.*

soft, pliable udder. A cow's udder that has a soft, spongy texture.

soft red winter wheat. A type of wheat (*q.v.*) that has a long kernel and whose germ is large; the suture is open, very deep and easily split. These wheats are lower in protein and higher in starch percentage than the hard winter types. They are grown in the more humid climates of the north central and eastern sections of the U.S.

soft ripe. Designating fruit that is fully ripe or soft. Also called *dead ripe.*

soft rot. Any of various bacterial and fungous decays of plant organs in which the major effect is a slimy softening of the plant tissues. See *brown rot of potato; brown rot of stone fruits; soft rot of vegetables; black leg: 2.*

soft rot of potato. See *brown rot of potato; black leg: 2.*

soft rot of stone fruits. See *brown rot of stone fruits.*

soft rot of sweet potato. A fungous disease of sweet potatoes caused by *Rhizopus nigricans,* family Mucoraceae, characterized by affected roots becoming very soft and watery and by water exuding when the skin is broken. The disease, progressing very rapidly, causes loss of sweet potatoes in the field, in storage, and in transit. Also called *mush rot; vinegar rot.*

soft rot of vegetables. A rapid, soft, wet rot of carrots, onions, celery, potatoes, and many other vegetables, caused chiefly by bacteria of the *Erwinia* spp. and *Pseudomonas* spp., notably by *Erwinia carotovora,* family Enterobacteriaceae.

soft-scald. Semiscald.

soft scald of apples. Apple soft scald.

soft scale. Any scale insect (*q.v.*) of the family Coccidae; an unarmored scale.

soft-shell egg. An egg with an incompletely calcified shell which may be due to one of several causes, including nutritional deficiencies of calcium, phosphorus or vitamin D, failure of the shell gland to function, or by violent peristaltic constrictions causing the egg to be forced through the uterus before completion of shell formation.

soft soap. A smeary, semiliquid soap that is usually made by treating an oil high in oleic acid with potash as the alkali. See *soap ashes.*

soft spot of cheese. A body defect of Cheddar cheese in which a portion of the cheese surface, most often near the center of the top or bottom, has decomposed, resulting in a soft, slimy, watery, foul-smelling area.

soft starch corn. A type of corn in which the kernels are composed almost entirely of soft starch. It is usually white, but all colors may be found. Also called *flour corn.*

soft thermopsis. *Thermopsis mollis,* family Leguminosae, a perennial herb grown for its clusters of yellow flowers. Native to North America. Also called *bush pea.*

soft wart. A small, solid growth arising from the skin or mucous membrane which is soft and not covered with a layer of horny cells. *Cf. hard wart.*

soft water. Water relatively free from various hardening compounds, particularly water free from a significant amount of carbonate of calcium, as distilled or rain water.

soft white wheat. A type of wheat (*q.v.*) whose

kernels are soft and chalky in appearance. They are softer in character than the hard winter wheats, possess a lower percentage of protein, are lower in gluten, and are used primarily for pastry flours and in breakfast food manufacture. They are often grown under irrigation in the western U.S., or on highly retentive soils in the north central U.S.

softwood. 1. Any wood of light texture which is easily cut. 2. In forestry, any tree having such wood, particularly a coniferous tree. 3. Immature, succulent wood, such as is used for a *softwood cutting q.v.*

softwood cutting. An immature, succulent stem taken from a plant which is used for rooting or grafting *(q.q.v.).* Also called *slip* (if used for rooting).

soggy body. Wet body.

soggy breakdown of apple. A storage rot occurring at temperatures below 38°F., which is closely associated with apple soft scald *(q.v.),* as a result of the invasion of rot-producing fungi, usually species of Alternaria, family Dematiaceae.

soggy potato. A potato with a relatively high percentage of water and a low percentage of starch which after cooking remains heavy and moist instead of light and flaky.

soil. 1. Comprehensively, any artificial or natural media in which plants grow, or can grow, as any of the possible variations in combinations of the four components: (a) mineral, or inorganic matter; (b) organic matter, both dead and living; (c) water; (d) gases. 2. In pedology, the surface or uppermost part of the zone of weathering which possesses a separate morphology, physical and chemical characteristics from the underlying parent material. The *soil* becomes a function of the variables: climate, parent material, organisms, relief, and time. The normal *soil* is the mature product of its environment. 3. In agronomy, the surface or humous layer, the plow *soil,* as distinguished from the underlying material or subsoil. 4. In engineering, especially in construction or excavation work, any geologic material other than indurated, or hard, rock. 5. Land, ground, earth, dirt. 6. The residue remaining on dishes, dairy utensils and equipment after their use. 7. To cover with soil. 8. To cut green forage and transport it to livestock in a confined area as a source of roughage.

soil acidity. Excess of hydrogen (H) ions over hydroxyl (OH) ion concentration in the soil, consisting of: (a) active, which is estimated by pH below 7.01, and (b) inactive or reserve, less successfully estimated by base exchange determinations, as opposed to soil alkalinity. *Soil acidity* is an important factor in the growth of plants both under agricultural and natural conditions. See *potential soil acidity.*

soilage. Freshly-cut green fodder fed to animals confined in an area.

soil aggregate. A single mass consisting of many soil particles held together, such as a clod, prism, crumb, or granule. Also called *aggregate.*

soil alkalinite. 1. Alkaline compounds capable of producing hydroxyl (OH) ions in the soil. 2. The excess of such materials over acid-forming materials.

soil alkalinity. Excess hydroxyl (OH) ions over hydrogen (H) ion concentration in the soil, as opposed to soil acidity. See *alkali soil.*

soil amendment. Any material which directly or indirectly improves soil conditions when used: *e.g.,* ground limestone, gypsum, peat moss, straw, cinders, etc.

soil analysis. Chemical, mechanical, and minerologic determinations of the separate components of a soil which may be complete or partial depending upon the objectives. Usually soil analysis is a laboratory procedure but some analyses may be made in the field also.

soil association. A group of soils with or without common characteristics, geographically associated in an individual pattern. It may constitute a natural land type or a geographical unit in soil mapping. A *soil association* is described in terms of the taxonomic units included, their relative proportions, and their pattern of association. Sometimes called *natural land type.* See *land type; soil complex.*

soil auger. A tool used for boring into the soil and withdrawing a sample for field or laboratory observation. Soil augers are of two general types: those with worm-type bits, and those having a hollow cylinder with a cutting edge in place of the worm bit.

soil bacteria. Two large groups of bacteria, based on their energy requirements: (a) the heterotrophic bacteria which obtain their energy and carbon from complex organic substances; (b) the autotrophic bacteria which can obtain their energy from the oxidation of inorganic elements or compounds, their carbon from carbon dioxide, and their nitrogen and other minerals from inorganic compounds. See *bacteria.*

soil ball. The rounded mass of soil left around the roots of trees or shrubs excavated for transplanting which prevents loss of moisture and reduces root loss and shock. It is usually covered with burlap and tightly bound. See *balled and burlapped.*

soil bank. 1. Surplus soil from the excavation of ditches. 2. A terrace with a vertical drop on the downhill side whose flat part is suitable for cultivation.

Soil Bank Plan. A U.S. federal program to (a) compensate farmers for planting soil-building crops on land poorly suited to cultivation; (b) reduce the amount of grain, fiber and food grown and to store the fertility in the soil rather than commodities in warehouses; (c) relieve producers of crops and livestock from excessive production from acres diverted from surplus crops. Suggestions for accomplishing the above include two main categories: (i) farmers to voluntarily divert acreage from allotment crops (corn, wheat, cotton, rice, peanuts, tobacco) for one or more years: no crop may be harvested from Soil Bank acres;

payment for these diverting acres could be in certificates negotiable for cash or surplus commodities: eligibility for price supports would still be dependent upon compliances with acreage allotments and marketing quotas; (ii) farmers to build up conservation reserve: this would be in addition to the acreage reserve and would encourage more desirable land use: using land poorly suited to tillage for forage, trees or water storage. Payments for a long period of time necessary to eliminate these acres from production are made to the farmers. The aim of the program is multifold: it would cut production and thereby raise farm prices; it diverts acres from grain, food and fiber production already overproduced in the country; it would help restore the fertility of the soil and normalize the water-table level.

soil binder. Any plant which forms a dense ground cover or holds soil in the mesh of its roots: *e.g.*, sod-forming grasses.

soil-blanched. Designating plants bleached by banking soil around them to exclude light, as in celery growing. See *hill: 4.*

soil blowing. The movement of dry soil by wind action. See *dust storm.*

soil buffer compounds. Clay, organic matter, and such compounds as carbonates and phosphates which enable the soil to resist appreciable change in pH value.

soil building. 1. The formation of soil from parent materials by natural processes. 2. Increasing the productive capacity of soil by manuring, fertilizing, liming, and growing soil-improving crops.

soil building plant. Any plant which has a high capacity for adding nutrients to the soil, especially those plants which have nodules of nitrogen-fixing bacteria in their roots: *e.g.*, alfalfa and red clover.

soil chemistry. That branch of chemistry which deals with investigations into the chemical nature of soil components and their interrelationships.

soil class. 1. Any of the soil groupings based on the relative proportion of soil separates or upon texture alone. 2. A group of soils which has a relatively definite range in some property or attribute, or a group in relation to environment and based on texture, structure, acidity, slope, natural drainage, etc.

soil classification. The systematic grouping of soils into categories from higher to lower. It may be done on a genetic basis, on the basis of the intrinsic properties of different soils, or a combination of the two. The smaller taxonomic units of a scientific system may be regrouped for various purposes, such as agriculture, highway construction, and forestry.

soil climate. The moisture and temperature conditions existing within the soil.

soil colloid. Inorganic and organic matter of very small size having a relatively large surface area per unit of mass. Under certain conditions, *soil colloids* form a more or less stable suspension or dispersion in water and are distinguishable from true solutions in that all particles have not dispersed to the molecular state. One micron, or 0.001 mm. has been arbitrarily set as the upper limit for size of a *soil colloid.*

soil complex. A mapping unit used in detailed soil surveys where two or more defined taxonomic units are so intimately associated geographically that they cannot be separated by boundaries on the scale used.

soil conditioner. Synthetic chemicals having the ability to stabilize soil aggregates and to improve soil structure or tilth. See *HPAN; IBMA; VAMA.*

soil conservation. 1. The preservation of soil resources; the maintenance of fertility and productivity of soil when in use. 2. The various regulatory measures and technical practices designed to preserve soils or land against deterioration.

Soil Conservation District. In the U.S., a legal subdivision of a state government created for developing and carrying out a soil conservation program within specified boundaries. It is governed by local land owners and occupiers.

soil conservationist. 1. One engaged in the practice or promotion of soil conservation. 2. A technical or administrative position in the Soil Conservation Service of the U.S. Department of Agriculture.

Soil Conservation Service (U.S.). A bureau of the U.S. Department of Agriculture established by the Soil Conservation Act of 1935. Its basic purpose is to aid in bringing about physical adjustments in land use and treatment that will conserve natural resources, establish a permanent and balanced agriculture, and reduce the hazards of floods and sedimentation. Its program is carried out principally under memoranda of understanding with soil conservation districts (*q.v.*). It has major responsibilities in watershed management, and it supervises the agricultural phase of certain specified water utilization programs in the western states. See *Soil Erosion Service.*

Soil Conservation Survey. A soil survey with added classification of slope, erosion, and other physical factors which affect the use of the soil. Usually land use is also mapped.

soil conserving. Descriptive prefix: preserving against deterioration, as *soil conserving* crops, *soil conserving* practices, *soil conserving* plans, *soil conserving* plants, *soil conserving* species, *soil conserving* studies.

soil consistence. The relative neutral attraction of the particles in the whole soil mass or their resistance to separation or deformation, as evidenced in cohesion and plasticity. *Soil consistence* is described by such general terms as loose or open; slightly, moderately, or very compact; mellow; friable; crumbly; plastic; sticky; soft; firm; hard; and cemented.

soil core tube. One of several types of soil sampling tools. It consists of a metal cylinder driven or pushed into the soil to obtain a core sample. See *King soil sampler: 1.*

soil creep. The very slow downward movement of soil on slopes under the force of gravity, as distinguished from landslides, slippage, earth flows, avalanches, and wash from rainfall runoff.

soil degradation. In pedology, the natural change of one type of mature soil into another, more leached type. Most frequently it refers to the change of a black Prairie or Chernozem type into a Podzol type.

soil depleting. Descriptive prefix: that which is conducive to the deterioration of soils, especially the removal or loss of essential plant food elements: *e.g., soil depleting* crops, *soil depleting* practices.

soil depleting crops. 1. Any field crop which is harvested completely leaving little or no residue in or on the soil. 2. Any crop which, under the prevailing farm practice, removes the greatest quantity of nutrients from the soil.

soil depletion. Decreased productivity of soil which results from loss of plant nutrients by crop removal. See *soil deterioration.*

soil depth. 1. Technically the thickness of the solum or soil proper, as distinguished from the C and D horizons of the profile. 2. The thickness of the humous layer or plowsoil.

soil deterioration. Loss of potential productive capacity of a soil which results from destructive processes accelerated by the activities of man: *e.g.,* through soil erosion, waterlogging and excessive accumulation of alkali.

soil development. The work of the natural evolutionary process of soil formation which leads to maturity or complete differentiation into soil horizons.

soil discharge. Discharge of water from soil or rock directly by evaporation.

soiled. Designating a product which has dirt on or under the surface.

soiled front. A soiling of the feathers in front of the neck and in the region of the crop of fowls which causes a damp, darkened, and permanently discolored appearance as a result of the escape of salivary and other fluids from the mouth. Such a condition is reported as a common symptom of the neural or fowl paralysis type of the avian leukosis complex (*q.v.*). See *fowl paralysis.*

soil erosion. Removal of soil material from a land surface by wind or water, including normal *soil erosion* and accelerated *soil erosion.* (Sometimes used loosely in reference to accelerated erosion only.) See *accelerated erosion; erosion; normal erosion.*

soil erosion process. Phenomena associated with soil movement by water, wind, and accompanying conditions.

Soil Erosion Service. In the U.S., an emergency agency first authorized in 1933 which was placed in the Department of Interior, and subsequently in 1935 transferred to the Department of Agriculure, as the Soil Conservation Service.

soil extract. The solution separated from a soil suspension or from a soil at a particular moisture content, by filtration, centrifugation, or displacement.

soil family. A category in soil classification between soil series and great soil group *(q.q.v.).* A taxonomic group of soils composed of soil series.

soil fertility. Conditions in a soil which are favorable for sustaining plant growth; not necessarily synonymous with crop productivity, inasmuch as a less fertile soil in nature may be made highly productive for crop use.

soil formation factor. The independent variables that define the soil system. Five main groups of soil formation factors are generally recognized by soil scientists: parent rock, climate, organisms, topography, and time.

soil fumigant. Any of a number of gases which, when injected into soil, destroy harmful organisms: *e.g.,* ethylene dibromide; hexachloropropene.

soil genesis. The mode of origin of the soil, particularly the processes which are responsible for the development of the solum from the parent material.

soil heaving. The raising of the ground surface by the formation of layers of ice columns in the soil which results from successive thawing and freezing. See *heaving.*

soil horizon. A layer of soil of a vertical section or the profile with distinctive characteristics that have been produced through the operation of soil-building processes. See *horizon A; horizon B; horizon C; soil profile.*

soil improvement. The practices for, or the results of, making the soil more productive for growing plants by the use of fertilizers, drainage, addition of organic matter, irrigation, etc.

soiling. In farm management, the cutting and bringing of green forage to livestock in place of allowing the animals to eat the green feed where it grows.

soiling crop. Green forage crops that are cut in the field and fed in fresh condition to livestock.

soil insecticide. Any insecticide applied to the soil to destroy obnoxious insects in the soil or to prevent them from inhabiting the soil.

soilless gardening. Growing plants in a nutrient solution; water culture. See *hydroponics.*

soil line. The soil surface; the line of contact of the soil with the atmosphere.

soil losses. Removal of physical or chemical portions of the soil by natural or artificial means which may include a change to a less desirable structure. Losses may result from elements of fertility removed in harvested crops, by erosion, or by leaching by excessive use of irrigation water.

soil management. The preparation and manipulation of soil for the production of crops. Continuous, profitable production involves the maintenance of the physical structure of the soil, the organic matter, the available nutrients, the biological activities, and the conservation of soil and water.

soil management groups. Groups of soil units of sufficient area and with similar adaptations or management requirements for one or more specific purposes, such as adapted crops or crop rotations, drainage practices, fertilization, forestry, highway engineering, etc.

soil map. A map designed to show the distribution of soil types or other soil mapping units in relation to the prominent physical and cultural features of the earth's surface. The following kinds of soil maps are recognized: (a) *detailed soil map,* a map on which the boundaries between all soil types that are significant to potential use (generally field-management systems) are shown. The scale of the map depends upon the purpose to be served, the intensity of land use, the pattern of the soils, and the scale of other cartographic materials available. Traverses are usually made at one-quarter mile or more frequent intervals. Commonly a scale of 4 ins. = 1 mi. (1:15,840) is now used for field mapping in the U.S.; (b) *detailed reconnaissance soil map,* a map in which some portions satisfy the specifications for *detailed soil maps* while other portions are *reconnaissance soil maps;* (c) *generalized soil map,* a small-scale map made to bring out the contrasts within large areas by generalization of more detailed maps. They vary from soil association maps of a county on a scale of 1 in. = 1 mi. (1:63,360) to maps of larger regions showing associations dominated by one or more great soil groups; (d) *reconnaissance soil map,* a map which is made by observation of an area at intervals, such that the complete land area is not examined, as is the case with detailed surveys. The intervals of traversing vary from about one-half mile to several miles. The units shown are soil associations. The maps are usually made for exploratory purposes to outline areas of soil suitable for more intensive development. The scale is usually smaller than for detailed maps.

soil maturity. In pedology, the natural development of a soil beginning with raw parent material and continuing to a state of maturity; a mature soil is one which expresses the full effect of environmental factors. Time is a factor although the time required for maturity varies.

soil microbiology. The branch of science which deals with the study of microorganisms in the soil, such as algae, fungi, bacteria, protozoa.

soil miller. Any of various types of soil-pulverizing, rotary hoes. See *rotary hoe.*

soil mixture. A mixture which consists of soil with varying proportions of one or more of the following: manure, peat moss, leaf mold, and sand used for growing seedlings, pot plants, plants in beds in greenhouses, and as a medium for the rooting of cuttings. Also called *potting soil.* See *standard soil mixture.*

soil moisture. The water in soil in any of the various forms and physical and chemical relationships which is normally expressed as a percentage of the weight of water to the oven-dry weight of soil: ordinarily, the water in soil which is available for plant use. See *holard.*

soil moisture capacity. The amount of water expressed in percentage of dry weight that a soil can retain against the pull of gravity. See *field capacity.*

soil morphology. The kinds and arrangements of horizons or separate natural layers in a soil profile.

soil mottling. In pedology, a variegation of color in a soil horizon, as mottles, spots, splotches, or veins of one color in a background mass of another. The different colors may be about equally divided in the mass.

soil mulch. 1. The natural cover of fragmental dead plant matter or litter on a forest soil. 2. The corresponding loose matter on the surface of a grassland soil, in contrast to an artificial mulch. See *mulch.*

soil nutrients. Nutrients in soil which are available for the use of growing plants.

soil particle. Disparate solid matter in the soil of any size which is not aggregated or cemented. A soil separate in the mechanical analysis of a soil.

soil phase. In soil classification and mapping, a subdivision or variant from the type. The phase may be based upon some slight difference, physical or chemical, from the normal character of the type. Often, though unscientifically, it is based upon some external factor in relation to use, such as, slope, degree of erosion, drainage, etc.

soil physics. That branch of science which is concerned with the application of the laws, knowledge, and techniques of physics to the study of soils.

soil porosity. The degree to which the soil is permeated with pores or cavities. It is expressed in percent of the volume of the soil unoccupied by soil particles.

soil productivity. The ability of a soil to produce crops under a particular condition of management. See *soil fertility.*

soil profile. A vertical section of a soil. The section, or face of an exposure made by a cut, may exhibit with depth a succession of separate layers although these may not be separated by sharp lines of demarcation. See *horizon A; horizon B; horizon C; soil horizon.*

Soil Province. In the early period of classification and mapping in the Soil Survey Reports of the U.S. Department of Agriculture, types and series of soils were correlated according to the Province in which they were located. The Province was a geographic division based mainly upon unity in geologic formations or upon physiography on the assumption that all soils located within such areas would have common characteristics and unity as agricultural land. (The term is not obsolete in pedologic terminology.)

soil reaction. The degree of acidity or alkalinity of the soil mass expressed in pH values.

soil renovation. The restoration of a badly deteriorated soil to a condition suitable for use.

soil rot. A bacterial disease of Irish and sweet potatoes, caused by *Actinomyces* spp., family Actinomycetaceae, characterized by roseate pocks on the tubers and roots.

soil-sampling tool. Any of various kinds of augers,

metal tubes (driven and pushed into the soil), cutting spades, post-hole diggers, narrow-bladed trowels, and other implements used for obtaining samples of soil for description, for laboratory determinations, and for exhibit.

soil-saving dam. A dam of earth, concrete, or other material which is placed across a gully or natural watercourse to impound runoff water and to collect eroded soil in a pool.

soil-saving dike. An earth dike constructed on the lower end of an irrigated field, usually with an adjustable outlet, for retaining sediment in the field.

soil science. The science which deals with the scientific study of soil as a natural body and an economic resource. See *pedology.*

soil scientist. A trained investigator or research worker who specializes in the scientific study of soils.

soils demonstration. A field or plot used to demonstrate proper soil fertilization, handling, or management.

soil separate. Any of a group of soil particles which has definite size limits. The names and sizes of separates recognized in the U.S. are: very coarse sand, 2.0–1.0 mm.; coarse sand, 1.0–0.05 mm.; medium sand, 0.05–0.25 mm.; fine sand 0.25–0.10 mm.; very fine sand 0.10–0.05 mm.; silt, 0.05–0.002 mm.; and clay, 0.002 mm. The classes of separates recognized by the International Society of Soil Science are: I, 2.0–0.2 mm.; II, 0.2–0.02 mm.; III, 0.02–0.002 mm.; IV, 0.002 mm.

soil series. A taxonomic group of soils, intermediate between family and type. As originally conceived, it was a group which could include a complete range in textural classes in the surface layers, or plow soil, without affecting the subsoil or parent rock.

soil shield. A device on a cotton planter which prevents the soil from falling back into the furrow opening before the seed is deposited.

soil shredder. A machine used in greenhouse culture for mixing soil ingredients, particularly for fining manure or organic matter.

soil sickness. Unproductiveness in soil frequently encountered in greenhouses which often results from the presence of parasitic soil organisms.

soil slaking. The phenomenon expressed by crumbling, flaking, or complete disintegration of a mass of dry soil when immersed in water. Also called *slacking.*

soil slip. A downward movement of a mass of soil on hillsides under conditions of complete saturation. The distance covered and the mass of the movement in a *soil slip* is less than in a landslide or avalanche.

soil slope. An incline of a land surface. See *slope.*

soil solution. 1. The liquid part of soil that surrounds soil particles and contains the elements in solution that are absorbed by plant roots. It is not constant in amount or composition. 2. Technically, the aqueous solution which exists in equilibrium with the soil particles at a particular moisture tension.

soil sterilization. 1. The incorporation in the soil of a material that is toxic to plants. 2. The use of steam or direct heat to destroy all plant and animal life in a soil. Practiced especially in greenhouses to rid soil of plant-disease organisms and insects.

soil structure. The natural arrangement of individual particles in soil into separate aggregates, various in form and size, designated as: granular, nuciform, crumb, prismatic, columnar, blocky, and platy. Also called *structural aggregate.*

soil survey. The systematic classification and mapping of soils in an area. *Soil surveys* are classified according to the intensity and purpose of field examinations. See *soil map.*

soil survey report. A written statement accompanying a soil map which describes the geography of the area surveyed, the characteristics and capabilities of the soil types or units shown on the map. It may also discuss the principal factors responsible for soil development.

soil temperature. The temperature in the soil mass which may vary according to depth and local climate. It is an important factor in the growth of plants.

soil testing. Any of the various laboratory and field examinations of soil samples to determine the amounts, kinds, and availability of plant nutrients. Tests may also be made in relation to plant growth, agriculture, highway and building construction, such as acidity, alkalinity, moisture, porosity, penetrability, etc.

soil testing set. A small, compact container and the various chemicals, equipment and instructions needed for making rapid qualitative and quantitative tests on soils. It is designed for use mainly in the field but it may be used in the laboratory. Also called *soil testing kit; soil testing outfit.*

soil texture. The relative proportion in a soil of the various size groups of individual soil grains. The coarseness or fineness of the soil depends on the predominance of one or the other of these groups, which are silt, clay, coarse sands, loam, etc. See *soil class; soil separate.*

soil tilth. Soil structure or physical condition favorable for germination and development of crop plants. It is usually produced by loosening the soil with plows, harrows, or other agricultural implements, and by other agronomic practices.

soil type. 1. A group of soils which have genetic horizons similar in differentiating characteristics (including texture and arrangement in the soil profile) and developed from a particular type of parent material. In a taxonomic classification, it is a category lower than family or series: a subdivision based primarily on soil texture. 2. In a general sense, any taxonomic order, group, or any kind of soil division. See *land type.*

soil-type name. In Soil Survey mapping by the U.S. Department of Agriculture, the geographic name given to a soil type, as a city, town, river, or other place-name representing the locality where the type was first recognized.

soil washing. Soil erosion which results from rainfall, snow melt, or irrigation runoff.

soil water. 1. Any or all of the various forms of water which may be contained in the soil, as free, capillary, combined. 2. The water below the ground surface in the zone of aeration which may be discharged as evaporation, plant transpiration, and seepage.

soja bean. Soybean.

solar wax-extractor. A glass-covered, insulated box used to melt beeswax by the heat of the sun.

solder top. An early type of container used for canning fruits and vegetables. A small opening in the top of the metal can allowed it to be filled, and a metal disk was then soldered over the opening.

sold in the seed. Designating cotton which is sold without ginning.

sole. 1. The bottom of the furrow on which the plow bottom slides. 2. The bottom of a plow. Also called *slade.* 3. The bottom of an animal's foot.

sole ownership. The owning of property, particularly real estate, in one person's name in fee simple.

solid color. A uniform color unmixed with any other shade; self color.

solid-colored flower. See *self-colored flower.*

solid, compact and close. Designating a desirable body and texture of cheese.

solid-cone nozzle. A spray nozzle similar to the hollow-cone nozzle (*q.v.*) except that a solid cone of spray is emitted. At short distances it deposits spray on the target in the form of a disk rather than of a ring; especially suited for close spraying, as with boom spraying of field crops.

solid disk. Round, smooth-edged, solid, circular pieces of metal used on certain types of disk harrows, as compared to the cut-away disks and the spading disks used on other types.

solidified honey. Granulated honey.

solid manure. Manure in solid form as contrasted to liquid manure *q.v.*

solid moldboard. A moldboard (*q.v.*) which has no perforations or cut-out sections which would break the smooth surface of the plow. *Cf. slat moldboard.*

solid-pack tomato. Whole, firm, evenly-colored tomatoes which are packed by hand into cans without the addition of juice or purée.

solid planting. 1. A planting of one strain, variety, or species. 2. Intensive broadcast or drill sowing such that the proximity of the subsequent plants prevents cultivation, as contrasted to row planting.

solids. All parts of liquid products or solutions, except the moisture, including the starch, fat, sugar, casein, and minerals.

solid seeding. Solid planting.

solid shield. A fender attachment for a cultivator which consists of a solid piece of metal, in contrast to one made of spaced rods or wire. See *shield.*

solids-not-fat. Everything in milk except the moisture and the fatty matter.

solifluction. The slow, mass movement down slope of a blanket of soil and rock weathering because of saturation with water. Most common in mountain regions.

solodized soil. A gray leached layer over the dark colored hard layer of a Solonetz (soil) *q.v.*

Solomon Islands ivyarum. *Scindapsus aureus,* family Araceae, a tender, climbing, perennial herb grown in the warm greenhouse for its habit and its foliage. Native to the Solomon Islands. Also called *hunter's robe; Colombo agent; devil's ivy.*

Solomonplume. Any plant of the genus Smilacina. Also called *false Solomon's-seal.* See *feather Solomonplume; starry Solomonplume.*

Solomon's lily. Blackcalla arum.

Solonchak (soil). An intrazonal group of soils which has a high concentration of soluble salts; usually light-colored and without characteristic structural form. It developed under grass or shrub vegetation in an arid, semiarid, or subhumid climate. It is approximately equivalent to saline soil.

Solonetz (soil). An intrazonal group of soils which has a variable surface horizon of friable soil underlaid by dark-colored, hard soil, ordinarily with columnar structure and usually highly alkaline. It developed under grass or shrub vegetation, mostly in a subhumid or semiarid climate.

Soloth (soil). An intrazonal group of soils which has a thin surface layer of brown, friable soil over a gray leached horizon resting on a brown or dark brown horizon; alkali soils developed in a semiarid or subhumid climate under shrubs, grasses, or a mixture of grass and trees.

soluble. A substance which is capable of being dissolved in another, usually a liquid. Feedstuffs containing *solubles* contain a higher percentage of vitamins, especially B complex, than those without *solubles.*

soluble ash. That portion of milk ash which is soluble in water.

soluble whey protein. Those fractions of milk proteins which are not precipitated by acids, thus passing into the whey, where they may be reclaimed by certain processes.

solum. The upper part of the soil profile above the parent material, in which the processes of soil formation have taken place. In mature soils this includes the A and B horizons which may be, and usually are, greatly unlike the parent material beneath. Living roots and life processes are largely confined to the *solum.* See *horizon A; horizon B; parent material.*

solute. Any substance which is dissolved in another.

solution. A mixture of two or more substances which is chemically and physically homogenous.

solution of formaldehyde. See *formalin.*

solvent. 1. A liquid that is used to dissolve a substance. 2. Designating an individual or business which is in good financial condition; *i.e.,* able to pay all debts and other liabilities.

solvent extracted cottonseed flakes. The product

obtained after extracting part of the oil from cottonseed by the use of solvents.

solvent extracted cottonseed meal. That cottonseed meal which has been manufactured by the solvent extraction process. It must be so designated and tagged at the time of sale. It must meet the quality specifications, as cottonseed meal, prime quality *q.v.*

solvent extracted cottonseed oil. That crude cottonseed oil which is produced by the solvent extraction process with or without pre-pressing. It is sold according to the grade specification it meets.

solvent process. The process of extracting oil from oil-bearing seeds, wool, etc., by a solvent.

soma. The body, in contrast with the germ or germ plasm.

somatic. 1. Designating body tissues. 2. Having chromosomes in pairs, one of each pair normally coming from the female parent and one from the male, as contrasted with germinal tissue which gives rise to germ cells.

somatoplasm. The body tissues.

song-bird. A singing bird, generally beneficial or relatively harmless on farms in contrast to nuisance birds, such as the crow, and game birds, such as quail and pheasants.

sonic boom. A compression wave in the lower atmosphere and the resulting explosive sound caused when a super-speed jet plane crosses the sound barrier. It may cause cattle, horses, etc., to stampede, or disrupt the routine of excitable animals, such as mink or fowl, and even cause them to injure themselves or their young.

sonic vibrator. A disk vibrating at a high speed to emit a penetrating sound: *e.g.*, a foghorn. A form of the *sonic vibrator* has been used in an attempt to homogenize milk by shattering the fat globules as the milk flows over the vibrating disk.

soot fertilizer. Chimney soot resulting from the burning of wood or coal and sometimes applied to soils for its fertilizing value. The effectiveness of the old practice of using a soot and common salt mixture for soil amendment is controversial.

sooty blotch of apple. See *sooty blotch of pome fruits.*

sooty blotch of clover. A leaf disease of clover caused by the fungus, *Cymadothea trifolii*, distributed throughout the temperate zones on alsike and white clover and less commonly on red and crimson clover. It appears as dark brown, angular, leaf blotches usually followed by minor defoliation. Also called *brown blotch.*

sooty blotch of pome fruits. A fungous disease of pome fruits caused by *Gloeodes pomigena*, family Leptostromataceae. Indefinite, dull black to gray blotches of various sizes appear on the fruit. Also called *cloud; sooty fungus; sooty mold; sooty smudge.*

sooty mold. A mold caused usually by the fungus *Meliola camelliae*, family Meliolaceae, or *Capnodium citri*, family Capnodiaceae, occurring on citrus and growing on the honeydew secreted by certain scale insects, whitefly larvae, and plant lice. The mold interferes with the physiological function of the leaves and produces blemishes on fruit.

sooty smudge of pome fruits. Sooty blotch of pome fruits.

sooty stripe of sorghum. A fungous disease of sorghum caused by *Ramulispora sorghi*, family Tuberculariaceae, characterized by lesions, first gray to tan in the center with a red margin, later sooty-colored.

Soprassata. Any sausage stuffed in crinkly hog middles. It consists mainly of coarsely chopped pork and contains black peppers as one of its condiments.

sore back. A common ailment with saddle horses, usually the result of a broken saddle gall.

sore head. Fowl pox.

sore mouth. Contagious ecthyma.

sore shin. A fungous disease of cotton seedlings caused by *Rhizoctonia solani*, or *Pellicularia filamentosa*, family Thelephoraceae, characterized by stem canker on the seedling near the soil line. The disease is generally distributed and commonly develops during periods of cold, wet weather. Also called *anthracnose seedling blight. Cf. cotton anthracnose.*

sore shin of tobacco. A fungous disease of tobacco caused by *Rhizoctonia solani*, characterized by dark-colored cankers and dwarfing and occasional breaking of plants.

sorgho. Sorgo.

sorghum. Any of the various plants of the genus Sorghum (syn. Holcus), family Gramineae. Species and varieties have important uses as fodder, forage, hay, grain, and as a source of sirup and broomcorn. See *broomcorn; durra; Johnsongrass; Kafir; milo; shallu; sorgo; Sudan grass.*

sorghum anthracnose. See *cereal anthracnose.*

sorghum charcoal rot. A fungous disease of sorghum and many other crops caused by *Macrophomina phaseoli*, family Sphaerioidaceae, characterized by poorly-filled heads, light-weight kernels, premature drying, lodging, and softening and discoloring of the stalks.

sorghum fodder. Fodder or stover from the sorghum plant cured in shocks in the same manner as corn.

sorghum gray leaf spot. A fungous disease of sorghums caused by *Cercospora sorghi*, family Dematiaceae, characterized by a spotting of the leaves. Endemic in humid and warm regions.

sorghum hay. Hay made from thickly planted sweet sorghum (sorgo). Also called *cane hay.*

sorghum head smut. See *head smut.*

sorghum leaf rust. A fungous disease of sorghum caused by *Puccinia purpurea*, family Pucciniaceae. Its development is promoted by warm, humid weather and it is often serious in humid sections of southeastern North America.

sorghum midge. *Contarinia sorghicola* (Coq.), family Itonididae (Cecidomyiidae), a midge that infests sorghums extracting the plant juices of the developing seeds and blasting or blighting the grain spikelets.

sorghum molasses. A type of sirup usually farm-produced from sorgo. It is common in the south-central U.S.

sorghum pan. A large, rectangular pan of copper, tin plate, or galvanized iron used for concentrating juice or sap of maple, cane, or sorghum to a sirupy consistency by boiling. Thin strips of metal divide it crosswise into sections which are connected in such a manner as to provide a zigzag path for the movement of the juice during the boiling process.

sorghum poisoning. A poisoning of livestock which results from ingestion of sorghum containing prussic acid (hydrocyanic acid), a deadly poison which develops in dangerous amounts under certain conditions in many of the sorghums.

sorghum root and stalk rot. Any of several root and stalk rots that attack sorghum, especially milo disease *q.v.*

sorghum silage. Silage made from either sweet or grain sorghums ensiled when the seeds are hard and ripe. It is somewhat inferior to well-eared corn silage in feed value.

sorghum sirup. A thick sirup made from the juice of sorgo or sweet sorghums by boiling the juice in shallow evaporating pans until it is concentrated.

sorghum stover. Sorghum fodder with the heads removed. It is similar to corn stover in composition and general value.

sorghum webworm. *Celama sorghiella* (Riley), family Noctuidae, a small moth whose larvae feed on the ripening kernels of grain sorghum and broomcorn.

sorgo. *Sorghum vulgare* var. *saccharatum,* family Gramineae, a tall, sweet herb with a juicy pith high in sugar, whose chief use formerly was for sirup, but now is primarily grown for forage. It is one of the four classes of sorghum commonly grown in the United States. Native to Europe and Asia. Also called *imphee; saccharine sorghum; sorgho; sweet sorghum; sugar sorghum.*

sori. The plural of sorus *q.v.*

sorrel. Certain plants of the genus Rumex, family Polygonaceae. See *sheep sorrel: 1; garden sorrel.* 2. A coat-color of some animals. It includes the red shades of chestnut.

sorrel rhubarb. *Rheum palmatum,* family Polygonaceae, a perennial herb grown for its striking foliage. Native to Asia.

sorrel tree. Sourwood.

sorter. A person or machine which separates or grades products according to quality, size, or other specifications.

sorting. 1. Separating or grading fruits, vegetables, or other products according to size or quality; often only for the removal of those items of inferior quality. 2. Dividing wool fleeces into various qualities (not grading).

sorting chute. A narrow alley with gates used for sorting or separating animals.

sorting rack. A rack used for sorting onions, so designed that dirt and small bulbs fall through the slats; the thicknecks, injured or decayed bulbs being picked out by hand. It is often made so that the bulbs can be divided into two or more grades as they pass over the rack.

sorus (plural: **sori**). 1. A fungus fruiting structure or spore mass, especially one of rusts and smuts. 2. A cluster of sporangia found on the underside of the leaves of ferns.

souari. Caryocar *q.v.*

sound. 1. In good condition; not damaged; acceptable. 2. Designating a horse free from unsoundness or blemishes of any kind. 3. Designating wool that has good strength.

sound cultivated oats. Designating cultivated oats that are not heat-damaged, badly ground-damaged, badly weather-damaged, sprouted, frosted, or otherwise materially damaged.

sound-cutting. In hardwood lumber, a piece that is free from rot and shake.

sound knot. A knot that is solid and is not decayed. See *knot: 4.*

sound-merchantable. Designating lumber not of an established grade which is salable for some specific purpose.

soundness. Freedom from defects or blemishes as: (a) a wool fleece that shows strong wool fiber and no weak spots; (b) a hog used for breeding purposes that is free from any noticeable defect; (c) a horse that shows good feet and legs.

sound-wormy. A recognized grade of hardwood lumber showing the effects of insect activity.

soup bone. The fore and hind shanks of a beef carcass. In modern meat cutting, many other bones of the meat carcass are trimmed out and sold as *soup bones.*

sour. In butter grading, designating butter with a sour flavor. Such butter is rated C.

sour aroma. Anything that is distinctly sour to the sense of smell: *e.g.,* sour cream, sauerkraut.

source. 1. The point of origin of a stream, river, etc. 2. A place or an object from which something may be obtained: *e.g.,* a *source* of supply; grapes are a *source* of wine.

source of inoculum. The place or object on or in which inoculum is produced.

sour cherry. *Prunus cerasus,* family Rosaceae, a deciduous tree, grown for its fruit, the red, tart cherry used in cooking, preserves, pies, etc. Important varieties include Montmorency, Early Richmond, and English Morello. Also called *red tart cherry.* The pitted, processed fruit is known as *r.s.d.* (red, sour, pitted) cherry in trade. See *cherry: 1.*

sour cherry ring spot. See *ring spot of stone fruits.*

sour cherry yellows. An important but not fully understood virus disease which occurs in the nursery and orchard mainly on sour cherries. Its symptoms are the appearance of yellow and green mottling of some of the leaves, followed by leaf shedding. Affected trees may eventually show a willowy type of twig growth and a much reduced spur system.

sourclover. Annual yellow sweetclover.

sour-cream butter. Any butter which is made from cream whose acidity has exceeded 0.20 percent.

sour-cream buttermilk. Any buttermilk which results from the churning of clean, ripened cream with an acidity of more than 0.20 percent.
sour crop. See *thrush: 2.*
sour egg. An egg with a sour odor which is usually associated with an early stage of white rot; an inedible egg.
sour flavor. A flavor defect most frequently associated with dairy products, such as milk, butter, and ice cream, which is characterized by a sour taste and odor resulting from the reduction of milk sugar to lactic acid by lactic acid producing bacteria. See *sour milk.*
sour-grass. Shore podgrass.
sour gum. See *black tupelo; water tupelo.*
sour gum casecutter. Tupelo leaf miner.
souring. The production of acid, chiefly lactic acid, by bacteria, such as Streptococcus. See *lactic acid culture.*
sour milk. Milk having a distinctly sour taste which results from bacterial decomposition producing lactic acid.
sour orange. *Citrus aurantium,* family Rutaceae, an evergreen tree cultivated in all subtropical regions for its flowers, leaves, and fruit yielding volatile oils (bigarade oils) of characteristic odor much prized for perfume. The fruits are used extensively for the making of marmalade and liqueurs. Native to Asia. Also called *Seville orange; bigarade.*
sour orange scab. A fungous disease of citrus caused by *Elsinoë fawcettii,* family Myriangiaceae, characterized by yellowish to reddish, corky or wart-like projections on fruits, leaves, and twigs. Leaves often become wrinkled, stunted and misshapen, and the fruits may be distorted. The disease occurs chiefly in humid regions and on susceptible species of citrus, particularly on the sour orange.
sour paspalum. *Paspalum conjugatum,* family Gramineae, a rapidly-spreading, troublesome weed pest (Hawaii). Also called *hilo grass.*
sour rot. A contact decay of citrus fruits, particularly of lemons in storage, caused by *Oospora citri-aurantii,* family Moniliaceae.
sour-sap. A winter injury, particularly of apricot, almond, and other stone fruits, characterized by stagnation and fermentation of the sap in the bark, cambium, and young wood. The affected tree dies suddenly as it is leafing out in the spring.
sour soil. 1. An acid soil *(q.v.).* 2. Loosely, any swampland soil or one on which sheep sorrel is an abundant weed, although such soils are not everywhere acid. *Cf. sweet soil.*
soursop. *Annona muricata,* family Annonaceae, a small, evergreen tree grown for its edible green, spiny, ovoid fruit used mostly for ices and beverages. It is also grown as a grafting stock for cherimoya. Native to tropical America. Also called *guanabana.*
sourweed. Sheep sorrel.
sour-whey. Designating a flavor defect of cheese which results from the high acidity of the milk from which the cheese was made and/or acid developed in the curd during manufacture.
sourwood. *Oxydendrum arboreum,* family Ericaceae, an evergreen tree grown in frost-free areas for its white flowers in panicles 8 to 10 ins. long. Native to North America. Also called *elk tree; sorrel tree; elk wood.*
souse. A food product made from pigs' feet, hearts, tongues, and shoulder hocks. The meat is cooked, cut into chunks, replaced in the broth in which it was cooked, seasoned with salt, pepper and vinegar, poured into pans, and set away to chill and jell. *Souse* is usually cut into slices and eaten cold. Also called *pan haus; pig souse.*
South American waxmallow. *Malvaviscus arboreus,* family Malvaceae, a shrub grown in warm areas and as a greenhouse plant for its foliage and showy, red flowers. Native to tropical America.
Southdale. A breed of sheep which resulted from cross-breeding the Southdown and Corriedale breeds.
South Devon. A dual-purpose breed of cattle developed in Devonshire, England.
Southdown. One of the smallest and oldest breeds of sheep which had its origin in the Southdown area of England and which is best known for its high quality mutton carcass. Mature ewes weigh from 125 to 160 lbs.
southeast decumaria. *Decumaria barbara,* family Saxifragaceae, a hardy, climbing, woody vine, grown as an ornamental climber for stone walls or as a creeping shrub for its habit and white flowers. Native to North America. Also called *climbing hydrangea.*
southerly buster. A strong, south wind with rain, snow, or sleet (New Zealand).
southern armyworm. *Prodenia eridania* (Cram.), family Noctuidae, an insect whose larvae infest vegetables and several other plants from Florida to South Carolina (U.S.).
southern bacterial wilt. Granville wilt.
southern balsam fir. Fraser balsam fir.
southern bayberry. *Myrica cerifera,* family Myricaceae, an evergreen shrub or small tree grown for its handsome foliage and grayish-white fruits. The source of American vegetable wax. Native to North America. Also called *bayberry; southern waxmyrtle; waxmyrtle.*
southern bean mosaic. A virus that causes a mottling of the leaves and a pronounced shiny appearance of the pods.
southern blight. A blight caused by the fungus *Sclerotium rolfsii,* which attacks a plant through the roots and underground parts of the stem at the soil line, and causes a rapid wilting of the plant. The disease is often serious on peanuts, tomatoes, peppers, and many other crops in the southeastern and Gulf states (U.S.).
southern bur clover. Spotted medic.
southern cabbageworm. *Pieris protodice* B. & L., family Pieridae, an insect pest of crucifers.
southern California miner's inch. The quantity of water flowing freely through an opening 1 sq. in.

in area, in a vertical wall under a pressure of 4 ins. measured from the center of the opening, which equals 1/50 cu. ft. per second. See *miner's inch.*

southern cane. Giant cane.

southern catalpa. *Catalpa bignonioides,* family Bignoniaceae, a deciduous, round-headed tree, grown for its white, yellow-striped flowers. Native to North America. Also called *Indian bean; bean tree.*

southern cattle fever. Cattle tick fever.

southern chicken flea. Sticktight flea.

southern chunk. A small draft horse which stands from 14.2 to 15.2 hands high and weighs 800 to 1100 lbs. It is adapted to work on southern plantations (U.S.). *Cf. southerner.*

southern corn-leaf blight. A fungous disease of corn caused by *Cochliobolus heterostrophus,* which kills the green tissue of the leaves and reduces yield. It occurs throughout the corn areas of the southern U.S. where the warm, humid climate favors its development.

southern corn-leaf spot. A fungous disease of corn caused by *Helminthosporium carbonum,* family Dematiaceae, which produces circular spots up to 1 in. in diameter on the leaves, sheaths, ears, and husks. The spots may become black with spores.

southern corn rootworm. *Diabrotica undecimpunctata howardi* (Barb.), family Chrysomelidae, an insect whose larvae often cause serious injury by burrowing in the roots, crown, and stems of young corn and many other plants, including peanuts, beans, and cucurbits. The adults eat the foliage of these plants. Also called *spotted cucumber beetle.*

southern corn rust. A rust of corn caused by the fungus, *Puccinia polysora,* family Pucciniaceae, which has the general characteristics of corn rust (*q.v.*). It is found from Massachusetts westward to Indiana (U.S.) and southward to Mexico and Central America.

southern cornstalk borer. *Diatraea crambidoides* (Grote.), family Crambidae, a straw-colored moth whose larvae feed on foliage but usually bore inside of stalks of corn and sorghum. It is distributed from Virginia south and west to Louisiana and Missouri, U.S.

southern cowpea weevil. See *cowpea weevil.*

southern crabapple. *Malus augustifolia,* family Rosaceae, a deciduous tree grown for its fragrant, rose or pink flowers and for its edible, yellow-green fruit used in jellies. Native to North America.

southern curled mustard. Potherb mustard.

southern cypress. Baldcypress.

southern dewberry. See *dewberry.*

southerner. A small, chunky type of horse standing 15 to 15.3 hands high weighing from 800 to 1300 lbs. which has been much used in southern U.S. by farmers for light farm work and for light delivery-wagon use. *Cf. southern chunk.*

southern fever tick. See *cattle tick.*

southern field pea. See *cowpea: 2.*

southern fire ant. *Solenopsis xyloni* (McCook), family Formicidae, a reddish ant with a fiery sting. It is most common in the southeastern states of U.S.

southern fox grape. Bullace grape.

southern grassworm. See *fall armyworm.*

southern green stink bug. *Nezara viridula* (L.), family Pentatomidae, an insect about ⅔ in. long which sucks sap from soybeans and cowpeas causing pods to drop; forming hardened, knotty areas; or producing stunted or distorted seed. In Florida, U.S., these bugs often cause severe loss of fruit and kill succulent shoots on young citrus trees.

southern hackberry. See *hackberry: 1.*

southern hogpeanut. *Amphicarpa bracteata,* family Leguminosae, an herbaceous, perennial vine, which has been grown in southern U.S. mainly as a feed for livestock. Native to North America. Also called *earth pea.*

southern house mosquito. *Culex quinquefasciatus* Say, family Culicidae, a blood-sucking insect that in some parts of the world is an important vector of a filarial (*q.v.*) parasite which lives in the blood and lymph, and of encephalitis in birds and horses.

southern leaf spot. A fungous disease of corn and teosinte caused by *Helminthosporium maydis,* family Dematiaceae, characterized by numerous, buff or reddish-brown, elongated spots between the veins. It is found world wide.

southern lyctus beetle. *Lyctus planicollis* Lec., family Lyctidae, a small beetle whose larvae eat dry, seasoned wood, chiefly the sapwood of hardwoods, often tunneling through timbers and sometimes boring out through the surface after lumber has been made into flooring, girders, and interior finish.

southern magnolia. *Magnolia grandiflora,* family Magnoliaceae, a magnificent, evergreen tree, grown in frost-free areas for its large, fragrant, cup-shaped, white flowers. Its wood is used for furniture, fixtures, Venetian blinds, interior finish, sash, doors, and millwork. Native to North America. Also called *evergreen magnolia; bull bay.*

southern maidenhair. *Adiantum capillus-veneris,* family Polypodiaceae, a tropical fern grown in the greenhouse, and outdoors along the more moist portions of the Pacific coast as an ornamental. Native to tropical areas of America. Also called *dudder grass; Venus hair.*

southern pea. Any of several edible cowpeas. See *cowpea: 2.*

southern pine beetle. *Dendroctonus frontalis* Zimm., family Scolytidae, an insect that infests pine and makes winding egg galleries in the inner bark. It kills healthy trees, as well as attacking weakened or felled specimens. It is a serious pest of pine from Pennsylvania to Florida and west to Texas and Oklahoma, U.S.

southern poplar. *Populus deltoides* var. *missouriensis,* family Salicaceae, a deciduous, hardy tree, grown as a shade tree and for its lumber used in pulp wood, crates, etc. Native to North America. Also called *cottonwood; balsam poplar.*

southern redcedar. *Juniperus silicicola,* family Pinaceae, an evergreen tree, grown for its valuable timber used for posts, chests, closet lining, millwork, pencil slats, woodenware, water buckets, etc. Native to North America.

southern red oak. *Quercus falcata,* family Fagaceae, a deciduous timber tree, attaining 100 ft. or more in height which is also grown as an ornamental or shade tree. Native to North America.

southern squash bug. See *horned squash bug.*

southern tobacco hornworm. See *tobacco hornworm.*

southern waxmyrtle. Southern bayberry.

southern white cedar. Whitecedar falsecypress.

southern wilt. Granville wilt.

southernwood. Oldman wormwood.

southern wool. Wool of bright color produced in southeastern U.S. See *bright wool.*

southwest. A geographic subdivision of the U.S., but one without any precisely fixed, or generally accepted boundaries. Reference is to the states of New Mexico and Arizona, but the western parts of Texas and Oklahoma, parts of Colorado and Nevada, and southern California may be included, although the latter is also designated as Pacific Northwest.

southwestern corn borer. *Diatraea grandiosella* Dyar., family Crambidae, a corn borer, introduced into the U.S. from Mexico, which very closely resembles the southern cornstalk borer *(q.v.),* but differs chiefly in its habit of internally girdling the stalk. It is found from Kansas, Nebraska, and Colorado southward to Texas, New Mexico, and Arizona.

southwestern flints. A group of drought-resistant varieties of flint and semiflint corn adapted to the semiarid regions of southwestern U.S. Similar strains have been cultivated since prehistoric times by the Indians of this area.

southwestern pine beetle. *Dendroctonus barberi* Hopk., family Scolytidae, a black insect that makes winding, transverse egg galleries in ponderosa pine in New Mexico, Arizona, Colorado, Utah, Nevada, and California, U.S.

sow. 1. A female swine, usually one that shows evidence of having produced pigs or one that is obviously pregnant. 2. To plant seeds by scattering either broadcast or by uneven distribution in a row.

sow belly. Salt pork; fat bacon which is not smoked.

sow broadcast. To scatter seeds widely upon a field by hand or by some mechanical device, in contrast to placing such seeds in rows or hills.

sowbug. *Porcellio laevis,* family Oniscidae, an insect-like creature, commonly found under flower pots or decaying boards, which is a scavenger on rotting plants and manure. In the greenhouse, it may cause some damage by feeding about the roots and on tender portions of plants near the ground. Also called *pillbug.*

sower. 1. One who sows or scatters seeds upon a field; one who operates a machine for this purpose. 2. A tool or mechanical device used to scatter seeds.

sowing. The age-old practice of scattering of seeds upon a seed bed or field in preparation for a future crop. This may be performed by hand or by any of many different types of mechanical equipment.

sow pig. A young female swine.

sow thistle. 1. Any of the several plants of the genus Sonchus, family Compositae. 2. S. *oleraceus,* an erect, coarse, leafy herb with smooth stems, cut leaves, and yellow, dandelion-like, flowers which is a weed pest. Also called *dindle.* See *field sow thistle.*

soya bean. Soybean.

soybean. *Glycine Max,* family Leguminosae, an annual, erect bushy herb, extensively grown for forage, as a cover crop, for green manuring, and particularly for its edible, nutritious, oil-bearing seeds which vary in color from green through yellow and brown to black. Over 70 percent of the American *soybean* acreage is grown for the seed, chiefly for the production of soybean oil and soybean oil meal. Native to Asia. Also called *soya bean; soja bean.* See *soybean oil; soybean oil meal.*

soybean anthracnose. There are two distinct fungi concerned with the two types of anthracnose occurring on soybeans: (a) *Glomerella glycines,* family Gnomoniaceae, of which the perithecia of the perfect stage and the acervuli of the conidial stage *(Colletotrichum glycines)* occur on seedlings and pods; (b) *Colletotrichum truncatum,* family Melanconiaceae, which is more abundant in the stems. Apparently these fungi attack both soybean *(Glycine max)* and Lima bean *(Phaseolus limensis).*

soybean hay. Hay made from the entire soybean plant which is cut and cured at any growth stage after the pods are formed and before the leaves turn yellow and fall. Well-cured *soybean hay* is palatable to livestock and is as rich in protein as alfalfa.

soybean hay meal. A livestock feedstuff which is the product obtained by grinding the entire soybean hay without the addition of any stems, straw, or foreign materials, or the abstraction of leaves or beans. It must be reasonably free from other crop plants and weeds and must not contain more than 33 percent of crude fiber.

soybean hull. The outer covering of the soybean.

soybean mill feed. A livestock feedstuff which is the by-product resulting from the manufacture of soybean flour or grits. It is composed of soybean hulls and the offal from the tail of the mill.

soybean mosaic. A virus disease of soybeans in which the symptoms appear as a distortion of the leaves: narrower than normal with margins turning downward. Affected plants produce misshapen pods and fewer seeds than normal plants. The disease is seed-borne and is commonly found over all the soybean areas of the United States.

soybean nematode. *Heterodera glycines,* family

Heteroderidae, a cyst-producing nematode, which has been reported from southeastern North Carolina (U.S.). The soybean roots are invaded by the larvae.

soybean oil. An oil obtained from the seeds of the soybean, used extensively in the preparation of many food and industrial products, including vegetable shortening, margarines, salad dressings, paints, soaps, plastics, and linoleum.

soybean oil cake. A livestock feedstuff which is the product remaining after the extraction of part of the oil from soybeans by pressure or solvents. If a name descriptive of the process of manufacture, such as expeller, hydraulic, or solvent extracted be used, the produce must correspond thereto.

soybean oil chips. One form of the product remaining after the extraction of the oil from soybeans. It is similar in composition to soybean oil cake *q.v.*

soybean oil meal. A livestock feedstuff consisting of ground soybean oil cake or oil chips. If a name descriptive of the process of manufacture, such as expeller-, hydraulic-, or solvent-extracted be used, the product must correspond thereto. *Soybean oil meal* has an average protein content of about 45 percent and ranks high as a protein supplement for livestock feeding.

soybean stem canker. A serious, fungous disease caused by the fungus *Diaporthe phaseolorum* var. *caulivora,* family Valsaceae, occurring only in the cornbelt area of the U.S. and the adjacent province of Ontario, Canada. The fungus is different from that organism causing pod and stem blight, which is *D. phaseolorum* var. *sojae,* with which it is sometimes confused. The canker organism can be inoculated into several legumes, but the natural mode of infestation into these and soybean is still not known. See *brown stem rot.*

sp. (spp.). The abbreviation, singular and plural respectively, for species (*q.v.*): *e.g.,* when one species of a genus of plant or animal is referred to, the name may be written *Canis* sp. meaning one species of dog; or *Canis* spp. meaning more than one species of the genus Canis.

space. 1. To plant seeds so as to provide a certain interval between each seed or group of seeds, as in planting corn. 2. To set out plants or to thin plants so that they are separated by a distance determined by the grower.

spaced row. A cultural system with strawberries in which individual plants are planted or are thinned to stand at a distance of 12 to 24 ins. apart in the row. *Cf. matted row system.*

spacing. The thinning of young plants in a row to the desired distance apart by using a hoe or other tool. It is important for best growth and production with such crops as cotton, sugar beets, etc.

spacing lever. On a cultivator, a lever used to change the distance between the gangs of shovels.

space spray. Insect spray and spray equipment used in insect control in open spaces, in a dairy barn, hog house, etc.

spade. A long, flat, sharp-bladed, earth-excavating, hand tool made in several shapes, which is pushed into the soil by foot pressure to turn over or excavate soil.

spading. The act of turning the soil over with a spade.

spading disk. A type of disk harrow in which the circular disks are serrated on the cutting edge to give them more cutting or penetrating action.

spading fork. A short-handled spade in which the spade blade is made of six, pointed, flat tines used for turning sandy or loose soil and/or for removing tubers from the hill or row, etc. Also called *fork; garden fork.*

Spalen. A very hard, cows' milk cheese named for the wooden containers, *spalen,* in which the cheeses are shipped.

spalt. In shingle cutting, the residue of a shingle bolt.

span. 1. A pair of animals usually harnessed together as a team. Also called *spann.* 2. The space between the top of the two walls of a ravine. 3. The length or extent of a bridge between supports.

spangle. A clearly defined marking of a distinctive color located at the end of a solid-color feather: *e.g.,* the *spangle* on the tip of the white feathers of the Silver-Spangled Hamburg chicken.

spangled. 1. Designating a feather of a solid color that has a clearly-defined, V-shaped, rounded at the top marking of a distinctive color on its tip. 2. Designating a variety of chicken whose plumage is so marked.

Spaniard. See *speargrass.*

Spanish. See *White-faced Black Spanish.*

Spanish bayonet. Yucca.

Spanish bluebell. Spanish squill.

Spanish broom. Weaversbroom.

Spanish button. Black centaurea.

Spanish cedar. Cigarbox cedrela.

Spanish chestnut. European chestnut.

Spanish clover. Spanishclover deervetch.

Spanishclover deervetch. *Lotus americanus,* family Leguminosae, an annual herb, with red or rose-colored flowers, found on dry soils of central U.S. which is of value as a wildlife plant. Native to the central U.S. Also called *birdsfoot trefoil; compass plant; prairie lotus.*

Spanish dagger. Yucca.

Spanish jasmine. *Jasminum officinale* var. *grandiflorum,* family Oleaceae, a shrub grown for its very fragrant, white flowers tinged with crimson and for its blossoms used in making perfumes. Native to India. Also called *Catalonian jasmine.*

Spanish lime. Honeyberry.

Spanish Merino. A breed of sheep developed and bred in Spain which is the ancestor of the present day Merino, a highly valued, fine-wool breed naturalized throughout the world.

Spanish mombin. Mombin.

Spanish moss. Treebeard tillandsia.

Spanishneedles. *Bidens bipinnata,* family Compositae, a common weed plant whose achenes (seeds) have two to four stiff bristles or awns which are barbed and cling to hair, fleece, or clothing. Native to eastern U.S.

Spanish oysterplant. *Scolymus hispanicus,* family Compositae, a biennial herb grown for its edible tap root with a flavor similar to vegetable-oyster salsify. Native to southern Europe. Also called *scolymus; Spanish salsify; golden thistle.*

Spanish plum. Mombin.

Spanish pony. A small horse common to western U.S., which is a descendant of the horses brought in by the early Spanish explorers. It is very similar to the other horses of this region called cayuses, mustangs, and Indian ponies.

Spanish red scale. Dictyospermum scale.

Spanish salsify. Spanish oysterplant.

Spanish squill. *Scilla hispanica,* family Liliaceae, a small, bulbous herb grown for its cheery, early-spring, blue, white, pink, and rose flowers. Native to the Iberian Peninsula. Also called *Spanish bluebell; Spanish jacinth.*

Spanishthyme coleus. *Coleus amboinicus,* family Labiatae, a fleshy herb, grown in the tropics for its aromatic shoots. Native to the East Indies. Also called *country borage; East India thyme.*

Spanish violet. Horned violet.

span-roof house. Even-span house.

spar. A moderately thick, lengthy pole, often used as a rafter or other form of support in barns, sheds, etc.

spareribs. A retail cut of pork made up of the rib bones and sternum removed from the belly or bacon side.

sparkleberry. Farkleberry.

sparse. Thinly scattered, not dense, as a *sparse* growth of plants.

spasm. An involuntary, sudden, and violent muscular contraction of a painful character.

spasmodic colic. Acute indigestion in horses.

spatterdock cow lily. *Nuphar advenum,* family Nymphaeaceae, an aquatic herb, grown in a pool for its globe-shaped, yellow flowers. Native to North America. Also called *dog lily; yellow pond lily; cowlily; beaver root.*

Spaulding rule. A diagram rule for estimating the lumber content of logs which disregards taper and allows for an 11/32 in. saw kerf, with size of slab varying with size of log. It is the statute rule for California and is widely used elsewhere on the Pacific Coast (U.S.). Also called *Columbia River Rule.*

spavin. A disease affecting the hock joint of a horse's hind leg, usually a bony growth on the inner, lower part of the hock that causes lameness. See *bone spavin; bog spavin; blood spavin; occult spavin.*

spay. To remove the ovaries of a female animal. *Cf. castrate.*

spayed heifer. A heifer which is unsexed by removal of ovaries. Such heifers make good feeder cattle.

spear. A sprout or shoot of a plant, as a blade of grass, a young stalk of asparagus, etc.

speargrass. Various species of Aciphylla, family Umbelliferae. Species are herbaceous plants with stiff, sharp leaves, growing out thickly from the ground in all directions. Each leaf terminates in a sharp spear that can penetrate clothing. Horses and other livestock carefully avoid the plant. All species are native to New Zealand except two originating in the mountains of Australia. Also called *Spaniard.*

spearmint. *Mentha spicata,* family Labiatae, an aromatic, perennial herb widely grown for its foliage used in flavoring and for its aromatic oil, the mint of commerce. Its flowers are a source of nectar for bees. Also called *garden mint.*

special. In marketing, designating fine-flavored butter that scores 93 points or above.

special breed milk. High quality milk of a particular breed, produced under rules prescribed by the breed association involved, such as Golden Guernsey milk.

specialized farming. Farming in which a major part of the income is normally derived from one enterprise.

specialized livestock ranch. See *cow ranch.*

special market. A limited market for a special product, either in kind or quality, such as certified milk, maple sirup, etc.

special-purpose. Designating an animal which is selected or used for a single purpose, such as a dairy cow kept only for milk production, or a horse kept only for riding.

special-purpose crop. A field crop planted or produced for a single purpose, such as rye sown as a cover crop, alfalfa for seed, peas for canning, etc.

special steamed bone meal. The dried, ground product, suitable for animal feeding, obtained by cooking dried bones under steam pressure in the process of obtaining gelatin or glue after the removal of grease and meat fiber.

specialty. 1. A special line of work or business. 2. A product to which special attention is given. 3. In ice cream manufacture, ice cream formed in individual molds, center and layer bricks, in pies, cakes, log rolls, etc.

specialty farmer. A farmer who specializes in a single crop or farm enterprise, such as a celery or tobacco grower, a broiler producer, etc.

species. A more or less continuous and interbreeding group of animals or plants which have in common one or more characters by which they may be differentiated from other *species; species* differences are similar in kind to breed differences, but usually more extreme, with the result that crosses between *species* are impossible or, if possible, are usually sterile; a subgroup of a genus.

specific. 1. A medicine that cures a particular disease. 2. Pertaining to a species. 3. Produced by a particular microorganism. 4. Restricted by nature to a special animal, thing, etc. 5. Exerting a peculiar influence over any part of the body.

specific gravity. In solids and liquids, the ratio between the weight of an object or substance and the weight of an equal volume of water or other substance adopted as a standard.

specific method. Any method that allows an unconfused detection (or an accurate determination) of the amount of an individual element or compound in a mixture.

specific retention. The volume percentage of water retained against the pull of gravity after saturation of soil or rock and drainage to a remote body of mobile water.

specific yield. The amount of water which a rock or earth will yield after being saturated and allowed to drain under the conditions specified for determining specific retention. It is expressed as percent of volume. The sum of specific retention and specific yield equals the porosity of the material drained.

specimen (tree, shrub, vine). 1. A plant of outstanding form, habit of growth, character, vigor, and health, grown by itself so as to display it to the best advantage or to allow it to assume interesting growth. 2. In nursery trade, a plant with a heavier than average stem or trunk.

speciosum lily. *Lilium speciosum,* family Liliaceae, a perennial, bulbous herb grown for its fragrant, drooping, white or pinkish, red-spotted flowers. Native to Japan. Also called *Japanese lily.*

speck. Pork, bacon, or other considerably fat meat.

speck-in-eye. A small scar on the eyeball of an animal, especially of a horse. It does not impair vision.

speckle. A small patch or dot of color.

speckled alder. 1. *Alnus incana,* family Betulaceae, a deciduous shrub or small tree of several varieties, grown as ornamentals. Native to North America, Europe, and Asia. Also called *hoary alder; European alder.* 2. See *hazel alder.*

speckled beadlily. *Clintonia umbellulata,* family Liliaceae, a perennial herb grown for its clusters of white flowers. Native to North America. Also called *dog plum; wild corn.*

speckled John. St. Johnswort.

speckled leaf blotch. A fungous disease of wheat, rye, and bluegrass caused by *Septoria tritici,* family Sphaerioidaceae, which is characterized by the death of the leaves during the late fall, winter, and early spring. Often confused with winter injury.

speckled snow mold. A slimy, gray rot of grasses and cereal leaves under the snow in mid or late winter, caused by the fungi *Typhula* spp., family Clavariaceae.

specklepod loco. *Astragalus lentiginosus,* family Leguminosae, a plant whose foliage, green or dried, is poisonous for livestock. Symptoms of poisoning are dullness, irregularity of gait, dragging of the feet, a solitary habit, loss of flesh, and a shaggy coat. As the animal ceases to eat, it dies. Found from Washington to Nevada, U.S. Also called *spotted loco.*

spectrophotometric. Pertaining to a spectrometer, an instrument which can separate a single ray of light of mixed wave lengths (colors) into its various components, thus allowing an estimation of the extent of absorption of lights of given wave lengths by a body or fluid or solid matter. *Spectrophotometric* methods are used to determine small amounts of both organic compounds and inorganic elements in plant and animal tissues, as trace amounts of DDT and other insecticides in tissues of animals.

speculator. A person who engages in a hazardous business expecting that the profits will be large enough to compensate for the risk involved.

speculum. A metal instrument or glass tube inserted into certain organs of the body, as the ear, mouth, nose, rectum, etc., for examination of the interior of the organ by means of a reflected light.

speculum method of insemination. A method of artificial insemination. A metal or glass speculum is inserted into the vagina, and the cervix is located by means of a pen flashlight or head light. When the cervix is located, the loaded insemination tube is then inserted and the semen discharged.

speed cone. One of the stepped cone pulleys attached in a pair and used to vary the speed of the shafts on a machine.

speed sprayer. A blower type spraying machine that employs a very large volume of air to distribute concentrated spray mixtures on trees. The typical model utilizes a low pressure, low volume pump which forces the spray material under low pressure to the fan where it is discharged as small spray droplets into the air stream which breaks them up into small particles and carries these particles to the leaf surface.

speedwell. Any plant of the genus Veronica, family Scrophulariaceae. Some species are hardy herbs suitable for the border or rock garden; some are attractively flowering, prostrate plants good for ground cover; others are common weedy herbs. See *bastard speedwell; brooklime; comb speedwell; creeping speedwell; drug speedwell; gentian speedwell; Hungarian speedwell; spike speedwell; woolly speedwell.*

speedy cutting. The striking of the inside of the hind cannon against the front foot as the hind foot is brought forward and passes the front foot on the outside in over-reaching, when a horse is running.

spelt. *Triticum spelta,* family Gramineae, a cereal which retains its glumes when threshed. Used for livestock feed but is of no value for flour making. Bearded and beardless varieties are grown both in spring and winter, but are of little commercial importance. Native to southern Europe and western Asia.

speltoid. Any mutants of the wheat, *Triticum vulgare,* possessing characters of spelt.

speltz. See *spelt.*

spent. Of animals, especially horses, completely exhausted.

spent bone black. Bone black or bone charcoal left after use in the refining of sugar. It is valued by fertilizer manufacturers chiefly because of its phosphorus content.

spent oxide. Gashouse tankage.

sperm. A mature male germ cell, a spermatozoon or a spermatozoid.

spermatheca. In insects and some other invertebrates, a small globular sac connected with the vagina by a small duct which serves as a reservoir for the vast number of spermatozoa to be used later for fertilizing eggs.

spermatic cord. The cord-like structure by which the testicle is suspended within the scrotum. It contains the vas deferens, the blood vessels, the nerves of the testicle, and a small muscle.

spermatids. The four cells produced from a single primary spermatocyte in spermatogenesis. By transformation in shape, *spermatids* become mature spermatozoa or sperms.

spermatium. A spore from a spermogonium (pycnium) in rusts. It acts as a male gamete (pycniospore).

spermatocyte. One of the various kinds of cells produced in the development of male gametes in animals. Primary *spermatocytes* are formed by ordinary mitotic cell divisions from spermatogonia. Meiotic divisions of primary *spermatocytes* result in the production of secondary *spermatocytes* which in turn give rise to spermatids by equational cell division.

spermatogenesis. The development of male gametes or sperm cells.

spermatogonia. Primordial male germ cells which by ordinary mitotic cell division give rise to the primary spermatocytes.

spermatophore. In spermogonium, the section in which the spores originate.

spermatophyta. Flowering and seed-bearing plants.

spermatozoa. Plural of spermatozoon *q.v.*

spermatozoon. A mature male sex cell of animals; a sperm cell.

sperm collector. See *semen collector.*

sperm-mother-cell. Spermatocyte.

spermogonium. Male sexual reproductive fruiting body in the rusts; pycnium.

spewing sickness. See *orange sneezeweed.*

sphagnum. Any plant of the genus Sphagnum, family Sphagnaceae. Species are mosses found in bogs from the polar region to the equator. The structure of the leaves is such that they have the power to absorb and retain large amounts of moisture. This gives value to *sphagnum* in horticultural work so that nurserymen and florists use it for various purposes. Also used for wound dressings. Also called *moss peat; sphagnum moss.*

sphagnum peat. See *peat; sphagnum.*

sphincter. A ring-shaped muscle which closes an orifice: *e.g., sphincter* muscles found in the lower end of a cow's teat. They act to press the teat tissues together and close the opening to the milk duct of the teat.

sphinx moth. Any of several moths, family Sphingidae, whose larvae are called hornworms. Several species are important pests of cultivated crops, deciduous trees, and shrubs. See *tobacco hornworm; tomato hornworm; catalpa sphinx.*

spiceberry. Checkerberry.

spicebush. *Calycanthus occidentalis,* family Calycanthaceae, a deciduous shrub grown for its light brown flowers. Native to California, U.S. Also called *California sweetshrub.*

Spiced. A cheese to which one or more spices are added during the making process.

spice tree. California laurel.

spicule. 1. Any of the points on the spore-bearing cells of certain fungi. 2. A small, floral spike.

spider. 1. Any of various Arachnids comprising the order Araneida. Individuals have eight legs and only two main body divisions. The abdomen has spinnerets for spinning silk threads used for constructing webs which serve as nests and as traps for prey. Most *spiders* are harmless to man and because of the insects they devour and their aid in pollination, they are exceedingly beneficial. 2. Anything suggestive of a *spider* in form, as various mechanical parts having radiating members. 3. A tumor or growth sometimes suspended in a cow's teat canal *(q.v.)* by strands of fibrous tissue which resembles a spider web (colloquial).

spiderflower. Brazilian glorybush.

spiderlily. Hymenocallis.

spider mite. Any mite of the family Tetranychidae. Species feed on the stems and leaves of plants.

spider teat. Any of various types of fibrous growths in the teat canal which interfere with the milking process (colloquial).

spiderweb houseleek. *Sempervivum arachnoideum,* family Crassulaceae, a tender, small, perennial herb, grown in the rock garden for its rosette foliage and red flowers. Native to southern Europe.

spiderwort. Any of the various plants of the genus Tradescantia, family Commelinaceae. Species have rose-purple, blue, or white flowers. See *Virginia spiderwort.*

spike. 1. A very large nail or anything shaped like a nail, as the pointed iron teeth of a spike-toothed harrow, etc. 2. A flower-cluster or type of inflorescence in which the individual flowers are sessile (stalkless or nearly so) on a central axis, as in the common plantain. 3. The spike-like panicle of wheat, barley, rye, certain grasses, and other plants. 4. A thick, upright stem which carries one or more flowers, as the stems of gladiolus, foxglove, and delphinium.

spike bentgrass. Big spike bentgrass.

spike crazyweed. *Oxytropis macouni,* family Leguminosae, a poisonous plant whose foliage, green or dried, is poisonous to livestock. Symptoms of poisoning are dullness, irregularity of gait, lack of appetite, dragging of the feet, a solitary habit, loss of flesh, and a shaggy coat. As the animal ceases to eat it dies. Native to North America; found from Montana to Colorado, U.S.

spiked alder. Summersweet clethra.

spiked flexible harrow. A later version of the chain harrow *(q.v.),* with sharp teeth at ends of links. Some models had two sets of teeth on each link. Used on cultivated lands.

spike flower. See *spike: 4.*

spike gayfeather. *Liatris spicata,* family Compositae, a perennial herb grown in the wild garden or in the informal border for its rose-purple flowers. Native to North America. Also called *prairie pine; devil's-bit.*

spike grass. See *broadleaf uniola.*

spike harrow. Spike-tooth harrow.

spikelet. Any small spike, the unit of inflorescence in grasses which is often provided with a small pedicel. Each *spikelet* may consist of one, but usually several florets.

spikelet rot of wheat. A bacterial disease of wheat caused by *Pseudomonas atrofaciens,* family Pseudomonadaceae, characterized by a light, brown bacterial rot at the base of the spikelet. When severe, it extends to the base of the kernels. Also called *basal glume rot.*

spike muhly. *Muhlenbergia wrighti,* family Gramineae, one of the important range grasses of the plains and slopes of Oklahoma, Colorado, Utah, New Mexico, Arizona, U.S., and northern Mexico. Native to North America.

spikenard. American spikenard.

spike redtop. Big spike bentgrass.

spike rush. Spikesedge.

spikesedge. *Eleocharis palustris,* family Cyperaceae, a watergrass growing in marshes and wet places from Labrador to Alaska southward to California and Pennsylvania. It is a weed pest in the rice fields of California, U.S. Native to North America.

spike speedwell. *Veronica spicata,* family Scrophulariaceae, a perennial herb grown for its dense, blue, flower clusters. Native to Europe. Also called *cat's-tail speedwell.*

spikespine coryphantha. *Coryphantha robustispina,* family Cactaceae, a small, ball-like cactus, sometimes grown for its 4 in., globular, pincushion shape and its salmon-pink flowers. Native to southern U.S. and Mexico. Also called *devil's pincushion; pineapple cactus.*

spike-tooth drag harrow. Spike-tooth harrow.

spike-tooth harrow. A device used to smooth and level the soil directly after plowing. It is made up of a number of wooden or steel teeth resembling long spikes attached to bars. Each unit, called a section, is 4 to 5 ft. wide and has 25 to 35 teeth. When guard rails protect the end of the bars, the harrow is called *closed-end harrow.*

spike top. 1. The pointed, dead tip of a living tree. 2. An unthrifty tree with a dead top.

spike trisetum. *Trisetum spicatum,* family Gramineae, a perennial bunchgrass which occurs mostly in scattered stands on alpine slopes and meadows in arctic America and southward to New England, and the western states in the U.S., which supplies considerable forage of good quality.

spike winterhazel. *Corylopsis spicata,* family Hamamelidaceae, an ornamental shrub grown for its early-spring, fragrant, yellow flowers. Native to Asia.

spiladophilous. Designating a plant or animal which lives in clay.

spile. A short tube or small spout inserted (a) into a hole bored in the outer body of a maple tree to aid in the collecting of the sap for maple sirup and sugar; (b) through a hole in a ditch levee to carry water to an irrigation furrow.

spillway. Any of several types of hydraulic structures used to convey water safely from an upper to a lower level, such as an overfall dam, an open channel chute, or a drop inlet tube.

spin. To twist or transform the fibers of wool, cotton, silk, flax, etc., into thread or yarn preparatory to making cloth or fabric.

spinach. *Spinacia oleracea,* family Chenopodiaceae, an annual herb whose basal, green leaves are a very important vegetable. Native to Asia.

spinach aphid. See *green peach aphid.*

spinach-beet. Leaf beet.

spinach blight. Cucumber mosaic.

spinach blue mold. See *downy mildew.*

spinach flea beetle. *Disonycha xanthomelas* (Dalm.), family Chrysomelidae, an insect whose larvae eat small holes in the leaves of spinach.

spinach leaf miner. *Pegomya hyoscyami* (Panz.), family Muscidae, a fly whose maggots mine the leaves making light-colored blotches filled with dark excrement on spinach, beets, and related plants. The leaf vegetables thus damaged are rendered unfit for consumption.

spinach yellows. A blight of spinach caused by the cucumber mosaic *(q.v.)* virus. Also called *blight.*

spinal. Referring to the spine or vertebral column.

spindle. 1. The fine threads of achromatic protoplasm arranged in a fusiform mass within the cell during mitosis. 2. In flower-bearing plants, to produce the stalks on which flowers grow.

spindletree. Euonymus.

spindle tuber. A virus disease of potatoes which causes the plant tops to be abnormally stiff and erect, the leaves smaller and darker green with terminal leaflets usually twisted. The tubers of affected plants are elongated and tend to be pointed especially at the stem end, resembling a sweet potato in shape.

spindle-up. Designating a plant with a sickly growth as a result of excessive amounts of alkali in the soil. Such a plant shows a lack of shoot and foliage development.

spindling. Designating a plant which grows into a long, slender stalk or stem; growing to stalk rather than to flower or fruit; growing weakly or disproportionately; leggy, often as a result of light deficiency or crowding.

spine. 1. A stiff, sharp-pointed outgrowth on a plant or animal. 2. The vertebral column; the backbone.

spineless horsebrush. See *gray horsebrush.*

spinner's waste. Waste fiber from the spinning room.

spinning count. Applied to wool, any of the different numbers indicating the fineness of the yarn which can be spun from it.

spinning dope. The liquid from which synthetic fiber is formed in the spinning process. It consists of the fiber-forming material dissolved in a solvent. The dope is forced through small holes in a spinneret just before being solidified as filaments by cooling, coagulation, or solvent evaporation.

spinose. Covered with, or containing, spines.

spiny elm caterpillar. The larva of the mourning-cloak butterfly *q.v.*

spiny mapleworm. See *green-striped mapleworm.*

spiny oakworm. *Anisota stigma* (F.), family Citheroniidae, a southern species much like the orange-striped oakworm *q.v.*

spiny rolling grass. *Spinifex hirsutus,* family Gramineae, a salt-tolerant grass with broad, hairy, silver-gray leaves and long, creeping rhizomes, which make it a valuable aid against sand erosion. The mature seed-heads are blown along the sand on their elastic spines, dropping seeds as they roll. Native to Australia. Also called *hairy spinifex; coast silver-grass.*

spirability. The ability of cotton fibers to be twisted, as in the manufacture of a yarn or thread. Under the microscope the spiralled cotton fibers give the appearance of long and irregularly twisted ribbons. This properly permits the fibers, when twisted together, to interlock better one with another.

spiral-auger conveyor. See *screw conveyor.*

spiral gravity separator. A seed separator consisting of a spiral incline down which seeds travel by gravity. As the seeds flow down, the heavier or rounder seeds move faster, swing to the outside of the chute, and are separated from the lighter seed. Also called *gravity separator.*

spiralism. An abnormal condition of plants in which normally straight stems produce an irregular, scraggly growth with a flat or open spiral form.

spiral stomach worm. *Dispharynx nasuta,* family Acuariidae, a short, white roundworm, often curved or even twisted into a spiral, occasionally found in the glandular stomach of chickens, turkeys, and pigeons. If present in considerable numbers they may affect the health of the bird. The sowbug and pillbug *(q.q.v.)* may function as intermediate hosts for the spiral roundworm.

spiral wildcelery. *Vallisneria spiralis,* family Amaryllidaceae, an aquatic plant of slow moving waters which is a valuable wildlife plant as food for the canvasback duck. Native to North America. Also called *wild celery; tape grass; eelgrass.*

spirea. Any plant of the genus Spiraea, family Rosaceae. Species are handsome, flowering, deciduous shrubs valuable as ornamentals. See *Billiard spirea; bridalwreath spirea; broadleaf meadowsweet spirea; bumalda spirea; Douglas spirea; garland spirea; germander spirea; hardhack spirea; Henry spirea; Japanese spirea; Japanese white spirea; Korean spirea; narrowleaf meadowsweet spirea; Margarita spirea; Mongolian spirea; Reeves spirea; striped spirea; Thunberg spirea; Vanhoutte spirea; Veitch spirea; willowleaf spirea.*

spirea aphid. *Aphis spiraecola* Patch, family Aphidae, an insect which infests spirea and citrus, curling and stunting leaves and twigs, deforming fruit, and covering everything with honeydew. Also called *green citrus aphid.*

spirit vinegar. The product made by the acetous fermentation of dilute distilled alcohol which contains not less than 4 gms. of acetic acid per 100 cc. at 20°C. Also called *distilled vinegar; grain vinegar.*

spirochete. Any microorganism that is spiral or wavy in form, highly flexible, and capable of contracting, as the organisms that cause leptospirosis and syphilis.

spirochetosis. 1. Any infectious disease caused by spirochetes. 2. An acute, highly fatal, septicemic disease of fowls characterized by the presence of the causal organism, *Borrelia anserina,* in the blood stream during the height of the disease.

spittlebug. Any of the small, sucking insects of the family Cercopidae, which are not true bugs but are closely related to leafhoppers. The young nymphs produce masses of whitish, frothy material in which they live. Many *spittlebugs* are harmless, but several are injurious to plants. Also called *froghoppers,* because of their leaping habit.

Spitkase. A small, spiced, skim milk cheese made in Germany from cows' milk.

splash. 1. The part of a liquid and material in suspension which is forced up or away from a surface by sudden impact, such as a raindrop splashing loose soil. 2. To produce a *splash.*

splash erosion. The disintegration of aggregates and the consequent dispersion of soil particles by the impact of falling rain. Drops falling on a sheet flow of soil-laden runoff water create a splashing and turbulence which helps to keep the dispersed soil in suspension with resultant sheet erosion.

splay-footed. Designating a flat-footed horse, cow, or other animal.

spleen. A large, glandlike, ductless organ in the upper part of the body cavity on the left side between the stomach and the diaphragm.

spleenwort. Any plant of the genus Asplenium, family Polypodiaceae. Species are widely distributed temperate and tropical ferns of great diversity of habit. Certain species are grown in greenhouses; others in the outdoor fernery or wild garden. See *birdsnestfern; ebony spleenwort; maidenhair spleenwort; mother spleenwort.*

splenic fever. Anthrax.

splice grafting. A simple method of grafting in which both the cion and the stock are cut with a diagonal cut; the two cut surfaces are fitted together, bound, and waxed. It is used for tender wood, which does not split readily, and for small shoots.

splint basket. A basket constructed largely or entirely of thin strips of wood. *Splint baskets* are made in various shapes and sizes, up to and larger than bushel capacity. They are used extensively in marketing as containers for fruits and vegetables.

splint bone. The small metacarpal or metatarsal bones found on each side of the cannon bone.

splints. A hard, bony enlargement on the splint bone which is located on the inside of the fore cannon bone of the leg of a horse. It may appear rarely on the outside of the front legs, but seldom on the hind legs. It is more often a blemish than an unsoundness. See *splint bone.*

splintwood. Sapwood.

split. 1. A deformed flower. See *splitting; splits.* 2. To separate thick hides into layers in the preparation of leather.

split clip. A small, metal staple made of fine wire which is put in the calyx of a carnation to pull the flower together when the calyx is split. Also called *carnation clip; Baur clip.* See *splitting.*

split comb. Of a chicken, a single comb with one or more of the points, or the blade, split or divided perpendicularly from front to rear with the two parts overlapping; a disqualification.

split crotch. Of a tree, a split or separation at the juncture of two large limbs.

split ditch. A lateral ditch used in the sugarcane system of surface land drainage in Louisiana, U.S. Such ditches are usually parallel and spaced 100 to 250 ft. apart with rows of sugar cane parallel to them.

split hoof. See *quarter crack; sand crack.*

split peas. 1. Peas hulled and dried so that the cotyledons usually separate. 2. A flour made from *split peas.* Also called *pease-meal.*

split-rail. 1. A long bar of timber split from a log or large limb and used largely in the construction of rail fences. 2. Any wooden bar that is split from a log, limb, bolt, or the like.

splits. Bulbs which are split from a mother-bulb, as those formed and naturally split within the mother-bulb of narcissus. See *split: 1.*

split shingle. A rough, flat, roofing shingle, split from a shingle bolt, usually by a broadax; a shake.

split stave. A stave split from a stave bolt by a froe, mallet, and wedges.

split stock. Material, such as shakes, ties, and grape stakes, rived from redwood logs.

split tail. A chicken's tail showing a decided gap between the top main-tail feathers at their base which results from improper placement of the quills; a general disqualification *q.v.*

splitting. A deformation of a flower which results from a vertical split between two segments of the calyx (as in carnation) or two segments of the perianth (as occasionally in lily). In carnation it is called carnation split and calyx split and such a flower is called a split.

split wing. A bird's wing which is abnormally formed and shows a decided gap between the primary and secondary wing feathers; a general disqualification *q.v.*

splitworm. Potato tuberworm.

spodium. Charcoal made from bone or animal material, used in manufacturing superphosphate.

spoil. To deteriorate by molding or rotting. See *spoilage.*

spoilage. 1. Hay or forage that has been improperly cured or stored. 2. Any objectionable change which has occurred in a food, feed, or material. 3. Putrefactive changes which occur in canned goods as a result of underprocessing causing the growth of vegetative cells, spores or organisms, or as a result of the growth of organisms entering the can after processing.

spoil bank. A small to large ridge of soil excavated from, and piled alongside, a drainage ditch. In good drainage practice, this spoil is leveled, *i.e.*, scattered over a considerable area alongside the ditch.

spoke. One of the set of bars or rods radiating from the hub of a wheel and supporting the felloe or rim.

sponge. The porous, elastic mass of interlacing fibers forming the internal skeleton of certain marine animals of tropical seas, and having great power of absorbing water. Gathered, treated and sold on a commercial basis, pieces of the above substance are used in various ways, as for washing, cleaning, etc.

spongospora scab. Powdery scab of potato.

spongy-bodied cheese. A cheese that has an undesirable spongy character and lacks quality in the texture of the body.

spongy body. A defect of butter, a weak body, as shown by its quick melting down or comparative softness when exposed to ordinary room temperatures.

spongy ice cream. A fluffy ice cream characterized by the presence of large air cells and an open texture throughout.

spontaneous combustion. An internal heating in a product or substance which reaches an ignition point without an external source of heat. In stored hay, temperatures within the hay of approximately 160° F. are considered to be reaching the danger point of internal ignition.

spool. 1. Of disk harrows, one of the spool-shaped, cast-iron parts in the disk gang assembly which holds the disks at equal distances apart. Also called *spreader spool.* 2. Any more or less round device on which wire, rope, etc., is rolled for easy handling.

sporadic. Occurring in scattered instances; isolated; appearing off and on.

sporadic disease. A disease which occurs in scattered or isolated instances.

sporangia. The plural of sporangium *q.v.*

sporangiophore. A thread-like structure (hypha) which bears a sporangium *q.v.*

sporangiospore. A spore produced in a sporangium, particularly by the black bread mold and related fungi.

sporangium (plural: **sporangia**). A fungous fruiting body that produces nonsexual spores (sporangiospores) within its external walls.

spore. The one to many-celled unit of a fern, fungus, bacterium, or protozoon that has entered the resting state and is capable of growth and reproduction when conditions become favorable. *Spores* in lower forms correspond to seeds in plants.

spore dust. A standardized powder containing spores and used in the biologic control of insect pests. The dust is placed at intervals in the soil, and the spores infect the larvae of certain insects, such as Japanese beetle, causing diseases which destroy them. See *milky disease.*

spore-forming. Designating bacteria or other organisms that reproduce by the formation of spores or seedlike bodies.

spore-mat. A spore-bearing layer of mycelium.

spore mother cell. The cell which undergoes one meiotic division in the development of microspores (or pollen grains).

sporidia. The plural of sporidium *q.v.*

sporidium (plural: **sporidia**). A spore which is produced on a fruiting body called a promycelium; the basidiospore of rust and smut fungi.

sporocyst. In flukes, or trematodes, the intermediate asexual generation or second larval stage and, in some cases, the third larval stage. In protozoa belonging to the group Coccidia, a *sporocyst* is a sac or cyst formed within an oöcyst; it contains one or more sporozoites *q.v.*

sporocyte. A spore mother cell of plants.

sporogenous. Producing spores; spore bearing.

sporophore. A fungous structure that produces or supports spores.

sporophyll. An adapted or modified leaf which produces sporangia (*q.v.*): *e.g.,* the fern. Plants which bear more than one type of spore (heterosporous) produce both microsporangia, which form microspores (*q.v.*), and megasporangia, which form megaspores *q.v.*

sporophyte. The spore-forming generation in the life-cycle of plants, normally diploid. The visible seed-bearing plant is the *sporophyte.*

Sporotrichum fruit rot of pome fruits. A fungous disease of pome fruits caused by *Sporotrichum malorum,* family Moniliaceae, characterized by sunken, light brown spots which may involve large areas of the fruit.

sporozoa. A class of protozoa which consists of many parasitic species with complicated life cycle.

sporozoite. A small, usually elongated, infective stage of sporozoan parasites, such as coccidia, plasmodia (malaria), etc.

sport. A genetic variation which is either induced or occurs spontaneously as a mutation (*q.v.*) in plants or animals.

sporulate. To produce spores.

sporulated. Designating that stage of development of certain plants and animals in which spores have developed.

sporulation. 1. In bacteria, the formation of spores within the body of the bacterium. The spores represent the inactive resting, or resistant, forms. 2. In coccidia, a kind of reproduction by which the fertilized cell within the oöcyst wall splits up into new individuals, called sporozoites. *Sporulation* of coccidial oöcysts usually occurs after the oöcyst has been discharged from the body of the host. 3. In plants and animals, the process of spore formation.

spot. 1. A mark or blotch on something; a small part of a surface differing from the rest in appearance or character, as a localized, discolored, or necrotic area on a leaf, fruit, or other plant part. 2. An area differing in color from the ground or main coat-color of an animal. 3. A definite locality or place; a small piece of ground. 4. To discover the condition of plants or animals. 5. To blaze a tree. 6. To transplant or prick off young plants.

spot basis. In marketing in the grain trade, the difference between the cash price and the futures price of a particular option in a specified market.

spot blotch. An important fungous disease of barley caused by *Helminthosporium sativum,* family Dematiaceae, which appears as brownish circular spots causing the leaves to turn brown. Infections also occur on sheaths, glumes, spikelets, and grain.

spot burning. A method of slash disposal in which only the heaviest accumulations of slash are fired, and the fire is not allowed to spread over the entire cut-over area.

spot checking. The inspection of an occasional bale of cotton, container of produce, etc., to see if the quality of the product is up to the intended grade or specification.

spot cotton. All cotton bought or sold for cash.

spot disease of strawberries. A fungous disease of strawberries caused by *Mycosphaerella fragariae,* family Mycosphaerellaceae, characterized at first by deep purple lesions and later by a white spot surrounded by a dark purple zone. When numerous, they may cause the death of the leaflet. Also called *leaf spot of strawberries.*

spot fire. A fire set in advance of, or away from, the main fire by flying sparks or embers.

spot infestation. Insect damage of a decidedly local nature which is confined to small areas.

spot market. The market for goods available for immediate delivery. *Cf. futures.*

spot necrosis of apples and prunes. See *drought spot: 1.*

spot plant. 1. To plant in small, open areas between established stands. 2. To plant at designated spacings or intervals.

spot price. The price at which the spot or cash commodity is selling. In grain trading, it is called the *cash* price.

spot rot. 1. Lenticel-centered brown spots on decaying apples in cold storage, caused by a fungus of the genus Botrytis (*q.v.*). 2. See *Jonathan spot.*

spotted. 1. Designating an animal, plant, product, etc., which has a part of its surface differing from the rest in appearance or character. 2. A common coat-color of some breeds of animals in which there are well-defined areas of red and white, or black and white colors in the body cast.

spotted asparagus beetle. *Crioceris duodecimpunctata* (L.), family Chrysomelidae, an insect that infests asparagus. The larvae bore into the developing berries and eat out the pulp and seeds. *Cf. asparagus beetle.*

spotted beebalm. *Monarda punctata,* family Labiatae, a perennial herb grown in masses for its yellowish, purple-spotted flowers. It is a good bee plant. Native to North America. Also called *horse mint; rignum.*

spotted beet webworm. *Hymenia perspectalis* (Hbn.), family Pyraustidae, an insect similar to but smaller than the Hawaiian beet webworm (*q.v.*). Also called *small beet webworm.*

spotted bellflower. *Campanula punctata,* family Campanulaceae, a perennial herb grown for its lilac-spotted, white flowers. Native to Asia.

spotted blister beetle. *Epicauta maculata* (Say), family Meloidae, an insect that infests many field and garden crops, being most abundant in the western U.S.

spotted bur clover. Spotted medic.

spotted catsear. *Hypochoeris radicata,* family Compositae, a perennial herb which is a common weed pest. Also called *California dandelion; cape weed; gosmore.*

spotted centaurea. *Centaurea maculosa,* family Compositae, an annual or biennial plant with thistle-like head, which is a common weed of roadsides, meadows, and waste places. Also called *spotted knapweed.*

spotted cowbane. Spotted waterhemlock.

spotted cucumber beetle. See *southern corn rootworm.*

spotted cutworm. *Amathes c-nigrum* (L.), family Noctuidae, an insect that feeds on many low-growing plants and cuts them off near the soil surface.

spotted deadnettle. See *deadnettle: 3.*

spotted eucalyptus. *Eucalyptus maculata,* family Myrtaceae, an aromatic tree grown as an ornamental. Native to Australia.

spotted euphorbia. *Euphorbia maculata,* family Euphorbiaceae, a common weed found in eastern and central U.S. Contact with the milky sap causes a dermatitis with blisters. Also called *milk purslane; spotted spurge.*

spotted geranium. *Geranium maculatum,* family Geraniaceae, a perennial herb grown for its rose-purple flowers. Native to North America. Also called *American alumroot; chocolate flower; wild geranium.*

spotted hemlock borer. See *hemlock borer.*

spotted knapweed. Spotted centaurea.

spotted lady beetle. See *two-spotted lady beetle.*

spotted lady's-thumb. *Polygonum persicaria,* family Polygonaceae, an erect, annual garden weed. Also called *peachwort; heartweed; spotted knotweed; persicary.*

spotted loco. Specklepod loco.

spotted medic. *Medicago arabica,* family Leguminosae, an annual herb grown extensively for permanent pastures in southern areas (U.S.) where winters are mild, as a cover crop, and for green manure. Also called *southern bur clover; spotted bur clover.* See *California bur clover.*

Spotted Poland China swine. A lard-type breed of swine which originated in the U.S., largely in Indiana, shortly before World War I. The crossing of Gloucester Old Spot swine with Poland China swine provided the foundation for the breed.

spotted snapweed. A species of snapweed *q.v.*

spotted spurge. Spotted euphorbia.

spotted tick fever. Rocky Mountain wood tick.

spotted tussock moth. *Halisidota maculata* (Harr.), family Arctiidae, a moth whose larvae feed on many trees in the northern U.S. It is not a serious defoliator.

spotted waterhemlock. *Cicuta maculata,* family Umbelliferae, a poisonous, fleshy-rooted, perennial herb, which occurs chiefly in wet meadows and pastures, and along ditches and streams, from eastern North America westward to the Great Plains. The plant, especially the roots, contains a violent poison which causes serious or fatal results if eaten by man or livestock. Native to North America. Also called *beaver poison; cowbane; musquash root; water hemlock.*

spotted wilt. A group of virus diseases classified by Holmes as family Lethaceae. Symptoms are foliage bronzing and spotting. Two strains are described for tomato. *Spotted wilt* is also designated for certain virus diseases of pea, celery and China aster.

spotted wilt of geranium. A disease of greenhouse geraniums in the U.S. Pacific Northwest in which the leaves are marked by bright yellow or white spots, often arc or ring-shaped; possibly a manifestation of the effects of the virus of spotted wilt.

spotting board. Dibble board.

spotting fire. A fire which spreads as a result of sparks or embers falling ahead and starting new fires.

spot treatment. A treatment which is applied only to certain parts of an area, or to certain plants, but not to all.

spotty stand. Any irregular or uneven stand of grain, grass, or other crop.

spot wool. Wool ready for immediate delivery.

spout. 1. A pipe, tube, or similar conduit, through which a liquid, grain, seed, or other loose material flows and is discharged. 2. A trough or chute by which grain, coal, sand, or the like is discharged from a conveyor or bin to a truck, etc. 3. A pipe or projection (as of a sprinkling can) by which a liquid is discharged or poured. 4. A spigot. 5. A spile which is used in the collection of sap. 6. To shoot out, as a liquid, etc., from a *spout;* to flow or shoot out with force.

spouty land. Spots, usually at the base of slopes, which are boggy, springy, and excessively wet from water from springs or seeps (colloquial).

spraddle leg. Slipped tendon.

sprain. 1. A severe wrench or strain of the parts around a joint which causes pain, swelling, and difficulty in moving. 2. See *internal brown spot.*

sprangletop. Any species of the genus Leptochloa, family Gramineae. The only species important is green sprangletop, *L. dubia,* a grass valuable for grazing and hay. Native to North and South America.

spray. 1. A small, growing or detached shoot, twig, or branch of a plant with leaves, flowers, berries, etc.; a horizontal arrangement of flowers or greens. 2. A solution or suspension of material, as insecticides, fungicides, etc., which is broken up into fine, liquid particles driven through the air by mechanical means, coating surfaces of plants or other objects that it contacts. 3. A jet of fine liquid particles forced under pressure from a spray nozzle of a spraying device. 4. A device for shooting out a jet or *spray*. See *sprayer*. 5. To scatter a liquid in fine particles. 6. To sprinkle or treat with a spray, as to direct a *spray* of fine particles of a liquid upon plants or other objects.

spray boom. On a spray rig, an assembly of several spray nozzles on a horizontal pipe used to extend the area of coverage of a single operation, as for spraying several rows of potatoes, etc. A similar arrangement on a vertical pipe is used for spraying trees.

spray cone. A jet of spray discharged in the form of a hollow cone. See *hollow-cone nozzle; solid-cone nozzle.*

spray cover. Distribution of spray material on a surface; the extent, thickness, etc., of coverage.

spray cultivation. The destruction of weeds in growing crops by the use of chemical weed-killers. See *chemical weed control.*

spray-dried skim milk. A milk product made by atomizing skim milk in a stream of dry, heated air removing thereby most of the moisture. *Cf. drum-dried skim milk.*

spray-drying process. A process which employs currents of heated air for the drying of fluid products: *e.g.*, drying fluid milk, skim milk, or buttermilk, or their condensed products by spraying them into a chamber of heated air which absorbs the moisture, the solids falling to the bottom in the powdered form being withdrawn and packaged. Eggs and certain other products are also dried by this process.

spray egg drier. See *spray-drying process.*

sprayer. A device or mechanism by means of which a solution or suspension of a material in liquid is broken up into droplets of a predetermined size range and distributed by hydraulics, by air blast, by gravity from an airplane, or by a combination of the above. It is used to sprinkle or treat plants, trees, etc., with a spray. See *spray: 2; spray machine.*

sprayer-duster. A spraying machine of the fan or blower type in which a fan, propeller, or turbine is used to produce a blast of air at great velocity. Fishtail outlets are often used to direct the air blast in a band which will cover the trees on one or both sides of the equipment. Most machines of this type may be used to apply either sprays or dusts, in some cases in combination with liquids.

spray gun. A short, hollow, hand-operated rod with a nozzle and cut-off usually adjustable by means of a twining handle, used for directing sprays with a wide fog for nearby foliage or a closely compacted stream to reach high tree tops, in contrast to a spray rod (*q.v.*) which has non-adjustable fog type nozzles that are held into the tree tops by the length of the rod.

spray helmet. A headpiece which covers the head and shoulders of the wearer and protects the eyes, nose, throat, etc., from drifting spray.

spray injury. Injury resulting from the application of a chemical compound, as a spray to the foliage or other parts of a plant. The injurious effects may include damage to leaves, blossoms, fruit, twigs, or to the entire plant, sometimes resulting in general necrosis and death. See, *e.g.*, *Bordeaux injury.*

spray irrigation. Irrigation by means of spray from pipes or pipe projections above the soil surface. Also called *sprinkler irrigation.* See *sprinkler irrigation system.*

spray machine. Any of several types of machines used for applying spray material to plants, etc. See *compressed-air sprayer; sprayer-duster; spray Venturi system.*

spray nozzle. The attachment on the end of a sprayer extension tube that breaks up spray liquids into fine drops so that the material may be spread evenly on plants or other surfaces. See *nozzle.*

spray pump. Any pump used for pressurizing spray material to deliver through the spray nozzle with force sufficient to achieve the desired break-up. See *compressed-air sprayer.*

spray rig. A complete, mobile spraying unit which consists of the necessary parts for distributing insecticides or other fluid material in the form of a spray. See *compressed-air sprayer; sprayer-duster.*

spray ring. A group of farmers or orchardists who join together to share the expense and use of a spray rig or who jointly contract for the services of a professional spray outfit.

spray rod. A slender pipe, usually with ¼ in. inside diameter, obtainable in various lengths, equipped with a spray nozzle at one end while the other end is connected through a shut-off valve to a supply hose. The pipe may be enclosed by a bamboo casing for rigidity. Used for extending the area of application in spraying. Also called *extension tube.*

spray russeting. Bordeaux injury.

spray Venturi system. A gravity-feed system, used in aerial spraying, in which several Venturi tubes (*q.v.*) are evenly spaced along the wings or underparts of the fuselage of an airplane for the dispersing of spray material.

spray washer. A jet or spray of water, or a series of jets or sprays, used for washing vegetables, poultry, or other products in preparation for marketing or processing.

spread. 1. A straddle; the difference in price between two delivery months in the same or different markets, or the sale of the one thing against a simultaneous purchase of the other. Straddling between a foreign and the domestic market is often referred to as *arbitrage.* 2. An extensive tract of land. 3. A ranch, including the buildings and the extent of land grazed by cattle or sheep (western U.S.). 4. The distribution of a disease. 5. Butter, peanut butter, margarine or other food mixtures used as a spread on bread to improve its flavor, palatability, and nutritive value. 6. To scatter, as to broadcast seed, fertilizer, etc. 7. To disseminate disease.

spreader. 1. A device used for distributing water uniformly in or from a channel. 2. A spreading agent, or spreader-sticker. 3. An instrument provided with hooks or flanges used to spread open the incision made in the body wall of a bird. 4. A device that scatters or spreads: *e.g.*, a manure *spreader*, a fertilizer *spreader* (*q.v.*). 5. An animal, etc., capable of acting as a parasite or disease vector. 6. A species of microorganism which tends to grow profusely over the entire surface of the culture medium.

spreader spool. See *spool.*

spreader-sticker. A spreading agent added to sprays that also results in better than normal adhesion of the suspended particles.

spreading agent. A material that reduces surface tension of a liquid and causes it to spread or disperse in a thin film over a surface instead of depositing as droplets, as in spraying to cover the surface of leaves and fruits.

spreading brood. 1. In beekeeping, putting a comb without brood between two combs of brood to induce the queen to lay in it. 2. Moving frames of brood to supers or to other colonies.

spreading cotoneaster. *Cotoneaster divaricata,* family Rosaceae, a shrub which bears bright red fruit, grown for its pinkish flowers. Native to China.

spreading crabgrass. *Digitaria decumbens,* family Gramineae, a perennial, creeping grass grown in warm areas as a pasture grass. Native to South Africa.

spreading decline. A disease of citrus in Florida, U.S., associated with the burrowing nematode.

spreading dogbane. *Apocynum androsaemifolium,* family Asclepiadaceae, a stout, branching, milky-juiced, perennial, poisonous herb, which, if ingested by cattle and horses, often proves fatal. Native to North America. Also called *milk ipecac; wild ipecac; honey-bloom; wandering milkweed.*

spreading pasqueflower. *Anemone patens* (*Pulsatilla p.*), family Ranunculaceae, a low-growing, anemone-like, poisonous, perennial herb grown for its early, large, bluish-purple flowers appearing before the leaves develop. Sheep on prairie pastures are reported to have been poisoned by eating the flowers and young plants. Native to North America. Also called *April fool pasqueflower.*

Sprengel tube. A U-shaped tube with capillary ends at right angles which is used for determining the specific gravity of milk and fats. The tube is weighed empty, filled with water and then with the fluid, whose specific gravity is given by the ratio of its weight to that of water. See *specific gravity.*

Sprenger asparagus. *Asparagus sprengeri,* family Liliaceae, a fernlike herb grown in the greenhouse for use as greens (*q.v.*) in bouquets, etc. Potted plants are often used in decorations for weddings, etc. Native to South Africa. Also called *asparagus fern; emerald feather; Sprengeri.*

Sprengeri. Sprenger asparagus.

sprig. 1. A shoot, twig, or spray of a plant. 2. A portion of stem and root of grass, used for transplanting. 3. To plant with *sprigs.*

sprig budding. Shield-budding.

sprig grafting. Shield graft(ing).

spring. 1. A natural flow of water from the ground. The source may be either shallow or deep-seated, and the *spring* may be at the base of a slope where a pervious stratum overlies a relatively impervious one, or it may be an artesian *spring* issuing at the surface through rock fracture or along fault lines, or a subterranean stream issuing through a solution opening as in a limestone terrain. 2. The season between winter and summer. 3. An elastic device or part which recovers its shape after being stretched, compressed, bent, etc., as the curved steel blades of a *spring*-tooth harrow. 4. The act of *springing* or leaping. 5. Referring to the *spring* season of the year in which a crop is harvested, or a flower blooms, as an early-*spring* crop, or a *spring* flower. 6. To indicate proximity of calving by a swelling of the udder, as in a cow.

spring barley. A variety of barley that is planted in the spring; a very common type grown in the U.S.

spring beauty. Any plant of the genus Claytonia, family Portulacaceae. Species are low-growing, perennial herbs with fibrous roots and slender, weak, narrow leaves. Also called *grassflower; Mayflower.* See *Virginia springbeauty.*

spring black stem of alfalfa. A fungous disease of alfalfa, sweetclover, and other clovers in North America and Europe, caused by *Ascochyta imperfecta, A. meliloti,* or *Mycosphaerella* and *Phoma trifolii;* characterized by dark brown to black lesions on the stems and petioles.

spring calf. A calf born in the spring of the year.

spring cankerworm. *Paleacrita vernata* (Peck), family Geometridae, a moth whose larvae feed on foliage and spin a silken thread by which they are often found suspended from a twig or branch of a tree. Cankerworms are serious defoliators of shade and fruit trees and are of general distribution east of the Rocky Mountains, U.S. *Cf. fall cankerworm.*

spring chicken. A young chicken, usually only a few months old. See *springer: 1.*

spring crop. Any crop that is planted or starts rapid growth in the early spring and is ready for harvest in the spring season, as rhubarb, lettuce, and radishes.

spring dwarf of strawberry. A microscopic, parasitic, nematode disease of strawberries caused by *Aphelenchoides fragariae,* family Aphelenchidae, characterized early in the season by small leaves, short petioles and crinkled and curled leaflets. See *summer dwarf of strawberry.*

springer. 1. In the live-poultry industry, a young chicken, larger than a broiler and smaller than a roaster, which commonly weighs from 3¼ to 4¼ lbs.; a fryer. 2. In marketing, a pregnant cow or heifer due to calve shortly. 3. A sealed can both ends of which are bulging, either one of which can be easily flattened. 4. A spoiled can.

spring-fall range. Those grazing areas of the western range which, because of grazing patterns, climatic factors, forage production, quantity of water, etc., are seasonally grazed only in spring and fall. Such grazing may not be necessarily obligatory but it is convenient in a grazing program where summer range *(q.v.)* at higher elevations is productive and available only during the summer season.

spring farrow. Pigs born from January 1 to June 30.

Springfield type of amortization. In the U.S., the terms of a farm real estate loan which provide for equal annual or semiannual payments on the principal over the period of the loan and interest payments on the remaining balances. See *amortization.*

spring grain aphid. Greenbug.

springhalt. See *stringhalt.*

spring heath. *Erica carnea,* family Ericaceae, a low, prostrate shrub, not over 12 ins. high, grown for its small flowers in varieties of red, rose, or white. Native to Europe.

springing cow (heifer). A pregnant cow or heifer which shows signs of the approaching birth of its young, as evidenced by a relaxation of the ligaments and muscles on either side of the tail-head and also by a slight elevation of the tail-head.

spring lamb. A new-crop lamb that comes to market in the spring of the year. (The term is used to distinguish the one from old-crop or fed lambs that were born the previous year. The term *spring* is usually dropped on July 1 and the term *lamb* is then used.)

spring lily. A native or naturalized spring-flowering lily. See *white fawnlily.*

spring migrant. An animal, insect, or plant that migrates or changes habitat in the spring of the year.

spring oats. Oats sown in the spring.

spring of ribs. In stock judging, the evaluation of the conformation of an animal. A good *spring of ribs* indicates a healthy, vigorous body conformation and a good lung and digestive capacity.

spring pig. A pig farrowed in the winter or early spring; usually one farrowed from January 1 to June 30.

spring planting. Planting in the spring season.

spring-plowing. The turning of the soil in the spring.

spring rabbitbrush. Littleleaf horsebrush.

spring rain. Rains of the spring season which are specially important in helping to warm the soil and to accelerate early plant growth, and which also may supply much of the moisture to be stored in the soil for later growth.

spring rye. A type of fast-growing rye which may be planted in the early spring and still make a mature crop the same season.

spring-sown. Designating a crop which is sown in the spring: *e.g., spring-sown* wheat.

spring squill. *Scilla verna,* family Liliaceae, a bulbous herb grown for its early-spring, blue, fragrant flowers. Native to Europe. Also called *sea onion; sea squill.*

springtails. Any one of a large number of minute, wingless insects of the order Collembola which has chewing or piercing mouth parts and which jumps long distances by means of a forked appendage, the furcula, folded forward under the abdomen when at rest. Found in damp places, some species infest seeds, seedlings, mushrooms, and other plants or plant parts close to the ground.

springtime. The season of the year following winter and preceding summer. For the North Temperate Zone it begins with the vernal equinox and ends with the summer solstice, and is generally recognized to be from March 21 to June 21.

spring tooth. One of the curved, spring-steel, soil-working parts of the spring-tooth harrow or spring-tooth cultivator *q.q.v.*

spring-tooth cultivator. Any cultivator designed to use spring teeth similar to those of a spring-tooth harrow, instead of cultivator shovels.

spring-tooth harrow. A type of field harrow whose teeth are narrow, curved, steel bands pointed on the contact end and placed in such a position as to penetrate deeply into the ground and stir the soil as the harrow moves forward. It is especially adapted to rough or stony ground. Also called *spring harrow.*

spring trip. On the shank of a cultivator, a spring device which allows the shovel to pivot back and pass over an obstruction without bending or breaking.

spring water. Water which flows from a spring *q.v.*

spring wheat. Wheat sown early in the spring and harvested in the summer of the same year.

spring-wheat region. The region in which spring wheat is commonly grown: North Dakota, South Dakota, Montana, and areas in bordering states to the south and east, U.S.

spring wood. The less dense, large-celled part of a growth layer formed in woody plants by rapid growth during the spring.

spring wool. Wool shorn in the spring as contrasted with fall wool, where sheep are shorn twice a year. It is preferred to fall wool as it is longer stapled.

springy pasterns. Pasterns *(q.v.)* that are strong but springy and elastic in action.

sprinkle. 1. A light rain. 2. To scatter water, sand, seeds, etc., in drops or particles. 3. To irrigate by a hose and nozzle.

sprinkler. 1. A device for spraying water on plants or lawns; a sprinkling unit attached to a water pipe or hose. 2. A sprinkling can.

sprinkler irrigation. Irrigation by means of above-ground applicators which project water outward through the air making it reach the soil in droplet form approaching rainfall. Applicators commonly used are rotary or fixed sprinklers, oscillating or perforated pipe.

sprinkler irrigation system. A system of irrigation in which water is pumped under pressure through pipes to above-ground applicators which distribute it through the air to soil. Under rare circumstances the pump has been eliminated, as where the water source is on a hillside, such as a spring or reservoir, and the water is piped to the applicator in the valley. The elevation difference provides the head for application pressure. *Sprinkler irrigation systems* are classified as fixed, semiportable, and portable.

sprinkling. Distributing water to plants and soil by a sprinkler or nozzle.

sprinkling can. A bucket equipped with a long spout terminating in a perforated sprinkling head. When tilted, it distributes water in numerous small jets. Used for watering flowers, seedlings in frames or small seed beds.

sprocket wheel. A wheel with cogs or projections so arranged as to engage the links of a chain, to transmit power as in grain combines, manure spreaders, etc.

sprout. 1. A new growth from a seed, rootstock, etc. 2. A perennial stem which grows from a stump, root, or tree; a sucker. 3. Something resembling a *sprout* in growth. 4. See *Brussels sprouts.* 5. To begin to grow or germinate; to put forth shoots, *sprouts,* buds, etc. 6. To cause plants, seeds, etc., to develop *sprouts* or shoots, as rain *sprouts* seed. 7. To remove *sprouts.*

sprouting broccoli. See *broccoli.*

spruce. Any tree of the genus Picea, family Pinaceae. Species are handsome, evergreen trees with dense crowns of conical form, widely used for timber and pulpwood and valuable for horticultural planting. Native to the cooler regions of the Northern Hemisphere. See *Alcock spruce; black spruce; Colorado spruce; Englemann spruce; Norway spruce; red spruce; Serbian spruce; silver spruce; Sitka spruce; tigertail spruce; white spruce; yeddo spruce.*

spruce aphid. *Aphis abietina* Wlk., family Aphidae, a dull green insect that is very destructive to Sitka spruce in Oregon and Washington, U.S.

spruce beetle. Any beetle infesting the spruce, especially any species of the genus Dendroctonus, family Scolytidae. The larvae feed and develop beneath the bark.

spruce borer. 1. The larva of any beetle of the families Buprestidae or Cerambycidae, developing beneath the bark of the spruce. 2. The larva of the spruce beetle *q.v.*

spruce bud scale. *Physokermes piceae* (Schr.), family Coccidae, an insect that occasionally infests the branchlets of Norway spruce. See *scale insect.*

spruce budworm. *Choristoneura fumiferana* (Clem.), family Tortricidae, a moth whose larvae infest conifers feeding on the terminal shoots which they have webbed together to form shelters. The tops of infested trees first appear as though scorched by fire, and if heavily infested the trees may die. The budworm is one of the most destructive insects in the forests of northern U.S. and southern Canada.

spruce gall aphid. Cooley spruce gall aphid.

spruce mite. Spruce spider mite.

spruce needle miner. *Taniva albolineana* (Kearf.), family Olethreutidae, a brown moth whose larvae bore into needles at the bases and make webs from entrance holes in needles to twigs on spruce from the eastern states to Colorado and Idaho, U.S.

spruce sawfly. *Neodiprion abietis,* family Diprionidae, a sawfly *(q.v.)* whose larvae defoliate spruce and pine trees, causing great destruction. In the 1930's, this fly reached epidemic proportions in Canadian forests, before being controlled by *spruce sawfly* parasites from Europe. Found first in Canada.

spruce spider mite. *Oligonychus ununguis* (Jac.), family Tetranychidae, a black mite which sucks sap from foliage and spins webbing between needles of evergreens. These mites may kill young spruces the first season and cause older trees to die progressively.

sprucetop grama. *Bouteloua chondrosioides,* family Gramineae, an erect, perennial bunchgrass, 1 to 3 ft. tall, with numerous, slender, flat leaves, and rather naked culms. Largely restricted to mesas and rocky hills of western Texas to southern Arizona, U.S., Mexico and Guatemala, it has a fair forage value.

sprung hock. See *curb: 1.*

spud. 1. A narrow, sharp spade-like tool usually fitted with a long handle used for undercutting and lifting out deep-rooted weeds, for removing bark from trees, for tree planting, etc. 2. A potato *q.v.* 3. In well drilling, to raise and lower the drilling tool.

spur. 1. A short, stubby shoot, as in some fruit trees where the *spurs* bear flowers for more than one year or as grapes are pruned to *spurs* of one to two buds each, etc. 2. A hollow, tubular projection from some part of a flower, very conspicuous in the columbine, usually secreting nectar to induce the visitation of insects. 3. A horn-like protuberance which grows from the inner side of the shank of a fowl. 4. A steel contrivance secured to a rider's heel and used to urge the horse by its pressure. See *spur rowel. Cf. bronc spur.*

spur blight. 1. Fire blight (bacterial) infections involving fruit spurs. See *fire blight.* 2. A fungous disease characterized by brown or purplish areas appearing on the canes of red raspberries at the nodes caused by *Didymella applanata* and resulting in defoliation and weakening of the canes. See *cane blight of raspberries.*

spur dike. A dike of rock or other material built from the bank into the channel for bank protection or for channel improvement. Also called *jetty.*

spurge. Certain plants of the genus Euphorbia, family Euphorbiaceae. See *floweringspurge euphorbia; spotted euphorbia; leafy euphorbia; toothed euphorbia.*

spurgelaurel daphne. *Daphne laureola,* family Thymelaeaceae, an erect, bushy shrub grown in warm areas for its evergreen leaves and yellow flowers. Native to southern Europe and western Asia. Also called *copse laurel.*

spurge nettle. Risky treadsoftly.

spur lupine. *Lupinus laxiflorus,* family Leguminosae, a perennial, poisonous herb which occurs from Montana and British Columbia southward to Arizona and California. Conditions of poisoning with sheep and other livestock are reported, with conspicuous symptoms of labored breathing, frothing at the mouth, and coma. The poisonous substance is most concentrated in the seed pod.

spurred bellflower. *Campanula alliariaefolia,* family Campanulaceae, a popular, perennial herb grown for its nodding, white flowers. Native to western Asia.

spur rowel. A small wheel with radiating points, attached at the end of a horseman's spur as an added goad.

spurry. Corn spurry.

spur strap. A leather strap which attaches a spur to a horseman's boot.

spur terrace. In erosion control, a short terrace which is used to collect and hold or divert runoff.

squab. A nestling pigeon raised for its flesh, which may be marketed when fully-feathered under the wings and just before it is ready to leave the nest, usually at 25 to 35 days of age when it may weigh from 12 to 24 ozs.

squab broiler. A small broiler (*q.v.*) that is usually under eight weeks of age, commonly of Leghorn or other light-weight breeds.

squab turkey. A young turkey which weighs under 4 or 5 lbs.

square. 1. Of flooring or roofing, an area of 100 sq. ft. 2. As applied to shingles, that number which will cover 100 sq. ft. of surface. 3. An L-shaped instrument used in carpentry to lay out or test square work, etc. 4. See *cotton square.*

square and blocky. In stock judging, designating a desirable body conformation for a beef animal of good meat type.

square bottle. A glass milk bottle with square sides and rounded corners. *Square bottles* fit together and require less space than round ones.

square dock. A sheep's deep, square, level rump with a high-set tailhead; a desirable characteristic.

square edge and sound. Designating square-edged lumber, or timber, which is free of bark and which has no lack of wood from any cause.

square-edged. Designating lumber that does not have bark or absence of wood on its squared edges. *Cf. waney.*

square-ended ram. See *square dock.*

square foot survey method. A standard technique for determining average density and composition of range vegetation by systematically located plots, usually of 100 sq. ft., on which the herbage of each species is estimated individually in square feet of ground completely covered.

square gaited. Designating the gait of a horse in which the action is straight on all four feet.

squarely placed. Designating a horse whose legs are set well apart and of good conformation. Such placement of legs is associated with surefootedness and ease of action.

square on four corners. Designating the walking gait of a horse in which the flat walk is *square on all four corners* and is bold, easy and graceful.

square planting. A system of arranging plantings of trees in an orchard in which trees are set out at the point of intersection of lines forming squares. See *hexagonal planting; triangular planting.*

squarepod deervetch. *Lotus tetragonolobus,* family Leguminosae, a prostrate, annual herb grown for its edible pods and seeds. Native to southern Europe. Also called *winged pea.*

square-quartered. Designating a beef animal that is rectangular in body conformation with thick, heavy, smooth, well-meated front and rear quarters.

square rump. An animal's long, level, broad rump with level pin bones and tailhead.

square-toe shoe. A shoe used on a horse to make the hoof breakover straight when the horse is in action.

square trot. A two-beat, diagonal gait of horses.

squash. The fruit of certain various tender, vine-like, annual plants of the genus Cucurbita, family Cucurbitaceae, which is used as a vegetable: *C. maxima; C. moschata,* and certain varieties of *C. pepo.* Also the vine or plant which bears such fruit. See *crookneck group; Hubbard group; pattypan group; pumpkin; summer squash; vegetable marrow: 1; winter squash.*

squash beetle. *Epilachna borealis* (F.), family Coccinellidae, an insect whose spiny larvae feed on foliage of squash and other cucurbits. Also called *squash ladybeetle.*

squash bug. *Anasa tristis* (DeG.), family Coreidae, an insect whose nymphs feed on cucurbits by sucking the sap. They inject a toxic substance into the vines causing wilting. (It is commonly called *stinkbug* because a disagreeable odor is given off when the insect is disturbed or crushed.)

squash ladybeetle. Squash beetle.

squash vine borer. *Melittia cucurbitae* (Harr.), family Aegeriidae, a wasp-like moth which lays its eggs on squash and related plants. The larvae tunnel into the stem causing the vine to wilt.

squat pack. A method of packing dressed poultry for export trade in which 12 birds are packed with their backs up in a single layer, six on a side with feet doubled under the breasts and the necks extended between the birds on the opposite side of the box. Also called *export pack. Cf. breast pack; double-layer side pack; single-layer side pack.*

squatter. 1. One who settles on land especially on public land without right or title (U.S.). 2. The term has had a speckled career in Australia, originally used to designate promoters of crimes, receivers of stolen property, vagrants, convicts holding tickets of leave, etc. But *squatter* did not always have, even when first used, dishonest implications. It was sometimes applied to all persons who placed themselves upon public lands without license and later extended to describe persons who occupied Crown land for sheep-raising under a lease or license, and eventually, to those who held their sheep run as freehold, the latter being a wealthy and highly-respected group for the most part. With the selection legislation of the mid-nineteenth century, attempts were made to open lands held by *squatters* for sale to farmers. These acts intensified a bitter class-war between selectors (*q.v.*) and *squatters,* gave no adequate protection to either group, and alienated a large proportion of land from the *squatters* without increasing agriculture very much. See *selection: 2.*

squatter's right. In the U.S., the right to occupancy of land created by virtue of long and undisturbed use in the absence of legal title or arrangement.

squaw bush. 1. Any dogwood. 2. American cranberrybush viburnum (*q.v.*). 3. Matrimonyvine (*q.v.*). 4. Skunkbush sumac *q.v.*

squawcarpet ceanothus. *Ceanothus prostratus,* family Rhamnaceae, an evergreen, prostrate shrub, grown for its foliage and blue flowers. Native to North America. Also called *Mahala mat.*

squaw corn. A strain of corn whose kernels, shaped like flint corn kernels, are composed almost entirely of soft starch. It is grown in all colors but white and blue are most common. Also called *flour corn; soft corn.*

squawroot. 1. *Conopholis americana,* family Orobanchaceae, a parasitic, scaly herb found on roots of oak and hemlock. Also called *American broomrape.* 2. *Carum gairdneri,* family Umbelliferae, a tuberous-rooted herb grown for its small, white flowers. Native to western North America. Also called *Indian potato; yamp; yampa.* 3. Blue cohosh (*q.v.*). 4. Purple trillium *q.v.*

squaw winter. A short spell of snow and cold weather in the early autumn.

squill. Any of the plants of the genus Scilla, family Liliaceae. Species are bulbous herbs cultivated for their early spring blooms. See *blue squill; Siberian squill; Spanish squill; spring squill; star-hyacinth squill.*

squirrel corn. *Dicentra canadensis,* family Fumariaceae, a rather weak herb, similar to Dutchman's-breeches (*q.v.*). Cattle are reported to have been poisoned by grazing on this plant, the symptoms being sudden trembling, frothing at the mouth, and convulsions. Native to eastern North America.

squirrel tail. 1. Any of the plants of the genus Sitanion, family Gramineae. Species are low to tall perennial bunchgrasses with bristly spikes. Their forage value is intermediate to low. See *Hansen squirreltail.* 2. A fowl's tail, any part of which extends forward beyond a perpendicular line drawn through the junction of tail and back. A disqualification in exhibits, except with Japanese bantams.

squirrel-tail grass. Foxtail barley.

squirting cucumber. *Ecballium elaterium,* family Cucurbitaceae, a hairy, perennial, herbaceous vine grown as a curiosity for its fruit which, when detached from its stalk, squirts out its brownish seeds with explosive force. Native to the Mediterranean areas of Europe and Asia.

stabilization. The point in the development of vegetation at which it is in equilibrium with the climate. Each stage of succession tends toward *stabilization,* but under natural conditions it is actually attained in only the last or climax stage. Thus, each group of invaders ushering in a new stage may increase up to a certain point and persist long enough to appear like a stable community, only to be replaced eventually by another stage until final *stabilization* is reached with a climax cover.

stabilized grade. The channel slope at which neither erosion nor silting occurs.

stabilizer. 1. A food additive to ice cream mix or other foods, such as gelatin or a sodium alginate base product, used to improve the texture and body quality of the product. 2. A substance added to any product to prevent change, such as rancidity in fats.

stable. 1. a. A building used for the lodging and feeding of horses, usually one fitted with stalls, mangers, feed boxes, and a place to keep harness or riding gear. b. A building equipped with stanchions, racks, etc., in which cattle, goats, etc., are kept: *e.g.,* a cow *stable.* 3. To put an animal into a *stable.* 4. Referring to things, chemical compounds, or conditions, not readily changed by action of designated agents or circumstances.

stable broom. A stiff broom, commonly a push broom, used to sweep out stables.

stabled. Having been put into the stable; kept in the stable, as the horses were *stabled* and fed; the cows were *stabled* during the rainy season.

stable fly. *Stomoxys calcitrans* (L.), family Muscidae, a small fly, similar to the house fly but with biting mouth parts that enable it to pierce the skin and suck the blood of animals. Due to its irritations animals lose weight, the yield of dairy cattle is reduced, and work animals become unmanageable. It occurs in all parts of the U.S. and throughout most of the world.

stable halter. A device, made of light rope or leather for leading or controlling horses. It consists of a head and nose piece with a lead or tying strap.

stable manure. The excreta of livestock intermixed with straw or other bedding material which commonly includes stalks and uneaten, coarse stems of forage, corn cobs, etc.; an important organic fertilizer. See *manure.*

stable price. 1. A price usually referred to as being firm and having the ability to withstand change. The *stable price* might be in terms of a fixed number of dollars to be exchanged for a given commodity or unit of a commodity. 2. A price which maintains a specified rate of exchange between two commodities or between one commodity and all other commodities.

stabling. The act of putting a horse or other animal into a stable; the quartering of livestock in a stable.

stack. 1. An orderly built weather resistant pile or rick of the sheaves or bundles of a grain crop before the grain is threshed, or the orderly piling together of the straw or cured hay into a pile or rick in such a manner that it will stand the weather and be preserved for future use. *Stacks* are either conical or rectangular and of many sizes. 2. A quantity of bales, crates, shipping containers, etc., arranged as a group in a pile. 3. To assemble a quantity of hay, straw, etc., into a pile or heap.

stacking. The act of placing hay, straw, bean vines, etc. into a pile.

stacking derrick. A form of derrick which lifts hay onto a stack, consisting of an upright mast with a swinging gaff or boom, usually swiveling completely around the mast.

stack silage. A method for preserving green crops by placing the material in a stack above ground and covering it with a water-proof and air-proof thatch of straw or sod. It is much used in Europe.

stack stand. A platform which supports stacks of hay or grain keeping them above the wet ground and free from vermin. Also called *staddles.*

staddle. 1. Anything which serves for a support; a prop. 2. A small tree, especially a forest tree.

staff. A heavy, stiff stick or rod, usually with a snap on one end, used in handling animals.

staff gauge. A graduated scale on a wooden staff, plank, metal-plate pier, wall, etc., by which the elevation of the water surface may be read.

staff of life. The cereal grains wheat and corn used in making bread (colloquial).

stafftree. American bittersweet.

stag. 1. In animals, a male castrated after reaching sexual maturity which shows pronounced sex development. 2. A horse which is thick and coarse in the throat latch and crest from late castration. 3. An imperfectly or recently castrated sheep or steer. 4. In marketing, an uncastrated male chicken with flesh slightly darkened and toughened and with comb and spur development showing the bird to be in a state of development between a roasting chicken and a cock. 5. A boar hog usually castrated after having passed breeding usefulness. 6. A wild mature male deer.

stag beef. Beef produced from bulls and stags that is usually strong-flavored and tough.

stag bush. Blackhaw viburnum.

stage. The elevation of a water surface above any chosen datum plane, often above an established low-water plane; gauge height.

stag-faced. In horses, referring to the face somewhat indented (concave) between forehead and muzzle; dish-faced.

staggerbush lyonia. *Lyonia mariana,* family Ericaceae, a slender, upright shrub with oval or elliptical leaves and white flowers, sometimes grown as an ornamental. Although seldom eaten by livestock, conditions of poisoning have been reported with sheep browsing on its herbage with observable symptoms of frothing at the mouth, convulsions and, later, paralysis of the limbs. Native to the U.S. Also called *wicopy.*

stagger grass. Crowpoison.

staggers. 1. A disease in horses, in the southeastern U.S., believed to result from eating the atamasco lily (*q.v.*). 2. See *gid.*

stagger weed. Squirrel corn.

staggerwort. See *sneezeweed: 2.*

stagheaded. 1. Designating a living tree with a part of the top dead. 2. Designating an animal with a large, coarse head.

staghorn fern. 1. Any fern of the genus Platycerium, family Polypodiaceae. 2. *P. bifurcatum,* a fern commonly grown in the conservatory and greenhouse as a decoration. Native to Australia and Polynesia. Also called *elkhorn fern.*

staghorn sumac. *Rhus typhina,* family Anacardiaceae, a deciduous shrub or small tree grown for its autumnal, brilliant red foliage. Native to North America.

Stagonospora leaf spot of grasses. A fungous disease caused by *Stagonospora arenaria,* family Sphaerioidaceae, characterized by the pycnidia being globose to flattened, dark-colored and immersed or partly projecting through the epidermis of the leaf.

stain. 1. A discoloration produced by foreign matter; a soiling, as of eggs by contact with damp litter, etc. 2. A dye or a mixture of dyes which renders minute or transparent structures visible or which differentiates tissue elements by coloring with selective *stains* for microscopic examination.

stained (egg). A market classification for an egg whose shell is stained or soiled with spots that together cover not over ⅛ of the shell surface with no adhering dirt.

stained-antigen, rapid, whole-blood test. A test to detect the presence of pullorum or fowl typhoid disease organisms in poultry which involves a drop of blood and a drop of stained, pullorum antigen. A drop of antigen (*q.v.*) is placed on a test plate; a drop of blood obtained with a wire loop from a pricked wing-vein is added and the mixture stirred with the loop. If the bird is a carrier of pullorum or typhoid organisms, crumb-like particles appear in the mixture in a few seconds. Also called *pullorum-typhoid test.* See *agglutination: 3.*

stained wool. Wool that has been stained by urine, etc., so that it cannot be scoured to a pure white color. It is subject to price discount.

stake. 1. A wooden or metal strip, usually pointed at one end for driving into the ground, used for marking a location in surveying, supporting a vine or plant, tethering an animal, etc. 2. To place *stakes* as markers. 3. To support plants by tying to *stakes.* 4. To tether an animal to a *stake.*

stake-body truck. A truck whose body sides and ends consist of removable panels that are affixed to vertical stakes. The ends of the stakes fit into sockets at the outer edge of the truck bed to hold the panels in an upright position. Locking clips and pins connect the panels together to complete the enclosure.

staking drainage ditches. Determination of the location and depth of a proposed drainage ditch in the field by placing stakes at regular intervals to indicate the center line of the ditch and by use of hub and guard stakes to indicate the exact depth of ditch in reference to the top of the hub stake which is usually driven flush with the ground surface. The guard stake indicates the location of the hub stake and the depth of cut is written in crayon on the guard stake, as C 4.24.

staking ground. A place to tether horses (southwestern U.S.).

staking pond (dam). The locating of a pond site including the dam and outlet, and the marking of the location of each with stakes.

staking terrace line. Laying out terrace lines and placing stakes along each as markers.

stalactite frost structure. Frozen soil which contains tiny icicles or columns of ice separating soil sheets; heaved soil. It is associated with partial thawing and refreezing.

stale. 1. A flavor defect commonly found in dairy products but also found in other food products, such as meats, bread, fruits, and vegetables. It is the result of oxidation and is most commonly associated with extended age of the product, poor storage, poor ventilation, and unsatisfactory sanitation. 2. Designating a horse which has difficulty in urinating, usually as a result of disease.

stale cream. A cream lacking freshness of flavor; one indicating slight chemical or bacterial decomposition.

stalk. 1. The stem or main axis of a plant, especially of an herbaceous plant. 2. Any stem or plant part which supports leaves, flowers or fruit; a petiole, pedicel, or peduncle.

stalk borer. *Papaipema nebris* (Guen.), family Noctuidae, an insect that may feed on any plant large enough and soft enough for its operations. It apparently prefers giant ragweed and corn in later stages, but is often a pest in flower and vegetable gardens where it tunnels into the stalks of dahlia, hollyhocks, aster, rhubarb, pepper, potato, tomato, tobacco, and other plants. It is distributed generally throughout the U.S. east of the Rocky Mountains.

stalk cutter. A field machine of which several types are available ranging from a free-rolling, trailing unit that rolls down and cuts up stalks, to power driven units that beat or cut and shred stalks of corn, sorghum, cotton and even light brush.

stalk rot. A rot of the stalks of corn, sorghum, and several other plants, caused by any of several species of fungi and bacteria. It may follow a period of drought, extreme heat, or other unfavorable conditions that weaken the plant. The disease is also favored by injuries to the stem, crown, or roots, caused by cultivation implements, insects, hail, or other agent that makes an opening for entrance of destructive fungi and bacteria.

stalk smut of rye. A fungous disease caused by *Urocystis occulta,* family Tilletiaceae, characterized by sori forming in the culm tissue and in the flowers, preventing the spike from emerging.

stall. The space in a barn which is occupied by a single animal, such as a dairy cow or horse, for feeding and handling. There are many types of *stalls* from the simple tie space to very elaborate mechanical devices.

stall barn. A barn used for sheltering dairy cattle and/or young stock where the adult animals are confined to stalls by means of stanchions, straps, halters, or chains during part of the year, as in the winter and for milking. Roughages and concentrates may be fed at the individual stalls. None, part, or all of feeds and bedding may be stored in the structure. Also called *stanchion barn.*

stall cock. A valve on the vacuum line of the milking machine which is placed at the stall of each cow for attaching the suction hose of the milking machine.

stallion. A male horse used for breeding purposes; it should be a highly selected animal with very superior breed characteristics.

stallion and three mares. In the show ring, a stallion and three mares which are owned by one exhibitor.

stallion barn. A separate barn of suitable construction and comfort where a stallion is kept away from the rest of the horses.

stallioner. A person who handles a stallion and is experienced in horse breeding.

stallion plague. See *dourine.*

stallion's court. a. A breeding establishment where a noted stallion is kept for public or selected service. b. The collection of mares held for breeding to a particular stallion.

stamen. The organ of a flower which bears the pollen (microspores) consisting of the stalk (filament) and the anther. See *flower.*

staminate. Designating a flower that has stamens but no pistil and hence is imperfect. *Cf. pistillate.*

stampede. A wild rush of cattle or horses as a result of fright.

stanchion. A tying or controlling device, usually of wood or metal, used in barns to control animals, usually dairy cows.

stanchion hose. A rubber hose with shaped ends used for connecting a milking machine to the vacuum line.

stand. 1. The proper number of uniformly distributed plants per acre. 2. The relative number of plants per area, as a poor, good, or medium *stand.* In forestry, fully-stocked, understocked, pure, mixed, or residual *stand.* 3. A hive of bees. 4. Density of game per acre, area, etc. 5. A stallion's court. 6. To cease walking or moving; to take or keep a certain position. 7. To rise to the feet. 8. Of a stallion, to be available for breeding purposes.

stand age. The average age of the trees which compose a stand. In practice, it is applied to even-aged stands by obtaining the average age of representative dominants.

standard. 1. A selected unit for comparison. 2. An upright piece of wood or metal, fitting into one of the spaces provided around the edge of a wagon or truck bed. 3. The narrow, and usually erect, unit of the inner whorl of the perianth of the flower of an iris. 4. The large, usually upright, petal in the blossoms of legumes. 5. In forestry, a seedling or healthy sprout selected in a coppice rotation. 6. A plant that has an upright, treelike stem or trunk and is not on a dwarfing stock or supported by a trellis or wall; specifically a fruit tree of natural size. 7. The measuring device used to determine the height of a horse in hands. 8. The part of a plow, extending downward from the beam, to which the soil-working parts are attached; also, the part of a disk harrow which extends downward from the frame and to which the gang assembly is attached. 9. Referring to a product that conforms to an established specification or standard.

standard baby beef. A good, choice, or prime steer or heifer of the beef breeds, 12 to 18 months of age, weighing 700 to 1,000 lbs.

standard barrel. A standardized container, of 7,056 cu. ins. capacity, used for marketing produce. For cranberries, the *standard barrel* contains 5,826 cu. ins.

standard bee(s). Bees which maintain their fundamental characteristics in their adaptability to various environments.

standard berry box. A standardized container, of half pint, one pint, and one quart sizes, used for marketing berries (U.S. standard).

Standardbred. A breed of horses, trotters and pacers, developed in the U.S. primarily for speed and endurance as roadsters and racers and named for their ability to trot or pace a mile in standard time. Its origin is attributed to a mixture of the following breeds: Thoroughbred, Norfolk Trotters, Narragansett Pacers, Arabian, and Morgan. In general, the *Standardbred* in appearance is rather rangy, leggy, deep-chested, narrow, and angular. The physical form is less important than speed in the selection. The breed holds the world record for a driving horse on a mile racing track.

Standard-bred poultry. An exhibition breed of domestic poultry. *Cf. production-bred poultry.*

standard cow day. A feeding index for a cow which is 16 lbs. total digestible nutrients per day from grazing.

standard crate. A standard western melon crate, of 12 x 12 x 22⅜ ins. inside measurement, used in packing and shipping melons.

standard-density bale. A flat bale of cotton with approximate dimensions of 56 x 28 x 18-20 ins. which weighs approximately 22-28 lbs. per cu. ft. and 500 lbs. per bale. The bale is usually covered with 2-pound jute bagging, 3 yards on top and 3 yards on the bottom, and is tied with 8 steel ties.

Standard D.H.I.A. plan. A uniform plan of recording milk and butterfat production for dairy cows. The tester, called a Dairy Herd Improvement Association supervisor, calls at each member's farm for two milkings a month: one on an unannounced evening and the other the following morning. He weighs the milk from each cow and takes a sample of the milk which is tested for butterfat content. Milk production, butterfat test, as well as feed costs per cow are entered in the herd book of the owner and a copy sent to the U.S. Department of Agriculture.

standard flat. A standard western melon crate used for packing and shipping melons which is 4½ x 13 x 22⅜ ins. inside measurement. See *standard crate.*

standard grade. A California, U.S., canning-fruit grade which represents fruit of good quality, reasonably good and uniform in color, free from blemishes, uniform in size, uniform in degree of ripeness, and symmetrical.

standard hamper. Any of 6 sizes of hampers which is accepted as legal under Federal U.S. standards. The standard sizes are ¼, ½, ⅝, ¾, 1, and 1½ bushel.

standardization. 1. Conformance to some standard by law or by an individual concern. 2. The act of selecting the most suitable dimensions or proportions for structural or mechanical parts, thus obtaining uniformity of practice. 3. Adjustment of the fat and/or other ingredients of a dairy product to a desired percentage.

standardization of milk. The practice of modifying normal cow's milk by the addition or subtraction of butterfat in the form of cream so that the modified milk conforms to a certain desired standard, such as a four percent milk.

standardized milk. Milk which has been so modified as to contain a definite amount of one or more of its constituents. Such standardization may be accomplished by separating and removing a definite amount of each of the skim milk and cream. Milk of known different fat contents from different cows or farms may also be standardized to any desired fat content.

standard knot. A knot that is sound and not larger than 1½ ins. in diameter. In hardwoods, it is not larger than 1¼ ins. in diameter. See *knot: 4.*

standard length. In softwood lumber, lengths of boards in multiples of even feet; in hardwood lumber, lengths from 3 to 20 ft., including both odd and even number of feet.

standard middlings. See *wheat standard middlings.*

standard milk. Milk which conforms to certain standards of composition for fats and total solids or for bacteriological requirements as prescribed by legal milk regulations or ordinances.

standard of living. Desired levels (rather than actual) of living which include both cultural and material possessions. (The term may refer to norms the members of a group apply to each other or to norms applied to the members by outside observers.) See *level of living.*

Standard of Perfection. See *American Standard of Perfection.*

standard pack. See *standard grade.*

standard planting. A planting made on land whose topography permits an orchard to be planted in straight rows with trees in the rows set a uniform distance apart.

standard pot. A flower pot that is as wide at the top as it is high. Such pots are found in various diameter sizes from 1 in. up to 14 and 16 ins.

Standard-prune constricting mosaic. A virus disease of the Standard variety of prune of the mosaic type, characterized by a constriction of the leaf blade in the mottled areas.

standard rod. In shop practice, a rod of well-seasoned yellow pine used as a basis and reference for accurate measurement in a workshop. Such rods are usually made from 5 to 10 ft. long, the section of the wood being about 2 by 2 ins. Each 12 in. division is marked on a brass inlaid plate.

standards. A term used in grading butter, signifying 90-score, which is available in car lots from one creamery.

standards box. A box which contains 12 individual samples of cotton for each of the grade standards, each sample differing slightly in grade factors, thus creating a range within each box. In determining the grade of a given lot of cotton, a grader may compare a sample with one of the small samples in the *standards box* for color; with another for foreign matter; and with a third for preparation. Each box carries the seal of the U.S. Department of Agriculture and the certificate and signature of the Secretary of Agriculture.

standard shorts. See *wheat brown shorts; wheat gray shorts; wheat red dog; wheat standard middlings; wheat white shorts.*

standard soil mixture. A mixture of three parts soil and either one part organic material, such as manure, peat moss, leaf mold, or sand which is used for rooting pot plants and in benches and beds in the greenhouse. See *soil mixture.*

standard specification. A model schedule or list of requirements, qualities, dimensions and tests which is proposed to be adhered to as a basis for the manufacture of similar articles, structures or mechanisms, designed to be made in large numbers.

standard wheat middlings. See *wheat standard middlings.*

standby crew. A group of men who are specially organized, trained, and placed for dependable and prompt action on forest fires. They may be assigned other duties during intervals between fire assignments.

stand density. In forestry, the density of stocking expressed in terms of number of trees or total basal area per acre.

stand-density index. In forestry, the number of trees per acre which a stand has, or will have, assuming that (a) the average diameter of a tree is 10 ins. and (b) the change in number of trees and average diameter progresses as in a normal, fully-stocked stand.

stand for service. 1. To be announced as being available to the public as a breeding sire, as a stallion or other male animal. Such an animal must meet certain standards of health, pedigree, and breed character. 2. A female animal that is willing to be served by a male.

standing corn. The corn crop as it stands in the row in a field.

standing crop. Any crop as it stands in the field before harvest, as corn, wheat, pasture grass, etc.

standing cypress. Texasplume.

standing grain. A crop of any of the cereal grains, such as wheat, oats, etc., as it stands in the field before harvest.

standing grass. See *standing crop.*

standing hay. See *standing crop.*

standing herd. Watching the herd (southwestern U.S.).

standing plant. 1. An unharvested plant. 2. A plant that is not broken over or lodged.

standing rib roast. A retail cut of beef for roasting. It usually consists of two or more ribs with adhering meat, taken from the upper part of the rib section immediately to the rear of the blade bone of a beef forequarter.

standing stalk. Any of the upright stalks in a field of corn, the ears usually being harvested from the *standing stalk.*

standing stall. A space in the barn where animals, especially horses, are individually confined when not in use or loose in yard or pasture. It should have a comfortable floor to stand on, be of adequate size and efficiently arranged for feeding and handling.

standing timber. Timbers still on the stump; uncut timber.

standing water. Any water that collects or stands for a period of time on the surface of land following a heavy rainfall or fast melting snow.

Standish honeysuckle. *Lonicera standishi,* family Caprifoliaceae, a hardy, partly-evergreen shrub grown in the shrub border and for its fragrant, creamy-white flowers and red fruit. Native to China.

stand squarely on legs. In stock judging, to have correct leg placement and character.

stand the winter. To endure, as field crops resistant to cold weather, and to carry over in the frozen ground during the winter, as winter wheat, hardy alfalfa, etc.

stand too close in front legs. In stock judging, a term used to describe an animal that is narrow-chested and deficient in the front quarters.

stand up. To be erect, as any plant that has a sturdy or stiff central stalk; a very important characteristic in plants to be harvested by machinery.

Stangenkase. A Limburger-type cheese similar to Backsteiner, made in Germany from partly skimmed cows' milk.

Stansbury cliffrose. *Cowania stansburiana,* family Rosaceae, a leafy, stiff-stemmed shrub or small tree which occurs in New Mexico and southern Colorado to Utah and California in woodland and ponderosa pine forest at elevations of 4,000 to 8,000 ft. Where abundant, it provides good-quality forage for cattle and sheep. Native to North America.

staphylococci. Disease-producing bacteria comprising the genus Staphylococcus, family Coccaceae. Spherical in shape, they occur in clusters, like a bunch of grapes.

staphylococcic mastitis. Mastitis (*q.v.*) caused by staphylococcic bacteria.

staple. 1. A bar, strip of iron, wire, etc., bent into the form of a U, or similar shape, and pointed at both ends so that it may be driven into a surface to hold wire fencing to a post, to hold a hasp or hook, to fasten together the parts of a splint basket, hamper or box, etc. 2. Fiber of such plants as cotton and flax, of wool, etc., especially with reference to length. Cotton with a *staple* of 1$\frac{1}{16}$ ins. in length or less is generally considered short *staple* and that with a *staple* length of 1$\frac{1}{8}$ ins. or more is called long *staple* or *staple* cotton. Wool is measured without stretching to remove the crimp. Fine *staple* wool from the western areas in the U.S. means the same as fine combing from other sections. 3. A lock or bunch of wool fibers as they exist in the fleece. 4. To judge the length of cotton fiber.

staple cotton. See *long staple.*

staple crop. A crop commonly produced and having a ready market.

staple length. The length of cotton or wool fiber as obtained by measuring the natural staple without stretching.

staple wool. Fine, strictly combing wool from the western half of the U.S.

stapling. The classing or grading of cotton or wool on the basis of length of fiber or staple.

staple puller. A heavy pliers designed for grasping and pulling staples, or an iron bar with a pointed projection at one end to be forced into the loop of the staple to pull it out of a fence post, etc.

star. A small white spot in the center of an animal's forehead, especially of a horse. Also called *starfaced.*

starapple. Any tree of the genus Chrysophyllum, family Sapotaceae. Species are grown in warm areas as ornamentals or fruit trees. Native to tropical America. See *cainito starapple.*

star astilbe. *Astilbe simplicifolia,* family Saxifragaceae, a perennial herb grown for its narrow clusters of white flowers. Native to Japan.

star-bloom. Pinkroot spigelia.

starch. A white, odorless, tasteless, granular or powdery complex carbohydrate which occurs widely in plants, especially in seeds, bulbs, and tubers. It is an important constituent of food and is used in making commercial glucose for stiffening fabrics and pastes.

star check. A heart check *(q.v.)* in which the separation extends in more than one direction from the pith.

starch hyacinth. See *grapehyacinth: 2.*

starch storage. The storage of carbohydrates in the form of starch in seeds, bulbs, tubers, roots, twigs, or other plant parts.

star feed. See *finger plate.*

starflower. *Trientalis borealis,* family Primulaceae, a hardy, herbaceous perennial grown for its star-shaped, white flowers. Native to North America. Also called *chickweed wintergreen.*

starglory. Any of several vines of the genus Quamoclit, family Convolvulaceae; native to tropical America. See *cypressvine starglory.*

star gooseberry. Otaheite gooseberry leafflower.

starhyacinth squill. *Scilla amoena,* family Liliaceae, a bulbous herb grown for its blue or whitish flowers. Native to Europe.

star ipomoea. Cypressvine starglory.

starjasmine. Chinese starjasmine.

starlily. *Leucocrinum montanum,* family Liliaceae, a bulbous herb grown for its funnel-shaped, fragrant, white flowers. Occasional livestock poisoning by its ingestion has been observed. Native to North America. Also called *sand lily.*

star magnolia. *Magnolia stellata,* family Magnoliaceae, a deciduous shrub or small tree grown for its fragrant, white flowers. Native to Japan.

star-of-Bethlehem. 1. Any bulbous herb of the genus Ornithogalum, family Liliaceae. See *Arabian star-of-Bethlehem; cape chinkerichee; milk star-of-Bethlehem.* 2. *O. umbellatum,* a bulbous herb grown for its clusters of star-like, white flowers borne on a flowering stem. Native to Europe. Also called *summer snowflake; sleepy Dick.*

star pruning. See *wedge-style pruning.*

starry Solomonplume. *Smilacina stellata,* family Liliaceae, a perennial herb grown for its greenish-white flowers in short racemes. Native to North America. Also called *false Solomon's-seal.*

starry Solomon's-seal. Starry Solomonplume.

start. To grow seedling plants, such as cabbage and tomatoes, in specially prepared beds from which they are later transplanted in the field when weather and soil conditions permit.

started-bud plant. See *started-eye plant.*

started chick. A chick which has been fed and watered, usually one that has been brooded at the hatchery for two or more weeks before shipping.

started-eye plant. A rose plant that is budded (usually in June) and from which the stock top is removed after about four weeks. The bud grows and the resulting shoot branches by the end of the first year. Such plants are produced in western U.S. and are used by florists for greenhouse forcing. Also called *started-bud plants.*

starter. 1. A prepared culture of desired organisms used for inoculating milk and cream in the making of various dairy products, such as butter, cheese, etc., chiefly to enhance the flavor. See *mother culture.* 2. The first food provided for young animals. See *calf starter; chick starter.*

starter fertilizer. A liquid or easily soluble dry fertilizer applied at planting or transplanting time.

starting battery. In poultry, a collection of brooding units, commonly wire cages arranged in a group with provisions for supplying heat, food, and water to young chicks, poults, etc., during their early growth period.

starting mash. In poultry, a specially prepared feed consisting of a mixture of water, ground cereal or cereal products together with the necessary protein, vitamin, and mineral supplements, which is designed to constitute the entire ration for young chicks, poults, etc., from four to six weeks of age.

star tulip. Mariposa.

starviolet dalibarda. *Dalibarda repens,* family Rosaceae, a perennial herb grown in shady places in the wild garden for its white flowers. Native to North America. Also called *false violet; dew drop.*

starve out. To hamper the growth of a plant so that it cannot grow and increase in a normal manner.

Repeated cuttings, planting other, more vigorous, competitive plants, and excessive cultural practices are the most common methods used.

starweed. Chickweed.

starwort. Any herb of the genus Stellaria, family Caryophyllaceae. See *Easterbell starwort; chickweed.*

stassanization. A heat treatment of milk, common in Europe, in which the product is heated to temperatures somewhat higher and for shorter periods than is commonly employed in the high-temperature, short-time process of heat treatment in the U.S.

state conservationist. In the U.S., the administrative officer in a state of the Soil Conservation Service.

State Department of Agriculture. In the U.S., a department of the state government established to administer regulatory laws relating to agricultural and consumer interests, primarily with matters within the state to which federal legislation does not apply.

state fair. A state government-sponsored exhibit of livestock, farm equipment, and farm products for the benefit of its citizens. In the U.S., nearly every state holds an annual *state fair.*

state-owned lands. Lands which belong to and are administered by a state.

state trading. A governmental trade restriction in which the government, instead of private organizations, does the international trading.

state veterinarian. A veterinary official of a State Department of Agriculture responsible for the proper enforcement of all regulatory laws regarding disease control, health and sanitation of livestock, and livestock products of the state.

static head. 1. The vertical distance the top of a column of fluid is above a reference level. 2. The total head without deduction for velocity head or losses; *e.g.*, the difference in elevation of head-water and tail-water of a power plant.

static load. Any load at rest; as, in hydraulics, the pressure on a pump piston due to static head.

static suction lift. In lift pumps, the vertical distance in feet from the centerline of pump to the free level of the liquid to be pumped.

station. 1. A standard of length, usually 100 ft., measured along the centerline of a road or along a survey line. 2. In Australia, a large sheep or cattle ranch which includes the owner's house, working sheds and the total grazing area. 3. The ideal pose or style, including height and reach, of a domestic fowl, especially in the game breeds of chicken. 4. See *experiment station.*

stationary baler. A stationary machine which bales hay and straw. See *pick-up baler.*

stationary mill. A sawmill or feed grinder, etc., which has a permanent location, as contrasted with a portable unit which may be moved at frequent intervals.

stationary phase. In a bacterial culture, the period of quiescence before active growth begins.

stationary spray plant. A spray outfit that remains in a fixed place. Generally the plant consists of a large-capacity tank, a power unit and pump of sufficient capacity to force spray liquids through underground pipes to all parts of an orchard.

stationary thresher. A threshing unit located in a fixed place to which harvested grain, beans, etc., are brought to be threshed, as contrasted to a combine harvester-thresher unit which threshes the crop as the unit moves over the field.

station hand. A man employed on a sheep or cattle station (Australia and New Zealand).

station manager. A person who directs the work of a farm or station, buys and sells the stock, employs the help, signs the checks, etc. (Australia and New Zealand).

status. 1. In anthropology, the rights and duties of an individual in reference to another individual where they stand in a reciprocal relation to each other. 2. In sociology, a hierarchy of positions in a society and a summation of a person's position in socio-economic terms.

statute inch. In northern California, U.S., a flow of 1.5 cubic feet per minute, or 1/40 cubic foot per second. See *miner's inch; southern California miner's inch.*

staty gum. Horncap eucalyptus.

stauntonvine. Japanese stauntonvine.

stave. Any long, narrow strip, usually of wood set edge to edge, used in making fences or enclosures, such as silos, tubs and buckets.

stave jointer. A machine tool used for turning the edges of barrel staves.

stave silo. A large, round bin or receptacle, made of narrow strips of wood or concrete staves, placed edge to edge and held together with steel bands, in which chopped green forage is stored. See *silo.*

stay. 1. A guy rope, cable, etc., that supports a vertical pole, mast, etc. 2. A prop or tie piece that supports or holds parts together, or contributes to stiffness, as a timber supporting the frame of a structure, the posts of a fence, etc.

stay bolt. A rod connecting opposite plates or parts to stiffen or prevent them from being shifted or bulged out.

stay chain. 1. The chain on a wagon or truck box which prevents the sides from bulging when the box is loaded. 2. A chain on each side of a wagon joining the loop near the end of the doubletree to the front axle to lessen the swing of the doubletree.

stay wire. Any of the several cross wires of woven wire fencing.

steady. 1. A command to a horse to calm him or to make him go more slowly. 2. Regular, constant, uniform; as a *steady* breeze.

steady market. A market in which there is no distinct tendency for a rise or fall of prices.

steady, true, and practical action. The highly satisfactory action of a horse.

steak. 1. A thick, usually cross-sectional, slice of meat, its name depending on the part of the carcass from which it is cut: *e.g.*, round *steak*, sirloin *steak*. It is usually, but not necessarily, from the rear quarters of an animal and may be cooked in any of several ways. If not qualified, the term refers to beef *steak.* 2. A thick, usually cross-sectional, slice of fish: *e.g.,* halibut *steak,* salmon *steak.*

steak knife. 1. A long, rather broad-bladed knife, upward curving at the end, used in farm slaughter for cutting up meat. 2. A knife used at the table for cutting up meat, particularly steak.

steam distillation. A distillation which is performed by boiling a mixture of water and immiscible liquid, or by blowing steam through a liquid (or wood chips) and collecting a distillate of water and substance.

steamed bone meal. See *bone meal (steamed).*

steamer. A device for obtaining beeswax from old combs in which the combs are suspended in a closed tank and treated with steam.

steam-heated pan. A pan, similar in principle to the sorghum pan, which consists of a long, shallow, open, wooden box, lined with copper or tin, in which juice is boiled by a closed steam coil (usually of copper) which extends the full length of the pan and is removable. It is used for concentration of juice, sap, etc., to a sirupy consistency.

steaming. A steaming treatment given lumber as part of a kiln drying process.

steaming hay. An old practice used in connection with the feeding of coarse hay, the main object of which is to increase the proportion of the hay that is consumed. (The reduction in the hay wasted, however, is not sufficient to justify the labor and the expense involved.)

steam jacket. A space formed in the walls of a vessel or chamber (as by an inner and outer casing) whereby steam may be circulated to heat the contents. Steam-jacketed kettles are commonly used by processors in the preparation of jelly, preserves, etc.

steam-jacketed kettle. See *steam jacket.*

steam tank. A chamber or vat heated by steam, used for various purposes, such as rendering fats, etc.

steapsin. Pancreatic lipase, a lipolytic enzyme secreted by the pancreas which has the power of hydrolizing fats to fatty acids and glycerol.

stearic acid. One of the fatty acids occurring in combined form in animal and vegetable oils; one of the major fatty acids in butter and most fats. Commercial *stearic acid* is used in large quantities in rubber compounding and in the preparation of soaps, greases, and chemicals.

stearine. The solid material of any fat which is obtained, in the process of refining, by filtration from an animal or vegetable oil after chilling or freezing.

steckling. 1. A root such as beet, carrot, turnip, etc., which is stored over winter and replanted the following spring for seed production purposes. Also called *stechling.* 2. A cutting or a transplanted seedling.

steckling bed. A bed in which beet roots for seed are grown. The beets are transplanted to the seed row after overwintering in storage. See *steckling: 1.*

steel. 1. A very hard steel rod very finely ground (usually fluted), and fitted with a handle, used for sharpening kitchen and butcher knives. 2. To sharpen with a *steel.*

steel-blue lady beetle. *Orcus chalybeus* (Bdv.), family Coccinellidae, a lady beetle which was introduced from Australia into California, U.S., in 1892 to prey on black, red, and purple scales.

steel gang plows. A plow constructed entirely of steel with two or more bottoms or plowshares. It is in common use with tractor power.

steel gray. The coat-color of a horse that is an even shade of gray over the body with a black or dark gray mane and tail.

steel plow. The common plow in which the share and moldboard are made of hardened steel. It is universal in the U.S. but still rare in many parts of the world.

steel squirrel. Giraffe.

steel tie. A flat band of steel used to tie cotton bales, etc.

steelyard. A weighing apparatus consisting of a beam or lever with two unequal arms so that an adjustable or sliding weight on the lower graduated arm can balance a much greater weight suspended from the shorter arm. The principle of the *steelyard* is employed in the construction of various weighing machines.

steeplebush. Hardhack spirea.

steeplechase. A horse race across country over a prescribed course which has artificial fences, water-jumps, and other obstacles.

steeplechaser. 1. One who rides in a steeplechase *(q.v.).* 2. A horse which is trained to run in a steeplechase.

steer. A male bovine castrated before the development of secondary sex characters.

steer character. The character of a bovine intermediate betwen the bull and the cow which is shown by a general trimness, compactness, and neatness.

steer in prime condition. A steer in the highest condition of fattening or finish.

steer-roping horse. A horse trained to follow a steer at close range until a lasso is thrown, to run ahead and to one side until the noose tightens and is held taut, and to stand while the steer is thrown to the ground.

steershead bleedingheart. *Dicentra uniflora,* family Fumariaceae, a perennial, poisonous herb occurring from Washington to Utah and California, U.S. Native to North America.

Steinbuscher. A soft, cows' milk cheese, similar to Romadur. First made in Steinbusch, Brandenburg, Germany.

stele. 1. In vascular plants, the central cylinder of the stems and roots. 2. In wood anatomy, that part of the stem which includes the vascular system (xylem and phloem).

stem. 1. The principal axis of a plant from which buds and shoots develop; a tree trunk; stock; stalk. 2. Underground *stem,* tuber, corm or rootstock often used for propagation. 3. The stalk supporting a leaf, flower, or fruit; a peduncle, pedicel, or petiole.

stem analysis. The measurement of stated cross sections of a tree to determine its growth and development at different periods of its life.

stem and graft canker. A fungous disease of cultivated roses caused by *Coniothyrium fuckelii*, family Sphaerioidaceae, characterized by the death of branches or the entire plant, depending on the severity of infection. The fungus enters through wounds. See *dieback.*

stem break. Cercosporella foot rot of cereals.

stem break of flax. A fungous disease of flax caused by *Polyspora lini*, family Melanconiaceae, characterized by light gray to brown, circular lesions with darkened margins on the cotyledons and gray to brown lesions on the leaves, capsules, and stems appearing late in the season.

stem cabbage. Kohlrabi.

stem canker. A fungous infection which causes discoloration and necrotic areas on the stems of various plants usually killing the section above the diseased area. See *blueberry cane canker; Rhizoctonia disease of potato; sore shin; stem and graft canker.*

stem check. See *brown check of celery.*

stem cutting. 1. Any part of a stem used for plant propagation, usually of three general types of wood: (a) softwood or green cutting taken from a plant while in leaf, the stem being succulent; (b) hardwood cutting, taken from the leafless, dormant plant; (c) semi-hardwood cutting, intermediate between softwood and hardwood cuttings. 2. In sugar-cane propagation, a part of a selected stalk or stem cut into three- or four-eye lengths to be planted, usually in August or September, in field furrows.

stem-end browning. A non-parasitic browning of the vascular ring of the potato. The cause is not fully established.

stem-end rot. Any rot of fruits, etc., beginning at the point of connection with the plant, as the decay of citrus fruit or Cucurbits caused by fungi of the genus Diplodia.

stem-end rot and leaf blotch of strawberries. A fungous disease of strawberries caused by *Gnomonia fragariae*, family Gnomoniaceae, characterized by the withering of the calyx and a light brown, firm rot at the base of the fruit. The disease results in considerable losses on susceptible varieties, such as Robinson. The infestation on the leaves is not well understood.

stem-end rot of citrus. Citrus stem-end rot.

stem-end rot of watermelon. Watermelon stem-end rot.

stem grass. Any of the grasses that normally grow as more or less isolated and single stems, as contrasted with bunch and sod grasses.

stemless gentian. *Gentiana acaulis*, family Gentianaceae, a perennial herb grown in the rock garden for its solitary, deep blue flowers sometimes spotted on the inside. Native to the Alps and the Pyrenees. See *gentian.*

stemless townsendia. *Townsendia exscapa*, family Compositae, a perennial herb grown in the rock garden for its flower heads with white or purplish rays, nestled in a rosette of very narrow leaves. Native to North America. Also called *Easter daisy.*

stemmed. 1. Having a stem or stems. 2. With the stem or stems removed, as of grapes, cherries, etc.

stemmer. A person who removes stems, as a fruit *stemmer*, a tobacco *stemmer*, etc.

stemmy. Having a high proportion of stems, as hay from which leaves have been shattered.

stem nematode. See *nematode.*

stemphylium leaf spot of clover. A fungous disease of clover caused by *Pleospora herbarum*, family Sphaeriaceae, characterized at first by small, irregular, dark brown, sunken spots and later by light and dark brown to sooty lesions.

stem pitting. 1. An infectious disease of citrus with symptoms similar to those of tristeza and blind pocket (*q.v.*). Common in commercial orchards in South Africa. 2. An apple virus disease associated with Virginia Crab rootstocks. The phloem extends into the xylem, showing pits in the wood when the bark is peeled. Affected trees lack vigor and productivity.

stem rot. 1. A rot of stems of various succulent or herbaceous plants caused by any of several species of bacteria or fungi. A wide variety of field, garden, and greenhouse plants including cereals, potato, tobacco, sweet potato, soybean, aster, carnation, geranium, chrysanthemum, and snapdragon are susceptible to *stem rot.* See *black leg; brown stem rot; foot rot of cereals; Fusarium rot; Rhizoctonia disease of potato; stem rot of rice; sweetpotato wilt.*

stem rot of rice. A fungous disease of rice caused by *Leptosphaeria Salvinii*, family Sphaeriaceae, characterized by a decay of the stem at the waterline.

stem rot of sweet potato. See *sweet potato wilt.*

stem rust. See *rusts.*

stem rust of barley. A disease of barley caused by one or more of three specialized fungus varieties: *Puccinia graminis tritici, P. graminis secalis*, and certain specialized races confined to barley. See *rusts; stem rust of wheat.*

stem rust of grain. See *stem rust of wheat; stem rust of oats; stem rust of rye.*

stem rust of grasses. Any of numerous species of rusts (*q.v.*) that infect the stems of grasses. *Puccinia graminis*, family Pucciniaceae, is a common species that attacks many grasses and is better known as a destructive disease of small grain.

stem rust of oats. A fungous disease caused by *Puccinia graminis avenae*, family Pucciniaceae, with symptoms similar to those in the stem rust of wheat (*q.v.*). See *rusts.*

stem rust of rye. A fungous disease caused by *Puccinia graminis secalis*, family Pucciniaceae, with symptoms similar to those in the stem rust of wheat (*q.v.*). See *rusts.*

stem rust of wheat. A fungous disease caused by *Puccinia graminis tritici*, family Pucciniaceae, characterized by two stages of rust, the red or uredial stage present on the leaves and culms of the maturing wheat plant, and the black or telial stage present just before and at maturity of the plant. See *rusts.*

stem smut. A disease infecting the stems of cereals and other grasses, caused by *Ustilago* spp., family Ustilaginaceae, characterized by the development of conspicuous, dusty, brown or black layers of smut around the internodes of the stems.

steppe. 1. A cows' whole milk cheese with color added, made in Germany, Austria, and Denmark. 2. A treeless grassland which exists under semiarid climates, such as the east-west zonal belt of plains in Russia, lying between the forest land of the north and deserts on the south, or the semiarid Great Plains of western U.S. lying east of the Rocky Mountains.

stepping pace. The slow pace of a saddle horse which is the modified form of the true pace.

sterile. 1. In animals, incapable of reproduction; unable to produce normal living young. 2. In soils, unproductive; barren; producing little or nothing, as a *sterile* soil. 3. In biological products, etc., free from contamination with living bacterial, fungal, or viral organisms; or designating an organism not capable of growing or multiplying.

sterile concentrated milk. Fresh whole milk from which most of the water has been removed, resulting in reduction of bulk and, at the same time, sterilization. When water is added to the concentrated product, the reconstituted milk simulates fresh, pasteurized, homogenized milk.

sterility. 1. The condition of being sterile, infertile, or incapable of reproducing. 2. The inability of a fruit plant to mature fruits with viable seeds. 3. The inability of an animal or fowl to produce offspring. *Cf. fertility.*

sterilization. 1. The act or process of sterilizing, of destroying all germ life. 2. The condition of being sterilized.

sterilize. 1. To remove or destroy all living organisms on or in an object or material, as by treating with heat, chemicals, or mechanically. 2. To render incapable of producing offspring, as by removing sex organs or inhibiting their function.

sterilizer. Any of several devices used for destroying all living organisms, such as germs, molds, and bacteria.

sterilizing solution. Any chemical solution capable of destroying bacteria or other organisms.

sternum. The breastbone of an animal.

sterol. Any of a group of solid cyclic alcohols, such as cholesterol and phytosterol, with wide distribution among animals and plants. The *sterols* are neutral and comparatively stable substances which occur partly in the free condition and partly esterified with higher fatty acids. They are of great biological importance since irradiation of some *sterols* leads to the formation of a form of vitamin D.

stevia. Fragrant piqueria.

Stewart's disease. Bacterial wilt of corn.

stewing chicken. A matured female chicken of any of the heavy breeds which is too mature for roasting.

stick. 1. A piece of wood, of indefinite size, that is long as compared with its diameter or cross section. 2. A thick, sirupy by-product obtained by evaporating the tank water produced in the processing of packing-house tankage. *Stick* is usually mixed with dried, low-grade tankage and used as a fertilizer, or is added to wet rendered tankage to be used as livestock feed. 3. To stab or cut with a knife, as in animal slaughter, to sever the jugular veins to cause free bleeding; also with poultry, to pierce the brain with a narrow-bladed knife to cause feather follicles to relax for ease in plucking feathers.

sticker. 1. A substance added to fungicides and insecticides to improve adherence to plant surfaces. 2. A strip of wood placed between the courses of lumber in a pile and at right angles to the boards to facilitate air circulation. 3. A person whose job is to stick animals in a slaughter house. See *stick: 3.*

sticker stain. a. A brown or blue stain on wood caused by fungi that occurs in air-seasoning of lumber where the stickers rest on the faces of the boards. b. A brown, chemical stain which occurs on and beneath the surface of portions of a board that are in contact with stickers.

sticking agent. See *sticker: 1.*

sticking knife. A knife used in the slaughtering of animals or fowls. It varies in size and shape depending upon the type of animals being slaughtered: *e.g.,* a poultry-sticking knife has a blade about ¼ in. wide and 3 or 4 ins. long.

stick lac. The natural form of the hard resinous substance, lac, which is secreted by a species of scale insect in India and Burma. *Stick lac* is collected from the branches and twigs of the trees and, after being purified by a melting process, is stretched into thin sheets. After drying, these sheets are broken into thin, small flakes to form the flake lac of commerce.

stickseed. Any plant of the genus Lappula, family Boraginaceae. Species are weed plants having bristly, adhesive fruit, and are found in abandoned fields and waste places in the North Temperate Zone.

stick-tight. See *beggarticks.*

sticktight flea. *Echidnophaga gallinacea* (Westw.), family Pulicidae, an insect pest of poultry in the southern part of the U.S. The fleas attach themselves to the comb, face, ear-lobes, and wattles of fowls by their mouth parts, engorging themselves with the blood of the host. They are found in clusters and, as their name implies, are removed with difficulty. Also called *southern chicken flea.*

sticky cheese. A body or texture defect of cheese in which cheese clings or sticks to the trier or cutting knife. It is caused by the presence of too high a moisture content. *Cf. crumbly: 4.*

sticky gypsophila. *Gypsophila viscosa,* family Caryophyllaceae, a perennial herb grown for its pink flowers. Native to western Asia.

sticky point. The moisture content at which a well-mixed, kneaded soil barely fails to adhere to a polished nickel or stainless steel surface when the shearing speed is 5 cm. per second.

stiff. Difficult to bend or stretch; inflexible; not pliant; rigid; firm; as *stiff* wood.

stiff bushpoppy. *Dendromecon rigida,* family Papaveraceae, an evergreen shrub grown in warm areas for its yellow flowers. Native to North America.

stiff-lamb disease. An ailment of young suckling lambs, generally occurring between one and eight weeks of age, characterized by stiffness, staggering, or inability to walk or to nurse. Because it is associated with a deficiency of vitamin E, supplements of this vitamin are frequently used as a preventive and treatment. Also called *white-muscle disease.*

stiff-leg derrick. In rigging, an implement for hoisting or lowering heavy loads equipped with a hoisting tackle suspended from the end of the boom to handle the load and a tackle connecting the boom and the mast for raising or lowering the boom. The mast is held in an upright position by inclined wood or steel struts, called stiff legs, extending from the top of the mast to the sill.

stifle. The joint next above the hock (tarsus) in the hind leg of animals; comparable to the knee joint of man. Also called *stifle joint.*

stifled. Of, or pertaining to, a horse which has the patella of the stifle out of place.

stifle joint. See *stifle.*

stigma. That part of the pistil (usually the summit) to which pollen adheres, effecting pollination.

stigmonose. A disease of leaves and other tissues (stems, fruits, etc.) exposed to the air in which there are produced small, discolored spots, mostly light green or yellow, sometimes strongly colored, caused by the reaction of the plant tissues to the salivary secretions of sucking insects (aphids, etc.). These reactions are to be distinguished from the effects of the transmission of virus diseases by the insects concerned. (The word was invented by Ernst A. Bessey.)

stilbestrol. See *diethylstilbestrol.*

stile. A set of steps, ladder, etc., which allows passage of a person over or through a fence or wall while at the same time forming a barrier for livestock.

still. A vessel in which a liquid or semisolid substance is submitted to distillation; *i.e.,* its more volatile constituents are evaporated and led to another vessel or apparatus in which they are condensed. Such devices are used by mint growers in the production of peppermint and spearmint oil. (Sometimes the term is used to include the whole apparatus.) See *distill; distillation.*

stillage. The spent mash resulting from the distillation of alcohol which is the material from which distillers' grains and distillers' dried solubles are produced.

still-born. Born lifeless; dead at birth.

still house. A building in which distillation is carried on; a distillery.

stilling basin. An open-top structure or excavation at the foot of a waterfall, drop structure, or rapids, used to dissipate the energy of the descending current.

stilling well. A protected, miniature well, beside and connected with a body of water, used for accurately determining the changes in water level. It is usually used in connection with the measurement of discharge water in a canal, stream, river, or flume.

Stilton. A hard, mild, blue-veined, cows' milk cheese which is considered by many people to be the finest English cheese.

sting. 1. The resulting wound or welt caused by the sharp-pointed organ of certain insects, scorpions, etc. Also called *stinger.* 2. The smart or irritation produced by touching certain plants: *e.g.,* the nettle. 3. The puncture or wound in the skin or rind of fruit, etc., made by the ovipositor or mouth parts of an insect.

stinging ant. See *fire ant; imported fire ant; little fire ant; southern fire ant.*

stinging nettle. See *great nettle; dog nettle: 1; risky treadsoftly.*

stink bug. See *squash bug; green stink bug.*

stinkgrass. *Eragrostis cilianensis,* family Gramineae, a weedy, annual grass with disagreeable odor when fresh, occurring in cultivated ground, fields and waste places from Maine to Washington, south throughout the U.S. and Mexico to Argentina. Horses have been reported poisoned by eating *stinkgrass* in large quantities over a considerable period of time.

stinking smut. See *bunt of wheat.*

stinking willie. Ragwort groundsel.

stinkweed. Field pennycress.

stion. A plant which consists of a stock and cion growing in combination.

stionic. Designating a plant which consists of a stock and cion growing in combination.

stipe. 1. The stalk-like support of a pistil; the leaf-stalk of a fern frond. 2. In mycology, the stem which supports a cap or pileus, as the stem of a mushroom or toadstool.

stipule. Either of the two appendages at the base of the leaves of many plants.

stirk. A 6- to 12-month-old heifer.

stirrup. An attachment to a riding saddle serving as an aid in mounting and as a foot rest for the rider, made of metal, wood, or heavy leather in many different styles.

stir the soil. The common practice in agriculture of turning or mixing the upper layers of the soil by some tool or implement, such as a spade or plow, for improving its physical conditions, to admit greater air and moisture penetration, and to control weeds.

stitch. 1. A balk or narrow ridge of land in a field, as a plowed or unplowed strip between two furrows. 2. A single insertion of the needle (in the pump method) to introduce a curing pickle into a ham. 3. To sew up a bag of seed, grain, flour, etc.

stock. 1. *Mathiola incana,* family Cruciferae, a

biennial or perennial herb grown for its blue, purple, reddish, or yellowish, fragrant flowers. Native to southern Europe. Also called *Brompton stock; castle gillyflower.* See *Grecian stock.* 2. A plant or plant part upon which a cion is inserted in propagation. See *nursery stock.* 3. Livestock; domesticated farm animals. 4. Material held for future use or distribution. 5. Material destined to be wrought into finished products, as crate *stock,* barrel *stock,* cider *stock,* etc. 6. A plant or plant part that furnishes cuttings for propagation. Also called *stock plant.* 7. The main stem or trunk of a plant. 8. A rootstock (rhizome). 9. The original type from which a group of plants or animals has been derived. 10. The base or handle of a whip. 11. The stump of a tree (after the tree is felled). 12. (plural) A small enclosure in which an animal is secured in a standing position during shoeing or an operation. 13. To provide or supply with livestock, as to *stock* a pasture or range with cattle, sheep, etc. 14. To assemble a supply of materials or commodities. 15. Of, or pertaining to, livestock, as *stock* barn, *stock* feed, etc.

stock barn. A building used primarily for the housing of livestock.

stock beet. Mangel wurzel.

stock bull. In marketing, an old or no longer useful breeding bull sent to the market for slaughter.

stock car. A slatted or open freight car for the handling and shipping of livestock. Some are double-decked for handling hogs and sheep.

stock cattle. Usually young steers or cows, light, thin, lacking in maturity and finish. Also called *stockers.*

stock country. Any region of extensive grasslands suitable for livestock grazing, as the western and southwestern range areas of the U.S.

stock cutting. See *stem cutting: 1.*

stock driveway. See *driveway: 1.*

stocked legs. A swelling of the legs of livestock, especially horses, below the knees and hocks, resulting from prolonged confinement and lack of exercise. Also called *stocking.*

stocked quadrat. A census plot (*q.v.*) in which members of a plant species are counted.

stocker. 1. In marketing, a meat-producing animal capable of additional growth and finish and usually considered to be thinner than a feeder. 2. Range cattle purchased for stocking a farm, in contrast to feeders.

stocker cattle. See *stocker.*

stock farm. An enterprise which is concerned chiefly with the production of livestock, and sometimes with the production of farm crops and pasture for the maintenance of such livestock.

stock feed. Any feed suitable for livestock, especially prepared feeds or supplementary feeds.

stock fence. Any fence that will confine livestock.

stock food. A mixture of different feeds, mineral, and medicinal compounds, fed to livestock to improve health and vitality.

stock fount. A watering device, usually consisting of a small receptacle or water cup, which is automatically kept full by a float valve, or filled when a valve is operated by the pressure of the muzzle or snout of the animal.

stock hog. A young hog capable of additional growth before being finished for market.

stock horse. 1. The common saddle horse of the cattle ranches of the western range (U.S.), which resulted from the crossing of many different breeds of improved horses with the Spanish horses brought into Mexico in 1519. A *stock horse* must be sturdy, agile, surefooted, and an easy keeper. See *steer-roping horse.* 2. A stallion used for breeding.

stock horse class. In the show ring, a classification for range saddle horses.

stockinet. A knitted, coarse, elastic, netlike type of cotton cloth, similar to stocking material, which is used as a cover for fresh-killed meat to protect it from dirt, from drying out, and from changing color. It is also used as a bandage material.

stocking. 1. The relative number of livestock per unit area for a specific time. In range management, the relative intensity of animal population, ordinarily expressed as the number of acres of range allowed for each animal for a specific period. 2. In wildlife management, the density of animal population in relation to carrying capacity. 3. In forestry, the density of a stand of trees, such as well-stocked, overstocked, partially stocked. 4. See *stocked legs.*

stocking leg. A white leg on a dark-colored horse.

stockman. A person who is engaged primarily in the production of livestock in contrast with a farmer.

stock melon. A variety of the watermelon (*q.v.*) grown, usually in cornfields, for livestock forage and for preserves. See *citronmelon.*

stock oil-emulsion. A fluid or paste containing more than 65 percent but less than 90 percent oil emulsified with a small amount of water. It is diluted in water for use as an insecticide. Also called *oil emulsion.* See *emulsion.*

stock pen. Any pen used for confining livestock. Commonly a pen at a shipping point or stockyard in which cattle, sheep, swine, etc., are kept prior to shipping or slaughtering.

stock penning. An old way of manuring land, practiced in Florida and southern Georgia (U.S.), in which livestock are kept in a small enclosure of an acre or two. When the land is sufficiently manured, the pen is moved to a new location.

stock pile. A quantity of some material kept for future sale or use.

stock plant. Any plant whose parts (slips, cuttings, etc.) are used for propagation. Also called *stock.*

stock rooting. In plant propagation, the rooting of cuttings which are taken from stock plants.

stock route. See *driveway: 1.*

stock-rustlers. Cattle and/or horse thieves.

stocks. See *stock: 12.*

stock saddle. Western saddle.

stock salt. Common sodium chloride, granulated, in blocks, or in cakes, used for salting livestock. See *rock salt.*

stock scale. A large platform scale, often provided with gates or panels, on which livestock or other objects can be weighed.

stock show. A display of livestock for public exhibition, for rating by judges, or for inspection by prospective buyers.

stock solution. A concentrated solution from which a portion is taken and diluted as needed, as a spray, etc.

stock tank. Any structure holding water for livestock. Commonly it is a deep trough, bowl or tank equipped with a float valve or other means of controlling the water level. Also called *basin.*

stock-tank warmer. A stove or other heating device used to raise the temperature of tank water to prevent freezing during the winter and thus to induce the animals to drink a normal amount.

stock tonic. Various medicinal preparations which are sold as condition powders, tonics, etc., and purported to improve the health and vitality of livestock. (Unless used carefully the results are often unsatisfactory.) See *conditioner: 1.*

stock water development. The development of new or improved sources of stock water supplies, such as wells, springs, and ponds, together with storage and delivery systems.

stock-watering tank. Stock tank.

stocky. Short and heavy bodied; sturdily built.

stockyard. 1. A yard for keeping or holding livestock. 2. A series of pens or yards where market animals are collected for sale. It may be only a pen or two along a railroad siding in a small town, or it may be as extensive as the great *stockyard* systems of Chicago.

stockyard company. The company that owns and controls the land, yards, and equipment for unloading and handling livestock at the larger livestock markets.

stockyard fever. See *equine influenza; hemorrhagic septicemia.*

stockyard pneumonia. Hemorrhagic septicemia.

Stoke's aster. Stokesia

stokesia. *Stokesia laevis,* family Compositae, a hardy, perennial herb grown for its interesting foliage and large, lavender-blue flowers. Native to America.

stolon. 1. Stems that, as they grow, bend over with their weight so that the tips touch the soil, often continuing growth along and on the soil. Roots may be produced at nodes touching the soil. 2. See *runner: 1.*

stoloniferous. Having or developing stolons; spreading by stolons.

stoma (plural **stomata**). An opening in the epidermal layer of plant tissues which leads to intercellular spaces. These small openings may open or close, depending upon climatic conditions, by means of guard or bulliform cells, etc., and are necessary to photosynthesis, transpiration, etc. Also called *breathing pore.*

stomach. In man and certain other vertebrate animals, a large saclike organ into which food passes from the esophagus or gullet for storage while undergoing the early stages of digestion. In true ruminants: *e.g.,* the bovine, the *stomach* is comprised of four compartments; the rumen (paunch), reticulum (honeycomb), omasum (manyplies), and abomasum (true *stomach*). In birds there are commonly two such compartments: the proventriculus (glandular fore-*stomach*) and the gizzard (muscular posterior-*stomach*). In most invertebrates, there are analogous parts that function as a *stomach.*

stomach ball. An indigestible mass of hair, fibrous material, or the like, which has been formed and is retained in the stomach or paunch of an animal, particularly in the paunch of cattle, sheep, and goats. See *hair ball; bezoar.*

stomach bloat. Acute indigestion in horses.

stomach-dirt. An accumulation of dirt in the stomach; a commonly occurring condition in horses and cattle grazing on loosely implanted vegetation.

stomach insecticide. Any insecticide that becomes effective only if taken into the alimentary tract. It is generally applied to control chewing insects.

stomach poison. Any material that when taken through the mouth into the alimentary tract or stomach causes sickness or the death of the insect or animal. See *stomach insecticide; rodenticide.*

stomach tube. A flexible, rubber tube which when passed through the esophagus or nasal passage into the stomach permits the introduction of treatments, or removal of gastric contents.

stomach worm. Any parasitic worm found in the stomach of animals.

stomata. The plural of stoma *q.v.*

stomatal transpiration. See *transpiration.*

stone. 1. In land description, a detached rock fragment on the surface of, or imbedded in, the soil. 2. In soil survey mapping, a rock which is greater than 10 ins. in diameter or more than 15 ins. in length. See *boulder.* 3. The hard seed of some pulpy fruits, as peach, plum, etc. See *clingstone; freestone.* 4. A unit of weight, 14 lbs. (avoirdupois, British). 5. See *testes.* 6. To take the *stone* or pit out of *stone* fruits.

stone boat. 1. A flat-bottomed, sledlike device used for short hauls and heavy loads over soft or wooded land not accessible to wagon or truck. It is commonly used for removing stones from fields. 2. A similar sled or drag on which stones of known weight are placed to test the pulling ability of a team of horses or oxen.

stone brood. A brood of bee larvae which has been killed by the fungus *Aspergillus flavus,* family Moniliaceae, and become very hard and mummified.

stone cell. In plant anatomy, a metamorphosed, parenchyma cell which is short, approximately isodiametric, and characterized by thick, heavily lignified walls and simple, commonly ramiform pits.

stone clover. See *rabbitfoot clover.*

stone crock. A cylindrical, thick-walled, glazed ceramic vessel available in capacities ranging from 1 pint to 20 gallons or more; used as a container for lard, butter, pickles, sauerkraut, meat-pickle solutions, etc.

stonecrop. Any plant of the genus Sedum, family Crassulaceae. Species are mostly low-growing, chiefly perennial, succulent herbs particularly adapted to the rock garden, but a few of them can be used in the flower border. *Stonecrop* and *sedum* are used interchangeably. See *goldmoss stonecrop; liveforever stonecrop; Texas stonecrop; white sedum.*

stone fence. A wall-like, stone enclosure of farm fields usually laid up without the use of mortar. The stones are commonly gathered from the fields to be enclosed. (In the earlier period of agricultural development in the U.S., it was a common form of fencing in areas of stony soils, especially in New England.)

stone fruit. A fruit with the seed or kernel surrounded by a hard endocarp (or stone) within the pulp or flesh, as a plum, peach, cherry. See *drupe.*

stone fruits bacterial canker. Bacterial canker of stone fruits.

stone fruits bacterial spot. Bacterial spot of stone fruits.

stone fruits brown rot. Brown rot of stone fruits.

stone fruits Coryneum blight. Coryneum blight of stone fruits.

stone fruits ring spot. Ring spot of stone fruits.

stone fruits virus. Any virus associated with stone fruits of the genus Prunus.

stone hammer. A hammer formed with a face at one end and a thick, blunt edge at the other; used for breaking stones.

stone lime. Calcium oxide (*q.v.*), known under several commercial names, as lime, burned lime, quicklime, caustic lime.

stone quarry. An excavation where stone is blasted or cut from bed rock for building or other purposes; a stone pit.

stones. See *concretion; urinary calculi.*

stone walling. 1. Building stone walls. 2. Wall of stone. 3. Material used for stone walls.

stoniness. An abnormal development of grit cells, as in the pear.

stony land. Land which contains enough stone, either detached fragments in or on the soil or rock outcrop, to interfere with or completely prevent tillage.

stony pit of pear. A virus disease of pear characterized by dark green areas just under the epidermis of the newly formed fruit; by deeply pitted and deformed, mature fruit; blisters on two-year old twigs; a ribbed appearance of the trunk; and a veinlet chlorosis of the leaves.

stook. A shock of corn or small grain.

stooked. Designating a grouping of four or more grain bundles set upright together into a shock or stook.

stool. 1. Fecal material; evacuation from the digestive tract. 2. To develop several stems from the crown of a plant or from a stump.

stooling. The development of more than one stem, or culm, on a single plant. Also called *tillering.*

stool layering. Mound layering.

stool out. Of a plant, to produce shoots from the base or crown.

stool shoot. A sprout produced from a bud developed in the tissues of a wound callus on a stump.

stoop labor. Hand labor used in planting, thinning, hoeing, or picking of certain crops, as sugar beets, cucumbers, strawberries, in which the worker has to work in a stooped position.

stopback. A stunting disease of young peach and pear trees caused by the tarnished plant bug (*q.v.*). The end bud of the primary shoot dies and the lateral shoots are thus forced to develop to an abnormal size.

stop bud. Crown bud.

stopping. The stoppage of terminal growth by pruning, disbudding, or pinching off terminal shoots.

stop planks. See *flashboard.*

storage. 1. Placing or holding in reserve in usable form and condition, as the *storage* of nutrients by plants in roots, for future use; the *storage* of honey by bees, during periods of abundance. 2. The holding of products for future sale or use, as in a granary, root cellar, cold *storage* warehouse, etc.

storage bin. A compartment, room, or small structure in which products are held for current or future use, as a grain bin, a potato bin.

storage eggs. Generally, eggs held under refrigeration for more than 30 days. (Usually, spring eggs of good quality are held until late summer or fall for price advance, commonly for about 6 months. Eggs held under refrigeration for 30 days or less are sold according to their grade and are not regarded as *storage eggs* but are sometimes known in the egg trade as *refrigerator eggs.)*

storage flavor. A stale or musty flavor resulting from poor ventilation, faulty temperature, or unsanitary conditions in which some food products are held.

storage house. Any building used for the holding of farm products, particularly a well ventilated building for the storage of fruits and vegetables.

storage organ. Any plant part in which elaborated food materials are stored.

storage-packed eggs. Selected eggs packed especially for storage: ordinarily, eggs of good quality, from which small, checked, and washed eggs have been sorted out, packed in new cases with new flats and fillers of the type prescribed by trade specifications.

storage pit. Cold pit.

storage scab. Black specks or round, irregular spots that may occur on apples in commercial storage at low temperatures due to infection by the apple scab fungus *Venturia inaequalis,* family Mycosphaerellaceae. *Storage scab* is favored by high humidity, high storage temperatures, and delay in harvesting and storing. The fungus occurs on the fruit in the orchard but is not apparent at the picking time.

storage spot. Certain shallow, sunken spots which appear on citrus fruits stored under unfavorable conditions of temperature and humidity. The spots gradually turn brown, especially on exposure to warmer temperature.

storage tank. A cylindrical tank of large size, commonly constructed of wood or stainless steel, used for storing water or liquid, dairy and food products, or a galvanized metal or wood tank used for storage of water on the farm.

storax. 1. A fragrant, dull-colored gum which exudes from the Oriental sweetgum, *Liquid-ambar orientalis,* family Hamamelidaceae. It is used in the manufacture of pharmaceuticals and perfumes. Also called *styrax.* 2. See *snowbell.*

store. To place products in some satisfactory place for safe keeping for future use, as to *store* grain in bins, goods in warehouses, etc.

store room. Room in a storehouse; a repository; a room in which articles are kept in reserve.

storing. The holding or keeping of excess products in reserve.

storksbill. See *heronbill; alfileria.*

storm. A violent disturbance of the lower atmosphere, frequently accompanied by wind, rain, snow, sleet, or hail and usually by lightning and thunder during summer.

storm cellar. An underground structure designed to protect people from tornadoes and severe storms.

stormflow. Runoff from a storm.

storm-proof. 1. Designating varieties of cotton whose open bolls hold cotton well during storms or periods of rainy, windy weather. See *storm resistance.* 2. Designating any structure, roof, etc., which withstands heavy weather with little or no damage to it or contents stored under or within.

storm resistance. A quality possessed by certain varieties of cotton which have moderately firm locks that cling together well and bolls that hang down so that the bur tends to form a roof over the locks, thus providing drainage and protection from damage by rain water. See *storm-proof: 1.*

storm sewer. A sewer which carries off only surface runoff water and is therefore different in function from a sanitary sewer.

stormy. 1. Designating an unsettled condition of the weather. 2. Designating a gassy fermentation in milk caused by spore-forming anaerobic bacteria.

stot. See *steer.*

stove. 1. Any device in which fuel may be burned for cooking or heating. 2. To expose to the fumes of burning sulfur in order to bleach damp skeins of yarn or other woolen material, etc.; a simple method used since earliest times.

stovehouse. 1. The warmest part of the greenhouse. 2. A greenhouse. Also called *warmhouse.*

stover. 1. The stem and leafy parts of corn fodder after the ears have been removed. 2. Sorghum forage.

Stracchino. Any of the several types of whole-milk cheese made in Italy. The best known is Stracchino di Gorgonzola.

straddle. 1. To stand with the legs, ends, or parts staggered, as the spokes of a wagon wheel where they join the hub. 2. To mount a horse. 3. To stand with one foot on either side of a fence, row, etc. 4. To cultivate on both sides of a row simultaneously. 5. See *spread: 1.*

straggler. 1. An animal that wanders or strays from a flock or herd. 2. A plant, branch, etc. that grows irregularly.

straight-away can washer. A mechanical type of automatic washing device for milk cans in which the cans are carried straight through the washing process by means of a ratchet drive.

straight bar bit. A single, straight, smooth, rounded bar of iron used as a bit on a horse bridle. It is the easiest on the horse of the many bits in use.

straightbred. Designating an animal with a straight line of ancestry or pedigree within a recognized breed, such as a registered Thoroughbred horse or a purebred Jersey cow.

straight cactus dahlia. A type of dahlia which has double flowers with the margins of most of the floral rays rolled toward the lower side for at least half their length and the rays nearly straight with their tips slightly incurved or recurved.

straight head of rice. A disease of rice in which heads of affected plants develop slowly, remain green longer than normal, stand erect when normal heads are drooping, and do not produce seed. It is apparently caused by soil organic material that in certain stages of decay interferes with normal root development.

straight hocks. An undesirable conformation in the hind legs of a horse which predisposes it to foot injury and malformation of the hock. Also called *upright hocks; post-legged.*

straight pasterns. An undesirable lower leg conformation of a horse which predisposes it to foot injury. Also called *upright pasterns; stilty.*

straight-run. In marketing, designating eggs which are sold without being graded as to size or quality.

straight-run chicks. Day-old chicks which are unseparated as to sex.

straight shoulders. An objectionable body conformation in horses in which the angle of the shoulder blade to the lower leg is straight. Also called *upright shoulders.*

straight topline. In livestock judging, the back of an animal that is level from front to rear. It is a desirable quality, especially in cattle and sheep.

straight underline. In livestock judging, the bottom line of an animal that is relatively level as viewed from the side; not cut up at front or rear. It is a desirable quality, especially in cattle.

strain. 1. A group of plants of common lineage which, although not taxonomically distinct from others of the species or variety, are distinguishable on the basis of productiveness, vigor, resistance to drought, cold, or diseases; or of other ecological or physiological characteristics. 2. A group of individuals within a breed which differ in one or more characters from the other members of the breed: *e.g.,* the Milking Shorthorns or Polled Herefords. 3. An organism or group of organisms which differs in origin or in minor respects from other organisms of the same species or variety. 4. A virus entity whose properties and behavior indicate relationship to a type virus and are sufficiently constant to enable the entity to be recognized whenever isolated. 5. A severe muscular effort on the part of a draft or other animal which may result in muscle, ligament, or other damage. 6. The condition of an overworked part operating above optimum load, such as a belt or motor operating under *strain.* 7. To filter a liquid and free it of impurities by passing it through some medium or fabric which can retain the solid matter and allow the liquid to pass.

strained. 1. Designating a liquid that has been passed through a screen or other medium to remove sediment or other solid matter. 2. Designating the muscular damage and other results following extreme muscular effort by animals.

strained honey. See *extracted honey.*

strained tomatoes. The product which is obtained by straining sound, ripe tomatoes, raw or cooked, through a screen to remove skins and seeds.

strainer. Any device through which any liquid is passed for purification or to separate it from solid matter, as a screen or cloth used to strain a liquid; a filter.

straining. 1. Infection in the genitalia of animals incidental to parturition, including subsequent infection. 2. Over-exertion on the part of animals. 3. Overworking motors, engines, etc., on heavy loads.

strake. An iron band fastening wheel fellies to each other; it differs from the tire in being made of separate pieces.

strand. 1. Each one of the separate units of fibers, threads, wires, etc., which, when twisted or woven together, may make up a rope, cable, fence wire, etc. 2. Any single filament.

stranger. A sheep of a neighbor's or another's run or range (Australia and New Zealand).

strangles. An acute, contagious, febrile disease of horses and mules characterized by catarrhal inflammation of the mucous membranes of the nasal passages and pharynx. Also called *avives; bastard strangles; distemper of horses; shipping fever of horses.*

strangleweed. Dodder.

strap. 1. A narrow piece of leather, or other flexible material which is used to retain something in place or to surround and hold things together. 2. A driving belt of leather, rubber, etc. 3. A short piece of rope spliced together at the ends, for use in slinging things together, for hoisting by a tackle. Also called *strop.* 4. A narrow plate of thin iron used to hold different wooden parts together, being secured to each piece by nails, bolts, or screws, as in securing the timbers of a roof, the sides of a box, or framing generally. 5. The leaf of a hinge. 6. The stirrup-shaped part of a clevis or shackle.

strap leaves. See *tobacco frenching.*

Strasburg acre. A measure of land which is about one-half of the English and American acre.

strata (singular: **stratum**). 1. In geology, layers or beds of rock which are nearly lithologic units. 2. In ecology, different vertical layers of vegetation, as ground cover, herbaceous growth, shrubs, and trees.

stratification. A method of storing seeds at a temperature between 35°F. and 45°F. in alternate layers or mixed in moist sand, peat moss or other medium, as a means of breaking the rest period.

stratified. Composed of, or arranged in, strata or layers, as *stratified* alluvium. (Strictly, the term is applied to geological deposits and not to soils proper.)

stratify. To expose seed to stratification *q.v.*

stratum. The singular of strata *q.v.*

straw. 1. The stems and leaves of grain crop plants which remain after threshing. 2. The fallen leaves of pines and similar trees.

strawberry. Any of the essentially stemless, perennial herbs of the genus Fragaria, family Rosaceae. The *strawberry* has trifoliate leaves, white flowers, and long slender runners by means of which it naturally propagates itself. In its many horticultural varieties, it is widely grown and much prized for its delicious red fruit which in a strict sense is a much enlarged, juicy, very fleshy receptacle bearing achenes. See *chiloe strawberry; European strawberry; hautbois strawberry; Virginia strawberry.*

strawberry barrel. A large barrel which is filled with rich soil and in which holes about 1½ ins. in diameter are cut in many places around the sides. In these holes strawberry plants are set. One such barrel will serve for two dozen or more plants and is useful as well as ornamental in a small garden or yard.

strawberry black root. A strawberry disease which primarily affects the cortex and later the vascular system. Several different fungi and nematodes have been associated with the condition. It is also aggrevated by winter injury. Also called *critical black root of strawberry.*

strawberry-blossom weevil. See *strawberry weevil.*

strawberry bud nematode. *Aphelenchoides fragariae,* family Aphelenchidae, a nematode which infests strawberries from Cape Cod to Virginia, U.S., causing strawberry cauliflower: small, narrow, twisted leaves and death of buds.

strawberry clover. *Trifolium fragiferum,* family Leguminosae, a perennial herb, similar to white clover, which has pinkish to reddish heads. It grows well in damp places, and in the U.S. is grown mainly in the northwestern states for forage and soil improvement. Native to Europe and Asia.

strawberry comb. A chicken's comb whose shape resembles the outline and surface of a strawberry.

strawberry crinkle. A virus disease of strawberry characterized by a crinkling or wrinkling of the leaves and the appearance of small, yellow spots which expand as the leaves grow, resulting in a flecked appearance. The affected plants are reduced in vigor and production.

strawberry crown borer. *Tyloderma fragariae* (Riley), family Curculionidae, a dark brown snout beetle which lays its eggs at the base of strawberry leaves. The yellowish grubs tunnel through the crowns, killing or stunting the plants.

strawberry crown miner. *Aristotelia fragariae*

Busck, family Gelechiidae, a moth, the larva of which burrows in crowns of plants, causing stunting and poor foliage.

strawberry crown mite. See *cyclamen mite.*

strawberry crown moth. *Ramosia bibionipennis* (Bdv.), family Aegeriidae, a wasplike moth whose larvae bore into the crowns of strawberry plants causing the foliage to turn yellow, often killing recent transplants and seriously weakening older plants.

strawberry fruitworm. See *omnivorous leaf tier.*

strawberry geranium. Strawberry saxifrage.

strawberry gray mold. A fungous disease of strawberries caused by *Botrytis chinerea,* family Moniliaceae, characterized by loss of buds and blossoms and a gray mold rot of the fruit.

strawberry groundcherry. *Physalis alkekengi,* family Solanaceae, a hardy, perennial herb grown for its showy, orange-red, inflated calyx. Native to Europe and Asia. Also called *winter cherry; Chinese lantern-plant.*

strawberry guara. Cattley guara.

strawberry leaf blight. A fungous disease caused by *Dendrophoma obscurans,* family Sphaerioidaceae, characterized by round, elliptical, or V-shaped lesions on the leaves. The fungus is also associated with the sepals. It is most noticeable in midsummer.

strawberry leaf roll. A minor, virus disease of strawberries which causes a downward rolling of the leaflets, most pronounced in the basal part. The leaf surface becomes ruffled, rugose, and pale green.

strawberry leaf roller. *Ancylis comptana fragariae* (W.&R.), family Olethreutidae, a small gray moth whose larvae fold and roll together the strawberry leaves, webbing them together and feeding within.

strawberry leaf scorch. A fungous disease caused by *Diplocarpon earliana,* family Dermateaceae, in which dark purple spots (which may later coalesce) appear not only on the leaf blades but also on the potioles, fruit pedicels, and on the sepals. Often large areas of the leaf take on a reddish or purplish hue, and badly infected leaves may curl and burn. The disease differs from strawberry leaf spot *(q.v.)* in that the spots never show the white central area of the leaf spot disease.

strawberry leaf spot. A fungous disease caused by *Mycosphaerella fragariae,* family Mycosphaerellaceae, which may first appear on the upper surface of leaves as small, deep purple, somewhat indefinite areas. As the spot enlarges, the central area becomes brown but later turns to a light brown or white spot surrounded by a dark purple zone, giving a bird's-eye effect. When severe the disease reduces yield and quality of berry. See *strawberry leaf scorch.*

strawberry leathery rot. A fungous disease caused by *Phytophthora cactorum,* family Pythiaceae. The fungus attacks berries in the field in all stages of growth and usually occurs only where the berries come in contact with the soil. On ripe fruit, the rot finally results in a mushy condition of the fruit.

strawberry madrone. *Arbutus unedo,* family Ericaceae, a broad-leaved, evergreen tree, which bears clusters of white or pink flowers and fleshy, orange-red, strawberry-like fruit, edible but tasteless. It is grown in warm regions as an ornamental. Native to Europe. Also called *strawberry tree.* See *Pacific madrone.*

strawberry nematode. Any of the extremely small nematodes *(q.v.)* which infests the strawberry, particularly *Aphelenchoides fragariae,* family Aphelenchidae, responsible for spring dwarf, and *A. besseyi* which causes summer dwarf of plants. The symptoms are very similar, the difference being mainly in time of attack. Affected plants may be recognized by the abnormal character of the crown: the bud is small and poorly developed, the development of new foliage is retarded, and the leaves are small, distorted and crinkled. Severely affected plants set no fruit and are worthless.

strawberry powdery mildew. A fungous disease of strawberry leaves caused by *Sphaerotheca macularis,* family Erysiphaceae, which is occasionally severely damaging but is generally of minor importance.

strawberry raspberry. *Rubus illecebrosus,* family Rosaceae, a hardy, near-herbaceous raspberry grown for its fragrant, white, long-sepaled flowers and large, scarlet, sour fruit, edible when cooked. Native to Japan. Also called *balloonberry.*

strawberry red core. Strawberry red stele.

strawberry red root rot. Strawberry red stele.

strawberry red stele. A root-rot disease caused by the fungus *Phytophthora fragariae* which is evidenced by early-spring stunting or dwarfing of the plant and wilting and dying of older leaves. In diseased plants, the small fibrous leaves are discolored, or have disappeared, leaving only the long rat-tail roots. When the central cylinder of the root is exposed by splitting, the core is found to be dark red. Also called *brown stele; strawberry red root rot.*

strawberry roan. The coat-color of a horse, predominately red or reddish-bay with white hairs rather uniformly intermixed.

strawberry root aphid. *Aphis forbesi* Weed, family Aphidae, a bluish-green aphid that feeds on the roots of strawberry plants, seriously affecting the vigor of the infested plant. The aphids hatch early in the spring and are carried to strawberry roots by the little brown cornfield ants which feed on the honeydew secreted by the aphids.

strawberry root rot. See *strawberry red stele.*

strawberry root weevil. *Brachyrhinus ovatus* (L.), family Curculionidae, a beetle with a short, blunt snout, the larvae of which feed on the fine roots and crowns of plants. In addition to strawberries, they also attack conifers, cane fruit, cucurbits, and various deciduous trees and shrubs. Also called *crown girdler.*

strawberry rootworm. *Paria canella* Wilcox, family Chrysomelidae, a dark brown beetle which feeds on the foliage of many fruit plants, riddling the leaves with small shot-holes. The larvae feed on strawberry crowns and roots. This insect is sometimes a serious pest in greenhouses.

strawberry sawfly. *Empria maculatus,* family Tenthredinidae, a small, dark-colored insect whose larval stage, the strawberry slug, damages the strawberry plant by eating the foliage.

strawberry saxifrage. *Saxifraga sarmentosa,* family Saxifragaceae, a trailing, perennial herb grown as a basket plant or pot plant for its varicolored leaves and white flowers. Native to Asia. Also called *strawberry geranium; Aaron's beard; beefsteak saxifrage; mother-of-thousands.*

strawberry shrub. 1. Any shrub of the genus Calycanthus, family Calycanthaceae, grown for their fragrant flowers. Also called *American allspice.* See *pale sweetshrub; spicebush.* 2. *C. floridus,* grown for its purplish-brown flowers. Native to southeastern U.S. Also called *Carolina allspice; common sweetshrub; sweetshrub.*

strawberry stem-end fruit rot. A fruit rot characterized by dead calyx caps. It is associated with infections of *Gnomonia fragariae,* family Gnomoniaceae, and *Dendrophoma obscurans,* family Sphaerioidaceae.

strawberry tan rot. A minor, fungous rot of strawberries caused by *Pezizella lythri,* family Helotiaceae, characterized by tan, conical, rotted portions.

strawberry tree. Strawberry madrone.

strawberry weevil. *Anthonomus signatus* Say, family Curculionidae, a snout beetle, the female of which punctures strawberry buds with her long beak to insert eggs, then crawls down and girdles the stem, causing it to wilt. Distributed mainly in eastern U.S. Also called *clipper.*

straw board. Thick paper board or mill board made chiefly or entirely from straw, generally wheat or rye straw. Used in bookbinding, etc.

straw boss. A subordinate foreman in a logging camp.

straw carrier. A straw elevator at the end of the thresher which lifts the straw onto a rack or into a mow.

strawed. Referring to straw, as a short-*strawed* variety of grain.

strawflower. *Helichrysum bracteatum,* family Compositae, an annual herb grown as an everlasting for its white, orange, red, pink, or yellow flowers. Native to Australia. Also called *everlasting flower; paper flower.*

straw itch mite. *Pyemotes ventricosus* (Newp.), family Pyemotidae, a small mite frequently encountered at harvest time in handling wheat and other small grains. On men they cause a hivelike eruption, often over much of the body, accompanied by a very severe itching. Also called *harvest mite.*

straw loft. A layer of straw supported by an open ceiling, usually wire mesh or slats, which provides insulation and allows air passage for ventilation in poultry or swine houses.

straw pile. A refuse pile which consists of straw and chaff left after threshing grain. If handled and formed into a weather-resisting unit, it is usually called a straw stack *q.v.*

straw rack. 1. A bunkerlike structure usually built in the open cattle yard in which straw may be stored for livestock feeding. 2. A shaker rack or carrier on a threshing machine which lifts the straw as it comes from the cylinder and discharges it at the tail of the machine. The lattice design of the carrier sifts the grain and chaff as it carries the straw over.

straw spreader. A mechanical device attached on the rear of a grain combine which distributes threshed straw over the ground behind the combine.

straw stack. A pile of straw usually stacked when threshing grain to help shed rain and preserve the contents for feeding and bedding livestock.

straw tying. See *green tying.*

straw wine. Any wine produced from grapes that have been partially dried to increase their sugar content, yielding a wine of high alcohol content which may be fortified by adding brandy, if desired. Such wines are produced to a limited extent in France and Spain.

strawy manure. Manure that contains a high proportion of straw, usually in an undecomposed or slightly decomposed state.

stray. An animal that has wandered away from the herd, flock, or farm to which it belongs.

streak. In turpentining, the incision made when a tree is chipped.

streak canal. The duct or canal at the lower end of a cow's teat through which milk is drained from the teat cistern in milking. The opening is closed by a sphincter muscle. Also called *teat canal.*

streak disease of corn. See *corn streak.*

streak disease of sugar cane. See *corn streak.*

streak mosaic. A virus disease of wheat characterized by severe, yellow spotting, streaking, and mottling of the leaves.

streak of brambles. A virus disease of brambles characterized by dark blue or violet-blue markings varying from dots to long streaks on the young shoots in early summer or midsummer. Also called *eastern blue stem; rosette.*

streak of grasses. A virus disease of grasses transmitted mechanically and by the grass mite, characterized by the appearance of greenish-yellow, narrow, linear, intermittent streaks on the leaves followed by chlorisis and a reduction in the yield of grain. Also called *brown leaf blight of grasses.*

streak of sweetpea. A disease of peas caused by several viruses, as spotted wilt and pea streak virus. Brown streaks appear on stems, petioles, and flower stalks.

stream. 1. Flowing water in a natural or artificial channel. It may range in volume from a small creek to a major river. 2. A jet of water, as from a nozzle. 3. A continuous flow or succession of anything: air, gas, liquids, light, electricity, persons, animals, materials, etc.

stream bank. Either of the usual boundaries (not flood boundaries) of a stream channel. See *river bank.*

streambank erosion control. Vegetative or mechanical protection of erodible stream banks. Frequently provided where valuable farm land or highways may be affected.

streambank management. Development or protection of vegetation along stream courses.

streambank wheatgrass. *Agropyron riparium,* family Gramineae, a forage grass found in dry or moist meadows and hills from North Dakota to Alberta and Washington, south to Oregon and Colorado in the U.S.; sometimes used in range reseeding. Native to North America. See *wheatgrass.*

stream bed. In any river or stream, the natural bottom over which water flows.

stream erosion. The erosion caused by a river or stream cutting into its natural banks, usually as a result of excess volume or shift in stream current.

stream flow. That portion of the precipitation which finds it way into natural or artificial runoff channels.

stream gauging. The quantitative measurement of stream flow by current meters, weirs, or other devices at selected locations.

street planting. The practice of planting trees and shrubs along the parkway of streets for shade and ornament.

street tree. Any tree planted along the side of a street for shade and ornament.

strength. 1. The capacity to resist force; solidity or toughness; the quality of bodies by which they endure the application of force without breaking or yielding. 2. Bodily or muscular power; force; vigor. 3. The potency or power of a liquid or other substance; intensity of active properties. 4. The firmness of a market price for a given commodity; a tendency to rise or remain firm in price.

strength of attack. The manpower and equipment with which a fire is attacked.

streptococci. Plural of streptococcus *q.v.*

streptococci mastitis. An udder inflammation which results from infection by streptococcal organisms, mainly by a single species, *Streptococcus agalactiae.* See *mastitis.*

streptococcus (plural: **streptococci**). Any bacterium of the genus Streptococcus, family Coccaceae. Several species, as S. *pyogenes,* S. *scarlatinae,* etc., are responsible for various diseases in man and animals. Certain others, as S. *kefir* and S. *cremoris,* are used in making cheese, etc.

streptococcus infection. 1. Any infection in which streptococci are the causative organisms. 2. The acute form of an infectious disease of chickens, due to infection by *Streptococcus gallinarum* organisms. Turkeys, pigeons, and other fowls are also susceptible. Although the disease is usually fatal, subacute and chronic forms, called sleeping sickness, also exist.

streptomycin. An antibiotic derived from the organism, *Streptomyces griseus.* It is beneficial in controlling certain human as well as plant diseases.

stretcher. A hand-operated mechanical device of several types used in stretching fence, such as barbed or woven wire, to a desired tautness. It consists principally of a stretcher bar, a ratchet device, stretcher arms and a stretcher clamp.

stretcher bar. The main iron bar of a fence wire stretcher.

stretcher clamp. The clamp which holds the wire in place while being stretched with a fence stretcher. See *stretcher.*

striations. In wood anatomy, fine linear streaks or markings in the cell wall.

strickle. Strike-off stick.

strict good middling. A grade of lint cotton which shall be American upland cotton which in color, leaf, and preparation is within the range represented by a set of samples, in the custody of the U.S. Department of Agriculture in the District of Columbia, in a container marked "Original official cotton standards of the United States, American upland, No. 2 or strict good middling, effective August 20, 1936."

strict good ordinary. A grade of lint cotton which shall be American upland cotton which in color, leaf, and preparation is within the range represented by a set of samples, in the custody of the U.S. Department of Agriculture in the District of Columbia, in a container marked "Original official cotton standards of the United States, American upland, No. 8 or strict good ordinary, effective August 20, 1936."

strict good ordinary extra white. A grade of lint cotton which shall be American upland cotton which in leaf and preparation is No. 8 or strict good ordinary, but which is whiter than strict good ordinary.

strict low middling. A grade of lint cotton which shall be American upland cotton which in color, leaf, and preparation is within the range represented by a set of samples, in the custody of the U.S. Department of Agriculture in the District of Columbia, in a container marked "Original official cotton standards of the United States, American upland, No. 6 or strict low middling, effective August 20, 1936."

strict low middling extra white. A grade of lint cotton which shall be American upland cotton which in leaf and preparation is No. 6 or strict low middling, but which is whiter than No. 6 or strict low middling.

strict low middling spotted. A grade of lint cotton which shall be American upland cotton which in leaf and preparation is No. 6 or strict low middling, but which in color is between No. 6 or strict low middling and No. 6 tinged or strict low middling tinged.

strict low middling tinged. A grade of lint cotton which shall be American upland cotton which in leaf and preparation is No. 6 or strict low middling, but which in color is within the range represented by a set of samples in the custody of the U.S. Department of Agriculture in the District of Columbia, in a container marked "Original official cotton standards of the United States, American upland, No. 6 tinged or strict low middling tinged, effective August 20, 1936."

strictly combing. Wool that has a staple length of at least 2½ ins. over the main part of the fleece (to be combed by the English system). The coarser the wool, the longer it must be to be classed as *strictly combing. Cf. French combing.*

strict middling. A grade of lint cotton which shall be American upland cotton which in color, leaf, and preparation is within the range represented by a set of samples in the custody of the U.S. Department of Agriculture in the District of Columbia, in a container marked "Original official cotton standards of the United States, American upland, No. 4 or strict middling, effective August 20, 1936."

strict middling extra white. A grade of lint cotton which shall be American upland cotton which in leaf and preparation is No. 4 or strict middling, but which is whiter than No. 4 or strict middling.

strict middling gray. A grade of lint cotton which shall be American upland cotton which in leaf and preparation is No. 4 or strict middling, but which is more gray in color than No. 4 or strict middling and no darker in color than the dullest bale in No. 4 or low middling.

strict middling spotted. A grade of lint cotton which shall be American upland cotton which in leaf and preparation is No. 4 or strict middling, but which in color is between No. 4 or strict middling and No. 4 tinged or strict middling tinged.

strict middling tinged. A grade of lint cotton which shall be American upland cotton which in leaf and preparation is No. 4 or strict middling, but which in color is within the range represented by a set of samples in the custody of the U.S. Department of Agriculture in the District of Columbia, in a container marked "Original official cotton standards of the United States, American upland, No. 4 tinged or strict middling tinged, effective August 20, 1936."

strict middling yellow stained. A grade of lint cotton which shall be American upland cotton which in leaf and preparation is No. 4 or strict middling, but which is deeper in color than No. 4 tinged or strict middling tinged.

stricture. An abnormal, localized contraction of any passage or duct of the body; a constriction.

stride. The distance from one foot print of a horse to the print of the same foot when it next comes fully to the ground.

Striga. *Striga asiatica,* family Scrophulariaceae, a weed that has newly invaded the U.S., so far being found only in North and South Carolina where it has caused severe injury to corn. It is an obligate parasite only on grasses, and affects corn, crabgrass, Japanese millet, Johnson grass, sorghum, Sudan grass, and others. The plants emerge after a period of feeding underground on the roots of the host. They grow 2 to 15 ins. tall, have red blossoms with a yellow eye, and produce enormous numbers of spore-like seeds. Native to Asia and South Africa. Also called *witchweed.*

strike-off stick. A straight, smooth, hardwood stick used in leveling off the grain in a grain measuring or testing bucket. Also called *strickle.*

strike root. Of a plant, to throw out lateral roots from the main root, crown, or stem, near the surface or just below the ground, which turn down and take root in the soil, bracing and giving strength as well as added nourishment to the plant. Corn is a typical example of a plant that *strikes root.*

striking pan. In sugar making, the last of a series of evaporating pans in which the more concentrated liquor collects. Also called *striking tache.*

string. 1. A cord or line of small diameter. 2. The several fumigating tents which are used by a single crew of men in fumigating orchards. 3. A group of partly broken horses which are assigned to a cowboy or horse trainer for his personal use or for further breaking or training. 4. A group of horses in a pack train. 5. To unroll wire prior to stretching and fastening to posts, etc.

string beans. Any immature beans used in the pod for food. See *kidney bean.*

string binder. A harvesting machine which uses a stout twine to tie the cut grain into bundles or sheaves. See *grain binder; corn binder.*

stringhalt. An abnormal condition in the horse characterized by involuntary, spasmodic lifting of one or both hind legs when walking or trotting. Also called *springhalt.*

stringhaltered. Afflicted with stringhalt *q.v.*

string leaf. Tobacco mosaic.

stringless bean. Any of a number of varieties of kidney bean *(q.v.)* whose unripe, tender pods are free or practically free from fibrous threads.

string out tile. The laying of a number of tiles end to end in the field, prior to placement in the ground; as in a drainage line.

stringy. 1. Resembling string or fiber; containing fibrous strands or elements, as certain vegetables, fruit, meat, etc.; viscid, containing or forming threadlike, slimy, or ropy, strands of milk when poured or ladled from a vessel. 2. Designating a horse afflicted with stringhalt.

stringy milk. Milk that contains tough, stringy clots as it is drawn from an infected udder. See *mastitis.*

strip. 1. A relatively narrow piece, as a *strip* of land. 2. Each of the squirts or streams of milk as taken from a cow's teat in milking. 3. To tear off, as the threads from a bolt or nut, the leaves, husks, suckers, from a plant or plant part, or the hide or pelt from a carcass. 4. To take the last of the milk from a cow's udder by hand milking after the machine milker has been removed.

strip a cow. To remove the last of the milk from each teat of the cow's udder by a stripping rather than a squeezing action. See *strip: 4.*

strip burning. Controlled use of fire to destroy surface material: *e.g.,* an area to be broadcast burned may be successively burned in strips or the method may simply involve the burning of inflammable material along railroads, highways, etc. See *slash disposal.*

strip cropping. The growing of crops in a systematic arrangement of strips or bands which serve as vegetative barriers to wind and water erosion.

strip cup. A small metal cup or vessel with a fine wire strainer or inner liner into which the first streams of milk from each teat are milked from the cow for examination to detect any indication of mastitis infection or any other abnormal condition of the milk or udder.

strip-cup test. See *strip cup.*

stripe. 1. A line or relatively narrow strip of color differing from the adjacent or ground color of a surface. 2. See *bacterial stripe; barley stripe.*

striped alder sawfly. *Hemichroa crocea* (Fourc.), an insect whose larvae feed on alder leaves, and occasionally on birch, eating everything but the midrib and larger veins.

striped blister beetle. *Epicauta vittata* (F.), family Meloidae, an insect which feeds on many vegetable crops and weeds. Its body contains cantharadin, which will blister tender skin.

striped crotalaria. *Crotalaria mucronata,* family Leguminosae, an erect-growing, tall, annual herb with woody stems and yellow flowers, grown in warm climates chiefly for soil-improvement purposes. Native to the tropical areas.

striped cucumber beetle. *Acalymma vittata* (F.), family Chrysomelidae, an insect that is a serious pest of cucumber and other cucurbits. The adults devour the leaves and stems and the larvae feed on the underground parts frequently destroying the root system. Found in central and eastern U.S.

striped feather. A feather having in the central part of the web a longitudinal stripe of color that differs from the color of its edge. Usually the strip extends through the web running parallel with the outer edges and tapering to a point near the lower extremity of the feather. Such feathers are found particularly in the hackle and saddle feathers of some parti-colored varieties of chickens.

striped flea beetle. *Phyllotreta striolata* (F.), family Chrysomelidae, a widely distributed insect pest of cabbage and many other garden plants.

striped garden caterpillar. *Polia legitima* (Grote), family Noctuidae, a generally-distributed cutworm, often abundant in late summer and early fall.

stripe disease of barley. See *barley stripe.*

striped maple. *Acer pensylvanicum,* family Aceraceae, a hardy, small tree, grown for its large, bright green, white-striped leaves and striped branches which are conspicuous in winter. Native to eastern North America. Also called *false dogwood; moosewood.*

striped marigold. *Tagetes tenuifolia,* family Compositae, an annual herb grown as an edging plant for its small size and orange flowers. Native to Mexico. Also called *Mexican marigold.*

striped pipsissewa. *Chimaphila maculata,* family Ericaceae, a low, evergreen herb grown for its leathery, striped, green leaves and occasional white flowers. Native to North America. Also called *dragon's-tongue; spotted wintergreen; rheumatism root.*

striped spirea. *Spiraea superba,* family Rosaceae, a hybrid, deciduous shrub, grown for its light rose or whitish-pink flowers.

striped tulip. A class of tulip with many varieties which are not self-colored but are variously marked, streaked or feathered with other colors, like the bizarre tulip and bybloemen tulip *q.q.v.*

stripe mosaic of wheat. A virus disease of wheat, seed-borne and also transmitted mechanically, characterized by chlorotic, and later by brown, mottled stripes in the leaf blade.

stripe rust. A fungous disease of grasses and cereals caused by *Puccinia glumarum,* family Pucciniaceae, characterized by yellow pustules arranged in long rows or stripes.

stripe smut of grasses. A fungous disease of grasses caused by *Ustilago striiformis,* family Ustilaginaceae, characterized by long, narrow, almost black stripes on the leaves and leaf sheaths. See *smut.*

strip farming. Farming on land where large fields and long slopes are protected from water erosion by plowing on the contour and planting alternate strips to grasses and row crops. See *contour farming.*

strip grazing. Rotational grazing.

stripper. 1. A device or machine used for stripping cotton from the plant at harvest time. See *cotton sled; stripping machine.* 2. A hand-operated implement or a power-driven machine used for gathering grass seed by combing or stripping the seed from the plant stems. 3. A machine for clearing new ground.

stripping. 1. Clearing the ground of brush or other undesirable cover preliminary to planting. 2. The cutting of all marketable timber without regard for future growth. 3. The removing of leaves and/or plant parts from a plant. 4. See *strip: 4.*

stripping machine. 1. A machine that combs or pulls the seed from standing grass in seed harvesting. 2. A machine that gathers cotton from standing plants.

strippings. In milking, the last milk drawn from the cow's udder. Also called *afterings.*

strip rotation. A crop rotation for contour strips within a field, using erosion resisting crops alternating with cultivated crops.

strip sodding. The laying of sod in strips separated by unsodded spaces.

strip steak. The steak cut from the loin strip of a beef carcass.

strip the crop. To remove the crop content of dressed poultry by a downward stroking and massaging of the crop of a bird that is suspended head downward. The crop material drains from the mouth.

strip (belt) transect. A rectangular area used in ecological studies of range vegetation. See *transect.*

strobila. The entire adult tapeworm which has a head (scolex), neck, and a chain of segments.

strobilus. A cone-like reproductive structure consisting of an axis bearing sporophylls; *i.e.*, spore-bearing leaves which are either microsporophylls (stamens) or macrosporophylls (carpels), the latter being open in the Gymnospermae, or closed so as to enclose the ovules in the Angiospermae. See *conifer*.

stroke. 1. The linear distance traveled in one motion by a piston or ram, whether in an engine or pump. 2. A sudden and severe attack, as of apoplexy or paralysis.

stroma. A mass of fungus hyphae, sometimes including host tissues, which contains or bears spores differing from a sclerotium in this respect.

strong. 1. Having a specific quality to a high degree, as *strong* flavor, etc. 2. Characterized by steady or advancing prices, as a market for a particular commodity.

strong and masculine. Indicating a vigorous, healthy male animal with the desired male characteristics.

strong back. Of livestock, the length, width, depth of flesh and back conformation desired in a good animal.

strong breed character. A character common to all members of a particular breed.

strong-caramel. An overheated condition in condensed milk in which a degree of caramelization of the milk sugar has taken place.

strong constitution. Good health and vigor, good size, well-sprung ribs and ample body conformation in the forequarter of an animal.

strong cotton. Hard-bodied cotton that will give good spinning results; not soft, weak, irregular or wasty.

strong fleece. A deep, even fleece over the entire body of the sheep.

strong flour. Flour prepared from hard wheat whose high gluten content makes it suitable for bread making.

strong land. Land composed of clay or silt soils. Such soils generally are durable under cultivation and their fertility is less quickly depleted by cropping than that of sandy soils.

strong loin. A wide and deeply-fleshed loin development, an important characteristic in meat and work animals.

strong muzzle. A muzzle characterized by large, open nostrils and strongly formed mouth, an indication of a good feeder in a cow or steer.

strong pastern. Of livestock, having pasterns of firm, dense bone well attached to adjacent parts. See *pastern*.

strong-rooted. Designating plants with an abundance of roots that penetrate deeply into the soil. See *deep-rooted; shallow-rooted*.

strong wool. See *robust: 1*.

strongyle. Any roundworm of the family Strongylidae. They are parasitic in the organs and tissues of man and various animals often causing severe injuries or death. Some species are specific to certain hosts: *e.g.*, the intestinal *strongyles* (palisade worms) of horses will not infest cattle, sheep, or swine.

strontium. Sr, a bivalent metallic element of the calcium group, present in soils and plants in extremely small amounts but not regarded as essential to normal plant growth. Radioactive *strontium*, a product of atomic and thermonuclear explosions may be present in fall-out left on vegetation which is consumed by man either directly or indirectly, as through cows' milk, thus becoming a possible serious menace to health.

structural profile. Soil profile.

structural timber. Heavy lumber suitable for load-bearing purposes.

structure of milk. Normal milk, secreted by the mammary gland which consists of a serum containing fat globules, lactose, salts, albumin, globulin, and solid particles in a colloidal or finely divided condition in turn consisting of casein and attached calcium phosphate. In addition, as milk comes from the udder it contains some tissue cells, an occasional leucocyte, and perhaps a few harmless bacteria.

Strumella canker of oak. A destructive, slow-growing, perennial canker *(q.v.)*, caused by the fungus *Strumella coryneoidea*, family Tuberculariaceae, primarily on forest oaks, especially the red oak group. Occasionally it affects ornamental plants and other trees, such as hickory, chestnut, beech, and maple. The fungus enters the trunk mainly through dead branch stubs.

strut. 1. A part of a machine or structure used chiefly to hold things apart. In general, any piece of a frame which resists thrust or pressure in the direction of its own length. *Cf. stay; tie*. 2. In an animal, exaggerated step or gait. 3. To walk with affected dignity or exaggerated step.

strutting. Designating the placement of teats on a cow's udder in which the teats are wide apart and hang at an angle to the body of the udder when it is distended.

strychnine. An extremely poisonous vegetable alkaloid which has an intensely bitter taste and is used as poison to kill mice, sparrows, pocket gophers, ground squirrels, etc. It is also used in medicine as a stimulant.

stub. 1. A short blunt projection, as a blunt projecting stem, branch, or root. 2. A tree with broken stem less than 20 ft. 3. The quill portion of a short feather appearing on the shanks or toes of otherwise clean-shanked birds.

stubbing. The heading back of a tree by severe pruning. Also called *dehorning*.

stubble. The basal portion of grain, forage, or some other plants, which remains after the top portion has been harvested.

stubble-berry. Black nightshade.

stubble crop. 1. A crop developed from the stubble of the previous season. 2. A crop sown on grain stubble after harvest for ploughing under as green manure the following spring.

stubble ground. A field from which grain or hay has been harvested. See *stubble*.

stubble moldboard. See *stubble plow*.

stubble mulch. A protective cover provided by leaving plant residues of any previous crop as a mulch on the soil surface when preparing for and planting the succeeding crop.

stubble pasture. A field from which a crop of wheat or other grain has been previously harvested and on which animals are placed to consume the crop refuse as well as weeds and annual grasses that may follow the grain crop.

stubble plow. A plow designed for turning stubble land. It has a short, highly-curved moldboard to bring about the maximum of soil pulverization.

stubble shearing. Shearing a short distance from the skin, as with show sheep.

stubborn disease. A virus disease of citrus characterized by deformed, acorn-shaped fruit, brush-like, stunted growth of twigs and foliage, and irregular blooming. Pink nose and a blue discoloration of the albedo are also associated with this disease.

stubby. Referring to the faulty foot action of a horse in which the toe strikes the ground first, resulting in a stubby step.

stubby root nematode. Any of several species of root nematodes of the genus Trichodorus or *Belonolaimus gracilis,* which feed on root tips of plants causing them to stop growing and to turn brown. Parts of the roots may then die probably because of attacks of secondary invaders. The final result is a reduced root system with many short root stubs. The attacks are particularly damaging to seedlings.

stub pole. A short pole (tongue) with fittings for attaching a wagon or trailer to a truck or tractor.

stub row. A short row of fruit trees occurring when trees are set on contour lines. In a planting, *stub rows* are usually avoided by not adhering strictly to contour lines.

stub-runner furrow opener. A device placed on a planter just ahead of the planting mechanism to open a furrow-like trench in the soil to receive the seed as it is dropped by the planter. It is suited to rough and stony ground. Its sloping forward edge causes residues to slide upward and off to either side.

stuck. Designating lumber that has been put into piles with narrow strips, stickers, between the layers.

stuck yolk. The condition of an egg in which the yolk adheres to the inside of the shell. *Stuck yolks* may occur in eggs that have deteriorated from long holding. Such eggs are classed as inedible.

stud. 1. A unit of selected animals kept for breeding purposes, usually applied to horses. 2. Abbreviation for *stud* horse, a stallion; or for *stud* ram, male sheep.

stud book. The official record book of a livestock registry association for horses, ponies, and jacks.

stud farm. A livestock farm or establishment sometimes under government supervision, that specializes in breeding and raising herd sires for sale or distribution to smaller breeders or farmers. Also a farm that specializes in stallions for stud purposes.

stud-flock. A flock of carefully selected breeding ewes kept for raising young rams for future herd sires.

stud groom. An individual in charge of a stallion who supervises the breeding operation.

stud mating. The breeding of an individual female fowl to a selected male for a desired combination of genetic qualities; used mainly to obtain exhibition individuals.

stud of mares. A unit of mares kept for breeding purposes.

stud ram. A male sheep kept for breeding purposes.

stud stock. Breeding sheep registered in the flock book and kept largely for the raising of young rams for flock improvement (New Zealand).

stuffed geese. Fat geese produced in sections of Europe by the feeding of special rations and by cramming the birds. See *noodling; cramming machine.*

stumbler. A horse with faulty knee action in the front legs which is often indicated by a blemish or patch of white hair on the knees; a very serious fault.

stump. The basal part of a tree or plant left after the bole and top have been cut.

stumpage. The amount of timber uncut in any area of woods; also its estimated value.

stump analysis. The measurement of the stump cross-section to determine basal growth and development at different periods in the life of a tree.

stump fence. A fence built of pulled tree stumps laid more or less on edge so that the prong-like horizontal roots interlock. (It was a common form of fencing for fields following the lumbering of the pine forests in parts of Michigan and some other northern states in the U.S.)

stumping. A method of rooting shoots of woody plants or shrubs. Old shoots are removed at the ground level and a mixture of peat and sand heaped over the stumps to a depth of 4 to 5 ins. The sand, through which the young shoots push, may be held in place by a board frame and must be kept moist to assure the successful rooting of the shoots.

stump puller. An appliance of several types, usually involving the use of a cable and winch; used for extracting stumps from the soil.

stump sucker. A shoot arising from a stump and developing into a tree.

stump wood. The resinous stump and roots of longleaf or slash pine used for extraction of turpentine and rosin.

stunning. Hitting an animal in the forehead with a heavy blunt hammer to render it senseless before slaughtering.

stunning ax. A heavy, blunt-nosed ax or sledge used in slaughter houses for stunning or killing animals.

stunt. To check or hinder the growth or development of an animal or plant.

stunted. Slowed, checked, or hindered growth or development; dwarfed.

stunt of blueberry. Blueberry stunt.

stunt of dahlia. A virus disease of dahlias characterized by dwarfing, mottled and distorted leaves, and malformed flowers. It is spread by aphids.

sturdy conformation. The body conformation of an animal having enough flesh, firm, strong bones and a substantially well-proportioned body.

sturdy cypresspine. *Callitris robusta,* family Pinaceae, a rapid-growing, pyramidal tree grown in warm climates for its fine foliage. Native to Australia.

sturgeon. Any of the large, long-lived, noncommunal, fresh- or salt-water fishes of the family Acipenseridae. The adults are caught as they enter shallow, coastal waters, rivers, or streams, to spawn. The flesh is edible, the air bladder used in the manufacture of isinglass, and the roe (*q.v.*) made into caviar. The latter is a fairly important industry in America, Germany, Norway, Russia, and Sweden. The common Atlantic *sturgeon* is *Acipenser sturio.* Native to the North Temperate Zone.

sty. A pen where swine are housed and fed. Usually it consists of a low shed and a dirt yard. Also called *pigsty.*

stylar end rot. A whitish to drab sunken area at the distal end of the rind of Persian limes and lemons resulting from the adverse effects of hot weather.

style. 1. In the pistil of a flower, the part between the ovary and the stigma; if the *style* is lacking, the stigma is sessile on the ovary. 2. The manner in which an animal displays itself while at rest or in action. 3. The manifestation of those characteristics which contribute to the general beauty, pleasant appearance, and attractiveness of an animal.

stylish. Designating an animal with attractive and correct body conformation, attractive bearing, and smooth and graceful action.

stylish carriage. The carriage of an animal of excellent body conformation, attractive bearing, and smooth and graceful action.

stylish head and neck. The head and neck of an animal with a very superior conformation and quality.

stylish heifer. A heifer of very superior quality and conformation.

stylish horse. A nicely conformed horse with attractive carriage and excellent leg action.

styrax. See *storax; snowbell.*

Styria. A cows' whole-milk, cylindrical cheese made in the Province of Styria, Austria.

subacute. A clinical condition intermediate between acute and chronic.

subalpine. The mountain zone just below cold timberline.

subbreed. An offshoot or subdivision of a major breed of livestock, such as Polled Jerseys.

subclass. A primary division of a class *q.v.*

subclinical. Designating early or mild stages of a disease before signs or symptoms are noticeable; symptomless or low-grade infections.

subclinical form. A form or stage of a disease without clinical symptoms. Also called *inapparent form of disease.*

subculture. A culture of bacteria, etc., prepared on a fresh medium by inoculation from a preexisting culture. See *culture: 3. a.*

subcutaneous. Situated or occurring beneath the skin. A *subcutaneous* injection is put just under the skin.

subcutaneous emphysema. A condition in which air or gas accumulates under the skin in the loose areolar tissue. Air can enter the tissues beneath the skin through various external wounds or by way of an injury to the lungs. Gas may accumulate subcutaneously due to the action of certain bacterial infections. *Subcutaneous emphysema* crackles (like crumpling tissue paper) when a hand is rubbed over the area.

subcutaneous tuberculin test. A test for the detection of tuberculosis in animals. It consists of first determining the normal temperature curve of the animal to be tested, then injecting subcutaneously the recommended amount of standardized tuberculin. A post-injection temperature curve of two degrees Fahrenheit higher than the highest preinjection temperature is considered to be a reaction indicating that the animal is tuberculous. See *tuberculosis.*

subcutis. The tissues beneath the skin attaching it to underlying parts.

subgrade. The final level to which an excavation or fill is made before applying a surfacing material.

subhumid. See *subhumid climate.*

subhumid climate. A climate intermediate between semiarid and humid which has sufficient precipitation to support a moderate to dense growth of tall and short grasses but in most instances insufficient to support a dense deciduous forest. Some subhumid areas, where the rainfall comes mostly during the growing season, have scattered deciduous trees with grass vegetation in between. The annual rainfall in *subhumid climates* may be as low as 20 ins. in cool regions and as high as 60 ins. in hot areas. The precipitation-effectiveness indexes are 32 to 48 for the dry subhumid and 48 to 64 for the moist subhumid.

subirrigation. Water supplied to the soil from ditches through underground tile lines, perforated pipe lines, or by natural subsoil moisture, in sufficient amounts to maintain a water table sufficiently close to the soil surface to supply adequate water quantities for crop needs. A low permeability layer in the soil profile is essential to the maintenance of the water table. Plants and planted seeds in containers, such as pots, etc., are frequently subirrigated by setting the container in another water-containing pot.

subject-matter specialist. In the extension service of a Land-Grant college or university in the U.S., a worker who has a specialized knowledge and experience in a particular field.

sublease. A lease from a tenant.

sublethal. Less than fatal in effect, as a *sublethal* dosage or application of a toxic substance, etc.

sublimate of mercury. See *corrosive sublimate.*

subluxation. A partial dislocation, a sprain of a tendon or ligament which occurs near a joint.

submarginal land. Land incapable of sustaining a certain use or ownership status economically. See *marginal land; super-marginal land.*

submaxillary. Situated below the lower jaw.

submerged weir. A weir (*q.v.*) which in use has the tail water (*q.v.*) level equal to or higher than the weir crest.

subnormal. Below the normal of average expectancy.

suborder. A category next below an order and above a family in a classification of animals and plants.

subregion. A minor division of a larger region. *Subregions* may be delineated on such bases as similarity of land features, climate, fertility, agricultural type, etc.

subscald. See *semiscald.*

subsidiary crop. A crop planted with a main crop, such as clover in small grain.

subsidy. A government grant of money to aid or encourage a private enterprise that serves to benefit the public. In agriculture, the *subsidy* may be granted to the total group or to one component group, such as the dairy farmer or the wheat grower. It may be effected through tariffs on an imported agricultural commodity, through price support programs, direct payments, or the providing of services and information.

subsistence. Means of support for existing; essential materials for maintaining or supporting animal life.

subsistence farmer. A farmer whose farm operations provide barely the living requirements of the operator and his family rather than an excess for sale on the market.

subsoil. 1. In common agricultural usage, that part of the profile which lies below the usual plow depth without any specific limitation in depth or kind of material; a B horizon, the first change with depth in texture or structure; *e.g.*, a clay layer underlying a sandy surface. 2. To plow deeply into the *subsoil.*

subsoil cultivator. See *subsoil plow.*

subsoiling. The breaking up of the dense subsurface layers of the soil and thin layers of hard pan.

subsoil plow. A special type of implement designed to penetrate hard or impervious soils so as to aid in water retention and better root penetration. There are several types consisting principally of one or more heavy, curved, chisel-like plow points which penetrate deeply into the soil without turning the surface. See *subsoil tillage.*

subsoil tillage. The use in arid or semiarid regions of specially designed implements, like the subsoil tillers and field cultivators with duck-foot sweeps, which stir the soil below the surface but do not materially disturb the surface or surface cover. See *subsurface tiller.*

subspecies. A group of individuals within a species, distinguished by certain common geographical or varietal characters. See *variety.*

substance. In judging horses, the quantity and quality of bone in the body conformation. Among some horsemen, general ruggedness.

substandard. Below the established standard.

substandard cream. A pasteurized, sweet cream product low in butterfat, which contains 10 to 12 percent fat. Also called *breakfast cream; cereal cream; half-and-half.*

substrate. The surface or material on or within which an organism or group of organisms grows.

substratum. The C or D horizon of a soil.

subsurface flow. Water which enters the ground, but which emerges as seepage into streams, ditches, rivers, or springs. *Cf. surface runoff.*

subsurface irrigation. Irrigation by means of underground, open-joint tile, perforated pipes, or open ditch, to maintain the watertable at a desired distance below the ground surface. See *subirrigation.*

subsurface tiller. A type of farm implement which is used in arid and semiarid regions to till and firm the soil beneath the surface and to conserve moisture by not disturbing the trash or crop residue left on the surface.

subsurface water. The water of precipitation which penetrates into the soil. It occupies the depth between the ground surface and the lower limit of water-bearing rock formations.

subterranean clover. *Trifolium subterraneum,* family Leguminosae, a decumbent, winter annual herb grown mostly for grazing. Also called *annual clover.*

subterranean clover mosaic. A virus disease of subterranean clover, transmitted by aphids, and characterized by mottling and distortion of the leaflets, dwarfing of the plants, and a reduction in seed production.

subtiller. Subsurface tiller.

subtropical. In part, tropical; designating the mesothermal temperature province.

suburban. Designating areas adjacent to cities or towns whose population is largely urban in outlook and habits, usually deriving its livelihood from the nearby cities or towns rather than from agriculture or local industry. See *exurban.*

suburbanite. A person who lives on a small plot of land near a city or other concentration of population and who enjoys the advantages and conveniences of both country and city living.

succade gourd. Yellowflowergourd.

succession. 1. The progressive development of vegetation towards its highest ecological expression, the climax. The replacement of one plant community by another. 2. Transfer of operation and of ownership of a farm from one generation to the next, as from father to son or son-in-law.

succession cropping. The growing of two or more crops one after the other on the same land in one growing season.

succory. Chicory.

succotash. A dish of American Indian origin which consists principally of fresh lima beans and kernels of fresh sweet corn boiled together with or without meat.

succulence. A condition of plants characterized by juiciness, freshness and tenderness, making them appetizing to animals.

succulent. 1. Juicy; having a high percentage of water, as plants having juicy tissues. 2. Any plant belonging to the genera Cactus, Crassula, etc.

succulent feed. Fresh or juicy feed of vegetable origin, as green grass, silage, roots, etc.

succulents. A group of plants which have ways of overcoming a deficiency of water as (a) the cacti and other plants that have fleshy leaves or stems or both and (b) yucca, agave and other plants that have leathery foliage, often reduced in area, so that they are able to survive long droughts.

succus. Juices or fluids extracted from or secreted by an organism.

suck. 1. Of an animal, to draw milk from the teat of the udder by application of suction by the mouth; to nurse. 2. To draw a liquid, air, etc., by vacuum-producing action.

sucker. 1. A secondary shoot which develops from the root, crown or stem of a plant. 2. A cow that sucks herself. 3. Of a plant, to produce secondary shoots *(suckers)*. 4. To remove *suckers* or shoots from a plant, as from tobacco, etc.

suckering. 1. The act or practice of removing suckers (*q.v.*) from a plant. 2. Of a plant, currently producing suckers.

sucking insects. Any insect which has piercing and sucking mouth parts, as lice, scale insects, chinch bugs, squash beetles, etc.

suckle. To nurse at the breast or udder.

suckling clover. *Trifolium dubium,* family Leguminosae, an annual, highly palatable herb grown as a permanent pasture plant. Native to Europe. Also called *least hop clover.*

suckling pig. A young pig still nursing its mother. When slaughtered at this stage it produces a small carcass for roasting whole.

sucklings. The young of mammals that are being nursed by the female; unweaned animals.

sucrose. $C_{12}H_{22}O_{11}$, cane or beet sugar: *i.e.,* common table sugar; a carbohydrate used in media for fermentation reactions.

suction dredge. Hydraulic dredge.

suction pump. A pump for drawing up water by suction created by a piston and valves in a vertical or horizontal cylinder. Water cannot be raised more than a theoretical height of 34 ft. by such means.

suction system. A system for drying various products in which air is moved through ducts at a pressure less than atmospheric pressure and is therefore pulled through the product to be dried. Also called *exhaust system; vacuum system.* See *pressure system.*

sudadero. Saddle fender.

Sudan colanut. *Cola acuminata,* family Sterculiaceae, a tree which yields the colanut containing a large amount of caffeine, and used in the soft drink industry. Native to tropical Africa. Also called *kolanut.*

Sudan grass. *Sorghum vulgare* var. *sudanense,* family Gramineae, an annual herb grown for summer pasture, hay, soiling, and silage. *Sudan grass* has been known to produce prussic acid poisoning (*q.v.*) but is much less likely to do so than other sorghums. Native to Africa.

sudden death. See *falling disease.*

suet. The hard, flaky, wax-like fat around the kidneys and loins of some animals, as cattle or sheep, used in cooking or for making tallow.

Suffolk. 1. A very popular early breed of draft horse originating in the county of Suffolk, England, which always has some shade of chestnut. 2. A large, Down breed of sheep characterized by jet black, polled head and ears, and black legs from the hocks and knees down. It is a mutton-type breed with mature ewes weighing from 160 to 225 lbs. The *Suffolk* originated in the county of Suffolk, England.

suffrutescent. Designating a plant which is partly woody and perennial at the base but with herbaceous, annual growth above, such as the carnation.

sugar. 1. Any of a group of carbohydrates characterized by being soluble in water, crystalline, and having a relatively sweet taste. 2. See *sucrose.*

sugar apple. *Annona squamosa,* family Annonaceae, a deciduous tree which bears small, greenish flowers, and a conical, pulpy, edible fruit which has a thick, scaly rind. In the U.S., it is grown only in southern Florida and to a very limited extent. Native to tropical America. Also called *sweetsop; cinnamon apple.*

sugar beet. *Beta saccharifera,* family Chenopodiaceae, a biennial herb grown as an annual for its large, fleshy roots containing large amounts of sugar, usually from 12 to 18 percent, a source of domestic sugar in the U.S. Native to Europe and Asia.

sugar-beet Cercospora leaf spot. A fungous disease of sugar beets caused by *Cercospora beticola,* family Dematiaceae, which blights the tops of plants so that root growth is dwarfed and the percentage of sucrose lowered.

sugar-beet curly top. See *curly top.*

sugar-beet drill. A two to six row drill that plants beet seed in rows at a depth of ½ to 1½ ins. It is often equipped with an attachment for applying fertilizer below or at the side of the seed.

sugar-beet Fusarium yellows. A fungous disease of sugar beets caused by *Fusarium conglutinans betae,* family Tuberculariaceae, characterized by the leaves becoming pale yellow and mottled, the central core of the root turning yellow or brown, and eventually black.

sugar-beet leafhopper. See *beet leafhopper.*

sugar-beet mosaic. A virus disease of sugar beets caused by *Marmor betae,* family Marmoriaceae, which produces mottling of the young leaves. The virus is transmitted by aphids.

sugar-beet root aphid. *Pemphigus betae* Doane, family Aphidae, an insect found on the roots of sugar beets, and many weeds. It reduces both the size and quality of the beets by sucking sap from the roots.

sugar-beet webworm. See *beet webworm.*

sugar-beet wireworm. *Limonius californicus* Mann, family Elateridae, one of several wireworms that may damage sugar beets.

sugar bush. A grove of sugar maple (*q.v.*) trees tapped for sap, which is made into maple sirup or sugar. Also called a *bush.*

sugar camp. A sugar maple tree grove, or a place in the forest where the maple sap is made into sugar or sirup. See *sugar bush.*

sugar cane. *Saccharum officinarum,* family Gramineae, a tall-growing plant, with big stalks and long, slender, coarse leaves. Chiefly valued for the sweet juices expressed from its stem, the plant is cultivated in tropical regions for sugar and its by-products. Also cultivated in the southern states, U.S. Usually propagated from stalk cuttings. Probably a cultigen.

sugarcane basal stem rot. A root and stem rot that affects the stem above and below ground level. The disease is attributed to fungi associated with the root-rot complex, especially *Marasmius sacchari* and species of Pythium.

sugarcane beetle. *Euetheola rugiceps* (Lec.), family Scarabaeidae, a beetle that infests sugar cane and corn, gnawing at the stems between the surface of the ground and the roots.

sugarcane borer. *Diatraea saccharalis* (F.), family Crambidae, a moth whose larvae bore into the interior of the stalks of sugar cane, corn, broom corn, sorghums and Sudan grass, killing the young canes and weakening the larger ones. It is a very close relative of the southern cornstalk borer *q.v.*

sugarcane chlorotic streak of grasses. A virus disease transmitted by leafhoppers, and characterized by white leaf stripes distributed uniformly over the leaves.

sugarcane gummosis. A bacterial disease, primarily of the vascular system of sugar cane, caused by *Xanthomonas vasculorum,* family Pseudomonadaceae. It receives its name from the slimy gum that oozes from the cut end of affected stalks. Diseased plants may have yellowish streaks, sometimes dotted with red or brownish spots on the leaves, usually near the tip. The disease is transmitted by infected cuttings used for propagation and by other means of physical contact.

sugarcane leafhopper. *Perkinsiella saccharicida* Kirk, family Delphacidae, a leafhopper (*q.v.*) that sucks sap from the leaves of sugar cane causing loss of vigor and stunting of the plant when present in large numbers.

sugarcane leaf scald. A bacterial disease caused by *Xanthomonas albilineans,* family Pseudomonadaceae, characterized by the appearance of sharply defined, white stripes which may extend over the entire length of the leaf and onto the sheath; sometimes the entire shoot is chlorotic to nearly white. Diseased plants have a characteristic appearance: the terminal whorl of leaves curves inward at the tips which are often dry and withered. The disease is highly infectious and is spread through infected cuttings or by physical contact.

sugarcane mealybug. Pink sugarcane mealybug.

sugarcane mosaic. A virus disease transmitted by aphids and mechanically. It is characterized by light green or yellow streaking and mottled appearance of the leaves. Some varieties of sugar cane are dwarfed severely, others show only leaf symptoms. In certain resistant varieties the symptoms are inconspicuous.

sugarcane mottled stripe. A minor, bacterial disease caused by *Xanthomonas rubrisubalbicans,* characterized by chlorotic, elongated blotches and stripes on the leaves.

sugarcane pineapple disease. A seed rotting disease caused by the fungus *Ceratostomella paradoxa,* family Ceratostomataceae. The rotting cuttings have a characteristic odor like decaying pineapple.

sugarcane pokkah boeng. A top-rotting disease of sugar cane caused by the fungus *Fusarium moniliforme,* family Tuberculariaceae. Injury varies from slight chlorosis and splitting of the base of young unfolding leaves to top rotting which may kill the growing point.

sugarcane Pythium root rot. A fungous disease of sugar cane caused by *Pythium arrhenomonas,* family Pythiaceae, characterized by the destruction of roots, severe wilting and yellowing of the leaves, stunting, and sometimes death of the plant. Deficient tillering is also characteristic. See *sugarcane basal stem rot.*

sugarcane red rot. A serious fungous disease of sugar cane in the southern U.S. causing a rot of cuttings that results in faulty stands and a loss of sucrose in mill cane. Infection of the stalks usually follows injury by the sugarcane borer (*q.v.*). The fungus that causes red rot is commonly known by its imperfect stage, *Colletotrichum falcatum,* family Melanconiaceae, although *Physalospora tucumanensis,* family Mycosphaerellaceae, is the perfect stage of the fungus.

sugarcane red stripe. A bacterial disease of sugar cane caused by *Xanthomonas rubrilineans,* family Pseudomonadaceae, characterized by narrow, sharply-defined, blood-red stripes which may be short or may extend nearly the length of the leaves, usually more prevalent on the younger unfolded leaves and often resulting in top rotting. The disease may extend into the mature portion of the stem to ground level. It is accompanied by a characteristic odor. Outbreaks of the disease are generally infrequent and localized.

sugarcane ring spot. A disease of sugar cane caused by the fungus *Leptosphaeria sacchari,* producing purple-bordered spots.

sugarcane root rot. See *sugarcane basal stem rot; sugarcane Pythium root rot.*

sugarcane sereh disease. A destructive virus disease of sugar cane caused by *Manus sacchari,* characterized by necrosis of the phloem and tops and general degeneration of the plant.

sugarcane smut. A disease of sugar cane caused by the fungus *Ustilago scitamintre,* family Ustilaginaceae, characterized by the spread of smut spores from the growing point of a long whiplike shoot.

sugarcane streak. A virus disease of sugar cane characterized by many narrow, elongated, sharply-defined, white streaks on the leaves resulting in reduced plant growth. Its insect vector is *Cicadulina mobila.*

sugar cure. Any method of curing meat which includes the use of sugar. The inclusion of sugar in a meat cure counteracts the astringent quality of the salt, makes meat juicier, enhances the flavor, and assists in improving the color of the cured meat. See *dry cure; sweet pickle cure.*

sugared honey. Granulated honey.

sugared oat mill feed. Oat mill feed (*q.v.*) to which molasses has been added.

sugar eucalyptus. *Eucalyptus cladocalyx,* family Myrtaceae, a semi-tropical, deciduous tree grown as a bee tree and as an ornamental for its flowers which grow in umbels of 6 to 12 blooms. Native to Australia. Also called *sugar gum.*

sugar-factory lime. Lime that has been used in sugar refining. On a dry basis it will usually test about 80 percent calcium carbonate ($CaCO_3$). It has use for liming soils and is usually sold by the cubic yard. Two yards have a lime equivalent of about one ton of ground limestone.

sugar gum. Sugar eucalyptus.

sugar hydrometer. Sirup hydrometer.

sugaring off. A social gathering of people about a maple sirup shed at a time when the sap of sugar maple trees is gathered and boiled down into maple sirup or sugar. It is similar to cane-grind parties in the South (U.S.).

sugar loaf. Douglas clematis.

sugar maple. *Acer saccharum,* family Aceraceae, a large, deciduous tree, chiefly of northern U.S. and Canada from Quebec to Pennsylvania, westward to Minnesota and Kansas, whose sweet sap may be collected in early spring and used to make sirup and sugar. The tree is sometimes planted for shade and for road-side plantings. Its wood is used in furniture, flooring, etc. Native to North America. See *maple; maple sirup; sentry maple.*

sugar-maple borer. *Glycobius speciosus* (Say), family Cerambycidae, the larva of a beetle which bores a wide channel several feet long in the inner bark and sapwood of sugar maple, causing cracks in the bark to which other injurious insects are attracted.

sugar mule. A market class of mules that ranges from 16 to 17 hands in height and 1,150 to 1,300 lbs. in weight. It is suitable for use in sugarcane fields.

sugar pack. Processed fruit with sugar added, packed for freezing. See *dry pack: 1.*

sugar palm. Any palm tree that can yield sugar, especially palms of the genus Arenga, family Palmaceae. See *gomuti sugarpalm.*

sugar pea. See *edible-podded pea; sugarpod gardenpea.*

sugar pine. *Pinus lambertiana,* family Pinaceae, a tall-growing, evergreen tree whose easily-worked wood is used for millwork, door, sash, interior and exterior trim, foundry patterns, etc. Native to North America.

sugar planter. One who operates a sugarcane plantation.

sugarpod gardenpea. *Pisum sativum* var. *macrocarpon,* family Leguminosae, an annual herb grown as a vegetable plant for its edible seed pod. Native to Europe and Asia.

sugar sirup. 1. A sweet, more or less viscid liquid made by concentrating the juices of sugar-producing plants, as sugar cane, sugar maple, and sorgo. 2. A sweet solution made by boiling, to varying degrees of concentration, a mixture of beet or cane sugar and water. Used as a sweetener in various food and beverage preparations, in the canning of fruit products, etc.

sugar-tree land. A type of agricultural land with a native growth of sugar maple (*Acer saccharum*). It was recognized by early settlers in U.S., especially in Indiana and Ohio, as first class land.

sugar weather. Weather (toward the beginning of spring) in which the temperature during the day rises above 32° F. and during the night drops below the freezing point causing the sap to start flowing in sugar maple trees and indicating that the time for making maple sugar is at hand.

sugi. Cryptomeria.

suine. A fatty material, obtained from pigs' lard, used as a substitute for butter.

suint. The water-soluble material resulting from perspiration naturally found in a wool fleece as a part of the yolk *q.v.*

suitcase farmer. A part-time farmer or one who maintains temporary residence on the land staying long enough to plant a crop and, on return, to harvest it. (Found mostly in the Great Plains region, U.S., and particularly in wheat farming.)

sulfadiazine. See *sulfa drug.*

sulfa drug. Any of a group of drugs that have a marked bacteriostatic action and used medicinally largely to enable the host to develop antibodies to combat the invading organisms. Included in the group of sulfa drugs are: sulfanilamide, sulfadiazine, sulfathiazole, sulfapyridine, sulfaguanidine, sulfamerazine, sulfamethazine, sulfaquinoxaline, and others. See *sulfonamides.*

sulfamate. A salt of sulfamic acid, used as an herbicide in weed control.

sulfate. A salt of sulfuric acid as ammonium *sulfate.*

sulfate of aluminum. Aluminum sulfate.

sulfate of ammonia. Ammonium sulfate.

sulfate of copper. Copper sulfate.

sulfate of iron. Ferrous sulfate.

sulfate of lime. Calcium sulfate.

sulfate of potash. Potassium sulfate.

sulfate of potash-magnesia. A double sulfate of potash and magnesia; a potash salt containing not less than 25 percent of potash (K_2O), nor less than 25 percent of sulfate of magnesia, and not more than 2.5 per cent of chlorine. Used as a fertilizer. Also called *double manure salt.*

sulfide spoilage. Spoilage of canned goods sometimes caused by *Clostridium nigrificans.* It is characterized by the production of hydrogen sulfide and blackening of the product. The can remains flat.

sulfide taste. An offensive taste and odor sometimes found in wine due to the presence of sulfur on the grapes or in the tanks.

sulfite. A salt of sulfurous acid. Certain *sulfites* or bisulfites are sometimes used to treat potatoes, carrots, etc., to retard oxidation changes in storage. See *sulfite wood pulp.*

sulfite wood pulp. Wood pulp obtained by the digestion of wood at high temperature and pressure with calcium bisulfite cooking acid.

sulfite yeast. The yeast grown on the waste liquor from the sulfite wood-pulping process. Produced under sanitary conditions and suitable for animal feeding, it contains not less than 40 percent protein.

sulfofication. The oxidation of sulfur, free from organic or inorganic compounds, largely through the action of microorganisms. In the soil, the oxides are quickly changed into sulfites and sulfates.

sulfonamides. A group of chemicals often used as drugs to suppress the normal growth and reproduction of many disease-producing bacteria, which includes sulfanilamide, sulfapyridine, and sulfathiazole. See *sulfa drug.*

sulfonated oil. An oil treated with sulfuric acid or closely allied substances to render it partly soluble in water. Certain *sulfonated oils* are used as spreaders in insect sprays.

sulfur. S, an elementary, yellow mineral substance, insoluble in water, easily fusible and inflammable. Also called *brimstone.* One of the minor but important elements in soil fertility and used in relatively large amounts by most plants, it is an important constituent of both protein and protoplasm. The powder form is an effective insecticide for many true bugs, some leafhoppers, mites, and other insects. The dust is used as a fungicide in the control of mildew, etc. When burned it forms sulfur dioxide, a gas which is highly toxic to insects and has long been used as a fumigant as well as a bleaching agent.

sulfur cinquefoil. *Potentilla recta,* family Rosaceae, a perennial herb growing 12 to 24 ins. tall, with soft hairy stems and branches terminated by clusters of yellow flowers. It is a weed pest east of the Rocky Mountains, U.S. Native to Europe.

sulfur dioxide. SO_2, a compound produced by burning sulfur; it has a suffocating odor and is used as a fumigant for the control of certain insects, for the prevention of molds on dried fruits, for bleaching wool, straw goods, etc.

sulfur dust. Dry, finely-powdered sulfur which is specially toxic to mites, such as chiggers and red spider mites, and to newly-hatched scale insects. Sometimes it is used in greenhouses by working the dust into the soil to control certain microorganisms. Also used as a fungicide, as in the control of mildew, etc.

sulfur eriogonum. *Eriogonum umbellatum,* family Polygonaceae, a wooly herb with golden-yellow flowers, sometimes grown as a bee plant. Native to the western U.S. Also called *sulfur-flower.*

sulfuric acid. H_2SO_4, a dense, heavy, exceedingly corrosive, oily liquid which can decompose animal and vegetable tissue and has a great affinity for water, giving out heat on combining with it. It is used in the manufacture of superphosphate fertilizers by converting the insoluble rock phosphate to a soluble and available form.

sulfur lime spray. A mixture of hydrated lime and 325-mesh sulfur made wettable by the addition of calcium caseinate. It was previously used on peaches as a fungicide substitute for self-boiled lime sulfur but is now superseded by wettable sulfur (*q.v.*). Also called *New Jersey dry mix sulfur lime.*

sulfurous acid. H_2SO_3, an acid resulting from the combination of sulfur dioxide and water, used as a bleaching agent, as a preservative in fruit juices, etc.

sulfur shelf-mushroom. An edible mushroom; the fruiting body of *Polyporus sulphureus,* family Polyporaceae.

sulking. 1. In commercial mushroom culture, a depression of growth and reduction in yield from an excessive accumulation of carbon dioxide and an unsaturated hydrocarbon gas, given off by the growing mushrooms, which are unable to escape due to insufficient ventilation in mushroom houses. 2. In horses, refusal to obey commands promptly.

sulky. 1. A very light, two-wheeled cart used in training and racing harness horses. 2. A heavy, two-wheeled vehicle used in breaking horses. 3. See *logging wheels.*

sulky disk plow. A one-bottom, disk plow (*q.v.*) provided with a seat for the driver.

sulky hayrake. See *dump rake.*

sulky plow. A one-bottom, horse-drawn plow mounted on wheels and provided with a seat upon which the operator rides.

sulky seat. A type of easy, comfortable seat suspended by springs to the frame of a plow or other implement on which the operator may ride.

sulla sweetvetch. See *red honeysuckle: 1.*

sulphur. See *sulfur.*

Sultans. A very old, ornamental breed of chickens of southeastern Europe, bred in Turkey under the name of Sultan's Fowl, which was introduced into England about 1850 and into the U.S. some years later. *Sultans* have large, globular, and compact crests, very full beards and white plumage.

sultan snapweed. *Impatiens sultani,* family Balsaminaceae, a tender, brittle, succulent herb, grown as a house plant and as a bedding plant for its solitary flowers originally scarlet but also in various colors in hybrids and sports. Native to eastern Africa. See *snapweed.*

sumac. A shrub or slender tree of the genus Rhus, family Anacardiaceae. See *flameleaf sumac; fragrant sumac; poison sumac; skunkbush sumac; smooth sumac; staghorn sumac.*

Sumatra. An old breed of game chickens which originated in Sumatra. The variety Black Sumatra, included in the Oriental class, has very lustrous, greenish-black plumage. The tail of the male is long and drooping with an abundance of long sickle and covert feathers. They are raised mainly for exhibition.

summer annual. Any plant which grows from seed each spring or summer, produces seed that same season, and does not survive the winter.

summer bedding. Any tender, ornamental plant grown in a flower bed or border during the summer, including tender annuals, some of which are grown in a greenhouse during winter, and other summer-blooming plants whose rootstocks or bulbs are stored over winter.

summer black stem of alfalfa. A fungous disease caused by *Cercospora zebrina,* family Dematiaceae, which usually occurs in warm weather and is characterized by the presence of a leaf spot and the blackening of the upper part of the stems.

summer crookneck squash. See *crookneck group.*

summercypress. Any plant of the genus Kochia, family Chenopodiaceae. See *Belvedere summercypress.*

summer dieback of apple. Withertip of apples.

summer dwarf of strawberry. A warm-weather disease caused by the nematode *Aphelenchoides besseyi,* family Aphelenchidae. The symptoms are the same as for spring dwarf of strawberry *q.v.* except that it occurs in summer.

summer fallow. The plowing of a field prior to or during the summer and cultivating enough to control weeds in preparation for a later crop.

summer grape. *Vitis aestivalis,* family Vitaceae, a woody, high-climbing vine, which bears edible, black fruit usually juicy and sweet but sometimes dry and sourish. Native to North America. Also called *bunch grape; pigeon grape.* See *grape.*

summer heliotrope. Valerian.

summer legume. A legume which grows from seed each spring or summer, produces seed that same season, and does not survive the winter: *e.g.,* as cowpeas, soybeans, etc.

summer oil spray. An oil emulsion spray suitable for use in insect control when trees are in foliage. Such sprays differ from dormant oil sprays in that they contain much more highly refined oils and are of necessity weaker, rendering them safe for most types of foliage. The oil may be the primary insecticide in a spray or may be added to make other materials more effective. Summer oils are commonly mixed with nicotine sulfate to combat sucking insects and may also be combined with arsenicals or with wetting agents.

summer phlox. *Phlox paniculata,* family Polemoniaceae, a perennial herb grown for its variously colored flowers. Native to North America. Also called *garden phlox; perennial phlox.*

summer range. 1. Range areas, usually at high elevations, which are necessarily grazed during the summer season because of the occurrence of snow in the other seasons. 2. Any range area which is grazed more conveniently in summer than in any other season.

summer-range shelter. A small, low, well-ventilated, portable structure which has a roof, wire sides, and floor. Commonly one of several such shelters is provided for growing chicks on summer range to furnish protection from heat and storms, and from animals that may prey on them.

summer sausage. A dry, uncooked sausage made of coarsely ground meat (usually beef and pork) and seasoning. The average mixture is cured for several days in a cool place below 40° F., then stuffed into casings and smoked. Since this sausage is consumed without cooking, it is necessary to have the inside temperature reach 142° F. during the smoking process to comply with federal (U.S.) regulations. Also called *cervelat; saveloy.*

summer savory. *Satureia hortensis,* family Labiatae, an annual, fragrant herb, which is one of the culinary herbs commonly included in herb gardening. The leafy portions of the plant are used in salads and dried for use in soups, stews, etc. Native to Europe.

summer snowflake. Star-of-Bethlehem.

summer spore. A spore that germinates without resting, often short-lived, associated with rapid fungus increase during a favorable season.

summer spray. Any of various sprays applied during the summer at the time the trees or other plants are in foliage. See *summer oil spray; dormant spray.*

summer squash. Any of several types or groups of squash commonly used as a vegetable soon after harvesting, rather than being kept through late fall and winter, in contrast to the winter squash or Hubbard group (*q.q.v.*). See *crookneck group; patty-pan group; squash.*

summersweet clethra. *Clethra alnifolia,* family Clethraceae, a hardy shrub grown for its very fragrant, white or pink flowers in erect clusters. Native to North America. Also called *spiked alder; sweet pepperbush; summer-sweet.*

summer tying. Green tying.

summer wood. The portion of the annual growth ring of a tree formed during the latter part of the yearly growth period. It is usually more dense and stronger mechanically than spring wood.

summer yellow cottonseed oil. Cottonseed oil that is sweet in flavor and odor, prime in color, clear and brilliant in appearance and free from water and settlings, and should contain not more than ⅛ of 1 percent free fatty acid according to the rules of the Interstate Cotton Seed Crushers' Association (U.S.).

sump. Usually a small depression or excavation in which water is collected for removal by pumping. In pump drainage, the walls of the *sump* are curbed to prevent sloughing.

sumpweed. Any plant of the genus Iva, family Compositae. Its several species are included in a group of plants whose pollen cause hayfever in the Rocky Mountain states (U.S.) and westward.

sun blotch. A virus disease of avocado whose typical symptoms are a yellow streaking of the green stems and branches and a yellow to red streak on the fruit. Affected trees tend to have a recumbent, willowy type of growth and some may be stunted. The disease is readily transmitted through budwood or graftwood.

sunburn. 1. Injury to the leaves, fruit, or other parts of plants, or to the skin of animals, as a result of exposure to intense sunlight. See *sunscald.* 2. Greening of potatoes, onions, and certain root crops due to exposure to sunlight. 3. A superficial inflammation of the skin of hogs (especially white hogs) on rape pasture, when they are wet from the dew on the plants and exposed to bright sunlight.

sun-cured hay. Hay cured (dehydrated) in direct sunlight (as in field curing) as distinguished from hay that is artificially dehydrated.

sundew. Roundleaf sundew.

sundial lupine. *Lupinus perennis,* family Leguminosae, a hardy, perennial herb grown for its white to blue flowers. Native to North America. Also called *Quaker bonnets.*

sundowner. An Australian bush-wanderer who made a habit of approaching a homestead at sundown with an appeal for work, and, work being impossible at that hour, thus obtained free rations for the night. He usually walked the same route, avoiding hill-climbing where possible, and seldom camped in the same place more than one night. The *sundowner* flourished mainly in the last two decades of the 19th century, and vanished as settlement and communications increased. Known as a *tussocker* in New Zealand. See *tramp; waltzing matilda. Cf. swagman.*

sun drying. The drying and preserving of a product, especially peaches, apricots and raisins, by placing on trays exposed to solar heat without controlled conditions of temperature, humidity, and air flow.

sun euphorbia. *Euphorbia helioscopia,* family Euphorbiaceae, an annual herb which is a weed in waste places in the northeastern U.S. and eastern Canada. Contact with the milky sap causes a dermatitis with blisters and inflammation in human beings. Native to Europe.

sunfishing. A violent form of bucking by horses, especially broncos.

sunflower. *Helianthus annuus,* family Compositae, a tall, annual herb growing up to 12 ft. or more in height, which has very large ovalish leaves and immense flower heads that are yellow except for the central disk. It is often planted in groups or rows for screening purposes or to provide shade for poultry on summer range. It is also sometimes grown for silage or for its seed which is used as a bird feed. The seed yields an edible oil. Native to North America. Also called *Russian sunflower; combflower.*

sunflower heliopsis. *Heliopsis helianthoides,* family Compositae, a perennial, sunflower-like herb grown for its yellow flower heads. Native to North America. Also called *false sunflower.*

sunflower-seed oil meal. A product obtained by grinding the sunflower-seed oil cake remaining after the extraction of oil from sunflower seed. It is used as a supplement in livestock feeds.

sunflower silage. Silage from sunflower plants which is usually made when the heads are just showing bloom; prepared chiefly in regions where the season is too short and cool for corn. *Sunflower silage* is somewhat less palatable and lower in feeding value than corn silage.

sunk fence. A ditch with a retaining wall used to divide lands without defacing the landscape.

sun moss. Portulaca.

sunn. See *sunn hemp.*

sunn hemp. *Crotolaria juncea,* family Leguminosae, a tall shrub with silky stems and bright yellow flowers whose branches yield a fiber, the commercial *sunn,* used in making ropes, sacking, etc. Native to India.

sunrose. 1. Any plant of the genus Helianthemum, family Cistaceae. 2. *H. nummularium,* a woody plant grown in warm areas for its yellow, roselike flowers. Native to southern Europe.

sunscald. Injury to plant leaves, stems, fruits, etc., which results from a combination of high temperature, high humidity, and intense sunshine. See *sunburn; winter sunscald (of trees).*

sun spurge. Sun euphorbia.

sunstroke. 1. Severe injury or killing of heat-sensitive plants by excessive heat during midsummer periods of cloudless skies with temperatures ranging from 95 to 100° F. or more. 2. See *heat prostration (stroke).*

super. A top or additional compartment added to a beehive in which bees deposit their nectar and honey.

super baby beef. A well-finished, good quality calf, ready for market at 650 to 850 lbs.

super cream. Cream with about 75 percent fat. The economy in storage and transportation makes it convenient for certain purposes. See *plastic cream.*

supercritical flow. A flow condition in an open channel in which the velocity is greater than critical velocity.

superfetation. Second impregnation of a female already pregnant.

superficial scab. A brownish abrasion of the skin of a potato. See *corky scab of potato.*

superficial scald. Scalded areas of plants later frequently invaded by rot-producing fungi which completely destroy the tissue.

superheated milk. Unsweetened condensed milk heated to a high temperature after condensing.

superior balance and quality. In the show ring, qualities of desirable conformation, flesh, or wool of sheep.

superior fleece in density, length, and covering. In the show ring, sheep showing desirable qualities of wool.

superior quality cottonseed. Cottonseed that by analysis contains less than ½ percent foreign matter, more than 8 but less than 10 percent moisture, and less than ¾ percent free fatty acids, in the oil in the seed.

supermarginal land. Highly productive land which is capable of sustained use with profit to the operator. See *marginal land; submarginal land.*

supermild mosaic. A very faint mosaic disease *(q.v.)* symptom.

superparasitism. The condition which results from

several different infestations or two or more species of primary parasites.

superphosphate. A fertilizer product which results from treating rock phosphate with sulfuric acid. The grade shows the available phosphoric acid: *e.g.*, 20 percent *superphosphate* contains 20 percent phosphoric acid. The highest grades of *superphosphate* are made by treating the phosphate rock with phosphoric acid resulting in products carrying 40 to 65 percent of the acid. Also called *acid phosphate.*

supersede. To rear a young queen to take the place of the mother queen while she is still in the hive.

supersedure cell. A special cell for rearing a young queen to supersede *(q.v.)* an old queen.

supplement. A feed or feed mixture that is relatively richer in a specific nutrient than the basic feed ingredients in a ration to which it is added. It may be used to supply a single nutrient or may contain a mixture of vitamins, proteins, minerals and other growth stimulants.

supplemental heat. In drying produce, the addition of a small amount of heat to bring about a limited rise of temperature, usually less than 20° F., which is used to dry the product within the maximum time permissible before spoilage.

supplemental irrigation. Irrigation carried on in areas where annual rainfall is normally high enough for good yields but is insufficient for some crops during part of a growing season.

supplemental substance. Material not primarily toxic which when added to sprays or dusts makes the sprays or dusts cover more economically.

supplemental (supplementary) feed. 1. Concentrates or hay fed to livestock on the range when forage is insufficient, of inferior quality, or unavailable. 2. See *supplement.*

supplementarity. The relationships between the outputs of two enterprises. If, between two enterprises, the output of one can be expanded for a given outlay without reducing the output of the other, by making it possible to utilize more fully certain fixed factors of production, the two enterprises exhibit *supplementarity.*

supplementary crops and livestock. Crops and livestock which do not compete at the same time for the use of a productive resource but supplement each other for fuller total use of the resource.

supplementary enterprise. A branch of the farm business which fits with another without serious conflict in the use of the farmer's productive resources. See *supplementarity.*

supply. The number of heads of livestock or the quantity of other produce on a particular market day given in figures. In the absence of actual figures, estimates are sometimes given relative to the receipts of the preceding marketing day.

support. See *price supports.*

suppuration. The formation of pus.

Surati. A cheese made from buffalo's milk. One of the best known of the few varieties of cheese made in India.

surcingle. A strap or girth around the body of a beast of burden for holding the saddle or load in place or used for throwing the animal.

surface color. 1. The red color of the skin of ripening and ripe fruit. Also called *overlying color. Cf. ground color.* 2. The color of that part of the plumage of a fowl that is visible when the feathers are in their normal, resting position.

surface fire. A fire which runs over the forest floor and burns only the surface litter, the loose debris, and the smaller vegetation.

surface incrustation. Hard, white spots which appear on the surface of ice cream or sherbet because of crystallization of some of the sucrose and the freezing of the liberated water.

surface inlet. An opening in the ground surface, usually at the lowest elevation in a field, which is curbed with tile, metal pipe or concrete and connected to a subsurface drain. It permits runoff water that would normally accumulate in this depressional area to be removed through this drain. Also called *catch basin.*

surface irrigation. Irrigation distribution of water over the soil surface by flooding or in furrows for storage in the soil for plant use.

surface measure. The area in square feet on one face of a board. When the boards are 1 in. thick, the surface measure is equivalent to board measure with board foot as its unit.

surface profile. The longitudinal profile assumed by the surface of a stream of water flowing in an open channel.

surface rot of sweet potato. A fungous disease of stored sweet potatoes caused by *Fusarium oxysporum,* family Tuberculariaceae, characterized by grayish-brown to black spots appearing after the sweet potatoes have been in storage two months or more.

surface runoff. The portion of runoff which flows to stream channels over the surface of the ground. It is quite distinct from the runoff absorbed by the soil, lost by evaporation, or retained as surface storage. The usual sources of *surface runoff* are rainfall occurring with intensities exceeding the rate of infiltration and snow upon the ground melting at rates exceeding the rate of infiltration. *Surface runoff* is the principal source of floodflow occurring in streams at times of rainstorm or melting snow. See *subsurface flow.*

surface sealing. Such orientation and packing of dispersed soil particles in the surface layer of soil as makes it almost impermeable to water.

surface soil. The upper part of soil commonly stirred by tillage implements. See *top soil.*

surface-sown. Designating seed sown on the surface of the ground.

surface storage. 1. The storage of water in surface reservoirs, such as constructed reservoirs, ponds, swamps and lakes to reduce flood flow, increase low-water flow, and retain runoff for irrigation. 2. The retention of water in minute depressions on the soil during a period of runoff. Also called *depression storage.*

surface taint. A very serious flavor defect of dairy

products, usually butter, associated with putrefaction and uncleanliness.

surface tension. Tendency of a liquid to contract until its area is the smallest possible for a given volume of liquid. This makes it appear to be covered by an elastic membrane.

surface tillage. Cultivating or working the soil to a shallow depth or barely below the surface.

surface to volume ratio. The ratio between the surface area and volume of bulk milk, in which the greater the volume, the lower the relative surface of the exposed milk and consequently the less contamination from the container per unit capacity.

surface-tube automatic watering. Surface-watering *(q.v.)* by means of perforated copper tubes, used to water benches in greenhouses.

surface water. Water on the surface, as lakes and rivers, in contrast to that underground. *Surface water* properly includes snow and ice.

surface-watering. Applying water to the surface of the soil as when watering benches and potted plants in greenhouses.

surfactant. A chemical product which can affect the surface relation of molecules or particles of other materials, as in sprays.

surging. In well drilling, the rapid up and down movement of a plunger to wash the fine sand, silt, and clay out of the water bearing formation which surrounds the screen. Also called *backwashing.*

surinam cherry. 1. *Malpighia glabra,* family Malpighiaceae, a shrub about 10 ft. tall, grown for its edible, aromatic fruit used in making jams and preserves. Native to tropical America. 2. Pitanga *q.v.*

Surinam cockroach. *Pycnoscelus surinamensis* (L.), family Blattidae, a large insect which eats seedlings, injures flowers, and girdles plants in the greenhouse.

surplus. Excess of a commodity making its supply greater than the demand for it in terms of returning a desirable price for it. In the U.S. *surpluses* are legally defined in price supports (*q.v.*). The level of price supports is determined by the estimated percentage the total supply is of the normal supply at the beginning of the market year.

surplus commodity. See *Commodity Credit Corporation.*

surplus honey. The honey removed from the hive over and above the amount the bees need for their own use.

surplus milk. The amount of market milk available in addition to the daily requirements for fluid milk in a particular market. The extra milk which cannot be utilized as beverage milk but must be manufactured into other products (U.S.).

surplus plan. See *Commodity Credit Corporation.*

surprise butter grading. See *pick-up contest.*

surprise contest. Pick-up contest.

surprise lily. Autumn lycoris.

surra. An infectious disease caused by the protozoan, *Trypanosoma evansi,* characterized by anemia, fever, edema, marked loss of weight, or the appearance of minor hemorrhages of the mucous and serous membranes. Usually fatal in horses, camels, and mules, it is not so serious in dogs or cattle. Endemic in India, the Philippines, and Africa. Also called *surrah.*

survey. An organized visual study of conditions associated with a problem, as a soil survey *q.v.*

survey party. A group of trained individuals engaged in a survey.

survey party chief. The leader in charge of a survey party. Also called *survey party leader.*

survival colony. The percentage of animals which continue to exist and persist in a group as compared with the number at a previous time, as immediately following stocking or breeding season.

suscept. Any plant or animal liable to infection by a given pathogen.

susceptibility. The state of being rather easily affected, sensitized, or capable of receiving infection or infestation.

susotoxin. A relatively poisonous ptomaine derived from the cultures of bacillus responsible for hog cholera. It is toxic only if ingested in fairly large amounts.

suspended milker. A type of milking machine suspended by a strap or surcingle on the body of a cow.

suspension. Solid particles mixed with but not dissolved in a fluid, which can be filtered off or removed by centrifugal force, as butterfat in milk.

Sussex. 1. A very old, general-purpose, English breed of chickens, valuable as a table fowl. The *Sussex* has a single comb, white skin and nonfeathered shanks. Mature cocks weigh about 9 lbs.; mature hens about 8 lbs. The egg-shell color is brown. Varieties are the Speckled, the Red, and the Light. 2. English beef cattle similar to, but larger than, Devon (*q.v.*), characterized by moderately dark red coloring and long, slightly-turned horns.

sustained yield management. Controlled exploitation of a forest unit in such a way that the annual or periodic yield of timber or other products can be maintained in perpetuity.

suture. 1. A line along which two things or parts are united. 2. In botany, a line of union or a line of dehiscence. 3. In horticulture, a longitudinal protuberance along one side of drupaceous fruits, as in the peach. 4. To unite or join by sewing.

Suworow sea-lavender. *Limonium suworow,* family Plumbaginaceae, a spiky herb, grown for its late winter and early spring, lavender flowers. Native to Asia. Also called *statice; Russian statice.*

Sveciaost. A Swedish, cows' milk cheese made principally for domestic consumption.

swage. A slightly in-curved metal tool used in blacksmith shops for shaping metal.

swagman. A man who tramps about in search of casual or seasonal employment; in general, decent men seeking work, however brief, who come from the ranks of shearers, farm-hands, fencers, or are just born wanderers (Australia). A number of colorful legends surround the *swagman* and the sundowner (*q.v.*) and their activities. See *tramp.*

Swainson pea. Any of the herbaceous plants of genus Swainsona, family Leguminosae, which is native to Australia. Also called *senna-pea.*

swale. A small, shallow, wet sag in land surface; a natural drainage way which does not necessarily contain a water course or permanent stream.

swallow bug. *Oeciacus vicarius* Horv., family Cimicidae, an insect that sucks the blood of poultry at night.

swallowtail. See *dovetail.*

swallowtail butterfly. Black swallowtail.

swamp. 1. A natural area which has standing water on the surface for all or a considerable period of the year and which has a dense cover of native vegetation. See *marsh; bog.* 2. To clear the ground of underbrush, fallen trees and other obstructions to facilitate subsequent logging operations, such as skidding.

swamp azalea. *Rhododendron viscosum,* family Ericaceae, a deciduous shrub, grown in warm areas for its white or pinkish, very fragrant flowers. Native to North America.

swamp blackberry. *Rubus hispidus,* family Rosaceae, a prostrate, vinelike shrub, which has value as a wildlife plant and is sometimes planted for ground cover. Native to North America. Also called *swamp dewberry.*

swamp blueberry. Highbush blueberry.

swamp chestnut oak. Chestnut oak.

swamper. A worker who clears away obstructions in logging.

swamp fever. See *equine infectious anemia.*

swamp hickory. Water hickory.

swamp honeysuckle. Swamp azalea.

Swamp Land Acts. Acts of U.S. Congress, 1849, 1850, 1860, in which wet lands, part of the Public Domain, were granted to certain states. The original purpose, rarely carried out, was to enable states to reclaim such lands by construction of levees and drains.

swamp laurel. Bog kalmia.

swamp locust. Waterlocust.

swamp milkweed. *Asclepias incarnata,* family Asclepiadaceae, a perennial, marsh herb, grown for its rose-purple, ball-like flower clusters. Native to North America. Also called *rose milkweed.*

swamp persicaria. *Polygonum coccineum,* family Polygonaceae, a strong-growing, perennial weed found in low, wet ground in the U.S.

swamp rose. *Rosa palustris,* family Rosaceae, a hardy shrub, grown for its pink flowers. Native to North America.

swamp smartweed. *Polygonum hydropiperoides,* family Polygonaceae, a weed plant which grows in shallow water and moist places in the U.S.

swamp sumac. Poison sumac.

swamp sunflower. *Helianthus angustifolius,* family Compositae, a perennial swamp herb sometimes grown for its showy, yellow flowers. Native to North America.

swamp white oak. *Quercus bicolor,* family Fagaceae, a hardy, deciduous tree, grown for its handsome gray bark and foliage, gray-green on the undersides. Native to North America.

Swan River daisy. *Brachycome iberidifolia,* family Compositae, an herbaceous annual grown for its blue, rose, or white flowers. Native to Australia.

sward. A closely grazed or mowed area in which the grass and other plant species are close-growing, making an almost complete ground cover.

swarm. 1. The simultaneous emergence or assembly in one place of large numbers of insects, especially bees, often for establishing a new colony. 2. To assemble or emerge simultaneously.

swarm box. A box containing combs of pollen and honey used by queen bee breeders to produce starting cells.

swarm cell. A specially constructed cell placed along the bottom of the comb to produce young queens for swarming.

swarm spore. An asexual reproductive cell which has motion by means of flagella. Also called *zoospore.*

swarm to issue. The time of swarming which usually is from 10 in the morning until 2 in the afternoon.

swath. The single strip cut by an implement or machine in cutting or harvesting hay or grain.

swath-cured. Designating hay or other crops left in the swath for curing.

sway-back. 1. A disease of new-born and young lambs characterized by a staggering gait or inability to walk, endemic in parts of the United Kingdom. 2. Designating a horse with a concave back.

sweat. 1. In animals, moisture exuded through the pores of the skin. 2. Moisture which is given off or collects on the surface of an object.

sweat fly. *Didea fasciata* Macq., family Syrphidae, a predaceous insect whose larva feeds on aphids.

sweat gland. A gland that secretes perspiration.

sweat room. A room in which fruit is artificially degreened with ethylene.

sweat shed (pen). A place to hold sheep so they will sweat before they are shorn to soften the yolk and make shearing easier.

Swede. Rutabaga.

Swedish bark gauge. An instrument used for measuring the thickness of bark on standing timber.

Swedish clover. Alsike clover.

Swedish gang mill. A gang saw mill for round logs which has a carriage permitting adjustment of sawing to the curvatures of the log.

sweeny. A wasting or atrophy of the shoulder muscles of a horse.

sweep. 1. A soil tillage device resembling a cultivator shovel which has attached wings; used for cultivating soil growing cotton, etc. 2. A gradual bend in a log, pole, or piling; a defect. 3. A horse-powered prime mover. Horses were hitched to a long level pole that was attached to a gear wheel which, through contact with other gears, transmitted power through a turning shaft to another machine, such as an elevator, corn sheller, etc., as the horses walked around in a circular path about the unit.

sweep rake. A hayrake with straight, horizontal wooden teeth, by means of which the hay may be raked from either the swath or windrow. Also called *hay rake.* See *rear-hitch sweep rake; side-hitch sweep rake.*

sweepstake prize. A prize awarded to a specimen or group of specimens winning the highest honor in a combination of classes.

sweep stock plow. A plow which has a highly arched or curved stock or beam.

sweet. In dairying, a nonacid, nonsoured and non-ripened product which is free from taint.

sweet acacia. *Acacia farnesiana,* family Leguminosae, a tropical shrub grown for its deep yellow, fragrant flowers used in the preparation of certain perfumes. Native to Australia. Also called *huisache.*

sweet alyssum. *Lobularia maritima,* family Cruciferae, an herb grown as an annual edging or bedding plant for its matting habit and its white or lilac flowers. Native to Europe and Asia. Also called *snowdrift.*

sweetanise. *Osmorhiza occidentalis,* family Umbelliferae, a mountain herb growing at altitudes up to 10,000 ft., palatable to stock and game animals. Native to North America. Also called *western cicely.*

sweetautumn clematis. *Clematis paniculata,* family Ranunculaceae, a hardy, woody vine grown on a trellis or fence for its fragrant, autumn-blossoming, white flowers. Native to Japan. Also called *Japanese clematis; Japanese virginsbower.*

sweet azalea. *Rhododendron arborescens,* family Ericaceae, a deciduous shrub grown for its handsome, fragrant, white or pinkish flowers. Native to North America. Also called *tree azalea.*

sweet balsam. Fragrant cudweed.

sweet basil. *Ocimum basilicum,* family Labiatae, an annual, aromatic, many-branched herb grown for its leaves used as a flavoring. Native to Europe and Asia.

sweetbay magnolia. *Magnolia virginiana,* family Magnoliaceae, a deciduous or half-evergreen tree grown for its very fragrant, white flowers. Native to North America. Also called *sweet bay; beaver tree; white bay; swamp bay; swamp laurel.*

sweetbell redpepper. *Capsicum frutescens* var. *grossum,* family Solanaceae, an annual herb grown for its large, bell-shaped, edible, furrowed, variously colored, mild-flavored fruit used in salads and as a flavoring. Native to America. Also called *bell pepper.*

sweet birch. *Betula lenta,* family Betulaceae, an aromatic tree grown for its dark, reddish-brown bark that glows in winter. The rose-colored wood is used for furniture and boat building. Native to North America. Also called *black birch; cherry birch; mountain mahogany.*

sweet bouvardia. *Bouvardia humboldti,* family Rubiaceae, a tropical shrub grown in the greenhouse for its showy, very fragrant, white flowers.

sweet braheapalm. *Brahea dulcis,* family Palmaceae, a medium-sized fan palm grown as an ornamental. Native to Mexico. Also called *palma dulce.*

sweetbread. The thymus and pancreas glands of an animal which are used as food.

sweetbrier rose. *Rosa eglanteria,* family Rosaceae, a many-branched shrub grown for its aromatic foliage and pink flowers. Native to Europe. Also called *eglantine.*

sweet buckeye. Yellow buckeye.

sweet butter. That butter which has no salt added to it in its manufacture.

sweet butterbur. *Petasites fragrans,* family Compositae, an evergreen, perennial herb grown as a cover for stony or unsightly garden banks. Native to the Mediterranean area. Also called *sweet coltsfoot.*

sweet cherry. *Prunus avium,* family Rosaceae, a deciduous tree of very ancient culture, yielding a globe-shaped fruit, the cherry of commerce. Native to Europe and Asia. See *bigarreau cherry; heart cherry; mazzard cherry.*

sweet cherry bacterial canker. A bacterial disease of sweet cherries and other stone fruits caused by *Pseudomonas syringae,* family Pseudomonadaceae, characterized by elongated, brown, water-soaked or gum-soaked lesions in the bark and outermost sapwood.

sweet cherry crinkle leaf. A nontransmissible disorder of Bing and Black Tartarian varieties of cherry accompanied by unfruitfulness. It has been responsible for great monetary losses to cherry growers in the western U.S.

sweet cherry deep suture. A nontransmissible disorder of sweet cherry characterized by rough leaves and deep sutures of fruits. Considerable variation of intensity has been observed.

sweet cherry mottle leaf. A mosaic virus disease of sweet cherry which causes chlorotic, mottled, or necrotic patterns in the leaves.

sweet cicely. Myrrh.

sweet claret. A sweet, unfortified red wine which is prepared from very ripe grapes in California.

sweet clematis. Plume clematis.

sweetclover. Any herb of the genus Melilotus, family Leguminosae. Species are Asiatic annual or biennial herbs which have been known in the Mediterranean region for 2,000 years and are grown in the U.S. for pasture, hay, silage, soiling, soil improvement and as a honey crop. See *annual yellow sweetclover; Hubam sweetclover; white sweetclover; yellow sweetclover.*

sweetclover disease of livestock. A livestock disease which results in a hemorrhage or series of hemorrhages that may be fatal. It is caused by ingestion of sweetclover with molds and mildews upon the natural coumarins in the plant. Eating such plants produces dicoumarol in the animal's body, which antagonizes vitamin K production and thereby prevents the formation of prothrombin needed for blood clotting.

sweetclover gooseneck. A fungous disease of sweetclover caused by *Ascochyta caulicola,* family Sphaerioidaceae, characterized by stem canker and hypertrophy of the stems, without the stem blackening.

sweetclover mosaic. A virus disease of sweetclover

caused by any of a number of viruses infecting other legumes, characterized by leaf mottling and chlorosis.

sweetclover Phytophthora root rot. A root rot of sweetclover caused by *Phytophthora cactorum,* family Pythiaceae. See *Phytophthora.*

sweetclover Pseudopeziza leaf spot. A fungous disease of sweetclover caused by *Pseudopeziza meliloti,* family Mollisiaceae, characterized by dark spots ranging in color from olive to reddish-brown, purple, or black, on either leaf surface.

sweetclover virus disease. Any of a number of virus diseases which attack sweetclover. Symptoms vary but usually include vein chlorosis, mottling, chlorotic patches between the veins and a reduction in vigor.

sweetclover weevil. *Sitona cylindricollis* Fahr., family Curculionidae, a weevil which feeds on sweetclover, often defoliating the plants. The larvae feed on the roots and damage the plant.

sweet coneflower. *Rudbeckia subtomentosa,* family Compositae, a perennial herb grown for its golden-yellow flowers. Native to North America.

sweet corn. *Zea mays* var. *saccharata,* a variety of corn characterized by edible kernels with a high sugar and low starch content which are wrinkled and translucent when dry. It is used as a fresh garden delicacy and a processed food. See *corn: 1.*

sweet corn smut. See *corn smut.*

sweet cream. See *cream.*

sweet cream butter. Butter made from cream in which little or no ripening or development of acidity has taken place. See *butter.*

sweet cream buttermilk. A by-product resulting from the churning of sweet cream with less than 0.20 percent acidity.

sweet cucumber pickle. A cucumber pickle prepared with a spiced, sweet vinegar.

sweet-curd. A U.S. cheese made from cows' milk by the usual Cheddar process, except that the milk is not ripened and the curd is cut, heated and dried without waiting for the development of acidity.

sweet-curd hole. A type of gas hole sometimes noted in Cheddar cheese which results from undesirable fermentation.

sweet curdling. Abnormal fermentation in milk, characterized by coagulation of the milk at very low acidity.

sweetened condensed skimmed milk. Condensed skim milk to which sugar has been added.

sweetened condensed whole milk. Condensed whole milk to which sugar has been added for preservation.

sweetened corn gluten feed. Corn gluten feed (*q.v.*) to which molasses has been added.

sweetened oat mill feed. Oat mill feed (*q.v.*) to which molasses has been added.

sweet everlasting. Fragrant cudweed.

sweet false chamomile. German chamomile.

sweet fennel. See *fennel.*

sweetfern. *Comptonia peregrina* (*C. asplenifolia*), family Myricaceae, an aromatic, deciduous shrub with fernlike leaves, grown as an ornamental ground cover in dry, sandy soil. Native to eastern North America. Also called *shrubby fern.*

sweet flag. *Acorus calamus,* family Araceae, a yellowish, perennial herb whose aromatic rootstock is used in medicines and tonics. Also called *calamus.*

sweet gale. *Myrica gale,* family Myriaceae, a bog shrub grown as an ornamental for its dense catkins. Native to Europe, Asia and North America. Also called *bog myrtle; Dutch myrtle; moor myrtle.*

sweetgrass. 1. Any plant of the genus Hierochloe, family Gramineae. 2. *H. odorata,* a perennial grass sometimes grown for its sweet odor. It is often placed on the floors of churches on saints' days. Native to Europe. Also called *holy grass; Seneca grass; vanilla grass.*

sweet gum. See *American sweetgum.*

sweet honeysuckle. *Lonicera caprifolium,* family Caprifoliaceae, a climbing, woody vine bearing orange-colored fruit, grown for its habit and for its yellowish-white flowers sometimes tinged with purple. Native to Europe and Asia. Also called *Italian honeysuckle.*

sweet iris. *Iris pallida,* family Iridaceae, a perennial herb grown for its lavender-blue flowers. Native to the Alps.

sweet lemon. A horticultural variety of the lemon, *Citrus limonia,* family Rutaceae, which has no commercial value, but is sometimes grown as a curiosity.

sweet marigold. *Tagetes lucida,* family Compositae, a tender, perennial herb grown as an annual for its fragrant, orange-yellow, dense clusters of flowers. Native to Mexico.

sweet marjoram. *Majorana hortensis,* family Labiatae, a very fragrant, perennial herb grown as an annual for its leaves used as a seasoning and garnish and for its essential oil. Also called *annual marjoram.*

sweet milk. Milk which is free from taint and developed acidity.

sweet mockorange. *Philadelphus coronarius,* family Saxifragaceae, a hardy, deciduous shrub grown for its fragrant, creamy-white flowers. Native to Europe and Asia. Also called *false syringa.*

sweet orange. *Citrus sinensis,* family Rutaceae, a broad-leaved, evergreen tree grown for its edible fruit, the orange of commerce. The plant has become naturalized in many tropical and subtropical regions where new varieties have been developed and commercial culture has become very important. Native to China.

sweet orange crinkly leaf. A virus disease of sweet orange characterized by puckering and cupping of the leaves.

sweet orange fruit scab. A fungous disease of sweet oranges caused by *Elsinoë australis,* family Myriangiaceae, characterized by irregular, or round, corky excrescences often covering a large part of the fruit.

sweet orange psorosis. A virus disease of sweet orange characterized by lesions and scaly bark. Also called *scaby bark.*

sweet orange tristeza. A virus disease of sweet orange characterized by a symptom known as quick decline *q.v.*

sweet osmanthus. *Osmanthus fragrans,* family Oleaceae, an evergreen, perennial shrub or small tree grown for its small, very fragrant, white flowers. Native to Asia.

sweetpea. *Lathyrus odoratus,* family Leguminosae, a very popular, viny, herbaceous, annual grown for its beautiful varicolored blossoms. Native to Sicily.

sweet pepper. Any variety of pepper which is mild and not hot. See *hot pepper; sweetbell redpepper.*

sweet pickle cure. A method of preserving meat in a brine or pickle which consists of salt, sugar, saltpeter, and water.

sweet potato. *Ipomoea batatas,* family Convolvulaceae, a tropical, perennial vine grown for its edible roots, the *sweet potato* of commerce. A very important food crop of the tropics, it is grown for human food, animal feed, and the manufacture of sirups, pectin, flour, and starch, in warm areas. The vines have antibiotic properties. See *yam.*

sweetpotato beetle. See *tortoise beetle.*

sweetpotato black rot. See *black rot of sweet potato.*

sweetpotato Ceratostomella. A fungous disease which attacks sweet potatoes in the field and in storage, caused by *Ceratostomella fimbriata,* family Ceratostomataceae. It produces moist, circular, dark brown or black spots, often covering the entire sweet potato, and black cankers on the stems.

sweetpotato flea beetle. *Chaetocnema confinis* Crotch, family Chrysomelidae, a garden insect that infests the sweet potato and numerous other plants. Long, narrow grooves are eaten in the leaves, wilting and browning them, and causing the plant to become stunted.

sweetpotato leaf beetle. *Typophorus viridicyaneus* (Crotch), family Chrysomelidae, a blue-green insect that eats leaves at the crown of the sweet potato. The larvae also burrow through the vine underground.

sweetpotato meal. A meal made from dried or dehydrated sweet potatoes which is of some value as a stockfeed. It can be included in mash for poultry.

sweetpotato pulp. The dried residue remaining as a by-product in the process of separation of starch from sweet potatoes; occasionally used as a feedstuff.

sweetpotato soft rot. A fungous disease of sweet potatoes caused by *Rhizopus nigricans,* family Mucoraceae. It is a black bread mold which causes the soft rot or ring rot of the roots in storage and can be highly destructive.

sweetpotato stem rot. See *sweetpotato wilt.*

sweetpotato weevil. *Cylas formicarius elegantulus* (Sum.), family Curculionidae, a slender-snout beetle which feeds on the sweet potato and closely related plants of the morning-glory family. The female lays eggs in the roots in the field, which hatch in the stored sweet potatoes and honeycombs them with white grubs. Of Asiatic origin, the beetle is now wide-spread in southern U.S.

sweetpotato whitefly. *Bemisia inconspicua* (Quaint.), family Aleyrodidae, an insect that attacks sweet potatoes in Florida, U.S. The larvae appear on the underside of the leaves and their honeydew encourages a black mold.

sweetpotato wilt. A stem rot of sweet potato caused by *Fusarium batatatis* and *F. hyperoxysporum,* family Tuberculariaceae. It is a major disease of the sweet potato and is characterized by death of the runners with browning of the vascular tissues.

sweet rocket. Dame's-rocket.

sweet sansevieria. *Sansevieria thyrsiflora,* family Liliaceae, a tender, perennial herb grown as a house plant for its long, slender, green leaves, banded with yellow, and its fragrant, greenish-white flowers. Native to South Africa. Also called *guinsensis; snake plant; mother-in-law's tongue; sword plant.*

sweet scabious. *Scabiosa atropurpurea,* family Dipsacaceae, an annual herb grown for its scented, white, rose, purple, or blue flowers. Native to Europe. Also called *mourning bride.*

sweet shrub. See *strawberry shrub: 2.*

sweet soil. A soil that is not acid; *i.e.,* one near a pH of 7. *Cf. sour soil.*

sweetsop. Sugar apple.

sweet sorghum. Any variety of sorghum which has stems filled with sweet juice and is grown for forage and sirup production, in contrast to the grain or nonsaccharine varieties. Also called *sorgo.*

sweet-sorghum hay. The hay or fodder from the sweet sorghum *(q.v.)* in contrast to that from the grain sorghums. The stalks are more nutritious than those of the latter. Also called *cane hay.*

sweet Spanish. Any of several varieties of onion of Spanish origin. They are large-sized, mild in flavor, and grown in some of the western states of the U.S.

sweet Sudan. A cross between a saccharine sorghum and Sudan grass; a fast-growing sorghum which is more resistant to disease and chinch bugs than sweet sorghum.

sweet syringa. Mock orange.

sweet unfortified wine. Red or white sweet wines which do not have wine brandy added.

sweet vernalgrass. *Anthoxanthum odoratum,* family Gramineae, a grass sometimes grown for its fragrance. Escaped it becomes a weed in the meadows. Native to Europe and Asia.

sweet violet. *Viola odorata,* family Violaceae, a perennial herb grown in shade in old-fashioned gardens for its fragrance and deep violet flowers. It is the parent of the florists' varieties of violets. Native to Europe, Africa, and Asia. Also called *English violet; florist's violet; garden violet; tufted pansy.*

sweet water. 1. Water from wells, springs, and surface water which is fresh in contrast to that which is salty or alkali. 2. In cooling systems, refrigerated fresh water in contrast to brine.

sweet white violet. *Viola blanda,* family Violaceae, a perennial herb grown in the wild garden for its white flowers with purple veins. Native to North America.

sweetwilliam. *Dianthus barbatus,* family Caryophylaceae, an herb grown as a biennial for its red, rose-purple, white or varicolored flowers. Native to Europe and Asia. Also called *bunch pink.*

sweet William phlox. *Phlox divaricata,* family Polemoniaceae, a perennial herb grown for its mauve flowers. Native to North America. Also called *wild sweetwilliam; blue phlox.*

sweetwilliam silene. *Silene armeria,* family Caryophyllaceae, an annual herb grown for its pink flowers. Native to southern Europe. Also called *none-so-pretty; sweet William catchfly.*

sweet wine. Wine in which alcoholic fermentation has been arrested and which contains in 100 cc. at 68°F. not less than one gram of sugar.

sweet woodruff. *Asperula odorata,* family Rubiaceae, a perennial herb grown for its white flowers and fragrant, dried foliage. Native to Europe and Asia.

swell. 1. In a saddle, the part in front of the seat which rises to the horn. Also called *fork; front.* 2. A tin can containing preserved foods with its ends bulging due to the formation of gas inside.

swell-butted. Designating a tree enlarged at the base. Also called *bottle-butted; churn-butted.*

swelled can. See *swell: 2.*

swell head. 1. A turkey disease which affects the sinuses or hollow spaces in the bony structure of the face. See *infectious sinusitis.* 2. A poisoning which occurs among sheep and goats grazing on certain plants, such as littleleaf horsebrush, *Tetradymia glabrata,* atamoscolily, *Zephyrantes atamasco,* etc. Also called *big head.* See *photosensitization.*

swelling. The expansion of wood or other materials caused by the absorption of water or other liquids.

swill. Liquid food for domestic animals, especially swine, consisting of ground feed mixed with water or milk or liquid garbage from the kitchen. Also called *slop.*

swine. Any mammal of the family Suidae. Domesticated species are grown for their edible flesh and fat, for their hides, and for their bristles. *Swine* are very important in the agricultural economy of the U.S. Also called *hog; pig.*

swine brucellosis. A disease of swine caused by the organism, *Brucella suis,* and closely related to brucellosis of cattle. It may be characterized by abortions, lameness, posterior paralysis, lowered fertility, or sterility.

swine dysentery. An infectious enteritis of swine attributed to *Salmonella* spp. Also called *infectious enteritis in pigs; necro in pigs; pig typhus; salmonellosis suis.* See *swine enteritis.*

swine enteritis. Inflammation of the intestines which may result from any of several microorganisms, improper feeding or overloading of the digestive tract, or chemical poisons, such as arsenic or mercury. The outstanding symptoms are diarrhea and debility.

swine erysipelas. An infectious bacterial disease of pigs characterized by high fever, reddish or purplish spots on the skin and by general debility and lameness. The causative organism is *Erysipelothrix rhusiopathiae.* Also called *diamond skin disease.*

swine fever. A febrile virus disease of hogs, highly infectious, contagious, and usually fatal. It is characterized by a high fever. More commonly called *hog cholera.*

swine index. In selection of pigs, swine and hogs, the list of a number of factors, such as conformation, rate of gain, feed efficiency, carcass value, prolificacy, etc.

swine influenza. A contagious viral disease of young pigs characterized by mild inflammation of the respiratory passages.

swine pellagra. A disease of swine, caused by a protein and niacin (Vitamin B complex) deficiency in feed and characterized by retarded growth, skin lesions and enteritis. It can be treated by making up the nutritional deficiency.

swine plague. A respiratory bacterial disease of swine caused by *Pasteurella multocida.* Also known as *hemorrhagic septicemia; pasteurellosis.*

swinepox. A pox disease of swine, very contagious, caused by *Variola porcina* or *V. suilla,* and characterized by local lesions of the skin of the abdomen, flanks, head and behind the ears.

swing around stacker. A hay stacker which can be raised and swung around to the side to drop hay or straw at any place desired.

swing (cattle). The main part of a traveling herd of cattle (western U.S.).

swing dingle. A single sled with wood-shod runners and a tongue with lateral play; used in hauling logs down steep slopes or on bare ground.

swinging drawbar. 1. A bar of steel with a pivot at one end attached to a tractor. The other end is attached to a plow, wagon, etc. 2. A wooden bar which may be removed from a fence and serves as a gate.

swing men. 1. In trail herding of cattle, the riders who ride behind the front rider and on the sides of the herd to keep the animals from spreading out and straying (western U.S.). 2. Dairy plant personnel whose function is to relieve regular employees when on vacation, etc.

swing milker. The milking unit suspended from the body of the cow by a strap. Also called *suspended milker.*

swing the gate. To shear sheep the fastest of any shearer in the shed (Australia and New Zealand).

swirl. An abnormal growth of hair on an animal in which the hair grows in a whorl *q.v.*

swirling-jet nozzle. A nozzle, used on knapsack sprayers for insect control, that emits a coarse, spattering spray.

Swiss. A large, hard, pressed-curd, cows' milk cheese with an elastic body and a mild, nutlike flavor which is best known because of the large holes or eyes that develop in the curd as the cheese ripens. Also called *Schweitzer.*

Swiss chard. See *leaf beet.*

Swiss-cheese plant. Ceriman.

Swiss eyes. The holes or openings characteristic of Swiss cheese.

Swiss harp. A specific wire-strung frame used in cutting curdled milk in the Swiss cheese making process. See *cheese knife; curd knives.*

Swiss holes. The large, uniformly distributed gas holes characteristic of Swiss cheese and found occasionally in Cheddar cheese. They result from a particular bacterial growth.

Swiss mountain pine. *Pinus mugo,* family Pinaceae, a hardy, evergreen, shrubby, nearly prostrate tree grown as an ornamental, often in massed plantings. Native to southern Europe.

Swiss steak. A cut of beef, usually from the round, 1 to 2 ins. thick, which is cooked with moist heat.

Swiss stone pine. *Pinus cembra,* family Pinaceae, a hardy, evergreen tree grown as an ornamental and for its edible seeds. Native to Europe and Asia.

switch. The brush of hair on the end of the tail of a dairy cow or other bovine animal.

switcher. A nervous animal that vigorously moves its tail.

switchgrass. *Panicum virgatum,* family Gramineae, a tall, perennial grass found growing in patches or bunches, which is widely distributed throughout the eastern U.S. west to Montana, Nevada, and Arizona. It is palatable as a range species when young. Native to North America.

swivel disk jointer. A concave disk swivel-mounted on a plow; a self-aligning jointer that swings to one side when an obstruction is encountered and automatically swings back after the obstruction has been passed. See *jointer.*

swivel plow. The hillside plow *(q.v.)* which is a reversible moldboard plow.

swollen legs. A condition of horses in which the legs swell or stock up especially during periods of inactivity. See *stocked legs.*

swollen nose. See *bull nose.*

swordgrass. Silvergrass.

sword leaves. See *tobacco frenching.*

sword plant. Sweet sansevieria.

sycamore. American plane-tree.

sycamore anthracnose. A destructive, fungous disease of the sycamore tree caused by *Gnomonia veneta,* family Gnomoniaceae, which attacks the leaves, producing leaf blotches, finally involving the entire leaf.

sycamore fig. See *sycomore fig.*

sycamore lace bug. *Corythucha ciliata* (Say), family Tingidae, an insect whose nymphs feed on the leaves of sycamore and other shade trees.

sycamore maple. Planetree maple.

sycamore plant bug. *Plagiognathus albatus,* Van D., a bug which sucks out the juices on the upper sides of sycamore leaves, presumably leaving a poisonous material, resulting in yellowish or reddish spots which produce holes in the leaves.

sycomore. Sycomore fig.

sycomore fig. *Ficus sycomorus (F. sycamorus),* family Moraceae, a tree grown in frost-free areas for shade and also for its fruit which is edible but of inferior quality. Native to the U.A.R. It is the sycamore of the Bible.

Sydney acacia. *Acacia longifolia,* family Leguminosae, a tall shrub or small tree grown for its golden-yellow flowers or as a street tree in warmer regions. Native to Australia. Also called *Sydney golden wattle; black wattle; feathery wattle; golden wattle.*

Sydney golden wattle. Sydney acacia.

sylvinite. A potassium salt, a mixture of the minerals sylvite and halite. It is a source of potassium chloride used in commercial fertilizers. Also called *hardsalt.*

symbiont. An organism which participates in symbiosis *q.v.*

symbiosis. The living together in intimate association of two dissimilar organisms with a resulting mutual benefit: *e.g.,* lichens with algae and fungi, and leguminous plants living in association with *Rhizobia* spp. See *nitrogen fixation parasite; parasitism.*

symbiotic. Referring to symbiosis *(q.v.)* and nitrogen fixation by legumes in soils.

symbiotic bacteria. Bacteria in nodules or tubercles growing on roots of legumes which have the power to fix free nitrogen from the air into forms that can be utilized by plants.

symmetry. In the show ring, balanced or harmonious development of all parts of the animal.

symphylid. Any small, white, rapidly-moving millipede, which lives on the small roots and root hairs of plants causing dwarfing and hardening. It is an occasional pest in greenhouses.

sympodial. 1. Designating the forking of plants in which one branch continues to develop and the other to abort. *Cf. monopodial.* 2. Designating a plant in which the main stem is a continuous development of the same secondary axis of one fork of a dichotomy. Used with grapevines.

symptom. A perceptible change in any part of the body which indicates disease or a stage of disease. A group of symptoms that, considered together, characterize a disease syndrome. A sign, mark, or indication.

symptomatic anthrax. See *blackleg: 4.*

symptom aggregate. Symptom picture.

symptomatology. The science of symptoms of disease.

symptom picture. The total effect of the individual symptoms of a disease in a given host.

synapsis. Meiosis.

syncarp. A collective or aggregate fruit, such as the mulberry and pineapple.

syndrome. The sum total of the symptoms or signs by which a disease is characterized or recognized.

synergist. Two or more agents or medications which, when combined, produce an effect greater than the additive effect of both when operating alone.

synergistic. Designating two agents or medications working together.

synergized. Designating an agent that has been made more efficient in its biological activity by the addition of another agent (synergist), which in itself may or may not have biological activity of the sort in question. Pyrethrum: *e.g.*, is *synergized* (has its insecticidal activity increased) by piperonyl butoxide, or piperonyl cyclonene, neither of which possesses marked insecticidal properties.

syngamy. Union of the gametes in fertilization.

synthetic. Artificially produced; produced by man's effort and design rather than naturally occurring; chemically manufactured.

synthetic cream. A cream-like material made by the emulsificaton of non-milk fats or hardened oils, such as whale, ground nut, etc., with dried egg, lecithin, soya, glyceryl, monosterate, etc. It is not to be confused with artificial cream *q.v.*

synthetic detergent. A non-soap chemical compound which has detergent properties.

synthetic manure. 1. Organic material, such as leaves, grass, straw, to which a mineral fertilizer and lime have been added to encourage decomposition. After decomposition it is spread on, or worked into, the soil in the manner of barnyard manure. Also called *artificial manure; compost.* 2. Synthetic chemical fertilizers, such as ammonia and urea.

Syrian globe daisy. *Globularia trichosantha,* family Globulariaceae, an erect, perennial herb grown for its blue flowers. Native to Asia.

Syrian juniper. *Juniperus drupacea,* family Cupressaceae, an evergreen tree grown as an ornamental. Native to the Near East.

syringa. See *mockorange; lilac.*

syrup. See *sirup.*

syrup jubea. *Jubaea spectabilis,* family Palmaceae, a palm grown in frost-free areas out-of-doors as an ornamental and in the north as a pot plant or in a tub for decoration. Native to Chile. Also called *coquito palm; wine palm; monkey coconut.*

systematic. Pertaining to or according to a system.

systemic. Pertaining to the body as a whole and not confined to one organ or part of the body, as a *systemic* infection.

systemic effects. Effects related to the body as a whole or those related to a series of parts or organs which unite in a common function.

systemic fungi. Fungi that grow throughout the body tissues.

systemic infection. An infection in which the microorganisms have spread throughout the body.

systemic insecticide. A substance which, when absorbed by plants, or injected or given orally to animals, renders the plants and animals toxic to insects feeding on them.

system of farming. The detailed organization, methods of operation and practices used on a particular farm.

Szechwan David lily. *Lilium davidi* var. *willmottiae,* family Liliaceae, a bulbous herb grown for its orange-red, dark-spotted flowers. Native to China.

szik soil. A saline or alkaline soil.

T

tabac. Tobacco; snuff (French).

tabasco redpepper. *Capsicum frutescens* var. *conoides,* family Solanaceae, a tropical, woody shrub grown as an annual for its pungent, red fruits used in condiments. Also grown as a pot plant. Also called *cone pepper.*

tabernaemontana. Any tree or shrub of the genus Tabernaemontana, family Apocynaceae. See *crapejasmine tabernaemontana.*

tabernaemontanus bulrush. *Scirpus tabernaemontani,* family Cyperaceae, a perennial, marsh herb grown for its flower spikes. Native to Europe.

tabidum. See *duff.*

table. To bend a plant over, as to *table* broomcorn when harvesting it.

table cream. See *coffee cream.*

table fern. See *brake.*

table grape. Any species of grape whose fresh, picked fruit is palatable.

table land. An elevated area or region of flat or undulating surface, bounded by steep slopes. See *mesa.*

table poultry. Poultry raised for meat production.

table scraps. Garbage or leftovers; uneaten food, often fed to chickens, swine, and dogs.

table sirup. Sirup prepared for human consumption.

table stock potatoes. Potatoes for human consumption as contrasted to potatoes raised for seed. See *seed potato.*

table system of sorting. A system of sorting tomatoes, etc., in small canneries, in which the fruit is placed on a table and hand sorted for size, blemishes, etc.

table use. Fit for human consumption and desirable enough to be served as food.

tabulated pedigree. In dairy cattle, a table or chart which presents ancestry, data of production, and other merit records, such as verified milk and butterfat records, show records, etc., of an individual animal.

tacamac poplar. Southern poplar.

tacca. Any herb of the genus Tacca, family Taccaceae. Species are grown as greenhouse plants. See *Fiji-arrowroot tacca.*

tacheture. In animals, a lack of pigmentation in the skin resulting from congenital causes, such as in the piebald horse.

tachina fly. A tachinid fly.

tachinid fly. Any fly of the family Tachinidae, whose larvae are good for insect control as they are parasitic on many noxious insects.

tack. 1. One handle of a scythe. 2. A strip which binds stems to some support, as to a wall. 3. Land or pasturage which is leased (Scottish). 4. Saddle, harness and other light equipment, used with saddle or running horses. 5. To prune (British).

Tackey pony. A breed of horses which are excellent riding horses for children. Native to southeastern U.S. Also called *marsh pony*.

tackling. Draft horse harness.

tack room. A room in a stable in which saddles, bridles, spurs, boots, harness, etc., are kept.

tacky. An animal in poor condition (southern U.S.).

tacso. See *passionflower; passionfruit*.

Taffelost. A sharp-flavored, Scandinavian whey cheese.

tag. 1. A dung-covered lock of wool. Also called *dag; daglock*. 2. A metal piece attached to an animal for identification, or a cardboard or cloth label attached to the container of a product, a feed or fertilizer, giving the content analysis, etc. 3. A lock of cotton fiber which adheres to the boll after picking. 4. To place a *tag* on a product, animal, etc., for identification.

tag alder. Hazel alder.

tagetes. Marigold.

tagged. Designating a product, animal, or plant on which an identification tag has been placed.

tagger. A device used to remove tags from sheep.

tagging. 1. The shearing of the dungy wool from a sheep's crotch. Also called *dagging*. 2. The process of attaching identification to animals. 3. The loss of bits of wool on weeds, fences, etc., by a sheep.

tagging chute. A chute *(q.v.)* used to hold cattle while they are being tagged.

taggy. Designating a sheep which has many tags.

tag lock. See *tag*.

tags. 1. The dung hanging on the wool around the crotch of a sheep. 2. The dirtied wool around the crotch of a sheep. Also called *dags; daglock*.

tag sore. The sticking of a sheep's tail to the tags due to fecal excoriation.

tag-tailed. Referring to a sheep with tags about its crotch. Also called *dag-tailed*.

Tahiti apple. Mombin.

Tahiti arrowroot. Fiji-arrowroot tacca.

Tahiti gooseberry. Otaheite gooseberry leafflower.

Tahiti mulberry. Papermulberry.

Tahiti walnut. Candlenuttree.

tail. 1. The posterior part of the vertebral column of animals. It is usually covered with hair, some of which may be quite long. Also called *brush*. 2. A fanlike row of rather stiff feathers on the posterior part of a bird. See *main tail feathers*. 3. The lowest grade of flour. See *tail of the mill*. 4. A wisp of hay not properly tucked into a bale. 5. The bottom layer of produce in a container. 6. The weakest and poorest sheep (Australia and New Zealand). 7. The woody cane of a plant which has been propagated by tip layerage. 8. To dock an animal's tail (Australia and New Zealand). 9. To keep sheep together in a flock (Australia and New Zealand). 10. To remove the tail from a carcass. 11. To assist an undernourished cow to its feet by pulling on the tail.

tail band. The crupper of a harness.

tailboard. 1. The gate at the rear end of a wagon or truck which can be let down or removed to facilitate loading or unloading. Also called *end board; end gate; shoveling board*. 2. See *breast chain: 2*.

tail chewing. The tendency of certain animals to gnaw at their own tails near the anus, sometimes caused by the animal's being infested with worms or afflicted with mange.

tail-coverts. Those curved feathers which are in front and at the sides of the tail of a bird.

tail down. 1. To roll logs down a skidway for loading. 2. In a rodeo and in a roundup, to twist and pull an animal's tail (especially cattle) so as to cause it to fall and to subdue it for branding, etc.

tailed. Designating an animal whose tail has been removed. See *docked*.

tailed pepper. Cubeb pepper.

tail ender. An unthrifty, runty animal.

tailer. On a threshing machine, a belt to carry straw away from the separating tables (obsolete).

tail feather. See *main tail feathers*.

tail gate. See *tailboard: 1*.

tail grape. Fragrant tail-grape.

tail head. The basal part of an animal's tail.

tail hook. A single, grappling hook used in logging.

tailing. 1. Material unthreshed after the first time through the cylinder and concaves of the thresher, and which is recirculated into the thresher by means of the tailing elevator. 2. The inferior or refuse material of the milling industry. 3. The twisting of a stubborn animal's tail to help move it.

tail male. The sire or top-line.

tail marker. In wildlife, a brightly-colored identification tag which is attached to a tail feather. Also called *tail bank*.

tail of the mill. The stock material left over at the end of the flour milling process which contains large portions of fluffy, fibrous material and finely ground bran. Some low-grade flour may be separated from this material, the remainder being mixed with separations of similar character from other parts of the process to make up the "shorts."

tailor. The preparation of gardenias for sale by removal of the flower from the stem at the point where the two vegetative branches join. Four or five leaves are sewed or stapled to a cardboard collar after which the collar is pushed up around the base of the calyx and a piece of wire is run through the ovary, bent down, and taped.

tailrace. A channel which conducts water away from a waterwheel; an after bay.

tail riders. Mounted cowboys who, during a drive, stayed behind the herd to keep the herd moving (western U.S.).

tail set. A device of leather and metal which, when attached to a horse, causes the horse to hold its tail high. See *tail-setting: 2*.

tail-setting. 1. The position and manner of attachment of the tail of an animal. 2. The nicking or slight cutting of the muscles of the underside of a horse's tail so as to cause the horse to hold its tail higher. See *tail set*.

tail test. See *intradermal tuberculin test*.

tail tree. In forestry power skidding, a tree at the end of a run to which the tackle is fastened.

tail up. To assist a weak or crippled animal to get on its feet by pulling on its tail.

tail water. The water just downstream from a structure.

taily. Designating a plant which grows tails.

taint. 1. A contamination or off-flavor of milk or other products. 2. To infect; contaminate.

tainui. Virescent Anzacwood.

tai-tai. A cultivated variety of the sour orange (*q.v.*) whose flower buds are fire-dried and used by the Chinese as a scenting material for tea.

taiwan hemp. Ramie.

Takane cherry. *Prunus nipponica,* family Rosaceae, a deciduous tree, grown for its white or pink flowers. Native to Japan.

take. 1. The uniting of a cion with a stock following grafting or budding. 2. To accept a male in coitus. 3. To heal on, as the uniting of a cion with the stock after grafting or budding. 4. To result in a mild infection after vaccination.

take a calf from a cow. To wean a calf.

take-all. 1. A fungous disease of wheat, rye, barley, and oats caused by *Ophiobolus graminis.* It is characterized by stunting, premature ripening and by black lesions at the base of the culms. 2. See *wheat mosaic-rosette.*

take-off. 1. The removal of the hide from a carcass. 2. See *power take-off*.

take root. To establish a healthy and normal root system.

take the gate. To divide sheep into classes by running them through a chute and forcing them into proper pens (Australia and New Zealand).

take up. 1. To secure a homestead (*q.v.*). 2. To gather together in one place, as weanlings are *taken up* for the night.

take-up box. In grain machinery, a device for taking up the slack on conveyer or elevator belts which is made in various patterns. The principle mechanism in each case is a shifting bearing for the shaft of a conveyer pulley which can be moved to and fro in its frame by means of a screw to alter the tension of the belt.

tal. The fiber derived from the palmyra palm (*Borassus flabellifer*).

talc. A finely-ground powder, from the mineral talc, a hydrated magnesium metasilicate, which is used as a carrier for poison dusts and in pharmaceutical products as a filler.

Taleggio. A soft, surface-ripened, cows' whole milk cheese with a moldy rind, made in Italy.

talk him out of it. To cause a horse to lose its desire to buck by talking quietly.

tall bilberry. Highbush blueberry.

tall blueberry. Highbush blueberry.

tall buttercup. *Ranunculus acris,* family Ranunculaceae, a perennial, fibrous-rooted herb grown for its yellow flowers. It is poisonous to livestock. Native to Europe. See *buttercup* for symptoms. Also called *blister flower; butter-and-eggs; buttercup; butter rose; field buttercup; tall crowfoot.*

tall butterwort. One species of butterwort *q.v.*

tall coneflower. Cutleaf coneflower.

tall-corn. Of, or pertaining to, the corn belt (humorous).

tall crowfoot. Tall buttercup.

tall cupflower. *Nierembergia frutescens,* family Solanaceae, a perennial plant grown in the flower border as a tender annual, as a greenhouse plant, and as a pot plant for its white, lilac- or blue-tinted flowers with yellow throats. Native to Chile.

tall field buttercup. Tall buttercup.

tall gayfeather. *Liatris scariosa,* family Compositae, a perennial herb grown for its tall flower stalks with bluish-purple flower heads. Native to North America. Also called *rattlesnake master; blue blazing star.*

tall grasses. An association of tall-stemmed native grasses found in prairie areas, such as the bluestem, *Andropogon* spp. See *mid grass; short grass.*

tall-grass prairie. A region in central North America, lying east of the 100th meridian in the Mississippi Valley, and some 250,000,000 acres in area. Its native vegetation is the tall grasses, but much of it is now farm land.

tall indigo. Hemp sesbania.

tall larkspur. Duncecap larkspur.

tall marsh grass. Prairie cordgrass.

tall meadow oat. Tall oatgrass.

tall meadowrue. *Thalictrum polygamum,* family Ranunculaceae, a tall, perennial herb grown in the wild garden for its graceful shape and its white flowers. Native to eastern North America. Also called *musquash weed.*

tall mullein. Clasping mullein.

tall oatgrass. *Arrhenatherum elatius,* family Gramineae, a tall, perennial grass sometimes grown for forage. Native to the Mediterranean region. Also called *tall meadow oat; tall meadow oat grass; meadow oat grass; evergreen grass.*

tallow. 1. The fat extracted from the adipose tissue of cattle and sheep. Used in candle making, soap manufacture, etc. See *suet*. 2. Designating various plants which yield flammable waxes, have a tallowy taste, form a fattening feed, or yield greasy substances resembling *tallow*: *e.g.,* Chinese *tallow*-tree; *tallow*-weed; wax myrtle; etc.

tallow bayberry. Wax myrtle.

tallower. An animal, as a cow or sheep, which produces tallow.

tallow gourd. Chinese waxgourd *q.v.*

tallowiness. Tallowy flavor *q.v.*

tallowing. Fattening of animals to make them produce tallow *q.v.*

tallow loaf. Drumhead cabbage.

tallow shrub. See *wax myrtle.*

tallow tree. See *Chinese tallowtree; candlenuttree.*

tallowy flavor. A flavor defect in milk or milk products suggesting the oily taste of old meat or tallow. It results from oxidation of some of the fatty constituents of milk.

tall pint. A can for ripe olives containing at least 9 oz. net.
tall quart. A can for ripe olives containing at least 19 oz. net.
tall sisymbrium. Tumblemustard.
tall speedwell. Culversphysic.
tall sunflower. Giant sunflower.
tall timber. Wild, uninhabited country wholly or partially covered with forest (colloquial).
tally. 1. A label or tag attached to a product or animal for identification. 2. The number of products or animals handled or produced. 3. 100 sheep (Australia and New Zealand). 4. The total sheep sheared by each shearer in one day (Australia and New Zealand). 5. To record the number of products handled. 6. To mark for identification.
tally book. A record book showing the number of sheep shorn by each shearer in a day (Australia and New Zealand).
tally man. One who keeps the record of the number of animals handled.
tally pen. A pen in which sheep are placed for counting after being shorn.
talon. 1. The hind part or heel of the foot or hoof of such animals as deer, swine, horses, etc. 2. A claw of a bird, usually of a bird of prey.
talus. The heap or fragments of rock and soil material which collects at the foot of a cliff or very steep slope chiefly as a result of rock weathering and gravitational force. See *scree.*
tamarack. 1. Foresters' name for the eastern larch (*q.v.*). Also called *American larch;* sometimes erroneously called *tamarisk.* 2. Lodgepole pine.
tamarack pine. Lodgepole pine.
tamarind. *Tamarindus indica,* family Leguminosae, a widely-distributed, tropical tree grown in frost-free areas for its edible fruit and as a shade tree. Also called *amli.*
tamarisk. 1. Any shrub or tree of the genus Tamarix, family Tamaricaceae. Also called *tamarix.* See *French tamarisk; smallflower tamarisk; fivestamen tamarisk.* 2. See *German falsetamarix; tamarack.*
tamarix. See *tamarisk: 1.*
tambac. 1. A variety of tumbak (*q.v.*). 2. *Aquilaria agallocha,* family Thymelaeaceae, a tree with resinous wood burnt as incense. Native to Indonesia. Also called *aloes* (Bible); *agalwood; agilwood; aguilawood; aloeswood; eaglewood.*
tambo. A corral (South America).
tame. 1. Domesticated. 2. Cultivated. 3. Designating an animal which has been made docile or tractable, as a wild horse is *tamed* or broken. 4. To domesticate. 5. To make docile.
tame hay. Hay produced from sown meadows, as contrasted with the forage from wild areas of native forage plants.
tame pasture. Lands cultivated and seeded and occupied largely by selected pasture plants used chiefly or entirely for the grazing of livestock.
Tamie. A cheese made by a secret process by the Trappist Monks in Savoy, France. It is similar to Tome de Beaumont.
tamp. 1. Any tool or device used for firming soil. 2. To firm soil which has been dug, as around a plant after transplanting.
tamperproof bottle closure. Tamperproof closure.
tamperproof closure. A special type of bottle cap or closure which cannot be removed and replaced without showing evidence of its removal. Also called *tamperproof bottle closure; sealed bottle closure.*
tampon. A plug made of cotton, sponge, or gauze, used for absorption of vaginal secretions.
Tamworth. A bacon-type English breed of swine which has a long neck, body, and legs and whose coat varies from golden-red to dark red. The oldest breed of hogs, it is large and rugged.
tan. 1. A Japanese unit of land measurement equal to 0.245 acre. 2. A coat-color of some animals, especially dogs. It is a yellowish-brown similar to the color of well-tanned leather, and varies in shade from light to dark. 3. To convert a hide into leather.
tanbark. The bark of hemlock, chestnut, oak, mimosa (wattle), or tanoak, etc., which is used as a source of tannin. The residue is often used as a ground covering in circus lots, race tracks, livestock pavilions, etc. Also called *tanner's bark.*
tanbark oak. 1. Any oak from which tanbark is obtained: as black oak (*q.v.*). 2. See *tanoak.* 3. Chestnut oak *q.v.*
tanbark tree. Canada hemlock.
tandem. One behind the other, as the hitching of horses or tractors one behind the other as a tandem team of animals.
tandem disk harrow. Double-disk harrow.
tandem hitch. See *tandem.*
tandem team. A pair of work animals hitched one ahead of the other rather than side by side.
tan disease. A physiological disorder of plants caused by excess soil moisture, characterized by either the roots or the aerial portions becoming swollen and by the outermost cork layers breaking and peeling off.
tangan. See *castorbean.*
tang bees. To beat on a pan, etc., to cause a swarm of bees to settle.
tangelo. *Citrus reticulata* x *C. paradisi,* family Rutaceae, an evergreen, hybrid tree grown for its delicate, juicy, edible fruit.
tangent saw. See *bastard-saw.*
tangerine. A horticultural variety of the mandarin orange (*Citrus nobilis* var. *deliciosa*) grown for its deep orange colored, loose-skinned, edible, sweet fruit. See *mandarin.*
Tangier peavine. *Lathyrus tingitanus,* family Leguminosae, an annual herb grown for its sweet-pea-like, odorless, dark, red-purple and red flowers. It is sometimes grown for forage. Native to the western Mediterranean. Also called *Tangier scarlet pea.*
tanglefoot. 1. Any sticky substance which can be applied to a tree trunk to catch and hold climbing insects to prevent their damaging the tree. 2. Broom deervetch.
tanglehead. *Heteropogon contortus,* family Gramineae, a perennial forage grass useful for reseed-

ing ranges and pastures, found in the warmer areas from Texas to Arizona, U.S. Native to North America.

tangleroot. A physiological abnormality of the pineapple in which the roots mass around the rootstock rather than growing outward.

tangletail. Goldmoss stonecrop.

tangor. *Citrus reticulata* x *C. sinensis,* family Rutaceae, a hybrid tree grown in warm areas for its sweet, edible fruit.

tania. Dasheen.

tanier. Dasheen.

tank. 1. A pond or pool, natural or artificial, which serves as a supply of drinking water for livestock (western U.S.). 2. A metal or wooden receptacle used to hold liquids, as the *tank* of a sprayer, a water *tank,* a gasoline *tank.*

tankage. 1. A fertilizer consisting of, and restricted to, the rendered, dried and ground meat and bone by-products of the carcasses of animals which have been slaughtered, or which have died from other causes. 2. See *digester tankage.* 3. An animal protein supplement feed.

tank agriculture. Hydroponics.

tanker. 1. A specialized truck which carries a very large, specially-built tank for the transporting of bulk milk. 2. A specialized truck with a large tank of water, a power pump, and base, used to fight forest fires.

tanker operator. The individual who drives the bulk milk tanker, collects the milk from the various producers, and transports it to the milk plant, receiving station, or buyer. Also called *tanker driver.*

tank gardening. Hydroponics.

tank heater. A stove specially designed to prevent freezing of drinking water for livestock.

tanking. 1. A method used to prepare livestock feedstuff from animal tissue in which the tissues are pressure cooked by steam. The liquids are drawn off and evaporated and the residue is pressed. The evaporated liquids are added to the pressed residue and the whole is dried and ground. Also called *wet rendering.* 2. Designating equipment used to prepare tankage. See *tankage: 1.*

tank-mix oil. Any spray which is prepared immediately before use by adding oil, the emulsifier, and spreader separately to the water in the spray tank and emulsifying by agitation and pumping.

tank room. The room in a meat-processing plant in which tankage is prepared. See *tankage: 1.*

tank sprayer. Any sprayer (*q.v.*) that is equipped with a supply tank or reservoir, as contrasted to the atomizer.

tank truck. See *tanker: 1.*

tannage. 1. The materials (chemicals, etc.) used in tanning. 2. The process of tanning.

tanner. One who practices the art and science of tanning leather. See *tan:* 3.

tanner's bark. Tanbark.

tannery. A place where the art or process of tanning or leather dressing is carried on.

tannic acid. Tannin; gallic acid, which is obtained from various sources: chestnut (*Castanea* spp.) wood and bark; redquebracho (*Schinopsis* spp.); oak (*Quercus* spp.) bark, acorns, and galls; tanoak (*Lithocarpus* spp.) bark; hemlock (*Tsuga* spp.) bark; American mangrove (*Rhizophora mangle*) bark; wattle (*Acacia* spp.) bark; dividivi (*Caesalpinia coriaria*) leaves; eucalyptus (*Eucalyptus* spp.) bark; pecan (*Carya illinoensis*) shells; sumac (*Rhus* spp.) leaves; cabbage palm (*Sabal palmetto*) roots. It is soluble in water and is capable of combining with the proteins in animal skins to convert the skins into leather.

tannin. Tannic acid.

tanning. The conversion of animal hides into leather.

tanoak. 1. Any evergreen tree of the genus Lithocarpus, family Fagaceae. 2. *L. densiflorus,* a broadleaved, evergreen tree grown as an ornamental and for its bark used in tanning. Native to North America. Also called *tanbark oak.*

tan rot of strawberries. A relatively unimportant, fungous disease of strawberries caused by *Pezizella lythri,* family Pezizaceae.

tansy. 1. Any herbaceous plant of the genus Tanacetum, family Compositae. 2. *T. vulgare,* a perennial, strong-scented, bitter-tasting herb sometimes grown for its yellow flowers and in the herb garden for its leaves used in seasoning and medicinally. Escaped, it becomes a weed pest. Native to Europe. Also called *English cost.*

tansy ragwort. Ragwort groundsel.

tanwood. Any of the various trees which yield tannin.

Tanzenberger. A Limburger-type cheese made in southern Austria.

tap. 1. In lumbering, a cut made from the inside of a log. 2. A wooden basket which is used for packing figs. 3. A faucet on a pipe or container containing liquid. 4. To remove a taproot from a plant. 5. To make a cut in a tree for obtaining sap as in the manufacture of maple sugar and turpentine.

tapadera. A leather hood which covers the front of a stirrup.

tape gage. See *chain gage.*

tape grass. Spiral wildcelery.

taper. The gradual reduction in diameter in the trunk of a tree from the base to the top. Also called *rise.*

taperpetal buglelily. *Watsonia angusta,* family Iridaceae, a bulbous herb grown for its scarlet flowers. Native to South Africa.

tapeworm. A parasitic intestinal worm of a flattened, tapelike form, order Cestoda, composed of separate parts or segments.

tapeworm plant. Ribbonbush.

tapioca plant. Cassava.

tapis vert. A formal garden.

tapnet. See *tap: 2.*

tapoun. A root segment.

tap(p)a. The bark of the papermulberry *q.v.*

tapping. 1. The boring of holes in the trees and the attaching of buckets to them to collect the sap, as in the manufacture of maple sugar, turpentine, natural rubber, etc.

tappoon. See *check dam.*

taproot. The primary descending root, usually conical in form, of a plant from which lateral branching roots may develop. Examples of crop plants which have *taproots* are carrots and alfalfa. See *primary root; root.*

taps. Stirrup coverings (southwest U.S.).

tap water. Water which comes from a water faucet, as contrasted with well water or rain water.

tar. 1. A black, liquid mixture of hydrocarbons and their derivatives obtained by distillation of wood, peat, coal, shale, etc. 2. *Tar* used as a disinfectant on a sheep accidentally cut by the shearer (Australia and New Zealand).

tara. See *taro.*

tarata. Tarata pittosporum.

tarata pittosporum. *Pittosporum eugenioides,* family Pittosporaceae, an evergreen shrub or small tree, bearing yellow-green flowers, grown in warm areas as an ornamental. Native to New Zealand. Also called *tarata; taratah; hedge laurel; turpentine tree; white mapau.*

tara vine. Bower actinidia.

tarboy. A person who daubs disinfectant on the wound of a sheep caused by a shearing accident (Australia and New Zealand).

tarbush. American tarbush.

tar distillate. The condensed distillate from tar, sometimes used as a constituent of dormant sprays for the control of aphids, oystershell scale, etc.

tare. 1. Weed seeds. 2. Deduction made in the weight of packaged products to allow for the weight of the container; gross weight minus the net weight. 3. Deduction made in the weight of sugar beets to allow for earth, crown tops, etc., which might adhere to the beets. 4. Any impurity in seed crops, such as dirt, chaff, weed seeds, broken seeds, etc. 5. See *vetch.*

tarefitch. See *vetch.*

tare room. The place at a sugar beet factory where the amount of earth, leaves, crown, etc., adhering to the beets is determined.

tares. Darnel ryegrass.

target spot. Early blight of potato.

target spot of red clover. Stemphylium leaf spot of clover.

target spot of sorghum. A fungous disease of Johnsongrass, Sudan grass, and sorghums, caused by *Helminthosporium sorghicola,* family Dematiaceae, characterized by small, round or oval lesions of alternate light brown and tan tissues on the plant leaves.

target spot of soybeans. A fungous disease of soybeans caused by *Phyllosticta* spp., family Sphaerioidaceae, characterized by a spotting of the petioles, stems, pods, and seeds. The spots vary from a pin-point to ½ in. in diameter, the larger ones suggesting a target after a dull green or yellow-green halo surrounding the spot.

target system. In the strategy of change, the system, group, or individual, such as a teacher, an agricultural extension agent, agricultural extension service.

Targhee sheep. A new breed of sheep developed at the U.S. Sheep Experiment at Dubois, Idaho. It is the result of crossbreeding the Rambouillet with Lincoln and certain other long-wool breeds, producing a type especially suited to northwestern range conditions. The *Targhee* is a polled, white-faced sheep of intermediate size with mature rams weighing about 200 lbs. and ewes about 130.

tariff. An excise tax levied by a government on imported products.

tariff wall. The imposition of an excise tax in a country which tends to reduce or prohibit the importation of certain products from another.

tarnished plant bug. *Lygus lineolaris* (P. de B.), family Miridae, a bug that infests many fruits and garden plants, releasing a toxin that sometimes causes a deformity or dwarfing. See *stopback.*

taro. 1. *Colocasia antiquorum,* family Araceae, a tuberous-rooted, perennial herb, grown as a summer bedding plant for its decorative foliage and for its large, edible, starchy root. Native to southeast Asia. Also called *elephant's-ear.* 2. See *dasheen.*

tar oil. See *tar distillate.*

tarp. Tarpaulin.

tarpan. A miniature horse which roamed wild over Europe for thousands of years and finally became extinct in the latter half of the 19th century. Also called *Tartar horse; Prejvalsky horse.*

tar paper. A heavy paper, coated or impregnated with a bituminous derivative, used as a temporary covering or an undercovering for siding and roofing, the top half of drainage-tile joints, or to cover a frame, as a *tar-paper* shack.

tarpaulian. See *tarpaulin.*

tarpaulin. A large, water-proof, canvas sheet, used to protect produce, etc., from inclement weather.

tarragon. *Artemisia dracunculus,* family Compositae, a perennial herb grown as a source of an aromatic, pungent, flavoring extract used in pickle manufacture and in the production of *tarragon* vinegar. Native to Europe and Asia. Also called *estragon.*

tarragon vinegar. Vinegar usually derived from cider, flavored with tarragon extract.

tarrify. See *treacle erysimum.*

tar spot. 1. A disease of grasses caused by *Phyllachora* spp., especially *P. graminis,* family Dothideaceae, characterized by black, sunken, glossy spots on the leaves. 2. A fungous disease of maples, willows, and oaks, caused by *Rhytisma* spp., family Phacidiaceae, characterized by black, thick, raised, tarlike spots up to ½ in. in diameter on the upper leaf surfaces.

tartar emetic. Potassium antimonyl tartrate, an insecticide used in sweetened bait sprays to control fruit flies and thrips, and as a constituent of ant poisons. It is also used as a medicinal ingredient in appetite stimulators for the bovine. Also called *tartarized antimony.*

Tartar horse. See *tarpan.*

Tartarian aster. Tatarian aster.
Tartarian buckwheat. Tartary buckwheat.
Tartarian dogwood. Tatarian dogwood.
Tartarian honeysuckle. *Lonicera tatarica,* family Caprifoliaceae, a hardy, vigorous shrub grown for its white or pink flowers and red fruit. Native to Asia. Also called *garden fly honeysuckle; bush honeysuckle.*
Tartarian oat. Side oat.
tartaric acid. An acid occurring widely in plants used in dyeing, photography, medicine, and jelly manufacture.
tartarized antimony. Tartar emetic.
Tartary buckwheat. *Fagopyrum tataricum,* family Polygonaceae, an annual herb grown for its seed sometimes used in poultry feeds. Native to India. Also called *India wheat; Kangra buckwheat; duck wheat; wildgoose; rye buckwheat; hull-less buckwheat; bloomless buckwheat.*
tarweed. Fireweed fiddleneck.
Tasmanian black spot. Apple scab.
Tasmanian bluebell. Australian bluebell-creeper.
Tasmanian cedar tree. See *swamp gum: 1.*
Tasmanian blue eucalyptus. *Eucalyptus globulus,* family Myrtaceae, a gigantic, evergreen tree, grown in semitropical and warm temperate areas as an ornamental, a bee tree, and for its valuable lumber. Native to Australia. Also called *blue gum.*
Tasmania stringy-bark. Messmate stringybark eucalyptus.
tassel. The male inflorescence of the corn plant.
tasselflower. 1. Any herb of the genus Emilia, family Compositae. See *earleaf tasselflower.* 2. See *prairie clover.* 3. See *love-lies-bleeding.*
tassel grapehyacinth. *Muscari comosum,* family Liliaceae, a bulbous herb grown in the rock garden and under shrubs for its purplish-blue or reddish-purple flowers. Native to Europe and Asia. Also called *feathered hyacinth; feather hyacinth; tassel hyacinth.*
tassel hyacinth. Tassel grapehyacinth.
tasseling time. In corn fields, an indefinite period during the growing season between tassel emergence and pollen shedding of corn plants. It generally occurs during July in the corn belt of the U.S. but may vary with season, time of planting, variety, etc., in other countries.
taste. 1. The flavor of a product as determined by placing the sapid substance in the mouth. 2. A small amount or sample. 3. In judging, to place a morsel or a few drops of a product in the mouth to savor, but often not to swallow.
taster. A person skilled in the act of tasting.
tat. See *tattoo.*
Tatarian aster. *Aster tataricus,* family Compositae, a tall, perennial herb, grown for its late-summer or autumnal, bluish-purple flowers. Native to Asia. Also called *Tartarian aster.*
Tatarian dogwood. *Cornus alba,* family Cornaceae, a hardy shrub, bearing whitish-blue fruit, grown for its bright red twigs, white flowers. Native to eastern Asia. Also called *Tartarian dogwood.*
Tatarian sea-lavender. *Limonium tataricum,* family Plumbaginaceae, a salt marsh, perennial herb, grown for its white and red, green-veined flowers. It can be used as an everlasting. Native to southern Europe.
tatter leaf. A stone-fruit virus disease of the ring spot type characterized by lacelike perforations in the leaves.
tattoo. 1. A letter, number, or other mark, pricked with indelible ink into the skin of an animal for identification. 2. To make such a mark.
tattoo-netted corky bark. A virus disease of citrus trees characterized by irregular, elongated areas, spots, and ridges of corky bark, all of which gives a tattooed effect. See *corky bark.*
tattoo oil. The pigment used in making a tattoo *q.v.*
taw. 1. To dress and prepare, as the skins of sheep, lambs, goats and kids, for gloves, etc., by imbuing with alum, salt, and other softening and bleaching agents. 2. To prepare hemp by beating.
tawhiwhi. See *tawhiwhi pittosporum.*
tawhiwhi pittosporum. *Pittosporum tenuifolium,* family Pittosporaceae, an evergreen tree grown in frost-free areas for its foliage and dark purple flowers. Native to New Zealand. Also called *black mapau kohuhu.*
tawkee. See *goldenclub; Virginia arrow arum.*
tawkin. See *goldenclub; Virginia arrow arum.*
tawny daylily. *Hemerocallis fulva,* family Liliaceae, a lily-like herb, grown for its rusty, orange-red flowers. Native to Europe and Asia.
tax. A compulsory charge which is levied by a governmental, national, state, or local unit against income or wealth for the common good.
tax deed land. Land which is held under a deed issued by a governmental unit pursuant to sale for nonpayment of taxes.
taxonomic. Referring to taxonomy, the science of classifying organisms.
taxonomist. A scientist who classifies plants and animals according to their morphological relationships.
taxonomy. The science of classification of plants and animals according to their morphological relationships and of the principles of such classification.
tayer. Sour, curdled milk (India).
T.B. Tuberculosis.
T-bone steak. A retail cut of beef for broiling which is taken from the loin section of the carcass. Characterized by its T-shaped bone, it is one of the most desirable cuts.
T-budding. See *shield budding.*
TCA. The sodium salt of trichloroacetic acid which is used as an herbicide to control quackgrass and annual weed grasses.
tche. *Cudrania tricuspidata,* family Moraceae, a shrub or small tree grown for its leaves fed to silkworms, and its edible, red fruit. It is also grown in hedges. Native to Asia.
T cut. See *shield budding.*
T. D. N. Total digestible nutrient.
tea. *Camellia sinensis* (*Thea sinensis*), family Theaceae, an evergreen tree or shrub grown in Asia for its leaves, the *tea* of commerce. It is also

grown in warm areas as an ornamental. Probably native to China.

teaberry. See *checkerberry; creeping pearlberry.*

tea broom. See *broom tea tree.*

tea garden. A plantation of tea trees.

teak. *Tectona grandis,* family Verbenaceae, a tall tree grown for its hard, durable, yellowish-brown lumber used in shipbuilding, furniture manufacture, etc. Native to the East Indies.

teakwood. Wood of teak *q.v.*

team. 1. Two or more horses, mules, etc., which are harnessed to the same vehicle, plow, etc. 2. Two or more specialists who are jointly investigating a problem. 3. Each drover's dogs (Australia and New Zealand). 4. Two or more judges. 5. In Pennsylvania, U.S., a single animal, harness and implement which is used in service or labor. 6. To hitch two or more draft animals to the same vehicle, plow, etc. 7. To match two or more draft animals to serve as a team.

team lines. The leather reins attached to the bridle bits of a team.

teamster. A person who drives a team.

tea-of-heaven. *Hydrangea serrata,* family Saxifragaceae, a shrub grown for its blue or white flowers. Native to Japan and Korea.

tea oil tree. Sasanqua camellia.

tea olive. See *osmanthus.*

tea plant. See *tea.*

tear. In wool marketing, the expressed proportion of top (*q.v.*) to noils (*q.v.*), as 7 to 1, meaning "7 lbs. of top to 1 lb. of noils."

tea rose. *Rosa odorata,* family Rosaceae, an evergreen or half-evergreen, partially climbing shrub, varieties of which are widely grown for their double, white, pink, or yellow flowers. It is too tender to survive outdoors except in very warm areas. Native to China.

tear out. 1. To exhaust the fertility of the soil, as by continuous growing of a single crop. 2. To remove shrubs or trees by removing most of the roots with the tops.

tear-stain. Citrus anthracnose.

teart of cattle. Molybdenum poisoning.

tea scrub. A dense vegetation composed of tea trees (Australia).

tease. 1. To stimulate an animal to accept coitus. 2. To vex or annoy. 3. Of fibrous materials such as wool or flax, to separate the strands of the fibers, or to prepare the fibers, by combing, for spinning. 4. To form a nap on cloth by stroking the loose fibers in one direction with comb-like, natural, or mechanical, teasels.

teasel. Any herb of the genus Dipsacus, family Dipsaceae. See *Venuscup teasel; fuller's teasel.*

teaselgourd. *Cucumis dipsaceus,* family Cucurbitaceae, a prickly stemmed, herbaceous vine grown as an ornamental or as a curiosity. Also called *hedgehog gourd.*

teaser. 1. An animal which is used to stimulate sexually one of the opposite sex. 2. Designating a male animal which is used to locate females of the same species in heat.

teasing pole. A pole used to separate a stallion from a mare to determine from the mare's action whether or not she will accept coitus with the stallion.

teasing stall. A stall in which a female teaser animal is placed for the purpose of stimulating a male sexually.

teat. The fleshy protruberance through which milk is drawn from the udder of a mammal. Also called *tit.* (*Teat* is used in formal discourse; *tit* is more common colloquially.) Also called *thelium.* See *nipple.*

teat canal. The passageway from the teat cistern to the outside. Also called *streak canal.*

teat cistern. The cavity within the teat proper.

teat cup. 1. The receptacles of a milking machine that hold the rubber inflations. 2. A cup used to apply remedial solutions to a cow's teat.

teat dilator. A plug placed in the teat canal either to dilate it or to keep it from closing if injured.

teat erosion. A lesion on the end of a cow's teat which is caused by constant irritation, such as from strong washing solutions or a milking machine.

teat eversion. The pulling out of the lining of the teat as a result of the teat cup being removed while there is still a vacuum.

teat expander. Teat dilator.

teat opener. Any instrument used to open the teat canal in diseased or injured condition.

teat patency. Insufficient closure of the teat sphincter. See *leaking teat.*

teat plug. Teat dilator.

tea tree. 1. Any of several plants of the genus Leptospermum, family Myrtaceae. See *broom teatree; Victoria teatree.* 2. See *tea.*

teat slitter. Bistoury.

teat spider remover. An instrument used for removal of growths (spiders) from within the teat canal.

teat stenosis. A narrowing or complete closure of the teat canal.

teat trim. The trimming away of the mammary glands from a bacon side.

teat tube. Milking tube.

technical. 1. Concerned with a particular science, industrial art, profession, sport, etc. 2. Practicing, using, or pertaining to the technique rather than the theory or underlying principles involved in the execution of a project. 3. Designating the grade of a commodity manufactured in the usual commercial manner. 4. Pertaining to or designating a market where prices are controlled by speculation or manipulation.

technique. An expert manner of executing practical details in a project.

technological change. Alterations in methods of production, distribution, and work organization, generally aimed at increasing economic effectiveness. See *social change.*

technology. 1. The branch of science dealing with the industrial arts. 2. Technical terms.

tecoma yew. Cabbage angelintree.

ted. To agitate and loosen hay in the swath with a tedder for drying purposes.

tedder. A device consisting of a wheel-mounted frame with a series of small forks attached to a crankshaft. It is used to stir and loosen hay in the swath for more even and quicker drying.

teepee-hill system. A system for training red raspberries. Five to eight canes, enough apart to make a base, with sides 15 ins. long are selected and drawn together and securely tied at the top. Another tie is made about 8 ins. lower.

tee scraper. A scraper shaped like the letter T, used on disk implements to keep the concave side of a disk clean.

Teeswater. 1. An old, English breed of cattle thought to be one of the progenitors of the Shorthorn. 2. An old, English breed of sheep.

teeth. 1. The tines of a rake, fork, etc. 2. The steel spikes on a spiked-toothed harrow. 3. See *tooth: 1.*

teff. *Eragrostis abyssinica,* family Gramineae, a fragrant grass grown for its cereal grain and as an ornamental. Native to North Africa.

teg. 1. A two-year-old sheep. 2. The fleece of a two-year-old sheep.

teil. European linden.

telegraph plant. Telegraph tickclover.

telegraph tickclover. *Desmodium gyrans,* family Leguminosae, a tropical, perennial herb grown as a curiosity because of the movements of its leaflets. Native to Asia.

Teleme. A pickled, white, slightly-cured, goats' or ewes' milk cheese which is made in the Balkan countries.

telltale thermometer. A special type of thermometer used in the canning industry. It operates like a clinical thermometer in that the mercury does not automatically descend from the maximum. It is used in cans during sterilization to record the highest temperatures reached.

temper. 1. The proper relative condition of moisture in grain preparatory to milling. 2. The relative hardness or softness of the metal in implements and tools. 3. The relative mildness or viciousness of an animal. 4. Milk of lime, etc., added to boiling sirup to clarify it.

temperament. Disposition of an animal.

temperate house. Cool greenhouse.

temperate rain forest. A woodland usually with a dominant species; found in a temperate zone in areas having little frost.

temperature. 1. The amount of heat or cold measured in degrees on different scales, as Fahrenheit or Centigrade. At sea level, water freezes at 32° F. and boils at 212° F. or 100° C. 2. The degree of heat in a living body. 3. Abnormal heat in a living body. Also called *fever.* See *thermometer.*

temperature inversion. A temperature condition in the atmosphere in which the temperature rises with increased elevation. A common condition when a radiation type frost forms.

temperature of holding. Holding temperature.

temperature test. Subcutaneous tuberculin test.

tempered. Designating a cheese which has been held in a room in which the temperature ranges from 50 to 60°F. (10 to 15.5°C.) for a sufficient time to secure a uniform temperature throughout its body.

tempering. 1. Accustoming planting stock gradually to a material change in temperature. 2. Treatment of tools and implements by blacksmiths, etc., to impart a degree of hardness or softness to the metal.

templet excavator. An excavating machine with a frame shaped to suit the width of the ditch. It consists of a bucket, with a cutting edge at either end, which moves back and forth on the underside, cutting out the earth, or an endless chain with buckets which cut out and remove the soil.

temporary dam. 1. A dam constructed of not very enduring material, such as brush, wire, or loose rock, for erosion control. 2. An easily removable dam made of metal, wood, or soil; used in irrigation to divert small flows of water.

temporary intermediate. One of four teeth of a horse which appears at about 5 weeks of age on either side and close by the first pair of teeth in the upper and lower jaws.

temporary meadow. An area on which hay or pasture plants occupy the land for a short period of time. An annual crop, such as Sudan grass or millet, may be considered a *temporary meadow* crop in contrast to perennial forage crops, such as alfalfa.

temporary pasture. A pasture that is used for grazing for a short period, usually not more than one season.

temporary permit. A license issued by the U.S. Forest Service to graze domestic livestock on a National Forest for a period of time not exceeding one year.

temporary silo. Any structure built of snow fence, baled straw, etc., within which the crops of one season may be ensiled. These structures usually last until the stored material is fed after which they are no longer useful. *Cf. trench silo.*

temporary structure. Any mechanical structure used to retard erosion or waterflow which is constructed on a temporary basis, but adequate enough to insure beneficial use over a short period, as one or two seasons.

temporary teeth. Milk teeth.

temporary unsoundness. In an animal, a condition temporarily preventing its full utilization: *e.g.,* lameness, disease. *Cf. breeding unsoundness.*

tenacious soil. A soil that shows a decided resistance to rupture. The soil mass coheres firmly. Sticky and *tenacious* are often used as synonyms, but in soil usage the former is taken to refer to adhesion, the latter to cohesion. Both may be present in a soil at the same time.

tenancy. The status of being a tenant. Farm *tenancy* as a study deals with (a) the conditions and circumstances under which land is rented and operated and (b) the rights and responsibilities of farm landlords and tenants.

tenancy-by-the-entirety. A joint-tenancy between husband and wife which differs from joint-tenancy (which can be between any relatives or non-relatives) in that neither husband nor wife can break the arrangement without the other's consent.

tenancy from year to year. A situation in which a landlord and a farm tenant agree that the tenant rents and operates the farm on a one-year basis, but with the understanding that the lease will be automatically extended for another year unless either party serves written notice on or before an agreed date, upon the other party, informing him of the termination of the lease at the end of the year.

tenancy in common. A legal arrangement in which each party owns an undivided share in property.

tenant. A person who rents or leases land from another person on a share basis, cash basis, or a combination of the two. See *renter; sharecropper.*

tenant farm. A farm which is rented or leased by the owner to another person for operation.

tenant farmer. A farmer who rents or leases land from another person, or one who works it on shares.

tenant-operator. A farmer who rents the farm he operates. *Cf. owner-operator.*

tend. To care for, as to *tend* a garden, to *tend* a flock of sheep.

tender. 1. Designating a plant which is unable to withstand a frost. See *hardy:* 2. 2. Sensitive; painful, as a *tender* abdomen. 3. Designating abnormally weak wool fibers.

tenderable. Designating a grade or class of a commodity which may be delivered in settlement of a future contract.

tender annual. Any annual plant which cannot withstand a frost and whose seeds will not live outdoors over winter in the regions where subfreezing temperatures prevail in winter.

tenderfoot. Originally applied to a newcomer to the ranching and mining regions of the Old West, U.S. Commonly applied to anyone new to any outdoor type of living.

tenderized. Designating a cut of meat which has been macerated or made tender by enzyme action, mechanical means, or by the action of certain chemical substances.

tenderloin. 1. The psoas muscle lying on either side of the backbone of meat animals and considered a very choice cut of meat. 2. See *oyster.*

tenderometer. A device which determines the amount of force necessary to shear peas, an indication of maturity.

tender wool. Abnormally weak wool. See *hunger fine.*

tendon. The strong tissue terminating a muscle and attached to a bone, for leverage purposes; usually a dense, cordlike structure of various thicknesses. Wide, flat tendons are called aponeuroses and fasciae.

tendon puller. A device, used in poultry slaughter, which consists of a hinged bar and a fixed notch. The feet of the birds are placed in the notch and the legs are affixed to the bar. By pulling at the bar the tendons in the legs are removed.

tendovaginitis. Inflammation of a tendon and its sheath.

tendril. A thin, leafless, thread-like organ of a plant which usually coils around whatever it grasps enabling the plant to climb, as the grape.

tendron. A young bud or tender shoot of a plant.

ten-eighty. Sodium fluoracetate *q.v.*

ten-frame hive. The most widely used type of beehive, containing 10 frames 9⅛ x 17⅝ ins. spaced ⅜ in. from center to center. Also called *standard Langstroth hive.*

ten-lined June beetle. *Polyphylla decem-lineata* (Say), family Scarabaeidae, a beetle, whose larva, a white grub, may girdle and destroy the roots of ornamental plants.

Tennessee red cedar. Eastern redcedar.

Tennessee Valley Authority (U.S.). A corporation created by an act of Congress, May 18, 1933, to take custody of the Wilson Dam and appurtenant plants at Muscle Shoals, Alabama, and to operate them in the interest of national defense and for the development of new types of fertilizers for use in agricultural programs, and for the development of the Tennessee River and its tributaries in the interest of navigation, the control of floods, and the generation and disposition of hydroelectric power.

Tennessee Walking Horse. A saddle-type breed of horse developed in central Tennessee, U.S., which stands about 15.2 hands high and weighs from 1,000 to 1,250 lbs. It has a good disposition and a natural running-walk gait. The colors are sorrel, chestnut, black, bay, gray, roan, yellow, and white.

ten-petal mentzelia. *Mentzelia decapetala,* family Loasaceae, a biennial, hardy herb grown for its large, white or yellow flowers. Native to North America. Also called *prairie lily; gumbo lily.*

tensiometer. Any of several types of devices which measures moisture, tension or condition of water in soil.

tent caterpillar. Certain insects of the genus Malacosoma. See *eastern tent caterpillar; California tent caterpillar; forest tent caterpillar; western tent caterpillar.*

tent worm. Tent caterpillar.

tenure. In rural sociology and agricultural economics, the nature of property rights under which land is held and utilized. (In the U.S., almost all property rights in land are lodged in the land owner through the system of ownership in fee simple. The state retains the right of eminent domain.)

teonacaztei. Sacred ear.

teosinte. Mexican teosinte.

tepary bean. *Phaseolus acutifolius* var. *latifolius,* family Leguminosae, an annual, tropical herb, which is an important vegetable crop grown in arid regions for its edible seeds. Native to North America.

TEPP. Tetraethyl pyrophosphate, an insecticide.

tequila. An alcoholic beverage, made in Mexico, obtained by the crude distillation of the sweet sap of tequila agave (*q.v.*) and other species of the genus Agave, family Amaryllidaceae. The alcohol content averages about 50%.

tequila agave. *Agave tequilana,* family Amaryllidaceae, a succulent, stemless plant, widely grown as a source of a liquor high in alcohol content. Native to Mexico. Also called *tequila mescal.*

tequila mescal. Tequila agave.

teratoma. 1. A plant tumor which has a mixture of different tissues along with incipient roots or shoots. 2. A tumor which contains congenitally derived, fetal remains, such as teeth, hair, etc.

term. The gestation period.

terminal. 1. A station for delivery or receipt of produce. 2. Designating growing or being located at the end of a branch, or stem. See *terminal bud.*

terminal bleach of cereals. A rapid whitening of the leaf tips of wheat, barley, and oats that have been growing under irrigation, when exposed to arid winds. The leaf tips first turn a dark gray-green and may be white to a distance of 1 to 3 ins. within 24 hrs.

terminal bud. The bud which develops at the end of a branch or stem.

terminal elevators. The reservoirs of the grain trade which make possible an orderly, year-round distribution of grain supplies from the farm to the table. They provide storage space where grain may be held between harvest time and the time it will be needed by industry, prepare grain for storage by conditioning and cleaning it (this conserves the value of the grain and improves its quality for milling, processing or feeding purposes), and act as wholesale distributors, prepared to sell grain in car lots and boat loads whenever buyers of grain need supplies.

terminalization. Movement of a chiasma between diplotene and metaphase of meiosis; so called because movement is usually toward the ends.

terminal market. A city or market into which large amounts of produce are brought for sale and distribution.

terminal moraine. The rock debris deposited by a glacier in front of the ice at any point where the glacial terminus remained stationary for a time. The topography is usually rolling to rough. See *moraine.*

termite. Any of numerous social insects of the order Isoptera. Also called *white ant.* See *eastern subterranean termite.*

termite shield. A shield, usually of sheet metal, which is placed in or on a foundation wall or other mass of masonry or around pipes to prevent the passage of termites.

term permits. Grazing rights or privileges, for national forests, issued to qualified applicants for ten year periods.

ternate Mexicanorange. *Choisya ternata,* family Rutaceae, an aromatic, evergreen shrub, grown in warm areas for its foliage and its fragrant, white flowers. Native to Mexico. Also called *Mexican orange.*

terpeney. Designating an odor or taste suggestive of turpentine.

terrace. 1. A long, low embankment or ridge of earth constructed across a slope with a flat or graded channel to control runoff and minimize soil erosion by scouring. Two general types are bench *terrace* and ridge *terrace.* Variations of these types are the horizontal bench *terrace,* sloping bench *terrace,* broad base *terrace,* narrow base *terrace,* Nichol's *terrace q.q.v.* 2. A raised, level area of earth supported on one side by a wall, bank, etc., as a terraced yard. Often there is one above another on a hillside or slope. 3. A natural land feature which is an elevated, relatively long, flat upland with a short scarp or embankment-like front. The plain may be narrow and the *terrace* steep or bench-like, or it may be wide and properly designated a *terrace* plain. *Terraces* represent former sea, lake, and river levels, and are locally significant in soil and land classification. 4. To construct a *terrace,* especially one across a slope to control erosion and runoff.

terrace channel. The depression along the upper side of the terrace ridge in which the water flows to the outlet, or is retained when the channel is not graded so that the water does not have an outlet other than through the soil.

terrace crown. The highest part of the terrace ridge; the top of the terrace.

terrace grade. The slope of the terrace channel in inches or feet per 100 ft. length. *Terrace grades* may be variable or uniform.

terrace height. The vertical difference in elevation between the bottom of the terrace channel and the crown of the terrace ridge at adjacent points.

terrace interval (horizontal). The horizontal distance in feet from the center of one terrace line, ridge, or channel to the same point on an adjacent terrace.

terrace interval (vertical). The vertical distance in feet from the center of one terrace line, ridge, or channel to the corresponding point on an adjacent terrace. (The interval of the first terrace is the vertical distance from the top of the hill to the terrace line of the terrace.)

terrace outlet. The water channel at the end of the terrace into which the flow from one or more terraces is discharged and conveyed from the field.

terrace oultet structure. A structure usually installed in or at the end of a terrace outlet to prevent gullying.

terracer. Any of several types of machines or devices that lift and roll, throw, scrape and push, or carry earth and mound it in a ridge. See *slip scraper; V-drag; grader: 2; carryall: 2; fresno scraper; plow.*

terrace ridge. That part of the raised embankment of earth which extends above the original surface of the soil and serves as a dike in diverting the runoff from its natural course.

terracette. A small hillside step or bench commonly believed to have been caused by livestock. It usually involves slumping on a small scale indicating gradual downhill movement of the entire surface. Also called *cattle terrace; sheep track; cat step.*

terrace width. The combined width of the terrace channel and ridge as measured horizontally from the edge of the cut slope on the upper edge of the channel to the foot of the lower slope of the terrace ridge.

terracing. A practice of constructing a ditch or channel, with a ridge below, across the slope at various vertical intervals to intercept runoff. The channel and ridge are usually constructed in a way that will permit contour operations with farm equipment.

terrain. 1. The general topographical aspect of a particular area or region. 2. See *terrane.*

terramycin. Oxytetrocycline hydrochloride, an antibiotic derived from the organism, *Streptomyces rimosus.*

terrane. Geologically, a region or extent of land over which a particular rock formation, or topographic aspect prevails: generally used in relation to fitness or unfitness of the land for some particular purpose. Also called *terrain.*

terrapin scale. *Lecanium nigrofasciatum* Perg., family Coccidae, a scale insect that infests common fruit trees and many ornamentals. Found throughout the eastern and southern U.S.

terrarium. A tightly-fitted, glass-enclosed, indoor garden, resembling an aquarium, in which plants are grown in earth for decoration, plant propagation, nature study, etc. Also called *Wardian case; bottle garden; crystal garden; glass garden.*

Terrell grass. Virginia wildrye.

terrestrial. 1. Referring to earth. 2. Designating a plant which lives in soil as contrasted to one which is epiphytic (*q.v.*) or one growing in water (aquatic or hydrophytic). 3. Designating a ground bird such as a pheasant, partridge, or chicken, as contrasted to an aerial bird.

terrestrial soil. Soil developed on land under relatively dry conditions, consisting mainly of mineral or inorganic matter.

terre tenant. A person in actual possession of the land.

territory wool. Domestic wool produced west of the 100th meridian in the U.S.

tertiary shoot. A shoot or branch that grows from a secondary shoot (*q.v.*) during the same year that the secondary shoot is produced.

test. 1. A means of trial or examination for performance, productivity, validity, quality, etc. 2. To examine for performance, productivity, validity, etc.

testa. Seed coat.

test cross. 1. In crossbreeding, a backcross to the recessive parental type of the cross. 2. A cross between two inbred lines to test for specific combining ability.

tester. 1. A special strain used to estimate the heredity of other strains by crossing them with test strain and studying their cross progenies. See *progeny test.* 2. One who tests, as milk *tester;* a unit personnel of a Dairy Herd Improvement Association. 3. A person who analyzes or grades, such as a taste *tester,* or a device, such as a seed *tester,* corn *tester,* etc.

testes. The glands that produce the male sex cells and hormones; testicles.

test feeds period. 1. The period during which the vector is fed upon the healthy test plant. 2. The duration during which animals are on a particular type of feed for experimental purposes.

test field. An area on which trials for performance, productivity, validity, etc., are conducted and data secured; any field where tests are run.

test fire. In forestry, a controlled fire set with or without the knowledge of the detection organization to check the alertness of lookout men or the effectiveness of any lookout station.

testicle. See *testes.*

testicle remover. In caponizing, a pincers-like device or a scoop remover to grasp and remove the testicle after the incision is made.

testicular cord. The spermatic cord.

testis. A testicle.

testosterone. The male sex hormone; a crystalline steroid, obtained as a by-product of packing houses from the testicles of bulls. It is also produced synthetically.

test pit. A hole dug in the ground to determine the character of the subsoil or substratum prior to ditching, laying footings, etc.

test weight per bushel. The actual weight per measured bushel of grain. Any given grain may vary in weight per bushel between the varieties depending on harvesting conditions, etc.

tetanic. Pertaining to severe muscular contractions.

tetanus. An infectious, serious disease of men and animals which results from the toxin of *Clostridium tetani,* family Bacillaceae. Also called *lockjaw.*

tetany. A condition in an animal in which there are localized, spasmodic, muscular contractions. See *grass tetany; transit tetany.*

tether. 1. A rope, chain, etc., used to tie an animal to prevent its straying while grazing. 2. To tie an animal with a rope, chain, etc., to prevent straying but allowing the animal to graze.

tether devil. See *bitter nightshade; matrimonyvine.*

tethering. Restraining an animal by tying it to a post, etc., with a long rope, chain, etc., so that restricted grazing is possible.

tethery. Designating tangled wool fibers.

tetrachloroethane. CH_2CCl_4, a chemical compound used as a soil fumigant, nemacide, and herbicide.

tetracycline. A broad-spectrum antibiotic produced by the organism *Streptomyces rimosus* and sold under various brand-names, as Tetracyn, Acromycin, Polycycline, etc. A number of other antibiotics are chemically derived from it.

tetradecenoic acid. An unsaturated, fatty acid which occurs in small quantities in milk fat.

tetraethyl pyrophosphate. See *TEPP.*

tetramere stomach worm. See *chicken stomach worm.*

tetramethylthiuram disulfide. See *thiram.*

tetraploid. An organism whose cells contain four haploid (monoploid) sets of chromosomes.

tetrasomic. Designating an organism whose cells

contain four members of a particular chromosome, the rest of the chromosome complement being diploid. (Chromosome formula: 2n + 2). See *trisomic.*

tetter berry. See *white bryony.*

tetterwort. Bloodroot.

Texan pride. Drummond phlox.

Texas bluebonnet. Texas lupine.

Texas bluegrass. *Poa arachnifera,* family Gramineae, a perennial, vigorous, forage grass, found in the U.S. from North Carolina to Texas. It is sometimes cultivated for winter pasture. Native to North America.

Texas brown-eyed Susan. Pinewoods coneflower.

Texas cattle fever. Cattle tick fever.

Texas croton. *Croton texensis,* family Euphorbiaceae, an annual herb found on dry sandy soils from Alabama to Wyoming, U.S., to Mexico. A poisonous plant, it is seldom grazed. Native to North America.

Texas fever. Cattle tick fever.

Texas itch. Common scab (mange) of cattle.

Texas leaf-cutting ant. *Atta texana* (Buckl.), family Formicidae, a serious insect pest of garden and field crops which cuts leaves off the plants and invades houses to steal seed and farinaceous foods. It is found in eastern and southern Texas and western Louisiana, U.S.

Texas loco. Woolly loco.

Texas longhorn. Cattle native to southwestern U.S., especially to Texas, which are descendants of cattle first brought into this region by the Spanish explorer, Coronado, in 1510. A special characteristic is the very large, long, curving horns. Also called *Longhorn.*

Texas lupine. *Lupinus subcarnosus,* family Leguminosae, an annual, hardy herb grown for its showy, blue flowers with white or yellow spots. Native to North America. Also called *Texas bluebonnet; bluebonnet.*

Texas millet. Texas panicum.

Texas Panhandle. The high plains region of northwest Texas, U.S., important agriculturally both as range country and for farm crops.

Texas panicum. *Panicum texanum,* family Gramineae, a weedy, annual grass sometimes used for hay.

Texas pea. *Astragalus austrinus,* family Leguminosae, a perennial, bushy herb grown for stock forage. Native to southwestern U.S. Also called *milkvetch.*

Texasplume. *Gilia rubra,* family Polemoniaceae, a perennial or biennial herb grown for its showy flower clusters with the corolla scarlet outside but yellow, red-dotted within. Native to North America. Also called *standing cypress; trailing fire; tree cypress.*

Texas root rot. Cotton root rot.

Texas sage. *Salvia coccinea,* family Labiatae, a perennial herb grown for its whorls of scarlet flowers. Native to North America.

Texas steer. Designating a steer hide which has a brand on the side.

Texas stock. A single-bottom, walking plow consisting of a beam, handles, and a shank whose handles are attached ahead of the shank on the beam so that braces go from the rear of the beam to the handles.

Texas stonecrop. *Sedum pulchellum,* family Crassulaceae, an evergreen, succulent, perennial herb grown in the rock garden for its foliage and purplish flowers. Native to North America. Also called *widow's cross; flowering moss.*

Texas thistle. Buffalobur nightshade.

Texas umbrella tree. Umbrella chinaberry.

Texel. A ewes' milk cheese made on an island off the coast of the Netherlands.

texture. 1. The physical properties of a product; that is, the structure and arrangement of the parts which make up the whole. See *soil texture.*

thalamus. See *receptacle: 1.*

thallium. A rodenticide, also used to control ants.

thallophyte. Designating a plant of the phylum Thallophyta. Lower forms of this phylum are unicellular and reproduce through asexual spores or vegetatively. In the higher forms, the plant bodies show no differentiation into distinct parts, and reproduce both sexually and asexually. *Cf. cormophyte.*

thapes. Gooseberry.

thatch. 1. A leaf, straw, reed, etc., covering for a roof, haystack, etc., for protection from the elements. 2. To cover with leaves, straw, reeds, etc., so as to protect from the elements.

thatchpalm. Florida silverpalm.

thaw. 1. A period of weather following ice and snow which is warm enough to cause melting. 2. To melt frozen products. 3. To melt ice, snow, and frost, in soil. See *January thaw.*

theave. A ewe which has not borne a lamb (British).

theelin. Estrone.

theelol. Estriol.

Thenay. A soft, whole milk, Camembert-like, French cheese which is cured about one month.

Theodore. Fiador.

therapeutic. Pertaining to the treatment of disease; curative.

therapeutic agent. Any medicament given to cure or arrest a disease.

therapy. The sum total of the treatment given to cure disease in plants, animals, and man.

thermal belt. A well-defined area on the sides of valleys, as on the eastern slope of the San Joaquin Valley in California, U.S., in which descending cool air is warmed by compression as it nears the lower elevation, while at the lowest elevation cold air accumulates. These belts are of great agricultural importance, for in them the danger of frost is minimized.

thermal death point. The amount of heat required to kill a particular organism.

thermal-fog applicator. Any spraying device in which oil solutions of insecticides are finely atomized by a blast of heated air or gas.

thermoperiodic adaptation. The adjustment of plants in their native or artificial habitat to specific temperature requirements.

thermoperiodicity. The response of plants to changes in day and night temperatures.

thermo-photoperiodic induction. Photothermal induction.

therophytes. Annual grasses and forbs that pass through the unfavorable season in the seed stage and complete their life cycle (from seed to seed) in one growing season.

thiamin. A member of the vitamin B complex; vitamin B_1.

thickened. Designating a common body and texture defect, as in condensed milk, characterized by a jellied formation.

thickener. A food adjunct, as gelatine, which is added to food to improve the viscosity or to give it a firm body.

thickening substance. Thickener.

thicket. A dense growth of shrubs or small trees.

thick-fleshed. Designating an animal with a great amount of muscle and fat.

thick-leaf phlox. Carolina phlox.

thick leg. An edematous condition in animals; elephantiasis.

thickneck. Designating an onion which has an abnormally thick stem.

thickset. 1. Designating a short, stout body. 2. Planted closely together.

thickspike wheatgrass. *Agropyron dasystachyum*, family Gramineae, a grass of fair forage value. Native to North America; found from Hudson Bay to Alaska, Lake Superior to southern Colorado.

thick streak. A retail cut of pork (England).

thick white. The viscid or jelly-like part of an egg white. See *thin white.*

thickwind. Any chronic, respiratory, horse ailment.

thief. 1. An open-ended glass tube used to obtain samples from bulk milk (British). 2. See *sucker.*

thief ant. *Solenopsis molesta* (Say), family Formicidae, a common house ant pest which shows a preference for protein foods. It may also damage germinating seeds of grains.

thigh. 1. The part of an animal's hind leg between the hock and the trunk. 2. A piece or cut of fowl from the leg between the drumstick and the body.

thill. Either of the shafts between which a horse or other animal is hitched to a vehicle.

thimble. Foxglove.

thimbleberry. Any of several species of the genus Rubus, family Rosaceae. Species are grown mainly for their ornamental thimble-shaped fruit. See *blackcap raspberry; fragrant thimbleberry; western thimbleberry; white thimbleberry.*

thimbleflower. See *selfheal; pinewoods coneflower; foxglove.*

thimble pot. The smallest standard flower pot (*q.v.*). Its inside diameter is 2 ins.

thin. 1. To reduce the number of plants in a row or area by hoeing, pulling, etc. 2. Designating an animal with little flesh. 3. Designating a pulse which is very feeble.

thin body. A body defect of dairy products, as buttermilk and ice cream, characterized by low viscosity.

thin fleshed. In stock judging, designating an animal which lacks muscling and finish.

thin land. Soil which is unproductive because of a lack of depth or fertility. *Cf. deep soil.*

thinleaf sunflower. *Helianthus decapetalus*, family Compositae, a perennial herb grown in the flower border for its large, yellow flowers. Native to North America. Also called *dahlia sunflower.*

thin market. A market so sensitive to volume that any substantial bids or offerings are likely to advance or depress the prices.

thinning. 1. The removal of plants in a stand to reduce crowding. 2. Removing certain flowers or clusters of flowers or individual fruits after fruit has set and natural dropping has occurred. 3. Removing live branches in a tree crown. 4. In woodland management, the cutting made in immature stands after the sapling stage to increase the rate of growth of the remaining trees.

thin out. See *thin: 1.*

thin rind hog. Hampshire swine.

thin-shell egg. An egg with a soft shell or no shell.

thin soil. Thin land.

thin stand. 1. A relatively smaller number of plants in an area than considered satisfactory. 2. A poor establishment of plants.

thin streak. A cut of pork (England).

thin super. A grade of foundation for frames for supers in beehives which is very clear, light-colored, and thin. Also called *thin surplus.*

thin surplus. Thin super.

thin white. The fluid part of an egg white as distinguished from the more viscous part, called thick white (*q.v.*). The white of an egg consists of four alternating envelopes of thin and thick white, the outermost envelope being made up of *thin white.*

tight soil. A soil that is compact, relatively impervious, and difficult to plow; usually a clay.

thiocyanate. A salt of thiocyanic acid, containing sulfur in combination with cyongen. Used as an insecticide and as a synergist.

thiodiphenylamine. See *phenothiazine.*

thionate. A salt of one or more of the thionic acids of the generic formula, H_2SnO_6, where n may be 2, 3, 4, or 5. It is used in agriculture as a pesticide dust.

thionate dust. A pesticide dust containing a compound of sulfur. See *ferbam.*

thiram. Tetramethylthiuram disulfide, a seed disinfectant.

thirds. Butter scoring below 84 points (obsolete).

thirty-six's (36's). A British grade of wool equal to United States braid (*q.v.*), to Australian and New Zealand Cotswold or Lincoln, to South American 6's.

thirty-two's (32's). A British grade of wool equal to United States braid *q.v.*

thistle. Prickly leaved herbs of the family Compositae, often of the genera Cirsium, Carduus, Cnicus, Onopordum.

thistle aphid. *Anuraphis cardui* (L.), family Aphidae, an insect pest of thistle that also infests chrysanthemums in the summer and apricots, plums, and prunes in the winter.

thole. A wood or metal pin or peg used to fasten the handle to a scythe (*q.v.*); an oar to the oar lock, etc.

Thomas phosphate. Basic slag. Also called *Thomas meal.*

Thomas slag. Basic slag.

thong. A strap of leather.

thoracic cavity. Cavity of the chest containing the heart and lungs.

thorax. The chest.

thorn. 1. A hard, sharp-pointed, leafless branch of a plant. 2. Any shrub or tree which bears *thorns,* as the hawthorn, buckthorn, Christ's thorn. 3. Any of the wild species of the genus Crataegus, family Rosaceae.

thornapple. Jimsonweed datura.

thornbush. See *thorn: 2.*

thorn-headed worm. A parasite belonging to the phylum Acanthocephala, characterized by a muscular snout or proboscis covered with large hooks or spines. These parasites have no free-living stage and larval development occurs in small arthropods, usually small crustacea or insects. The only species of importance, *Macracanthorhynchus hirudinaceus,* occurs in the intestines of swine. At each place of its attachment there results a nodule or swelling. It contributes to the general unthriftiness of a hog.

thornlocust. Honeylocust.

thorn tree. See *thorn: 2.*

thorny. Abounding with thorns.

thorny acacia. Honeylocust.

thorny ononis. Cammock.

Thoroughbred. A breed of saddle horses developed in the United Kingdom, which stand about 16 hands and weigh about 1,000 lbs. The *Thoroughbred* is bay, brown, black, chestnut, tan, gray, or roan and is essentially a race horse. The term is sometimes confused with purebred *q.v.*

thoroughpin. A soft swelling which appears behind and on the upper part of the hock and in front of the large tendon of the horse. See *blood spavin.*

thoroughstem. Boneset.

thoroughwax. Boneset.

thoroughwort. Boneset.

thousand-headed-cabbage sprouts. See *Brussels sprouts.*

thousand-leaf. Yarrow.

thrash. See *thresh.*

thread blight. A fungous disease of citrus and coffee caused by *Corticium koleroga* or *C. stevensii,* family Thelephoraceae, characterized by minute pustules on the upper surfaces of the leaves. As the spots coalesce, the leaf turns smoky gray. It results in the death of twigs and leaves.

threading. Separating small trees from a bundle preparatory to transplanting.

threadleaf groundsel. *Senecio longilobus,* family Compositae, a perennial herb found on dry range land from Colorado to Utah, U.S., and south to Mexico, which is poisonous to livestock which graze on it. Symptoms of poisoning are the same as for ragwort groundsel (*q.v.*). Native to U.S. and Mexico.

threadleaf sedge. *Carex filifolia,* family Cyperaceae, a grasslike plant, found from Texas, U.S., to the Yukon between 3,000 and 12,000 ft. altitude, which is rated as good forage. Native to western North America. Also called *niggerwool.*

thread-necked strongyles. Various species of roundworms (nematodes) of the genus Nemotodirus, family Trichostrongylidae, parasites in the small intestine of domestic and wild ruminants, as well as certain other mammals. When large numbers are present in the small intestine of a host, they may cause parasitic dysentery.

threadworms. Small roundworm (nematode) parasites of the genus Strongyloides, which live in the small intestine of man and various domestic and wild animals. Some of the important species are *S. stercoralis* (in man, dog); *S. papillasus* (cattle, sheep, goats); *S. ransomi* (swine); *S. westeri* (equines). Infection results from infective larvae penetrating the hosts' skin or by being swallowed. Larvae may cause a dermatitis, and adults (female only) in the small intestine may cause parasitic dysentery, anemia, etc.

thready. Designating a rapid pulse which is scarcely perceptible.

three-and-five-gaited horse. A class of light horses *(q.v.)* developed for riding. See *three-gaited horse; five-gaited saddle horse.*

three A (3A) standards. A list of detailed requirements, specifications, recommendations, etc.; formulated by the International Association of Milk and Food Sanitarians, the U.S. Public Health Service, and the Dairy Industry Committee to cover most pieces of dairy machinery.

three-banded leafhopper. *Erythroneura tricincta* Fitch, family Cicadellidae, an insect that infests grapes in central U.S. The leaves of infested plants are spotted followed by a yellowing and death.

three bulls. In stock shows for beef cattle, the best three bulls owned by an exhibitor.

three-cornered alfalfa hopper. *Spissistilus festinus,* (Say), family Membracidae, an insect that infests alfalfa, beans, cowpeas, tomatoes, melons, and certain trees. It punctures the stem of the plant to feed or lay eggs. Also called *alfalfa treehopper.*

three-eighths blood. A U.S. grade of wool for length, denoting a fiber of 3 ins. which is equal to English 56's; Australian and New Zealand Halfbred and South American 1's.

three ewe lambs. In stock shows, a pen of three female lambs.

three-field system. A system of fallowing in which land tracts are divided into three fields, with one field fallowed each year. The practice dates back to the Middle Ages in Europe.

three-gaited horse. A saddle horse which has the following gaits: walk, trot, and canter.

three-leaved ivy. Poison ivy.

three-lined potato beetle. *Lema trilineata* (Oliv.),

family Chrysomelidae, an insect that feeds on the foliage of potatoes and related plants.

three off. Designating a horse between 3 and 3½ years of age.

three-quarter bred. Designating animals from the same mother but whose sires are half-brothers by the same father.

three-quarter pot. A flower pot *(q.v.)* with a diameter from 6 to 12 ins. but whose depth is only ¾ the regulation depth (or ¾ the diameter at the top). Also called *azalea pot.*

three ram lambs. In stock shows, a pen of three male lambs.

three-row lister. A lister *(q.v.)* that plows three rows at a time.

three-row lister cultivator. A lister cultivator *(q.v.)* that cultivates three rows at once.

three-row lister planter. A lister planter *(q.v.)* that plants three rows at a time.

three-tined pitchfork. A pitchfork having three tines which is designed for use in handling hay, grain bundles, etc.

three-tined weeder. A flower garden implement with a short handle which has three sharply curved tines or prongs designed for removing weeds.

three-way cross. 1. A cross involving three inbred lines, strains or breeds. 2. In swine, a system of rotation breeding involving 3 boars of different breeds.

three-wheel sweep rake. A sweep rake *(q.v.)* that has a seat mounted on a rear truck consisting of one wheel which casters in any direction. A lever raises and lowers the teeth.

three-wheel terracer. A grader *(q.v.)* that has one wheel at the front of the frame, two wheels at the rear. The blade is raised or lowered by levers.

three-year-old. In stock shows, a horse which was foaled on or after January 1 and on or before December 31 of the third preceding year.

threonine. See *amino acids.*

thresh. To beat and separate the grain or seed from the rest of the plant of wheat, oats, barley, etc., by means of a grain threshing machine or combine *q.q.v.*

thresher. Grain threshing machine.

thresherman. A man employed in operating a grain thresher.

threshing. The process by which grain or seed is separated from the straw. See *thresh.*

threshing cylinder. Cylinder.

threshing floor. A hardened earthen, wooden or other floor on which grain is threshed or beaten out of the straw by hand (obsolete in most areas).

threshing machine. A combination machine in which grain of wheat or other cereals is separated from straw and chaff. See *grain threshing machine.*

threshing mask. A respirator worn by a person engaged in threshing.

threshing ring. 1. A group of farmers who own a combine or grain thresher cooperatively. 2. A group of farmers who cooperate at threshing time. They may or may not own the threshing equipment.

threshing time. The time of year during which grain is threshed.

threshold period. Minimum acquisition feeding period.

thrift. 1. Any plant of the genus Armeria, family Plumbaginaceae, summer-blooming, perennial herbs. They are useful in the flower garden border. 2. *A. maritima,* a perennial herb grown for its pink, purple, or white flowers. Native to Europe. 3. A healthy condition in plants or animals.

thriftless. 1. Designating a plant or animal which is not thriving. 2. Designating poor soil.

thrifty. 1. Designating a plant or animal which is thriving. 2. Designating productive soil.

thrips. Very small insects of the order Thysanoptera, for the most part very injurious to plants. See *banded greenhouse thrips; chrysanthemum thrips; citrus thrips; flower thrips; gladiolus thrips; greenhouse thrips; onion thrips; pear thrips; tobacco thrips.*

thrips injury. A disorder of plants resulting from infestation by thrips. A non-parasitic disease of plants, in which infestation by thrips damages the plant. Often the terminal bud is destroyed, and the leaves appear torn, cupped, or distorted. Infestation may reduce the vigor of the plant and make it susceptible to invasion by disease.

thrive. To be healthy and vigorous.

throat. 1. The front part of the neck through which the windpipe and esophagus pass. 2. An entrance or a passageway into a machine, as the throat of an ensilage cutter. 3. See *horn neck; throatlatch.*

throat bot fly. *Gasterophilus nasalis* (L.), family Gasterophilidae, an insect that lays its eggs on the hair under the horse's jaw. The larvae crawl into the mouth and eventually pass into the stomach and duodenum. Also called *chin fly.*

throatlatch. The strap on a bridle or halter which passes under the horse's throat.

throat of a plow. The space behind and above the share up to the horizontal portion of the beam that gives room for handling the furrow slice and trash.

throatstrap. Throatlatch.

throatwort. 1. *Trachelium caeruleum,* family Campanulaceae, a perennial herb grown for its blue or white, dense, flower cluster. Native to the Mediterranean region. 2. See *Coventry bells; foxglove; spike gayfeather.*

throck. The standard of a plow (British dialectal).

throe. A tool used for splitting timber into shingles or clapboards. Also called *froe; frow.*

thromboplastin. A substance existing in the tissue which causes clotting of the blood. It is available commercially as a packing-house by-product.

thropple. See *throttle: 1.*

throttle. 1. The throat of an animal. 2. The control which regulates the flow of fuel to an engine. 3. To choke or cut off the oxygen supply to an animal.

through. One complete coverage of all of the middles or areas between the rows in an entire field by a cultivating implement. Once *through* the corn field is a complete cultivation of all the middles in the field.

through check. A crack in lumber which extends from one side of the piece to the other.

throw. 1. In horseback riding, the violent ejection of the rider from the horse's back by action of the horse. 2. To cause an animal, as a horse or cow, to fall to the ground before branding, treating, etc. 3. To wield a lasso.

throw back. 1. In breeding, a reversion to an ancestral type or ancestral characteristic. 2. To revert to an ancestral type or characteristic.

throwboard. An extra sideboard placed on one side of a wagon or truck used in harvesting corn by hand to prevent the ears from being thrown over and beyond the wagon or truck. The thrown ears ricochet from the board into the wagon.

throw-in. A turning inward of the disks of a disk cultivator so as to cause the soil to be moved toward the planted row. *Cf. throw-out: 1.*

thrown. 1. Designating a person ejected from an animal's back as a result of the animal's action. 2. Designating an animal which has been forced to a prone position.

throw-out. 1. A turning outward of the disks of a disk cultivator so as to cause the soil to be moved away from the planted row. *Cf. throw-in.* 2. A market classification for inferior, coarse, rough hogs of all weights. 3. A device used to remove the carcass of a hog from a vat of scalding water. 4. Any product which does not meet certain minimum standards. 5. An animal discarded for failure to meet certain requirements or specifications.

throw up. 1. To hill earth or soil around the crowns of plants. 2. To vomit.

thrumwort. Love-lies-bleeding.

thrush. 1. A foul-smelling, degenerative condition involving the frog in the sole of a horse's foot. Predisposing causes are generally considered to be unhygienic conditions such as standing in mud and filth. 2. A fungous disease of the digestive tract of poultry caused by *Candida albicans.* About the only clinical signs are depression and emaciation. Also called *crop mycosis; intestinal mycosis.*

thrust hoe. See *scuffle hoe.*

thumb capping. The seating of the cardboard milk bottle closure by hand instead of machine.

thumb feather. See *index feather.*

thumb-marks. A disfiguring depression on the sides of a chicken's single comb *q.v.*

thumb pot. The next to the smallest of the standard flower pots *(q.v.)* that has an inside diameter of 2½ ins.

thumps. An animal ailment resembling hiccoughs in man which is seen in baby pig anemia, swine influenza, and in verminous pneumonia.

Thunberg daylily. *Hemerocallis thunbergi,* family Liliaceae, a lilylike herb grown for its fragrant, lemon-yellow flowers. Native to Japan.

Thunberg kudzubean. *Pueraria thunbergiana,* family Leguminosae, a perennial, climbing vine grown for its habit and its clusters of purple flowers. Because of its rapid growth it is used in warm areas for erosion control and for forage. Native to Asia. Also called *ko.*

Thunberg lespedeza. *Lespedeza thunbergi,* family Leguminosae, an herblike shrub grown for its autumnal, rose purple, flower clusters. Native to Asia.

Thunberg pine. Japanese black pine.

Thunberg spirea. *Spiraea thunbergi,* family Rosaceae, a deciduous shrub, grown for its numerous, pure white flowers. Native to China and Japan.

thunderflower. See *corn poppy; evening campion.*

thundergodvine. Regels threewingnut.

thunderstorm. A rain storm, usually intense but of short duration, accompanied by lightning and thunder.

thunderwood. Poison sumac.

thunk. A wartlike growth in a cow's teat canal.

thurberia weevil. *Anthonomus grandis thurberiae* Pierce, a variety of the boll weevil *(q.v.)* that infests cotton in the southwestern U.S.

Thuringer. A cured summer sausage which consists largely of beef with a small amount of pork.

thurl. The region between the pinbone and the hip joint on a dairy cow. Its desirable confirmation is that it should be high, wide, and level with the rest of the hip region.

Thurlow weeping willow. *Salix elegantissima,* family Salicaceae, a hardy, deciduous tree, grown as an ornamental for its drooping branches. Native to Japan.

thwarter. A disease of sheep, and occasionally of pigs, caused by the tick *Ixodes ricinus,* family Ixodidae, characterized by abnormal conformation of the neck, loss of muscular control, paralysis, and often death. Occurs in Scotland and northern England. Also called *thwarter-ill; thorter-ill; louping-ill.*

thyme. 1. Any aromatic, woody, perennial herb of the genus Thymus, family Labiatae. Species are grown as ornamentals and for their aromatic foliage used in seasoning. 2. *T. vulgaris,* an herb cultivated in the herb garden as a source of an essential oil and for its leaves used in seasoning. Native to southern Europe.

thyme dodder. Clover dodder.

thymeleaf speedwell. *Veronica serpyllifolia,* family Scrophulariaceae, a small perennial weed, found in lawns, fields, and waste places. Also called *Pauls betony.*

thymus gland. A ductless, glandular body situated in the thorax which reaches its maximum development early in life. Colored grayish-red, it usually has two longitudinal lobes. The animal *thymus gland,* used as food, is called *sweetbread.*

thyroid gland. A ductless gland situated in the neck which secretes thyroxin, a hormone that controls the rate of metabolism and has a profound influence on growth and production.

ti. *Cordyline terminalis,* family Liliaceae, a plant with large elliptical leaves in a cluster near the top and white or greenish flowers giving rise to pea-sized, red berries. Many horticultural varieties are grown in greenhouses and in warm, subtropical, or tropical climates out-of-doors, especially because of their brilliant leaf colors. Its leaves are used in place of plates by indigenous populations in some parts of the world. Native to Indonesia.

Tibet. A hard, grating, ewes' milk cheese made in Tibet.

Tibet honeysuckle. *Lonicera thibetica,* family Caprifoliaceae, a hardy, low, deciduous shrub grown for its fragrant, rosy-purple flowers and ornamental, red fruit. Native to China.

tick. Any of various blood-sucking arachnids which fasten themselves to warm blooded animals. Some are important vectors of diseases.

tickborne. Transmitted by ticks, as cattle tick fever.

tick farcy. Mycotic lymphangitis.

tick fever. See *piroplasmosis; cattle tick fever.*

tick infested. Designating an animal or area of land which abounds in ticks.

ticking. Specks or small dots of color on an animal's coat or on the plumage of a fowl.

tickle grass. Witch grass.

tick paralysis. A disease of sheep, cattle and man, caused by the ticks *Ixodes* spp. or *Dermacentor* spp., characterized by a paralysis first affecting the legs, next the chest and neck, and by the ultimate death resulting from the heart becoming involved.

tick quarantine. A legal prohibition against shipping livestock from one area to another in an attempt to prevent the spread of ticks.

tick-seed. 1. Beggarticks. 2. Designating a plant with a sticky seed. 3. Any plant of the genus Coreopsis, family Compositae. Species are grown as ornamentals for their yellow, crimson, or brown heads. 4. One of the various plants of the genus Desmochinus, family Leguminosae. Species are characterized by trifoliate leaves and sticky seeds.

tidal flat. An area of nearly flat, barren mud periodically covered by tidal waters.

tidal marsh. Low, flat marshland traversed by interlacing channels and tidal sloughs, inundated by high tides. It supports a cover of salt marsh vegetation, usually low-growing plants, with areas of rushes and reeds.

tideland. The area adjacent to the sea or ocean which is covered during floodtide. The shallow sea bottom (in some instances several miles from shore).

tideland spruce. Sitka spruce.

tidewater cypress. Baldcypress.

tie. 1. A string, rope, metal band, or wire which encircles a bale of cotton, hay, etc., holding the material in a limited space and form and making for ease of handling, storage, and transportation. 2. A rope, twine, etc., for fastening purposes. 3. A stanchion for securing an animal. 4. See *railroad tie.* 5. Depression, usually in the middle of the back of an animal, caused by the skin adhering to the backbone. 6. To fasten, secure, with a rope, string, wire, etc.

tie down by the tail. To twist an animal's tail in such a fashion as to prevent its rising.

tie hard and fast. To tie the end of the lariat or rope to the saddle horn so as to prevent escape of an animal or loss of the rope.

tie-in chains. Chains used in multihitches along with buckback straps so that each individual animal is held in place and the entire team can be handled by two lines.

tie-out chain. A chain used to fasten an animal to a stake, tree, post, etc.

tie out the bale. To fasten baling wire around a bale of hay.

tier. 1. In grafting and budding, one who ties the bud or cion to the stem or stock. 2. One of the horizontal layers in the vertical stacking of boards in a well-defined pile for seasoning. 3. A layer of lumber, boxes, crates, bags, etc., in storage.

tierce. A cask intermediate in size between a barrel and a hogshead.

ti-es. See *canistel lucuma; eggfruit.*

tie stall. 1. A stanchion to which an animal is fastened. 2. A stall in which an animal is fastened by a halter or chain.

tigerflower. *Tigrida pavonia,* family Iridaceae, a tender, cormous herb grown in the flower garden or in the greenhouse for its cup-shaped, red, yellow, or orange-spotted flowers. Native to Mexico and Guatemala. Also called *flower of Tigris.*

tiger lily. *Lilium tigrinum,* family Liliaceae, a bulbous herb grown for its large, orange-red or salmon-red, black-spotted flowers. Native to Japan and China.

tiger swallowtail. *Papilio glaucus* L., family Papilionidae, a butterfly whose larva feeds on apple, ash, birch, cherry, lilac, poplar and some other trees. It is not particularly destructive.

tigertail spruce. *Picea polita,* family Pinaceae, a tall, hardy, slow-growing, evergreen tree grown as an ornamental. Native to Japan.

tight cooperage. Wooden barrels so constructed that they will hold liquid products. *Cf. slack cooperage.*

tight in the boll. Designating seed cotton which can be pulled from the boll only in pieces.

tight side. The right side of a beef carcass. *Cf. loose side.*

tight soil. A compact clay soil difficult to plow and till and permitting a very slow permeation of water.

tight-wooled. Designating (a) a Rambouillet lamb with heavy wrinkles on the neck and body, (b) any breed of sheep with short, dense, compact fleece.

tigre. Appaloosa.

Tignard. A hard, blue-veined, ewes' or goats' milk cheese made in France.

tikor. a. *Curcuma angustifolia,* family Zingiberaceae, an herb having tubers from which a starch is made. Native to Indonesia. b. The starch derived from *tikor.* Also called *East India arrowroot; East Indian arrowroot.*

tikouka. Giant dracena.

tile. 1. A short length of pipe made of fired shale, clay, or concrete, used to carry away excess water, to distribute sewage from the septic tank, etc. Also called *drainage tile.* 2. To drain an area by means of *tile.*

tile bat. A broken piece of tile.

tiled. Designating an area of land which has been under-drained with tile, as contrasted to open ditches.

tile drain(age) system. Any of several different arrangements of tile drains for removing excess water from the land, as contrasted to drainage by means of open ditches.

tile drained. Tiled.

tile hook. An L-shaped rod which is used to lower a tile into a trench and to place it adjacent to, and in line with, the previously laid tile.

tile-horned prionus. *Prionus imbricornis* (L.), family Cerambycidae, the larva of a beetle which bores into the roots of oak and chestnut in southeastern U.S.

tile line. The line of tile, after placement and covering with soil, to provide for the removal of excess subsurface water.

tiler. A person who lays tile for drainage.

tile silo. A silo constructed of vitreous tile (in the form of curved plates).

tile spade. A spade used by tilers which has a long, narrow blade and a short handle.

till. 1. A deposit or mixture of earth, sand, gravel, and boulders which has been transported by, and deposited under, glaciers. *Till* is generally but not everywhere unstratified. Also called *boulder clay q.v.* 2. To cultivate, plow, fit and sow land.

tillable acres. That part of the farm land that can be used for crops without additional drainage, clearing, or irrigation.

tillage. The practice of working the soil to bring about more favorable conditions for plant growth.

tillage equipment. Field tools and machinery designed for efficient and satisfactory handling of the soil: *e.g.,* plows, harrows, disks, and cultivators.

tillage for impounding water. A regular or special tillage operation conducted in such a way as to confine a large portion of rainfall so that it may soak into the soil.

till basket. A rectangular or square basket composed of two sheets of wood veneer placed at right angles to each other and folded so as to form a flat-bottomed container to hold berries, cherries, etc., for handling and retail sale. Its capacity is usually one pint or one quart.

tiller. 1. A shoot which rises from the crown of a plant. 2. A sprout which rises from a stump of a tree. 3. To put forth shoots from the crown of a plant; to stool.

tillering. The production of shoots from the crown of a plant; stooling.

tillering stage. See *growth stages.*

tiller plow. One-way disk plow.

till plain. A level or undulating land surface covered by till. See *till: 1.*

Tilsit(er). A cow's milk, cured, firm, slightly yellow cheese. It has a mild to medium-sharp flavor similar to Limburger and is made in East Prussia.

tilted udder. An udder of a cow which is not attached squarely.

tilth. The artificial condition of soil structure brought about by tillage or cultivation. (Terms indicating the condition of *tilth* are the same as those indicating the soil structure under the broken or disturbed conditions, as mealy, mellow, pulverulent, granular, crumb, buckshot, mulch, nut-structure, cloddy.)

tilting lever. Leveling lever.

timber. 1. Forest stands. 2. Sawed lumber 5 x 5 ins. or more. 3. Sawed lumber 4½ x 6 ins. or more (U.K.). 4. Wood in form for heavy construction. 5. A warning cry that a tree or branch is falling.

Timber Culture Act. An act of the U.S. Congress, first passed in 1873, and subsequently amended, which entitled a settler to a certificate of ownership of 160 acres of land in the Great Plains upon proof of planting and survival of a certain number of trees on 40 acres.

timber danthonia. *Danthonia intermedia,* family Gramineae, a bunchgrass found from Quebec, Canada, to New Mexico, U.S., in the spruce belt, which is of good forage value. Native to North America. Also called *timber oatgrass.*

timbered. Designating a forested area.

timber frame. Post frame.

timberland. A forested area.

timber line. The highest extension of woody or forest growth on mountains above which further tree growth is prevented because of environmental influences. The elevation at which the *timber line* occurs varies with latitude.

timber marking. The selection and indication, usually by blaze or paint spot, of trees which are to be cut or retained in a cutting operation.

timber milkvetch. Timber poisonvetch.

timber oatgrass. Timber danthonia.

timber poisonvetch. *Astragalus convallarius,* family Leguminosae, a poisonous plant found on mountain slopes from British Columbia, Canada, to Arizona, U.S. Characteristics of poisoning in animals are dullness, irregularity in gait, solitary habit, loss of nervous sensibility, loss of flesh, shaggy coat, loss of appetite, and may result in death.

timber rot. A fungous rot of lumber and timber caused by *Poria incrassata,* family Polyporaceae.

timber stand improvement. All cuttings, not a part of a major harvest felling, made during the life of a forest stand for improving the stand as regards composition, condition, or rate of growth.

timber wheels. Logging wheels.

timbo. Any woody plant of the genus Lonchocarpus, family Leguminosae, from whose roots is extracted rotenone, the chief and most effective of several insecticides contained in these plants. Native to South America. The crude product is sometimes called *derris root.* Also called *cubé.* See *rotenone.*

time of one-half response. The time required to obtain a change in moisture content of grains and produce from the original moisture content half of the way to equilibrium, usually expressed in hours.

time pinch. The removal of the terminal buds of chrysanthemums at a fixed date instead of according to plant height: *e.g.*, for a variety maturing October 15, the bud should be pinched on July 8.

timothy. *Phleum pratense,* family Gramineae, one of the best known hay grasses; easy to establish and easy to grow. The common name *herd's grass* has been applied to *timothy* in some parts of the U.S., especially New England. Native to Europe and Asia.

tin. A thin metal can (British).

tine. One of the long, slender projections of a rake, pitchfork, etc. Also called *tooth; prong; spike.*

tinged. One of the six groups (next to the lowest) of upland cotton *(q.v.)* based on color in the official classification of the U.S. Department of Agriculture.

tinned food. Canned food (British).

tinted egg. 1. Any poultry egg whose shell is not white or brown. 2. An Easter egg.

tiny white specks. A defect of cheese characterized by specks resembling grains of salt scattered throughout the body.

tip. 1. The exposed part of the wool fiber in the fleece. 2. The end of a branch, twig, etc. 3. A place where refuse is dumped (England). 4. To remove the ends of branches, stems, twigs, etc.

ti-palm. See *ti.*

tipburn. Any disease of plants caused by environmental or other factors in which the margins and tips of the leaves burn or turn brown. On potatoes, it is caused by feeding of flea beetles. *Cf. burn of potato; hopper burn.*

tip cutting. Softwood cutting.

tip deterioration. Endoxerosis.

tip-dwarf mite. *Eriophyes thujae* Gar., family Eriophyidae, a mite which feeds on the eastern arborvitae, causing a yellowish discoloration of the terminal growth. It may kill young trees. Also called *western cedar mite.*

tip layering. A method of plant reproduction especially useful for black raspberries. The tips of the branches are covered with soil so that roots may be developed from the nodes at the end of the branch. When the roots have developed, the stem is cut away. Also called *arching layers.*

tip moth. Any moth of the genus Rhyacionia, whose larvae bore into the terminal buds and tips of pines.

tipped. Designating the web end of a feather different in color from the rest of the feather.

tipping. Topping.

tippy wool. A crusted fleece of wool.

tip roast. A retail cut of roast beef from the thin end of the loin.

tire. The outer covering of rubber on a wheel of a vehicle.

tired soil. A soil which, for want of nourishment, is incapable of producing a satisfactory growth of a particular crop. Also called *exhausted soil.*

tisswood. 1. The wood of the redbay persea (*q.v.*). 2. See *mountain silverbell.*

tit. See *teat.*

titanium dioxide. TiO_2, one of the more common minerals in the earth's crust. It is present in soils, especially in clay soil, but its role in plant growth, if any, is undetermined.

titer. The minimum quantity of a substance required to produce a specific reaction with a given amount of another substance. The measure used to designate the relative amount of antibody present in body fluids, such as blood serum, that causes clumping of a standard suspension of bacteria. The dilutions showing complete and incomplete agglutination may be used to describe the *titer* of a serum.

tithymal grass. Leafy euphorbia.

title. The legal interest of an owner or owners in property, usually real estate. It may vary from fee simple, the highest type of ownership, to tenancy in common, joint tenancy, or contingent interests. The *title* of ownership is transferred from one party to another by means of a deed. The history and present status of the *title* is shown by its abstract.

title insurance. A guaranty made by a title insurer or by a governmental unit of a state that the title to a particular piece of real estate is vested in a named person at a certain date.

titular group leader. The elected leader of an organized and formal group.

toad. Any of several tailless amphibians of the genus Bufo or similar genera, of the family Bufonidae. Species are generally terrestrial and insectivorous.

toadflax. Any plant of the genera Linaria and Kickxia, specifically butter-and-eggs toadflax *q.v.*

toad lily. See *hairy toadlily; American waterlily.*

toad's mouth. See *snapdragon: 2.*

toadstool. The fruiting structure of a fungus of class Basidiomycetes, order Agaricales. The vegetative part of the fungus consists of fine, branching, usually white threads of cells which grow in the soil, in wood (often causing decay in it), in manure, or in other vegetable organic substances. The fruit, produced in the air, is usually fleshy, more rarely tough or even woody consisting of a cap (pileus), the underside of which has radiating plates (gills) on which the spores are produced. More often the cap may sit on a central stalk (stipe) but this may be lateral or wanting in those *toadstools* that are attached by one edge. Popularly, but not scientifically, the name *toadstool* is applied to poisonous species, and mushroom *(q.v.)* to edible ones.

toadstool disease. Mushroom root rot.

toad trillium. *Trillium sessile,* family Liliaceae, a hardy, perennial herb grown in the wild garden for its early-spring, purplish-red flowers. Native to North America. Also called *red trillium; nosebleed; bloody butchers.*

tobacco. *Nicotiana tabacum,* family Solanaceae, an annual, herbaceous plant which yields the *tobacco* of commerce. Although the acreage planted is relatively small, it is one of the important money crops of the U.S. Probably native to tropical regions in America.

tobacco angular leaf spot. A sporadic, bacterial, necrotic leaf disease of tobacco caused by *Pseudomonas angulata.*

tobacco barn. A barn specially designed for curing tobacco. See *air cure; open-fire curing; flue cure.*

tobacco beetle. Cigarette beetle.

tobacco blue mold. Downy mildew of tobacco.

tobacco-brown hay. Hay which, cocked, stacked, or baled, has undergone some oxidation accompanied by considerable heat, turns brown, and has an odor similar to cured tobacco.

tobacco budworm. *Heliothis virescens* (F.), family Noctuidae, a caterpillar that infests tobacco, eating into the buds and unfolded leaves. It also infests cotton, geranium, ageratum, and solanaceous plants.

tobacco bug. *Cyrtopeltis minimus* (Uhl.), family Miridae, a small green and black insect which damages tobacco leaves by sucking the juices from them. Also called *suckfly.*

tobacco dust. A powdered, insect repellent of unspecified nicotine content, which is prepared from dry roots, stems, and tobacco factory waste. See *nicotine dust.*

tobacco flea beetle. *Epitrix hirtipennis* (Melsh.), family Chrysomelidae, a serious pest of tobacco. The adult beetles chew small, round holes in the leaves and the larvae feed on the roots and tunnel into the stalk. It also infests other solanaceous plants.

tobacco fly. See *tobacco hornworm.*

tobacco frenching. A nitrogen deficiency disease of tobacco in which the chlorotic foliage is deformed, the leaves narrowed, thickened, and increased in number. See *frenching: 2.*

tobacco-hoeing attachment. An attachment for a cultivator which consists of two gangs each having three points for close-in cultivation of tobacco plants.

tobacco hollow stalk. A bacterial soft rot of tobacco stalks caused by *Erwinia aroideae,* family Enterobacteriaceae, which may spread and continue under wet conditions in the barns during the curing process.

tobacco hornworm. *Protoparce sexta* (Johan.), family Sphingidae, a green and white worm up to 4 ins. long which infests tobacco, eating the foliage. It is also a serious pest of tomato and eggplant. See *tomato hornworm.*

tobacco leaf powder. Tobacco dust.

tobacco mildew. A disease which infects tobacco plants while still in the field, caused by the fungus *Peronospora hyoscyami.*

tobacco mosaic. A virus disease of tobacco, and numerous other plants outside the family Solanaceae, characterized in young plants by a downward curling and distortion of the youngest leaves along with a chlorosis. Dark green spots appear on the leaves as they grow, developing into irregular, crumpled, blister-like areas as the rest of the leaf becomes chlorotic. Old leaves may have scattered, necrotic spots, and there is a stunting of the plant. Also called *brindle; calico.*

tobacco moth. *Ephestia elutella* (Hbn.), family Phycitidae, an insect pest of tobacco in storage, which also infests stored foods.

tobacco necrosis. A virus disease of tobacco and a wide range of plants, characterized by stunting. In tulips, it is characterized by severe necrosis and death of the plant.

tobacco pole rot. A rot of tobacco leaves during the curing process in the barns which is prevalent when wet, humid weather occurs during the growing and harvesting season. The rots may be initiated in the field. Several species of fungi and bacteria, such as Pythium, Sclerotinia, Erwinia, etc., have been associated with the rotting. Also called *tobacco pole burn; tobacco barn rot.*

tobacco powder. Tobacco dust.

tobacco rattle. A virus disease of tobacco characterized by gray, necrotic spots on the leaves, downward leaf curling, enlargement and drying of the lesions. The dried tissue makes a rattling sound as the tissue breaks.

tobacco ring spot. A seed-borne virus disease of tobacco, legumes, and soybeans, characterized by concentric light green to yellow patterns on the leaves, and irregularly parallel lines along the veins. The necrotic tissue dries and new rings form around the older ones.

tobacco ring spot virus. A widespread, sap-transmissible virus of which there are several strains that vary in symptoms produced on differential host plants. The virus has a wide host range, infecting many plant families. In tobacco, the symptoms consist of well-defined, necrotic rings on leaves, with considerable necrosis and stunting of the plant, followed by a disappearance of symptoms. In other plants, symptoms vary including rings, yellow areas, and stunting. There are other types of *tobacco ring spot viruses* that are not considered to be related, such as tobacco Bergerac ring spot virus, tobacco broad ring spot virus and tobacco broken ring spot virus.

tobacco sauces. Tobacco extracts, containing 1 to 10 percent nicotine, formerly used extensively in Europe for flavoring smoking and chewing tobacco.

tobacco sick soil. Soil infested with the root-rot fungus which attacks the tobacco plant. The sickness is commonly ascribed to depletion of available plant nutrients in the soil by continued cultivation of tobacco.

tobacco spear. A spear-like device fitted over the end of a lath or stick to aid in piercing the butt ends of tobacco plants and in spacing them on the lath before hanging in the barn for curing.

tobacco splitworm. Potato tuberworm.

tobacco stems. Waste tobacco products which are ground up to be used as a fertilizer material for nitrogen and potash. The nicotine may or may not be removed.

tobacco stick. A lath on which tobacco leaves or plants are hung for curing.

tobacco streak. A virus disease of tobacco and leguminous plants characterized by necrosis, puckering, leaf curl of young leaves, darkening of the areas adjacent to veins, and necrotic streaks along the midrib. The plant soon recovers and the symptoms disappear.

tobacco thrips. *Frankliniella fusca* (Hinds), family Thripidae, an insect pest of many plants but especially important on cigar-wrapper tobacco. Both adults and young feed along the veins of the leaves, defacing the leaves and reducing their market value.

tobacco water. The resultant liquid of tobacco infused in boiling water, used as an insecticide and in veterinary medicine. Also called *tobacco liquor.*

tobacco webworm. Corn root webworm.

tobacco wilt. Granville wilt.

tobacco wireworm. *Conoderus vespertinus* (F.), family Elateridae, a common tobacco insect which eats the seed and underground plant parts.

tobacco worm. See *tomato hornworm.*

tobira pittosporum. *Pittosporum tobira,* family Pittosporaceae, an evergreen, winter-flowering shrub grown in warmer climates, and in conservatories, for its clusters of fragrant, white or greenish flowers. Native to China and Japan. Also called *Japanese pittosporum.*

tobosa. *Hilaria mutica,* family Gramineae, a range grass which is especially well adapted to the dry plains and sand hills of southwestern U.S., and whose palatability and forage rating are good. Native to North America.

toddy. a. The fresh sap from any of several East Indian palm trees of the genus Caryota, family Palmaceae, used as a drink in many tropical countries. b. The fermented sap from such trees, an intoxicating liquor.

toddy fishtailpalm. *Caryota urens,* family Palmaceae, a tall palm which is grown for ornament in Florida, U.S. It yields a fiber, Kittul, similar to horsehair. Native to the East Indies. Also called *wine palm; jaggery palm.*

toddy ladle. Centuryplant agave.

toe. 1. Any one of the digits on the foot. 2. The lowest downstream edge of a dam. 3. The bottom of a seed furrow opened by the furrow opener of a grain drill. 4. The tuberous roots of dahlia, used for propagation.

toe calk. A bar of steel welded to the under surface side of the toe of a horseshoe to attain more adequate traction.

toe clip. A small piece of steel drawn up from the toe of the horseshoe. When the shoe is in place on the foot, the *toe clip* is pressed against the toe of the foot to keep the shoe from slipping back.

toe crack. A crack in the front wall of a horse's hoof which often results in lameness. See *sand crack; quarter crack.*

toe feathering. The growth of feathers on the shanks and toes of fowls.

toe-in. Front gather.

toe mark. To punch holes in the membrane between the toes of fowls for identification purposes.

toe out. To walk with the feet pointed outward; splay-footed; slew-footed.

toe sack. Gunny sack.

toe scraper. A metal scraper used to remove the dirt from the convex side of a disk of a single-disk furrow opener. See *tee scraper.*

toe weight shoe. A weighted horseshoe designed to make a horse pick his feet up higher and extend them further forward than he otherwise would.

Toggenburg. A breed of milk goats, perhaps the most popular and most widely used of any in the U.S., varying in color from light brown to chocolate. Its ears are white with a dark spot in the middle, and there are two white streaks on the face. The legs are white from the knees down. Bucks weigh 160 lbs. or more and the does 125 lbs. or over.

toggled. See *toggle system.*

toggle system. A system for treating new born kids in which the young goats are individually staked 10 ft. apart to a rope about 15 ins. long. Each rope is fastened to a swivel which in turn is fastened to a leather loop attached to the kids fetlock joint. The kids are kept thus for about 10 days or until they can follow the does to the range.

toggling. See *toggle system.*

to hurt for rain. To suffer greatly for lack of water, as a crop.

Tokay grape. A Hungarian variety of the grape *Vitis vinifera,* family Vitaceae, cultivated for its brilliant red berries from which a sweet, rich wine is made. Native to Hungary. Also grown in California, U.S.

tolerance. 1. The ability of a tree to grow in the shade of other trees and in competition with them. 2. The acquired ability of an animal or insect to take poison without ill effects. 3. The ability of a species to develop and survive under unfavorable environmental conditions. 4. A percentage of off-grade or off-sized produce which is allowed in a particular grade, or containers in a lot, that fail to meet grade specifications, especially of fruits and vegetables. 5. The amount of toxic residue permitted by federal law to be present on or in food and forage products that are to be marketed for consumption. 6. The accuracy limits specified in the construction of a machine part as plus or minus .005.

tolerance of fillage. The allowable variations of measurement in the filling of a milk bottle.

tolerant. Capable of sustaining disease without serious injury or crop loss. See *tolerance.*

toll. 1. The part of the flour or meal a miller keeps as payment for grinding the grain. 2. The fee paid for use of a bridge, ferry, road, etc.

Tolmie mariposa. *Calochortus tolmei,* family Liliaceae, a cormous herb grown for its lilac-tinged flowers with purple and white hairs. Native to

North America. Also called *giant star tulip; Tolmie pussyears.*

Tolmie pussyears. Tolmie mariposa.

tom. 1. The male turkey. 2. The male of certain animals, as the *tom cat.*

tomatillo groundcherry. *Physalis ixocarpa,* family Solanaceae, an annual herb grown for its inflated calyx with purple veins. Escaped, it can become a weed. Native to Mexico. Also called *Mexican groundcherry.*

tomato. *Lycopersicon esculentum,* family Solanaceae, a tender, annual herb, one of the most important vegetable crops, grown for its delicious, edible fruits eaten raw or cooked. Native to South America. Also called *apple of love.*

tomato anthracnose. The most serious of the tomato fruit rots, caused by the fungus *Colletotrichum villosum,* family Melanconiaceae. It is characterized by cankers on the fruit, rarely on the leaves.

tomato apples. An apple fruit malformation which results from winter injury to the apple blossom bud.

tomato aspermy. A virus disease of tomatoes, chrysanthemums, tobacco, and other plants. In tomatoes, it is characterized by the fruits failing to set after infection.

tomato black rot. Nailhead spot.

tomato curly top. See *curly top.*

tomato cyclone. See *cyclone.*

tomato eggplant. Ethiopian eggplant.

tomato foot rot. A foot rot of plants (*q.v.*) caused by *Phytophthora cryptogea,* family Pythiaceae, which attacks potato, tomato, petunia, China aster, and wallflower.

tomato fruit rot. A fungous disease of tomato caused by *Phoma destructiva,* family Sphaerioidaceae, characterized by brown or blackish lesions, 1 to 3 cms. in diameter, on green or ripe fruit. On the fruit in storage or in transit the fungus may form a growth of dirty-gray mycelium and ruin it.

tomato fruitworm. See *corn earworm.*

tomato hornworm. *Protoparce quinquemaculata* (Haw.), family Sphingidae, an insect whose larva feeds ravenously on tomato, tobacco, eggplant, pepper, and potato. Found throughout the U.S. and Canada. See *tobacco hornworm.*

tomato juice. The unconcentrated, pasteurized product consisting of the liquid, with a substantial portion of the pulp expressed from ripe tomatoes with or without application of heat or addition of salt.

tomato ketchup. See *catchup.*

tomato leaf blight. See *Septoria leaf spot.*

tomato leaf mold. A fungous leaf mold of tomatoes caused by *Cladosporium fulvum,* family Dematiaceae, which is troublesome in greenhouses.

tomato mosaic. See *tobacco mosaic.*

tomato paste. See *paste.*

tomato paste from trimming stock. Strained tomatoes from trimming stock, concentrated by evaporation.

tomato pinworm. *Keiferia lycopersicella* (Busck), family Gelechiidae, a serious insect pest of tomato, eggplant, and potato in the U.S. The young larvae mine the leaves, and the older ones puncture the fruit.

tomato psyllid. *Paratrioza cockerelli* (Sulc), family Psyllidae, an insect pest of potato, tomato, and other plants of the nightshade family. Infestation results in psyllid yellows *q.v.*

tomato purée. The product obtained by the evaporation of strained tomatoes, with or without the addition of salt, which contains not more than 10 to 12 percent of tomato solids determined by drying in a vacuum at 70°C. See *purée.*

tomato root rot. See *tomato foot rot.*

tomato russet mite. *Vasates lycopersici* (Massee), family Eriophyidae, an insect pest of tomato which sometimes causes severe defoliation.

tomato sauce. A semi-fluid preparation of cooked tomatoes, whole or strained, served with a meal. See *tomato purée; catchup; paste.*

tomato wilt. A widespread fungous disease of tomato caused by *Fusarium lycopersici,* family Tuberculariaceae, or by *Verticillium albo-atrum,* or *Pseudomonas solanacearum,* family Bacteriaceae, characterized by a yellowing, upward rolling, and a wilting of the leaves.

tomato worm. See *tomato hornworm.*

tomato yellows. See *curly top.*

tombac. See *tumbak.*

Tome de Beaumont. A cows' milk, French cheese, cured for about 6 weeks.

Tom Thumb. Designating certain dwarf species or varieties.

ton. A unit of weight, generally equal to 2,000 lbs. The short *ton* is 2,000 lbs. and the long *ton* is 2,240 lbs.

tone. The condition of normal balance between an organism and its environment.

tongs. A tool consisting of two flat, curved bars pivoted about a common center, commonly used for grasping objects.

tongue. 1. A part of the mouth, composed chiefly of muscle, on which the taste buds lie. It is one of the organs of taste, and moves the food being chewed around in the mouth and back to the throat for swallowing. 2. A retail cut of beef, veal, mutton, and lamb, consisting of the whole *tongue* of an animal. 3. The wooden or metal pole which acts as a guide and is fastened to the front end of a vehicle or plow. It separates the two members of a team of draft animals. The eveners for pulling the implement may also be attached to the *tongue.* 4. To make a slit in a plant's stem in grafting.

tongue fern. Japanese feltfern.

tongue grafting. See *whip grafting.*

tongue grass. Chickweed.

tongue loller. An animal whose tongue hangs out of its mouth; usually a horse or cow.

tongue truck. A small wheel attached to a plow or two-wheeled vehicle at or about the point of hitch. It is used to help support the weight of the plow, etc.

tonic. 1. A medicinal preparation to increase the strength or tone of the body system. 2. Designating a plant exhibiting normal reactions to external influences. 3. Referring to muscular contraction, as *tonic* convulsions. See *conditioner.*

tonka bean. *Dipteryx odorata (Coumarouna o.),* family Leguminosae, a large, tropical tree whose black, almond-like seed, the *tonka bean* of commerce, is used in flavoring, perfumery, and to scent snuffs. Native to South America.

tonkabean wood. *Alyxia buxifolia,* family Apocynaceae, a small, straggling, evergreen shrub sometimes grown for its fragrant, tonkabean-like odor. Native to Tasmania. Also called *scentwood; boxleaf alyxia.*

tonnage. The total weight in tons of any particular produce.

too dark. Designating a color defect of chocolate ice cream in which the product appears almost black.

too deep color. A mild color defect of strawberry ice cream in which there is too much red.

too high, vivid. Designating a deep yellow coloring of vanilla ice cream; a defect.

tool. 1. Any manually-operated implement or instrument, as an ax, hoe, saw, lathe, etc. 2. Any of the machinery on the farm including both hand and mechanical *tools.*

tool bar. 1. A heavy frame on a tractor to which tools can be attached for various tillage operations. 2. See *arbor:* 3.

tool gang. See *gang.*

too light. Designating the color defect of a product, as chocolate and strawberry ice cream, pale in color and lacking appeal.

tool shed. A shed used for repairing farm implements and for storage of tools.

too much overlap. A finish and appearance defect in cheese resulting from the bandage being too long for the cheese and overlapping more than 1 in.

toona. 1. *Toona ciliata,* family Meliaceae, a tall, nearly evergreen tree useful for its flowers, which yield a dye, and its soft, reddish wood valuable for making furniture. Native to the Himalayas. Also called *Indian mahogany.* 2. Panama gumtree.

too pale, chalky, lacking. Designating the color defect of vanilla ice cream characterized by too much whiteness.

too rigid. Designating a body and texture defect of cultured cream in which the cream stands up rigidly, resembling jelly. Angular edges of spooned cultured cream tend to retain their angularity.

too sweet. Designating the flavor defect of a food product in which there is an excess of sweetness.

tooth. 1. In animals, one of the hard, bone-like appendages of the jaws used to tear and to masticate food. 2. See *tine.* 3. A cog. 4. Any of the piercing projections on the cutting edge of a saw. 5. One of the projecting or piercing points or prongs of a comb or a card.

toothache tree. Pricklyash.

tooth bar. In a spiked-toothed harrow *(q.v.)* the arbor to which the spikes are fixed.

toothed burclover. California bur clover.

toothed euphorbia. *Euphorbia dentata,* family Euphorbiaceae, a plant which secretes a poisonous, milky juice. The effects of poisoning are swelling of the eyes and mouth, abdominal pains, collapse, excessive scours, and death. Native to North America. Also called *toothed spurge.*

toothed flea beetle. *Chaetocnema denticulata* (Ill.), family Chrysomelidae, an insect that eats small portions of the leaves of corn causing the plant to appear bleached. Infestation also retards growth and causes wilting of the leaves even in wet weather.

toothed spurge. Toothed euphorbia.

tooth float. A device for filing horses' teeth.

toothleaf jacaranda. *Jacaranda cuspidifolia,* family Bignoniaceae, a tree grown in very warm areas for its 2-foot long leaves and for its bluish-violet flowers. Native to Brazil and Argentina.

top. 1. The above-ground parts of certain plants, especially the leaves, as beet *tops,* turnip *tops.* See *crown: 1.* 2. Scoured, combed, long wool from which the noils have been removed. 3. The highest price paid for a product within a certain period of time. 4. The upper, branchy portion of a felled tree. It is sold as firewood, charcoal source, etc. 5. To cut off the crown and leaves, as of the sugar beet. 6. To remove the upper portion of the crown of a tree. 7. To place the best articles of produce, as eggs, or fruit, on the top layer, in a container so that the whole container appears to include articles of higher quality than is actually the case. 8. To sort out animals that have reached a certain stage of development or finish.

topato. Pomato.

top chuck. A retail cut of beef taken from the neck, down to the blade bone, and rolled with sheets of cod fat.

top crop. The bolls of cotton, growing at the top of the plant. They are the last to be picked late in the season.

top cross. 1. In corn-breeding, a cross *(q.v.)* in which an inbred line is used as the pollen parent and a commercial variety as the seed plant. 2. A cross between purebred males and grade females.

top diameter. The diameter at the uppermost end of a log or salable piece of timber.

top door freezer. The chest-type freezer.

top-dress. To apply fertilizer or manure on top of the ground without working it into the soil.

top dressing. A surface application of manure or fertilizer which is not worked into the soil.

toper's plant. Small burnet.

Topfkase. A sour, cows' milk, cooked curd cheese made in Germany.

top fuller. The forging tool with which a blacksmith makes furrows in a piece of red hot iron when drawing it down. It is usually held by means of a twisted osier or hazel rod and is struck by the sledge.

top grade. A market grade of produce which is either the best or the next best after fancy *q.v.*

top grafting. Grafting performed on the branches of a plant, as contrasted to root grafting, crown grafting. Also called *top working.*

top growth. The above-ground portions of a plant.

topiary work. The clipping and training of shrubs into ornamental but unnatural shapes.

topinambou. Jerusalemartichoke sunflower.

top kill. 1. A portion of the top of a tree which is dead while the remainder of the tree is alive. See *staghead.* 2. To kill the top growth of a plant with chemicals.

topknot. A usually decorative crest of feathers on the head of a bird or a tuft of hair on the head of an animal.

topline. The line along an animal's backbone, stretching from the head to the tailhead.

top milk. That portion of unhomogenized bottled milk which rises to the top of the bottle and contains the larger proportion of the butter fat.

top necrosis. The relatively rapid killing of a bud, branch, or entire top of a plant. Also called *acronecrosis.*

topographic map. A scale representation, by means of conventional signs, of a part of the earth's surface, showing the culture, relief, elevations above a datum, hydrography, and, frequently, the vegetation.

topography. Representation of a surface and its features, such as relief, streams, lakes, woods, etc.

top onion. *Allium cepa* var. *viviparum,* family Liliaceae, a botanical variety of the garden onion, which produces small bulbs in place of some or all of the flowers. The small bulbs are planted for green onions for spring use. Also called *Egyptian onion; tree onion; perennial tree onion; winter onion.*

toposequence. A sequence or chain of soils whose properties vary with topographic position. Theoretically, the slope is the variable, and other factors of soil formation remain the same. See *catena (soil).*

top pad. A pad, usually leather, or leather-lined, at the inside top of a horse collar, which rests on the horse's neck. Also called *neck pad.*

topped. 1. Designating livestock which are removed from a group of animals: *e.g.,* when a large number of animals are fed for market (or for another purpose), individuals will reach the desired weight or finish at different times. The herd will be *topped* by removing or sorting out the animals that have attained the desired point (cut, finish, etc.). This process will be followed at intervals until all of the animals are sold. The better animals are being continually removed. 2. Referring to vegetables in which the vegetative portions have been removed from the root, bulb, etc., as with beets, carrots, onions, etc. 3. Referring to cutting off the top of a tree either before or after felling.

topping. 1. In propagation by budding, the removing of the stock above the inserted bud by cutting to cause growth of the bud. 2. In pruning, the removal of the ends of branches, as a hedge is sheared. 3. The breaking off of the top of a plant. 4. The creamy mixture used to decorate ice cream, cakes, molds, specials, etc.

top primrose. *Primula obconica,* family Primulaceae, a tender, perennial herb grown in the greenhouse for its lilac or pink flowers. Native to China.

top quality. Designating generally, but not specifically, produce of the very best sort.

top round. A retail cut of beef which is a steak cut from the hind quarter. It is the choicest cut from the round. See *bottom round.*

tops. High, mountain, grazing country, sometimes above timber line (New Zealand).

top set. Small bulbs (bulbels) which are produced in place of some or all of the flowers of the top onion *(q.v.).* They are used as sets.

top sickness in tobacco. A boron deficiency disease of tobacco.

top soil. 1. The surface portion of the soil, including the average plow depth, or the surface horizon where this is deeper than plow depth. It cannot be precisely defined as to depth or productivity except in reference to a particular soil type. 2. Horizon A (*q.v.*). 3. Presumably fertile soil material used to topdress roadbanks, gardens, lawns, etc.

top spit. Top soil.

top supering. The placing of one super of drawn comb on top of another on beehives as the bees are storing nectar and honey.

top the stand. 1. To cut grain for collecting seed by setting the cutting bar so that just enough straw is cut to make bundles. 2. Cutting the top portion of the plants in a forage field.

top working. Top grafting.

top yeast. Yeast *(q.v.)* which forms fermentation on top of a substance.

torchlily. a. Any one of the several plants of the genus Kniphofia, family Liliaceae. b. *K. uvaria,* a tall, herbaceous plant grown for its scarlet flowers. Native to Africa. Also called *red-hot poker.*

Toringo crabapple. *Malus sieboldii,* family Rosaceae, a deciduous tree grown for its pink or deep rose flowers, and for its edible fruit used in jellies. Native to Asia.

tornado. A violent whirlwind which rotates in a counter clockwise direction with a destructive diameter usually ranging from 50 to 200 yards. The average forward speed is 25-40 m.p.h. It is usually associated with cold fronts, squall lines and severe thunderstorms. In the U.S., the paths are most frequently from southwest to northeast, and from south to north along the Atlantic coast. Not to be confused with cyclone. Also called *twister.*

torn bandage. A defect in the wrapping of cheeses in which the cloth bandage covering the exterior is torn.

Torrens system. A system of registration of title to land in which the government guarantees the title. It is in effect in some states in the U.S., in

Australia, South Africa, to some extent in Canada, and in Central Europe.

torrential rain. A violent downpour of rain.

torreya. Any tree of the genus Torreya, family Taxaceae. Species are interesting evergreens sometimes grown as ornamentals. Because of their nutmeg-like fruits, they are sometimes called *false nutmeg.* See *California torreya; Japanese torreya.*

tortoise beetle. An insect of the family Chrysomelidae. Also called *gold bug.* See *argus tortoise beetle; black-legged tortoise beetle; golden tortoise beetle.*

tortoise flower. White turtlehead.

tortoise-shell color. A coat-color pattern of swine in which black is restricted to spots on a white or red background.

torula. 1. A group of wild yeasts which are of considerable importance in the fermentation of fruit juices. 2. Lactose fermenting yeasts which produce undesirable changes in dairy products.

torus. See *receptacle: 1.*

Toscanello. A hard, ewes' milk, grating cheese made in Italy.

total acres. All land which is included in one farm.

total digestible nutrients. A standard evaluation of the usefulness of a particular feed for livestock which includes all the digestible organic nutrients: protein, fiber, nitrogen-free extract, and fat (the latter being multiplied by 2.25 because its energy value for animals is approximately 2.25 times that of protein or carbohydrates). Also called *T.D.N.*

total farm capital. The value of land with such permanent improvements as are ordinarily transferred with the title to the land: livestock, machinery, equipment, feed, crops, and supplies constituting together a farm operating unit.

total farm expense. The total farm expense includes the following: (a) cash expenditures for hired labor, feed purchases, crop expense, livestock expense, taxes, machine hire, and other cash farm expenses; (b) the net decreases in power and machinery, farm improvements, orchard, and in feed, crops and supplies. A net decrease occurs in such accounts where the sum of the beginning inventory plus the purchases exceeds the sum of the ending inventory plus the sales. In the feed, crops and supplies account, however, feed purchases and crop expenses are not here included in computing net decreases or net increases in the account; and (c) a charge at current hired wage rates, for farm work performed by unpaid family labor other than the operator, and a charge for the cash cost of board furnished to hired labor. For some uses in determining the *total farm expense* for making a farm business analysis, a charge for the farm operator's labor is also included. By so doing, it is possible to compute the total labor expense per tillable acre, the total power and machinery cost per tillable acre, etc. On a rented farm, the landlord's portion of the expense is also included in order to have the total expense for the complete farm.

total shorts. Wheat gray shorts.

total solids. Dry matter.

total supply. The national, indicated production of a product, plus carry-over from the previous year or years, plus the estimated imports.

Touareg. A skimmed milk, dry, hard cheese made by the Berbers in North Africa.

touch-me-not. See *snapweed: 1.*

tough. 1. Designating market grain which contains moisture in excess of 13.5 to 14.5 percent. 2. Designating plant or animal products that are not brittle, crisp, or tender.

Touloumisio. A Feta-like cheese made in Greece.

Toulouse. A heavy-weight breed of geese, grown primarily for meat production, whose plumage is dark gray with a narrow edging of lighter gray. It is almost white on the abdomen. The gander weighs up to 26 lbs. and the goose up to 20 lbs.

tournant oil. A low grade olive oil derived from the expressed marc (from the residue) of the first pressing after fermentation.

Tourne bacterium. In wine-making, a trade name for a bacterial organism that affects the fermentation process. It grows in the absence of air, producing lactic acid. Infected wine turns cloudy brown or purple, tiny bubbles appear on the surface, and it may taste musty or watered. Sulfur dioxide has been found to inhibit the bacteria and pasteurization destroys it.

Tourne disease of wine. See *Tourne bacterium.*

tous-les-mois. Edible canna.

tow. 1. The short fibers of flax and hemp after they have been combed out of the stalk. 2. The hatchelled parts of flax, jute and hemp when made ready for spinning. 3. A strong rope.

towbug. Cigarette beetle.

towcock. Cowpea.

towdie. A pullet which has not laid an egg (Scottish).

towelgourd. Any plant of the genus Luffa, family Cucurbitaceae, native of the tropics. The unripened fruit is eaten in some countries. In North America it is raised as an ornamental for its green-colored fruits. The dry fibrous interior of the fruit is often used as a sponge or cloth. Also called *Egyptian bath sponge; dishcloth gourd; vegetable sponge; rag-gourd; loofa.*

tower. 1. A hydraulically operated, telescoping mast capable of elevating to a height of 26 or 30 ft. in which there are furnace, heating pipes, and cabinets for trays of produce to be dehydrated. The trays are stacked one on top of the other. 2. A platform attached to and above a sprayer from which a man sprays the tops of tall trees.

tower excavator. For the construction of levees, etc., a device consisting of two towers mounted on rails or caterpillars which is separated by the borrow pit and connected by cables. A drag bucket is pulled to the levee by the cables and dumped, whereupon it is pulled back to the other tower.

towerman. In forestry, a lookout man stationed at a tower in a forest to watch for forest fires.

tower silo. A cylindrical tower made of wood, concrete, tile, metal, etc., used for storage of silage. It is the most common type of silo *(q.v.)* in the U.S. Also called *upright silo.*

tower sprayman. The person who sprays in orchards from a tower mounted on a power sprayer. See *tower: 2.*

town cress. Gardencress peppergrass.

township. 1. A unit of land in the survey of the public domain by the General Land Office of the U.S. 2. Land survey measurements north and south from the base line as from the meridian *(q.v.)* are in units of 6 miles and are designated as Towns (*townships*), while those running east and west are designated as Ranges. A description "T4N, R4E" designates a *township* which is the fourth town north of the base line and in the fourth range east of the meridian. A *township* is approximately 16 miles square and contains 36 square miles or sections. 3. A small unit of local government in some sections of the U.S. (In some New England states this unit is called a town). 4. A townsite (Australia). 5. A subdivision of certain provinces in Canada.

toxaphene. A chlorinated hydrocarbon insecticide.

toxemia. A generalized blood poisoning which results from absorption of bacterial or other poisons. In this condition only the poisons or toxins are found in the blood stream, and the bacteria or other poison producers are confined to a wound or other affected area.

toxemic. Pertaining to the presence in the blood of toxic or poisonous products, such as those formed by bacterial infection.

toxic. Poisonous; caused by poison.

toxicant. Any poison or poisonous substance.

toxicarol. A rotenoid obtained from the roots of *Tephrosia toxicaria* (fishdeath tephrosia); a poisonous compound.

toxicity. State or degree of being poisonous.

toxicology. The science which deals with poisons, antidotes, toxins, effects of poisons, and the recognition of poisons.

toxic residue. A poisonous residue left on plants, in the soil, or on animals, by a spray or dust.

toxin. A poison produced by animal or vegetable cells. It causes antibodies to form when it is injected into animals. Among the various *toxins* are those developed by snakes (venom), insects, and bacteria, as in tetanus, botulism, and diphtheria. See *toxic; toxicity.*

toy. Designating any small or dwarf variety, as a *toy* spaniel.

toyon. Christmasberry.

trace. 1. Either of the two leather straps or chains, attached at one end to the hames clamped to the horse collar or breast band, and at the other to tugs on a whippletree. See *tug chain.* 2. The force to pull the vehicle, etc., as applied to traces. 3. See *trace element.* 4. To follow the course of nutrient elements in plants or animals, as by use of radioisotopes. 5. Designating an artificial flavoring material with little or no true flavoring.

trace bud. Latent bud.

trace element. Any of certain chemical elements necessary in minute quantities for optimum growth and development of plants and animals. *Cf. essential element.*

trace high. See *kilt.*

trace mineral. See *trace element.*

trace of precipitation. Precipitation in amounts too small to be measured. Less than .005 in. for rain and less than .05 in. for snow.

tracer. A radioisotope mixed with a stable substance, by means of which a material can be traced as it undergoes physical and chemical changes. In agricultural research: *e.g.*, radioactive isotopes of phosphorus in a chemical fertilizer can be traced through the plant as it is absorbed by the roots.

trachea. The windpipe; in mammals, it extends from the throat to the bronchi.

tracheid. Any of the water-conducting, tube-like cells in the xylem of a plant. They also act as a support.

track roller. An overhead track upon which a movable hoist operates to facilitate the movement of milk cans to and from the cooler.

tract. 1. An area of land of any size, but bigger than a lot. 2. An anatomical structure as the digestive *tract,* cerebellospinal *tract.*

tractable. Docile; easily controlled.

traction duster. A dusting machine which derives the power for blowing or spreading dust from the ground wheels on which it is mounted.

traction sprayer. A sprayer (*q.v.*) which receives the power to compress the air in the tank from the wheels on which it is mounted. See *power sprayer.*

tractor. 1. A vehicle with an automotive or diesel type engine used to supply power to other machines in one of 5 ways: (a) pulling at the drawbar; (b) belt power from the belt pulley; (c) rotary power from the power take-off shaft; (d) hydraulic power for the operation of hydraulic cylinders or other positioning mechanisms; (e) electric power when a generator is mounted on the tractor. 2. Designating an implement pulled or powered by a tractor.

tractor binder. A grain binder (*q.v.*) drawn by a tractor.

trade-center community. A trade center and all the area surrounding it from which residents come to obtain supplies and services. See *community.*

trade egg. In marketing, a shell egg which is small or slightly dirty but of reasonably good quality.

trade in futures. To buy and sell futures *q.v.*

trader. A person who buys and sells produce.

trading center. A city with an important market for a particular kind of produce. See *trade-center community.*

trading limit. 1. Prices above or below which trading is not allowed during any one day. 2. In the U.S., the maximum amount of contracts an individual is allowed to hold at any time in certain commodities, under Commodity Exchange Act.

traffic. 1. Buying and selling. 2. The movement of produce by rail, truck, or ship. 3. The movement of units of vehicles on a highway.

trail. 1. A pathway made either by repeated passage of men or animals or constructed for easier passage. It is not usable for vehicles. 2. To move cattle or sheep over a long distance to a pasture, a market, or shipping point.

trail blazer. A pioneer.

trail boss. A foreman in charge of a herd being taken up a trail for grazing or marketing.

trail broken. Designating cattle or horses accustomed to the trail.

trail builder. See *bulldozer.*

trail cutter. A man inspecting a trail herd and cutting out any stock which does not belong to the driver.

trail drive. To drive a herd along the trail to pasture or market.

trail driver. One who drives the herd along a trail.

trailer. A vehicle with one, two, or four wheels, designed to be pulled by another vehicle. For agricultural purposes, it is used to carry heavy loads.

trailerette. A one-wheeled trailer.

trailer hitch. A device for attaching a trailer to a pulling vehicle. Usually, it consists of a ball and socket joint with a set-screw or other safety devices affixed to it to prevent a loosening of the hitch.

trailer mower. A ground-driven mower designed for attachment to the tractor drawbar; hence, it operates behind the tractor.

trail hand. A puncher (*q.v.*) or herder who helps in driving livestock over the land.

trail herd. A herd of cattle which is taken by trail to market or to a distant range.

trail herding. Directing and controlling the movement of a group of livestock on restricted overland routes.

trailing. 1. The driving of livestock from place to place. 2. The voluntary wandering of livestock about a range, usually in search of forage, water, or salt. 3. Designating a plant which puts forth long, recumbent stems. See *creeping.*

trailing arbutus. 1. Any plant of the genus Epigaea, family Ericaceae. 2. *E. repens,* an evergreen, very fragrant wild flower which may be grown in the wild garden for its white or pinkish flowers. Native to North America. Also called *ground laurel; winter pink; Mayflower.*

trailing azalea. Alpine azalea.

trailing begonia. Begonia treebine.

trailing blackberry. Cutleaf blackberry.

trailing fire. Texasplume.

trailing fuchsia. *Fuchsia procumbens,* family Onagraceae, a trailing shrub grown in a hanging basket for its habit and for its orange-purple calyx. Native to New Zealand.

trailing lantana. *Lantana sellowiana,* family Verbenaceae, a vine-like shrub grown for its trailing habit and its pinkish-lilac flowers. Native to South America. Also called *weeping lantana.*

trailing mahonia. Oregongrape.

trailing myrtle. Periwinkle.

trailing phlox. *Phlox nivalis,* family Polemoniaceae, a perennial herb grown for its white or pink flowers. Native to North America. Also called *camla.*

trailing sumac. Poison ivy.

trailing yew. Canada yew.

trailjackal. Any man who by dishonest means lived off the trail herds or who preyed on them (southwestern U.S.).

train. 1. To direct the growth of a plant to a desired shape by pruning while young or fastening its stem and branches to a support. 2. To instruct and drill an animal so as to teach it certain desired habits.

trained horse. A horse trained to do specific jobs on the range or farm.

trainer. 1. An intermediate or overtopped tree which by its shading or abrasive actions hastens the natural pruning of crop trees. 2. A person who trains animals.

tramp. 1. To compact soil usually around a transplant by treading on it or by pounding it with a tool. See *tamp.* 2. To compact silage in the silo by walking on the material. 3. A man who wanders about the country working only when necessary (U.S. and England). See *sundowner; swagman.*

tramper. A heavy weight, usually hydraulically operated, which loosely presses lint cotton preparatory to baling.

tramper side. Either of two press boxes in a cotton gin, arranged side by side and supported by a center column, which can be revolved. When lint cotton is falling into a box it is the *tramper side.*

trampling. 1. Treading under feet; the damage to plants or soil brought about by congested movements of livestock, including mechanical injury to tree reproduction and ground cover in woods. 2. Compacting soil in earthen dams and reservoirs by livestock to make the dam or reservoir impervious to water (largely replaced by machine compaction).

transect. A cross-section of an area which is used as a sample for recording, mapping, or studying the vegetation and its use. A *transect* may be a series of plots, a belt or strip, or merely a line, depending on the purpose of use.

transfer hose. A flexible hose, usually of plastic, which is used to convey the milk from the bulk milk tank to the pump of the bulk milk truck.

transgressive segregation. The appearance in the F_2 (or later) generations of individuals showing a more extreme development of a character than either parent, assumed to be due to cumulative and complementary effects of genes contributed by the parents of the original hybrid. Adequate testing of variation in the parents is required to establish its occurrence.

transient. Having only a brief existence; short-lived.

transition. A short conduit uniting two other conduits having different hydraulic elements. It is designed for a minimum of hydraulic disturbance.

transitional parity prices. In the change from old parity prices to the new parity prices, a price provision which prevents the drop in price of any commodity from being more than 5 percent per year. See *parity prices.*

transitional soil. A soil which somewhat resembles two different soils and genetically is related to them.

transition cell. A cell in a beehive larger than a worker cell, but smaller than a drone cell, where a change from the worker- to the drone-size bee occurs.

transition soil. Transitional soil.

transitory. Of brief duration; passing quickly.

transit rot. Any rot of produce which occurs while the produce is in transit.

transit tetany. A condition likely to occur in very warm weather to an animal which is transported over very long distances in confined quarters. Symptoms consist of distress, rapid respiration, high and hard pulse, sweating, spasms of muscles of the neck and shoulders.

translocate. To transfer the products of metabolism, etc., from one part of a plant to another.

translocated herbicide. An herbicide which, when sprayed or applied to one portion of a plant, travels to another part of the plant.

translocated spray. See *translocated herbicide.*

translocation. 1. In genetics, change in position of a segment of a chromosome to another part of the same chromosome or of a different chromosome. 2. See *translocate.*

transpiration. The process by which water vapor leaves the living plant body and enters the atmosphere. It includes cuticular *transpiration,* or direct evaporation into the atmosphere from moist membranes through the cuticle; and stomatal *transpiration,* or outward diffusion into the atmosphere through the stomata and lenticels of water vapor previously evaporated from imbibed membranes into gas spaces within the plant. See *evaporation.*

transpiration ratio. The ratio of the weight of water transpired by a crop in its growth to the dry weight of the crop.

transpire. To give off moisture as water vapor from a living plant. See *transpiration.*

transplant. 1. A seedling which has been moved one or more times. 2. To transfer seedlings from the seedbed and set them in the ground.

transplanter. See *dibble; transplanting machine.*

transplant flat. See *flat: 1.*

transplanting machine. A wheeled machine which has a furrow opener, a tank for water supply, disks for closing the furrow, and seats low to the ground for a person to sit on while placing the transplants in the furrow.

transportation tank. A tank used for over-the-road transportation of milk, fluid milk products and fruits, such as cherries in water from the orchard to the processing plant.

transported soil. A soil formed by the consequent or subsequent weathering of materials transported and deposited by some agency, such as water, air, or ice.

transport truck. 1. A hand propelled vehicle having two or more wheels which is used to haul or transport material. 2. A large truck for hauling loads on the highway.

transversal. See *side step.*

trap. 1. A two-wheeled, one-horse cart. 2. Any of various devices used for catching living insects, animals, birds, etc. 3. A device which consists of a U-shaped pipe or chamber so that liquid flowing through it always forms a seal against a flow-back of gas. 4. A chamber to collect sediment flowing in a pipe.

trap crop. Any crop planted for the purpose of attracting insects, fungi, etc., away from other nearby crops.

trap efficiency. A percentage index which shows the effectiveness of a reservoir in removing silt or sediment from a stream which enters it.

trap nest. A fowl nest so arranged that the hen may enter it, but not leave until released by an attendant. It is used for checking egg production.

Trappist. A mild, pale yellow, semisoft, cured cheese which is made from cows' whole milk to which ewes' or goats' milk may or may not be added. It is ripened five or six weeks. The cheese was made originally by Trappist monks in Europe.

trappy. Designating a horse's gait which is quick and high with a comparatively short stride.

trap row. See *trap crop.*

trap tree. A felled or killed tree used as a lure for insects. See *trap crop.*

trash. 1. Leaves, stalks, husks, etc., left on the ground after harvest. 2. Anything worthless or useless, as *trash* in seed cotton.

trash farming. Stubble mulch farming.

trash guard. An attachment, usually a rod or wire, for a moldboard plow, which drags trash into the furrow for covering.

trash mover. A storm which produces sufficient runoff to move trash.

trash shield. A part of a plow: a piece of sheet metal so shaped as to form a hood over the top side of the furrow slice as it is turned over.

trashy cultivation. Tillage which leaves the major portion of crop residues on the soil surface. Also called *trashy tillage.*

trauma. A wound or injury; a deep emotional shock.

traumatic heartwood. In trees, pathological heartwood (*q.v.*) that has developed in the vicinity of wounds. See *heartwood.*

traumatic pericarditis. Inflammation of the outer coat of the heart sac, which may be caused by a foreign body.

travelersjoy. *Clematis vitalba,* family Ranunculaceae, a hardy, woody, high-climbing vine grown on a trellis, fence, etc., for its almond-scented, small, late-summer, white flowers. Native to Europe. Also called *old-man's-beard; withywind.*

Traveling Stock Reserve. Lands in Australia which are reserved by the government to be used as routes for the movement of stock.

travel stain. Undesirable markings which are made on honeycomb by bees repeatedly crawling across the comb.
travel-time map. In forestry, a map that shows the time required to reach various parts of a unit by the initial action fireman or crew from specified positions.
traverse time. The time which elapses from the instant a particle of air enters a bulk of grain until it has traversed a particular path and leaves the grain, usually expressed in seconds.
travis. The partition or half-wall dividing individual stalls in a barn. Also called *traverse.*
Travnik. A soft cheese made from ewes' milk to which a small portion of goats' milk is added.
travois. See *crotch: 2.*
tray. 1. A wide, flat-bottomed, topless, shallow container used for picking, carrying, handling, drying, or storing produce. 2. A short piece of heavy wrapping paper on which seedless grapes are sun-dried.
tray agriculture. Hydroponics.
tray gardening. Hydroponics.
treacleberry. Feather Solomonplume.
treacle erysimum. *Erysimum cheiranthoides,* family Cruciferae, an annual herb which is a weed pest among annual plants and in waste spaces. Also called *treacle mustard; wormseed mustard; tarrify.*
tread. 1. Any injury to the front legs of a horse caused by its overreaching with a hind foot as it runs. 2. That part of a wheel or tire which rests or bears on the ground. 3. That endless belt on which a crawler tractor moves. Usually called *track.*
treader. A rotary hoe pulled backwards to give more compaction of the surface soil than when it is pulled forward. The same effect may be produced by reversing the wheels of the rotary hoe and pulling it in a forward direction.
treadle guide. Parallel shift.
treasureflower gazania. *Gazania rigens,* family Compositae, a perennial herb grown as a tender annual for its handsome flowers which have orange rays with a white-eyed black spot near the base. Native to South Africa.
treasure-vine. Climbing dahlia.
treat. 1. To care for a sick animal or diseased plant by giving it proper attention and medication. 2. To subject a product to an action or process to improve it in some manner. 3. To subject plants, animals or soil to various chemicals, practices, etc., in order to learn which are beneficial or harmful.
treatment. 1. The care, attention, and medication given a sick animal or diseased plant. 2. Any specific application of a chemical or any particular practice which may be applied to plants, soils, animals, etc., to bring about a change or to learn what may be beneficial or harmful.
treatment of choice. The best of several drugs or systems of medication available for a specific purpose, when all are comparatively evaluated from standpoints of safety, efficiency, simplicity, and cost.
Trebizond date. The fruit of the Russian olive *q.v.*
treble superphosphate. Double superphosphate.
trebletree. An evener (*q.v.*) in a draft pulled by three horses.
tree. Any woody, perennial plant which normally has one well-defined stem and a definitely formed crown. It is usually considered to have a minimum mature height of 8 to 10 ft. See *bush; fruit tree; nut tree; shade tree; shrub; street tree.*
tree age. 1. The length of time since the germination of a seed or from the initial budding or vegetative propagation. 2. In horticulture, the length of time since a tree was set in an orchard.
tree aralia. Devil's-walking-stick.
tree azalea. Sweet azalea.
tree basin. See *basin: 1.*
treebeard tillandsia. *Tillandsia usneoides,* family Bromeliaceae, an epiphytic plant growing wild on trees from Virginia, U.S., southward. It is prized for its drooping habit and is used for decoration. Native to tropical and subtropical America. Also called *long moss; Spanish moss.*
treebine. Any of the tender, softwood vines of the genus Cissus, family Vitaceae. Species are grown in the greenhouse or as house plants except in the subtropical areas where they are grown out-of-doors. See *begonia treebine; evergreen treebine; ivy treebine.*
tree caliper. 1. A caliper-like device used to measure diameters of tree trunks and logs. 2. The diameter of a tree trunk.
Treece. A small, plastic-curd, Italian cheese which is eaten fresh.
tree class. One of the groups into which all forest trees may be divided on the basis of vigor, insect vulnerability, value, crown class, age class, size, form, diameter, etc.
tree cotton. Asiatic tree cotton.
tree count. In forestry, a rapid counting of trees without regard to species, diameter, height, etc.
tree cypress. Texasplume.
tree expert. A person thoroughly trained in the care of trees. See *forester.*
tree facer. A turpentining tool with a strong, concave, cutting head for removing bark preparatory to seating the cup.
tree farm. A tract of land on which trees, especially forest species, are grown as a managed crop.
tree fern. Any fern of the family Cyatheaceae. Species are notable for their upright woody stems, from a few ft. to 60 ft. or more in height. They are tropical, usually growing in humid locations. The important genera are: *Alsophila,* mainly Australasian; *Cibotium,* Australia and Hawaii; *Cyathea,* West Indies, South America, Africa, New Zealand; *Dicksonia,* Australasia. In large greenhouses the commoner species used are *Cyathea arborea,* from the West Indies, and *Alsophila australis,* from Australia.
tree fruit. Fruit borne by trees, as the peach, apple, cherry, plum, etc., as contrasted to the fruit of brambles, strawberries, etc. See *small fruit.*
tree heath. *Erica arborea,* family Ericaceae, a shrub or small tree grown in the warm regions

for its white, fragrant flower clusters. In Europe, it is grown as a source for briar pipe bowls. Native to the Mediterranean region.

treehopper. Any insect of the family Membracidae. It injures trees and shrubs by sucking sap. See *buffalo treehopper; two-marked treehopper; three-cornered alfalfa hopper.*

tree injection. The introduction of a chemical into the sap stream of a tree (a) to kill it, (b) to prevent disease, (c) to prevent chlorosis, (d) to color the wood prior to sawing, or (e) to destroy insects.

tree lifter. A mechanical device used for loosening the soil prior to pulling out seedlings and transplants to avoid damage to the roots.

tree line. Timber line.

tree lupine. *Lupinus arboreus,* family Leguminosae, an herbaceous, perennial, shrubby herb grown in the flower border for its fragrant, sulfur-yellow flowers. It is also useful as a sand binder. Native to California, U.S. Also called *yellow lupine.*

tree man. A person commercially or scientifically interested in trees. See *nurseryman.*

tree mildew. A powdery mildew of trees and shrubs caused by *Phyllactinia corylea.*

tree myrtle. Feltleaf ceanothus.

tree of chastity. Chastetree.

tree of heaven. Tree of heaven ailanthus.

tree of heaven ailanthus. *Ailanthus altissima,* family Simarubaceae, a medium-sized, deciduous tree grown as a street tree for its ability to withstand smoke and city conditions. Native to China. Also called *tree of heaven; false varnish tree.*

tree onion. Top onion.

tree orchid. Any epiphytic orchid.

tree paint. Any one of several types of wound dressing for trees, such as orange shellac, asphaltum paints, creosote paints, grafting waxes, house paint, Bordeaux paste, commercial tree paints, etc. Ideally, it should disinfect, prevent fungus entrance, prevent checking, permit callus growth, be toxic to insects, be easily applied, allow excess moisture to evaporate, and should not crack on drying, or injure the tissues.

tree peony. *Paeonia suffruticosa,* family Ranunculaceae, a small shrubby, perennial herb grown for its spring, rose-red or white flowers. Native to China.

tree-perching. Designating an epiphytic *q.v.*

tree planting. To grow trees, shrubs, vines, or cuttings to prevent excessive runoff and soil loss. It includes plantings for gully control, post lots, sand dune fixation, bank protection, reinforcement of existing woodlands, windbreaks, etc. Another purpose of *tree planting* is to obtain wood products or trees to sell for ornament, as Christmas trees, etc.

tree poppy. Stiff bushpoppy.

tree primrose. Evening primrose.

tree protector. Any of the several kinds of material for wrapping loosely around the trunk of a tree to protect it from sunburn or injury by rodents.

tree pruner. 1. A long rod with an attached sharpened hook, blade, and rope. The blade is movable and is fixed to the hook. When the hook is placed over a small limb and the rope attached to the blade is pulled, the limb is cut off. 2. One who prunes trees.

tree ripe. Designating a fruit which has ripened on a tree, as contrasted to one which has been picked and allowed to ripen after harvesting.

tree root rot. See *mushroom root rot.*

tree scraper. A short-handled tool with a sharply-curved blade which is used to remove loose bark from trees harboring insects.

tree silktassel. *Garrya elliptica,* family Garryaceae, an evergreen shrub grown for its attractive foliage and catkins. Native to North America. Also called *silk tassel tree; tassel tree; quinine bush.*

tree stool. A stump.

tree sugar. Maple sugar.

tree surgeon. A person skilled in pruning, spraying, fertilizing, and repairing of injury to trees, such as cleaning and filling of cavities.

tree surgery. The after-care of trees which includes pruning, repair of injury, spraying, saving trees from change in grade, and fertilizing when necessary. See *tree surgeon.*

tree tobacco. *Nicotiana glauca,* family Solanaceae, a perennial, treelike, tropical herbaceous plant grown for its stately habit, blue-green foilage, and yellow flowers. Native to South America.

treetomato. *Cyphomandra betacea,* family Solanaceae, a tender shrub which bears tomato-like, acid fruit, grown as an ornamental in the U.S. Native to South America.

tree top. See *crown: 1.*

tree trunk. The main stem of a tree; the bole.

tree wrap. Specially prepared paper wrapped spirally around a tree trunk to protect it from sunburn, etc.

tree yucca. Joshua tree.

tref. Designating a food which does not meet the dietary laws for Orthodox Jews. See *Kosher.*

trefoil. Clover.

trellis. A latticework frame or wires supported by posts on which vines or flowers are trained. It may be constructed as a bower, arbor, etc.

trematode. Parasitic flatworm of the class Trematoda, such as the sheep liver fluke.

trembles. A disease of cattle and sheep which results from ingestion of white snakeroot, jimmyweed, or burrowweed, all of which contain tremetol, an unsaturated alcohol. It is characterized by a trembling in the muscles about the eyes and legs, depression, inactivity, constipation, nausea, rapid and labored breathing, weakness, prostration, and sometimes, death. Humans who drink milk from affected cows may contract this disease. Also called *milk sickness.*

trembling in pigs. See *shivering in little pigs.*

trembling poplar. See *quaking aspen; European aspen.*

tremor. A trembling of the voluntary muscles.

tremulous air cell. An enlarged air cell in poultry eggs which causes the membranes to split apart beyond the edges of the cell. The cell moves slightly as the egg is turned from one side to another. Such an egg is graded C, or the lowest edible grade.

trench. 1. A long, relatively narrow excavation; a ditch. 2. To dig a ditch; to make a long, relatively narrow excavation.

trencher. Any of several types of machines which digs trenches.

trenching. A method used by gardeners for renewing and deepening the soil. Trenches 12 to 24 ins. wide and 10 to 20 ins. deep are excavated with a spade, and the soil thickness separated into 2 or 3 layers. The old surface layer is buried and the subsurface layer placed on top. Manure may be placed in the bottom of the trench or between separate layers. The method is a laborious one and not much used in the U.S. but it is popular in the United Kingdom.

trench plow. 1. A long-winged plow used to start the excavation or trench for placing tile for subsurface drainage. Sometimes called a *ditcher.* 2. A plow with a U-shaped blade to loosen the soil in trenches. See *ditching plow.*

trench silo. A trench excavated in a hillside or on firm ground, usually lined with wood or concrete retaining walls. Commonly about 15 to 25 ft. wide, 6 to 8 ft. deep and as long as the capacity desired. A *trench silo* must have good drainage and its use is largely limited to arid or semiarid climates.

trespass. To enter unlawfully on a property.

tret. An allowance formerly made to a purchaser of a product to allow for waste due to transportation, miscounting, etc., which consisted of giving 4 lbs. of the product free for every purchase of 100 lbs. after deducting the tare.

triage. A marketing classification of broken coffee berries. It is the lowest grade (English).

trial run. A preliminary test to estimate the performance of a machine, procedure, or of any device involving motion.

triangle. See *rattle.*

triangle pot roast. A retail cut of beef for roasting taken from the chuck *q.v.*

triangular. Designating a flexible bracing system for trees with weak or split crotches, or for heavy lateral limbs.

triangular planting. An arrangement for planting trees in which four trees are planted so as to form a rectangle. The fifth tree is planted at the point where the diagonals from the corners intersect. Also called *quincunx.*

triangular weir. A contracted measuring weir notch with sides that form an angle with its apex downward, the crest is the apex of the angle; a V-shaped weir.

tribasic copper sulfate. $CuSO_4 \cdot 3\ Cu(OH)_2 \cdot H_2O$, a copper salt used as a fungicide.

tribasic phosphate of soda. Trisodium phosphate, a sodium salt which in a 2.5 percent solution has an appreciable disinfecting value.

tribe. 1. In botany, a subdivision of a family, roughly equivalent to subfamily. 2. In animal breeding, a group or combination of animals descended through the female line from a particular progenitrix or foundress.

tribulosis. A disease of sheep, also called big head (*q.v.*), that results from their feeding on puncturevine (*Tribulus terrestris*).

tric. See *bovine genital trichomoniasis.*

tricalcium phosphate. $Ca_3(PO_4)_2$, a salt of phosphoric acid, present in rock and bones. It is the source of almost all phosphate and phorphorus-containing substances. Reaction of *tricalcium phosphate* and sulfuric acid yields the widely-used fertilizer marketed as superphosphate. Also called *bone phosphate; B.P.L.; calcium phosphate; phosphate of lime.*

trichiasis. A turning inward of the eyelashes.

trichina. See *trichinosis.*

trichina worm. See *trichinosis.*

trichinosis. A parasitic disease of animals and man caused by the nematode *Trichinella spiralis.* In severe cases the disease is characterized by diarrhea, anorexia, fever, muscular pains, stiffness, difficulty in breathing and chewing, and loss of weight. Infection in animals occurs by ingestion of tissue (namely muscle) of animals previously infected. In man the disease is contracted by eating raw or improperly cooked pork from pigs harboring the infection. Detection of the disease in swine is not always possible at the time of slaughter, therefore it is assumed that all pork may be infected with trichinae (*trichinosis*) and should not be consumed until adequately cooked. Also called *trichiniasis.*

trichloroacetonitrile. A non-inflammable insect fumigant.

trichloroethylene. A non-inflammable insect fumigant.

trichobezoar. Hair ball.

Trichoderma rot. A rot of lemon fruit in storage caused by *Trichoderma lignorum,* family Moniliaceae, characterized by a whitish growth which turns greenish-yellow, yellow and bluish-green.

trichomonad. A microscopic, single-celled animal of the genus Trichomonas, which moves by the action of 3 to 5 whip-like flagella at the front end, and a membrane which undulates along one side of the body. See *trichomoniasis.*

trichomonal antigen. An extract from a certain type of one-celled animal (trichomonad) which, when injected into the body of a higher animal, causes it to produce substances (anti-bodies) to combat the trichomonad.

trichomoniasis. An infection with trichomonads. Common types in domestic animals are bovine genital *trichomoniasis* and *trichomoniasis* of the upper and lower digestive tracts of poultry *q.q.v.*

trichomoniasis of the lower digestive tract. A disease of the lower digestive tract of chickens, guineas, etc., but especially of turkeys caused by the protozoa *Trichomonas gallinarum,* characterized by droopiness, loss of weight, diarrhea with yellowish fluid droppings, and death (for young birds 4–8 days after the symptoms are evident; for older fowls after a longer interval).

trichomoniasis of the upper digestive tract. A disease of the upper digestive tract of chickens, pigeons, but especially of turkeys caused by the protozoa *Trichomonas gallinae,* characterized by droopiness, loss of appetite, drooping wings, emaciation, drooling, and death (in young birds as early as one day after the symptoms are apparent, for older birds the interval is longer).

trichophytobezoar. Hair and food which an animal has ingested and which has gathered in the rumen or the stomach in the form of balls. See *hair ball.*

trick-horse. A horse trained to perform different feats.

tricresol. See *cresols.*

trier. A probe used to obtain a representative, small sample of produce without cutting or untying the container, as a cheese or grain *trier.*

Trieste mustard. *Brassica juncea,* family Cruciferae, an annual herb, one form of which, called Southern Curled, is grown for greens. Native to Europe and Asia. Also called *India mustard.*

trifoliate orange. *Poncirus trifoliata,* family Rutaceae, a spiny, deciduous tree grown as an almost impenetrable hedge tree, as grafting stock for tender citrus, and for its white flowers. Native to China.

trigeneric hybrid. A hybrid which results from crossing plants of three different genera.

trigonella. Any plant of the genus Trigonella, family Leguminosae. See *fenugreek trigonella.*

trihybrid. a. In genetics, a cross which involves three pairs of genes. b. An individual which is heterozygous for three pairs of genes.

trillium. Any plant of the genus Trillium, family Liliaceae. Species are early-spring flowering and are useful in the wild garden in shady places. See *dwarf trillium; nodding trillium; Pacific trillium; painted trillium; purple trillium; toad trillium.*

trim. To remove unwanted or undesirable portions of a product, plant, etc., as to *trim* fat from a ham or to *trim* a tree or hedge.

trim a sample. To remove bagging-stained or damaged cotton from the outer surface of a sample. Also called *dress a sample.*

trim carcass. Any animal that dresses out well, showing good fleshing and finish, without excess or gobby fat or coarse bone.

trimmed loin. A wholesale cut of beef or pork loin from which the flank has been removed and the kidney and suet or leaf fat taken out.

trimmed short loin. A wholesale cut of beef short loin which has the kidney and suet removed.

trimming. 1. The cutting and dressing of an animal coat or feet, as *trimming* a horse's mane, tail, and hooves. 2. The removal of unwanted or undesirable portions from a product. 3. The part so removed. 4. The pruning of a plant.

trimming saws. In lumber sawing, a series of swing saws which are suspended over transfer chains somewhat similar to those used in the slasher. The saws are spaced at such intervals that any two of them will saw stock to commercial lengths.

trimness. In stock judging, the condition of an animal's being in good flesh without having excess weight.

trinity lily. Dwarf trillium.

trio. 1. A group consisting of three objects or organisms. 2. A male and two female birds of the same variety which are shown as a unit in exhibitions.

trip. 1. A safety device which allows an automatic release, as a *trip* on the shank of a cultivator. 2. To release suddenly, as to *trip* a sling and release hay.

tripe. A retail cut of beef consisting of the walls of the rumen and the reticulum.

tripe-hound. A sheep dog (Australia and New Zealand).

trip hook. A part of the binding and tying mechanism on a grain binder. When sufficient grain has been collected, the pressure against the *trip hook* forces it back releasing the trip dog to start the operation of the binding mechanism. The *trip hook* can be adjusted to increase or decrease the size of bundle.

triple barrel pump. A pump having three barrels connected with a common suction pipe. The pistons are operated by a three-throw crank. The cranks are at an angle of 120°, so that each piston is always at a different part of the stroke from the others. The pump, being double acting, produces an almost uniform flow. Also called *triplex pump.*

triple superphosphate. See *double superphosphate.*

triplet lily. Grassnut brodiaea.

tripletree. See *evener.*

trip lever. A lever which, when operated, will cause a sudden release or emptying.

triple-yolk egg. An egg which contains three yolks.

tripod stacker. A device used for elevating hay to the top of a stack which consists of three poles arranged as a tripod with a pulley suspended from the top.

trip rope. Rope, string, chain, etc., attached to a trip which is used to activate the trip. See *trip: 1.*

trisomic. Referring to an organism having three chromosomes of one type (Chromosome formula 2n/1).

tristeza. A virus disease of citrus characterized by failure and dying of sweet oranges, tangerines, and grapefruit grown on certain root stocks, such as on sour orange stock. It is caused by destruction of food conducting vessels in bark at the bud union, resulting in starvation and death of roots.

tritoma. Torchlily.

triturated. Ground or rubbed into fine particles.

triumph tulip. A class or type of tulip with many varieties which originated by crossing early, Darwin, breeder, and cottage tulips. They have sturdy growth and large flowers and blossom earlier than the Darwin tulips.

trocar. A sharp-pointed instrument used by veterinarians for piercing a cavity, such as the paunch of a cow, which is generally used in conjunction with a cannula, a hollow tube. The *trocar* fits inside the cannula and punctures the cavity wall. It is then withdrawn and the cannula is left in place to allow for escape of gas or fluid.

trockentorf. A relatively undecomposed, peatlike deposit found on the surface of well-drained soils under forest cover. It is composed of the remains of leaves and fragments of wood.

Trojan iris. *Iris trojana,* family Iridaceae, a perennial herb grown for its blue and red-purple flowers. It is a bearded iris. Native to Turkey.

trompillo. Silverleaf nightshade.

trona. $Na_2CO_3 \cdot 2H_2O$, a glistening black mineral, which is the principal constituent of black alkali of arid soil.

tropical. Designating a plant, disease, etc., which flourishes in the tropics or where conditions are made to resemble the tropics in temperature, humidity, length of day, etc.

tropical almond terminalia. *Terminalia catappa,* family Combretaceae, a tropical tree grown in warm areas as a street tree. Native to Asia. Also called *Indian almond.*

tropical carpetgrass. *Axonopus compressus,* family Gramineae, a stoloniferous, perennial grass sometimes used as a turf grass. Its forage rating is good. Native to the U.S., found from Florida to Louisiana. Also called *true carpet grass.*

tropical flint. Flint corn (*q.v.*) that is grown in the tropics. In the U.S., it is grown in the Gulf States. See *Creole flint.*

tropical fowl mite. *Ornithonyssus bursa* (Berl.), family Dermanyssidae, a mite that infests domestic fowls and wild birds. These mites congregate around the vent resulting in bloody scabs. Severe infestations can cause death. It is widely distributed throughout the northern, eastern, and central U.S. Also called *feather mite.* See *northern fowl mite.*

tropical lilac. Creeping skyflower.

tropics. The area of the earth's surface lying between the Tropic of Cancer and the Tropic of Capricorn. It is considered generally to be an area of high temperature often accompanied by high humidity.

tropism. A growth reaction of a plant to various external or internal stimuli. Of the former may be mentioned phototropism, the increased growth toward or away from light, geotropism, growth with respect to the stimuli of gravity, chemotropism, hydrotropism, thigmotropism, etc. Probably in all these cases the external stimuli affect the production of growth hormones which induce specific growth.

trot. A horse's two-beat gait, in which the diagonal fore and hind legs act together. Speed varies from horse to horse. The road horse *trot* is a fast-stepping *trot,* characterized by long and rapid strides executed with an extreme degree of extension. The heavy harness horse *trot* and the hackney pony *trot* are high-stepping *trots,* characterized by height and spring of stride, the horse setting himself, going collectedly, and executing each step with an extreme degree of flexion and the utmost precision.

trots. A diarrheal or abnormally loose condition of the bowels. Sometimes also called *skitters.*

trotter. 1. A horse used for sulky racing. See *light-harness horse; Standardbred.* 2. Lamb shank.

trotting plate. Running plates.

trough. 1. A long, relatively narrow, open-topped vessel used to hold water or feed for livestock. 2. A shallow conduit, usually of V or U section, open at the top.

trough hopper. A hopper used to furnish a continuous supply of feed into a trough *q.v.*

trough scraper. A hoe-shaped tool used to clean feed troughs.

trout flower. Fawnlily.

troutlily. Fawnlily.

Trouville. A soft, ripened, whole milk cheese made in France.

trowel. 1. A short-handled tool having a relatively narrow, concave and pointed blade. It is used for transplanting, planting bulbs, etc. 2. A hand tool used in concrete work and handling mortar in bricklaying.

truck. 1. A strong vehicle, especially one with four wheels, a wagon or an automotive vehicle used for transporting machinery, freight, and other heavy articles. 2. A low vehicle similar to a wheelbarrow but having two small, stout wheels in place of one, and a forward lip, used for moving barrels, boxes, etc., by hand; a barrow-*truck.* 3. A swiveling carriage, having two, four, or six wheels, placed under a locomotive or car. 4. Vegetables raised for sale. 5. To transport produce by means of an automotive *truck.*

truck crop. A vegetable crop usually raised on a relatively large acreage, under intensive methods of farming, at a place comparatively distant from the markets upon which it is sold. See *market garden.*

truckee pine. Jeffrey pine.

trucker dealer. An entrepreneur who buys produce from the farmer at the farm and hauls it to a market for sale.

truck farm. A farm producing one or several kinds of vegetables which are shipped to and sold on distant markets. *Cf. market garden.*

truck farmer. A farmer engaged primarily in the production of truck crops *q.v.*

truck gardener. Truck farmer.

truck gardening. The growing of a truck crop *q.v.*

truck grower. Truck farmer.

truck horse. Draft horse.

trucking. 1. The transportation of produce by automotive trucks, as contrasted to rail, water, or air transportation. 2. See *truck gardening.*

truck-ins. Livestock transported to the yards in trucks, as contrasted to that brought in by rail or driven in. *Cf. drive-ins.*

true. Like the parental type, without change, as a variety which breeds *true.*

true aloe. *Aloe vera,* family Liliaceae, a succulent herb grown in a pot or tub as a specimen plant for its grayish-green leaves and yellow flowers.

The dehydrated sap of the leaves of *true aloe* and several other species of the genus Aloe is called bitter aloe. It has a medicinal value as a tonic, a purgative and an agent promoting menstrual discharge. Native to the Mediterranean region. Also called *Barbados aloe.*

true bay. Grecian laurel.

true-breeding. Designating varieties which conform to the parental type with respect to certain characteristics, such as color, disease resistance, etc.

true daisy. English daisy.

true flavor. Natural flavor.

true forget-me-not. *Myosotis scorpioides,* family Boraginaceae, a perennial herb grown for its blue flowers with yellow, pink, or white centers. Native to Europe and Asia.

true lavender. *Lavandula officinalis,* family Labiatae, an undershrub which is an old flower garden favorite. It is grown for its fragrant flower buds dried and used in perfumery, sachets, etc. Native to the Mediterranean region. Also called *lavender; white lavender; French lavender.*

truelove. 1. Purple trillium. 2. *Paris quadrifolia,* family Liliaceae, a rhizomatous herb with yellowish-green flowers. Native to Europe. Also called *herb Paris.*

true mistletoe. American-mistletoe.

true mountain mahogany. *Cercocarpus montanus,* family Rosaceae, a tender, evergreen shrub grown as an ornamental for its 3 to 5 in. long, feathery-tailed fruit. It is found on dry ridges at altitudes of 4,000 to 10,000 ft. from northern Montana to New Mexico, U.S., and furnishes a large amount of high-quality browse.

true myrtle. *Myrtus communis,* family Myrtaceae, an evergreen, aromatic shrub grown in warm regions for its handsome foliage. Also grown by florists in greenhouses as a pot plant. Native to the Mediterranean region. Also called *Dutch myrtle.*

trueness. 1. In wool grading, the uniformity in diameter of the fiber from its base to tip. 2. See *true.*

true prairie. See *prairie: 2.*

true shield budding. Budding (*q.v.*) of citrus by means of bark bud-shields from the cion, without wood.

truestar anisetree. *Illicium verum,* family Magnoliaceae, an evergreen shrub or small tree grown in frost-free areas for its purplish-red flowers and its star-shaped, carminative, anise-scented fruit. Native to China. Also called *badian; star anise; Chinese anise.*

true stomach. Abomasum.

true to type. See *true.*

truffle. The subterranean fruit of a fungus of the family Tuberaceae, which is highly prized as a food or condiment. Mostly, the genus concerned is Tuber, although a few *truffles* belong to other, but closely related genera. The vegetative body of the fungus consists of slender branching filaments (hyphae) of usually white cells which permeate the soil and encase and penetrate the roots of the woody plants from which their nourishment is obtained. Very often these trees are species of oaks, sometimes of birch. The *truffles* are neatly harvested by the aid of trained pigs or dogs that detect them in the soil by their odor and dig them up. This manner of harvesting is employed in the countries of southern Europe and northern Africa bordering on the Mediterranean Sea. Some *truffles* occur in California and here and there throughout the U.S. but not in sufficient quantity to warrant commercial exploitation. The species mostly harvested in Europe are *Tuber aestivum, T. brumale, T. magnatum, T. melanosporum,* etc. Those of *T. magnatum* possess a garlic flavor.

trumpet ash. Trumpetcreeper.

trumpetbush. Trumpetcreeper.

trumpetcreeper. 1. Any plant of the genus Campsis, family Bignoniaceae. Species are rampant growing vines. See *Chinese trumpetcreeper.* 2. *C. radicans,* a stout, woody vine climbing as high as 30 ft., which is grown for its orange-scarlet flowers and as a covering for unsightly buildings, etc. Native to North America. Also called *trumpetvine; trumpetflower; fox-glove.*

trumpetflower. See *cross vine; trumpetcreeper; trumpet honeysuckle.*

trumpet honeysuckle. *Lonicera sempervirens,* family Caprifoliaceae, a climbing, perennial vine, evergreen in warm areas, which is grown on a fence, trellis, etc., for its bright orange or red flowers, yellow inside. Native to North America. Also called *coral honeysuckle.*

trumpet lily. See *Easter lily: 1.*

trumpet narcissus. Daffodil.

trumpet pitcherplant. *Sarracenia flava,* family Sarraceniaceae, an insectivorous, perennial plant, which is sometimes grown as a curiosity and as a bog plant. The pitcher-like leaves are sometimes 3 ft. long. Native to North America. Also called *Eve's-cup.*

trumpetvine. Trumpetcreeper.

truncated soil profile. A soil profile from which one or more of the upper horizons normally present have been removed by natural or accelerated erosion or by other means. The profile may have lost part or all of the A and sometimes the B horizon leaving as soil only the poor, undeveloped parent material or C horizon. A comparison of eroded soil profiles of the same area, soil type, and slope conditions, indicates the degree of truncation.

trunk. The main unbranched body, stalk, or stem of a vine or tree. See *stem: 1.*

trunk scald. See *sunscald.*

trussing. The tying compactly of a dressed bird to conserve juiciness and flavor in roasting.

trustee. A person who holds property in trust; a person or agent to whom legal title has been transferred and who has the responsibility of managing the property for the benefit of a second party. The second party is entitled to all the rents or benefits. In equity, the second party is the real owner.

truth-in-fabric laws. Those laws which require the manufacturer of woolens and wool products to label each product to show the percentage of virgin wool used.

try. To determine if a female is in heat by bringing her and a male in close proximity to each other. They are usually kept separated by a fence, wall, or teasing pole *q.v.*

tryer. 1. A grade stallion which is used for the purpose of determining if a mare is in heat. See *try.* 2. See *trier.*

Tryon's scab. A disease of citrus in Australia and New Zealand caused by *Ramularia scabiosa,* family Moniliaceae, characterized by corky projections on the fruits, a distinct distortion of the fruit and crater-like lesions on the leaves. Also called *Australian citrus scab; citrus scab.*

trypanosome. A microscopic, free-swimming, tailed protozoan parasite of the genus Trypanosoma, family Trypanosomatidae. Many species are not injurious, but a few cause serious diseases of animals and man, such as *T. brucei* and *T. gambiense,* both of which gain entry to the body by the bite of the tsetse fly *q.v.*

trypsin. One of the principal proteolytic enzymes of the pancreatic secretion.

tryptophan. One of a group of chemical substances (amino acids) which constitute fundamental units from which proteins are built and which are needed for the building and repair of tissues; an essential amino acid.

Tschernozem (Tshernosem, Tchernayzem). Black soil, particularly that of the Russian steppes. See *Chernozem.*

Tschil. A cows' or ewes' sour, skimmed milk cheese made in the Middle East.

tsetse fly. *Glossina mortisans,* family Muscidae, a fly of central and southern Africa which is a carrier of the parasite (*Trypanosoma brucei*) that causes the disease nagana in various animals, as horses, cattle, or goats. The parasite is conveyed by the fly's bite. The widespread presence of the fly, during certain seasons, renders some districts completely unsuitable for habitation by such domestic animals. A related species (*G. palpalis*) carries the trypanosome, *T. gambiense,* that causes sleeping sickness in man.

T-super. Air-way super.

T-tin. A metal support, shaped like the letter T, used to hold the comb honey section in a super.

tub. A sales package for wholesale butter which holds from 10 to 64 lbs. The prevailing size is 63 lbs.

tuba. Fermented sap from different palm genera from which alcohol is manufactured in the Philippines.

tubaroot jewelvine. *Derris elliptica,* family Leguminosae, a tropical plant grown for its rootstock, the principal source of rotenone *q.v.*

tub butter. Butter placed on the market in tub lots and sold by being ladled from the tub in the amount desired by the customer.

tuber. 1. A swollen, fleshy portion of an underground stem or shoot which bears minute, scale-like leaves with buds or eyes, as in the potato. *Cf. corm; flowering bulb.* 2. Loosely but incorrectly, any plant with a bulbous stem or root.

tubercel. See *legume.*

tubercle. 1. A small tuber. 2. The small nodules ascribed to the action of symbiotic organisms especially on the roots of legumes. 3. A nodule or small eminence in or on soft tissue. 4. A rough, rounded eminence on a bone.

tubercle bacilli. Microorganisms that cause tuberculosis.

tuber crop. See *root crops; tuber.*

tubercular. Referring to an animal which has tuberculosis; susceptible to tuberculosis.

tubercule. See *tubercle.*

tuberculin. A biological agent derived from the growth and further processing of the tubercle bacilli which is used for the detection or diagnosis of tuberculosis in animals and man.

tuberculosis. A chronic infectious, contagious disease of man and practically all warm-blooded animals which is caused by *Mycobacterium tuberculosis.*

tuberculosis of bovine. An infectious disease, especially of milch cows, caused by the bovine type of bacillus *Mycobacterium tuberculosis.* It affects the animals' liver, lymph nodes, spleen, and udder. The milk from diseased cattle can infect humans also.

tuberculosis of chickens. See *avian tuberculosis.*

tuberculosis of hogs. A disease caused by bovine, avian, and occasionally human types of the bacillus *Mycobacterium tuberculosis,* which produces ulceration in the affected animal's intestines. Also called *swine tuberculosis.*

tuberculosis of plants. Certain bacterial diseases of plants characterized by abnormal swellings or hollow pockets.

tuberculous. Referring to tuberculosis or infection by tuberculosis.

Tubergen desert candle. *Eremurus tubergeni,* family Liliaceae, a hybrid, perennial herb grown for its yellow flowers on scapes 8 to 10 ft. high.

tuber goldfields. *Baeria macrantha,* family Compositae, a hardy, annual herb grown for its profuse, yellow flowers. Native to North America.

tuber indexing. Growing a seed piece of a potato under conditions suitable for virus determination. According to the reaction, the remainder of the tubers produced by that plant are either condemned as unfit for planting or certified as disease-free. Also called *hill indexing.* See *virus indexing.*

tube roller. A land roller constructed of short sections of pipe mounted on an axle, with cast-iron heads in which are bolted U-shaped cross-pieces. This type of roller crushes clods but does not leave a smooth surface.

tuberose. *Polianthes tuberosa,* family Amaryllidaceae, a tender, tuberous, perennial herb grown for its wax-white, fragrant flowers. Native to Mexico.

tuberous. Designating any plant which bears tubers.

tuberous begonia. Any of a group of summer-blooming begonias having fleshy tuberous roots which are derived from several species of the genus Begonia. Also called *tuberous-rooted begonia.*

tuberous root. 1. An enlarged root which tapers toward both ends as in dahlia and sweet potato. 2. The legal designation of the enlarged, underground parts of the dahlia. The law allows patenting of plants propagated by roots and refuses patents to plants propagated by tubers.

tuberous waterlily. Magnolia waterlily.

tuber rot. See *early blight of potato.*

tuber-unit method. A system used in the production of certified seed potatoes in which seed pieces from each potato are planted consecutively in the row. The hills are marked off in units of 4 hills. If a diseased hill is found, the entire unit of four is destroyed.

tuber waterhemlock. *Cicuta vagans,* family Umbelliferae, a perennial, poisonous herb found on edges of ponds or in bogs. Symptoms of poisoning in cattle are stomach pain, nausea, violent vomiting, diarrhea, dilated pupils, difficult breathing, frothing at the mouth, thready pulse, convulsions, and death. It is perhaps one of the most poisonous plants found in the U.S. Native to North America. Also called *Oregon waterhemlock.*

tuberworm. See *potato tuberworm.*

tube well. A small well for obtaining water from shallow strata which consists of a pointed pipe driven into the ground without boring. See *driven well.*

tub gardening. A method of growing certain flowers, fruits, and vegetables in small quantities in a wooden barrel in which 4-in. holes no closer than 12 ins. apart are bored in the sides. The barrel is filled with rich soil and plants are set in the holes. See *strawberry barrel.*

tubman. A person who tends tubs of scalding water for the dipping of hog carcasses in slaughter houses.

tubular roller. See *cast-iron drum roller.*

tub-washed. Designating wool which has been washed after shearing (obsolete).

Tuckahoe Indian bread. The large, edible underground fungus, *Poria cocos,* formerly eaten by a tribe of American Indians.

tucker. 1. A device on a hay baler which squares the ends and prevents the bale from having tails. 2. Food; board (Australia and New Zealand).

tucket. Unripened corn.

tuck in. To cause a horse to hold its head high with its chin toward its chest.

tufa. See *tuff.*

tuff. Fine volcanic detritus in various states of stratification and consolidation. *Tufa* applies to similar rocks, but more especially to a kind of porous rock formed as a deposit from springs or streams; usually applied to calcareous deposits as travertine and calcareous tufa. See *detritus.*

tuffaceous. Tufflike.

tuft. A small cluster, as of grass, with the roots intertwined.

tufted. Designating plants growing in a tuft.

tufted columbine. Columbine meadowrue.

tufted hairgrass. *Deschampsia caespitosa,* family Gramineae, a perennial bunchgrass found from Greenland to Alaska, south to New Jersey and west to California, U.S., in bogs and wet places. Its forage rating is good and it is sometimes cut for hay.

tufted pansy. See *horned violet; sweet violet.*

tufted sandwort. Moss sandwort.

tufting. The practice of planting sod or grasses in tufts or small bunches.

tuft tree. Giant dracena.

tug chain. In a harness, a small length of chain one of whose ends is attached to the trace (*q.v.*) and the other to one end of the whiffletree.

tularemia. A disease caused by the microorganism *Bacterium tularenese,* common in southwestern U.S. among rabbits and squirrels, and transmitted to other animals and man by the deer fly and other insects. It is occasionally transmitted to man by handling infected animals, causing an irregular fever lasting for several weeks. In animals, it is frequently fatal. Also called *rabbit fever.*

tule bulrush. *Scirpus acutus,* family Cyperaceae, an aquatic or marshland, perennial, grass-like herb grazed to some extent. It may also be a weed pest. Native to North America.

tule lands. Marshy land which is occupied chiefly by bulrushes of the genus Scirpus, especially such land in the lower part of the Sacramento Valley in California, U.S. Such land, when reclaimed by drainage, has been highly productive agricultural land. Also called *tulare.*

tule potato. Arrowhead.

tulip. Any plant of the genus Tulipa, family Liliaceae. Species are bulbous herbs and are popular for their early spring bloom, especially, *T. gesneriana,* the parental form of most of the flower garden varieties. See *tulip classes (or types).*

tulip blight. Botrytis blight of tulip.

tulip bulb aphid. *Anuraphis tulipae* (Fonsc.), family Aphidae, a sucking insect which infests plants at or beneath the surface of the soil causing shriveling of bulbs and stunted growth. It attacks tulips, iris, freesia, crocus, and gladiolus. Also called *iris root aphid.*

tulip classes (or types). See *bizarre tulip; breeder tulip; bybloemen tulip; cottage tulip; Darwin tulip; early double tulip; early single tulip; lily-flowering tulip; Mendel tulip; parrot tulip; Rembrandt tulip; triumph tulip.*

tulip dropper. See *dropper:* 2.

tulip fire. Botrytis blight of tulip.

tulip mold. Botrytis blight of tulip.

tulip orchid. *Cattleya citrina,* family Orchidaceae, an ephiphytic plant widely hybridized among orchid fanciers and grown in the greenhouse for sale as a corsage plant for its yellow, white-lipped flowers. Native to tropical America.

tulip poplar. See *tulip tree.*

tulip poppy. *Papaver glaucum,* family Papaveraceae, an annual herb grown for its numerous cup-shaped, scarlet flowers. Native to the Near East.

tulip root. A disease of oats caused by infestation of the nematode *Ditylenchus dipsaci.*

tulip root rot. A rot of roots and basal stem resulting in dwarfing and reduced vigor. It is associated with the fungus *Pythium debaryanum* var. *pelargonii,* family Pythiaceae.

tulip scale. See *tuliptree scale.*

tulip tree. *Liriodendron tulipifera,* family Magnoliaceae, a deciduous, magnificent, broadly pyramidal tree, up to 150 ft. tall, grown for its size, shape, and for its lovely white, tulip-like flowers. Its wood has a number of commercial uses including veneer, cabinet-making, boxes, etc. Native to North America. Also called *yellow poplar* (forestry name); *white wood; tulip poplar.*

tuliptree aphid. *Macrosiphum liriodendri* (Monell), family Aphidae, an insect that infests the leaves of the tulip tree and broadleaved evergreens.

tuliptree scale. *Toumeyella liriodendri* (Gmel.), family Coccidae, an insect that infests the tulip tree, magnolia, and linden. It is one of the largest scale insects and is widely distributed in the U.S. east of the Rocky Mountains.

tulipwood. The wood of the tulip tree *q.v.*

tumbak. A coarse tobacco of Iran. Also called *tumbaki; tumbek; tumbeki.*

tumblemustard. *Sisymbrium altissimum,* family Cruciferae, a weed pest found from Nova Scotia to the Pacific, U.S. Feeding on it in considerable amounts may cause chronic enteritis, hemorrhagic diarrhea, colic, abortion, nephritis, and paralysis in livestock. Also called *tumbling mustard.*

tumbler. The pupa of the mosquito.

tumble ringworm. *Cycloloma atriplicifolium,* family Chenopodiaceae, an annual herb which is a weed pest of the tumble-weed sort. It is found west and south of Minnesota, U.S.

tumble weed. Any weed that breaks away from the ground, usually at or near the surface, at the end of the growing season and because of its form is often blown considerable distances, scattering seeds en route.

tumbleweed amaranth. *Amaranthus graecizans,* family Amaranthaceae, a coarse, annual weed which is a fairly common weed pest. Also called *tumble pigweed.*

tumble windmillgrass. *Chloris verticillata,* family Gramineae, a perennial, weedy grass, the inflorescence of which breaks away at maturity and rolls as a tumble weed; however it is sometimes grazed by livestock early in the season. Also called *windmill grass.*

tumbling bay. Weir.

tumbling mustard. Tumblemustard.

tumbling pigweed. Tumbleweed amaranth.

tumor. A swelling; a new growth of cells or tissues governed by factors independent of the laws of growth of the host. It may be either benign or malignant.

tumor gall of tobacco. A virus disease of tobacco characterized by tumors on the roots and stem.

tumulus scab. A type of corky scab of potato (*q.v.*) in which the scabs are elevated and warty with sloping sides.

tun. A measure of 252 gallons; four hogsheads.

tuna oil. A livestock feed supplement which is the product obtained by extraction of part of the oil from the cannery refuse of tuna.

Tundra. A great soil group in which the profile consists of dark brown peaty layers over either grayish or brownish horizons resting on ever-frozen substrata. Native vegetation consists of mosses, lichens, and shrubs. Climate is cold, humid; soil development process is gleization with medium natural fertility for crop plants. Agricultural use is mostly pasture and some very short season crops. Hunting and trapping are associated enterprises.

tung. See *tung oil tree.*

tung oil. A vegetable drying oil expressed or extracted with solvents from the nutmeats of trees of the genus Aleurites. Also called *Chinawood oil; wood oil.*

tung oil pomace. A fertilizer consisting of the ground residue remaining after the extraction of tung oil from the seeds of the tung oil tree.

tung oil tree. *Aleurites fordi,* family Euphorbiaceae, a small, tropical tree grown for its nutlike fruit, the seeds of which yield the tung oil, used in dyes, varnishes, waterproofing agents, etc. Native to central Asia. Also called *China wood oil tree; tung.*

tunic flower. Saxifrage tunicflower.

Tunis. A minor, hardy, medium-sized breed of sheep originating in North Africa which has a brownish face and rather coarse wool.

Tunisgrass. *Sorghum virgatum,* family Gramineae, an annual, forage grass with awned spikelets on panicles up to 2 ft. long. Native to North Africa.

Tunisian bee. *Apis nigra,* a ferocious honeybee. Also called *Punic bee.*

tunnel. 1. A gallery. 2. To burrow, make a passageway, in plant stems, leaves, roots, etc., as by larvae of certain insects.

tup. A ram.

tupelo. Any tree of the genus Nyssa, family Cornaceae. Species are cultivated for their fine, autumnal foliage. See *black tupelo; water tupelo.*

tupelo leaf miner. *Antispila nysaefoliella* Clem., family Heliozelidae, an insect whose larva infests the tupelo, mining the leaves. On reaching maturity, the larva cuts an oval case out of a leaf and falls to earth along with it. Also called *sour-gum casecutter.*

tup hog. A male lamb from weaning time until it is shorn.

tup yeld. Designating a barren ewe.

turban buttercup. Persian buttercup.

turban group. A group of varieties of winter squash, which has a turban-shaped button at the blossom end. See *banana group; Hubbard group.*

turbary rights. The rights to cut turf on another person's land. See *turf: 2.*

turbinates. Scroll-shaped bones in the nasal cavity.

turbine pump. A multiple centrifugal pump, with several impellers in series, which is suitable for pumping against high heads. Also called *series pump; multistage centrifugal pump.*

turbulent velocity. The velocity of a liquid in a particular conduit above which turbulent flow will always exist, and below which the flow may be either turbulent or laminar, depending on circumstances.

turf. 1. A close-growing, well-knit, usually fine-leaved growth of a grass, mixture of grasses, or other plant species, which is best maintained by mowing, fertilizing, and watering so as to present a pleasing appearance. It is useful for lawns, golf courses, horse racing tracks, athletic fields, etc. 2. A slab of peat used for fuel (British). 3. The surface layer of a peat soil. 4. To produce a *turf* by the seeding, sodding or other vegetative propagation of grass or other plant species.

turfing. The removal of turf and the extraction of roots from it before returning it to the soil.

turfingdaisy. *Matricaria tchihatchewi,* family Compositae, a low, carpet-like, perennial herb grown as a bedding plant for its white, daisy-like flowers. Native to Turkey.

turgidity. See *turgor.*

turgor. The distension of the cell wall and protoplasmic layer of plants by fluids. It is essential to growth.

turicata. *Ornithodorus turicata* (Dugés), family Argasidae, a fever tick, which pesters hogs, cattle, and occasionally man, causing extreme irritation. Found in Mexico and southwestern U.S. Also called *relapsing fever tick.*

turion. See *sucker.*

Turk. An old breed of horses related to the Arabians whose influence is shown today in all light breeds.

turken. The female offspring of the turkey cock and domestic hen.

Turkestan alfalfa. A strain of *Medicago sativa* var. *falcata,* family Leguminosae, which comes from the area of Turkestan, Russia. It is smaller, shorter-lived, and less hardy than alfalfa *q.v.*

Turkestan bluestem. *Andropogon ischaemum,* family Gramineae, a perennial bunchgrass which is of limited use as a pasture and hay crop in the U.S. Native to Asia.

turkey. *Meleagris gallopavo,* a large, native American bird, now largely domesticated, which is raised for its delicious meat. Roast *turkey* is traditionally served in the U.S. at Thanksgiving and Christmastime. See *Bronze; Broad Breasted Bronze; Beltsville Small White turkey; Bourbon Red; Narragansett; Slate; White Holland; Black turkey.*

turkey berry. Indiancurrant coralberry.

turkey breeder. A person who breeds and raises turkeys for market.

turkey bur. Great burdock.

turkey corn. Squirrel corn.

turkey gnat. *Simulium meridionale* Riley, family Simuliidae, a small, blackish fly that congregates about the eyes, ears, and nostrils of chicken and turkeys, greedily sucking blood from the host. The bite may result in symptoms similar to mastoiditis (*q.v.*). It may be a vector of the organism causing Onchocerciosis in humans.

turkey house. A verminproof, weatherproof roosting shelter for turkeys.

turkey louse. *Lipeurus caponis* (L.), family Philopteridae, a louse, parasitic on chickens and turkeys, which latches on to the bird and eats its feathers.

turkeymullein. *Eremocarpus setigerus,* family Euphorbiaceae, an annual, weedy plant which yields blackish seeds that are nutritionally fattening for the turkeys. Sheep have been killed from grazing on this plant for the hairs covering the stems and leaves are indigestible. Native to western U.S. Also called *dove-weed.*

turkey oak. 1. *Quercus laevis,* family Fagaceae, a deciduous tree sometimes grown in dry sandy soils as an ornamental. Native to the southern U.S. 2. See *European Turkey oak.* 3. See *southern red oak.*

turkey patty. A meat patty prepared by a special machine by knitting together the desired proportions of light and dark meat of turkey to form a convenient portion for use in sandwiches or as a meat serving.

turkey poult. A young turkey before its sex can be determined. Also called *poult.*

turkey steak. A steak cut from a turkey carcass.

turkey sticker. A short-handled, narrow-bladed knife used to pierce the brain of a turkey in killing it for dry picking.

Turkish bean. Scarlet runner bean.

Turkish filbert. *Corylus colurna,* family Betulaceae, a deciduous tree grown as an ornamental and for its edible nuts. Native to southeastern Europe and western Asia.

Turkish walnut. Persian walnut.

Turk's-cap lily. *Lilium superbum,* family Liliaceae, a showy, bulbous, erect herb grown in the bog garden for its orange-red, dark-spotted flowers. Native to North America. Also called *American martagon.*

Turkscap meloncactus. *Melocactus intortus,* family Cactaceae, a tropical, cylindrical cactus, with yellow or brown spines and pinkish flowers; grown as a curiosity. Native to the West Indies.

turmeric. *Curcuma longa,* family Zingiberaceae, a tropical, aromatic, perennial herb grown for its rootstock, source of commercial *tumeric,* used as a dye, medicine, and mostly as a condiment, chiefly in India. Native to India.

tumeric root. 1. Bloodroot. 2. Goldenseal.

turn. 1. To change the position of an egg in an incubator or in the nest where the hen is incubating it. 2. To restrain an animal, as a fence *turns* a cow. 3. To plow, so that the lower part of the soil is brought to the surface and the former surface part is covered. 4. To begin to show signs of ripening, as a fruit. 5. To direct an animal to pasture. See *turn in.* 6. To sour, as milk. 7. To change, as leaves in the autumn.

turnabout. A cattle disease caused by inflammation of the brain, evidenced by shaking, grinding of teeth, vertigo, staggering, and coma. Also called *turn-about sickness.*

turn away. See *settle: 2.*

turned-in toes. Designating the badly placed forefeet of an animal making it predisposed to winging or paddling in the action of the forefeet. Such an animal is also called *pigeon-toed.*

turned-out toes. Designating the badly placed forefeet of an animal, causing interference in foot action. Sometimes called *dishing; splay-footed.*

turn-furrow. The moldboard of a plow (obsolete).

turn in (to). To direct livestock into a pasture or field.

turning. 1. See *turn: 3.* 2. Designating milk which is beginning to sour.

turning plow. Moldboard plow.

turning shovel. Moldboard hiller.

turning space. An area or space left uncultivated in a field or orchard and used to reverse the direction of a tractor or team working the field, etc. See *head land: 2.*

turning sweet. See *chilling.*

turning under. The covering or burying of the surface soil, trash, a green manure crop, etc., by plowing.

turnip. *Brassica rapa,* family Cruciferae, a biennial herb grown as an annual for its root and its leaves. Widely used as a vegetable. Native to Europe, Asia and Africa. See *turnip greens.*

turnip aphid. *Rhopalosiphum pseudobrassicae* (Davis), family Aphidae, an insect that infests various cruciferous plants. Infested leaves curl, and, if not killed, the plant is dwarfed and forms small heads slowly. Also called *turnip louse.*

turnip cabbage. Kohlrabi.

turnip garden parsley. *Petroselinum crispum* var. *radicosum,* family Umbelliferae, a biennial or perennial herb grown for its edible, thickened root. Native to Europe and western Asia. Also called *hamburg parsley; turnip-rooted parsley.*

turnip grass. Bulb panicum.

turnip greens. The leaves of the turnip (*q.v.*) used as a potherb.

turnip maggot. *Hylemya floralis* (Fall), family Muscidae, an insect that tunnels in the roots of cruciferous plants. Infested seedlings wilt, turn yellow, and die.

turniproot chervil. *Chaerophyllum bulbosum,* family Umbelliferae, a biennial herb grown for its edible, fleshy root. Native to Europe. Also called *bulbous chervil.*

turnip-rooted cabbage. See *rutabaga; kohlrabi.*

turnip-rooted celery. Celeriac.

turnip-rooted parsley. Turnip garden parsley.

turnip sawfly. *Athalia spinarum,* family Tenthredinidae, a sawfly which, by means of its ovipositor, makes cracks in the edges of turnip leaves to deposit its eggs therein. The larvae, known as *black niggers, black-jacks,* or *turnip niggers,* eat the leaves when they hatch. Native to Europe.

turnip-stemmed cabbage. Kohlrabi.

turnip tops. Turnip greens.

turnipy. Designating an off-flavor of milk or its products suggesting cooked turnips.

turnout. 1. A horse complete with saddle and bridle, especially for buying or selling (New Zealand). 2. To produce.

turnpike geranium. Jerusalemoak goosefoot.

turn plow. Moldboard plow.

turn row. A relatively narrow strip of uncultivated land at the margin of a field on which the plow or other equipment may be turned. See *head-land: 2.*

turn sickness. Gid.

turn spout. A discharging spout for grain elevators, etc. It is mounted on a swiveling socket or support so that it may be slued to any desired angle.

turn under. See *turn: 3.*

turnway. A grazed or vegetative strip at the ends of a terrace or strip-cropped field, used for turning during the process of tillage.

turnwrest plow. A plow with a shifting moldboard so designed that the furrow slice is always thrown the same way whatever the direction of movement (Scotland and England). Also called *Kentish plough.* See *reversible plow.*

turpene. A flavor defect of ices and sherbets suggesting turpentine.

turpentine. 1. Oleoresin. 2. The essential oil derived from the distillation of oleoresin *q.v.* 3. To tap a tree to obtain oleoresin.

turpentine hack. A tool used for barking and cutting the face on the pine trees to collect the crude resin.

turpentine orchard. A stand of coniferous trees from which oleoresin is taken.

turpentine tree. Tarata pittosporum.

turpentining. Conducting the collection of oleoresin in such a way as to avoid severe damage to timber stands of turpentine pines so that a profitable, sustained collection can be maintained along with timber conservation.

turtle back. Virginia dayflower.

turtle backing. Preparation of a wide bed, in the form of a low rounded ridge, bordered on each side by furrows; a tillage method used in the planting of sugar cane on flat, wet lands.

tush. A tooth located between the incisors and molars; the eye tooth; tusk.

tussock. A dense, heavy tuft or matted growth of grass or sedge which forms a small hillock.

tussocker. See *sundowner.*

tussock grass. Alkali sacaton.

tussock-grassland. A range on which the grass is heavily tufted (New Zealand).

tussock moth. See *hickory tussock moth; spotted tussock moth; western tussock moth; white-marked tussock moth.*

Twdr Sir. A ewes' skimmed milk, cured cheese made in Serbia.

twibil. 1. A mattock whose blade has one end like an ax, the other like an adz. 2. A mortising tool. 3. A reaping hook used for cutting beans and peas (England). Also called *twibill; twybill.*

twig. A small branch or shoot of a woody plant of no particular size or length. See *spur: 1.*

twig blight. Any sudden dying of the twigs or terminal branches of woody plants which may result from a variety of causes.

twig drop. The shedding or dropping of twigs or terminal branches of woody plants which may result from any of a number of reasons.

twig girdler. *Oncideres cingulata* (Say), family Cerambycidae, a beetle which neatly cuts off twigs and branches of hickory, pecan, persimmon, oak, poplar, tupelo, honey locust, and other trees in the eastern U.S.

twig graft. A graft made with a twig. See *graft.*

twiggy. Abounding with twigs.

twig pruner. *Elaphidion villosum* (F.), family Cerambycidae, a beetle whose larvae cut twigs from oak, hickory, pecan, maple, locust, hockberry, elm, walnut, sweetgum, some fruits trees and some shade trees. Also called *oak twig pruner.*

twin. 1. Either one of two young born at one birth by an animal whose species normally bears only one at a birth. 2. To bear two young at a birth when only one is usual.

twinberry. Bearberry honeysuckle.

twine binder. 1. That part of a binder which does the actual tying of the grain or fodder bundles. 2. A binder which uses twine as the tying material. See *binder: 1.*

twining button flower. *Hibbertia volubilis,* family Dilleniaceae, a high-climbing, woody vine grown to cover a trellis, etc., and for its golden flowers. Native to Australia. Also called *button flower; Guineagold-vine.*

twining milkweed. Grecian silkvine.

twinning. The production of two young at one birth by an animal which normally produces only one young at a birth.

twin sisters. American twinflower.

twist. 1. A distortion of a board so that the 4 corners of any face are not in the same plane. 2. The depth and fullness between the upper part of the hind legs of an animal. 3. A fungous disease of wheat and rye caused by *Dilophospora alopecuri,* and/or the eelworm, *Tylenchus tritici.* The symptoms are much like those of nematode disease of wheat *q.v.*

twisted comb. A rooster's single comb (*q.v.*) that is irregularly shaped.

twisted eglantine. Woodbine honeysuckle.

twisted feather. A fowl's feather which has a twisted quill or shaft.

twisted heath. *Erica cinerea,* family Ericaceae, a small shrub grown as a pot plant for its winter bloom of rose-purple flowers. Native to Europe. Also called *Scotch heath.*

twisted leaf. A virus disease reported from western U.S. and Canada which is of economic importance in but a few orchards. Symptoms on Bing cherry are shortened internodes with compact tufts of twisted leaves and necrosis of midribs or lateral veins.

twisted shovel. A cultivator shovel used for general work. The shovel is cut off higher on one side than the other with the result that most of the soil is shed off the high side. The shovels on a gang are mounted at a slight angle so that the soil shed by one shovel covers the trough cut by the preceding shovel.

twisted stomach worm. *Haemonchus contortus,* a stomach worm which infests sheep, cattle and goats. Infestation is characterized by a severe anemia, pallid skin, visible mucous membranes, weakness, standing still, a swelling (bottle jaw), and hard dry feces. Also called *twisted wire worm; barber pole worm; large stomach worm; common stomach worm.*

twister. Tornado.

twist him down. To force a steer or cow to the ground by turning or twisting its head.

twitch. A device used for pinching or squeezing the nose (upper lip) of a horse to take the animal's attention away from some manipulation or minor operation. It is also used as a control measure on some unruly horses. The device may be a loop of rope or chain or a clamp, or may be accomplished with the hand. In some localities, especially where mules are common, the device may be placed on the ear. Also called *twist.*

twitch-grass. Quack grass.

twitch logs. To pull logs out of the forest by means of an ox team.

two bulls. In stock shows for beef cattle, the best two bulls bred and owned by an exhibitor.

two-eye stub cutting. A cutting of a plant for propagation which comprises a short section of the stem (not terminal) on which there are two buds.

two females. In stock shows for beef cattle, the best two females bred and owned by an exhibitor.

two-field system. An early system of cropping in England in which a year of wheat was followed by a year of fallow.

2,4-D. A selective herbicide.

2,4-dinitroanisole. An ovicide.

two-gang disk harrow. Double-disk harrow.

two-grooved loco. *Astragalus bisulcatus,* family Leguminosae, a poisonous plant found from Manitoba, Canada, to New Mexico, U.S. For symptoms of poisoning see *blue loco.*

two-grooved milkvetch. Two-grooved loco.

two-horse. Designating a vehicle, plow, cultivator, wagon, etc., drawn by a team of 2 horses.

two-lined chestnut borer. *Agrilus bilineatus* (Web.), family Buprestidae, the larva of a beetle which infests oak, chestnut, beech, and ironwood. It mines in the cambium, girdles the trunk and branches. Weakened trees may die rapidly.

two-litter system. A swine production program in which each sow farrows a spring and fall litter of pigs.

two-man. Designating a tool, implement, etc., which is operated by two men, as a *two-man* saw.

two-marked treehopper. *Enchenopa binotata* (Say), family Membracidae, an insect that infests butternut, locust, bittersweet, sycamore, hickory, willow, redbud, and hoptree.

2-phenylcyclohexanol. A complex organic chemical compound which, diluted with alcohol, is used to repel chiggers, flies, ticks, and mosquitoes.
two-price system. Multiple-price system.
two-rib roast. A retail cut of beef for roasting which is a rib roast consisting of the fifth and sixth ribs.
Tworog. A fresh, sour milk cheese made in Russia.
two-row. Designating a farm implement which works two rows at a time, as a *two-row* corn planter.
two-rowed barley. *Hordeum distichon,* family Gramineae, a cereal plant grown for its grain used as livestock feed, for minor use in pearling, and for flour. Native to Asia.
two-run process. A process used in the manufacture of vinegar in which the freshly fermented liquid is passed through a generator to become as acetified as desired. See *one-run process.*
two-sided gait. Any of several horse gaits in which the horse travels from the ground with its right fore- and hind-legs and with its left fore- and hind-legs alternately, such as a pace.
two-spotted lady beetle. *Adalia bipunctata* (L.), family Coccinellidae, a beneficial lady beetle *q.v.*
two-spotted mite. Two-spotted spider mite.
two-spotted spider mite. *Tetranychus telarius* (L.), family Tetranychidae, an insect that infests greenhouse plants and, out-of-doors, evergreens, hydrangeas, roses, and clematis.
two-stage dam. A dam which has two distinct drops.
two-storied forest. A stand of trees in which two heights or age classes of considerable difference are developed or are intended to be maintained, as upper story or overwood and lower story or underwood. (The term is not applicable to forest under reproduction, in which the appearance of two stories is the temporary result of an incomplete process, but to those forests of which the two stories of growth are a permanent feature.)
two-story Langstroth hive. Ten-frame hive.
two-story tree. 1. A pruned fruit tree which develops a water-spout limb strong enough to become a leader and to develop another or second top. Also called *sucker top.* 2. A fruit tree which has two distinct levels of framework branches.
two-striped grasshopper. *Melanoplus bivittatus* (Say), family Acrididae, an insect that is a widely distributed pest of grain, fruit trees, and cultivated crops.
two-toothed sheep. 1. A sheep as soon as it is shorn for the first time. 2. A one-year-old sheep.
two-trunk Kniffin system. A system of training grape vines. One trunk reaches to the lower wire, the other to the top wire and a cane is tied to the right and to the left along each wire. See *single-trunk, four-cane Kniffin system; six-cane Kniffin system.*
two-way livestock. Livestock suitable either for slaughter or feeding.
two-way sulky. A sulky plow which has both a left-hand and right-hand bottom. Only one plow is used at a time. The furrows are turned in the same direction across the field to avoid dead furrows.
two-wheel. Designating a vehicle, plow, etc., which is supported by and moves on two wheels on one axle.
two-wing silverbell. *Halesia diptera,* family Styracaceae, a tender shrub or small tree grown for its white flowers. Native to North America. Also called *snowdrop tree; silverbell tree.*
two-wire horizontal trellis. A system for managing bramble fruits in which 1 to 2 ft. crosspieces are fastened to posts, and wires are fastened to the ends of the crosspieces. The canes are trained between the wires.
two year old and over. In stock judging, any animal 2 or more years old.
tying in. The use of tie-in-chains (*q.v.*) in multihitches.
tympanites. Distention of the abdomen of an animal which results from the presence of gas or air in the intestines or peritoneal cavity. See *wind colic.*
tympany of the crop. A condition of a bird in which gas collects in and distends the crop.
type designation. A legend used to indicate the principal vegetative cover and forage species of a range or subtype. It is composed of the type numbers followed by principal species symbols.
type host. A plant or animal from which a parasite was first taken for the purpose of describing: *e.g.,* if a flea is first described from a red fox, and later the same flea is found on the gray fox, coyote, dog, and man, the *type host* is the red fox.
type locality. 1. The particular place at which a plant specimen was collected. 2. The place where a kind of soil or a particular geological formation is typically displayed.
type map. A map which shows plant types (associations) and their designations and/or other range-management information.
type of farming. The form of organization of a farm business, primarily the kinds and amounts of the different crop and livestock enterprises found therein.
type of farming area. An area where most farmers follow a similar type of farming. The transportation and market facilities, soil types, topography and climate are usually also similar.
type of livestock. An ideal which presumably embodies all the characters contributory to the species, breed, variety, etc., valued for some certain purpose or purposes: *e.g.,* a beef *type.*
type of soil. Soil type.
type penstemon. *Penstemon hirsutus,* family Scrophulariaceae, a perennial herb grown for its long, toothed, purple or violet flowers. Native to North America.
type species. A species which is designated as typical for a genus of plants, fungi, etc.
type writeup. The field notes taken for a range type or subtype in making a range survey. It includes estimates of density and composition, record of proper use factors, appraisals of range and

soil conditions, and other information pertinent to management.

Typhula blight. A fungous disease of winter cereals and grasses caused by *Typhula itoana* or *T. idahoensis,* following periods of heavy snow or cloudy, cold weather. It is characterized by a felty, white mycelial mat over the plants and soil. The affected plants fade, wither, and turn brown. The disease disappears as the weather turns warm.

Typhula snow mold of grasses. Typhula blight.

typical. In appraising, that which most frequently occurs or exists in the particular situation under consideration.

typy. Designating an animal which closely approximates or resembles the accepted type. See *type of livestock.*

tyrothricin. A combination of the two antibiotics, gramicidin and tyrocidine, derived from the organism *Bacillus brevis.*

tyrotoxicon. A ptomaine poison produced in milk, cheese, cream, or ice cream.

U

udder. The encased group of mammary glands provided with teats or nipples, as in a mare, ewe, sow or cow. Also called *bag.*

udder attachment. The manner in which the udder of a dairy cow is attached to the body. The attachment should be long in the rear extending well up behind with good width, and in front should join the abdomen well forward and be well up, with the floor of the udder level.

udder balm. Any patent medicine which is supposed to have a beneficial effect when applied to a cow's udder.

udder cannula. A small, plastic, metal, or hard rubber tube designed for the injection of suffusing solutions into the udder through the teat canal.

udder clap. Mastitis.

udder felon. Mastitis.

udder support. A heavy, cloth support for a cow's udder.

udder tube. A device similar to an udder cannula (*q.v.*), except that it does not have an outer case through which the point of the injecting instrument is thrust.

udo. *Aralia cordata,* family Araliaceae, a perennial herb grown for its young, edible shoots. Native to Japan.

Uganda cherry-orange. East African cherry-orange.

ugly-nest caterpillar. *Archips cerasivorana* (Fitch), family Tortricidae, an insect that infests cherry, and sometimes apple trees. It is found in the northern U.S. Also called *cherry-tree Tortrix.*

ulcer. An open sore that discharges pus and is difficult to heal.

ulcerate. To break down into an ulcer, or to become affected with ulcers.

ulcerative cellulitis. Ulcerative lymphangitis.

ulcerative cloacitis. Cloacitis.

ulcerative enteritis. An inflammatory condition of the intestines in which the lining of the intestine has a variable number of crater-like erosions (ulcers). There is a marked tendency toward very loose or watery evacuations.

ulcerative lymphangitis. A contagious disease of a horse's skin caused by *Corynebacterium pseudotuberculosis,* characterized by pustules which, on breaking, discharge a yellow pus that dries, leaving a yellow crust. It involves the limbs primarily and resembles cutaneous glanders. Also called *Canadian horse pox; ulcerative cellulitis.*

ule. Panama gumtree.

ullage. In commercial law, the amount by which a cask, upon being gauged, is found not to be completely full.

ultimate analysis. The determination of the elements in a substance or mixture, as a soil, in contrast to a proximate analysis in which the compounds composing the mixture are determined.

ultra baby beef. Super baby beef.

ultraviolet light. That portion of the spectrum which is composed of light waves just shorter than violet light. It is used in irradiation, disinfection, and sterilization.

umbellifer. Any plant of the family Umbelliferae, as carrot, parsley.

umbilical cord. The cord, consisting of two arteries, a vein and a urachus which connects the placenta to the fetus at the navel in mammals.

umbilical hernia. A hernia (*q.v.*) in which the bowels or omentum protrude from the abdominal cavity through the umbilical ring.

umbilicus. The navel.

umbrella catalpa. Umbrella southern catalpa.

umbrella chinaberry. *Melia azedarach* var. *umbraculiformis,* family Meliaceae, a tree grown for its drooping foliage. Also called *Texas umbrella tree.*

umbrella cover. A metal, milk-can cover which is rounded on the top so that trash, water, etc., will roll off and not catch. See *plug cover.*

umbrella elm. Camperdown elm.

umbrella fir. Umbrellapine.

umbrella flatsedge. *Cyperus alternifolius,* family Cyperaceae, a grass-like herb, grown as a pot plant for its interesting, umbrella-shaped cluster of leaves. Escaped, it can become a weed pest. Native to Africa. Also called *umbrella plant.*

umbrella Kniffin system. A system of training grapes in which a two-wire trellis is used. The trunk is carried to or almost to the top wire, and the canes are carried obliquely from the head to the lower wire. Also called *umbrella system.*

umbrella leaf. American umbrellaleaf.

umbrella magnolia. *Magnolia tripetala,* family

Magnoliaceae, a small tree grown for its large, fragrant, white flowers, and as an understock. Native to North America. Also called *elk wood; umbrella tree.*

umbrella palm. Canterbury umbrellapalm.

umbrellapine. *Sciadopitys verticillata,* family Taxodiaceae, an evergreen tree widely grown as a slow-growing ornamental. Native to Japan.

umbrella plant. Umbrella flatsedge.

umbrella southern catalpa. A horticultural variety of the southern catalpa, *Catalpa bignonioides,* family Bignoniaceae, a dwarfed, globular-shaped tree grown as an accent plant.

umbrella system. Umbrella Kniffin system.

umbrella tree. Umbrella magnolia.

umkokolo. Keiapple.

unarmed. Designating a plant which is thornless.

unavailable. Designating those compounds or chemical elements in the soil or feed which are not in a form available for use by a plant or animal.

unbalanced udder. The udder of a cow which is unevenly attached to her body.

unbroken. 1. In marketing, designating an egg free from cracks or breaks in the shell. 2. Designating an untrained horse. See *unbroke(n) horse.* 3. Designating soil which has not been plowed.

unbroke(n) horse. A horse which has not been trained to the saddle or the harness.

uncap. To remove the caps from the cells of a honeycomb for extracting the honey.

uncapping box. A box or tank into which cappings from a honeycomb are placed for draining preparatory to extraction.

uncapping tank. Uncapping box.

unclean flavor. A flavor defect of milk or its products suggesting dirt, filth, etc.

uncolored. Designating a product to which no artificial coloring has been added.

unconfined ground water. Ground water not under hydrostatic pressure; it is under atmospheric pressure and the water table rises or falls as the volume of stored water changes.

uncontrolled burning. See *wildfire: 2.*

uncured. Designating a product which has not been cured. See *cure: 1.*

underbrush. Woody plants, usually short in structure, growing in a forest underneath tree species.

under color. The color of the down, or downy feathers, of a bird. It is not evident when the feathers are in the normal, folded position. See *ground color: 1.*

under cultivation. Cultivated.

undercut. 1. A notch cut in the trunk of a standing tree below the level of the major cut and on the side to which the tree is to fall. It determines the direction of falling. 2. A saw cut made on the underside of a large branch beyond the point of severance, prior to making the actual primary cut, to prevent splitting or tearing. 3. The harvesting of less timber from a stand than that budgeted. 4. The tenderloin muscle of beef (British).

undercutting. 1. Cutting away at the base or underpart of a steep slope or cliff, as by stream, glacier, wind, or wave erosion, thereby steepening the slope or producing an overhanging cliff. 2. The removal of declining lower branches of a tree to let the upper and newer parts replace them.

underdrain. Drain tiles placed in trenches and covered over with soil to such a depth that the surface soil can be cultivated and the profile adequately drained.

underdrainage. That drainage consisting of drain tiles placed in trenches deep enough to allow the covering soil to be cultivated and the profile adequately drained.

underfeed. To engage in poor feeding practices, usually providing limited or restricted energy intake.

underflow. 1. The water flowing beneath the dry beds of rivers, especially in arid regions. 2. The rate of flow or discharge of subsurface water.

under glass. In a greenhouse.

undergrades. In marketing, eggs inferior to those of the standard grade.

undergrazing. An intensity of grazing which fails to use fully the forage available for consumption under a system of conservation range management.

underground runoff. Water which flows toward stream channels after infiltration into the ground. See *subsurface flow.*

underground stem. A true subterranean stem which occurs in the form of a rhizome, tuber, etc.

underground water. See *ground water.*

undergrowth. Underbrush, seedlings, shoots, and small saplings.

under irrigation. 1. Supplied with water for crop production entirely or in part by means of irrigation. See *subirrigation.* 2. Inadequate irrigation.

under lath. In a lath house.

underline. The line through the middle of the chest and belly to the rear legs of an animal.

underlying color. Ground color.

undermilled rice. Rice with only the bran removed.

underpinning. An animal's legs.

underplant. To set out young trees, or sow seeds, within an existing stand.

under reamer. In well boring, a tool having two legs or cutters spread open by a spring, used to enlarge the borehole underneath the casing.

underrun. The amount by which mill-cut merchantable lumber falls short of the estimate.

under runner. A grinding mill in which only the lower of the two stones revolves. This form is especially useful in grinding oats, corn, chalk, clay, etc. The material rests on the revolving stone, and is moved out from the center by centrifugal force.

under saddle. Designating a horse trained to work with a saddle on its back.

undersetting. The operation of supporting earth in a cutting beneath a rock by building a retaining wall against the face of the earth bank. Also called *underpinning.*

undershot jaw. A protruding lower jaw. Also called *bulldog-jaw; prognathism.*

undershrub. Any low-growing shrub which can be planted beneath a tree or in front of higher shrubs.
understock. See *rootstock.*
understocked stand. A stand of trees in which the growing space is not effectively occupied by crop trees.
under-stocking. Pasturing or grazing a strength of livestock less than the carrying capacity of a pasture or range.
understory. In a forest, that part of the trees which is below the overstory *q.v.*
undertenant. A tenant who leases from one who is himself a tenant.
underwood. See *understory.*
undesirable flavor. Any flavor that is foreign to a product of good commercial quality, such as sour, bitter, yeasty, fermented and fruity, or other off-flavors in cheese.
undeveloped flavor. A flavor defect of cultured milk which has a slighly unclean odor and a weak or mild acid taste, both of which produce a repulsive sensation in the mouth.
undifferentiated soil groups. Soil mapping units in which two or more similar taxonomic soil units occur but not in a regular geographic association: *e.g.*, the steep phases of two or more similar soils might be shown as a unit on a map because topography so dominates the properties. See *soil association; soil complex.*
undisseminated. In a flower show, designating a variety which can be exhibited only by the originator or his agent.
undrawn. Designating killed poultry with the feathers removed but with the viscera unremoved. Also called *New York dressed.*
undried oat groats. See *hulled oats.*
undulant fever. An acute, febrile disease of man with a tendency to become chronic, caused by the species of Brucella: *Brucella melitensis, Brucella abortus,* and *Brucella suis.* Also called *Malta-fever; Mediterranean fever.*
undulating. Designating a relief, landscape aspect, or lay-of-the-land, characterized by successive rolls, or rounded elevations, and depressions; not flat, but with less relief and less broken than hilly land.
uneven. Designating a market when there are many sales out of the general trend.
uneven-aged. Designating a stand of trees in which there are considerable differences in the ages of various trees and in which three or more age classes are represented. *Cf. even-aged.*
uneven bandage. A bandage defect of cheese characterized by the bandage being longer along one side than the other.
uneven body. A body and texture defect of fermented milk in which the upper portion of the bottled milk is quite viscous, whereas the lower portion is lacking in viscosity.
uneven edge. A finish and appearance defect in cheese due to poor workmanship in which a narrow, raised rim is formed around the outer edge of the cheese.
uneven size. An appearance and finish defect in cheese in which the cheese is not within the weight tolerance for that particular style of cheese.
uneven-span house. A greenhouse having one side or span of the roof longer than the other, used largely for side-hill construction. See *even-span house.*
uneven, waxy color. A color defect of cheese in which the color in different parts of the product varies.
unfenced. Designating an area having no fences to impede the movement of livestock; often refers to the open range country in the western U.S.
unfermented acidophilus milk. Highly pasteurized, sweet milk with many *Lactobacillus acidophilus* cells suspended in it. See *acidophilus milk.*
unfertile. Barren, not fertile.
unfertilized. Not fertilized.
unfree water. See *bound water.*
unfruitful flower. A flower which does not develop a fruit.
ungulate. Referring to a hoofed quadruped, as a cow.
unhardy. Tender.
unhitch. To disconnect a horse or team from a vehicle, or to disconnect a tractor from a vehicle, etc.
unhulled peanut oil feed. A livestock feedstuff which is the ground residue obtained after the extraction of part of the oil from whole peanuts. Also called *peanut meal and hulls.*
unhulled seed. Any type of seed, normally covered by the ovary wall, bracts, or other parts of the floret, that has not been processed down to the naked caryopsis.
unifoliate. Designating a plant with single leaflets.
uniform flow (hydraulics). A constant flow of discharge whose mean velocity is also constant, an ideal condition that can be only approximated in practice. If the velocity of the constant discharge varies, the flow is *steady non-uniform flow.* Also called *steady uniform flow.*
uniform white sapwood rot. A heart rot of hardwood, maple, and beech, caused by the fungus *Hydnum septentrionale,* family Hydnaceae.
unilateral horizontal cordon. A method of training vinifera grapes and dwarf fruit trees in which one horizontal arm or lateral branch is developed from a trunk about 2 ft. above the ground. The horizontal arm extends on a wire from the top of the trunk. See *bilateral horizontal cordon; vertical cordon.*
unimproved land. Land that has not been improved or changed from its natural conditions for agricultural use. See *scrub: 1.*
unimproved wool. That wool which contains no blood of Merino or English sheep and other wools not finer than 40's.
union. 1. The proper, healthy uniting, jointing, of a stock and cion in a graft. 2. The place on the stem of such a uniting.
union drive. A log drive which belongs to several

different owners all of whom share expenses pro rata.

uniparous. Producing only one egg or one offspring at a time.

unipera. An animal that regularly produces only one offspring at a parturition.

unisexual. Designating a flower which has either stamens or pistils but not both.

unit. 1. A single thing or item of produce. 2. A recognized measure of weight, volume, or distance, such as a bushel, gallon, or mile.

unit character. A character believed to be inherited according to Mendel's law, as determined by the allelomorphs at a single gene locus (obsolete).

unit cleaner. A cleaner that cleans only the cotton for one gin stand.

unit-cooler system. A method of distributing refrigeration in cold-storage rooms in which air is forced through coils containing a refrigerant. The coils are continuously defrosted by a brine spray or intermittently by water spray. The air is discharged at the top of the unit and returned at the bottom.

U(nited) S(tates) Approved. An official designation for breeding flocks, hatching eggs and chicks qualifying under the rules and standards of the U(nited) S(tates) National Poultry Improvement Plan (*q.v.*). For *U(nited) S(tates) Approved* classification, all males and females in breeding flocks are selected by authorized state agency. See *U(nited) S(tates) Certified for Eggs.*

U(nited) S(tates) Certified for Eggs. An official designation for breeding flocks, hatching eggs and chicks qualifying under the rules and standards of the U(nited) S(tates) National Poultry Improvement Plan (*q.v.*). For *U(nited) S(tates) Certified for Eggs* classification, the breeding flock specifications are: (a) all males R. O. P. See *U(nited) S(tates) Record of Performance,* or (b) all males and females from U(nited) S(tates) Performance Tested Parent Stock (*q.v.*), mated in the same combination as used in the qualifying parent stock.

U(nited) S(tates) Certified for Meat. An official designation for breeding flocks, hatching eggs and chicks qualifying under the rules and standards of the U(nited) S(tates) National Poultry Improvement Plan (*q.v.*). For *U(nited) S(tates) Certified for Meat* classification, all male and female breeding stock to be from U(nited) S(tates) Performance Tested Parent Stock (*q.v.*), mated in the same combination as used in qualifying parent stock.

U(nited) S(tates) National Poultry Improvement Plan (N.P.I.P.). A plan established in 1935 with approval of the U.S. Secretary of Agriculture and an appropriation by Congress. Objectives of the plan are: (a) to improve production and meat qualities of poultry; (b) reduce losses from hatchery disseminated diseases; (c) establish uniform terminology for grades; and (d) control advertising of poultry products. The U.S. Department of Agriculture cooperates with state authorities in the administration of health and production regulations for the poultry industry. Participation in the plan is optional with the states, and with the hatcheries and their flocks within each state.

U(nited) S(tates) Performance Tested Parent Stock. An official classification under the U(nited) S(tates) National Poultry Improvement Plan (*q.v.*) for poultry flocks which qualify through the performance of their progeny in an official random sample test *q.v.*

U(nited) S(tates) Pullorum-Typhoid Clean. Under the U(nited) S(tates) National Poultry Improvement Plan (*q.v.*), an official classification for poultry flocks in which no pullorum or typhoid reactors were found on the first official blood-test, provided that if a reactor or reactors were found on the first test, the flock may be qualified with two consecutive official negative tests at least 21 days apart.

U(nited) S(tates) Pullorum-Typhoid Passed. Under the U(nited) S(tates) National Poultry Improvement Plan (*q.v.*), an official classification for poultry flocks in which no pullorum or typhoid reactors were found on the last official test. Birds must be at least 5 months old when tested and must be tested within 12 months of time hatching eggs are to be used. The interval between tests must be at least 21 days.

U(nited) S(tates) Record of Performance (R.O.P.). An official designation, under the U(nited) S(tates) National Poultry Improvement Plan (*q.v.*) for females that qualify on individual or family trapnest records. Males qualify by being produced from a single-male mating of qualified *R.O.P.* birds, and by being reasonable representatives of breed or variety and by showing health and vigor and no serious defects.

unit-livestock-price method. The valuation of inventoried animals in which the number of animals is multiplied by an average unit cost of raising the animal to a specified age.

unit requirements. The quantity of the factors of production used under given conditions for a given unit of farm production or the operation of a unit of farm power.

univalent. Designating a chromosome unpaired at meiosis. See *bivalent chromosome.*

unleached wood ashes. A fertilizer composed of wood ashes that have had no part of their plant food removed and that contain 4 percent or more of water soluble potash (K_2O). *Cf. leached wood ashes.*

unmarketable. Designating produce that cannot be sold because of its poor quality, or lack of demand.

unmerchantable wool. Wool which has been washed but which is not yet sufficiently clean to be classed as standard washed wool (obsolete).

unmottled curly dwarf. A virus disease of the potato characterized by a dwarfing of the plant, rugosity of the leaves, leaf rolling, and cracking of the tubers.

unnatural strawberry color. A color defect of

strawberry ice cream which lacks the desired hue of a strawberry color.

unnatural Swiss cheese flavor. A flavor defect of Swiss cheese which has a clean taste but one foreign to Swiss cheese.

unnecessary overlapping. Uneven bandage.

unpaid family labor. Labor used in conducting the farm business which is furnished by the members of the farm family, other than the operator, for which no direct wage is paid. Its value is determined by the amount of additional labor the operator would have had to hire at current wages to carry on the same size business had the family labor not been available.

unpolished rice. The whole rice grain which retains its outer natural color.

unpublished pedigree. Closed pedigree.

unripe cotton. Dead cotton.

unripened-cream butter. Butter made either from pasteurized sweet cream or pasteurized sour cream to which no culture has been added.

unrodded. Designating a building which has no lightning rods.

unsalted butter. Butter which contains no added salt.

unsaturated fats. Fats of vegetable origin such as olive oil or cottonseed oil.

unscoured wool. Wool which has not been scoured. See *scoured wool.*

unsettled market. A market in which supply, demand, and prices are subject to rapid changes.

unsex. To remove the ovaries of a female or the testes of a male animal or bird; to castrate; caponize.

unsexed. Designating an animal which has been spayed, caponized, or castrated.

unshod. Designating a horse which has no horseshoes on its hooves.

unslaked lime. See *burnt lime; calcium oxide.*

unsound horse. A horse which has a defect that interferes with the complete utilization of the animal's ability to perform a service. See *breeding unsoundness; hereditary unsoundness; temporary unsoundness.*

unsound knot. A knot that is not solid across the face or which is not as solid as the surrounding wood due to decay or to mechanical separation. See *knot: 4.*

untenderable. Not tenderable *q.v.*

unthrifty. Not thrifty *q.v.*

unwashed wool. Wool that is in its original, greasy condition.

up. 1. In the show ring, to announce the name of a rider: *e.g.,* Mr. John Doe *up.* 2. Denoting a plant which has sprouted, as the corn is *up.*

upbreeding. The use of a superior breed upon an inferior or mongrel one in order to improve the offspring of the inferior breed.

up-budding. Shield budding.

up-country. The country which lies inland from the sea-coast; the interior or backwoods.

up-grading. Upbreeding.

uphand sledge. In blacksmithing, the lifting of the sledge over the work.

up-headed. Designating an animal which holds its head high.

upkeep. The cost of maintenance.

upland. 1. Highland, hill-land, or an elevated plain in association with and in contrast to a valley plain or lowland. 2. See *upland cotton.* 3. Designating crops or crop varieties which are grown on upland in contrast to lowland areas.

upland cotton. *Gossypium hirsutum,* family Malvaceae, a tropical, woody herb grown as an annual for its fiber, the source of much of the commercial cotton in the U.S. It is also important for the edible oil derived from the seed, and for the stockfeed residue from the seed after extraction of the oil. It is one of the most important cash crops of the U.S. Native to America. Also called *American upland cotton.*

U. S. Department of Agriculture Grades of Upland Cotton	
No. 1	middling fair
No. 2	strict good middling
No. 3	good middling
No. 4	strict middling
No. 5	middling
No. 6	strict low middling
No. 7	low middling
No. 8	strict good ordinary
No. 9	good ordinary

upland cress. See *bitter wintercress; gardencress peppergrass.*

upland game. Birds and small animals which are hunted, particularly for sport.

upland rice. Any rice variety grown without irrigation. See *lowland: 3.*

upland weed. In cranberry culture, a non-aquatic or non-waterloving weed.

upper elevator canvas. Elevator canvas.

upright. An erect branch attached to a prostrate stem of a trailing plant. Fruits of cranberry are borne on *uprights.*

upright method. Any of several methods of training grapes in which two or more arms are laid off from a vertical main stem or trunk to the right and left on a horizontal wire. As the shoots grow upward they are tied to wires above.

upright storage. Vertical storage.

upright (vertical) silo. A vertical structure used for preserving chopped, green feed silage. Usually it is made of wooden staves, concrete or tile blocks or metal sheets held together by bands or reinforcements. Diameter and height vary with the capacity desired.

upright tree. A fastigiate tree (*q.v.*); a tree which holds its limbs upright like a Lombardy poplar.

upset. 1. To enlarge by blows upon the end, as a

blacksmith *upsets* a bar of heated iron by bumping its end upon the anvil. 2. To cause any change in the normal functions of plants or animals.

upstanding. Designating an animal which stands well on its legs.

upstream. Towards the head of the river or stream from the point of reference. *Cf. downstream.*

uramon. $(NH_2)_2CO$, a nitrogen fertilizer which contains about 42 percent nitrogen.

uranium. U, a heavy, grayish element of the chromium group. It is radioactive, rare, and found only in combination, as in uranite and pitchblende. The radioactivity is measured by the Geiger-Müller Counter which consists of a conducting cylinder, filled with certain gases and having a coaxial, thin wire which detects the passage of radioactive waves through the cylinder walls by their ionizing effects on the gases.

urao. $K_2CO_3 \cdot NaHCO_3 \cdot 2H_2O$, a hydrous sodium carbonate. The trona (*q.v.*) of arid regions of the western U.S. is an impure form of *urao.*

urate. A substance formed by the combination of certain chemical elements with uric acid.

urban. 1. Belonging to or residing in a town or city rather than the countryside. 2. In the U.S., designating any of the following residence categories, used by the Bureau of the Census: (a) places having 2,500 persons or more, incorporated as cities, boroughs, and villages; (b) incorporated towns of 2,500 persons or more, except in New England, New York, and Wisconsin where towns are simply minor civil divisions; (c) the densely populated areas, incorporated and unincorporated, around cities with a population of 50,000 or more; and (d) unincorporated places with a population of 2,500 or more outside any urban fringe. See *rural-farm; rural-nonfarm.*

urban fringe. That area on the periphery of cities and their more densely populated suburbs which has lost its original distinctive rural character because of urban decentralization. See *rurban.*

urd. *Phaseolus mungo,* family Leguminosae, an annual, herbaceous plant grown for forage, green manure, or its small, elongated, dark seeds used for feed. Probably native to Indonesia. Also called *black gram; mungo.*

urea. 1. A non-protein, organic compound of nitrogen, made synthetically by a combination of ammonia and carbon dioxide, and used in fertilizers and as a livestock feed supplement. 2. The chief compound of nitrogen in the urine of mammals.

urea-ammonia liquors. Any of several solutions of urea in crude aqueous ammonia, used in ammoniating superphosphate.

urediospore. Repeating spore.

uremia. Accumulation of urinary constituents in the blood, and the toxic condition that results.

uremic poisoning. Uremia.

ureter. The tube that conveys the urine from the kidney to the bladder.

urethra. In most mammals, the canal or tube which carries the urine away from the bladder and serves as a duct for the passage of the male's semen.

Uri. A hard-boiled, cows' milk cheese made in Switzerland.

uric acid. A white, crystalline compound, derived from guano (*q.v.*). It contains about 33 percent available nitrogen.

urinary calculi. Stones of phosphates, urates, oxalates, and lime salts which lodge in some part of the urinary tract, as the kidney, ureter, bladder, urethra. Symptoms of the presence of these stones are obstruction to the flow of urine, passage of a few drops of blood, and convulsive pain when urinating

urine. In man and higher animals, a fluid consisting of certain waste products of metabolism (*q.v.*). Extracted from the blood by kidneys, it is conveyed to and stored in the bladder, from where it is expelled through the urethra at intervals.

urodeum. In the cloaca of a bird, that part into which the genital ducts and the ureters empty.

urolithiasis. The formation of urinary calculi.

uropygial gland. The preen gland.

Urseren. A mild-flavored, cured, cows' milk cheese made in Switzerland.

usage factor. In bulk milk tanks, a numerical value which represents the portion used in relation to the total tank capacity. The *usage factor* for a 300 gallon tank in which 200 gallons were cooled would be ⅔ or 66.7 percent.

usambara violet. African violet.

U.S.D.A. United States Department of Agriculture.

useful life. Of a fixed asset, the normal period of usefulness.

use hazard. In land use, any serious inconvenience or danger incident to occupancy, as poisonous plants, landslides, floods, avalanches, insects, or dust.

utensil flavor. A flavor defect of milk or its products suggesting the odor of poorly-washed dairy utensils and equipment. Also called *dishrag flavor.*

uterine capsule. A medication in capsule form which is placed in the uterus of an animal after parturition. It aids in preventing putrefaction of the retained afterbirth, assists in loosening the afterbirth attachments, and checks various types of uterine infections.

uterine insemination. In artificial insemination, the practice of placing the semen in the uterus, as contrasted to placing it in the vagina or in the cervix.

uterus. The womb; in female mammals, an organ in which the young develops before birth.

utility. 1. The next to the lowest grade of veal and the fourth lowest grade of beef. 2. An organization providing services, such as gas or electricity. 3. A mediocre grade of fruit.

utility building. A farm building designed to house miscellaneous equipment and supplies.

utility class. In stock shows, gaited and harness horses.

utilization. 1. The extent to which grazing animals have cropped the current growth within their reach. It may be expressed as the percentage in height or weight of forage removed, or in such general terms as excessive, close, moderate, or light. 2. The extent to which a ration, feed or nutrient is used.

utilization cut. A percentage reduction or other expressions of type values applied to the total forage acres to account for areas partially or wholly unusable by livestock because of such factors as down timber, rocks, steepness of slope, critical watershed values, game, recreation, erosion, or poor range condition.

utilization value. In forest planting, an expression denoting the usefulness of a seed and determined by its purity and germinative energy. *Utilization value* percent = purity percent × germinative energy present ÷ 100.

uva. 1. A soft, indehiscent, grapelike fruit; a raisin or grape. 2. See *bearberry.*

uvala. A natural land feature in a karst region. It is characterized by depression larger than a sinkhole, often formed by the coalescence of several sink holes.

V

vacant land. Designating unclaimed and unoccupied land which may be partially cultivated.

vaccinal titer. A measurable amount of agglutinin or antibody which results from vaccination.

vaccinate. To inoculate with a bacteria or virus so as to produce immunity against a disease.

vaccination. An inoculation made with bacteria or virus so as to produce immunity against a disease.

vaccine. A suspension of a living, disease-producing agent sometimes attenuated, reduced, or inactivated, such as bacteria or a virus used in vaccination for disease control.

vachette clasp. A short, curved piece of steel wire, with sharp pointed ends used to bind a toe crack or quarter crack of a horse's hoof.

vacreation. 1. The process of removing odors in milk and its fluid products by direct steam injection and volatilization under vacuum. 2. Treatment of fluid dairy products with a vacreator.

vacreator. A multiple unit consisting of steam jets and vacuum chambers designed to remove undesirable odor from milk and/or cream.

vacuolated cytoplasm. The living substance of cells, exclusive of the nucleus, when filled with or containing bubble-like structures.

vacuum dropper. See *separator dropper.*

vacuum line. A tube, part of a milking machine, that is attached to the teat cup and the vacuum pump. As the vacuum pump exhausts the air in the *vacuum line,* milk is drawn by suction from the cow.

vacuum-packed whole kernel. A type of canned corn which is whole kernels with the addition of a liquid.

vacuum pan. Any of the several devices used to produce evaporated or condensed milk, maple sirup, etc., under partial vacuum. The unit consists usually of a jacket or body, a heating surface, a condenser and trap, and a vacuum pump.

vadose (water). Water held in soil, or other surficial geological formations, above the level of permanent ground water.

vagabond bee. Any worker bee which lacks an individual foraging area.

vagabond crambus. *Crambus vulgivagellus* Clem., family Crambidae, the larva of a moth which prepares a silk-lined tunnel near the surface of the soil and feeds normally on grass roots. It sometimes infests corn. Also called *blackheaded sod webworm.*

vagina. The canal in female mammals extending from the uterus to the vulva.

vagina cylinder. See *vagina shell; artificial vagina.*

vaginal collection. One of the oldest methods of obtaining semen for artificial insemination in which semen is removed from the vagina of a female just after coitus by means of a syringe or vaginal spoon.

vaginal jelly. A water-soluble, often a vegetable, compound used to lubricate the artificial vagina to prevent injury or pain to the penis in collecting semen for artificial insemination. It may also be applied to instruments for insertion into an animal's vagina to prevent injury and to make entrance easier.

vaginal spoon. A long-handled spoon used to recover semen from the vagina of a female immediately following coitus. See *vaginal collection.*

vagina shell. The rigid, outer covering of the artificial vagina.

vaginitis. Vaginal inflammation. See *granular vaginitis.*

vale. A small or large valley (a poetic term).

Valencia rind spot. A spotting and pitting, usually on the stem end, of Valencia oranges in the late summer and early fall. It seems to be caused in the mature fruit by hot weather and high humidity.

valerian. 1. Any of the several herbs or shrubs of the genus Valeriana, family Valerianaceae. 2. *V. officinalis,* a perennial herb grown for its fragrant, pink, white, or lavender flowers. Native to Europe and Asia. Also called *garden heliotrope; cherry pie; false heliotrope; summer heliotrope.*

valine. See *amino acids.*

valley. 1. An elongated, erosional depression usually occupied by a stream which has a downward slope conforming to the direction of flow of the occupying stream, and includes both bottomland and slopes. Some features designated as *valleys* may be primarily structural, due to rock folding and faulting rather than erosional; some may be dry or without streams; some in arid regions may be basins which do not have the usual form of valleys of humid regions. 2. The drainage area of some rivers, as the Mississippi *Valley,* or a combination of lowland and mountainous ridges as the Great Appalachian *Valley* of eastern U.S.

valley land. The lower lying or smoother part of a valley, as differentiated from hill, ridge, or mountain land.

valley lily. Lily-of-the-valley.

valley mahogany. True mountain mahogany.

valley oak. California white oak.

valley train. Deposits of comminuted rock material which were carried in prehistoric times by streams originating from the melting ice of glaciers. The deposits are confined to valleys and thus are distinguished from outwash plains. (A physical feature of parts of the glacial region of the U.S.) The *valley train* often constitutes a distinctive land type of agricultural significance.

valley white oak. California white oak.

vallota. *Vallota speciosa,* family Amaryllidaceae, a bulbous herb grown for its large, funnel-shaped, scarlet flowers borne in terminal clusters on long, hollow, leafless, flowering stalks. Native to South America. Also called *Scarborough lily.*

valone. A chemical used as a constituent of a lousicide.

valonia oak. *Quercus macrolepis,* family Fagaceae, an evergreen tree sometimes grown as an ornamental in warm areas. Native to Europe and Asia. Also called *aegilops; egilops.*

valorization. The establishment and maintenance of an arbitrary price for a product by governmental action through subsidies, loans, government purchase, etc.

Valparaiso oak. Canyon live oak.

valuable consideration. Any service, goods, etc., given in payment or partial payment. (The term is often used in a deed, as "one dollar and other *valuable considerations,*" to prevent disclosure of the full price.)

value of crops produced per tillable acre. A value determined by multiplying the total production of each crop on a farm by a representative farm price, as of December 15 or a similar period. Typical or average prices for the region are used. The value of the total crop production is divided by the number of tillable acres in the farm for the determination of the value per tillable acre.

value of farm operator's labor. An allowance used primarily in computing the rate of return on the farm capital investments for the services of the farm operator for his labor at the rate at which he would have to pay another man to perform these functions. It is exclusive of the family living from the farm. It is used in computing the rate earned on the capital invested in the farm business and the total and per tillable acre charge for farm labor.

value of farm production. The gross (farm) income (*q.v.*) and the value of farm products furnished to the farm family.

value orientation. A generalized disposition on the part of a human group or society to select or favor one of several alternative modes or action and/or belief.

valve-controlled hydrant. In irrigation, a hydrant fitted with a small valve and connected by a short riser to the underground lateral.

valve rope knife. An emergency tool used in well boring by the cable system, consisting of a cylinder with a knife balanced in it like a valve in a pump. The tool is lowered around the rope until the desired depth is reached, when an upward pull engages the knife and severs the rope.

VAMA. A synthetic chemical, modified vinyl acetate maleic acid, which has been used experimentally as a soil conditioner.

van. A large, covered truck or wagon.

vandalroot. Valerian.

vane. 1. The thin web part of a bird's feather. 2. A thin plate or strip of metal pivoted on the top of a spire or mast to show the direction of the wind; a weathercock. 3. A blade of a windmill, fan, centrifugal pump, or similar apparatus.

vane churn. One of several types of churns used in the commercial production of butter.

Vanhoutte spirea. *Spiraea vanhouttei,* family Rosaceae, a hybrid shrub, able to withstand city conditions very well, and grown for its lovely habit and numerous pure white flowers. Also called *bridal wreath.*

vanilla. 1. Any one of the several plants of the genus Vanilla, family Orchidaceae. 2. See *Mexican vanilla.* 3. The extract derived from the *vanilla* pods which is extensively used in confections.

vanilla bean. The long, bean-like capsule of the vanilla *q.v.*

vanilla extract. The flavoring extract prepared from vanilla beans, with or without sugar or glycerine, which contains in 100 cc. the soluble materials from not less than 10 grams of the vanilla bean.

vanilla grass. Sweetgrass.

vanilla leaf. Vanilla trilisa.

vanilla trilisa. *Trilisa odoratissima,* family Compositae, an aromatic, perennial herb grown for its rose-purple flowers. Native to North America. Also called *Carolina vanilla; vanilla leaf.*

vanilla-wafer flavor. A flavor defect of vanilla ice cream which suggests the taste of vanilla wafers.

vanillin. A sparkling, white, fuzzy crystalline substance which has a characteristic vanilla-like odor.

vapor drift. The movement of herbicidal vapors from the area of application to adjacent areas.

vaquero. A Mexican cowboy.

var. The abbreviation for the word variety: *e.g.,* as in *Sarcoptes scabiei* var. *bovis.*

vara. A Spanish-American measure of length, equal to about 33 ins., which was used in description of Mexican land grants (southwestern U.S.).

variable. Designating a plant or animal which is not true to type.

variable costs. Farming expenses which change with output: *e.g.*, hired labor, tractor fuel, repairs, feed, fertilizer, breeding fees, feeder stock, seed, and other crop expenses. *Cf. fixed cost.*

variable-depth attachment. A planter attachment for planting seeds at variable depths below the surface of the soil.

variable drop. A device on a check-row planter which varies the speed on the plate in the hopper and thus changes the number of kernels in the hills.

variable factor of production. An input that is subject to the control of an entrepreneur.

variable grade. A terrace channel having a variable slope which decreases as the distance from the outlet increases: *e.g.*, a terrace 1,200 ft. long may be built in four 300-ft. sections each having different grades, the outlet end having 0.40 ft. fall per 100 ft.; the next section, 0.30; the third, 0.20, and the fourth section, 0.10 ft. per 100 ft.

variable hen louse. Turkey louse.

variable oak leaf caterpillar. *Heterocampa manteo* (Dbldy.), family Notodontidae, a caterpillar which voraciously feeds on the foliage of the oak, linden, walnut, birch, elm, hawthorn, and persimmon.

variable payment loan. In farming, a loan in which the payments of the principal are scheduled to fluctuate with crop yields or farm income.

variation. In biology, the occurrence of differences among the individuals of the same species.

variegated. 1. Designating leaves of plants which have a pattern caused by the alternation of scattered areas or stripes on leaf margins of different colors. 2. Designating a flower each petal of which is streaked or striped with a color or colors different from the basic color. See *self-colored flower.*

variegated alfalfa. A group of alfalfa (*q.v.*) perhaps a hybrid of the purple-flowered and the yellow-flowered species. It is more resistant to cold and drought than other commercially-grown alfalfa. See *Turkestan alfalfa; yellow-flowered alfalfa.*

variegated cutworm. *Peridroma margaritosa* (Haw.), family Noctindae, the larva of a moth, probably the most widely distributed and destructive pests of garden crops, fruit trees, and vines. It eats the foliage, buds, and fruits of the plants.

variegated leafcroton. *Codiaeum variegatum*, family Euphorbiaceae, a tropical, foliage plant which, in many varieties, is cultivated widely for its gorgeously-colored leaves. Native to Australia and Polynesia. Also called *croton; leafcroton.*

variegated spurge. See *snow-on-the-mountain euphorbia.*

variegation. See *marginal variegation; sectional variegation.*

varietal. Concerning, or designating a variety.

variety. a. A group of related plants or animals which differs from other similar groups by characters too trivial or inconstant to be recognized as a species; often any category of lower rank than a species. See *cultivar.* b. In domesticated animals, a subdivision of a breed based on some minor character such as color, etc.

variety hybrid. A cross between two varieties of the same species.

variola. Cow-pox.

varnish. See *desert varnish.*

varnish sumac. Flameleaf sumac.

varnish tree. Candlenuttree.

varve clay. Alternating dark and light bands or layers in a clay deposit. Each pair, or *varve*, represents one year's accumulation, the light, thicker, and coarser layer being laid in the summer and the darker, thinner, and finer-textured layer in winter. Most *varve clay* deposits were formed at the close of the last ice age when glacial drainage entered standing bodies of water.

vascular. 1. Concerning blood vessels. 2. Concerning the conducting system of plants (xylem, phloem).

vascular plant. A plant that has special fibrovascular tissue through which water and the dissolved organic foods are conducted in the xylem and the phloem portions, respectively, of the fibrovascular bundles. These plants consist of the ferns, their allies, and the seed-producing plants, such as the gymnosperms and the angiosperms.

vase. Designating a plant that has been pruned and/or trained to be wider at the top than at the bottom.

vasectomize. To emasculate a male animal by removing, severing, or ligating the vas deferens, the excretory spermatic duct. The testes are not removed in this operation.

vase pruning. See *head pruning.*

Vaseygrass. *Paspalum urvillei*, family Gramineae, a grass grown for hay in the southeastern U.S. It has a fair forage rating. Native to South America.

Västgotäost. A Swedish, medium firm, mild, sweet, nutty-flavored, cured cheese made from partly-skimmed cows' milk.

vat. Any large vessel, tub, cistern, etc., used for holding liquids, as a vessel used in large-scale brewing.

vat pasteurization. A method of milk pasteurization employing a definite unit of milk; commonly called *holder pasteurization.*

V-drag. A type of horse-drawn scraper which consists of two heavy timbers fastened together so as to resemble the letter V. The hitch is usually made at or near the apex. It is used in the construction of small terraces and shallow water channels.

veal. 1. The flesh obtained from a *veal* calf. 2. To feed a young calf by hand, or allow it to suckle a cow other than its mother, and to supplement its milk ration with grain gruel, etc., to prepare it for slaughter.

veal bird. A retail cut of veal which consists of several small pieces of veal fastened on to a skewer and shaped to resemble a drumstick. See *mock chicken; city chicken.*

veal calf. In marketing, a calf not over 3 months old which weighs from 100 to 200 lbs. See *deacon; bob: 1.*

veal chuck. See *veal shoulder.*
veal chuck steak. A retail cut of veal for broiling made from the first, second, and third ribs.
veal crown roast. A crown roast (*q.v.*) of veal.
veal cutlet. A retail cut of veal for broiling made from the round *q.v.*
vealer. Veal calf.
veal foreshank. A retail cut of veal made from the shank of the front leg.
veal hindshank. A retail cut of veal made from the shank of the hind leg.
vealing. See *veal: 2.*
veal rack. See *lamb rack.*
veal rib chop. A retail cut of veal for braising made from the ribs. It includes a rib and the eye muscle.
veal riblet. A retail cut of veal for braising or stewing which comes from the veal breast.
veal rib roast. A retail cut of veal for roasting which consists of two or more ribs and the eye muscle.
veal round roast. A retail cut of veal for braising or roasting which comes from the round.
veal round steak. Veal cutlet.
veal rump roast. A retail cut of veal for roasting which comes from the top of the round.
veal shoulder. A wholesale cut of veal which comes from the shoulder and the first through the sixth ribs.
veal stew. A retail cut of small pieces of veal for stewing which comes from the breast.
vealy in color. Designating a meat carcass which is gray.
Veatch silktassel. See *Veitch silktassel.*
vector. A living transmitter of disease organisms; an arthropod or other species that usually is biologically or mechanically an essential agent in transmitting a particular disease or pathogenic organism.
vedalia. *Rodolia cardinalis* (Muls.), family Coccinellidae, an insect of great importance, used to control cottony-cushion scale (*q.v.*). Also called *Australian lady beetle; cardinal ladybird.*
vega. 1. A fertile, grass-covered tract or extensive plain (southwestern U.S.). 2. Irrigated land from which a single crop per year is produced. *Cf. huerta.*
vegetable. 1. The edible part of an herbaceous plant. 2. Any plant part that is eaten either cooked or raw during the principal part of a meal rather than as a dessert.
vegetable antimony. Boneset.
vegetable bezoar. A bezoar (*q.v.*) that is formed from the hairs of clover, awns of oats, etc.
vegetable bin. A storage space or bin for vegetables, often potatoes, carrots, beets, etc. See *cold cellar.*
vegetable butter. See *oleomargarine.*
vegetable buttermilk. A drink, for those who cannot consume solid foods, which consists of a vegetable purée (15 percent), buttermilk (70 percent), water (15 percent), and sugar to taste.
vegetable crop. Any plant crop from which vegetables are harvested.
vegetable egg. Eggplant.
vegetable fat. Vegetable oil.
vegetable flavor. A flavor defect of milk or its products which suggests raw or cooked vegetables.
vegetable forcing. The use of artificial heat, light, etc., to cause vegetables to mature more quickly or at a different season than normal. See *forcing.*
vegetable garden. A relatively small plot of land used for growing vegetables for home consumption.
vegetable gardening. 1. The art and science of growing vegetables. 2. The growing of vegetables at home.
vegetable gold. Saffron.
vegetable horsehair. Mediterraneanpalm.
vegetable lamb. Cotton.
vegetable lemon. Mango melon.
vegetable marrow. 1. A group of squashes which are varieties of the pumpkin, *Cucurbita pepo.* Members of the group are oblong; have a smooth skin, hard outer covering, and range in color from cream to dark green. 2. See *American avocado.*
vegetable matter. Any material which is, or only recently was, plant tissue.
vegetable mold. See *humus.*
vegetable oil. Any oil derived from cottonseed, flaxseed, soybeans, peanuts, olives, corn, castor beans, etc., as contrasted to animal fats.
vegetable orange. Mango melon.
vegetable oyster. *Tragopogon porrifolius,* family Compositae, a biennial, tap-rooted herb grown for its edible root whose flavor suggests oysters. Also called *salsify; oyster plant.*
vegetable pear. Chayote.
vegetable shortening. Edible vegetable fat which has been homogenized and solidified for use as a lard substitute.
vegetable show. An exhibition for display of specimen vegetables.
vegetable spermaceti. Chinese insect wax.
vegetable sponge. Towelgourd.
vegetable tallow. Chinese tallowtree.
vegetable wax. A waxy secretion from a plant, as from the bayberry.
vegetable weevil. *Listroderes costirostris obliquus* (Klug), family Curculionidae, a beetle found in the Gulf states and California (U.S.) which, along with its larvae, feeds voraciously on beet, cabbage, carrot, cauliflower, lettuce, mustard, onion, radish, potato, turnip, tomato, Swiss chard, and spinach. It defoliates the whole plant with the exception of stems and leaf midribs.
vegetal. Of, designating, pertaining to, or like, a vegetable.
vegetal discharge. The water of soil and substrata which is absorbed by roots of plants and discharged in the process of transpiration.
vegetarian. 1. One who, because of cultural reasons or personal conviction, abstains from eating meat (in the strictest sense, also milk, butter, and eggs). 2. An herbivorous (*q.v.*) animal.
vegetate. To produce stems and leaves from seeds or buds.

vegetated. Abounding in, or covered with plants.
vegetated channel. A vegetated, natural, or artificially constructed waterway used to conduct the accumulated runoff from cultivated land, terraced land, diversions, or from fields in a contour strip-crop system.
vegetation. Any group or association of plants; the sum of vegetable life; plants in general.
vegetation cover. The entire canopy of vegetation on an area. Also called *vegetal cover; vegetative cover.*
vegetation subtype. See *association: 1.*
vegetative. 1. Designating life and growth of leaves and stems in a plant as contrasted to fruiting or seed production. 2. In soils, referring to the productive power of a soil.
vegetative branch. A branch that bears leaves but no flowers or fruit.
vegetative control. Any method of effecting soil and moisture conservation by the use of vegetation.
vegetative cover. Any plant growth occupying the land.
vegetative growth. Above-ground parts of a plant, such as leaves and stems.
vegetative litter. The accumulation of plant material made up of dead and cast-off parts of plants.
vegetative mantle. The complete coverage of the soil by vegetation, especially perennial ground-cover vegetation.
vegetative reproduction. Reproduction of plants by plant parts other than seed of the parent plant. See *asexual.*
vehicle. A car, carriage, wagon, truck, trailer, or other conveyance.
veil. A light, metal screen or diaphanous cloth material draped about the head when working with bees to prevent being stung on the face or neck.
vein. 1. One of the systems of branching tubes, etc., which carry blood back to the heart. 2. One of the fibrovascular bundles which forms the framework of a leaf.
veinbanding. A common symptom of virus diseases of plants in which a broad band of chlorotic tissue develops along the veins or in which there is chlorosis or necrosis in the interveinal parenchyma.
vein blight. See *bacterial blight of cotton.*
vein clearing. A symptom of a virus disease of plants in which there is a clearing or chlorosis of the tissue in or adjacent to the veins.
veining. The pattern or system of veins.
Veitch fir. *Abies veitchi,* family Pinaceae, an evergreen tree grown as an ornamental. Native to Japan.
Veitch silktassel. *Garrya veitchi,* family Garryaceae, an evergreen shrub, growing on dry sites in western U.S., which is of minor importance as a browse plant.
Veitch spirea. *Spiraea veitchi,* family Rosaceae, a deciduous shrub grown for its dense clusters of white flowers. Native to China.
veld(t). South African grassland with few or no shrubs or trees.
velocity. The speed of flow in feet per minute. The *velocity* of freed air through stored products varies greatly, but the average *velocity* is greater than the apparent *velocity,* and is the apparent *velocity* divided by the void space.
velvet bean. Any plant of the genus Stizolobium, family Leguminosae. Species are grown in warm areas for feed, soil building, green manure, pasturing, and sometimes for ornament. Also called *banana bean.*
velvetbean caterpillar. *Anticarsia gemmatalis* Hbn., family Noctuidae, a caterpillar that feeds on soybean, velvetbean, cowpeas, peanuts, kudzubean, and the black locust.
velvet bean meal. A livestock feedstuff which is ground velvetbeans containing only an unavoidable trace of hulls or pods.
velvet bentgrass. *Agrostis canina,* family Gramineae, a perennial herb used as a turfgrass. Native to North America. Also called *brown bent; dog bent; dog's grass.*
velvet dock. Flannel mullein.
velvet flower. Love-lies-bleeding.
velvetgrass. *Holcus lanatus,* family Gramineae, an ornamental, perennial grass whose leaves are covered with soft, velvet hairs. Native to Europe. Also called *dart grass; mesquite grass.*
velvetleaf. See *chingma abutilon; flannel mullein.*
velvet mimosa. Velvetpod mimosa.
velvet plant. See *flannel mullein; Java velvetplant.*
velvetpod mimosa. *Mimosa dysocarpa,* family Leguminosae, a plant with bipinnate foliage, found on the semiarid ranges from Texas to Arizona and Mexico, whose forage rating is fair. Native to U.S. and Mexico.
velvet sumac. Staghorn sumac.
velvety. Designating a product whose surface is soft and smooth.
vena contracta. The most contracted section area of a stream, jet, or nappe, just beyond the plane of the orifice or notch, from which it issues.
venation. The arrangement of the veins in a leaf.
vendee. A person who buys something, especially a land buyer.
vender. Vendor.
Vendôme. A soft, French, ripened cheese, cured in a cool, moist cellar or buried in ashes. It is similar to Camembert *q.v.*
vendor. A person who sells something; a seller. Also called *vender.*
veneer band. A thin sheet of wood which forms an open-ended container. It is placed in a flat, filled with soil to receive a plant for the first transplanting or seeds. Seeds of some vegetables are planted in veneer bands in flats, and later set out in gardens or fields. The band is then set out with the transplant and not removed.
veneer grafting. A useful form of grafting, especially for ornamentals, in which a thin, wedge-shaped section, about 1 in. long, is cut from the stock. The lower end of the cut extends about

⅓ the diameter of the stock. The cion is then cut to match and the union is tied and waxed.

venereal disease. A disease which may be spread by coitus.

Veneto. A granular, greenish-yellow or straw-colored, Italian, Parmesan-type cheese which is cured from 1 to 2 years.

Venezuela treebine. *Cissus rhombifolia,* family Vitaceae, a vine grown as a house vine. Native to South America.

venison. The edible flesh of deer.

venom. Poisonous matter secreted by such creatures as snakes, bees, scorpions.

venomous. Designating a reptile, insect, etc., which secretes venom.

vent. 1. A small opening or passage, as an opening for ventilation. 2. The opening of the cloaca; the anus of a bird. 3. A brand mark made to indicate that the animal no longer belongs to the original owner. 4. To cancel an old brand on an animal to show change in ownership.

vent gleet. Cloacitis.

ventilate. To produce a free circulation of air for heating, ventilation, etc.

ventilated bale. A bale which has one or more perforations through it in the long axis direction to provide ventilation for drying.

ventilated T-super. Air-way super.

ventilating fan. An electric fan so placed or installed that it brings about an exchange of air in a given space or in the material being ventilated.

ventilating flues. The passages or conduits used to conduct air in a ventilating system.

ventilating rim. A cap for the super of a beehive which has screened openings to allow easier ventilation.

ventilation. The movement of air which brings about an exchange of air in the material or space being ventilated.

ventilation front. The loci of all points of equal travel time of the circulating air in the product being conditioned.

ventilator. An adjustable or fixed opening used to regulate the flow of air through a space or material. Louvers or other devices are used to keep rainfall out of the openings. The air flow may result from wind pressure, convection, or fan forces.

vent picking. A vice of poultry which consists of one or several chickens picking the everted vent of another. See *cannibalism.*

ventral. Designating the abdominal or under surface of the body.

vent shield. A shield used to cover a chicken's vent to prevent picking.

Venturi flume. A type of open flume with a contracted throat causing a drop in the hydraulic grade line, used for measuring flow of liquids. See *Parshall flume.*

Venturi meter. A proprietary measuring device which consists essentially of a Venturi tube and some form of flow-registering device. (Patented by the late Clemens Herschel.) See *Venturi tube.*

Venturi tube. A closed conduit which is gradually contracted to a throat causing a reduction of pressure head by which the velocity through the throat may be determined. The contraction is generally followed, but not necessarily, by gradual enlargement to original size. Piezometers connected to the pipe above the contracting section and at the throat indicate the drop in the pressure head, which is an index of flow.

Venuscup teasel. *Dipsacus sylvestris,* family Dipsaceae, a thistle-like weed, which has become naturalized in eastern North America. Native to Europe. Also called *English thistle; teasel.*

Venusflytrap. *Dionaea muscipula,* family Droseraceae, a perennial herb grown for its remarkable habit of catching and digesting insects.

Venus hair. Southern maidenhair.

Venuslookingglass. 1. Any plant of the genus Specularia, family Campanulaceae. 2. S. *speculum-veneris,* an annual, erect herb grown for its deep blue or white flowers. Native to the Mediterranean area.

Venus's chariot. Aconite monkshood.

veratrine. The crude mixture of alkaloids derived from the drug sabadilla *q.v.*

verbena. Any plant of the genus Verbena, family Verbenaceae. Species are popular for the flower garden and for pot plants. See *blue verbena; moss verbena; rose verbena.*

verbena bud moth. *Endothenia hebesana* (Wlk.), family Olethrentidae, an insect that infests verbena and lionsheart.

verdant. Designating fields or forests which are green with growing vegetation.

verge. 1. In a city or town, the strip of grass lying between the street and the sidewalk. 2. The edge of a flower bed.

verjuice. The sour juice of unripe fruit.

vermicelli. A form of spaghetti having a small diameter.

vermicide. Any substance which kills internal, parasitic worms; an anthelmintic. *Cf. vermifuge.*

vermiculite. A mineral, or minerals, classified with the micas, which with treatment at high temperatures, expands into worm-like scales and becomes a loose, absorbent mass. It is recognized as a clay-mineral constituent of some clays, but its exact chemical composition is not known. Its chief use is in insulation. Commercial *vermiculite* is used as a mulch for seed beds, as a medium for rooting plant cuttings, and in potting plants.

vermifuge. A drug or chemical that expels worms from animals; an anthelmintic. *Cf. vermicide.*

vermilion globe sour orange. A variety of the sour orange (*q.v.*) used as a rootstock for sweet orange grafting.

vermin. Any noxious animal; insect, acarid, rodent, etc.

verminous. 1. Due to, or pertaining to, worms. 2. Abounding in or infested with vermin.

vermorel nozzle. A spraying nozzle with a small orifice to which the liquid comes directly by a screw thread, and is dispersed as a fine, conical spray at low pressure.

verna barberry. *Berberis vernae,* family Berberidaceae, a deciduous shrub grown for its yellow flower clusters and its pale red fruit. Native to China.

vernal clematis. *Clematis cirrhosa,* family Ranunculaceae, a woody, climbing vine, grown in warm areas for its early, large, yellowish-white flowers. Native to Europe and Asia. Also called *evergreen clematis.*

vernalization. Seed treatment in which the germinating seed is held in artificial darkness, at low temperatures, for a fixed period to induce early flowering and fruiting when sown.

verruca vulgaris. See *bleeding wart.*

verrucosis. See *sour orange scab.*

Verschaffelt coleus. *Coleus blumei* var. *verschaffelti,* family Labiatae, a tender herb grown in the flower garden and as a pot plant for its variegated leaves. Native to Indonesia. Also called *foliage plant.*

vert. In forestry, any plant, etc., which grows and bears green leaves.

vertebra. Any of the bony segments which form the spinal column or backbone.

vertical cordon. A seldom-used system of training the vinifera grape so that spurs are spaced along an upright trunk four to six feet in height.

vertical discharge pipe. A type of installation for airplane spraying in which the spray is emitted from a pipe carried vertically downward for 18 ins. below the fuselage. The forward velocity of the liquid as it leaves the pipe is great enough to cause it to atomize as it strikes the comparatively still air.

vertical drainage. The disposal of excess water through surface inlets or wells into a pervious subsoil or substratum.

vertical farming. The combination of production, processing, and marketing of a farm product as a single business.

vertical grain. Designating lumber that has been sawed lengthwise into quarters, which are further cut into planks of various sizes. Such boards show the grain beautifully.

vertical integration. The control by a single firm of two or more stages in the chain of production, processing and distribution. This chain extends from the supply of production resources to the point at which the commodity reaches the consumer. Recent developments in agriculture include extensive *integration* in the broiler industry and with some *integration* in laying flocks, poultry, hog fattening, and cattle feeding.

vertical interval. In terrace farming, the vertical distance in feet from the center of one terrace line, ridge, or channel to the corresponding point on an adjacent terrace. The interval of the first terrace is the vertical distance from the top of the hill to the staked channel of the terrace.

vertical lift. A lift for a cutter bar on a mowing machine by which the operator, without moving from his seat, raises the cutter bar to a vertical position to clear obstructions and to turn corners by using hand and foot levers. Also called *plain lift.*

vertical mulching. A recent, experimental agronomic practice in which stubble, straw, cornstalks, and other plant residues are buried in a deep, narrow channel in the soil. An implement similar to the shank of a mole plow cuts a deep channel about 6 ins. wide at the top and narrowing at the bottom, and a chopper blows the plant matter in the channel. The practice is purported to promote more rapid infiltration of water, to lessen soil erosion, provide drainage, and other benefits.

vertical storage. A storage structure whose height is greater than its width or diameter. See *silo; bin; tank.*

vertical suction. In plowing, the underside curvature from the point of the share to the heel of the landside.

vertical T budding. A bud graft (*q.v.*) in which a vertical cut about 1½ ins. long is made through the bark of the stock and another cut is made at right angles to the first so as to form a cut similar to the letter T. The bud is placed inside the vertical cut and tied. Also called *T-budding.* See *inverted T.*

vertical training. See *vertical cordon.*

vertical zonality of soils. The distribution of different great soil groups on mountain slopes in which each group occupies a definite climatic and vegetation zone.

Verticillium wilt. Wilt produced on many plants by the fungus *Verticillium albo-atrum,* family Moniliaceae, which attacks plant roots and vascular bundles of the stem, discoloring and warping them and often killing the infected plant. Host plants are found mainly in the dicotyledons, in various families, frequently in plants with woody stems. Among the tree species attacked are avocado, ash, apricot, ailanthus, basswood, almond, peach, plum, maple, elm, catalpa, persimmon, black locust; among woody shrubs, cotton, raspberry, blackberry, sumac, barberry, elderberry, viburnum, rose, currant; and herbaceous plants such as tomato, potato, capsicum, eggplant, muskmelon, cucumber, watermelon, squash, beet, aster, chrysanthemum, peppermint, pumpkin, and a host of others. Also called *blue stripe wilt.*

vertigo. Staggers; a condition of disturbed equilibrium in which, because of dizzyness, an animal is unable to maintain its balance.

vervain. Verbena.

vervain mallow. Hollyhock mallow.

very coarse sand. See *soil separates.*

very fine sand. See *soil separates.*

very heavy breeds. The chickens of large breeds: Brahma, Langhan, Cochin, Orpington, Cornish, and Jersey Giant.

vesicle. A small bladder or sac which contains liquid, as a blister.

vesicular exanthema. In hogs, an acute, highly infectious, virus disease which causes the formation of vesicles or blisters on the snout, mucous membranes of the mouth, the feet between the toes, the soles, and dew claws.

vesicular mole. A fleshy mass or tumor formed in the uterus of animals that is cystic and resembles a bunch of grapes.

vesicular stomatitis. A virus disease of cattle, swine, and horses which resembles foot-and-mouth disease and vesicular exanthema. It is very common in the southeastern U.S.

vesicular structure. A soil structure characterized by round or egg-shaped cavities or vesicles.

vessel. 1. A tube or duct in a body which circulates blood, lymph, etc., within it. 2. A tubelike structure within a plant for the transportation of water and nutrients. 3. A container, as a flower pot, watering can, etc.

vessel segment. See *member vessel.*

vesture of land. Legally, all vegetation, except the trees, on the surface of the land which clothes it externally.

vetch. 1. Any of the various plants of the genus Vicia, family Leguminosae. Species are grown as cover crops, as green manure, and for pasture and forage. See *hairy vetch.* 2. *V. sativa,* an annual or biennial herb grown as a cover crop, as green manure, and for forage in the South and on the Pacific Coast, U.S. Native to Europe. Also called *fodder vetch; spring vetch; tare.*

vetch anthracnose. A common disease of vetch, caused by the fungus *Colletotrichum villosum,* family Melanconiaceae, characterized by lesions on the stems, leaves and pods. Also called *anthracnose of vetch.*

vetch bruchid. *Bruchus brachialis* Fahr., family Bruchidae, an insect which lays its eggs on the seeds of *Vicia* spp. The larva enters the seed and feeds on it. However, the infestation is not serious enough to materially affect production for forage.

vetchling. Grass peavine.

vetchworm. Corn earworm.

veteran. Any old tree; often an old tree which remains from a former stand.

veterinarian. An authorized practitioner of veterinary science *q.v.*

veterinary. Concerning veterinary science. (Often misused for *veterinarian.*)

veterinary inspector. A veterinarian employed by a municipal, county, state or federal agency, for public health work, meat or milk inspection, border and portal entry inspection, disease eradication, or one of many other professional duties.

veterinary medicine. Veterinary science.

veterinary practitioner. A duly licensed veterinarian actively engaged in diagnosing and treating diseases of domestic and pet animals.

veterinary science. A branch of knowledge, dealing mainly with domestic animals, that encompasses anatomy, physiology, breeding (including breed improvement), nutrition, animal diseases and treatment (including diseases transmissible to man), man's use of animals, etc. The first recognized veterinary school was established in 1761 in France; in 1791 England had such a school with others to follow; in North America several schools flourished after 1852, which, however, did not survive when standards were raised after World War 1, requiring a high school diploma and four years of college. To that was added two years of preprofessional training, bringing the total up to six years of study. At present schools of *veterinary science* in the U.S. and Canada are, for the most part, allied with large universities.

veterinary surgeon. A veterinarian who performs surgery. Broadly, any veterinary practitioner.

vetkousie. *Mesembryanthemum pomeridianum,* family Aizoaceae, an annual herb grown for its yellow flowers and as a potherb. Native to South Africa.

vetticoat. European cornsalad.

vetticost. European cornsalad.

vi(apple). Mombin.

viability. 1. Ability to live (immediately after birth or hatching). 2. The capacity of seeds to germinate. 3. The state of being alive.

viable. 1. Living. 2. Specifically in regard to organisms or agents of disease, able to cause infection in animals. 3. Capable of living. 4. Capable of germinating, as seeds.

vibrionic abortion. See *vibriosis.*

vibriosis. A contagious and infectious disease of cattle and sheep, usually spread venereally, characterized by abortion and infertility. The causative organism is *Vibrio fetus.*

viburnum. Any plant of the genus Viburnum, family Caprifoliaceae. Species are deciduous shrubs cultivated as ornamentals. See *American cranberrybush viburnum; blackhaw viburnum; European cranberrybush viburnum; hobblebush viburnum; Kentucky viburnum; Koreanspice viburnum; laurestinus viburnum; mapleleaf viburnum; nannyberry viburnum; snowball European viburnum; wayfaringtree viburnum; witherod viburnum.*

viburnum aphid. *Anuraphis viburniphila* (Patch), family Aphidae, an insect that infests the snowball European viburnum causing a downward curling of the leaves.

vice. Any seriously undesirable habit of an animal, such as vent picking and cannibalism in poultry; balking, kicking, shying, etc., in horses.

viceroy. *Limenitis archippus* (Cram.), family Oecophoridae, a butterfly whose larvae feed on the poplar and willow.

Victoria. A minor, medium-sized, breed of white hogs, which originated in the U.S.

Victoria blight of oats. See *Helminthosporium blight of oats.*

Victoria lily. Royal waterplatter.

Victorian bower spinach. New Zealand-spinach.

Victorian-box. Orangeberry pittosporum.

Victorian hazel. Virescent Anzacwood.

Victoria teatree. *Leptospermum laevigatum,* family Myrtaceae, an evergreen shrub grown in warm regions for reclamation of moving sands. Native to Australia. Also called *teatree; Australian teatree.*

vicuña. An expensive fabric made from the delicate wool of the wild South American ruminant *Lama vicugna.*

Vieth's ratio. The ratio between lactose, protein, and ash in milk (12.7: 9:1.2).
vigor. Animal or vegetative energy.
vilitis. A condition of a horse's hoof in which the soft part of the wall becomes inflamed.
village form of settlement. A form of agricultural settlement in which farmers reside in a center surrounded by their lands. *Cf. isolated form of settlement; line village form of settlement.*
villi. Small, finger-like processes of the inner lining of the small intestine and in the attachment of placenta to uterus.
Villiers. A square, soft cheese made in France.
vinaigrette. Vinegar sauce.
vinca. See *periwinkle.*
vine. 1. The grape. 2. Any woody or herbaceous plant which trails, climbs, or creeps as contrasted to those which stand without support. 3. A more or less prostrate vegetable garden plant, as potato *vine,* tomato *vine.*
vine bark louse. The grape scale.
vine beetle. Any of the several beetles of the genera Colaspis and Anomala, very harmful to grapevine leaves and branches. See *e.g., grape colaspis.* See also *grape leaf skeletonizer; grape rootworm.*
vine borer. A beetle whose larvae burrow in grapevine wood or pith.
vine bower. Italian clematis.
vine cactus. Ocotillo.
vine cuttings. Sections of a plant cut off to be planted for vegetative reproduction. See *slip: 4.*
vine fruit. Any fruit grown on woody vines: grapes, muscadines, etc., as contrasted to tree fruits, bush fruits, bramble fruits.
vinegar. A condiment made from various sugary and starchy materials with subsequent acetic fermentation, which must contain as least 4 percent acetic acid. See *apple cider vinegar; cider vinegar; distilled vinegar; grape vinegar; spirit vinegar; wine vinegar.*
vinegar bluecurls. *Trichostema lanceolatum,* family Labiatae, a woody herb which is a very important bee plant from California to Oregon, U.S. Native to North America. Also called *camphorweed.*
vinegar dried grains. Yeast dried grains.
vinegar eel. *Anguillula aceti,* an organism which frequently occurs in vinegar generators. Generators may become so badly infested that acetification is interfered with.
vinegar fly. Any fly of the family Drosophilidae, which breeds in fermenting fruits, or in crevices around a vinegar generator. It is a widespread pest of vinegar factories.
vinegar rot. Sweetpotato soft rot.
vinegar tree. Staghorn sumac.
vinegar weed. Vinegar bluecurls.
vine hawk moth. Any of several moths whose larvae feed on grapevine leaves. See *achemon sphinx.*
vine leaf folder. Grape leaf folder.
vineless. Designating varieties of the sweet potato which have short, upright branches. *Cf. viney.*
vinelilac. Any plant of the genus Hardenbergia, family Leguminosae. Species are grown as outdoor vines in warm areas and in a cool greenhouse. See *Compton vine lilac; singleleaf vinelilac.*
vine maple. *Acer circinatum,* family Aceraceae, a deciduous, small tree grown for its handsome, autumnal, red and orange foliage. Native to North America.
vine-mesquite. *Panicum obtusum,* family Gramineae, a perennial grass of the southwestern U.S. whose forage rating is fair. Native to North America.
vine mosaic of grapes. A virus disease of grapes characterized by the development of bright yellow leaves on the spring growth, some of which develop a little green coloring. Later leaves may be only slightly affected. Certain varieties fail to set fruit.
vine peach. Mango melon.
viner. 1. A machine used to shell canning peas which consists of a metal cylinder inside which are beaters attached to a central axis. While the cylinder revolves slowly, the beaters revolve rapidly hulling the peas. Also called *vining machine.* 2. A machine to harvest peas.
vine turner and shaker. An attachment for a potato digger *(q.v.)* that shakes the dirt off the harvested potatoes and drops them along with the vines in a neat row.
vine wilt. Sweetpotato wilt.
viney. Designating any plant species which has vine-like branches. *Cf. vineless.*
vineyard. A plantation, farm, or land on which grapes are grown.
vineyardist. A person who cultivates or grows grapes.
vineyard plow. A plow specially designed for use in vineyards. It has adjustable handles to allow the plow to work close to the vines without injuring them.
V-ing. The plowing of a furrow in a vineyard on the north side of a row so that trays of harvested grapes being sun-dried may be tilted to the south and receive the full rays of the sun.
vinhatico. Madeira Bay persea.
vinifera grape. European grape.
vining. The mechanical shelling of canning peas.
vining machine. Viner.
vining station. A covered, portable, or fixed shed in a production area where one or more viners *(q.v.)* operates. The use of the *vining station* eliminates the need for hauling the vines and pea pods to the cannery.
vin ordinaire. Low-priced, common, bulk and table wines of France.
vintage. 1. The wine production for one year. 2. Designating wines from a particularly good year.
viny. Having or producing many vines.
viola. 1. Any of several varieties of perennial flowering plants most of which are descendants of crosses between the horned violet, *Viola cornuta,* the European yellow violet, *V. lutea,* and possibly other species. They have flowers intermediate in size between those of pansy and violet, are valu-

able as edging plants, and are available in white, lavender, purple, yellow, apricot, and wine-red colored flowers. 2. See *violet.*

violet. Any one of the several species of the genus Viola, family Violaceae, mostly perennial herbs. This genus also includes the garden pansy. The florists *violets* are varieties of the sweet *violet (q.v.), V. odorata.* Also called *viola.* See *African violet; American dog violet; birdsfoot violet; bogbice violet; Canada violet; dame's-rocket; horned violet; palmate violet; sweet white violet.*

violet aphid. *Micromyzus violae* (Perg.), family Aphidae, an insect that infests violets both in and out-of-doors.

violet bloom. Matrimonyvine.

violet boltonia. *Boltonia latisquama,* family Compositae, a tall, perennial, aster-like herb grown for its blue-violet flowers. There is a dwarf *violet boltonia (B. latisquama* var. *nana)* sometimes called *dwarf pinkray boltonia.* Native to North America.

violet cress. Portugal diamondflower.

violet root rot. A disease of alfalfa, beets, and other crops caused by the imperfect fungus *Rhizoctonia crocorum,* whose rare perfect stage is *Helicobasidium purpureum,* family Auriculariaceae. It occurs in limited patches in the fields, under these crops, producing a purplish growth over the base of the stem and the roots.

violet sage. *Salvia nemorosa,* family Labiatae, a perennial undershrub grown for its purplish-violet flowers. Native to Europe and Asia.

violet sawfly. *Ametastegia pallipes* (Spin.), family Tenthredinidae, an insect whose larvae skeletonize the leaves of *Viola* spp. eating holes or cutting in from the margins of the leaves. Sometimes the entire infested plant is defoliated. Found in eastern U.S.

violet sorrel. Violet woodsorrel oxalis.

violet wheatgrass. Slender wheatgrass.

violet woodsorrel oxalis. *Oxalis violacea,* family Oxalidaceae, a perennial herb grown in the shade in the wild garden for its rose-purple, or occasionally pinkish-white, flowers. Native to North America. Also called *violet sorrel; purple wood sorrel.*

viosterol. Irradiated ergosterol; vitamin D_2.

vipersbugloss. *Echium vulgare,* family Boraginaceae, a poisonous, biennial herb common in the New England states and frequently west of the Mississippi River, U.S. Physical contact with the bristles on the leaves and stems produces a severe dermatitis accompanied by inflammation and itching. Native to Europe and Asia. Also called *blue thistle; blue devil; vipers bugloss; blueweed.*

viper's grass. Blacksalsify serpentroot.

viral. Having the nature of a virus; pertaining to a virus; like a virus.

viremia. The presence of a virus in the blood.

virescent Anzacwood. *Pomaderris apetala,* family Rhamnaceae, a small tree grown in warm regions for its greenish-white flowers. Native to Australia and New Zealand. Also called *tainui; Victorian hazel.*

virgate. 1. In botany, referring to a plant, (a) with many branches, or (b) rodlike in appearance. 2. An old English land measure of varying extent which was usually equal to 30 acres.

virgin. 1. Any female which has not had coitus, as a *virgin* queen bee. 2. In turpentining, the face or cut the first year a tree is bled. 3. Undisturbed, unplowed, uncut.

virginalis mockorange. *Philadelphus virginalis,* family Saxifragaceae, a polybrid, deciduous shrub grown as an ornamental and for forcing for its double, white flowers.

virgin dip. In turpentining, the oleoresin obtained the first year a tree is bled.

virgin forest. A mature or overmature forest which has grown entirely uninfluenced by human activity.

Virginia arrow arum. *Peltandra virginica,* family Araceae, a stemless, perennial, aquatic herb grown for its attractive foliage in the bog garden at the edges of pools. Native to North America. Also called *tawkee; tawkin; green arrow arum.*

Virginia bluebells. *Mertensia virginica,* family Boraginaceae, a perennial, shade-enduring herb grown in the wild garden for its blue flowers which are purple in bud. It is also used by florists for forcing in pots. Native to North America. Also called *Virginia cowslip; American lungwort; tree lungwort.*

Virginia box. A standard apple container 1.13 x 1.13 x 1.56 ft. which holds 1⅛ bushels. See *Michigan box; northwestern box.*

Virginia cedar. Eastern redcedar.

Virginia cowslip. Virginia bluebells.

Virginia crab. A variety of crabapple used as a hardy interstock.

Virginia creeper. *Parthenocissus quinquefolia,* family Vitaceae, a tall-growing, woody vine grown for its handsome foliage and for its habit of clinging to stone, brick, or wooden surfaces. Native to North America. Also called *American joy; woodbine; American ivy; false grape.*

Virginia date plum. Persimmon.

Virginia dayflower. *Commelina virginica,* family Commelinaceae, a perennial, troublesome weed pest found in the southern states of the U.S. Also called *turtle back.*

Virginia fence. The zigzag rail fence which consists of rails split from logs and crossed at the ends when laid to form the fence. Also called *worm fence; rail fence.*

Virginia goat's rue. Virginia tephrosia.

Virginia grape fern. Rattlesnake fern.

Virginia ham. a. A ham which has been chilled throughout and rubbed thoroughly with fine table salt. The hock end is packed with salt, and the entire ham is covered with a layer of salt. It is then left 7 weeks to cure in a cool, dry place. After curing, the ham is rubbed with a mixture of New Orleans molasses, brown sugar, black pepper, cayenne pepper, and saltpeter, and is left in a cool place for 2 weeks. Afterwards it is hung hock down in a meat house for 30 days. b. A ham which is rubbed on the face end with saltpeter. This is followed by a liberal rubbing with fine salt and curing for 3 to 5 days. After curing the ham is washed and rubbed with black pepper. It is then smoked intermittently for a month, peppered again, and aged 10 months in a bag. It is a famous delicacy in the U.S.

Virginia juniper. Eastern redcedar.

Virginia lionsheart. *Physostegia virginiana,* family Labiatae, a hardy, perennial herb grown for its clusters of purple-red, rose-pink, or lilac flowers. Native to North America. Also called *obedient plant; false dragonhead.*

Virginia lungwort. Virginia bluebells.

Virginia mountainmint. *Pycnanthemum virginianum,* family Labiatae, an herbaceous, aromatic perennial grown for its white or purplish flowers. Native to North America.

Virginia penthorum. *Penthorum sedoides,* family Saxifragaceae, a perennial herb which is a weed pest. Native to North America. Also called *ditch stonecrop.*

Virginia pepperweed. *Lepidium virginicum,* family Cruciferae, an annual herb which is a weed pest. When eaten by dairy cows, it causes an objectionable flavor in the milk.

Virginia pine. *Pinus virginiana,* family Pinaceae, a bushy, evergreen tree characteristic of barren and dry, sandy soils in eastern U.S. Native to North America. Also called *scrub pine; Jersey pine; spruce pine; poverty pine.*

Virginia poke. Pokeberry.

Virginia red cedar. Eastern redcedar.

Virginia sap-pine. Loblolly pine.

Virginia saxifrage. *Saxifraga virginiensis,* family Saxifragaceae, a small, perennial herb grown in dry, rocky, and partly shady places for its white flowers. Native to North America. Also called *early saxifrage; rock saxifrage.*

Virginia snakeroot. *Aristolochia serpentaria,* family Aristolochiaceae, a perennial herb grown for its rootstock used medicinally. Native to North America.

Virginia spiderwort. *Tradescantia virginiana,* family Commelinaceae, a perennial herb grown for its violet-purple or bluish flowers. Native to North America. Also called *spiderwort.*

Virginia springbeauty. *Claytonia virginica,* family Portulacaceae, a perennial herb grown in partial shade for its early-spring, white flowers sometimes tinged or streaked with pink. Native to North America. Also called *Mayflower; springbeauty; grass-flower; good-morning spring.*

Virginia stewartia. *Stewartia malacodendron,* family Theaceae, a tender shrub grown for its large, white or bluish flowers. Native to North America. Also called *silkie camellia.*

Virginia stock. *Malcomia maritima,* family Cruciferae, an annual, spreading herb grown for its reddish to white flowers. Native to the Mediterranean area.

Virginia strawberry. *Fragaria virginiana,* family Rosaceae, a perennial herb which is rarely cultivated but is prized in the wild for its small, red, very sweet, edible fruit. Native to North America. See *strawberry.*

Virginia sumac. Staghorn sumac.

Virginia sun-cured. A type of tobacco, grown in north-central Virginia, which is a recognized type in manufacturing and for export.

Virginia sweetspire. *Itea virginica,* family Saxifragaceae, a shrub grown for its showy, white flowers. Native to North America. Also called *chase shrub; Virginia willow; sweet spire.*

Virginia tea. Virginia sweetspire.

Virginia tephrosia. *Tephrosia virginiana,* family Leguminosae, a perennial herb grown for its showy, yellowish-purple flowers. Its root is a source of rotenone (*q.v.*). Native to North America. Also called *goat's-rue; catgut; wild sweet pea; rabbit's pea.*

Virginia thorn. Washington hawthorn.

Virginia trumpet flower. See *trumpetcreeper.*

Virginia wake robin. Virginia arrow arum.

Virginia waterleaf. *Hydrophyllum virginianum,* family Hydrophyllaceae, a perennial herb grown for its white or violet-purple flowers. Native to North America. Also called *waterleaf; Indian salad; Shawnee salad.*

Virginia wildrye. *Elymus virginicus,* family Gramineae, a tall, coarse, perennial grass, which is abundant east of the 100th meridian in the U.S., where it is used mainly to stabilize soil. It is also a fairly palatable forage plant. Native to North America. Also called *Terrel grass.*

Virginia willow. Virginia sweetspire.

virgin land development. 1. Generally, the reclamation of any land area hitherto not under cultivation for crop production: *e.g.,* marsh lands. 2. In the U.S.S.R., a project, initiated by Central Committee Decree, March 2, 1954, to increase agricultural production by putting idle (virgin) lands under cultivation. Specified were land areas in Kazakhstan, Siberia, the Urals, the Volga, and some parts of the north Caucasus.

virgin oil. The oil obtained by the first pressing of olives, considered superior in flavor and quality to subsequent pressings.

virgin queen. An unmated queen bee.

virginsbower. *Clematis virginiana,* family Ranunculaceae, a hardy, climbing or sprawling, woody vine grown for its habit and white flowers. Native to North America. Also called *woodbine; lovevine; old-man's beard.*

virgin sod. Natural sod (*q.v.*), especially that of the prairies of midwestern and western U.S., in contrast to that produced by grasses seeded by man.

virgin soil. Soil in its natural state as distinguished from soil or land which has been plowed or otherwise altered by man for cultivated crops or other uses.

virgin timber. Virgin forest.

virgin wool. Wool which has been used in the manufacture of fabrics for the first time.

virogenic. Designating a disease caused by a virus.

virologist. A person skilled in the science of virology.

virology. The study of viruses and virus diseases.

virosis. A disease caused by a virus.

virostatic. Possessing the ability to arrest multiplication of a virus.

virucide. A chemical or physical agent that kills or inactivates viruses; a disinfectant.

virucidal. Possessing the ability completely to inactivate a virus.

virulence. The disease-producing power of an organism.

virulent. Highly pathogenic; having great disease-producing capacity; deadly; very poisonous or harmful.

virulent culture. A culture or growth of a microorganism (bacteria or virus) which is alive and not attenuated or modified, and which is capable of multiplication and producing a reaction or possible disease in susceptible hosts.

virulent latent. A form of latent microorganism which differs from the common latent in its increased virulence and in the greater production of lesions.

viruliferous. Virus-carrying; particularly designating insect vectors.

virus (plural **viruses**). A self-reproducing agent that is considerably smaller than a bacterium and can multiply only within the living cells of a suitable host. Most *viruses* are too small to be seen even with the aid of the ordinary microscope but can be photographed with the aid of the electron microscope. *Viruses* usually are considered to be living agents or microorganisms but some have characteristics of non-living matter. They are protein-containing bodies of high molecular weight capable of multiplying and acting like living organisms when in living tissue. They are the cause of many animal and plant diseases, such as smallpox, measles, tobacco mosaic, etc. Recovery from some viral diseases confers lasting immunity.

virus antagonism. See *virus interference.*

virus competition. See *virus interference.*

virus indexing. The determination of the presence or absence of viruses in symptomless growing plants, dormant plants, or seeds by: (a) inoculation to virus-sensitive plants, by rubbing with expressed juice, insect feeding techniques, or by some form of grafting; (b) growing divisions or portions of the symptomless plants, such as buds, sprouts, or samples of seed, under conditions favorable for symptom expression. See *tuber indexing.*

virus indicator plant. A species or plant variety that reacts visibly to specific virus diseases and that may be used to demonstrate the presence or absence of such viruses in less sensitive plants. See *virus indexing.*

virus interaction. The action of a virus in altering the normal development of other viruses or virus strains which is expressed by partial or complete suppression, by synergistic association, by modification of the type of symptoms in the host plant, or by abnormal increase in concentration of one virus.

virus interference. The action of one virus, strain, or isolate which causes partial or complete suppression of other viruses, strains, or isolates.

viscera. The internal organs of the body, particularly in the chest and abdominal cavities, such as the heart, lungs, liver, intestines, and kidneys.

visceral lymphomatosis. One type of the avian leukosis complex *(q.v.)* characterized by the presence of tumorous masses which involve some of the viscera. Diarrhea is frequently present but often the fowl has no outward symptoms.

viscogen. A thickening agent for cream prepared from lime (CaO), water and sucrose which increases the whipping property of cream.

viscolising. 1. The treatment of milk or cream by special high pressure equipment to break up the fat globules and make the product homogeneous. 2. A questionable procedure of enhancing the depth of cream layer on milk by partial homogenization of a portion of the milk and remixing.

viscometer. An instrument used for measuring the viscosity of liquids either by noting the time it takes drops to traverse an inclined glass or porcelain slope or by observing the time taken by a measured quantity to run out of a standard orifice into another vessel.

viscosity. The resistance of a fluid to flow; thickness of a fluid.

visibility meter. An instrument used for measuring the dependable range of distance at which a standard smoke column can be detected by the eye under existing conditions of haze. Also called *haze meter.*

visible supply. In U.S. grain marketing, a statistical compilation of the amounts of the different kinds of grain in elevators east of the Rocky Mountains which receive and ship grain without converting it into manufactured products. It indicates the amount of grain that is potentially in merchandising channels.

vitamin. An organic substance which performs specific and necessary functions in relatively small concentrations in an organism. Required for normal growth and maintenance, *vitamins* are not utilized as building units for the structure of the organism and do not furnish energy, but are essential for the transformation of energy and for the regulation of the metabolism of the structural units.

vitamin A. Antixerophthalmic factor. It is mainly concerned with the maintenance of healthy epithelial structures.

vitamin A & D feeding oil. A vitamin supplement for feedstuffs, such as fish or fish-liver oil or a blend of two or more of the following: Vitamin A and/or D concentrate, synthetic vitamin D fish oil, fish oil, marine animal oil, or edible vegetable oil. The vitamin potence should be stated in A. O. A. C. chick units of vitamin D and U. S. P. units of vitamin A per pound (or units per gram).

vitamin A feeding oil. A livestock feedstuff vitamin supplement which is either fish or fish-liver oil or a blend of two or more of the following: vitamin A concentrate, fish-liver oil, fish oil, marine animal oil, or edible vegetable oil. The vitamin potency should be stated in U. S. P. units of vitamin A per pound (or units per gram).

vitamin B. The original name given to the crude preparations that relieved beriberi and which were later found to contain many factors. Also called *water-soluble B.* See *vitamin B complex.*

vitamin B complex. Factors isolated from yeasts, liver, and other sources of the original "water-soluble B" which includes the following vitamins: (a) **Biotin**, a factor which cures alopecia (baldness) and "spectacle eye" in rats fed a biotin-deficient diet supplemented by raw egg whites. Also called *anti-egg white injury factor; vitamin H.* (b) **Cholin,** a factor required for certain metabolism processes. (c) **Folic acid,** a factor required for growth and blood formation. Also called *vitamin M; vitamin B_{10}; vitamin B_{11}.* (d) **Inositol,** a factor which prevents alopecia (baldness) and "spectacle eye" in mice and other animals. (e) **Niacin,** a factor which prevents pellagra. Also called *anti-black-tongue factor; nicotinamide; niacinamide; nicotinic acid amide; PP factors.* (f) **P-aminobenzoic acid,** a factor which prevents gray hair in rats and promotes growth in chicks. Also called *PABA.* (g) **Pantothenic acid,** a factor which prevents dermatitis in chicks. Also called *chick antidermatitis factor; antigray hair factor.* (h) **Pyridoxine,** a factor which prevents a specific dermatitis in young rats and in man. Also called *vitamin B_6; vitamin Y.* (i) **Riboflavin,** a component of Warbow's "yellow enzyme" which is essential for cellular oxidation and reduction. Also called *vitamin G.* (j) **Thiamin,** an antiberiberi factor. Also called *aneurine; antineuritic factor; vitamin B_1; vitamin F.* (k) **Vitamin B_{12},** an antipernicious anemia factor. Also called *cobalamin; erythrocyte maturation factor.*

vitamin C. Ascorbic acid, a factor which is essential in the prevention of scurvy. Also called *antiscorbutic factor; avitamic acid; cevitamic acid.*

vitamin D. The antirachitic vitamin. It is essential in the metabolism of calcium. Several forms of vitamin D are recognized as D_2 found in plant products and D_3 the animal form found in fish oils and in the blood after irradiation.

vitamin deficiency. Any disease which is caused by the lack of one or more vitamins in the diet: *e.g.*, scurvy.

vitamin D feeding oil. A vitamin, livestock-feeding supplement which is either fish or fish-liver oil or a blend of two or more of the following: vitamin D concentrate, synthetic vitamin D, fish-liver oil, marine animal oil, or edible vegetable oil. The vitamin potency should be stated in A. O. A. C. chick units of vitamin D per pound (or units per gram).

vitamin D milk. Milk whose vitamin D content has been increased above that of normal milk by special processes in accordance with public health regulation by feeding cows irradiated yeast, by irradiation of milk or by addition of vitamin D concentrate to the milk. Also called *vitamin-fortified milk.*

vitamin E. A factor which is important in the stabilization of vitamin A. It prevents muscular dystrophy in certain animals and is a sterility factor in experimental animals. Also called *antisterility factor; the tocopherols.* See *stiff-lamb disease.*

vitamin G. See *vitamin B complex: (i).*

vitamin H. See *vitamin B complex: (a).*

vitaminized milk. See *vitamin D milk.*

vitamin K. Blood clotting factor. It is essential for the formation of prothrombin. Also called *koagulations vitamin* (Danish name); *antihemorrhagic factor.*

vitamins. See *vitamin.*

vitelline membrane. A membrane enclosing the yolk of an egg.

viticulture. The art and science of growing grapes.

vitreous. Resembling glass in hardness or brittleness.

vitrified tile. Clay or shale tile which has been so burned as to produce a glassy or vitreous product.

vitriol. Sulfuric acid or some of its compounds. Blue *vitriol* is hydrous copper sulfate; green *vitriol,* copperas; red *vitriol* is either a sulfate of cobalt or a ferric sulfate; white *vitriol* is hydrated zinc sulfate.

vitriolated tartar. Potassium sulfate.

vivarium. A glass box resembling an aquarium which is used to keep or raise animals or plants. See *terrarium.*

vivary. See *vivarium.*

vives. Strangles *q.v.*

vivianite. A hydrous ferrous phosphate, a mineral, and a number of oxidation forms of it, which is found in soils especially those in swamps and bogs. Also called *blue ochre.*

viviparous. 1. Giving birth to living young. *Cf. oviparous.* 2. Producing organs of reproduction while yet attached to the parent plant.

Vize. A hard, Greek, ewes' milk cheese similar to Romano.

vly. A swamp or marsh; or a creek (a localism of Dutch origin, New York, U.S.).

vocational agricultural education. In the U.S., a program of systematic instruction which is given to persons who have entered upon, or are preparing to enter upon, the work of the farm. It usually refers to a public-school program for youth in high school and for adults and other out-of-school persons engaged in farming. Also called *agricultural education; vocational agriculture.*

voetganger. *Locustana pardalina,* family Locustidae, a South African species of locust, the wingless larvae of which gather together in vast armies, swarming out and laying waste an entire countryside.

void. To evacuate feces and/or urine.

Void. A soft, cows' milk, French cheese which is similar to Limburger.

void space. The space between kernels in a bulk of grain which is usually expressed as percent of total volume.

volatile. Easily evaporated.

volatilization. Vaporization; the conversion of a liquid or solid into vapors.

volcanic ash. The finely comminuted ejecta of volcanic eruptions which in thick deposits, as the parent material of soil, occupy only an extremely small surface area. The ash or dust carried by winds is widely distributed and incorporated in soils.

volcanic soil. A soil which is derived from or consists of volcanic rocks especially ash and other fragmentary ejecta.

vole. *Microtus* spp., a small rodent resembling a mouse which has a blunt nose, short tail and ears, and a stout body. It often does much damage to crops. Also called *meadow mouse; field mouse.*

Volga wildrye. See *mammoth wildrye.*

voltine. The annual brood-producing frequency of silkworms. A monovoltine silkworm produces a single brood in a season; a polyvoltine (multivoltine) silkworm produces more than one.

volume regulation. A method of controlling the annual or periodic cut from a forest on the basis of growing stock and increment, using formulae and disregarding area.

volume table. A tabular statement which shows, for a given species of trees, the average log content of trees of different sizes. The usual *volume table* is based on diameter and height.

volumetric standard. A container used to calibrate and test the bulk milk tank and other volumetric measuring devices. The containers are usually checked and sealed by weights and measures agencies to deliver a precise volume of liquid. Two types in common use are the standard with a graduated neck, and the slicer-plate type.

volunteer. 1. Any plant which grows from self-sown seed. 2. Designating a plant which grows from self-sown seed.

volunteer seeding. See *self-seeding.*

volute. The water chamber in a centrifugal pump which has a spiral casing.

vomit. 1. Material which is ejected from the stomach through the mouth or nostrils. 2. To eject matter from the stomach through the mouth or nostrils; to throw up.

Vorarlberg Sour Milk. A hard, cows' milk cheese made from whole or skim milk which is ripened in a cool cellar. It is greasy in texture and has a strong odor.

V-shaped comb. The comb of a chicken which consists of two well-defined, horn-shaped combs, as in the Houdan.

vulture-hock. Stiff quill feathers growing on the thighs and extending backward behind the hocks of certain fowls. The effect resembles the feathering of a vulture. A disqualification with most breeds of chickens but a requirement for a few sorts, such as the Sultans and the Mille Fleur Bantams.

vulva. The external genital (reproductive) organs of the female: the opening to the vagina. In insects and other invertebrates, the external opening of the oviduct.

W

waddy. See *cowboy.*

wadi. A desert water course which contains water only immediately after a heavy rain (Sahara and Arabian countries).

wafer ash. See *Baldwin hoptree; hoptree.*

waffle steak. A retail cut of beef usually from a tough piece that is tenderized by macerating in a waffling machine.

wage(s). The remuneration for labor paid in money or in other valuable considerations.

wagon. See *farm wagon.*

wagon bed. The main body of a wagon, not including the wheels; the part of a wagon in which the load is placed. Also called *wagon box.*

wagon box. Wagon bed.

wagon drag. A skid, shoe, or drag used to check the speed of a wagon on a downgrade.

wagoner. 1. One who has charge of a wagon. 2. A teamster in the business of transportation by wagons.

wagon gear. See *farm truck: 1.*

wagon hammer. The pin which attaches the doubletree to the tongue of a wagon and serves as a wheel wrench. Also called *evener pin.*

wagon horse. In marketing, an active, undersized draft horse which stands 15.2 to 16.2 hands high and weighs from 1,350 to 1,650 lbs. It must trot well.

wagon load. An indefinite amount of produce, hay, manure, etc., which is sufficient to fill a wagon rack or box.

wagon pole. See *tongue: 3.*

wagon road. Prior to the advent of the automobile, a road sufficiently improved to permit use for wagon transportation.

wagon seat. The wooden seat, with or without springs, which is used on a wagon.

wagon yard. An old fashioned hostelry which consisted of a hotel of sorts, and a stable for horses, at which a person could spend the night and have his animals cared for (obsolete).

wahoo. Eastern wahoo.

waif. A stray animal, specifically a lost sheep.

wain. A carriage for the transportation of goods on wheels; a wagon (obsolete).

wainable. Designating that which may be plowed or manured or is tillable.

wainscot oak. European Turkey oak.

wainwright. One who does all the wood work in wagon making. Also called *wagonwright.*

wainy lumber. See *waney.*

waist carrier. A device consisting of a strap to be tied around a person's waist and one or two baskets to which the strap is attached which is used to carry produce, such as raspberries, during the harvesting.

waiver. A relinquishment of a claim; *e.g.*, a landlord may waive his lien on a tenant's crop in favor of a lender who wants to make the tenant a loan.

wakerobin. 1. Trillium. 2. Cuckoopint.

Walachian sheep. *Ovis strepsiceros*, a remarkable, domestic breed of sheep with extremely long, erect, spiral horns. Native to southeastern Europe and western Asia. Also called *Zackelschaf.*

walk. A slow, flat-footed, four-beat, horse's gait which is one of the most useful gaits whether in harness or under saddle, if executed with snap and animation.

walk-in cooler. A room-size refrigerator.

walking. Designating any farm implement whose use requires the operator to walk behind or along side: *e.g.*, the *walking* cultivator.

walking a farm. The procedure followed in the preliminary stages of appraising the value of a farm in which the appraiser walks over all parts of the farm to observe and record the physical features including land, drainage, crops, and woodlot, which affect the value of the farm.

walking class. In stock shows, a Tennessee Walking Horse (*q.v.*) that has a good running walk, a good flat-footed walk, and a slow, easy canter.

walking dredge. A type of dry-land dipper dredge (*q.v.*) mounted on several large sledges which are moved forward individually as the machine moves forward (obsolete).

walkingfern. *Camptosorus rhizophyllus*, family Polypodiaceae, a hardy fern, the long, pointed fronds of which root at the tip, grown in partial shade in the wild garden or in a fernery. Native to North America. Also called *walking leaf.*

walking horse. 1. Any horse trained to do the running walk, foxtrot, and canter. See *plantation gait.* 2. Tennessee Walking Horse.

walking horse class. See *walking class.*

walking leaf. Walkingfern.

walkingstick. *Diapheromera femorata* (Say), family Phasmatidae, a long, thin insect resembling a twig, found from New England to the Rocky Mountains, U.S., which feeds on a variety of woody plants. Also called *northern walkingstick.*

walk in lanes. Wedge-style pruning.

walk-in storage unit. See *walk-in cooler.*

walk-through. Designating a building or milking room so designed that the traffic moves in one door or gate and out another, rather than in and out the same door.

walk-trot-canter horse. Three-gaited horse.

walkway. A levee which is constructed through a marsh range to permit cattle to travel into and across the land, for improving accessibility, and for more uniform use of the range.

walleye. See *amaurosis.*

wall fern. See *polypody.*

wallflower. 1. Any herb of the genus Cheiranthus, family Cruciferae. 2. *C. cheiri*, a perennial herb grown in the flower garden or rock garden for its fragrant, yellow or orange-brown flowers. Native to the Himalayas.

wall gillyflower. See *wallflower.*

wall grass. Goldmoss stonecrop.

wall hopper. A hopper which is built or set against a wall.

wall moss. Goldmoss stonecrop.

wallow. 1. A mud hole in which hogs lie. 2. To roll, lie, move lazily in a mud hole, as a hog.

wall pepper. Goldmoss stonecrop.

wall rockcress. *Arabis albida*, family Cruciferae, a perennial herb grown in rock gardens for its fragrant, white flowers. Native to the Caucasus.

wall tree. An espaliered tree.

walnut. Any tree of the genus Juglans, family Juglandaceae. Species are tall, deciduous, valuable nut and timber trees. See *butternut: 1; eastern black walnut; Persian walnut.*

walnut anthracnose. A destructive disease of walnuts very similar to blotch of Persian walnuts *q.v.*

walnut aphid. *Chromaphis juglandicola* (Kltb.), family Aphidae, an aphid that infests the Persian walnut feeding on the underside of the leaves. A sooty mold grows on the aphid's honeydew.

walnut blight. A blight of walnut caused by *Pseudomonas juglandis*, family Pseudomonadaceae.

walnut blister mite. *Aceria erinea* (Nal.), family Eriophyidae, a mite which infests walnuts causing yellow or brown felt-like galls on the undersides of leaves.

walnut blotch. Walnut anthracnose.

walnut caterpillar. *Datana integerrima* G. & R., family Notodontidae, a moth whose larvae feed in colonies on walnut, hickory, and pecans in the U.S. from Maine to Florida and west to Kansas.

walnut comb. A type of chicken comb resulting from a cross of a rose comb with a pea comb variety of chickens, which is a low comb with ridges suggesting the exterior of a walnut shell.

walnut datana. Walnut caterpillar.

walnut husk fly. *Rhagoletis completa* Cress., family Trypetidae (Tephritidae), a fly infesting walnuts which lays eggs on the husks. The larvae tunnel in the husk resulting in staining of the shell and reducing of the value. Found in central and western U.S.

walnut scale. *Aspidiotus juglans-regiae* Comst., family Diaspididae, a round scale insect found on walnut, apple, ash, cherry, poplar, currant, dogwood, elm, grape, sweetgum, holly, hockberry, linden, locust, maple, oak, peach, pear, plum, prune, and rose.

walnut-shell. An organic flour prepared from walnut shells which is used as a dust carrier in the application of poison dusts.

waltzing matilda. An Australian term, of unknown origin, used to designate the wanderings of the sundowner and the swagman (*q.q.v.*), the matilda being the bedroll or swag carried by these men. Also called *walking matilda.*

wandering bee. See *vagabond bee.*

wandering dune. An unstablized sand dune; one that is being shifted about by wind action.

wandering Jew. *Tradescantia fluminensis*, family Commelinaceae, a prostrate, perennial herb grown in a hanging basket as a house plant for its habit and its white flowers. Native to Brazil. See *spiderwort.*

wandering Jew zebrina. *Zebrina pendula,* family Commelinaceae, a trailing herb grown in the greenhouse for its foliage and reddish-purple flowers. Native to Mexico.

wandering milkweed. Spreading dogbane.

wandflower. Any of the perennial, herbaceous plants of the genus Sparaxis, family Iridaceae. Species are grown for ornament. See *big wandflower.*

waney. Designating lumber which has bark or a lack of wood on a square-edged piece.

wanigan. A floating cookhouse used by lumbermen when driving logs down a stream in the spring log drive.

wapatoo. See *arrowhead.*

warble fly. Any of the various flies of the family Hypodermatidae, as the ox warble fly, whose larvae live beneath the integument of cattle and certain other mammals and thus lower the quality and value of the meat and hide. See *common cattle grub; northern cattle grub.*

war bridle. Any of several types of halters, hackamores, or bridles with a part of the rope passing through the horse's mouth in such a way that a powerful leverage is placed on the horse's jaw or head. It is used on vicious or unbroken horses.

Wardian case. See *terrarium.*

warehouse receipt. A negotiable certificate issued by a warehouse, listing goods or produce stored there by an individual or firm, which must be surrendered to receive delivery of the goods.

warehouse truck. A manual, two-wheeled truck which consists of a frame of two stout pieces of wood, braced with metal crosspieces, having wheels at one end and handles at the other. It is used to transport heavy objects for short distances.

warm-blooded animal. Any animal, such as a bird or mammal, whose body temperature is warmer than its surrounding medium, as contrasted with the cold-blooded animals, such as reptiles.

warm-blooded stallion. A male horse free from draft horse blood: *e.g.,* an Arabian, Thoroughbred, etc.

warm-climate. Designating a tender (*q.v.*) plant.

warm greenhouse. See *warm house.*

warm house. A house (*q.v.*) of a range of greenhouses which is maintained at a day temperature of about 75° and night temperature of about 60°. Rose, poinsettia, gardenia, and many other plants that do well in these temperatures are propagated in a *warm house.* Also called *warm greenhouse.* See *cool house.*

warming room. A room in or attached to a sheep barn which can be heated for caring for early lambs.

warm season. Designating a plant which thrives best when the temperature is regularly quite high: *e.g.,* okra.

warm shower. A rain whose sensible temperature is warm to the touch.

warm test. A germination test in which seed is put on a germinating medium and then placed in a temperature conducive to rapid germination.

warm up. 1. To feed feeder cattle on concentrates for a short time so that they show the effects of feed but are not finished off. 2. To operate the engine of a tractor for a short while before placing it under load to bring the engine up to a satisfactory operating temperature.

warm weather. See *warm season.*

warp. 1. Yarn which runs the long way in a woven fabric. 2. In lumber, any variation of a board from a plane surface often caused by a too rapid loss of moisture in the curing process. 3. To vary from a plane surface, as a board.

warping. Fertilizing land by the addition of alluvial muds, as by flooding with river water or tidewater carrying a large amount of sediment in suspension.

warp wool. See *warp: 1.*

warrant. In land allotments, a certificate authorizing ownership of public tracts of land to the holder. Also called *land warrants.* In some presettlement land surveys in Pennsylvania, U.S., the land was subdivided into blocks of 1,000 acres called *warrants.*

warranty deed. A legal document which passes the title to real estate from person to person and warrants the title against encumbrances of any nature.

warren. An area, place, or enclosure in which rabbits are bred and raised. Also called *rabbit warren.*

warren hoe. A garden hoe with a heart-shaped blade, especially used for making furrows.

warrior grasshopper. Clear winged grasshopper.

wart. 1. A tumor on the skin or mucous membrane of an animal composed of fibrous tissue covered over with epithelial cells similar to those of the part of the body on which they are located. For the most part, they are painless and do not interfere with the function of an animal. See *hard wart; soft wart.* 2. An unorganized proliferation of plant cells with an appearance like an animal wart. 3. One of the effects of cucumber mosaic *q.v.* 4. See *potato wart.*

wart of peach. A virus disease of peaches characterized by the production of warty fruit, or sometimes by dwarfed, misshapen, or cracked fruit with a gummy exudate.

warty barberry. *Berberis verruculosa,* family Berberidaceae, a low, evergreen shrub grown as an ornamental for its foliage, large, golden-yellow flowers and bluish-black fruit. Native to China.

warty disease. See *potato wart.*

warty philodendron. *Philodendron verrucosum,* family Araceae, a tropical plant grown in the greenhouse for its pale green, salmon-lined leaves. Native to tropical America.

Warwickshire. A hard, sweet-cured, cows' whole milk, English cheese, aged three or four months. It is quite similar to Derby.

wash. 1. In the arid regions, especially of western U.S., the channel of an occasional stream in the alluvium in the bottom of a canyon or in that filling a basin. It may be enclosed by vertical banks or may be wide and flat in cross section. 2. A ravine or gully on arid slopes, especially in western U.S. 3. The colluvial or alluvial deposit, at the base of a slope, resulting from erosion. 4. A shallow part of a river or arm of the sea. 5. A fermented wort (*q.v.*) from which spirit is distilled. 6. To cleanse by using water and/or other cleaning agents. 7. To move material, such as sand, soil, etc., as water *washes* soil.

washboard. Designating a corrugated surface, as on a road or a piece of lumber.

Washed-curd. A semisoft to slightly firm, cows' milk cheese similar to Cheddar except that the curd is washed with water before it is salted. Also called *Soaked-curd.*

washed wool. Wool that has been washed on the sheep's back in cold water (almost obsolete).

washer. Any of many various mechanical devices to cleanse produce, such as potatoes, with water and brushes or by causing a turbulence in the water in which the produce is placed and to which a cleansing agent may be added.

washing. 1. The act of cleansing something with water to which often a soap or other cleaning agent is added. 2. Soil movement by erosion.

washing powder. Any of numerous dry, pulverized detergents or dry, ground soaps.

Washington cedar. Giant arborvitae.

Washington hawthorn. *Crataegus phaenopyrum*, family Rosaceae, an armored, dense, round-headed, deciduous tree grown as an ornamental for its white flowers and its brightly colored, autumnal fruit, and as an almost impenetrable hedge plant. Native to North America.

Washington lily. *Lilium washingtonianum*, family Liliaceae, a bulbous herb, 4 to 6 ft. high, grown for its white or purple-spotted flowers. Native to western North America.

Washington lupine. *Lupinus polyphyllus*, family Leguminosae, a hardy, perennial, herbaceous, poisonous plant which with its many varieties is grown for its white, yellow, or blue flowers. Found wild on the West Coast, U.S., it is especially toxic to sheep. Symptoms of poisoning are nervousness, labored breathing, convulsions, and frothing at the mouth.

Washington plant. Carolina fanwort.

Washington('s) bower. Matrimonyvine.

Washington thorn. Washington hawthorn.

wash land. Any land which is periodically overflowed.

Washoe canary. A burro, or donkey (applied by miners in western U.S.). Also called *Rocky Mountain canary.*

washout. 1. The erosion of a portion of a levee, railroad bed, road, etc., by water. 2. Designating a channel made by erosion in the deposit of a certain sediment and filled by fresh material. 3. The flushing out of a pipe, etc.

wash plain. See *alluvial fan.*

washy coupled. Of, or pertaining to a horse that is loose or long coupled.

waste. 1. Any wood, lumber, or other material, which may have merchantable value but which is not utilized. 2. Remnants of cloth, hulls, etc., used in cleaning, wiping, etc. 3. Any part of produce not utilized by the main industry concerned with the product, such as the waste from a retail cut of meat, slabs at a lumber mill, cinders at a furnace, etc. 4. See *wasty wool.*

waste bank. The ridge of excavated earth parallel to a drainage ditch. Also called *spoil bank.*

waste gate. A water gate in a dam or reservoir for the discharge of excess water.

wasteland. Land essentially incapable at present of producing anything of value (not used to describe an idle farm or forest land).

waste lime. Any industrial waste or by-product containing calcium or calcium and magnesium in forms that will neutralize acids. It may be designated by a prefix suggesting the name of the industry or process by which it is produced, as gas-house *lime;* tanner's *lime,* etc. Also called *by-product lime.* See *lime: 1.*

wasteway. The channel used to convey water discharged into it from a spillway; an escape or sluice; a spillway.

waste weir. A weir (*q.v.*) used for the escape of excess water. See *waste gate.*

wasting ears. Harvested, immature, field corn used as a vegetable. Also called *green corn; roasting ears.*

wasty. 1. In stock or meat judging, designating an animal with a large amount of fat in proportion to the lean meat, or one with a large middle indicating a low dressing percentage. 2. Designating wool or lint cotton in which there is much undesirable or unsalable fiber.

wasty wool. Wool with fibers short, weak, and tangled in the lock.

watch tower. In forestry, a tower usually constructed on a hill top or prominence for a fire warden or ranger to observe the surrounding area and keep a lookout for fires.

water. H_2O, hydrogen oxide, although the liquid may contain associated molecules, H_4O_2, H_6O_3, etc. The melting and freezing point is 0°C. (32°F.) and the boiling point 100°C. (212°F.). The most valuable natural resource and the most limiting factor in crop production. See *natural water.*

water apple. Bullocksheart custardapple.

water application efficiency. The percentage of water applied that can be accounted for as moisture increase in the soil occupied by the rooting system of the crop.

water application rate. The rate in inches per hour that irrigation water is applied to fields.

water arum. Wild calla.

water bath. A cleansing, soaking, heating, etc., by means of water in a tub, vat, boiler, etc.

water bean. American lotus.

water beech. See *American hornbeam; American planetree.*

water-belly of steers. Urinary calculi.

water birch. Any of various North American birches *(Betula fontinalis; B. nigra; B. piperi),* family Betulaceae, found in moist or wet areas. See *river birch.*

water blister. A collection of watery fluid beneath the epidermis (skin or mucous membrane) which usually results from excessive pressure or from irritants.

water bloom poisoning. A poisoning of livestock resulting from drinking pond or stream water which is contaminated with certain blue-green algae. This poisoning is rare but very lethal.

water breaker. See *breaker: 1.*

water brush. Eastern baccharis.

water buffalo. *Bos bubalus,* family Boridae, a large animal, 10 ft. long including a 2 ft. tail and up to 6 ft. high at the shoulder, found wild in the jungles from India to the Philippines at elevations from sea level to 6,000 ft. The coat color varies from a dark gray to black. The animal has long horns, curving back over the body, which sometimes measure as much as 12 ft. along the curve. Widely domesticated, the *water buffalo* is raised as a beast of burden, for its rich milk, and its hide from which a good quality leather is manufactured. Also called *carabao* (Philippines); *gamoose, jemoose* (Egypt).

water buttercup. Marshmarigold.

water cabbage. American waterlily.

water calamint. Field mint.

water canna. Powdered thalia.

water carrier. A worker bee which gathers water and transports it to the hive.

water chamber. Volute.

waterchestnut. *Trapa natans,* family Onagraceae, an aquatic herb grown in a pond or pool for its beautiful foliage and edible fruit. Native to Europe and Asia. Also called *Chinese water chestnut; waternut.*

water chinquapin (chinkapin). American lotus.

water closet. A small room or compartment provided with a device for defecation and relief. See *privy.*

water conditioner. Any of several devices or materials which change the composition of water to a desirable state, as a water softener, an acidifier, etc.

water conservation. The saving or recovery of water for beneficial uses.

water content. 1. The water of the soil or habitat; (physiological) the available water supply; (physical) the total amount of soil water. 2. The amount of water in a material in relation to dry weight.

watercore. A wetwood *(q.v.)* condition in certain western conifers.

water core of apples. A physiological disorder of apples caused by unfavorable water relations, characterized by glassy or watery areas in the flesh or pulp near the core or extending out toward the surface.

watercourse. 1. A channel, usually natural, through which a stream of water flows and which may be dry during unusual drought periods. 2. Legally, a natural stream flowing in a well-defined channel (the stream still retains its character as a *watercourse* where it floods its banks). Water flowing underground in a known and well defined channel also constitutes a *watercourse* and is governed by law applicable to surface streams.

watercress. *Nasturtium officinale,* family Cruciferae, a hardy, perennial, aquatic herb grown as a salad plant for its edible leaves. Native to Europe.

watercress leaf beetle. *Phaedon aeruginosus* Suffr., family Chrysomelidae, a beetle which infests watercress.

watercress sowbug. *Lirceus brachyurus* (Harger), family Asellidae, an insect that infests watercress.

water crop. Pendulous crop.

water culture. Hydroponics.

water cushion. A pool of water maintained to break the impact of water outflowing from a dam, chute, or other spillway structure.

water daisy. Marshmarigold.

water-dilution system. An obsolete system of cream separation in which cold water was added directly to whole milk on the theory that the milk would become less viscid and thus facilitate the rising of the fat globules.

water disposal pattern. The water routing system of a given area, as a system of grassed waterways or cultivated land for the removal of concentrated flow of surface runoff.

water dock. Goldenclub.

water elevator. 1. Any mechanism used for raising water, especially to a considerable height above ground, as a windmill operating a pump. 2. A vertical or inclined conveyor whose buckets dip up water from a source of supply and deliver it into a conduit or elevated receptacle, thus fulfilling the functions of a pump. Under the name of chain pumps, they are much used in smaller sizes.

water elm. American elm.

waterer. Any of numerous devices, such as watering cans, water fountains, etc., used to furnish water to plants or animals.

water erosion. Erosion *(q.v.)* by water. See *splash erosion; sheet erosion; rill erosion; gully erosion.*

water fallow. The maintenance of a rice field under a cover of water for 1 or 2 years during which a crop of fish may be produced. It is practiced in rice growing in Arkansas, U.S.

water farm. 1. The cultivation of fish in farm ponds. 2. See *hydroponics.*

water feather. Chile parrotsfeather.

water finder. Dowser.

water fountain. 1. Any of several devices which automatically furnish drinking water to animals or birds. 2. A decorative, artificially produced, jet of water usually issuing from a piece of sculpture in or surrounded by a pool.

waterfront. Land, with or without buildings, which fronts on a body of water.

waterfront zone. A strip of land of varying width along a lake, pond, stream, ocean, or other water body, on which the scenic values are safeguarded from impairment by excluding or restricting the utilization of natural resources, commercial development, and occupancy.

water furrow. A shallow trench or depression between two raised soil beds in which irrigation water or surface drainage water can flow.

watergall. Eastern baccharis.

water gauge. 1. A staff gauge used to indicate the water level in an open channel above a specified datum. 2. A pressure gauge used to indicate the pressure in a closed water system. 3. Any device used to indicate the water level in an open channel.

water gang. A constructed channel used for drainage or irrigation.

water gap. 1. A gorge cut through a ridge by a stream. 2. The gap below a fence where it crosses a small stream.

water garden. A pool used in landscaping to grow aquatic plants.

water garden plant. A plant such as the waterlily and lotus, grown in a pool. See *aquatics: 1.*

water gate. See *waste gate.*

water glass. A solution of sodium or potassium silicate made by fusing quartz sand with potash or sodium hydrate. It is a heavy liquid used as an adhesive, protective coating, and fire-proofing agent.

water goggles. Marshmarigold.

water gowan. Marshmarigold.

water-grass. Barnyardgrass.

water ground. 1. Designating corn which has been soaked in warm water and passed through crushing mills so regulated that the hulls, endosperms, and embryos are mechanically separated without damage. 2. Designating corn meal that is ground at a water grist mill in contrast to that ground at a mill.

water gum. Water tupelo.

waterhawthorn. Any plant of the genus Aponogeton, family Aponogetonaceae. Species are aquatic plants for greenhouse pools or aquaria. See *cape waterhawthorn; lace waterhawthorn.*

water head. 1. The source of a stream. 2. A pond created to furnish irrigation for a small plot, or to furnish limited power for a mill. 3. The pressure created by the depth of water in such a pond.

waterhemlock. Any plant of the genus Cicuta, family Umbelliferae. Species are among the most poisonous of plants. See *Bolander waterhemlock; bulb waterhemlock; California waterhemlock; Douglas waterhemlock; spotted waterhemlock; tuber waterhemlock; western waterhemlock.*

water hickory. *Carya aquatica,* family Juglandaceae, a deciduous tree grown for its valuable wood used in tool handles, sporting goods, ladders, furniture, implements, woodenware, and for fuel and smoking meat. Native to North America. Also called *bitter pecan; swamp hickory.*

waterhole. 1. A natural depression in which water collects or stands, often in the dry bed of a stream; a spring in the desert. 2. A natural spring or pool on the open range where cattle drink (western U.S.).

water holly. Oregongrape.

waterhyacinth. 1. Any plant of the genus Eichhornia, family Pontederiaceae. Species are aquatic herbs grown in ponds. See *sawpetal waterhyacinth.* 2. *E. crassipes,* an aquatic herb grown for its blue and yellow spotted, violet flowers. Escaped, it can be a very serious pest. Native to tropical America.

water in. To thoroughly water a transplant.

watering. 1. Furnishing water for the consumption of plants or animals. 2. Designating a device for furnishing water to plants or animals. See *sprinkling can.*

watering cup. An automatic device consisting of a cup and a spring-operated faucet connected to a water system, so constructed that as an animal nudges the spring plate in the cup a valve is opened and a slow stream of drinking water delivered.

watering place. A place on a farm or ranch at which there is a natural or artificial source of drinking water for livestock.

watering trough. A fixed or portable, open, elongated container or conduit, which holds or carries drinking water for livestock and poultry.

water injury. Any of various types of injury to plants which results from an excessive amount of water in or on the soil.

water intake. 1. The ingestion of water. 2. A point or place in a pond, reservoir, etc., where the water flows in. 3. The point where water flows into a structure.

water jacket. A metal outercovering for a kettle, etc., which is filled with heated water. See *steam jacket.*

water jump. A trench or ditch which serves as a fence or boundary for a field. Also called *ha-ha.*

waterlemon. *Passiflora laurifolia,* family Passifloraceae, a climbing vine grown in frost-free areas for its edible fruit. Native to the tropics. Also called *yellow granadilla; Jamaica honeysuckle; passion fruit; pomme liane.*

water level. 1. The surface of water which is used in measurement of surface or stream water depth. 2. The free water level in the soil. 3. See *water table.*

waterlily. 1. Any of several aquatic plants of the family Nymphaeaceae, which includes the genera Brasenia, Euryale, Nelumbium, Nuphar, Nymphaea, and Victoria. 2. Any plant of the genus Nymphaea. See *American waterlily; cape waterlily; European white waterlily; yellow Mexican waterlily.*

waterlily aphid. *Rhopalosiphum nymphaeae* (L.), family Aphidae, an aphid which infests waterlily, waterplantain, cattail, almond, apricot, and peach.

waterlily leaf beetle. *Galerucella nymphaeae* (L.), family Chrysomelidae, a beetle which feeds on the upper surfaces of the leaves of aquatic plants.

waterlily leaf cutter. *Nymphula obliteralis* (Wlk.), family Pyraustidae, the aquatic larva of a moth which prepares a case made of part of a leaf fastened together with silk. It infests the foliage of aquatic plants causing the leaves to become ragged and rotten.

water line. See *water table.*

water lobelia. *Lobelia dortmanna,* family Lobeliaceae, a poisonous, perennial herb. Ingestion results in nausea, exhaustion, prostration, dilation of the pupils, stupor, convulsions, and death. Native to Europe.

waterlocust. *Gleditsia aquatica,* family Leguminosae, an armored, deciduous, tender tree grown for its feathery leaves and greenish flowers. Native to North America. Also called *swamp locust.*

waterlogged. 1. Designating lands where the ground water stands at a level that is detrimental to plants. It may result from over-irrigation, flooding, seepage, or lack of drainage. 2. Designating a product saturated with water.

waterlogging. A soil condition resulting from water saturation which may be permanent or for considerable periods of time. *Waterlogging* may occur naturally, as in poorly-drained flat lands or in swamps, or may be man-made, as from excessive use of water in irrigation. *Cf. rise of alkali.*

water maize. Royal waterplatter.

water mallow. Rose mallow.

waterman. A person officially designated to supervise or actually control the release of irrigation water from ditches, mains, etc., usually for irrigation in irrigation districts. Also called *watermaster.* See *ditch rider.*

water maple. See *red maple; silver maple.*

water mark. 1. Any mark, such as stain from suspended sediment, which indicates the level to which surface water has risen. 2. A water-soaked spot on citrus fruit resulting from a moderate freezing temperature.

watermark disease. A parasitic disease of willow, caused by *Bacterium salicis,* characterized by a wilting of the leaves in spring or early summer, the wilted leaves remaining on the branches for some time, and by a colorless exudate which turns brown upon long exposure to air.

watermaster. Waterman.

watermelon. *Citrullus vulgaris,* family Cucurbitaceae, a tropical, tender, annual, herbaceous vine of almost innumerable varieties, grown as an important cash crop for its very popular fruit, the *watermelon* of commerce. Native to South Africa. Also called *anguria.* See *citronmelon; stock melon.*

watermelon stem-end rot. A rot induced by the fungus *Diplodia natalensis,* family Sphaerioidaceae. It progresses rapidly, affecting the fruit at all stages of development. The first indication is a browning and shriveling of the stem followed by a decay of the fruit, starting at the stem end. The disease is responsible for a high percentage of the watermelons spoiled in transit.

watermelon wilt. A widespread, fungous disease of watermelons caused by *Fusarium oxysporum f. niveum,* family Tuberculariaceae, characterized by yellowing, wilting, and dieback.

water milfoil. Canada parrotsfeather.

water nut. 1. *Eleocharis tuberosa,* family Cyperaceae, a rush-like, aquatic plant grown for its edible corms rich in sugar and proteins. Native to eastern Asia. Also called *Chinese water chestnut.* 2. Waterchestnut.

water nymph. American waterlily.

water oak. *Quercus nigra,* family Fagaceae, a deciduous tree which is easily transplanted and grown as a street tree. Native to North America.

water of hydration. 1. Moisture lost and never regained by wool dried at a temperature high enough to turn it brown. 2. Synthetically prepared water.

water out. To furnish sufficient drinking water at one time to animals or fowls.

water-packed cotton. 1. Cotton in a bale into which water has penetrated during the baling process, causing damage to the fiber. 2. A bale of cotton apparently dry on the exterior but, through exposure to the weather or through other means, damaged by water in its interior.

water pan. A shallow container used to hold drinking water for fowls.

water parsnip. Hemlock waterparsnip.

water pimpernell. Scarlet pimpernell.

water pine. Chinacypress.

water place. See *watering place.*

water plant. See *aquatic.*

water plantain. American waterplantain.

waterplatter. Any plant of the genus Victoria, family Nymphaeaceae. See *royal waterplatter.*

water poplar. Southern poplar.

waterpoppy. *Hydrocleis nymphoides,* family Butomaceae, a tender, aquatic herb grown in an aquarium, greenhouse pool, or in an out-of-doors pool in very warm regions for its light yellow flowers. Native to Brazil.

waterproof. To apply a covering or coating which will shed water and reduce water absorption to a minimum.

water raking. A method of harvesting cranberries. A shallow flood is put on the bog and the berries raked off as they float on or near the surface of the water.

water repellent. Designating a cloth or product which, due to an application of certain substances, sheds most of the water that falls on it.

water requirement. The quantity of water, regardless of its source, required by a crop in a given period of time for its normal growth under field conditions. It includes surface evaporation and other economically unavoidable wastes. Usually it is expressed as depth (volume per unit area) for a given time. See *consumptive use; evapotranspiration; irrigation requirement.*

water reserve. Land set aside by the government as an area for maintaining stream flow (Australia).

water resources. All surface and underground water available in a given area or drainage basin.

water retained. Field capacity.

water retting. See *ret.*

water rice. Annual wildrice.

water right. A legal right to use the water of a natural stream, furnished by a canal for general or for specific purposes. It may entitle a person to use the canal to full capacity, to a measured extent, or for a definite period of time, and to change the place of diversion, storage, or use of water as long as it does not infringe upon the rights of other people. In some states, in the U.S., the rights to water may be sold and transferred separately from the land on which it has been used. See *riparian doctrine.*

water rot. An after-effect of water spot *(q.v.)* in which blue and green molds invade the water spot on a citrus fruit if wet weather persists. If dry weather quickly follows, the injured tissue dries to a brown, sunken area.

water runoff. See *runoff.*

water rush. See *rush.*

water seedbox. *Ludwigia natans,* family Onagraceae, an evergreen, aquatic or marsh herb grown in the aquarium or greenhouse pool for its foliage and yellow flowers. Native to South America. Also called *false loosestrife.*

watershed. The total runoff from a region which supplies the water of a river, or lake; a catchment area, or drainage basin. (Strictly the *watershed* is the drainage divide.)

watershed management. The administration and regulation of the aggregate resources of a drainage basin for the production of water and the control of erosion, streamflow, and floods.

watershield. See *Carolina fanwort; Schreber watershield.*

water shoot. See *water sprout.*

water sick. Designating land which has become unproductive because of various residual salts from irrigation water.

water skegs. Yellowflag iris.

water-slaked lime. Hydrated lime.

water snake. a. Any of the snakes of the genus Natrix, especially *N. sipedon* of North America, and other allied genera, inhabiting fresh water and feeding on aquatic animals like fish. Severe in appearance and temperament, their sting, though painful, is not generally poisonous. A poisonous species, moccasin *water snake* or water moccasin, *Agkistrodon piscivorous,* family Crotalidae, with a hollow depression between the eyes and nostril, is found in southern U.S. b. A second group of *water snakes,* family Homalopsinae, characterized by a tapered tail, back teeth, and valve-like nostrils on the snout, are found in Australia and Polynesia.

water snowflake. India floatingheart.

water-soaked. 1. Designating lesions on plant parts which appear to have a clear liquid within the lesion. 2. Designating any product which has an excess of moisture.

waters of the United States. All waters within the U.S. which are navigable for the purpose of commerce or whose navigation successfully aids commerce.

water-soluble. Designating a substance which is soluble in water.

water spot. A nonparasitic breakdown of the rind of citrus fruits probably caused by imbibition of external water by the white portion of the rind. It produces a raised, blister-like area on the surface of the rind. Minute cracks rupture the cuticle and several layers of underlying cells so that the water spreads. The spots appear at old wounds, at the navel, and at the collar portion around the old button of the fruit.

water spreading. 1. The artificial application of water to lands for the purpose of storing it in the ground for subsequent withdrawal by pumps for crops. 2. Irrigation by surplus waters out of cropping season. 3. The diversion of runoff water from gullies or water courses and its distribution on adjacent, gently sloping, grazing lands needing additional water. The volume of water flowing down the channels is reduced and the moisture absorbed by the spreading area increases the growth of vegetation.

watersprout. A vigorous, long shoot which grows from the trunk, crown, or a large branch of a tree. Also called *sucker; water shoot.*

water spruce. See *black spruce; Colorado spruce; Canada hemlock.*

water stain. A yellowish to blackish surface discoloration of light-colored woods which results from dripping water during seasoning.

water stress. See *wilting.*

water suffocation. The killing of a plant by an excessive amount of soil water. See *waterlogging.*

water supply. Stored water either in tanks, ponds, or underground reservoirs which is available as needed.

water swivel. In well-boring, a combined universal joint and hoist coupling which form the connection between the water supply pipe and the drill rods, permitting complete rotation of the tools.

water system. The well, the pump and its parts including the pipe in the well, the storage tank, and the distributing pipes.

water table. The upper boundary of the zone of saturation except where it is bounded by an impervious stratum. A surface where the hydrostatic pressure is equal to the atmospheric pressure, as determined by the water level in open wells that barely penetrate the zone of saturation *q.v.*

water tank. An elevated water reservoir.

water-tight grain. Wood that is sound and tight-knotted.

water-tolerant. Designating a plant which will continue to thrive in soil in which there is considerable moisture.

water trough. A long, shallow container from which horses, cattle, and other kinds of livestock drink water.

water tupelo. *Nyssa aquatica,* family Cornaceae, a tender, swamp tree grown for its handsome, autumnal foliage. Native to North America. Also called *tupelo gum; bay poplar; cotton gum; sour gum.*

water user. A person who irrigates land for crop production (western U.S.).

waterway. A natural or artificially constructed course for the concentrated flow of water.

waterweed. See *Canada waterweed.*

water weevil. Rice water weevil.

waterwillow. *Decodon verticillatus,* family Lythraceae, a woody herb grown in the bog garden for its pinkish-purple flowers. Native to North America. Also called *swamp loosestrife; wild oleander.*

water witching. To use a divining rod *(q.v.)* for locating water. See *dowser.*

watery. 1. In meats, designating a carcass that is soft and underfinished with an apparent high moisture content. 2. See *pasty.*

water year. A special grouping of the periods of the year to facilitate water supply studies. The U.S. Geological Survey uses October 1 to September 30 as the *water year.* See *climate year.*

watery egg. A hen's egg with its albumen less viscid than normal.

water yield. 1. The total outflow of a drainage basin through either surface channels and/or subsurface aquifers. 2. The surplus of precipitation over infiltration and canopy interception.

watery rot. Watery soft rot.

watery soft rot. A storage and transit disease of practically all vegetable crops caused by *Sclerotinia sclerotiorum,* family Sclerotiniaceae, characterized by a soft, watery breakdown, lack of odor and a cottony mold. Also called *drop; melters; leak.*

watery white. See *watery egg.*

Watson magnolia. *Magnolia watsoni,* family Magnoliaceae, a deciduous, small tree or shrub grown for its large, sweet-scented, white, crimson-centered flowers. Native to Japan.

wattle. 1. The fleshy processes of the skin pendent from the throat or the head of a bird or reptile. 2. The flap of skin which hangs from the throat of a goat or some swine (England). 3. Material made up of interwoven twigs bound together with mud or plaster. 4. Loose flesh pendent from the mouth of some fishes. 5. Certain species of the genus Acacia, family Leguminosae (Mimosaceae), especially *A. penninervis, A. pycnantha,* and *A. saligna,* which are raised for their bark used in tanning. The bark, stripped in the spring, yields up to 45 percent more tannin than the other tanbarks. Native to Australia. Also called *mimosa.*

wattle edema. Edema of the wattles.

wattle test. Any of several tests, for various bird diseases, in which an injection is made into a bird's wattles. Swelling of the wattle is usually indicative that the bird has the disease tested for.

Waukegan creeping juniper. *Juniperus horizontalis* var. *douglasi,* family Cupressaceae, a hardy, trailing, prostrate, evergreen shrub grown as foundation planting. Its blue-gray leaves turn grayish-purple in the fall. Native to North America. Also called *Waukegan juniper.*

wave action erosion. Erosion on shore lines of oceans and lakes from the action of waves.

wave wash. See *wave action erosion.*

wavyleaf plantain-lily. *Hosta undulata,* family Liliaceae, a perennial herb grown for its funnel-shaped, pale lavender flowers. Native to Asia.

wax. 1. Any substance similar to beeswax *(q.v.)* in composition and use, as carnauba wax, candelilla wax, ouricury wax, bayberry wax, Japan wax, sugarcane wax, ucuuba wax, flax wax, cotton wax, esparto wax, cauassu wax, murumuru wax, palm wax, spermaceti, lanolin, Chinese insect wax, shellac wax, ceresin wax, montan wax, and paraffin. 2. To apply *wax* to the surface of a fruit, vegetable, foliage, or flower for preservation.

wax bean. A type of either the kidney bean, *Phaseolus vulgaris,* or the bush kidney bean, *P. vulgaris* var. *humilis,* family Leguminosae, important vegetable plants grown for their golden-yellow seed pods harvested, before the seeds mature, for immediate consumption or processing. Also called *butter bean.* See *string bean.*

wax berry. See *snowberry: 2.*

wax breaker. An automatic picking machine fitted with rubber fingers on a revolving drum to knock off solidified wax previously applied on the carcasses of birds. The wax carries the feathers with it and so removes the feathers from the birds.

wax-dipping tank. A tank of liquid wax into which killed birds are dipped to facilitate the removal of feathers.

wax dolls. Drug fumitory.

wax flower. Madagascar stephanotis.

waxflower pyrola. *Pyrola elliptica,* family Pyrolaceae, a hardy, low-growing, evergreen herb grown in the wild garden for its olive-green foliage and fragrant, waxy, greenish-white flowers. Native to North America. Also called *lesser wintergreen.*

wax gland. A wax-secreting gland of the worker bee.

wax gourd. Any plant of the genus Benincasa, family Cucurbitaceae. See *Chinese waxgourd.*

wax miller. Greater wax moth.

wax moth. Greater wax moth.

wax myrtle. 1. Any of several plants of the genus Myrica, family Myricaceae, which yields fragrant and relatively dripless wax for making candles. Also called *tallow bayberry; tallow shrub; wax shrub.* See *bayberry; northern bayberry; southern waxmyrtle; sweet gale.*

wax-picking. A method of picking poultry. The bird carcasses are coated with hot, melted wax. When cold, the wax is removed, the feathers and pin feathers being carried with it.

wax pink. Portulacca.

waxplant. *Hoya carnosa,* family Asclepiadaceae, a fleshy, tropical vine grown in the greenhouse and out-of-doors in frost-free areas for its fragrant, white, pink-eyed, waxy flowers. Native to China and Australia.

wax-pluck. To remove the feathers from a bird by wax-picking *q.v.*

wax-plucking. Wax-picking.

wax press. An appliance used to extract beeswax from the honeycomb.

wax scale. 1. The separate pieces of wax secreted by the worker bee. 2. Florida wax scale.

wax shrub. Wax myrtle.

waxwork. American bittersweet.

wax worm. Greater wax moth.

waxy body. Designating a desirable smoothness in the body and texture of a cheese, as in one which responds to pressure of thumb and fingers like cold butter.

waxy corn. An agricultural variety of corn having a waxy kernel and grown as a substitute for tapioca.

waxy maize. See *waxy corn.*

wayfaringtree viburnum. *Viburnum lantana,* family Caprifoliaceae, a hardy, deciduous, treelike shrub grown for its white flowers, and ornamental red fruit turning black.

way-going crop. The crop which is taken from the land the year a tenant leaves a farm.

way thistle. Canada thistle.

waythorn. Buckthorn.

way wort. Scarlet pimpernel.

weak. 1. Designating a defect of cultured cream, butter, and cheese characterized by lack of viscosity or strength in body and texture. 2. Designating a market condition in which the prices tend to go downward.

weak and wasty cotton. In marketing, (of little or no standardization but reflecting a buyer's opinion) a poor grade of cotton.

weak and watery. Designating an egg white that is thin and generally lacking in viscosity.

weak cotton. See *weak and wasty cotton.*

weak flour. Flour prepared from soft wheat.

weak neck of sorghum. A nonpathogenic disorder of sorghum characterized by the culm breaking below the head and poorly developed heads with light-weight seed.

weak, watery body. A body defect of ice cream characterized by a low melting point, low viscosity, and a small proportion of solids. Weak ice cream is compressed easily.

wean. 1. To make a young animal cease to depend on its mother's milk. 2. To accustom partly grown birds to do without artificial heat.

weaned. Designating a young animal which has stopped or has been stopped from suckling its mother.

weaner. 1. Any of many devices used to prevent a young animal's suckling its mother. 2. A weaned lamb, 5 to 12 months old, before it grows its permanent teeth (Australia). 3. Any lamb from the time it is weaned until it is sheared. 4. A weaned range calf.

weanling. An animal that has recently been weaned or taken from its mother.

wear(ing) plate. Metal plates which support the rear side of the knife on the cutter bar of a mower or binder.

wear out. 1. To deplete farm land by exploitative cropping or other improper practices. 2. To destroy through use, as to *wear out* a combine.

weather. 1. The atmospheric conditions prevailing at any specified time, or those prevailing during any particular period, as shown by meteorological observations and records of air temperature, barometric pressure, wind velocity, humidity, clouds, and precipitation. 2. Bad weather. 3. To be subjected to the influences of atmospheric conditions, as the abrasive action of rainfall, the disintegration due to frost, the deteriorating influences of oxygen and other gases contained in the atmosphere, etc.

weatherbird. An untrained person who is observant of weather conditions and who often forecasts changes in weather accurately.

weather board. See *siding: 1.*

weathered. 1. Designating the physical and chemical disintegration or decomposition of rocks under natural conditions. 2. Designating a product which has been exposed to a variety of weather conditions and has changed as a result.

weather hazard. A chance that bad weather will occur at an inappropriate time.

weathering. 1. The action of all natural forces on a product left exposed to the weather. 2. The deterioration of the unprotected surface layer of wood which has been exposed to the weather. 3. Atmospheric action on rock surfaces producing decomposition, disintegration, or alteration of rocks at or close to the earth's surface. *Weathering* implies no removal of material other than that in solution in drainage waters.

weatherman. A meteorologist, publicly or privately employed, who forecasts weather.

weather plant. Jequirity rosarypea.

weatherproof. Constructed or protected so that exposure to the weather will not interfere with successful operation, storage, preservation, etc.

weather stain. A grayish-brown surface discoloration of lumber or other wood products, resulting from exposure to the weather.

weather vane. Any of several devices used to show the direction from which the wind is blowing. It is often a decoration for a barn or house roof.

weave. In horses, to move the body, neck and head, from side to side of the stall in a restless manner. See *jig.*

weaver. A horse that sways backward and forward in its stall.

weaversbroom. *Spartium junceum,* family Leguminosae, an almost leafless shrub grown for its abundance of yellow blossoms. Native to the Mediterranean area.

weaving system. A system of training trailing bramble fruit canes on a two-wire vertical trellis in which 2 or 3 canes are taken together and passed from upper to lower wires and back again so as to form a circle or wreath.

web. 1. The blade of a saw. 2. That portion of an ordinary anvil, between the head and the base, which is of reduced size. 3. The cotton disk used in filtering milk. 4. The fat which surrounds the paunch and intestines of sheep. 5. The disk of a rolling colter. 6. The membrane which unites the fingers and toes, especially of amphibians and water birds. 7. To tie plant parts together with silk or a *web,* as an insect. 8. See *web of feather.* 9. See *cylinder hay loader.*

Webber's brown fungus. *Aegerita webberi* Faw., a fungus used in biological control of the citrus whitefly, the cloudy-winged whitefly, the woolly whitefly, and the citrus blackfly.

web fat. See *web: 4.*

web-footed. Designating any of certain water fowls whose toes are joined together by a membrane, as a duck.

web halter. A halter made from heavy cloth similar to canvas.
web hay loader. Cylinder hay loader.
web of feather. The flat part of a fowl's feather which consists of a series of barbs on both sides of the shaft.
web of feet. See *web-footed.*
web up. To enclose or tie together plant parts with silk or a web, as does the bagworm.
web worm. See *greater wax moth; bagworm.*
wedding flight. Mating flight.
wedge. A piece of hard metal or wood, tapering from the butt to the front end, which can be driven or forced into a narrow opening. It is used to split logs, wood, etc., and to hold a tree off the saw in felling. A *wedge* may also be used to tighten a loose connection or to reinforce a weak joint in wood or metal.
wedge bone sirloin. A retail cut of meat for broiling, taken from the rump end of the loin.
wedgeleaf ceanothus. Buckbrush ceanothus.
wedge-style pruning. A little-used system of pruning bearing apple trees in which four channels or wedges extending toward the center of the tree are pruned out. It facilitates spraying and harvesting. (*Wedge-style pruning* differs from the four-leaf clover system of pruning in that in the latter the channels or wedges are established when the tree is young.) Also called *clover-leaf style pruning; star pruning; walk in lanes.*
weed. 1. A plant out of place. Thus, rye growing in a field of wheat is a *weed.* 2. More popularly, an herbaceous plant which takes possession of fallow fields or finds its way into lawns or planted fields, crowding the vegetation planted there and robbing it of moisture and nutrients. 3. A plant whose usefulness is not recognized or which is undesirable because of odor, spines, prickles, or poisonous characteristics. 4. A tree of inferior value in a forest, or one growing in a street or lawn, where it is not wanted, like the seedlings of the boxelder. 5. Excessive vegetative growth. 6. Any plant that harbors insects, fungi, or viruses that may spread to nearby crop plants. 7. A horse or other domestic animal that is undersized or a misfit. 8. To remove *weeds* from a lawn, field, or other places where they are not wanted, usually manually but also by machinery. 9. Designating a flavor of dairy products, especially milk, which resembles that of certain vegetation (onion, wild garlic, leeks, etc.).
weed and brush bar. A specially constructed cutter bar on a mowing machine, with stub guards and extra-heavy knives for cutting heavy weeds and underbrush.
weed attachment. A wheel placed at the outer end of the cutter bar to carry it some 6 to 12 ins. off the ground so that the weeds may be cut without undue strain upon the mower parts.
weed bar. Weed and brush bar.
weed boom. A boom on a spray rig designed for the application of herbicides. See *boom sprayer.*
weed competition. The competition between weeds and desirable plants. See *competitive plant.*
weed control. Any device or system used for the destruction, checking of growth and spreading of weeds by cultivation, by the use of herbicides, cultural practices, etc.
weeder. 1. A light spring-tooth harrow (*q.v.*) which has long, flexible teeth. 2. A hand tool consisting of a short handle and several curved prongs, used to remove weeds from flower beds. 3. Any machine used to go over an entire field, regardless of rows, in the early stages of development of both crops and weeds. Specifically a light, spring-tined machine which cultivates lightly over the entire field, regardless of rows, when weeds are small and easily dislodged without undue damage to crop plants. 4. A rotary hoe.
weeder hoe. See *sickle: 1.*
weeder tooth. A long, flexible harrow tooth.
weed free. 1. Designating a standing crop, pasture, lawn, etc., in which there are relatively few weeds. 2. Designating crop seeds from which all weed seeds have been separated in the cleaning process.
weed hook. A rod attached to a plow beam, extending out to the front and side of the plow bottom, to bend weeds over in such a fashion that they will be completely buried in the furrow.
weediness. The state or condition of a field, flower bed, lawn, etc., in which there is an abundance of weeds.
weeding. The removal of undesirable plants from a field, flower bed, lawn, etc.
weeding hoe. A special tooth for a harrow used for weeding.
weed killer. See *herbicide.*
weed knife. A curved tooth for a harrow, used to cut weeds below the soil surface. Also called *knife harrow.*
weed patch. An area of indeterminate size on which weeds are growing profusely.
weed spray. See *herbicide.*
weed sprayer. See *sprayer.*
weed tree. Any undesirable species of tree which interferes with the development of crop trees, or one which grows where it is not wanted.
weedy. 1. Designating an off-flavor of milk or its products which suggests the odor or taste of certain weeds. 2. Designating an abundance of weeds in a field, flower bed, lawn, etc. 3. Designating growth of high vigor of a plant, usually associated with fruiting. 4. Designating a thin or undesirable animal.
weep. 1. To turn sirupy, as a jelly sometimes does. 2. To exude sap, as a cut or broken grapevine. 3. To exude moisture from a plant surface.
weephole. A hole through an abutment or retaining wall to provide for the passage of accumulated seepage.
weeping. Designating a tree with pendent branches: *e.g.,* Babylon *weeping* willow.
weeping bulrush. *Scirpus cernuus,* family Cyperaceae, a perennial, marsh herb grown as a pot plant for its drooping habit. Native to Europe.
weeping eczema. An eczema (*q.v.*) that has a serous discharge.

weeping forsythia. *Forsythia suspensa,* family Oleaceae, a deciduous shrub, with long, arching limbs, grown for its profuse, early-spring, yellow blossoms. Native to China. Also called *weeping golden bell.*

weeping golden bell. Weeping forsythia.

weeping lantana. Trailing lantana.

weeping lovegrass. *Eragrostis curvula,* family Gramineae, a perennial grass suitable for seeding semidesert lowlands. Native to North America.

weeping willow. See *Babylon weeping willow; Thurlow weeping willow; Wisconsin weeping willow.*

weevil. Any of the several beetles of the family Curculionidae, which has mouth parts at the end of a more or less prolonged snout. They are injurious plant or seed eaters.

weevily. Infested with weevils.

weichselwood. The wood of the mahaleb cherry *q.v.*

weigela. Any of the several plants of the genus Weigela, family Caprifoliaceae. Species are hardy shrubs grown for their late spring blossoms. See *crimson weigela; old fashioned weigela.*

weigh-a-day-a-month DHIA plan. A plan for determining milk production of dairy cows in which the owner weighs the milk from each cow on one day a month. A herd summary is calculated in a central computing office at the end of the year and returned to the owner. This is the least expensive of the three systems of records under the DHIA system. See *Dairy Herd Improvement Association.*

weigh beam. The beam of a scale on which the sliding poise operates and on which are graduations to show weight.

weighbridge. The platform of a scale, flush with the road or street, on which vehicles can drive to be weighed; or a smaller platform used for weighing sacks of grain or cans of cream.

weigher. 1. A mechanism on a grain thresher, combine, etc., which automatically weighs a quantity of grain and then automatically records bushels and dumps the grain into a spout for bagging, or into a truck box. 2. One who weighs a crop (produce), usually for a buyer, as a wheat *weigher.*

weighing attachment. See *weigher: 1.*

weighmaster. An appointed, public official whose duty is to weigh and certify net weight of produce.

weight. 1. A piece of metal attached to the end of a calf's horns to cause the horn to grow in a desirable, downward-curved pattern. 2. To attach *weights* to a calf's horns.

weight box. A shallow pan attached to a disk harrow, on which heavy stones, etc., can be placed so that the disks will penetrate the soil to a greater depth than otherwise.

weight method. A method of arriving at the volume of forage or herbage on a range area by clipping and weighing samples or by estimating weight.

weigh-ticket. The ticket issued by a weighmaster which shows the gross, tare, and net weight.

weight pan. Weight box.

weight with standard butterfat. A basis for purchasing milk to be processed and sold as market milk in which there is a differential for milk with fat over or below standard.

Weil's disease. Leptospirosis.

weir. A notch of regular form through which water flows. See *measuring weir.*

weir basin. Approach channel.

weir head. The depth of water flowing over the weir crest as measured at a point at least 2.5 and preferably 4 times the depth of water flowing over the weir, upstream from the weir. The zero point of the measuring scale should be placed at the exact level of the weir crest.

weir notch. The opening in a weir for the passage of water.

Weisslacker. A soft, ripened, cows' milk cheese made in Bavaria.

well. 1. A hole in the earth down to a supply of water. See *artesian well; drilled well; driven well; dug well; horizontal well; radial well.* 2. A spring or natural flow of water.

well-bred. Designating a plant or animal which has a good pedigree.

well-broken horse. A horse which is well trained to the saddle or harness.

well casing. The wooden, stone, brick, metal, concrete, etc., lining of a well.

well-centered yolk. A yolk that occupies the center of the egg.

well cress. See *watercress.*

well curbing. The platform which is a part of the raised, enclosing frame for a well.

well-cured. Designating a desired product in cheese ripening.

well-developed. 1. Designating a characteristic, desirable, and pronounced flavor of a cheese. 2. Designating a satisfactory condition of animal or plant development in whole or in part.

well developing. The pumping, backwashing, and surging of a new well to pump out the fine sand, silt, and clay so as to achieve maximum flow through the remaining coarser material.

well-drained. Designating a soil (in humid regions) that normally contains only capillary moisture, in contrast to a swampy or waterlogged condition.

well-dressed. 1. Designating a quality and condition for dressed poultry which does not show any dressing defect or slight dressing defect (*q.q.v.*). It may have one broken wing if the break is above the wing tip, if the bone does not protrude, and if there is no appreciable bruise or blood clot. 2. Designating a cheese which has a properly fitted bandage.

well driller. A man who bores or drills wells.

well drilling. The making of a drilled well *q.v.*

well-fatted. Designating a fattened, finished bird or animal.

well-finished. Designating an animal or bird that has a deep covering of natural fleshing intermingled and covered with a layer of firm fat.

well-fleshed. 1. Designating a quality and condition for dressed poultry in which the bird has a well-developed breast, with some fat covering over thighs and back, and with a good covering of flesh over all bones. 2. Designating a well-muscled animal carrying a limited amount of fat or finish.

well grass. See *watercress.*

well house. A small house, shed, or the like which is built over a well.

well-kept. Designating a farm, farmstead, etc., which is maintained in an excellent condition.

well point. A perforated pipe covered with a sand screen which is sunk into sand to permit the pumping of ground water and the exclusion of sand.

well sinker. A person who bores or drills wells.

well-sprung. Designating an animal with a good spring of ribs.

well water. Water which is drawn or pumped from a well, in contrast to ditch water, creek water, rain water, spring water, etc.

Welsh. A breed of horses of the saddle-pony type, originating in Wales, which stands about 12 hands high and weighs somewhat less than 500 lbs. The most usual colors are chestnut, bay, gray, and black.

Welsh bean. See *kidney bean.*

Welsh Black. An English breed of dual-purpose cattle. They are black and weigh up to 1350 lbs.

Welsh onion. Chibol.

Welsh Pony. See *Welsh.*

Welshpoppy. *Meconopsis cambrica,* family Papaveraceae, a perennial herb grown for its large, yellow flowers. Native to western Europe. Also called *English poppy; Welchpoppy.*

wen. An encysted skin tumor.

Wensleydale. A cured, medium-hard, blue-veined, cylindrical, 10 to 12 lb. cows' whole milk cheese made in England.

Werder. A semisoft, German cheese, made from cows' whole or partly skimmed milk, ripened by bacteria that produce a red tint on the surface. Also called *Elbinger.*

Wessex Saddleback swine. An English breed of swine, characterized by a black ground color, with a white belt extending from the back over the shoulders to the ventral side of the body; similar to the American Hampshire swine.

West. 1. That portion of the U.S. lying west of the 100th meridian. Sometimes arbitrarily defined as that part west of the 100th meridian, that part west of the Mississippi River, or the Rocky Mountains region and westward. The geographic location of where the *West* begins has been variable according to the time in history and according to context. See *Far West; Pacific Northwest; Southwest; Wild West.* 2. Western Australia.

West Coast fir. Douglas fir.

West Coast hemlock. Pacific hemlock.

western arborvitae. Giant arborvitae.

western azalea. *Rhododendron occidentale,* family Ericaceae, a deciduous shrub grown for its white or pinkish, yellow-blotched flowers and yellow or scarlet, autumnal foliage. It may be poisonous to sheep and goats. Native to western North America.

western balsam bark beetle. *Dryocoetes confusus* Swaine, family Scolytidae, a beetle that infests spruce, fir, and pine in western and northwestern U.S. It is capable of killing the trees by constructing a nuptial chamber and radiating egg galleries beneath the bark.

western barley. A market classification of barley which includes all white barley grown west of the U.S. Great Plains, and may not include more than 10 percent barley of other classes.

western birch. Water birch.

western black flea beetle. *Phyllotreta pusilla* Horn, family Chrysomelidae, a beetle whose larvae mine the leaves and feed on the roots of cabbage, cauliflower, horseradish, mustard, radish, watercress, turnip, sugar beet, corn, wallflower, and rockcress.

western bleeding-heart. Pacific bleeding-heart.

western blight. Curly top.

western blue joint. See *bluestem: 1.*

western blue stem of brambles. See *Verticillium wilt.*

western bracken. *Pteridium aquilinum,* family Polypodiaceae, a fern which is poisonous to livestock. After about 2 months of feeding on it cattle become sick with very high temperature, difficult breathing, salivation, nosebleed and hemorrhage. Native to western North America. Also called *brake fern; hog brake; bracken.*

western brome grass. Mountain brome.

western catalpa. Northern catalpa.

western cedar. Sierra juniper.

western cedar borer. *Trachykele blondeli* Marseul, family Buprestidae, the larva of a beetle which makes galleries in Rocky Mountain juniper, California incensecedar, cypress, and related genera.

western cedar mite. Tip-dwarf mite.

western celery mosaic. A virus disease of celery in Colorado and California, U.S., characterized by white-colored, translucent, sunken spots on the outer petioles, followed by necrotic spotting. It also affects carrots, celeriac, dill, caraway, and coriander.

western chicken flea. *Ceratophyllus niger* Fox, family Ceratophyllidae, a blood-sucking insect which infests birds in Maine and on the Pacific Coast, U.S.

western chokecherry. *Prunus virginiana* var. *demissa,* family Rosaceae, a deciduous tree whose leaves are poisonous. It is found in the U.S. from the Dakotas west to California on prairies and dry slopes. Symptoms of poisoning are uneasiness, staggering, convulsions, difficulty in breathing, rolling eyes, and tongue hanging out. Then the animal becomes quiet, bloats, and dies within an hour after ingestion.

western clematis. Western virginsbower.

western coffee. Cascara buckthorn.

western corn rootworm. *Diabrotica virgifera* Lec., family Chrysomelidae, a beetle whose larva is sometimes injurious to corn and other grain crops in Colorado, U.S. Also called *Colorado corn rootworm.*

western crab(apple). Prairie crabapple.
western cucumber mosaic. A minor, virus disease of celery.
western dogwood. Pacific dogwood.
western dropwort. Bowman's root.
western duck. In marketing, a duck of Pekin, Roven, or mixed breeds which is produced on general farms in midwestern U.S. The birds are firm-fleshed, creamy-yellow skinned and weigh 3 to 5 lbs.
western equine encephalomyelitis. Equine encephalomyelitis.
western false hellebore. California false hellebore.
western fescue. *Festuca occidentalis,* family Gramineae, a grass found on dry, rocky, woody slopes from Michigan, U.S., to British Columbia, Canada, whose herbage is highly palatable to stock. Its forage rating is good.
western field wireworm. *Limonius infuscatus* Mots., family Elateridae, an insect that is injurious to plants on the U.S. West Coast.
western fir. Douglas fir.
western grape leaf skeletonizer. *Harrisina brillians* B. & McD., family Zygaenidae, an insect which occurs in southwestern U.S., infesting wild and cultivated grapes in compact colonies in late summer.
western grape rootworm. *Adoxus obscurus* (L.), family Chrysomelidae, a rootworm that infests grapes and fireweed in California, U.S. The beetle eats chain-like strips in leaves and holes in the berries. The larva feeds on smaller roots and the bark of the bigger roots.
western hemlock. Pacific hemlock.
western juniper. Sierra juniper.
western larch. *Larix occidentalis,* family Pinaceae, a deciduous tree, important as a source of wood used as ties, poles, in building construction, planing-mill products, boxes, crates, etc. Native to North America. Also called *western tamarack.*
western monkshood. Columbia monkshood.
western needlegrass. *Stipa occidentalis,* family Gramineae, a perennial bunchgrass which occurs abundantly in and west of the Rocky Mountains, U.S., and whose forage value is good.
western paper birch. Water birch.
western peach tree borer. *Sanninoidea exitiosa graefi* (Hy. Edw.), family Aegeriidae, a destructive insect pest of peach, plum, prune, apricot, and sometimes of apple and almond. The larva tunnels the trunks from 2 to 3 ins. below the soil to about 10 ins. above.
western pine beetle. *Dendroctonus brevicomis* Lec., family Scolytidae, a most important insect pest of ponderosa pine and Coulter pine along the U.S. West Coast, causing great losses each year. The adults construct egg galleries between the bark and sapwood.
western pitch pine. Ponderosa pine.
western potato flea beetle. *Epitrix subcrinita* (Lec.), family Chrysomelidae, a serious insect pest which infests potato, tomato, and related plants, as well as apple, ash, arbutus, bean, beet, cabbage, carrot, celery, clover, corn, cucumber, holly, honeysuckle, lettuce, maple, muskmelon, pumpkin, phlox, radish, rhubarb, spinach, sunflower, viburnum, violet, and watermelon. The young adult feeds on the foliage riddling the leaves. Especially serious in western U.S.
western ragweed. *Ambrosia psilostachya,* family Compositae, a perennial herb which is a weed pest and whose pollen often causes hay fever. Native to North America. Also called *perennial ragweed.*
western range. The native livestock grazing areas of the Great Plains, the Rocky Mountains, the Intermountain and the Pacific Coast regions of the U.S.
western raspberry fruitworm. *Byturus bakeri* Barber, family Byturidae, a moth which infests raspberries and loganberries, feeding on the buds, blossoms, and young leaves. The larva bores its way into fruit rendering it unfit for human consumption.
western raspberry rust. A rust of raspberries caused by *Phragmidium rubi-idaei,* family Pucciniaceae. Also called *leaf rust; cane rust.*
western red. A market subclass of soft red winter wheat grown west of the Great Plains, U.S.
western red cedar. See *Rocky Mountain juniper; giant arborvitae.*
western rust. Western raspberry rust.
western saddle. A heavy, deep-seated saddle with a distinctly raised horn and cantle, as contrasted to the English saddle, which is light and comparatively flat. It is used for working livestock on the ranches in western U.S. Also called *stock saddle.*
western sand cherry. Bessey cherry.
western sneezeweed. Orange sneezeweed.
western snowberry. *Symphoricarpos occidentalis,* family Caprifoliaceae, a deciduous shrub which is a fair browse plant for sheep, goats and cattle on the western range, and is also grown as an ornamental. Native to North America. Also called *buck brush; wolfberry.*
western spotted cucumber beetle. *Diabrotica undecimpunctata* Mann., family Chrysomelidae, a beetle which feeds on almost every plant except conifers. The larva feeds on roots of corn, sweetpea, and grass, and the adults mine holes in the ripening fruit.
western spruce. Sitka spruce.
western spruce box. A type of box used for shipping dressed poultry. It is made of strong veneer, with solid ends and a cleated top.
western striped cucumber beetle. *Acalymma trivittata* (Mann.), family Chrysomelidae, a beetle whose adult feeds on foliage of cucumbers, muskmelons, pumpkin, squash, etc., while the larva feeds on their roots. It is found on the West Coast and Arizona, U.S.
western striped flea beetle. *Phyllotreta ramosa* (Crotch), family Chrysomelidae, a flea beetle *(q.v.)* that infests crucifers in California, U.S.
western swordfern. *Polystichum munitum,* family Polypodiaceae, a hardy, evergreen fern grown for ornament. Native to western North America. Also called *giant holly fern.*

western tamarack. Western larch.

western tent caterpillar. *Malacosoma pluviale* (Dyar.), family Lasiocampidae, a tent caterpillar *(q.v.)* that feeds on alder, apple, hawthorn, cherry, currant, and rose.

western thimbleberry. *Rubus parviflorus,* family Rosaceae, an erect, unarmed shrub grown occasionally for its white flowers. Native to North America.

western tomato blight. See *curly top.*

western tussock moth. *Hemerocampa vetusta* (Bdv.), family Lymantriidae, a moth whose caterpillar feeds on the leaves and fruit of apple, almond, apricot, blackberry, Brazil peppertree, California buckthorn, cherry, hawthorn, oak, pear, plum, prune, walnut, and willow. Sometimes called *California tussock moth.*

western virginsbower. *Clematis ligusticifolia,* family Ranunculaceae, a climbing, woody vine grown on a trellis, fence, etc., for its habit and for its white flowers. Native to western North America. Also called *hill clematis.*

western waterhemlock. *Cicuta occidentalis,* family Umbelliferae, an extremely poisonous, perennial herb. Symptoms of poisoning in stock are frothing at the mouth, uneasiness, pain, convulsions, kicking, throwing back of the head, rigid legs, bellowing, and spasmodic contractions of the diaphragm. Death follows ingestion rather rapidly. Native to western North America.

western wheatgrass. *Agropyron smithi,* family Gramineae, a tall, coarse, smooth-stemmed grass bearing whitish flowers; widely cultivated as a forage crop. Native to western U.S. Also called *bluestem; bluestem wheatgrass.*

western white fir. White fir.

western white pine. *Pinus monticola,* family Pinaceae, an evergreen tree important for its wood used for matches, boxes, construction, millwork, fixtures, etc. Native to western Canada and U.S.

western white wheat. A market type of white wheat.

western X-disease. A virus disease of stone fruits, generally distributed throughout western U.S. On peach, it is characterized by red and yellow coloration of the leaves followed by progressive shedding from base to tip; on chokecherry by early red and yellow coloration of leaves; on sweet and sour cherry by small, delayed ripening fruits or wilt and decline. Symptoms may vary with different strains or due to a complex mixture of viruses. Similar to the eastern form of X-disease (*q.v.*) except that nearby diseased chokecherries are not necessary for spread in orchards. A group of somewhat similar diseases, as wilt and decline, little cherry, western X little cherry, buckskin of sweet cherry, Green Valley strain, Napa Valley strain of buckskin, may result from other associated virus or virus strains. See *buckskin: 4; decline: 3; little cherry; wilt.*

western yarrow. *Achillea lanulosa,* family Compositae, a perennial herb of fair forage value. It is grazed somewhat on the western ranges, U.S.

western yellow blight. See *curly top.*

western yellow pine. Ponderosa pine.

western-yellow-pine mistletoe. Dwarf mistletoe.

western yellow rust of raspberries. A fungous disease caused by *Phragmidium rubi-idaei,* family Pucciniaceae, characterized by small, light yellow pustules on the leaves, and by elliptical, cankerlike lesions on the stems. Usually occurs in the Pacific coast of northwestern U.S.

West Friesian. A soft, uncured, cows' skim milk cheese made in Holland.

West Highland cattle. A small, hardy breed of beef cattle with thick, shaggy hair, thick mane, widely-spaced, long horns, heavy dewlap, and sturdy frame. The coat color may vary from a mixture of yellow, black, brindle, and red to red and black. Although the *West Highland cattle* are quite small and slow-maturing, the quality of their beef is excellent. Native to the western uplands of Scotland. Also called *Kyloes.*

West India gherkin. West Indian gherkin.

West India gooseberry. Otaheite gooseberry leafflower.

West Indian arrowroot. Bermuda arrowroot.

West Indian cabbagetree. Cabbage angelintree.

West Indian cedar. Cigarbox cedrela.

West Indian fruit fly. *Anastrepha mombinpraeoptans* Seín, family Tephritidae, an insect pest of grapefruit and sour orange in tropical America.

West Indian gherkin. *Cucumis anguria,* family Cucurbitaceae, a trailing vine grown as a curiosity and for its fruit used in pickles. Native to tropical America. Also called *anguria; bur gherkin.*

West Indian locust. *Hymenaea courbaril,* family Leguminosae, a large, hardwood tree, bearing white flowers and woody pods. The brown wood is used for construction, the pods yield an edible pulp, and a resin obtained from the tree has medicinal and wood-finishing value. Native to tropical America. Also called *algarroba; courbaril.*

West Indian peach scale. See *white peach scale.*

West Indies mahogany. *Swietenia mahagoni,* family Meliaceae, an evergreen tree grown for its hard, red wood, used especially in furniture. Native to tropical America.

Westphalian ham. A ham prepared by rubbing with salt and saltpeter and cured for two weeks. It is then placed in a brine solution for 18 days; next, ripened in a cool cellar for 4 weeks after which it is soaked in luke-warm water for 12 hours. Finally, it is smoked over a beechwood fire on which juniper twigs and berries are thrown throughout the smoking period.

Westphalia Sour-milk. A small, sour-milk, surface-ripened, cows' milk German cheese which is ripened in a cool, moist cellar.

wet. 1. To dampen; to sprinkle; to furnish water to a plant. 2. Characterized by rain, as *wet* weather. 3. Moist, covered, soaked, or saturated with water.

wetback. A term originally indicating a Mexican laborer who illegally entered the U.S. by swimming the Rio Grande river. Commonly, a derogatory term for any Mexican migrant worker employed in the U.S. They constitute an important part of the farm labor in the Southwest and other sections of the U.S. Also called *spick.*

wet band. A group of ewes or nannies accompanied by nursing young. Also called *wet herd.*

wet beet pulp. A livestock feedstuff obtained in the manufacture of beet sugar. It consists of the residue left after sugar is extracted and consists of about 90 percent of moisture. It is usually fed in the wet condition but may be stored temporarily in large silos prior to hauling and feeding. It is low in protein and phosphorus.

wet body. A body defect of ice cream characterized by the shiny surface appearance of inadequately frozen ice cream as it is drawn from the freezer. Also called *soggy body.*

wet brewers' grains. In beer manufacture, the residue left after the wort is extracted and from which brewers' dried grains (*q.v.*) are prepared.

wet butt. A mottling and weakening of the tissues of tobacco as a result of freezing before curing has been finished.

wet climate. The climate in which precipitation effectiveness is such that rain-forest vegetation prevails. The Thornthwaite precipitation-effectiveness index is about 128. See *precipitation-effectiveness.*

wet down. 1. To water a plant so that the soil surrounding its roots is soaked. 2. To bring soil to field capacity by rain, as it *wet down* six ins.

wet-dry climate. Any climate which has alternating wet and dry seasons, such as a wet summer and a dry winter, or the reverse. Sometimes the climatic condition in the trade-wind belt where daily showers are interspersed with dry sunny weather is also considered to be wet-dry. The term is used more commonly, however, to apply to alternating wet and dry seasons.

wet feet. 1. A condition in plants which results when the watering is in excess of drainage. 2. A tree or other plant which has its roots in an excessively wet soil.

wet-hand milking. An unhygienic method of hand milking in which the initial squirts of milk are used to wet the fingers and hands to lubricate the teats. Also called *wet milking.*

wet hay. Hay not sufficiently cured for safe storage. See *tobacco-brown hay.*

wether hog(g). 1. A wether lamb from the time of weaning until it is shorn. See *tup hog; wether lamb: 1.* 2. A second season wether (England).

wether lamb. 1. A male sheep castrated before it reaches maturity or develops male characteristics. 2. The fleece from any sheep after the first fleece has been removed (England).

wet land. 1. Land that is swampy or marshy in contrast to naturally well-drained or dry land. 2. In an agronomic sense, land that is temporarily excessively wet.

wet mash. Any mash to which water, milk, or other liquid has been added for feeding purposes. *Cf. dry food.*

wet method. A method of setting cranberry cuttings in which mowed vines are broadcast on thoroughly soaked peat or muck and shallowly cut into the soft soil; the water table is held high for most of the first year.

wet milking. See *fatmilking; wet-hand milking.*

wet-mix feed. Wet mash.

wet pocket. A low area in a field, woods, etc., in which water collects and remains for considerable periods of time.

wet pox. The diphtheritic form of fowl pox (*q.v.*) characterized by yellowish, raised, membranous patches in the mouth and throat.

wet-process milling. A process used in milling corn or sorghum grain, involving softening, coarse grinding, removal of the germ, hull fiber, starch, glutin, and other solubles, by a variety of processes. The main product, starch, may be treated to produce corn sugar, the germ solvent extracted to produce cooking oil, and a protein fraction, "zein," removed as a parent material for artificial fiber. By-products are animal feeds and a steep-liquor used in the manufacture of penicillin. Also called *wet-process steeping.*

wet-process steeping. Wet-process milling.

wet rendering. See *tanking: 1.*

wet rot. A symptom of late blight and rot of potato (*q.v.*) in which tubers in heavy, damp soil, decay before harvest.

wet season. 1. The season or time of year during which most of the precipitation occurs. *Cf. dry season.* 2. A season in which the precipitation is greatly in excess of normal.

wet sheep. Ewes with suckling lambs. See *dry sheep; wet band.*

wet, soggy, mushy body. A body defect of Devonshire cream characterized by a very soft, almost slushy texture entirely unlike the relatively firm body and granular texture desired in clotted cream.

wet soil. 1. Any soil which is wet or inadequately drained for short or long periods of time. 2. Marshy, boggy soil.

wet spot. 1. A dark, moist area on the bark of locust trees caused by sap flow from wounds made by the feeding of young larvae of the locust borer, *Megacyllene robiniae* (Forst.), in early spring. 2. See *wet pocket.*

wettable. Designating a powder or dust which will readily mix with water.

wettable powder (dust). Any material manufactured in the form of a powder or dust that can be mixed with or dissolved readily in water.

wettable sulfur. That sulfur so treated as to be readily miscible in water. It is widely used in spray mixtures, especially for its fungicidal effect.

wetting agent. A chemical product that lowers the surface tension of another substance.

wetting power. The ability of a detergent (*q.v.*) to lower surface and interfacial tension of water and immersed objects, aiding wetting and penetration of the material to be cleaned.

wet-weather blight. A sporadic cotton disease caused by the fungus, *Ascochyta gossypii*, family Sphaeropsidaceae, which occurs in serious proportions in cool, moist weather. It is characterized by small, circular, white spots on the cotyledons and leaves. The lesions enlarge, coalesce, turn brown, and become rough. The spots then fall out leaving a ragged appearance and sometimes there is defoliation. Often the lesions girdle the stem and kill the terminal buds, resulting in the loss of the stand. In the summer the lesions are from ½ to 1 in. long, occurring at the branch axils and are ragged at the edges. Also called *Ascochyta blight; Ascochyta canker; wet-weather canker; Ascochyta leaf spot.*

wet weather canker. Wet-weather blight.

wet wing. See *favus: 1.*

wetwood. Wood with an abnormally high water content and a translucent or water-soaked appearance which develops only in living trees and does not originate through soaking logs or lumber in water.

wet year. A year in which the amount of precipitation is considerably above normal.

Weymouth bit. A type of bit for a bridle which has a bar with a U-shaped curve in it and which is so constructed that considerable pressure can be brought to bear on the horse's mouth.

Weymouth pine. Eastern white pine.

whale guano. A fertilizer which is the residue from the flesh and fat of the whale after it is rendered to extract the oil. Also called *whale meat; whale tankage.*

whale meal. A livestock feedstuff prepared from the clean, dried, undecomposed flesh of the whale after part of the oil has been extracted. If it contains more than 3 percent of salt (NaCl), the amount of salt must constitute a part of the brand name, provided that in no case shall the salt content of this product exceed 7 percent.

whale meat. Whale guano.

whale oil. A lubricating oil obtained principally from the fat or blubber of the Greenland or right whale, which is largely used for oiling wools during combing, in batching flax and other fibers, in currying, and in dressing chamois leathers. The oil of commerce may be obtained from the blubber of any whale or dolphin. Also called *train oil.*

whale's back. A hill which rises sharply and drops off sharply (Australia and New Zealand).

whale tankage. Whale guano.

whang. Tough leather adapted for strings, thongs, and belt laces which is commonly of calf skin.

wharf rat. Brown rat.

wheat. 1. Any of the several plants of the genus Triticum, family Gramineae. Only a few species, however, are important. 2. *T. aestivum* (*sativum*), an annual herb which is one of the most important crop plants of the world, furnishing grain, one of the principal sources of human food, animal feed, and manufactured products, such as alcohol and glucose. The *wheat* grain is not commonly fed to livestock in the U.S. but the by-products of milling are extensively used. Place of origin unknown, presumably western Asia. 3. In marketing, any grain which (before the removal of dockage) consists of 50 percent or more wheat and not more than 10 percent of other grains for which standards have been established under the provisions of the U.S. Grain Standards Act. See *club wheat; durum wheat; hard red spring wheat; hard red winter wheat; mixed wheats; Polish wheat; poulard wheat; red durum wheat; soft red winter wheat; soft white wheat; white wheat.*

wheat anthracnose. See *cereal anthracnose.*

wheat aphid. Any of the various plant lice of the family Aphidae, as the English grain aphid (*q.v.*), which suck the juices from wheat still in the field.

wheat beetle. 1. *Silvanus surinamensis*, a small, reddish-brown insect whose larvae feed on cereal crops, especially wheat. 2. Drugstore beetle. 3. Any grain or flour beetle.

Wheat Belt. A region in which wheat is the principal crop. In the U.S., it extends from northern Texas and western Oklahoma to North Dakota and the Canadian border. The southern part of the U.S. *Wheat Belt* is the winter *wheat belt* and the northern part is the spring *wheat belt.*

wheat binder. See *grain binder.*

wheat bran. A livestock feedstuff which is the coarse outer covering of the wheat kernel as separated from cleaned and scoured wheat in the usual process of commercial milling. Also called *bran.*

wheat bran and wheat standard middlings. A livestock feedstuff which consists of the two commodities, wheat bran and wheat standard middlings (*q.v.*) mixed in the proportions obtained in the usual process of commercial milling.

wheat brown shorts. A livestock feedstuff which consists mostly of the fine particles of wheat bran, wheat germ, and very little of the fibrous material obtained from the tail of the mill. It is obtained in the commercial milling of wheat and contains less than 7.5 percent of crude fiber. Also called *brown shorts; wheat red shorts.*

wheat bulb worm. See *wheat stem maggot.*

wheat country. A Wheat Belt.

wheat cutworm. Larva of any lepidopterous insect, as the fall armyworm (*q.v.*), that kills wheat by cutting off the stalk at the ground.

wheat distillers' dried grains. A livestock feedstuff which is the dried residue left in the manufacture of alcohol and distilled liquors from wheat or from a grain mixture in which wheat predominates.

wheat distillers' dried grains with solubles. A livestock feedstuff which is the dried residue left in the manufacture of alcohol and distilled liquors from wheat or from a grain mixture in which wheat predominates and which contains the major portion of the condensed screened stillage dried therewith.

wheat flour. See *flour.*

wheat flour middlings. A livestock feedstuff that consists of wheat standard middlings and wheat red dog (*q.q.v.*) proportionately combined in the process of milling and containing less than 6.0 percent of crude fiber.

wheat germ. The embryo of the wheat seed which is rich in vitamin B_6.

wheat germ oil cake. A livestock feedstuff which

is obtained after oil has been partially expressed from commercial wheat germs. It contains more than 29 percent of protein.

wheat germ oil meal. A livestock feedstuff which is the ground wheat germ oil cake *q.v.*

wheatgrass. Any of the perennial bunchgrasses of the genus Agropyron, family Gramineae. In many sections of the northwestern and the Rocky Mountain regions of the U.S. the *wheatgrasses* constitute the most important forage species. See *bearded wheatgrass; beardless bluebunch wheatgrass; quack grass; slender wheatgrass; thickspike wheatgrass.*

wheat gray middlings. Wheat gray shorts.

wheat gray shorts. A livestock feedstuff which consists of the fine particles of the outer bran, the inner bran or bee-wing bran, the germ, and the offal or fibrous material obtained from the tail of the mill. It is obtained in the commercial milling of wheat and contains less than 6.0 percent of crude fiber. Also called *total shorts.*

wheat head armyworm. *Faronta diffusa* (Wlk.), family Noctuidae, a caterpillar which infests wheat, grasses, and other small grains.

wheat jointworm. *Harmolita tritici* (Fitch), family Eurytomidae, a larva of a fly that is a serious pest of wheat in the U.S. The fly lays eggs in the stem tissues of a wheat plant near a joint. The larvae develop and feed inside the stem weakening it and making it apt to lodge.

wheat land. Land ordinarily planted to wheat or which is productive when planted to wheat.

wheatland plow. One-way disk plow.

wheat leaf miner. Corn blotch leaf miner.

wheat low-grade feed flour. A livestock feedstuff which is a by-product obtained in commercial flour milling. It consists principally of wheat flour with small quantities of aleurone and fine wheat bran particles and contains less than 1.5 percent of crude fiber.

wheat middlings. See *wheat flour middlings; wheat standard middlings; wheat white shorts.*

wheat midge. *Sitodiplosis mosellana* (Gehin), family Itonididae, a small insect which lays eggs among the bracts of wheat. The larvae feed on the kernels, causing wilting, discoloration, and lodging. Destructive to wheat in Europe and the U.S.

wheat mildew. See *downy mildew; powdery mildew.*

wheat mixed feed. A livestock feedstuff which consists of wheat bran and the wheat gray or total shorts proportionately combined as obtained in the usual process of commercial milling. It contains less than 8.5 percent of crude fiber. Also called *mill-run wheat feed.*

wheat mosaic-rosette. A virus disease of wheat, barley, and rye, characterized by dwarfing, an excessively dark-green color of leaves in the fall, a dying of the outer leaves and the autumn tillers, excessive development of the spring tillers, a yellow streaking, and sometimes death of the plant. There are two types of this disease: (1) The green type, which shows a light green pattern on a normal green background. (2) The yellow type, which shows light yellow patterns. Also called *rosette of wheat; take-all.*

wheat moth. Angoumois grain moth.

wheat poisoning. A condition seen in cattle grazing on lush wheat pasture wherein the animals show a posterior paralysis, tend to drop down, and become semicomatose. It is due to an improper phosphorus-calcium balance in the blood as a result of eating wheat high in phosphorus content.

wheat red dog. A livestock feedstuff which is a by-product obtained in commercial flour milling. It consists principally of aleurone *(q.v.)* with small quantities of wheat flour and fine wheat bran particles and contains less than 4.0 percent of crude fiber. Also called *red dog flour; wheat red dog flour.*

wheat red dog flour. Wheat red dog.

wheat red shorts. Wheat brown shorts.

wheat region. Wheat Belt.

wheat rust. See *crown rust; leaf rust; leaf rust of cereals; rusts; stripe rust.*

wheat sawfly. 1. The wheat stem sawfly. 2. Any of various small sawflies of the genus Dolerus, whose larvae damage both the wheat stems and heads. Native to America. 3. The grass sawfly *(Pachynematus extensicornis),* whose larvae damage wheat by feeding on the blades.

wheat scab. See *scab: 3.*

wheat shorts. Wheat standard middlings.

wheat standard middlings. A livestock feedstuff which consists mostly of fine particles of wheat bran, wheat germ, and very little of the fibrous material obained from the tail of the mill. Obtained in the usual commercial process of milling, it contains not more than 9.5 percent of crude fiber.

wheat states. Wheat Belt.

wheat stem maggot. *Meromyza americana* Fitch, family Chloropidae, a maggot which infests wheat, rye, barley, oats, bluegrass, and timothy. It tunnels in the stems causing the heads to dry out and whiten while the lower stem and leaves are still green.

wheat stem rust. Stem rust of wheat.

wheat stem sawfly. *Cephus cinctus* Nort., family Cephidae, a serious insect pest of wheat which also infests spring rye, barley, spelt, timothy, and some grasses. It tunnels the stem causing excessive lodging.

wheat straw. The dried, mature wheat stem and leaves used for roughage, bedding in barns, and as poultry litter.

wheat straw-worm. *Harmolita grandis* (Riley), family Eurytomidae, an insect pest of wheat, found west of the Mississippi, U.S. Passing the winter in the larval and pupal stages in the stubble or straw, it emerges in April or May, laying an egg in the stem wall just above the ground. The first generation larva works its way into the stem, eats off the developing head and prevents tiller development. The second generation larva, becoming active as the plant is starting to joint, has a stunting effect on the growth and weakens the straw even though a head may be produced.

wheat streak mosaic. A virus disease of wheat in the western U.S., characterized by severe, yellow spotting, streaking, or mottling; or by mild, light green or weak yellow streaking and mottling. It also attacks corn, oats, barley, winter-annual bromegrasses, Canada bluegrass, etc.

wheat striate mosaic. A virus disease of wheat characterized by fine, light green or yellow lines in a streaked pattern. In some plants, there is severe stunting.

wheat take-all. See *take-all.*

wheat total shorts. Wheat gray shorts.

wheat white middlings. Wheat white shorts.

wheat white shorts. A livestock feedstuff which consists of a small portion of the fine wheat bran particles, the wheat germ, and a large portion of the fibrous offal obtained from the tail of the mill. Obtained in the usual process of flour milling, it contains not more than 3.5 percent of crude fiber.

wheat wireworm. *Agriotes mancus* (Say), family Elateridae, a beetle whose larva infests wheat, generally feeding on roots of mature plants. Sometimes it eats the seed in the ground or the underground parts of the small plant, causing it to wither and die.

wheel. 1. A disk or circular frame with a hub in the center. An axle, for the loads to rest on, goes into or through the hub. Power, if any, is often furnished the *wheel* by the axle. Ground-driven *wheels* supply power to the axle for operation of other machine components. 2. (cap.) In the U.S., any of various political organizations set up during the latter part of the nineteenth century for the benefit primarily of farmers: *e.g.*, the National Agricultural *Wheel.* 3. A cylinder of cheese. See *drum.*

wheelbarrow. 1. A box placed on a light horizontal type of A-frame and supported at one end by one or two wheels. Handles at the other end of the frame are lifted to utilize the mechanical advantage of the wheel to move light loads of earth or other materials. 2. Designating implements built or transported on a *wheelbarrow,* as the *wheelbarrow* sprayer.

wheel colter. Rolling colter.

wheeler. Wheel horse.

Wheeler bluegrass. *Poa nervosa,* family Gramineae, a perennial grass of good forage value, found in open woods, dry meadows, and on old burned-over areas at medium altitudes. Native to western North America.

wheel excavator. Any of several different kinds of excavators which consist generally of a frame, a power unit, and a large wheel. On the rim of the wheel are fastened buckets with cutters which dig a trench as the wheel is turned. It is commonly used in trenching for tile drain installation.

wheel harrow. A wheeled farm implement which is dragged over the ground for leveling or pulverizing the soil or covering seed, as distinguished from the harrow which slides on the ground.

wheel-hoe cultivator. See *garden cultivator.*

wheel horse. In a team of horses in a tandem hitch, the horse nearest the wheels of a vehicle.

wheel plow. Any plow set up on wheels.

wheel track planting. Planting a crop, as corn, on plowed fields, in the tractor wheel tracks. The weight of the wheels crush and firm the soil. The rough surface left reduces the amount of potential soil erosion.

whey fermentation product. A livestock feedstuff obtained by the concentration or dehydration of the liquid product resulting from the action of the ferment on the basic medium of whey, wherein there is a synthesis of nutritionally significant quantities of one or more vitamins.

wheyfers. A confection which consists of sweetened condensed whey, precooked dry cereal, and nuts.

whey fudge. A confection which consists of sweetened condensed whey, sugar, corn sirup, invert sirup, and chocolate.

wheying off. A body and texture defect of cultured cream, cultured milk, and ice cream, characterized by a seeping out of the whey in the product.

whey sauce. A cream-style sauce for vegetables which consists of whey, cream, starch or flour, and salt.

whey soluble. See *condensed whey solubles; dried whey solubles.*

whey taint. A flavor defect of cheese which has a mildy acid taste and an aroma characteristic of fermented whey. See *acid flavor.*

whiffletree. Singletree.

whin. Gorse.

whinny. The gentle, soft cry of a horse.

whip. 1. Any of several kinds of instruments for lashing animals, usually consisting of a handle and a lash. 2. An unbranched shoot of a woody plant, particularly the first year's growth. 3. A tall, slender tree in the upper crown colony. 4. To lash an animal. 5. To beat cream, egg white, etc., so that it becomes a froth and holds shape. See *whipping cream.*

whip-and-tongue graft. Whip graft(ing).

whip break. To train a horse by use of a whip. (The whip is used sparingly, and never out of anger, so that the horse learns that when he does certain things he is punished and when he does certain other things he is rewarded.)

whip graft(ing). A method of grafting, usually at the collar or on the roots of small seedling trees as in the propagation of nursery stock, or in joining small branches as in top-working. The stock is cut off with a long slanting cut in which a vertical slit, forming a kind of tongue, is made. The cuts on the lower end of the cion are made to complement those on the stock so that it is possible to insert the tongue of one in the slit of the other. Tying and waxing may or may not be done. Also called *whip-and-tongue graft.* See *root grafting.*

whipped. Designating cream or egg whites, etc., which have been beaten into a froth.

whipped cream flower. A decoration for a confection made from whipped cream in the form of a flower.

whipping cream. Raw or pasteurized sweet cream containing 30 to 40 percent fat, which has been aged for sufficient time to enhance its whipping qualities. Also called *double cream.*

whipping of the tongue. The sudden, lashing, sidewise movement of the tongue of a wagon, etc., as the vehicle moves over rough terrain.

whippletree. Singletree.

whippoorwill cowpea. A variety of the cowpea *q.v.*

whippoorwill's boots. Pitcherplant.

whip socket. A socket on or near the dashboard of a buggy or other horse-drawn vehicle into which the butt of the whip is placed when the whip is not in use.

whipstalk. See *whip: 2.*

whipstitch. To plow the land so that it is ridged (England). See *rafter: 2.*

whiptail. A molybdenum-deficiency disease of cauliflower characterized by malformation of the leaves and unsatisfactory curd formation. The disease is thought to be more prevalent on acid soils.

whip tree. Singletree.

whipworm. Any worm of the genus Trichuri and related genera, which live internally in the intestines of man and various domestic animals.

whirlbone. 1. The upper part of the hip bone. 2. The knee bone.

whirldry. A piece of equipment used to remove honey from wax cappings by centrifugal force.

whirlwind. A windstorm, a rapidly rotating inward and upward spiral motion of the air, covering a very small area. It is often seen on hot, summer days, especially on open or prairie land. Also called *devil duster.*

whirlwind seeder. A device for the broadcast sowing of small seeds and light grains which consists of a fabric hopper, a disk with prominent ridges, and a crank. As the crank is turned, the disk revolves rapidly and throws the seed, dropping on it from the hopper, in all directions.

whirlwind terracer. A terracer *(q.v.)* that consists of a shortened moldboard plow to cut and lift a furrow slice up into a rapidly revolving vertical power-driven auger which throws the soil to the side.

whiskered beets. Curly top.

whiskers. The grayish-white fungous threads (properly, sporangiophores) of the bread-mold fungus, *Rhizopus nigricans,* family Mucoraceae, which grow out of starchy, vegetable matter such as bread, sweet potatoes, spoiling squash, or pumpkin, etc. At their tips, they bear the round, dark-colored sporangia, or spore cases, within which the numerous spores are produced.

whiskey. A distilled alcoholic beverage made from cereal grain by conversion of the starch into sugar, fermentation of the sugar, concentration of the alcohol and associated products by distillation, and improvement of quality by aging in oak barrels. See *Scotch whisky.*

whispering bells. Yellow whispering bells.

whistler. A horse which makes loud whistling noises through the nostrils when breathing.

whistling. The noise made by a horse afflicted by roaring when breathing.

white. 1. A coat-color of an animal, or the color of a bird's plumage, which is characterized by the absence of any color. 2. The albumen of an egg. 3. In marketing, any egg whose shells are pure white. 4. In marketing, a type of Spanish sweet orange. See *blood oval.* 5. A color designation for upland cotton next in whiteness to extra white. 6. See *white wheat.*

white alder. Winterberry.

white alkali soil. See *alkali soil.*

white ant. A termite.

white apple. American potatobean.

white apple leafhopper. *Typhlocyba pomaria* McA., family Cicadellidae, an insect that feeds on the apple, both as an adult and as a nymph, by sucking the plant juices on the undersides of the leaves. Infested trees in late summer have pale foliage with specks of greenish-white showing through from the underside. The new foliage is curled, sometimes burned, and the fruit is spotted with excrement.

white arsenic. As_2O_3, a chemical compound, arsenic *(q.v.)* that is one of the cheapest of the arsenicals. It is used in sprays, poison baits, etc., as an insecticide.

white ash. *Fraxinus americana,* family Oleaceae, a large, deciduous tree grown for its lumber used for handles, furniture, vehicle parts, sporting goods (bats, oars, paddles, etc.), and other articles. It is one of the better street trees in places where the soil is favorable. Native to eastern North America.

white asp. White poplar.

white asparagus. Garden asparagus with blanched shoots.

white bachelor's button. Evening campion.

white balsam. Fragrant cudweed.

white baneberry. *Actea alba,* family Ranunculaceae, a perennial herb grown in the wild garden for its white flowers and white fruit. Its berries are poisonous, and ingestion results in rapid pulse, dizziness, and colicky pains. Native to North America. Also called *white cohosh.*

white-barked canoe birch. See *European white birch; paper birch.*

whitebark scapbloom. Chaparral whitehorn ceanothus.

white bead. White baneberry.

whitebeam mountainash. *Sorbus aria,* family Rosaceae, a deciduous, hardy tree which bears orange- or scarlet-colored berries, grown for its handsome foliage. Native to Europe. Also called *chess-apples.*

white bedstraw. *Galium mollugo,* family Rubiaceae, a perennial herb grown in the flower border or rock garden for its tiny, white flowers. Native to Europe.

white beech. See *American beech; European hornbeam.*

white beet. Leaf beet.

white bent (grass). See *redtop.*

white berry. White baneberry.

white birch. 1. *Betula alba,* family Betulaceae, a common birch with white or light gray bark. Native to Europe. 2. Paper birch. 3. Gray birch.

white blight. A fungous disease of sweetpeas caused by *Erostrotheca multiformis,* characterized by tan to buff-colored, angular spots on the leaves; enlarging of the spots to involve the whole leaf; and premature leaf drop.

white boltonia. *Boltonia asteroides,* family Compositae, a perennial herb grown for its white, violet, or purple flowers. Native to North America. Also called *false aster; false starwort.*

white bottle. Cowbell silene.

white break. A virus disease of gladiolus characterized by the production of short spikes, fewer florets than normal, and a bleaching of the petals of colored varieties.

white brittlebush. *Encelia farinosa,* family Compositae, a desert, shrubby plant grown for its clusters of yellow flowers. Native to North America.

white broom. Portuguese broom.

whitebrush. Privet lippia.

white bryony. *Bryonia alba,* family Cucurbitaceae, an herbaceous plant from whose perennial roots a cathartic is sometimes extracted. Native to Europe and western Asia.

white bud disease. A zinc-deficiency disease of corn characterized by yellow streaks on old leaves which become necrotic. Young leaves unfolding in the bud may be white or yellow.

white Burley. A type of tobacco for manufacturing and export classification which is thin-bodied, light-colored, and cured.

white bush. English hawthorn.

white buttonwood. American planetree.

white campion. Evening campion.

white cedar. See *California incensecedar; Chinaberry; eastern arborvitae; giant arborvitae; Lawson falsecypress; MacNab cypress; whitecedar falsecypress.*

whitecedar falsecypress. *Chamaecyparis thyoides,* family Pinaceae, an evergreen, swamp tree whose wood is used for poles, shingles, woodenware, planing-mill products, water tanks, boats, boxes, fences, etc. Native to eastern U.S. Also called *white cedar; Atlantic white cedar; southern white cedar.*

White (cheese). A summer, cows' skim milk, uncured, French cheese.

white clematis. Western virginsbower.

white clover. *Trifolium repens,* family Leguminosae, a perennial herb which spreads rapidly by stolons and is seldom unduly injured by mowing or grazing. It is grown as an important pasture plant and is used in lawn mixtures. Native to Europe. Also called *Dutch clover; white Dutch clover.*

white club wheat. A subclass of white wheat which includes the Sonora variety or the white club variety, either singly or in combination. It shall not include more than 10 percent of common white wheat other than Sonora, either singly or in combination.

white cockle. Evening campion.

white cohosh. White baneberry.

white comb. See *favus: 1.*

white corallita. 1. Sometimes the name applied to Christmasvine porana *(q.v.)* in Florida, U.S. 2. Any of a small group of tendril-climbing herbs of the genus Antigonon, family Polygonaceae, grown in warm regions for their ornamental foliage and numerous flowers of various colors.

white corn. 1. Corn with white kernels. See *yellow corn.* 2. A marketing class of corn which should not contain more than 2 percent corn of colors other than white. 3. Wheat, oats, barley (England).

white cornel. Flowering dogwood.

white-cup. White cupflower.

white cupflower. *Nierembergia rivularis,* family Solanaceae, a perennial, flowering herb, treated as an annual. It resembles a small petunia and has white flowers sometimes tinged with blue or rose. Its drooping habit adapts it to rock walls. Native to Argentina. Also called *white-cup.*

white cypress. Baldcypress.

white daisy. Oxeye daisy.

white deadnettle. See *deadnettle: 4.*

white deal. Norway spruce.

white diarrhea. Pullorum disease.

white Dutch. A white-flowered variety of the scarlet runner bean *q.v.*

white Dutch clover. White clover.

white Dutch runner. White Dutch.

white eardrop. Dutchman's-breeches.

white-edge morning-glory. *Ipomoea nil,* family Nymphaeaceae, a perennial vine grown on a trellis, fence, etc., in warm areas for its large, purple, blue, or rose-pink flowers. Native to the Tropics. Also called *Japanese morning-glory.*

white eggs. In marketing, eggs with white shells, as contrasted to brown-shelled eggs.

white Egyptianlotus. *Nymphaea lotus,* family Nymphaeaceae, an aquatic plant grown in a pool for its night-blooming, white flowers with pinkish outer petals. Native to Egypt. Also called *Egyptian lotus; Egyptian white lotus; white lotus of Egypt.*

white elm. American elm.

white emperor of grapes. A minor, virus disease of Emperor variety of grapes, characterized by greenish-yellow or pink fruit instead of bright red berries; a reduction of sugar content of the fruit; and dark, thick, wrinkled, downward-curling leaves. The leaves become reddish or bronze-colored along the veins and yellowish between them. The virus is the same which causes grape leafroll in Europe.

white eye. See *glass eye; ocular lymphomatosis.*

white face. 1. A coat color characteristic of certain breeds of animals, such as Hereford cattle, who have a red body, a white face, and white markings on the brisket, underline and switch of the tail. See *Hereford; Rambouillet.* 2. In the U.S. western range, sheep of fine wool breeding (obsolete).

White-faced Black Spanish. One of the oldest

Mediterranean breeds of chicken characterized by an unusual enamel-white face, lustrous, greenish-black plumage, dark slate shanks and white skin. Its egg color is white, and it is raised mainly for exhibition. Also called *Black Spanish.*

white-faced crossbred lambs. A type of crossbred feeder lambs characterized by white face.

white fairylantern. White mariposa.

white falsehellebore. *Veratrum album,* family Liliaceae, a hardy, perennial herb sometimes grown in a shady border or in a wild garden for its greenish flowers white inside. Young livestock and chickens may be fatally poisoned by eating the plant. Symptoms of poisoning are salivation, retching, purging, weakness, general paralysis, spasms, rapid thread-like pulse, lowered temperature, shallow respiration, cold skin, blindness and death. Native to Europe and northern Asia.

white fawnlily. *Erythronium albidum,* family Liliaceae, a bulbous herb grown in the rock garden for its white or pinkish-purple flowers. Native to North America. Also called *fawn lily; spring lily.*

white-felt blight of conifers. A rare blight of coniferous trees, caused by the fungus, *Acanthostigma parasiticum,* family Sphaeriaceae.

white fir. *Abies concolor,* family Pinaceae, an evergreen tree grown as an ornamental. Native to North America. Also called *Colorado fir; California white fir; concolor fir; western white fir.*

white flag. Orrisroot iris.

whiteflies. See *whitefly.*

white flour. Flour.

white-flowered gourd. Calabash gourd.

white-flowering dogwood. Flowering dogwood.

whitefly. A minute, sucking insect of the family Aleyrodidae, which is, for the most part, tropical. See *citrus blackfly; citrus whitefly, cloudy-winged whitefly; rhododendron whitefly; woolly whitefly.*

white frangipani. *Plumeria alba,* family Apocynaceae, a tropical tree widely grown for its showy, fragrant, white flowers. Native to the West Indies. Also called *white trumpet flower.*

white-fringed beetles. *Graphognathus* spp., family Curculionidae, insect pests which are general feeders, having been observed on over 385 different species of plants. They infest a plant in vast numbers on the lower part of the stem or taproot causing the plant to yellow, wilt, and die.

white fringed orchis. The white fringeorchid.

white-fringe fungus. A fungus, *Fusarium aleyrodis* Petch, used in the biological control of citrus whitefly, cloudy-winged whitefly, purple scale, and perhaps other scale insects.

white fringeorchid. *Habenaria blephariglottis,* family Orchidaceae, a bog orchid grown in the wild garden for its small, pure-white flower. Native to North America. Also called *white fringed orchis.*

white fringetree. *Chionanthus virginicus,* family Oleaceae, a shrub or tree grown for its showy, white flowers. Native to North America. Also called *old-man's-beard.*

white globe lily. White mariposa.

white goldenrod. Silver goldenrod.

white goosefoot. Lambsquarters goosefoot.

white gourd (of India). Chinese waxgourd.

white grain sorghum. A marketing grade for grain sorghum which includes all varieties of white grain sorghums and may include not more than 10 percent of grain sorghums of other colors.

white grapehyacinth. *Muscari botryoides* var. *album,* family Liliaceae, a bulbous, perennial herb grown for its early-spring, white flowers. Native to southern Europe.

white grapewort. White baneberry.

white grass. Velvetgrass.

white gray. Designating a coat color of some animals characterized by a grayness, very light or almost devoid of colored hair. In some cases also called *flea-bitten.*

white grease. The fat, except that from the intestines, obtained from swine which have died before slaughter. It is used in manufacturing processes but not for food.

white grubs. White, curled larvae of May beetles, family Scarabaeidae, which are found in the soil. They feed on the roots of various plants, including forest trees.

white gum. American sweetgum.

white-head. Take-all.

white-headed scale fungi. Any of three species of fungi, used in the biological control of scale insects: (a) *Podonectria coccicola* has been found on purple scale insects, Glover's scale insects, and chaff scale insects; (b) *P. aurantii* has been found on citrus scale insects in Brazil and Formosa; (c) *P. echinata* has been observed in Ceylon on scale insects.

white heart. See *aster yellows; Dutchman's-breeches.*

whiteheart hickory. Mockernut hickory.

white heart rot (of deciduous trees). A fungous disease of deciduous trees caused by *Fomes igniarius,* family Polyporaceae, characterized by yellowish-white areas in the heartwood surrounded by a zone of yellowish-green to brownish-black. The decayed portions turn light and become soft. The fruiting bodies of the fungus appear on the surface as hoof-shaped brackets at knots or wounds.

white heath. Tree heath.

white heath aster. Heath aster.

white heifer disease. A condition frequently found in white Shorthorn heifers in which an imperforate hymen, poorly developed genitals, or absence of part of the genital organs causes sterility. A genetic factor, it is associated with color character. The condition is occasionally found in other breeds regardless of color.

white hellebore. See *American falsehellebore; white falsehellebore.*

White Holland. A breed of turkey whose plumage is pure white with a deep black beard. The tom weighs up to 33 lbs., the hen up to 18 lbs.

white holly. American holly.

white honeyflower. Black locust.

white horse nettle. Silverleaf nightshade.

white ironbark. White ironbark eucalyptus.

white ironbark eucalyptus. *Eucalyptus leucoxylon,* family Myrtaceae, a tall tree, with smooth, pale to dark gray bark, which is of value in California, U.S., as a lumber tree, bee tree, and an ornamental. Native to Australia. Also called *fat cake; white ironbark.*

white jasmine. *Jasminum nudiflorum,* family Oleaceae, a deciduous shrub grown for its showy, yellow flowers which appear in late winter or early spring before the leaves. Native to China; naturalized in the U.S. from New York southward. Also called *naked jasmine; winter jasmine.*

white laurel. See *sweetbay magnolia; western azalea.*

white lavender. True lavender.

white leaf. White poplar.

whiteleaf Japanese magnolia. *Magnolia obovata,* family Magnoliaceae, a deciduous tree grown for its fragrant, white flowers. Native to Japan.

white lilyturf. *Ophiopogon jaburan,* family Liliaceae, a half-hardy, evergreen, perennial herb grown as an ornamental ground cover for its foliage, its white flowers, and violet-blue fruit. Native to Japan. Also called *jaburan; mondo grass.*

white lime. Silver linden.

white linden. Silver linden.

white line. The strip of horn on the sole of a horse's hoof which is lighter in color and more plastic than the horn of the sole but less dense than the horn of the wall. It is into this part that nails are driven in shoeing the horse.

white-lined sphinx. *Celerio lineata* (F.), family Sphingidae, the larva of a moth which feeds on apple, rhododendron, beet, collard, currant, elm, fuchsia, gooseberry, grape, melon, pear, plum, portulaca, prune, tomato, turnip, and other plants. It is common in western U.S.

white locoweed. Lambert crazyweed.

white lotus of Egypt. White Egyptianlotus.

white lupine. *Lupinus albus,* family Leguminosae, an annual herb grown to some extent for soil improvement. Native to Europe and Asia.

whiteman's foot. Rippleseed plantain.

white mapau. Tarata pittosporum.

white mariposa. 1. *Calochortus albus,* family Liliaceae, a bulbous herb grown for its globe-like, white flowers. Native to North America. Also called *white fairylantern; white globe lilly.* 2. See *charming mariposa.*

white-marked tussock moth. *Hemerocampa leucostigma* (J. E. Smith), family Lymantriidae, a moth whose larva feeds on the foliage of deciduous trees: elm, linden, maple, horsechestnut, poplar, sycamore, buckeye, willow, apple, pear, quince, plum, etc., skeletonizing the leaves and scarring the fruit.

white meat. The meat from the breast of domesticated chickens and turkeys which is white-colored when cooked. *Cf. dark meat.*

white melilot. White sweetclover.

white middlings. Wheat flour middlings.

white mosaic. A type of tobacco mosaic *q.v.*

white mulberry. *Morus alba,* family Moraceae, a hardy, deciduous tree grown in Asia for its leaves used as food for silkworms and in the U.S. as an ornamental. Native to Asia.

white-muscle disease. Stiff-lamb disease.

white mustard. *Brassica hirta,* family Cruciferae, an annual herb grown to some extent for its edible leaves used in salads and as a potherb. Escaped, it can become a weed pest. Native to Europe and Asia.

white myrtle. California laurel.

white Norway. Norway spruce.

white oak. 1. *Quercus alba,* family Fagaceae, a roundheaded, important deciduous, timber tree, which is one of the slowest growing and longest lived trees in the eastern U.S. It is a magnificent tree, but it is not grown ordinarily as an ornamental due to difficulty in transplanting. The commercial uses of its wood are extensive. 2. See *bur oak; post oak; swamp white oak.*

white oats. A market classification of oats based on color of the lemma and palea of matured grain. It includes the yellow varieties.

white peach scale. *Pseudaulacaspis pentagona* (Targ.), family Diaspididae, an armored scale insect found on the U.S. East Coast, which infests ceanothus, alder, poplar, dogwood, shadbush, and tuliptree. Also called *West Indian peach scale.*

White Pekin. Pekin.

white pigweed. Lambsquarters goosefoot.

white pine. Eastern white pine.

white-pine aphid. *Cinara strobi* (Fitch), family Aphidae, an aphid which infests the eastern white pine east of the Mississippi, sometimes killing small trees. A sooty mold develops in the honeydew secreted by the aphids.

white-pine blister rust. A destructive, fungous disease of white pine and other five-needled pines, caused by the rust, *Cronartium ribicola,* family Melampsoraceae. Hyphal-threads of the fungus penetrate the sapwood and inner bark, causing white blisters to appear on the bark. The uredial and telial stages of this rust occur on currants and gooseberries. Occurs in eastern and Pacific U.S. and Europe. Also called *blister rust of pine.*

white-pine cone beetle. *Conophthorus corniperda,* (Schw.), family Scolytidae, a beetle whose larva is destructive to cones of the eastern white pine, feeding on the scales, seeds, and other parts of the cone. Occurs in Canada and northern U.S.

white-pine rust. See *white-pine blister rust.*

white-pine sawfly. *Neodiprion pinetum* (Nort.), family Diprionidae, the larva of an insect that infests the eastern white pine and sometimes other pines in northeastern U.S.

white-pine weevil. *Pissodes strobi* (Peck), family Curculionidae, an insect that infests eastern white pine and other pines and spruces in eastern U.S. The leader on an infested plant becomes brown in contrast to the rest of the tree as a result of girdling by the weevils.

white pitting. A virus disease of gladiolus characterized by short, stiff, brittle leaves, a white pitting of the petals, and a white or brown streaking of the stems and leaves.

whitepopinac leadtree. *Leucaena glauca,* family Leguminosae, an evergreen, tropical tree grown as an ornamental in frost-free areas. Native to America.

white poplar. *Populus alba,* family Salicaceae, a hardy, deciduous, quick-growing, short-lived tree grown as an ornamental. Native to Europe. Also called *abele; Dutch beech; silver-leaved poplar.*

white potato. See *potato.*

white robin. Evening campion.

whiteroot. Eurasian Solomonseal.

white root rot of apples. A fungous disease of apples caused by *Corticium galactinum,* family Thelephoraceae, characterized by a white or cream-colored layer of mycelia on the root surface, and by a knotting or gnarling of the roots.

white rosemary. Heath aster.

white rot. 1. A fungous disease of onions, garlic, shallot, leak, Welsh onion, and wild garlic, caused by *Sclerotium cepivorum,* characterized by a yellowing and dying back of the leaves. A semiwatery decay affects the roots. The young plants may wilt and die rapidly. 2. A rotting of the contents of an egg which results from the invasion of bacteria. To the candler, the egg appears watery. Also called *addled egg.*

white rusts. Minor fungous diseases of plants caused by various species of the genus Albugo, family Albuginaceae. The presence of the abundant mycelium (*q.v.*) within the leaves or floral organs may cause them to become deformed or discolored. On the leaves and other green portions the fungus forms shining, white blisters under the epidermis. When the epidermis is ruptured by the presence of the developing conidia, the latter are distributed by the wind and rain and germinate to infect other host plants. The following host plants and the fungous species causing white rust are important: amaranth, *A. bliti;* crucifers (cabbage, cauliflower, radish, horseradish, wallflower, etc.) *A. candida;* purslane, *A. portulacae;* salsify, *A. tragopogonis;* spinach, *A. occidentalis;* and sweet potato, *A. ipomoeae-panduratae.*

white sage. *Salvia apiana,* family Labiatae, a shrubby herb grown for its white-hairy, dense foliage. It is an important bee plant. Native to California, U.S. Also called *greasewood.*

white sandalwood. *Santalum album,* family Santalaceae, a partially parasitic tree whose fragrant wood is used for ornamental carvings and cabinet work. Native to India.

white sanicle. White snakeroot.

white sapote. *Casimiroa edulis,* family Rutaceae, a tropical tree with leathery leaves and small, greenish flowers, grown in warm areas for its sweet, persimmon-like gray or yellowish-green, edible fruit. Native to Central America. Also called *Mexican apple.*

white scours. See *calf scours.*

white sedum. *Sedum album,* family Crassulaceae, a perennial, evergreen herb with a creeping habit, which is grown in the rock garden for its habit, foliage, and white flowers. Native to the Mediterranean area. Also called *worm grass.*

Whiteside test. A test for abnormal milk in which 2 parts by volume of normal sodium hydroxide is added to 10 parts of milk and stirred. A viscid mass is formed if the milk has a higher than normal leucocyte count which is indicative of mastitis.

white smut. Any fungous disease caused by the parasitic fungi of the genus Entyloma, family Tilletiaceae, in which the smut spores remain enclosed in the tissues of the host plants and do not escape as a black powdery mass. In addition, numerous conidiophores push out through the epidermis and form a white layer with numerous conidia.

white smut of dahlia. See *white smut.*

white smut of spinach. See *white smut.*

white snakeroot. *Eupatorium rugosum* (*E. urticaefolium*), family Compositae, a perennial herb grown for its branched clusters of white flowers which last from midsummer to frost. A poisonous plant, it is found wild from eastern North America to Minnesota and Texas. Symptoms of poisoning in livestock are trembling of the legs, depression, inactivity, constipation, nausea, rapid breathing, and collapse. The poison may be transmitted to humans by milk from a cow which has eaten the plant. Also called *richweed; white sanicle; white snakeweed; Indian sanicle.* See *milk sickness.*

white specks. A minor color defect of American Cheddar cheese characterized by tiny, white specks throughout.

white spongy rot. White heart rot (of deciduous trees).

white spot. A whitish lesion on the stems of conifer seedlings, just above the ground, which results from high temperatures at the soil line. It may be fatal.

white spot of alfalfa. A physiological disease of alfalfa which results from an oversupply of water either by rain or by irrigation. There may be localized spotting or marginal injury.

white spruce. *Picea glauca,* family Pinaceae, a hardy, evergreen tree, which, with its varieties, is grown as an ornamental. Its wood is used in pulp, cooperage, boxes, general building construction, planing-mill products, furniture, ladders, etc. Native to North America. Also called *catpine; skunk spruce.*

whitestem gooseberry. *Ribes inerme,* family Saxifragaceae, an almost spineless shrub, occurring in the Rocky Mountains and northwestern U.S., at 2,000 to 9,500 ft. Sheep, goats, and cattle browse on it when more palatable species are not available. Its food value rating is fair.

white swallowwort. *Cynanchum vincetoxicum,* family Asclepiadaceae, a perennial herb which is a weed in northeastern U.S. Cattle, horses, and sheep are reported to have been poisoned from eating it. Native to Europe.

white swamp azalea. Swamp azalea.

white sweetclover. *Melilotus alba,* family Leguminosae, a biennial herb, which has a wide range of adaptation and which is grown for pasture, hay, silage, green manure, and as a bee plant. Native to Europe. Also called *biennial white sweet clover; Bokhara clover.*

white thimbleberry. *Rubus odoratus* var. *albidus,* family Rosaceae, a shrub occasionally grown for its white flowers. Native to North America. See *fragrant thimbleberry.*

whitethorn. Hawthorn.

white tip of rice. A chlorosis of the rice plant attributed to a deficiency of magnesium.

white top. See *pepperweed whitetop; daisy fleabane; redtop.*

white trumpet. Easter lily.

white-trumpet lily. Easter lily.

white trunk rot. White heart rot (of deciduous trees).

white turtlehead. *Chelone glabra,* family Scrophulariaceae, a perennial herb grown in partial shade for its white flowers. Native to North America. Also called *snakehead; shell-flower; tortoise flower.*

white upland aster. *Aster ptarmicoides,* family Compositae, a perennial herb grown for its white, branching clusters of flowers. Native to North America.

white walnut. See *butternut: 1.*

whitewash. 1. A mixture of slaked lime, water, and other ingredients as desired, used inside chicken coops, hen houses, on tree trunks, etc., as an insecticide, miticide, repellent, etc. It is sometimes used as a paint for fences, shacks, barns, and tenant houses. 2. To apply *whitewash.*

white-water. Braxy.

white waterlily. American waterlily.

white weed. Oxeye daisy.

white wheat. A marketing classification of wheat which has a lack of red pigment in the grain. It is used for cake and pastry flour, crackers, and breakfast foods. See *hard white wheat; soft white wheat; white club wheat; western white wheat.*

white willow. *Salix alba,* family Salicaceae, a deciduous tree, grown as an ornamental. Native to Europe, Asia, and North Africa.

white wintergreen. Creeping pearlberry.

whitewood. See *basswood; cattlebush; Norway spruce; tulip tree.*

white wood aster. *Aster divaricatus,* family Compositae, a fall-flowering, perennial herb grown for its white flowers. Native to North America. Also called *early aster.*

white-wool. Designating those breeds of sheep whose wool is creamy-white.

white yam. Winged yam.

whitly tree. European mountainash.

Whitney godetia. *Godetia grandiflora,* family Onagraceae, a popular, summer-flowering, annual herb, grown for its large, purple, rose to white flowers. Native to North America.

Whitson gillyflower. Dame's-rocket.

whoa. A command given to a horse to stop.

whole grain. 1. Cereal grain which is not cracked or chopped. 2. Those grains which retain their natural form and all their natural constituents.

whole-kernel corn. See *Maryland-style corn.*

whole milk. Milk, especially cows' milk, which has not had any portion of the fat removed.

whole pressed cottonseed. A livestock feedstuff which is the product that results from subjecting whole undecorticated cottonseed to the expeller or screw process for extraction of oil. It includes the entire cottonseed less the oil extracted and the linters removed.

whole pressed cottonseed, prime quality. Whole pressed cottonseed (*q.v.*) that must be produced from sound cottonseed, shall be firm in texture, must not have a sour, musty or burnt odor, and shall contain not less than 22 percent of crude protein. It shall be sold according to its protein content.

whole pressed peanuts. See *unhulled peanut oil feed.*

whole-root graft. Root grafting (*q.v.*) in which a whole root is used for the stock instead of a piece of root. See *piece-root graft; nurse-root-graft.*

wholesale cut. A large cut of meat, used for ready distribution, from which the retail cuts are taken.

whole-wheat. Whole-wheat flour.

whole-wheat flour. The food product made by grinding wheat. It contains all the constituents of the cleaned grain in their natural proportions. Also called *whole-wheat; entire-wheat flour; graham flour; Arlington meal.*

whorl. 1. A swirl, cowlick, in the hair on an animal's coat. 2. Three or more twigs, flowers, leaves, floral parts, arranged in a circle at one point on a plant.

whorled milkweed. Poison milkweed.

Wichura rose. *Rosa wichuraiana,* family Rosaceae, an evergreen or half-evergreen prostrate shrub grown for hybridizing and as an ornamental ground cover for its habit and its white, fragrant flowers. Native to eastern Asia. Also called *memorial rose.*

wick. 1. A thick, heavy cloth, fiber glass, etc., placed in the bottom of a flower pot with one end in a supply of water, which furnishes a constant supply of water to the plant. 2. A loosely-woven cloth which, by capillary attraction, draws up a proper amount of oil, in a lamp, or tallow, in a candle, for illumination purposes. 3. A farm, especially a dairy farm (English). Also called *wica.*

wickerwork. Basket work made by the interlacing of wickers, otherwise osiers or withes, the twigs of certain willows.

wicket. 1. A small gate set within a larger one; a half door. 2. A small sluice or opening for letting the water out of a canal lock. 3. A valve or throttle in the chute of a water wheel for regulating the flow of water.

wicket dam. A movable barrier made of shutters revolving about a central axle (obsolete).

wicopy. Atlantic leatherwood.

wide. Designating a thick-bodied animal.

wideleaf sea-lavender. *Limonium latifolium,* family Plumbaginaceae, a perennial, sea or marsh herb grown for its white and blue flowers used for everlastings. Native to Europe and Asia.

wide loop. A promiscuous collecting of cattle by less than honest means, without regard to their actual ownership (southwestern U.S.).

wide nutritive ration. A feed or ration which has little protein in proportion to its non-nitrogenous nutrients. See *narrow nutritive ration.*

wide ration. See *wide nutritive ration.*

widespread device. A device attached to the rear end of a manure spreader consisting of curved or spiral, steel blades which rotate rapidly to broadcast manure. One-half of the device is set to throw to the right, the other to the left.

wide-sprung. Designating an animal with well-arched ribs and a broad, deep chest.

wide-track seeder. A type of broadcast seeder which has the wheels set far apart at each end of the seed box. See *narrow-track seeder.*

widow flower. Sweet scabious.

widow's-cross. Texas stonecrop.

width of carcass. The distance from the chine bones to the outside line or curve of the carcass.

wiener. A sausage made from pork and beef sometimes also with veal which is smoked and cooked. Also called *wienerwurst; wienie.*

Wiesenboden. Blackish soils high in organic matter which developed on wet sites mostly supporting grasses and sedges. The soil development process is gleization in humid to subhumid climates, and the surface organic layer is underlain by a grayish or glei horizon usually at a depth of 6 to 30 ins. The soils may be fairly productive for agriculture if properly drained and managed. See *humic-gley soil; glei horizon; meadow soil.*

wigging. Clipping the wool around the eyes of a sheep (Australia).

wiggletail. See *seat guide.*

wild. 1. Uncultivated; not cultivated by man; growing without the care of man. 2. Not domesticated, as a *wild* animal.

wild alfalfa. Sickle alfalfa.

wild areas. Areas of forest lands similar to wilderness area (*q.v.*) which are subject to the same restrictions and regulations but are smaller in size, namely less than 100,000 but more than 5,000 acres.

wild ash. American mountain ash.

wild balsam apple. Wild mockcucumber.

wild barley. Foxtail barley.

wild bean. See *American potatobean; lupine.*

wild bees. Bees living in hollow trees or other abodes not prepared for them by man, as distinguished from the bees raised on a farm.

wildbergamot beebalm. *Monarda fistulosa,* family Labiatae, a perennial herb, grown for its lilac or purple flowers. Native to North America.

wild black cherry. Black cherry.

wild bleeding-heart. Fringed bleedingheart.

wild blueberry. Any of the several unnamed and uncultivated species of blueberry, genus Vaccinium, family Ericaceae, which grow in acid, heathland soils.

wild blue phlox. Sweetwilliam phlox.

wild brier. Sweetbrier rose.

wild buckwheat. Dullseed cornbind.

wild bugloss. *Anchusa arvensis,* family Boraginaceae, a small, bristly plant.

wild calla. *Calla palustris,* family Araceae, an aquatic herb grown along the edge of a pond in mud or shallow water for its oval or heart-shaped leaves and green flowers. It is poisonous for livestock. Symptoms of poisoning are an intense burning feeling in the mucous membranes of the mouth and throat. Native to the North Temperate Zone. Also called *water arum.* See *callalily.*

wild chamomile. See *German chamomile.*

wild carrot. *Daucus carota,* family Umbelliferae, an annual or biennial herb, which is a serious weed pest but also a beautiful flower plant. Eaten by cows, it imparts an objectionable flavor to the milk. Native to Europe and Asia. Also called *bird's nest; lace-flower; devil's-plague; bee's-nest plant; Queen Anne's lace.*

wild celery. Any plant of the genus Vallisneria, family Hydrocharitaceae. Species are submerged, aquatic herbs. See *spiral wildcelery.*

wild cherry poisoning. See *black cherry.*

wild corn. The speckled beadlily.

wild cucumber. The wild mockcucumber.

wilderness. In the early period of settlement of the U.S., primitive forested region, which was uninhabited by the white man (except possibly by the trapper and trader), as contrasted with established settlements and the frontier.

wilderness area. An area of land providing isolation from the sights and sounds associated with modern living; having the general appearance of being unaltered from its natural state by commercial, industrial, or agricultural activities and characterized by primitive conditions of transportation and the absence of permanent habitations.

wildfire. 1. A bacterial leaf-spot disease of tobacco and soybeans caused by *Pseudomonas tabaci,* family Pseudomonadaceae, characterized on tobacco by yellow lesions with white spots of dead tissue in the center. In soybeans, it is characterized by light brown to dark necrotic spots each surrounded by a yellow, sometimes indistinct, halo. Severe infestation can cause severe shedding. 2. The promiscuous, indiscriminate, or accidental burning of vegetation. It involves no responsibility for damage to property resulting from escape of the flames.

wild flag. The blueflag iris.

wildflooding. Irrigation water released in a field without controlled distribution.

wild flower. Any plant, with attractive or interesting blossoms, leaves, habits, etc., which flourishes untended and uncultivated in its natural habitat.

wild fowl. 1. Wild birds collectively. 2. Wild, aquatic birds; sometimes only the game species, such as ducks and geese.

wild garlic. Field garlic.

wild geranium. The spotted geranium.

wildginger. Any plant of the genus Asarum, family Aristolochiaceae. See *Canada wildginger.*

wildgoose. Tartary buckwheat.

wildgoose plum. *Prunus munsoniana,* family Rosaceae, a variety of plum with fruit yellowish or red with a slight bloom; it is of little commercial importance. Native to south central and southeastern U.S.

wild grass. Any grass, family Gramineae, which has not been sown and which grows untended on ranges, pastures, meadows, etc.

wild hay. Hay made from native or wild, uncultivated grasses and plants. See *tame hay.*

wild hoarhound. *Eupatorium aromaticum,* family Compositae, a perennial herb, grown for its clusters of white flowers. Native to North America. Also called *pool-root.*

wild hollyhock. Foothill checkermallow.

wild honeysuckle. Scarlet gaura.

wild horse. An untamed and/or undomesticated horse. In the U.S., the *wild horses,* roaming the ranges, have descended from the once domesticated horses. Generally, very similar to the common domestic breeds they are distinct from the *wild horse* of Africa and are captured and used as a source of bucking broncos in rodeos, as riding ponies, and for slaughter.

wild horse runner. A person who catches wild horses (western U.S.).

wild hyacinth. The blue squill.

wild hydrangea. Smooth hydrangea.

wildindigo. Any plant of the genus Baptisia, family Leguminosae. Also called *false indigo; wild indigo.* See *blue wildindigo; yellow wildindigo.*

wilding. Wildling.

wild ipecac. The spreading dogbane.

wild iris. Blueflag iris.

wild land. Uncultivated land maintained primarily for its protective vegetation cover, wood, forage production, recreation, or wildlife.

wild lantana. Pink sandverbena.

wild leek. *Allium tricoccum,* family Liliaceae, a bulbous, perennial herb, which is a weed pest. Cows feeding on it give milk which has an onion off-flavor. Also called *ramps.*

wild lettuce. Prickly lettuce.

wild licorice. 1. *Glycyrrhiza lepidota,* family Leguminosae, a perennial herb growing to 5 ft., with small, yellow flowers and prickle-covered pods. Unlike licorice *(q.v.),* it is not grown commercially, though its roots have similar properties. Native to western North America. Also called *American licorice.* 2. The three species of the genus Galium, family Rubiaceae, which are slender, bedstraw herbs with sweet roots. Native to Europe, Asia and North America. See *white bedstraw.* 3. Buttonbush. 4. Jequirity rosarypea.

wildlife. The complete fauna, exclusive of domesticated animals. In common usage, the mammals and birds of rural and wilderness areas.

wildlife area improvement. Planting, fencing, ditching, control of grazing, fire, and water level, etc., that will provide food, cover, and water for wildlife in such areas as marshes, odd areas, fence rows, hedges, gullies, streambanks, borrow pits, and water-logged land.

wildlife area management. Management of an area, particularly of its vegetation and water, for the benefit of wildlife.

wildlife border. Strips of herbaceous and/or woody plants developed around the edges of fields or woodland for the encouragement and benefit of wildlife, and sometimes for the additional purpose of checking erosion.

wildlife conservation. 1. Preservation of rare or vanishing species. 2. Reasonable use and management of environmental factors so as to allow maximum breeding populations and cropping of game, furbearers, etc.

wildlife land. a. Land, whether or not suitable for other uses, which is used only or mainly for producing or maintaining wildlife. b. Land of such character that its only practical use is for wildlife habitat or production.

wildlife management. Any planned activity aimed at increasing or decreasing numbers of any wildlife species, restricting damage or increasing beneficial activities of wildlife.

wild lilac. Ceanothus.

wildling. 1. An escape; a cultivated plant which has become wild and thrives without cultivation. 2. Any wild plant or animal. 3. In forestry, a seedling produced naturally outside of a nursery.

wild madder. White bedstraw.

wild mockcucumber. *Echinocystis lobata,* family Cucurbitaceae, an annual, herbaceous, rapidly-growing vine grown as an ornamental for quick cover. It is harmful because it is often a host for the virus of cucumber mosaic (*q.v.*). Native to North America. Also called *wild balsam apple; mockcucumber.*

wild monkshood. Clambering monkshood.

wild morning-glory. See *European glorybind; hedge glorybind.*

wild mustard. Charlock.

wild oat. 1. *Avena fatua,* family Gramineae, an annual grass, distributed over most of the U.S., and which is a weed pest in small grains. It is grazed on the western ranges and is sometimes used for hay. Its general forage rating is good when pastured or harvested in the immature stages. Native to Europe and Asia. 2. See *cheatgrass brome; little merrybells.*

wild oleander. Waterwillow.

wild oleaster. Silver buffaloberry.

wild onion. See *Canada garlic; field garlic.*

wild-orange. Carolina laurelcherry.

wild ox. *Bos primigenius,* the early wild ox of Europe.

wild pansy. *Viola tricolor,* family Violaceae, a very hardy, lovely, temperate zone, perennial herb of straggling habit that is the wild stock from which the garden pansy, *V.t. hortensis,* grown for its almost year-round, blue, white, and yellow, solitary flowers is derived. Also called *hearts-ease; Johnny-jump-up.*

wild passionflower. Maypop passionflower.

wild pasture. A pasture of native or volunteer grasses and plants.

wild pea. Arrow crotalaria.

wild pepper-grass. Virginia pepperweed.

wild pink. Arethusa.

wild-plantain. Carib heliconia.

wild plum. American plum.

wild pumpkin. Calabazilla.

wild radish. *Raphanus raphanistrum,* family Cruciferae, an annual herb which is a weed pest in grain fields. Stock feeding on large quantities of the seed mixed with grain may become poisoned. Characteristics of the illness are enteritis, hemorrhagic diarrhea, colic, abortion, nephritis, apathy, and paralysis of heart and respiratory organs.

wild raisin. Witherod viburnum.

wild red oat. Animated oat.

wild rice. Annual wildrice.

wild rosemary. Crystaltea ledum *q.v.*

wild rye. Any of the tall, perennial grasses of the genus Elymus, family Gramineae. See *Canada wildrye; giant wildrye; blue wildrye.*

wilds. Country which has not been cleared and cultivated: woods, uninhabited prairies, remote areas, etc.

wild sage. Fringed sagebrush.

wild sarsaparilla. *Aralia nudicaulis,* family Araliaceae, a wild, perennial herb sometimes grown in the shade in the wild garden for its ball-like cluster of greenish-yellow flowers. Native to North America. Also called *American sarsaparilla.*

wild senna. *Cassia marilandica,* family Leguminosae, a hardy, perennial herb grown for its June-blooming, yellow flowers. Native to North America. Also called *American senna.*

wild snapdragon. Butter-and-eggs toadflax.

wild spikenard. Feather Solomonplume.

wild stonecrop. Mountain stonecrop.

wild sunflower. Giant sunflower.

wild sweet crabapple. *Malus coronaria,* family Rosaceae, a small tree, grown for its early-spring, pale pink, fragrant blossoms. It bears a small, yellow, hard fruit. Native to North America. Also called *American crab; sweet crabapple; garland crabapple.*

wild sweet pea. Virginia tephrosia.

wild sweet potato. Bigroot morningglory.

wild sweetwilliam. Sweetwilliam phlox.

wild tansy. Ragweed.

wild thyme. Mother-of-thyme.

wild tobacco. See *Aztec tobacco; coyote tobacco.*

wild tomato. Carolina horsenettle.

Wild West. A term applied to the western areas of the U.S. during early settlement when law and order was almost nonexistent. It suggests a fabulous, sometimes almost mythical, West of the cowboys, Indians, fur traders, gold and silver miners, remittance men, squatters, rustlers, land sharks, sod huts, dancehall girls, gamblers, saloon keepers, Mexicans, pony express, stage coaches, cattle barons, bad men; of the Alamo, of Dodge City, of Cheyenne, of Denver, of Central City; of the gold rush, of the Santa Fe trail, etc.

wild wormwood. Canada sagebrush.

will. A legal document, signed by a person and two unrelated witnesses, containing instructions for the disposition of his property after his death.

Willamette Valley. The drainage basin of the Willamette River and its tributaries in western Oregon, U.S. It is mainly a flat alluvial plain partly covered by hills of basalt rock. Part of the land is excessively wet but the arable soils are very productive, yielding many crops, notably hops.

willow. 1. Designating a shrub or tree of the genus Salix, family Salicaceae. Species are quick-growing and often brittle. See *Babylon weeping willow; basket willow; black willow; goat willow; laurel willow; pussy willow; Thurlow weeping willow; white willow; Wisconsin weeping willow.* 2. See *desertwillow; waterwillow; willow oak.*

willow flea weevil. *Rhynchaenus rufipes* (Lec.), family Curculionidae, a beetle, found from Maine to Oregon, Colorado, New Mexico, and California, U.S., whose larva mines the leaves of Lombardy poplar, aspen, and river birch.

willow-herb. See *willow-weed; purple lythrum.*

willow leaf beetle. The imported willow leaf beetle.

willowleaf spirea. *Spiraea salicifolia,* family Rosaceae, a deciduous shrub grown for its rose-pink flowers. Native to southern Europe and Asia. Also called *queen-of-the-meadow; bridewort.*

willow oak. *Quercus phellos,* family Fagaceae, a deciduous tree grown for shade and ornament in warm areas. Native to North America.

willow pittosporum. *Pittosporum phillyraeoides,* family Pittosporaceae, a small, evergreen tree grown as an ornamental for its drooping limbs and yellow flowers. Native to Australia. Also called *narrow-leaved pittosporum; butter bush.*

willow sawfly. *Nematus ventralis* Say, family Tenthredinidae, the larva of a sawfly which infests willows and poplars, feeding on the leaves. Trees are sometimes defoliated. Also called *yellow-spotted willow slug.*

willow shoots. Slender, much-branched, fairly erect, wiry shoots, appearing most frequently on larger limbs of peaches infested with peach yellows (*q.v.*). They have small, yellow leaves sometimes spotted with red. Also called *broom shoots.*

willow twig. A virus disease of peach characterized by the death and shedding of lateral buds and the consequent bare growth. Reported from California, U.S.

willowwort. Golden loosestrife.

willow-weed. Any plant of the genus Epilobium, family Onagraceae. Also called *epilobe; willow-herb; epilobium.*

Wilson barberry. *Berberis wilsonae,* family Berberidaceae, a half-evergreen shrub grown for its clusters of golden-yellow flowers and coral-red or salmon-colored, ornamental fruit. It is sometimes rust infected. Native to China.

Wilstermarsch. A cows' milk cheese made in Germany. Similar to Tilsiter, it has a mild acid flavor, and is cured three to four weeks.

wilt. 1. Loss of freshness and a drooping of the foliage of a plant due to an inadequate supply of moisture, excessive transpiration by the plant, and as a result of a disease interfering with the utilization of water by the plant. 2. Of a plant, to lose freshness of the foliage and droop, or to become flaccid as a result of inadequate soil moisture, disease, or its being cut.

wilt disease. See *stem rot; Fusarium (wilt diseases); Verticillium wilt.*

wilted. Designating a plant whose foliage has lost its freshness and is drooping as a result of a lack of moisture in the soil, a disease, or its being cut.

wilting. The drooping and loss of freshness of foliage due to water deficiency, commonly as a result of the blocking of the conductive system of the plant or of a deficiency of available soil moisture.

wilting coefficient. The moisture, in percentage of dry weight, which remains in the soil within the root-feeding zone as plants reach a condition of permanent wilting. An approximation to the permanent wilting point is obtained by dividing the moisture equivalent by 1.84. Also called *wilting point; wilting range.*

wilting method. A method of making silage from legumes, grasses, and small grains in which the cut plants are allowed to lose moisture until the moisture content is around 65 percent before storage. See *acidified silage.*

wilting percentage. Permanent wilting point.

wilting point. See *wilting coefficient; permanent wilting point.*

wilt of aster. See *Verticillium wilt.*

wilt of beans. Bacterial wilt of beans.

wilt of begonias. See *Verticillium wilt.*

wilt of cabbage. Cabbage yellows.

wilt of citrus. Citrus blight.

wilt of cotton. See *Verticillium wilt.*

wilt of eggplant. See *Verticillium wilt.*

wilt of elm. Dothiorella wilt.

wilt of flax. See *Fusarium wilt (diseases).*

wilt of maple. See *Verticillium wilt.*

wilt of sweet potato. See *Fusarium wilt (diseases).*

wilt of tomato. See *Fusarium wilt (diseases).*

wilt-resistant. Designating a plant variety which does not readily become infected with a wilt disease.

wilts. Plant diseases in which the plant becomes limp and droops, generally because invading organisms obstruct the flow of sap. *Erwinia tracheiphila* is the cause of cucumber wilt, *Xanthomonas campestris* of cabbage wilt, and *Bacterium stewartii* of corn wilt.

Wiltshire. A hard, sweet-curd, cows' milk, English cheese, similar to Derby.

Wiltshire bacon. Wiltshire side.

Wiltshire side. An English cut of pork which is half of a dressed hog carcass with the backbone, aitchbone, shoulder blade, and feet removed. Also called *Wiltshire bacon.*

winch. 1. A hand crank for turning a machine which revolves. See *reel.* 2. A drumlike device, turned manually or mechanically by a crank, on which rope is wound; used for hauling or hoisting.

Winchester bushel. Once a standard bushel measure in Pennsylvania, it was 18.5 ins. in diameter and 8 ins. deep, or 2150.42 cu. ins. in capacity.

wind. 1. The horizontal movement of air on and above the earth's surface. (The best velocity for driving windmills is from 15 to 20 miles per hour.) 2. The blast of air from a blower or fan. 3. The breathing of a horse. See *heaves; roaring.*

wind and work. A common guarantee, given by the seller of a horse at an auction, which denotes some unsoundness but warrants good wind with ability and willingness to work.

wind blast. The current of air forced out by a blower or fan.

wind blower. 1. An attachment to a grain thresher which consists of a large fan funnel, intake device, and a discharge pipe for forcing the straw away from the thresher up onto a stack. See *wind stacker.* 2. A mechanical device used in modern hay barns for forcing air through newly stored hay to hasten drying.

wind-borne. Designating diseases, seeds, spores, etc., distributed by air movements. Also called *air-borne.*

windbreak. 1. Any object which serves as an obstacle to free movement of surface winds. In forestry, trees serve such a purpose. Tree *windbreaks* are classified according to their general arrangement as rows and hedgerows, belts and shelterbelts of three or more rows, and groves, or in the most extensive case, forests. They may be of natural or artificial origin. 2. A vegetative barrier used to reduce or check the force of the wind.

wind burn. 1. A withered, seemingly blasted, appearance of foliage which results from the action of violent wind on a plant. Fruit may also exhibit mechanical damage from exposure to wind while being shipped, or from dehydration. 2. The bronzed condition of some evergreen foliage caused by an inadequate water supply usually during late winter. It is associated with high transpiration rate and frozen soil or low soil-water content.

wind colic. A form of colic (*q.v.*) in which there is a distention of the bowels with gas so that a drumlike sound is heard when the animal's side is slapped. Also called *flatulent colic; tympanites.*

wind dunes. See *dune.*

wind erosion. Erosion of soil and rock formations by wind action.

windfall. 1. A tree uprooted or broken off by wind. 2. An area on which the trees have been blown down by wind. 3. A fruit that falls from a tree before the crop is harvested, usually associated with wind. Also called *drop.*

wind-firm. Designating a tree which withstands a heavy wind.

windflower. Anemone.

wind-gall. A puff or bursal distention at the side of the tendon at or above the fetlock of a horse.

wind gap. A natural land feature: a narrow gorge, or notch in a ridge, which may have been cut by a stream, but in which the stream is no longer present.

wind indicator. A device that indicates the direction and velocity of the surface wind. See *vane.*
winding. In a horse's gait, a twisting of the striding leg around in front of the supporting leg in much the same manner as in paddling.
wind injury. See *wind burn.*
wind mantle. A screen of trees and woody shrubs, commonly used to delimit the border of a stand or forest, which prevents penetration of wind into the interior.
windmill. A device which consists of propeller-like blades, sometimes called sails, attached to a gear head and connected to a central shaft and a rear vane to guide the wheel (fan) into the wind. The action of the wind causes the blades to turn the shaft. The energy of the wind is used most frequently for pumping water and generating electricity.
windmill governor. A coiled, steel spring located in the base of the vane frame of a modern windmill which is so adjusted that it holds the disk or fan wheel in the wind under ordinary pressures, but yields when the pressure becomes that of a storm.
windmill grass. 1. The stiff, spreading-radiating, spiked species of the genus Chloris, family Gramineae, some of which are weeds but have a fair forage rating. 2. See *tumble windmill grass.*
windmill palm. Fortunes windmill palm.
windmill pump. A well pump of the reciprocating type actuated through a crank driven by a windmill. A reservoir or supply tank is generally erected adjacent to the windmill derrick, a float in this tank throwing the pump out of gear when full, and engaging it once more as the water level falls.
windmill sails. The propeller-like blades of a windmill which are usually approximately rectangular with a length of 5 times their breadth, the total amount of surface being about ¼ the area of the circle described by their motion.
window box. A flower box (*q.v.*) usually located outside a window.
window gardening. The growing of plants in window boxes *q.v.*
wind-pollinated. Designating those plants which have the pollen carried by the wind instead of by insects, as the pines, grasses, etc.
wind poppy. Flaming poppy.
wind power. The force of the atmosphere in motion exerted upon the vanes or sails of windmills and various other motors.
wind pressure. The crushing power or force of the wind against some obstacle or opposing force.
wind-puff. Air under the skin.
windrow. 1. A long, relatively low, ridge of hay, sheaves of grain, etc., which is made to effect drying, curing, etc. 2. Accumulations of slash left in rows by loggers. 3. To rake hay or to place sheaves in a relatively low, long row for drying and curing.
windrow baler. A hay baler (*q.v.*) that moves across a field, picking up hay from the windrow, baling it, and delivering the bales to a truck or onto the ground.
windrower. A mechanical device used for taking the cut hay or grain from the swath and turning it into a windrow ready for further handling with the hay loader, field chopper, hay baler, or combine.
windrowing attachment. An attachment for a cutter bar, consisting of a number of metal bars, each curved upward at the rear end. They decrease in length from about 8 ft. to 3 ft. The purpose is to windrow harvested crops.
windrow loader. Cylinder hay loader.
windrow pick-up attachment. An attachment for a combine which consists of a series of metal fingers on an endless belt. The fingers pick up grain which has been windrowed for combining.
windrow pickup baler. Windrow baler.
windrow turner. Any of several different devices which turns over windrowed hay to assist in curing.
wind shield. The vertical board or canvas covered frame that is mounted behind the platform canvas on a grain binder to keep the wind from affecting the fall of cut grain onto the platform canvas.
wind stacker. An attachment on a grain thresher which consists of a large fan or blower and a tubelike stacking pipe. The fan receives the straw and blows it out of the pipe. The pipe can be extended and moved in both the vertical and horizontal plane to aid in stacking the straw.
windstorm. Strong to violent winds which are usually not accompanied by precipitation.
wind strip cropping. The production of crops in long, relatively narrow strips which are placed crosswise of the direction of the prevailing winds without regard to the contour of the land.
wind sucking. See *cribber.*
wind tight. Designating a structure, etc., which is so tight as to prevent the passing through of wind.
wind valley. Wind gap.
windward. 1. Designating the point from which the wind blows. 2. Situated toward the point from which the wind blows.
windy. Of, or pertaining to, a roarer *q.v.*
wine. 1. The product made by the normal alcoholic fermentation of the juice of sound ripe grapes containing not less than 7 percent or more than 16 percent of alcohol by volume and in 100 cubic centimeters (20° C.), not more than 0.1 gram of sodium chloride nor more than 0.2 gram of potassium sulfate; and for red *wine* not more than 0.14 gram and for white *wine* not more than 0.12 gram volatile acids produced by fermentation and calculated as acetic acid. 2. Any fermented alcoholic beverage which is made from miscellaneous fruits, as blackberry *wine,* elderberry *wine,* etc.
wineberry. See *wine raspberry; mako mako wineberry.*
wine grape. European grape.
wine palm. Toddy fishtailpalm.
wine plant. Garden rhubarb.

wine raspberry. *Rubus phoenicolasius,* family Rosaceae, a hardy, prickly shrub grown for its pinkish or white flowers and as a hedge plant. Native to China and Japan. Also called *wineberry.*

winery. An establishment which manufactures wine.

Winesap. A widely known variety of apple bearing a large, red, round-oblate, good-flavored fruit which keeps well.

wine stone. Cream of tartar which crystallizes on the walls of wine tanks.

wine taster. A judge of wines who bases his decisions on the flavor and aroma of the wine.

wine vinegar. A vinegar made from grape wine. Usually quite high in acidity, it may contain as much as 7 percent acid, calculated as acetic. *Wine vinegar* may be prepared by running wine through generators, or by simply allowing it to stand in barrels from which the bung has been removed. See *grape vinegar.*

wing. 1. A leaf-like, dry or membranous expansion or appendage of a plant part, such as along some stems and petioles and of samaras and some capsules. 2. Either of the lateral petals of a pea-like flower. See *standard: 4; keel: 2.* 3. An organ of flight for birds, insects, bats, etc. 4. A piece of bird meat consisting of the organ of flight with the feathers removed. 5. The outside corner of the cutting edge of a plow. 6. Fan or blower blade. 7. An extension on a building. 8. The vane of a windmill.

wing band. A metal or plastic device placed around the web cord of the wing for marking and identifying a bird.

wing-band pliers. A pliers-like tool used to fasten a wing band *q.v.*

wing bar. On parti-colored chickens, the stripe or bar of color extending across the middle of the wing. It is formed by the color or markings of the double row of secondary wing coverts.

wing bay. The triangular part of the wing (of chickens), below the wing bar, formed by the exposed ends of the secondary wing feathers when the wing is folded.

wing bearing. The flat surface on the wing of the plow share. Up to 1.5 ins. wide, it is the portion that remains in contact with the ground and carries some of the downward thrust.

wing bow. The upper part of the wing of a chicken below the shoulder and wing-front and above the wing bar.

wing coverts. The small, close-fitting feathers which cover the bend of bird's wing and the quills of the major (primary and secondary) wing-feathers.

wing dam. A water-deflecting structure, such as a wall, crib, dike, row of piles, or other barrier which projects streamward from the shore; a spur dike.

winged euonymus. *Euonymus alatus,* family Celastraceae, a deciduous shrub grown for its bright red, autumnal foliage. Native to China and Japan. Also called *winged spindle-tree.*

winged fruit. A fruit having a wing. See *wing: 1.*

winged lythrum. *Lythrum alatum,* family Lythraceae, a perennial herb grown for its purple flowers. Native to North America. Also called *milk willow-herb.*

winged pea. Squarepod deervetch.

winged pigweed. Tumble ringwing.

winged shoulders. A set of shoulders having a winglike appearance, peculiar to some cattle, particularly Guernseys.

winged spindle-tree. Winged euonymus.

winged tobacco. *Nicotiana alata,* family Solanaceae, a flowering, perennial herb grown as an annual in the flower border for its night-scented, violet flowers, white inside. Native to South America. Also called *nicotiana; flowering tobacco.*

winged-wheel feed. A type of a feeding device for fertilizer distributors consisting of a series of horizontal disks with prominent ridges. The disks revolve at high speeds, and broadcast the fertilizer as it falls on them.

winged yam. *Dioscorea alata,* family Dioscoreaceae, an herbaceous vine widely grown in the Orient for its edible tuber and often in the U.S. as an ornamental. Native to eastern Asia. Also called *white yam.*

wing everlasting. *Ammobium alatum,* family Compositae, a bushy perennial herb grown as an annual for its yellow flowers which, when dried, hold their colors almost indefinitely. Native to Australia. See *everlasting: 1.*

wing feather. In the wing section of a bird, any feather, especially one of the large primary or secondary feathers, which forms part of the principal wing surface in flight. Also called *remiges.*

wing finisher. A mechanical device for removing feathers from the wings of birds. It consists of a revolving drum fitted with rubber fingers.

wing front. The upper, front edge of a bird's wing at the shoulder.

wing hiller. A type of cultivator attachment used for hilling crops, etc.

winging. Of a horse, paddling or throwing of the feet outward when in motion.

wing louse. Turkey louse.

wingnut. Any tree of the genus Pterocarya, family Juglandaceae. Species are deciduous trees grown as ornamentals for their attractive leaves and winged fruits. Also called *false walnut.* See ***Chinese** wingnut.*

wingseed. Any seed having a wing. See *wing: 1.*

wing shovel. See *sweep: 1.*

wingstem passionflower. *Passiflora alata,* family Passifloraceae, a climbing vine with red and white flowers, grown for its many-seeded, edible berries from which jams, sherbets, and soft drinks are made in many tropical countries. Native to tropical America.

wing-tipped. See *wingseed.*

wink. To clip the wool away from a sheep's eyes (New Zealand).

winnow. To separate grain, etc., from the husks by means of an air current or by wind.

winnowing. The process of separating threshed

grain from dust, chaff, etc., by a combination of sifting and blowing machines.

winter. 1. The cold season of the year in the temperate and arctic zones. 2. To spend the *winter* and retain viability or vigor throughout the *winter* period.

winter aconite. *Eranthis hyemalis,* family Ranunculaceae, a tuberous herb grown for its very early-spring, yellow flowers. Native to Europe and Asia.

winter annual. An annual plant *(q.v.)* whose seed germinates in the fall or early winter, over-winters in a vegetative condition, often in a rosette form, completes vegetative growth, matures a crop of seed, and then dies in the late spring or early summer.

winter apple. Any variety of apple that matures late in the season and can be held in refrigerated storage for a minimum of 3 months, as Jonathan, Delicious, Winesap, etc.

winter barley. A winter annual *(q.v.)* barley. *Cf. spring barley.*

winterberry. 1. *Ilex verticillata,* family Aquifoliaceae, a shrub whose branches with red berries are gathered and used locally for Christmas decorations. Native to North America. Also called *black alder; Michigan holly; dogberry; false alder.* 2. See *inkberry.*

winterberry euonymus. *Euonymus bungeanus,* family Celastraceae, a shrub or small tree grown for its foliage, its yellow flowers, and its persistent pinkish fruits. Native to China.

winter blight of stone fruits. Coryneum blight of stone fruits.

winter bluegrass. Mutton bluegrass.

winter browning. A browning or bronzing of evergreens in the winter which results from a condition in which transpiration is fairly rapid whereas the roots in the cold or frozen soil cannot replace the water deficit. Also called *wind burn.*

winter capemarigold. *Dimorphotheca aurantiaca,* family Compositae, a perennial, somewhat woody herb grown as an annual for its white, bluish, or red flowers. Native to South Africa. Also called *African daisy.*

winter cauliflower. Broccoli.

winter cherry. Strawberry groundcherry.

winter cluster. The massing together of bees in a hive, as in a ball, to generate and conserve heat for wintering.

winter cover crop. Any of several plants, such as winter rye, wheat, or oats, planted in the early autumn. It makes sufficient growth before winter comes to aid in protecting the soil from wind and water erosion. In the spring it is usually plowed under and used as green manure.

wintercreeper. Bigleaf wintercreeper euonymus.

wintercreeper euonymus. *Euonymus fortunei* var. *radicans,* family Celastraceae, a trailing or climbing, evergreen, woody vine widely grown for its foliage. Native to Asia.

wintercress. Any plant of the genus Barbarea, family Cruciferae. Species are sometimes used as substitutes for watercress. See *early wintercress; bitter wintercress.*

winter crop. A crop plant which is a winter annual *q.v.*

winter daffodil. Fall daffodil.

winter dormancy. Dormancy *(q.v.)* that occurs during winter.

winter-dropped lamb. A lamb born in the winter.

winter drying. Winter browning.

winter-fallow. Designating land which is fallow in winter.

winter fat. *Eurotia lanata,* family Chenopodiaceae, a low shrub found from Texas, U.S. and Manitoba, Canada to the West Coast. One of the outstanding browse plants on the western range, its forage rating is excellent.

winter feed(ing). Feed for, or feeding of livestock during the winter months.

winter grafting. Grafting *(q.v.)* performed in the winter. See *dormant grafting.*

winter grain. Any cereal plant which is a winter annual *q.v.*

winter grape. Frost grape.

winter grass. See *annual bluegrass; mutton bluegrass.*

wintergreen. Any of several plants of the genus Gaultheria, family Ericaceae. See *checkerberry.*

wintergreen barberry. *Berberis julianae,* family Berberidaceae, a very hardy, evergreen shrub, grown as an ornamental for its foliage. Native to China.

wintergreen oil. Oil of wintergreen.

winter hardiness. The physiological adjustment within a plant that enables it to endure the rigors of winter (snow, freezing, ice) and to retain its vigor the next season.

winter-hardy. Designating a plant which successfully endures the rigors of winter. See *winter hardiness.*

winterhazel. Any shrub of the genus Corylopsis, family Hamamelidaceae. See *spike winterhazel; buttercup winterhazel.*

winter heath. Spring heath.

winter honeysuckle. *Lonicera fragrantissima,* family Caprifoliaceae, a hardy, deciduous shrub, grown for its very fragrant, creamy-white flowers. Native to China.

wintering. The care, feeding, and maintenance of livestock through the winter months. See *winter feed(ing).*

winter injury. Any injury to plants which results from the rigors of winter; the death of plant parts due to drying during cold weather, especially when soil is frozen deeply.

winter irrigation. The application of water to soil in late fall or early spring for subsequent use by plants.

winter jasmine. White jasmine.

winterkill. Of plants, to die as a result of the rigors of the winter, as by low temperatures, snow, frost, etc.

winterkilling. The dying of plants as a result of the rigors of winter.

winter lettuce. Endive.

winter melon. *Cucumis melo* var. *inodorus,* family Cucurbitaceae, a tender, annual herb which is a variety of muskmelon *(q.v.),* grown for its sweet, edible fruit. It keeps well in storage. Perhaps native to Turkey.

winter mulch. Mulch *(q.v.)* applied in the autumn for the protection of plants or plant parts during the winter months.

winter oats. Oats sown in the autumn in areas where winter weather is less severe to produce a crop the following spring or early summer.

winter onion. Top onion.

winter over. See *winter: 2.*

winter pause. In poultry, the stoppage in egg production in winter.

winter pink. See *trailing arbutus: 2.*

winter plowing. The plowing of farm land during the winter months to conserve moisture and to obtain a better tilth for the next spring. See *fall plowing.*

winter plum. Persimmon.

winter purslane. Miner's lettuce.

winter range. The range, usually at low altitudes, used for pasture in the winter months in western U.S. See *spring-fall range; summer range; yearlong range grazing.*

winter rape. *Brassica napus,* family Cruciferae, an annual herb grown in Europe for its seeds which yield the colza oil or rape oil of commerce, and in the U.S. as a cover crop and as a temporary pasture crop. It provides excellent forage over a long season, requiring no cultivation. Native to Europe and Asia.

winter rye. Rye planted in the late summer or early autumn, which endures winter weather and produces a crop the next spring or early summer.

winter savory. *Satureia montana,* family Labiatae, a small shrub grown for its leaves used as a flavoring. It is less delicate than summer savory *(q.v.).* Native to Europe.

Winterbark drimys. *Drimys winteri,* family Magnoliaceae, an evergreen tree grown in warm areas for its jasmine-scented flowers and aromatic foliage. Native to Central and South America.

winter sets in. Winter begins.

winter squash. Any of the several varieties of the squash, *Cucurbita maxima,* family Cucurbitaceae, which have hard, outer shells and keep well in winter storage. *Cf. summer squash.*

winter storage. 1. Storage of produce during the winter. 2. Storage of nutrients within a plant for winter maintenance and/or for the following growing season. 3. See *cold cellar.*

winter sun scald (of trees). Bark injury on the sun-exposed side of tree trunks resulting from periodic freezing of the tissues. It is characterized by loose, brown, dead bark, sometimes splitting and cracking. Eventually the bark peels away to produce an open wound or dead spot on the trunk.

winter sweet. See *origanum.*

winter sweet pea. Goat's-rue swainsonpea.

winter turnip. Rutabaga.

winter tying. The training of grape vines on a trellis by tying with cords during the dormant season. *Cf. green tying.*

winter vegetable. 1. A vegetable produced during the winter months, especially one grown in the warm winter sections of the country as contrasted to a hothouse vegetable. 2. A vegetable which stores well and is used during the winter.

winter vetch. Hairy vetch.

winterweed. Chickweed.

winter wheat. The type of wheat that is planted in the fall, carries over the winter as a young seedling, and matures in the early summer of the following year.

wire. 1. A slender thread of metal, usually flexible, which may be used for such purposes as to carry electric current, tie bales, construct fences, make temporary repairs on machinery, etc. 2. See *barbed wire.* 3. See *trash guard.*

wire basket. A basket or pail-shaped container made from wire which is used to gather eggs or other produce.

wire cob rake. An endless, eccentric-sprocket-agitated belt of wire mesh which is a part of a corn sheller. Shelled corn cobs, husks, and silks drop onto it and the corn passes through to the cleaning sieves below. The cobs and foreign matter pass over the cob rake to the hopper of the cob stacker.

wire crawl. The undesirable, sidewise movement of the check wire *(q.v.)* in planting corn or other crops with a check-row planter. See *cross-over.*

wire cut. In animals, a wound made by barbed wire or other wire fencing.

wire cutter. 1. A pliers-like tool used to cut wire. 2. A cowboy on the western range in the U.S. who cut down or destroyed the fences of farmers in the early days of settlement.

wire dam. A temporary dam which has a head wall constructed of mesh wire.

wire fence. Any fence in which wire is used as the barrier material, such as barbed wire, woven wire, mesh, etc.

wire gauze strainer. A milk strainer usually made with wire gauze bottoms and stiff sides of tin plate.

wire grass. Any of the several grasses with thin, long, wiry leaves. See *Bermudagrass; Canada bluegrass; goose grass; pineland threeawn.*

wire grip. A heavy metal hook fastened to a stretcher clamp on the one end and to a block and tackle on the other, in order to stretch a fence wire.

wire grub. A wireworm.

wire plant. Wirevine.

wire rush. Baltic rush.

wire shield. A special attachment for a cultivator, composed of a frame and seven or eight stiff wires parallel to the ground, used to prevent the covering of young plants by soil during cultivation.

wire snipper. See *wire cutter: 2.*

wire stem. A stem canker stage of damping-off in cabbage when the fungous infections occur on older seedlings.

wirestem muhly. See *wood grass: 1.*

wire-stretcher. Any of several devices used to draw fence wire taut before the fence is fastened to posts.

wire travel. See *wire crawl.*

wirevine. *Muehlenbeckia complexa,* family Polygonaceae, a woody, vinelike plant, grown in a hanging basket for its drooping, wiry stems. Native to New Zealand. Also called *maidenhair vine.*

wireworm. The larva of the click beetle, family Elateridae, which eats seed, underground stems, small roots, and burrows into potatoes, carrots, etc.

Wisconsin curd test. A test used to detect milk which is unsatisfactory for use in cheesemaking.

Wisconsin false blossom. False blossom.

Wisconsin rake. A long-handled scoop which has teeth so arranged that a worker can comb cranberry or blueberry vines, take off the berries, and drop them in a box-like compartment behind the teeth. Also called *Wisconsin scoop.*

Wisconsin scoop. Wisconsin rake.

Wisconsin weeping willow. *Salix blanda,* family Salicaceae, a hardy, hybrid, deciduous tree grown for its pendulous branches and foliage.

Wisconsin white hemlock. Canada hemlock.

Wisconsin white pine. Eastern white pine.

wishbone. The forked or V-shaped bone located at the front of the breast of fowls; the clavicle. The tender meat which surrounds it is a popular piece when cooked.

wisteria. Any plant of the genus Wisteria (Wistaria), family Leguminosae. Species are woody vines grown for their beautiful flowers. See *Japanese wisteria; Kentucky wisteria; Chinese wisteria.*

witch-alder. Fothergilla.

witchbells. See *bluebell: 1.*

witches'-broom. An abnormal type of growth of an herbaceous plant or of a branch of a woody plant in which buds normally remaining dormant begin to grow, forming a bushy, broomlike growth, usually with the leaves reduced in size.

witch grass. *Panicum capillare,* family Gramineae, an annual grass widespread throughout the U.S., found in waste places, old fields, barnyards, etc. It is a weed pest. Also called *old witch grass; tickle grass; fool-hay.*

witchhazel. 1. Any plant of the genus Hamamelis, family Hamamelidaceae. Species are shrubs used for the shrub border. 2. *H. virginiana,* a coarse shrub 15 ft. in height, grown for its bright yellow flowers. Native to North America. 3. See *fothergilla.*

witching wand. Divining rod (stick, wand).

witchweed. Striga.

Withania. An East Indian cheese made from milk coagulated with rennet obtained from withania berries (*Withania coagulans*).

with calf. Designating a cow that is pregnant.

withe. A slender, rather long switch, shoot, or branch of a woody plant, especially of a willow.

wither. Of a plant, to become dehydrated, drooped, and limp.

withe-rod. Witherod viburnum.

witherod viburnum. *Viburnum cassinoides,* family Caprifoliaceae, a large, upright, deciduous shrub grown for its yellowish-pink fruits turning blue-black. Native to North America. Also called *withe-rod; false Paraguay tea; Appalachian tea; wild raisin.*

withers. The ridge at the apex of the shoulder bones of a horse, cow, sheep, etc.

withertip. See *leaf and twig blight.*

withertip of apples. A copper-deficiency disease of apples in Australia characterized by brown spots on the terminal leaves that generally wither and fall.

withertip of citrus. See *citrus anthracnose; lime anthracnose.*

withertip of flax. A calcium deficiency disease of flax in Australia characterized by a sudden bending and loss of turgidity of the stem about 2 to 4 ins. below the tip, necrosis, and death of the tip.

with return. A condition under which a stallion is bred to a mare in which he must serve her in the next heat period if she does not become pregnant after the first service. See *no return.*

withy. The willow or osier used in basket making, wickerwork, etc. Also called *with; withes.* See *willow; basket willow.*

withywind. Travelersjoy.

witloof. French endive.

witness corner. A marker set on a property line leading to a corner, used where it would be impracticable to maintain a monument at the corner itself.

witness mark (stake). A mark or stake set to indicate the position (approximate or exact) of a propery corner, instrument station, or other survey point.

witness tree. A living tree marked by a land surveyor to indicate the location of a corner of a tract of land. For the public domain, surveyed by the U.S. General Land Office, the section and quarter-section quarters were so marked. Also called *bearing tree.*

woadwaxen. *Genista tinctoria,* family Leguminosae, a hardy, small shrub grown for its foliage and yellow flowers. Formerly it was used as a source of a blue dye. Native to Europe and Asia. Also called *base broom; dyer's greenweed.*

wold. Rolling upland, or hill country (England).

wolf. 1. Certain large, dog-like, carnivorous animals of the genus Canis, especially *C. occidentalis,* the gray or timber wolf. At one time in the U.S. it was a serious predator of domesticated animals, particularly sheep. 2. See *common cattle grub; northern cattle grub.*

wolfberry. 1. Any plant of the genus Lycium, family Solanaceae. Also called *desertthorn.* See *matrimonyvine.* 2. See *western snowberry.* 3. See *silverberry.*

wolf grape. See *bitter nightshade; matrimonyvine.*

wolfsbane. Aconite monkshood.

wolf's teeth. The first premolar tooth situated

just in front of the first well-developed tooth; it is a much-reduced vestige and is usually shed soon after erupting. Also called *wolf tooth.*

wolftail. 1. *Lycurus phleoides,* family Gramineae, a slender, perennial grass found from west Texas to Arizona, U.S., and south to Mexico at elevations of 4,000 to 8,000 ft.; its forage rating is good. 2. An old-time imaginary and fictitious disease of cattle. Also called *wolf-in-the-tail.* See *hollow horn.*

wolf tree. A tree which occupies more space than its silvicultural value warrants, thus curtailing the space of better species. (Usually applied to a broad-crowned, short-stemmed reserve, or a weed tree of large size which has been growing for some time and overtops others or takes valuable space.)

wolf willow. Silver buffaloberry.

wolverine. *Gulo gulo,* family Mustelidae, a weasel-like, carnivorous mammal with shaggy, blackish-brown fur, a bushy tail and short legs, which causes great annoyance to trappers by cunningly stealing both bait and trapped animals. Its pelt is of little commercial value, though often used by northern tribes. Native to the northern, arctic and subarctic regions of the world. Also called *glutton.*

womb. Uterus.

wonga-wonga. Jasmine pandorea.

wood. 1. Fuel, as a winter's supply of *wood;* firewood. 2. The hard, fibrous substances beneath the bark of a tree, or shrub; xylem. See *wood fiber; heartwood; sapwood.* 3. Designating a product made of *wood;* wooden. 4. See *forest; lumber; timber; woodland.*

wood-and-wire excluder. A type of queen excluder made of strips of wood and round wire bars spaced alternately.

wood anemone. American wood anemone.

wood ant. Allegheny mound ant.

wood ashes. The residue which remains after burning wood. It has a value as a fertilizer and was used by early settlers in the home manufacture of lye. Sometimes it is used as a source of calcium in livestock feeds, especially for hogs. See *unleached wood ashes; leached wood ashes.*

woodbark. 1. The bark of trees. 2. A dust carrier prepared from pulverized bark.

woodbine. A common name for many plants. See *Carolinajessamine; hedge glorybind; Virginia creeper; virginsbower; woodbine honeysuckle.*

woodbine honeysuckle. *Lonicera periclymenum,* family Caprifoliaceae, a hardy, woody, climbing vine grown for its fragrant, red flowers and as a cover for unsightly stumps, etc. Native to Europe and Asia. Also called *woodbine; twisted eglantine; Dutch woodbine; European honeysuckle.*

woodboat. In forestry, a single sled which has two skids attached to the bunk by their forward ends, with the rear ends dragging. It is used to haul cordwood from steep or rocky slopes.

wood borer. Any of the larva of insects that burrows or tunnels in wood.

wood cellulose. The residue which remains after chemical treatment of wood to remove extractives, lignin, and hemicelluloses.

wood charcoal. A carbon fuel prepared by heating or charring wood out of contact with air, either in retorts or in stacks in which a retarded combustion is effected. The carbon thus obtained is nearly pure; it is light and extremely porous possessing the power of absorbing gases. It is used as a fuel and is fed to animals, especially hogs, because it absorbs gases in passing through the intestines. It is also fed to poultry as an additive to mash or scratch feeds.

woodchuck. A marmot, *Marmota monax,* an animal which occasionally becomes a nuisance on farms because of its burrows in pastures, orchards, and fields. Also called *groundhog.*

wood creosote. See *creosote.*

wood distillation. The decomposition of wood in a closed retort by heat so that wood alcohol, acetone, acetic acid, water, and tar are driven off, and only charcoal remains in the retort.

wooded. Concerning, or designating an area full of trees.

wooden follower. In brick cheesemaking, a wooden board shaped to fit a wooden form. After the curd is placed in the form, the *wooden follower* is placed on it and weighted down with bricks to compress the cheese.

wood engraver. Bark beetle.

wood evil. See *blackleg: 4.*

wooden stave silo. A type of vertical silo made from wooden staves and steel hoops to form a circular structure.

wooden tongue. 1. A tongue of a wagon, mower, etc., made of wood. 2. See *actinobacillosis; actinomycosis.*

woodfern. 1. Any of several ferns of the genus Dryopteris, family Polypodiaceae. Also called *fern meadowbur; sweet fern.* 2. *D. intermedia,* a hardy fern grown in the outdoor fern garden and sold by florists as an ornamental. Native to North America. Also called *fancyfern.*

wood fiber. Much elongated cells, with attenuated ends and thickened walls, empty of living contents at maturity, whose chief function is support or protection of the parts of the plant. The walls are lignified. The wood fibers may make up a considerable portion of the cells of the wood of dicotyledons but may be found also in herbaceous stems to which they give part of their stiffness. See *fiber-tracheid.*

wood fretter. A wood borer.

woodfringe. See *mountainfringe.*

Woodgate rust. A rust of the Scotch pine caused by the aecial stage of a form of *Cronartium quercuum,* which produces large, round galls on the woody twigs and young branches of the plant. Occurs in New Hampshire, New York, and Michigan, U.S.

wood grass. 1. *Muhlenbergia mexicana,* family Polygonaceae, a grass used for binding soil and as hay. Native to North America. 2. See *yellow Indiangrass.*

wood gum. A yellow, sticky substance extracted

from wood by dilute alkali. It is often composed of xylan with smaller amounts of mannan, araban, and galactan.

wood-hoop silo. See *wooden stave silo.*

wood horsetail. Field horsetail.

wood hyacinth. Blue squill.

woodland. 1. Any area whose vegetation is composed essentially of woody plants. See *farm woodland; woodlot; forest.* 2. Pinyon-juniper, and chaparral forest types (southwestern U.S.). 3. Open stands of oaks and Digger pines (Pacific Coast states of the U.S.).

woodland management. The management of existing woodlands and plantations that have passed the establishment stage including all measures designed to improve the quality and quantity of woodland growing stock and to maintain litter and herbaceous ground cover for soil and water conservation. It includes all such measures as planting, improvement-cutting, thinning, pruning, slash disposal, fire protection, and grazing control.

woodland pasture. 1. Farm woodlands used for grazing. 2. Wooded areas with grass and other grazing plants growing in open spaces among trees.

woodland primrose. Cowslip primrose.

woodlandstar. 1. Any herb of the genus Lithophragma, family Saxifragaceae. 2. *L. affinis,* a perennial, tuberous-rooted herb, grown in the wild garden or rock garden for its cluster of white flowers. Native to North America.

wood laurel. Spurgelaurel daphne.

wood lice. 1. The larvae of the greater wax moth *(q.v.).* 2. Any of the larvae of several insects, as the termites, that live in or under the wood and damage it.

wood lily. 1. *Lilium philadelphicum,* family Liliaceae, a bulbous herb grown for its orange-red, dark-spotted flowers. Native to North America. 2. See *lily-of-the-valley; trillium.*

woodlot. A tract of woodland, a part of a farm and usually only a few acres in size, which may have a single or multiple use, as a source of wood fuel and logs for lumber, as a woodland pasture, and as a means of providing food and protective cover for wildlife.

wood merrybells. *Uvularia perfoliata,* family Liliaceae, a hardy, perennial herb sometimes grown in the border or in the wild garden for its bell-shaped, pale yellow flowers. Native to North America. Also called *mohawkweed; mealy bellwort.*

wood mint. *Blephilia ciliata,* family Labiatae, a perennial herb with bluish-purple flowers. It is a valuable bee plant. Native to North America.

woodnettle. *Laportea canadensis,* family Urticaceae, a perennial herb found in woodlands. On contact with skin, it produces a severe itching. Native to North America.

wood-oil. The drying oil, similar to linseed oil, obtained from the seeds of several species of the genus Aleurites, family Euphorbiaceae, especially the tung oil tree and Japan wood oil tree *q.q.v.*

wood pocket of lemon. A disease of Lisbon lemons, probably caused by a virus, characterized by a defect or crack in the bark, by a discoloration of wood beneath the crack, and by a variegated appearance of the leaves. Also called *lignocortosis of lemon.*

wood preservative. Any of several chemical compounds used to prevent decay in wood by sealing the exposed surfaces or by penetrating the wood: *e.g.,* creosote.

wood pulp. See *pulp: 1.*

wood rat. Any rodent of the genus Neotoma. Species are soft-furred rats sometimes destructive to citrus trees. Mostly found in California, U.S. and southern North America.

wood rot. The decay of wood caused by a fungus or fungi living on or within the wood.

woods. See *forest; farm woodland; woodlot; woodland.*

wood sage. The American germander.

Woods ceropegia. *Ceropegia woodi,* family Asclepiadaceae, a tropical, trailing or prostrate, greenhouse plant, grown for its heart-shaped leaves and its pinkish-purple flowers. Native to South America. Also called *hearts-on-strings; rosaryvine.*

woods cull. Butts and other merchantable sections of trees commonly left in the woods.

wood shavings. Very thin strips of wood resulting from planing boards. Sometimes used as bedding for animals and for kindling.

woodsorrel oxalis. *Oxalis acetosella,* family Oxalidaceae, a stemless, perennial herb, grown in the wild garden for a cushion effect and for its white, pinkish-veined flowers. Native to Europe. Also called *sleeping-beauty; sheep sorrel; shamrock.*

wood-stave silo. Wooden-stave silo.

wood strawberry. European strawberry.

wood substance. In wood anatomy, the solid, fused material of which woody tissue is composed; specifically, it is the cell wall consisting of cellulose, hemicellulose, and lignin.

wood sugar. Sugar made from the carbohydrate part of wood by hydrolysis. It is chiefly glucose, containing some xylose and possibly small amounts of galactose, manose, and arabanose.

woodsy. Designating an area which resembles or suggests a woodland or forest.

wood-tar creosote. A distillate oil derived from wood tar, a by-product in the destructive distillation of either hardwoods or softwoods. It is a valuable timber preserver.

wood technology. The study of wood in all its aspects; the science of wood, including its anatomy, chemistry, physical properties, treatment, etc.

wood tick. American dog tick.

wood turpentine. The essential oil obtained from pine stumps or other resinous wood by destructive or steam distillation.

wood wax(en). Woadwaxen.

wood-wool. A wood product resembling wool, made from waste pulp or fine hair-like shavings. Used primarily as insulating and packing material.

woody. 1. Designating a plant which has some characteristics of wood. See *wood: 2.* 2. Designating a forested or woodland area. 3. Designating a plant with a high content of lignin *q.v.*

woody aster. *Aster xylorrhiza,* family Compositae, a perennial herb which is poisonous to sheep grazing on it. Symptoms of poisoning are impairment of vision, abdominal pain, depraved appetite, and paralysis. Found on the western range in Wyoming, Utah and Nevada, U.S.

woody flavor. A flavor defect of butter suggesting the piny odor of a new churn or butter tub.

woody nightshade. Bitter nightshade.

woody plant. Any shrub, tree, or certain vines, as distinguished from herbaceous (*q.v.*) plants, which produces wood and has buds surviving above ground during the winter.

woody tongue. See *actinomycosis; actinobacillosis.*

wool. The fibers constituting the soft, curly coats of sheep, camel, yak, etc. After cotton, sheep *wool* is the most important source of natural fiber used in apparel, upholstery, carpets, etc.

wool-ball. Wool ingested by nursing lambs which have been sucking at tags (*q.v.*) instead of at the teats. It forms a ball in the stomach and may cause a serious digestive disturbance. See *hairball.*

wool bin. In a barn or shearing shed, an open compartment where sheared wool is stacked by class or grade until it is pressed (Australia and New Zealand).

wool blind(ness). A condition in which the wool around a sheep's eyes is so heavy and thick that it interferes with vision. See *wink.*

wool book. A ledger in which a record is kept of the number, description, and weight of each bale of wool as it is pressed (Australia and New Zealand).

wool brand. An identification device stenciled on the coat of a sheep (Australia and New Zealand).

wool card. Card.

wool carding. The process of disentangling and separating the fibers of wool and delivering them in a parallel condition suitable for spinning. This was formerly done by means of two hand combs, one of which was fixed in a stationary frame. To facilitate the work, the combs were warmed in a charcoal pot. At the present time the operation of carding is confined to the ordinary stapled or Merino wools which possess the property of felting and is effected by cylinders warmed with steam covered with carding cloth.

wool classer. One who prepares wool for the market by segregating the fleeces in the shearing pen according to strict trade standards (Australia and New Zealand).

wool classing. The grading of skirted fleeces and pieces before packing them in the shearing sheds (Australia and New Zealand).

wool cleaner. A machine used for beating, shaking, and cleaning wool prior to scouring and dyeing.

wool clip. The annual wool crop.

wool comb. See *card.*

wool combing. The process of dealing with long stapled wools which are to be used in worsted manufacture. The operations are like carding and are effected with a gilling machine.

wooled. Abounding with, covered with, wool.

woolen fabric. See *woolens.*

woolen mill. Any mill which makes fabric from wool. See *gin: 1.*

woolens. Those fabrics manufactured in whole or in part of woolen yarn; *i.e.,* from uncombed wool. *Cf. worsteds.*

woolen wool. Short wools adapted to the manufacture of woolens as distinguished from worsteds.

woolen yarn. Yarn manufactured from uncombed wool.

wool fat. The waxy, greasy substance that exudes from the sheep's skin and clings to the wool fibers. After refining, it is sold in the drug trade as lanolin. In the crude form it is often known in commerce as degras. Also called *yolk.*

woolfell. A sheepskin from a slaughtered or dead sheep from which the wool has not been removed.

wool grader. Wool classer.

wool grades. Various methods of grading wool: *e.g.,* according to type, use, fineness of fiber, amount of yarn 1 lb. of wool will spin (worsted count system), etc. Classification systems vary from country to country and in each system skilled judgment is of great importance, since the grading of wool is not an entirely objective operation. See *American grades of wool; Australian grades of wool; B. A. 1-6.*

wool grease. Wool fat.

wool grower. A rancher or farmer who maintains a flock of sheep principally for shearing.

wool having a weak spot. Wool with a break.

wool house. Pullery.

wool in the grease. Grease wool.

woolly. 1. Abounding with, covered with, wool. 2. Designating a lanate plant; one covered with soft, loose, somewhat matted hairs which resemble wool.

woolly alder aphid. *Prociphilus tessellatus* (Fitch), family Aphidae, a bluish-black aphid that infests the alder. Also called *alder blight aphid; maple leaf aphid.*

woolly apple aphid. *Eriosoma lanigerum* (Hausm.), family Aphidae, a widely distributed aphid which infests apple, pear, hawthorn, elm, and mountain-ash. Infestation is characterized by white cottony masses over masses of purplish aphids in wounds on the trunk and branches. Seriously infested trees may be stunted, retarded, or killed.

woolly Asiaglory. *Argyreia speciosa,* family Convolvulaceae, a large, climbing plant grown in the greenhouse and outdoors in warm areas for its showy, rose-colored, and funnel-shaped flowers. Native to Indonesia. Also called *elephant creeper; woolly morning-glory.*

woolly bear. Salt-marsh caterpillar.

woolly betony. *Stachys lanata,* family Labiatae, a hardy, perennial herb grown for its dense clusters of purple flowers. Native to Iran and the Caucasus. Also called *lamb's ears.*

woolly crazyweed. Woolly loco.

woolly croton. *Croton capitatus,* family Euphorbiaceae, an annual herb which, as hay, is sometimes poisonous to stock. Native to North America.

woolly elm aphid. *Eriosoma americanum* (Riley), family Aphidae, an aphid which infests the American elm. Infested trees are characterized by curled or rolled leaves.

woolly foot (grama). Black grama.

woolly groundsel. Threadleaf groundsel.

woolly knot. See *hairy root: 1.*

woollyleaf loco. *Astragalus leucophyllus,* family Leguminosae, a perennial herb which is poisonous to livestock. For symptoms of poisoning see *woolly loco.*

woolly loco. *Astragalus mollissimus,* family Leguminosae, a poisonous plant, found from South Dakota to Arizona, U.S. Both its green and dried foliage are poisonous to livestock. Symptoms of poisoning are dullness, irregularity of gait, lack of appetite, dragging of the feet, a solitary habit, loss of flesh, and shaggy coat. As the animal ceases to eat, it dies. Also called *purple locoweed; Texas loco; woolly crazyweed.*

woolly manzanita. *Arctostaphylos tomentosa,* family Ericaceae, an evergreen shrub grown for its foliage, white-felty beneath, and its pink or white flowers. Native to North America.

woolly morning-glory. Woolly Asiaglory.

woolly mother-of-thyme. *Thymus serpyllum* var. *lanuginosus,* family Labiatae, a woody, perennial herb widely grown for its trailing habit; its gray, hairy stems; and its small, purplish flowers. Native to the Mediterranean area. Also called *woolly thyme.*

woolly pear aphid. *Eriosoma pyricola* Baker, family Aphidae, an aphid which infests pear trees and has similar characteristics to the woolly apple aphid *q.v.*

woollypod milkweed. *Asclepias eriocarpa,* family Asclepiadaceae, a perennial herb found in California, U.S., which is very poisonous to sheep and cattle. Symptoms of poisoning are depression, loss of appetite, diarrhea, low temperature, and rapid pulse. Also called *broadleaved milkweed.*

woolly root. Curly top.

woolly speedwell. *Veronica incana,* family Scrophulariaceae, an herb grown as an edging plant for its low growth habit and its blue flowers. Native to northern Asia.

woolly streak. A disease of apples which results from excessive water in the soil and which is characterized by a dry, reddish-brown cell formation in the core.

woolly thyme. Woolly mother-of-thyme.

woolly whitefly. *Aleurothrixus floccosus* (Mask.), family Aleyrodidae, an insect pest of citrus which has been controlled by internal parasites.

wool maggot fly. Black blow fly.

wool pack. The cotton, wool, or other cloth material used to wrap a bale of wool.

wool pool. A grouping of the wool of many producers into a single total amount so it can be sold on the market as a single unit. Such commodity pools are very helpful in securing a more satisfactory market than the individual producer could command for his small quantity. Wool, cotton, tobacco, are frequently pooled.

wool roller. The assistant to the wool classer (*q.v.*) who skirts and rolls the fleece (Australia and New Zealand).

wool shed. The building in which sheep are shorn (Australia and New Zealand).

wool sorter. One who sorts wool according to certain grades or classifications.

woolsorter's disease. Anthrax (*q.v.*) in man.

wool sorting. The process of separating fleeces into parts according to fineness and length. See *fleece: 1.*

wool stapler. Wool sorter.

wool top. In wool combing, the fibers which are combed and rolled into a rope-like strand of yarn. See *top: 2; noils.*

wool tree. Kapok ceiba.

wool type. Any breed of sheep produced primarily for wool as contrasted to mutton sheep *q.v.*

wool washing. The process necessary to remove the suint, or natural grease, and dirt from the fleece. It is treated first in alkaline liquor which saponifies the grease and then is washed in fresh water.

wool washings and waste. A fertilizer derived from the washings from cleaning wool which contains as much as 1 to 5 percent of potash.

wool waste. Wool washings and waste.

wool wax. Wool fat.

wool with a break. A defect of wool fiber resulting in lack of strength, characterized by a small strand or lock's breaking squarely across when stretched or pulled.

Wooton loco. *Astragalus wootoni,* family Leguminosae, a poisonous plant, found in southern New Mexico, and western Texas, U.S., and in northern Mexico. Both its green and dried foliage are poisonous. Symptoms of poisoning are dullness, irregularity of gait, lack of appetite, dragging of the feet, a solitary habit, loss of flesh, and shaggy coat. As the animal ceases to eat, it dies. Also called *locoweed.*

Worcestershire sauce. A food flavoring product which is prepared from a mixture of spiced vinegar, soy sauce, walnut catsup, lime juice, lime oil, sugar, salt, spices, water, and a stabilizer.

work. 1. The performance of any productive operation, act, etc. 2. The productive operation, act, matter, etc., upon which a person expends energy. 3. (plural) The inner machinery, as the *works* of a combine. 4. The capacity for performing productive operations, etc., as the horse has a lot of *work* left in him. 5. To perform a productive operation, act, etc. 6. To expend energy in any productive act, etc. 7. To operate in a machine, as the mower *works.* 8. To engage in a vocation or profession, as he *works* in Extension. 9. To proceed or move slowly, as water *works* down into the soil. 10. To respond to exertion of energy, as sandy land *works* well. 11. To cause to labor, as to *work* horses. 12. To cultivate the soil. 13. To bud or to graft. 14. To ferment, as cider *works.* 15. To drive livestock on horseback or by means of a dog.

workability. The state or quality of being shaped, worked, etc.

work and wind test. A test for horses to be offered for sale in which it is determined if the horse is well broken and works well in harness, and if its breathing is normal.

work animal. Any domestic animal primarily used for work purposes: horses, mules and oxen.

work day. Working day.

worked ground. Cultivated land.

worker. The female bee, other than the queen, whose organs of reproduction are only partially developed, and which gathers nectar and pollen, tends the brood, brings in water, protects the hive, etc. See *queen bee; drone.*

worker bee. Worker.

worker cell. One of the cells in a beehive in which a worker is reared.

work horse. 1. Any horse used as a source of power, especially one which pulls vehicles, plows, etc. *Cf. saddle horse; riding horse.* 2. In racing, a horse which is run with the race horse in training for competition.

working. 1. The dimensional changes of lumber as a result of moisture changes. 2. Fermenting. 3. See *work.*

working assets. Working farm capital.

working circle. The primary unit of forest management, with well-defined boundaries, usually based on topography, which is large enough to furnish a sustained yield of forest products sufficient to support dependent industries or communities.

working day. The day considered for purposes of work, such as an eight or ten hour *working day.*

working dog. 1. A dog trained to drive sheep, cattle, etc., in such a manner that the animals are not harmed. 2. A dog trained to run a treadmill, pull a sled or cart, etc.

working farm capital. The livestock, machinery, equipment, feed, other supplies, and cash, used in the operation of a farm business.

working group. In forestry, the parts of a forest which have similar silviculture management and rotation.

working man. Any person who engages in productive activity, especially a manual worker.

workman. A person who performs manual labor.

work mule. Any mule used for work, such as pulling a vehicle, plow, etc. See *pack mule.*

work of digestion. The loss of energy which occurs in chewing, digesting and assimilating the food; heat increment.

work only. A condition of a sale of a horse in which the seller guarantees only the willingness and ability of the horse to work.

work over. To bud or graft an individual tree or group of trees in an orchard to other horticultural varieties.

workshop. A place or building on farms where repairs to machinery or new pieces of equipment are made.

work simplification area. An area of study and research dedicated to the objective of increasing labor efficiency on farms through doing the job quicker and/or easier. Time studies are usually involved and the job is broken down into separate units or steps in search of new and better work methods.

work stock. The total number of animals used for work, as horses, mules, etc., on a farm.

work unit. The amount of work for pay usually done by a man working with average labor efficiency in a ten-hour day on crops, productive livestock, and other work on the farm.

worm. 1. Any small, soft-bodied, usually limbless animal, such as a larva, grub, maggot, earthworm, silkworm, etc. 2. To rid an animal of internal, parasitic *worms.*

worm cast. See *cast: 2.*

wormed. Designating an animal which has been given a vermifuge or anthelmintic.

worm fence. Virginia fence.

worm-grass. See *pinkroot spigelia; white sedum.*

worm nest. In cattle, a protuberance or swollen area on the flank or lower portion of the chest which contains filarial worms.

wormseed goosefoot. *Chenopodium ambrosioides,* family Chenopodiaceae, an annual herb grown as a source of the chenopodium oil used in medicines and vermifuges. Escaped, it is a weed pest and, if ingested, poisons livestock. Native to tropical America. Also called *Mexican tea.*

wormseed oil. Oil expressed from the wormseed goosefoot (*q.v.*) that is used as a component of certain vermifuges. Also called *American wormseed oil; chenopodium oil.*

worm stage. A stage of development of certain insects in which the insect is a larva.

wormwood. 1. Any of the several plants of the genus Artemisia, family Compositae. Species are cultivated for their aromatic qualities. 2. *A. absinthium,* a woody, perennial herb grown for its oil which is an ingredient of absinthe. The oil is also used in decoctions for sprains and bruises, and the plant is credited with having other medicinal uses as a tonic, diuretic, and anthelmintic. In the U.S., it is a specialty crop grown on very small acreages in Michigan and a few other states. Native to Europe.

wormy. Designating plants, animals, or products which are infested with worms.

wormy trap. A small enclosure, containing a few acres, where screwworm-infested animals are confined for treatment.

worn-out land. Land which has lost most or all of its fertility by cropping or erosion.

worsteds. Fabrics made in whole or part from a combed wool yarn. Also called *worsted fabric. Cf. woolens.*

worsted yarn. Yarn made from combed wool.

wort. The sweet admixture of malt that produces alcohol and carbon dioxide in the manufacture of beer. It is the filtrate from a mixture of malted barley and grain in which amylase converts the starch into maltose and glucose, and in which other enzymes convert the grain proteins into peptides and phosphates, along with other minerals and accessory growth factors. Its pH is 4.3 to 4.5 and at 30° C. it is an ideal medium for the growth of yeasts.

-wort. A suffix denoting a plant, as a butter*wort.*

wound. Any violently caused disruption of the continuity of an internal or external tissue.

wound dressing. Any substance applied to wounds to protect them and promote healing, and, with plants, to prevent the entrance of decay organisms.

wound gum. See *gummosis.*

wound parasite. Any parasite which is not able to enter a host except through wounds or injured tissue.

wound stain. A discoloration which originates in or is developed as a result of a wound in a tree.

wound tumor disease of legumes. A minor, virus disease of legumes which develops at swellings or tumors at places previously punctured by leafhoppers.

wrangler. A herder or handler of wild or range horses on ranches.

wrap. 1. Any sheet material used to enclose produce, as polyethylene, paper, cellophane, etc. 2. To enclose produce in a *wrap.*

wrapper leaf. In marketing, a type of tobacco used for the outer covering in the manufacture of cigars.

wreath system. See *weaving system.*

wriggler. The larva of a mosquito.

wring the neck. To kill a chicken by breaking its neck.

wrinkled. Designating a bandage defect of cheese characterized by a fold appearing to be pressed into the cheese.

wry neck. 1. A twisting of the neck and head of fowls into an unnatural position, possibly as a symptom of paralysis and other diseases. 2. An animal with its neck distorted due to its having been carried in a faulty position in the uterus.

wry tail. The tail of a fowl or mammal (usually cattle), when it is permanently carried to one side.

Wyandotte. A medium-weight, dual-purpose breed of chickens of the American class. The *Wyandotte* has a rose comb, yellow skin, nonfeathered shanks, smooth-fitting plumage and a characteristic curvilinear breed type. Mature cocks weigh about 8½ lbs. and mature hens about 6½ lbs. The egg-shell color is brown. The varieties are the Silver-Laced, Golden-Laced, White, Black, Buff, Partridge, Silver-Penciled and the Columbian. There are also bantam sorts of most of the above varieties.

wyche elm. Scotch elm.

wych hazel. 1. Witchhazel. 2. Scotch elm.

wyomingite. See *leucite.*

Wyomings. In marketing, sheep which originated in Wyoming, U.S.

X

X. An obsolete grade of washed wool from Merino sheep. (The more X's, as XXX, the finer the wool.)

xanthone. $C_{13}H_8O_2$, an insecticide and ovicide used in the control of the codling moth (*q.v.*) and various orchard mites.

xanthophyll. $C_{40}H_{56}O_2$, a dark brown compound, usually associated with chlorophyll and carotene in plants, which imparts a golden-yellow to ivory color depending on concentration. *Xanthophyll,* or a related compound, is responsible for the color of yellow flower petals. See *anthoxanthin.*

xanthosis. See *yellows: 1.*

xarque. Jerked beef.

X-disease. 1. A disease of cattle characterized by thickening and wrinkling of the skin on the neck, withers, and sometimes over most of the body, by salivation, lacrymation, depression, and intermittent diarrhea. It is caused by the animal's eating toxic substances (chlorinated naphthalene) frequently contained in machine oils, lubricants, and certain wood preservatives. Also called *disease X; hyperkeratosis.* 2. See *avian monocytosis.*

X-disease (of stone fruits). A virus disease which was first noticed on peach and later on chokecherry, *Prunus virginiana,* and other *Prunus* spp. The virus is transmissable by grafting or insects *(Colladonus geminatus)* to cultivated stone fruits. It is also artificially transmitted to various herbaceous plants. On peach, the oldest leaves progressively turn yellow and red and defoliate. Early season red and yellow coloring is the prominent symptom of chokecherry. Immature fruit, wilt, and decline are symptoms on cherry. Also called *yellow-red disease; yellow-red virosis; eastern X-disease.* See *western X-disease; buckskin: 4; little cherry.*

xenia. The immediate effect of pollen on the endosperm due to the phenomenon of double fertilization which may allow the male parent to affect the appearance of the seed.

xeric. Dry; as applied to soils, those which normally have a very low content of moisture, and to plants, those tolerant to a very dry soil or arid conditions.

xerophthalmia. A condition in animals and fowls, caused by a vitamin A deficiency, characterized by a roughened coat and red, sore, puffed eyelids, and an abnormally dry and lusterless condition of the eyeball.

xerophytes. Plants which habitually grow where the evaporation stress is high and the water supply low.

xerophytic vegetation. Vegetation which is characteristic of the desert regions, such as thorny brush, cacti, shrubs, and small flowering annual and perennial plants.

xochinacaztli. Sacred ear.

xylem. That portion of the fibrovascular system of a vascular plant including the non-living tracheids and vessels, and the immediately associated living parenchyma cells as well as the supporting and protective wood fibers. In plants with woody stems, the *xylem* lies inside the cambium (*q.v.*). *Xylem* is the tissue which makes up most if not all of the lumber, wood, and wood products of commerce. *Cf. phloem.*

xylometer. A tank into which wood can be submerged to measure its volume by the amount of water it displaces.

Y

yabbie. A small, destructive, fresh-water crayfish, mainly species of the genus Cherax, family Parastacidae, which is an industrious burrower. *Yabbies* often cause collapse of earthen banks in bore-water drains, irrigation channels, and stock dams, by tunneling, and have even been known to attack concrete walls of large dams. Found in Australia.

yak. *Bos grunniens,* family Bovidae, a large, wild or domesticated bovine characterized by short hair on the back and long, wavy hair on the underparts and tail. It is blackish-brown, or sometimes black and white (domesticated variety), in color. Domesticated, the *yak* provides milk, hide, and hair, which is woven into fabrics, and is an important beast of burden in elevated regions of central Asia. Its tail is often used as a fly-flapper (called a *chowry*) and for decoration. Native to Tibet and central Asia. See *zho.*

yam. 1. Any of the several plants of the genus Dioscorea, family Dioscoreaceae. Some species are cultivated for food and some for ornament. See *winged yam; cinnamonvine; cushcush yam.* 2. See *sweet potato.*

yamp. See *squawroot: 2.*

Yangtao actinidia. *Actinidia chinensis,* family Dilleniaceae, a hardy, woody vine grown as an ornamental covering for a fence, arbor, etc., but especially for a trellis. It produces an edible fruit which may be used in preserves. Native to China.

yapon. Yaupon.

yard. 1. A unit of linear measurement equal to 3 ft. or 36 ins. 2. The grounds which immediately surround a dwelling. 3. An enclosed area in which stock, fowls, etc., are kept, as a chicken *yard.* It may be used as a suffix, as in stock*yard.*

yardage. The fee charged by a transit company or stockyard for the use of its storage yards.

yard-land. In old English law, a measure of land, of uncertain area, apparently ranging from 20 to 50 acres.

yardlong bean. Yardlong cowpea.

yardlong cowpea. *Vigna sesquipedalis,* family Leguminosae, an herbaceous vine grown in warm areas as a cover crop and green manure. Probably native to Asia. Also called *yardlong bean; asparagus bean.*

yard lumber. Lumber which has been air-dried in a yard.

yard trader. A person who buys and sells livestock at a market for profit. See *commission company.*

yarovization. Vernalization.

yarrow. *Achillea millefolium,* family Compositae, a hardy, perennial herb, a weed pest in unworked fields, roadsides, etc., generally in North America. It is grazed moderately by stock, and its forage rating is fair. Native to Europe, Asia, and North America. Also called *milfoil; thousand-leaf.*

yate eucalyptus. *Eucalyptus cornuta,* family Myrtaceae, an aromatic tree grown for its foliage and greenish-yellow flowers. Native to Australia. Also called *yate tree.*

yaupon. 1. *Ilex vomitoria,* family Aquifoliaceae, an evergreen shrub with smooth leaves and scarlet berries, grown as an ornamental and wildlife plant. The leaves are sometimes used as a tea-substitute, and were formerly used by Indians in preparing a black drink used for medicine and as a ceremonial beverage. Native to North America. Also called *cassena; emetic holly; yapon; youpon; yupon.* 2. Dahoon.

yautia malanga. *Xanthosoma sagittaefolium,* family Araceae, a tropical herb grown for its edible rootstock. Native to America. Also called *tania; taniers.*

yean. To give birth to young, especially by goats and sheep.

yeanling. 1. A young sheep or goat. 2. Referring to a newborn or young of sheep or goats.

yearling. 1. Generally, an animal more than one but less than two years of age. 2. In stock shows, a horse foaled on or after January 1 and on or before December 31 of the preceding year. 3. In stock shows, a sheep born on or after January 1 and on or before December 31 of the preceding year.

year-long range grazing. The practice of maintaining animals on the same pasture throughout the season or year.

year-long range land. A range suitable for grazing at any or all times in the year. See *spring-fall range; summer range land; winter range.*

yeast. 1. A yellowish substance composed of microscopic, unicellular fungi of family Saccharomycetaceae, which induces fermentation in juices, worts, doughs, etc. The fungi may reproduce asexually by the division and separation of the cells into two equal parts, as in the fission *yeasts,* or, more frequently, in the budding *yeasts,* by the budding out of small outgrowths which enlarge until they attain full size, and then separate from the parent cell. Most *yeasts* produce enzymes which bring about the fermentation of carbohydrates to produce alcohols and carbon dioxide. The most common of such *yeasts* is *Saccharomyces cerevisiae* whose fermentary process produces the carbon dioxide which leavens bread dough, as well as the alcohol used for beverages (beer, wines, etc.). This is the so-called bread-yeast or brewers'-yeast. In yeast inoculation of a substance, the surface or top yeast ferments it at a faster rate than the sedimentary yeast at the bottom. 2. The asporogenous yeasts, family Torulopsidaceae, in which the production of internal spores (ascospores) is lacking. Some of them (especially in the genus Candida) are parasites of man or other animals and may cause various diseases, such as thrush. In recent years, especially in Germany, some yeasts have been cultivated on sugar-containing derivatives of sawdust and have become valuable sources of animal feed. Also called *pseudo-yeast.*

yeast dried grains. A livestock feedstuff which is the properly dried residue from the mixture of cereals, malt and malt sprouts (sometimes cottonseed meal) obtained in the manufacture of yeast or vinegar. Also called *vinegar dried grains.*

yeast hole. An undesirable texture defect of Cheddar cheese in which there are large gas holes. See *fish eyes.*

yeasty flavor. A flavor defect of cheese, butter, and some fruit juices which suggests a taste and aroma of yeast fermentation.

Yeatman's pea. Cowpea.

yeddo euonymus. *Euonymus yedoensis,* family Celastraceae, a shrub grown for its brilliant scarlet, autumnal foliage. Native to Asia.

yeddo spruce. *Picea jezoensis,* family Pinaceae, a hardy, evergreen tree grown as an ornamental. Native to Asia.

yellow. In marketing, a yellow chroma in cotton. See *blue.*

yellow adder's tongue. Fawnlily.

yellow alstroemeria. *Alstroemeria aurantiaca,* family Amaryllidaceae, an herb grown for its yellow, green-tipped, brown-spotted flowers. Native to Chile.

yellow ash. American yellowwood.

yellow asphodel. Jacobsrod.

yellow azalea. Flame azalea.

yellowband iris. *Iris ochroleuca,* family Iridaceae, a perennial herb grown for its tall flower stalk bearing white, golden-yellow marked flowers. It is a beardless iris. Native to Turkey.

yellow-bark oak. Black oak.

yellow basswood. See *basswood.*

yellow bean mosaic. A virus disease of beans, sweetclover, red clover, and gladiolus, characterized by dark green areas near the veins of infected leaves, a puckering of the leaves, and a development of blisterlike lesions.

yellow bedstraw. *Galium verum,* family Rubiaceae, a perennial herb grown as a bedding plant and for its small, yellow flowers. Native to Europe. Also called *cheese rennet; cheese running; curdwort; Our Lady's bedstraw.*

yellow bell. Fawnlily.

yellow belly. See *yellow berry of wheat.*

yellow berry of wheat. A nitrogen deficiency disease in red wheat characterized by the normally hard and flinty grains becoming partially or entirely starchy in composition and light yellow in color.

yellow birch. *Betula lutea,* family Betulaceae, a hardy, deciduous tree which is important for its timber used in veneer, hardwood distillation, railroad ties, slack cooperage, furniture, planing-mill products, woodenware, scientific instruments, etc. Native to North America.

yellow blight. See *watery soft rot.*

yellow body. A temporary endocrine body which remains on the surface of the ovary of mammals following the extrusion of an egg. It generally remains fully-developed as long as pregnancy exists. As long as it persists, the female does not come into heat. Also called *corpus luteum.*

yellow bristlegrass. *Setaria lutescens,* family Gramineae, an annual grass which is a weed pest in cultivated fields throughout the U.S. Native to Europe. Also called *yellow foxtail -grass; yellow pigeon-grass.*

yellow buckeye. *Aesculus octandra,* family Hippocastanaceae, a hardy, deciduous tree grown for shade and in lawns. Its leaves are poisonous to stock and its nutlike seeds have poisoned children. Symptoms of poisoning are inflammation of the mucous membranes, vomiting, depression, weakness, and paralysis. Native to North America. Also called *sweet buckeye.*

yellow bud mosaic. A virus disease of peach characterized by yellow-colored buds sprouting for only a few millimeters with retarded growth. Older leaves are blotched along veins without distortion and the fruit is deformed on some varieties.

yellow butterflypalm. *Chrysalidocarpus lutescens,* family Palmaceae, a palm widely used by florists as a tub or pot palm in the greenhouse for interior decoration. Native to Madagascar. Also called *areca; cane palm.*

yellow butterwort. *Pinguicula lutea,* family Lentibulariaceae, a small herb with golden-yellow leaves, sometimes grown as a novelty. Native to southeastern U.S.

yellow calla. Golden callalily.

yellow camomile. Golden camomile.

yellow cedar. See *Sierra juniper; Nootka falsecypress; eastern arborvitae.*

yellow centaurea. Star thistle.

yellow cherry. Cherry tomato.

yellow chestnut oak. Chinkapin oak.

yellow copper oxide. See *copper oxide.*

yellow corn. a. Corn which has yellow kernels in contrast to the other common color, white. The yellow corn contains carotenoid pigments which can be converted into vitamin A in the animal body. b. A marketing class for corn which may not contain more than five percent grains of any color other than yellow.

yellow cosmos. *Cosmos sulphureus,* family Compositae, an annual herb, up to 7 ft. high, grown for its height and orange-yellow flowers. Native to Mexico.
yellow cottonwood. Southern poplar.
yellow cypress. Baldcypress.
yellow daisy. Blackeyedsusan.
yellow deal. The wood of the red pine, shortleaf pine, or Scotch pine.
yellow dock. Curly dock.
yellow dun. A coat-color of a horse that is light, lemon-yellow.
yellow dwarf mosaic of grasses. A virus disease of barley and certain other cereals and grasses characterized by bright yellow blotches developing from the leaf tip throughout the leaf blade, stunting and producing excessive tillering in the plant, which causes reduced root and spike development, and reduced kernel formation.
yellow dwarf of onion. A virus disease named after the two conspicuous symptoms which it produces: yellowing and dwarfing of the plant. It is transmitted by aphids and leafhoppers and not by seed.
yellow dwarf of potato. A virus disease characterized by dwarfing, yellowing of the foliage, rugosity and rolling upward of the leaves, and stem, leaf, and tuber necrosis.
yellow elder. Florida yellowtrumpet.
yellow flag. Yellowflag iris.
yellowflag iris. *Iris pseudacorus,* family Iridaceae, a perennial herb grown in a pool, bog garden and on a bank of a stream. It forms a large clump producing many rich yellow flowers. Native to Turkey, Europe, and North Africa. Also called *European yellowflag iris; false sweet flag; yellow flag; daggers.*
yellow flat. A virus disease of lilies, especially of the Easter lily, characterized by the top leaves of the plant turning light green and curling downward, assuming a cylindrical rather than pyramidal shape, and by stunting.
yellow-flowered alfalfa. One of the commonly cultivated groups or varieties of alfalfa *(q.v.)* which has yellow flowers.
yellow-flowered biennial sweetclover. See *yellow sweetclover.*
yellow flower gourd. *Cucurbita pepo* var. *ovifera,* family Cucurbitaceae, a variety of the pumpkin *(q.v.)* grown for its ornamental, yellow fruits, which are otherwise inedible. Also called *succade gourd.*
yellow foxtail-grass. Yellow bristlegrass.
yellow fungus of citrus whitefly. *Aschersonia goldiana,* family Zythiaceae, a fungus that destroys the citrus whitefly in Florida and the West Indies. Its pure culture in sweet potato strips is used to introduce the fungus into infested groves.
yellow galt. In animals, an inflammation of the udder.
yellow gentian. *Gentiana lutea,* family Gentianaceae, a perennial herb grown for its bitter, tonic rootstock. Native to southern Europe and western Asia.
yellow goatsbeard. Meadow salsify.
yellow grain sorghum. A market classification of grain sorghum which includes all varieties with yellow and salmon-pink grains. It may not include more than 10 percent of grain sorghums of other colors.
yellow granadilla. Watermelon.
yellow grease. Yellow fat from a hog carcass, unfit for human consumption, which is used in manufacturing soap, etc.
yellow hard winter wheat. A marketing classification for a subclass of hard red winter wheat *(q.v.)* which may contain not more than 25 percent of dark, hard, and vitreous kernels.
yellow-headed fireworm. *Acleris minuta* (Rob.), family Tortricidae, a caterpillar pest of cranberries which webs leaves together and feeds inside them. Also called *cranberry worm.*
yellow-headed spruce sawfly. *Pikonema alaskensis* (Rob.), family Tenthredinidae, a sawfly whose larvae feed on the needles of the spruces of the northern states.
yellow-heliotrope. Orange streptosolen.
yellow honeysuckle. *Lonicera flava,* family Caprifoliaceae, a hardy, twining, woody vine grown on a trellis, fence, etc., for its habit and for its fragrant, orange-yellow flowers. Native to North America.
yellow hornpoppy. *Glaucium flavum,* family Papaveraceae, a biennial herb grown for its blue flowers. Native to southern Europe.
yellow Indiangrass. *Sorghastrum nutans,* family Gramineae, a tall, perennial grass found in the U.S. from the Atlantic Coast to Montana, and in Mexico, on open prairies, in open woods, and on dry slopes. Its forage rating is excellent.
yellowing. The loss of green color of a plant's foliage due to degeneration or disorganization of chlorophyll or chloroplasts. See *chlorosis.*
yellow iris. Yellowflag iris.
Yellowish-Brown Lateritic soil. A zonal group of soils characterized by yellow-brown friable and granular surface horizons high in clay content over yellow or reddish-yellow friable clay material. The overlying parent materials are usually not strongly mottled. It developed under tropical forest in a hot, humid to subhumid, wet-dry climate.
yellow jessamine. See *Carolinajessamine; jasmine.*
yellow late rust. Cane and leaf rust.
yellow leaf. Cherry leaf spot.
yellow leaf blight. See *potash hunger.*
yellow leaf blotch. Alfalfa yellow leaf blotch.
yellow leaf of cherry. Cherry leaf spot.
yellow leaf roll of peach. A virus disease of peach characterized by a yellowing and inward rolling of the leaf edges, accompanied by reddish or brown coloring and premature defoliation.
yellow leaf spot of wheat. A minor, fungous disease of wheat in soft red and hard red winter wheat regions, caused by the fungus, *Helminthosporium tritici-vulgaris,* family Dematiaceae.
yellow locust. Black locust.
yellow lupine. Tree lupine.

yellow mariposa. *Calochortus luteus,* family Liliaceae, a cormous herb grown for its yellow or orange flowers, brown-spotted and brown-striped. Native to North America. Also called *butterfly tulip; mariposa tulip; mariposa lily.*

yellow mealworm. *Tenebrio molitor* L., family Tenebrionidae, the larva of a beetle which infests stored grain. It may also infest squabs, eating the skin at the neck and vent. The adult beetle affects setting hens by feeding on the feet.

yellow melitot. Yellow sweetclover.

yellow mercuric oxide. HgO, a chemical used as a fungicide. Also called *yellow oxide.*

yellow Mexican waterlily. *Nymphaea mexicana,* family Nymphaeaceae, a tropical, herbaceous plant, grown for its canary-yellow flowers. Native to tropical America.

yellow mombin. See *mombin.*

yellow monkshood. *Aconitum lutescens,* family Ranunculaceae, a perennial herb found from Idaho to New Mexico, U.S., which may be poisonous to stock feeding on it. For symptoms of poisoning see *Columbia monkshood.*

yellow mosaic of bean. Usually a more virulent form of mosaic than the common bean mosaic which differs in not being transmissible in the seed. Also, it occurs on some varieties of bean that are resistant to the common mosaic. Both are transmissible by aphids.

yellow mosaic of raspberry. Probably the most common virus disease of the black varieties. The foliage of diseased plants is distinctly yellow and the mottling on the leaves is rather diffused, seldom showing the dark green blisters of green mosaic. Both green and yellow mosaics are spread by the plant louse, *Amphorophora rubi,* family Aphidae.

yellow mosaic of rose. A rather widespread, mild type of mosaic of the rose, especially of the varieties used as stocks for budding or grafting.

yellow-necked caterpillar. *Datana ministra* (Drury), family Notodontidae, a caterpillar, generally distributed over the U.S., which infests apple, pear, cherry, quince, and shade and forest trees. The young caterpillars emerging from egg masses on the undersides of leaves, skeletonize the leaf and spread to other leaves to feed. Also called *datana caterpillar.*

yellow nut grass. Chufa flatsedge.

yellow oats. A class of oats which is based on the yellow color of the seed.

yellow oleander. Luckynut thevetia.

yellow oxalis. Creeping oxalis.

yellow oxide (of mercury). Yellow mercuric oxide.

yellow paintbrush. Field hawkweed.

yellow phosphorus. The toxic component of many commercial rat poisons.

yellow pigeon-grass. Yellow bristlegrass.

yellow pine-flax. New Mexico flax.

Yellow Podzolic soil. A taxonomic, zonal group of soils, characterized by a thin layer of organic matter over a grayish leached layer underlain by a yellow B horizon. It developed under forest, in humid warm climates and shows some laterization. The fertility level is low for crop plants. In more recent soil classification this group is combined with Red-Yellow Podzolic soils *q.v.*

yellow pond lily. Spatterdock cow lily.

yellow poplar. See *cucumbertree magnolia; tulip tree.*

yellow-red virosis. X-disease (of stone fruits).

yellowrocket. Bitter wintercress.

yellowroot. 1. *Xanthorhiza (Zanthorhiza) simplicissima,* family Ranunculaceae, an herb grown for its brownish-purple flowers. Also called *shrub yellowroot.* 2. Common goldthread.

yellow rot. A bacterial, greenhouse disease of hyacinths caused by *Xanthomonas hyacinthi (Pseudomonas hyacinthi),* characterized by yellow stripes on the leaves which turn brown and by a rotting of the bulb.

yellow rust. Stripe rust.

yellow rust of brambles. Western yellow rust of raspberries.

yellows. 1. Any plant disease in which yellowing (or chlorosis) and stunting are principal symptoms. Also called *xanthosis.* See *aster yellows; cabbage yellows; carnation yellows; carrot yellows; peach yellows; sour cherry yellows.* 2. Jaundice (*q.v.*) in domestic animals, as cattle, horses, sheep, etc.

yellow sage. Lantana.

yellow scale. *Aonidiella citrina* (Coq.), family Diaspididae, a major scale insect pest of citrus.

yellows of walnut. A zinc deficiency disease of Persian walnuts characterized by a yellowish-green mottle between the lateral veins of the leaves; a narrow, small leaf; bunching of the leaves; a tendency of the leaves to stand erect; and death of twigs and small branches.

yellow soybean. A market class of soybeans based on the yellow color of the seed.

yellow spot. A minor fungous disease of sugar cane caused by *Cercospora koepkei,* family Dematiaceae.

yellow-spotted willow slug. Willow sawfly.

yellow stage. Endoxerosis.

yellow star thistle. Star thistle.

yellowstem white willow. *Salix alba* var. *vitellina,* family Salicaceae, a hardy, deciduous tree, grown for its bright yellow twigs. Native to Europe and Asia. Also called *golden osier; golden willow.*

yellow stonecrop. Jenny stonecrop.

yellow stripe. Yellow spot.

yellow-striped armyworm. *Prodenia ornithogalli* Guen., family Noctuidae, a cutworm that infests cotton by feeding on young plants, boring into squares and bolls. However, it feeds generally on many different plants. Also called *cotton cutworm.*

yellow sweetclover. Either of two similar species of sweet-smelling herbs of the genus Melilotus *(M. altissima* and *M. officinalis),* family Leguminosae, with small, pea-like, yellow flowers. Growing to 8 ft., it is sometimes used as a forage crop. Native to Europe and Asia; naturalized in North America. Also called *King's clover; yellow melilot.*

yellow teat disease. Anaplasmosis.

yellow tips. Endoxerosis.

yellow toadflax. Butter-and-eggs toadflax.

yellow top. Alfalfa yellows.

yellow trefoil. Black medic.

yellowtuft. *Alyssum argenteum,* family Cruciferae, a perennial herb grown for its deep yellow, long-blooming flowers. Native to Europe.

yellow vegetable. A vegetable which has a yellow colored edible part, as the carrot.

yellow violet. Downy violet.

yellow-weed. Orange sneezeweed.

yellow whispering bells. *Emmenanthe penduliflora,* family Hydrophyllaceae, an annual herb grown for its bell-shaped, cream-yellow flowers. Native to North America.

yellow wildindigo. *Baptisia tinctoria,* family Leguminosae, a perennial, tumbleweed-type herb grown for its bright yellow flowers. Stock feeding on it in the wild have been poisoned. Native to North America. Also called *clover broom; wild indigo; false indigo.*

yellowwood. Any deciduous tree of the genus Cladrastis, family Leguminosae. Species are grown as ornamentals. See *American yellowwood; Osage orange.*

yellow wood sorrel. Creeping oxalis.

yellow woollybear. *Diacrisia virginica* (F.), family Arctiidae, a woolly, yellowish caterpillar, up to 2 ins. long, which feeds voraciously on field and garden plants in summer and autumn. It is distributed generally over the U.S.

yeoman. A farmer who owned a small estate (England, obsolete).

yerba. See *herb; plant: 1.*

yerbabuena. *Micromeria chamissonis,* family Labiatae, a trailing, perennial herb grown for its fragrant foliage. Native to North America.

yerbadepasmo baccharis. *Baccharis pteronioides,* family Compositae, a low shrub found in the U.S. from Texas and Arizona southward. It is poisonous to cattle browsing it.

yerba santa. Any one of the several shrubs of the genus Eriodictyon, family Hydrophyllaceae. Species are aromatic and are used as ornamentals where native. Native to western North America. Also called *eriodictyon.*

yew. Any one of the several evergreen, hardy shrubs or trees of the genus Taxus, family Taxaceae. Species are valuable, slow-growing ornamentals and are among the most popular of all evergreens. The wood, bark, leaves, and seed are very poisonous. See *English yew; Canada yew; Japanese yew.*

yew pine. Black spruce.

yield. 1. The quantity of or aggregate products resulting from cultivation or growth, as a *yield* of 30 bushels of wheat per acre. 2. The percentage of clean wool remaining in a lot after scouring. 3. The ratio of carcass weight to live weight. 4. To produce products as a result of cultivation or growth, as a tree *yields* fruits.

yielder. A plant or animal which produces products, often as a good or poor *yielder.*

yield of clean content. The clean scoured wool of the fleece.

yield per acre. The actual amount of a product produced per acre. The product of the average yield per individual plant times the number of individual plants per acre, which is determined by random sampling and taking the actual yield from smaller areas of given size, such as row segments and computing the acre yield from these samples.

ymer. A soured cows' milk, yoghurt-like dairy beverage made in Denmark.

yodogava azalea. *Rhododendron yedoense,* family Ericaceae, a deciduous or half-evergreen, low shrub, grown for its double, purple-spotted, rosy-lilac flowers. Native to Asia. Also called *Korean azalea.*

yoghurt. A semisolid, fermented milk product obtained by inoculating milk at body temperature with the *yoghurt* bacteria *Lactobacillus bulgaricus, Streptococcus lactis, Thermobacterium yoghurti,* etc., and then incubating it overnight (8 hours) at the same temperature. The bacteria convert the milk sugar and milk proteins to lactic acid, and peptones and amino acids, which produces the characteristic tangy taste of the product. *Yoghurt* contains almost all the known vitamins of the B-complex group, has a low caloric value, and is easily digested, which makes it an ideal food for people of all ages and tastes. In India, *yoghurt* (locally called *dahin*) is churned and, after the addition of cold water, the resulting butter-solids separated and heated to yield ghee *q.v.*

yoke. 1. A variously-shaped wooden frame which is placed on the necks of two oxen to work them together as a team. 2. A clamp which unites two pieces. 3. A team, as a *yoke* of oxen. 4. A wood or steel bar placed at the end of a tongue connected by rings and leather straps to the hames of the harness of a pair of horses. It holds up the tongue and aids in holding back the wagon on a downhill slope (obsolete).

yoke elm. European hornbeam.

yokel. A country bumpkin, or slow-witted rustic (rare).

Yokohama velvetbean. *Stizolobium hassjoo,* family Leguminosae, a tropical, annual vine grown in warm areas for its dark purple flowers and as a forage plant. Native to Japan.

yolk. 1. The yellow part of a fowl's egg that has a germinal disk located on its outer edge from which the embryo develops. The remainder of the yolk, together with the white, provides the nourishment on which the embryo makes its growth. 2. In wool scouring, all the substances, such as degras and perspiration, present in the fleece of a sheep when it is sheared.

Yorkshire. A bacon-type breed of swine originating in England. The hair coat is white, the ears erect, and the face generally dish-shaped. Boars weigh up to 700 lbs. and sows up to 600. In England three types of the breed are recognized: Large White, Middle White, and Small White, but no distinction is made in the U.S.

Yorkshire Coach. A breed of light coach horses of a bay color which was popular in the Old West in the U.S.

Yorkshire pudding. A yellow-colored pudding made from batter similar to pancake batter. Flavored with meat drippings, it is baked in a flat-based pan usually directly beneath and often with the roasting meat. Originating in the county of Yorkshire, England, it is eaten with the meat (usually beef) dish.

York spot. See *cork: 3.*

Yoshino cherry. *Prunus yedoensis,* family Rosaceae, a deciduous tree grown for its white to pink flowers. Native to Japan.

young cheese. A Cheddar cheese which has been cured from 10 days to 3 months.

young fustic. Smoketree.

young goose. A gosling.

young hen. A pullet.

young herd. In stock shows, one boar pig and three gilts no older than senior pigs.

young soil. Recently deposited or exposed soil material on which the soil profile development has begun but is in the initial stages.

young tom. A male turkey less than 1 year old.

youpon. See *yaupon.*

youth and old age. See *zinnia.*

Ysabella. A basic coat-color of some horses having silver manes and tails. There are three shades: the red *Ysabella,* palomino, and palomino-albino.

yucca. Any plant of the genus Yucca, family Liliaceae. Species are semidesert plants sometimes grown for ornament. Native to the U.S., Mexico, and the West Indies. Also called *dagger plant.* See *Adam's needle yucca; Joshua tree; soaptree yucca.*

yucca moth. *Tegeticula yuccasella* (Riley), family Prodoxidae, an insect which pollinates yuccas in southwestern U.S.

yucca plant bug. *Halticotoma valida* Reut., family Miridae, a plant bug that infests yucca. Infested plant leaves are stippled with excrement.

Yulan magnolia. *Magnolia denudata,* family Magnoliaceae, a deciduous tree grown for its fragrant, cup-shaped, white flowers. Native to China.

Yunnan meadowrue. *Thalictrum dipterocarpum,* family Ranunculaceae, a perennial, flowering plant, grown for its rosy-lilac flowers. Native to China.

yupon. See *yaupon.*

Z

zaman. Rain tree.

zambac. Arabian jasmine.

zanja. An irrigation ditch (southwestern U.S.).

zapater spoilage. A spoilage of olives during fermentation in which the olives develop a sage-like odor and flavor.

zebra caterpillar. *Ceramica picta* (Harr.), family Noctuidae (Phalaenidae), a caterpillar that feeds generally in late summer on truck, garden, and fruit crops, trees, flowers, etc.

Zebu. A strain of cattle originating in India; widely domesticated throughout India, China, and East Africa, they are used as beasts of burden, meat animals, and for their milk. The *Zebu* has a large hump over the shoulders, short horns, pendulous ears, and a large dewlap. Although not an ideal beef animal, it is widely used in breeding heat-tolerant beef cattle in Africa, Australia, southern U.S., etc. The Brahman, an alleged variety of the *Zebu,* was developed in the U.S. However, the names are used interchangeably. See *ranch cattle.*

zein. A prolamin, an alcohol-soluble protein which is obtained from corn.

zeolites. A group of light-colored, hydrous silicates containing potash and soda. They are soluble and could play an important base exchange role in soils. Their presence in soils has not been definitely proven, and in recent years less importance has been attached to *zeolites* as soil minerals. They are used in some systems of water softeners.

zephyr-flower. Zephyrlily.

zephyrlily. Any species of the genus Zephyranthes, family Amaryllidaceae. Species are grown as ornamentals. Also called *fairy lily.* See *atamasco-lily; rosepink zephyrlily.*

zerlate. See *ziram.*

zero grazing. Green chopping.

zero margin diameter limit. The diameter below which on the average, trees or logs will not yield material of enough value to cover the conversion costs.

zero pasture. Cutting green forage and hauling it to stock in lieu of pasturing. Sometimes called *zero grazing.* See *green chopping.*

Zeyher mockorange. *Philadelphus zeyheri,* family Saxifragaceae, a hybrid, deciduous shrub grown for its white flowers.

zho. An animal of the ox family, resulting from a cross between a bull yak *(q.v.)* and a common cow; used as a domestic animal in northern India. Native to Tibet. Also called *zobo; zobu* (males); *zhomo; jomo* (females).

zibeb. See *raisin.*

Ziegel. An Austrian cheese made from cows' whole milk with or without the addition of extra cream; cured for a month.

Ziger. A German whey cheese made from cows' milk which is very similar to Ricotta *q.v.*

zigzag clover. *Trifolium medium,* family Leguminosae, a perennial herb with strong, creeping rootstocks, which develops seed sparingly, and is not grown commercially. Native to Europe.

zigzag rail fence. Virginia fence.

zimate. See *ziram.*

Zimmerman pine moth. *Dioryctria zimmermani*

(Grote), family Phycitidae, a moth which lays its eggs on bark of pines in western U.S. The larvae bore into the trunks and branches. Infested trees show brown tips and sometimes broken tops.

zinc. Zn, a metallic chemical element, one of the so-called trace elements in soils, essential for both plants and animals.

zinc arsenate. $Zn_3 (AsO_4)_2$, a compound used as a component of sprays for control of leaf chewing insects, cabbage worms, the codling moth, etc.

zinc arsenite. $Zn (AsO_2)_2$, a compound used as a very effective component of insecticides.

zincate. See *ziram.*

zinc chloride. $ZnCl_2$, a compound used as a fungicide and wood preservative.

zinc dimethyl dithiocarbamate. See *ziram.*

zinc oxide. ZnO, a compound used as a fungicide, as a seed fumigant, and as source of zinc in livestock rations.

zinc phosphide. A compound, Zn_3P_2, which is a component of raticides and insecticides.

zinc trichlorophenate. A compound which is used as a fungicide, especially for seed treatment.

zinnia. 1. Any plant of the genus Zinnia, family Compositae. Species are grown as ornamentals. 2. Z. *elegans,* an annual herb grown for its attractive flowers. Its many varieties produce practically all the flower colors. Native to Mexico.

ziram. $ZnC_6H_{12}N_2S_4$, a compound used as a component of fungicides.

zit-kwa. Chinese waxgourd.

Zizania. A genus of marsh grasses, family Gramineae. Species are cultivated for their narrow, rice-like grains and as ornamentals. *Z. aquatica* (annual wildrice) is the source of the wild rice of commerce.

Zomma. A Turkish, plastic-curd cheese which is high in fat content.

zonal soil. Any one of the great groups of soil which has a well-developed soil profile that reflects the dominant influence of climate and vegetation in its formation.

zonate eyespot of grasses. A fungous disease of grasses caused by *Helminthosporium giganteum,* family Dematiaceae, which is characterized by an elongated eyespot with a brown halo and a tan center on the leaves.

Zondek-Aschheim test. A pregnancy test in which a small amount of urine from the female tested is injected into the blood stream of a virgin rabbit. Pregnancy is established if certain changes occur in the ovaries of the rabbit.

zone line. A narrow, orange, brown, or black line of hyphae, or decomposition products, which is formed within wood in the process of decay by certain wood-destroying fungi. *Zone lines* delimit small, irregular bodies of decay.

zone of aeration. In hydrology, the upper member of the zone of rock fracture. It includes the belt of soil water, water of the intermediate belt, and water of the capillary fringe.

zone of rock fracture. The zone of the lithosphere from the ground surface to the lower limits of porous, water bearing rock formations.

zone of saturation. The lower member of the zone of rock fracture. The zone of ground water, the upper surface of which is the water table. Water from this zone feeds springs, streams and wells. Generally, it is beneath the depth of penetration of plant roots.

zone of weathering. That depth, or thickness, of the surface crust of the earth in which weathering, or subaerial decay of rocks, has taken place. It includes soil and the parent material of soil.

zoning act. The division of the land of political subdivision into districts and setting up regulations pertaining to use of land; height, area, bulk, and use of buildings, and yard requirements; and in some states density of population. Zoning is effected by local ordinance under the police power of the state granted by specific legislation generally termed an enabling act. Zoning may apply to rural as well as urban areas. See *rural zoning.*

zoology. The science which deals with the study of animals.

zoonosis. 1. Any animal disease that can be transmitted from animals to other animals and to man. 2. A disease due to animal parasites.

zoophagous parasite. Any parasite that thrives in or on animals.

zoospore. An asexual, motile reproductive cell, which, by repeated divisions of the protoplast, develops into a new plant; a means of reproduction of many green algae, lower fungi, and some olive-brown algae.

zootechny. The scientific breeding and/or domestication of animals.

zygote. 1. The cell which is produced by the union of male and female gametes in reproduction; a fertilized egg cell. 2. The individual developing from such a cell.

zymogenic flora. The plants which develop in a soil as a result of the addition of an excessive quantity of readily decomposed organic matter or manure.

zymurgy. A branch of applied chemistry dealing with fermentation.

APPENDIX A

Reference List

AGRICULTURE (general)

"The Agricultural Education Magazine," Vol. XXIII, Nos. 1-4, 7-11, 1950-1951

BADING, H., *Wörterbuch der Landwirtschaft.* Hamberg: Park-Verlag, Claassen-Goverts G.M.-B.H., 1947

BARNHART, CLARENCE L., *The American College Dictionary.* New York: Harper, 1948

"Better Farming Methods," April, May, November, 1950

BLACK, H. C., *Black's Law Dictionary.* St. Paul, Minn.: 4th ed., 1951

BURNHAM, CHARLES R., "Glossary of Genetic Terms," University of Minnesota, in ms.

CARLISLE, ROSE JEANNE, "A Southwestern Dictionary," Albuquerque, N. M.: University of New Mexico, M.A. Thesis unpub., 1939

CARPENTER, J. RICHARD, *An Ecological Glossary.* New York: Hafner Publishing Co., 1956

CHISOLM, ALEC H., ed. *The Australian Encyclopaedia.* East Lansing, Michigan: Michigan State University Press, 1958, 10 Vols.

Climate and Man, Yearbook of Agriculture, Washington, D.C.: United States Department of Agriculture, 1941

COOK, G. C., *Handbook on Teaching Vocational Agriculture.* Danville, Ill.: Interstate, 1936

Eastern States Farmers Handbook. West Springfield, Mass.: Eastern States Farmers' Exchange, 1950

EMERSON, G., *American Farmers Encyclopedia.* New York: A. O. Moore, 1858

FAIRCHILD, HENRY PRATT, *Dictionary of Sociology.* New York: Philosophical Library, 1944

Farm Policy Dictionary. Agricultural Extension Service, Columbus, Ohio: Ohio State University, 1956?

Food and Life, Yearbook of Agriculture. Washington, D.C.: United States Department of Agriculture, 1939

GARDNER, D. P., *Farmers Dictionary.* New York: Harper & Bro., 1846

GAYLORD, CLINTON G., "A Few Farming Terms Used in the United States," United States Department of Agriculture Ext. Serv. No. 624, Washington, D.C., May, 1951

"Glossary of Agricultural Terms," Unpublished, Michigan State University, East Lansing

HAYES, H. K., IMMER, F. R., *Methods of Plant Breeding.* New York: McGraw-Hill, 1942

JACOBS, MORRIS B., GERSTEIN, MAURICE J., WALTER, WILLIAM G., *Dictionary of Microbiology.* Princeton, N.J.: D. Van Nostrand Co., Inc., 1957

KOHLER, ERIC L., *A Dictionary for Accountants.* New York: Prentice-Hall, Inc., 1952

LOUDON, JOHN C., *Loudon's Encyclopedia of Agriculture.* London: Longman, Orme, Brown, Green, and Longmans, 4th ed., 1839

MATHEWS, M. M., *The Dictionary of Americanisms.* Chicago, Ill.: University of Chicago Press, 1951

MCFARLANE, N. L., WINRIGHT, G. L., "Desert Agriculture," Berkeley: University of California Ext. Serv. Circ. No. 176, June 1951

Midwest Farm Handbook. Ames: Iowa State College Press, 2nd ed., 1951

NASCO "Catalog No. 14," Fort Atkinson, Wis.: National Agricultural Supply Company, 1950-1951

The New International Dictionary. Springfield, Mass.: G. & G. Merriam Co., 1944

RHAM, W., *Dictionary of the Farm.* London: Routledge, revised edition, 1853

"Service Letter to Teachers of Agriculture in Michigan," East Lansing: Michigan State University Division of Education, Series XIII, Nos. 1 & 2, Nov. 8, 1950 and Feb. 20, 1951

SLOANE, MRS. T. W., personal letter, Conway Co., Ark., Feb. 17, 1951

SPECTOR, WILLIAM S., ed., *Handbook of Biological Data.* Springfield, Ohio: Carpenter Lith. & Printing Co., 1956

"Teaching Techniques and Instructional Planning for High-School Classes in Vocational

Agriculture," Lansing, Mich.: The State Board of Control for Vocational Education, Bulletin No. 290, 1948

TRESSLER, D. K., EVERS, C. F., *The Freezing Preservation of Foods.* New York: A. V. I. Publishing Company, 2nd ed., 1947

TWEENEY, C. F., HUGHES, L. E. C., *Chambers Technical Dictionary.* New York: Macmillan Company, 1954

WILCOX, EARLEY V., *Modern Farmers' Cyclopedia of Agriculture.* New York: Orange Judd Publishing Co., 1946

WHITE, JOHN M., *The Farmer's Handbook.* Norman: University of Oklahoma Press, 1948

WOLFE, L. S., *Farm Glossary.* Orangeburg, S. C.: L. S. Wolfe, 1948

AGRICULTURAL CHEMISTRY

FREAR, D. E. H., *Agricultural Chemistry.* New York: Van Nostrand, 1950, 2 vols.

HANA, L. W., *Hana's Handbook of Agricultural Chemicals.* Forest Grove, Oregon: L. W. Hana, 1952

SHEPHARD, H. H., *The Chemistry and Action of Insecticides.* New York: McGraw-Hill, 1951

AGRICULTURAL ECONOMICS

BERRY, RUSSELL L., HILL, ELTON B., HENDERSON, SIDNEY, "How to Keep Your Farm in the Family," East Lansing, Mich.: Michigan State University Experiment Station Spec. Bull. No. 357, Apr. 1949

DUMMEIER, E. F., HEFLEBOWER, R. B., NORMAN, T., *Economics with Applications to Agriculture.* New York: McGraw-Hill, 3rd ed., 1950

EDWARDS, A. C., JR., GOULD, H. P., "Production and Marketing Calendar of American Fruits," Washington, D.C.: United States Department of Agriculture Reports FS-55, Apr. 1932

"Farm Management Terms and Definitions," *Doane's Agricultural Digest.* St. Louis, Missouri: Doane Agriculture Service, Inc., 1946

HILL, ELTON B., HARRIS, MARSHALL, "Family Farm-Operating Agreements," East Lansing, Mich.: Michigan State University Experiment Station Spec. Bull. No. 368, North Central Regional Publication No. 17, Jan., 1951

MULLINS, TROY, SLUSHER, M. W., "Comparison of Farming Systems for Large Rice Farms in Arkansas," Fayetteville, Arkansas: University of Arkansas, Experiment Station, Bull. No. 509, June, 1951

RENNE, R. R., *Land Economics.* New York: Harper, 1947

ROBERTSON, L. S., WOODS, R. H., *Farm Business Management.* New York: Lippincott, rev. ed., 1951

SLUSHER, M. W., MULLINS, TROY, "Production and Costs for Enterprises on Rice Farms," Fayetteville, Arkansas: University of Arkansas, Experiment Station Bull. No. 489, October, 1949

THOMSEN, F. L., *Agricultural Marketing.* New York: McGraw-Hill, 1951

U. S. Government Organization Manual 1957-58. Washington, D.C.: Federal Register Division National Archives and Records Services, 1957

AGRICULTURAL ENGINEERING

Agricultural Engineering. XXIX, Nos. 1-12, 1948

BARGER, E. L., CARLETON, W. M., MCKIBBEN, E. G., BAINER, R., "Farm Tractors and Power Units," mimeo, nd

"Barn Equipment and Hay Unloading Tools," Catalog No. 45-3, Chicago: H. D. Hudson Mfg. Co., 1950

BARRE, H. J., SAMMET, L. L., *Farm Structures.* New York: Wiley, 1950

"Bull. 800-49, Supplement to General Catalog No. 800," Hoopeston, Ill.: Sprague-Sells Div. of Food Machinery and Chemical Corp., 1949

"Catalog 1950-1951," Chicago, Ill.: Seedburo Equipment Co., May, 1950

"Catalog No. 28," St. Johns, Mich.: The F. C. Mason Co., 1947

COOK, G. C., SCRANTON, L. L., MCCOLLY, H. F., *Farm Mechanics Text and Handbook.* Danville, Ill.: Interstate, rev. ed., 1951

"Dictionary of Standard Terms for Use in Describing Containers and Loading and Bracing Methods for Fruits, Fresh, and Vegetables, Fresh or Green," New York: Bull. No. 45, Assn. of Am. R. R., Freight Container Bureau, Dec. 15, 1944

Ditching with Dynamite. Wilmington, Del.: E. I. du Pont de Nemours & Co., 2nd ed., 1945

"Farm Equipment Catalog," Chicago, Ill.: Montgomery Ward, 1950

Farm Planner's Engineering Handbook Upper Mississippi Region III. Milwaukee: United States Department of Agriculture Soil Conservation Service, 2nd ed., 1952

"General Sales Catalog," Chicago, Ill.: International Harvester, Aug., 1949

ISRAELSEN, O. W., *Irrigation Principles and Practices.* New York: Wiley, 2nd ed., 1950

LEWIS, M. R., "Practical Irrigation," Washington, D.C.: United States Department of Agriculture Farmers' Bull. No. 1922, Jan., 1943

MEREDITH, MAMIE, "The Nomenclature of American Pioneer Fences," *Southern Folklore Quarterly,* XVI (June, 1951), 109-151

PICKELS, G. W., *Drainage and Flood-Control Engineering.* New York: McGraw-Hill, 1st ed., 1925

ROE, H. B., NEAL, J. H., "Farm Drainage Practices," St. Paul, Minn.: University of Minnesota, Ext. Bull. 149, Oct., 1938

SCHAENZER, J. P., *Rural Electrification.* New York: Bruce Publ. Co., 1935

SMITH, H. P., *Farm Machinery & Equipment.* New York: McGraw-Hill, 3rd ed., 1948

"Special Equipment Catalog," No. 32, Chicago, Ill.: The Creamery Package Mfg. Co.
"Sprayers and Dusters, Catalog No. 51," Elkhart, Ind., Dobbins Mfg. Co., nd
STAEBNER, F. E., "Supplemental Irrigation," Washington, D.C.: United States Department of Agriculture Farmers' Bulletin No. 1846, Oct., 1940
WOOLEY, J. C., *Farm Buildings.* New York: McGraw-Hill, 2nd ed., 1946

ANIMAL HUSBANDRY

ACLAND, L. G. D., *The Early Canterbury Runs.* Christchurch, N.Z.: Whitcombe & Tombs, 1951
American Hampshire Herdsman. Vol. 25, No. 11, Nov. 1950
ANDERSON, A. L., *Introductory Animal Husbandry.* New York: Macmillan, rev. ed., 1951
"The Berkshire News," Vol. 15, No. 12; Vol. 16, No. 1, October, November, 1950
BLACK, W. H., "Feeding Cattle for Beef," Washington, D.C.: United States Department of Agriculture Farmers' Bull. No. 1549, rev. Apr., 1940
BLACK, W. H., MCCOMAS, E. W., "Beef Production on the Farm," Washington, D.C.: United States Department of Agriculture Farmers' Bull. No. 1592, rev. ed., Apr. 1940
BRIGGS, H. M., *Modern Breeds of Livestock.* New York: Macmillan, 1949
BROWN, GEORGE, "Selection and Care of the Farm Sheep Flock," East Lansing, Mich.: M.S.U. Ext. Bull. No. 242, October, 1942
CARROLL, W. E., KRIDER, J. L., *Swine Production.* New York: McGraw-Hill, 1950
The Cattleman. Vol. XXXVI, No. 8, Jan., 1950
COFFEY, W. C., *Productive Sheep Husbandry.* rev. by W. G. Kammlade. Philadelphia: Lippincott, 1937
"The Duroc News," Vol. 24, No. 10, October, 1950
ENSMINGER, M. E., *Animal Science.* Danville, Ill.: Interstate, 2nd ed., 1951
GORMAN, J. A., *The Western Horse.* Danville, Ill.: Interstate, 1949
"Hog Breeder," Vol. 25, No. 7, July, 1950
HULTZ, F. S., HILL, J. A., *Range Sheep and Wool.* New York: Wiley, 1931
KAMMLADE, W. G., *Sheep Science.* Philadelphia: Lippincott, 1947
KELLEY, R. B., *Native and Adapted Cattle.* Sydney, Australia: Angus and Robertson, Ltd., 1959
MCMILLEN, W. N., "Profitable Pork Production," East Lansing, Mich.: M.S.U. Ext. Bull. No. 299, April, 1949
MEISTER, JACK A., "Horseshoeing Terms," in ms., 1952
NORDBY, J. E., BEESON, W. M., FOURT, D. L., *Livestock Judging Handbook.* Danville, Ill.: Interstate, 1947
PERRY, E. J., ed., *The Artificial Insemination of Farm Animals.* New Brunswick, N. J.: Rutgers University Press, rev. ed., 1947
PETERS, W. H., *Livestock Production.* New York: McGraw-Hill, 1942
"The Poland China Breeder," W. C. Williams, Vol. 2, No. 6, Jan., 1950
"The Poland China World," March, 1950
RICE, V. A., ANDREWS, F. N., *Breeding and Improvement of Farm Animals.* New York: McGraw-Hill, 4th ed., 1951
RUSSELL, E. Z., "Swine Production," rev. by J. H. Seller, Washington, D.C.: United States Department of Agriculture Farmers' Bull. No. 1437, 1940
SMITH, W. W., *Pork Production.* New York: Macmillan, 3rd ed., 1952
SNAPP, R. R., *Beef Cattle.* New York: Wiley, 4th ed., 1952
WALL, J. F., *A Horseman's Handbook on Practical Breeding.* Camden, S.C.: Thoroughbred Bloodline, 3rd ed. rev., 1950
WINTERS, L. M., *Animal Breeding.* New York: Wiley, 4th ed., 1949

BEES

GROUT, R. A., ed., *The Hive and the Honey Bee.* Hamilton, Ill.: Dadant & Sons, rev. ed., 1949

DAIRY

ASSOCIATES OF ROGERS, *Fundamentals of Dairy Science.* New York: Reinhold, 1935
BARTLETT, R. W., *The Milk Industry.* New York: Ronald Press, 1946
BURKE, A. D., *Practical Manufacture of Cultured Milks and Kindred Products.* Milwaukee, Wis.: Olsen Publishing Co., 1938
DAVIS, J. G., *A Dictionary of Dairying.* London: Leonard Hill, 1950
DEYOE, G. P., BALTZER, A. C., and others, "Improvement Projects in Dairying," Lansing, Michigan: State Board of Control for Voc. Ed. Bull. No. 260, nd
ECKLES, C. H., *Dairy Cattle and Milk Production.* rev. by E. L. Anthony. New York: Macmillan, 4th ed., 1950
ECKLES, C. H., COMBS, W. B., MACY, H., *Milk and Milk Products.* New York: McGraw-Hill, 4th ed., 1951
ELDRIDGE, E. F., *Industrial Waste Treatment Practice.* New York: McGraw-Hill, 1942
FARRALL, A. W., *Dairy Engineering.* New York: Wiley, 1942
HALL, C. W., "Terms and Definitions of Bulk Milk Handling," in ms. East Lansing, Mich.: Michigan State University, 1957
HAMMER, B. W., *Dairy Bacteriology.* New York: Wiley, 3rd ed., 1948
HENDERSON, H. O., LARSON, C. W., PUTNEY, F. S., *Dairy Cattle Feeding and Management.* New York: Wiley, 3rd ed., 1950

HUNZIKER, O. F., *The Butter Industry*. La Grange, Ill.: Hunziker, 3rd ed., 1940

——, *Condensed Milk and Milk Powder*. La Grange, Ill.: Hunziker, 7th ed. rev., 1949

Laboratory Manual. Washington, D.C.: Milk Industry Foundation, 2nd ed., 1949

MCCULLOCH, E. C., *Disinfection and Sterilization*. Philadelphia: Lea & Febiger, 1945

NELSON, J. A., TROUT, G. M., *Judging Dairy Products*. Milwaukee, Wis.: Olsen Publishing Co., 3rd ed. rev., 1951

OLSON, T. M., *Elements of Dairying*. New York: Macmillan, 1938

PETERSEN, W. E., *Dairy Science*. Philadelphia: Lippincott, 2nd ed., 1950

SANDERS, GEORGE P., *Cheese Varieties and Descriptions*. United States Department of Agriculture Handbook No. 54, 1953

SOMMER, H. H., *Market Milk and Related Products*. Madison: published by the author, 3rd ed., 1952

TURNBOW, G. D., TRACY, P. H., RAFFETTO, L. A., *The Ice Cream Industry*. New York: Wiley, 2nd ed., 1949

VAN SLYKE, L. L., PRICE, W. V., *Cheese*. New York: Orange-Judd, 2nd ed., 1952

WHITTIER, E. O., WEBB, B. H., *By-Products from Milk*. New York: Reinhold, 1950

WIECHERS, S. G., GOEDE, B. D., *Continous Butter Making*. Amsterdam: North-Holland Publishing Co., 1950

FEEDS

BLACK, W. H., "Feeding Cattle for Beef," Washington, D.C.: United States Department of Agriculture Farmers' Bull. No. 1549, rev. ed., 1940

The Feed Bag Red Book. (Buyers' Guide for 1948), Milwaukee, Wis.: Editorial Service Co., 1948

Feed Trade Manual. Chicago: American Miller Publishing, 5th ed., 1947

Flour & Feed. Vol. 51, No. 4, September 1950

MAYNARD, L. A., *Animal Nutrition*. New York: McGraw-Hill, 3rd ed., 1951

MORRISON, F. B., *Feeds & Feeding*. Ithaca, N. Y.: Morrison Publishing Co., 8th ed., 1951

SEIDEN, RUDOLPH, PFANDER, W. H., *The Handbook of Feedstuffs*. New York: Springer, 1957

FIBER CROPS

"Better Cottons," Beltsville, Md.: United States Department of Agriculture Research Administration, September, 1947

BROWN, H. B., *Cotton*. New York: McGraw-Hill, 2nd ed., 1938

"The Classification of Cotton," Washington, D.C.: Bureau of Agricultural Economics, United States Department of Agriculture Misc. Pub. No. 310, May, 1938

DILLMAN, A. C., "Flaxseed Production in the North Central States," Washington, D.C.: United States Department of Agriculture Farmers' Bull. No. 1747, rev., March, 1942

"National Cottonseed Products Association Rules Governing Transactions Between Members," New Orleans: 1954

ROBINSON, B., "Flax-Fiber Production," Washington, D.C.: United States Department of Agriculture Farmers' Bull. No. 1728, July, 1940

FIELD CROPS

Agronomy Journal, XLI, Nos. 1-12, 1949

Agronomy Journal, XLII, Nos. 1-10, 1950

BAYLES, B. B., TAYLOR, J. W., "Wheat Production in the Eastern U.S.," Washington, D.C.: United States Department of Agriculture Farmers' Bull. No. 2006, September, 1951

DICKINSON, JAMES G., *Diseases of Field Crops*. New York: McGraw-Hill, 1956

HUTCHESON, T. B., WOLFE, T. K., KNIPPS, M. S., *The Production of Field Crops*. New York: McGraw-Hill, 3rd ed., 1948

"Important Facts About the Grain Exchange," Chicago, Ill.: The Board of Trade of the City of Chicago

MCNEAL, XZIN, "Artificial Drying of Combined Rice," Fayetteville, Ark.: University of Arkansas Expt. Sta. Bull. No. 487, June, 1949

——, "Effect of Combine Adjustment on Harvest Losses of Rice," Fayetteville, Ark.: University of Arkansas Expt. Sta. Bull. No. 500, December, 1950

——, "When to Harvest Rice for Best Milling Quality and Germination," Fayetteville, Ark.: University of Arkansas Expt. Sta. Bull. No. 504, December, 1950

RATHER, H. C., *Field Crops*. New York: McGraw-Hill, 1942 (Copy used was one for 2nd rev. by Carter M. Harrison)

RICHEY, FREDERICK D., "The What and How of Hybrid Corn," Washington, D.C.: United States Department of Agriculture Farmers' Bull. No. 1744, January, 1935

SIMMONS, CHARLES F., "Rice Production and Riceland Uses in Arkansas," Fayetteville, Ark.: University of Arkansas Ext. Serv. Circular No. 424, June, 1940

WALLACE, H. A., and BRESSMAN, E. N., *Corn and Corn Growing*. New York: Wiley, 5th ed., rev. by J. J. Newlin, Edgar Anderson, Earl N. Bressman, 1949

——, and BROWN, W. L., *Corn and Its Early Fathers*. East Lansing, Michigan: Michigan State University Press, 1956

WILSON, H. K., *Grain Crops*. New York: McGraw-Hill, 1948

FORAGE CROPS

Grass, Yearbook of Agriculture. Washington, D.C.: United States Department of Agriculture, 1948

HITCHCOCK, A. S., "Manual of the Grasses of the United States," Washington, D.C.: United States Department of Agriculture Misc. Pub. 200, 1935

Research and Farming, IX, Raleigh, N. C.: Agri. Expt. Station, progress report No. 1, July, 1950

SAMPSON, A. W., *Range Management.* New York: Wiley, 1952

FORESTRY

BROWN, H. P., PANCHIN, A. J., FORSAITH, C. C., *Textbook of Wood Technology.* New York: McGraw-Hill, I, 1949

HAWLEY, R. C., *Forestry Terminology.* Washington, D.C.: Society of American Foresters, nd

HORTICULTURE

"Abstracts of Lecturers," East Lansing, Mich.: M. S. U. Continuing Education Service, Fruit Growers Conference, Jan. 18, 19, 20, 1950

"American Fruit Grower," Cleveland, Ohio: Vol. 70, No. 2, Vol. 70, No. 7, Feb., July, 1950

ANDERSON, HARRY W., *Diseases of Fruit Crops.* New York: McGraw-Hill, 1956

BAILEY, LIBERTY HYDE, *Cyclopedia of Horticulture.* London: Macmillan, 1914, 6 vols.; and BAILEY, ETHEL ZOE, compilers, *Hortus Second.* New York: The Macmillan Company, 1947

BARTHOLOMEW, E. T., "Endoxerosis, or Internal Decline of Lemon Fruits," Berkeley: University of California Expt. Sta. Bull. No. 605, April, 1937

____, SINCLAIR, WALTON B., HORSPOOL, R. P., "Freeze Injury and Subsequent Seasonal Changes in Valencia Oranges & Grapefruit," Berkeley, California: University of California Expt. Sta. Bull. No. 719, October, 1950

____, SINCLAIR, WALTON B., TURRELL, F. M., "Granulation of Valencia Oranges," Berkeley, California: University of California Expt. Sta. Bull. 647, February, 1941

CARYL, R. E., "Citrus Culture in California," Berkeley, California: University of California Ext. Serv. Circ. No. 114, rev. by J. C. Johnson, February, 1946

COONY, J. J., and others, "Avocado Irrigation," Berkeley: University of California Ext. Serv., May, 1951

CRUESS, W. V., *Commercial Fruit & Vegetable Products.* New York: McGraw-Hill, 3rd ed., 1948

FERNALD, M. L., ed. *Gray's Manual of Botany.* New York: American Book Company, 8th ed., 1950

FROST, HOWARD B., CAMERON, JAMES W., "Frua and Dweet," Berkeley: University of California Expt. Sta. Bull. No. 721, February, 1951

GARDNER, V. R., BRADFORD, F. C., HOOKER, H. D., JR., *The Fundamentals of Fruit Production.* New York: McGraw-Hill, 2nd ed., 1939

GOODALL, GEORGE E., and others, "Planning Your New Avocado Grove," Berkeley: University of California Ext. Serv., April, 1951

HIRT, RAY R., "Common Arboricultural Terms," *Arborist's News,* Vol. XIV, No. 10, October, 1949

HOTTES, ALFRED C., *Climbers and Ground Covers.* New York: A. T. De La Mare Co., 1949

____, *The Book of Shrubs.* New York: A. T. De La Mare Co., 1952

____, *The Book of Trees.* New York: A. T. De La Mare Co., 1952

____, *The Book of Perennials.* New York: A. T. De La Mare Co., 1952

LAURIE, A., KIPLINGER, D. C., *Commercial Flower Forcing.* Philadelphia: Blakiston, 5th ed., 1950

"Market Growers Journal," Akron, Ohio: Jan., Feb., Mar., 1952

"The Papaya," Berkeley: University of California Agr. Ext. Serv., mimeo copy, nd

PARKER, WILLIS "Navels Must Be Good," *Citrus Leaves,* November, 1950

PIRONE, P. P., *Maintenance of Shade and Ornamental Trees.* New York: Oxford Press, 2nd ed., 1948

POST, K. *Florist Crop Production and Marketing.* New York: Orange-Judd, 1952

PUFFER, R. E. and others, "Harvesting & Marketing Your Avocados," Berkeley: University of California Ext. Serv., June, 1951

REHDER, ALFRED, *Manual of Cultivated Trees and Shrubs.* New York: Macmillan Co., 2nd ed., 1956

RYERSON, KNOWLES, "The Edible Passion Fruit," Berkeley, Calif.: mimeo, nd

____, "The Feijoa in California," Berkeley: University of California Ext. Serv., mimeo, nd

____, "The Jujube in California," Berkeley: University of California Ext. Serv., mimeo, nd

SHOEMAKER, J. S., *Small Fruit Culture.* Philadelphia: Blakiston, 2nd ed., 1950

SMOYER, K. M. and others, "Why Prune Avocado Trees," Berkeley: University of California Ext. Serv., January, 1951

Standardized Plant Names. Harrisburg, Pa.: J. Horace McFarland Company, 1942

TAYLOR, NORMAN, *Taylor's Encyclopedia of Gardening.* Boston: The American Garden Guild, Inc., and Houghton Mifflin, 1948

THOMPSON, HOMER C., *Vegetable Crops.* New York: McGraw-Hill, 4th ed., 1949

WEBBER, H. J., BATCHELOR, L. D. and others, *The Citrus Industry.* Berkeley: University of California Press, 1948, 2 Vols.

WHITE, F. A. and others, "Start Right with Your Young Avocado Trees," Berkeley: University of California Ext. Serv., November, 1950

WRIGHT, W. J., *Greenhouses.* New York: Orange-Judd, rev. ed., 1952

INSECTS

Borror, Donald J., DeLong, Dwight M., *An Introduction to the Study of Insects.* New York: Rinehart & Co., 1954

Brown, A. W. A., *Insect Control by Chemicals.* New York: Wiley, 1951

de la Torre-Bueno, J. R., *A Glossary of Entomology.* Brooklyn, N. Y.: Brooklyn Entomological Society, 1950

Entomological Society of America, *Common Names of Insects Approved by the Entomological Society of America.* Washington, D.C.: The Society, December 1955 (Volume I, No. 4 of the *Bulletin of the Entomological Society of America*)

Insects, Yearbook of Agriculture. Washington, D.C.: United States Department of Agriculture, 1952

Journal of Economic Entomology. Menashi, Wis., American Ass'n. of Econ. Entomologists, Vol. 43, No. 1, February, 1950

Metcalf, C. L. and Flint, W. P., *Destructive and Useful Insects.* New York: McGraw-Hill, 3rd ed. rev. by R. L. Metcalf, 1951

Peairs, L. M., *Insect Pests of Farm, Garden, and Orchards.* New York: Wiley, 4th ed., 1948

Quayle, H. J., "Control of Citrus Insects and Mites," Berkeley: University of California Ext. Serv. Circ. No. 123, October, 1941

Wescott, Cynthia, *The Gardener's Bug Book.* Garden City, N.Y.: The American Garden Guild and Doubleday, 1956

MEAT PRODUCTS

Brown, George A., "The Home Meat Supply," East Lansing, Mich.: M.S.U. Ext. Serv. Bull. No. 151, December, 1945

"Meat: U.S. Standards for Grade Issued by the U.S. Department of Agriculture," (ms. of article prepared for the *Encyclopedia Americana.* Sent to the editors by J. C. Pierce, Jr., Chief Standardization Section, Standardization & Grading Division, United States Department of Agriculture, Washington, D.C., March 5, 1951)

"Official U.S. Standards for Grades of Carcass Beef," Washington, D.C.: United States Department of Agriculture Production & Marketing Adm. Service & Regulatory Announcement No. 99, December, 1950

"Official U.S. Standards for Grades of Lamb Carcasses, Yearling Mutton, and Mutton Carcasses," Washington, D.C.: Bureau of Ag. Econ., Service & Regulatory Announcement No. 123, March, 1931

"Official U.S. Standards for Grades of Slaughter Cattle," Washington, D.C.: United States Department of Agriculture Production & Marketing Adm. Serv. & Regulatory Announcement No. 112, January, 1951

"Official U.S. Standards for Grades of Veal and Calf Carcasses," Washington, D.C.: United States Department of Agriculture Prod. & Marketing Adm. Serv. & Reg. Announcement No. 114, February, 1947

"Official U.S. Standards for Grades of Vealers and Slaughter Calves," Washington, D.C.: United States Department of Agriculture Bureau of Ag. Econ., Serv. and Reg. Announcements No. 113, September, 1928

"Proposed Standards for Grades of Slaughter Barrows & Gilts," Washington, D.C.: United States Department of Agriculture Prod. & Marketing Adm. (mimeo), February 15, 1950

"Rules & Regulations of the Secretary of Agriculture Governing the Grading and Certification of Meats, Prepared Meats, and Meat Products," Washington, D.C.: United States Department of Agriculture Prod. & Marketing Adm. Serv. & Reg. Announcement No. 98, rev., April 9, 1948

Tucker, H. Q. and others; Bratzler, L. J., editor, *A Cross Sectional Muscle Nomenclature of the Beef Carcass.* East Lansing, Michigan: Michigan State University Press, 1952

Ziegler, P. T., *The Meat We Eat.* Danville, Ill.: Interstate, 2nd ed., 1949

PLANT DISEASES

Ainsworth, G. C., Bisby, G. R., *A Dictionary of the Fungi.* Kew, Surrey: The Commonwealth Mycological Institute, 4th ed., 1954

Bessey, E. A., *Morphology & Taxonomy of Fungi.* Philadelphia: Blakiston, 1950

Chester, K. S., *Nature & Prevention of Plant Diseases.* Philadelphia: Blakiston, 2nd ed., 1947

Dykstra, T. P., "Potato Diseases and Their Control," Washington, D.C.: United States Department of Agriculture Farmers' Bull. No. 1881, rev. ed., February, 1948

Heald, F. D., *Manual of Plant Diseases.* New York: McGraw-Hill, 2nd ed., 1933

Index of Plant Diseases in the U.S. Issued by the Plant Disease Survey, U.S. Department of Agriculture

Klotz, L. J., Fawcett, H. S., *Color Handbook of Citrus Diseases.* Berkeley: University of California Press, 2nd ed. rev., 1948

Klotz, L. J. and others, "Water Spot of Navel Oranges: Studies of the Problem to 1948," Berkeley: University of California Expt. Sta., 1948

Plant Diseases, Yearbook of Agriculture. Washington, D.C.: United States Department of Agriculture, 1953

Snell, Walter H., Dick, Esther A., *A Glossary of Mycology.* Cambridge, Mass.: Harvard University Press, 1957

Terminology of Plant Virus Diseases. Canadian Phytopathological Society, Terminology Committee, nd

WALKER, J. C., *Plant Pathology*. New York: McGraw-Hill, 1950

WESCOTT, CYNTHIA, *Plant Disease Handbook*. New York: D. Van Nostrand Co., 1950

POULTRY HUSBANDRY

The American Standard of Perfection. Davenport, Iowa: Am. Poultry Association, rev. and ed. to 1947, 1947

BENJAMIN, E. W., PIERCE, H. C., TERMOHLEN, W. D., *Marketing Poultry Products*. New York: Wiley, 4th ed., 1949

Catalog No. 50. Newport, Ky.: Nat'l. Band & Tag Co., March, 1949

SEIDEN, R., *Poultry Handbook*. New York: Van Nostrand, 2nd ed., 1952

SEEDS

"Seed Trade Buyers Guide," Chicago: Seed World Publications, 1949

"The Seed World," Barrington, Illinois: March 18, 1949, September 15, 1950

SOILS AND FERTILIZERS

"Fertilizer Review," National Fertilizer Association, Washington, D.C.: July-Aug., 1948, Nov.-Dec., 1948, Jan.-Feb., 1949, May-June, 1949, Jul.-Aug.-Sept., 1949

"Glossary of Pedologic (Soils) and Landform Terminology for Soil Engineers." Washington, D.C.: National Academy of Science, Special Report 25, 1957

"Glossary of Soil Terms," *Doane's Agricultural Digest*. St. Louis, Missouri: Doane Agriculture Service, Inc., 645-648, 1943

"Glossary of Terms Used in the Field of Soil and Water Conservation," (tentative), Soil Conservation Soc. of America, December, 1947

JACKS, G. V., "Multilingual Vocabulary of Soil Science," Italy: Food and Agriculture Organization of United Nations, 1955

MILLAR, C.E., TURK, L. M., *Fundamentals of Soil Science*. New York: Wiley, 2nd ed., 1951

MEINZER, OSCAR E., "Outline of Ground-Water Hydrology," Washington, D.C.: Washington Government Printing Office, Water-Supply Paper 494, 1923

MOORE, H. C., *Dictionary of Fertilizer Materials and Terms*. Philadelphia: Ware Bros., rev. ed., 1941

MOORE, W. G., *Dictionary of Geography*. Harmondsworth, Middlesex, England: Penguin Books, 1952

PARKER, E. R., JONES, W. W., "Effects of Fertilizers upon the Yields, Size and Quality of Orange Fruits," Berkeley: University of California Expt. Sta. Bull. No. 722, March, 1951

Plant Food Journal. Washington, D.C.: Vol. 4, No. 2, April-May-June, 1951

RICE, C. M., *Dictionary of Geological Terms*. Ann Arbor, Michigan: Edwards Bros., 1952

Soil, Yearbook of Agriculture, Washington, D.C.: United States Department of Agriculture, 1957

Soils and Men, Yearbook of Agriculture, Washington, D.C.: U.S. Government Pr. Off., United States Department of Agriculture, 1938

Soil and Water Conservation Glossary. Soil Conservation of America, reprinted from the Journal, VII, 1, 2 & 3, 1952

Soil Science Society of America Proceedings, Vol. XX, No. 3, 1956

VEATCH, J. O., *Soils and Lands of Michigan*. East Lansing, Michigan: Michigan State University Press, 1953

Water, Yearbook of Agriculture, Washington, D.C.: United States Department of Agriculture, 1955

VETERINARY MEDICINE

Animal Diseases. Yearbook of Agriculture, Washington, D.C.: United States Department of Agriculture, 1956

BARGER, E. H., CARD, L. E., *Diseases and Parasites of Poultry*. Philadelphia: Lea & Febiger, 4th ed., 1949

DUKES, H. H., *Physiology of Domestic Animals*. Ithaca, N.Y.: Comstock, 6th ed. rev., 1947

DYKSTRA, R. R., *Animal Sanitation and Disease Control*. Danville, Ill.: Interstate, rev. ed., 1949

Keeping Livestock Healthy, Yearbook of Agriculture, Washington, D.C.: United States Department of Agriculture, 1942

MILLER, WILLIAM C., WEST, GEOFFREY P., *Black's Veterinary Dictionary*. London: Black, 3rd ed., 1953

WEEDS AND KILLERS

DARLINGTON, E. T., BESSEY, E. A., MEGEE, C. R., GRIGSBY, B. H., "Some Important Michigan Weeds," East Lansing, Mich.: M.S.U. Expt. Sta. Spec. Bull. No. 304, 2nd rev., May, 1951

KAPP, L. C., BARTHOLOMEW, R. P., "The Effect of Controlling Weeds by Irrigation and Seeding Practices on Rice Yields," Fayetteville, Ark.: University of Arkansas Expt. Sta. Bull. No. 467, June, 1947

MUENSCHER, W. C., *Poisonous Plants of the United States*. New York: Macmillan, rev. ed., 1951

ROBBINS, W. W., CRAFTS, A. S., RAYNOR, R. N., *Weed Control*. New York: McGraw-Hill, 1942

APPENDIX B

EDITOR-IN-CHIEF

WINBURNE, JOHN, Assistant Dean of the University College and Professor of American Thought and Language, Michigan State University. A.B. 1934, M.A. 1938, University of Colorado; Ph.D. 1951, Michigan State University.

ASSOCIATE EDITORS

ANTHONY, ERNEST L., Dean Emeritus, College of Agriculture, Michigan State University. B.S. 1912, University of Missouri; M.S. 1914, Pennsylvania State College.

BARRETT, FRANK N., Associate Professor Emeritus of Poultry Husbandry, Michigan State University. B.S. 1917, M.S. 1929, University of Illinois.

BESSEY, ERNST A., Dean Emeritus, Advanced School of Graduate Studies, Michigan State University. A.B. 1896, B.S. 1897, M.A. 1898, University of Nebraska; Ph.D. 1904, University of Halle, Germany.

MEGEE, C. R., Assistant Dean Emeritus, College of Agriculture, Michigan State University. B.S. 1914, M.A. 1916, University of Missouri; Ph.D. 1935, Michigan State University.

VEATCH, J. O., Professor Emeritus of Soil Science, Michigan State University. B.S. 1904, University of Missouri.

EDITORIAL BOARD

BERG, HERBERT A., Assistant Director of Cooperative Extension Service, Michigan State University. B.S. 1922, M.S. 1925, University of Illinois.

CARLETON, WALTER M., United States Department of Agriculture. B.S. 1938, Kansas State University; M.S. 1946, Iowa State University; Ph.D. 1948, Michigan State University.

CATION, DONALD P., Associate Professor of Botany, Michigan State University. B.S. 1925, M.S. 1928, Iowa State University.

CONNER, GABEL H., Professor of Veterinary Surgery and Medicine, Michigan State University. B.S. 1940, D.V.M. 1941, Washington State University; M.S. 1943, Iowa State University; Ph.D. 1953, University of Minnesota.

HARRISON, CARTER M., Professor of Farm Crops, Michigan State University. B.S 1926, M.S. 1927, University of Wisconsin; Ph.D. 1930, University of Chicago.

HILL, ELTON B., Professor of Agricultural Economics, Michigan State University. B.S. 1915, M.A. 1922, Michigan State University; M.S. 1927, Cornell University.

HOEFER, J. A., Professor of Animal Husbandry, Michigan State University. B.S. 1937, M.S. 1940, Ph.D. 1946, Purdue University.

KIDDER, ERNEST H., Associate Professor of Agricultural Engineering, Michigan State University. B.Ag.E. 1935, University of Minnesota; M.S. 1947, University of Illinois.

MARSHALL, ROY E., Assistant Director Emeritus, Agricultural Experiment Station, Michigan State University. B.S. 1913, University of Nebraska; M.S. 1915, Oregon State College; Ph.D. 1931, University of Minnesota.

RUSSELL, C. E., Associate Professor Emeritus of Horticulture, Michigan State University. B.S. 1926, Michigan State University; M.S. 1928, Oregon State University.

TROUT, G. MALCOLM, Professor of Food Science, Michigan State University. B.S. 1923, M.S. 1924, Iowa State University; Ph.D. 1936, Cornell University.

EDITORIAL RESEARCH STAFF

A. ROBIN BOWERS
AVIS BUTLER
NIRMAL SINGH DHESI
SURJIT SINGH DULAI
JOHN S. L. HOLLAND
PATRICIA MAGAW
SUSAN H. MCLEAN
LARRY NICHOLSON
LINDA L. PECKHAM
JULIE POWERS RICHARDS
JOHN THOMPSON
KATHLEEN WOLCOTT